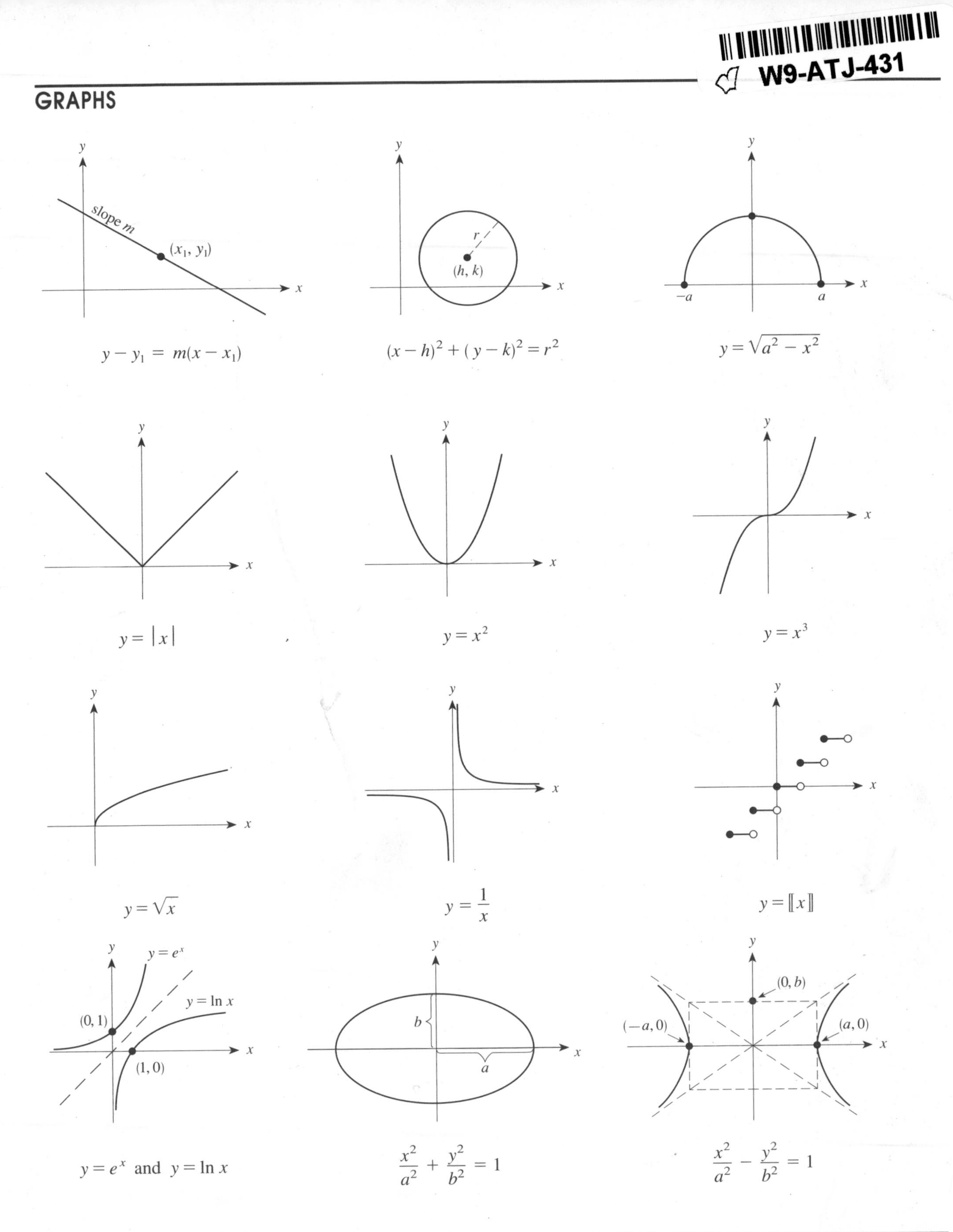
W9-ATJ-431
GRAPHS
slope m
(x_1, y_1)
$y - y_1 = m(x - x_1)$
r
(h, k)
$(x - h)^2 + (y - k)^2 = r^2$
$-a$
a
$y = \sqrt{a^2 - x^2}$
$y = |x|$
$y = x^2$
$y = x^3$
$y = \sqrt{x}$
$y = \frac{1}{x}$
$y = [\![x]\!]$
$y = e^x$
$y = \ln x$
(0, 1)
(1, 0)
$y = e^x$ and $y = \ln x$
b
a
$\frac{x^2}{a^2} + \frac{y^2}{b^2} = 1$
$(0, b)$
$(-a, 0)$
$(a, 0)$
$\frac{x^2}{a^2} - \frac{y^2}{b^2} = 1$

PRECALCULUS

Fifth Edition

PRECALCULUS

A PROBLEMS-ORIENTED APPROACH

Fifth Edition

DAVID COHEN
Department of Mathematics
University of California
Los Angeles

WEST PUBLISHING COMPANY
Minneapolis/St. Paul New York Los Angeles San Francisco

Copyediting: Susan Gerstein
Text Design: Janet Bollow
Technical Illustrations: TECHarts, Boulder, Colorado
Composition: Parkwood Composition
Cover Image: Scott A. Burns, Urbana, IL
The image is obtained in analyzing a numerical method of solution for a system of nonlinear equations arising in an engineering design problem. Iteration of functions, a topic new to this edition of *Precalculus,* plays a key role in generating the solutions and the resulting image.

WEST'S COMMITMENT TO THE ENVIRONMENT

In 1906, West Publishing Company began recycling materials left over from the production of books. This began a tradition of efficient and responsible use of resources. Today, 100% of our legal bound volumes are printed on acid-free, recycled paper consisting of 50% new paper pulp and 50% paper that has undergone a de-inking process. We also use vegetable-based inks to print all of our books. West recycles nearly 27,700,000 pounds of scrap paper annually—the equivalent of 229,300 trees. Since the 1960s, West has devised ways to capture and recycle waste inks, solvents, oils, and vapors created in the printing process. We also recycle plastics of all kinds, wood, glass, corrugated cardboard, and batteries, and have eliminated the use of polystyrene book packaging. We at West are proud of the longevity and the scope of our commitment to the environment.

West pocket parts and advance sheets are printed on recyclable paper and can be collected and recycled with newspapers. Staples do not have to be removed. Bound volumes can be recycled after removing the cover.

Production, Prepress, Printing and Binding by West Publishing Company

British Library Cataloguing-in-Publication Data. A catalogue record for this book is available from the British Library.

610 Opperman Drive
P.O. Box 64526
St. Paul, MN 55164-0526

Printed in the United States of America
04 03 02 01 00 99 98 8 7 6 5 4

LIBRARY OF CONGRESS CATALOGING-IN-PUBLICATION DATA

Cohen, David, 1942 Sept. 28–
Precalculus : a problems-oriented approach / David Cohen. -- 5th ed.
p. cm.
Includes index.
ISBN 0-314-06921-6 (hc : alk. paper)
1. Mathematics. I. Title.
QA39.2.C64 1997
512′.1--dc20 95-47495
CIP

FOR MY SON ANDY

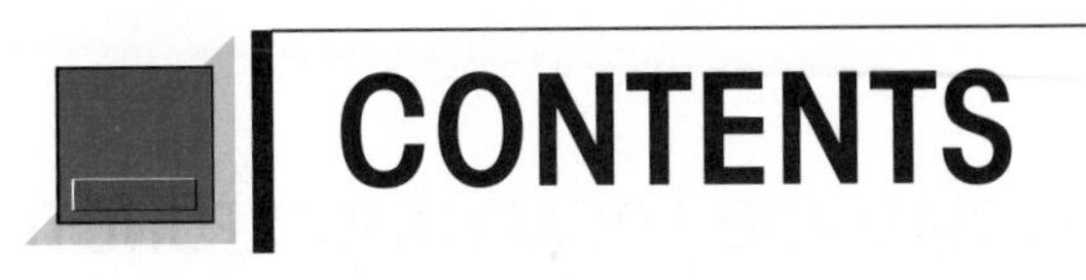

CONTENTS

10 SYSTEMS OF EQUATIONS 600

11 ANALYTIC GEOMETRY 672

12 ROOTS OF POLYNOMIAL EQUATIONS 752

ADDITIONAL TOPICS 822

APPENDIXES A-1

PREFACE TO THE FIFTH EDITION

This text is for students who are preparing to take calculus or other courses requiring a background in precalculus mathematics. As in the earlier editions, my goal has been to create a book that is *accessible* to the student. The presentation is student-oriented in three specific ways. First, I've tried to talk to, rather than lecture at, the student. Second, examples are consistently used to introduce, to explain, and to motivate concepts. And third, all of the initial exercises for each section are carefully correlated with the worked examples in that section.

AUDIENCE In writing *Precalculus,* I have assumed that the students have been exposed to intermediate algebra, but that they have not necessarily mastered that subject. Also, for many precalculus students, there may be a gap of several years between their last mathematics course and the present one. For these reasons, the review material in parts of the first two chapters is unusually thorough.

CURRICULUM REFORM This new edition of *Precalculus* reflects several of the major themes that have been developed in the curriculum reform movement of this decade. Overall, there is an increased emphasis upon graphs and visualization, and there is a tighter linkage between the graphical, numerical, and algebraic viewpoints. (See, for example, Section 4.3, *More on Iteration. Quadratics and Population Growth;* or Section 5.5, *Equations and Inequalities with Logs and Exponents.*)

TECHNOLOGY In the following discussion and throughout this text, the term *graphing utility* refers to either a graphics calculator or a computer with software for graphing and analyzing functions.

Over the past five to ten years, all of us in the mathematics teaching community have become increasingly aware of the *graphing utility* and its potential for making a positive impact in our teaching. While it is clear that more reforms lie ahead, many of the specific details have already evolved in workshops and in classrooms across the country. At present, however, even within a given school, some instructors teach the course with the graphing utility, while others do not. Consequently, this 1997 edition of *Precalculus* allows for both approaches to instruction. In addition to the regular exercise sets at the end of every section, there are also graphing utility exercises for the appropriate sections in Chapters 2 through 12.

FEATURES

1. *Word problems and applications.* Word problems and strategies for solving them are explained and developed throughout the book. Maximum-minimum problems relating to quadratic functions are discussed in detail in Section

4.5. The preceding section introduces some strategies for approaching these problems. To ensure that precalculus students gain appropriate practice and experience with these important strategies, the initial exercises in Section 4.4 make specific references to the corresponding worked examples in the text. In general, applications are integrated throughout the text. Two complete sections (5.6 and 5.7) are devoted to applications of the exponential function.

2. *Emphasis on graphing.* Graphs and techniques for graphing are developed throughout the text, and graphs are used to explain and reinforce algebraic concepts. See, for example, Sections 2.6 (inequalities), 3.3 (graphing techniques), and 7.4 (sine and cosine).
3. *Calculator exercises.* There are two broad categories of calculator exercises in this text:

   (i) GRAPHING UTILITY EXERCISES
   Following the regular exercise sets, there are graphing utility exercises for the appropriate sections in Chapters 2 through 12. These exercises reinforce and supplement the core material. Many of the exercises are divided into parts and ask the student to investigate a question from each of the three perspectives—graphical, numerical, and algebraic. (See, for example, Exercises 21 and 22 on pages 208–209.)

   (ii) (ORDINARY) CALCULATOR EXERCISES
   As in the previous edition, there are numerous calculator exercises integrated throughout the regular exercise sets. Just as with the graphing utility exercises, these exercises reinforce or supplement the core material. Where appropriate, real-life data is used. See, for example, Exercises 30 and 33 on page 184. Some of the calculator exercises contain surprising results that motivate subsequent theoretical questions; and a few of these exercises demonstrate that the use of a calculator cannot replace thinking or the need for mathematical proofs.

4. *End-of-chapter material.* Each chapter concludes with a detailed chapter summary, a *Writing Mathematics* section, an extensive Chapter Review exercise set, and a Chapter Test. In the *Writing Mathematics* questions, the student is asked to organize his or her thoughts and respond in complete sentences. Some of the questions are simply true-or-false questions that can be explained in just a sentence or two. In other cases, some study, perhaps some group work, and a more elaborate written response is required. For example, in Chapter 1 (pages 30–32), the *Writing Mathematics* section describes ancient Babylonian, Greek, and Arab techniques for solving quadratic equations; the student is then asked to demonstrate and explain the method in her or his own words. As another example, in Chapter 9, the *Writing Mathematics* section describes a new proof for the law of cosines from the *College Mathematics Journal,* and the student is asked to fill in the details. (See page 594.)

CHANGES IN THIS EDITION

Comments and suggestions from students, instructors, and reviewers have helped me to revise this text in a number of ways that I believe will make the book more useful to the instructor and more accessible to the student. The major changes occur in the following areas.

1. The material on inequalities, an important subject for both precalculus and calculus, has been moved from Chapter 1 to Chapter 2 (Sections 2.5 and

2.6). This allows the use of graphs to help explain and interpret the solutions of inequalities. This is one aspect of the increased emphasis on graphs and visualization in this new edition.

2. A new feature, the Graphical Perspective begins in Chapter 2. (See, for example, pages 41 and 50.) These graphs supplement the text discussion and help the student with visualization. Of course, any graph in any math book can be said to do this. But the Graphical Perspectives are designed so that students who have a graphing utility can (and should be encouraged to) produce their own versions of the picture. Experience shows that this involves the students in a more active way in reading the text. (For students without the technology, there is no problem; they simply get a book with more pictures and more emphasis on graphical interpretation of algebraic results.)
3. As described earlier, beginning in Chapter 2 there are graphing utility exercises, at the ends of appropriate sections, following the regular exercise sets. In contrast to the previous edition, these graphing utility exercises are not tied to a specific brand or model of calculator. (Keystroke-level instructions with additional exercises are, however, provided in the *Graphing Utilities Manual* that is available with this text. Additionally, Appendix A.1 at the end of this book provides a brief overview of the use of graphing utilities in the context of precalculus.)
4. In Chapter 3 (*Functions*), Section 3.3 has been expanded to include a discussion of the order in which translations and reflections are to be carried out. In Section 3.4, text and exercises on iteration of functions have been added. Graphical iteration (the "cobweb diagram") is explained in detail. This forms the background for a major new section in Chapter 4, *More on Iteration. Quadratics and Population Growth.* This section (4.3) introduces fixed points, both algebraically and geometrically, and uses them in a very informal graphical introduction to population growth and dynamical systems. For students (and instructors) who wish to pursue these topics further, but still at the precalculus level, I recommend the excellent paperback by Robert L. Devaney, *Chaos, Fractals, and Dynamics* (Menlo Park, California: Addison-Wesley, 1990).
5. In Chapter 5, a new section (5.5) has been added, *Equations and Inequalities with Logs and Exponents.* (A portion of this material on equations was covered only briefly in the previous edition.) Graphs and Graphical Perspectives are used extensively in this section to interpret the examples and solutions.

   In Section 5.7, the data in the exercises on exponential growth and decay has been extensively updated and (as in Chapter 4) there is an increased emphasis on modeling real-life data (with linear, quadratic, and exponential functions). There are new exercises asking the student to compare the results obtained by using different models for the same data set. Exercises on the logistic growth curve have also been added to the section on exponential growth and decay.
6. Chapter 7 in the previous edition (*Trigonometric Functions of Real Numbers*) is now split into two chapters: Chapter 7, *Graphs of the Trigonometric Functions;* and Chapter 8, *Analytical Trigonometry.*

   In Chapter 7, the initial section on radian measure from the previous edition has been expanded slightly and divided into two sections. Section 7.1 now contains material on the definition of radian measure and the evaluation of the trigonometric functions only. The geometric material on arc length, sector area, and angular speed is now covered in Section 7.2. Note that none of these three topics involves the trigonometric functions in its definition. Thus, moving these topics out of the key Section 7.1 sharpens the focus of 7.1. Additionally, the geometry in Section 7.2 is enriched with the discussion of the area of a segment of a circle and exercises involving elements of gothic architecture and Hippocrates' results on the area of a lune. (See Exercises 29–36 on pages 423–426.)

The introductory section on the graphs of sine and cosine (7.3 in the previous edition, 7.4 in the present one) now includes an extended example on solving trigonometric equations of the form $\cos x = k$. Classroom experience has shown that this is an excellent way to emphasize the properties of the cosine graph (and, likewise in exercises, the sine graph) and to integrate the graphical and algebraic viewpoints. This material also helps to prepare the student for sections in the next chapter on trigonometric equations and inverse functions.

As an application in analyzing functions of the form $y = A\sin(Bx - C)$ and $y = A\cos(Bx - C)$, there is a new section (7.5) on simple harmonic motion.

7. Chapter 8, *Analytical Trigonometry,* incorporates the addition formulas for sine, cosine, and tangent in one section (8.1). Section 8.3 on the product-to-sum and the sum-to-product formulas is new.
8. Two new sections on partial fractions have been added in Chapter 12 (*Roots of Polynomial Equations*). Together, they form a comprehensive treatment of the subject, more complete than that offered in many calculus texts. If the instructor wishes, however, only the first section can be covered; it is self-contained and omits the theory.

SUPPLEMENTARY MATERIALS

1. The *Student's Solutions Manual,* by Ross Rueger, contains complete solutions for the odd-numbered exercises and for the test questions at the end of each chapter.
2. The *Instructor's Solutions Manual,* by Ross Rueger, contains answers or solutions for every even-numbered exercise in *Precalculus.*
3. The *Graphing Utilities Manual* covers each of the six most popular graphing calculators and the following three computer programs: Mathematica, MAPLE, and DERIVE. It also provides documentation and laboratory exercises for *GraphToolz* (described in No. 8).
4. The *Computer-Generated Testing Program,* Westest 3.2, is available to schools adopting *Precalculus.* (There are versions for both Macintosh and Windows.)
5. The *Test Bank,* by Charles Heuer, is a package that contains three different chapter tests, as well as two multiple-choice versions, for each chapter in *Precalculus.*
6. West's *Math Tutor* software, by Mathens, for DOS, Windows, and Macintosh, is an algorithmically based tutorial, keyed to the text, that allows students to work problems on the computer. The software has been completely revised based on extensive reviews, making it much easier to use.
7. *Cohen's Precalculus Video Series* are video tapes available to qualified adopters. Each tape is keyed to a particular section of *Precalculus.*
8. *Graph Toolz,* by Tom Saxton, is a software program for graphing and evaluating functions. This easy-to-use software is free to qualified adopters. Doc-

umentation and laboratory exercises are found in the *Graphing Utilities Manual.* (It is available for the Macintosh family of computers.)

9. *Transparency masters* for many of the key figures or tables appearing in the text are available to schools adopting *Precalculus.*

ACKNOWLEDGMENTS Many students and teachers from both colleges and high schools have made useful constructive suggestions about the text and exercises, and I thank them for that. Special thanks go to Professor Charles Heuer for his careful work in checking the text and the exercise solutions for accuracy. In preparing and revising the manuscript, I received valuable suggestions and comments from the following reviewers.

Curtis Chapman
Oakland University
Rochester, MI

Greg Dietrich
Florida Community College—Kent Campus
Jacksonville, FL

Harvey Greenwald
California Polytechnic and State University
San Luis Obispo, CA

Christine Loritsch
Pasadena City College
Pasadena, CA

Mary Moynihan
Cape Cod Community College
West Barnstable, MA

The following individuals helped to shape this revision by responding to a survey dealing with content issues.

Lois Bearden
Schoolcraft College

Edward Bradley
California State University—Sacramento

Judy Cain
Tompkins-Cortland Community College

Maurice Chabot
University of South Maine
Math Department

James Eckerman
American River College

Gillion Zoe Elston
University of California

Joe Feidler
California State University

Penelope Fowler
Tennessee Wesleyan College

Ray Glenn
Tallahassee Community College

Madelyn Gould
DeKalb College—Central Campus

Gail Greene
Montreat-Anderson College

Robert Holman
Minot State University

Terry Jenkins
Clarke College

Giles Maloof
Boise State University

Lucy Michal
El Paso Community College

Richard Pilgrim
University of California—San Diego

Lisa Taylor
California State University—Sacramento

Jan Vandever
South Dakota State University

Donald Vick
Minot State University

I would like to thank Professor Ross Rueger, who wrote the supplementary manuals and created the answer section for the text; and Susan Gerstein, who copyedited the manuscript. Finally, to Peter Marshall, Jane Bass, Christine Hurney, Stacy Lenzen and their staffs at West Publishing Company, thank you for all your work and help in bringing this manuscript into print.

David Cohen
Lunada Bay, California, 1997

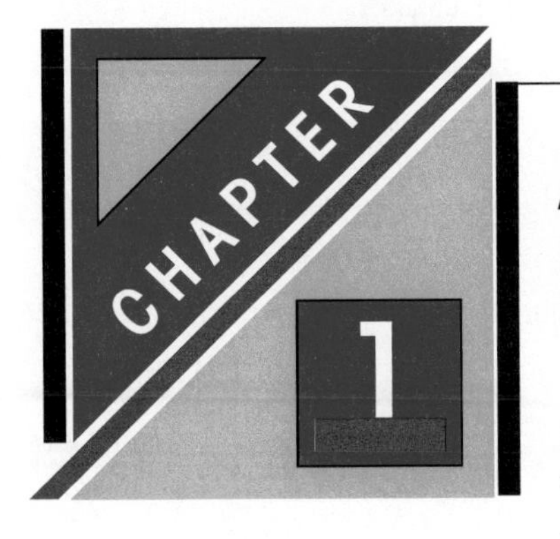

ALGEBRA BACKGROUND FOR PRECALCULUS

There are topics [in algebra] *whose consideration prepares a student for a deeper understanding.*

Leonhard Euler (1707–1783) in *Introduction to Analysis of the Infinite,* translated by John D. Blanton (New York: Springer-Verlag, 1988)

Perhaps Pythagoras was a kind of magician to his followers because he taught them that nature is commanded by numbers. There is a harmony in nature, he said, a unity in her variety, and it has a language: numbers are the language of nature.

Jacob Bronowski in *The Ascent of Man* (Boston: Little, Brown and Co., 1973)

INTRODUCTION

In Chapters 1 and 2 we review several key topics from algebra and coordinate geometry that form the foundation for our work in precalculus. Although you are probably already familiar with some of this material from previous courses, don't be lulled into a false sense of security. Now, in your second exposure to these topics, you really have the opportunity to master them. Take advantage of this opportunity; it will pay great dividends both in this course and in calculus.

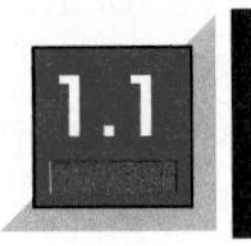

SETS OF REAL NUMBERS

What secrets lie hidden in decimals?

Stephen P. Richards in *A Number for Your Thoughts* (New Providence, NJ: S. P. Richards, 1982)

Natural numbers have been used since time immemorial; fractions were employed by the ancient Egyptians as early as 1700 B.C.; and the Pythagoreans, in ancient Greece, about 400 B.C., discovered numbers, like $\sqrt{2}$, which cannot be fractions.

Stefan Drobot in *Real Numbers* (Englewood Cliffs, NJ: Prentice-Hall, Inc., 1964)

Here, as in your previous mathematics courses, most of the numbers we deal with are *real numbers.* These are the numbers used in everyday life, in the sciences, in industry, and in business. Perhaps the simplest way to define a real number is this: A **real number** is any number that can be expressed in decimal form. Some examples of real numbers are

$$7(=7.000\ldots) \qquad -\tfrac{2}{3}(=-0.\overline{6}) \qquad \sqrt{2}\,(=1.4142\ldots)$$

(The bar above the 6 in the decimal $-0.\overline{6}$ indicates that the 6 repeats indefinitely.)

Certain sets of real numbers are referred to often enough to be given special names. These are summarized in the box that follows.

PROPERTY SUMMARY SETS OF REAL NUMBERS

NAME	DEFINITION AND COMMENTS	EXAMPLES
Natural numbers	These are the ordinary counting numbers: 1, 2, 3, and so on.	1, 4, 29, 1066
Integers	These are the natural numbers along with their negatives and zero.	$-26, 0, 1, 1776$
Rational numbers	As the name suggests, these are the real numbers that are *ratios* of two integers (with nonzero denominators, of course). It can be proved that a real number is rational if and only if its decimal expansion terminates (e.g., 3.15) or repeats (e.g., $2.\overline{43}$).	$4\ (=\frac{4}{1}), -\frac{2}{3}$, $1.7\ (=\frac{17}{10})$, $4.\overline{3}, 4.1\overline{73}$
Irrational numbers	These are the real numbers that are not rational. Section A.4 of the Appendix contains a proof of the fact that the number $\sqrt{2}$ is irrational. The proof that π is irrational is more difficult. The first person to prove that π is irrational was the Swiss mathematician J. H. Lambert (1728–1777).	$\sqrt{2}, 3+\sqrt{2}, 3\sqrt{2}$, $\pi, 4+\pi, 4\pi$

As you've seen in previous courses, the real numbers can be represented as points on a **number line,** as shown in Figure 1. As indicated in Figure 1, the point associated with the number zero is referred to as the **origin.**

Origin

$-4 \quad -3 \quad -2 \quad -1 \quad 0 \quad 1 \quad 2 \quad 3 \quad 4$

FIGURE 1

The fundamental fact here is that there is a **one-to-one correspondence** between the set of real numbers and the set of points on the line. This means that each real number is identified with exactly one point on the line; conversely, with each point on the line we identify exactly one real number. The real number associated with a given point is called the **coordinate** of the point. As a practical matter, we're usually more interested in relative locations than precise locations on a number line. For instance, since π is approximately 3.1, we show π slightly to the right of 3 in Figure 2. Similarly, since $\sqrt{2}$ is approximately 1.4, we show $\sqrt{2}$ slightly less than halfway from 1 to 2 in Figure 2.

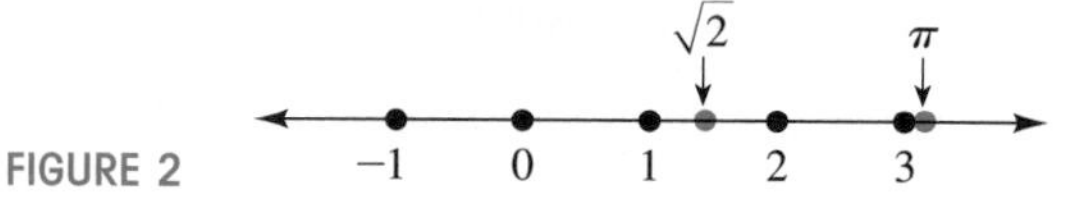

FIGURE 2

It is often convenient to use number lines that show reference points other than the integers used in Figure 2. For instance, Figure 3(a) displays a number line with reference points that are multiples of π. In this case, it is the integers that we then locate approximately. For example, in Figure 3(b) we show the approximate location of the number 1 on such a line.

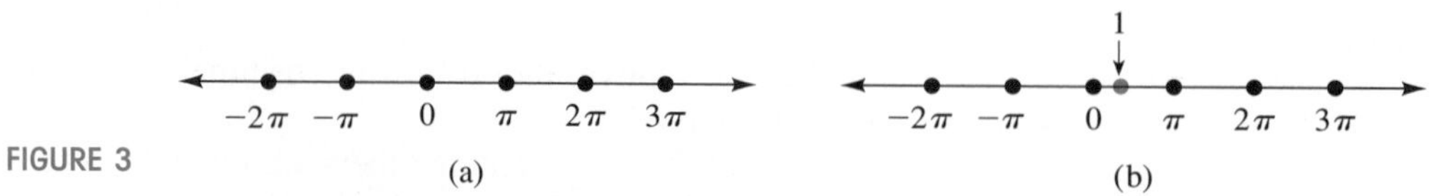

FIGURE 3

Two of the most basic relations for real numbers are **less than** and **greater than,** symbolized by $<$ and $>$, respectively. For ease of reference, we review these and two related symbols in the box that follows.

PROPERTY SUMMARY NOTATION FOR LESS THAN AND GREATER THAN

NOTATION	DEFINITION	EXAMPLES
$a < b$	a is less than b. On a number line, oriented as in Figure 1, 2, or 3, the point a lies to the left of b.	$2 < 3$; $-4 < 1$
$a \leqslant b$	a is less than or equal to b.	$2 \leqslant 3$; $3 \leqslant 3$
$b > a$	b is greater than a. On a number line oriented as in Figure 1, 2, or 3, the point b lies to the right of a. [$b > a$ is equivalent to $a < b$.]	$3 > 2$; $0 > -1$
$b \geqslant a$	b is greater than or equal to a.	$3 \geqslant 2$; $3 \geqslant 3$

In general, relationships involving real numbers and any of the four symbols $<$, $\leqslant$, $>$, and $\geqslant$ are called **inequalities.** One of the simplest uses of inequalities is in defining certain sets of real numbers called *intervals.* Roughly speaking, any uninterrupted portion of the number line is referred to as an **interval.** In the definitions that follow, you'll see notations such as $a < x < b$. This means that *both* of the inequalities $a < x$ and $x < b$ hold; in other words, the number x is between a and b.

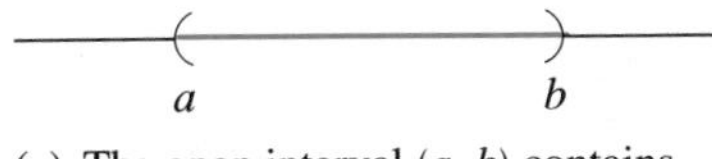
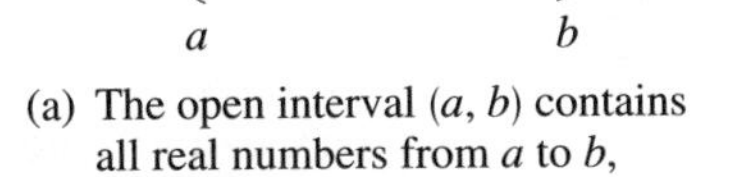

(a) The open interval (a, b) contains all real numbers from a to b, excluding a and b.

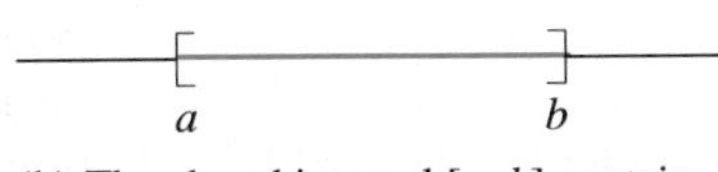

(b) The closed interval $[a, b]$ contains all real numbers from a to b, including a and b.

FIGURE 4

DEFINITION: Open Intervals and Closed Intervals

The **open interval** (a, b) consists of all real numbers x such that $a < x < b$. See Figure 4(a).
The **closed interval** $[a, b]$ consists of all real numbers x such that $a \leqslant x \leqslant b$. See Figure 4(b).

Notice that the brackets in Figure 4(b) are used to indicate that the numbers a and b are included in the interval $[a, b]$, whereas the parentheses in Figure 4(a) indicate that a and b are excluded from the interval (a, b). At times you'll see notation such as $[a, b)$. This stands for the set of all real numbers x such that $a \leqslant x < b$. Similarly, $(a, b]$ denotes the set of all numbers x such that $a < x \leqslant b$.

EXAMPLE 1 Show each interval on a number line, and specify inequalities describing the numbers x in each interval.

$$[-1, 2] \qquad (-1, 2) \qquad (-1, 2] \qquad [-1, 2)$$

Solution See Figure 5.

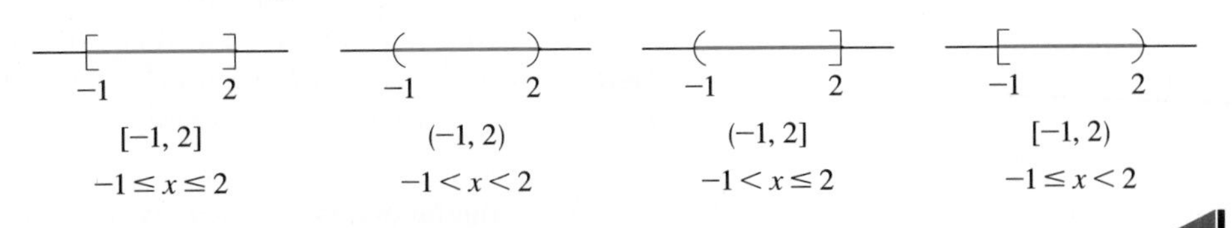

FIGURE 5

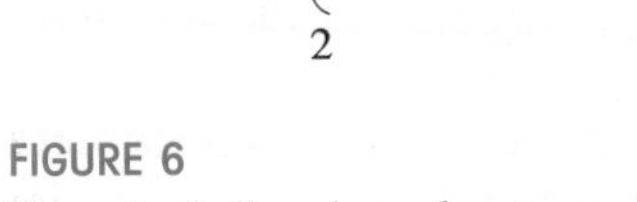

FIGURE 6
The set of all real numbers x such that $x > 2$.

In addition to the four types of intervals shown in Figure 5, we can also consider **unbounded intervals.** These are intervals that extend indefinitely in one direction or the other, as shown, for example, in Figure 6. We also have a convenient notation for unbounded intervals; for example, we indicate the unbounded interval in Figure 6 with the notation $(2, \infty)$.

COMMENT AND CAUTION The symbol ∞ is read *infinity*. It is not a real number, and its use in the context $(2, \infty)$ is only to indicate that the interval has no right-hand boundary. In the box that follows, we define the five types of unbounded intervals. Note that the last interval, $(-\infty, \infty)$, is actually the entire real-number line.

PROPERTY SUMMARY **UNBOUNDED INTERVALS**

NOTATION	DEFINING INEQUALITY	EXAMPLE
(a, ∞)	$x > a$	$(2, \infty)$; 2
$[a, \infty)$	$x \geq a$	$[2, \infty)$; 2
$(-\infty, a)$	$x < a$	$(-\infty, 2)$; 2
$(-\infty, a]$	$x \leq a$	$(-\infty, 2]$; 2
$(-\infty, \infty)$		$(-\infty, \infty)$; 2

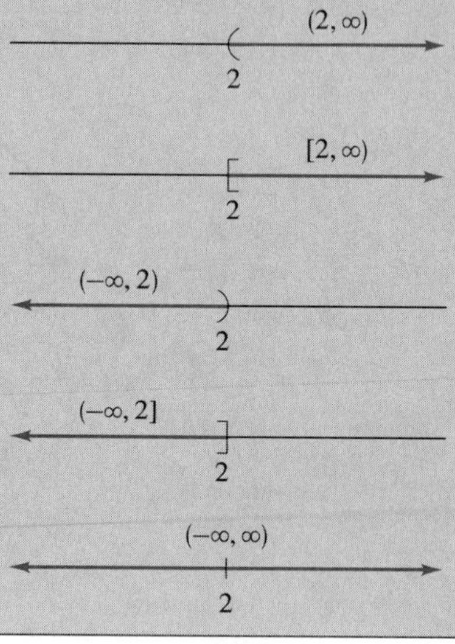

EXAMPLE 2 Indicate each set of real numbers on a number line: **(a)** $(-\infty, 4]$; **(b)** $(-3, \infty)$.

Solution

(a) The interval $(-\infty, 4]$ consists of all real numbers that are less than or equal to 4; see Figure 7.

(b) The interval $(-3, \infty)$ consists of all real numbers that are greater than -3; see Figure 8.

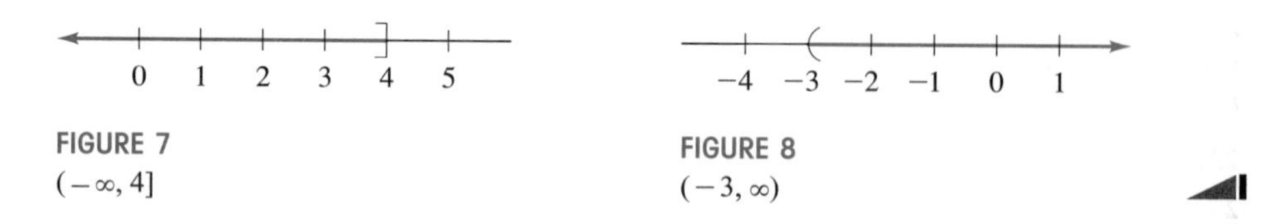

FIGURE 7
$(-\infty, 4]$

FIGURE 8
$(-3, \infty)$

We conclude this section by mentioning that our treatment of the real-number system has been rather informal, and we have not derived any of the rules of arithmetic and algebra using the most basic properties of the real numbers. However, we do list those basic properties and derive some of their consequences in Section A.3 of the Appendix. (For example, in the Appendix we prove that the product of two negative numbers is positive.)

1-11 odd
13, 21, 23, 26
35-57 odd
62, 63, 64

EXERCISE SET 1.1

A

In Exercises 1–12, determine whether the number is a natural number, an integer, a rational number, or an irrational number. (Some numbers fit in more than one category.) The following facts will be helpful in some cases: Any number of the form $\sqrt{n}$, where n is a natural number that is not a perfect square, is irrational. Also, the sum, difference, product, and quotient of an irrational number and a nonzero rational are all irrational. (For example, the following four numbers are irrational: $\sqrt{6}$, $\sqrt{10}-2$, $3\sqrt{15}$, and $-5\sqrt{3}/2$.)

1. **(a)** 7 **(b)** -7

2. **(a)** -203 **(b)** $204/2$

3. **(a)** $27/4$ **(b)** $\sqrt{27/4}$

4. **(a)** $\sqrt{7}$ **(b)** $4\sqrt{7}$

5. **(a)** 10^6 **(b)** $10^6/10^7$

6. **(a)** 8.7 **(b)** $8.\overline{7}$

7. **(a)** 8.74 **(b)** $8.\overline{74}$

8. **(a)** $\sqrt{99}$ **(b)** $\sqrt{99}+1$

9. $3\sqrt{101}+1$ **10.** $(3-\sqrt{2})+(3+\sqrt{2})$

11. $\dfrac{\sqrt{5}+1}{4}$ **12.** $\dfrac{0.1234}{0.5677}$

In Exercises 13–22, draw a number line similar to the one shown in Figure 1. Then indicate the approximate location of the given number. Where necessary, make use of the approximations $\sqrt{2}\approx 1.4$ and $\sqrt{3}\approx 1.7$. (The symbol $\approx$ means "is approximately equal to."*)*

13. $11/4$ **14.** $-7/8$ **15.** $1+\sqrt{2}$

16. $1-\sqrt{2}$ **17.** $\sqrt{2}-1$ **18.** $-\sqrt{2}-1$

19. $\sqrt{2}+\sqrt{3}$ **20.** $\sqrt{2}-\sqrt{3}$ **21.** $(1+\sqrt{2})/2$

22. $(2\sqrt{3}+1)/2$

In Exercises 23–32, draw a number line similar to the one shown in Figure 3(a). Then indicate the approximate location of the given number.

23. $\pi/2$ **24.** $3\pi/2$ **25.** $\pi/6$ **26.** $7\pi/4$

27. -1 **28.** 3 **29.** $\pi/3$ **30.** $3/2$

31. $2\pi+1$ **32.** $2\pi-1$

33. Use a calculator to show the approximate location of the irrational number $(\sqrt{139}-5)/3$ on a number line.

34. Use a calculator to show the approximate location of the irrational number $\sqrt{\dfrac{3-\sqrt{5}}{2}}$ on a number line.

In Exercises 35–44, say whether the statement is TRUE *or* FALSE. *(In Exercises 41–44, do not use a calculator or tables; use instead the approximations $\sqrt{2}\approx 1.4$ and $\pi\approx 3.1$.)*

35. $-5<-50$ **36.** $0<-1$ **37.** $-2\leqslant -2$

38. $\sqrt{7}-2\geqslant 0$ **39.** $\frac{13}{14}>\frac{15}{16}$ **40.** $0.\overline{7}>0.7$

41. $2\pi<6$ **42.** $2\leqslant(\pi+1)/2$ **43.** $2\sqrt{2}\geqslant 2$

44. $\pi^2<12$

In Exercises 45–58, show the given interval on a number line.

45. $(2, 5)$ **46.** $(-2, 2)$ **47.** $[1, 4]$

48. $[-\frac{3}{2}, \frac{1}{2}]$ **49.** $[0, 3)$ **50.** $(-4, 0]$

51. $(-3, \infty)$ **52.** $(\sqrt{2}, \infty)$ **53.** $[-1, \infty)$

54. $[0, \infty)$ **55.** $(-\infty, 1)$ **56.** $(-\infty, -2)$

57. $(-\infty, \pi]$ **58.** $(-\infty, \infty)$

B

From grade school on, we all acquire a good deal of experience in working with rational numbers. Certainly we can tell when two rational numbers are equal, although it may require some calculation, as, for example, in the case of $\frac{129}{31}=\frac{2193}{527}$. In Exercises 59 and 60, you'll see that in working with irrational numbers, it's sometimes hard to recognize when two numbers are equal. For each equation in Exercises 59 and 60, use a calculator to verify that the quantities on both sides agree, to six decimal places. (In each case it can be shown that the two quantities are indeed equal.)

59. **(a)** $\sqrt{6}+\sqrt{2}=2\sqrt{2+\sqrt{3}}$

(b) $\sqrt{3+\sqrt{5}}+\sqrt{3-\sqrt{5}}=\sqrt{10}$

(c) $\sqrt{\sqrt{6+4\sqrt{2}}}=\sqrt{2+\sqrt{2}}$

60. **(a)** $\sqrt{7}-\sqrt{8-2\sqrt{7}}=1$

(b) $\sqrt[3]{2+\sqrt{5}}+\sqrt[3]{2-\sqrt{5}}=1$

(c) $\sqrt[5]{176+80\sqrt{5}}=1+\sqrt{5}$

61. The value of the irrational number π, correct to ten decimal places (without rounding), is 3.1415926535. By using a calculator, determine to how many decimal places each of the following quantities agrees with π.

(a) $(4/3)^4$ [This is the value used for π in the Rhind papyrus, an ancient Babylonian text written about 1650 B.C.]

(b) $22/7$ [Archimedes (287–212 B.C.) showed that $223/71<\pi<22/7$. The use of the approximation $22/7$ for π was introduced to the western world through the writings of Boethius (ca 480–520), a Roman philosopher, mathematician, and statesman. Among all fractions with numerators and denominators less than 100, the fraction 22/7 is the best approximation to π.]

(c) 355/113 [This approximation of π was obtained in fifth-century China by Zu Chong-Zhi (430–501) and his son. According to David Wells in *The Penguin Dictionary of Curious and Interesting Numbers* (New York: Viking Penguin, Ltd., 1986), "This is the best approximation of any fraction below 103993/33102."]

(d) $\dfrac{63}{25}\left(\dfrac{17+15\sqrt{5}}{7+15\sqrt{5}}\right)$ [This approximation for π was obtained by the Indian mathematician Scrinivasa Ramanujan (1887–1920)].

Remark: A simple approximation that agrees with π through the first 14 decimal places is $\dfrac{355}{113}\left(1-\dfrac{0.0003}{3533}\right)$. (This approximation was also discovered by Ramanujan.) For a fascinating account of the history of π, see the book by Petr Beckmann, *A History of* π, 4th ed. (New York: Golem Press, 1977), and for a very contemporary look at π, see Richard Preston's article, "The Mountains of Pi" in *The New Yorker* (March 2, 1992; pp. 36–67).

In Exercises 62–64, give an example of irrational numbers a and b such that the indicated expression is **(a)** *rational;* **(b)** *irrational.*

62. $a+b$ **63.** ab **64.** a/b

65. **(a)** Give an example in which the result of raising a rational number to a rational power is an irrational number.

(b) Give an example in which the result of raising an irrational number to a rational power is a rational number.

66. Can an irrational number raised to an irrational power yield an answer that is rational? This problem shows that the answer here is "yes." (However, if you study the following solution very carefully, you'll see that even though we've answered the question in the affirmative, we've not pinpointed the specific case in which an irrational number raised to an irrational power is rational.)

(a) Let $A=(\sqrt{2})^{\sqrt{2}}$. Now, either A is rational or A is irrational. If A is rational, we are done. Why?

(b) If A is irrational, we are done. Why?
Hint: Consider $A^{\sqrt{2}}$.

Remark: For more about this problem and related questions, see the article "Irrational Numbers," by J. P. Jones and S. Toporowski in *American Mathematical Monthly,* vol. 80 (1973), pp. 423–424.

1.2 ABSOLUTE VALUE

There has been a real need in analysis for a convenient symbolism for "absolute value" . . . and the two vertical bars introduced in 1841 by Weierstrass, as in $|z|$, have met with wide adoption; . . .

Florian Cajori in *A History of Mathematical Notations,* Vol. 1 (La Salle, Ill.: The Open Court Publishing Co., 1928)

As an aid in measuring distances on the number line, we review the concept of *absolute value.* We begin with a definition of absolute value that is geometric in nature. Then, after you have developed some familiarity with the concept, we explain a more algebraic approach that is often useful in analytical work.

DEFINITION: Absolute Value (geometric version)

The **absolute value** of a real number x, denoted by $|x|$, is the distance from x to the origin.

For instance, because the numbers 5 and -5 are both five units from the origin, we have $|5|=5$ and $|-5|=5$. Here are three more examples:

$$|17|=17 \qquad \left|-\frac{2}{3}\right|=\frac{2}{3} \qquad |0|=0$$

In dealing with an expression such as $|-5+3|$, the convention is to compute the quantity $-5+3$ first, then to take the absolute value. We therefore have, in this case,

$$|-5+3| = |-2| = 2$$

EXAMPLE 1 Evaluate each expression.

(a) $5 - |6-7|$ **(b)** $||-2| - |-3||$

Solution

(a) $5 - |6-7| = 5 - |-1| = 5 - 1 = 4$ **(b)** $||-2| - |-3|| = |2-3| = |-1| = 1$

As we said at the beginning of this section, there is an equivalent, more algebraic way to define absolute value. According to this equivalent definition, the value of $|x|$ is x itself when $x \geqslant 0$, and the value of $|x|$ is $-x$ when $x < 0$. We can write this symbolically as follows:

DEFINITION: Absolute Value (algebraic version)

$$|x| = \begin{cases} x & \text{when } x \geqslant 0 \\ -x & \text{when } x < 0 \end{cases}$$

EXAMPLE

$|-7| = -(-7) = 7$

By looking at examples with specific numbers, you should be able to convince yourself that both definitions yield the same results. We use the algebraic definition of absolute value in Examples 2 and 3.

EXAMPLE 2 Rewrite each expression in a form that does not contain absolute values:

(a) $|\pi - 4| + 1$; **(b)** $|x-5|$, given that $x \geqslant 5$; **(c)** $|t-5|$, given that $t < 5$.

Solution

(a) The quantity $\pi - 4$ is negative (since $\pi \approx 3.14$), therefore its absolute value is equal to $-(\pi - 4)$. In view of this, we have

$$|\pi - 4| + 1 = -(\pi - 4) + 1 = -\pi + 5$$

(b) Since $x \geqslant 5$, the quantity $x - 5$ is nonnegative, therefore its absolute value is equal to $x - 5$ itself. Thus, we have

$$|x-5| = x - 5 \qquad \text{when } x \geqslant 5$$

(c) Since $t < 5$, the quantity $t - 5$ is negative. Therefore its absolute value is equal to $-(t-5)$, which in turn is equal to $5 - t$. In view of this, we have

$$|t-5| = 5 - t \qquad \text{when } t < 5$$

EXAMPLE 3 Simplify the expression $|x-1| + |x-2|$, given that x is in the open interval $(1, 2)$.

Solution Since x is greater than 1, the quantity $x - 1$ is positive and, consequently,

$$|x-1| = x - 1$$

On the other hand, we are also given that x is less than 2. Therefore the quantity $x-2$ is negative, and we have

$$|x-2| = -(x-2) = -x+2$$

Putting things together now, we can write

$$\begin{aligned} |x-1| + |x-2| &= (x-1) + (-x+2) \\ &= -1+2 = 1 \end{aligned}$$

The solutions in Examples 2 and 3 use the algebraic version of the definition of absolute value. In the error box that follows, the first two items indicate possible misuses of this definition. *Suggestion:* Before reading further, cover up the columns headed "Correction" and "Comment" in the box and try to decide for yourself where each error lies.

ERRORS TO AVOID

ERROR	CORRECTION	COMMENT
$\lvert 3-\sqrt{2}\rvert \not= -(3-\sqrt{2}) = -3+\sqrt{2}$	$\lvert 3-\sqrt{2}\rvert = 3-\sqrt{2}$	Because the quantity $3-\sqrt{2}$ is positive, the appropriate formula is $\lvert x\rvert = x$, not $\lvert x\rvert = -x$.
$\lvert x-y\rvert \not= -(x-y) = -x+y$	$\lvert x-y\rvert = \begin{cases} x-y & \text{if } x-y \geq 0 \\ -(x-y) & \text{if } x-y < 0 \end{cases}$	The error arises in assuming that the quantity $x-y$ is negative. Unless additional information is available, $\lvert x-y\rvert$ cannot be simplified.
$\lvert a+b\rvert \not= \lvert a\rvert + \lvert b\rvert$	$\lvert a+b\rvert = \begin{cases} a+b & \text{if } a+b \geq 0 \\ -(a+b) & \text{if } a+b < 0 \end{cases}$	The absolute value of a sum is, in general, not equal to the sum of the individual absolute values. For example, if $a=-1$ and $b=2$, then $\lvert a+b\rvert = \lvert -1+2\rvert = \lvert 1\rvert = 1$, but $\lvert a\rvert + \lvert b\rvert = \lvert -1\rvert + \lvert 2\rvert = 3$.

In the box that follows, we list several basic properties of the absolute value. Each of these properties can be derived from the definitions. (With the exception of the *triangle inequality,* we shall omit the derivations. For a proof of the triangle inequality, see Exercise 71.)

PROPERTY SUMMARY **PROPERTIES OF ABSOLUTE VALUE**

1. For all real numbers x, we have
   (a) $|x| \geq 0$;
   (b) $x \leq |x|$ and $-x \leq |x|$;
   (c) $|x|^2 = x^2$.
2. For all real numbers a and b, we have
   (a) $|ab| = |a|\,|b|$ and $|a/b| = |a|/|b|$ $(b \neq 0)$;
   (b) $|a+b| \leq |a| + |b|$ (the triangle inequality).

EXAMPLE 4 Write the expression $|-2-x^2|$ in an equivalent form that does not contain absolute values.

Solution We have

$$\begin{aligned} |-2-x^2| &= |-1(2+x^2)| \\ &= |-1|\,|2+x^2| \qquad \text{using Property 2(a)} \\ &= 2+x^2 \end{aligned}$$

The last equality follows from the fact that x^2 is nonnegative for any real number x and, consequently, the quantity $2+x^2$ is positive.

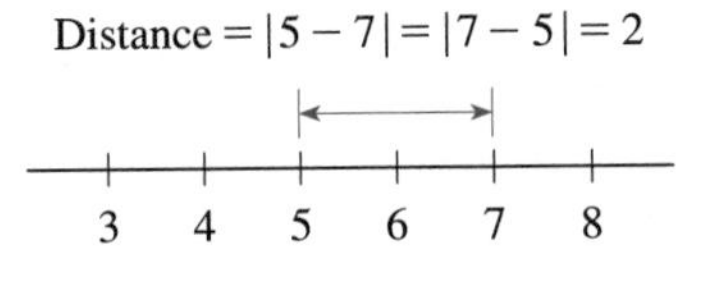

FIGURE 1

If we think of the real numbers as points on a number line, the distance between two numbers a and b is given by the absolute value of their difference. For instance, as indicated in Figure 1, the distance between 5 and 7, namely 2, is given by either $|5-7|$ or $|7-5|$. For reference, we summarize this simple but important fact as follows.

PROPERTY SUMMARY **DISTANCE ON A NUMBER LINE**

The **distance** between a and b is $|a-b| = |b-a|$.

EXAMPLE 5 Rewrite each of the following statements using absolute values.

(a) The distance between 12 and -5 is 17.
(b) The distance between x and 2 is 4.
(c) The distance between x and 2 is less than 4.
(d) The point t is more than five units from the origin.

Solution
(a) $|12-(-5)| = 17$ or $|-5-12| = 17$

(b) $|x-2| = 4$ or $|2-x| = 4$

(c) $|x-2| < 4$ or $|2-x| < 4$

(d) $|t| > 5$

EXAMPLE 6 In each case, the set of real numbers satisfying the given inequality is one or more intervals on the number line. Show the interval(s) on a number line.

(a) $|x| < 2$ **(b)** $|x| > 2$ **(c)** $|x-3| < 1$ **(d)** $|x-3| \geq 1$

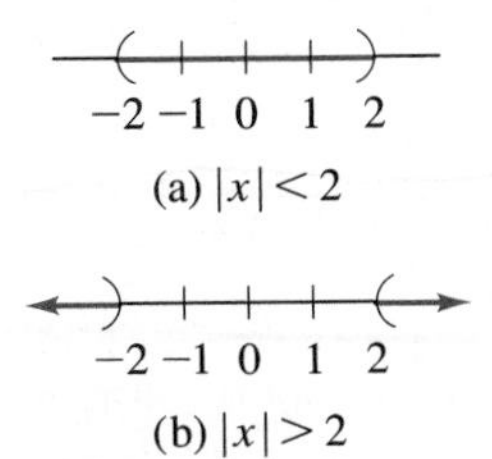

FIGURE 2

Solution
(a) The given inequality tells us that the distance from x to the origin is less than two units. So, as indicated in Figure 2(a), the point x must lie in the open interval $(-2, 2)$.
(b) The condition $|x| > 2$ means that x is more than two units from the origin. Thus, as indicated in Figure 2(b), the point x lies either to the right of 2 or to the left of -2.

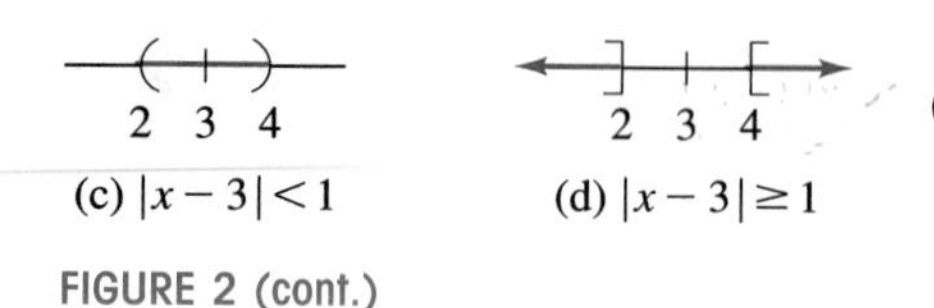

(c) $|x-3|<1$ (d) $|x-3|\geq 1$

FIGURE 2 (cont.)

(c) The given inequality tells us that x must be less than one unit away from 3 on the number line. Looking one unit to either side of 3, then, we see that x must lie between 2 and 4 and x cannot equal 2 or 4. See Figure 2(c).

(d) The inequality $|x-3| \geqslant 1$ says that x is at least one unit away from 3 on the number line. This means that either $x \geqslant 4$ or $x \leqslant 2$, as shown in Figure 2(d). [Here's an alternative way of thinking about this: The numbers satisfying the given inequality are precisely those numbers that *do not* satisfy the inequality in part (c). So for part (d), we need to shade that portion of the number line that was not shaded in part (c).]

EXERCISE SET 1.2

11, 13, 15, 21, 25–37 odd, 51–61 odd, 68

A

In Exercises 1–16, evaluate each expression.

1. $|3|$
2. $3+|-3|$
3. $|-6|$
4. $-6-|-6|$
5. $|-1+3|$
6. $|-6+3|$
7. $\left|-\frac{4}{5}\right|-\frac{4}{5}$
8. $\left|\frac{4}{5}\right|-\frac{4}{5}$
9. $|-6+2|-|4|$
10. $|-3-4|-|-4|$
11. $\big||-8|+|-9|\big|$
12. $\big||-8|-|-9|\big|$
13. $\left|\dfrac{27-5}{5-27}\right|$
14. $\dfrac{|27-5|}{|5-27|}$
15. $|7(-8)|-|7|\,|-8|$
16. $|(-7)^2|+|-7|^2-(-|-3|)^3$

In Exercises 17–24, evaluate each expression, given that $a=-2$, $b=3$, and $c=-4$.

17. $|a-b|^2$
18. $a^2-|bc|$
19. $|c|-|b|-|a|$
20. $|b+c|-|b|-|c|$
21. $|a+b|^2-|b+c|^2$
22. $\dfrac{|a|+|b|+|c|}{|a+b+c|}$
23. $\dfrac{a+b+|a-b|}{2}$
24. $\dfrac{a+b-|a-b|}{2}$

In Exercises 25–38, rewrite each expression in a form that does not contain absolute values.

25. $|\sqrt{2}-1|-1$
26. $|1-\sqrt{2}|+1$
27. $|x-3|$, where $x \geqslant 3$
28. $|x-3|$, where $x<3$
29. $|t^2+1|$
30. $|x^4+1|$
31. $|-\sqrt{3}-4|$
32. $|-\sqrt{3}-\sqrt{5}|$
33. $|x-3|+|x-4|$, where $x<3$
34. $|x-3|+|x-4|$, where $x>4$
35. $|x-3|+|x-4|$, where $3<x<4$
36. $|x-3|+|x-4|$, where $x=4$
37. $|x+1|+4|x+3|$, where $-\frac{5}{2}<x<-\frac{3}{2}$
38. $|x+1|+4|x+3|$, where $x<-3$

In Exercises 39–50, rewrite each statement using absolute values, as in Example 5.

39. The distance between x and 4 is 8.
40. The distance between x and 4 is at least 8.
41. The distance between x and 1 is 1/2.
42. The distance between x and 1 is less than 1/2.
43. The distance between x and 1 is at least 1/2.
44. The distance between x and 1 exceeds 1/2.
45. The distance between y and -4 is less than 1.
46. The distance between x^3 and -1 is at most 0.001.
47. The number y is less than three units from the origin.
48. The number y is less than one unit from the number t.
49. The distance between x^2 and a^2 is less than M.
50. The sum of the distances of a and b from the origin is greater than or equal to the distance of $a+b$ from the origin.

In Exercises 51–62, the set of real numbers satisfying the given inequality is one or more intervals on the number line. Show the intervals on a number line.

51. $|x|<4$
52. $|x|<2$
53. $|x|>1$
54. $|x|>0$
55. $|x-5|<3$
56. $|x-4|<4$
57. $|x-3| \leqslant 4$
58. $|x-1| \leqslant 1/2$
59. $\left|x+\frac{1}{3}\right|<\frac{3}{2}$
60. $\left|x+\frac{\pi}{2}\right|>1$
61. $|x-5| \geqslant 2$
62. $|x+5| \geqslant 2$

B

63. In parts (a) and (b), sketch the interval or intervals corresponding to the given inequality.
    (a) $|x-2|<1$ **(b)** $0<|x-2|<1$
    (c) In what way do your answers in (a) and (b) differ? (The distinction is important in the study of *limits* in calculus.)

64. Given two real numbers a and b, the notation **max(*a, b*)** denotes the larger of the two numbers. For instance, if $a = 1$ and $b = 2$, then $\max(1, 2) = 2$. In cases where $a = b$, then $\max(a, b)$ denotes the common value of a and b. It can be shown (see Exercise 69) that $\max(a, b)$ can be expressed in terms of absolute value as follows:

$$\max(a, b) = \frac{a + b + |a - b|}{2}$$

Verify this equation in each of the following cases:
(a) $a = 1$ and $b = 2$ **(b)** $a = -1$ and $b = -5$
(c) $a = b = 10$

65. Given two real numbers a and b, the notation **min(*a, b*)** denotes the smaller of the two numbers. In cases where $a = b$, then $\min(a, b)$ denotes the common value of a and b. It can be shown (see Exercise 70) that $\min(a, b)$ can be expressed in terms of absolute value as follows:

$$\min(a, b) = \frac{a + b - |a - b|}{2}$$

Verify this equation in each of the following cases:
(a) $a = 6$ and $b = 1$ **(b)** $a = 1$ and $b = -6$
(c) $a = b = -6$

66. Show that for all real numbers a and b, we have

$$|a| - |b| \leq |a - b|$$

Hint: Beginning with the identity $a = (a - b) + b$, take the absolute value of each side and then use the triangle inequality.

67. Show that

$$|a + b + c| \leq |a| + |b| + |c|$$

for all real numbers a, b, and c. *Hint:* The left-hand side can be written $|a + (b + c)|$. Now use the triangle inequality.

68. Explain why there are no real numbers that satisfy the equation $|x^2 + 4x| = -12$.

C

69. (As background for this exercise, you'll need to have worked Exercise 64.) Prove that

$$\max(a, b) = \frac{a + b + |a - b|}{2}$$

Hint: Consider three separate cases: $a = b$; $a > b$; and $b > a$.

70. (As background for this exercise, you'll need to have worked Exercise 65.) Prove that

$$\min(a, b) = \frac{a + b - |a - b|}{2}$$

71. Complete the following steps to prove the triangle inequality.
(a) Let a and b be real numbers. Which property in the summary box on page 8 tells us that $a \leq |a|$ and $b \leq |b|$?
(b) Add the two inequalities in part (a) to obtain $a + b \leq |a| + |b|$.
(c) In a similar fashion, add the two inequalities $-a \leq |a|$ and $-b \leq |b|$ and deduce that $-(a + b) \leq |a| + |b|$.
(d) Why do the results in parts (b) and (c) imply that $|a + b| \leq |a| + |b|$?

1.3 POLYNOMIALS AND FACTORING

The Cartesian use of letters near the beginning of the alphabet for parameters and those near the end as unknown quantities, the adaptation of exponential notation to these, and the use of the Germanic symbols + and − all combined to make Descartes' algebraic notation look like ours, for, of course, we took ours from him.

Carl B. Boyer, *A History of Mathematics,* 2nd ed., revised by Uta C. Merzbach (New York: John Wiley & Sons, Inc., 1991)

As background for our work on polynomials, we first review the terms *constant* and *variable.* By way of example, consider the familiar expression for the area of a circle of radius r, namely, πr^2. Here π is a constant; its value never changes throughout the discussion. On the other hand, r is a variable; we can substitute any positive number for r to obtain the area of a particular circle. More generally, by a **constant** we mean either a particular number (such as π, or -17, or $\sqrt{2}$) or a letter whose value remains fixed (although perhaps unspecified) throughout a given discussion. In contrast, a **variable** is a letter for which we can substitute any number selected from a given set of numbers. The given set of numbers is called the **domain** of the variable.

Some expressions will make sense only for certain values of the variable. For instance, $1/(x - 3)$ will be undefined when x is 3 (for then the denominator is zero). So in this case we would agree that the domain of the variable x consists of all real numbers except $x = 3$. Similarly, throughout this chapter we adopt the following convention:

THE DOMAIN CONVENTION

The domain of a variable in a given expression is the set of all real-number values of the variable for which the expression is defined.

In algebra it's customary (but certainly not mandatory) to use letters near the end of the alphabet for variables; letters from the beginning of the alphabet are generally used for constants. So, for example, in the expression $ax + b$, the letter x is the variable and a and b are constants.

EXAMPLE 1 Specify the variable, the constants, and the domain of the variable for each expression:

(a) $3x + 4$; **(b)** $\dfrac{1}{(t-1)(t+3)}$; **(c)** $ay^2 + by + c$; **(d)** $4x + 3x^{-1}$.

Solution

	VARIABLE	CONSTANTS	DOMAIN
(a) $3x + 4$	x	3, 4	the set of all real numbers
(b) $\dfrac{1}{(t-1)(t+3)}$	t	$1, -1, 3$	the set of all real numbers except $t = 1$ and $t = -3$
(c) $ay^2 + by + c$	y	a, b, c	the set of all real numbers
(d) $4x + 3x^{-1}$	x	4, 3	the set of all real numbers except $x = 0$ (Recall that $x^{-1} = 1/x$.)

The expressions in parts (a) and (c) of Example 1 are *polynomials*. By a **polynomial in *x*** we mean an expression of the form

$$a_n x^n + a_{n-1}x^{n-1} + \cdots + a_1 x + a_0$$

where n is a nonnegative integer and $a_n \neq 0$. The individual expressions $a_k x^k$ making up the polynomial are called **terms.** In this chapter, the **coefficients** a_k will always be real numbers. For example, the terms of the polynomial $x^2 - 7x + 3$ are x^2, $-7x$ and 3; the coefficients are 1, -7, and 3. In writing a polynomial, it's customary (but not mandatory) to write the terms in order of decreasing powers of x. For instance, we would usually write $x^2 - 7x + 3$ rather than $x^2 + 3 - 7x$. The highest power of x in a polynomial is called the **degree** of the polynomial. For example, the degree of the polynomial $x^2 - 7x + 3$ is 2. In the case of a polynomial consisting of only a nonzero constant a_0, we say that the degree is zero (because $a_0 = a_0 x^0$). No degree is defined for the polynomial whose only term is zero.

Some additional terminology is useful in describing polynomials that have only a few terms. A polynomial with only one term (such as $3x^2$) is a **monomial;** a polynomial with only two terms (such as $2x + 3$) is a **binomial;** and a polynomial with only three terms (such as $5x^2 - 6x + 3$) is a **trinomial.** The table that follows provides examples of this terminology.

Expression	Polynomial?	If Polynomial: Degree	Terms	Coefficients
$2x^3 - 3x^2 + 4x - 1$	yes	3	$2x^3, -3x^2, 4x, -1$	$2, -3, 4, -1$
$t^2 + 1$	yes	2	$t^2, 1$	$1, 1$
-12	yes	0	-12	-12
$2x^{-4} + 5$	no			
$\dfrac{1}{2x-3}$	no			
$\sqrt{4x^2+1}$	no			
$\sqrt{2}x + 1$	yes	1	$\sqrt{2}x, 1$	$\sqrt{2}, 1$

In your previous algebra courses, you learned to add, subtract, and multiply polynomials. In the multiplication of polynomials, several particular types of products occur so frequently that it is well worth your time to memorize the results. In the box that follows, we display these special products. (Exercises 9 and 10 will ask you to verify these formulas.) As you memorize these formulas, keep in mind that it's the *form,* or *pattern,* that is important—not the specific choice of letters.

PROPERTY SUMMARY — SPECIAL PRODUCTS

1. $(A - B)(A + B) = A^2 - B^2$
2. **(a)** $(A + B)^2 = A^2 + 2AB + B^2$
   (b) $(A - B)^2 = A^2 - 2AB + B^2$
3. **(a)** $(A + B)^3 = A^3 + 3A^2B + 3AB^2 + B^3$
   (b) $(A - B)^3 = A^3 - 3A^2B + 3AB^2 - B^3$
4. **(a)** $(A + B)(A^2 - AB + B^2) = A^3 + B^3$
   (b) $(A - B)(A^2 + AB + B^2) = A^3 - B^3$ } Here

EXAMPLE 2 Use the Special Products to compute each product.

(a) $(5xy - 4)(5xy + 4)$ **(b)** $(3\sqrt{xy} - z)(3\sqrt{xy} + z)$
(c) $(4x^3 - 3)^2$ **(d)** $(2x + 5)^3$
(e) $(x - 2)(x^2 + 2x + 4)$

Solution

(a) $(5xy - 4)(5xy + 4) = (5xy)^2 - 4^2 = 25x^2y^2 - 16$ using Special Product 1

(b) $(3\sqrt{xy} - z)(3\sqrt{xy} + z) = (3\sqrt{xy})^2 - z^2 = 9xy - z^2$ using Special Product 1

(c) $(4x^3 - 3)^2 = (4x^3)^2 - 2(4x^3)(3) + 3^2 = 16x^6 - 24x^3 + 9$ Special Product 2(b)

(d) $(2x + 5)^3 = (2x)^3 + 3(2x)^2(5) + 3(2x)(5)^2 + (5)^3 = 8x^3 + 60x^2 + 150x + 125$ Special Product 3(a)

(e) $(x - 2)(x^2 + 2x + 4) = x^3 - 2^3 = x^3 - 8$ Special Product 4(b)

There are many cases in algebra in which the process of *factoring* simplifies the work at hand. To **factor** a polynomial means to write it as a product of two or more nonconstant polynomials. For instance, a factorization of $x^2 - 9$ is given by

$$x^2 - 9 = (x - 3)(x + 3)$$

In this case, $x - 3$ and $x + 3$ are the **factors** of $x^2 - 9$.

There is one convention that we need to agree on at the outset. If the polynomial or expression that we wish to factor contains only integer coefficients, then the factors (if any) should involve only integer coefficients. For example, according to this convention, we will not consider the following type of factorization in this section:

$$x^2 - 2 = (x - \sqrt{2})(x + \sqrt{2})$$

because it involves coefficients that are irrational numbers. (We should point out, however, that factorizations such as this are useful at times. We'll make use of this type of factorization in Section 12.8.) As it happens, $x^2 - 2$ is an example of a polynomial that cannot be factored using integer coefficients. We say in such a case that the polynomial is **irreducible over the integers.**

If the coefficients of a polynomial are rational numbers, then we do allow factors with rational coefficients. For instance, the factorization of $y^2 - \frac{1}{4}$ over the rational numbers is given by

$$y^2 - \frac{1}{4} = \left(y - \frac{1}{2}\right)\left(y + \frac{1}{2}\right)$$

We'll consider five techniques for factoring in this section. (These techniques will be used throughout the text when we simplify expressions and when we solve equations.) In Table 1, we summarize these techniques. Notice that three of the formulas in the table are just restatements of Special Product formulas. Remember, it is the form or pattern in the formula that is important—not the specific choice of letters.

TABLE 1 Basic Factoring Techniques

Technique	Example or Formula	Remark
Common factor	$3x^4 + 6x^3 - 12x^2 = 3x^2(x^2 + 2x - 4)$ $4(x^2 + 1) - x(x^2 + 1) = (x^2 + 1)(4 - x)$	In any factoring problem, the first step always is to look for the common factor of highest degree.
Difference of squares	$x^2 - a^2 = (x - a)(x + a)$	There is no corresponding formula for a sum of squares; $x^2 + a^2$ is irreducible over the integers.
Trial and error	$x^2 + 2x - 3 = (x + 3)(x - 1)$	In this example, the only possibilities, or trials, are: **(a)** $(x - 3)(x - 1)$ **(b)** $(x - 3)(x + 1)$ **(c)** $(x + 3)(x - 1)$ **(d)** $(x + 3)(x + 1)$ By inspection or by carrying out the indicated multiplications, we find that only case (c) checks.
Difference of cubes Sum of cubes	$x^3 - a^3 = (x - a)(x^2 + ax + a^2)$ $x^3 + a^3 = (x + a)(x^2 - ax + a^2)$	Verify these formulas for yourself by carrying out the multiplications. Then memorize the formulas.
Grouping	$x^3 - x^2 + x - 1 = (x^3 - x^2) + (x - 1)$ $= x^2(x - 1) + (x - 1)\cdot 1$ $= (x - 1)(x^2 + 1)$	This is actually an application of the common factor technique.

The idea in factoring is to use one or more of these techniques until each of the factors obtained is irreducible. The examples that follow show how this works in practice.

EXAMPLE 3 Factor: **(a)** $x^2 - 49$; **(b)** $(2a - 3b)^2 - 49$.

Solution
(a) $x^2 - 49 = x^2 - 7^2$
$= (x - 7)(x + 7)$ difference of squares

(b) Notice that the pattern or form is the same as in part (a): it's a difference of squares. We have

$$\begin{aligned}(2a - 3b)^2 - 49 &= (2a - 3b)^2 - 7^2 \\ &= [(2a - 3b) - 7][(2a - 3b) + 7] \quad \text{difference of squares} \\ &= (2a - 3b - 7)(2a - 3b + 7)\end{aligned}$$

EXAMPLE 4 Factor: **(a)** $2x^3 - 50x$; **(b)** $3x^5 - 3x$.

Solution
(a) $2x^3 - 50x = 2x(x^2 - 25)$ common factor
$= 2x(x - 5)(x + 5)$ difference of squares

(b) $3x^5 - 3x = 3x(x^4 - 1)$ common factor
$= 3x[(x^2)^2 - 1^2]$
$= 3x(x^2 - 1)(x^2 + 1)$ difference of squares
$= 3x(x - 1)(x + 1)(x^2 + 1)$ difference of squares, again

EXAMPLE 5 Factor: **(a)** $x^2 - 4x - 5$; **(b)** $(a + b)^2 - 4(a + \text{b}) - 5$.

Solution
(a) $x^2 - 4x - 5 = (x - 5)(x + 1)$ trial and error

(b) Note that the form of this expression is

$$(\quad)^2 - 4(\quad) - 5$$

This is the same *form* as the expression in part (a). So we need only replace x with the quantity $a + b$ in the solution for part (a). This yields

$$\begin{aligned}(a + b)^2 - 4(a + b) - 5 &= [(a + b) - 5][(a + b) + 1] \\ &= (a + b - 5)(a + b + 1)\end{aligned}$$

EXAMPLE 6 Factor: $2z^4 + 9z^2 + 4$.

Solution We use trial and error to look for a factorization of the form $(2z^2 + ?)(z^2 + ?)$. There are three possibilities:

(i) $(2z^2 + 2)(z^2 + 2)$ **(ii)** $(2z^2 + 1)(z^2 + 4)$ **(iii)** $(2z^2 + 4)(z^2 + 1)$

Each of these yields the appropriate first term and last term, but (after checking) only possibility (ii) yields $9z^2$ for the middle term. The required factorization is then $2z^4 + 9z^2 + 4 = (2z^2 + 1)(z^2 + 4)$.

QUESTION Why didn't we consider any possibilities with subtraction signs in place of addition signs?

EXAMPLE 7 Factor: **(a)** $x^2 + 9$; **(b)** $x^2 + 2x + 3$.

Solution

(a) The expression $x^2 + 9$ is irreducible over the integers. (This can be discovered by trial and error.) If the given expression had instead been $x^2 - 9$, then it could have been factored as a difference of squares. Sums of squares, however, cannot in general be factored over the integers.

(b) The expression $x^2 + 2x + 3$ is irreducible over the integers. (Check this for yourself by trial and error.)

EXAMPLE 8 Factor: $x^2 - y^2 + 10x + 25$.

Solution Familiarity with the special products suggests that we try grouping the terms this way:

$$(x^2 + 10x + 25) - y^2$$

Then we have

$$\begin{aligned} x^2 - y^2 + 10x + 25 &= (x^2 + 10x + 25) - y^2 \\ &= (x+5)^2 - y^2 \\ &= [(x+5) - y][(x+5) + y] \qquad \text{difference of squares} \\ &= (x - y + 5)(x + y + 5) \end{aligned}$$

EXAMPLE 9 Factor: $ax + ay^2 + bx + by^2$.

Solution We factor a from the first two terms and b from the second two, to obtain

$$ax + ay^2 + bx + by^2 = a(x + y^2) + b(x + y^2)$$

We now recognize the quantity $x + y^2$ as a common expression that can be factored out. We have then

$$a(x + y^2) + b(x + y^2) = (x + y^2)(a + b)$$

The required factorization is therefore

$$ax + ay^2 + bx + by^2 = (x + y^2)(a + b)$$

As you may wish to check for yourself, this factorization can also be obtained by trial and error.

EXAMPLE 10 Factor: **(a)** $t^3 - 125$; **(b)** $8 + (a-2)^3$.

Solution

(a)
$$\begin{aligned} t^3 - 125 &= t^3 - 5^3 \\ &= (t-5)(t^2 + 5t + 25) \qquad \text{difference of cubes} \end{aligned}$$

(b)
$$\begin{aligned} 8 + (a-2)^3 &= 2^3 + (a-2)^3 \\ &= [2 + (a-2)][2^2 - 2(a-2) + (a-2)^2] \qquad \text{sum of cubes} \\ &= a(4 - 2a + 4 + a^2 - 4a + 4) \\ &= a(a^2 - 6a + 12) \end{aligned}$$

As you can check now, the expression $a^2 - 6a + 12$ is irreducible over the integers. Therefore the required factorization is $a(a^2 - 6a + 12)$.

For some calculations (particularly in calculus), it's helpful to be able to factor an expression involving fractional exponents. For instance, suppose (as in the next example), we want to factor the expression

$$x(2x-1)^{-1/2}+(2x-1)^{3/2}$$

The common expression to factor out here is $(2x-1)^{-1/2}$; the technique is to *choose the expression with the smaller exponent*. *Note:* For a review of fractional exponents, see Section A.6 in the Appendix.

EXAMPLE 11 Factor: $x(2x-1)^{-1/2}+(2x-1)^{3/2}$.

Solution

$$\begin{aligned} x(2x-1)^{-1/2}+(2x-1)^{3/2} &= (2x-1)^{-1/2}[x+(2x-1)^{3/2+1/2}] \\ &= (2x-1)^{-1/2}[x+(2x-1)^{2}] \\ &= (2x-1)^{-1/2}(4x^2-3x+1) \end{aligned}$$

We conclude this section with examples of four common errors to avoid in factoring. The first two errors (in the box that follows) are easy to detect; simply multiplying out the supposed factorizations shows that they do not check. The third error can result from a lack of familiarity with the basic factoring techniques listed on page 14. The fourth error indicates a misunderstanding of what is required in factoring, namely, the final quantity must be expressed as a *product* of terms or expressions, not a sum.

ERRORS TO AVOID

ERROR	CORRECTION	COMMENT
$x^2+6x+9 \neq (x-3)^2$	$x^2+6x+9=(x+3)^2$	Check the middle term.
$x^2+64 \neq (x+8)(x+8)$	x^2+64 is irreducible over the integers.	A sum of squares is, in general, irreducible over the integers. (A difference of squares, however, can always be factored.)
x^3+64 is irreducible over the integers.	$x^3+64=(x+4)(x^2-4x+16)$	Although a sum of squares is irreducible, a sum of cubes can be factored.
x^2-2x+3 factors as $x(x-2)+3$	x^2-2x+3 is irreducible over the integers.	The polynomial x^2-2x+3 is the *sum* of the expression $x(x-2)$ and the constant 3. But, by definition, the factored form of a polynomial must be a *product* (of two or more nonconstant polynomials).

EXERCISE SET 1.3

A

In Exercises 1–6, specify the domain of each variable.

1. **(a)** $2x^2-3x+5$ **(b)** $2x^{1/2}-3x+5$
2. **(a)** $y-1$ **(b)** $2y/(y-1)$
3. **(a)** $ax+b$ **(b)** $ax^{1/3}+b$
4. **(a)** $x+x^{-1}$ **(b)** $t^{-1}+2t^{-2}$
5. **(a)** $4\sqrt{t}+t^2$ **(b)** $\dfrac{4}{(t-1)\sqrt{t}}$
6. **(a)** $\dfrac{1}{x}$ **(b)** $\dfrac{1}{(x-1)(x+2)(x-3)}$

For Exercises 7 and 8, specify the degree and the (nonzero) coefficients of each polynomial.

7. **(a)** 4 **(b)** $4x^3$ **(c)** $x^6 + 4x^3 - x^2 + 2$

8. **(a)** 0 **(b)** $1 - x + 6x^5$ **(c)** $\sqrt{2}x^2 - 2\sqrt{2}$

In Exercises 9 and 10, you are asked to verify the special products in the box on page 13. As an example, here is a verification for Special Product 4(a).

EXAMPLE

$$\begin{aligned}(A+B)(A^2-AB+B^2) &= A(A^2-AB+B^2) \\ &\quad + B(A^2-AB+B^2) \\ &= A^3 - A^2B + AB^2 \\ &\quad + A^2B - AB^2 + B^3 \\ &= A^3 + (AB^2 - AB^2) \\ &\quad + (A^2B - A^2B) + B^3 \\ &= A^3 + 0 + 0 + B^3 = A^3 + B^3\end{aligned}$$

9. Verify the Special Products 1, 2(a), and 2(b) given on page 13.

10. Verify the Special Products 3(a), 3(b), and 4(b) given on page 13.

For Exercises 11–22, use the Special Products formulas to compute the products.

11. **(a)** $(x - y)(x + y)$ **(b)** $(x^2 - 5)(x^2 + 5)$

12. **(a)** $(3y - a^2)(3y + a^2)$
(b) $(3\sqrt{y} - a^2)(3\sqrt{y} + a^2)$

13. **(a)** $(A - 4)(A + 4)$
(b) $[(a + b) - 4][(a + b) + 4]$

14. **(a)** $(\sqrt{ab} + \sqrt{c})(\sqrt{ab} - \sqrt{c})$
(b) $(a^{1/2} + b^{1/2})(a^{1/2} - b^{1/2})$

15. **(a)** $(x - 8)^2$ **(b)** $(2x^2 - 5)^2$

16. **(a)** $(2^m + 1)^2$ **(b)** $(a^m - a^{-m})^2$

17. **(a)** $(\sqrt{x} + \sqrt{y})^2$ **(b)** $(\sqrt{x + y} - \sqrt{x})^2$

18. **(a)** $(2x + y)^3$ **(b)** $(x - 2y)^3$

19. **(a)** $(a + 1)^3$ **(b)** $(3x^2 - 2a^2)^3$

20. **(a)** $(x - y)(x^2 + xy + y^2)$
(b) $(x^2 - y^2)(x^4 + x^2y^2 + y^4)$

21. **(a)** $(x + 1)(x^2 - x + 1)$
(b) $(x^2 + 1)(x^4 - x^2 + 1)$

22. **(a)** $(5 - y)(25 + 5y + y^2)$
(b) $(3 + x)(9 - 3x + x^2)$

In Exercises 23–82, factor each polynomial or expression. If a polynomial is irreducible, state this. (In Exercises 23–28, the factoring techniques are specified.)

23. (Common factor and difference of squares)
(a) $x^2 - 64$ **(b)** $7x^4 + 14x^2$
(c) $121z - z^3$ **(d)** $a^2b^2 - c^2$

24. (Common factor and difference of squares)
(a) $1 - t^4$ **(b)** $x^6 + x^5 + x^4$
(c) $u^2v^2 - 225$ **(d)** $81x^4 - x^2$

25. (Trial and error)
(a) $x^2 + 2x - 3$ **(b)** $x^2 - 2x - 3$
(c) $x^2 - 2x + 3$ **(d)** $-x^2 + 2x + 3$

26. (Trial and error)
(a) $2x^2 - 7x - 4$ **(b)** $2x^2 + 7x - 4$
(c) $2x^2 + 7x + 4$ **(d)** $-2x^2 - 7x + 4$

27. (Sum or difference of cubes)
(a) $x^3 + 1$ **(b)** $x^3 + 216$
(c) $1000 - 8x^6$ **(d)** $64a^3x^3 - 125$

28. (Grouping)
(a) $x^4 - 2x^3 + 3x - 6$
(b) $a^2x + bx - a^2z - bz$

29. **(a)** $144 - x^2$
(b) $144 + x^2$
(c) $144 - (y - 3)^2$

30. **(a)** $4a^2b^2 + 9c^2$
(b) $4a^2b^2 - 9c^2$
(c) $4a^2b^2 - 9(ab + c)^2$

31. **(a)** $h^3 - h^5$
(b) $100h^3 - h^5$
(c) $100(h + 1)^3 - (h + 1)^5$

32. **(a)** $x^4 - x^2$
(b) $3x^4 - 48x^2$
(c) $3(x + h)^4 - 48(x + h)^2$

33. **(a)** $x^2 - 13x + 40$
(b) $x^2 - 13x - 40$

34. **(a)** $x^2 + 10x + 16$
(b) $x^2 - 10x + 16$

35. **(a)** $x^2 + 5x - 36$
(b) $x^2 - 13x + 36$

36. **(a)** $x^2 - x + 6$
(b) $x^2 + x - 6$

37. **(a)** $3x^2 - 22x - 16$
(b) $3x^2 - x - 16$

38. **(a)** $x^2 - 4x + 1$
(b) $x^2 - 4x - 3$

39. **(a)** $6x^2 + 13x - 5$
(b) $6x^2 - x - 5$

40. **(a)** $16x^2 + 18x - 9$
(b) $16x^2 - 143x - 9$

41. **(a)** $t^4 + 2t^2 + 1$
(b) $t^4 - 2t^2 + 1$
(c) $t^4 - 2t^2 - 1$

42. **(a)** $t^4 - 9t^2 + 20$
(b) $t^4 - 19t^2 + 20$
(c) $t^4 - 19t^2 - 20$

43. **(a)** $4x^3 - 20x^2 - 25x$
(b) $4x^3 - 20x^2 + 25x$

44. **(a)** $x^3 + x^2 + x$
(b) $x^3 + 2x^2 + x$

45. **(a)** $ab - bc + a^2 - ac$
(b) $(u + v)x - xy + (u + v)^2 - (u + v)y$

46. **(a)** $3(x + 5)^3 + 2(x + 5)^2$
(b) $a(x + 5)^3 + b(x + 5)^2$

47. $x^2z^2 + xzt + xyz + yt$

48. $a^2t^2 + b^2t^2 - cb^2 - ca^2$

49. $a^4 - 4a^2b^2c^2 + 4b^4c^4$

50. $A^2 - B^2 + 16A + 64$

51. $A^2 + B^2$

52. $x^2 + 64$

53. $x^3 + 64$

54. $27 - (a - b)^3$

55. $(x + y)^3 - y^3$

56. $(a + b)^3 - 8c^3$

57. $x^3 - y^3 + x - y$

58. $8a^3 + 27b^3 + 2a + 3b$

59. **(a)** $p^4 - 1$ **(b)** $p^8 - 1$

60. **(a)** $p^4 + 4$ **(b)** $p^4 - 4$

61. $x^3 + 3x^2 + 3x + 1$

62. $-1 + 6x - 12x^2 + 8x^3$

63. $x^2 + 16y^2$

64. $4u^2 + 25v^2$

65. $(25/16) - c^2$

66. $(81/4) - y^2$

67. $z^4 - (81/16)$

68. $\dfrac{(a+b)^2}{4} - \dfrac{a^2b^2}{9}$

69. $\dfrac{125}{m^3n^3} - 1$

70. $\dfrac{x^3}{8} - \dfrac{512}{x^3}$

71. $\frac{1}{4}x^2 + xy + y^2$

72. $x^2 + x + 1$

73. $64(x - a)^3 - x + a$

74. $64(x - a)^4 - x + a$

75. $ax^2 + (a + b)x + b$

76. $(5a^2 - 11a + 10)^2 - (4a^2 - 15a + 6)^2$

77. $(x + 1)^{1/2} - (x + 1)^{3/2}$

78. $(x^2 + 1)^{3/2} + (x^2 + 1)^{7/2}$

79. $(x + 1)^{-1/2} - (x + 1)^{-3/2}$

80. $(x^2 + 1)^{-2/3} + (x^2 + 1)^{-5/3}$

81. $(2x + 3)^{1/2} - \frac{1}{3}(2x + 3)^{3/2}$

82. $(ax + b)^{-1/2} - \sqrt{ax + b}/b$

B

In Exercises 83–87, factor each expression.

83. $A^3 + B^3 + 3AB(A + B)$

84. $p^3 - q^3 - p(p^2 - q^2) + q(p - q)^2$

85. $2x(a^2 + x^2)^{-1/2} - x^3(a^2 + x^2)^{-3/2}$

86. $\frac{1}{2}(x - a)^{-1/2}(x + a)^{-1/2} - \frac{1}{2}(x + a)^{1/2}(x - a)^{-3/2}$

87. $y^4 - (p + q)y^3 + (p^2q + pq^2)y - p^2q^2$

88. **(a)** Factor: $x^4 + 2x^2y^2 + y^4$.
(b) Factor: $x^4 + x^2y^2 + y^4$. *Hint:* Add and subtract a term. [Keep part (a) in mind.]
(c) Factor $x^6 - y^6$ as a difference of squares.
(d) Factor $x^6 - y^6$ as a difference of cubes. [Use the result in part (b) to obtain the same answer as in part (c).]

89. The expression $\dfrac{x^2 - 16}{x - 4}$ is undefined when $x = 4$. In this exercise we investigate the values of the expression when x is very close to 4.
(a) Complete the tables at the top of the next column.

x	$\dfrac{x^2-16}{x-4}$
3.9	
3.99	
3.999	
3.9999	
3.99999	

x	$\dfrac{x^2-16}{x-4}$
4.1	
4.01	
4.001	
4.0001	
4.00001	

(b) On the basis of the tables in part (a), what value does the expression $\dfrac{x^2 - 16}{x - 4}$ seem to be approaching as x gets closer and closer to 4? This "target value" is referred to as the *limit* of $\dfrac{x^2 - 16}{x - 4}$ as x approaches 4. (The notion of a limit is made more precise in calculus.)
(c) How could factoring have been used to obtain this limit without the work in part (a)?

90. The expression $\dfrac{x^3 - 8}{x - 2}$ is undefined when $x = 2$. In this exercise, we investigate the values of the expression when x is very close to 2.
(a) Complete the following tables.

x	$\dfrac{x^3-8}{x-2}$
1.9	
1.99	
1.999	
1.9999	
1.99999	

x	$\dfrac{x^3-8}{x-2}$
2.1	
2.01	
2.001	
2.0001	
2.00001	

(b) On the basis of the tables in part (a), what "target value," or limit, does the expression $\dfrac{x^3 - 8}{x - 2}$ seem to be approaching as x gets closer and closer to 2?
(c) How could factoring have been used to obtain this limit without the work in part (a)?

C

In Exercises 91–96, factor the expressions.

91. $x^4 + 64$ *Hint:* Add and subtract the term $16x^2$.

92. $x^4 - 15x^2 + 9$ *Hint:* Add and subtract a term.

93. $(x + y)^2 + (x + z)^2 - (z + t)^2 - (y + t)^2$

94. $(a - a^2)^3 + (a^2 - 1)^3 + (1 - a)^3$

95. $(b - c)^3 + (c - a)^3 + (a - b)^3$

96. $(a + b + c)^3 - a^3 - b^3 - c^3$

1.4 QUADRATIC EQUATIONS

The formula for the quadratic equation seems first to have been discovered by the Moslems around 900 A.D., although quadratic equations had been solved by the Babylonians 3000 years earlier.

Charles Robert Hadlock in *Field Theory and Its Classical Problems* (Washington, D.C.: Mathematical Association of America, 1978)

An equation that can be written in the form

$$ax^2 + bx + c = 0, \qquad \text{where } a, b, \text{ and } c \text{ are real numbers and } a \neq 0$$

is called a **quadratic equation.** Examples of quadratic equations are

$$x^2 - 8x - 9 = 0 \qquad 2y^2 - 5 = 0 \qquad 3m^2 = 1 - m$$

Recall that a **solution,** or **root,** of an equation is a value for the unknown that makes the equation a true statement. For example, the value $x = -1$ is a solution of the quadratic equation $x^2 - 8x - 9 = 0$, because when x is replaced by -1 in this equation, we have

$$\begin{aligned}(-1)^2 - 8(-1) - 9 &= 0\\ 1 + 8 - 9 &= 0\\ 0 &= 0 \qquad \text{True}\end{aligned}$$

When applicable, one of the simplest techniques for solving quadratic equations involves factoring. This method relies on the following familiar and important property of the real-number system.

PROPERTY SUMMARY — ZERO-PRODUCT PROPERTY OF REAL NUMBERS

$$pq = 0 \quad \text{if and only if} \quad p = 0 \text{ or } q = 0 \quad \text{(or both)}$$

For example, to solve the equation $x^2 - 2x - 3 = 0$ by factoring, we have

$$\begin{aligned}x^2 - 2x - 3 &= 0\\ (x - 3)(x + 1) &= 0\end{aligned}$$

$$x - 3 = 0 \quad\Big|\quad x + 1 = 0$$
$$x = 3 \quad\Big|\quad x = -1$$

As you can easily check, the values $x = 3$ and $x = -1$ both satisfy the given equation.

The zero-product property remains valid when we have a product of three or more factors equal to zero. We use this fact in the next example.

EXAMPLE 1 Solve each equation.

(a) $6x^3 - x^2 - 2x = 0$ **(b)** $y^3 + 3y^2 - y - 3 = 0$

Solution

(a) We can factor out the common term x from the left-hand side of the given equation. This yields

$$\begin{aligned}x(6x^2 - x - 2) &= 0\\ x(2x + 1)(3x - 2) &= 0\end{aligned}$$

$$x = 0 \quad\Big|\quad 2x + 1 = 0 \quad\Big|\quad 3x - 2 = 0$$
$$\quad\Big|\quad x = -1/2 \quad\Big|\quad x = 2/3$$

The solutions of the equation are then $x = 0$, $x = -1/2$, and $x = 2/3$.

(b)

$$\begin{aligned} y^3 + 3y^2 - y - 3 &= 0 \\ y^2(y+3) - (y+3) &= 0 \\ (y+3)(y^2-1) &= 0 \\ (y+3)(y-1)(y+1) &= 0 \end{aligned}$$

$y + 3 = 0$	$y - 1 = 0$	$y + 1 = 0$
$y = -3$	$y = 1$	$y = -1$

As you can now easily check, each of the three values $y = -3$, $y = 1$, and $y = -1$ satisfies the given equation.

In Example 1 we solved two equations by using the elementary factoring techniques from Section 1.3. Of course, not all equations are so readily solved by factoring. Consider, for example, the quadratic equation $x^2 - 2x - 4 = 0$. There are only three possible factorizations with integer coefficients, but none yields the appropriate middle term, $-2x$, when multiplied out:

$$(x-4)(x+1) \qquad (x+4)(x-1) \qquad (x-2)(x+2)$$

Moreover, it is not always obvious whether a particular equation, such as $x^2 + 156x + 5963 = 0$, can be solved by factoring. (We will come back to this equation in Exercise 39.)

The technique of *completing the square* provides a systematic approach for solving quadratic equations. We'll demonstrate this technique by solving the equation $x^2 - 2x - 4 = 0$. First, we rewrite the equation in the form

$$x^2 - 2x = 4 \tag{1}$$

with the x-terms isolated on the left-hand side of the equation. To complete the square, we follow these two steps:

STEP 1 Take half of the coefficient of x and square it.

STEP 2 Add the number obtained in Step 1 to both sides of the equation.

For equation (1), the coefficient of x is -2. Taking half of -2 and then squaring it gives us $(-1)^2$, or 1. Now, as directed in Step 2, we add 1 to both sides of equation (1). This yields

$$\begin{aligned} x^2 - 2x + 1 &= 4 + 1 \\ (x-1)^2 &= 5 \\ x - 1 &= \pm\sqrt{5} \\ x &= 1 \pm \sqrt{5} \end{aligned}$$

We have now obtained two solutions, $1 + \sqrt{5}$ and $1 - \sqrt{5}$. (Exercise 6 will ask you to verify that these values indeed satisfy the original equation.) In the box that follows, we summarize the technique of completing the square, and we indicate how the process got its name.

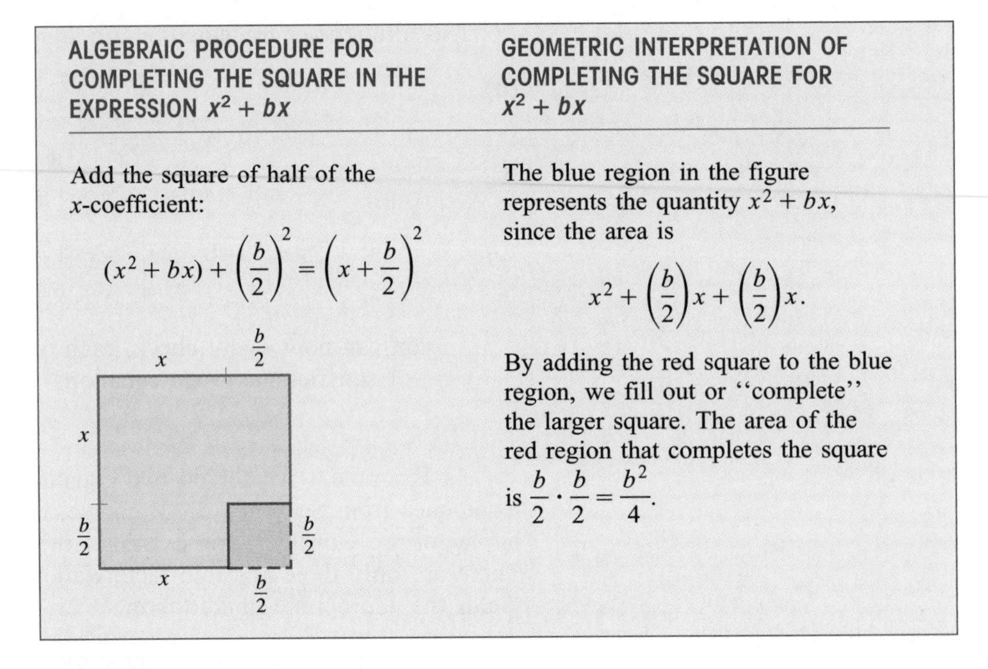

ALGEBRAIC PROCEDURE FOR COMPLETING THE SQUARE IN THE EXPRESSION $x^2 + bx$

Add the square of half of the x-coefficient:

$$(x^2 + bx) + \left(\frac{b}{2}\right)^2 = \left(x + \frac{b}{2}\right)^2$$

GEOMETRIC INTERPRETATION OF COMPLETING THE SQUARE FOR $x^2 + bx$

The blue region in the figure represents the quantity $x^2 + bx$, since the area is

$$x^2 + \left(\frac{b}{2}\right)x + \left(\frac{b}{2}\right)x.$$

By adding the red square to the blue region, we fill out or "complete" the larger square. The area of the red region that completes the square is $\frac{b}{2} \cdot \frac{b}{2} = \frac{b^2}{4}$.

Rather than repeat the process of completing the square for each new equation to be solved, we can instead use a general formula for solving any quadratic equation. The derivation of this **quadratic formula** runs as follows. We start with the general quadratic equation $ax^2 + bx + c = 0$ $(a \neq 0)$. In order to use the procedure for completing the square described in the preceding box, we first divide both sides of the given equation by a, so that the coefficient of x^2 is 1:

$$x^2 + \frac{b}{a}x + \frac{c}{a} = 0$$

Subtracting c/a from both sides yields

$$x^2 + \frac{b}{a}x = -\frac{c}{a}$$

Now, to complete the square, we add $[\frac{1}{2}(b/a)]^2$, or $b^2/4a^2$, to both sides. That gives us

$$x^2 + \frac{b}{a}x + \frac{b^2}{4a^2} = \frac{b^2}{4a^2} - \frac{c}{a}$$

$$\left(x + \frac{b}{2a}\right)^2 = \frac{b^2 - 4ac}{4a^2}$$

$$x + \frac{b}{2a} = \pm\sqrt{\frac{b^2 - 4ac}{4a^2}} = \pm\frac{\sqrt{b^2 - 4ac}}{2|a|}$$

$$= \pm\frac{\sqrt{b^2 - 4ac}}{2a}$$

This last equality follows from the fact that for any real number $a \neq 0$, the expressions $\pm 2|a|$ and $\pm 2a$ both represent the same two numbers. We now conclude that the solutions are

$$x = -\frac{b}{2a} + \frac{\sqrt{b^2 - 4ac}}{2a} \quad \text{and} \quad x = -\frac{b}{2a} - \frac{\sqrt{b^2 - 4ac}}{2a}$$

These solutions are usually written in the more compact form displayed in the following box.

THE QUADRATIC FORMULA

The solutions of the equation $ax^2 + bx + c = 0$ $(a \neq 0)$ are given by

$$x = \frac{-b \pm \sqrt{b^2 - 4ac}}{2a}$$

EXAMPLE 2 Use the quadratic formula to solve $2x^2 = 4 - x$.

Solution We first rewrite the equation $2x^2 + x - 4 = 0$ so that it has the form $ax^2 + bx + c = 0$. By comparing these last two equations, we see that $a = 2$, $b = 1$, and $c = -4$. Therefore,

$$x = \frac{-b \pm \sqrt{b^2 - 4ac}}{2a} = \frac{-1 \pm \sqrt{1^2 - 4(2)(-4)}}{2(2)} = \frac{-1 \pm \sqrt{33}}{4}$$

Thus, the two solutions are $\frac{-1 + \sqrt{33}}{4}$ and $\frac{-1 - \sqrt{33}}{4}$.

EXAMPLE 3 Figure 1 shows a right circular cylinder with radius r and height h. The formula for the total surface area S of the cylinder is

$$S = 2\pi r^2 + 2\pi rh \tag{2}$$

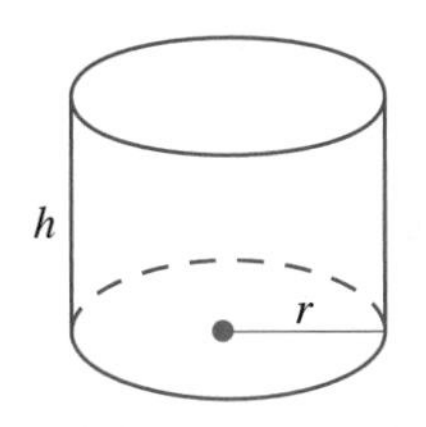

FIGURE 1

Solve equation (2) for r in terms of h and S.

Solution By rewriting equation (2) in the following form, we can see that the equation is quadratic in r, and we can determine what values to use for a, b, and c in the quadratic formula:

$$(2\pi)r^2 + (2\pi h)r - S = 0$$

We have $a = 2\pi$, $b = 2\pi h$, and $c = -S$, therefore

$$r = \frac{-2\pi h \pm \sqrt{4\pi^2h^2 - 4(2\pi)(-S)}}{4\pi} = \frac{-2\pi h \pm 2\sqrt{\pi^2h^2 + 2\pi S}}{4\pi}$$
$$= \frac{-\pi h \pm \sqrt{\pi^2h^2 + 2\pi S}}{2\pi}$$

This last expression gives us two solutions to equation (2). In the present context, however, the radius r cannot be negative, so we discard the "minus" option within the plus-or-minus sign, since that would lead to a negative value for r. Thus we have

$$r = \frac{-\pi h + \sqrt{\pi^2h^2 + 2\pi S}}{2\pi}$$

QUESTION How do we know that this last expression is indeed positive?

In Examples 2 and 3 we found that each quadratic equation had two real roots (although in Example 3 we were interested in only the positive real root). As the next example indicates, however, it is also possible for quadratic equations to have only one real root, or even no real roots. [In Example 4(b) we make use of the complex number system; if you need a review of this topic, see Section 12.1.]

EXAMPLE 4 Use the quadratic formula to solve each equation:

(a) $4x^2 - 12x + 9 = 0$; **(b)** $x^2 + x + 2 = 0$.

Solution

(a) Here $a = 4$, $b = -12$, and $c = 9$, so

$$x = \frac{12 \pm \sqrt{(-12)^2 - 4(4)(9)}}{2(4)} = \frac{12 \pm \sqrt{144 - 144}}{8}$$

$$= \frac{12 \pm 0}{8} = \frac{12}{8} = \frac{3}{2}$$

In this case our only solution is 3/2. We refer to 3/2 as a **double root.** *Note:* We've used the quadratic formula here only for an illustration; in this case we can solve the given equation more efficiently by factoring. This yields

$$(2x - 3)(2x - 3) = 0$$

Both factors lead us directly to the root $x = 3/2$. (This is why 3/2 is called a double root.)

(b) Since $a = 1$, $b = 1$, and $c = 2$, we have

$$x = \frac{-1 \pm \sqrt{(1)^2 - 4(1)(2)}}{2(1)} = \frac{-1 \pm \sqrt{-7}}{2}$$

Now, if we confine ourselves to the real-number system, the equation has no solutions, because the expression $\sqrt{-7}$ is undefined within the real-number system. However, if we take a broader point of view and work within the complex number system, then there are indeed two solutions. As you'll see in Section 12.1 (or as you perhaps already know from a previous course) these two nonreal complex solutions can be written

$$\frac{-1 + i\sqrt{7}}{2} \quad \text{and} \quad \frac{-1 - i\sqrt{7}}{2}$$

The quantity $b^2 - 4ac$ that appears under the radical sign in the quadratic formula is called the *discriminant.* In Examples 2 and 3, the discriminant was positive and, consequently, each equation had two real solutions. In Example 4(a), in which the discriminant was zero, we obtained only one (real) solution. In Example 4(b), the discriminant was negative and, consequently, we obtained two solutions involving the imaginary unit i. These observations are generalized in the box that follows.

THE DISCRIMINANT $b^2 - 4ac$

Consider the quadratic equation $ax^2 + bx + c = 0$, where a, b, and c are real numbers and $a \neq 0$. The expression $b^2 - 4ac$ is called the **discriminant.**

1. If $b^2 - 4ac > 0$, then the equation has two distinct real roots.
2. If $b^2 - 4ac = 0$, then the equation has exactly one real root.
3. If $b^2 - 4ac < 0$, then the equation has no real root. There are two (nonreal) complex-number roots. (These two roots are *complex conjugates,* that is, they have the form $A \pm Bi$, where A and B are real numbers and $B \neq 0$.)

EXAMPLE 5

(a) Compute the discriminant to determine how many real solutions there are for the equation $x^2 + x - 1 = 0$.

(b) Find a value for k such that the quadratic equation $x^2 + \sqrt{2}x + k = 0$ has exactly one real solution.

Solution

(a) Here $a = 1$, $b = 1$, and $c = -1$. Therefore,

$$b^2 - 4ac = 1^2 - 4(1)(-1) = 5$$

Since the discriminant is positive, the equation has two distinct real solutions.

(b) For the equation to have exactly one real solution, the discriminant must be zero, that is,

$$\begin{aligned} b^2 - 4ac &= 0 \\ (\sqrt{2})^2 - 4(1)(k) &= 0 \\ 2 - 4k &= 0 \\ k &= 1/2 \end{aligned}$$

The required value for k is $1/2$.

The techniques presented in this section can often be used to solve equations that contain radicals. As a simple example, consider the equation

$$x - 2 = \sqrt{x}$$

By squaring both sides, we obtain

$$\begin{aligned} x^2 - 4x + 4 &= x \\ x^2 - 5x + 4 &= 0 \\ (x-4)(x-1) &= 0 \end{aligned}$$

Therefore $\quad x = 4 \quad$ or $\quad x = 1$

The value $x = 4$ checks in the original equation, but the value $x = 1$ does not. (Verify this.) We say in this case that the value $x = 1$ is an **extraneous root** or an **extraneous solution** of the original equation. [This extraneous root arises from the fact that if $a^2 = b^2$, it is not necessarily true that $a = b$. For instance, $(-3)^2 = 3^2$, but certainly $-3 \neq 3$.] In summary, whenever we square both sides of an equation (or raise both sides to an even integral power), there is a possibility of introducing extraneous roots. For this reason, it is essential to check all candidates for solutions obtained in this manner.

Our final example in this section involves an equation containing several square-root expressions. Before squaring both sides in such cases, it is usually a good idea to see if we can first isolate one of the radical expressions on one side of the equation.

EXAMPLE 6 Solve: $\sqrt{9+x}+\sqrt{1+x}-\sqrt{x+16}=0.$

Solution

$$\sqrt{9+x}+\sqrt{1+x}=\sqrt{x+16} \qquad \text{adding } \sqrt{x+16} \text{ to both sides in preparation for squaring}$$

$$(\sqrt{9+x}+\sqrt{1+x})^2=(\sqrt{x+16})^2$$

$$9+x+2\sqrt{9+x}\,\sqrt{1+x}+1+x=x+16$$

$$2\sqrt{9+x}\,\sqrt{1+x}=6-x$$

$$(2\sqrt{9+x}\,\sqrt{1+x})^2=(6-x)^2$$

$$4(9+10x+x^2)=36-12x+x^2$$

$$3x^2+52x=0$$

$$x(3x+52)=0$$

$$x=0 \quad \Big| \quad 3x+52=0$$

$$x=-52/3$$

We have found two possible solutions: 0 and $-52/3$. Since these were obtained by squaring both sides of the equation, we need to check to see if they satisfy the original equation. With $x=0$, the original equation becomes $\sqrt{9}+\sqrt{1}-\sqrt{16}=0$, or $3+1-4=0$, which is certainly true. On the other hand, with $x=-52/3$, the quantities underneath the radicals in the original equation are negative and, consequently, the (real) square roots are undefined. So $-52/3$ is an extraneous solution, and $x=0$ is the only solution of the given equation.

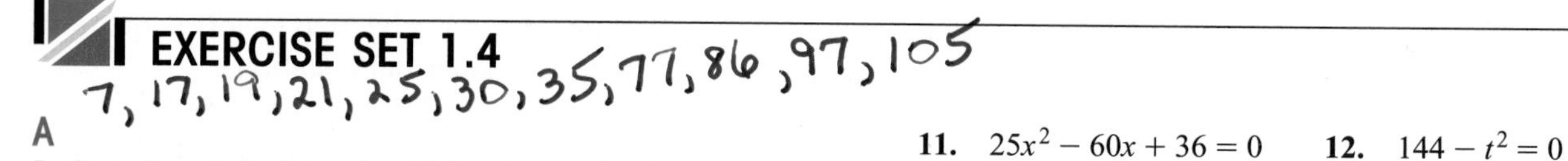

EXERCISE SET 1.4

A

In Exercises 1–5, determine if the given value is a solution of the equation.

1. $2x^2-6x-36=0;\ x=-3$
2. $(y-4)(y-5)=0;\ y=-1$
3. $4x^2-1=0;\ x=-1/4$
4. $m^2+m-\frac{5}{16}=0;\ m=1/4$
5. $x^2-2x-6=0;\ x=-1-\sqrt{7}$
6. Verify that the numbers $1+\sqrt{5}$ and $1-\sqrt{5}$ both satisfy the equation $x^2-2x-4=0$.

In Exercises 7–24, solve each equation by factoring.

7. $x^2-5x-6=0$
8. $x^2-5x=-6$
9. $x^2-100=0$
10. $4y^2+4y+1=0$
11. $25x^2-60x+36=0$
12. $144-t^2=0$
13. $10z^2-13z-3=0$
14. $3t^2-t-4=0$
15. $(x+1)^2-4=0$
16. $x^2+3x-40=0$
17. $x(2x-13)=-6$
18. $x(3x-23)=8$
19. $x^3+x^2-156x=0$
20. $8x^3-18x^2-5x=0$
21. $8y^3-4y^2-6y+3=0$
22. $y^4+2y^3-9y^2-18y=0$
23. $-y^3+y^2+2y=2$
24. $y^6-9y^4-4y^2=-36$

In Exercises 25–36, use the quadratic formula to solve the equation.

25. $x^2+10x+9=0$
26. $x^2-10x+25=0$
27. $x^2-x-5=0$
28. $x^2-x+5=0$
29. $2x^2+3x-4=0$
30. $4x^2-3x-9=0$

31. $12x^2 + 32x + 5 = 0$
32. $10x^2 - x - 1 = 0$
33. $2x^2 = x + 5$
34. $3 - 2x = -4x^2$
35. $-6x^2 + 12x = -1$
36. $-\sqrt{2}x^2 + x = -\sqrt{2}$

In Exercises 37 and 38, use the quadratic formula and a calculator to solve for x. Round each answer to two decimal places. Note: After completing Exercises 37 and 38, compare the two equations and their solutions. These exercises show that a slight change in one of the coefficients can sometimes radically alter the nature of the solutions.

37. $x^2 + 3x + 2.249 = 0$
38. $x^2 + 3x + 2.251 = 0$

In Exercises 39 and 40, use the quadratic formula and a calculator to solve for x. (These exercises provide examples of equations that could be solved by factoring, but for which the quadratic formula is clearly more efficient.)

39. $x^2 + 156x + 5963 = 0$
40. $52x^2 - 165x + 108 = 0$

In Exercises 41–52, solve the equations using any method you choose.

41. $(3x - 2)^2 = 3x - 2$
42. $3(x + 1)^2 - 4(x + 1) = 0$
43. $3x^2 + 4x - 3 = 0$
44. $13x^2 = 52$
45. $x^2 = 24$
46. $x^2 + 16x = 0$
47. $x(x - 1) = 1$
48. $x(x - 1) = -4$
49. $\frac{1}{2}x^2 - x - \frac{1}{3} = 0$
50. $x^2 + 34x + 288 = 0$
51. $2\sqrt{5}x^2 - x - 2\sqrt{5} = 0$
52. $\sqrt{2}x^2 + x = 10\sqrt{2}$

In Exercises 53–58, solve for x in terms of the other letters.

53. $2y^2x^2 - 3yx + 1 = 0 \quad (y \neq 0)$
54. $(ax + b)^2 - (bx + a)^2 = 0 \quad (a \neq \pm b)$
55. $(x - p)^2 + (x - q)^2 = p^2 + q^2$
56. $21x^2 - 2kx - 3k^2 = 0 \quad (k > 0)$
57. $12x^2 = ax + 20a^2 \quad (a > 0)$
58. $3Ax^2 - 2Ax - 3Bx + 2B = 0 \quad (A \neq 0)$

In Exercises 59–64, solve for the indicated letter.

59. $2\pi r^2 + 2\pi rh = 20\pi$; for r
60. $2\pi y^2 + \pi yx = 12$; for y
61. $-16t^2 + v_0t = 0$; for t
62. $-\frac{1}{2}gt^2 + v_0t + h_0 = 0$; for t
63. $x^3 + bx^2 - 2b^2x = 0$; for x
64. $\dfrac{1}{y^3} + \dfrac{b}{y^2} - \dfrac{2b^2}{y} = 0$; for y

In Exercises 65–72, use the discriminant to determine how many real roots the equation has.

65. $x^2 - 12x + 16 = 0$
66. $2x^2 - 6x + 5 = 0$
67. $4x^2 - 5x - \frac{1}{2} = 0$
68. $4x^2 - 28x + 49 = 0$
69. $x^2 + \sqrt{3}x + \frac{3}{4} = 0$
70. $\sqrt{2}x^2 + \sqrt{3}x + 1 = 0$
71. $y^2 - \sqrt{5}y = -1$
72. $\dfrac{m^2}{4} - \dfrac{4m}{3} + \dfrac{16}{9} = 0$

In Exercises 73–76, find values for k such that the equation has exactly one real root.

73. $x^2 + 12x + k = 0$
74. $3x^2 + (\sqrt{2k})x + 6 = 0$
75. $x^2 + kx + 5 = 0$
76. $kx^2 + kx + 1 = 0$

In Exercises 77–88, determine all of the real number solutions for the equation.

77. **(a)** $\sqrt{x - 8} = 4$ **(b)** $\sqrt{x - 8} = 2x + 1$
78. **(a)** $\sqrt{3x + 1} = 4$ **(b)** $\sqrt{3x + 1} = 2x - 1$
79. $\sqrt{1 - 3x} = 2$
80. $\sqrt{x^2 + 5x - 2} = 2$
81. $\sqrt{x^4 - 13x^2 + 37} = 1$ Hint: Let $x^2 = t$ and $x^4 = t^2$
82. $\sqrt{y + 2} = y - 4$
83. $\sqrt{1 - 2x} + \sqrt{x + 5} = 4$
84. $\sqrt{x - 5} - \sqrt{x + 4} + 1 = 0$
85. $\sqrt{3 + 2t} + \sqrt{-1 + 4t} = 1$
86. $\sqrt{2t + 5} - \sqrt{8t + 25} + \sqrt{2t + 8} = 0$
87. $\sqrt{2x + 1} + \sqrt{x + 4} = 1$
88. $\sqrt{2x + 6} + \sqrt{x + 4} - \sqrt{8x + 9} = 0$

B

In Exercises 89–98, find all the real solutions of the equation. You'll first want to eliminate the fractions by multiplying both sides by the least common denominator. [For example, in Exercise 89, first multiply both sides by $x(x + 5)$.] Be sure to check your answers for these exercises, because multiplying both sides of an equation by an expression involving the variable may introduce extraneous roots.

89. $\dfrac{3}{x + 5} + \dfrac{4}{x} = 2$
90. $\dfrac{5}{x + 2} - \dfrac{2x - 1}{5} = 0$
91. $1 - x - \dfrac{2}{6x + 1} = 0$
92. $\dfrac{x^2 - 3x}{x + 1} = \dfrac{4}{x + 1}$
93. $\dfrac{3x^2 - 6x - 3}{(x + 1)(x - 2)(x - 3)} + \dfrac{5 - 2x}{x^2 - 5x + 6} = 0$
94. $\dfrac{6}{x^2 - 1} + \dfrac{x}{x + 1} = \dfrac{3}{2}$
95. $\dfrac{2x}{x^2 - 1} - \dfrac{1}{x + 3} = 0$
96. $\dfrac{3}{x^2 - x - 2} - \dfrac{4}{x^2 + x - 6} = \dfrac{1}{3x + 3}$
97. $\dfrac{x - 1}{x + 1} - \dfrac{x + 1}{x + 3} + \dfrac{4}{x^2 + 4x + 3} = 0$
98. $\dfrac{x}{x - 2} + \dfrac{x}{x + 2} = \dfrac{8}{x^2 - 4}$

In Exercises 99–102, find all the real solutions of the equation.

99. $\sqrt{\sqrt{x}+\sqrt{a}}+\sqrt{\sqrt{x}-\sqrt{a}}=\sqrt{2\sqrt{x}+2\sqrt{b}}$

100. $x=\sqrt{3x+x^2-3\sqrt{3x+x^2}}$

101. $\dfrac{\sqrt{x}-a}{\sqrt{x}}-\dfrac{\sqrt{x}+a}{\sqrt{x}-b}=0 \quad (a>0,\ b>0)$

Suggestion: Let $\sqrt{x}=t$.

102. $x-\sqrt{x^2-x}=\sqrt{x} \quad (x>0)$

Suggestion: If $x \neq 0$, then you can divide through by $\sqrt{x}$ to obtain a simpler equation.

103. In the accompanying figure, the point P is located on line segment $\overline{AB}$ in such a way that

$$\frac{AB}{AP}=\frac{AP}{PB}$$

(In words: P divides $\overline{AB}$ into two segments such that the ratio of AB to the longer segment is equal to the ratio of the longer segment to the shorter segment.)

(a) Show that this ratio is equal to $\frac{1}{2}(1+\sqrt{5})$.
Hints: Let $AP=x$ and $PB=y$. Obtain an equation involving x and y, and then use the quadratic formula to solve for y in terms of x.

(b) The ratio determined in part (a) is denoted by the Greek letter *phi*: ϕ. That is,

$$\phi=\frac{1+\sqrt{5}}{2}$$

The letter ϕ is used in honor of the ancient Greek sculptor Phidias, who is said to have used this ratio in his work and who may have strongly influenced the architects of the Parthenon in Athens. [See, for example, page 63 in H. E. Huntley's *The Divine Proportion* (New York: Dover Publications, 1970.]

Verify that the number ϕ satisfies the following algebraic properties.

(i) ϕ is the positive root of the quadratic equation $x^2-x-1=0$.
(ii) $\phi^2=\phi+1$
(iii) $\phi^3=2\phi+1$ *Hint:* $\phi^3=\phi(\phi^2)$
(iv) $\phi^{-1}=\phi-1$
(v) $\phi^{-2}=-\phi+2$

104. In the figure (at the top of the next column), $ABCD$ is a square and each side has length 1. The point M is the midpoint of $\overline{DC}$, and the arc is a portion of a circle with center M and radius MB.

(a) Find MB.
(b) Explain why $MF=MB$.

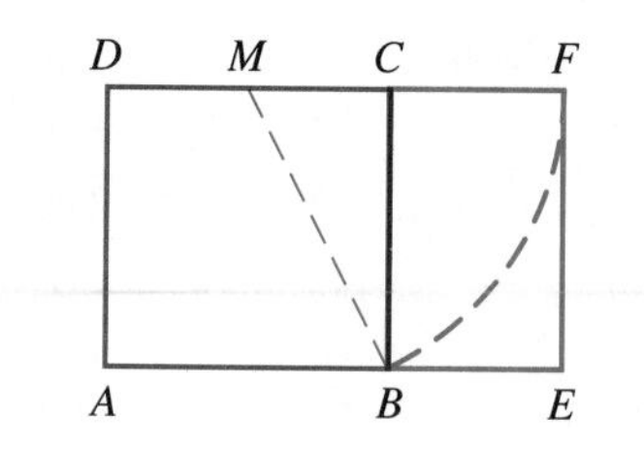

(c) Use the results in parts (a) and (b) to show $DF=\phi$. (The number ϕ is defined in Exercise 103.)
(d) Show that $CF=1/\phi$.
(e) Show that $DF/CF=\phi^2$.

105. For a certain right circular cylinder, the height is 3 m and the total surface area is $8\pi\ \text{m}^2$. Find the radius of the cylinder.

106. A ball is thrown straight upward. Suppose that the height of the ball at time t is $h=-16t^2+96t$, where h is in feet and t is in seconds, with $t=0$ corresponding to the instant that the ball is first tossed.
(a) How long does it take for the ball to land?
(b) At what time is the height 80 ft? Why does this question have two answers?

107. During a flu epidemic in a small town, a public health official finds that the total number of people P who have caught the flu after t days is closely approximated by the formula

$$P=-t^2+26t+106 \quad (1 \leq t \leq 13)$$

(a) How many have caught the flu after 10 days?
(b) After approximately how many days will 250 people have caught the flu?

108. The radius of a circle is r units. By how many units should the radius be increased so that the area increases by b square units?

109. A piece of wire L in. long is cut into two pieces. Each piece is then bent to form a square. If the sum of the areas of the two squares is $5L^2/128$, how long is each piece of wire?

110. **(a)** Show that the sum of the roots of the equation $x^2+px+q=0$ is $-p$.
(b) Show that the product of the roots of the equation $x^2+px+q=0$ is q.

111. In this section we solved quadratic equations by factoring, by completing the square, and by using the quadratic formula. This exercise shows how to solve a quadratic equation by the *method of substitution.* As an example, we use the equation

$$x^2+x-1=0 \qquad (1)$$

(a) In equation (1), make the substitution $x=y+k$. Show that the resulting equation can be written

$$y^2+(2k+1)y=1-k-k^2 \qquad (2)$$

(b) Find a value for k so that the coefficient of y in equation (2) is 0. Then, using this value of k, show that equation (2) becomes $y^2 = 5/4$.

(c) Solve the equation $y^2 = 5/4$. Then use the equation $x = y + k$ to obtain the solutions of equation (1).

112. Use the substitution method (explained in Exercise 111) to solve the quadratic equation $2x^2 - 3x + 1 = 0$.

113. For which values of A and B will the roots of the equation $x^2 + Ax + B = 0$ be A and B?

114. For which values of k will the roots of the equation $x^2 = 2x(3k + 1) - 7(2k + 3)$ be equal?

115. Let α and β be the roots of the quadratic equation $ax^2 + bx + c = 0$. Verify each of the following statements.

(a) $\alpha + \beta = -b/a$

(b) $\alpha\beta = c/a$

(c) $\alpha^2 + \beta^2 = \dfrac{b^2 - 2ac}{a^2}$ *Suggestion:* Use (a) and (b) and the fact that $\alpha^2 + \beta^2 = (\alpha + \beta)^2 - 2\alpha\beta$.

(d) $\dfrac{1}{\alpha^2} + \dfrac{1}{\beta^2} = \dfrac{b^2 - 2ac}{c^2}$ *Suggestion:* Add the fractions $1/\alpha^2$ and $1/\beta^2$.

116. Here is an outline for a slightly different derivation of the quadratic formula. The advantage to this method is that fractions are avoided until the very last step. Fill in the details.

(a) Beginning with $ax^2 + bx = -c$, multiply both sides by $4a$. Then add b^2 to both sides.

(b) Now factor the resulting left-hand side and take square roots.

CHAPTER ONE SUMMARY OF PRINCIPAL TERMS AND NOTATION

TERMS OR NOTATIONS	PAGE REFERENCE	COMMENTS
1. Natural numbers, integers, rational numbers, and irrational numbers	2	The box on page 2 provides both definitions and examples. Also note the theorem in the box, which explains how to distinguish between rationals and irrationals in terms of their decimal representations.
2. $a < b$ $b > a$	3	a is less than b. b is greater than a.
3. (a, b) $[a, b]$	3	The open interval (a, b) consists of all real numbers between a and b, excluding a and b. The closed interval $[a, b]$ consists of all real numbers between a and b, including a and b.
4. (a, ∞)	4	The unbounded interval (a, ∞) consists of all real numbers x such that $x > a$. The infinity symbol, ∞, does not denote a real number. It is used in the context (a, ∞) to indicate that the interval has no right-hand boundary. For the definitions of the other types of unbounded intervals, see the box on page 14.
5. $\lvert x \rvert$	6, 7	The absolute value of a real number x is the distance of x from the origin. This is equivalent to the following algebraic definition: $\lvert x \rvert = \begin{cases} x & \text{if } x \geq 0 \\ -x & \text{if } x < 0 \end{cases}$
6. Constant, variable, and the domain convention	11, 12	By a constant we mean either a particular number (such as -8 or π) or a letter whose value remains fixed (although perhaps unspecified) throughout a given discussion. In contrast, a variable is a letter for which we can substitute any number from a given set of numbers. The given set is called the domain of the variable. According to the domain convention, the domain of a variable in a given expression consists of all real numbers for which the expression is defined.

TERMS OR NOTATIONS	PAGE REFERENCE	COMMENTS
7. Polynomial	12	A polynomial is an expression of the form $a_nx^n + a_{n-1}x^{n-1} + \cdots + a_1x + a_0$. The individual expressions a_kx^k making up the polynomial are called terms. The numbers a_k are called coefficients.
8. Degree of a polynomial	12	For a polynomial that is not simply a constant, the degree is the highest power (that is, the exponent) of x that appears. The degree of a nonzero constant polynomial is zero. Degree is undefined for the zero polynomial.
9. Special products	13	The box on page 13 lists four basic types of products that you should memorize because they occur so frequently.
10. Factoring techniques	14	The techniques for factoring listed in the box on page 14 are fundamental for simplifying expressions and for solving equations.
11. Quadratic equation	20	A quadratic equation is an equation that can be written in the form $ax^2 + bx + c = 0$, where $a \neq 0$.
12. The zero-product property	20	This property of real numbers (and of complex numbers in general) can be stated as follows. If $pq = 0$, then $p = 0$ or $q = 0$; conversely, if $p = 0$ or $q = 0$, then $pq = 0$. (We used this property in solving quadratic equations by factoring.)
13. The quadratic formula: $x = \dfrac{-b \pm \sqrt{b^2 - 4ac}}{2a}$	23	This is the quadratic formula; it provides the solutions of the quadratic equation $ax^2 + bx + c = 0$. The formula is derived on pages 22–23 by means of the useful technique of completing the square.
14. Discriminant	24	The discriminant of the quadratic equation $ax^2 + bx + c = 0$ is the number $b^2 - 4ac$. As indicated in the box on page 25, the discriminant provides information about the roots of the equation.
15. Extraneous solution (or extraneous root)	25	In solving equations, certain processes (such as squaring both sides) can lead to answers that do not check in the original equation. These numbers are called extraneous solutions (or extraneous roots).

WRITING MATHEMATICS

1. More than 3000 years ago, the ancient Babylonian mathematicians solved quadratic equations. A method they used is demonstrated (using modern notation) in the following example.

   To find the positive root of $x^2 + 8x = 84$:

   Rewrite the equation as $x(x + 8) = 84$, and let $y = x + 8$. Then the equation to be solved becomes $xy = 84$. Now take half the coefficient of x in the original equation, which is 4, and define another variable t by $t = x + 4$. Then we have $x = t - 4$ and $y = t + 4$, and therefore

$$(t - 4)(t + 4) = 84$$
$$t^2 = 100$$
$$t = 10$$

   With $t = 10$, we get $x = 10 - 4 = 6$, the required positive root.

On your own or with a group of classmates, work through the above example, filling in any missing details if necessary. Then (strictly on your own), use the Babylonian method to find the positive root of each of the following equations. Write out your solutions in detail, as if you were explaining the method to another student who'd not seen it before. This will involve a combination of English composition and algebra. Also, in part (b), check your answer by using the quadratic formula.

(a) $x^2 + 14x = 72$ **(b)** $x^2 + 2Ax = B \quad (B > 0)$

2. The ancient Greek mathematicians (2500 years ago) used geometric methods to solve quadratic equations or, rather, to construct line segments whose lengths were the roots of the equations. According to historian Howard Eves in *An Introduction to the History of Mathematics,* 6th ed. (Philadelphia: Saunders College Publishing, 1990):

> Imbued with the representation of a number by a length and completely lacking any adequate algebraic notation, the early Greeks devised ingenious geometrical processes for carrying out algebraic operations.

One of the methods used by the ancient Greeks to solve a quadratic equation is described in the following example (using modern algebraic notation).

> To construct a line segment whose length is equal to the (positive) root of the equation $x^2 + 8x = 84$:
>
> Begin with a line segment $\overline{AB}$ of length 8. At B, construct a line segment $\overline{BC} \perp \overline{AB}$ such that the length of $\overline{BC}$ is $\sqrt{84}$. Next, let M be the midpoint of $\overline{AB}$. With M as center, draw a circular arc of radius $\overline{MC}$ intersecting $\overline{AB}$ (extended) at P. Then the length of $\overline{AP}$ is the required root.

On your own or with a group of classmates, work through this construction. That is, sketch the appropriate figure and verify that the construction indeed yields the positive root of the equation. Then, on your own, use the Greek method to determine the positive root of each of the following equations. Write out your work in detail, as if you were explaining it to a student who had not seen it before. Be sure to include the appropriate geometric figures and an explanation of why the method works.

(a) $x^2 + 14x = 72$ **(b)** $x^2 + 2cx = d \quad (c, d > 0)$

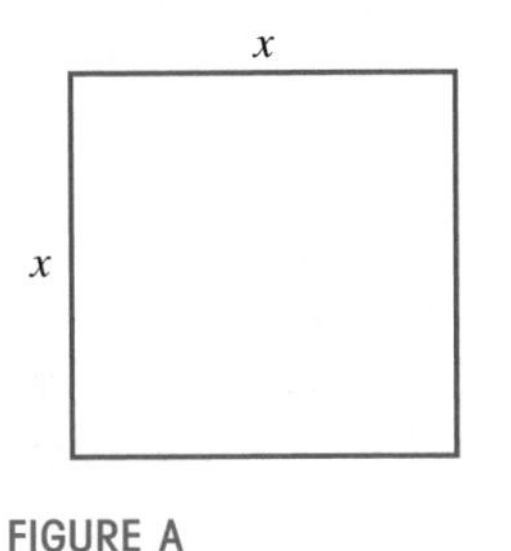

FIGURE A

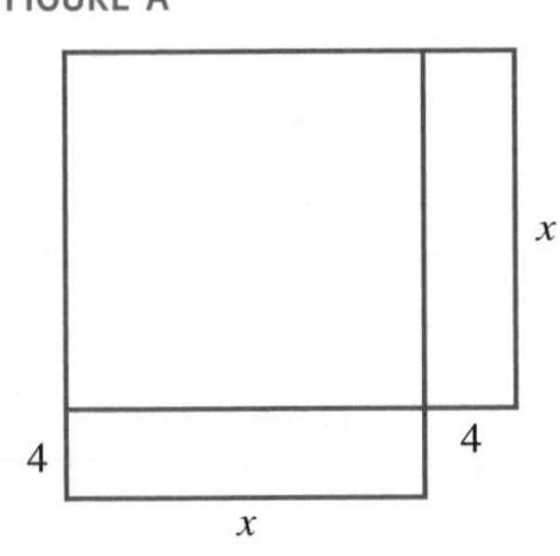

FIGURE B

3. The example in Exercise 1 demonstrates a Babylonian method for solving a quadratic equation. According to Professor Victor J. Katz in *A History of Mathematics* (New York: HarperCollins College Publishers, 1993), "Whatever the ultimate origin of this method, a close reading of the wording of the [ancient clay] tablets seems to indicate that the scribe had in mind a geometric procedure [completing the square] . . . ."

   One of the earliest *explicit* uses of the completing-the-square technique to solve a quadratic is due to the ninth-century Arab mathematician and astronomer Muhammed al-Khwārizmī. The following example demonstrates al-Khwārizmī's use of completing the square to solve the equation $x^2 + 8x = 84$.

> To find the positive root of $x^2 + 8x = 84$:
>
> Begin with a square of side x, as in Figure A. Take half of the coefficient of x: this is one-half of 8, or 4. Now form two rectangles, each with dimensions 4-by-x, and adjoin them to the square, as indicated in Figure B. Then, as in

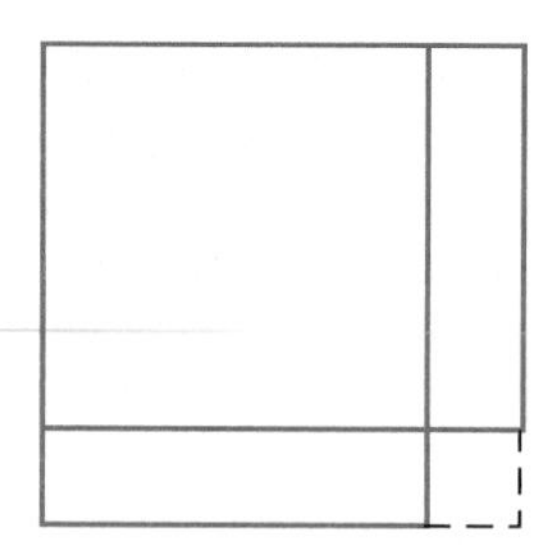

FIGURE C

Figure C, draw the dashed lines to complete the (outer) square. In Figure B, the combined area of the square and the two rectangles is $x^2 + 8x$. But, from the equation we wish to solve, $x^2 + 8x$ is equal to 84. In Figure C, the area of the small square in the lower right-hand corner is 16. Thus the area of the entire outer square is $16 + 84$, or 100. But the area of the outer square is also $(x + 4)^2$. Therefore

$$(x + 4)^2 = 100 \quad \text{and consequently} \quad x + 4 = 10$$

The positive root we are looking for is therefore $x = 6$.

On your own or with a group of classmates, work through this example, filling in the missing details as necessary. Then, on your own, use this completing-the-square process to find the positive root of each of the following equations. Write out your solutions in detail, as if you were explaining the method to another student who'd not seen it before. Be sure to include the appropriate geometric figures.

(a) $x^2 + 14x = 72$ **(b)** $x^2 + 2Ax = B \quad (B > 0)$

CHAPTER ONE REVIEW EXERCISES

Factor each expression in Exercises 1–20.

1. $a^2 - 16b^2$
2. $x^2 + ax + yx + ay$
3. $8 - (a + 1)^3$
4. $x^2 - 18x + 81$
5. $a^2x^3 + 2ax^2b + b^2x$
6. $8a^2x^2 + 16a^3$
7. $8x^2 + 6x + 1$
8. $12x^2 - 2x - 4$
9. $a^4x^4 - x^8a^8$
10. $2x^2 - 2bx + ax - ab$
11. $8 + 12a + 6a^2 + a^3$
12. $(x^2 + 2x - 8)^2 - (2x + 1)^2$
13. $4x^2y^2z^3 - 3xyz^3 - z^3$
14. $(x + y - 1)^2 - (x - y + 1)^2$
15. $1 - x^6$
16. $12x^3 + 44x^2 - 16x$
17. $a^2x^2 + 2abx + b^2 - 4a^2b^2x^2$
18. $a^2 + 2ab + b^2 + a + b$
19. $a^2 - b^2 + ac - bc + a^2b - b^2a$
20. $5(a + 1)^2 + 29(a + 1) - 144$

Use the Special Products formulas in Section 1.3 to compute the products in Exercises 21–34.

21. $(9x - y)(9x + y)$
22. $(3x + 1)^2$
23. $(2ab - 3)^2$
24. $(ax + b)^2(ax - b)^2$
25. $(2a + 3)^3$
26. $(x^2 - 2)^3$
27. $(1 - 3a)(1 + 3a + 9a^2)$
28. $(x^n + y^n)(x^{2n} - x^ny^n + y^{2n})$
29. $(x^{1/2} - y^{1/2})(x^{1/2} + y^{1/2})$
30. $(\sqrt{x} - \sqrt{3})(\sqrt{x} + \sqrt{3})$
31. $(x^{1/3} + y^{1/3})^3$
32. $(x^{1/3} + y^{1/3})(x^{2/3} - x^{1/3}y^{1/3} + y^{2/3})$
33. $(x^2 - 3x + 1)^2$ *Hint:* Rewrite the expression as $[x^2 - (3x - 1)]^2$.
34. $(x^2 + x - 1)(x^2 + x + 1)$

For Exercises 35–39, rewrite the statements using absolute values and inequalities or equalities.

35. The distance between x and 6 is 2.
36. The distance between x and a is less than 1/2.
37. The distance between a and b is 3.
38. The distance between x and -1 is 5.
39. The distance between x and 0 exceeds 10.
40. What can you say about x if $|x - 5| = 0$?

Rewrite each of the expressions in Exercises 41–46 in a form that does not contain absolute values.

41. $|\sqrt{6} - 2|$
42. $|2 - \sqrt{6}|$
43. $|x^4 + x^2 + 1|$
44. **(a)** $|x - 3|$, if $x < 3$ **(b)** $|x - 3|$, if $x > 3$
45. $|x - 2| + |x - 3|$, if: **(a)** $x < 2$ **(b)** $2 < x < 3$ **(c)** $x > 3$
46. $|x + 2| + |x - 1|$, if: **(a)** $x < -2$ **(b)** $-2 < x < 1$ **(c)** $x > 1$

In Exercises 47–52, sketch each interval on a number line.

47. $(3, 5)$
48. $(3, 5]$

49. The set of all negative real numbers that are in the interval $[-5, 2]$

50. $(-\infty, 4)$

51. $[-1, \infty)$

52. The set of real numbers that belongs to either of the intervals $(0, \sqrt{2}]$ or $(\sqrt{2}, 2]$

In Exercises 53–57, sketch the intervals described by the given inequalities.

53. $|x-6|<3$

54. $|x-\frac{1}{2}|<1$

55. $|x+1|\geq 1$

56. $|x|\geq 5$

57. **(a)** $0<|x-4|<5$ **(b)** $|x-4|<5$

58. Polynomials can be used to approximate more complicated expressions. For example, in calculus it is shown that when x is close to zero, the expression $\sqrt{1+x}$ can be approximated by the polynomial $1+\frac{1}{2}x$. Complete the following table to see evidence of this. (Use a calculator to compute $\sqrt{1+x}$; report the values to six decimal places without rounding.)

x	$1+\frac{x}{2}$	$\sqrt{1+x}$
0.1		
0.01		
0.001		

59. Using calculus, it can be shown that when x is close to zero, the expression $1/\sqrt{1-x^2}$ can be approximated by the polynomial $1+\frac{1}{2}x^2+\frac{3}{8}x^4$. Complete the table to see evidence of this. (Use a calculator and round your answers to six significant digits.)

x	$1+\frac{x^2}{2}+\frac{3x^4}{8}$	$1/\sqrt{1-x^2}$
0.1		
0.01		

60. **(a)** Use a calculator to evaluate the expression

$$\frac{\sqrt{3-\sqrt{5}}}{\sqrt{2}+\sqrt{7-3\sqrt{5}}}.$$

(b) Use a calculator to evaluate the expression $\sqrt{5}/5$. If you do the calculator work carefully, your results in parts (a) and (b) will suggest (but not prove!) that the two expressions are equal. Follow steps (c), (d), and (e) to prove that the two expressions are indeed equal.

(c) Multiply the expression in part (a) by $\dfrac{\sqrt{2}-\sqrt{7-3\sqrt{5}}}{\sqrt{2}-\sqrt{7-3\sqrt{5}}}$, which equals 1. After carrying out the indicated operations, you should obtain $\dfrac{\sqrt{6-2\sqrt{5}}-2\sqrt{9-4\sqrt{5}}}{-5+3\sqrt{5}}$.

(d) Verify that $\sqrt{6-2\sqrt{5}}=\sqrt{5}-1$. Also verify that $\sqrt{9-4\sqrt{5}}=\sqrt{5}-2$.

(e) Use the results in part (d) to simplify the expression in part (c). The final result, after rationalizing the denominator, should be $\sqrt{5}/5$.

In Exercises 61–78, find all the real solutions of each equation.

61. $12x^2+2x-2=0$

62. $4y^2-21y=18$

63. $\frac{1}{2}x^2+x-12=0$

64. $x^2+\frac{13}{2}x+10=0$

65. $\dfrac{x}{5-x}=\dfrac{-2}{11-x}$

66. $\dfrac{x^2}{(x-1)(x+1)}=\dfrac{4}{x+1}+\dfrac{4}{(x-1)(x+1)}$

67. $\dfrac{1}{3x-7}-\dfrac{2}{5x-5}-\dfrac{3}{3x+1}=0$

68. $4x^2+x-2=0$

69. $x^5-2x^3-2x=0$

70. $x^3-6x^2+7x=0$

71. $\sqrt{4-3x}=5$

72. $\sqrt{x}=\sqrt{x+27}-1$

73. $\sqrt{4x+3}=\sqrt{11-8x}-1$

74. $2-\sqrt{3\sqrt{2x-1}+x}=0$

75. $\dfrac{2}{\sqrt{x^2-36}}+\dfrac{1}{\sqrt{x+6}}-\dfrac{1}{\sqrt{x-6}}=0$

76. $\sqrt{x-\sqrt{1-x}}+\sqrt{x}-1=0$

77. $\sqrt{x+7}-\sqrt{x+2}=\sqrt{x-1}-\sqrt{x-2}$

78. $\dfrac{\sqrt{x}-4}{\sqrt{x}-18}=3$

For Exercises 79–82, find the value(s) of k for which the equation has exactly one real root.

79. $kx^2-6x+5=0$

80. $x^2+x+k^2=0$

81. $kx(x+2)=-1$

82. $x^2+(k+1)x+2k=0$

If an object is thrown vertically upward from a height of h_0 feet with an initial speed of v_0 ft/sec, then its height h (in feet) after t seconds is given by

$$h=-16t^2+v_0t+h_0$$

Make use of this formula in working Exercises 83–85.

83. A ball is thrown vertically upward from ground level with an initial speed of 64 ft/sec.
(a) At what time will the height of the ball be 15 ft? (Two answers.)
(b) At what time t does the ball hit the ground?

84. One ball is thrown vertically upward from a height of 50 ft with an initial speed of 40 ft/sec. At the same instant, another ball is thrown vertically upward from a height of 100 ft with an initial speed of 5 ft/sec. Which ball hits the ground first?

85. An object is projected vertically upward. Suppose that its height is H ft at t_1 sec and again at t_2 sec. Express the initial speed in terms of t_1 and t_2.

Answer: $16(t_1 + t_2)$

Exercise 86 (including the clever solution that is outlined) appears in the book The USSR Olympiad Problem Book *by D. O. Shklarsky et al. (San Francisco: W. H. Freeman and Co., 1962).*

86. Follow steps (a)–(d) to solve the equation

$$\sqrt{a - \sqrt{a + x}} = x \quad (a \geq 1)$$

(a) By squaring as usual, obtain the equation

$$x^4 - 2ax^2 - x + a^2 - a = 0.$$

(b) Although the equation obtained in part (a) is a fourth-degree equation in x, it is only a quadratic in a. Solve for a in terms of x to obtain $a = x^2 + x + 1$ or $a = x^2 - x$.
(c) Solve each of the equations in part (b) for x in terms of a.
(d) Check your results in part (c) to eliminate any extraneous roots.

CHAPTER ONE TEST

In Exercises 1–4, factor each expression. (If an expression is irreducible over the integers, say so.)

1. **(a)** $64 - (x - 2)^3$ **(b)** $w^2y^2 + 16$
2. **(a)** $2x^2 + 3x - 8$ **(b)** $6x^2y^2z + xyz - z$
3. $6x(1 - x^2)^{-1/2} - 3x^2(1 - x^2)^{-3/2}$
4. $9ax + 6bx - 6ay - 4by$
5. Use one of the Special Product formulas to compute $(2a + 3b^2)^3$.
6. Simplify the expression $|x - 6| + |x - 7|$, given that $6 < x < 7$.
7. Specify the intervals that are described by the inequalities.
   (a) $|x - 4| < 1/10$ **(b)** $x \geq -2$

In Exercises 8–11, find all of the real solutions of each equation.

8. **(a)** $x^2 + 4x = 5$ **(b)** $x^2 + 4x = 1$
9. $\sqrt{5 - 2x} - \sqrt{2 - x} - \sqrt{3 - x} = 0$
10. $2x^3 + x^2 - 3x = 0$
11. $y^3 + 3y^2 - 8y - 24 = 0$
12. Find a positive value for k such that the equation

$$x^2 + 2kx + 3(k + 1) = 0$$

has exactly one real-number solution. Give two forms for your answer: one involving a radical and the other a calculator approximation rounded to two decimal places.

COORDINATES, GRAPHS, AND INEQUALITIES

The mathematics I was taught in my freshman year was algebra (old stuff), trigonometry (too much of it), and analytical geometry (a revelation).

Paul R. Halmos in *I Want To Be a Mathematician* (N.Y.: Springer-Verlag, 1985)

. . . [Descartes and Fermat] *were moved by the same spirit: they wanted to show how the Renaissance algebra of Cardan and his successors could be applied to the geometry of the Greeks, . . .*

From *A Source Book in Mathematics, 1200–1800,* D. J. Struik, ed. (Princeton: Princeton University Press, 1986)

INTRODUCTION

In Chapter 1 we saw an important connection between algebra and geometry: The points on a line can be labeled using the real numbers. Just as each point on a line can be labeled with a unique real number, each point in a *plane* can be labeled with a unique *pair* of real numbers. The details of this are reviewed in Section 2.1, Rectangular Coordinates. These ideas are then used throughout the chapter to graph equations and solve inequalities. The basic theme in the work on graphing equations is this: We start with a given equation relating two variables, say, x and y. By choosing values for x, corresponding y-values are then determined. The resulting number pairs (x, y) (called *ordered pairs*) are then interpreted as points in the plane. This provides another basic connection between algebra and geometry: We can associate a geometric picture—a graph—with an equation in two variables. Sometimes it is easier to work directly from a graph, other times from the equation. But in any event, having the latitude to think either algebraically or geometrically (or both) will clarify many problems later on.

2.1 RECTANGULAR COORDINATES

The name coordinate *does not appear in the work of Descartes. This term is due to Leibniz, and so are* abscissa *and* ordinate *(1692).*

David M. Burton in *The History of Mathematics, an Introduction,* 2nd ed. (Dubuque, Iowa: Wm. C. Brown Publishers, 1991)

FIGURE 1

In previous courses you learned to work with a rectangular coordinate system such as that shown in Figure 1. In this section we review some of the most basic formulas and techniques that are useful here.

The point of intersection of the two perpendicular number lines, or **axes,** is called the **origin** and is denoted by the letter O. The horizontal and vertical axes are often labeled the ***x*-axis** and the ***y*-axis,** respectively, but any other variables will do just as well for labeling the axes. For instance, Figure 2 shows

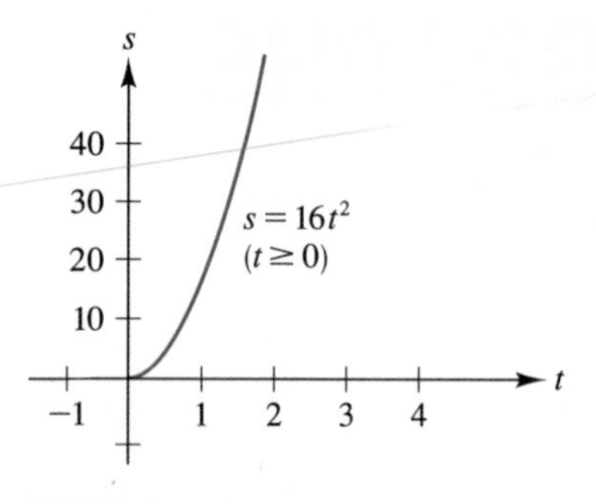

FIGURE 2
A graph of the formula $s = 16t^2$ in a t-s coordinate system. [The formula relates the distance s (in feet) and the time t (in seconds) for an object falling in a vacuum.]

a t-s coordinate system. (We'll discuss curves, or graphs, like the one in Figure 2 in later sections.)

Notice that in Figures 1 and 2 the axes divide the plane into four regions, or **quadrants,** labeled I through IV, as shown in Figure 1. Unless indicated otherwise, we assume that the same unit of length is used on both axes. In Figure 1, the same scales are used on both axes; in Figure 2, the scale used on the s-axis is different than the scale on the t-axis.

Now look at the point P in Figure 3(a). Starting from the origin O, one way to reach P is to move three units in the positive x-direction and then two units in the positive y-direction. That is, the location of P relative to the origin and the axes is "right 3, up 2." We say that the **coordinates** of P are $(3, 2)$. The first number within the parentheses conveys the information "right 3," and the second number conveys the information "up 2." We say that the **x-coordinate** of P is 3 and the **y-coordinate** of P is 2. Likewise, the coordinates of point Q in Figure 3(a) are $(-2, 4)$. With this coordinate notation in mind, observe in Figure 3(b) that $(3, 2)$ and $(2, 3)$ represent different points; that is, the order in which the two numbers appear within the parentheses affects the location of the point. Figure 3(c) displays various points with given coordinates; you should check for yourself that the coordinates correspond correctly to the location of each point.

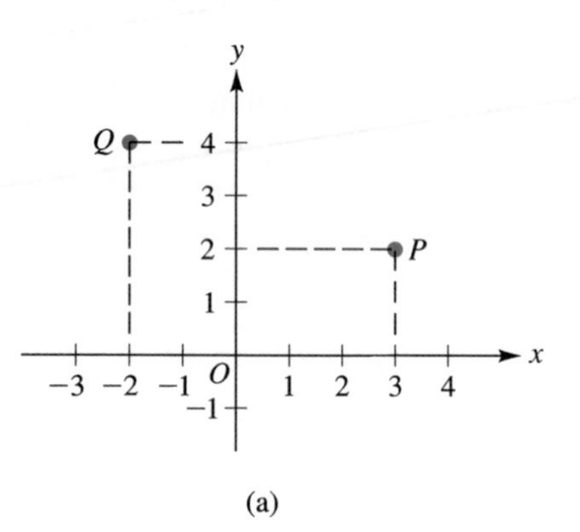

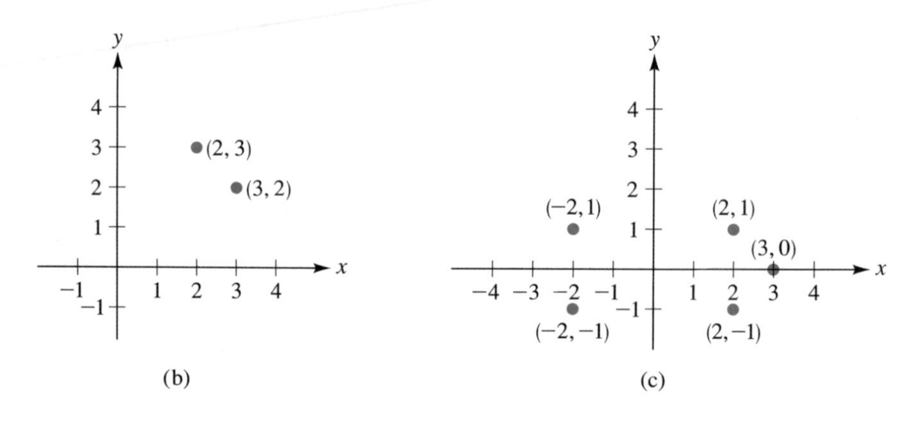

FIGURE 3

Some terminology and notation: The x-y coordinate system that we have described is often called a **Cartesian coordinate system.** The term *Cartesian* is used in honor of René Descartes, the seventeenth-century French philosopher and mathematician. The coordinates (x, y) of a point P are referred to as an **ordered pair.** Recall, for example, that $(3, 2)$ and $(2, 3)$ represent different points; that is, the order of the numbers matters. The x-coordinate of a point is sometimes referred to as the **abscissa** of the point; the y-coordinate is the **ordinate.** The notation $P(x, y)$ means that P is a point whose coordinates are (x, y). At times, we abbreviate the phrase *the point whose coordinates are* (x, y) to simply *the point* (x, y).

The remainder of our work in this section depends on a key result from elementary geometry, the Pythagorean theorem. For reference, we state this theorem and its converse in the box that follows. (See Exercises 62 and 63 for two proofs of the theorem.)

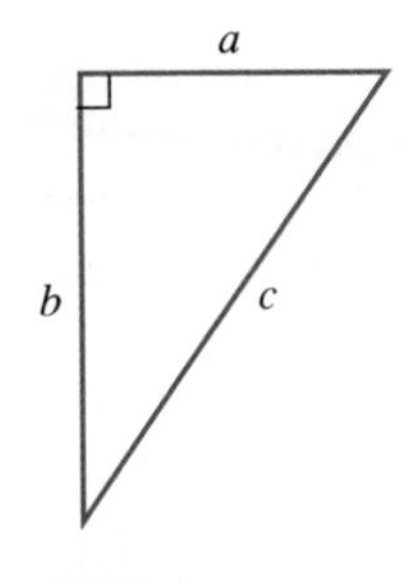

FIGURE 4

THE PYTHAGOREAN THEOREM AND ITS CONVERSE

1. The Pythagorean theorem
(See Figure 4.) In a right triangle, the lengths of the sides are related by the equation

$$a^2 + b^2 = c^2$$

where a and b are the lengths of the sides forming the right angle (the *legs*) and c is the length of the *hypotenuse* (the side opposite the right angle).

2. Converse
If the lengths of the sides of a triangle are related by an equation of the form $a^2 + b^2 = c^2$, then the triangle is a right triangle, and c is the length of the hypotenuse.

EXAMPLE 1 Use the Pythagorean theorem to calculate the distance d between the points (2, 1) and (6, 3).

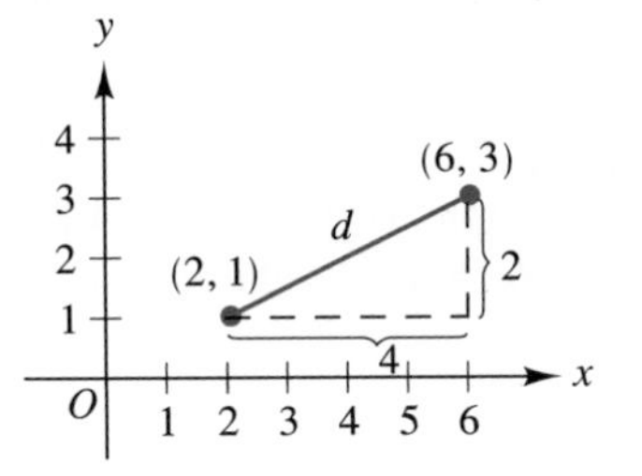

FIGURE 5

Solution We plot the two given points and draw a line connecting them, as shown in Figure 5. Then we draw the dashed lines parallel to the axes, as shown, and apply the Pythagorean theorem to the right triangle that is formed. The base of the triangle is four units long. You can see this simply by counting spaces or by using absolute value, as discussed in Section 1.2: $|6-2| = 4$. The height of the triangle is found to be two units, either by counting spaces or by computing the absolute value: $|3-1| = 2$. Thus we have

$$d^2 = 4^2 + 2^2 = 20$$
$$d = \sqrt{20} = \sqrt{4}\sqrt{5} = 2\sqrt{5}$$

We can apply the method used in Example 1 to derive a general formula for the distance d between any two points (x_1, y_1) and (x_2, y_2): see Figure 6. Just as before, we draw in the right triangle and apply the Pythagorean theorem. We have

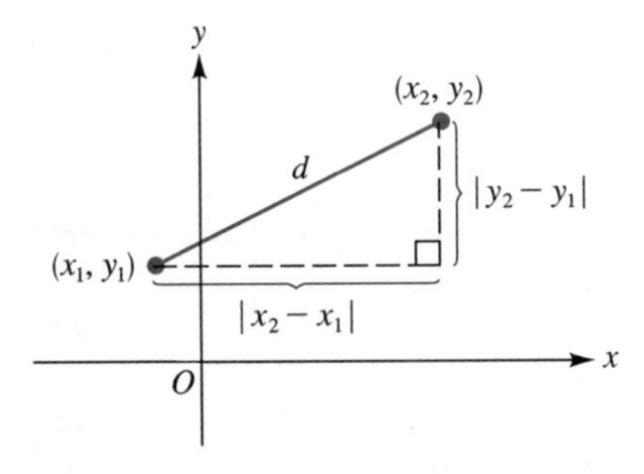

FIGURE 6

$$d^2 = |x_2 - x_1|^2 + |y_2 - y_1|^2$$
$$= (x_2 - x_1)^2 + (y_2 - y_1)^2 \qquad \text{(Why?)}$$

and therefore

$$d = \sqrt{(x_2 - x_1)^2 + (y_2 - y_1)^2}$$

This last equation is referred to as the **distance formula.** For reference, we restate it in the box that follows.

THE DISTANCE FORMULA

The distance d between the points (x_1, y_1) and (x_2, y_2) is given by

$$d = \sqrt{(x_2 - x_1)^2 + (y_2 - y_1)^2}$$

Examples 2 and 3 demonstrate some simple calculations involving the distance formula. *Note:* In computing the distance between two given points, it does not matter which one you treat as (x_1, y_1) and which as (x_2, y_2), because quantities such as $x_2 - x_1$ and $x_1 - x_2$ are negatives of each other and so their squares are equal.

EXAMPLE 2 Calculate the distance between the points $(2, -6)$ and $(5, 3)$.

Solution Substituting $(2, -6)$ for (x_1, y_1) and $(5, 3)$ for (x_2, y_2) in the distance formula, we have

$$\begin{aligned} d &= \sqrt{(5-2)^2 + [3-(-6)]^2} \\ &= \sqrt{3^2 + 9^2} = \sqrt{90} \\ &= \sqrt{9}\sqrt{10} = 3\sqrt{10} \end{aligned}$$

You should check for yourself that the same answer is obtained using $(2, -6)$ as (x_2, y_2) and $(5, 3)$ as (x_1, y_1).

EXAMPLE 3 Is the triangle with vertices $D(-2, -1)$, $E(4, 1)$, and $F(3, 4)$ a right triangle?

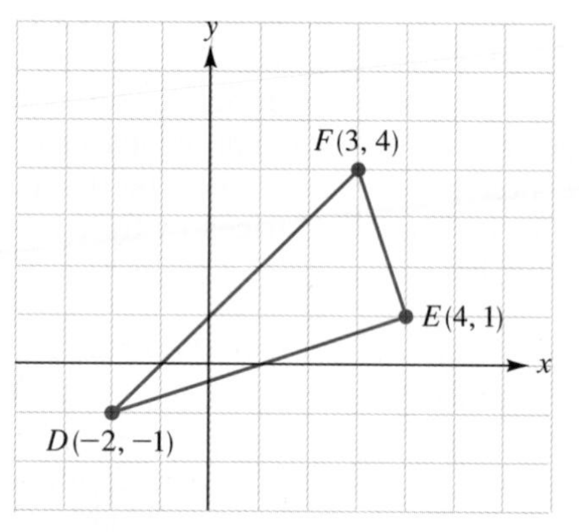

FIGURE 7

Solution First we sketch the triangle in question; see Figure 7. From the sketch it appears that angle E could be a right angle, but certainly this is not a proof. We need to use the distance formula to calculate the lengths of the three sides and then check whether any relation of the form $a^2 + b^2 = c^2$ holds. The calculations are as follows:

$$\begin{aligned} DE &= \sqrt{[4-(-2)]^2 + [1-(-1)]^2} = \sqrt{36+4} = \sqrt{40} \\ EF &= \sqrt{(4-3)^2 + (1-4)^2} = \sqrt{1+9} = \sqrt{10} \\ DF &= \sqrt{[3-(-2)]^2 + [4-(-1)]^2} = \sqrt{25+25} = \sqrt{50} \end{aligned}$$

Because $(\sqrt{40})^2 + (\sqrt{10})^2 = (\sqrt{50})^2$, we are indeed guaranteed that $\triangle DEF$ is a right triangle. (In Section 2.3, you'll see that this result can be obtained more efficiently using the concept of "slope.")

We can use the distance formula to obtain the equation of a circle. Figure 8 shows a circle with center (h, k) and radius r. By definition, a point (x, y) is on this circle if and only if the distance from (x, y) to (h, k) is r. Thus we have

$$\sqrt{(x-h)^2 + (y-k)^2} = r$$

or, equivalently,

$$(x-h)^2 + (y-k)^2 = r^2 \tag{1}$$

(These last two equations are equivalent because two nonnegative quantities are equal if and only if their squares are equal.)

The work in the previous paragraph tells us two things. First, if a point (x, y) lies on the circle in Figure 8, then x and y together satisfy equation (1). Second, if a pair of numbers x and y satisfies equation (1), then the point (x, y) lies on the circle in Figure 8. (Does it sound to you as if the previous two sentences say the same thing? They don't! Think about it.) Equation (1) is called **the standard form for the equation of a circle.** For reference, we record the result in the box that follows.

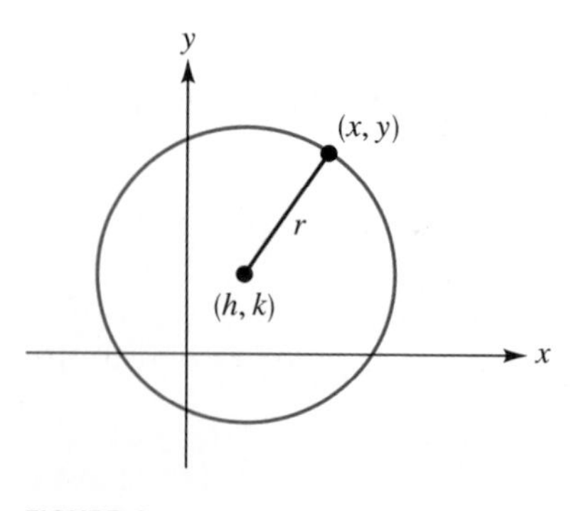

FIGURE 8

THE EQUATION OF A CIRCLE IN STANDARD FORM

The equation of a circle with center (h, k) and radius r is

$$(x-h)^2+(y-k)^2=r^2$$

EXAMPLE 4

(a) Write the equation of the circle with center $(-2, 1)$ and radius 3.
(b) Does the point $(-4, 3)$ lie on this circle?

Solution
(a) In the equation $(x-h)^2+(y-k)^2=r^2$, we substitute the given values $h=-2$, $k=1$, and $r=3$. This yields

$$[x-(-2)]^2+(y-1)^2=3^2$$

or

$$(x+2)^2+(y-1)^2=9 \tag{2}$$

This is the standard form for the equation of the given circle.
(b) To find out if the point $(-4, 3)$ lies on the circle, we check to see if the coordinates $x=-4$ and $y=3$ satisfy equation (2). We have

$$(-4+2)^2+(3-1)^2 \stackrel{?}{=} 9$$
$$(-2)^2+2^2 \stackrel{?}{=} 9$$
$$8=9 \qquad \text{FALSE}$$

This shows that the values $x=-4$ and $y=3$ do not satisfy equation (2). Consequently, the point $(-4, 3)$ does not lie on the circle.

In the example just concluded, we found the standard form for the equation of a given circle. An alternative form of this equation is found by carrying out the indicated algebra in equation (2) to obtain

$$x^2+4x+4+y^2-2y+1=9$$

and therefore

$$x^2+4x+y^2-2y-4=0$$

The disadvantage to this alternative form is that information regarding the center and radius is no longer readily visible.

When the equation of a circle is not in standard form, the technique of completing the square (discussed in Section 1.4) can be used to convert the equation to standard form. Example 5 shows how this is done.

EXAMPLE 5 Determine the center and radius of the circle given by

$$4x^2-24x+4y^2+16y+51=0$$

Solution First we divide through by 4 so that the coefficients of x^2 and y^2 are both 1:

$$x^2-6x+y^2+4y+\frac{51}{4}=0$$

Now, in preparation for completing the squares, we write

$$(x^2 - 6x + ____) + (y^2 + 4y + ____) = -\frac{51}{4}$$

To complete the square in x, we need to add $\left(-\frac{6}{2}\right)^2$, or 9. To complete the square in y, we need to add $\left(\frac{4}{2}\right)^2$, or 4. Thus we have

$$(x^2 - 6x + 9) + (y^2 + 4y + 4) = -\frac{51}{4} + 9 + 4$$

or

$$(x - 3)^2 + (y + 2)^2 = -\frac{51}{4} + \frac{52}{4} = \frac{1}{4}$$

$$(x - 3)^2 + (y + 2)^2 = \left(\frac{1}{2}\right)^2$$

This is the equation in standard form. The center of the circle is $(3, -2)$; the radius is $1/2$.

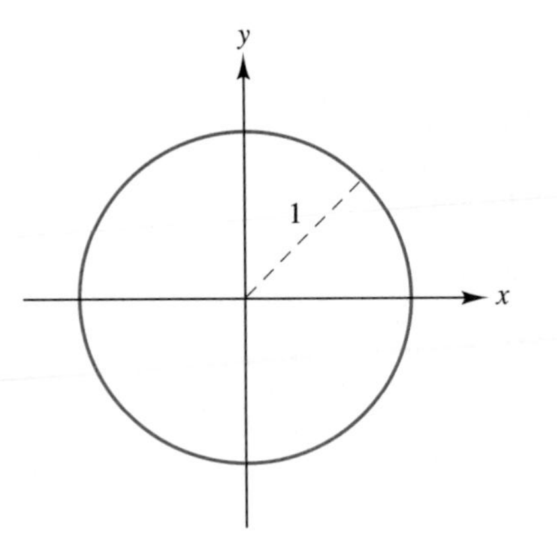

FIGURE 9
The unit circle $x^2 + y^2 = 1$.

For a large portion of our work in trigonometry later in this book, we will refer to the **unit circle.** This is the circle with center $(0, 0)$ and radius 1 (see Figure 9). To obtain an equation for this circle, we substitute the values $h = 0$, $k = 0$, and $r = 1$ in the standard form for the equation of a circle. This gives us $(x - 0)^2 + (y - 0)^2 = 1^2$ or, more simply,

$$x^2 + y^2 = 1$$

This is the equation for the unit circle; we will use this equation in the next example.

EXAMPLE 6 As indicated in Figure 10, the point P lies in the fourth quadrant, on the unit circle. If the x-coordinate of P is $1/2$, determine the y-coordinate of P.

Solution Since P lies on the unit circle, the coordinates of P satisfy the equation $x^2 + y^2 = 1$. We are given that the x-coordinate of P is $1/2$. Substituting the value $x = 1/2$ in the equation $x^2 + y^2 = 1$ gives us

$$\left(\frac{1}{2}\right)^2 + y^2 = 1$$

$$y^2 = 1 - \frac{1}{4} = \frac{3}{4} \quad \text{and therefore} \quad y = \pm\frac{\sqrt{3}}{2}$$

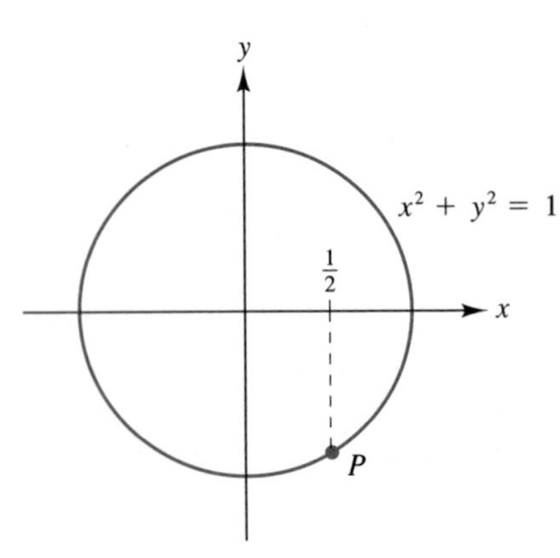

FIGURE 10

We want to choose the negative root here because P lies in Quadrant IV (in which all y-coordinates are negative). Thus, the y-coordinate of P is $-\sqrt{3}/2$.

There is another way to write the equation of a circle that is useful in work with graphing calculators and computers. This involves solving the equation for y. As an example, we'll use equation (2) from Example 4(a):

$$(x + 2)^2 + (y - 1)^2 = 9$$

GRAPHICAL PERSPECTIVE

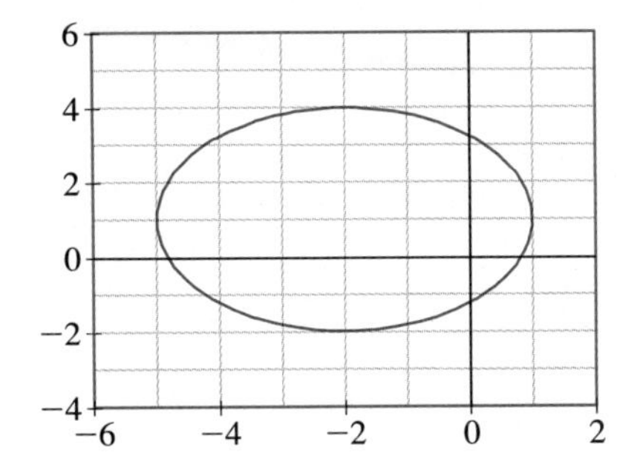

FIGURE 12
A distorted view of the circle $(x+2)^2+(y-1)^2=9$ results when the unit of length or scale on the x-axis differs from that on the y-axis. In this particular view, the unit of length on the y-axis is two-thirds that on the x-axis. (Specific techniques for obtaining an undistorted view are discussed in the Graphics Calculator Supplement.)

We solve this equation for y as follows:

$$(y-1)^2 = 9-(x+2)^2$$
$$y-1 = \pm\sqrt{9-(x+2)^2}$$
$$y = 1\pm\sqrt{9-(x+2)^2}$$

The equation of the circle with center $(-2, 1)$ and radius 3, is therefore equivalent to the *pair* of equations $y=1+\sqrt{9-(x+2)^2}$ and $y=1-\sqrt{9-(x+2)^2}$. In work with a graphing calculator or computer, both of these equations are graphed to yield the required circle. See Figure 11 (below) and the Graphical Perspective in Figure 12. In Figure 11, note that the equation with the positive square root represents the upper semicircle, and the equation with the negative square root represents the lower semicircle.

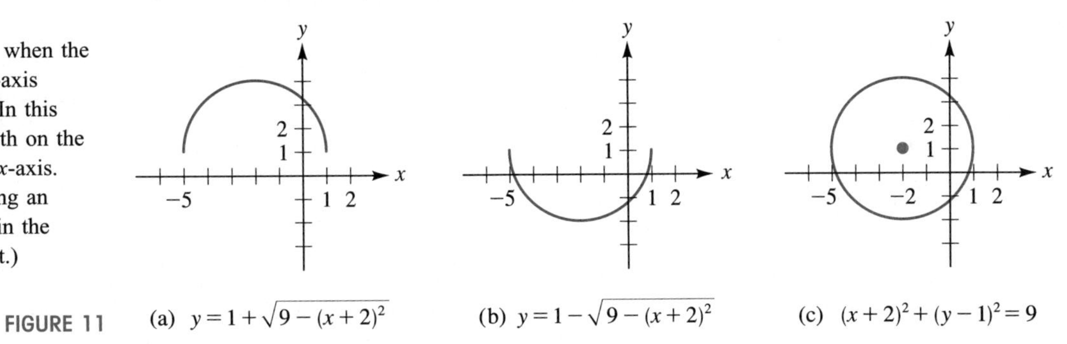

FIGURE 11 (a) $y=1+\sqrt{9-(x+2)^2}$ (b) $y=1-\sqrt{9-(x+2)^2}$ (c) $(x+2)^2+(y-1)^2=9$

In Example 7 we use a simple result that you may recall from previous courses: the *midpoint formula*. This result is summarized in the box that follows. (For a proof of the formula, see Exercise 61.)

THE MIDPOINT FORMULA

The midpoint of the line segment joining the points $P(x_1, y_1)$ and $Q(x_2, y_2)$ is

$$\left(\frac{x_1+x_2}{2}, \frac{y_1+y_2}{2}\right)$$

EXAMPLE

The midpoint of the line segment joining $(2, -15)$ and $(4, 5)$ is

$$\left(\frac{2+4}{2}, \frac{-15+5}{2}\right)=(3, -5)$$

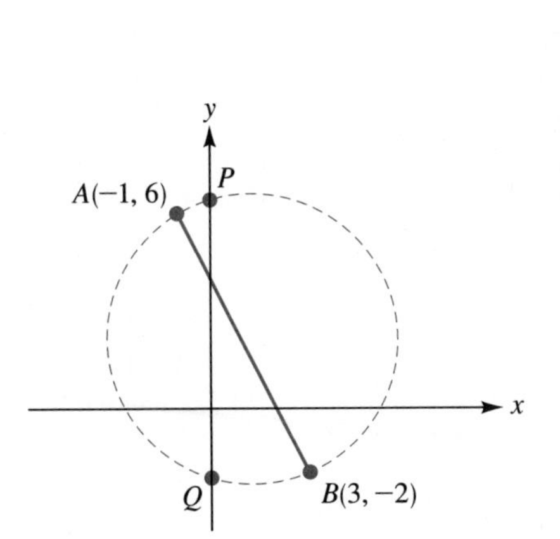

FIGURE 13
$\overline{AB}$ is a diameter. What are the y-intercepts of the circle?

EXAMPLE 7 Points $A(-1, 6)$ and $B(3, -2)$ are the endpoints of a diameter of a circle, as shown in Figure 13. Determine the coordinates of the points P and Q where the circle intersects the y-axis. (The y-coordinates of the points P and Q are called the *y-intercepts* of the circle. Both y-intercepts and x-intercepts are discussed more generally in the next section.)

Solution The x-coordinates of both P and Q are zero. So, our strategy here is as follows: If we knew the equation of the circle, then we could find the required y-coordinates by setting $x=0$ in the equation of the circle. To write the equation of the circle, we need to know the center and the radius. We can find these by using the midpoint formula and the distance formula.

The center of the circle is the midpoint of diameter $\overline{AB}$. Therefore, the coordinates of the center are

$$x = \frac{-1+3}{2} = 1 \quad \text{and} \quad y = \frac{6+(-2)}{2} = 2$$

So the center is the point $(1, 2)$. Since the radius r is the distance from the center to point $A(-1, 6)$, we have

$$r = \sqrt{[1-(-1)]^2 + (2-6)^2} = \sqrt{4+16} = \sqrt{20}$$

Therefore,

$$r^2 = 20$$

It follows now that the equation of the circle is

$$(x-1)^2 + (y-2)^2 = 20$$

For the y-intercepts, we set $x = 0$ in this last equation to obtain

$$(y-2)^2 = 19$$
$$y - 2 = \pm\sqrt{19}$$
$$y = 2 \pm \sqrt{19}$$

Thus, the y-intercepts are $2+\sqrt{19}$, which is positive, and $2-\sqrt{19}$, which is negative. Consequently, the coordinates of P and Q are $P(0, 2+\sqrt{19})$ and $Q(0, 2-\sqrt{19})$.

EXERCISE SET 2.1

8, 21, 23, 33, 35, 43, 45

A

1. Plot the points $(5, 2)$, $(-4, 5)$, $(-4, 0)$, $(-1, -1)$, and $(5, -2)$.
2. Draw the square $ABCD$ whose vertices (corners) are $A(1, 0)$, $B(0, 1)$, $C(-1, 0)$, and $D(0, -1)$.
3. **(a)** Draw the right triangle PQR with vertices $P(1, 0)$, $Q(5, 0)$, and $R(5, 3)$.
   (b) Use the formula for the area of a triangle, $A = \frac{1}{2}bh$, to find the area of triangle PQR in part (a).
4. **(a)** Draw the trapezoid $ABCD$ with vertices $A(0, 0)$, $B(7, 0)$, $C(6, 4)$, and $D(4, 4)$.
   (b) Compute the area of the trapezoid. (See the inside front cover of this book for the appropriate formula.)

In Exercises 5–10, calculate the distance between the given points.

5. **(a)** $(0, 0)$ and $(-3, 4)$ **(b)** $(2, 1)$ and $(7, 13)$
6. **(a)** $(-1, -3)$ and $(-5, 4)$ **(b)** $(6, -2)$ and $(-1, 1)$
7. **(a)** $(-5, 0)$ and $(5, 0)$ **(b)** $(0, -8)$ and $(0, 1)$
8. **(a)** $(-5, -3)$ and $(-9, -6)$
   (b) $(\frac{9}{2}, 3)$ and $(-2\frac{1}{2}, -1)$
9. $(1, \sqrt{3})$ and $(-1, -\sqrt{3})$
10. $(-3, 1)$ and $(374, -335)$ (Use a calculator.)
11. Which point is farther from the origin?
    (a) $(3, -2)$ or $(4, \frac{1}{2})$
    (b) $(-6, 7)$ or $(9, 0)$
12. Use the distance formula to show that, in each case, the triangle with given vertices is an isosceles triangle.
    (a) $(0, 2)$, $(7, 4)$, $(2, -5)$
    (b) $(-1, -8)$, $(0, -1)$, $(-4, -4)$
    (c) $(-7, 4)$, $(-3, 10)$, $(1, 3)$
13. In each case, determine if the triangle with the given vertices is a right triangle. *Hint:* Find the lengths of the sides and then use the *converse* of the Pythagorean theorem.
    (a) $(7, -1)$, $(-3, 5)$, $(-12, -10)$
    (b) $(4, 5)$, $(-3, 9)$, $(1, 3)$
    (c) $(-8, -2)$, $(1, -1)$, $(10, 19)$
14. **(a)** Two of the three triangles specified in Exercise 13 are right triangles. Find their areas.

(b) Calculate the area of the remaining triangle in Exercise 13 by using the following formula for the area A of a triangle with vertices (x_1, y_1), (x_2, y_2), and (x_3, y_3):

$$A = \frac{1}{2}|x_1y_2 - x_2y_1 + x_2y_3 - x_3y_2 + x_3y_1 - x_1y_3|$$

(The derivation of this formula is given in Exercise 64.)

(c) Use the formula giveh in part (b) to check your answers in part (a).

15. Use the formula given in Exercise 14(b) to calculate the area of the triangle with vertices $(1, -4)$, $(5, 3)$, and $(13, 17)$. Conclusion?

16. The coordinates of points A, B, and C are $A(-4, 6)$, $B(-1, 2)$, and $C(2, -2)$.
(a) Show that $AB = BC$ by using the distance formula.
(b) Show that $AB + BC = AC$ by using the distance formula.
(c) What can you conclude from parts (a) and (b)?

In Exercises 17–24, determine the center and the radius for the circle. Also, find the y-coordinates of the points (if any) where the circle intersects the y-axis.

17. $(x - 3)^2 + (y - 1)^2 = 25$

18. $x^2 + (y + 1)^2 = 20$

19. $x^2 + y^2 = \sqrt{2}$

20. $x^2 + y^2 - 10x + 2y + 17 = 0$

21. $x^2 + y^2 + 8x - 6y = -24$

22. $4x^2 - 4x + 4y^2 - 63 = 0$

23. $9x^2 + 54x + 9y^2 - 6y + 64 = 0$

24. $3x^2 + 3y^2 + 5x - 4y = 1$

In Exercises 25–32, you are given either the x- or the y-coordinate of a point P on the unit circle, and you are given the quadrant in which P is located. Find the remaining coordinate (as in Example 6). In Exercises 31 and 32, use a calculator and round to three decimal places.

25. **(a)** $x = 3/5$; I
(b) $x = 3/5$; IV

26. **(a)** $x = -5/13$; II
(b) $x = -5/13$; III

27. **(a)** $y = -\sqrt{2}/2$; III
(b) $y = -\sqrt{2}/2$; IV

28. **(a)** $y = \sqrt{3}/2$; II
(b) $y = -\sqrt{3}/2$; III

29. **(a)** $x = -2/3$; II
(b) $x = -2/3$; III

30. **(a)** $y = 2/3$; I
(b) $y = 23/$; II

31. **(a)** $x = 0.652$; I
(b) $y = 0.652$; I

32. **(a)** $x = -0.293$; II
(b) $y = -0.293$; III

In Exercises 33 and 34, determine the equation of the circle in standard form, given the coordinates of the diameter $\overline{PQ}$.

33. $P(-4, -2)$ and $Q(6, 4)$

34. $P(1, -3)$ and $Q(-5, -5)$

In Exercises 35 and 36, find the midpoint of the line segment joining points P and Q.

35. **(a)** $P(3, 2)$ and $Q(9, 8)$
(b) $P(-4, 0)$ and $Q(5, -3)$
(c) $P(3, -6)$ and $Q(-1, -2)$

36. **(a)** $P(12, 0)$ and $Q(12, 8)$
(b) $P(\frac{3}{5}, -\frac{2}{3})$ and $Q(0, 0)$
(c) $P(1, \pi)$ and $Q(3, 3\pi)$

37. The coordinates of A and B are $A(-1, 2)$ and $B(5, -3)$. If B is the midpoint of line segment $\overline{AC}$, what are the coordinates of C?

38. The coordinates of the points S and T are $S(4, 6)$ and $T(10, 2)$. If M is the midpoint of $\overline{ST}$, find the midpoint of $\overline{SM}$.

39. **(a)** Sketch the parallelogram with vertices $A(-7, -1)$, $B(4, 3)$, $C(7, 8)$, and $D(-4, 4)$.
(b) Compute the midpoints of the diagonals $\overline{AC}$ and $\overline{BD}$.
(c) What conclusion can you draw from part (b)?

40. The vertices of $\triangle ABC$ are $A(1, 1)$, $B(9, 3)$, and $C(3, 5)$.
(a) Find the perimeter of $\triangle ABC$.
(b) Find the perimeter of the triangle that is formed by joining the midpoints of the three sides of $\triangle ABC$.
(c) Compute the ratio of the perimeter in part (a) to the perimeter in part (b).
(d) What theorem from geometry provides the answer for part (c) without using the results in (a) and (b)?

41. This exercise refers to $\triangle ABC$ whose vertices are specified in Exercise 40.
(a) Compute the sum of the squares of the lengths of the sides of $\triangle ABC$.
(b) Compute the sum of the squares of the lengths of the three medians. (A **median** is a line segment drawn from a vertex to the midpoint of the opposite side.)
(c) Check that the ratio of the answer in part (a) to that in part (b) is 4/3. *Remark:* As Exercise 58 asks you to show, this ratio is 4/3 for any triangle.

42. **(a)** Sketch the circle of radius 1 centered at the origin.
(b) Write the equation for this circle.
(c) Does the point $(\frac{3}{5}, \frac{4}{5})$ lie on this circle?
(d) Does the point $(-1/2, \sqrt{3}/2)$ lie on this circle?

In Exercises 43 and 44, an equation of a circle is given. Solve the equation for y (as on page 41), and then graph the resulting semicircles, as in Figures 11(a) and 11(b) in the text.

43. $(x + 5)^2 + (y + 3)^2 = 4$

44. $(x - 4)^2 + (y - 1)^2 = 9$

B

45. The center of a circle is the point (3, 2). If the point (−2, −10) lies on this circle, find the equation of the circle.

46. Find the equation of the circle tangent to the x-axis and with center (3, 5). *Hint:* First draw a sketch.

47. Find the equation of the circle tangent to the y-axis and with center (3, 5).

48. Find the equation of the circle passing through the origin and with center (3, 5).

49. Use the Pythagorean theorem to find the length a in the following figure. Then find b, c, d, e, f, and g.

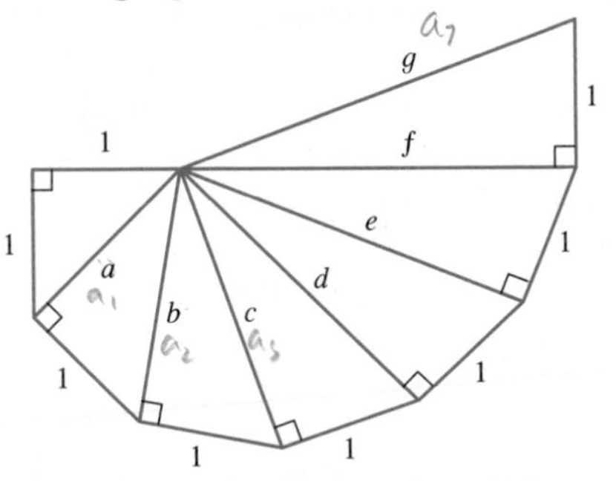

Remark: This figure provides a geometric construction for the irrational numbers $\sqrt{2}, \sqrt{3}, \ldots, \sqrt{n}$ where n is a nonsquare natural number. According to Boyer's *A History of Mathematics,* 2nd ed. (New York: John Wiley & Sons, Inc., 1991), "... Plato ... says that his teacher Theodorus of Cyrene ... was the first to prove the irrationality of the square roots of the nonsquare integers from 3 to 17, inclusive. It is not known how he did this, nor why he stopped with $\sqrt{17}$." One plausible reason for Theodorus stopping with $\sqrt{17}$ may have to do with the figure shown here. Theodorus may have known that the figure begins to overlap itself at the stage where $\sqrt{18}$ would be constructed.

50. (A numerologist's delight) Using the Pythagorean theorem and your calculator, compute the area of a right triangle in which the lengths of the hypotenuse and one leg are 2045 and 693, respectively.

In Exercises 51 and 52, determine the equation of the circle that satisfies the given conditions. Write the equation in standard form.

51. The circle passes through the origin and is concentric with the circle $x^2 - 6x + y^2 - 4y + 4 = 0$.

52. The circle passes through (−4, 1) and its center is the midpoint of the line segment joining the centers of the two circles $x^2 + y^2 - 6x - 4y + 12 = 0$ and $x^2 + y^2 - 14x + 47 = 0$.

53. Find a value for t such that points (0, 2) and (12, t) are 13 units apart. *Hint:* By the distance formula, $13 = \sqrt{(12-0)^2 + (t-2)^2}$. Now square both sides and solve for t.

54. (a) Find values for t such that the points (−2, 3) and (t, 1) are six units apart.
(b) Can you find a real number t such that the points (−2, 3) and (t, 1) are one unit apart? Why or why not?

55. The diagonals of a parallelogram bisect each other. Steps (a), (b), and (c) outline a proof of this theorem.
(a) In the parallelogram $OABC$ shown in the figure, check that the coordinates of B must be $(a + b, c)$.

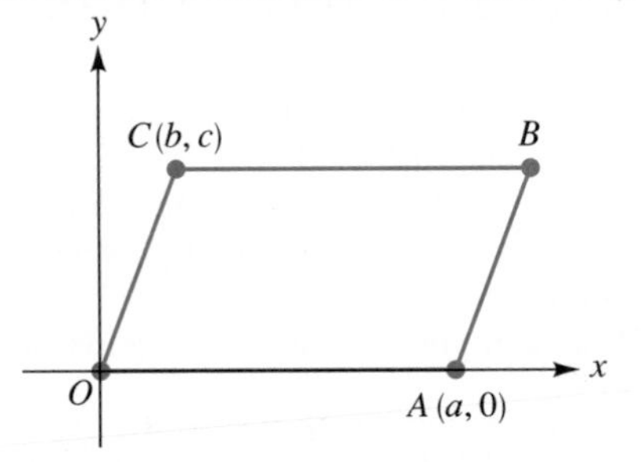

(b) Use the midpoint formula to calculate the midpoints of diagonals $\overline{OB}$ and $\overline{AC}$.
(c) The two answers in part (b) are identical. This shows that the two diagonals do indeed bisect each other, as we wished to prove.

56. Prove that in a parallelogram, the sum of the squares of the lengths of the diagonals equals the sum of the squares of the lengths of the four sides. (Use the figure in Exercise 55.)

57. Use the figure in Exercise 55 to prove the following theorem: If the lengths of the diagonals of a parallelogram are equal, then the parallelogram is a rectangle. *Hint:* Equate the two expressions for the lengths of the diagonals and conclude that $ab = 0$. The case in which $a = 0$ can be discarded.

58. The sum of the squares of the sides of any triangle equals 4/3 of the sum of the squares of the medians. Prove this, taking the vertices to be (0, 0), (2, 0), and (2a, 2b). *Note:* The expression *squares of the sides* means *squares of the lengths of the sides.*

59. In $\triangle ABC$, let M be the midpoint of side $\overline{BC}$. Prove that $AB^2 + AC^2 = 2(BM^2 + AM^2)$. *Hint:* Let the coordinates be $A(0, 0)$, $B(2, 0)$, and $C(2a, 2b)$.

60. If the point (x, y) is equidistant from the points (−3, −3) and (5, 5), show that x and y satisfy the equation $x + y = 2$. *Hint:* Use the distance formula.

61. Suppose that the coordinates of points P, Q, and M are

$$P(x_1, y_1) \qquad Q(x_2, y_2) \qquad M\left(\frac{x_1 + x_2}{2}, \frac{y_1 + y_2}{2}\right)$$

Follow steps (a) and (b) to prove that M is the midpoint of the line segment from P to Q.

(a) By computing both of the distances PM and MQ, show that $PM = MQ$. (This shows that M lies on the perpendicular bisector of line segment $\overline{PQ}$, but it does not show that M actually lies *on* $\overline{PQ}$.)

(b) Show that $PM + MQ = PQ$. (This shows that M does lie on $\overline{PQ}$.)

62. This problem outlines one of the shortest proofs of the Pythagorean theorem. The proof was discovered by the Hindu mathematician Bhāskara (1114–ca 1185). (For other proofs, see the next exercise and Exercise 112 on page 99.) In the figure we are given a right triangle ACB with the right angle at C, and we want to prove that $a^2 + b^2 = c^2$. In the figure, $\overline{CD}$ is drawn perpendicular to $\overline{AB}$.

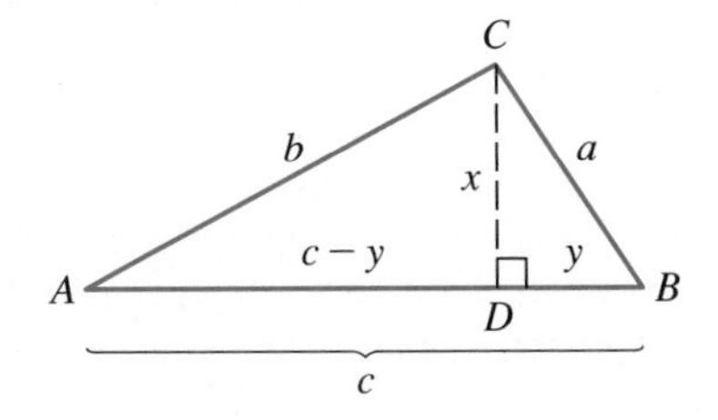

(a) Check that $\angle CAD = \angle DCB$ and that $\triangle BCD$ and $\triangle BAC$ are similar.

(b) Use the result in part (a) to obtain the equation $a/y = c/a$, and conclude that $a^2 = cy$.

(c) Show that $\triangle ACD$ is similar to $\triangle ABC$, and use this to deduce that $b^2 = c^2 - cy$.

(d) Combine the two equations deduced in parts (b) and (c) to arrive at $a^2 + b^2 = c^2$.

63. One of the oldest and simplest proofs of the Pythagorean theorem is found in the ancient Chinese text *Chou Pei Suan Ching.* This text was written during the Han period (206 B.C.–A.D. 222), but portions of it may date back to 600 B.C. The proof in *Chou Pei Suan Ching* is based on the following diagram.

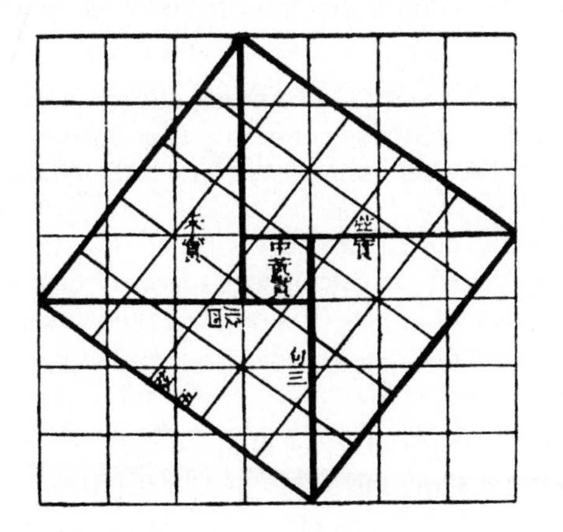

A diagram accompanying a proof of the ''Pythagorean'' theorem in the ancient Chinese text *Chou Pei Suan Ching* [from *Science and Civilisation in China, vol. 3,* by Joseph Needham (Cambridge: Cambridge University Press, 1959.)]

In this exercise we explain the details of the proof.

(a) Starting with the right triangle in Figure A, we make four replicas of this triangle and arrange them to form the pattern shown in Figure B. Explain why the outer quadrilateral in Figure B is a square.

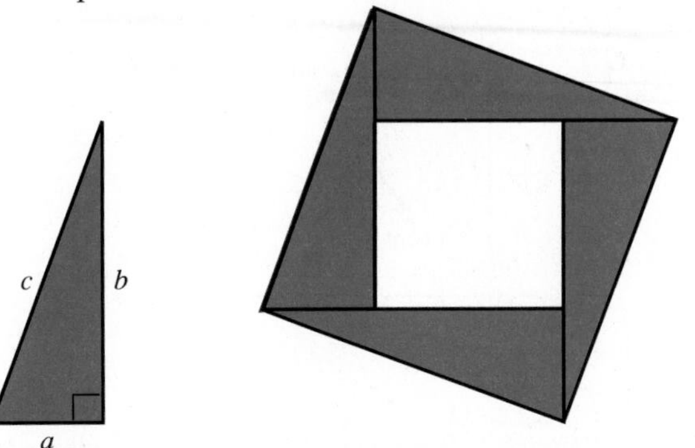

FIGURE A FIGURE B

(b) The unshaded region in the center of Figure B is a square. What is the length of each side?

(c) The area of the outer square in Figure B is $(\text{side})^2 = c^2$. This area can also be computed by adding up the areas of the four right triangles and the inner square. Compute the area in this fashion. After simplifying, you should obtain $a^2 + b^2$. Now conclude that $a^2 + b^2 = c^2$, since both expressions represent the same area.

C

64. This problem indicates a method for calculating the area of a triangle when the coordinates of the three vertices are given.

(a) Calculate the area of $\triangle ABC$ in the figure.
Hint: First calculate the area of the rectangle enclosing $\triangle ABC$, then subtract the areas of the three right triangles.

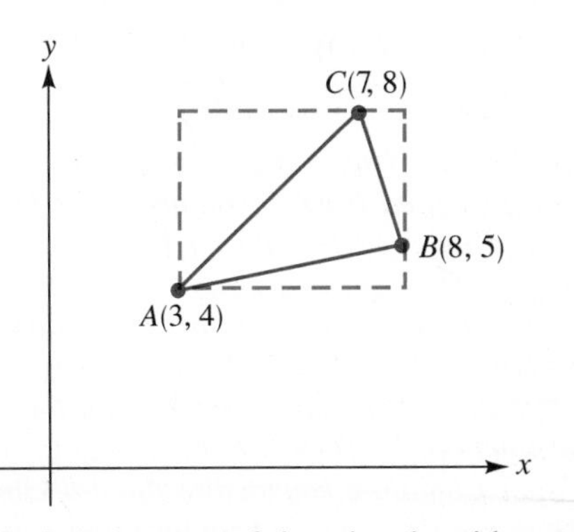

(b) Calculate the area of the triangle with vertices (1, 3), (4, 1), and (10, 4). *Hint:* Work with an enclosing rectangle and three right triangles, as in part (a).

(c) Using the same technique that you used in parts (a) and (b), show that the area of the triangle in the following figure is given by

$$A = \frac{1}{2}(x_1y_2 - x_2y_1 + x_2y_3 - x_3y_2 + x_3y_1 - x_1y_3)$$

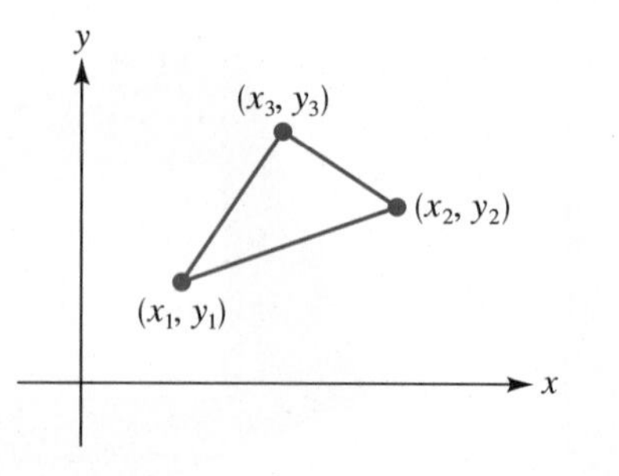

Remark: If we use absolute value signs instead of the parentheses, then the formula will hold regardless of the relative positions or quadrants of the three vertices. Thus, the area of a triangle with vertices (x_1, y_1), (x_2, y_2), (x_3, y_3) is given by

$$A = \frac{1}{2}|x_1y_2 - x_2y_1 + x_2y_3 - x_3y_2 + x_3y_1 - x_1y_3|$$

65. Refer to the following figure.

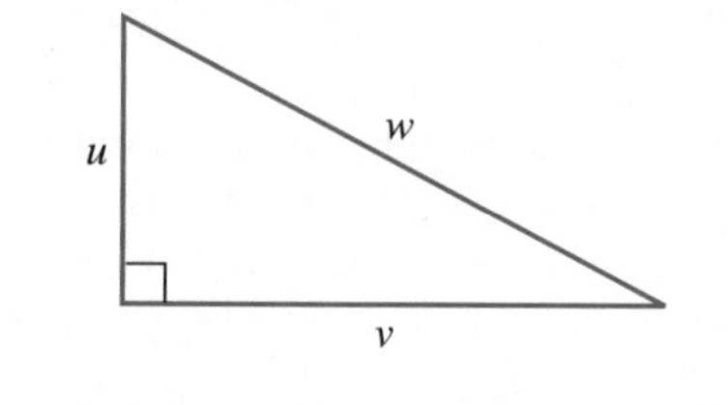

(a) If $u = \dfrac{2(m+n)}{n}$ and $v = \dfrac{4m}{m-n}$, show that

$$w = \frac{2(m^2+n^2)}{(m-n)n}$$

(b) Show that $\frac{1}{2}uv = u + v + w$ (that is, the perimeter is numerically equal to the area).

(c) Give an example of a right triangle in which the length of each side is an integer and the perimeter is numerically equal to the area.

GRAPHING UTILITY EXERCISES FOR SECTION 2.1

In Exercises 1–8, graph the circles using the technique explained in the text. In each case, choose the viewing window so that all of the circle is visible. Also, be sure that the same size unit is used on both axes so that the circle is not distorted.

1. $x^2 + y^2 = 36$

2. $x^2 + y^2 = 144$

3. $x^2 + y^2 = 2$

4. $x^2 + y^2 = 0.1$

5. $(x+2)^2 + (y-4)^2 = 9$

6. $(x-1)^2 + (y-3)^2 = 4$

7. $(x-12)^2 + (y+10)^2 = 90$

8. $x^2 + (y+7)^2 = 75$

In Exercises 9–14, assume that P is a point on the unit circle $x^2 + y^2 = 1$. In each case, you are given both the quadrant for P and the x-coordinate for P.

(a) *Use a graphing utility to graph the circle and to approximate (to one decimal place) the corresponding y-coordinate of P.*

(b) *Use algebra to compute an expression for the exact y-coordinate of P. Then use the calculator to give a decimal approximation, rounded to three decimal places. Check to see that your calculator approximation is consistent with the graphical estimate obtained in part (a).*

9. Quadrant I; $x = 2/3$

10. Quadrant II; $x = -1/4$

11. Quadrant II; $x = -3/5$

12. Quadrant IV; $x = 5/8$

13. Quadrant IV; $x = 0.7$

14. Quadrant III; $x = -0.5$

You know from the text that the graph of the equation $x^2 + y^2 = 9$ is a circle with center (0, 0) and radius 3. In Exercises 15–21, use a graphing utility to investigate what happens when the equation is changed slightly. (As in graphing circles, you'll first need to solve the given equations for y.)

15. (a) $x^2 + 2y^2 = 9$ (The resulting curve is an *ellipse*. We'll study this curve in a later chapter.)

(b) On the same set of axes, graph the ellipse $x^2 + 2y^2 = 9$ and the circle $x^2 + y^2 = 9$. At which points do the two curves appear to intersect?

(c) Carry out the calculations to verify that the points indicated in part (b) are indeed the exact intersection points.

16. $2x^2 + y^2 = 9$

17. (a) $x^2 - y^2 = 9$ (The resulting curve is a *hyperbola*. We'll study this curve in a later chapter.)

(b) $x^2 - y^2 = 0$

(c) On the same set of axes, graph the equations given in parts (a) and (b). Are there any intersection points?

18. $x^2 - y^3 = 9$

19. $x^3 - y^2 = 9$

20. **(a)** $x^3 + y^3 = 9$ **(b)** $x^5 + y^5 = 9$

(c) On the same set of axes, graph the equations given in parts (a) and (b). Based on the pattern you see, what do you think the graph of the equation $x^7 + y^7 = 9$ will look like? Draw a sketch (by hand) to show your prediction. Then use a graphing utility to see how accurate your prediction is.

21. **(a)** $x^4 + y^4 = 9$ **(b)** $x^6 + y^6 = 9$

(c) On the same set of axes, graph the equations given in parts (a) and (b). Based on the pattern you see, what do you think the graph of the equation $x^8 + y^8 = 9$ will look like? Draw a sketch (by hand) to show your prediction. Then use a graphing utility to see how accurate your prediction is.

In Exercises 22–24, graph the equations. Hint: Make use of the zero-product property from Section 1.4.

22. $(x^2 + y^2 - 1)(x^2 + y^2 - 4) = 0$

23. $(x^2 - y^2 - 1)(x^2 - y^2) = 0$

24. $[(x - 1)^2 + y^2 - 1][(x + 1)^2 + y^2 - 1] = 0$

2.2 GRAPHS AND EQUATIONS, A SECOND LOOK

He [Pierre de Fermat (1601–1665)] *invented analytic geometry in 1629 and described his ideas in a short work entitled* Introduction to Plane and Solid Loci, *which circulated in manuscript form from early 1637 on but was not published during his lifetime. The credit for this achievement has usually been given to Descartes on the basis of his* Geometry, *which was published late in 1637. . . . However, nothing that we would recognize as analytic geometry can be found in Descartes's essay, except perhaps the idea of using algebra as a language for discussing geometric problems.*

George F. Simmons in *Calculus Gems, Brief Lives and Memorable Moments* (New York: McGraw-Hill, Inc.: 1992)

In this section we look at the connection between two basic skills that you've worked on in previous courses:

- graphing equations
- solving equations

We also present a list of six basic graphs to memorize. These graphs will provide handy examples when we discuss functions in Chapter 3. We begin by recalling the definition of the *graph* of an equation.

DEFINITION: The Graph of an Equation

The **graph** of an equation in two variables is the set of all points whose coordinates satisfy the equation.

TABLE 1 $y = 3x - 2$

x	y
0	−2
1	1
2	4
3	7
−1	−5
−2	−8

Suppose, for example, that we want to graph the equation $y = 3x - 2$. We begin by noting that the domain of the variable x in the expression $3x - 2$ is the set of all real numbers. Now we choose various values for x and in each case compute the corresponding y-value from the equation $y = 3x - 2$. For example, if $x = 0$, then $y = 3(0) - 2 = -2$. Table 1 summarizes the results of some of these calculations. The first line in Table 1 tells us that the point with coordinates $(0, -2)$ is on the graph of $y = 3x - 2$. Reading down the table, we see that some other points on the graph are $(1, 1)$, $(2, 4)$, $(3, 7)$, $(-1, -5)$, and $(-2, -8)$.

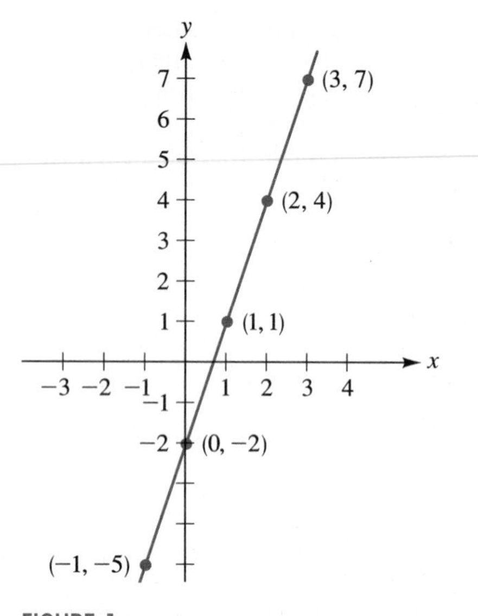

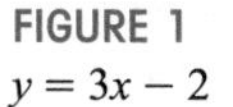

FIGURE 1
$y = 3x - 2$

We now plot (locate) the points we have determined and note, in this example, that they all appear to lie on a straight line. We draw the line indicated; this is the graph of $y = 3x - 2$ (see Figure 1). In Examples 1 and 2, we ask simple questions that will test your understanding of the process we've just described and of graphing in general.

EXAMPLE 1 Does the point $(1\frac{2}{3}, 2\frac{2}{3})$ lie on the graph of $y = 3x - 2$?

Solution Looking at the graph in Figure 1, we certainly see that the point $(1\frac{2}{3}, 2\frac{2}{3})$ is very close to the line, but from this visual inspection alone, we cannot be certain that this point actually lies on the line. To settle the question, then, we check to see if the values $x = 1\frac{2}{3} = \frac{5}{3}$ and $y = 2\frac{2}{3} = \frac{8}{3}$ together satisfy the equation $y = 3x - 2$. We have

$$\frac{8}{3} \stackrel{?}{=} 3\left(\frac{5}{3}\right) - 2$$

$$\frac{8}{3} \stackrel{?}{=} 3 \qquad \text{No!}$$

Since the coordinates $(1\frac{2}{3}, 2\frac{2}{3})$ evidently do not satisfy the equation $y = 3x - 2$, we conclude that the point $(1\frac{2}{3}, 2\frac{2}{3})$ is not on the graph. [The calculation further tells us that the point $(1\frac{2}{3}, 3)$ *is* on the graph.]

EXAMPLE 2 Find the value of the constant a, given that the graph of the equation $y = a(x - 2)^2 + 3$ passes through the origin.

Solution Since the graph passes through the origin, the values $x = 0$, $y = 0$ satisfy the given equation. So we have

$$0 = a(0 - 2)^2 + 3$$
$$0 = 4a + 3$$
$$-4a = 3$$
$$a = -3/4$$

The required value of a is $-3/4$. (Thus the graph of $y = -\frac{3}{4}(x - 2)^2 + 3$ passes through the origin; see Figure 2 for a Graphical Perspective.)

GRAPHICAL PERSPECTIVE

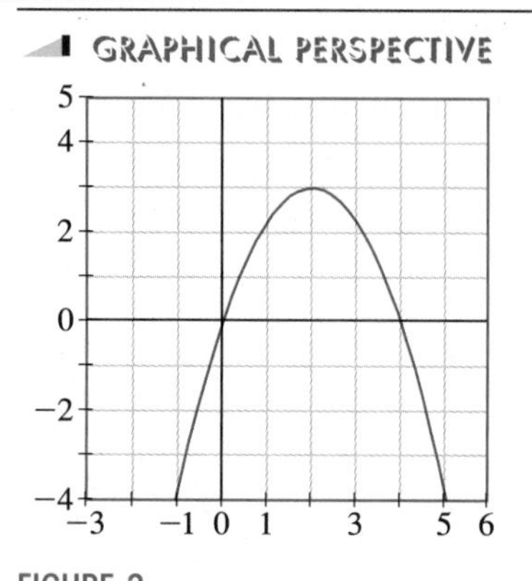

FIGURE 2
The graph of $y = -\frac{3}{4}(x - 2)^2 + 3$ passes through the origin.

The next example illustrates a type of graph that often appears in beginning calculus courses (in connection with the notion of a *limit*).

EXAMPLE 3 Graph the equation: $y = \dfrac{3x^2 - 5x + 2}{x - 1}$.

Solution We need to do two things here before we set up a table of values. First, note that the domain of the expression on the right-hand side of the given equation consists of all real numbers except $x = 1$. Thus, whatever the appearance of the resulting graph, it cannot contain a point with x-coordinate 1. Next, can we simplify the given expression? (If so, that would save work when we make the calculations for our table.) In this case, we can simplify things as follows:

$$y = \frac{3x^2 - 5x + 2}{x - 1} = \frac{(3x - 2)(x - 1)}{x - 1} = 3x - 2$$

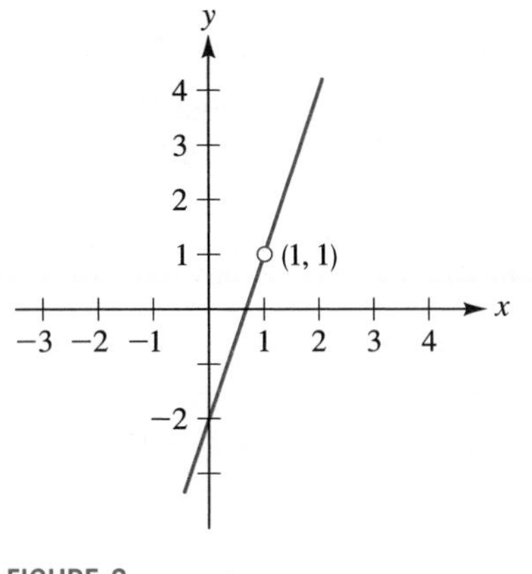

FIGURE 3
$y = \dfrac{3x^2 - 5x + 2}{x - 1}$

So, to graph the given equation we need only graph the simpler equation $y = 3x - 2$, remembering that the point with an x-coordinate of 1 must be excluded from the graph. We've previously done the work for $y = 3x - 2$ (see Figure 1). Thus, the required graph is obtained by deleting the point (1, 1) from the graph in Figure 1. This is shown in Figure 3. [The open circle in Figure 3 is used to indicate that the point (1, 1) does not belong to the graph.]

When we graph an equation, it's helpful to know where the curve or line crosses the x- or y-axis. By a ***y*-intercept** of a graph, we mean the y-coordinate of a point where the graph crosses the y-axis: For instance, returning to Figure 1, the y-intercept is -2. Notice that this y-intercept can be obtained algebraically just by setting x equal to zero in the given equation $y = 3x - 2$. In a similar fashion, an ***x*-intercept** of a graph is the x-coordinate of a point where the graph crosses the x-axis. In Figure 1 we can see that the x-intercept is a number between 0 and 1. To determine this x-intercept, we set y equal to zero in the equation $y = 3x - 2$. That yields $0 = 3x - 2$ and, consequently, $x = 2/3$. So the x-intercept of the graph in Figure 1 is $2/3$.

EXAMPLE 4 Figure 4 shows a graph of the equation $y = 2x^3 - 2x^2 - x$. Determine the x-intercepts of the graph.

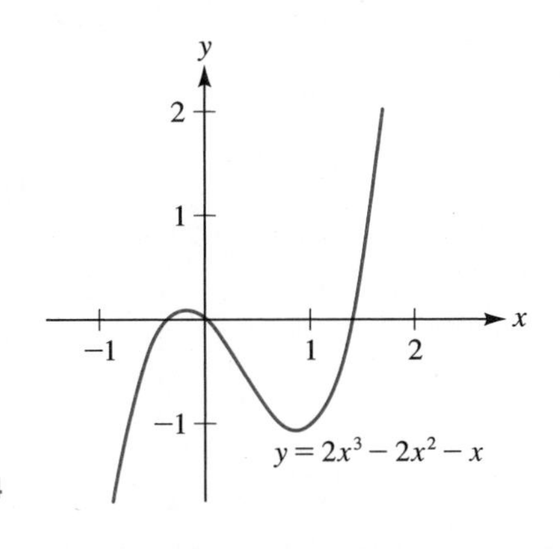

FIGURE 4

Solution At the points where the graph crosses the x-axis, the y-coordinates are zero. Setting y equal to zero in the given equation yields

$$2x^3 - 2x^2 - x = 0$$
$$x(2x^2 - 2x - 1) = 0$$

Thus, $x = 0$ or $2x^2 - 2x - 1 = 0$. This last equation can be solved by using the quadratic formula. As you should check for yourself, the roots are $(1 \pm \sqrt{3})/2$. We've now found the three x-intercepts of the graph in Figure 3: they are 0, $(1 + \sqrt{3})/2$, and $(1 - \sqrt{3})/2$. *Suggestion:* Use your calculator to check that these last two values are consistent with Figure 4.

Whether one is graphing equations "by hand" or using a graphing calculator or computer, a question often arises as to whether or not all of the essential features of the graph are clear. The next example indicates how our work on intercepts is helpful here.

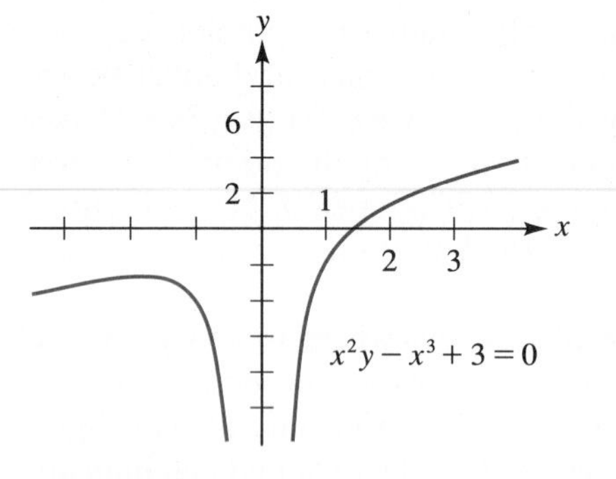

FIGURE 5
Does the graph ever cross the y-axis?

EXAMPLE 5 Figure 5 shows a graph of the equation $x^2y - x^3 + 3 = 0$.

(a) From the portion of the graph shown in Figure 5, it appears that there is no y-intercept. Use the given equation to show that this is indeed the case.

(b) As indicated in Figure 5, there is an x-intercept between 1 and 2. Find this intercept. Give two forms for the answer: an exact expression and a calculator approximation rounded to two decimal places.

Solution

(a) If there were a y-intercept (perhaps somewhere below the portion of the graph displayed in Figure 5), we could find it by setting x equal to zero in the given equation. But with $x = 0$, the given equation becomes $3 = 0$, which is clearly impossible. We thus conclude that there is no y-intercept; the graph never does cross the y-axis.

(b) The x-intercept of the graph is determined by setting y equal to zero in the given equation. That gives us $x^2(0) - x^3 + 3 = 0$, and therefore $x^3 = 3$. To solve this equation for x, we take the cube root of both sides to obtain

$$x = \sqrt[3]{3}$$
$$\approx 1.44 \quad \text{using a calculator}$$

Notice that the value $x = 1.44$ is consistent with the graph shown in Figure 5. Also, see the Graphical Perspective in Figure 6.

GRAPHICAL PERSPECTIVE

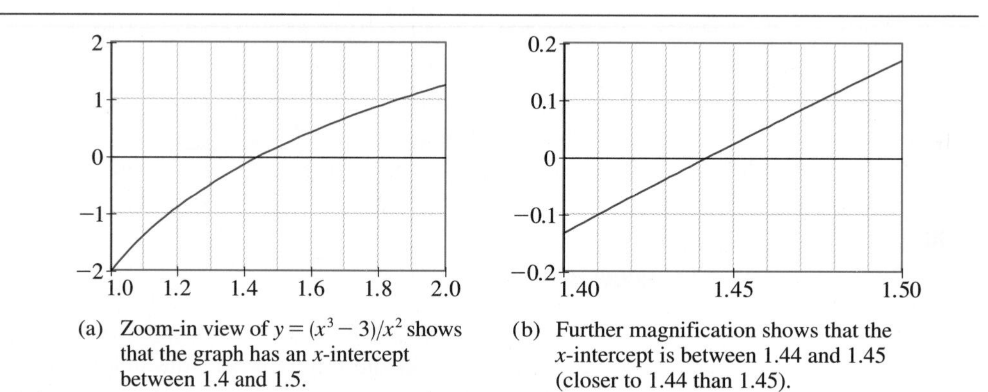

(a) Zoom-in view of $y = (x^3 - 3)/x^2$ shows that the graph has an x-intercept between 1.4 and 1.5.

(b) Further magnification shows that the x-intercept is between 1.44 and 1.45 (closer to 1.44 than 1.45).

FIGURE 6
To graph $x^2y - x^3 + 3 = 0$, first solve for y:

$$x^2y = x^3 - 3$$
$$y = (x^3 - 3)/x^2$$

When graphing equations "by hand," how many points do we need to plot before we're sure about the essential features of the graph? In later sections of this text, you'll see that there are a number of techniques and concepts that make it unnecessary to plot a large number of points. For now, however, we point out that some equations arise frequently enough to make it worth memorizing the basic shapes and features of their graphs. Figure 7 displays six such equations and their graphs. Exercises 35 and 36 at the end of this section ask you to set up tables and verify for yourself that the graphs in Figure 7 are indeed correct. From now on (with the exception of Exercises 35 and 36), if you need to sketch a graph of one of these basic equations, you should do so from memory.

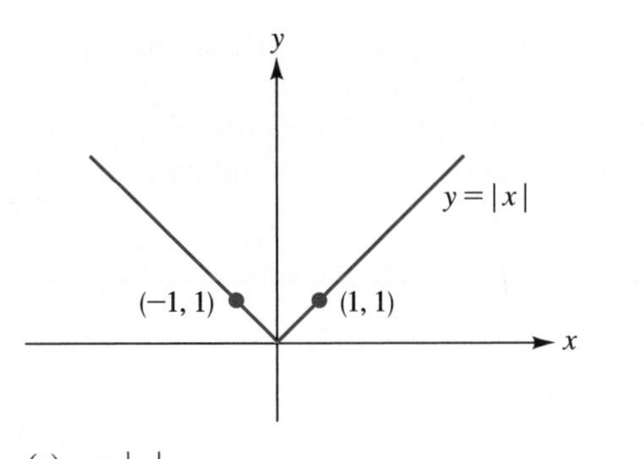

(a) $y = |x|$
The domain of the variable x is $(-\infty, \infty)$.

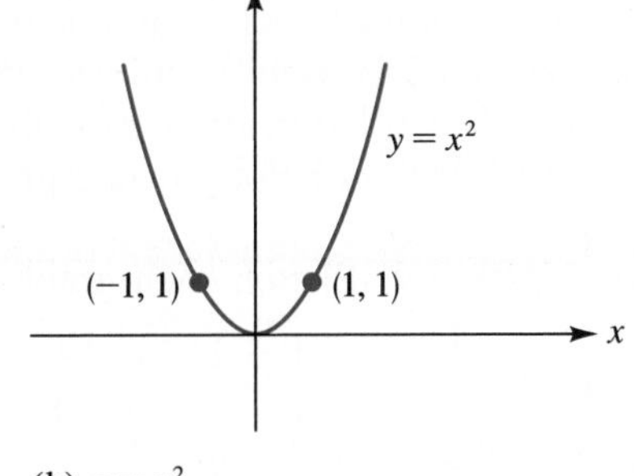

(b) $y = x^2$
The domain of the variable x is $(-\infty, \infty)$.

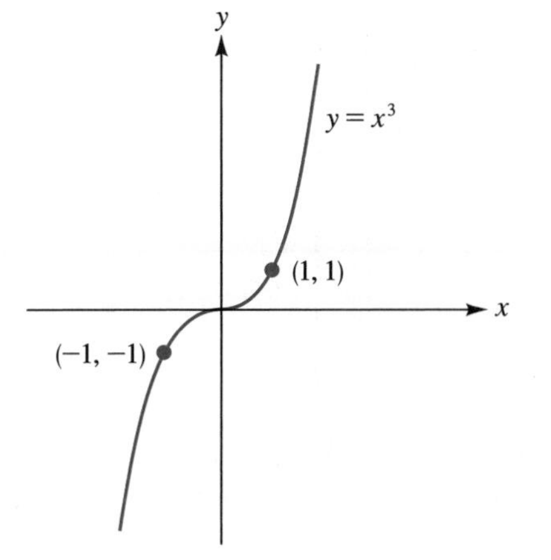

(c) $y = x^3$
The domain of the variable x is $(-\infty, \infty)$.

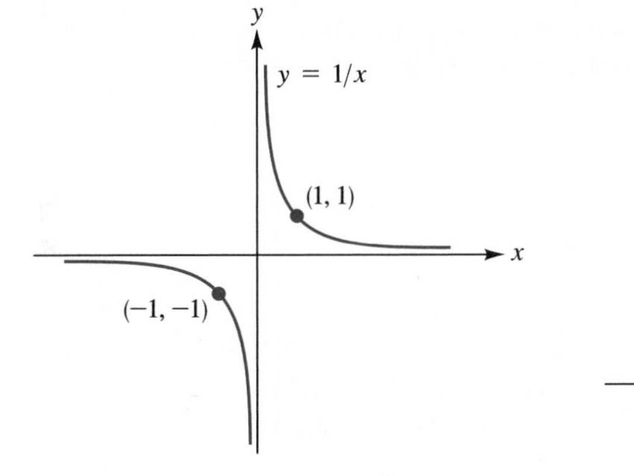

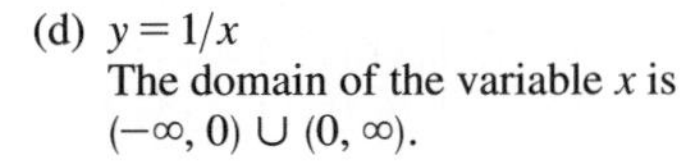
(d) $y = 1/x$
The domain of the variable x is
$(-\infty, 0) \cup (0, \infty)$.

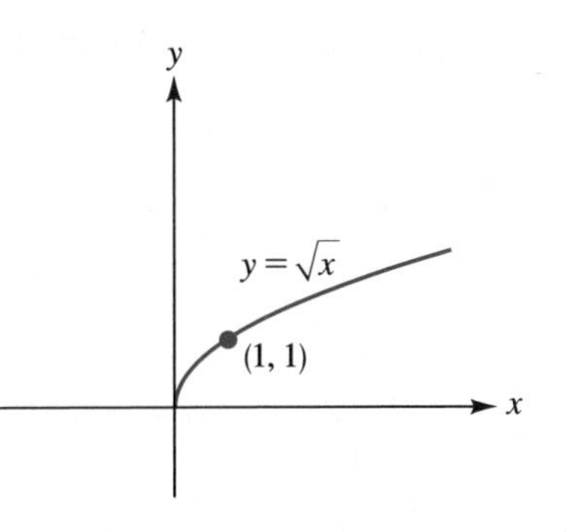

(e) $y = \sqrt{x}$
The domain of the variable x is $[0, \infty)$.

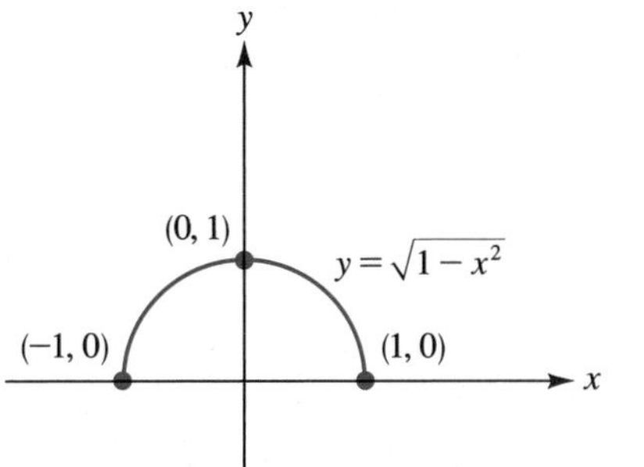

(f) $y = \sqrt{1 - x^2}$
The domain of the variable x is $[-1, 1]$.

FIGURE 7

EXAMPLE 6 The variable y is defined in terms of the variable x as follows:

$$y = \begin{cases} x^2 & \text{if } x < 2 \\ 1/x & \text{if } x \geqslant 2 \end{cases}$$

(a) Find y when $x = 3/2$, when $x = 2$, and when $x = 3$.

(b) Sketch the graph of $y = \begin{cases} x^2 & \text{if } x < 2; \\ 1/x & \text{if } x \geqslant 2. \end{cases}$

Solution

(a) To find y when $x = 3/2$, do we use the equation $y = x^2$ or the equation $y = /x$? According to the given inequalities, we should use the equation $y = x^2$ whenever the x-values are less than 2. So, for $x = 3/2$ (which surely is less than 2), we have $y = (3/2)^2 = 9/4$. On the other hand, the inequalities tell us to use the equation $y = 1/x$ whenever the x-values are greater than or equal to 2. So when x is 2, we have $y = 1/2$, and when x is 3, we have $y = 1/3$.

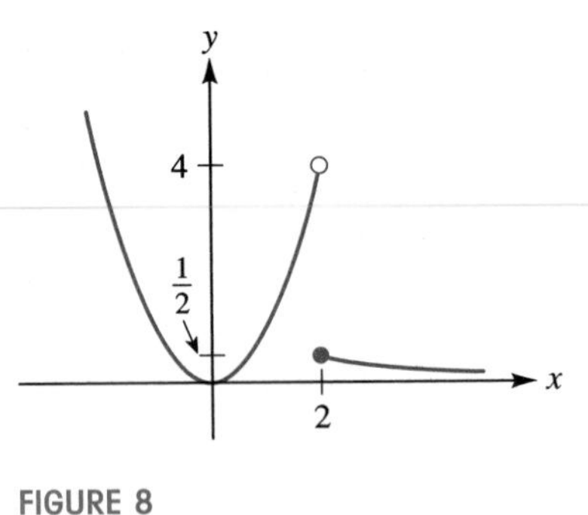

FIGURE 8

A graph of $y = \begin{cases} x^2, & x<2 \\ \frac{1}{x}, & x \geq 2 \end{cases}$

(b) For the graph, we look back at Figures 7(b) and (d) and choose the appropriate portion of each. The result is displayed in Figure 8. The open circle in the figure is used to indicate that the point (2, 4) does not belong to the graph. The filled-in circle, on the other hand, is used to indicate that the point $(2, \frac{1}{2})$ does belong to the graph.

EXAMPLE 7 Sketch each graph:

(a) $y = \begin{cases} |x| & \text{if } -1 \leq x \leq 1, \\ \sqrt{x} & \text{if } \quad 1 \leq x \leq 4; \end{cases}$

(b) $y = \begin{cases} |x| & \text{if } -1 \leq x < 1, \\ \sqrt{x} & \text{if } \quad 1 < x \leq 4. \end{cases}$

Solution

(a) We want to "paste together" portions of the graphs of $y = |x|$ and $y = \sqrt{x}$. The appropriate portions of each graph are indicated in color in Figures 9(a) and 9(b). Using Figures 9(a) and 9(b) as a guide, we sketch the required graph in Figure 9(c).

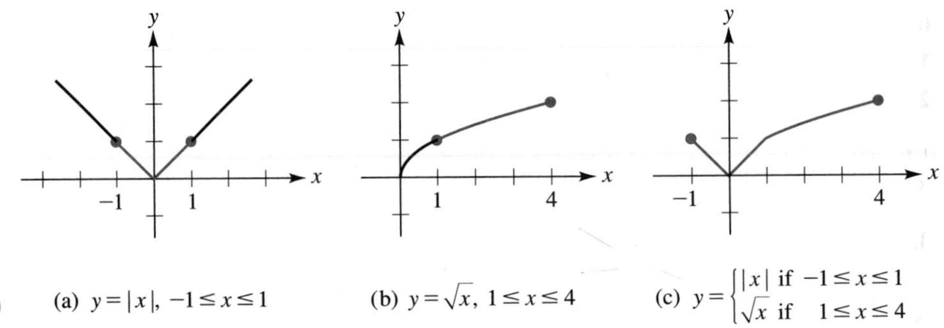

FIGURE 9 (a) $y = |x|, -1 \leq x \leq 1$ (b) $y = \sqrt{x}, 1 \leq x \leq 4$ (c) $y = \begin{cases} |x| \text{ if } -1 \leq x \leq 1 \\ \sqrt{x} \text{ if } \quad 1 \leq x \leq 4 \end{cases}$

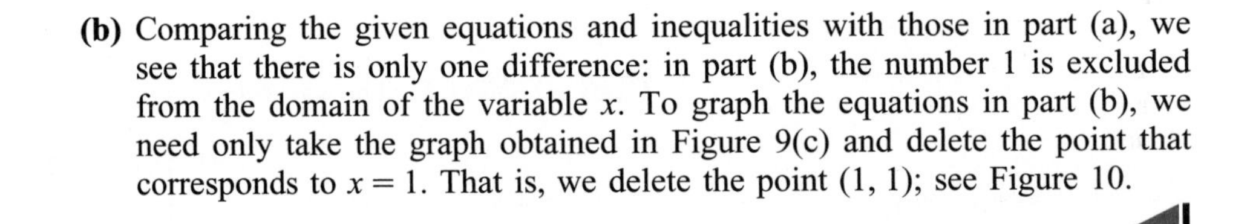

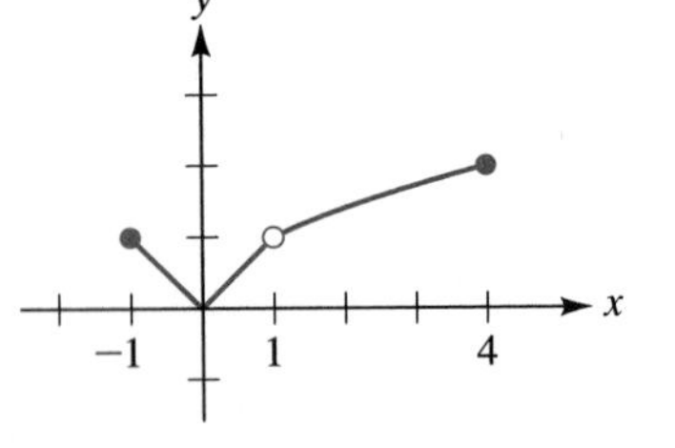

FIGURE 10

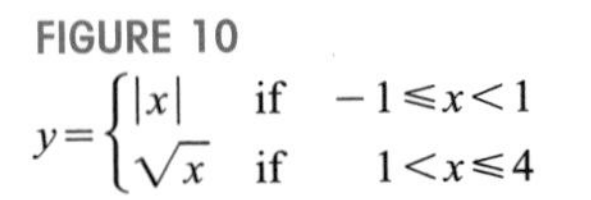

$y = \begin{cases} |x| & \text{if } -1 \leq x < 1 \\ \sqrt{x} & \text{if } \quad 1 < x \leq 4 \end{cases}$

(b) Comparing the given equations and inequalities with those in part (a), we see that there is only one difference: in part (b), the number 1 is excluded from the domain of the variable x. To graph the equations in part (b), we need only take the graph obtained in Figure 9(c) and delete the point that corresponds to $x = 1$. That is, we delete the point (1, 1); see Figure 10.

We conclude this section by comparing the graphs in Figures 8, 9(c), and 10. Each graph was obtained by combining portions of the basic graphs in Figure 7. In Figures 8 and 10 the graphs have a break or gap in them. We say that the graph (or more precisely, the *function*) in Figure 8 is *discontinuous* when $x = 2$. (The two pieces do not meet at $x = 2$.) The function in Figure 10 is discontinuous when $x = 1$. (There is a gap or missing point when $x = 1$.) In Figure 9(c), however, the two portions of the graph form a graph with no break or gap. We say that the function in Figure 9(c) is *continuous* when $x = 1$. A rigorous definition of continuity is properly a subject for calculus. However, even at the intuitive level at which we've presented the idea here, you'll find that the concept is useful in helping you to organize your thoughts about graphs.

Do 3, 11, 17, 19, 28, 29, 31, 33, 39, 43

EXERCISE SET 2.2

A

In Exercises 1–6, determine whether the given point lies on the graph of the equation, as in Example 1. Note: You're not asked to draw the graph.

1. $(8, 6)$; $y = \frac{1}{2}x + 3$
2. $\left(\frac{3}{5}, -\frac{17}{5}\right)$; $y = -\frac{2}{3}x - 3$
3. $(4, 3)$; $3x^2 + y^2 = 52$
4. $(4, -2)$; $3x^2 + y^2 = 52$
5. $(a, 4a)$; $y = 4x$
6. $(a - 1, a + 1)$; $y = x + 2$

In Exercises 7–12, you are given an equation and a point. In each case, find the value of the constant a so that the graph of the equation passes through the given point.

7. $y = ax + 6$; $(2, -3)$ **8.** $y = -5x - a$; $(-4, 1)$
9. $y = a(x - 6)^2 - 2$; $(-5, 2)$
10. $y^2 = 4ax$; $(4, 2)$
11. $x^2 + ay^2 = 28$; $(1, -3)$
12. $ax^2 - y^2 = 12$; $(4, 6)$

In Exercises 13–18, specify the domain of the variable x and then graph the equation, as we did in Example 3.

13. $y = \dfrac{x^2 + x - 20}{x - 4}$ **14.** $y = \dfrac{6x^2 + 11x + 4}{3x + 4}$

15. $y = \dfrac{x^2 - 9}{x - 3}$ **16.** $y = \dfrac{x - x^2}{x}$

17. $\dfrac{-x^3 + 4x^2 + x - 4}{x^2 - 1}$ **18.** $\dfrac{x^2 - 4}{(x + 2)(2 - x)}$

In Exercises 19–24, the graph of the equation is a straight line. Graph the equation after finding the x- and y-intercepts.

19. $3x + 4y = 12$ **20.** $3x - 4y = 12$
21. $y = 2x - 4$ **22.** $x = 2y - 4$
23. $x + y = 1$ **24.** $2x - 3y = 6$

In Exercises 25–28, determine any x- or y-intercepts for the graph of the equation. Note: You're not asked to draw the graph.

25. (a) $y = x^2 + 3x + 2$ **(b)** $y = x^2 + 2x + 3$
26. (a) $y = x^2 - 4x - 12$ **(b)** $y = x^2 - 4x + 12$
27. (a) $y = x^2 + x - 1$ **(b)** $y = x^2 + x + 1$
28. (a) $y = 6x^3 + 9x^2 + x$ **(b)** $y = 9x^3 + 6x^2 + x$

In Exercises 29–34, each figure shows a graph of an equation. Find the x- and y-intercepts of the graph. If an intercept involves a radical, give both the radical form of the answer and a calculator approximation rounded to two decimal places. (Check to see that your calculator answer is consistent with the given figure.)

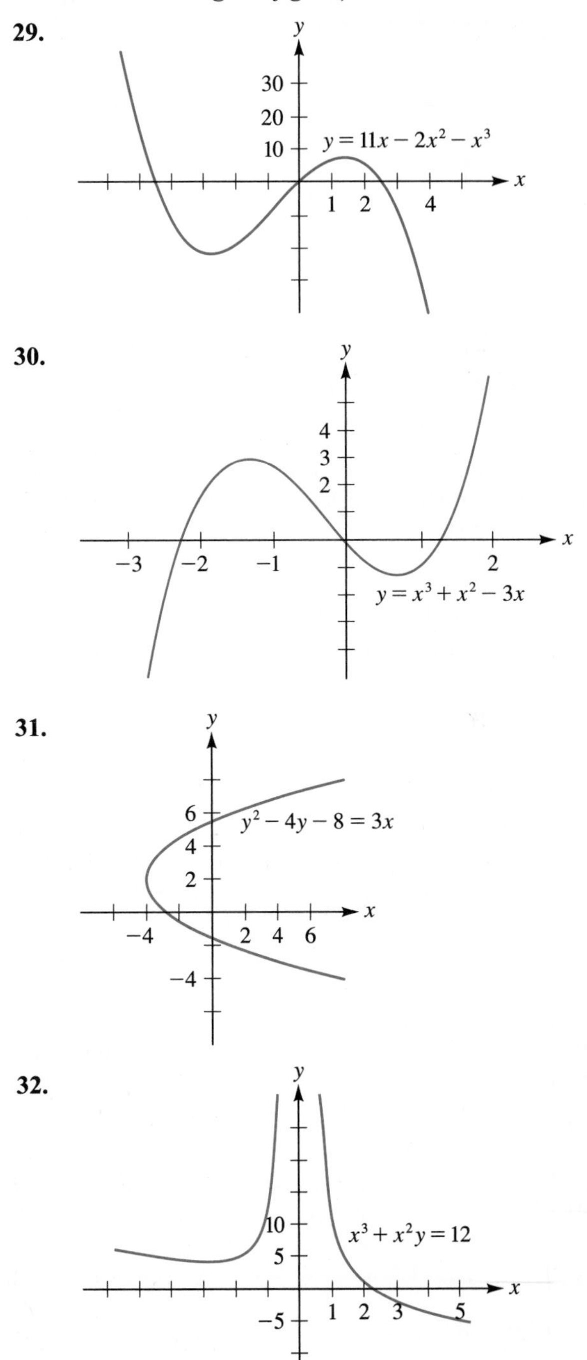

33. **34.**

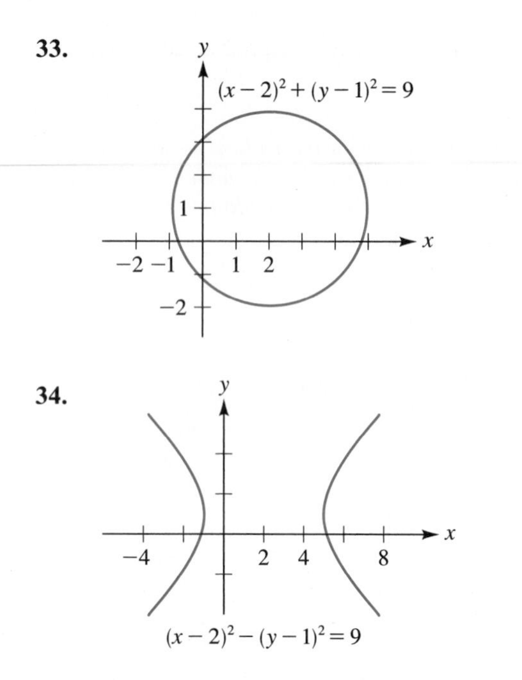

In Exercises 35 and 36, set up a table and graph each equation. [In Exercises 36(b) and (c), use a calculator to compute the square roots.]

35. **(a)** $y=|x|$ **(b)** $y=x^2$ **(c)** $y=x^3$

36. **(a)** $y=1/x$ **(b)** $y=\sqrt{x}$ **(c)** $y=\sqrt{1-x^2}$

In Exercises 37–42, sketch the graphs, as in Examples 6 and 7.

37. $y=\begin{cases} |x| & \text{if } x\leq 0 \\ x^2 & \text{if } x>0 \end{cases}$

38. **(a)** $y=\begin{cases} x^3 & \text{if } x\leq 0 \\ |x| & \text{if } x>0 \end{cases}$

(b) $y=\begin{cases} |x| & \text{if } x\leq 0 \\ x^3 & \text{if } x>0 \end{cases}$

39. **(a)** $y=\begin{cases} \sqrt{x} & \text{if } 0\leq x\leq 1 \\ 1/x & \text{if } 1<x<2 \end{cases}$

(b) $y=\begin{cases} \sqrt{x} & \text{if } 0\leq x<1 \\ 1/x & \text{if } 1<x<2 \end{cases}$

40. **(a)** $y=\begin{cases} \sqrt{1-x^2} & \text{if } -1\leq x\leq 0 \\ x^2 & \text{if } 0<x\leq 2 \end{cases}$

(b) $y=\begin{cases} \sqrt{1-x^2} & \text{if } -1\leq x<0 \\ x^2 & \text{if } 0<x\leq 2 \end{cases}$

41. $y=\begin{cases} x^3 & \text{if } -1\leq x<0 \\ \sqrt{x} & \text{if } 0\leq x<1 \\ 1/x & \text{if } 1\leq x\leq 3 \end{cases}$

42. $y=\begin{cases} |x| & \text{if } -2\leq x\leq -1 \\ \sqrt{1-x^2} & \text{if } -1<x<1 \\ |x| & \text{if } 1\leq x\leq 2 \end{cases}$

B

43. The figure shows the graph of the equation $F=\frac{9}{5}C+32$, which relates temperature on the Fahrenheit scale to temperature on the Celsius scale.

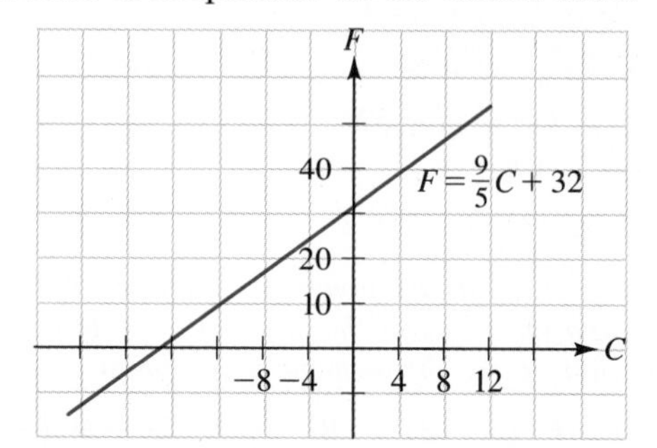

(a) Use the graph to estimate the Celsius temperature corresponding to 0°F.

(b) Use the equation to find the exact Celsius temperature corresponding to 0°F. Check that your answer is consistent with your estimate in part (a).

44. Graph the following equations on the same set of axes and in each case include only the portion of the graph between $x=0$ and $x=1$. Draw the graphs carefully enough to ensure that you can make accurate comparisons among them.

(a) $y=x$ **(b)** $y=x^2$ **(c)** $y=x^3$ **(d)** $y=x^4$

What is the pattern here? What would you say that $y=x^{100}$ must look like in this interval?

45. The accompanying figure shows the graph of $y=\sqrt{x}$. Use this graph to estimate the following quantities (to one decimal place).

(a) $\sqrt{2}$ **(b)** $\sqrt{3}$

(c) $\sqrt{6}$ *Hint:* $\sqrt{ab}=\sqrt{a}\sqrt{b}$

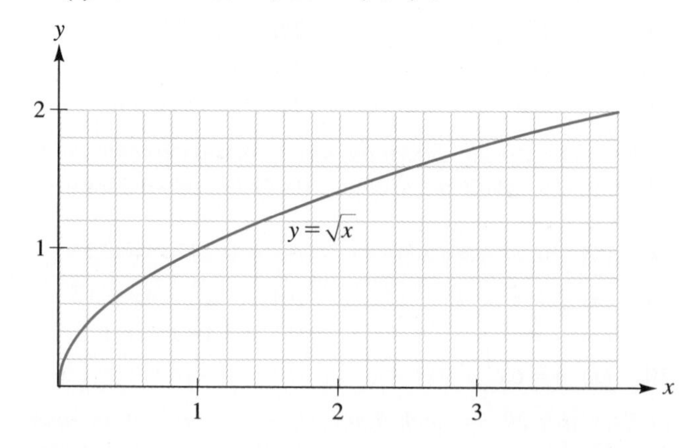

46. The following figure shows the graph of $y = \sqrt[3]{x}$. Use the graph to estimate the following quantities. In each case, compare your estimate with a value obtained from a calculator.

(a) $\sqrt[3]{2}$ **(b)** $\sqrt[3]{5}$ **(c)** $\sqrt[3]{6.5}$

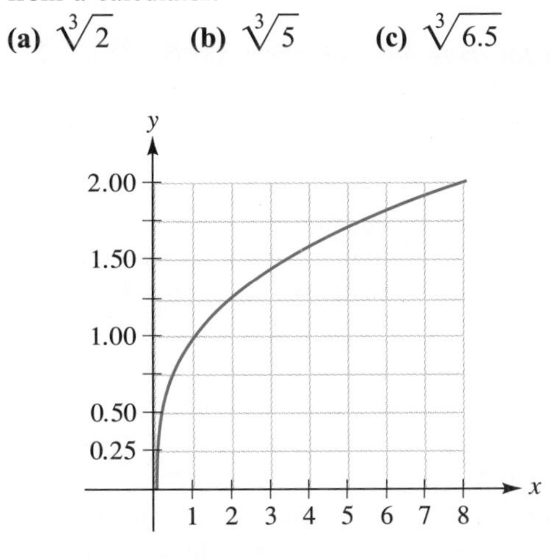

47. In a certain biology experiment, the number N of bacteria increases with time t as indicated in the accompanying figure.

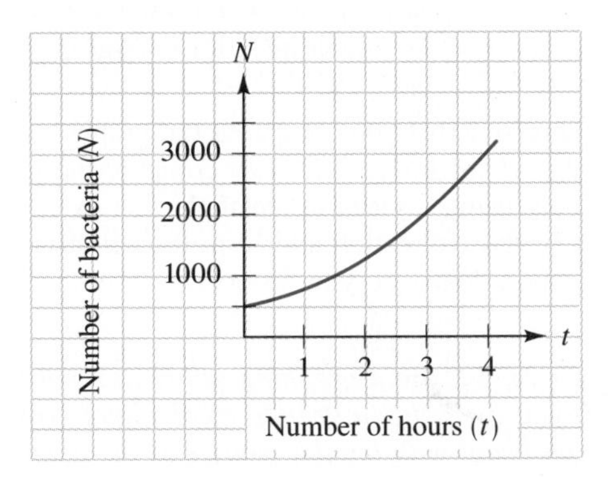

(a) How many bacteria are initially present when $t = 0$?

(b) Approximately how long does it take for the original colony to double in size?

(c) For which value of t is the population approximately 2500?

(d) During which time interval does the population increase more rapidly, between $t = 0$ and $t = 1$ or between $t = 3$ and $t = 4$?

In Exercises 48 and 49, determine the x-coordinates of the points A and B. (Each dashed line is parallel to one of the coordinate axes.)

48.

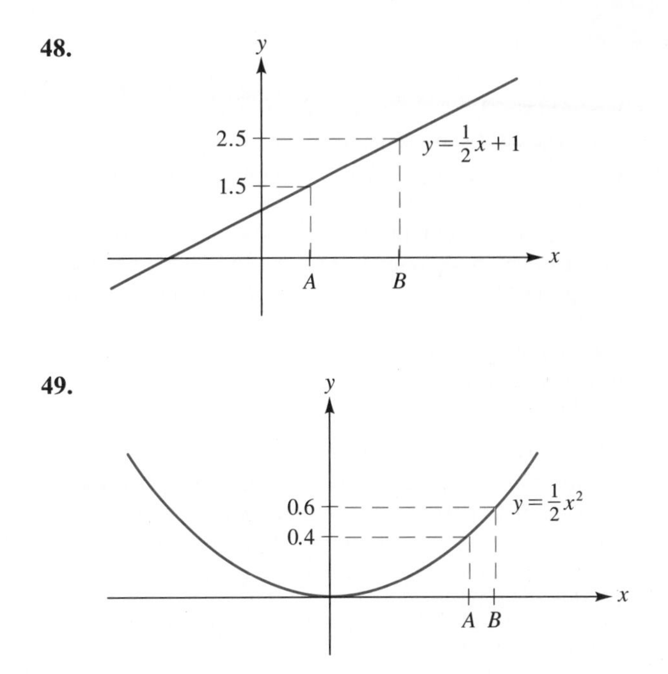

C

50. Suppose that the circle $x^2 + 2Ax + y^2 + 2By = C$ has two x-intercepts, x_1 and x_2, and two y-intercepts, y_1 and y_2. Prove each statement.

(a) $\dfrac{x_1 + x_2}{y_1 + y_2} = \dfrac{A}{B}$ **(b)** $x_1x_2 - y_1y_2 = 0$

(c) $x_1x_2 + y_1y_2 = -2C$

51. In the accompanying figure, the x-coordinates of the points A and B are a and $1/a$, respectively. (Assume that $0 < a < 1$.) If P denotes an arbitrary point on the semicircle, show that

$$PA/PB = a$$

What is remarkable about this result?

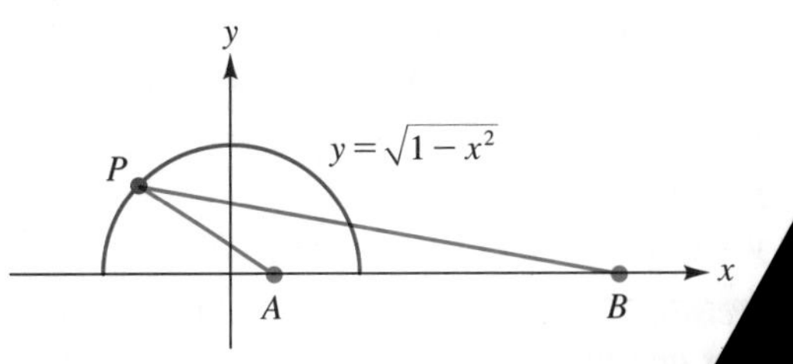

GRAPHING UTILITY EXERCISES FOR SECTION 2.2

1. According to Example 2, the graph of the equation $y = -\frac{3}{4}(x-2)^2 + 3$ passes through the origin. Use a graphing utility to verify this fact.
2. In Example 3, we saw that the graph of the equation $y = (3x^2 - 5x + 2)/(x - 1)$ is a line with one point missing. Most graphing utilities, however, will ignore this distinction and simply plot the line. (One notable exception: *Analyzer* * software, published by Addison-Wesley, will indicate the missing point with a small open circle, just as in Figure 3 in the text.) Find out how your graphing utility operates in this respect by graphing the equation $y = (3x^2 - 5x + 2)/(x - 1)$. Does the point (1, 1) appear to be a part of the graph?

In Exercises 3–6, the graph of each equation is a line with one or two points missing. Graph the equations and specify the coordinates of the missing points. In Exercises 5 and 6, you will need to factor the numerators and the denominators to determine the missing points.

3. $y = (2x^2 + 9x + 10)/(x + 2)$
4. $y = (4x^2 - 9)/(2x + 3)$
5. $y = (2x^3 - 11x^2 + 10x + 8)/(x^2 - 6x + 8)$
6. $y = (3x^3 + 5x^2 - 27x - 45)/(3x^2 - 4x - 15)$

In Exercises 7–14: **(a)** *Use a graphing utility to graph the equations and to approximate the x-intercepts. In approximating the x-intercepts, use a "solve" key or a sufficiently magnified view to ensure that the values you give are correct in the first two decimal places.* **(b)** *Determine exact expressions for the x-intercepts, as in Examples 4 and 5(b). Then use a calculator to evaluate the expressions. Round to three decimal places and use the results to check your answers in part (a).*

7. $y = x^2 - 2x - 2$
8. $y = 2x^2 + x - 5$

$\cdot = 2x^3 - 5x$

10. $y = 3x^3 + 5x^2 + x$

$^3 + 9x^2 - 28x - 63$

ᵗor by grouping.

$21x - 9$

$^2 - 3x + 3)$

')

''ty to graph the
cepts. In approxi-
key or a sufficiently
ues you give are cor-
s. Remark: None of the

x-intercepts for these four equations can be obtained using factoring techniques.

15. $y = x^3 - 3x + 1$
16. $y = 8x^3 - 6x - 1$
17. $y = x^5 - 6x^4 + 3$
18. $y = 2x^5 - 5x^4 + 5$
19. Graph each of the equations $y = |x|$, $y = x^2$, and $y = x^3$. Check to see that the graphs that you obtain are consistent with those shown in Figures 7(a), 7(b), and 7(c).
20. Graph each of the equations $y = 1/x$, $y = \sqrt{x}$, and $y = \sqrt{1 - x^2}$. Check to see that the graphs that you obtain are consistent with those shown in Figures 7(d), 7(e), and 7(f).
21. **(a)** Graph $y = x^2$ using a viewing rectangle that extends from -2 to 2 in the x-direction and from 0 to 4 in the y-direction. Notice how flat the graph appears to be near the origin.
    (b) Follow part (a) using $y = x^3$ instead of $y = x^2$.
    (c) Using the viewing rectangle in part (a), graph both $y = x^2$ and $y = x^3$ on the same set of axes. Where do the graphs intersect? For x-values in the open interval (0, 1), which quantity is larger, x^2 or x^3?
22. In Exercise 21(c) you compared the graphs of $y = x^2$ and $y = x^3$ for $0 < x < 1$. Now compare the graphs for $x > 1$ by using a viewing rectangle that extends from 0 to 10 in the x-direction and from 0 to 1000 in the y-direction. What do you observe?

In Exercises 23–32, use a graphing utility to obtain the graphs. (If you are using a graphing calculator, you may need to experiment with the dot mode and the connected mode to see which produces the more useful graph in a given case.)

23. $y = \begin{cases} x^2 - 1 & \text{if } x \leq 2 \\ 5 - x & \text{if } x > 2 \end{cases}$

24. $y = \begin{cases} -x^2 & \text{if } -2 < x \leq 0 \\ \frac{1}{2}x & \text{if } 0 < x \leq 4 \end{cases}$

25. $y = \begin{cases} x^2 & \text{if } 0 \leq x < 1 \\ 1/x & \text{if } 1 \leq x \leq 4 \end{cases}$

26. $y = \begin{cases} |x| & \text{if } -4 \leq x < 0 \\ x^2 & \text{if } 0 \leq x < 1 \\ 1/x & \text{if } 1 \leq x \leq 4 \end{cases}$

27. $y = \begin{cases} \sqrt{1-x^2} & \text{if } -1 \leqslant x < 0 \\ \sqrt{x} & \text{if } \quad 0 \leqslant x \leqslant 4 \end{cases}$

28. $y = \begin{cases} x^3 & \text{if } -1 \leqslant x < 0 \\ \sqrt{x} & \text{if } \quad 0 \leqslant x \leqslant 4 \end{cases}$

29. $y = \begin{cases} -1/x & \text{if } x < -1 \\ x^2 & \text{if } -1 \leqslant x < 1 \\ 1/x & \text{if } x \geqslant 1 \end{cases}$

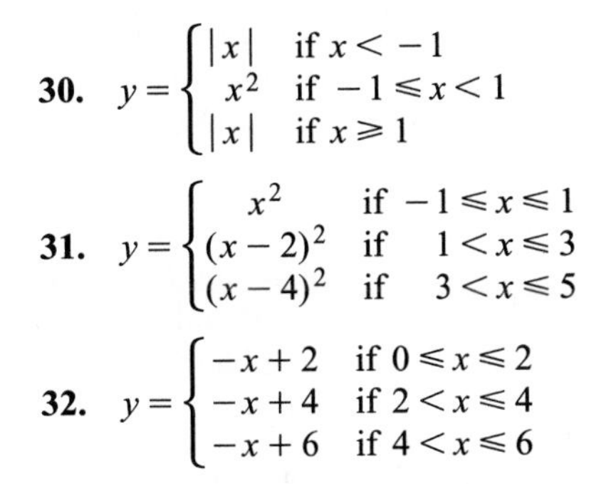

30. $y = \begin{cases} |x| & \text{if } x < -1 \\ x^2 & \text{if } -1 \leqslant x < 1 \\ |x| & \text{if } x \geqslant 1 \end{cases}$

31. $y = \begin{cases} x^2 & \text{if } -1 \leqslant x \leqslant 1 \\ (x-2)^2 & \text{if } \quad 1 < x \leqslant 3 \\ (x-4)^2 & \text{if } \quad 3 < x \leqslant 5 \end{cases}$

32. $y = \begin{cases} -x+2 & \text{if } 0 \leqslant x \leqslant 2 \\ -x+4 & \text{if } 2 < x \leqslant 4 \\ -x+6 & \text{if } 4 < x \leqslant 6 \end{cases}$

2.3 EQUATIONS OF LINES

He [Pierre de Fermat (1601–1665)] *introduced perpendicular axes and found the general equations of straight lines and circles and the simplest equations of parabolas, ellipses, and hyperbolas; and he further showed in a fairly complete and systematic way that every first- or second-degree equation can be reduced to one of these types.*

George F. Simmons in *Calculus Gems, Brief Lives and Memorable Moments* (New York: McGraw-Hill Book Company, 1992)

In this section we take a systematic look at equations of lines and their graphs. We begin by recalling the concept of *slope,* which you've seen in previous courses. The slope of a nonvertical line is a number that measures the slant or direction of the line; it is defined as follows.

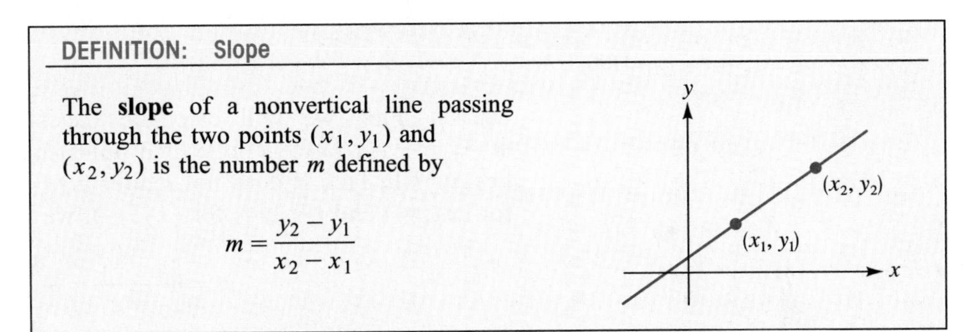

DEFINITION: Slope

The **slope** of a nonvertical line passing through the two points (x_1, y_1) and (x_2, y_2) is the number m defined by

$$m = \frac{y_2 - y_1}{x_2 - x_1}$$

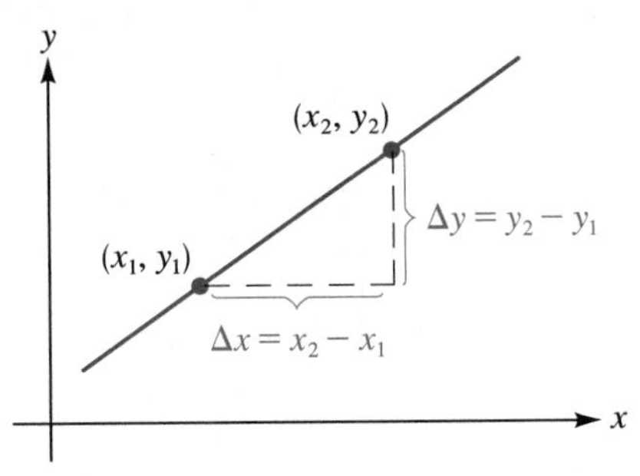

FIGURE 1

Note that the quantity $x_2 - x_1$ appearing in the definition of slope is the amount by which x changes as we move from (x_1, y_1) to (x_2, y_2) along the line. We denote this change in x by the symbol Δx (read *delta* x). Thus $\Delta x = x_2 - x_1$ (see Figure 1). Similarly, the symbol Δy is defined to mean the change in y: $\Delta y = y_2 - y_1$. Using these ideas, we can rewrite our definition of slope as $m = \Delta y / \Delta x$.

The slope of a line does not depend on which two particular points on the line are used in the calculation. To see why this is so, consider Figure 2. The

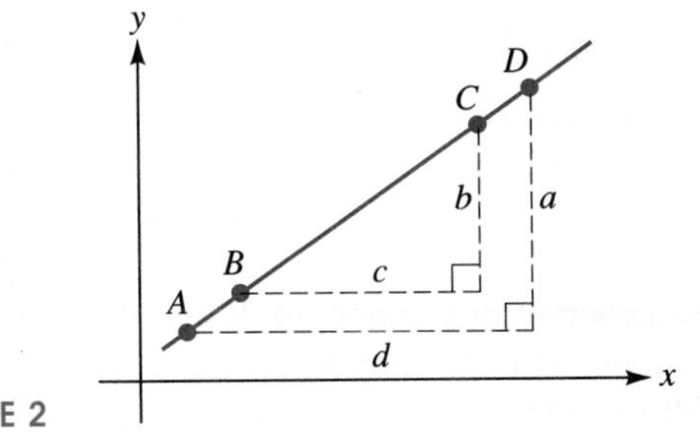

FIGURE 2

two right triangles are similar (because the corresponding angles are equal). This implies that the corresponding sides of the two triangles are proportional, and so we have

$$\frac{a}{d} = \frac{b}{c}$$

Now notice that the left-hand side of this equation represents the slope $\Delta y/\Delta x$ calculated using the points A and D, and the right-hand side represents the slope calculated using the points B and C. Thus the values we obtain for the slope are indeed equal.

EXAMPLE 1 Compute and compare the slopes of the three lines shown in Figure 3.

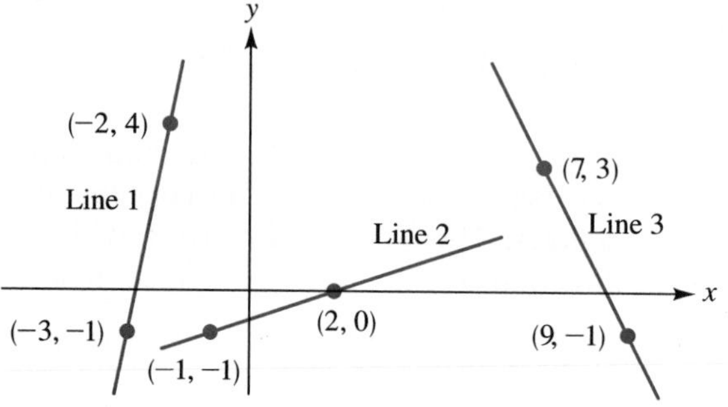

FIGURE 3

Solution First we will calculate the slope of Line 1, using the formula $m = (y_2 - y_1)/(x_2 - x_1)$. Which point should we use as (x_1, y_1) and which as (x_2, y_2)? In fact, it does not matter how we label our points. Using $(-3, -1)$ for (x_1, y_1) and $(-2, 4)$ for (x_2, y_2), we have

$$m = \frac{y_2 - y_1}{x_2 - x_1} = \frac{4 - (-1)}{-2 - (-3)} = 5$$

So the slope of Line 1 is 5. If, instead, we had used $(-2, 4)$ for (x_1, y_1) and $(-3, -1)$ for (x_2, y_2), then we'd have $m = (-1 - 4)/(-3 + 2) = 5$, the same result. This is not accidental because, in general,

$$\frac{y_2 - y_1}{x_2 - x_1} = \frac{y_1 - y_2}{x_1 - x_2}$$

(Exercise 63 asks you to verify this identity.) Next, we calculate the slopes of Lines 2 and 3.

$$\text{Slope of line 2: } \frac{0 - (-1)}{2 - (-1)} = \frac{1}{3} \qquad \text{Slope of line 3: } \frac{3 - (-1)}{7 - 9} = -2$$

Lines 1 and 2 both have positive slopes and slant upward to the right. Note that Line 1 is steeper than Line 2 and correspondingly has the larger slope, 5. Line 3 has a negative slope, -2, and slants downward to the right.

The observations made in Example 1 are true in general. Lines with a positive slope slant upward to the right, the steeper line having the larger slope. Likewise, lines with a negative slope slant downward to the right; see Figure 4. Figure 5 shows for comparison some values of m for various lines.

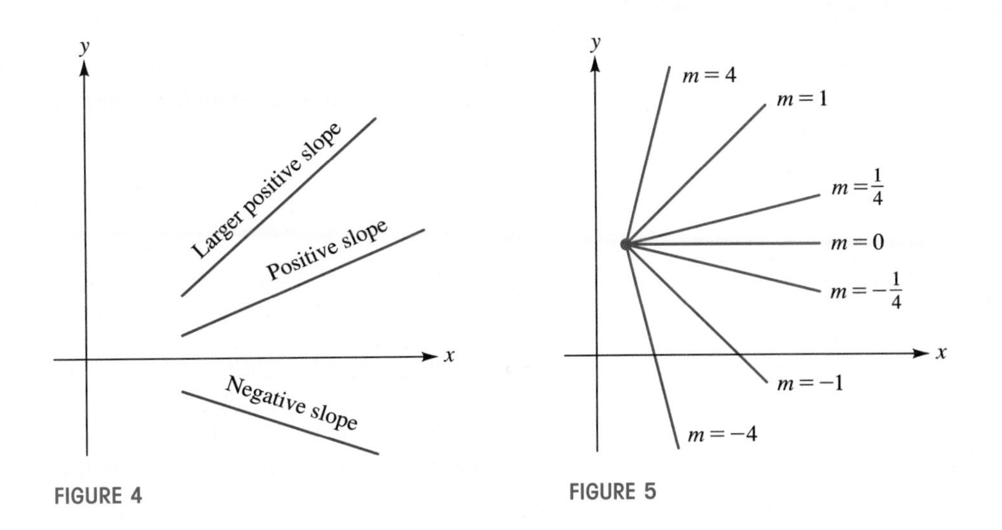

FIGURE 4

FIGURE 5

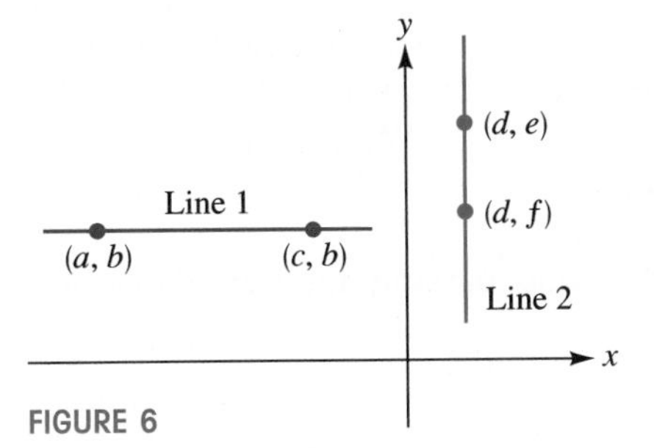

FIGURE 6

We have yet to mention slopes for horizontal or vertical lines. In Figure 6, Line 1 is horizontal; it passes through the points (a, b) and (c, b). Note that the two y-coordinates must be the same in order for the line to be horizontal. Line 2 in Figure 6 is vertical; it passes through (d, e) and (d, f). Note that the two x-coordinates must be the same in order for the line to be vertical. For the slope of Line 1 we have $m = \dfrac{b-b}{c-a} = \dfrac{0}{c-a} = 0$ (provided $a \neq c$). Thus the slope of Line 1, a horizontal line, is zero. For the vertical line in Figure 6, the calculation of slope begins with writing $\dfrac{e-f}{d-d}$, but the denominator is zero, and since division by zero is undefined, we conclude that slope is undefined for vertical lines. We summarize these results in the box that follows.

HORIZONTAL AND VERTICAL LINES

1. The slope of a horizontal line is zero.
2. Slope is not defined for vertical lines.

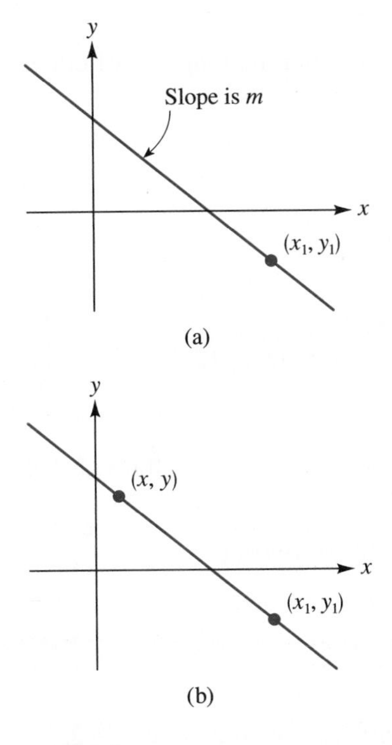

FIGURE 7

We can use the concept of slope to find the equation of a line. Suppose we have a line that has slope m and passes through the point (x_1, y_1), as shown in Figure 7(a). Let (x, y) be any other point on the line, as in Figure 7(b). Then the slope of the line is given by $m = (y - y_1)/(x - x_1)$, therefore, $y - y_1 = m(x - x_1)$. Note that the given point (x_1, y_1) also satisfies this last equation, because in that case we have $y_1 - y_1 = m(x_1 - x_1)$, or $0 = 0$. The equation $y - y_1 = m(x - x_1)$ is called the **point–slope formula.** We have shown that any point on the line satisfies this equation. (Conversely, it can be shown that if a point satisfies this equation, then the point does lie on the given line.)

THE POINT–SLOPE FORMULA

The equation of a line that has slope m and passes through the point (x_1, y_1) is

$$y - y_1 = m(x - x_1)$$

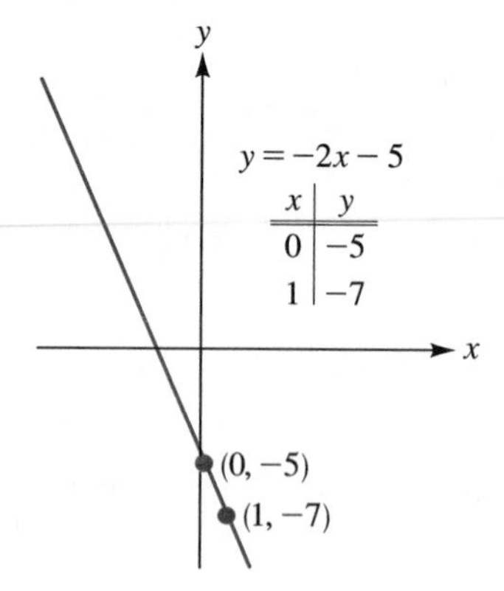

FIGURE 8
$y = -2x - 5$

EXAMPLE 2 Write the equation of the line that passes through $(-3, 1)$ and has a slope of -2. Sketch a graph of the line.

Solution Since the slope and a point are given, we use the point–slope formula:

$$
\begin{aligned}
y - y_1 &= m(x - x_1) \\
y - 1 &= -2[x - (-3)] \\
y - 1 &= -2x - 6 \\
y &= -2x - 5
\end{aligned}
$$

This is the required equation. One way to graph this line is to pick values for x and then compute the corresponding y-values, as done in Section 2.2. The table and graph are displayed in Figure 8. Notice that if we know ahead of time that the graph is a line, a table as brief as that in Figure 8 is sufficient.

There is another way to go about graphing the line $y = -2x - 5$ in Example 2. Since slope is (change in y)/(change in x), a slope of -2 (or $-2/1$) can be interpreted as telling us that if we start at $(-3, 1)$ and let x increase by one unit, then y must decrease by two units to bring us back to the line. Following this path in Figure 9 takes us from $(-3, 1)$ to $(-2, -1)$. We now draw the line through these two points, as shown in Figure 9.

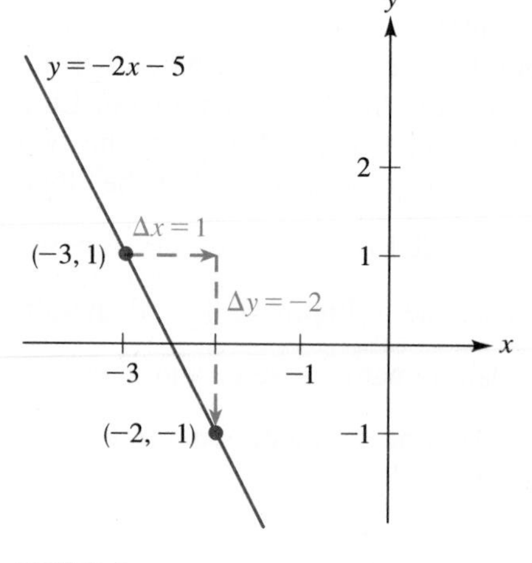

FIGURE 9

EXAMPLE 3 Find the equation of the line passing through the points $(-2, -3)$ and $(2, 5)$.

Solution The slope of the line is

$$m = \frac{y_2 - y_1}{x_2 - x_1} = \frac{5 - (-3)}{2 - (-2)} = \frac{8}{4} = 2$$

Knowing the slope, we can apply the point–slope formula, making use of either $(-2, -3)$ or $(2, 5)$. Using the point $(2, 5)$ as (x_1, y_1), we have

$$
\begin{aligned}
y - y_1 &= m(x - x_1) \\
y - 5 &= 2(x - 2) \\
y - 5 &= 2x - 4 \\
y &= 2x + 1
\end{aligned}
$$

Thus the required equation is $y = 2x + 1$. You should check for yourself that the same answer is obtained using the point $(-2, -3)$ instead of $(2, 5)$ in the last set of calculations.

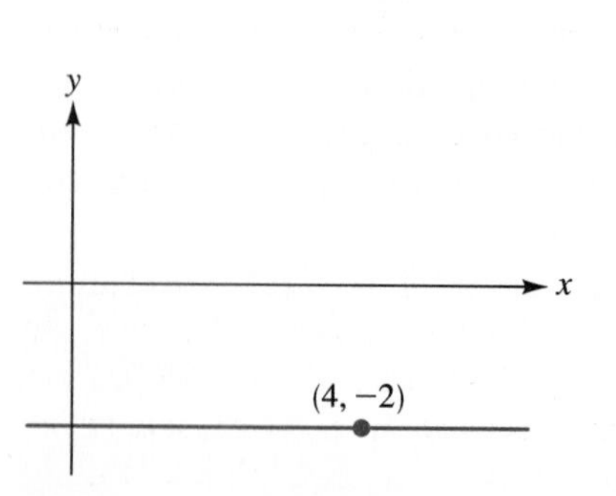

FIGURE 10

EXAMPLE 4 Find the equation of the horizontal line passing through the point $(4, -2)$; see Figure 10.

Solution Since the slope of a horizontal line is zero, we have

$$
\begin{aligned}
y - y_1 &= m(x - x_1) \\
y - (-2) &= 0(x - 4) \\
y &= -2
\end{aligned}
$$

Thus the equation of the horizontal line passing through $(4, -2)$ is $y = -2$.

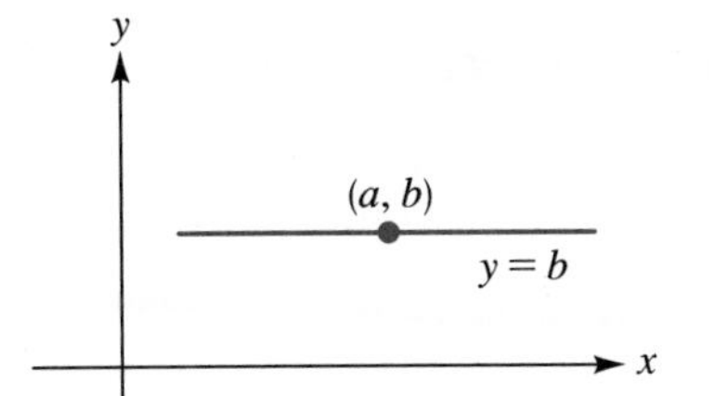

FIGURE 11

By using the point–slope formula exactly as we did in Example 4, we can show more generally that the equation of the horizontal line in Figure 11 is $y = b$. What about the equation of the vertical line in Figure 12 passing through the point (a, b)? Because slope is not defined for vertical lines, the point–slope formula is not applicable this time. However, note that as we move along the vertical line, the x-coordinate is always a; it is only the y-coordinate that varies. The equation $x = a$ expresses exactly these two facts; it says that x must always be a and it places no restrictions on y.

In the box that follows, we summarize our results concerning horizontal and vertical lines.

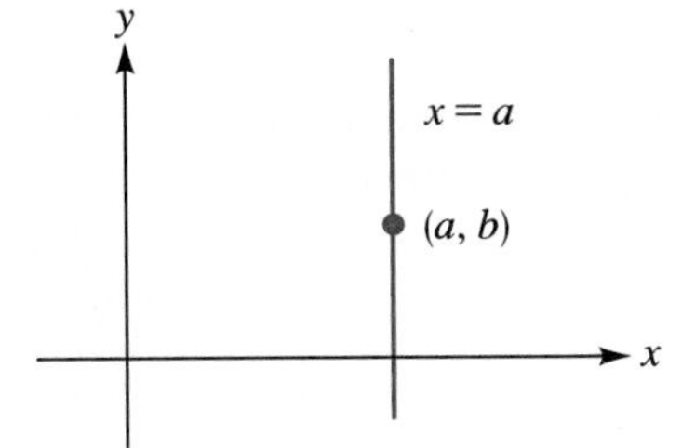

FIGURE 12

EQUATIONS OF HORIZONTAL AND VERTICAL LINES

1. The equation of a horizontal line through the point (a, b) is $y = b$; see Figure 11.
2. The equation of a vertical line through the point (a, b) is $x = a$; see Figure 12.

Another basic form for the equation of a line is the *slope–intercept form.* We are given the slope m and y-intercept b, as shown in Figure 13, and we want to find the equation of the line. To say that the line has a y-intercept of b is the same as saying that the line passes through $(0, b)$. The point–slope formula is applicable now, using the slope m and the point $(0, b)$. We have

$$y - y_1 = m(x - x_1)$$
$$y - b = m(x - 0)$$
$$y = mx + b$$

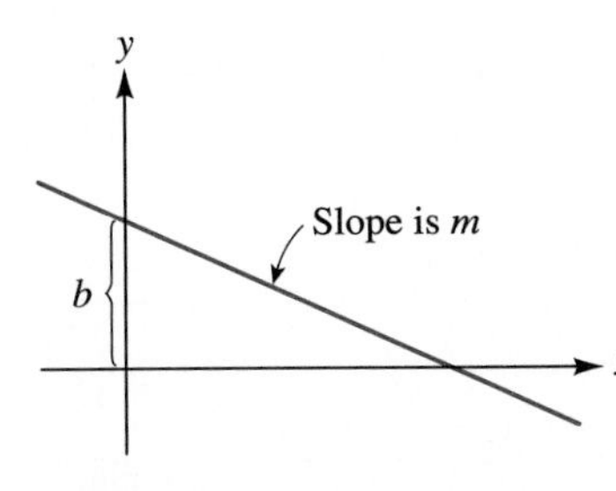

FIGURE 13

This last equation is called the **slope–intercept formula.**

THE SLOPE–INTERCEPT FORMULA

The equation of a line with slope m and y-intercept b is

$$y = mx + b$$

EXAMPLE 5 Write the equation of a line that has slope 4 and a y-intercept of 1. Graph the line.

Solution Substituting $m = 4$ and $b = 1$ in the equation $y = mx + b$ yields

$$y = 4x + 1$$

This is the required equation. We could draw the graph by first setting up a simple table, but for purposes of emphasis and review we proceed as we did just after Example 2. Starting from the point $(0, 1)$, we interpret the slope of 4 as saying that if x increases by 1, then y increases by 4. This takes us from $(0, 1)$ to the point $(1, 5)$, and the line can now be sketched as in Figure 14.

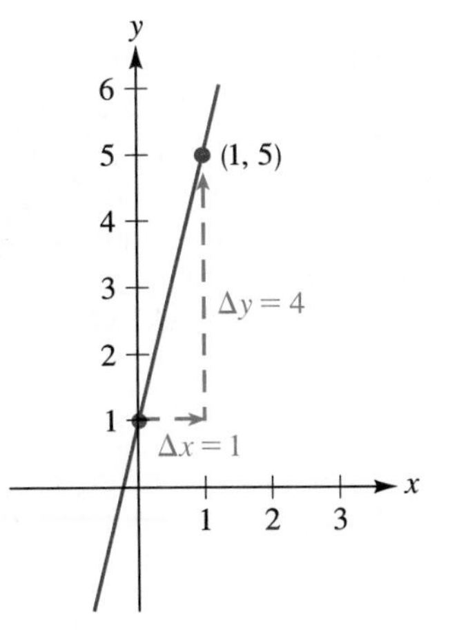

FIGURE 14
$y = 4x + 1$

EXAMPLE 6 Find the slope and y-intercept of the line $3x - 5y = 15$.

Solution First we solve for y so that the equation is in the form $y = mx + b$:

$$3x - 5y = 15$$
$$-5y = -3x + 15$$

and therefore

$$y = \frac{3}{5}x - 3$$

The slope m and the y-intercept b can now be read directly from the equation: $m = 3/5$ and $b = -3$.

As a consequence of our work up to this point, we can say that the graph of any **linear equation** $Ax + By + C = 0$, where A and B are not both zero, is a line, since an equation of this type can always be rewritten in one of the following three forms: $y = mx + b$, or $x = a$, or $y = b$. In the box that follows, we summarize our basic results on equations of lines.

PROPERTY SUMMARY EQUATIONS OF LINES

EQUATION	COMMENT
$y - y_1 = m(x - x_1)$ (the point–slope formula)	This is the equation of a line that has slope m and passes through the point (x_1, y_1).
$y = mx + b$ (the slope–intercept formula)	This is the equation of a line with slope m and y-intercept b.
$Ax + By + C = 0$	The graph of any equation of this form (where A and B are not both zero) is a line. Special cases: If $A = 0$, then the equation can be written $y = -C/B$, which is the equation of a horizontal line with y-intercept $-C/B$. If $B = 0$, then the equation can be written $x = -C/A$, which represents a vertical line with x-intercept $-C/A$.

We conclude this section by discussing two useful relationships regarding the slopes of parallel lines and the slopes of perpendicular lines. First, nonvertical parallel lines have the same slope; see Figure 15. This should seem reasonable if you recall that slope is a number indicating the direction or slant of a line. The relationship concerning slopes of perpendicular lines is not so obvious. The slopes of two nonvertical perpendicular lines are negative reciprocals of each other. That is, if m_1 and m_2 denote the slopes of the two perpendicular lines, then $m_1 = -1/m_2$ or, equivalently, $m_1 m_2 = -1$. For example, if a line has a slope of $2/3$, then any line perpendicular to it must have a slope of $-3/2$. For reference we summarize these facts in the box that follows. (Proofs of these facts are outlined in detail in Exercises 61 and 62.)

GRAPHICAL PERSPECTIVE

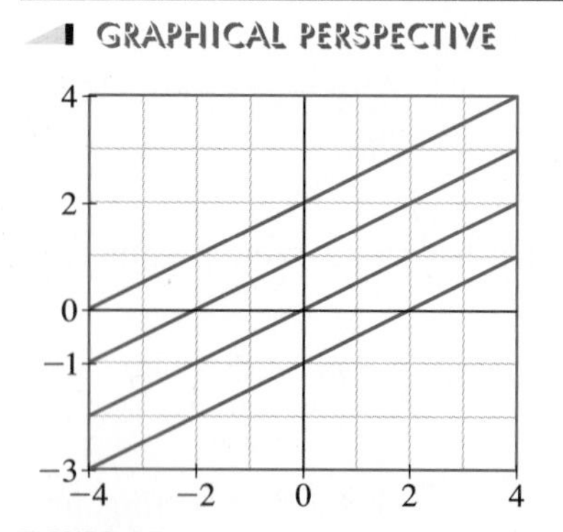

FIGURE 15
Parallel lines have the same slope. Each line here has a slope of 0.5. The equations of the lines, from top to bottom, are $y = 0.5x + 2$, $y = 0.5x + 1$, $y = 0.5x$, and $y = 0.5x - 1$.

PARALLEL AND PERPENDICULAR LINES

Let m_1 and m_2 denote the slopes of two nonvertical lines. Then:

1. the lines are parallel if and only if $m_1 = m_2$; and
2. the lines are perpendicular if and only if $m_1 = -1/m_2$.

EXAMPLE 7 Are the lines $3x - 6y - 8 = 0$ and $2y = x + 1$ parallel?

Solution By solving each equation for y, we can see what the slopes are:

$$\begin{aligned} 3x - 6y - 8 &= 0 \\ -6y &= -3x + 8 \\ y &= \frac{-3}{-6}x + \frac{8}{-6} \\ y &= \frac{1}{2}x - \frac{4}{3} \end{aligned} \qquad \Bigg| \qquad \begin{aligned} 2y &= x + 1 \\ y &= \frac{1}{2}x + \frac{1}{2} \end{aligned}$$

From this we see that both lines have the same slope, namely, $m = 1/2$. It follows therefore that the lines are parallel.

EXAMPLE 8 Find the equation of the line that is parallel to $5x + 6y = 30$ and passes through the origin.

Solution First we find the slope of the line $5x + 6y = 30$:

$$\begin{aligned} 5x + 6y &= 30 \\ 6y &= -5x + 30 \\ y &= -\frac{5}{6}x + 5 \end{aligned}$$

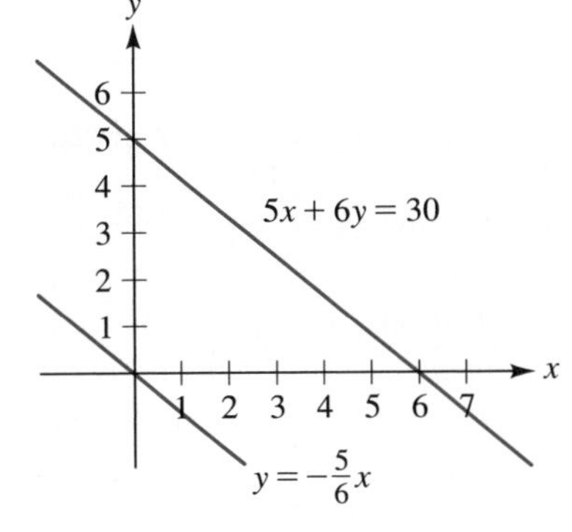

FIGURE 16

The slope is the x-coefficient in the last equation: $m = -5/6$. A parallel line will have the same slope. Since the required line is to pass through $(0, 0)$, we have

$$\begin{aligned} y - y_1 &= m(x - x_1) \\ y - 0 &= -\frac{5}{6}(x - 0) \\ y &= -\frac{5}{6}x \end{aligned}$$

This is the equation of the line that is parallel to $5x + 6y = 30$ and that passes through the origin, as required; see Figure 16.

EXAMPLE 9 Find the equation of the line that is tangent to the circle $x^2 + y^2 = 25$ at the point $(3, -4)$; see Figure 17. Write the answer in the form $y = mx + b$.

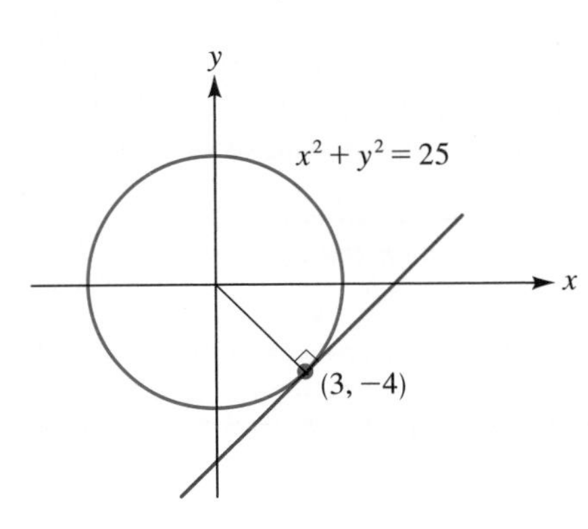

FIGURE 17

Solution We have one point on the tangent line, namely, $(3, -4)$. If we knew the slope, then we could apply the point–slope formula. From elementary geometry, we know that the tangent line is perpendicular to the radius. Since the radius passes through $(0, 0)$ and $(3, -4)$, its slope is

$$\frac{-4 - 0}{3 - 0} = -\frac{4}{3}$$

and, consequently, the slope of the tangent is $3/4$. The point–slope formula is now applicable:

$$y - y_1 = m(x - x_1)$$

$$y - (-4) = \frac{3}{4}(x - 3)$$

Check now for yourself that this can be simplified to $y = \frac{3}{4}x - \frac{25}{4}$, which is the required equation of the tangent line.

EXERCISE SET 2.3

DO 1, 5, 11, 13, 15, 23, 26, 31, 33, 35, 39, 51, 55

A

In Exercises 1–3, compute the slope of the line passing through the two given points. In Exercise 3, include a sketch with your answers.

1. **(a)** $(-3, 2), (1, -6)$ **(b)** $(2, -5), (4, 1)$
(c) $(-2, 7), (1, 0)$ **(d)** $(4, 5), (5, 8)$

2. **(a)** $(-3, 0), (4, 9)$ **(b)** $(-1, 2), (3, 0)$
(c) $(\frac{1}{2}, -\frac{3}{5}), (\frac{3}{2}, \frac{3}{4})$ **(d)** $(\frac{17}{3}, -\frac{1}{2}), (-\frac{1}{2}, \frac{17}{3})$

3. **(a)** $(1, 1), (-1, -1)$ **(b)** $(0, 5), (-8, 5)$
(c) $(-1, 1), (1, -1)$ **(d)** $(a, b), (b, a)$
(Assume $a \neq b$.)

4. Compute the slope of the line in the following figure using each pair of points indicated.
(a) A and B **(b)** B and C **(c)** A and C
The principle involved here is that no matter which pair of points you choose, the slope is the same.

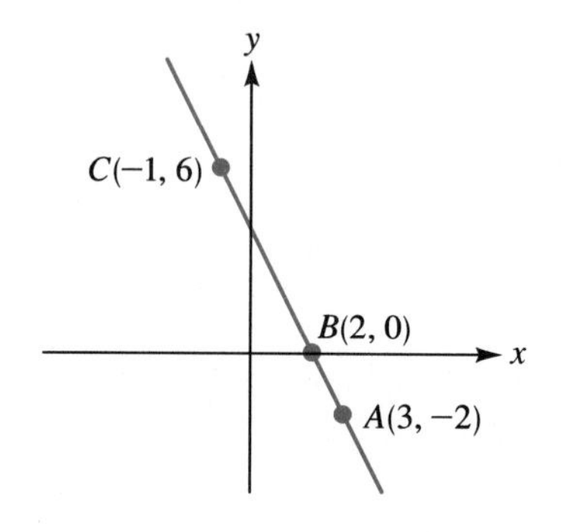

5. The slopes of four lines are indicated in the figure. List the slopes m_1, m_2, m_3, m_4 in order of increasing value.

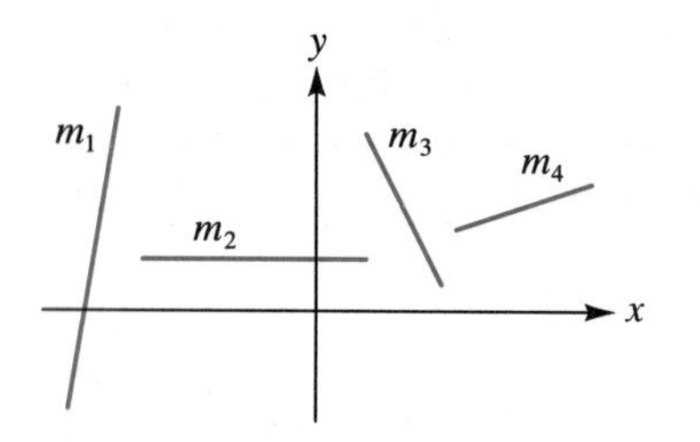

6. Refer to the accompanying figure.

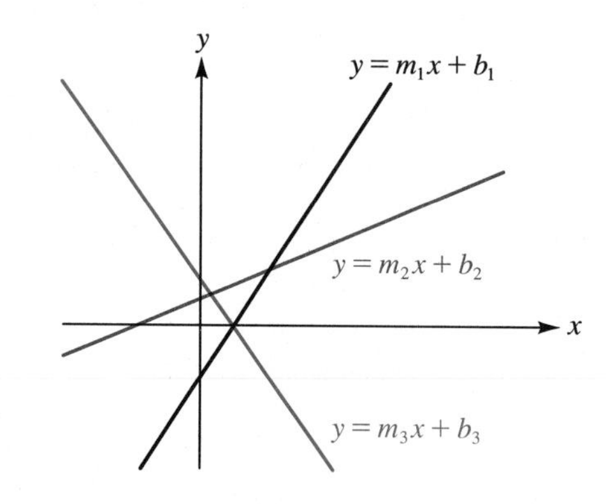

(a) List the slopes m_1, m_2, and m_3 in order of increasing size.
(b) List the numbers b_1, b_2, and b_3 in order of increasing size.

In Exercises 7–9, three points A, B, and C are specified. Determine if A, B, and C are collinear (lie on the same line) by checking to see whether the slope of $\overline{AB}$ equals the slope of $\overline{BC}$.

7. $A(-8, -2)$, $B(2, \frac{1}{2})$, $C(11, -1)$

8. $A(4, -3)$, $B(-1, 0)$, $C(-4, 2)$

9. $A(0, -5)$, $B(3, 4)$, $C(-1, -8)$

10. If the area of the "triangle" formed by three points is zero, then the points must in fact be collinear. Use this observation, along with the formula in Exercise 14(b) of Exercise Set 2.1, to rework Exercise 9.

In Exercises 11 and 12, find the equation for the line having the given slope and passing through the given point. Write your answers in the form $y = mx + b$.

11. **(a)** $m = -5$; $(-2, 1)$ **(b)** $m = 4$; $(4, -4)$
(c) $m = 1/3$; $(-6, -2/3)$ **(d)** $m = -1$; $(0, 1)$

12. **(a)** $m = 22$; $(0, 0)$
(b) $m = -222$; $(0, 0)$
(c) $m = \sqrt{2}$; $(0, 0)$

In Exercises 13 and 14, find the equation for the line passing through the two given points. Write your answer in the form $y = mx + b$.

13. **(a)** $(4, 8)$ and $(-3, -6)$ **(b)** $(-2, 0)$ and $(3, -10)$
(c) $(-3, -2)$ and $(4, -1)$

14. **(a)** $(7, 9)$ and $(-11, 9)$ **(b)** $(5/4, 2)$ and $(3/4, 3)$
(c) $(12, 13)$ and $(13, 12)$

In Exercises 15 and 16, write the equation of a vertical line passing through the given point. In Exercises 17 and 18, write the equation of a horizontal line passing through the given point.

15. $(-3, 4)$ **16.** $(8, 5)$ **17.** $(-3, 4)$ **18.** $(8, 5)$

19. Is the graph of the line $x = 0$ the x-axis or the y-axis?

20. Is the graph of the line $y = 0$ the x-axis or the y-axis?

In Exercises 21 and 22, find the equation of the line with the given slope and y-intercept.

21. **(a)** slope -4; y-intercept 7
(b) slope 2; y-intercept $3/2$
(c) slope $-4/3$; y-intercept 14

22. **(a)** slope 0; y-intercept 14
(b) slope 14; y-intercept 0

In Exercises 23–28, find an equation for the line that is described, and sketch the graph. For Exercises 23–25, write the final answer in the form $y = mx + b$; for Exercises 26–28, write the answer in the form $Ax + By + C = 0$.

23. **(a)** Passes through $(-3, -1)$ and has slope 4
(b) Passes through $(5/2, 0)$ and has slope $1/2$
(c) Has x-intercept 6, y-intercept 5
(d) Has x-intercept -2, slope $3/4$
(e) Passes through $(1, 2)$ and $(2, 6)$

24. **(a)** Passes through $(-7, -2)$ and $(0, 0)$
(b) Passes through $(6, -3)$ and has y-intercept 8
(c) Passes through $(0, -1)$ and has the same slope as the line $3x + 4y = 12$
(d) Passes through $(6, 2)$ and has the same x-intercept as the line $-2x + y = 1$
(e) Has x-intercept -6, y-intercept $\sqrt{2}$

25. Has the same x- and y-intercepts as the circle $x^2 + y^2 + 4x - 4y + 4 = 0$

26. Passes through $(3, -5)$ and through the center of the circle $4x^2 + 8x + 4y^2 - 24y + 15 = 0$

27. Passes through $(-3, 4)$ and is parallel to the x-axis

28. Passes through $(-3, 4)$ and is parallel to the y-axis

In Exercises 29 and 30, find the x- and y-intercepts of the line, and find the area and perimeter of the triangle formed by the given line and the axes.

29. **(a)** $3x + 5y = 15$ **(b)** $3x - 5y = 15$

30. **(a)** $5x + 4y = 40$ **(b)** $2x + 4y = \sqrt{2}$

31. Determine whether each pair of lines is parallel, perpendicular, or neither.
(a) $3x - 4y = 12$; $4x - 3y = 12$
(b) $y = 5x - 16$; $y = 5x + 2$
(c) $5x - 6y = 25$; $6x + 5y = 0$
(d) $y = -\frac{2}{3}x - 1$; $y = \frac{3}{2}x - 1$
(e) $-2x - 5y = 1$; $y - \frac{2}{5}x - 4 = 0$
(f) $x = 8y + 3$; $4y - \frac{1}{2}x = 32$

32. Are the lines $y = x + 1$ and $y = 1 - x$ parallel, perpendicular, or neither?

In Exercises 33–38, find an equation for the line that is described. Write the answer in the two forms $y = mx + b$ and $Ax + By + C = 0$.

33. Is parallel to $2x - 5y = 10$ and passes through $(-1, 2)$

34. Is parallel to $4x + 5y = 20$ and passes through $(0, 0)$

35. Is perpendicular to $4y - 3x = 1$ and passes through $(4, 0)$

36. Is perpendicular to $x - y + 2 = 0$ and passes through $(3, 1)$

37. Is parallel to $3x - 5y = 25$ and has the same y-intercept as the line $6x - y + 11 = 0$

38. Is perpendicular to $9y - 2x = 3$ and has the same x-intercept as the circle $x^2 + y^2 - 12x - 2y = -36$

39. **(a)** Sketch the circle $x^2 + y^2 = 25$.
(b) Find the equation of the line tangent to this circle at the point $(-4, -3)$. Include a sketch.

40. **(a)** Tangents are drawn to the circle $x^2 + y^2 = 169$ at the points $(5, 12)$ and $(5, -12)$. Find the equations of the tangents.
(b) Find the coordinates of the point where these two tangents intersect.

B

41. Show that the slope of the line passing through the two points $(3, 9)$ and $(3 + h, (3 + h)^2)$ is $6 + h$.

42. Show that the slope of the line passing through the two points (x, x^2) and $(x + h, (x + h)^2)$ is $2x + h$.

43. Show that the slope of the line passing through the two points (x, x^3) and $(x + h, (x + h)^3)$ is $3x^2 + 3xh + h^2$.

44. Show that the slope of the line passing through the points $(x, \sqrt{x})$ and $(x + h, \sqrt{x + h})$ can be written

$$1/(\sqrt{x + h} + \sqrt{x})$$

45. Show that the slope of the line passing through the two points $\left(x, \dfrac{1}{x}\right)$ and $\left(x + h, \dfrac{1}{x + h}\right)$ is $\dfrac{-1}{x(x + h)}$.

46. Why is the letter m used to denote slope? Look up the verb *monter* in a French-to-English dictionary.

47. Sketch the curve $y = x^2$ and indicate the point $T(3, 9)$ on the graph. In each case that follows, you're given the x-coordinate of a point P on the curve. Find the corresponding y-coordinate, and then calculate the slope of the line through P and T. Display your results in a table.
(a) $x = 2.5$ **(b)** $x = 2.9$ **(c)** $x = 2.99$
(d) $x = 2.999$ **(e)** $x = 2.9999$
As P gets closer to T, what numerical value does the slope of $\overline{PT}$ seem to approach? What would you estimate to be the slope of the line that is tangent to the curve $y = x^2$ at point T?

48. Follow the instructions for Exercise 47 using the curve $y = \sqrt{x}$, the point $T(1, 1)$, and the following sequence of x-values.
(a) $x = 1.1$ **(b)** $x = 1.01$ **(c)** $x = 1.001$
(d) $x = 1.0001$ **(e)** $x = 1.00001$
(Round your answers to six decimal places.)

49. Find a point P on the curve $y = x^3$ such that the slope of the line passing through P and $(1, 1)$ is $3/4$.

50. Find a point P on the curve $y = \sqrt{x}$ such that the slope of the line through P and $(1, 1)$ is $1/4$.

51. Find a point P on the curve $y = 1/x$ such that the slope of the line through P and $(2, 1/2)$ is $-1/16$.

52. Find a point P on the circle $x^2 + y^2 = 20$ such that the slope of the radius drawn from $(0, 0)$ to P is 2. (You will get two answers.)

53. A line with a slope of -5 passes through the point $(3, 6)$. Find the area of the triangle in the first quadrant formed by this line and the coordinate axes.
Hint: First find the equation of the line.

54. The y-intercept of the line in the figure is 6. Find the slope of the line if the area of the shaded triangle is 72 square units.

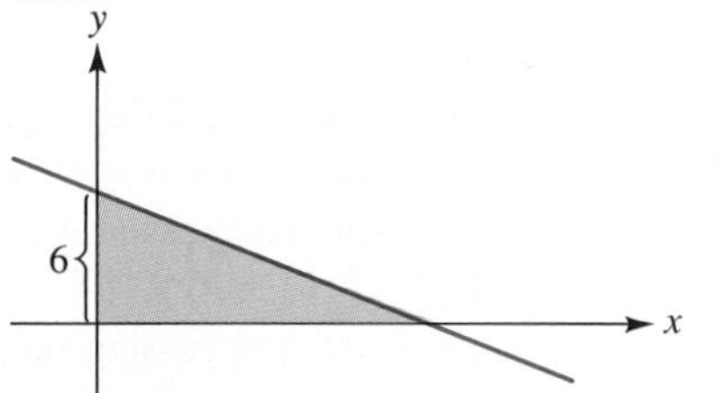

55. **(a)** Sketch the line $y = \frac{1}{2}x - 5$ and the point $P(1, 3)$. Follow parts (b)–(d) to calculate the perpendicular distance from point $P(1, 3)$ to the line.
(b) Find the equation of the line that passes through $P(1, 3)$ and is perpendicular to the line $y = \frac{1}{2}x - 5$.
(c) Find the coordinates of the point where these two lines intersect.
(d) Use the distance formula to find the perpendicular distance from $P(1, 3)$ to the line $y = \frac{1}{2}x - 5$.

56. **(a)** Show that the equation of the line tangent to the circle $x^2 + y^2 = a^2$ at the point (x_1, y_1) on the circle is

$$x_1 x + y_1 y = a^2$$

(b) Use the result in part (a) to rework Exercises 39(b) and 40(a).

57. Follow the instructions in parts (a)–(e) to determine the slope m of the line in the figure.

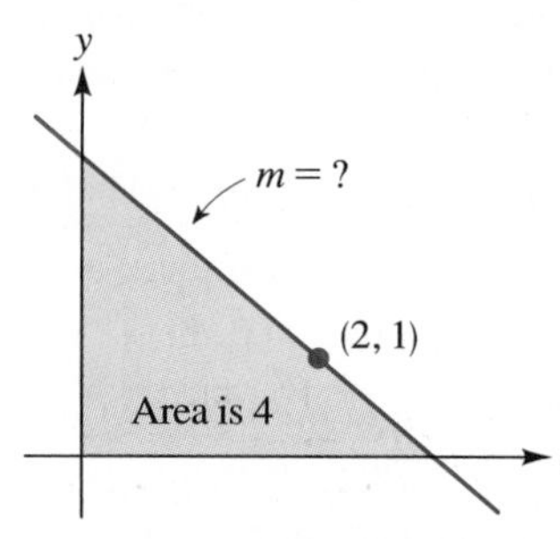

(a) Use the point–slope formula to show that the equation of the line is $y - 1 = m(x - 2)$.
(b) Show that the y-intercept for the line is $-2m + 1$. Show that the x-intercept is $(2m - 1)/m$.
(c) Show that the area of the triangle, which is one-half base times height, is $\frac{1}{2}\left(\frac{2m-1}{m}\right)(-2m + 1)$.
(d) Set the expression for area in part (c) equal to 4, as given in the figure, and simplify the resulting equation to obtain $4m^2 + 4m + 1 = 0$.
(e) Solve the equation for m. You should obtain $m = -1/2$.

58. A line with a negative slope passes through the point $(3, 1)$. The area of the triangle bounded by this line and the axes is 6 square units. Find the possible slopes of the line. *Hint:* See Exercise 57.

59. By analyzing sales figures, the accountant for College Stereo Company knows that 280 units of a CD player can be sold each month when the price is $P = \$195$/unit. The figures also show that for each \$15 hike in the price, 10 fewer units are sold monthly.
(a) Let x denote the number of units sold per month and P the price per unit. Find an equation that expresses x in terms of P, assuming that this relationship is linear. *Hint:* $\Delta x/\Delta P = -10/15 = -2/3$.
(b) Use the equation that you found in part (a) to determine how many units can be sold in a month when the price is \$270/unit.
(c) What should the price be to sell 205 units/month?

60. Imagine that you own a grove of orange trees, and suppose that from past experience you know that when 100 trees are planted, each tree will yield approximately 240 oranges/yr. Furthermore, you've noticed that when additional trees are planted in the grove, the yield per tree decreases. Specifically, you have noted that the yield per tree decreases by about 20 oranges for each additional tree planted.

(a) Let y denote the yield per tree when x trees are planted. Find a linear equation relating x and y.

(b) Use the equation in part (a) to determine how many trees should be planted to obtain a yield of 400 oranges/tree.

(c) If the grove contains 95 trees, what yield can you expect from each tree?

61. This exercise outlines a proof of the fact that two nonvertical lines are parallel if and only if their slopes are equal. The proof relies on the following observation for the given figure: The lines $y = m_1x + b_1$ and $y = m_2x + b_2$ are parallel if and only if the two vertical distances AB and CD are equal. (In the figure, the points C and D both have x-coordinate 1.)

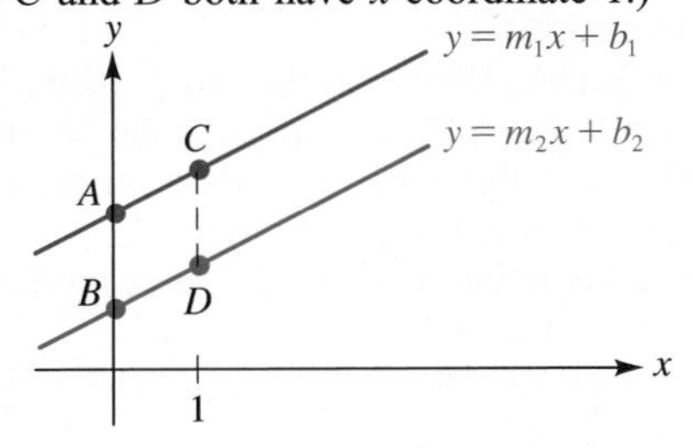

(a) Verify that the coordinates of A, B, C, and D are

$A(0, b_1)$ $\quad$ $B(0, b_2)$

$C(1, m_1 + b_1)$ $\quad$ $D(1, m_2 + b_2)$

(b) Using the coordinates in part (a), check that $AB = b_1 - b_2$ and $CD = (m_1 + b_1) - (m_2 + b_2)$.

(c) Use part (b) to show that the equation $AB = CD$ is equivalent to $m_1 = m_2$.

62. This exercise outlines a proof of the fact that two nonvertical lines with slopes m_1 and m_2 are perpendicular if and only if $m_1m_2 = -1$. In the accompanying figure, we've assumed that our two nonvertical lines $y = m_1x$ and $y = m_2x$ intersect at the origin. [If they did not intersect there, we could just as well work with lines parallel to these that do intersect at (0, 0), recalling that parallel lines have the same slope.] The proof relies on the following geometric fact:

$$\overline{OA} \perp \overline{OB} \quad \text{if and only if} \quad (OA)^2 + (OB)^2 = (AB)^2$$

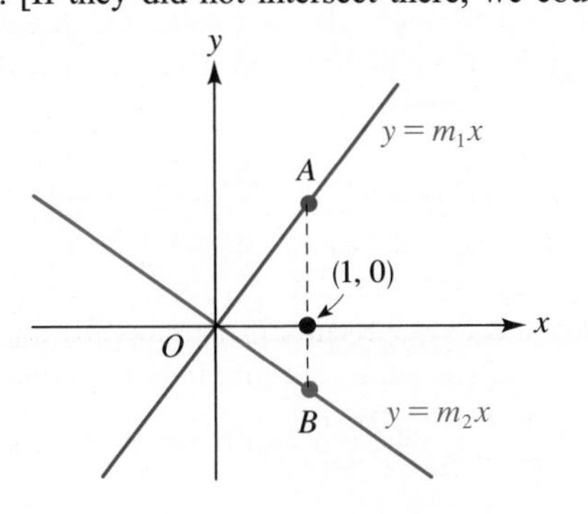

(a) Verify that the coordinates of A and B are $A(1, m_1)$ and $B(1, m_2)$.

(b) Show that

$$OA^2 = 1 + m_1^2$$
$$OB^2 = 1 + m_2^2$$
$$AB^2 = m_1^2 - 2m_1m_2 + m_2^2$$

(c) Use part (b) to show that the equation $OA^2 + OB^2 = AB^2$ is equivalent to $m_1m_2 = -1$.

63. Verify the identity

$$(y_2 - y_1)/(x_2 - x_1) = (y_1 - y_2)/(x_1 - x_2)$$

What does this identity tell you about calculating slope?

64. The following figure shows two tangent lines drawn from the point (a, b) to the circle $x^2 + y^2 = R^2$. Follow steps (a)–(d) to show that the equation of the line passing through the two points of tangency is

$$ax + by = R^2$$

(This problem, along with the following clever solution, appears in the classic text by Isaac Todhunter, *A Treatise on Plane Co-ordinate Geometry,* first published in 1855.)

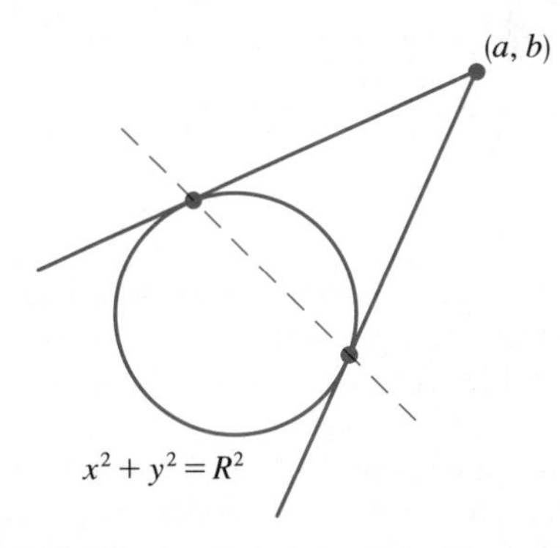

(a) Let (x_1, y_1) be one of the points of tangency. Show that the equation of the tangent line through this point is $x_1x + y_1y = R^2$.

(b) Using the result in part (a), explain why $x_1a + y_1b = R^2$.

(c) In a similar fashion, explain why $x_2a + y_2b = R^2$, where (x_2, y_2) is the other point of tangency.

(d) The equation $ax + by = R^2$ represents a line. Explain why this line must pass through (x_1, y_1) and (x_2, y_2).

C

65. Given two lines L_1 and L_2 passing through the origin, as shown in the figure, a third line L is said to bisect the area between L_1 and L_2 provided that for each point P on L, the area of region A equals the area of region B. (In the figure, each dashed line is parallel to one of the coordinate axes.)

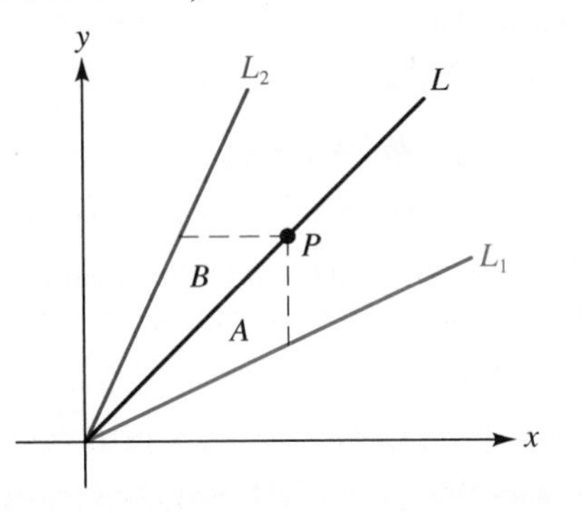

(a) Suppose that the equation of L_1 is $y = \frac{1}{2}x$ and the equation of L_2 is $y = 8x$. Find the equation of L.

(b) Suppose that the equations of L_1 and L_2 are $y = nx$ and $y = mx$, respectively. Find the equation of L.

66. The vertices of a triangle are $A(-4, 0)$, $B(2, 0)$, and $C(0, 6)$. Let M_1, M_2, and M_3 be the midpoints of $\overline{AB}$, $\overline{BC}$, and $\overline{AC}$, respectively. Let H_1, H_2, and H_3 be the feet of the altitudes on sides $\overline{AB}$, $\overline{BC}$, and $\overline{AC}$, respectively.

(a) Find the equation of the circle passing through M_1, M_2, and M_3.
Answer: $3x^2 + 3y^2 + 3x - 11y = 0$

(b) Find the equation of the circle passing through H_1, H_2, and H_3.
Answer: $3x^2 + 3y^2 + 3x - 11y = 0$

(c) Find the point P at which the three altitudes intersect. *Answer:* $(0, 4/3)$

(d) Let N_1, N_2, and N_3 be the midpoints of $\overline{AP}$, $\overline{BP}$, and $\overline{CP}$, respectively. Show that the circle obtained in parts (a) and (b) passes through N_1, N_2, and N_3. *Note:* This circle is called the *nine-point circle* of $\triangle ABC$.

(e) The *circumcircle* of $\triangle ABC$ is the circle passing through the three points A, B, and C. Show that the coordinates of the center Q of the circumcircle are $Q(-1, 7/3)$. *Hint:* Use the result from geometry that says that the perpendicular bisectors of the sides of the triangle intersect at the center of the circumcircle.

(f) For $\triangle ABC$, show that the radius of the nine-point circle is one-half the radius of the circumcircle.

(g) Show that the midpoint of line segment $\overline{QP}$ is the center of the nine-point circle.

(h) Find the point G where the three medians of $\triangle ABC$ intersect.

(i) Show that G lies on line segment $\overline{QP}$ and that $PG = 2(GQ)$.

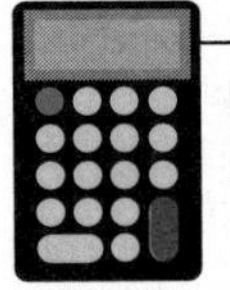

GRAPHING UTILITY EXERCISES FOR SECTION 2.3

1. **(a)** What are the equations of the four lines passing through the origin with slopes 1, 2, 3, and 10? Graph the four lines. As the slopes increase, what happens to the graphs?

(b) What are the equations of the four lines passing through the origin with slopes -1, -2, -3, and -10? Graph the four lines. As the slopes decrease from -1 to -10, what happens to the graphs?

2. **(a)** The following four lines are parallel: $y = 2x - 1$, $y = 2x$, $y = 2x + 1$, and $y = 2x + 2$. Without drawing the graphs, how do you know this to be true?

(b) On the same set of axes, graph the four lines in part (a).

3. Graph the following set of parallel lines on the same set of axes: $y - 1 = -0.5(x - 2)$, $y - 1 = -0.5(x + 2)$, $y - 1 = -0.5(x - 4)$.

4. In each case, graph the pair of perpendicular lines on the same set of axes.

(a) $y = 2x$ and $y = -\frac{1}{2}x$

(b) $y = \frac{8}{3}x$ and $y = -\frac{3}{8}x$

(c) $y = 3x - 4$ and $y = -\frac{1}{3}x - 4$

5. In parts (a) through (d), first solve the equation for y. Then graph the equation and make a note of the x- and y-intercepts.

(a) $\dfrac{x}{2} + \dfrac{y}{3} = 1$ **(b)** $\dfrac{x}{-2} + \dfrac{y}{-3} = 1$

(c) $\dfrac{x}{6} + \dfrac{y}{5} = 1$ **(d)** $\dfrac{x}{-6} + \dfrac{y}{-5} = 1$

(e) Based on your results in parts (a) through (d), describe, in general, the graph of the equation $\dfrac{x}{a} + \dfrac{y}{b} = 1$, where a and b are constants.

6. First, with paper and pencil, find the equation of the line that is tangent to the circle $x^2 + y^2 = 25$ at the point $(3, -4)$. Then graph the line and the circle. Check your picture by comparing it to Figure 17.

In Exercises 7–10, you are given the equation of a circle and a point on the circle. Determine the equation of the tangent to the circle at the given point; then check your result by graphing the circle and the tangent line.

7. $x^2 + y^2 = 20$; $(-2, 4)$
8. $x^2 + y^2 = 74$; $(5, -7)$
9. $(x - 10)^2 + (y + 10)^2 = 289$; $(15, 8)$
10. $x^2 + y^2 = 4$; $(-\sqrt{3}, -1)$
11. Find the equation of the line that passes through the origin and is parallel to the line $3x + 4y = 12$. Graph the lines to check your answer. (Be sure to use the same size unit and same scale on both axes so that perpendicular lines indeed appear to meet at right angles.)
12. Find the equation of the line that passes through the origin and is perpendicular to the line $4x - 5y = 20$. Graph the lines to check your answer.

For Exercises 13 and 14, you'll need to recall the following definitions and results from elementary geometry. In a triangle, a line segment drawn from a vertex to the midpoint of the opposite side is called a ***median.*** *The three medians of a triangle are* ***concurrent;*** *that is, they intersect in a single point. This point of intersection is called the* ***centroid*** *of the triangle. A line segment drawn from a vertex perpendicular to the opposite side is an* ***altitude.*** *The three altitudes of a triangle are concurrent; the point where the altitudes intersect is the* ***orthocenter*** *of the triangle.*

13. This exercise provides an example of the fact that the medians of a triangle are concurrent.
    (a) The vertices of $\triangle ABC$ are as follows:

    $$A(-4, 0) \qquad B(2, 0) \qquad C(0, 6)$$

    Use a graphing utility to draw $\triangle ABC$. (Since $\overline{AB}$ coincides with the x-axis, you won't need to draw a line segment for this side.) *Note:* If the graphing utility you use does not have a provision for drawing line segments, you will need to determine an equation for the line in each case and then graph the line.
    (b) Find the coordinates of the midpoint of each side of the triangle, then add the three medians to your picture in part (a). Note that the three medians do appear to intersect in a single point. Use the graphing utility to estimate the coordinates of the centroid.
    (c) Using paper and pencil, find the equation of the medians from A to $\overline{BC}$ and from B to $\overline{AC}$. Then (solving simultaneous equations, as you did in intermediate algebra), determine the exact coordinates of the centroid. How do these numbers compare with your estimates in part (b)?
14. This illustrates the fact that the altitudes of a triangle are concurrent. Again, we'll be using $\triangle ABC$ with vertices $A(-4, 0)$, $B(2, 0)$, and $C(0, 6)$. Note that one of the altitudes of this triangle is just the portion of the y-axis extending from $y = 0$ to $y = 6$; thus, you won't need to graph this altitude; it will already be in the picture.
    (a) Using paper and pencil, find the equations for the three altitudes. (Actually, you're finding equations for the lines that coincide with the altitude segments.)
    (b) Use a graphing utility to draw $\triangle ABC$ along with the three altitude lines that you determined in part (a). Note that the altitudes appear to intersect in a single point. Use the graphing utility to estimate the coordinates of this point.
    (c) Using simultaneous equations (as in intermediate algebra), find the exact coordinates of the orthocenter. Are your estimates in part (b) close to these values?

2.4 SYMMETRY AND GRAPHS

Symmetry is a working concept. If all the object is symmetrical, then the parts must be halves (or some other rational fraction) and the amount of information necessary to describe the object is halved (etc.).

Alan L. Macay, Department of Crystallography, University of London

As you will see throughout this text, there are a number of techniques that help us to understand the essential features of a graph. In the box that follows, we introduce three types of *symmetry* that are useful in graphing equations.

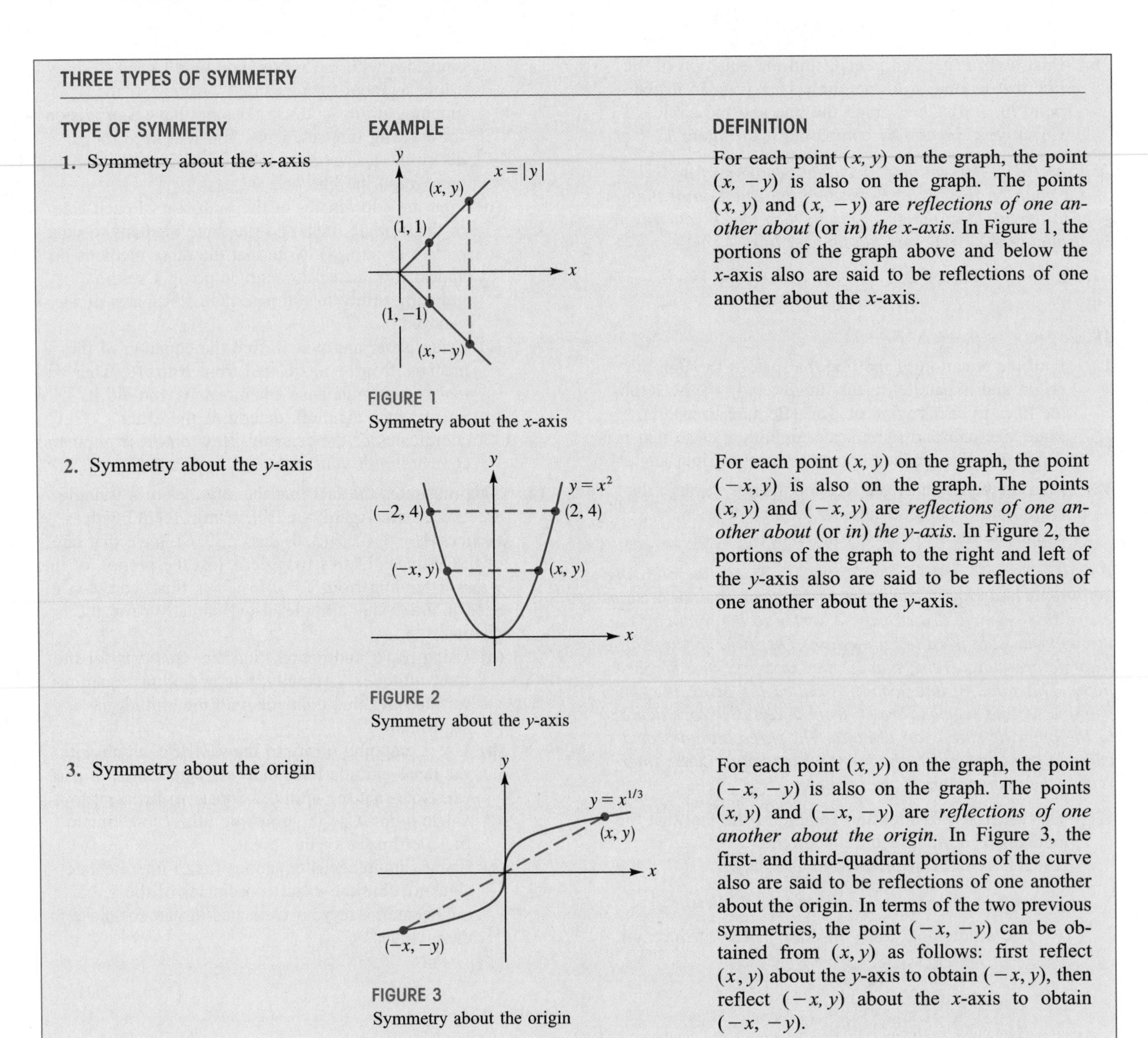

THREE TYPES OF SYMMETRY

TYPE OF SYMMETRY	EXAMPLE	DEFINITION
1. Symmetry about the x-axis	FIGURE 1 Symmetry about the x-axis	For each point (x, y) on the graph, the point $(x, -y)$ is also on the graph. The points (x, y) and $(x, -y)$ are *reflections of one another about* (or *in*) *the x-axis.* In Figure 1, the portions of the graph above and below the x-axis also are said to be reflections of one another about the x-axis.
2. Symmetry about the y-axis	FIGURE 2 Symmetry about the y-axis	For each point (x, y) on the graph, the point $(-x, y)$ is also on the graph. The points (x, y) and $(-x, y)$ are *reflections of one another about* (or *in*) *the y-axis.* In Figure 2, the portions of the graph to the right and left of the y-axis also are said to be reflections of one another about the y-axis.
3. Symmetry about the origin	FIGURE 3 Symmetry about the origin	For each point (x, y) on the graph, the point $(-x, -y)$ is also on the graph. The points (x, y) and $(-x, -y)$ are *reflections of one another about the origin.* In Figure 3, the first- and third-quadrant portions of the curve also are said to be reflections of one another about the origin. In terms of the two previous symmetries, the point $(-x, -y)$ can be obtained from (x, y) as follows: first reflect (x, y) about the y-axis to obtain $(-x, y)$, then reflect $(-x, y)$ about the x-axis to obtain $(-x, -y)$.

EXAMPLE 1 A line segment $\mathscr{L}$ has endpoints $(1, 2)$ and $(5, 3)$. Sketch the reflection of $\mathscr{L}$ about: **(a)** the x-axis; **(b)** the y-axis; and **(c)** the origin.

Solution

(a) First reflect the endpoints $(1, 2)$ and $(5, 3)$ about the x-axis to obtain the new endpoints $(1, -2)$ and $(5, -3)$, respectively. Then join these two points as indicated in Figure 4(a), in Quadrant IV.

(b) Reflect the given endpoints about the y-axis to obtain the new endpoints $(-1, 2)$ and $(-5, 3)$; then join these two points as shown in Figure 4(a), in Quadrant II.

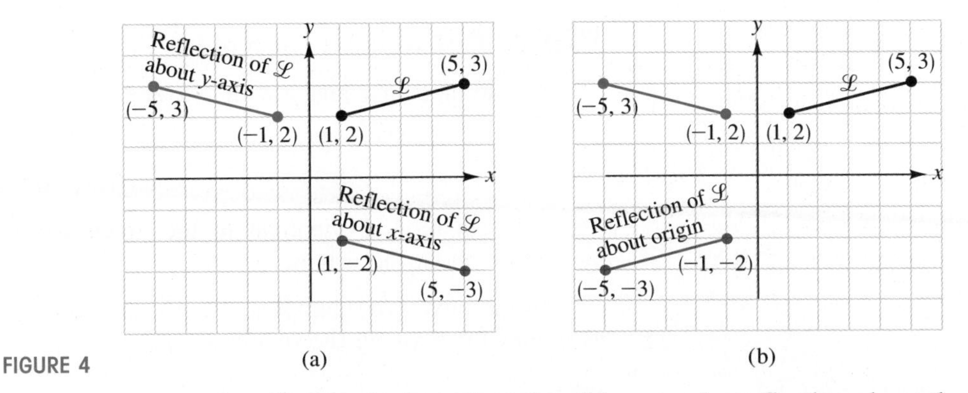

FIGURE 4

(c) As described in the box preceding this example, reflection about the origin can be carried out in two steps: first reflect about the y-axis, then reflect about the x-axis. In part (b) we obtained the reflection of $\mathscr{L}$ in the y-axis. So now we need only reflect the line segment obtained in part (b) about the x-axis. See Figure 4(b).

In the box that follows, we list three rules for testing whether the graph of an equation possesses any of the types of symmetry we've been discussing. (The validity of each rule follows directly from the definitions of symmetry.)

THREE TESTS FOR SYMMETRY

1. The graph of an equation is symmetric about the y-axis if replacing x with $-x$ yields an equivalent equation.
2. The graph of an equation is symmetric about the x-axis if replacing y with $-y$ yields an equivalent equation.
3. The graph of an equation is symmetric about the origin if replacing x and y with $-x$ and $-y$, respectively, yields an equivalent equation.

EXAMPLE 2

(a) Test for symmetry about the y-axis: $y = x^4 - 3x^2 + 1$.

(b) Test for symmetry about the x-axis: $x = \dfrac{y^2}{y-1}$.

Solution

(a) Replacing x with $-x$ in the original equation gives us

$$y = (-x)^4 - 3(-x)^2 + 1 \qquad \text{or} \qquad y = x^4 - 3x^2 + 1$$

Since this last equation is the same as the original equation, we conclude that the graph is symmetric about the y-axis.

(b) Replacing y with $-y$ in the original equation gives us

$$x = \frac{(-y)^2}{(-y) - 1} \qquad \text{or} \qquad x = \frac{y^2}{-y - 1}$$

Since this last equation is not equivalent to the original equation, we conclude that the graph is not symmetric about the x-axis.

EXAMPLE 3 Is the graph of $y = 2x^3 - x$ symmetric about the origin?

Solution Replacing x and y with $-x$ and $-y$, respectively, gives us

$$-y = 2(-x)^3 - (-x) = -2x^3 + x$$
$$y = 2x^3 - x \qquad \text{multiplying through by } -1$$

This last equation is identical to the given equation. Therefore the graph is symmetric about the origin.

In the next two examples, we use the notions of symmetry and intercepts as guides for drawing the required graphs.

EXAMPLE 4 Graph the equation: $y = -x^2 + 5$.

Solution The domain of the variable x in the expression $-x^2 + 5$ is the set of all real numbers. However, since the graph must be symmetric about the y-axis (why?), we need to be able to sketch the graph only to the right of the y-axis; then the portion to the left will be the mirror image of this (with the y-axis as the mirror). The x- and y-intercepts (if any) are computed as follows:

x-INTERCEPTS	y-INTERCEPTS
$-x^2 + 5 = 0$	$y = -(0)^2 + 5$
$x^2 = 5$	$y = 5$
$x = \pm\sqrt{5}\ (\approx \pm 2.2)$	

In Figure 5(a) we've set up a short table of values and sketched the graph for $x \geq 0$. The complete graph is then obtained by reflection about the y-axis, as shown in Figure 5(b). [The curve in Figure 5(b) is a *parabola*. This type of curve will be considered in detail in later chapters.]

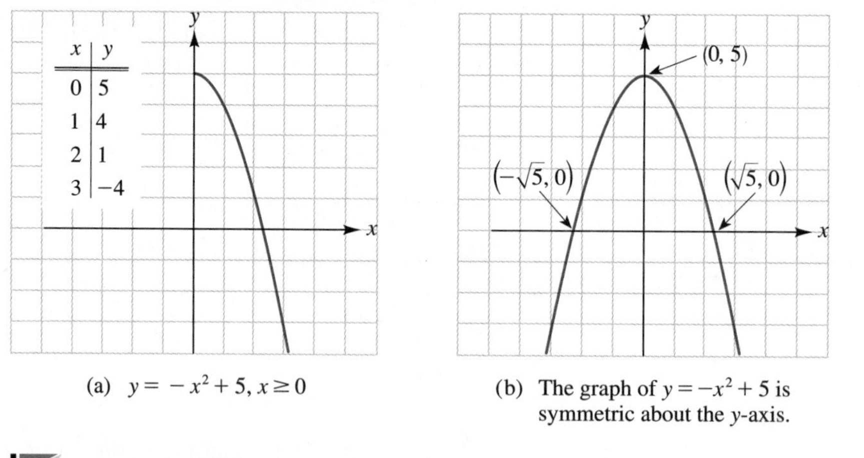

FIGURE 5 (a) $y = -x^2 + 5, x \geq 0$ (b) The graph of $y = -x^2 + 5$ is symmetric about the y-axis.

EXAMPLE 5 Graph: $y = -4/x$.

Solution Before doing any calculations, note that the domain of the variable x consists of all real numbers other than zero. So, however the resulting graph may look, it cannot contain a point with an x-coordinate of zero. In other words, the graph cannot cross the y-axis. (Check for yourself that the graph cannot cross the x-axis either.) Now, since the graph is symmetric about the origin (why?), we need to be able to sketch the graph only for $x > 0$; the portion corresponding to $x < 0$ can then be obtained by reflection about the origin. In Figure 6(a) we've

x	1	2	3	4	5	6	7	8	$\frac{1}{2}$	$\frac{1}{3}$
y	-4	-2	$-\frac{4}{3}$	-1	$-\frac{4}{5}$	$-\frac{2}{3}$	$-\frac{4}{7}$	$-\frac{1}{2}$	-8	-12

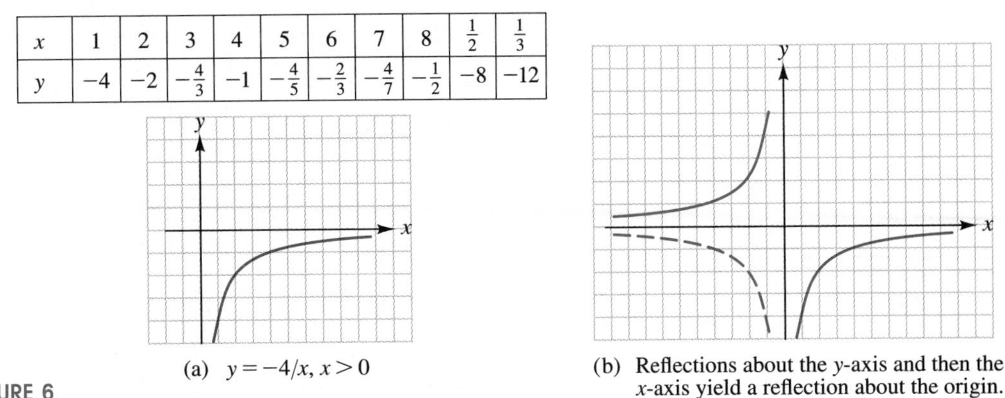

(a) $y=-4/x, x>0$

(b) Reflections about the y-axis and then the x-axis yield a reflection about the origin.

FIGURE 6

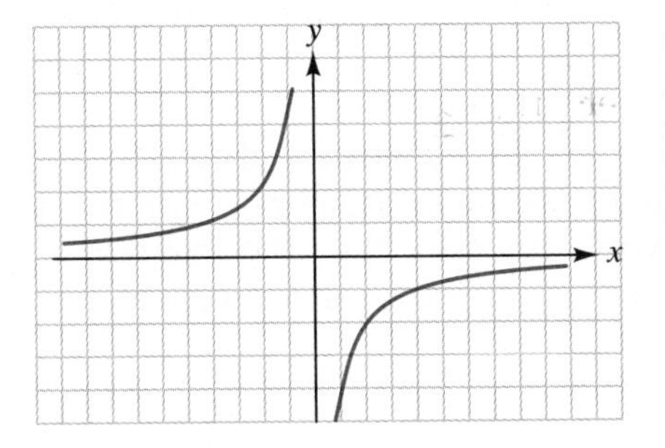

FIGURE 7
The graph of $y = -4/x$ is symmetric about the origin.

set up a table of values and sketched the graph for $x>0$. (Note the fractional x-values at the end of the table. Since y is undefined when x is zero, it's informative to pick x-values near zero and look at the corresponding y-values.) In Figure 6(b) we have reflected the first-quadrant portion of the graph first about the y-axis and then about the x-axis to obtain the required reflection about the origin. Figure 7 shows our final graph of $y = -4/x$.

We can use the graph in Figure 6(a) to introduce the idea of an *asymptote.* A line is an **asymptote** for a curve if the distance between the line and the curve approaches zero as we move out farther and farther along the line. So, for the curve in Figure 6(a), both the x-axis and the y-axis are asymptotes. (We'll return to this idea several times later in the text. For other pictures of asymptotes, see Figure 2 in Section 4.7.)

Symmetry about the x-axis and symmetry about the y-axis are both examples of *symmetry about a line.* In these two cases, the axis of symmetry is one of the coordinate axes. Now we'll look at another example of this type of symmetry: symmetry about the line $y=x$. This kind of symmetry will arise when we analyze inverse functions in the next chapters.

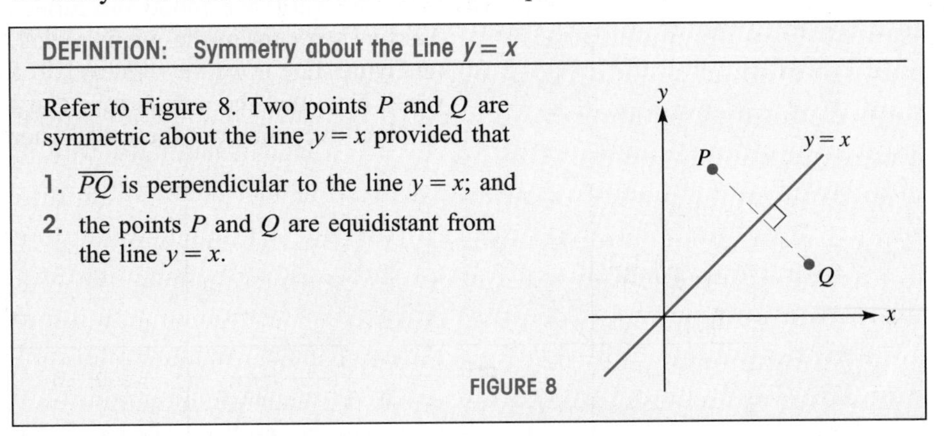

DEFINITION: Symmetry about the Line $y=x$

Refer to Figure 8. Two points P and Q are symmetric about the line $y=x$ provided that

1. $\overline{PQ}$ is perpendicular to the line $y=x$; and
2. the points P and Q are equidistant from the line $y=x$.

FIGURE 8

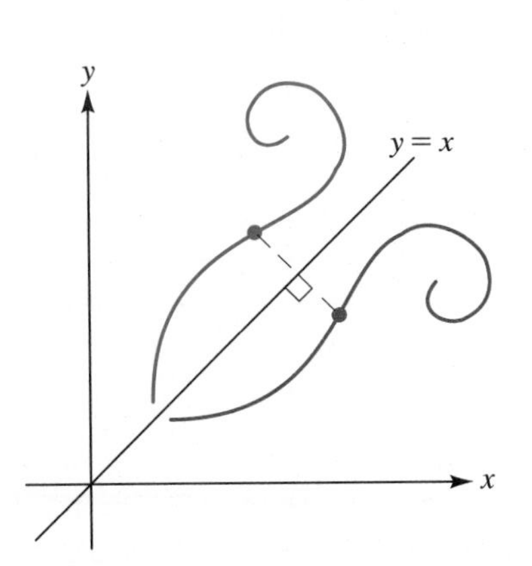

FIGURE 9
The two curves are symmetric about the line $y=x$.

This definition says that P and Q are symmetric about the line $y=x$ if $y=x$ is the perpendicular bisector of line segment $\overline{PQ}$. In Figure 8, we say that the two points P and Q are **reflections** of each other about the line $y=x$ and that $y=x$ is the **axis of symmetry.** In addition, we say that two curves are symmetric about $y=x$ if each point on one curve is the reflection of a corresponding point on the other curve, and vice versa; see Figure 9.

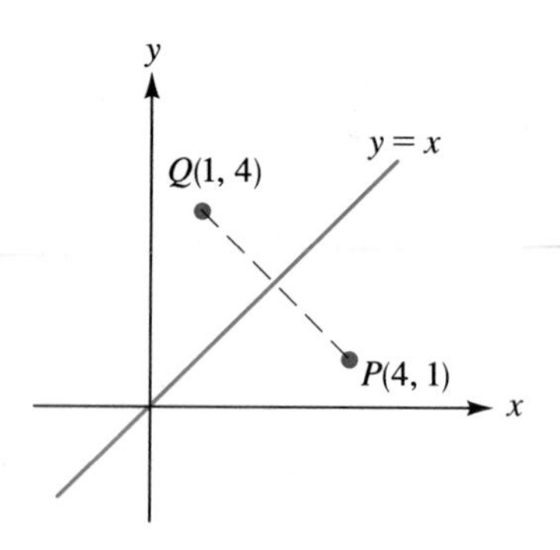

FIGURE 10

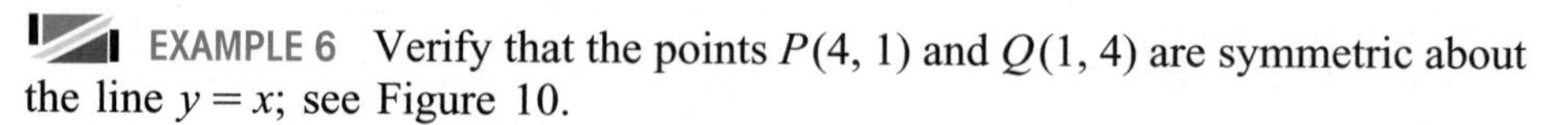

EXAMPLE 6 Verify that the points $P(4, 1)$ and $Q(1, 4)$ are symmetric about the line $y = x$; see Figure 10.

Solution We have to show that the line $y = x$ is the perpendicular bisector of the line segment $\overline{PQ}$. Now, the slope of $\overline{PQ}$ is

$$m = \frac{4 - 1}{1 - 4} = -1$$

We know that the slope of the line $y = x$ is 1. Since these two slopes are negative reciprocals, we conclude that the line $y = x$ is perpendicular to $\overline{PQ}$. Next, we must show that the line $y = x$ passes through the midpoint of $\overline{PQ}$. But (as you should check for yourself), the midpoint of $\overline{PQ}$ is $\left(\frac{5}{2}, \frac{5}{2}\right)$, which does lie on the line $y = x$. In summary, then, we've shown that the line $y = x$ passes through the midpoint of $\overline{PQ}$ and is perpendicular to $\overline{PQ}$. Thus P and Q are symmetric about the line $y = x$, as we wished to show.

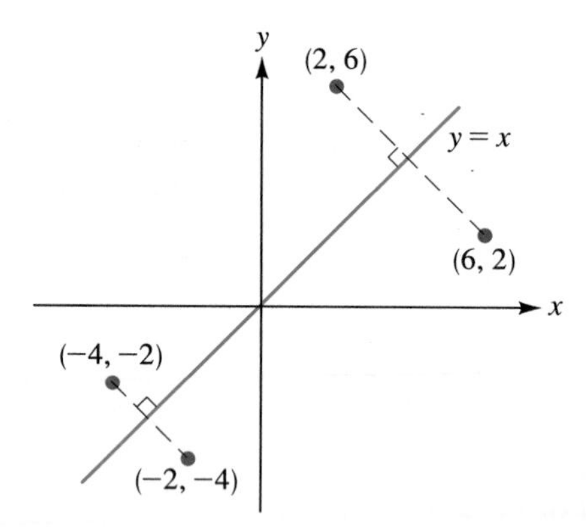

FIGURE 11
Two examples of the fact that (a, b) and (b, a) are symmetric about $y = x$.

Just as we've shown that (4, 1) and (1, 4) are symmetric about the line $y = x$, we can also show that in general, (a, b) and (b, a) are always symmetric about the line $y = x$. Figure 11 displays two examples of this, and Exercise 31 at the end of this section asks you to supply a proof for the general situation.

EXAMPLE 7 A line segment $\mathscr{L}$ has endpoints $(-4, 1)$ and $(-1, 2)$. Sketch the line segments that result from **(a)** a reflection in the line $y = x$; **(b)** a reflection in the line $y = x$ followed by a reflection in the y-axis; and **(c)** a reflection in the y-axis followed by a reflection in the line $y = x$. Are the results in parts (b) and (c) the same?

Solution

(a) First reflect the endpoints $(-4, 1)$ and $(-1, 2)$ in the line $y = x$ to obtain the new endpoints $(1, -4)$ and $(2, -1)$, respectively. Then join these new endpoints as shown in Figure 12(a).

(b) In Figure 12(a) we found the reflection of $\mathscr{L}$ in the line $y = x$. Reflecting this line segment in the y-axis yields the result shown in Figure 12(b).

(c) In Figure 12(c), we've reflected $\mathscr{L}$ first in the y-axis and then in the line $y = x$. Note that the results here and in part (b) are not the same. The order in which we carry out these operations makes a difference.

FIGURE 12

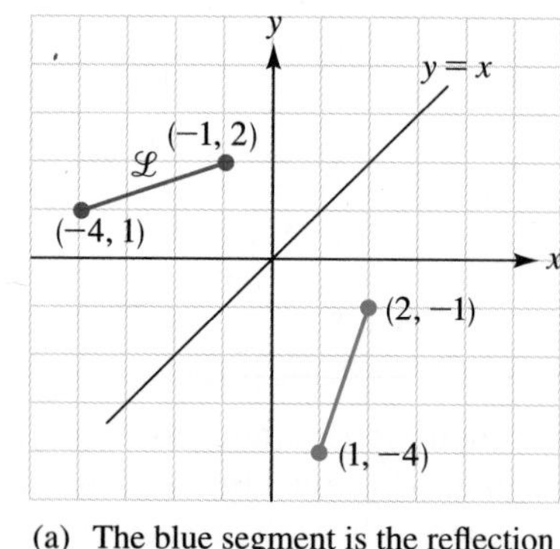

(a) The blue segment is the reflection of $\mathscr{L}$ in the line $y = x$.

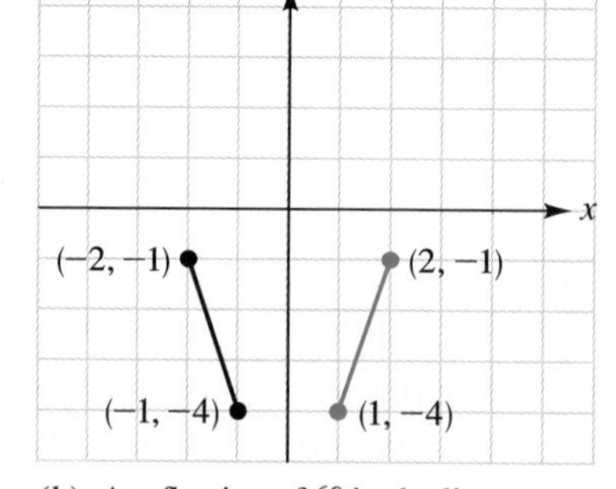

(b) A reflection of $\mathscr{L}$ in the line $y = x$ followed by a reflection in the y-axis yields the black segment.

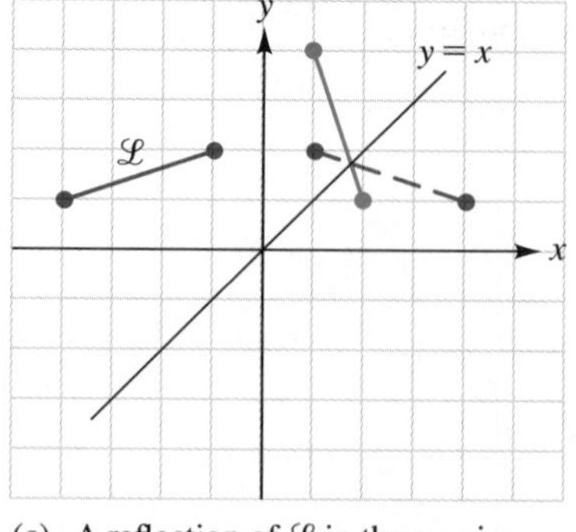

(c) A reflection of $\mathscr{L}$ in the y-axis followed by a reflection in the line $y = x$ yields the blue segment.

EXERCISE SET 2.4

A

In Exercises 1–6, the endpoints of a line segment $\overline{AB}$ are given. Sketch the reflection of $\overline{AB}$ about **(a)** *the x-axis;* **(b)** *the y-axis;* **(c)** *the origin; and* **(d)** *the line $y = x$.*

1. $A(1, 4)$ and $B(3, 1)$

2. $A(-1, -2)$ and $B(-5, -2)$

3. $A(-2, -3)$ and $B(2, -1)$

4. $A(-3, -3)$ and $B(-3, -1)$

5. $A(0, 1)$ and $B(3, 1)$

6. $A(-2, -2)$ and $B(0, 0)$

In Exercises 7 and 8, fill in the entries in the tables with YES *or* NO *to indicate whether the graph of the equation possesses the given symmetry.*

7.

	Symmetric about the x-axis	Symmetric about the y-axis	Symmetric about the origin
$y = x^2$			
$y = x^3$			
$y = \sqrt{x}$			

8.

	Symmetric about the x-axis	Symmetric about the y-axis	Symmetric about the origin
$y = \lvert x \rvert$			
$y = 1/x$			
$y = \sqrt{1 - x^2}$			

In Exercises 9–26, graph the equation after determining the x- and y-intercepts and whether the graph possesses any of the three types of symmetry described on page 70.

9. $y = 4 - x^2$

10. $y = -x^3$

11. $y = -1/x$

12. $x = y^2 - 1$

13. $y = -x^2$

14. $y = 1/x^2$

15. $y = -1/x^3$

16. $y = |x| - 2$

17. $y = \sqrt{x^2}$

18. $y = x + 1$

19. $y = x^2 - 2x + 1$

20. $x = y^3 - 1$

21. $y^2 = 2x - 4$

22. $|y| = 2x - 4$

23. $y = 2x^2 + x - 4$

24. $y = 2x^2$

25. **(a)** $x + y = 2$
(b) $|x| + y = 2$

26. **(a)** $x + |y| = 2$
(b) $|x + y| = 2$

B

27. The accompanying figure shows the graphs of $y = \frac{3}{4}x - 2$ and $y = |\frac{3}{4}x - 2|$.

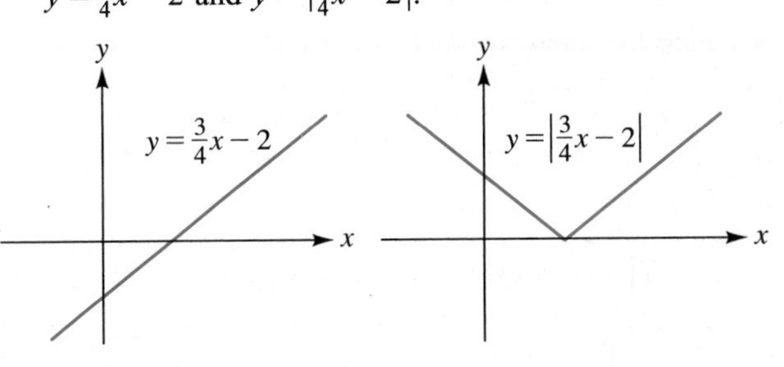

(a) Determine the x- and y-intercepts for each graph.
(b) Which portions of the two graphs are identical? (Give your answer in terms of an interval along the x-axis.)
(c) Explain how the graph of $y = |\frac{3}{4}x - 2|$ can be obtained from that of $y = \frac{3}{4}x - 2$ by means of reflection.

28. The accompanying figure shows the graphs of the two equations $y = x^2 - 6x + 8$ and $y = |x^2 - 6x + 8|$.

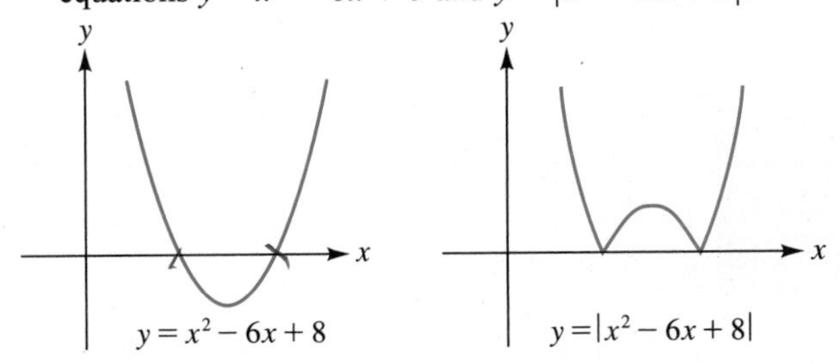

(a) Determine the x- and y-intercepts for each graph.
(b) Which portions of the two graphs are identical? (Give your answer in terms of intervals along the x-axis.)
(c) Explain how the graph of $y = |x^2 - 6x + 8|$ can be obtained from that of $y = x^2 - 6x + 8$ by means of reflection.

29. Reread the definition of symmetry about the line $y = x$ given on page 73. Then use the method shown in Example 6 to show that the points $(7, -1)$ and $(-1, 7)$ are symmetric about the line $y = x$.

30. Use the method shown in Example 6 to show that $(5, 2)$ and $(2, 5)$ are symmetric about $y = x$.

31. Use the method of Example 6 to show that the points (a, b) and (b, a) are symmetric about the line $y = x$.

32. Points P and Q are reflected in the line $y = x$ to obtain points P' and Q', respectively. Does the distance from P to Q equal the distance from P' to Q'?

33. Pick any three points on the line $y = \frac{1}{2}x + 3$. Reflect each point in the line $y = x$. Show that the three reflected points all lie on one line. What is the equation of that line?

34. Determine the intercepts and symmetry for the equation $y = x^{2/3}$. Then graph the equation. (Use a calculator.)

35. Find the intercepts and symmetry for $y = x^3 - 27x$. Then use a calculator (as necessary) to set up a table of x- and y-values, with x running from 0 to 6 in increments of $1/2$. Finally, graph the equation.

36. Consider the equation $y = 4x^2/(1 + x^2)$.
(a) Use a calculator to set up a table with x-values running from 0 to 5 in increments of $1/2$.
(b) Graph the equation using your table of values and symmetry.

C

In Exercises 37–40, use the following definition of symmetry about a line: Two points P and Q are ***symmetric about a line*** $\mathcal{L}$ *provided that:*

1. $\overline{PQ}$ *is perpendicular to* $\mathcal{L}$; and
2. *points P and Q are equidistant from* $\mathcal{L}$.

In other words, P and Q are symmetric about the line $\mathcal{L}$ *if* $\mathcal{L}$ *is the perpendicular bisector of line segment* $\overline{PQ}$. *We say that* $\mathcal{L}$ *is* ***the axis of symmetry*** *and that P and Q are* ***reflections of each other through*** $\mathcal{L}$.

37. Determine values for m and b so that the points $(8, 2)$ and $(4, 8)$ are symmetric about the line $y = mx + b$.

38. Reflect the point $(2, 1)$ in the line $y = 3x$. What are the coordinates of the point you obtain?

39. **(a)** If the point (a, b) is reflected in the line $y = 3x$, show that the coordinates of the reflected point are

$$\left(\frac{3b - 4a}{5}, \frac{3a + 4b}{5}\right)$$

(b) If the point (a, b) is reflected in the line $y = mx$, what are the coordinates of the reflected point?

40. **(a)** Find the reflection of the point (a, b) in the line $y = -x$. *Answer:* $(-b, -a)$
(b) Begin with the point (a, b). Reflect this point about the x-axis to obtain a point Q. Then reflect Q about the y-axis to obtain a point R. Then reflect R about the line $y = x$ to obtain a point S. Show that S is the reflection of (a, b) in the line $y = -x$.

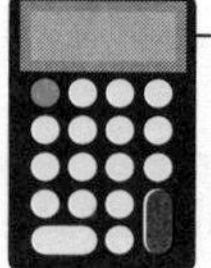

GRAPHING UTILITY EXERCISES FOR SECTION 2.4

In Exercises 1–14, graph each equation. Does the graph appear to possess any of the three types of symmetry defined on page 70? In cases where your answer is YES, use the symmetry tests on page 71 to verify your answer.

In each case, begin with a ***standard viewing rectangle,*** *which is the default setting for your graphing utility. (For most graphing calculators, for example, the standard viewing rectangle extends from −10 to 10 in both the x- and the y-directions.) Then, if necessary, use a zoom-in view for a second look. In Exercises 1–6, suggestions for the second view are provided.*

1. $y = x^2 - 3x$ (second view: x from -2 to 5; y from -4 to 10)

2. $y = x^3 - 3x$ (second view: x from -2 to 5; y from -4 to 10)

3. $y = 2^x$ (second view: x from -3 to 3; y from 0 to 8)

4. $y = 2^{|x|}$ (second view: x from -3 to 3; y from 0 to 8)

5. $y = 1/(x^2 - x)$ (second view: x from -2 to 2; y unchanged)

6. $y = 1/(x^3 - x)$ (second view: x from -2 to 2; y unchanged)

7. $y = x^2 - 0.2x - 15$ *Hint:* Look closely at the x-intercepts.

8. $y = x^2 - 2x^3$

9. $y = \sqrt{|x|}$

10. $y = \sqrt{|x|^3}$

11. $y = 2x - x^3 - x^5 + x^7$

12. $y = |2x - x^3 - x^5 + x^7|$

13. $y = x^4 - 10x^2 + \frac{1}{4}x$ *Suggestion:* For the second view, try zooming out in the y-direction.

14. $y = x^4 - 10x^2 + \frac{1}{4}$

15. In the text we said that a line is an *asymptote* for a curve if the distance between the line and the curve approaches zero as we move farther and farther out along the line. In terms of graphing, this means that as we zoom out, the curve and the line eventually appear to be indistinguishable. In this exercise, we'll demonstrate this using the curve $y = -4/x$ (which we graphed in Figure 7). As indicated in the text, both the x- and the y-axes are asymptotes for this curve. First, graph

$y = -4/x$ using a viewing rectangle that extends from -5 to 5 in both the x- and the y-directions. Then take a second look using a viewing rectangle that extends from -30 to 30 in both the x- and the y-directions. At this scale, you'll see that the curve is virtually indistinguishable from an asymptote when either $|x| > 8$ or $|y| > 8$.

16. **(a)** Graph the equation $y = 20/x$ using a standard viewing rectangle.
(b) Although both the x- and the y-axes are asymptotes for this curve, the graph in part (a) does not show this clearly. Take a second look using a viewing rectangle that extends from -100 to 100 in both the x- and the y-directions. Note that the curve indeed appears indistinguishable from an asymptote when either $|x|$ or $|y|$ is sufficiently large.

17. In Exercises 15 and 16, the asymptotes were the x- and the y-axes. In this exercise we look at a curve with asymptotes that are lines other than the axes.
(a) Graph the curve $y = (3x - 6)/(x + 1)$. (Use a standard viewing rectangle.)
(b) Adjust the viewing rectangle so that x extends from -100 to 100. Note that the line $y = 3$ appears to be a horizontal asymptote for the curve.
(c) Complete the following two tables to obtain numerical evidence that the line $y = 3$ is an asymptote for the curve. Round your answers to five decimal places.

x	10	100	1000	10000	100000
$\frac{3x-6}{x+1}$					

x	-10	-100	-1000	-10000	-100000
$\frac{3x-6}{x+1}$					

(d) Look at the graphs of $y = (3x - 6)/(x + 1)$ and the asymptote $y = 3$ using a viewing rectangle that extends from -10 to 10 in the x-direction and -50 to 50 in the y-direction. As the picture suggests, the curve has a second asymptote: the line $x = -1$ is a vertical asymptote. If the graphing utility you are using has a provision for drawing vertical lines, add the line $x = -1$ to the picture.

INEQUALITIES

The fundamental results of mathematics are often inequalities *rather than* equalities.

E. Beckenbach and R. Bellman in *An Introduction to Inequalities* (New York: Random House, 1961)

If we replace the equal sign in an equation with any one of the four symbols $<$, $\leqslant$, $>$, or $\geqslant$, we obtain an **inequality.** As with equations in one variable, a real number is a **solution** of an inequality if we obtain a true statement when the variable is replaced by the real number. For example, the value $x = 5$ is a solution of the inequality $2x - 3 < 8$, because when $x = 5$ we have

$$2(5) - 3 < 8$$
$$7 < 8 \qquad \text{which is true}$$

We also say in this case that the value $x = 5$ **satisfies** the inequality. To **solve** an inequality means to find all of the solutions. The set of all solutions of an inequality is called (naturally enough) the **solution set.**

Recall that two equations are said to be equivalent if they have exactly the same solutions. Similarly, two inequalities are **equivalent** if they have the same solution set. Most of the procedures used for solving inequalities are similar to those for equalities. For example, adding or subtracting the same number on both sides of an inequality produces an equivalent inequality. We need to be careful, however, in multiplying or dividing both sides of an inequality by the same nonzero number. For instance, suppose that we start with the inequality $2 < 3$ and multiply both sides by 5. That yields $10 < 15$, which is certainly true. But if we multiply both sides of the inequality $2 < 3$ by -5, we obtain $-10 < -15$, which is false. Multiplying both sides of an inequality by the same *positive* number preserves the inequality, whereas multiplying by a *negative* number reverses the inequality. In the following box, we list some of the principal properties of inequalities. In general, whenever we use Property 1 or 2 in solving an inequality, we obtain an equivalent inequality. Also, note that each property can be rewritten to reflect the fact that $a < b$ is equivalent to $b > a$. For example, Property 3 can just as well be written this way: If $b > a$ and $c > b$, then $c > a$.

PROPERTY SUMMARY — PROPERTIES OF INEQUALITIES

PROPERTY	EXAMPLE
1. If $a < b$, then $a + c < b + c$ and $a - c < b - c$.	If $x - 3 < 0$, then $(x - 3) + 3 < 0 + 3$ and, consequently, $x < 3$.
2. **(a)** If $a < b$ and c is positive, then $ac < bc$ and $a/c < b/c$.	If $\frac{1}{2}x < 4$, then $2(\frac{1}{2}x) < 2(4)$ and, consequently, $x < 8$.
(b) If $a < b$ and c is negative, then $ac > bc$ and $a/c > b/c$.	If $-\frac{x}{5} < 6$, then $(-5)\left(-\frac{x}{5}\right) > (-5)(6)$ and, consequently, $x > -30$.
3. The transitive property: If $a < b$ and $b < c$, then $a < c$.	If $a < x$ and $x < 2$, then $a < 2$.

EXAMPLE 1 Solve each of the following inequalities:

(a) $2x - 3 < 5$; **(b)** $5t + 8 \leqslant 7(1 + t)$.

Solution

(a) Our work follows the same pattern that we would use to solve the equation $2x - 3 = 5$. We have

$$2x - 3 < 5$$

$$2x < 8 \quad \text{adding 3 to both sides}$$

$$x < 4 \quad \text{dividing both sides by 2}$$

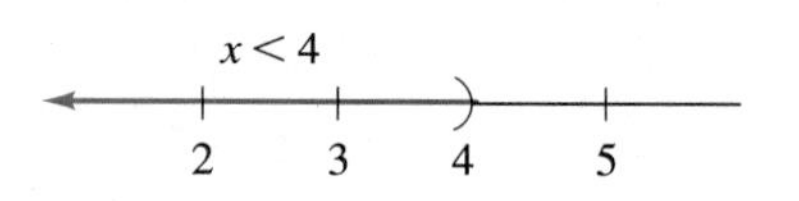

FIGURE 1

The solution set is therefore $(-\infty, 4)$; see Figure 1.

(b) Again, we follow the pattern that we would use to solve the corresponding equation.

$$5t + 8 \leqslant 7 + 7t$$

$$-2t \leqslant -1 \quad \text{subtracting } 7t \text{ and 8 from both sides}$$

$$t \geqslant \frac{1}{2} \quad \text{dividing both sides by } -2\text{, which reverses the inequality}$$

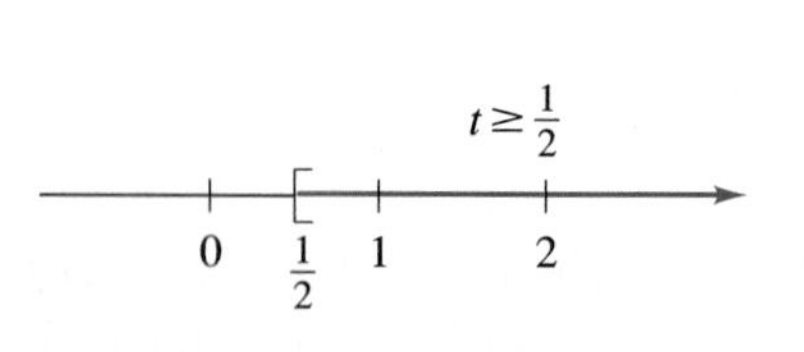

FIGURE 2

The solution set is therefore $[1/2, \infty)$; see Figure 2.

In Section 2.2 we saw that there is a close connection between solving equations and graphing. (Example: The roots of the equation $x^2 - 4 = 0$ are the x-intercepts of the graph of $y = x^2 - 4$.) It's also true that solving inequalities and graphing are closely related. Consider, for instance, the inequality $2x - 3 < 5$ that we solved in Example 1(a). To interpret this inequality geo-

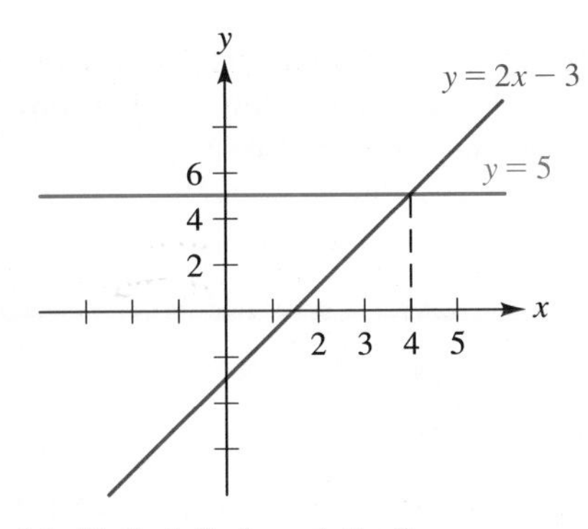

(a) To the left of $x = 4$, the line $y = 2x - 3$ is below the line $y = 5$.

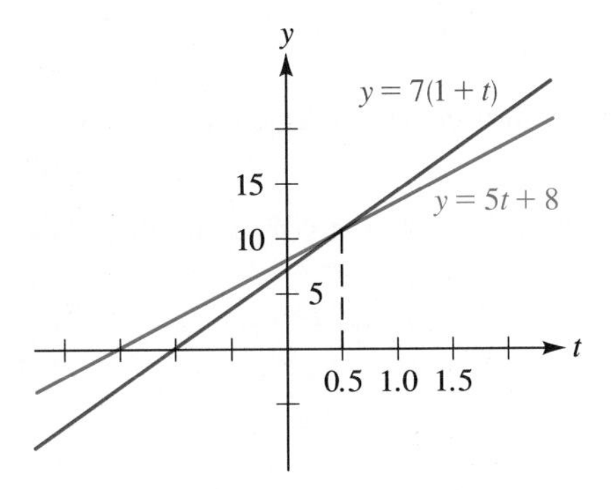

(b) The graph indicates that if $t \geq 1/2$ then $5t + 8 \leq 7(t + 1)$.

FIGURE 3

metrically, we graph the two equations $y = 2x - 3$ and $y = 5$. Then, as indicated in Figure 3(a), the graph of $y = 2x - 3$ is below the line $y = 5$ as long as x is to the left of 4. The previous sentence translates algebraically into

$$2x - 3 < 5 \quad \text{provided} \quad x < 4$$

This confirms the result in Example 1(a).

Similarly, for a geometric interpretation of the inequality in Example 1(b), we graph the two lines $y = 5t + 8$ and $y = 7(1 + t)$ in a t-y coordinate system. As indicated in Figure 3(b), the line $y = 5t + 8$ is below the line $y = 7(1 + t)$ provided t is to the right of $1/2$. Furthermore, at $t = 1/2$, the lines intersect. That is, for $t = 1/2$, the quantities $5t + 8$ and $7(1 + t)$ are equal. Algebraically, we can restate this information by saying

$$5t + 8 \leqslant 7(1 + t) \quad \text{provided} \quad t \geqslant 1/2$$

This confirms the result in Example 1(b).

In the next example, we solve the inequality

$$-\frac{1}{2} < \frac{3 - x}{-4} < \frac{1}{2}$$

By definition, this is equivalent to the pair of inequalities

$$-\frac{1}{2} < \frac{3 - x}{-4} \quad \text{and} \quad \frac{3 - x}{-4} < \frac{1}{2}$$

One way to proceed here would be first to determine the solution set for each inequality. Then the set of real numbers common to both solution sets would be the solution set for the original inequality. However, the method shown in Example 2 is more efficient.

EXAMPLE 2 Solve: $-\dfrac{1}{2} < \dfrac{3 - x}{-4} < \dfrac{1}{2}$.

Solution We begin by multiplying through by -4. Remember, this will reverse the inequalities:

$$2 > 3 - x > -2$$

Next, with a view toward isolating x, we first subtract 3 to obtain

$$-1 > -x > -5$$

Finally, multiplying through by -1, we have

$$1 < x < 5$$

The solution set is therefore the open interval $(1, 5)$.

The next example refers to the Celsius and Fahrenheit scales for measuring temperature.* The formula relating the temperature readings on the two scales is

$$F = \frac{9}{5}C + 32$$

*The Celsius scale was devised in 1742 by the Swedish astronomer Anders Celsius. The Fahrenheit scale was first used by the German physicist Gabriel Fahrenheit in 1724.

EXAMPLE 3 Over the temperature range $32° \leq F \leq 39.2°$ on the Fahrenheit scale, water contracts (rather than expands) with increasing temperature. What is the corresponding temperature range on the Celsius scale?

Solution

$$32 \leq F \leq 39.2 \qquad \text{given}$$

$$32 \leq \frac{9}{5}C + 32 \leq 39.2 \qquad \text{substituting } \frac{9}{5}C + 32 \text{ for } F$$

$$0 \leq \frac{9}{5}C \leq 7.2 \qquad \text{subtracting 32}$$

$$0 \leq C \leq \frac{5}{9}(7.2) \qquad \text{multiplying by } \frac{5}{9}$$

$$0 \leq C \leq 4$$

Thus, a range of 32°F to 39.2°F on the Fahrenheit scale corresponds to 0°C to 4°C on the Celsius scale.

In the next three examples, we solve inequalities that involve absolute values. The following theorem is very useful in this context.

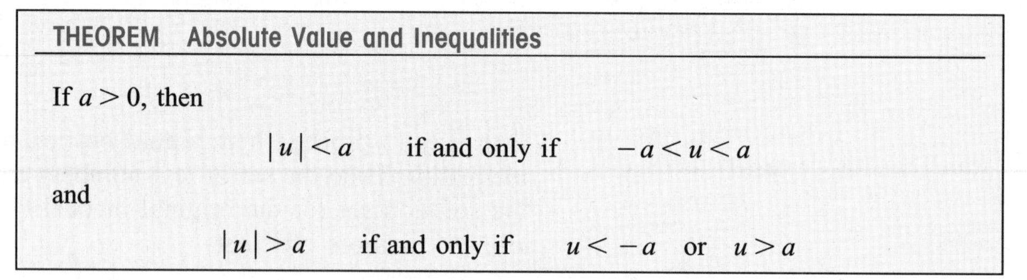

THEOREM Absolute Value and Inequalities

If $a > 0$, then

$$|u| < a \quad \text{if and only if} \quad -a < u < a$$

and

$$|u| > a \quad \text{if and only if} \quad u < -a \quad \text{or} \quad u > a$$

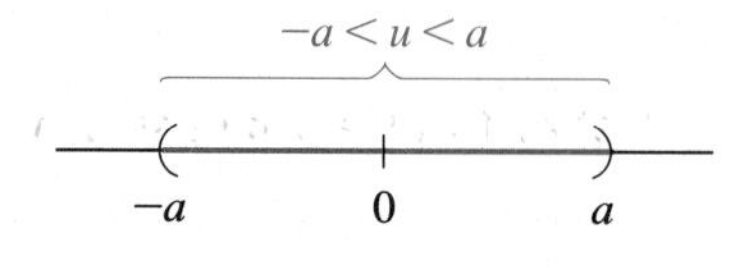

FIGURE 4

You can see why this theorem is valid if you think in terms of distance and position on a number line. The condition $-a < u < a$ means that u lies between $-a$ and a, as indicated in Figure 4. But this is the same as saying that the distance from u to zero is less than a, which in turn can be written $|u| < a$. (The second part of the theorem can be justified in a similar manner.)

EXAMPLE 4 Solve: **(a)** $|x| < 1$; **(b)** $|x| \geq 1$.

Solution

(a) By the theorem we just discussed, the condition $|x| < 1$ is equivalent to

$$-1 < x < 1$$

The solution set is therefore the open interval $(-1, 1)$.

(b) In view of the second part of the theorem, the inequality $|x| \geq 1$ is satisfied when x satisfies either of the inequalities

$$x \leq -1 \quad \text{or} \quad x \geq 1$$

The solution sets for these last two inequalities are $(-\infty, -1]$ and $[1, \infty)$, respectively. Consequently, the solution set for the given inequality $|x| \geq 1$ consists of the two intervals $(-\infty, -1]$ and $[1, \infty)$.

In Example 4(b), we found that the solution set consisted of two intervals on the number line. We have a convenient notation for describing such sets.

Given any two sets A and B, we define the set $A \cup B$ (read ***A* union *B***) to be the set of all elements that are in A or in B (or in both). For example, if $A = \{1, 2, 3\}$ and $B = \{4, 5\}$, then $A \cup B = \{1, 2, 3, 4, 5\}$. As another example, the union of the two closed intervals [3, 5] and [4, 7] is given by

$$[3, 5] \cup [4, 7] = [3, 7]$$

because the numbers in the interval [3, 7] are precisely those numbers that are in [3, 5] or [4, 7] (or in both). Using this notation, we can write the solution set for Example 4(b) as

$$(-\infty, -1] \cup [1, \infty)$$

EXAMPLE 5 Solve: $|x - 3| < 1$.

Solution We'll show two methods.

FIRST METHOD We use the theorem preceding Example 4. With $u = x - 3$ and $a = 1$, the theorem tells us that the given inequality is equivalent to

$$-1 < x - 3 < 1$$

or (by adding 3)

$$2 < x < 4$$

The solution set is therefore the open interval (2, 4).

ALTERNATIVE METHOD The given inequality tells us that x must be less than one unit away from 3 on the number line. Looking one unit to either side of 3, then, we see that x must lie strictly between 2 and 4. The solution set is therefore (2, 4), as obtained using the first method.

EXAMPLE 6 Solve: $\left|1 - \dfrac{t}{2}\right| > 5$.

Solution Referring again to the theorem, we use the fact that $|u| > a$ means that either $u < -a$ or $u > a$. So, in the present example, the given inequality means that either

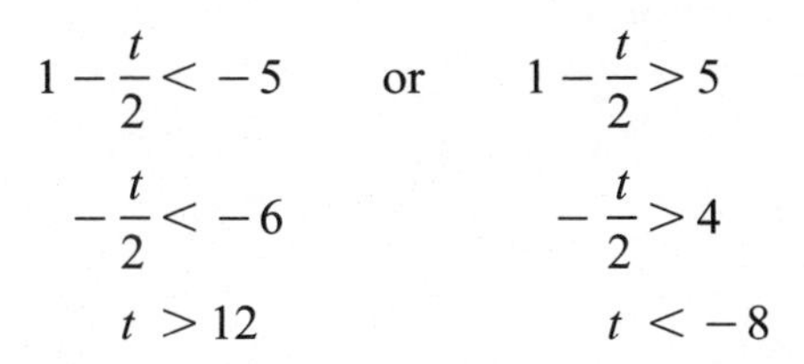

$$1 - \frac{t}{2} < -5 \quad \text{or} \quad 1 - \frac{t}{2} > 5$$

$$-\frac{t}{2} < -6 \qquad\qquad -\frac{t}{2} > 4$$

$$t > 12 \qquad\qquad t < -8$$

This tells us that the given inequality is satisfied when t is in either of the intervals $(12, \infty)$ or $(-\infty, -8)$. Thus the solution set is $(-\infty, -8) \cup (12, \infty)$.

Earlier in this section we saw that there is a close connection between inequalities and graphs. The next example emphasizes this point again.

EXAMPLE 7 The graph of the equation $y = |14x - 2| - 3$ has a V shape, as shown in Figure 5.

(a) Use the graph to estimate the solution set for the inequality

$$|14x - 2| - 3 \leq 0$$

GRAPHICAL PERSPECTIVE

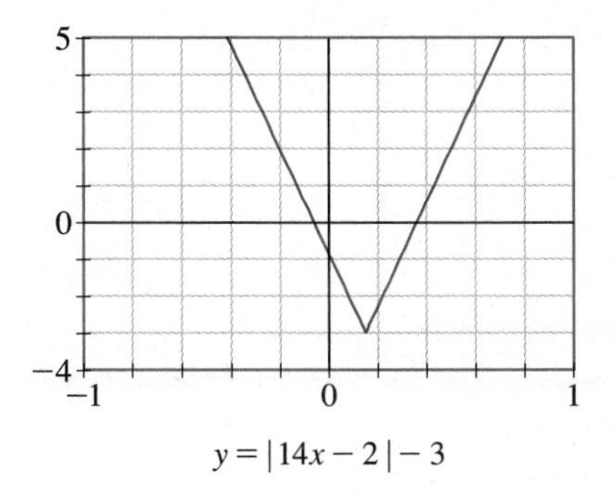

FIGURE 5

(b) Solve the inequality algebraically to determine the solution set precisely. Compare the results with those in part (a).

Solution

(a) As indicated in Figure 5, the graph has two x-intercepts. At each intercept, the value of the quantity $|14x - 2| - 3$ is zero, while for x-values between these two intercepts, the quantity $|14x - 2| - 3$ is negative (because the graph is *below* the x-axis). Let's estimate the two intercepts. The smaller of the two is approximately halfway between 0 and -0.2, call it -0.1. The larger x-intercept is between 0.3 and 0.4, call it 0.35. In summary, the two x-intercepts are approximately -0.1 and 0.35. From the graph, then, the solution set for the given inequality is a closed interval $[a, b]$, where $a \approx -0.1$ and $b \approx 0.35$.

(b) To obtain the precise values for a and b, we first write the given inequality as $|14x - 2| \leq 3$. Then we have

$$-3 \leq 14x - 2 \leq 3$$
$$-1 \leq 14x \leq 5$$
$$-\frac{1}{14} \leq x \leq \frac{5}{14}$$

So, the solution set is the closed interval $[-\frac{1}{14}, \frac{5}{14}]$. Using a calculator (or dividing by hand!), we find that $-\frac{1}{14} \approx -0.071$, and $\frac{5}{14} \approx 0.357$. Note that these numbers are indeed consistent with the estimates that we obtained graphically in part (a).

EXERCISE SET 2.5

Do 7, 10, 15, 26, 29, 47

A

In Exercises 1–34, solve the inequality and specify the answer using interval notation.

1. $x + 5 < 4$

2. $2x - 7 < 11$

3. $1 - 3x \leq 0$

4. $6 - 4x \leq 22$

5. $4x + 6 < 3(x - 1) - x$

6. $2(t - 1) - 3(t + 1) \leq -5$

7. $1 - 2(t + 3) - t \leq 1 - 2t$

8. $t - 4[1 - (t - 1)] > 7 + 10t$

9. $\frac{3x}{5} - \frac{x - 1}{3} < 1$

10. $\frac{2x + 1}{2} + \frac{x - 1}{3} < x + \frac{1}{2}$

11. $\frac{x - 1}{4} - \frac{2x + 3}{5} \leq x$

12. $\frac{x}{2} - \frac{8x}{3} + \frac{x}{4} > \frac{23}{6}$

13. $-2 \leq x - 6 \leq 0$

14. $-3 \leq 2x + 1 \leq 5$

15. $-1 \leq \frac{1 - 4t}{3} \leq 1$

16. $\frac{2}{3} \leq \frac{5 - 3t}{-2} \leq \frac{3}{4}$

17. $0.99 < \frac{x}{2} - 1 < 0.999$

18. $\frac{9}{10} < \frac{3x - 1}{-2} < \frac{91}{100}$

19. **(a)** $|x| \leq 1/2$ **(b)** $|x| > 1/2$

20. **(a)** $|x| > 2$ **(b)** $|x| \leq 2$

21. **(a)** $|x| > 0$ **(b)** $|x| < 0$

22. **(a)** $|t| \geq 0$ **(b)** $|t| \leq 0$

23. **(a)** $x - 2 < 1$ **(b)** $|x - 2| < 1$ **(c)** $|x - 2| > 1$

24. **(a)** $x - 4 \geq 4$ **(b)** $|x - 4| \geq 4$ **(c)** $|x - 4| \leq 4$

25. **(a)** $1 - x \leq 5$ **(b)** $|1 - x| \leq 5$ **(c)** $|1 - x| > 5$

26. **(a)** $3x + 5 < 17$ **(b)** $|3x + 5| < 17$ **(c)** $|3x + 5| > 17$

27. **(a)** $a - x < c$ **(b)** $|a - x| < c$ **(c)** $|a - x| \geq c$

28. (Assume $b < c$ throughout this exercise.)
(a) $|x - a| + b < c$ **(b)** $|x + a| + b > c$
(c) $|x + a| + b < c$

29. $\left|\frac{x - 2}{3}\right| < 4$

30. $\left|\frac{4 - 5x}{2}\right| > 1$

31. $\left|\frac{x+1}{2}-\frac{x-1}{3}\right|<1$

32. $\left|\frac{3(x-2)}{4}+\frac{4(x-1)}{3}\right| \leqslant 2$

33. **(a)** $|(x+h)^2-x^2|<3h^2$, where $h>0$
(b) $|(x+h)^2-x^2|<3h^2$, where $h<0$

34. **(a)** $|3(x+2)^2-3x^2|<1/10$
(b) $|3(x+2)^2-3x^2|<\varepsilon$, where $\varepsilon>0$

For Exercises 35–38: **(a)** *use the graph to determine which of the following general forms describes the solution set for the given inequality:*

$[a, b] \qquad (-\infty, a) \cup (b, \infty) \qquad [a, \infty) \qquad (a, \infty)$

(b) *use the graph to estimate the values of a and b;*
(c) *solve the inequality algebraically to determine the exact values of a and b, then write the solution set using interval notation.*

35. $7x-2 \geqslant 0$

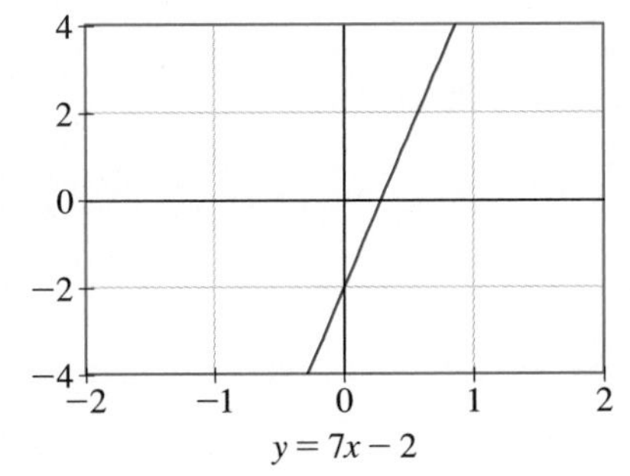

$y=7x-2$

36. $6-13x<0$

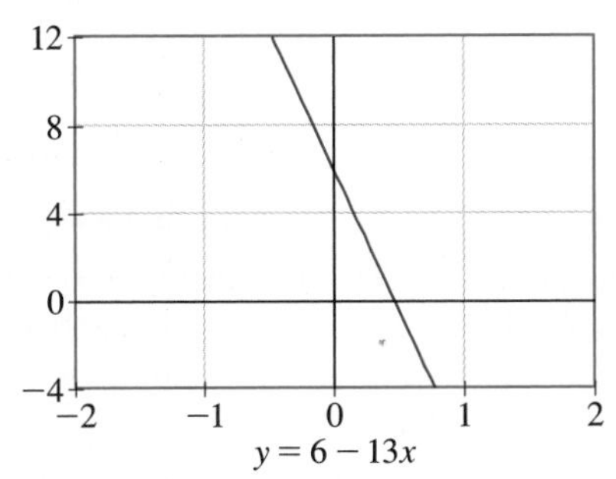

$y=6-13x$

37. $|8x-3|-2 \leqslant 0$

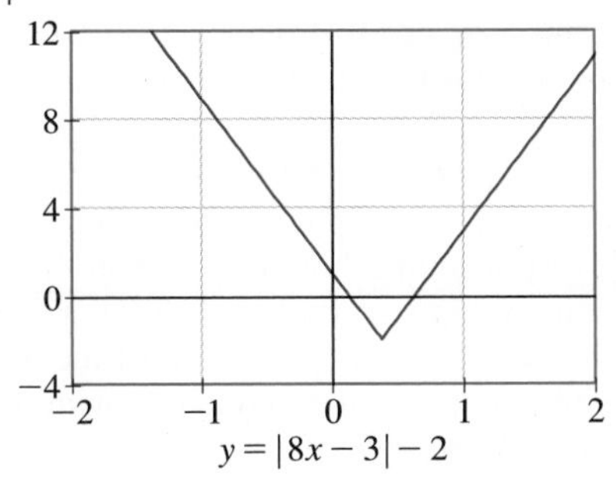

$y=|8x-3|-2$

38. $1-|15x-3|<0$

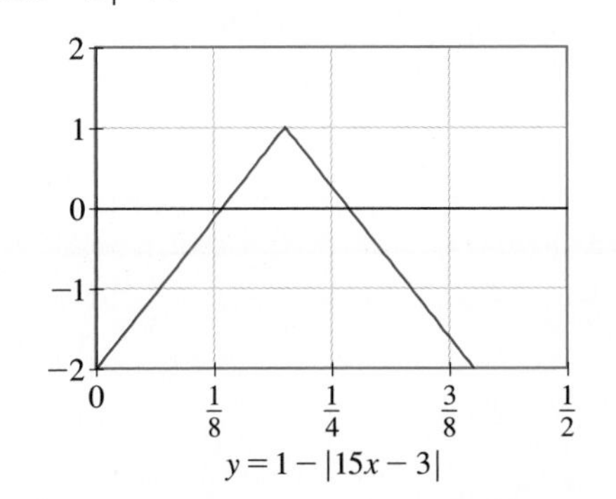

$y=1-|15x-3|$

39. The surface temperature of the moon varies over the interval $-280° \leqslant F \leqslant 260°$ on the Fahrenheit scale. What is the corresponding interval on the Celsius scale? (Round the values that you obtain to the nearest 5°C.)

40. Data from the *Apollo 11* moon mission in July 1969 showed temperature readings on the lunar surface varying over the interval $-183° \leqslant C \leqslant 112°$ on the Celsius scale. What is the corresponding interval on the Fahrenheit scale? (Round the numbers you obtain to the nearest integers.)

41. From the cloud tops of Venus to the planet's surface, temperature readings range over the interval $-25° \leqslant C \leqslant 475°$ on the Celsius scale. What is the corresponding range on the Fahrenheit scale?

42. Measurements sent back to Earth from the *Viking I* spacecraft in June 1976 indicated temperature readings on the surface of Mars in the range $-139° \leqslant C \leqslant -28°$ on the Celsius scale. What is the corresponding range on the Fahrenheit scale? (Round the numbers you obtain to the nearest integers.)

43. The temperature of the variable star Delta Cephei varies over the interval $5100° \leqslant C \leqslant 6500°$ on the Celsius scale. What is the corresponding range on the Fahrenheit scale? (Round the numbers in your answer to the nearest 100°F.)

44. Data from the *Mariner 10* spacecraft (launched November 3, 1973) indicate that the surface temperature on the planet Mercury varies over the interval $-170° \leqslant C \leqslant 430°$ on the Celsius scale. What is the corresponding interval on the Fahrenheit scale? (Round the values that you obtain to the nearest 10°F.)

B

45. Given two positive numbers a and b, we define the **geometric mean** and the **arithmetic mean** as follows:

$$\text{G.M.} = \sqrt{ab} \qquad \text{A.M.} = \frac{a+b}{2}$$

(a) Complete the following table, using a calculator as

necessary so that the entries in the third and fourth columns are in decimal form.

a	b	$\sqrt{ab}$ (G.M.)	$(a+b)/2$ (A.M.)	Which is larger, G.M. or A.M.?
1	2			
1	3			
1	4			
2	3			
3	4			
5	10			
9	10			
99	100			
999	1000			

(b) Prove that the following inequality is valid for all positive numbers a and b:

$$\sqrt{ab} \leqslant \frac{a+b}{2}$$

In words: *The geometric mean is less than or equal to the arithmetic mean.* *Hint:* Use the following property of inequalities. If x and y are positive, then the inequality $x \leqslant y$ is equivalent to $x^2 \leqslant y^2$.

46. Given two positive numbers a and b, we define the **root mean square** as follows:

$$\text{R.M.S.} = \sqrt{\frac{a^2+b^2}{2}}$$

(a) Complete the following table, using a calculator as necessary so that the entries in the third and fourth columns are in decimal form.

a	b	$(a+b)/2$ (A.M.)	$\sqrt{(a^2+b^2)/2}$ (R.M.S.)	Which is larger, A.M. or R.M.S.?
1	2			
1	3			
1	4			
2	3			
3	4			
5	10			
9	10			
99	100			
999	1000			

(b) Prove that the following inequality is valid for all positive numbers a and b:

$$\frac{a+b}{2} \leqslant \sqrt{\frac{a^2+b^2}{2}}$$

In words: *The arithmetic mean is less than or equal to the root mean square.* [*Use the hint in Exercise 45(b).*]

47. Solve the inequality $|x-1|+|x-2|<3$. *Hint:* Begin by considering three cases: $x<1$; $1 \leqslant x<2$; $x \geqslant 2$.

48. Solve the inequality $6-|x+3|-|x-2|<0$. (Adapt the hint in Exercise 47.)

49. (a) Complete the following table; use a calculator and round to four decimal places.

x	y	$\frac{x}{y}+\frac{y}{x}$	True or False: $\frac{x}{y}+\frac{y}{x} \geqslant 2$
1	1		
2	3		
3	5		
4	7		
5	9		
9	10		
49	50		
99	100		

(b) Prove that for all positive numbers x and y, we have $\frac{x}{y}+\frac{y}{x} \geqslant 2$. *Hint:* Use the inequality in Exercise 45(b). Take $a=x/y$ and $b=y/x$.

50. Suppose that $|x-2|<0.1$ and $|y-3|<0.01$. Show that $|xy-6| \leqslant 0.321$. *Hint:* First check that $xy-6=x(y-3)+3(x-2)$. Then use the triangle inequality (page 8).

C

51. Let a and b be positive numbers. In Exercises 45 and 46, you were asked to show that

$$\sqrt{ab} \leqslant \frac{a+b}{2} \leqslant \sqrt{\frac{a^2+b^2}{2}}$$

This exercise shows how to establish these important inequalities using geometric, rather than algebraic, methods. In the figure at the top of the next column, $\overline{CF}$ is the diameter of a semicircle with center E. Let $CD=a$ and $DF=b$.

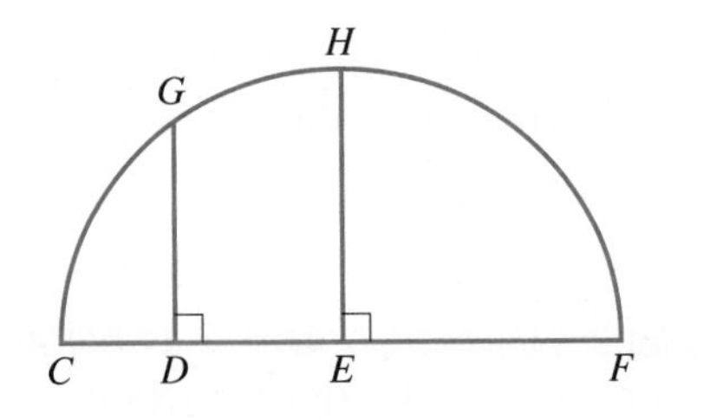

(a) Explain why $CE = EH = (a+b)/2$.
(b) Show that $DG = \sqrt{ab}$ and $DH = \sqrt{(a^2+b^2)/2}$.
(c) Explain why $DG \leqslant EH \leqslant DH$, and conclude from this that $\sqrt{ab} \leqslant (a+b)/2 \leqslant \sqrt{(a^2+b^2)/2}$.

52. Let a, b, c, and d be positive real numbers. Use the inequality in Exercise 45(b) to prove each of the following inequalities.
(a) $\sqrt{abcd} \leqslant (ab+cd)/2$
(b) $\sqrt{ab} + \sqrt{cd} \leqslant \sqrt{(a+c)(b+d)}$

53. Let x denote the width of a rectangle with perimeter 30 ft.
(a) Show that the area A (in square feet) of the rectangle is given by $A = x(15-x)$.
(b) Use the inequality in Exercise 45(b) to show that $A \leqslant 225/4$.
(c) For which value of x is $A = 225/4$? What are the dimensions of the rectangle in this case?

54. Let x denote the width of a rectangle with area 25 ft^2.
(a) Show that the perimeter P (in feet) of the rectangle is given by $P = 2x + (50/x)$.
(b) Use the inequality in Exercise 45(b) to show that $P \geqslant 20$.
(c) For which value of x is $P = 20$? What are the dimensions of the rectangle in this case?

55. Let s, t, u, and v be four nonnegative numbers. Show that

$$\sqrt[4]{stuv} \leqslant \frac{s+t+u+v}{4}$$

Hint: Write $\sqrt[4]{stuv} = \sqrt{\sqrt{st}\sqrt{uv}}$, and use the arithmetic-geometric mean inequality.

56. Let a, b, and c be nonnegative numbers. Follow Steps (a) through (c) to show that

$$\sqrt[3]{abc} \leqslant \frac{a+b+c}{3}$$

(a) Let A, B, and C be real numbers. Verify the following identity by carrying out the indicated operations on the right-hand side of the equation.

$$3ABC = A^3 + B^3 + C^3 - \tfrac{1}{2}(A+B+C) \times [(A-B)^2 + (B-C)^2 + (C-A)^2]$$

(b) Now suppose that A, B, and C are nonnegative numbers. Use the identity in part (a) to explain why

$$3ABC \leqslant A^3 + B^3 + C^3$$

(c) In the identity in part (b), make the substitutions $A^3 = a$, $B^3 = b$, and $C^3 = c$. Show that the result can be written $\sqrt[3]{abc} \leqslant \frac{1}{3}(a+b+c)$, as required.

MORE ON INEQUALITIES

It is rather surprising to think that the man who surveyed and mapped Virginia was one of the founders of algebra as we know the subject today. Such however is the case, for Thomas Harriot was sent by Sir Walter Raleigh to accompany Sir Richard Grenville (1585) to the New World, where he made the survey of that portion of American territory.

David Eugene Smith in *History of Mathematics,* Vol. I (New York: Ginn and Co., 1923)

The symbols of inequality > and < were introduced by [Thomas] *Harriot* [1560–1621]. *The signs ⩾ and ⩽ were first used about a century later by the Parisian hydrographer Pierre Bouguer.*

Florian Cajori in *A History of Mathematics,* 4th ed. (New York: Chelsea Publishing Company, 1985)

The last example in the previous section emphasized the connection between inequalities and graphs. This is our starting point for the present section. We begin with an example in which polynomial inequalities are solved by relying on a graph. Then we'll use the ideas there to explain a method for solving inequalities even when a graph is not given.

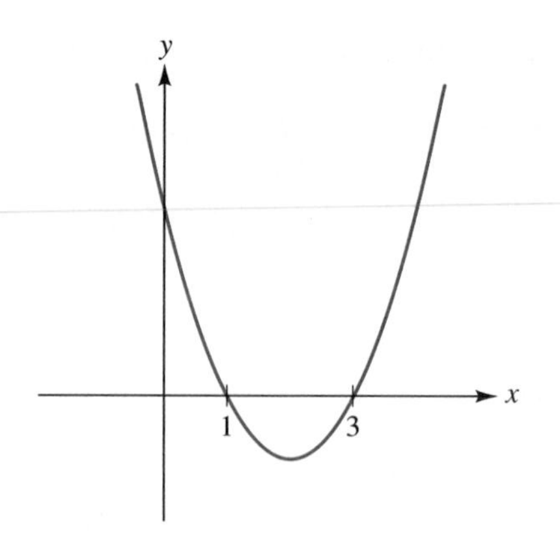

FIGURE 1
$y = x^2 - 4x + 3$

EXAMPLE 1 Figure 1 shows the graph of the equation $y = x^2 - 4x + 3$. Use the graph to solve each of the following inequalities:

(a) $x^2 - 4x + 3 < 0$; **(b)** $x^2 - 4x + 3 \geqslant 0$.

Solution

(a) As indicated in Figure 1, the graph of $y = x^2 - 4x + 3$ is below the x-axis when (and only when) the x-values are in the open interval $(1, 3)$. This is the same as saying that the quantity $x^2 - 4x + 3$ is negative when (and only when) the x-values are in the open interval $(1, 3)$. Or, rewording once more, the solution set for the inequality $x^2 - 4x + 3 < 0$ is the open interval $(1, 3)$.

(b) The graph in Figure 1 is above the x-axis when (and only when) the x-values are in either of the two intervals $(-\infty, 1)$ or $(3, \infty)$. This is the same as saying that the quantity $x^2 - 4x + 3$ is positive when (and only when) the x-values are in the set $(-\infty, 1) \cup (3, \infty)$. Furthermore, the graph shows that the quantity $x^2 - 4x + 3$ is equal to 0 when $x = 1$ and when $x = 3$. In summary, the solution set for the inequality $x^2 - 4x + 3 \geqslant 0$ is the set $(-\infty, 1] \cup [3, \infty)$.

We can summarize the essential features of our work in Example 1 as follows. The x-intercepts of the graph in Figure 1 divide the number line into three intervals, namely, $(-\infty, 1)$, $(1, 3)$, and $(3, \infty)$. Within each of these intervals, the algebraic sign of $x^2 - 4x + 3$ stays the same. [For example, to the left of $x = 1$, that is, on the interval $(-\infty, 1)$, the quantity $x^2 - 4x + 3$ is always positive.] More generally, it can be shown (using calculus) that this same type of behavior—regarding *persistence of sign*—occurs with all polynomials and, indeed, with quotients of polynomials as well. This important fact, along with the definition of a *key number,* is presented in the box that follows.

PROPERTY SUMMARY — KEY NUMBERS AND PERSISTENCE OF SIGN

Let P and Q be polynomials with no common factors (other than, possibly, constants), and consider the following four inequalities:

$$\frac{P}{Q} < 0 \qquad \frac{P}{Q} \leqslant 0 \qquad \frac{P}{Q} > 0 \qquad \frac{P}{Q} \geqslant 0$$

The **key numbers** for each of these inequalities are the real numbers for which $P = 0$ or $Q = 0$. (Geometrically speaking, the key numbers are the x-intercepts for the graphs of the equations $y = P$ and $y = Q$.) It can be proved (using calculus) that the algebraic sign of P/Q does not change within each of the intervals determined by the key numbers.

EXAMPLES

1. The key numbers for the inequality $(x - 7)/(x - 8) \geqslant 0$ are 7 and 8.
2. The key numbers for the inequality $(x - 3)(x + 3) < 0$ are ± 3. (In this case, the polynomial Q is the constant 1.)

We'll show how this result is applied by solving the polynomial inequality

$$x^3 - 2x^2 - 3x > 0$$

First, using factoring techniques from Section 1.3, we rewrite the inequality in the equivalent form

$$x(x + 1)(x - 3) > 0$$

The key numbers, then, are the solutions of the equation $x(x+1)(x-3)=0$. That is, the key numbers are $x=0$, $x=-1$, and $x=3$. Next, we locate these numbers on a number line. As indicated in Figure 2, this divides the number line into four distinct intervals.

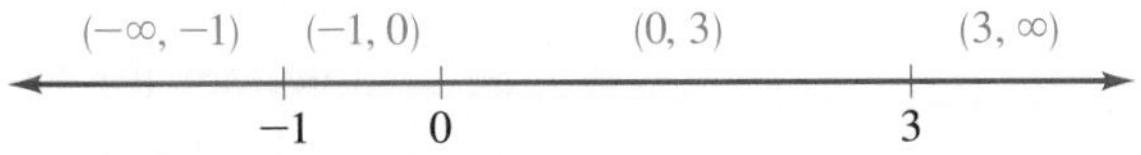

$(-\infty, -1)$ $(-1, 0)$ $(0, 3)$ $(3, \infty)$

-1 0 3

FIGURE 2

Now, according to the result stated in the box just prior to this example, no matter what x-value we choose in the interval $(-\infty, -1)$, the resulting sign of x^3-2x^2-3x $[=x(x+1)(x-3)]$ will always be the same. Thus, to see what that sign is, we first choose any convenient *test number* in the interval $(-\infty, -1)$, say, $x=-2$. Then, using $x=-2$, we determine the sign of $x(x+1)(x-3)$ simply by considering the sign of each factor, as indicated in Table 1.

TABLE 1

Interval	Test Number	x	$x+1$	$x-3$	$x(x+1)(x-3)$
$(-\infty, -1)$	-2	neg.	neg.	neg.	neg.

On the interval $(-\infty, -1)$, the sign of x^3-2x^2-3x $[=x(x+1)(x-3)]$ is negative because it is the product of three negative factors.

Looking at Table 1, we conclude that the values of x^3-2x^2-3x are negative *throughout* the interval $(-\infty, -1)$ and, consequently, no number in this interval satisfies the given inequality. Next, we carry out similar analyses for the remaining three intervals, as shown in Table 2. (You should verify for yourself that the entries in the table are correct.)

TABLE 2

Interval	Test Number	x	$x+1$	$x-3$	$x(x+1)(x-3)$
$(-1, 0)$	$-\frac{1}{2}$	neg.	pos.	neg.	pos.
$(0, 3)$	1	pos.	pos.	neg.	neg.
$(3, \infty)$	4	pos.	pos.	pos.	pos.

On the interval $(-1, 0)$, the product $x(x+1)(x-3)$ is positive because it has two negative factors and one positive factor. On $(0, 3)$, the product is negative because it has two positive factors and one negative factor. And on $(3, \infty)$, the product is positive because all the factors are positive.

Looking at the two tables now, we conclude that the sign of $x(x+1)(x-3)$ is positive throughout both of the intervals $(-1, 0)$ and $(3, \infty)$ and, consequently, all the numbers in these intervals satisfy the given inequality. Moreover, our work also shows that the other two intervals that we considered are not part of the solution set. As you can readily check, the key numbers themselves do not satisfy the given inequality for this example. In summary, then, the solution set for the inequality $x^3-2x^2-3x>0$ is $(-1, 0) \cup (3, \infty)$.

Furthermore, it is important to notice that the work we just carried out also provides us with three additional pieces of information.

The solution set for $x^3 - 2x^2 - 3x \geq 0$ is $[-1, 0] \cup [3, \infty)$.

The solution set for $x^3 - 2x^2 - 3x < 0$ is $(-\infty, -1) \cup (0, 3)$.

The solution set for $x^3 - 2x^2 - 3x \leq 0$ is $(-\infty, -1] \cup [0, 3]$.

In the box that follows, we summarize the steps for solving polynomial inequalities.

STEPS FOR SOLVING POLYNOMIAL INEQUALITIES

1. If necessary, rewrite the inequality so that the polynomial is on the left-hand side and zero is on the right-hand side.
2. Find the key numbers for the inequality and locate them on a number line.
3. List the intervals determined by the key numbers.
4. From each interval, choose a convenient test number. Then use the test number to determine the sign of the polynomial throughout the interval.
5. Use the information obtained in the previous step to specify the required solution set. [Don't forget to take into account whether the original inequality is strict ($<$ or $>$) or nonstrict ($\leq$ or $\geq$).]

EXAMPLE 2 Solve $x^4 \leq 14x^3 - 48x^2$.

Solution First, we rewrite the inequality so that zero is on the right-hand side. Then we factor the left-hand side as follows:

$$x^4 - 14x^3 + 48x^2 \leq 0$$
$$x^2(x^2 - 14x + 48) \leq 0$$
$$x^2(x - 6)(x - 8) \leq 0$$

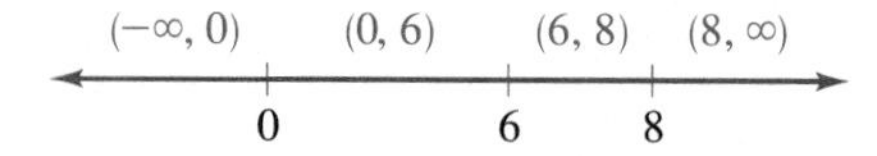

FIGURE 3

From this last line, we see that the key numbers are $x = 0$, 6, and 8. As indicated in Figure 3, these numbers divide the number line into four distinct intervals. We need to choose a test number from each interval and see whether the polynomial is positive or negative on the interval. This work is carried out in the following table.

Interval	Test Number	x^2	$x-6$	$x-8$	$x^2(x-6)(x-8)$
$(-\infty, 0)$	-1	pos.	neg.	neg.	pos.
$(0, 6)$	1	pos.	neg.	neg.	pos.
$(6, 8)$	7	pos.	pos.	neg.	neg.
$(8, \infty)$	9	pos.	pos.	pos.	pos.

The table shows that the quantity $x^2(x - 6)(x - 8)$ is negative only for x-values in the interval $(6, 8)$. Also, as noted at the start, the quantity is equal to zero when $x = 0$, 6, or 8. Thus, the solution set consists of the numbers in the closed interval [6, 8], along with the number 0. We can write this set

$$[6, 8] \cup \{0\}$$

where $\{0\}$ denotes the set that has zero as its only member.

GRAPHICAL PERSPECTIVE

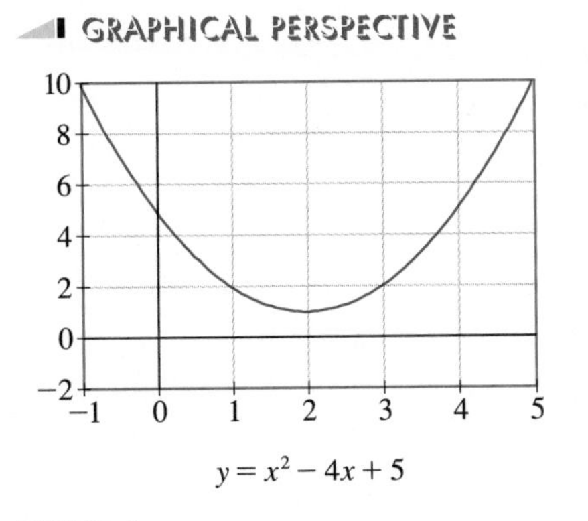

$y = x^2 - 4x + 5$

FIGURE 4
The graph of $y = x^2 - 4x + 5$ lies entirely above the x-axis. Algebraically this means that the quantity $x^2 - 4x + 5$ is positive for all values of x. Consequently, the solution set for $x^2 - 4x + 5 > 0$ is the set of all real numbers. Equivalently, there are no real numbers satisfying $x^2 - 4x + 5 < 0$.

EXAMPLE 3 Solve: **(a)** $x^2 - 4x + 5 > 0$; **(b)** $x^2 - 4x + 5 < 0$.

Solution

(a) The equation $x^2 - 4x + 5 = 0$ has no real solution (because the discriminant of the quadratic is -4, which is negative), so there is no key number. Consequently, the polynomial $x^2 - 4x + 5$ never changes sign. To see what that sign is, choose the most convenient test number, namely, $x = 0$, and evaluate the polynomial: $0^2 - 4(0) + 5 > 0$. So the polynomial is positive for every value of x, and the solution set is $(-\infty, \infty)$, the set of all real numbers.

(b) Our work in part (a) shows that no real number satisfies the inequality $x^2 - 4x + 5 < 0$. See Figure 4 for a graphical perspective on this and the result in part (a).

The technique used in the previous examples can also be used to solve inequalities involving quotients of polynomials. For these cases recall that the definition of a key number includes the x-values for which the denominator is zero. For example, the key numbers for the inequality $\dfrac{x+3}{x-4} \geqslant 0$ are -3 and 4.

EXAMPLE 4 Solve: $\dfrac{x+3}{x-4} \geqslant 0$.

Solution The key numbers are -3 and 4. As indicated in Figure 5, these numbers divide the number line into three intervals. In the table that follows, we've chosen a test number from each interval and determined the sign of the quotient $(x + 3)/(x - 4)$ for each interval.

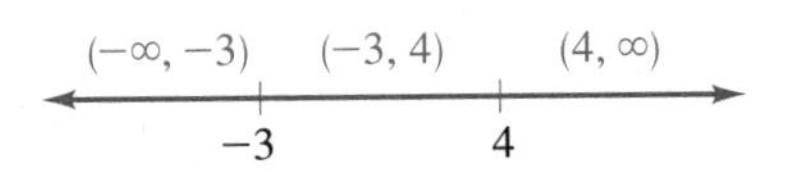

FIGURE 5

Interval	Test Number	$x+3$	$x-4$	$\dfrac{x+3}{x-4}$
$(-\infty, -3)$	-4	neg.	neg.	pos.
$(-3, 4)$	0	pos.	neg.	neg.
$(4, \infty)$	5	pos.	pos.	pos.

From these results, we conclude that the solution set for $\dfrac{x+3}{x-4} \geqslant 0$ contains the two intervals $(-\infty, -3)$ and $(4, \infty)$. However, we still need to consider the two endpoints, -3 and 4. As you can easily check, the value $x = -3$ does satisfy the given inequality, but $x = 4$ does not. In summary, then, the solution set is $(-\infty, -3] \cup (4, \infty)$. See Figure 6 (on the next page) for a graphical interpretation of this result. [In Figure 6, the graph of the equation $y = (x + 3)/(x - 4)$ has two distinct pieces, or *branches,* much like the graph of $y = 1/x$, which we looked at in Section 2.2. The vertical line $x = 4$ is an asymptote for the graph. Graphs such as these are studied in Section 4.7, *Rational Functions*.]

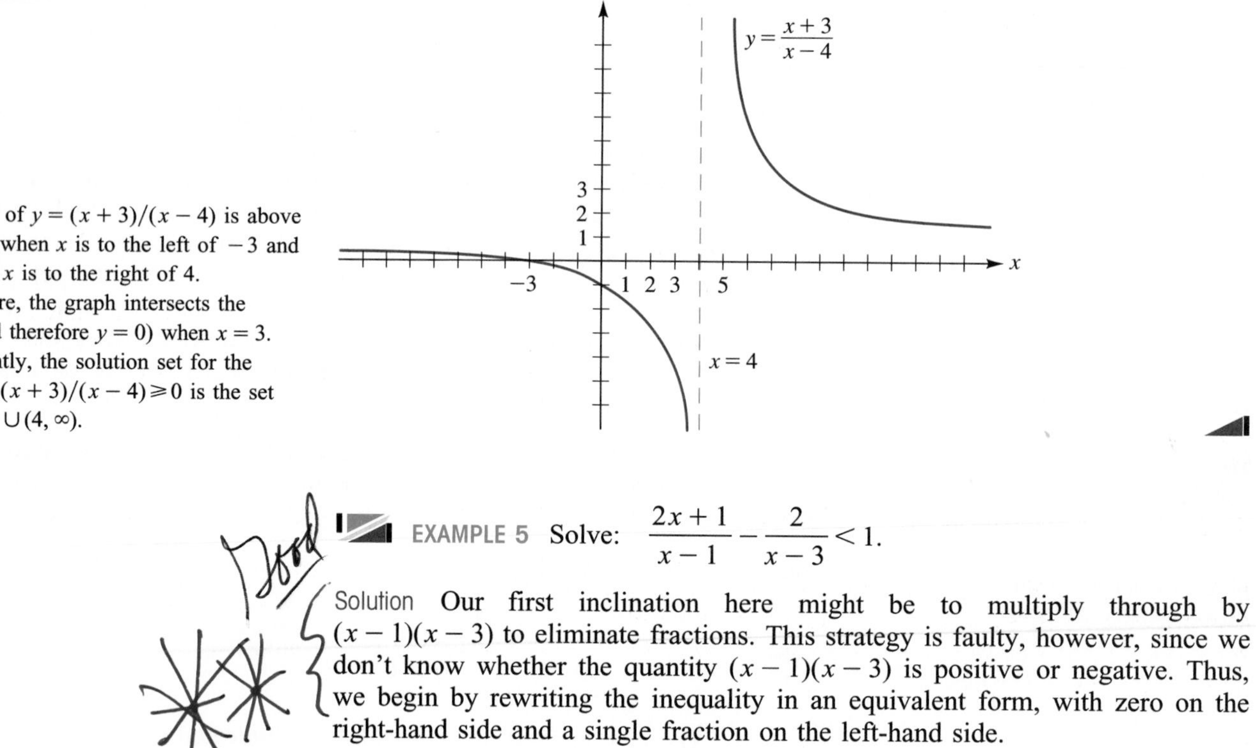

FIGURE 6
The graph of $y = (x + 3)/(x - 4)$ is above the x-axis when x is to the left of -3 and also when x is to the right of 4. Furthermore, the graph intersects the x-axis (and therefore $y = 0$) when $x = 3$. Consequently, the solution set for the inequality $(x + 3)/(x - 4) \geq 0$ is the set $(-\infty, -3] \cup (4, \infty)$.

EXAMPLE 5 Solve: $\dfrac{2x+1}{x-1} - \dfrac{2}{x-3} < 1$.

Solution Our first inclination here might be to multiply through by $(x - 1)(x - 3)$ to eliminate fractions. This strategy is faulty, however, since we don't know whether the quantity $(x - 1)(x - 3)$ is positive or negative. Thus, we begin by rewriting the inequality in an equivalent form, with zero on the right-hand side and a single fraction on the left-hand side.

$$\frac{2x+1}{x-1} - \frac{2}{x-3} - 1 < 0$$

$$\frac{(2x+1)(x-3) - 2(x-1) - 1(x-1)(x-3)}{(x-1)(x-3)} < 0$$

$$\frac{x^2 - 3x - 4}{(x-1)(x-3)} < 0 \qquad \text{Check the algebra!}$$

$$\frac{(x+1)(x-4)}{(x-1)(x-3)} < 0 \qquad (1)$$

The key numbers are those x-values for which the denominator or the numerator is zero. By inspection, then, we see that these numbers are -1, 4, 1, and 3. As Figure 7 indicates, these numbers divide the number line into five distinct intervals.

$(-\infty, -1)$ $(-1, 1)$ $(1, 3)$ $(3, 4)$ $(4, \infty)$

−1 1 3 4

FIGURE 7

Now, just as in the previous examples, we choose a test number from each interval and determine the sign of the quotient for that interval. (You should check each entry in the following table for yourself.)

Interval	Test Number	$(x+1)(x-4)$	$(x-1)(x-3)$	$\dfrac{(x+1)(x-4)}{(x-1)(x-3)}$
$(-\infty, -1)$	-2	pos.	pos.	pos.
$(-1, 1)$	0	neg.	pos.	neg.
$(1, 3)$	2	neg.	neg.	pos.
$(3, 4)$	7/2	neg.	pos.	neg.
$(4, \infty)$	5	pos.	pos.	pos.

From these results, we can see that the quotient on the left-hand side of inequality (1) is negative (as required) on the two intervals $(-1, 1)$ and $(3, 4)$. Now we need to check the endpoints of these intervals. When $x = -1$ or $x = 4$, the quotient is zero, and so, in view of the original inequality, we exclude these two x-values from the solution set. Furthermore, the quotient is undefined when $x = 1$ or $x = 3$, so we must also exclude those two values from the solution set. In summary, then, the solution set is $(-1, 1) \cup (3, 4)$.

EXERCISE SET 2.6

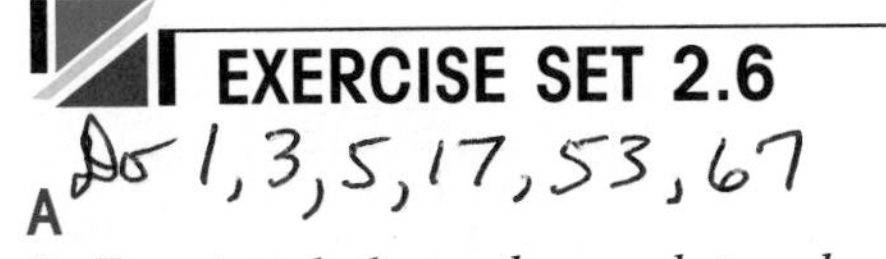

A

In Exercises 1–6, use the graph to solve each inequality. (Assume that each figure shows all of the essential features of the graph of the equation; that is, that there are no surprises "out of camera range.")

1. **(a)** $x^2 - 3x - 4 \geq 0$
   (b) $x^2 - 3x - 4 \leq 0$

$y = x^2 - 3x - 4$

2. **(a)** $-\frac{1}{2}x^2 - \frac{7}{2}x - 5 > 0$
   (b) $-\frac{1}{2}x^2 - \frac{7}{2}x - 5 < 0$

$y = -\frac{1}{2}x^2 - \frac{7}{2}x - 5$

3. **(a)** $x^4 - 4x^3 + 6x^2 - 4x + 2 < 0$
   (b) $x^4 - 4x^3 + 6x^2 - 4x + 2 > 0$

$y = x^4 - 4x^3 + 6x^2 - 4x + 2$

4. **(a)** $-3(x+3)^2 \leq 0$
   (b) $-3(x+3)^2 < 0$

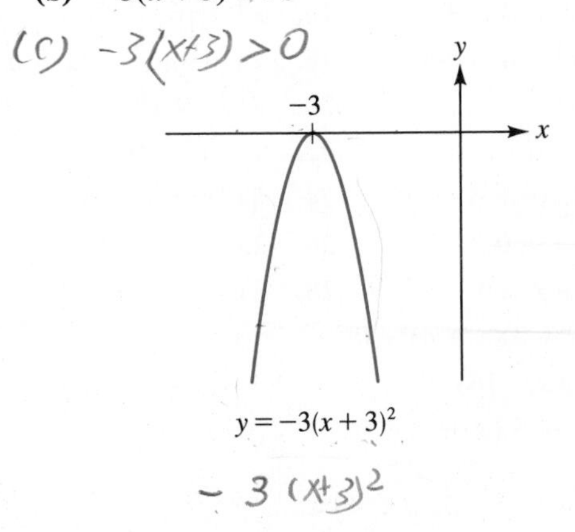

$y = -3(x+3)^2$

Don't clear out the fractions! See example 5.

5. **(a)** $x^3 - 3x^2 - x + 3 \geq 0$
(b) $x^3 - 3x^2 - x + 3 < 0$

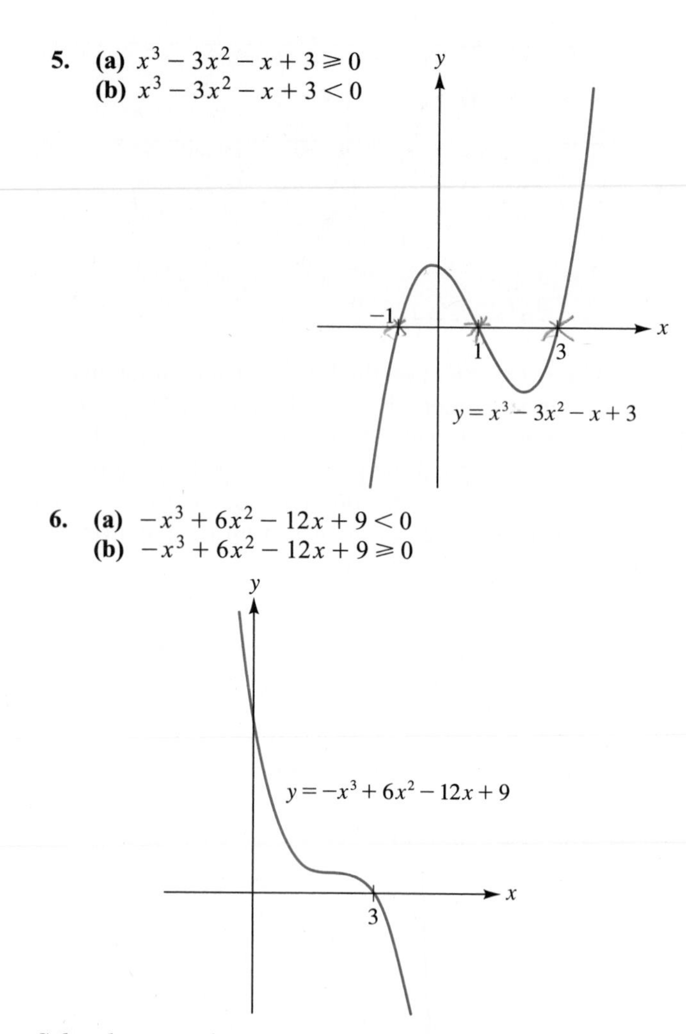

6. **(a)** $-x^3 + 6x^2 - 12x + 9 < 0$
(b) $-x^3 + 6x^2 - 12x + 9 \geq 0$

Solve the inequalities in Exercises 7–68. Suggestion: A calculator is useful for approximating the key numbers in Exercises 25, 26, 43, 52, and 54.

7. $x^2 + x - 6 < 0$
8. $x^2 + 4x - 32 < 0$
9. $x^2 - 11x + 18 > 0$
10. $2x^2 + 7x + 5 > 0$
11. $9x - x^2 \leq 20$
12. $3x^2 + x \leq 4$
13. $x^2 - 16 \geq 0$
14. $24 - x^2 \geq 0$
15. $16x^2 + 24x < -9$
16. $x^4 - 16 < 0$
17. $x^3 + 13x^2 + 42x > 0$
18. $2x^3 - 9x^2 + 4x \geq 0$
19. $225x \leq x^3$
20. $81x^2 \leq x^6$
21. $2x^2 + 1 \geq 0$
22. $1 + x^2 < 0$
23. $12x^3 + 17x^2 + 6x < 0$
24. $8x^4 < x^2 - 2x^3$
25. $x^2 + x - 1 > 0$
26. $2x^2 + 9x - 1 > 0$
27. $x^2 - 8x + 2 \leq 0$
28. $3x^2 - x + 5 \leq 0$
29. $(x-1)(x+3)(x+4) \geq 0$
30. $x^4(x-2)(x-16) \geq 0$
31. $(x+4)(x+5)(x+6) < 0$
32. $(x - \frac{1}{2})(x + \frac{1}{2})(x + \frac{3}{2}) < 0$
33. $(x-2)^2(3x+1)^3(3x-1) > 0$
34. $(2x-1)^3(2x-3)^5(2x-5) > 0$
35. $(x-3)^2(x+1)^4(2x+1)^4(3x+2) \leq 0$
36. $x^4 - 25x^2 + 144 \leq 0$
37. $20 \geq x^2(9 - x^2)$
38. $x^2(3x^2 + 11) \geq 4$
39. $9(x-4) - x^2(x-4) < 0$
40. $(x+1)^2 - 5(x+1) > 14$
41. $4(x^2 - 9) - (x^2 - 9)^2 > -5$
42. $x(1 - x^2)^4 + (x+3)(1 - x^2)^4 \geq 0$
43. $(x-4)(2x^2 - 6x - 1) < 0$
44. $(x+2)^3(x^2 - 4x - 2) < 0$
45. $x^3 + 2x^2 - x - 2 > 0$
46. $2x^4 + x^3 - 16x - 8 > 0$
47. $\dfrac{x-1}{x+1} \leq 0$
48. $\dfrac{x+4}{2x-5} \leq 0$
49. $\dfrac{2-x}{3-2x} \geq 0$
50. $\dfrac{x^2-1}{x^2+8x+15} \geq 0$
51. $\dfrac{x^2-8x-9}{x} < 0$
52. $\dfrac{x^2-3x+1}{1-x} < 0$
53. $\dfrac{2x^3+5x^2-7x}{3x^2+7x+4} > 0$
54. $\dfrac{x^2-x-1}{x^2+x-1} > 0$
55. $\dfrac{x}{x+1} > 1$
56. $\dfrac{2x}{x-2} < 3$
57. $\dfrac{1}{x} \leq \dfrac{1}{x+1}$
58. $\dfrac{2}{x} < \dfrac{x}{2}$
59. $\dfrac{x+2}{x+5} \leq 1$
60. $\dfrac{2}{x-4} - \dfrac{1}{x-3} \geq \dfrac{2}{3}$
61. $\dfrac{1}{x-2} - \dfrac{1}{x-1} \geq \dfrac{1}{6}$
62. $\dfrac{2x}{x+5} + \dfrac{x-1}{x-5} < \dfrac{1}{5}$
63. $\dfrac{1+x}{1-x} - \dfrac{1-x}{1+x} < -1$
64. $\dfrac{x+1}{x+2} > \dfrac{x-3}{x+4}$
65. $\dfrac{3-2x}{3+2x} > \dfrac{1}{x}$
66. $\dfrac{x}{x-2} - \dfrac{3}{x+1} \geq 2$
67. $1 + \dfrac{1}{x} \geq \dfrac{1}{1+x}$
68. $x - \dfrac{10}{x-1} \geq 4$

For Exercises 69 and 70, determine the domain of the variable in each expression. (Recall the domain convention from Section 1.3: The domain is the set of all real-number values for the variable for which the expression is defined.)

69. **(a)** $\sqrt{x^2 - 4x - 5}$ **(b)** $\sqrt{1/(x^2 - 4x - 5)}$

70. **(a)** $\sqrt{\dfrac{x+2}{x-4}}$ **(b)** $\sqrt[3]{\dfrac{x+2}{x-4}}$

B

71. For which values of b will the equation $x^2 + bx + 1 = 0$ have real solutions?

72. The sum of the first n natural numbers is given by

$$1 + 2 + 3 + \cdots + n = \frac{n(n+1)}{2}$$

For which values of n will the sum be less than 1225?

73. For which values of a is $x = 1$ a solution of the inequality $\frac{2a + x}{x - 2a} < 1$?

74. Solve: $\frac{ax + b}{\sqrt{x}} > 2\sqrt{ab}$, where $a > 0$, $b > 0$.

75. The two shorter sides in a right triangle have lengths x and $1 - x$, where $x > 0$. For which values of x will the hypotenuse be less than $\sqrt{17}/5$?

76. A piece of wire 12 cm long is cut into two pieces. Denote the lengths of the two pieces by x and $12 - x$. Both pieces are then bent into squares. For which values of x will the combined areas of the squares exceed 5 cm^2?

77. Let V and S denote the volume and total surface area, respectively, of a right circular cylinder of radius r and height 1 unit. For which r-values will the ratio V/S be less than 1/3?

78. Let V and S denote the volume and total surface area, respectively, of a right circular cone of radius r and height 1 unit. For which r-values will the ratio V/S be less than 4/27?

79. **(a)** Use a calculator to complete the following table. (Round the results to two decimal places.)

x	1	2	5	10	100	200	10,000
$\frac{x^2 + 1000}{2x^2 + x}$							

(b) For which positive real numbers x will we have $\frac{x^2 + 1000}{2x^2 + x} < 0.5$?

(c) What is the smallest *natural number* n for which $\frac{n^2 + 1000}{2n^2 + n} < 0.5$?

C

80. Find a nonzero value for c so that the solution set for the inequality

$$x^2 + 2cx - 6c < 0$$

is the open interval $(-3c, c)$.

81. Solve: $(x - a)^2 - (x - b)^2 > (a - b)^2/4$, where $a > b$.

82. Solve: $x^3 + (1/x^3) \geq 3$. (Use a calculator to approximate the key numbers.)

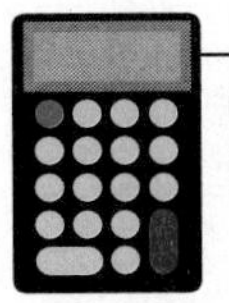

GRAPHING UTILITY EXERCISES FOR SECTION 2.6

1. In the text (pages 86–87) we solved the inequality $x^3 - 2x^2 - 3x > 0$. Graph the equation $y = x^3 - 2x^2 - 3x$ and explain or describe (in complete sentences) the relationship between the graph and the solution set of the given inequality.

2. In Example 2 in the text, we solved the inequality $x^4 \leq 14x^3 - 48x^2$. Graph the equation $y = x^4 - 14x^3 + 48x^2$ and explain or describe (in complete sentences) the relationship between the graph and the solution set of the given inequality.

3. **(a)** Graph the equation $y = x^2 - 4x + 5$.
(b) Use the graph in part (a) to solve the inequality $x^2 - 4x + 5 > 0$. Check that your answer agrees with the result obtained in Example 3(a) in the text.
(c) Use the graph in part (a) to solve the inequality $x^2 - 4x + 5 < 0$. Check that your answer agrees with the result obtained in Example 3(b) in the text.

4. **(a)** Using a viewing rectangle that extends from -5 to 5 in both the x- and the y-directions, graph the equation $y = \frac{2x + 1}{x - 1} - \frac{2}{x - 3} - 1$.
(b) Using the picture obtained in part (a), make a preliminary estimate for the solution set of the inequality $\frac{2x + 1}{x - 1} - \frac{2}{x - 3} - 1 > 0$.
(c) Actually, the viewing rectangle used in part (a) misses one of the essential portions of the graph of the given equation. Change the viewing rectangle

so that x extends from -10 to 10 and y extends from -10 to 20. Now use the new picture to estimate the solution set of the inequality given in part (b).

(d) Use the graph obtained in part (c) to estimate the solution set for the inequality $\frac{2x+1}{x-1} - \frac{2}{x-3} < 1$. Then check that your estimate is consistent with the answer obtained in Example 5 in the text.

In Exercises 5–24, use a graph to estimate the solution set for each inequality. For the endpoints of the intervals in the solution set (that is, for the x-intercepts of the graph), use a "solve" key, or zoom in until you are sure of the first three decimal places in each case. Also, for Exercises 5–16, use an algebraic technique to obtain exact values for the endpoints of the intervals. Check that the values are consistent with the results obtained graphically.

5. $x^2 - 5x + 3 \leq 0$

6. $x^2 + x - 4 \leq 0$

7. $2x^2 - 5x \geq 1$

8. $2x^2 + 5x \geq 1$

9. $0.25x^2 - 6x - 2 < 0$

10. $0.25x^3 - 6x^2 - 2x < 0$

11. $x^4 - 2x^2 - 1 > 0$

12. $x^4 - 2x^2 + 1 > 0$

13. $(x^2 - 5)/(x^2 + 1) \leq 0$

14. $(x^3 - 5)/(x^2 + 1) \leq 0$

15. $(x^2 + 1)/(x^2 - 5) \leq 0$

16. $(x^2 + 1)/(x^3 - 5) \leq 0$

17. $x^3 + 2x \geq -1$

18. $x^3 - 2x \geq -1$

19. $x^4 - 2x > -1$

20. $x^5 - 2x > -1$

21. $x - \frac{x^3}{3!} + \frac{x^5}{5!} - \frac{x^7}{7!} < 0$ *Note:* For natural numbers n, the symbol $n!$ denotes the product of the first n natural numbers. For instance, $3! = 1 \times 2 \times 3 = 6$.

22. $1 - \frac{x^2}{2!} + \frac{x^4}{4!} - \frac{x^6}{6!} + \frac{x^8}{8!} < 0$

23. $(x^2 - 5x - 5)/(x^2 + 5x + 5) \geq 0$

24. $(x^3 - 4x^2 - x - 5)/(x^4 - 2x - 1) \leq 0$

CHAPTER TWO SUMMARY OF PRINCIPAL TERMS AND FORMULAS

TERMS OR FORMULAS	PAGE REFERENCE	COMMENTS
1. The Pythagorean theorem $a^2 + b^2 = c^2$	37	In a right triangle, the lengths of the sides are related by this equation, where c is the length of the hypotenuse. Conversely, if the lengths of the sides of a triangle are related by an equation of the form $a^2 + b^2 = c^2$, then the triangle is a right triangle, and c is the length of the hypotenuse.
2. $d = \sqrt{(x_2 - x_1)^2 + (y_2 - y_1)^2}$	37	This is the distance d between the two points (x_1, y_1) and (x_2, y_2).
3. $(x - h)^2 + (y - k)^2 = r^2$	38	This is the equation of a circle with center at (h, k) and radius r.
4. The unit circle	40	This refers to the circle with center (0, 0) and radius 1. The equation of the unit circle is $x^2 + y^2 = 1$.
5. $\left(\frac{x_1 + x_2}{2}, \frac{y_1 + y_2}{2}\right)$	41	This is the midpoint of the line segment joining (x_1, y_1) and (x_2, y_2).
6. Graph of an equation	47	The graph of an equation in the variables x and y is the set of all points (x, y) with coordinates that satisfy the equation.
7. x-intercept and y-intercept	49	An x-intercept of a graph is the x-coordinate of a point where the graph intersects the x-axis. Similarly, a y-intercept is the y-coordinate of a point where the graph intersects the y-axis.
8. $m = \frac{y_2 - y_1}{x_2 - x_1}$	57	This is the slope m of a nonvertical line passing through the points (x_1, y_1) and (x_2, y_2).
9. $y - y_1 = m(x - x_1)$	59	This is the point–slope form for the equation of a line that passes through the point (x_1, y_1) and has slope m.
10. $y = mx + b$	61	This is the slope–intercept form for the equation of a line with slope m and y-intercept b.
11. $m_1 = m_2$	62	Nonvertical parallel lines have the same slope.

TERMS OR FORMULAS	PAGE REFERENCE	COMMENTS
12. $m_1 = -1/m_2$	62	Two nonvertical lines are perpendicular if and only if the slopes are negative reciprocals of one another.
13. Symmetry about the x-axis	70	A graph is symmetric about the x-axis if, for each point (x, y) on the graph, the point $(x, -y)$ is also on the graph. The points (x, y) and $(x, -y)$ are *reflections* of each other about (or in) the x-axis.
14. Symmetry about the y-axis	70	A graph is symmetric about the y-axis if, for each point (x, y) on the graph, the point $(-x, y)$ is also on the graph. The points (x, y) and $(-x, y)$ are *reflections* of each other about (or in) the y-axis.
15. Symmetry about the origin	70	A graph is symmetric about the origin if, for each point (x, y) on the graph, the point $(-x, -y)$ is also on the graph. The points (x, y) and $(-x, -y)$ are *reflections* of each other about the origin.
16. Symmetry tests	71	**(i)** The graph of an equation is symmetric about the y-axis if replacing x with $-x$ yields an equivalent equation. **(ii)** The graph of an equation is symmetric about the x-axis if replacing y with $-y$ yields an equivalent equation. **(iii)** The graph of an equation is symmetric about the origin if replacing x and y with $-x$ and $-y$, respectively, yields an equivalent equation.
17. Symmetry about the line $y = x$	73	Two points P and Q are symmetric about the line $y = x$ provided that $\overline{PQ}$ is perpendicular to $y = x$ and the points P and Q are equidistant from $y = x$. Two points of the form (a, b) and (b, a) are always symmetric about the line $y = x$, and we say that these two points are *reflections* of each other in the line $y = x$.
18. The solution set of an inequality	77	This is the set of numbers that satisfy the inequality.
19. $A \cup B$	81	The set $A \cup B$ consists of all elements that are in at least one of the two sets A and B.
20. Key numbers of an inequality	86	Let P and Q denote polynomials and consider the following four inequalities: $\frac{P}{Q} < 0 \quad \frac{P}{Q} \leq 0 \quad \frac{P}{Q} > 0 \quad \frac{P}{Q} \geq 0$ The key numbers for each of these inequalities are those real numbers for which $P = 0$ or $Q = 0$. It can be proved that the algebraic sign of P/Q is constant within each of the intervals determined by these key numbers.

WRITING MATHEMATICS

1. The following geometric method for solving quadratic equations of the form $x^2 - ax + b = 0$ is due to the British writer Thomas Carlyle (1795–1881):

> To solve the equation $x^2 - ax + b = 0$, plot the points $A(0, 1)$ and $B(a, b)$. Then draw the circle with $\overline{AB}$ as diameter. The x-intercepts of this circle are the roots of the equation.

Use the techniques of this chapter to verify for yourself that this method indeed yields the roots for the equation $x^2 - 6x + 5 = 0$. (You need to find the equation of the circle, determine the x-intercepts, and then check that the numbers you obtain are the roots of the given equation.) After you have done this, carefully write out your verification in detail (using complete sentences), as if you were explaining the method and why it works to a classmate. Be sure to make explicit reference to any formulas or equations that you use from the chapter.

2. In Section 2.3 we developed the point–slope formula, $y - y_1 = m(x - x_1)$, and the slope–intercept formula, $y = mx + b$. Although these two formulas look quite different from one another, both are merely restatements of the definition of slope. Write a paragraph (or two, at the most) to explain this last sentence. Make use of the following two figures.

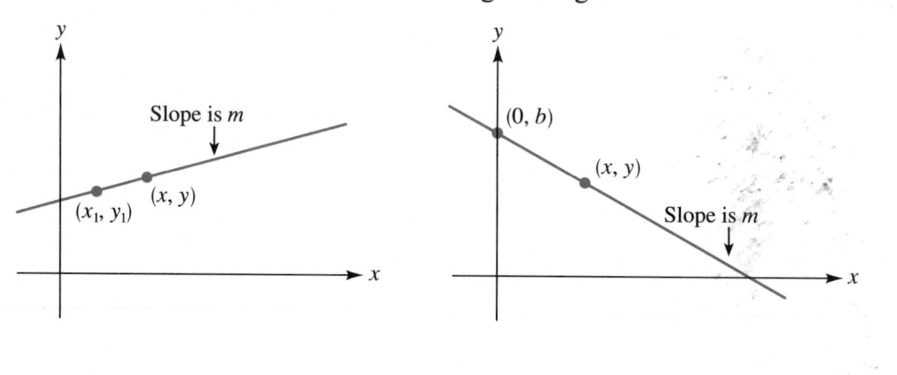

CHAPTER TWO REVIEW EXERCISES

In Exercises 1–18, find an equation for the line satisfying the given conditions. For Exercises 1–9, write the answer in the form $y = mx + b$; in Exercises 10–18, write the answer in the form $Ax + By + C = 0$.

1. Passes through $(-4, 2)$ and $(-6, 6)$
2. $m = -2$; y-intercept 5
3. $m = 1/4$; passes through $(-2, -3)$
4. $m = 1/3$; x-intercept -1
5. x-intercept -4; y-intercept 8
6. $m = -10$; y-intercept 0
7. y-intercept -2; parallel to the x-axis
8. Passes through $(0, 0)$ and is parallel to $6x - 3y = 5$
9. Passes through $(1, 2)$ and is perpendicular to the line $x + y + 1 = 0$
10. Passes through $(1, 1)$ and the center of the circle $x^2 - 4x + y^2 - 8y + = 0$
11. Passes through the centers of the circles $x^2 + 4x + y^2 + 2y = 0$ and $x^2 - 4x + y^2 - 16y = 0$
12. $m = 3$; has the same x-intercept as the line $3x - 8y = 12$
13. Passes through the origin and the midpoint of the line segment joining the points $(-2, -3)$ and $(6, -5)$
14. Is tangent to the circle $x^2 + y^2 = 20$ at the point $(-2, 4)$
15. Is tangent to the circle $x^2 - 6x + y^2 + 8y = 0$ at the point $(0, 0)$
16. Passes through $(2, 4)$; the y-intercept is twice the x-intercept
17. Passes through $(2, -1)$; the sum of the x- and y-intercepts is 2 (There are two answers.)
18. Passes through $(4, 5)$; has no x-intercept
19. Find the distance between the points $(-1, 2)$ and $(4, -10)$.
20. **(a)** Find the perimeter of the triangle with vertices $A(3, 1)$, $B(7, 4)$, and $C(-2, 13)$.
    (b) Find the perimeter of the triangle formed by joining the midpoints of the sides of the triangle in part (a).
21. Which point is farther from the origin, $(15, 6)$ or $(16, 2)$?
22. Find the midpoint of the line segment joining
    (a) $(4, 9)$ and $(10, -2)$; **(b)** $(\frac{2}{3}, -1)$ and $(-\frac{3}{4}, -\frac{1}{3})$.

In Exercises 23–32, test each equation for symmetry about the x-axis, the y-axis, and the origin.

23. $y = x^4 - 2x^2$

24. $y = x^4 - 2x^2 + 1$

25. $y = x^3 + 5x$

26. $y = x^3 + 5x + 1$

27. $y^2 = (x + y)^4$

28. $y = 2/(1 - x^2)$

29. $y = 3x - (1/x)$

30. $y = 2^x + 2^{-x}$

31. $y = 2^x - 2^{-x}$

32. $x^2 + y^2 + 2x = 2\sqrt{x^2 + y^2}$

For Exercises 33–40, tell whether each graph is symmetric about the x-axis, the y-axis, the origin, or the line $y = x$.

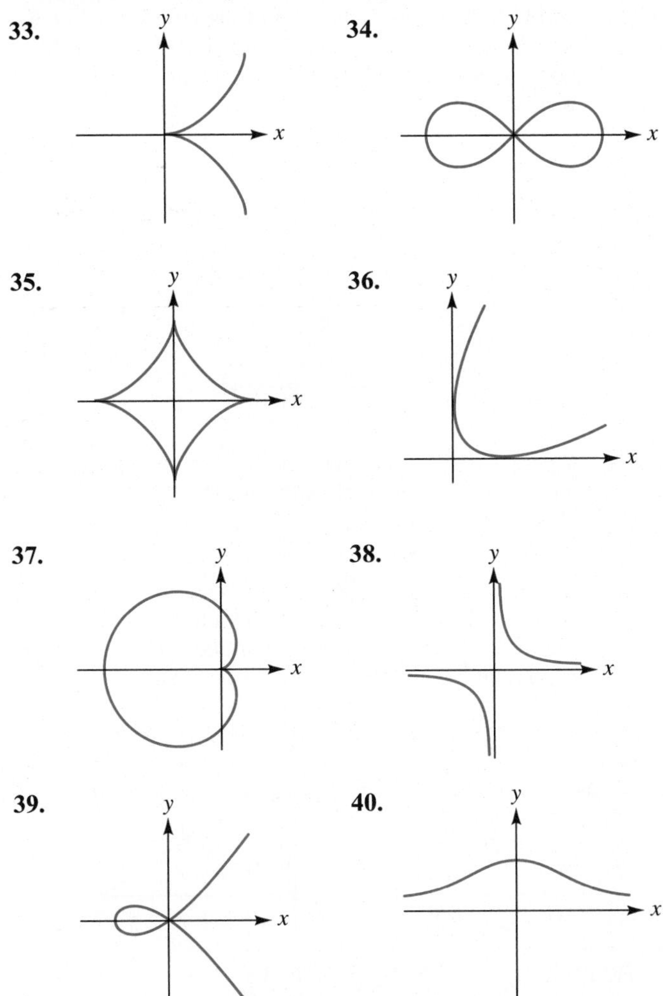

In Exercises 41–44, determine the coordinates of the point (or points) P satisfying the given conditions.

41. The point P lies on the line $y = -\frac{2}{3}x + 1$ and the distance from P to the origin is $\sqrt{10}$ units.

42. The point P lies on the line $y = 17 - x$ and the distance from P to the origin is 13 units.

43. The point P lies on the curve $y = x^2$ and the sum of the x- and y-coordinates of P is 2.

44. The point P lies on the curve $y = \sqrt{x}$ and the product of the x- and y-coordinates is $27/64$.

Graph the equations in Exercises 45–62.

45. $x = 9 - y^2$

46. $y = -2x + 6$

47. $x^2 + y^2 = 1$

48. $(x + 2)^2 + y^2 = 1$

49. $y = |x| + 1$

50. $y = 1 - |x|$

51. $y = x^2 - 4x + 4$

52. $y = -x$

53. $y = |x - 2| + 2$

54. $y = |x - 2| - 2$

55. $3x + y = 0$

56. $|x| + |y| = 1$

57. $x^2 - 4x + y^2 + 6y = 0$

58. $x^2 + 2x + y^2 + 8y = 8$

59. $y = \dfrac{x^2 - 16}{x - 4}$

60. $y = 1 + \dfrac{1}{x}$

61. $(4x - y + 4)(4x + y - 4) = 0$

62. $(y - x)(y - x + 2)(y - x - 2) = 0$

63. Find a value for t such that the slope of the line passing through $(2, 1)$ and $(5, t)$ is 6.

64. Express in terms of x the slope of the line passing through (x, x^2) and $(-3, 9)$.

65. Express in terms of x the slope of the line passing through $(-2, -8)$ and (x, x^3).

66. The vertices of a parallelogram are $(0, 0)$, $(5, 2)$, $(8, 7)$, and $(3, 5)$. Find the midpoint of each diagonal. Observation?

67. The vertices of a right triangle are $A(0, 0)$, $B(0, 2b)$, and $C(2c, 0)$. Let M be the midpoint of the hypotenuse. Compute the three distances MA, MB, and MC. What do you observe?

68. The vertices of parallelogram $ABCD$ are $A(-4, -1)$, $B(2, 1)$, $C(3, 3)$, and $D(-3, 1)$.
(a) Compute the sum of the squares of the two diagonals.
(b) Compute the sum of the squares of the four sides. What do you observe?

In Exercises 69–72, two points are given. In each case compute **(a)** *the distance between the two points,* **(b)** *the slope of the line segment joining the two points, and* **(c)** *the midpoint of the line segment joining the two points.*

69. $(2, 5)$ and $(5, -6)$

70. $(3/2, 1)$ and $(7/2, 9)$

71. $(\sqrt{3}/2, 1/2)$ and $(-\sqrt{3}/2, -1/2)$

72. $(\sqrt{3}, \sqrt{2})$ and $(\sqrt{2}, \sqrt{3})$

73. A line passes through the points $(1, 2)$ and $(4, 1)$. Find the area of the triangle bounded by this line and the coordinate axes.

74. Let $P_1(x_1, y_1)$ and $P_2(x_2, y_2)$ be two given points. Let Q be the point $(\frac{1}{3}x_1 + \frac{2}{3}x_2, \frac{1}{3}y_1 + \frac{2}{3}y_2)$

(a) Show that the points P_1, Q, and P_2 are collinear. *Hint:* Compute the slope of $\overline{P_1Q}$ and the slope of $\overline{QP_2}$.

(b) Show that $P_1Q = \frac{2}{3}P_1P_2$. (In other words, Q is on the line segment $\overline{P_1P_2}$ and two-thirds of the way from P_1 to P_2.)

75. (a) Let the vertices of $\triangle ABC$ be $A(-5, 3)$, $B(7, 7)$, and $C(3, 1)$. Find the point on each median that is two-thirds of the way from the vertex to the midpoint of the opposite side. (Recall that a median of a triangle is a line segment drawn from a vertex to the midpoint of the opposite side.) What do you observe? *Hint:* Use the result in Exercise 74.

(b) Follow part (a) but take the vertices to be $A(0, 0)$, $B(2a, 0)$, and $C(2b, 2c)$. What do you observe? What does this prove?

76. Suppose that the circle $(x-h)^2 + (y-k)^2 = r^2$ has two x-intercepts, x_1 and x_2, and two y-intercepts, y_1 and y_2. Compute the quantity $x_1x_2 - y_1y_2$.

In Exercises 77–99, solve the inequalities.

77. $4x - 5 \leq 23$

78. $1 - 3x < 2 + 3x$

79. $-1 < \dfrac{1-2(1+x)}{3} < 1$

80. $3 < \dfrac{x-1}{-2} < 4$

81. $|x| \leq 1/2$

82. $|t-2| < 1$

83. $|x+4| < 1/10$

84. $|3-5x| < 2$

85. $|2x-1| \geq 5$

86. $x^2 - 21x + 108 \leq 0$

87. $x^2 + 3x - 40 < 0$

88. $x^2 \geq 15x$

89. $x^2 - 6x - 1 < 0$

90. $(x+12)(x-1)(x-8) < 0$

91. $x^4 - 34x^2 + 225 < 0$

92. $\dfrac{x+12}{x-5} > 0$

93. $\dfrac{(x-7)^2}{(x+2)^3} \geq 0$

94. $\dfrac{(x-6)^2(x-8)(x+3)}{(x-3)^2} \leq 0$

95. $\dfrac{x^2-10x+9}{x^3+1} \leq 0$

96. $\dfrac{3x+1}{x-4} < 1$

97. $\dfrac{1-2x}{1+2x} \leq \dfrac{1}{2}$

98. $x^2 + \dfrac{1}{x^2} > 3$ *Suggestion:* Use a calculator to evaluate the key numbers.

99. $\sqrt{x} - \dfrac{5}{\sqrt{x}} \leq 4$

For Exercises 100–102, find the values of k for which the roots of the equations are real numbers.

100. $kx^2 - 6x + 5 = 0$

101. $x^2 + x + k^2 = 0$

102. $kx(x+2) = -1$

103. The point $(1, -2)$ is the midpoint of a chord of the circle $x^2 - 4x + y^2 + 2y = 15$. Find the length of the chord.

104. In the accompanying figure, points P and Q trisect the hypotenuse in $\triangle ABC$. Prove that the square of the hypotenuse is equal to 9/5 the sum of the squares of the distances from the trisection points to the vertex A of the right angle. *Hint:* Let the coordinates of B and C be $(0, 3b)$ and $(3c, 0)$, respectively. Then the coordinates of P and Q are $(c, 2b)$ and $(2c, b)$, respectively.

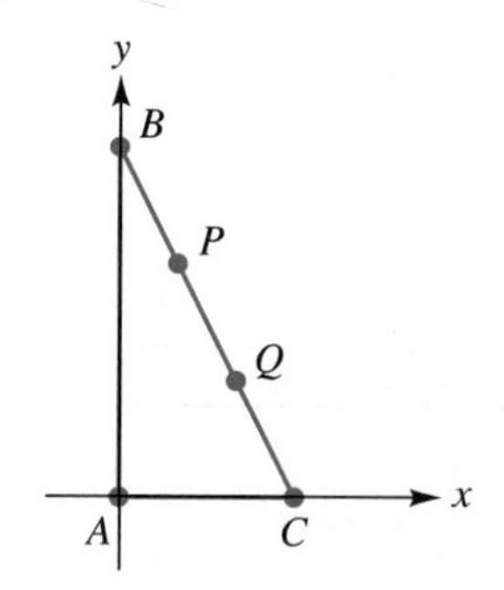

105. Figure A shows a triangle with sides of lengths s, t, and u and a median of length m. Prove that

$$m^2 = \frac{1}{2}(s^2 + t^2) - \frac{1}{4}u^2$$

Hint: Set up a coordinate system as indicated in Figure B. Then each of the quantities m^2, s^2, t^2, and u^2 can be computed in terms of a, b, and c.

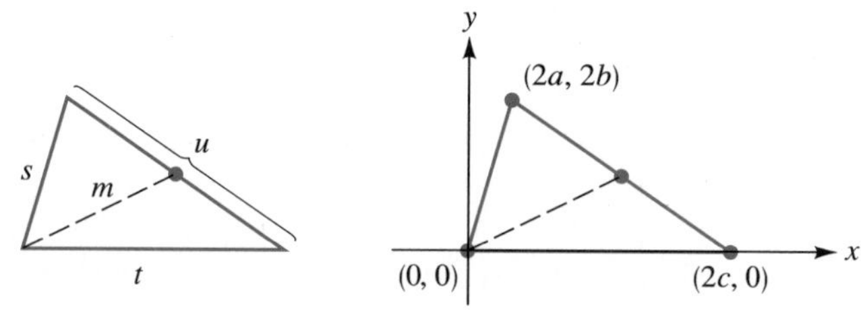

FIGURE A FIGURE B

106. In this exercise you'll derive a useful formula for the (perpendicular) distance d from the point (x_0, y_0) to the line $y = mx + b$. The formula is

$$d = \frac{|y_0 - mx_0 - b|}{\sqrt{1+m^2}}$$

(a) Refer to the figure (on the next page).

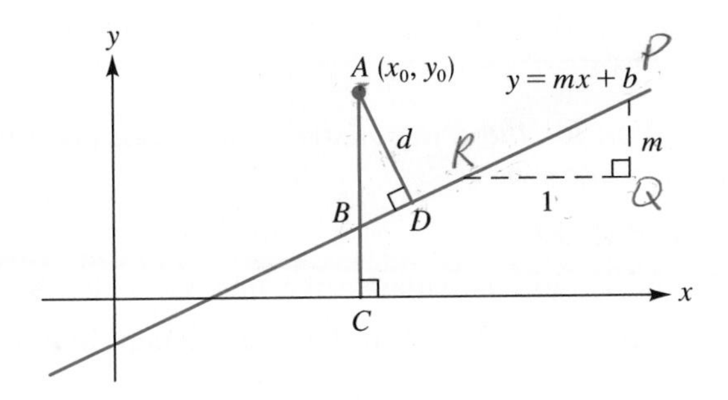

Use similar triangles to show that

$$\frac{d}{AB} = \frac{1}{\sqrt{1+m^2}}$$

Therefore, $d = AB/\sqrt{1+m^2}$.

(b) Check that $AB = AC - BC = y_0 - mx_0 - b$.

(c) Conclude from parts (a) and (b) that $d = (y_0 - mx_0 - b)/\sqrt{1+m^2}$. For the general case (in which the point and line may not be situated as in our figure), we need to use the absolute value of the quantity in the numerator to assure that AB and d are nonnegative.

107. Use the formula given in Exercise 106 to find the distance from the point (1, 2) to the line $y = \frac{1}{2}x - 5$.

108. Use the formula given in Exercise 106 to demonstrate that the distance from the point (x_0, y_0) to the line $Ax + By + C = 0$ is

$$d = \frac{|Ax_0 + By_0 + C|}{\sqrt{A^2 + B^2}}$$

109. Use the formula given in Exercise 108 to find the distance from the point $(-1, -3)$ to the line $2x + 3y - 6 = 0$.

110. Find the equation of the circle with center (2, 3) that is tangent to the line $x + y - 1 = 0$. *Hint:* Use the formula given in Exercise 108 to find the radius.

111. In the figure (at the top of the next column), the circle is tangent to the x-axis, to the y-axis, and to the line $3x + 4y = 12$.

(a) Find the equation of the circle. *Suggestion:* First decide what the relationships must be between h, k, and r in the equation $(x-h)^2 + (y-k)^2 = r^2$. Then find a way to apply the formula given in Exercise 106.

(b) Let S, T, and U denote the points where the circle touches the x-axis, the line $3x + 4y = 12$, and the y-axis, respectively. Find the equation of the line through A and T; through B and U; through S and C.

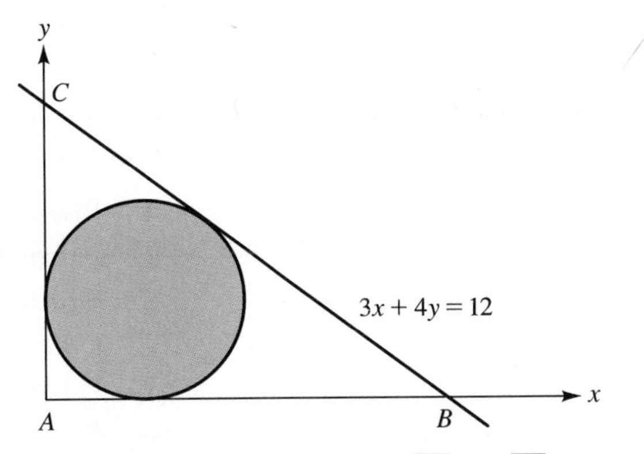

(c) Where do the line segments $\overline{AT}$ and $\overline{CS}$ intersect? Where do the line segments $\overline{AT}$ and $\overline{BU}$ intersect? What do you observe? *Remark:* The point determined here is called the *Gergonne point of the triangle ABC,* so-named in honor of its discoverer, French mathematician Joseph-Diez Gergonne (1771–1859).

112. This exercise outlines a proof of the Pythagorean theorem that was discovered by James A. Garfield, the twentieth President of the United States. Garfield published the proof in 1876, when he was Republican leader in the House of Representatives. We start with a right triangle with legs of length a and b and hypotenuse of length c. We want to prove that $a^2 + b^2 = c^2$.

(a) Take two copies of the given triangle and arrange them as shown in Figure A. Explain why the angle marked θ is a right angle. (θ is the Greek letter "theta.")

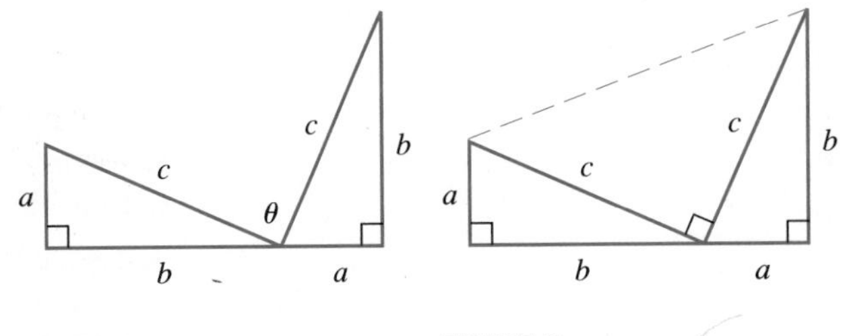

FIGURE A FIGURE B

(b) Draw the line segment indicated in Figure B. Notice that the outer quadrilateral in Figure B is a trapezoid. (Two sides are parallel.) The area of this trapezoid can be computed in two distinct ways: using the formula for the area of a trapezoid (given on the inside front cover of this book) or adding the areas of the three right triangles in Figure B. Compute the area of the trapezoid in each of these two ways. When you equate the two answers and simplify, you should obtain $a^2 + b^2 = c^2$, as required.

CHAPTER TWO TEST

In Problems 1–4, find an equation for the line satisfying the given conditions. Write the answer in the form $y = mx + b$.

1. Passes through $(1, -2)$ and $(3, 8)$
2. Passes through $(2, -1)$ and is perpendicular to the line $5x + 6y = 30$
3. Passes through the origin and the midpoint of the line segment joining the points $(-5, -2)$ and $(3, 8)$
4. Is tangent to the circle $x^2 - 8x + y^2 + 10y = 0$ at the origin
5. Which point is farther from the origin, $(3, 9)$ or $(5, 8)$?
6. Test each equation for symmetry about the x-axis, the y-axis, and the origin: **(a)** $y = x^3 + 5x$; **(b)** $y = 3^x + 3^{-x}$; **(c)** $y^2 = 5x^2 + x$.

In Problems 7–9, graph the equations and specify all x- and y-intercepts.

7. $3x - 5y = 15$
8. $2x^2 - 10x + 2y^2 + 6y = 17$
9. $y = |x + 1| + 2$
10. Find a value for t such that the slope of the line passing through the points $(-4, 3)$ and $(t, 13)$ is 2.
11. Find the point (or points) on the graph of the equation $y = x^2$ where the sum of the x- and y-coordinates is 12.
12. Does the point $(-1/2, 5)$ lie on the graph of the equation $y = 4x^2 - 8x$?
13. Sketch each graph.

    (a) $y = \dfrac{4x^2 - 1}{2x + 1}$ **(b)** $y = \begin{cases} \sqrt{1 - x^2} & \text{if } -1 \leq x \leq 0 \\ |x| & \text{if } \quad 0 < x \leq 2 \end{cases}$

    (c) $y = \begin{cases} \sqrt{x} & \text{if } 0 \leq x \leq 1 \\ 1/x & \text{if } 1 < x \leq 4 \end{cases}$
14. The endpoints of line segment $\overline{AB}$ are $A(3, -1)$ and $B(-1, 2)$. Sketch the reflection of $\overline{AB}$ about **(a)** the x-axis; **(b)** the y-axis; **(c)** the origin; and **(d)** the line $y = x$.
15. A line passes through the points $(6, 2)$ and $(1, 4)$. Find the area of the triangle bounded by this line and the x- and y-axes.

In Problems 16–20, solve the inequalities.

16. $4(1 + x) - 3(2x - 1) \geq 1$
17. $\dfrac{3}{5} < \dfrac{3 - 2x}{-4} < \dfrac{4}{5}$
18. $(x - 4)^2(x + 8)^3 \geq 0$
19. $\dfrac{1}{x} + \dfrac{1}{x + 1} + \dfrac{1}{x + 2} \geq 0$
20. $|3x - 10| - 2 \leq 0$

FUNCTIONS

From the beginning of modern mathematics in the 17th century the concept of function has been at the very center of mathematical thought.

Richard Courant and Fritz John in *Introduction to Calculus and Analysis* (New York: Wiley-Interscience, 1965)

The word ''function'' was introduced into mathematics by Leibniz, who used the term primarily to refer to certain kinds of mathematical formulas. It was later realized that Leibniz's idea of function was much too limited in scope, and the meaning of the word has since undergone many steps of generalization.

Tom M. Apostol in *Calculus,* 2nd ed. (New York: John Wiley & Sons, 1967)

The prodigious output of Leonhard Euler (1707–1783) contains many great achievements, among the most important of which must be the reformulation of the calculus around the ideas of function and variable, and his creation of a flexible and powerful algebraic language for handling functions.

John Fauvel and Jeremy Gray (editors), *The History of Mathematics: A Reader* (London: Macmillan Education, Ltd., 1987)

INTRODUCTION

We begin the study of functions in this chapter. The first section deals with matters of definition and notation. In Sections 3.2 and 3.3 we discuss graphs of functions; several techniques are developed that allow us to graph functions with a minimum of calculation. In Sections 3.4 and 3.5 we find that two functions can be combined in various ways to produce new functions. One of these ways of combining two functions is known as *composition of functions.* As you'll see, this is the unifying theme between Sections 3.4 and 3.5.

3.1 THE DEFINITION OF A FUNCTION

There are numerous instances in mathematics and its applications in which one quantity corresponds to or depends on another according to some definite rule. Consider, for example, the equation $y = 3x - 2$. Each time that we select an x-value, a corresponding y-value is determined, in this case according to the rule *multiply by* 3, *then subtract* 2. In this sense, the equation $y = 3x - 2$ is an example of a *function.* It is a *rule* specifying a y-value corresponding to each x-value. It is useful to think of the x-values as inputs and the corresponding y-values as outputs. The function, or rule, then tells us what output results from a given input. This is indicated schematically in Figure 1. As another example, the area A of a circle depends on the radius r according to the rule or function $A = \pi r^2$. For each value of r there is a corresponding value for A obtained by the rule *square r and multiply the result by* π. In this case the inputs are the values of r and the outputs are the corresponding values of A.

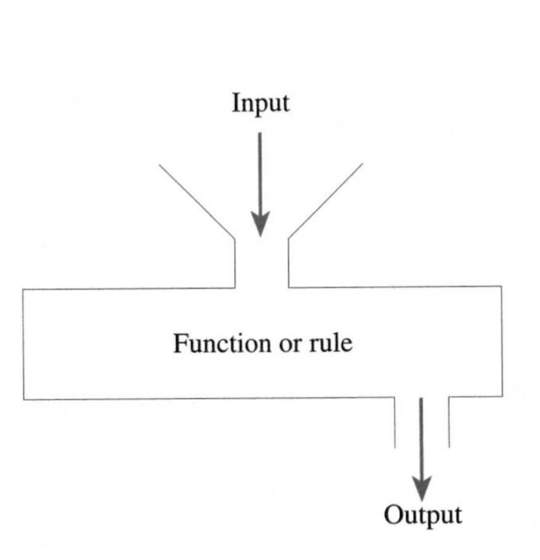

FIGURE 1

For most of the functions studied in this text (and in beginning calculus), the inputs and outputs are real numbers and the function or rule is specified by means of an equation. This was the case with the two examples we have given; however, this need not always be the case. A function can be specified by means

TABLE 1

Mercury	0
Venus	0
Earth	1
Mars	2
Jupiter	16
Saturn	18
Uranus	15
Neptune	8
Pluto	1

of a table or a graph. Consider, for example, the correspondences in Table 1, which indicate the number of moons each planet in the solar system was known to have as of 1996. If we think of the inputs as the planets listed in the left-hand column of Table 1 and the outputs as the numbers in the right-hand column, then these correspondences constitute a function. The rule for this function can be stated, *Assign to each planet in the solar system the number of moons it was known to have in 1996.*

Table 2 provides another example of a function that is specified by means of a table rather than a formula. The table indicates the (approximate) number of farms in the United States over the years 1850–1990. If we think of the years as the inputs and the number of farms as the outputs, we can plot the data from this function in a rectangular coordinate system, as shown in Figure 2. Notice that even a quick glance at the graph lets you immediately see the trends in the data. We'll study graphs of functions in detail in Sections 3.2 and 3.3.

TABLE 2 Number of Farms in the United States, 1850–1990

Year	1850	1860	1870	1880	1890	1900	1910	1920	1930	1940	1950	1960	1970	1980	1990
Farms (thousands)	1449	2044	2660	4009	4565	5740	6366	6454	6295	6102	5388	3962	2954	2440	2143

Source: *The Universal Almanac 1994,* John W. Wright, ed. (Kansas City, Missouri: Andrews and McMeel, 1993)

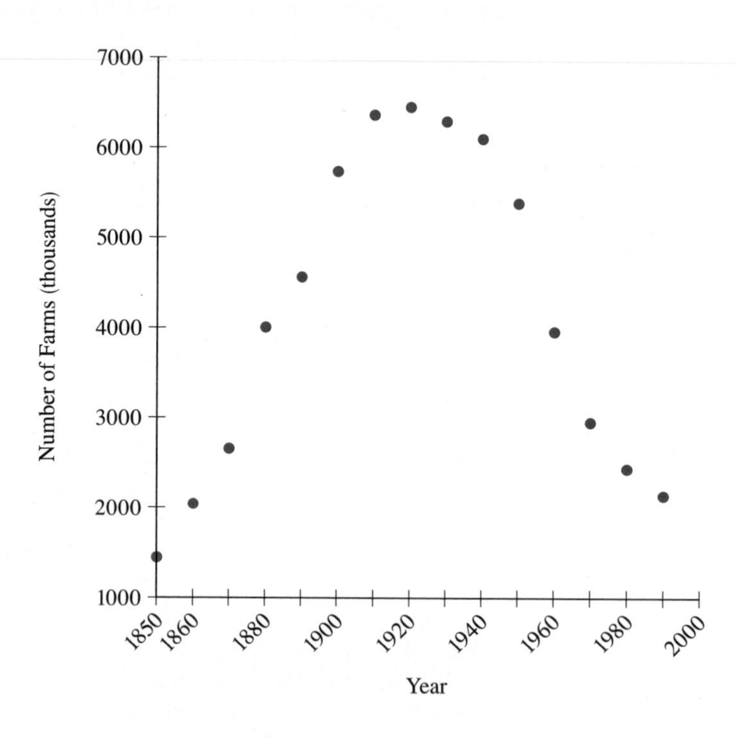

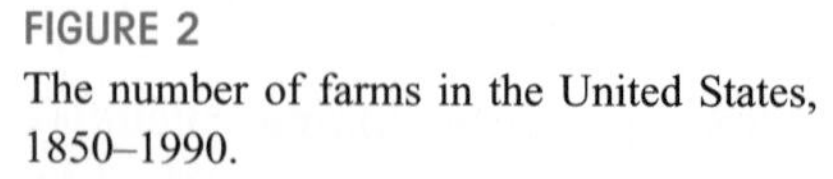

FIGURE 2
The number of farms in the United States, 1850–1990.

A definition of the term "function" is given in the box that follows. As you will see, the definition is broad enough to encompass all the examples we have just seen.

DEFINITION: Function

Let A and B be two nonempty sets. A **function** from A to B is a rule of correspondence that assigns to each element in A exactly one element in B.

The set A in the definition just given is called the **domain** of the function. Think of the domain as the set of all possible inputs. The set of all outputs, on the other hand, is called the **range** of the function. When a function is defined by means of an equation, the letter representing elements from the domain (that is, the inputs) is called the **independent variable.** For example, in the equation $y = 3x - 2$, the independent variable is x. The letter representing elements from the range (that is, the outputs) is called the **dependent variable.** In the equation $y = 3x - 2$, the dependent variable is y; its value *depends* on x. This is also expressed by saying that *y is a function of x.*

For functions defined by equations, we'll agree on the following convention regarding the domain: Unless otherwise indicated, the domain is assumed to be the set of all real numbers that lead to unique real-number outputs. (This is essentially the domain convention described in Section 1.3.) Thus, the domain of the function defined by $y = 3x - 2$ is the set of all real numbers, whereas the domain of the function defined by $y = 1/(x - 5)$ is the set of all real numbers except 5. [The expression $1/(x - 5)$ is undefined when $x = 5$ because the denominator is then zero.]

EXAMPLE 1 Find the domain of the function defined by each equation: **(a)** $y = \sqrt{2x + 6}$; **(b)** $s = 1/(t^2 - 6t - 7)$.

Solution

(a) The quantity under the radical sign must be nonnegative, so we have

$$\begin{aligned} 2x + 6 &\geqslant 0 \\ 2x &\geqslant -6 \\ x &\geqslant -3 \end{aligned}$$

The domain is therefore the interval $[-3, \infty)$.

(b) Since division by zero is undefined, the domain of this function consists of all real numbers t except those for which the denominator is zero. Thus to find out which values of t to exclude, we solve the equation $t^2 - 6t - 7 = 0$. We have

$$t^2 - 6t - 7 = 0$$
$$(t - 7)(t + 1) = 0$$
$$t - 7 = 0 \quad \Big| \quad t + 1 = 0$$
$$t = 7 \quad \Big| \quad t = -1$$

It follows now that the domain of the function defined by $s = 1/(t^2 - 6t - 7)$ is the set of all real numbers except $t = 7$ and $t = -1$.

EXAMPLE 2 Find the domain of the function defined by each equation:

(a) $y = \sqrt{\dfrac{x + 3}{x - 4}}$; **(b)** $y = \sqrt[3]{\dfrac{x + 3}{x - 4}}$.

Solution

(a) The quantity underneath the square root sign must be nonnegative, so we require

$$\frac{x+3}{x-4} \geqslant 0$$

This inequality can be solved using the techniques of Section 2.6. In fact, we did solve exactly this inequality in Example 4 of Section 2.6. As you can check (preferably on your own, without looking back), the solution set is $(-\infty, -3] \cup (4, \infty)$. This is the domain of the given function.

(b) As opposed to the situation with square roots, cube roots are defined for all real numbers. Thus, the only trouble spot for $y = \sqrt[3]{(x+3)/(x-4)}$ occurs when the denominator is zero, that is, when x is 4. Consequently, the domain of the given function consists of all real numbers except $x = 4$.

EXAMPLE 3 Find the range of the function defined by $y = \dfrac{x+2}{x-3}$.

Solution The range of this function is the set of all outputs y. One way to see what restrictions the given equation imposes on y is to solve the equation for x as follows:

$$\begin{aligned} y(x-3) &= x+2 \qquad \text{multiplying by } x-3 \\ xy - 3y &= x+2 \\ xy - x &= 3y+2 \\ x(y-1) &= 3y+2 \\ x &= \frac{3y+2}{y-1} \end{aligned}$$

From this last equation we see that the value of y cannot be 1. (The denominator is zero when $y = 1$.) The range therefore consists of all real numbers except $y = 1$.

We often use single letters to name functions. If f is a function and x is an input for the function, then the resulting output is denoted by $f(x)$. This is read *f of x* or *the value of f at x*. As an example of this notation, suppose that f is the function defined by

$$f(x) = x^2 - 3x + 1 \qquad (1)$$

Then $f(-2)$ denotes the output that results when the input is -2. To calculate this output, just replace x with -2 throughout equation (1). This yields

$$\begin{aligned} f(-2) &= (-2)^2 - 3(-2) + 1 \\ &= 4 + 6 + 1 = 11 \end{aligned}$$

That is, $f(-2) = 11$. Figure 3 summarizes this result and the notation.

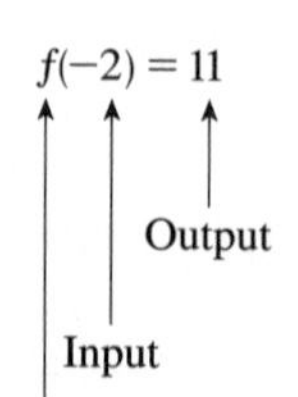

FIGURE 3

EXAMPLE 4 Let $f(x) = 1/(x-1)$. Compute: **(a)** $f(0)$; **(b)** $f(t)$; **(c)** $f(x-1)$.

Solution

(a) $f(0) = \dfrac{1}{0-1} = -1$

(b) $f(t) = 1/(t-1)$

(c) Replace x with the quantity $x-1$ throughout the given equation. This yields

$$f(x-1) = \frac{1}{(x-1)-1} = \frac{1}{x-2}$$

EXAMPLE 5 Let $g(x) = 1 - x^2$. Compute $g(x-1)$.

Solution In the equation $g(x) = 1 - x^2$, we substitute the quantity $x-1$ in place of each occurrence of x. This gives us

$$\begin{aligned} g(x-1) &= 1 - (x-1)^2 \\ &= 1 - (x^2 - 2x + 1) = -x^2 + 2x \end{aligned}$$

Thus, $g(x-1) = -x^2 + 2x$.

Here is a slightly different perspective on function notation that we can apply in Example 5 and in similar problems. Instead of writing $g(x) = 1 - x^2$, we can write $g(\quad) = 1 - (\quad)^2$, with the understanding that whatever quantity goes in the parentheses on the left-hand side of the equation must also be placed in the parentheses on the right-hand side. In particular, if we want $g(x-1)$, we simply write $x-1$ inside each set of parentheses:

$$g(\quad) = 1 - (\quad)^2 \qquad \text{and therefore} \qquad g(x-1) = 1 - (x-1)^2$$

From here on, the algebra is the same as in Example 5. We again obtain $g(x-1) = -x^2 + 2x$.

In the box that follows, we indicate several errors to avoid when using function notation. The first error indicated in the box involves confusion between a function and a number. For any real numbers r, a, and b, it is always true that $r(a+b) = ra + rb$; this is the familiar *distributive law* for real numbers. If f is a function, however, it is not in general true that $f(a+b) = f(a) + f(b)$.

ERRORS TO AVOID

ERROR	EXAMPLE USING $f(x) = x + 1$, $a = 2$, $b = 3$
$f(a+b) \neq f(a) + f(b)$	$f(2+3) = f(5) = 6$; $f(2) + f(3) = 3 + 4 = 7$; therefore $f(2+3) \neq f(2) + f(3)$
$f(ab) \neq f(a) \cdot f(b)$	$f(2 \cdot 3) = f(6) = 7$; $f(2) \cdot f(3) = 3 \cdot 4 = 12$; therefore $f(2 \cdot 3) \neq f(2) \cdot f(3)$
$f\left(\dfrac{1}{a}\right) \neq \dfrac{1}{f(a)}$	$f\left(\dfrac{1}{2}\right) = \dfrac{1}{2} + 1 = \dfrac{3}{2}$; $\dfrac{1}{f(2)} = \dfrac{1}{2+1} = \dfrac{1}{3}$; therefore $f\left(\dfrac{1}{2}\right) \neq \dfrac{1}{f(2)}$
$\dfrac{f(a)}{f(b)} \neq \dfrac{a}{b}$	$\dfrac{f(2)}{f(3)} = \dfrac{2+1}{3+1} = \dfrac{3}{4} \neq \dfrac{2}{3}$

The next two examples using function notation involve calculations with a **difference quotient.** This is an expression of the form

$$\frac{f(x+h) - f(x)}{h} \qquad \text{or} \qquad \frac{f(x) - f(a)}{x - a}$$

For now we'll concentrate on the algebra used in calculating these quantities. In the next section you'll see some applications; difference quotients are used to study the *average rate of change* of a function.

EXAMPLE 6 Let $f(x) = x^2 + 3x$. Compute $\dfrac{f(x) - f(2)}{x - 2}$.

Solution

$$\frac{f(x)-f(2)}{x-2} = \frac{(x^2+3x)-[2^2+3(2)]}{x-2} = \frac{x^2+3x-10}{x-2} = \frac{(x-2)(x+5)}{x-2} = x+5$$

The difference quotient is $x + 5$.

EXAMPLE 7 Let $G(x) = 2/x$. Find $\dfrac{G(x+h) - G(x)}{h}$.

Solution

$$\frac{G(x+h)-G(x)}{h} = \frac{[2/(x+h)]-(2/x)}{h}$$

An easy way to simplify this last expression is to multiply it by $\dfrac{(x+h)x}{(x+h)x}$, which equals 1. This yields

$$\frac{G(x+h)-G(x)}{h} = \frac{(x+h)x}{(x+h)x} \cdot \frac{\dfrac{2}{x+h} - \dfrac{2}{x}}{h} = \frac{2x-2(x+h)}{h(x+h)x} = \frac{2x-2x-2h}{h(x+h)x} = \frac{-2h}{h(x+h)x} = \frac{-2}{(x+h)x}$$

In Examples 1 through 7 we've considered functions that are defined by means of equations. It is important to understand, however, that not all equations and rules define functions. For example, consider the equation $y^2 = x$ and the input $x = 4$. Then we have $y^2 = 4$ and, consequently, $y = \pm 2$. So we have *two* outputs in this case, whereas the definition of a function requires that there be *exactly one* output. Example 8 provides some additional perspective on this situation.

EXAMPLE 8 Let $A = \{b, g\}$ and $B = \{s, t, u, z\}$. Which of the four correspondences in Figure 4 represent functions from A to B? For those correspondences that do represent functions, specify the range in each case.

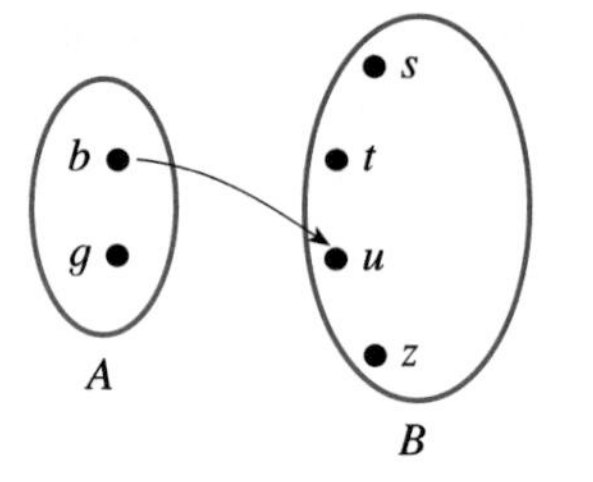

(a)

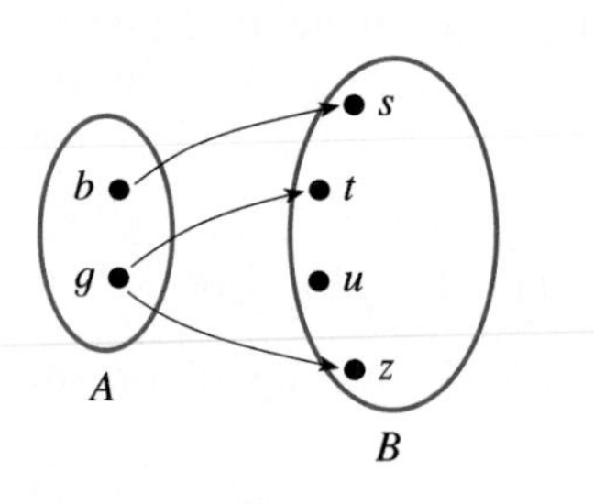

(b)

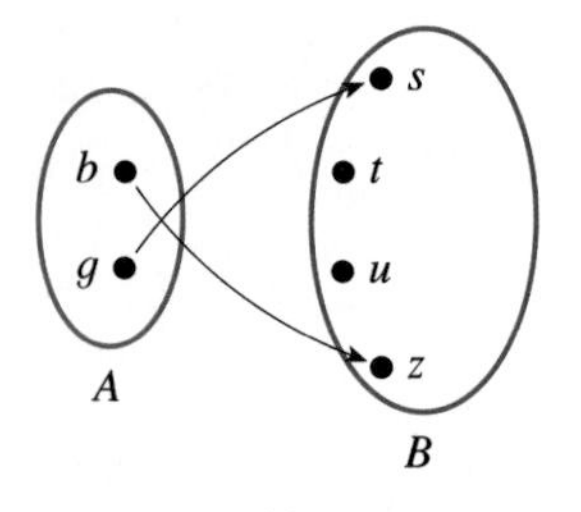

(c)

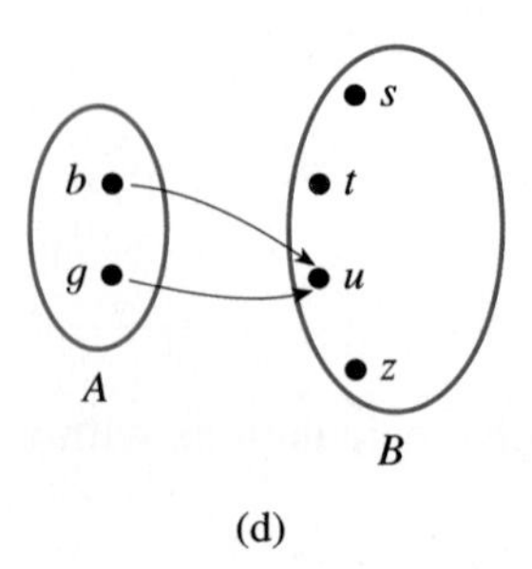

(d)

FIGURE 4

Solution

(a) This is not a function. The definition requires that *each* element in A be assigned an element in B. The element g in this case has no assignment.

(b) This is not a function. The definition requires *exactly one* output for a given input. In this case there are two outputs for the input g.

(c) and **(d)** Both of these rules qualify as functions from A to B. For each input there is exactly one output. (Regarding the function in (d) in particular, notice that nothing in the definition of a function prohibits two different inputs from producing the same output.) For the function in (c), the outputs are s and z, and so the range is the set $\{s, z\}$. For the function in (d), the only output is u, consequently the range is the set $\{u\}$.

Because the function concept is so central to the rest of the material in this text, we conclude this section with a review of the terminology introduced so far.

PROPERTY SUMMARY — TERMINOLOGY FOR FUNCTIONS

TERM	DEFINITION AND COMMENTS	EXAMPLE
Function	Given two nonempty sets A and B, a function f from A to B is a rule that assigns to each element x in A exactly one element $f(x)$ in B. We think of each element x in A as an input and the corresponding element $f(x)$ in B as an output.	The equation $f(x) = x^2$ defines a function. The rule in this case is: *For each real number x, compute its square.* Given the input $x = 3$, for example, the output is $f(3) = 9$.
Domain and range	The domain of a function is the set of all inputs; this is the set A in the definition. The range is the set of all outputs; see Figure 5.	The domain and the range of the function defined by $f(x) = x^2$ are the sets $(-\infty, \infty)$ and $[0, \infty)$, respectively.
Independent and dependent variables	The letter used to represent the elements in the domain of a function (the inputs) is the independent variable. The letter used for elements in the range (the outputs) is the dependent variable.	For the function $y = x^2$, the independent variable is x and the dependent variable is y. The value of y *depends* on x.

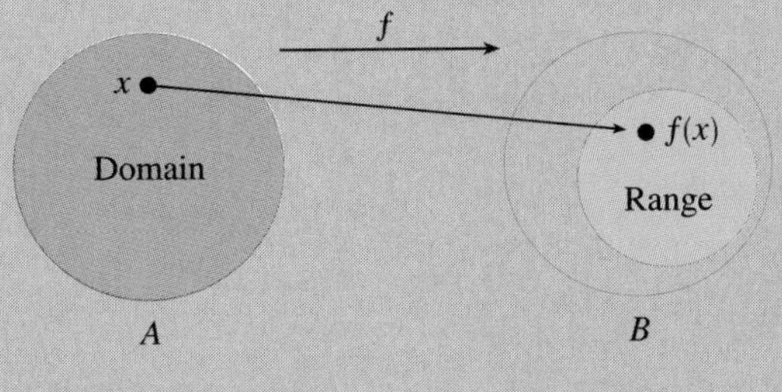

f
x
f(x)
Domain
Range
A
B

FIGURE 5
A function f from A to B.

EXERCISE SET 3.1

A Do 1, 5, 17, 31, 37

In Exercises 1–12, determine the domain of each function.

1. **(a)** $y = -5x + 1$
(b) $y = 1/(-5x + 1)$
(c) $y = \sqrt{-5x + 1}$
(d) $y = \sqrt[3]{-5x + 1}$

2. **(a)** $s = 3t + 12$
(b) $s = 1/(3t + 12)$
(c) $s = \sqrt{3t + 12}$
(d) $s = \sqrt[3]{3t + 12}$

3. **(a)** $f(x) = x^2 - 9$
(b) $g(x) = 1/(x^2 - 9)$
(c) $h(x) = \sqrt{x^2 - 9}$
(d) $k(x) = \sqrt[3]{x^2 - 9}$

4. **(a)** $F(t) = t^2 + 4t$
(b) $G(t) = 1/(t^2 + 4t)$
(c) $H(t) = \sqrt{t^2 + 4t}$
(d) $K(t) = \sqrt[3]{t^2 + 4t}$

5. **(a)** $f(t) = t^2 - 8t + 15$
(b) $g(t) = 1/(t^2 - 8t + 15)$
(c) $h(t) = \sqrt{t^2 - 8t + 15}$
(d) $k(t) = \sqrt[3]{t^2 - 8t + 15}$

6. **(a)** $F(x) = 2x^2 + x - 6$
(b) $G(x) = 1/(2x^2 + x - 6)$
(c) $H(x) = \sqrt{2x^2 + x - 6}$
(d) $K(x) = \sqrt[3]{2x^2 + x - 6}$

7. **(a)** $f(x) = (x - 2)/(2x + 6)$
(b) $g(x) = \sqrt{(x - 2)/(2x + 6)}$
(c) $h(x) = \sqrt[3]{(x - 2)/(2x + 6)}$

8. **(a)** $F(t) = (3t - 4)/(7 - 2t)$
(b) $G(t) = \sqrt{(3t - 4)/(7 - 2t)}$
(c) $H(t) = \sqrt[3]{(3t - 4)/(7 - 2t)}$

9. **(a)** $y = 2(x - 1)(x + 2)(x - 3)$
(b) $y = \sqrt[3]{2(x - 1)(x + 2)(x - 3)}$
(c) $y = \sqrt[4]{2(x - 1)(x + 2)(x - 3)}$

10. **(a)** $z = 5(x - 4)(5 - x)(x + 1)$
(b) $z = \sqrt[5]{5(x - 4)(5 - x)(x + 1)}$
(c) $z = \sqrt[6]{5(x - 4)(5 - x)(x + 1)}$

11. $f(x) = (4x + 1)/(x^3 - 1)$

12. $g(x) = (x^2 - 1)/(x^3 + 27)$

In Exercises 13–24, find the domain and the range of each function.

13. $y = 4x - 5$

14. $y = 125 - 12x$

15. $y = 4x^3 - 5$

16. $y = 125 - 12x^3$

17. $g(x) = \dfrac{4x - 20}{3x - 18}$

18. $f(x) = \dfrac{1 - x}{x}$

19. **(a)** $f(x) = \dfrac{x + 3}{x - 5}$ **(b)** $F(x) = \dfrac{x^3 + 3}{x^3 - 5}$

20. **(a)** $g(x) = \dfrac{2x - 7}{3x + 24}$
(b) $G(x) = \dfrac{2x^3 - 7}{3x^3 + 24}$

21. $s = t^2 + 4$

22. $s = 2t^2 - 10$

23. $H(u) = (au + b)/u$, where a and b are constants, and $b \neq 0$

24. $K(v) = cv^2 + d$, where c and d are constants, $c \neq 0$

25. Let $A = \{x, y, z\}$ and $B = \{1, 2, 3\}$. Which of the rules displayed in the figure represent functions from A to B?

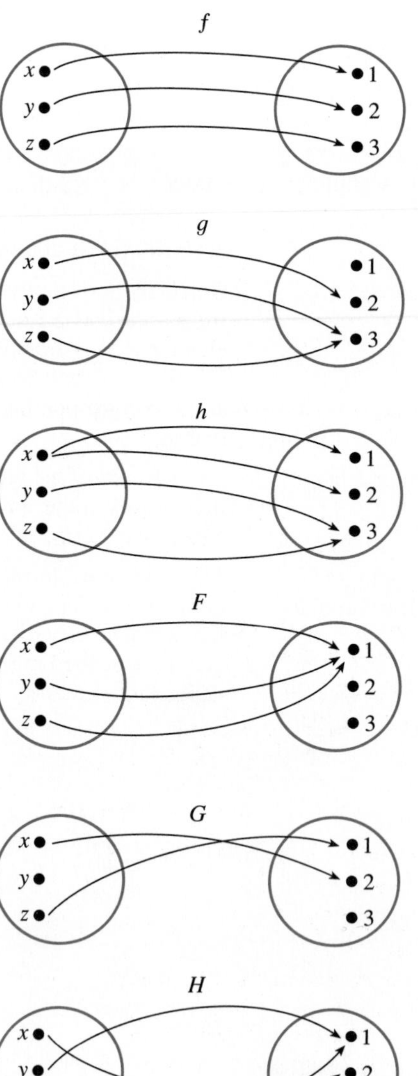

26. Let $D = \{a, b\}$ and $C = \{i, j, k\}$. Which of the rules displayed in the following figure represent functions from D to C?

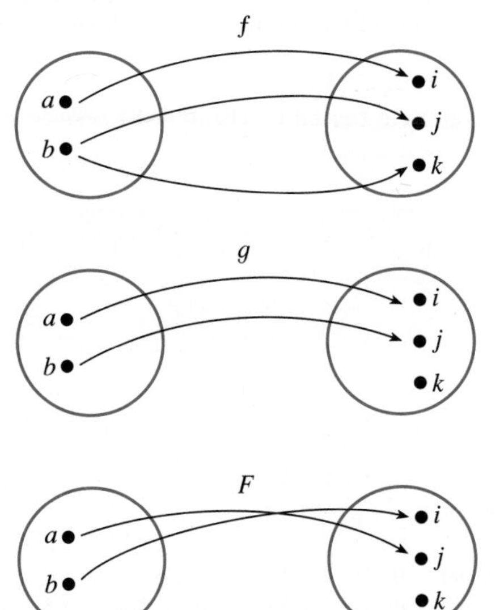

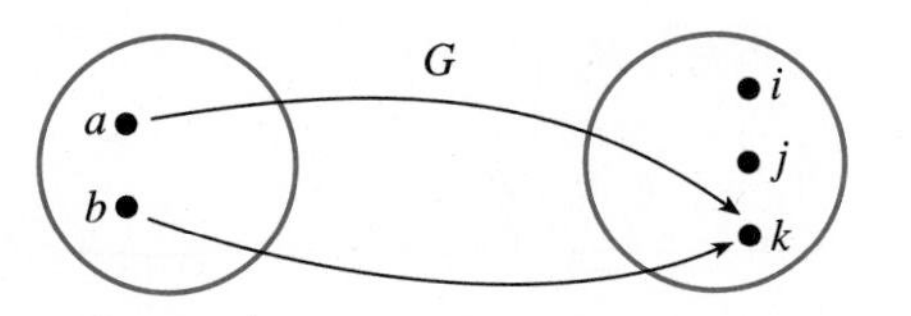

27. **(a)** Specify the range for each rule that represents a function in Exercise 25.
(b) Specify the range for each rule that represents a function in Exercise 26.

28. **(a)** Suppose that in Exercise 25 all the arrows were reversed. Which, if any, of the new rules would be functions from B to A?
(b) Suppose that in Exercise 26 all the arrows were reversed. Which, if any, of the new rules would be functions from C to D?

29. Each of the following rules defines a function with domain the set of all real numbers. Express each rule in the form of an equation.

EXAMPLE
The rule *For each real number, compute its square* can be written $y = x^2$.

(a) For each real number, subtract 3 and then square the result.
(b) For each real number, compute its square and then subtract 3 from the result.
(c) For each real number, multiply it by 3 and then square the result.
(d) For each real number, compute its square and then multiply the result by 3.

30. Each of the following rules defines a function with domain equal to the set of all real numbers. Express each rule in words.
(a) $y = 2x^3 + 1$ **(b)** $y = 2(x+1)^3$
(c) $y = (2x+1)^3$ **(d)** $y = (2x)^3 + 1$

31. Let $f(x) = x^2 - 3x + 1$. Compute the following.
(a) $f(1)$ **(b)** $f(0)$ **(c)** $f(-1)$
(d) $f(3/2)$ **(e)** $f(z)$ **(f)** $f(x+1)$
(g) $f(a+1)$ **(h)** $f(-x)$ **(i)** $|f(1)|$
(j) $f(\sqrt{3})$ **(k)** $f(1+\sqrt{2})$ **(l)** $|1 - f(2)|$

32. Let $H(x) = 1 - x + x^2 - x^3$.
(a) Which number is larger, $H(0)$ or $H(1)$?
(b) Find $H(\frac{1}{2})$. Does $H(\frac{1}{2}) + H(\frac{1}{2}) = H(1)$?

33. Let $f(x) = 3x^2$. Find the following.
(a) $f(2x)$ **(b)** $2f(x)$ **(c)** $f(x^2)$
(d) $[f(x)]^2$ **(e)** $f(x/2)$ **(f)** $f(x)/2$
For checking: No two answers are the same.

34. Let $f(x) = 4 - 3x$. Find the following.
(a) $f(2)$ **(b)** $f(-3)$ **(c)** $f(2) + f(-3)$
(d) $f(2+3)$ **(e)** $f(2x)$ **(f)** $2f(x)$
(g) $f(x^2)$ **(h)** $f(1/x)$ **(i)** $f[f(x)]$
(j) $x^2f(x)$ **(k)** $1/f(x)$ **(l)** $f(-x)$
(m) $-f(x)$ **(n)** $-f(-x)$

35. Let $H(x) = 1 - 2x^2$. Find the following.
(a) $H(0)$ **(b)** $H(2)$
(c) $H(\sqrt{2})$ **(d)** $H(5/6)$
(e) $H(x+1)$ **(f)** $H(x+h)$
(g) $H(x+h) - H(x)$ **(h)** $\dfrac{H(x+h) - H(x)}{h}$

36. **(a)** If $f(x) = 2x + 1$, does $f(3+1) = f(3) + f(1)$?
(b) If $f(x) = 2x$, does $f(3+1) = f(3) + f(1)$?
(c) If $f(x) = \sqrt{x}$, does $f(3+1) = f(3) + f(1)$?

37. Let $R(x) = (2x-1)/(x-2)$. Find the following.
(a) domain and range of R **(b)** $R(0)$
(c) $R(1/2)$ **(d)** $R(-1)$
(e) $R(x^2)$ **(f)** $R(1/x)$
(g) $R(a)$ **(h)** $R(x-1)$

38. Let $g(x) = 2$, for all x. Find each output.
(a) $g(0)$ **(b)** $g(5)$ **(c)** $g(x+h)$

39. Let $d(t) = -16t^2 + 96t$.
(a) Compute $d(1)$, $d(3/2)$, $d(2)$, and $d(t_0)$.
(b) For which values of t is $d(t) = 0$?
(c) For which values of t is $d(t) = 1$?

40. Let $A(x) = |x^2 - 1|$. Compute $A(2)$, $A(1)$, and $A(0)$.

41. Let $g(t) = |t - 4|$. Find $g(3)$. Find $g(x+4)$.

42. Let $f(x) = x^2/|x|$.
(a) What is the domain of f?
(b) Find $f(2)$, $f(-2)$, $f(20)$, and $f(-20)$.
(c) What is the range of f? *Hint:* Look over your results in part (b).

In Exercises 43–50, compute the difference quotient $\dfrac{f(x+h)-f(x)}{h}$ *for each function.*

43. $f(x) = 8x - 3$

44. $f(x) = -2x + 9$

45. **(a)** $f(x) = x^2$
(b) $f(x) = 2x^2$
(c) $f(x) = 2x^3$

46. **(a)** $f(x) = -3x^2$
(b) $f(x) = -3x^2 + 4$
(c) $f(x) = -3x^3 + 4$

47. $f(x) = 2x^2 - 3x + 1$

48. $f(x) = -x^2 + 6x - 2$

49. $f(x) = 3/x$

50. $f(x) = -2/x^2$

In Exercises 51–58, compute the difference quotient $\dfrac{g(x)-g(a)}{x-a}$ *for each function.*

51. $g(x) = 4x^2$

52. $g(x) = -2x^2 + 9$

53. $g(x) = x^2 - 2x + 4$

54. $g(x) = 2x^2 - x + 1$

55. $g(x) = 4x^3$

56. $g(x) = -x^3 + 2$

57. $g(x) = 5/x^2$

58. $g(x) = -3/x^2$

59. Let $f(x) = x/(x-1)$. Compute each difference quotient.
(a) $\dfrac{f(x)-f(a)}{x-a}$
(b) $\dfrac{f(x)-f(3)}{x-3}$
(c) $\dfrac{f(x+h)-f(x)}{h}$
(d) $\dfrac{f(3+h)-f(3)}{h}$

60. Let $g(t) = 1/(3t-5)$. Compute each difference quotient.
(a) $\dfrac{g(t)-g(a)}{t-a}$
(b) $\dfrac{g(t)-g(2)}{t-2}$
(c) $\dfrac{g(t+h)-g(t)}{h}$
(d) $\dfrac{g(2+h)-g(2)}{h}$

61. In each case a pair of functions f and g are given. Find all real numbers x_0 for which $f(x_0) = g(x_0)$.
(a) $f(x) = 4x - 3;\ g(x) = 8 - x$
(b) $f(x) = x^2 - 4;\ g(x) = 4 - x^2$
(c) $f(x) = x^2;\ g(x) = x^3$
(d) $f(x) = 2x^2 - x;\ g(x) = 3$

In Exercises 62–66, use a calculator for the computations.

62. Let $h(x) = \sqrt{x}$.
(a) Compute $\dfrac{h(5)-h(1)}{5-1}$ and $\dfrac{1}{h(5)+1}$. Round both answers to three decimal places.
(b) Why are the two answers in part (a) the same?

63. When $1000 is deposited in a savings account at an annual rate of 12% *compounded quarterly,* the amount in the account after t years is given by

$$A(t) = 1000\left(1 + \frac{0.12}{4}\right)^{4t}$$

(a) Compute $A(1) - A(0)$. This is the amount the account will grow in the first year.
(b) Compute $A(10) - A(9)$. This is the amount the account will grow in the tenth year.

64. Let $f(n) = [1 + (1/n)]^n$.
(a) Complete the table. (Round the results to three decimal places.)

n	1	2	5	10	15	20
$f(n)$						

(b) By trial and error, find the smallest natural number n such that $f(n) > 2.7$.
(c) Using your calculator, can you find a number n such that $f(n) \geqslant 2.8$?

65. Let $g(n) = n^{1/n}$.
(a) Complete the table. (Round the results to four decimal places.)

n	2	3	4	5	6	7	8
$g(n)$							

(b) By trial and error, find the smallest natural number n such that $g(n) < 1.2$.

66. Consider the function f defined by

$$f(x) = x^2 + \frac{2}{x^2} \quad (x > 0)$$

(a) Complete the table. (Round the results to four decimal places.)

x	1	1.05	1.10	1.15	1.20	1.25
$f(x)$						

(b) Which x-value in the table yields the smallest value for $f(x)$?
(c) It can be shown using calculus that the input x yielding the smallest possible output $f(x)$ for this function is $x = 2^{1/4}$. Which x-value in the table is closest to this?
(d) Compute $f(2^{1/4})$. Which value of $f(x)$ in the table is closest to $f(2^{1/4})$?

In Exercises 67–70, refer to the following table. The left-hand column of the table lists four errors to avoid in working with function notation. In each case, use the function $f(x) = x^2 - 1$ and give a numerical example showing that the expressions on each side of the equation are not equal.

	Errors to Avoid	Numerical example showing that the equation is not, in general, valid
67.	$f(a+b) = f(a) + f(b)$	
68.	$f(ab) = f(a) \cdot f(b)$	
69.	$f\left(\frac{1}{a}\right) = \frac{1}{f(a)}$	
70.	$\frac{f(a)}{f(b)} = \frac{a}{b}$	

B

71. Let $f(n) = 1 - n$. Simplify $f(f(n))$.

72. Let $f(n) = 1 - n$. Simplify $f(f(n+2) + 2)$.

For Exercises 73–78, two functions s and c are defined as follows:

$$s(x) = \frac{2x}{1+x^2} \qquad c(x) = \frac{1-x^2}{1+x^2}$$

73. Show that the functions s and c possess the following properties.
(a) $s(0) = 0$; $s(1) = 1$; $s(-x) = -s(x)$
(b) $c(0) = 1$; $c(1) = 0$; $c(-x) = c(x)$
(c) $[s(x)]^2 + [c(x)]^2 = 1$

74. What does the identity $s(-x) = -s(x)$ [from Exercise 73(a)] tell you about the graph of $y = s(x)$? *Hint:* Refer to the symmetry tests in Section 2.4.

75. What does the identity $c(-x) = c(x)$ [from Exercise 73(b)] tell you about the graph of $y = c(x)$?

76. Are there any inputs x for which $s(x) = c(x)$?

77. **(a)** Show that $s(x) \leq 1$ for all x. *Hint:* Use the techniques of Section 2.6 to solve the inequality $s(x) \leq 1$.
(b) Show that $s(x) \geq -1$ for all x.

78. Follow Exercise 77, but use $c(x)$ rather than $s(x)$.

79. Let $f(x) = (x - a)/(x + a)$.
(a) Find $f(a)$, $f(2a)$, and $f(3a)$. Is it true that $f(3a) = f(a) + f(2a)$?
(b) Show that $f(5a) = 2f(2a)$.

80. Let $M(x) = (x - a)/(x + a)$. Compute $M(1/x)$.

81. Let $\phi(y) = 2y - 3$. Show that $\phi(y^2) \neq [\phi(y)]^2$.

82. Let $k(x) = 5x^3 + \frac{5}{x^3} - x - \frac{1}{x}$. Show that $k(x) = k(1/x)$.

83. Let $f(x) = 2x + 3$. Find values for a and b such that the equation $f(ax + b) = x$ is true for all values of x.

84. If $p(x) = 2^x$, verify each identity:
(a) $2p(x) = p(x+1)$; **(b)** $p(a+b) = p(a) \cdot p(b)$;
(c) $p(x) \cdot p(-x) - 1 = 0$.

85. Let $f(t) = (t - x)/(t + y)$. Show that

$$f(x+y) + f(x-y) = \frac{-2y^2}{x^2 + 2xy}$$

86. Let $f(z) = \frac{3z-4}{5z-3}$. Find $f\left(\frac{3z-4}{5z-3}\right)$.

87. Let $F(x) = \frac{ax+b}{cx-a}$. Show that $F\left(\frac{ax+b}{cx-a}\right) = x$.
(Assume that $a^2 + bc \neq 0$.)

88. If $f(x) = -2x^2 + 6x + k$ and $f(0) = -1$, find k.

89. If $g(x) = x^2 - 3xk - 4$ and $g(1) = -2$, find k.

90. Let $h(x) = x^2 - 4x - c$. Find a nonzero value for c such that $h(c) = c$.

91. A function doesn't always have to be given by an algebraic formula. For example, let the function L be defined by the following rule: *$L(x)$ is the exponent to which 2 must be raised to yield x.* (For the moment, we won't concern ourselves with the domain and range.) Then $L(8) = 3$, for example, since the exponent to which 2 must be raised to yield 8 is 3 (that is, $8 = 2^3$). Find the following outputs.
(a) $L(1)$ **(b)** $L(2)$
(c) $L(4)$ **(d)** $L(64)$
(e) $L(1/2)$ **(f)** $L(1/4)$
(g) $L(1/64)$ **(h)** $L(\sqrt{2})$
The function L is called a *logarithm function.* The usual notation for $L(x)$ in this example is $\log_2 x$. Logarithm functions will be studied in Chapter 5.

92. Let $f(x) = ax^2 + bx + c$. Show that

$$\frac{f(x+h) - f(x)}{h} = 2ax + ah + b$$

93. Let $q(x) = ax^2 + bx + c$. Evaluate

$$q\left(\frac{-b + \sqrt{b^2 - 4ac}}{2a}\right)$$

94. By definition, a **fixed point** for the function f is a number x_0 such that $f(x_0) = x_0$. For instance, to find any fixed points for the function $f(x) = 3x - 2$, we write $3x_0 - 2 = x_0$. On solving this last equation, we find that $x_0 = 1$. Thus, 1 is a fixed point for f.

Calculate the fixed points (if any) for each function.
(a) $f(x) = 6x + 10$ **(b)** $g(x) = x^2 - 2x - 4$
(c) $S(t) = t^2$ **(d)** $R(z) = (z + 1)/(z - 1)$

95. Let $f(x) = \dfrac{3x - 4}{x - 3}$.
(a) Find $f[f(x)]$.
(b) Find $f[f(22/7)]$. Try not to do it the hard way!

96. Consider the two rules F and G, where F is the rule that assigns to each person his or her mother, and G is the rule that assigns to each person his or her aunt. Explain why F is a function but G is not.

*In Exercises 97–100, use this definition: A **prime number** is a positive whole number with no factors other than itself and 1. For example, 2, 13, and 37 are primes, but 24 and 39 are not. By convention, 1 is not considered prime, so the list of the first few primes is as follows:*

$$2, 3, 5, 7, 11, 13, 17, 19, 23, 29, \ldots$$

97. Let G be the rule that assigns to each positive integer the nearest prime. For example, $G(8) = 7$, since 7 is the prime nearest 8. Explain why G is not a function. How could you alter the definition of G to make it a function?

98. Let f be the function that assigns to each natural number x the number of primes that are less than or equal to x. For example, $f(12) = 5$ because, as you can easily check, five primes are less than or equal to 12. Similarly, $f(3) = 2$ because two primes are less than or equal to 3. Find $f(8)$, $f(10)$, and $f(50)$.

99. If $P(x) = x^2 - x + 17$, find $P(1)$, $P(2)$, $P(3)$, and $P(4)$. Can you find a natural number x for which $P(x)$ is not prime?

100. If $P(x) = x^2 - x + 41$, find $P(1)$, $P(2)$, $P(3)$, and $P(4)$. Can you find a natural number x for which $P(x)$ is not prime?

101. If $g(a) = b$ and $g(b) = a$, simplify $g(g(a))$ and $g(g(g(a)))$.

102. Let $f(x) = 2 - 2|x - 1|$. Find two distinct numbers a and b such that $f(a) = b$ and $f(b) = a$.
Hint: Assume $a < 1$ and $b > 1$. (*Remark:* Exercise 104 asks you to solve this same problem but without using the added hypotheses $a < 1$ and $b > 1$.)

C

103. Let $f(n) = n^2 - n + 2$. Find a value for k such that the equation

$$f(n^2 + k) = f(n) \cdot f(n + 1)$$

holds for all values of n.

104. Solve Exercise 102 but without using the assumptions $a < 1$ and $b > 1$.

105. Let $f(x) = ax^2 + bx + c$, where $a < 0$. Show that

$$\frac{f(x_1) + f(x_2)}{2} \leq f\left(\frac{x_1 + x_2}{2}\right)$$

*In Exercises 106–108, use the following definitions. Two positive integers are said to be **relatively prime** if they have no common factor other than 1. For example, 4 and 7 are relatively prime, but 4 and 18 are not relatively prime, nor are 5 and 10. The **Euler ϕ-function** is then defined as follows:*
(i) The domain of ϕ is the set of positive integers;
(ii) $\phi(1) = 1$;
(iii) for $m > 1$, $\phi(m)$ is the number of positive integers that are less than m and relatively prime to m.
For example, to find $\phi(14)$, first make a list of the positive integers that are less than 14 and are relatively prime to 14: 1, 3, 5, 9, 11, 13. Now count how many numbers there are in this list. There are six numbers in the list, and therefore $\phi(14) = 6$.

106. (a) For the following tables, first verify for yourself that the completed entries are correct. Then complete the remaining entries.

m	1	2	3	4	5	6	7	8	9	10	11
$\phi(m)$	1	1	2	2	4						

m	12	13	14	15	16	17	18	19	20
$\phi(m)$			6						8

(b) Use the results in part (a) to decide whether each of the following statements is true or false.
(i) $\phi(2) \cdot \phi(5) = \phi(2 \cdot 5)$
(The dots denote multiplication.)
(ii) $\phi(2) \cdot \phi(6) = \phi(2 \cdot 6)$
(iii) $\phi(2) \cdot \phi(7) = \phi(2 \cdot 7)$
(iv) $\phi(3) \cdot \phi(4) = \phi(3 \cdot 4)$
(v) $\phi(3) \cdot \phi(5) = \phi(3 \cdot 5)$
(vi) $\phi(3) \cdot \phi(6) = \phi(3 \cdot 6)$

107. (a) The factors of the number 6 are 1, 2, 3, and 6. Use the table in Exercise 106(a) to verify that $\phi(1) + \phi(2) + \phi(3) + \phi(6) = 6$.
(b) The factors of the number 15 are 1,3, 5, and 15. Use the tables in Exercise 106(a) to verify that $\phi(1) + \phi(3) + \phi(5) + \phi(15) = 15$.

108. Suppose that p is a prime number. Explain (in complete sentences) why $\phi(p) = p - 1$.

109. Let $S(x) = (3^x - 3^{-x})/2$ and $C(x) = (3^x + 3^{-x})/2$. Show that the functions S and C possess the given properties.
(a) $S(0) = 0$; $C(0) = 1$; $S(1) = 4/3$; $C(1) = 5/3$
(b) $[C(x)]^2 - [S(x)]^2 = 1$
(c) $S(-x) = -S(x)$ and $C(-x) = C(x)$
(d) $S(x+y) = S(x)C(y) + C(x)S(y)$
(e) $C(x+y) = C(x)C(y) + S(x)S(y)$
(f) $S(2x) = 2S(x)C(x)$ and $C(2x) = [C(x)]^2 + [S(x)]^2$
(g) $S(3x) = 3S(x) + 4[S(x)]^3$
(h) $[S(x) + C(x)]^2 = S(2x) + C(2x)$

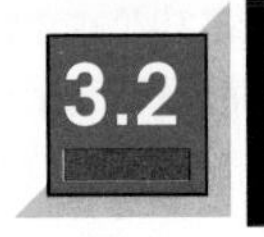

3.2 THE GRAPH OF A FUNCTION

In my own case, I got along fine without knowing the name of the distributive law until my sophomore year in college; meanwhile I had drawn lots of graphs.

Professor Donald E. Knuth (one of the world's preeminent computer scientists) in *Mathematical People* (Boston: Birkhäuser, 1985)

When the domain and range of a function are sets of real numbers, we can graph the function in the same way we graphed equations in Chapter 2. In graphing functions, the usual practice is to reserve the horizontal axis for the independent variable and the vertical axis for the dependent variable. The function or rule then tells you how you must pick your y-coordinate, once you have selected an x-coordinate.

DEFINITION: Graph of a Function

The **graph** of a function f in the x-y plane consists of those points (x, y) such that x is in the domain of f and $y = f(x)$. See Figure 1.

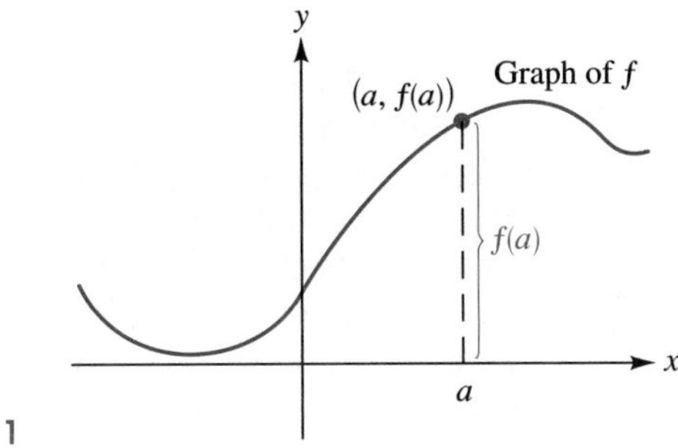

FIGURE 1

 **EXAMPLE 1** Figure 2 shows portions of the graphs of the functions $f(x) = \sqrt{x}$ and $y = x$. What are the coordinates of the points P, Q, and R in Figure 2? (Each dashed line is parallel to one of the coordinate axes.) For those coordinates involving radicals, also supply a calculator approximation rounded to two decimal places.

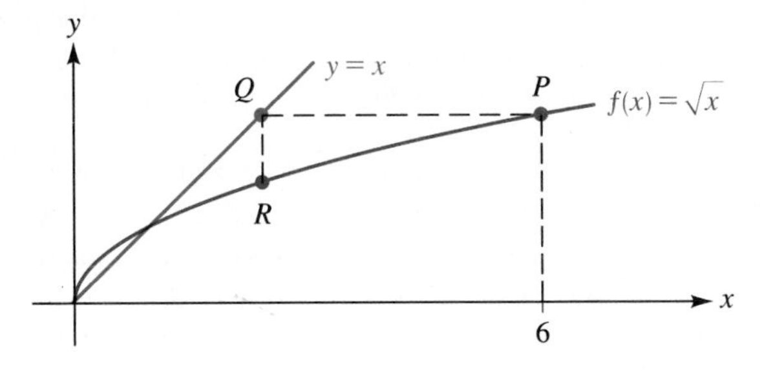

FIGURE 2
What are the coordinates of P, Q, and R?

Solution In Figure 2, the point P lies on the graph of $f(x)=\sqrt{x}$. Since the x-coordinate of P is 6, the y-coordinate is $f(6)$, which, according to the rule for f, is $\sqrt{6}$. Thus, the coordinates of P are $(6, \sqrt{6})$. Next, we want the coordinates of Q. Figure 2 shows that the points P and Q have the same y-coordinate. So, the y-coordinate of Q must also be $\sqrt{6}$. What about the x-coordinate of Q? Since Q lies on the graph of $y=x$, the x-coordinate of Q must be the same as its y-coordinate, namely, $\sqrt{6}$. Thus, the coordinates of Q are $(\sqrt{6}, \sqrt{6})$.

Finally, let us determine the coordinates of the point R. According to Figure 2, the x-coordinate of R is the same as that of Q, which we know to be $\sqrt{6}$. The y-coordinate of R is then found by taking the square root of this x-coordinate, to obtain $\sqrt{\sqrt{6}}$. (We've used the fact that R lies on the graph of $f(x)=\sqrt{x}$.)

Summarizing now, the required coordinates are as follows. (You should check for yourself that the calculator values given are correct.)

$$P:\quad (6, \sqrt{6}) \approx (6, 2.45)$$
$$Q:\quad (\sqrt{6}, \sqrt{6}) \approx (2.45, 2.45)$$
$$R:\quad (\sqrt{6}, \sqrt{\sqrt{6}}) \approx (2.45, 1.57)$$

EXAMPLE 2 Graph the functions f and g defined as follows:

(a) $f(x)=x^2-1$; **(b)** $g(x)=|x^2-1|$.

Solution

(a) The graph of this function is by definition the graph of the equation $y=x^2-1$. [Whether we label the vertical axis y or $f(x)$ is immaterial.] As preparation for drawing the graph, we first determine the domain of f, the x- and y-intercepts of the graph, and any symmetries the graph may have. As you can readily check, the results are as follows.

domain: $(-\infty, \infty)$	x-intercepts: $1, -1$
symmetry: about the y-axis	y-intercept: -1

The required graph is then sketched as in Figure 3(a).

(b) The graph of g is shown in Figure 3(b). It is obtained from the graph of f in Figure 3(a) as follows. First, for $x \geqslant 1$ or $x \leqslant -1$, the graph of g is identical to that of f because in this case, we have

$$g(x)=|x^2-1|=x^2-1=f(x)$$

Second, on the interval $(-1, 1)$, the graph of g is obtained by reflecting the graph of f in the x-axis because the quantity x^2-1 is negative on this interval and therefore

$$g(x)=|x^2-1|=-(x^2-1)=-f(x)$$

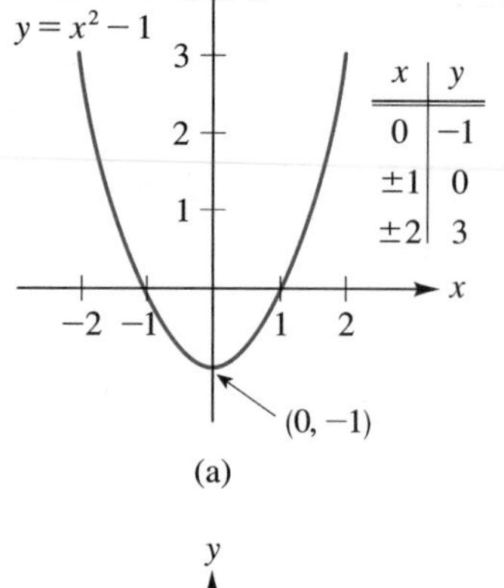

(a)

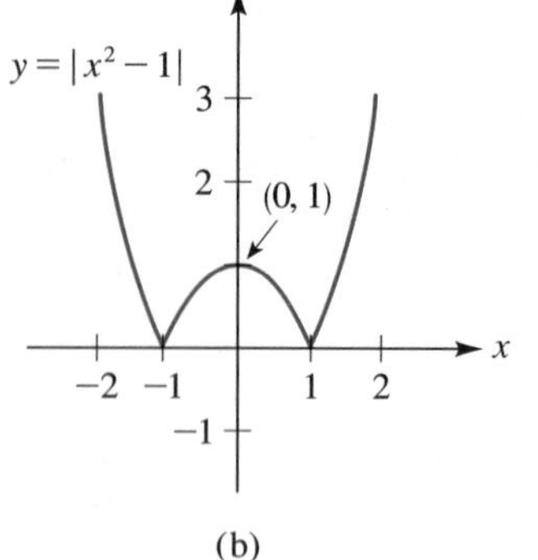

(b)

FIGURE 3

EXAMPLE 3 Specify the domain and the range of the function g graphed in Figure 4(a).

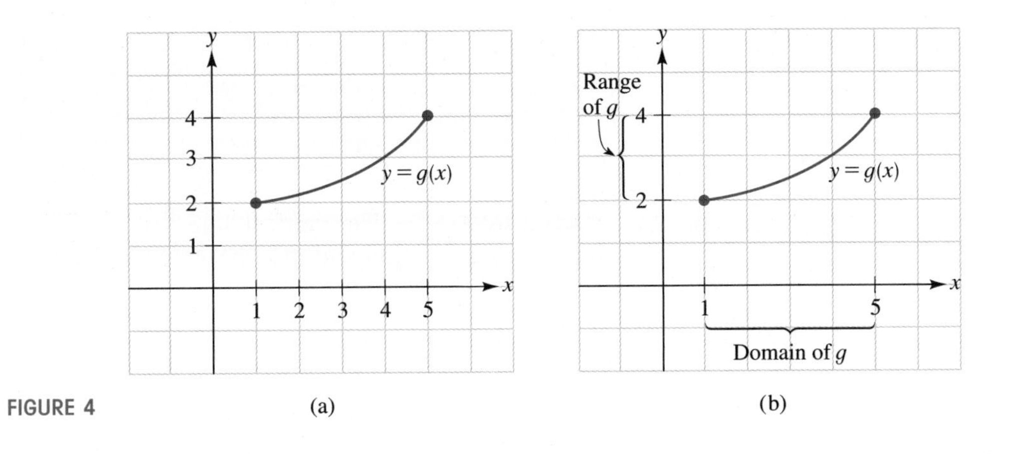

FIGURE 4

Solution The domain of g is just that portion of the x-axis (the inputs) utilized in graphing g. As Figure 4(b) indicates, this amounts to all real numbers x from 1 to 5, inclusive: $1 \leq x \leq 5$. Recall from Section 1.1 that this set of numbers is denoted by $[1, 5]$. To find the range of g, we need to check which part of the y-axis is utilized in graphing g. As Figure 4(b) indicates, this is the set of all real numbers y between 2 and 4, inclusive: $2 \leq y \leq 4$. Our shorthand notation for this interval of numbers is $[2, 4]$.

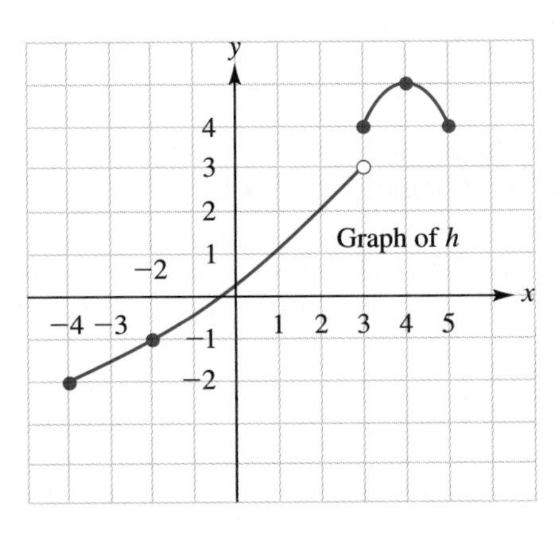

FIGURE 5

EXAMPLE 4 The graph of a function h is shown in Figure 5. The open circle in the figure is used to indicate that the point $(3, 3)$ does not belong to the graph of h.

(a) Specify the domain and the range of the function h.
(b) Determine each value: **(i)** $h(-2)$; **(ii)** $h(3)$; **(iii)** $|h(-4)|$.

Solution

(a) The domain of h is the interval $[-4, 5]$. The range is the set $[-2, 3) \cup [4, 5]$.
(b) **(i)** The function notation $h(-2)$ stands for the y-coordinate of that point on the graph of h whose x-coordinate is -2. Since the point $(-2, -1)$ is on the graph of h, we conclude that $h(-2) = -1$.
(ii) We have $h(3) = 4$ because the point $(3, 4)$ lies on the graph of h. Note that h would not be considered a function if the point $(3, 3)$ were also part of the graph. (Why?)
(iii) Since the point $(-4, -2)$ lies on the graph of h, we write $h(-4) = -2$. Thus $|h(-4)| = |-2| = 2$.

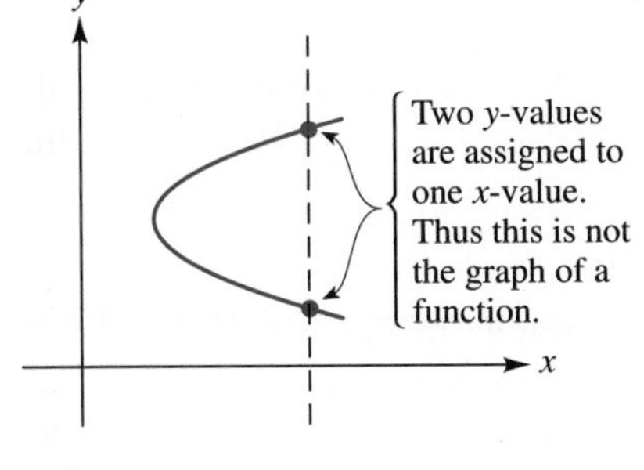

FIGURE 6

Most of the graphs that we looked at in Chapter 2 are graphs of functions. However, it's important to understand that not every graph represents a function. Consider, for example, the graph in Figure 6.

Figure 6 shows a vertical line intersecting the graph in two distinct points. The specific coordinates of the two points are unimportant. What the vertical line helps us to see is that two different y-values have been assigned to the same x-value, and therefore the graph cannot be the graph of a function $y = f(x)$. These remarks are generalized in the box that follows.

> **VERTICAL LINE TEST**
>
> A graph in the x-y plane represents a function $y = f(x)$ provided that any vertical line intersects the graph in at most one point.

EXAMPLE 5 The vertical line test implies that the graph in Figure 7(a) represents a function and that the graph in Figure 7(b) does not.

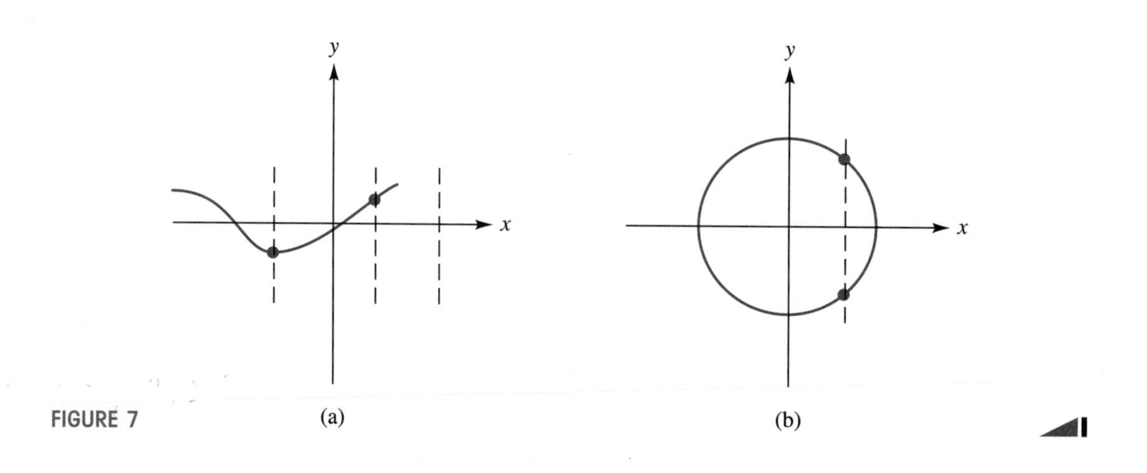

FIGURE 7

EXAMPLE 6 Figure 8 shows three closely related graphs. (Notice that the graphs are constructed from two of the six basic equations described in Section 2.2.) Which of the graphs represent functions?

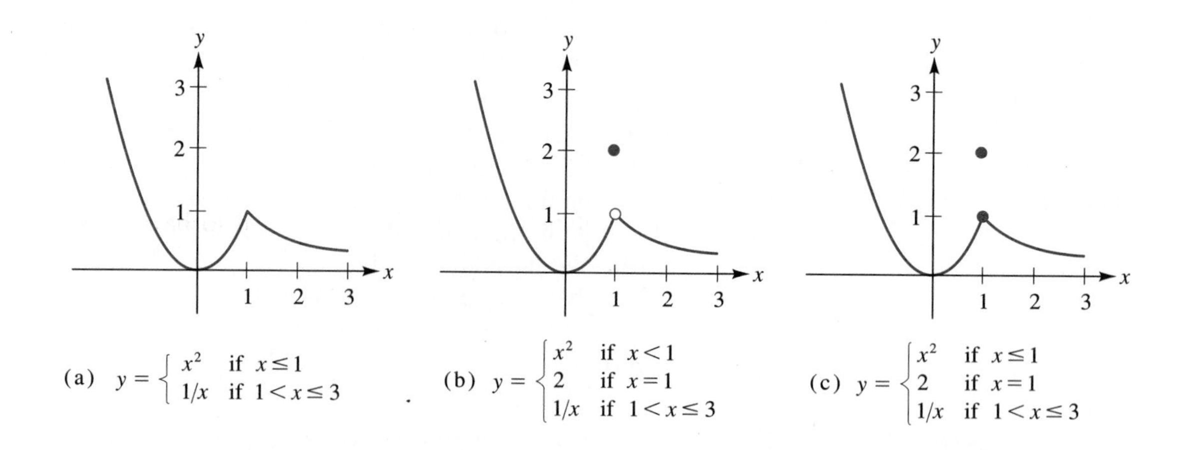

FIGURE 8

Solution The vertical line test tells us that the graphs in Figures 8(a) and 8(b) represent functions, but that the graph in Figure 8(c) does not.

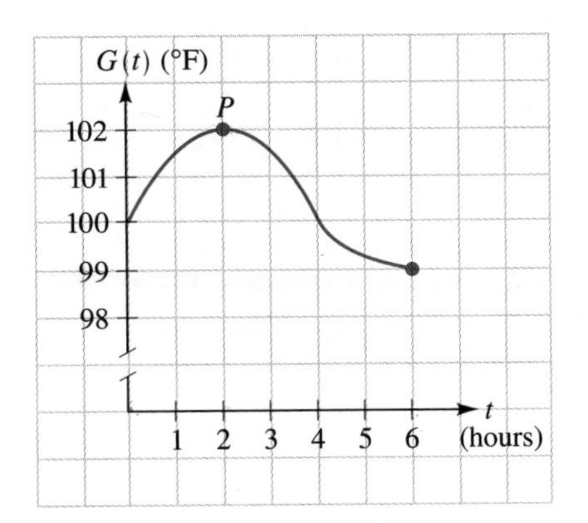

FIGURE 9
The graph of the function G is a fever graph, where $G(t)$ is a patient's temperature t hours after 12 noon, $0 \leq t \leq 6$.

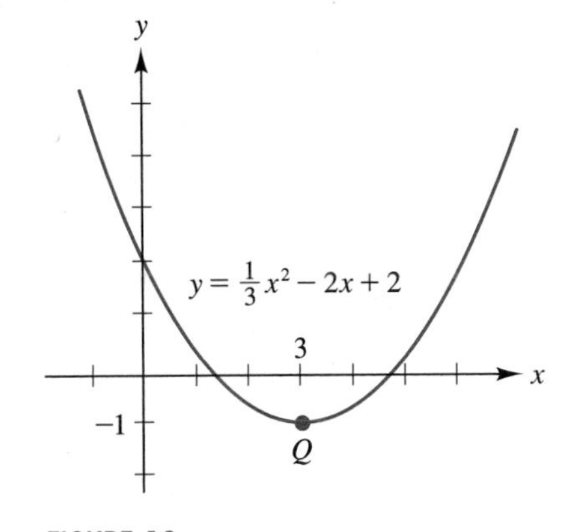

FIGURE 10
The point Q is a turning point. The minimum value of the function is -1. The function is decreasing on $(-\infty, 3)$ and increasing on $(3, \infty)$.

One of the most basic questions regarding a function is this: What are the highest and the lowest points on the graph? We will use Figures 9 and 10 to introduce some terminology that is useful here.

The points P and Q in Figures 9 and 10 are called *turning points*. At a **turning point,** the graph changes from rising to falling, or vice versa. In Figure 9, the highest point on the graph of the function G is $P(2, 102)$. We say that the *maximum value* of the function G is 102; and it occurs at $t = 2$. More generally, we say that $f(x_0)$ is the **maximum value** of a function f if the inequality $f(x_0) \geq f(x)$ holds for every x in the domain of f. Minimum values are defined similarly: $f(x_0)$ is the **minimum value** for a function f if the inequality $f(x_0) \leq f(x)$ holds for every x in the domain of f. Assuming that the domain of the function in Figure 9 is $[0, 6]$, the minimum value of the function G is 99, and it occurs when $t = 6$. In Figure 10, the minimum value of the function is -1, and it occurs when $x = 3$. Not every function has a maximum or minimum value. For example, the function $y = x^3$ possesses neither a maximum nor a minimum value. (Picture the graph.)

The function G in Figure 9 is said to be *increasing* on the open interval $(0, 2)$ and *decreasing* on the interval $(2, 6)$. In terms of the temperature interpretation for Figure 9, the patient's temperature is rising between noon and two o'clock and falling between two and six o'clock. In Figure 10, the function is decreasing on the interval $(-\infty, 3)$ and increasing on the interval $(3, \infty)$. For theoretical work, the terms increasing and decreasing can be defined in terms of inequalities. A function f is **increasing** on an interval if the following condition holds: If x_1 and x_2 are in the interval and $x_1 < x_2$, then $f(x_1) < f(x_2)$. Similarly, a function f is said to be **decreasing** on an interval if the following condition holds: If x_1 and x_2 are in the interval and $x_1 < x_2$, then $f(x_1) > f(x_2)$.

In the previous section we discussed difference quotients, and in this section we're working with graphs of functions. The following definition ties together these two concepts.

DEFINITION: The Average Rate of Change of a Function

Refer to Figure 11. The **average rate of change** of a function f on the interval $[a, b]$ is the slope of the line segment joining the two points $(a, f(a))$ and $(b, f(b))$. If $y = f(x)$, the average rate of change is denoted by $\Delta y/\Delta x$ or $\Delta f/\Delta x$, and we have

$$\frac{\Delta y}{\Delta x} = \frac{f(b) - f(a)}{b - a}$$

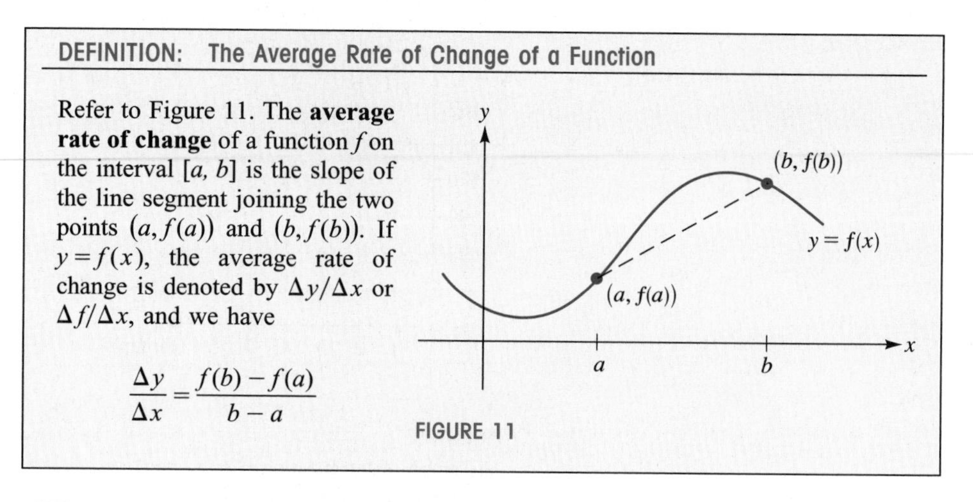

FIGURE 11

EXAMPLE 7 Compute and compare the average rates of change for the function $f(x) = x^2$ on the following intervals: **(a)** $[-2, 0]$; **(b)** $[0, 1]$; **(c)** $[1, 2]$.

Solution

(a) On $[-2, 0]$: $\dfrac{\Delta f}{\Delta x} = \dfrac{f(0) - f(-2)}{0 - (-2)} = \dfrac{0^2 - (-2)^2}{2} = -2$

(b) On $[0, 1]$: $\dfrac{\Delta f}{\Delta x} = \dfrac{f(1) - f(0)}{1 - 0} = \dfrac{1^2 - 0^2}{1} = 1$

(c) On $[1, 2]$: $\dfrac{\Delta f}{\Delta x} = \dfrac{f(2) - f(1)}{2 - 1} = \dfrac{2^2 - 1^2}{1} = 3$

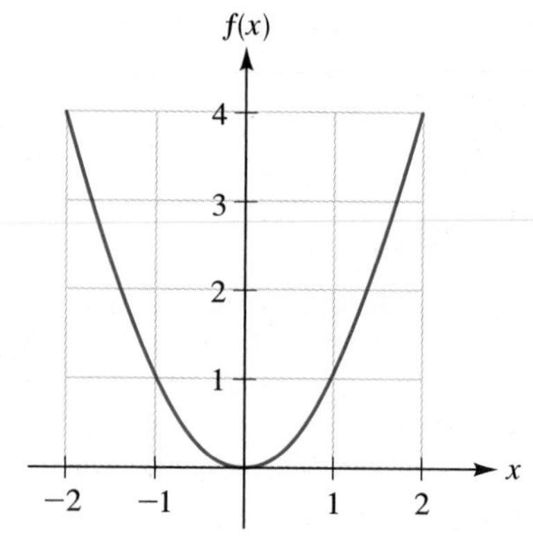

FIGURE 12
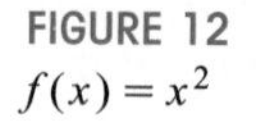
$f(x) = x^2$

(Refer to Figure 12.) On the interval $[-2, 0]$, the function is decreasing and the average rate of change is negative. On both of the intervals $[0, 1]$ and $[1, 2]$, the function is increasing and the average rates of change are positive. The graph rises more steeply on the interval $[1, 2]$ than on $[0, 1]$ and, correspondingly, the average rate of change is greater for the interval $[1, 2]$ than for $[0, 1]$.

A WORD OF CAUTION Like other types of averages occurring in arithmetic and statistics, the average rate of change can mask or hide some details. For instance, just because the average rate of change is positive on an interval, that does not guarantee that the function is increasing throughout that interval. For example, look back at Figure 11; for the function shown there, the average rate of change on $[a, b]$ is positive (because the slope of the dashed line is positive). The function, however, is increasing only on a portion of $[a, b]$, not throughout the entire interval.

In applications, it's informative to keep track of the units associated with the average rate of change. Consider, for instance, the patient's fever graph in Figure 9 (on page 117). Let's compute the average rate of change from $t = 2$ hr to $t = 6$ hr (that is, from 2 P.M. to 6 P.M.). We have

$$\frac{\Delta G}{\Delta t} = \frac{G(6) - G(2)}{6 - 2} = \frac{99° - 102°}{4 \text{ hr}} = \frac{-3°}{4 \text{ hr}} = -\frac{3}{4} \text{ degree/hr}$$

Interpretation: Between 2 P.M. and 6 P.M., the patient's temperature is dropping at an average rate of $3/4$ of a degree per hour.

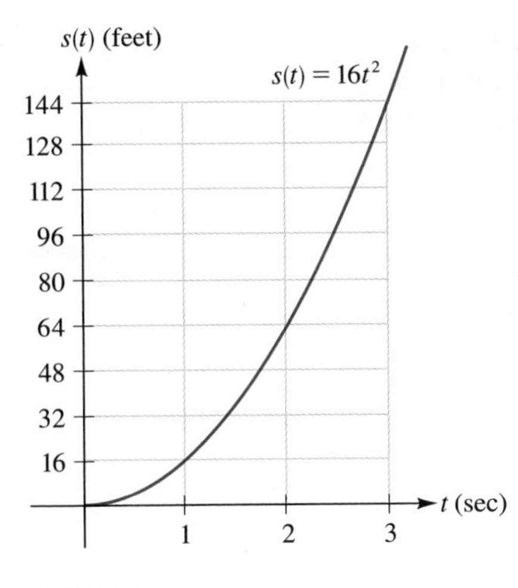

FIGURE 13

As another example, consider the function in Figure 13: $s(t) = 16t^2$, where $t \geq 0$. This function relates the distance $s(t)$ and the time t for an object falling in a vacuum. Here t is measured in seconds, $s(t)$ is in feet, and $t = 0$ corresponds to the instant that the object begins to fall. For instance, after 1 second, the object falls a distance of $s(1) = 16(1^2) = 16$ ft. And after 2 seconds, the total distance will be $s(2) = 16(2^2) = 64$ ft. To see how the units work here, let's calculate the average rate of change $\Delta s/\Delta t$ from $t = 1$ to $t = 3$. We have

$$\frac{\Delta s}{\Delta t} = \frac{s(3) - s(1)}{3 \text{ sec} - 1 \text{ sec}} = \frac{16(3^2) \text{ ft} - 16(1^2) \text{ ft}}{2 \text{ sec}} = \frac{128 \text{ ft}}{2 \text{ sec}} = 64 \text{ ft/sec}$$

Notice that the units here have the form *distance per unit time.* So in this case, $\Delta s/\Delta t$ gives us the *average velocity* of the object. More generally, whenever we have a function expressing distance $s(t)$ in terms of time t, the **average velocity** over an interval is defined to be the average rate of change $\Delta s/\Delta t$ over that interval.

EXAMPLE 8 Refer to the distance function in Figure 13, $s(t) = 16t^2$.

(a) Find a general expression for the average velocity $\Delta s/\Delta t$ over the interval $[t_1, t_2]$. (Assume that $t_1 \neq t_2$.)

(b) Use the result in part (a) to compute the average velocity over the following intervals: **(i)** $t = 1$ sec to $t = 2$ sec; **(ii)** $t = 1$ sec to $t = 1.5$ sec; **(iii)** $t = 1$ sec to $t = 1.1$ sec.

Solution

(a) By definition, the average velocity or average rate of change is

$$\begin{aligned} \frac{\Delta s}{\Delta t} &= \frac{s(t_2) - s(t_1)}{t_2 - t_1} \\ &= \frac{16t_2^2 - 16t_1^2}{t_2 - t_1} = \frac{16(t_2 - t_1)(t_2 + t_1)}{t_2 - t_1} && \text{factoring} \\ &= 16(t_2 + t_1) && \text{simplifying} \end{aligned}$$

So, we have $\Delta s/\Delta t = 16(t_2 + t_1)$. As explained previously, the units here are feet per second. Question: Where did we use the given condition $t_1 \neq t_2$?

(b) From $t = 1$ to $t = 2$: $\Delta s/\Delta t = 16(t_2 + t_1) = 16(2 + 1) = 48$ ft/sec
From $t = 1$ to $t = 1.5$: $\Delta s/\Delta t = 16(t_2 + t_1) = 16(1.5 + 1) = 40$ ft/sec
From $t = 1$ to $t = 1.1$: $\Delta s/\Delta t = 16(t_2 + t_1) = 16(1.1 + 1) = 33.6$ ft/sec

NOTE Exercise 59 asks you to carry out these calculations several steps further, using shorter and shorter time intervals. In calculus, this procedure leads to the concept of *instantaneous* (as opposed to *average*) *velocity.*

The definition and the examples that we've been using for the average rate of change have involved difference quotients of the form $\dfrac{f(b) - f(a)}{b - a}$. What about $\dfrac{f(x + h) - f(x)}{h}$, the other form of difference quotient that we worked with in the previous section? As indicated in Figure 14, this difference quotient also represents an average rate of change. In this case, it represents the average rate of change for a function f on the interval $[x, x + h]$.

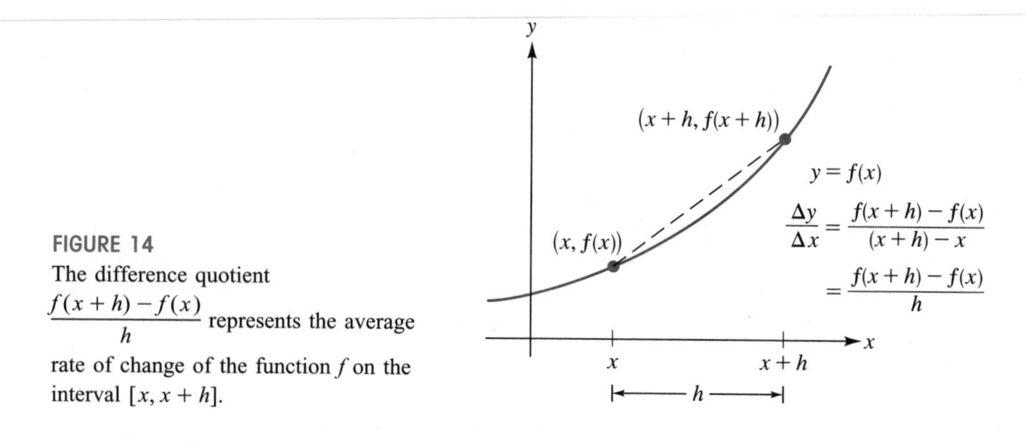

FIGURE 14
The difference quotient $\frac{f(x+h)-f(x)}{h}$ represents the average rate of change of the function f on the interval $[x, x + h]$.

EXERCISE SET 3.2

9, 11, 13, 17, 25

A

In Exercises 1–4, specify the y-coordinate of the point P on the given function. In each case, give an exact expression and also a calculator approximation rounded to three decimal places.

1.

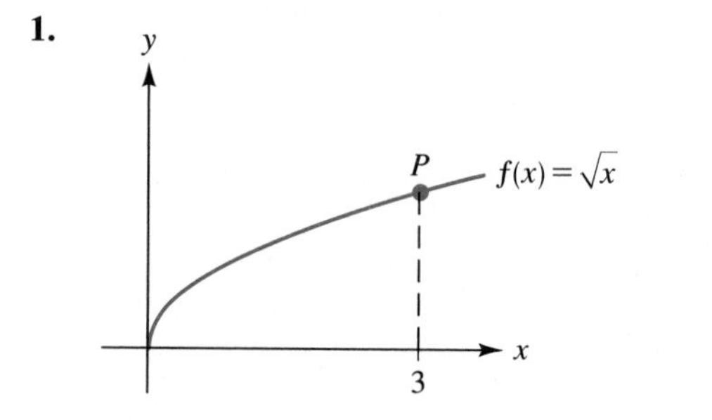

2. **3.**

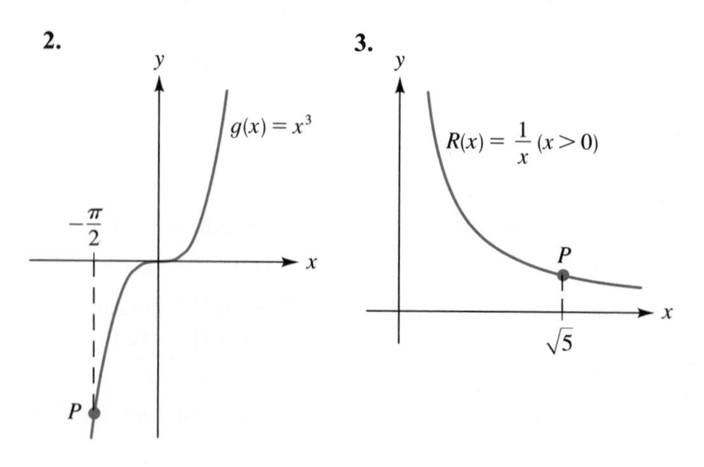

4.

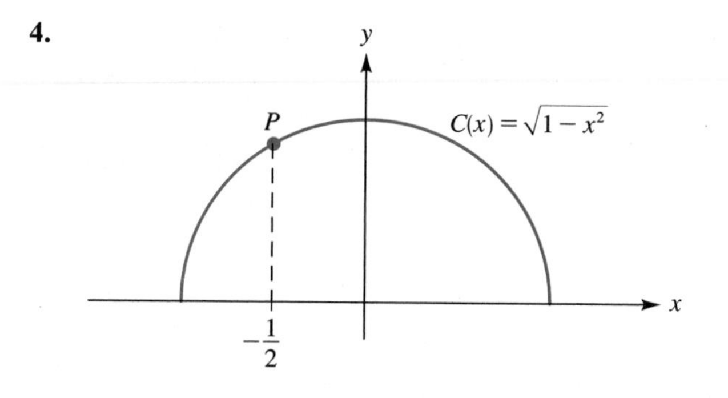

In Exercises 5–8, determine the coordinates of the points P, Q, and R in each figure; give an exact expression and also a calculator approximation rounded to three decimal places. Assume that each dashed line is parallel to one of the coordinate axes. (In Exercise 8, note that the line is $y = x/2$ rather than $y = x$.)

5.

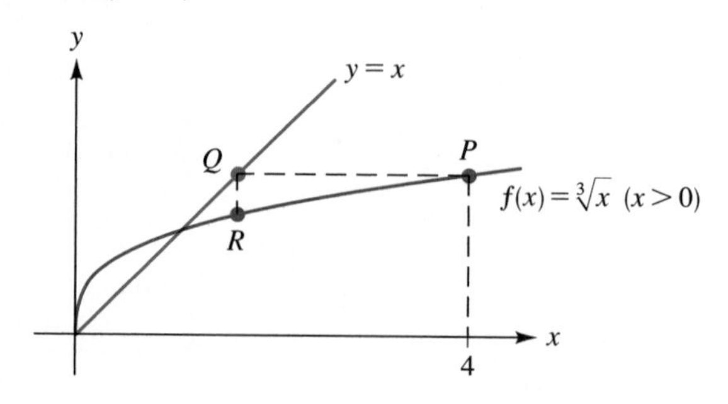

6.

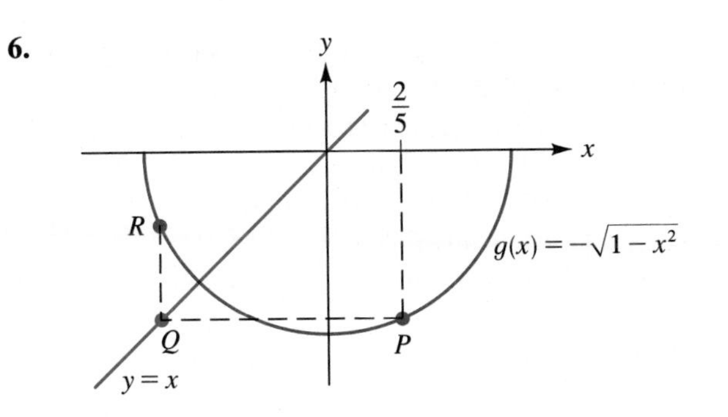

7.

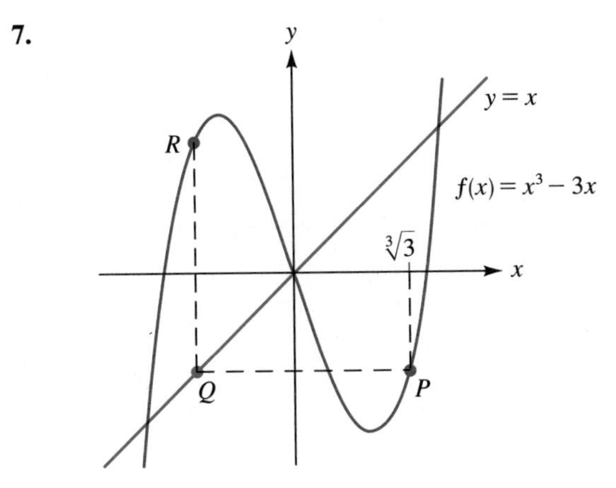

8.

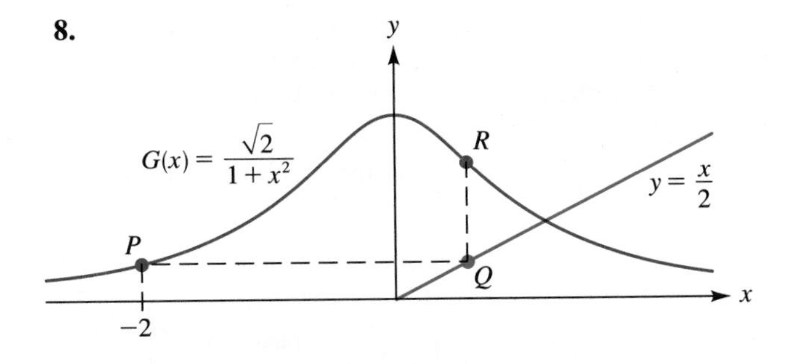

In Exercises 9–16, the graph of a function is given. In each case, specify the domain and the range of the function. (The axes are marked off in one-unit intervals.)

9.

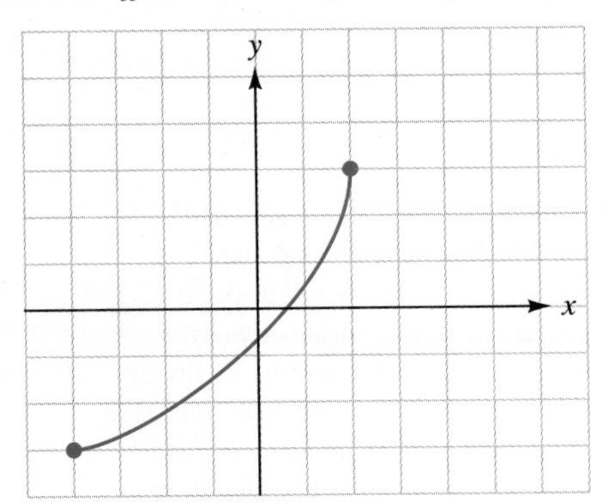

10.

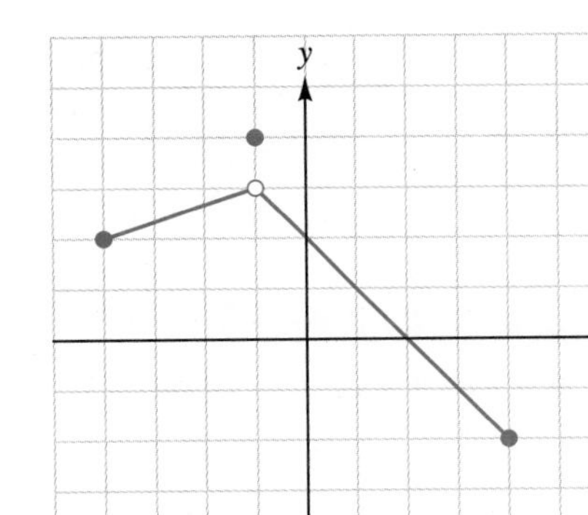

11.

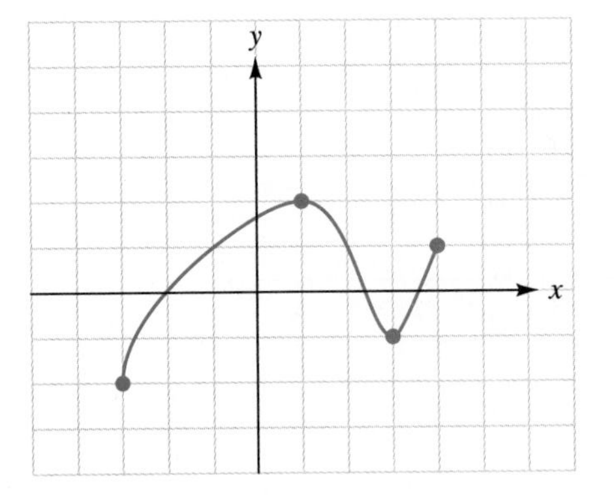

12.

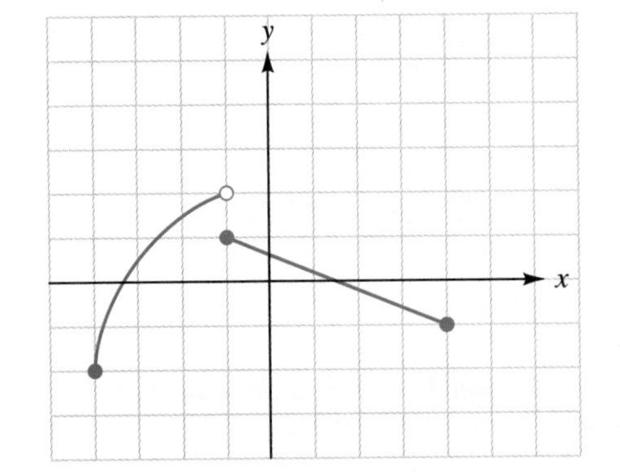

13.

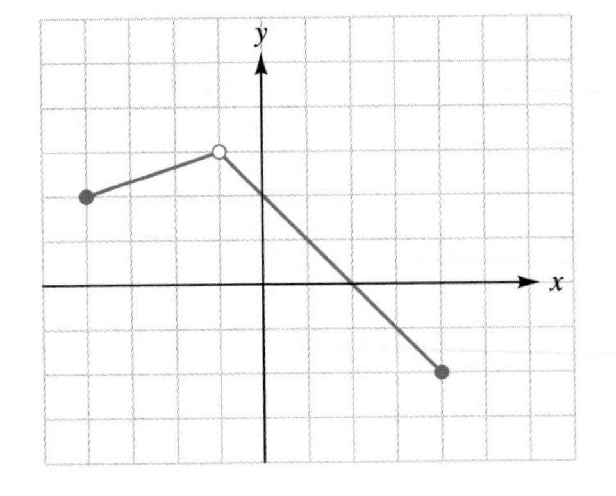

14.

15.

16.

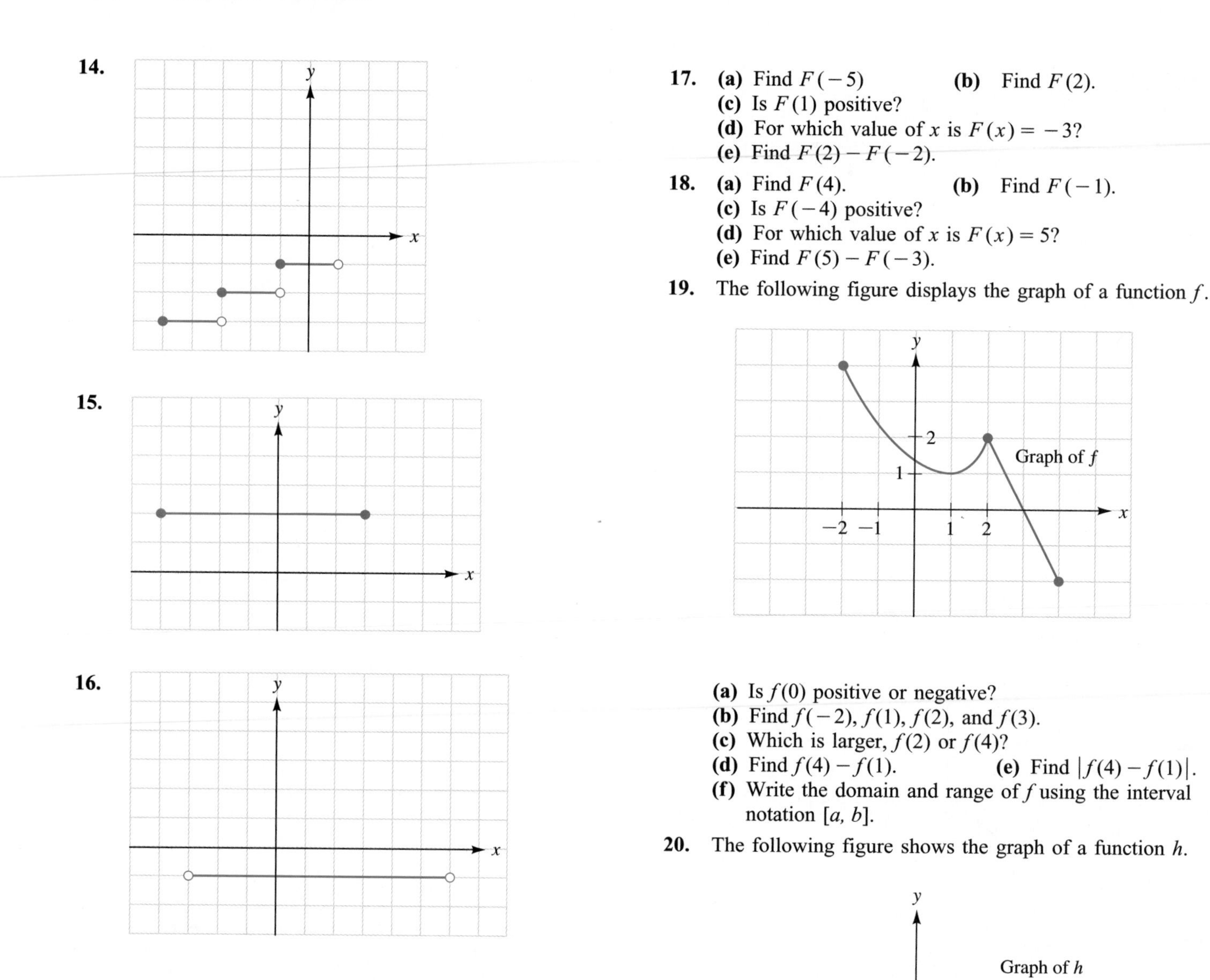

In Exercises 17 and 18, refer to the graph of the function F in the figure. (Assume that the axes are marked off in one-unit intervals.)

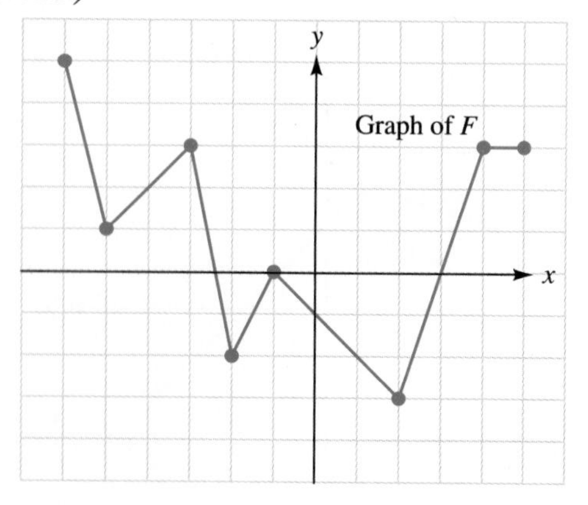

17. **(a)** Find $F(-5)$ **(b)** Find $F(2)$.
(c) Is $F(1)$ positive?
(d) For which value of x is $F(x) = -3$?
(e) Find $F(2) - F(-2)$.

18. **(a)** Find $F(4)$. **(b)** Find $F(-1)$.
(c) Is $F(-4)$ positive?
(d) For which value of x is $F(x) = 5$?
(e) Find $F(5) - F(-3)$.

19. The following figure displays the graph of a function f.

(a) Is $f(0)$ positive or negative?
(b) Find $f(-2)$, $f(1)$, $f(2)$, and $f(3)$.
(c) Which is larger, $f(2)$ or $f(4)$?
(d) Find $f(4) - f(1)$. **(e)** Find $|f(4) - f(1)|$.
(f) Write the domain and range of f using the interval notation $[a, b]$.

20. The following figure shows the graph of a function h.

(a) Find $h(a)$, $h(b)$, $h(c)$, and $h(d)$.
(b) Is $h(0)$ positive or negative?
(c) For which values of x does $h(x) = 0$?
(d) Which is larger, $h(b)$ or $h(0)$?
(e) As x increases from c to d, do the corresponding values of $h(x)$ increase or decrease?
(f) As x increases from a to b, do the corresponding values of $h(x)$ increase or decrease?

In Exercises 21 and 22, refer to the graphs of the functions f and g in the figure. Assume that the domain of each function is $[-3, 3]$ *and that the axes are marked off in one-unit intervals.*

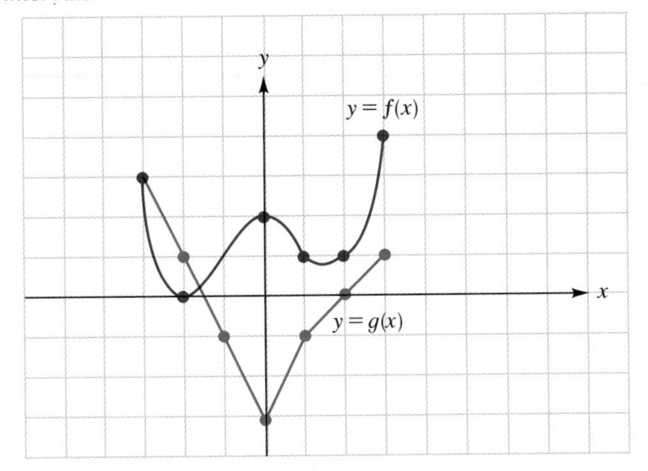

21. **(a)** Which is larger, $f(-2)$ or $g(-2)$?
(b) Compute $f(0) - g(0)$.
(c) Which among the following three quantities is the smallest?

$$f(1) - g(1) \qquad f(2) - g(2) \qquad f(3) - g(3)$$

(d) For which value(s) of x does $g(x) = f(1)$?
(e) Is the number 4 in the range of f or in the range of g?

22. **(a)** For the interval $[0, 3]$, is the quantity $g(x) - f(x)$ positive or negative?
(b) For the interval $(-3, -2)$, is the quantity $g(x) - f(x)$ positive or negative?
(c) Compute $\dfrac{f(x) - f(2)}{x - 2}$ when $x = 3$.
(d) Compute $\dfrac{g(x) - g(-2)}{x + 2}$ when $x = -3$.

23. Complete the following table.

Function	$\lvert x \rvert$	x^2	x^3
Domain			
Range			
Turning point			
Maximum value			
Minimum value			
Interval(s) where increasing			
Interval(s) where decreasing			

24. Set up and complete a table like the one in Exercise 23 for the three functions $1/x$, $\sqrt{x}$, and $\sqrt{1 - x^2}$.

In Exercises 25–28, you are given functions with domain $[0, 4]$. *Specify:* **(a)** *the range of each function;* **(b)** *the maximum value of the function;* **(c)** *the minimum value of the function;* **(d)** *interval(s) where the function is increasing; and* **(e)** *interval(s) where the function is decreasing.*

25.

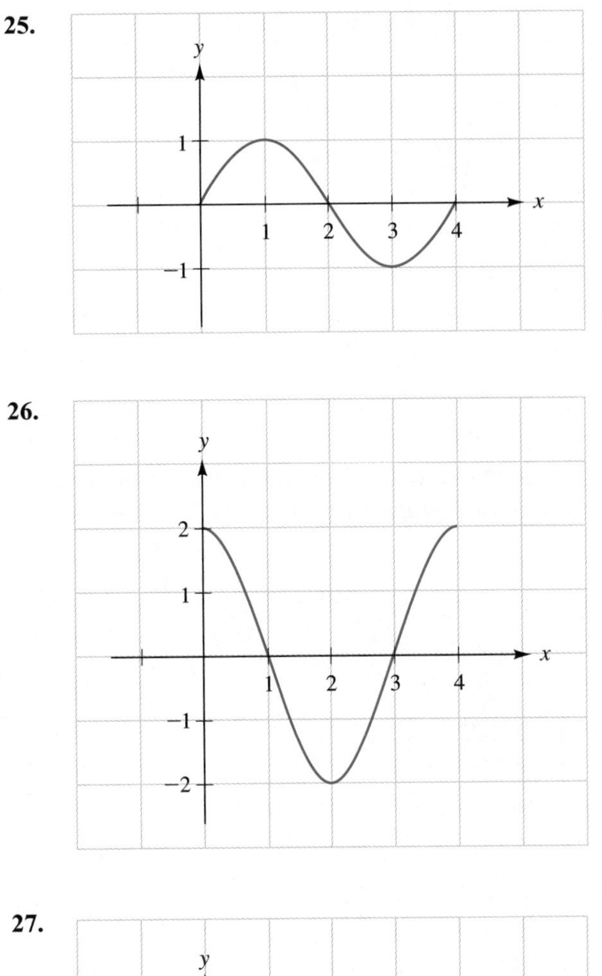

26.

27.

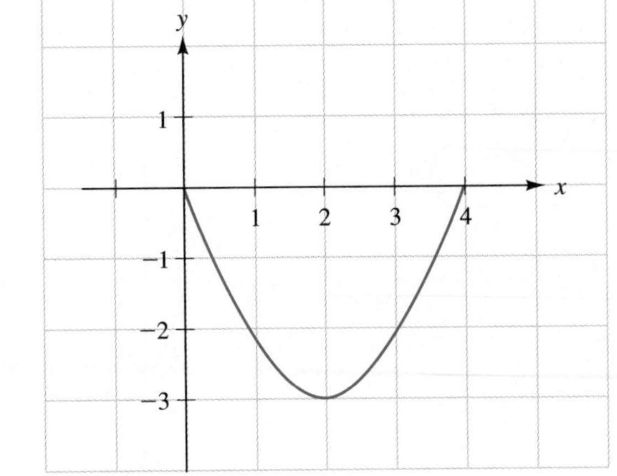

28.

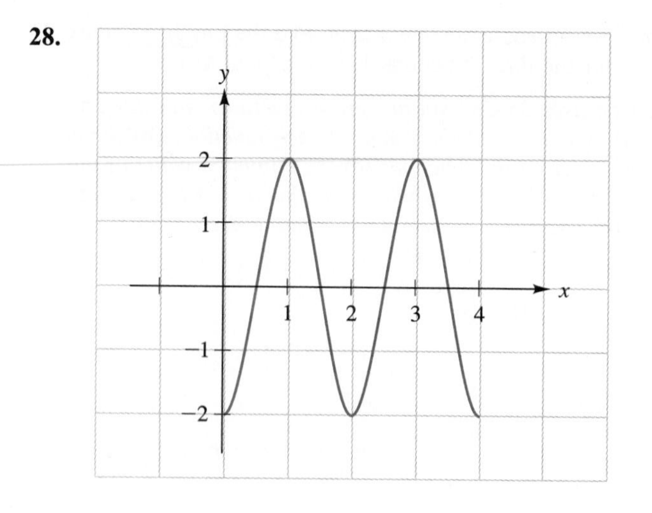

30.

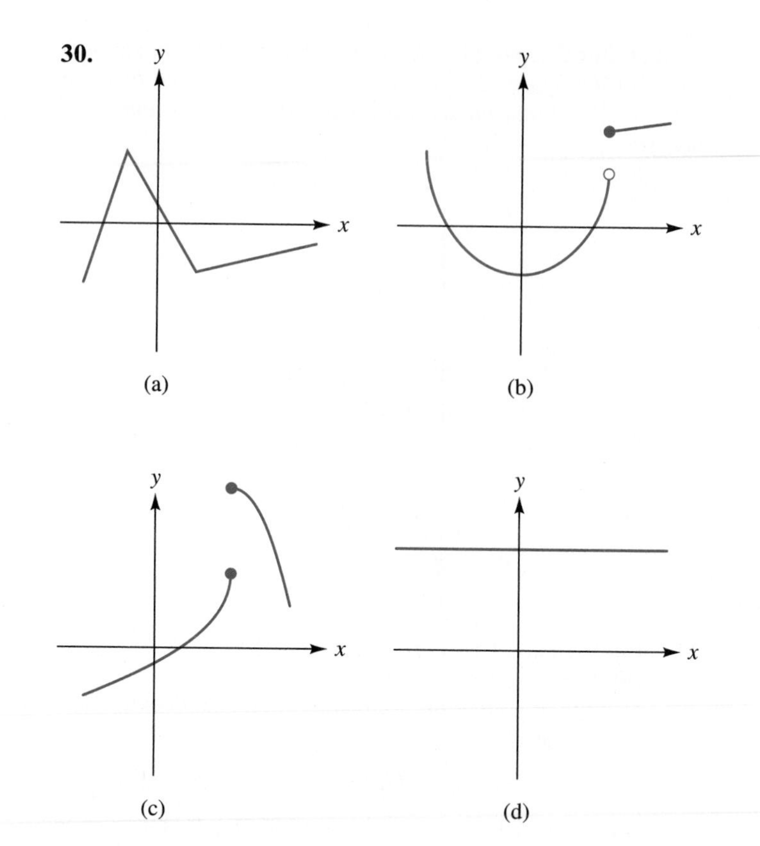

In Exercises 29 and 30, use the vertical line test to determine whether each graph represents a function $y = f(x)$.

29.

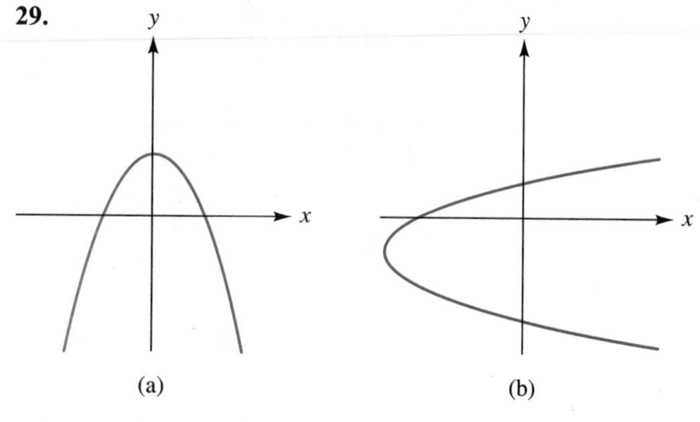

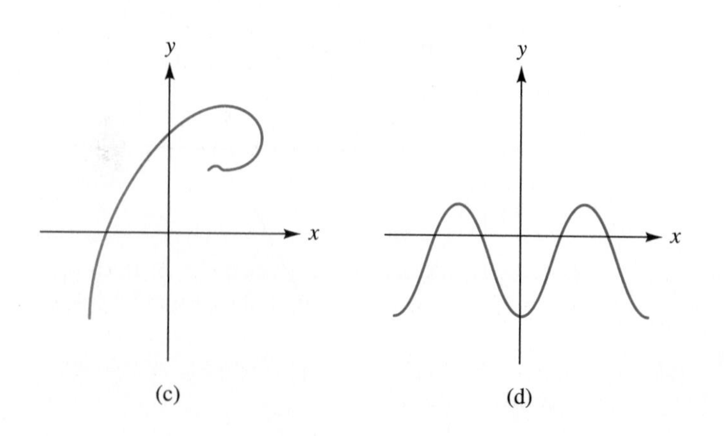

In Exercises 31–40, graph the function defined by the given rules.

31. $A(x) = \begin{cases} x^3 & \text{if } -2 \leqslant x \leqslant -1 \\ x^2 & \text{if } x > -1 \end{cases}$

32. $B(x) = \begin{cases} \sqrt{1 - x^2} & \text{if } -1 \leqslant x < 1 \\ 1/x & \text{if } x \geqslant 1 \end{cases}$

33. $C(x) = \begin{cases} x^3 & \text{if } x < 1 \\ \sqrt{x} & \text{if } x > 1 \end{cases}$

34. **(a)** $f(x) = \begin{cases} x^2/|x| & \text{if } x \neq 0 \\ 0 & \text{if } x = 0 \end{cases}$

(b) $F(x) = \begin{cases} x^2/|x| & \text{if } x \neq 0 \\ 1 & \text{if } x = 0 \end{cases}$

35. **(a)** $g(x) = \begin{cases} x/|x| & \text{if } x \neq 0 \\ 0 & \text{if } x = 0 \end{cases}$

(b) $G(x) = \begin{cases} x/|x| & \text{if } x \neq 0 \\ 1 & \text{if } x = 0 \end{cases}$

36. $U(x) = \begin{cases} 1 & \text{if } x \leq -2 \\ -1 & \text{if } x > -2 \end{cases}$

37. $f(x) = \begin{cases} 1/x & \text{if } x < -1 \\ x & \text{if } -1 \leq x \leq 1 \\ 1/x & \text{if } x > 1 \end{cases}$

38. $g(x) = \begin{cases} 1/x & \text{if } x < -\frac{1}{2} \\ 1 & \text{if } -\frac{1}{2} \leq x \leq 1 \\ x^3 & \text{if } x > 1 \end{cases}$

39. $y = \begin{cases} x & \text{if } 0 \leq x < 1 \\ x-1 & \text{if } 1 \leq x < 2 \\ x-2 & \text{if } 2 \leq x < 3 \end{cases}$

40. $T(x) = \begin{cases} 2x+2 & \text{if } -1 \leq x < 0 \\ -2x+2 & \text{if } 0 \leq x < 1 \\ 2x-2 & \text{if } 1 \leq x < 2 \\ -2x+6 & \text{if } 2 \leq x \leq 3 \end{cases}$

41. Specify the domain and the range of the function whose graph is the horizontal line $y = 3$.

42. Specify the domain and the range of the function whose graph is the horizontal line $y = -\sqrt{2}$.

43. Let $f(x) = x^2$. Find the slope of the line passing through the points $(3, f(3))$ and $(4, f(4))$. What does this have to do with the average rate of change of f on the interval [3, 4]?

44. Let $g(x) = 1/x$. Find the slope of the line passing through the points $(2, g(2))$ and $(6, g(6))$. What does this have to do with the average rate of change of g on the interval [2, 6]?

In Exercises 45–54, compute the average rate of change of the function on the given interval. In Exercises 51–54, simplify your answer as much as possible.

45. $f(x) = x^2 + 2x$ on [3, 5]
46. $f(x) = \sqrt{x}$ on [4, 9]
47. $g(x) = 2x^2 - 4x$ on $[-1, 3]$
48. $g(x) = x^3 - x$ on [1, 2]
49. $h(t) = 2t - 6$ on [5, 12]
50. $h(t) = 16 - 7t$ on $[-\sqrt{5}, \sqrt{2}]$
51. $f(x) = 3/x$ on $[a, b]$
52. $f(x) = 3x^2$ on $[a, b]$
53. $F(x) = -2x^3$ on $[a, b]$
54. $F(x) = 4/(x-1)$ on $[a, b]$
55. The following graph shows the temperature $G(t)$ of a solution during the first 8 minutes of a chemistry experiment.

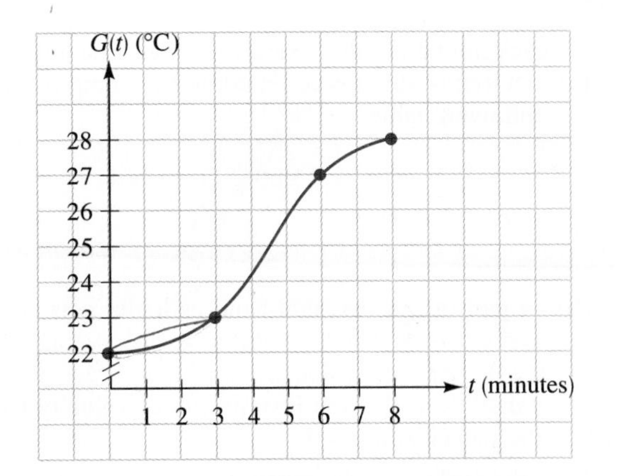

Compute the average rate of change of temperature, $\Delta G/\Delta t$, over the following intervals. (Be sure to specify the units as part of each answer.)
(a) $t = 0$ min to $t = 3$ min **(b)** $t = 3$ min to $t = 6$ min
(c) $t = 6$ min to $t = 8$ min

56. Iodine-131 is a radioactive substance. The accompanying graph shows how an initial one-gram sample decays over a 32-day period.

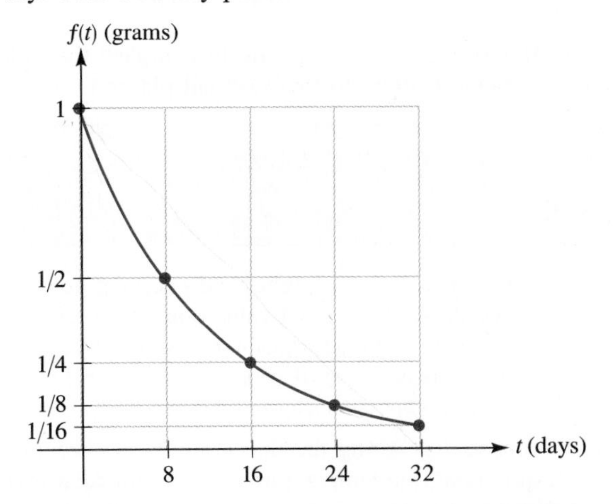

(a) Compute $\Delta f/\Delta t$ over each of the following intervals: $t = 0$ days to $t = 8$ days; $t = 8$ days to $t = 16$ days; $t = 16$ days to $t = 24$ days; $t = 24$ days to $t = 32$ days. (Be sure to specify the units with each answer.)
(b) Find the average of the four answers in part (a).
(c) Compute $\Delta f/\Delta t$ over the interval from $t = 0$ to $t = 32$. Is the answer the same as that obtained in part (b)?

B

57. Let $f(x) = 1/x$. Find a number b such that the average rate of change of the function f on the interval $[1, b]$ is $-1/5$.

58. Let $f(x) = \sqrt{x}$. Find a number b such that $\Delta f/\Delta x$ on the interval $[1, b]$ is $1/7$.

59. As background for this exercise, reread Example 8.
(a) Let $s(t) = 16t^2$ (as in Example 8). Complete the following table.

Interval	[1, 1.1]	[1, 1.01]	[1, 1.001]	[1, 1.0001]	[1, 1.00001]
$\Delta s/\Delta t$					

(b) In part (a), as the right-hand endpoint gets closer and closer to 1, what value does the average velocity $\Delta s/\Delta t$ seem to be approaching? This "target value" is called the **instantaneous velocity** (or just the **velocity**) at $t = 1$.

60. Near the surface of the moon, the distance traveled by a falling object is given by the formula $s(t) = \frac{8}{3}t^2$, where t is measured in seconds, $s(t)$ is in feet, and $t = 0$ corresponds to the instant that the object begins to fall.
(a) Show that the average velocity over the interval $[t_1, t_2]$ is given by

$$\frac{\Delta s}{\Delta t} = \frac{8}{3}(t_2 + t_1)$$

(b) Use the result in part (a) to complete the following table. (Round to four decimal places.)

Interval	[2, 2.1]	[2, 2.01]	[2, 2.001]	[2, 2.0001]	[2, 2.00001]
$\Delta s/\Delta t$					

(c) In part (b), as the right-hand endpoint gets closer and closer to 2, what value does the average velocity seem to be approaching? (As mentioned in the previous exercise, this target value is called the *instantaneous velocity* (or *velocity*) when $t = 2$.)

61. Suppose that during the first 4 hours of a laboratory experiment, the temperature of a certain substance is given by the formula $f(t) = t^3 - 6t^2 + 9t$, where t is measured in hours, with $t = 0$ corresponding to the time the experiment begins, and $f(t)$ is the temperature (degrees Fahrenheit) of the substance at time t hours. Calculate the average rate of change of temperature between the following times.
(a) $t = 0$ and $t = 1$ **(b)** $t = 1$ and $t = 2$
(c) $t = 0$ and $t = 3$ **(d)** $t = 0$ and $t = 4$

62. **(a)** Let f be the temperature function given in Exercise 61. Compute $\Delta f/\Delta t$ on the interval $[0, t_1]$.
(b) Use the result in part (a) to complete the following table. (Round to four decimal places.)

Interval	[0, 0.1]	[0, 0.01]	[0, 0.001]	[0, 0.0001]
$\Delta f/\Delta t$				

(c) In part (b), as the right-hand endpoint gets closer and closer to 0, what value does the average rate of change of temperature seem to be approaching? (This target value tells us the rate at which the temperature is changing at the instant that the experiment begins.)

63. The accompanying graph and table show gasoline consumption in the United States over the years 1982–1992.

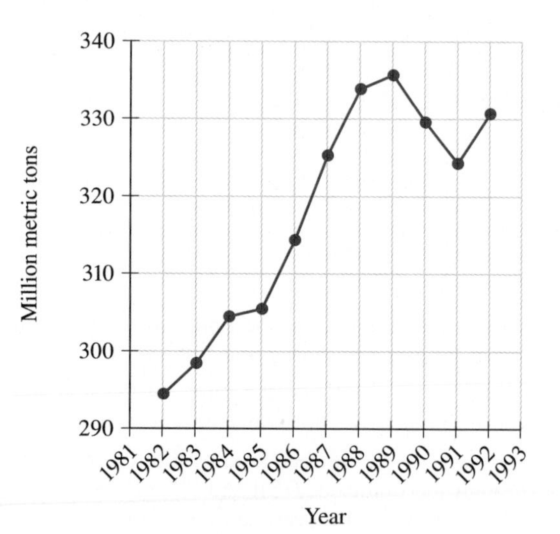

Year	Gasoline Consumption (million metric tons)
1982	294.4
1983	298.4
1984	304.4
1985	305.4
1986	314.3
1987	325.2
1988	333.8
1989	335.6
1990	329.5
1991	324.2
1992	330.6

Gasoline Consumption in the U.S., 1982–1992. Source: *BP Statistical Review of World Energy, 1993*

(a) According to the graph, is the average rate of change of consumption greater over the period 1983–1984 or over the period 1984–1985?
(b) Use the values in the table to compute the average rate of change of consumption for each of the periods mentioned in part (a). Check that your results are consistent with your answer in part (a).
(c) According to the graph, is the average rate of change of consumption positive or negative over the period 1989–1991?

(d) Use the values in the table to compute the average rate of change of consumption over the period 1989–1991. Check that the result is consistent with your answer in part (c).

(e) Compute the average rates of change of consumption over the periods 1982–1992 and 1991–1992. Which is greater?

64. The accompanying graph and table show the number of U.S. households with video cassette recorders (VCRs) over the years 1982–1992. Let y denote the number of households with VCRs in the year t. (So, for example, according to the first entry in the table, if $t = 1978$, then $y = 200{,}000$ VCRs.)

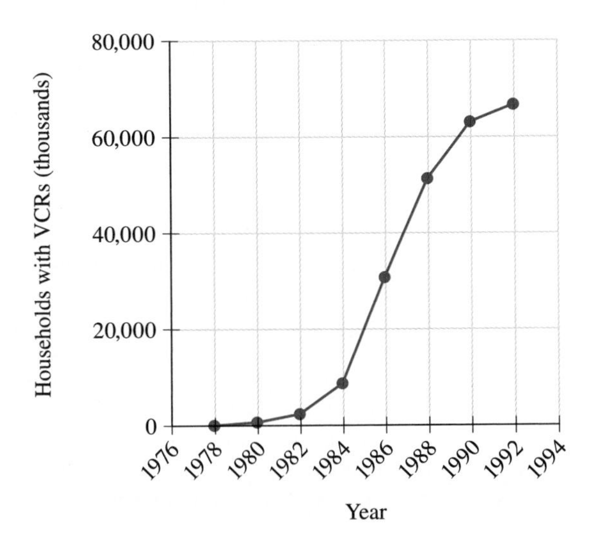

Year	Households with VCRs (thousands)
1978	200
1980	840
1982	2530
1984	8880
1986	30920
1988	51390
1990	63180
1992	66776

Number of U.S. Households with Video Cassette Recorders, 1978–1992.
Source: *Nielsen Media Research,* 1993

(a) According to the graph, is $\Delta y/\Delta t$ greater over the period 1988–1990 or over the period 1990–1992?

(b) Use the values in the table to compute $\Delta y/\Delta t$ for each of the periods mentioned in part (a). (Round to four decimal points.) Use the results to check your answer in part (a).

(c) Use the graph to decide whether the number of households with VCRs is growing faster over the period 1984–1986 or over 1978–1992. *Hint:* For the period 1978–1992, use a ruler to draw a line segment connecting the appropriate points on the graph. Then compare slopes.

(d) Use the table to compute $\Delta y/\Delta t$ over each of the two periods mentioned in part (c). Use the results to check your answer in part (c).

65. (As background for this exercise, you need to have read the discussion following Example 8 in the text.) Let $f(x) = x^2$.

(a) Find and simplify a general expression for the average rate of change, $\dfrac{f(x+h) - f(x)}{h}$.

(b) Use the result in part (a) to complete the following tables.

h	2	1	0.1	0.01	0.001	0.0001
$\Delta f/\Delta x$ on the interval $[1, 1+h]$						

h	-2	-1	-0.1	-0.01	-0.001	-0.0001
$\Delta f/\Delta x$ on the interval $[1, 1+h]$						

(c) Based on the results in part (b), what value would you say $\Delta f/\Delta x$ is approaching as h approaches zero? This "target value" is called the instantaneous rate of change of $f(x) = x^2$ at $x = 1$.

(d) The accompanying figure shows the graph of $f(x) = x^2$ and the line that is tangent to the curve at $x = 1$. The slope of this tangent line is defined as the instantaneous rate of change of $f(x) = x^2$ at $x = 1$. [You found this number in part (c).] Use the point–slope formula to find the equation of this tangent line. Write the final answer in the form $y = mx + b$.

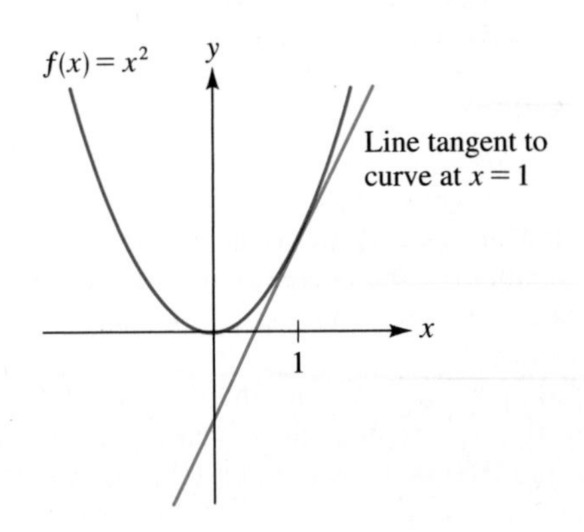

66. Let $f(x) = 1/x$.

(a) Find and simplify a general expression for the average rate of change $\Delta f/\Delta x$ on the interval $[2, 2+h]$.

(b) Use the result in part (a) to complete the following tables.

h	1	0.1	0.01	0.001	0.0001
$\Delta f/\Delta x$ on the interval $[2, 2+h]$					

h	-1	-0.1	-0.01	-0.001	-0.0001
$\Delta f/\Delta x$ on the interval $[2, 2+h]$					

(c) Based on the results in part (b), what value would you say $\Delta f/\Delta x$ is approaching as h approaches zero? This "target value" is called the instantaneous rate of change of $f(x) = 1/x$ at $x = 2$.

(d) The accompanying figure shows the graph of $f(x) = 1/x$ $(x > 0)$ and the line that is tangent to the curve at $x = 2$. The slope of this tangent line is defined as the instantaneous rate of change of $f(x) = 1/x$ at $x = 2$. [You found this number in part (c).] Use the point–slope formula to find the equation of this tangent line. Write the final answer in the form $y = mx + b$.

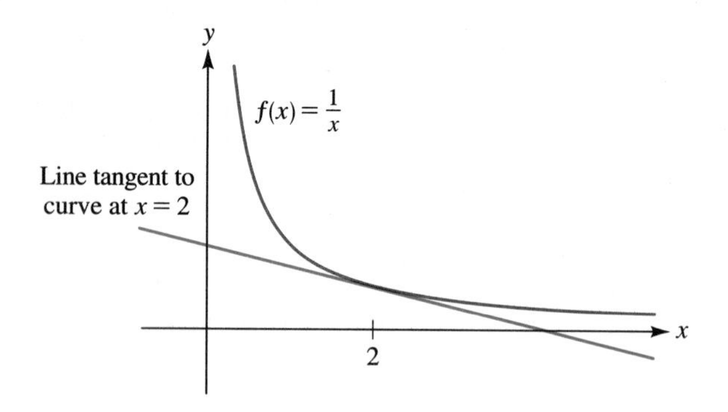

(e) Find the area of the triangle in the first quadrant bounded by the tangent line and the coordinate axes.

67. A bug travels counterclockwise around the square in the figure, beginning from the point (1, 0). As indicated in the figure, t denotes the distance the bug has traveled, and $P(t)$ denotes the bug's location when it has traveled the distance t. For example, when $t = 0$, $P(t)$ is the point (1, 0), and when $t = 1$, $P(t)$ is the point (1, 1).

(a) A function S is defined as follows: $S(t)$ is the y-coordinate of the point $P(t)$. For example, if $t = 1/2$, then $P(t)$ is the point (1, 1/2) and therefore $S(t) = 1/2$. Set up a table showing the values of $S(t)$ with t running from 0 to 3 in increments of 1/4. Then graph the function S on the interval $0 \leq t \leq 3$.

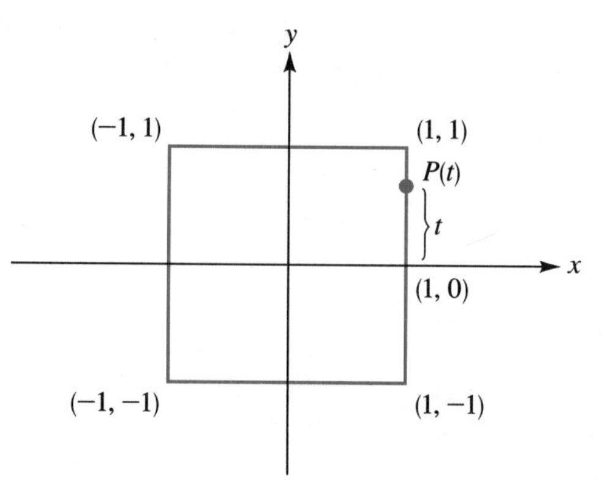

(b) Set up a table showing the values of $S(t)$ with t running from 3 to 5 in increments of 1/4. Then graph the function S on the interval $3 \leq t \leq 5$.

(c) Graph the function S for $0 \leq t \leq 8$. This corresponds to one complete trip around the square by the bug.

(d) Graph the function S for $8 \leq t \leq 16$. How does your result relate to the graph in part (c)?

68. Let $P(t)$ be defined as in Exercise 67. Define a function C as follows: $C(t)$ is the x-coordinate of the point $P(t)$. Graph the function C for $0 \leq t \leq 8$.

69. The notation $[x]$ denotes the greatest integer that is less than or equal to x. For instance, $[3] = 3$; $[4\frac{1}{2}] = 4$; and $[-4/3] = -2$.

(a) Complete the following table.

x	0	0.1	0.5	0.9	1.0
$[x]$					

(b) Graph the **greatest integer function** $y = [x]$ on the interval $-2 \leq x < 3$.

(c) The domain of the greatest integer function is $(-\infty, \infty)$. What is the range?

70. (Continuation of Exercise 69) Set up tables and graph each of the following for $-2 \leq x \leq 2$: **(a)** $y = ([x])^2$; **(b)** $y = [x^2]$.

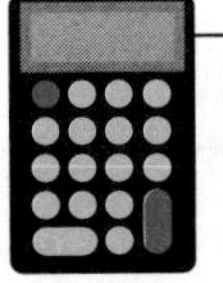

GRAPHING UTILITY EXERCISES FOR SECTION 3.2

In general, the techniques of calculus are required to obtain the precise coordinates of the turning points of a function. In Exercises 1–8, you'll find the approximate coordinates of turning points using a graphing utility.

1. Let $f(x) = x^3 - x + 1$.
(a) Graph this function in the standard viewing rectangle. How many turning points do there appear to be?
(b) For a better view of the turning points, graph the function using a viewing rectangle that extends from -2 to 2 in both the x- and the y-directions.
(c) Zoom in repeatedly on each turning point, estimating its x-coordinate at each step. Continue this process for each turning point until you are certain about the first three decimal places in the x-coordinate. (If you are using a graphics calculator such as a Texas Instruments or a Casio, the TRACE key is useful here.)
(d) Using calculus, it can be shown that the precise x-coordinates of the turning points are $\pm 1/\sqrt{3}$: use your calculator to evaluate these numbers. Compare your best estimates in part (c) with these exact x-coordinates. To how many decimal places do they agree?
(e) Use the graph in part (a) and the information given in part (d) to specify the intervals on which f is increasing and the intervals on which f is decreasing.

2. Let $f(x) = x^4 - x + 1$.
(a) Graph this function in the standard viewing rectangle and note that there appears to be only one turning point. Then check that there are no surprises "out of camera range" by looking at the graph in a viewing rectangle that extends from -30 to 30 in both the x- and the y-directions.
(b) Zoom in repeatedly on the turning point, estimating its x-coordinate at each step. Continue the process until you are certain about the first three decimal places in the x-coordinate.
(c) Using calculus, it can be shown that the precise x-coordinate of the turning point is $\sqrt[3]{0.25}$; use your calculator to evaluate this number. Compare your best estimate in part (c) with this exact x-coordinate. To how many decimal places do they agree?
(d) Use the graph in part (a) and the information given in part (c) to specify the interval on which f is increasing and the interval on which f is decreasing.

In Exercises 3–8, follow a procedure similar to that used in Exercises 1 and 2 to estimate the turning points and to specify where the function is increasing and where it is decreasing. The values given after each function are the exact x-coordinates of the turning points, obtained using calculus.

3. $f(x) = x^3(x-2)^2$; $x = 2$; $x = 6/5$

4. $f(x) = x/(2x^2+1)$; $x = \pm\sqrt{2}/2$

5. $f(x) = x/(2x^3+1)$; $x = \sqrt[3]{0.25}$

6. $f(x) = x^3 - 3x^2 - 12x - 2$; $x = 1 \pm \sqrt{5}$

7. $y = x + \dfrac{1}{2x^3}$; $x = \pm\sqrt[4]{1.5}$

8. $y = x^2 + \dfrac{1}{2x^3}$; $x = \sqrt[5]{0.75}$

9. (a) What is the domain of the function $f(x) = \sqrt[3]{x}$? (Answer this without relying on a graph.)
(b) Use your graphing utility to graph the function $f(x) = \sqrt[3]{x}$. If your graph is not consistent with the answer that you gave in part (a), try entering the function as

$$f(x) = \begin{cases} \sqrt[3]{x} & \text{if } x \geq 0 \\ -\sqrt[3]{-x} & \text{if } x < 0 \end{cases}$$

(This should get you around a shortcoming of some graphing utilities: they require nonnegative inputs in computing roots, even cube roots.)

10. Follow the instructions preceding Exercise 3 for each of the following functions. (As background for this problem, you need to have worked Exercise 9.)
(a) $y = \sqrt[3]{x} - x^2$; $x = 1/\sqrt[5]{216}$; $x = 0$
(b) $y = \sqrt[3]{x} - x$; $x = \pm\sqrt{3}/9$; $x = 0$

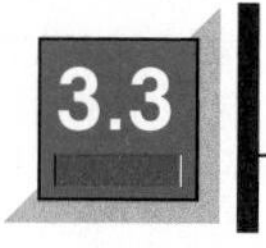

3.3 TECHNIQUES IN GRAPHING

. . . geometrical figures are graphic formulas.

David Hilbert (1862–1943)

The simple geometric concepts of reflection and translation can be used to great advantage in graphing. We discussed the idea of reflection in the x-axis and in the y-axis in Section 2.4. By a **translation** of a graph, we mean a shift in its location such that every point of the graph is moved the same distance in the same direction. The size and the shape of a graph are unchanged by a translation.

EXAMPLE 1 The graph of a function G is the quarter-circle shown in Figure 1. In each case, sketch the resulting graph after the following operations are carried out on the graph of G:

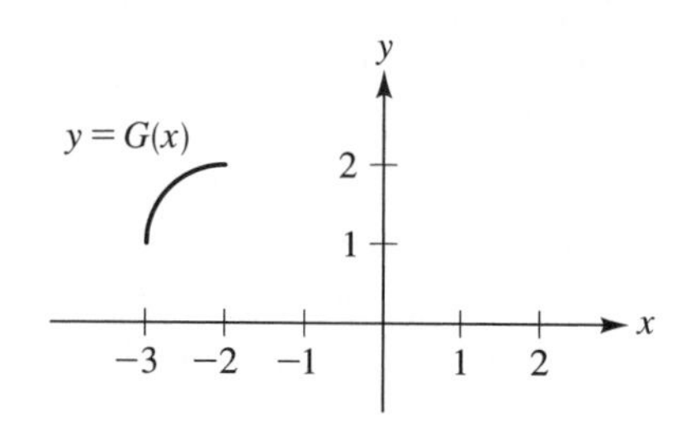

FIGURE 1

(a) a translation of four units to the right followed by a translation of one unit vertically downward;
(b) a translation of three units to the right followed by a reflection in the y-axis;
(c) a reflection in the y-axis followed by a translation of three units to the right. [Note that these are the same two operations used in part (b), but here the order is reversed.]

Solution
(a) Figure 2 shows the result of translating the graph of G four units to the right and then one unit down. (As you can check by drawing a sketch, the same end result is obtained if we first translate one unit down and then four units to the right.) *Fact:* If several translations are to be carried out in succession, the end result will be the same, no matter in what order the individual translations are carried out.

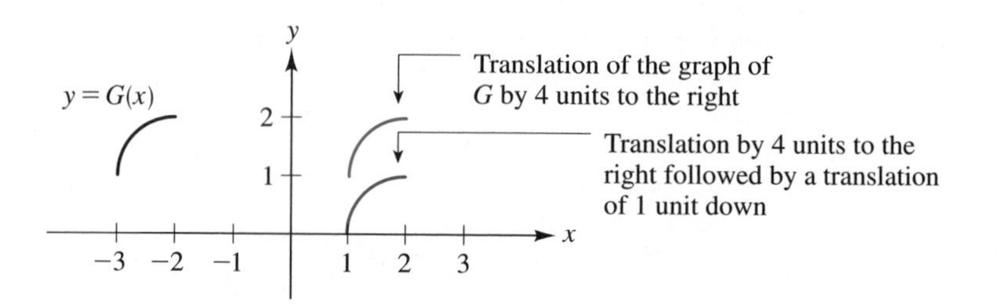

FIGURE 2
The red curve is the end result when the graph of G is translated four units to the right and then one unit down.

(b) Figure 3(a) shows the graph of G after a translation of three units to the right. When this is followed by a reflection in the y-axis, we obtain the result shown in Figure 3(b).

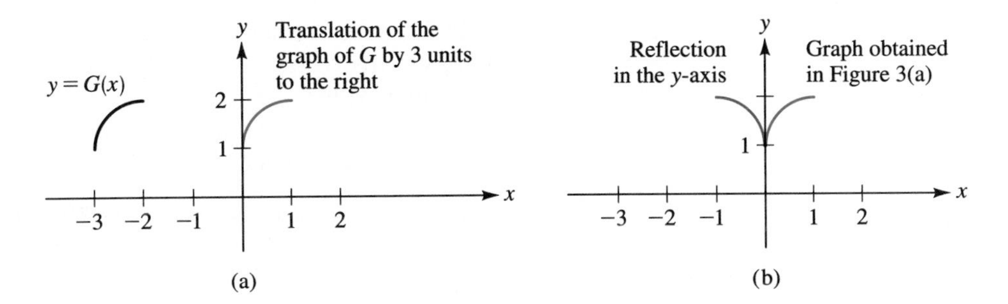

FIGURE 3
Figure (a) shows the result of translating the graph of G three units to the right. When this is followed by a reflection in the y-axis, we obtain the red curve in Figure (b).

(c) Figure 4 shows what happens to the graph of G if we use the same two operations that were used in part (b) but reverse the order in which they are carried out. Note that the end result is quite different from that obtained in part (b).

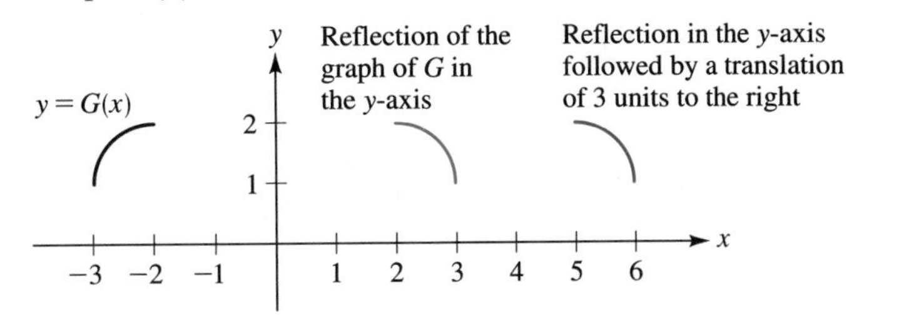

FIGURE 4
The reflection of the graph of G in the y-axis followed by a translation of three units to the right.

Given the graph of a function f, it is useful to know how to sketch efficiently the graphs of the following closely related functions. (In the following equations, c denotes a positive constant.)

$$y=f(x)+c \qquad y=f(x+c) \qquad y=-f(x)$$
$$y=f(x)-c \qquad y=f(x-c) \qquad y=f(-x)$$

Each of these can be graphed by translating or reflecting the graph of $y=f(x)$. In the box that follows, we summarize the techniques that are involved. (Exercises 45–48 will help you to see why these techniques are valid, and Section 11.2 will explain these techniques from a more general point of view.)

PROPERTY SUMMARY **TRANSLATIONS AND REFLECTIONS**

(In Items 1–4, the letter c denotes a positive constant.)

EQUATION	HOW TO OBTAIN THE GRAPH FROM THAT OF $y=f(x)$
1. $y=f(x)+c$	Translate c units vertically upward.
2. $y=f(x)-c$	Translate c units vertically downward.
3. $y=f(x+c)$	Translate c units to the left.
4. $y=f(x-c)$	Translate c units to the right.
5. $y=-f(x)$	Reflect in the x-axis.
6. $y=f(-x)$	Reflect in the y-axis.

In Figure 5 we show examples of the first four techniques in the box, the techniques involving translation. Figure 6 displays examples of the remaining two techniques, the ones involving reflection. You should look over these figures carefully before going on to Examples 2 through 5, in which the various graphing techniques are combined.

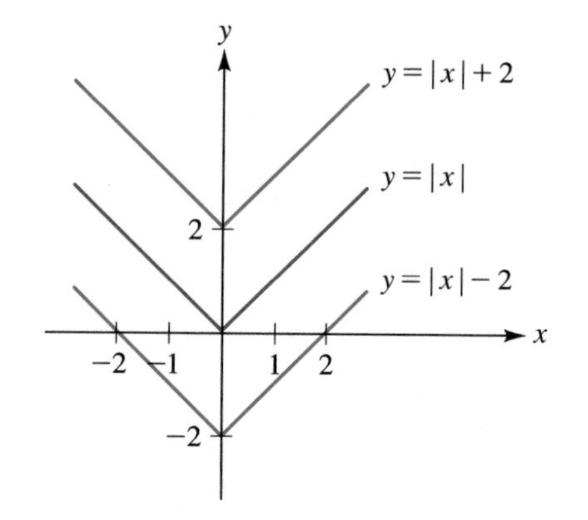

(a) To graph $y = |x| + 2$, translate $y = |x|$ up 2 units. To graph $y = |x| - 2$, translate $y = |x|$ down 2 units.

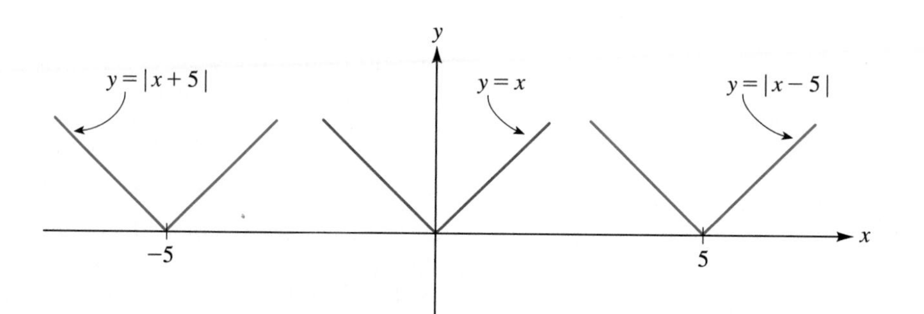

FIGURE 5

(b) To graph $y = |x + 5|$, translate $y = |x|$ to the left 5 units. To graph $y = |x - 5|$, translate $y = |x|$ to the right 5 units.

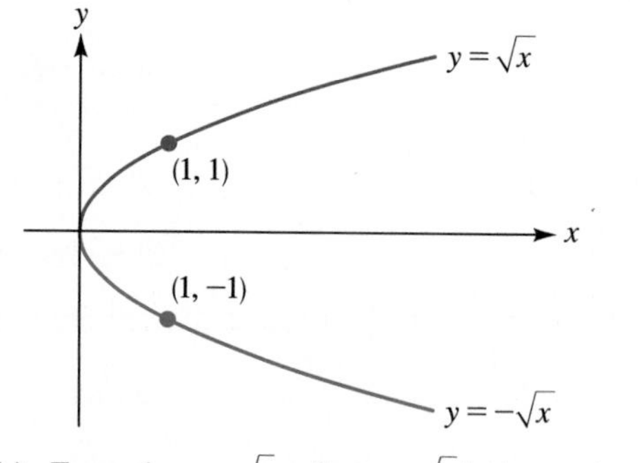

(a) To graph $y = -\sqrt{x}$, reflect $y = \sqrt{x}$ in the x-axis. More generally, to graph $y = -f(x)$, reflect the graph of $y = f(x)$ in the x-axis.

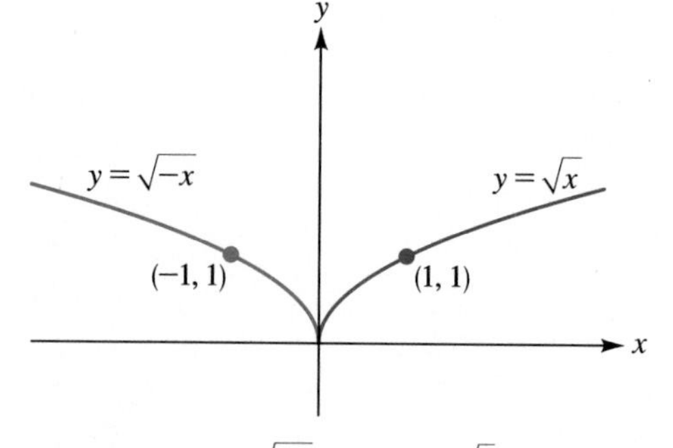

(b) To graph $y = \sqrt{-x}$, reflect $y = \sqrt{x}$ in the y-axis. More generally, to graph $y = f(-x)$, reflect the graph of $y = f(x)$ in the y-axis.

FIGURE 6

EXAMPLE 2 Graph $y = (x - 2)^2 + 1$.

Solution We begin with the graph of $y = x^2$ in Figure 7(a), then we move the curve two units to the right to obtain the graph of $y = (x - 2)^2$; see Figure 7(b). Next we move the curve in Figure 7(b) up one unit to obtain the graph of $y = (x - 2)^2 + 1$; see Figure 7(c).

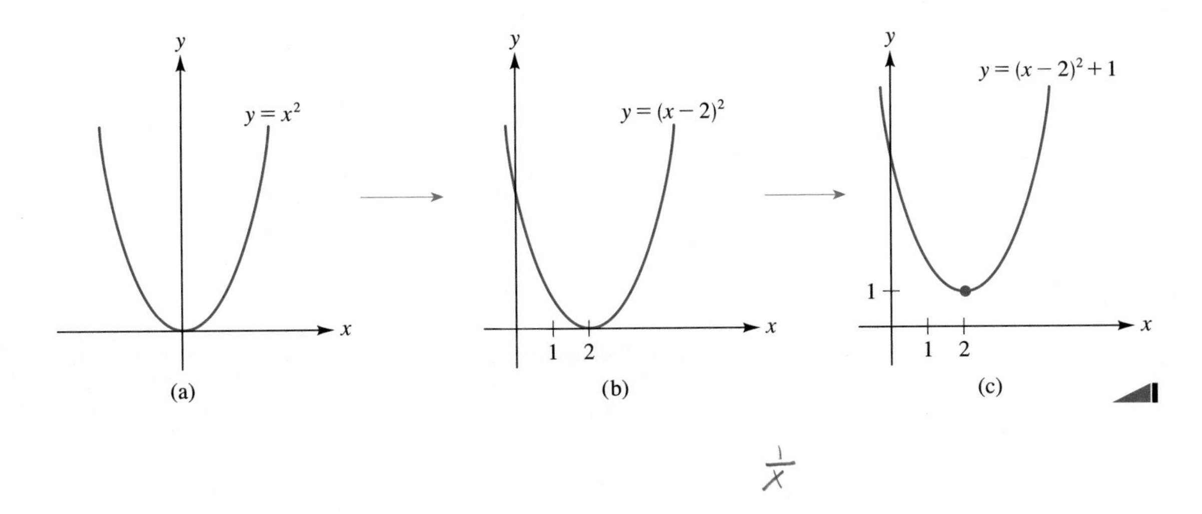

FIGURE 7

$\frac{1}{x}$

EXAMPLE 3 Graph the functions $y = \dfrac{1}{x - 1}$ and $y = \dfrac{1}{x - 1} + 1$.

Solution We begin with the graph of $y = 1/x$ in Figure 8(a). The x- and y-axes are asymptotes for this graph. Moving this graph to the right one unit yields the graph of $y = \dfrac{1}{x - 1}$, shown in Figure 8(b). Note that the vertical asymptote moves one unit to the right also, but the horizontal asymptote is unchanged. Next, we move the graph in Figure 8(b) up one unit (why?) to obtain the graph of $y = \dfrac{1}{x - 1} + 1$, as shown in Figure 8(c). [You should verify for yourself that the y-intercepts in Figures 8(b) and 8(c) are -1 and 0, respectively.]

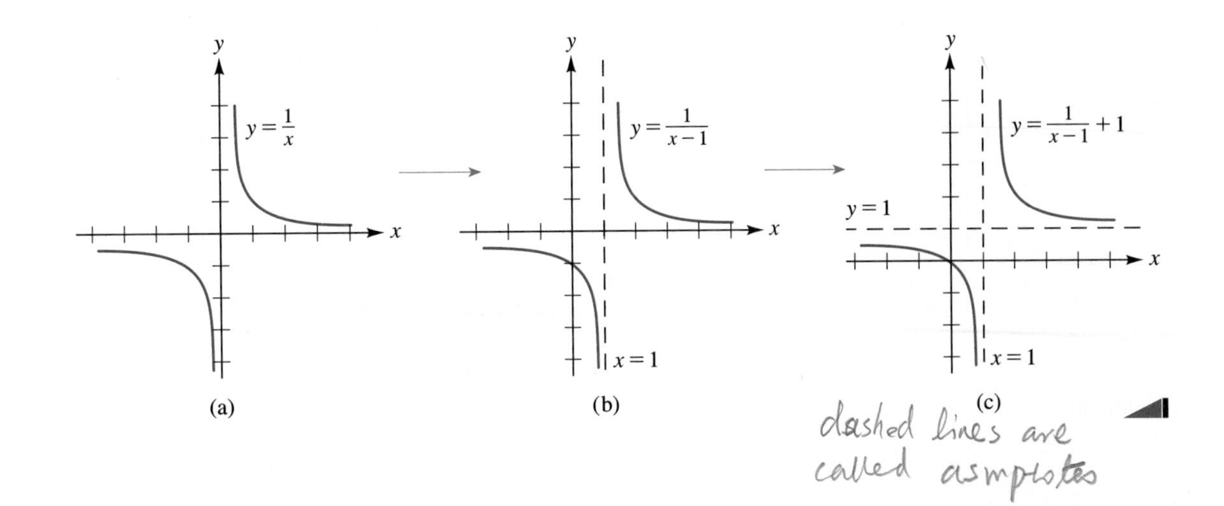

FIGURE 8

EXAMPLE 4 Graph each equation:
(a) $y = -|x|$; **(b)** $y = -|x - 2| + 3$.

Solution See Figure 9.

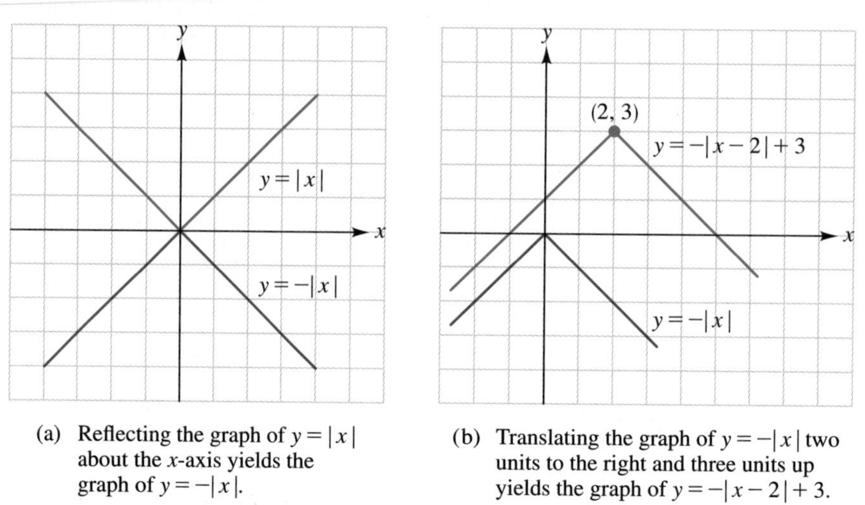

FIGURE 9

(a) Reflecting the graph of $y = |x|$ about the x-axis yields the graph of $y = -|x|$.

(b) Translating the graph of $y = -|x|$ two units to the right and three units up yields the graph of $y = -|x - 2| + 3$.

EXAMPLE 5 The graph of a function f is a line segment joining the points $(-3, 1)$ and $(2, 4)$. Graph each of the following functions:

(a) $y = f(-x)$; **(b)** $y = -f(x)$; **(c)** $y = -f(-x)$;
(d) $y = f(x + 1)$ **(e)** $y = -f(x + 1)$ **(f)** $y = f(1 - x)$.

Solution See Figure 10.

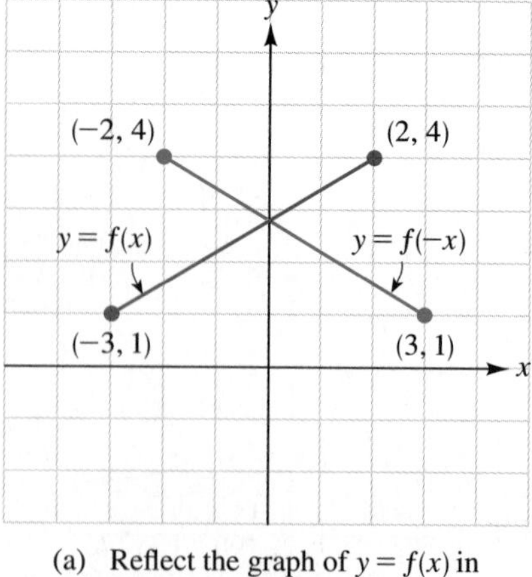

(a) Reflect the graph of $y = f(x)$ in the y-axis to obtain the graph of $y = f(-x)$.

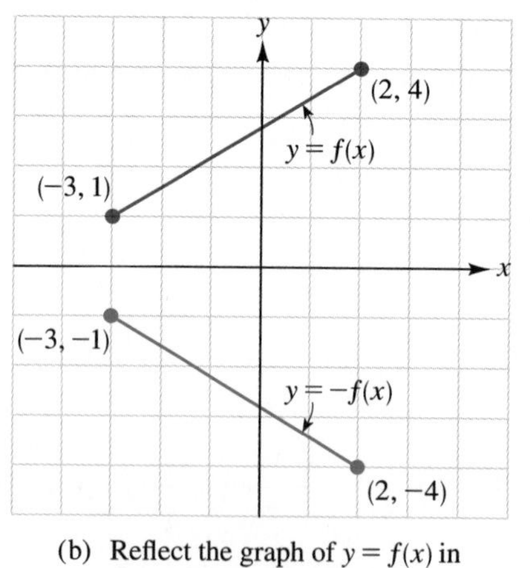

(b) Reflect the graph of $y = f(x)$ in the x-axis to obtain the graph of $y = -f(x)$.

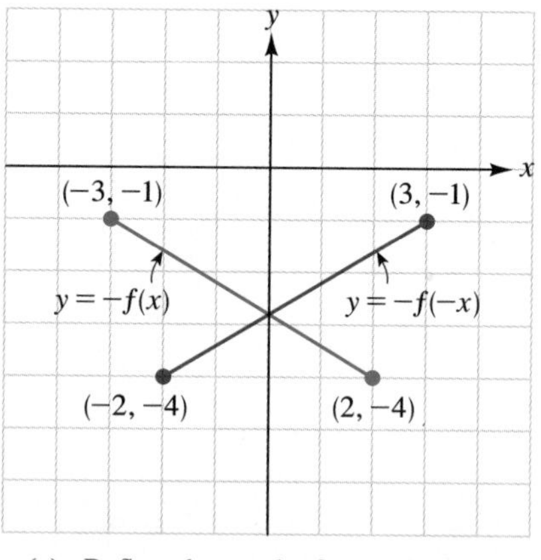

(c) Reflect the graph of $y = -f(x)$ in the y-axis to obtain the graph of $y = -f(-x)$. [Or, reflect the graph of $y = f(-x)$ in the x-axis.]

FIGURE 10

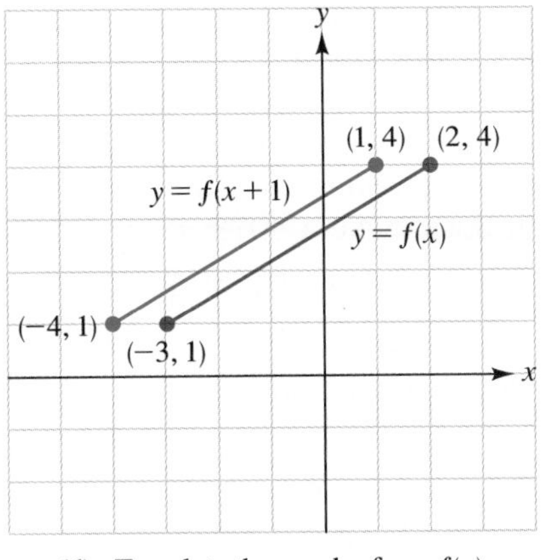

(d) Translate the graph of $y = f(x)$ one unit to the left to obtain the graph of $y = f(x + 1)$.

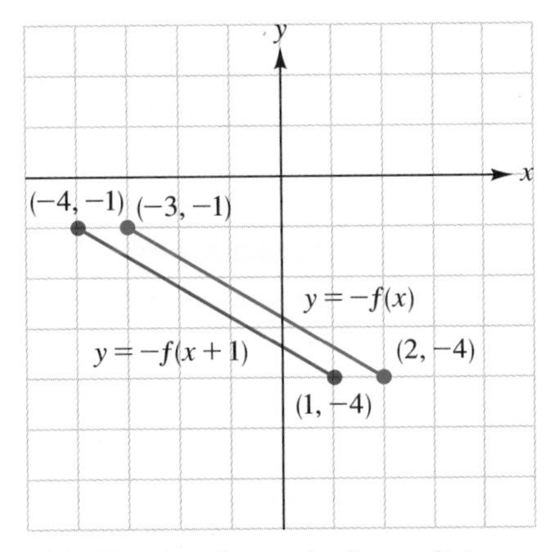

(e) Translate the graph of $y = -f(x)$ one unit to the left to obtain the graph of $y = -f(x + 1)$. [Or, reflect the graph of $y = f(x + 1)$ in the x-axis.]

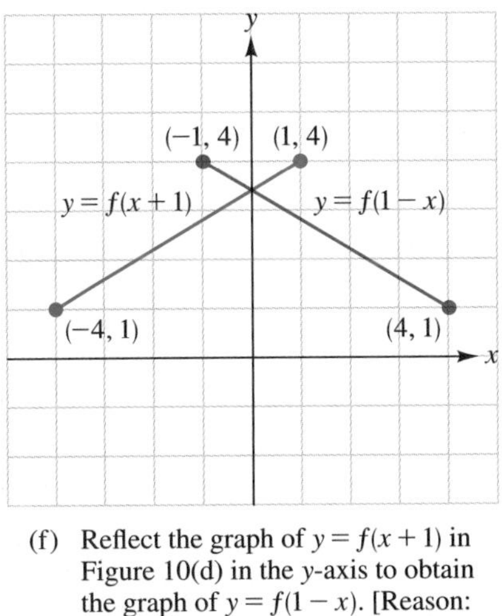

(f) Reflect the graph of $y = f(x + 1)$ in Figure 10(d) in the y-axis to obtain the graph of $y = f(1 - x)$. [Reason: Replacing x with $-x$ in $y = f(x + 1)$ yields $y = f(1 - x)$.]

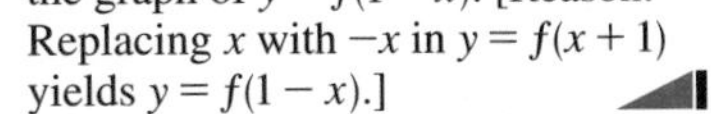

FIGURE 10 (continued)

We conclude this section by outlining what we have done and by pointing out some potential trouble spots that students sometimes encounter. Beginning on page 132, we've used translations and reflections as aids in graphing. The Property Summary Box on page 131 lists the basic techniques. The first examples in the use of these techniques are given in Figures 5 and 6. Go back and review these figures for a moment. As you do, note that each equation that is graphed involves but one application of one basic technique.

What about graphing equations that involve more than just one of the basic techniques? To be specific, what about equations involving a translation and a reflection, or two translations, or two reflections? Does it make a difference which operation you carry out first? [We've graphed such equations in Examples 2 through 5 but, with the exception of Example 5(c), this issue hasn't been explicitly raised.] As indicated in Table 1, there are four cases where it doesn't matter which of the two required operations is carried out first.

TABLE 1 Pairs of Operations on Graphs for Which the Order Does Not Affect the Final Result

Pairs of Operations for Which the Order Doesn't Affect the Final Result	Example	Comment
1. Two translations	$y = \sqrt{x - 2} + 1$	This involves translating $y = \sqrt{x}$ in both the x- and the y-directions. Exercise 49 shows that the order in which these translations are carried out does not affect the final result.
2. Two reflections (one in x-axis, one in y-axis)	$y = -\sqrt{-x}$	This involves reflecting $y = \sqrt{x}$ in both the x- and the y-axes. Exercise 51 shows that the order in which these reflections are carried out does not affect the final result.
3. Translation in x-direction and reflection in x-axis	$y = -\sqrt{x - 2}$	Beginning with the graph of $y = \sqrt{x}$, there are two options: translate first and then reflect; or reflect first and then translate. Exercise 53 shows that both options lead to the same final result.
4. Translation in y-direction and reflection in y-axis	$y = \sqrt{-x} + 1$	Beginning with the graph of $y = \sqrt{x}$, there are two options: translate first and then reflect; or reflect first and then translate. Exercise 55 shows that both options lead to the same final result.

Notice the similarity in Items 3 and 4 of Table 1. In each case, the translation is parallel to the axis of reflection. Specifically, Item 3 involves a translation in the x-direction and a reflection in the x-axis; Item 4 involves a translation in the y-direction and a reflection in the y-axis. As we've said, for these cases it does not matter which of the two operations you carry out first. There are two related cases, however, where it does make a difference:

Case 1: a translation in the y-direction coupled with a reflection in the x-direction

Case 2: a translation in the x-direction coupled with a reflection in the y-direction

[Actually, you have already seen an instance documenting Case 2; go back and reread parts (b) and (c) of Example 1 on pages 130–131.]

Cases 1 and 2 are potential trouble spots that we referred to earlier. As an example of Case 1, consider the equation $y = -\sqrt{x} + 2$. The two operations here are a translation of 2 units up in the y-direction and a reflection in the x-axis. As indicated in Figure 11, the correct procedure here is *first* to reflect $y = \sqrt{x}$ in the x-axis to obtain $y = -\sqrt{x}$; *then* to translate the graph of $y = -\sqrt{x}$ up 2 units to obtain $y = -\sqrt{x} + 2$. In summary for this case, "x-axis reflection before y-translation." Exercise 57 shows that an incorrect result is obtained by first translating up 2 units and then reflecting in the x-axis. To avoid this error, think along the following lines: The "+2" in the equation $y = -\sqrt{x} + 2$ indeed says to translate up 2 units; but what is it that's being translated? According to the given equation, the graph to be translated up 2 units is $y = -\sqrt{x}$. So in this case you need to do the reflection before the translation.

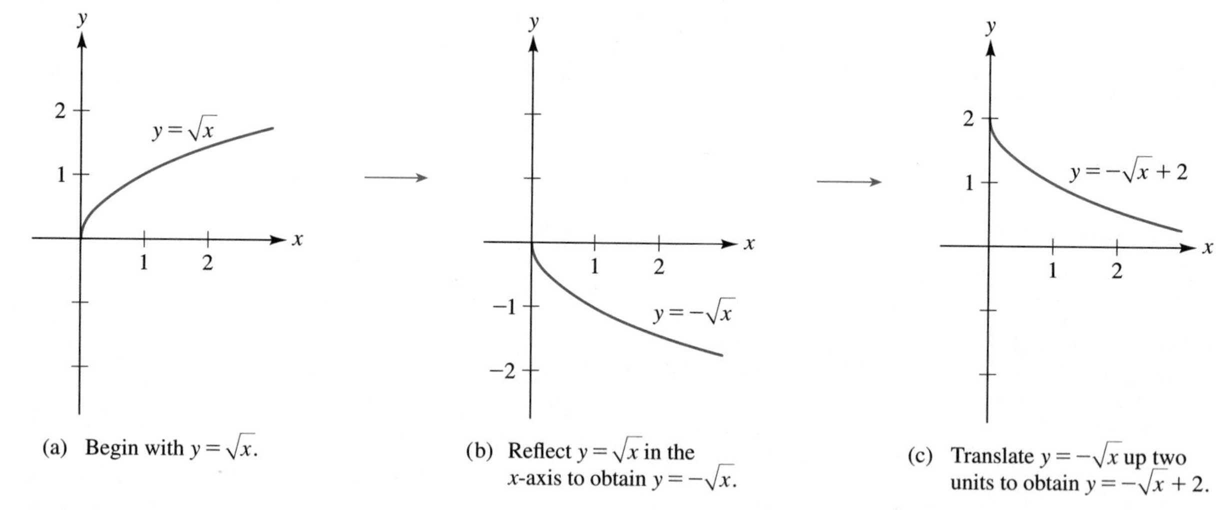

(a) Begin with $y = \sqrt{x}$.

(b) Reflect $y = \sqrt{x}$ in the x-axis to obtain $y = -\sqrt{x}$.

(c) Translate $y = -\sqrt{x}$ up two units to obtain $y = -\sqrt{x} + 2$.

FIGURE 11
Steps in graphing $y = -\sqrt{x} + 2$; x-axis reflection before y-translation.

A good example for exploring Case 2 (a translation in the x-direction coupled with a reflection in the y-direction) is given by the equation

$$y = \sqrt{-x + 2}$$

How would *you* graph this equation? Evidently, a reflection in the y-axis is called for at some point (because of the "$-x$" in the equation), and it appears that a translation of 2 units in the x-direction will be required. A common error that is sometimes made here is to apply the translation rule for graphing $y=f(x+c)$. But that rule does not apply: there is no "$-x$" in the generic equation $y=f(x+c)$. So, rather than introduce yet another rule, one for graphing equations of the form $y=f(-x+c)$ or $y=f(-x-c)$, we explain how to proceed using the material you've already learned. The general principle we'll rely on is this: replacing x with $-x$ throughout an equation reflects the graph in the y-axis. To graph $y=\sqrt{-x+2}$, first ignore the negative sign directly to the left of x, and graph the simpler equation $y=\sqrt{x+2}$, as shown in Figure 12(a). Then, for the graph of $y=\sqrt{-x+2}$, just reflect the graph of $y=\sqrt{x+2}$ in the y-axis, as shown in Figure 12(b). (For additional perspective on this, see Exercise 59.)

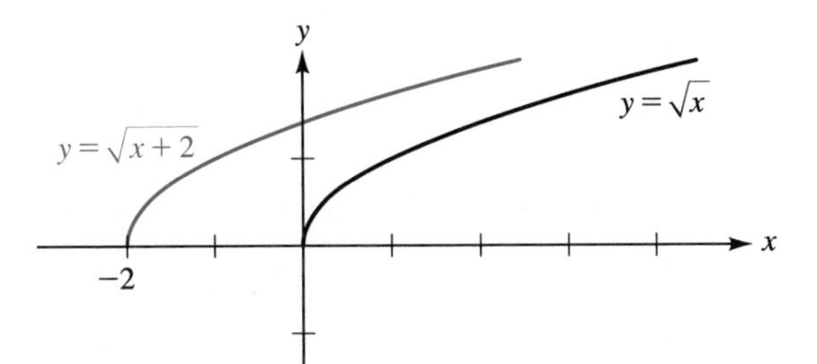

(a) Translate the graph of $y=\sqrt{x}$ to the left 2 units to obtain the graph of $y=\sqrt{x+2}$.

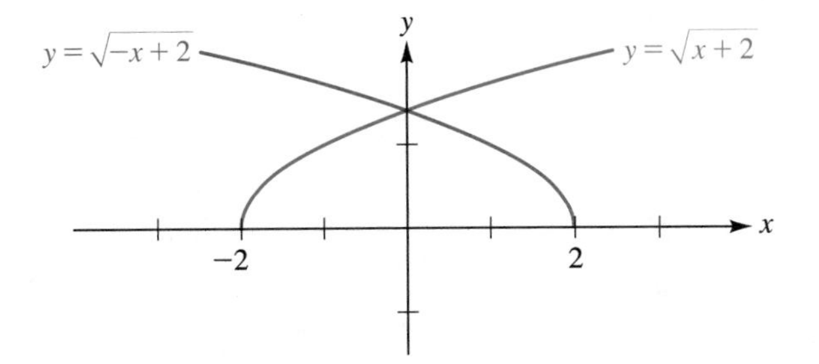

(b) Reflect the graph of $y=\sqrt{x+2}$ in the y-axis to obtain the graph of $y=\sqrt{-x+2}$.

FIGURE 12

Steps in graphing $y=\sqrt{-x+2}$: first graph $y=\sqrt{x+2}$; then reflect in the y-axis to obtain the graph of $y=\sqrt{-x+2}$.

EXERCISE SET 3.3

Do 3, 11, 17, 21, 22, 29, 41, 45

A

In Exercises 1 and 2, the right-hand column contains instructions for translating and/or reflecting the graph of $y=f(x)$. Match each equation in the left-hand column with an appropriate set of instructions in the right-hand column.

1.

(a) $y=f(x-1)$

(b) $y=f(x)-1$

(c) $y=f(x)+1$

(d) $y=f(x+1)$

(e) $y=f(-x)+1$

(f) $y=f(-x)-1$

(g) $y=-f(x)+1$

(h) $y=-f(x+1)$

(i) $y=-f(x)-1$

(j) $y=f(1-x)+1$ *Hint:* See part (f) of Example 5.

(k) $y=-f(-x)+1$

(A) Translate left 1 unit.

(B) Reflect in the x-axis, then translate left 1 unit.

(C) Translate right 1 unit.

(D) Reflect in the x-axis, then translate up 1 unit.

(E) Reflect in the x-axis, then translate down 1 unit.

(F) Translate down 1 unit.

(G) Reflect in the x-axis, reflect in the y-axis, then translate up 1 unit.

(H) Translate left 1 unit, then reflect in the y-axis, then translate up 1 unit.

(I) Translate up 1 unit.

(J) Reflect in the y-axis, then translate up 1 unit.

(K) Reflect in the y-axis, then translate down 1 unit.

2.

(a) $y=f(x+2)+3$

(b) $y=f(x+3)+2$

(c) $y=f(x-2)+3$

(d) $y=f(x-2)-3$

(e) $y=f(x+2)-3$

(f) $y=f(x-3)+2$

(g) $y=f(x-3)-2$

(A) Translate left 2 units, then translate down 3 units.

(B) Translate left 3 units, then translate up 2 units.

(C) Translate right 3 units, then translate up 2 units.

(D) Translate left 3 units, then translate down 2 units.

(E) Translate right 3 units and down 2 units.

(F) Reflect in the y-axis, then translate up 2 units.

(G) Reflect in the x-axis, then translate right 2 units.

(h) $y = f(x+3) - 2$ **(H)** Reflect in the x-axis, then translate left 2 units.

(i) $y = -f(x+2)$ **(I)** Translate left 2 units, then reflect in the y-axis.

(j) $y = -f(x-2)$ **(J)** Translate right 2 units, then translate up 3 units.

(k) $y = f(2-x)$ **(K)** Translate left 2 units, then translate up 3 units.

(l) $y = f(-x) + 2$ **(L)** Translate right 2 units and down 3 units.

In Exercises 3–24, sketch the graph of the function.

3. $y = x^3 - 3$
4. $y = x^2 + 3$
5. $y = (x+4)^2$
6. $y = (x+4)^2 - 3$
7. $y = (x-4)^2$
8. $y = (x-4)^2 + 1$
9. $y = -x^2$
10. $y = -x^2 - 3$
11. $y = -(x-3)^2$
12. $y = -(x-3)^2 - 3$
13. $y = \sqrt{x-3}$
14. $y = \sqrt{x-3} + 1$
15. $y = -\sqrt{x+1}$
16. $y = -\sqrt{x+1} + 1$
17. $y = \dfrac{1}{x+2} + 2$
18. $y = \dfrac{1}{x-3} - 1$
19. $y = (x-2)^3$
20. $y = (x-2)^3 + 1$
21. $y = -x^3 + 4$
22. $y = -(x-1)^3 + 4$
23. **(a)** $y = |x+4|$
(b) $y = |4-x|$
(c) $y = -|4-x| + 1$
24. **(a)** $y = \sqrt{x+2}$
(b) $y = \sqrt{2-x}$
(c) $y = -\sqrt{2-x}$

In Exercises 25–40, sketch the graph of the function, given that f, F, and g are defined as follows:

$$f(x) = |x| \qquad F(x) = 1/x \qquad g(x) = \sqrt{1-x^2}$$

25. $y = f(x-5)$
26. $y = -f(x-5)$
27. $y = f(5-x)$
28. $y = -f(5-x)$
29. $y = 1 - f(x-5)$
30. $y = f(-x)$
31. $y = F(x+3)$
32. $y = F(x) + 3$
33. $y = -F(x+3)$
34. $y = F(-x) + 3$
35. $y = g(x-2)$
36. $y = -g(x-2)$
37. $y = 1 - g(x-2)$
38. $y = g(-x)$
39. $y = g(2-x)$
40. $y = -g(2-x)$

In Exercises 41 and 42, the graph of a function $y = f(x)$ is a line segment joining the points $(-4, 1)$ and $(-1, 2)$. In each case, sketch the graph of the given equation.

41. **(a)** $y = -f(x)$
(b) $y = f(-x)$
(c) $y = -f(-x)$
(d) $y = -f(x-1)$
(e) $y = -f(1-x)$
(f) $y = 1 - f(1-x)$

42. **(a)** $y = f(x) - 2$
(b) $y = |f(x) - 2|$
(c) $y = f(x-4) - 2$
(d) $y = |f(x-4) - 2|$
(e) $y = |f(x) - 4| - 2$
(f) $y = 2 - |f(x) - 4|$

In Exercises 43 and 44, refer to the graph of the function g shown in the figure in order to graph each function.

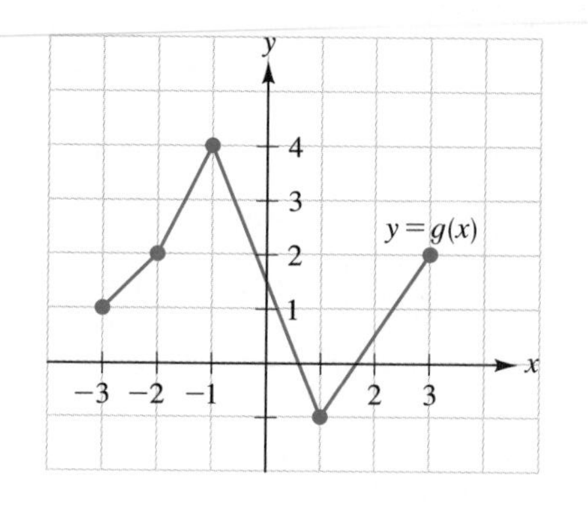

43. **(a)** $y = g(-x)$
(b) $y = -g(x)$
(c) $y = -g(-x)$

44. $y = -g(x-3) - 1$

45. **(a)** Complete the following table.

x	x^2	$x^2 - 1$	$x^2 + 1$
0			
±1			
±2			
±3			

(b) Using the results in the table, graph the functions $y = x^2$, $y = x^2 - 1$, and $y = x^2 + 1$ on the same set of axes. How are the graphs related?

46. **(a)** Complete the given table.

x	x^2	$(x-1)^2$	$(x+1)^2$
0			
1			
2			
3			
−1			
−2			
−3			

(b) Using the results in the table, graph the functions $y = x^2$, $y = (x-1)^2$, and $y = (x+1)^2$ on the same set of axes. How are the graphs related?

47. **(a)** Complete the following table. (Use a calculator where necessary.)

x	$\sqrt{x}$	$-\sqrt{x}$
0		
1		
2		
3		
4		
5		

(b) Using the results in the table, graph the functions $y = \sqrt{x}$ and $y = -\sqrt{x}$ on the same set of axes. How are the two graphs related?

48. **(a)** Complete the given tables. (Use a calculator where necessary.)

x	0	1	2	3	4	5
$\sqrt{x}$						

x	0	-1	-2	-3	-4	-5
$\sqrt{-x}$						

(b) Using the tables, graph the functions $y = \sqrt{x}$ and $y = \sqrt{-x}$ on the same set of axes. How are the graphs related?

B

As background for Exercises 49–56, you should review Table 1 on page 135.

49. This exercise relates to a comment made in Item 1 of Table 1.

(a) Translate the graph of $y = \sqrt{x}$ to the right 2 units, then translate the resulting graph 1 unit up. For reference, label this final graph G_1.

(b) Translate the graph of $y = \sqrt{x}$ up 1 unit, then translate the resulting graph 2 units to the right. For reference, label this final graph G_2. After you have done this, note that G_1 and G_2 are the same. That is, the order in which these translations are carried out does not affect the final result.

50. Let P be a point with coordinates (a, b), and assume that c and d are positive numbers. (The condition that c and d are positive isn't really necessary in this problem, but it will help you to visualize things.)

(a) Translate the point P by c units in the x-direction to obtain a point Q, then translate Q by d units in the y-direction to obtain a point R. What are the coordinates of the point R?

(b) Translate the point P by d units in the y-direction to obtain a point S, then translate S by c units in the x-direction to obtain a point T. What are the coordinates of the point T?

(c) Compare your answers in parts (a) and (b). What have you demonstrated? (Answer in complete sentences.)

51. This exercise relates to a comment made in Item 2 of Table 1.

(a) Reflect the graph of $y = \sqrt{x}$ in the x-axis, then reflect the resulting graph in the y-axis. For reference, label this final graph G_1.

(b) Reflect the graph of $y = \sqrt{x}$ in the y-axis, then reflect the resulting graph in the x-axis. For reference, label this final graph G_2. After you have done this, note that G_1 and G_2 are the same. That is, the order in which the x-axis and y-axis reflections are carried out does not affect the final result.

52. Let P be a point with coordinates (a, b).

(a) Reflect P in the x-axis to obtain a point Q, then reflect Q in the y-axis to obtain a point R. What are the coordinates of the point R?

(b) Reflect P in the y-axis to obtain a point S, then reflect S in the x-axis to obtain a point T. What are the coordinates of the point T?

(c) Compare your answers in parts (a) and (b). What have you demonstrated? (Answer in complete sentences.)

53. This exercise relates to a comment made in Item 3 of Table 1.

(a) Translate the graph of $y = \sqrt{x}$ to the right 2 units, then reflect the resulting graph in the x-axis. For reference, label this final graph G_1.

(b) Reflect the graph of $y = \sqrt{x}$ in the x-axis, then translate the resulting graph 2 units to the right. Label this final graph G_2. After you have done this, note that G_1 and G_2 are the same. That is, the order in which these two operations are carried out does not affect the final result.

54. Let P be a point with coordinates (a, b).

(a) Translate P by c units in the x-direction to obtain a point Q, then reflect Q in the x-axis to obtain a point R. What are the coordinates of the point R?

(b) Reflect P in the x-axis to obtain a point S, then translate S by c units in the x-direction to obtain a point T. What are the coordinates of the point T?

(c) Compare your answers in parts (a) and (b). What have you demonstrated? (Answer in complete sentences.)

55. This exercise relates to a comment made in Item 4 of Table 1.

(a) Translate the graph of $y = \sqrt{x}$ up 1 unit in the y-direction, then reflect the resulting graph in the y-axis. For reference, label this final graph G_1.

(b) Reflect the graph of $y = \sqrt{x}$ in the y-axis, then translate the resulting graph up 1 unit. Label this final graph G_2. After you have done this, note that G_1 and G_2 are the same. That is, the order in which these two operations are carried out does not affect the final result.

56. Let P be a point with coordinates (a, b).

(a) Translate P by c units in the y-direction to obtain a point Q, then reflect Q in the y-axis to obtain a point R. What are the coordinates of the point R?

(b) Reflect P in the y-axis to obtain a point S, then translate S by c units in the y-direction to obtain a point T. What are the coordinates of the point T?

(c) Compare your answers in parts (a) and (b). What have you demonstrated? (Answer in complete sentences.)

57. As background for this exercise, review Figure 11, which shows how to graph the equation $y = -\sqrt{x} + 2$.

(a) Translate the graph of $y = \sqrt{x}$ up 2 units in the y-direction, then reflect the resulting graph in the x-axis. You should obtain a picture similar to the following.

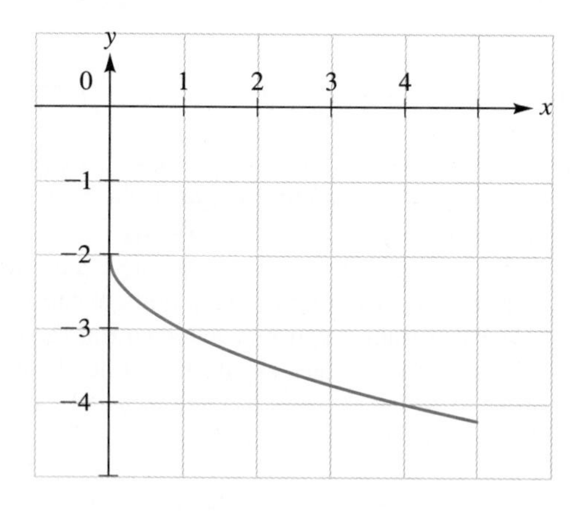

(b) The graph shown in part (a) is *not* the graph of $y = -\sqrt{x} + 2$. Verify this as follows. First, note that the graph shown in part (a) passes through the point $(1, -3)$. Now check that the pair of numbers $x = 1$ and $y = -3$ does *not* satisfy the equation $y = -\sqrt{x} + 2$.

(c) What is the equation for the graph obtained in part (a)?

58. Let P be a point with coordinates (a, b), and let c denote a nonzero real number.

(a) Translate P by c units in the y-direction to obtain a point Q, then reflect Q in the x-axis to obtain a point R. What are the coordinates of the point R?

(b) Reflect P in the x-axis to obtain a point S, then translate S by c units in the y-direction to obtain a point T. What are the coordinates of the point T?

(c) By comparing y-coordinates, show that the points R and T are not the same point. (You'll need to use the assumption $c \neq 0$.) This shows that when an x-axis reflection is coupled with a y-direction translation, the order *does* affect the final result.

59. As background for this exercise, review Figure 12, which shows how to graph the equation $y = \sqrt{-x + 2}$.

(a) Reflect the graph of $y = \sqrt{x}$ in the y-axis, then translate the resulting graph 2 units to the left. Your final graph should be similar to the following.

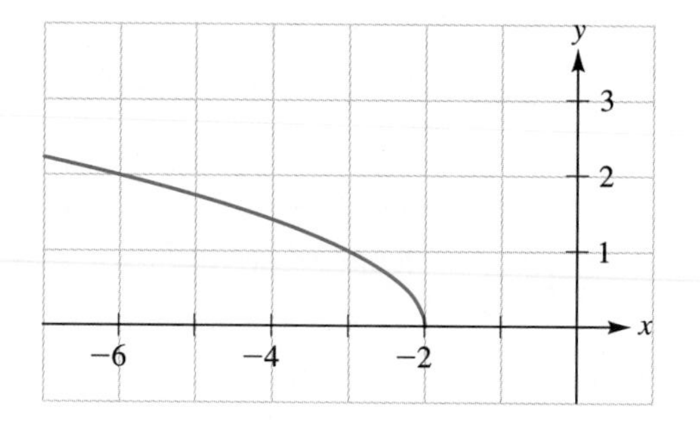

(b) The graph shown in part (a) is *not* the graph of $y = \sqrt{-x + 2}$. Verify this as follows. First, note that the graph shown in part (a) passes through the point $(-6, 2)$. Now check that the pair of numbers $x = -6$ and $y = 2$ does *not* satisfy the equation $y = \sqrt{-x + 2}$.

(c) What is the equation for the graph obtained in part (a)?

For Exercises 60 and 61, assume that (a, b) is a point on the graph of $y = f(x)$, and specify the corresponding point on the graph of each equation. [For example, the point that corresponds to (a, b) on the graph of $y = f(x - 1)$ is $(a + 1, b)$.]

60. **(a)** $y = f(x - 3)$ **(b)** $y = f(x) - 3$ **(c)** $y = f(x - 3) - 3$ **(d)** $y = -f(x)$ **(e)** $y = f(-x)$ **(f)** $y = -f(-x)$ **(g)** $y = f(3 - x)$ **(h)** $y = -f(3 - x) + 1$

61. **(a)** $y = f(-x) + 2$ **(b)** $y = -f(-x) + 2$ **(c)** $y = -f(x - 3)$ **(d)** $y = 1 - f(x + 1)$ **(e)** $y = f(1 - x)$ **(f)** $y = -f(1 - x) + 1$

62. Show that $\frac{x}{x-1} = \frac{1}{x-1} + 1$, provided that $x \neq 1$. Then use this fact to graph the function g defined by $g(x) = x/(x-1)$.

63. Verify that $-x^2 + 2x + 2 = -(x-1)^2 + 3$. Then use this fact to graph the function h defined by $h(x) = -x^2 + 2x + 2$.

64. What is the range of the function f defined by $f(x) = \frac{x^2}{1 - \sqrt{1-x^2}}$? *Hint:* The function is easy to graph after rationalizing the denominator.

65. **(a)** A function f is said to be **even** if the equation $f(-x) = f(x)$ is satisfied by all values of x in the domain of f. Explain why the graph of an even function must be symmetric about the y-axis.
(b) Show that each function is even by computing $f(-x)$ and then noting that $f(x)$ and $f(-x)$ are equal.
(i) $f(x) = x^2$ **(ii)** $f(x) = 2x^4 - 6$
(iii) $f(x) = 3x^6 - \frac{4}{x^2} + 1$

66. **(a)** A function f is said to be **odd** if the equation $f(-x) = -f(x)$ is satisfied by all values of x in the domain of f. Show that if (x, y) is a point on the graph of an odd function f, then the point $(-x, -y)$ is also on the graph. (This implies that the graph of an odd function must be symmetric about the origin.)
(b) Show that each function is odd by computing $f(-x)$ as well as $-f(x)$ and then noting that the two expressions obtained are equal.
(i) $f(x) = x^3$ **(ii)** $f(x) = -2x^5 + 4x^3 - x$
(iii) $f(x) = |x|/(x + x^7)$

67. Is each function odd, even, or neither? (See Exercises 65 and 66 for definitions.)
(a) $f(x) = \frac{1-x^2}{2+x^2}$ **(b)** $g(x) = \frac{x - x^3}{2x + x^3}$
(c) $h(x) = x^2 + x$ **(d)** $F(x) = (x^2 + x)^2$
(e) $G(x) = \begin{cases} 1 & \text{if } x > 0 \\ 0 & \text{if } x = 0 \\ -1 & \text{if } x < 0 \end{cases}$
Suggestion for part (e): Look at the graph.

68. Suppose that the function f is increasing on the interval $(2, 4)$ and decreasing on the intervals $(-\infty, 2)$ and $(4, \infty)$. On what interval(s) is each function increasing?
(a) $y = f(-x)$ **(b)** $y = -f(x)$

69. Suppose that the function g is decreasing on $(-\infty, 1)$ and increasing on $(1, \infty)$. On what interval(s) is the function $y = -g(-x)$ increasing?

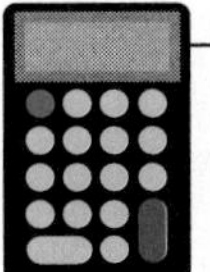

GRAPHING UTILITY EXERCISES FOR SECTION 3.3

In Exercises 1–3, you will use a graphing utility to provide examples of the six graphing techniques listed in the box on page 131.

1. (As background for this exercise, you should first work Exercise 9 in the Graphing Utility Exercises for the previous section.)
(a) Graph the two functions $f(x) = x^{2/3}$ and $g(x) = x^{2/3} + 3$. Observe that the graph of g is obtained by translating the graph of f *up* 3 units. This illustrates Property 1 on page 131.
(b) Graph the two functions $f(x) = x^{2/3}$ and $h(x) = x^{2/3} - 3$. Observe that the graph of h is obtained by translating the graph of f *down* 3 units. This illustrates Property 2 on page 131.

2. On the same set of axes, graph the three functions

$$F(x) = \sqrt{1 - x^3}$$
$$L(x) = \sqrt{1 - (x+2)^3}$$
$$R(x) = \sqrt{1 - (x-2)^3}$$

(*Suggestion:* Use a viewing rectangle in which x extends from -5 to 3 and y extends from 0 to 5.) Observe that the graph of L is obtained by translating the graph of F to the *left* 2 units. Also, the graph of R is obtained by translating the graph of F to the *right* 2 units. This illustrates Properties 3 and 4 on page 131.

3. (A viewing rectangle that extends from -4 to 4 in both the x- and y-directions is appropriate for this exercise.)
(a) On the same set of axes, graph the two functions $f(x) = 2^x$ and $g(x) = -2^x$. [*Algebra reminder:* -2^x means $-(2^x)$, not $(-2)^x$]. Observe that the graph of g is obtained by reflecting the graph of f in the x-axis. This illustrates Property 5 on page 131.
(b) On the same set of axes graph the two functions $f(x) = 2^x$ and $h(x) = 2^{-x}$. Observe that the graph of h is obtained by reflecting the graph of f in the y-axis. This illustrates Property 6 on page 131.

4. (A viewing rectangle that extends from -5 to 5 in both the x- and the y-directions is appropriate for this exercise.)
(a) Graph the function $y = x/(x-1)$.
(b) From the screen display, it appears that the graph may be a translation of the basic $y = 1/x$ graph. Prove that this is indeed the case by using algebra (*not your calculator*) to show that

$$\frac{x}{x-1} = \frac{1}{x-1} + 1$$

What translations are involved here?
(c) Verify the identity in part (b) visually. That is, on the same set of axes, graph $y = \dfrac{x}{x-1}$ and $y = \dfrac{1}{x-1} + 1$. Observe that the resulting graphs appear to be identical.

For Exercises 5–10, tell how the graph in part (a) should be translated and/or reflected to obtain the remaining graphs. Then use a graphing utility to draw the graphs and check your statements.

5. **(a)** $y = x^2$ **(b)** $y = (x+3)^2$ **(c)** $y = -(x+3)^2$ **(d)** $y = (-x+3)^2$

6. **(a)** $y = x^3$ **(b)** $y = (x+2)^3$ **(c)** $y = -(x+2)^3$ **(d)** $y = (-x+2)^3$

7. **(a)** $y = |x|$ **(b)** $y = |x+4|$ **(c)** $y = |-x+4|$ **(d)** $y = |-x+4| - 2$

8. **(a)** $F(x) = \frac{1}{2}x^3 - \frac{3}{2}x^2$ **(b)** $y = F(-x)$ **(c)** $y = -F(x)$ **(d)** $y = -F(-x)$

9. **(a)** $G(x) = 2/(1+x^2)$ **(b)** $y = G(x+3)$ **(c)** $y = G(x-3)$ **(d)** $y = -G(x) + 2$

10. **(a)** $y = \sqrt{x+2}$ **(b)** $y = \sqrt{2-x}$ **(c)** $y = 3 - \sqrt{2-x}$

3.4 METHODS OF COMBINING FUNCTIONS. ITERATION

The notation ϕx to indicate a function of x was introduced by . . . [John Bernoulli] *in 1718, . . . but the general adoption of symbols like f, F, ϕ, and ψ . . . to represent functions, seems to be mainly due to Euler and Lagrange.*

W. W. Rouse Ball in *A Short Account of the History of Mathematics* (New York: Dover Publications, 1960)

The composition of functions is basic to the study of iteration and chaos.

Hartmut Jürgens *et al., Fractals for the Classroom: Strategic Activities,* Vol. 2 (New York: Springer-Verlag, 1992)

Two given numbers a and b can be combined in various ways to produce a third number. For instance, we can form the sum $a + b$ or the difference $a - b$ or the product ab. Also, if $b \neq 0$, we can form the quotient a/b. Similarly, two functions can be combined in various ways to produce a third function. Suppose, for example, that we start with the two functions $y = x^2$ and $y = x^3$. It seems natural to define their sum, difference, product, and quotient as follows:

$$\begin{aligned} \text{sum:}\quad & y = x^2 + x^3 \\ \text{difference:}\quad & y = x^2 - x^3 \\ \text{product:}\quad & y = x^2x^3 = x^5 \\ \text{quotient:}\quad & y = \frac{x^2}{x^3} = \frac{1}{x} \end{aligned}$$

Indeed, this is the idea behind the formal definitions we now give.

DEFINITIONS: Arithmetical Operations with Functions

Let f and g be two functions. Then the **sum** $f + g$, the **difference** $f - g$, the **product** fg, and the **quotient** f/g are functions defined by the following equations:

$$(f+g)(x) = f(x) + g(x) \tag{1}$$
$$(f-g)(x) = f(x) - g(x) \tag{2}$$
$$(fg)(x) = f(x) \cdot g(x) \tag{3}$$
$$(f/g)(x) = f(x)/g(x) \qquad \text{provided } g(x) \neq 0 \tag{4}$$

For the functions defined by equations (1), (2), and (3), the domain is the set of all inputs x belonging to both the domain of f and the domain of g. For the quotient function in equation (4), we impose the additional restriction that the domain exclude all inputs x for which $g(x) = 0$.

EXAMPLE 1 Let $f(x) = 3x + 1$ and $g(x) = x - 1$. Compute each of the following: **(a)** $(f + g)(x)$; **(b)** $(f - g)(x)$; **(c)** $(fg)(x)$; **(d)** $(f/g)(x)$.

Solution

(a) $(f + g)(x) = f(x) + g(x) = (3x + 1) + (x - 1) = 4x$

(b) $(f - g)(x) = f(x) - g(x) = (3x + 1) - (x - 1) = 2x + 2$

(c) $(fg)(x) = [f(x)][g(x)] = (3x + 1)(x - 1) = 3x^2 - 2x - 1$

(d) $(f/g)(x) = \dfrac{f(x)}{g(x)} = \dfrac{3x + 1}{x - 1}$, where $x \neq 1$

For the remainder of this section, we are going to discuss a method of combining functions known as **composition of functions.** As you will see, this method is based on the familiar algebraic process of substitution. Suppose, for example, that f and g are two functions defined by

$$f(x) = x^2 \qquad g(x) = 3x + 1$$

Choose any number in the domain of g, say, $x = -2$. We can compute $g(-2)$:

$$g(-2) = 3(-2) + 1 = -5$$

Now let's use the output -5 that g has produced as an *input* for f. We obtain

$$f(-5) = (-5)^2 = 25$$

Consequently,

$$f[g(-2)] = 25$$

So, beginning with the input -2, we've successively applied g and then f to obtain the output 25. Similarly, we could carry out this same procedure for any other number in the domain of g. Here is a summary of the procedure:

1. Start with an input x and calculate $g(x)$.
2. Use $g(x)$ as an input for f; that is, calculate $f[g(x)]$.

We use the notation $f \circ g$ to denote the function, or rule, that tells us to assign the output $f[g(x)]$ to the initial input x. In other words, $f \circ g$ denotes the rule consisting of two steps: *First apply g; then apply f.* We read the notation $f \circ g$ as *f circle g* or *f composed with g.* In Figure 1, we summarize these ideas.

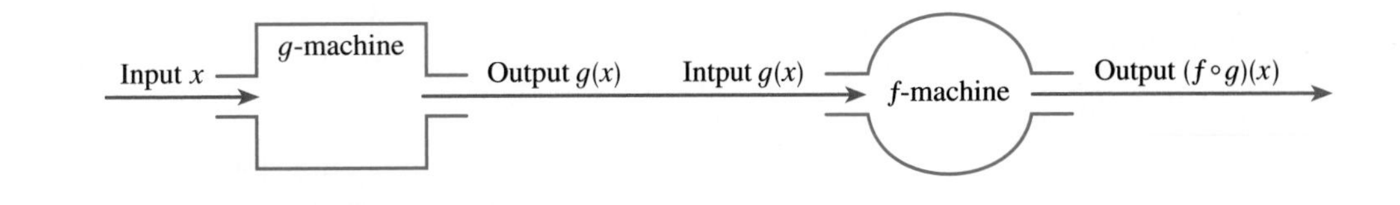

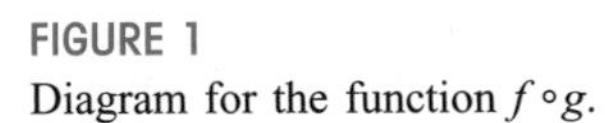
FIGURE 1
Diagram for the function $f \circ g$.

When we write $g(x)$, we assume that x is in the domain of the function g. Likewise, for the notation $f[g(x)]$ to make sense, the outputs $g(x)$ must themselves be acceptable inputs for the function f. Our formal definition, then, for the composite function $f \circ g$ is as follows.

$f \circ g$

DEFINITION: Composition of Functions: $f \circ g$

Given two functions f and g, the function $f \circ g$ is defined by

$$(f \circ g)(x) = f[g(x)]$$

The domain of $f \circ g$ consists of those inputs x (in the domain of g) for which $g(x)$ is in the domain of f.

EXAMPLE 2 Let $f(x) = x^2$ and $g(x) = 3x + 1$. Compute $(f \circ g)(x)$ and $(g \circ f)(x)$.

Solution

$$\begin{aligned} (f \circ g)(x) &= f[g(x)] && \text{definition of } f \circ g \\ &= f(3x+1) && \text{definition of } g \\ &= (3x+1)^2 && \text{definition of } f \\ &= 9x^2 + 6x + 1 \end{aligned}$$

$$\begin{aligned} (g \circ f)(x) &= g[f(x)] && \text{definition of } g \circ f \\ &= g(x^2) && \text{definition of } f \\ &= 3(x^2) + 1 && \text{definition of } g \\ &= 3x^2 + 1 \end{aligned}$$

Notice that the two results obtained in Example 2 are not the same. This shows that, in general, $f \circ g$ and $g \circ f$ represent different functions.

EXAMPLE 3 Let f and g be defined as in Example 2: $f(x) = x^2$ and $g(x) = 3x + 1$. Compute $(f \circ g)(-2)$.

Solution We will show two methods.

FIRST METHOD Using the formula for $(f \circ g)(x)$ developed in Example 2, we have

$$(f \circ g)(x) = 9x^2 + 6x + 1$$

and, therefore,

$$\begin{aligned} (f \circ g)(-2) &= 9(-2)^2 + 6(-2) + 1 \\ &= 36 - 12 + 1 \\ &= 25 \end{aligned}$$

ALTERNATIVE METHOD Working directly from the definition of $f \circ g$, we have

$$\begin{aligned} (f \circ g)(-2) &= f[g(-2)] \\ &= f[3(-2) + 1] \\ &= f(-5) \\ &= (-5)^2 = 25 \end{aligned}$$

In Examples 2 and 3, the domain of both f and g is the set of all real numbers. And as you can easily check, the domain of both $f \circ g$ and $g \circ f$ is also the set of all real numbers. In Example 4, however, some care needs to be taken in describing the domain of the composite function.

EXAMPLE 4 Let f and g be defined as follows:

$$f(x) = x^2 + 1 \qquad g(x) = \sqrt{x}$$

Compute $(f \circ g)(x)$. Find the domain of $f \circ g$ and sketch its graph.

Solution $(f \circ g)(x) = f[g(x)]$

$$= f(\sqrt{x}) = (\sqrt{x})^2 + 1 = x + 1$$

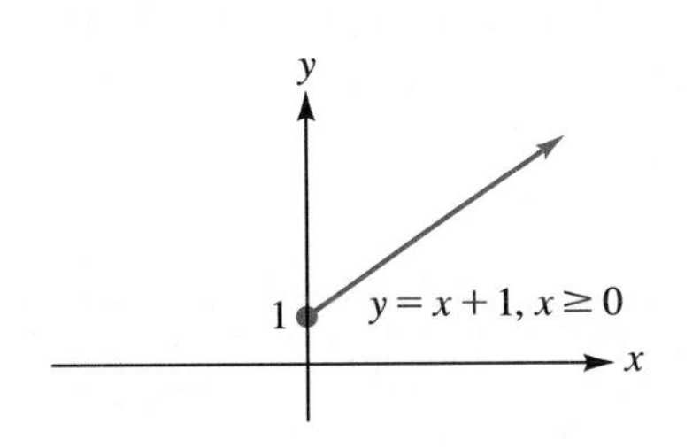

FIGURE 2
The graph of the function $f \circ g$ in Example 4.

So we have $(f \circ g)(x) = x + 1$. Now, what about the domain of $f \circ g$? Our first inclination may be to say (incorrectly!) that the domain is the set of all real numbers, since any real number can be used as an input in the expression $x + 1$. However, the definition of $f \circ g$ on page 144 tells us that the inputs for $f \circ g$ must first of all be acceptable inputs for g. Given the definition of g, then, we must require that x be nonnegative. On the other hand, for any nonnegative input x, the number $g(x)$ will be an acceptable input for f. (Why?) In summary, then, the domain of $f \circ g$ is the interval $[0, \infty)$. The graph of $f \circ g$ is shown in Figure 2.

One reason for studying the composition of functions is that it lets us express a given function in terms of simpler functions. This is often useful in calculus. Suppose, for example, that we wish to express the function C defined by

$$C(x) = (2x^3 - 5)^2$$

as a composition of simpler functions. That is, we want to come up with two functions f and g so that the equation

$$C(x) = (f \circ g)(x)$$

holds for every x in the domain of C.

We begin by thinking what we would do to compute $(2x^3 - 5)^2$ for a given value of x. First, we would compute the quantity $2x^3 - 5$, then we would square the result. Therefore, recalling that the rule $f \circ g$ tells us to do g *first,* we let $g(x) = 2x^3 - 5$. And then, since the next step is squaring, we let $f(x) = x^2$. Now let's see if these choices for f and g are correct; that is, let us calculate $(f \circ g)(x)$ and see if it really is the same as $C(x)$.

Using $f(x) = x^2$ and $g(x) = 2x^3 - 5$, we have

$$(f \circ g)(x) = f[g(x)] = f(2x^3 - 5) = (2x^3 - 5)^2 = C(x)$$

This shows that our choices for f and g were indeed correct, and we have expressed C as a composition of two simpler functions.*

Note that in expressing C as $f \circ g$, we chose g to be the "inner" function, that is, the quantity inside the parentheses: $g(x) = 2x^3 - 5$. This observation is used in Example 5.

EXAMPLE 5 Let $s(x) = \sqrt{1 + x^4}$. Express the function s as a composition of two simpler functions f and g.

Solution Let g be the "inner" function; that is, let $g(x)$ be the quantity inside the radical:

$$g(x) = 1 + x^4$$

And let's take f to be the square root function:

$$f(x) = \sqrt{x}$$

*Other answers are possible, too. For instance, if $F(x) = (x - 5)^2$ and $G(x) = 2x^3$, then (as you should verify for yourself) $C(x) = F[G(x)]$.

Now we need to verify that these are the appropriate choices for f and g; that is, we need to check that the equation $(f \circ g)(x) = s(x)$ is true for every x in the domain of s. We have

$$(f \circ g)(x) = f[g(x)] = f(1 + x^4) = \sqrt{1 + x^4} = s(x)$$

Thus, $(f \circ g)(x) = s(x)$, as required.

We conclude this section by describing the process of *iteration* for a function. This simple process is of fundamental importance in modern mathematics and science in the study of fractals and chaos.* The process of iteration is sequential; it proceeds step-by-step. We begin with a function f and an initial input x_0. In the first step, we compute the output $f(x_0)$. For the second step, we use the number $f(x_0)$ as an *input* for f and compute $f(f(x_0))$. The process then continues in this way: at each step, we use the output from the previous step as the new input.

For example, suppose that we start with the function $f(x) = x/2$ and the input $x_0 = 6$. Then, for the first three steps in the iteration process, we have

$$f(6) = \frac{6}{2} = 3$$

$$f(3) = \frac{3}{2} = 1.5$$

$$f(1.5) = \frac{1.5}{2} = 0.75$$

After this, a calculator becomes convenient. Check for yourself that the iterations run as follows:

$$6 \to 3 \to 1.5 \to 0.75 \to 0.375 \to 0.1875 \to 0.09375 \to 0.046875 \to \cdots$$

This list of numbers is called the *orbit* of 6 under the function f. In the list or orbit, the number 3 is the *first iterate* of 6, the number 1.5 is the *second iterate* of 6, and so on. In the box that follows, we summarize these ideas and introduce a useful notation for the iterates.

DEFINITION: Iterates

Given a function f and an input x_0, the **iterates** of x_0 are the numbers $f(x_0)$, $f(f(x_0))$, $f(f(f(x_0)))$, and so on. The number $f(x_0)$ is called the **first iterate,** the number $f(f(x_0))$ is called the **second iterate,** and so on. Subscript notation is used to denote the iterates as follows.

$x_1 = f(x_0)$ the first iterate

$x_2 = f(f(x_0))$ the second iterate

$x_3 = f(f(f(x_0)))$ the third iterate

$\vdots$

The **orbit** of x_0 under the function f is the list of numbers $x_0, x_1, x_2, x_3, \ldots$.

*See, for example, pp. 314–363 in *The Nature and Power of Mathematics* by Donald M. Davis (Princeton: Princeton University Press, 1993). For a wider view, see the bestseller by James Gleick: *Chaos: Making a New Science* (New York: Penguin Books, 1988).

Notice that the subscripts in this notation tell you how many times to apply the function f. For example, x_1 indicates that f is applied once to obtain $f(x_0)$, and x_2 indicates that the function is applied twice to obtain $f(f(x_0))$.

EXAMPLE 6 Compute the first four iterates in each case. Use a calculator as necessary. In part (b), round the final answers to three decimal places.

(a) $f(x) = x^2$, $x_0 = -2$; **(b)** $g(x) = \sqrt{x}$, $x_0 = 0.1$

Solution

(a)
$$x_1 = (-2)^2 = 4$$
$$x_2 = (4)^2 = 16$$
$$x_3 = (16)^2 = 256$$
$$x_4 = (256)^2 = 65{,}536$$

(b) We use the square root key (or keys) on the calculator repeatedly. As you should check for yourself, this yields

$$x_1 = \sqrt{0.1} = 0.316227\ldots \approx 0.316 \quad \text{rounding to three decimal places}$$
$$x_2 = \sqrt{0.316227\ldots} = 0.562341\ldots \approx 0.562$$
$$x_3 = \sqrt{0.562341\ldots} = 0.749894\ldots \approx 0.750$$
$$x_4 = \sqrt{0.749989\ldots} = 0.865964\ldots \approx 0.866$$

The process of iterating a function can be visualized in an x-y coordinate system. We sketch the function $y = f(x)$ along with the line $y = x$, as in Figure 3. Then, starting with the initial input x_0 on the x-axis, we follow the horizontal and vertical segments to determine points P, Q, and R on the graph of f. As the next paragraph explains, the y-coordinates of the points P, Q, and R are the first, second, and third iterates, respectively, of x_0. Additional iterates are obtained by continuing this pattern. In Figure 4 we show how this looks in a specific case. The y-coordinates of the five points P, Q, R, S, and T are the first five iterates of $x_0 = 0.1$ under the function $g(x) = -4x^2 + 4x$. Although Figure 4 at

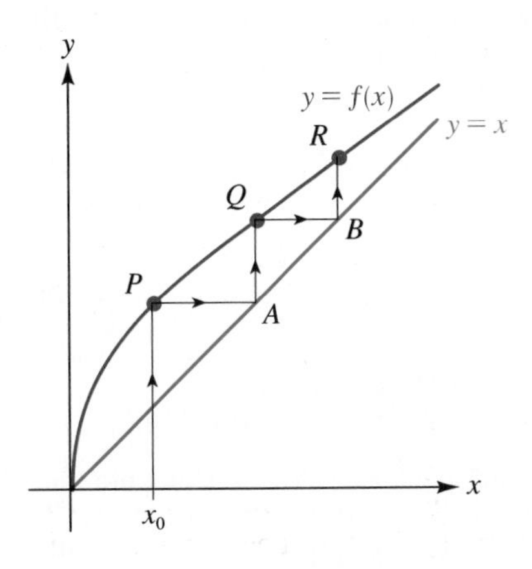

FIGURE 3
The y-coordinates of the points P, Q, and R are, respectively, the first three iterates of x_0 under the function f.

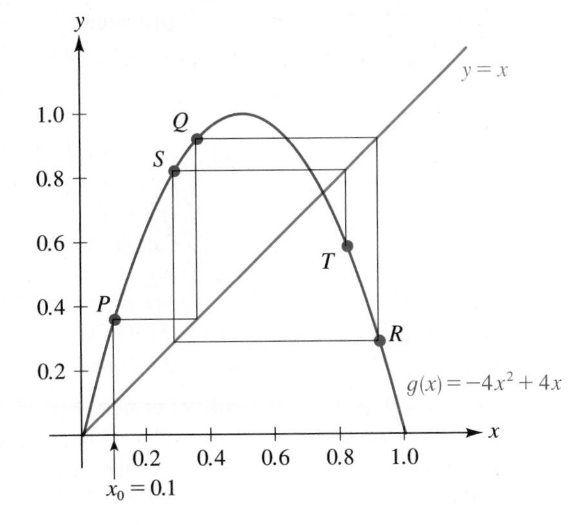

FIGURE 4
Graphical iteration of $g(x) = -4x^2 + 4x$ with initial input $x_0 = 0.1$. The y-coordinates of the five points P, Q, R, S, and T are, respectively, the first five iterates of x_0.

first appears more complicated than Figure 3, it is important to understand that the pattern is formed in exactly the same way. We start with the initial input on the x-axis and travel vertically to determine the point P on the curve. After that, each successive point on the curve is generated in the same way: Go horizontally to the line $y = x$ and then vertically to the curve.

As we've said, this paragraph explains why the y-coordinates of the three points P, Q, and R in Figure 3 are, respectively, the first three iterates of x_0 under the function f. In Figure 3, the x-coordinate of P is given to be x_0. So (according to the definition of the graph of a function), the y-coordinate of P is $f(x_0)$. In other words, the y-coordinate of P is the first iterate of x_0. Next in Figure 3, we move horizontally from the point P to the point A. Because the movement is horizontal, the y-coordinate doesn't change; it's the y-coordinate of P, or $f(x_0)$. Now, what about the x-coordinate of A? Since the point A lies on the line $y = x$, the x- and y-coordinates of A must be identical. Consequently, the x-coordinate of A must be the number $f(x_0)$. Finally, we move vertically in Figure 3 from A to Q. The point Q has the same x-coordinate as the point A, namely, $f(x_0)$. Now, since Q lies on the graph of the function f, we conclude that the y-coordinate of Q is $f(f(x_0))$. That is, the y-coordinate of Q is indeed the second iterate of x_0. The reasoning to show that the y-coordinate of R is the third iterate of x_0 is entirely similar, so we omit giving the details here.

EXERCISE SET 3.4

Do 1, 11, 15, 19, 23

A

In Exercises 1–10, compute each expression, given that the functions f, g, h, k, and m are defined as follows:

$f(x) = 2x - 1$ $\quad$ $k(x) = 2$, *for all x*

$g(x) = x^2 - 3x - 6$ $\quad$ $m(x) = x^2 - 9$

$h(x) = x^3$

1. **(a)** $(f + g)(x)$ **(b)** $(f - g)(x)$ **(c)** $(f - g)(0)$
2. **(a)** $(fh)(x)$ **(b)** $(h/f)(x)$ **(c)** $(f/h)(1)$
3. **(a)** $(m - f)(x)$ **(b)** $(f - m)(x)$
4. **(a)** $(fg)(x)$ **(b)** $(fg)(1/2)$
5. **(a)** $(fk)(x)$ **(b)** $(kf)(x)$ **(c)** $(fk)(1) - (kf)(2)$
6. **(a)** $(g + m)(x)$ **(b)** $(g + m)(x) - (g - m)(x)$
7. **(a)** $(f/m)(x) - (m/f)(x)$ **(b)** $(f/m)(0) - (m/f)(0)$
8. **(a)** $[h \cdot (f + m)](x)$ *Note:* h and $(f + m)$ are two functions; the notation $h \cdot (f + m)$ denotes the product function. **(b)** $(hf)(x) + (hm)(x)$
9. **(a)** $[m \cdot (k - h)](x)$ **(b)** $(mk)(x) - (mh)(x)$ **(c)** $(mk)(-1) - (mh)(-1)$
10. **(a)** $(g + g)(x)$ **(b)** $(g - g)(x)$ **(c)** $(kg)(x)$ **(d)** $(g + g)(-3) - (kg)(-3)$
11. Let $f(x) = 3x + 1$ and $g(x) = -2x - 5$. Compute the following.
    (a) $(f \circ g)(x)$ **(b)** $(f \circ g)(10)$ **(c)** $(g \circ f)(x)$ **(d)** $(g \circ f)(10)$
12. Let $f(x) = 1 - 2x^2$ and $g(x) = x + 1$. Compute the following.
    (a) $(f \circ g)(x)$ **(b)** $(f \circ g)(-1)$ **(c)** $(g \circ f)(x)$ **(d)** $(g \circ f)(-1)$ **(e)** $(f \circ f)(x)$ **(f)** $(g \circ g)(-1)$
13. Compute $(f \circ g)(x)$, $(f \circ g)(-2)$, $(g \circ f)(x)$, and $(g \circ f)(-2)$ for each pair of functions.
    (a) $f(x) = 1 - x$; $g(x) = 1 + x$
    (b) $f(x) = x^2 - 3x - 4$; $g(x) = 2 - 3x$
    (c) $f(x) = x/3$; $g(x) = 1 - x^4$
    (d) $f(x) = 2^x$; $g(x) = x^2 + 1$
    (e) $f(x) = x$; $g(x) = 3x^5 - 4x^2$
    (f) $f(x) = 3x - 4$; $g(x) = (x + 4)/3$
14. Let $h(x) = 4x^2 - 5x + 1$, $k(x) = x$, and $m(x) = 7$ for all x. Compute the following.
    (a) $h[k(x)]$ **(b)** $k[h(x)]$ **(c)** $h[m(x)]$ **(d)** $m[h(x)]$ **(e)** $k[m(x)]$ **(f)** $m[k(x)]$

15. Let $F(x) = \dfrac{3x-4}{3x+3}$ and $G(x) = \dfrac{x+1}{x-1}$. Compute the following.
(a) $(F \circ G)(x)$ **(b)** $F[G(t)]$ **(c)** $(F \circ G)(2)$
(d) $(G \circ F)(x)$ **(e)** $G[F(y)]$ **(f)** $(G \circ F)(2)$

16. Let $f(x) = (1/x^2) + 1$ and $g(x) = 1/(x-1)$.
(a) Compute $(f \circ g)(x)$.
(b) What is the domain of $f \circ g$?
(c) Graph the function $f \circ g$.

17. Let $M(x) = (2x-1)/(x-2)$.
(a) Compute $M(7)$ and then $M[M(7)]$.
(b) Compute $(M \circ M)(x)$.
(c) Compute $(M \circ M)(7)$, using the formula you obtained in part (b). Check that your answer agrees with that obtained in part (a).

18. Let $F(x) = (x+1)^5$, $f(x) = x^5$, and $g(x) = x+1$. Which of the following is true for all x?

$$(f \circ g)(x) = F(x) \qquad \text{or} \qquad (g \circ f)(x) = F(x)$$

19. Refer to the graphs of the functions f, g, and h to compute the required quantities. Assume that all the axes are marked off in one-unit intervals.

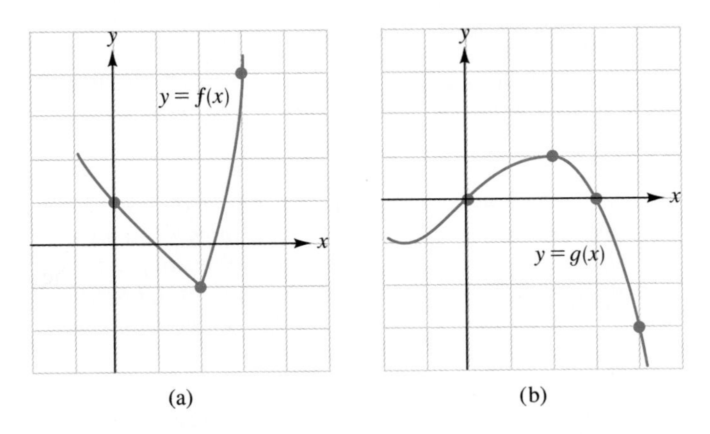

(a) (b)

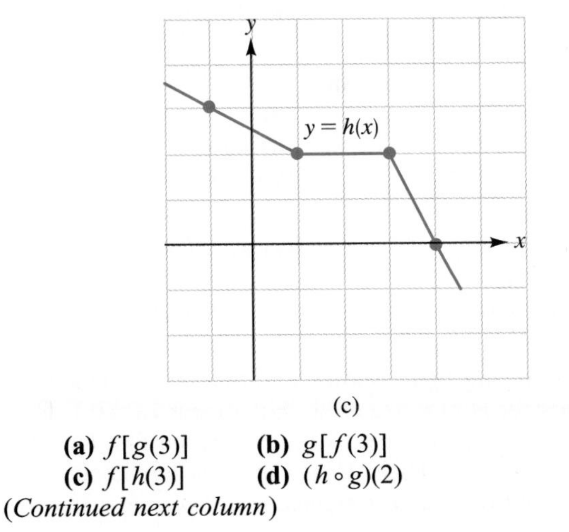

(c)

(a) $f[g(3)]$ **(b)** $g[f(3)]$
(c) $f[h(3)]$ **(d)** $(h \circ g)(2)$
(Continued next column)

(e) $h\{f[g(3)]\}$
(f) $(g \circ f \circ h \circ f)(2)$ *Note:* This means first do f, then h, then f, then g.

20. Let $f(x) = 2x - 1$. Find $f[f(x)]$. If $z = f[f(x)]$, find $f(z)$.

21. **(a)** Let $T(x) = 4x^3 - 3x^2 + 6x - 1$ and $I(x) = x$. Find $(T \circ I)(x)$ and $(I \circ T)(x)$.
(b) Let $G(x) = ax^2 + bx + c$ and $I(x) = x$. Find $(G \circ I)(x)$ and $(I \circ G)(x)$.
(c) What general conclusion do you arrive at from the results of parts (a) and (b)?

22. The figure shows the graphs of two functions, F and G. Use the graphs to compute each quantity.
(a) $(G \circ F)(1)$ **(b)** $(F \circ G)(1)$
(c) $(F \circ F)(1)$ **(d)** $(G \circ G)(-3)$
(e) $(G \circ F)(5)$ **(f)** $(F \circ F)(5)$
(g) $(F \circ G)(-1)$ **(h)** $(G \circ F)(2)$

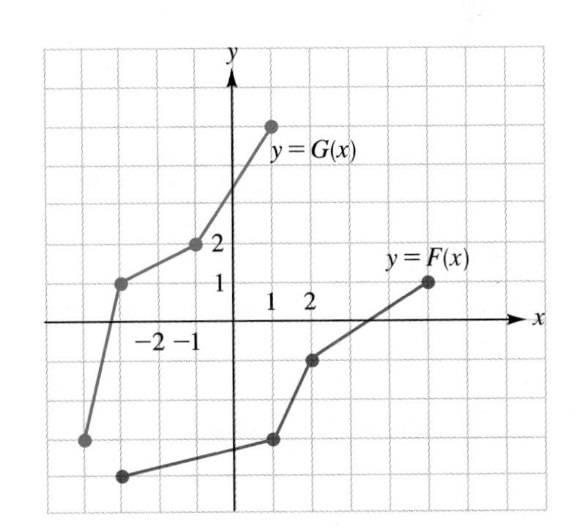

23. The domain of a function f consists of the numbers -1, 0, 1, 2, and 3. The following table shows the output that f assigns to each input.

x	-1	0	1	2	3
$f(x)$	2	2	0	3	1

The domain of a function g consists of the numbers 0, 1, 2, 3, and 4. The following table shows the output that g assigns to each input.

x	0	1	2	3	4
$g(x)$	3	2	0	4	-1

Use this information to complete the following tables for $f \circ g$ and $g \circ f$. *Note:* Two of the entries will be undefined.

x	0	1	2	3	4
$(f \circ g)(x)$					

x	-1	0	1	2	3	4
$(g \circ f)(x)$						

24. The following two tables show certain pairs of inputs and outputs for functions f and g.

x	0	$\pi/6$	$\pi/4$	$\pi/3$	$\pi/2$
$f(x)$	0	1/2	$\sqrt{2}/2$	$\sqrt{3}/2$	1

y	0	1/4	$\sqrt{2}/4$	1/2	$\sqrt{2}/2$	3/4	$\sqrt{3}/2$	1
$g(y)$	$\pi/2$	π	0	$\pi/3$	$\pi/4$	0	$\pi/6$	0

Use this information to complete the following table of values for $(g \circ f)(x)$.

x	0	$\pi/6$	$\pi/4$	$\pi/3$	$\pi/2$
$(g \circ f)(x)$					

25. Let $f(x) = 2x + 1$ and $g(x) = 3x - 4$.
(a) Find $(f \circ g)(x)$ and graph the function $f \circ g$.
(b) Find $(g \circ f)(x)$ and graph the function $g \circ f$.

26. Let $F(x) = x^2$ and $G(x) = x - 4$.
(a) Find $F[G(x)]$ and graph the function $F \circ G$.
(b) Find $G[F(x)]$ and graph the function $G \circ F$.

27. Let $g(x) = \sqrt{x} - 3$ and $f(x) = x - 1$.
(a) Sketch a graph of g. Specify the domain and range.
(b) Sketch a graph of f. Specify the domain and range.
(c) Compute $(f \circ g)(x)$. Graph the function $f \circ g$ and specify its domain and range.
(d) Find a formula for $g[f(x)]$. Which values of x are acceptable inputs here? That is, what is the domain of $g \circ f$?
(e) Use the results of part (d) to sketch a graph of the function $g \circ f$.

28. Let $F(x) = -x^2$ and $G(x) = \sqrt{x}$. Determine the domains of $F \circ G$ and $G \circ F$.

29. Let $C(x) = (3x - 1)^4$. Express C as a composition of two simpler functions.

30. Let $C(x) = (1 + x^2)^3$. Find functions f and g so that $C(x) = (f \circ g)(x)$ is true for all values of x.

31. Express each function as a composition of two functions.
(a) $F(x) = \sqrt[3]{3x + 4}$ **(b)** $G(x) = |2x - 3|$
(c) $H(x) = (ax + b)^5$ **(d)** $T(x) = 1/\sqrt{x}$

32. Let $a(x) = x^2$, $b(x) = |x|$, and $c(x) = 3x - 1$. Express each of the following functions as a composition of two of the given functions.
(a) $f(x) = (3x - 1)^2$ **(b)** $g(x) = |3x - 1]$
(c) $h(x) = 3x^2 - 1$

33. Let $a(x) = 1/x$, $b(x) = \sqrt[3]{x}$, $c(x) = 2x + 1$, and $d(x) = x^2$. Express each of the following functions as a composition of two of the given functions.
(a) $f(x) = \sqrt[3]{2x + 1}$ **(b)** $g(x) = 1/x^2$
(c) $h(x) = 2x^2 + 1$ **(d)** $K(x) = 2\sqrt[3]{x} + 1$
(e) $l(x) = \dfrac{2}{x} + 1$ **(f)** $m(x) = \dfrac{1}{2x + 1}$
(g) $n(x) = x^{2/3}$

34. Express $G(x) = 1/(1 + x^4)$ as a composition of two simpler functions, one of which is $f(x) = 1/x$.

In Exercises 35–40, use the given function and compute the first six iterates of each initial input x_0. In cases where a calculator answer contains four or more decimal places, round the final answer to three decimal places. (However, during the calculations, work with all of the decimal places that your calculator affords.)

35. $f(x) = 2x$
(a) $x_0 = 1$
(b) $x_0 = 0$
(c) $x_0 = -1$

36. $f(x) = \frac{1}{4}x$
(a) $x_0 = 16$
(b) $x_0 = 0$
(c) $x_0 = -16$

37. $g(x) = 2x + 1$
(a) $x_0 = -2$
(b) $x_0 = -1$
(c) $x_0 = 1$

38. $g(x) = \frac{1}{4}x + 3$
(a) $x_0 = 3$
(b) $x_0 = 4$
(c) $x_0 = 5$

39. $F(x) = x^2$
(a) $x_0 = 0.9$
(b) $x_0 = 1$
(c) $x_0 = 1.1$

40. $G(x) = x^2 + 0.25$
(a) $x_0 = 0.4$
(b) $x_0 = 0.5$
(c) $x_0 = 0.6$

41. The following figure shows a portion of the iteration process for $f(x) = \sqrt{x}$ with initial input $x_0 = 0.1$. Use the figure to estimate (to within 0.05) the first four iterates of $x_0 = 0.1$. Then, use a calculator to compute the four iterates. (Round to three decimal places, where appropriate.) Check that your calculator results are consistent with the estimates obtained from the graph.

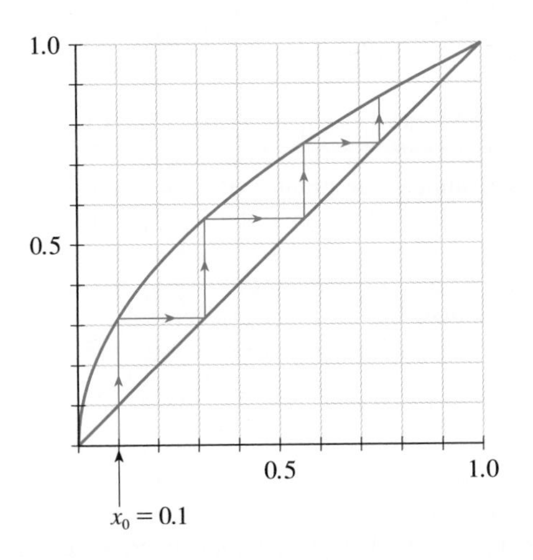

42. The following figure shows a portion of the iteration process for $f(x) = 3.6(x - x^2)$ with initial input $x_0 = 0.3$. Use the figure to estimate (to within 0.1, or closer if it seems appropriate) the first seven iterates of $x_0 = 0.3$. Then, use a calculator to compute the seven iterates. (Round the final answers to three decimal places.) Check that your calculator results are consistent with the estimates obtained from the graph.

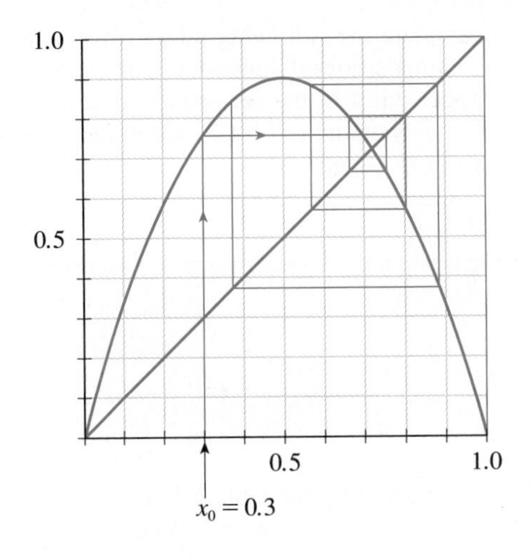

B

43. The circumference of a circle with radius r is given by $C(r) = 2\pi r$. Suppose that the circle is shrinking in size and that the radius at time t is given by

$$r = f(t) = \frac{1}{t^2 + 1}$$

Assume that t is measured in minutes and that r, or $f(t)$, is in feet. Find $(C \circ f)(t)$ and use this to find the circumference when $t = 3$ min.

44. A spherical weather balloon is being inflated in such a way that the radius is given by

$$r = g(t) = \tfrac{1}{2}t + 2$$

Assume that r is in meters and t is in seconds, with $t = 0$ corresponding to the time that inflation begins. If the volume of a sphere of radius r is given by

$$V(r) = \tfrac{4}{3}\pi r^3$$

compute $V[g(t)]$ and use this to find the time at which the volume of the balloon is $36\pi\ \text{m}^3$.

45. Suppose that a manufacturer knows that the daily production cost to build x bicycles is given by the function C, where

$$C(x) = 100 + 90x - x^2 \qquad (0 \leq x \leq 40)$$

That is, $C(x)$ represents the cost in dollars of building x bicycles. Furthermore, suppose that the number of bicycles that can be built in t hr is given by the function f, where

$$x = f(t) = 5t \qquad (0 \leq t \leq 8)$$

(a) Compute $(C \circ f)(t)$.
(b) Compute the production cost on a day that the factory operates for $t = 3$ hr.
(c) If the factory runs for 6 hr instead of 3 hr, is the cost twice as much?

46. Suppose that in a certain biology lab experiment, the number of bacteria is related to the temperature T of the environment by the function

$$N(T) = -2T^2 + 240T - 5400 \qquad (40 \leq T \leq 90)$$

Here, $N(T)$ represents the number of bacteria present when the temperature is T degrees Fahrenheit. Also, suppose that t hr after the experiment begins, the temperature is given by

$$T(t) = 10t + 40 \qquad (0 \leq t \leq 5)$$

(a) Compute $N[T(t)]$.
(b) How many bacteria are present when $t = 0$ hr? When $t = 2$ hr? When $t = 5$ hr?

47. Let $g(x) = 4x - 1$. Find $f(x)$, given that the equation $(g \circ f)(x) = x + 5$ is true for all values of x.

48. Let $g(x) = 2x + 1$. Find $f(x)$, given that $(g \circ f)(x) = 10x - 7$.

49. Let $f(x) = -2x + 1$ and $g(x) = ax + b$. Find a and b so the equation $f[g(x)] = x$ holds for all values of x.

50. Let $f(x) = (3x - 4)/(x - 3)$.
(a) Compute $(f \circ f)(x)$.
(b) Find $f[f(113/355)]$. (Try not to do it the hard way.)

51. Let $f(x) = x^2$ and $g(x) = 2x - 1$.
(a) Compute $\dfrac{f[g(x)] - f[g(a)]}{g(x) - g(a)}$.
(b) Compute $\dfrac{f[g(x)] - f[g(a)]}{x - a}$.

52. Suppose that $y = f(x)$ and $g(y) = x$. Show that $(g \circ f)(x) - (f \circ g)(y) = x - y$.

53. Given three functions f, g, and h, we define the function $f \circ g \circ h$ by the formula

$$(f \circ g \circ h)(x) = f\{g[h(x)]\}$$

For example, if $f(x) = x^2$, $g(x) = x + 1$, and $h(x) = x/2$, we evaluate $(f \circ g \circ h)(x)$ as

$$\begin{aligned}(f \circ g \circ h)(x) &= f\{g[h(x)]\} \\ &= f\left[g\left(\frac{x}{2}\right)\right] = f\left(\frac{x}{2} + 1\right) \\ &= \left(\frac{x}{2} + 1\right)^2 = \frac{x^2}{4} + x + 1\end{aligned}$$

Using $f(x) = x^2$, $g(x) = x + 1$, and $h(x) = x/2$, compute each composition.
(a) $(g \circ h \circ f)(x)$
(b) $(h \circ f \circ g)(x)$
(c) $(g \circ f \circ h)(x)$
(d) $(f \circ h \circ g)(x)$
(e) $(h \circ g \circ f)(x)$
Hint for checking: No two answers are the same.

54. Let $F(x) = 2x - 1$, $G(x) = 4x$, and $H(x) = 1 + x^2$. Compute each composition.
(a) $(F \circ G \circ H)(x)$
(b) $(G \circ F \circ H)(x)$
(c) $(F \circ H \circ G)(x)$
(d) $(H \circ F \circ G)(x)$
(e) $(H \circ G \circ F)(x)$
(f) $(G \circ H \circ F)(x)$

55. Let $f(x) = x^2$, $g(x) = 1 - x$, and $h(x) = 3x$. Express each function as a composition of f, g, and h.
(a) $p(x) = 1 - 9x^2$
(b) $q(x) = 3 - 3x^2$
(c) $r(x) = 1 - 6x + 9x^2$
(d) $s(x) = 3 - 6x + 3x^2$

56. Suppose that three functions f', g, and g' are defined as follows (f' is read *f prime*):

$$f'(x) = 3x^2 \qquad g(x) = 2x^2 + 5 \qquad g'(x) = 4x$$

(a) Find $f'[g(x)] \cdot g'(x)$. (The dot denotes multiplication, not composition.)
(b) Evaluate $f'[g(-1)] \cdot g'(-1)$.
(c) Find $f'[g(t)] \cdot g'(t)$.

57. *The $3x + 1$ conjecture* Define a function f, with domain the positive integers, as follows:

$$f(x) = \begin{cases} 3x + 1 & \text{if } x \text{ is odd} \\ x/2 & \text{if } x \text{ is even} \end{cases}$$

(a) Compute $f(1)$, $f(2)$, $f(3)$, $f(4)$, $f(5)$, and $f(6)$.
(b) Compute the first three iterates of $x_0 = 1$.
(c) Compute the iterates of $x_0 = 3$ until you obtain the value 1. (After this, the iterates will recycle through the simple pattern obtained in part (b).)
(d) *The $3x + 1$ conjecture* asserts that for any positive integer x_0, the iterates eventually return to the value 1. Verify that this conjecture is valid for each of the following values of x_0: 2, 4, 5, 6, and 7.

Remark: At present, the $3x + 1$ conjecture is indeed a conjecture, not a theorem. No one yet has found a proof that the assertion is valid for *every* positive integer. Computer checks, however, have verified the conjecture on a case-by-case basis for values of x_0 ranging into the billions. For a history of this problem and references, see the article, "The $3x + 1$ Problem and Its Generalizations" by Jeffrey C. Lagarias in *The American Mathematical Monthly* (vol. 92, Jan. 1985, pp. 3–23).

C

58. Let $i(x) = x$, $a(x) = -x$, $b(x) = 1/x$, and $c(x) = -1/x$. Assume that the domain of each function is $(0, \infty)$.
(a) Complete the following table, where the operation is composition of functions. For example, c is the proper entry in the second row, third column, since if you compute $(a \circ b)(x)$ you'll get $-1/x$, which is $c(x)$.

$\circ$	i	a	b	c
i				
a				
b				
c				

(b) According to your table, does $a \circ b = b \circ a$?
(c) Use your table to find a^2 (meaning $a \circ a$), b^2, c^2, and c^3.
(d) Use your table to find out if $(a \circ b) \circ c = a \circ (b \circ c)$.
(e) Let G denote the set consisting of the four functions you've worked with in this exercise. Now define a function f, with domain G, as follows: $f(t) = t^2$ for each element t in the set G. [For example, $f(a) = a^2 = a \circ a = i$.] What is the range of f?

(f) Verify that $f(a \circ b) = [f(a)] \circ [f(b)]$.
(g) Find out if $f(a \circ b \circ c) = [f(a)] \circ [f(b)] \circ [f(c)]$.

59. Let p and q be two numbers that add to 1. Define the function F as follows (next column):

$$F(x) = p - \frac{1}{x+q}$$

Show that $F\{F[F(x)]\} = x$.

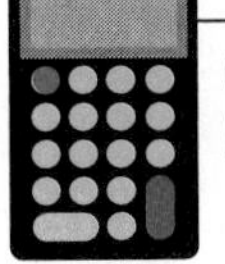

GRAPHING UTILITY EXERCISES FOR SECTION 3.4

In Exercises 1–4, complete each table of values. In Exercise 4, round your answers to three decimal places. [Before proceeding, you may want to check the instruction manual for your graphing utility or the Graphics Calculator Supplement for this text to see how your machine operates with respect to composition of functions. For instance, for the TI-82 graphics calculator, you can enter two functions f and g and then evaluate $f(g(x_0))$ without first obtaining an explicit formula for $f(g(x))$.] As you complete the tables in Exercises 1–4, note that, in general, $f(g(x)) \neq g(f(x))$.

1. $f(x) = 2x^3,\ g(x) = 5 - 4x$

x	0.1	0.2	0.3
$f(g(x))$			
$g(f(x))$			

2. $f(x) = 4x - 1,\ g(x) = -x^2 + 5x + 1$

x	1.4	1.5	1.6
$f(g(x))$			
$g(f(x))$			

3. $f(x) = 3x^2 - x + 4,\ g(x) = 5 - 4x$

x	7	8	9
$f(g(x))$			
$g(f(x))$			

4. $f(x) = \sqrt{x} - 4,\ g(x) = 2x^2 - x$

x	10	11	12
$f(g(x))$			
$g(f(x))$			

5. Suppose that an oil spill in a lake covers a circular area and that the radius of the circle is increasing according to the formula $r = f(t) = 15 + t^{1.65}$, where t represents the number of hours since the spill was first observed and the radius r is measured in meters. (Thus, when the spill was first discovered, $t = 0$ hr and the initial radius was $r = f(0) = 15 + 0^{1.65} = 15$ m.)
(a) Since the area $A(r)$ of any circle is πr^2, the area of the oil spill after t hours is given by the composite function $(A \circ f)(t)$. Compute a table of values for this composite function with t running from 0 to 5 in increments of 0.5. (Round each output to the nearest integer.) Then use the table to answer the questions that follow in parts (b) through (d).
(b) After one hour, what is the area of the spill (rounded to the nearest 10 m^2)?
(c) Initially, what was the area of the spill (when $t = 0$)? Approximately how many hours does it take for this area to double?
(d) Compute the average rate of change of the area of the spill from $t = 0$ to $t = 2.5$ and from $t = 2.5$ to $t = 5$. Over which of the two intervals is the area increasing faster?

Exercises 6–8 indicate how iteration is used in finding roots of numbers and roots of equations. (The functions that are given in each exercise were determined using Newton's method, *a process studied in calculus.)*

6. Let $f(x) = 0.5\left(x + \dfrac{3}{x}\right)$.
(a) Compute the first ten iterates of $x_0 = 1$ under the function f. What do you observe?
(b) Use your calculator to evaluate $\sqrt{3}$ and compare the answer to your results in part (a). What do you observe?
(c) It can be shown that for any positive number x_0, the iterates of x_0 under the function $f(x) = 0.5(x + 3/x)$ always approach the number $\sqrt{3}$. (You'll see the reasons for this in Section 4.3.)

Looking at your results in parts (a) and (b), which is the first iterate that agrees with $\sqrt{3}$ through the first three decimal places? Through the first eight decimal places?

(d) Compute the first ten iterates of $x_0 = 50$ under the function f, then answer the questions presented in part (c).

7. Let $f(x) = (2x^3 + 7)/3x^2$.

(a) Compute the first ten iterates of $x_0 = 1$ under the function f. What do you observe?

(b) Evaluate the expression $\sqrt[3]{7}$ and compare the answer to your results in part (a). What do you observe?

(c) It can be shown that for any positive number x_0, the iterates of x_0 under the function $f(x) = (2x^3 + 7)/3x^2$ always approach the number $\sqrt[3]{7}$. Looking at your results in parts (a) and (b), which is the first iterate that agrees with $\sqrt[3]{7}$ through the first three decimal places? Through the first eight decimal places?

8. The following figure shows the graph of the function $y = x^5 - 4x + 2$. According to the figure, the graph has an x-intercept between $x = 1$ and $x = 2$. In other words, the equation $x^5 - 4x + 2 = 0$ has a root between 1 and 2. Call the root r, as indicated in the figure. An interesting fact about this root, this real number r, is that it is an irrational number that *cannot* be written in terms of radicals. (Another, more familiar example of an irrational number that cannot be expressed in terms of radicals is π.)

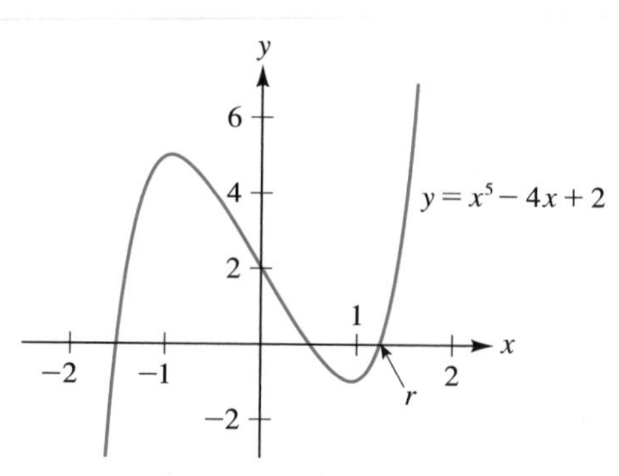

(a) Although the real number r cannot be expressed in terms of radicals, it can be determined to as many decimal places as we please by iterating the function $f(x) = (4x^5 - 2)/(5x^4 - 4)$, with initial input $x_0 = 2$. Compute the first eight iterates, keep a record of the results, and then go on to part (b).

(b) The value of the root r, correct through ten decimal places is 1.2435963905. Which is the first iterate to agree with r through the first three decimal places? Through the first eight decimal places?

3.5 INVERSE FUNCTIONS

PROBLEM
SOLUTION

If anybody ever told me why the graph of $y = x^{1/2}$ is the reflection of the graph of $y = x^2$ in a 45° line, it didn't sink in. To this day, there are textbooks that expect students to think that it is so obvious as to need no explanation. This is a pity, if only because [in calculus] *it is such a common practice to define the natural logarithm first and then define the exponential function as its inverse.*

Professor Ralph P. Boas, recalling his student days in the article "Inverse Functions" in *The College Mathematics Journal,* vol. 16(1985), p. 42.

We shall introduce the idea of inverse functions through an easy example. Consider the function f defined by $f(x) = 2x$. A short table of values for f and the graph of f are displayed in Figure 1 on the next page.

Now we ask this question: What happens if we take the table for f (in Figure 1) and interchange the entries in the x and y columns to obtain a new table that looks like the following?

x	y
−4	−2
−2	−1
0	0
2	1
4	2

← New table formed by interchanging the inputs and outputs for f

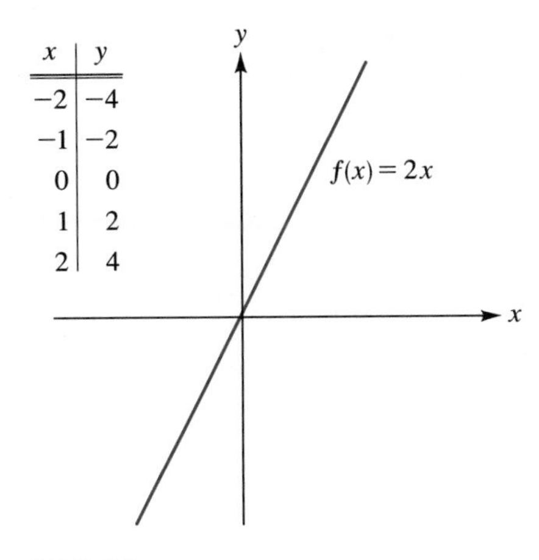

FIGURE 1

Several observations are in order. Our new table does itself represent a function (in this case): for each input x there is exactly one output y. Let's call this new function g. Note that each output is one-half of the corresponding input. So the function g can be described by the formula $g(x) = x/2$. In summary, we began with the "doubling function," $f(x) = 2x$; then, by interchanging the inputs with the outputs, we obtained the "halving function," $g(x) = x/2$.

The doubling function f and the halving function g are in a certain sense opposites; each reverses, or undoes, the effect of the other. That is, if you begin with x and then calculate $g[f(x)]$, you get x again; similarly, $f[g(x)]$ is also equal to x. (Verify these last two statements for yourself.)

The two functions f and g that we've been discussing are an example of a pair of *inverse functions.* The general definition for inverse functions is given in the box that follows.

> **DEFINITION: Inverse Functions**
>
> Two functions f and g are **inverses** of one another provided that
>
> $$f[g(x)] = x \qquad \text{for each } x \text{ in the domain of } g$$
>
> and
>
> $$g[f(x)] = x \qquad \text{for each } x \text{ in the domain of } f$$

EXAMPLE 1 Verify that the functions f and g are inverses, where

$$f(x) = \frac{1}{3}x + 2 \qquad \text{and} \qquad g(x) = 3x - 6$$

Solution In view of the definition, we must check that $f[g(x)] = x$ and $g[f(x)] = x$. We have

$$\begin{aligned} f[g(x)] &= f(3x - 6) \\ &= \tfrac{1}{3}(3x - 6) + 2 = x \qquad \text{(Check the algebra.)} \end{aligned}$$

Thus, $f[g(x)] = x$. Now we still need to check that $g[f(x)] = x$. We have

$$\begin{aligned} g[f(x)] &= g\left(\tfrac{1}{3}x + 2\right) \\ &= 3\left(\tfrac{1}{3}x + 2\right) - 6 = x \qquad \text{(Again, check the algebra.)} \end{aligned}$$

Having shown that $f[g(x)] = x$ and $g[f(x)] = x$, we conclude that f and g are indeed inverse functions.

EXAMPLE 2 Suppose that f and g are a pair of inverse functions. If $f(2) = 3$, what is $g(3)$?

Solution We can use either of two methods.

FIRST METHOD The quantities 3 and $f(2)$ are declared to be equal. Thus, whether we use 3 or $f(2)$ as an input for g, the result must be the same. We then have

$$\begin{aligned} 3 &= f(2) \\ g(3) &= g[f(2)] \\ g(3) &= 2 \qquad \text{using the fact that } g[f(x)] = x \text{ for any } x \text{ in the domain of } f \end{aligned}$$

Thus $g(3) = 2$.

ALTERNATIVE METHOD Think of a table of values for the function f. Since $f(2) = 3$, one entry in the table must look like this:

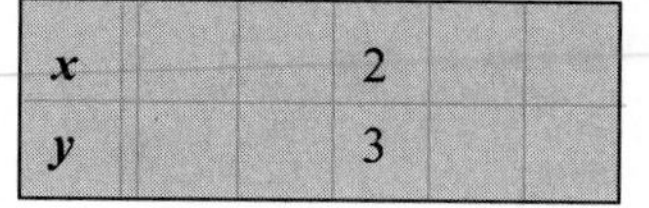

x			2		
y			3		

Now, g reverses the roles of x and y, so in the table for g, one entry must look like this:

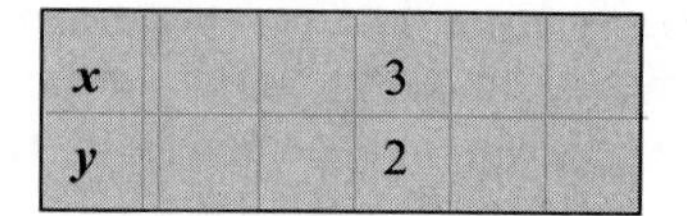

x			3		
y			2		

Thus, $g(3) = 2$, as obtained previously. In Figure 2 we show a graphic interpretation of this method.

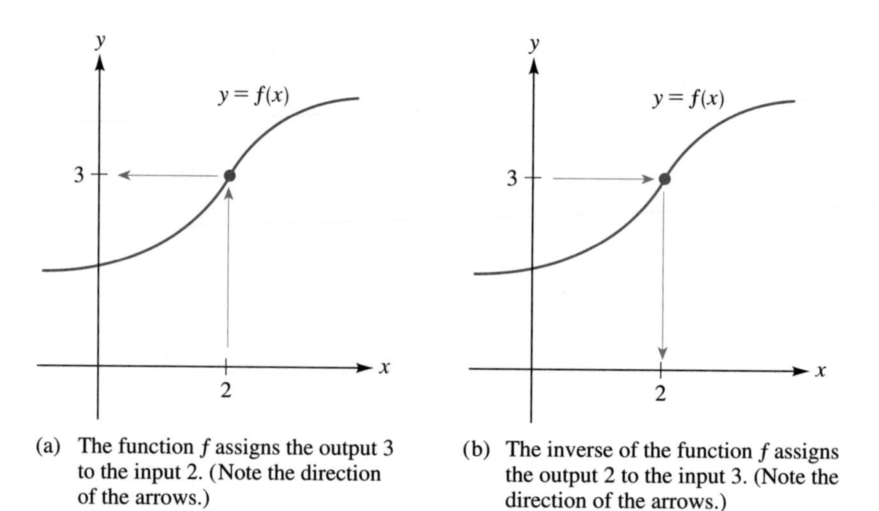

FIGURE 2

(a) The function f assigns the output 3 to the input 2. (Note the direction of the arrows.)

(b) The inverse of the function f assigns the output 2 to the input 3. (Note the direction of the arrows.)

It is customary to use the notation f^{-1} (read *f inverse*) for the function that is the inverse of f. So in Example 1, for instance, we have

$$f(x) = \frac{1}{3}x + 2 \quad \text{and} \quad f^{-1}(x) = 3x - 6 \tag{1}$$

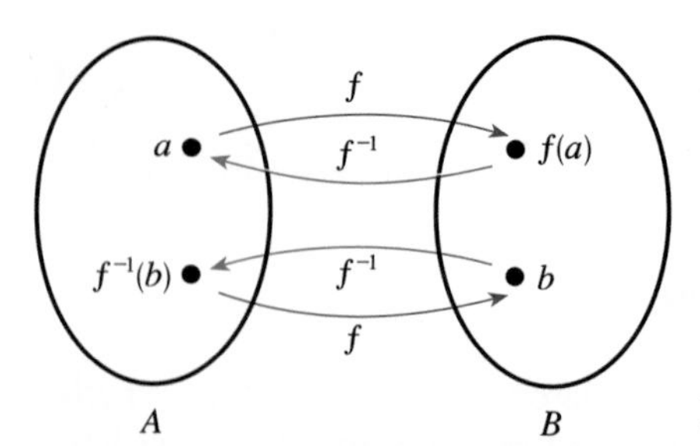

FIGURE 3
The action of inverse functions: The set A in the figure is both the domain of f and the range of f^{-1}. The set B is both the domain of f^{-1} and the range of f.

CAUTION In the context of functions, $f^{-1}(x)$ does not in general mean $\dfrac{1}{f(x)}$. For instance, regarding the pair of inverse functions in equations (1), it's certainly not true in general that $3x - 6 = \dfrac{1}{\frac{1}{3}x + 2}$. (For other examples, see Exercises 7 and 8 at the end of this section.) For reference now, in the box that follows, we rewrite the defining equations for inverse functions using the f^{-1} notation. Figure 3 provides a graphic summary of these equations.

$$f[f^{-1}(x)] = x \qquad \text{for each } x \text{ in the domain of } f^{-1}$$

and

$$f^{-1}[f(x)] = x \qquad \text{for each } x \text{ in the domain of } f$$

EXAMPLE 3 Suppose $f(x) = 4x^3 + 7$. Find $f[f^{-1}(5)]$. (Assume that f^{-1} exists and that 5 is in its domain.)

Solution By definition, $f[f^{-1}(x)] = x$. Substituting $x = 5$ on both sides of this identity directly gives us

$$f[f^{-1}(5)] = 5 \qquad \text{as required}$$

Note that we didn't need to make use of the formula $f(x) = 4x^3 + 7$, nor did we need to find a formula for $f^{-1}(x)$.

EXAMPLE 4 Solve the following equation for x, given that the domain of both f and f^{-1} is $(-\infty, \infty)$ and that $f(1) = -2$:

$$3 + f^{-1}(x - 1) = 4$$

Think $f(x)$ then $f^{-1}(x)$ then $f^{-1}(x-1)$. Do not start with $f(x-1)$!!

Solution

$$\begin{aligned}
3 + f^{-1}(x - 1) &= 4 \\
f^{-1}(x - 1) &= 1 \\
f[f^{-1}(x - 1)] &= f(1) && \text{applying } f \text{ to both sides} \\
x - 1 &= f(1) && \text{definition of inverse function} \\
x &= 1 + f(1) = 1 + (-2) = -1
\end{aligned}$$

We'll postpone a discussion until the end of this section as to which functions have inverses and which do not. For functions that do have inverses, however, there's a simple method that can often be used to determine those inverses. This method is illustrated in Examples 5 and 6.

EXAMPLE 5 Let $f(x) = 3x - 4$. Find $f^{-1}(x)$.

Solution We begin by rewriting the given equation as

$$y = 3x - 4$$

We know that f^{-1} interchanges the inputs and outputs of f, so to determine f^{-1}, we need only switch the x's and y's in the equation $y = 3x - 4$. That gives us $x = 3y - 4$. Now we solve for y as follows:

$$\begin{aligned}
x &= 3y - 4 \\
x + 4 &= 3y \\
\frac{x + 4}{3} &= y
\end{aligned}$$

Thus, the inverse function is $y = (x + 4)/3$. We can also write this as

$$f^{-1}(x) = \frac{x + 4}{3}$$

Actually, we should call this result just a *candidate* for f^{-1}, since certain technical matters remain to be discussed. However, if you compute $f[f^{-1}(x)]$ and $f^{-1}[f(x)]$ and find that they both do equal x, then the matter is settled, by definition. Exercise 17 at the end of this section asks you to carry out those calculations.

Here is a summary of our procedure for calculating $f^{-1}(x)$.

TO FIND $f^{-1}(x)$ FOR THE FUNCTION $y=f(x)$

1. Interchange x and y in the equation $y=f(x)$.
2. Solve the resulting equation for y.

EXAMPLE 6 Let $f(x)=\dfrac{2x-1}{3x+5}$. Find $f^{-1}(x)$.

Solution

STEP 1 Write the given function as $y=\dfrac{2x-1}{3x+5}$. Interchange x and y to get

$$x=\frac{2y-1}{3y+5}$$

STEP 2 Solve the resulting equation for y:

$$\begin{aligned} x&=\frac{2y-1}{3y+5} \\ 3xy+5x&=2y-1 && \text{multiplying by } 3y+5 \\ 3xy-2y&=-5x-1 && \text{rearranging} \\ y(3x-2)&=-5x-1 \\ y&=\frac{-5x-1}{3x-2} \end{aligned}$$

Thus, the inverse function is given by

$$f^{-1}(x)=\frac{-5x-1}{3x-2}$$

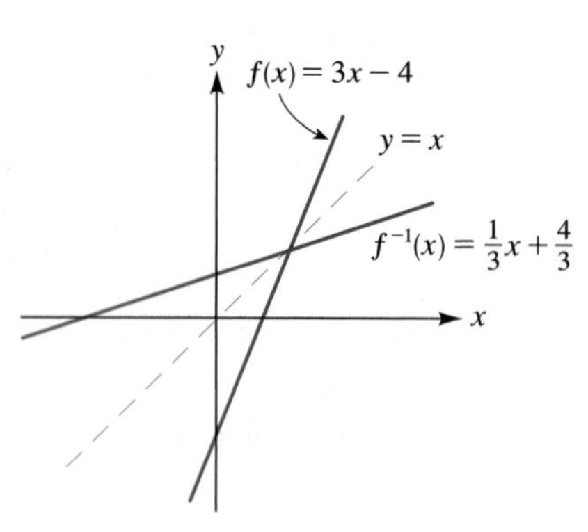

FIGURE 4

A certain symmetry always occurs when we graph a function and its inverse. As an example, consider the functions f and f^{-1} from Example 5:

$$f(x)=3x-4 \qquad f^{-1}(x)=\frac{1}{3}x+\frac{4}{3}$$

Figure 4 shows the graphs of f, f^{-1}, and the line $y=x$. Note that the graphs of f and f^{-1} are symmetric about the line $y=x$. This kind of symmetry, in fact, always occurs for the graphs of any pair of inverse functions.

Why are the graphs of f and f^{-1} always mirror images of each other about the line $y=x$? First, recall that the function f^{-1} switches the inputs and outputs of f. Thus, a point (a, b) is on the graph of f if and only if (b, a) is on the graph

$$f[f^{-1}(x)] = x \qquad \text{for each } x \text{ in the domain of } f^{-1}$$

and

$$f^{-1}[f(x)] = x \qquad \text{for each } x \text{ in the domain of } f$$

EXAMPLE 3 Suppose $f(x) = 4x^3 + 7$. Find $f[f^{-1}(5)]$. (Assume that f^{-1} exists and that 5 is in its domain.)

Solution By definition, $f[f^{-1}(x)] = x$. Substituting $x = 5$ on both sides of this identity directly gives us

$$f[f^{-1}(5)] = 5 \qquad \text{as required}$$

Note that we didn't need to make use of the formula $f(x) = 4x^3 + 7$, nor did we need to find a formula for $f^{-1}(x)$.

EXAMPLE 4 Solve the following equation for x, given that the domain of both f and f^{-1} is $(-\infty, \infty)$ and that $f(1) = -2$:

$$3 + f^{-1}(x - 1) = 4$$

Think $f(x)$ then $f^{-1}(x)$ then $f^{-1}(x-1)$. Do not start with $f(x-1)$!!

Solution

$$\begin{aligned} 3 + f^{-1}(x-1) &= 4 \\ f^{-1}(x-1) &= 1 \\ f[f^{-1}(x-1)] &= f(1) && \text{applying } f \text{ to both sides} \\ x - 1 &= f(1) && \text{definition of inverse function} \\ x &= 1 + f(1) = 1 + (-2) = -1 \end{aligned}$$

We'll postpone a discussion until the end of this section as to which functions have inverses and which do not. For functions that do have inverses, however, there's a simple method that can often be used to determine those inverses. This method is illustrated in Examples 5 and 6.

EXAMPLE 5 Let $f(x) = 3x - 4$. Find $f^{-1}(x)$.

Solution We begin by rewriting the given equation as

$$y = 3x - 4$$

We know that f^{-1} interchanges the inputs and outputs of f, so to determine f^{-1}, we need only switch the x's and y's in the equation $y = 3x - 4$. That gives us $x = 3y - 4$. Now we solve for y as follows:

$$\begin{aligned} x &= 3y - 4 \\ x + 4 &= 3y \\ \frac{x+4}{3} &= y \end{aligned}$$

Thus, the inverse function is $y = (x + 4)/3$. We can also write this as

$$f^{-1}(x) = \frac{x+4}{3}$$

Actually, we should call this result just a *candidate* for f^{-1}, since certain technical matters remain to be discussed. However, if you compute $f[f^{-1}(x)]$ and $f^{-1}[f(x)]$ and find that they both do equal x, then the matter is settled, by definition. Exercise 17 at the end of this section asks you to carry out those calculations.

Here is a summary of our procedure for calculating $f^{-1}(x)$.

TO FIND $f^{-1}(x)$ FOR THE FUNCTION $y=f(x)$

1. Interchange x and y in the equation $y=f(x)$.
2. Solve the resulting equation for y.

EXAMPLE 6 Let $f(x)=\dfrac{2x-1}{3x+5}$. Find $f^{-1}(x)$.

Solution

STEP 1 Write the given function as $y=\dfrac{2x-1}{3x+5}$. Interchange x and y to get

$$x=\frac{2y-1}{3y+5}$$

STEP 2 Solve the resulting equation for y:

$$x=\frac{2y-1}{3y+5}$$

$$3xy+5x=2y-1 \qquad \text{multiplying by } 3y+5$$

$$3xy-2y=-5x-1 \qquad \text{rearranging}$$

$$y(3x-2)=-5x-1$$

$$y=\frac{-5x-1}{3x-2}$$

Thus, the inverse function is given by

$$f^{-1}(x)=\frac{-5x-1}{3x-2}$$

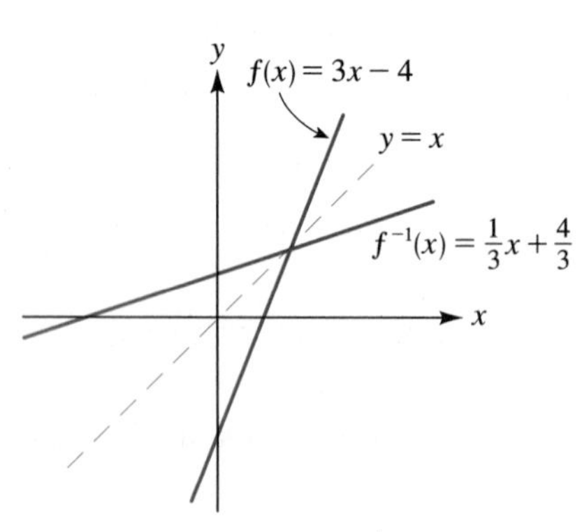

FIGURE 4

A certain symmetry always occurs when we graph a function and its inverse. As an example, consider the functions f and f^{-1} from Example 5:

$$f(x)=3x-4 \qquad f^{-1}(x)=\frac{1}{3}x+\frac{4}{3}$$

Figure 4 shows the graphs of f, f^{-1}, and the line $y=x$. Note that the graphs of f and f^{-1} are symmetric about the line $y=x$. This kind of symmetry, in fact, always occurs for the graphs of any pair of inverse functions.

Why are the graphs of f and f^{-1} always mirror images of each other about the line $y=x$? First, recall that the function f^{-1} switches the inputs and outputs of f. Thus, a point (a, b) is on the graph of f if and only if (b, a) is on the graph of f^{-1}. But (as we know from Section 2.4) the points (a, b) and (b, a) are mirror images in the line $y = x$. It follows, therefore, that the graphs of f and f^{-1} are reflections of each other about the line $y = x$. For reference, we record this useful fact in the box that follows.

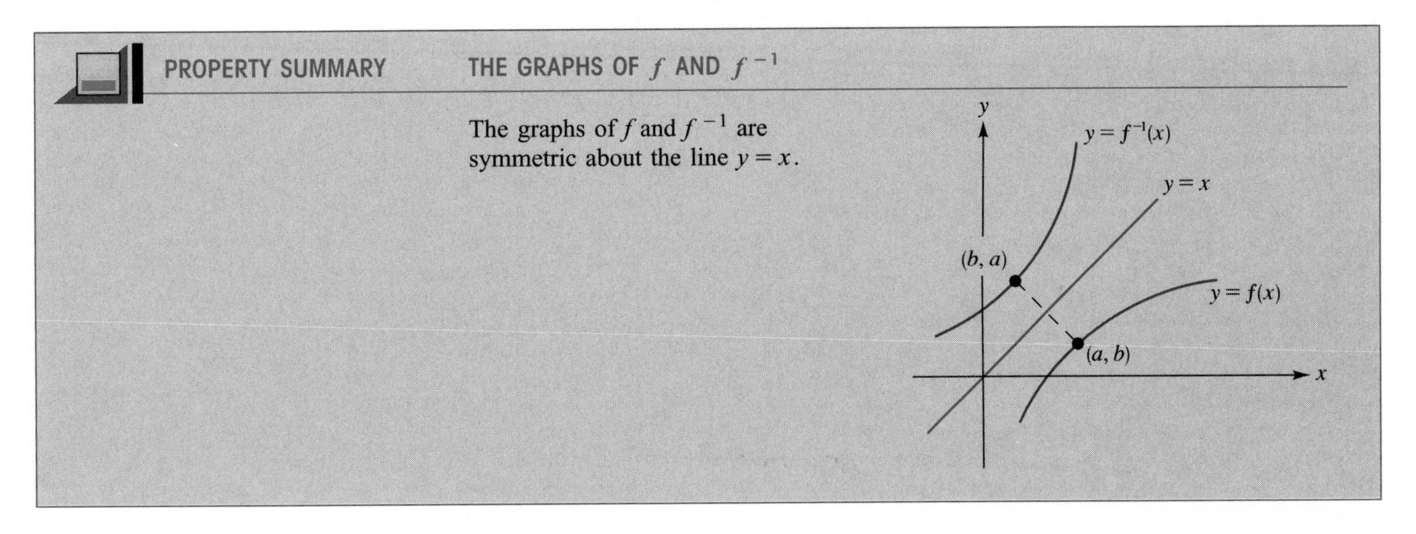

EXAMPLE 7 The graph of a function f consists of the line segment joining the points $(-2, -3)$ and $(-1, 4)$, as shown in Figure 5. Sketch a graph of f^{-1}.

Solution The graph of f^{-1} is obtained by reflecting the graph of f about the line $y = x$. The reflection of $(-2, -3)$ is $(-3, -2)$, and the reflection of $(-1, 4)$ is $(4, -1)$. To graph f^{-1}, then, we plot the reflected points $(-3, -2)$ and $(4, -1)$ and connect them with a line segment, as shown in Figure 6. For reference, Figure 6 also shows the graphs of f and $y = x$.

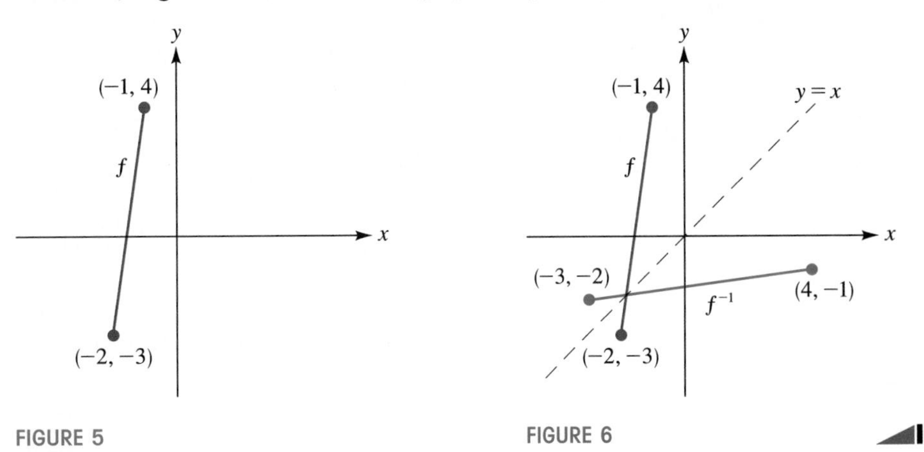

FIGURE 5

FIGURE 6

EXAMPLE 8 Let $g(x) = x^3$. Find $g^{-1}(x)$, and then, on the same set of axes, sketch the graphs of g, g^{-1}, and $y = x$.

Solution We begin by writing $y = x^3$, then switching x and y and solving for y. We first have

$$x = y^3$$

To solve this equation for y, we take the cube root of both sides to obtain

$$\sqrt[3]{x} = y$$

So the inverse function is $g^{-1}(x) = \sqrt[3]{x}$. We could graph the function g^{-1} by plotting points, but the easier way is to reflect the graph of $g(x) = x^3$ about the line $y = x$; see Figure 7.

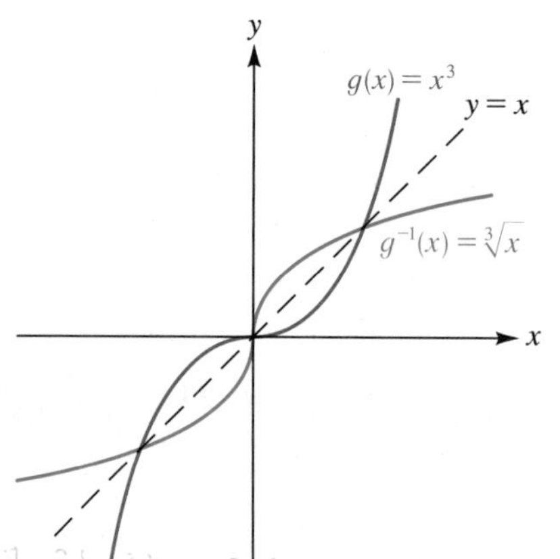

FIGURE 7

TABLE 1

(a) $F(x) = x^2$

x	y
1	1
2	4
3	9
−1	1
−2	4
−3	9

(b) Reversing the inputs and outputs

x	y
1	1
4	2
9	3
1	−1
4	−2
9	−3

Earlier in this section, we mentioned that not every function has an inverse function. For instance, Tables 1(a) and (b) indicate what happens if we begin with the function defined by $F(x) = x^2$ and interchange the inputs and outputs. The resulting table, Table 1(b), does not define a function. For instance, the input 4 in Table 1(b) produces two distinct outputs, 2 and -2, but the definition of a function requires exactly one output for a given input.

Why didn't this difficulty arise when we looked for the inverse of $f(x) = 2x$ at the beginning of this section? The answer is that there is an essential difference between the functions defined by $f(x) = 2x$ and $F(x) = x^2$. With $f(x) = 2x$, different inputs never yield the same output. Or, to put it another way, for the function $f(x) = 2x$, the only time it can happen that $f(a) = f(b)$ is when $a = b$. This condition guarantees that interchanging the inputs and outputs of f will yield a function. With $F(x) = x^2$, however, it *can* happen that $F(a) = F(b)$ and yet $a \neq b$. For instance,

$$F(2) = F(-2) \qquad \text{because } 2^2 = 4 = (-2)^2$$

but, of course, $2 \neq -2$.

It's useful to have a name for functions such as $f(x) = 2x$, in which distinct inputs always yield distinct outputs. We call these types of functions *one-to-one* functions.

DEFINITION: One-to-One

A function f is **one-to-one** provided that the following condition holds for all a and b in the domain of f:

$$\text{If } f(a) = f(b) \quad \text{then } a = b$$

EXAMPLES

$f(x) = 2x$ is one-to-one.

$F(x) = x^2$ is not one-to-one because $F(2) = F(-2)$, yet $2 \neq -2$.

EXAMPLE 9 **(a)** Let $f(x) = 3x - 4$. Show that f is one-to-one.
(b) Let $g(x) = \sqrt{2x^2 + 5}$. Show that g is not one-to-one.

Solution

(a) To show that f is one-to-one, we assume that $f(a) = f(b)$, and then we must show that $a = b$. From the condition $f(a) = f(b)$ we have

$$3a - 4 = 3b - 4 \quad \text{using the definition of } f$$
$$3a = 3b \quad \text{adding 4 to both sides}$$
$$a = b \quad \text{dividing by 3}$$

This shows that the function f is one-to-one.

(b) To show that g is not one-to-one, all we need to do is to demonstrate one specific case where the condition $g(a) = g(b)$ fails to imply that $a = b$. Using $a = 1$ and $b = -1$, we have

$$g(1) = \sqrt{2(1)^2 + 5} = \sqrt{7} \qquad \text{and} \qquad g(-1) = \sqrt{2(-1)^2 + 5} = \sqrt{7}$$

So, $g(1) = g(-1) = \sqrt{7}$, but certainly $1 \neq -1$. This shows that the function g is not one-to-one.

Using graphs, there is an easy way to tell which functions are one-to-one:

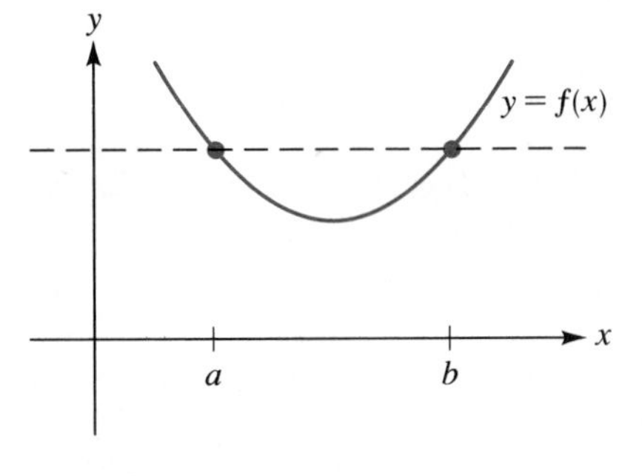

FIGURE 8

HORIZONTAL LINE TEST

A function f is one-to-one if and only if each horizontal line intersects the graph of $y = f(x)$ in at most one point.

Figure 8 will help you to see why the horizontal line test is valid. The figure shows a horizontal line intersecting the graph of a function f at two distinct points with x-coordinates a and b. From the graph, we see that $f(a) = f(b)$, even though $a \neq b$. Thus, the function represented in Figure 8 is not one-to-one.

EXAMPLE 10 Use the horizontal line test to determine which functions in Figure 9 are one-to-one.

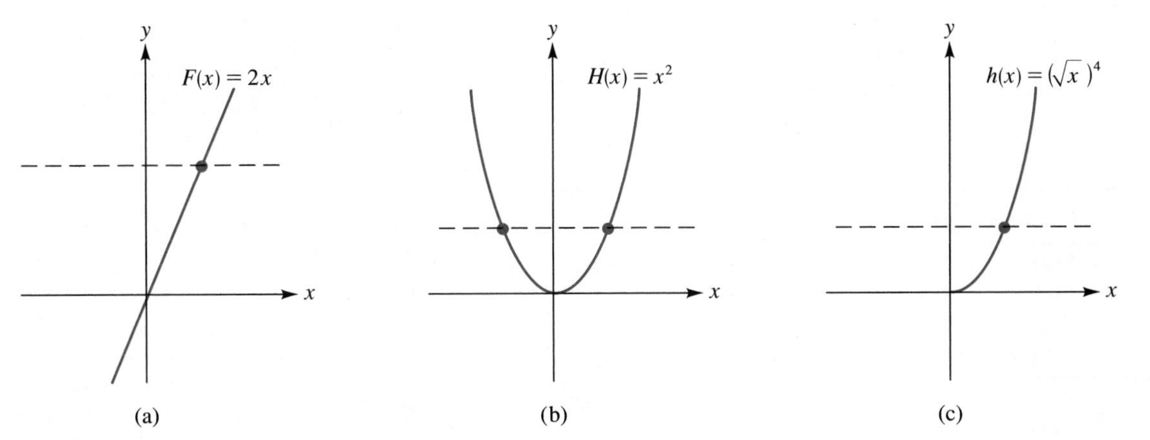

FIGURE 9

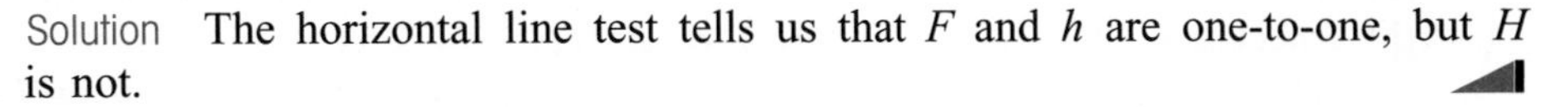

Solution The horizontal line test tells us that F and h are one-to-one, but H is not.

The following theorem tells us that the functions with inverses are precisely the one-to-one functions. (The proof of the theorem isn't difficult, but we'll omit it here.)

THEOREM

A function f has an inverse function if and only if f is one-to-one.

EXAMPLE 11 Which of the three functions in Figure 9 have inverse functions?

Solution According to our theorem, the functions with inverses are those that are one-to-one. In Example 10, we saw that F and h were one-to-one but H was not. So F and h each have inverses but H does not.

EXERCISE SET 3.5

Do 1, 3, 5, 9, 11, 13, 17, 21, 25, 45

A

1. Verify that the given pairs of functions are inverse functions.
   (a) $f(x) = 3x;\ g(x) = x/3$
   (b) $f(x) = 4x - 1;\ g(x) = \frac{1}{4}x + \frac{1}{4}$
   (c) $g(x) = \sqrt{x};\ h(x) = x^2$ [Assume that the domain of both g and h is $[0, \infty)$.]
2. Which pairs of functions are inverses?
   (a) $f(x) = -3x + 2;\ g(x) = \frac{2}{3} - \frac{1}{3}x$
   (b) $F(x) = 2x + 1;\ G(x) = \frac{1}{2}x - 1$
   (c) $G(x) = x^3;\ H(x) = 1 - x^3$
   (d) $f(t) = t^3;\ g(t) = \sqrt[3]{t}$

In Exercises 3 and 4, suppose that f and g are a pair of inverse functions.

3. If $f(7) = 12$, what is $g(12)$? (If you need a hint, reread Example 2.)
4. If $g(-2) = 0$, what is $f(0)$?

In Exercises 5 and 6, let $f(x) = x^3 + 2x + 1$, and assume that the domain of f^{-1} is $(-\infty, \infty)$. Simplify each expression (as in Example 3).

5. **(a)** $f[f^{-1}(4)]$ **(b)** $f^{-1}[f(-1)]$
   (c) $(f \circ f^{-1})(\sqrt{2})$ **(d)** $f[f^{-1}(t + 1)]$
6. **(a)** $f(0)$
   (b) $f^{-1}(1)$ *Hint:* Use the result in part (a).
   (c) $f(-1)$ **(d)** $f^{-1}(-2)$
7. Let $f(x) = 2x + 1$.
   (a) Find $f^{-1}(x)$.
   (b) Calculate $f^{-1}(5)$ and $1/f(5)$. Are your answers the same?
8. Let $g(t) = \dfrac{1}{t} + 1$.
   (a) Find $g^{-1}(t)$.
   (b) Calculate $g^{-1}(2)$ and $1/g(2)$, and note that the answers are not the same.
9. Let $f(x) = 3x - 1$.
   (a) Compute $f^{-1}(x)$.
   (b) Verify that $f[f^{-1}(x)] = x$ and that $f^{-1}[f(x)] = x$.
   (c) On the same set of axes, sketch the graphs of f, f^{-1}, and the line $y = x$. Note that the graphs of f and f^{-1} are symmetric about the line $y = x$.
10. Follow Exercise 9, but use $f(x) = \frac{1}{3}x - 2$.
11. Follow Exercise 9, but use $f(x) = \sqrt{x - 1}$. [The domain of f^{-1} will be $[0, \infty)$.]
12. Follow Exercise 9, but use $f(x) = 1/x$.
13. Let $f(x) = (x + 2)/(x - 3)$.
    (a) Find the domain and range of the function f.
    (b) Find $f^{-1}(x)$.
    (c) Find the domain and range of the function f^{-1}. What do you observe?
14. Let $f(x) = (2x - 3)/(x + 4)$. Find $f^{-1}(x)$. Find the domain and range for f and f^{-1}. What do you observe?
15. Let $f(x) = 2x^3 + 1$. Find $f^{-1}(x)$.
16. In our discussion at the beginning of this section, we considered the functions $f(x) = 2x$ and $g(x) = x/2$. Verify that $f[g(x)] = x$ and that $g[f(x)] = x$. On the same set of axes, sketch the graphs of f, g, and $y = x$.
17. This exercise refers to the comments made at the end of Example 5. Compute $f[f^{-1}(x)]$ and $f^{-1}[f(x)]$ using the functions $f(x) = 3x - 4$ and

$f^{-1}(x) = (x+4)/3$. By actually carrying out the calculations, you can see in each case that the answer is indeed x. Then, on the same set of axes, sketch the graphs of f, f^{-1}, and the line $y = x$.

18. Let $f(x) = (x-3)^2$ and take the domain to be $[3, \infty)$.
(a) Find $f^{-1}(x)$. Give the domain of f^{-1}.
(b) On the same set of axes, graph both f and f^{-1}.

19. Let $f(x) = (x-3)^3 - 1$.
(a) Compute $f^{-1}(x)$.
(b) On the same set of axes, sketch the graphs of f, f^{-1}, and the line $y = x$. *Hint:* Sketch f using the ideas of Section 4.3. Then reflect f to get f^{-1}.

20. The following figure shows the graph of a function f. (The axes are marked off in one-unit intervals.) Graph each of the following functions.

(a) $y = f^{-1}(x)$ **(b)** $y = f(x-2)$
(c) $y = f(x) - 2$ **(d)** $y = f(-x)$
(e) $y = -f(x)$ **(f)** $y = f^{-1}(x-2)$
(g) $y = f^{-1}(x) - 2$ **(h)** $y = f^{-1}(x-2) - 2$
(i) $y = f^{-1}(-x)$ **(j)** $y = -f^{-1}(x)$

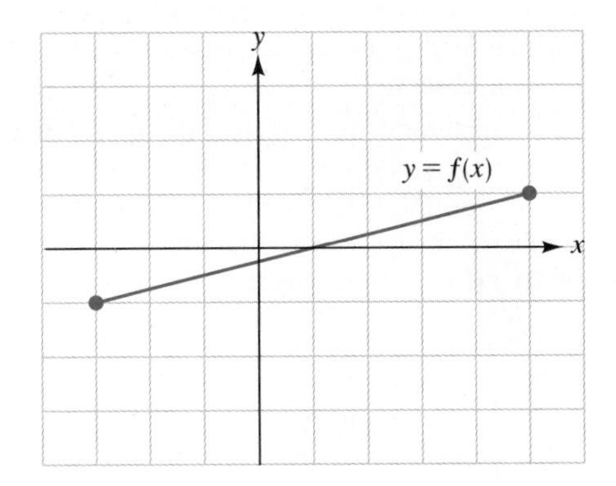

21. The following figure shows the graph of a function g. (The axes are marked off in one-unit intervals.) Graph each of the following functions.

(a) $y = g^{-1}(x)$ **(b)** $y = g^{-1}(x) - 1$
(c) $y = g^{-1}(x-1)$ **(d)** $y = g^{-1}(-x)$
(e) $y = -g^{-1}(x)$ **(f)** $y = -g^{-1}(-x)$

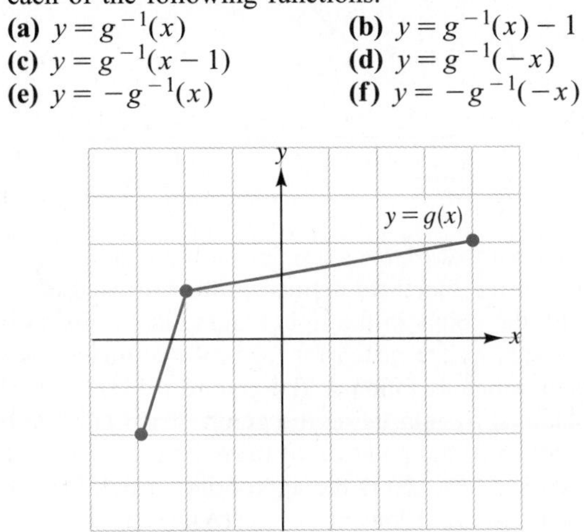

22. The following figure shows the graph of a function $y = h(x)$.
(a) Sketch a graph that is the reflection of h about the y-axis. What is the equation of the graph you obtain?
(b) On the same set of axes, sketch graphs of h, h^{-1}, and the line $y = x$.
(c) Let k be the function whose graph is the reflection of h in the x-axis. Graph the function k^{-1}.

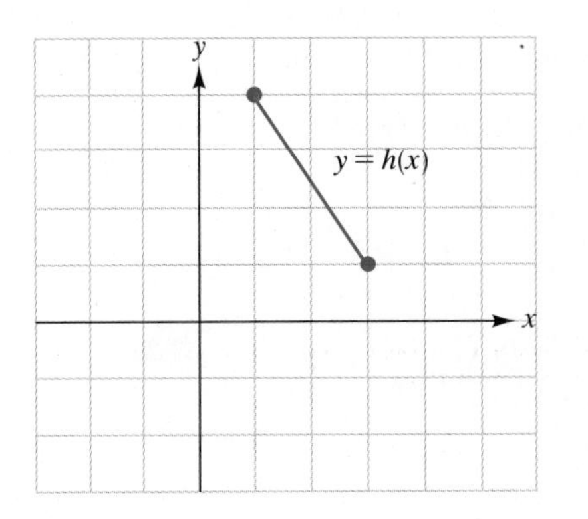

In Exercises 23–34, use the horizontal line test to determine whether the function is one-to-one (and therefore has an inverse).

23. $y = x^2 + 1$ **24.** $y = \sqrt{x}$
25. $f(x) = 1/x$ **26.** $g(x) = |x|$
27. $y = x^3$ **28.** $y = 1 - x^3$
29. $y = \sqrt{1 - x^2}$ **30.** $y = 3x$
31. $g(x) = 5$ (for all x)
32. $y = mx + b$ $(m \neq 0)$

33. $f(x) = \begin{cases} x^2 & \text{if } -1 \leq x \leq 0 \\ x^2 + 1 & \text{if } x > 0 \end{cases}$

34. $g(x) = \begin{cases} x^2 & \text{if } -1 \leq x < 0 \\ x^2 + 1 & \text{if } x \geq 0 \end{cases}$

In Exercises 35–40, use the method of Example 9(a) to show that the function is one-to-one.

35. $f(x) = 4x - 9$ **36.** $g(x) = \dfrac{1}{x} + 3$
37. $g(x) = \sqrt{2x + 1}$ **38.** $f(x) = \sqrt[3]{2 - 5x}$
39. $f(t) = \dfrac{t + 4}{2t - 7}$ **40.** $g(t) = \dfrac{6 - 7t}{1 + 3t}$

In Exercises 41 and 42, give an example to show that the function is not one-to-one [as in Example 9(b)].

41. $f(x) = 2x^2 - 3x$ **42.** $f(x) = \sqrt{x^2 - 6}$

43. Let $f(x) = (3x - 2)/(5x - 3)$. Show that $f[f(x)] = x$. This says that f is its own inverse.

44. Let $f(x) = (ax + b)/(cx - a)$. Show that f is its own inverse. (Assume that $a^2 + bc \neq 0$.)

In Exercises 45–48, assume that the domain of f and f^{-1} is $(-\infty, \infty)$. Solve the equation for x or for t (whichever is appropriate) using the given information.

45. **(a)** $7 + f^{-1}(x - 1) = 9$; $f(2) = 6$
(b) $4 + f(x + 3) = -3$; $f^{-1}(-7) = 0$

46. **(a)** $f^{-1}(2x + 3) = 5$; $f(5) = 13$
(b) $f(1 - 2x) = -4$; $f^{-1}(-4) = -5$

47. $f^{-1}\left(\dfrac{t+1}{t-2}\right) = 12$; $f(12) = 13$

48. $f\left(\dfrac{1-2t}{1+2t}\right) = 7$; $f^{-1}(7) = -3$

B

49. Let $f(x) = \sqrt{x}$.
(a) Find $f^{-1}(x)$. What is the domain of f^{-1}? [The domain is not $(-\infty, \infty)$.]
(b) In each case, determine whether the given point lies on the graph of f or f^{-1}.
(i) $(4, 2)$ **(ii)** $(2, 4)$ **(iii)** $(5, \sqrt{5})$
(iv) $(\sqrt{5}, 5)$ **(v)** $(a, f(a))$, where $a \geq 0$
(vi) $(f(a), a)$ **(vii)** $(b, f^{-1}(b))$, where $b \geq 0$
(viii) $(f^{-1}(b), b)$

50. Let $F(x) = x^3 + 7x - 5$. Find $F^{-1}(-5)$. [Assume that the domain of F^{-1} is $(-\infty, \infty)$.]

51. In the following figure, determine the coordinates of the points A, B, C, and D. Express your answers in terms of the function f and the number a. (Each dashed line through A is parallel to an axis.)

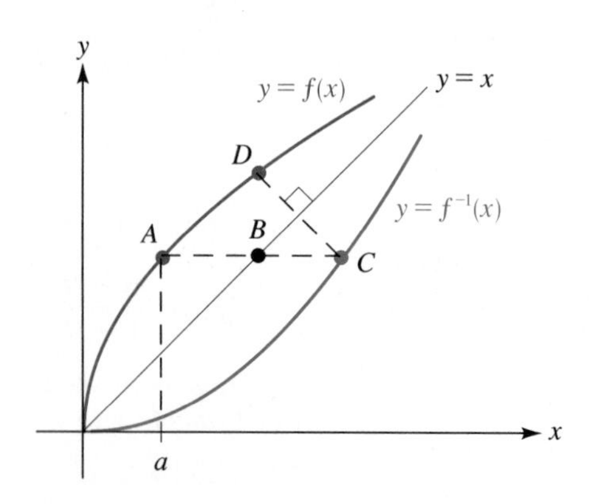

52. In the following figure, determine the coordinates of the points A, B, C, D, E, and F. Specify the answers in terms of the functions f and f^{-1} and the number b.

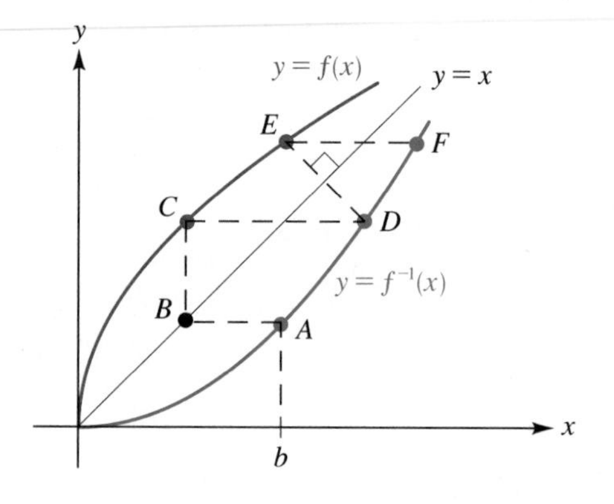

53. Let $f(x) = 3/(x - 1)$.
(a) Find the average rate of change of f on the interval $[4, 9]$.
(b) Find $f^{-1}(x)$ and then compute the average rate of change of f^{-1} on the interval $[f(4), f(9)]$. What do you observe?

54. The accompanying figure shows the graph of a function f on the interval $[a, b]$. If the average rate of change of f on the interval $[a, b]$ is k, show that the average rate of change of the function f^{-1} on the interval $[f(a), f(b)]$ is $1/k$, where $k \neq 0$.

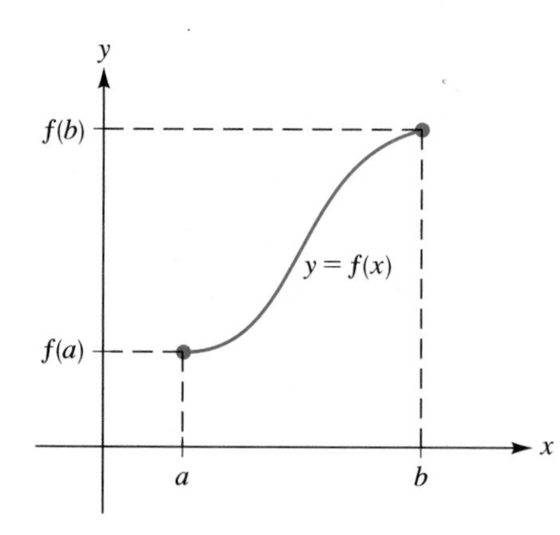

55. Suppose that (a, b) is a point on the graph of $y = f(x)$. Match the functions defined in the left-hand column with the points in the right-hand column. For example, the appropriate match for (a) in the left-hand column is determined as follows. The graph of $y = f(x) + 1$ is obtained by translating the graph of $y = f(x)$ up one unit. Thus, the point (a, b) moves up to $(a, b + 1)$ and, consequently, (E) is the appropriate match for (a).
(a) $y = f(x) + 1$ **(A)** $(-a, b)$

(b) $y = f(x+1)$	**(B)** (b, a)
(c) $y = f(x-1)+1$	**(C)** $(a-1, b)$
(d) $y = f(-x)$	**(D)** $(-b, a+1)$
(e) $y = -f(x)$	**(E)** $(a, b+1)$
(f) $y = -f(-x)$	**(F)** $(1-a, b)$
(g) $y = f^{-1}(x)$	**(G)** $(-a, -b)$
(h) $y = f^{-1}(x)+1$	**(H)** $(-b, 1-a)$
(i) $y = f^{-1}(x-1)$	**(I)** $(b, -a)$
(j) $y = f^{-1}(-x)+1$	**(J)** $(a, -b)$
(k) $y = -f^{-1}(x)$	**(K)** $(b+1, a)$
(l) $y = -f^{-1}(-x)+1$	**(L)** $(a+1, b+1)$
(m) $y = 1-f^{-1}(x)$	**(M)** $(b, a+1)$
(n) $y = f(1-x)$	**(N)** $(b, 1-a)$

C

56. As background for this problem you will need to have worked Exercise 51.

(a) In working Exercise 51 for a group of students, one of the author's teaching assistants assumed that the two distances AB and BC (in the figure accompanying Exercise 51) are equal. Demonstrate that this general assumption is not valid by using $f(x) = \sqrt{x}, f^{-1}(x) = x^2$, where $x \geq 0$, and completing the following table.

a	AB	BC	$AB = BC$? (yes or no)
1/2			
1/4			
3/4			

(b) As in part (a), let $f(x) = \sqrt{x}$ and $f^{-1}(x) = x^2$, where $x \geq 0$. Is there any value for a, where $0 < a < 1$, for which the two distances AB and BC are equal? If so, find it; if not, explain why not.

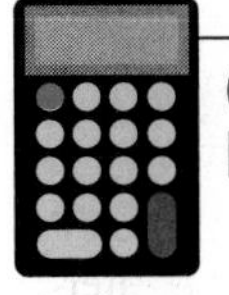

GRAPHING UTILITY EXERCISES FOR SECTION 3.5

In Exercises 1–6, the two given functions are inverse functions. In each case, graph the pair of inverse functions on the same set of axes along with the line $y = x$. Note the symmetry. [Be sure to use the same size unit and the same scale on both axes. (Why?)]

1. $y = 4x + 8$
$y = \frac{1}{4}x - 2$

2. $y = 3x - 5$
$y = (x+5)/3$

3. $f(x) = \sqrt{x-2}, \quad x \geq 2$
$g(x) = x^2 + 2, \quad x \geq 0$

4. $F(x) = \sqrt{2x+3}, \quad x \geq -1.5$
$G(x) = (x^2 - 3)/2, \quad x \geq 0$

5. $y = 1/\sqrt{x}$
$y = 1/x^2, \quad x > 0$

6. $y = 1/(2x-3)$
$y = (3x+1)/2x$

In Exercises 7–13, graph each function and then apply the horizontal line test to see whether the function is one-to-one.

7. $y = x^2 + 2x$

8. $y = x^3 + 2x$

9. $y = 2x^3 + x^2$ *Suggestion:* Begin with the standard viewing rectangle and then use a zoom-in view. (First looks can be deceiving.)

10. $y = 0.01x^4 - 1$

11. $y = 2x^5 + x - 1$

12. **(a)** $f(x) = x^3 + x^2 + x$
(b) $g(x) = x^3 - x^2 + x$
(c) $h(x) = x^3 - x^2 - x$

13. **(a)** $F(x) = x^x, \quad x > 1$
(b) $G(x) = x^x, \quad x > 0$
(c) $H(x) = x^{1/x}, \quad x > 0$

14. Some turning points are difficult to detect with a graphing utility. Here's an example: Graph the function $y = (x^2 + x - 2)/x^2$ in the standard viewing rectangle. Do you see a turning point? Using calculus, it can be shown that there *is* a turning point at $x = 4$. See if you can find a viewing rectangle to support this fact. Is the function $y = (x^2 + x - 2)/x^2$ one-to-one on the interval $(2, \infty)$?

15. Let $F(x) = \sqrt{1 - x^2}$, with domain $-1 \leq x \leq 0$. Find $F^{-1}(x)$ and then graph both functions on the same set of axes. (Use the same size unit and same scale on both axes.) Use the picture to check your algebra. That is, are the two graphs symmetric about the line $y = x$?

16. In this exercise you'll investigate the inverse of a composite function. In parts (b) and (c), which involve graphing, be sure to use the same size unit and scale on both axes so that symmetry about the line $y = x$ can be checked visually.

(a) Let $f(x) = 2x + 1$ and $g(x) = \frac{1}{4}x - 3$. Compute each of the following:
(i) $f(g(x))$ **(ii)** $g(f(x))$

(iii) $f^{-1}(x)$ **(v)** $f^{-1}(g^{-1}(x))$
(iv) $g^{-1}(x)$ **(vi)** $g^{-1}(f^{-1}(x))$

(b) On the same set of axes, graph the two answers that you obtained in (i) and (v) of part (a). Note that the graphs are *not* symmetric about $y = x$. The conclusion here is that the inverse function for $f(g(x))$ is not $f^{-1}(g^{-1}(x))$.

(c) On the same set of axes, graph the two answers that you obtained in (i) and (vi) of part (a); also put the line $y = x$ into the picture. Note that the two graphs *are* symmetric about the line $y = x$. The conclusion here is that the inverse function for $f(g(x))$ is $g^{-1}(f^{-1}(x))$. In fact, it can be shown that this result is true in general. For reference, then, we summarize this fact about the inverse of a composite function in the box that follows.

$$(f \circ g)^{-1} = g^{-1} \circ f^{-1}$$

17. Let $f(x) = \dfrac{3x}{1 + |x|}$.

(a) Graph this function. Note that, according to the horizontal line test, the function is one-to-one.

(b) Since the function is one-to-one, it has an inverse. Compute this inverse. *Hint:* Consider two cases. First, if $x \geqslant 0$, then the given equation is $y = 3x/(1 + x)$. Second, if $x < 0$, the given equation is equivalent to $y = 3x/(1 - x)$. (Why?) When you think you have found f^{-1}, graph it and f on the same set of axes and check that the graphs are symmetric about $y = x$.

Answer: For the inverse, you should obtain

$$y = \begin{cases} x/(3 - x) & \text{if } 0 \leqslant x < 3 \\ x/(3 + x) & \text{if } -3 < x < 0 \end{cases}$$

18. The curve in the figure that follows is known as the *Trisectrix of Maclaurin.* [The curve, studied by the Scottish mathematician Colin Maclaurin (1698–1746), can be used to trisect an angle.]

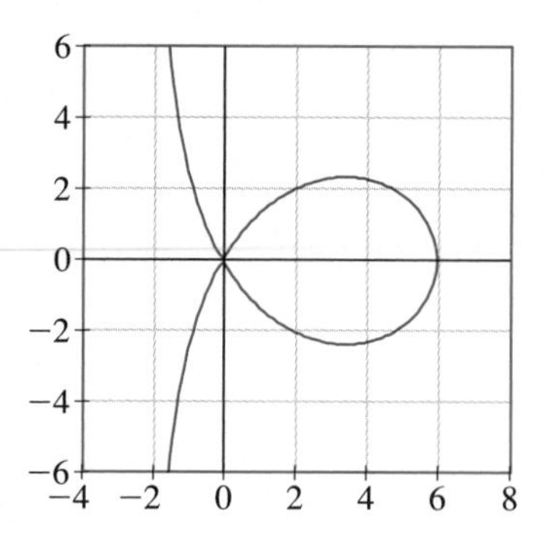

(a) The curve is composed of the graphs of two functions f and g:

$$f(x) = \sqrt{\frac{6x^2 - x^3}{2 + x}} \qquad g(x) = -\sqrt{\frac{6x^2 - x^3}{2 + x}}$$

Graph these two functions on the same set of axes to obtain the Trisectrix of Maclaurin.

(b) For the rest of this exercise we focus on the first-quadrant portion of the curve. To do this, delete the graph of g from your picture and change the viewing rectangle so that x extends from 0 to 7 and y extends from 0 to 3.

(c) Let $y = G(x)$ denote the function that you just graphed in part (b). Apply the horizontal line test to the graph to determine if the function G is one-to-one on the interval $[0, 2]$.

(d) Is G one-to-one on the interval $[0, 5]$?

(e) Use your graphing utility to estimate the largest value of c for which the function G is one-to-one on the interval $[0, c]$. (This is also the input for which the function G is maximum. Thus, for a good estimate, you'll want to zoom in several times on the turning point.)

(f) Using calculus, it can be shown that the exact value of c is $2\sqrt{3}$. Use a calculator to evaluate this expression. To how many decimal places does your estimate in part (e) agree with this number?

CHAPTER THREE SUMMARY OF PRINCIPAL TERMS AND FORMULAS

TERMS OR FORMULAS	PAGE REFERENCE	COMMENTS
1. Function	101	Given two nonempty sets A and B, a function from A to B is a rule of correspondence that assigns to each element of A exactly one element in B.
2. Domain	103	The domain of a function is the set of all inputs for that function. If f is a function from A to B, then the domain is the set A.
3. Range	103	The range of a function from A to B is the set of all elements in B that are actually used as outputs.

TERMS OR FORMULAS	PAGE REFERENCE	COMMENTS
4. $f(x)$	104	Given a function f, the notation $f(x)$ denotes the output that results from the input x.
5. The graph of a function	113	The graph of a function f consists of those points (x, y) such that x is in the domain of f and $y = f(x)$.
6. Vertical line test	116	A graph in the x-y plane represents a function $y = f(x)$ provided that any vertical line intersects the graph in at most one point.
7. Turning point	117	A turning point on a graph is a point where the graph changes from rising to falling, or vice versa. See, for example, Figures 9 and 10 on page 117.
8. Maximum value and minimum value	117	An output $f(a)$ is a maximum value of the function f if $f(a) \geq f(x)$ for every x in the domain of f. An output $f(b)$ is a minimum value if $f(b) \leq f(x)$ for every x in the domain of f.
9. Increasing function	117	A function f is increasing on an interval if the following condition holds: If a and b are in the interval and $a < b$, then $f(a) < f(b)$. Geometrically, this means that the graph is rising as we move in the positive x-direction.
10. Decreasing function	117	A function f is decreasing on an interval if the following condition holds: If a and b are in the interval and $a < b$, then $f(a) > f(b)$. Geometrically, this means that the graph is falling as we move in the positive x-direction.
11. Average rate of change	118	The average rate of change, $\Delta f/\Delta x$, of a function f on an interval $[a, b]$ is the slope of the line segment through the two points $(a, f(a))$ and $(b, f(b))$: $\dfrac{\Delta f}{\Delta x} = \dfrac{f(b) - f(a)}{b - a}$.
12. Techniques for graphing functions	131	The box on page 131 lists six techniques involving translation and reflection. For a comprehensive example, see Example 5 on pages 134–135.
13. $f \circ g$	144	The composition of two functions f and g: $(f \circ g)(x) = f[g(x)]$.
14. Iterates	146	Given a function f and an input x_0, the iterates of x_0 are the numbers $f(x_0), f(f(x_0)), f(f(f(x_0))), \ldots$. The number $f(x_0)$ is called the first iterate, the number $f(f(x_0))$ is called the second iterate, and so on. The orbit of x_0 under the function f is the list of numbers $x_0, f(x_0), f(f(x_0)), \ldots$
15. Inverse functions	155	Two functions f and g are inverses of one another provided that the following two conditions are met. First, $f[g(x)] = x$ for each x in the domain of g. Second, $g[f(x)] = x$ for each x in the domain of f.
16. f^{-1}	156	f^{-1} denotes the inverse function for f. *Note:* In this context, f^{-1} does not mean $1/f$.
17. One-to-one function	160	A function f is said to be one-to-one provided that distinct inputs always yield distinct outputs. Geometrically, this means that every horizontal line intersects the graph of $y = f(x)$ in at most one point. Algebraically, this means that f satisfies the following condition: If a and b are in the domain of f and $f(a) = f(b)$, then $a = b$. The relationship between inverse functions and one-to-one functions is this: A function f has an inverse if and only if f is one-to-one.

WRITING MATHEMATICS

1. Section 3.5 on inverse functions begins with the following quotation from Professor R. P. Boas:

> If anybody ever told me why the graph of $y = x^{1/2}$ is the reflection of the graph of $y = x^2$ $[x \geqslant 0]$ in a 45° line, it didn't sink in. To this day, there are textbooks that expect students to think that it is so obvious as to need no explanation.

Show your instructor that you are not in the dark about inverse functions. Write out an explanation of why the graphs of $y = x^{1/2}$ and $y = x^2$ $(x \geqslant 0)$ are indeed reflections of one another in a 45° line. (As preparation for your writing, you'll probably need to look over Sections 2.4 and 3.5 and make a few notes.)

2. Decide which of the following rules are functions. Write out your reasons in complete sentences. (In each case, assume that the domain is the set of students in your school.)
   (a) F is the rule that assigns to each person his or her brother.
   (b) G is the rule that assigns to each person his or her aunt.
   (c) H is the rule that assigns to each person his or her mother.
   (d) K is the rule that assigns to each person his or her mother or father.

3. Given the graph of two functions f and g, points on the graph of the composite function $g \circ f$ can be constructed geometrically. Use the following figure to explain (in complete sentences) how this works. *Hint:* Determine in succession the coordinates of A, B, C, and D.

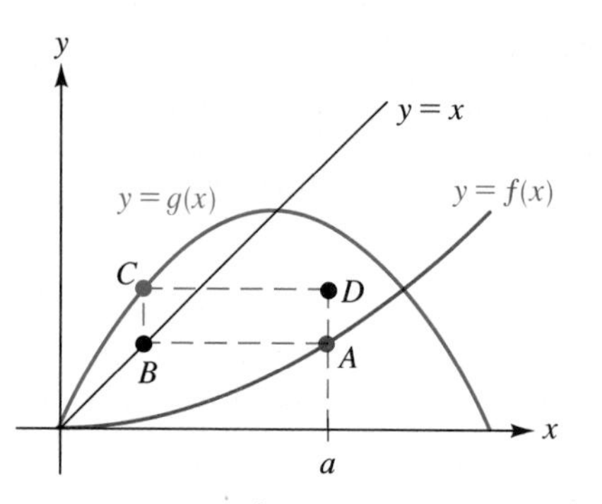

CHAPTER THREE REVIEW EXERCISES

1. **(a)** Find the domain of the function defined by $f(x) = \sqrt{15 - 5x}$.
   (b) Find the range of the function defined by $g(x) = (3 + x)/(2x - 5)$.

2. Let $f(x) = 3x^2 - 4x$ and $g(x) = 2x + 1$. Compute each of the following: **(a)** $(f - g)(x)$; **(b)** $(f \circ g)(x)$; **(c)** $f[g(-1)]$.

3. A **linear function** is a function defined by an equation of the form $F(x) = ax + b$, where a and b are real numbers. Suppose f and g are linear functions. Is the function $f \circ g$ a linear function?

4. A **quadratic function** is a function defined by an equation of the form $F(x) = ax^2 + bx + c$, where a, b, and c are real numbers and $a \neq 0$.
   (a) Give an example of quadratic functions f and g for which $f + g$ is also a quadratic function.
   (b) Give an example of quadratic functions f and g for which $f + g$ is not a quadratic function.

(c) Suppose f and g are quadratic functions. Is the function $f \circ g$ a quadratic function?

5. **(a)** Compute $\dfrac{F(x) - F(a)}{x - a}$ given that $F(x) = 1/x$.

(b) Compute $\dfrac{g(x+h) - g(x)}{h}$ for the function defined by $g(x) = x - 2x^2$.

6. The y-intercept for the graph of $y = f(x)$ is 4. What is the y-intercept for the graph of $y = -f(x) + 1$?

7. **(a)** Find $g^{-1}(x)$ given that $g(x) = (1 - 5x)/3x$.

(b) The figure displays the graph of a function f. Sketch the graph of f^{-1}.

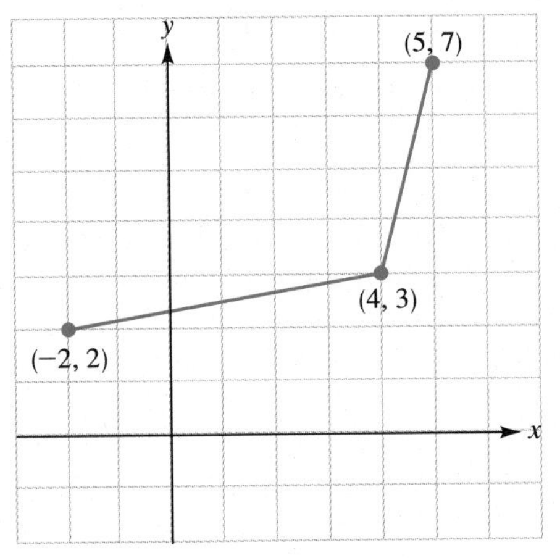

8. Refer to the function f in Exercise 7(b).

(a) Compute $\Delta f/\Delta x$ on $[-2, 5]$.

(b) Find the average rate of change of f^{-1} on $[2, 7]$.

9. Graph each function and specify the intercepts.

(a) $y = |x + 2| - 3$ **(b)** $y = \dfrac{1}{x+2} - 1$

10. Refer to the following graph of $y = g(x)$. The domain of g is $[-5, 2]$.

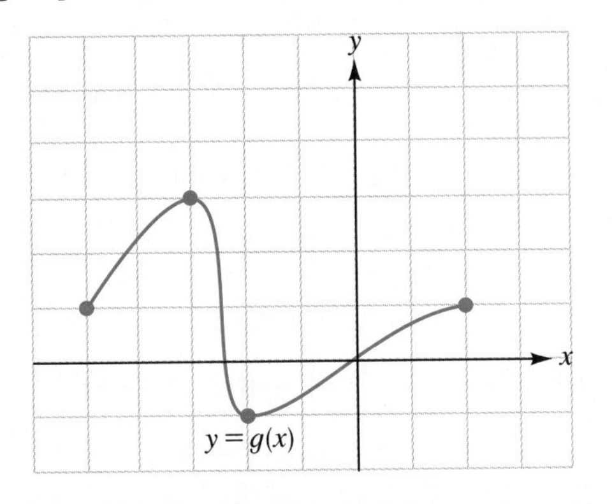

(a) What are the coordinates of the turning point(s)?

(b) What is the maximum value of g?

(c) Which input yields a minimum value for g?

(d) On which interval(s) is g increasing?

(e) Compute $\Delta g/\Delta x$ for each of the following intervals: $[-5, -3]$; $[-3, -2]$; $[-5, 2]$.

11. Let $f(x) = 3x^2 - 2x$. Find **(a)** $f(-1)$ and **(b)** $f(1 - \sqrt{2})$.

12. Graph the function G defined by

$$G(x) = \begin{cases} \sqrt{1 - x^2} & \text{if } -1 \leq x < 0 \\ \sqrt{x} & \text{if } x \geq 0 \end{cases}$$

13. Express the slope of a line passing through the points $(5, 25)$ and $(5 + h, (5 + h)^2)$ as a function of h.

14. Graph the function $y = |9 - x^2|$. Specify symmetry and intercepts.

15. The following figure shows the graph of a function $y = f(x)$. Sketch the graph of $y = f(-x)$.

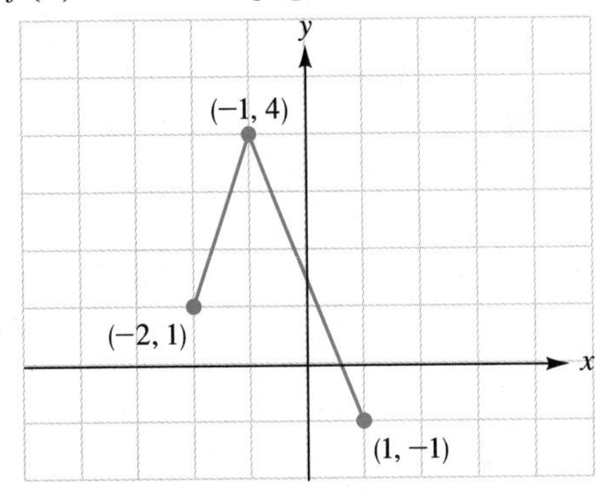

16. Given that the domains of f and f^{-1} are both $(-\infty, \infty)$ and that $f^{-1}(1) = -4$, solve the following equation for t: $2 + f(3t + 5) = 3$.

17. Let $f(x) = \frac{1}{2}x + 2$.

(a) Compute the first three iterates of $x_0 = 1$.

(b) Refer to the following figure. Explain why the y-coordinates of the points A, B, and C are the iterates that you computed in part (a).

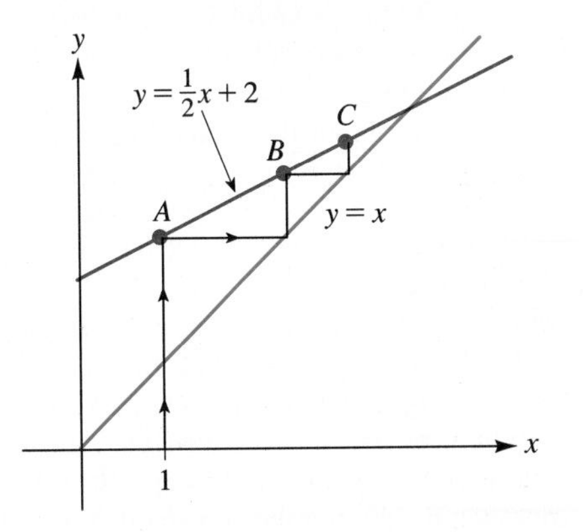

(c) Determine the y-coordinate of the point where the lines $y = x$ and $y = \frac{1}{2}x + 2$ intersect.

(d) If we continue the pattern of dashed lines in the figure, it appears that the iterates of $x_0 = 1$ will approach 4. Verify this empirically by using your calculator to compute the fourth through tenth iterates. (Round your final answers to three decimal places.)

18. **(a)** Let $f(x) = \frac{1}{2}x + 2$ (as in Exercise 17). Compute the first ten iterates of $x_0 = 5$. (After the third iterate, round the final answers to three decimal places.) What number do the answers seem to be approaching?

(b) Let $g(x) = 2x + 1$. Compute the first ten iterates of $x_0 = 1$. Do you observe a pattern similar to the one in part (a)?

19. Let $f(x) = \sqrt{x+1}$.

(a) Use the accompanying figure to estimate (to the nearest two-tenths) the first three iterates of $x_0 = -0.8$.

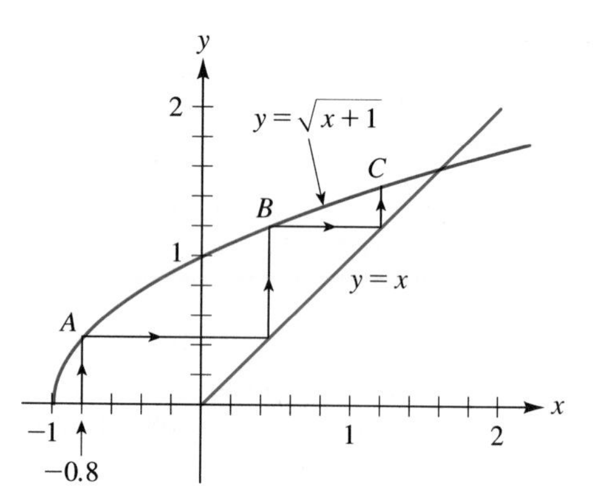

(b) Use a calculator to compute the first three iterates of $x_0 = -0.8$. Round the final results to two decimal places. Check that your results are consistent with the estimates in part (a).

(c) Determine the y-coordinate of the point where the line $y = x$ meets the curve $y = \sqrt{x+1}$.
Hint: First obtain the x-coordinate by solving the equation $x = \sqrt{x+1}$.

(d) If we continue the pattern in the figure, it appears that the iterates of $x_0 = -0.8$ will approach the number determined in part (c). Verify this empirically as follows: First, compute the first twelve iterates of $x_0 = -0.8$, rounding the final results to three decimal places. Next, use a calculator to evaluate the answer from part (c). Note that the iterates indeed appear to be approaching the number obtained in part (c). (This number is known as the *golden ratio*; see Exercises 103 and 104 in Exercise Set 1.4, page 28.)

In Exercises 20–42, sketch the graph and specify any x- or y-intercepts.

20. $y = 4 - x^2$

21. $y = -(x-1)^2 + 2$

22. $y = \dfrac{1}{x} + 1$

23. $f(x) = \dfrac{1}{x+1}$

24. $y = \dfrac{1}{x+1} + 1$

25. $y = |x+3|$

26. $g(x) = -\sqrt{x-4}$

27. $h(x) = \sqrt{1-x^2}$

28. $f(x) = \sqrt{1-x^2} + 1$

29. $y = 1 - (x+1)^3$

30. $y = 4 - \sqrt{-x}$

31. $y = (\sqrt{x})^2$

32. $f \circ g$, where $g(x) = x + 3$ and $f(x) = x^2$

33. $f \circ g$, where $g(x) = \sqrt{x-1}$ and $f(x) = -x^2$

34. $f(x) = \begin{cases} \sqrt{1-x^2} & \text{if } -1 \leq x \leq 0 \\ \sqrt{x} + 1 & \text{if } x > 0 \end{cases}$

35. $F(x) = \begin{cases} -\sqrt{1-x^2} & \text{if } -1 \leq x < 0 \\ \sqrt{x} & \text{if } x \geq 0 \end{cases}$

36. $y = \begin{cases} |x-1| & \text{if } 0 \leq x \leq 2 \\ |x-3| & \text{if } 2 < x \leq 4 \end{cases}$

37. $y = \begin{cases} 1/x & \text{if } 0 < x \leq 1 \\ 1/(x-1) & \text{if } 1 < x \leq 2 \end{cases}$

38. $(y + |x| - 1)(y - |x| + 1) = 0$

39. f^{-1}, where $f(x) = \frac{1}{2}(x+1)$

40. g^{-1}, where $g(x) = \sqrt[3]{x+2}$

41. $f \circ f^{-1}$, where $f(x) = \sqrt{x-2}$

42. $f^{-1} \circ f$, where $f(x) = \sqrt{x-2}$

In Exercises 43–52, find the domain of the function.

43. $y = 1/(x^2 - 9)$

44. $y = x^3 - x^2$

45. $y = \sqrt{8 - 2x}$

46. $y = x/(6x^2 + 7x - 3)$

47. $y = \sqrt{|2 - 5x|}$

48. $y = (25 - x^2)/\sqrt{x^2 + 1}$

49. $y = \sqrt{x^2 - 2x - 3}$

50. $y = \sqrt{5 - x^2}$

51. $y = x + \dfrac{1}{x}$

52. $z = \dfrac{1}{t - \sqrt{t+2}}$

In Exercises 53–58, determine the range of the function.

53. $y = \dfrac{x+4}{3x-1}$

54. $y = \dfrac{2x-3}{x-2}$

55. $f \circ g$, where $f(x) = 1/x$ and $g(x) = 3x + 4$

56. $g \circ f$, where $f(x) = \dfrac{x+2}{x-1}$ and $g(x) = \dfrac{x+1}{x+4}$

57. f^{-1}, where $f(x) = \dfrac{x}{3x-6}$

58. f^{-1}, where $f(x) = \dfrac{5-x}{1+x}$

In Exercises 59–66, express the function as a composition of two or more of the following functions:

$f(x) = 1/x \qquad g(x) = x - 1 \quad F(x) = |x| \quad G(x) = \sqrt{x}$

59. $a(x) = \dfrac{1}{x-1}$

60. $b(x) = \dfrac{1}{x} - 1$

61. $c(x) = \sqrt{x-1}$

62. $d(x) = \sqrt{x} - 1$

63. $A(x) = (1/\sqrt{x}) - 1$

64. $B(x) = |x - 2|$

65. $C(x) = \sqrt[4]{x} - 1$

66. $D(x) = 1/\sqrt{x-3}$

For Exercises 67–100, compute the indicated quantity using the functions f, g, and F defined as follows:

$$f(x) = x^2 - x \qquad g(x) = 1 - 2x \qquad F(x) = \frac{x-3}{x+4}$$

67. $f(-3)$

68. $f(1 + \sqrt{2})$

69. $F(3/4)$

70. $f(t)$

71. $f(-t)$

72. $g(2x)$

73. $f(x - 2)$

74. $g(x + h)$

75. $g(2) - g(0)$

76. $f(x) - g(x)$

77. $|f(1) - f(3)|$

78. $f(x + h) - f(x)$

79. $f(x^2)$

80. $f(x)/x \quad (x \neq 0)$

81. $[f(x)][g(x)]$

82. $f[f(x)]$

83. $f[g(x)]$

84. $g[f(3)]$

85. $(g \circ f)(x)$

86. $(g \circ f)(x) - (f \circ g)(x)$

87. $(F \circ g)(x)$

88. $\dfrac{g(x+h) - g(x)}{h}$

89. $\dfrac{f(x+h) - f(x)}{h}$

90. $\dfrac{F(x) - F(a)}{x - a}$

91. $F^{-1}(x)$

92. $F[F^{-1}(x)]$

93. $F^{-1}[F(x)]$

94. $F^{-1}(0) - \dfrac{1}{F(0)}$

95. $(g \circ g^{-1})(x)$

96. $g^{-1}(x)$

97. $g^{-1}(-x)$

98. $\dfrac{g^{-1}(x+h) - g^{-1}(x)}{h}$

99. $F^{-1}[F(22/7)]$

100. $T^{-1}(x)$, where $T(x) = f(x)/x \quad (x \neq 0)$

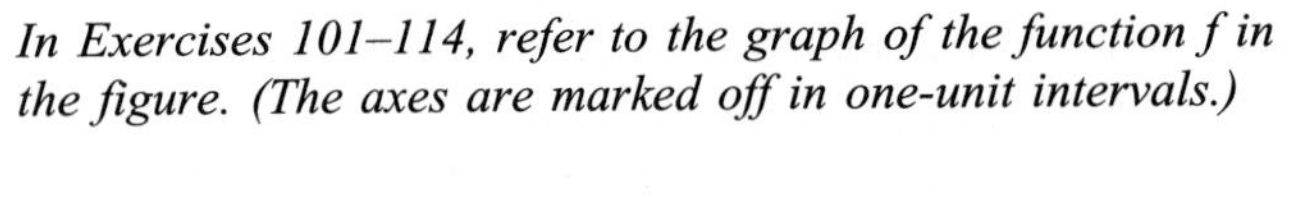

In Exercises 101–114, refer to the graph of the function f in the figure. (The axes are marked off in one-unit intervals.)

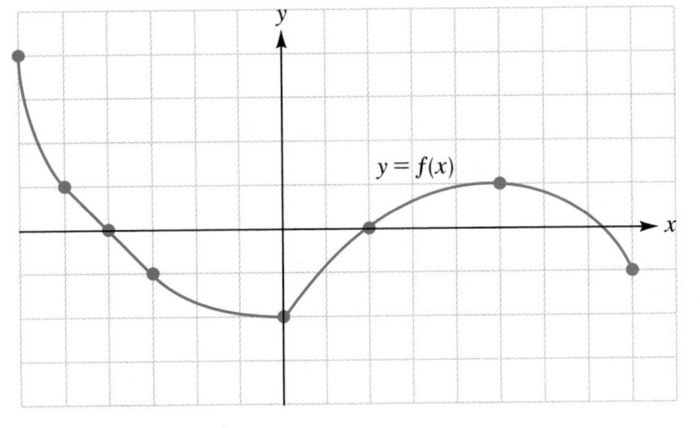

101. Is $f(0)$ positive or negative?

102. Specify the domain and range of f.

103. Find $f(-3)$.

104. Which is larger, $f(-5/2)$ or $f(-1/2)$?

105. Compute $f(0) - f(8)$.

106. Compute $|f(0) - f(8)|$.

107. Specify the coordinates of the turning points.

108. What are the minimum and the maximum values of f?

109. On which interval(s) is f decreasing?

110. For which x-values is it true that $1 \leq f(x) \leq 4$?

111. What is the largest value of $f(x)$ when $|x| \leq 2$?

112. Is f a one-to-one function?

113. Does f possess an inverse function?

114. Compute $f[f(-4)]$.

For Exercises 115–130, refer to the graphs of the functions f and g in the figure. Assume that the domain of each function is [0, 10].

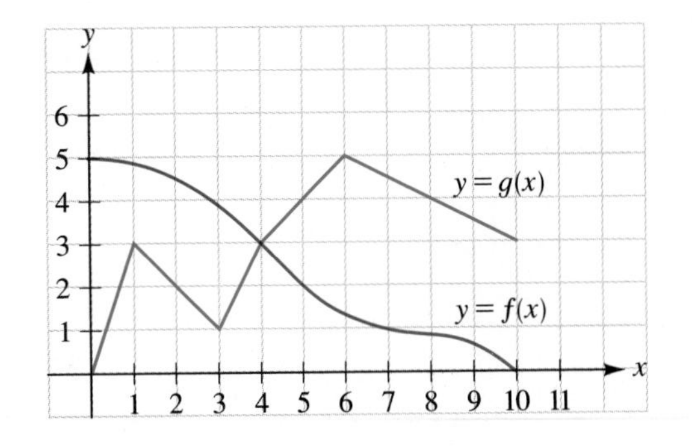

115. For which x-value is $f(x) = g(x)$?

116. For which x-values is it true that $g(x) \leq f(x)$?

117. **(a)** For which x-value is $f(x) = 0$?
(b) For which x-value is $g(x) = 0$?

118. Compute $f(0) + g(0)$.

119. Compute each of the following.
(a) $(f + g)(8)$ **(b)** $(f - g)(8)$
(c) $(fg)(8)$ **(d)** $(f/g)(8)$

120. Compute each of the following.
(a) $g[f(5)]$ **(b)** $f[g(5)]$
(c) $(g \circ f)(5)$ **(d)** $(f \circ g)(5)$

121. Which is larger, $(f \circ f)(10)$ or $(g \circ g)(10)$?

122. Compute $g[f(10)] - f[g(10)]$.

123. For which x-values is it true that $f(x) \geq 3$?

124. For which x-values is it true that $|f(x) - 3| \leq 1$?

125. What is the largest number in the range of g?

126. Specify the coordinates of the highest point on the graph of each of the following equations.
(a) $y = g(-x)$ **(b)** $y = -g(x)$
(c) $y = g(x - 1)$ **(d)** $y = f(-x)$
(e) $y = -f(x)$ **(f)** $y = -f(-x)$

127. On which intervals is the function g decreasing?

128. What are the coordinates of the turning points of g?

129. For which values of x in the interval (4, 7) is the quantity $\dfrac{f(x) - f(5)}{x - 5}$ negative?

130. For which values of x in the interval (0, 5) is the quantity $\dfrac{f(x) - f(2)}{x - 2}$ positive?

CHAPTER THREE TEST

1. Find the domain of the function defined by $f(x) = \sqrt{x^2 - 5x - 6}$.
2. Find the range of the function defined by $g(x) = (2x - 8)/(3x + 5)$.
3. Let $f(x) = 2x^2 - 3x$ and $g(x) = 2 - x$. Compute each of the following:
(a) $(f - g)(x)$; **(b)** $(f \circ g)(x)$; **(c)** $f[g(-4)]$.
4. Let $f(t) = 2/t$. Compute $\dfrac{f(t) - f(a)}{t - a}$.
5. Let $g(x) = 2x^2 - 5x$. Compute $\dfrac{g(x + h) - g(x)}{h}$.
6. Find $g^{-1}(x)$ given that $g(x) = -4x/(6x + 1)$.
7. The graph of a function f is a line segment joining the two points $(-3, 1)$ and (5, 6). Determine the slope of the line segment that results from graphing the equation $y = -f^{-1}(x)$.
8. Graph each function and specify the intercepts:
(a) $y = -|x - 3| + 1$; **(b)** $y = \dfrac{1}{x + 3} - 2$.
9. The figure shows the graph of a function g with domain $[-4, 2]$.

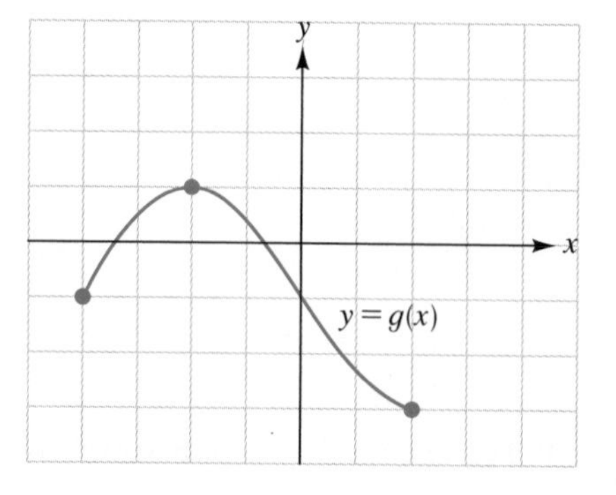

(a) Specify the range of g.
(b) What are the coordinates of the turning point?
(c) What is the minimum value of g?
(d) Which input yields a maximum value for g? What is that maximum value?
(e) On which interval is g decreasing?
(f) Compute $\Delta g/\Delta x$ on the interval $[-4, 2]$.

10. Let $f(x) = x^2 - 3x - 1$. Compute each of the following:
(a) $f(-3/2)$; **(b)** $f(\sqrt{3} - 2)$.

11. Graph the function F defined by

$$F(x) = \begin{cases} |x+1| & \text{if } x < -1 \\ -x^2 + 1 & \text{if } x > -1 \end{cases}$$

What is the domain of this function?

12. In the accompanying figure, each dashed line is parallel to one of the coordinate axes. Determine the y-coordinates of the points P and Q. Give two forms for each answer: an exact answer (involving one or more radicals) and a calculator approximation rounded to two decimal places.

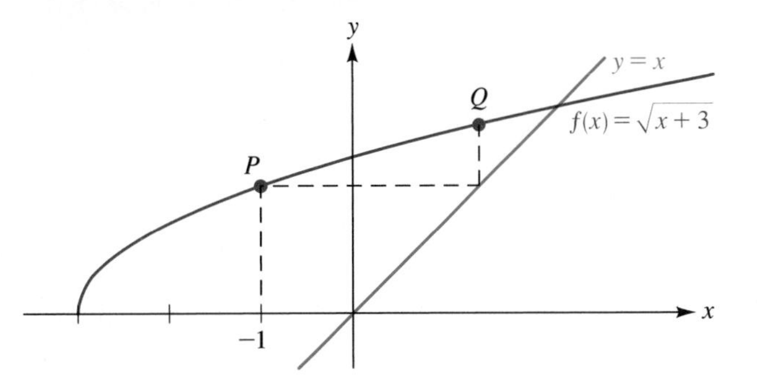

13. **(a)** Let $f(x) = 3 - 2x$. Show that f is one-to-one.
(b) Let $g(x) = \sqrt{1 + 4x^2}$. Show that g is not one-to-one.

14. In the following figure, the graph of $f(x) = 1 - |2x - 1|$ is shown in red.

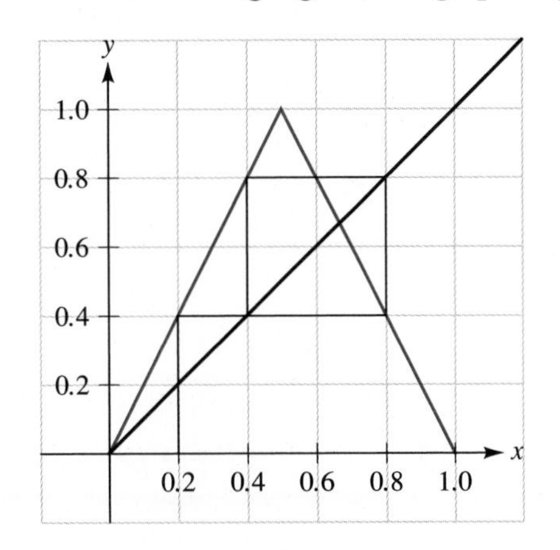

(a) Use the figure to specify the first six iterates of $x_0 = 0.2$.

(b) Use a calculator to compute the first six iterates of $x_0 = 0.2$. Are your results consistent with those in part (a)?

15. The graph of f is a line segment joining the points $(1, 3)$ and $(5, -2)$. Sketch the graph of $y = f(-x)$.
16. Given that the domains of both f and f^{-1} are $(-\infty, \infty)$ and that $f^{-1}(-3) = 1$, solve the equation $5 + f(4t - 3) = 2$ for t.
17. Let $f(x) = 1/x$. Find a value for b so that on the interval $[1, b]$, we have $\Delta f/\Delta x = -1/10$.
18. During the first five hours of a lab experiment, the temperature of a solution is given by

$$F(t) = 0.16t^2 - 1.6t + 35 \quad (0 \leqslant t \leqslant 5)$$

where t is measured in hours, $t = 0$ corresponds to the time that the experiment begins, and $F(t)$ is the temperature (in degrees Celsius) at time t.

(a) Find (and simplify) a formula for $\Delta F/\Delta t$ over the interval $[t_1, 5]$.
(b) Use the result in part (a) to compute the average rate of change of temperature during the period from $t = 4$ hr to $t = 5$ hr. (Be sure to include the units as part of the answer.)

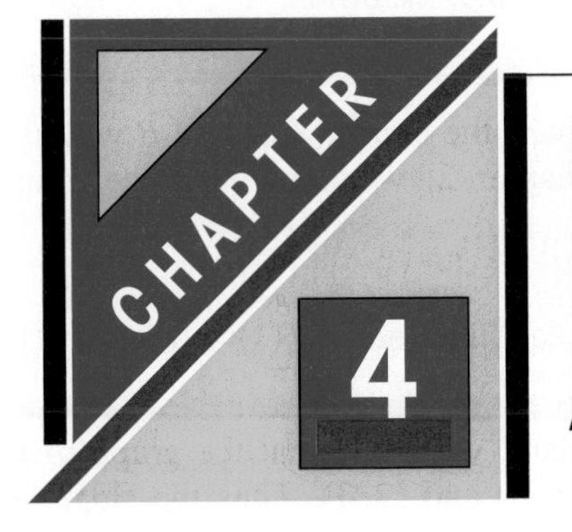

POLYNOMIAL AND RATIONAL FUNCTIONS: APPLICATIONS TO ITERATION AND OPTIMIZATION

INTRODUCTION

In the previous chapter we studied some rather general rules for working with functions and their graphs. Now we focus our attention on just a few particular types of functions. In Section 4.1 we discuss linear functions. As you'll see, the applications here are quite diverse, ranging from business and economics to physics and statistics. Section 4.2 contains a discussion of quadratic functions. In Section 4.3 we learn more about iteration of functions (first discussed in Section 3.4). An example in Section 4.3 shows how iteration of quadratic functions is used in the study of population growth. Section 4.4 could aptly be titled ''Translating English into Algebra.'' In that section we develop the background skills for solving the maximum and minimum problems in Section 4.5. If a friend of yours happens to be taking calculus, you'll find that you can solve some of your friend's maximum and minimum problems without using calculus. In the last two sections of the chapter, 4.6 and 4.7, we present a number of techniques for graphing polynomial and rational functions. Although these graphs can be obtained by using a graphing calculator, our goal here is to *understand* why the graphs look as they do.

LINEAR FUNCTIONS

All decent functions are practically linear.

Professor Andrew Gleason

I had a moment of mixed joy and anguish, when my mind took over. It raced well ahead of my body and drew my body compellingly forward. I felt that the moment of a lifetime had come.

Dr. Roger Bannister, first person to run the mile in under four minutes, recalling the last lap of his record-breaking run at the Ilffley Road Track, Oxford, England, May 6, 1954. The quotation is from Roger Bannister, *The Four Minute Mile* (New York: Dodd, Mead & Company, 1955).

By a **linear function** we mean a function defined by an equation of the form

$$f(x) = Ax + B$$

where A and B are constants. In this chapter, the constants A and B will always be real numbers. From our work in Chapter 2, we know that the graph of $y = Ax + B$ is a straight line.

EXAMPLE 1 Suppose that f is a linear function. If $f(1) = 0$ and $f(2) = 3$, find an equation defining f.

Solution From the statement of the problem, we know that the graph of f is a straight line passing through the points $(1, 0)$ and $(2, 3)$. Thus the slope of the line is

$$m = \frac{y_2 - y_1}{x_2 - x_1} = \frac{3 - 0}{2 - 1} = 3$$

Now we can use the point–slope formula using the point $(1, 0)$ to find the required equation. We have

$$y - y_1 = m(x - x_1)$$
$$y - 0 = 3(x - 1)$$
$$y = 3x - 3$$

This is the equation defining f. If we wish, we can rewrite it using function notation: $f(x) = 3x - 3$.

One basic application of linear functions that occurs in business and economics is *linear* or *straight-line depreciation.* In this situation we assume that the value V of an asset (such as a machine or an apartment building) decreases linearly over time t; that is, $V = mt + b$, where the slope m is negative.

EXAMPLE 2 A factory owner buys a new machine for \$8000. After ten years, the machine has a salvage value of \$500.

(a) Assuming linear depreciation (as indicated in Figure 1), find a formula for the value V of the machine after t years, where $0 \leq t \leq 10$.
(b) Use the formula derived in part (a) to find the value of the machine after five years.

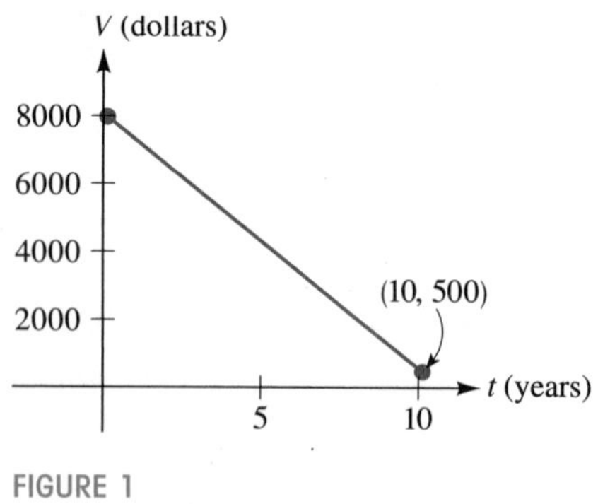

FIGURE 1

Solution

(a) We need to determine m and b in the equation $V = mt + b$. From Figure 1 we see that the V-intercept of the line segment is $b = 8000$. Furthermore, since the line segment passes through the two points $(10, 500)$ and $(0, 8000)$, the slope m is

$$\frac{8000 - 500}{0 - 10} = -750$$

So we have $m = -750$ and $b = 8000$, and the required equation is

$$V = -750t + 8000 \qquad (0 \leq t \leq 10)$$

(b) Substituting $t = 5$ in the equation $V = -750t + 8000$ yields

$$V = -750(5) + 8000 = -3750 + 8000 = 4250$$

Thus the value of the machine after five years is \$4250.

It is instructive to keep track of the units associated with the slope in Example 2. Repeating the slope calculation and keeping track of the units, we have

$$m = \frac{\$8000 - \$500}{0 \text{ years} - 10 \text{ years}} = -\$750/\text{year} \tag{1}$$

Thus, slope is a *rate of change.* In this example, the slope represents the rate of change of the value of the machine. Notice that the calculations in (1) are exactly those that we would have used in the previous chapter to compute the *average* rate of change $\Delta V/\Delta t$. In the context of linear functions, we often suppress the word ''average'' and refer simply to the *rate of change,* because no matter what two points we choose on the graph of a linear function, the slope will always be the same.

As a second example of slope as a rate of change, let us suppose that a small manufacturer of handmade running shoes knows that her total cost in dollars, $C(x)$, for producing x pairs of shoes each business day is given by the linear function

$$C(x) = 10x + 50$$

Figure 2(a) displays a table and the graph for this function. Actually, since x represents the daily number of pairs of shoes, x can assume only whole-number values. So, technically, the graph that should be given is the one in Figure 2(b). However, the graph in Figure 2(a) turns out to be useful in practice, and we shall follow this convention.

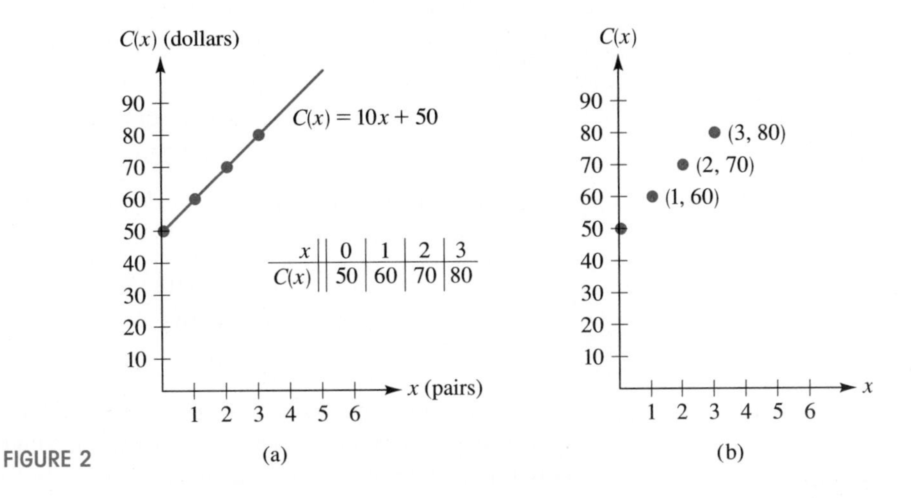

FIGURE 2

The slope of the line $C(x) = 10x + 50$ in Figure 2(a) is 10, the coefficient of x. But to understand the units involved, let's calculate the slope using two of the points in Figure 2(a). Using the points (0, 50) and (1, 60) and keeping track of the units, we have

$$m = \frac{\$60 - \$50}{1 \text{ pair} - 0 \text{ pairs}} = \frac{\$10}{1 \text{ pair}} = \$10/\text{pair}$$

Again the slope is a rate. In this case, the slope represents the rate of increase of cost; each additional pair of shoes produced costs the manufacturer \$10.

In the study of economics, an equation that gives the cost $C(x)$ for producing x units of a commodity is called a **cost equation** or **cost function.** When the graph of the cost equation is a line, we define the **marginal cost** as the additional cost to produce one more unit. Thus, in the preceding example, the marginal cost is \$10 per pair, and we see that the slope of the line in Figure 2(a) represents this marginal cost.

EXAMPLE 3 Suppose that the cost $C(x)$ in dollars of producing x bicycles is given by

$$C(x) = 625 + 45x$$

(a) Find the cost of producing 10 bicycles.
(b) What is the marginal cost?
(c) Use the answers in parts (a) and (b) to find the cost of producing 11 bicycles. Then check the answer by evaluating $C(11)$.

Solution
(a) Using $x = 10$ in the cost equation, we have

$$C(10) = 625 + 45(10) = 1075$$

Thus, the cost of producing ten bicycles is \$1075.
(b) Since C is a linear function, the marginal cost is the slope and we have

$$\text{marginal cost} = \$45 \text{ per bicycle}$$

(c) According to the result in part (b), each additional bicycle costs \$45. Therefore, we can compute the cost of 11 bicycles by adding \$45 to the cost for 10 bicycles:

$$\begin{aligned}\text{cost of 11 bicycles} &= \text{cost of 10 bicycles} + \text{marginal cost}\\ &= \$1075 + \$45 = \$1120\end{aligned}$$

So, the cost of producing 11 bicycles is \$1120. We can check this result by using the cost function $C(x) = 625 + 45x$ to compute $C(11)$ directly:

$$\begin{aligned}C(11) &= 625 + 45(11)\\ &= 625 + 495\\ &= \$1120 \qquad \text{as obtained previously}\end{aligned}$$

FIGURE 3
The slope of the line is the rate of change of distance with respect to time.

Interpreting slope as a rate of change is not restricted to applications in business or economics. Suppose, for example, that you are driving a car at a steady rate of 50 mph. Using the distance formula from elementary mathematics,

$$\text{distance} = \text{rate} \times \text{time}$$

or

$$d = rt$$

we have, in this case, $d = 50t$, where d represents the distance traveled (in miles) in t hours. In Figure 3, we show a graph of the linear function $d = 50t$. The slope of this line is 50, the coefficient of t. But 50 is also the given rate of speed, in miles per hour. So again, slope is a rate of change. In this case, slope is the **velocity,** or *rate of change of distance with respect to time.*

We conclude this section by indicating one way that linear functions are used in statistical applications. Tables 1 and 2 show the evolution of the record for the one-mile run during the years 1911–1993. In case you're wondering why Table 1 begins with 1911, John Paul Jones was the first twentieth-century runner to break the previous century's record of 4:15.6, set in 1895. Table 1 ends with the year 1954, the first year in which the "four-minute barrier" was broken. (See the quote by Roger Bannister at the beginning of this section.)

TABLE 1 EVOLUTION OF THE RECORD FOR THE MILE RUN, 1911–1954

x (year)	y (time)	
1911	4:15.4	(John Paul Jones, U.S.)
1913	4:14.6	(John Paul Jones, U.S.)
1915	4:12.6	(Norman Taber, U.S.)
1923	4:10.4	(Paavo Nurmi, Finland)
1931	4:09.2	(Jules Ladoumegue, France)
1933	4:07.6	(Jack Lovelock, New Zealand)
1934	4:06.8	(Glen Cunningham, U.S.)
1937	4:06.4	(Sidney Wooderson, Great Britain)
1942	4:06.2	(Gunder Haegg, Sweden)
1942	4:06.2	(Arne Andersson, Sweden)
1942	4:04.6	(Gunder Haegg, Sweden)
1943	4:02.6	(Arne Andersson, Sweden)
1944	4:01.6	(Arne Andersson, Sweden)
1945	4:01.4	(Gunder Haegg, Sweden)
1954	3:59.4	(Roger Bannister, Great Britain)
1954	3:58.0	(John Landy, Australia)

TABLE 2 EVOLUTION OF THE RECORD FOR THE MILE RUN THROUGH 1993 (continued from Table 1)

x (year)	y (time)	
1957	3:57.2	(Derek Ibbotson, Great Britain)
1958	3:54.5	(Herb Elliott, Australia)
1962	3:54.4	(Peter Snell, New Zealand)
1964	3:54.1	(Peter Snell, New Zealand)
1965	3:53.6	(Michel Jazy, France)
1966	3:51.3	(Jim Ryun, U.S.)
1967	3:51.1	(Jim Ryun, U.S.)
1975	3:51.0	(Filbert Bayi, Tanzania)
1975	3:49.4	(John Walker, New Zealand)
1979	3:49.0	(Sebastian Coe, Great Britain)
1980	3:48.8	(Steve Ovett, Great Britain)
1981	3:48.53	(Sebastian Coe, Great Britain)
1981	3:48.40	(Steve Ovett, Great Britain)
1981	3:47.33	(Sebastian Coe, Great Britain)
1985	3:46.32	(Steve Cram, Great Britain)
1993	3:44.39	(Noureddine Morceli, Algeria)

Let's begin by looking at the data in Table 1 graphically. In Figure 4(a), we've plotted the (x, y) pairs given in the table. The resulting plot is called a **scatter diagram.**

A striking feature of the scatter diagram in Figure 4(a) is that the records do not appear to be leveling off; rather, they seem to be decreasing in an approximately linear fashion. Using the *least-squares technique* from the field of statistics, it can be shown that the linear function that best fits the data in Table 1 is

$$f(x) = -0.370x + 962.041 \tag{2}$$

The graph of this line is shown in Figure 4(b). The line itself is referred to as the **regression line,** or the **least-squares line.** (The formulas for determining this line are given in Exercise 51.)

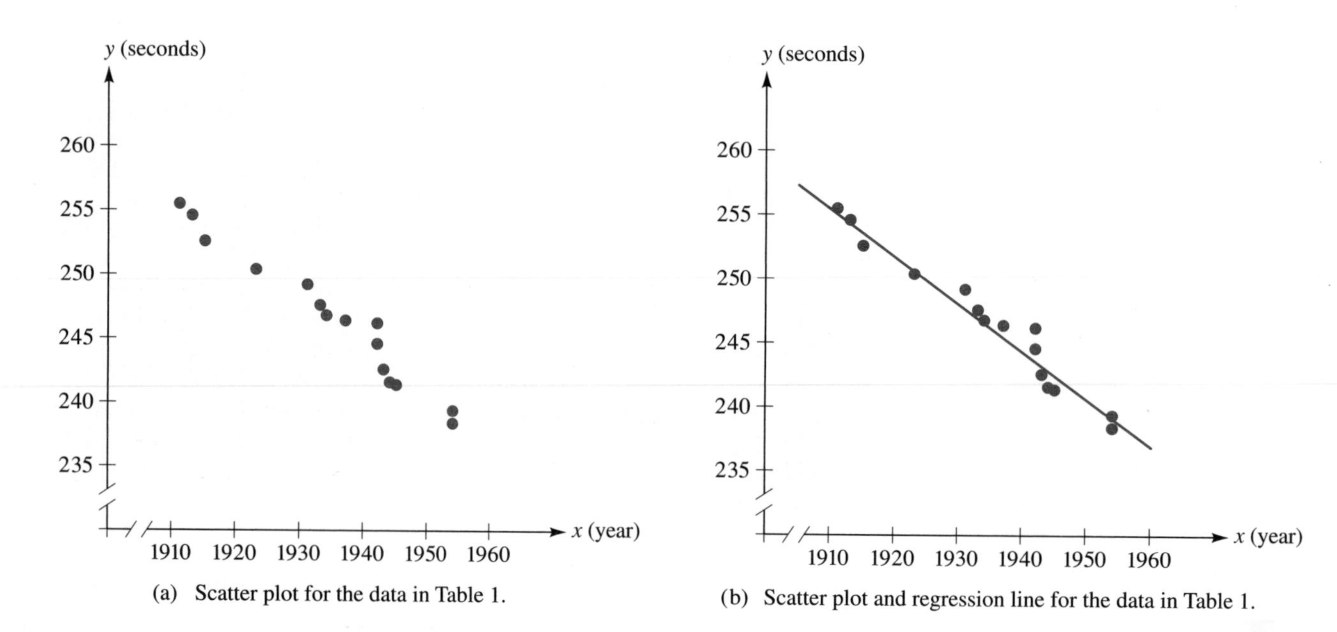

(a) Scatter plot for the data in Table 1.

(b) Scatter plot and regression line for the data in Table 1.

FIGURE 4

As an intuitive check on the least-squares line, let's use it to estimate what the record for the mile might have been in 1993; then we'll check our prediction against the actual record (from Table 2). Using $x = 1993$ in equation (2), we have

$$\begin{aligned} f(1993) &= -0.370(1993) + 962.041 \\ &\approx 224.6 \text{ seconds} \end{aligned}$$

Thus, after converting seconds to minutes, our least-squares estimate for the 1993 record is 3:44.6. This agrees favorably with the actual record of 3:44.39 set by Noureddine Morceli of Algeria. Exercises 37 and 38 at the end of this section ask you to carry out similar calculations for some of the other mile records.

EXAMPLE 4

(a) The June 1976 issue of *Scientific American* contained an article entitled "Future Performance in Footracing" by H. W. Ryder, H. J. Carr, and

P. Herget. According to the authors of this article, "It appears likely that within 50 years the record [for the mile] will be down to 3:30." Use the regression line defined in equation (2) to make a projection for the mile time in the year 2026 (which will be 50 years after the article appeared). Is your projection close to the one given in the *Scientific American* article?

(b) The regression line defined in equation (2) is based only on the data from Table 1. If we use all the data from both Table 1 and Table 2, then the least-squares technique can be used to derive the following regression line:

$$f(x) = -0.400x + 1019.472 \qquad (3)$$

Use equation (3) to make a projection for the mile time in the year 2026. Which projection is closer to the value predicted in the *Scientific American* article, this one or the one calculated in part (a)?

Solution

(a) Setting $x = 2026$ in equation (2) gives us

$$f(2026) = -0.370(2026) + 962.041 = 212.421 \text{ seconds}$$

Converting this result to minutes and then rounding to one decimal place, our projection for the mile time in 2026 is 3:32.4. This is fairly close to the projection in *Scientific American;* our projected time is about $2\frac{1}{2}$ seconds slower.

(b) Using equation (3), which is based on the data in both Tables 1 and 2, we have

$$f(2026) = -0.400(2026) + 1019.472 = 209.072 \text{ seconds}$$

Converting seconds to minutes and then rounding to one decimal place, we obtain a projected mile time of 3:29.1. Although our projection in part (a) agreed fairly well with the figure in *Scientific American,* this result is closer still; it's only about one second faster.

EXERCISE SET 4.1

A Do 1, 5, 6, 9, 13, 17

In Exercises 1–10, find the linear functions satisfying the given conditions.

1. $f(-1) = 0$ and $f(5) = 4$
2. $f(3) = 2$ and $f(-3) = -4$
3. $g(0) = 0$ and $g(1) = \sqrt{2}$
4. The graph passes through the points (2, 4) and (3, 9).
5. $f(\frac{1}{2}) = -3$ and the graph of f is a line parallel to the line $x - y = 1$.
6. $g(2) = 1$ and the graph of g is perpendicular to the line $6x - 3y = 2$.
7. The graph of f is a horizontal line that passes through the larger of the two y-intercepts of the circle $x^2 - 2x + y^2 - 3 = 0$.
8. The x- and y-intercepts of the graph of g are 1 and 4, respectively.
9. The graph of the inverse function passes through the points $(-1, 2)$ and $(0, 4)$.
10. The x- and y-intercepts of the inverse function are 5 and -1, respectively.
11. Let $f(x) = 3x - 4$ and $g(x) = 1 - 2x$. Determine whether the function $f \circ g$ is linear.
12. Explain why there is no linear function with a graph that passes through all three of the points $(-3, 2)$, $(1, 1)$, and $(5, 2)$.
13. A factory owner buys a new machine for \$20,000. After eight years, the machine has a salvage value of \$1000. Find a formula for the value of the machine after t years, where $0 \leq t \leq 8$.

14. A manufacturer buys a new machine costing \$120,000. It is estimated that the machine has a useful lifetime of ten years and a salvage value of \$4000 at that time.
(a) Find a formula for the value of the machine after t years, where $0 \leq t \leq 10$.
(b) Find the value of the machine after eight years.

15. A factory owner installs a new machine costing \$60,000. Its expected lifetime is five years, and at the end of that time the machine has no salvage value.
(a) Find a formula for the value of the machine after t years, where $0 \leq t \leq 5$.
(b) Complete the following depreciation schedule.

End of Year	Yearly Depreciation	Accumulated Depreciation	Value V
0	0	0	60,000
1			
2			
3			
4			
5		60,000	0

16. Let x denote a temperature on the Celsius scale, and let y denote the corresponding temperature on the Fahrenheit scale.
(a) Find a linear function relating x and y; use the facts that 32°F corresponds to 0°C, and 212°F corresponds to 100°C. Write the function in the form $y = Ax + B$.
(b) What Celsius temperature corresponds to 98.6°F?
(c) Find a number z for which z°F = z°C.

17. Suppose that the cost $C(x)$, in dollars, of producing x electric fans is given by $C(x) = 450 + 8x$.
(a) Find the cost to produce 10 fans.
(b) Find the cost to produce 11 fans.
(c) Use your answers in parts (a) and (b) to find the marginal cost. (Then check that your answer is the slope of the line.)

18. Suppose that the cost to a manufacturer of producing x units of a certain motorcycle is given by $C(x) = 220x + 4000$, where $C(x)$ is in dollars.
(a) Find the marginal cost.
(b) Find the cost of producing 500 motorcycles.
(c) Use your answers in parts (a) and (b) to find the cost of producing 501 motorcycles.

19. Suppose that the cost $C(x)$, in dollars, of producing x units of a portable compact disc player is given by $C(x) = 400 + 50x$.
(a) Compute $C(n + 1) - C(n)$.
(b) What is the marginal cost?
(c) What is the relationship between the answers in parts (a) and (b)?

20. Suppose that the cost $C(x)$, in dollars, of producing x compact discs is given by $C(x) = 0.5x + 500$.
(a) Graph the given equation.
(b) Compute $C(150)$.
(c) If you add the marginal cost to the answer in part (b), you obtain a certain dollar amount. What does this amount represent?

21. The following graphs each relate distance and time for a moving object. Determine the velocity in each case.

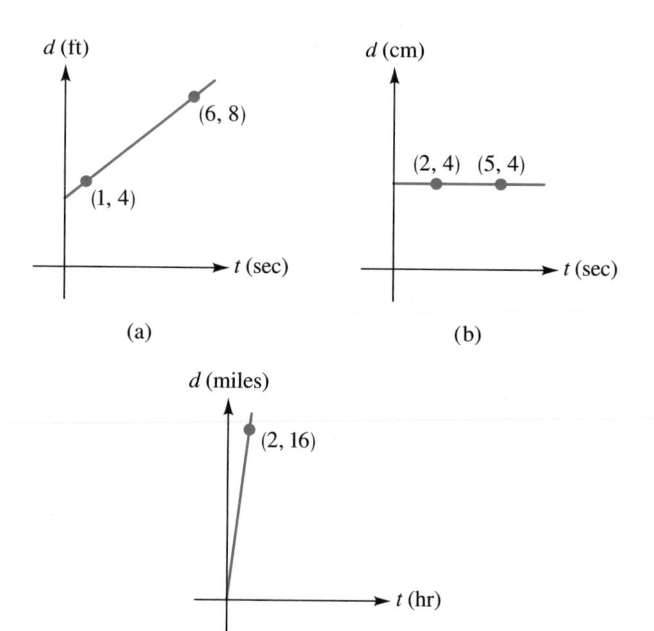

22. The distance d, in feet, covered by a particular object in t sec is given by the equation $d = 5t$.
(a) Find the velocity of the object.
(b) Find the distance covered in 15 sec.
(c) Use your answers in parts (a) and (b) to find the distance covered in 16 sec. Check your answer by substituting $t = 16$ in the given distance formula.

23. Two points A and B move along the x-axis. After t sec, their positions are given by the equations

$$A: \quad x = 3t + 100$$
$$B: \quad x = 20t - 36$$

(a) Which point is traveling faster, A or B?
(b) Which point is farther to the right when $t = 0$?
(c) At what time t do A and B have the same x-coordinate?

24. A point moves along the x-axis, and its x-coordinate after t seconds is $x = 4t + 10$. (Assume that x is in centimeters.)
(a) What is the velocity?
(b) What is the x-coordinate when $t = 2$ sec?
(c) Use your answers in parts (a) and (b) to find the x-coordinate when $t = 3$ sec. Check your answer by letting $t = 3$ in the given equation.

In Exercises 25–28, the population data is from The Universal Almanac 1994, *John W. Wright, ed. (Kansas City: Andrews and McMeel, 1993)*

25. The following table gives the population of California in 1970 and in 1980.

x (Year)	y (Population)
1970	19,971,069
1980	23,667,764

(a) Find the equation of the linear function whose graph passes through the two (x, y) points given in the table.
(b) Use the linear function determined in part (a) to make a projection for the population of California in 1990. (Round your answer to the nearest 1000.)
(c) The actual California population for 1990 was 29,760,021. Does the linear function yield a projection that is too high or too low?
(d) Use the following formula to compute the percent error in your projection. (Round your final answer to one decimal place.)

$$\text{percent error} = \frac{|\text{actual value} - \text{projected value}|}{\text{actual value}} \times 100$$

26. Follow Exercise 25 using the following data for Texas.

x (Year)	y (Population)
1970	11,198,655
1980	14,225,513

In 1990, the population of Texas was 16,986,510.

27. Follow Exercise 25 using the following data for Florida.

x (Year)	y (Population)
1970	6,791,418
1980	9,746,961

In 1990, the population of Florida was 12,937,926.

28. Follow Exercise 25 using the following data for the state of Hawaii. In part (b), round to the nearest 100 rather than the nearest 1000.

x (Year)	y (Population)
1970	769,913
1980	964,691

In 1990, the population of the state of Hawaii was 1,108,229.

In general, the growth of plants or animals does not follow a linear pattern. For relatively short intervals of time, however, a linear function may provide a reasonable description of the growth. Exercises 29 and 30 provide examples of this.

29. **(a)** In an experiment with sunflower plants, H. S. Reed and R. H. Holland measured the height of the plants every seven days for several months. [The experiment is reported in *Proceedings of the National Academy of Sciences,* vol. 5 (1919), p. 140.] The following data are from this experiment.

x (Number of days)	21	49
y [Average height (cm) of plants after x days]	67.76	205.50

Find the linear function $y = f(x)$ whose graph passes through the two (x, y) points given in the table. (Round each number in the answer to two decimal places.)
(b) Use the linear function determined in part (a) to estimate the average height of the plants after 28 days. (Round your answer to two decimal places.)
(c) In the experiment, Reed and Holland found that the average height after 28 days was 98.10 cm. Is your estimate in part (b) too high or too low? Compute the percent error in your estimate, using the formula given in Exercise 25(d).
(d) Follow parts (b) and (c) using $x = 14$ days. For the computation of percent error, you need to know that Reed and Holland found that the average height after 14 days was 36.36 cm.
(e) As you've seen in parts (c) and (d), your estimates are quite close to the actual values obtained in the experiment. This indicates that for a relatively short interval, the growth function is nearly linear. Now repeat parts (b) and (c) using $x = 84$ days, which is a longer interval of time. You'll find that the linear function does a poor job in describing the growth

of the sunflower plants. To compute the percent error, you need to know that Reed and Holland determined the average height after 84 days to be 254.50 cm.

30. **(a)** The biologist R. Pearl measured the population of a colony of fruit flies *(Drosophila melanogaster)* over a period of 39 days. [The experiment is discussed in Pearl's book, *The Biology of Population Growth* (New York: Alfred Knopf, 1925).] Two of the measurements made by Pearl are given in the following table.

x **(Number of days)**	12	18
y **(Population)**	105	225

Find the linear function $y = f(x)$ whose graph passes through the two (x, y) points given in the table.

(b) Use the linear function determined in part (a) to estimate the population after 15 days. Then compute the percent error in your estimate, given that the actual population after 15 days, as found by Pearl, was 152. The formula for percent error is given in Exercise 25(d).

(c) Use the linear function determined in part (a) to estimate the population after 9 days. Then compute the percent error in your estimate, given that after 9 days, Pearl found the actual population to be 39.

(d) As you've seen in parts (b) and (c), your estimates are quite close to the actual values obtained in the experiment. This indicates that for relatively short intervals of time, the growth is nearly linear. Now repeat part (b) using $x = 39$ days, which covers a longer interval of time. You'll find that the linear function does a poor job in describing the population growth. For the computation of percent error, you need to know that Pearl determined the population after 39 days to be 938.

31. **(a)** On graph paper, plot the following points: $(1, 0)$, $(2, 3)$, $(3, 6)$, $(4, 7)$.

(b) In your scatter diagram from part (a), sketch a line that best seems to fit the data. Estimate the slope and the y-intercept of the line.

(c) The actual regression line in this case is $y = 2.4x - 2$. Add the graph of this line to your sketch from parts (a) and (b).

32. **(a)** On graph paper, plot the following points: $(1, 2)$, $(3, 2)$, $(5, 4)$, $(8, 5)$, $(9, 6)$.

(b) In your scatter diagram from part (a), sketch a line that best seems to fit the data.

(c) Using your sketch from part (b), estimate the y-intercept and the slope of the regression line.

(d) The actual regression line is $y = 0.518x + 1.107$. Check to see if your estimates in part (c) are consistent with the actual y-intercept and slope. Graph this line along with the points given in part (a). (In sketching the line, use the approximation $y = 0.5x + 1.1$.)

33. The table in the following figure shows the population of Los Angeles over the years 1930–1990. The accompanying graph shows the corresponding scatter plot and regression line. The equation of the regression line is

$$f(x) = 37{,}546.068x - 71{,}238{,}863.429$$

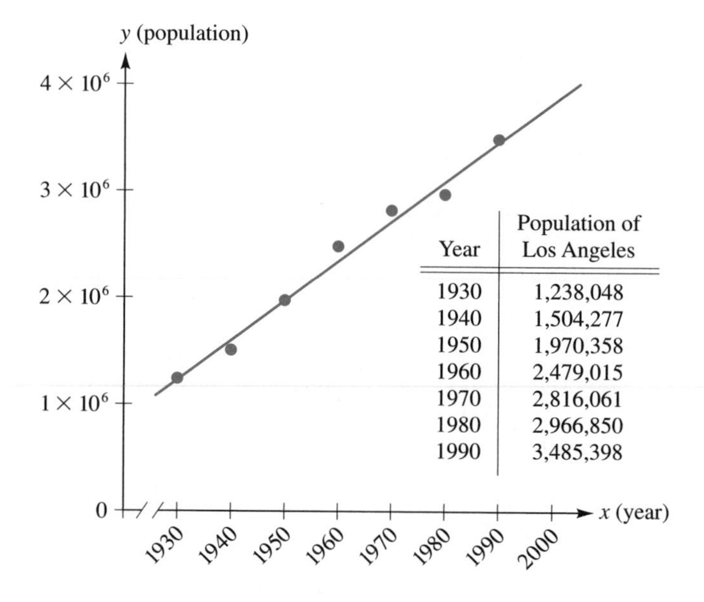

Year	Population of Los Angeles
1930	1,238,048
1940	1,504,277
1950	1,970,358
1960	2,479,015
1970	2,816,061
1980	2,966,850
1990	3,485,398

(a) Predict whether the population of Los Angeles might reach 3.8 million by the year 2000.

(b) Find $f^{-1}(x)$.

(c) Use your answer in part (b) to estimate the year in which the population of Los Angeles might reach 4 million. *Hint:* For the function f, the inputs are years and the outputs are populations; for f^{-1}, the inputs are populations and the outputs are years.

34. In this exercise we use a linear function to summarize and predict sales (actually, shipments) of compact discs (CDs) in the United States. The compact discs we are referring to are audio or music CDs, rather than those used with a personal computer. The data in this exercise are taken from *The Universal Almanac 1994*, John W. Wright, ed. (Kansas City: Andrews and McMeel, 1993). According to *The Universal Almanac*, "Less than 100,000 CDs were sold in 1983, but the format outsold LPs in 1988 and overtook cassette tapes in 1992.'' The table in the following figure shows the number of CDs shipped in the U.S. over the years 1986–1991. (These are net figures; returns from the retailers have been accounted for.) The graph in the figure shows the corresponding scatter plot and regression line. The equation of the regression line is

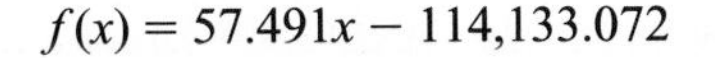

$$f(x) = 57.491x - 114{,}133.072$$

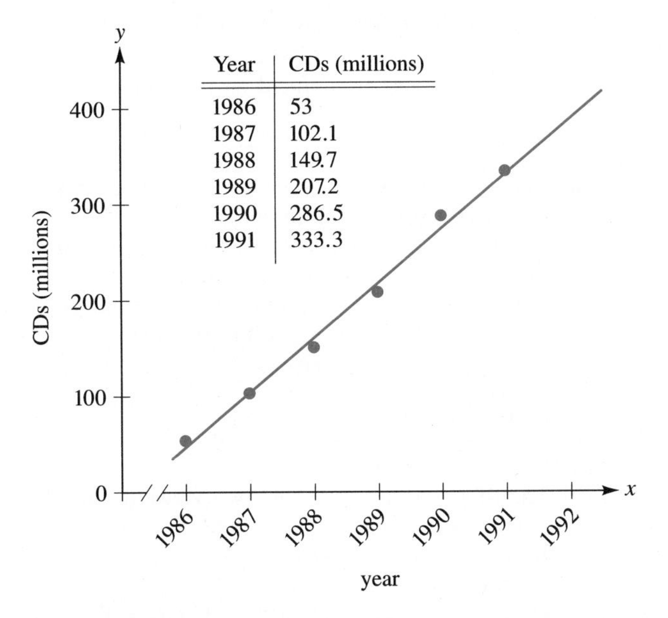

Year	CDs (millions)
1986	53
1987	102.1
1988	149.7
1989	207.2
1990	286.5
1991	333.3

(a) Use the regression line to predict the number of CDs shipped in 1992. Then say whether your estimate is too high or too low, given that the actual number of shipments for 1992 was 407.5 million.

(b) Use the regression line to estimate the number of CDs that might be shipped in the year 2000.

(c) Find $f^{-1}(x)$.

(d) Use your answer in part (c) to predict the year in which the number of CDs shipped might reach 1 billion (= 1,000 million). *Hint:* For the function f, you input a year and the output is the number of CDs for that year; for the function f^{-1}, you input the number of CDs and the function outputs a corresponding year.

Afterword: One might argue that the growth in CD sales cannot continue indefinitely in this same linear fashion; at some point sales and shipments should level off. In the next chapter we do consider functions with graphs that ''level off.''

35. Many companies in the United States manufacture or assemble their products abroad to save on labor costs. For example, the toy industry in the United States imports approximately 30% of its products from China. (GI Joe and Teenage Mutant Ninja Turtles are assembled in China.) The following table shows the dollar value of these imports during the period 1986–1990, while the graph displays the scatter plot and regression line for the data. The equation of the regression line is

$$f(x) = (0.3428)x - 680.6114$$

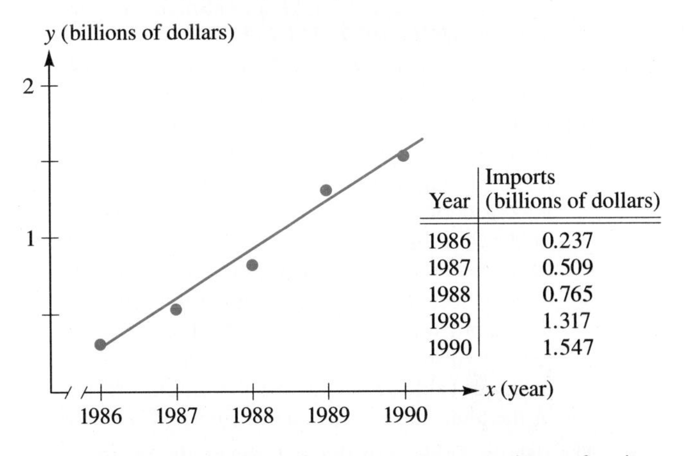

Year	Imports (billions of dollars)
1986	0.237
1987	0.509
1988	0.765
1989	1.317
1990	1.547

(a) Use the regression line to make an estimate for the value of the toy imports from China in 1996.

(b) Predict whether the value of the toy imports will exceed 5 billion dollars by the year 2000.

36. This exercise illustrates one of the pitfalls that can arise in using a regression line: large discrepancies can occur when the regression line is used to make long-term projections. The following table shows the population of California during the years 1860–1900.
The accompanying graph displays the scatter plot of the data; the equation of the regression line is

$$f(x) = (28632.69)x - 52928780$$

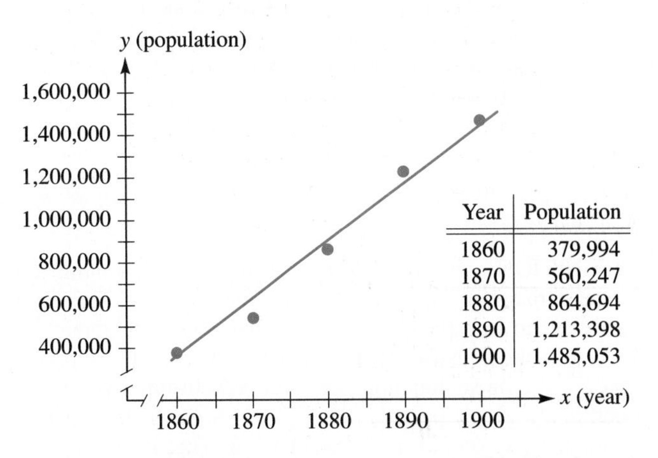

Year	Population
1860	379,994
1870	560,247
1880	864,694
1890	1,213,398
1900	1,485,053

As you can see from the scatter plot, the population growth over this period is very nearly linear.

(a) Use the regression line to estimate the population of California in 1990.

(b) According to the U.S. Bureau of the Census, the population of California in 1990 was 29,839,250. Is your estimate in part (a) close to this figure?

37. The regression line defined by equation (2) in the text is based on the data in Table 1 covering the years 1911–1954.

(a) Use equation (2) to make an estimate for the mile record in 1957. Then check Table 2 to see if your estimate is close. Is your estimate too high or too low?

(b) Use equation (2) to estimate the mile record for 1958. When you check the actual record in Table 2 for 1958, you'll find that your estimate in part (a) is considerably higher. This isn't surprising in view of the fact that, in setting the record in 1958, Herb Elliott bettered the then-existing record by more than $2\frac{1}{2}$ seconds, the largest single improvement of the record for the entire century. (For a fascinating look at Herb Elliott and many of the other runners listed in Tables 1 and 2, see the article "Milers" in the June 27, 1994, issue of *Sports Illustrated.*)

38. The data in Table 2 of the text covers the years 1957–1993. The equation of the regression line for this data is

$$y = -0.318x + 858.955$$

(a) Use this equation to estimate what the mile record might have been for the year 1954. Then check Table 1 to see the actual record for 1954, which was set by John Landy. Is your estimate too high or too low? Compute the percent error in your estimate using the formula given in Exercise 25(d).

(b) Use the given equation to estimate what the mile record might have been back in 1911. Check your estimate against the actual record as shown in Table 1, and compute the percent error. How does the percent error here compare to that in part (a)? Why is this to be expected?

B

In Exercises 39–41, you'll need to use the following definition and ideas.

DEFINITION: A **fixed point** of a function f is a number x in the domain of f with the property that $f(x) = x$. There are two ways to find fixed points. The first way is algebraic: just solve the equation $f(x) = x$ for x. For example, suppose that f is the linear function $f(x) = \frac{1}{3}x + 2$. In this case the condition $f(x) = x$ becomes $\frac{1}{3}x + 2 = x$. Solving this equation yields (as you should now check) $x = 3$. That is, 3 is a fixed point for the function f. The other way to find the fixed point is geometric. Draw careful graphs of $f(x) = \frac{1}{3}x + 2$ and $y = x$; the x-coordinate of the intersection point is the required fixed point. Try this on a sheet of graph paper so that you get an accurate sketch. You should find that the x-coordinate of the intersection point is 3, which agrees with the answer that was obtained algebraically.

39. Let $f(x) = -\dfrac{3}{2}x + \dfrac{15}{2}$.

(a) Find a fixed point of f using the algebraic method (as described in the definition).

(b) Use the geometric (graphical) method to confirm that your answer in part (a) is correct.

(c) What are the first five iterates of 3 under the function f? (Iterates are defined on p. 146.)

40. Let $g(x) = 8 - \dfrac{1}{3}x$.

(a) Find a fixed point of g using the algebraic method (as described in the definition).

(b) Use the geometric (graphical) method to confirm that your answer in part (a) is correct.

(c) Let b denote the fixed point that you found in part (a). Simplify each of the following expressions: $f(b)$; $f(f(b))$; $f(f(f(b)))$.

41. Let $f(x) = \dfrac{1}{2}x + \dfrac{3}{2}$.

(a) Find a fixed point of f.

(b) The following figures show two views of the graphs of the function f and the line $y = x$. Use the figures to find the first five iterates of $x = 1$ under the function f. (Use Figure A for the first three iterates; use the zoom-in view in Figure B for the remaining iterates.)

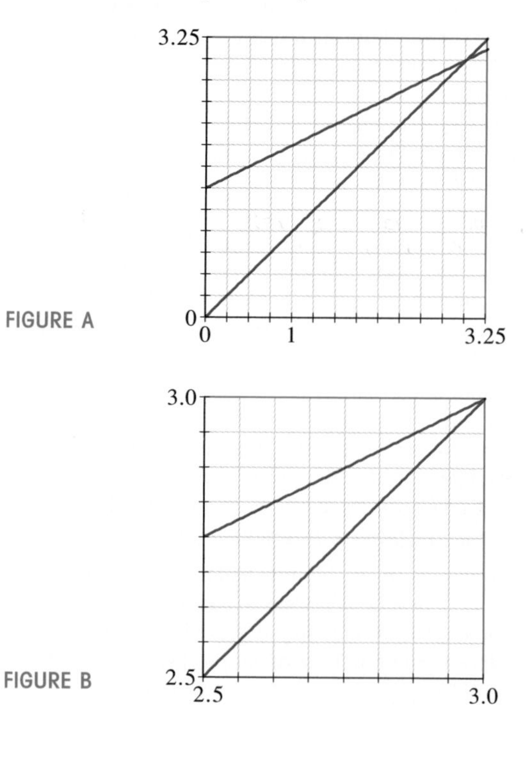

FIGURE A

FIGURE B

(c) Use a calculator to compute the first twelve iterates of $x = 1$ under the function f. (Where appropriate, round the final answers to four decimal places.) Check that your first five iterates agree with the values you found in part (b).

(d) Use the results in parts (c) and (a) to fill in the blanks in the following sentence. The iterates are approaching the number ____, which is the ____________ (two words) of the function f.

42. **(a)** The accompanying figure shows the beginning of the iteration process for the function $f(x) = 2x - 1$ with initial input $x = 2$. Use the figure to read off the first three iterates of $x = 2$.

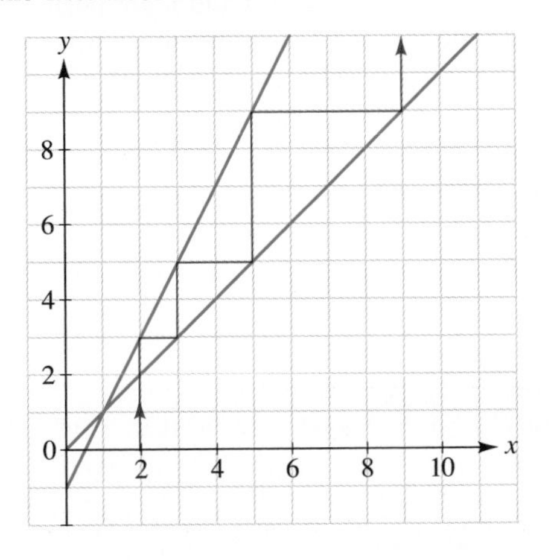

(b) As indicated by the pattern emerging in the figure, the iterates of $x = 2$ will grow larger and larger, without bound. By calculating, find the first iterate to exceed 100 and the first iterate to exceed 1000.

In Exercises 43–46, let f and g be the linear functions defined by

$$f(x) = Ax + B \quad (A \neq 0) \quad \text{and} \quad g(x) = Cx + D \quad (C \neq 0)$$

In each case, compute the average rate of change of the given function on the interval $[a, b]$.

43. **(a)** f **(b)** $f \circ f$ **(c)** $g \circ f$ **(d)** $f \circ g$

44. **(a)** f^{-1} **(b)** g^{-1}

45. **(a)** $(f \circ g)^{-1}$ **(b)** $(g \circ f)^{-1}$

46. **(a)** $f^{-1} \circ g^{-1}$ **(b)** $g^{-1} \circ f^{-1}$

47. Show that the linear function $f(x) = mx$ satisfies the following identities:
(a) $f(a + b) = f(a) + f(b)$; **(b)** $f(ax) = af(x)$.

48. Let f be a linear function such that

$$f(a + b) = f(a) + f(b)$$

for all real numbers a and b. Show that the graph of f passes through the origin. *Hint:* Let $f(x) = Ax + B$ and show that $B = 0$.

49. Find a linear function $f(x) = mx + b$ such that m is positive and $(f \circ f)(x) = 9x + 4$.

50. **(a)** Let f be a linear function. Show that

$$f\left(\frac{x_1 + x_2}{2}\right) = \frac{f(x_1) + f(x_2)}{2}$$

(In words: The output of the average is the average of the outputs.)

(b) Show (by using an example) that the equation in part (a) does not hold for the function $f(x) = x^2$.

51. This exercise shows how to compute the slope and the y-intercept of the regression line. As an example, we'll work with the simple data set given in Exercise 31.

x	1	2	3	4
y	0	3	6	7

(a) Let Σx denote the sum of the x-coordinates in the data set, and let Σy denote the sum of the y-coordinates. Check that $\Sigma x = 10$ and $\Sigma y = 16$.

(b) Let Σx^2 denote the sum of the squares of the x-coordinates, and let Σxy denote the sum of the products of the corresponding x- and y-coordinates. Check that $\Sigma x^2 = 30$ and $\Sigma xy = 52$.

(c) The slope m and the y-intercept b of the regression line satisfy the following pair of simultaneous equations [in the first equation, n denotes the number of points (x, y) in the data set]:

$$\begin{cases} nb + (\Sigma x)m = \Sigma y \\ (\Sigma x)b + (\Sigma x^2)m = \Sigma xy \end{cases}$$

So, in the present example, these equations become

$$\begin{cases} 4b + 10m = 16 \\ 10b + 30m = 52 \end{cases}$$

Solve this pair of equations for m and b, and check that your answers agree with the values in Exercise 31(c).

In Exercises 52–55, use the method described in Exercise 51 to find the equation of the regression line for the given data sets.

52.

x	2	4	8	10
y	−7	−5	−2	−1

53.

x	1	2	3	4	5
y	2	3	9	9	11

54.

x	1	2	3	4	5
y	16	13.1	10.5	7.5	2

55.

x	520	740	560	610	650
y	81	98	83	88	95

C

56. Suppose that f is a linear function satisfying the condition $f(kx) = kf(x)$ for all real numbers k. Prove that the graph of f passes through the origin.

57. **(a)** Find all linear functions f satisfying the identity $f(f(x)) = 2x + 1$. (For your answer, rationalize any denominators containing radicals.)

(b) Find all linear functions f satisfying the identity

$$f(f(f(x))) = 2x + 1$$

(Rationalize any denominators containing radicals.)

58. Let a and b be real numbers, and suppose that the inequality $ax + b \leq 0$ has no solutions. What can you say about the linear function $f(x) = ax + b$? Answer in complete sentences and justify what you say.

59. Are there any linear functions f satisfying the identity $f^{-1}(x) = f(f(x))$? If so, list them; if not, explain why not.

60. Let $f(x) = ax + b$, where a and b are positive numbers, and assume that the following equation holds for all values of x: $f(f(x)) = bx + a$. Show that $a + b = 1$.

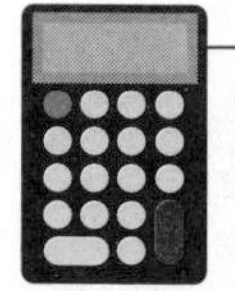

GRAPHING UTILITY EXERCISES FOR SECTION 4.1

1. **(a)** Reread Example 2. Then graph the function $f(x) = -750x + 8000$. Adjust the viewing rectangle (and the scales on the axes) so that your picture is similar to Figure 1 on page 176.

(b) Use your graphing utility to estimate the y-coordinate of the point on the graph corresponding to $x = 5$. How close is your estimate to the answer \$4250 obtained in the text?

2. Linear functions can be used to approximate more complicated functions. This is one of the meanings or implications of the quotation by Professor Gleason on page 175. This exercise illustrates that idea.

(a) Using calculus, it can be shown that the equation of the line that is tangent to the curve $y = x^2$ at the point (1, 1) is $y = 2x - 1$. Verify this visually by graphing the two functions $y = x^2$ and $y = 2x - 1$ on the same set of axes. (*Suggestion:* Use a viewing rectangle that extends from -2 to 3 in the x-direction and from -3 to 4 in the y-direction.) Note that the tangent line is virtually indistinguishable from the curve in the immediate vicinity of the point (1, 1).

(b) For numerical rather than visual evidence of how well the linear function $y = 2x - 1$ approximates the function $y = x^2$ in the immediate vicinity of (1, 1), complete the following tables.

x	0.9	0.99	0.999
x^2			
$2x - 1$			

x	1.1	1.01	1.001
x^2			
$2x - 1$			

3. Follow Exercise 2 using the curve $y = 1/x$ and the line $y = -0.25x + 1$, which is tangent to the curve at the point $(2, 1/2)$. (*Suggestion:* First use the standard viewing rectangle. Then, for a better view, use a viewing rectangle extending from 0 to 5 in the x-direction and from 0 to 2 in the y-direction.) For part (b), complete the following tables.

x	1.9	1.99	1.999
$1/x$			
$-0.25x + 1$			

x	2.1	2.01	2.001
$1/x$			
$-0.25x + 1$			

As preparation for Exercises 4 and 5, you may need to check the owner's manual for your graphing utility to see if there is a convenient option for computing the equation of a regression line. [If not, it's still possible (but not convenient) to work these exercises by making use of the technique indicated in Exercise 51 in Exercise Set 4.1.]

4. The following table shows world grain production for selected years over the period 1975–1990.

Year	World Grain Production (Million Tons)
1975	1250
1976	1363
1977	1337
1978	1467
1979	1428
1980	1447
1981	1499
1982	1550
1990	1780

Source: Lester R. Brown et al. in *Vital Signs 1994* (New York: W. W. Norton & Company, 1994)

(a) Compute the equation of the regression line using the data for the years 1975–1980, inclusive. (In the x-y data pairs, use x for the year and y for the grain production.)

(b) Use the equation of the regression line to make estimates for the world grain production in each of the years 1981, 1982, and 1990.

(c) How close are your estimates in part (b) to the actual values given in the table for the years 1981, 1982, and 1990? Answer this by computing the percent error for each estimate. For percent error, use the formula

$$\text{percent error} = \frac{|(\text{actual value}) - (\text{estimated value})|}{\text{actual value}} \times 100$$

Which of the three estimates is the best? Which is the worst?

(d) Use the regression line to estimate the world grain production for the year 1993. You'll find that your estimate is much higher than the actual 1993 production, which was approximately 1682 million tons. What is the percent error in your estimate? (According to Lester Brown's *Vital Signs, 1994,* the drop in world grain production for 1993 was due almost entirely to the effects of poor weather on the U.S. corn crop that year.)

5. The following table shows world natural gas production over the years 1989–1992. The abbreviation tcf stands for *trillion cubic feet.*

Year	1989	1990	1991	1992
Natural Gas Production (tcf)	69.231	71.081	72.034	72.516

Source: BP Statistical Review of World Energy 1994 (British Petroleum Company, London)

(a) Compute the equation of the regression line for this data. (In the x-y data pairs, use x for the year and y for the natural gas production.)

(b) Use the regression line to estimate natural gas production for the year 1993.

(c) According to *BP Statistical Review of World Energy 1994,* world natural gas production for 1993 was 74.025 tcf. Is your estimate in part (b) close to this figure? Is your estimate too high or too low?

(d) The data from the table and from part (c) indicate that world natural gas production is increasing. However, for purposes of making a conservative estimate, assume for the moment that the production of natural gas levels off at about 74 tcf per year. According to *BP Statistical Review of World Energy 1994,* the 1993 proved world reserves of natural gas were 5016.2 tcf. Carry out the following computation and interpret your answer.
Hint: Keep track of the units.

$$\frac{5016.2 \text{ tcf}}{74 \text{ tcf/year}} = \ldots$$

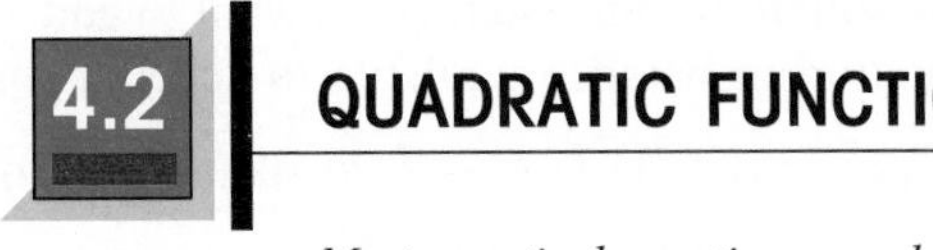

4.2 QUADRATIC FUNCTIONS

Most practical questions can be reduced to problems of largest and smallest magnitudes. . . .

Pafnuti Lvovich Chebyshev (Russian mathematician, 1821–1894). According to Professor George F. Simmons in *Calculus Gems* (New York: McGraw-Hill, 1992), ". . . [Chebyshev] was the most eminent Russian mathematician of the nineteenth century . . . and is considered the founder of the great school of mathematics that has been flourishing in Russia for the past century."

After the linear functions, the next simplest functions are the **quadratic functions,** which are defined by equations of the form

$$f(x) = ax^2 + bx + c \qquad (a \neq 0)$$

where a, b, and c are constants and a is not zero. In this chapter the constants a, b, and c will always be real numbers. We will see that the graph of any quadratic function is a curve called a **parabola,** which is similar or identical in shape to the graph of $y = x^2$. Figure 1 displays the graphs of two typical parabolas.

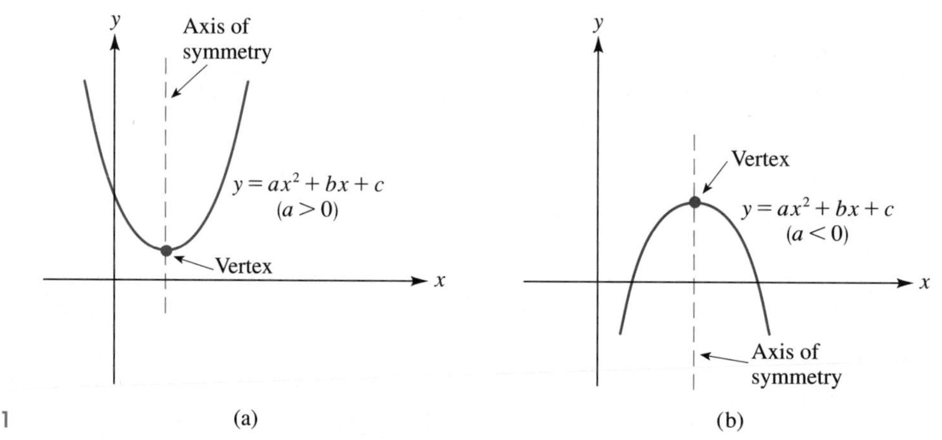

FIGURE 1 (a) (b)

Subsequent examples will demonstrate that the parabola $y = ax^2 + bx + c$ opens upward when $a > 0$ and downward when $a < 0$. As Figure 1 indicates, the turning point on the parabola is called the **vertex.** The **axis of symmetry** of the parabola $y = ax^2 + bx + c$ is the vertical line passing through the vertex.

The methods that we will use for dealing with quadratic functions have already been developed in Chapters 1 and 3. In particular, the following two topics are prerequisites for understanding the examples in this section:

1. completing the square; for a review, see Section 1.4;
2. translations and reflections; for a review, see Section 3.3.

EXAMPLE 1 Graph the function $y = x^2 - 2x + 3$.

Solution The idea here is to use the technique of completing the square; this will enable us to obtain the required graph simply by shifting the basic $y = x^2$ graph. We begin by writing the given equation

$$y = x^2 - 2x \qquad +3$$

To complete the square for the x-terms we want to add 1. (Check this.) Of course, to keep the equation in balance we have to account for this by writing

$$y = (x^2 - 2x + 1) + 3 - 1 \qquad \text{adding zero to the right side}$$

or

$$y = (x - 1)^2 + 2$$

Now, as we know from Section 3.3, the graph of this last equation is obtained by moving the parabola $y = x^2$ one unit in the positive x-direction and two units in the positive y-direction. This shifts the vertex from the origin to the point $(1, 2)$. See Figure 2.

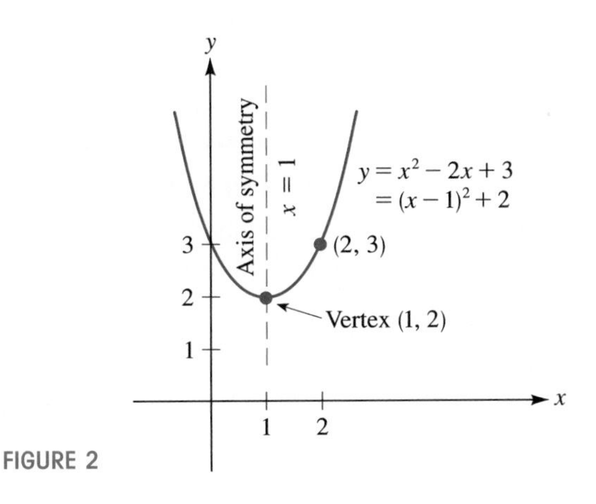

FIGURE 2

NOTE As a guide to sketching the graph, you'll want to know the y-intercept. To find the y-intercept, substitute $x = 0$ in the given equation to obtain $y = 3$. Then, given the vertex $(1, 2)$ and the point $(0, 3)$, a reasonably accurate graph can be quickly sketched. [Actually, once you find that $(0, 3)$ is on the graph, you also know that the reflection of this point about the axis of symmetry is on the graph. This is why the point $(2, 3)$ is shown in Figure 2.]

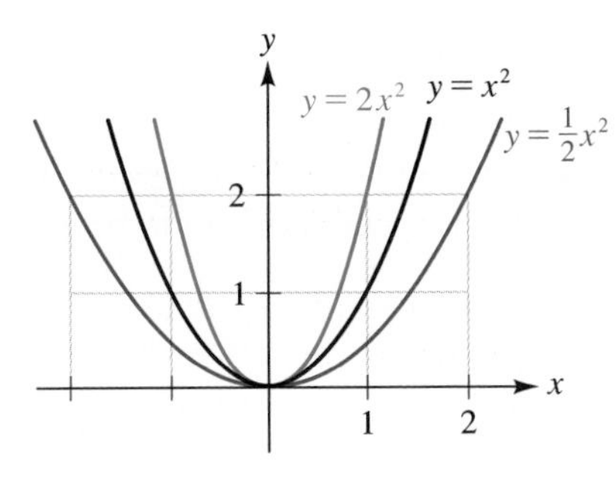

FIGURE 3

Now we want to compare the graphs of $y = x^2$, $y = 2x^2$, and $y = \frac{1}{2}x^2$. The last two equations were not specifically discussed in the previous chapters, so for the moment we can graph them by first setting up tables. Figure 3 displays the graphs. All three graphs are parabolas that open upward, but the shapes are not identical. The parabola $y = 2x^2$ is narrower than $y = x^2$, while $y = \frac{1}{2}x^2$ is wider than $y = x^2$. These observations about shape would also apply to $y = -x^2$, $y = -2x^2$, and $y = -\frac{1}{2}x^2$; in these cases, though, the parabolas open downward rather than upward. The observations that we have just made are generalized in the box that follows.

PROPERTY SUMMARY **THE GRAPH OF $y = ax^2$**

1. The graph of $y = ax^2$ is a parabola with vertex at the origin. It is similar in shape to $y = x^2$.
2. The parabola $y = ax^2$ opens upward if $a > 0$, downward if $a < 0$.
3. The parabola $y = ax^2$ is narrower than $y = x^2$ if $|a| > 1$, wider than $y = x^2$ if $|a| < 1$.

EXAMPLE 2 Sketch the graph of $y = 3(x - 1)^2$.

Solution Because of the $x - 1$, we shift the basic parabola $y = x^2$ one unit to the right. The factor of 3 in the given equation tells us that we want to draw a parabola that is narrower than $y = x^2$ but that has the same vertex, $(1, 0)$. To see exactly how narrow to draw $y = 3(x - 1)^2$, we need to know another point on

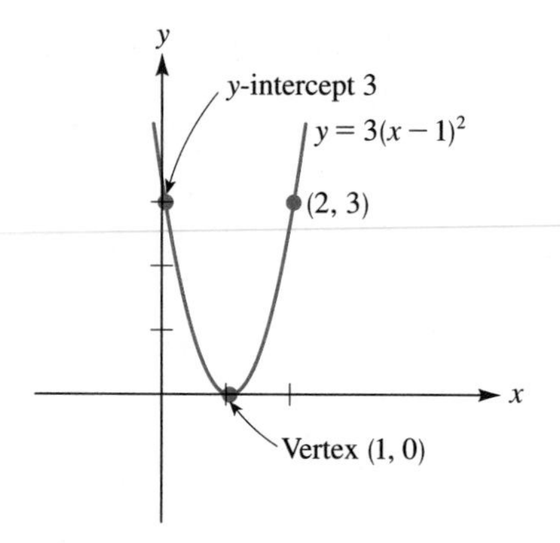

FIGURE 4

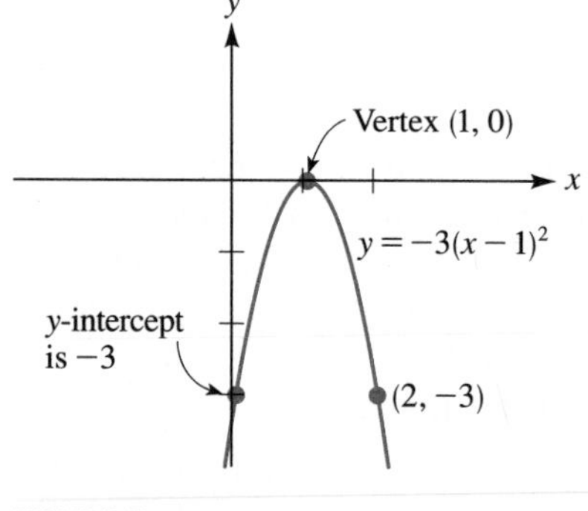

FIGURE 5

the graph other than the vertex, (1, 0). An easy point to obtain is the y-intercept. Setting $x = 0$ in the equation yields

$$y = 3(0-1)^2 = 3$$

Now that we know the vertex, (1, 0), and the y-intercept, 3, we can sketch a reasonably accurate graph; see Figure 4.

EXAMPLE 3 Sketch the graph of $y = -3(x-1)^2$.

Solution In Example 2 we sketched the graph of $y = 3(x-1)^2$. By reflecting that graph in the x-axis, we obtain the graph of $y = -3(x-1)^2$. See Figure 5.

EXAMPLE 4 Graph the function $f(x) = -2x^2 + 12x - 16$ and specify the vertex, axis of symmetry, maximum or minimum value of f, and x- and y-intercepts.

Solution The idea is to complete the square, as in Example 1. We have

$$
\begin{aligned}
y &= -2x^2 + 12x \qquad - 16 \\
&= -2(x^2 - 6x \qquad) - 16 \\
&= -2(x^2 - 6x + 9) - 16 + 18 && \text{adding (2)(9) = 18 to keep the equation in balance} \\
&= -2(x-3)^2 + 2 && (1)
\end{aligned}
$$

From equation (1), we see that the required graph is obtained simply by shifting the graph of $y = -2x^2$ "right 3, up 2," so that the vertex is (3, 2). As a guide to sketching the graph, we want to compute the intercepts. The y-intercept is -16. (Why?) For the x-intercepts, we replace y with 0 in equation (1) to obtain

$$
\begin{aligned}
-2(x-3)^2 &= -2 \\
(x-3)^2 &= 1 \\
x - 3 &= \pm 1 \\
x &= 4 \quad \text{or} \quad 2
\end{aligned}
$$

Thus the x-intercepts are $x = 4$ and $x = 2$. Knowing these intercepts and the vertex, we can sketch the graph as in Figure 6. You should check for yourself that the information accompanying Figure 6 is correct.

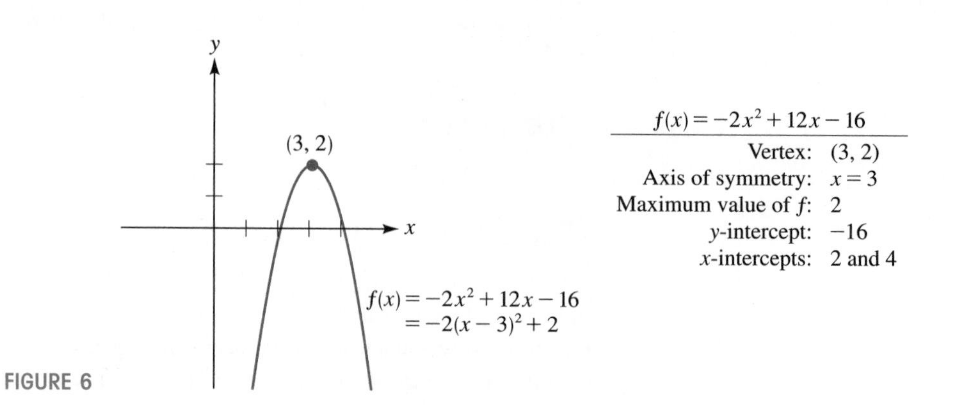

FIGURE 6

The next three examples involve maximum and minimum values of functions. As background for this, we first summarize our basic technique for graphing parabolas.

PROPERTY SUMMARY **THE GRAPH OF THE PARABOLA $y = ax^2 + bx + c$**

By completing the square, the equation of the parabola $y = ax^2 + bx + c$ can always be rewritten in the form

$$y = a(x-h)^2 + k$$

In this form, the vertex of the parabola is (h, k) and the axis of symmetry is the line $x = h$. The parabola opens upward if $a > 0$ and downward if $a < 0$.

EXAMPLE 5 Among all possible inputs for the function

$$g(x) = 3x^2 - 2x - 6$$

which yields the smallest output? What is this minimum output?

Solution Think in terms of a graph. Since the graph of this function is a parabola opening upward, the input we want is the x-coordinate of the vertex. By completing the square (as shown in Example 4), we find that the given equation can be rewritten

$$g(x) = 3\left(x - \frac{1}{3}\right)^2 - \frac{19}{3}$$

(You should verify this for yourself, using Example 4 as a model.) Thus, the vertex of the parabola is $(1/3, -19/3)$. From this, and from the fact that the parabola opens upward, we conclude that the input $x = 1/3$ produces the smallest output, and this minimum output is $g(1/3) = -19/3$.

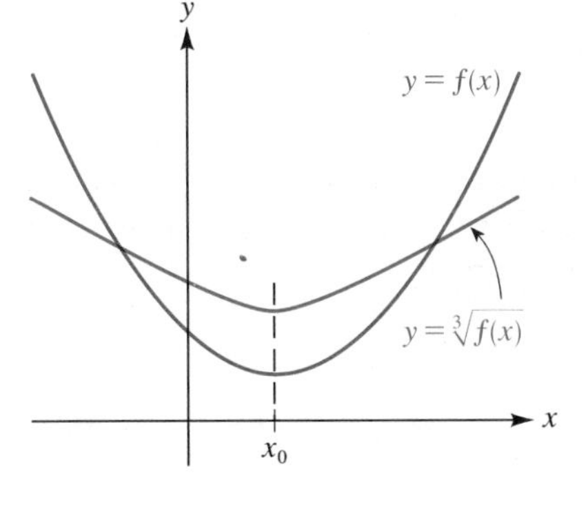

FIGURE 7
The minimum value occurs at x_0 for both of the functions $y = \sqrt[3]{f(x)}$ and $y = f(x)$.

For the previous two examples, keep in mind that we were able to find the maximum or minimum easily only because the functions were quadratics. In contrast, you cannot expect to find the minimum of $y = x^4 - 8x$ using the method of Examples 4 and 5 because it is not a quadratic function. In general, the techniques of calculus are required to find maxima and minima for functions other than quadratics. There are some cases, however, in which our present method can be adapted to functions that are closely related to quadratics. For instance, in the next example, we look for an input that minimizes a function of the form $y = \sqrt[3]{f(x)}$, where f is a quadratic function. As indicated in Figure 7, we need only find the input that minimizes the quadratic function $y = f(x)$, because this same input also minimizes $y = \sqrt[3]{f(x)}$. (The outputs of the two functions are, of course, different, but the point here is that the same *input* does the job for both functions.)

EXAMPLE 6 Let $f(x) = \sqrt[3]{x^2 - x + 1}$. Which x-value yields the minimum value for the function f? What is this minimum value?

Solution According to the remarks just prior to this example, the x-value that minimizes the function $f(x) = \sqrt[3]{x^2 - x + 1}$ will be the same x-value that minimizes the function $y = x^2 - x + 1$. To find this x-value we complete the square, just as we've done previously in this section. We have

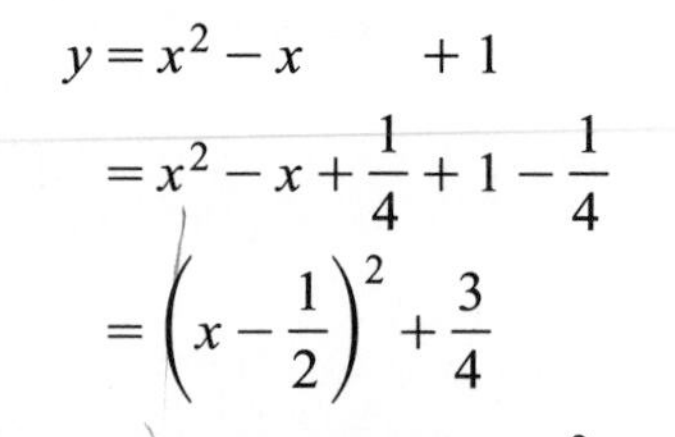

$$\begin{aligned} y &= x^2 - x \qquad + 1 \\ &= x^2 - x + \frac{1}{4} + 1 - \frac{1}{4} \\ &= \left(x - \frac{1}{2}\right)^2 + \frac{3}{4} \end{aligned}$$

This shows that the vertex of the parabola $y = x^2 - x + 1$ is $(1/2, 3/4)$. Since this parabola opens upward, we conclude that the input $x = 1/2$ will produce the minimum value for $y = x^2 - x + 1$ and also for $f(x) = \sqrt[3]{x^2 - x + 1}$. For the minimum value of the function f, we compute

$$\begin{aligned} f(1/2) &= \sqrt[3]{(1/2)^2 - (1/2) + 1} \\ &= \sqrt[3]{3/4} \approx 0.91 \end{aligned}$$

In summary: The minimum value of the function f occurs when $x = 1/2$. This minimum value is $\sqrt[3]{3/4}$, which is approximately 0.91. See Figure 8 for a graphical perspective.

GRAPHICAL PERSPECTIVE

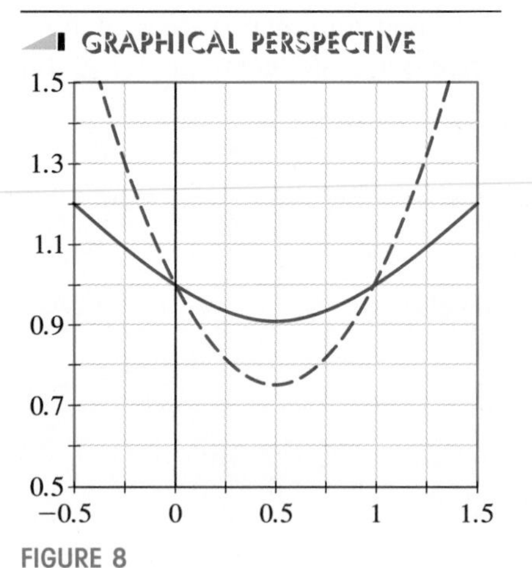

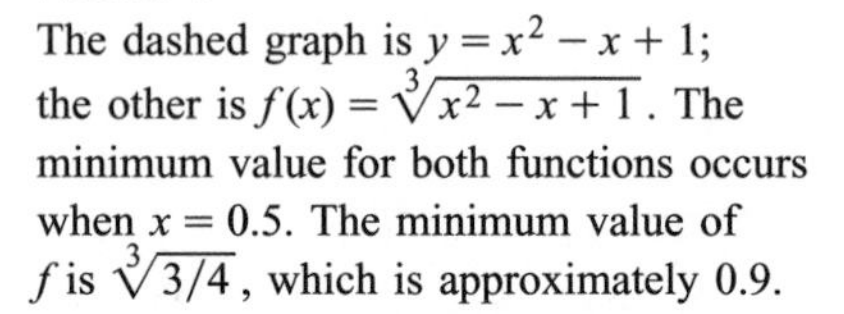

FIGURE 8
The dashed graph is $y = x^2 - x + 1$; the other is $f(x) = \sqrt[3]{x^2 - x + 1}$. The minimum value for both functions occurs when $x = 0.5$. The minimum value of f is $\sqrt[3]{3/4}$, which is approximately 0.9.

EXAMPLE 7 As indicated in Figure 9, the maximum value of the function $y = -t^4 + 3t^2$ occurs when t is between 1.0 and 1.5 and also when t is between -1.0 and -1.5. Determine these t-values exactly, then obtain calculator approximations rounded to three decimal places.

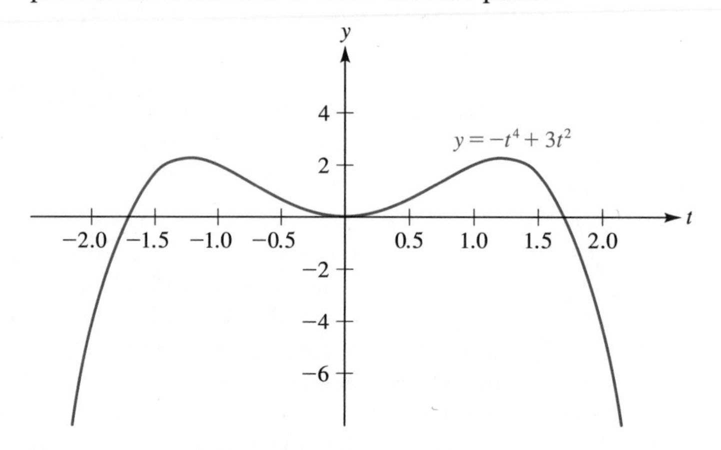

FIGURE 9

Solution The substitution $t^2 = x$ will reduce this question to one about a quadratic function. If $t^2 = x$, then $t^4 = x^2$ and we have

$$\begin{aligned} y &= -t^4 + 3t^2 \\ &= -x^2 + 3x \\ &= -\left(x - \frac{3}{2}\right)^2 + \frac{9}{4} \qquad \text{completing the square, as we've done throughout this section (Check the algebra!)} \end{aligned}$$

The graph of this last equation is a parabola that opens downward, and the maximum occurs when x is $3/2$. So we have

$$t^2 = x = 3/2$$

and therefore

$$t = \pm\sqrt{\frac{3}{2}} = \pm\frac{\sqrt{3}}{\sqrt{2}} = \pm\frac{\sqrt{6}}{2} \approx \pm 1.225$$

Note that these calculator values are consistent with Figure 9.

EXERCISE SET 4.2

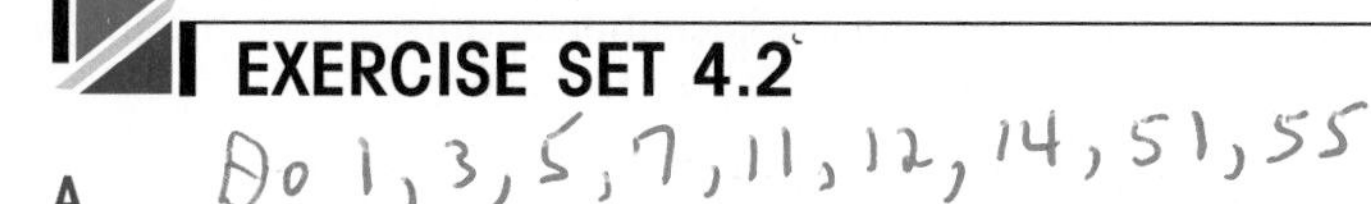

A

In Exercises 1–18, graph the quadratic function. Specify the vertex, axis of symmetry, maximum or minimum value, and intercepts.

1. $y = (x+2)^2$
2. $y = -(x+2)^2$
3. $y = 2(x+2)^2$
4. $y = 2(x+2)^2 + 4$
5. $y = -2(x+2)^2 + 4$
6. $y = x^2 + 6x - 1$
7. $f(x) = x^2 - 4x$
8. $F(x) = x^2 - 3x + 4$
9. $g(x) = 1 - x^2$
10. $y = 2x^2 + \sqrt{2}x$
11. $y = x^2 - 2x - 3$
12. $y = 2x^2 + 3x - 2$
13. $y = -x^2 + 6x + 2$
14. $y = -3x^2 + 12x$
15. $s = 16t^2$
16. $s = -\frac{1}{2}t^2 - t$
17. $s = 2 + 3t - 9t^2$
18. $s = -\frac{1}{4}t^2 + t - 1$

For Exercises 19–24, determine the input that produces the largest or smallest output (whichever is appropriate). State whether the output is largest or smallest.

19. $y = 2x^2 - 4x + 11$
20. $f(x) = 8x^2 + x - 5$
21. $g(x) = -6x^2 + 18x$
22. $s = -16t^2 + 196t + 80$
23. $f(x) = x^2 - 10$
24. $h(x) = x^2 - 10x$

In Exercises 25–30, find the maximum or minimum value for each function (whichever is appropriate). State whether the value is a maximum or minimum.

25. $y = x^2 - 8x + 3$
26. $y = \frac{1}{2}x^2 + x + 1$
27. $y = -2x^2 - 3x + 2$
28. $y = -\frac{1}{3}x^2 - 2x$
29. $f(t) = -12t^2 + 1000$
30. $g(t) = 400t^2$
31. How far from the origin is the vertex of the parabola $y = x^2 - 6x + 13$?
32. Find the distance between the vertices of the parabolas $y = -\frac{1}{2}x^2 + 4x$ and $y = 2x^2 - 8x - 1$.

For Exercises 33–38, the functions f, g, and h are defined as follows:

$$f(x) = 2x - 3 \qquad g(x) = x^2 + 4x + 1 \qquad h(x) = 1 - 2x^2$$

In each exercise, classify the function as linear, quadratic, or neither.

33. $f \circ g$
34. $g \circ f$
35. $g \circ h$
36. $h \circ g$
37. $f \circ f$
38. $h \circ h$

In Exercises 39 and 40, determine the inputs that yield the minimum values for each function. Compute the minimum value in each case.

39. **(a)** $f(x) = \sqrt{x^2 - 6x + 73}$
    (b) $g(x) = \sqrt[3]{x^2 - 6x + 73}$
    (c) $h(x) = x^4 - 6x^2 + 73$
40. **(a)** $F(x) = (4x^2 - 4x + 109)^{1/2}$
    (b) $G(x) = (4x^2 - 4x + 109)^{1/3}$
    (c) $H(x) = 4x^4 - 4x^2 + 109$

In Exercises 41 and 42, determine the maximum value for each function.

41. **(a)** $f(x) = \sqrt{-x^2 + 4x + 12}$
    (b) $g(x) = \sqrt[3]{-x^2 + 4x + 12}$
    (c) $h(x) = -x^4 + 4x^2 + 12$
42. **(a)** $F(x) = (27x - x^2)^{1/2}$
    (b) $G(x) = (27x - x^2)^{1/3}$
    (c) $H(x) = 27x^2 - x^4$

B

43. Let $f(x) = ax^2 + bx + c$. Compute the average rate of change $\Delta f/\Delta x$ for the interval $[x_1, x_2]$.
44. Let $f(x) = ax^2 + bx + c$. Show that
$$\frac{f(x+h) - f(x)}{h} = 2ax + ah + b$$
45. By completing the square, show that the coordinates of the vertex of the parabola $y = ax^2 + bx + c$ are $(-b/2a, -D/4a)$, where $D = b^2 - 4ac$.
46. Compute the average of the two x-intercepts of the graph of $y = ax^2 + bx + c$. (Assume $b^2 - 4ac > 0$.) How does your answer relate to the result in Exercise 45?

47. Let $f(x) = ax^2 + bx + c$.

(a) Show that $f\left(\frac{-b}{2a} + h\right) = ah^2 - \frac{D}{4a}$, where D denotes the discriminant $b^2 - 4ac$.

(b) Show that $f\left(\frac{-b}{2a} - h\right) = ah^2 - \frac{D}{4a}$.

Remark: The results in parts (a) and (b) prove that the parabola is symmetric about the line $x = -b/2a$. Why?

48. Find the x-coordinate of the vertex of the parabola $y = (x - a)(x - b)$, where a and b are constants.

As background for Exercises 49 and 50, read the definition of an iterate on page 146.

49. The figure shows portions of the parabola $f(x) = x^2 - 0.6$ and the line $y = x$.

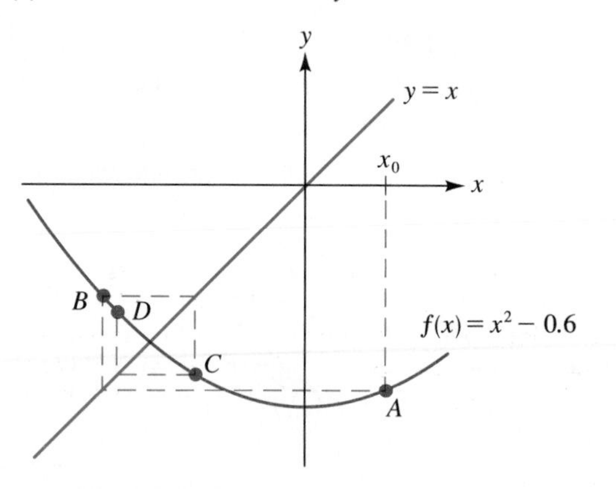

(a) Explain why the y-coordinates of A, B, C, and D are the first four iterates of x_0 for the function $f(x) = x^2 - 0.6$.

(b) Suppose that $x_0 = 0.2$. Compute the first four iterates. (For the third and fourth iterates, round the final answers to three decimal places.)

50. The following figure shows portions of the parabola $f(x) = x^2 - 0.3$ and the line $y = x$.

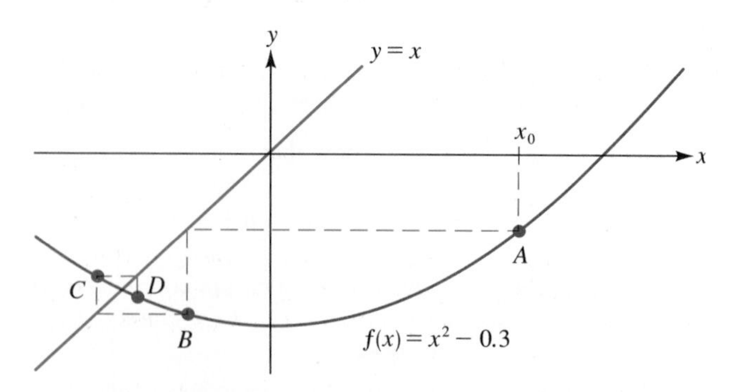

(a) Explain why the y-coordinates of A, B, C, and D are the first four iterates of x_0 for the function $f(x) = x^2 - 0.3$.

(b) Suppose that $x_0 = 0.4$. Compute the first four iterates. For the third and fourth iterates, round the final answers to four decimal places.

(c) Use your calculator to compute the fifth through tenth iterates of $x_0 = 0.4$. Round the final answers to four decimal places.

(d) Use the quadratic formula and your calculator to find the y-coordinate of the point in the third quadrant where the line $y = x$ intersects the parabola $f(x) = x^2 - 0.3$. Round the answer to four decimal places. Note how close your answer is to the tenth iterate in part (c). (As the picture suggests, if the iteration process is continued indefinitely, the iterates approach this y-coordinate.)

In Exercises 51–54, find quadratic functions satisfying the given conditions.

51. The graph passes through the origin, and the vertex is the point (2, 2).

52. The graph is obtained by translating $y = x^2$ four units in the negative x-direction and three units in the positive y-direction.

53. The vertex is $(3, -1)$, and one x-intercept is 1.

54. The axis of symmetry is the line $x = 1$. The y-intercept is 1. There is only one x-intercept.

55. Find the equation of the circle that passes through the origin and is centered at the vertex of the parabola $y = 2x^2 + 12x + 14$.

56. For which value of c will the minimum value of the function $f(x) = x^2 + 2x + c$ be $\sqrt{2}$?

C

57. Consider the quadratic function $y = px^2 + px + r$, where $p \neq 0$.

(a) Show that if the vertex lies on the x-axis, then $p = 4r$.

(b) Show that if $p = 4r$, then the vertex lies on the x-axis.

58. Let $g(x) = x^2 + 2(a + b)x + 2(a^2 + b^2)$, where a and b are constants.

(a) Show that the coordinates of the vertex are

$$(-(a + b), (a - b)^2)$$

(b) Use the result in part (a) to explain why the graph of g has no x-intercepts unless $a = b$, in which case there is only one x-intercept.

59. Let $f(x) = ax^2 + bx$, where $a \neq 0$ and $b \neq 0$. Find a value for b such that the equation $f(f(x)) = 0$ has exactly three real roots.

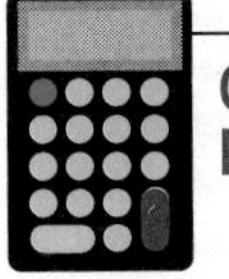

GRAPHING UTILITY EXERCISES FOR SECTION 4.2

Exercises 1 and 2 show how the shape of the parabola $y = ax^2$ changes as a changes. For Exercise 1, use a viewing rectangle extending from -3 to 3 in the x-direction and from 0 to 10 in the y-direction. For Exercise 2, use a viewing rectangle extending from -10 to 10 in the x-direction and from 0 to 10 in the y-direction.

1. **(a)** On the same set of axes, graph the four parabolas $y = x^2$, $y = 2x^2$, $y = 3x^2$, and $y = 8x^2$. As the coefficient of x^2 increases from 1 to 8, how do the resulting graphs change?
   (b) Where do you think the graph of $y = 50x^2$ would fit in with the graphs in part (a)? After answering, use the graphing utility to see if you are right.

2. **(a)** On the same set of axes, graph the four parabolas $y = x^2$, $y = 0.5x^2$, $y = 0.25x^2$, and $y = 0.125x^2$. As the coefficient of x^2 decreases from 1 to 0.125, how do the resulting graphs change?
   (b) Where do you think the graph of $y = 0.02x^2$ would fit in with the graphs in part (a)? After answering, use the graphing utility to see if you are right.

3. **(a)** Graph the two functions $y = x^2 - x + 1$ and $f(x) = \sqrt[3]{x^2 - x + 1}$. (Use a viewing rectangle extending from -3 to 3 in the x-direction and from 0 to 4 in the y-direction. As indicated in Example 6, the minimum value for both functions occurs when $x = 0.5$. Check that your graphs are consistent with this fact.
   (b) Zoom in repeatedly on the turning point for the function $f(x) = \sqrt[3]{x^2 - x + 1}$. See if you can determine the minimum value correct to three decimal places. (For the precise value of the minimum, see Example 6.)

4. **(a)** Graph the two functions $y = x^2 + 4x$ and $y = x^2 - 4x$. How are the two graphs related (in terms of symmetry)?
   (b) Follow part (a) using the two functions $y = -2x^2 + 3x + 4$ and $y = -2x^2 - 3x + 4$.
   (c) Which one of the graphing techniques from Section 3.3 relates to what you have observed in parts (a) and (b)?

In Section 4.1, you saw how a linear function can be used to summarize a data set. Quadratic functions can also be used for this purpose. In this latter case, you need an option on your graphing utility for quadratic *(or* second-degree polynomial*) regression. The next exercise involves both linear and quadratic regression.*

5. The following table indicates estimates for the number of AIDS cases reported worldwide for selected years covering the period 1986–1993.

Year	1986	1987	1988	1989	1990	1993
AIDS Cases (thousands)	279	493	813	1275	1872	4820

Source: Global AIDS Policy Coalition, Harvard School of Public Health, as reported in *Vital Signs 1994,* Lester Brown et al. (New York: W. W. Norton & Co., 1994)

   (a) Find the equation of the regression line for the years 1986–1989. (For convenience, let the numbers 1986–1989 correspond to 0–3, respectively.)
   (b) Use the equation of the regression line to make projections for the number of reported AIDS cases in 1990 and in 1993. Then compare your projections with the values given in the table. Are your projections too high or too low?
   (c) Find the quadratic regression for the years 1986–1989. Then use the resulting quadratic function to make projections for the number of reported AIDS cases in 1990 and in 1993. Which of the two projections, 1990 or 1993, comes closer to the corresponding value in the table? Are your projections more accurate or less accurate than those obtained in part (b)? Are your projections too high or too low?

6. **(a)** Use a graphing utility to find the equation of the line passing through the two points $(6, -2)$ and $(10, 0)$. *Hint:* The regression line is the line you want.
   (b) Graph the line determined in part (a) to check that the two given points indeed appear to lie on the line.
   (c) Use a graphing utility to find the equation of the parabola passing through the three points $(-1, 19)$, $(1, 3)$, and $(2, -2)$. *Hint:* A quadratic regression yields the required function.
   (d) Graph the parabola determined in part (c) and check visually that each of the three given points appears to lie on the parabola. (Adjust the viewing rectangle as appropriate.)

7. **(a)** The following table shows world energy production for the years 1988–1990.

World Energy Production 1988–1990 (from coal, natural gas, crude oil, nuclear power, and hydroelectric power)

Year	1988	1989	1990
World Energy Production (quadrillion Btu)	320.16	332.72	340.20

Source: Annual Energy Review 1993 (United States Energy Information Administration)

Figure A shows a scatter plot and regression line for this data, as studied in Section 4.1. Figure B shows the same scatter plot, but this time a quadratic function, rather than a linear function, is used to model the data. Use a graphing utility (or computer software) to determine equations for these two functions.

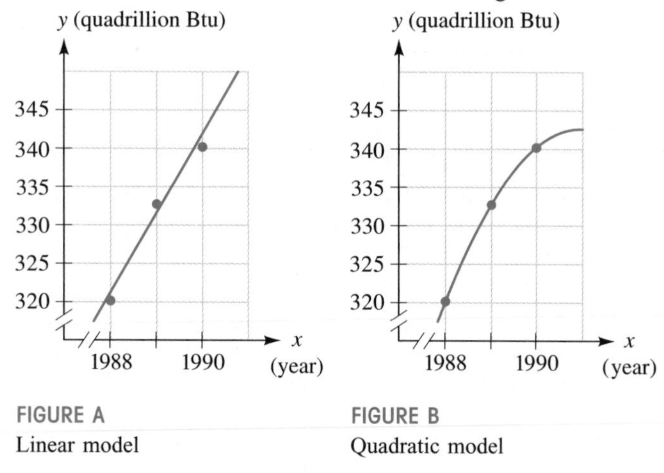

FIGURE A
Linear model

FIGURE B
Quadratic model

(b) Complete the following table using the equations that you determined in part (a). Check that your projections for 1991 are consistent with the visual information in Figures A and B.

Year	1991	1992
Estimate for World Energy Production (quadrillion Btu) Using Linear Model		
Estimate for World Energy Production (quadrillion Btu) Using Quadratic Model		

(c) According to the United States Energy Information Administration, the world energy production for 1991 was 344.41 quadrillion Btu. Looking at the results in part (b), which model gives the closer estimate?

(d) According to the United States Energy Information Administration, the world energy production for 1992 was 346.33 quadrillion Btu. Looking at the results in part (b), which model gives the closer estimate?

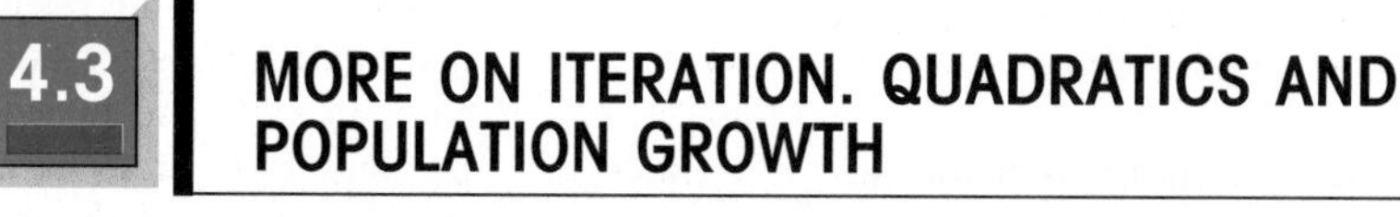

MORE ON ITERATION. QUADRATICS AND POPULATION GROWTH

Although I shall henceforth adopt the habit of referring to the variable X as "the population," there are countless situations outside population biology where . . . [iteration of functions] *applies. . . . Examples in economics include models for the relationship between commodity quantity and price, for the theory of business cycles, and for the temporal sequences generated by various other economic quantities.*

. . . I would therefore urge that people be introduced to, say, [the iteration process for $f(x) = kx(1 - x)$] *early in their mathematical education. This equation can be studied phenomenologically by iterating it on a calculator, or even by hand. Its study does not involve as much conceptual sophistication as does elementary calculus. Such study would greatly enrich the student's intuition about nonlinear systems.*

Biologist Robert M. May, "Simple mathematical models with very complicated dynamics," *Nature,* vol. 261 (1976), pp. 459–467.

In this section we continue the work on iteration that we began in Section 3.4. The prerequisites for our discussion are the definitions in the box back on page

146, and also the explanation of graphical iteration on pages 147–148. The following (review) example will help you to recall the notation for iterates.

EXAMPLE 1 Let $f(x) = 3x - x^2$ and $x_0 = 4$. Compute the first two iterates, x_1 and x_2, of x_0.

Solution Using the notation from Section 3.4, we have

$$\begin{aligned} x_1 &= f(x_0) \\ &= f(4) = 3(4) - 4^2 = -4 \end{aligned}$$

Now we use the value just obtained for x_1 to compute x_2:

$$\begin{aligned} x_2 &= f(x_1) \\ &= f(-4) = 3(-4) - (-4)^2 = -28 \end{aligned}$$

If we start with a function f and an input x, it's usually not the case that $f(x)$ turns out to be the same as x itself. That is, usually, the output is not the same as the input. But sometimes this does happen. Take, for example, the function $f(x) = 3x - 2$ and the input $x = 1$. Then we have

$$f(1) = 3(1) - 2 = 1$$

So, for this particular function, the input $x = 1$ is an instance where "input = output." The input $x = 1$ in this case is called a *fixed point* of the function $f(x) = 3x - 2$. In the box that follows, we give the general definition of a fixed point.

DEFINITION: Fixed Point of a Function

A fixed point of a function f is an input x in the domain of f such that

$$f(x) = x$$

EXAMPLE

Both 0 and 1 are fixed points for $f(x) = x^2$ because $f(0) = 0^2 = 0$ and $f(1) = 1^2 = 1$.

EXAMPLE 2 Find the fixed points (if any) for each function:

(a) $f(x) = 1 - x$; **(b)** $g(x) = 1 + x$; **(c)** $h(x) = x^2 - x - 3$.

Solution

(a) We're looking for a number x such that $f(x) = x$. In view of the definition of f, this last equation becomes

$$1 - x = x$$

or

$$2x = 1$$

and therefore

$$x = 1/2$$

This shows that the function f has one fixed point; it is $x = 1/2$.

(b) If x is a fixed point of $g(x) = 1 + x$, we have

$$1 + x = x$$

But then subtracting x from both sides of this last equation yields $1 = 0$, which is impossible. We conclude from this that there is no fixed point for the function $g(x) = 1 + x$.

(c) The fixed points (if any) are the solutions of the quadratic equation $x^2 - x - 3 = x$. Subtracting x from both sides of this equation, we have

$$x^2 - 2x - 3 = 0$$
$$(x - 3)(x + 1) = 0$$

Looking at this last equation, we can see that there are two roots: 3 and -1. Each of these numbers is a fixed point for the given function $h(x) = x^2 - x - 3$. That is, $h(3) = 3$ and $h(-1) = -1$. [You should verify each of these last two equations for yourself by actually computing $h(3)$ and $h(-1)$.]

A fixed point of a function can be interpreted geometrically: it is the x-coordinate of a point where the given function intersects the line $y = x$. Figure 1 shows the fixed points for the functions in the example that we've just completed.

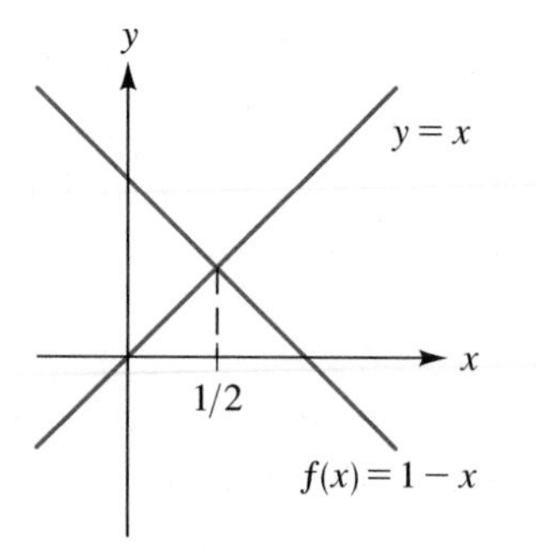

(a) The fixed point of f is $1/2$.

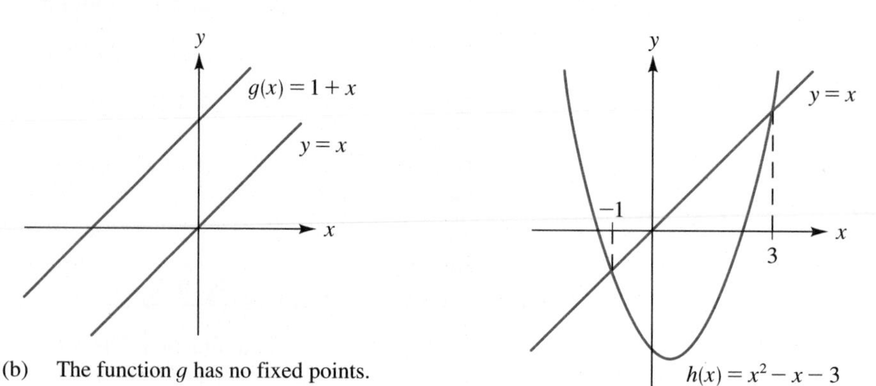

(b) The function g has no fixed points.

(c) The fixed points of h are -1 and 3.

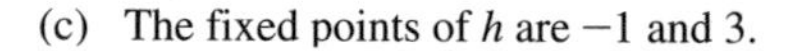

FIGURE 1
Fixed points for the functions in Example 2.

Fixed points are related to the iteration process in several ways. Suppose that a number a is a fixed point of the function f. Then by definition we have $f(a) = a$, which says that the first iterate of a is equal to a itself. Similarly, all of the subsequent iterates of the fixed point a will be equal to a. For instance, for the second iterate, we have

$$\begin{aligned} f(f(a)) &= f(a) && \text{substituting } a \text{ for } f(a) \text{ on the left-hand side} \\ &= a && \text{again because } f(a) = a \end{aligned}$$

This shows that the second iterate is equal to a. The same type of calculation will show that any subsequent iterate of the fixed point a is equal to a.

Another connection between fixed points and iteration is this: for some functions, a fixed point can be a "target value" for other iterates. We'll explain this using Figure 2 (on the next page).

Figure 2 shows the first four steps in the iteration process for $f(x) = \frac{1}{2}x + 2$ with $x_0 = 1$. (See Section 3.4 if you need to review graphical iteration.) As indicated in Figure 2, the input 4 is a fixed point for the function f, and the iteration process follows a staircase pattern that approaches the point (4, 4). We

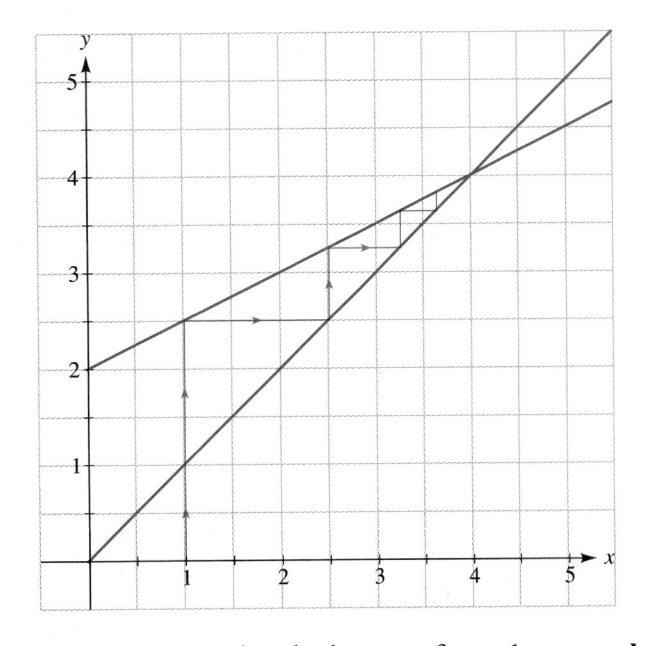

FIGURE 2
The first four steps in the iteration process for $f(x) = \frac{1}{2}x + 2$ with $x_0 = 1$.

say in this case that the iterates of $x_0 = 1$ **approach** the fixed point 4 and that this target value 4 is an **attracting fixed point** of the function f. Table 1 gives you a more numerical look at what is meant by saying that the iterates approach the target value 4.

In the table, notice, for example, that

x_5 differs from 4 by less than 0.1;

x_{10} differs from 4 by less than 0.01;

and x_{15} differs from 4 by less than 0.0001.

What's important here is that the differences between the iterates and 4 can be made as small as we please by carrying out the iteration process sufficiently far. The idea of a target value or *limit* is made more precise in calculus. But, for our purposes, Figure 2 and Table 1 will certainly give you an intuitive understanding of the idea and what we mean by saying that the iterates approach 4.

TABLE 1 THE ITERATES APPROACH 4.

x_1	2.5
x_2	3.25
x_3	3.625
x_4	3.8125
x_5	3.906 . . .
x_6	3.953 . . .
x_7	3.976 . . .
x_8	3.988 . . .
x_9	3.9941 . . .
x_{10}	3.9970 . . .

(a) x_1 through x_{10}

x_{11}	3.9985 . . .
x_{12}	3.99926 . . .
x_{13}	3.99963 . . .
x_{14}	3.99981 . . .
x_{15}	3.999908 . . .
x_{20}	3.9999971 . . .
x_{25}	3.999999910 . . .

(b) x_{11} through x_{15} along with x_{20} and x_{25}

Figures 3 and 4 show two more ways that the iteration process may relate to fixed points. In Figure 3, there is an attracting fixed point for the iterates of $x_0 = -0.1$, but this time the iteration process approaches the fixed point through a spiral pattern rather than a staircase pattern. To find the fixed point (and thereby determine the number that the iterates are approaching), we need to solve the quadratic equation $x^2 - 0.5 = x$. As you should check for yourself by means of the quadratic formula and then a calculator, the relevant root here is $(1-\sqrt{3})/2 \approx -0.366$. Notice that this value is consistent with Figure 3. In Figure 4, the fixed point 1 is a **repelling fixed point** for the iterates of $x_0 = 1.25$. As Figure 4 indicates, the iterates of $x_0 = 1.25$ move farther and farther away from the value 1. Indeed, as you can check with a calculator, the first five iterates of 1.25 are as follows (we're rounding to two decimal places):

$$x_1 \approx 1.56 \qquad x_2 \approx 2.44 \qquad x_3 \approx 5.96 \qquad x_4 \approx 35.53 \qquad x_5 \approx 1262.18$$

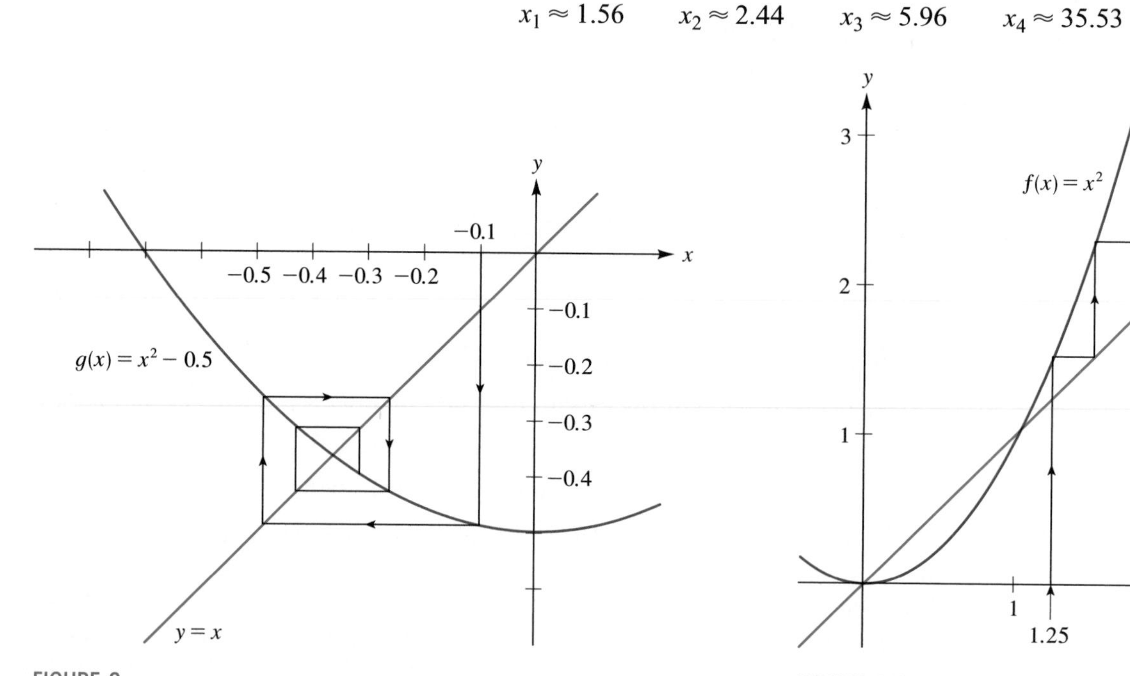

FIGURE 3
The iterates of $x_0 = -0.1$ under the function $g(x) = x^2 - 0.5$ approach the attracting fixed point $\frac{1}{2}(1-\sqrt{3}) \approx -0.366$.

FIGURE 4
The iterates of $x_0 = 1.25$ under the function $f(x) = x^2$ move away from the repelling fixed point 1.

The iteration process for functions is often applied in the study of population growth. The word "population" here is used in a general sense. It needn't refer only to human populations. For instance, biological or ecological studies may involve animal, insect, or bacterial populations. (Also, see the quotation at the beginning of this section.) The following equation defines one type of quadratic function that has been studied extensively in this context:

$$f(x) = kx(1-x) \tag{1}$$

In using this idealized model, we assume that the population size is measured by a number between 0 and 1, where 1 corresponds to the maximum possible population size in the given environment and 0 corresponds to the case where the population has become extinct. We start with a given input x_0 $(0 \leq x_0 \leq 1)$ that represents the fraction of the maximum population size that is initially present. For instance, if the maximum possible population of catfish in a pond is 100 and initially there were 70 catfish, then we would have $x_0 = 70/100 = 0.7$.

The next basic assumption in using equation (1) to model population size is that the iterates of x_0 represent the fraction of the maximum possible population present after each successive time interval. That is,

x_1 is the (fraction of the maximum) population after the first time interval;

x_2 is the (fraction of the maximum) population after the second time interval;

and, in general,

x_n is the (fraction of the maximum) population after the nth time interval.

In a given study, the time intervals might be measured, for example, in years, in months, or in breeding seasons. The constant k in equation (1) is the *growth parameter;* it is related to the rate of growth of the particular population being studied. Science writer James Gleick has described k this way: "In a pond, it might correspond to the fecundity of the fish, the propensity of the population not just to boom but also to bust. . . ." [*Chaos: Making a New Science* (New York: Viking Penguin, Inc., 1988)]

EXAMPLE 3 In the Mississippi Delta region, many farmers have replaced unproductive cotton fields with catfish ponds. Suppose that a farmer has a catfish pond with a maximum population size of 500 and that initially the pond is stocked with 50 catfish. Also, assume that the growth parameter for this population is $k = 2.9$, so that equation (1) becomes

$$f(x) = 2.9x(1 - x) \tag{2}$$

Finally, assume that the time intervals here are breeding seasons.

(a) What is the value for x_0?
(b) Use Figure 5 (on the next page) to estimate the iterates x_1 through x_5. Then use a calculator to compute these values. Round the final answers to three decimal places. Interpret the results.
(c) As indicated in Figure 5, the iteration process is spiraling in on a fixed point of the function. (Figure 6 at the end of this section will demonstrate this in greater detail.) Find this fixed point and interpret the result.

Solution

(a) $x_0 = \dfrac{\text{initial population}}{\text{maximum population}} = \dfrac{50}{500} = 0.1$

(b) Looking at Figure 5, it appears that x_1, the first iterate of x_0, is between 0.25 and 0.30, much closer to the former number than the latter. As an estimate, let's say that x_1 is about 0.26. This and the other estimates are given in Table 2. *Suggestion:* Make the estimates for yourself before looking at the estimates we give. Some slight discrepancies are okay. In the bottom row of Table 2 are the values of the iterates obtained using a calculator. You should verify these for yourself. (There should be no discrepancies here.)

The results in Table 2 tell us what is happening to the population through the first five breeding seasons. From an initial population of 50

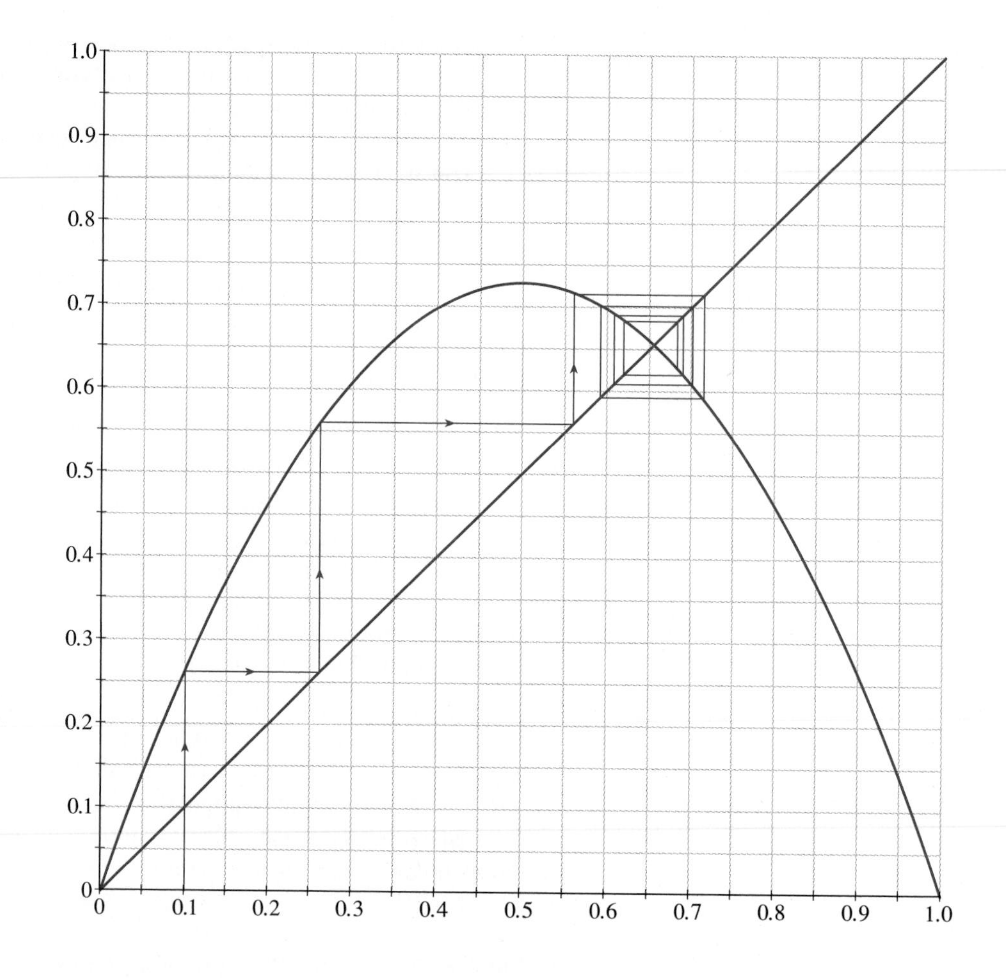

FIGURE 5
The first ten iterations of $x_0 = 0.1$ under $f(x) = 2.9x(1 - x)$.

TABLE 2 ITERATES OF $x_0 = 0.1$ UNDER THE FUNCTION $f(x) = 2.9x(1 - x)$. (Calculator values rounded to three decimal places.)

	x_1	x_2	x_3	x_4	x_5
From graph	0.26	0.56	0.71	0.58	0.70
From calculator	0.261	0.559	0.715	0.591	0.701

catfish, the population size steadily increases through the first three breeding seasons (the numbers in the table are getting bigger). The population size then drops after the fourth breeding season, and goes back up after the fifth season. (These facts can be deduced from Figure 5, as well as from Table 2.) To compute the actual numbers of fish, we need to multiply each number in the bottom row of Table 2 by 500. (Remember, the iterates in Table 2 represent fractions of the maximum possible population size 500.) For example, to compute the actual number of fish at the end of the third breeding season, we multiply the maximum population size of 500 by x_3:

$$\begin{aligned} x_3 \times 500 &\approx 0.715 \times 500 \\ &\approx 358 \text{ catfish} \end{aligned}$$

The number of catfish at the end of each of the other breeding seasons is obtained in the same manner. See Table 3; use your calculator to check each of the entries in the table.

TABLE 3 CATFISH POPULATION

n	0	1	2	3	4	5
Number of fish after n breeding seasons	50	131	280	358	296	351

(c) The fixed point we are looking for occurs when the parabola in Figure 5 intersects the line $y = x$. So, following the method in Example 2, we need to solve the equation $2.9x(1 - x) = x$. We have

$$\begin{aligned} 2.9x(1 - x) &= x \\ -2.9x^2 + 1.9x &= 0 \\ x(-2.9x + 1.9) &= 0 \end{aligned}$$

Therefore, $x = 0$ or $-2.9x + 1.9 = 0$, i.e.,

$$x = \frac{-1.9}{-2.9} = 0.655 \qquad \text{using a calculator and rounding to three decimal places}$$

This shows that there are two fixed points for the function f, namely, 0 and 19/29. Looking at Figure 5, we know that 19/29 is the fixed point that we are interested in here, not 0. So the iterates of $x_0 = 0.1$ are approaching the value 19/29, which is approximately 0.655.

We now summarize and interpret the results. There were initially 50 catfish in a pond that could hold at most 500. As we saw in part (b), the population size increases over the first three breeding seasons. After this, as Figure 5 shows, the population oscillates up and down but draws closer and closer to an equilibrium population corresponding to the fixed point $x = 19/29$. This equilibrium population is

$$\frac{19}{29} \times 500 \approx 328 \text{ catfish}$$

We conclude this section with some pictures indicating only three of the many possibilities that can arise in the iteration of quadratic functions of the form $f(x) = kx(1 - x)$. Figure 6 (on the next page) concerns the function with growth parameter $k = 2.9$ that we used in the catfish example: $f(x) = 2.9x(1 - x)$. In Figure 6(a), we've carried out the iteration of $x_0 = 0.1$ through the twentieth iterate. (In the catfish example, Figure 5 goes only as far as the tenth iterate.) Figure 6(a) indicates quite clearly that the iterates are indeed approaching a fixed point of the function. Figure 6(b) presents another way to visualize the long-term behavior of the iterates. Values of n are marked on the horizontal axis, values of the iterates x_n are marked on the vertical axis, and the points with coordinates (n, x_n) are then plotted. For example, since $x_0 = 0.1$, we plot the point $(0, 0.1)$; and since $x_1 = 0.261$, we plot the point $(1, 0.261)$. The line segments in Figure 6(b) are drawn in only to help the eye see the pattern that is emerging. Three facts that can be inferred from Figure 6(b) are as follows: after the first few iterates, the iterates oscillate up and down; the magnitude of the oscillations is decreasing; and, in the long run, the iterates are approaching a number between 0.6 and 0.7. (In Example 3 we found this value to be approximately 0.655.)

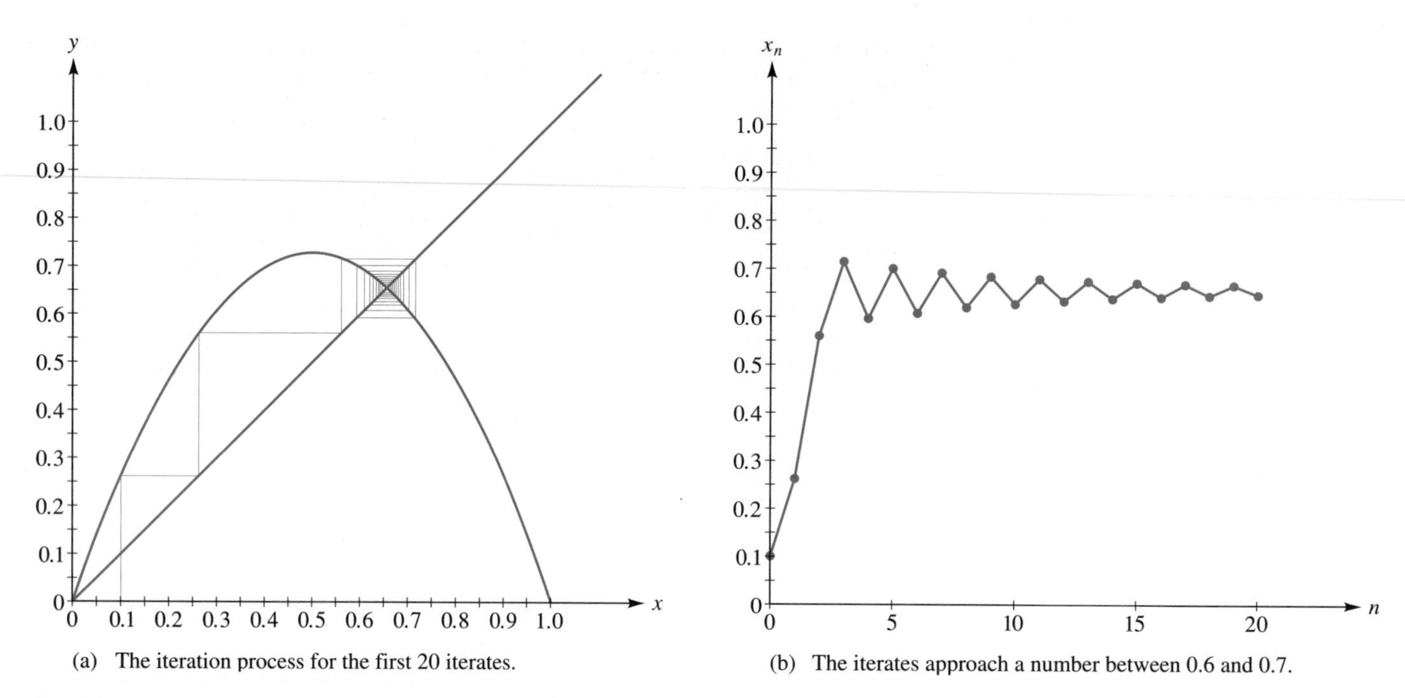

(a) The iteration process for the first 20 iterates.

(b) The iterates approach a number between 0.6 and 0.7.

FIGURE 6
The iteration of $f(x) = kx(1 - x)$ with growth parameter $k = 2.9$ and $x_0 = 0.1$.

Unlike the iteration pictured in Figure 6(a), Figure 7(a) shows a case in which the iteration process is spiraling away from, rather than toward, a fixed point. Again, we've used the initial input $x_0 = 0.1$, but this time the growth parameter is $k = 3.2$. As indicated in Figure 7(b), the long-term behavior of the iterates becomes quite predictable: they alternate between two values. Figure 7(b) shows that the smaller of these two values is between 0.5 and 0.6, while

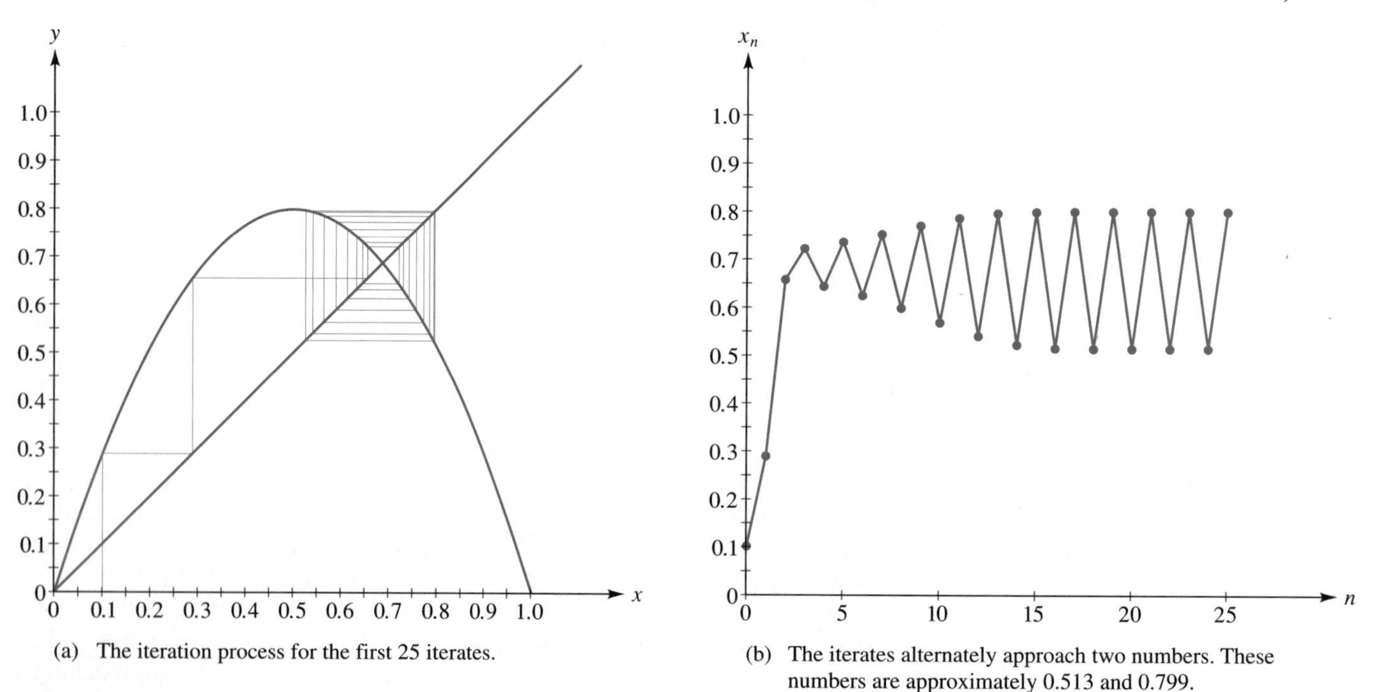

(a) The iteration process for the first 25 iterates.

(b) The iterates alternately approach two numbers. These numbers are approximately 0.513 and 0.799.

FIGURE 7
The iteration of $f(x) = kx(1 - x)$ with growth parameter $k = 3.2$ and $x_0 = 0.1$.

the larger is approximately 0.8. Exercise 34 gives you formulas for computing these two limiting values for the iterates. (They turn out to be, approximately, 0.513 and 0.799.)

In Figure 8, once again we take the initial input to be $x_0 = 0.1$ but this time the growth parameter is $k = 3.9$. Now the iterates seem to fluctuate widely with no apparent pattern, in sharp contrast to the previous two figures, where there were clear patterns. Phenomena such as this are the subject of **chaos theory,** a new branch of twentieth-century mathematics with wide application. [For a readable and nontechnical introduction to this relatively new subject, see James Gleick's best-seller, *Chaos: Making a New Science* (New York: Viking Penguin, Inc., 1988). For a little more detail on the mathematics, see the paperback by Donald M. Davis, *The Nature and Power of Mathematics* (Princeton, N.J.: Princeton University Press, 1993), pp. 314–363.]

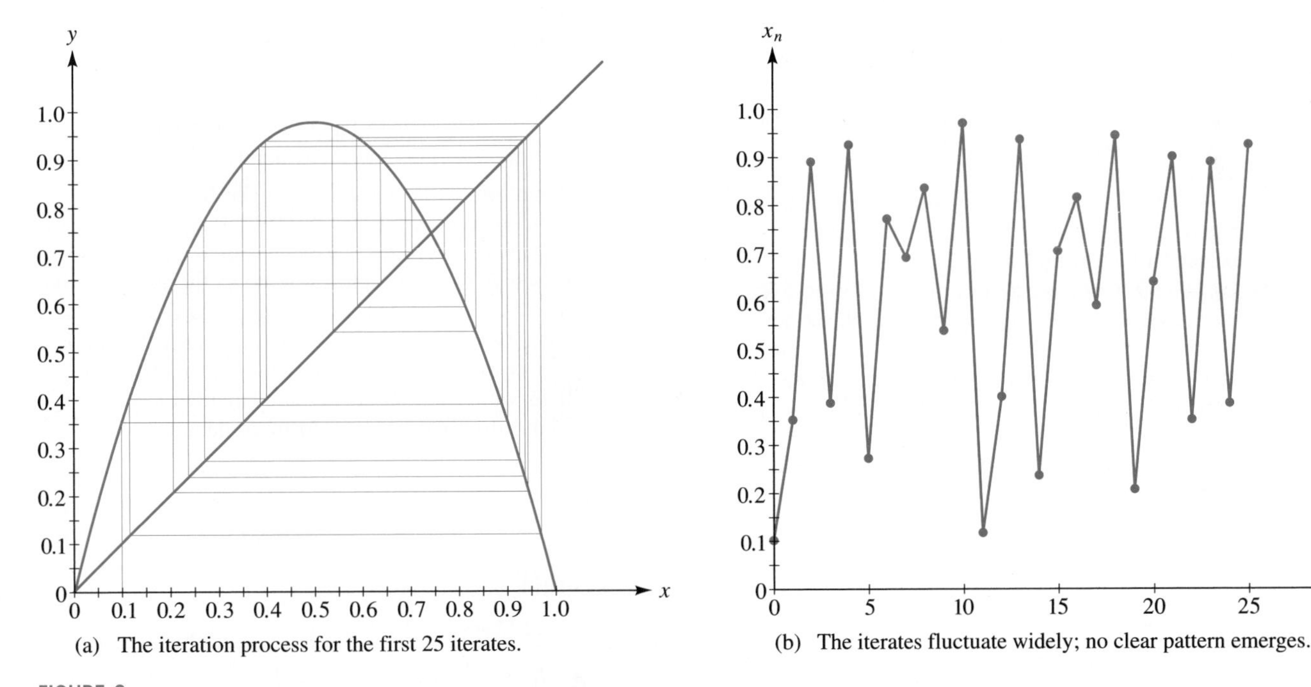

FIGURE 8
The iteration of $kx(1-x)$ with growth parameter $k = 3.9$ and $x_0 = 0.1$.

EXERCISE SET 4.3

A

In Exercises 1–18, find all real numbers (if any) that are fixed points for the given functions.

1. $f(x) = -4x + 5$
2. $g(x) = 3x - 14$
3. $G(x) = \frac{1}{2} + x$
4. $F(x) = (7 - 2x)/8$
5. $h(x) = x^2 - 3x - 5$
6. $H(t) = 3t^2 + 18t - 6$
7. $f(t) = t^2 - t + 1$
8. $F(t) = t^2 - t - 1$
9. $k(t) = t^2 - 12$
10. $K(t) = t^2 + 12$
11. $T(x) = 1.8x(1 - x)$
12. $T(y) = 3.4y(1 - y)$
13. $g(u) = 2u^2 + 3u - 4$
14. $G(u) = 3u^2 - 4u - 2$
15. $f(x) = 7 + \sqrt{x - 1}$
16. $f(x) = \sqrt{10 + 3x} - 4$
17. $G(x) = (x - 3)/(x + 5)$
18. $M(x) = (x - 36)/(x + 13)$

19. This exercise refers to the function $g(x) = x^2 - 0.5$ in Figure 3 in the text.

(a) Use the quadratic formula to verify that one of the fixed points of this function is $(1 - \sqrt{3})/2$, then use your calculator to check that this is approximately -0.366.

(b) According to the text, the iterates of -0.1 approach the value determined in part (a). Use your calculator: which is the first iterate to have the digit 3 in the first decimal place?

(c) Use your calculator: which is the first iterate to have the digit 6 in the second decimal place?

20. This exercise refers to the function $f(x) = x^2$ in Figure 4 in the text. According to the text, the iterates of 1.25 move farther and farther away from the fixed point 1. In this exercise you'll see that iterates of other points even closer to the fixed point 1 nevertheless are still "repelled" by 1.

(a) Let $x_0 = 1.1$. Use your calculator: which is the first iterate to exceed 10? Which is the first iterate to exceed one million?

(b) Let $x_0 = 1.001$. Use your calculator: which is the first iterate to exceed 2? Which is the first iterate to exceed one million?

(c) Let $x_0 = 0.99$. Compute x_1 through x_{10} to see that the iterates are indeed moving farther and farther away from the fixed point 1. What value are the iterates approaching? Is this value a fixed point of the function?

21. The accompanying figure shows the first eight steps in the iteration process for $f(x) = -0.7x + 2$, with $x_0 = 0.4$.

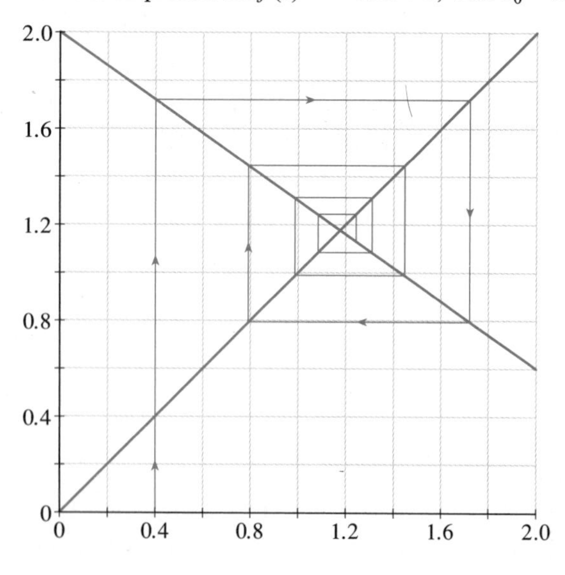

(a) Complete the following table. For the values obtained from the graph, estimate to the nearest one-tenth; for the calculator values, round the final answers to three decimal places.

	x_1	x_2	x_3	x_4	x_5	x_6	x_7	x_8
From graph								
From calculator								

(b) The figure shows that the iterates are approaching a fixed point of the function. Determine the exact value of this fixed point and then give a calculator approximation rounded to three decimal places.

(c) In the table for part (a), you used a calculator to compute the first eight iterates. Which of these iterates is the first to have the same digit in the first decimal place as the fixed point?

22. The figure on the next page shows the first six steps in the iteration process for $f(x) = 2.9x(1 - x)$, with $x_0 = 0.2$.

(a) Complete the following table. For the values obtained from the graph, estimate to the nearest 0.05, or closer if you can. For instance, to the nearest 0.05, the first iterate is 0.45. But, since the graph shows the iterate a bit above 0.45, the estimate 0.46 would be better. For the calculator work, round the final answers to three decimal places.

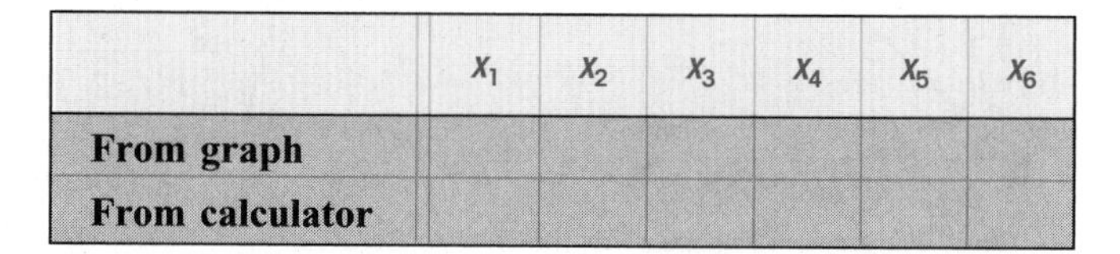

	x_1	x_2	x_3	x_4	x_5	x_6
From graph						
From calculator						

(b) The figure shows that the iterates are approaching a fixed point of the function. Use the figure to estimate, to the nearest 0.05, a value for this fixed point. Then, determine the exact value of this fixed point; then also give a calculator approximation rounded to three decimal places.

(c) In the table for part (a), you used a calculator to compute the first six iterates. Which of these iterates is the first to have the same digit in the first decimal place as the fixed point? (*Remark:* Agreement in the second decimal place doesn't occur until the 26th iterate.)

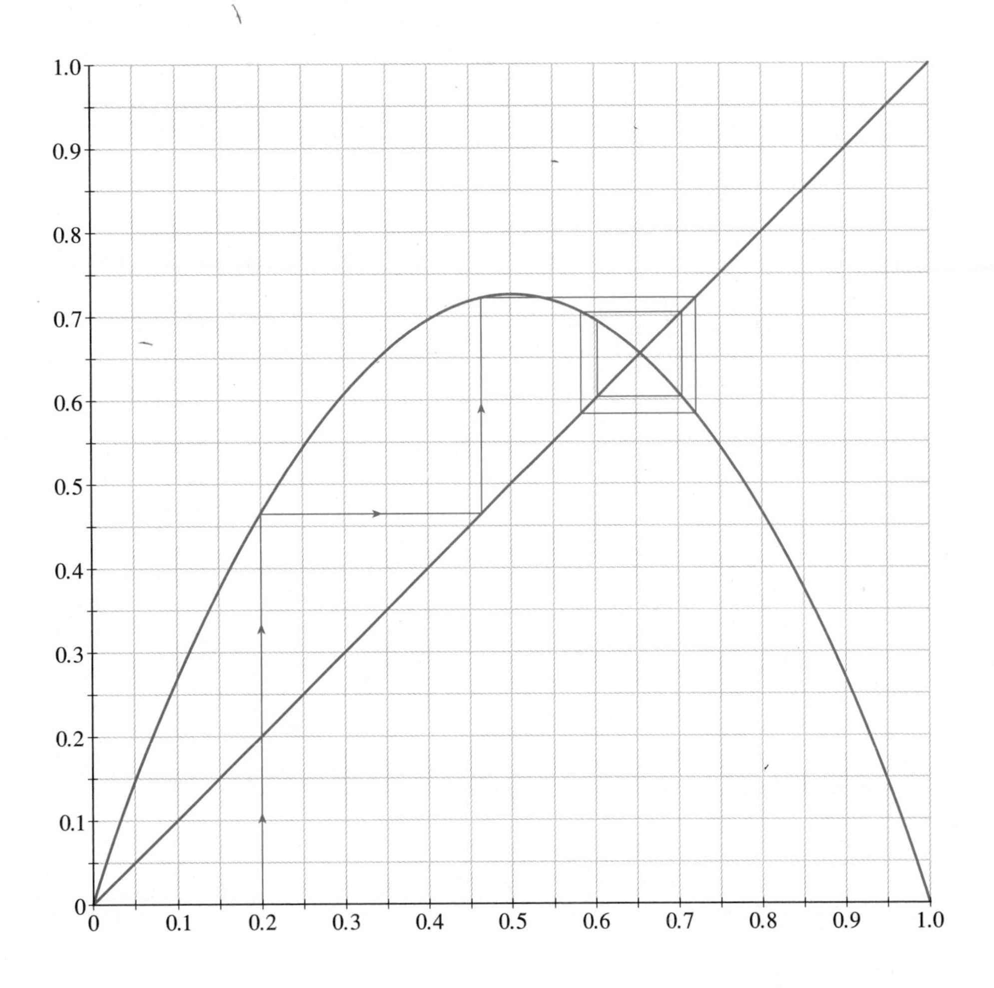

FIGURE FOR EXERCISE 22

23. The figure on the top of page 210 shows the first nine steps in the iteration process for $f(x) = 4x(1 - x)$, with $x_0 = 0.9$. (Qualitatively, note that this figure is quite different from those in the previous two exercises or in Figures 2 through 7 in the text. Here, no clear pattern in the iterates seems to emerge.)

(a) Complete the following table. For the values obtained from the graph, estimate to the nearest 0.05, or closer if you can. For instance, to the nearest 0.05, the first iterate is 0.35. But, since the graph shows the iterate a bit above 0.35, the estimate 0.36 would be better. For the calculator work, round the final answers to three decimal places.

	x_1	x_2	x_3	x_4	x_5	x_6	x_7	x_8	x_9
From graph									
From calculator									

(b) In this part of the exercise, we again work with the function $f(x) = 4x(1 - x)$. Furthermore, we will use an initial input that is very close to the one in part (a). As you will see, however, the results will be remarkably different from those in part (a). (This phenomenon is called **sensitivity to initial conditions;** it is one of the characteristic behaviors studied in chaos theory.) For x_0 we use the value

$$x_0 = \frac{5 + \sqrt{5}}{8} \approx 0.9045 \ldots$$

[Note that this input differs from the one used in part (a) by less than 0.01. (Exercise 33 shows how this seemingly "off-the-wall" input was obtained.)] Using algebra (and not a calculator!), determine exact expressions for x_1 and x_2. What do you observe? What are x_3 and x_4? What is the general pattern here? *Note:* If you were to use a calculator rather than "old-fashioned algebra" for

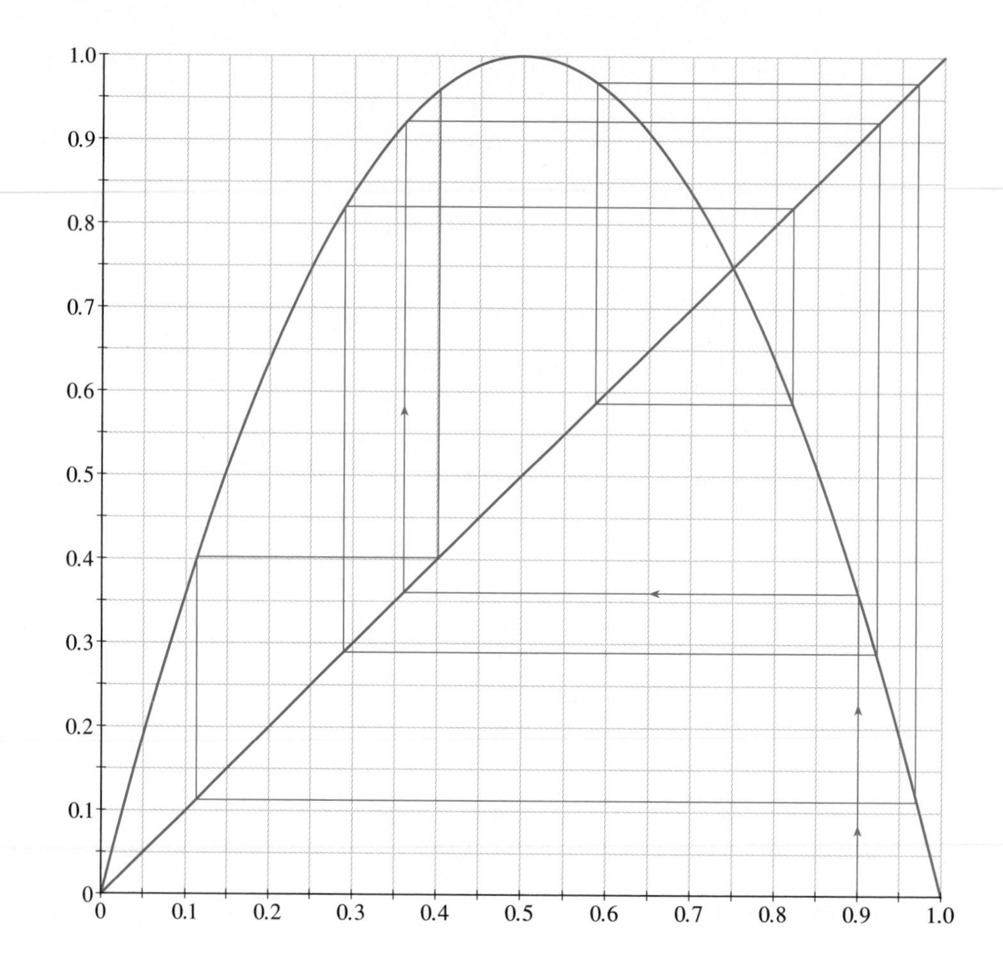

FIGURE FOR EXERCISE 23

all of this, due to rounding errors you might miss seeing the patterns.

(c) Take $x_0 = 0.905$ and use your calculator to compute the iterates x_1 through x_{10}. Round the final results to three decimal places. Is the behavior of the iterates more like that in part (a) or in part (b)?

24. Figure A shows the iteration process for $f(x) = 0.4x(1 - x)$, using the initial input $x_0 = 0.18$. Figure B shows a zoom-in view of the bottom-left square in Figure A. As indicated in both figures, the iterates approach the fixed point 0. Parts (a) and (b) provide some numerical perspective on this.

(a) Use Figure A to complete the following table. For the iterates that you read from the graph, estimate to the nearest 0.01. For the calculator values, round the final answers to three decimal places.

	x_1	x_2	x_3
From Figure A			
From calculator			

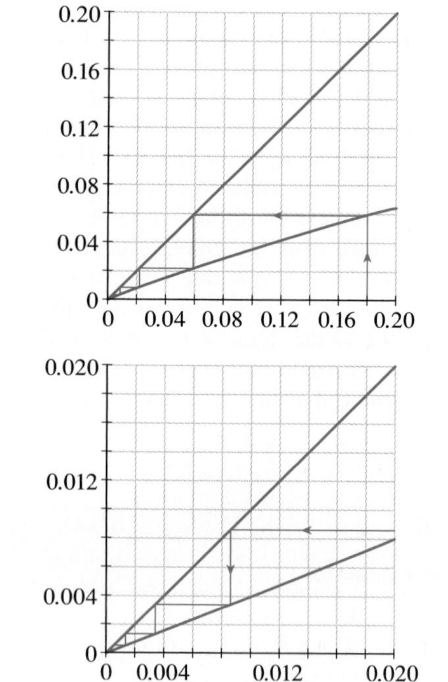

FIGURE A

FIGURE B

(b) Use Figure B to complete the table in the next column. For the iterates that you read from the graph, estimate at least to the nearest 0.001. (Some estimates can be made to the nearest 0.0005.) For the calculator values, round the final answers to four decimal places.

	x_4	x_5	x_6
From Figure B			
From calculator			

(c) Use Figure B to improve the graphical estimate for x_3 that you gave in part (a). Estimate to the nearest 0.0005. Check your estimate using a calculator.

25. The accompanying figure shows the iteration process for the function

$$f(x) = \begin{cases} 2x & \text{if } x \leq 1/2 \\ 2 - 2x & \text{if } x > 1/2 \end{cases}$$

with initial input $x_0 = 0.7$. (This function is sometimes called a *tent function* or a *hat function* because of the shape of the graph.)

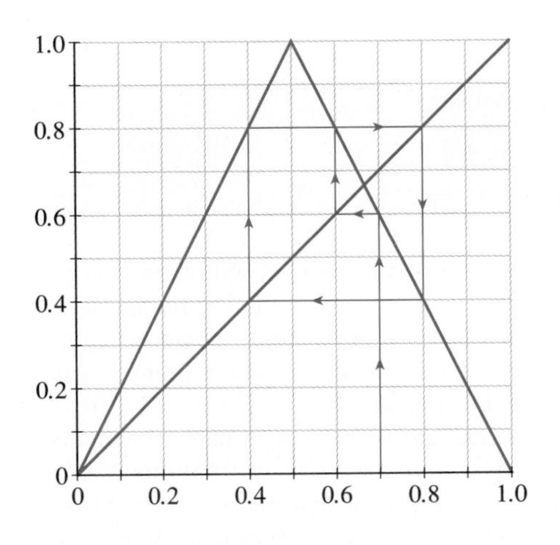

(a) Use the figure to list the iterates x_1 through x_8. Then use a calculator (or paper-and-pencil arithmetic) to check your answers. In one or two complete sentences, describe the pattern that emerges. (The iterate $x_0 = 0.7$ in this case is said to be *eventually periodic.*)

(b) Compute the first eight iterates of $x_0 = 0.75$. In a complete sentence, describe the pattern that emerges. (The iterate $x_0 = 0.75$ in this case is said to be an *eventually fixed point* of the function f.)

(c) Compute the first ten iterates of $x_0 = 2$. Describe the pattern that emerges.

26. The following figure shows a portion of the graphs of $y = x$ and the function f defined by

$$f(x) = \begin{cases} 2x & \text{if } x \leq 1/2 \\ 2 - 2x & \text{if } x > 1/2 \end{cases}$$

The grid lines in the figure are marked off in one-thirteenth-unit intervals.

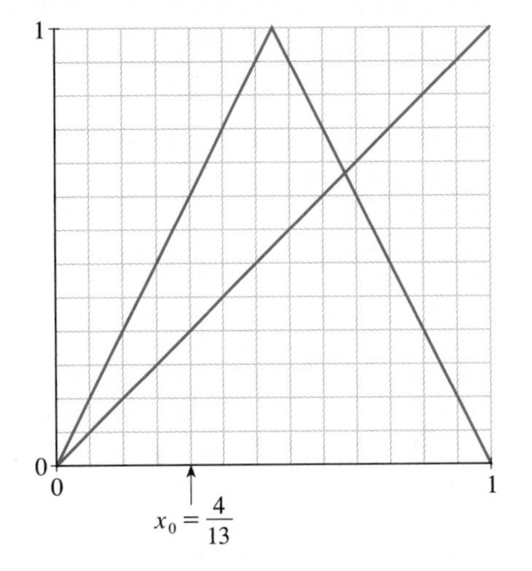

(a) Use the figure (not a calculator) to complete the following table for the first twelve iterates of $x_0 = 4/13$ under the given function f. (Unlike the case for most of the previous figures, here you'll need to draw in the iteration path for yourself.)

x_0	x_1	x_2	x_3	x_4	x_5	x_6	x_7	x_8	x_9	x_{10}	x_{11}	x_{12}
4/13												

(b) What is the pattern for these iterates? Based on the pattern, what are the values for x_{60}, x_{61}, and x_{62}?

As background for Exercises 27–30, you need to have read Example 3 in this section. As in Example 3, assume that there is a catfish pond with a maximum population size of 500 catfish. Also assume, unless stated otherwise, that the initial population size is 50 catfish (so that $x_0 = 0.1$).

27. (a) In Example 3, we used a growth parameter of $k = 2.9$ and we computed the first five iterates of $x_0 = 0.1$. Now (under these same assumptions) compute x_{21} through x_{25}, given that $x_{20} \approx 0.64594182$. Round your final answers to four decimal places.

(b) Use the results in part (a) to complete the following table. [Compare your table to Table 3 in the text; note that the population size continues to oscillate up and down, but now the sizes of the oscillations are much less. This provides additional evidence that the population size is approaching an equilibrium value. (In Example 3 we determined this equilibrium size to be about 328 catfish.)]

n	20	21	22	23	24	25
Number of fish after n breeding seasons						

(c) In parts (a) and (b), and in Example 3, we worked with a growth parameter of 2.9, and we found that the iterates of $x_0 = 0.1$ were approaching a fixed point of the function. Now assume instead that the growth parameter is $k = 0.75$ (but, still, that $x_0 = 0.1$). Determine the iterates x_1 through x_5. (Round to five decimal places.) Are the iterates approaching a fixed point of the function $f(x) = 0.75x(1 - x)$? Interpret your results.

28. As in Example 3, take $x_0 = 0.1$, but now assume that the growth parameter is $k = 3$, so that equation (1) in the text becomes $f(x) = 3x(1 - x)$.

(a) Complete the following tables. (Round the final answers for the iterates to three decimal places.) Notice that after the third breeding season, the population oscillates up and down, as in Example 3.

n	0	1	2	3	4	5
x_n	0.1					
Number of fish after n breeding seasons	50					

n	6	7	8	9	10
x_n					
Number of fish after n breeding seasons					

(b) Complete the following table, given that $x_{100} \approx 0.643772529$. Round your final answers for the iterates to four decimal places. In your results, note that the population continues to oscillate up and down, but that the sizes of the oscillations are less than those observed in part (a).

n	101	102	103	104	105	106
x_n						
Number of fish after n breeding seasons						

(c) The iterates that you computed in parts (a) and (b) are approaching a fixed point of the function $f(x) = 3x(1 - x)$. Find this fixed point and the corresponding equilibrium population size.

29. (a) As in Example 3, take $x_0 = 0.1$, but now assume that the growth parameter is $k = 3.1$, so that equation (1) in the text becomes $f(x) = 3.1x(1 - x)$. Complete the following three tables. For the third table, use the fact that $x_{20} \approx 0.56140323$. Round your final answers for the iterates to four decimal places.

n	0	1	2	3	4	5
x_n	0.1					
Number of fish after n breeding seasons	50					

n	6	7	8	9	10
x_n					
Number of fish after n breeding seasons					

n	21	22	23	24	25	26
x_n						
Number of fish after n breeding seasons						

(b) Your results in part (a) will show that the population size is oscillating up and down, but that the iterates don't seem to be approaching a fixed point. Indeed, determine the (nonzero) fixed point of the function $f(x) = 3.1x(1 - x)$. Then note that the successive iterates in part (a) actually move farther and farther away from this fixed point.

(c) It can be shown that the long-term behavior of the iterates in this case resembles the pattern in Figure 7(b) on page 206; that is, the iterates alternately approach two values. Use the following formulas, with $k = 3.1$, to determine these two values a and b that the iterates are alternately approaching. (The formulas are developed in Exercise 34.) Round the answers to four decimal places. Check to see that your answers are consistent with the results in part (a).

$$a = \frac{1 + k + \sqrt{(k - 3)(k + 1)}}{2k}$$

$$b = \frac{1 + k - \sqrt{(k - 3)(k + 1)}}{2k}$$

(d) What are the two populations corresponding to these two numbers a and b?

30. As in Example 3, assume that the maximum population size of the pond is 500 catfish, but now suppose that the growth parameter is $k = 0.6$.

(a) Suppose that the initial population is again 50 catfish, so that $x_0 = 0.1$. Compute the first ten iterates. Do they appear to be approaching a fixed point? Interpret the results.

(b) Follow part (a), but assume that the initial population is 450, so that $x_0 = 450/500 = 0.9$.

(c) What relationship do you see between the iterates in part (a) and in part (b)?

B

31. **(a)** Suppose that c and d are inputs for a function g and that $g(c) = d$ and $g(d) = c$. List the first six iterates of c and the first six iterates of d. What is the pattern? The set $\{c, d\}$ is said to be a **2-cycle** for the function g.

(b) The following figure shows the iteration process for the function $T(x) = 1 - |2x - 1|$, with initial input $x_0 = 0.4$. Use the figure to list the first six iterates of 0.4 under the function T.

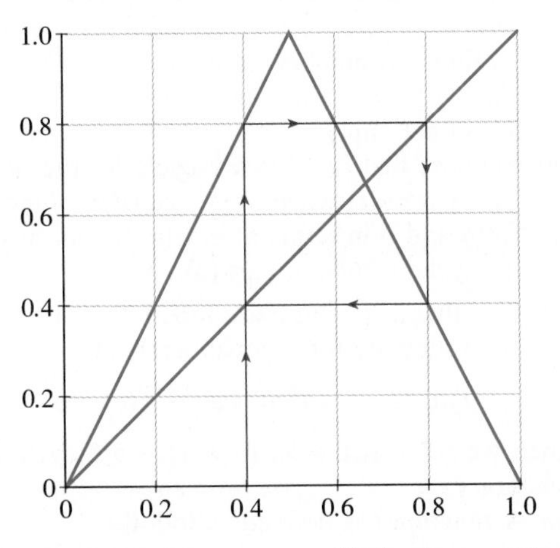

(c) In the following sentence, fill in the two blank spaces with numbers. The work in part (b) shows that {__, __} is a 2-cycle for the function T.

(d) In part (b) you used a graph to list the first six iterates of $x_0 = 0.4$ under the function $T(x) = 1 - |2x - 1|$. Now, using a calculator (or just simple arithmetic), actually compute x_1 and x_2 and thereby check your results.

32. As background for this exercise, you need to have worked part (a) in the previous exercise so that you know the definition of a 2-cycle.

(a) The curve in the following figure is the graph of the function $Q(x) = x^2 - 7$. Use the figure to list the first six iterates of -3 under the function Q. Also, list the first six iterates of 2.

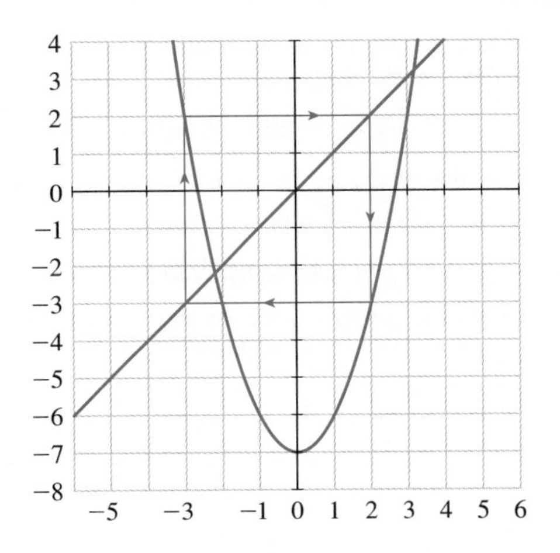

(b) In the following sentence, fill in the two blank spaces with numbers. The work in part (a) shows that {__, __} is a 2-cycle for the function Q.

(c) In part (a) you used a graph to list the first six iterates of -3 under the function Q. Now, using the formula $Q(x) = x^2 - 7$, actually compute x_1 and x_2 and thereby check your results.

(d) Use your calculator to compute the first six iterates of $x_0 = -2.99$ under the function $Q(x) = x^2 - 7$. Note that the behavior of the iterates is vastly different than that observed in part (a), even though the initial inputs differ by only 0.01. (As pointed out in Exercise 23, this type of behavior is referred to as *sensitivity to initial conditions.*)

33. Let $f(x) = 4x(1 - x)$. In this exercise we find distinct inputs a and b such that $f(a) = b$ and $f(b) = a$. As indicated in Exercise 31, the set $\{a, b\}$ is called a **2-cycle** for the function f.

(a) From the equation $f(a) = b$ and the definition of f, we have

$$4a(1 - a) = b \tag{1}$$

Likewise, from the equation $f(b) = a$ and the definition of f, we have

$$4b(1 - b) = a \tag{2}$$

Subtract equation (2) from equation (1) and show that the resulting equation can be written

$$4(b - a)(b + a - 1) = b - a \tag{3}$$

(b) Divide both sides of equation (3) by the quantity $b - a$. (The quantity $b - a$ is nonzero because we are assuming that a and b are distinct.) Then solve the resulting equation for b in terms of a. You should obtain

$$b = \frac{5}{4} - a \tag{4}$$

(c) Use equation (4) to substitute for b in equation (1). After simplifying, you should obtain

$$16a^2 - 20a + 5 = 0 \tag{5}$$

(d) Use the quadratic formula to solve equation (5) for a. You should obtain

$$a = (5 \pm \sqrt{5})/8$$

(e) Using the positive root for the moment, suppose $a = (5 + \sqrt{5})/8$. Use this expression to substitute for a in equation (4). Show that the result is $b = (5 - \sqrt{5})/8$. Now check that these values of a and b satisfy the conditions of the problem. That is, given that $f(x) = 4x(1 - x)$, show that

$$f\left[\frac{1}{8}(5 + \sqrt{5})\right] = \frac{1}{8}(5 - \sqrt{5})$$

and

$$f\left[\frac{1}{8}(5 - \sqrt{5})\right] = \frac{1}{8}(5 + \sqrt{5})$$

Note: If we'd begun part (e) by using the other root of the quadratic, namely, $a = (5 - \sqrt{5})/8$, then we would have found $b = (5 + \sqrt{5})/8$, so no new information would have been obtained. In summary, the 2-cycle for the function f consists of the two numbers that are the roots of equation (5).

34. Let $f(x) = kx(1 - x)$ and assume that $k > 0$. Follow the method of Exercise 33 to show that the values of a and b for which $f(a) = b$ and $f(b) = a$ are given by the formulas

$$a = \frac{1 + k + \sqrt{(k-3)(k+1)}}{2k}$$

and

$$b = \frac{1 + k - \sqrt{(k-3)(k+1)}}{2k}$$

Note that for $k > 3$, both of these expressions represent real numbers (because the quantity beneath the radical sign is then positive). In summary: For all values of k greater than 3, the function $f(x) = kx(1 - x)$ has a unique 2-cycle $\{a, b\}$, where a and b are given by the preceding formulas.

35. **(a)** Let $f(x) = 3.5x(1 - x)$. Use the formulas in Exercise 34 to find values for a and b such that $\{a, b\}$ is a 2-cycle for this function [that is, so that $f(a) = b$ and $f(b) = a$]. Exact answers are required, not calculator approximations.

(b) The following figure shows the iteration process for the 2-cycle determined in part (a). Use the answers in part (a) to specify the coordinates of the four points P, Q, R, and S.

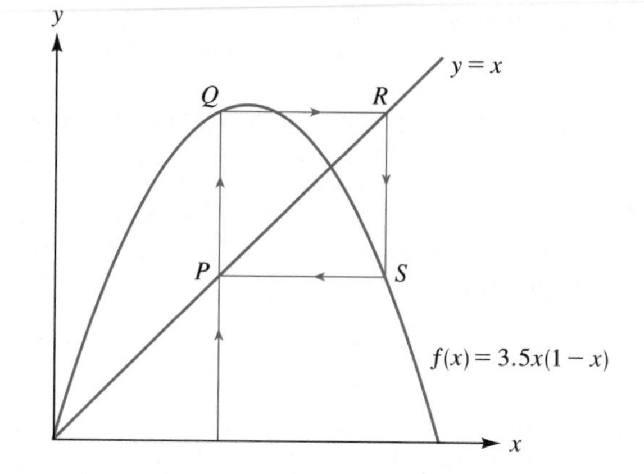

36. Let $f(x) = 3.2x(1 - x)$. Use the formulas in Exercise 34 to determine the values of a and b such that $\{a, b\}$ is a 2-cycle for this function. Use a calculator to evaluate the answers and round to three decimal places. [You'll find that these are the two numbers referred to in the caption for Figure 7(b) in this section.]

37. **(a)** Find a 2-cycle for the function $f(x) = x^2 - 3$. That is, find real numbers a and b such that $f(a) = b$ and $f(b) = a$. *Caution:* The formulas in Exercise 34 do not apply.

(b) Suppose that $\{a, b\}$ is a 2-cycle for the function $g(x) = x^2 - c$, where c is a constant. Find formulas for a and b in terms of c. Check your answers by using the results in part (a).

38. Suppose that a, b, and c are inputs for a function f and that the following three equations hold:

$$f(a) = b \qquad f(b) = c \qquad f(c) = a$$

Then we say that the set $\{a, b, c\}$ is a **3-cycle** for the function f.

(a) A function f is defined as follows:

$$f(x) = \begin{cases} 2x & \text{if } x \leq 1/2 \\ 2 - 2x & \text{if } x > 1/2 \end{cases}$$

Find a 3-cycle $\{a, b, c\}$ for the function f, given that a is less than $1/2$ while b and c are each greater than $1/2$.

(b) Find a 3-cycle $\{a, b, c\}$ for the function f in part (a), given that a and b are each less than $1/2$ and c is greater than $1/2$.

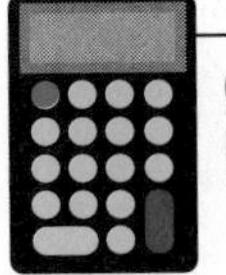

GRAPHING UTILITY EXERCISES FOR SECTION 4.3

In Exercises 1–12: **(a)** *Graph each function along with the line $y = x$. Use the graph to determine how many (if any) fixed points there are for the given function.* **(b)** *For those cases in which there are fixed points, use the zoom-in capability of the graphing utility to estimate the fixed point. (In each case, continue the zoom-in process until you are sure about the first three decimal places.)*

1. $f(x) = x^3 + 3x + 2$

2. $g(x) = x^3 - 3x + 2$

3. $h(x) = x^3 - 3x - 3.07$

4. $k(x) = x^3 - 3x - 3.08$

5. $s(t) = t^4 + 3t - 2$

6. $u(t) = t^4 + 3t + 2$

7. $m(t) = t^4 - 3t^2 + 3.5$

8. $n(t) = t^4 - 3t^2 + 3.6$

9. $F(t) = t^4 - 3t^2 + 1.06$

10. $G(t) = t^4 - 3t^2 + 1.07$

11. $f(x) = x^4 - 3x^3$

12. $g(x) = 1/(x^4 - 3x^3)$

13. (a) Graph the function $g(x) = x^2 - 0.5$ and the line $y = x$. (Use a viewing rectangle that extends from -1 to 1 in both the x- and the y-directions.)

(b) From the graph, note that the function $g(x) = x^2 - 0.5$ has a fixed point that lies between -0.4 and -0.3. According to the text (page 202), this fixed point is $(1 - \sqrt{3})/2$. Evaluate this expression, and write down the first eight decimal places.

(c) Answer the following questions by using a graphing utility to compute the iterates of $x_0 = -0.1$ under the function $g(x) = x^2 - 0.5$.

(i) Which is the first iterate to agree with the fixed point in the first two decimal places? Does the next iterate after this also agree with the fixed point in the first two decimal places?

(ii) Which is the first iterate to agree with the fixed point in the first four decimal places? Does the next iterate after this also agree with the fixed point in the first four decimal places?

14. (a) According to Example 3 in the text, the iterates of $x_0 = 0.1$ under the function $f(x) = 2.9x(1 - x)$ approach the fixed point 19/29. Using a calculator, write down the first five decimal places of this fixed point.

(b) Which is the first iterate to agree with the fixed point in the first two decimal places? For this iterate, what is the corresponding number of fish in the pond?

(c) Which is the first iterate to agree with the fixed point in the first three decimal places? For this iterate, what is the corresponding number of fish in the pond?

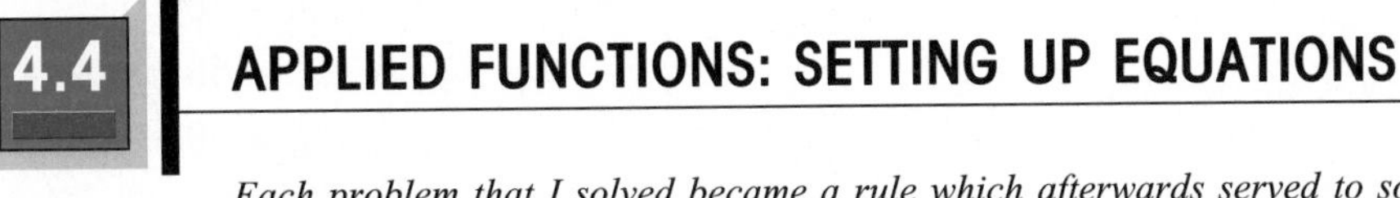

4.4 APPLIED FUNCTIONS: SETTING UP EQUATIONS

Each problem that I solved became a rule which afterwards served to solve other problems.

René Descartes (1596–1650)

I hope that I shall shock a few people in asserting that the most important single task of mathematical instruction in the secondary schools is to teach the setting up of equations to solve word problems.

George Polya (1887–1985)

One of the first steps in problem solving often involves defining a function. The function then serves to describe or summarize a given situation in a way that is both concise and (one hopes) revealing. In this section we practice setting up equations that define such functions. In Section 4.5 we will use this skill in solving an important class of applied problems involving quadratic functions.

For many of the examples in this section, we'll rely on the following four-step procedure (on the next page) to set up the required equation. You may eventually want to modify this procedure to fit your own style. The important point, however, is that it is possible to approach these problems in a systematic manner. A word of advice: You're accustomed to working mathematics problems in which the answers are numbers. In this section, the answers are functions (or, more precisely, equations defining functions); you'll need to get used to this.

STEPS FOR SETTING UP EQUATIONS

STEP 1 After reading the problem carefully, draw a picture that conveys the given information.

STEP 2 State in your own words, as specifically as you can, what the problem is asking for. (This usually requires rereading the problem.) Now, assuming that the problem asks you to find a particular quantity (or a formula for a particular quantity), assign a variable to denote that key quantity.

STEP 3 Label any other quantities in your figure that appear relevant. Are there equations relating these quantities?

STEP 4 Find an equation involving the key variable that you identified in Step 2. (Some people prefer to do this right after Step 2.) Now, as necessary, substitute in this equation using the auxiliary equations from Step 3 to obtain an equation involving only the required variables.

EXAMPLE 1 The perimeter of a rectangle is 100 cm. Express the area of the rectangle in terms of the width x.

Solution Let's follow our four-step procedure.

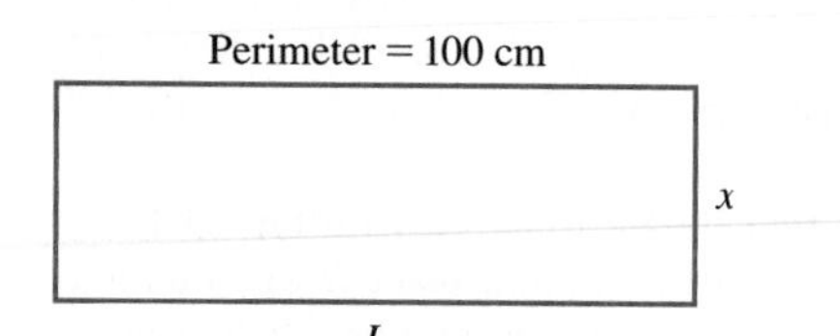

FIGURE 1

STEP 1 See Figure 1. The figure conveys the given information that the perimeter is 100 cm and that the width is x. (For the moment, ignore the label L at the base of the rectangle; it enters the picture in Step 3, but it's not part of the given information.)

STEP 2 We want to express the area of the rectangle in terms of x, the width. Let A stand for the area of the rectangle.

STEP 3 Call the length of the rectangle L, as indicated in Figure 1. Then, since the perimeter is given as 100 cm, we have

$$2x + 2L = 100$$
$$x + L = 50$$
$$L = 50 - x \qquad (1)$$

STEP 4 The area of a rectangle equals width times length:

$$A = x \cdot L$$
$$= x(50 - x) \qquad \text{substituting for } L \text{ using equation (1)}$$
$$= 50x - x^2 \qquad (2)$$

This is the required equation expressing the area of the rectangle in terms of the width x.

To emphasize this dependence of A on x, we can use function notation to rewrite equation (2):

$$A(x) = 50x - x^2$$

The domain of this area function is the open interval (0, 50). To see why this is so, first note that $x > 0$ because x denotes a width. Furthermore, in view of equation (1), we must have $x < 50$, otherwise L, the length, would be zero or negative.

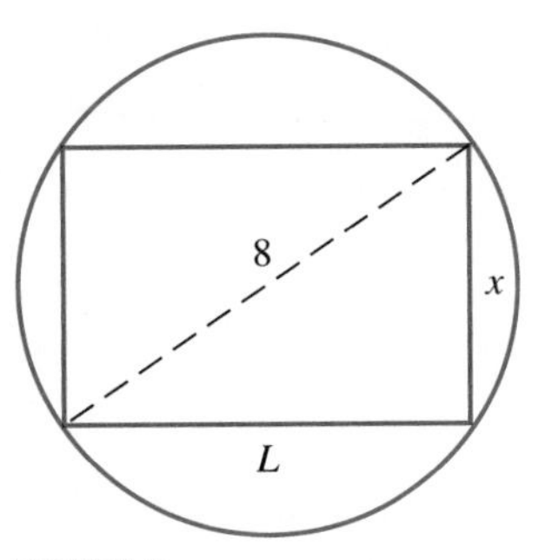

FIGURE 2

EXAMPLE 2 A rectangle is inscribed in a circle of diameter 8 cm. Express the perimeter of the rectangle as a function of its width x.

Solution We follow our four-step procedure.

STEP 1 See Figure 2. Notice that the diagonal of the rectangle is a diameter of the circle.

STEP 2 The problem asks us to come up with a formula or a function that gives us the perimeter of the rectangle in terms of x, the width. Let P denote the perimeter.

STEP 3 Let L denote the length of the rectangle, as shown in Figure 2. Then, by the Pythagorean theorem, we have

$$L^2 + x^2 = 8^2 = 64$$

This equation relates the length L and the width x. Rather than leaving the equation in this form, however, we'll solve for L in terms of x (because the instructions for the problem mention x, not L):

$$L^2 = 64 - x^2$$
$$L = \sqrt{64 - x^2} \qquad (3)$$

STEP 4 The perimeter of the rectangle is the sum of the lengths of the four sides. Thus,

$$P = x + x + L + L = 2x + 2L \qquad (4)$$

This equation expresses P in terms of x and L. However, the problem asks for P in terms of just x. Using equation (3) to substitute for L in equation (4), we have

$$P = 2x + 2\sqrt{64 - x^2} \qquad (5)$$

This is the required equation. It expresses the perimeter P as a function of the width x, so if you know x, you can calculate P.

To emphasize this dependence of P on x, we can employ function notation to rewrite equation (5):

$$P(x) = 2x + 2\sqrt{64 - x^2} \qquad (6)$$

Before leaving this example, we need to specify the domain of the perimeter function in equation (6). An easy way to do this is to look back at Figure 2. Since x represents a width, we certainly want $x > 0$. Furthermore, Figure 2 tells us that $x < 8$, because in any right triangle, a leg is always shorter than the hypotenuse. Putting these observations together, we conclude that the domain of the perimeter function in equation (6) is the open interval $(0, 8)$.

NOTE Although it would make sense *algebraically* to use an input such as $x = -1$ in equation (6), it does not make sense in our *geometric* context, where x denotes the width.

EXAMPLE 3 Let $P(x, y)$ be a point on the curve $y = \sqrt{x}$. Express the distance from P to the point $(1, 0)$ as a function of x.

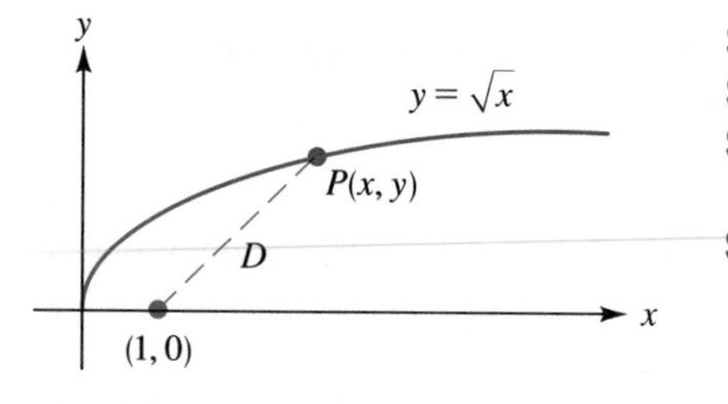

FIGURE 3

Solution

STEP 1 See Figure 3.

STEP 2 We want to express the length of the dashed line in Figure 3 in terms of x. Call this length D.

STEP 3 There are no other quantities in Figure 3 that need labeling. But don't forget that we are given

$$y = \sqrt{x} \tag{7}$$

STEP 4 By the distance formula, we have

$$\begin{aligned} D &= \sqrt{(x-1)^2 + (y-0)^2} \\ &= \sqrt{x^2 - 2x + 1 + y^2} \end{aligned} \tag{8}$$

Now we can use equation (7) to eliminate y in equation (8):

$$\begin{aligned} D(x) &= \sqrt{x^2 - 2x + 1 + (\sqrt{x})^2} \\ &= \sqrt{x^2 - 2x + 1 + x} \\ &= \sqrt{x^2 - x + 1} \end{aligned} \tag{9}$$

Equation (9) expresses the distance as a function of x, as required. What about the domain of this distance function? Since the x-coordinate of a point on the curve $y = \sqrt{x}$ can be any nonnegative number, the domain of the distance function is $[0, \infty)$.

EXAMPLE 4 A point $P(x, y)$ lies in the first quadrant on the parabola $y = 16 - x^2$, as indicated in Figure 4. Express the area of the triangular region in Figure 4 as a function of x.

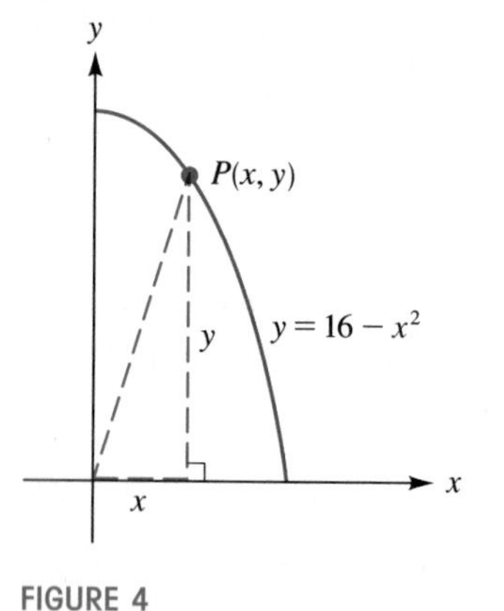

FIGURE 4

Solution

STEP 1 See Figure 4.

STEP 2 We want to express the area of the shaded triangle in terms of x. Let A denote the area of this triangle.

STEP 3 Since the coordinates of P are (x, y), the base of our triangle is x and the height is y. Also, x and y are related by the given equation

$$y = 16 - x^2 \tag{10}$$

STEP 4 The area of a triangle equals $\frac{1}{2}$(base)(height):

$$A = \tfrac{1}{2}(x)(y)$$

so

$$\begin{aligned} A(x) &= \tfrac{1}{2}(x)(16 - x^2) && \text{substituting for } y \text{ using equation (10)} \\ &= 8x - \tfrac{1}{2}x^3 \end{aligned}$$

This last equation expresses the area of the triangle as a function of x, as required. (Exercise 46 at the end of this section will ask you to specify the domain of this area function.)

EXAMPLE 5 A piece of wire x inches long is bent into the shape of a circle. Express the area of the circle in terms of x.

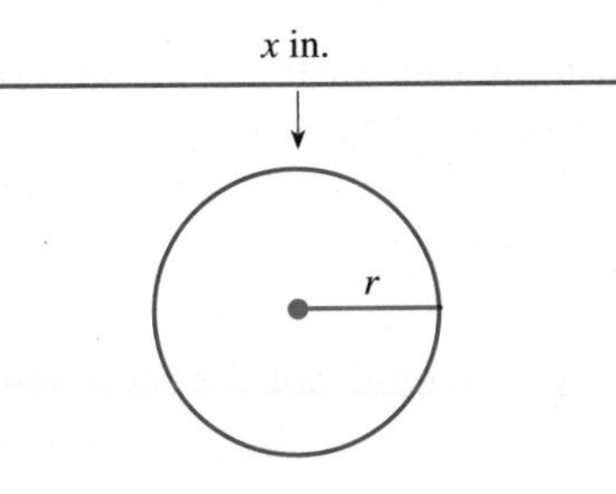

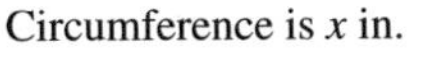

FIGURE 5

Solution

STEP 1 See Figure 5.

STEP 2 We are supposed to express the area of the circle in terms of x, the circumference. Let A denote the area.

STEP 3 The general formula for circumference in terms of radius is $C = 2\pi r$. Since in our case the circumference is given as x, our equation becomes

$$x = 2\pi r \qquad \text{or} \qquad r = \frac{x}{2\pi} \tag{11}$$

STEP 4 The general formula for the area of a circle in terms of the radius is

$$A = \pi r^2 \tag{12}$$

This expresses A in terms of r, but we want A in terms of x. So we replace r in equation (12) by the quantity given in equation (11). This yields

$$A(x) = \pi\left(\frac{x}{2\pi}\right)^2 = \frac{\pi x^2}{4\pi^2} = \frac{x^2}{4\pi}$$

Thus, we have

$$A(x) = \frac{x^2}{4\pi}$$

This is the required equation. It expresses the area of the circle in terms of the circumference x. In other words, if we know the length of the piece of wire that is to be bent into a circle, we can use that length to calculate what the area of the circle will be.

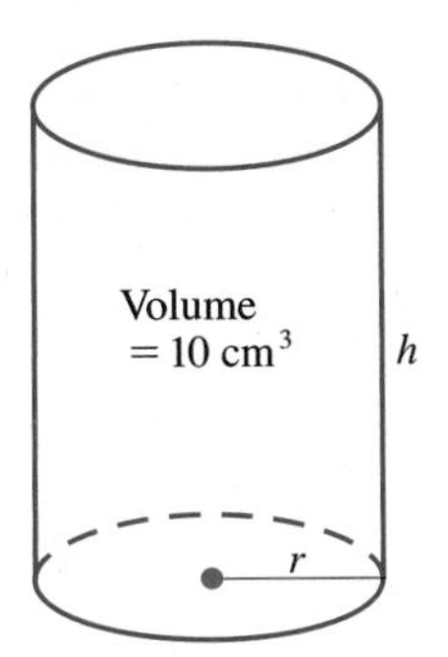

$V = \pi r^2 h$
$S = 2\pi r^2 + 2\pi r h$

FIGURE 6

EXAMPLE 6 Figure 6 displays a right circular cylinder, along with the formulas for its volume V and total surface area S. Given that the volume is 10 cm^3, express the surface area S as a function of r, the radius of the base.

Solution

STEP 1 See Figure 6.

STEP 2 We are given a formula that expresses the surface area S in terms of both r and h. We want to express S in terms of just r.

STEP 3 We are given that $V = 10$ and also that $V = \pi r^2 h$. Thus,

$$\pi r^2 h = 10$$

and consequently, expressing h in terms of r,

$$h = \frac{10}{\pi r^2} \tag{13}$$

STEP 4 We take the given formula for S, namely,

$$S = 2\pi r^2 + 2\pi r h$$

and replace h with the quantity given in equation (13). We get

$$S(r) = 2\pi r^2 + 2\pi r\left(\frac{10}{\pi r^2}\right)$$

$$= 2\pi r^2 + \frac{20}{r}$$

This is the required equation. It expresses the total surface area in terms of the radius r. Since the only restriction on r is that it be positive, the domain of the area function is $(0, \infty)$.

We have followed the same four-step procedure in Examples 1 through 6. Of course, no single method can cover all possible cases. As usual, common sense and experience are often necessary. Also, you should not feel compelled to follow this procedure at any cost. Keep this in mind as you study the last two examples in this section.

EXAMPLE 7 Two numbers add up to 8. Express the product P of these two numbers in terms of a single variable.

Solution If we call the two numbers x and $8-x$, then their product P is given by

$$P(x) = x(8-x) = 8x - x^2$$

That's it. This last equation expresses the product as a function of the variable x. Since there are no restrictions on x (other than its being a real number), the domain of this function is $(-\infty, \infty)$.

EXAMPLE 8 In economics, the revenue R generated by selling x units at a price of p dollars per unit is given by

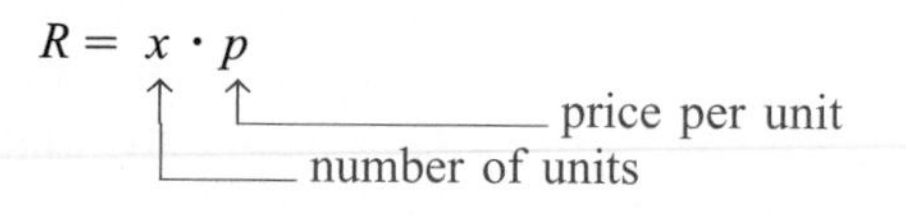

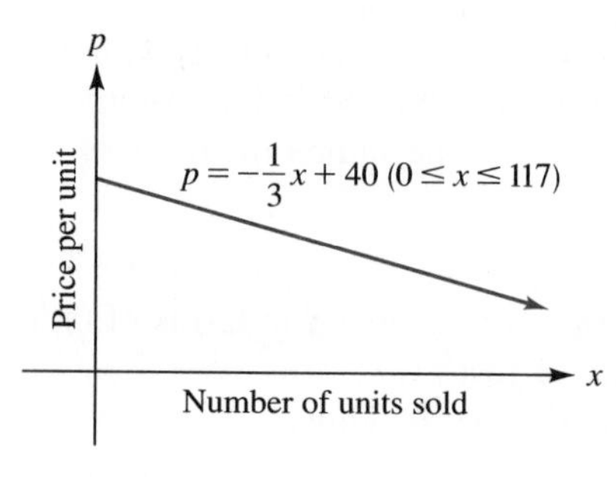

FIGURE 7

In Figure 7, we are given a hypothetical function relating the selling price of a certain item to the number of units sold. Such a function is called a **demand function.** Express the revenue as a function of x.

Solution More than anything else, this problem is an exercise in reading. After reading the problem several times, we find that it comes down to this:

given: $R = x \cdot p$ and $p = -\frac{1}{3}x + 40 \quad (0 \leq x \leq 117)$

find: an equation expressing R in terms of x

In view of this, we write

$$R = x \cdot p = x\left(-\frac{1}{3}x + 40\right)$$

Thus we have

$$R(x) = -\frac{1}{3}x^2 + 40x \qquad (0 \leq x \leq 117)$$

This is the required function. It allows us to calculate the revenue when we know the number of units sold. Note that this revenue function is a quadratic function.

QUESTION FOR REVIEW AND ALSO PREVIEW How would you find the maximum revenue in this case?

EXERCISE SET 4.4

A

1. **(a)** The perimeter of a rectangle is 16 cm. Express the area of the rectangle in terms of the width x. *Suggestion:* First reread Example 1.

(b) The area of a rectangle is 85 cm^2. Express the perimeter as a function of the width x.

2. A rectangle is inscribed in a circle of diameter 12 in.

(a) Express the perimeter of the rectangle as a function of its width x. *Suggestion:* First reread Example 2.

(b) Express the area of the rectangle as a function of its width x.

3. A point $P(x, y)$ is on the curve $y = x^2 + 1$.

(a) Express the distance from P to the origin as a function of x. *Suggestion:* First reread Example 3.

(b) In part (a), you expressed the length of a certain line segment as a function of x. Now express the slope of that line segment in terms of x.

4. A point $P(x, y)$ lies on the curve $y = \sqrt{x}$, as shown in the figure.

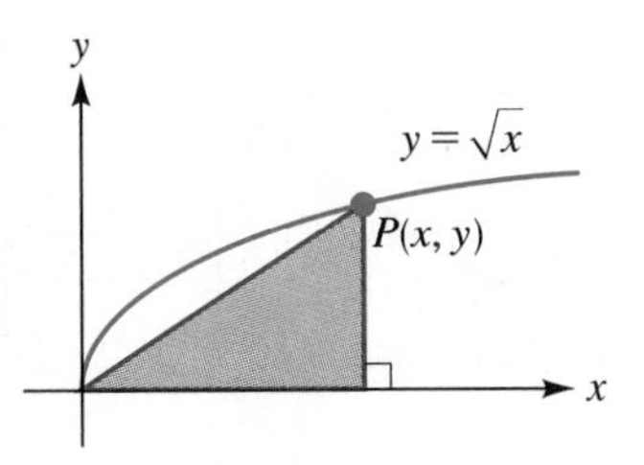

(a) Express the area of the shaded triangle as a function of x. *Suggestion:* First reread Example 4.

(b) Express the perimeter of the shaded triangle in terms of x.

5. A piece of wire πy inches long is bent into a circle.

(a) Express the area of the circle as a function of y. *Suggestion:* First reread Example 5.

(b) If the original piece of wire were bent into a square instead of a circle, how would you express the area in terms of y?

6. The volume of a right circular cylinder is 20 $in.^3$ Express the total surface area of the cylinder as a function of r, the radius of the base. *Suggestion:* First reread Example 6.

7. Two numbers add to 16.

(a) Express the product of the two numbers in terms of a single variable. *Suggestion:* First reread Example 7.

(b) Express the sum of the squares of the two numbers in terms of a single variable.

(c) Express the difference of the cubes of the two numbers in terms of a single variable. (There are two answers.)

(d) What happens when you try to express the average of the two numbers in terms of one variable?

8. The product of two numbers is 16. Express the sum of the squares of the two numbers as a function of a single variable.

9. Given a demand function $p = -\frac{1}{4}x + 8$, express the revenue as a function of x. (See Example 8 for terminology and definitions.)

10. In Example 1, we considered a rectangle with perimeter 100 cm. We found that the area of such a rectangle is given by $A(x) = 50x - x^2$, where x is the width of the rectangle. Compute the numbers $A(1)$, $A(10)$, $A(20)$, $A(25)$, and $A(35)$. Which width x seems to yield the largest area $A(x)$?

11. In Example 2, we considered a rectangle inscribed in a circle of diameter 8 cm. We found that the perimeter of such a rectangle is given by

$$P(x) = 2x + 2\sqrt{64 - x^2}$$

where x is the width of the rectangle.

(a) Use a calculator to complete the following table, rounding each answer to two decimal places.

x	1	2	3	4	5	6	7
$P(x)$							

(b) In your table, what is the largest value for $P(x)$? What is the width x in this case?

(c) Using calculus, it can be shown that among all possible widths x, the width $x = 4\sqrt{2}$ cm yields the largest possible perimeter. Use a calculator to compute the perimeter in this case, and check to see that the value you obtain is indeed larger than all of the values obtained in part (a).

12. In economics, the demand function for a given commodity tells us how the unit price p is related to the number of units x that are sold. Suppose we are given a demand function

$$p = 5 - \frac{x}{4} \quad (p \text{ in dollars})$$

(a) Graph this demand function.

(b) How many units can be sold when the unit price is \$3? Locate the point on the graph of the demand function that conveys this information.
(c) To sell 12 items, how should the unit price be set? Locate the point on the graph of the demand function that conveys this information.
(d) Find the revenue function corresponding to the given demand function. (Use the formula $R = x \cdot p$.) Graph the revenue function.
(e) Find the revenue when $x = 2$, when $x = 8$, and when $x = 14$.
(f) According to your graph in part (d), which x-value yields the greatest revenue? What is that revenue? What is the corresponding unit price?

13. Let $2s$ denote the length of the side of an equilateral triangle.
(a) Express the height of the triangle as a function of s.
(b) Express the area of the triangle as a function of s.
(c) Use the function you found in part (a) to determine the height of an equilateral triangle, each side of which is 8 cm long.
(d) Use the function you found in part (b) to determine the area of an equilateral triangle, each side of which is 5 in. long.

14. If x denotes the length of a side of an equilateral triangle, express the area of the triangle as a function of x.

15. The height of a right circular cylinder is twice the radius. Express the volume as a function of the radius.

16. Using the information given in Exercise 15, express the radius as a function of the volume.

17. The volume of a right circular cylinder is 12π in.^{3}
(a) Express the height as a function of the radius.
(b) Express the total surface area as a function of the radius.

18. The total surface area of a right circular cylinder is 14 in.^{2} Express the volume as a function of the radius.

19. The volume V and the surface area S of a sphere of radius r are given by the formulas $V = \frac{4}{3}\pi r^3$ and $S = 4\pi r^2$. Express V as a function of S.

20. The base of a rectangle lies on the x-axis, while the upper two vertices lie on the parabola $y = 10 - x^2$. Suppose that the coordinates of the upper right vertex of the rectangle are (x, y). Express the area of the rectangle as a function of x.

21. The hypotenuse of a right triangle is 20 cm. Express the area of the triangle as a function of the length x of one of the legs.

22. **(a)** Express the area of the shaded triangle in the accompanying figure as a function of x.
(b) Express the perimeter of the shaded triangle as a function of x.

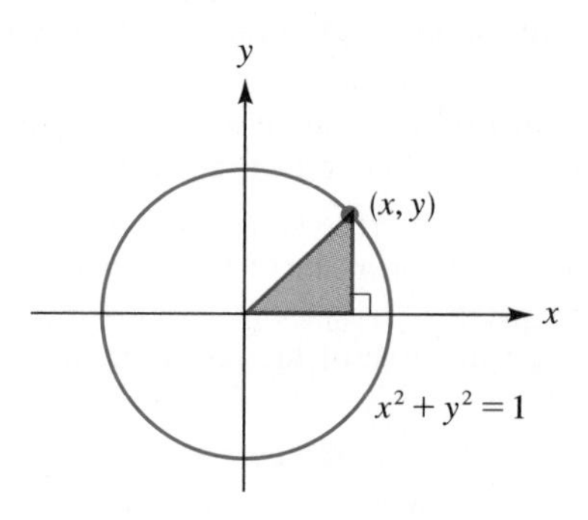

23. For the following figure, express the length AB as a function of x. *Hint:* Note the similar triangles.

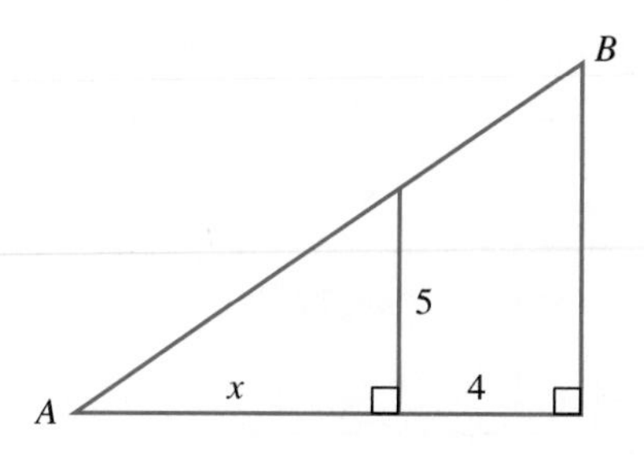

24. Five hundred feet of fencing is available to enclose a rectangluar pasture alongside a river, which serves as one side of the rectangle (so only three sides require fencing—see the figure). Express the area of the rectangular pasture as a function of x.

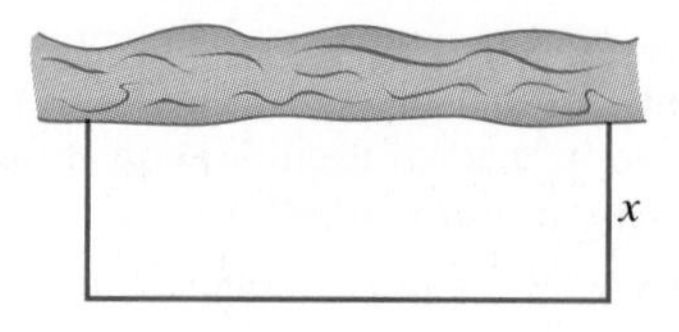

25. After rereading Example 1, complete the following table. Which x-value in the table yields the largest area A? What is the corresponding value of the length L in that case?

x	5	10	20	24	24.8	24.9	25	25.1	25.2	45
$A(x)$										

26. After rereading Example 3, complete the following three tables. For each table, specify the x-value that yields the smallest distance D. (For Table 2, round the values of D to two significant digits. For Table 3, round the values of D to seven significant digits.)

TABLE 1

x	1	2	3	4	5
D					

TABLE 2

x	0.25	0.50	0.75
D			

TABLE 3

x	0.498	0.499	0.500	0.501	0.502
D					

27. **(a)** After rereading Example 4, complete the following three tables. For each table, specify the x-value that yields the largest area A. [For Tables 2 and 3, round the values of A to six significant digits. In part (b), you'll see why we are asking for that many digits.]

TABLE 1

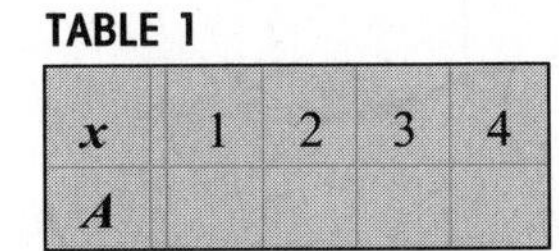

x	1	2	3	4
A				

TABLE 2

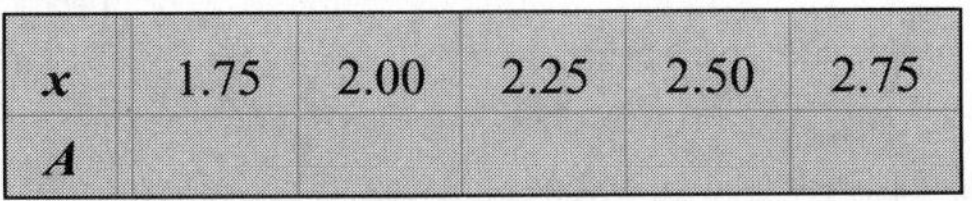

x	1.75	2.00	2.25	2.50	2.75
A					

TABLE 3

x	2.15	2.20	2.25	2.30	2.35
A					

(b) Using calculus, it can be shown that the x-value yielding the largest area A is $x = 4\sqrt{3}/3$. Which x-value in the tables is closest to this x-value? To six significant digits, what is the area A when $x = 4\sqrt{3}/3$?

B

28. A piece of wire 4 m long is cut into two pieces, then each piece is bent into a square. Express the combined area of the two squares in terms of one variable.

29. A piece of wire 3 m long is cut into two pieces. Let x denote the length of the first piece and $3 - x$ the length of the second. The first piece is bent into a square and the second into a rectangle in which the width is half the length. Express the combined area of the square and the rectangle as a function of x. Is the resulting function a quadratic function?

In Exercises 30–33, refer to the following figure, which displays a right circular cone along with the formulas for the volume V and the lateral surface area S.

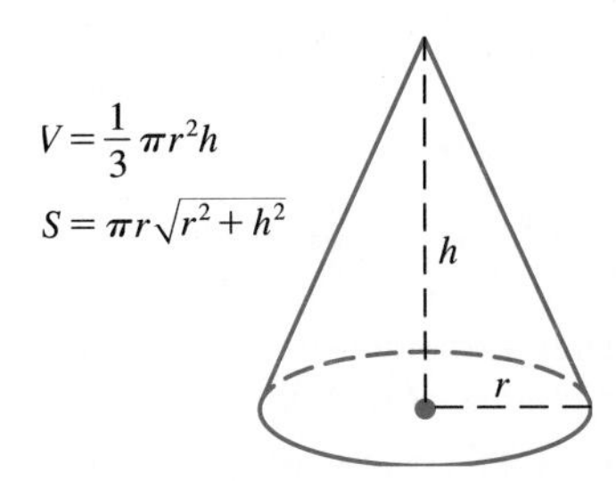

30. The volume of a right circular cone is 12π cm.^{3}
(a) Express the height as a function of the radius.
(b) Express the radius as a function of the height.

31. Suppose that the height and radius of a right circular cone are related by the equation $h = \sqrt{3}r$.
(a) Express the volume as a function of r.
(b) Express the lateral surface area as a function of r.

32. The volume of a right circular cone is 2 ft.^{3} Show that the lateral surface area as a function of r is given by

$$S = \frac{\sqrt{\pi^2 r^6 + 36}}{r}$$

33. In a certain right circular cone, the volume is numerically equal to the lateral surface area.
(a) Express the radius as a function of the height.
(b) Express the height as a function of the radius.

34. A line is drawn from the origin O to a point $P(x, y)$ in the first quadrant on the graph of $y = 1/x$. From point P, a line is drawn perpendicular to the x-axis, meeting the x-axis at B.
(a) Draw a figure of the situation described.
(b) Express the perimeter of $\triangle OPB$ as a function of x.
(c) Try to express the area of $\triangle OPB$ as a function of x. What happens?

35. A piece of wire 14 in. long is cut into two pieces. The first piece is bent into a circle, the second into a square. Express the combined total area of the circle and the square as a function of x, where x denotes the length of the wire that is used for the circle.

36. A wire of length L is cut into two pieces. The first piece is bent into a square, the second into an equilateral triangle. Express the combined total area of the square and the triangle as a function of x, where x denotes the length of wire used for the triangle. (Here, L is a constant, not another variable.)

37. An athletic field with a perimeter of $\frac{1}{4}$ mile consists of a rectangle with a semicircle at each end, as shown in the figure. Express the area of the field as a function of r, the radius of the semicircle.

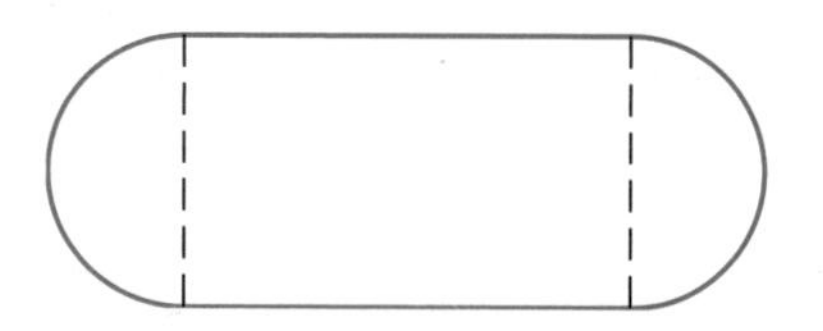

38. A square of side x is inscribed in a circle. Express the area of the circle as a function of x.

39. An equilateral triangle of side x is inscribed in a circle. Express the area of the circle as a function of x.

40. **(a)** Refer to the accompanying figure. An offshore oil rig is located at point A, which is 10 miles out to sea. An oil pipeline is to be constructed from A to a point C on the shore and then to an oil refinery at point D, farther up the coast. If it costs \$8000 per mile to lay the pipeline in the sea and \$2000 per mile on land, express the cost of laying the pipeline in terms of x, where x is the distance from B to C.

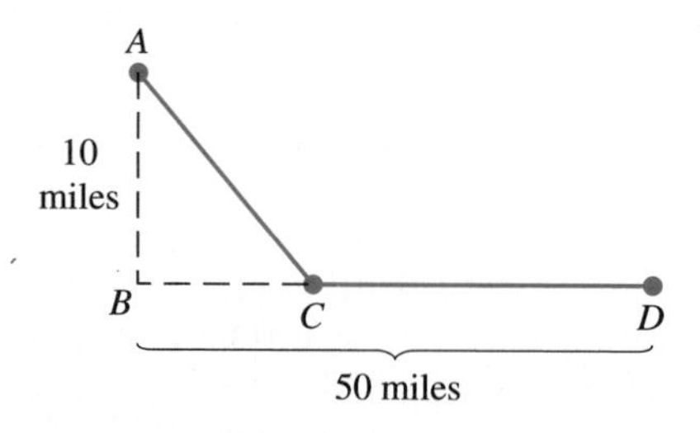

(b) Use a calculator and your result in part (a) to complete the following table, rounding the answers to the nearest 100 dollars.

x (miles)	0	10	20	30	40	50
Cost (dollars)						

(c) Based on the results in part (b), does it appear that the cost function is increasing, decreasing, or neither on the interval $0 \leqslant x \leqslant 50$? Which of the x-values in the table yields the lowest cost?

(d) Use a calculator and your result in part (a) to complete the following table, rounding the answers to the nearest 100 dollars.

x (miles)	0	4	8	12	16	20
Cost (dollars)						

(e) Again, answer the questions in part (c), but now take into account the results in part (d) also.

41. **(a)** An open-top box is constructed from a 6-by-8-inch rectangular sheet of tin by cutting out equal squares at each corner and then folding up the flaps, as shown in the figure. Express the volume of the box as a function of x, the length of the side of each cutout square.

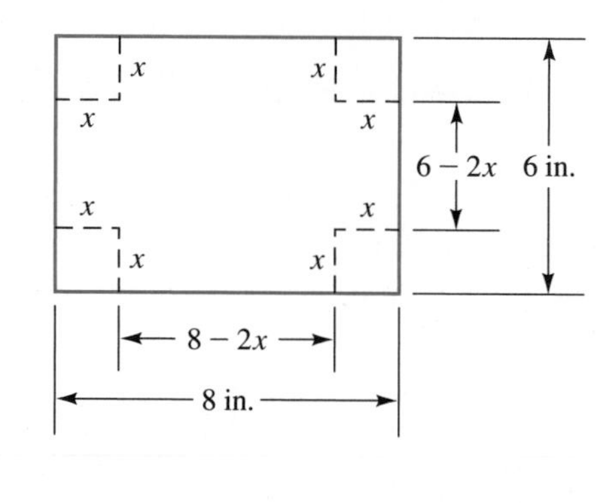

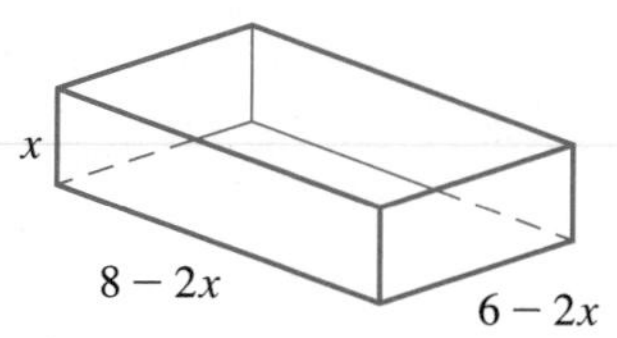

(b) Use a calculator and your result in part (a) to complete the following table. Round the answers to the nearest 0.5 in.^{3}

x (in.)	0	0.5	1.0	1.5	2.0	2.5	3.0
Volume (in.^{3})							

(c) Based on the results in part (b), which x-value in the interval $0 \leqslant x \leqslant 3$ appears to yield the largest volume for the box?

(d) Use a calculator and your result in part (a) to complete the following table. Round off the answers to the nearest 0.1 in.^{3}

x (in.)	0.8	0.9	1.0	1.1	1.2	1.3	1.4
Volume (in.^{3})							

(e) Again, answer the questions in part (c), but now take into account the results in part (d), too.

42. Follow Exercise 41(a), but assume that the original piece of tin is a square, 12 in. on each side.

43. A Norman window is in the shape of a rectangle surmounted by a semicircle, as shown in the figure. Assume that the perimeter of the window is 32 ft.

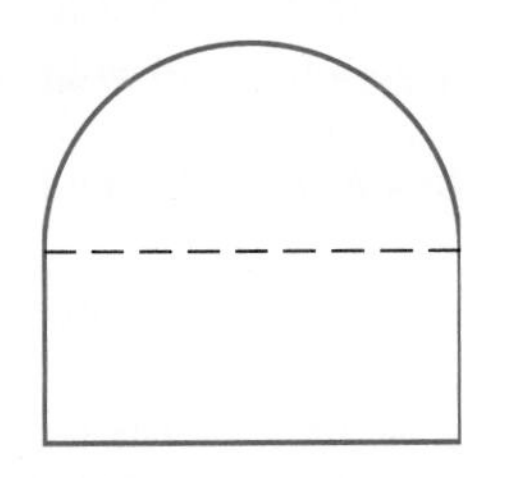

(a) Express the area of the window as a function of r, the radius of the semicircle.
(b) The function you were asked to find in part (a) is a quadratic function, so its graph is a parabola. Does the parabola open upward or downward? Does it pass through the origin? Show that the vertex of the parabola is

$$\left(\frac{32}{\pi+4}, \frac{512}{\pi+4}\right)$$

44. Refer to the following figure. Express the lengths CB, CD, BD, and AB in terms of x.
Hint: Recall the theorem from geometry stating that in a 30°-60°-90° right triangle, the side opposite the 30° angle is half the hypotenuse.

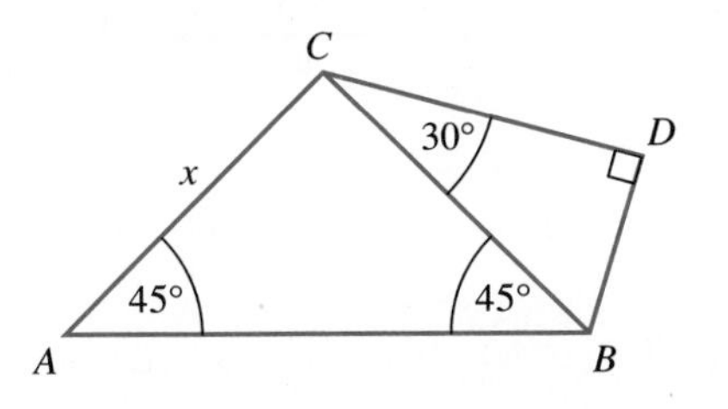

45. Refer to the following figure; let s denote the ratio of y to z.
(a) Express y as a function of s.
(b) Express s as a function of y.
(c) Express z as a function of s.
(d) Express s as a function of z.

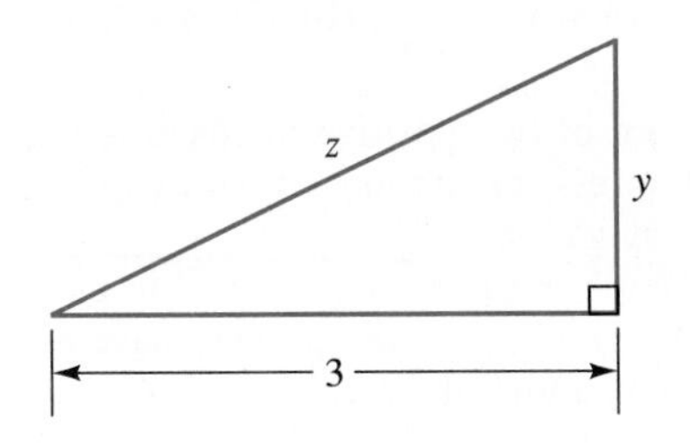

46. Refer to Example 4.
(a) What is the x-intercept of the curve $y = 16 - x^2$ in Figure 4?
(b) What is the domain of the area function in Example 4, assuming that the point P does not lie on the x- or y-axis?

47. The following figure shows the parabola $y = x^2$ and a line segment $\overline{AP}$ drawn from the point $A(0, -1)$ to the point $P(a, a^2)$ on the parabola.

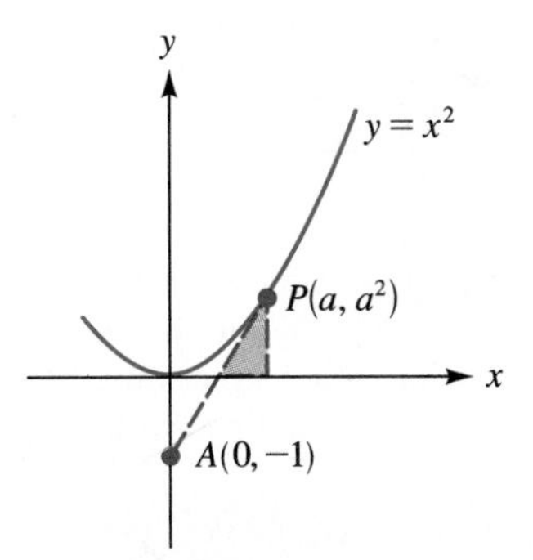

(a) Express the slope of $\overline{AP}$ in terms of a.
(b) Show that the area of the shaded triangle in the figure is given by

$$\text{area} = \frac{a^5}{2(a^2+1)}$$

48. A rancher who wishes to fence off a rectangular area finds that the fencing in the east-west direction will require extra reinforcement owing to strong prevailing winds. Fencing in the east-west direction will therefore cost \$12 per (linear) yard, as opposed to a cost of \$8 per yard for fencing in the north-south direction. Given that the rancher wants to spend \$4800 on fencing, express the area of the rectangle as a function of x, its width. The required function is in fact a quadratic, so its graph is a parabola. Does the parabola open upward or downward? By considering this graph, find which width x yields the rectangle of largest area. What is this maximum area?

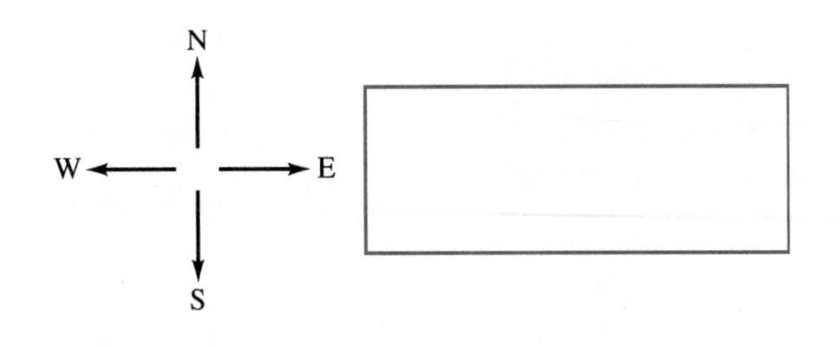

C

49. The following figure shows two concentric squares. Express the area of the shaded triangle as a function of x.

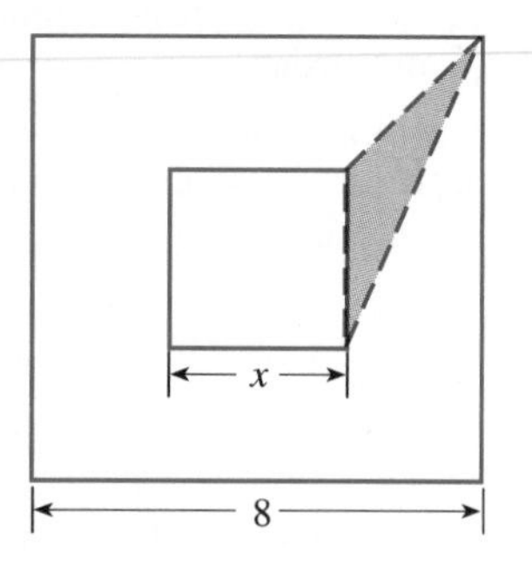

50. The following figure shows two concentric circles of radii r and R. Let A denote the area within the larger circle but outside the smaller one. Express A as a function of x (where x is defined in the figure).

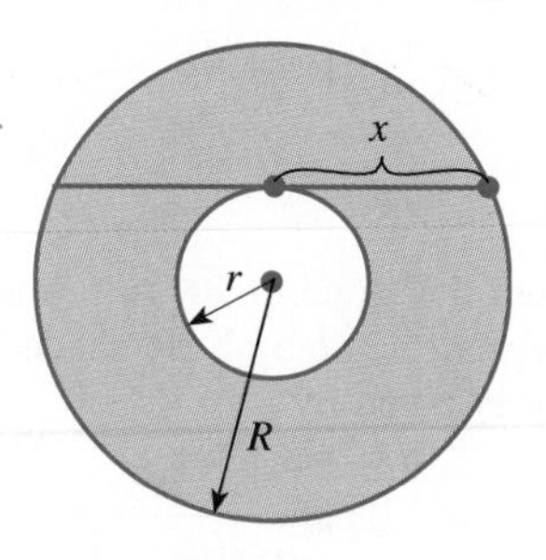

51. A straight line with slope m $(m < 0)$ passes through the point $(1, 2)$ and intersects the line $y = 4x$ at a point in the first quadrant. Let A denote the area of the triangle bounded by $y = 4x$, the x-axis, and the given line of slope m. Express A as a function of m.

52. A line with slope m $(m < 0)$ passes through the point (a, b) in the first quadrant and intersects the line $y = Mx$ $(M > 0)$ at another point in the first quadrant. Let A denote the area of the triangle bounded by $y = Mx$, the x-axis, and the given line with slope m. Express A in terms of m, M, a, and b.

Answer: $A = \dfrac{M(am - b)^2}{2m(m - M)}$

53. A line with slope m $(m < 0)$ passes through the point (a, b) in the first quadrant. Express the area of the triangle bounded by this line and the axes in terms of m.

54. One corner of a page of width a is folded over and just reaches the opposite side, as indicated in the figure. Express L, the length of the crease, in terms of x and a.

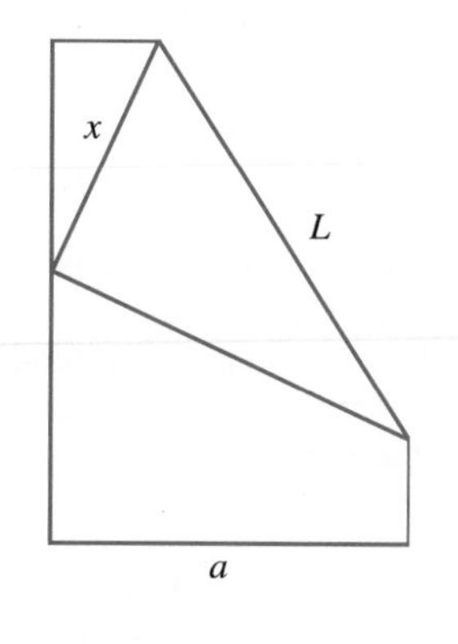

GRAPHING UTILITY EXERCISES FOR SECTION 4.4

In Exercises 1–8, graph the indicated function and answer the following questions: **(a)** *Is the function a quadratic function?* **(b)** *How many turning points are there?* **(c)** *Is there a maximum value for the function?* **(d)** *Is there a minimum value?*

1. $A(x) = 50x - x^2$, $0 < x < 50$ (This is the function that we obtained in Example 1.)
2. $P(x) = 2x + 2\sqrt{64 - x^2}$, $0 < x < 8$ (from Example 2)
3. $D(x) = \sqrt{x^2 - x + 1}$, $x \geq 0$ (from Example 3)
4. $A(x) = 8x - \frac{1}{2}x^3$, $0 < x < 4$ (from Example 4)
5. $S(r) = 2\pi r^2 + 20/r$, $r > 0$ (from Example 6)
6. $R(x) = x\left(-\frac{1}{3}x + 40\right)$, $x \geq 0$ (from Example 8)
7. **(a)** $f(x) = x + \dfrac{1}{x}$, $x > 0$; **(b)** $g(x) = x^2 + \dfrac{1}{x}$, $x > 0$
8. **(a)** $F(x) = \sqrt{x} + \dfrac{1}{x}$; **(b)** $G(x) = \sqrt{x} - \dfrac{1}{x}$
9. The sum of two positive numbers is $\sqrt{11}$.
   (a) Express the product of the two numbers in terms of a single variable.
   (b) Graph the function obtained in part (a). Based on the graph, does the product have a minimum or maximum value?

10. The product of two positive numbers is $\sqrt{11}$.
(a) Express the sum of the two numbers in terms of a single variable.
(b) Graph the function obtained in part (a). Based on the graph, does the sum have a minimum or maximum value?

4.5 MAXIMUM AND MINIMUM PROBLEMS

. . . problems on maxima and minima, although new features in an English textbook, stand so little in need of apology with the scientific public that I offer none.

G. Chrystal in his preface to *Textbook of Algebra* (1886)

In daily life it is constantly necessary to choose the best possible (optimal) solution. A tremendous number of such problems arise in economics and in technology. In such cases it is frequently useful to resort to mathematics.

V. M. Tikhomirov in *Stories about Maxima and Minima* (translated from the Russian by Abe Shenitzer) (Providence, Rhode Island: The American Mathematical Society, 1990)

You have already seen several examples of maximum and minimum problems in the second section of this chapter. Before reading further in the present section, you should first review Examples 5 through 7 on pages 193–195.

Actually, we begin this section's discussion of maximum and minimum problems at a more intuitive level. Consider, for example, the following question: If two numbers add to 9, what is the largest possible value of their product? To gain some insight here, we carry out a few preliminary calculations (see Table 1).

TABLE 1

x	y	$x+y$	xy
−2	11	9	−22
−1	10	9	−10
0	9	9	0
1	8	9	8
2	7	9	14
3	6	9	18
4	5	9	(20)

We have circled the 20 in the right-hand column of Table 1 because it *appears* to be the largest product. We say "appears" because our table is incomplete. For instance, what if we allowed x- and y-values that are not whole numbers? Might we get a product exceeding 20? Table 2 shows the results of some additional calculations along these lines.

TABLE 2

x	y	$x+y$	xy
1	8	9	8
1.5	7.5	9	11.25
2	7	9	14
2.5	6.5	9	16.25
3	6	9	18
3.5	5.5	9	19.25
4	5	9	20
4.5	4.5	9	20.25

As you can see from Table 2, there is a product exceeding 20, namely, 20.25. Now the question is, if we further expand our tables, can we find an even larger product, one exceeding 20.25? And here we have come about as far as we want to go using this approach involving arithmetic and tables. For no matter what candidate we come up with for the largest product, there will always be the question of whether we might do still better using a larger table.

Nevertheless, this approach was useful, for it showed us what is really at the heart of a typical maximum or minimum problem. Essentially, we are trying to sort through an infinite number of possible cases and pick out the required extreme case. In the example at hand, there are infinitely many pairs of numbers x and y adding to 9. We are asked to look at the products of all of these pairs and see which (if any!) is the largest. Example 1 shows how to apply our knowledge of quadratic functions to solve this problem in a definitive manner.

EXAMPLE 1 Two numbers add to 9. What is the largest possible value for their product?

Solution Call the two numbers x and $9 - x$. Then their product P is given by

$$P = x(9 - x) = 9x - x^2$$

The graph of this quadratic function is the parabola in Figure 1. Note the accompanying calculations for the vertex. As Figure 1 and the accompanying calculations show, the largest value of the product P is 20.25. This is the required solution. (Note from the graph that there is no smallest value of P.)

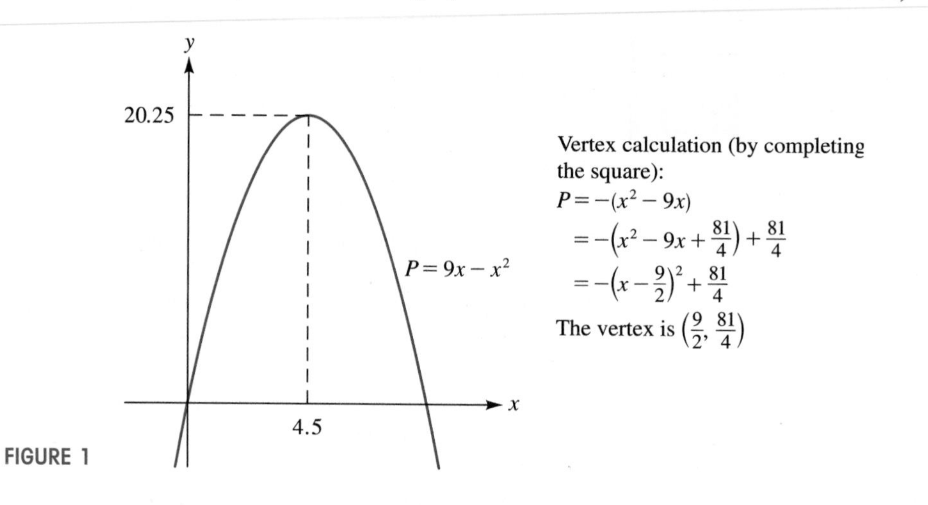

FIGURE 1

Example 1 illustrates the general strategy we will follow for solving the maximum and minimum problems in this section.

STRATEGY FOR SOLVING THE MAXIMUM AND MINIMUM PROBLEMS IN THIS SECTION

1. Express the quantity to be maximized or minimized in terms of a single variable. For instance, in Example 1, we found $P = 9x - x^2$. In setting up such functions, you'll want to keep in mind the four-step procedure used in the previous section.
2. Assuming that the function you have determined is a quadratic, note whether its graph, a parabola, opens upward or downward. Check whether this is consistent with the requirements of the problem. For instance, the parabola in Figure 1 opens downward; so it makes sense to look for a *largest,* not a smallest, value of P. Now complete the square to locate the vertex. (If the function is not a quadratic but is closely related to a quadratic, these ideas may still apply. See, for instance, Examples 6 and 7 in Section 4.2.)
3. After you have determined the vertex, you must relate that information to the original question. In Example 1, for instance, we were asked for the product P, not for x.

EXAMPLE 2 Among all rectangles having a perimeter of 10 ft, find the dimensions (length and width) of the one with the greatest area.

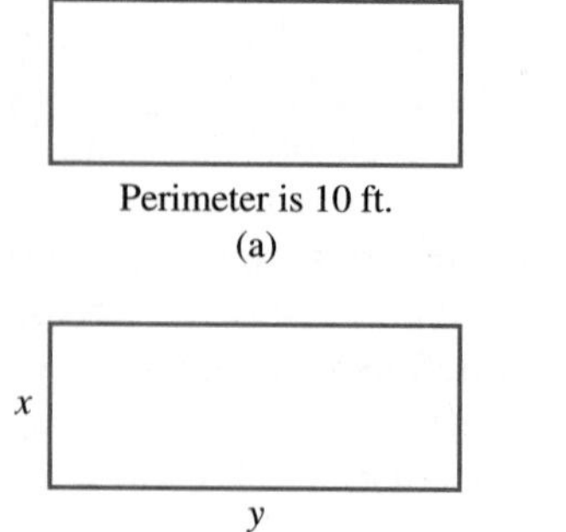

FIGURE 2

Solution First we want to set up a function that expresses the area of the rectangle in terms of a single variable. In doing this, we'll be guided by the four-step procedure that we used in Section 4.4. Figure 2(a) displays the given information. Our problem is to determine the dimensions of the rectangle that has the greatest area. As Figure 2(b) indicates, we can label the dimensions x

and y. Since the perimeter is given as 10 ft, we have

$$2x + 2y = 10$$
$$x + y = 5$$
$$y = 5 - x \tag{1}$$

Letting A denote the area of the rectangle, we can write

$$A = xy$$
$$A(x) = x(5 - x) \quad \text{substituting for } y \text{ using equation (1)}$$
$$= 5x - x^2$$

This expresses the area of the rectangle in terms of the width x. Since the graph of this quadratic function is a parabola that opens downward, it does make sense to talk about the maximum. We can find the vertex of the parabola by completing the square:

$$A(x) = -(x^2 - 5x \qquad)$$
$$= -\left(x^2 - 5x + \frac{25}{4}\right) + \frac{25}{4} \quad \text{adding zero}$$
$$= -\left(x - \frac{5}{2}\right)^2 + \frac{25}{4}$$

GRAPHICAL PERSPECTIVE

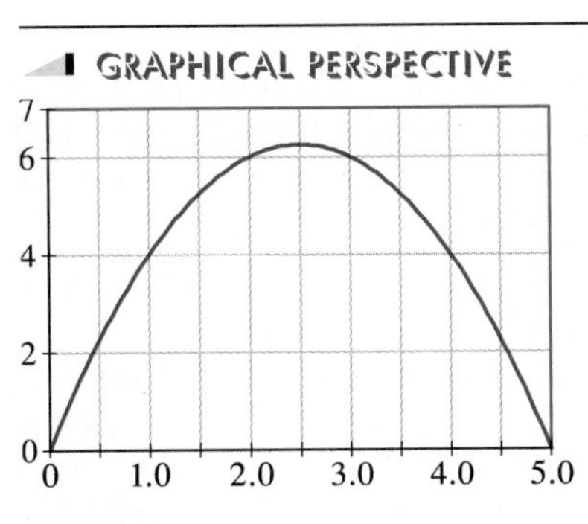

FIGURE 3
The graph of the area function, $A(x) = 5x - x^2$, is a parabola that opens downward. The maximum area occurs when $x = 2.5$ ft.

So the vertex is $\left(\frac{5}{2}, \frac{25}{4}\right)$ and, in particular, the width $x = \frac{5}{2}$ ft yields the maximum area. The problem asks us for the length and width of this rectangle, not its area. To compute the length y, we again use equation (1):

$$y = 5 - x = 5 - \frac{5}{2} = \frac{5}{2}$$

Thus, among all rectangles having a perimeter of 10 ft, the one with the greatest area is actually the square with dimensions of $2\frac{1}{2}$ ft by $2\frac{1}{2}$ ft. See Figure 3 for a graphical view of this result.

EXAMPLE 3 Suppose that a baseball is tossed straight up and that its height as a function of time is given by the formula

$$h = -16t^2 + 64t + 6$$

In this formula, h is measured in feet and t in seconds, with $t = 0$ corresponding to the instant that the ball is released. What is the maximum height of the ball? When does the ball reach this height?

Solution The given function tells us how the height h of the ball depends on the time t. We want to know the largest possible value for h. Since the graph of the given function is a parabola opening downward, we can determine the largest value of h just by finding the vertex of the parabola. After completing the square, we find that the original equation can be rewritten.

$$h = -16(t - 2)^2 + 70 \tag{2}$$

(You should verify this for yourself. If you get stuck, follow the model for completing the square in Example 4 of Section 4.2.) From equation (2) we see that

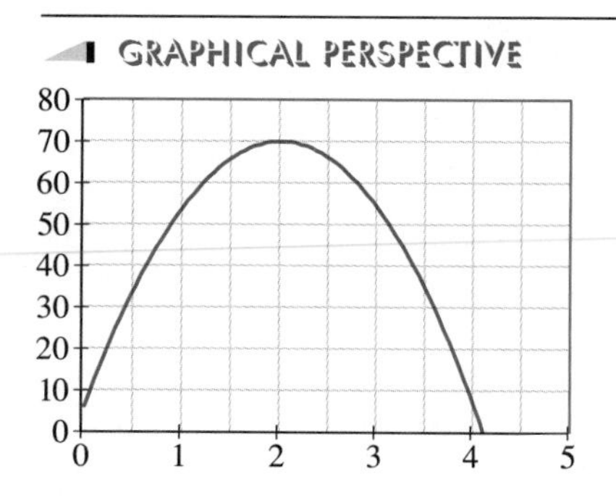

FIGURE 4
The graph of the height function $h = -16t^2 + 64t + 6$ is a parabola that opens downward. The maximum height is 70 ft, and this occurs when $t = 2$ sec.

the vertex of the parabola $h = -16t^2 + 64t + 6$ is (2, 70). Therefore the maximum height of the ball is 70 ft, and the ball reaches this height at $t = 2$ sec. The graph of the parabola given by equation (2) is shown in Figure 4. *Caution:* The path of the baseball is *not* the parabola. Why?

In the next example, we look for an input to minimize a function of the form $y = \sqrt{f(x)}$, where f is a quadratic function with values that are always nonnegative. (There was an example similar to this in Section 4.2.) As indicated in Figure 5, we need only find the input that minimizes the quadratic function $y = f(x)$, because this same input also minimizes $y = \sqrt{f(x)}$. (The *outputs* of the two functions are different; but the same *input* serves to minimize both functions.)

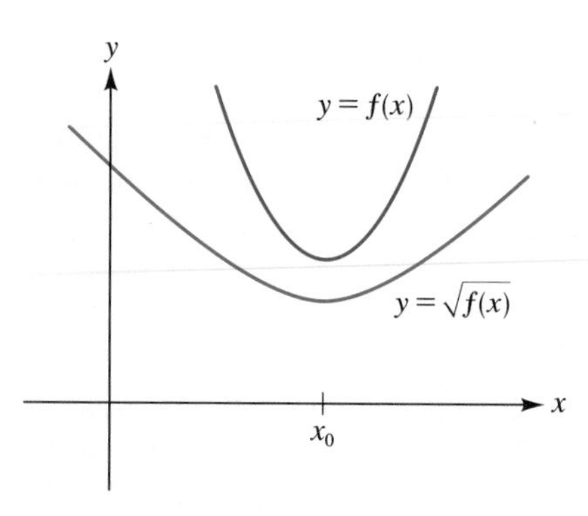

FIGURE 5
The same input x_0 minimizes both functions $y = f(x)$ and $y = \sqrt{f(x)}$ [provided $f(x)$ is nonnegative for all x].

EXAMPLE 4 Which point on the curve $y = \sqrt{x}$ is closest to the point (1, 0)?

Solution In Figure 6, we let D denote the distance from a point (x, y) on the curve to the point (1, 0). We are asked to find out exactly which point (x, y) will make the distance D as small as possible. Using the distance formula, we have

$$\begin{aligned} D &= \sqrt{(x-1)^2 + (y-0)^2} \\ &= \sqrt{x^2 - 2x + 1 + y^2} \end{aligned} \tag{3}$$

This expresses D in terms of both x and y. To express D in terms of x alone, we use the given equation $y = \sqrt{x}$ to substitute for y in equation (3). This yields

$$\begin{aligned} D(x) &= \sqrt{x^2 - 2x + 1 + (\sqrt{x})^2} \\ &= \sqrt{x^2 - x + 1} \end{aligned}$$

Now, according to the remarks just prior to this example, the x-value that minimizes the function $D(x) = \sqrt{x^2 - x + 1}$ is the same x-value that minimizes $y = x^2 - x + 1$. (Actually, to apply those remarks, we need to know that the expression $x^2 - x + 1$ is nonnegative for all inputs x. See Exercise 43 for this.) To determine the x-value that minimizes $y = x^2 - x + 1$, we complete the square, as usual.

$$\begin{aligned} y &= x^2 - x + 1 \\ &= \left(x - \frac{1}{2}\right)^2 + \frac{3}{4} \qquad \text{Check the algebra.} \end{aligned}$$

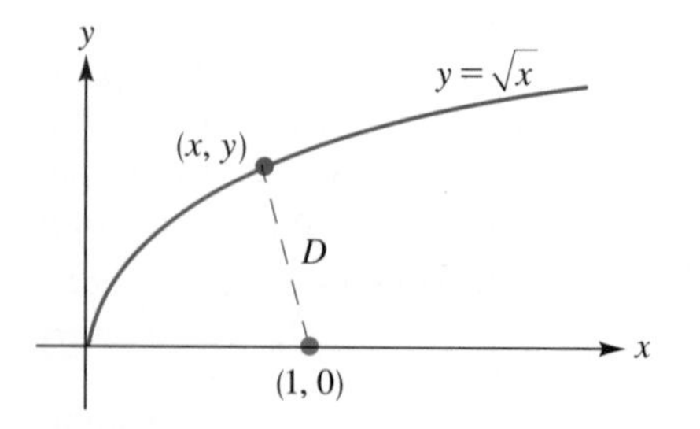

FIGURE 6

From this last equation we conclude that the quantity $x^2 - x + 1$ will be smallest when $x = 1/2$.

Now that we have x, we can find the required point on the curve $y = \sqrt{x}$. [The problem did not ask us to specify the distance $D(x)$, so we needn't calculate it.] Substituting $x = 1/2$ in the equation $y = \sqrt{x}$, we have

$$y = \sqrt{\frac{1}{2}} = \frac{1}{\sqrt{2}} \cdot \frac{\sqrt{2}}{\sqrt{2}} = \frac{\sqrt{2}}{2}$$

Thus, among all points (x, y) on the curve $y = \sqrt{x}$, the point closest to $(1, 0)$ is $(1/2, \sqrt{2}/2)$. This is the required solution.

QUESTION Why would it have made no sense to ask instead for the point *farthest* from $(1, 0)$?

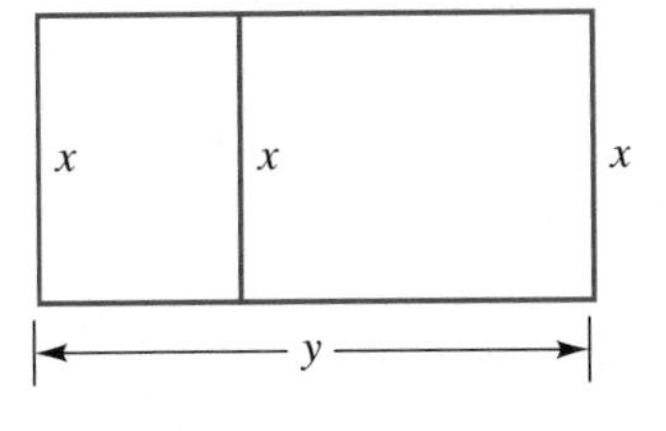

FIGURE 7

EXAMPLE 5 Suppose that you have 600 m of fencing with which to build two adjacent rectangular corrals. The two corrals are to share a common fence on one side, as shown in Figure 7. Find the dimensions x and y so that the total enclosed area is as large as possible.

Solution You have 600 m of fencing to be set up as shown in Figure 7. The question is how to choose x and y so that the total area is maximum. Since the total length of fencing is 600 m, we can relate x and y by writing

$$\begin{aligned} x + x + x + y + y &= 600 \\ 3x + 2y &= 600 \\ 2y &= 600 - 3x \\ y &= 300 - \frac{3}{2}x \end{aligned} \tag{4}$$

Letting A denote the total area, we have

$$\begin{aligned} A &= xy \\ A(x) &= x\left(300 - \frac{3}{2}x\right) \qquad \text{using equation (4) to substitute for } y \\ &= 300x - \frac{3}{2}x^2 \end{aligned}$$

This last equation expresses the area as a function of x. Note that the graph of this function is a parabola opening downward, so it does make sense to talk about a maximum. We complete the square, as usual, to locate the vertex. As you can check, the result is

$$A(x) = -\frac{3}{2}(x - 100)^2 + 15000$$

Thus the x-coordinate of the vertex is 100; this is the x-value that maximizes the area. The corresponding y-value can now be calculated using equation (4):

$$y = 300 - \frac{3}{2}x = 300 - \frac{3}{2}(100) = 150$$

Thus, by choosing x to be 100 m and y to be 150 m, the total area in Figure 7 will be as large as possible. Incidentally, note that the exact location of the fence dividing the two corrals does not influence our work or the final answer.

EXAMPLE 6 Suppose that the following function relates the selling price, p, of an item to the quantity sold, x:

$$p = -\frac{1}{3}x + 40 \qquad (p \text{ in dollars})$$

For which value of x will the corresponding revenue be a maximum? What are the maximum revenue and the unit price in this case?

Solution First, recall the formula for revenue given on page 220.

$$R = \text{number of units} \times \text{price per unit}$$

Using this, we have

$$R = x \cdot p$$

$$R(x) = x\left(-\frac{1}{3}x + 40\right) = -\frac{1}{3}x^2 + 40x \qquad (5)$$

We want to know which value of x yields the largest revenue R. Since the graph of the revenue function in equation (5) is a parabola opening downward, the required x-value is the x-coordinate of the vertex of the parabola. We complete the square to locate the vertex:

$$\begin{aligned} R(x) &= -\frac{1}{3}(x^2 - 120x \quad) \\ &= -\frac{1}{3}(x^2 - 120x + 3600) + 1200 \qquad \text{adding } 0\ (= -1200 + 1200) \\ &= -\frac{1}{3}(x - 60)^2 + 1200 \end{aligned}$$

The vertex is therefore (60, 1200). Consequently, the revenue is a maximum when $x = 60$ items, and this maximum revenue is \$1200. For the corresponding unit price, we have

$$p(x) = -\frac{1}{3}x + 40$$

$$p(60) = -\frac{1}{3}(60) + 40 = 20$$

Thus the unit price corresponding to the maximum revenue is \$20.

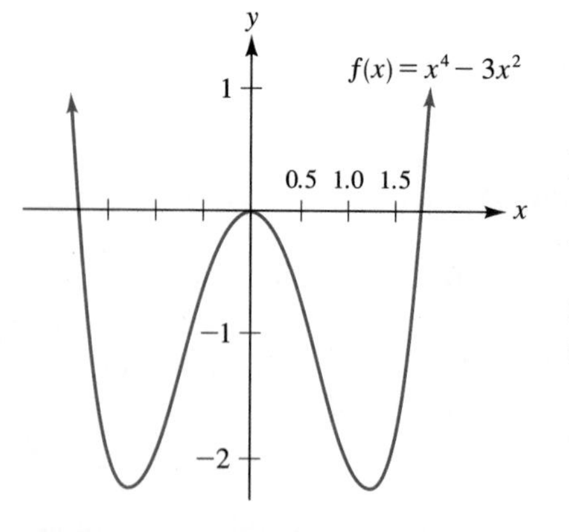

FIGURE 8

EXAMPLE 7 Figure 8 shows a graph of the function

$$f(x) = x^4 - 3x^2$$

From the graph, it is clear that the minimum of the function is slightly less than -2. Find the exact value of this minimum and the corresponding x-values at which it occurs.

Solution Although this is not a quadratic function, we can nevertheless use the technique of completing the square in this particular case. We have

$$\begin{aligned} f(x) &= x^4 - 3x^2 + \frac{9}{4} - \frac{9}{4} \\ &= \left(x^2 - \frac{3}{2}\right)^2 - \frac{9}{4} \\ &= -\frac{9}{4} + \left(x^2 - \frac{3}{2}\right)^2 \end{aligned}$$

From this last equation we see that $f(x)$ is never less than $-9/4$ (because the quantity being added to $-9/4$ is nonnegative). Furthermore, $f(x)$ does attain the value $-9/4$ when $x^2 = 3/2$; that is, when $x = \pm\sqrt{3/2}$. Thus the minimum value of f is $-9/4$ and the corresponding x-values are

$$x = \pm\frac{\sqrt{3}}{\sqrt{2}} = \pm\frac{\sqrt{6}}{2} \quad \text{rationalizing the denominator}$$

NOTE Use your calculator to approximate these x-values, and then check that they are consistent with the graph in Figure 8.

EXERCISE SET 4.5

Do 1, 3, 11, 17, 5

A

1. Two numbers add to 5. What is the largest possible value of their product?
2. Find two numbers adding to 20 such that the sum of their squares is as small as possible.
3. The difference of two numbers is 1. What is the smallest possible value for the sum of their squares?
4. For each quadratic function, state whether it would make sense to look for a highest or a lowest point on the graph. Then determine the coordinates of that point.
   (a) $y = 2x^2 - 8x + 1$ **(b)** $y = -3x^2 - 4x - 9$
   (c) $h = -16t^2 + 256t$ **(d)** $f(x) = 1 - (x + 1)^2$
   (e) $g(t) = t^2 + 1$ **(f)** $f(x) = 1000x^2 - x + 100$
5. Among all rectangles having a perimeter of 25 m, find the dimensions of the one with the largest area.
6. What is the largest possible area for a rectangle with a perimeter of 80 cm?
7. What is the largest possible area for a right triangle in which the sum of the lengths of the two shorter sides is 100 in.?
8. The perimeter of a rectangle is 12 m. Find the dimensions for which the diagonal is as short as possible.
9. Two numbers add to 6.
   (a) Let T denote the sum of the squares of the two numbers. What is the smallest possible value for T?
   (b) Let S denote the sum of the first number and the square of the second. What is the smallest possible value for S?
   (c) Let U denote the sum of the first number and twice the square of the second number. What is the smallest possible value for U?
   (d) Let V denote the sum of the first number and the square of twice the second number. What is the smallest possible value for V?
10. Suppose that the height of an object shot straight up is given by $h = 512t - 16t^2$. (Here h is in feet and t is in seconds.) Find the maximum height and the time at which the object hits the ground.
11. A baseball is thrown straight up, and its height as a function of time is given by the formula $h = -16t^2 + 32t$ (where h is in feet and t is in seconds).
    (a) Find the height of the ball when $t = 1$ sec and when $t = 3/2$ sec.
    (b) Find the maximum height of the ball and the time at which that height is attained.
    (c) At what times is the height 7 ft?
12. Find the point on the curve $y = \sqrt{x}$ that is nearest to the point (3, 0).
13. Which point on the curve $y = \sqrt{x - 2} + 1$ is closest to the point (4, 1)? What is this minimum distance?
14. Find the coordinates of the point on the line $y = 3x + 1$ closest to (4, 0).
15. **(a)** What number exceeds its square by the greatest amount?
    (b) What number exceeds twice its square by the greatest amount?
16. Suppose that you have 1800 m of fencing with which to build three adjacent rectangular corrals, as shown in the figure. Find the dimensions so that the total enclosed area is as large as possible.

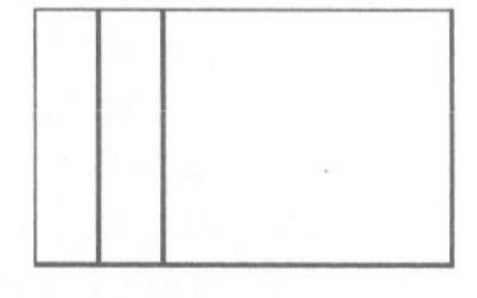

17. Five hundred feet of fencing is available for a rectangular pasture alongside a river, the river serving as one

side of the rectangle (so only three sides require fencing). Find the dimensions yielding the greatest area.

18. Let $A = 3x^2 + 4x - 5$ and $B = x^2 - 4x - 1$. Find the minimum value of $A - B$.

19. Let $R = 0.4x^2 + 10x + 5$ and $C = 0.5x^2 + 2x + 101$. For which value of x is $R - C$ a maximum?

20. Suppose that the revenue generated by selling x units of a certain commodity is given by $R = -\frac{1}{5}x^2 + 200x$. Assume that R is in dollars. What is the maximum revenue possible in this situation?

21. Suppose that the function $p = -\frac{1}{4}x + 30$ relates the selling price p of an item to the number of units x that are sold. Assume that p is in dollars. For which value of x will the corresponding revenue be a maximum? What is this maximum revenue and what is the unit price?

22. The action of sunlight on automobile exhaust produces air pollutants known as *photochemical oxidants.* In a study of cross-country runners in Los Angeles, it was shown that running performances can be adversely affected when the oxidant level reaches 0.03 part per million. Suppose that on a given day, the oxidant level L is approximated by the formula

$$L = 0.059t^2 - 0.354t + 0.557 \qquad (0 \leq t \leq 7)$$

where t is measured in hours, with $t = 0$ corresponding to 12 noon, and L is in parts per million. At what time is the oxidant level L a minimum? At this time, is the oxidant level high enough to affect a runner's performance?

23. **(a)** Find the smallest possible value of the quantity $x^2 + y^2$ under the restriction that $2x + 3y = 6$.
(b) Find the radius of the circle whose center is at the origin and that is tangent to the line $2x + 3y = 6$. How does this answer relate to your answer in part (a)?

B

24. Through a type of chemical reaction known as *autocatalysis,* the human body produces the enzyme trypsin from the enzyme trypsinogen. (Trypsin then breaks down proteins into amino acids, which the body needs for growth.) Let r denote the rate of the chemical reaction in which trypsin is formed from trypsinogen. It has been shown experimentally that $r = kx(a - x)$, where k is a positive constant, a is the initial amount of trypsinogen, and x is the amount of trypsin produced (so x increases as the reaction proceeds). Show that the reaction rate r is a maximum when $x = a/2$. In other words, the speed of the reaction is greatest when the amount of trypsin formed is half the original amount of trypsinogen.

25. **(a)** Let $x + y = 15$. Find the minimum value of the quantity $x^2 + y^2$.
(b) Let C be a constant and $x + y = C$. Show that the minimum value of $x^2 + y^2$ is $C^2/2$. Then use this result to check your answer in part (a).

26. Suppose that A, B, and C are positive constants and that $x + y = C$. Show that the minimum value of $Ax^2 + By^2$ occurs when $x = BC/(A + B)$ and $y = AC/(A + B)$.

27. The following figure shows a square inscribed within a unit square. For which value of x is the area of the inner square a minimum? What is the minimum area? *Hint:* Denote the lengths of the two segments that make up the base of the unit square by t and $1 - t$. Now use the Pythagorean theorem and congruent triangles to express x in terms of t.

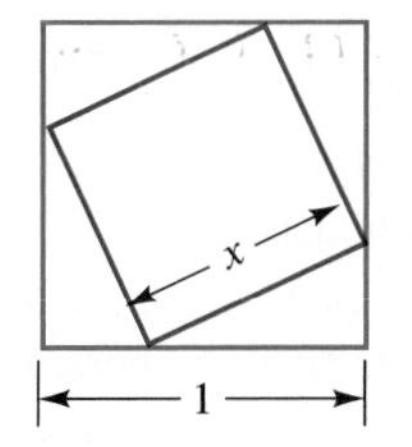

28. **(a)** Find the coordinates of the point on the line $y = mx + b$ that is closest to the origin

Answer: $\left(\dfrac{-mb}{1+m^2}, \dfrac{b}{1+m^2}\right)$

(b) Show that the perpendicular distance from the origin to the line $y = mx + b$ is $|b|/\sqrt{1+m^2}$. *Suggestion:* Use the result in part (a).
(c) Use part (b) to show that the perpendicular distance from the origin to the line $Ax + By + C = 0$ is $|C|/\sqrt{A^2+B^2}$.

29. The point P lies in the first quadrant on the graph of the line $y = 7 - 3x$. From the point P, perpendiculars are drawn to both the x-axis and the y-axis. What is the largest possible area for the rectangle thus formed?

30. Show that the largest possible area for the shaded rectangle shown in the figure is $-b^2/4m$. Then use this to check your answer to Exercise 29.

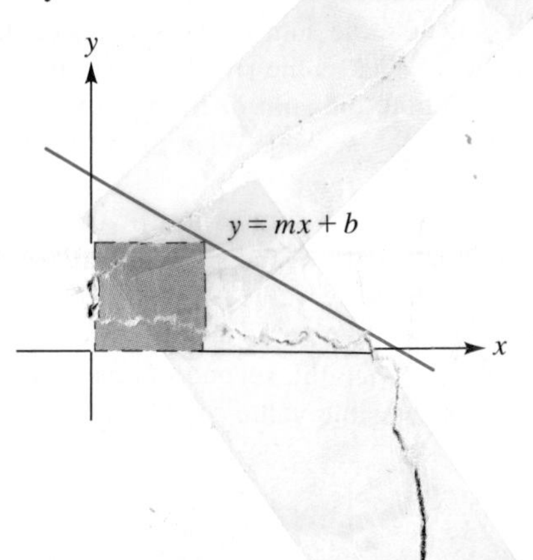

31. Show that the maximum possible area for a rectangle inscribed in a circle of radius R is $2R^2$.
Hint: Maximize the square of the area.

32. An athletic field with a perimeter of $\frac{1}{4}$ mile consists of a rectangle with a semicircle at each end, as shown in the figure. Find the dimensions x and r that yield the greatest possible area for the rectangular region.

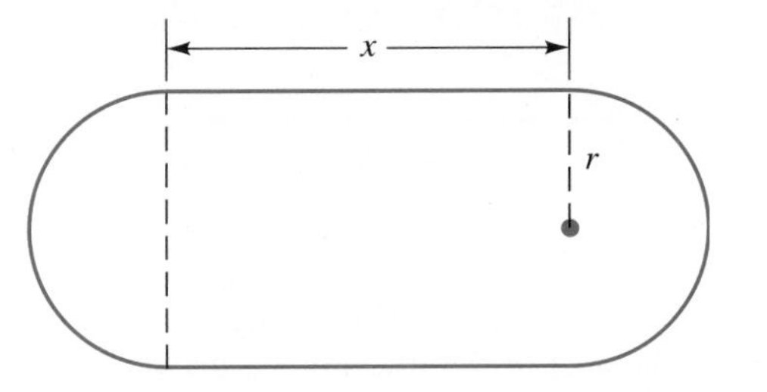

33. A rancher who wishes to fence off a rectangular area finds that the fencing in the east-west direction will require extra reinforcement owing to strong prevailing winds. Because of this, the cost of fencing in the east-west direction will be \$12 per (linear) yard, as opposed to a cost of \$8 per yard for fencing in the north-south direction. Find the dimensions of the largest possible rectangular area that can be fenced for \$4800.

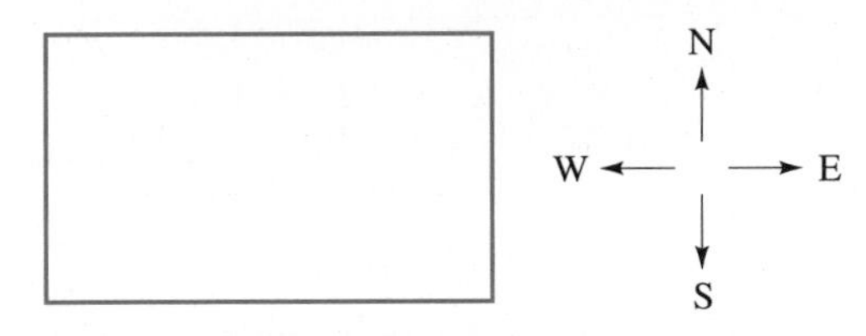

34. Let $f(x) = (x-a)^2 + (x-b)^2 + (x-c)^2$, where a, b, and c are constants. Show that $f(x)$ will be a minimum when x is the average of a, b, and c.

35. Let $y = a_1(x-x_1)^2 + a_2(x-x_2)^2$, where a_1, a_2, x_1, and x_2 are all constants. In addition, suppose that a_1 and a_2 are both positive. Show that the minimum of this function occurs when

$$x = \frac{a_1x_1 + a_2x_2}{a_1 + a_2}$$

36. Among all rectangles with a given perimeter P, find the dimensions of the one with the shortest diagonal.

37. By analyzing sales figures, the economist for a stereo manufacturer knows that 150 units of a compact disc player can be sold each month when the price is set at $p = \$200$ per unit. The figures also show that for each \$10 hike in price, five fewer units are sold each month.
(a) Let x denote the number of units sold per month and let p denote the price per unit. Find a linear function relating p and x.
Hint: $\Delta p/\Delta x = 10/(-5) = -2$
(b) The revenue R is given by $R = xp$. What is the maximum revenue? At what level should the price be set to achieve this maximum revenue?

38. Let $f(x) = x^2 + px + q$, and suppose that the minimum value of this function is 0. Show that $q = p^2/4$.

39. For which numbers t will the value of the function $f(t) = t^2 - t^4$ be as large as possible?

40. Find the minimum value of the function $f(t) = t^4 - 8t^2$.

41. Among all possible inputs for the function $f(t) = -t^4 + 6t^2 - 6$, which ones yield the largest output?

42. Let $f(x) = x - 3$ and $g(x) = x^2 - 4x + 1$.
(a) Find the minimum value of $g \circ f$.
(b) Find the minimum value of $f \circ g$.
(c) Are the results in parts (a) and (b) the same?

43. This exercise completes a detail mentioned in Example 4. In that example, we used the result that the input minimizing the function $D(x) = \sqrt{x^2 - x + 1}$ is the same input that minimizes the function $y = x^2 - x + 1$. For this result to be applicable, we need to know that the quantity $x^2 - x + 1$ is nonnegative for all x. Parts (a) and (b) each suggest a way to verify that this is indeed the case. [Actually, parts (a) and (b) both show that the quantity $x^2 - x + 1$ is positive for all x.]
(a) Use the fact that $x^2 - x + 1$ can be written as $\left(x - \frac{1}{2}\right)^2 + \frac{3}{4}$ to explain (in complete sentences) why the quantity $x^2 - x + 1$ is positive for all inputs x.
(b) Compute the discriminant of the quadratic $x^2 - x + 1$ and explain why the graph of $y = x^2 - x + 1$ has no x-intercepts. Then, use this fact to explain why the quantity $x^2 - x + 1$ is always positive. *Hint:* Is the graph of the parabola $y = x^2 - x + 1$ U-shaped up or U-shaped down?

44. Let $f(x) = (x-1)^2 - 4$.
(a) Sketch the graph of the function f and note that the minimum value occurs when x is 1.
(b) Although the minimum value of the function f occurs when $x = 1$, the minimum value of the function $y = \sqrt{f(x)}$ does not occur when $x = 1$. Explain why this does not contradict the statement in the caption to Figure 5 on page 230.
(c) Find the value(s) of x that minimize the function $y = \sqrt{f(x)}$. *Hint:* What is the domain of $y = \sqrt{f(x)}$?

In Exercises 45–52, use the following result to lessen the amount of computation required: The x-coordinate of the vertex of the parabola $y = ax^2 + bx + c$ is given by the formula $x = -b/(2a)$. (As indicated in Exercise 45 in Section 4.2, this result can be derived by completing the square.)

45. A piece of wire 16 in. long is to be cut into two pieces. Let x denote the length of the first piece and $16 - x$ the

length of the second. The first piece is to be bent into a circle and the second piece into a square.
(a) Express the total combined area A of the circle and the square as a function of x.
(b) For which value of x is the area A a minimum?
(c) Using the x-value that you found in part (b), find the ratio of the lengths of the shorter to the longer piece of wire. *Answer:* $\pi/4$

46. A 30-inch piece of string is to be cut into two pieces. The first piece will be formed into the shape of an equilateral triangle and the second piece into a square. Find the length of the first piece if the combined area of the triangle and the square is to be as small as possible.

C

47. Repeat Exercise 45, but assume that the length of the wire is L inches.

48. Let $f(x) = x^2 + bx + 1$. Find a positive value for b such that the distance from the origin to the vertex of the parabola is as small as possible.

49. The next figure shows a rectangle inscribed in a given triangle of base b and height h. Find the ratio of the area of the triangle to the area of the rectangle when the area of the rectangle is maximum.

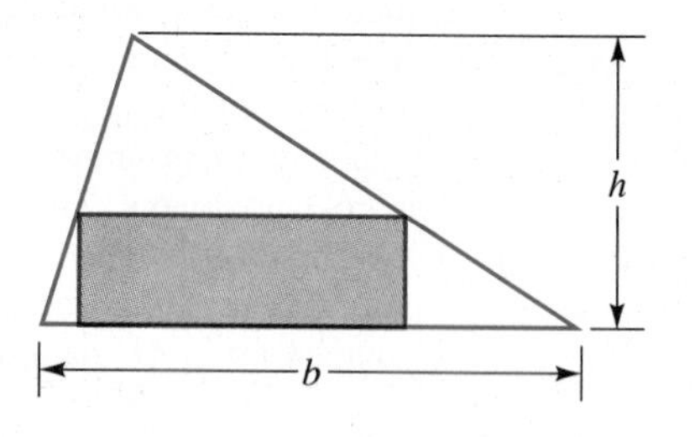

50. A Norman window is in the shape of a rectangle surmounted by a semicircle, as shown in the figure. Assume that the perimeter of the window is P, a constant. Show that the area of the window is a maximum when both x and r are equal to $P/(\pi + 4)$. Show that this maximum area is $\frac{1}{2}P^2/(\pi + 4)$.

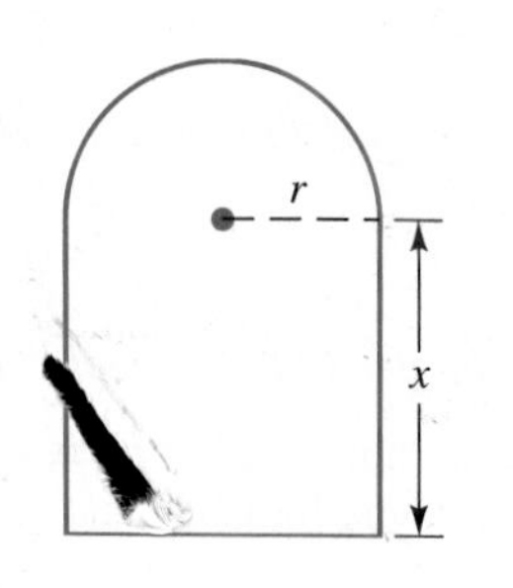

51. A triangle is inscribed in a semicircle of diameter $2R$, as shown in the figure. Show that the smallest possible value for the area of the shaded region is $(\pi - 2)R^2/2$.

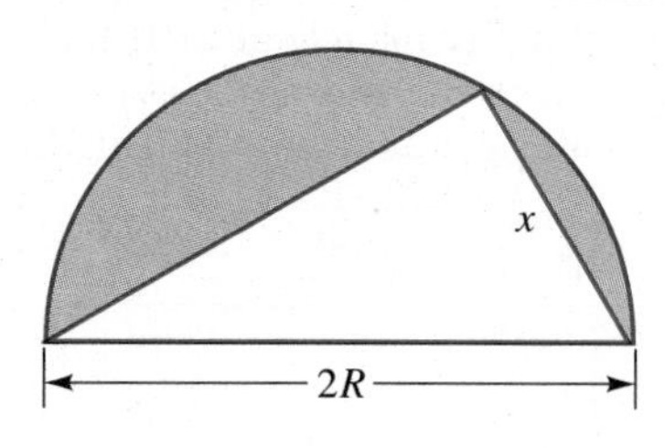

Hint: The area of the shaded region is a minimum when the area of the triangle is a maximum. Find the value of x that maximizes the *square* of the area of the triangle. This will be the same x that maximizes the area of the triangle.

52. **(a)** Complete the following table. Which x-y pair in the table yields the smallest sum $x + y$?

x	0.5	1	1.5	2	2.5	3	3.5
y							
xy	12	12	12	12	12	12	12
$x + y$							

(b) Find two positive numbers with a product of 12 and as small a sum as possible.
Hint: The quantity that you need to minimize is $x + (12/x)$, where $x > 0$. But

$$x + \frac{12}{x} = \left(\sqrt{x} - \sqrt{\frac{12}{x}}\right)^2 + 2\sqrt{12}$$

This last expression is minimized when the quantity within parentheses is zero. Why?
(c) Use a calculator to verify that the two numbers obtained in part (b) produce a sum that is smaller than any of the sums obtained in part (a).

53. What is the smallest possible value for the sum of a positive number and its reciprocal? *Hint:* After setting up the appropriate function, adapt the hint given in Exercise 52(b).

54. Let $f(x) = \dfrac{2x^2 - 4x + 1}{2x}$ where $x > 0$.
(a) Complete the following two tables. In both cases, specify the x-value that yields the smallest output. (Round the outputs in the tables to six decimal places.)

x	0.4	0.5	0.6	0.7	0.8	0.9
$f(x)$						

x	0.68	0.69	0.70	0.71	0.72	0.73
$f(x)$						

(b) Among all positive inputs for the given function f, find the one for which $f(x)$ is a minimum.

Hint: Write $f(x)$ as $\left(x + \dfrac{1}{2x}\right) - 2$, and then adapt the hint given in Exercise 52(b).

(c) Use a calculator to verify that the x-value obtained in part (b) produces an output that is smaller than any of the outputs obtained in part (a).

55. Let $G(x) = (a + x)(b + x)/x$, where a and b are positive constants and $x > 0$. Show that the minimum value of G is $(\sqrt{a} + \sqrt{b})^2$ and that this minimum value occurs when $x = \sqrt{ab}$.

Hint: Show that $G(x) = (a + b) + \left(\dfrac{ab}{x} + x\right)$, then adapt the hint in Exercise 52(b).

56. A line with a positive slope m passes through the point $(-2, 1)$.

(a) Show that the area of the triangle bounded by this line and the axes is $\left(2m + \dfrac{1}{2m}\right) + 2$.

(b) For which value of m will the area be a minimum? *Hint:* Adapt the hint given in Exercise 52(b).

57. Let $f(x) = 1/(2x^2 - x + 1)$.

(a) Experiment with your calculator; what is the largest output you can find for the function f?

(b) Prove that for every value of x, we have $f(x) \leq 8/7$.

58. Find the largest value of the function $f(x) = 1/(x^4 + 2x^2 + 1)$.

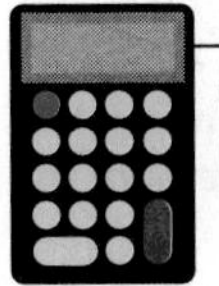

GRAPHING UTILITY EXERCISES FOR SECTION 4.5

Exercises 1–12 are maximum-minimum problems quoted from the following calculus textbooks with permission of the publishers.

EXERCISES	TEXT
1, 2	*Calculus,* 5th ed., Howard Anton (New York: John Wiley & Sons, Inc., 1995)
3, 4	*Calculus,* 5th ed., Roland E. Larson, Robert P. Hostetler, Bruce H. Edwards (Lexington, Mass: D.C. Heath and Co., 1994)
5, 6	*Calculus and Analytic Geometry,* 5th ed., Sherman K. Stein, Anthony Barcellos (New York: McGraw-Hill, Inc., 1992)
7, 8	*Calculus with Analytic Geometry,* 4th ed., C.H. Edwards, Jr., David E. Penney (Englewood Cliffs, N.J.: Prentice-Hall, 1994)
9, 10	*Calculus: One and Several Variables,* 7th ed., Saturnino L. Salas, Einar Hille, revised by Garret J. Etgen (New York: John Wiley & Sons, Inc., 1995)
11, 12	*Calculus,* 3rd ed., James Stewart (Pacific Grove, Calif.: Brooks/Cole Publishing Co. 1995)

For Exercises 1–12: **(a)** *first set up the appropriate function to be maximized or minimized, just as you have practiced in Sections 4.4 and 4.5;* **(b)** *use a graphing utility to graph the function and to determine (at least approximately) the number or numbers that the problem asks for; and* **(c)** *if the function obtained in part (a) is a quadratic or is closely related to a quadratic, use the techniques from Section 4.5 to solve the problem and thereby check your answer in part (b).*

1. A rectangular area of 3200 ft^2 is to be fenced off. Two opposite sides will use fencing costing \$1 per foot and the remaining sides will use fencing costing \$2 per foot. Find the dimensions of the rectangle of least cost.

2. A cylindrical can, open at the top, is to hold 500 cm^3 of liquid. Find the height and radius that minimize the amount of material needed to manufacture the can.

3. A page is to contain 30 square inches of print. The margins on each side are 1 inch. Find the dimensions of the page so that the least paper is used.

4. A power station is on one side of a river that is 1/2 mile wide, and a factory is 6 miles downstream on the other side. It costs \$6 per foot to run power lines overland and \$8 per foot to run them underwater. Find the most economical path for the transmission line from the power station to the factory.

5. What point on the parabola $y = x^2$ is closest to the point (3, 0)?

6. **(a)** How should one choose two nonnegative numbers whose sum is 1 in order to maximize the sum of their squares?
(b) To minimize the sum of their squares?

7. A rectangle of fixed perimeter 36 is rotated about one of its sides, thus sweeping out a figure in the shape of a right circular cylinder. [Refer to the accompanying figure.] What is the maximum possible volume of that cylinder?

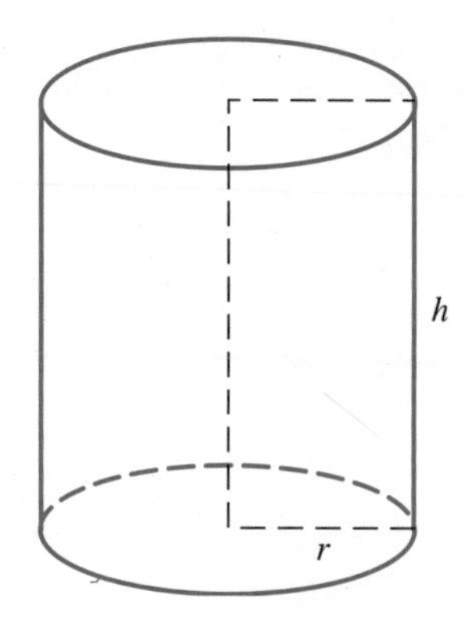

8. A farmer has 600 yd of fencing with which to build a rectangular corral. Some of the fencing will be used to construct two internal divider fences, both parallel to the same two sides of the corral. [Refer to the accompanying figure.] What is the maximum possible total area of such a corral?

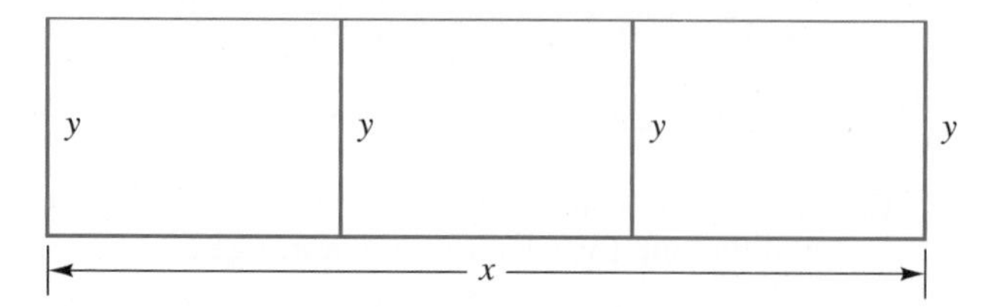

9. A rectangular warehouse will have 5000 square feet of floor space and will be separated into two rectangular rooms by an interior wall. The cost of the exterior walls is \$150 per linear foot and the cost of the interior wall is \$100 per linear foot. Find the dimensions that will minimize the cost of building the warehouse.

10. Let ABC be a triangle with vertices $A = (-3, 0)$, $B = (0, 6)$, $C = (3, 0)$. Let P be a point on the line segment that joins B to the origin. Find the position of P that minimizes the sum of the distances between P and the vertices.

11. A Norman window has the shape of a rectangle surmounted by a semicircle. (Thus the diameter of the semicircle is equal to the width of the rectangle.) If the perimeter of the window is 30 ft, find the dimensions of the window so that the greatest possible amount of light is admitted.

12. A fence 8 ft tall runs parallel to a tall building at a distance of 4 ft from the building. What is the length of the shortest ladder that will reach from the ground over the fence to the wall of the building?

4.6 POLYNOMIAL FUNCTIONS

We can rephrase the definitions of linear and quadratic functions using the terminology for polynomials given in Section 1.3.

A function f is **linear** if $f(x)$ is a **polynomial of degree 1:**

$$f(x) = a_1x + a_0 \tag{1}$$

A function f is **quadratic** if $f(x)$ is a **polynomial of degree 2:**

$$f(x) = a_2x^2 + a_1x + a_0 \tag{2}$$

In view of equation (1), linear functions are sometimes called **polynomial functions of degree 1.** Similarly, in view of equation (2), quadratic functions are **polynomial functions of degree 2.** More generally, by a **polynomial function of degree *n*,** we mean a function defined by an equation of the form

$$f(x) = a_nx^n + a_{n-1}x^{n-1} + \cdots + a_1x + a_0 \tag{3}$$

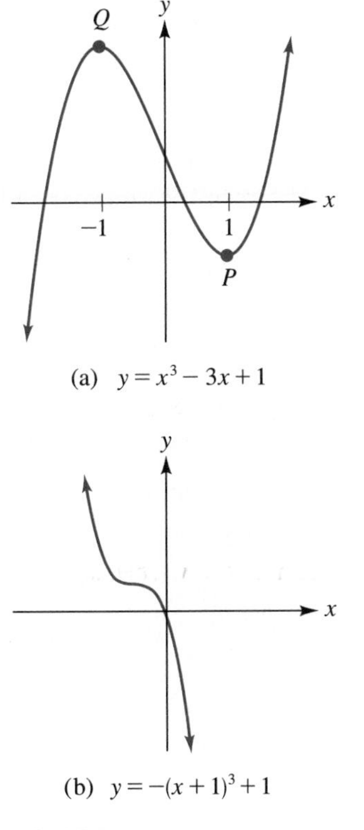

FIGURE 1

where n is a nonnegative integer and $a_n \neq 0$. Throughout the remainder of this chapter, the coefficients a_k will always be real numbers.

In principle, we can obtain the graph of any polynomial function by setting up a table and plotting a sufficient number of points. Indeed, this is just the way a graphing calculator or computer operates. However, to *understand* why the graphs look as they do, we want to discuss some additional methods for graphing polynomial functions.

There are three facts that we shall need. By way of example, look at the graphs of the polynomial functions in Figure 1. First, notice that both graphs are unbroken, smooth curves with no "corners." As is shown in calculus, this is true for the graph of every polynomial function. In contrast, the graphs in Figures 2 and 3 cannot represent polynomial functions. The graph in Figure 2 has a break in it, and the graph in Figure 3 has what's called a **cusp.**

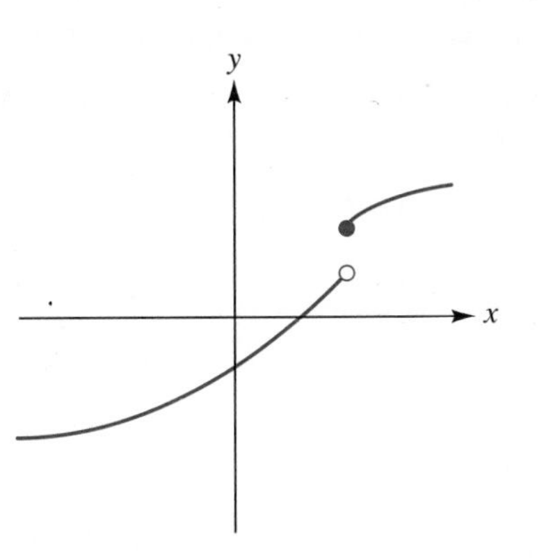

FIGURE 2
Since the graph has a break, it cannot represent a polynomial function.

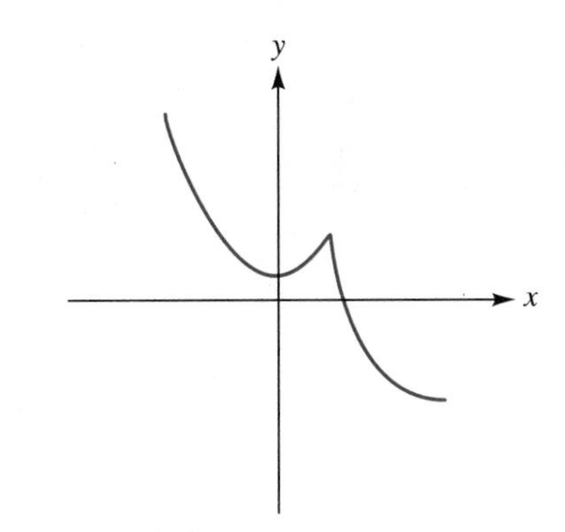

FIGURE 3
Since the graph has a cusp, it cannot represent a polynomial function.

Now look back at the graph in Figure 1(a). Recall (from Section 3.2) that points such as P and Q are called **turning points.** These are points where the graph changes from rising to falling, or vice versa. It is a fact (proved in calculus) that the graph of a polynomial function of degree n has *at most* $n - 1$ turning points. For instance, in Figure 1(a), there are two turning points, while the degree of the polynomial is 3. However, as Figure 1(b) indicates, we needn't have any turning points at all.

A third property of polynomial functions concerns their behavior when $|x|$ is very large. We'll illustrate this property using the function $y = x^3 - 3x + 1$, graphed in Figure 1(a). Now, in Figure 1(a), the x-values are relatively small; for instance, the x-coordinates of P and Q are 1 and -1, respectively. In Figure 4, however, we show the graph of this same function using units of 100 on the x-axis. On this scale, the graph appears indistinguishable from that of $y = x^3$. In particular, note that as $|x|$ gets very large, $|y|$ grows very large.

It's easy to see why the function $y = x^3 - 3x + 1$ resembles $y = x^3$ when $|x|$ is very large. First, let's rewrite the equation $y = x^3 - 3x + 1$ as

$$y = x^3\left(1 - \frac{3}{x^2} + \frac{1}{x^3}\right)$$

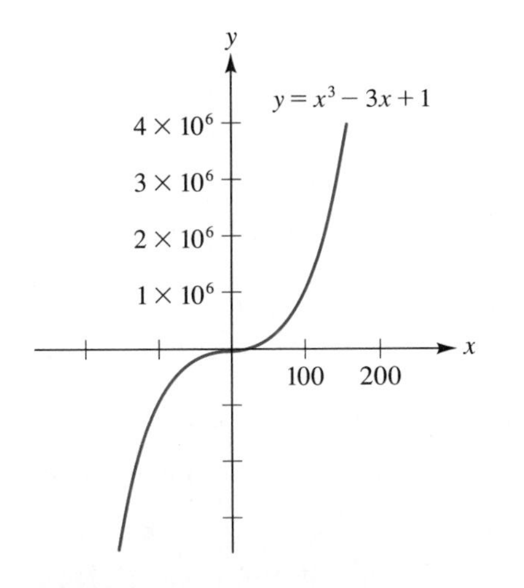

FIGURE 4
When $|x|$ is very large, the graph of $y = x^3 - 3x + 1$ appears indistinguishable from that of $y = x^3$.

Now, when $|x|$ is very large, both $3/x^2$ and $1/x^3$ are close to zero. So we have

$$y \approx x^3(1 - 0 + 0)$$
$$\approx x^3 \qquad \text{when } |x| \text{ is very large}$$

The same technique that we've just used in analyzing $y = x^3 - 3x + 1$ can be applied to any (nonconstant) polynomial function. The result is summarized in item 3 in the following box.

PROPERTY SUMMARY **GRAPHS OF POLYNOMIAL FUNCTIONS**

1. The graph of a polynomial function of degree 2 or greater is an unbroken smooth curve. (For degrees 1 and 0, the graph is a line.)
2. The graph of a polynomial function of degree n has at most $n - 1$ turning points.
3. For the graph of any polynomial function (other than a constant function), as $|x|$ gets very large, $|y|$ grows very large. If

$$f(x) = a_n x^n + a_{n-1}x^{n-1} + \cdots + a_1 x + a_0 \qquad (a_n \neq 0)$$

then

$$f(x) \approx a_n x^n \qquad \text{when } |x| \text{ is very large.}$$

EXAMPLE 1 A function f is defined by

$$f(x) = -x^3 + x^2 + 9x + 9$$

Which of the graphs in Figure 5 might represent this function?

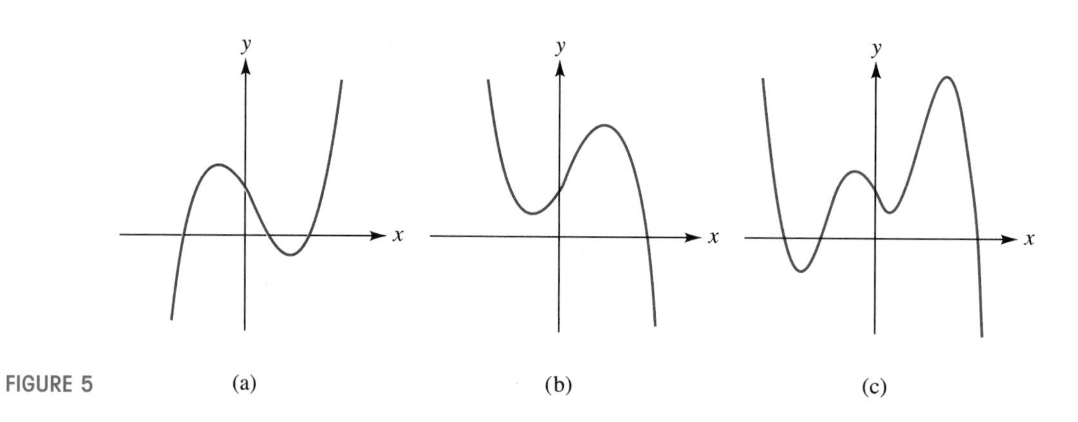

FIGURE 5

Solution When $|x|$ is very large, $f(x) \approx -x^3$. This rules out the graph in Figure 5(a). The graph in Figure 5(c) can also be ruled out, but for a different reason. That graph has four turning points, whereas the graph of the cubic function f can have at most two turning points. The graph in Figure 5(b), on the other hand, does have two turning points; furthermore, that graph does behave like $y = -x^3$ when $|x|$ is very large. The graph in Figure 5(b) might be (in fact, it is) the graph of the given function f.

The simplest polynomial functions to graph are those of the form $y = x^n$. From earlier work, we are already familiar with the graphs of $y = x$, $y = x^2$, and $y = x^3$. For reference, these are shown in Figure 6.

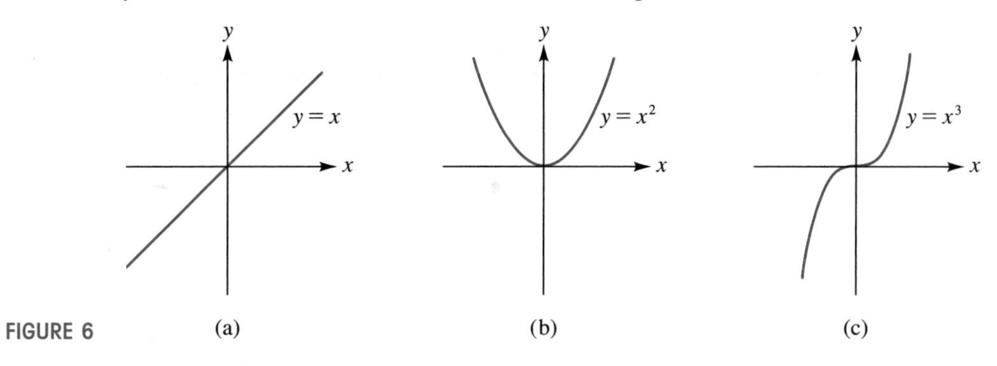

FIGURE 6

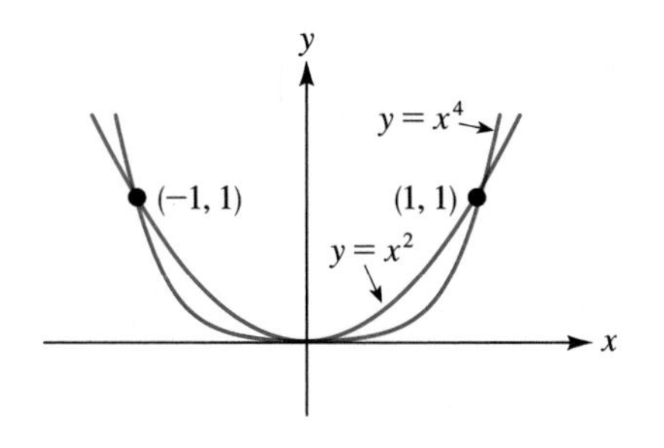

FIGURE 7

When n is greater than 3, the graph of $y = x^n$ resembles the graph of $y = x^2$ or $y = x^3$, depending on whether n is even or odd. Consider, for instance, the graph of $y = x^4$, shown in Figure 7 along with the graph of $y = x^2$. Just as with $y = x^2$, the graph of $y = x^4$ is a symmetric, U-shaped curve passing through the three points $(-1, 1)$, $(1, 1)$, and $(0, 0)$. However, in the interval $-1 < x < 1$, the graph of $y = x^4$ is flatter than that of $y = x^2$. Similarly, the graph of $y = x^6$ in this interval would be flatter still. The data in Table 1 show why this is true. Figure 8 displays the graphs of $y = x^2$, $y = x^4$, and $y = x^6$ for the interval $0 \leq x \leq 1$.

TABLE 1

x	0.2	0.4	0.6	0.8	1.0
x^2	0.04	0.16	0.36	0.64	1.0
x^4	0.0016	0.0256	0.1296	0.4096	1.0
x^6	0.000064	0.004096	0.046656	0.262144	1.0

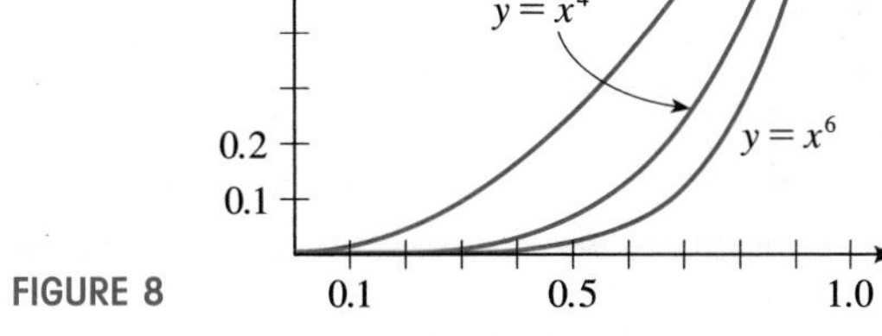

FIGURE 8

Incidentally, Figure 8 indicates one of the practical difficulties you may encounter in trying to draw an accurate graph of $y = x^n$. Suppose, for instance, that you want to graph $y = x^6$, and the lines you draw are 0.01 cm thick. Also suppose that you use the same scale on both axes, taking the common unit to be 1 cm. Then, in the first quadrant, your graph of $y = x^6$ will be indistinguishable from the x-axis when $x^6 < 0.01$, or $x < \sqrt[6]{0.01} \approx 0.46$ cm (using a calculator). This explains why sections of the graphs in Figure 8 appear horizontal.

For $|x| > 1$, the graph of $y = x^4$ rises more rapidly than that of $y = x^2$. Similarly, the graph of $y = x^6$ rises still more rapidly. This is shown in Figures 9(a) and (b). (Note the different scales used on the y-axes in the two figures.)

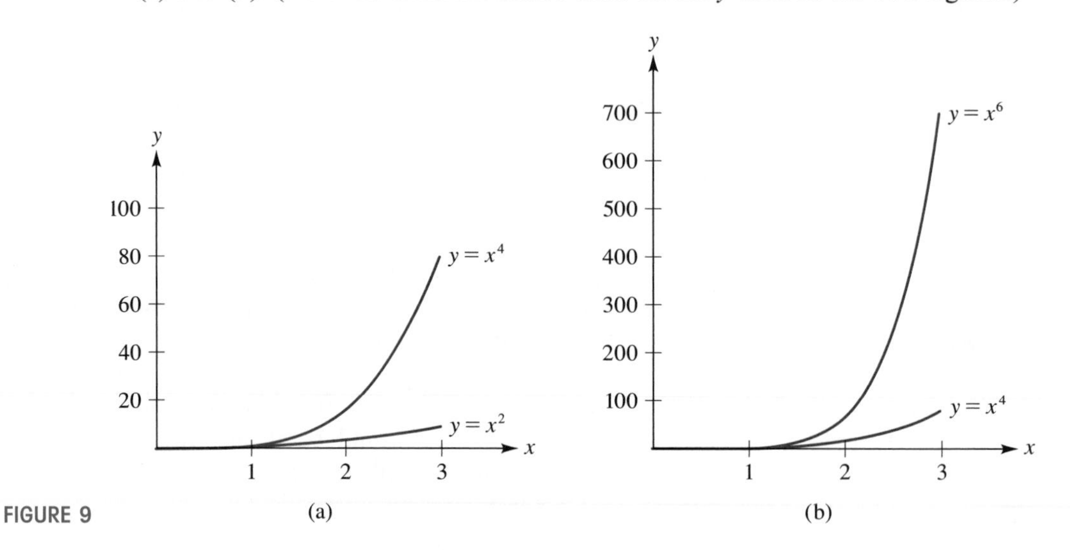

FIGURE 9

As mentioned earlier, when n is odd, the graph of $y = x^n$ resembles that of $y = x^3$. In Figure 10 we compare the graphs of $y = x^3$ and $y = x^5$. Notice that both curves pass through (0, 0), (1, 1), and $(-1, -1)$. For reasons similar to those explained for even n, the graph of $y = x^5$ is flatter than that of $y = x^3$ in the interval $-1 < x < 1$, and the graph of $y = x^7$ is flatter still. For $|x| > 1$, the graph of $y = x^5$ is steeper than that of $y = x^3$, and $y = x^7$ is steeper still.

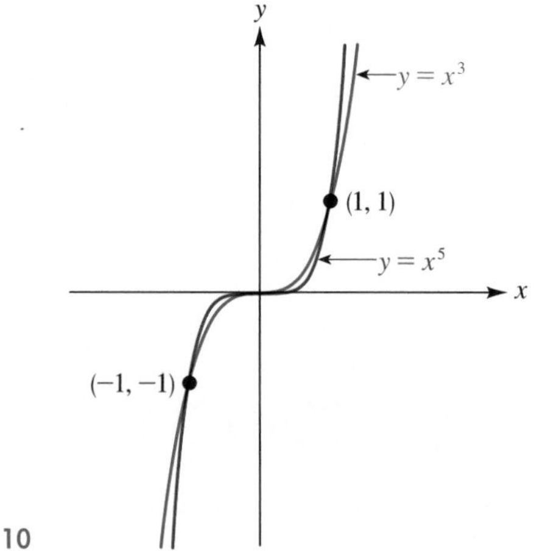

FIGURE 10

In Section 4.2, we observed the effect of the constant a on the graph of $y = ax^2$. Those same comments apply to the graph of $y = ax^n$. For instance, the graph of $y = \frac{1}{2}x^4$ is wider than that of $y = x^4$, while the graph of $y = -\frac{1}{2}x^4$ is obtained by reflecting $y = \frac{1}{2}x^4$ in the x-axis.

EXAMPLE 2 Sketch the graph of $y = (x+2)^5$ and specify the y-intercept.

Solution The graph of $y = (x+2)^5$ is obtained by moving the graph of $y = x^5$ two units to the left. As a guide to drawing the curve, we recall that $y = x^5$ passes through the points $(1, 1)$ and $(-1, -1)$. Thus, $y = (x+2)^5$ must pass through $(-1, 1)$ and $(-3, -1)$, as shown in Figure 11.

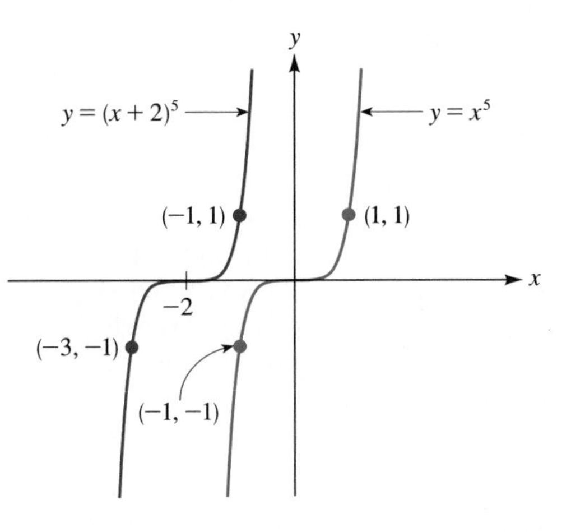

FIGURE 11

Although the curve rises and falls very sharply, it is important to realize that it is never really vertical. For instance, the curve eventually crosses the y-axis. To find the y-intercept, we set $x = 0$ to obtain $y = 2^5 = 32$. Thus, the y-intercept is 32.

EXAMPLE 3 Graph the function $y = -2(x-3)^4$.

Solution We begin with the graph of $y = -2x^4$, in Figure 12(a). The points $(1, -2)$ and $(-1, -2)$ are obtained by substituting $x = 1$ and $x = -1$, respectively, in the equation $y = -2x^4$. Now if we replace x with $x - 3$ in the equation $y = -2x^4$, we have $y = -2(x-3)^4$, which we can graph by translating the graph in Figure 12(a) to the right 3 units; see Figure 12(b).

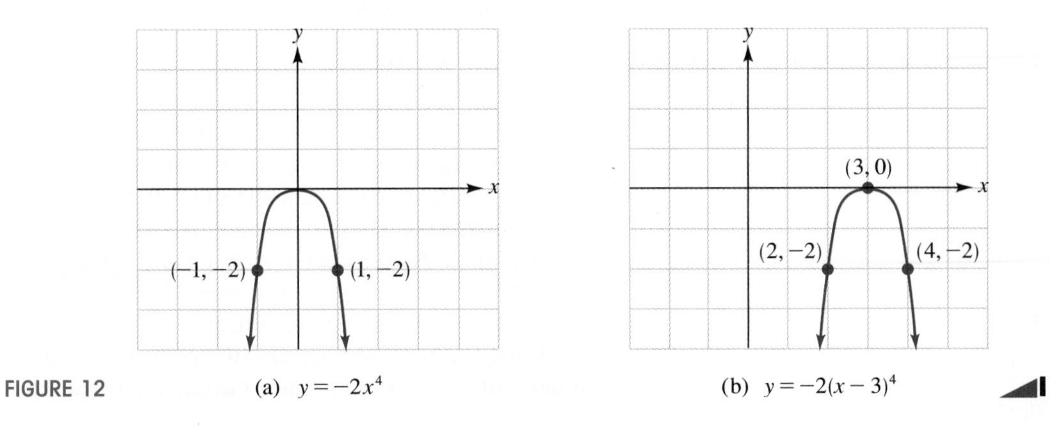

FIGURE 12 (a) $y = -2x^4$ (b) $y = -2(x-3)^4$

We can use our work on solving inequalities (in Section 2.6) to graph polynomial functions that are in factored form. Consider, for example, the function

$$f(x) = x(x+1)(x-3)$$

First of all, by inspection, we see that $f(x) = 0$ when $x = 0$, $x = -1$, or $x = 3$. These are the x-intercepts for the graph. Also, note that the y-intercept is 0. (Why?) Next, we want to know what the graph looks like in the intervals between the x-intercepts. To do this, we solve the two inequalities $f(x) > 0$ and $f(x) < 0$ using the technique in Section 2.6. Table 2 shows the results. (You should check these results for yourself; if you need a review, the details of this very example are worked out on pages 86–88.)

TABLE 2

Interval	$x(x+1)(x-3)$
$(-\infty, -1)$	negative
$(-1, 0)$	positive
$(0, 3)$	negative
$(3, \infty)$	positive

Now we interpret the results in Table 2 graphically. When x is in either of the intervals $(-\infty, -1)$ or $(0, 3)$, the graph lies below the x-axis (because $f(x) < 0$); and when x is in either of the intervals $(-1, 0)$ or $(3, \infty)$, the graph lies above the x-axis (because $f(x) > 0$). This information is summarized in Figure 13(a). The three dots in the figure indicate the x-intercepts of the graph. The shaded regions are the **excluded regions** through which the graph cannot pass. The graph must pass only through the unshaded regions (and through the three x-intercepts). In Figure 13(b) we have drawn a rough sketch of a curve satisfying these conditions.

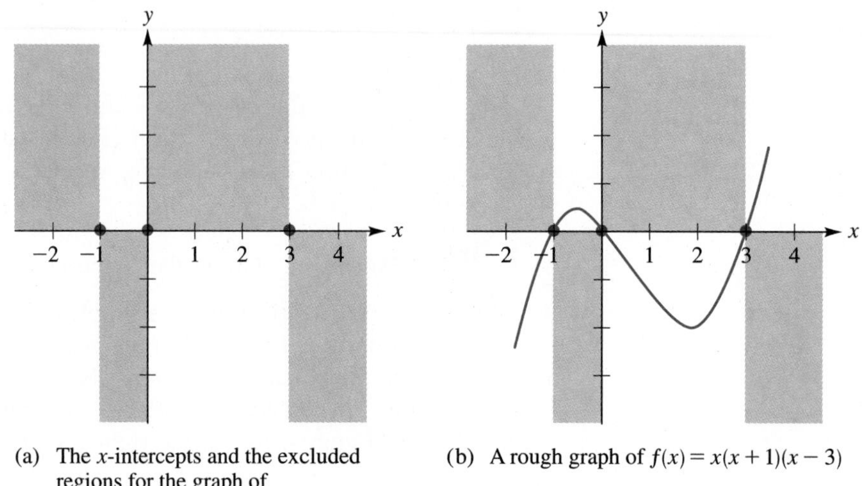

FIGURE 13 (a) The x-intercepts and the excluded regions for the graph of $f(x) = x(x+1)(x-3)$ (b) A rough graph of $f(x) = x(x+1)(x-3)$

Notice that to draw a smooth curve satisfying the conditions of Figure 13(a), we need at least two turning points: one between $x = -1$ and $x = 0$ and another between $x = 0$ and $x = 3$. On the other hand, since the degree of $f(x)$ is 3, there can be no more than two turning points. Thus Figure 13(b) has exactly two turning points. As another check on our rough sketch, we note that for large values of $|x|$, the graph indeed resembles that of $y = x^3$. (We are using the third property in the summary box on page 240.)

While the precise location of the turning points is a matter for calculus, we can nevertheless improve upon the sketch in Figure 13(b) by computing $f(x)$ for

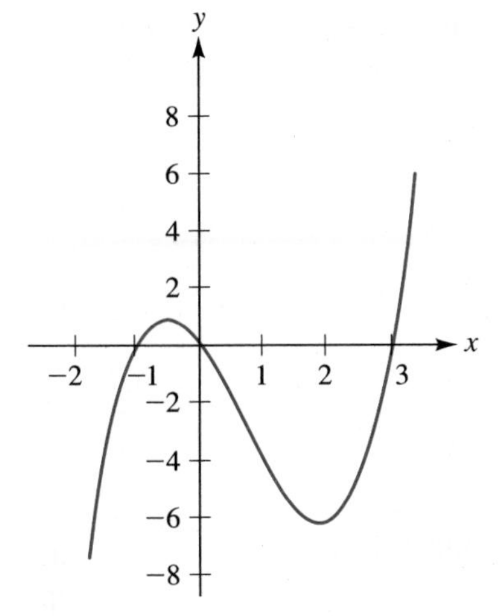

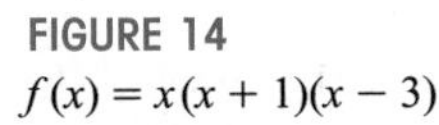
FIGURE 14
$f(x) = x(x + 1)(x - 3)$

some specific values of x. Some reasonable choices in this case are the inputs -2, $-\frac{1}{2}$, 1, 2, and 4. As you can check, the resulting points on the graph are $(-2, -10)$, $\left(-\frac{1}{2}, \frac{7}{8}\right)$, $(1, -4)$, $(2, -6)$, and $(4, 20)$. We can now sketch the graph, as shown in Figure 14.

In the example just concluded, the polynomial $f(x) = x(x + 1)(x - 3)$ has no *repeated factors*. That is, none of the factors is squared or cubed or raised to a higher power. Now let us look at two examples in which there are repeated factors. The observations we make will help us in determining the general shape of a graph without the need for plotting a large number of individual points. In Figure 15(a), we show the graph of $g(x) = x(x + 1)(x - 3)^2$. Notice that in the immediate vicinity of the intercept at $x = 3$, the graph has the same general shape as that of $y = A(x - 3)^2$.

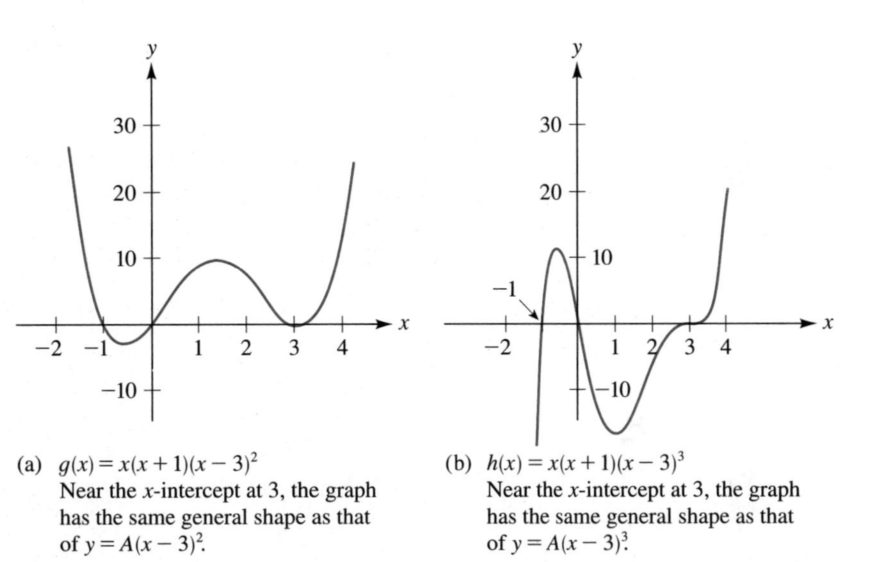

FIGURE 15 (a) $g(x) = x(x + 1)(x - 3)^2$
Near the x-intercept at 3, the graph has the same general shape as that of $y = A(x - 3)^2$.

(b) $h(x) = x(x + 1)(x - 3)^3$
Near the x-intercept at 3, the graph has the same general shape as that of $y = A(x - 3)^3$.

Similarly, in Figure 15(b), we show the graph of $h(x) = x(x + 1)(x - 3)^3$; notice that in the immediate vicinity of $x = 3$, the graph has the same general shape as that of $y = A(x - 3)^3$. In the box that follows, we state the general principle underlying these observations. (The principle can be justified using calculus.)

THE BEHAVIOR OF A POLYNOMIAL FUNCTION NEAR AN x-INTERCEPT

Let $f(x)$ be a polynomial and suppose that $(x - a)^n$ is a factor of $f(x)$. [Furthermore, assume that none of the other factors of $f(x)$ contains $(x - a)$.] Then, in the immediate vicinity of the x-intercept at a, the graph of $y = f(x)$ closely resembles that of $y = A(x - a)^n$.

The principle that we have just stated is easy to apply because we already know how to graph functions of the form $y = A(x - a)^n$. The next example shows how this works.

EXAMPLE 4 Describe the behavior of each function in the immediate vicinity of the indicated x-intercept.

(a) $f(x) = \frac{1}{2}(x-3)(x-1)^3$; intercept: $x = 1$
(b) $g(x) = (x+1)(x+4)(x+3)^2$; intercept: $x = -3$

Solution

(a) We make the following observation:

$$f(x) = \frac{1}{2}(x-3)(x-1)^3$$

When x is close to 1, this factor [$(x-3)$] is close to $1-3$, or -2.

So if x is very close to 1, we have the approximation

$$f(x) \approx \frac{1}{2}(1-3)(x-1)^3 = \frac{1}{2}(-2)(x-1)^3 = -(x-1)^3$$

Thus, in the immediate vicinity of $x = 1$, the graph of f closely resembles $y = -(x-1)^3$. Notice the technique we used to obtain this result. We retained the factor corresponding to the intercept $x = 1$, and we approximated the remaining factor using the value $x = 1$. See Figure 16.

(b) We use the approximation technique shown in part (a).

$$g(x) = (x+1)(x+4)(x+3)^2$$

When x is close to -3, this factor [$(x+1)$] is close to $-3+1$, or -2.

When x is close to -3, this factor [$(x+4)$] is close to $-3+4$, or 1.

Thus, when x is close to -3, we have the approximation

$$g(x) \approx (-3+1)(-3+4)(x+3)^2 = -2(x+3)^2$$

This tells us that in the immediate vicinity of $x = -3$, the graph of g resembles $y = -2(x+3)^2$. See Figure 17.

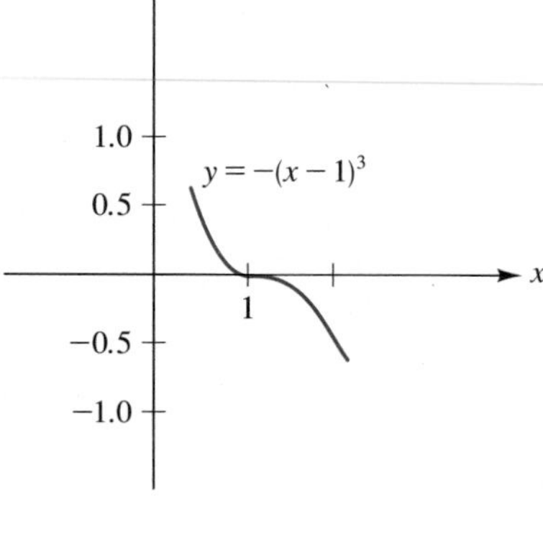

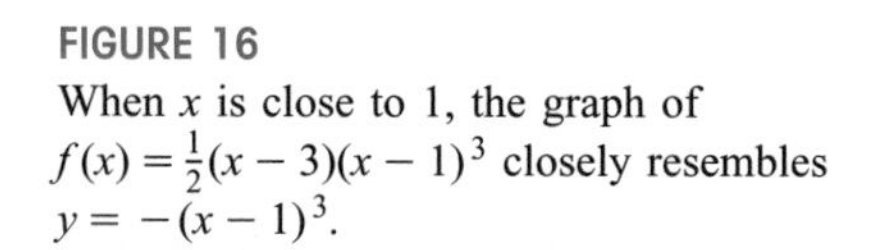

FIGURE 16
When x is close to 1, the graph of $f(x) = \frac{1}{2}(x-3)(x-1)^3$ closely resembles $y = -(x-1)^3$.

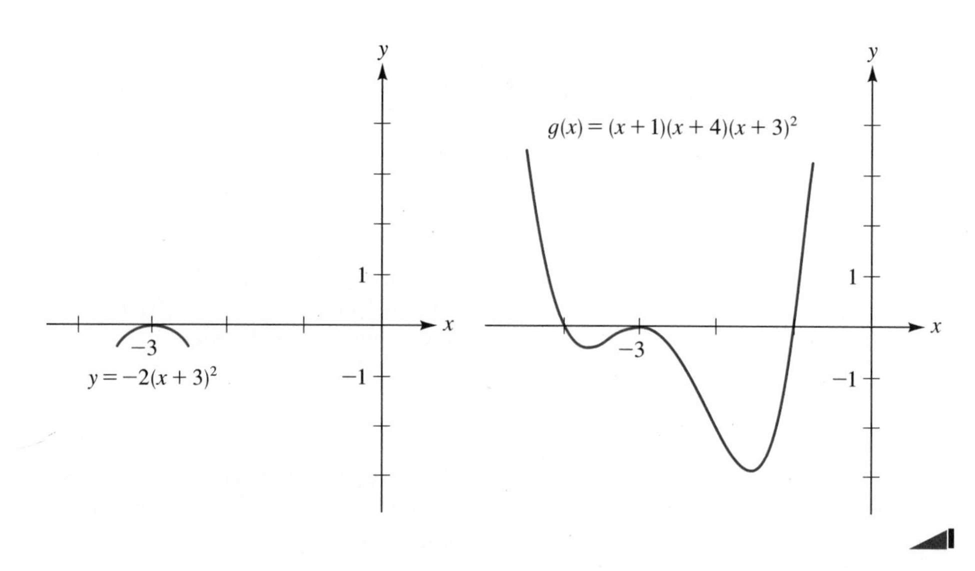

FIGURE 17
When x is close to -3, the graph of $g(x) = (x+1)(x+4)(x+3)^2$ closely resembles $y = -2(x+3)^2$.

EXERCISE SET 4.6

A

In Exercises 1–4, give a reason (as in Example 1) why each graph cannot represent a polynomial function of degree 3.

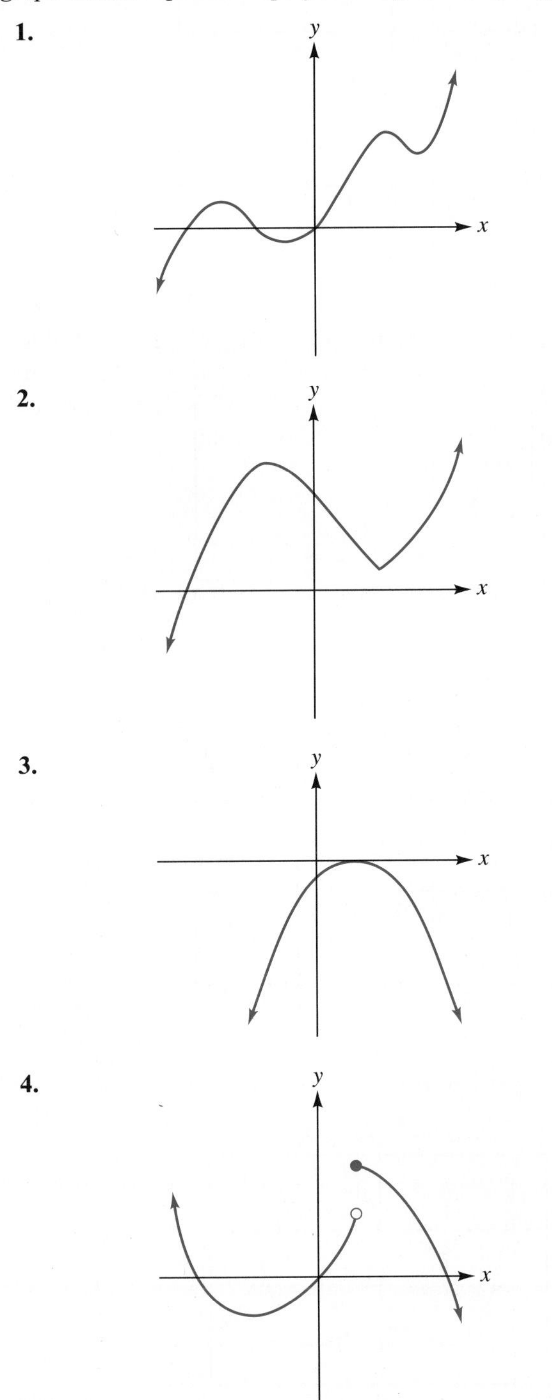

In Exercises 5–8, give a reason why each graph cannot represent a polynomial function that has highest-degree term $2x^5$.

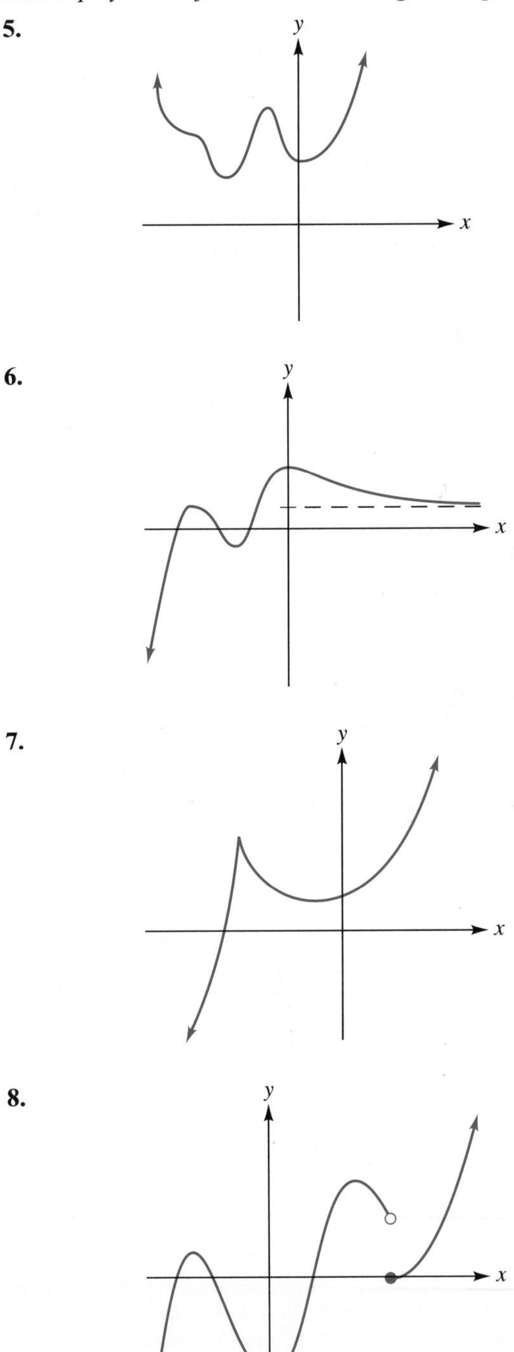

In Exercises 9–20, sketch the graph of each function and specify all x- and y-intercepts.

9. $y = (x-2)^2 + 1$
10. $y = -3x^4$
11. $y = -(x-1)^4$
12. $y = -(x+2)^3$
13. $y = (x-4)^3 - 2$
14. $y = -(x-4)^3 - 2$
15. $y = -2(x+5)^4$
16. $y = -2x^4 + 5$
17. $y = (x+1)^5/2$
18. $y = \frac{1}{2}x^5 + 1$
19. $y = -(x-1)^3 - 1$
20. $y = x^8$

In Exercises 21–28, **(a)** *determine the x- and y-intercepts and the excluded regions for the graph of the given function. Specify your results using a sketch similar to Figure 13(a). In Exercises 25–28, you will first need to factor the polynomial.* **(b)** *Graph each function.*

21. $y = (x-2)(x-1)(x+1)$
22. $y = (x-3)(x+2)(x+1)$
23. $y = 2x(x-2)(x-1)$
24. $y = (x-3)(x-2)(x+2)$
25. $y = x^3 - 4x^2 - 5x$
26. $y = x^3 - 9x$
27. $y = x^3 + 3x^2 - 4x - 12$
28. $y = x^3 - 5x^2 - x + 5$

In Exercises 29–38, **(a)** *determine the x- and y-intercepts and the excluded regions for the graph of the given function. Specify your results using a sketch similar to Figure 13(a).* **(b)** *Describe the behavior of the function at each x-intercept that corresponds to a repeated factor. Specify your results using a sketch similar to the left-hand portion of Figure 17.* **(c)** *Graph each function.*

29. $y = x^3(x+2)$
30. $y = (x-1)(x-4)^2$
31. $y = 2(x-1)(x-4)^3$
32. $y = (x-1)^2(x-4)^2$
33. $y = (x+1)^2(x-1)(x-3)$
34. $y = x^2(x-4)(x+2)$
35. $y = -x^3(x-4)(x+2)$
36. $y = 4(x-2)^2(x+2)^3$
37. $y = -4x(x-2)^2(x+2)^3$
38. $y = -3x^3(x+1)^4$

B

In Exercises 39–44, determine the x-intercepts of the polynomial functions. If an intercept involves a radical, give two forms for the answer: a simplified radical expression and a calculator approximation rounded to three decimal places. In all cases, be sure that your answers are consistent with the graph that is given.

39.

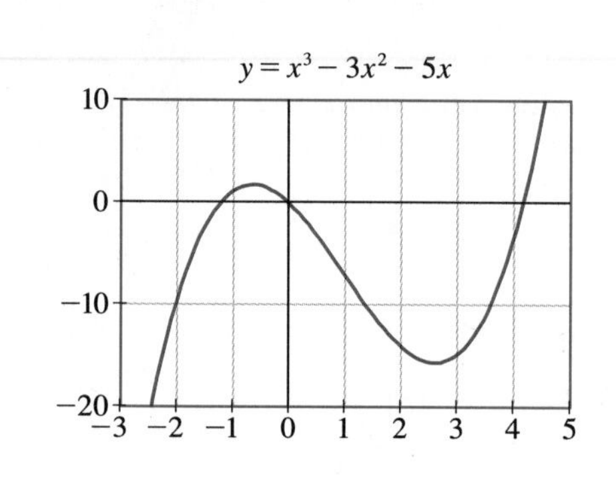

40.

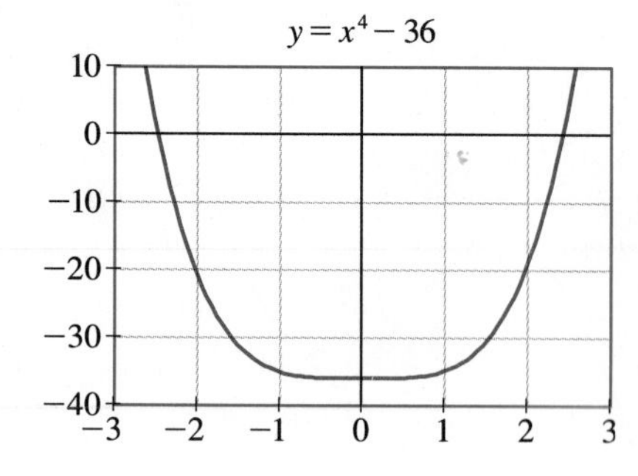

41.

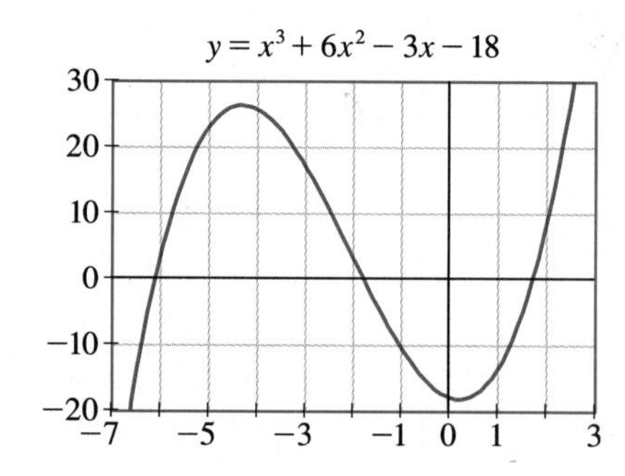

42.

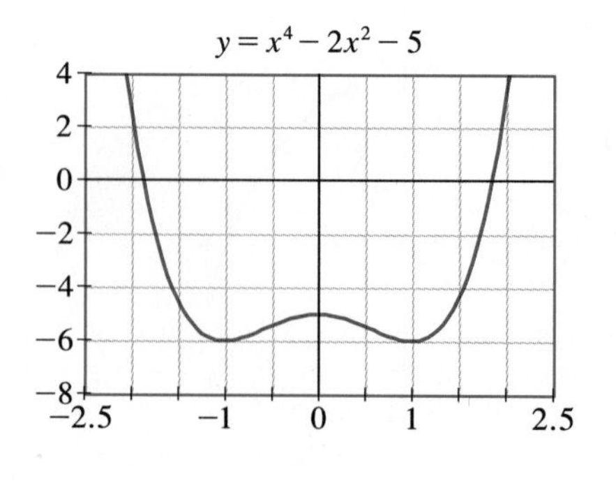

43.

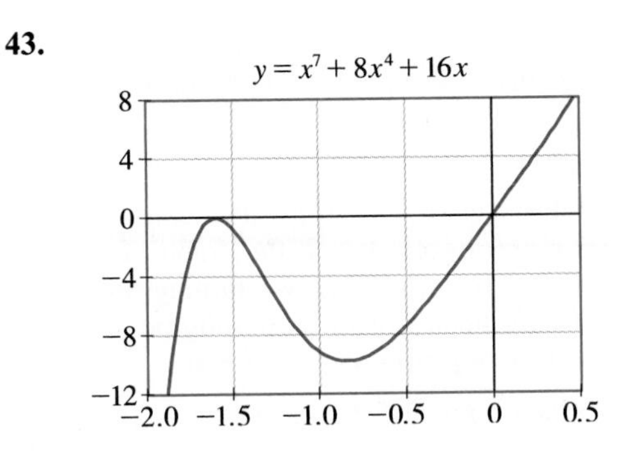

44.

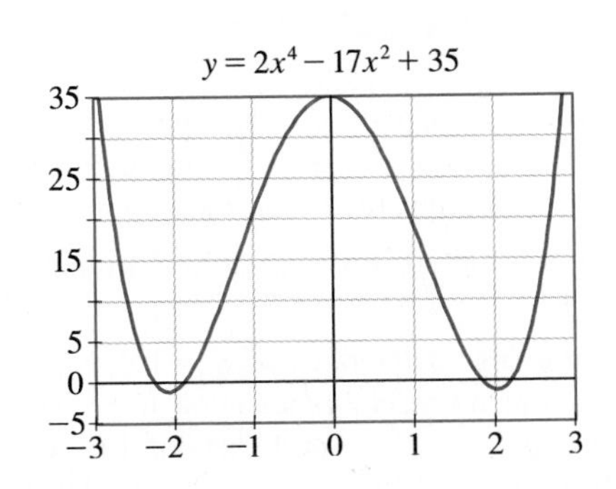

In Exercises 45–50, six functions are defined as follows:

$$f(x) = x \qquad g(x) = x^2 \qquad h(x) = x^3$$
$$F(x) = x^4 \qquad G(x) = x^5 \qquad H(x) = x^6$$

Refer also to the following figure.

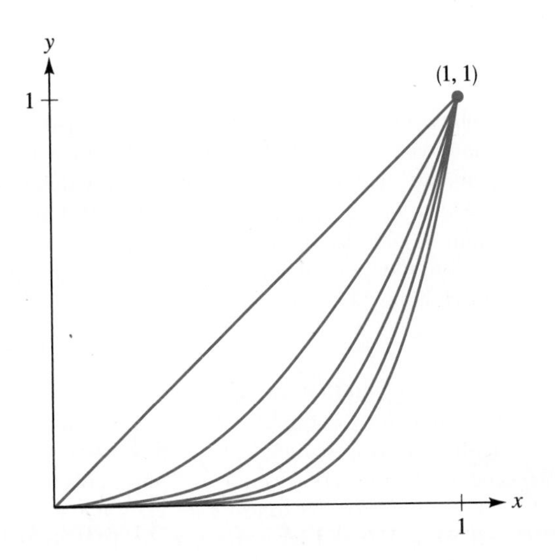

45. The six graphs in the figure are the graphs of the six given functions for the interval [0, 1], but the graphs are not labeled. Which is which?

46. For which x-values in [0, 1] will the graph of g lie strictly below the horizontal line $y = 0.1$? Use a calculator to evaluate your answer. Round off the result to two significant figures.

47. Follow Exercise 46, using the function H instead of g.

48. Find a number t in [0, 1] such that the vertical distance between $f(t)$ and $g(t)$ is $1/4$.

49. Is there a number t in [0, 1] such that the vertical distance between $g(t)$ and $F(t)$ is 0.26?

50. Find all numbers t in [0, 1] such that $F(t) = G(t) + H(t)$.

51. Do the graphs of $y = x$ and $y = x^2/100$ intersect anywhere other than at the origin? (See the following figure.)

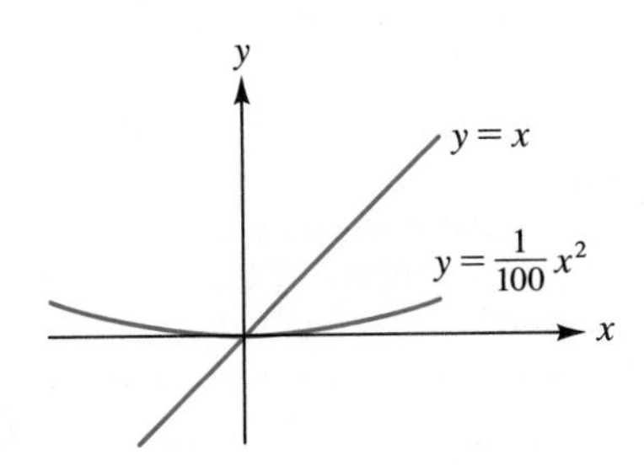

52. **(a)** Graph the function $y = 4x^2 - x^4$.
(b) Find the coordinates of the turning points. *Hint:* See Example 7 in Section 4.2.

53. **(a)** Graph the function $D(x) = x^2 - x^4$.
(b) Find the turning points of the graph. *Hint:* See Example 7 in Section 5.2.
(c) On the same set of axes, sketch the graphs of $y = x^2$ and $y = x^4$ for $0 \leqslant x \leqslant 1$. What is the maximum vertical distance between the graphs?

54. **(a)** An open-top box is to be constructed from a 6-by-8-inch rectangular sheet of tin by cutting out equal squares at each corner and then folding up the resulting flaps. Let x denote the length of the side of each cutout square. Show that the volume $V(x)$ is

$$V(x) = x(6 - 2x)(8 - 2x)$$

(b) What is the domain of the volume function in part (a)? [The answer is *not* $(-\infty, \infty)$.]
(c) Graph the volume function.
(d) Use your graph to estimate the maximum possible volume for the box.

55. A cylinder is inscribed in a sphere of radius 6 cm. Let r denote the radius and h the height of the cylinder, as shown in the figure. We are going to estimate the maximum possible volume for the cylinder.
(a) Show that $h = 2\sqrt{36 - r^2}$.
(b) Let V denote the volume of the cylinder. Check that $V = 2\pi r^2\sqrt{36 - r^2}$.

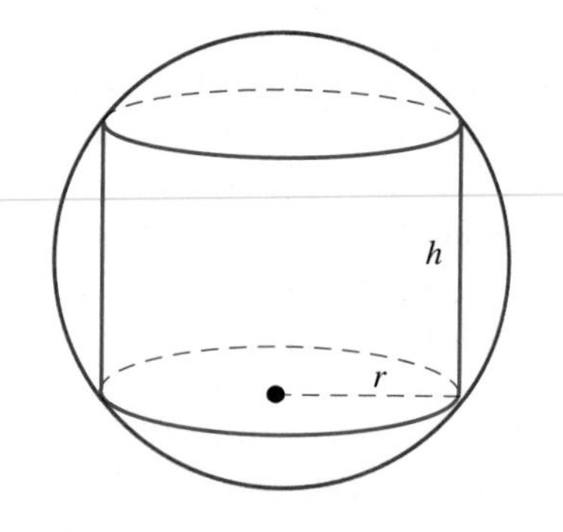

(c) What is the domain of the volume function in part (b)?

(d) The expression for V in part (b) is not a polynomial but the square of the expression is, so we'll work with the function $f(r) = 4\pi^2r^4(36 - r^2)$. At the end, we can convert back to V by taking the square root of $f(r)$. Graph the function $f(r) = 4\pi^2r^4(36 - r^2)$. (To fine-tune your graph, use a calculator to compute $f(r)$ for r running from 0 to 6 in increments of 0.5.) From the graph, estimate the maximum value of $f(r)$. Now estimate the maximum possible volume of the cylinder.

56. **(a)** Use a symmetry test from Section 2.4 to show that the graph of $y = 2x - x^3$ is symmetric about the origin.

(b) By factoring, show that $2x - x^3$ can be rewritten as $-x(x - \sqrt{2})(x + \sqrt{2})$. Then graph the function $y = 2x - x^3$ using the techniques for factored polynomials in this section.

(c) Use the following recipe (from calculus) to fine-tune your graph in part (b): If the graph of a function $f(x) = ax + bx^3$ has two turning points, then the x-coordinates of those turning points are the roots of the equation $a + 3bx^2 = 0$.

(d) Find the fixed points of the function $y = 2x - x^3$.

(e) Use a calculator to compute the first six iterates of $x_0 = 1.6$. (Round the final answers to three decimal places.)

(f) Given that the iterates of $x_0 = 1.6$ approach a fixed point of the function, use your results in part (e) to say which fixed point the iterates are approaching.

(g) Repeat parts (e) and (f) using the initial input $x_0 = 1.7$.

(h) Compute the first six iterates of $x_0 = 1.8$. Note that the results bear no resemblance at all to either of the patterns obtained for the iterates of 1.6 or 1.7. Use your graph from part (b) to explain why the iterates of $x_0 = 1.8$ behave as they do.

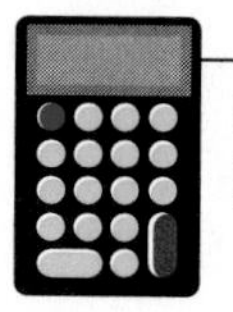

GRAPHING UTILITY EXERCISES FOR SECTION 4.6

1. According to Property 3 on page 240, the graph of the polynomial function

$$f(x) = a_nx^n + a_{n-1}x^{n-1} + \cdots + a_0$$

resembles the graph of $y = a_nx^n$ when $|x|$ is large. In this exercise we look at some examples of this property.

(a) Look at the graph of $y = x^2 + 6x + 2$ in the standard viewing rectangle. According to Property 3, this graph resembles that of $y = x^2$ when $|x|$ is large. To see a vivid demonstration of this, change the viewing rectangle so that x extends from -1000 to 1000 and y extends from $-100{,}000$ to 100,000. (If your graphing utility has scale settings, choose the size of unit on the x-axis to be 100 and that on the y-axis to be 10,000.) As you can see, the resulting graph indeed resembles the familiar parabola $y = x^2$. Indeed, add the graph of $y = x^2$ to the picture. What do you observe?

(b) Follow the general procedure in part (a) for the function $y = -0.25x^3 + 3x + 4$. (Begin with the standard viewing rectangle and then switch to a view in which x runs from -100 to 100 and y runs from -1000 to 1000.)

(c) Follow the general procedure in part (a) for the polynomial function

$$y = 2x^5 - 16x^4 - 26x^3 + 352x^2 - 72x - 1140$$

(You will find that the standard viewing rectangle does not show enough of the graph. Begin instead with a viewing rectangle extending from -6 to 8 in the x-direction and from -2000 to 2000 in the y-direction. For the long-range view, let x run from -250 to 250 and let y run from -10^{11} to 10^{11}.)

2. This exercise demonstrates the behavior of a polynomial function near its x-intercepts. Let

$$g(x) = x(x + 1)(x - 3)^2$$

(a) Graph the function g in the standard viewing rectangle. Then, to help focus on the x-intercepts, change the viewing rectangle so that x extends from -2 to 4. (Retain the previous settings for y.)

(b) Next (without a graphing utility), use the estimation technique explained in Example 4 to verify the following statements.

(i) Near $x = 0$, the graph of g resembles the line $y = 9x$.

(ii) Near $x = -1$, the graph of g resembles the line $y = -16(x + 1)$.

(iii) Near $x = 3$, the graph of g resembles the parabola $y = 12(x - 3)^2$.

(c) Add the graph of the line $y = 9x$ to your picture. What do you observe?

(d) Remove $y = 9x$ from the picture and add the graph of the line $y = -16(x + 1)$. What do you observe?

(e) Remove the line $y = -16(x + 1)$ from the picture and add the graph of the parabola $y = 12(x - 3)^2$. What do you observe?

Problems 3–5 are maximum-minimum problems, just as you practiced in Section 4.5, but here the function that you set up will turn out to be a polynomial of degree greater than 2. For each problem, first set up the appropriate function to be maximized or minimized, then use a graphing utility to graph the function and to determine (at least approximately) the number or numbers that the problem asks for.

3. Which point on the curve $y = x^3$ is closest to the point (4, 5)? *Remark:* If you work with the *square* of the distance, then a polynomial function is directly involved. But with the graphing utility, in fact, it is just as easy to work with the distance function itself.

4. The following figure shows a rectangle with its base on the x-axis and its upper two vertices on the curve $y = 1 - x^4$. What is the maximum possible area for such a rectangle?

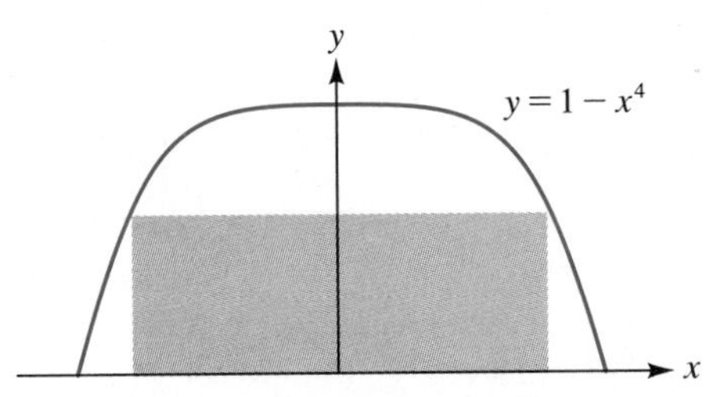

5. An open-top box is constructed from a 9-in.-by-12-in. rectangular sheet of tin by cutting out the same sized squares at each corner and then folding up the flaps. Find the largest possible volume for the box.

RATIONAL FUNCTIONS

We shall not attempt to explain the numerous situations in life sciences where the study of such graphs is important nor the chemical reactions which give rise to the rational functions r(x). . . . Let it suffice to mention that, to the experimental biochemist, theoretical results concerning the shapes of rational functions are of considerable interest.

W. G. Bardsley and R. M. W. Wood in "Critical Points and Sigmoidicity of Positive Rational Functions," *The American Mathematical Monthly,* vol. 92 (1985), pp. 37–42.

After the polynomial functions, the next simplest functions are the **rational functions.** These are functions defined by equations of the form

$$y = \frac{f(x)}{g(x)}$$

where $f(x)$ and $g(x)$ are polynomials. In general, throughout this section, when we write a function such as $y = f(x)/g(x)$, we assume that $f(x)$ and $g(x)$ contain no common factors (other than constants). (Exercises 36 and 37 ask you to consider several cases in which $f(x)$ and $g(x)$ do contain common factors.) Also, for each of the examples that we discuss, the degree of $f(x)$ is less than or equal to the degree of $g(x)$. (A case in which the degree of $f(x)$ exceeds the degree of $g(x)$ is developed in the exercises.)

EXAMPLE 1 Specify the domain of the rational function defined by

$$y = \frac{3x - 2}{x^2 - 1}$$

Also, find the x- and y-intercepts (if any) for the graph of this function.

Solution By factoring the denominator, we can rewrite the given equation

$$y = \frac{3x - 2}{(x - 1)(x + 1)}$$

Since the denominator is zero when $x = 1$ and when $x = -1$, it follows that the domain of the given function consists of all real numbers except 1 and -1. To determine the x-intercepts of the graph, we set $y = 0$ (as usual) in the given equation to obtain

$$\frac{3x-2}{x^2-1} = 0 \qquad (x \neq 1, x \neq -1)$$

$$3x - 2 = 0 \qquad \text{multiplying both sides of the equation by the nonzero quantity } x^2 - 1 \text{, since } x \neq \pm 1$$

$$x = 2/3$$

Thus, the only x-intercept is $x = 2/3$. For the y-intercept, we set $x = 0$ (as usual) in the given equation. As you can readily check, this yields $y = 2$. In summary then, the x- and y-intercepts are $2/3$ and 2, respectively.

In the box that follows, we summarize the ideas used in Example 1 for determining the domain and the x-intercepts of a rational function.

Let R be a rational function defined by

$$R(x) = \frac{f(x)}{g(x)}$$

where $f(x)$ and $g(x)$ are polynomials with no common factors (other than constants). Then the domain of R consists of all real numbers for which $g(x) \neq 0$; the x-intercepts (if any) are the real solutions of the equation $f(x) = 0$.

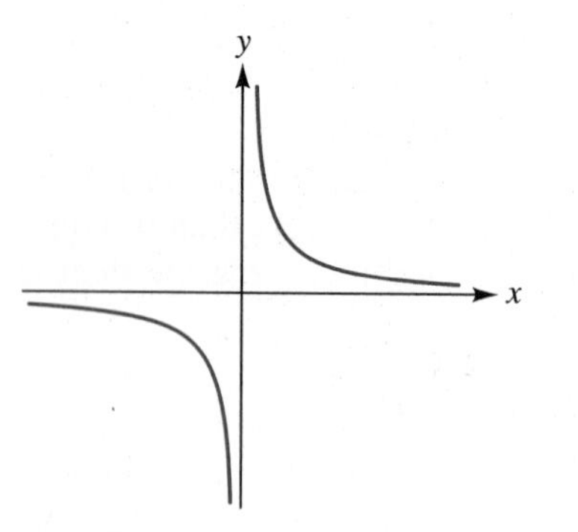

FIGURE 1
The graph of $y = 1/x$.

You are already familiar with the graph of one rational function, because in Section 2.2 we graphed $y = 1/x$. For convenience, this graph is shown again in Figure 1. The graph in Figure 1 differs from the graph of every polynomial function in two important aspects. First, the graph has a break in it; it is composed of two distinct pieces or **branches.** (Recall from the previous section that the graph of a polynomial function never has a break in it.) In general, the graph of a rational function has one more branch than the number of distinct real values for which the denominator is zero. For instance, referring to the function in Example 1, we can expect the graph of $y = (3x - 2)/(x^2 - 1)$ to have three branches.

The second way that the graph in Figure 1 differs from that of a polynomial function is related to *asymptotes.* A line is an **asymptote** for a curve if the distance between the line and the curve approaches zero as we move out farther and farther along the line. Thus, the x-axis is a horizontal asymptote for the graph in Figure 1, while the y-axis is a vertical asymptote. It can be shown that the graph of a polynomial function never has an asymptote. Figure 2 displays additional examples of curves with asymptotes.

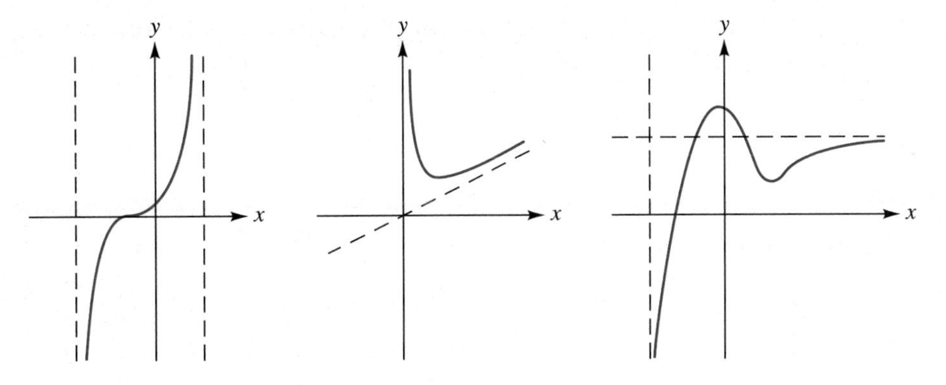

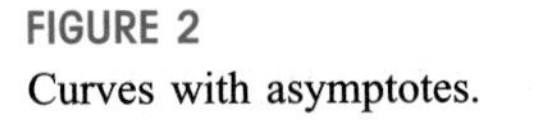

FIGURE 2
Curves with asymptotes.

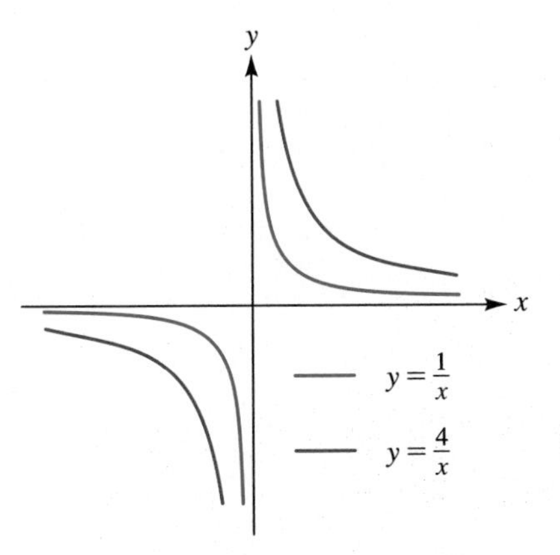

FIGURE 3

If k is a positive constant, the graph of $y = k/x$ resembles that of $y = 1/x$. Figure 3 shows the graphs of $y = 1/x$ and $y = 4/x$. Once we know about the graph of $y = k/x$, we can graph any rational function of the form

$$y = \frac{ax + b}{cx + d}$$

The next three examples show how this is done using the translation and reflection techniques explained in Section 3.3.

EXAMPLE 2 Graph $y = 1/(x - 1)$.

Solution By translating the graph of $y = 1/x$ one unit to the right, we obtain the graph of $y = 1/(x - 1)$, shown in Figure 4. Notice that the horizontal asymptote is unchanged by the translation. However, the vertical asymptote moves one unit to the right. (The y-intercept is -1; why?)

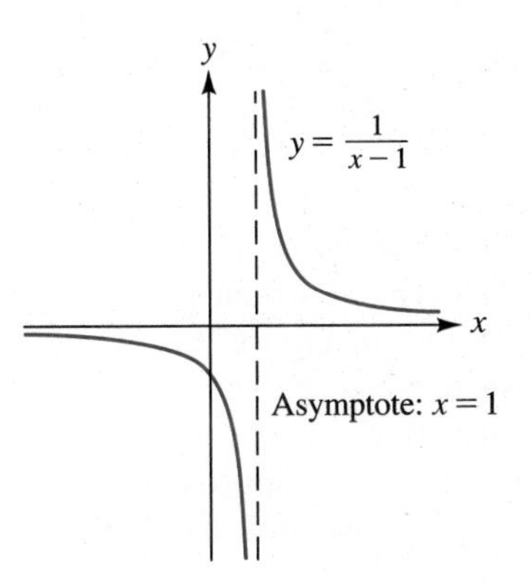

FIGURE 4
The graph of $y = 1/(x - 1)$.

EXAMPLE 3 Graph $y = 2/(x - 1)$ and $y = -2/(x - 1)$.

Solution The graph of $y = 2/(x - 1)$ has the same basic shape and location as the graph of $y = 1/(x - 1)$ shown in Figure 4. As a further guide to sketching $y = 2/(x - 1)$, we can pick several convenient x-values on either side of the asymptote $x = 1$ and then compute the corresponding y-values. After doing this, we obtain the graph shown in Figure 5(a). By reflecting this graph about the x-axis, we obtain the graph of $y = -2/(x - 1)$, which is shown in Figure 5(b).

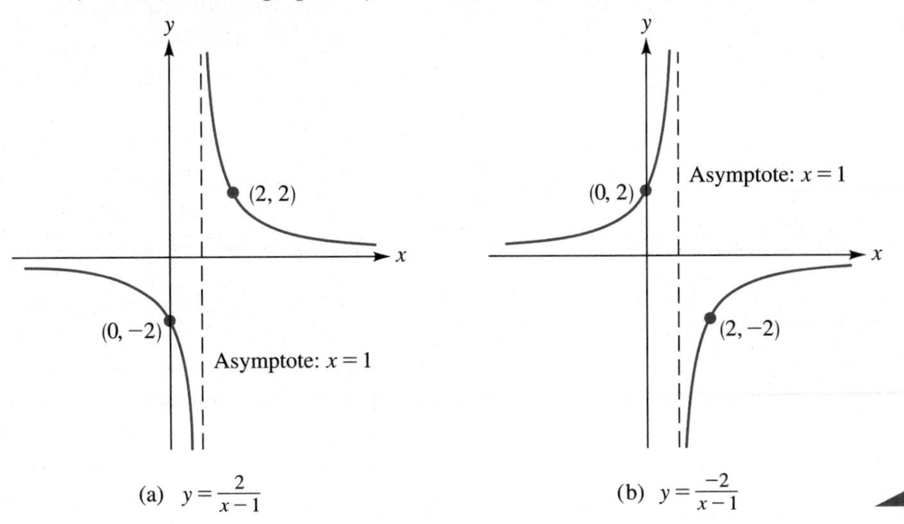

FIGURE 5 (a) $y=\frac{2}{x-1}$ (b) $y=\frac{-2}{x-1}$

EXAMPLE 4 Graph $y = \dfrac{4x-2}{x-1}$.

Solution First, as you can readily check, the x- and y-intercepts are $1/2$ and 2, respectively. Next, using long division, we find that

$$\frac{4x-2}{x-1} = 4 + \frac{2}{x-1}$$

(If you need a review of the long division process from basic algebra, see Section 12.2.) We conclude that the required graph can be obtained by moving the graph of $y = 2/(x-1)$ up 4 units in the y-direction; see Figure 6. Notice that the vertical asymptote is still $x = 1$, but the horizontal asymptote is now $y = 4$ instead of the x-axis.

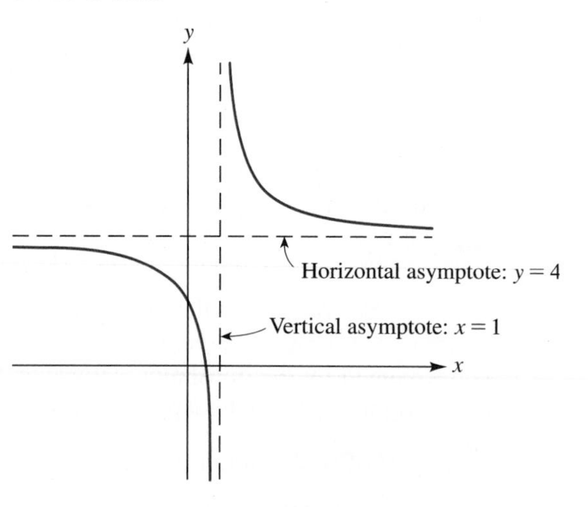

FIGURE 6
The graph of $y = \dfrac{4x-2}{x-1}$.

Now let's look at rational functions of the form $y = 1/x^n$. First we'll consider $y = 1/x^2$. As with $y = 1/x$, the domain consists of all real numbers except $x = 0$. For $x \neq 0$, the quantity $1/x^2$ is always positive. This means that the graph will always lie above the x-axis. Furthermore, the graph will be symmetric about the y-axis. (Why?) As $|x|$ becomes very large, the quantity $1/x^2$ approaches zero; this is true whether x itself is positive or negative. So when $|x|$ is very large, we expect the graph of $y = 1/x^2$ to look as shown in Figure 7(a). On the other hand, when x is a very small fraction, close to zero, either negative or positive, the quantity $1/x^2$ is very large. For instance, if $x = 1/10$, we find that $y = 100$, and if $x = 1/100$, we find that $y = 10{,}000$. Thus, as x approaches zero, from the right or from the left, the graph of $y = 1/x^2$ must look as shown in Figure 7(b).

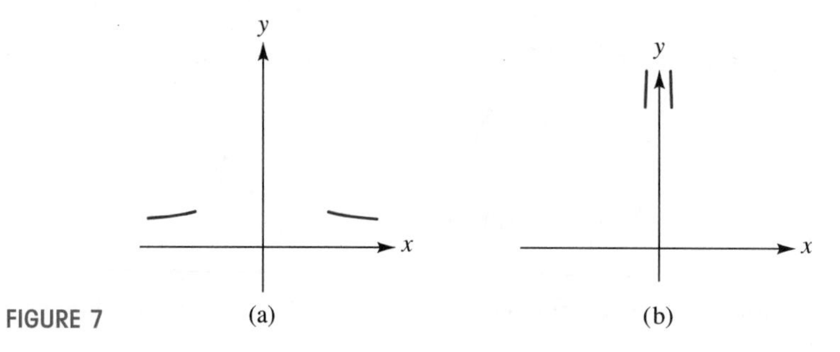

FIGURE 7

Now, by plotting several points and taking Figures 7(a) and 7(b) into account, we obtain the graph of $y = 1/x^2$, shown in Figure 8. Also, by following a similar line of reasoning, we find that the graph of $y = 1/x^3$ looks as shown in Figure 9. (Note that $y = 1/x^3$ is symmetric about the origin.)

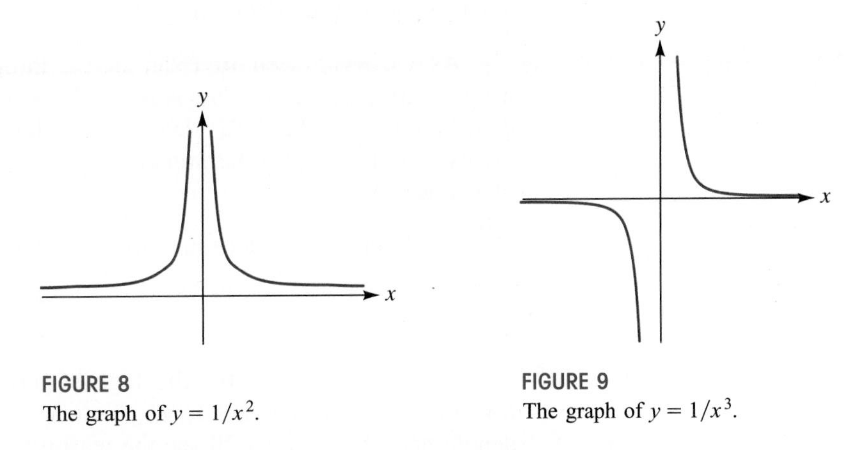

FIGURE 8
The graph of $y = 1/x^2$.

FIGURE 9
The graph of $y = 1/x^3$.

In general, when n is an even integer greater than 2, the graph of $y = 1/x^n$ resembles that of $y = 1/x^2$. When n is an odd integer greater than 3, the graph of $y = 1/x^n$ resembles that of $y = 1/x^3$.

EXAMPLE 5 Graph $y = -1/(x+3)^2$.

Solution Refer to Figure 10. Begin with the graph of $y = 1/x^2$. By moving the graph three units to the left, we obtain the graph of $y = 1/(x+3)^2$. Then by reflecting the graph of $y = 1/(x+3)^2$ about the x-axis, we get the graph of $y = -1/(x+3)^2$.

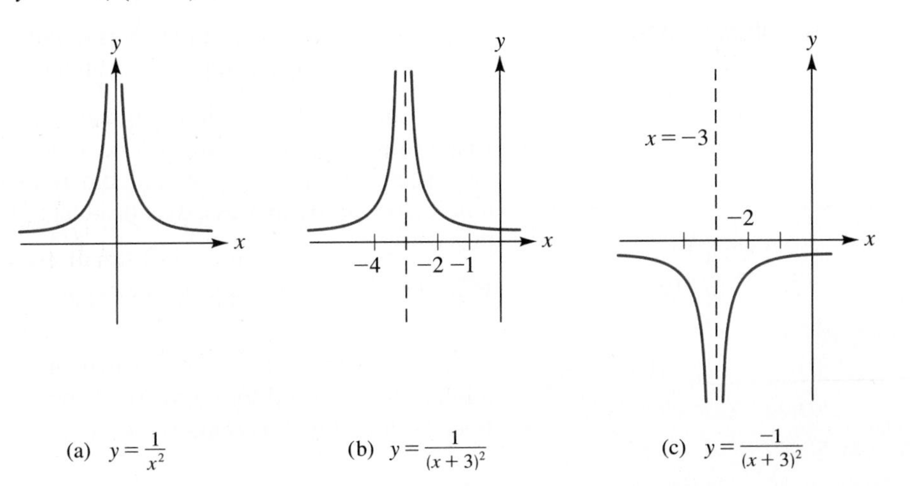

FIGURE 10

QUESTION What are the y-intercepts in Figures 10(b) and (c)?

EXAMPLE 6 Graph $y = 4/(x-2)^3$.

Solution Moving the graph of $y = 1/x^3$ two units to the right gives us the graph of the function $y = 1/(x-2)^3$. The graph of $y = 4/(x-2)^3$ will have the same

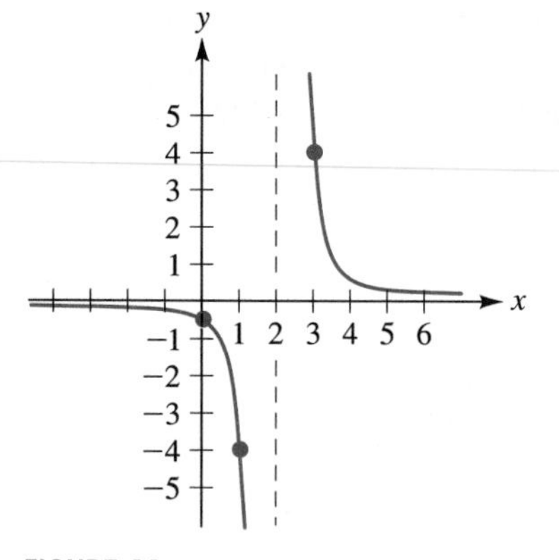

FIGURE 11
The graph of $y = 4/(x-2)^3$.

basic shape and location. As a further guide to sketching the required graph, we can pick several convenient x-values near the asymptote $x = 2$ and compute the corresponding y-values. Using $x = 0$, $x = 1$, and $x = 3$, we find that the points $(0, -\frac{1}{2})$, $(1, -4)$, and $(3, 4)$ are on the graph. With this information, the graph can be sketched as in Figure 11.

As we have seen in Examples 2 through 6, the vertical asymptotes of $y = f(x)/g(x)$ are found by solving the equation $g(x) = 0$. In Example 7 we demonstrate a method for determining the horizontal asymptote. The method involves dividing both the numerator and the denominator by the highest power of x that appears.

EXAMPLE 7 Determine the horizontal asymptote for the graph of

$$y = \frac{3x^2}{x^2 - 4x + 1}$$

Solution First we look for the highest power of x that appears in the given fractional expression; this is x^2. Now we divide both the numerator and the denominator by x^2. (You'll see the reason for doing this in a moment.) We have

$$y = \frac{3x^2}{x^2 - 4x + 1} = \frac{\frac{3x^2}{x^2}}{\frac{x^2 - 4x + 1}{x^2}} = \frac{3}{1 - \frac{4}{x} + \frac{1}{x^2}}$$

When $|x|$ grows very large, the two fractions $4/x$ and $1/x^2$ both approach zero. We therefore have

$$y \approx \frac{3}{1 - 0 + 0} = 3 \qquad \text{when } |x| \text{ is very large}$$

This tells us that the line $y = 3$ is a horizontal asymptote for the graph of the given function. See Table 1 for a numerical perspective on this.

TABLE 1 AS x GROWS LARGER AND LARGER, THE VALUES OF $3x^2/(x^2 - 4x + 1)$ APPROACH 3.

x	$\frac{3x^2}{x^2 - 4x + 1}$
100	3.12 . . .
1000	3.012 . . .
10,000	3.0012 . . .
100,000	3.00012 . . .
1,000,000	3.000012 . . .
10,000,000	3.0000012 . . .
100,000,000	3.00000012 . . .

In the previous section you learned a technique for determining the general shape of a polynomial function near an x-intercept. That same technique is also very useful in graphing rational functions in which both the numerator and denominator can be expressed as products of linear factors. Example 8 shows how this works.

EXAMPLE 8 Graph the function $y = \dfrac{(x-3)(x+2)}{(x+1)(x-2)}$.

Solution By inspection, we see that the x-intercepts are $x = 3$ and $x = -2$, the y-intercept is 3, and the vertical asymptotes are the lines $x = -1$ and $x = 2$. To find the horizontal asymptote, we write

$$y = \frac{x^2 - x - 6}{x^2 - x - 2} = \frac{1 - \frac{1}{x} - \frac{6}{x^2}}{1 - \frac{1}{x} - \frac{2}{x^2}} \qquad \text{dividing numerator and denominator by } x^2$$

$$\approx \frac{1 - 0 - 0}{1 - 0 - 0} \qquad \text{when } |x| \text{ is large}$$

This shows that the line $y = 1$ is a horizontal asymptote.

Now we want to see how the graph looks in the immediate vicinity of the x-intercepts and the vertical asymptotes. To do this, we use the approximation technique explained in the previous section for polynomial functions. Let's start with the x-intercept at $x = 3$. As in Section 4.6, we'll retain the factor $x - 3$ and approximate the remaining factors using $x = 3$. So, for x near 3, we have

$$y = \frac{(x-3)(x+2)}{(x+1)(x-2)} \approx \frac{(x-3)(3+2)}{(3+1)(3-2)} = \frac{5}{4}(x-3)$$

Thus, in the immediate vicinity of the x-intercept $x = 3$, the required graph will closely resemble the line $y = \frac{5}{4}(x - 3)$. The remaining calculations for approximating the graph near the other x-intercept and near the two vertical asymptotes are carried out in exactly the same manner. As Exercise 35 will ask you to verify, the results are:

x near 3: $y \approx \frac{5}{4}(x-3)$

x near -2: $y \approx -\frac{5}{4}(x+2)$

x near -1: $y \approx \frac{4/3}{x+1}$

x near 2: $y \approx \frac{-4/3}{x-2}$

We summarize these results as follows. As the graph passes through the points $(3, 0)$ and $(-2, 0)$, it resembles the lines $y = \frac{5}{4}(x - 3)$ and $y = -\frac{5}{4}(x + 2)$, respectively. Near the vertical asymptote $x = -1$, the graph has the same basic shape as $y = 1/(x + 1)$, and near the vertical asymptote $x = 2$, the graph resembles $y = -1/(x - 2)$; see Figure 12.

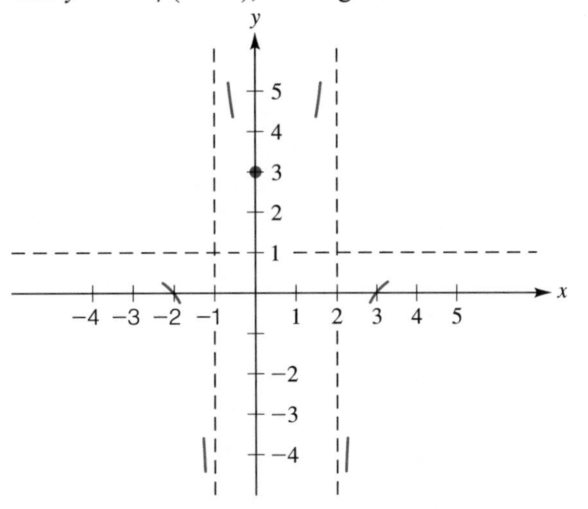

FIGURE 12

The graph of $y = \frac{(x-3)(x+2)}{(x+1)(x-2)}$ near its x-intercepts and vertical asymptotes.

Finally, we want to find out how the graph approaches the horizontal asymptote $y = 1$. Perhaps the simplest way is to do some calculations. In Table 2

TABLE 2 WHEN $|x|$ IS LARGE, y IS LESS THAN 1.

x	y
10	0.95
100	0.99
−10	0.96
−100	0.9996

we have computed the outputs for some relatively large values of $|x|$. From the table, we see that when $|x|$ is large, y is less than 1. Graphically, this means that the curve approaches the asymptote $y = 1$ from below. In Figure 13 we summarize what we have discovered up to this point. Then, using Figure 13 as a guide, we can sketch the required graph, as shown in Figure 14.

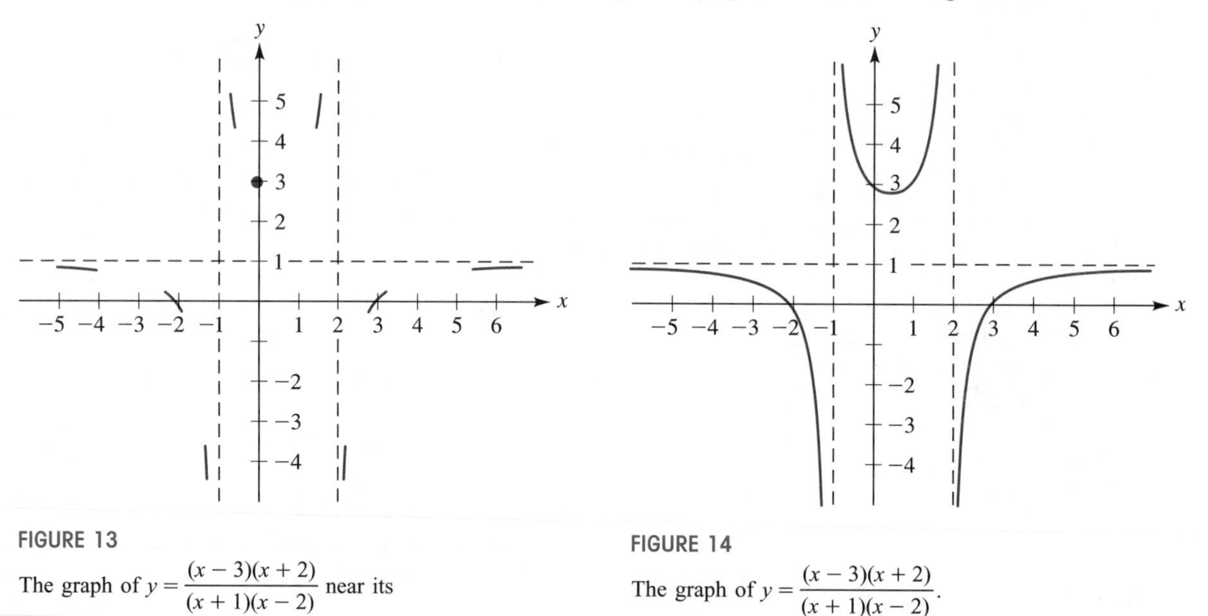

FIGURE 13

The graph of $y = \dfrac{(x-3)(x+2)}{(x+1)(x-2)}$ near its x-intercepts and its horizontal and vertical asymptotes.

FIGURE 14

The graph of $y = \dfrac{(x-3)(x+2)}{(x+1)(x-2)}$.

NOTE A general method for finding the coordinates of the lowest point on the middle branch of the graph is studied in calculus. Those coordinates can also be found, however, by applying algebraic techniques. (See Exercise 38.)

EXERCISE SET 4.7

A

In Exercises 1–6, find the domain and the x- and y-intercepts for each rational function.

1. $y = (3x + 15)/(4x - 12)$

2. $y = (x + 6)(x + 4)/(x - 1)^2$

3. $y = (x^2 - 8x - 9)/(x^2 - x - 6)$

4. $y = (2x^2 + x - 5)/(x^2 + 1)$

5. $y = (x^2 - 4)(x^3 - 1)/x^6$

6. $y = (x^5 - 2x^4 - 9x + 18)/(8x^3 + 2x^2 - 3x)$

In Exercises 7–30, sketch the graph of each rational function. Specify the intercepts and the asymptotes.

7. $y = 1/(x + 4)$

8. $y = -1/(x + 4)$

9. $y = 3/(x + 2)$

10. $y = -3/(x + 2)$

11. $y = -2/(x - 3)$

12. $y = (x - 1)/(x + 1)$

13. $y = (x - 3)/(x - 1)$

14. $y = 2x/(x + 3)$

15. $y = (4x - 2)/(2x + 1)$

16. $y = (3x + 2)/(x - 3)$

17. $y = 1/(x - 2)^2$

18. $y = -1/(x - 2)^2$

19. $y = 3/(x + 1)^2$

20. $y = -3/(x + 1)^2$

21. $y = 1/(x+2)^3$

22. $y = -1/(x+2)^3$

23. $y = -4/(x+5)^3$

24. $y = x/[(x+1)(x-3)]$

25. $y = -x/[(x+2)(x-2)]$

26. $y = 2x/(x+1)^2$

27. **(a)** $y = 3x/[(x-1)(x+3)]$
(b) $y = 3x^2/[(x-1)(x+3)]$

28. **(a)** $y = (4x^2+1)/(x^2-1)$
(b) $y = (4x^2+1)/(x^2+1)$

29. **(a)** $f(x) = (x-2)(x-4)/[x(x-1)]$
(b) $g(x) = (x-2)(x-4)/[x(x-3)]$
[Compare the graphs you obtain in parts (a) and (b). Notice how a change in only one constant can radically alter the nature of the graph.]

30. **(a)** $f(x) = (x-1)(x+2.75)/[(x+1)(x+3)]$
(b) $g(x) = (x-1)(x+3.25)/[(x+1)(x+3)]$
[Compare the graphs you obtain in parts (a) and (b). Notice how a relatively small change in one of the constants can radically alter the graph.]

B

In Exercises 31–34, graph the functions. Note: Each graph intersects its horizontal asymptote once. Find the intersection point before sketching your final graph.

31. $y = \dfrac{(x-4)(x+2)}{(x-1)(x-3)}$

32. $y = \dfrac{(x-1)(x-3)}{(x+1)^2}$

33. $y = \dfrac{(x+1)^2}{(x-1)(x-3)}$

34. $y = \dfrac{2x^2-3x-2}{x^2-3x-4}$

35. (This exercise refers to Example 8.) Let

$$y = \frac{(x-3)(x+2)}{(x+1)(x-2)}$$

Verify each of the following approximations.
(a) When x is close to -2, then $y \approx -\frac{5}{4}(x+2)$.
(b) When x is close to -1, then $y \approx \dfrac{4/3}{x+1}$.
(c) When x is close to 2, then $y \approx \dfrac{-4/3}{x-2}$.

In Exercises 36 and 37, graph the functions. Notice in each case that the numerator and denominator contain at least one common factor. Thus, you can simplify each quotient; but don't lose track of the domain of the function as it was initially defined.

36. **(a)** $y = \dfrac{x+2}{x+2}$ **(b)** $y = \dfrac{x^2-4}{x-2}$
(c) $y = \dfrac{x-1}{(x-1)(x-2)}$

37. **(a)** $y = \dfrac{x^2-9}{x+3}$ **(b)** $y = \dfrac{x^2-5x+6}{x^2-2x-3}$
(c) $y = \dfrac{(x-1)(x-2)(x-3)}{(x-1)(x-2)(x-3)(x-4)}$

38. This exercise shows you how to determine the coordinates of the lowest point on the middle branch of the curve in Figure 14. The basic idea is as follows. Suppose that the required coordinates are (h, k). Then the horizontal line $y = k$ is the unique horizontal line intersecting the curve in one and only one point. (Any other horizontal line intersects the curve either in two points or not at all.) In steps (a) through (c) that follow, we use these observations to determine the point (h, k).
(a) Given any horizontal line $y = k$, its intersection with the curve in Figure 14 is determined by solving the following pair of simultaneous equations:

$$\begin{cases} y = k \\ y = \dfrac{x^2-x-6}{x^2-x-2} \end{cases}$$

In the second equation of the system, replace y with k and show that the resulting equation can be written

$$(k-1)x^2 - (k-1)x + (6-2k) = 0 \qquad (1)$$

(b) If k is indeed the required y-coordinate, then equation (1) must have exactly one real solution. Set the discriminant of the quadratic equation equal to zero to obtain

$$(k-1)^2 - 4(k-1)(6-2k) = 0$$

and deduce from this that $k = 1$ or $k = 25/9$. The solution $k = 1$ can be discarded. (To see why, look at Figure 14.)
(c) Using the value $y = 25/9$, show that the corresponding x-coordinate is $1/2$. Thus, the required point is $\left(\frac{1}{2}, \frac{25}{9}\right)$.

39. Graph the function $y = x/(x-3)^2$. Use the technique explained in Exercise 38 to find the coordinates of any turning points on the graph.

40. Graph the function $y = 2/(x - x^2)$. Use the technique explained in Exercise 38 to find the coordinates of any turning points on the graph.

An asymptote that is neither horizontal nor vertical is called a ***slant*** *or* ***oblique asymptote.*** *For example, as indicated in the following figure (next page), the line $y = x$ is a slant asymptote for the graph of $y = (x^2+1)/x$. To understand why the line $y = x$ is an asymptote, we carry out the indicated division and write the function in the form*

$$y = x + \frac{1}{x} \qquad (2)$$

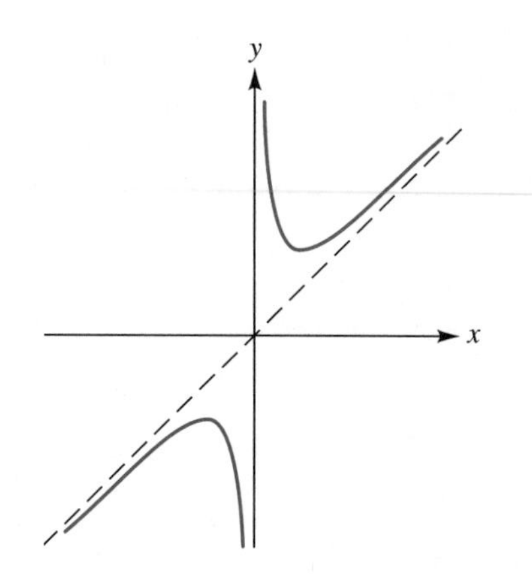

From equation (2), we see that if $|x|$ is very large then $y \approx x + 0$; that is, $y \approx x$, as we wished to show. Equation (2) actually tells us more than this. When $|x|$ is very close to zero, equation (2) yields $y \approx 0 + (1/x)$. In other words, as we approach the y-axis, the curve looks more and more like the graph of $y = 1/x$. In general, if we have a rational function $f(x)/g(x)$ in which the degree of $f(x)$ is 1 greater than the degree of $g(x)$, then the graph has a slant asymptote that is obtained as follows. Divide $f(x)$ by $g(x)$ to obtain an equation of the form

$$\frac{f(x)}{g(x)} = (mx + b) + \frac{h(x)}{g(x)}$$

where the degree of $h(x)$ is less than the degree of $g(x)$. Then the equation of the slant asymptote is $y = mx + b$. (For instance, using the previous example, we have $y = mx + b = x$ and $h(x)/g(x) = 1/x$.) In Exercises 41–43, you are asked to graph functions that have slant asymptotes.

41. Let $y = F(x) = \dfrac{x^2 + x - 6}{x - 3}$.

(a) Use long division to show that

$$\frac{x^2 + x - 6}{x - 3} = (x + 4) + \frac{6}{x - 3}$$

(b) The result in part (a) shows that the line $y = x + 4$ is a slant asymptote for the graph of the function F. Verify this fact empirically by completing the following two tables. (Use a calculator.)

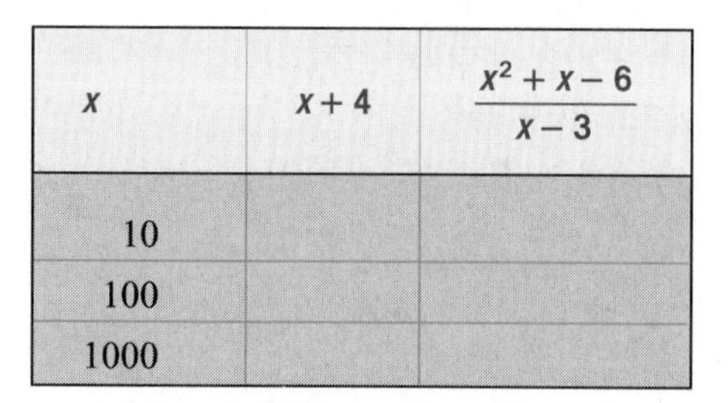

x	$x + 4$	$\dfrac{x^2 + x - 6}{x - 3}$
10		
100		
1000		

x	$x + 4$	$\dfrac{x^2 + x - 6}{x - 3}$
−10		
−100		
−1000		

(c) Determine the vertical asymptote and the x- and y-intercepts for the graph of F.

(d) Graph the function F. (Use the techniques in this section along with the fact that $y = x + 4$ is a slant asymptote.)

(e) Use the technique explained in Exercise 38 to find the coordinates of the two turning points on the graph of F.

42. **(a)** Show that the line $y = x - 2$ is a slant asymptote for the graph of $F(x) = x^2/(x + 2)$.

(b) Sketch the graph of F.

43. Show that the line $y = -x$ is a slant asymptote for the graph of $y = (1 - x^2)/x$. Then sketch the graph of this function.

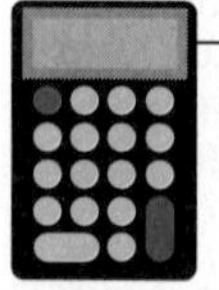

GRAPHING UTILITY EXERCISES FOR SECTION 4.7

Problems 1–7 are maximum-minimum problems, just as you practiced in Section 4.5, but now the function that you set up will turn out to be a rational function rather than a quadratic. To find the required maximum or minimum and/or input, first set up the appropriate function. Then use a graphing utility to graph the function and to approximate the required values.

1. The figure on the next page shows a line with slope m passing through the point (2, 1). Find the slope of this line so that the area of the shaded triangle is a minimum. *Hint:* You should obtain the following function for the area A as a function of the slope: $A = \dfrac{-4m^2 + 4m - 1}{2m}$

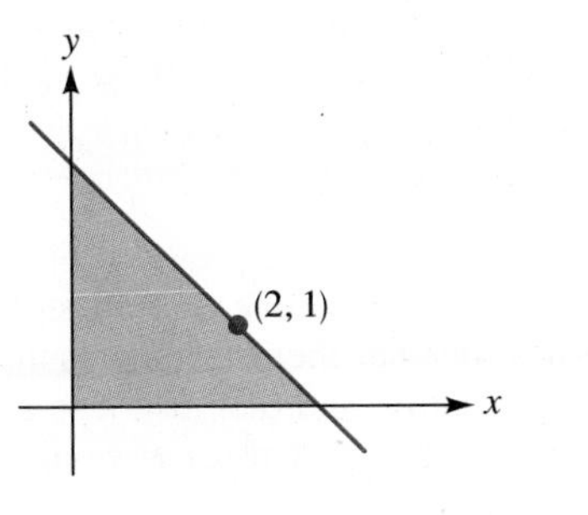

2. The product of two positive numbers is 6.
   (a) Express the sum S of the two numbers as a function of one variable. *Answer:* $S = (x^2 + 6)/x$
   (b) For which value of x will the sum be as small as possible? What is the other number in this case, and what is the corresponding minimum sum?
   Hint for checking: Using calculus, it can be shown that the exact value for x is $\sqrt{6}$. Evaluate this using a calculator. How close were you to this value?

3. Find the point P on the first-quadrant portion of the parabola $y = x^2$, with the property that the slope of the line through $(0, -1)$ and P is a minimum. [After you've finished this problem, use the graphing utility to draw the parabola and the line through P and $(0, -1)$. What do you observe?]

4. What is the smallest possible value for the sum of a positive number and its reciprocal?

5. The area of a rectangle is 6 m^2.
   (a) Express the perimeter of the rectangle as a function of the width x.
   (b) For which value of x is the perimeter a minimum? What is the length in this case?

6. A printer designing a small rectangular poster decides to make the area of the poster 480 in.^{2} The margins on the top and bottom of the poster are to be 3 in. and 4 in., respectively. The margins on the sides are both to be 2.5 in.
   (a) Express the area of the printed portion of the poster as a function of x, the width of the entire poster.
   (b) For which width x will the printed area be a maximum? What is the corresponding length in this case?
   (c) Using calculus, it can be shown that the exact value for x is $20\sqrt{42}/7$. To how many decimal places does your value for x in part (b) agree with this answer?

7. The sum of two positive numbers a and b is 1. Find the minimum possible value for the quantity $(a + a^{-1})^2 + (b + b^{-1})^2$.

In Section 4.7 we studied rational functions in which the degree of the numerator was less than or equal to the degree of the denominator. Exercises 8 and 9 examine cases in which the degree of the numerator is greater than that of the denominator.

8. Let $f(x) = (x^3 + 2x^2 + 1)/(x^2 + 2x)$.
   (a) Graph the function f using a viewing rectangle extending from -5 to 5 in both the x- and y-directions.
   (b) Add the graph of the line $y = x$ to your picture in part (a). Note that to the right of the origin, as x increases, the graph of f begins to look more and more like the line $y = x$. This also occurs to the left of the origin as x decreases.
   (c) The results in part (b) suggest that the line $y = x$ may be an asymptote for the graph of f. Verify this visually by changing the viewing rectangle so that it extends from -20 to 20 in both the x- and the y-directions. What do you observe?
   (d) Using algebra, verify the identity

$$\frac{x^3 + 2x^2 + 1}{x^2 + 2x} = \frac{1}{x^2 + 2x} + x.$$

   Then explain why, for large values of $|x|$, the graph of f looks more and more like the line $y = x$. *Hint:* Substitute some large numbers (such as 100 or 1000) into the expression $1/(x^2 + 2x)$. What happens?

9. Let $f(x) = (x^5 + 1)/x^2$.
   (a) Graph the function f using a viewing rectangle that extends from -4 to 4 in the x-direction and from -8 to 8 in the y-direction.
   (b) Add the graph of the curve $y = x^3$ to your picture in part (a). Note that as $|x|$ increases (that is, as x moves away from the origin), the graph of f looks more and more like the curve $y = x^3$. For additional perspective, first change the viewing rectangle so that y extends from -20 to 20. (Retain the x-settings for the moment.) Describe what you see. Next, adjust the viewing rectangle so that x extends from -10 to 10 and y extends from -100 to 100. Summarize your observations.
   (c) In the text we said that a line is an asymptote for a curve if the distance between the line and the curve approaches zero as we move further and further out along the curve. The work in part (b) shows that a *curve* can be an asymptote for another curve. In particular, part (b) shows that the distance between the curve $y = x^3$ and the graph of the given function f approaches zero as we move further and further out along the graph of f. That is, the curve $y = x^3$ is an asymptote for the graph of the given function f.

Complete the following two tables for a numerical perspective on this. In the tables, d denotes the vertical distance between the curve $y = x^3$ and the graph of f:

$$d = \left| \frac{x^5 + 1}{x^2} - x^3 \right|$$

x	5	10	50	100	500
d					

x	-5	-10	-50	-100	-500
d					

(d) Parts (b) and (c) have provided both a graphical and a numerical perspective. For an algebraic perspective that ties together the previous results, verify the following identity, and then use it to explain why the results in parts (b) and (c) were inevitable:

$$\frac{x^5 + 1}{x^2} = x^3 + \frac{1}{x^2}$$

CHAPTER FOUR SUMMARY OF PRINCIPAL TERMS

TERMS	PAGE REFERENCE	COMMENTS
1. Linear function	175	A linear function is a function of the form $f(x) = Ax + B$. The graph of a linear function is a straight line. An important idea that arose in several of the examples is that the slope of a line can be interpreted as a rate of change. Two instances of this are marginal cost and velocity.
2. Quadratic function	190	A quadratic function is a function of the form $f(x) = ax^2 + bx + c$. The graph of a quadratic function is a parabola. See the boxes on pages 191 and 193.
3. Fixed point	199	A fixed point of a function f is an input x in the domain of f such that $f(x) = x$. For example, $x = 1$ is a fixed point for $f(x) = \sqrt{x}$ because $f(1) = \sqrt{1} = 1$.
4. Polynomial function	238	A polynomial function of degree n is a function of the form $$f(x) = a_n x^n + a_{n-1} x^{n-1} + \cdots + a_1 x + a_0$$ where n is a nonnegative integer and $a_n \neq 0$. Three basic properties of polynomial functions are summarized in the box on page 240.
5. Rational function	251	A rational function is a function of the form $y = f(x)/g(x)$, where $f(x)$ and $g(x)$ are polynomials.
6. Asymptote	252	A line is said to be an asymptote for a curve if the distance between the line and the curve approaches zero as we move out farther and farther along the line. See, for example, Figure 2 on page 253, in which the dashed lines are asymptotes.

WRITING MATHEMATICS

Consider the following two problems.

(A) The perimeter of a rectangle is 2 m.
  (i) Express the area of the rectangle as a function of the width w.
  (ii) Find the maximum possible area for such a rectangle.

(B) The area of a rectangle is 2 m^2.
  (i) Express the perimeter of the rectangle as a function of the width w.
  (ii) Find the minimum possible perimeter for such a rectangle.

1. After working out Problem A for yourself, write out the solution in complete sentences, as if you were explaining it to a classmate or to your instructor.

Be sure to let the reader know where you are headed and why each of the main steps is necessary.

2. After working out part (i) of Problem B for yourself, write out the solution in complete sentences.
3. Explain why the methods of Section 4.5 are not applicable for solving the second part of Problem B.
4. The following result is known as the *arithmetic–geometric mean inequality:*

   For all nonnegative real numbers a and b, we have $a + b \geq 2\sqrt{ab}$, with equality holding when and only when $a = b$.

   By working with a classmate or your instructor, find out how to apply this result to the second part of Problem B. Then write out your solution in complete sentences. How does your final answer here compare with that in Problem A?

CHAPTER FOUR REVIEW EXERCISES

1. Find $G(0)$ if G is a linear function such that $G(1) = -2$ and $G(-2) = -11$.
2. **(a)** Let $f(x) = 3x^2 + 6x - 10$. For which input x is the value of the function a minimum? What is that minimum value?
   (b) Let $g(t) = 6t^2 - t^4$. For which input t is the value of the function a maximum?
3. Suppose the function $p = -\frac{1}{8}x + 100 \quad (0 \leq x \leq 12)$ relates the selling price p of an item to the quantity x that is sold. Assume that p is in dollars. What is the maximum revenue possible in this situation?
4. Graph the function $y = -1/(x + 1)^3$.
5. Graph the function $y = (x - 4)(x - 1)(x + 1)$.
6. Graph the function $f(x) = x^2 + 4x - 5$. Specify the vertex, the x- and y-intercepts, and the axis of symmetry.
7. A factory owner buys a new machine for \$1000. After five years, the machine has a salvage value of \$100. Assuming linear depreciation, find a formula for the value V of the machine after t years, where $0 \leq t \leq 5$.
8. Graph the function $y = 2(x - 3)^4$. Does the graph cross the y-axis? If so, where?
9. Graph the function $y = (3x + 5)/(x + 2)$. Specify all intercepts and asymptotes.
10. What is the largest area possible for a right triangle in which the sum of the lengths of the two shorter sides is 12 cm?
11. Graph the function $x/[(x + 2)(x - 4)]$.
12. Let $P(x, y)$ be a point [other than $(-1, -1)$] on the graph of $f(x) = x^3$. Express the slope of the line passing through the points P and $(-1, -1)$ as a function of x. Simplify your answer as much as possible.
13. A rectangle is inscribed in a circle. The circumference of the circle is 12 cm. Express the perimeter of the rectangle as a function of its width w.
14. Give a reason why each of the following two graphs cannot represent a polynomial function with highest-degree term $-\frac{1}{3}x^3$.

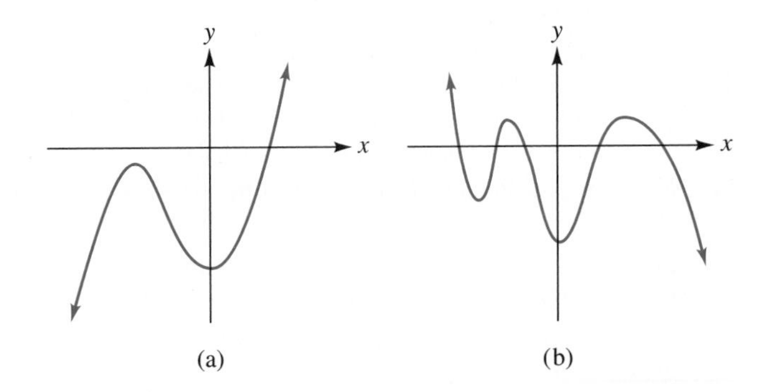

(a) (b)

In Exercises 15–20, find equations for the linear functions satisfying the given conditions. Write each answer in the form $f(x) = mx + b$.

15. $f(3) = 5$ and $f(-2) = 0$
16. $f(1) = 5$ and the graph of f passes through the origin.
17. $f(4) = -1$ and the graph of f is parallel to the line $3x - 8y = 16$.
18. The graph passes through (6, 1) and the x-intercept is twice the y-intercept.

19. $f(-3) = 5$, and the graph of the inverse function passes through (2, 1).

20. The graph of f passes through the vertices of the two parabolas $y = x^2 + 4x + 1$ and $y = \frac{1}{2}x^2 + 9x + \frac{81}{2}$.

In Exercises 21–26, graph the quadratic functions. In each case, specify the vertex and the x- and y-intercepts.

21. $y = x^2 + 2x - 3$

22. $f(x) = x^2 - 2x - 15$

23. $y = -x^2 + 2\sqrt{3}x + 3$

24. $f(x) = 2x^2 - 2x + 1$

25. $y = -3x^2 + 12x$

26. $f(x) = -4x^2 + 16x$

27. Find the distance between the vertices of the two parabolas $y = x^2 - 4x + 6$ and $y = -x^2 - 4x - 5$.

28. Find the value of a, given that the maximum value of the function $f(x) = ax^2 + 3x - 4$ is 5.

29. The sum of two numbers is $\sqrt{3}$. Find the largest possible value for their product.

30. The sum of two numbers is 2/3. What is the smallest possible value for the sum of their squares?

31. Suppose that an object is thrown vertically upward (from ground level) with an initial velocity of v_0 ft/sec. It can be shown that the height h (in feet) after t sec is given by the formula $h = v_0t - 16t^2$.
(a) At what time does the object reach its maximum height? What is that maximum height?
(b) At what time does the object strike the ground?

32. Let $f(x) = 4x^2 - x + 1$ and $g(x) = (x - 3)/2$.
(a) For which input will the value of the function $f \circ g$ be a minimum?
(b) For which input will the value of $g \circ f$ be a minimum?

33. **(a)** Let P be a point on the parabola $y = x^2$. Express the distance from P to the point (0, 2) in terms of x.
(b) Which point in the second quadrant on the parabola $y = x^2$ is closest to the point (0, 2)?

34. Find the coordinates of the point on the line $y = 2x - 1$ closest to $(-5, 0)$.

35. Find all values of b such that the minimum distance from the point (2, 0) to the line $y = \frac{4}{3}x + b$ is 5.

36. What number exceeds one-half its square by the greatest amount?

37. Suppose that $x + y = \sqrt{2}$. Find the minimum value of the quantity $x^2 + y^2$.

38. For which numbers t will the value of $9t^2 - t^4$ be as large as possible?

39. Find the maximum area possible for a right triangle with a hypotenuse of 15 cm. *Hint:* Let x denote the length of one leg. Show that the area is $A = x\sqrt{225 - x^2}/2$. Now work with A^2.

40. For which point (x, y) on the curve $y = 1 - x^2$ is the sum $x + y$ a maximum?

41. Let $f(x) = x^2 - (a^2 + 2a)x + 2a^3$, where $0 < a < 2$. For which value of a will the distance between the x-intercepts of f be a maximum?

42. Suppose that the revenue R (in dollars) generated by selling x units of a certain product is given by $R = 300x - \frac{1}{4}x^2 \quad (0 \leq x \leq 1200)$. How many units should be sold to maximize the revenue? What is that maximum revenue?

43. Suppose that the function $p = 160 - \frac{1}{5}x$ relates the selling price p of an item to the quantity x that is sold. Assume that p is in dollars. For which value of x will the revenue R be a maximum? What is the selling price p in this case?

44. A piece of wire 16 cm long is cut into two pieces. Let x denote the length of the first piece and $16 - x$ the length of the second. The first piece is formed into a rectangle in which the length is twice the width. The second piece of wire is also formed into a rectangle, but with the length three times the width. For which value of x is the total area of the two rectangles a minimum?

45. The following figure shows the graph of $f(x) = 3\sqrt{x - 1}$ and the first five steps in the iteration process for $x_0 = 2$.

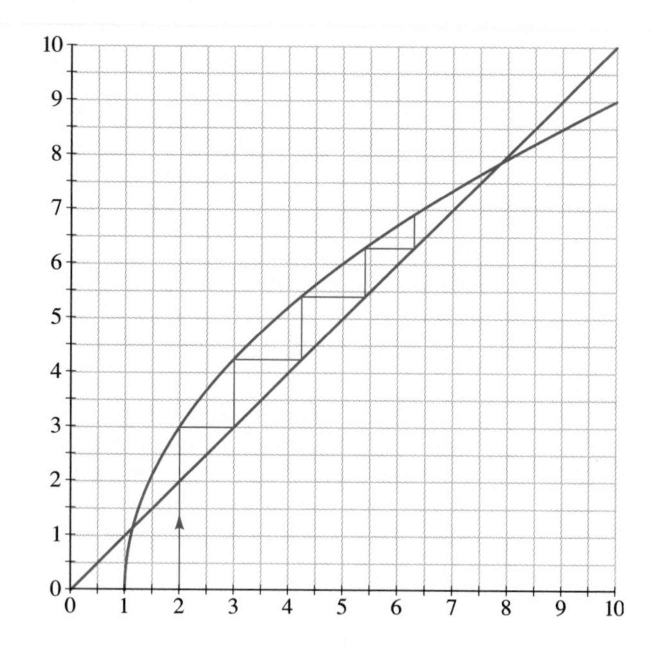

(a) Use the figure to estimate the iterates x_1 through x_5. (Estimate to the nearest one-half or one-fourth, as appropriate.) Then use a calculator to compute the iterates. Round the final answers to three decimal places. Check that the results are consistent with your graphical estimates.

(b) As indicated in the figure, the iteration process follows a staircase pattern that approaches an intersection point of the curve $f(x) = 3\sqrt{x-1}$ and the line $y = x$. The x-coordinate of this point is a fixed point for the function f. Determine this fixed point, and thus find the number that the iterates are approaching.

46. Find all real numbers that are fixed points of the given functions.
(a) $f(x) = x^2 - 8$ **(b)** $g(x) = x^2 + 8$
(c) $y = 4x - x^3$ **(d)** $y = 8x^2 - x - 15$

47. In the following figure, $PQRS$ is a square with sides parallel to the coordinate axes. The coordinates of points A and B are $A(a, 0)$ and $B(b, 0)$. Show that $f(a) = b$ and $f(b) = a$.

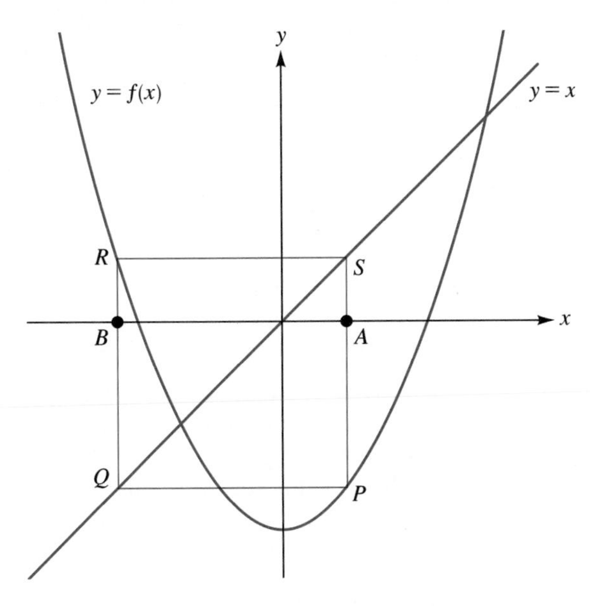

48. Let $f(x) = x^2 - 2$. Find two numbers a and b $(a \neq b)$ such that $f(a) = b$ and $f(b) = a$.

In Exercises 49–64, graph each function and specify the x- and y-intercepts and asymptotes, if any.

49. $y = (x+4)(x-2)$ **50.** $y = (x+4)(x-2)^2$
51. $y = -(x+5)^3$ **52.** $y = -x(x+1)$
53. $y = -x^2(x+1)$ **54.** $y = -x^3(x+1)$
55. $y = x(x-2)(x+2)$
56. $y = (x-3)(x+1)(x+5)$
57. $y = (3x+1)/x$ **58.** $y = (1-2x)/x$
59. $y = -1/(x-1)^2$ **60.** $y = (x+1)/(x+2)$
61. $y = (x-2)/(x-3)$ **62.** $y = x/(x-2)(x+4)$
63. $y = \dfrac{x^2 - 2x + 1}{x^2 - 4x + 4}$ **64.** $y = \dfrac{x(x-2)}{(x-4)(x+4)}$

65. Let $f(x) = x^2 + 2bx + 1$.
(a) If $b = 1$, find the distance from the vertex of the parabola to the origin.
(b) If $b = 2$, find the distance from the vertex of the parabola to the origin.
(c) For which real numbers b will the distance from the vertex to the origin be as small as possible?

66. Find the range of the function $f(x) = -2x^2 + 12x - 5$. *Hint:* Look at the graph.

67. The range of the function $y = x^2 - 2x + k$ is the interval $[5, \infty)$. Find the value of k.

68. Find a value for b such that the range of the function $f(x) = x^2 + bx + b$ is the interval $[-15, \infty)$.

69. Find the range of the function $y = \dfrac{(x-1)(x-3)}{x-4}$.
Hint: Solve the equation for x in terms of y using the quadratic formula. If you're careful with the algebra, you will find that the expression under the resulting radical sign is $y^2 - 8y + 4$. The range of the given function can then be found by solving the inequality $y^2 - 8y + 4 \geq 0$.

70. In the following figure, triangle OAB is equilateral and $\overline{AB}$ is parallel to the x-axis. Find the length of a side and the area of the triangle OAB.

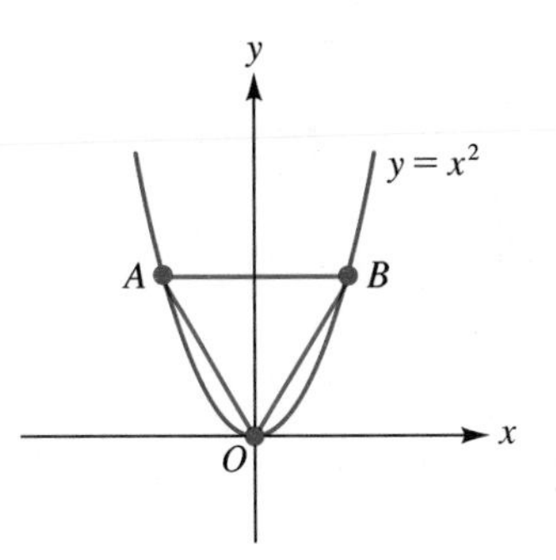

71. Let A denote the area of the right triangle in the first quadrant that is formed by the y-axis and the lines $y = mx$ and $y = m$. (Assume $m > 0$.) Express the area of the triangle as a function of m.

72. **(a)** Express the distance from the origin to the vertex of the parabola $y = x^2 - 2bx$ $(b > 0)$ as a function of b.
(b) Let V denote the vertex of the parabola in part (a), and let A and B denote the points where the curve meets the x-axis. Express the area of $\triangle VAB$ as a function of b.
(c) A circle is drawn with center V and passing through A and B. Then, through the smaller y-intercept of the circle, a tangent line is drawn. Express the x-intercept of this tangent line as a function of b.

73. In the accompanying figure, the radius of the circle is $OC = 1$. Express the area of $\triangle ABC$ as a function of x.

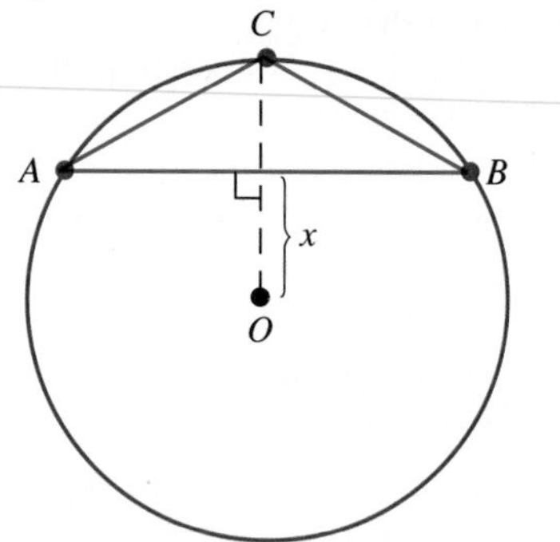

74. **(a)** Factor the expression $x^3 - 3x^2 + 4$.
Hint: Subtract and add 1, then factor by grouping.
(b) Use the factorization from part (a) to graph the function $y = x^3 - 3x^2 + 4$.

75. **(a)** Let $f(x) = ax^2 + bx + c \quad (a \neq 0)$. Find a constant x_0 such that the following equation is an identity:

$$f(x_0 + x) = f(x_0 - x)$$

(b) What is the geometric significance of your answer in part (a)?

76. In Section 4.2, we saw how the sign of the constant a influences the graph of the parabola $f(x) = ax^2 + bx + c$. In this exercise you'll see how the sign of b affects the graph.
(a) On the same set of axes, graph the parabolas $y = x^2 + 4x + 1$ and $y = x^2 - 4x + 1$. What type of symmetry do you observe?
(b) Let $f(x) = ax^2 + bx + c$. Compute $f(-x)$.
(c) Use the result in part (b) to describe what happens to the graph of $f(x) = ax^2 + bx + c$ when the sign of b is reversed.

CHAPTER FOUR TEST

1. A linear function L satisfies the conditions $L(-2) = -4$ and $L(5) = 1$. Find $L(0)$.

2. **(a)** Let $F(x) = 4x - 2x^2$. What is the maximum value for this function? On which interval is this function increasing?
(b) Let $G(t) = 9t^4 + 6t^2 + 2$. For which value of t is this function a minimum?

3. Let $f(x) = (x - 3)(x + 4)^2$.
(a) Determine the intercepts and the excluded regions for the graph of f.
(b) Determine the behavior of f when x is very close to -4.
(c) Sketch the graph of f.

4. Graph each function:
(a) $y = -1/(x - 3)^3$;
(b) $y = -1/(x - 3)^2$.

5. Graph the function $y = -x^2 + 7x + 6$. Specify the intercepts, the axis of symmetry, and the coordinates of the turning point.

6. Suppose that the function $p = -\frac{1}{6}x + 80 \quad (0 \leq x \leq 400)$ relates the selling price of an item to the quantity x that is sold. Assume that p is in dollars. What is the maximum revenue possible in this situation? Which price p generates this maximum revenue?

7. The price of a new machine is \$14,000. After ten years, the machine has a salvage value of \$750. Assuming linear depreciation, find a formula for the value of the machine after t years, where $0 \leq t \leq 10$.

8. Find the real numbers (if any) that are fixed points of the given functions.
   (a) $f(x) = 2x(1 - x)$
   (b) $g(x) = (4x - 1)/(3x + 6)$

9. The following figure shows the first six steps in the iteration process for the function $f(x) = 0.6 - x^2$ with initial input $x_0 = 0.2$.

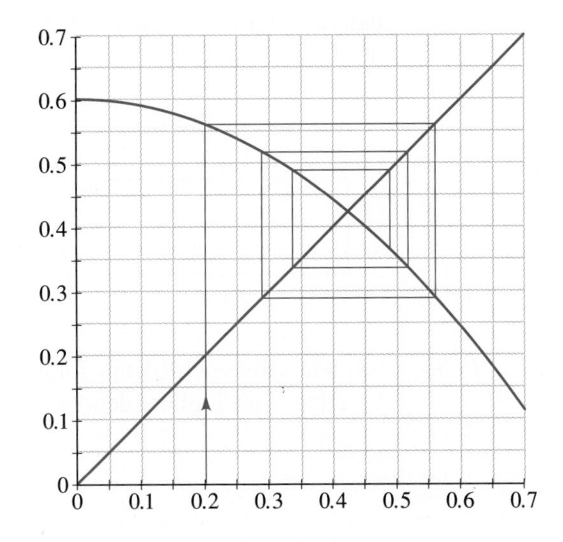

   (a) Use the figure and your calculator to complete the following table. For your answers from the graph, try to estimate to the nearest 0.02 (or 0.01). For the calculator answers, round to three decimal places.

	x_1	x_2	x_3	x_4	x_5	x_6
From graph						
From calculator						

   (b) Compute the two fixed points of the function $f(x) = 0.6 - x^2$. Give two forms for each answer: an exact expression involving a radical and a calculator approximation rounded to four decimal places. Which fixed point do the iterates in part (a) approach?
   (c) Use your calculator to compute the first six iterates of $x_0 = 1$. Do the iterates seem to approach either of the fixed points determined in part (b)?
   (d) Follow part (c) using $x_0 = 2$.

10. Graph the function $y = -\frac{1}{2}(3 - x)^3$. Specify the intercepts. *Hint:* First graph $y = -\frac{1}{2}(x + 3)^3$.

11. Graph the function $y = (2x - 3)/(x + 1)$. Specify the intercepts and asymptotes.

12. **(a)** Suppose that $P(x, y)$ is a point on the line $y = 3x - 1$ and Q is the point $(-1, 3)$. Express the length PQ as a function of x.
    (b) For which value of x is the length PQ a minimum?

13. Let $f(x) = x(x - 2)/(x^2 - 9)$.

(a) Find the vertical and the horizontal asymptotes.
(b) Use a sketch to show the behavior of f near the x-intercepts.
(c) Use a sketch to show the behavior of f near the asymptotes.
(d) Sketch the graph of f.

14. A rectangle is inscribed in a semicircle of diameter 8 cm. (See the accompanying figure.) Express the area of the rectangle as a function of the width w of the rectangle.

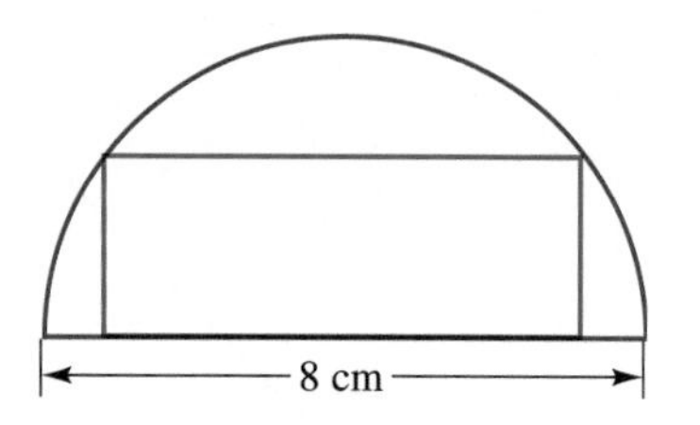

15. Explain why each of the following graphs cannot represent a polynomial function with highest-degree term $-x^3$.

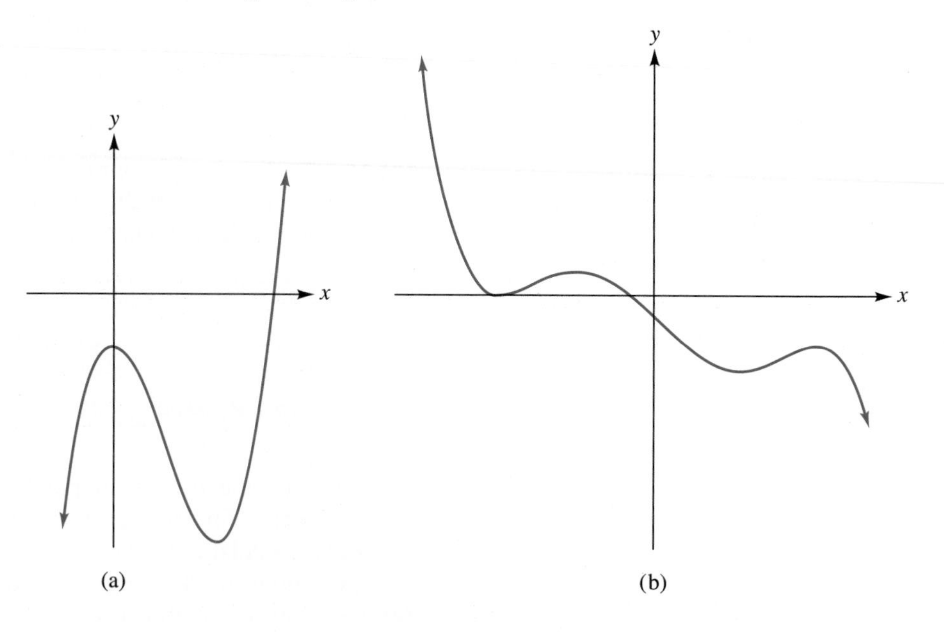

(a) (b)

16. For both political and economic reasons, the global arms trade has declined dramatically since 1987. The following table shows how worldwide exports of conventional weapons (measured in billion 1991 dollars) fell over the period 1987–1990.

Year	1987	1988	1989	1990
Exports of conventional weapons (billion 1991 dollars)	68.7	61.9	54.3	39.8

Worldwide exports of conventional weapons, 1987–1990. Source: Lester R. Brown, et al., *Vital Signs 1994* (New York: W. W. Norton, 1994)

(a) Draw a scatter diagram (i.e., plot the points) for the data in the table. Then sketch in the line that you feel best seems to fit the data.

(b) Using formulas from statistics, it can be shown that the equation for the regression line is $y = -9.430x + 18807.730$. Use this equation to estimate the worldwide exports of conventional weapons for the year 1991. (The actual estimate for 1991 from the afore-mentioned publication was 25.5 billion dollars. Is your estimate too high or too low?)

EXPONENTIAL AND LOGARITHMIC FUNCTIONS

The exponential growth of population and its attendant assault on the environment is so recent that it is difficult for people to appreciate how much damage is being done.

Nathan Keyfitz in "The Growing Human Population," *Scientific American,* vol. 261 (September 1989), 119–126

... in 1958 ... C. D. Keeling of The Scripps Institution of Oceanography ... began a series of painstaking measurements of CO_2 concentration on a remote site at what is now the Mauna Loa Observatory in Hawaii. His observations, continued to the present, show an exponential growth of atmospheric carbon dioxide.... Keeling's observations have been duplicated at other stations in various parts of the world over shorter periods of time. In all sets of observations, the exponential increase is clear....

Gordon MacDonald in "Scientific Basis for the Greenhouse Effect," from *The Challenge of Global Warming,* Dean Edwin Abrahamson, ed. (Washington, D.C.: Island Press, 1989)

INTRODUCTION

In the previous chapter we saw that linear functions can be used to describe or *model* certain data sets. For instance, when we looked at the records for the one-mile run, we found that the times decreased in a linear fashion. In this chapter we consider two other important types of functions that are often useful as models: the *exponential functions* and their inverses, the *logarithmic functions.* Figure 1 provides an example. Figure 1(a) shows the population of the United States over the years 1800–1900 along with a scatter plot of the data. The scatter plot makes it clear that the population did *not* increase in a linear fashion. Figure 1(b) shows the graph of a function that does fit this data very closely; this function is an exponential function. (The actual equation for this particular function is $y = ab^t$, where $a \approx 4.33 \times 10^{-15}$ and $b \approx 1.027$.)

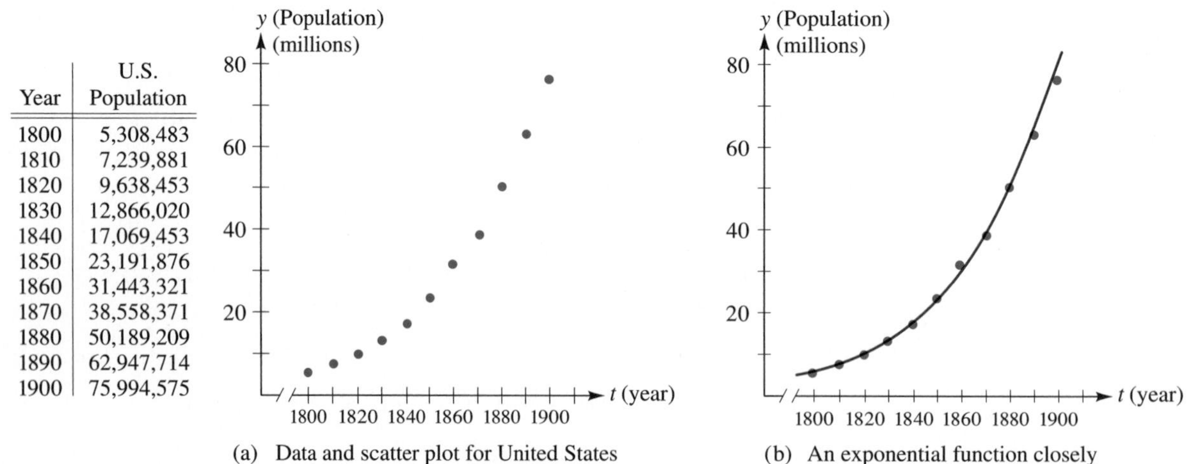

Year	U.S. Population
1800	5,308,483
1810	7,239,881
1820	9,638,453
1830	12,866,020
1840	17,069,453
1850	23,191,876
1860	31,443,321
1870	38,558,371
1880	50,189,209
1890	62,947,714
1900	75,994,575

FIGURE 1

(a) Data and scatter plot for United States population over the years 1800-1900.

(b) An exponential function closely models the population growth.

As you will see in Sections 5.6 and 5.7, the applications of exponential functions are quite diverse and not restricted to population growth. For instance, in studying the greenhouse effect, scientists obtain graphs like the one in Figure 2. This graph shows the concentration of carbon dioxide in the atmosphere over the years 1800–1995. Again, the growth can be described or modeled very closely by an exponential function.

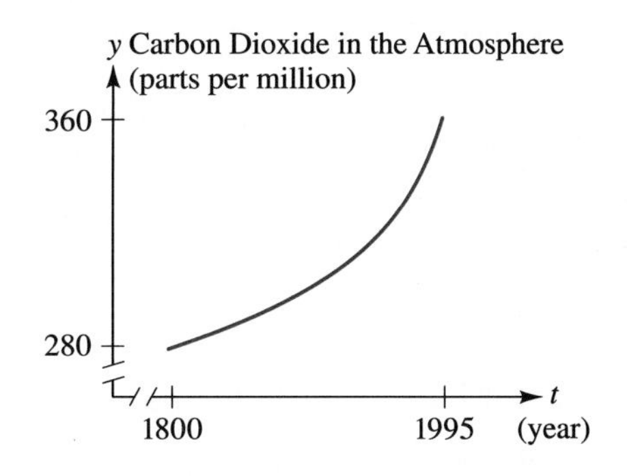

FIGURE 2
The exponential growth of carbon dioxide levels in the atmosphere.

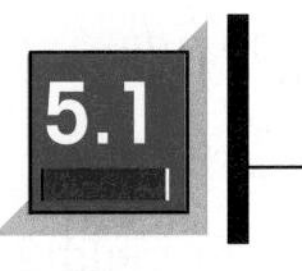

EXPONENTIAL FUNCTIONS

We begin with an example. Suppose that your mathematics instructor, in an effort to improve classroom attendance, offers to pay you each day for attending class! Suppose you are to receive 2¢ on the first day you attend class, 4¢ the second day, 8¢ the third day, and so on, as shown in Table 1. How much money will you receive for attending class on the 30th day?

TABLE 1

x (Day Number)	y (Amount Earned That Day)
1	2¢ $(=2^1)$
2	4¢ $(=2^2)$
3	8¢ $(=2^3)$
4	16¢ $(=2^4)$
5	32¢ $(=2^5)$
⋮	⋮
x	2^x

As you can see by looking at Table 1, the amount y earned on day x is given by the rule, or *exponential function,*

$$y = 2^x$$

Thus on the 30th day (when $x = 30$), you will receive

$$y = 2^{30} \text{ cents}$$

If you use a calculator, you will find this amount to be well over 10 million dollars! The point here is simply this: Although we begin with a small amount, $y = 2$¢, repeated doubling quickly leads to a very large amount. In other words, the exponential function grows very rapidly.

Before leaving this example, we mention a simple method for quickly estimating numbers such as 2^{30} (or any power of two) in terms of the more familiar powers of ten. Begin by observing that

$$2^{10} \approx 10^3 \qquad \text{(a useful coincidence, worth remembering)}$$

Now just cube both sides to obtain

$$(2^{10})^3 \approx (10^3)^3 \qquad \text{or} \qquad 2^{30} \approx 10^9$$

Thus 2^{30} is about one billion. To convert this number of cents to dollars, we divide by 100 or 10^2 to obtain

$$\frac{10^9}{10^2} = 10^7 \text{ dollars}$$

which is 10 million dollars, as mentioned before.

EXAMPLE 1 Estimate 2^{40} in terms of a power of 10.

Solution Take the basic approximation $2^{10} \approx 10^3$ and raise both sides to the fourth power. This yields

$$(2^{10})^4 \approx (10^3)^4$$

or

$$2^{40} \approx 10^{12}, \qquad \text{as required}$$

EXAMPLE 2 Estimate the power to which 10 must be raised to yield 2.

Solution We begin with our approximation

$$10^3 \approx 2^{10}$$

Raising both sides to the power $1/10$ yields

$$(10^3)^{1/10} \approx (2^{10})^{1/10}$$

and, consequently,

$$10^{3/10} \approx 2$$

Thus the power to which 10 must be raised to yield 2 is approximately $3/10$.

In Appendix A.6 we define the expression b^x, where x is a rational number. We also state that if x is irrational, then b^x can be defined so that the usual properties of exponents continue to hold. Although a rigorous definition of irrational exponents requires concepts from calculus, we can nevertheless convey the basic idea by means of an example. (We need to do this before we give the general definition for exponential functions.)

How shall we assign a meaning to $2^{\sqrt{2}}$, for example? The basic idea is to evaluate the expression 2^x successively by using rational numbers x that are closer and closer to $\sqrt{2}$. Table 2 displays the results of some calculations along these lines.

TABLE 2 Values of 2^x for Rational Numbers x Approaching $\sqrt{2}$ $(= 1.41421356\ldots)$

x	1.4	1.41	1.414	1.4142	1.41421	1.414213
2^x	2.6...	2.65...	2.664...	2.6651...	2.66514...	2.665143...

The data in the table suggests that as x approaches $\sqrt{2}$, the corresponding values of 2^x approach a unique real number, call it t, with a decimal expansion that begins as 2.665. Furthermore, by continuing this process we can obtain (in theory, at least) as many places in the decimal expansion of t as we wish. The value of the expression $2^{\sqrt{2}}$ is then defined to be this number t. The following results (stated here without proof) summarize this discussion and also pave the way for the definition we will give for exponential functions.

PROPERTY SUMMARY **REAL NUMBER EXPONENTS**

Let b denote an arbitrary positive real number. Then:

1. For each real number x, the quantity b^x is a unique real number.
2. When x is irrational, we can approximate b^x as closely as we wish by evaluating b^r, where r is a rational number sufficiently close to the number x.
3. The properties of rational exponents continue to hold for irrational exponents.
4. If $b^x = b^y$ and $b \neq 1$, then $x = y$.

EXAMPLE 3 Use the properties of exponents to simplify each expression: **(a)** $(3^{\sqrt{2}})^{\sqrt{2}}$; **(b)** $(3^{\sqrt{2}})^2$.

Solution

(a) $(3^{\sqrt{2}})^{\sqrt{2}} = 3^{\sqrt{2} \times \sqrt{2}} = 3^2 = 9$

(b) $(3^{\sqrt{2}})^2 = (3^2)^{\sqrt{2}}$ (Why?)

$= 9^{\sqrt{2}}$

EXAMPLE 4 Solve the equation $4^x = 8$.

Solution First let's estimate x, just to get a feeling for what kind of answer to expect. Since $4^1 = 4$, which is less than 8, and $4^2 = 16$, which is more than 8, we know that our final answer should be a number between 1 and 2. To obtain this answer, we take advantage of the fact that both 4 and 8 are powers of 2. Using this fact, we can write the given equation as

$$(2^2)^x = 2^3$$
$$2^{2x} = 2^3$$
$$2x = 3 \quad \text{using Property 4 in the Property Summary Box}$$
$$x = 3/2$$

Note that the answer $x = 3/2$ is indeed between 1 and 2, as we estimated.

For the remainder of this section, b denotes an arbitrary positive constant other than 1. In the box that follows, we define the exponential function with base b.

DEFINITION: The Exponential Function with Base b

Let b denote an arbitrary positive constant other than 1. The **exponential function with base b** is defined by the equation

$$y = b^x$$

Note In many scientific applications, functions of the form $y = ab^{kx}$, where a, b, and k are constants, are also referred to as exponential functions.

EXAMPLES

1. The equations $y = 2^x$ and $y = 3^x$ define the exponential functions with bases 2 and 3, respectively.
2. The equation $y = (1/2)^x$ defines the exponential functions with base $1/2$.
3. The equations $y = x^2$ and $y = x^3$ do not define exponential functions.

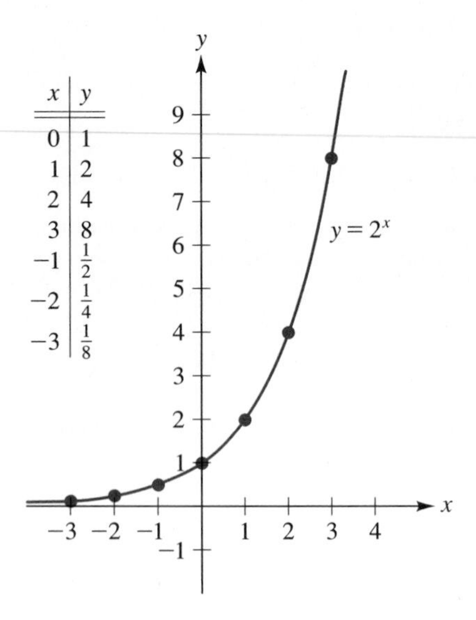

FIGURE 1

To help with our analysis of exponential functions, let's set up a table and use it to graph the exponential function $y = 2^x$. This is done in Figure 1. In drawing a smooth and unbroken curve, we are actually relying on the results in the Property Summary Box on the previous page. The key features of the exponential function $y = 2^x$ and its graph are:

1. The domain of $y = 2^x$ is the set of all real numbers. The range is the set of all *positive* real numbers.
2. The y-intercept of the graph is 1. The graph has no x-intercept.
3. For $x > 0$, the function increases or grows very rapidly. For $x < 0$, the graph rapidly approaches the x-axis; the x-axis is a horizontal asymptote for the graph. (Recall from Section 4.7 that a line is an **asymptote** for a curve if the separation distance between the curve and the line approaches zero as we move farther and farther out along the line.)

You should memorize the basic shape and features of the graph of $y = 2^x$ so that you can sketch it as needed without first setting up a table. The next example shows why this is useful.

EXAMPLE 5 Graph each of the following functions. In each case specify the domain, the range, the intercept(s), and the asymptote.

(a) $y = -2^x$ **(b)** $y = 2^{-x}$ **(c)** $y = (1/2)^x$ **(d)** $y = 2^{-x} - 2$

Solution

(a) Recall that -2^x means $-(2^x)$, not $(-2)^x$. The graph of $y = -2^x$ is obtained by reflecting the graph of $y = 2^x$ in the x-axis. See Figure 2(a).

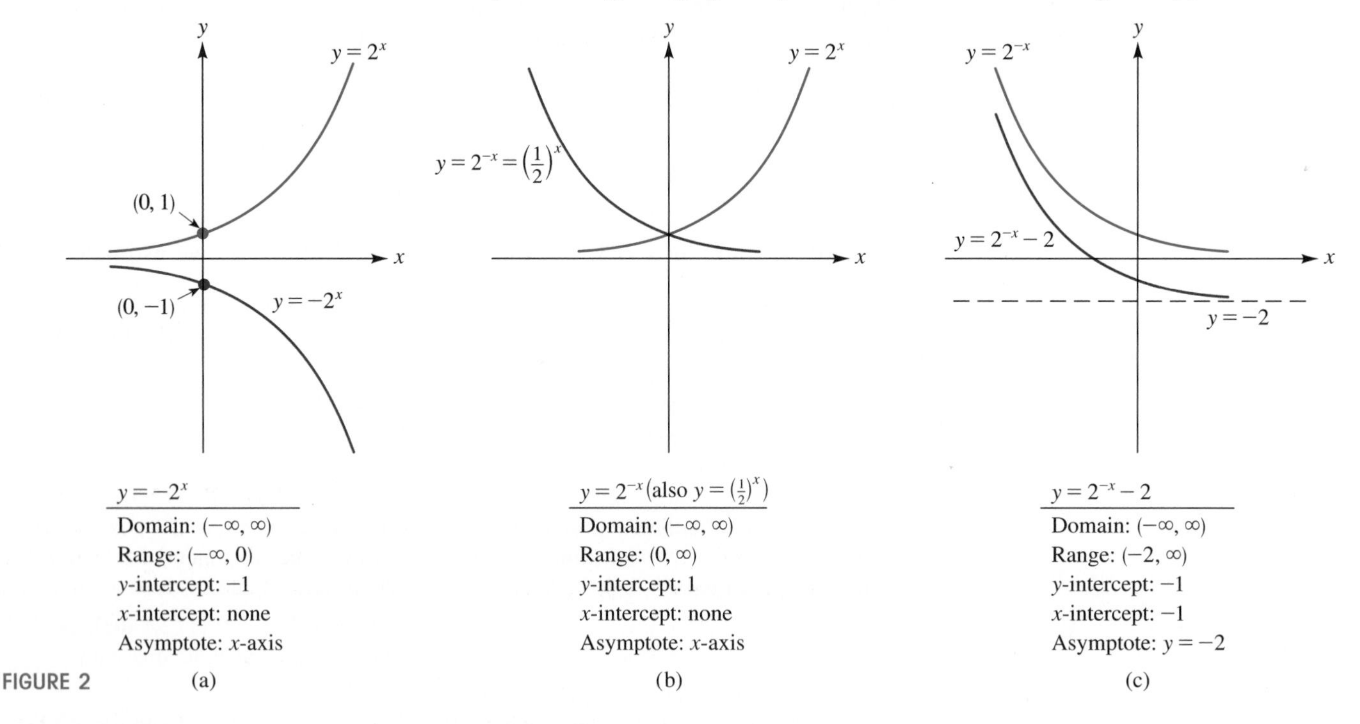

FIGURE 2

(b) Similarly, the graph of $y = 2^{-x}$ is obtained from the graph of $y = 2^x$ by reflection in the y-axis. See Figure 2(b).

(c) Next, regarding $y = (1/2)^x$, observe that

$$\left(\frac{1}{2}\right)^x = \frac{1^x}{2^x} = \frac{1}{2^x} = 2^{-x}$$

In other words, $y = (\frac{1}{2})^x$ is really the same function as $y = 2^{-x}$, which we already graphed in Figure 2(b).

(d) Finally, to graph $y = 2^{-x} - 2$, take the graph of $y = 2^{-x}$ in Figure 2(b) and move it two units in the negative y-direction, as shown in Figure 2(c). Note that the asymptote and y-intercept will also move down two units. To find the x-intercept, we set $y = 0$ in the given equation to obtain

$$\begin{aligned} 2^{-x} - 2 &= 0 \\ 2^{-x} &= 2^1 \\ -x &= 1 \qquad \text{using Property 4 on page 273} \\ x &= -1 \end{aligned}$$

Thus, the x-intercept is -1.

In the next example, we apply our knowledge about the graph of $y = 2^x$ to solve an equation. In particular, we use the fact that the graph of $y = 2^x$ always lies above the x-axis; for no value of x is 2^x ever zero.

EXAMPLE 6 Solve the equation $x^2 2^x - 2^x = 0$.

Solution First we factor the left-hand side of the equation; the common term is 2^x. This gives us

$$2^x(x^2 - 1) = 0$$

Since 2^x is never zero, we can now divide both sides of this last equation by 2^x (without losing a solution) to obtain

$$\begin{aligned} x^2 - 1 &= 0 \\ (x - 1)(x + 1) &= 0 \end{aligned}$$

and, consequently,

$$x = 1 \qquad \text{or} \qquad x = -1$$

Thus the required solutions are $x = 1$, $x = -1$. (You should check for yourself that each of these values satisfies the original equation.)

Now what about exponential functions with bases other than 2? As Figure 3 indicates, the graphs are similar to $y = 2^x$. The graph of $y = 4^x$ rises more rapidly than $y = 2^x$ when x is positive. For negative x-values, the graph of $y = 4^x$ is below that of $y = 2^x$. You can see why this happens by taking $x = -1$, for example, and comparing the values of 4^x and 2^x. If $x = -1$, then

$$2^x = 2^{-1} = \frac{1}{2} \qquad \text{but} \qquad 4^x = 4^{-1} = \frac{1}{4}$$

Therefore $4^x < 2^x$ when $x = -1$. Notice also in Figure 3 that all three graphs have the same y-intercept of 1. This follows from the fact that $b^0 = 1$ for any positive number b.

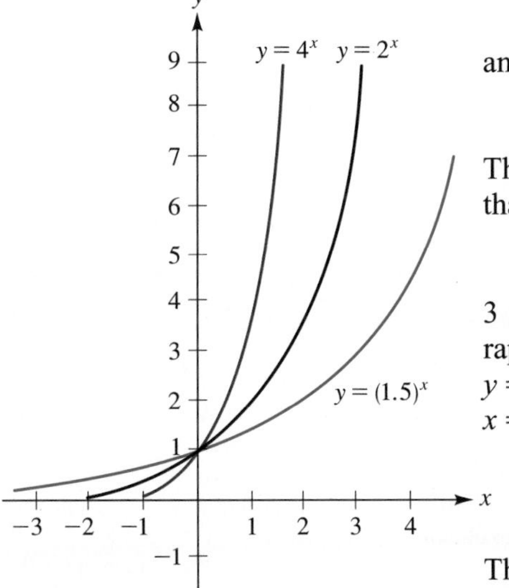

FIGURE 3

The exponential functions in Figure 3 (on the previous page) all have bases larger than 1. To see examples in which the bases are in the interval $0 < b < 1$, we need only reflect the graphs in Figure 3 in the y-axis. For instance, the reflection of $y = 4^x$ in the y-axis gives us the graph of $y = (1/4)^x$. (We discussed the idea behind this in Example 5; see Figure 2(b), for instance.)

In the box that follows, we summarize what we've learned up to this point regarding the exponential function $y = b^x$.

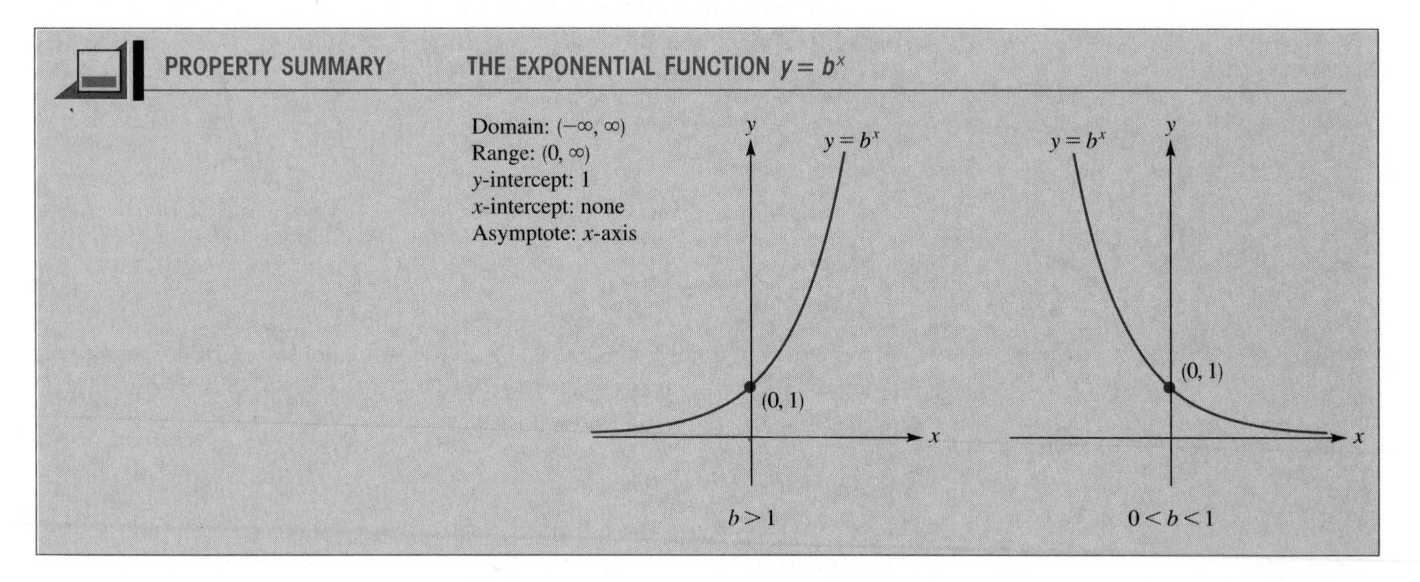

EXAMPLE 7 Graph the function $y = -3^{-x} + 1$.

Solution The required graph is obtained by reflecting and translating the graph of $y = 3^x$, as shown in Figure 4.

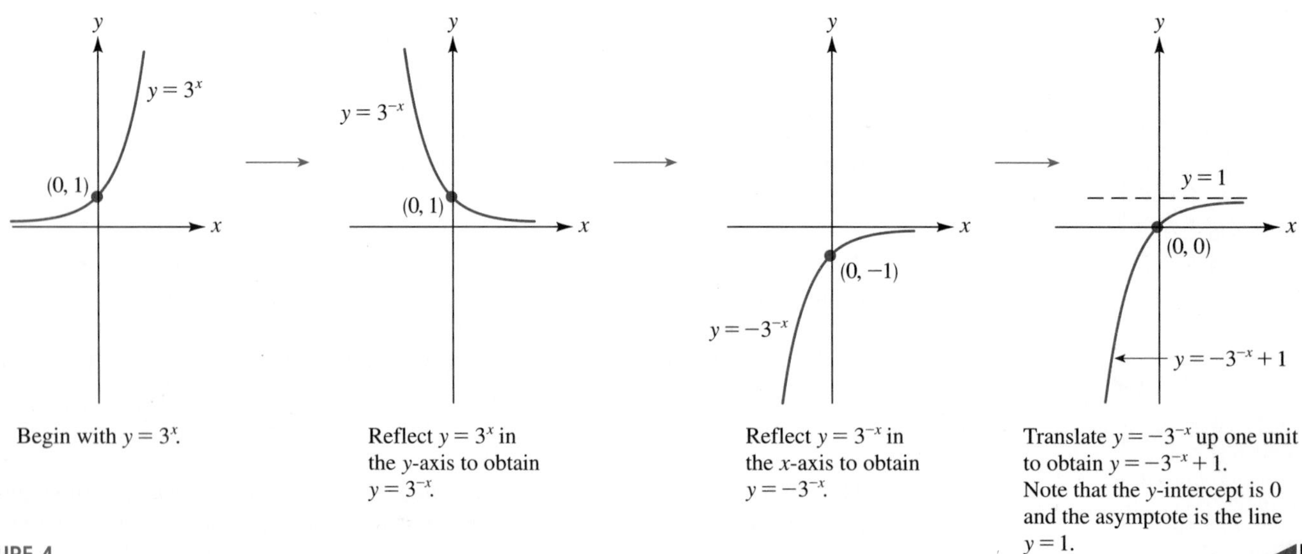

Begin with $y = 3^x$.

Reflect $y = 3^x$ in the y-axis to obtain $y = 3^{-x}$.

Reflect $y = 3^{-x}$ in the x-axis to obtain $y = -3^{-x}$.

Translate $y = -3^{-x}$ up one unit to obtain $y = -3^{-x} + 1$. Note that the y-intercept is 0 and the asymptote is the line $y = 1$.

FIGURE 4

EXERCISE SET 5.1

Do 1, 3, 5, 11, 15, 25, 27

A

In Exercises 1 and 2, estimate each quantity in terms of powers of ten, as in Example 1.

1. (a) 2^{30} (b) 2^{50}

2. (a) 2^{90} (b) 4^{50}

In Exercises 3–10, use the properties of exponents to simplify each expression. In Exercises 9 and 10, write the answers in the form b^n, where b and n are real numbers.

3. $(5^{\sqrt{3}})^{\sqrt{3}}$

4. $(\sqrt{2}^{\sqrt{2}})^{\sqrt{2}}$

5. $(4^{1+\sqrt{2}})(4^{1-\sqrt{2}})$

6. $(3^{2+\sqrt{5}})(3^{2-\sqrt{5}})$

7. $\dfrac{2^{4+\pi}}{2^{1+\pi}}$

8. $\dfrac{10^{\pi+2}}{10^{\pi-2}}$

9. $(\sqrt{5}^{\sqrt{2}})^2$

10. $[(\sqrt{3})^{\pi}]^4$

In Exercises 11 and 12, solve each equation, as in Example 4.

11. **(a)** $3^x = 27$ **(b)** $9^t = 27$ **(c)** $3^{1-2y} = \sqrt{3}$ **(d)** $3^z = 9\sqrt{3}$

12. **(a)** $2^x = 32$ **(b)** $2^t = 1/4$ **(c)** $2^{3y+1} = \sqrt{2}$ **(d)** $8^{z+1} = 32\sqrt{2}$

In Exercises 13–16, specify the domain of the function.

13. $y = 2^x$

14. $y = 1/2^x$

15. $y = 1/(2^{x-1})$

16. $y = 1/(2^x - 1)$

In Exercises 17–24, graph the pair of functions on the same set of axes.

17. $y = 2^x$; $y = 2^{-x}$

18. $y = 3^x$; $y = 3^{-x}$

19. $y = 3^x$; $y = -3^x$

20. $y = 4^x$; $y = -4^x$

21. $y = 2^x$; $y = 3^x$

22. $y = (1/3)^x$; $y = 3^x$

23. $y = (1/2)^x$; $y = (1/3)^x$

24. $y = (1/2)^{-x}$; $y = (1/3)^{-x}$

For Exercises 25–32, graph the function and specify the domain, range, intercept(s), and asymptote.

25. $y = -2^x + 1$

26. $y = -3^x + 3$

27. $y = 3^{-x} + 1$

28. $y = 3^{-x} - 3$

29. $y = 2^{x-1}$

30. $y = 2^{x-1} - 1$

31. $y = 3^{x+1} + 1$

32. $y = 1 - 3^{x-1}$

For Exercises 33–36, solve the equation.

33. $3x(10^x) + 10^x = 0$

34. $4x^2(2^x) - 9(2^x) = 0$

35. $3(3^x) - 5x(3^x) + 2x^2(3^x) = 0$

36. $\dfrac{(x+4)10^x}{x-3} = 2x(10^x)$

B

37. For this exercise, recall from Section 4.3 that a fixed point of a function f is an input x in the domain of f such that $f(x) = x$.

(a) On the same set of axes, sketch the graphs of $y = 2^{-x}$ and $y = x$.

(b) Your graph in part (a) shows that the function $y = 2^{-x}$ has exactly one fixed point. Between what two consecutive integers does this fixed point lie? [To determine the fixed point of the function $y = 2^{-x}$ precisely, we would need to solve the equation $2^{-x} = x$. However, this equation cannot be solved by using any of our basic techniques from Chapter 1. Go on to part (c) to find out about an approximate solution.]

(c) The following figure shows the curve $y = 2^{-x}$ and the parabola $y = \frac{1}{8}x^2 - \frac{5}{8}x + 1$. Notice that, when x is between 0 and 2, the curves appear virtually indistinguishable from one another. Thus, to find an approximate solution (between 0 and 2) of the equation $2^{-x} = x$, solve the *quadratic* equation $\frac{1}{8}x^2 - \frac{5}{8}x + 1 = x$. [Use the answer from part (b) to decide which of the two roots is appropriate.] Give two forms for the final answer: an exact expression involving a radical and a calculator approximation rounded to three decimal places.

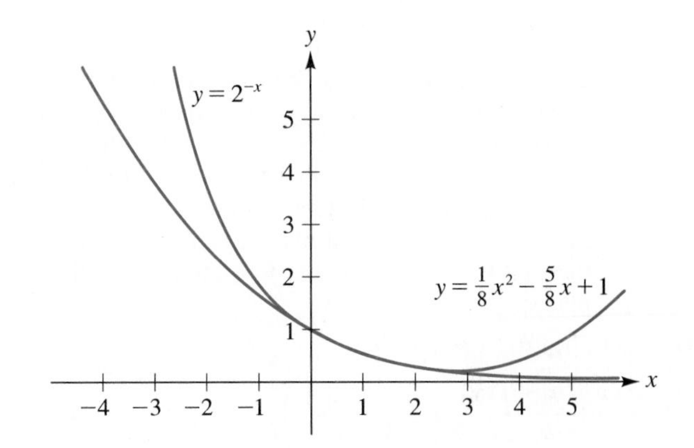

(d) Using either techniques from calculus or a graphics calculator, it can be shown that the fixed point of the function $y = 2^{-x}$ is 0.641179 (rounded to six decimal places). To how many decimal places does your approximation in part (c) agree with this value?

38. The following figure shows the first five steps in the iteration process for $f(x) = 2^{-x}$ with initial input $x_0 = 0.1$.

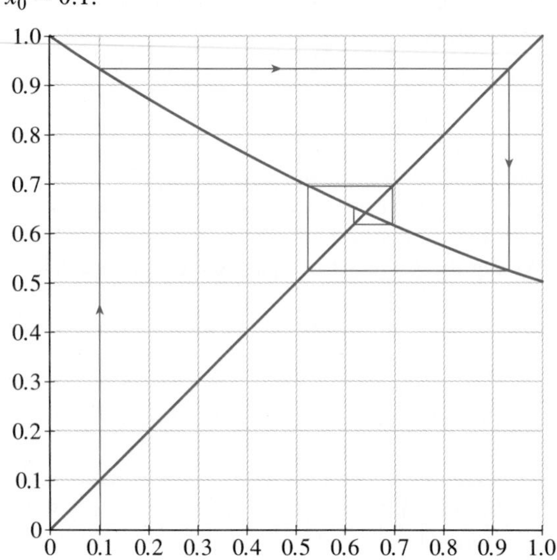

(a) Use the figure and your calculator to complete the following table. For the values that you read from the graph, try to estimate to within 0.02 (or 0.01 if it seems appropriate). For the calculator values, round the final answers to four decimal places.

	x_1	x_2	x_3	x_4	x_5
From Graph					
From Calculator					

(b) The preceding figure shows that the iterates of 0.1 are approaching a fixed point of the function. As mentioned in Exercise 37(d), this fixed point is approximately 0.641179. Using your calculator, find out which iterate is the first to have a 6 in the first decimal place; which iterate is the first to have 64 in the first two decimal places; and which iterate is the first to have 641 in the first three decimal places.

39. Let $f(x) = 2^x$. Show that

$$\frac{f(x+h) - f(x)}{h} = 2^x\left(\frac{2^h - 1}{h}\right).$$

40. Let $\phi(t) = 1 + a^t$. Show that $\dfrac{1}{\phi(t)} + \dfrac{1}{\phi(-t)} = 1$.

41. Let $f(x) = 2^x$ and let g denote the function that is the inverse of f.

(a) On the same set of axes, sketch the graphs of f, g, and the line $y = x$.

(b) Using the graph you obtained in part (a), specify the domain, range, intercept, and asymptote for the function g.

42. Let $S(x) = (2^x - 2^{-x})/2$ and $C(x) = (2^x + 2^{-x})/2$. Compute $[C(x)]^2 - [S(x)]^2$.

For Exercises 43–50, refer to the following figure, which shows portions of the graphs of $y = 2^x$, $y = 3^x$, and $y = 5^x$. In each case, ***(a)*** *use the figure to estimate the indicated quantity, and* ***(b)*** *use a calculator to compute the indicated quantity, rounding the result to two decimal places.*

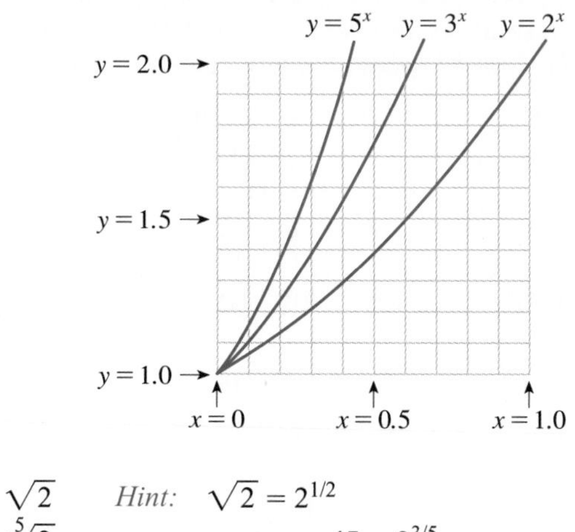

43. $\sqrt{2}$ *Hint:* $\sqrt{2} = 2^{1/2}$

44. $\sqrt[5]{2}$

45. $2^{3/5}$

46. $\sqrt[5]{8}$ *Hint:* $8^{1/5} = 2^?$

47. $\sqrt{3}$

48. $\sqrt[3]{3}$

49. $5^{3/10}$

50. $\sqrt[4]{5}$

In Exercises 51–54, refer to the following graph of the exponential function $y = 10^x$. Use the graph to estimate (to the nearest tenth) the solution of each equation. (After we've studied logarithmic functions later in the chapter, we will be able to obtain more precise solutions.)

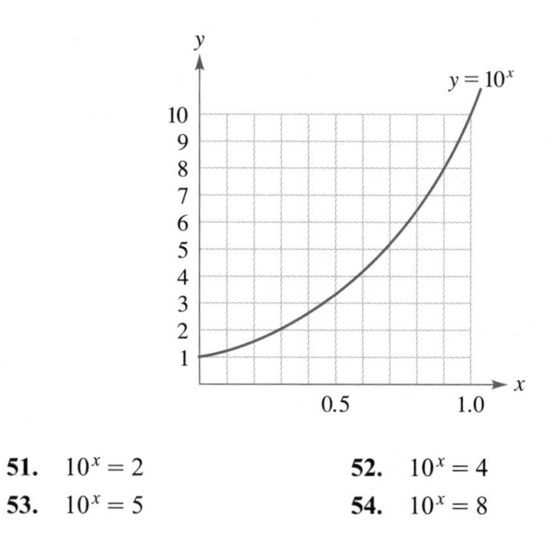

51. $10^x = 2$

52. $10^x = 4$

53. $10^x = 5$

54. $10^x = 8$

C

55. This exercise serves as a preview for the work on logarithms in Section 5.3. Follow Steps (a)–(f) to complete the table. (Notice that one entry in the table is already filled in. Reread Example 2 in the text to see how that entry was obtained.)

(a) Fill in the entries in the right-hand column corresponding to $x = 1$ and $x = 10$.

(b) Note that 4 and 8 are powers of 2. Use this information along with the approximation $10^{0.3} \approx 2$ to find the entries in the table corresponding to $x = 4$ and $x = 8$.

(c) Find the entry corresponding to $x = 5$.
Hint: $5 = 10/2 \approx 10/10^{0.3}$.

(d) Find the entry corresponding to $x = 7$.
Hint: $7^2 \approx 50 = 5 \times 10$; now make use of your answer in part (c).

(e) Find the entry corresponding to $x = 3$.
Hint: $3^4 \approx 80 = 8 \times 10$.

(f) Find the entries corresponding to $x = 6$ and $x = 9$. *Hint:* $6 = 3 \times 2$ and $9 = 3^2$.

Remark: This table is called a *table of logarithms to the base* 10. We say, for example, that the logarithm of 2 to the base 10 is (about) 0.3. We write this symbolically as $\log_{10} 2 \approx 0.3$.

x	Exponent to Which 10 Must Be Raised to Yield x
1	
2	≈0.3
3	
4	
5	
6	
7	
8	
9	
10	

GRAPHING UTILITY EXERCISES FOR SECTION 5.1

1. Reread Example 4 in Section 5.1. Then, for a graphical perspective, graph the exponential function $y = 4^x$ along with the line $y = 8$. By zooming in, estimate the x-coordinate of the intersection point. How close do you come to the solution $x = 3/2$ obtained in the text?

2. **(a)** Graph the function $y = 2^x$ in the standard viewing rectangle. Check that your result is similar to that shown in Figure 1 in the text.

(b) In the view obtained in part (a), note that the curve is virtually indistinguishable from the x-axis for x-values that are less than -2. According to the text, however, the curve never actually touches the x-axis. Change the y-boundaries in your viewing rectangle so that y extends from -1 to 1. Does any part of the curve appear to touch the x-axis? What if you change the y-boundaries so that y extends from -0.1 to 0.1?

3. Graph the equation $y = x^2 2^x - 2^x$ in the standard viewing rectangle. What are the x-intercepts? How does this relate to Example 6 in the text?

4. For this exercise use a viewing rectangle extending from -2 to 3 in the x-direction and from -1 to 20 in the y-direction.

(a) In the same picture, graph the functions $y = 2^x$, $y = 3^x$, and $y = 4^x$.

(b) Where do you think the graph of $y = 75^x$ would fit into your picture from part (a)? After answering, add the graph of $y = 75^x$ to your picture and see if you're right.

For Exercises 5–10: **(a)** *use pencil and paper to determine the intercepts and asymptotes for the graph of each function; and* **(b)** *use a graphing utility to graph the function (choose an appropriate viewing rectangle so that the essential features of the graph are clear), then check to see if your answers in part (a) are consistent with the graph.*

5. $y = -3^{x-2}$

6. $y = -3^{x-2} + 1$

7. $y = 4^{-x} - 4$

8. $y = 4 + (1/4)^x$

9. $y = 10^{x-1}$

10. $y = -10^{x-1} + 0.5$

11. If you add two quadratic functions, the result is again a quadratic function (assuming that the x^2-terms don't add to zero). For example, if $f(x) = x^2 + 2x$ and $g(x) = 2x^2 - 1$, then the sum is $(f + g)(x) = 3x^2 + 2x - 1$, another quadratic. But this is not the case for exponential functions. To explore a visual example, consider the function $y = 2^x + 2^{-x}$. This is the sum of the two exponentials $y = 2^x$ and $y = (1/2)^x$.

(a) Graph the function $y = 2^x + 2^{-x}$ in the standard viewing rectangle. As you can see, the resulting graph is U-shaped.

(b) For comparison, add the graph of $y = x^2$ to your picture. Note that (at this scale) the graphs are very similar for $|x| > 2$.

(c) Actually, the graph of $y = 2^x + 2^{-x}$ rises much more steeply than does $y = x^2$. To demonstrate this, change the viewing rectangle so that x extends from -100 to 100 and y extends from 0 to 100,000.

12. Follow the general procedure in Exercise 11, but use the two functions $y = 2^x - 2^{-x}$ and $y = x^3$.

13. As background for this exercise, reread Example 2 in the text. (The example shows how to estimate the value of x for which $10^x = 2$.)

(a) In the same picture, graph the two functions $y = 10^x$ and $y = 2$. Use a viewing rectangle in which x runs from 0 to 1 and y runs from 0 to 10.

(b) Explain how your picture in part (a) supports the answer 3/10 that was obtained in Example 2.

5.2 THE EXPONENTIAL FUNCTION $y = e^x$

This is undoubtedly the most important function in mathematics.

Walter Rudin in *Real and Complex Analysis* (New York: McGraw-Hill, 1966)

From the standpoint of calculus and scientific applications, one particular base for exponential functions is by far the most useful. This base is a certain irrational number that lies between 2 and 3 and is denoted by the letter e. For purposes of approximation, you'll need to know that

$$e \approx 2.7$$

[The Swiss mathematician Leonhard Euler (1707–1783) introduced the letter e to denote this number. To six decimal places, the value of e is 2.718281....]

At the precalculus level, it's hard to escape the feeling that $y = 2^x$ or $y = 10^x$ is by far more simple and more natural than $y = e^x$. So as we work with e and e^x in this chapter, you will need to take it on faith that, in the long run, the number e and the function $y = e^x$ make life simpler, not more complex.

There are several different (but equivalent) ways that the number e can be defined. One way involves investigating the values of the expression

$$\left(1 + \frac{1}{x}\right)^x$$

as x becomes larger and larger. See Table 1.

TABLE 1

x	$\left(1 + \frac{1}{x}\right)^x$
10	2.5937424...
100	2.7048138...
1000	2.7169239...
10^4	2.7181459...
10^5	2.7182682...
10^6	2.7182804...

The data in Table 1 suggest that the quantity $\left(1 + \frac{1}{x}\right)^x$ gets closer and closer to e as x becomes larger and larger. Indeed, in many calculus books, the number e is defined as the *limiting value* or the *limit* of the quantity $\left(1 + \frac{1}{x}\right)^x$ as x grows ever larger. Now admittedly, we have not defined here the meaning of *limiting value* or *limit*—that is a topic for calculus. Nevertheless, Table 1 should give you a reasonable, if intuitive, appreciation of the idea. In Figure 1 we give a graphical interpretation.

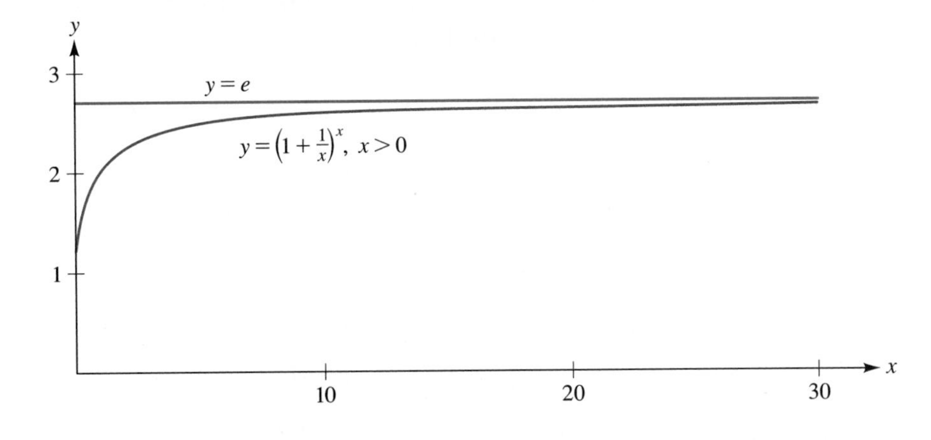

FIGURE 1
The line $y = e$ is an asymptote for the graph of $y = \left(1 + \frac{1}{x}\right)^x$, $x > 0$. As x increases without bound, the values of $\left(1 + \frac{1}{x}\right)^x$ approach the number e.

One more comment about Table 1 and the number e: In Section 5.6, you'll see that the data in Table 1 can be interpreted in terms of banking and compound interest. For this reason, the constant e is sometimes called "the banker's constant."

In calculus the number e is sometimes introduced in a way that involves slopes of lines. You know that the graph of each exponential function $y = b^x$ passes through the point (0, 1). Figure 2(a) shows the exponential function $y = 2^x$ along with a line that is tangent to the curve at the point (0, 1). By carefully measuring rise and run, it can be shown that the slope of this tangent line is about 0.7. Figure 2(b) shows a similar situation with the curve $y = 3^x$. Here the slope of the tangent line through (0, 1) is approximately 1.1. Now,

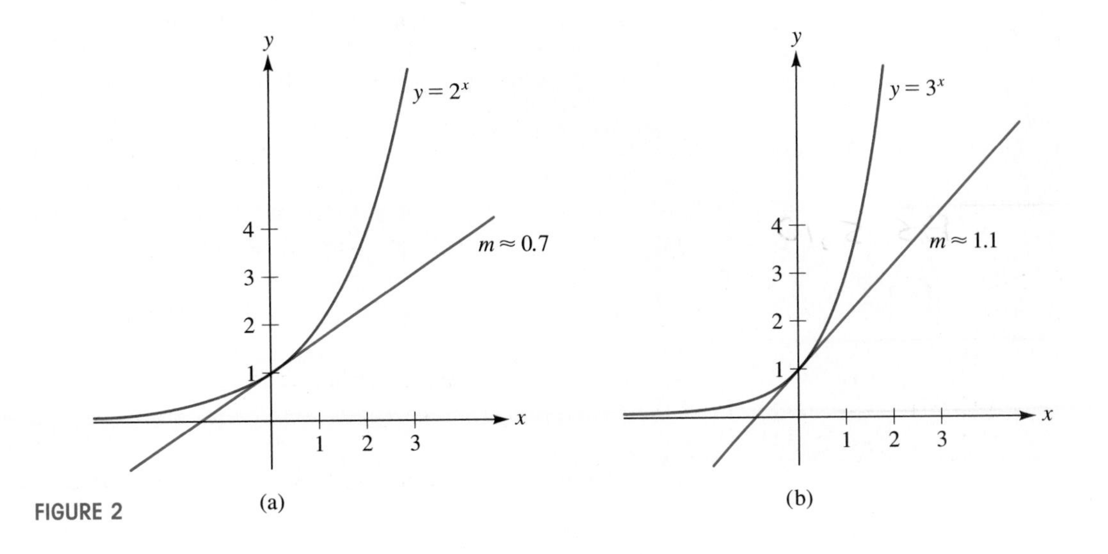

FIGURE 2

since the slope of the tangent to $y = 2^x$ is a bit less than 1, while that for $y = 3^x$ is a bit more than 1, it seems reasonable to suppose that there is a number between 2 and 3, call it e, with the property that the slope of the tangent through (0, 1) is exactly 1. See Figure 3 and the Property Summary Box that follows.

REMARK It's certainly not obvious that this definition for e is equivalent to the definition involving the expression $(1 + 1/x)^x$. The methods of calculus are required to demonstrate that the definitions are indeed equivalent. (For an informal justification, however, see Exercise 46.)

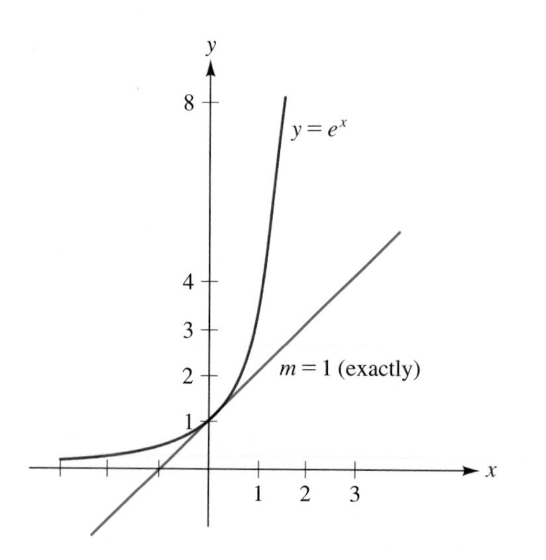

FIGURE 3
The slope of the tangent to the curve $y = e^x$ at the point (0, 1) is $m = 1$.

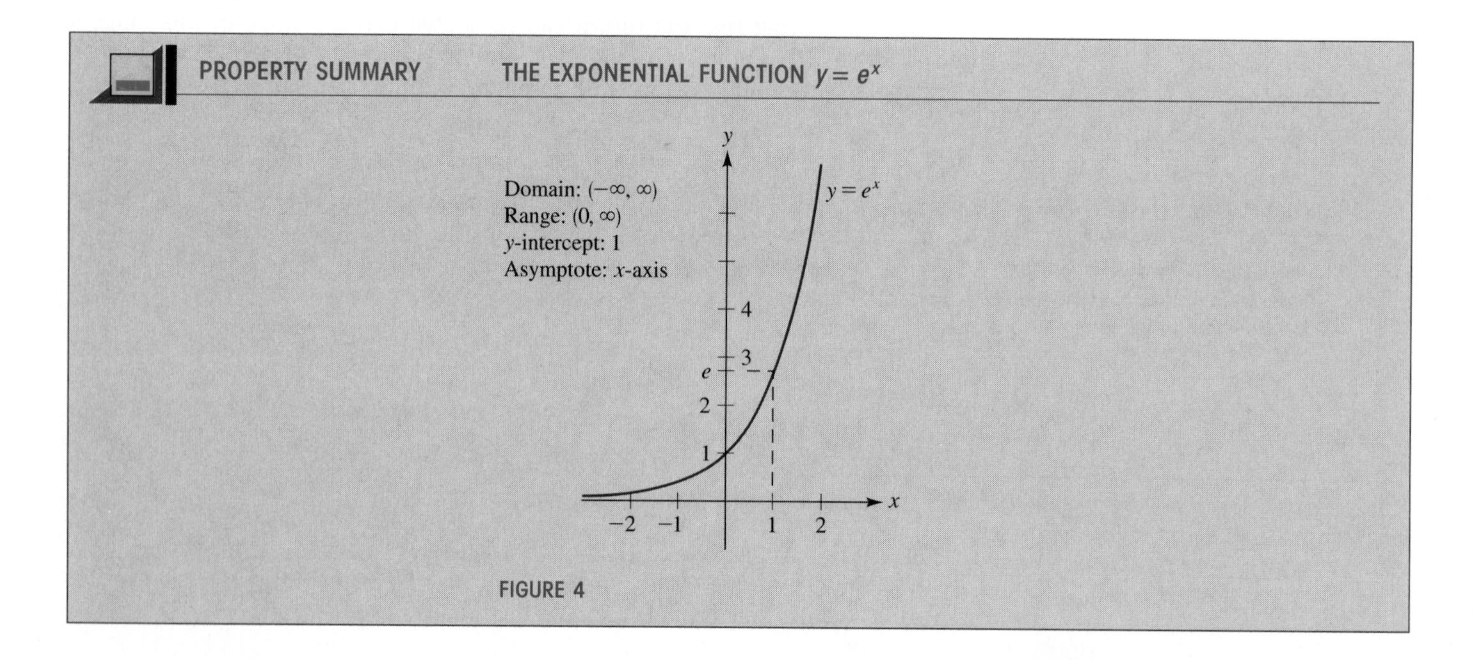

FIGURE 4

EXAMPLE 1 Graph each of the following functions, specifying the domain, range, intercept, and asymptote: **(a)** $y = e^{x-1}$; **(b)** $y = -e^{x-1}$; **(c)** $y = -e^{x-1} + 1$.

Solution

(a) We begin with the graph of $y = e^x$ (on the previous page). Moving the graph to the right one unit yields the graph of $y = e^{x-1}$, shown in Figure 5(a). The x-axis is still an asymptote for this translated graph, but the y-intercept will no longer be 1. To find the y-intercept, we replace x with 0 in the given equation to obtain

$$y = e^{-1} = \frac{1}{e} \approx 0.37 \qquad \text{using a calculator}$$

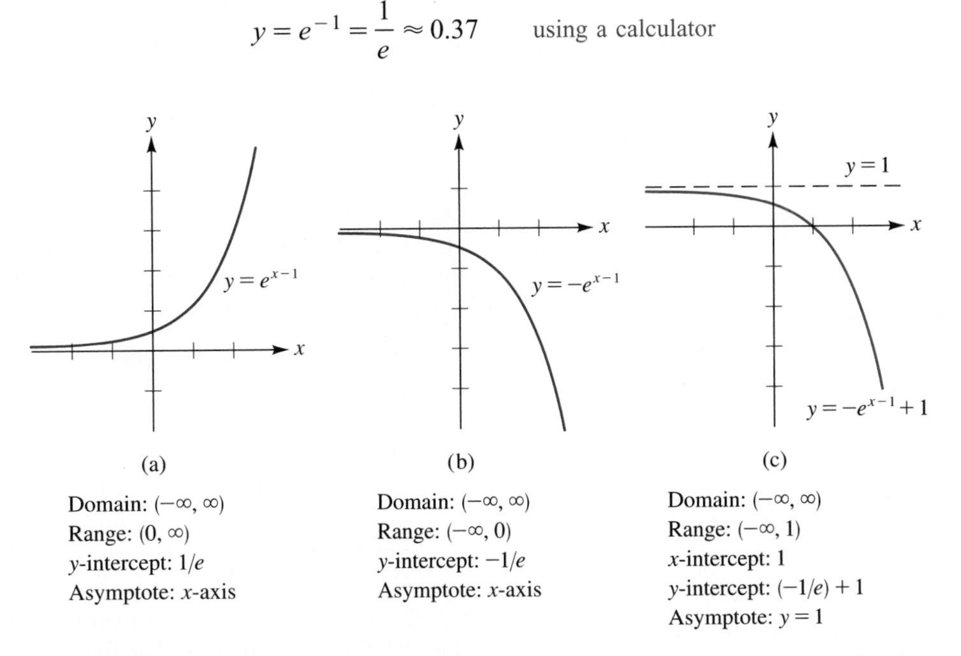

(a) Domain: $(-\infty, \infty)$; Range: $(0, \infty)$; y-intercept: $1/e$; Asymptote: x-axis

(b) Domain: $(-\infty, \infty)$; Range: $(-\infty, 0)$; y-intercept: $-1/e$; Asymptote: x-axis

(c) Domain: $(-\infty, \infty)$; Range: $(-\infty, 1)$; x-intercept: 1; y-intercept: $(-1/e) + 1$; Asymptote: $y = 1$

FIGURE 5

(b) Reflecting the graph from part (a) in the x-axis yields the graph of $y = -e^{x-1}$; see Figure 5(b). Note that under this reflection, the y-intercept moves from $1/e$ to $-1/e$.

(c) Translating the graph in Figure 5(b) up one unit produces the graph of $y = -e^{x-1} + 1$, shown in Figure 5(c). Under this translation, the asymptote moves from $y = 0$ (the x-axis) to $y = 1$. Also, the y-intercept moves from $-1/e$ to $(-1/e) + 1$ (≈ 0.63). The x-intercept in Figure 5(c) is obtained by setting $y = 0$ in the equation $y = -e^{x-1} + 1$. This yields

$$\begin{aligned} 0 &= -e^{x-1} + 1 \\ e^{x-1} &= e^0 \\ x - 1 &= 0 \qquad \text{(Why?)} \\ x &= 1 \end{aligned}$$

One of the characteristic features of exponential functions is their rapid rate of growth. You will see instances of this in the next example, in the exercises, and in the applications sections later in this chapter.

EXAMPLE 2 Let $f(x)=x^2$, $g(x)=2^x$, and $h(x)=e^x$. Compute and compare the average rates of change for these functions on the following intervals: **(a)** [0, 1]; **(b)** [9, 10].

Solution

(a) On the interval [0, 1]:

$$\frac{\Delta f}{\Delta x}=\frac{f(1)-f(0)}{1-0}=\frac{1^2-0^2}{1-0}=1$$

$$\frac{\Delta g}{\Delta x}=\frac{g(1)-g(0)}{1-0}=\frac{2^1-2^0}{1-0}=1$$

$$\frac{\Delta h}{\Delta x}=\frac{h(1)-h(0)}{1-0}=\frac{e^1-e^0}{1-0}\approx 1.7$$

On the interval [0, 1], the average rate of change for the functions $f(x)=x^2$ and $g(x)=2^x$ is 1. The average rate of change for $h(x)=e^x$ on this interval is slightly larger; it is approximately 1.7.

(b) On the interval [9, 10]:

$$\frac{\Delta f}{\Delta x}=\frac{f(10)-f(9)}{10-9}=\frac{100-81}{1}=19$$

$$\frac{\Delta g}{\Delta x}=\frac{g(10)-g(9)}{10-9}=\frac{2^{10}-2^9}{1}=512$$

$$\frac{\Delta h}{\Delta x}=\frac{h(10)-h(9)}{10-9}=\frac{e^{10}-e^9}{1}\approx 13{,}923 \qquad \text{using a calculator}$$

On the interval [9, 10], the average rate of change for $g(x)=2^x$ is more than 25 times as great as that for $f(x)=x^2$. The average rate of change for $h(x)=e^x$ is a whopping 733 times greater than that for $f(x)=x^2$. (Check these last two statements with your calculator.)

EXERCISE SET 5.2

DS 5, 10

A

In Exercises 1–12, graph the function and specify the domain, range, intercept(s), and asymptote.

1. $y=e^x$

2. $y=e^{-x}$

3. $y=-e^x$

4. $y=-e^{-x}$

5. $y=e^x+1$

6. $y=e^{x+1}$

7. $y=e^{x+1}+1$

8. $y=e^{x-1}-1$

9. $y=-e^{x-2}$

10. $y=-e^{x-2}-2$

11. $y=e-e^x$

12. $y=e^{-x}-e$

13. On the same set of axes, graph the functions $y=2^x$, $y=e^x$, and $y=3^x$.

14. On the same set of axes, graph the functions $y=2^{-x}$, $y=e^{-x}$, and $y=3^{-x}$.

In Exercises 15—22, answer TRUE *or* FALSE. *You do not need a calculator for these exercises. Rather, use the fact that e is approximately* 2.7.

15. $e < 1/2$

16. $e < 5/2$

17. $\sqrt{e} < 1$

18. $e^2 < 4$

19. $e^2 < 9$

20. $e^3 < 27$

21. $e^{-1} < 0$

22. $e^0 = 1$

In Exercises 23–30, refer to the following graph of $y = e^x$. In each case, use the graph to estimate the indicated quantity to the nearest tenth (or closer, if it seems appropriate). Also, use a calculator to obtain a second estimate. Round the calculator values to three decimal places.

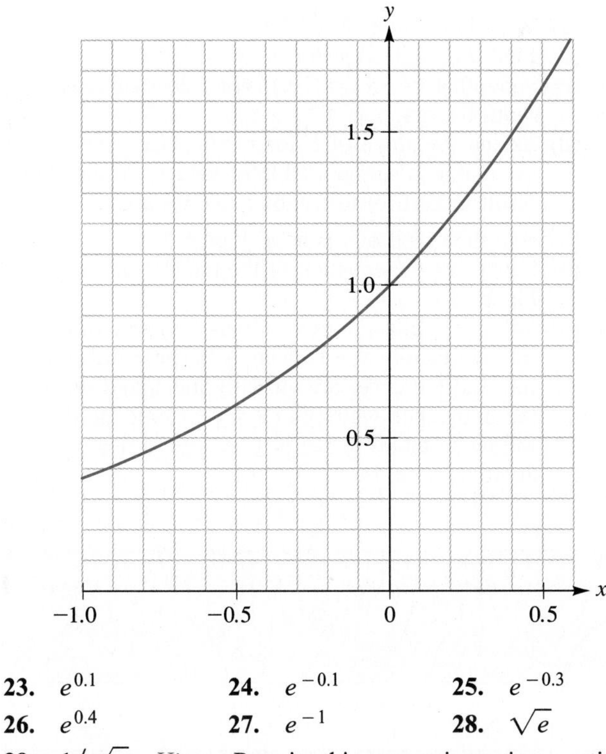

23. $e^{0.1}$

24. $e^{-0.1}$

25. $e^{-0.3}$

26. $e^{0.4}$

27. e^{-1}

28. $\sqrt{e}$

29. $1/\sqrt{e}$ *Hint:* Rewrite this expression using a rational exponent, that is, $1/\sqrt{e} = e^?$.

30. $\sqrt[5]{e}$

31. **(a)** Use the graph preceding Exercise 23 to estimate the value of x for which $e^x = 1.5$.

(b) One of the keys on your calculator will allow you to solve the equation in part (a) for x; it is the "ln" key. Use your calculator to compute ln(1.5). Round your answer to three decimal places and check to see that it is consistent with the value obtained graphically in part (a). *Remark:* The "ln" function, which is called the *natural logarithm* function, is discussed in detail in the next section.

32. Follow Exercise 31 using the equation $e^x = 0.6$ (rather than $e^x = 1.5$).

33. Follow Exercise 31 using the equation $e^x = 1.8$.

34. Follow Exercise 31 using the equation $e^x = 0.4$.

35. Let $f(x) = x^4$, $g(x) = 2^x$, and $h(x) = e^x$. Compute and compare (as in Example 2) the average rates of change for these functions on the following intervals:
(a) [2, 3]; **(b)** [8, 9]; **(c)** [14, 15].

36. Let $f(x) = x^6$, $g(x) = 2^x$, and $h(x) = e^x$. Compute and compare (as in Example 2) the average rates of change for these functions on the following intervals:
(a) [1, 2]; **(b)** [16, 17]; **(c)** [27, 28].

37. The following figure shows portions of the graphs of the exponential functions $y = e^x$ and $y = \pi^x$.

(a) Use the figure (not your calculator) to determine which number is larger, e^π or π^e.

(b) Use your calculator to check your answer in part (a).

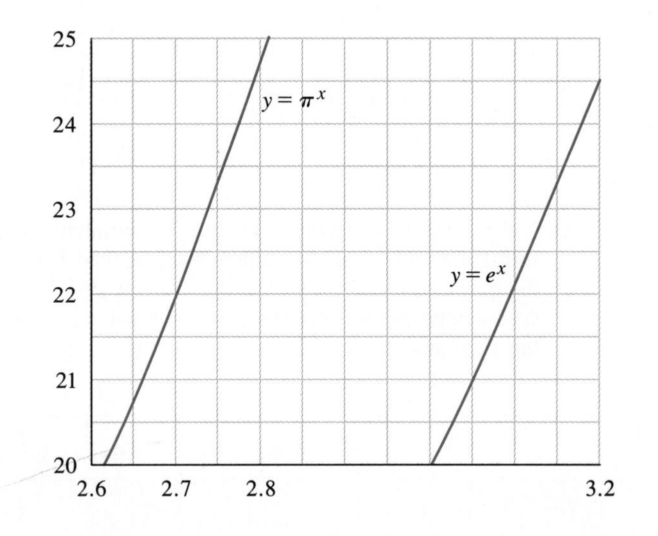

38. **(a)** Use your calculator to approximate the numbers $e\pi$ and $e + \pi$. *Remark:* It is not known whether these numbers are rational or irrational.

(b) Use your calculator to approximate $\left(e^{\pi\sqrt{163}} - 744\right)^{1/3}$. *Remark:* Contrary to the empirical evidence on your calculator, it is known that this number is irrational.

(c) Use your calculator to compute 878/323. (The result agrees with e through the fourth decimal place. The fraction 878/323 is the best rational approximation for e using numerators and denominators less than 1000.)

B

39. The following figure shows the first five steps in the iteration process for $f(x) = e^{-x}$, with initial input $x_0 = 1$.

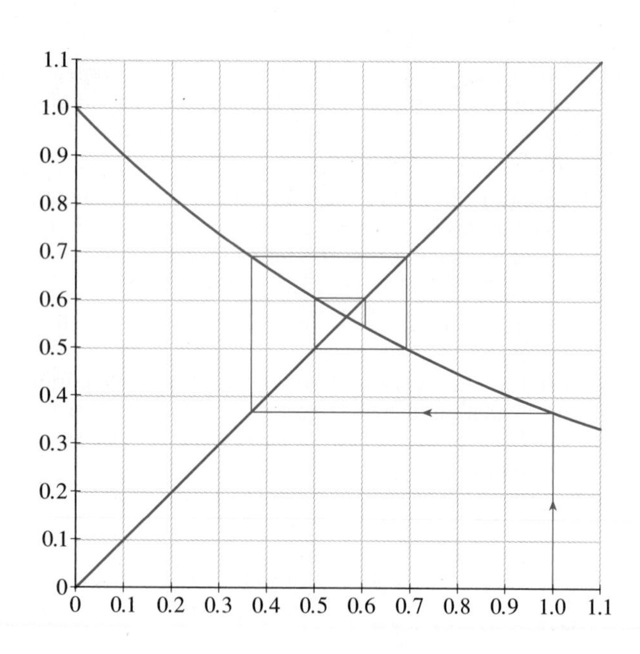

(a) Use the figure and your calculator to complete the following table. For the values that you read from the graph, estimate to within 0.05 (or closer if it seems appropriate). For the calculator values, round the final answers to four decimal places.

	x_1	x_2	x_3	x_4	x_5
From Graph					
From Calculator					

(b) The iteration figure shows that the iterates of 1 are approaching a fixed point that is between 0.5 and 0.6. The actual value of this fixed point, rounded to six decimal places, is 0.567143. Use your calculator to find out which iterate is the first to have 56 in the first two decimal places. Does the iterate after that also begin with 56 in the first two decimal places? What about the iterate after that?

(c) Experiment with your calculator or with sketches: Can you find an initial input for which the iterates do not approach the fixed point in part (b)?

40. The **hyperbolic sine function S** is defined by

$$S(x) = \frac{e^{x} - e^{-x}}{2}$$

(a) What is the domain of S?
(b) Find the y-intercept for the graph of S.
(c) Show that $S(-x) = -S(x)$. What does this say about the graph of S?
(d) Sketch the graph of S for $x \geq 0$. (Use your calculator to set up a table of values.) Then use the result in part (c) to complete the graph.

41. The **hyperbolic cosine function C** is defined by

$$C(x) = \frac{e^{x} + e^{-x}}{2}$$

(a) What is the domain of C?
(b) Find the y-intercept for the graph of C.
(c) Show that $C(-x) = C(x)$. What does this say about the graph of C?
(d) Sketch the graph of C for $x \geq 0$. (Use your calculator to set up a table of values.) Then use the result in part (c) to complete the graph.

42. Refer to the graph of $y = e^{x}$ in Figure 3.
(a) Show that the equation of the tangent line in Figure 3 is $y = x + 1$.
(b) When x is close to zero, e^{x} can be approximated by the quantity $x + 1$. (This is because the tangent line and the curve are virtually indistinguishable in the immediate vicinity of $x = 0$.) Complete the following tables to see just how good this approximation is.

x	0.3	0.2	0.1	0.01	0.001	0.0001
$x+1$						
e^{x}						

x	−0.3	−0.2	−0.1	−0.01	−0.001	−0.0001
$x+1$						
e^{x}						

43. **(a)** Use a calculator to complete the following table. Round your answers to six decimal places.

x	1000	10^{4}	10^{5}	10^{6}	10^{7}
$(1 + 2/x)^{x}$					

(b) Use a calculator to compute e^2. What do you observe?

44. **(a)** Use a calculator to complete the following table. Round your answers to six decimal places.

x	1000	10^4	10^5	10^6	10^7
$(1 + 3/x)^x$					

(b) Use a calculator to compute e^3. What do you observe?

45. Let $f(x) = e^x$. Let L denote the function that is the inverse of f.

(a) On the same set of axes, sketch the graphs of f and L. *Hint:* You do not need the equation for $L(x)$.

(b) Specify the domain, range, intercept, and asymptote for the function L and its graph.

(c) Graph each of the following functions. Specify the intercept and asymptote in each case.

(i) $y = -L(x)$
(ii) $y = L(-x)$
(iii) $y = L(x - 1)$

46. The text showed two distinct ways to define the number e. One involved the expression $(1 + 1/x)^x$ and the other involved tangent lines. In this exercise you'll see that these two approaches are, in fact, related. For convenience, we'll write the expression $(1 + 1/x)^x$ using the letter n rather than x (so we can use x for something else in a moment). Then we have

$$\left(1 + \frac{1}{n}\right)^n \approx e \quad \text{as } n \text{ becomes larger and larger, without bound} \qquad (1)$$

(a) Define x by the equation $x = 1/n$. As n becomes larger and larger without bound, what happens to x? Complete the following table.

n	100	1000	10^6	10^9
x				

(b) Substitute $x = 1/n$ in approximation (1) to obtain

$$(1 + x)^{1/x} \approx e \qquad \text{as } x \text{ approaches zero}$$

Next, raise both sides to the power x to obtain

$$(1 + x) \approx e^x \qquad \text{as } x \text{ approaches zero} \qquad (2)$$

Now complete the following table to see just how well the quantity $1 + x$ approximates e^x as x approaches zero.

x	$1 + x$	e^x	$e^x - (1 + x)$
0.1			
0.01			
0.001			

This suggests (but does not *prove*) that the line $y = 1 + x$ is tangent to the curve $y = e^x$ at $x = 0$.

In Exercises 47–54, decide which of the following properties apply to each function. (More than one property may apply to a function.)

A. *The function is increasing for* $-\infty < x < \infty$.
B. *The function is decreasing for* $-\infty < x < \infty$.
C. *The function has a turning point.*
D. *The function is one-to-one.*
E. *The graph has an asymptote.*
F. *The function is a polynomial function.*
G. *The domain of the function is* $(-\infty, \infty)$.
H. *The range of the function is* $(-\infty, \infty)$.

47. $y = e^x$

48. $y = e^{-x}$

49. $y = -e^x$

50. $y = e^{|x|}$

51. $y = x + e$

52. $y = x^2 + e$

53. $y = x/e$

54. $y = e/x$

C

55. Let S and C denote the functions defined in Exercises 40 and 41, respectively. Prove each of the following identities. [The identities in parts (b) and (c) have already appeared in Exercises 40 and 41, but they are included here for the sake of completeness.]

(a) $[C(x)]^2 - [S(x)]^2 = 1$
(b) $S(-x) = -S(x)$
(c) $C(-x) = C(x)$
(d) $S(x + y) = S(x)C(y) + C(x)S(y)$
(e) $C(x + y) = C(x)C(y) + S(x)S(y)$
(f) $S(2x) = 2S(x)C(x)$
(g) $C(2x) = [C(x)]^2 + [S(x)]^2$

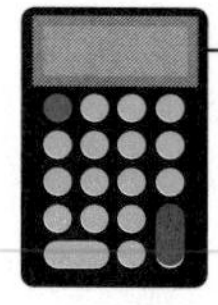

GRAPHING UTILITY EXERCISES FOR SECTION 5.2

1. In this exercise we compare three functions:

$$y = 2^x, \qquad y = e^x, \qquad y = 3^x.$$

(a) Begin by graphing all three functions in the standard viewing rectangle. The picture confirms three facts that you know from the text: on the positive x-axis, the functions increase very rapidly; on the negative x-axis, the graphs approach the asymptote (which is the x-axis) as you move to the left; the y-intercept in each case is 1.

(b) To compare the functions for positive values of x, use a viewing rectangle in which x extends from 0 to 3 and y extends from 0 to 10. Note that the graph of e^x is bounded between the graphs of 2^x and 3^x, just as the number e is between 2 and 3. In particular, the picture that you obtain demonstrates the following fact: For positive values of x,

$$2^x < e^x < 3^x$$

(c) To see the graphs more clearly when x is negative, change the viewing rectangle so that x extends from -3 to 0 and y extends from 0 to 1. Again, note that the graph of e^x is bounded between the graphs of 2^x and 3^x, but now the graph of 3^x is the bottom (rather than the top) curve in the picture. This demonstrates the following fact: For negative values of x,

$$3^x < e^x < 2^x$$

2. (a) According to the text, the line $y = x + 1$ is tangent to the curve $y = e^x$ at the point (0, 1) on the curve. Verify this visually by graphing both the curve and the line in the standard viewing rectangle. Note that the curve and the line are virtually indistinguishable in the immediate vicinity of the point (0, 1).

(b) Zoom in on the point (0, 1); use a viewing rectangle in which x extends from -0.05 to 0.05 and y extends from 0.95 to 1.05. Again, note that the line and the curve are virtually indistinguishable in the immediate vicinity of the point (0, 1). For a numerical look at this, complete the following table. In the columns for e^x and $e^x - (x + 1)$, round each entry to four decimal places. When you are finished, observe that the closer x is to 0, the closer the agreement between the two quantities $x + 1$ and e^x.

x	$x+1$	e^x	$e^x-(x+1)$
-0.05			
-0.04			
-0.03			
-0.02			
-0.01			
0.00			
0.01			
0.02			
0.03			
0.04			
0.05			

In Problems 3–8, first tell what translations or reflections are required to obtain the graph of the given function from that of $y = e^x$. Then graph the given function along with $y = e^x$ and check that the picture is consistent with your ideas.

3. $y = e^{-x}$

4. $y = -e^x$

5. $y = -e^{-x}$

6. (a) $y = e^x - 1$ **(b)** $y = e^{-x} - 1$

7. (a) $y = e^{x-1}$ **(b)** $y = e^{-x-1}$

8. (a) $y = e^{x-1} - 1$ **(b)** $y = e^{-x-1} - 1$

9. In this exercise you'll look at functions of the form $y = e^{x/k}$, where k is a constant. Set the viewing rectangle so that both x and y extend from 0 to 5.

(a) In the same picture, graph the following five functions:

$$y = e^x \qquad y = e^{x/3} \qquad y = e^{x/5} \qquad y = e^{x/7} \qquad y = e^{x/9}$$

What is the y-intercept for each graph? What is the general pattern that emerges? That is, as k increases, what happens to the appearance of the graph of $y = e^{x/k}$?

(b) Based on your answer in part (a), what do you think the graph of $y = e^{x/1000}$ will look like in the viewing rectangle that you're using? After answering, add the graph of $y = e^{x/1000}$ to your picture and see if you're right.

(c) Despite the appearance of the graph of $y = e^{x/1000}$ that you obtained in part (b), the long-term behavior of the function $y = e^{x/1000}$ is quite different. Graph this function using a viewing rectangle that extends from 0 to 15,000 in the x-direction and from 0 to 100,000 in the y-direction. Now what do you observe?

In Section 5.2, we saw that the number e can be defined as the "limit" of the expression $(1 + 1/x)^x$ as x becomes indefinitely large. We write this fact symbolically as follows:

$$\lim_{x\to\infty}\left(1+\frac{1}{x}\right)^x = e$$

Actually, this is a particular instance of the following more general result: For each real number a, the limiting value (as x becomes indefinitely large) of $\left(1+\frac{a}{x}\right)^x$ *is* e^a. *In symbols,* $\lim_{x\to\infty}\left(1+\frac{a}{x}\right)^x = e^a$. *In Exercises 10 and 11, you'll use the graphing utility to draw graphs that illustrate this result.*

10. Begin this exercise with a viewing rectangle extending from 0 to 50 in the x-direction and from 0 to 10 in the y-direction.
 (a) Graph the line $y = e^2$ along with the function $y = \left(1+\frac{2}{x}\right)^x$. Note that the line appears to be an asymptote for the curve.
 (b) For more convincing evidence that the line is an asymptote, change the viewing rectangle so that x extends from 0 to 1000. The picture that you obtain supplies evidence for the following fact (which is proved in calculus): $\lim_{x\to\infty}\left(1+\frac{2}{x}\right)^x = e^2$.

11. Begin this exercise with a viewing rectangle extending from 0 to 100 in the x-direction and from 0 to 30 in the y-direction.
 (a) Graph the line $y = e^3$ along with the function $y = \left(1+\frac{3}{x}\right)^x$. Note that the line appears to be an asymptote for the curve.
 (b) For more convincing evidence that the line is an asymptote, change the viewing rectangle so that x extends from 0 to 1000. The picture that you obtain supplies evidence for the following fact (which is proved in calculus): $\lim_{x\to\infty}\left(1+\frac{3}{x}\right)^x = e^3$.

In Section 4.1, you saw how a linear function can be used to model a data set. The next exercise shows a case in which an exponential function provides a much better model. For this exercise you will need options on your graphing utility for linear regression (as in Section 4.1) and for exponential regression.

12. **(a)** In the chapter introduction, there is a table showing United States population over the years 1800–1900. Enter this data into your graphing utility and obtain a scatter plot. (Adjust the viewing rectangle so that your picture is similar to the version of the scatter plot shown in the chapter introduction.)
 (b) Use your graphing utility to perform a linear regression for the data. Add the graph of the regression line to your scatter plot. Note that the regression line reflects the general fact that the population is increasing. However, in detail, the regression line misses most of the data points. More importantly, the line fails to convey much important information. For example, from 1800 to about 1820, the population grows relatively slowly. After that, the population as well as the growth rate of the population is increasing. This information is lost in using the regression line as a model.
 (c) Now use the graphing utility to perform an exponential regression, that is, find a function of the form $y = ab^x$ that fits the data. Graph the resulting equation [along with the scatter plot from part (a)]. Note that the curve appears to fit the data much more closely than the linear function in part (b). (In the text and Exercises for Section 5.7, you'll see more examples in which functions of the form $y = ab^x$ are used to model population growth.)

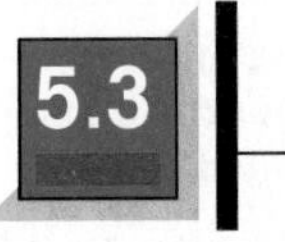

5.3 LOGARITHMIC FUNCTIONS

For the number whose logarithm is unity, let e be written,...

Leonhard Euler in a letter written in 1727 or 1728

[John Napier] *hath set my head and hands a work with his new and admirable logarithms. I hope to see him this summer, if it please God, for I never saw book that pleased me better, or made me more wonder.*

Henry Briggs, March 10, 1615

It has been thought that the earliest reference to the logarithmic curve was made by the Italian Evangelista Torricelli in a letter of the year 1644, but Paul Tannery made it practically certain that Descartes knew the curve in 1639.

Florian Cajori in *A History of Mathematics*, 4th ed. (New York: Chelsea Publishing Co., 1985)

In the previous two sections we studied exponential functions. Now we consider functions that are inverses of exponential functions. These inverse functions are called *logarithmic functions.*

Having said this, let's back up for a moment to review briefly some of the basic ideas behind inverse functions (as discussed in Section 3.5). We start with a given function F, say, $F(x) = 3x$ for example, that is one-to-one. (That is, for each output there is but one input.) Then, by interchanging the inputs and outputs, we obtain a new function, the so-called inverse function. In the case of $F(x) = 3x$, it's easy to find an equation defining the inverse function. We just interchange x and y in the equation $y = 3x$ to obtain $x = 3y$. Solving for y in this last equation then gives us $y = \frac{1}{3}x$, which defines the inverse function. Using function notation, we can summarize the situation by writing $F(x) = 3x$ and $F^{-1}(x) = \frac{1}{3}x$.

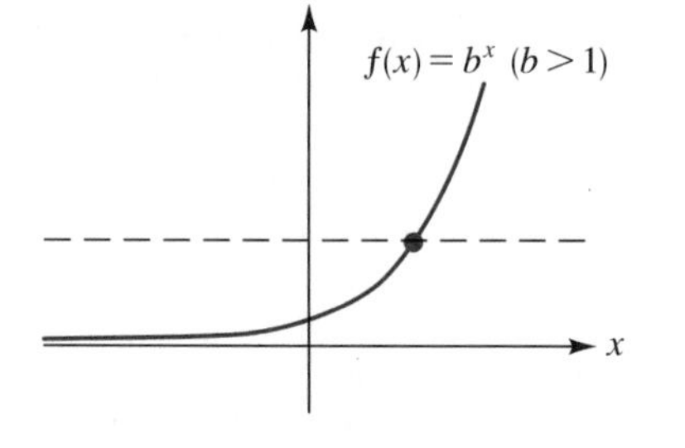

In the preceding paragraph we saw that a particular linear function had an inverse, and we found a formula for that inverse. Now let's repeat that same reasoning beginning with an exponential function. First, we must make sure the exponential function f defined by $f(x) = b^x$ is one-to-one. We can see this by applying the horizontal line test, as indicated in Figure 1. Next, since $f(x) = b^x$ is one-to-one, it has an inverse function. Let's study this inverse.

We begin by writing the exponential function $f(x) = b^x$ in the form

$$y = b^x \tag{1}$$

Then, in order to obtain an equation for f^{-1}, we interchange x and y in equation (1). This gives us

$$x = b^y \tag{2}$$

The crucial step now is to express equation (2) in words:

y is the exponent to which b must be raised to yield x. (3)

Statement (3) defines the function that is the inverse of $y = b^x$. Now we introduce a notation that will allow us to write this statement in a more compact form.

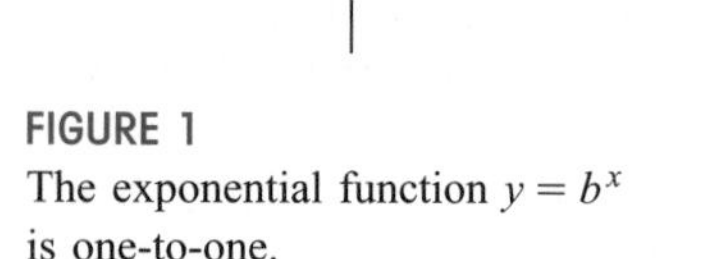

FIGURE 1
The exponential function $y = b^x$ is one-to-one.

DEFINITION: $\log_b x$

We define the expression $\log_b x$ to mean "the exponent to which b must be raised to yield x." ($\log_b x$ is read *log base b of x* or *the logarithm of x to the base b.*)

EXAMPLES

(a) $\log_2 8 = 3$, since 3 is the exponent to which 2 must be raised to yield 8.
(b) $\log_{10}(1/10) = -1$, since -1 is the exponent to which 10 must be raised to yield 1/10.
(c) $\log_5 1 = 0$, since 0 is the exponent to which 5 must be raised to yield 1.

Using this notation, statement (3) becomes

$$y = \log_b x$$

Since (2) and (3) are equivalent, we have the following important relationship.

$y = \log_b x$ is equivalent to $x = b^y$

TABLE 1

Exponential Form of Equation	Logarithmic Form of Equation
$8 = 2^3$	$\log_2 8 = 3$
$\frac{1}{9} = 3^{-2}$	$\log_3 \frac{1}{9} = -2$
$1 = e^0$	$\log_e 1 = 0$
$a = b^c$	$\log_b a = c$

We say that the equation $y = \log_b x$ is in **logarithmic form** and that the equivalent equation $x = b^y$ is in **exponential form.** Table 1 displays some examples.

Let us now summarize our discussion up to this point.

1. According to the horizontal line test, the function $f(x) = b^x$ is one-to-one and therefore possesses an inverse. This inverse function is written
$$f^{-1}(x) = \log_b x$$
2. $\log_b a = c$ means that $a = b^c$.

To graph the function $y = \log_b x$, we recall from Section 3.5 that the graph of a function and its inverse are reflections of one another about the line $y = x$. Thus, to graph $y = \log_b x$, we need only reflect the curve $y = b^x$ about the line $y = x$. This is shown in Figure 2.

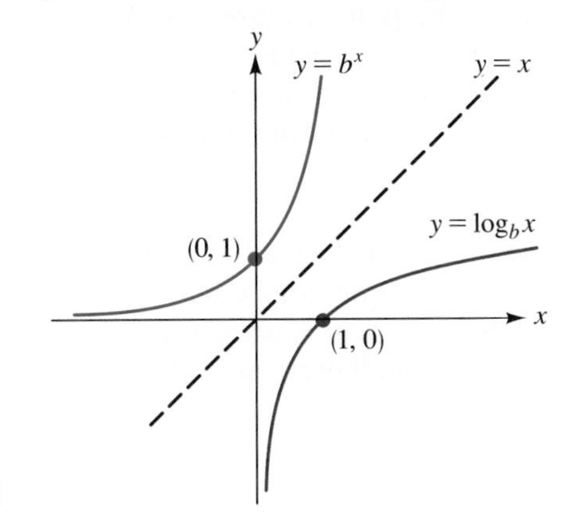

FIGURE 2

NOTE For the rest of this chapter we assume that the base b is greater than 1 when we use the expression $\log_b x$.

With the aid of Figure 2, we can make the following observations about the function $y = \log_b x$.

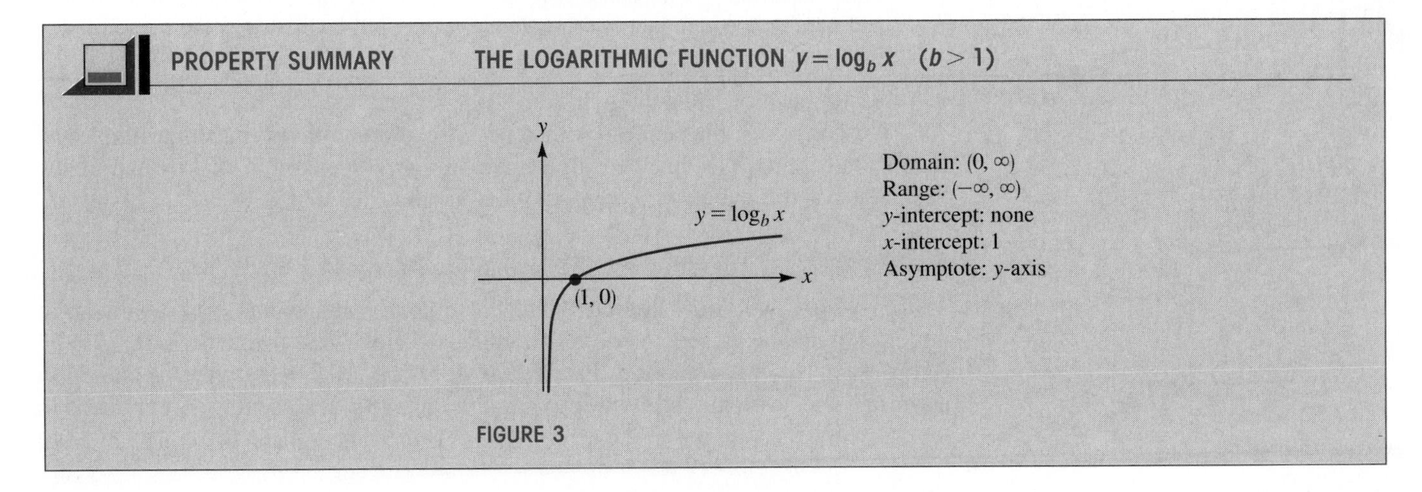

FIGURE 3

One aspect of the function $y = \log_b x$ may not be immediately apparent to you from Figures 2 and 3. The function grows or increases *very* slowly. Consider

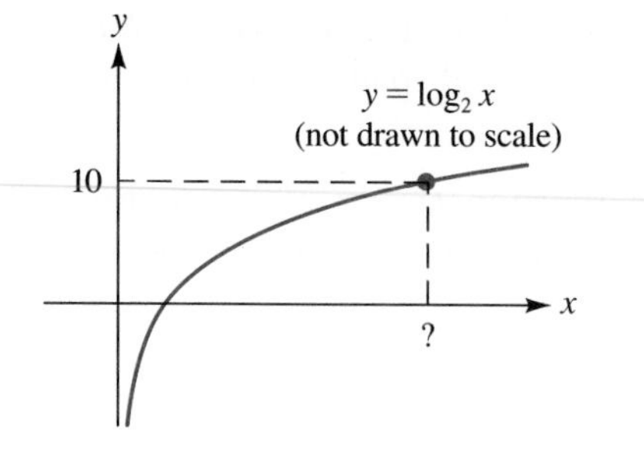

FIGURE 4

$y = \log_2 x$, for example. Let's ask how large x must be before the curve reaches the height $y = 10$ (see Figure 4).

To answer this question, we substitute $y = 10$ in the equation $y = \log_2 x$:

$$10 = \log_2 x$$

Writing this equation in exponential form yields

$$x = 2^{10} = 1024$$

In other words, we must go out beyond 1000 on the x-axis before the curve $y = \log_2 x$ reaches a height of 10 units. Exercise 55 at the end of this section asks you to show that the graph of $y = \log_2 x$ doesn't reach a height of 100 until x is greater than 10^{30}. (Numbers as large as 10^{30} rarely occur in any of the sciences. For instance, the distance in inches to the Andromeda galaxy is less than 10^{24}.) The point we are emphasizing here is this: The graph of $y = \log_2 x$ (or $\log_b x$) is always rising, but very slowly.

Before going on to consider some numerical examples, let's pause for a moment to think about the notation we've been using. In the expression

$$\log_b x$$

the name of the function is $\log_b$ and the input is x. To emphasize this, we might be better off writing $\log_b x$ as $\log_b(x)$, so that the similarity to the familiar $f(x)$ notation is clear. However, for historical reasons,* the convention is to suppress the parentheses, and we follow that convention here. In the box that follows, we indicate some errors that can occur upon forgetting that $\log_b$ is the name of a function, not a number.

ERRORS TO AVOID

ERROR	CORRECTION	COMMENT
$\dfrac{\log_2 8}{\log_2 4} \not= \dfrac{8}{4}$	$\dfrac{\log_2 8}{\log_2 4} = \dfrac{3}{2}$	$\log_2$ is the name of a function. It is not a factor that can be "cancelled" from the numerator and denominator.
$\dfrac{\log_2 16}{16} \not= \log_2$	$\dfrac{\log_2 16}{16} = \dfrac{4}{16} = \dfrac{1}{4}$	This "equation" is nonsense. On the left-hand side there is a quotient of two real numbers, which is a real number. On the right-hand side is the name of a function.

We conclude this section with a set of examples involving logarithms and logarithmic functions. In one way or another, every example makes use of the key fact that the equation $\log_b a = c$ is equivalent to $b^c = a$.

EXAMPLE 1 Which quantity is larger: $\log_3 10$ or $\log_7 40$?

Solution First we estimate $\log_3 10$. This quantity represents the exponent to which 3 must be raised to yield 10. Since $3^2 = 9$ (less than 10), but $3^3 = 27$ (more than 10), we conclude that the quantity $\log_3 10$ lies between 2 and 3. In a similar way we can estimate $\log_7 40$; this quantity represents the exponent to which 7 must be raised to yield 40. Since $7^1 = 7$ (less than 40) while $7^2 = 49$

*The notation *Log* was introduced in 1624 by the astronomer Johannes Kepler (1571–1630). In Leonhard Euler's text *Introduction to Analysis of the Infinite,* first published in 1748, appears the statement, "It has been customary to designate the logarithm of y by the symbol $\log y$."

(more than 40), we conclude that the quantity $\log_7 40$ lies between 1 and 2. It now follows from these two estimates that $\log_3 10$ is larger than $\log_7 40$.

EXAMPLE 2 Evaluate $\log_4 32$.

Solution Let $y = \log_4 32$. The exponential form of this equation is $4^y = 32$. Now, since both 4 and 32 are powers of 2, we can rewrite the equation $4^y = 32$ using the same base on both sides:

$$\begin{aligned}(2^2)^y &= 2^5 \\ 2^{2y} &= 2^5 \\ 2y &= 5 && \text{using Property 4 on page 273} \\ y &= 5/2 && \text{as required}\end{aligned}$$

EXAMPLE 3 Graph the following equations:

(a) $y = \log_{10} x$; **(b)** $y = -\log_{10} x$.

Solution

(a) The function $y = \log_{10} x$ is the inverse function for the exponential function $y = 10^x$. Thus we obtain the graph of $y = \log_{10} x$ by reflecting the graph of $y = 10^x$ in the line $y = x$; see Figure 5(a).

(b) To graph $y = -\log_{10} x$, we reflect $y = \log_{10} x$ in the x-axis; see Figure 5(b).

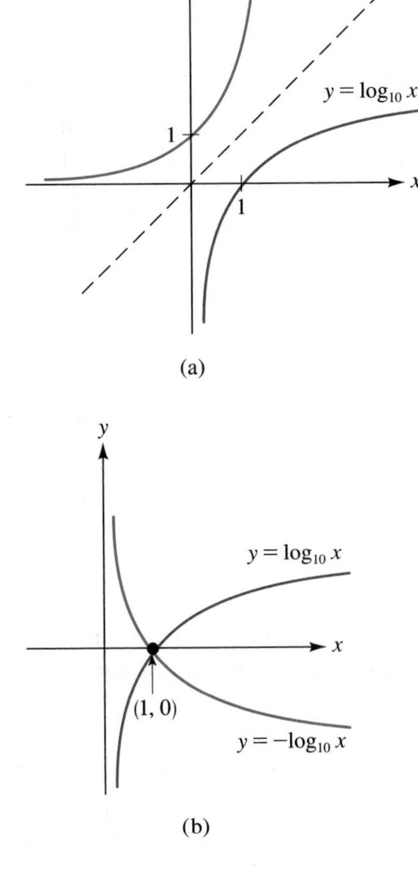

FIGURE 5

EXAMPLE 4 Find the domain of the function $f(x) = \log_2(12 - 4x)$.

Solution As you can see by looking back at Figure 3 on page 291, the inputs for the logarithmic function must be positive. So, in the case at hand, we require that the quantity $12 - 4x$ be positive. Consequently, we have

$$\begin{aligned}12 - 4x &> 0 \\ -4x &> -12 \\ x &< 3\end{aligned}$$

Therefore the domain of the function $f(x) = \log_2(12 - 4x)$ is the interval $(-\infty, 3)$.

The next example concerns the exponential function $y = e^x$ and its inverse function, $y = \log_e x$. Many books, as well as calculators, abbreviate the expression $\log_e x$ by $\ln x$, read *natural log of* x.* For reference and emphasis, we repeat this fact in the following box. (Incidentally, on most calculators, "log" is an abbreviation for $\log_{10}$.)

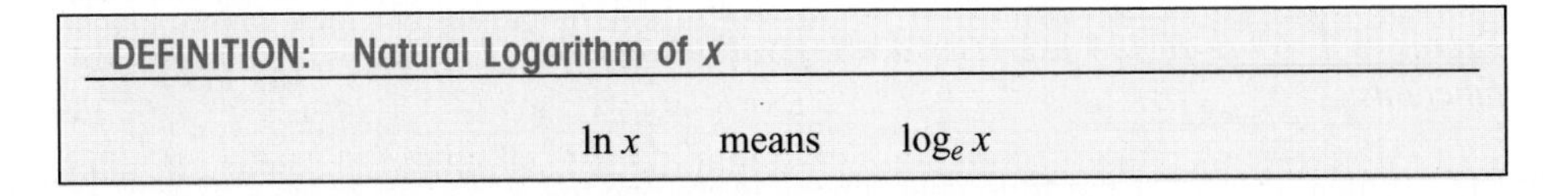

DEFINITION: Natural Logarithm of x

$\ln x$ means $\log_e x$

*According to the historian Florian Cajori, the notation $\ln x$ was used by (and perhaps first introduced by) Irving Stringham in his text *Uniplanar Algebra* (San Francisco: University Press, 1893).

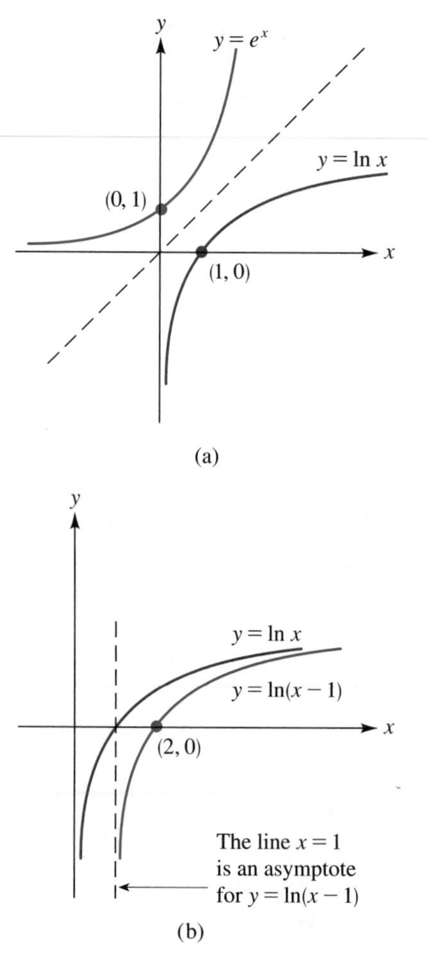

FIGURE 6

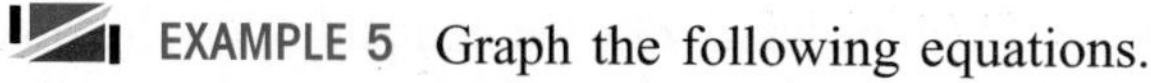
EXAMPLE 5 Graph the following equations.
(a) $y = \ln x$; **(b)** $y = \ln(x - 1)$.

Solution

(a) The function $y = \ln x$ ($= \log_e x$) is the inverse of $y = e^x$. Thus its graph is obtained by reflecting $y = e^x$ in the line $y = x$, as in Figure 6(a).

(b) To graph $y = \ln(x - 1)$, we take the graph of $y = \ln x$ and move it one unit in the positive x-direction; see Figure 6(b).

EXAMPLE 6 Simplify each expression: **(a)** $\ln e$; **(b)** $\ln 1$.

Solution

(a) The expression $\ln e$ denotes the exponent to which e must be raised to yield e; this exponent is 1. Thus $\ln e = 1$.

(b) Similarly, $\ln 1$ denotes the exponent to which e must be raised to yield 1. Since $e^0 = 1$, zero is the required exponent. Thus $\ln 1 = 0$.

EXAMPLE 7 Solve the equations: **(a)** $10^{2x} = 200$; **(b)** $e^{3t-1} = 2$.

Solution

(a) We write the equation $10^{2x} = 200$ in logarithmic form to obtain

$$2x = \log_{10} 200$$
$$x = \frac{\log_{10} 200}{2}$$

Without a calculator or tables, we leave the answer in this form. Alternatively, using a calculator we obtain $x \approx 1.15$.

(b) To solve $e^{3t-1} = 2$, we write the equation in logarithmic form:

$$3t - 1 = \log_e 2$$

that is,

$$3t - 1 = \ln 2$$
$$3t = 1 + \ln 2$$
$$t = \frac{1 + \ln 2}{3}$$

Without a calculator or tables, this is the final answer. On the other hand, using a calculator we obtain $t \approx 0.56$.

EXERCISE SET 5.3

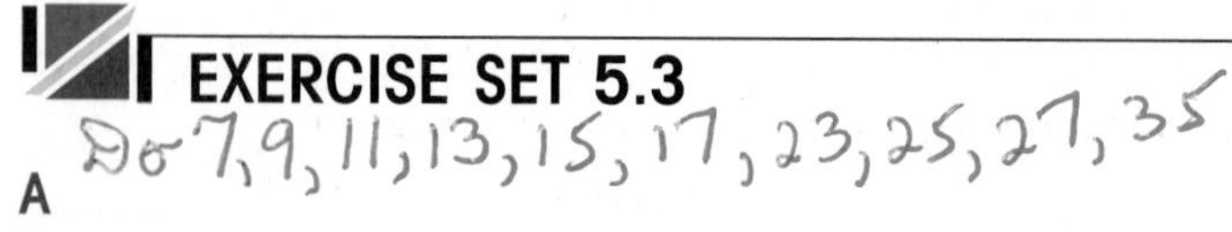

A

Exercises 1–6 are review exercises dealing with inverse functions.

1. Which, if any, of the following functions are one-to-one and therefore possess an inverse?
(a) $y = x^2 + 1$ **(b)** $y = 3x$
(c) $y = (x + 1)^3$

2. Does each of the following functions have an inverse?

(a) $f(x) = \begin{cases} x^2 & \text{if } -1 \leqslant x \leqslant 0 \\ x^2 + 1 & \text{if } x > 0 \end{cases}$

(b) $g(x) = \begin{cases} x^2 & \text{if } -1 \leqslant x < 0 \\ x^2 + 1 & \text{if } x \geqslant 0 \end{cases}$

3. Let $f(x) = (2x - 1)/(3x + 4)$. Find each quantity.
(a) $f^{-1}(x)$ **(b)** $1/f(x)$ **(c)** $f^{-1}(0)$ **(d)** $1/f(0)$

4. Let $f(x) = x^3 + 2x + 1$. Evaluate $f[f^{-1}(5)]$. [Assume that the domain of f^{-1} is $(-\infty, \infty)$.]

5. The graph of $y = f(x)$ is a line segment joining the two points $(3, -2)$ and $(-1, 5)$. What are the corresponding endpoints for the graph of $y = f^{-1}(x-1)$?

6. Which (if either) of the following conditions tells us that a function is one-to-one?
(a) For each input there is exactly one output.
(b) For each output there is exactly one input.

In Exercises 7 and 8, write each equation in logarithmic form.

7. **(a)** $9 = 3^2$ **(b)** $1000 = 10^3$
(c) $7^3 = 343$ **(d)** $\sqrt{2} = 2^{1/2}$

8. **(a)** $1/125 = 5^{-3}$ **(b)** $e^0 = 1$
(c) $5^x = 6$ **(d)** $e^{3t} = 8$

9. Write each equation in exponential form.
(a) $\log_2 32 = 5$ **(b)** $\log_{10} 1 = 0$
(c) $\log_e \sqrt{e} = 1/2$ **(d)** $\log_3(1/81) = -4$
(e) $\log_t u = v$

10. Complete the tables.
(a)

x	1	10	10^2	10^3	10^4	10^{-1}	10^{-2}	10^{-3}
$\log_{10} x$								

(b)

x	1	e	e^2	e^3	e^4	e^{-1}	e^{-2}	e^{-3}
$\ln x$								

11. Which quantity is larger, $\log_5 30$ or $\log_8 60$?

12. Which quantity is larger?
(a) $\log_{10} 90$ or $\log_e e^5$ **(b)** $\log_2 3$ or $\log_3 2$

In Exercises 13 and 14, evaluate each expression.

13. **(a)** $\log_9 27$ **(b)** $\log_4(1/32)$ **(c)** $\log_5 5\sqrt{5}$

14. **(a)** $\log_{25}(1/625)$ **(b)** $\log_{16}(1/64)$
(c) $\log_{10} 10$ **(d)** $\log_2 8\sqrt{2}$

In Exercises 15 and 16, solve each equation for x by converting to exponential form. In Exercises 15(b) and 16, give two forms for each answer: one involving e and the other a calculator approximation rounded to two decimal places.

15. **(a)** $\log_4 x = -2$ **(b)** $\ln x = -2$

16. **(a)** $\log_5 x = e$ **(b)** $\ln x = -e$

In Exercises 17 and 18, find the domain of each function.

17. **(a)** $y = \log_4 5x$ **(b)** $y = \log_{10}(3-4x)$
(c) $y = \ln(x^2)$ **(d)** $y = (\ln x)^2$
(e) $y = \ln(x^2 - 25)$

18. **(a)** $y = \ln(2 - x - x^2)$ **(b)** $y = \log_{10} \dfrac{2x+3}{x-5}$

19. In the accompanying figure, what are the coordinates of the four points A, B, C, and D?

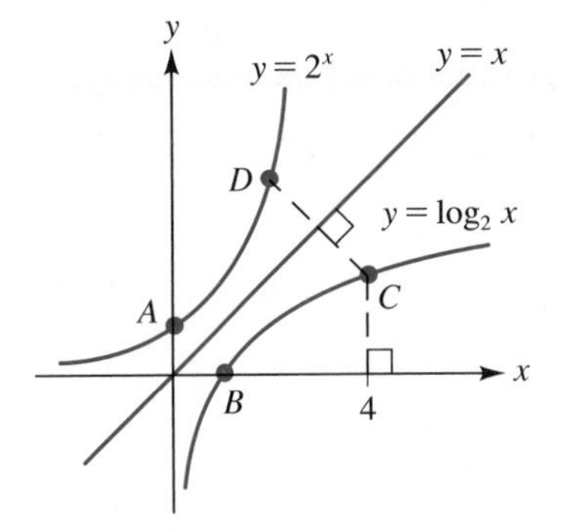

20. In the accompanying figure, what are the coordinates of the points A, B, C, and D?

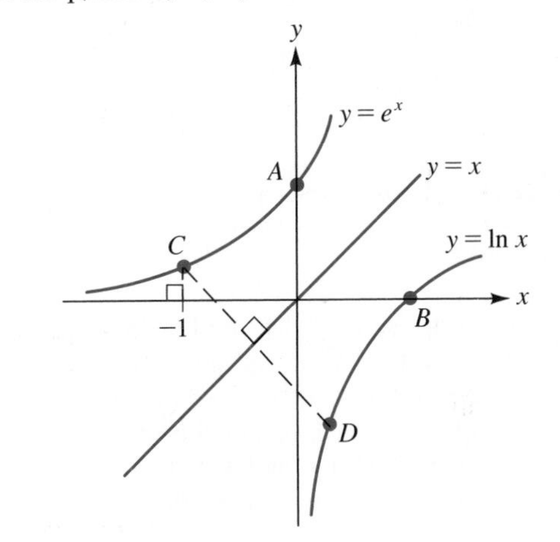

In Exercises 21–26, graph each function and specify the domain, range, intercept(s), and asymptote.

21. **(a)** $y = \log_2 x$ **(b)** $y = -\log_2 x$
(c) $y = \log_2(-x)$ **(d)** $y = -\log_2(-x)$

22. **(a)** $y = \ln x$ **(b)** $y = -\ln x$
(c) $y = \ln(-x)$ **(d)** $y = -\ln(-x)$

23. $y = -\log_3(x-2) + 1$ **24.** $y = -\log_{10}(x+1)$

25. $y = \ln(x + e)$ **26.** $y = \ln(-x) + e$

In Exercises 27 and 28, simplify each expression.

27. **(a)** $\ln e^4$ **(b)** $\ln(1/e)$ **(c)** $\ln\sqrt{e}$

28. **(a)** $\ln e$ **(b)** $\ln e^{-2}$ **(c)** $(\ln e)^{-2}$

In Exercises 29–36, find all the real-number solutions for each equation. In each case, give both the exact value of the answer and a calculator approximation rounded to two decimal places.

29. $10^x = 25$ **30.** $10^x = 145$

31. $10^{x^2} = 40$

32. $(10^x)^2 = 40$

33. $e^{2t+3} = 10$

34. $e^{t-1} = 16$

35. $e^{1-4t} = 12.405$

36. $e^{3x^2} = 112$

37. In this exercise, the notation $\log x$ means $\log_{10} x$. The following figure shows the first two steps in the iteration of $x_0 = 0.6$ under the function $f(x) = -\log x$.

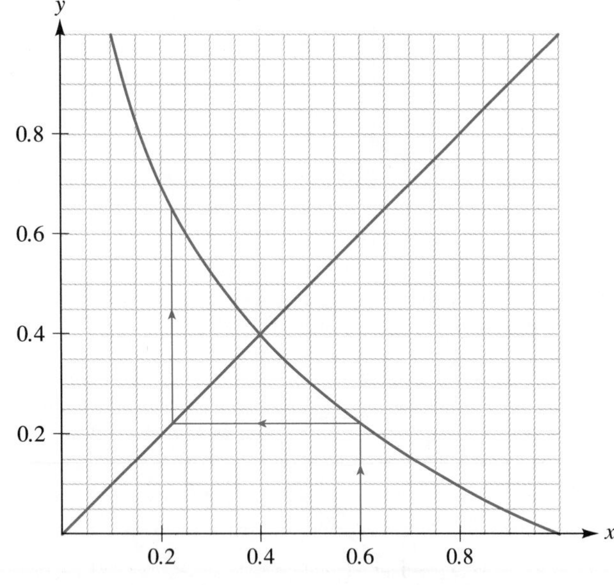

(a) The figure indicates that the function $f(x) = -\log x$ has a fixed point very close to 0.4. Is the fixed point exactly 0.4? Use your calculator to compute $-\log(0.4)$. Conclusion?

(b) Using a sharp pencil and a ruler, continue the iteration process in the figure out through the sixth iterate. (You may want to make a photocopy of the figure and work from that.) Then complete the following table. For the values obtained from the graph, estimate to the nearest 0.05 (or closer, where it seems appropriate). For the calculator work, round the final answers to four decimal places.

	x_1	x_2	x_3	x_4	x_5	x_6
From Graph						
From Calculator						

(c) Use your calculator to compute the iterates x_7 through x_9. (Again, round the final answers to four decimal places.) Based on your results here and in part (b), do the iterates appear to be approaching the fixed point of the function?

(d) Use your calculator to compute x_{10}, and note that an error message is displayed on the calculator screen. Explain (What is the domain of the logarithm function?)

38. The following figure shows a portion of the graph of $y = 10^x$.

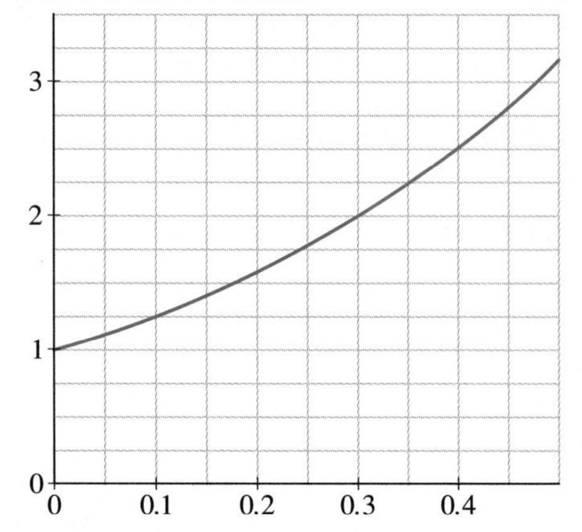

Use the graph to estimate (to the nearest 0.05) each of the following quantities. Also, use a calculator to evaluate the expressions; round the answers to three decimal places. As usual, check to see that the calculator values are consistent with the values obtained from the graph.

(a) $\log_{10}(1.25)$ **(b)** $\log_{10}(2)$ **(c)** $\log_{10}(1.125)$
(d) $\log_{10}(2.5)$ **(e)** $\log_{10}(2.25)$ **(f)** $\log_{10}(3.125)$

39. The accompanying figure shows a portion of the graph of $y = e^x$ for $1.5 \leq x \leq 1.7$. Using the figure, estimate the quantity ln 5 to two decimal places. Then use a calculator to determine the percent error in your approximation. *Hint:* Write the equation $x = \ln 5$ in exponential form.

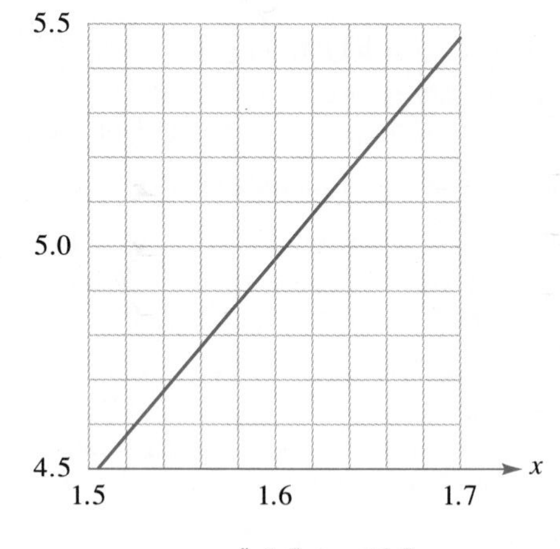

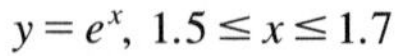
$y = e^x,\ 1.5 \leq x \leq 1.7$

40. **(a)** Use the symmetry tests from Section 2.4 to determine if the graph of $y = \ln|x|$ is symmetric about the x-axis, the y-axis, or the origin.

(b) Graph the equation $y = \ln|x|$. *Hint:* First graph the equation for $x > 0$; then use symmetry.

In Exercises 41 and 42, refer to the following figure. The equation of the curve in the figure is $y = \ln|x|$. In each case, use a calculator to determine the coordinates of the indicated point. Round the final answers to three decimal places.

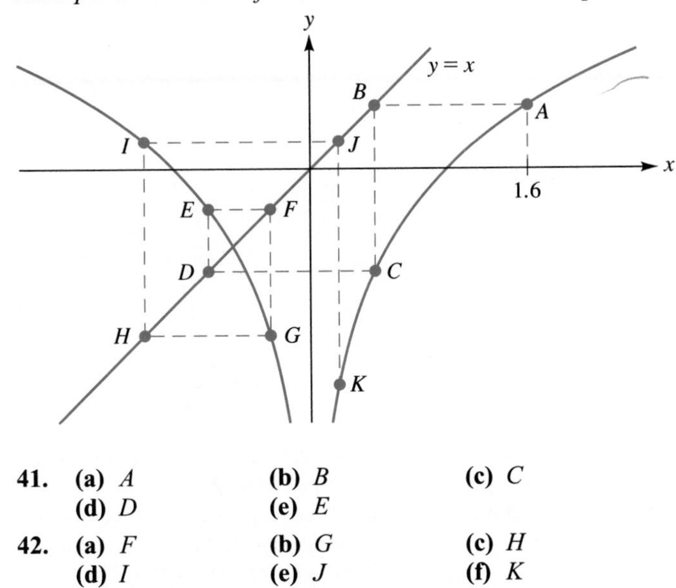

41. **(a)** A **(b)** B **(c)** C
(d) D **(e)** E

42. **(a)** F **(b)** G **(c)** H
(d) I **(e)** J **(f)** K

In Exercises 43–46, use a calculator to match the equations with the graphs.

43. $f(x) = \sqrt{x}$; $g(x) = \ln x$

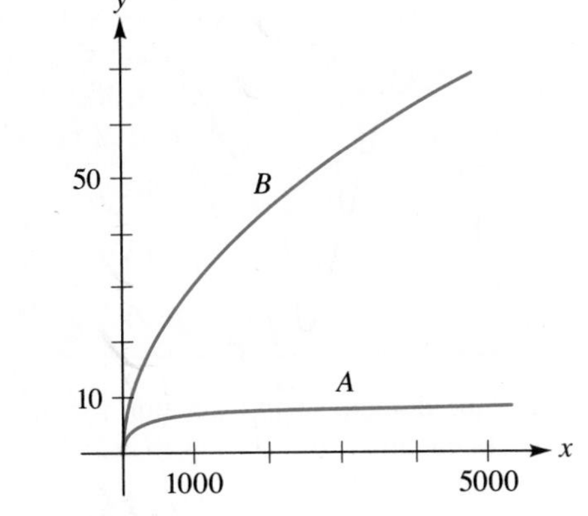

44. $f(x) = \sqrt[10]{x}$; $g(x) = \ln x$

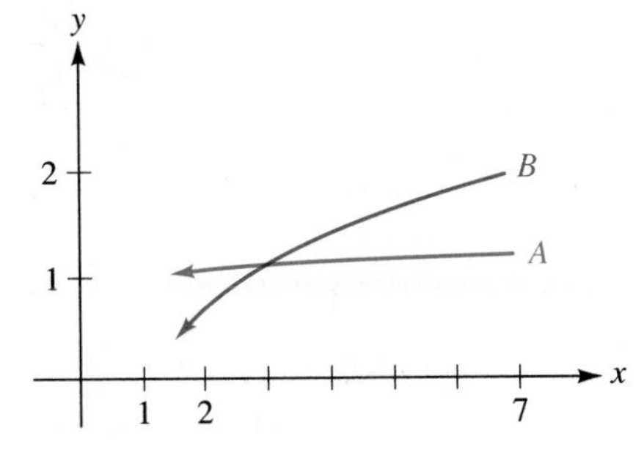

45. $f(x) = \sqrt[10]{x}$; $g(x) = \ln x$

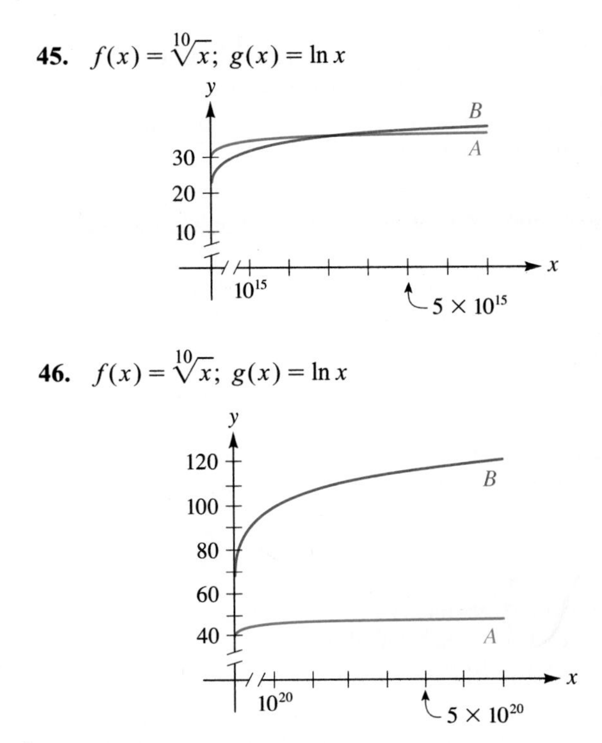

46. $f(x) = \sqrt[10]{x}$; $g(x) = \ln x$

B

47. The accompanying figure shows portions of the graphs of $y = \log_2 x$ and $y = \log_3 x$. Which is which? Give your reasons in complete sentences.

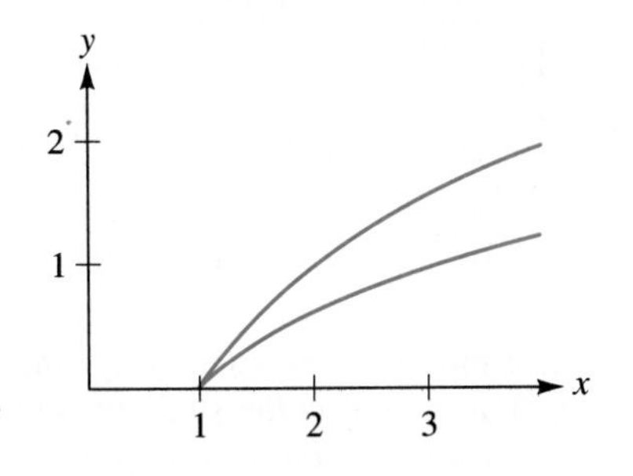

48. The accompanying figure shows portions of the graphs of $y = \log_3 x$ and $y = \log_6 x$. Which is which? Explain your reasons in complete sentences.

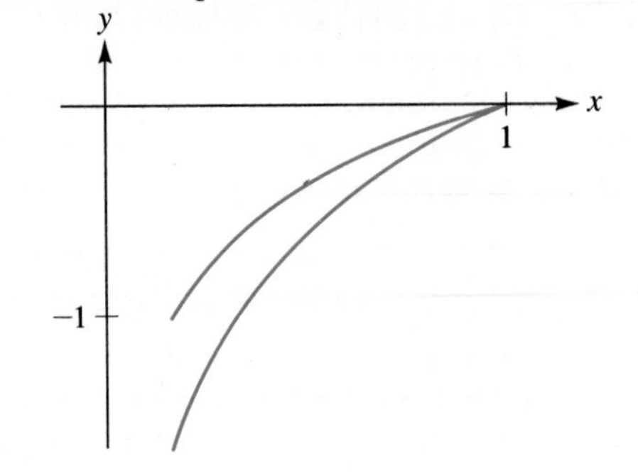

49. Let $f(x) = e^{x+1}$. Find $f^{-1}(x)$ and sketch its graph. Specify any intercept or asymptote.

50. Let $g(t) = \ln(t-1)$. Find $g^{-1}(t)$ and draw its graph. Specify any intercept or asymptote.

51. Sketch the region bounded by $y = e^x$, $y = e^{-x}$, the x-axis, and the vertical lines $x = \pm 1$. Why must the area of this region be less than two square units?

52. Solve $2^{2x} - 2^{x+1} - 15 = 0$ for x. *Hint:* First show the equation is equivalent to $(2^x)^2 - 2(2^x) - 15 = 0$.

53. Solve $e^{2x} - 5e^x - 6 = 0$ for x.

54. Solve $e^x - 3e^{-x} = 2$ for x. *Hint:* Multiply both sides by e^x.

55. Estimate a value for x such that $\log_2 x = 100$. Use the approximation $10^3 \approx 2^{10}$ to express your answer as a power of 10. *Answer:* 10^{30}

56. **(a)** How large must x be before the graph of $y = \ln x$ reaches a height of $y = 100$?
(b) How large must x be before the graph of $y = e^x$ reaches a height of **(i)** $y = 100$?
(ii) $y = 10^6$?

In Exercises 57 and 58, use the following information on pH. Chemists define pH by the formula pH $= -\log_{10}[H^+]$, *where* $[H^+]$ *is the hydrogen ion concentration measured in moles per liter. For example, if* $[H^+] = 10^{-5}$, *then pH* $= 5$. *Solutions with a pH of 7 are said to be* neutral; *a pH below 7 indicates an* acid; *and a pH above 7 indicates a* base.

57. **(a)** For some fruit juices, $[H^+] = 3 \times 10^{-4}$. Determine the pH and classify these juices as acid or base.
(b) For sulfuric acid, $[H^+] = 1$. Find the pH.

58. An unknown substance has a hydrogen ion concentration of 3.5×10^{-9}. Classify the substance as acid or base.

59. This exercise indicates one of the ways the natural logarithm function is used in the study of prime numbers. Recall that a prime number is a natural number greater than 1 with no factors other than itself and 1. For example, the first ten prime numbers are 2, 3, 5, 7, 11, 13, 17, 19, 23, and 29.
(a) Let $P(x)$ denote the number of prime numbers that do not exceed x. For instance, $P(6) = 3$, since there are three prime numbers (2, 3, and 5) that do not exceed 6. Compute $P(10)$, $P(18)$, and $P(19)$.
(b) According to the **prime number theorem,** $P(x)$ can be approximated by $x/(\ln x)$ when x is large and, in fact, the ratio $\dfrac{P(x)}{x/(\ln x)}$ approaches 1 as x grows larger and larger. Verify this empirically by completing the following table. Round your results to three decimal places. (These facts were discovered by Carl Friedrich Gauss in 1792, when he was 15 years old. It was not until 1896, more than 100 years later, that the prime number theorem was formally proved by the French mathematician J. Hadamard and also by the Belgian mathematician C. J. de la Vallée-Poussin.)

x	$P(x)$	$\dfrac{x}{\ln x}$	$\dfrac{P(x)}{x/\ln x}$
10^2	25		
10^4	1229		
10^6	78498		
10^8	5761455		
10^9	50847534		
10^{10}	455052512		

(c) In 1808, A. M. Legendre found that he could improve upon Gauss's approximation for $P(x)$ by using the expression $x/(\ln x - 1.08366)$ rather than $x/(\ln x)$. Complete the following table to see how well Legendre's expression approximates $P(x)$. Round your results to four decimal places.

x	$P(x)$	$\dfrac{x}{\ln x - 1.08366}$	$\dfrac{P(x)}{x/(\ln x - 1.08366)}$
10^2	25		
10^4	1229		
10^6	78498		
10^8	5761455		
10^9	50847534		
10^{10}	455052512		

60. **(a)** What is the domain of the function $y = \sqrt{\ln x}$?
(b) Solve the equation $\sqrt{\ln x} = (1 + \ln x)/2$.
Hint: Let $t = \ln x$.

In Exercises 61–68, decide which of the following properties apply to each function. (More than one property may apply to a function.)

A. *The function is increasing for* $-\infty < x < \infty$.
B. *The function is decreasing for* $-\infty < x < \infty$.
C. *The function has a turning point.*
D. *The function is one-to-one.*
E. *The graph has an asymptote.*
F. *The function is a polynomial function.*
G. *The domain of the function is* $(-\infty, \infty)$.
H. *The range of the function is* $(-\infty, \infty)$.

61. $y = \ln x$

62. $y = \ln(-x)$

63. $y = -\ln x$

64. $y = \ln|x|$

65. $y = \ln x + e$

66. $y = x + \ln e$

67. $y = -\ln(-x)$

68. $y = (\ln x)^2$

In Exercises 69 and 70, use logarithms (and a calculator) to find a natural number n satisfying the given conditions.

69. $e^n < 10^{12} < e^{n+1}$

70. $10^n < e^{100} < 10^{n+1}$

71. Using, as usual, x for the inputs and y for the outputs, graph the function defined by the equation $\ln y = x$.

C

72. **(a)** Find the domain of the function f defined by $f(x) = \ln(\ln x)$.
(b) Find $f^{-1}(x)$ for the function f in part (a).
(c) Find the domain of the function g defined by $g(x) = \ln[\ln(\ln x)]$.

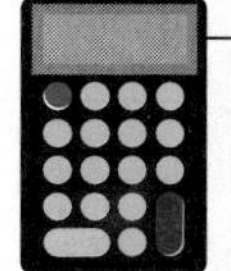

GRAPHING UTILITY EXERCISES FOR SECTION 5.3

1. **(a)** According to the text, what kind of symmetry would you see if you were to graph $y = e^x$ and $y = \ln x$ in the same picture?
(b) Confirm your answer in part (a) by graphing the two functions along with the line $y = x$. (Start with a viewing rectangle that extends from -5 to 5 in both the x- and y-directions. Then make any adjustments that are necessary to ensure that the same size unit and the same scale are used on both axes.)

2. Follow Exercise 1 using the two functions $y = 10^x$ and $y = \log_{10} x$.

3. Let $F(x) = e^x$, $G(x) = \ln x$, $H(x) = x - 2$, and $K(x) = x^2$. Specify the domain of each of the following functions and then check your answers by graphing. *Hint:* The answers for (c) and (d) are not the same.

(a) F **(b)** G **(c)** $G \circ F$ **(d)** $F \circ G$
(e) $G \circ H$ **(f)** $H \circ G$ **(g)** $G \circ K$ **(h)** $K \circ G$

4. Use graphs to help answer the following questions. (You will need to experiment with the settings for the viewing rectangle.)
(a) For which x-values is $\ln x < \log_{10} x$?
(b) For which x-values is $\ln x > \log_{10} x$?
(c) For which x-values is $e^x < 10^x$?
(d) For which x-values is $e^x > 10^x$?

5. This exercise demonstrates the very slow growth of the natural logarithm function $y = \ln x$. We consider the following question: How large must x be before the graph of $y = \ln x$ reaches a height of 10?
(a) Graph the function $y = \ln x$ using a viewing rectangle that extends from 0 to 10 in the x-direction and 0 to 12 in the y-direction. Note how slowly the graph rises. Use the graphing utility to estimate the height of the curve (the y-coordinate) when $x = 10$.
(b) Since we are trying to see when the graph of $y = \ln x$ reaches a height of 10, add the horizontal line $y = 10$ to your picture. Next, adjust the viewing rectangle so that x extends from 0 to 100. Now use the graphing utility to estimate the height of the curve when $x = 100$. [As both the picture and the y-coordinate indicate, we're still not even half way to 10. Go on to part (c).]
(c) Change the viewing rectangle so that x extends to 1000, then estimate the y-coordinate corresponding to $x = 1000$. (You'll find that the height of the curve is almost 7. We're getting closer.)
(d) Repeat part (c) with x extending to 10,000. (You'll find that the height of the curve is over 9. We're almost there.)
(e) The last step: change the viewing rectangle so that x extends to 100,000, then use the graphing utility to estimate the x-value for which $\ln x = 10$. As a check on your estimate, rewrite the equation $\ln x = 10$ in exponential form, and evaluate the expression that you obtain for x.

6. This exercise uses the natural logarithm and a regression line (i.e., a linear function) to find a simple formula relating the following two quantities:

x: the average distance of a planet from the Sun

y: the period of a planet (i.e., the time required for a planet to make one complete revolution around the sun)

(a) Complete the following table. (In the table, the abbreviation AU stands for *astronomical unit,* a unit of distance. By definition, one AU is the average distance from the Earth to the Sun.

Planet	x Average Distance from the Sun (AU)	y Period (years)	ln x	ln y
Mercury	0.387	0.241		
Venus	0.723	0.615		
Earth	1.000	1.000		
Mars	1.523	1.881		
Jupiter	5.202	11.820		

(b) Use a graphing utility to obtain a scatter plot and a regression equation $Y = AX + B$ for the pairs of numbers $(\ln x, \ln y)$. After some rounding, you should obtain $A = 1.5$ and $B = 0$. Thus the quantities $\ln x$ and $\ln y$ are related as follows:

$$\ln y = 1.5 \ln x$$

(c) Show that the equation obtained in part (b) can be simplified to

$$y = x^{1.5} \tag{1}$$

(d) Equation (1) was obtained using data for the first five planets from the Sun. Complete the following table to see how well equation (1) fits the data for the remaining planets.

Planet	x (AU)	y (years) [calculated from eqn. (1)]	y (years) (from astronomical observations)
Saturn	9.555		29.46
Uranus	19.22		84.01
Neptune	30.11		164.79
Pluto	39.44		248.50

Remark: Equation (1) is *Kepler's third law of planetary motion.* According to George F. Simmons in *Calculus Gems* (New York: McGraw-Hill, Inc., 1992), "Kepler had struggled for more than twenty years to find this connection between a planet's *distance* from the sun and the *time* required to complete its orbit. He published his discovery in 1619 in a work entitled *Harmonices Mundi* (The Harmonies of the World)."

7. This exercise uses the natural logarithm and a regression line to find a simple formula relating the following two quantities:

x: the average distance of a planet from the Sun

y: the average orbital velocity of the planet

(a) Complete the following table.

Planet	x Average Distance from the Sun (millions of miles)	y Average Orbital Velocity (miles/sec)	ln x	ln y
Mercury	35.98	29.75		
Venus	67.08	21.76		
Earth	92.96	18.51		
Mars	141.64	14.99		

(b) Use a graphing utility to obtain a scatter plot and a regression equation $Y = AX + B$ for the pairs of numbers $(\ln x, \ln y)$. You will find that $A = -0.500$ and $B = 5.184$. Thus the quantities $\ln x$ and $\ln y$ are related as follows:

$$\ln y = -0.500 \ln x + 5.184$$

(c) Show that the equation obtained in part (b) can be rewritten in the form

$$y = \frac{e^{5.184}}{\sqrt{x}} \tag{2}$$

Hint: Convert from logarithmic to exponential form.

(d) Equation (2) was obtained using data for the first four planets from the Sun. Complete the following table to see how well equation (2) fits the data for the remaining planets. (Round the answers to two decimal places.)

Planet	x Average Distance from the Sun (millions of miles)	y Average Orbital Velocity (miles/sec) [calculated from eqn. (2)]	y Average Orbital Velocity (miles/sec) (from astronomical observations)
Jupiter	483.63		8.12
Saturn	888.22		5.99
Uranus	1786.55		4.23
Neptune	2799.06		3.38
Pluto	3700.75		2.95

PROPERTIES OF LOGARITHMS

... he [John Napier] *invented the word "logarithms," using two Greek words,*
... arithmos, *"number" and* logos, *"ratio." It is impossible to say exactly what he had in mind when making up this word.*

Alfred Hooper in *Makers of Mathematics* (New York: Random House, Inc., 1948)

The teaching of logarithms as a computing device is vanishing from the schools....

The logarithmic function, however, will never die, for the simple reason that logarithmic and exponential variations are a vital part of nature and of analysis.

Howard Eves in *An Introduction to the History of Mathematics,* 6th ed. (Philadelphia: Saunders College Publishing, 1990)

A few basic properties of logarithms are used repeatedly. Our procedure in this section will be to state these properties, discuss their proofs, and then look at examples.

Properties of Logarithms

1. **(a)** $\log_b b = 1$ **(b)** $\log_b 1 = 0$
2. $\log_b PQ = \log_b P + \log_b Q$
   The log of a product is the sum of the logs of the factors.
3. $\log_b(P/Q) = \log_b P - \log_b Q$
   The log of a quotient is the log of the numerator minus the log of the denominator.
   As a useful particular case, we have $\log_b(1/Q) = -\log_b Q$.
4. $\log_b P^n = n \log_b P$
5. $b^{\log_b P} = P$

NOTE P and Q are assumed to be positive in Properties 2–5.

In essence, each of these properties follows from the equivalence of the two equations $y = \log_b x$ and $b^y = x$. For instance, the equivalent exponential forms for Properties 1(a) and 1(b) are $b^1 = b$ and $b^0 = 1$, respectively, both of which are certainly valid.

To prove Property 2, we begin by letting $x = \log_b P$. The equivalent exponential form of this equation is

$$P = b^x \tag{1}$$

Similarly, we let $y = \log_b Q$. The exponential form of this equation is

$$Q = b^y \tag{2}$$

If we multiply equation (1) by equation (2), we get

$$PQ = b^x b^y$$

and, therefore,

$$PQ = b^{x+y} \tag{3}$$

Next we write equation (3) in its equivalent logarithmic form. This yields

$$\log_b PQ = x + y$$

But, using the definitions of x and y, this last equation is equivalent to

$$\log_b PQ = \log_b P + \log_b Q$$

That completes the proof of Property 2.

The proof of Property 3 is quite similar to the proof given for Property 2. Exercise 72(a) asks you to carry out the proof of Property 3.

We turn now to the proof of Property 4. We begin by letting $x = \log_b P$. In exponential form, this last equation becomes

$$b^x = P \qquad (4)$$

Now we raise both sides of equation (4) to the power n. This yields

$$(b^x)^n = b^{nx} = P^n$$

The logarithmic form of this last equation is

$$\log_b P^n = nx$$

or (from the definition of x)

$$\log_b P^n = n \log_b P$$

The proof of Property 4 is now complete.

Property 5 is again just a restatement of the meaning of $\log_b P$. To derive this property, let $x = \log_b P$. Therefore,

$$b^x = P$$

Now, in this last equation, we simply replace x with $\log_b P$. The result is

$$b^{\log_b P} = P \qquad \text{as required}$$

Now let's see how these properties are used. To begin with, we display some simple numerical examples in the box that follows.

PROPERTY SUMMARY **PROPERTIES OF LOGARITHMS**

PROPERTY	EXAMPLE
$\log_b P + \log_b Q = \log_b PQ$	Simplify: $\log_{10} 50 + \log_{10} 2$ Solution: $\log_{10} 50 + \log_{10} 2 = \log_{10}(50 \cdot 2) = \log_{10} 100 = 2$
$\log_b P - \log_b Q = \log_b \frac{P}{Q}$	Simplify: $\log_8 56 - \log_8 7$ Solution: $\log_8 56 - \log_8 7 = \log_8 \frac{56}{7} = \log_8 8 = 1$
$\log_b P^n = n \log_b P$	Simplify: $\log_2 \sqrt[5]{16}$ Solution: $\log_2 \sqrt[5]{16} = \log_2(16^{1/5}) = \frac{1}{5} \log_2 16 = \frac{4}{5}$
$b^{\log_b P} = P$	Simplify: $3^{\log_3 7}$ Solution: $3^{\log_3 7} = 7$

The examples in the box showed how we can simplify or shorten certain expressions involving logarithms. The next example is also of this type.

EXAMPLE 1 Express as a single logarithm with a coefficient of 1:

$$\tfrac{1}{2}\log_b x - \log_b(1 + x^2)$$

Solution

$$\begin{aligned}\tfrac{1}{2}\log_b x - \log_b(1 + x^2) &= \log_b x^{1/2} - \log_b(1 + x^2) && \text{using Property 4 on page 301}\\ &= \log_b \frac{x^{1/2}}{1 + x^2} && \text{using Property 3}\end{aligned}$$

This last expression is the required answer.

Property 2 says that the logarithm of a product of two factors is equal to the sum of the logarithms of the two factors. This can be generalized to any number of factors. For instance, with three factors we have

$$\begin{aligned}\log_b(ABC) &= \log_b[A(BC)]\\ &= \log_b A + \log_b BC && \text{using Property 2}\\ &= \log_b A + \log_b B + \log_b C && \text{using Property 2 again}\end{aligned}$$

The next example makes use of this idea.

EXAMPLE 2 Express as a single logarithm with a coefficient of 1:

$$\ln(x^2 - 9) + 2\ln\frac{1}{x + 3} + 4\ln x \qquad (x > 3)$$

Solution

$$\begin{aligned}\ln(x^2 - 9) + 2\ln\frac{1}{x + 3} + 4\ln x &= \ln(x^2 - 9) + \ln\left[\left(\frac{1}{x + 3}\right)^2\right] + \ln x^4\\ &= \ln(x^2 - 9) + \ln\frac{1}{(x + 3)^2} + \ln x^4\\ &= \ln\left[(x^2 - 9)\cdot\frac{1}{(x + 3)^2}\cdot x^4\right]\\ &= \ln\frac{(x^2 - 9)x^4}{(x + 3)^2}\end{aligned}$$

This last expression can be simplified still further by factoring the quantity $x^2 - 9$ as $(x - 3)(x + 3)$. Then a factor of $x + 3$ can be divided out of the numerator and denominator of the fraction. The result is

$$\begin{aligned}\ln(x^2 - 9) + 2\ln\frac{1}{x + 3} + 4\ln x &= \ln\frac{(x - 3)x^4}{x + 3}\\ &= \ln\frac{x^5 - 3x^4}{x + 3}\end{aligned}$$

In the examples we have considered so far, we've used the properties of logarithms to shorten given expressions. We can also use these properties to expand an expression. (This is useful in calculus.)

EXAMPLE 3 Write each of the following quantities as sums and differences of simpler logarithmic expressions. Express each answer in such a way that no logarithm of products, quotients, or powers appears.

(a) $\log_{10}\sqrt{3x}$ **(b)** $\log_{10}\sqrt[3]{\dfrac{2x}{3x^2+1}}$ **(c)** $\ln\dfrac{x^2\sqrt{2x-1}}{(2x+1)^{3/2}}$

Solution

(a) $\log_{10}\sqrt{3x} = \log_{10}(3x)^{1/2} = \frac{1}{2}\log_{10}(3x)$
$= \frac{1}{2}(\log_{10} 3 + \log_{10} x)$

(b) $$\log_{10}\sqrt[3]{\frac{2x}{3x^2+1}} = \log_{10}\left[\left(\frac{2x}{3x^2+1}\right)^{1/3}\right]$$
$$= \frac{1}{3}\log_{10}\frac{2x}{3x^2+1}$$
$$= \frac{1}{3}[\log_{10} 2x - \log_{10}(3x^2+1)]$$
$$= \frac{1}{3}[\log_{10} 2 + \log_{10} x - \log_{10}(3x^2+1)]$$

(c) $$\ln\frac{x^2\sqrt{2x-1}}{(2x+1)^{3/2}} = \ln\frac{x^2(2x-1)^{1/2}}{(2x+1)^{3/2}}$$
$$= \ln x^2 + \ln(2x-1)^{1/2} - \ln(2x+1)^{3/2}$$
$$= 2\ln x + \frac{1}{2}\ln(2x-1) - \frac{3}{2}\ln(2x+1)$$

EXAMPLE 4 Given that $\log_b 2 = A$ and $\log_b 6 = B$, express each of the following in terms of A and/or B.

(a) $\log_b 8$ **(b)** $\log_b\sqrt{6}$ **(c)** $\log_b 12$
(d) $\log_b 3$ **(e)** $\log_{\sqrt{b}} 2$ **(f)** $\log_b(b/36)$

Solution

(a) $\log_b 8 = \log_b 2^3 = 3\log_b 2 = 3A$
(b) $\log_b\sqrt{6} = \log_b 6^{1/2} = \frac{1}{2}\log_b 6 = \frac{1}{2}B$
(c) $\log_b 12 = \log_b(2\times 6) = \log_b 2 + \log_b 6 = A + B$
(d) $\log_b 3 = \log_b(6/2) = \log_b 6 - \log_b 2 = B - A$
(e) Let $\log_{\sqrt{b}} 2 = x$. (So we want to express x in terms of A and/or B.) Writing this equation in exponential form, we have $(\sqrt{b})^x = 2$, and therefore

$$(b^{1/2})^x = 2$$
$$b^{x/2} = 2$$

Now we rewrite this last equation in logarithmic form to obtain $\log_b 2 = x/2$, and therefore

$$x = 2\log_b 2 = 2A$$

(f) $\log_b(b/36) = \log_b b - \log_b 36$
$= 1 - \log_b 6^2 = 1 - 2\log_b 6 = 1 - 2B$

EXAMPLE 5 Figure 1 shows the graph of $y = 2^x - 3$. (The axes are marked off in one-unit intervals.) As indicated in the figure, the x-intercept of the curve is between 1 and 2. Determine both an exact expression for this x-intercept and a calculator approximation rounded to two decimal places.

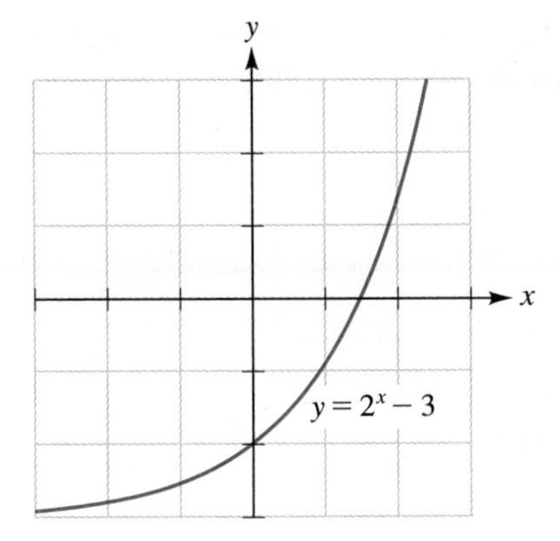

FIGURE 1

GRAPHICAL PERSPECTIVE

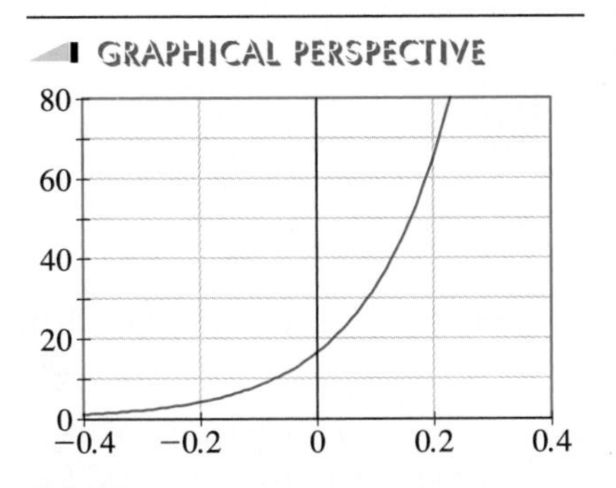

FIGURE 2
The curve $y = 4^{5x+2}$ intersects the line $y = 70$ at a point whose x-coordinate is slightly greater than 0.2. Example 6 shows that this x-coordinate is 0.213 (rounded to three decimal places).

Solution Setting y equal to zero in the given equation yields $2^x - 3 = 0$, or $2^x = 3$. Notice that the unknown, x, appears in the exponent. To solve for x, we take the logarithm of both sides of the equation. (Base e logarithms are used in the following computations; Exercise 71 asks you to check that the use of base 10 logarithms produces an equivalent answer.) We have

$$\begin{aligned} \ln 2^x &= \ln 3 \\ x \ln 2 &= \ln 3 && \text{using Property 4 on page 301} \\ x &= \frac{\ln 3}{\ln 2} \approx 1.58 && \textit{Caution: } \frac{\ln 3}{\ln 2} \neq \ln 3 - \ln 2 \end{aligned}$$

EXAMPLE 6 Solve for x: $4^{5x+2} = 70$. Use a calculator to evaluate the final answer; round to three decimal places.

Solution The fact that the unknown appears in the exponent suggests that Property 4 (the property that "brings down" the exponent) may be useful. To put Property 4 into play, we'll take the logarithm of both sides of the given equation. Either base ten or base e logarithms can be used here (because they're both on the calculator). Since we used base e logarithms in the previous example, let's use base ten here just for the sake of variety. (You can check for yourself at the end that base e logarithms will yield the same numerical answer.)

Taking the base ten logarithm of both sides of the given equation, we have

$$\begin{aligned} \log_{10} 4^{5x+2} &= \log_{10} 70 \\ (5x+2)\log_{10} 4 &= \log_{10} 70 && \text{using Property 4} \\ 5x + 2 &= (\log_{10} 70)/(\log_{10} 4) \\ x &= \frac{[(\log_{10} 70)/(\log_{10} 4)] - 2}{5} \\ x &\approx 0.213 \end{aligned}$$

The required solution is approximately 0.213. Figure 2 shows a graphical interpretation of this result.

It is sometimes necessary to convert logarithms in one base to logarithms in another base. After the next example, we will state a formula for this. However, as the next example indicates, it is easy to work this type of problem from the basics, without relying on a formula.

EXAMPLE 7 Express $\log_2 5$ in terms of base 10 logarithms.

Solution Let $z = \log_2 5$. The exponential form of this equation is

$$2^z = 5$$

We now take the base 10 logarithm of each side of this equation to obtain

$$\begin{aligned} \log_{10} 2^z &= \log_{10} 5 \\ z \log_{10} 2 &= \log_{10} 5 && \text{using Property 4 on page 301} \\ z &= \frac{\log_{10} 5}{\log_{10} 2} \end{aligned}$$

Given our definition of z, this last equation can be written

$$\log_2 5 = \frac{\log_{10} 5}{\log_{10} 2}$$

This is the required answer.

The method shown in Example 7 can be used to convert between any two bases. Exercise 72(b) at the end of this section asks you to follow this method, using letters rather than numbers, to arrive at the following general formula.

PROPERTY SUMMARY — CHANGE OF BASE FOR LOGARITHMS

FORMULA	EXAMPLES	
$\log_a x = \dfrac{\log_b x}{\log_b a}$	$\log_2 3 = \dfrac{\log_{10} 3}{\log_{10} 2} = \dfrac{\ln 3}{\ln 2}$	$\log_{10} e = \dfrac{\ln e}{\ln 10} = \dfrac{1}{\ln 10}$

EXAMPLE 8 Simplify the product $(\log_2 10)(\log_{10} 2)$.

Solution We'll use the change-of-base formula so that both factors in the given product involve logarithms with the same base. Expressing the first factor, $\log_2 10$, in terms of base ten logarithms, we have

$$\log_2 10 = \frac{\log_{10} 10}{\log_{10} 2} = \frac{1}{\log_{10} 2}$$

and therefore $$(\log_2 10)(\log_{10} 2) = \frac{1}{\log_{10} 2}(\log_{10} 2) = 1$$

So, the given product is equal to 1. [Using the same method, one can show that, in general, $(\log_a b)(\log_b a) = 1$.]

Examples 1–8 have dealt with applications of the five properties of logarithms. However, you also need to understand what the properties *don't* say. For instance, Property 3 does not apply to an expression such as $\dfrac{\log_{10} 5}{\log_{10} 2}$. [Property 3 *would* apply if the expression were $\log_{10}(5/2)$.] In the box that follows, we list some errors to avoid in working with logarithms.

ERRORS TO AVOID

ERROR	CORRECTION	COMMENT
$\log_b(x+y) \neq \log_b x + \log_b y$	$\log_b(xy) = \log_b x + \log_b y$	In general, there is no simple identity involving $\log_b(x+y)$. This is similar to the situation for $\sqrt{x+y}$ (which is not equal to $\sqrt{x}+\sqrt{y}$).
$\dfrac{\log_b x}{\log_b y} \neq \log_b x - \log_b y$	$\log_b\left(\dfrac{x}{y}\right) = \log_b x - \log_b y$	In general, there is no simple identity involving the quotient $(\log_b x)/(\log_b y)$.
$(\ln x)^3 \neq 3 \ln x$	$(\ln x)^3 = (\ln x)(\ln x)(\ln x)$	$(\ln x)^3$ is not the same as $\ln(x^3)$. Regarding the latter, we do have $\ln(x^3) = 3 \ln x$, where $x > 0$.
$\ln \dfrac{x}{2} \neq \dfrac{\ln x}{2}$	$\ln \dfrac{x}{2} = \ln x - \ln 2$ for $x > 0$ and $\dfrac{\ln x}{2} = \dfrac{1}{2} \ln x = \ln\sqrt{x}$ for $x > 0$	The confusion between $\ln \dfrac{x}{2}$ and $\dfrac{\ln x}{2}$ is sometimes due simply to careless handwriting.

EXERCISE SET 5.4

Do 1-9 odd, 13, 21, 43, 49, 53, 57

A

In Exercises 1–10, simplify the expression by using the definition and properties of logarithms.

1. $\log_{10} 70 - \log_{10} 7$
2. $\log_{10} 40 + \log_{10}(5/2)$
3. $\log_7 \sqrt{7}$
4. $\log_9 25 - \log_9 75$
5. $\log_3 108 + \log_3(3/4)$
6. $\ln e^3 - \ln e$
7. $-\frac{1}{2} + \ln\sqrt{e}$
8. $e^{\ln 3} + e^{\ln 2} - e^{\ln e}$
9. $2^{\log_2 5} - 3\log_5 \sqrt[3]{5}$
10. $\log_b b^b$

In Exercises 11–19, write the expression as a single logarithm with a coefficient of 1.

11. $\log_{10} 30 + \log_{10} 2$
12. $2\log_{10} x - 3\log_{10} y$
13. $\log_5 6 + \log_5(1/3) + \log_5 10$
14. $p\log_b A - q\log_b B + r\log_b C$
15. (a) $\ln 3 - 2\ln 4 + \ln 32$
    (b) $\ln 3 - 2(\ln 4 + \ln 32)$
16. (a) $\log_{10}(x^2 - 16) - 3\log_{10}(x+4) + 2\log_{10} x$
    (b) $\log_{10}(x^2 - 16) - 3[\log_{10}(x+4) + 2\log_{10} x]$
17. $\log_b 4 + 3[\log_b(1+x) - \frac{1}{2}\log_b(1-x)]$
18. $\ln(x^3 - 1) - \ln(x^2 + x + 1)$
19. $4\log_{10} 3 - 6\log_{10}(x^2+1) + \frac{1}{2}[\log_{10}(x+1) - 2\log_{10} 3]$

In Exercises 20–26, write the quantity using sums and differences of simpler logarithmic expressions. Express the answer so that logarithms of products, quotients, and powers do not appear.

20. (a) $\log_{10}\sqrt{(x+1)(x+2)}$
    (b) $\ln\sqrt{\dfrac{(x+1)(x+2)}{(x-1)(x-2)}}$
21. (a) $\log_{10}\dfrac{x^2}{1+x^2}$ (b) $\ln\dfrac{x^2}{\sqrt{1+x^2}}$
22. (a) $\log_b\dfrac{\sqrt{1-x^2}}{x}$ (b) $\ln\dfrac{x\sqrt[3]{4x+1}}{\sqrt{2x-1}}$
23. (a) $\log_{10}\sqrt{9-x^2}$ (b) $\ln\dfrac{\sqrt{4-x^2}}{(x-1)(x+1)^{3/2}}$
24. (a) $\log_b\sqrt[3]{\dfrac{x+3}{x}}$ (b) $\ln\dfrac{1}{\sqrt{x^2+x+1}}$
25. (a) $\log_b\sqrt{x/b}$
    (b) $2\ln\sqrt{(1+x^2)(1+x^4)(1+x^6)}$
26. (a) $\log_b\sqrt[3]{\dfrac{(x-1)^2(x-2)}{(x+2)^2(x+1)}}$
    (b) $\ln\left(\dfrac{e-1}{e+1}\right)^{3/2}$

In Exercises 27–36, suppose b is a positive constant greater than 1, and let A, B, and C be defined as follows:

$$\log_b 2 = A \qquad \log_b 3 = B \qquad \log_b 5 = C$$

In each case, use the properties of logarithms to evaluate the given expression in terms of A, B, and/or C. (In Exercises 31–36, use the change-of-base formula.)

27. (a) $\log_b 6$ (b) $\log_b(1/6)$
    (c) $\log_b 27$ (d) $\log_b(1/27)$
28. (a) $\log_b 10$ (b) $\log_b 100$
    (c) $\log_b 0.01$ (d) $\log_b 0.3$
29. (a) $\log_b(5/3)$ (b) $\log_b 0.6$
    (c) $\log_b(5/9)$ (d) $\log_b(5/16)$
30. (a) $\log_b\sqrt{5}$ (b) $\log_b\sqrt{15}$
    (c) $\log_b\sqrt[3]{0.4}$ (d) $\log_b\sqrt[4]{60}$
31. (a) $\log_3 b$ (b) $\log_3(10b)$
32. (a) $\log_{b^2} 5$ (b) $\log_{\sqrt{b}} 2$
33. (a) $\log_{3b} 2$ (b) $\log_{3b} 15$
34. (a) $\log_{5b} 1.2$ (b) $\log_{5b} 2.5$
35. (a) $(\log_b 5)(\log_5 b)$ (b) $(\log_b 6)(\log_6 b)$
36. (a) $\log_{2b} 6 + \log_{2b}(1/6)$
    (b) $\log_{18}(1/b)$

In Exercises 37 and 38, suppose that $\log_{10} A = a$, $\log_{10} B = b$, and $\log_{10} C = c$. Express the following logarithms in terms of a, b, and c.

37. (a) $\log_{10} AB^2C^3$ (b) $\log_{10} 10\sqrt{A}$
    (c) $\log_{10}\sqrt{10ABC}$ (d) $\log_{10}(10A/\sqrt{BC})$
38. (a) $\log_{10} A + 2\log_{10}(1/A)$
    (b) $\log_{10}(A/10)$
    (c) $\log_{10}\dfrac{100A^2}{B^4\sqrt[3]{C}}$
    (d) $\log_{10}\dfrac{(AB)^5}{C}$

In Exercises 39 and 40, suppose that $\ln x = t$ and $\ln y = u$. Write each expression in terms of t and u.

39. (a) $\ln(ex)$ (b) $\ln xy - \ln(x^2)$
    (c) $\ln\sqrt{xy} + \ln(x/e)$ (d) $\ln(e^2 x\sqrt{y})$

40. **(a)** $\ln(e^{\ln x})$ **(b)** $e^{\ln(\ln xy)}$
(c) $\ln\left(\frac{ex}{y}\right) - \ln\left(\frac{y}{ex}\right)$ **(d)** $\dfrac{(\ln x)^3 - \ln(x^4)}{\left(\ln \frac{x}{e^2}\right)\ln(xe^2)}$

In Exercises 41 and 42, graph the equations and determine the x-intercepts (as in Example 5).

41. **(a)** $y = 2^x - 5$
(b) $y = 2^{x/2} - 5$

42. **(a)** $y = 3^{x-1} - 2$
(b) $y = 3^{1-x} - 4$

In Exercises 43–47, solve the equations. Express the answers in terms of natural logarithms.

43. $5 = 2e^{2x-1}$ *Suggestion:* First divide by 2.

44. $3e^{1+t} = 2$ **45.** $2^x = 13$

46. $5^{3x-1} = 27$ **47.** $10^x = e$

48. $10^{2x+3} = 280$

In Exercises 49 and 50, solve the equations. Give two forms for each answer: one involving base ten logarithms and the other a calculator approximation rounded to three decimal places.

49. $3^{x^2-1} = 12$ **50.** $2^{9-x^2} = 430.5389$

In Exercises 51–56, express the quantity in terms of base 10 logarithms.

51. $\log_2 5$ **52.** $\log_5 10$ **53.** $\ln 3$
54. $\ln 10$ **55.** $\log_b 2$ **56.** $\log_2 b$

In Exercises 57–61, express the quantity in terms of natural logarithms.

57. $\log_{10} 6$ **58.** $\log_2 10$ **59.** $\log_{10} e$
60. $\log_b 2$, where $b = e^2$ **61.** $\log_{10}(\log_{10} x)$

62. Give specific examples showing that each statement is *false*.
(a) $\log(x+y) = \log x + \log y$
(b) $(\log x)/(\log y) = \log x - \log y$
(c) $(\log x)(\log y) = \log x + \log y$
(d) $(\log x)^k = k \log x$

63. True or false?
(a) $\log_{10} A + \log_{10} B - \frac{1}{2}\log_{10} C = \log_{10}(AB/\sqrt{C})$
(b) $\log_e \sqrt{e} = 1/2$ **(c)** $\ln\sqrt{e} = 1/2$
(d) $\ln x^3 = \ln 3x$
(e) $\ln x^3 = 3 \ln x$
(f) $\ln 2x^3 = 3 \ln 2x$
(g) $\log_a c = b$ means $a^b = c$.
(h) $\log_5 24$ is between 5^1 and 5^2.
(i) $\log_5 24$ is between 1 and 2.
(j) $\log_5 24$ is closer to 1 than to 2.
(k) The domain of $g(x) = \ln x$ is the set of all real numbers.
(l) The range of $g(x) = \ln x$ is the set of all real numbers.
(m) The function $g(x) = \ln x$ is one-to-one.

Use a calculator for Exercises 64–66.

64. **(a)** Check Property 2 using the values $b = 10$, $P = \pi$, and $Q = \sqrt{2}$.
(b) Let $P = 3$ and $Q = 4$. Show that $\ln(P+Q) \neq \ln P + \ln Q$.
(c) Check Property 3 using the values $b = 10$, $P = 2$, and $Q = 3$.
(d) If $P = 10$ and $Q = 20$, show that $\ln(PQ) \neq (\ln P)(\ln Q)$.
(e) Check Property 3 using natural logarithms and the values $P = 17$ and $Q = 76$.
(f) Show that $(\log_{10} 17)/(\log_{10} 76) \neq \log_{10} 17 - \log_{10} 76$.

65. **(a)** Check Property 4 using the values $b = 10$, $P = \pi$, and $n = 7$.
(b) Using the values given for b, P, and n in part (a), show that $\log_b P^n \neq (\log_b P)^n$.
(c) Verify Property 5 using the values $b = 10$ and $P = 1776$.
(d) Verify that $\ln 2 + \ln 3 + \ln 4 = \ln 24$.
(e) Verify that
$$\log_{10} A + \log_{10} B + \log_{10} C = \log_{10}(ABC),$$
using the values $A = 11$, $B = 12$, and $C = 13$.
(f) Let $f(x) = e^x$ and $g(x) = \ln x$. Compute $f[g(2345.6)]$.
(g) Let $f(x) = 10^x$ and $g(x) = \log_{10} x$. Compute $g[f(0.123456)]$.

66. Evaluate $[\frac{1}{\pi}\ln(640320^3 + 744)]^2$. *Remark:* Contrary to the empirical evidence from your calculator, it is known that this number is irrational.

B

67. *The approximation* $\ln(1+x) \approx x$: As indicated in the accompanying graph, the values of $\ln(1+x)$ and x are very close to one another for small positive values of x.

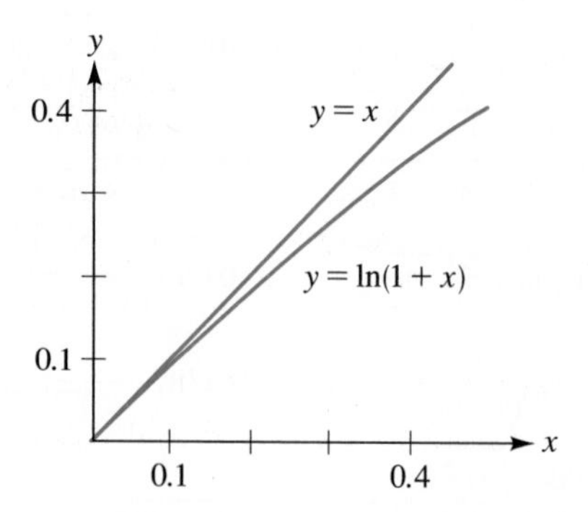

Using your calculator, complete the table to obtain numerical evidence of this. For your answers, report the first six decimal places of the calculator display (don't round off).

x	0.1	0.05	0.005	0.0005
$\ln(1+x)$				

68. As indicated in the figure, the graph of $y = ae^{bx}$ passes through the two points (2, 10) and (8, 80). Follow steps (a) through (c) to determine the values of the constants a and b.

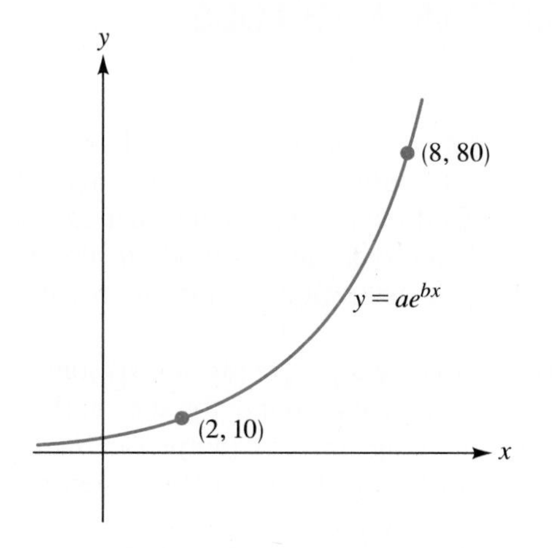

(a) Since the point (8, 80) lies on the graph, the pair of values $x = 8$ and $y = 80$ satisfies the equation $y = ae^{bx}$, that is,

$$80 = ae^{8b} \qquad (1)$$

Similarly, since the point (2, 10) lies on the graph, we have

$$10 = ae^{2b} \qquad (2)$$

Now use equations (1) and (2) to show that $b = (\ln 8)/6$.

(b) In equation (2), substitute for b using the expression obtained in part (a). Show that the resulting equation can be written $10 = a(e^{\ln 8})^{1/3}$.

(c) Use one of the properties of logarithms to simplify the right-hand side of the equation in part (b), then solve for a. (You should obtain $a = 5$.)

In Exercises 69 and 70, you are given the coordinates of two points on the graph of the curve $y = ae^{bx}$. In each case, determine the values of a and b. Hint: *Use the method explained in Exercise 68.*

69. $(-2, 324)$ and $(1/2, 4/3)$ **70.** $(1, 2)$ and $(4, 8)$

71. **(a)** Use base 10 logarithms to solve the equation $2^x - 3 = 0$.

(b) Use your calculator to evaluate the answer in part (a). Also use the calculator to evaluate $(\ln 3)/(\ln 2)$.

(c) Prove that $(\log_{10} 3)/(\log_{10} 2) = (\ln 3)/(\ln 2)$.

72. **(a)** Prove that $\log_b(P/Q) = \log_b P - \log_b Q$.
Hint: Study the proof of Property 2 in the text.

(b) Prove the **change of base formula:**

$$\log_a x = \frac{\log_b x}{\log_b a}$$

Hint: Use the method of Example 7 in the text.

73. Show that $\log_b \dfrac{\sqrt{3}+\sqrt{2}}{\sqrt{3}-\sqrt{2}} = 2\log_b(\sqrt{3}+\sqrt{2})$.

74. **(a)** Show that $\text{lob}_b(P/Q) + \log_b(Q/P) = 0$.
(b) Simplify: $\log_a x + \log_{1/a} x$.

75. Simplify: $b^{3\log_b x}$.

76. Is there a constant k such that the equation $e^x = 2^{kx}$ holds for all values of x?

77. Prove that $\log_b a = 1/(\log_a b)$.

78. Simplify: $(\log_2 3)(\log_3 4)(\log_4 5)$.

79. **(a)** Without using your calculator, show that $\dfrac{1}{\log_2 \pi} + \dfrac{1}{\log_5 \pi} > 2$. *Hint:* Convert to base π logarithms.

(b) Without using your calculator, show that $\log_\pi 2 + \dfrac{1}{\log_\pi 2} > 2$. *Hint:* First explain, in complete sentences, how you know that the quantity $\log_\pi 2$ is positive. Then apply the inequality given in Exercise 45(b) in Section 2.5.

(c) Use your calculator (and the change-of-base formula) to find out which of the two quantities is larger: $\dfrac{1}{\log_2 \pi} + \dfrac{1}{\log_5 \pi}$ or $\log_\pi 2 + \dfrac{1}{\log_\pi 2}$.

80. A function f with domain $(1, \infty)$ is defined by the equation $f(x) = \log_x 2$.
(a) Find a value for x such that $f(x) = 2$.
(b) Is the number that you found in part (a) a fixed point of the function f?

C

81. Prove that $(\log_a x)/(\log_{ab} x) = 1 + \log_a b$.

82. Simplify a^x when $x = \log_b(\log_b a)/\log_b a$.

83. If $a^2 + b^2 = 7ab$, where a and b are positive, show that

$$\log[\tfrac{1}{3}(a+b)] = \tfrac{1}{2}(\log a + \log b)$$

no matter which base is used for the logarithms (but it is understood that the same base is used throughout).

84. Let $y = \ln[\ln(\ln x)]$. First, complete the following table using a calculator. After doing that, disregard the evidence in your table and prove (without a calculator, of course) that the range of the given function is actually the set of all real numbers.

x	100	1000	10^6	10^{20}	10^{50}	10^{99}
y						

85. Let $f(x) = \ln\left(1 - \frac{1}{x^2}\right)$.

(a) Use the properties of logarithms (and some algebra) to show that

$$f(2) + f(3) + f(4) = \ln(5/8)$$

(b) Use a calculator to check the result in part (a).

5.5 EQUATIONS AND INEQUALITIES WITH LOGS AND EXPONENTS

Initially, [John] *Napier* [1550–1617] *called logarithms "artificial numbers" but later coined the term* logarithm, *meaning "number of the ratio."*

Ronald Calinger, ed., *Classics of Mathematics* (Englewood Cliffs, New Jersey: Prentice Hall, 1995)

... we would like to give a value for z, such that $a^z = y$. This value of z, insofar as it is viewed as a function of y, it is called the LOGARITHM of y.

Leonhard Euler (1707–1783), *Introductio in analysin infinitorum* (Lausanne: 1748) [This classic text has been translated by John D. Blanton (Berlin: Springer-Verlag, 1988).]

In Examples 2 through 9 in this section we illustrate some of the more common approaches for solving equations and inequalities involving exponential and logarithmic expressions. Although there is no single technique that can be used to solve every equation or inequality of this type, the methods in this section do have much in common: they all rely on the definition and the basic properties of logarithms.

Actually, in the previous two sections, we've already solved some equations involving exponential expressions. So, as background for the work to follow, you might want to review the following three examples. *Study suggestion:* Try solving for yourself the following three equations before you turn back to the text's solution.

EXAMPLE	EQUATION
Example 7 in Section 5.3	$10^{2x} = 200$
Example 5 in Section 5.4	$2^x - 3 = 0$
Example 6 in Section 5.4	$4^{5x+2} = 70$

EXAMPLE 1 Consider the following two equations:

$$(\ln x)^2 = 2 \ln x \qquad (1)$$

$$\ln(x^2) = 2 \ln x \qquad (2)$$

(a) Is either one of the values $x = 1$ or $x = 2$ a root of equation (1)?
(b) Is either one of the values $x = 1$ or $x = 2$ a root of equation (2)?

Solution

(a) To see if the value $x = 1$ satisfies equation (1), we have

$$(\ln 1)^2 = 2 \ln 1$$
$$0^2 = 2(0) \qquad \text{True}$$

Thus $x = 1$ is a root of equation (1). To see if $x = 2$ is a root, we write

$$(\ln 2)^2 = 2 \ln 2$$
$$\ln 2 = 2 \qquad \text{dividing through by } \ln 2\ (\neq 0)$$

This last equation is not valid. You can see this by using a calculator, or, more directly, by rewriting the equation in exponential form to obtain

$e^2 = 2$, which is clearly false. (Why?) Consequently, the value $x = 2$ is not a solution of equation (1).

(b) You've seen equation (2) before, or at least one very much like it. It's an example of one of the basic properties of logarithms that we studied in the previous section. That is, we know from the previous section that the equation $\ln(x^2) = 2 \ln x$ holds for every value of x in the domain of the function $y = \ln x$. In particular, then, since both $x = 1$ and $x = 2$ are in the domain of the function $y = \ln x$, we can conclude immediately, without any work required, that both of the values 1 and 2 satisfy equation (2).

The example that we have just completed serves to remind us of the difference between a *conditional equation* and an *identity*. An identity is true for all values of the variable in its domain. For example, the equation $\ln(x^2) = 2 \ln x$ is an identity; it is true for every positive real number x. In contrast to this, a conditional equation is true only for some (or perhaps even none) of the values of the variable. Equation (1) in Example 1 is a conditional equation; we saw that it is true when $x = 1$ and false when $x = 2$. The equation $2^x = -1$ is an example of a conditional equation that has no real-number root. (Why?) Most of the equations that we solve in this section are conditional equations; but watch for a few identities to pop up in the examples and exercises. It can make your work much easier.

GRAPHICAL PERSPECTIVE

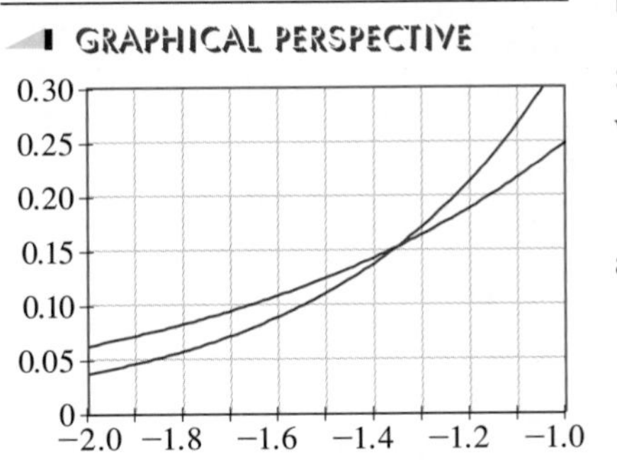

FIGURE 1
The curves $y = 4^x$ and $y = 3^{2x+1}$ intersect at a point whose x-coordinate is between -1.4 and -1.3. In Example 2 we found this x-value to be -1.355 (rounded to three decimal places). (Question: Which of the two graphs in the figure represents $y = 4^x$?)

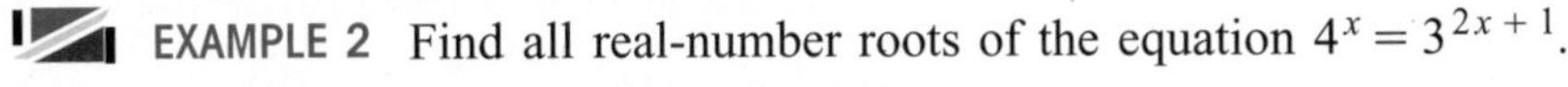

EXAMPLE 2 Find all real-number roots of the equation $4^x = 3^{2x+1}$.

Solution Taking the natural logarithm of both sides of the given equation, we have

$$\ln(4^x) = \ln(3^{2x+1})$$

and therefore

$$\begin{aligned} x \ln 4 &= (2x + 1)\ln 3 \\ x \ln 4 &= 2x \ln 3 + \ln 3 \\ x(\ln 4 - 2 \ln 3) &= \ln 3 \\ x &= \frac{\ln 3}{\ln 4 - 2 \ln 3} \end{aligned}$$

Using a calculator now, we find that x is approximately -1.355. In Figure 1 we show a graphical check of this solution. Exercise 77 asks you to check the solution algebraically.

EXAMPLE 3 Find all real-number roots of the following equations:

(a) $\ln(\ln x) = 2$; **(b)** $e^{\ln x} = -2$; **(c)** $e^{\ln x} = 2$.

Solution

(a) For convenience, we let t stand for $\ln x$. Then the given equation becomes

$$\ln t = 2$$

Rewriting this equation in exponential form, we have

$$t = e^2$$

or, in view of the definition of t,

$$\ln x = e^2$$

Now we rewrite this last equation in exponential form to obtain

$$\begin{aligned} x &= e^{e^2} \\ &\approx 1618.2 \quad \text{using a calculator} \end{aligned}$$

As Exercise 78 asks you to verify, the value $x = e^{e^2}$ indeed checks in the original equation.

(b) Think for a moment about the graph of the basic exponential function $y = e^x$. The graph is always above the x-axis. This tells us that e raised to any exponent is never negative. Consequently, the equation $e^{\ln x} = -2$ has no real-number solution.

(c) From Section 5.3, we know that $e^{\ln x} = x$ for all positive numbers x, so the given equation $e^{\ln x} = 2$ becomes simply $x = 2$, and we are done.

In the next example we solve equations involving logarithms by converting them to exponential form. For part (b) of the example, you need to recall the quadratic formula.

EXAMPLE 4 Find all real-number roots of the following equations:

(a) $\log_{10}(x^2 - 2) = 3$; **(b)** $\log_{10}(x^2 - 2x) = 3$.

Solution

(a) Converting to exponential form, we have

$$x^2 - 2 = 10^3$$

and therefore

$$x^2 = 1002$$

or

$$x = \pm\sqrt{1002} \approx \pm 31.65$$

As Exercise 79(a) asks you to check, both of the values $\pm\sqrt{1002}$ satisfy the given equation.

(b) Again we begin by converting to exponential form. This yields

$$x^2 - 2x = 1000$$

and therefore

$$x^2 - 2x - 1000 = 0$$

Now we apply the quadratic formula:

$$\begin{aligned} x = \frac{-b \pm \sqrt{b^2 - 4ac}}{2a} &= \frac{2 \pm \sqrt{(-2)^2 - 4(1)(-1000)}}{2} \\ &= \frac{2 \pm \sqrt{4004}}{2} = \frac{2 \pm \sqrt{4(1001)}}{2} \\ &= \frac{2 \pm 2\sqrt{1001}}{2} = 1 \pm \sqrt{1001} \end{aligned}$$

As Exercise 79(b) asks you to verify, both of the numbers $1+\sqrt{1001}$ and $1-\sqrt{1001}$ satisfy the given equation. Using a calculator, we find that these two values are approximately 32.6 and -30.6, respectively. In Figure 2 we show a graphical check of the solutions that we've determined in this example.

GRAPHICAL PERSPECTIVE

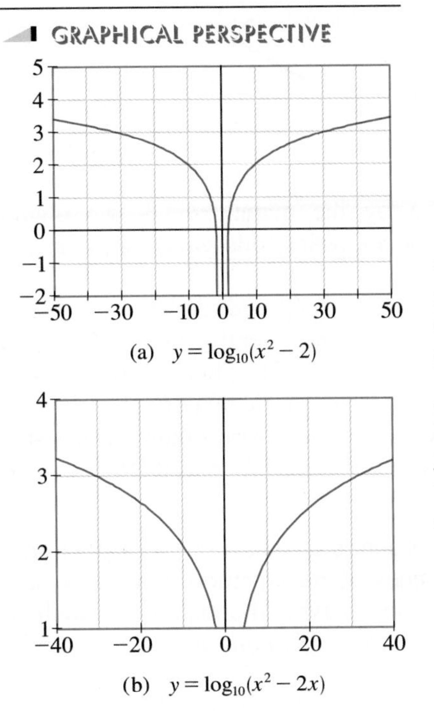

FIGURE 2
In (a), the figure shows that the curve $y=\log_{10}(x^2-2)$ intersects the horizontal line $y=3$ at two points whose x-coordinates are approximately 30 and -30. In part (a) of the example, we found these two numbers to be 31.65 and -31.65, respectively (rounded to two decimal places). Similarly, in (b) the figure shows that the curve $y=\log_{10}(x^2-2x)$ intersects the horizontal line $y=3$ at two points, one with an x-coordinate slightly greater than 30, the other with an x-coordinate very close to -30. In part (b) of the example, we found these two x-coordinates to be 32.6 and -30.6, respectively (rounded to one decimal place).

If you look over both parts of the example just completed, you'll see that we used the same general procedure in both cases: convert from logarithmic to exponential form, then solve the resulting quadratic equation. The next example works in the other direction: first we solve a quadratic equation, then we isolate the variable by converting from exponential to logarithmic form.

EXAMPLE 5 As indicated in Figure 3, the graph of the function $y=2(3^{2x})-3^x-3$ has an x-intercept between 0.25 and 0.50. Find the exact value of this intercept and also a calculator approximation rounded to three decimal places.

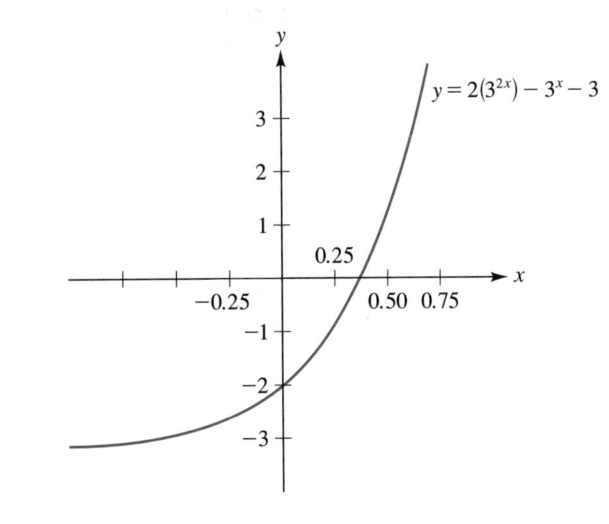

FIGURE 3

Solution We need to solve the equation $2(3^{2x})-3^x-3=0$. The observation that helps here is that 3^{2x} can be written as $(3^x)^2$. This lets us rewrite the equation as

$$2(3^x)^2-3^x-3=0$$

Now, for convenience, we let $t=3^x$ so that the equation becomes

$$2t^2-t-3=0$$

or

$$(2t-3)(t+1)=0 \qquad \text{Check the factoring.}$$

$$2t-3=0 \quad \Big| \quad t+1=0$$
$$t=3/2 \quad \Big| \quad t=-1$$

Using the value $t=3/2$ in the equation $t=3^x$ gives us

$$3^x=3/2$$

We can solve this last equation by taking the logarithm of both sides. Using base e logarithms, we have

$$\ln(3^x) = \ln 1.5$$

and therefore $\quad x \ln 3 = \ln 1.5$

Dividing both sides of this last equation by the quantity $\ln 3$, we obtain $x = (\ln 1.5)/(\ln 3)$. As Exercise 80 asks you to verify, this value for x indeed checks in the original equation.

Now, what about the other value for t that we found? With $t = -1$, we have $3^x = -1$, but that is impossible. (Why?) Consequently, there is but one root of the given equation, namely, $x = (\ln 1.5)/(\ln 3)$. This is the required x-intercept for the graph shown in Figure 3. Using a calculator, we find that $\ln(1.5)/(\ln 3) \approx 0.369$. Note that this value is consistent with Figure 3, in which the x-intercept appears to be roughly halfway between 0.25 and 0.50.

When we solved equations in earlier chapters of this text, we learned that some techniques, such as squaring both sides of an equation, may introduce extraneous roots. The next example shows another type of situation in which an extraneous root may be introduced. We'll return to this point after the example.

EXAMPLE 6 Solve for x: $\log_3 x + \log_3(x+2) = 1$

Solution Using Property 2 (on page 301), we can write the given equation as

$$\log_3[x(x+2)] = 1 \quad \text{or} \quad \log_3(x^2 + 2x) = 1$$

Writing this last equation in exponential form yields

$$\begin{aligned} x^2 + 2x &= 3^1 \\ x^2 + 2x - 3 &= 0 \\ (x+3)(x-1) &= 0 \end{aligned}$$

Thus, we have

$$x + 3 = 0 \quad \text{or} \quad x - 1 = 0$$

and, consequently,

$$x = -3 \quad \text{or} \quad x = 1$$

Now let's check these values in the original equation to see if they are indeed solutions.

If $x = -3$, the equation becomes

$$\log_3(-3) + \log_3(-1) \stackrel{?}{=} 1$$

Neither expression is defined, since the domain of the logarithm function does not contain negative numbers.

If $x = 1$, the equation becomes

$$\log_3 1 + \log_3 3 \stackrel{?}{=} 1$$

$$0 + 1 \stackrel{?}{=} 1 \qquad \text{True}$$

Thus, the value $x = 1$ is a solution of the original equation, but $x = -3$ is not.

In the example we just concluded, an extraneous solution ($x = -3$) was generated along with the correct solution, $x = 1$. How did this happen? It happened because in the second line of the solution, we used the property $\log_b PQ = \log_b P + \log_b Q$ which is valid only when both P and Q are positive. In the box that follows, we generalize this remark about extraneous solutions.

PROPERTY SUMMARY **EXTRANEOUS SOLUTIONS FOR LOGARITHMIC EQUATIONS**

Using the properties of logarithms (on page 301) to solve logarithmic equations may introduce extraneous solutions that do not check in the original equation. (This occurs because the logarithm function requires positive inputs, but in solving an equation, we may not know ahead of time the sign of an input involving a variable.) Therefore, it is always necessary to check any candidates for solutions that are obtained in this manner.

Earlier in this book, in Sections 2.5 and 2.6, we learned how to solve inequalities involving polynomials. Those same techniques, along with the properties of logarithms, are often useful in solving inequalities involving logarithmic and exponential expressions. In the box that follows, we list some additional facts that are helpful in solving these types of inequalities.

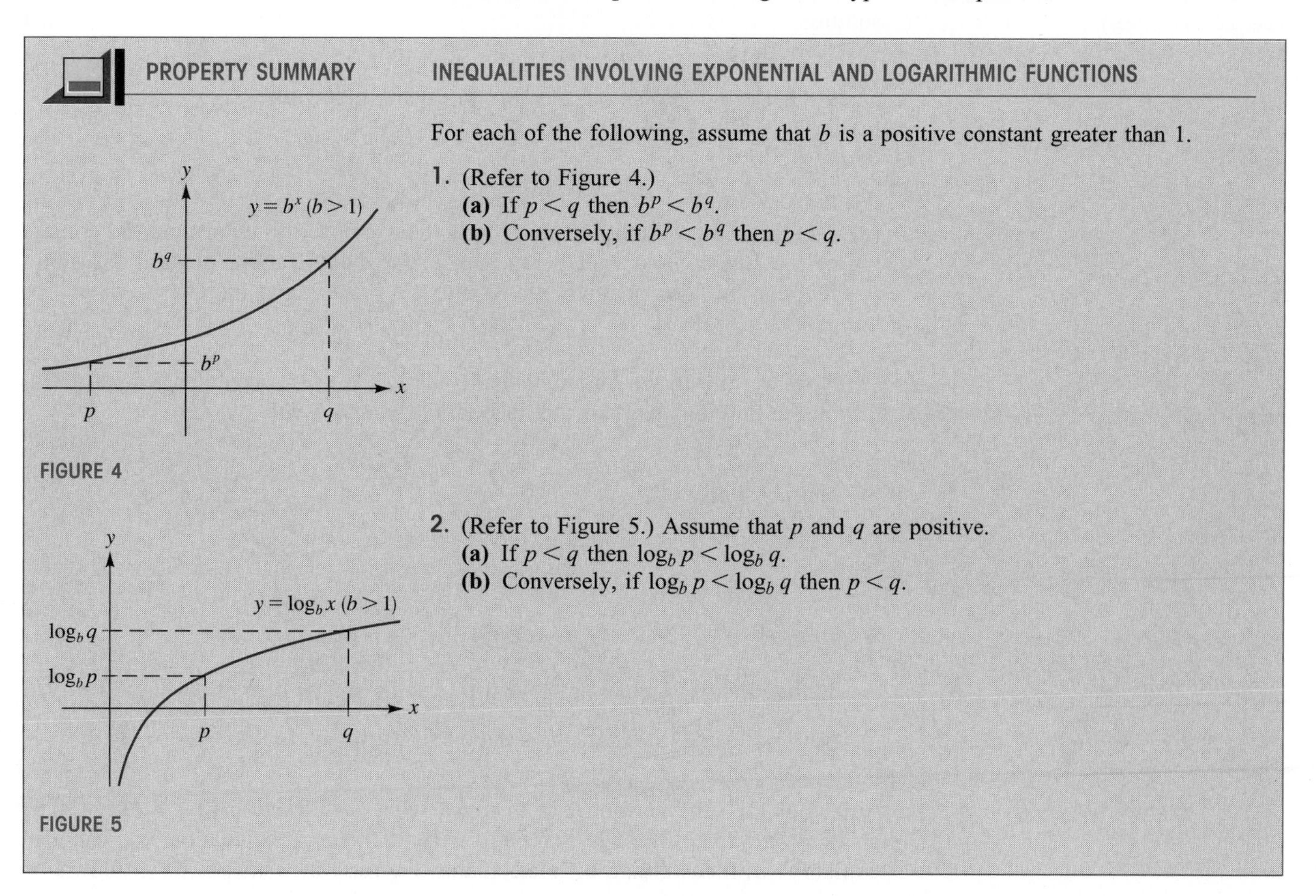

PROPERTY SUMMARY **INEQUALITIES INVOLVING EXPONENTIAL AND LOGARITHMIC FUNCTIONS**

For each of the following, assume that b is a positive constant greater than 1.

1. (Refer to Figure 4.)
   (a) If $p < q$ then $b^p < b^q$.
   (b) Conversely, if $b^p < b^q$ then $p < q$.

FIGURE 4

2. (Refer to Figure 5.) Assume that p and q are positive.
   (a) If $p < q$ then $\log_b p < \log_b q$.
   (b) Conversely, if $\log_b p < \log_b q$ then $p < q$.

FIGURE 5

GRAPHICAL PERSPECTIVE

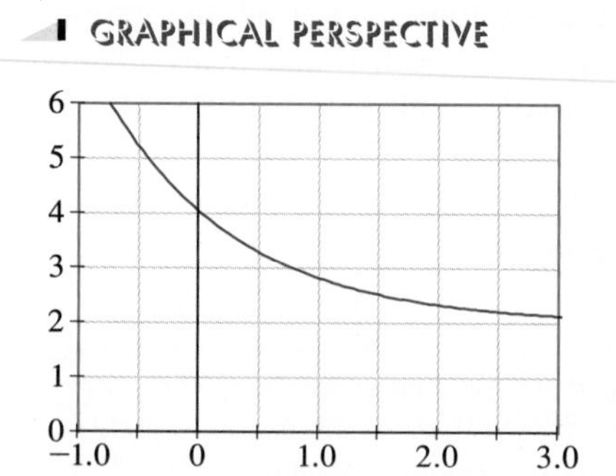

FIGURE 6
The graph of the equation $y = 2(1 + 0.4^x)$ indicates that the solution set of the inequality $2(1 + 0.4^x) < 5$ is an interval of the form (a, ∞), where a is approximately -0.5. In Example 7 we found that this value of a is $(\ln 1.5)/(\ln 0.4) \approx -0.443$ (rounded to three decimal places).

EXAMPLE 7 Solve the inequality $2(1 + 0.4^x) < 5$.

Solution Dividing both sides by 2, we have

$$\begin{aligned} 1 + 0.4^x &< 5/2 \\ 0.4^x &< 3/2 \\ \ln(0.4^x) &< \ln(3/2) \qquad \text{using Property 2(a) on page 315} \\ x \ln(0.4) &< \ln(3/2) \end{aligned}$$

Now, to isolate x, we want to divide both sides of this last inequality by the quantity $\ln(0.4)$, which is negative. (Without relying on a calculator, how do you know that this quantity is negative?) Since dividing both sides by a negative quantity reverses the inequality, we obtain

$$x > \frac{\ln(3/2)}{\ln(0.4)}$$

So the solution set consists of all real numbers greater than $\dfrac{\ln(3/2)}{\ln(0.4)}$.

Using a calculator to evaluate this last expression, we find that it is approximately -0.443. With this approximation, we can use interval notation to write the solution set as $(-0.443, \infty)$. See Figure 6 for a graphical view of this solution.

EXAMPLE 8 Solve each of the following inequalities:
(a) $\ln(2 - 3x) \leq 1$;
(b) $e^{2-3x} \leq 1$.

Solution
(a) As a preliminary but necessary first step, we need to determine the domain of the function $y = \ln(2 - 3x)$. Since the inputs for the natural logarithm function must be positive, we require $2 - 3x > 0$, and therefore

$$-3x > -2 \qquad \text{or} \qquad x < 2/3$$

Now we turn to the given inequality $\ln(2 - 3x) \leq 1$. Applying Property 1(a) on the previous page to this inequality, we can write

$$e^{\ln(2-3x)} \leq e^1$$

and consequently

$$\begin{aligned} 2 - 3x &\leq e \\ -3x &\leq e - 2 \\ x &\geq \frac{e-2}{-3} = \frac{2-e}{3} \end{aligned}$$

Putting things together now, we want x to be greater than or equal to $(2 - e)/3$ but less than $2/3$. Thus, the solution set is the interval $\left[\dfrac{2-e}{3}, \dfrac{2}{3}\right)$.

NOTE For this interval notation to make sense, the fraction $(2 - e)/3$ must be smaller than $2/3$. Although you can verify this using a calculator, you can also explain it without relying on a calculator. Exercise 81 asks you to do this.

(b) Since the domain of the exponential function is the set of all real numbers, there is no preliminary restriction on x, as there was in part (a). Taking the natural logarithm of both sides of the given inequality [as in Property 2(a) on page 315], we obtain

$$\ln(e^{2-3x}) \leq \ln(1)$$

and therefore

$$\begin{aligned} 2 - 3x &\leq 0 \\ -3x &\leq -2 \\ x &\geq 2/3 \end{aligned}$$

The solution set of the given inequality is the interval $[2/3, \infty)$.

EXAMPLE 9

(a) Determine the domain of the function $f(x) = \log_{10} x + \log_{10}(x - 2)$.
(b) Solve the inequality $\log_{10} x + \log_{10}(x - 2) \leq \log_{10} 24$.

Solution

(a) As you know, the logarithm function requires positive inputs. So for the expression $\log_{10} x$ we require $x > 0$, while for the expression $\log_{10}(x - 2)$ we require

$$x - 2 > 0 \qquad \text{and therefore} \qquad x > 2$$

In summary, then, we want x to be greater than zero and, at the same time, we need to have x greater than 2. Notice now that if x is greater than 2, it's automatically true that x is greater than zero. Thus, the domain of the given function f consists of all real numbers greater than 2.

(b) On the left-hand side of the given inequality, we can use one of the basic properties of logarithms to write

$$\log_{10} x + \log_{10}(x - 2) = \log_{10}[x(x - 2)] = \log_{10}(x^2 - 2x)$$

With this result, the given inequality becomes

$$\log_{10}(x^2 - 2x) \leq \log_{10} 24$$

and therefore

$$x^2 - 2x \leq 24 \qquad \text{using Property 2(b) on page 315}$$

or

$$x^2 - 2x - 24 \leq 0 \tag{1}$$

Inequality (1) is one of the types of inequalities that you learned to solve earlier in this text (in Section 2.6). As Exercise 82 at the end of this section asks you to show, the solution set for inequality (1) is the closed interval $[-4, 6]$. This closed interval, however, is not the solution set for the original inequality; we need to take into account the result in part (a). Putting things together, then, we want just those numbers in the interval $[-4, 6]$ that are greater than 2. Consequently, the solution set for the given inequality is the interval (2, 6]. Figure 7 shows a graphical view of this result.

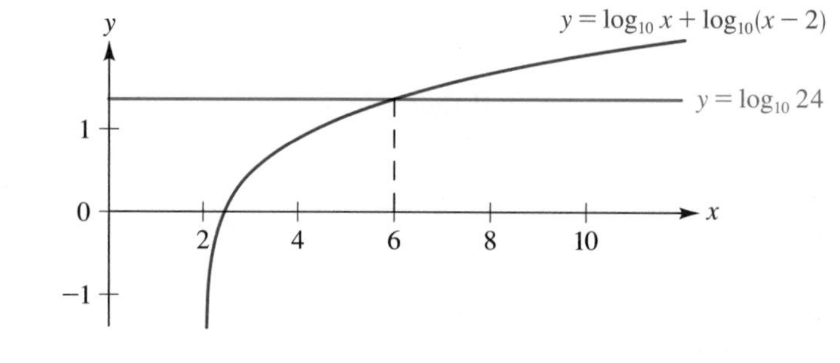

FIGURE 7
The curve $y = \log_{10} x + \log_{10}(x - 2)$ is below the horizontal line $y = \log_{10} 24$ when $2 < x < 6$. At $x = 6$ the curve and the line intersect. Thus, the solution set of the inequality $\log_{10} x + \log_{10}(x - 2) \leq \log_{10} 24$ is the interval (2, 6].

EXERCISE SET 5.5

Do 1, 7, 29, 31, 39, 47, 55, 73

A

To help you get started, Exercises 1–10 correlate directly with Examples 2–6 as shown in the chart. (Thus, if you need help in any of Exercises 1–10, first consult the indicated example.)

Exercise	1, 2	3, 4	5, 6	7, 8	9, 10
Based on Example No.	2	3(a)	4(a)	4(b)	5

In Exercises 1–41, find all the real-number roots of each equation. In each case, give an exact expression for the root and also (where appropriate) a calculator approximation rounded to three decimal places.

1. $5^x = 3^{2x-1}$
2. $7^{-4x} = 2^{1+3x}$
3. $\ln(\ln x) = 1.5$
4. $\log_3[\log_3(2x)] = -2$
5. $\log_{10}(x^2 + 36) = 2$
6. $\log_2(2x^2 - 4) = 5$
7. $\log_{10}(2x^2 - 3x) = 2$
8. $\log_9(x^2 + x) = 0.5$
9. $10^{2x} + 3(10^x) - 10 = 0$
10. $3(2^{2x}) - 11(2^x) - 4 = 0$
11. **(a)** $\ln(x^3) = 3 \ln x$
(b) $(\ln x)^3 = 3 \ln x$
12. **(a)** $(\log_{10} x)^2 = 2 \log_{10} x$
(b) $\log_{10}(x^2) = 2 \log_{10} x$
13. **(a)** $\log_3 6x = \log_3 6 + \log_3 x$
(b) $\log_3 6x = 6 \log_3 x$
14. **(a)** $\ln x = (\log_{10} x)/(\log_{10} e)$
(b) $\ln x = (\log_{10} e)/(\log_{10} x)$
15. $7^{\log_7 2x} = 2x$
16. $7^{\log_7 2x} = 7$
17. $\log_2(\log_3 x) = -1$
18. $\ln[\ln(\ln x)] = 1$
19. $\ln 4 - \ln x = (\ln 4)/(\ln x)$
20. $\log_5 x - \log_5 10 = \log_5(x/10)$
21. $\ln(3x^2) = 2 \ln(3x)$
22. $\ln(3x^2) = 2 \ln(\sqrt{3}x)$
23. $\log_{16} \dfrac{x+3}{x-1} = \dfrac{1}{2}$
24. $\log_{\sqrt{2}} \dfrac{1-4x}{1+4x} = 4$
25. **(a)** $e^{2x} + 2e^x + 1 = 0$
(b) $e^{2x} - 2e^x + 1 = 0$
(c) $e^{2x} - 2e^x - 3 = 0$
(d) $e^{2x} - 2e^x - 4 = 0$
26. **(a)** $4e^{6x} - 12e^{3x} + 9 = 0$
(b) $4e^{6x} + 12e^{3x} + 9 = 0$
(c) $4e^{6x} - 16e^{3x} - 9 = 0$
(d) $e^{6x} - 12e^{3x} - 9 = 0$
27. $e^x - e^{-x} = 1$ *Hint:* Multiply both sides by e^x.
28. $e^x + e^{-x} = 2$
29. $2^{5x} = 3^x(5^{x+3})$
30. $e^{3x} = 10^{2x}(2^{1-x})$
31. $\log_6 x + \log_6(x + 1) = 1$
32. $\log_6 x + \log_6(x + 1) = 0$
33. $\log_9(x + 1) = \frac{1}{2} + \log_9 x$
34. $\log_2(x + 4) = 2 - \log_2(x + 1)$
35. $\log_{10}(2x + 4) + \log_{10}(x - 2) = 1$
36. $\ln x + \ln(x + 1) = \ln 12$
37. $\log_{10}(x + 3) - \log_{10}(x - 2) = 2$
38. $\ln(x + 1) = 2 + \ln(x - 1)$
39. $\log_{10}(x + 1) = 2 \log_{10}(x - 1)$
40. $\log_2(2x^2 + 4) = 5$
41. $\log_{10}(x - 6) + \log_{10}(x + 3) = 1$
42. Solve for x in terms of a: $\log_2(x + a) - \log_2(x - a) = 1$.
43. Solve for x in terms of y:
(a) $\log_{10} x - y = \log_{10}(3x - 1)$;
(b) $\log_{10}(x - y) = \log_{10}(3x - 1)$.
44. Solve for x in terms of b: $\log_b(1 - 3x) = 3 + \log_b x$.

In Exercises 45–60, solve the inequalities.

45. $3(2 - 0.6^x) \leq 1$

46. $6(5 - 1.6^x) \geq 13$

47. $4(10 - e^x) \leq -3$

48. $\frac{2}{3}(1 - e^{-x}) \leq -3$

49. $\ln(2 - 5x) > 2$

50. $3 \log_{10}(4x + 3) < 1$

51. $e^{2+x} \geq 100$

52. $4^{5-x} > 15$

53. $2^x > 0$

54. $\log_2 x \geq 0$

55. $\log_2 \frac{2x - 1}{x - 2} < 0$

56. $\ln \frac{3x - 2}{4x + 1} > \ln 4$

57. $e^{x^2 - 4x} \geq e^5$

58. $10^{-x^2} \leq 10^{-12}$

59. $e^{(1/x) - 1} > 1$

60. $e^{1/(x-1)} > 1$

61. **(a)** Specify the domain of the function $y = \ln x + \ln(x - 4)$.
(b) Solve the inequality $\ln x + \ln(x - 4) \leq \ln 21$.

62. **(a)** Specify the domain of the function $y = \ln x + \ln(x + 2)$.
(b) Solve the inequality $\ln x + \ln(x + 2) \leq \ln 35$.

63. Solve the inequality $\log_2 x + \log_2(x + 1) - \log_2(2x + 6) < 0$.

64. Solve the inequality $\log_{10}(x^2 - 6x - 6) > 0$.

65. **(a)** Specify the domain of each function: $y = \log_2(2x - 1)$; $y = \log_4 x$.
(b) Solve the inequality $\log_2(2x - 1) > \log_4 x$.

66. **(a)** Specify the domain of each function: $y = \log_2(3x + 2)$; $y = \log_4 25x$.
(b) Solve the inequality $\log_2(3x + 2) > \log_4 25x$.

B

In Exercises 67–76, solve each equation. In Exercises 71–76, solve for x in terms of the other letters.

67. $3(\ln x)^2 - \ln(x^2) - 8 = 0$

68. $x^{1 + \log_x 16} = 4x^2$

69. $\log_6 x = \dfrac{1}{\dfrac{1}{\log_2 x} + \dfrac{1}{\log_3 x}}$

70. $\dfrac{\ln(\sqrt{x + 4} + 2)}{\ln\sqrt{x}} = 2$

71. $\alpha \ln x + \ln \beta = 0$

72. $3 \ln x = \alpha + 3 \ln \beta$

73. $y = Ae^{kx}$

74. $\beta = 10 \log_{10}(x/x_0)$

75. $y = a/(1 + be^{-kx})$

76. $T = T_1 + (T_0 - T_1)e^{-kx}$

In Exercises 77–80, you are given an equation and a root that was obtained in an example in the text. In each case: **(a)** *verify (algebraically) that the root indeed satisfies the equation; and* **(b)** *use a calculator to check that the root satisfies the equation.*

77. *(From Example 2)*
$4^x = 3^{2x+1}$; $x = (\ln 3)/(\ln 4 - 2 \ln 3)$

78. *[From Example 3(a)]* $\ln(\ln x) = 2$; $x = e^{e^2}$

79. **(a)** *[From Example 4(a)]*
$\log_{10}(x^2 - 2) = 3$; $x = \pm\sqrt{1002}$
(b) *[From Example 4(b)]*
$\log_{10}(x^2 - 2x) = 3$; $x = 1 \pm \sqrt{1001}$

80. *(From Example 5)*
$2(3^{2x}) - 3^x - 3 = 0$; $x = (\ln 1.5)/\ln 3$

81. *[From Example 8(a)]* Explain, in one or two complete sentences, how you know (without using a calculator) that the fraction $(2 - e)/3$ is less than $2/3$.

82. *[From Example 9(b)]* Solve the inequality $x^2 - 2x - 24 \leq 0$. You should find that the solution set is the closed interval $[-4, 6]$.

C

In Exercises 83 and 84, solve the inequalities.

83. $\log_\pi[\log_4(x^2 - 5)] < 0$

84. $\dfrac{1}{\log_2 x} + \dfrac{1}{\log_3 x} + \dfrac{1}{\log_4 x} > 2$

In Exercises 85 and 86, solve the equations.

85. $x^{(x^x)} = (x^x)^x$

86. $(\pi x)^{\log_{10} \pi} = (ex)^{\log_{10} e}$

87. Solve for x (assuming that $a > b > 0$):
$$(a^4 - 2a^2b^2 + b^4)^{x-1} = (a - b)^{2x}(a + b)^{-2}$$

Answer: $x = \dfrac{\ln(a - b)}{\ln(a + b)}$

88. Let $f(x) = \ln(x + \sqrt{x^2 + 1})$. Find $f^{-1}(x)$.

89. Suppose that $\log_{10} 2 = a$ and $\log_{10} 3 = b$. Solve for x in terms of a and b:
$$6^x = \frac{10}{3} - 6^{-x}$$

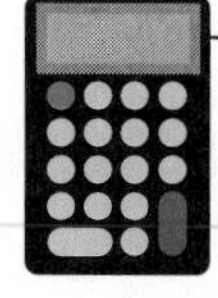

GRAPHING UTILITY EXERCISES FOR SECTION 5.5

1. (a) In Example 1(a) we found that $x = 1$ is a root of the equation $(\ln x)^2 = 2 \ln x$. Verify this visually by graphing the two functions $y = (\ln x)^2$ and $y = 2 \ln x$ and noting that the graphs indeed appear to intersect at $x = 1$. (Use a viewing rectangle in which x extends from 0 to 10 and y extends from -10 to 10.)
   (b) Your picture in part (a) shows that there is another point where the graphs intersect. In other words, there is a second root of the equation $(\ln x)^2 = 2 \ln x$. Using the picture from part (a), estimate this root to the nearest 0.5. Then zoom in repeatedly on the intersection point until you are certain of the first two decimal places in the root.
   (c) Using pencil and paper, solve the equation $(\ln x)^2 = 2 \ln x$. Use one of your answers to check your estimation in part (b).
2. Graph the two functions $y = \ln(x^2)$ and $y = 2 \ln x$ in the same picture. What do you observe? What does this have to do with the statements made in the text in the solution of Example 1(b)?
3. As background for this exercise, reread Example 2.
   (a) Figure 1 in the text offers a graphical perspective on the solution of the equation $4^x = 3^{2x+1}$. Use a graphing utility to create your own version of Figure 1. Then zoom in on the intersection point of the two curves to estimate the x-coordinate. This x-coordinate is a root of the given equation. (Continue zooming in until you are certain about the first three decimal places of the root.)
   (b) According to Example 2, the root of the equation $4^x = 3^{2x+1}$ is $(\ln 3)/(\ln 4 - 2 \ln 3)$. Evaluate this expression and use it to check your estimate in part (a).
   (c) For a different graphical perspective on Example 2, rewrite the given equation in the equivalent form $4^x - 3^{2x+1} = 0$. Now graph the function $y = 4^x - 3^{2x+1}$. The x-intercept of this graph is a root of the given equation $4^x = 3^{2x+1}$. Use your graphing utility to estimate this x-intercept; check that your answer is consistent with the results in part (a).
4. As background for this exercise, reread Example 5.
   (a) Figure 3 in the text shows that the equation $2(3^{2x}) - 3^x - 3 = 0$ has a root between 0.25 and 0.50. Use a graphing utility to create your own version of Figure 1. Then, by zooming in on the x-intercept of the graph, estimate the root. (Continue zooming in until you are certain about the first three decimal places of the root.)
   (b) According to Example 5, the exact value of the root is $(\ln 1.5)/(\ln 3)$. Evaluate this expression and use the result to check your estimate in part (a).
5. (a) In the standard viewing rectangle, graph the function $y = \ln(2 - 3x)$ and the line $y = 1$. Note that the curve and the line intersect at a point in the second quadrant. For a closer look, switch to a viewing rectangle in which x extends from -1 to 1 and y extends from -2 to 2.
   (b) What is the connection between the intersection point observed in part (a) and the solution to Example 8(a) in the text? (Explain in complete sentences.)
   (c) Estimate the x-coordinate of the intersection point by zooming in (Continue zooming in until you are certain about the first two decimal places.)
   (d) According to Example 8(a), the exact value of the x-coordinate of the intersection point is $(2 - e)/3$. Evaluate this expression and use the result to check your estimate in part (c).
6. (a) Graph the two functions $f(x) = (\ln x)/(\ln 3)$ and $g(x) = \ln x - \ln 3$. (Use a viewing rectangle in which x extends from 0 to 10 and y extends from -5 to 5.) Why aren't the two graphs identical? That is, doesn't one of the basic log identities say that $(\ln a)/(\ln b) = \ln a - \ln b$?
   (b) Your picture in part (a) indicates that
   $$\frac{\ln x}{\ln 3} > \ln x - \ln 3 \quad (0 < x \leq 10)$$
   Find a viewing rectangle in which $(\ln x)/(\ln 3) \leq \ln x - \ln 3$.
   (c) Use the picture that you obtain in part (b) to estimate the value of x for which $(\ln x)/(\ln 3) = \ln x - \ln 3$.
   (d) Solve the equation $(\ln x)/(\ln 3) = \ln x - \ln 3$ algebraically and use the result to check your estimate in part (c).

COMPOUND INTEREST

$S = Pe^{rt}$. This result is remarkable both because of its simplicity and the occurrence of e. (Who would expect that number to pop up in finance theory?)

Philip Gillett in *Calculus and Analytic Geometry,* 3rd ed. (Lexington, Mass: D.C. Heath & Company, 1988)

We begin this section by considering how money accumulates in a savings account. Eventually, this will lead us back to the number e and to functions of the form $y = ae^{bx}$.

The following idea from arithmetic is a prerequisite for our discussion. To increase a given quantity by, say, 15%, we multiply the quantity by 1.15. For instance, suppose that we want to increase \$100 by 15%. The calculations can be written

$$100 + 0.15(100) = 100(1 + 0.15) = 100(1.15)$$

Similarly, to increase a quantity by 30%, we would multiply by 1.30, and so on. The next example displays some calculations involving percentage increase. The results may surprise you unless you're already familiar with this topic.

EXAMPLE 1 An amount of \$100 is increased by 15% and then the new amount is increased by 15%. Is this the same as an overall increase of 30%?

Solution To increase \$100 by 15%, we multiply by 1.15 to obtain \$100(1.15). Now to increase this new amount by 15%, we multiply it by 1.15 to obtain

$$[(\$100)(1.15)](1.15) = \$100(1.15)^2 = \$132.25$$

Alternatively, if we increase the original \$100 by 30%, we obtain

$$\$100(1.30) = \$130$$

Comparing, we see that the result of two successive 15% increases is greater than the result of a single 30% increase.

Now let's look at another example and use it to introduce some terminology. Suppose that you place \$1000 in a savings account at 10% interest *compounded annually.* This means that at the end of each year, the bank contributes to your account 10% of the amount that is in the account at that time. Interest compounded in this manner is called **compound interest.** The original deposit of \$1000 is called the **principal,** denoted P. The interest rate, expressed as a decimal, is denoted by r. Thus, $r = 0.10$ in this example. The variable A is used to denote the **amount** in the account at any given time. The calculations displayed in Table 1 show how the account grows.

TABLE 1

Time Period	Algebra	Arithmetic
After 1 year	$A = P(1 + r)$	$A = 1000(1.10)$ $= \$1100$
After 2 years	$A = [P(1 + r)](1 + r)$ $= P(1 + r)^2$	$A = 1100(1.10)$ $= \$1210$
After 3 years	$A = [P(1 + r)^2](1 + r)$ $= P(1 + r)^3$	$A = 1210(1.10)$ $= \$1331$

We can learn several things from Table 1. First, consider how much interest is paid each year.

Interest paid for first year: $\$1100 - \$1000 = \$100$
Interest paid for second year: $\$1210 - \$1100 = \$110$
Interest paid for third year: $\$1331 - \$1210 = \$121$

Thus the interest earned each year is not a constant; it increases each year.

If you look at the algebra in Table 1, you can see what the general formula should be for the amount after t years.

COMPOUND INTEREST FORMULA (INTEREST COMPOUNDED ANNUALLY)

Suppose that a principal of P dollars is invested at an annual rate r that is compounded annually. Then the amount A after t years is given by

$$A = P(1 + r)^t$$

EXAMPLE 2 Suppose that \$2000 is invested at $7\frac{1}{2}\%$ interest compounded annually. How many years will it take for the money to double?

Solution In the formula $A = P(1 + r)^t$, we use the given values $P = \$2000$ and $r = 0.075$. We want to find how long it will take for the money to double; that is, we want to find t when $A = \$4000$. Making these substitutions in the formula, we obtain

$$4000 = 2000(1 + 0.075)^t$$

and, therefore,

$$2 = 1.075^t \qquad \text{dividing by 2000}$$

We can solve this exponential equation by taking the logarithm of both sides. We use base e logarithms. (Base 10 would also be convenient here.) This yields

$$\ln 2 = \ln 1.075^t$$

and, consequently,

$$\ln 2 = t \ln 1.075 \quad \text{(Why?)}$$

To isolate t, we divide both sides of this last equation by ln 1.075. This yields

$$t = \frac{\ln 2}{\ln 1.075} \approx 9.6 \text{ years}$$

Now, assuming that the bank computes the compound interest only at the end of the year, we must round the preliminary answer of 9.6 years and say that when $t = 10$ years, the initial \$2000 will have *more than* doubled. Table 2 adds some perspective to this. The table shows that after 9 years, something less than \$4000 is in the account; whereas after 10 years, the amount exceeds \$4000.

TABLE 2 $A = 2000(1.075)^t$

t (years)	A (dollars)
9	3834.48
10	4122.06

In Example 2, the interest was compounded annually. In practice, though, interest is usually computed more often. For instance, a bank may advertise a rate of 10% per year compounded semiannually. This means that after half a year, the interest is compounded at 5%, and then after another half year, the interest is again compounded at 5%. If you review Example 1, you'll see that two compoundings at a rate of r is not the same as one compounding at a rate of $2r$. The formula $A = P(1 + r)^t$ can be generalized to cover such cases where interest is compounded more than once each year.

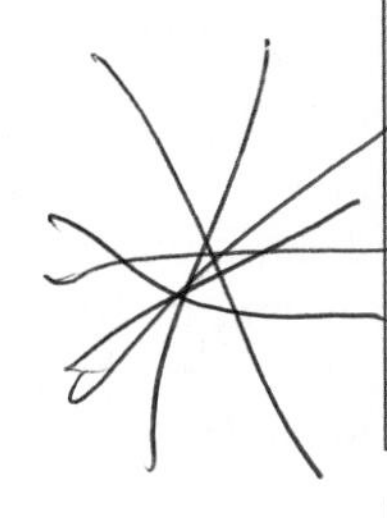

COMPOUND INTEREST FORMULA (INTEREST COMPOUNDED n TIMES PER YEAR)

Suppose that a principal of P dollars is invested at an annual rate r that is compounded n times per year. Then the amount A after t years is given by

$$A = P\left(1 + \frac{r}{n}\right)^{nt}$$

EXAMPLE 3 Suppose that \$1000 is placed in a savings account at 10% per annum. How much is in the account at the end of one year if the interest is **(a)** compounded once each year ($n = 1$)? **(b)** compounded quarterly ($n = 4$)?

Solution
We use the formula $A = P[1 + (r/n)]^{nt}$.

(a) For $n = 1$, we obtain

$$\begin{aligned} A &= 1000\left(1 + \frac{0.10}{1}\right)^{(1)(1)} \\ &= 1000(1.1) \\ &= \$1100 \end{aligned}$$

(b) For $n = 4$, we obtain

$$\begin{aligned} A &= 1000\left(1 + \frac{0.10}{4}\right)^{4(1)} \\ &= 1000(1.025)^4 \\ &= \$1103.81 \quad \text{using a calculator} \end{aligned}$$

Notice that compounding the interest quarterly rather than annually yields the greater amount. This is in agreement with our observations in Example 1.

The results in Example 3 will serve to illustrate some additional terminology used by financial institutions. In that example, the interest for the year under quarterly compounding was

$$\$1103.81 - \$1000 = \$103.81$$

Now, \$103.81 is 10.381% of \$1000. We say in this case that the **effective rate** of interest is 10.381%. The given rate of 10% per annum compounded once a year is called the **nominal rate.** (The nominal rate and the effective rate are also referred to as the *annual rate* and the *annual yield,* respectively.) The next example further illustrates these ideas.

EXAMPLE 4 A bank offers a nominal interest rate of 12% per annum for certain accounts. Compute the effective rate if interest is compounded monthly.

Solution Let P denote the principal, which earns 12% ($r = 0.12$) compounded monthly ($n = 12$). Then with $t = 1$, our formula yields

$$A = P\left(1 + \frac{0.12}{12}\right)^{12(1)} = P(1.01)^{12}$$

Using a calculator to approximate the quantity $(1.01)^{12}$, we obtain

$$A \approx P(1.12683)$$

This shows that the effective interest rate is about 12.68%.

There are two rather natural questions to ask upon first encountering compound interest calculations:

QUESTION 1 For a fixed period of time (say, one year), does more and more frequent compounding of interest continue to yield greater and greater amounts?

QUESTION 2 Is there a limit on how much money can accumulate in a year when interest is compounded more and more frequently?

The answer to both of these questions is yes. If you look back over Example 1, you'll see evidence for the affirmative answer to Question 1. For additional evidence, and for the answer to Question 2, let's do some calculations. To keep things as simple as possible, suppose a principal of \$1 is invested for 1 year at the nominal rate of 100% per annum. (More realistic figures could be used here, but the algebra becomes more cluttered.) With these data, our formula becomes

$$A = 1\left(1 + \frac{1}{n}\right)^{n(1)} \qquad \text{or} \qquad A = \left(1 + \frac{1}{n}\right)^{n}$$

Now, as we know from Section 5.2, the expression $\left(1 + \frac{1}{n}\right)^{n}$ approaches the value e as n grows larger and larger; see Table 3.

Table 3 shows that the amount does increase with the number of compoundings. But assuming that the bank rounds to the nearest penny, Table 3 also shows that there is no difference between compounding hourly, compounding

TABLE 3 Results of Compounding Interest More and More Frequently

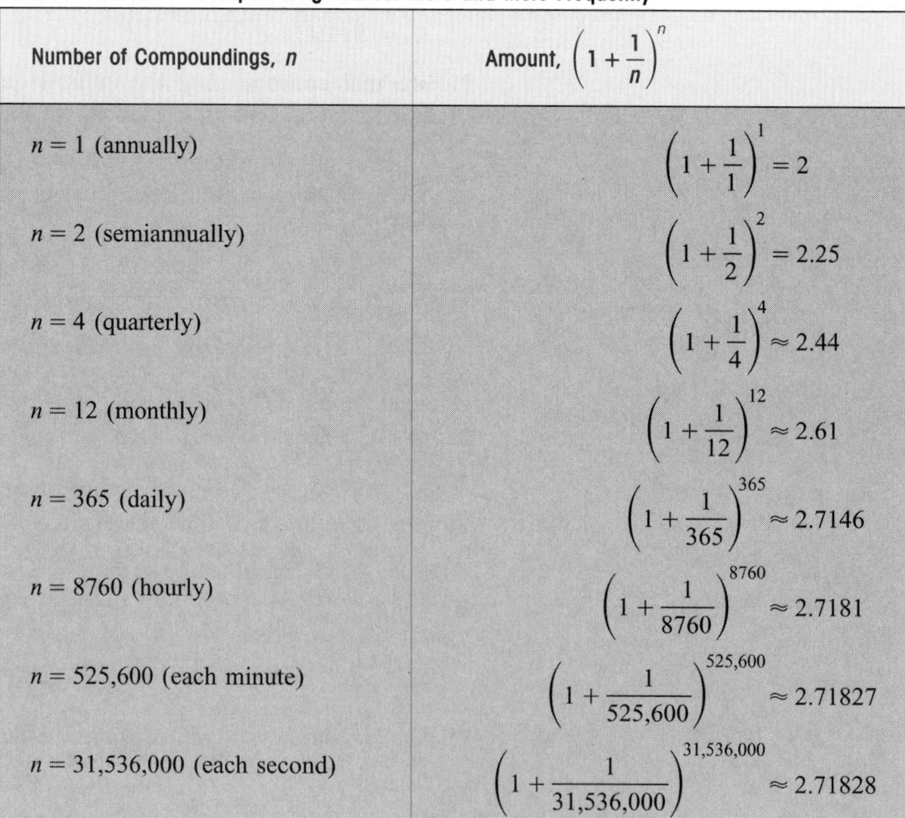

Number of Compoundings, n	Amount, $\left(1 + \frac{1}{n}\right)^{n}$
$n = 1$ (annually)	$\left(1 + \frac{1}{1}\right)^{1} = 2$
$n = 2$ (semiannually)	$\left(1 + \frac{1}{2}\right)^{2} = 2.25$
$n = 4$ (quarterly)	$\left(1 + \frac{1}{4}\right)^{4} \approx 2.44$
$n = 12$ (monthly)	$\left(1 + \frac{1}{12}\right)^{12} \approx 2.61$
$n = 365$ (daily)	$\left(1 + \frac{1}{365}\right)^{365} \approx 2.7146$
$n = 8760$ (hourly)	$\left(1 + \frac{1}{8760}\right)^{8760} \approx 2.7181$
$n = 525{,}600$ (each minute)	$\left(1 + \frac{1}{525{,}600}\right)^{525{,}600} \approx 2.71827$
$n = 31{,}536{,}000$ (each second)	$\left(1 + \frac{1}{31{,}536{,}000}\right)^{31{,}536{,}000} \approx 2.71828$

each minute, and compounding each second. In each case, the rounded amount is \$2.72.

Some banks advertise interest compounded not monthly, daily, or even hourly, but *continuously*—that is, at each instant. The formula for the amount earned under continuous compounding of interest is as follows.

COMPOUND INTEREST FORMULA (INTEREST COMPOUNDED CONTINUOUSLY)

Suppose that a principal of P dollars is invested at an annual rate r that is compounded continuously. Then the amount A after t years is given by

$$A = Pe^{rt}$$

EXAMPLE 5 A sum of \$100 is placed in a savings account at 5% per annum compounded continuously. Assuming no subsequent withdrawals or deposits, when will the balance reach \$150?

Solution Substitute the values $A = 150$, $P = 100$, and $r = 0.05$ in the formula $A = Pe^{rt}$ to obtain

$$150 = 100e^{0.05t}$$

and, therefore,

$$1.5 = e^{0.05t}$$

To solve this last equation for t, we rewrite it in its equivalent logarithmic form:

$$0.05t = \ln 1.5$$

$$t = \frac{\ln 1.5}{0.05} \text{ years}$$

Using a calculator, we find that $t \approx 8.1$ years. In other words, it will take slightly more than 8 years 1 month for the balance to reach \$150.

In the next example, we compare the nominal rate with the effective rate under continuous compounding of interest.

EXAMPLE 6

(a) Given a nominal rate of 8% per annum compounded continuously, compute the effective interest rate.

(b) Given an effective rate of 8% per annum, compute the nominal rate.

Solution

(a) With the values $r = 0.08$ and $t = 1$, the formula $A = Pe^{rt}$ yields

$$A = Pe^{0.08(1)}$$

$$A \approx P(1.08329) \quad \text{using a calculator}$$

This shows that the effective interest rate is approximately 8.33% per year.

(b) We now wish to compute the nominal rate r, given an effective rate of 8% per year. An effective rate of 8% means that the initial principal P grows to

$P(1.08)$ by the end of the year. Thus, in the formula $A = Pe^{rt}$, we make the substitutions $A = P(1.08)$ and $t = 1$. This yields

$$P(1.08) = Pe^{r(1)}$$

Dividing both sides of this last equation by P, we have

$$1.08 = e^r$$

To solve this equation for r, we rewrite it in its equivalent logarithmic form:

$$r = \ln(1.08)$$
$$r \approx 0.07696 \qquad \text{using a calculator}$$

Thus, a nominal rate of about 7.70% per annum yields an effective rate of 8%. Table 4 summarizes these results.

TABLE 4 Comparison of Nominal and Effective Rates in Example 6

Nominal Rate (% per annum)	Effective Rate (% per annum)
8	8.33
7.70	8

We now have two basic formulas for compound interest:

$$A = P\left(1 + \frac{r}{n}\right)^{nt} \qquad \text{for interest compounded } n \text{ times per year}$$

and

$$A = Pe^{rt} \qquad \text{for interest compounded continuously}$$

At first glance, these formulas appear to be very different from one another: the second formula involves e, the first does not. However, appearances can be deceiving. As the next example shows, the equation $A = P\left(1 + \frac{r}{n}\right)^{nt}$ can be rewritten in a form that is entirely similar to $A = Pe^{rt}$.

EXAMPLE 7 Find a constant k such that

$$P\left(1 + \frac{r}{n}\right)^{nt} = Pe^{kt} \tag{1}$$

Solution We want to solve for k in equation (1). After dividing both sides of the equation by P, we have

$$\left(1 + \frac{r}{n}\right)^{nt} = e^{kt}$$

Writing this last equation in logarithmic form yields

$$kt = \ln\left(1 + \frac{r}{n}\right)^{nt} = nt \ln\left(1 + \frac{r}{n}\right) \qquad \text{(Why?)}$$

$$k = n \ln\left(1 + \frac{r}{n}\right) \qquad \text{dividing by } t \tag{2}$$

This shows that the formula $A = P\left(1 + \frac{r}{n}\right)^{nt}$ can be rewritten $A = Pe^{kt}$, where k is given by equation (2).

Now we come to one of the remarkable and characteristic features of compound interest and of growth governed by the formula $A = Pe^{rt}$. By the **doubling**

time we mean, as the name implies, the amount of time required for a given principal to double. The surprising fact here is that the doubling time does not depend on the principal P. To see why this is so, we begin with the formula

$$A = Pe^{rt}$$

We are interested in the time t at which $A = 2P$. Replacing A by $2P$ in the formula yields

$$2P = Pe^{rt}$$
$$2 = e^{rt}$$
$$rt = \ln 2$$
$$t = \frac{\ln 2}{r}$$

Denoting the doubling time by T_2, we have the following formula.

$$\text{Doubling time} = T_2 = \frac{\ln 2}{r}$$

As you can see, the formula for the doubling time T_2 does not involve P, but only r. Thus, at a given rate under continuous compounding, \$2 and \$2000 would both take the same amount of time to double. (This idea takes some getting used to.)

EXAMPLE 8 Compute the doubling time T_2 when a sum is invested at an interest rate of 4% per annum compounded continuously.

Solution $T_2 = \frac{\ln 2}{r} = \frac{\ln 2}{0.04} \approx 17.3$ years using a calculator

There is a convenient approximation that allows us to estimate doubling times easily. Using a calculator, we see that

$$\ln 2 \approx 0.7$$

Using this approximation, we have the following rule of thumb for estimating doubling time:

$$T_2 \approx \frac{0.7}{r}$$

Let's use this rule to rework Example 8. With $r = 0.04$, we obtain

$$T_2 \approx \frac{0.7}{0.04} = \frac{70}{4} = 17.5 \text{ years}$$

Notice that this estimation is quite close to the actual doubling time obtained in Example 8.

The idea of doubling time is useful in graphing the function $A = Pe^{rt}$. In this discussion, we assume that P and r are constants, so that the amount A is a function of the time t. Suppose, for example, that a principal of \$1000 is invested at 10% per annum compounded continuously. Then the function we wish to graph is

$$A = 1000e^{0.1t}$$

The doubling time in this situation is

$$T_2 \approx \frac{0.7}{r} = \frac{0.7}{0.1} = 7 \text{ years}$$

Table 5 shows the results of doubling a principal of \$1000 every 7 years.

TABLE 5

t (years)	A (dollars)
0	1000
7	2000
14	4000
21	8000
28	16000

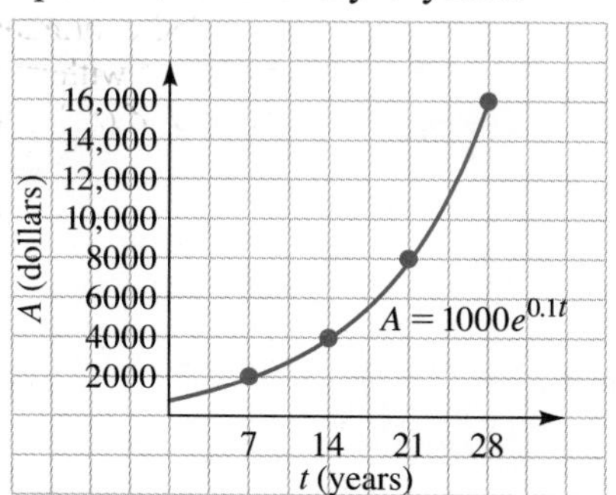

FIGURE 1
An exponential growth function.

We'll use the data in Table 5 to graph the function $A = 1000e^{0.1t}$. We'll mark off units on the t-axis in multiples of 7; on the A-axis, we'll use multiples of 2000. Figure 1 shows the result of plotting the points from Table 5 and then joining them with a smooth curve. Notice that the domain in this context is $[0, \infty)$.

EXERCISE SET 5.6

1. You invest \$800 at 6% interest compounded annually. How much is in the account after 4 years, assuming that you make no subsequent withdrawal or deposit?
2. A sum of \$1000 is invested at an interest rate of $5\frac{1}{2}$% compounded annually. How many years will it take until the sum exceeds \$2500? (First find out when the amount equals \$2500; then round off as in Example 2.)
3. At what interest rate (compounded annually) will a sum of \$4000 grow to \$6000 in 5 years?
4. A bank pays 7% interest compounded annually. What principal will grow to \$10,000 in 10 years?
5. You place \$500 in a savings account at 5% compounded annually. After 4 years you withdraw all your money and take it to a different bank, which advertises a rate of 6% compounded annually. What is the balance in this new account after 4 more years? (As usual, assume that no subsequent withdrawal or deposit is made.)
6. A sum of \$3000 is placed in a savings account at 6% per annum. How much is in the account after 1 year if the interest is compounded **(a)** annually? **(b)** semiannually? **(c)** daily?
7. A sum of \$1000 is placed in a savings account at 7% per annum. How much is in the account after 20 years if the interest is compounded **(a)** annually? **(b)** quarterly?
8. Your friend invests \$2000 at $5\frac{1}{4}$% per annum compounded semiannually. You invest an equal amount at the same yearly rate, but compounded daily. How much larger is your account than your friend's after 8 years?

9. You invest \$100 at 6% per annum compounded quarterly. How long will it take for your balance to exceed \$120? (Round your answer up to the next quarter.)

10. A bank offers an interest rate of 7% per annum compounded daily. What is the effective rate?

11. What principal should you deposit at $5\frac{1}{2}\%$ per annum compounded semiannually so as to have \$6000 after 10 years?

12. You place a sum of \$800 in a savings account at 6% per annum compounded continuously. Assuming that you make no subsequent withdrawal or deposit, how much is in the account after 1 year? When will the balance reach \$1000?

13. A bank offers an interest rate of $6\frac{1}{2}\%$ per annum compounded continuously. What principal will grow to \$5000 in 10 years under these conditions?

14. Given a nominal rate of 6% per annum, compute the effective rate under continuous compounding of interest.

15. Suppose that under continuous compounding of interest, the effective rate is 6% per annum. Compute the nominal rate.

16. You have two savings accounts, each with an initial principal of \$1000. The nominal rate on both accounts is $5\frac{1}{4}\%$ per annum. In the first account, interest is compounded semiannually. In the second account, interest is compounded continuously. How much more is in the second account after 12 years?

17. You want to invest \$10,000 for 5 years, and you have a choice between two accounts. The first pays 6% per annum compounded annually. The second pays 5% per annum compounded continuously. Which is the better investment?

18. Suppose that a certain principal is invested at 6% per annum compounded continuously.
(a) Use the rule $T_2 \approx 0.7/r$ to estimate the doubling time.
(b) Compute the doubling time using the formula $T_2 = (\ln 2)/r$.
(c) Do your answers in (a) and (b) differ by more than 2 months?

19. A sum of \$1500 is invested at 5% per annum compounded continuously.
(a) Estimate the doubling time.
(b) Compute the actual doubling time.
(c) Let d_1 and d_2 denote the actual and estimated doubling times, respectively. Define d by

$$d = |d_1 - d_2|$$

What percent is d of the actual doubling time?

20. A sum of \$5000 is invested at 10% per annum compounded continuously.
(a) Estimate the doubling time.
(b) Estimate the time required for the \$5000 to grow to \$40,000.

21. After carrying out the calculations in this problem, you'll see one of the reasons why the government imposes inheritance taxes and why laws are passed to prohibit savings accounts from being passed from generation to generation without restriction. Suppose that a family invests \$1000 at 8% per annum compounded continuously. If this account were to remain intact, being passed from generation to generation, for 300 years, how much would be in the account at the end of those 300 years?

22. A principal of \$500 is invested at 7% per annum compounded continuously.
(a) Estimate the doubling time.
(b) Sketch a graph similar to the one in Figure 1, showing how the amount increases with time.

23. A principal of \$7000 is invested at 5% per annum compounded continuously.
(a) Estimate the doubling time.
(b) Sketch a graph showing how the amount increases with time.

24. In one savings account, a principal of \$1000 is deposited at 5% per annum. In a second account, a principal of \$500 is deposited at 10% per annum. Both accounts compound interest continuously.
(a) Estimate the doubling time for each account.
(b) On the same set of axes, sketch graphs showing the amount of money in each account over time. Give the (approximate) coordinates of the point where the two curves meet. In financial terms, what is the significance of this point? (In working this problem, assume that the initial deposits in each account were made at the same time.)
(c) During what period of time does the first account have the larger balance?

25. A principal of \$2000 is deposited in a savings account (Account 1) at 4% per annum, compounded quarterly. At the same time, a principal of \$2000 is deposited in another savings account (Account 2) at 4% per annum compounded continuously.
(a) Complete the following table to compare the results. (Round the values to the nearest dollar.)

t **(years)**	1	2	3	4	10
A **(Account 1)**					
A **(Account 2)**					

(b) Compute the doubling time for each account.

(c) Compute the average rate of change $\Delta A/\Delta t$ for Account 1 over the period $t = 2$ years to $t = 3$ years.

(d) Compute the average rate of change $\Delta A/\Delta t$ for Account 2 over the period $t = 2$ years to $t = 3$ years.

26. The following table shows the **public debt** for the United States for the decade of the 1980s. [The *public debt* consists of the national (federal) debt plus the debts owed by state and local governments.]

Year	Public Debt (trillion $)
1980	0.907
1981	0.998
1982	1.142
1983	1.377
1984	1.572
1985	1.823
1986	2.125
1987	2.350
1988	2.602
1989	2.857

Source: The World Almanac (Mahwah, New Jersey: Funk & Wagnalls, 1993)

Both of the accompanying figures show scatter plots for this data. In the scatter plots, and in the work that follows, we take $t = 0$ to correspond to the year 1980. (So, for example, $t = 9$ represents 1989 and $t = 11$ represents 1991.) In Figure A, an exponential regression function is used to model the data; in Figure B, a quadratic function is used. The equations defining these two functions are as follows.

Exponential model (Figure A): $y = 0.9044e^{0.1338t}$

Quadratic model (Figure B): $y = 0.0084t^2 + 0.1509t + 0.8579$

(a) Use your calculator and the equations just given to complete the following table. Round your answers to two decimal places.

	1990	1991	1992
Public Debt (trillion $) (exponential model)			
Public Debt (trillion $) (quadratic model)			
Public debt (trillion $) (actual)	3.233	3.665	4.065

(b) Look over the results that you obtained in part (a). Of the two models, which one gives a consistently closer estimate for the public debt?

(c) Use the exponential model to make a projection for the public debt in the year 2000. Next, use the quadratic model to make the projection. Are the two estimates relatively close to one another?

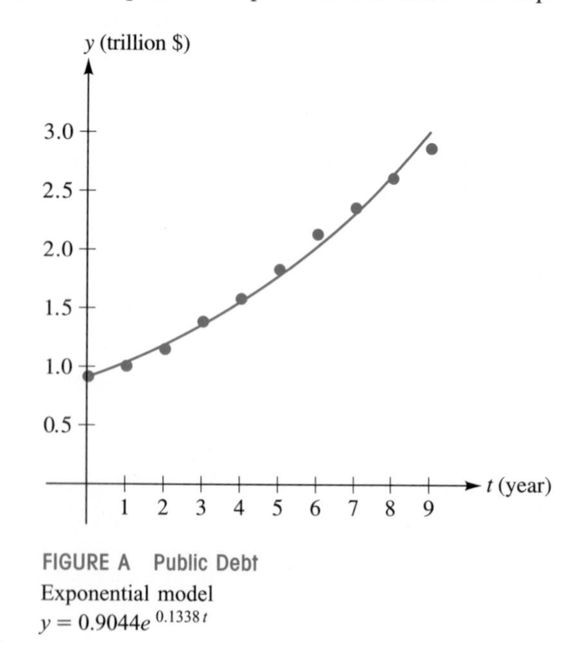

FIGURE A Public Debt
Exponential model
$y = 0.9044e^{0.1338t}$

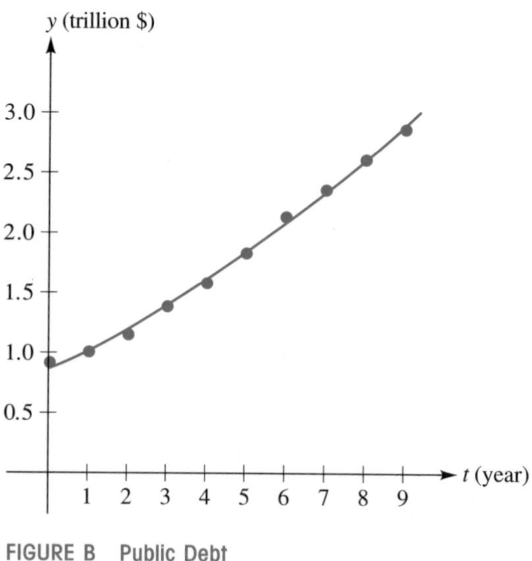

FIGURE B Public Debt
Quadratic model
$y = 0.0084t^2 + 0.1509t + 0.8579$

27. Economists use the **producer price index** to measure the average change in the price of wholesale goods. (So, roughly speaking, the larger the producer price index for a given year, the more that wholesale prices have jumped that year.) The following table lists the producer price index for the years, 1975, 1980, 1985 and 1990.

Year	1975	1980	1985	1990
Producer Price Index	58.4	89.8	103.2	116.3

Source: U.S. Bureau of Labor Statistics, *Producer Price Indices*

In each of the figures below we show a scatter plot for the data along with a particular type of regression function. Figure A uses a *logarithmic function* to model the data; Figure B uses a *quadratic function;* and Figure C uses a *power function* (that is, a function of the form $y = ax^b$). In the scatter plots, and in the work that follows, we take $t = 0$ to correspond to the year 1975. (So, for example, $t = 5$ represents 1980 and $t = 10$ represents 1985.)

(a) After looking at Figures A, B, and C, say which model appears to fit the four data points best. (That is, in your view, which curve comes closest to passing through all four of the given points?)

(b) Use your calculator and the equations supplied in the captions below the figures to complete the following table. Round your answers to one decimal place.

	1991	1992
Producer Price Index (logarithmic model)		
Producer Price Index (quadratic model)		
Producer Price Index (power model)		
Producer Price Index (actual)	116.5	117.2

(c) Which of the three models provides the best estimate for the actual 1991 producer price index?

(d) Which of the three models provides the best estimate for the actual 1992 producer price index?

(e) Use each of the three models to make projections for the producer price index for the year 2000. Which model yields the highest figure? the lowest?

(f) Of the three models, which one do you think is least likely to be useful for making long-range predictions about the producer price index? There is no single "right" answer here, but you should supply a reason (using complete sentences) for the answer you do give.

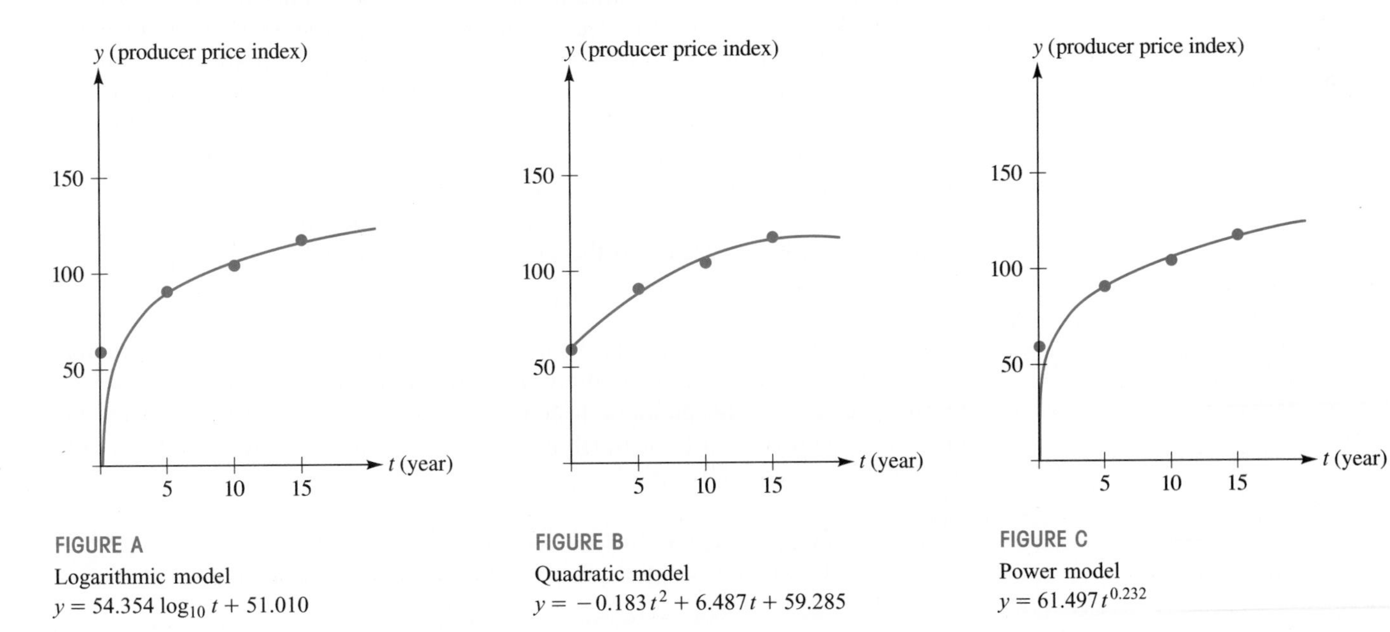

FIGURE A
Logarithmic model
$y = 54.354 \log_{10} t + 51.010$

FIGURE B
Quadratic model
$y = -0.183t^2 + 6.487t + 59.285$

FIGURE C
Power model
$y = 61.497t^{0.232}$

5.7 EXPONENTIAL GROWTH AND DECAY

Drawing by Professor Ann Jones, University of Colorado, Boulder. From the cover of *The Physics Teacher*, vol. 14, no. 7 (October 1976).

In newspapers and in everyday speech, the term *exponential growth* is used rather loosely to describe any situation involving rapid growth. In the sciences, however, **exponential growth** refers specifically to growth governed by functions of the form $y = ae^{bx}$, where a and b are positive constants. For example, since the function $A = Pe^{rt}$ (discussed in the previous section) has this general form, we say that money grows exponentially under continuous compounding of interest. Similarly, in the sciences, **exponential decay** refers specifically to decrease or decay governed by functions of the form $y = ae^{bx}$, where a is positive and b is negative. As examples in this section, we shall consider population growth, global warming, and radioactive decay.

Under ideal conditions involving unlimited food and space, a colony of bacteria increases according to the *growth law*

$$\mathcal{N} = \mathcal{N}_0 e^{kt}$$

In this formula, $\mathcal{N}$ is the population at time t and k is a positive constant related to the growth rate of the population. The number $\mathcal{N}_0$ is also a constant; it represents the size of the population at time $t = 0$. Sometimes, to emphasize the fact that $\mathcal{N}$ depends on (is a function of) t, we use function notation to rewrite the growth law as $\mathcal{N}(t) = \mathcal{N}_0 e^{kt}$.

EXAMPLE 1 Suppose that at the start of an experiment in a biology laboratory, 1200 bacteria are present in a colony. Four hours later, the population is found to be 2200.

(a) Determine the *growth constant k.*
(b) How many bacteria were there three hours after the experiment began? (Round to the nearest 50.)
(c) When will the population reach 3000?

Solution

(a) The initial population is 1200, so we have

$$\mathcal{N} = 1200e^{kt} \tag{1}$$

We are also given that $\mathcal{N} = 2200$ when $t = 4$. Substituting these values in equation (1) yields $2200 = 1200e^{4k}$, and therefore $e^{4k} = 11/6$. Rewriting this last equation in logarithmic form, we have

$$4k = \ln(11/6)$$

and therefore

$$k = \frac{\ln(11/6)}{4} \tag{2}$$

(b) Substituting $t = 3$ in equation (1) yields

$\mathcal{N} = 1200e^{3k}$ where k is given by equation (2)

$\mathcal{N} \approx 1900$ bacteria using a calculator and rounding to the nearest fifty

(c) We want to find t when $\mathcal{N} = 3000$. Substituting $\mathcal{N} = 3000$ in equation (1) yields $3000 = 1200e^{kt}$, and therefore $e^{kt} = 5/2$. Rewriting this equation in logarithmic form, we have

$kt = \ln(5/2)$

$t = \dfrac{\ln(5/2)}{k}$ where k is given by equation (2)

$t \approx 6$ hours using a calculator and rounding to the nearest hour

The population reaches 3000 approximately six hours after the start of the experiment.

In the next example we predict the population of the world in the year 2000, assuming that the population grows exponentially. In this context, it can be shown that the constant k in the formula $\mathcal{N} = \mathcal{N}_0 e^{kt}$ represents the **relative growth rate** of the population $\mathcal{N}$.* Algebraically, we will work with k just as we did in the calculations for compound interest in the previous section, where k represented an interest rate.

EXAMPLE 2 Statistical projections from the United Nations indicate that the relative growth rate for the world's population during the years 1992–2000 will be approximately 1.6% per year. (This is down from the all-time high of 2% per year in 1970.) Use the formula $\mathcal{N} = \mathcal{N}_0 e^{kt}$ to predict the world population in the year 2000, given that the population in 1995 was 5.702 billion people.

Solution Let $t = 0$ correspond to the year 1995. Then the year 2000 corresponds to the value $t = 5$. Our given data therefore are $k = 0.016$, $t = 5$, and $\mathcal{N}_0 = 5.702$

*The *relative growth rate* is $(\Delta\mathcal{N}/\mathcal{N})/\Delta t$; it is the fractional change in $\mathcal{N}$ per unit time. This is distinct from (but sometimes confused with) the *average growth rate*, $\Delta\mathcal{N}/\Delta t$.

(in units of one billion). Using these values in the formula $\mathcal{N} = \mathcal{N}_0 e^{kt}$, we obtain

$$\mathcal{N} = 5.702e^{(0.016)5}$$

$$\mathcal{N} \approx 6.177 \quad \text{using a calculator and rounding to three decimal places}$$

Thus, our prediction for the world's population in the year 2000 is 6.177 billion people.

EXAMPLE 3 The following statements about the increasing levels of carbon dioxide in the atmosphere appear in the book *The Challenge of Global Warming,* edited by Dean Edwin Abrahamson (Washington, D.C.: Island Press, 1989).

> Carbon dioxide, the single most important greenhouse gas, accounts for about half of the warming that has been experienced as a result of past emissions* and also for half of the projected future warming. The present [1988] concentration is now about 350 parts per million (ppm) and is increasing about 0.4% . . . per year. . . .

Assuming that the concentration of carbon dioxide in the atmosphere continues to increase exponentially at 0.4% per year, estimate when the concentration might reach 600 ppm. (This would be roughly twice the level estimated to exist prior to the Industrial Revolution.)

Solution Let $t = 0$ correspond to the year 1988. In the formula $\mathcal{N} = \mathcal{N}_0 e^{kt}$, we want to determine the time t when $\mathcal{N} = 600$. Using the values $\mathcal{N}_0 = 350$, $\mathcal{N} = 600$, and $k = 0.004$, we have

$$600 = 350e^{0.004t} \quad \text{or} \quad \frac{600}{350} = e^{0.004t}$$

and, consequently,

$$\ln(600/350) = 0.004t$$

$$t = \frac{\ln(600/350)}{0.004} \approx 135 \quad \text{using a calculator and rounding to the nearest whole number}$$

Now, 135 years beyond the base year of 1988 is 2123. Rounding once more, we summarize our result this way: If the carbon dioxide levels continue to increase exponentially at 0.4% per year, then the concentration will reach 600 ppm by approximately 2125.

In Examples 1 and 2, we used the function $\mathcal{N} = \mathcal{N}_0 e^{kt}$ to describe population growth. In Example 3 we used this function to model the increasing levels of carbon dioxide in the atmosphere. It is a remarkable fact that the same basic formula, but now with $k < 0$, describes radioactive decay. How do scientists know that this is the approximate formula here? We mention two reasons—one practical, one theoretical. As early as 1902, E. Rutherford and F. Soddy carried out experiments in England to measure radioactive decay. They found that an equation of the form $\mathcal{N} = \mathcal{N}_0 e^{kt}$, with $k < 0$, indeed aptly described their data. Alternatively, on theoretical grounds, it can be argued that the decay rate at each

*The two principal sources of these emissions are the burning of fossil fuels and the burning of vegetation in the tropics.

instant must be proportional to the amount $\mathcal{N}$ present at that instant. Under this condition, calculus can be used to show that the law of exponential decay applies. (See Exercise 63 for some of the details.)

In discussing radioactive decay, it is convenient to introduce the term *half-life.* (As you'll see, this is analogous to the concept of *doubling time,* which we discussed in the previous section.)

DEFINITION: Half-life

The **half-life** of a radioactive substance is the time required for half of a given sample to disintegrate. The half-life is an intrinsic property of the substance; it does not depend on the given sample size.

EXAMPLE

Iodine-131 is a radioactive substance with a half-life of 8 days. Suppose that 2 g are present initially. Then:

at $t = 0$, 2 g are present;
at $t = 8$ days, 1 g is left;
at $t = 16$ days, $\frac{1}{2}$ g is left;
at $t = 24$ days, $\frac{1}{4}$ g is left;
at $t = 32$ days, $\frac{1}{8}$ g is left.

TABLE 1

t (days)	$\mathcal{N}$ (amount)
0	$\mathcal{N}_0$
8	$\mathcal{N}_0/2$
16	$\mathcal{N}_0/4$
24	$\mathcal{N}_0/8$
32	$\mathcal{N}_0/16$

Just as we used the idea of doubling time to graph an exponential growth function in the previous section, we can use the half-life to graph an exponential decay function. Consider, for example, the radioactive substance iodine-131, which has a half-life of eight days. Table 1 shows what fraction of an initial amount remains at eight-day intervals. Using the data in this table, we can draw the graph of the decay function $\mathcal{N} = \mathcal{N}_0 e^{kt}$ for iodine-131 (see Figure 1). Notice that we are able to construct this graph without specifically evaluating the decay constant k. (The next example shows how to determine k.)

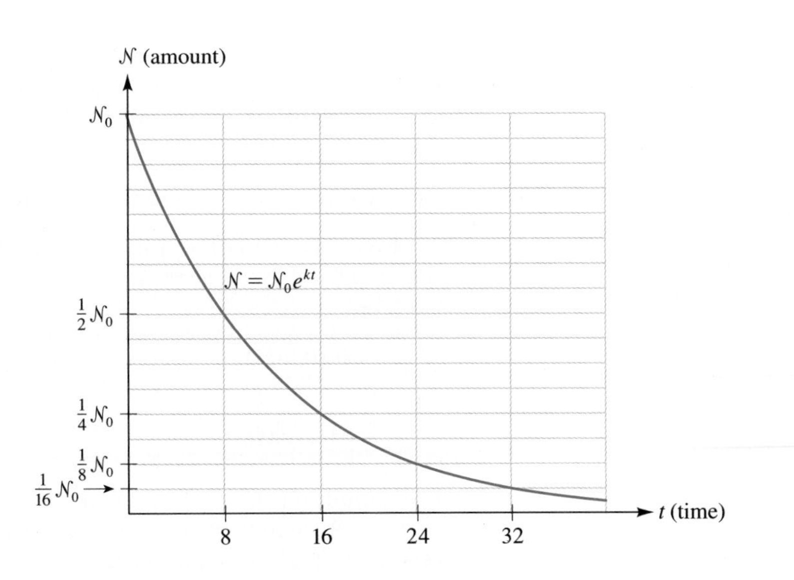

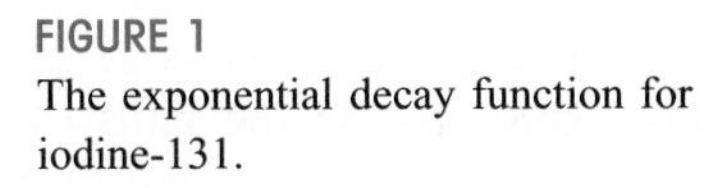
FIGURE 1
The exponential decay function for iodine-131.

EXAMPLE 4 Hospitals utilize the radioactive substance iodine-131 in the diagnosis of conditions of the thyroid gland. The half-life of iodine-131 is eight days.

(a) Determine the *decay constant* k for iodine-131.
(b) If a hospital acquires 2 g of iodine-131, how much of this sample will remain after 20 days?
(c) How long will it be until only 0.01 g remains?

Solution

(a) The half-life is eight days. This means that when t is 8, the value of $\mathcal{N}$ is $\frac{1}{2}\mathcal{N}_0$. Using these values for t and $\mathcal{N}$ in the formula $\mathcal{N} = \mathcal{N}_0 e^{kt}$, we obtain

$$\frac{1}{2}\mathcal{N}_0 = \mathcal{N}_0 e^{k8} \qquad \text{and therefore} \qquad \frac{1}{2} = e^{8k}$$

We solve for k by writing this last equation in logarithmic form:

$$8k = \ln(1/2) \qquad \text{and therefore} \qquad k = \frac{\ln(1/2)}{8}$$

If a numerical value for k is required, we can use a calculator (or tables) to obtain $k \approx -0.0866$.

(b) We are given that $\mathcal{N}_0 = 2$ g, and we want to find the value of $\mathcal{N}$ when $t = 20$ days. First, before using algebra and a calculator, let's estimate the answer to get a feeling for the situation. The half-life is eight days. This means that after eight days, 1 g remains; after 16 days, 0.5 g remains; and after 24 days, 0.25 g remains. Since 20 is between 16 and 24, it follows that after 20 days, the amount remaining will be something between 0.5 g and 0.25 g. Now for the actual calculations: Substitute the values $\mathcal{N}_0 = 2$ and $t = 20$ in the decay law $\mathcal{N} = \mathcal{N}_0 e^{kt}$. This yields

$$\mathcal{N} = 2e^{20k} \qquad \text{where } k \text{ has the value determined in part (a)}$$
$$\mathcal{N} \approx 0.354 \text{ g} \qquad \text{using a calculator}$$

Thus, after 20 days, approximately 0.35 g of the iodine-131 remains. Note that this figure is indeed between 0.5 g and 0.25 g, as we first estimated.

(c) We are given that $\mathcal{N}_0 = 2$ g and we want to find the time t at which $\mathcal{N}$ is 0.01 g. Substituting the values $\mathcal{N}_0 = 2$ and $\mathcal{N} = 0.01$ into the decay law $\mathcal{N} = \mathcal{N}_0 e^{kt}$ yields

$$0.01 = 2e^{kt} \qquad \text{where } k \text{ has the value determined in part (a)}$$

To solve this equation for t, we first divide both sides by 2 to obtain $0.005 = e^{kt}$. The logarithmic form of this last equation is $kt = \ln(0.005)$. Therefore

$$t = \frac{\ln(0.005)}{k}$$
$$t \approx 61.2 \text{ days} \qquad \text{using a calculator and the value of } k \text{ from part (a)}$$

In Example 4(a), we found that the decay constant for iodine-131 is given by $k = (\ln \frac{1}{2})/8$. Notice that the denominator in this expression is the half-life for iodine-131. By following the same reasoning used in Example 4(a), we find that the decay constant for any radioactive substance is always $\ln\frac{1}{2}$ divided by the half-life. For reference, this useful fact is restated in the box that follows.

PROPERTY SUMMARY — A FORMULA FOR THE DECAY CONSTANT k

FORMULA	EXAMPLE
$k = \dfrac{\ln\frac{1}{2}}{\text{half-life}}$	The half-life of strontium-90 is 28 years, and therefore $k = (\ln\frac{1}{2})/28$.

EXAMPLE 5 An article on nuclear energy appeared in the January 1976 issue of *Scientific American.* The author of the article was Hans Bethe (1906–1992), a Nobel prize winner in physics. At one point in the article, Professor Bethe discussed the disposal (through burial) of radioactive waste material from a nuclear reactor. The particular waste product under discussion was plutonium-239.

> . . . Plutonium-239 has a half-life of nearly 25,000 years, and 10 half-lives are required to cut the radioactivity by a factor of 1000. Thus, the buried wastes must be kept out of the biosphere for 250,000 years.

(a) Supply the detailed calculations to support the statement that 10 half-lives are required before the radioactivity is reduced by a factor of 1000.

(b) Show how the figure of 10 half-lives can be obtained by estimation, as opposed to detailed calculation.

Solution

(a) Let $\mathcal{N}_0$ denote the initial amount of plutonium-239 at time $t = 0$. Then the amount $\mathcal{N}$ present at time t is given by $\mathcal{N} = \mathcal{N}_0 e^{kt}$. We wish to determine t when $\mathcal{N} = \frac{1}{1000}\mathcal{N}_0$. First, since the half-life is 25,000 years, we have $k = (\ln\frac{1}{2})/25{,}000$. Now, substituting $\frac{1}{1000}\mathcal{N}_0$ for $\mathcal{N}$ in the decay law gives us

$$\frac{1}{1000}\mathcal{N}_0 = \mathcal{N}_0 e^{kt} \qquad \text{where } k \text{ is } (\ln\tfrac{1}{2})/25{,}000$$

$$\frac{1}{1000} = e^{kt}$$

$$kt = \ln\frac{1}{1000} \qquad \text{converting from exponential to logarithmic form}$$

$$t = \frac{\ln\frac{1}{1000}}{k} = \frac{\ln\frac{1}{1000}}{(\ln\frac{1}{2})/25{,}000} = \frac{(\ln\frac{1}{1000})(25{,}000)}{\ln\frac{1}{2}}$$

Using a calculator, we obtain the value 249,114.6 years; however, given the time scale involved, it would be ludicrous to announce the answer in this form. Instead, we round off the answer to the nearest thousand years and say that after 249,000 years, the radioactivity will have decreased by a factor of 1000. Notice that this result confirms Professor Bethe's ballpark estimate of 10 half-lives, or 250,000 years

(b)

After 1 half-life: $\mathcal{N} = \dfrac{\mathcal{N}_0}{2}$

After 2 half-lives: $\mathcal{N} = \dfrac{1}{2}\left(\dfrac{\mathcal{N}_0}{2}\right) = \dfrac{\mathcal{N}_0}{2^2}$

After 3 half-lives: $\mathcal{N} = \dfrac{1}{2}\left(\dfrac{\mathcal{N}_0}{2^2}\right) = \dfrac{\mathcal{N}_0}{2^3}$

Following this pattern, we see that after 10 half-lives, we should have

$$\mathcal{N} = \frac{\mathcal{N}_0}{2^{10}}$$

But as we noted in the first section of this chapter, 2^{10} is approximately 1000. Therefore, we have

$$\mathcal{N} \approx \frac{\mathcal{N}_0}{1000} \qquad \text{after 10 half-lives}$$

This is in agreement with Professor Bethe's statement.

Do 1,21

EXERCISE SET 5.7

In Exercises 1 and 2, assume that the populations grow exponentially, that is, according to the law $\mathcal{N} = \mathcal{N}_0 e^{kt}$.

1. At the start of an experiment, 2000 bacteria are present in a colony. Two hours later, the population is 3800.
(a) Determine the growth constant k.
(b) Determine the population five hours after the start of the experiment.
(c) When will the population reach 10,000?

2. At the start of an experiment, 2×10^4 bacteria are present in a colony. Eight hours later, the population is 3×10^4.
(a) Determine the growth constant k.
(b) What was the population two hours after the start of the experiment?
(c) How long will it take for the population to triple?

3. The figure in the next column shows the graph of an exponential growth function $\mathcal{N} = \mathcal{N}_0 e^{kt}$. Determine the values of $\mathcal{N}_0$ and k.

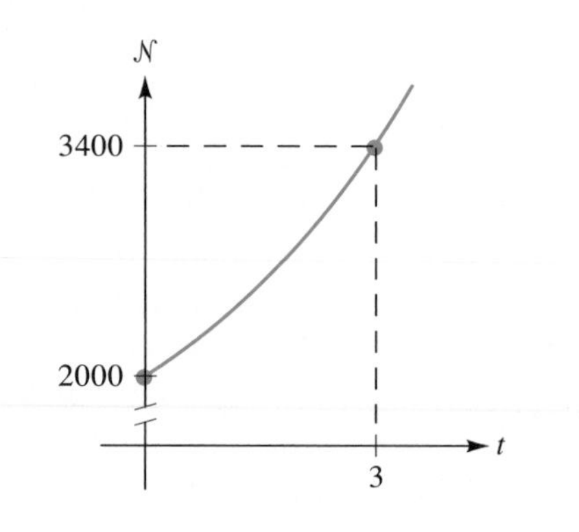

4. Suppose that you are helping a friend with his homework on the growth law, $\mathcal{N} = \mathcal{N}_0 e^{kt}$, and on his paper you see the equation $e^k = -0.75$. How do you know at this point that your friend must have made an error?

The statistics in Exercises 5–7 are taken from the 1995 World Population Data Sheet *(Washington, D.C.: Population Reference Bureau, Inc., 1995). In Exercises 5–7(a), complete the tables, assuming that the populations grow exponentially and that the indicated relative growth rates are valid through the year 2000.*

5.

Region	1995 Population (billions)	Percent of Population in 1995	Relative Growth Rate (percent per year)	Year 2000 Population (billions)	Percent of World Population in 2000
World	5.702	100	1.5	?	?
More developed regions	1.169	?	0.2	?	?
Less developed regions	4.533	?	1.9	?	?

6.

Country	1995 Population (millions)	Relative Growth Rate (percent per year)	Year 2000 Population (millions)	Percent Increase in Population
United States	263.2	0.7	?	?
People's Republic of China	1218.8	1.1	?	?
Mexico	93.7	2.2	?	?

7. **(a)**

Country	1995 Population (millions)	Relative Growth Rate (percent per year)	Year 2000 Population (millions)	Percent Increase in Population
Iraq	20.6	3.7	?	?
United Kingdom	58.6	0.2	?	?

(b) Iraq has one of the highest population growth rates in the world, and United Kingdom has one of the lowest. Use the data given in part (a) to compute an estimate for the year in which the two populations could be equal. (Round your final answer to the nearest five years.)
Hint: You need to solve the equation $20.6e^{0.037t} = 58.6e^{0.002t}$

8. In 1992, the countries Chad and Denmark each had populations of about 5.2 million. However, the relative growth rate for Chad was 2.5% per year, while that for Denmark was 0.1% per year.

(a) Assuming exponential growth and the given growth rates, make projections for the population of each nation in the year 2010.

(b) When might the population of Chad reach 10 million? What would the population of Denmark be at this time?

9. In 1992, Cyprus and Gaza each had populations of about 0.7 million. However, the relative growth rate for Cyprus was 1.1% per year, while that for Gaza was 4.6% per year.

(a) Assuming exponential growth and the given growth rates, make projections for the populations in the year 2000.

(b) When might the population of Cyprus reach 1 million? What would the population of Gaza be at this time?

10. The countries of Georgia and Tajikistan (former members of the Soviet Union) each had 1992 populations of 5.5 million. However, the relative growth rate for Georgia was 0.9% per year, while that for Tajikistan was 3.2% per year.

(a) Assuming exponential growth and the given growth rates, make projections for the population of each nation in the year 2010.

(b) In what year might the population of Georgia reach 10 million? What would the population of Tajikistan be at this time?

11. In 1994 the population of Kuwait was 1.3 million.

(a) Assuming exponential growth, what value for k would lead to a population of 2 million in 16 years (that is, by the year 2010)? *Remark:* The answer you obtain is, in fact, less than the current growth rate for Kuwait, which is about 3.3% per year.

(b) What value for k would lead to a population of 2 million in the year 2100?

12. According to the United States Bureau of the Census, the population of the United States grew most rapidly during the period 1800–1810 and least rapidly during the period 1930–1940. Use the following data to compute the relative growth rate (percent/year) for each of these two periods. In 1800, the population was 5,308,483; in 1810, it was 7,239,881. In 1930, the population was 123,202,624; in 1940, it was 132,164,569.

13. According to the U.S. Bureau of the Census, in the year 1850 the population of the United States was 23,191,876; in 1900, the population was 62,947,714.

(a) Assuming that the population grew exponentially during this period, compute the growth constant k.

(b) Assuming continued growth at the same rate, predict the 1950 population.

(c) The actual population for 1950, according to the Bureau of the Census, was 150,697,361. How does this compare to your prediction in part (b)? Was the actual growth over the period 1900–1950 faster or slower than exponential growth with the growth constant k determined in part (a)?

In Exercises 14–16, use the population statistics given in the following table, provided by the U.S. Bureau of the Census.

State	1930	1940	1950
California	5,677,251	6,907,387	10,586,223
New York	12,588,066	13,479,142	14,830,192
North Dakota	680,845	641,935	619,636

14. **(a)** Assume that the population of California grew exponentially over the period 1930–1940. Compute the growth constant k and express it as a percent (per year).

(b) Assume that the population of California grew exponentially over the period 1940–1950. Compute the growth constant k and express it as a percent.

(c) What would the 1990 California population have been if the relative growth rate obtained in part (b) had remained in effect throughout the period 1950–1990? *Hint:* Let $t = 0$ correspond to 1950. Then find $\mathcal{N}$ when $t = 40$.

(d) How does your prediction in part (c) compare to the actual 1990 population of 29,279,000?

15. Repeat Exercise 14 for New York. *Note:* The actual 1990 population of New York was 17,627,000.

16. Repeat Exercise 14 for North Dakota. *Note:* The actual 1990 population of North Dakota was 634,000.

17. Lester R. Brown, in his book *State of the World 1985* (New York: W. W. Norton and Co., 1985), makes the following statement:

> The projected [population] growth for North America, all of Europe, and the Soviet Union is less than the additions expected in either Bangladesh or Nigeria.

In this exercise, you are asked to carry out the type of calculations that could be used to support Lester Brown's projection for the period 1990–2025. The source of the data in the following table is the Population Division of the United Nations.

Region	1990 Population (millions)	Projected Relative Growth Rate (%/year)	2025 Projected Population (millions)
North America	275.2	0.7	?
(Former) Soviet Union	291.3	0.7	?
Europe	499.5	0.2	?
Nigeria	113.3	3.1	?

(a) Complete the table, assuming that the populations grow exponentially and that the indicated growth rates are valid over the period 1990–2025.

(b) According to the projections in part (a), what will the net increase in Nigeria's population be over the period 1990–2025?

(c) According to the projections in part (a), what will the net increase in the combined populations of North America, the (former) Soviet Union, and Europe be over the period 1990–2025?

(d) Compare your answers in parts (b) and (c). Do your results support or contradict Lester Brown's projection?

18. As of 1995, ten countries in the world had populations exceeding 100 million. These countries are listed (in order of population, from largest to smallest) in the table. According to *World Population Profile: 1987,* published by the United States Bureau of the Census, "The latest projections suggest that India's population may surpass China's in less than 60 years, or before today's youngsters in both countries reach old age." Using the data in the table, estimate the populations of the ten countries in the year 2050. (Assume that the indicated growth rates are valid over the period 1995–2050). Tabulate your final answers, listing the countries in order of population, from largest to smallest. Do your results for China and India support the projection by the Census Bureau?

Country	1995 Population (millions)	Relative Growth Rate (%/year)
1. China	1218.8	1.1
2. India	930.6	1.9
3. United States	263.2	0.7
4. Indonesia	198.4	1.6
5. Brazil	157.8	1.7
6. Russia	147.5	−0.6
7. Pakistan	129.7	2.9
8. Japan	125.2	0.3
9. Bangladesh	119.2	2.4
10. Nigeria	101.2	3.1

Source: 1995 World Population Data Sheet (Washington, D.C.: Population Reference Bureau, Inc.)

In Exercises 19 and 20, use the half-life information to complete each table. (The formula $\mathcal{N} = \mathcal{N}_0 e^{kt}$ is not required.)

19. **(a)** Uranium-228: half-life = 550 seconds

t (seconds	0	550	1100	1650	2200
$\mathcal{N}$ (grams)	8				

(b) Uranium-238: half-life $= 4.9 \times 10^9$ years

t (years)	0				
$\mathcal{N}$ (grams)	10	5	2.5	1.25	0.625

20. **(a)** Polonium-210: half-life $= 138.4$ days.

t (days)	0	138.4	276.8	415.2	
$\mathcal{N}$ (grams)	0.4				0.025

(b) Polonium-214: half-life $= 1.63 \times 10^{-4}$ second

t (seconds)	0				6.52×10^{-4}
$\mathcal{N}$ (grams)	0.1	0.05	0.025	0.0125	

21. The half-life of iodine-131 is eight days. How much of a one-gram sample will remain after seven days?

22. The half-life of strontium-90 is 28 years. How much of a 10 g sample will remain after **(a)** 1 year? **(b)** 10 years?

23. The radioactive isotope sodium-24 is used as a tracer to measure the rate of flow in an artery or vein. The half-life of sodium-24 is 14.9 hours. Suppose that a hospital buys a 40 g sample of sodium-24.
(a) How much of the sample will remain after 48 hours?
(b) How long will it be until only one gram remains?

24. The radioactive isotope carbon-14 is used as a tracer in medical and biological research. Compute the half-life of carbon-14 given that the decay constant k is -1.2097×10^{-4}. (The units for k here are such that your half-life answer will be in years.)

25. **(a)** The half-life of radium-226 is 1620 years. Draw a graph of the decay function for radium-226, similar to that shown in Figure 1.
(b) The half-life of radium-A is 3 min Draw a graph of the decay function for radium-A.

26. **(a)** The half-life of thorium-232 is 1.4×10^{10} years. Draw a graph of the decay function.
(b) The half-life of thorium-A is 0.16 sec. Draw a graph of the decay function.

27. The half-life of plutonium-241 is 13 years.
(a) How much of an initial 2 g sample remains after 5 years?
(b) Find the time required for 90% of the 2 g sample to decay. *Hint:* If 90% has decayed, then 10% remains.

28. The half-life of radium-226 is 1620 years.
(a) How much of a 2 g sample remains after 100 years?
(b) Find the time required for 80% of the 2 g sample to decay.

29. The half-life of thorium-229 is 7340 years.
(a) Compute the time required for a given sample to be reduced by a factor of 1000. Show detailed calculations, as in Example 5(a).
(b) Express your answer in part (a) in terms of half-lives.
(c) As in Example 5(b), estimate the time required for a given sample of thorium-229 to be reduced by a factor of 1000. Compare your answer with that obtained in part (b).

30. The Chernobyl nuclear explosion (in the former Soviet Union, on April 25, 1987) released large amounts of radioactive substances into the atmosphere. These substances included cesium-137, iodine-131, and strontium-90. Although the radioactive material covered many countries, the actual amount and intensity of the fallout varied greatly from country to country, due to vagaries of the weather and the winds. One area that was particularly hard hit was Lapland, where heavy rainfall occurred just when the Chernobyl cloud was overhead.
(a) Many of the pastures in Lapland were contaminated with cesium-137, a radioactive substance with a half-life of 33 years. If the amount of cesium-137 was found to be ten times the normal level, how long would it take until the level returns to normal? (That is, compute the time required for the amount of cesium-137 to be reduced by a factor of 10.)
(b) Follow part (a), but assume that the amount of cesium-137 was 100 times the normal level.
Remark: Several days after the explosion, it was reported that the level of cesium-137 in the air over Sweden was 10,000 times the normal level. Fortunately there was little or no rainfall.

31. Strontium-90, with a half-life of 28 years, is a waste product from nuclear fission reactors. One of the reasons great care is taken in the storage and disposal of this substance stems from the fact that strontium-90 is, in some chemical respects, similar to ordinary calcium. Thus, strontium-90 in the biosphere, entering the food chain via plants or animals, would eventually be absorbed into our bones. (In fact, everyone already has a measurable amount of strontium-90 in their bones as a result of fallout from atmospheric nuclear tests.)
(a) Compute the decay constant k for strontium-90.
(b) Compute the time required if a given quantity of strontium-90 is to be stored until the radioactivity is reduced by a factor of 1000.

(c) Using half-lives, estimate the time required for a given sample to be reduced by a factor of 1000. Compare your answer with that obtained in (b).

32. **(a)** Suppose that a certain country violates the ban against above-ground nuclear testing and, as a result, an island is contaminated with debris containing the radioactive substance iodine-131. A team of scientists from the United Nations wants to visit the island to look for clues in determining which country was involved. However, the level of radioactivity from the iodine-131 is estimated to be 30,000 times the safe level. Approximately how long must the team wait before it is safe to visit the island? The half-life of iodine-131 is 8 days.

(b) Rework part (a), assuming instead that the radioactive substance is strontium-90 rather than iodine-131. The half-life of strontium-90 is 28 years. Assume, as before, that the initial level of radioactivity is 30,000 times the safe level. (This exercise underscores the difference between a half-life of 8 days and one of 28 years.)

33. The *mean life* μ of a radioactive particle or nucleus is defined by $\mu = -1/k$, where k is the decay constant. (It can be shown that the mean life is the average time that a particle survives before it decays.)

(a) Show that $\mu = \dfrac{\text{half-life}}{\ln 2}$.

(b) Krypton-91 has a half-life of 10 sec. Compute the mean life of krypton-91.

(c) How much of a one-gram sample of krypton-91 remains after a period of three mean lives?

34. **(a)** Show that after a period of three mean lives, approximately 5% of an initial sample $\mathcal{N}_0$ will remain.

(b) Show that after a period of four mean lives, less than 2% of an initial sample $\mathcal{N}_0$ will remain.

35. An article that appeared in the August 13, 1994, *New York Times* reported

> German authorities have discovered . . . a tiny sample of weapons-grade nuclear material believed to have been smuggled out of Russia to interest foreign governments or terrorist groups that might want to build atomic bombs. . . . [the police] said they had seized the material, .028 ounces of highly enriched uranium-235, in June in . . . Bavaria . . . and have since arrested . . . [six] suspects. . . .

Suppose that the suspects, in an attempt to avoid arrest, had thrown the 0.028 ounces of uranium-235 into the Danube River, where it would sink to the bottom. How many ounces of the uranium-235 would still be in the river after 1000 years? The half-life of uranium-235 is 7.1×10^8 years.

36. Using function notation, the decay law for radioactive substances is $\mathcal{N}(t) = \mathcal{N}_0 e^{kt}$. Compute the ratio $\mathcal{N}(t+1)/\mathcal{N}(t)$. What do you observe?

37. In 1969 the United States National Academy of Sciences issued a report entitled *Resources and Man.* One conclusion in the report is that a world population of 10 billion "is close to (if not above) the maximum that an intensively managed world might hope to support with some degree of comfort and individual choice." (The figure "10 billion" is sometimes referred to as the *carrying capacity* of the Earth.)

(a) When the report was issued in 1969, the world population was about 3.6 billion, with a relative growth rate of 2% per year. Assuming continued exponential growth at this rate, estimate the year in which the Earth's carrying capacity of 10 billion might be reached.

(b) Repeat the calculations in part (a) using the following more recent data: In 1995, the world population was about 5.7 billion, with a relative growth rate of 1.5% per year. How does your answer compare with that in part (a)?

38. The following extract is from an article by Kim Murphy that appeared in the *Los Angeles Times* on September 14, 1994.

> CAIRO—Over a chorus of reservations from Latin America and Islamic countries still troubled about abortion and family issues, nearly 180 nations adopted a wide-ranging plan Tuesday on global population, the first in history to obtain partial endorsement from the Vatican.
>
> The plan, approved on the final day of the U.N. population conference here, for the first time tries to limit the growth of the world's population by preventing it from exceeding 7.2 billion people over the next two decades.

(a) As of 1995, the world population was 5.702 billion, with an annual growth rate of 1.5%/year. Assuming continued exponential growth at this rate, make a projection for the world population in the year 2020. Round off the answer to one decimal place. How does your answer compare to the target value of 7.2 billion mentioned in the article?

(b) Suppose that over the years 1995–2020, the value of the growth constant k for world population is as follows:

Period	k (%/year)
1995–2000	1.5
2000–2010	1.4
2010–2020	1.2

Using this scenario, make a projection for the world population in the year 2020. Round the answer to one decimal place. Does your answer exceed the target value of 7.2 billion mentioned in the article? (For the purposes of this exercise, ignore the inconsistency that this table supplies two different values of k for the year 2000 and two different values for the year 2010.)

(c) As in parts (a) and (b), assume that in 1995, the world population was 5.702 billion. Determine a value for the growth constant k so that exponential growth throughout the years 1995–2020 leads to a world population of 7.2 billion in the year 2020.

39. The following table gives the dollar values of China's exports to the United States over the years 1985–1991. The accompanying graph shows a scatter plot for the data along with an exponential regression curve that has been determined using statistical methods. In the scatter plot and in the equation for the curve, $t = 0$ corresponds to 1985, $t = 1$ corresponds to 1986, and so on.

Year	1985	1986	1987	1988	1989	1990	1991
China Exports to U.S. (millions of U.S. $)	2336	2633	3030	3399	4414	5314	6192

Source: Los Angeles Times, May 17, 1994

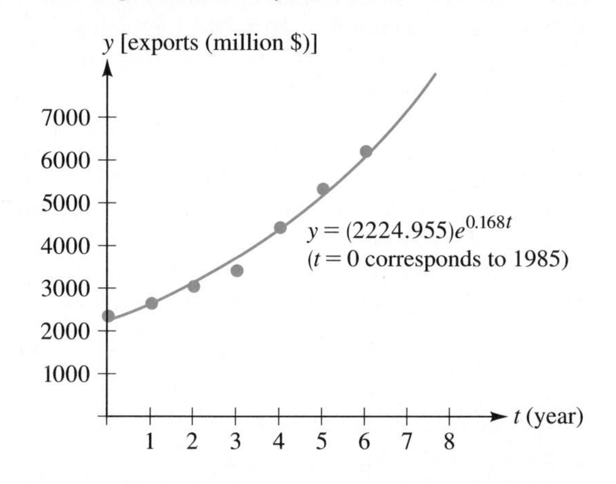

(a) Use the equation of the exponential curve, $y = (2224.955)e^{0.168t}$, to estimate China's exports to the U.S. for the year 1992 ($t = 7$). Is your estimate higher or lower than the actual figure for 1992, which was 8590 million dollars?

(b) Assuming that the exports to the U.S. continue to increase exponentially, make an estimate for China's exports to the U.S. in the year 2000.

(c) The exponential equation given in part (a) was obtained using techniques from the field of statistics. However, you can obtain a reasonable approximation to this equation using ideas from this chapter, rather than from statistics. Look again at the scatter plot for this data. You can see that the exponential curve seems to pass through the point corresponding to $t = 1$, and it very nearly passes through the point corresponding to $t = 4$. Referring back to the table now, notice that the coordinates of these two points in the scatter plot are $(1, 2633)$ and $(4, 4414)$. Thus, your assignment is to find constants a and b so that the curve $y = ae^{bt}$ will pass through the two points $(1, 2633)$ and $(4, 4414)$. (If you need help, a method for finding a and b is described in Exercise 68 in Section 5.4.)

(d) Use the equation that you determined in part (c) to rework part (a). Which of the equations comes closer to predicting the actual 1992 value?

40. In this exercise, we compare exponential functions with some other functions that are often useful in modeling (i.e., describing) real-world data. The data we'll look at concerns the *gross national product* (or *GNP*, for short). The GNP of a country is defined as the total market value of all the goods and services produced by the country during a given period. The following table gives the GNP for the United States during the years 1960, 1970, and 1980.

Year	1960	1970	1980
U.S. GNP (billions of $)	515.3	1015.5	2732.0

Source: Bureau of Economic Analysis, U.S. Department of Commerce

Figures A through D, which follow, show scatter plots for this data and functions that have been chosen to

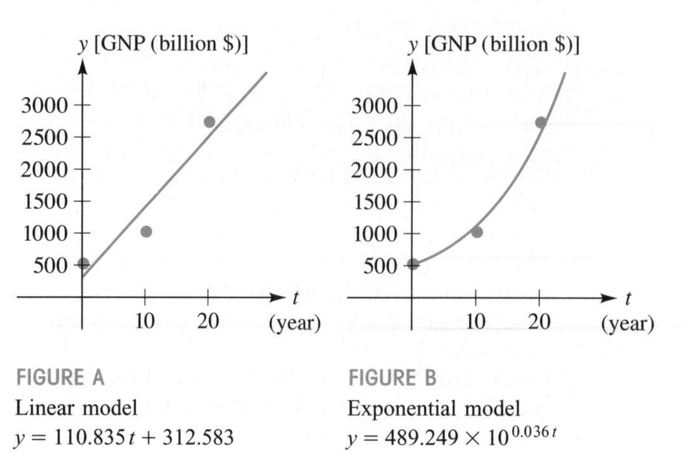

FIGURE A
Linear model
$y = 110.835t + 312.583$

FIGURE B
Exponential model
$y = 489.249 \times 10^{0.036t}$

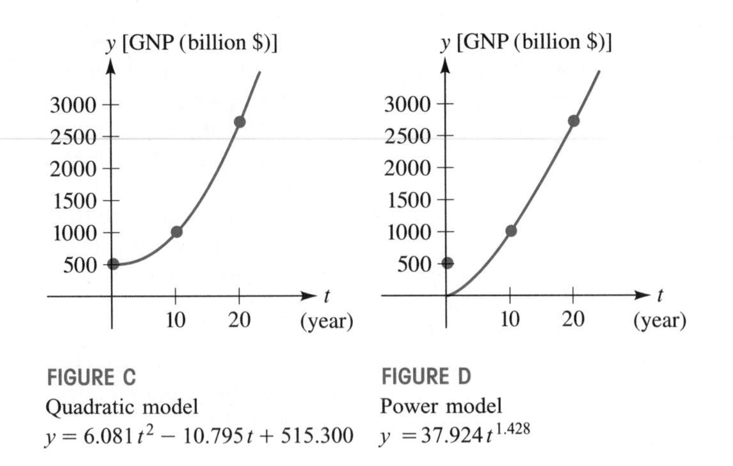

FIGURE C
Quadratic model
$y = 6.081t^2 - 10.795t + 515.300$

FIGURE D
Power model
$y = 37.924t^{1.428}$

model the data. (The equations for the functions were obtained using statistical techniques.) Figure A displays a linear function; this is the regression line that we studied in Section 4.1. Figure B uses an exponential function to model the data. Figure C employs a quadratic function. And Figure D uses a power function, which is a function of the form $y = at^b$, where a and b are constants.

In the scatter plots and in the equations, $t = 0$ corresponds to the year 1960, $t = 10$ corresponds to 1970, and $t = 20$ corresponds to 1980.

(a) Use the equations given in the figure captions along with your calculator to complete the following table. Round the answers to one decimal place.

	1990
GNP (billion $) (linear model)	
GNP (billion $) (exponential model)	
GNP (billion $) (quadratic model)	
GNP (billion $) (power model)	

(b) The actual value for the GNP in 1990 was 5524.5 billion dollars. Of the four models in part (a), which comes closest to this value? Which is the next closest?

(c) Repeat parts (a) and (b) for the year 1992. The actual value for the GNP in 1992 was 5961.9 billion dollars.

In Exercises 41–46, in cases where the data is not attributed, it has been compiled from the following sources.

BP Statistical Review of World Energy (London: British Petroleum Company, 1994)

Information Please Almanac [1994] (New York: Houghton Mifflin Company, 1993)

The Universal Almanac 1994 (Kansas City, Missouri: Andrews and McMeel, 1993)

The World Almanac and Book of Facts 1994 (Mahwah, New Jersey: Funk and Wagnalls, 1993)

World Resources 1992–93 (Oxford: Oxford University Press, 1992)

41. *Depletion of Nonrenewable Resources:* Suppose that the world population grows exponentially. Then, as a first approximation, it is reasonable to assume that the use of nonrenewable resources, such as petroleum and coal, also grows exponentially. Under these conditions, the following formula can be derived (using calculus):

$$A = \frac{A_0}{k}\left(e^{kT} - 1\right)$$

where A is the amount of the resource consumed from time $t = 0$ to $t = T$, the quantity A_0 is the amount of the resource consumed during the year $t = 0$, and k is the relative growth rate of annual consumption.

(a) Show that solving the formula for T yields

$$T = \frac{\ln[(Ak/A_0) + 1]}{k}$$

This formula gives the "life expectancy" T for a given resource. In the formula, A_0 and k are as previously defined, and A represents the total amount of the resource available.

(b) In 1987, worldwide consumption of oil was estimated at 3.090 billion metric tons, with a relative growth rate of 1.6%/year. Proven world oil reserves at that time were estimated to be 124 billion metric tons. Compute T, the life expectancy for oil, under these conditions.

(c) In the 1990s (up to the time of this writing), worldwide oil consumption has actually tended to level off, rather than to increase exponentially. The graph (at the top of the next column), for example, shows worldwide oil consumption over the years 1990–1993, inclusive. As the graph indicates, oil consumption seems to be stabilizing between 3.1 and 3.2 billion metric tons per year.

Assume for the moment that worldwide oil consumption stabilizes at 3.15 billion metric tons per year. Given that the proven world reserves of

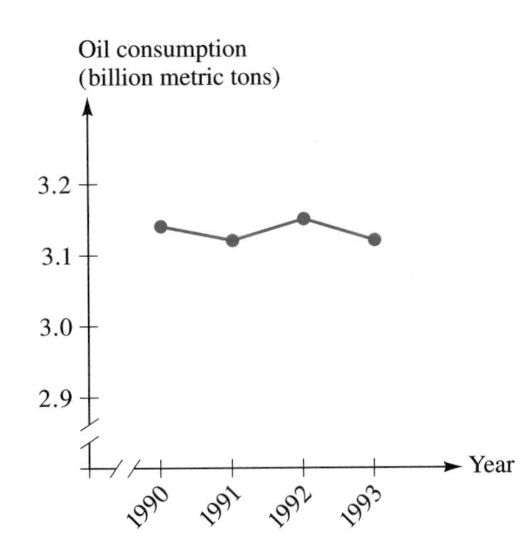

oil as of 1993 were estimated at 136.7 billion metric tons, compute the quantity

$$\frac{136.7 \text{ billion metric tons}}{(3.15 \text{ billion metric tons})/\text{year}}$$

to determine the life expectancy for oil under this scenario. [You'll find that this gives a longer life expectancy than that in part (a), but still the news is not encouraging.]

42. **(a)** During the years 1976–1979, worldwide consumption of natural gas was increasing exponentially. A United States Government study, *Global 2000 Report to the President,* which was carried out for President Jimmy Carter during that time, estimated that the 1976 worldwide consumption of natural gas was approximately 50 tcf (trillion cubic feet), while the total remaining gas ultimately available was 8493 tcf. Use these figures and the formula given in Exercise 41(a) to compute the life expectancy for natural gas, assuming that $k = 2\%$ per year.

(b) In the 1990s (up to the time of this writing), worldwide natural gas consumption has actually tended to level off, rather than increase exponentially. The following graph, for example, shows worldwide natural gas consumption over the years 1990–1993, inclusive. As the graph indicates, natural gas consumption seems to be stabilizing at about 70 tcf per year.

Assume for the moment that worldwide natural gas consumption remains at 70 tcf/year. Given that the proven world reserves of natural gas as of 1993 were estimated at 5016.2 tcf, compute the quantity

$$\frac{5016.2 \text{ tcf}}{(70 \text{ tcf})/\text{year}}$$

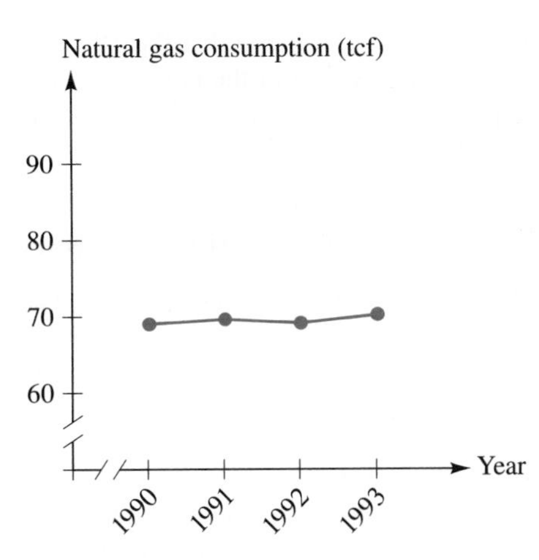

to determine the life expectancy for natural gas under this scenario. [As in the previous exercise with oil consumption, you'll find that this scenario gives a longer life expectancy than that in part (a), but again the news is not encouraging.]

In Exercises 43–46, refer to the following table, which shows world reserves and related data for selected minerals. The term "world reserves" refers to known resources that can be recovered under existing economic conditions using current technology.

Mineral	1990 World Reserves (1000 metric tons)	1990 Consumption (1000 metric tons)	Relative Growth Rate (percent per year)
Copper	321,000	10,773.2	2.2
Aluminum (in bauxite)	21,800,000	17,877.9	2.5
Lead	70,000	5544.5	0.4
Nickel	48,988	842.6	1.7

43. **(a)** Assume that the use of the minerals listed in the table is growing exponentially. Use the formula for T (given in Exercise 41) to compute the life expectancy for copper and the corresponding *depletion date* $(= T + 1990)$.

(b) What if the world reserves of copper were actually twice that listed in the table, and the relative growth rate were actually half that in the table? Compute the depletion date for copper under this scenario. (Assume that the 1990 consumption is as given in the table.)

(c) What if, from 1990 onward, consumption of copper stopped increasing exponentially and actually stabilized at the 1990 level of 10,773.2 thousand metric tons per year? (Assume that the 1990 world reserves are as given in the table.) Determine the life expectancy T for copper under this scenario by computing the quantity

$$\frac{321{,}000 \text{ thousand metric tons}}{(10{,}773.2 \text{ thousand metric tons})/\text{year}}$$

What is the corresponding depletion date for copper?

44. Follow Exercise 43 for aluminum.

45. Follow Exercise 43 for lead.

46. Follow Exercise 43 for nickel.

47. In this exercise, the term ''nonrenewable energy resources'' refers collectively to oil, natural gas, coal, and uranium. In the *International Energy Annual* (Washington, D.C.: U.S. Government Printing Office, 1992), it was estimated that total worldwide consumption of these nonrenewable resources for the year 1991 amounted to 346.83 quadrillion Btu.

(a) Suppose that yearly consumption increases exponentially. Use the equation $\mathcal{N} = \mathcal{N}_0 e^{kt}$ to estimate how long it will take until the yearly consumption is 600 quadrillion Btu, assuming that **(i)** $k = 1\%/\text{year}$; **(ii)** $k = 2\%/\text{year}$.

(b) Suppose, for this exercise, that the total nonrenewable energy resources ultimately available as of 1991 was 225,000 quadrillion Btu. Use the formula for T in Exercise 41 to compute the life expectancy for these nonrenewable resources if **(i)** $k = 1\%/\text{year}$; **(ii)** $k = 2\%/\text{year}$.

(c) Suppose, for this portion of the exercise, that world energy consumption remains at the 1991 level of 346.83 quadrillion Btu/year and that, as of 1991, the total nonrenewable energy resources ultimately available were 155,000 quadrillion Btu. Use your calculator to compute the life expectancy for the planet's nonrenewable energy resources under this scenario.

Remark: In Part (b), the given estimate for the total nonrenewable energy resources ultimately available as of 1991 was 225,000 quadrillion Btu. In fact, this figure is probably overly optimistic. Over two decades ago, in the *Global 2000 Report,* the U.S. Government estimated that in 1976, the remaining nonrenewable energy resources were about 160,000 quadrillion Btu. The estimate used in part (c) is closer to this figure. Also, the supposition of continued exponential growth [in parts (a) and (b)] may be unwarranted. According to the *BP Statistical Review of World Energy* (London: British Petroleum Co., 1994), ''Growth in world energy demand has stalled since 1990, mainly because of declining consumption in Non-OECD Europe [Eastern Europe].''

48. In 1956, scientists Beno Gutenberg and Charles Richter developed a formula to estimate the amount of energy E released in an earthquake. The formula is

$$\log_{10} E = 11.4 + 1.5M$$

where E is the energy in ergs and M is the Richter magnitude.

(a) An earthquake that occurred in the Los Angeles area in January 1994 had a magnitude of 6.8. Compute the corresponding energy E.

(b) The following table gives the magnitudes for two large earthquakes that occurred on the same day in the South Pacific in 1993.

Date	Location	M (Magnitude)
March 6, 1993	Solomon Islands	6.5
March 6, 1993	Fiji	6.7

Compute the energy of each earthquake. Then compute the ratio of the energies (larger to smaller). For the ratio, round the answer to the nearest integer.

(c) Solve the formula $\log_{10} E = 11.4 + 1.5M$ for E. Then use the result to check your answer in part (a).

(d) The following table gives the magnitudes for two large earthquakes.

Date	Location	M (Magnitude)
March 28, 1964	Alaska	8.6
January 16, 1993	Japan	7.0

Use the formula that you derived in part (c) to compute the ratio of the energies of the two quakes (larger to smaller). Round your answer to the nearest integer.

49. The age of some rocks can be estimated by measuring the ratio of the amounts of certain chemical elements within the rock. The method known as the *rubidium–strontium method* will be discussed here. This method has been used in dating the moon rocks brought back on the Apollo missions.

Rubidium-87 is a radioactive substance with a half-life of 4.7×10^{10} years. Rubidium-87 decays into the substance strontium-87, which is stable (nonradioactive). We are going to derive the following formula for the age of a rock:

$$T = \frac{\ln[(\mathcal{N}_s/\mathcal{N}_r) + 1]}{-k}$$

where T is the age of the rock, k is the decay constant for rubidium-87, $\mathcal{N}_s$ is the number of atoms of strontium-87 now present in the rock, and $\mathcal{N}_r$ is the number of atoms of rubidium-87 now present in the rock.

(a) Assume that initially, when the rock was formed, there were $\mathcal{N}_0$ atoms of rubidium-87 and none of strontium-87. Then, as time goes by, some of the rubidium atoms decay into strontium atoms, but the total number of atoms must still be $\mathcal{N}_0$. Thus, after T years, we have

$$\mathcal{N}_0 = \mathcal{N}_r + \mathcal{N}_s$$

or, equivalently,

$$\mathcal{N}_s = \mathcal{N}_0 - \mathcal{N}_r \qquad (1)$$

However, according to the law of exponential decay for the rubidium-87, we must have $\mathcal{N}_r = \mathcal{N}_0 e^{kT}$. Solve this equation for $\mathcal{N}_0$ and then use the result to eliminate $\mathcal{N}_0$ from equation (1). Show that the result can be written

$$\mathcal{N}_s = \mathcal{N}_r e^{-kT} - \mathcal{N}_r \qquad (2)$$

(b) Solve equation (2) for T to obtain the formula given at the beginning of this exercise.

50. (Continuation of Exercise 49)

(a) The half-life of rubidium-87 is 4.7×10^{10} years. Compute the decay constant k.

(b) Analysis of lunar rock samples taken on the Apollo 11 mission showed the strontium–rubidium ratio to be

$$\frac{\mathcal{N}_s}{\mathcal{N}_r} = 0.0588$$

Estimate the age of the these lunar rocks.

51. (Continuation of Exercise 49) Analysis of the so-called genesis rock sample taken on the Apollo 15 mission revealed a strontium–rubidium ratio of 0.0636. Estimate the age of this rock.

52. *Radiocarbon Dating:* Because rubidium-87 decays so slowly, the technique of rubidium–strontium dating is generally considered effective only for objects older than 10 million years. In contrast, archeologists and geologists rely on the *radiocarbon dating* method in assigning ages ranging from 500 to 50,000 years.

Two types of carbon occur naturally in our environment—carbon-12, which is nonradioactive, and carbon-14, which has a half-life of 5730 years. All living plant and animal tissue contains both types of carbon, always in the same ratio. (The ratio is one part carbon-14 to 10^{12} parts carbon-12.) As long as the plant or animal is living, this ratio is maintained. When the organism dies, however, no new carbon-14 is absorbed, and the amount of carbon-14 begins to decrease exponentially. Since the amount of carbon-14 decreases exponentially, it follows that the level of radioactivity also must decrease exponentially. The formula describing this situation is

$$\mathcal{N} = \mathcal{N}_0 e^{kT}$$

where T is the age of the sample, $\mathcal{N}$ is the present level of radioactivity (in units of disintegrations per hour per gram of carbon), and $\mathcal{N}_0$ is the level of radioactivity T years ago, when the organism was alive. Given that the half-life of carbon-14 is 5730 years and that $\mathcal{N}_0 = 920$ disintegrations per hour per gram, show that the age T of a sample is given by

$$T = \frac{5730 \ln(\mathcal{N}/920)}{\ln(1/2)}$$

In Exercises 53–58, use the formula derived in Exercise 52 to estimate the age of each sample. Note: *Some technical complications arise in interpreting such results. Studies have shown that the ratio of carbon-12 to carbon-14 in the air (and therefore in living matter) has not in fact been constant over time. For instance, air pollution from factory smokestacks tends to increase the level of carbon-12. In the other direction, nuclear bomb testing increases the level of carbon-14.*

53. Prehistoric cave paintings were discovered in the Lascaux cave in France. Charcoal from the site was analyzed and the level of radioactivity was found to be $\mathcal{N} = 141$ disintegrations per hour per gram. Estimate the age of the paintings.

54. Before radiocarbon dating was used, historians estimated that the tomb of Vizier Hemaka, in Egypt, was constructed about 4900 years ago. After radiocarbon dating became available, wood samples from the tomb were analyzed, and it was determined that the radioactivity level was 510 disintegrations per hour per gram. Estimate the age of the tomb on the basis of this reading and compare your answer to the figure already mentioned.

55. Before radiocarbon dating, scholars believed that agriculture (farming, as opposed to the hunter-gatherer existence) in the Middle East began about 6500 years ago. However, when radiocarbon dating was used to study an ancient farming settlement at Jericho, the radioactivity level was found to be in the range $\mathcal{N} = 348$ disintegrations per hour per gram carbon. Based on this evidence, estimate the age of the site, and compare your answer to the figure mentioned at the beginning of this exercise. Round your answer to the nearest 100 years. (Similar analyses in Iraq, Turkey, and other countries have since shown that agriculture was firmly established at least 9000 years ago.)

56. **(a)** Analyses of some ancient campsites in the Western Hemisphere reveal a carbon-14 radioactivity level of $\mathcal{N} = 226$ disintegrations per hour per gram carbon. Show that this implies an age of 11,500 years, to the nearest 500 years. *Remark:* An age of 11,500 years corresponds to the last Ice Age, when sea level was significantly lower than it is today. According to the *Clovis hypothesis,* this was the time humans first entered the Western Hemisphere, across what would have been a land bridge extending over what is now the Bering Strait. (The name "Clovis" refers to Clovis, New Mexico. In 1933, a spearpoint and bones were discovered there that were subsequently found to be about 11,500 years old.)

(b) In the article "Coming to America" (*Time Magazine,* May 3, 1993), Michael D. Lemonick describes some of the evidence indicating that the Clovis hypothesis is not valid:

> A team led by University of Kentucky archeologist Tom Dillehay discovered indisputable traces · · · [in Monte Verde, in southern Chile] of a human settlement that was inhabited between 12,800 and 12,300 years ago. Usually all scientists can find from that far back are stones and bones. In this case, thanks to a peat layer that formed during the late Pleistocene era, organic matter [to which radiocarbon dating can be applied] was mummified and preserved as well.

What range for the carbon-14 radioactivity level $\mathcal{N}$ corresponds to the ages mentioned in this article? (That is, compute the values of $\mathcal{N}$ corresponding to $T = 12{,}800$ and $T = 12{,}300$.)

57. The Dead Sea Scrolls are a collection of ancient manuscripts discovered in caves along the west bank of the Dead Sea. (The discovery occurred by accident when an Arab herdsman of the Taamireh tribe was searching for a stray goat.) When the linen wrappings on the scrolls were analyzed, the carbon-14 radioactivity level was found to be 723 disintegrations per hour per gram. Estimate the age of the scrolls using this information. Historical evidence suggests that some of the scrolls date back somewhere between 150 B.C. and A.D. 40. How do these dates compare with the estimate derived using radiocarbon dating?

58. **(a)** According to Exercise 52, the formula for the age T of a sample in terms of the radioactivity level $\mathcal{N}$ is $T = \dfrac{5730 \ln(\mathcal{N}/920)}{\ln(1/2)}$. Solve this equation for $\mathcal{N}$ and show that the result can be written $\mathcal{N} = 920(2^{-T/5730})$.

(b) The famous prehistoric stone monument Stonehenge is located 8 miles north of Salisbury, England. Excavations and radiocarbon dating have led anthropologists and archeologists to distinguish three periods in the building of Stonehenge: period I, 2800 years ago; period II, 2100 years ago; and period III, 2000 years ago. Use the formula in part (a) to compute the carbon-14 radioactivity levels that led the scientists to assign these ages.

Exercises 59–62 introduce a model for population growth that takes into account limitations on food and the environment. This is the **logistic growth model,** *named and studied by the nineteenth century Belgian mathematician and sociologist Pierre Verhulst. (The word "logistic" has Latin and Greek origins meaning "calculation" and "skilled in calculation," respectively.) In the logistic model that we'll study, the initial population growth resembles exponential growth. But then, at some point, due perhaps to food or space limitations, the growth slows down and eventually levels off, and the population approaches an equilibrium level. The basic equation that we'll use for logistic growth is*

$$N = \frac{P}{1 + ae^{-bt}}$$

where N is the population at time t, P is the equilibrium population (or the upper limit for population), and a and b are positive constants.

59. The following figure shows the graph of the logistic function $N(t) = 4/(1 + 8e^{-t})$. Note that in this equation the equilibrium population P is 4 and that this corresponds to the asymptope $N = 4$ in the graph.

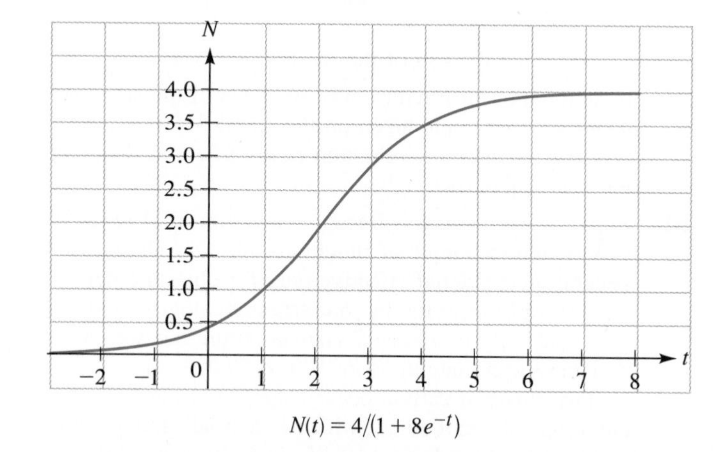

$N(t) = 4/(1 + 8e^{-t})$

(a) Use the graph and your calculator to complete the table at the top of the next column. For the values that you read from the graph, estimate to the nearest 0.25. For the calculator values, round to three decimal places.

	$N(-1)$	$N(0)$	$N(1)$	$N(4)$	$N(5)$
From Graph					
From Calculator					

(b) As indicated in the graph, the line $N = 4$ appears to be an asymptote for the curve. Confirm this empirically by computing $N(10)$, $N(15)$, and $N(20)$. Round each answer to eight decimal places.

(c) Use the graph to estimate, to the nearest integer, the value of t for which $N(t) = 3$.

(d) Find the exact value of t for which $N(t) = 3$. Evaluate the answer using a calculator, and check that it is consistent with the result in part (c).

(e) If $N(t) = 4/(1 + 8e^{-t})$, find $N^{-1}(t)$.

(f) Use your result from part (e) to solve for t in the equation $N(t) = 3$. Check your answer using the result in part (d).

60. The following figure shows the graph of a logistic function $N(t) = P/(1 + ae^{-bt})$. As indicated in the figure, the graph passes through the two points $(0, 1)$ and $(1, 2)$, and the asymptote is $N = 5$. In this exercise, we determine the values of the constants P, a, and b.

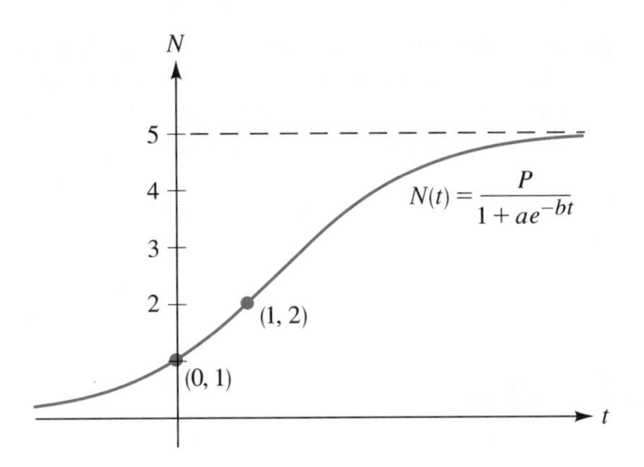

(a) As indicated in the figure, the logistic curve approaches an equilibrium population of 5. By definition, this is the value of the constant P, so the equation becomes

$$N(t) = 5/(1 + ae^{-bt})$$

Now use the fact that the graph passes through the point $(0, 1)$ to obtain an equation that you can solve for a. Solve the equation; you should obtain $a = 4$.

(b) With $P = 5$ and $a = 4$, the logistic equation becomes $N(t) = 5/(1 + 4e^{-bt})$. Use the fact that the graph of this equation passes through the point $(1, 2)$ to show that $b = \ln(8/3) \approx 0.9808$.

61. Biologist H. G. Thornton carried out an experiment in the 1920s to measure the growth of a colony of bacteria in a closed environment. As is common in the biology lab, Thornton measured the area of the colony, rather than count the number of individuals. (The reasoning is that the actual population size is directly proportional to the area, and the area is much easier to measure.) The following table and scatter plot summarize Thornton's measurements. Notice that the data points in the scatter plot suggest a logistic growth curve.

Days	0	1	2	3	4	5
Area of Colony (cm^2)	0.24	2.78	13.53	36.30	47.50	49.40

Source: H. G. Thornton, *Annals of Applied Biology,* 1922, p. 265

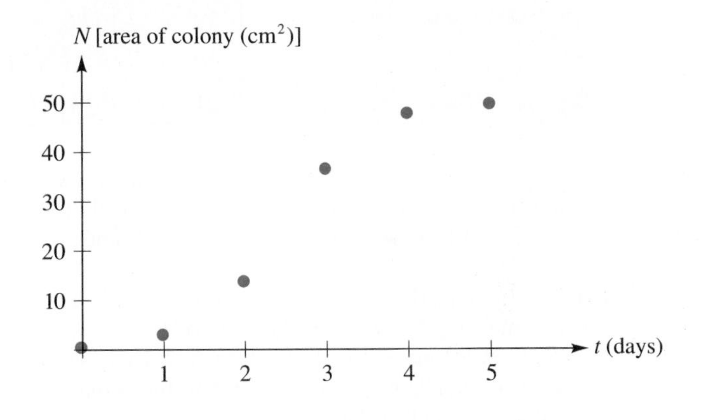

(a) Use the technique shown in Exercise 60 to determine a logistic function $N(t) = P/(1 + ae^{-bt})$ with a graph that passes through Thornton's two points $(0, 0.24)$ and $(2, 13.53)$. Assume that the equilibrium population is $P = 50$. Round the final values of a and b to two decimal places.

(b) Use the logistic function that you determined in part (a) to estimate the area of the colony after 1, 3, 4, and 5 days. Round your estimates to one decimal place. If you compare your estimates to the results that Thornton obtained in the laboratory, you'll see there is good agreement.

(c) Use the logistic function that you determined in part (a) to estimate the time at which the area of the colony was 10 cm^2. Express your answer in terms of days and hours, and round to the nearest half-hour.

62. The following data on the growth of *Lupinus albus,* a plant in the pea family, are taken from an experiment that was summarized in the classic text by Sir D'Arcy Wentworth Thompson, *On Growth and Form* (New York: Dover Publications, Inc., 1992). (The book was first published in 1917.) The accompanying figure shows a scatter plot for the data.

GROWTH OF *LUPINUS ALBUS*

Day	Length (mm)	Day	Length (mm)
4	**10.5**	**13**	**120.1**
5	16.3	14	132.3
6	23.3	15	140.6
7	32.5	16	149.7
8	42.2	17	155.6
9	58.7	18	158.1
10	77.9	19	160.6
11	93.7	20	161.4
12	107.4	21	161.6

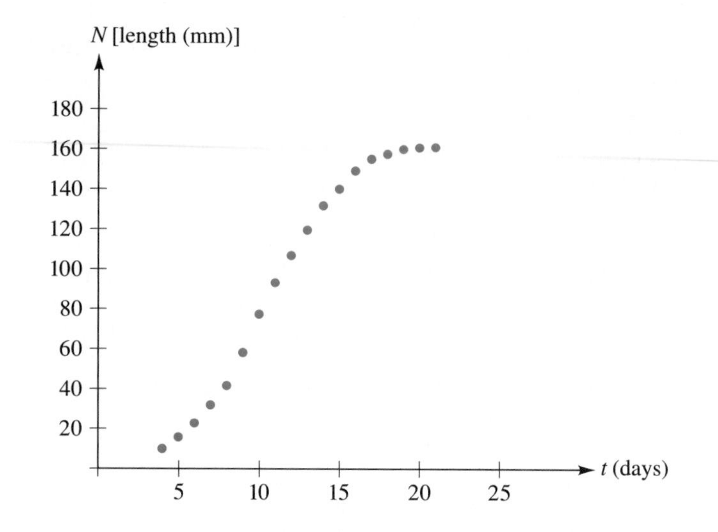

(a) Follow Exercise 60 to determine the equation of a logistic curve $N = P/(1 + ae^{-bt})$ that models this growth. Assume $N(0) = 1$, $N(10) = 77.9$, and $P = 162$.

(b) Use the equation that you determined in part (a) to compute $N(4)$, $N(8)$, $N(12)$, and $N(16)$. Round the answers to one decimal place. In each case, say if your answer is higher or lower than the actual value obtained in the experiment.

63. Let $\mathcal{N} = \mathcal{N}_0 e^{kt}$. In this exercise we show that if Δt is very small, then $\Delta\mathcal{N}/\Delta t \approx k\mathcal{N}$. In other words, over very small intervals of time, the average rate of change of $\mathcal{N}$ is proportional to $\mathcal{N}$ itself.

(a) Show that the average rate of change of the function $\mathcal{N} = \mathcal{N}_0 e^{kt}$ on the interval $[t, t + \Delta t]$ is given by

$$\frac{\Delta\mathcal{N}}{\Delta t} = \frac{\mathcal{N}_0 e^{kt}(e^{k\Delta t} - 1)}{\Delta t} = \frac{\mathcal{N}(e^{k\Delta t} - 1)}{\Delta t}$$

(b) In Section 6.2 we saw that $e^x \approx 1 + x$ when x is close to zero. Thus, if Δt is sufficiently small, we have $e^{k\Delta t} \approx 1 + k\Delta t$. Use this approximation and the result in part (a) to show that

$$\frac{\Delta\mathcal{N}}{\Delta t} \approx k\mathcal{N} \qquad \text{when } \Delta t \text{ is sufficiently close to zero}$$

CHAPTER FIVE SUMMARY OF PRINCIPAL TERMS AND FORMULAS

TERMS OR FORMULAS	PAGE REFERENCE	COMMENTS
1. Exponential function with base b	273	The exponential function with base b is defined by the equation $y = b^x$. It is understood here that the base b is a positive number other than 1.
2. Asymptote	274	A line is an asymptote for a curve if the separation distance between the curve and the line approaches zero as we move out farther and farther along the line.
3. The number e	280	The irrational number e is one of the basic constants in mathematics, as is the irrational number π. To five decimal places, the value of e is 2.71828. In calculus, e is the base most commonly used for exponential functions. The graph of the exponential function $y = e^x$ is shown in the box on page 282.

TERMS OR FORMULAS	PAGE REFERENCE	COMMENTS
4. $\log_b x$	290	The expression $\log_b x$ denotes the exponent to which b must be raised to yield x. The equation $\log_b x = y$ is equivalent to $b^y = x$.
5. $\ln x$	293	The expression $\ln x$ means $\log_e x$. Logarithms to the base e are known as *natural logarithms*. For the graph of $y = \ln x$, see Figure 6(a) in Section 5.3.
6. $\log_a x = \dfrac{\log_b x}{\log_b a}$	306	This is the change-of-base formula for converting logarithms from one base to another.
7. $A = P\left(1 + \dfrac{r}{n}\right)^{nt}$	323	This formula gives the amount A that accumulates after t years when a principal of P dollars is invested at an annual rate r that is compounded n times per year.
8. $A = Pe^{rt}$	325	This formula gives the amount A that accumulates after t years when a principal of P dollars is invested at an annual rate r that is compounded continuously.
9. $\mathcal{N} = \mathcal{N}_0 e^{kt}$ $(k > 0)$	332	This is the formula for exponential growth. (It can be shown that if a quantity $\mathcal{N}$ grows at a rate that is proportional to $\mathcal{N}$ itself, then $\mathcal{N}$ must follow this law of exponential growth.) In the formula, $\mathcal{N}$ is a function of the time t and $\mathcal{N}_0$ denotes the value of $\mathcal{N}$ at time $t = 0$. The positive number k is the *growth constant;* it is related to (but not equal to) the rate at which $\mathcal{N}$ increases. *Optional note:* It can be shown that k is the fractional change in $\mathcal{N}$ per unit time: $k = \dfrac{\Delta\mathcal{N}/\mathcal{N}}{\Delta t}$.
10. $\mathcal{N} = \mathcal{N}_0 e^{kt}$ $(k < 0)$	334	This is the decay law for radioactive substances. In the formula, $\mathcal{N}$ denotes the amount of the substance present at time t, and $\mathcal{N}_0$ is the amount present at time $t = 0$. The negative number k is the *decay constant;* it is related to (but not equal to) the rate at which $\mathcal{N}$ decreases. (See the optional note in item 9.)
11. Half-life	335	The half-life of a radioactive substance is the time required for half of a given sample to disintegrate. Notice that the half-life is an amount of time, not an amount of the substance.

WRITING MATHEMATICS

1. In Section 5.3, we defined logarithmic functions in terms of inverse functions.
   (a) Explain in general what is meant by a pair of inverse functions.
   (b) Assuming a knowledge of the function $f(x) = 2^x$, explain how to define and then graph the function $g(x) = \log_2 x$.

2. A student who was trying to simplify the expression $\dfrac{\ln(e^3)}{\ln(e^2)}$ wrote

$$\frac{\ln(e^3)}{\ln(e^2)} = \ln(e^3) - \ln(e^2) = 3 - 2 = 1$$

   (a) What is the error here? What do you think is the most probable reason for making this error?
   (b) Give the correct solution.

3. A student who wanted to simplify the expression $\dfrac{100}{\log_{10} 100}$ wrote

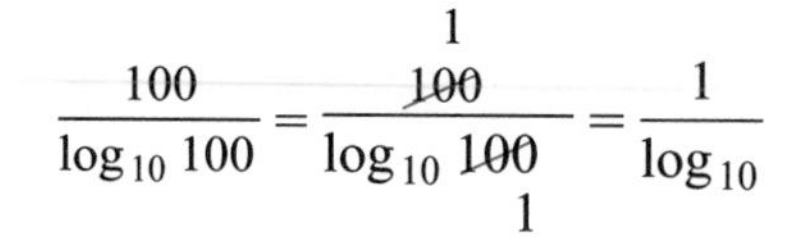

Explain why this is nonsense, and then indicate the correct solution.

4. Look over the following method for solving the equation $\ln x^2 = 6$:

$$2 \ln x = 6$$
$$\ln x = 3$$
$$x = e^3$$

As you can check, e^3 is a root of the given equation. The value $-e^3$, however, is also a root. Why is it that the demonstrated method fails to produce this root? Find a method that does produce both roots.

CHAPTER FIVE REVIEW EXERCISES

1. Which is larger, $\log_5 126$ or $\log_{10} 999$?
2. Graph the function $y = 3^{-x} - 3$. Specify the domain, range, intercept(s), and asymptote.
3. Suppose that the population of a colony of bacteria increases exponentially. If the population at the start of an experiment is 8000, and 4 hours later it is 10,000, how long (from the start of the experiment) will it take for the population to reach 12,000? (Express the answer in terms of base e logarithms.)
4. Express $\log_{10} 2$ in terms of base e logarithms.
5. Let f be the function defined by
$$f(x) = \begin{cases} 2^{-x} & \text{if } x < 0 \\ x^2 & \text{if } x \geq 0 \end{cases}$$
Sketch the graph of f and then use the horizontal line test to determine whether f is one-to-one.
6. Estimate 2^{60} in terms of an integral power of 10.
7. Solve for x: $\ln(x + 1) - 1 = \ln(x - 1)$.
8. Suppose that \$5000 is invested at 8% interest compounded annually. How many years will it take for the money to double? Make use of the approximations $\ln 2 \approx 0.7$ and $\ln 1.08 \approx 0.08$ to obtain a numerical answer.
9. On the same set of axes, sketch the graphs of $y = e^x$ and $y = \ln x$. Specify the domain and range for each function.
10. Solve for x: $xe^x - 2e^x = 0$.
11. Simplify: $\log_9(1/27)$.
12. Given that $\ln A = a$, $\ln B = b$, and $\ln C = c$, express $\ln [(A^2\sqrt{B})/C^3]$ in terms of a, b, and c.
13. The half-life of plutonium-241 is 13 years. What is the decay constant? Use the approximation $\ln(1/2) \approx -0.7$ to obtain a numerical answer.
14. Express as a single logarithm with a coefficient of 1: $3 \log_{10} x - \log_{10}(1 - x)$.
15. Solve for x, leaving your answer in terms of base e logarithms: $5e^{2-x} = 12$.
16. Let $f(x) = e^{x+1}$. Find a formula for $f^{-1}(x)$ and specify the domain of f^{-1}.
17. Suppose that in 1995 the population of a certain country was 2 million and increasing with a relative growth rate of 2%/year. Estimate the year in which the population will reach 3 million. (Use the approximation $\ln 1.5 \approx 0.40$ to obtain a numerical answer.)
18. Simplify: $\ln e + \ln\sqrt{e} + \ln 1 + \ln e^{\ln 10}$.
19. A principal of \$1000 is deposited at 10% per annum, compounded continuously. Estimate the doubling time and then sketch a graph that shows how the amount increases with time.
20. Simplify: $\ln(\log_8 56 - \log_8 7)$.

In Exercises 21–32, graph the function and specify the asymptote(s) and intercept(s).

21. $y = e^x$ **22.** $y = -e^{-x}$

23. $y = \ln x$ **24.** $y = \ln(x + 2)$

25. $y = 2^{x+1} + 1$ **26.** $y = \log_{10}(-x)$

27. $y = (1/e)^x$ **28.** $y = (1/2)^{-x}$

29. $y = e^{x+1} + 1$ **30.** $y = -\log_2(x + 1)$

31. $y = \ln(e^x)$ **32.** $y = e^{\ln x}$

In Exercises 33–49, solve the equation for x. (When logarithms appear in your answer, leave the answer in that form, rather than using a calculator.)

33. $\log_4 x + \log_4(x - 3) = 1$

34. $\log_3 x + \log_3(2x + 5) = 1$

35. $\ln x + \ln(x + 2) = \ln 15$ **36.** $\log_6 \dfrac{x + 4}{x - 1} = 1$

37. $\log_2 x + \log_2(3x + 10) - 3 = 0$

38. $2 \ln x - 1 = 0$ **39.** $3 \log_9 x = 1/2$

40. $e^{2x} = 6$ **41.** $e^{1-5x} = 3\sqrt{e}$

42. $2^x = 100$

43. $\log_{10} x - 2 = \log_{10}(x - 2)$

44. $\log_{10}(x^2 - x - 10) = 1$

45. $\ln(x + 2) = \ln x + \ln 2$

46. $\ln(2x) = \ln 2 + \ln x$ **47.** $\ln x^4 = 4 \ln x$

48. $(\ln x)^4 = 4 \ln x$ **49.** $\log_{10} x = \ln x$

50. Solve for x: $(\ln x)/(\ln 3) = \ln x - \ln 3$. Use a calculator to evaluate your result and round to the nearest integer. *Answer:* 206765

In Exercises 51–66, simplify the expression without using a calculator.

51. $\log_{10}\sqrt{10}$ **52.** $\log_7 1$

53. $\ln\sqrt[5]{e}$ **54.** $\log_3 54 - \log_3 2$

55. $\log_{10} \pi - \log_{10} 10\pi$ **56.** $\log_2 2$

57. 10^t, where $t = \log_{10} 16$ **58.** $e^{\ln 5}$

59. $\ln(e^4)$ **60.** $\log_{10}(10^{\sqrt{2}})$

61. $\log_{12} 2 + \log_{12} 18 + \log_{12} 4$

62. $(\log_{10} 8)/(\log_{10} 2)$ **63.** $(\ln 100)/(\ln 10)$

64. $\log_5 2 + \log_{1/5} 2$ **65.** $\log_2 \sqrt[7]{16\sqrt[3]{2\sqrt{2}}}$

66. $\log_{1/8}(1/16) + \log_5 0.02$

In Exercises 67–70, express the quantity in terms of a, b, and c, where $a = \log_{10} A$, $b, = \log_{10} B$, and $c = \log_{10} C$.

67. $\log_{10} A^2B^3\sqrt{C}$ **68.** $\log_{10}\sqrt[3]{AC/B}$

69. $16 \log_{10}\sqrt{A}\sqrt[4]{B}$ **70.** $6 \log_{10}[B^{1/3}/(A\sqrt{C})]$

In Exercises 71–76, find consecutive integers n and n + 1 such that the given expression lies between n and n + 1. Do not use a calculator.

71. $\log_{10} 209$ **72.** $\ln 2$ **73.** $\log_6 100$

74. $\log_{10}(1/12)$ **75.** $\log_{10} 0.003$ **76.** $\log_3 244$

77. **(a)** On the same set of axes, graph the curves $y = \ln(x + 2)$ and $y = \ln(-x) - 1$. According to your graph, in which quadrant do the two curves intersect?
(b) Find the x-coordinate of the intersection point.

78. The curve $y = ae^{bt}$ passes through the point (3, 4) and has a y-intercept of 2. Find a and b.

79. A certain radioactive substance has a half-life of T years. Find the decay constant (in terms of T).

80. Find the half-life of a certain radioactive substance if it takes T years for one-third of a given sample to disintegrate. (Your answer will be in terms of T.)

81. A radioactive substance has a half-life of M minutes. What percentage of a given sample will remain after $4M$ minutes?

82. At the start of an experiment, a colony of bacteria has initial population a. After b hours, there are c bacteria present. Determine the population d hours after the start of the experiment.

83. The half-life of a radioactive substance is d days. If you begin with a sample weighing b grams, how long will it be until c grams remain?

84. Find the half-life of a radioactive substance if it takes D days for P percent of a given sample to disintegrate.

In Exercises 85–90, write the expression as a single logarithm with a coefficient of 1.

85. $\log_{10} 8 + \log_{10} 3 - \log_{10} 12$

86. $4 \log_{10} x - 2 \log_{10} y$

87. $\ln 5 - 3 \ln 2 + \ln 16$

88. $\ln(x^4 - 1) - \ln(x^2 - 1)$

89. $a \ln x + b \ln y$

90. $\ln(x^3 + 8) - \ln(x + 2) - 2 \ln(x^2 - 2x + 4)$

In Exercises 91–98, write the quantity using sums and differences of simpler logarithmic expressions. Express the answer so that logarithms of products, quotients, and powers do not appear.

91. $\ln\sqrt{(x - 3)(x + 4)}$ **92.** $\log_{10} \dfrac{x^2 - 4}{x + 3}$

93. $\log_{10} \dfrac{3}{\sqrt{1 + x}}$ **94.** $\ln \dfrac{x^2\sqrt{2x + 1}}{2x - 1}$

95. $\log_{10}\sqrt[3]{x/100}$

96. $\log_{10}\sqrt{\dfrac{x^2+8}{(x^2+9)(x+1)}}$

97. $\ln\left(\dfrac{1+2e}{1-2e}\right)^3$

98. $\log_{10}\sqrt[3]{x\sqrt{1+y^2}}$

99. Suppose that A dollars are invested at $R\%$ compounded annually. How many years will it take for the money to double?

100. A sum of \$2800 is placed in a savings account at 9% per annum. How much is in the account after two years if the interest is compounded quarterly?

101. A bank offers an interest rate of 9.5% per annum, compounded monthly. Compute the effective interest rate.

102. A sum of D dollars is placed in an account at $R\%$ per annum, compounded continuously. When will the balance reach E dollars?

103. Compute the doubling time for a sum of D dollars invested at $R\%$ per annum, compounded continuously.

104. Your friend invests D dollars at $R\%$ per annum, compounded semiannually. You invest an equal amount at the same yearly rate, but compounded daily. How much larger is your account than your friend's after T years?

105. **(a)** You invest \$660 at 5.5% per annum, compounded quarterly. How long will it take for your balance to reach \$1000? Round off your answer to the next quarter of a year.
(b) You invest D dollars at $R\%$ per annum, compounded quarterly. How long will it take for your balance to reach nD dollars? (Assume that $n>1$.)

In Exercises 106–110, find the domain of each function.

106. **(a)** $y = \ln x$ **(b)** $y = e^x$

107. **(a)** $y = \log_{10}\sqrt{x}$ **(b)** $y = \sqrt{\log_{10} x}$

108. **(a)** $f \circ g$, where $f(x) = \ln x$ and $g(x) = e^x$
(b) $g \circ f$, where f and g are as in part (a)

109. **(a)** $y = \log_{10}|x^2 - 2x - 15|$
(b) $y = \log_{10}(x^2 - 2x - 15)$

110. $y = \dfrac{2+\ln x}{2-\ln x}$

111. Find the range of the function defined by $y = \dfrac{e^x+1}{e^x-1}$.

In Exercises 112–122, use the graph and the properties of logarithms to estimate the quantity to the nearest tenth.

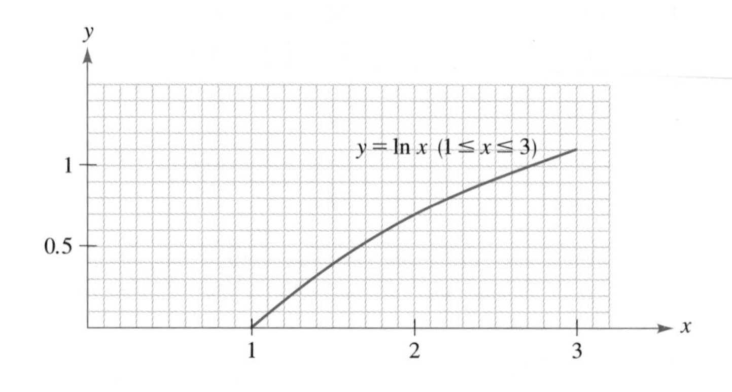

112. $\ln 2$ **113.** $\ln 0.5$ **114.** $\ln 3$

115. $\ln(1/9)$ **116.** $\ln 6$ **117.** $\ln 72$

118. $\ln(40.5)$ *Hint:* $40.5 = 81/2$

119. e *Hint:* If $x = e$, then $\ln x = 1$.

120. $e^{1.1}$ *Hint:* If $x = e^{1.1}$, then $\ln x = 1.1$.

121. $\log_2 3$ **122.** $\log_3 6$

123. The figure shows a portion of the graph of $y = e^x$, for $1.0970 \le x \le 1.1000$.
(a) Use the graph to estimate $\ln 3$ to four decimal places.
(b) Using your calculator, compute the percent error in the approximation in part (a).
(c) Using the approximation in part (a), estimate the quantities $\ln\sqrt{3}$, $\ln 9$, and $\ln(1/3)$.

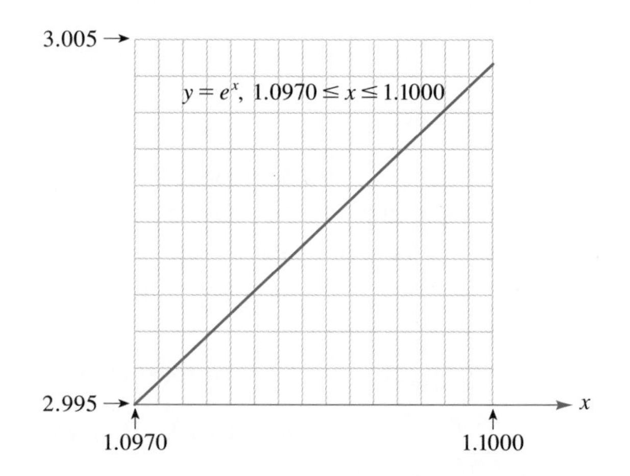

124. Given two positive numbers a and b, we define the geometric mean (G.M.) and the arithmetic mean (A.M.) as follows:

$$\text{G.M.} = \sqrt{ab} \qquad \text{A.M.} = (a+b)/2$$

If $a > b > 0$, we define the logarithmic mean (L.M.) by

$$\text{L.M.} = \frac{a-b}{\ln a - \ln b}$$

In Exercise 45 in Section 2.5, we saw that the geometric mean is always less than or equal to the arithmetic mean. In this exercise you will see that the logarithmic mean always lies between the geometric mean and the arithmetic mean. In symbols:

$$\text{G.M.} < \text{L.M.} < \text{A.M.}$$

(a) Complete the following table. Use a calculator and round to three decimal places in all rows except for the first and the last. In the first row round to four decimal places, and in the last round to five decimal places.

a	b	$\sqrt{ab}$ (G.M.)	$\frac{a-b}{\ln a - \ln b}$ (L.M.)	$\frac{a+b}{2}$ (A.M.)
1.1	1.01			
2	1			
3	1			
10	5			
10	9			
10	9.9			

(b) In each of the rows that you completed in part (a), notice that the geometric mean is less than the logarithmic mean, which is in turn less than the arithmetic mean. To demonstrate that this relationship holds in general, we'll rely on the following inequalities, which hold for all values of x greater than 1:

$$\sqrt{x} < \frac{x-1}{\ln x} < \frac{x+1}{2} \qquad (1)$$

Although a rigorous proof of these inequalities requires calculus, the following figure suggests that the inequalities are correct.

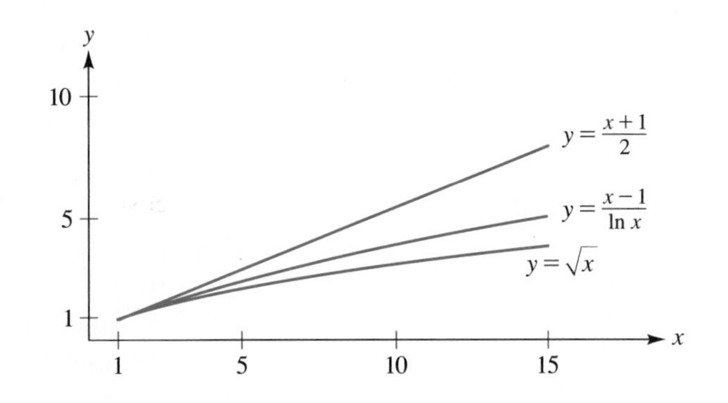

In the inequalities in (1), make the substitution $x = a/b$ and show that the resulting inequalities can be written in the form

$$\sqrt{ab} < \frac{a-b}{\ln a - \ln b} < \frac{a+b}{2} \qquad \text{as required}$$

Hint: After making the indicated substitution, multiply through by b.

CHAPTER FIVE TEST

1. Graph the function $y = 2^{-x} - 3$. Specify the domain, range, intercepts, and asymptote.
2. Suppose that the population of a colony of bacteria increases exponentially. At the start of an experiment there are 6000 bacteria, and one hour later the population is 6200. How long (from the start of the experiment) will it take for the population to reach 10,000? Give two forms for the answer: one in terms of base e logarithms and the other a calculator approximation rounded to the nearest hour.
3. Which is larger, $\log_2 17$ or $\log_3 80$? Explain the reasons for your answer.
4. Express $\log_2 15$ in terms of base e logarithms.
5. Estimate 2^{40} in terms of an integral power of 10.

6. Suppose that \$9500 is invested at 6% interest, compounded annually. How long will it take for the amount in the account to reach \$12,000?
7. **(a)** For which values of x is the identity $e^{\ln x} = x$ valid?
   (b) On the same set of axes, graph $y = e^x$ and $y = \ln x$.
8. Given that $\log_{10} A = a$ and $\log_{10} B = b$, express $\log_{10}(A^3/\sqrt{B})$ in terms of a and b.
9. Simplify each expression: **(a)** $\log_5(1/\sqrt{5})$; **(b)** $\ln e^2 + \ln 1 - e^{\ln 3}$.
10. Solve for x: $x^3e^x - 4xe^x = 0$.
11. The half-life of a radioactive substance is four days.
    (a) Find the decay constant.
    (b) How much of an initial 2 g sample will remain after 10 days?
12. Express as a single logarithm with a coefficient of 1: $2 \ln x - \ln\sqrt[3]{x^2 + 1}$.
13. Solve for x, leaving your answer in terms of base e logarithms: $-2e^{3x-1} = 9$.
14. Let $g(x) = \log_{10}(x - 1)$. Find a formula for $g^{-1}(x)$ and specify the range of g^{-1}.
15. A principal of \$12,000 is deposited at 6% per annum, compounded continuously.
    (a) Estimate the doubling time.
    (b) Sketch a graph showing how the amount increases with time.
16. **(a)** For which values of x is it true that $\ln(x^2) = 2 \ln x$?
    (b) For which values of x is it true that $(\ln x)^2 = 2 \ln x$?
17. Let $f(x) = \begin{cases} e^x & \text{if } x < 0, \\ \sqrt{1 - x^2} & \text{if } 0 \leqslant x < 1. \end{cases}$
    (a) Sketch the graph of f. **(b)** Is the function f one-to-one?
18. Solve for x: $\frac{1}{2} - \log_{16}(x - 3) = \log_{16} x$.
19. Solve for x: $6e^{2x} - 5e^x = 6$.
20. Solve each inequality: **(a)** $5(4 - 0.3^x) > 12$;
    (b) $\ln x + \ln(x - 3) \leqslant \ln 4$

TRIGONOMETRIC FUNCTIONS OF ANGLES

Trigonometry, as we know the subject today, is a branch of mathematics that is linked with algebra. As such it dates back only to the eighteenth century.

When treated purely as a development of geometry, however, it goes back to the time of the great Greek mathematician–astronomers. . . .

Alfred Hooper in *Makers of Modern Mathematics* (New York: Random House, Inc., 1948)

It is certain that the division of the ecliptic into 360 degrees . . . [was] *adopted by the Greeks from Babylon. . . .* [But] *It was Hipparchus* [ca 180–ca 125 B.C.] *who first divided the circle in general into 360 parts or degrees, and the introduction of this division coincides with his invention of trigonometry.*

Sir Thomas Heath in *A History of Greek Mathematics,* vol. II (London: Oxford University Press, 1921)

The origins of trigonometry are obscure. There are some problems in the Rhind papyrus [ca 1650 B.C.] *that involve the cotangent of the dihedral angles at the base of a pyramid, and . . . the Babylonian cuneiform tablet Plimpton 322 essentially contains a remarkable table of secants. It may be that modern investigations into the mathematics of ancient Mesopotamia will reveal an appreciable development of practical trigonometry.*

Howard Eves in *An Introduction to the History of Mathematics,* 6th ed. (Philadelphia: Saunders College Publishing, 1990)

INTRODUCTION

In general, there are two approaches to trigonometry at the precalculus level. In a sense, these two approaches correspond to the two historical roots of the subject mentioned in the first of our quotations in the margin. One approach centers around the study of triangles. Indeed, the word ''trigonometry'' is derived from two Greek words, *trigonon,* meaning triangle, and *metria,* meaning measurement. It is with this triangle approach that we begin the chapter. The second approach, which is in a sense a generalization of the first, we will take up in Chapters 7 and 8. Before getting down to specifics, we offer the following bird's-eye view of Chapters 6, 7, and 8 in terms of functions. As you'll see in subsequent chapters, a real familiarity with both approaches to trigonometry is necessary.

	INPUTS FOR THE FUNCTIONS	OUTPUTS	EMPHASIS
Chapter 6	angles	real numbers	The trigonometric functions are used to study triangles.
Chapters 7 and 8	real numbers	real numbers	The objects of study are the trigonometric functions themselves.

6.1 TRIGONOMETRIC FUNCTIONS OF ACUTE ANGLES

. . . [the English clergyman and mathematician] *William Oughtred* [1574–1660] *introduced a vast array of characters into mathematics . . . . In the* Trigonometria *(1657), he employed* ∠ *for angle. . . .*

Florian Cajori in *A History of Mathematical Notations,* vol. I (La Salle, Illinois: The Open Court Publishing Company, 1928)

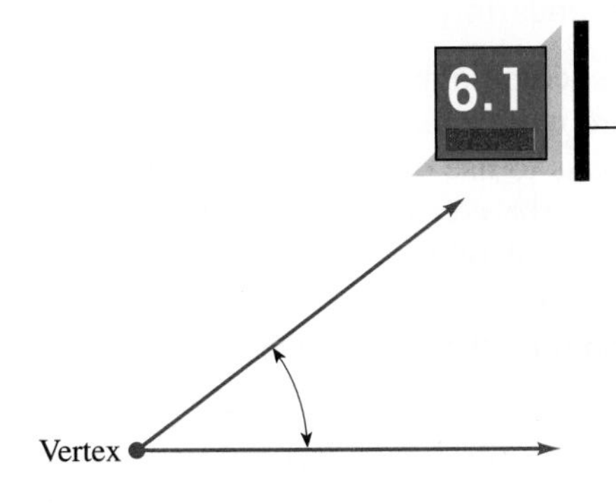

FIGURE 1

In elementary geometry, an **angle** is a figure formed by two rays with a common endpoint. As indicated in Figure 1, the common endpoint is called the **vertex**

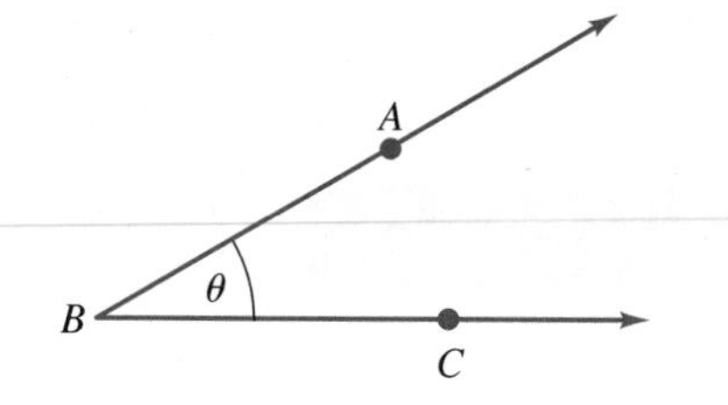

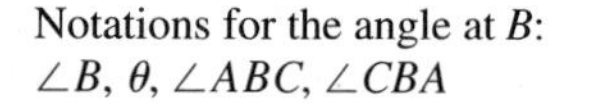
Notations for the angle at B:
$\angle B$, θ, $\angle ABC$, $\angle CBA$

FIGURE 2

of the angle. Recall that there are several conventions used in naming angles; Figure 2 indicates some of these. In Figure 2, the symbol θ is the lowercase Greek letter *theta.* Greek letters are often used to name angles. For reference, the Greek alphabet is given in the endpapers at the back of this book. The $\angle$ symbol that you see in Figure 2 stands for the word "angle." When three letters are used in naming an angle, as with $\angle ABC$ in Figure 2, the middle letter always indicates the vertex of the angle.

There are several ways to indicate the size of (i.e., the amount of rotation in) an angle. In this chapter we use the familiar units of *degrees.* Recall that 360 degrees is the measure of an angle obtained by rotating a ray through one complete circle. The symbol for degrees is °. For comparison, Figure 3 displays angles of various degrees. In this section, we will not make any distinction whether the rotation generating an angle is clockwise or counterclockwise; in a subsequent section this distinction will be important. The angles in Figures 3(a) and 3(b) are **acute angles;** these are angles with degree measure θ satisfying $0° < \theta < 90°$. The angle in Figure 3(d) is an **obtuse angle,** that is, an angle with degree measure θ satisfying $90° < \theta < 180°$.

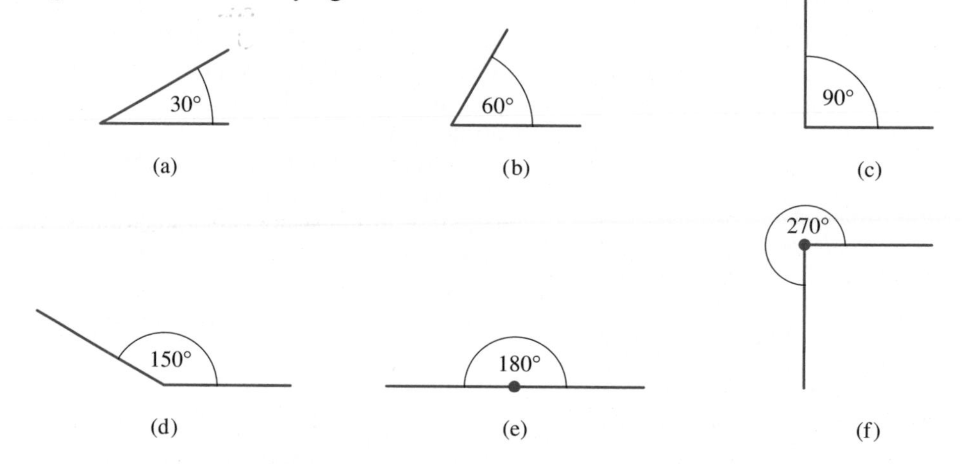

FIGURE 3

The degree is subdivided into two smaller units: the **minute,** which is 1/60 of a degree; and the **second,** which is 1/60 of a minute. The abbreviations for minutes and seconds are ′ and ″, respectively. Thus we have

$$1° = 60' \qquad \text{There are 60 minutes in one degree.}$$

and

$$1' = 60'' \qquad \text{There are 60 seconds in one minute.}$$

Until modern handheld calculators became commonplace (in the 1970s and 1980s), these subdivisions of the degree were widely used. Nowadays, however, the convention in most scientific applications is to use ordinary decimal notation for the fractional parts of a degree, and this is the convention we follow in this text. (We mention in passing that minutes and seconds are still widely used in some portions of astronomy and in navigation.)

One more convention: Suppose, for example, that the measure of an angle, say angle B, is 70°. We can write this as

$$\text{measure } \angle B = 70° \qquad \text{or, more concisely,} \qquad m\,\angle B = 70°$$

For ease of notation and speech, however, we will often simply write $\angle B = 70°$.

We are going to define six functions, the six *trigonometric functions.* As indicated by the title of this section, the inputs for these functions (in this section) will be acute angles. The outputs will be real numbers. In a later section we'll make a transition so that angles of any size can serve as inputs, and in Chapter 7 we'll make another transition, this time to real-number inputs. But for now, as we have said, the inputs will be acute angles.

For reasons that are more historical* than mathematical, the trigonometric functions have names that are words rather than single letters such as f. The use of words rather than single letters for naming functions is by no means peculiar to trigonometry. For instance, in almost every computer-programming language, the absolute value function is ABS. So, for example, we have

$$\text{ABS}(5) = 5 \qquad \text{and} \qquad \text{ABS}(-5) = 5$$

The names of the six trigonometric functions, along with their abbreviations, are as follows.

NAME OF FUNCTION	ABBREVIATION
cosine	cos
sine	sin
tangent	tan
secant	sec
cosecant	csc
cotangent	cot

Now, let θ be an acute angle in a right triangle, as shown in Figure 4. We define the six trigonometric functions of θ in the box that follows.

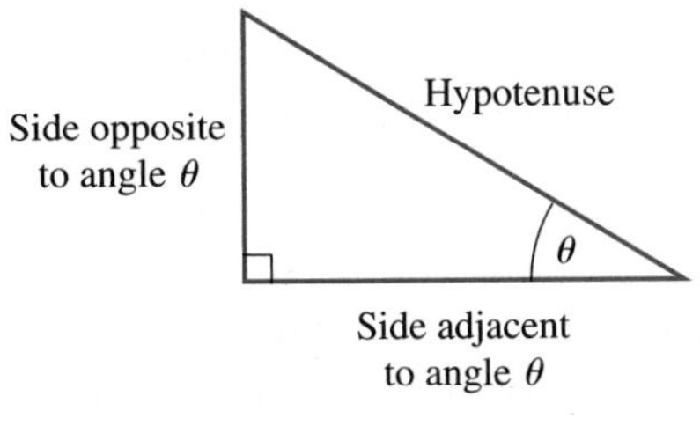

FIGURE 4

RIGHT TRIANGLE DEFINITIONS

(Refer to Figure 4.)

$$\cos\theta = \frac{\text{length of side adjacent angle } \theta}{\text{length of hypotenuse}} \qquad \sec\theta = \frac{\text{length of hypotenuse}}{\text{length of side adjacent angle } \theta}$$

$$\sin\theta = \frac{\text{length of side opposite angle } \theta}{\text{length of hypotenuse}} \qquad \csc\theta = \frac{\text{length of hypotenuse}}{\text{length of side opposite angle } \theta}$$

$$\tan\theta = \frac{\text{length of side opposite angle } \theta}{\text{length of side adjacent angle } \theta} \qquad \cot\theta = \frac{\text{length of side adjacent angle } \theta}{\text{length of side opposite angle } \theta}$$

Our work for the next several sections will be devoted to exploring these definitions and their consequences. Two preliminary observations that will help you to understand and memorize the definitions are these.

1. An expression such as $\cos\theta$ really means $\cos(\theta)$, where cos or cosine is the name of the function and θ is an input. It is for historical rather than mathematical reasons that the parentheses are suppressed.
2. There are three pairs of reciprocals in the definitions: $\cos\theta$ and $\sec\theta$; $\sin\theta$ and $\csc\theta$; and $\tan\theta$ and $\cot\theta$.

*For a discussion of the names of the trigonometric functions, see Howard Eves, *An Introduction to the History of Mathematics,* 6th ed., pp. 236–237 (Philadelphia: Saunders College Publishing, 1990).

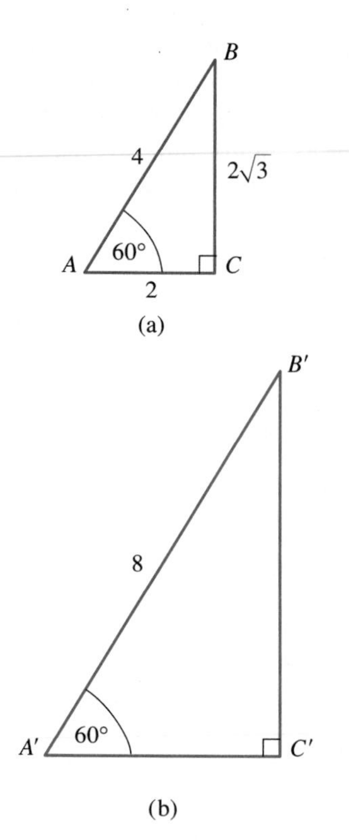

FIGURE 5

The definitions that we have just given for the trigonometric functions involve ratios of sides in a right triangle. The fact that *ratios* are involved is important. We'll use two examples to explain and emphasize this. Figure 5 shows two right triangles; both contain an angle of 60 degrees. In Figure 5(a) all three sides of the triangle are specified, while in Figure 5(b), only one side is specified. Using the data in Figure 5(a), we have

$$\tan 60° = \frac{\text{opposite}}{\text{adjacent}} = \frac{BC}{AC} = \frac{2\sqrt{3}}{2} = \sqrt{3}$$

So, from Figure 5(a), we have $\tan 60° = \sqrt{3}$. Now, the question is, will Figure 5(b) give us the same value for tan 60°? [It had better, otherwise tan will not qualify as a function. (Why?)] Indeed, Figure 5(b) does give us the same value for tan 60°, and it is not even necessary to use the fact that we are given a specific length in Figure 5(b). Elementary geometry* tells us that $\triangle ABC$ is similar to $\triangle A'B'C'$. Then, because corresponding sides of similar triangles are proportional, we have

$$\frac{BC}{AC} = \frac{B'C'}{A'C'} \tag{1}$$

Now look at the ratio on the right-hand side of equation (1); it represents tan 60° as we would calculate it in Figure 5(b). The left-hand side of the equation, as we've already seen from Figure 5(a), represents $\tan 60° = \sqrt{3}$. In summary, similar triangles and equation (1) tell us that we obtain the same value for tan 60° whether we use Figure 5(a) or 5(b).

The point of the example we've just concluded is this: For a given acute angle θ, the values of the trigonometric functions depend on the ratios of the lengths of the sides but not on the size of the particular right triangle in which θ resides. For emphasis, the next example makes this same point.

EXAMPLE 1 Figure 6 shows two right triangles. The first right triangle has sides 5, 12, and 13. The second right triangle is similar to the first (the angles are the same), but each side is 10 times longer than the corresponding side in the first triangle. Calculate and compare the values of $\sin\theta$, $\cos\theta$, and $\tan\theta$ for both triangles.

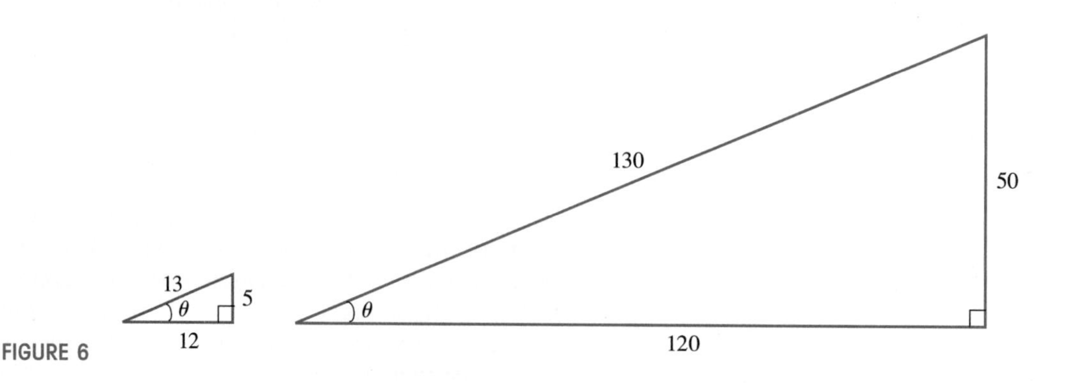

FIGURE 6

*If two angles of one triangle are congruent to two corresponding angles of another triangle, then the triangles are similar.

Solution

SMALL TRIANGLE

$$\sin\theta = \frac{\text{opposite}}{\text{hypotenuse}} = \frac{5}{13}$$

$$\cos\theta = \frac{\text{adjacent}}{\text{hypotenuse}} = \frac{12}{13}$$

$$\tan\theta = \frac{\text{opposite}}{\text{adjacent}} = \frac{5}{12}$$

LARGE TRIANGLE

$$\sin\theta = \frac{50}{130} = \frac{5}{13}$$

$$\cos\theta = \frac{120}{130} = \frac{12}{13}$$

$$\tan\theta = \frac{50}{120} = \frac{5}{12}$$

OBSERVATION The corresponding values for $\sin\theta$, $\cos\theta$, and $\tan\theta$ are the same for both triangles.

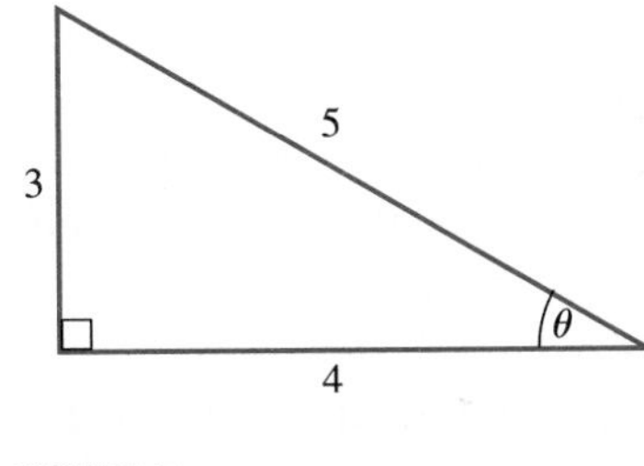

FIGURE 7

EXAMPLE 2 Let θ be the acute angle indicated in Figure 7. Determine the six quantities $\cos\theta$, $\sin\theta$, $\tan\theta$, $\sec\theta$, $\csc\theta$, and $\cot\theta$.

Solution We use the definitions:

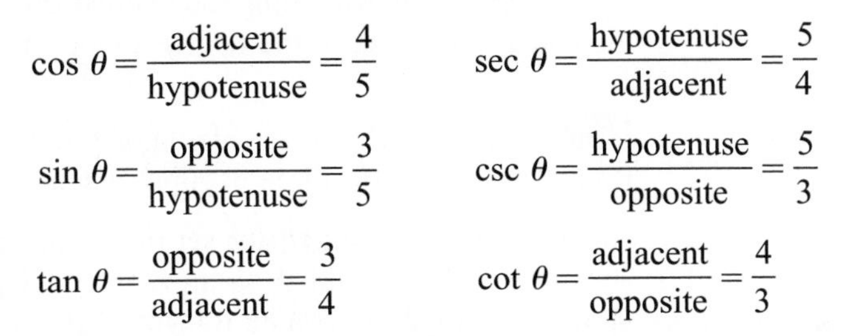

$$\cos\theta = \frac{\text{adjacent}}{\text{hypotenuse}} = \frac{4}{5} \qquad \sec\theta = \frac{\text{hypotenuse}}{\text{adjacent}} = \frac{5}{4}$$

$$\sin\theta = \frac{\text{opposite}}{\text{hypotenuse}} = \frac{3}{5} \qquad \csc\theta = \frac{\text{hypotenuse}}{\text{opposite}} = \frac{5}{3}$$

$$\tan\theta = \frac{\text{opposite}}{\text{adjacent}} = \frac{3}{4} \qquad \cot\theta = \frac{\text{adjacent}}{\text{opposite}} = \frac{4}{3}$$

Note the pairs of answers that are reciprocals. (As mentioned before, this helps in memorizing the definitions.)

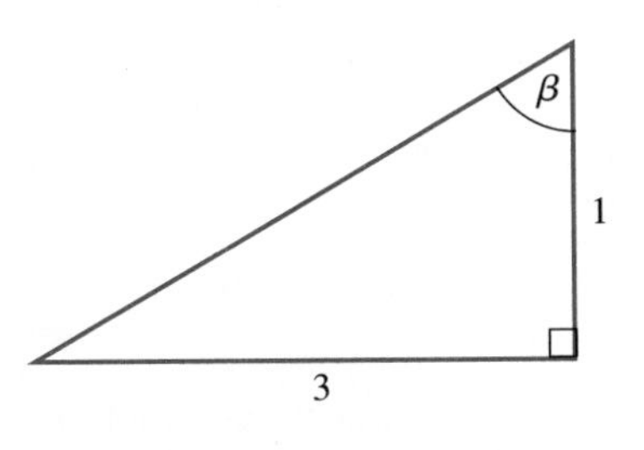

FIGURE 8

EXAMPLE 3 Let β be the acute angle indicated in Figure 8. Find $\sin\beta$ and $\cos\beta$.

Solution In view of the definitions, we need to know the length of the hypotenuse in Figure 8. If we call this length h, then by the Pythagorean theorem we have

$$h^2 = 3^2 + 1^2 = 10$$
$$h = \sqrt{10}$$

Therefore

$$\sin\beta = \frac{\text{opposite}}{\text{hypotenuse}} = \frac{3}{\sqrt{10}} = \frac{3\sqrt{10}}{10}$$

and

$$\cos\beta = \frac{\text{adjacent}}{\text{hypotenuse}} = \frac{1}{\sqrt{10}} = \frac{\sqrt{10}}{10}$$

In the next example we use a calculator to evaluate the trigonometric functions for a particular acute-angle input. All scientific calculators allow for two common units of angle measurement: the degree and the *radian*. Radian measure is discussed in Chapter 7; for the present chapter, though, we are using degree measure exclusively. So, before working through the next example on your own, you need to check that your calculator is set in the "degree mode" rather than the "radian mode." If necessary, consult the owner's manual for your calculator on this point.

One more calculator-related detail. For some calculators, when you want to evaluate a function, first you enter the input, then you press the appropriate function key. For example, on this type of calculator, the keystrokes for evaluating sin 40° are as follows:

40 (sin) input before name of function

For other types of calculators, you enter the name of the function before the input. On these calculators, the keystrokes for evaluating sin 40° are as follows:

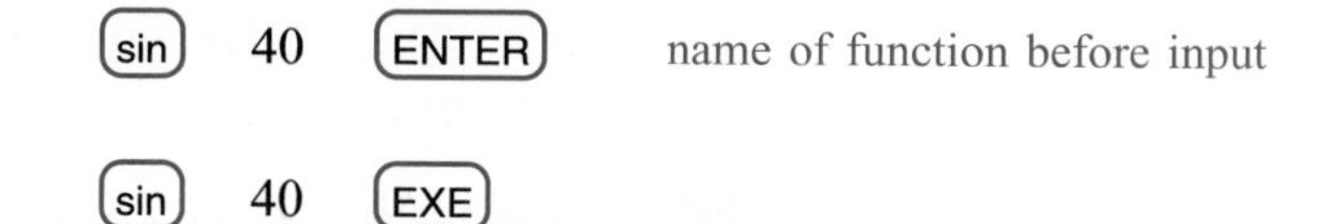

Before going on to read the next example, be sure you know which of the two types of calculators you have: *input before name of function* or *name of function before input.* Try evaluating the expression sin 40°. You should obtain (assuming the calculator is set in the degree mode) a display that begins 0.6427 . . . .

EXAMPLE 4 Use a calculator to evaluate cos 40°, sin 40°, and tan 40°. Round the results to three decimal places.

Solution With the calculator set in the degree mode, the sequence of keystrokes and the results (rounded to three decimal places) are as follows. You should verify each of these results for yourself.

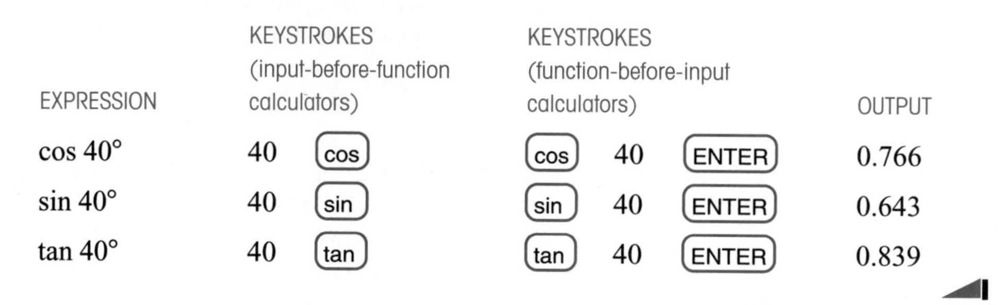

EXPRESSION	KEYSTROKES (input-before-function calculators)	KEYSTROKES (function-before-input calculators)	OUTPUT
cos 40°	40 (cos)	(cos) 40 (ENTER)	0.766
sin 40°	40 (sin)	(sin) 40 (ENTER)	0.643
tan 40°	40 (tan)	(tan) 40 (ENTER)	0.839

In the next example, we use a calculator to evaluate the three functions secant, cosecant, and cotangent for a particular input. These three functions are referred to as the **reciprocal functions.** This is because, from the definitions on page 359, it follows that

$$\sec\theta = \frac{1}{\cos\theta} \qquad \csc\theta = \frac{1}{\sin\theta} \qquad \cot\theta = \frac{1}{\tan\theta}$$

So, for example, although most calculators do not have a key labeled "sec," we can nevertheless evaluate the secant function by using the cosine key (cos) and then the reciprocal key (1/x) or (x^{-1}).

EXAMPLE 5 Use a calculator to evaluate sec 10°, csc 10°, and cot 10°. Round the results to three decimal places.

Solution As in the previous example, we first check that the calculator is in the degree mode. Then the sequence of keystrokes and the results are as follows (again, you should verify each of these results for yourself):

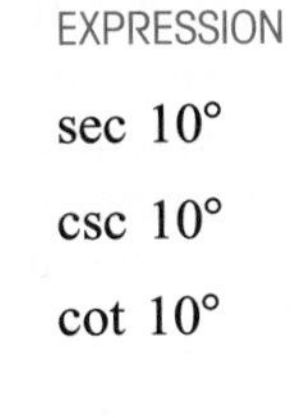
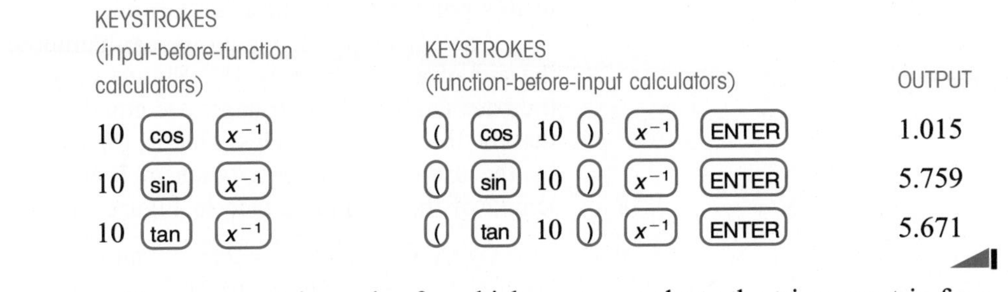

EXPRESSION	KEYSTROKES (input-before-function calculators)	KEYSTROKES (function-before-input calculators)	OUTPUT
sec 10°	10 [cos] [x^{-1}]	[(] [cos] 10 [)] [x^{-1}] [ENTER]	1.015
csc 10°	10 [sin] [x^{-1}]	[(] [sin] 10 [)] [x^{-1}] [ENTER]	5.759
cot 10°	10 [tan] [x^{-1}]	[(] [tan] 10 [)] [x^{-1}] [ENTER]	5.671

There are certain angles for which we can evaluate the trigonometric functions without using a calculator. Three such angles are 30°, 45°, and 60°. We focus on these particular angles for now, not because they are somehow more fundamental than others, but because a ready knowledge of their trigonometric values will provide a useful source of examples in our subsequent work.

In order to obtain the trigonometric values for angles of 30° and 60°, we rely on the following theorem from geometry. (For a proof of this theorem, see Exercise 51.)

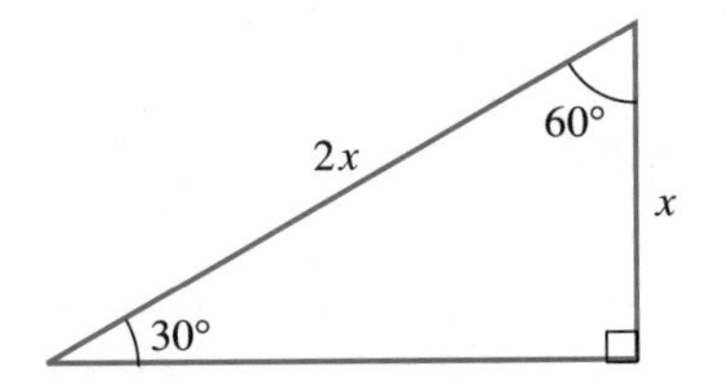

FIGURE 9

THEOREM: The 30°-60° Right Triangle

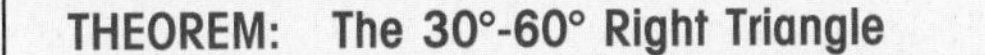

In a 30°-60° right triangle, the length of the side opposite the 30° angle is half the length of the hypotenuse. (See Figure 9.)

If we let y denote the length of the unmarked side in Figure 9, then (according to the Pythagorean theorem) we have $y^2 + x^2 = (2x)^2$ and, consequently,

$$y^2 = 4x^2 - x^2 = 3x^2$$
$$y = \sqrt{3}x$$

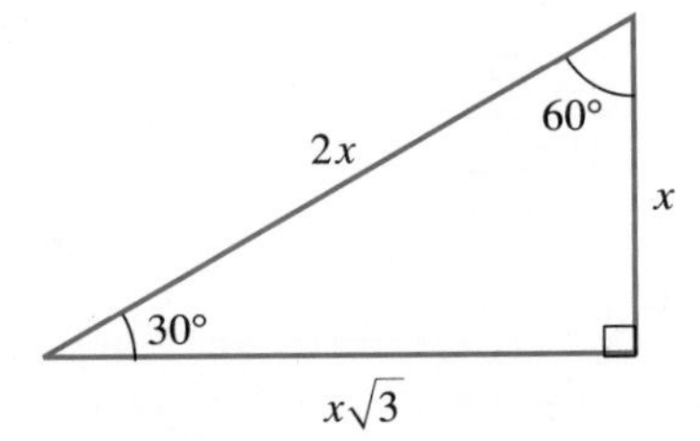

FIGURE 10

Our 30°-60° right triangle can now be labeled as in Figure 10. From this figure, the required trigonometric values can be written down by inspection. We will list only the values of the cosine, sine, and tangent, since the remaining three values are just the reciprocals of these.

$$\cos 30° = \frac{x\sqrt{3}}{2x} = \frac{\sqrt{3}}{2} \qquad \cos 60° = \frac{x}{2x} = \frac{1}{2}$$

$$\sin 30° = \frac{x}{2x} = \frac{1}{2} \qquad \sin 60° = \frac{x\sqrt{3}}{2x} = \frac{\sqrt{3}}{2}$$

$$\tan 30° = \frac{x}{x\sqrt{3}} = \frac{1}{\sqrt{3}} = \frac{\sqrt{3}}{3} \qquad \tan 60° = \frac{x\sqrt{3}}{x} = \sqrt{3}$$

There are two observations to be made here. First, the final answers do not involve x. Once again this shows that for a given angle, it is the *ratios* of the sides that determine the trigonometric values, not the size of the triangle. Second, notice that

$$\sin 30° = \cos 60° \qquad \text{and} \qquad \sin 60° = \cos 30°$$

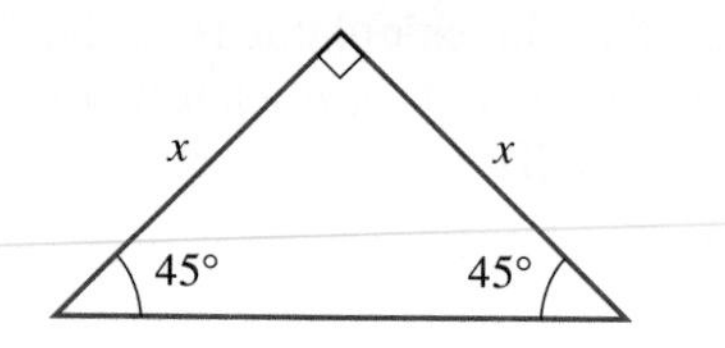

FIGURE 11

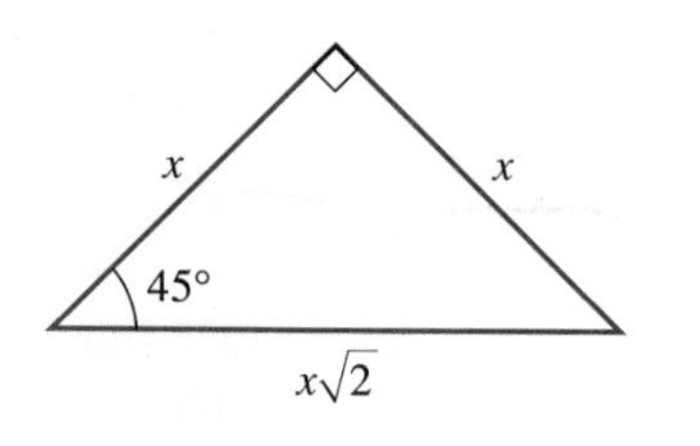

FIGURE 12

It is easy to see that this is no coincidence. The terms ''opposite'' and ''adjacent'' are relative ones. As a glance at Figure 10 shows, the side opposite to one acute angle is automatically adjacent to the other acute angle. We'll return to this point again in the next section.

Another angle for which we can compute the trigonometric values without the aid of a calculator is 45°. Figure 11 shows a 45°-45° right triangle. Since the base angles of this triangle are equal, it follows from a theorem of geometry that the triangle is isosceles. That is, the triangle has two equal sides, as indicated by the use of the letter x twice in Figure 11. If we now use y to denote the length of the unmarked side in Figure 11, we have

$$y^2 = x^2 + x^2 = 2x^2 \qquad \text{and, consequently,} \qquad y = \sqrt{2}x$$

Our 45°-45° right triangle can now be labeled as in Figure 12, and the trigonometric values can then be obtained by inspection. As before, we list only the values for cosine, sine, and tangent, since the remaining three are just the reciprocals of these.

$$\cos 45° = \frac{x}{x\sqrt{2}} = \frac{1}{\sqrt{2}} = \frac{\sqrt{2}}{2}$$

$$\sin 45° = \frac{x}{x\sqrt{2}} = \frac{1}{\sqrt{2}} = \frac{\sqrt{2}}{2}$$

$$\tan 45° = \frac{x}{x} = 1$$

Table 1 summarizes the results we've now obtained for angles of 30, 45, and 60 degrees. These results should be memorized.

TABLE 1

θ	$\sin\theta$	$\cos\theta$	$\tan\theta$
30°	$\frac{1}{2}$	$\frac{\sqrt{3}}{2}$	$\frac{\sqrt{3}}{3}$
45°	$\frac{\sqrt{2}}{2}$	$\frac{\sqrt{2}}{2}$	1
60°	$\frac{\sqrt{3}}{2}$	$\frac{1}{2}$	$\sqrt{3}$

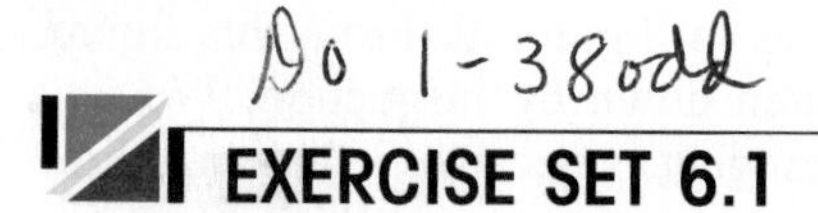

EXERCISE SET 6.1

A

In Exercises 1–4, use the definitions (as in Example 2) to evaluate the six trigonometric functions of (a) θ and (b) β. In cases where a radical occurs in a denominator, rationalize the denominator.

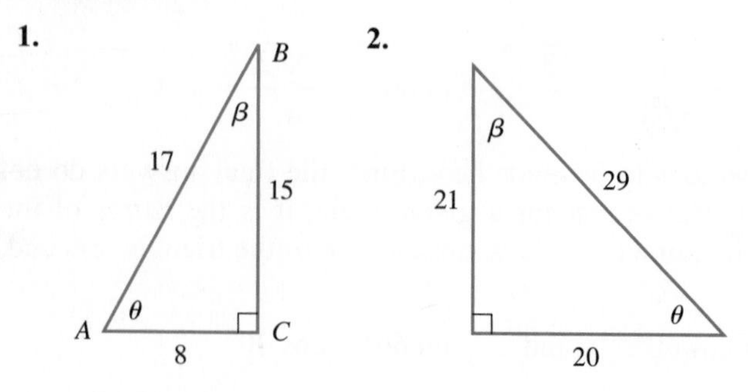

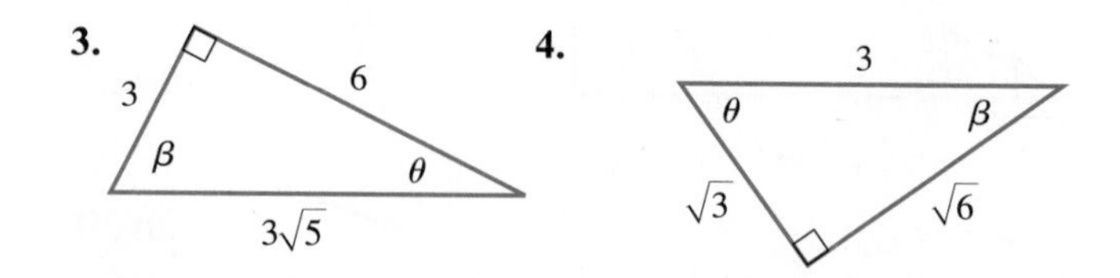

In Exercises 5–12, suppose that $\triangle ABC$ is a right triangle with $\angle C = 90°$.

5. If $AC = 3$ and $BC = 2$, find the following quantities: **(a)** $\cos A$, $\sin A$, $\tan A$; **(b)** $\sec B$, $\csc B$, $\cot B$.
6. If $AC = 6$ and $BC = 2$, find the following quantities: **(a)** $\cos A$, $\sin A$, $\tan A$; **(b)** $\sec B$, $\csc B$, $\cot B$.
7. If $AB = 13$ and $BC = 5$, compute the values of the six trigonometric functions of angle B.

8. If $AB = 3$ and $AC = 1$, compute the values of the six trigonometric functions of angle A.

9. If $AC = 1$ and $BC = 3/4$, compute each quantity: **(a)** $\sin B$, $\cos A$; **(b)** $\sin A$, $\cos B$; **(c)** $(\tan A)(\tan B)$.

10. If $AC = BC = 4$, compute the following: **(a)** $\sec A$, $\csc A$, $\cot A$; **(b)** $\sec B$, $\csc B$, $\cot B$; **(c)** $(\cot A)(\cot B)$.

11. If $AB = 25$ and $AC = 24$, compute each of the required quantities: **(a)** $\cos A$, $\sin A$, $\tan A$; **(b)** $\cos B$, $\sin B$, $\tan B$; **(c)** $(\tan A)(\tan B)$.

12. If $AB = 1$ and $BC = \sqrt{3}/2$, compute the following: **(a)** $\cos A$, $\sin B$; **(b)** $\tan A$, $\cot B$; **(c)** $\sec A$, $\csc B$.

In Exercises 13–18, use a calculator to compute $\cos\theta$, $\sin\theta$, *and* $\tan\theta$ *for the given value of* θ. *Round each result to three decimal places.*

13. $\theta = 65°$ **14.** $\theta = 21°$ **15.** $\theta = 38.5°$

16. $\theta = 12.4°$ **17.** $\theta = 80.06°$ **18.** $\theta = 0.99°$

In Exercises 19–24, use a calculator to compute $\sec\theta$, $\csc\theta$, *and* $\cot\theta$ *(as in Example 5) for the given value of* θ. *Round each result to three decimal places.*

19. $\theta = 20°$ **20.** $\theta = 40°$ **21.** $\theta = 17.5°$

22. $\theta = 18.5°$ **23.** $\theta = 1°$ **24.** $\theta = 89.9°$

In Exercises 25–36, verify that each equation is correct by evaluating each side. Do not use a calculator. The purpose of Exercises 25–36 is twofold. First, doing the problems will help you to review the values in Table 1 of this section. Second, the exercises serve as an algebra review. Note: *In the exercises, notation such as* $\sin^2\theta$ *stands for* $(\sin\theta)^2$. *This common notational convention will be discussed in the next section.*

25. $\cos 60° = \cos^2 30° - \sin^2 30°$

26. $\cos 60° = 1 - 2\sin^2 30°$

27. $\sin^2 30° + \sin^2 45° + \sin^2 60° = 3/2$

28. $\sin 30° \cos 60° + \cos 30° \sin 60° = 1$

29. $2 \sin 30° \cos 30° = \sin 60°$

30. $2 \sin 45° \cos 45° = 1$

31. $\sin 30° = \sqrt{(1 - \cos 60°)/2}$

32. $\cos 30° = \sqrt{(1 + \cos 60°)/2}$

33. $\tan 30° = \dfrac{\sin 60°}{1 + \cos 60°}$ **34.** $\tan 30° = \dfrac{1 - \cos 60°}{\sin 60°}$

35. $1 + \tan^2 45° = \sec^2 45°$ **36.** $1 + \cot^2 60° = \csc^2 60°$

37. **(a)** Use a calculator to evaluate cos 30° and cos 45°. Give your answers to as many decimal places as your calculator allows.
(b) In Table 1, there are expressions for cos 30° and cos 45°. Use your calculator to evaluate these expressions, and check to see that the results agree with those in part (a).

38. **(a)** Use a calculator to evaluate tan 30° and tan 60°. Give your answers to as many decimal places as your calculator allows.
(b) In Table 1, there are expressions for tan 30° and tan 60°. Use your calculator to evaluate these expressions, and check to see that the results agree with those in part (a).

B

For Exercises 39–46, four functions S, C, T, and D are defined as follows:

$$\left.\begin{aligned} S(\theta) &= \sin\theta \\ C(\theta) &= \cos\theta \\ T(\theta) &= \tan\theta \\ D(\theta) &= 2\theta \end{aligned}\right\} \quad 0° < \theta < 90°$$

In each case, use the values in Table 1 (which you are supposed to memorize) to decide if the statement is true or false. A calculator is not required.

39. $2[S(30°)] = S(60°)$ **40.** $T(45°) - (C \circ D)(30°) > 0$

41. $(T \circ D)(30°) > 1$ **42.** $T(60°) = 2[T(30°)]$

43. $S(45°) - C(45°) = 0$ **44.** $S(45°) - C(45°) = 0°$

45. $(C \circ D)(30°) = S(30°)$ **46.** $(T \circ D)(15°) - C(30°) < 0$

In Exercises 47 and 48, refer to the following figure. In the figure, the arc is a portion of a circle with center O and radius r.

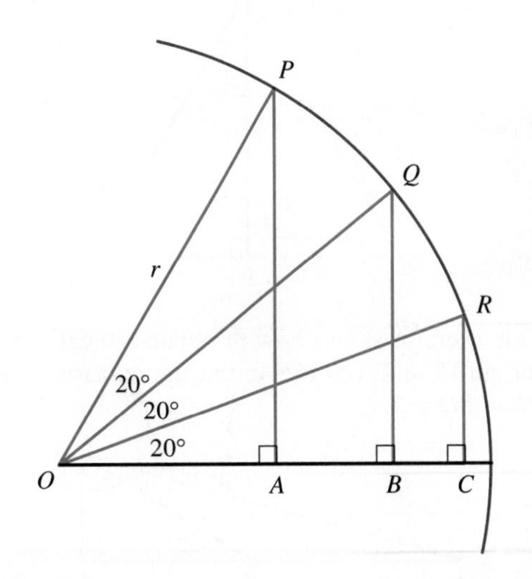

47. **(a)** Use the figure and the right-triangle definition of sine to explain (in complete sentences) why $\sin 20° < \sin 40° < \sin 60°$.

(b) Use a calculator to verify that $\sin 20° < \sin 40° < \sin 60°$.

48. **(a)** Use the figure and the right-triangle definition of cosine to explain (in complete sentences) why $\cos 20° > \cos 40° > \cos 60°$.

(b) Use a calculator to verify that $\cos 20° > \cos 40° > \cos 60°$.

For Exercises 49 and 50, refer to the following figure. In each case, say which of the two given quantities is larger. If the quantities are equal, say so.

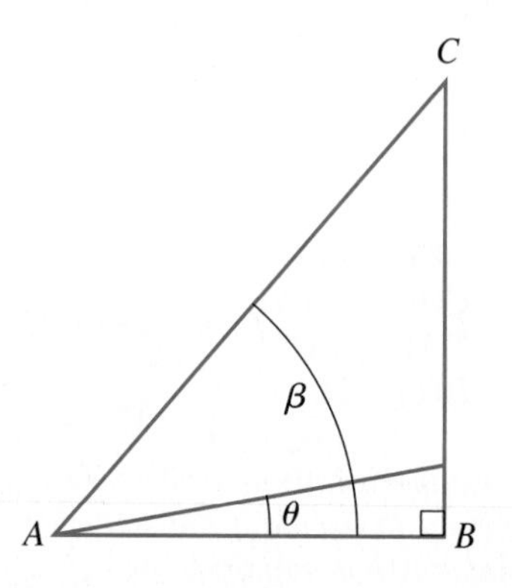

49. **(a)** $\cos\theta$, $\cos\beta$
(b) $\sec\theta$, $\sec\beta$

50. **(a)** $\tan\theta$, $\tan\beta$
(b) $\cot\theta$, $\cot\beta$

51. The following figure shows a 30°-60° right triangle, $\triangle ABC$. Prove that $AC = 2AB$. *Suggestion:* Construct $\triangle DBC$ as shown, congruent to $\triangle ABC$. Then note that $\triangle ADC$ is equilateral.

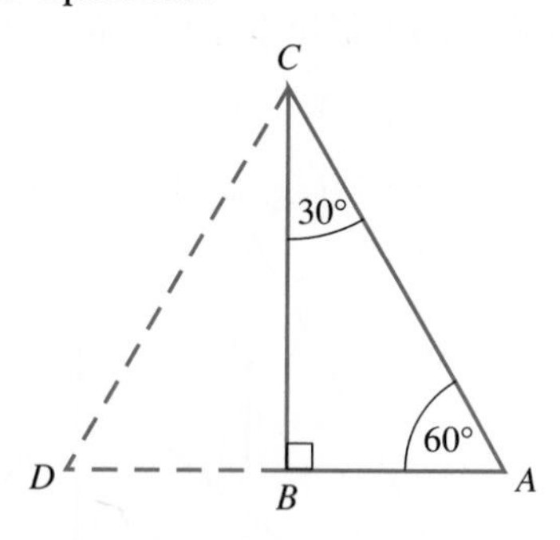

52. This exercise shows how to obtain radical expressions for $\sin 15°$ and $\cos 15°$. In the figure, assume that $AB = BD = 2$.

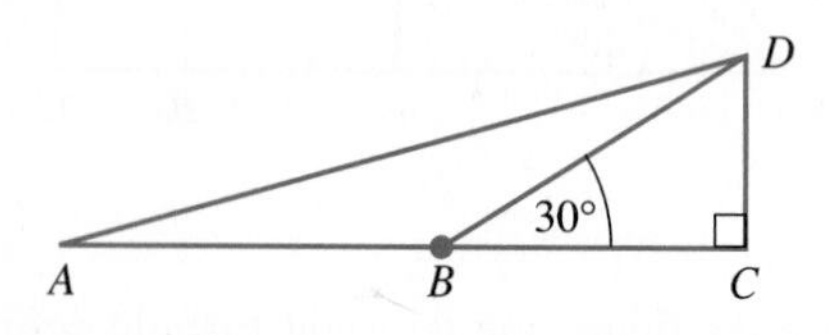

(a) In the right triangle BCD, note that $DC = 1$ because $\overline{DC}$ is opposite the 30° angle and $BD = 2$. Use the Pythagorean theorem to show that $BC = \sqrt{3}$.

(b) Use the Pythagorean theorem to show that $AD = 2\sqrt{2+\sqrt{3}}$.

(c) Show that the expression for AD in part (b) is equal to $\sqrt{6}+\sqrt{2}$. *Hint:* Two nonnegative quantities are equal if and only if their squares are equal.

(d) Explain why $\angle BAD = \angle BDA$.

(e) According to a theorem from geometry, an exterior angle of a triangle is equal to the sum of the two nonadjacent interior angles. Apply this to $\triangle ABD$ with exterior angle $DBC = 30°$, and show that $\angle BAD = 15°$.

(f) Using the figure and the values that you have obtained for the lengths, conclude that

$$\sin 15° = \frac{1}{\sqrt{6}+\sqrt{2}} \qquad \cos 15° = \frac{2+\sqrt{3}}{\sqrt{6}+\sqrt{2}}$$

(g) Rationalize the denominators in part (f) to obtain

$$\sin 15° = \frac{\sqrt{6}-\sqrt{2}}{4} \qquad \cos 15° = \frac{\sqrt{6}+\sqrt{2}}{4}$$

(h) Use your calculator to check the results in part (g).

53. If an angle θ is an integral multiple of 3°, then the real number $\sin\theta$ is either rational or expressible in terms of radicals. You've already seen examples of this with the angles 30°, 45°, and 60°. The accompanying table gives the values of $\sin\theta$ for some multiples of 3°. Use your calculator to check the entries in the table. *Remark:* If θ is an integral number of degrees, but not a multiple of 3°, then the real number $\sin\theta$ cannot be expressed in terms of radicals within the real-number system.

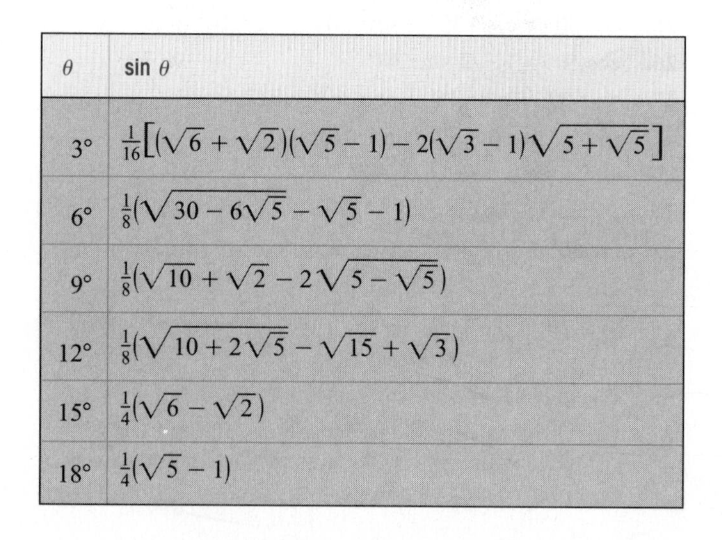

θ	$\sin\theta$
3°	$\frac{1}{16}\left[(\sqrt{6}+\sqrt{2})(\sqrt{5}-1) - 2(\sqrt{3}-1)\sqrt{5+\sqrt{5}}\right]$
6°	$\frac{1}{8}(\sqrt{30-6\sqrt{5}} - \sqrt{5} - 1)$
9°	$\frac{1}{8}(\sqrt{10}+\sqrt{2} - 2\sqrt{5-\sqrt{5}})$
12°	$\frac{1}{8}(\sqrt{10+2\sqrt{5}} - \sqrt{15}+\sqrt{3})$
15°	$\frac{1}{4}(\sqrt{6}-\sqrt{2})$
18°	$\frac{1}{4}(\sqrt{5}-1)$

54. **(a)** Use the expression for sin 18° given in Exercise 53 to show that the number sin 18° is a root of the quadratic equation $4x^2 + 2x - 1 = 0$.

(b) Use the expression for sin 15° given in Exercise 53 to show that the number sin 15° is a root of the equation $16x^4 - 16x^2 + 1 = 0$.

ALGEBRA AND THE TRIGONOMETRIC FUNCTIONS

Perhaps the first use of abbreviations for the trigonometric [functions] . . . *goes back to the* [Danish] *physician and mathematician, Thomas Finck* [1561–1646] . . . .

Florian Cajori in *A History of Mathematical Notations,* vol. II (Chicago: The Open Court Publishing Company, 1929)

Albert Girard (1595–1632), who seems to have lived chiefly in Holland, . . . interested himself in spherical trigonometry and trigonometry. In 1626, he published a treatise on trigonometry that contains the earliest use of our abbreviations sin, tan, *and* sec *for* sine, tangent, *and* secant.

Howard Eves in *An Introduction to the History of Mathematics,* 6th ed. (Philadelphia: Saunders College Publishing, 1990)

In this section, we are going to concentrate on the algebra involved in working with the trigonometric functions. This will help to pave the way both for the applications in the next section and for some of the more analytical parts of trigonometry in the next chapter. We begin by listing some common notational conventions.

NOTATIONAL CONVENTIONS

1. The quantity $(\sin \theta)^n$ is usually written $\sin^n \theta$. For example, $(\sin \theta)^2$ is written $\sin^2 \theta$. The same convention also applies to the other five trigonometric functions.*
2. Parentheses are often omitted in multiplication. For example, the product $(\sin \theta)(\cos \theta)$ is usually written $\sin \theta \cos \theta$. Similarly, $2(\sin \theta)$ is written $2 \sin \theta$.
3. An expression such as $\sin \theta$ really means $\sin(\theta)$, where sin or sine is the name of the function and θ is an input. It is for historical rather than mathematical reasons that the parentheses are suppressed. An exception to this, however, occurs in expressions such as $\sin(A + B)$; here the parentheses are necessary.

EXAMPLE 1

(a) Combine like terms: $3 \sin^2 \theta \cos \theta - 5 \sin^2 \theta \cos \theta$.
(b) Carry out the multiplication: $(2 \sin \theta - 3 \cos \theta)^2$.

Solution

(a) We do this in the same way we would simplify the algebraic expression $3S^2C - 5S^2C$. Since

$$3S^2C - 5S^2C = -2S^2C$$

we have

$$3 \sin^2 \theta \cos \theta - 5 \sin^2 \theta \cos \theta = -2 \sin^2 \theta \cos \theta$$

NOTE If the original expression had been $3 \sin^2 \theta \cos \theta - 5 \sin \theta \cos^2 \theta$, then there would have been no like terms to combine.

(b) We do this in the same way that we would expand $(2S - 3C)^2$. Since

$$(2S - 3C)^2 = 4S^2 - 12SC + 9C^2$$

we have

$$(2 \sin \theta - 3 \cos \theta)^2 = 4 \sin^2 \theta - 12 \sin \theta \cos \theta + 9 \cos^2 \theta$$

*A single exception to this convention occurs when $n = -1$. The meaning of expressions such as $\sin^{-1} x$ will be explained in Sections 4 and 5 in Chapter 8.

EXAMPLE 2 Factor: $\tan^2 A + 5 \tan A + 6$.

Solution

PRELIMINARY SOLUTION To help us focus on the algebra that is actually involved, let's replace each occurrence of the quantity $\tan A$ by the letter T. Then

$$T^2 + 5T + 6 = (T + 3)(T + 2)$$

ACTUAL SOLUTION $\tan^2 A + 5 \tan A + 6 = (\tan A + 3)(\tan A + 2)$

NOTE After you are accustomed to working with trigonometric expressions, you should be able to eliminate the preliminary step.

EXAMPLE 3 Add $\sin \theta + \dfrac{1}{\cos \theta}$.

Solution

PRELIMINARY SOLUTION

$$S + \frac{1}{C} = \frac{S}{1} + \frac{1}{C} = \frac{S}{1} \cdot \frac{C}{C} + \frac{1}{C}$$
$$= \frac{SC}{C} + \frac{1}{C} = \frac{SC + 1}{C}$$

ACTUAL SOLUTION

$$\sin \theta + \frac{1}{\cos \theta} = \frac{\sin \theta}{1} + \frac{1}{\cos \theta} = \frac{\sin \theta}{1} \cdot \frac{\cos \theta}{\cos \theta} + \frac{1}{\cos \theta}$$
$$= \frac{\sin \theta \cos \theta}{\cos \theta} + \frac{1}{\cos \theta} = \frac{\sin \theta \cos \theta + 1}{\cos \theta}$$

There are numerous identities involving the trigonometric functions. Recall that an **identity** is an equation that is satisfied by all relevant values of the variables concerned. Two examples of identities are $(x - y)(x + y) = x^2 - y^2$ and $x^3 = x^4/x$. The first of these is true no matter what real numbers are used for x and y; the second is true for all real numbers except $x = 0$. For now, in the context of right-triangle trigonometry, we'll consider only a few of the most basic identities.

PROPERTY SUMMARY BASIC RIGHT-TRIANGLE IDENTITIES

IDENTITY	EXAMPLES
1. $\sin^2 \theta + \cos^2 \theta = 1$	1. $\sin^2 30° + \cos^2 30° = 1$ $\sin^2 19° + \cos^2 19° = 1$
2. $\dfrac{\sin \theta}{\cos \theta} = \tan \theta$	2. $\dfrac{\sin 60°}{\cos 60°} = \tan 60°$
3. $\sin(90° - \theta) = \cos \theta$; $\cos(90° - \theta) = \sin \theta$	3. $\sin 70° = \cos 20°$; $\sin 20° = \cos 70°$
4. $\sec \theta = \dfrac{1}{\cos \theta}$; $\csc \theta = \dfrac{1}{\sin \theta}$; $\cot \theta = \dfrac{1}{\tan \theta}$	4. $\cot 40° = \dfrac{1}{\tan 40°}$

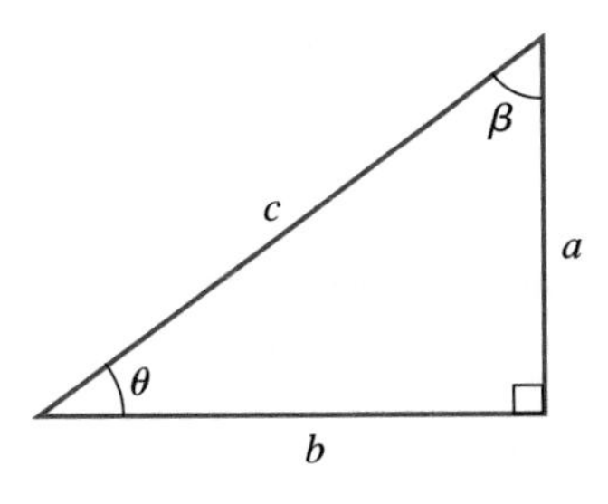

FIGURE 1

The fourth set of identities in the box follows directly from the definitions on page 359. For the proofs of the other identities, we refer to Figure 1.

PROOF THAT $\sin^2\theta + \cos^2\theta = 1$. Looking at Figure 1, we have $\sin\theta = a/c$ and $\cos\theta = b/c$. Thus

$$\begin{aligned}\sin^2\theta + \cos^2\theta &= \left(\frac{a}{c}\right)^2 + \left(\frac{b}{c}\right)^2 = \frac{a^2}{c^2} + \frac{b^2}{c^2} = \frac{a^2+b^2}{c^2}\\ &= \frac{c^2}{c^2} \qquad \text{using } a^2+b^2=c^2\text{, the Pythagorean theorem}\\ &= 1 \qquad \text{as required}\end{aligned}$$

PROOF THAT $(\sin\theta)/(\cos\theta) = \tan\theta$. Again with reference to Figure 1, we have, by definition, $\sin\theta = a/c$, $\cos\theta = b/c$ and $\tan\theta = a/b$. Therefore

$$\begin{aligned}\frac{\sin\theta}{\cos\theta} &= \frac{a/c}{b/c} = \frac{a}{c} \times \frac{c}{b}\\ &= \frac{a}{b} = \tan\theta \qquad \text{as required}\end{aligned}$$

PROOF THAT $\sin(90° - \theta) = \cos\theta$. First of all, since the sum of the angles in any triangle is 180°, we have

$$\begin{aligned}\theta + \beta + 90° &= 180°\\ \beta &= 90° - \theta\end{aligned}$$

Then $\sin(90° - \theta) = \sin\beta = b/c$. But also (by definition) $\cos\theta = b/c$. Thus

$$\sin(90° - \theta) = \cos\theta$$

since both expressions equal b/c. This is what we wanted to prove.

The proof that $\cos(90° - \theta) = \sin\theta$ is entirely similar and we shall omit it. We can conveniently summarize these last two results by recalling the notion of complementary angles. Two acute angles are said to be **complementary** provided their sum is 90°. Thus, the two angles θ and $90° - \theta$ are complementary. In view of this, we can restate the last two results as follows:

> If two angles are complementary, then the sine of (either) one equals the cosine of the other.

Incidentally, this result gives us an insight into the origin of the term *cosine:* it is a shortened form of the phrase *complement's sine.*

EXAMPLE 4 Suppose that B is an acute angle and $\cos B = \frac{2}{5}$. Find $\sin B$ and $\tan B$.

Solution

We'll show two methods. The first uses the identities $\sin^2 B + \cos^2 B = 1$ and $(\sin B)/(\cos B) = \tan B$. The second makes direct use of the Pythagorean theorem.

FIRST METHOD Replace $\cos B$ with $\frac{2}{5}$ in the identity $\sin^2 B + \cos^2 B = 1$. This yields

$$\sin^2 B + \left(\frac{2}{5}\right)^2 = 1$$
$$\sin^2 B = 1 - \frac{4}{25} = \frac{21}{25}$$
$$\sin B = \sqrt{\frac{21}{25}} = \frac{\sqrt{21}}{5}$$

Notice that we've chosen the positive square root here. This is because the values of the trigonometric functions of an acute angle are by definition positive. **CAUTION** In a later section, we'll extend the definitions of the trigonometric functions to include angles that are not acute. In those cases, some of the trigonometric values will not be positive; we will then need to pay more attention to the matter of signs.

For $\tan B$, we have

$$\tan B = \frac{\sin B}{\cos B} = \frac{\sqrt{21}/5}{2/5} = \frac{\sqrt{21}}{2}$$

ALTERNATE METHOD Since $\cos B = \frac{2}{5} =$ adjacent/hypotenuse, we can work with a right triangle labeled as in Figure 2. Using the Pythagorean theorem, we have $2^2 + x^2 = 5^2$, from which it follows that $x = \sqrt{21}$. Consequently,

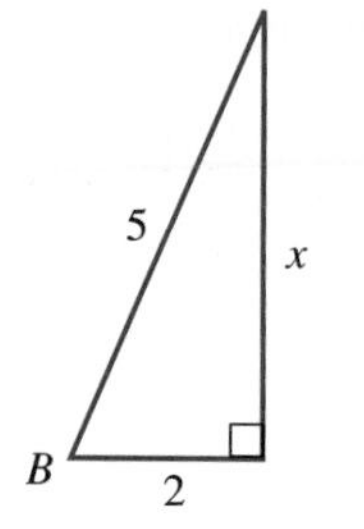

FIGURE 2

$$\sin B = \frac{\text{opposite}}{\text{hypotenuse}} = \frac{x}{5} = \frac{\sqrt{21}}{5}$$

and

$$\tan B = \frac{\text{opposite}}{\text{adjacent}} = \frac{x}{2} = \frac{\sqrt{21}}{2} \qquad \text{as obtained previously}$$

One of the many uses for the identities in this section is in simplifying trigonometric expressions. In the example that follows, we use the identity $\sin^2 A + \cos^2 A = 1$.

EXAMPLE 5 Combine and simplify: $\dfrac{\sin A}{\cos A} + \dfrac{\cos A}{\sin A}$.

Solution The common denominator is $\cos A \sin A$. Therefore we have

$$\begin{aligned}\frac{\sin A}{\cos A} + \frac{\cos A}{\sin A} &= \frac{\sin A}{\cos A} \cdot \frac{\sin A}{\sin A} + \frac{\cos A}{\sin A} \cdot \frac{\cos A}{\cos A} \\ &= \frac{\sin^2 A}{\cos A \sin A} + \frac{\cos^2 A}{\sin A \cos A} \\ &= \frac{\sin^2 A + \cos^2 A}{\cos A \sin A} \\ &= \frac{1}{\cos A \sin A}\end{aligned}$$

This is the required result. If we wish, we can write this answer in an alternative form that doesn't involve fractions:

$$\frac{1}{\cos A \sin A} = \frac{1}{\cos A} \cdot \frac{1}{\sin A} = \sec A \csc A$$

One of the most useful techniques for simplifying a trigonometric expression is first to rewrite it in terms of sines and cosines and then to carry out the usual algebraic simplifications. This is demonstrated in Examples 6 through 8.

EXAMPLE 6 Simplify the expression $\csc A \tan A \cos^2 A$.

Solution We use the basic identities to express everything in terms of sines and cosines.

$$\begin{aligned}
\csc A \tan A \cos^2 A &= \frac{1}{\cancel{\sin A}_{\,1}} \cdot \frac{\cancel{\sin A}^{\,1}}{\cos A} \cdot \cos^2 A \\
&= \frac{\cos^2 A}{\cos A} = \cos A
\end{aligned}$$

EXAMPLE 7 Simplify the expression $\dfrac{1 - \tan^2 B}{1 + \tan^2 B}$.

Solution

$$\frac{1 - \tan^2 B}{1 + \tan^2 B} = \frac{1 - \dfrac{\sin^2 B}{\cos^2 B}}{1 + \dfrac{\sin^2 B}{\cos^2 B}}$$

Since $\tan B = (\sin B)/(\cos B)$, it follows that $\tan^2 B = (\sin^2 B)/(\cos^2 B)$.

$$= \frac{(\cos^2 B)\left(1 - \dfrac{\sin^2 B}{\cos^2 B}\right)}{(\cos^2 B)\left(1 + \dfrac{\sin^2 B}{\cos^2 B}\right)}$$

multiplying both the numerator and the denominator by $\cos^2 B$

$$= \frac{\cos^2 B - \sin^2 B}{\cos^2 B + \sin^2 B}$$

$$= \frac{\cos^2 B - \sin^2 B}{1} = \cos^2 B - \sin^2 B$$

EXAMPLE 8 Express in terms of sines and cosines and simplify: $\dfrac{\tan(90° - A)}{\cos A \sin A}$.

Solution

$$\frac{\tan(90^\circ - A)}{\cos A \sin A} = \frac{\dfrac{\sin(90^\circ - A)}{\cos(90^\circ - A)}}{\cos A \sin A} \quad \text{applying the identity } \tan\theta = (\sin\theta)/(\cos\theta)$$

$$= \frac{\dfrac{\cos A}{\sin A}}{\cos A \sin A} \quad \text{(Why?)}$$

$$= \frac{\cos A}{\sin A} \cdot \frac{1}{\cos A \sin A} = \frac{1}{\sin^2 A}$$

The given expression is therefore equal to $1/\sin^2 A$. We can also write this as $\csc^2 A$.

EXERCISE SET 6.2

A Do 1, 3, 5, 7, 11, 13, 15, 19, 29, 35, 41, 49, 55

In Exercises 1–12, carry out the indicated operations.

1. **(a)** $-SC + 12SC$
(b) $-\sin\theta\cos\theta + 12\sin\theta\cos\theta$

2. **(a)** $10SC + 4SC - 16SC$
(b) $10\sin\theta\cos\theta + 4\sin\theta\cos\theta - 16\sin\theta\cos\theta$

3. **(a)** $4C^3S - 12C^3S$
(b) $4\cos^3\theta\sin\theta - 12\cos^3\theta\sin\theta$

4. **(a)** $-C^2S^2 + (2SC)^2$
(b) $-\cos^2\theta\sin^2\theta + (2\sin\theta\cos\theta)^2$

5. **(a)** $(1+T)^2$
(b) $(1+\tan\theta)^2$

6. **(a)** $(3-2T)^2$
(b) $(3-2\tan\theta)^2$

7. **(a)** $(T+3)(T-2)$
(b) $(\tan\theta+3)(\tan\theta-2)$

8. **(a)** $(S^2-3)(S^2+3)$
(b) $(\sec^2\theta-3)(\sec^2\theta+3)$

9. **(a)** $\dfrac{S-C}{C-S}$
(b) $\dfrac{\sin\theta-\cos\theta}{\cos\theta-\sin\theta}$

10. **(a)** $\dfrac{5-2T}{2T-5}$
(b) $\dfrac{5-2\tan\theta}{2\tan\theta-5}$

11. **(a)** $C+\dfrac{2}{S}$
(b) $\cos A+\dfrac{2}{\sin A}$

12. **(a)** $\dfrac{1}{S}-\dfrac{3}{C}$
(b) $\dfrac{1}{\sin A}-\dfrac{3}{\cos A}$

In Exercises 13–18, factor each expression.

13. **(a)** T^2+8T-9
(b) $\tan^2\beta+8\tan\beta-9$

14. **(a)** $3S^2+2S-8$
(b) $3\sec^2\beta+2\sec\beta-8$

15. **(a)** $4C^2-1$
(b) $4\cos^2 B-1$

16. **(a)** $16S^3-9S^2$
(b) $16\sin^3 B-9\sin^2 B$

17. **(a)** $9S^2T^3+6ST^2$
(b) $9\sec^2 B\tan^3 B+6\sec B\tan^2 B$

18. **(a)** $5C^2c^2-15Cc$
(b) $5\csc^2 B\cot^2 B-15\csc B\cot B$

In Exercises 19–30, use the given information to determine the values of the remaining five trigonometric functions. (Assume that all of the angles are acute.) When radicals appear in a denominator, rationalize the denominator. In Exercises 19–24, use either of the methods shown in Example 4. In Exercises 25–30, use the second method shown in Example 4.

19. $\sin\theta = 3/4$
20. $\sin\theta = 2/5$
21. $\cos\beta = \sqrt{3}/5$
22. $\cos\beta = \sqrt{7}/3$
23. $\sin A = 5/13$
24. $\cos A = 8/17$
25. $\tan B = 4/3$
26. $\tan B = 5$
27. $\sec C = 3/2$
28. $\csc C = \sqrt{5}/2$
29. $\cot\alpha = \sqrt{3}/3$
30. $\cot\alpha = \sqrt{3}/2$

In Exercises 31–34, use a calculator to determine the values of the remaining five trigonometric functions. (In each case, θ denotes an acute angle.) Round the final answers to three decimal places.

31. $\cos\theta = 0.4626$
32. $\sin\theta = 0.5917$
33. $\tan\theta = 1.1998$
34. $\sec\theta = 2.3283$

In Exercises 35–56, simplify each expression.

35. $\dfrac{\sin^2 A - \cos^2 A}{\sin A - \cos A}$ *Hint:* Factor the numerator.

36. $\dfrac{\sin^4 A - \cos^4 A}{\cos A - \sin A}$

37. $\sin^2 \theta \cos \theta \csc^3 \theta \sec \theta$

38. $\sin \theta \csc \theta \tan \theta$

39. $\cot B \sin^2 B \cot B$

40. $\dfrac{3 \sin \theta + 6}{\sin^2 \theta - 4}$

41. $\dfrac{\cos^2 A + \cos A - 12}{\cos A - 3}$

42. $\dfrac{\cos A - 2 \sin A \cos A}{\cos^2 A - \sin^2 A + \sin A - 1}$

43. $\dfrac{\tan \theta}{\sec \theta - 1} + \dfrac{\tan \theta}{\sec \theta + 1}$

44. $\cot \theta + \dfrac{1 - 2 \cos^2 \theta}{\sin \theta \cos \theta}$

45. $\sec A \csc A - \tan A - \cot A$

46. $(\sec A + \tan A)(\sec A - \tan A)$

47. $\dfrac{\cot^2 \theta}{\csc^2 \theta} + \dfrac{\tan^2 \theta}{\sec^2 \theta}$

48. $\dfrac{\tan \theta + \tan \theta \sin \theta - \cos \theta \sin \theta}{\sin \theta \tan \theta}$

49. $\dfrac{\cos(90° - \theta)}{\cos \theta}$

50. $\sin^2(90° - \beta) + \cos^2(90° - \beta)$

51. $\dfrac{\cos^2(90° - A)}{\sin^2(90° - A)} - \dfrac{1}{\cos^2 A}$

52. $\dfrac{\cos(90° - A)}{\csc A} + \dfrac{\sin(90° - A)}{\sec A}$

53. $1 - \sin(90° - \theta)\cos \theta$

54. $\dfrac{\cos A \tan A}{\tan(90° - A)} - \dfrac{1}{\sin(90° - A)}$

55. $\dfrac{\dfrac{\cos \theta + 1}{\cos \theta} + 1}{\dfrac{\cos \theta - 1}{\cos \theta} - 1}$

56. $\dfrac{\dfrac{\cot \beta + 1}{\cot \beta} + 1}{\dfrac{\cot \beta - 1}{\cot \beta} - 1}$

B

57. Suppose that

$$A \sin \theta + \cos \theta = 1 \quad \text{and} \quad B \sin \theta - \cos \theta = 1$$

Show that $AB = 1$. *Hint:* Solve the first equation for A, the second for B, and then compute AB.

58. If $\sin \alpha + \cos \alpha = a$ and $\sin \alpha - \cos \alpha = b$, show that

$$\tan \alpha = \frac{a + b}{a - b}$$

59. If $a \sin^2 \theta + b \cos^2 \theta = 1$, show that

$$\sin^2 \theta = \frac{1 - b}{a - b} \quad \text{and} \quad \tan^2 \theta = \frac{b - 1}{1 - a}$$

60. This exercise shows how to obtain radical expressions for sin 18° and cos 18°, using the following figure.

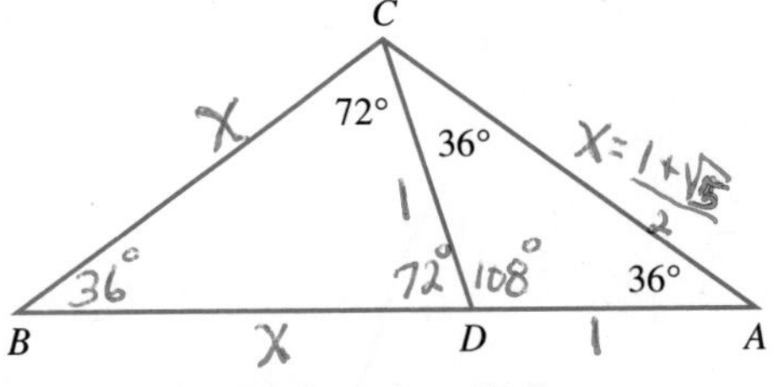

(a) Find $\angle B$, $\angle BDC$, and $\angle ADC$.

(b) Why does $AC = BC = BD$?

For the rest of this problem; assume $AD = 1$.

(c) Why does $CD = 1$?

(d) Let x denote the common lengths AC, BC, and BD. Use similar triangles to deduce that $x/(1 + x) = 1/x$. Then show that $x = (1 + \sqrt{5})/2$.

(e) In $\triangle BDC$, draw an altitude from B to $\overline{DC}$, meeting $\overline{DC}$ at F. Use right triangle BFC to conclude that $\sin 18° = 1/(1 + \sqrt{5})$.

(f) Rationalize the denominator in part (e) to obtain $\sin 18° = \frac{1}{4}(\sqrt{5} - 1)$.

(g) Use the identity $\sin^2 \theta + \cos^2 \theta = 1$, along with part (f), to show that

$$\cos 18° = \tfrac{1}{4}\sqrt{10 + 2\sqrt{5}}$$

(h) Use your calculator to check the results in parts (f) and (g).

61. Suppose that β is an acute angle and

$$\sin \beta = \frac{m^2 - n^2}{m^2 + n^2} \quad (m > n > 0)$$

Show that

$$\cos \beta = \frac{2mn}{m^2 + n^2} \quad \text{and} \quad \tan \beta = \frac{m^2 - n^2}{2mn}$$

6.3 RIGHT-TRIANGLE APPLICATIONS

We began this chapter in Section 6.1 by defining the six trigonometric functions of an acute angle θ. Then, in Section 6.2, we looked at some of the algebra that is involved in working with the trigonometric functions. Now we are ready to apply these ideas in solving some basic problems involving right triangles.

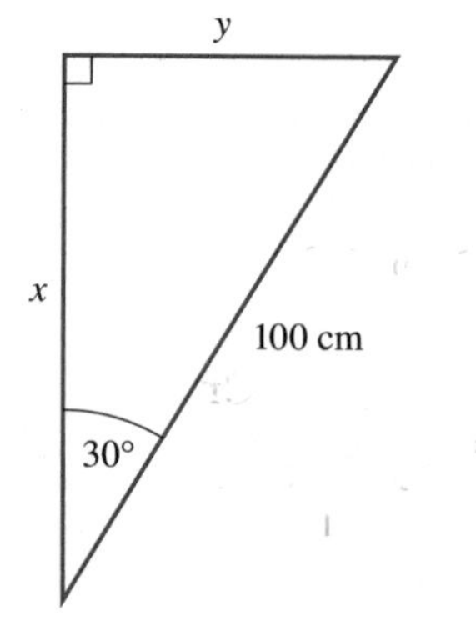

FIGURE 1

EXAMPLE 1 Use one of the trigonometric functions to find x in Figure 1.

Solution Relative to the given 30° angle, x is the adjacent side. The length of the hypotenuse is 100 cm. Since the adjacent side and the hypotenuse are involved, we use the cosine function here:

$$\cos 30^\circ = \frac{\text{adjacent}}{\text{hypotenuse}} = \frac{x}{100}$$

Consequently,

$$x = 100 \cos 30^\circ = (100) \cdot \left(\sqrt{3}/2\right) = 50\sqrt{3} \text{ cm}$$

This is the result we are looking for.

We used the cosine function in Example 1 because the adjacent side and the hypotenuse were involved. We could instead use the secant. In that case, again with reference to Figure 1, the calculations look like this:

$$\begin{aligned} \sec 30^\circ &= \frac{\text{hypotenuse}}{\text{adjacent}} = \frac{100}{x} \\ \frac{2}{\sqrt{3}} &= \frac{100}{x} \\ 2x &= 100\sqrt{3} \\ x &= \frac{100\sqrt{3}}{2} = 50\sqrt{3} \text{ cm} \qquad \text{as obtained previously} \end{aligned}$$

EXAMPLE 2 Find y in Figure 1.

Solution As you can see from Figure 1, the side of length y is opposite the 30° angle. Furthermore, we are given the length of the hypotenuse. Since the opposite side and the hypotenuse are involved, we use the sine function. This yields

$$\sin 30^\circ = \frac{\text{opposite}}{\text{hypotenuse}} = \frac{y}{100}$$

and therefore

$$y = 100 \sin 30^\circ = (100) \cdot (1/2) = 50 \text{ cm}$$

This is the required answer. Actually, we could have obtained this particular result much faster by recalling that in the 30°-60° right triangle, the side opposite the 30° angle, namely y, is half of the hypotenuse. That is, $y = 100/2 = 50$ cm, as obtained previously.

EXAMPLE 3 A ladder that is leaning against the side of a building forms an angle of 50° with the ground. If the foot of the ladder is 12 ft from the base of the building, how far up the side of the building does the ladder reach? See Figure 2.

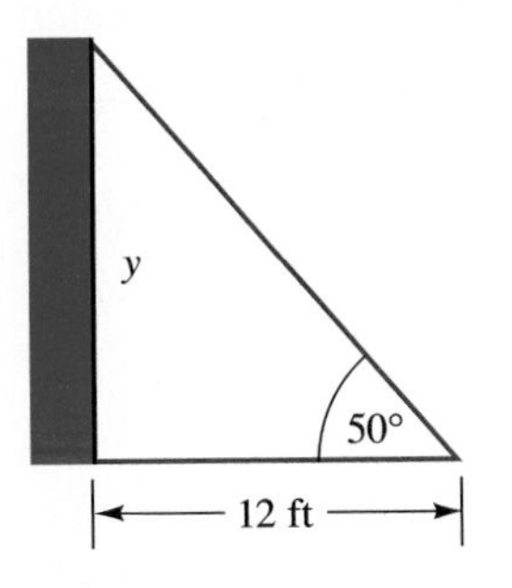

FIGURE 2

Solution In Figure 2 we have used y to denote the required distance. Notice that y is opposite the 50° angle, while the given side is adjacent to that angle. Since the opposite and adjacent sides are involved, we'll use the tangent function. (The cotangent function could also be used.) We have

$$\tan 50° = \frac{y}{12} \qquad \text{and therefore} \qquad y = 12 \tan 50°$$

Without the use of a calculator or tables, this is our final answer. On the other hand, using a calculator we find that $y = 14$ ft, to the nearest foot.

In the next example we derive a useful formula for the area of a triangle. The formula can be used when we are given two sides and the included angle but not the height.

EXAMPLE 4 Show that the area of the triangle in Figure 3 is given by

$$A = \tfrac{1}{2}ab \sin \theta$$

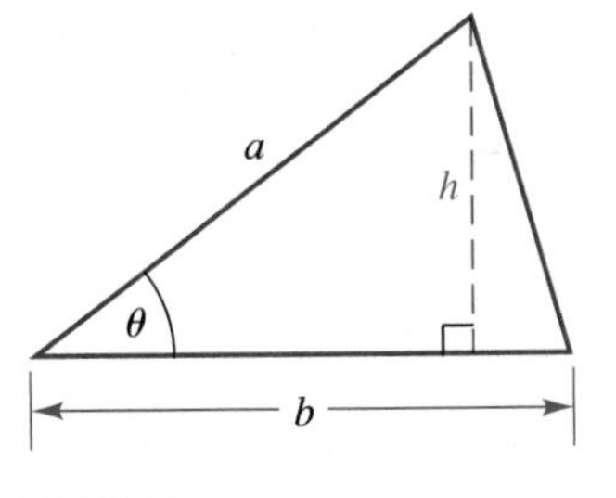

FIGURE 3

Solution In Figure 3, if h denotes the length of the altitude, then we have

$$\sin \theta = \frac{h}{a} \qquad \text{and therefore} \qquad h = a \sin \theta$$

This value for h can now be used in the usual formula for the area of a triangle:

$$A = \tfrac{1}{2}bh = \tfrac{1}{2}b(a \sin \theta) = \tfrac{1}{2}ab \sin \theta$$

The formula that we just derived in Example 4 is worth remembering. In words, the formula states that *the area of a triangle is equal to half the product of the lengths of two of the sides times the sine of the included acute angle.* In Section 6.4, we will see that the formula remains valid even when the included angle is not an acute angle.

EXAMPLE 5 Find the area of the triangle in Figure 4.

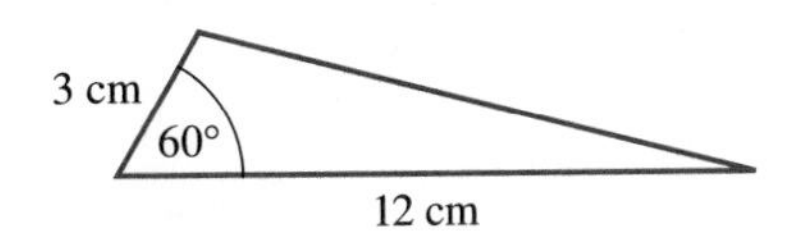

FIGURE 4

Solution In the formula $A = \frac{1}{2}ab \sin \theta$, we let $a = 3$ cm, $b = 12$ cm, and $\theta = 60°$. Then

$$\begin{aligned} A &= \frac{1}{2}(3)(12) \sin 60° = (18)\frac{\sqrt{3}}{2} \\ &= 9\sqrt{3} \text{ cm}^2 \end{aligned}$$

This is the required area.

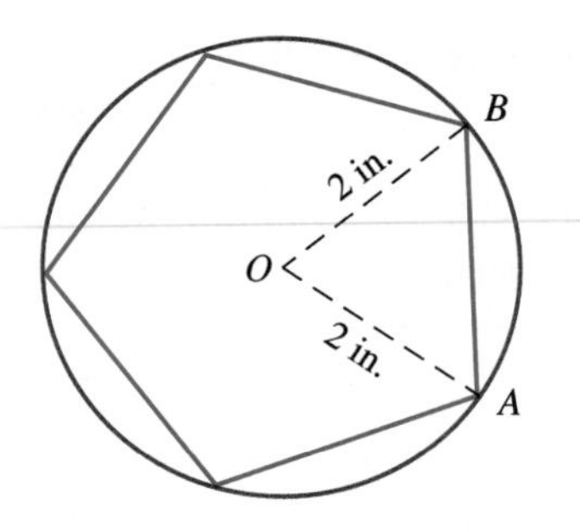

FIGURE 5

EXAMPLE 6 Figure 5 shows a regular pentagon inscribed in a circle of radius 2 in. (*Regular* means that all of the sides are equal and all of the angles are equal.) Find the area of the pentagon.

Solution The idea here is first to find the area of triangle BOA by using the area formula from Example 4. Then, since the pentagon is composed of five such identical triangles, the area of the pentagon will be five times the area of triangle BOA. We will make use of the result from geometry that, in a regular n-sided polygon, the central angle is $360°/n$. In our case, we therefore have

$$\angle BOA = \frac{360°}{5} = 72°$$

We can now find the area of triangle BOA:

$$\begin{aligned} \text{area} &= \tfrac{1}{2}ab \sin \theta \\ &= \tfrac{1}{2}(2)(2)\sin 72° = 2 \sin 72° \text{ in}^2 \end{aligned}$$

The area of the pentagon is five times this, or $10 \sin 72°$ in.^{2} (Using a calculator, this is about 9.51 in.^{2})

Now we introduce some terminology that will be used in the next two examples. Suppose that a surveyor sights an object at a point above the horizontal, as indicated in Figure 6(a). Then the angle between the line of sight and the horizontal is called the **angle of elevation.** The **angle of depression** is similarly defined for an object below the horizontal, as shown in Figure 6(b).

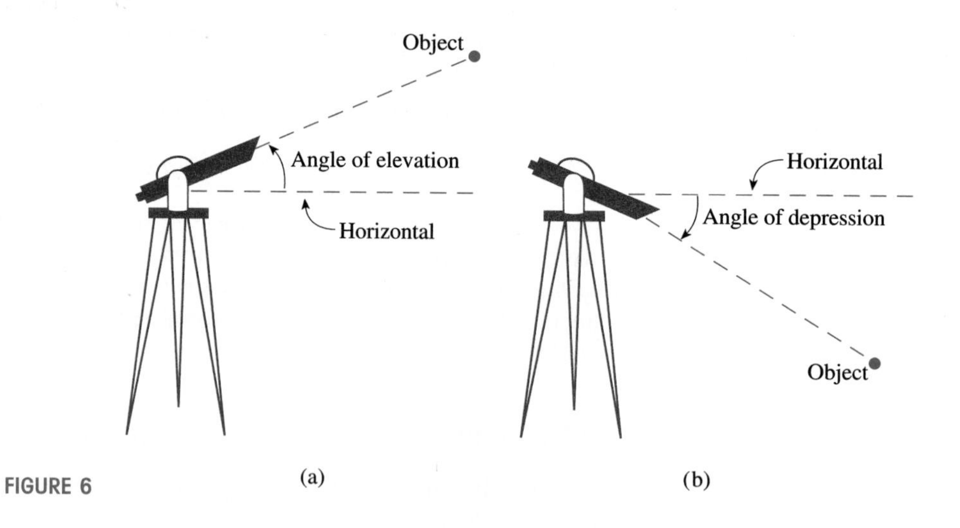

FIGURE 6

EXAMPLE 7 A helicopter hovers 800 ft directly above a small island that is off the California coast. From the helicopter, the pilot takes a sighting to a point P directly ashore on the mainland, at the water's edge. If the angle of depression is 35°, how far off the coast is the island? See Figure 7.

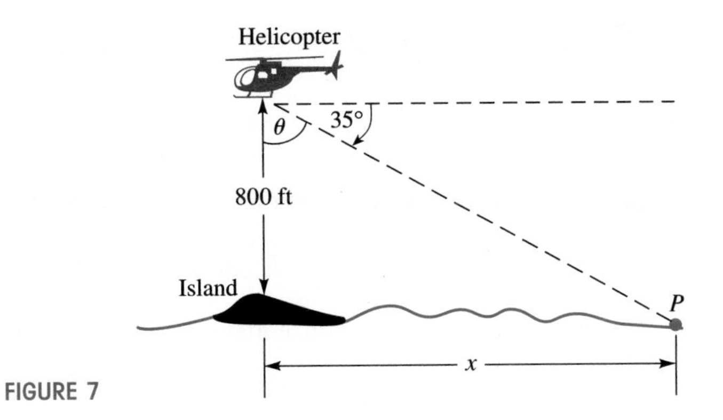

FIGURE 7

Solution Let x denote the distance from the island to the mainland. Then, as you can see from Figure 7, we have $\theta + 35° = 90°$, from which it follows that $\theta = 55°$. Now we can write

$$\tan 55° = \frac{x}{800}$$

or

$$x = 800 \tan 55° \approx 1150 \text{ ft}$$ using a calculator and rounding to the nearest 50 feet

EXAMPLE 8 Two satellite tracking stations, located at points A and B in California's Mojave Desert, are 200 miles apart. At a prearranged time, both stations measure the angle of elevation of a satellite as it crosses the vertical plane containing A and B. (See Figure 8.) If the angles of elevation from A and from B are α and β, respectively, express the altitude h of the satellite in terms of α and β.

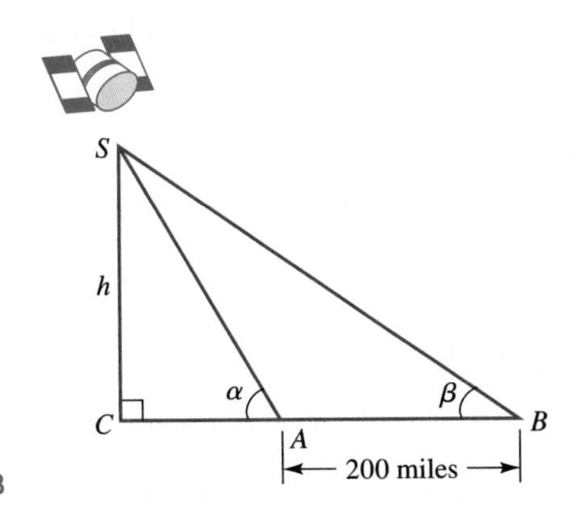

FIGURE 8

Solution We want to express the length $h = SC$ in Figure 8 in terms of the angles α and β. Note that $\overline{SC}$ is a side of both of the right triangles SCA and SCB.

Working first in the right triangle SCB, we have

$$\cot\beta = \frac{CA + 200}{h}$$

or

$$h\cot\beta = CA + 200 \qquad (1)$$

We can eliminate CA from equation (1) as follows. Looking at right triangle SCA, we have

$$\cot\alpha = \frac{CA}{h} \qquad \text{and thus} \qquad CA = h\cot\alpha$$

Using this last equation to substitute for CA in equation (1), we obtain

$$\begin{aligned} h\cot\beta &= h\cot\alpha + 200 \\ h\cot\beta - h\cot\alpha &= 200 \\ h(\cot\beta - \cot\alpha) &= 200 \\ h &= \frac{200}{\cot\beta - \cot\alpha} \text{ miles} \qquad \text{as required} \end{aligned}$$

Also, $h = \dfrac{-200\tan\alpha\tan\beta}{\tan\beta - \tan\alpha}$

EXAMPLE 9 The arc shown in Figure 9 is a portion of the unit circle, $x^2 + y^2 = 1$. Express the following quantities in terms of θ.

(a) OA **(b)** AB **(c)** OC **(d)** The area of $\triangle OAC$

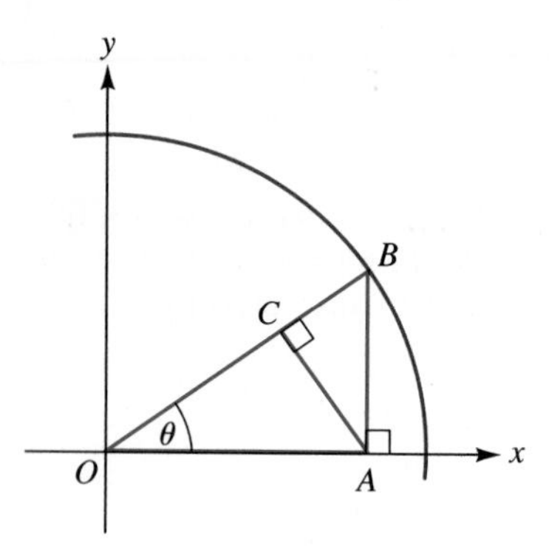

FIGURE 9

Solution

(a) In right triangle OAB:

$$\cos\theta = \frac{OA}{OB} = \frac{OA}{1}$$

$$OA = \cos\theta$$

(b) In right triangle OAB:

$$\sin\theta = \frac{AB}{OB} = \frac{AB}{1}$$

$$AB = \sin\theta$$

(c) In right triangle OAC, we have $\cos\theta = OC/OA$, and therefore

$$OC = OA\cos\theta = (\cos\theta)(\cos\theta) = \cos^2\theta$$

(d) area $\triangle OAC = \frac{1}{2}(OA)(OC)(\sin\theta)$ using the formula area $= \frac{1}{2}ab\sin\theta$

$= \frac{1}{2}(\cos\theta)(\cos^2\theta)(\sin\theta)$ using the results from parts (a) and (c)

$= \frac{1}{2}\cos^3\theta\sin\theta$

Do 1, 7, 10, 13, 15, 19, 25, 33

EXERCISE SET 6.3

A

For Exercises 1–6, refer to the following figure. (However, each problem is independent of the others.)

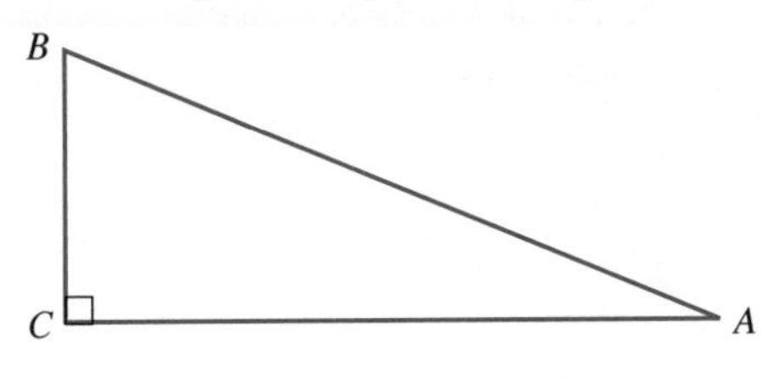

1. If $\angle A = 30°$ and $AB = 60$ cm, find AC and BC.
2. If $\angle A = 60°$ and $AB = 12$ cm, find AC and BC.
3. If $\angle B = 60°$ and $AC = 16$ cm, find BC and AB.
4. If $\angle B = 45°$ and $AC = 9$ cm, find BC and AB.
5. If $\angle B = 50°$ and $AB = 15$ cm, find BC and AC. (Round your answers to one decimal place.)
6. If $\angle A = 25°$ and $AC = 100$ cm, find BC and AB. (Round your answers to one decimal place.)
7. A ladder 18 ft long leans against a building. The ladder forms an angle of 60° with the ground.
   (a) How high up the side of the building does the ladder reach? [Here and in part (b), give two forms for your answers: one with radicals and one (using a calculator) with decimals, rounded to two places.]
   (b) Find the horizontal distance from the foot of the ladder to the base of the building.
8. From a point level with and 1000 ft away from the base of the Washington Monument, the angle of elevation to the top of the monument is 29.05°. Determine the height of the monument to the nearest half foot.
9. Refer to the following figure. At certain times, the planets Earth and Mercury line up in such a way that $\angle EMS$ is a right angle. At such times, $\angle SEM$ is found to be 21.16°. Use this information to estimate the distance MS of Mercury from the Sun. Assume that the distance from the Earth to the Sun is 93 million miles. (Round your answer to the nearest million miles. Because Mercury's orbit is not exactly circular, the actual distance of Mercury from the Sun varies from about 28 million miles to 43 million miles.)

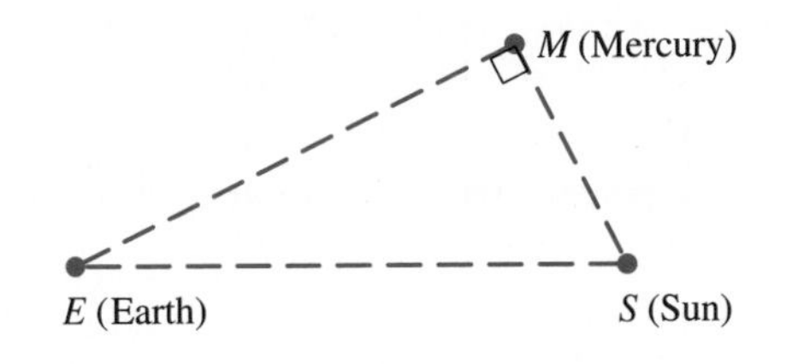

10. Determine the distance AB across the lake shown in the figure, using the following data: $AC = 400$ m, $\angle C = 90°$, and $\angle CAB = 40°$. Round the answer to the nearest meter.

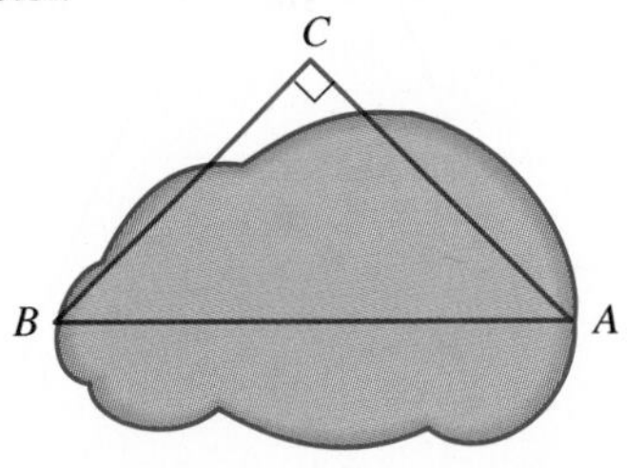

11. A building contractor wants to put a fence around the perimeter of a flat lot that has the shape of a right triangle. One angle of the triangle is 41.4°, and the length of the hypotenuse is 58.5 m. Find the length of fencing required. Round the answer to one decimal place.
12. Suppose that the contractor in Exercise 11 reviews his notes and finds that it is not the hypotenuse that is 58.5 m but rather the side opposite the 41.4° angle. Find the length of fencing required in this case. Again, round the answer to one decimal place.

For Exercises 13 and 14, refer to the following diagram for the roof of a house. In the figure, x is the length of a rafter measured from the top of a wall to the top of the roof; θ is the acute angle between a rafter and the horizontal; and h is the vertical distance from the top of the wall to the top of the roof.

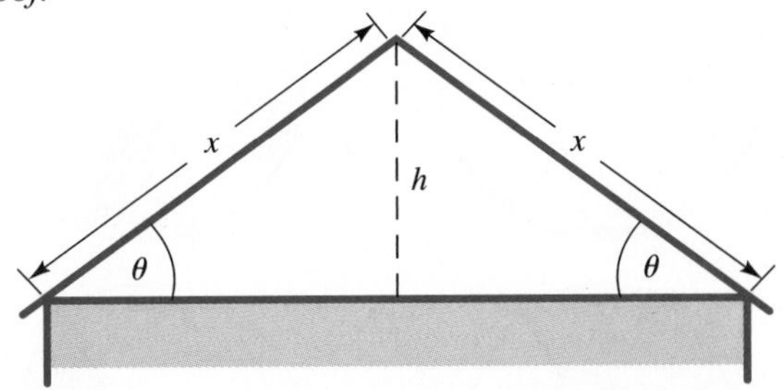

13. Suppose that $\theta = 39.4°$ and $x = 43.0$ ft.
   (a) Determine h. Round the answer to one decimal place.
   (b) The *gable* is the triangular region bounded by the rafters and the attic ceiling. Find the area of the gable. Round the final answer to one decimal place.
14. Suppose that $\theta = 34°$ and $h = 36.5$ ft.
   (a) Determine x. Round the answer to one decimal place.

(b) Find the area of the gable. Round the final answer to one decimal place. [See Exercise 13(b) for the definition of gable.]

In Exercises 15 and 16, find the area of the triangle. In Exercise 16, use a calculator and round the final answer to two decimal places.

15. **16.**

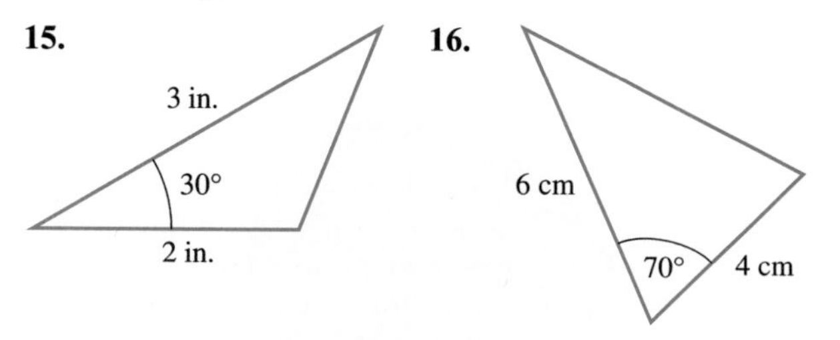

In Exercises 17–22, determine the area of the shaded region, given that the radius of the circle is 1 unit and the inscribed polygon is a regular polygon. Give two forms for each answer: an expression involving radicals or the trigonometric functions; a calculator approximation rounded to three decimal places.

17. **18.** **19.** **20.** **21.** **22.**

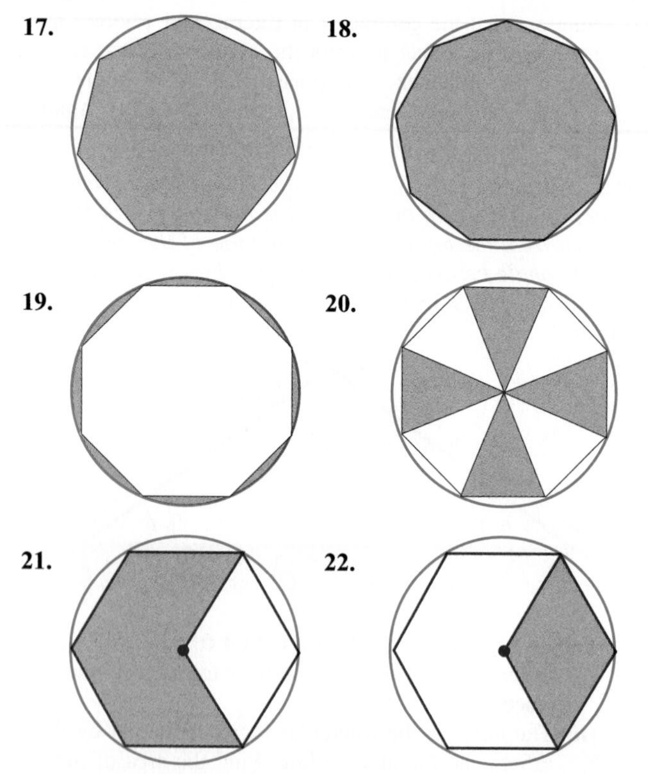

23. Show that the perimeter of the pentagon in Example 6 is $20 \sin 36°$. *Hint:* In Figure 5, draw a perpendicular from O to $\overline{AB}$.

24. In triangle OAB, lengths $OA = OB = 6$ in. and $\angle AOB = 72°$. Find AB. *Hint:* Draw a perpendicular from O to $\overline{AB}$. Round the answer to one decimal place.

25. The accompanying figure shows two ships at points P and Q, which are in the same vertical plane as an airplane at point R. When the height of the airplane is 3500 ft, the angle of depression to P is 48° and that to Q is 25°. Find the distance between the two ships. Round the answer to the nearest 10 feet.

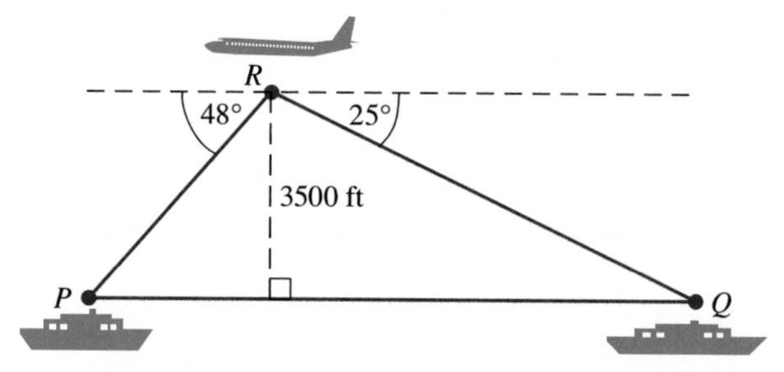

26. An observer in a lighthouse is 66 ft above the surface of the water. The observer sees a ship and finds the angle of depression to be 0.7°. Estimate the distance of the ship from the base of the lighthouse. Round the answer to the nearest 5 feet.

27. From a point on ground level, you measure the angle of elevation to the top of a mountain to be 38°. Then you walk 200 m farther away from the mountain and find that the angle of elevation is now 20°. Find the height of the mountain. Round the answer to the nearest meter.

28. A surveyor stands 30 yd from the base of a building. On top of the building is a vertical radio antenna. Let α denote the angle of elevation when the surveyor sights to the top of the building. Let β denote the angle of elevation when the surveyor sights to the top of the antenna. Express the length of the antenna in terms of the angles α and β.

29. In $\triangle ACD$, you are given $\angle C = 90°$, $\angle A = 60°$, and $AC = 18$ cm. If B is a point on $\overline{CD}$ and $\angle BAC = 45°$, find BD. Express the answer in terms of a radical (rather than using a calculator).

30. The radius of the circle in the following figure is 1 unit. Express the lengths OA, AB, and DC in terms of α.

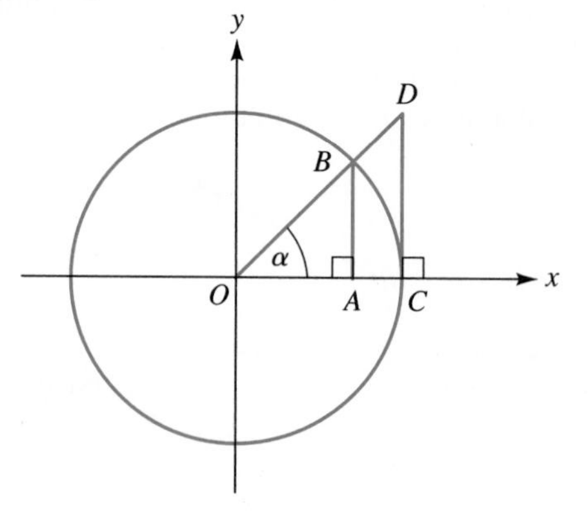

31. The arc in the next figure is a portion of the unit circle, $x^2 + y^2 = 1$.

(a) Express the following angles in terms of θ: $\angle BOA$, $\angle OAB$, $\angle BAP$, $\angle BPA$. (Assume θ is in degrees.)

(b) Express the following lengths in terms of $\sin\theta$ and $\cos\theta$: AO, AP, OB, BP.

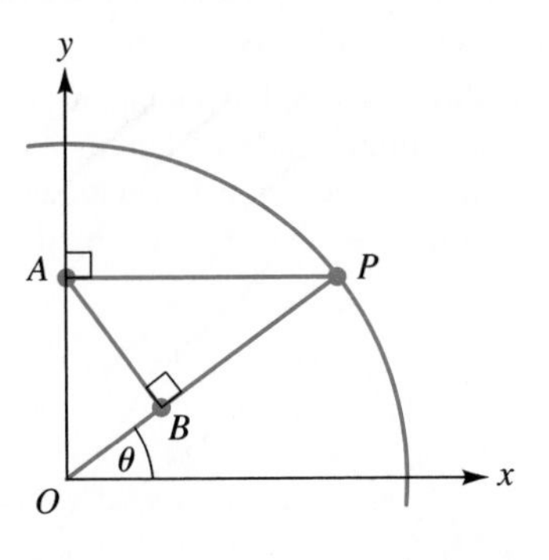

32. In the following figure, suppose that $AD = 1$. Find the length of each of the other line segments in the figure. When radicals appear in an answer, leave the answer in that form, rather than using a calculator.

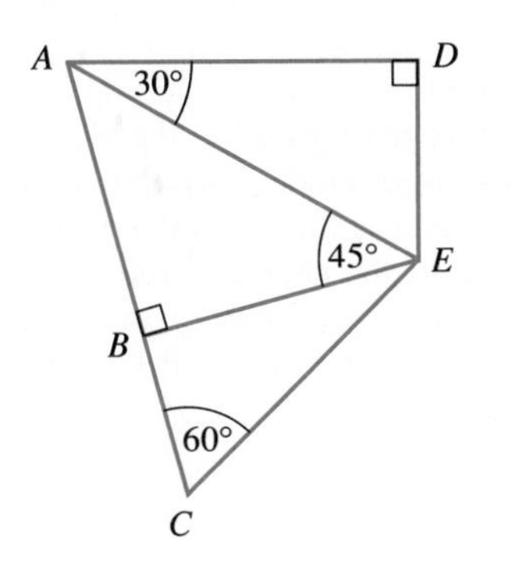

B

33. Refer to the figure. Express each of the following lengths as a function of θ: **(a)** BC; **(b)** AB; **(c)** AC.

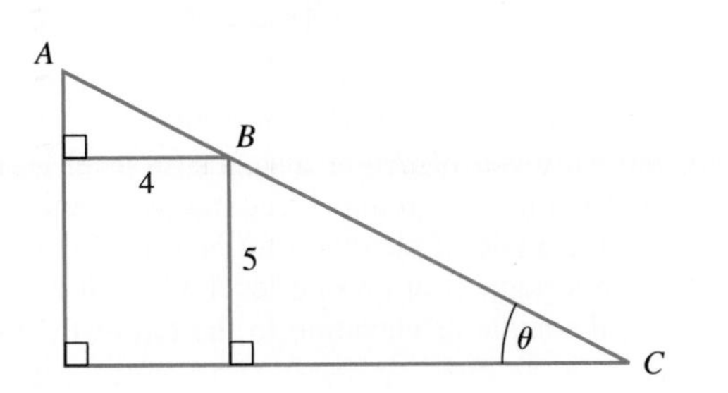

34. In the following figure, $AB = 8$ in. Express x as a function of θ. *Hint:* First work Exercise 33.

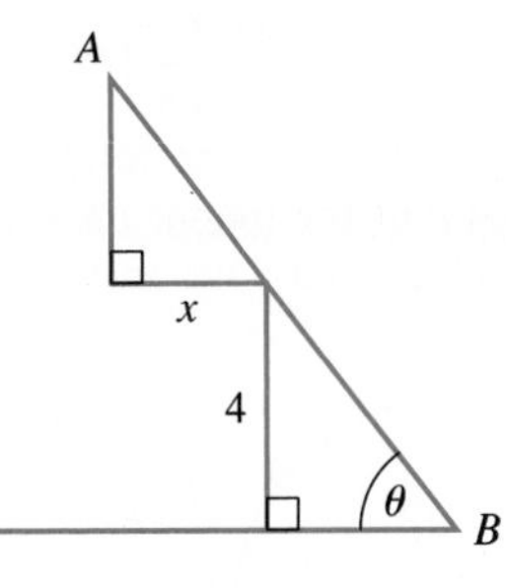

35. In the figure, line segment $\overline{BA}$ is tangent to the unit circle at A. Also, $\overline{CF}$ is tangent to the circle at F. Express the following lengths in terms of θ.

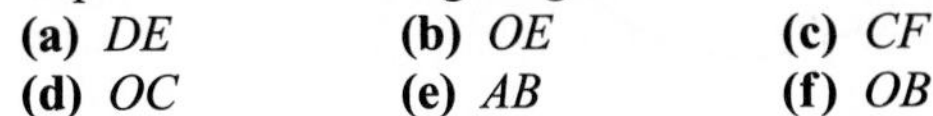

(a) DE	**(b)** OE	**(c)** CF
(d) OC	**(e)** AB	**(f)** OB

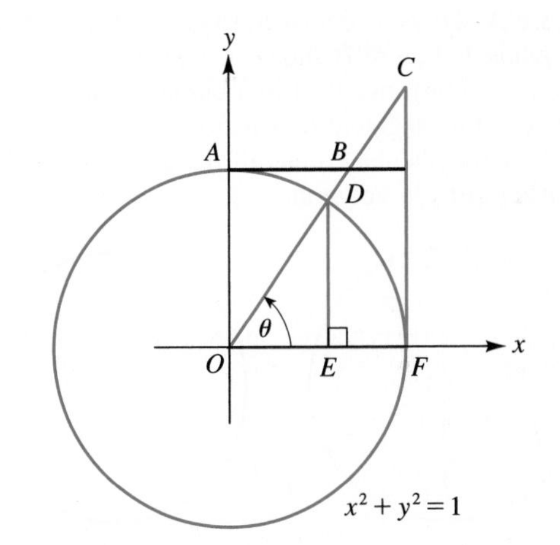

36. At point P on the earth's surface, the moon is observed to be directly overhead, while at the same time at point T, the moon is just visible. See the figure below.

(a) Show that $MP = \dfrac{OT}{\cos\theta} - OP$.

(b) Use a calculator and the following data to estimate the distance MP from the earth to the moon: $\theta = 89.05°$ and $OT = OP = 4000$ miles. Round your answer to the nearest thousand miles. (Because the moon's orbit is not really circular, the actual distance varies from about 216,400 miles to 247,000 miles.)

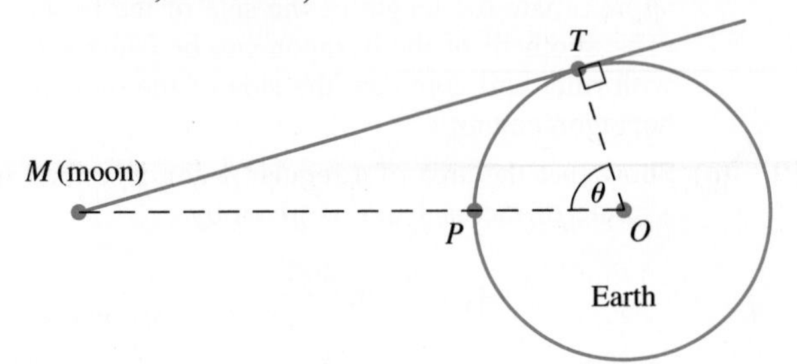

37. Refer to the accompanying figure. Let r denote the radius of the moon.

(a) Show that $r = \left(\dfrac{\sin \theta}{1 - \sin \theta}\right) PS$.

(b) Use a calculator and the following data to estimate the radius r of the moon: $PS = 238{,}857$ miles and $\theta = 0.257°$. Round your answer to the nearest 10 miles.

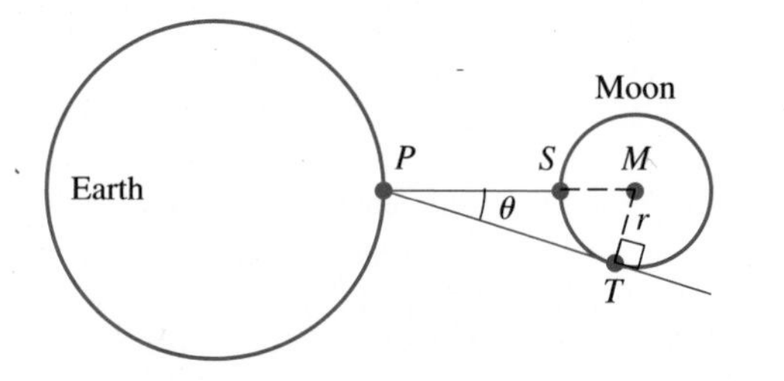

38. Figure A shows a regular hexagon inscribed in a circle of radius 1. Figure B shows a regular heptagon (seven-sided polygon) inscribed in a circle of radius 1. In Figure A, a line segment drawn from the center of the circle perpendicular to one of the sides is called an **apothem** of the polygon.

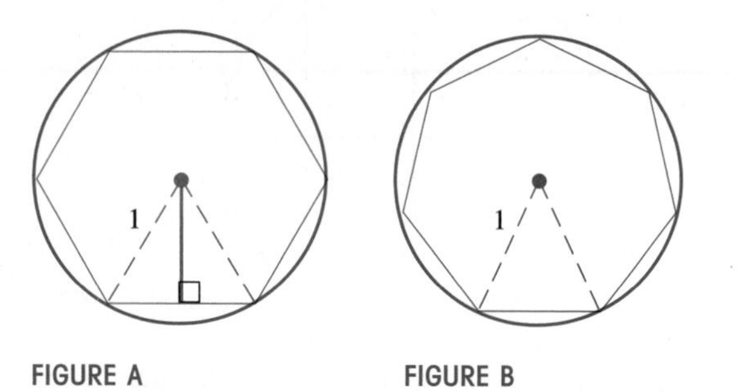

FIGURE A FIGURE B

(a) Show that the length of the apothem in Figure A is $\sqrt{3}/2$.

(b) Show that the length of one side of the heptagon in Figure B is $2 \sin(180°/7)$.

(c) Use a calculator to evaluate the expressions in parts (a) and (b). Round each answer to four decimal places, and note how close the two values are. Approximately two thousand years ago, Heron of Alexandria made use of this coincidence when he used the length of the apothem of the hexagon to approximate the length of the side of the heptagon. (The apothem of the hexagon can be constructed with ruler and compass; the side of the regular heptagon cannot.)

39. **(a)** Show that the area of a regular n-gon inscribed in a circle of radius 1 unit is given by

$$A_n = \frac{n}{2} \sin \frac{360°}{n}$$

(b) Use a calculator and the formula in part (a) to complete the following table.

n	5	10	50	100	1000	5000	10,000
A_n							

(c) Explain why the successive values of A_n in your table get closer and closer to π.

40. The figure shows a regular pentagon and a regular hexagon, with a common side of length 2 cm. Compute the area within the hexagon but outside of the pentagon. Round the answer to two decimal places.

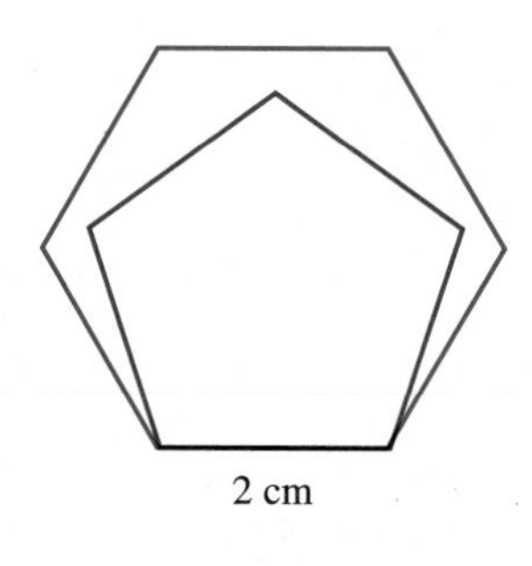

C

41. In the accompanying figure, the smaller circle is tangent to the larger circle. Ray PQ is a common tangent and ray PR passes through the centers of both circles. If the radius of the smaller circle is a and the radius of the larger circle is b, show that $\sin \theta = (b - a)/(a + b)$. Then, using the identity $\sin^2\theta + \cos^2\theta = 1$, show that $\cos \theta = 2\sqrt{ab}/(a + b)$.

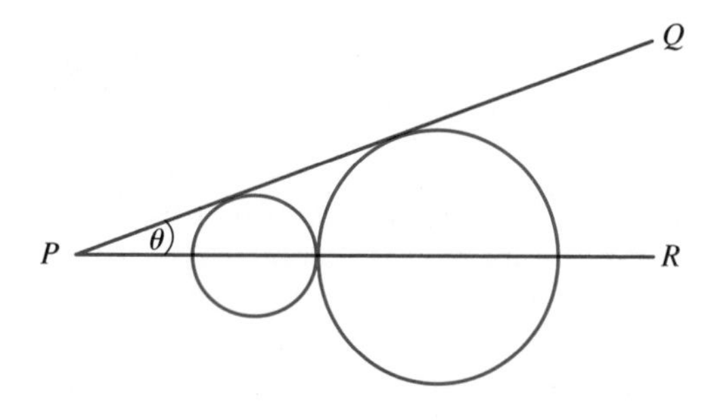

42. A vertical tower of height h stands on level ground. From a point P at ground level and due south of the tower, the angle of elevation to the top of the tower is θ. From a point Q at ground level and due west of the tower, the angle of elevation to the top of the tower is

β. If d is the distance between P and Q, show that

$$h = \frac{d}{\sqrt{\cot^2 \theta + \cot^2 \beta}}$$

43. The following problem is taken from *An Elementary Treatise on Plane Trigonometry,* 8th ed., by R. D. Beasley (London: Macmillan and Co., 1884):

> The [angle of] elevation of a tower standing on a horizontal plane is observed; a feet nearer it is found to be 45°; b feet nearer still it is the complement of what it was at the first station; shew that the height of the tower is $ab/(a-b)$ feet.

6.4 TRIGONOMETRIC FUNCTIONS OF ANGLES

Be clear about the signs of the functions. Your calculator will show them to you correctly, but you still need to be able to figure them out for yourself, quadrant by quadrant.

David Halliday and Robert Resnick in *Fundamentals of Physics,* 3rd ed. (New York: John Wiley and Sons, 1988)

Recall that the definitions of the trigonometric functions given in Section 6.1 apply only to acute angles. That is, as defined on page 359, the domains of trigonometric functions consist of all angles θ such that $0° < \theta < 90°$. In this section, we want to expand our definitions of the trigonometric functions to include angles of any size. To do this, we will need to be more explicit than we have been about angles and their size or measure.

For analytical purposes, we think of the two rays that form an angle as having been originally coincident; then while one ray is held fixed, the other is rotated to create the given angle. As Figure 1 indicates, the fixed ray is called the **initial side** of the angle, and the rotated ray is called the **terminal side.** By convention, we take the measure of an angle to be **positive** if the rotation is counterclockwise and **negative** if the rotation is clockwise. For example, the measure of the angle in Figure 2(a) is *positive* thirty degrees because the rotation is counterclockwise, whereas the measure of the angle in Figure 2(b) is *negative* thirty degrees because the rotation is clockwise. Figure 2(c) shows an angle of 390°.

FIGURE 1

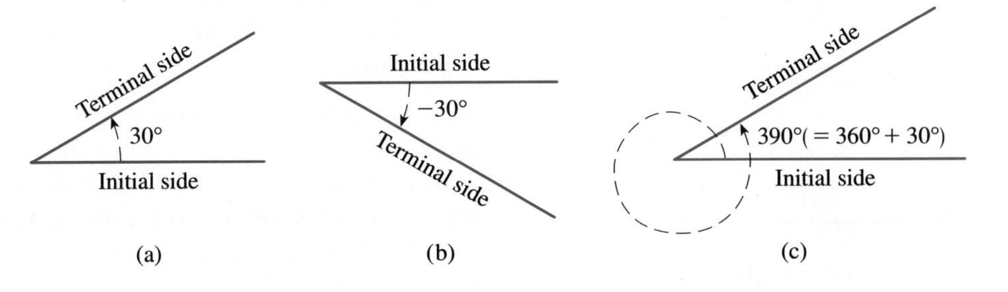

FIGURE 2
Angles generated by a counterclockwise rotation have positive measure. Angles generated by a clockwise rotation have negative measure.

In our development of trigonometry, it will be convenient to have a *standard position* for angles. In a rectangular coordinate system, an angle is in **standard position** if the vertex is located at (0, 0) and the initial side of the angle lies along the positive horizontal axis. Figure 3 (on the next page) shows examples of angles in standard position.

Now we are ready to extend our definitions of the trigonometric functions to accommodate angles of any size. We begin by placing the angle θ in standard

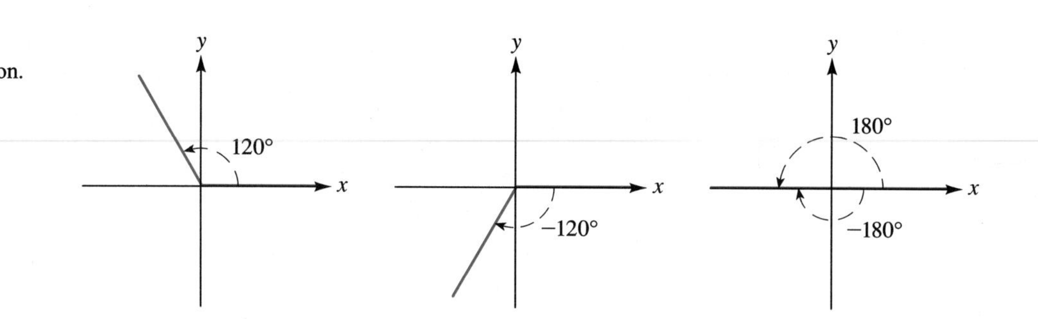

FIGURE 3
Examples of angles in standard position.

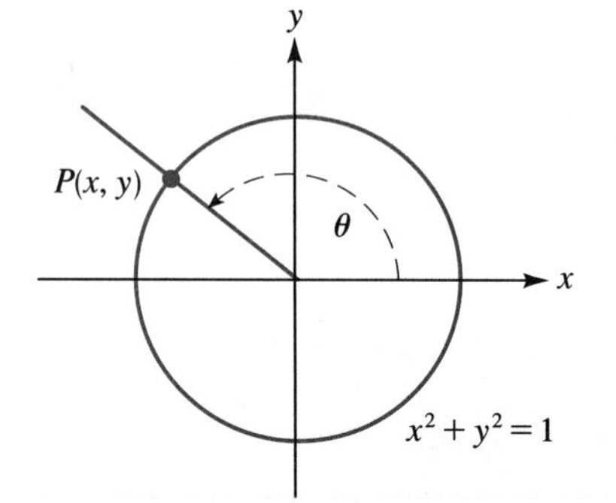

FIGURE 4
$P(x,y)$ denotes the point where the terminal side of angle θ intersects the unit circle.

position and drawing in the **unit circle** $x^2 + y^2 = 1$, as shown in Figure 4. (Recall from Chapter 2 that the equation $x^2 + y^2 = 1$ represents a circle of radius 1, with center at the origin.) Notice the notation $P(x, y)$ in Figure 4; this stands for the point P, with coordinates (x, y), where the terminal side of angle θ intersects the unit circle. With this notation, we define the six trigonometric functions of θ as follows.

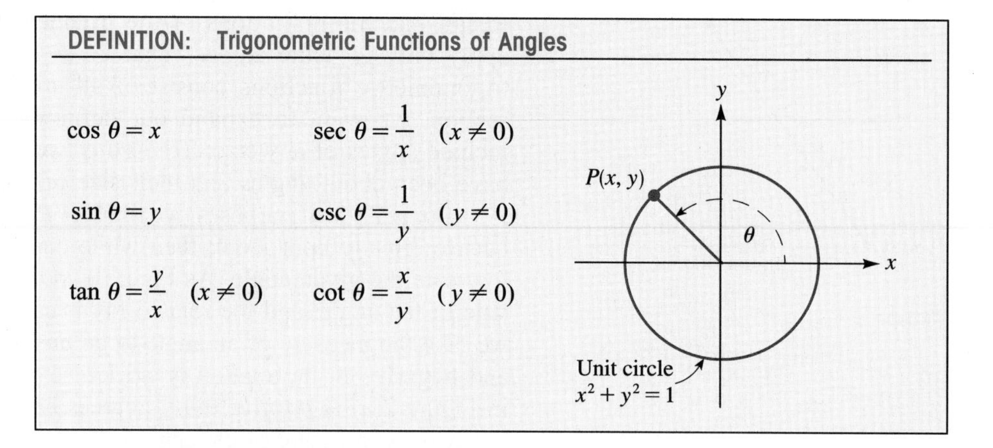

DEFINITION: Trigonometric Functions of Angles

$\cos\theta = x$ $\qquad \sec\theta = \frac{1}{x} \quad (x \neq 0)$

$\sin\theta = y$ $\qquad \csc\theta = \frac{1}{y} \quad (y \neq 0)$

$\tan\theta = \frac{y}{x} \quad (x \neq 0)$ $\qquad \cot\theta = \frac{x}{y} \quad (y \neq 0)$

Even before turning to some examples, we make three initial observations concerning our new definitions. The first two observations will help you to memorize the definitions. The third observation will help you to see why these new definitions are consistent with our previous work on right-triangle trigonometry.

1. $\cos\theta$ is the first coordinate of the point where the terminal side of angle θ intersects the unit circle; $\sin\theta$ is the second coordinate. (You can remember this by noting that, alphabetically, cosine comes before sine.)
2. Just as with the right-triangle definitions, we have the same three pairs of reciprocals: $\cos\theta$ and $\sec\theta$; $\sin\theta$ and $\csc\theta$; $\tan\theta$ and $\cot\theta$.
3. For the cases in which the angle θ is acute, these definitions are really equivalent to the original right-triangle definitions. Consider, for instance, the acute angle θ in Figure 5. According to our "new" definition of sine, we have

$$\sin\theta = y$$

On the other hand, applying the original right-triangle definition to Figure 5 yields

$$\sin\theta = \frac{\text{opposite}}{\text{hypotenuse}} = \frac{y}{1} = y$$

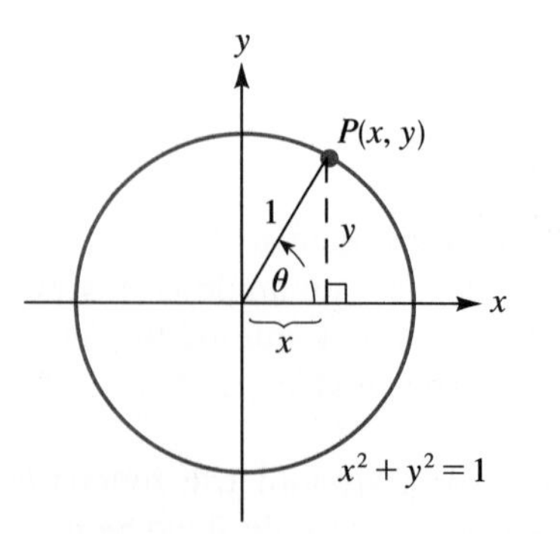

FIGURE 5

Thus, in both cases we obtain the same result. Using Figure 5, you should check for yourself that the same agreement also occurs for the other trigonometric functions.

FIGURE 6

EXAMPLE 1 Compute cos 90°, sin 90°, tan 90°, sec 90°, csc 90°, and cot 90°.

Solution We place the angle $\theta = 90°$ in standard position. Then, as Figure 6 indicates, the terminal side of the angle meets the unit circle at the point (0, 1). Now we apply the definitions:

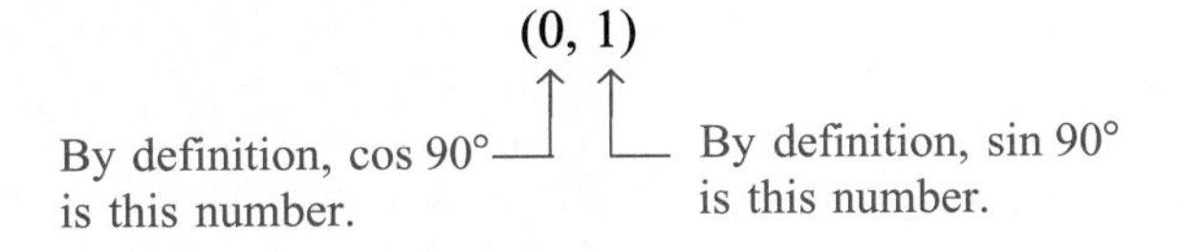

Thus,

$$\cos 90° = 0 \quad \text{and} \quad \sin 90° = 1$$

For the remaining trigonometric functions of 90°, we have

$$\tan 90° = \frac{y}{x} = \frac{1}{0} \quad \text{undefined}$$

$$\sec 90° = \frac{1}{x} = \frac{1}{0} \quad \text{undefined}$$

$$\csc 90° = \frac{1}{y} = \frac{1}{1} = 1$$

$$\cot 90° = \frac{x}{y} = \frac{0}{1} = 0$$

These are the required results.

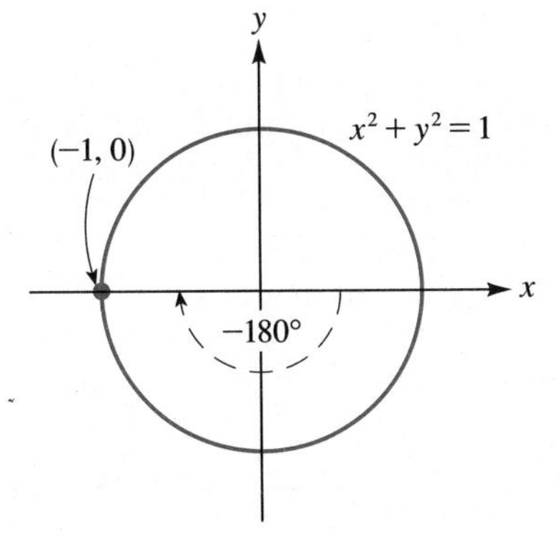

FIGURE 7

EXAMPLE 2 Evaluate the trigonometric functions of $-180°$.

Solution The results can be read off from Figure 7.

$$\cos(-180°) = x = -1 \qquad \sec(-180°) = \frac{1}{x} = \frac{1}{-1} = -1$$

$$\sin(-180°) = y = 0 \qquad \csc(-180°) = \frac{1}{y} = \frac{1}{0} \quad \text{undefined}$$

$$\tan(-180°) = \frac{y}{x} = \frac{0}{-1} = 0 \qquad \cot(-180°) = \frac{x}{y} = \frac{-1}{0} \quad \text{undefined}$$

In the two examples we have just concluded, we evaluated the trigonometric functions for $\theta = 90°$ and $\theta = -180°$. In the same manner, the trigonometric functions can be evaluated just as easily for any angle that is an integral multiple of 90°. Table 1 shows the results of such calculations. Exercise 14 at the end of this section asks you to make these calculations for yourself.

TABLE 1

θ	$\cos\theta$	$\sin\theta$	$\tan\theta$	$\sec\theta$	$\csc\theta$	$\cot\theta$
0°	1	0	0	1	undefined	undefined
90°	0	1	undefined	undefined	1	0
180°	−1	0	0	−1	undefined	undefined
270°	0	−1	undefined	undefined	−1	0
360°	1	0	0	1	undefined	undefined

EXAMPLE 3 Use Figure 8 (on the next page) to approximate the following trigonometric values to within successive tenths. Then use a calculator to check your answers. Round the calculator values to two decimal places.

(a) $\cos 160°$ and $\sin 160°$ **(b)** $\cos(-40°)$ and $\sin(-40°)$

Solution

(a) Figure 8 shows that (the terminal side of) an angle of 160° in standard position meets the unit circle at a point in Quadrant II. Letting (x, y) denote the coordinates of that point, we have (from Figure 8)

$$-1.0 < x < -0.9 \quad \text{and} \quad 0.3 < y < 0.4$$

But, by definition, $\cos 160° = x$ and $\sin 160° = y$ and, consequently,

$$-1.0 < \cos 160° < -0.9 \quad \text{and} \quad 0.3 < \sin 160° < 0.4$$

The corresponding calculator results are $\cos 160° \approx -0.94$ and $\sin 160° \approx 0.34$. Note that the estimations we obtained from Figure 8 are consistent with these calculator values. (When you check these calculator values for yourself, be sure that your calculator is set in the degree mode.)

(b) As you can verify using Figure 8, the terminal side of an angle of $-40°$ intersects the unit circle at the same point as does the terminal side of an angle of 320°. Letting (x, y) denote the coordinates of that point, we have (from Figure 8)

$$0.7 < x < 0.8 \quad \text{and} \quad -0.7 < y < -0.6$$

and, consequently,

$$0.7 < \cos(-40°) < 0.8 \quad \text{and} \quad -0.7 < \sin(-40°) < -0.6$$

The corresponding calculator values here are $\cos(-40°) \approx 0.77$ and $\sin(-40°) \approx -0.64$. Again, note that the estimations we obtain from Figure 8 are consistent with these calculator values.

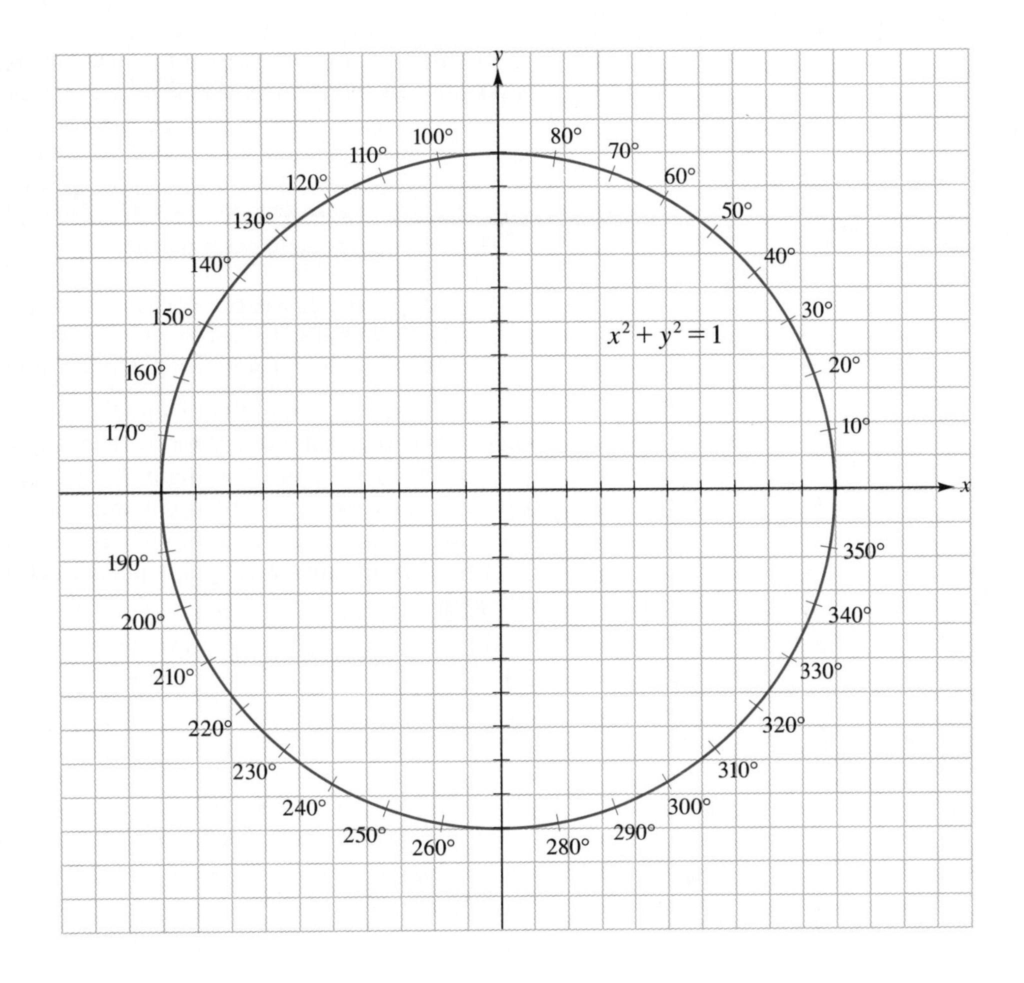

FIGURE 8
Degree measure on the unit circle for angles in standard position.

EXAMPLE 4 In each case, use Figure 8 to determine which quantity is larger. Then use a calculator to check the answer.

(a) sin 80° or cos 80° **(b)** sin 110° or sin 290°
(c) cos 20° or cos 40°

Solution

(a) In Figure 8, look at the point on the unit circle corresponding to 80°. The y-coordinate (which is sin 80°) is nearly 1, but the x-coordinate (cos 80°) is less than 0.2. So, $\sin 80° > \cos 80°$. Indeed, a calculator shows that $\sin 80° \approx 0.98$ and $\cos 80° \approx 0.17$.

(b) In Figure 8, look at the points on the unit circle corresponding to 110° and 290°. The y-coordinate for 110° is positive, while the y-coordinate for 290° is negative. In other words, sin 110° is positive and sin 290° is negative. So (without even worrying about the decimal places) we have $\sin 110° > \sin 290°$. A calculator check shows that $\sin 110° \approx +0.94$ and $\sin 290° \approx -0.94$.

(c) Look at the points on the unit circle corresponding to 20° and 40°. The x-coordinate for 20° is larger than the x-coordinate for 40°. Therefore, $\cos 20° > \cos 40°$. A calculator check confirms this: $\cos 20° \approx 0.94$ and $\cos 40° \approx 0.77$.

In evaluating the trigonometric functions, we'd like to take advantage, as much as possible, of the symmetry in Figure 8. To do this, we introduce the concept of a *reference angle.* (Even when we are using a calculator with the trigonometric functions, there are times, nevertheless, when we will need the reference angle concept. For instance, see Example 2(c) in Section 7.4.)

> **DEFINITION: The Reference Angle**
>
> Let θ be an angle in standard position, and suppose that θ is not a multiple of 90°. The **reference angle** associated with θ is the acute angle (with positive measure) formed by the x-axis and the terminal side of the angle θ.

In Figure 9 we show four examples of angles and their respective reference angles. The first part of Figure 9 shows how to find the reference angle for $\theta = 135°$. First we place the angle $\theta = 135°$ in standard position. Then we find the acute angle between the x-axis and the terminal side of θ. As you can see in this case, this acute angle is 45°; that is the reference angle associated with $\theta = 135°$. In the same way, you should work through the remaining three parts of Figure 9.

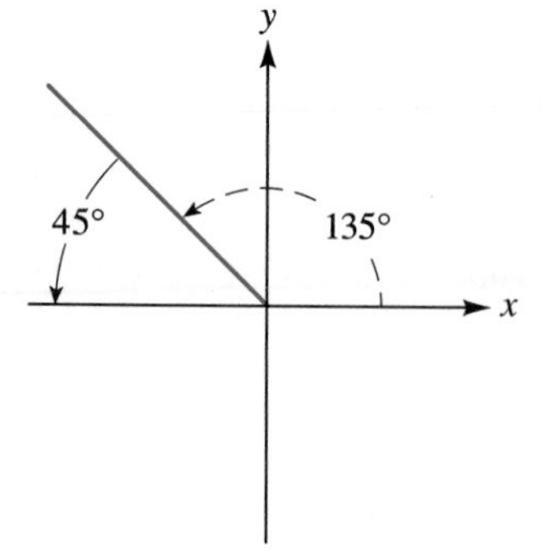

The reference angle for 135° is 45°. [180° − 135° = 45°]

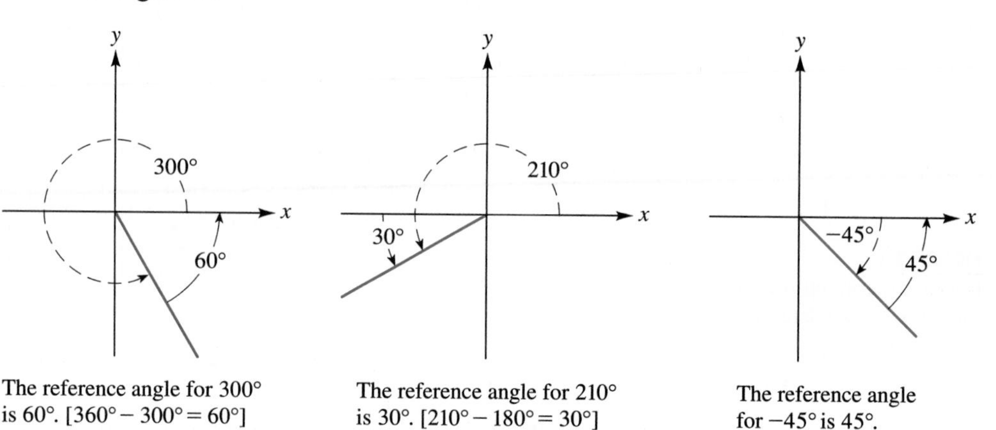

The reference angle for 300° is 60°. [360° − 300° = 60°]

The reference angle for 210° is 30°. [210° − 180° = 30°]

The reference angle for −45° is 45°.

FIGURE 9

Now let's look at an example to see how reference angles are used in evaluating the trigonometric functions. Suppose that we want to evaluate cos 150°. In Figure 10(a) we've placed the angle $\theta = 150°$ in standard position. As you can see, the reference angle for 150° is 30°.

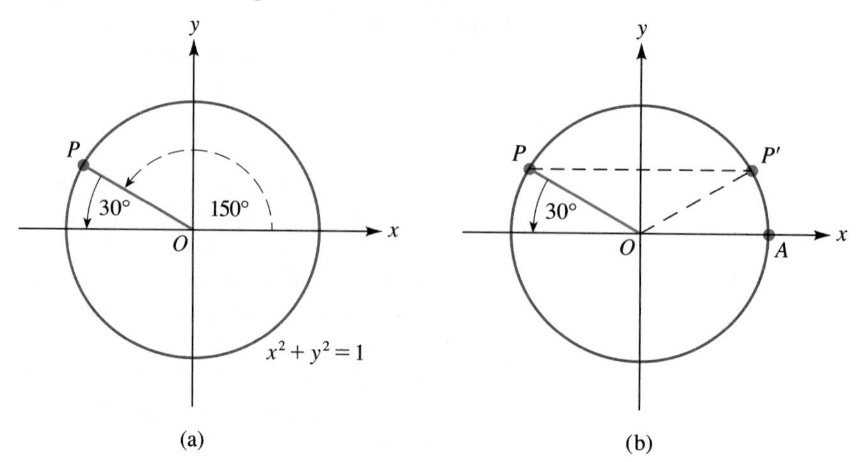

FIGURE 10 (a) (b)

By definition, the value of $\cos 150°$ is the x-coordinate of the point P in Figure 10(a). To find this x-coordinate, we reflect the line segment $\overline{OP}$ in the y-axis; the reflected line segment is the segment $\overline{OP'}$ in Figure 10(b). Since $\angle P'OA = 30°$, the x-coordinate of the point P' is by definition $\cos 30°$, or $\sqrt{3}/2$. The x-coordinate of P is then the negative of this, that is, the x-coordinate of P is $-\sqrt{3}/2$. It follows now, again by definition, that the value of $\cos 150°$ is $-\sqrt{3}/2$.

The same method that we have just used to evaluate $\cos 150°$ can be used to evaluate any of the trigonometric functions when the angles are not multiples of 90°. The following three steps summarize this method.

STEP 1 Determine the reference angle associated with the given angle.
STEP 2 Evaluate the given trigonometric function using the reference angle for the input.
STEP 3 Affix the appropriate sign to the number found in Step 2.

The next two examples illustrate this procedure.

EXAMPLE 5 Evaluate the following quantities: **(a)** $\sin 135°$; **(b)** $\cos 135°$; **(c)** $\tan 135°$.

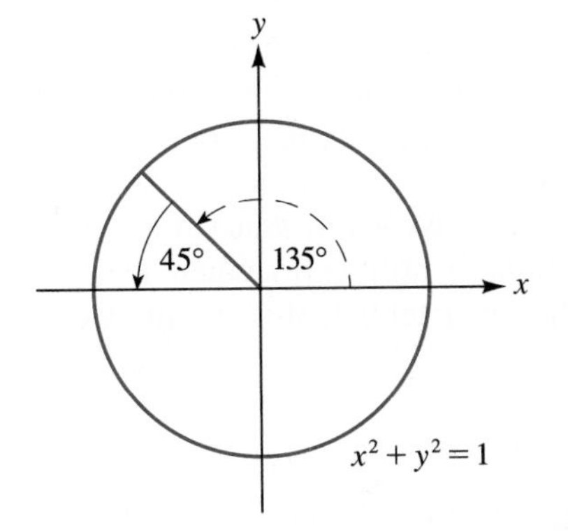

FIGURE 11

Solution As Figure 11 indicates, the reference angle associated with 135° is 45°.

(a) **STEP 1** The reference angle is 45°.
STEP 2 $\sin 45° = \sqrt{2}/2$
STEP 3 $\sin \theta$ is the y-coordinate. In Quadrant II, y-coordinates are positive. Thus, $\sin 135°$ is positive, since the terminal side of $\theta = 135°$ lies in the second quadrant. We therefore have

$$\sin 135° = \sqrt{2}/2$$

(b) **STEP 1** The reference angle is 45°.
STEP 2 $\cos 45° = \sqrt{2}/2$
STEP 3 $\cos \theta$ is the x-coordinate. In Quadrant II, x-coordinates are negative. Thus, $\cos 135°$ is negative, since the terminal side of $\theta = 135°$ lies in Quadrant II. We therefore have

$$\cos 135° = -\sqrt{2}/2$$

(c) **STEP 1** The reference angle is 45°.
STEP 2 $\tan 45° = 1$
STEP 3 By definition, $\tan \theta = y/x$. Now, the terminal side of $\theta = 135°$ lies in Quadrant II, in which y is positive and x is negative. Thus, $\tan 135°$ is negative. We therefore have

$$\tan 135° = -1$$

EXAMPLE 6 Evaluate the following quantities: **(a)** $\cos(-120°)$; **(b)** $\cot(-120°)$; **(c)** $\sec(-120°)$.

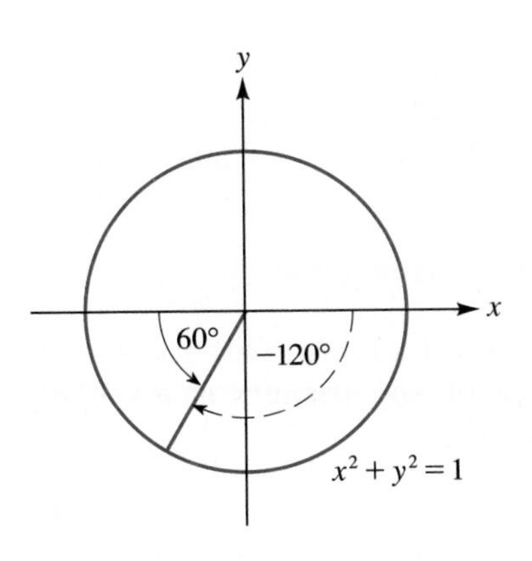

FIGURE 12

Solution As Figure 12 shows, the reference angle for $-120°$ is 60°.

(a) **STEP 1** The reference angle for $-120°$ is 60°.
STEP 2 $\cos 60° = 1/2$

STEP 3 $\cos\theta$ is the x-coordinate. In Quadrant III, the x-coordinates are negative. Thus, $\cos(-120°)$ is negative, since the terminal side of $\theta = -120°$ lies in Quadrant III. It now follows that

$$\cos(-120°) = -1/2$$

(b) **STEP 1** The reference angle for $-120°$ is $60°$.
STEP 2 $\cot 60° = \sqrt{3}/3$
STEP 3 By definition, $\cot\theta = x/y$. Now, the terminal side of $\theta = -120°$ lies in Quadrant III, in which both x and y are negative. So x/y is positive and we have

$$\cot(-120°) = \sqrt{3}/3$$

(c) We could follow our three-step procedure here, but in this case there is a faster method. In part (a) of this example, we found that $\cos(-120°) = -1/2$. Therefore $\sec(-120°) = -2$, because (according to the unit-circle definitions) $\sec\theta$ is the reciprocal of $\cos\theta$.

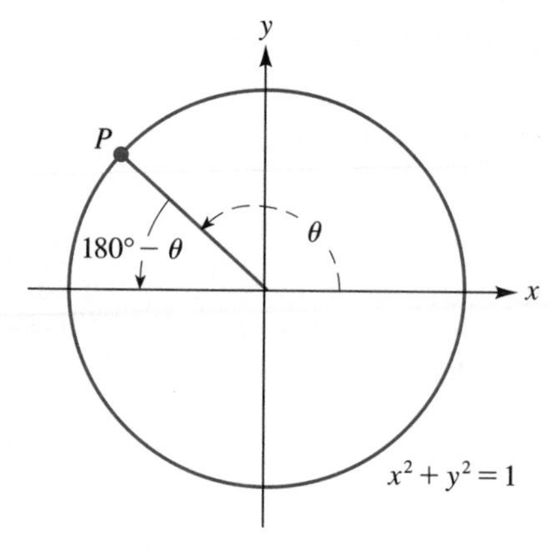

FIGURE 13

In this section we have used the unit circle to generalize our definitions of the trigonometric functions to accommodate angles of any size. We conclude this section now with an application that relies on both the unit-circle definitions and the right-triangle definitions. Recall that in the previous section we found that the area of a triangle is given by the formula $A = \frac{1}{2}ab\sin\theta$, where a and b are the lengths of two sides and θ is the angle included between those two sides. When we derived this result in the previous section, we were assuming that θ was an acute angle. In fact, however, the formula is still valid when θ is an obtuse angle (an angle between 90° and 180°). To establish this result, we'll need to rely on the following identity:

$$\sin(180° - \theta) = \sin\theta$$

We will use Figure 13 to establish this identity for the case in which θ is an obtuse angle. (Exercise 61 shows you how to prove this identity when θ is an acute angle, and Section 8.1 discusses identities such as this from a more general point of view.) Figure 13 shows an obtuse angle θ in standard position. Let us apply our three-step procedure to determine $\sin\theta$. (As you'll see, the end result will be the required identity.)

STEP 1 As indicated in Figure 13, the reference angle for θ is $180° - \theta$.
STEP 2 The sine of the reference angle is $\sin(180° - \theta)$.
STEP 3 $\sin\theta$ is positive because the terminal side of θ lies in Quadrant II. Therefore we have

$$\sin\theta = \sin(180° - \theta) \qquad \text{as required}$$

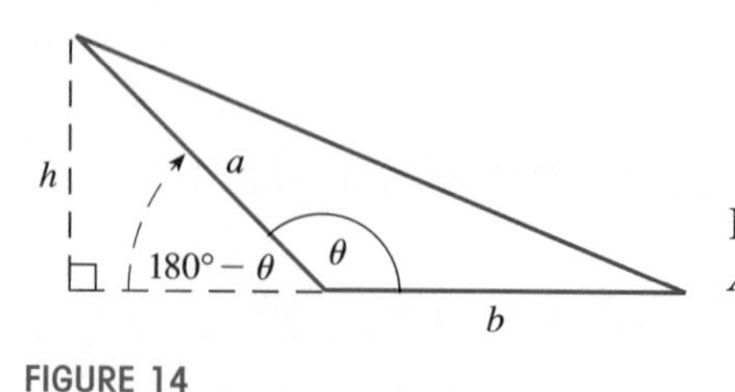

FIGURE 14

We are now prepared to show that the area A of the (red) triangle in Figure 14 is given by $A = \frac{1}{2}ab\sin\theta$. The area of any triangle is given by $A = \frac{1}{2}(\text{base})(\text{height})$. Thus, referring to Figure 14, we have

$$A = \tfrac{1}{2}bh \tag{1}$$

Also from Figure 14, we have

$$\sin(180° - \theta) = \frac{\text{opposite}}{\text{hypotenuse}} = \frac{h}{a}$$

$$h = a \sin(180° - \theta)$$

$$h = a \sin\theta \qquad \text{using the identity } \sin(180° - \theta) = \sin\theta \qquad (2)$$

Now we can use equation (2) to substitute for h in equation (1). This yields

$$A = \tfrac{1}{2}b(a \sin\theta) = \tfrac{1}{2}ab \sin\theta$$

The formula we've just derived is a useful one. We will use it in later work to prove the *law of sines.* For reference, we summarize the result in the box that follows.

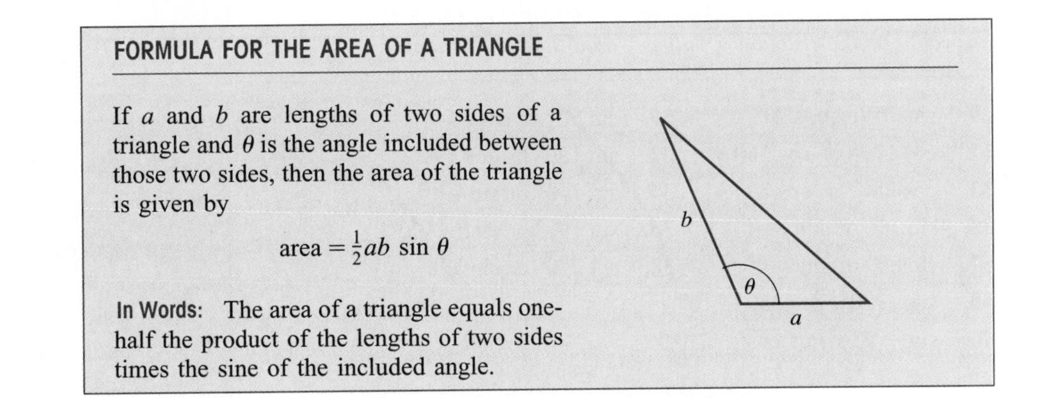

FORMULA FOR THE AREA OF A TRIANGLE

If a and b are lengths of two sides of a triangle and θ is the angle included between those two sides, then the area of the triangle is given by

$$\text{area} = \tfrac{1}{2}ab \sin\theta$$

In Words: The area of a triangle equals one-half the product of the lengths of two sides times the sine of the included angle.

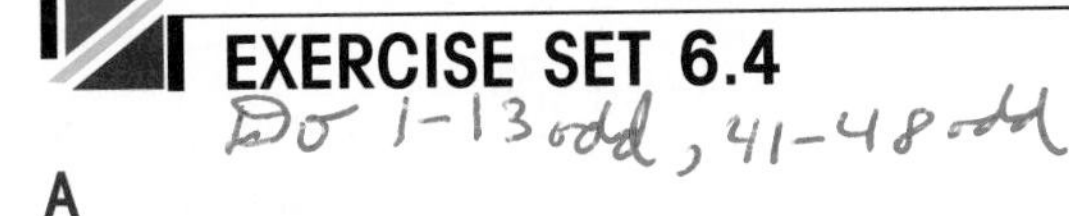

EXERCISE SET 6.4

Do 1-13 odd, 41-48 odd

A

In Exercises 1–8, sketch each angle in standard position and specify the reference angle.

1. (a) 110° (b) −110°
2. (a) 160° (b) −160°
3. (a) 200° (b) −200°
4. (a) 225° (b) −225°
5. (a) 300° (b) −300°
6. (a) 325° (b) −325°
7. (a) 60° (b) −60°
8. (a) 460° (b) −460°

In Exercises 9–13, use the definitions (not a calculator) to evaluate the six trigonometric functions of the angle. If a value is undefined, state this.

9. 270°
10. 450°
11. −270°
12. −630°
13. 810°
14. Use the definitions of the trigonometric functions to complete the table. (When you are finished, check your results against the values shown in Table 1.)

θ	cos θ	sin θ	tan θ	sec θ	csc θ	cot θ
0°						
90°						
180°						
270°						
360°						

In Exercises 15–28, use Figure 8 to approximate within successive tenths the given trigonometric values. Then use a calculator to approximate the values to the nearest hundredth.

15. sin 10° and sin(−10°)
16. cos 10° and cos(−10°)
17. cos 80° and cos(−80°)
18. sin 80° and sin(−80°)

19. sin 120° and sin(−120°)
20. cos 120° and cos(−120°)
21. sin 150° and sin(−150°)
22. cos 150° and cos(−150°)
23. cos 220° and cos(−220°)
24. sin 220° and sin(−220°)
25. cos 310° and cos(−310°)
26. sin 310° and sin(−310°)
27. sin(40° + 360°)
28. cos(40° + 360°)

In Exercises 29–40, use Figure 8 (on page 387) to determine which quantity is larger. Then use a calculator to check the answer. For the calculator values, round to two decimal places.

29. sin 70° or cos 70°
30. sin 60° or sin 70°
31. cos 170° or cos 160°
32. sin 190° or cos 190°
33. cos 280° or cos 290°
34. sin 160° or sin 170°
35. sin 10° or sin(−10°)
36. sin 190° or sin(−190°)
37. sin 80° or sin 110°
38. sin 230° or sin 320°
39. sin(−80°) or cos(−110°)
40. sin(−230°) or cos(−230°)

In Exercises 41–50, evaluate each expression using the method shown in Examples 5 and 6.

41. **(a)** cos 315° **(b)** cos(−315°) **(c)** sin 315° **(d)** sin(−315°)
42. **(a)** cos 240° **(b)** cos(−240°) **(c)** sin 240° **(d)** sin(−240°)
43. **(a)** cos 300° **(b)** cos(−300°) **(c)** sin 300° **(d)** sin(−300°)
44. **(a)** cos 150° **(b)** cos(−150°) **(c)** sin 150° **(d)** sin(−150°)
45. **(a)** cos 210° **(b)** cos(−210°) **(c)** sin 210° **(d)** sin(−210°)
46. **(a)** cos 585° **(b)** cos(−585°) **(c)** sin 585° **(d)** sin(−585°)
47. **(a)** cos 390° **(b)** cos(−390°) **(c)** sin 390° **(d)** sin(−390°)
48. **(a)** cos 405° **(b)** cos(−405°) **(c)** sin 405° **(d)** sin(−405°)
49. **(a)** sec 600° **(b)** csc(−600°) **(c)** tan 600° **(d)** cot(−600°)
50. **(a)** sec 330° **(b)** csc(−330°) **(c)** tan 330° **(d)** cot(−330°)

In Exercises 51 and 52, complete the tables.

51.

θ	sin θ	cos θ	tan θ
0°			
30°			
45°			
60°			
90°			
120°			
135°			
150°			
180°			

52.

θ	sin θ	cos θ	tan θ
180°			
210°			
225°			
240°			
270°			
300°			
315°			
330°			
360°			

In Exercises 53–56, use the given information to determine the area of each triangle.

53. Two of the sides are 5 cm and 7 cm, and the angle between those sides is 120°.

54. Two of the sides are 3 m and 6 m, and the included angle is 150°.

55. Two of the sides are 21.4 cm and 28.6 cm, and the included angle is 98.5°. (Round the final answer to one decimal place.)

56. Two of the sides are 5.98 cm and 8.05 cm, and the included angle is 107.1°. (Round the answer to one decimal place.)

57. The following figure shows an equilateral triangle inscribed in a circle of radius 12 cm. Use the method of Example 6 in Section 6.3 to compute the area of the triangle. Give two forms for your answer: one with a square root and the other a calculator approximation rounded to two decimal places.

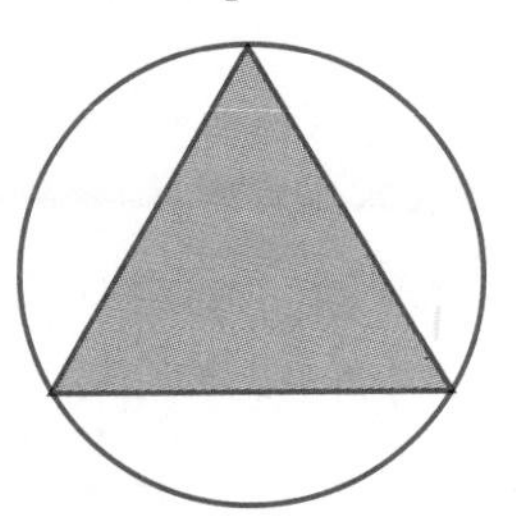

58. An equilateral triangle is inscribed in a circle of radius 8 cm. Compute the area of the shaded region in each of the following figures. Round your final answers to three decimal places.

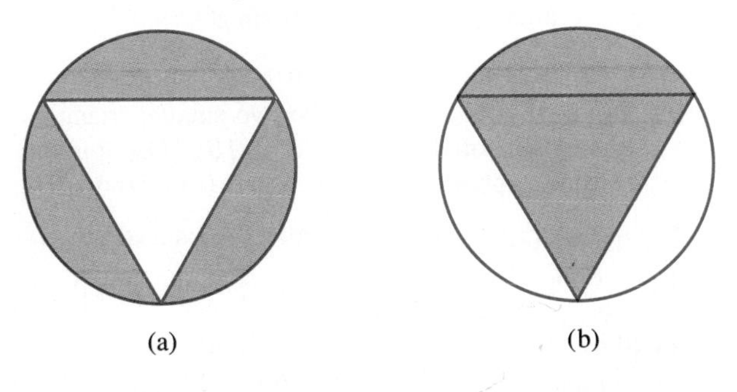

(a) (b)

B

59. **(a)** Complete the following table, using the words "positive" or "negative" as appropriate.

	Terminal Side of Angle θ Lies in			
	Quadrant I	**Quadrant II**	**Quadrant III**	**Quadrant IV**
cos θ and sec θ	positive	negative		
sin θ and csc θ				
tan θ and cot θ				

(b) The mnemonic (memory device) ASTC (*a*ll *s*tudents *t*ake *c*alculus) is sometimes used to recall the signs of the trigonometric values in each quadrant:

A ....All are positive in Quadrant I.

S ....Sine is positive in Quadrant II.

T ....Tangent is positive in Quadrant III.

C ....Cosine is positive in Quadrant IV.

Check the validity of this mnemonic against your chart in part (a).

60. The value of tan θ is undefined when $\theta = 90°$. Use a calculator to complete the following tables to investigate the behavior of tan θ as θ approaches 90°. Round each value to the nearest integer.

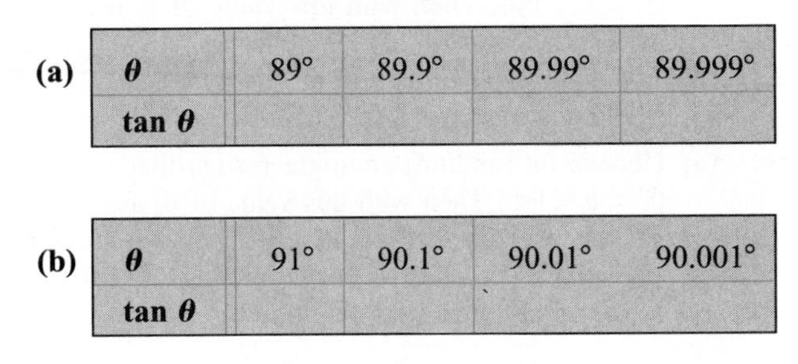

(a)

θ	89°	89.9°	89.99°	89.999°
tan θ				

(b)

θ	91°	90.1°	90.01°	90.001°
tan θ				

61. In the text we derived the identity $\sin(180° - \theta) = \sin \theta$ in the case when θ is an obtuse angle. Now we'll derive the identity when θ is an acute angle.
Note: If θ is an acute angle, then $180° - \theta$ is an obtuse angle.

(a) Sketch a figure showing the obtuse angle $180° - \theta$ in standard position. Note that the terminal side of the angle $180° - \theta$ lies in the second quadrant.

(b) What is the reference angle for the angle $180° - \theta$?

(c) Use Steps (a) and (b) to conclude that $\sin(180° - \theta) = \sin \theta$.

62. Check that the identity $\sin(180° - \theta) = \sin \theta$ is valid in the three cases $\theta = 0°$, $\theta = 90°$, and $\theta = 180°$.

For Exercises 63 and 64, refer to the following figure.

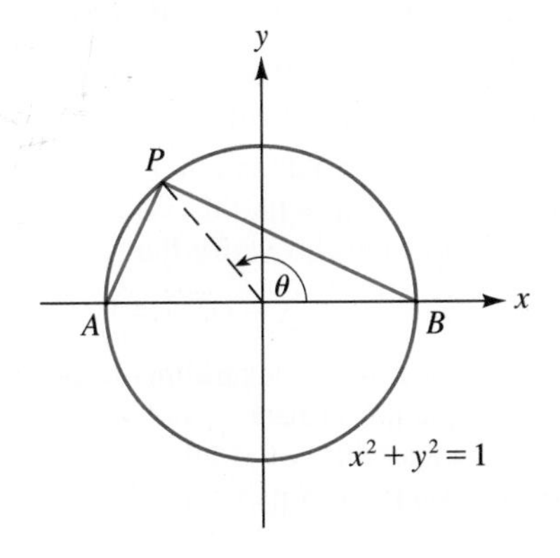

63. Express the area of $\triangle ABP$ as a function of θ.

64. **(a)** Express the slope of $\overline{PB}$ as a function of θ.

(b) Express the slope of $\overline{PA}$ as a function of θ.

(c) Using the expressions obtained in parts (a) and (b), compute the product of the two slopes.

(d) What can you conclude from your answer in part (c)?

As background for Exercises 65–70, you need to have studied logarithms in Sections 5.3 and 5.4.

65. **(a)** Choose (at random) an angle θ such that $0° < \theta < 90°$. Then with this value of θ, use your calculator to verify that $\log_{10}(\sin^2 \theta) = 2 \log_{10}(\sin \theta)$.
(b) For which values of θ in the interval $0° \leq \theta \leq 180°$ is the equation in part (a) valid?

66. **(a)** Choose (at random) an angle θ such that $0° < \theta < 90°$. Then with this value of θ, use your calculator to verify that $\log_{10}(\cos^2 \theta) = 2 \log_{10}(\cos \theta)$.
(b) For which values of θ in the interval $0° \leq \theta \leq 180°$ is the equation in part (a) valid?

For Exercises 67 and 68, let

$$\left.\begin{aligned} S(\theta) &= \sin \theta \\ C(\theta) &= \cos \theta \end{aligned}\right\} \quad 0° \leq \theta \leq 360°$$
$$L(x) = \ln x$$

67. What is the domain of the function $(L \circ S)(\theta)$?

68. What is the domain of the function $(L \circ C)(\theta)$?

69. **(a)** Choose (at random) an angle θ such that $0° < \theta < 90°$. Then with this value of θ, use your calculator to verify that

$$\ln\sqrt{1 - \cos \theta} + \ln\sqrt{1 + \cos \theta} = \ln(\sin \theta).$$

(b) Use the properties of logarithms to prove that if $0° < \theta < 90°$, then

$$\ln\sqrt{1 - \cos \theta} + \ln\sqrt{1 + \cos \theta} = \ln(\sin \theta).$$

(c) For which values of θ in the interval $0° \leq \theta \leq 360°$ is the equation in part (b) valid?

70. **(a)** Choose (at random) an angle θ such that $0° < \theta < 90°$. Then with this value of θ, use your calculator to verify that

$$\ln\sqrt{1 - \sin \theta} + \ln\sqrt{1 + \sin \theta} = \ln(\cos \theta).$$

(b) Use the properties of logarithms to prove that the equation in part (a) holds for all acute angles $(0° < \theta < 90°)$.
(c) Does the equation in part (a) hold if $\theta = 90°$? If $\theta = 0°$.
(d) For which values of θ in the interval $0° \leq \theta \leq 360°$ is the equation in part (a) valid?

71. *Formula for* $\sin(\alpha + \beta)$ In the following figure, $\overline{AD} \perp \overline{BC}$ and $AD = 1$.

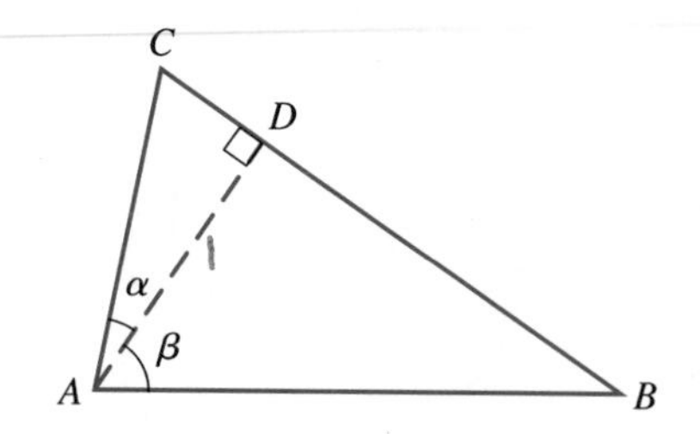

(a) Show that $AC = \sec \alpha$ and $AB = \sec \beta$.
(b) Show that

$$\text{area } \triangle ADC = \tfrac{1}{2} \sec \alpha \sin \alpha$$
$$\text{area } \triangle ADB = \tfrac{1}{2} \sec \beta \sin \beta$$
$$\text{area } \triangle ABC = \tfrac{1}{2} \sec \alpha \sec \beta \sin(\alpha + \beta)$$

(c) The sum of the areas of the two smaller triangles in part (b) equals the area of $\triangle ABC$. Use this fact and the expressions given in part (b) to show that

$$\sin(\alpha + \beta) = \sin \alpha \cos \beta + \cos \alpha \sin \beta$$

(d) Use the formula in part (c) to compute $\sin 75°$.
Hint: $75° = 30° + 45°$
(e) Show that $\sin 75° \neq \sin 30° + \sin 45°$.
(f) Compute $\sin 105°$ and then check that $\sin 105° \neq \sin 45° + \sin 60°$.

C

72. Use the figure to prove the following theorem: The area of a quadrilateral is equal to one-half of the product of the diagonals times the sine of the included angle. *Hint:* Find the areas of each of the four triangles that make up the quadrilateral. Show that the sum of those areas is $\frac{1}{2}(\sin \theta)[qs + rq + rp + ps]$. Then factor the quantity within the brackets.

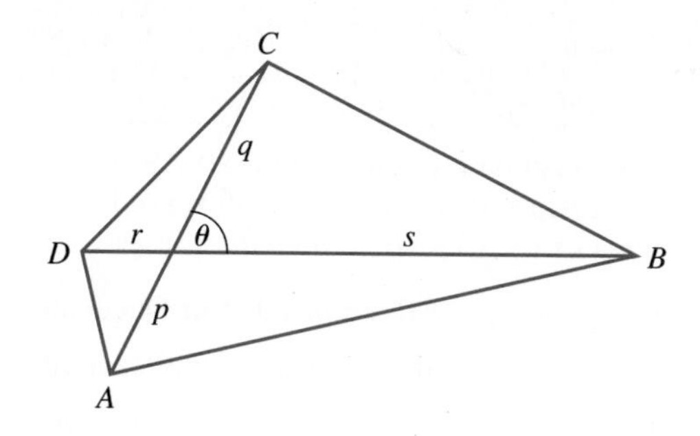

6.5 TRIGONOMETRIC IDENTITIES

A special name for the function which we call the sine is first found in the works of Āryabhata (c. 510) . . . .

It is further probable, from the efforts made to develop simple tables, that the Hindus were acquainted with the principles which we represent by the . . . [formula] $\sin^2 \phi + \cos^2 \phi = 1$. . . .

David Eugene Smith in *History of Mathematics,* vol. II (New York: Ginn and Company, 1925)

As we mentioned in Section 6.2, there are many identities involving the trigonometric functions. In fact, that's one reason why these functions are so useful. In the box that follows, we list some of the same basic identities that we studied in Section 6.2. Keep in mind, however, that now the identities will be true for all angles for which the expressions are defined, rather than for acute angles only. This means that we will need to pay attention to the signs of the trigonometric values. (You'll see how this works in Examples 1 and 2.)

PROPERTY SUMMARY **BASIC TRIGONOMETRIC IDENTITIES**

1. $\sin^2 \theta + \cos^2 \theta = 1$
2. $\dfrac{\sin \theta}{\cos \theta} = \tan \theta$
3. $\sec \theta = \dfrac{1}{\cos \theta}$; $\csc \theta = \dfrac{1}{\sin \theta}$; $\cot \theta = \dfrac{1}{\tan \theta}$

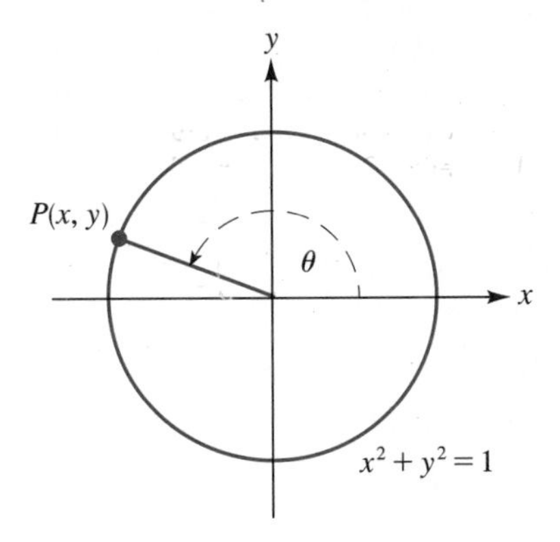

FIGURE 1

The second and the third identities in the box are immediate consequences of the unit-circle definitions of the trigonometric functions. To see why the first identity in the box is valid, consider Figure 1. Since the point $P(x, y)$ lies on the unit circle, we have

$$x^2 + y^2 = 1$$

But, by definition, $x = \cos \theta$ and $y = \sin \theta$. Therefore, we have

$$\cos^2 \theta + \sin^2 \theta = 1$$

which is essentially what we wished to show. Incidentally, you should also become familiar with the equivalent forms of this identity:

$$\cos^2 \theta = 1 - \sin^2 \theta \qquad \text{and} \qquad \sin^2 \theta = 1 - \cos^2 \theta$$

EXAMPLE 1 Given $\sin \theta = 2/3$ and $90° < \theta < 180°$, find $\cos \theta$ and $\tan \theta$.

Solution Substituting $\sin \theta = 2/3$ into the identity $\cos^2 \theta = 1 - \sin^2 \theta$ yields

$$\cos^2 \theta = 1 - \left(\frac{2}{3}\right)^2 = \frac{5}{9}$$

$$\cos \theta = \pm\sqrt{\frac{5}{9}} = \frac{\pm\sqrt{5}}{3}$$

To decide whether to choose the positive or the negative value, note that the given inequality $90° < \theta < 180°$ tells us that the terminal side of θ lies in the second quadrant. Since x-coordinates are negative in Quadrant II, we choose the negative value here. Thus

$$\cos\theta = \frac{-\sqrt{5}}{3}$$

For $\tan\theta$, we have

$$\tan\theta = \frac{y}{x} = \frac{\sin\theta}{\cos\theta} = \frac{2/3}{-\sqrt{5}/3}$$
$$= -\frac{2}{\sqrt{5}} = -\frac{2\sqrt{5}}{5}$$

EXAMPLE 2 Suppose that

$$\cos\theta = t/2$$

where $270° < \theta < 360°$ and $t > 0$. Express the other five trigonometric values as functions of t.

Solution Replacing $\cos\theta$ with the quantity $t/2$ in the identity $\sin^2\theta = 1 - \cos^2\theta$ yields

$$\sin^2\theta = 1 - \left(\frac{t}{2}\right)^2 = 1 - \frac{t^2}{4} = \frac{4 - t^2}{4}$$

and, consequently,

$$\sin\theta = \pm\frac{\sqrt{4 - t^2}}{2}$$

To decide whether to choose the positive or the negative value here, note that the given inequality $270° < \theta < 360°$ tells us that the terminal side of θ lies in the fourth quadrant. Since y-coordinates are negative in Quadrant IV, we choose the negative value here. Thus

$$\sin\theta = -\frac{\sqrt{4 - t^2}}{2}$$

To obtain $\tan\theta$, we use the identity $\tan\theta = (\sin\theta)/(\cos\theta)$. This yields

$$\tan\theta = \frac{-\sqrt{4 - t^2}/2}{t/2} = -\frac{\sqrt{4 - t^2}}{t}$$

We can now find the remaining three values simply by taking reciprocals:

$$\sec\theta = \frac{1}{\cos\theta} = \frac{2}{t}$$

$$\csc\theta = \frac{1}{\sin\theta} = -\frac{2}{\sqrt{4 - t^2}} = -\frac{2\sqrt{4 - t^2}}{4 - t^2}$$ rationalizing the denominator

$$\cot\theta = \frac{1}{\tan\theta} = -\frac{t}{\sqrt{4 - t^2}} = -\frac{t\sqrt{4 - t^2}}{4 - t^2}$$ rationalizing the denominator

In the next five examples, we are asked to show that certain trigonometric equations are, in fact, identities. The procedures here will be very much like those in our work on simplifying trigonometric expressions in Section 6.2, but now we have an advantage. In each case we are given an answer toward which to work. The identities in the next five examples should not be memorized; they are too specialized. Instead, concentrate on the proofs themselves, noting where the fundamental identities (such as $\sin^2\theta + \cos^2\theta = 1$) come into play.

EXAMPLE 3 Prove that the equation $\csc A \tan A \cos A = 1$ is an identity.

Solution We begin with the left-hand side and express each factor in terms of sines or cosines:

$$\csc A \tan A \cos A = \frac{1}{\cancel{\sin A}_{1}} \cdot \frac{\cancel{\sin A}^{1}}{\cancel{\cos A}} \cdot \cancel{\cos A}^{1}$$

$$= 1 \qquad \text{as required}$$

EXAMPLE 4 Prove that $1 - \dfrac{\cot^2\theta}{\csc^2\theta} = \sin^2\theta$.

Solution We'll work with the left-hand side, expressing everything in terms of sines and cosines and then simplifying (as in Section 6.2).

$$1 - \frac{\cot^2\theta}{\csc^2\theta} = 1 - \frac{(\cos^2\theta)/(\sin^2\theta)}{1/\sin^2\theta}$$

Since $\tan\theta = (\sin\theta)/(\cos\theta)$, it follows that $\tan^2\theta = (\sin^2\theta)/(\cos^2\theta)$ and consequently $\cot^2\theta = (\cos^2\theta)/(\sin^2\theta)$.

$$= 1 - \frac{\cos^2\theta}{1}$$

Multiplying both numerator and denominator of the fraction by $\sin^2\theta$

$$= \sin^2\theta$$

EXAMPLE 5 Prove that $\cos^2 B - \sin^2 B = \dfrac{1 - \tan^2 B}{1 + \tan^2 B}$.

Solution We begin with the right-hand side this time; it is the more complicated expression. As in the previous examples, we express everything in terms of sines and cosines.

$$\frac{1 - \tan^2 B}{1 + \tan^2 B} = \frac{1 - \dfrac{\sin^2 B}{\cos^2 B}}{1 + \dfrac{\sin^2 B}{\cos^2 B}} = \frac{\cos^2 B\left(1 - \dfrac{\sin^2 B}{\cos^2 B}\right)}{\cos^2 B\left(1 + \dfrac{\sin^2 B}{\cos^2 B}\right)}$$

$$= \frac{\cos^2 B - \sin^2 B}{\cos^2 B + \sin^2 B} = \frac{\cos^2 B - \sin^2 B}{1} = \cos^2 B - \sin^2 B$$

EXAMPLE 6 Prove that $\dfrac{\cos\theta}{1 - \sin\theta} = \dfrac{1 + \sin\theta}{\cos\theta}$.

Solution The suggestions given in the previous examples are not applicable here. Everything is already in terms of sines and cosines. Furthermore, neither side appears more complicated than the other. A technique that does work here is to begin with the left-hand side and multiply numerator *and* denominator by the same quantity, namely, $1 + \sin\theta$. Doing so gives us

$$\frac{\cos\theta}{1-\sin\theta} = \frac{\cos\theta}{1-\sin\theta}\cdot\frac{1+\sin\theta}{1+\sin\theta}$$
$$= \frac{(\cos\theta)(1+\sin\theta)}{1-\sin^2\theta}$$
$$= \frac{(\cos\theta)(1+\sin\theta)}{\cos^2\theta} = \frac{1+\sin\theta}{\cos\theta}$$

The general strategy for each of the proofs in Examples 3 through 6 was the same. In each case we worked with one side of the given equation, and we transformed it (into equivalent expressions) until it was identical to the other side of the equation. This is not the only strategy that can be used. For instance, an alternate way to establish the identity in Example 6 is as follows. The given equation is equivalent to

$$\frac{\cos\theta}{1-\sin\theta} - \frac{1+\sin\theta}{\cos\theta} = 0 \qquad (1)$$

Now we show that equation (1) is an identity. To do this, we combine the two fractions on the left-hand side, using the common denominator $(\cos\theta)(1-\sin\theta)$:

$$\frac{\cos\theta}{1-\sin\theta} - \frac{1+\sin\theta}{\cos\theta} = \frac{\cos^2\theta - (1+\sin\theta)(1-\sin\theta)}{(\cos\theta)(1-\sin\theta)}$$
$$= \frac{\cos^2\theta - (1-\sin^2\theta)}{(\cos\theta)(1-\sin\theta)}$$
$$= \frac{\cos^2\theta + \sin^2\theta - 1}{(\cos\theta)(1-\sin\theta)} = \frac{1-1}{(\cos\theta)(1-\sin\theta)} = 0$$

This shows that equation (1) is an identity, and thus the equation given in Example 6 is an identity.

Another strategy that can be used in establishing trigonometric identities is to work independently with each side of the given equation until a common expression is obtained. This is the strategy used in Example 7.

EXAMPLE 7 Prove that $\dfrac{1}{1-\cos\beta} + \dfrac{1}{1+\cos\beta} = 2 + 2\cot^2\beta$.

Solution

$$\text{Left-hand side} = \frac{1(1+\cos\beta) + 1(1-\cos\beta)}{(1-\cos\beta)(1+\cos\beta)}$$
$$= \frac{2}{1-\cos^2\beta}$$

$$\text{Right-hand side} = 2 + \frac{2\cos^2\beta}{\sin^2\beta}$$
$$= \frac{2\sin^2\beta + 2\cos^2\beta}{\sin^2\beta}$$
$$= \frac{2(\sin^2\beta + \cos^2\beta)}{1-\cos^2\beta}$$
$$= \frac{2}{1-\cos^2\beta}$$

We've now established the required identity by showing that both sides are equal to the same quantity.

Do 1, 5, 7, 15, 17, 23, 25, 27, 35

EXERCISE SET 6.5

A

In Exercises 1–8, use the given information to determine the remaining five trigonometric values. Rationalize any denominators that contain radicals.

1. **(a)** $\sin\theta = 1/5,\ 90° < \theta < 180°$
(b) $\sin\theta = -1/5,\ 180° < \theta < 270°$

2. **(a)** $\cos\theta = -3/4,\ 90° < \theta < 180°$
(b) $\cos\theta = 3/4,\ 270° < \theta < 360°$

3. **(a)** $\cos\theta = 5/13,\ 0° < \theta < 90°$
(b) $\cos\theta = -5/13,\ 180° < \theta < 270°$

4. **(a)** $\sin\theta = \sqrt{3}/6,\ 0° < \theta < 90°$
(b) $\sin\theta = -\sqrt{3}/6,\ 270° < \theta < 360°$

5. $\csc A = -3,\ 270° < A < 360°$
Hint: First find $\sin A$.

6. $\csc A = 6/5,\ 90° < A < 180°$

7. $\sec B = -3/2,\ 180° < B < 270°$

8. $\sec B = 25/24,\ 270° < B < 360°$

In Exercises 9–14, use the given information to express the remaining five trigonometric values as functions of t or u. Assume that t and u are positive. Rationalize any denominators that contain radicals.

9. $\cos\theta = t/3,\ 270° < \theta < 360°$

10. $\cos\theta = -2t/5,\ 90° < \theta < 180°$

11. $\sin\theta = -3u,\ 180° < \theta < 270°$

12. $\sin\theta = -u^3,\ 270° < \theta < 360°$

13. $\cos\theta = u/\sqrt{3},\ 0° < \theta < 90°$

14. $\sin\theta = 2u/\sqrt{7},\ 0° < \theta < 90°$

In Exercises 15–36, prove that the equations are identities.

15. $\sin\theta\cos\theta\sec\theta\csc\theta = 1$

16. $\tan^2 A + 1 = \sec^2 A$

17. $(\sin\theta\sec\theta)/(\tan\theta) = 1$

18. $\tan\beta\sin\beta = \sec\beta - \cos\beta$

19. $(1 - 5\sin x)/\cos x = \sec x - 5\tan x$

20. $\dfrac{1}{\sin\theta} - \sin\theta = \cot\theta\cos\theta$

21. $(\cos A)(\sec A - \cos A) = \sin^2 A$

22. $\dfrac{\sin\theta}{\csc\theta} + \dfrac{\cos\theta}{\sec\theta} = 1$

23. $(1 - \sin\theta)(\sec\theta + \tan\theta) = \cos\theta$

24. $(\cos\theta - \sin\theta)^2 + 2\sin\theta\cos\theta = 1$

25. $(\sec\alpha - \tan\alpha)^2 = \dfrac{1 - \sin\alpha}{1 + \sin\alpha}$

26. $\dfrac{\sin B}{1 + \cos B} + \dfrac{1 + \cos B}{\sin B} = 2\csc B$

27. $\sin A + \cos A = \dfrac{\sin A}{1 - \cot A} - \dfrac{\cos A}{\tan A - 1}$

28. $(1 - \cos C)(1 + \sec C) = \tan C\sin C$

29. $\csc^2\theta + \sec^2\theta = \csc^2\theta\sec^2\theta$

30. $\cos^2\theta - \sin^2\theta = 1 - 2\sin^2\theta$

31. $\sin A\tan A = \dfrac{1 - \cos^2 A}{\cos A}$

32. $\dfrac{\cot A - 1}{\cot A + 1} = \dfrac{1 - \tan A}{1 + \tan A}$

33. $\cot^2 A + \csc^2 A = -\cot^4 A + \csc^4 A$

34. $\dfrac{\cot^2 A - \tan^2 A}{(\cot A + \tan A)^2} = 2\cos^2 A - 1$

35. $\dfrac{\sin A - \cos A}{\sin A} + \dfrac{\cos A - \sin A}{\cos A} = 2 - \sec A\csc A$

36. $\tan A\tan B = \dfrac{\tan A + \tan B}{\cot A + \cot B}$

B

37. Only one of the following two equations is an identity. Decide which equation this is, and give a proof to show that it is, indeed, an identity. For the other equation, give an example showing that it is not an identity. (For example, to show that the equation $\sin\theta + \cos\theta = 1$ is not an identity, let $\theta = 30°$. Then the equation becomes $\frac{1}{2} + \frac{\sqrt{3}}{2} = 1$, which is false.)
(a) $\dfrac{\csc^2\alpha - 1}{\csc^2\alpha} = \cos\alpha$
(b) $(\sec^2\alpha - 1)(\csc^2\alpha - 1) = 1$

38. Follow the directions given in Exercise 37.
(a) $(\csc\beta - \cot\beta)^2 = \dfrac{1 + \cos\beta}{1 - \cos\beta}$
(b) $\cot\beta + \dfrac{\sin\beta}{1 + \cos\beta} = \csc\beta$

39. Prove the identity $\dfrac{\sin\theta}{1 - \cos\theta} = \dfrac{1 + \cos\theta}{\sin\theta}$ in two ways.
(a) Adapt the method of Example 6.
(b) Begin with the left-hand side and multiply numerator and denominator by $\sin\theta$.

In Exercises 40–44, prove that the equations are identities.

40. $\dfrac{2\sin^3\beta}{1 - \cos\beta} = 2\sin\beta + 2\sin\beta\cos\beta$
Hint: Write $\sin^3\beta$ as $(\sin\beta)(\sin^2\beta)$.

41. $\dfrac{\sec\theta - \csc\theta}{\sec\theta + \csc\theta} = \dfrac{\tan\theta - 1}{\tan\theta + 1}$

42. $1 - \dfrac{\sin^2\theta}{1 + \cot\theta} - \dfrac{\cos^2\theta}{1 + \tan\theta} = \sin\theta\cos\theta$

43. $(\sin^2\theta)(1 + n\cot^2\theta) = (\cos^2\theta)(n + \tan^2\theta)$

44. $(r\sin\theta\cos\phi)^2 + (r\sin\theta\sin\phi)^2 + (r\cos\theta)^2 = r^2$

45. **(a)** Factor the expression $\cos^3\theta - \sin^3\theta$.
(b) Prove the identity

$$\frac{\cos\phi\cot\phi - \sin\phi\tan\phi}{\csc\phi - \sec\phi} = 1 + \sin\phi\cos\phi$$

46. Prove the following identities. (These two identities, along with $\sin^2\theta + \cos^2\theta = 1$, are known as the *Pythagorean identities*. They will be discussed in the next chapter.)
(a) $\tan^2\theta = \sec^2\theta - 1$ **(b)** $\cot^2\theta = \csc^2\theta - 1$

47. If $\tan\alpha\tan\beta = 1$ and α and β are acute angles, show that $\sec\alpha = \csc\beta$. *Hint:* Make use of the identities in Exercise 46.

48. Refer to the figure. Express the slope m of the line as a function of θ. *Hint:* What are the coordinates of the point P?

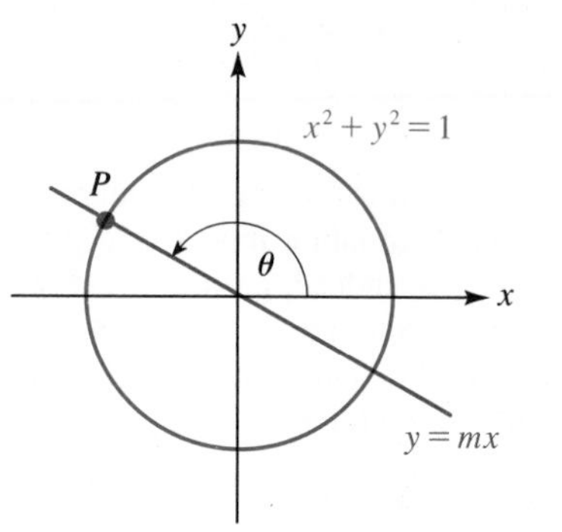

49. Suppose that $\sin\theta = (p - q)/(p + q)$, where p and q are positive and $90° < \theta < 180°$. Show that

$$\tan\theta = \frac{q - p}{2\sqrt{qp}}$$

50. In this exercise, you will use the unit-circle definitions of sine and cosine, along with the identity $\sin^2\theta + \cos^2\theta = 1$ to prove a surprising geometric result. In the figure, we show an equilateral triangle inscribed in the unit circle $x^2 + y^2 = 1$. The vertices of the equilateral triangle are $A\left(-\frac{1}{2}, -\frac{\sqrt{3}}{2}\right)$, $B(1, 0)$, and $C\left(-\frac{1}{2}, \frac{\sqrt{3}}{2}\right)$. Prove that for any point P on the unit circle, the sum of the squares of the distances from P to the three vertices is 6. *Hint:* Let the coordinates of P be $(\cos\theta, \sin\theta)$.

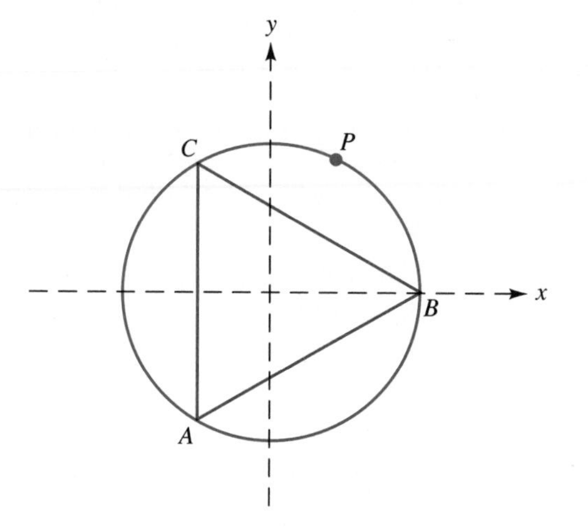

CHAPTER SIX SUMMARY OF PRINCIPAL TERMS AND FORMULAS

FORMULAS OR NOTATIONS	PAGE REFERENCE	COMMENTS
1. $\cos\theta$ $\sin\theta$ $\tan\theta$ $\sec\theta$ $\csc\theta$ $\cot\theta$	359	When θ is an acute angle in a right triangle, the six trigonometric functions of θ are defined as follows. (In these definitions, the word "adjacent" means "the length of the side adjacent to angle θ." The same conventions apply to the words "opposite" and "hypotenuse.") $\cos\theta = \dfrac{\text{adjacent}}{\text{hypotenuse}}$ $\sec\theta = \dfrac{\text{hypotenuse}}{\text{adjacent}}$ $\sin\theta = \dfrac{\text{opposite}}{\text{hypotenuse}}$ $\csc\theta = \dfrac{\text{hypotenuse}}{\text{opposite}}$ $\tan\theta = \dfrac{\text{opposite}}{\text{adjacent}}$ $\cot\theta = \dfrac{\text{adjacent}}{\text{opposite}}$

FORMULAS OR NOTATIONS	PAGE REFERENCE	COMMENTS
2. $\sin^n \theta$	367	$\sin^n \theta$ means $(\sin \theta)^n$. The same convention also applies to the other five trigonometric functions.
3. $\sin^2 \theta + \cos^2 \theta = 1$ $(\sin \theta)/(\cos \theta) = \tan \theta$ $\sin(90° - \theta) = \cos \theta$ $\cos(90° - \theta) = \sin \theta$ $\sec \theta = 1/\cos \theta$ $\csc \theta = 1/\sin \theta$ $\cot \theta = 1/\tan \theta$	368, 395	These are some of the most fundamental trigonometric identities. (We'll see many more in the following chapter.) With the exception of the third and fourth identities in this list, we discussed these identities in the context of both right-triangle and unit-circle trigonometry. Although the third and fourth identities were discussed only in the context of right triangles, you'll see in Chapter 8 that these two identities indeed hold for all angles.
4. $A = \frac{1}{2}ab \sin \theta$	375, 390	The area of a triangle equals one-half of the product of the lengths of two sides times the sine of the included angle.
5. Initial side of an angle Terminal side of an angle	383	We think of the two rays that form an angle as having been originally coincident; then, while one ray is held fixed, the other is rotated to create the given angle. The fixed ray is the initial side of the angle, and the rotated ray is the terminal side. (See Figure 1 on page 383). The measure of an angle is *positive* if the rotation is counterclockwise and *negative* if the rotation is clockwise. (See Figure 2 on page 383.)
6. Standard position	383	In a rectangular coordinate system, an angle is in standard position if the vertex is located at (0, 0) and the initial side of the angle lies along the positive horizontal axis. For examples, see Figure 3 on page 384.
7. $\cos \theta$ $\sin \theta$ $\tan \theta$ $\sec \theta$ $\csc \theta$ $\cot \theta$	384	If θ is an angle in standard position and $P(x, y)$ is the point where the terminal side of the angle meets the unit circle, then the six trigonometric functions of θ are defined as follows: $\cos \theta = x$ $\sec \theta = \frac{1}{x} \quad (x \neq 0)$ $\sin \theta = y$ $\csc \theta = \frac{1}{y} \quad (y \neq 0)$ $\tan \theta = \frac{y}{x} \quad (x \neq 0)$ $\cot \theta = \frac{x}{y} \quad (y \neq 0)$
8. Reference angle	388	Let θ be an angle in standard position in an *x-y* coordinate system, and suppose that θ is not a multiple of 90°. The reference angle associated with θ is the acute angle (with positive measure) formed by the *x*-axis and the terminal side of the angle θ.
9. The three-step procedure for evaluating the trigonometric functions	389	The following three-step procedure can be used to evaluate the trigonometric functions for angles that are not multiples of 90°. **STEP 1** Determine the reference angle associated with the given angle. **STEP 2** Evaluate the given trigonometric function using the reference angle for the input. **STEP 3** Affix the appropriate sign to the number found in Step 2. (See Examples 5 and 6 on pages 389–390.)

WRITING MATHEMATICS

1. A student who wanted to simplify the expression $\dfrac{30^\circ}{\sin 30^\circ}$ wrote

$$\frac{30^\circ}{\sin 30^\circ} = \frac{\cancel{30^\circ}}{\sin \cancel{30^\circ}} = \frac{1}{\sin}$$

Explain why this is nonsense and then indicate the correct solution.

2. Determine if each statement is TRUE or FALSE. In each case, write out your reason or reasons in complete sentences. If you draw a diagram to accompany your writing, be sure that you clearly label any parts of the diagram to which you refer in the writing.

   (a) If θ is an angle in standard position and $180^\circ < \theta < 270^\circ$, then $\tan \theta < \sin \theta$.

   (b) If θ is an acute angle in a right triangle, then $\sin \theta < 1$.

   (c) If θ is an acute angle in a right triangle, then $\tan \theta < 1$.

   (d) For all angles θ, we have $\sin \theta = \sqrt{1 - \cos^2 \theta}$.

   (e) For all acute angles θ, we have $\sin \theta = \sqrt{1 - \cos^2 \theta}$.

   (f) The formula for the area of a triangle, $A = \frac{1}{2}ab \sin \theta$, is not valid when θ is a right angle.

   (g) Every angle θ satisfying $0^\circ < \theta < 180^\circ$ is an allowable input for the sine function.

   (h) Every angle θ satisfying $0^\circ < \theta < 180^\circ$ is an allowable input for the tangent function.

3. A student who was asked to simplify the expression $\sin^2 33^\circ + \sin^2 57^\circ$ wrote $\sin^2 33^\circ + \sin 57^\circ = \sin^2(33^\circ + 57^\circ) = \sin^2 90^\circ = 1^2 = 1$.

   (a) Where is the error?

   (b) What is the correct answer?

4. **(a)** Use a calculator to evaluate the quantity $4 \sin 18^\circ \cos 36^\circ$.

   (b) The following *very* short article, with the accompanying figure, appeared in *The Mathematical Gazette,* Vol. XXIII (1939) p. 211. On your own or with a classmate, study the article and fill in the missing details. Then, strictly on your own, rewrite the article in a paragraph (or two at the most). Write as if you were explaining to a friend or classmate why $4 \sin 18^\circ \cos 36^\circ = 1$.

   "*To prove that* $4 \sin 18^\circ \cos 36^\circ = 1$.
   In the figure, $\sin 18^\circ = \frac{1}{2}a/b$, $\cos 36^\circ = \frac{1}{2}b/a$. Multiply."

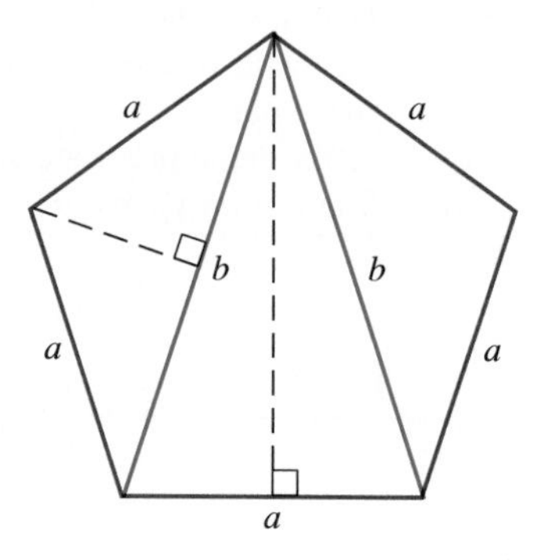

CHAPTER SIX REVIEW EXERCISES

For Exercises 1–16, evaluate each expression. (Don't use a calculator.)

1. $\sin 135^\circ$ **2.** $\cos(-60^\circ)$ **3.** $\tan(-240^\circ)$
4. $\sin 450^\circ$ **5.** $\csc 210^\circ$ **6.** $\sec 225^\circ$
7. $\sin 270^\circ$ **8.** $\cot(-330^\circ)$ **9.** $\cos(-315^\circ)$
10. $\cos 180^\circ$ **11.** $\cos 1800^\circ$ **12.** $\sec 120^\circ$
13. $\csc 240^\circ$ **14.** $\csc(-45^\circ)$ **15.** $\sec 780^\circ$
16. $\sin^2 33^\circ + \sin^2 57^\circ$

17. Simplify: $\ln(\sin^2 17^\circ + \cos^2 17^\circ) - \cos(\ln e^{\pi})$

18. If $\theta = 45^\circ$, find each of the following.
(a) $\cos\theta$ **(b)** $\cos^3\theta$ **(c)** $\cos 2\theta$
(d) $\cos 3\theta$ **(e)** $\cos(2\theta/3)$ **(f)** $(\cos 3\theta)/3$
(g) $\cos(-\theta)$ **(h)** $\cos^3 5\theta$

In Exercises 19–31, the lengths of the three sides of a triangle are denoted a, b, and c; the angles opposite these sides are A, B, and C, respectively. In each exercise, use the given information to find the required quantities.

	GIVEN	FIND
19.	$B = 90^\circ$, $A = 30^\circ$, $b = 1$	a and c
20.	$B = 90^\circ$, $A = 60^\circ$, $a = 1$	c and b
21.	$B = 90^\circ$, $\sin A = 2/5$, $a = 7$	b
22.	$B = 90^\circ$, $\sec C = 4$, $c = \sqrt{2}$	b
23.	$B = 90^\circ$, $\cos A = 3/8$	$\sin A$ and $\cot A$
24.	$B = 90^\circ$, $b = 1$, $\tan C = \sqrt{5}$	a
25.	$b = 4$, $c = 5$, $A = 150^\circ$	area of $\triangle ABC$
26.	$A = 120^\circ$, $b = 8$, area $\triangle ABC = 12\sqrt{3}$	c
27.	$c = 4$, $a = 2$, $B = 90^\circ$	$\sin^2 A + \cos^2 B$
28.	$B = 90^\circ$, $2a = b$	A
29.	$A = 30^\circ$, $B = 120^\circ$, $b = 16$	a
30.	$a = 7$, $b = 8$, $\sin C = 1/4$	area of $\triangle ABC$
31.	$a = b = 5$, $\sin(\frac{1}{2}C) = 9/10$	c and area of $\triangle ABC$

32. For the following figure, show that

$$y = x[\tan(\alpha + \beta) - \tan\beta]$$

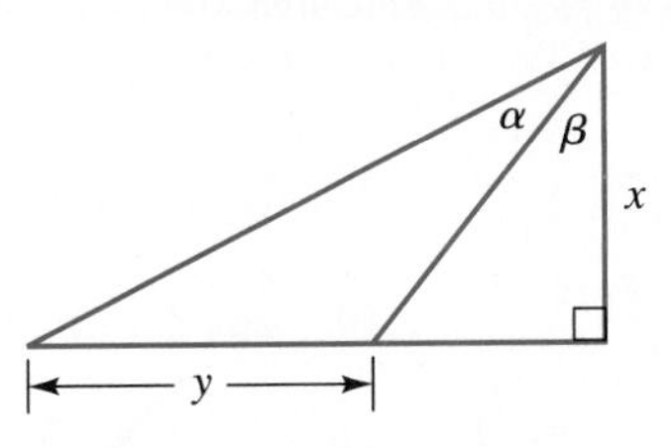

33. For the following figure, show that

$$y = \frac{x}{\cot\alpha - \cot\beta}$$

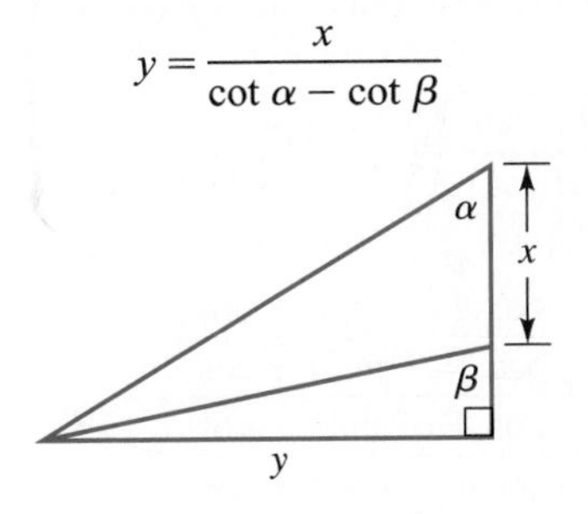

34. For the following figure, show that

$$\cot\theta = \frac{a}{b} + \cot\alpha$$

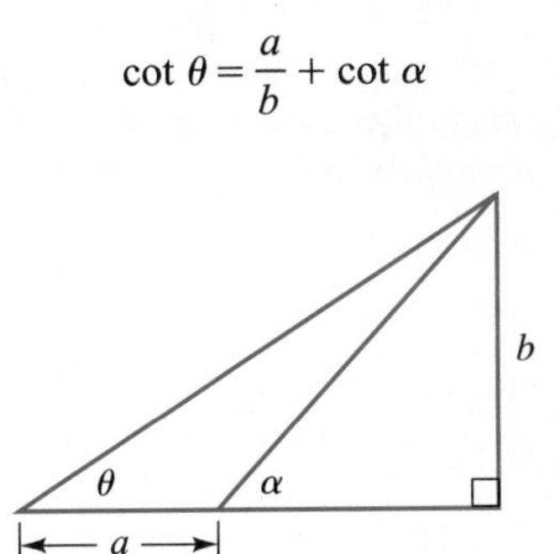

35. Suppose that θ is an acute angle in a right triangle and $\sin\theta = \dfrac{2p^2q^2}{p^4 + q^4}$. Find $\cos\theta$ and $\tan\theta$.

36. This problem is adapted from the text *An Elementary Treatise on Plane Trigonometry,* by R. D. Beasley, first published in 1884.
(a) Prove the following identity, which will be used in part (b): $1 + \tan^2\alpha = \sec^2\alpha$.
(b) Suppose that α and θ are acute angles and $\tan\theta = \dfrac{1 + \tan\alpha}{1 - \tan\alpha}$. Express $\sin\theta$ and $\cos\theta$ in terms of α. *Hint:* Draw a right triangle, label one of the angles θ, and let the lengths of the sides

opposite and adjacent to θ be $1 + \tan\alpha$ and $1 - \tan\alpha$, respectively.

Answer: $\sin\theta = (\cos\alpha + \sin\alpha)/\sqrt{2}$
$\cos\theta = (\cos\alpha - \sin\alpha)/\sqrt{2}$

In Exercises 37–46, convert each expression into one involving only sines and cosines and then simplify. (Leave your answers in terms of sines and/or cosines.)

37. $\dfrac{\sin A + \cos A}{\sec A + \csc A}$

38. $\dfrac{\csc A \sec A}{\sec^2 A + \csc^2 A}$

39. $\dfrac{\sin A \sec A}{\tan A + \cot A}$

40. $\cos A + \tan A \sin A$

41. $\dfrac{\cos A}{1 - \tan A} + \dfrac{\sin A}{1 - \cot A}$

42. $\dfrac{1}{\sec A - 1} \div \dfrac{1}{\sec A + 1}$

43. $(\sec A + \csc A)^{-1}[(\sec A)^{-1} + (\csc A)^{-1}]$

44. $\dfrac{\dfrac{\tan^2 A - 1}{\tan^3 A + \tan A}}{\dfrac{\tan A + 1}{\tan^2 A + 1}}$

45. $\dfrac{\dfrac{\sin A + \cos A}{\sin A - \cos A} - \dfrac{\sin A - \cos A}{\sin A + \cos A}}{\dfrac{\sin A + \cos A}{\sin A - \cos A} + \dfrac{\sin A - \cos A}{\sin A + \cos A}}$

46. $\dfrac{\sin A \tan A - \cos A \cot A}{\sec A - \csc A}$

In Exercises 47–50, refer to the following figure. This figure shows a highly magnified view of the point P, where the terminal side of an angle of 10° (in standard position) meets the unit circle.

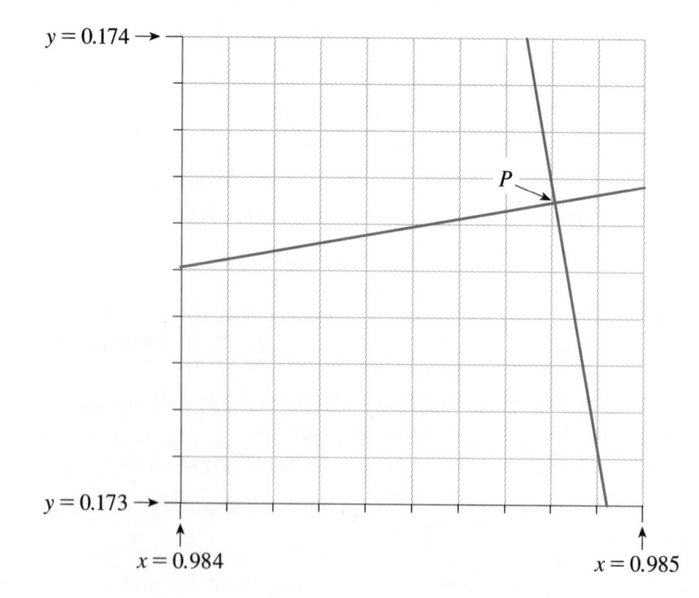

47. Using the figure, estimate the value of cos 10° to four decimal places. (Then use a calculator to check your estimate.)

48. Use your estimate in Exercise 47 to evaluate each of the following.
(a) cos 170° **(b)** cos 190° **(c)** cos 350°

49. Using the figure, estimate the value of sin 10° to four decimal places. (Then use a calculator to check your estimate.)

50. Use your estimate in Exercise 49 to evaluate each of the following.
(a) sin(−10°) **(b)** sin(−190°) **(c)** sin(−370°)

In Exercises 51–60, use the given information to find the required quantities.

	GIVEN	FIND
51.	$\cos\theta = 3/5$, $0° < \theta < 90°$	$\sin\theta$ and $\tan\theta$
52.	$\sin\theta = -12/13$, $180° < \theta < 270°$	$\cos\theta$
53.	$\sec\theta = 25/7$, $270° < \theta < 360°$	$\tan\theta$
54.	$\cot\theta = 4/3$, $0° < \theta < 90°$	$\sin\theta$
55.	$\csc\theta = 13/12$, $0° < \theta < 90°$	$\cot\theta$
56.	$\sin\theta = 1/5$, $0° < \theta < 90°$	$\cos(90° - \theta)$
57.	$\cos\theta = 5t$, $0° < \theta < 90°$	$\tan(90° - \theta)$
58.	$\sin\theta = -\sqrt{5}u$, $90° < \theta < 270°$	$\sec\theta$
59.	$\tan\theta = \dfrac{\sqrt{2} - 1}{\sqrt{2} + 1}$, $0° < \theta < 90°$	$\sin\theta$
60.	$\cos\theta = \dfrac{2u}{u^2 + 1}$, $0° < \theta < 90°$	$\tan\theta$

61. Suppose θ is an acute angle and $\tan\theta + \cot\theta = 2$. Show that $\sin\theta + \cos\theta = \sqrt{2}$.

62. A 100-ft vertical antenna is on the roof of a building. From a point on the ground, the angles of elevation to the top and the bottom of the antenna are 51° and 37°, respectively. Find the height of the building.

63. In an isosceles triangle, the two base angles are each 35°, and the length of the base is 120 cm. Find the area of the triangle.

64. Find the perimeter and the area of a regular pentagon inscribed in a circle of radius 9 cm.

65. In triangle ABC, let h denote the length of the altitude from A to $\overline{BC}$. Show that $h = a/(\cot B + \cot C)$.

66. The length of each side of an equilateral triangle is $2a$. Show that the radius of the inscribed circle is $a/\sqrt{3}$ and the radius of the circumscribed circle is $2a/\sqrt{3}$.

67. From a helicopter h ft above the sea, the angle of depression to the pilot's horizon is θ. Show that $\cot \theta = R/\sqrt{2Rh + h^2}$, where R is the radius of the Earth.

68. In triangle ABC, angle C is a right angle. If a, b, and c denote the lengths of the sides opposite angles A, B, and C, respectively, show that

$$\frac{\sin A - \sin B}{\sin A + \sin B} = \frac{a - b}{a + b}$$

In Exercises 69–78, show that each equation is an identity.

69. $\sin^2(90° - \theta) \csc \theta - \tan^2(90° - \theta) \sin \theta = 0$

70. $\dfrac{1 - \sin \theta \cos \theta}{(\cos \theta)(\sec \theta - \csc \theta)} \times \dfrac{\sin^2 \theta - \cos^2 \theta}{\sin^3 \theta + \cos^3 \theta} = \sin \theta$

71. $\cos A \cot A = \csc A - \sin A$

72. $\sec A - 1 = (\sec A)(1 - \cos A)$

73. $\dfrac{\cot A - 1}{\cot A + 1} = \dfrac{\cos A - \sin A}{\cos A + \sin A}$

74. $\dfrac{\sin A}{\csc A - \cot A} = 1 + \cos A$

75. $\dfrac{1}{1 - \cos A} + \dfrac{1}{1 + \cos A} = 2 + 2 \cot^2 A$

76. $\dfrac{1}{\csc A - \cot A} - \dfrac{1}{\csc A + \cot A} = 2 \cot A$

77. $$\frac{1}{1 + \sin^2 A} + \frac{1}{1 + \cos^2 A} + \frac{1}{1 + \sec^2 A} + \frac{1}{1 + \csc^2 A} = 2$$

78. $\dfrac{\sec A - \tan A}{\sec A + \tan A} = \left(\dfrac{\cos A}{1 + \sin A}\right)^2$

79. The radius of the circle in the given figure is one unit. The line segment $\overline{PT}$ is tangent to the circle at P, and $\overline{PN}$ is perpendicular to $\overline{OT}$. Express each of the following lengths in terms of θ.
(a) PN **(b)** ON **(c)** PT
(d) OT **(e)** NA **(f)** NT

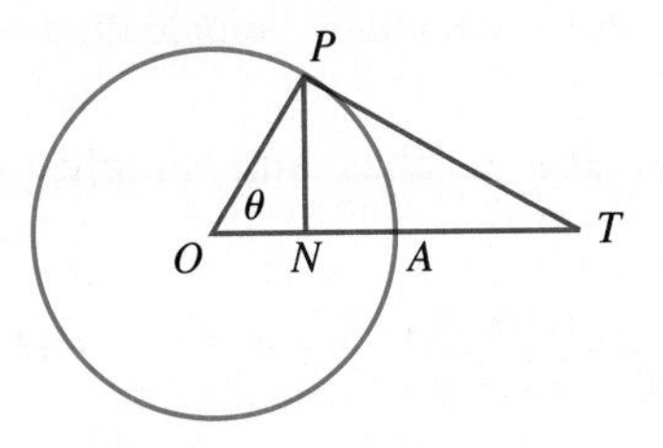

80. Refer to the figure accompanying Exercise 79.
(a) Show that the ratio of the area of triangle ONP to the area of triangle NPT is $\cot^2 \theta$.
(b) Show that the ratio of the area of triangle NPT to the area of triangle OPT is $\sin^2 \theta$.

81. Refer to the following figure. From a point R, two tangent lines are drawn to a circle of radius 1. These tangents meet the circle at the points P and Q. At the midpoint of the circular arc $\widehat{PQ}$, a third tangent line is drawn, meeting the other two tangents at T and U, as shown. Express the area of triangle RTU as a function of θ (where θ is defined in the figure).

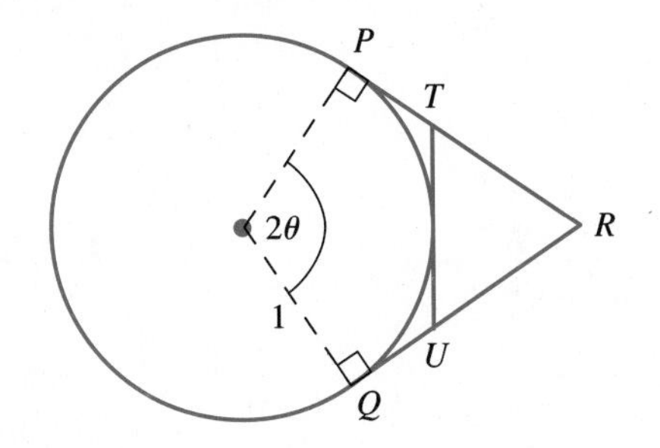

82. Suppose that in $\triangle ABC$, angle C is a right angle. Let a, b, and c denote the lengths of the sides opposite angles A, B, and C, respectively. Use the right-triangle definitions of sine and cosine to prove that

$$\frac{a^2 \cos^2 B - b^2 \cos^2 A}{a^2 - b^2} = 1$$

83. The following figure shows a semicircle with radius $AO = 1$.

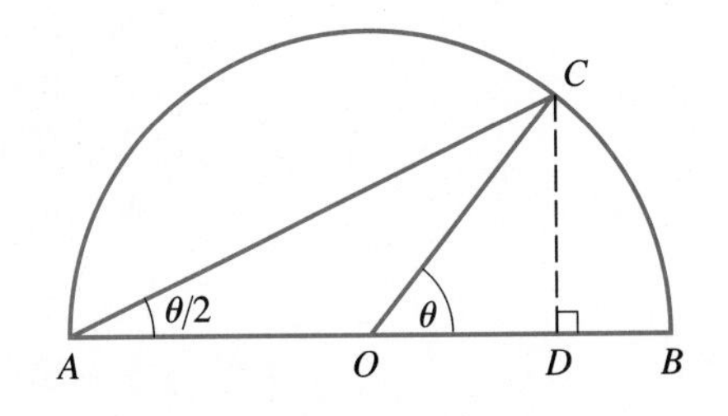

(a) Use the figure to derive the formula

$$\tan \frac{\theta}{2} = \frac{\sin \theta}{1 + \cos \theta}$$

Hint: Show that $CD = \sin \theta$ and $OD = \cos \theta$. Then look at right triangle ADC to find $\tan(\theta/2)$.

(b) Use the formula developed in part (a) to show that
(i) $\tan 15° = \dfrac{1}{2 + \sqrt{3}} = 2 - \sqrt{3}$;
(ii) $\tan 22.5° = \sqrt{2} - 1$.

84. This exercise outlines geometric derivations of formulas

$$\sin 2\theta = 2 \sin \theta \cos \theta \qquad \cos 2\theta = 2 \cos^2 \theta - 1$$

Refer to the figure, which shows a semicircle with diameter $\overline{SV}$ and radius $UV = 1$. In each case, supply the reasons for the given statement.

(a) $TU = \cos 2\theta$ **(b)** $TV = 1 + \cos 2\theta$

(c) $WT = \sin 2\theta$

(d) $WV = 2 \cos \theta$ *Hint:* Use right triangle SWV.

(e) $SW = 2 \sin \theta$

(f) $\sin \theta = \dfrac{\sin 2\theta}{2 \cos \theta}$ *Hint:* Use $\triangle WTV$ and the results in parts (c) and (d).

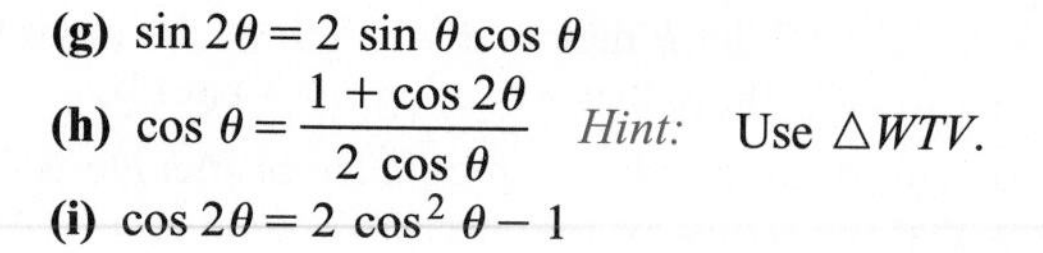

(g) $\sin 2\theta = 2 \sin \theta \cos \theta$

(h) $\cos \theta = \dfrac{1 + \cos 2\theta}{2 \cos \theta}$ *Hint:* Use $\triangle WTV$.

(i) $\cos 2\theta = 2 \cos^2 \theta - 1$

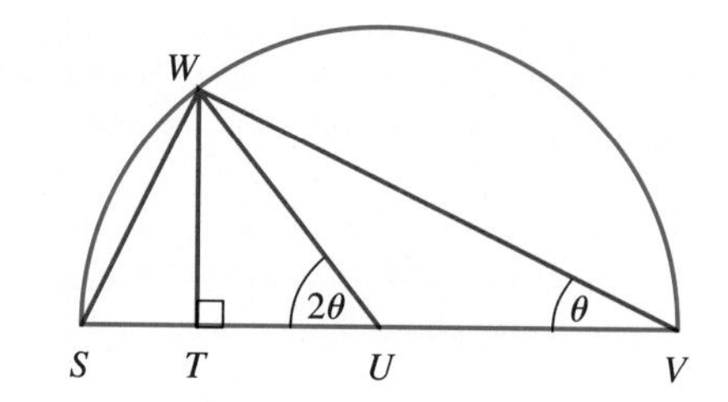

CHAPTER SIX TEST

1. Specify a value for each of the following expressions. (Exact values are required, not calculator approximations.)
(a) $\tan 30°$ **(b)** $\sec 45°$ **(c)** $\sin^2 25° + \cos^2 25°$ **(d)** $\sin 53° - \cos 37°$

2. Suppose that angle θ is in standard position and the terminal side of the angle is in the third quadrant. Which of the following are positive?
(a) $\sec \theta$ **(b)** $\csc \theta$ **(c)** $\cot \theta$

3. Evaluate each of the following: **(a)** $\sin(-270°)$; **(b)** $\cos 180°$; **(c)** $\tan 720°$.

4. Factor: $2 \cot^2 \theta + 11 \cot \theta + 12$.

5. Find the area of a triangle in which the lengths of two sides are 8 cm and 9 cm and the angle included between those two sides is 150°.

6. Evaluate each of the following (exact values required, not calculator approximations): **(a)** $\sin(-225°)$; **(b)** $\tan 330°$; **(c)** $\sec 120°$.

7. If $\sin \theta = -\sqrt{5}/5$ and $180° < \theta < 270°$, find $\cos \theta$ and $\tan \theta$.

8. If $\sin \theta = -u/3$ and $270° < \theta < 360°$, express $\cos \theta$ and $\cot \theta$ as functions of u.

9. Simplify: $$\frac{\dfrac{\cos \theta + 1}{\cos \theta} + 1}{\dfrac{\cos \theta - 1}{\cos \theta} - 1}.$$

10. A regular nine-sided polygon is inscribed in a circle of radius 3 m. Find the area of the polygon. Give two forms of the answer: one in terms of a trigonometric function and the other a calculator approximation rounded to three decimal places.

11. Express each term using sines and cosines and simplify as much as possible:
$$\frac{\cot^2 \theta}{\csc^2 \theta} + \frac{\tan^2 \theta}{\sec^2 \theta}$$

12. In the figure, $\angle DAB = 25°$, $\angle CAB = 55°$, and $AB = 50$ cm. Find CD. Leave your answer in terms of the trigonometric functions (rather than using a calculator).

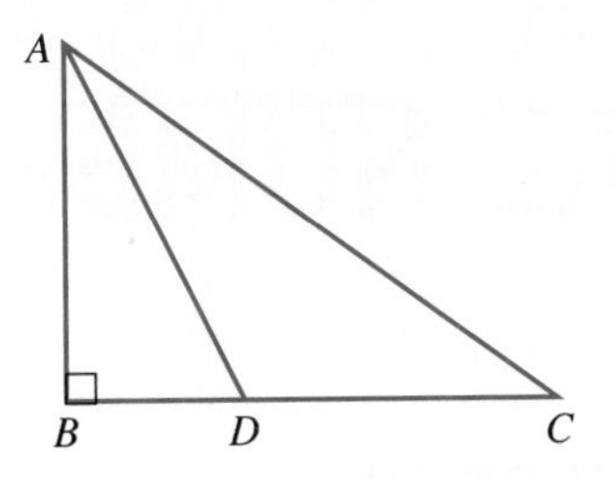

13. Without using a calculator, determine in each case which of the two trigonometric values is the larger. State your reasons. *Hint:* Sketch each pair of angles in standard position.
(a) sin 5° or sin 85° **(b)** sin 5° or cos 5° **(c)** tan 175° or tan 185°

14. In $\triangle ABC$ (shown in the figure), $\angle A = 20°$, $AB = 3.25$ cm, and $AC = 2.75$ cm. Complete steps (a) and (b) to determine the length CB.
(a) Find AD, DB, and CD. Leave your answers in terms of the trigonometric functions.
(b) Find CB by applying the Pythagorean theorem in $\triangle CDB$. For your final answer, use a calculator and round to two decimal places.

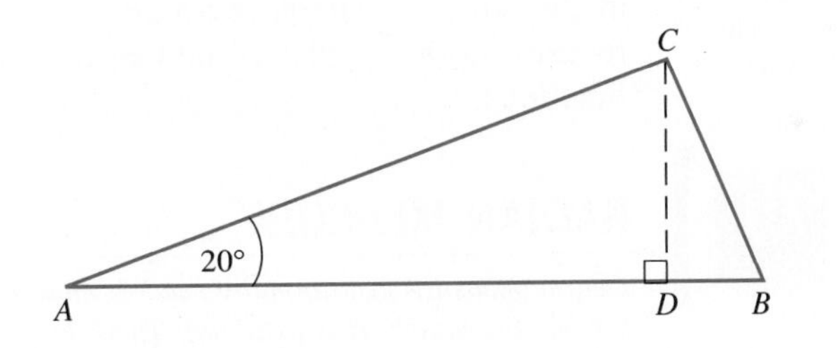

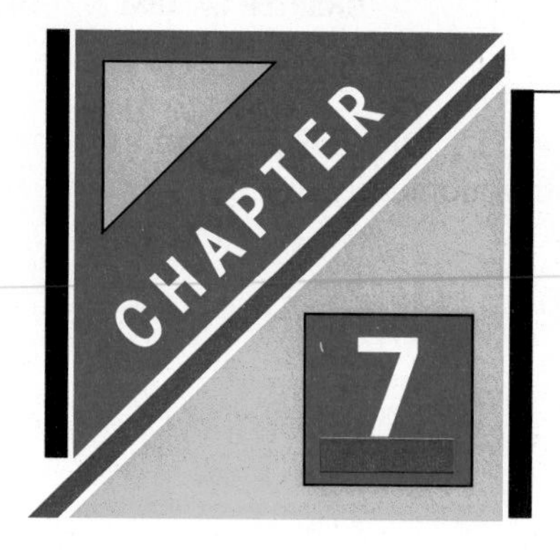

TRIGONOMETRIC FUNCTIONS OF REAL NUMBERS

Trigonometric functions were originally introduced and tabulated in calculations in astronomy and navigation and were used extensively in surveying. What is more important [now], *they occur whenever we want to describe a periodic process, like the vibration of a string, the tides, planetary motion, alternating currents, or the emission of light by atoms.*

From *The Mathematical Sciences,* edited by the Committee on Support of Research in the Mathematical Sciences with the collaboration of George Boehm (Cambridge, Mass.: The M.I.T. Press, 1969)

INTRODUCTION

As the opening quotation indicates, the trigonometric functions play an important role in many of the modern applications of mathematics. In these applications, the inputs for the functions are real numbers rather than angles. In order to make the transition from angle inputs to real-number inputs, we introduce *radian measure* for angles in Sections 7.1 and 7.2. Then, in Section 7.3, we use radian measure to restate the definitions of the trigonometric functions in such a way that the domains are indeed sets of real numbers. Among other advantages, this allows us to analyze the trigonometric functions using the graphing techniques from Chapter 3. Graphs relating to the sine and cosine functions are discussed in Sections 7.4 and 7.5. The next section, 7.6, shows how these graphs are used in the study of *simple harmonic motion.* The last section of the chapter, 7.7, focuses on the graphs of the tangent, the cotangent, the secant, and the cosecant functions.

RADIAN MEASURE

Degree measure, traditionally used to measure the angles of a geometric figure, has one serious drawback. It is artificial. There is no intrinsic connection between a degree and the geometry of a circle. Why 360 degrees for one revolution? Why not 400? or 100?

Calculus, One and Several Variables, 7th ed., by S. L. Salas and Einar Hille, revised by Garret J. Etgen (New York: John Wiley and Sons, Inc., 1995)

In elementary geometry, angles are measured using the familiar units of degrees. However, for the more analytical portions of trigonometry and for calculus, *radian measure* is most often used. The radian measure of an angle is defined as follows.

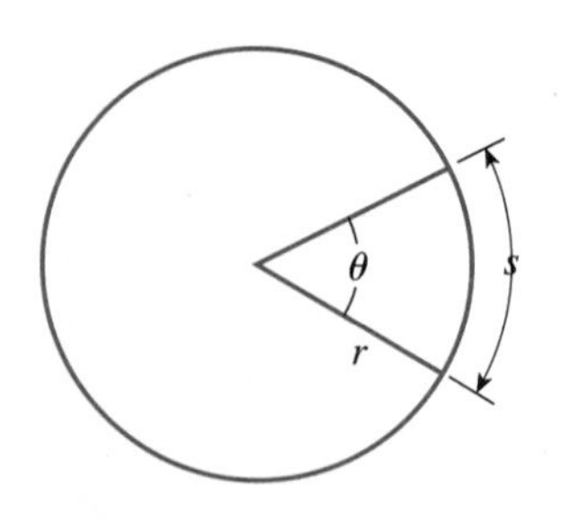

FIGURE 1
The radian measure θ is defined by the equation $\theta = s/r$.

DEFINITION: The Radian Measure of an Angle

Place the vertex of the angle at the center of a circle of radius r. Let s denote the length of the arc intercepted by the angle, as indicated in Figure 1. The **radian measure** θ of the angle is the ratio of the arc length s to the radius r. In symbols,

$$\theta = \frac{s}{r} \tag{1}$$

In this definition, it is assumed that s and r have the same linear units.

EXAMPLE 1 Determine the radian measure for each angle in Figure 2.

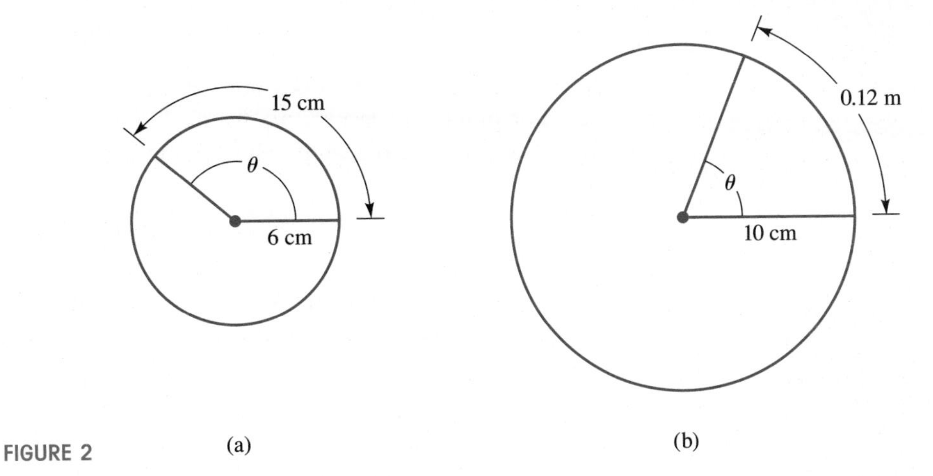

FIGURE 2

Solution In Figure 2(a) we have $r = 6$ cm, $s = 15$ cm, and therefore

$$\theta = \frac{s}{r} = \frac{15 \text{ cm}}{6 \text{ cm}} = \frac{5}{2}$$

So, for the angle in Figure 2(a), we obtain $\theta = 5/2$ radians. In the calculations just completed, notice that although both s and r have the dimensions of length, the resulting radian measure s/r is simply a real number with no dimensions. (The dimensions "cancel out.")

Before carrying out the calculations for the radian measure in Figure 2(b), we have to arrange matters so that the same unit of length is used for the radius and the arc length. Converting the arc length s to centimeters, we have $s = 12$ cm and, consequently,

$$\theta = \frac{s}{r} = \frac{12 \text{ cm}}{10 \text{ cm}} = \frac{6}{5}$$

Thus, in Figure 2(b), we have $\theta = 6/5$ radians. Again, note that the radian measure s/r is dimensionless. (Check for yourself that the same answer is obtained if we use meters rather than centimeters for the common unit of measurement.)

At first it may appear to you that the radian measure depends on the radius of the particular circle that we use. But as you will see, this is not the case. To gain some experience in working with the definition of radian measure, let's calculate the radian measure of the right angle in Figure 3. We begin with the formula $\theta = s/r$. Now, since θ is a right angle, the arc length s is one-quarter of the entire circumference. Thus,

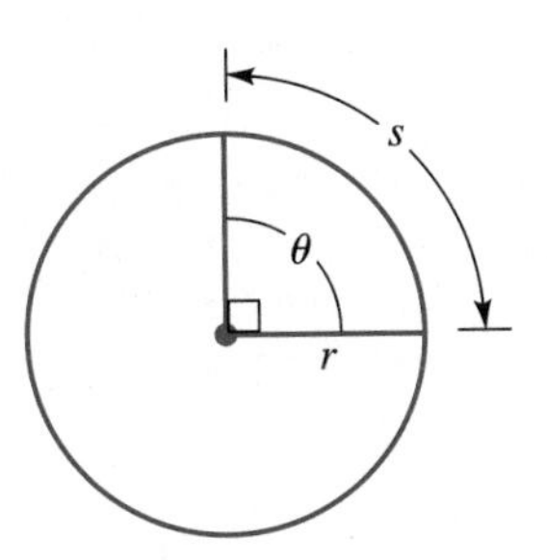

FIGURE 3

$$s = \frac{1}{4}(2\pi r) = \frac{\pi r}{2} \tag{2}$$

Using equation (2) to substitute for s in equation (1), we get

$$\theta = \frac{\pi r/2}{r} = \frac{\pi r}{2} \times \frac{1}{r} = \frac{\pi}{2}$$

In other words,

$$90° = \frac{\pi}{2} \text{ radians} \tag{3}$$

Notice that the radius r does not appear in our answer.

For practical reasons, we would like to be able to convert rapidly between degree and radian measure. Multiplying both sides of equation (3) by 2 yields

$$180° = \pi \text{ radians} \tag{4}$$

Equation (4) is useful and should be memorized. For instance, dividing both sides of equation (4) by 6 yields

$$\frac{180°}{6} = \frac{\pi}{6} \text{ radians}$$

or

$$30° = \frac{\pi}{6} \text{ radians}$$

Similarly, dividing both sides of equation (4) by 4 and 3, respectively, yields

$$45° = \frac{\pi}{4} \text{ radians} \qquad \text{and} \qquad 60° = \frac{\pi}{3} \text{ radians}$$

And, multiplying both sides of equation (4) by 2 gives us

$$360° = 2\pi \text{ radians}$$

Figure 4 summarizes some of these results.

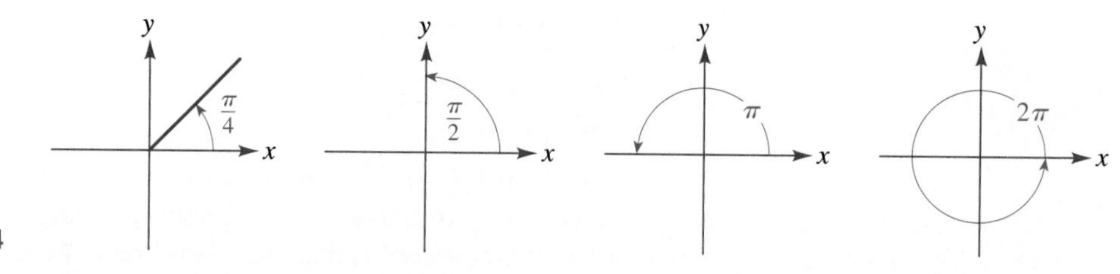

FIGURE 4

EXAMPLE 2

(a) Express 1° in radian measure.
(b) Express 1 radian in terms of degrees.

Solution

(a) Dividing both sides of equation (4) by 180 yields

$$1° = \frac{\pi}{180} \text{ radian } (\approx 0.017 \text{ radian})$$

(b) Dividing both sides of equation (4) by π yields

$$\frac{180°}{\pi} = 1 \text{ radian}$$

In other words, 1 radian is approximately 180°/3.14, or 57.3°.

From the results in Example 2, we have the following rules for converting between radians and degrees.

> To convert from degrees to radians, multiply by $\pi/180°$. To convert from radians to degrees, multiply by $180°/\pi$.

EXAMPLE 3 Convert 150° to radians.

Solution

$$150°\left(\frac{\pi}{180°}\right) = \frac{5\pi}{6} \qquad \text{reducing the fraction}$$

Thus,

$$150° = \frac{5\pi}{6} \text{ radians}$$

EXAMPLE 4 Convert $11\pi/6$ radians to degrees.

Solution

$$\frac{11\pi}{6}\left(\frac{180°}{\pi}\right) = \frac{(11)(180°)}{6} = 11(30°) = 330°$$

So, $11\pi/6$ radians $= 330°$.

We saw in Example 2(b) that 1 radian is approximately 57°. It is also important to be able to visualize an angle of 1 radian without thinking in terms of degree measure. This is done as follows. In the equation $\theta = s/r$, we let $\theta = 1$. This yields $1 = s/r$ and consequently $r = s$. In other words, in a circle, 1 radian is the central angle that intercepts an arc equal in length to the radius of the circle. In the box that follows we summarize this result. Figure 5 (on the next page) then displays angles of 1, 2, and 3 radians.

PROPERTY SUMMARY **RADIAN MEASURE**

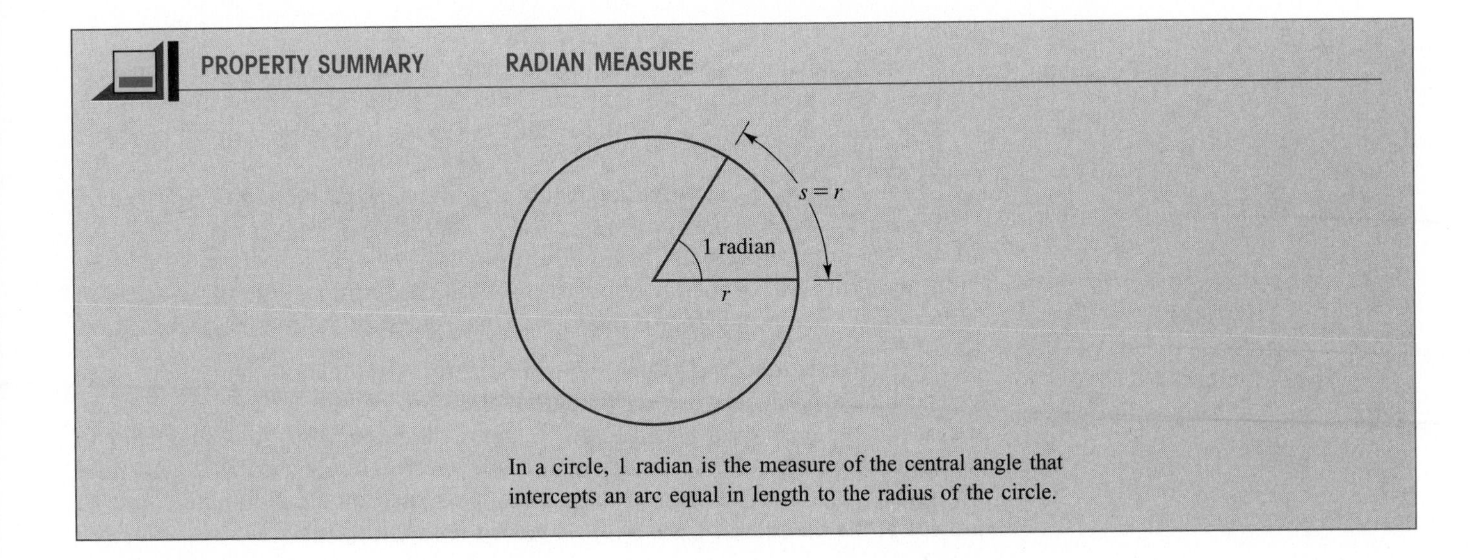

In a circle, 1 radian is the measure of the central angle that intercepts an arc equal in length to the radius of the circle.

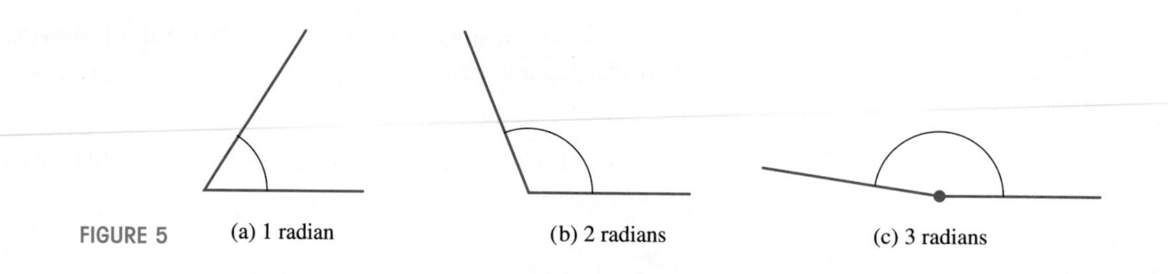

FIGURE 5 (a) 1 radian (b) 2 radians (c) 3 radians

From now on, when we specify the measure of an angle, we will assume that the units are radians unless the degree symbol is explicitly used. (This convention is also used in calculus.) For instance, the equation $\theta = 2$ means that θ is 2 radians. Similarly, the expression $\sin(\pi/6)$ refers to the sine of $\pi/6$ radians.

EXAMPLE 5 Evaluate the following quantities:

(a) $\sin \dfrac{\pi}{6}$; **(b)** $\cos 2\pi$.

Solution

(a) $\dfrac{\pi}{6}$ radians $= 30°$, therefore

$$\sin \frac{\pi}{6} = \sin 30° = \frac{1}{2} \qquad \text{because } \sin 30° = \frac{1}{2}$$

(b) 2π radians $= 360°$, therefore

$$\cos 2\pi = \cos 360° = 1 \qquad \text{because } \cos 360° = 1$$

EXAMPLE 6 Without using a calculator, determine whether the following values are positive or negative:

(a) $\cos 3$; **(b)** $\sin 1$; **(c)** $\tan 6$.

Solution

(a) Since π radians is 180°, we estimate that 3 radians is slightly less than 180° (because 3 is slightly less than π). Thus, in standard position, the terminal side of an angle of 3 radians lies in the second quadrant. Therefore, $\cos 3$ is negative.

(b) Since 1 radian is approximately 60° [as we saw in Example 2(b) and in Figure 5], $\sin 1$ is positive.

(c) We know that 6 radians is slightly less than 2π radians, which is one revolution, or 360°. Thus, in standard position, the terminal side of an angle of 6 radians lies in the fourth quadrant. Consequently, $\tan 6$ is negative.

As we mentioned in Chapter 6, all scientific calculators have keys with which you can indicate whether you are working in degrees or radians. In preparation for the next example, refer to the owner's manual for your calculator to make sure you know how it operates in this respect. [Here's a quick way to find out if you've properly set the calculator in the radian mode. Compute $\sin(\pi/6)$ and see if the calculator reads 0.5, as indeed it should.]

EXAMPLE 7 Use Figure 6 to approximate to within successive tenths the following trigonometric values. Then use a calculator to check your answers. Round off the calculator values to two decimal places:

(a) $\cos 3$ and $\sin 3$; **(b)** $\cos(-3)$ and $\sin(-3)$.

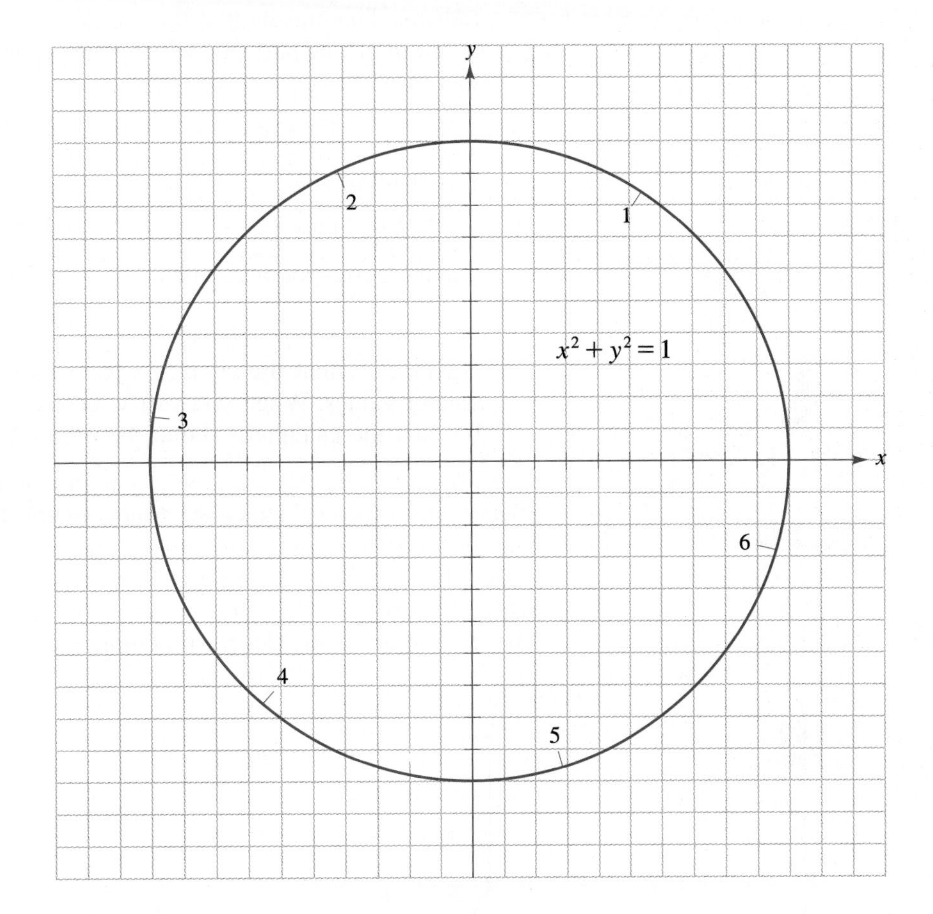

FIGURE 6
Radian measure on the unit circle for angles in standard position.

Solution

(a) Figure 6 shows that (the terminal side of) an angle of 3 radians in standard position meets the unit circle at a point in the second quadrant. Letting (x, y) denote the coordinates of that point, we have (using the grid in Figure 6)

$$-1.0 < x < -0.9 \quad \text{and} \quad 0.1 < y < 0.2$$

But, by definition, $\cos 3 = x$ and $\sin 3 = y$, so, consequently,

$$-1.0 < \cos 3 < -0.9 \quad \text{and} \quad 0.1 < \sin 3 < 0.2$$

The corresponding calculator values here are

$$\cos 3 \approx -0.99 \quad \text{and} \quad \sin 3 \approx 0.14$$

Note that the estimates we obtained from Figure 6 are consistent with these calculator values. (When you verify these calculator values for yourself, be sure that the calculator is set in the radian mode.)

(b) In Figure 6, we want to know where the terminal side of an angle of -3 radians intersects the unit circle. There is no marker for this point in Figure 6. But we can locate the point, nevertheless, by using symmetry. Let (x, y) denote the point corresponding to 3 radians, determined in part (a). By reflecting the point in the x-axis, we obtain the point corresponding to -3 radians. This means that the required point will have the same x-coordinate as the point in part (a), while the y-coordinate will be just the negative of that in part (a). So, given that

$$-1.0 < \cos 3 < -0.9 \qquad \text{and} \qquad 0.1 < \sin 3 < 0.2$$

we conclude that

$$-1.0 < \cos(-3) < -0.9 \qquad \text{and} \qquad -0.2 < \sin(-3) < -0.1$$

For the corresponding calculator values here, we obtain

$$\cos(-3) \approx -0.99 \qquad \text{and} \qquad \sin(-3) \approx -0.14$$

Again, note that the estimates from Figure 6 are consistent with these calculator values. When you verify these calculator values for yourself, remember that the calculator should be in the radian mode.

We conclude this section by pointing out that our three-step procedure (from Section 6.4) for evaluating the trigonometric functions is applicable whether we are working in degrees or radians. As background for this, Figure 7 displays examples of reference-angle calculations that use radian measure. You should carefully compare this with Figure 9 on page 388 where the same calculations are carried out using degree measure.

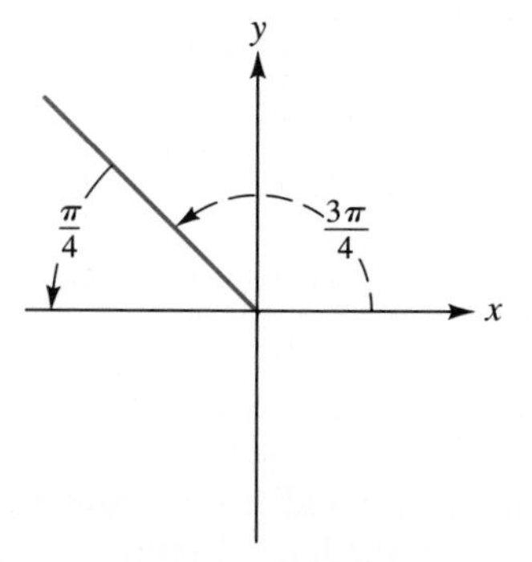

(a) The reference angle for $3\pi/4$ is $\pi/4$. $[\pi - (3\pi/4) = \pi/4]$

(b) The reference angle for $5\pi/3$ is $\pi/3$. $[2\pi - (5\pi/3) = \pi/3]$

(c) The reference angle for $7\pi/6$ is $\pi/6$. $[(7\pi/6) - \pi = \pi/6]$

(d) The reference angle for $-\pi/4$ is $\pi/4$.

FIGURE 7

EXAMPLE 8 Use the three-step procedure to evaluate each of the following expressions.

(a) $\cos\dfrac{3\pi}{4}$; **(b)** $\tan\dfrac{7\pi}{6}$; **(c)** $\csc\left(-\dfrac{\pi}{4}\right)$.

Solution

(a) **STEP 1** As indicated in Figure 7(a), the reference angle for $3\pi/4$ is $\pi/4$.
STEP 2 $\cos(\pi/4) = \sqrt{2}/2$

STEP 3 $\cos\theta$ is the x-coordinate. In Quadrant II, x-coordinates are negative, thus $\cos(3\pi/4)$ is negative, since the terminal side of $\theta = 3\pi/4$ lies in Quadrant II. Consequently we have $\cos(3\pi/4) = -\sqrt{2}/2$.

(b) STEP 1 As indicated in Figure 7(c), the reference angle for $7\pi/6$ is $\pi/6$.

STEP 2 $\tan(\pi/6) = \sqrt{3}/3$ (How do we know this?)

STEP 3 By definition, $\tan\theta = y/x$. Now, the terminal side of $\theta = 7\pi/6$ lies in Quadrant III, in which both x and y are negative. Thus the quotient y/x is positive, and we conclude that $\tan(7\pi/6) = \sqrt{3}/3$.

(c) STEP 1 As indicated in Figure 7(d), the reference angle for $-\pi/4$ is $\pi/4$.

STEP 2 $\csc(\pi/4) = \dfrac{1}{\sin(\pi/4)} = \dfrac{1}{1/\sqrt{2}} = \sqrt{2}$

STEP 3 By definition, $\csc\theta = 1/y$. Now, the terminal side of $\theta = -\pi/4$ lies in Quadrant IV, in which y is negative. Consequently $1/y$ is negative, and we have $\csc(-\pi/4) = -\sqrt{2}$.

Do 1, 5, 9, 11, 15, 19, 31, 35, 41

EXERCISE SET 7.1

A

In Exercises 1–4, use the definition $\theta = s/r$ to determine the radian measure of each angle.

1\. 2.

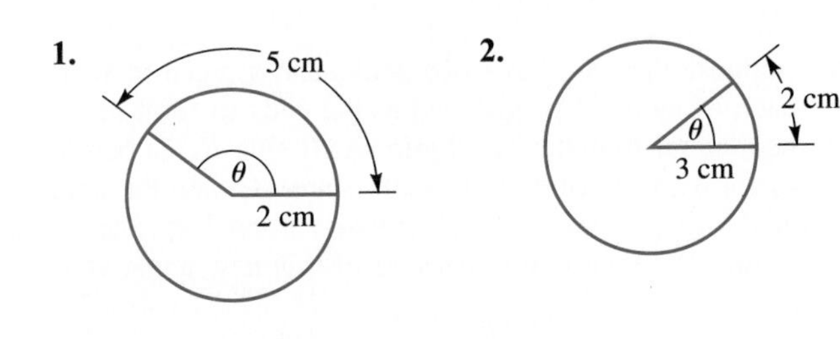

3\.

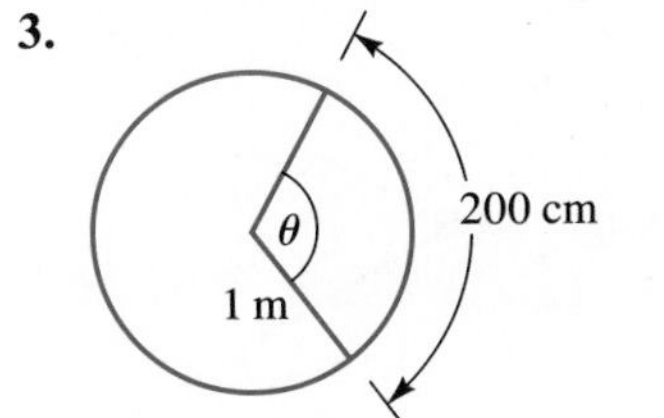

4\.

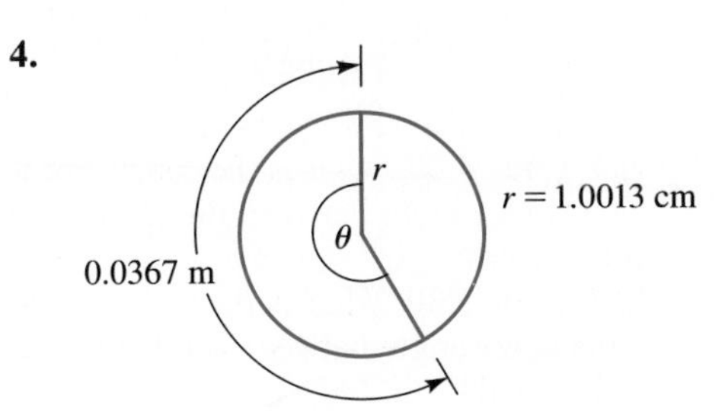

In Exercises 5–8, convert to radian measure. Express your answers both in terms of π and as decimal approximations rounded to two decimal places.

5\. **(a)** 45° **(b)** 90° **(c)** 135°

6\. **(a)** 30° **(b)** 150° **(c)** 300°

7\. **(a)** 0° **(b)** 360° **(c)** 450°

8\. **(a)** 36° **(b)** 35° **(c)** 720°

In Exercises 9–12, convert the radian measures to degrees.

9\. **(a)** $\pi/12$ **(b)** $\pi/6$ **(c)** $\pi/4$

10\. **(a)** 3π **(b)** $3\pi/2$ **(c)** 2π

11\. **(a)** $\pi/3$ **(b)** $5\pi/3$ **(c)** 4π

12\. **(a)** $5\pi/6$ **(b)** $11\pi/6$ **(c)** 0

In Exercises 13 and 14, convert the radian measures to degrees. Round the answers to two decimal places.

13\. **(a)** 2 **(b)** 3 **(c)** π^2

14\. **(a)** 1.32 **(b)** 0.96 **(c)** $1/\pi$

15\. Suppose that the radian measure of an angle is 3/2. Without using a calculator or tables, determine if this angle is larger or smaller than a right angle.
Hint: What is the radian measure of a right angle?

16\. Two angles in a triangle have radian measure $\pi/5$ and $\pi/6$. What is the radian measure of the third angle?

For Exercises 17 and 18, refer to the following figure, which shows all of the angles from 0° to 360° that are multiples of 30° or 45°.

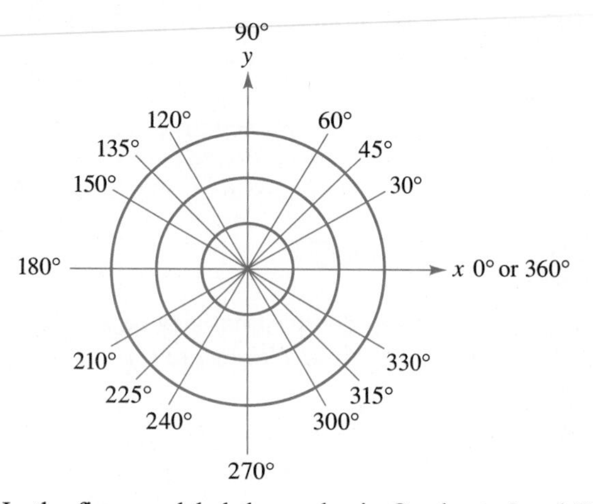

17. In the figure, relabel the angles in Quadrants I and II using radian measure.

18. In the figure, relabel the angles in Quadrants III and IV using radian measure.

In Exercises 19 and 20, complete the tables.

19.

θ	$\sin\theta$	$\cos\theta$	$\tan\theta$
0			
$\pi/2$			
π			
$3\pi/2$			
2π			

20.

θ	$\csc\theta$	$\sec\theta$	$\cot\theta$
0			
$\pi/2$			
π			
$3\pi/2$			
2π			

In Exercises 21–30, use Figure 6 to approximate the given trigonometric values to within successive tenths. Then use a calculator to compute the values to the nearest hundredth.

21. $\cos 1$ and $\sin 1$

22. $\cos 2$ and $\sin 2$

23. $\cos(-1)$ and $\sin(-1)$

24. $\cos(-2)$ and $\sin(-2)$

25. $\cos 4$ and $\sin 4$

26. $\cos 5$ and $\sin 5$

27. $\cos(-4)$ and $\sin(-4)$

28. $\cos(-5)$ and $\sin(-5)$

29. $\sin(1+2\pi)$

30. $\sin(2+2\pi)$

In Exercises 31–40, evaluate each expression (as in Example 8).

31. **(a)** $\cos(5\pi/6)$ **(b)** $\cos(-5\pi/6)$ **(c)** $\sin(5\pi/6)$ **(d)** $\sin(-5\pi/6)$

32. **(a)** $\cos(5\pi/3)$ **(b)** $\cos(-5\pi/3)$ **(c)** $\sin(5\pi/3)$ **(d)** $\sin(-5\pi/3)$

33. **(a)** $\cos(4\pi/3)$ **(b)** $\cos(-4\pi/3)$ **(c)** $\sin(4\pi/3)$ **(d)** $\sin(-4\pi/3)$

34. **(a)** $\cos(2\pi/3)$ **(b)** $\cos(-2\pi/3)$ **(c)** $\sin(2\pi/3)$ **(d)** $\sin(-2\pi/3)$

35. **(a)** $\cos(5\pi/4)$ **(b)** $\cos(-5\pi/4)$ **(c)** $\sin(5\pi/4)$ **(d)** $\sin(-5\pi/4)$

36. **(a)** $\cos(17\pi/6)$ **(b)** $\cos(-17\pi/6)$ **(c)** $\sin(17\pi/6)$ **(d)** $\sin(-17\pi/6)$

37. **(a)** $\sec(4\pi/3)$ **(b)** $\csc(-4\pi/3)$ **(c)** $\tan(4\pi/3)$ **(d)** $\cot(-4\pi/3)$

38. **(a)** $\sec(7\pi/4)$ **(b)** $\csc(-7\pi/4)$ **(c)** $\tan(7\pi/4)$ **(d)** $\cot(-7\pi/4)$

39. **(a)** $\sec(17\pi/6)$ **(b)** $\csc(-17\pi/6)$ **(c)** $\tan(17\pi/6)$ **(d)** $\cot(-17\pi/6)$

40. **(a)** $\sec(3\pi/4)$ **(b)** $\csc(-3\pi/4)$ **(c)** $\tan(3\pi/4)$ **(d)** $\cot(-3\pi/4)$

B

41. Suppose that you have two sticks and a piece of wire, each of length 1 ft, fastened at the ends to form an equilateral triangle; see Figure A. If side $\overline{BC}$ is bent out to form an arc of a circle with center A, then the angle at A will decrease from 60° to something less. See Figure B. What is the measure of this new angle at A?

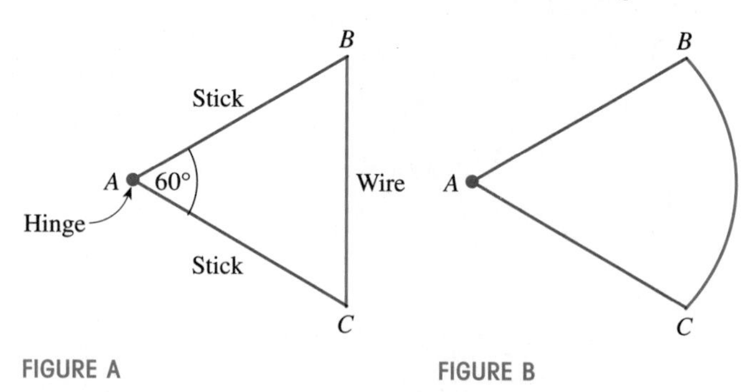

FIGURE A FIGURE B

42. **(a)** When a clock reads 4:00, what is the radian measure of the (smaller) angle between the hour hand and the minute hand?

(b) When a clock reads 5:30, what is the radian measure of the (smaller) angle between the hour hand and the minute hand?

43. Are there any real numbers x with the property that x degrees equals x radians? If so, find them; if not, explain why not.

44. Are there any real numbers x with the property that x degrees equals $2x$ radians? If so, find them; if not, explain why not.

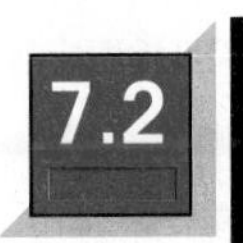

7.2 RADIAN MEASURE AND GEOMETRY

. . . I wrote to him [to Alexander J. Ellis, in 1874], *and he declared at once for the form "radian," on the ground that it could be viewed as a contraction for "radial angle. . . ."*

Thomas Muir, in a letter appearing in the April 7, 1910, issue of *Nature*

I shall be very pleased to send Dr. Muir a copy of my father's examination questions of June, 1873, containing the word radian. . . . *It thus appears that* radian *was thought of independently by Dr. Muir and my father, and, what is really more important than the exact form of the name, they both independently thought of the necessity of giving a name to the unit-angle.*

James Thomson, in a letter appearing in the June 16, 1910, issue of *Nature*

One of the advantages in using radian measure in precalculus and calculus is that many formulas then take on especially simple forms. The two basic formulas in the box that follows are examples of this.

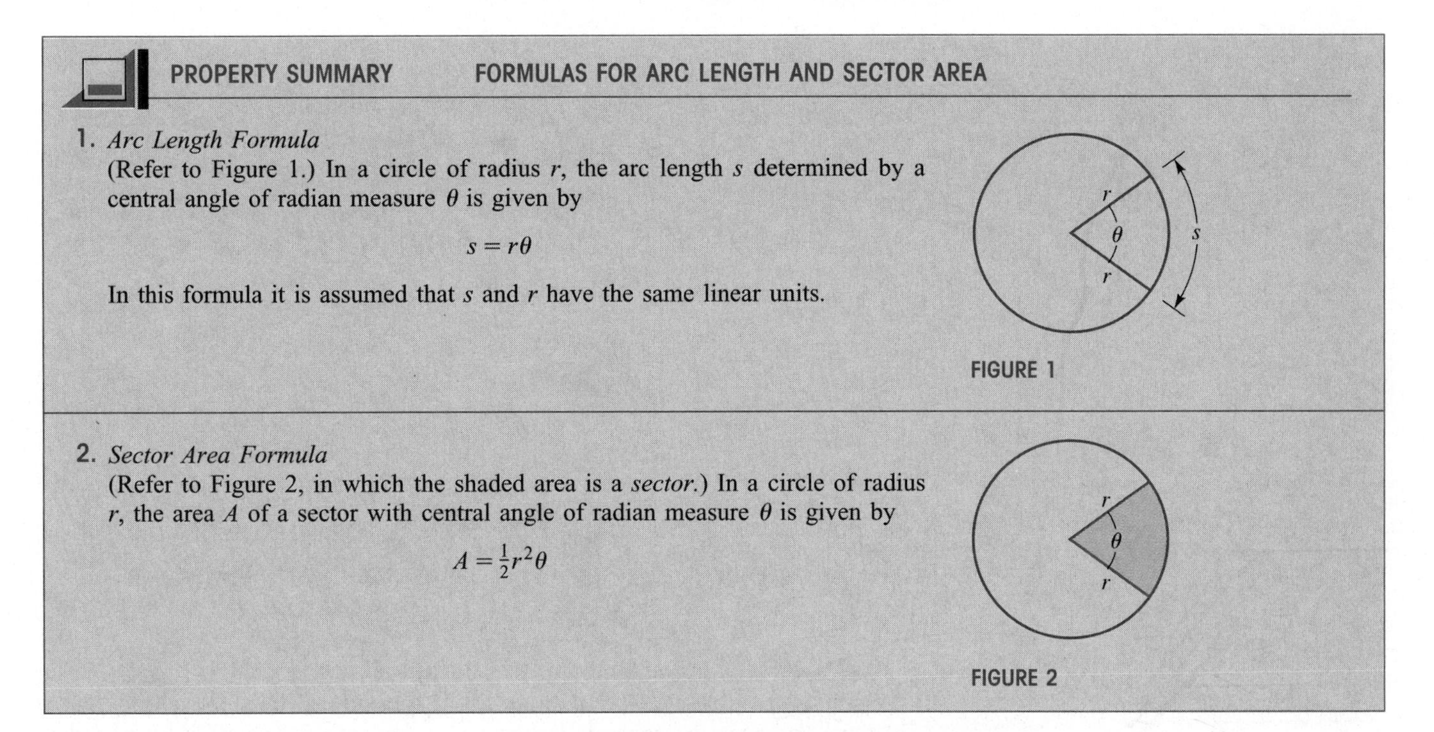

PROPERTY SUMMARY FORMULAS FOR ARC LENGTH AND SECTOR AREA

1. *Arc Length Formula*
(Refer to Figure 1.) In a circle of radius r, the arc length s determined by a central angle of radian measure θ is given by

$$s = r\theta$$

In this formula it is assumed that s and r have the same linear units.

FIGURE 1

2. *Sector Area Formula*
(Refer to Figure 2, in which the shaded area is a *sector.*) In a circle of radius r, the area A of a sector with central angle of radian measure θ is given by

$$A = \frac{1}{2}r^2\theta$$

FIGURE 2

The arc length formula is a direct consequence of the definition of radian measure. Recall from the previous section that the defining equation for radian measure is $\theta = s/r$. Multiplying both sides of this equation by r yields $\theta r = s$; that is, $s = r\theta$, which is the arc length formula.

To derive the sector area formula, we'll rely on the following two facts from elementary geometry. First, the area of a circle of radius r is πr^2. Second, the area A of a sector is directly proportional to the measure θ of its central angle. (For example, if you double the angle, the area of the sector is doubled.) The statement that the area A is directly proportional to the angle θ is expressed algebraically as

$$A = k\theta \tag{1}$$

To determine the value of the constant k, we use the fact that when $\theta = 2\pi$, the sector is actually a full circle with area πr^2. Substituting the values $\theta = 2\pi$ and $A = \pi r^2$ in equation (1) yields

$$\pi r^2 = k(2\pi)$$

and therefore

$$k = \tfrac{1}{2}r^2$$

Using this last value for k in equation (1), we obtain $A = \frac{1}{2}r^2\theta$, as required. The next three examples indicate how the arc length formula and the sector area formula are used in calculations.

EXAMPLE 1 Compute the indicated arc lengths in Figures 3(a) and 3(b). Express the answers both in terms of π and as decimal approximations rounded to two decimal places.

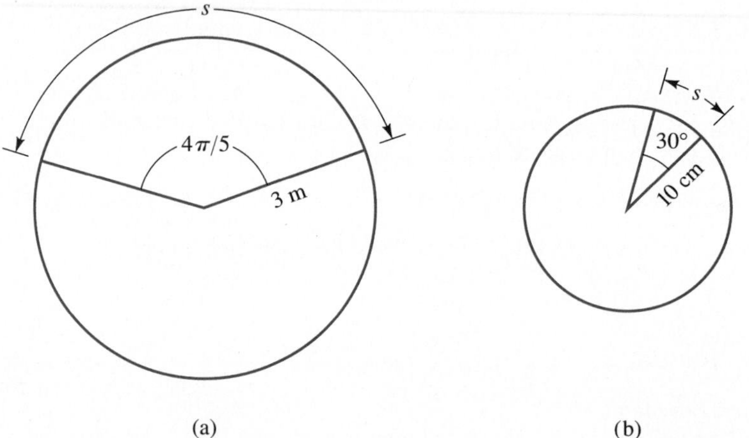

FIGURE 3

Solution

(a) We use the formula $s = r\theta$ with $r = 3$ m and $\theta = 4\pi/5$. This yields

$$s = (3 \text{ m})\left(\tfrac{4\pi}{5}\right) = \tfrac{12\pi}{5} \text{ m}$$

Using a calculator, we find that this is approximately 7.54 m.

(b) To apply the formula $s = r\theta$, the angle measure θ must be expressed in radians (because the formula is just a restatement of the definition of radian measure). We've seen previously that $30° = \pi/6$ radians. So we have

$$s = r\theta = (10 \text{ cm})\left(\tfrac{\pi}{6}\right) = \tfrac{5\pi}{3} \text{ cm}$$

Before picking up the calculator, note that we can obtain a quick approximation here: since $\pi \approx 3$, we have $s \approx 5(3/3) = 5$ cm. With a calculator now, we obtain $s = \frac{5\pi}{3}$ cm ≈ 5.24 cm.

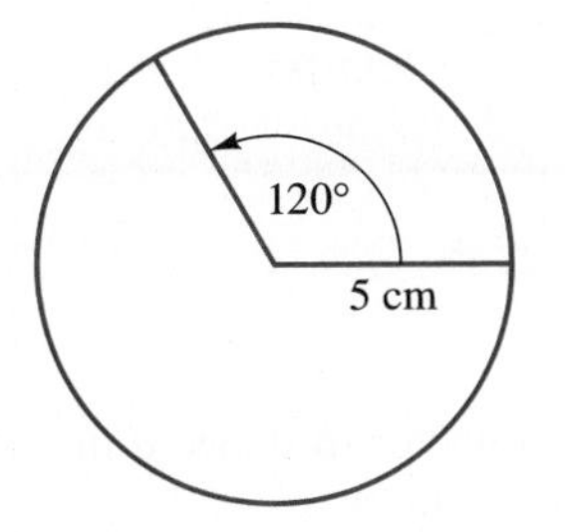

FIGURE 4

EXAMPLE 2 Compute the area of the sector in Figure 4.

Solution We first convert 120° to radians:

$$120°\left(\frac{\pi}{180°}\right) = \frac{2\pi}{3} \text{ radians}$$

We then have

$$\begin{aligned} A &= \tfrac{1}{2}r^2\theta \\ &= \tfrac{1}{2}(5 \text{ cm})^2\left(\tfrac{2\pi}{3}\right) = \tfrac{25\pi}{3} \text{ cm}^2 \ (\approx 26.18 \text{ cm}^2) \end{aligned}$$

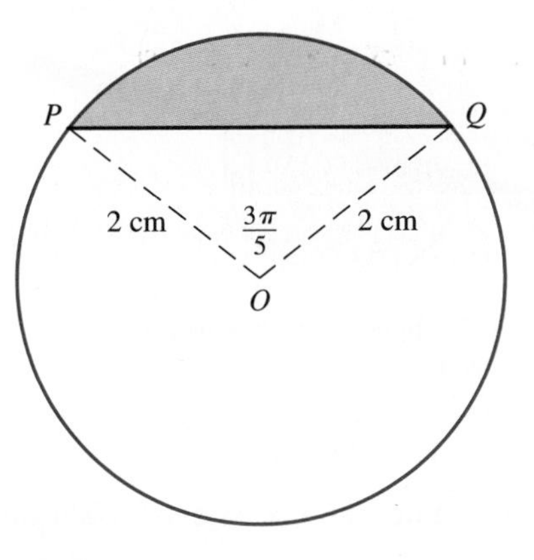

FIGURE 5

EXAMPLE 3 A **segment** of a circle is a region bounded by an arc of the circle and its chord. Compute the area of the segment (the shaded region) in Figure 5. Give two forms for the answer: an exact expression involving π and a calculator approximation rounded to two decimal places.

Solution Although we have not developed an explicit formula for the area of a segment, notice in Figure 5 that $\triangle OPQ$ and the shaded segment, taken together, form a sector of the circle. So we have

$$\text{area of segment} = (\text{area of sector } OPQ) - (\text{area of } \triangle OPQ)$$

For the area of the sector, we compute

$$\begin{aligned} \text{area of sector } OPQ &= \frac{1}{2}r^2\theta = \frac{1}{2}(2^2)\left(\frac{3\pi}{5}\right) \\ &= \frac{6\pi}{5} \text{ cm}^2 \end{aligned}$$

For the area of the triangle, we use the formula $A = \frac{1}{2}ab \sin\theta$ (from Section 6.4):

$$\text{area of } \triangle OPQ = \frac{1}{2}(2)(2)\sin\frac{3\pi}{5} = 2\sin\frac{3\pi}{5} \text{ cm}^2$$

Putting things together now, we obtain

$$\begin{aligned} \text{area of segment} &= \left(\frac{6\pi}{5} - 2\sin\frac{3\pi}{5}\right) \text{cm}^2 \\ &\approx 1.87 \text{ cm}^2 \end{aligned}$$

The next example indicates how radian measure is used in the study of rotating objects. As a prerequisite for this example, we first define the terms *angular speed* and *linear speed.*

DEFINITION: Angular Speed and Linear Speed

Suppose that a wheel rotates about its axis at a constant rate.

1. Refer to Figure 6(a). If a radial line turns through an angle θ in time t, then the **angular speed** of the wheel [denoted by the Green letter ω *(omega)*] is defined to be

$$\text{angular speed} = \omega = \frac{\theta}{t}$$

2. Refer to Figure 6(b). If a point P on the rotating wheel travels a distance d in time t, then the **linear speed** of P, denoted by v, is defined to be

$$\text{linear speed} = v = \frac{d}{t}$$

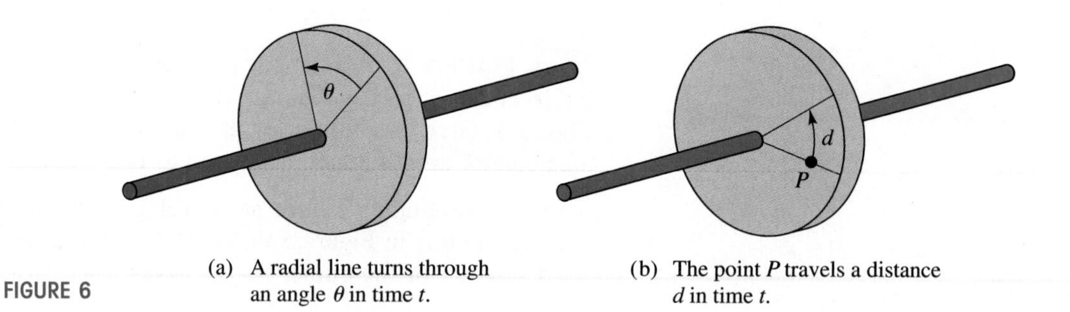

FIGURE 6

(a) A radial line turns through an angle θ in time t.

(b) The point P travels a distance d in time t.

EXAMPLE 4 A circular gear in a motor rotates at the rate of 100 rpm (revolutions per minute).

(a) What is the angular speed of the gear in radians per minute?
(b) Find the linear speed of a point on the gear 4 cm from the center.

Solution

(a) Each revolution of the gear is 2π radians. So in 100 revolutions, there are

$$\theta = 100(2\pi) = 200\pi \text{ radians}$$

Consequently, we have

$$\omega = \frac{\theta}{t} = \frac{200\pi \text{ radians}}{1 \text{ min}} = 200\pi \text{ radians/min}$$

(b) We can use the formula $s = r\theta$ to find the distance traveled by the point in 1 minute. Using $r = 4$ cm and $\theta = 200\pi$, we obtain

$$s = r\theta = 4(200\pi) = 800\pi \text{ cm}$$

The linear speed is therefore

$$v = \frac{d}{t} = \frac{800\pi \text{ cm}}{1 \text{ min}}$$
$$= 800\pi \text{ cm/min } (\approx 2513 \text{ cm/min})$$

There is a simple equation relating linear speed and angular speed. [Actually, as you'll see, the explanation we are about to give is just a repetition of the reasoning used in Example 4(b).] We refer back to Figure 6, in which the radial line turns through an angle θ in time t. Now we make two additional assumptions. First, we assume that the angle θ is measured in radians (so that we can apply the formula $s = r\theta$). Second, we assume that the point P is a distance r from the center of the circle. That is, the radius of the circular path is r. Then, according to our arc length formula, the distance d in Figure 6(b) is given by $d = r\theta$. So we have

$$\begin{aligned} v = \frac{d}{t} = \frac{r\theta}{t} &= r \times \frac{\theta}{t} \\ &= r\omega \qquad \text{using the definition of } \omega \end{aligned}$$

We have now shown that $v = r\omega$. For reference, we record this result in the box that follows.

LINEAR SPEED AND ANGULAR SPEED

Suppose that an object travels at a constant rate along a circular path of radius r. Then the linear speed v and the angular speed ω are related by the equation

$$v = r\omega$$

In this equation, the angular speed ω must be expressed in radians per unit time.

We can use the formula $v = r\omega$ to rework Example 4(b) in a more concise fashion. Recall that we wanted to find the linear speed v of a point 4 cm from the center. To do this, we need only substitute the values $r = 4$ cm and $\omega = 200\pi$ radians/min [from part (a) of the example] in the formula $v = r\omega$. This yields

$$\begin{aligned} v &= r\omega \\ &= (4 \text{ cm})(200\pi \text{ radians/min}) && (2) \\ &= 800\pi \text{ cm/min} \qquad \text{(as obtained previously)} && (3) \end{aligned}$$

Notice how the units work in equations (2) and (3). Although radians appear in equation (2), they do not appear in the final answer in equation (3); the units for the linear speed are simply cm/min. We drop the radians from the final result because, as pointed out in Section 7.1, radian measure is unitless.*

*Many physics texts gloss over this point in the calculations. One text that does provide sufficient emphasis here is *Physics,* 7th ed., by John D. Cutnell and Kenneth W. Johnson (New York: John Wiley & Sons, Inc., 1995). In that text, the authors state, "In calculations, therefore, the radian is treated as a unitless number and has no effect on other units that it multiplies or divides."

Do 1, 9, 11, 23

EXERCISE SET 7.2

A

In Exercises 1–4, find the arc length s in each case.

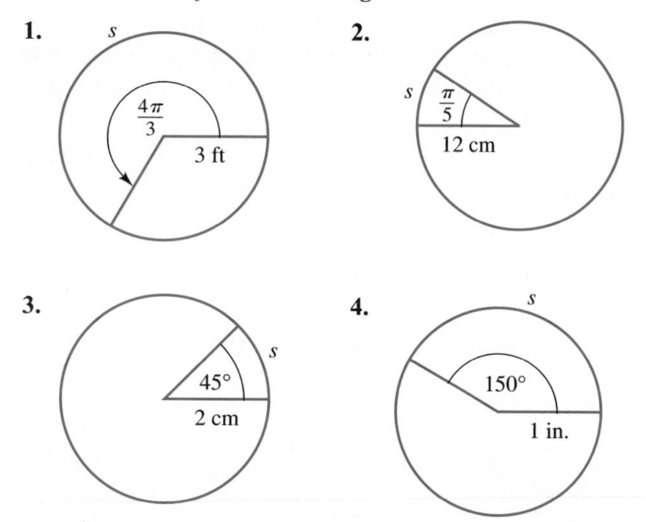

In Exercises 5 and 6, find the area of the sector determined by the given radius r and central angle θ. Express the answer both in terms of π and as a decimal approximation rounded to two decimal places.

5. **(a)** $r = 6$ cm; $\theta = 2\pi/3$ **(b)** $r = 5$ m; $\theta = 80°$
(c) $r = 24$ m; $\theta = \pi/20$ **(d)** $r = 1.8$ cm; $\theta = 144°$

6. **(a)** $r = 4$ cm; $\theta = \pi/10$ **(b)** $r = 16$ m; $\theta = 5°$
(c) $r = 21$ ft; $\theta = 11\pi/6$ **(d)** $r = 4.2$ in.; $\theta = 170°$

7. In a circle of radius 1 cm, the area of a certain sector is $\pi/5$ cm^2. Find the radian measure of the central angle. Express the answer in terms of π rather than as a decimal approximation.

8. In a circle of radius 3 m, the area of a certain sector is 20 m^2. Find the degree measure of the central angle. Round the answer to two decimal places.

In Exercises 9 and 10: **(a)** *find the perimeter of the sector; and* **(b)** *find the area of the sector. In each case, use a calculator to evaluate the answer and round to two decimal places.*

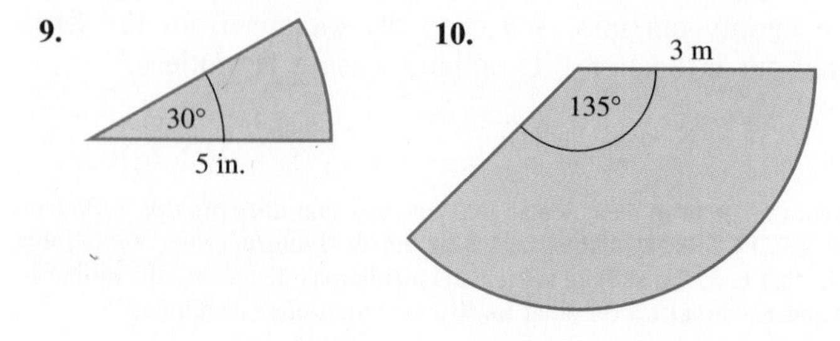

In Exercises 11 and 12, compute the area of the shaded segment of the circle. Give two forms for each answer: an exact expression and a calculator approximation rounded to two decimal places. In the figure for Exercise 11, recall the following convention (mentioned in Section 7.1): If degree measure is not explicitly indicated, assume that radian measure is used.

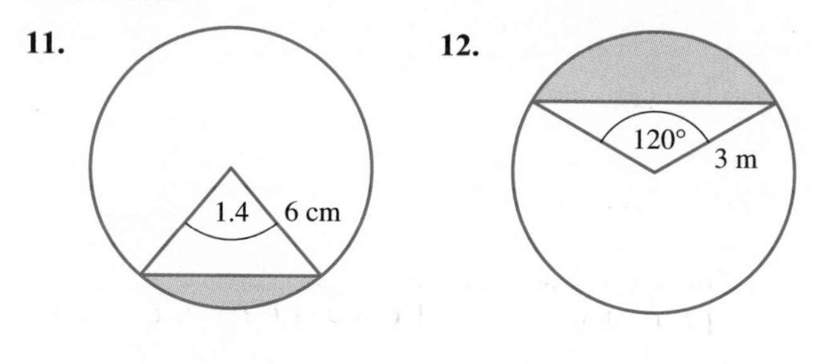

In Exercises 13–18, you are given the rate of rotation of a wheel as well as its radius. In each case, determine the following:
(a) *the angular speed, in units of radians/sec;*
(b) *the linear speed, in units of cm/sec, of a point on the circumference of the wheel;*
(c) *the linear speed, in cm/sec, of a point halfway between the center of the wheel and the circumference.*

13. 6 revolutions/sec; $r = 12$ cm

14. 15 revolutions/sec; $r = 20$ cm

15. 1080°/sec; $r = 25$ cm

16. 2160°/sec; $r = 60$ cm

17. 500 rpm; $r = 45$ cm

18. 1250 rpm; $r = 10$ cm

19. For this problem, assume that the earth is a sphere with a radius of 3960 miles and a rotation rate of 1 revolution per 24 hours.
(a) Find the angular speed. Express your answer in units of radians/sec, and round to two significant digits.
(b) Find the linear speed of a point on the equator. Express the answer in units of miles per hour, and round to the nearest 10 mph.

20. A wheel 3 ft in diameter makes x revolutions. Find x, given that the distance traveled by a point on the circumference of the wheel is 22619 ft. (Round your answer to the nearest whole number.)

B

In Exercises 21 and 22, suppose that a belt drives two wheels of radii r and R, as indicated in the figure.

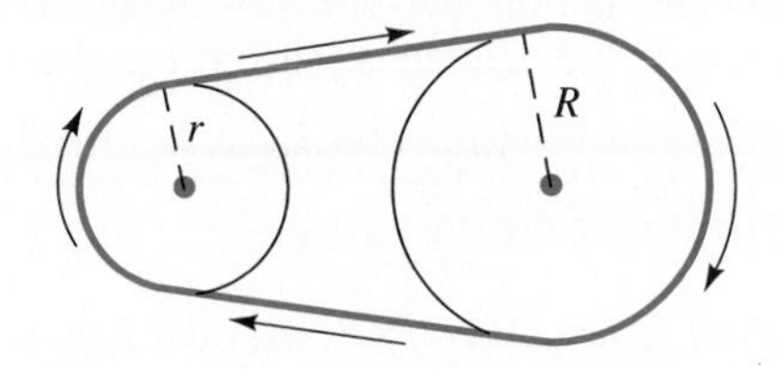

21. If $r = 6$ cm, $R = 10$ cm, and the angular speed of the larger wheel is 100 rpm, determine each of the following:
(a) the angular speed of the larger wheel in radians per minute;
(b) the linear speed of a point on the circumference of the larger wheel;
(c) the angular speed of the smaller wheel in radians per minute. *Hint:* Because of the belt, the linear speed of a point on the circumference of the larger wheel is equal to the linear speed of a point on the circumference of the smaller wheel.
(d) The angular speed of the smaller wheel in rpm.

22. Follow Exercise 21, assuming that $r = 5$ cm, $R = 15$ cm, and the angular speed of the larger wheel is 1800 rpm.

The latitude of a point P on the surface of the Earth is specified by means of the angle θ in the figure. For instance, the latitude of Paris, France, is 48°52′ N. The letter N is used here to indicate that the location is north of, rather than south of, the equator. (Recall that the notation 52′ indicates 52/60 of one degree.) In Exercises 23–28, use the arc length formula (and your calculator) to determine the distance $\widehat{PE}$ from the given location P to the equator. Assume that the Earth is a sphere with radius OP = OE = 3960 miles. Round each answer to the nearest 10 miles.

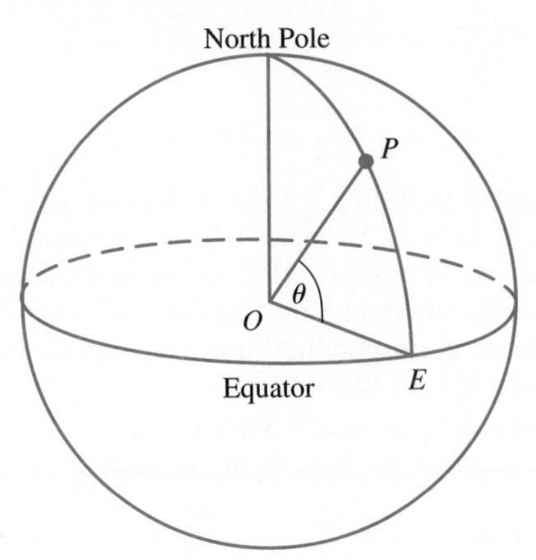

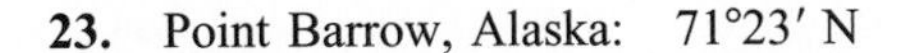

23. Point Barrow, Alaska: 71°23′ N

24. Singapore: 1°17′ N

25. Honolulu: 21°19′ N

26. Lagos: 6°27′ N

27. Washington, D.C.: 38°54′ N

28. Fairbanks, Alaska: 64°51′ N

29. Many of the window designs used in gothic architecture involve circles, sectors, and segments of circles. The **equilateral arch** in Figure A is an example of this. Figure B shows how the arch is designed. Starting with the equilateral triangle ABC, circular arc $\widehat{AC}$ is drawn with center B and radius AB. Similarly, circular arc $\widehat{BC}$ is drawn with center A and radius AB.

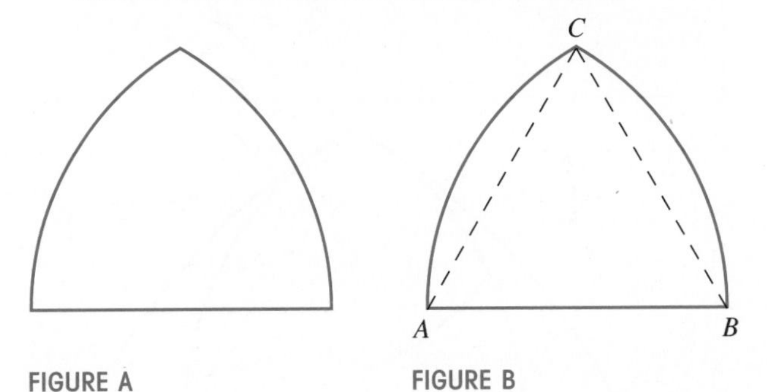

FIGURE A FIGURE B

(a) Let s denote the length of a side of the equilateral triangle in Figure B. Express the perimeter of the equilateral arch in terms of s.
(b) Express the area of the equilateral arch in terms of s.

30. Closely related to the equilateral arch in Exercise 29 is the **equilateral curved triangle** shown in Figure C. Just as with the equilateral arch in Exercise 29, the design begins with the equilateral triangle ABC. Circular arcs are then constructed on each side of the triangle, following the method explained in Exercise 29.

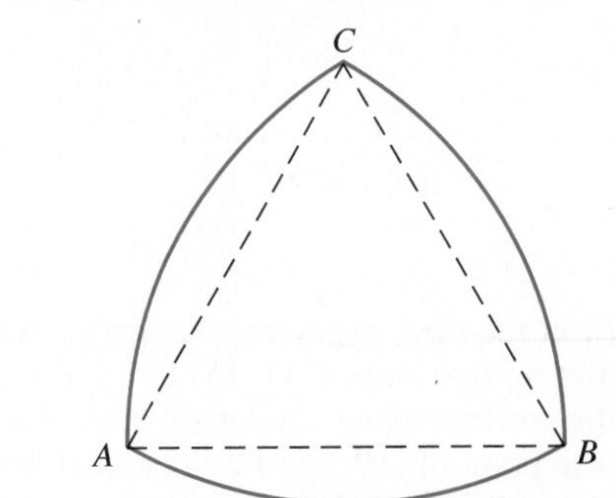

FIGURE C

(a) In Figure C, let s denote the length of a side of the equilateral triangle ABC. Express the perimeter and the area of the equilateral curved triangle in terms of s.
(b) Show that the area of the equilateral triangle ABC is approximately 61% of the area of the equilateral curved triangle ABC.

31. Figure D shows one of the gothic window designs used in Wells Cathedral, England. (The cathedral was constructed in the mid-thirteenth century.) Figure E indicates how the design is formed. We start with the equilateral $\triangle ABC$ and construct the equilateral arch *ABC*. Next, the midpoints of the three sides of $\triangle ABC$ are joined to create four smaller equilateral triangles. The two equilateral triangles *ADF* and *FEC* are then used to construct the two smaller equilateral arches shown in Figure E. And finally, the equilateral triangle *DBE* is used to construct the equilateral curved triangle within the top half of the figure.

(a) Let s denote the length of a side of the equilateral triangle *ABC*. Express the area of each of the equilateral arches *ABC* and *ADF* in terms of s. Also, find the ratio of the area of arch *ADF* to arch *ABC*.

(b) Express the area of the equilateral curved triangle *DBE* in terms of s.

(c) Express the area of the curved figure *DEF* in terms of s. (By "the curved figure *DEF*" we mean the region bounded by the circular arcs $\widehat{DE}$, $\widehat{EF}$, and $\widehat{FD}$.)

(d) Express the area of the curved figure *BEC* in terms of s.

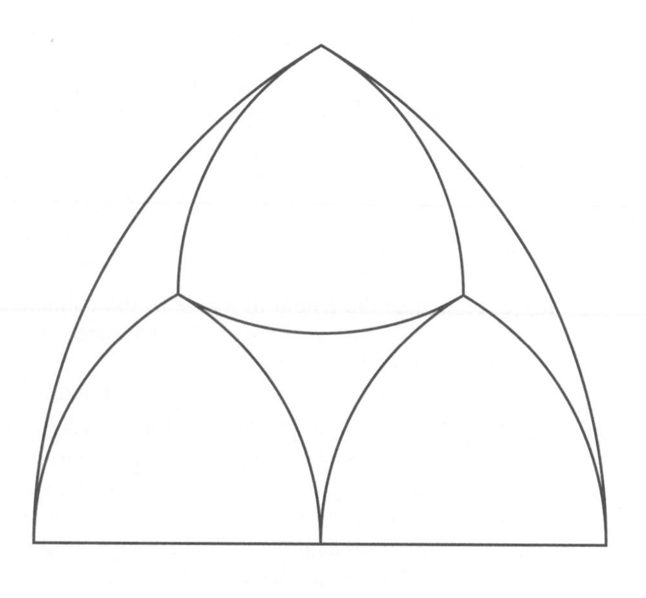

FIGURE D

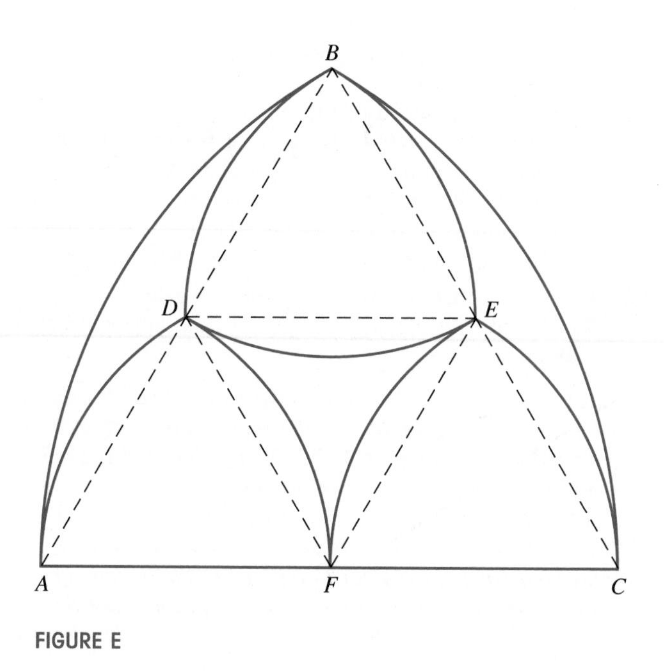

FIGURE E

32. Figure F (on the next page) shows a gothic window design from the cathedral at Reims, France. (The cathedral was constructed during the years 1211–1311.) Figure G indicates how the design is formed. Triangle *ABC* is equilateral and arch *ABC* is the corresponding equilateral arch. Equilateral triangle *GID*, congruent to triangle *ABC*, is constructed such that *D* is the midpoint of $\overline{AB}$, and $\overline{GI}$ is parallel to $\overline{AB}$. The points *E* and *F* are the midpoints of segments $\overline{DG}$ and $\overline{DI}$, respectively. The two small arches at the bottom of the figure are equilateral arches corresponding to the equilateral triangles *GHE* and *HIF*. For the circle in the figure, the center is *D* and the radius is $AD\ (= ED = FD)$.

(a) Let s denote the length of a side in each of the two equilateral triangles *ABC* and *GID*. Find the area and the perimeter (in terms of s) of the curved figure *AJBCA*. *Hint:* For the area, subtract the area of the semicircle *AJB* from the area of the equilateral arch *ABC*.

(b) Express (in terms of s) the area and the perimeter of equilateral arch *GHE*.

(c) Show that the area of the curved figure *EHF* is $s^2(2\sqrt{3} - \pi)/8$.

(d) Express (in terms of s) the perimeter and the area of the curved figure *FBI*.

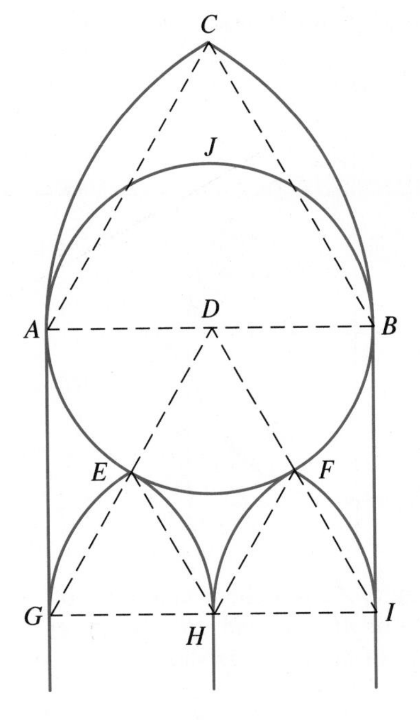

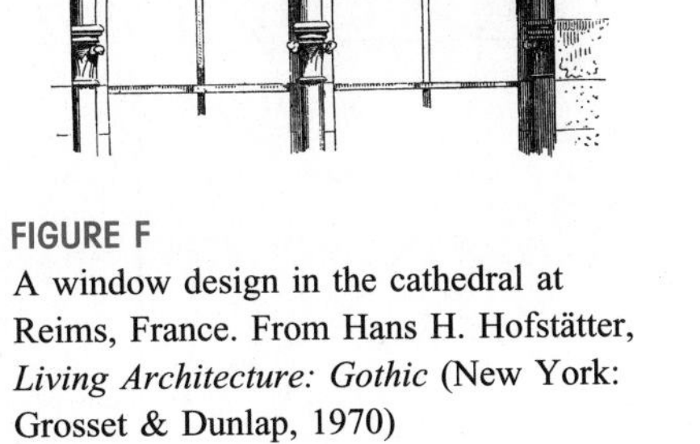

FIGURE F
A window design in the cathedral at Reims, France. From Hans H. Hofstätter, *Living Architecture: Gothic* (New York: Grosset & Dunlap, 1970)

FIGURE G

---

33. **(a)** The following figure shows a segment with central angle α in a circle of radius r. (Assume α is in radians.) Show that the area A of the segment is given by

$$A = \frac{1}{2}r^2(\alpha - \sin \alpha).$$

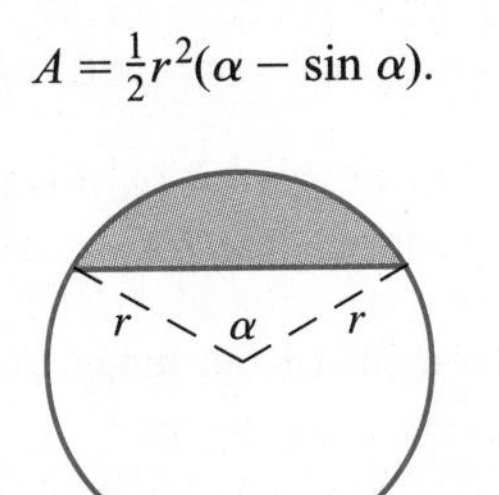

(b) In the following figure, O is the center of the circle, and the radius of the circle is 1 unit. Use the formula in part (a) to find the sum of the areas of the two shaded segments. Give two forms for the answer: an exact expression involving π and a calculator approximation rounded to two decimal places.

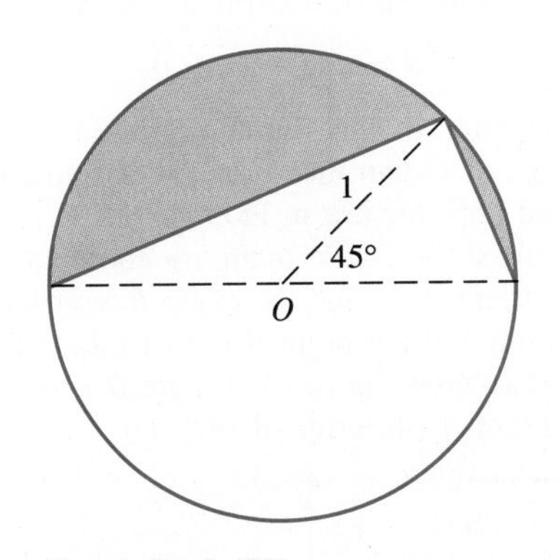

Figure for Exercise 33(b)

34. In the following figure (on the next page), arc ABC is a portion of a circle with center $D(0, -1)$ and radius $\overline{DC}$. The shaded crescent-shaped region is called a **lune.** Verify the following result, which was discovered (and proved) by the Greek mathematician Hippocrates of Chios approximately 2500 years ago. The area of the lune is equal to the area of the square $OCED$.
Hint: In computing the area of the lune, make

use of the formula given in Exercise 33(a) for the area of a segment.

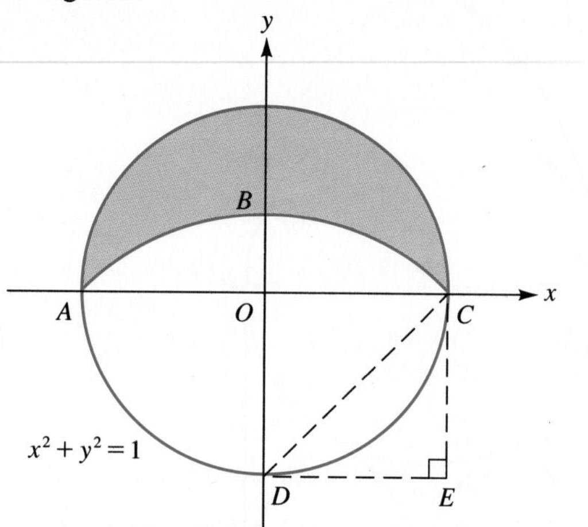

35. In the following figure, $\overline{AB}$ is a chord in a circle of radius 1. The length of $\overline{AB}$ is d, and $\overline{AB}$ subtends an angle θ at the center of the circle, as shown.

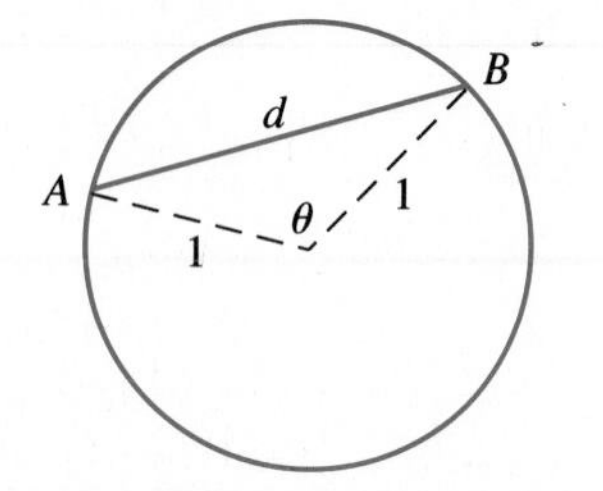

In this exercise we derive the following formula for the length d of the chord in terms of the angle θ:

$$d = \sqrt{2 - 2\cos\theta}$$

(The derivation of this formula does not require any knowledge of radian measure. The formula is developed here for use in Exercise 36.)

(a) We place the figure in an x-y coordinate system and orient it so that the angle θ is in standard position and the point B is located at (1, 0). (See the following figure.) What are the coordinates of the point A (in terms of θ)?

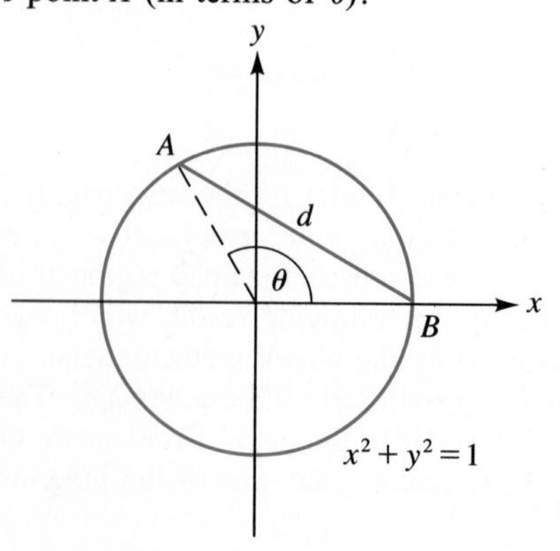

(b) Use the formula for the distance between two points to show that $d = \sqrt{2 - 2\cos\theta}$.

36. In the following figure, arc ABC is a semicircle with diameter $\overline{AC}$, and arc CDE is a semicircle with diameter $\overline{CE}$.

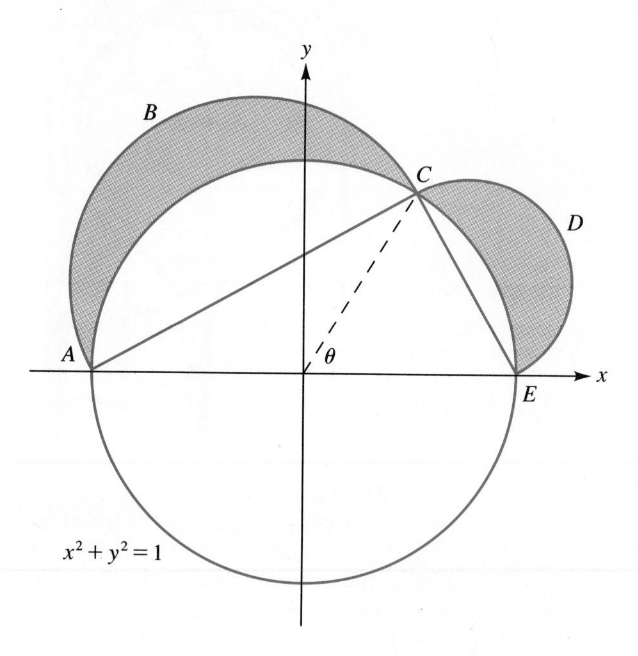

(a) Find the area of the lune CDE assuming that $\theta = 60°$. Give two forms for the answer: an exact expression involving π and a calculator approximation rounded to three decimal places. *Hint:* Use the formulas in Exercises 33(a) and 35.

(b) Follow part (a) for lune ABC, again assuming that $\theta = 60°$.

(c) Find the area of $\triangle ACE$ (still assuming that $\theta = 60°$). Then use the results in parts (a) and (b) to verify that the area of $\triangle ACE$ is equal to the sum of the areas of the lunes CDE and ABC.

REMARK In this exercise we used the value $\theta = 60°$. Actually, the result in part (c) holds in general: For any angle θ between 0° and 180°, the sum of the areas of the two lunes is equal to the area of the triangle. As with the result in Exercise 34, this fact was discovered and proved by the ancient Greek mathematician Hippocrates of Chios. According to Professor George F. Simmons in his text *Calculus Gems* (New York: McGraw-Hill Book Co., 1992), these results appear "to be the earliest precise determination of the area of a region bounded by curves."

37. The accompanying figure shows a circular sector with radius r cm and central angle θ (radian measure). The perimeter of the sector is 12 cm.

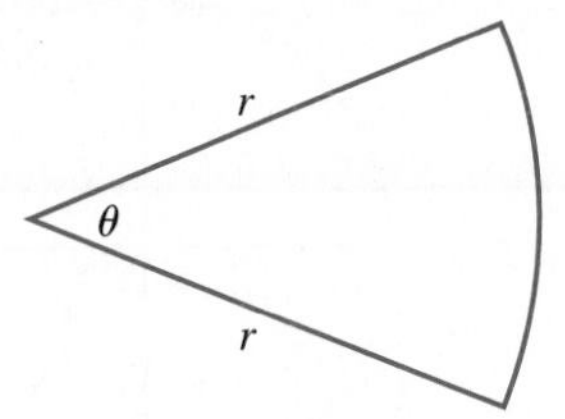

(a) Express r as a function of θ.
(b) Express the area A of the sector as a function of θ. Is this a quadratic function?
(c) Express θ as a function of r.
(d) Express the area A of the sector as a function of r. Is this a quadratic function?
(e) For which value of r is the area A a maximum? What is the corresponding value of θ in this case?

38. The following figure shows a semicircle of radius 1 unit and two adjacent sectors, AOC and COB.
(a) Show that the product P of the areas of the two sectors is given by

$$P = \frac{\pi\theta}{4} - \frac{\theta^2}{4}$$

Is this a quadratic function?
(b) For what value of θ is P a maximum?

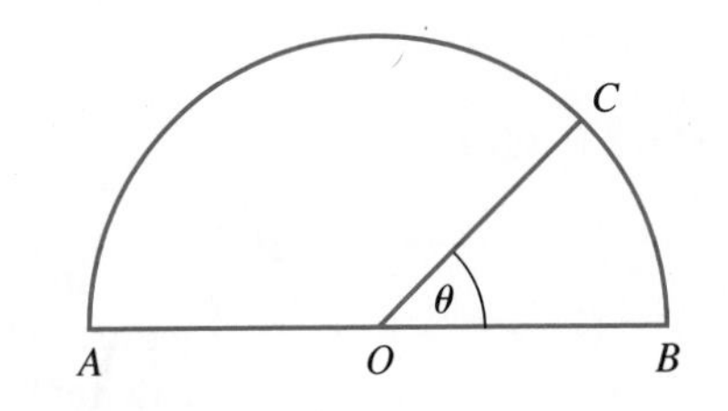

7.3 TRIGONOMETRIC FUNCTIONS OF REAL NUMBERS

In calculus and in the sciences, many of the applications of the trigonometric functions require that the inputs be real numbers, rather than angles. By making a small but crucial change in our viewpoint, we can define the trigonometric functions in such a way that the inputs are real numbers. As background for our discussion, recall that radian measure is defined by the equation $\theta = s/r$. In this definition, s and r are assumed to have the same linear units (for example, centimeters) and therefore the ratio s/r is a real number with no units. For example, in Figure 1 the radius is 3 cm, the arc length is 6 cm, and so the radian measure θ is given by

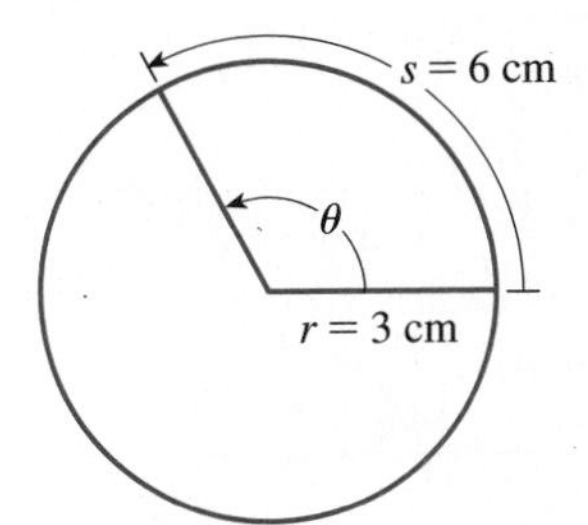

FIGURE 1

$$\theta = \frac{6 \text{ cm}}{3 \text{ cm}} = 2 \qquad \text{(a real number)}$$

Now, in the previous chapter (Section 6.4), the definitions of the trigonometric functions are based on the unit circle, so let's look at radian measure in that context. In the equation $\theta = s/r$, we set $r = 1$. This yields

$$\theta = \frac{s}{1} = s \qquad \text{(just as before, a real number)}$$

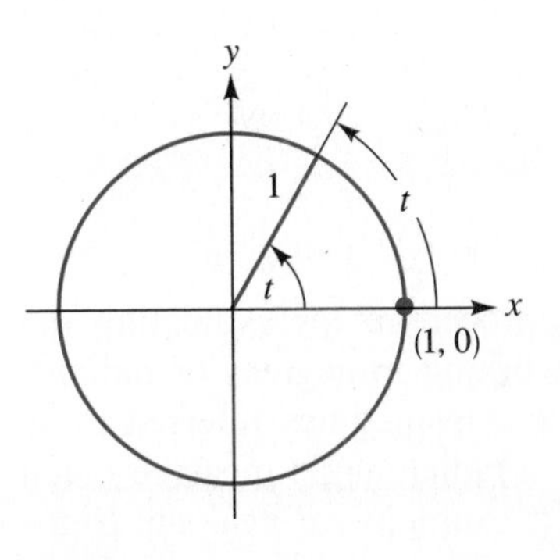

FIGURE 2
In the unit circle, the radian measure of an angle equals the measure of the intercepted arc.

Thus, in the unit circle, the radian measure of the angle is equal to the measure of the intercepted arc. For reference, we summarize this important observation in Figure 2. (We've used the letter t, rather than our usual θ, to emphasize the fact that both the radian measure of the angle and the measure of the intercepted arc are given by the same *real number*.)

The conventions regarding the measurement of arc length on the unit circle are the same as those introduced in the previous chapter for angles. As Figure 3 (on the next page) indicates, we measure from the point $(1, 0)$ and we assume that the positive direction is counterclockwise.

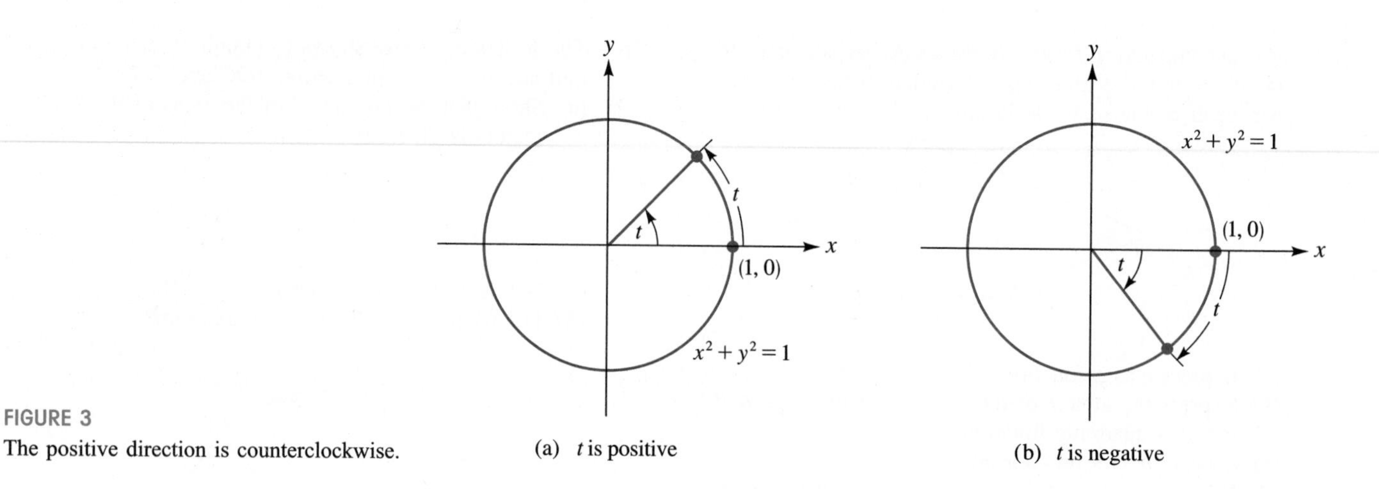

FIGURE 3
The positive direction is counterclockwise.

(a) t is positive

(b) t is negative

In the definitions that follow, you may think of t as either the measure of an arc or the radian measure of an angle. But in both cases, and this is the point, t denotes a real number.

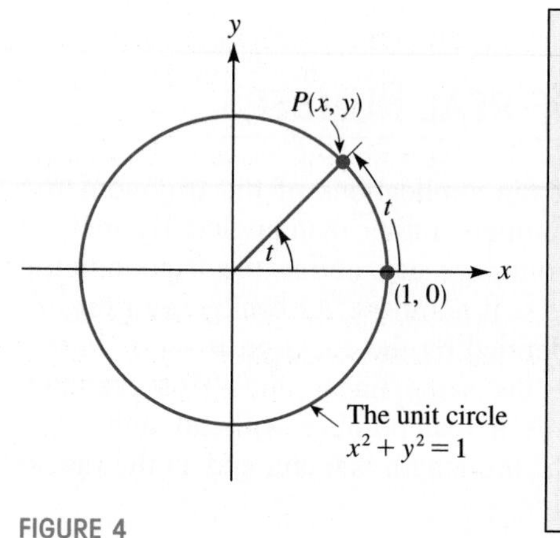

FIGURE 4

DEFINITION: Trigonometric Functions of Real Numbers

(Refer to Figure 4.) Let $P(x, y)$ denote the point on the unit circle that has arc length t from (1, 0). Or equivalently, let $P(x, y)$ denote the point where the terminal side of the angle with radian measure t intersects the unit circle. Then the six trigonometric functions of the real number t are defined as follows.

$$\cos t = x \qquad \sec t = \frac{1}{x} \quad (x \neq 0)$$

$$\sin t = y \qquad \csc t = \frac{1}{y} \quad (y \neq 0)$$

$$\tan t = \frac{y}{x} \quad (x \neq 0) \qquad \cot t = \frac{x}{y} \quad (y \neq 0)$$

Note that there is nothing essentially new here as far as evaluating the trigonometric functions. For instance, $\sin(\pi/2)$ is still equal to 1. What is different now is that the inputs are real numbers:

$$\sin \frac{\pi}{2} = 1$$

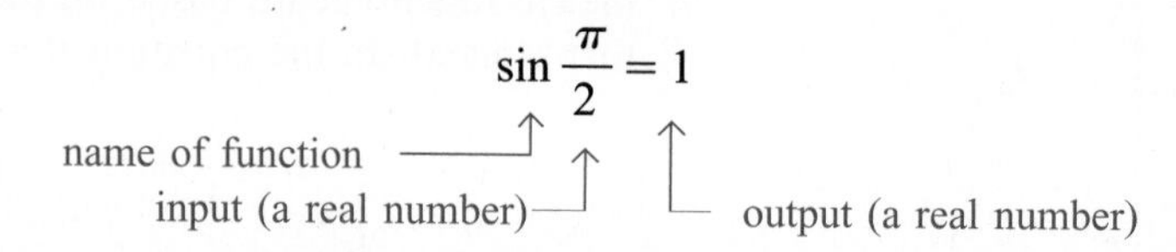

As explained in Section 7.1, the three-step procedure for evaluating the trigonometric functions is valid whether we are working in degrees or radians. When radian measure is used, the reference angle is sometimes referred to as the **reference number** to emphasize the fact that a radian angle measure (such as $\pi/4$ or 3) is indeed a real number. In the next example we evaluate trigonometric functions with real number inputs. As you'll see, however, there are no new techniques for you to learn in this context. As we've said, the new definition amounts to a change in viewpoint more than anything else.

EXAMPLE 1 Evaluate each of the following expressions:

(a) $\cos(2\pi/3)$; **(b)** $\sec(2\pi/3)$.

Solution

(a) The cosine of the real number $2\pi/3$ is, by definition, the cosine of $2\pi/3$ radians, which we can evaluate by means of our three-step procedure.

STEP 1 The reference number for $2\pi/3$ is $\pi/3$ [because $\pi - (2\pi/3) = \pi/3$].

STEP 2 $\cos(\pi/3) = 1/2$

STEP 3 $\cos t$ is the x-coordinate, and in Quadrant II, x-coordinates are negative. Thus, since the terminal side of $t = 2\pi/3$ is in Quadrant II, $\cos(2\pi/3)$ is negative. (Equivalently, an arc of length $2\pi/3$ terminates in Quadrant II.) Consequently, we have

$$\cos(2\pi/3) = -1/2$$

(b) We could use the three-step procedure to evaluate $\sec(2\pi/3)$ but in this case, it is much more direct simply to note that $\sec(2\pi/3)$ is the reciprocal of $\cos(2\pi/3)$, which we evaluated in part (a). Thus we have

$$\sec(2\pi/3) = -2$$

For the remainder of this section we discuss some of the more fundamental identities for the trigonometric functions of real numbers. As we said in Section 6.2, an **identity** is an equation that is satisfied by all values of the variable in its domain. As a demonstration of what this definition really means, consider the equation

$$\sin^2 t + \cos^2 t = 1$$

As we'll prove in a moment (and as you should already expect from the previous chapter), this equation is an identity. The domain of the variable t in this case is the set of all real numbers. So if we pick a real number at random, say, $t = 17$, then it follows that the equation

$$\sin^2 17 + \cos^2 17 = 1$$

is true. You should take a minute now to confirm this empirically with your calculator. That is, actually compute the left-hand side of the preceding equation and check that the result is indeed 1. The point is, no matter what real number we choose for t, the resulting equation will be true.

As an example of an equation that is not an identity, consider the *conditional equation*

$$2 \sin t = 1 \tag{1}$$

As you can check (without a calculator), this equation is true when $t = \pi/6$ but false when $t = \pi/3$. Thus, equation (1) is *not* satisfied by *all* values of the variable in its domain. In other words, equation (1) is not an identity. In a later section, we'll work with conditional equations such as equation (1). When we do that, we will be trying to solve these equations; that is, we'll want to find just which values of the variable (if any) do satisfy the equation.

QUESTION FOR REVIEW AND PREVIEW Can you find some real numbers, other than $\pi/6$, that also satisfy equation (1)?

There are five identities that are immediate consequences of the unit-circle definitions of the trigonometric functions. For example, using the unit-circle definitions, we have

$$\tan t = \frac{y}{x} \qquad \text{definition of } \tan t$$

$$= \frac{\sin t}{\cos t} \qquad \text{definitions of } \sin t \text{ and } \cos t$$

Thus, we have $\tan t = (\sin t)/(\cos t)$. This identity and four others are listed in the box that follows.

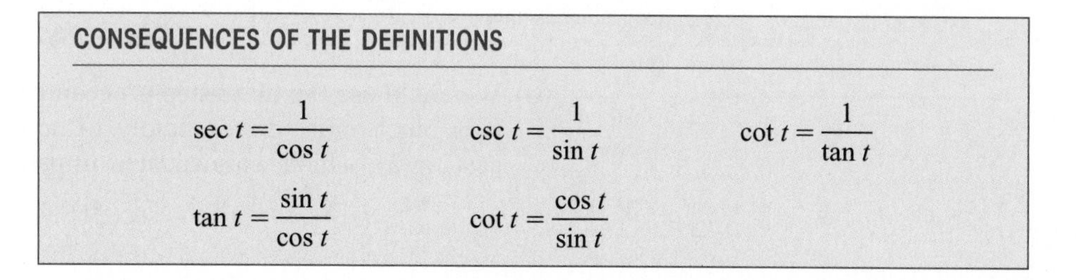

CONSEQUENCES OF THE DEFINITIONS

$$\sec t = \frac{1}{\cos t} \qquad \csc t = \frac{1}{\sin t} \qquad \cot t = \frac{1}{\tan t}$$

$$\tan t = \frac{\sin t}{\cos t} \qquad \cot t = \frac{\cos t}{\sin t}$$

The next identities that we consider are the three *Pythagorean identities.* The term ''Pythagorean'' is used here because, as you'll see, the proofs of the identities rely on the unit-circle definitions of the trigonometric functions; and the equation of the unit circle essentially is derived from the Pythagorean theorem.

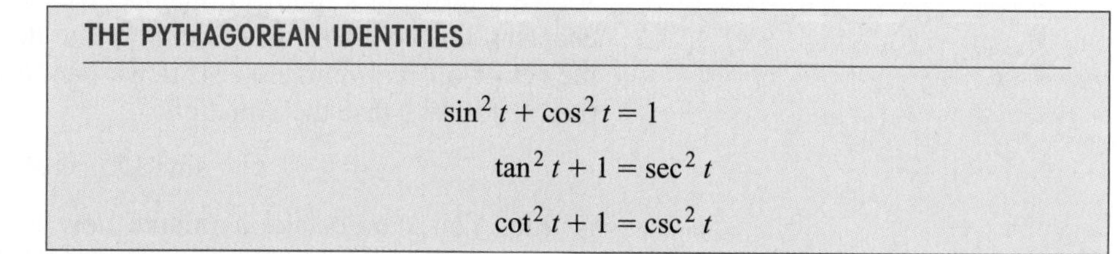

THE PYTHAGOREAN IDENTITIES

$$\sin^2 t + \cos^2 t = 1$$

$$\tan^2 t + 1 = \sec^2 t$$

$$\cot^2 t + 1 = \csc^2 t$$

To establish the identity $\sin^2 t + \cos^2 t = 1$, we think of the real number t, for the moment, as the radian measure of an angle in standard position, as indicated in Figure 5. Now we proceed exactly as in the previous chapter, when we proved this identity in the context of angles. Since the point (x, y) in Figure 5 lies on the unit circle, we have

$$x^2 + y^2 = 1$$

Now, replacing x by $\cos t$ and y by $\sin t$ gives us

$$\cos^2 t + \sin^2 t = 1$$

which is what we wished to show. You should also be familiar with two other ways of writing this identity:

$$\cos^2 t = 1 - \sin^2 t \qquad \text{and} \qquad \sin^2 t = 1 - \cos^2 t$$

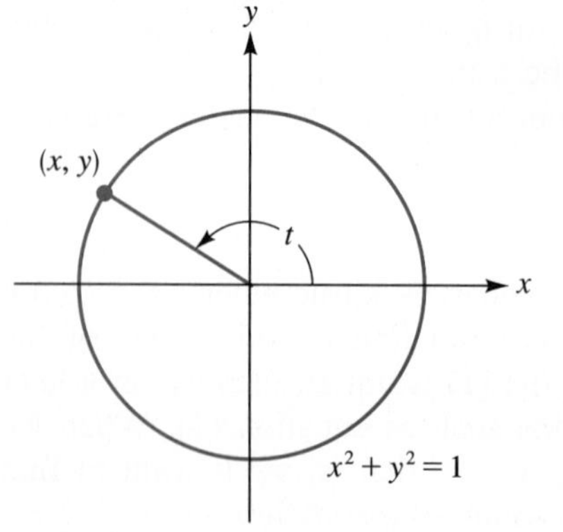

FIGURE 5
A real number t interpreted as the radian measure of an angle in standard position.

To prove the second of the Pythagorean identities, we begin with the identity $\sin^2 t + \cos^2 t = 1$ and divide both sides by the quantity $\cos^2 t$ to obtain

$$\frac{\sin^2 t}{\cos^2 t} + \frac{\cos^2 t}{\cos^2 t} = \frac{1}{\cos^2 t} \qquad \text{assuming } \cos t \neq 0$$

and, consequently,

$$\tan^2 t + 1 = \sec^2 t \qquad \text{as required}$$

Since the proof of the third Pythagorean identity is similar, we omit it here.

EXAMPLE 2 If $\sin t = 2/3$ and $\pi/2 < t < \pi$, compute $\cos t$ and $\tan t$.

Solution

$$\begin{aligned} \cos^2 t &= 1 - \sin^2 t && \text{using the first Pythagorean identity} \\ &= 1 - \left(\frac{2}{3}\right)^2 && \text{substituting} \\ &= 1 - \frac{4}{9} = \frac{5}{9} \end{aligned}$$

Consequently,

$$\cos t = \sqrt{5}/3 \qquad \text{or} \qquad \cos t = -\sqrt{5}/3$$

Now, since $\pi/2 < t < \pi$, it follows that $\cos t$ is negative. (Why?) Thus

$$\cos t = -\sqrt{5}/3 \qquad \text{as required}$$

To compute $\tan t$, we use the identity $\tan t = (\sin t)/(\cos t)$ to obtain

$$\tan t = \frac{2/3}{-\sqrt{5}/3} = \frac{2}{3} \times \frac{3}{-\sqrt{5}} = -\frac{2}{\sqrt{5}}$$

If required, we can rationalize the denominator (by multiplying by $\sqrt{5}/\sqrt{5}$) to obtain

$$\tan t = -2\sqrt{5}/5$$

EXAMPLE 3 If $\sec t = -5/3$ and $\pi < t < 3\pi/2$, compute $\cos t$ and $\tan t$.

Solution Since $\cos t$ is the reciprocal of $\sec t$, we have $\cos t = -3/5$. We can compute $\tan t$ using the second Pythagorean identity as follows:

$$\begin{aligned} \tan^2 t &= \sec^2 t - 1 \\ &= \left(-\frac{5}{3}\right)^2 - 1 = \frac{16}{9} \end{aligned}$$

Therefore,

$$\tan t = \frac{4}{3} \qquad \text{or} \qquad \tan t = -\frac{4}{3}$$

Since t is between π and $3\pi/2$, $\tan t$ is positive. (Why?) Thus

$$\tan t = \frac{4}{3}$$

The next example shows how certain radical expressions can be simplified through an appropriate trigonometric substitution. (This technique is often useful in calculus.)

EXAMPLE 4 In the expression $u/\sqrt{u^2-1}$, make the substitution $u = \sec\theta$ and show that the resulting expression is equal to $\csc\theta$. (Assume that $0 < \theta < \pi/2$.)

Solution Replacing u by $\sec\theta$ in the given expression yields

$$\frac{u}{\sqrt{u^2-1}} = \frac{\sec\theta}{\sqrt{\sec^2\theta - 1}}$$

$$= \frac{\sec\theta}{\sqrt{\tan^2\theta}} \qquad \text{using the second Pythagorean identity}$$

$$= \frac{\sec\theta}{\tan\theta} = \frac{1/\cos\theta}{\sin\theta/\cos\theta}$$

$$= \frac{1}{\sin\theta} = \csc\theta$$

QUESTION Where did we use the condition $0 < \theta < \pi/2$?

As we saw in Chapter 6, one technique for simplifying trigonometric expressions involves first writing everything in terms of sines and cosines. However, this is not necessarily the most efficient method. As the next example indicates, the second and third Pythagorean identities can be quite useful in this context.

EXAMPLE 5 Simplify the expression $\dfrac{\csc t + \csc t\cot^2 t}{\sec^2 t - \tan^2 t}$.

Solution

$$\frac{\csc t + \csc t\cot^2 t}{\sec^2 t - \tan^2 t} = \frac{(\csc t)(1+\cot^2 t)}{1} \qquad \text{factoring out the common term in the numerator and applying the second Pythagorean identity in the denominator}$$

$$= (\csc t)(\csc^2 t) \qquad \text{applying the third Pythagorean identity}$$

$$= \csc^3 t$$

As background for the next set of identities, we define the terms *even function* and *odd function.* A function f is said to be an **even function** provided

$$f(-t) = f(t) \qquad \text{for every value of } t \text{ in the domain of } f$$

The function defined by $f(t) = t^2$ serves as a simple example of an even function. Whether we use t or $-t$ for the input, the output is the same. For instance,

$$f(5) = 5^2 = 25 \qquad \text{and also} \qquad f(-5) = (-5)^2 = 25$$

The graph of an even function is always symmetric about the y-axis. (This follows directly from the symmetry tests in Chapter 2 on page 71.) For example, in Figure 6 we show the graph of the even function $f(t) = t^2$.

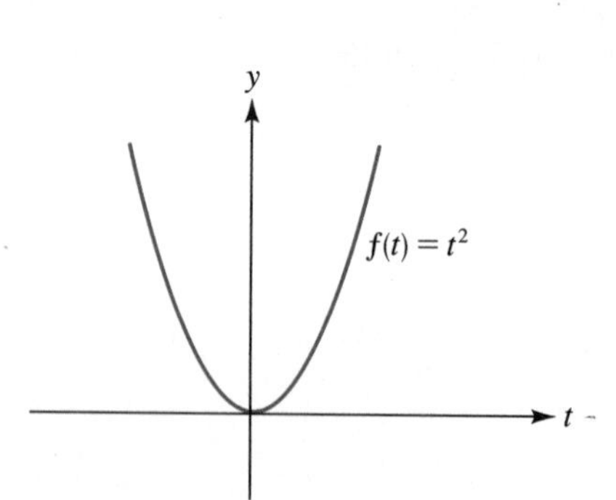

FIGURE 6
The graph of an even function is symmetric about the y-axis.

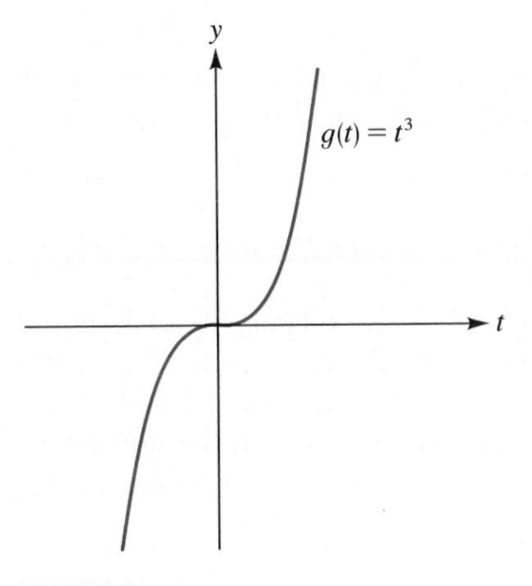

FIGURE 7
The graph of an odd function is symmetric about the origin.

A function f is said to be an **odd function** provided

$$f(-t) = -f(t) \qquad \text{for every value of } t \text{ in the domain of } f$$

The function defined by $g(t) = t^3$ is an odd function. For instance, if you compute $g(5)$ and $g(-5)$, you'll find that the outputs are the negatives of one another. (Check this.) The graph of an odd function is always symmetric about the origin. (Again, this follows from the symmetry tests in Chapter 2.) In Figure 7 we display the graph of the odd function $g(t) = t^3$.

Now let's return to our development of trigonometric identities. The three identities in the box that follows can be interpreted using the terminology we've just been discussing. These identities, called the opposite-angle identities, tell us that the cosine function is even and that the sine and tangent functions are odd. (We'll see the graphical aspects of this in the following sections.)

THE OPPOSITE-ANGLE IDENTITIES

$$\cos(-t) = \cos t$$
$$\sin(-t) = -\sin t$$
$$\tan(-t) = -\tan t$$

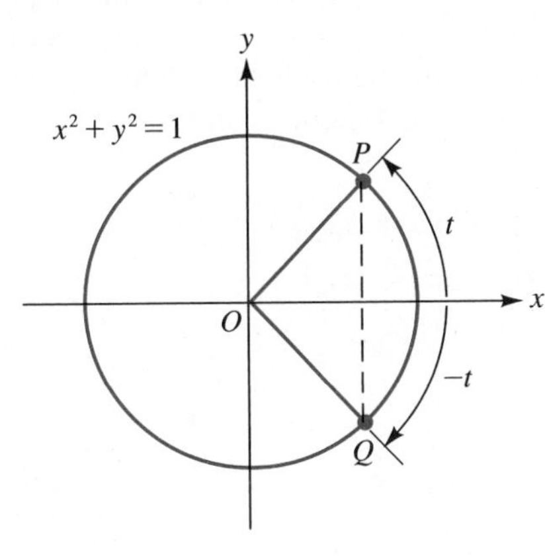

FIGURE 8

To see why the first two of these identities are true, consider Figure 8. (Although Figure 8 shows an arc of length t terminating in Quadrant I, the argument we use will work no matter where the arc terminates.) By definition, the coordinates of the points P and Q in Figure 8 are as follows:

$$P\colon\ (\cos t, \sin t)$$
$$Q\colon\ (\cos(-t), \sin(-t))$$

However, as you can see by looking at Figure 8, the x-coordinates of P and Q are the same, while the y-coordinates are negatives of each other. Thus,

$$\cos(-t) = \cos t$$

and

$$\sin(-t) = -\sin t \qquad \text{as we wished to show}$$

Now we can establish the third identity, involving $\tan(-t)$, as follows:

$$\tan(-t) = \frac{\sin(-t)}{\cos(-t)}$$
$$= \frac{-\sin t}{\cos t} = -\frac{\sin t}{\cos t} = -\tan t$$

EXAMPLE 6

(a) If $\sin t = -0.76$, find $\sin(-t)$.
(b) If $\cos s = -0.29$, find $\cos(-s)$.
(c) If $\sin u = 0.54$, find $\sin^2(-u) + \cos^2(-u)$.

Solution

(a) $\sin(-t) = -\sin t$
$$= -(-0.76) = 0.76$$

(b) $\cos(-s) = \cos s = -0.29$

(c) The opposite-angle identities are unnecessary here, and so too is the given value of $\sin u$. For, according to the first Pythagorean identity, we have

$$\sin^2(-u) + \cos^2(-u) = 1$$

The final identities that we are going to discuss in this section are simply consequences of the fact that the circumference C of the unit circle is 2π. (Substituting $r = 1$ in the formula $C = 2\pi r$ gives us $C = 2\pi$.) Thus, if we begin at any point P on the unit circle and travel a distance of 2π units along the perimeter, we return to the same point P. In other words, arc lengths of t and $t + 2\pi$ [measured from (1, 0), as usual] yield the same terminal point on the unit circle. Since the trigonometric functions are defined in terms of the coordinates of that point P, we obtain the following identities.

PERIODICITY OF SINE AND COSINE

$$\sin(t + 2\pi) = \sin t$$
$$\cos(t + 2\pi) = \cos t$$

These two identities are true for all real numbers t. As you'll see in the next section, they provide important information about the graphs of the sine and cosine functions; the graphs of both functions repeat themselves at intervals of 2π. Similar identities hold for the other trigonometric functions in their respective domains:

$$\tan(t + 2\pi) = \tan t \qquad \csc(t + 2\pi) = \csc t$$
$$\cot(t + 2\pi) = \cot t \qquad \sec(t + 2\pi) = \sec t$$

EXAMPLE 7 Evaluate $\sin \dfrac{5\pi}{2}$.

Solution First, simply as a matter of arithmetic, we observe that $\frac{5\pi}{2} = \frac{\pi}{2} + 2\pi$. Thus, in view of our earlier remarks, we have

$$\sin \tfrac{5\pi}{2} = \sin \tfrac{\pi}{2} = 1$$

The preceding set of identities can be generalized as follows. If we start at a point P on the unit circle and make two complete counterclockwise revolutions, the arc length we travel is $2\pi + 2\pi = 4\pi$. Similarly, for three complete revolutions, the arc length traversed is $3(2\pi) = 6\pi$. And, in general, if k is any integer, the arc length for k complete revolutions is $2k\pi$. (When k is positive, the revolutions are counterclockwise; when k is negative, the revolutions are clockwise.) Consequently, we have the following identities.

For any real number t and any integer k, the following identities hold:

$$\sin(t + 2k\pi) = \sin t \quad \text{and} \quad \cos(t + 2k\pi) = \cos t$$

EXAMPLE 8 Evaluate $\cos(-17\pi)$.

Solution $\cos(-17\pi) = \cos(\pi - 18\pi) = \cos \pi = -1$

EXERCISE SET 7.3

A

In Exercises 1–8, evaluate each expression (as in Example 1).

1. **(a)** $\cos(11\pi/6)$ **(b)** $\cos(-11\pi/6)$ **(c)** $\sin(11\pi/6)$ **(d)** $\sin(-11\pi/6)$
2. **(a)** $\cos(2\pi/3)$ **(b)** $\cos(-2\pi/3)$ **(c)** $\sin(2\pi/3)$ **(d)** $\sin(-2\pi/3)$
3. **(a)** $\cos(\pi/6)$ **(b)** $\cos(-\pi/6)$ **(c)** $\sin(\pi/6)$ **(d)** $\sin(-\pi/6)$
4. **(a)** $\cos(13\pi/4)$ **(b)** $\cos(-13\pi/4)$ **(c)** $\sin(13\pi/4)$ **(d)** $\sin(-13\pi/4)$
5. **(a)** $\cos(5\pi/4)$ **(b)** $\cos(-5\pi/4)$ **(c)** $\sin(5\pi/4)$ **(d)** $\sin(-5\pi/4)$
6. **(a)** $\cos(9\pi/4)$ **(b)** $\cos(-9\pi/4)$ **(c)** $\sin(9\pi/4)$ **(d)** $\sin(-9\pi/4)$
7. **(a)** $\sec(5\pi/3)$ **(b)** $\csc(-5\pi/3)$ **(c)** $\tan(5\pi/3)$ **(d)** $\cot(-5\pi/3)$
8. **(a)** $\sec(7\pi/4)$ **(b)** $\csc(-7\pi/4)$ **(c)** $\tan(7\pi/4)$ **(d)** $\cot(-7\pi/4)$
9. **(a)** List four positive real-number values of t for which $\cos t = 0$.
   (b) List four negative real-number values of t for which $\cos t = 0$.
10. **(a)** List four positive real numbers t such that $\sin t = 1/2$.
    (b) List four positive real numbers t such that $\sin t = -1/2$.
    (c) List four negative real numbers t such that $\sin t = 1/2$.
    (d) List four negative real numbers t such that $\sin t = -1/2$.

In Exercises 11–14, use a calculator to evaluate the six trigonometric functions using the given real-number input. (Round the results to two decimal places.)

11. **(a)** 2.06 **(b)** -2.06
12. **(a)** 0.55 **(b)** -0.55
13. **(a)** $\frac{\pi}{6}$ **(b)** $\frac{\pi}{6} + 2\pi$
14. **(a)** 1000 **(b)** $1000 - 2\pi$

In Exercises 15–22, check that both sides of the identity are indeed equal for the given values of the variable t. For part (c) of each problem, use your calculator.

15. $\sin^2 t + \cos^2 t = 1$
    (a) $t = \pi/3$
    (b) $t = 5\pi/4$
    (c) $t = -53$
16. $\tan^2 t + 1 = \sec^2 t$
    (a) $t = 3\pi/4$
    (b) $t = -2\pi/3$
    (c) $t = \sqrt{5}$
17. $\cot^2 t + 1 = \csc^2 t$
    (a) $t = -\pi/6$
    (b) $t = 7\pi/4$
    (c) $t = 0.12$
18. $\cos(-t) = \cos t$
    (a) $t = \pi/6$
    (b) $t = -5\pi/3$
    (c) $t = -4$
19. $\sin(-t) = -\sin t$
    (a) $t = 3\pi/2$
    (b) $t = -5\pi/6$
    (c) $t = 13.24$
20. $\tan(-t) = -\tan t$
    (a) $t = -4\pi/3$
    (b) $t = \pi/4$
    (c) $t = 1000$
21. $\sin(t + 2\pi) = \sin t$
    (a) $t = 5\pi/3$
    (b) $t = -3\pi/2$
    (c) $t = \sqrt{19}$
22. $\cos(t + 2\pi) = \cos t$
    (a) $t = -5\pi/3$
    (b) $t = \pi$
    (c) $t = -\sqrt{3}$

In Exercises 23 and 24, show that the equation is not an identity by evaluating both sides using the given value of t and noting that the results are unequal.

23. $\cos 2t = 2 \cos t$; $t = \pi/6$
24. $\sin 2t = 2 \sin t$; $t = \pi/2$
25. If $\sin t = -3/5$ and $\pi < t < \frac{3\pi}{2}$, compute $\cos t$ and $\tan t$.
26. If $\cos t = 5/13$ and $\frac{3\pi}{2} < t < 2\pi$, compute $\sin t$ and $\cot t$.
27. If $\sin t = \sqrt{3}/4$ and $\frac{\pi}{2} < t < \pi$, compute $\tan t$.
28. If $\sec s = -\sqrt{13}/2$ and $\sin s > 0$, compute $\tan s$.
29. If $\tan \alpha = 12/5$ and $\cos \alpha > 0$, compute $\sec \alpha$, $\cos \alpha$, and $\sin \alpha$.
30. If $\cot \theta = -1/\sqrt{3}$ and $\cos \theta < 0$, compute $\csc \theta$ and $\sin \theta$.
31. In the expression $\sqrt{9 - x^2}$, make the substitution $x = 3 \sin \theta$ $\left(0 < \theta < \frac{\pi}{2}\right)$, and show that the result is $3 \cos \theta$.
32. Make the substitution $u = 2 \cos \theta$ in the expression $1/\sqrt{4 - u^2}$, and simplify the result. (Assume that $0 < \theta < \pi$.)
33. In the expression $1/(u^2 - 25)^{3/2}$, make the substitution $u = 5 \sec \theta$ $\left(0 < \theta < \frac{\pi}{2}\right)$, and show that the result is $(\cot^3 \theta)/125$.
34. In the expression $1/(x^2 + 5)^2$, replace x by $\sqrt{5} \tan \theta$ and show that the result is $(\cos^4 \theta)/25$.
35. In the expression $1/\sqrt{u^2 + 7}$, let $u = \sqrt{7} \tan \theta$, where $0 < \theta < \frac{\pi}{2}$, and simplify the result.
36. In the expression $\sqrt{x^2 - a^2}/x$ $(a > 0)$, let $x = a \sec \theta$ $\left(0 < \theta < \frac{\pi}{2}\right)$, and simplify the result.
37. **(a)** If $\sin t = 2/3$, find $\sin(-t)$.
    (b) If $\sin \phi = -1/4$, find $\sin(-\phi)$.
    (c) If $\cos \alpha = 1/5$, find $\cos(-\alpha)$.
    (d) If $\cos s = -1/5$, find $\cos(-s)$.

38. **(a)** If $\sin t = 0.35$, find $\sin(-t)$.
(b) If $\sin \phi = -0.47$, find $\sin(-\phi)$.
(c) If $\cos \alpha = 0.21$, find $\cos(-\alpha)$.
(d) If $\cos s = -0.56$, find $\cos(-s)$.

39. If $\cos t = -1/3$ $\left(\frac{\pi}{2} < t < \pi\right)$, compute the following:
(a) $\sin(-t) + \cos(-t)$; **(b)** $\sin^2(-t) + \cos^2(-t)$.

40. If $\sin(-s) = 3/5$ $\left(\pi < s < \frac{3\pi}{2}\right)$, compute: **(a)** $\sin s$; **(b)** $\cos(-s)$; **(c)** $\cos s$; **(d)** $\tan s + \tan(-s)$.

In Exercises 41 and 42, use one of the identities $\cos(t + 2\pi k) = \cos t$ *or* $\sin(t + 2\pi k) = \sin t$ *to evaluate each expression.*

41. **(a)** $\cos\left(\frac{\pi}{4} + 2\pi\right)$ **(b)** $\sin\left(\frac{\pi}{3} + 2\pi\right)$
(c) $\sin\left(\frac{\pi}{2} - 6\pi\right)$

42. **(a)** $\sin(17\pi/4)$ **(b)** $\sin(-17\pi/4)$
(c) $\cos 11\pi$ **(d)** $\cos(53\pi/4)$
(e) $\tan(-7\pi/4)$ **(f)** $\cos(7\pi/4)$
(g) $\sec\left(\frac{11\pi}{6} + 2\pi\right)$ **(h)** $\csc\left(2\pi - \frac{\pi}{3}\right)$

In Exercises 43–46, use the Pythagorean identities to simplify the given expressions.

43. $\dfrac{\sin^2 t + \cos^2 t}{\tan^2 t + 1}$ **44.** $\dfrac{\sec^2 t - 1}{\tan^2 t}$

45. $\dfrac{\sec^2 \theta - \tan^2 \theta}{1 + \cot^2 \theta}$ **46.** $\dfrac{\csc^4 \theta - \cot^4 \theta}{\csc^2 \theta + \cot^2 \theta}$

In Exercises 47–54, prove that the equations are identities.

47. $\csc t = \sin t + \cot t \cos t$

48. $\sin^2 t - \cos^2 t = \dfrac{1 - \cot^2 t}{1 + \cot^2 t}$

49. $\dfrac{1}{1 + \sec s} + \dfrac{1}{1 - \sec s} = -2 \cot^2 s$

50. $\dfrac{1 + \tan s}{1 - \tan s} = \dfrac{\sec^2 s + 2 \tan s}{2 - \sec^2 s}$

51. $\cot \theta + \tan \theta + 1 = \dfrac{\cot \theta}{1 - \tan \theta} + \dfrac{\tan \theta}{1 - \cot \theta}$

52. $\dfrac{\sec s + \cot s \csc s}{\cos s} = \csc^2 s \sec^2 s$

53. $(\tan \theta)(1 - \cot^2 \theta) + (\cot \theta)(1 - \tan^2 \theta) = 0$

54. $(\cos \alpha \cos \beta - \sin \alpha \sin \beta)(\cos \alpha \cos \beta + \sin \alpha \sin \beta) = \cos^2 \alpha - \sin^2 \beta$

55. If $\sec t = 13/5$ and $\frac{3\pi}{2} < t < 2\pi$, evaluate

$$\frac{2 \sin t - 3 \cos t}{4 \sin t - 9 \cos t}$$

56. If $\sec t = (b^2 + 1)/2b$ and $\pi < t < \frac{3\pi}{2}$, find $\tan t$ and $\sin t$.

B

57. Use the accompanying figure to explain why the following four identities are valid. (The identities can be used to provide an algebraic foundation for the "reference angle technique" that we've used to evaluate the trigonometric functions.)

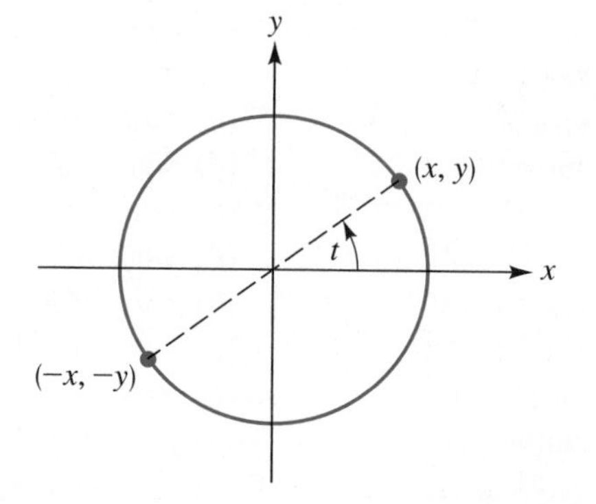

(i) $\sin(t + \pi) = -\sin t$
(ii) $\sin(t - \pi) = -\sin t$
(iii) $\cos(t + \pi) = -\cos t$
(iv) $\cos(t - \pi) = -\cos t$

58. Use two of the results in Exercise 57 to verify the identity $\tan(t + \pi) = \tan t$. (You'll see the graphical aspect of this identity in Section 7.7.)

59. In the equation $x^4 + 6x^2y^2 + y^4 = 32$, make the substitutions

$$x = X \cos \tfrac{\pi}{4} - Y \sin \tfrac{\pi}{4} \quad \text{and} \quad y = X \sin \tfrac{\pi}{4} + Y \cos \tfrac{\pi}{4}$$

and show that the result simplifies to $X^4 + Y^4 = 16$.

60. Suppose that $\tan \theta = 2$ and $0 < \theta < \pi/2$.
(a) Compute $\sin \theta$ and $\cos \theta$.
(b) Using the values obtained in part (a), make the substitutions

$$x = X \cos \theta - Y \sin \theta \quad \text{and} \quad y = X \sin \theta + Y \cos \theta$$

in the expression $7x^2 - 8xy + y^2$, and simplify the result.

61. In this exercise, we are going to find the minimum value of the function

$$f(t) = \tan^2 t + 9 \cot^2 t, \qquad 0 < t < \pi/2$$

(a) Set your calculator in the radian mode and complete the table. Round the values you obtain to two decimal places.

t	0.2	0.4	0.6	0.8	1.0	1.2	1.4
$f(t)$							

(b) Of the seven outputs you calculated in part (a), which is the smallest? What is the corresponding input?

(c) Prove that $\tan^2 t + 9 \cot^2 t = (\tan t - 3 \cot t)^2 + 6$.

(d) Use the identity in part (c) to explain why $\tan^2 t + 9 \cot^2 t \geqslant 6$.

(e) The inequality in part (d) tells us that $f(t)$ is never less than 6. Furthermore, in view of part (c), $f(t)$ will equal 6 when $\tan t - 3 \cot t = 0$. From this last equation, show that $\tan^2 t = 3$, and conclude that $t = \pi/3$. In summary, the minimum value of f is 6, and this occurs when $t = \pi/3$. How do these values compare with your answers in part (b)?

62. Let $f(\theta) = \sin \theta \cos \theta \quad \left(0 \leqslant \theta \leqslant \frac{\pi}{2}\right)$.

(a) Set your calculator in the radian mode and complete the table. Round the results to two decimal places.

θ	0	$\frac{\pi}{10}$	$\frac{\pi}{5}$	$\frac{\pi}{4}$	$\frac{3\pi}{10}$	$\frac{2\pi}{5}$	$\frac{\pi}{2}$
$f(\theta)$							

(b) What is the largest value of $f(\theta)$ in your table in part (a)?

(c) Show that $\sin \theta \cos \theta \leqslant 1/2$ for all real numbers θ in the interval $0 \leqslant \theta \leqslant \frac{\pi}{2}$. *Hint:* Use the inequality $\sqrt{ab} \leqslant (a+b)/2$ [given in Exercise 45(b) on page 84], with $a = \sin \theta$ and $b = \cos \theta$.

(d) Does the inequality $\sin \theta \cos \theta \leqslant 1/2$ hold for all real numbers θ?

63. Consider the equation

$$2 \sin^2 t - \sin t = 2 \sin t \cos t - \cos t$$

(a) Evaluate each side of the equation when $t = \pi/6$.

(b) Evaluate each side of the equation when $t = \pi/4$.

(c) Is the given equation an identity?

64. Suppose that

$$f(t) = (\sin t \cos t)(2 \sin t - 1)(2 \cos t - 1)(\tan t - 1)$$

(a) Compute each of the following: $f(0)$, $f(\pi/6)$, $f(\pi/4)$, $f(\pi/3)$, and $f(\pi/2)$.

(b) Is the equation $f(t) = 0$ an identity?

In Section 7.2, we pointed out that one of the advantages in using radian measure is that many formulas then take on particularly simple forms. Another reason for using radian measure is that the trigonometric functions can be closely approximated by very simple polynomial functions. To see examples of this, complete the tables in Exercises 65–68. Round (or, for exact values, simply report) the answers to six decimal places. In Exercises 67 and 68, note that the higher-degree polynomial provides the better approximation. Note: The approximating polynomials in Exercises 65–68 are known as ***Taylor polynomials,*** *after the English mathematician Brook Taylor (1685–1731). The theory of Taylor polynomials is developed in calculus.*

65.

t	$1 - \frac{1}{2}t^2$	$\cos t$
0.02		
0.05		
0.1		
0.2		
0.3		

66.

t	$t - \frac{1}{6}t^3$	$\sin t$
0.02		
0.05		
0.1		
0.2		
0.3		

67.

x	$\frac{1}{3}x^3 + x$	$\frac{2}{15}x^5 + \frac{1}{3}x^3 + x$	$\tan x$
0.1			
0.2			
0.3			
0.4			
0.5			

68.

x	$x^2 + x$	$\frac{1}{3}x^3 + x^2 + x$	$e^x \sin x$
0.1			
0.2			
0.3			
0.4			
0.5			

69. The accompanying figure shows two x-y coordinate systems. (The same unit of length is used on all four axes.) In the coordinate system on the left, the curve is a portion of the unit circle

$$x^2 + y^2 = 1$$

and A is the point $(1, 0)$. The points B, C, D, E, and F are located on the circle according to the information in the following table.

arc	$\overset{\frown}{AB}$	$\overset{\frown}{AC}$	$\overset{\frown}{AD}$	$\overset{\frown}{AE}$	$\overset{\frown}{AF}$
length	$\pi/12$	$\pi/6$	$\pi/4$	$\pi/3$	$5\pi/12$

Determine the y-coordinates of the points P, Q, R, S, and T. Give an exact expression for each answer and, where appropriate, a calculator approximation rounded to three decimal places.

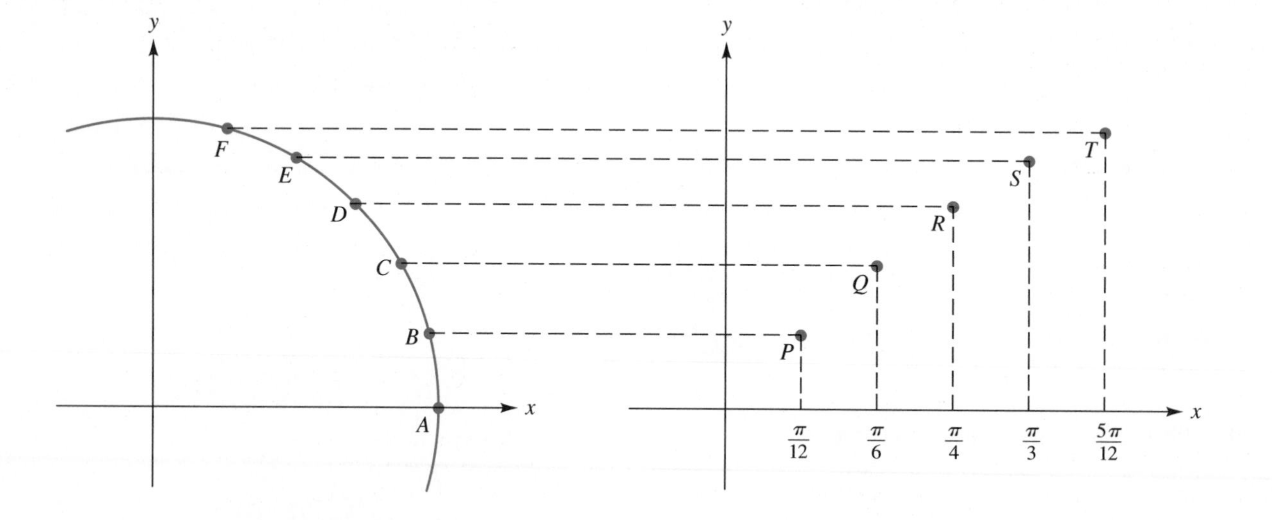

7.4 GRAPHS OF THE SINE AND COSINE FUNCTIONS

Our focus in this section is on the sine and cosine functions. As preparation for the discussion, we want to understand what is meant by the term *periodic function.* By way of example, both of the functions in Figure 1 are **periodic.** That is, their graphs display patterns that repeat themselves at regular intervals.

In Figure 1(a), the graph of the function f repeats itself every six units. We say that *the period of f is* 6. Similarly, the period of g in Figure 1(b) is 2π. In both cases, notice that the period represents the minimum number of units that we must travel along the horizontal axis before the graph begins to repeat itself. Because of this, we can state the definition of a periodic function as follows.

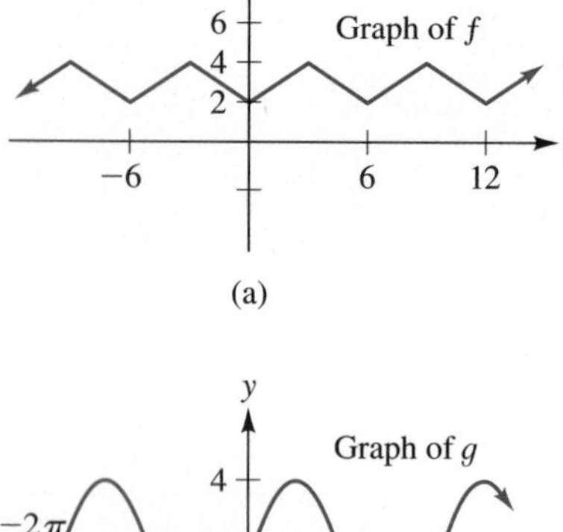

(a)

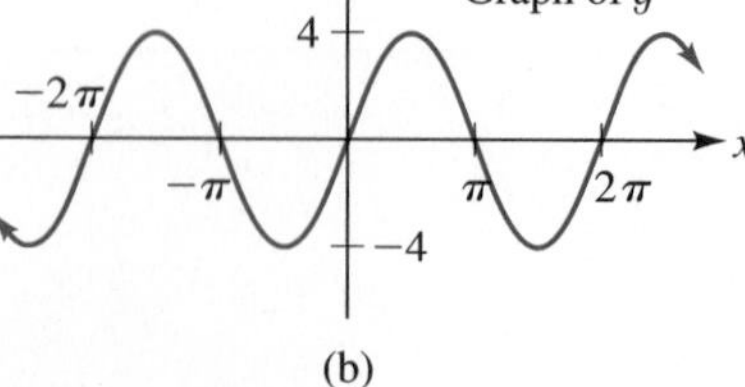

(b)

FIGURE 1

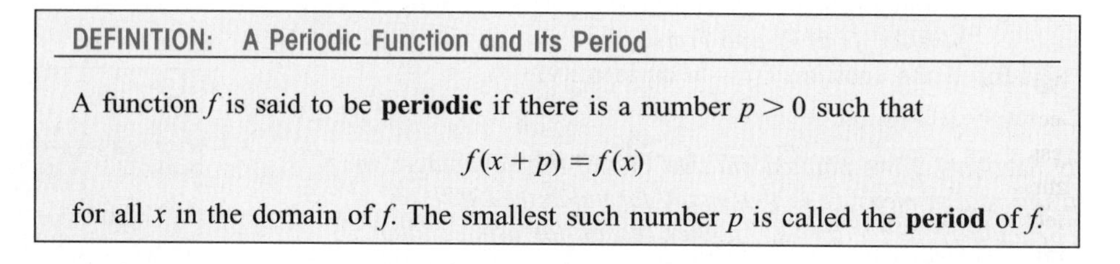

DEFINITION: A Periodic Function and Its Period

A function f is said to be **periodic** if there is a number $p > 0$ such that

$$f(x + p) = f(x)$$

for all x in the domain of f. The smallest such number p is called the **period** of f.

We also want to define the term *amplitude* as it applies to periodic functions. For a function such as g in Figure 1(b), in which the graph is centered about the horizontal axis, the amplitude is simply the maximum height of the graph above the horizontal axis. Thus, the amplitude of g is 4. More generally, we define the amplitude of any periodic function as follows.

DEFINITION: Amplitude

Let f be a periodic function and let m and M denote, respectively, the minimum and maximum values of the function. Then the **amplitude** of f is the number

$$\frac{M-m}{2}$$

For the function g in Figure 1(b), this definition tells us that the amplitude is $\frac{4-(-4)}{2} = 4$, which agrees with our previous value. Check for yourself now that the amplitude of the function in Figure 1(a) is 1.

Periodic functions are used throughout the sciences to analyze or describe a variety of phenomena ranging from the vibrations of an electron to the variations in the size of an animal population as it interacts with its environment. Figures 2 through 5 display some examples of periodic functions.

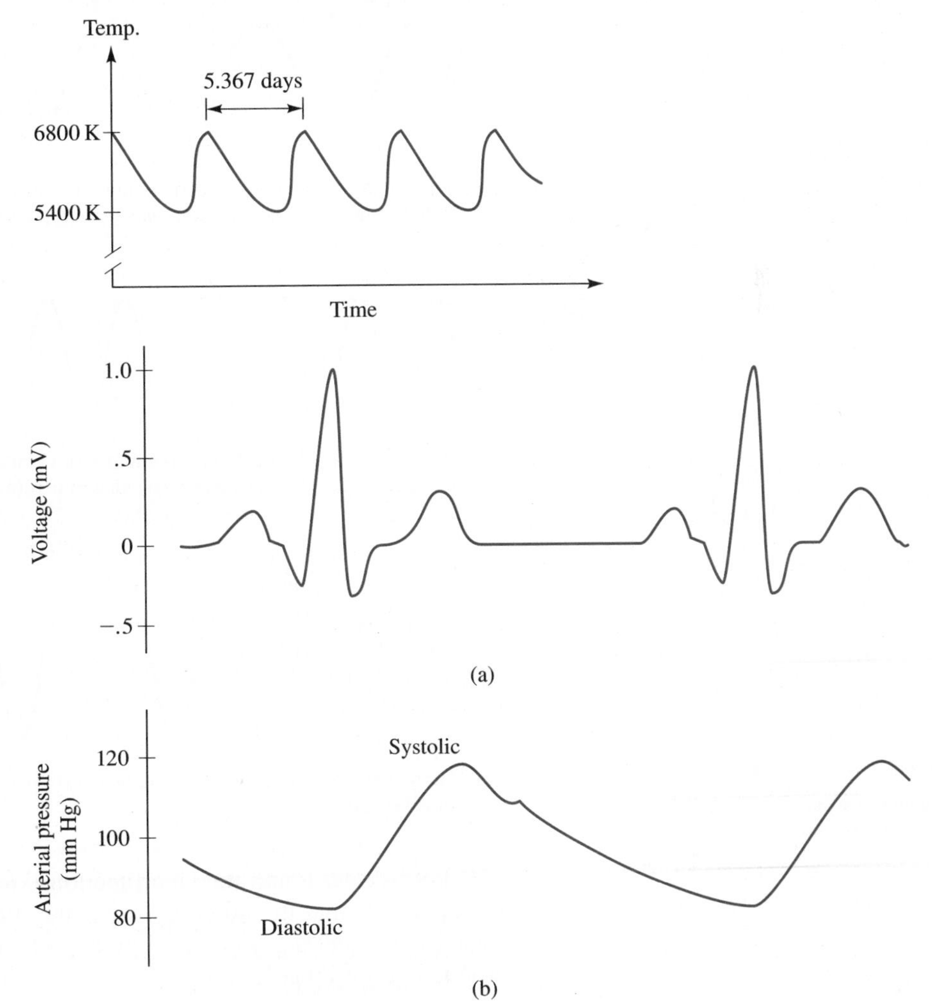

FIGURE 2
The surface temperature of the star Delta Cephei is a periodic function of time. The period is 5.367 days. The amplitude is $(6800 - 5400)/2 = 700$K.

FIGURE 3
Electrical activity of the heart and blood pressure as periodic functions of time. The figure shows (a) a typical ECG (electrocardiogram) and (b) the corresponding graph of arterial blood pressure [adapted from *Physics for the Health Sciences* by C. R. Nave and B. C. Nave (Philadelphia: W. B. Saunders Co., 1975)]

FIGURE 4
The sound wave generated by a note played on the bamboo flute is a periodic function. The sound wave is recorded on an oscilloscope. (Photograph by Professor Vern Ostdiek)

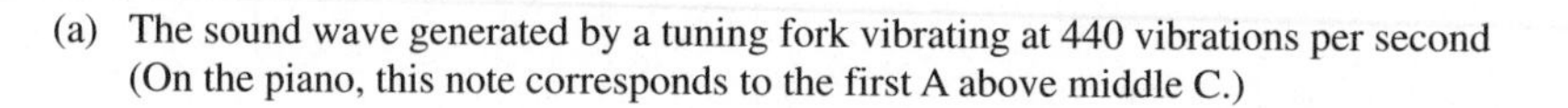

(a) The sound wave generated by a tuning fork vibrating at 440 vibrations per second (On the piano, this note corresponds to the first A above middle C.)

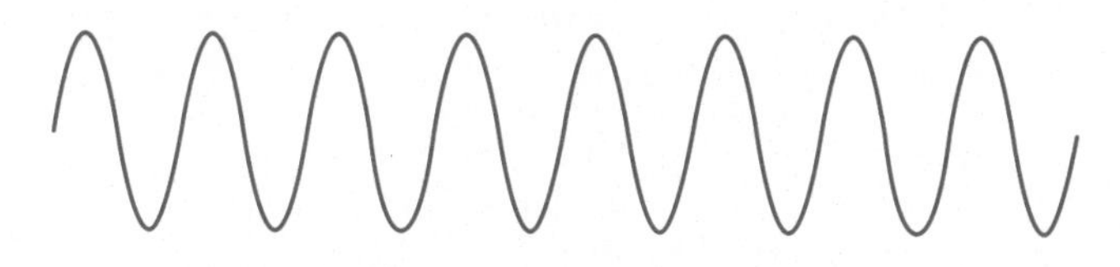

(b) The sound wave generated by a tuning fork vibrating at 880 vibrations per second [This note is an A one octave above that in part (a).]

(c) The sound wave generated by simultaneously striking the tuning forks in parts (a) and (b).

FIGURE 5
Sound waves.

Let us now graph the sine function $f(\theta) = \sin\theta$. We are assuming that θ is in radians, so that the domain of the sine function is the set of all real numbers. Even before making any calculations, we can gain strong intuitive insight into how the graph must look by carrying out the following experiment. After drawing the unit circle, $x^2 + y^2 = 1$, place your fingertip at the point (1, 0) and then move your finger counterclockwise around the circle. As you do this, keep track of what happens to the y-coordinate of your fingertip. (The y-coordinate is $\sin \theta$.) Figure 6 indicates the general results.

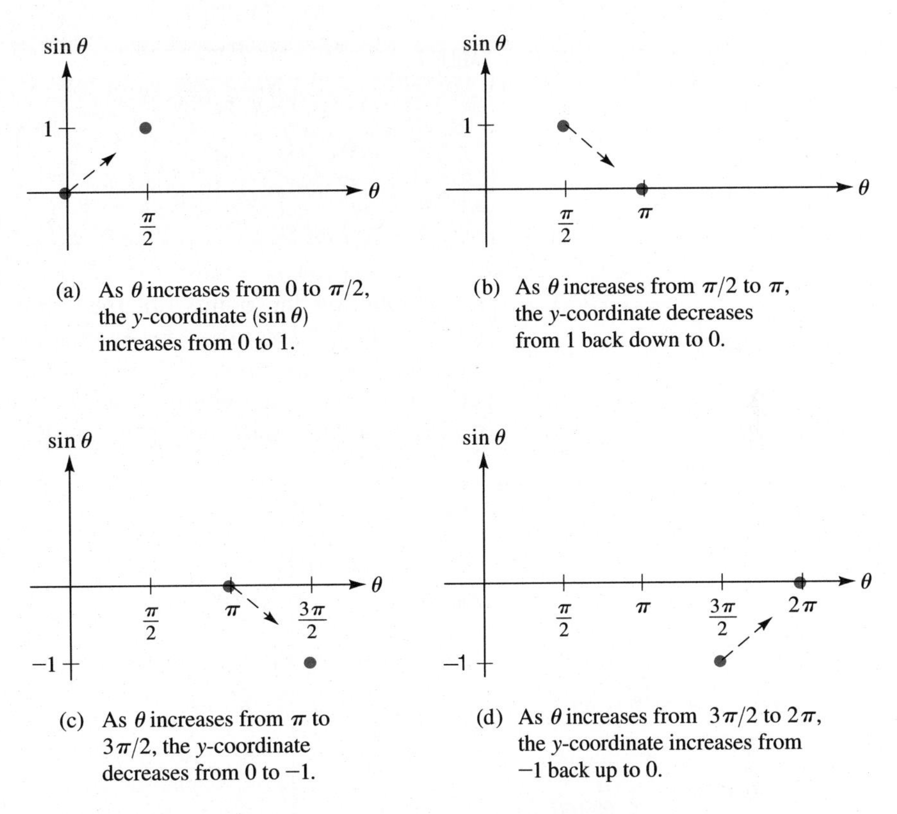

FIGURE 6

Figure 6 tells us a great deal about the sine function: where the function is increasing and decreasing, where the graph crosses the x-axis, and where the high and low points of the graph occur. Furthermore, since at $\theta = 2\pi$ we've returned to our starting point (1, 0), additional counterclockwise trips around the unit circle will just result in repetitions of the pattern established in Figure 6. In other words, the period of the function $y = \sin \theta$ is 2π.

As informative as Figure 6 is, however, there is not enough information there to tell us the precise shape of the required graph. For instance, all three of the curves in Figure 7 fit the specifications described in Figure 6.

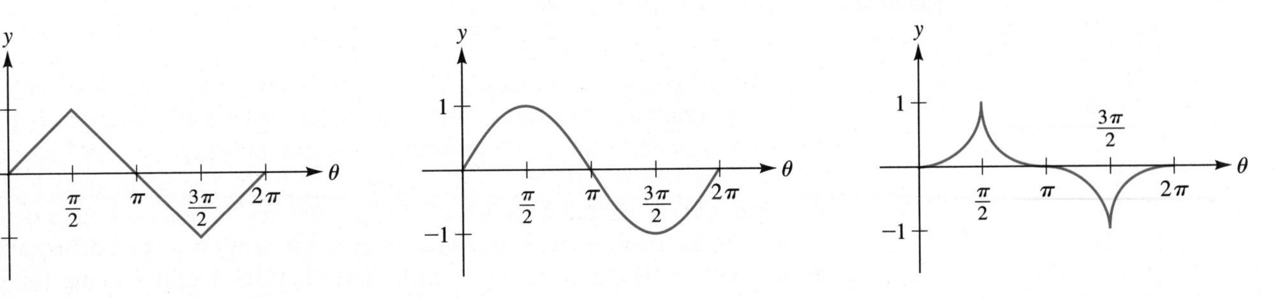

FIGURE 7

Now let's set up a table so that we can accurately sketch the graph of $y = \sin\theta$. Since the sine function is periodic (as we've just observed) with period 2π, our table need contain only values of θ between 0 and 2π, as shown in Table 1. This will establish the basic pattern for the graph.

TABLE 1

θ	0	$\frac{\pi}{6}$	$\frac{\pi}{3}$	$\frac{\pi}{2}$	$\frac{2\pi}{3}$	$\frac{5\pi}{6}$	π	$\frac{7\pi}{6}$	$\frac{4\pi}{3}$	$\frac{3\pi}{2}$	$\frac{5\pi}{3}$	$\frac{11\pi}{6}$	2π
$\sin\theta$	0	$\frac{1}{2}$	$\frac{\sqrt{3}}{2}$	1	$\frac{\sqrt{3}}{2}$	$\frac{1}{2}$	0	$-\frac{1}{2}$	$-\frac{\sqrt{3}}{2}$	-1	$-\frac{\sqrt{3}}{2}$	$-\frac{1}{2}$	0

In plotting the points obtained from Table 1, we use the approximation $\sqrt{3}/2 \approx 0.87$. Rather than approximating π, however, we mark off units on the horizontal axis in terms of π. The resulting graph is shown in Figure 8. By continuing this same pattern to the left and right, we obtain the complete graph of $f(\theta) = \sin\theta$, as indicated in Figure 9.

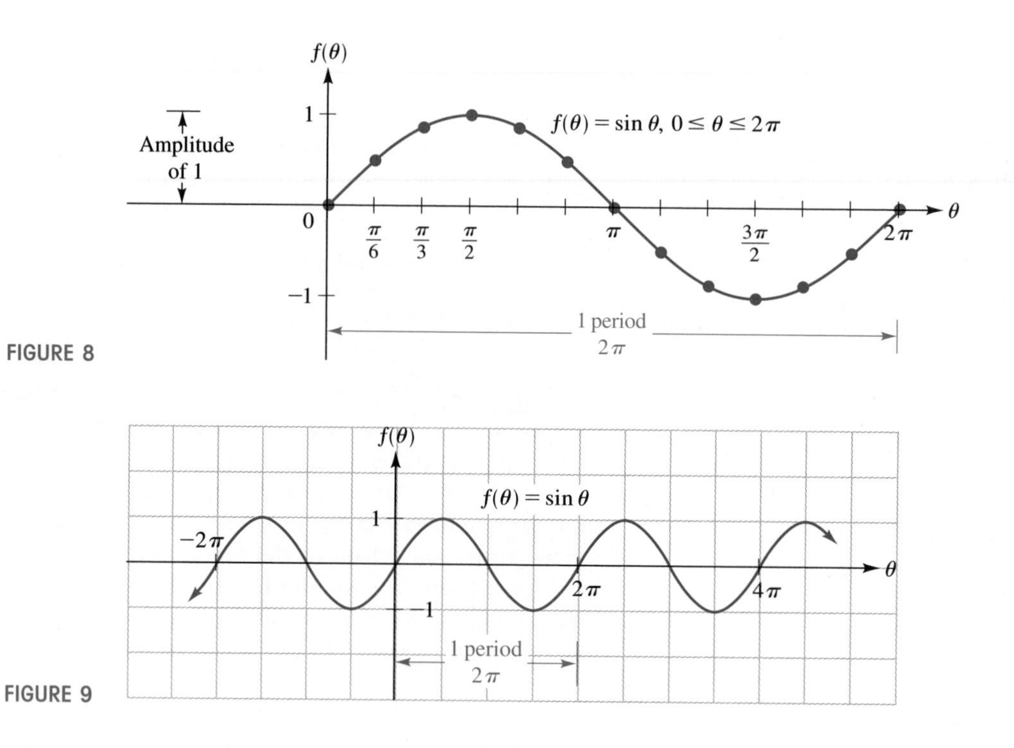

FIGURE 8

FIGURE 9

Before going on to analyze the sine function or to study the graphs of the other trigonometric functions, we're going to make a slight change in the notation we've been using. To conform with common usage, we will use x instead of θ on the horizontal axis, and we will use y for the vertical axis. The sine function is then written simply as $y = \sin x$, where the real number x denotes the radian measure of an angle or, equivalently, the length of the corresponding arc on the unit circle. For reference, we redraw Figures 8 and 9 using this familiar x-y notation, as shown in Figure 10.

You should memorize the graph of the sine curve in Figure 10(b) so that you can sketch it without first setting up a table. Of course, once you know the

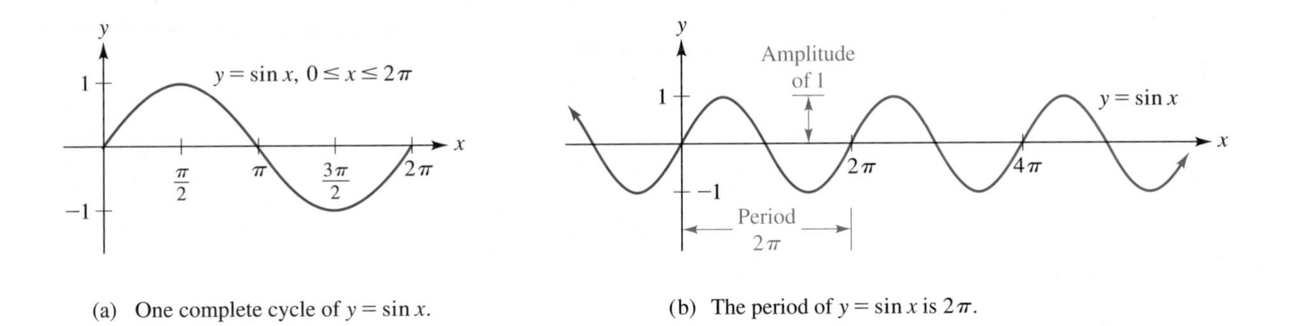

(a) One complete cycle of $y = \sin x$.

(b) The period of $y = \sin x$ is 2π.

FIGURE 10

shape and the location of the basic cycle shown in Figure 10(a), you automatically know the graph of the full sine curve in Figure 10(b).

We can use the graphs in Figure 10 to help us list some of the key properties of the sine function.

PROPERTY SUMMARY

THE SINE FUNCTION: $y = \sin x$

1. The domain of the sine function is the set of all real numbers. The range of the sine function is the closed interval $[-1, 1]$, and we have

$$-1 \leq \sin x \leq 1, \qquad \text{for all } x$$

2. The sine function is periodic, with period 2π. The amplitude is 1.

3. The graph of $y = \sin x$ consists of repetitions of the **basic sine wave** shown in Figure 11. The basic sine wave crosses the x-axis at the beginning, middle, and end of the cycle. The curve reaches its highest point one-quarter of the way through the cycle and its lowest point three-quarters of the way through the cycle.

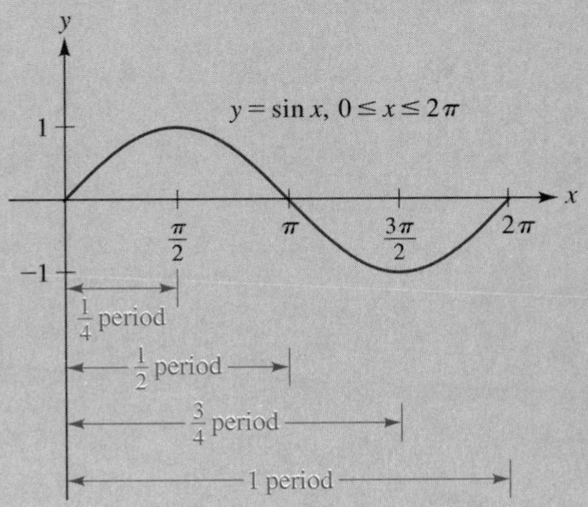

FIGURE 11
The basic sine wave.

EXAMPLE 1 Figure 12 shows a portion of the graph of $y = \sin x$.

(a) What are the coordinates of the three turning points A, B, and C? Here, and in part (b), report the answers both in terms of π and as calculator approximations rounded to three decimal places.
(b) What are the x-intercepts at D and E?

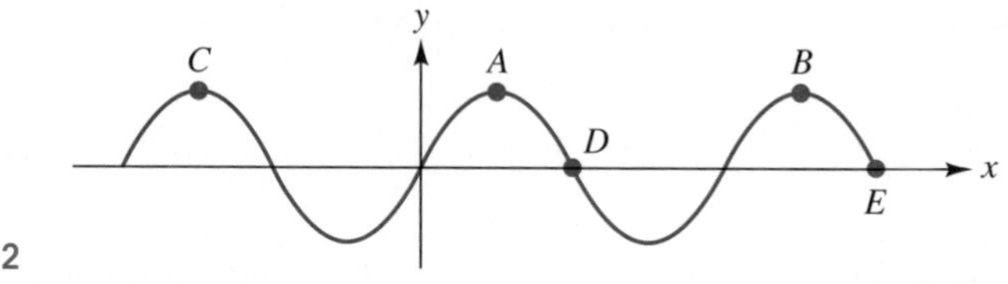

FIGURE 12

Solution

(a) From Figure 11 (which you should also memorize), we know that the coordinates of the point A are $(\pi/2, 1)$. Therefore, since the period of the sine function is 2π, the coordinates of B and C are as follows:

$$\text{Coordinates of } B: \quad \left(\frac{\pi}{2} + 2\pi, 1\right) \quad \text{or} \quad \left(\frac{5\pi}{2}, 1\right)$$

$$\text{Coordinates of } C: \quad \left(\frac{\pi}{2} - 2\pi, 1\right) \quad \text{or} \quad \left(-\frac{3\pi}{2}, 1\right)$$

As you should now check for yourself, the calculator approximations here are

$$A(1.571, 1) \qquad B(7.854, 1) \qquad C(-4.712, 1)$$

(b) From Figure 11, we also know the x-coordinate at D; it is π. Since the point E is 2π units to the right of D, the x-coordinate of E is $\pi + 2\pi = 3\pi$. Using a calculator, we have

x-intercept at D: 3.142

x-intercept at E: 9.425

We could obtain the graph of the cosine function by setting up a table, just as we did with the sine function. A more interesting and informative way to proceed, however, is to use the identity

$$\cos x = \sin\left(x + \frac{\pi}{2}\right) \tag{1}$$

(Exercise 49 at the end of this section outlines a geometric proof of this identity. Also, after studying Section 8.1, you'll see a simple way to prove this identity algebraically.) From our work on graphing functions in Chapter 3, we can interpret equation (1) geometrically. Equation (1) tells us that the graph of $y = \cos x$ is obtained by translating the sine curve $\pi/2$ units to the left. The result is shown in Figure 13(a). Figure 13(b) displays one complete cycle of the cosine curve, from $x = 0$ to $x = 2\pi$.

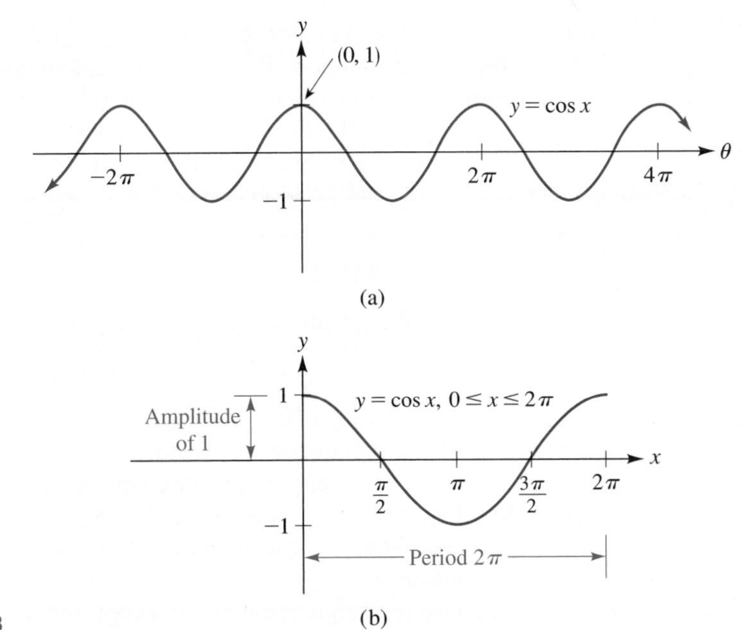

FIGURE 13

As with the sine function, you should memorize the graph and the basic features of the cosine function, which are summarized in the box that follows.

PROPERTY SUMMARY **THE COSINE FUNCTION: $y = \cos x$**

1. The domain of the cosine function is the set of all real numbers. The range of the cosine function is the closed interval $[-1, 1]$, and we have

$$-1 \leq \cos x \leq 1, \qquad \text{for all } x$$

2. The cosine function is periodic, with period 2π. The amplitude is 1.

3. The graph of $y = \cos x$ consists of repetitions of the **basic cosine wave** shown in Figure 14. At the beginning and end of the cycle, the basic cosine wave reaches its highest point. At the middle of the cycle, the curve reaches its lowest point. The x-intercepts occur one-quarter and three-quarters of the way through the cycle.

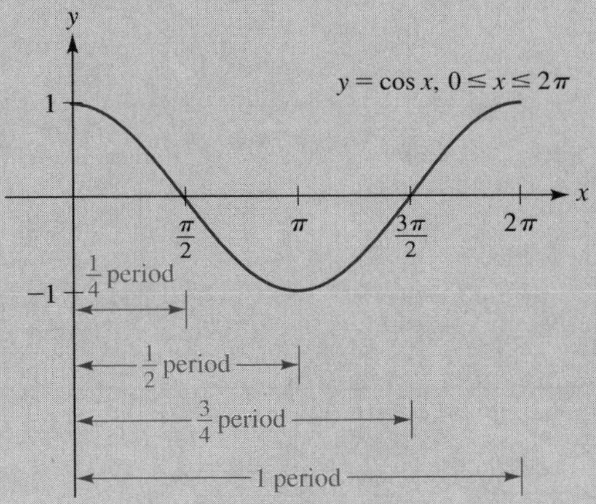

FIGURE 14
The basic cosine wave.

In the next example, we use a graph to estimate roots of equations involving the cosine function. We also use a calculator to obtain more accurate results. As you'll see, the calculator portion of the work involves more than just button pushing. You'll need to use the reference-angle concept, and you'll need to apply the following property of inverse functions (from Section 3.5).

$$f^{-1}(f(x)) = x \qquad \text{for every } x \text{ in the domain of } f$$

EXAMPLE 2

(a) Use the graph in Figure 15 to estimate a root of the equation

$$\cos x = 0.8$$

in the interval $0 \leq x \leq \pi/2$.

(b) Use a calculator to obtain a more accurate value for the root in part (a). Round the answer to three decimal places.

(c) Use the reference-angle concept to find all solutions to the equation $\cos x = 0.8$ in the interval $0 \leq x \leq 2\pi$. Round the answers to three decimal places.

(d) Use the reference-angle concept and a calculator to find all solutions of the equation $\cos x = -0.8$ within the interval $0 \leq x \leq 2\pi$. Again, round the answer to three decimal places.

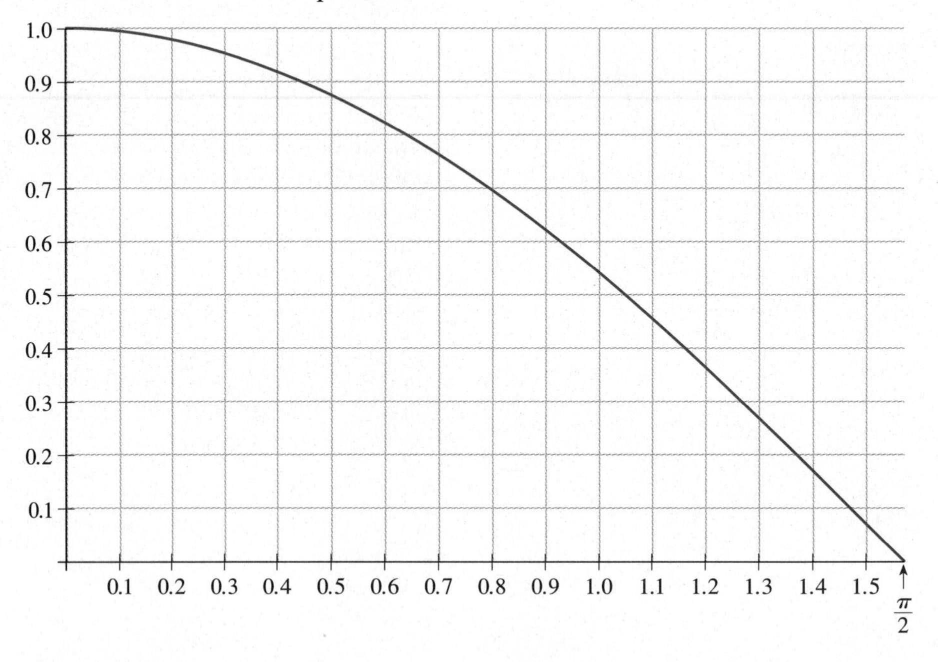

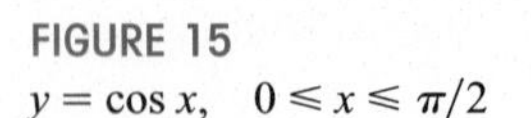
FIGURE 15
$y = \cos x, \quad 0 \leq x \leq \pi/2$

Solution

(a) In Figure 15, look at the point where the horizontal line $y = 0.8$ intersects the curve $y = \cos x$. The x-coordinate of this point, call it x_1, is the root we are looking for. (Why?) Using Figure 15, we estimate that x_1 is approximately halfway between $x = 0.6$ and $x = 0.7$. So, our estimate for the root is $x_1 \approx 0.65$.

(b) We can determine the root x_1 by using the *inverse cosine function.* Just as the notation f^{-1} denotes the inverse of a function f, so the notation $\cos^{-1}$

is often used to denote the inverse cosine function. Starting with the equation $\cos x_1 = 0.8$, we apply the inverse cosine to both sides to obtain

$$\cos^{-1}(\cos x_1) = \cos^{-1}(0.8)$$

and therefore

$$x_1 = \cos^{-1}(0.8)$$

We use a calculator, set in the radian mode, to evaluate the expression $\cos^{-1}(0.8)$. The keystrokes are as follows. (As in previous sections, we show keystrokes for the two basic types of calculators: input-before-function calculators and function-before-input calculators. If these keystrokes don't seem to apply to your particular brand of calculator, be sure to consult the instruction manual for the calculator.)

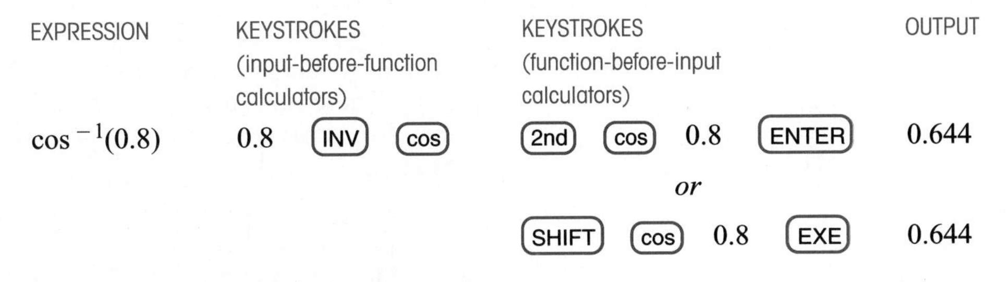

EXPRESSION	KEYSTROKES (input-before-function calculators)	KEYSTROKES (function-before-input calculators)	OUTPUT
$\cos^{-1}(0.8)$	0.8 [INV] [cos]	[2nd] [cos] 0.8 [ENTER]	0.644
		or	
		[SHIFT] [cos] 0.8 [EXE]	0.644

Note that the value 0.644 is consistent with the estimate $x_1 \approx 0.65$ that was obtained graphically in part (a).

(c) To find another root of the equation, we use the fact that the cosine is positive in Quadrant IV as well as in Quadrant I. The root $x_1 \approx 0.644$ is the radian measure of a first-quadrant angle (because $0.644 < \pi/2$). As indicated in Figure 16, we let x_2 denote the radian measure of the fourth-quadrant angle that has x_1 for its reference angle. Then we have $\cos x_2 = \cos x_1 = 0.8$. That is, x_2 is also a solution of the equation $\cos x = 0.8$. The calculations for x_2 now run as follows:

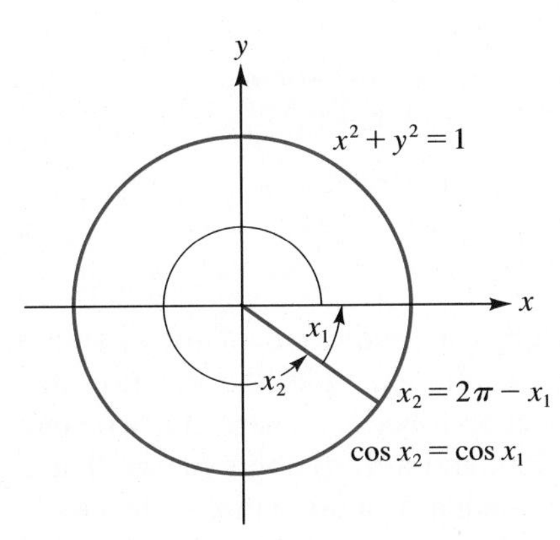

FIGURE 16

$$\begin{aligned} x_2 &= 2\pi - x_1 \\ &= 2\pi - \cos^{-1}(0.8) \\ &\approx 5.640 \quad \text{using a calculator (radian mode) and rounding to three decimal places} \end{aligned}$$

We've now determined two roots of the equation $\cos x = 0.8$ in the interval $0 \leq x \leq 2\pi$. There are no other roots in this interval because the cosine is negative in Quadrants II and III.

(d) In part (b) we determined a first-quadrant angle x_1 satisfying the equation $\cos x = 0.8$. Now we want solutions for the equation $\cos x = -0.8$. Since the cosine is negative in Quadrants II and III, we therefore want angles in Quadrants II and III that have x_1 for the reference angle. As indicated in Figure 17 (on the next page), these angles are $\pi - x_1$ and $\pi + x_1$. Computing, we have

$$\pi - x_1 = \pi - \cos^{-1}(0.8) \approx 2.498$$

and

$$\pi + x_1 = \pi + \cos^{-1}(0.8) \approx 3.785$$

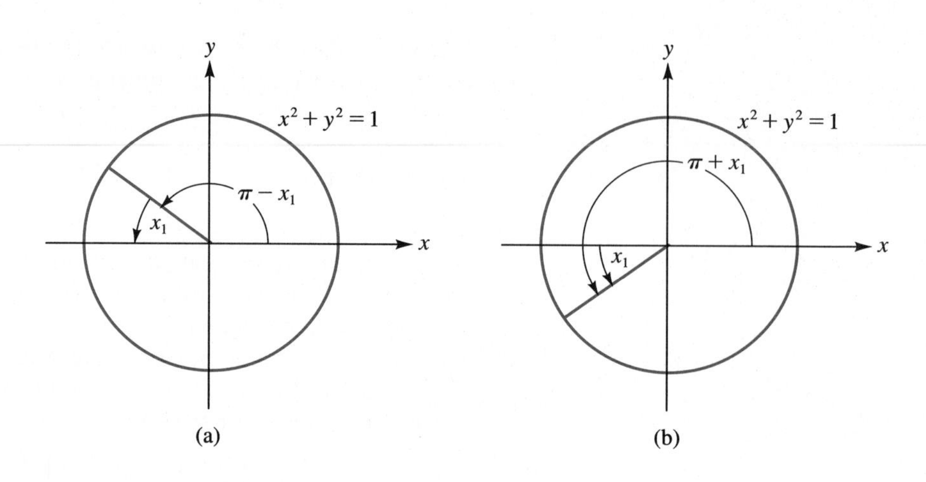

FIGURE 17
The reference angle for both $\pi - x_1$ and $\pi + x_1$ is x_1.

In summary, the equation $\cos x = -0.8$ has two roots in the closed interval $[0, 2\pi]$. These roots are approximately 2.498 and 3.785. In Figure 18, which summarizes this example, these two roots are denoted by x_3 and x_4, respectively.

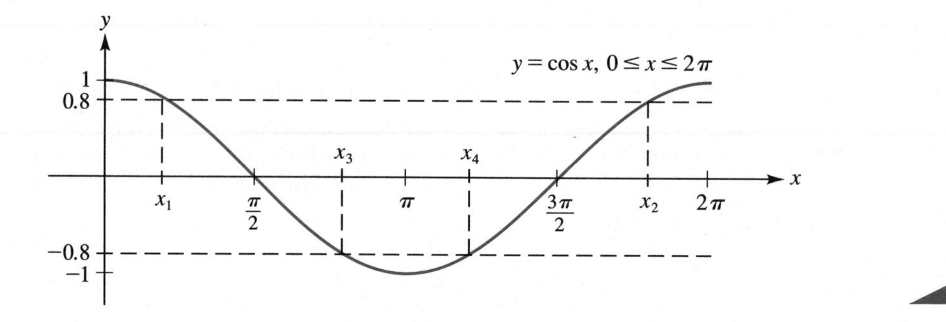

FIGURE 18
Summary of Example 2.
The roots of the equation $\cos x = 0.8$ in the interval $[0, 2\pi]$ are $x_1 \approx 0.644$ and $x_2 \approx 5.640$. The roots of the equation $\cos x = -0.8$ in the interval $[0, 2\pi]$ are $x_3 \approx 2.498$ and $x_4 \approx 3.785$.

In part (b) of the example just concluded, we used a calculator to find a number x such that $\cos x = 0.8$. As indicated in the example, the function that outputs such a number is called the inverse cosine function. There are two standard abbreviations for the name of this function, and a third abbreviation that is used on some types of calculators. The two standard abbreviations are $\cos^{-1}$ and arccos (read *arc cos*). And, as you saw in Example 2, the notation INV COS is used on some calculators. In the next chapter, we will analyze the inverse cosine function in some detail. But for our present purposes, you need only know the following definition:

$\cos^{-1}(x)$ denotes the number in the interval $[0, \pi]$ whose cosine is x.

As examples of this notation, we have

$\cos^{-1}(1/2) = \pi/3$	because $\cos\frac{\pi}{3} = \frac{1}{2}$ and $0 < \frac{\pi}{3} < \pi$
$\cos^{-1}(1/2) \neq 5\pi/3$	because although $\cos\frac{5\pi}{3} = \frac{1}{2}$, the number $\frac{5\pi}{3}$ is not in the required interval $[0, \pi]$
$\cos^{-1}(-1/2) = 2\pi/3$	because $\cos\frac{2\pi}{3} = -\frac{1}{2}$ and $0 < \frac{2\pi}{3} < \pi$
$\cos^{-1}(0.8) \approx 0.644$	as we saw in Example 2

In Example 2, Figure 18 tells us that the number x_1 is in the interval $[0, \pi]$ but x_2 is not. That is why the keystrokes on page 447 gave us the value for x_1 rather than x_2. (We then used the reference-angle or reference-number concept to obtain x_2.)

For some of the exercises in this section, you will need to use the *inverse sine function,* rather than the inverse cosine function that we have just been discussing. For reference, we define both functions in the box that follows. (These functions are discussed at greater length in the next chapter.)

DEFINITION: Inverse Cosine Function and Inverse Sine Function

NAME OF FUNCTION	ABBREVIATION	DEFINITION
Inverse cosine	$\cos^{-1}$	$\cos^{-1}(x)$ is the (unique) number in the interval $[0, \pi]$ whose cosine is x.
Inverse sine	$\sin^{-1}$	$\sin^{-1}(x)$ is the (unique) number in the interval $[-\pi/2, \pi/2]$ whose sine is x.

EXERCISE SET 7.4

Do 2, 19, 22, 37, 47

A

In Exercises 1–8, specify the period and amplitude for each function.

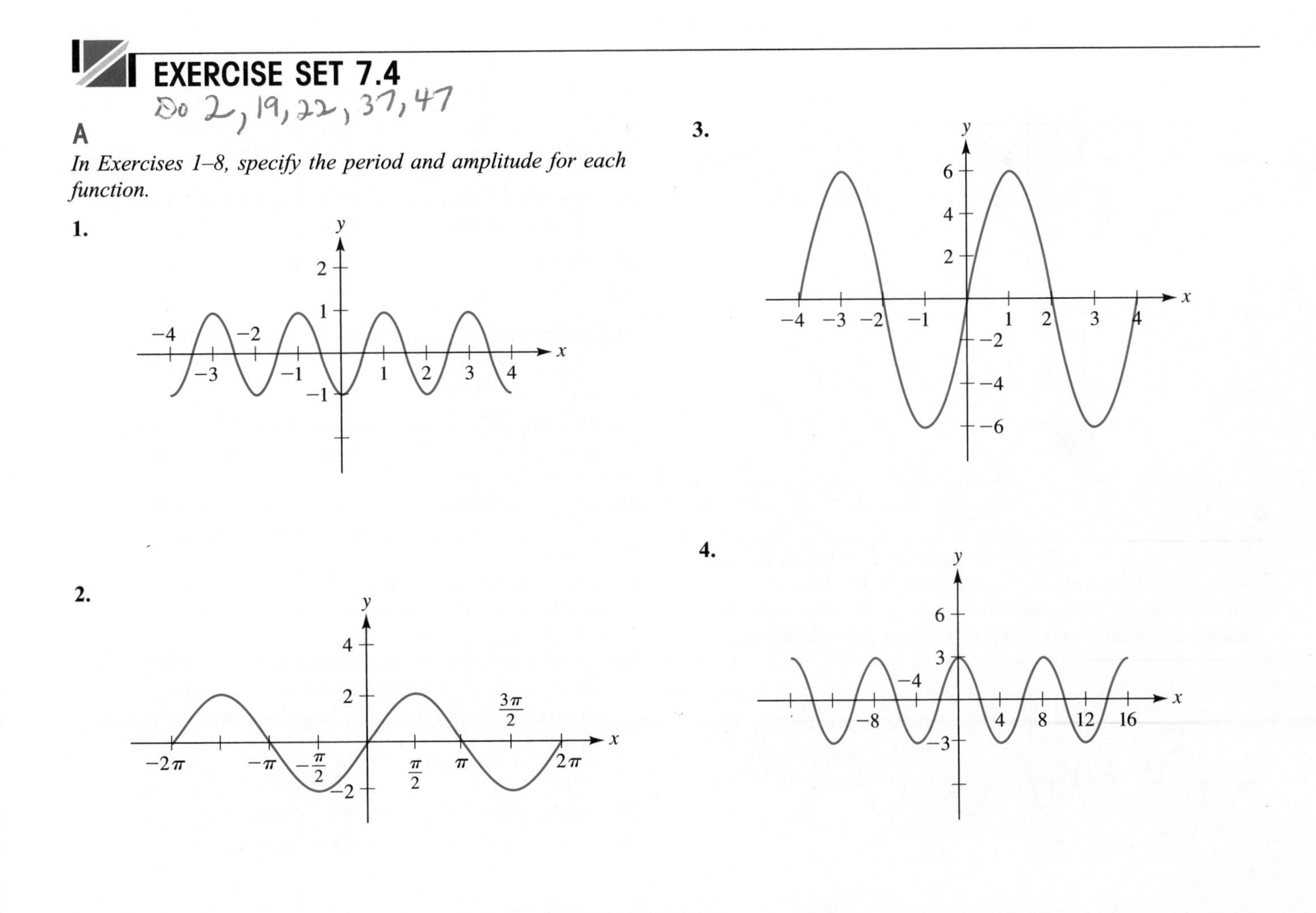

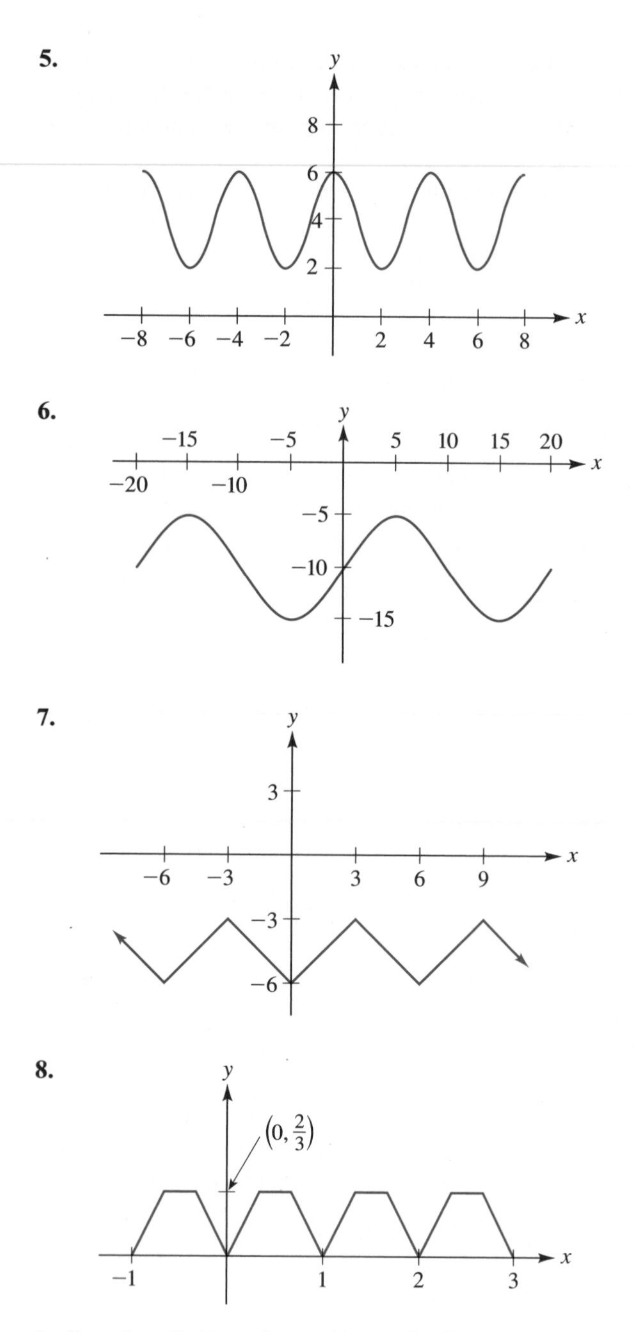

In Exercises 9–18, refer to the graph of $y = \sin x$ in the following figure. Specify the coordinates of the indicated points. Give the x-coordinates both in terms of π and as calculator approximations rounded to three decimal places.

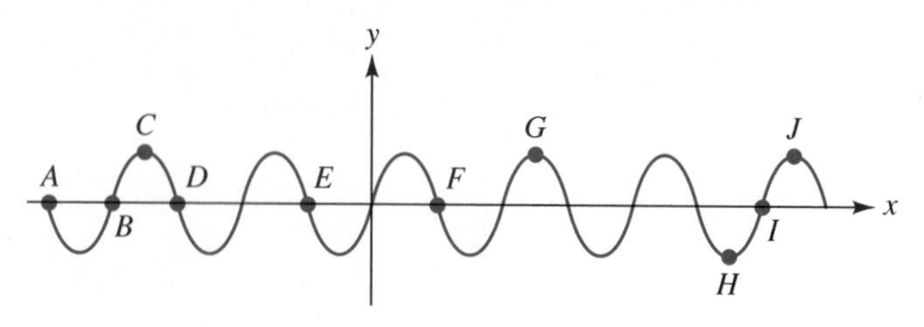

9. *C* **10.** *F* **11.** *G* **12.** *A*
13. *B* **14.** *J* **15.** *D* **16.** *H*
17. *E* **18.** *I*

In Exercises 19–22, state whether the function $y = \sin x$ is increasing or decreasing on the given interval. (The terms increasing *and* decreasing *are explained on page 117.)*

19. $\frac{3\pi}{2} < x < 2\pi$ **20.** $-\frac{\pi}{2} < x < \frac{\pi}{2}$

21. $\frac{5\pi}{2} < x < \frac{7\pi}{2}$ **22.** $-\frac{5\pi}{2} < x < -2\pi$

In Exercises 23–32, refer to the graph of $y = \cos x$ in the following figure. Specify the coordinates of the indicated points. Give the x-coordinates both in terms of π and as calculator approximations rounded to three decimal places.

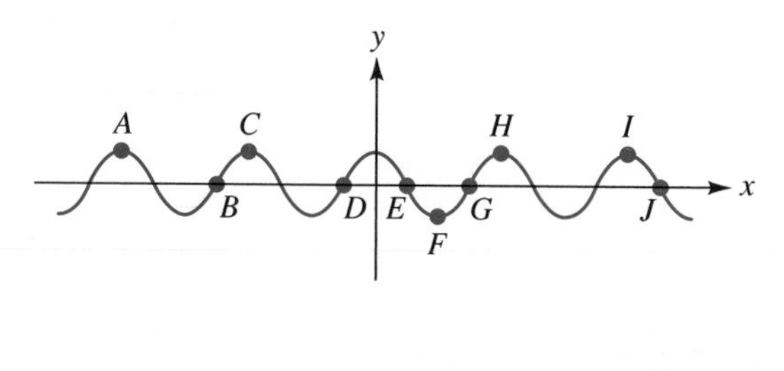

23. *J* **24.** *H* **25.** *A* **26.** *G*
27. *E* **28.** *D* **29.** *I* **30.** *F*
31. *B* **32.** *C*

In Exercises 33–36, state whether the function $y = \cos x$ is increasing or decreasing on the given interval.

33. $0 < x < \pi$ **34.** $6\pi < x < 7\pi$

35. $-\frac{\pi}{2} < x < 0$ **36.** $-\frac{5\pi}{2} < x < -2\pi$

In Exercises 37–42, you are given an equation of the form

$$\cos x = k, \quad \text{where } k \geq 0$$

(a) *Use the graph in Figure 15 (on page 446) to estimate (to the nearest 0.05) a root of the equation within the interval $0 \leq x \leq \pi/2$.*

(b) *Use a calculator to obtain a more accurate value for the root in part (a). Round the answer to four decimal places.*

(c) *Use the reference-angle concept and a calculator to find another root of the equation (within the interval $0 \leq x \leq 2\pi$).*

(d) *Use the reference-angle concept and a calculator to find all solutions of the equation $\cos x = -k$ within the interval $0 \leq x \leq 2\pi$. Again, round the answers to four decimal places.*

37. $\cos x = 0.7$ **38.** $\cos x = 0.9$

39. $\cos x = 0.4$ **40.** $\cos x = 0.3$

41. $\cos x = 0.6$ **42.** $\cos x = 0.55$

In Exercises 43–48, you are given an equation of the form

$$\sin x = k, \quad \text{where } k > 0$$

(a) *Use the graph in Figure A below to estimate (to the nearest 0.05) a root of the equation within the interval* $0 \leq x \leq \pi/2$.
(b) *Use a calculator (with the inverse sine function) to obtain a more accurate value for the root in part (a). Round the answer to four decimal places.*
(c) *Use the reference-angle concept and a calculator to find another root of the equation (within the interval* $0 \leq x \leq 2\pi$*). Round the answer to four decimal places.*
(d) *Use the reference-angle concept and a calculator to find all solutions of the equation* $\sin x = -k$ *within the interval* $0 \leq x \leq 2\pi$*. Round the answer to four decimal places.*

43. $\sin x = 0.1$
44. $\sin x = 0.6$
45. $\sin x = 0.4$
46. $\sin x = 0.2$
47. $\sin x = 0.85$
48. $\sin x = 0.7$

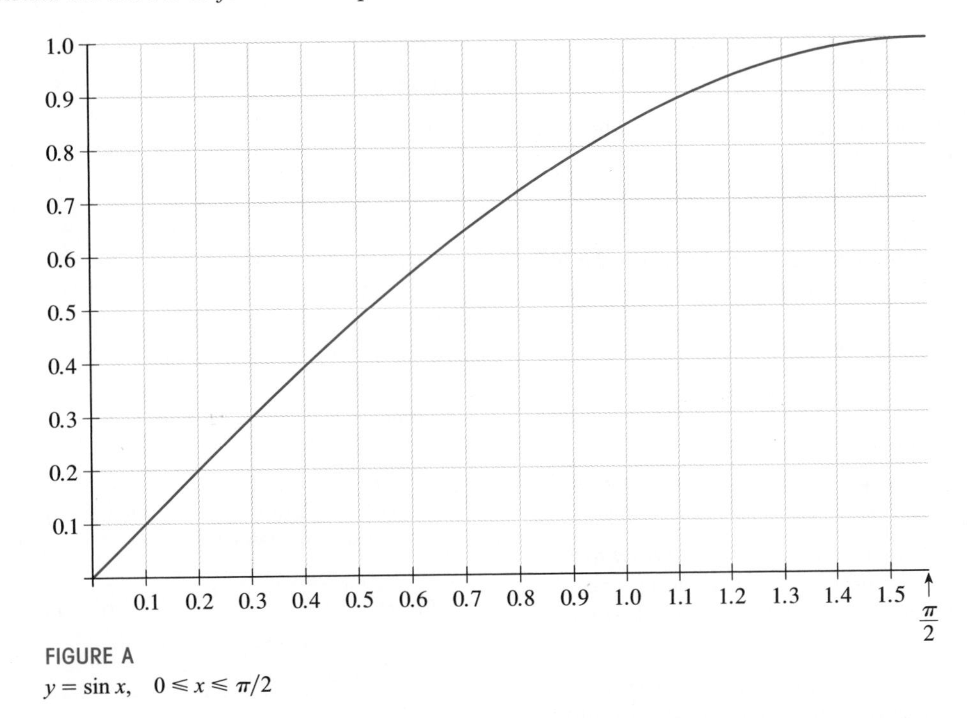

FIGURE A
$y = \sin x, \quad 0 \leq x \leq \pi/2$

B

49. In the text we used the identity $\cos \theta = \sin(\theta + \pi/2)$ in obtaining the graph of $y = \cos x$ from that of $y = \sin x$. In this exercise you'll derive this identity. Refer to the following figure. (Although the figure shows the angle of radian measure θ in the first quadrant, the proof can be easily carried over for the other quadrants as well.)

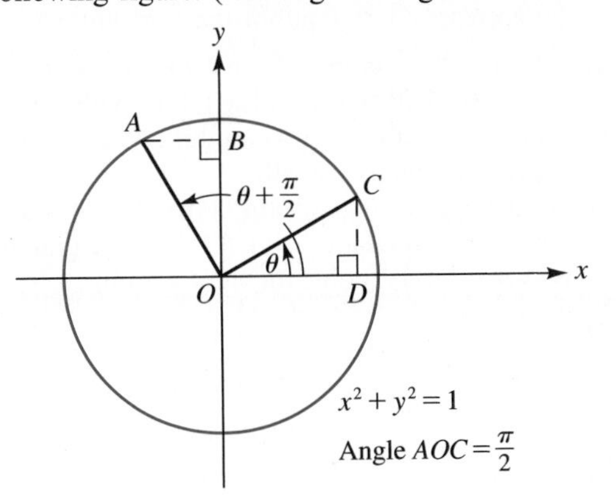

(a) What are the coordinates of C?

(b) Show that $\triangle AOB$ and $\triangle COD$ are congruent.

(c) Use the results in parts (a) and (b) to show that the coordinates of A are $(-\sin\theta, \cos\theta)$.

(d) Since the radian measure of $\angle DOA$ is $\theta + \frac{\pi}{2}$, the coordinates of A (by definition) are $\left(\cos\left(\theta + \frac{\pi}{2}\right), \sin\left(\theta + \frac{\pi}{2}\right)\right)$. Now explain why

$$\cos\theta = \sin\left(\theta + \frac{\pi}{2}\right)$$

and

$$-\sin\theta = \cos\left(\theta + \frac{\pi}{2}\right)$$

In Exercises 50–54, refer to the following figure, which shows the graphs of $y = \sin x$ *and* $y = \cos x$ *on the closed interval* $[0, 2\pi]$.

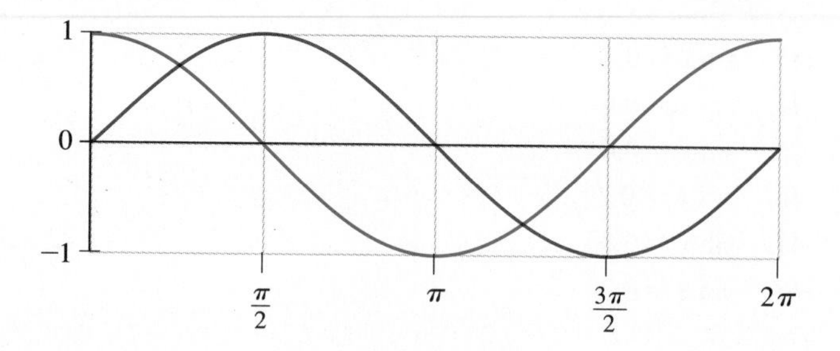

50. **(a)** When the value of $\sin x$ is a maximum, what is the corresponding value of $\cos x$?
(b) When the value of $\cos x$ is a maximum, what is the corresponding value of $\sin x$?

51. Follow Exercise 50, but replace the word "maximum" with "minimum."

52. **(a)** For which x-values in the interval $0 \leq x \leq 2\pi$ is $\sin x = \cos x$? *Hint:* Refer to Table 1 on page 364.
(b) For which x-values in the interval $0 \leq x \leq 2\pi$ is $\sin x < \cos x$?

53. Specify an open interval in which both the sine and cosine functions are decreasing.

54. Specify an open interval in which the sine function is decreasing but the cosine function is increasing.

As a prerequisite for Exercises 55 and 56, you need to have studied Section 4.3 on iteration and population growth. And in particular, you should be familiar with Example 3 on pages 203–205.

55. As in Example 3 in Section 4.3, suppose a farmer has a fishpond with a maximum population size of 500, and suppose that initially the pond is stocked with 50 fish. Unlike Example 3, however, assume that the growth equation is

$$f(x) = \cos x \quad (0 \leq x \leq 1)$$

Finally, as in Example 3, assume that the time intervals are breeding seasons.

(a) Use the accompanying iteration diagram and your calculator to complete the following table. For the values obtained from the graph, estimate to the nearest 0.05 (or closer, if it seems appropriate). For the calculator values, round the final answers to five decimal places. As usual, check to see that the calculator results are consistent with the estimates obtained from the graph.

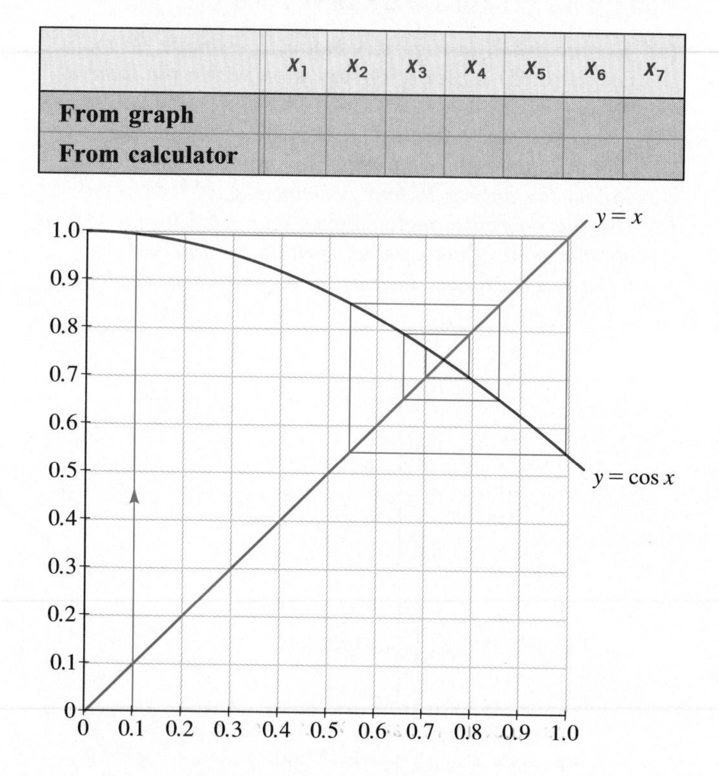

	x_1	x_2	x_3	x_4	x_5	x_6	x_7
From graph							
From calculator							

(b) Use the calculator results in part (a) to complete the following table.

n	0	1	2	3	4	5	6	7
Number of fish after n breeding seasons	50							

(c) As indicated in the figure accompanying part (a), the iteration process is spiraling in on a fixed point between 0.7 and 0.8. Using calculus (or simply a calculator), it can be shown that this fixed point is 0.7391 (rounded to four decimal places). What is the corresponding equilibrium population?

56. Suppose that in Exercise 55, instead of $f(x) = \cos x$, we use $g(x) = \sin x$ $(0 \leq x \leq 1)$ for the growth function.
(a) Complete a table similar to the one in Exercise 55(b) (assuming $x_0 = 50$).
(b) What do you think would be the long-term behavior of this population? *Hint:* Think graphically. Is there a fixed point for the function $g(x) = \sin x$?

GRAPHING UTILITY EXERCISES FOR SECTION 7.4

In Exercises 1–3, you'll use a graphing utility to obtain several different views of $y = \sin x$ and $y = \cos x$. If you are using a graphics calculator, make sure it is set for the radian mode (rather than the degree mode).

1. **(a)** Graph $y = \sin x$ using a viewing rectangle that extends from -7 to 7 in the x-direction and from -3 to 3 in the y-direction. Note that there is an x-intercept between 3 and 4. Using your knowledge of the sine function (and not the graphing utility), what is the exact value for this x-intercept?

(b) Refer to the graph that you obtained in part (a). How many turning points do you see? Note that one of the turning points occurs when x is between 1 and 2. What is the exact x-coordinate for this turning point?

(c) Add the graph of $y = \cos x$ to your picture. How many turning points do you see for $y = \cos x$? Note that one of the turning points occurs between $x = 6$ and $x = 7$. What is the exact x-coordinate for this turning point?

(d) Your picture in part (c) indicates that the graphs of $y = \sin x$ and $y = \cos x$ are just translates of one another. By what distance would we have to shift the graph of $y = \sin x$ to the left in order for it to coincide with the graph of $y = \cos x$?

(e) Most graphing utilities have an option that will let you mark off the units on the x-axis in terms of π. Check your instruction manual if necessary, and then, for the picture that you obtained in part (c), change the x-axis units to multiples of $\pi/2$. Use the resulting picture to confirm your answers to the questions in parts (a) through (c) regarding x-intercepts and turning points.

2. This exercise presents an interesting fact about the graph of $y = \sin x$ that is useful in numerical work.

(a) In the standard viewing rectangle, graph the function $y = \sin x$ along with the line $y = x$. Notice that the two graphs appear to be virtually identical in the vicinity of the origin. Actually, the only point that the two graphs have in common is (0, 0), but very near the origin, the distance between the two graphs is less than the thickness of the lines or dots that your graphing utility draws. To underscore this fact, zoom in on the origin several times. What do you observe?

(b) The work in part (a) can be summarized as follows. (Recall that the symbol $\approx$ means "is approximately equal to.")

$$\sin x \approx x \quad \text{when } |x| \text{ is close to } 0$$

To see numerical evidence that supports this result, complete the following tables.

x	0.253	0.0253	0.00253	0.000253
sin x				
$\lvert x - \sin x\rvert$				

x	-0.253	-0.0253	-0.00253	-0.000253
sin x				
$\lvert x - \sin x\rvert$				

3. **(a)** Graph the function $y = \cos x$ in the standard viewing rectangle, and look at the arch-shaped portion of the curve between $-\pi/2$ and $\pi/2$. This portion of the cosine curve has the general shape of a parabola. Could it actually be a portion of a parabola? Go on to part (b).

(b) Calculus shows that the answer to the question raised in part (a) is "no." But calculus also shows that there is a parabola, $y = 1 - 0.5x^2$, that closely resembles the cosine curve in the vicinity of $x = 0$. To see this, graph the parabola $y = 1 - 0.5x^2$ and the curve $y = \cos x$ in the standard viewing rectangle. Describe, in a complete sentence or two, what you observe.

(c) Complete the following table to see numerical evidence that strongly supports your observations in part (b).

x	1	0.5	0.1	0.01	0.001
cos x					
$1 - 0.5x^2$					

4. In Example 2 in Section 7.4, we determined two roots x_1 and x_2 of the equation $\cos x = 0.8$ in the interval $0 \leq x \leq 2\pi$. The values were $x_1 \approx 0.644$ and $x_2 \approx 5.640$. Check these results graphically as follows. Using a viewing rectangle extending from $x = 0$ to $x = 2\pi$, graph the curve $y = \cos x$ along with the horizontal line $y = 0.8$. Then, use the graphing utility (with repeated zooms) to estimate the x-coordinates of the two points where the curve $y = \cos x$ intersects the line $y = 0.8$. What do you observe?

5. In Example 2(d), we determined two roots x_3 and x_4 of the equation $\cos x = -0.8$ in the interval $0 \leqslant x \leqslant 2\pi$. The values were $x_3 \approx 2.498$ and $x_4 \approx 3.785$. Use the procedure indicated in the previous exercise to check these results graphically.

In Exercises 6 and 7, use graphs (as in Example 2) to estimate the roots of each equation for $0 \leqslant x \leqslant 2\pi$. Zoom in close enough on the intersection points until you are sure about the first two decimal places in each root. Then, use a calculator, as in Example 2, to determine more accurate values for the roots. Round the calculator values to four decimal places.

6. **(a)** $\cos x = 0.351$
   (b) $\cos x = -0.351$

7. **(a)** $\sin x = 0.687$
   (b) $\sin x = -0.687$

7.5 GRAPHS OF $y = A \sin(Bx - C)$ AND $y = A \cos(Bx - C)$

The graphs of $y = \sin x$ and $y = \cos x$ are the building blocks we need for graphing functions of the form

$$y = A \sin(Bx - C) \qquad \text{and} \qquad y = A \cos(Bx - C)$$

As a first example, consider $y = 2 \sin x$. To obtain the graph of $y = 2 \sin x$ from that of $y = \sin x$, we multiply each y-coordinate on the graph of $y = \sin x$ by 2. As indicated in Figure 1, this changes the amplitude from 1 to 2, but it does not affect the period, which remains 2π. As a second example, Figure 2 shows the graphs of $y = \cos x$ and $y = \frac{1}{2} \cos x$. Note that the amplitude of $y = \frac{1}{2} \cos x$ is $\frac{1}{2}$ and the period is, again, 2π. More generally, graphs of functions of the form $y = A \sin x$ and $y = A \cos x$ always have an amplitude of $|A|$ and a period of 2π.

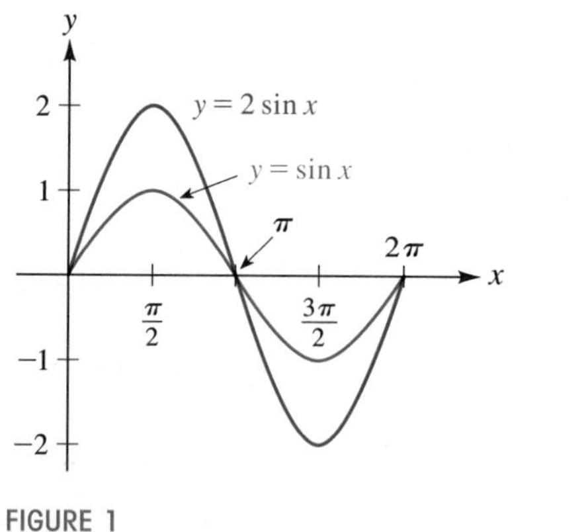

FIGURE 1
The amplitude of $y = 2 \sin x$ is 2. Both $y = \sin x$ and $y = 2 \sin x$ have a period of 2π.

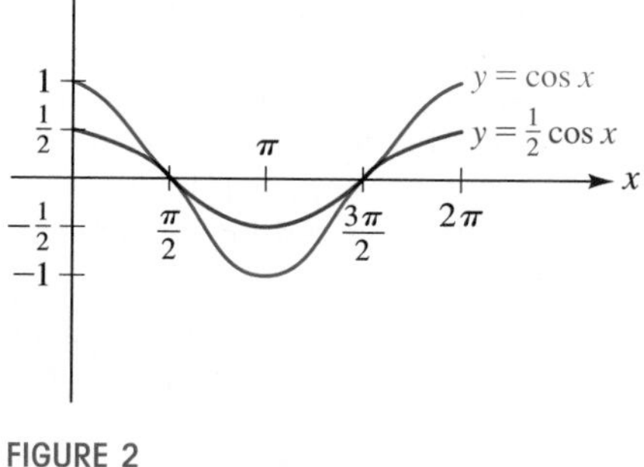

FIGURE 2
The amplitude of $y = \frac{1}{2} \cos x$ is $\frac{1}{2}$. Both $y = \cos x$ and $y = \frac{1}{2} \cos x$ have a period of 2π.

EXAMPLE 1 Graph the function $y = -2 \sin x$ over one period. On which interval(s) is the function decreasing?

Solution In Section 3.3 we saw that the graph of $y = -f(x)$ is obtained from that of $y = f(x)$ by reflection about the x-axis. Thus, we need only take the graph

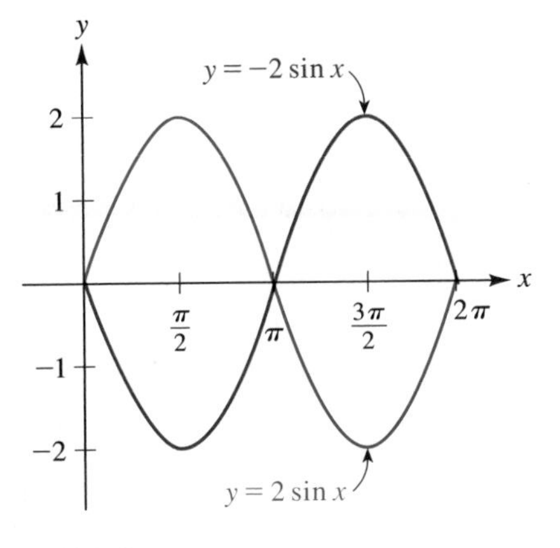

FIGURE 3

of $y = 2\sin x$ from Figure 1 and reflect it about the x-axis; see Figure 3. Note that both functions have an amplitude of 2 and a period of 2π. From the graph in Figure 3, we can see that the function $y = -2\sin x$ is decreasing on the intervals $(0, \pi/2)$ and $(3\pi/2, 2\pi)$.

We have just seen that functions of the form $y = A\sin x$ and $y = A\cos x$ have an amplitude of $|A|$ and a period of 2π. The next two examples show how to analyze functions of the form $y = A\sin Bx$ and $y = A\cos Bx$ $(B > 0)$. As you'll see, these functions have an amplitude of $|A|$ and a period of $2\pi/B$.

EXAMPLE 2 Graph the function $y = \cos 3x$ over one period.

Solution We know that the cosine curve $y = \cos x$ begins its basic pattern when $x = 0$ and completes that pattern when $x = 2\pi$. Thus, $y = \cos 3x$ will begin its basic pattern when $3x = 0$, and it will complete that pattern when $3x = 2\pi$. From the equation $3x = 0$ we conclude that $x = 0$, and from the equation $3x = 2\pi$ we conclude that $x = 2\pi/3$. Thus, the graph of $y = \cos 3x$ begins its basic pattern at $x = 0$ and completes the pattern at $x = 2\pi/3$. This tells us that the period is $2\pi/3$. Next, in preparation for drawing the graph, we divide the period into quarters, as shown in Figure 4(a). In Figure 4(b) we've plotted the points with x-coordinates shown in Figure 4(a). (We've also plotted the point on the curve corresponding to $x = 0$, where the basic pattern is to begin.) From Figure 4(b), we can see that the amplitude is going to be 1. Now, with the points in Figure 4(b) as a guide, we can sketch one cycle of $y = \cos 3x$, as shown in Figure 4(c).

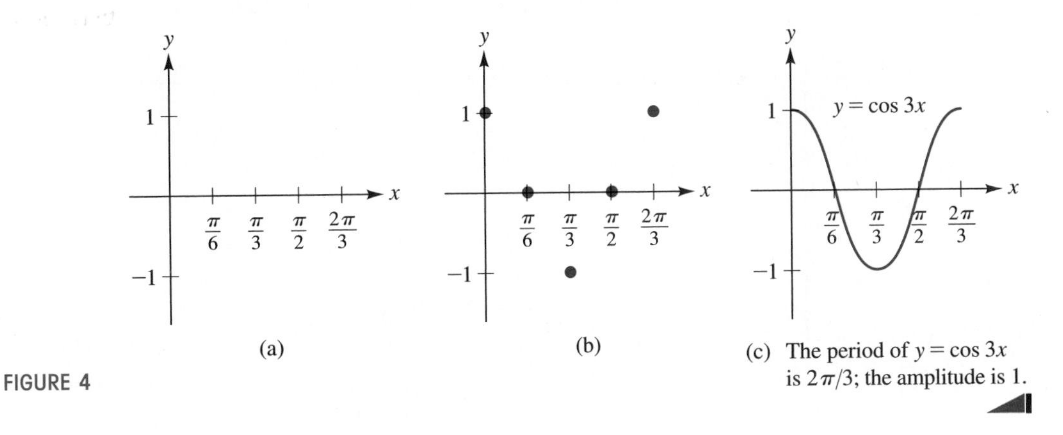

FIGURE 4

EXAMPLE 3 Graph each function over one period:

(a) $y = \frac{1}{2}\cos 3x$; **(b)** $y = -\frac{1}{2}\cos 3x$.

Solution

(a) In Example 2 we graphed $y = \cos 3x$. To obtain the graph of $y = \frac{1}{2}\cos 3x$ from that of $y = \cos 3x$, we multiply each y-coordinate on the graph of $y = \cos 3x$ by $1/2$. As indicated in Figure 5(a) (on the next page), this changes the amplitude from 1 to $1/2$, but it does not affect the period, which remains $2\pi/3$.

(b) The graph of $y = -\frac{1}{2}\cos 3x$ is obtained by reflecting the graph of $y = \frac{1}{2}\cos 3x$ about the x-axis, as indicated in Figure 5(b). Both functions have a period of $2\pi/3$ and an amplitude of $1/2$.

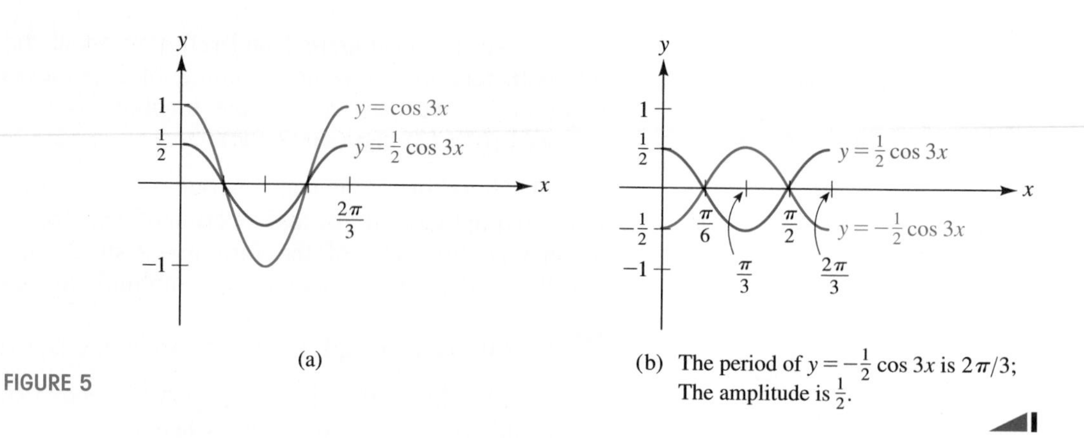

FIGURE 5

(b) The period of $y = -\frac{1}{2}\cos 3x$ is $2\pi/3$; The amplitude is $\frac{1}{2}$.

Before looking at more examples, let's take a moment to summarize where we are. Our work in Examples 2 and 3(a) shows how to graph $y = \frac{1}{2}\cos 3x$. The same technique we used in those examples can be applied to any function of the form $y = A\cos Bx$ or $y = A\sin Bx$. As indicated in the box that follows, for both functions, the amplitude is $|A|$ and the period is $2\pi/B$. (Exercise 39 asks you to use the method of Example 2 to show that the period is indeed $2\pi/B$.)

PROPERTY SUMMARY **THE GRAPHS OF $y = A\sin Bx$ AND $y = A\cos Bx$ $(B > 0)$**

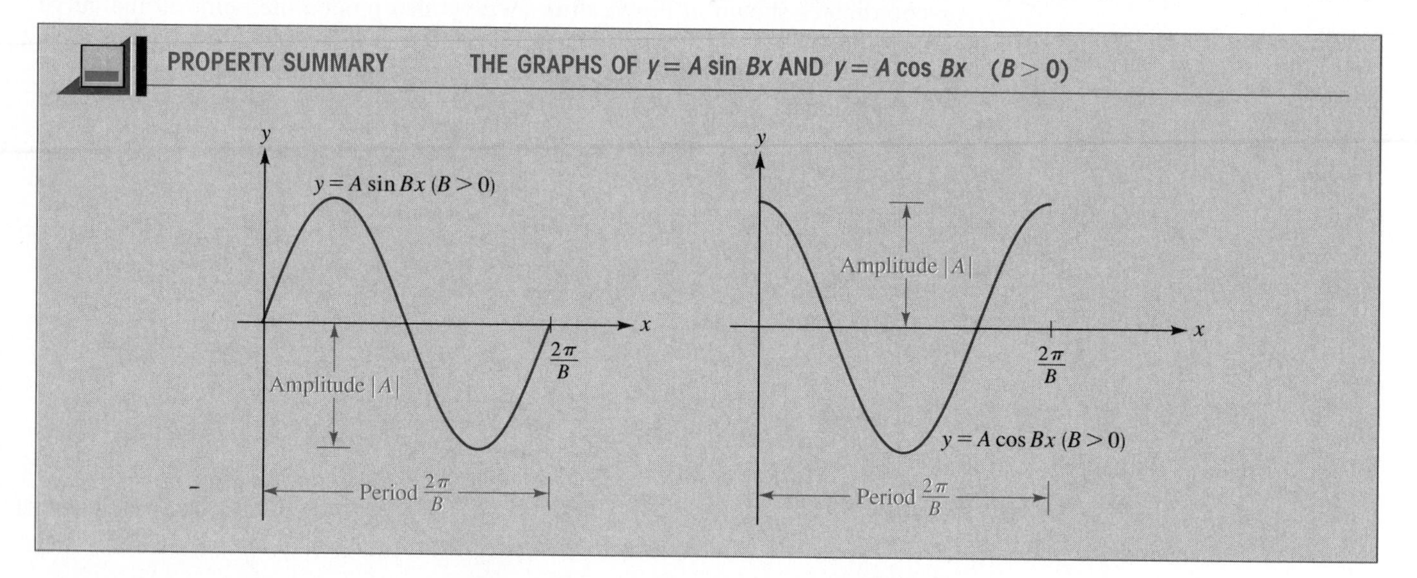

EXAMPLE 4 In Figure 6, a function of the form $y = A\sin Bx$ $(B > 0)$ is graphed for one period. Determine the values of A and B.

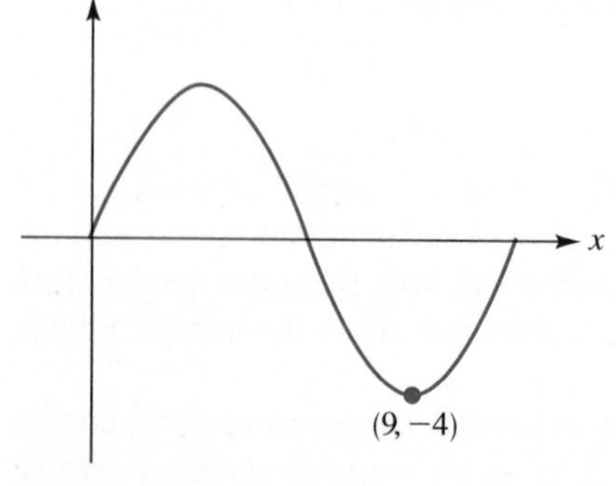

FIGURE 6

Solution From the figure, we see that the amplitude is 4. Also from the figure, we know that three-fourths of the period is 9, so

$$\frac{3}{4}\left(\frac{2\pi}{B}\right) = 9$$

$$\frac{\pi}{2B} = 3$$

$$B = \frac{\pi}{6} \qquad \text{(Check the algebra in the last two lines.)}$$

In summary, we have $A = 4$ and $B = \pi/6$; the equation of the curve is $y = 4\sin(\pi x/6)$.

EXAMPLE 5 Graph the function $y = 4\sin\left(2x - \frac{2\pi}{3}\right)$ over one period.

Solution The technique here is to factor the quantity within parentheses so that the coefficient of x is 1. We'll then be able to graph the function using a simple translation, as in Chapter 3. We have

$$y = 4\sin\left(2x - \frac{2\pi}{3}\right)$$

$$= 4\sin\left[2\left(x - \frac{\pi}{3}\right)\right] \qquad (1)$$

Now, note that equation (1) is obtained from $y = 4\sin 2x$ by replacing x with $x - \frac{\pi}{3}$. Thus, the graph of equation (1) is obtained by translating the graph of $y = 4\sin 2x$ a distance of $\pi/3$ units to the right. Figure 7(a) shows the graph of $y = 4\sin 2x$ over one period. By translating this graph $\pi/3$ units to the right, we obtain the required graph, as shown in Figure 7(b). Note that the translated graph has the same amplitude and period as the original graph. Also, as a matter of arithmetic, you should check for yourself that each of the x-coordinates shown in Figure 7(b) is obtained simply by adding $\pi/3$ to a corresponding x-coordinate in Figure 7(a). For example in Figure 7(a), the cycle ends at $x = \pi$; in Figure 7(b) the cycle ends at $\pi + \frac{\pi}{3} = \frac{4\pi}{3}$.

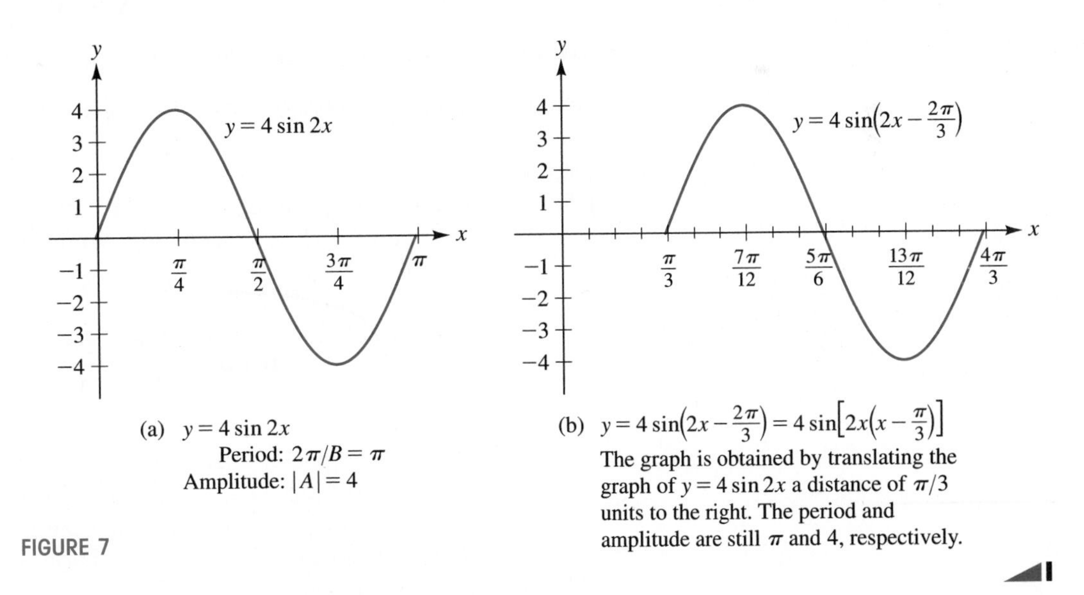

(a) $y = 4\sin 2x$
Period: $2\pi/B = \pi$
Amplitude: $|A| = 4$

(b) $y = 4\sin\left(2x - \frac{2\pi}{3}\right) = 4\sin\left[2x\left(x - \frac{\pi}{3}\right)\right]$
The graph is obtained by translating the graph of $y = 4\sin 2x$ a distance of $\pi/3$ units to the right. The period and amplitude are still π and 4, respectively.

FIGURE 7

In the example we just completed, we used translation to graph the function $y = 4\sin\left(2x - \frac{2\pi}{3}\right)$. In particular, we translated the graph of $y = 4\sin 2x$ so that the starting point of the basic cycle was shifted from $x = 0$ to $x = \pi/3$. The number $\pi/3$ in this case is called the *phase shift* of the function. In the box that follows, we define phase shift and we generalize the results of the graphing technique used in Example 5.

PROPERTY SUMMARY $y = A\sin(Bx - C)$ AND $y = A\cos(Bx - C)$ $(B > 0,\ C \neq 0)$

The graphs of $y = A\sin(Bx - C)$ and $y = A\cos(Bx - C)$ are obtained by horizontally translating the graphs of $y = A\sin Bx$ and $y = A\cos Bx$, respectively, so that the starting point of the basic cycle is shifted from $x = 0$ to $x = C/B$. The number C/B is called the **phase shift** for each of the functions $y = A\sin(Bx - C)$ and $y = A\cos(Bx - C)$. The amplitude and the period for these functions are $|A|$ and $2\pi/B$, respectively.

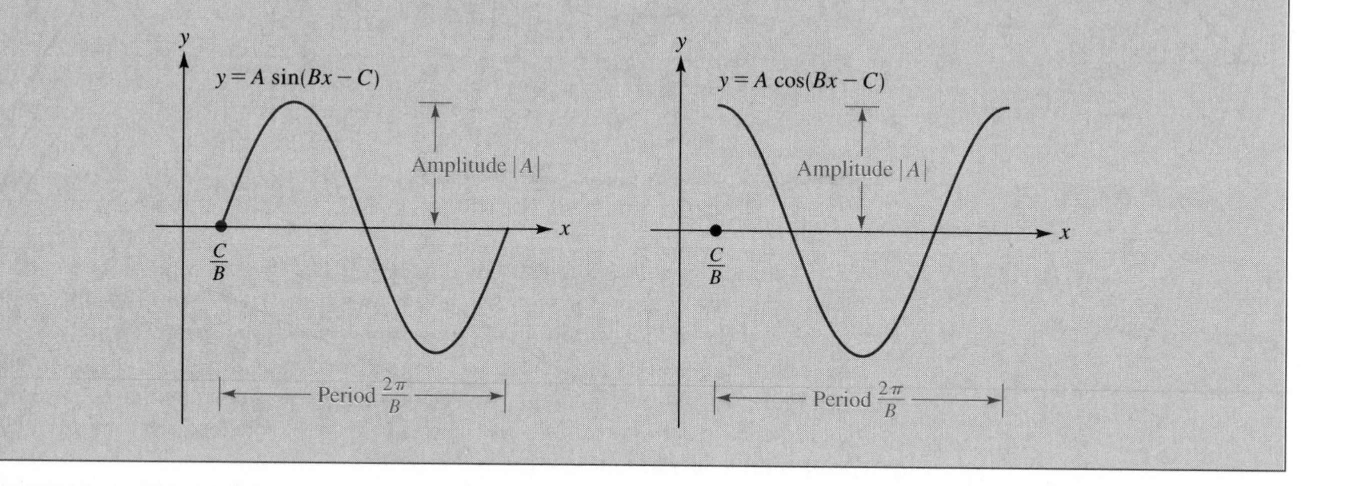

EXAMPLE 6 Specify the amplitude, period, and phase shift for each function: **(a)** $f(x) = 3\cos(4x - 5)$; **(b)** $g(x) = -2\cos\left(\pi x + \dfrac{2\pi}{3}\right)$.

Solution

(a) By comparing the given equation with $y = A\cos(Bx - C)$, we see that $A = 3$, $B = 4$, and $C = 5$. Consequently, we have

$$\text{amplitude} = |A| = 3$$

$$\text{period} = \frac{2\pi}{B} = \frac{2\pi}{4} = \frac{\pi}{2}$$

$$\text{phase shift} = \frac{C}{B} = \frac{5}{4}$$

For purposes of review, let's also calculate the phase shift without explicitly relying on the expression C/B. In the equation $f(x) = 3\cos(4x - 5)$, we can factor a 4 out of the parentheses to obtain

$$f(x) = 3\cos\left[4\left(x - \frac{5}{4}\right)\right]$$

This last equation tells us that we can obtain the graph of f by translating the graph of $y = 3\cos 4x$. In particular, the translation shifts the starting point of the basic cycle from $x = 0$ to $x = 5/4$. The number $5/4$ is the phase shift, as obtained previously.

(b) We have $A = -2$, $B = \pi$, and $C = -2\pi/3$, and therefore

$$\text{amplitude} = |A| = 2$$

$$\text{period} = \frac{2\pi}{B} = \frac{2\pi}{\pi} = 2$$

$$\text{phase shift} = \frac{C}{B} = \frac{-2\pi/3}{\pi} = -\frac{2}{3}$$

EXAMPLE 7 Graph the following function over one period:

$$g(x) = -2\cos\left(\pi x + \frac{2\pi}{3}\right)$$

Solution Our strategy is first to obtain the graph of $y = 2\cos\left(\pi x + \frac{2\pi}{3}\right)$. The graph of g can then be obtained by a reflection about the x-axis. We have

$$y = 2\cos\left(\pi x + \frac{2\pi}{3}\right) = 2\cos\left[\pi\left(x + \frac{2}{3}\right)\right]$$

Now, the graph of this last equation is obtained by translating the graph of $y = 2\cos\pi x$ a distance of 2/3 unit to the left. Figures 8(a) and 8(b) show the graphs of $y = 2\cos\pi x$ and $y = 2\cos\left[\pi\left(x + \frac{2}{3}\right)\right]$. By reflecting the graph of this last equation about the x-axis, we obtain the graph of g, as required. See Figure 8(c).

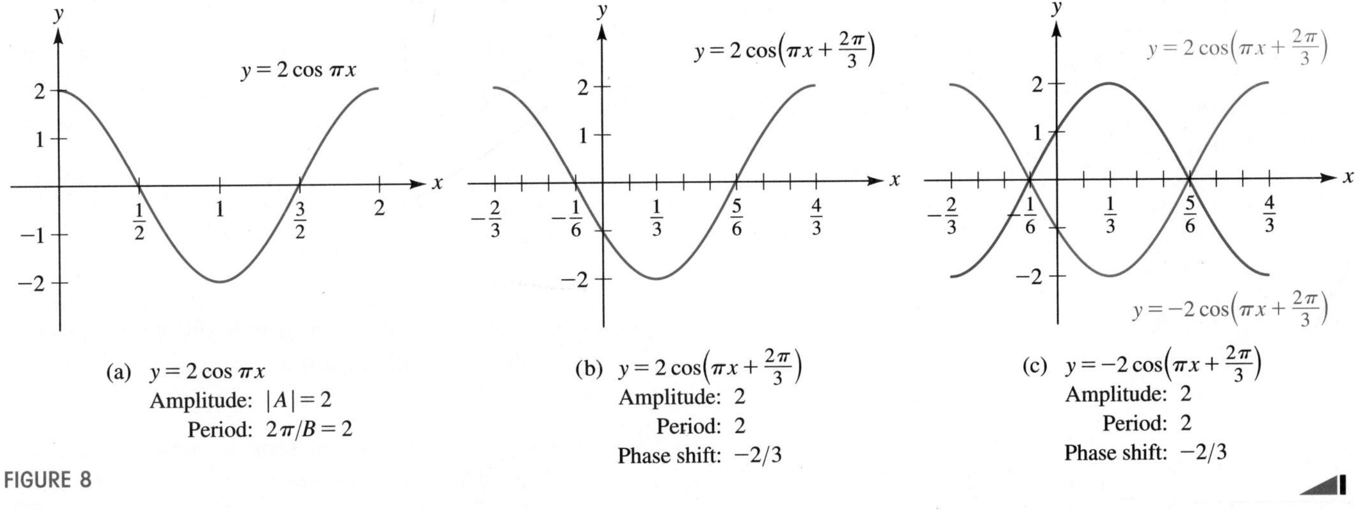

(a) $y = 2\cos\pi x$
Amplitude: $|A| = 2$
Period: $2\pi/B = 2$

(b) $y = 2\cos\left(\pi x + \frac{2\pi}{3}\right)$
Amplitude: 2
Period: 2
Phase shift: $-2/3$

(c) $y = -2\cos\left(\pi x + \frac{2\pi}{3}\right)$
Amplitude: 2
Period: 2
Phase shift: $-2/3$

FIGURE 8

EXERCISE SET 7.5

Do 1, 17, 23, 27

A

In Exercises 1–12, graph the functions for one period. In each case, specify the amplitude, period, x-intercepts, and interval(s) on which the function is increasing.

1. **(a)** $y = 2\sin x$
(b) $y = -\sin 2x$

2. **(a)** $y = 3\sin x$
(b) $y = \sin 3x$

3. **(a)** $y = \cos 2x$
(b) $y = 2\cos 2x$

4. **(a)** $y = \cos(x/2)$
(b) $y = -\frac{1}{2}\cos(x/2)$

5. **(a)** $y = 3\sin(\pi x/2)$
(b) $y = -3\sin(\pi x/2)$

6. **(a)** $y = 2\sin\pi x$
(b) $y = -2\sin\pi x$

7. **(a)** $y = \cos 2\pi x$
(b) $y = -4\cos 2\pi x$

8. **(a)** $y = -2\cos(x/4)$
(b) $y = -2\cos(\pi x/4)$

9. $y = 1 + \sin 2x$ **10.** $y = \sin(x/2) - 2$

11. $y = 1 - \cos(\pi x/3)$ **12.** $y = -2 - 2\cos 3\pi x$

In Exercises 13–28, determine the amplitude, period, and phase shift for the given function. Graph the function over one period. Indicate the x-intercepts and the coordinates of the highest and lowest points on the graph.

13. $f(x) = \sin\left(x - \frac{\pi}{6}\right)$ **14.** $g(x) = \cos\left(x + \frac{\pi}{3}\right)$

15. $F(x) = -\cos\left(x + \frac{\pi}{4}\right)$ **16.** $G(x) = -\sin(x + 2)$

17. $y = \sin\left(2x - \frac{\pi}{2}\right)$ **18.** $y = \sin\left(3x + \frac{\pi}{2}\right)$

19. $y = \cos(2x - \pi)$ **20.** $y = \cos\left(x - \frac{\pi}{2}\right)$

21. $y = 3\sin\left(\frac{1}{2}x + \frac{\pi}{6}\right)$ **22.** $y = -2\sin(\pi x + \pi)$

23. $y = 4\cos\left(3x - \frac{\pi}{4}\right)$ **24.** $y = \cos(x + 1)$

25. $y = \frac{1}{2}\sin\left(\frac{\pi x}{2} - \pi^2\right)$ **26.** $y = \cos\left(2x - \frac{\pi}{3}\right) + 1$

27. $y = 1 - \cos\left(2x - \frac{\pi}{3}\right)$ **28.** $y = 3\cos\left(\frac{2x}{3} + \frac{\pi}{6}\right)$

In Exercises 29–34, determine whether the equation for the graph has the form $y = A\sin Bx$ or $y = A\cos Bx$ (with $B > 0$) and then find the values of A and B.

29.

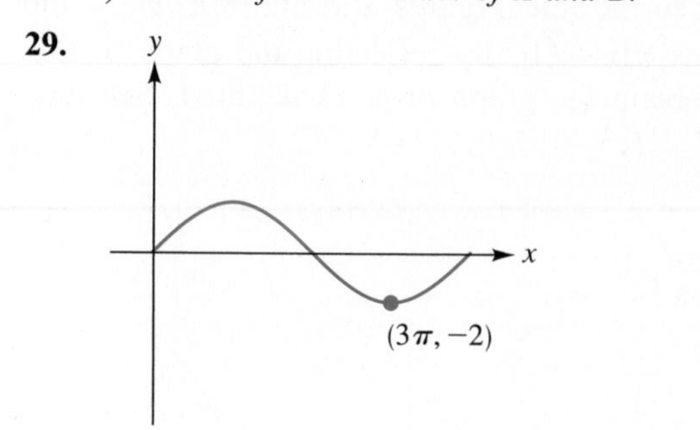

30.

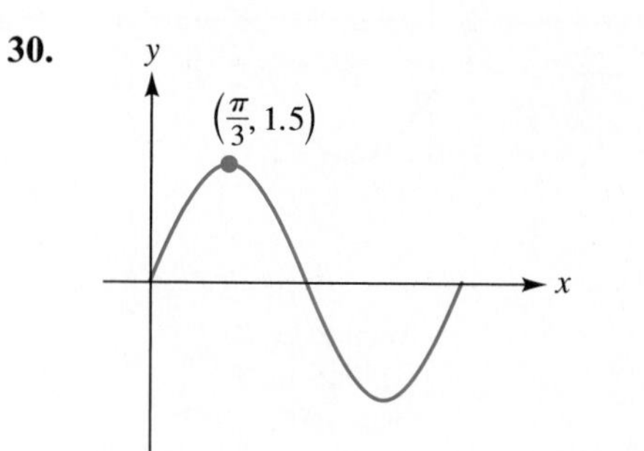

31.

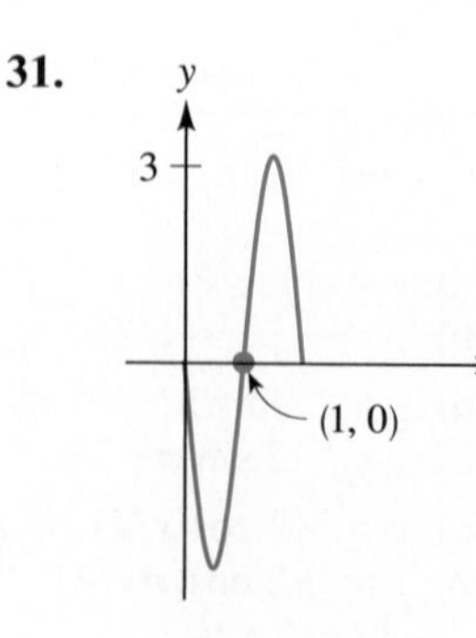

32. **33.** **34.**

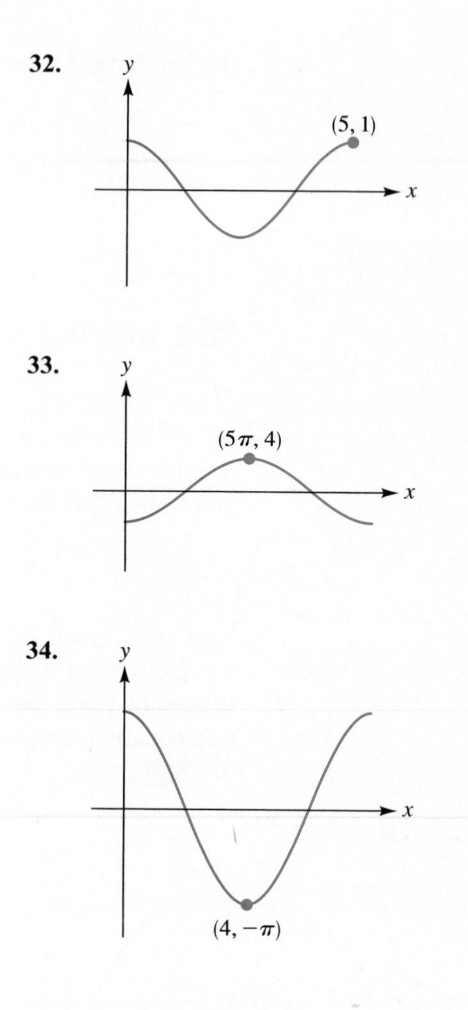

B

35. In Section 8.2 you'll see the identity $\sin^2 x = \frac{1}{2} - \frac{1}{2}\cos 2x$. Use this identity to graph the function $y = \sin^2 x$ for one period.

36. In Section 8.2 you'll see the identity $\cos^2 x = \frac{1}{2} + \frac{1}{2}\cos 2x$. Use this identity to graph the function $y = \cos^2 x$ for one period.

37. In Section 8.2 we derive the identity $\sin 2x = 2\sin x\cos x$. Use this to graph $y = \sin x\cos x$ for one period.

38. In Section 8.2 we derive the identity $\cos 2x = \cos^2 x - \sin^2 x$. Use this to graph $y = \cos^2 x - \sin^2 x$ for one period.

39. In Example 2 we showed that the period of $y = \cos 3x$ is $2\pi/3$. Use the same method to show that the period of $y = A\cos Bx$ is $2\pi/B$.

GRAPHING UTILITY EXERCISES FOR SECTION 7.5

1. Set the viewing rectangle so that it extends from 0 to 2π in the x-direction and from -4 to 4 in the y-direction. On the same set of axes, graph the four functions $y = \sin x$, $y = 2\sin x$, $y = 3\sin x$, and $y = 4\sin x$. What is the amplitude in each case? What is the period?
2. **(a)** Without using a graphing utility, specify the amplitude and the period for each of the following four functions: $y = \cos x$, $y = 2\cos x$, $y = 3\cos x$, and $y = 4\cos x$.
   (b) Check your answers in part (a) by graphing the four functions. (Use the viewing rectangle specified in Exercise 1.)
3. **(a)** Without using a graphing utility, specify the amplitude and the period for $y = 2\sin \pi x$ and for $y = \sin 2\pi x$.
   (b) Check your answers in part (a) by graphing the two functions. (Use a viewing rectangle that extends from 0 to 2 in the x-direction and from -2 to 2 in the y-direction.)

Instructions for Exercises 4–11.

(a) *Using pencil and paper, not a graphing utility, determine the amplitude, period, and (where appropriate) phase shift for each function.*

(b) *Use a graphing utility to graph each function for two complete cycles. [In choosing an appropriate viewing rectangle, you will need to use the information obtained in part (a).]*

(c) *Use the graphing utility to estimate the coordinates of the highest and the lowest points on the graph.*

(d) *Use the information obtained in part (a) to specify the exact values for the coordinates that you estimated in part (c).*

4. $y = -2.5\cos(3x + 4)$
5. $y = -2.5\cos(3\pi x + 4)$
6. $y = -2.5\cos\left(\frac{1}{3}x + 4\right)$
7. $y = -2.5\cos\left(\frac{1}{3}\pi x + 4\right)$
8. $y = \sin(0.5x - 0.75)$
9. $y = \sin(0.5x + 0.75)$
10. $y = 0.02\cos(100\pi x - 4\pi)$
11. $y = 0.02\cos(0.01\pi x - 4\pi)$

12. Let $F(x) = \sin x$, $G(x) = x^2$, and $H(x) = x^3$. Which, if any, of the following four composite functions have graphs that do not go below the x-axis? First, try to answer without using a graphing utility, then use the graphing utility to check yourself. (You will learn more this way than if you were to draw the graphs immediately.)

$$y = G(F(x)) \qquad y = F(G(x))$$
$$y = F(H(x)) \qquad y = H(F(x))$$

13. **(a)** Graph the two functions $y = \sin x$ and $y = \sin(\sin x)$ in the standard viewing rectangle. Then for a closer look, switch to a viewing rectangle extending from 0 to 2π in the x-direction and from -1 to 1 in the y-direction. Compare the two graphs; write out your observations in complete sentences.
    (b) Use the graphing utility to estimate the amplitude of the function $y = \sin(\sin x)$.
    (c) Using your knowledge of the sine function, explain why the amplitude of the function $y = \sin(\sin x)$ is the number $\sin 1$. Then evaluate $\sin 1$ and use the result to check your approximation in part (b).
14. Let $f(x) = e^{x/20}(\sin x)$.
    (a) Graph the function f using a viewing rectangle that extends from -5 to 5 in both the x- and the y-directions. Note that the resulting graph resembles a sine curve.
    (b) Change the viewing rectangle so that x extends from 0 to 50 and y extends from -10 to 10. Describe what you see. Is the function periodic?
    (c) Add the graphs of the two functions $y = e^{x/20}$ and $y = -e^{x/20}$ to your picture in part (b). Describe what you see.
15. For this exercise, use the standard viewing rectangle.
    (a) Graph the function $y = \ln(\sin^2 x)$.
    (b) Graph the function $y = \ln(1 - \cos x) + \ln(1 + \cos x)$.
    (c) Explain why the two graphs are identical.

SIMPLE HARMONIC MOTION

Amongst the most important classes of motions which we have to consider in Natural Philosophy, there is one, namely, Harmonic Motion, *which is of such immense use, not only in ordinary kinetics, but in the theories of sound, light, heat, etc., that we make no apology for entering here into considerable detail regarding it.*

Sir William Thomson (Lord Kelvin) and Peter Tait in *Treatise on Natural Philosophy* (Cambridge University Press, 1879)

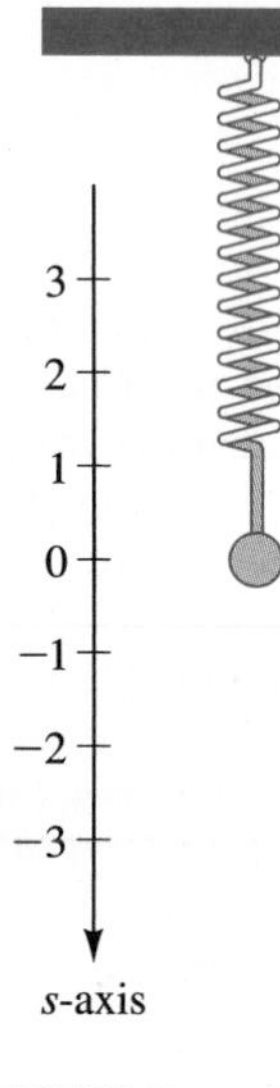

FIGURE 1

Figure 1 shows a mass attached to a spring hung from the ceiling. If we pull the mass down a bit and then release it, the resulting up-and-down oscillations are referred to as **simple harmonic motion.** (We are neglecting the effects of friction and air resistance here.) Other examples of simple harmonic motion (or combinations of simple harmonic motions) include the vibration of a string or of a column of air in a musical instrument and the vibration of an atom in a solid. Furthermore, the mathematics used in describing or analyzing these mechanical oscillations is the same as that used in studying electromagnetic oscillations (such as radio waves, microwaves, or the alternating electrical current in your house).

In order to analyze the motion of a mass attached to a spring, we set up a coordinate system as shown in Figure 1, in which the equilibrium position of the mass (the position before we pull it down) corresponds to $s = 0$. Now suppose that we pull the mass down to $s = -2$ and release it. If we take a sequence of "snapshots" at one-second time intervals, we will obtain the type of result shown in Figure 2.

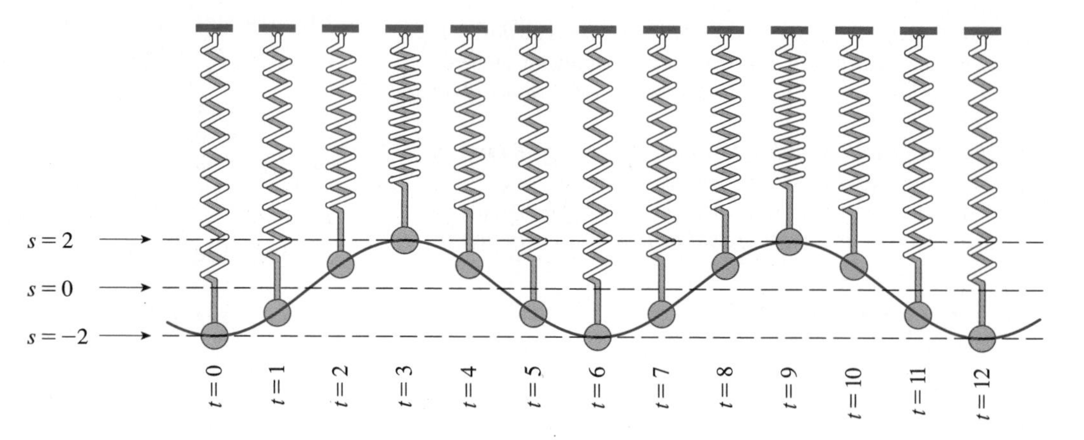

FIGURE 2
Snapshots of a mass–spring system taken at one-second intervals.

The curve in Figure 2 shows how the coordinate s of the mass changes over time. As you can see, the curve resembles the graphs of $y = A\sin(Bx - C)$ and $y = A\cos(Bx - C)$ that we considered in the previous section. Indeed, using calculus, it can be shown that simple harmonic motion is characterized by either one of these types of equations. For example, for the simple harmonic motion depicted in Figure 2, it can be shown that an appropriate equation relating the position s and the time t is

$$s = -2\cos\frac{\pi t}{3} \qquad (1)$$

Using equation (1) and Figure 2, we can give a physical interpretation to the term *period.* From Figure 2 we see that it takes six seconds for the mass to return to its starting position at $s = -2$. Then in the next six seconds, the same motion is repeated, and so on. We say that the **period** of the motion is six seconds. That is, it takes six seconds for the motion to go through one complete cycle. Notice that this agrees numerically with the period we calculate using equation (1) and the expression $2\pi/B$ from Section 7.5:

$$\text{period} = \frac{2\pi}{B} = \frac{2\pi}{\pi/3} = 6$$

In simple harmonic motion, the **frequency** f is the number of complete cycles per unit time, and it is given by

$$\text{frequency} = f = \frac{1}{\text{period}}$$

For instance, for the motion in Figure 2, we have

$$f = \frac{1}{\text{period}} = \frac{1}{6} \text{ cycles per second}$$

We mention in passing that 1 cycle per second (cps) is known as a *hertz,* abbreviated Hz. This unit is named after the German physicist Heinrich Hertz (1857–1894), who was the first person to produce and study radio waves.* Although you may not have realized it, you have probably heard this unit mentioned (implicitly, at least) many times on the radio. For example, when a radio station advertises itself as ''98.1 on the FM dial,'' this refers to radio waves with a frequency of 98.1 million hertz.

From Figure 2, we can see that the mass moves back and forth between $s = -2$ and $s = 2$. In other words, the maximum displacement of the mass from its equilibrium position (at $s = 0$) is two units. We say that the **amplitude** of the motion is two units. Notice that this agrees with the amplitude we would calculate using equation (1).

EXAMPLE 1 A mass on a smooth tabletop is attached to a spring, as shown in Figure 3. The coordinate system has been chosen so that the equilibrium position of the mass corresponds to $s = 0$. Assume that the mass moves in simple harmonic motion described by

$$s = 5 \cos \frac{\pi t}{4}$$

where s is in centimeters and t is in seconds.

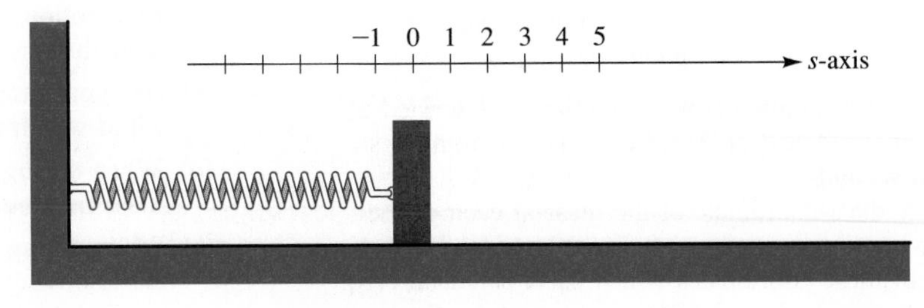

FIGURE 3

*For background and details, see the interesting article ''Heinrich Hertz'' by Philip and Emily Morrison in the December 1957 issue of *Scientific American.*

(a) Graph the function $s = 5\cos(\pi t/4)$ over the interval $0 \leq t \leq 16$. Specify the amplitude, the period, and the frequency of the motion.
(b) Use the graph to determine the times in this interval at which the mass is farthest from the origin.
(c) When during this interval of time is the mass passing through the equilibrium position?

Solution

(a) Using the techniques developed in Section 7.5, we obtain the graph shown in Figure 4. From the graph (or from the given equation), we have

$$\text{amplitude} = 5 \text{ cm}$$
$$\text{period} = 8 \text{ sec}$$
$$\text{frequency} = \frac{1}{8} \text{ cycles/sec}$$

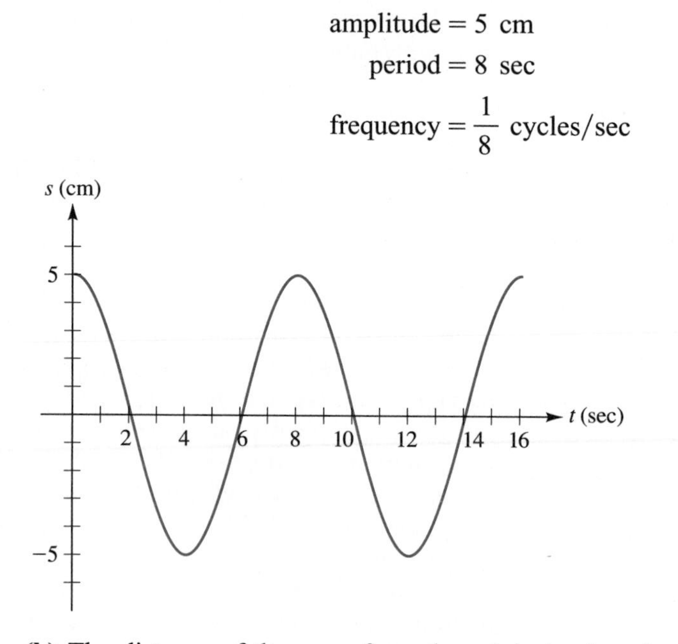

FIGURE 4

(b) The distance of the mass from the origin is given by $|s|$. According to Figure 4, the maximum value of $|s|$ is 5, and this occurs when $t = 0$ sec, 4 sec, 8 sec, 12 sec, and 16 sec.
(c) We are given that the equilibrium position of the mass is $s = 0$. From Figure 4, we see that s is zero when $t = 2$ sec, 6 sec, 10 sec, and 14 sec.

EXERCISE SET 7.6

A

For Exercises 1 and 2, suppose that we have a mass–spring system as shown in Figure 3 on page 463.

1. Assume that the simple harmonic motion is described by the equation $s = 4\cos(\pi t/2)$, where s is in centimeters and t is in seconds.
(a) Specify the s-coordinate of the mass at each of the following times: $t = 0$ sec, 0.5 sec, 1 sec, and 2 sec. (One of these coordinates will involve a radical sign; for this case, use a calculator and round the final answer to two decimal places.)
(b) Find the amplitude, the period, and the frequency of this motion. Sketch the graph of $s = 4\cos(\pi t/2)$ over the interval $0 \leq t \leq 8$.
(c) Use your graph to determine the times in this interval at which the mass is farthest from the origin.
(d) When during the interval of time $0 \leq t \leq 8$ is the mass passing through the origin?
(e) When during the interval of time $0 \leq t \leq 8$ is the mass moving to the right? *Hint:* The mass is moving to the right when the s-coordinate is increasing. Use the graph to see when s is increasing.

2. Assume that the simple harmonic motion is described by the equation $s = -6\cos(\pi t/6)$, where s is in centimeters and t is in seconds.

(a) Complete the table. (For coordinates that involve radical signs, use a calculator and round the result to two decimal places.)

t (sec)	1	2	3	4	5	6	7	8	9	10	11	12
s (cm)												

(b) Find the amplitude, period, and frequency of this motion. Sketch the graph of $s = -6\cos(\pi t/6)$ over the interval $0 \leq t \leq 24$.

(c) Use your graph to determine the times in this interval at which the mass is farthest from the equilibrium position.

(d) When during the interval of time $0 \leq t \leq 24$ is the mass passing through the origin?

(e) When during the interval of time $0 \leq t \leq 24$ is the mass moving to the left? *Hint:* The mass is moving to the left when the s-coordinate is decreasing. Use the graph to see when s is decreasing.

In Exercises 3 and 4, suppose that we have a mass–spring system, as shown in Figure 1 on page 462.

3. Assume that the simple harmonic motion is described by the equation $s = -3\cos(\pi t/3)$, where s is in feet, t is in seconds, and the equilibrium position of the mass is $s = 0$.

(a) Specify the amplitude, period, and frequency for this simple harmonic motion, and sketch the graph of the function $s = -3\cos(\pi t/3)$ over the interval $0 \leq t \leq 12$.

(b) When during the interval of time $0 \leq t \leq 12$ is the mass moving upward? *Hint:* The mass is moving upward when the s-coordinate is increasing. Use the graph to see when s is increasing.

(c) When during the interval of time $0 \leq t \leq 12$ is the mass moving downward? *Hint:* The mass is moving downward when the s-coordinate is decreasing. Use the graph to see when s is decreasing.

(d) For this harmonic motion, it can be shown (using calculus) that the velocity v of the mass is given by

$$v = \pi \sin(\pi t/3)$$

where t is in seconds and v is in ft/sec. Graph this velocity function over the interval $0 \leq t \leq 12$.

(e) Use your graph to find the times during this interval when the velocity is zero. At these times, where is the mass? (That is, what are the s-coordinates?)

(f) Use the graph to find the times when the velocity is maximum. Where is the mass at these times?

(g) Use the graph to find the times when the velocity is minimum. Where is the mass at these times?

(h) On the same set of axes, graph the velocity function $v = \pi \sin(\pi t/3)$ and the position function $s = -3\cos(\pi t/3)$ for $0 \leq t \leq 12$.

4. Assume that the simple harmonic motion is described by the equation $s = 4\cos(2t/3)$, where s is in feet, t is in seconds, and the equilibrium position of the mass is $s = 0$.

(a) Specify the amplitude, period, and frequency for this simple harmonic motion, and sketch the graph of the function $s = 4\cos(2t/3)$ over the interval $0 \leq t \leq 6\pi$.

(b) When during the interval of time $0 \leq t \leq 6\pi$ is the mass moving upward? *Hint:* The mass is moving upward when the s-coordinate is increasing. Use the graph to see when s is increasing.

(c) When during the interval of time $0 \leq t \leq 6\pi$ is the mass moving downward? *Hint:* The mass is moving downward when the s-coordinate is decreasing. Use the graph to see when s is decreasing.

(d) For this harmonic motion, it can be shown (using calculus) that the velocity v of the mass is given by $v = -\frac{8}{3}\sin(2t/3)$, where t is in seconds and v is in ft/sec. Graph this velocity function over the interval $0 \leq t \leq 6\pi$.

(e) Use your graph to find the times during this interval when the velocity is zero. At these times, where is the mass? (That is, what are the s-coordinates?)

(f) Use the graph to find the times when the velocity is maximum. Where is the mass at these times?

(g) Use the graph to find the times when the velocity is minimum. Where is the mass at these times?

5. The voltage in a household electrical outlet is given by

$$V = 170\cos(120\pi t)$$

where V is measured in volts and t in seconds.

(a) Specify the amplitude and the frequency for this oscillation.

(b) Graph the function $V = 170\cos(120\pi t)$ for two complete cycles beginning at $t = 0$.

(c) For which values of t in part (b) is the voltage maximum?

B

6. The following figure (on the next page) shows a simple pendulum consisting of a string with a weight attached at one end and the other end suspended from a fixed point. As indicated in the figure, the angle between the vertical and the pendulum is denoted by θ (where θ is in radians). Suppose that we pull the pendulum out from the equilibrium position (where $\theta = 0$) to a position $\theta = \theta_0$. Now we release the pendulum so that it swings back and forth. Then (neglecting friction and assuming θ_0

is a small angle), it can be shown that the angle θ at time t is very closely approximated by

$$\theta = \theta_0 \cos\left(t\sqrt{g/L}\right)$$

where L is the length of the pendulum, g is a constant (the acceleration due to gravity), and t is the time in seconds.

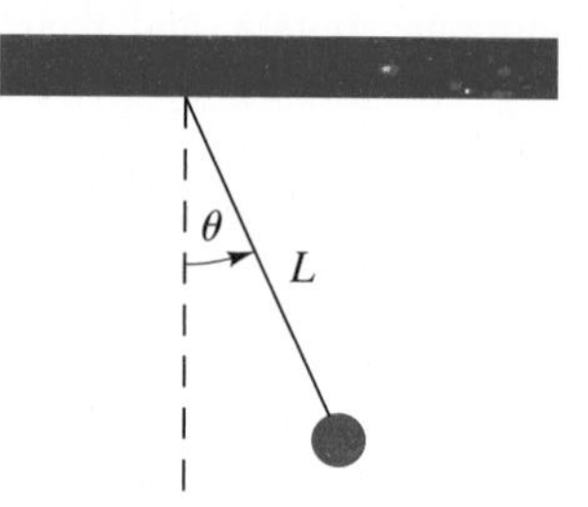

(a) What are the amplitude, the period, and the frequency for the motion defined by the equation $\theta = \theta_0 \cos\left(t\sqrt{g/L}\right)$? Assume that t is in seconds, L is in meters, and g is in m/sec^2.

(b) Use your results in part (a) to answer these two questions. Does the period of the pendulum depend upon the amplitude? Does the period depend upon the length L?

(c) Graph the function $\theta = \theta_0 \cos\left(t\sqrt{g/L}\right)$ for two complete cycles beginning at $t = 0$ and using the following values for the constants:

$$\theta_0 = 0.1 \text{ radian} \qquad L = 1 \text{ m} \qquad g = 9.8 \text{ m/sec}^2$$

(d) For which values of t during these two cycles is the weight moving to the right? *Hint:* The weight is moving to the right when θ is increasing; use your graph in part (c) to see when this occurs.

(e) The velocity V of the weight as it oscillates back and forth is given by

$$V = -\theta_0\sqrt{g/L}\,\sin\left(t\sqrt{g/L}\right)$$

where V is in m/sec. Graph this function for two complete cycles using the values of the constants given in part (c).

(f) At which times during these two cycles is the velocity maximum? What is the corresponding value of θ in each case?

(g) At which times during these two cycles is the velocity minimum? What is the corresponding value of θ in each case?

(h) At which times during these two cycles is the velocity zero? What is the corresponding value of θ in each case?

7. Refer to Figure A. Suppose that the point P travels counterclockwise around the unit circle at a constant angular speed of $\pi/3$ radians/sec. Assume that at time $t = 0$ sec, the location of P is $(1, 0)$.

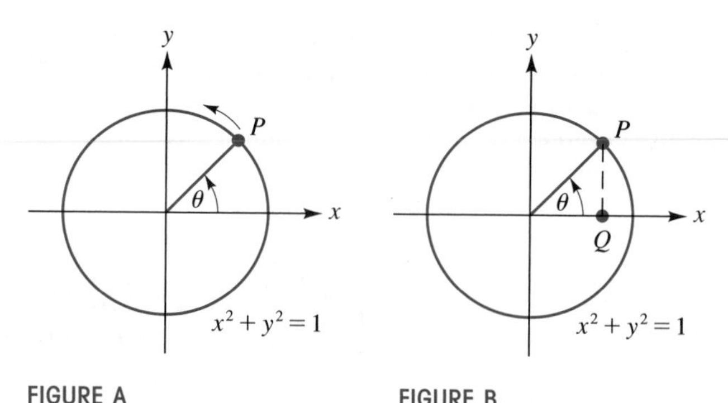

FIGURE A FIGURE B

(a) Complete the table.

t (sec)	0	1	2	3	4	5	6	7
θ (radians)								

(b) Now (for each position of the point P) suppose that we draw a perpendicular from P to the x-axis, meeting the x-axis at Q, as indicated in Figure B. The point Q is called the *projection* of the point P on the x-axis. As the point P moves around the circle, the point Q will move back and forth along the diameter of the circle. What is the x-coordinate of the point Q at each of the times listed in the table for part (a)?

(c) Draw a sketch showing the location of the points P and Q when $t = 1$ sec.

(d) Draw sketches as in part (c) for $t = 2$, 3, and 4 sec.

(e) The x-coordinate of Q is always equal to $\cos\theta$. Why?

(f) If you look back at your table in part (a), you'll see that the relation between θ and t is $\theta = \pi t/3$. Thus, the x-coordinate of Q at time t is given by

$$x = \cos(\pi t/3)$$

This tells us that as the point P moves around the circle at a constant angular speed, the point Q oscillates in simple harmonic motion along the x-axis. Graph this function for two complete cycles, beginning at $t = 0$. Specify the amplitude, the period, and the frequency for the motion.

(g) Using calculus, it can be shown that the velocity of the point Q at time t is given by

$$V = -\frac{\pi}{3}\sin\frac{\pi t}{3}$$

Graph this function for two complete cycles, beginning at $t = 0$.

(h) Use the graph in part (g) to determine the times (during the first two cycles) when the velocity of the point Q is zero. What are the corresponding x-coordinates of Q in each case?

(i) Use the graph in part (g) to determine the times (during the first two cycles) when the velocity of the point Q is maximum. Where is Q located at these times?

(j) Use the graph in part (g) to determine the times (during the first two cycles) when the velocity of the point Q is minimum. Where is Q located at these times?

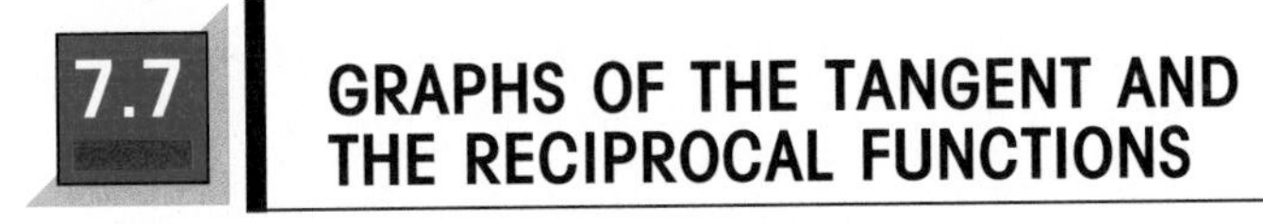

7.7 GRAPHS OF THE TANGENT AND THE RECIPROCAL FUNCTIONS

A third . . . function, the tangent of θ, or tan θ, *is of secondary importance, in that it is not associated with wave phenomena. Nevertheless, it enters into the body of analysis so prominently that we cannot ignore it.*

Samuel E. Urner and William B. Orange in *Elements of Mathematical Analysis* (Boston: Ginn and Co., 1950)

We have seen in the previous sections that the sine and cosine functions are periodic. The remaining four trigonometric functions are also periodic, but their graphs differ significantly from those of sine and cosine. The graphs of $y = \tan x$, $\cot x$, $\csc x$, and $\sec x$ all possess asymptotes.

We'll obtain the graph of the tangent function by a combination of both point-plotting and symmetry considerations. Table 1 displays a list of values for $y = \tan x$ using x-values in the interval $[0, \pi/2)$.

TABLE 1
As x increases from 0 toward $\pi/2$, the values of $\tan x$ increase slowly at first, then more and more rapidly.

x	0	$\frac{\pi}{6}$	$\frac{\pi}{4}$	$\frac{\pi}{3}$	$\frac{5\pi}{12}$ (= 75°)	$\frac{17\pi}{36}$ (= 85°)	$\frac{89\pi}{180}$ (= 89°)	$\frac{\pi}{2}$
tan x	0	$\frac{\sqrt{3}}{3} \approx 0.58$	1	$\sqrt{3} \approx 1.73$	3.73	11.43	57.29	undefined

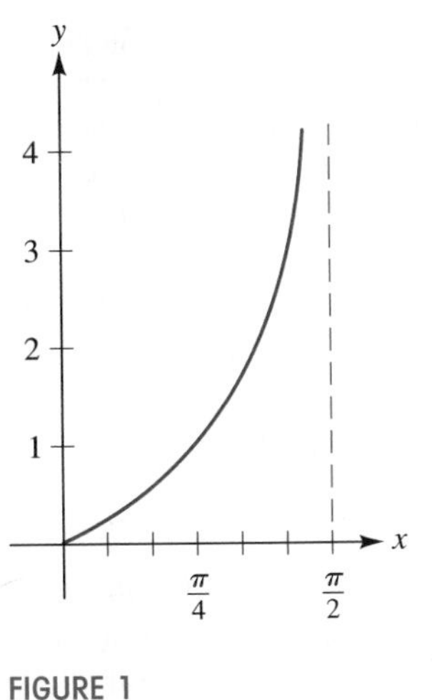

FIGURE 1
$y = \tan x, \quad 0 \leq x < \pi/2$

As indicated in Table 1, $\tan x$ is undefined when $x = \pi/2$. This follows from the identity

$$\tan x = \frac{\sin x}{\cos x} \tag{1}$$

When $x = \pi/2$, the denominator in this identity is zero. Indeed, when x is equal to any odd integral multiple of $\pi/2$ (e.g., $\pm 3\pi/2$, $\pm 5\pi/2$), the denominator in equation (1) will be zero and, consequently, $\tan x$ will be undefined.

Because $\tan x$ is undefined when $x = \pi/2$, we want to see how the graph behaves as x gets closer and closer to $\pi/2$. This is why the x-values $5\pi/12$, $17\pi/36$, and $89\pi/180$ are used in Table 1. In Figure 1, we've used the data in Table 1 to draw the graph of $y = \tan x$ for $0 \leq x \leq \pi/2$. As the figure indicates, the vertical line $x = \pi/2$ is an asymptote for the graph.

The graph of $y = \tan x$ can now be completed without further need for tables or a calculator. First, the identity $\tan(-x) = -\tan x$ (from Section 7.3) tells us that the graph of $y = \tan x$ is symmetric about the origin. So, after reflecting the graph in Figure 1 about the origin, we can draw the graph of $y = \tan x$ on the interval $\left(-\frac{\pi}{2}, \frac{\pi}{2}\right)$, as shown in Figure 2.

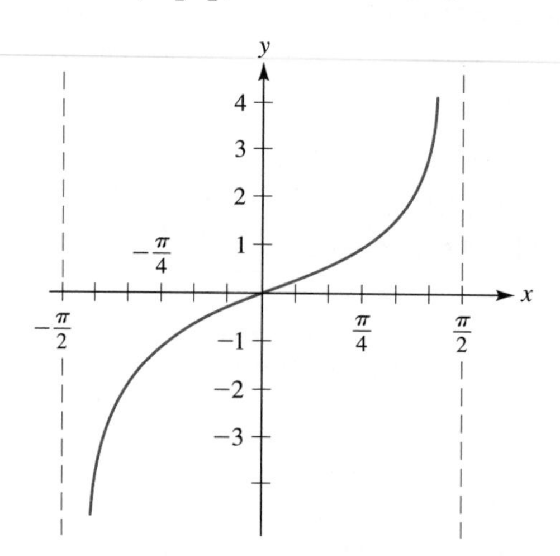

FIGURE 2
$y = \tan x, \quad -\frac{\pi}{2} < x < \frac{\pi}{2}$
The graph is symmetric about the origin.
The lines $x = \pm\pi/2$ are asymptotes.

Now, to complete the graph of $y = \tan x$, we use the identity

$$\tan(s + \pi) = \tan s \tag{2}$$

Looking at Figure 3, we can see why this identity is valid. By definition, the coordinates of P and Q are

$$P(\cos s, \sin s) \qquad \text{and} \qquad Q\big(\cos(s + \pi), \sin(s + \pi)\big)$$

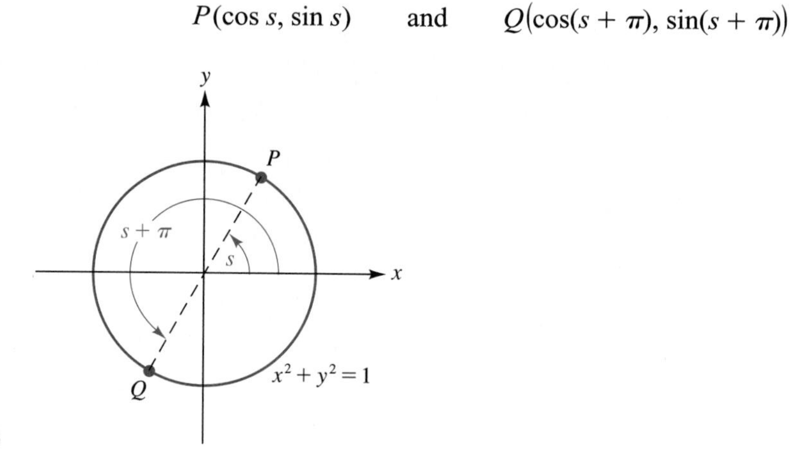

FIGURE 3

On the other hand, the points P and Q are symmetric about the origin, so the coordinates of Q are just the negatives of the coordinates of P. That is,

$$\cos(s + \pi) = -\cos s \qquad \text{and} \qquad \sin(s + \pi) = -\sin s$$

Consequently, we have

$$\tan(s + \pi) = \frac{\sin(s + \pi)}{\cos(s + \pi)} = \frac{-\sin s}{-\cos s}$$

$$= \frac{\sin s}{\cos s} = \tan s \qquad \text{as required}$$

[Although Figure 3 shows the angle with radian measure s terminating in Quadrant I, our proof is valid for the other quadrants as well. (Draw a figure for yourself and verify this.)]

Identity (2) tells us that the graph of $y = \tan x$ must repeat itself at intervals of length π. Taking this fact into account, along with Figure 2, we conclude that the period of $y = \tan x$ is exactly π. Our final graph of $y = \tan x$ is shown in the Property Summary Box that follows.

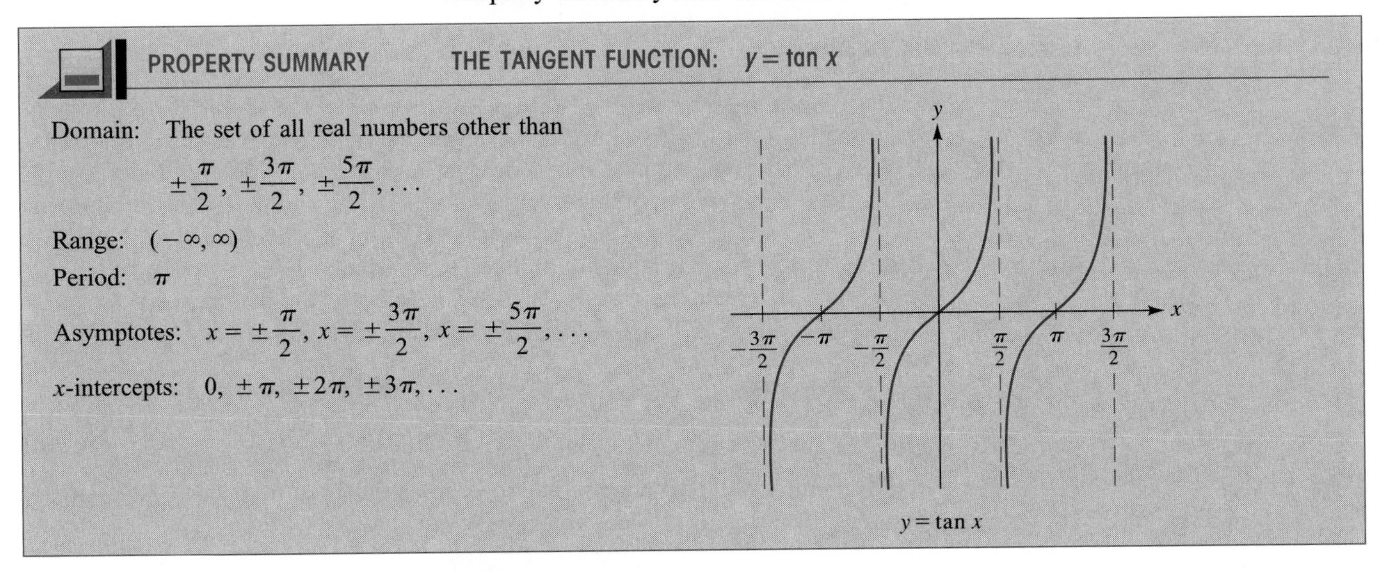

PROPERTY SUMMARY **THE TANGENT FUNCTION: $y = \tan x$**

Domain: The set of all real numbers other than

$\pm\frac{\pi}{2}, \pm\frac{3\pi}{2}, \pm\frac{5\pi}{2}, \ldots$

Range: $(-\infty, \infty)$

Period: π

Asymptotes: $x = \pm\frac{\pi}{2}, x = \pm\frac{3\pi}{2}, x = \pm\frac{5\pi}{2}, \ldots$

x-intercepts: $0, \pm\pi, \pm 2\pi, \pm 3\pi, \ldots$

$y = \tan x$

EXAMPLE 1 Graph the following functions for one period. In each case, specify the period, the asymptotes, and the intercepts:

(a) $y = \tan\left(x - \frac{\pi}{4}\right)$; **(b)** $y = -\tan\left(x - \frac{\pi}{4}\right)$.

Solution We begin with the graph of one period of $y = \tan x$, as shown in Figure 4(a). By translating this graph $\pi/4$ units to the right, we obtain the graph of $y = \tan\left(x - \frac{\pi}{4}\right)$, in Figure 4(b). With this translation, note that the left asymptote shifts from $x = -\pi/2$ to $x = -\frac{\pi}{2} + \frac{\pi}{4} = -\frac{\pi}{4}$; the right asymptote shifts from $x = \pi/2$ to $x = \frac{\pi}{2} + \frac{\pi}{4} = \frac{3\pi}{4}$; and the x-intercept shifts from 0 to $\pi/4$.

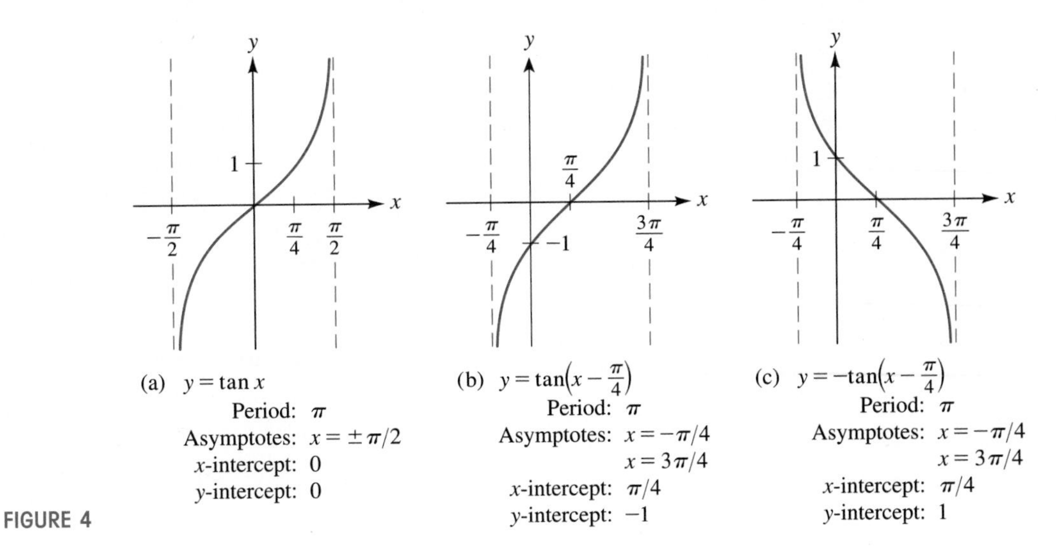

(a) $y = \tan x$
Period: π
Asymptotes: $x = \pm\pi/2$
x-intercept: 0
y-intercept: 0

(b) $y = \tan\left(x - \frac{\pi}{4}\right)$
Period: π
Asymptotes: $x = -\pi/4$
$x = 3\pi/4$
x-intercept: $\pi/4$
y-intercept: -1

(c) $y = -\tan\left(x - \frac{\pi}{4}\right)$
Period: π
Asymptotes: $x = -\pi/4$
$x = 3\pi/4$
x-intercept: $\pi/4$
y-intercept: 1

FIGURE 4

For the y-intercept of $y = \tan\left(x - \frac{\pi}{4}\right)$, we replace x with 0 in the equation. This yields

$$y = \tan\left(0 - \frac{\pi}{4}\right) = \tan\left(-\frac{\pi}{4}\right) = -\tan\frac{\pi}{4} = -1$$

Finally, for the graph of $y = -\tan\left(x - \frac{\pi}{4}\right)$, we need only reflect the graph in Figure 4(b) about the x-axis; see Figure 4(c).

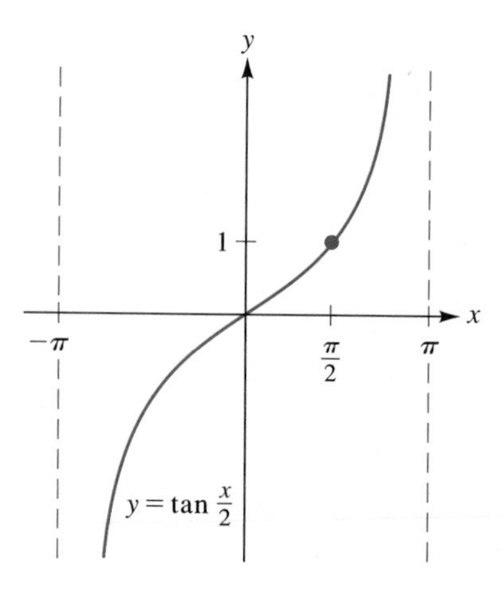

FIGURE 5

EXAMPLE 2 Graph the function $y = \tan(x/2)$ for one period.

Solution First we refer back to Figure 2, which shows the basic pattern for one period of $y = \tan x$. In this basic pattern, the asymptotes occur when the radian measure of the angle equals $-\pi/2$ or $\pi/2$. Consequently, for $y = \tan(x/2)$, the asymptotes occur when $x/2 = -\pi/2$ and when $x/2 = \pi/2$. From the equation $x/2 = -\pi/2$ we conclude that $x = -\pi$, and from the equation $x/2 = \pi/2$ we conclude that $x = \pi$. Thus, the asymptotes for $y = \tan(x/2)$ are $x = -\pi$ and $x = \pi$. The distance between these asymptotes, namely 2π, is the period of $y = \tan(x/2)$. This is twice the period of $y = \tan x$, so basically, we want to draw a curve with the same general shape as $y = \tan x$ but twice as wide; see Figure 5. Note that the graph in Figure 5 passes through the origin, since when $x = 0$, we have

$$y = \tan(0/2) = \tan 0 = 0$$

The graph of the cotangent function can be obtained from that of the tangent function by means of the identity

$$\cot x = -\tan\left(x - \frac{\pi}{2}\right) \tag{3}$$

(Exercise 37 shows how to derive this identity.) According to identity (3), the graph of $y = \cot x$ can be obtained by first translating the graph of $y = \tan x$ to the right $\pi/2$ units and then reflecting the translated graph about the x-axis. When this is done, we obtain the graph shown in the box that follows.

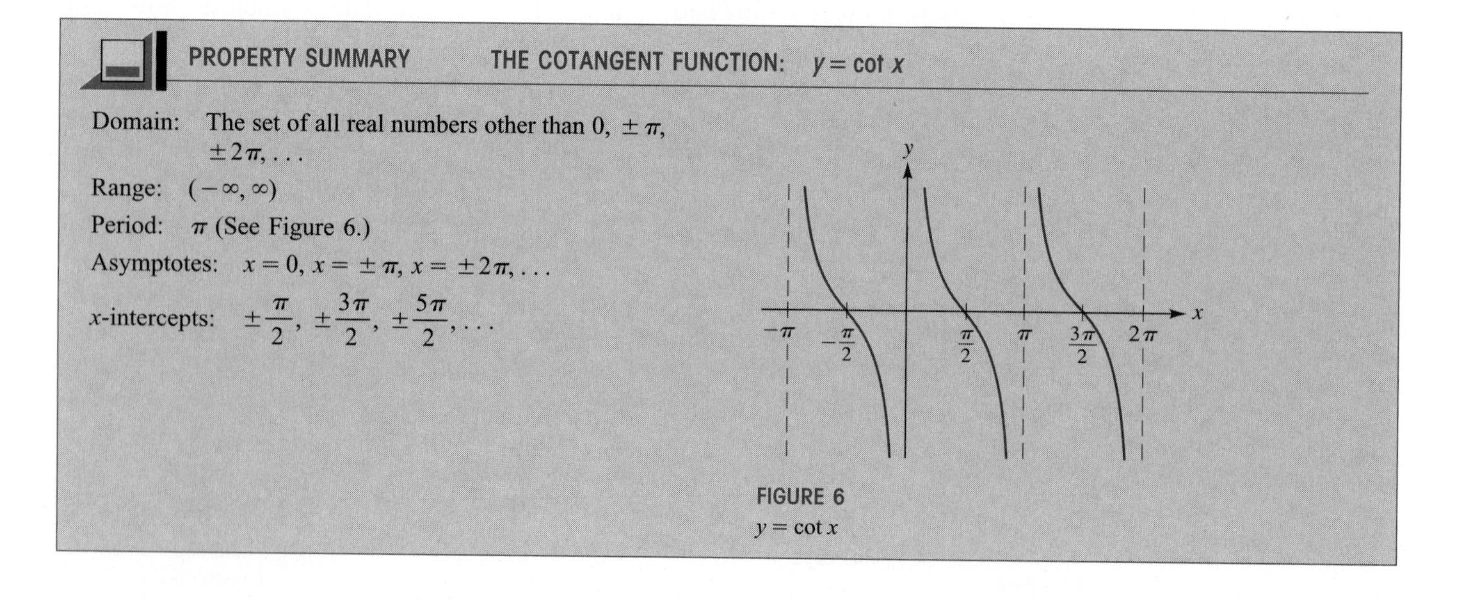

PROPERTY SUMMARY **THE COTANGENT FUNCTION: $y = \cot x$**

Domain: The set of all real numbers other than $0, \pm\pi, \pm 2\pi, \ldots$

Range: $(-\infty, \infty)$

Period: π (See Figure 6.)

Asymptotes: $x = 0, x = \pm\pi, x = \pm 2\pi, \ldots$

x-intercepts: $\pm\frac{\pi}{2}, \pm\frac{3\pi}{2}, \pm\frac{5\pi}{2}, \ldots$

FIGURE 6
$y = \cot x$

EXAMPLE 3 Graph each of the following functions for one period:

(a) $y = \cot \pi x$; **(b)** $y = \frac{1}{2} \cot \pi x$; **(c)** $y = -\frac{1}{2} \cot \pi x$.

Solution

(a) Looking at the graph of $y = \cot x$ in Figure 6, we see that one complete pattern or cycle of the graph is bounded by the asymptotes $x = 0$ and $x = \pi$. Now, for the function we are given, x has been replaced by πx. Thus, the corresponding asymptotes occur when $\pi x = 0$ and when $\pi x = \pi$; in other words, when $x = 0$ and when $x = 1$; see Figure 7(a).

(b) The graph of $y = \frac{1}{2} \cot \pi x$ will have the same general shape as that of $y = \cot \pi x$, but each y-coordinate on $y = \frac{1}{2} \cot \pi x$ will be one-half of the corresponding coordinate on $y = \cot \pi x$; see Figure 7(b).

(c) The graph of $y = -\frac{1}{2} \cot \pi x$ is obtained by reflecting the graph of $y = \frac{1}{2} \cot \pi x$ about the x-axis; see Figure 7(c).

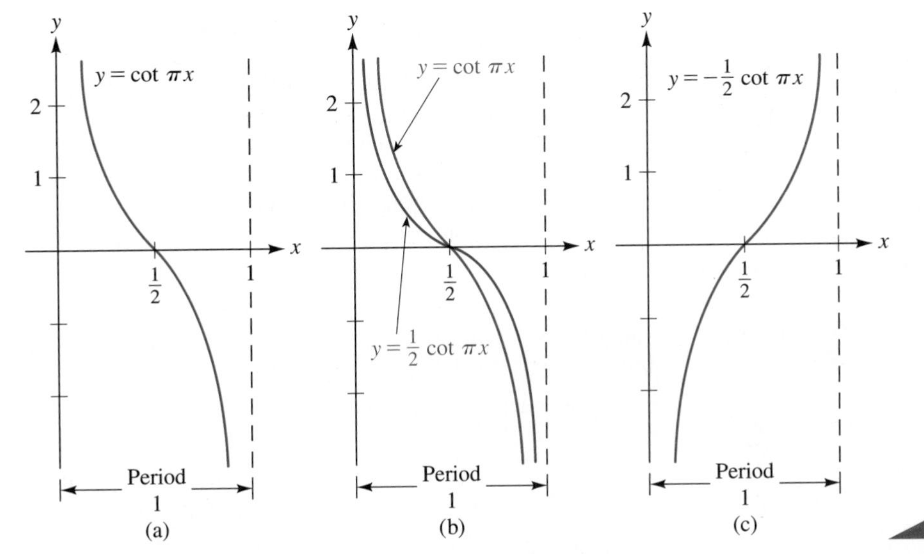

FIGURE 7

We conclude this section with a discussion of the graphs of $y = \csc x$ and $y = \sec x$. We will obtain these graphs in a series of easy steps, relying on the ideas of symmetry and translation. First consider the function $y = \csc x$. In Figure 8(a), we've set up a table and used it to sketch the graph of $y = \csc x$ on the interval $[\pi/2, \pi)$. Note that $\csc x$ is undefined when $x = \pi$. (Why?)

x	$\csc x$
$\pi/2$	1
$2\pi/3$	≈ 1.2
$5\pi/6$	2
$11\pi/12$	≈ 3.9

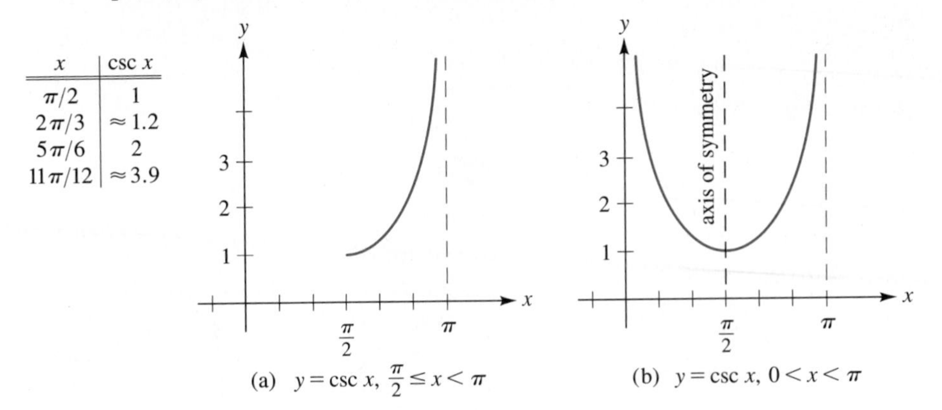

(a) $y = \csc x$, $\frac{\pi}{2} \le x < \pi$

(b) $y = \csc x$, $0 < x < \pi$

FIGURE 8

As Figure 8(a) indicates, the vertical line $x = \pi$ is an asymptote for the graph. The graph in Figure 8(a) can be extended to the interval $(0, \pi)$ by means of the identity

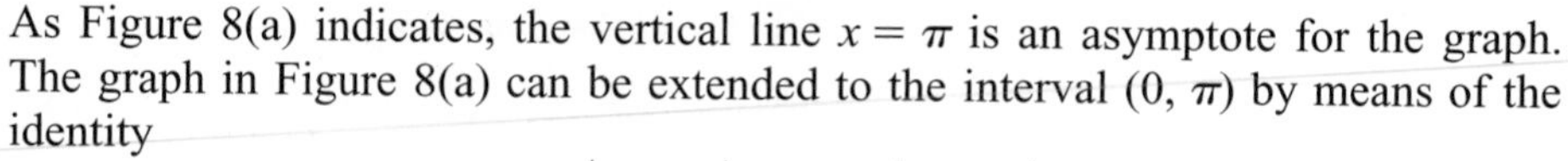

$$\csc\left(\frac{\pi}{2} + s\right) = \csc\left(\frac{\pi}{2} - s\right)$$

(Exercise 38 shows you how to verify this identity.) This identity tells us that, starting at $x = \pi/2$, whether we travel a distance s to the right or a distance s to the left, the value of $y = \csc x$ is the same. In other words, the graph of $y = \csc x$ is symmetric about the line $x = \pi/2$. In view of this symmetry, we can sketch the graph of $y = \csc x$ on the interval $(0, \pi)$, as shown in Figure 8(b).

The next step in obtaining the graph of $y = \csc x$ is to use the fact that the graph is symmetric about the origin. To verify this, we need to check that $\csc(-x) = -\csc x$. We have

$$\csc(-x) = \frac{1}{\sin(-x)} = \frac{1}{-\sin x} = -\csc x \qquad \text{as required}$$

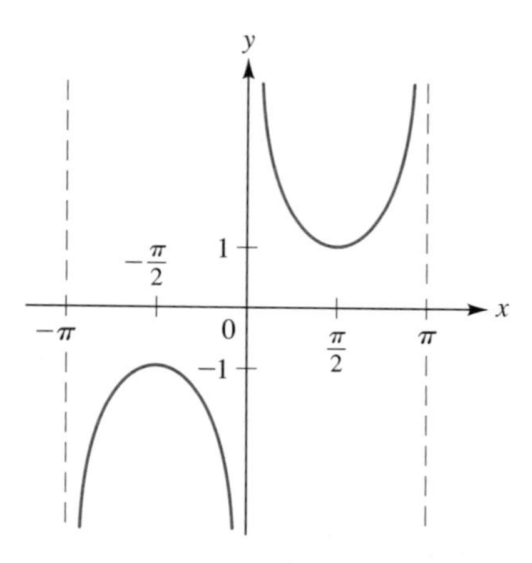

FIGURE 9

$y = \csc x, \quad -\pi < x < \pi$

The graph is symmetric about the origin.

Now, taking into account this symmetry about the origin and the portion of the graph that we've already obtained in Figure 8(b), we can sketch the graph of $y = \csc x$ over the interval $(-\pi, \pi)$, as shown in Figure 9.

To complete the graph of $y = \csc x$, we observe that the values of $\csc x$ must repeat themselves at intervals of 2π. This is because $\csc x = 1/(\sin x)$, and the sine function has a period of 2π. In view of this, we can draw the graph of $y = \csc x$ as shown in the box that follows.

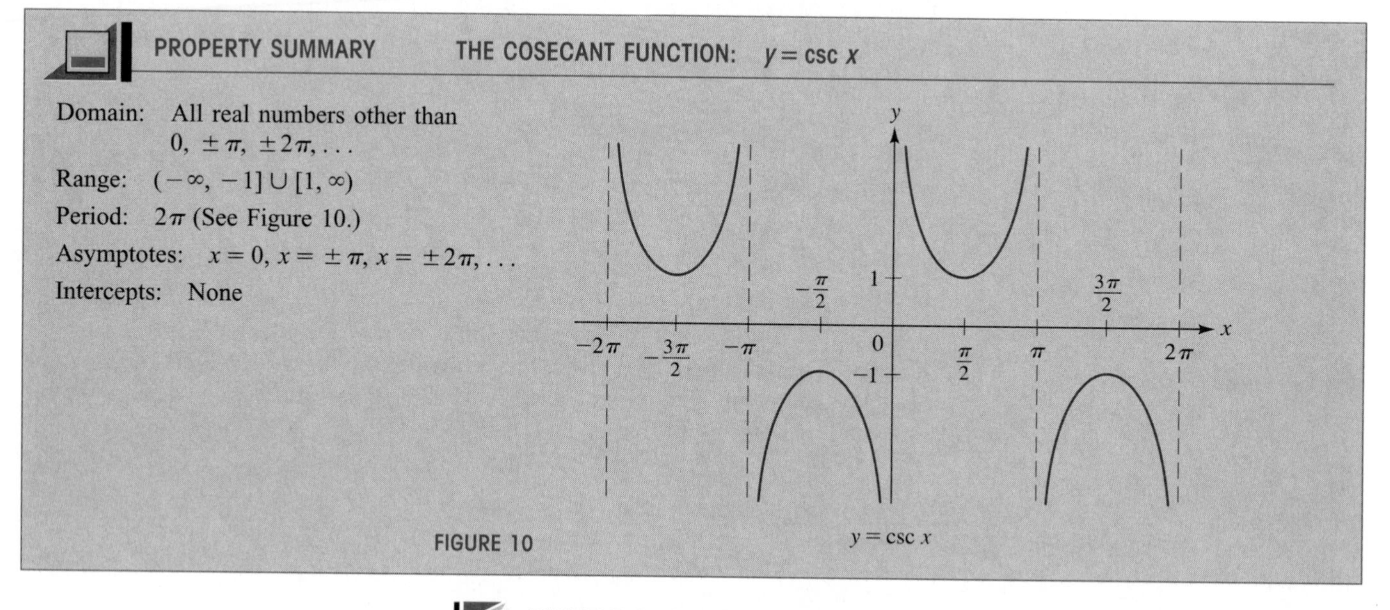

PROPERTY SUMMARY THE COSECANT FUNCTION: $y = \csc x$

Domain: All real numbers other than $0, \pm\pi, \pm 2\pi, \ldots$

Range: $(-\infty, -1] \cup [1, \infty)$

Period: 2π (See Figure 10.)

Asymptotes: $x = 0, x = \pm\pi, x = \pm 2\pi, \ldots$

Intercepts: None

FIGURE 10 $y = \csc x$

EXAMPLE 4 Graph the function $y = \csc(x/3)$ for one period.

Solution Since $\csc(x/3) = 1/\sin(x/3)$, it will be helpful first to graph one period of $y = \sin(x/3)$. This is done in Figure 11(a) using the techniques of Section 7.5. Note that the period of $y = \sin(x/3)$ is 6π. This is also the period of $y = \csc(x/3)$, because $\csc(x/3)$ and $\sin(x/3)$ are just reciprocals. The asymptotes for $y = \csc(x/3)$ occur when $\sin(x/3) = 0$. From Figure 11(a), we see that $\sin(x/3) = 0$ when $x = 0$, when $x = 3\pi$, and when $x = 6\pi$. These asymptotes are sketched in Figure 11(b). The colored points in Figure 11(b) indicate where the value of $\sin(x/3)$ is

1 or -1; the graph of $y = \csc(x/3)$ must pass through these points. (Why?) Finally, using the points and the asymptotes in Figure 11(b), we can sketch the graph of $y = \csc(x/3)$, as shown in Figure 11(c).

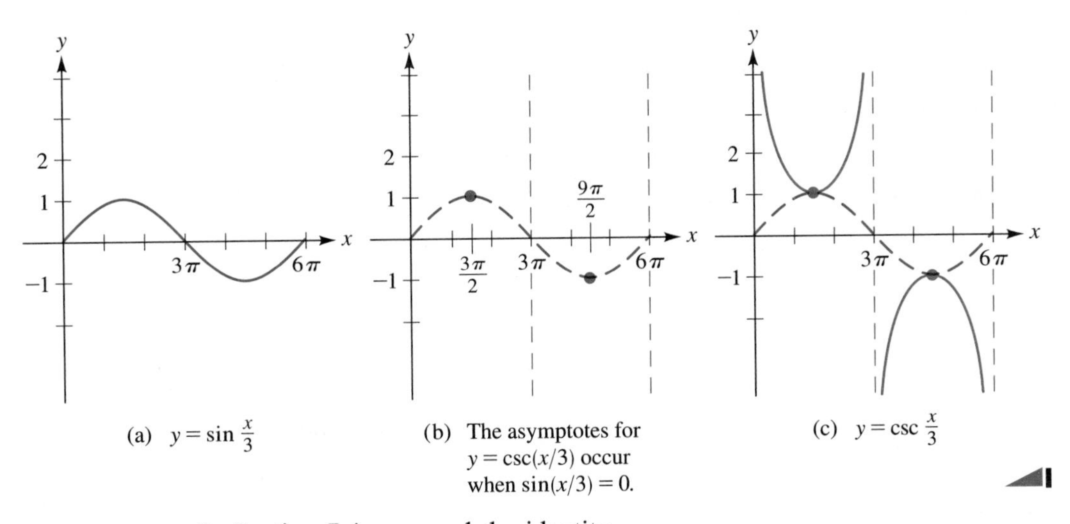

FIGURE 11

(a) $y = \sin \frac{x}{3}$

(b) The asymptotes for $y = \csc(x/3)$ occur when $\sin(x/3) = 0$.

(c) $y = \csc \frac{x}{3}$

In Section 7.4, we used the identity

$$\cos x = \sin\left(x + \frac{\pi}{2}\right) \tag{4}$$

to graph the cosine function. This identity tells us that the graph of $y = \cos x$ can be obtained by translating the graph of $y = \sin x$ to the left by $\pi/2$ units. Now, from identity (4), it follows that

$$\sec x = \csc\left(x + \frac{\pi}{2}\right)$$

This last identity tells us that the graph of $y = \sec x$ can also be obtained by a translation. In this case, by translating the graph of $y = \csc x$ a distance of $\pi/2$ units to the left, we obtain the graph of $y = \sec x$, as shown in the box that follows.

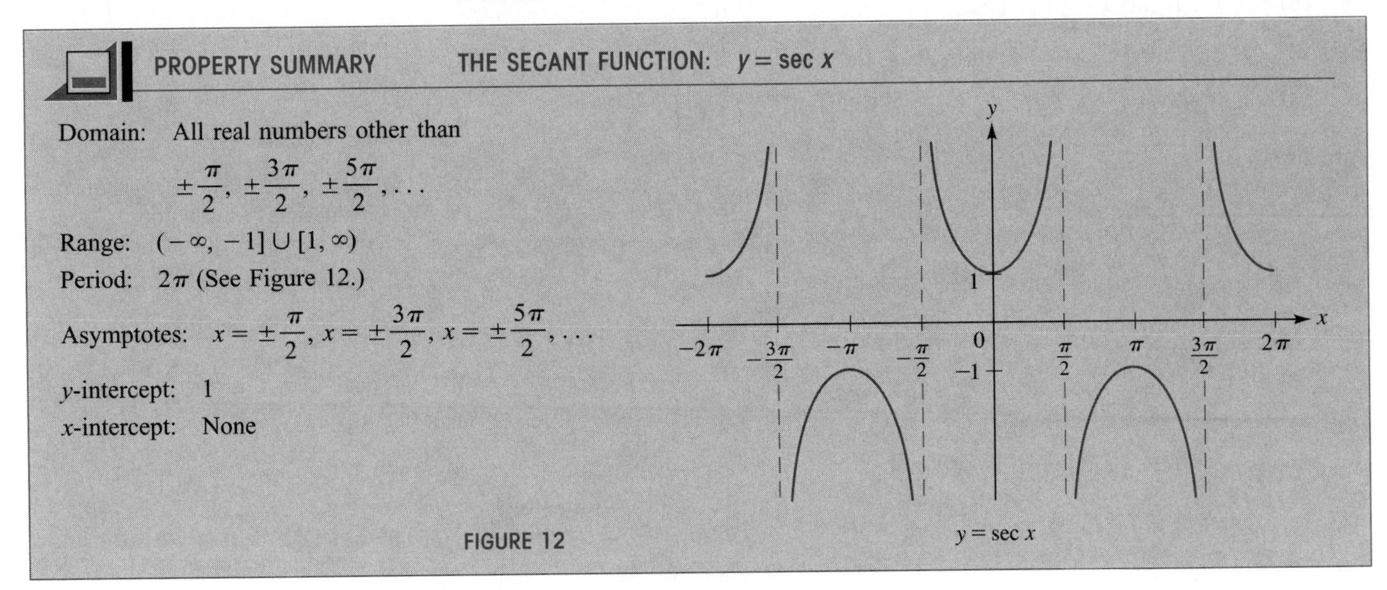

PROPERTY SUMMARY **THE SECANT FUNCTION: $y = \sec x$**

Domain: All real numbers other than

$$\pm\frac{\pi}{2}, \pm\frac{3\pi}{2}, \pm\frac{5\pi}{2}, \ldots$$

Range: $(-\infty, -1] \cup [1, \infty)$

Period: 2π (See Figure 12.)

Asymptotes: $x = \pm\frac{\pi}{2}, x = \pm\frac{3\pi}{2}, x = \pm\frac{5\pi}{2}, \ldots$

y-intercept: 1

x-intercept: None

FIGURE 12

$y = \sec x$

EXERCISE SET 7.7

A

In Exercises 1–24, graph each function for one period, and show (or specify) the intercepts and asymptotes.

1. **(a)** $y = \tan\left(x + \frac{\pi}{4}\right)$
   (b) $y = -\tan\left(x + \frac{\pi}{4}\right)$
2. **(a)** $y = \tan\left(x - \frac{\pi}{3}\right)$
   (b) $y = -\tan\left(x - \frac{\pi}{3}\right)$
3. **(a)** $y = \tan(x/3)$
   (b) $y = -\tan(x/3)$
4. **(a)** $y = 2 \tan \pi x$
   (b) $y = -2 \tan \pi x$
5. $y = \frac{1}{2}\tan(\pi x/2)$
6. $y = -\frac{1}{2}\tan 2\pi x$
7. $y = \cot(\pi x/2)$
8. $y = \cot 2\pi x$
9. $y = -\cot\left(x - \frac{\pi}{4}\right)$
10. $y = \cot\left(x + \frac{\pi}{6}\right)$
11. $y = \frac{1}{2}\cot 2x$
12. $y = -\frac{1}{2}\cot(x/2)$
13. $y = \csc\left(x - \frac{\pi}{4}\right)$
14. $y = \csc\left(x - \frac{\pi}{6}\right)$
15. $y = -\csc(x/2)$
16. $y = 2 \csc x$
17. $y = \frac{1}{3}\csc \pi x$
18. $y = -\frac{1}{2}\csc 2\pi x$
19. $y = -\sec x$
20. $y = -2 \sec x$
21. $y = \sec(x - \pi)$
22. $y = \sec(x + 1)$
23. $y = 3 \sec(\pi x/2)$
24. $y = -2 \sec(\pi x/3)$

In Exercises 25–28, graph each function for two periods. Specify the intercepts and the asymptotes.

25. **(a)** $y = 2 \sin\left(3\pi x - \frac{\pi}{6}\right)$
    (b) $y = 2 \csc\left(3\pi x - \frac{\pi}{6}\right)$
26. **(a)** $y = -\frac{1}{2}\sin\left(\pi x + \frac{\pi}{3}\right)$
    (b) $y = -\frac{1}{2}\csc\left(\pi x + \frac{\pi}{3}\right)$
27. **(a)** $y = -3 \cos\left(2\pi x - \frac{\pi}{4}\right)$
    (b) $y = -3 \sec\left(2\pi x - \frac{\pi}{4}\right)$
28. **(a)** $y = \cos\left(3x + \frac{\pi}{3}\right)$
    (b) $y = \sec\left(3x + \frac{\pi}{3}\right)$

B

For Exercises 29–32, six functions are defined as follows:

$$f(x) = \sin x \qquad g(x) = \csc x \qquad h(x) = \pi x - \frac{\pi}{6}$$
$$F(x) = \cos x \qquad G(x) = \sec x \qquad H(x) = \pi x + \frac{\pi}{4}$$

In each case, graph the indicated function for one period.

29. **(a)** $f \circ h$
    (b) $g \circ h$
30. **(a)** $F \circ H$
    (b) $G \circ H$
31. **(a)** $f \circ H$
    (b) $g \circ H$
32. **(a)** $F \circ h$
    (b) $G \circ h$

For Exercises 33–36, four functions are defined as follows:

$$f(x) = \csc x \qquad T(x) = \tan x$$
$$g(x) = \sec x \qquad A(x) = |x|$$

In each case, graph the indicated function over the interval $[-2\pi, 2\pi]$.

33. $A \circ T$
34. $A \circ g$
35. $A \circ f$
36. $f \circ A$
37. In the text, we used the identity
$$\cot s = -\tan\left(s - \frac{\pi}{2}\right)$$
to obtain the graph of $y = \cot x$ from that of $y = \tan x$. The following steps show one way to derive this identity. (Although the accompanying figure shows the angle with radian measure s terminating in the first quadrant, the proof is valid no matter where the angle terminates.) In the figure, $\overline{PO} \perp \overline{QO}$.

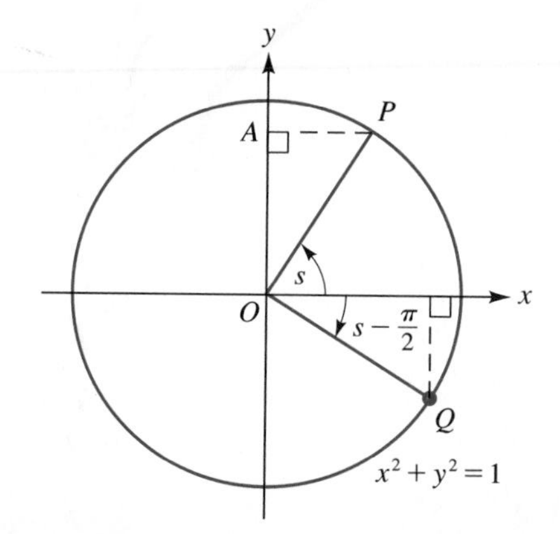

(a) Why are the coordinates of P and Q as follows?
$$P(\cos s, \sin s) \quad \text{and} \quad Q\left(\cos\left(s - \frac{\pi}{2}\right), \sin\left(s - \frac{\pi}{2}\right)\right)$$
(b) Using congruent triangles [and without reference to part (a)], explain why the y-coordinate of Q is the negative of the x-coordinate of P, and the x-coordinate of Q equals the y-coordinate of P.
(c) Use the results in parts (a) and (b) to conclude that
$$\sin\left(s - \frac{\pi}{2}\right) = -\cos s \quad \text{and} \quad \cos\left(s - \frac{\pi}{2}\right) = \sin s$$
(d) Use the result in part (c) to show that
$$\cot s = -\tan\left(s - \frac{\pi}{2}\right)$$

38. In this exercise we verify the identity $\csc\left(\frac{\pi}{2} + s\right) = \csc\left(\frac{\pi}{2} - s\right)$. Refer to the following figure, in

which $\angle AOB = \angle BOC = s$ radians. [Although the figure shows s in the interval $0 < s < \frac{\pi}{2}$, a similar proof will work for other intervals as well. (A proof that does not depend upon a picture can be given using the formula for $\sin(s+t)$, which is developed in Section 8.1.)]

(a) Show that the triangles AOE and COD are congruent and, consequently, that $AE = CD$.

(b) Explain why the y-coordinates of the points C and A are $\sin\left(\frac{\pi}{2} - s\right)$ and $\sin\left(\frac{\pi}{2} + s\right)$, respectively.

(c) Use parts (a) and (b) to conclude that $\sin\left(\frac{\pi}{2} + s\right) = \sin\left(\frac{\pi}{2} - s\right)$. It follows from this that $\csc\left(\frac{\pi}{2} + s\right) = \csc\left(\frac{\pi}{2} - s\right)$, as required.

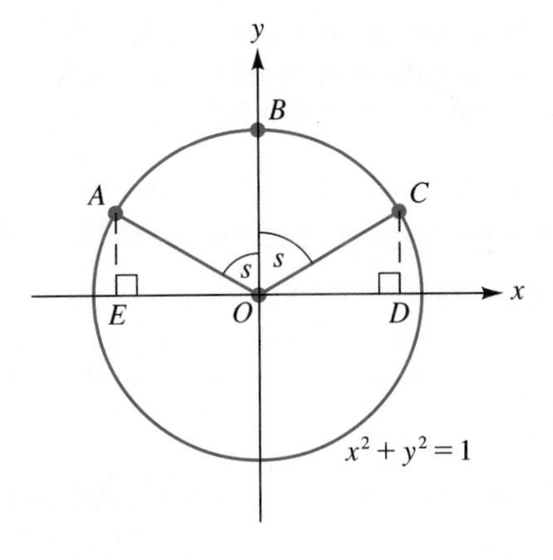

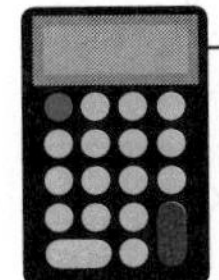

GRAPHING UTILITY EXERCISES FOR SECTION 7.7

1. Graph the function $y = \tan x$. Use a viewing rectangle that extends from -5 to 5 in both the x- and the y-directions. What are the exact values for the x-intercepts shown in your graph?

2. In graphing the tangent function in the text, we used the identity $\tan(x + \pi) = \tan x$. Check this identity by graphing the two equations $y = \tan x$ and $y = \tan(x + \pi)$ and noting that the graphs indeed appear to be identical.

3. In the text, we obtained the graph of the cotangent function from that of the tangent by means of the identity $\cot x = -\tan\left(x - \frac{\pi}{2}\right)$. Verify this identity graphically by graphing the two functions $y = \cot x$ and $y = -\tan\left(x - \frac{\pi}{2}\right)$ and noting that the graphs indeed appear to be identical.

4. In this exercise we compare the graphs of $f(x) = \tan x$ and $g(x) = x^3$ on the open interval $-\pi/2 < x < \pi/2$. On the same set of axes, graph these two functions for $-\pi/2 < x < \pi/2$. Then answer the following questions.

(a) Which of the two graphs appears to be horizontal as it passes through the origin?

(b) When x is in the open interval $(0, \pi/2)$, which quantity is larger, $\tan x$ or x^3?

(c) When x is in the open interval $(-\pi/2, 0)$, which quantity is larger, $\tan x$ or x^3?

In Exercises 5–10, graph each function. Adjust the viewing rectangle as necessary so that the graph is shown for at least two periods.

5. **(a)** $y = \tan(x/4)$
(b) $y = \tan(4x)$

6. **(a)** $y = \cot(2x)$
(b) $y = \cot(x/2)$

7. **(a)** $y = 0.5 \tan \pi x$
(b) $y = 0.5 \tan\left(\pi x + \frac{\pi}{3}\right)$
(c) $y = 0.5 \tan(\pi x + 1)$

8. **(a)** $y = 0.25 \cot x$
(b) $y = 0.25 \cot\left(x + \frac{\pi}{4}\right)$
(c) $y = -0.25 \cot\left(x + \frac{\pi}{4}\right)$

9. **(a)** $y = 0.4 \tan(x/2)$
(b) $y = 0.4 \tan(x/3)$
(c) $y = 0.4 \tan(x/5)$

10. **(a)** $y = 0.2 \cot(3x)$
(b) $y = -0.2 \cot(4x)$
(c) $y = -0.2 \cot(12x)$

11. Graph the functions $y = \sin x$ and $y = \csc x$ in the standard viewing rectangle. [For $\csc x$, use $1 \div (\sin x)$.] Observe that $|\sin x| \leq 1$, while $|\csc x| \geq 1$. At which points in the picture do we have $\sin x = \csc x$? At which points do we have $\sin x = -\csc x$?

12. Graph the functions $y = \cos x$ and $y = \sec x$ in the standard viewing rectangle. [For $\sec x$, use $1 \div (\cos x)$.] Observe that $|\cos x| \leq 1$, while $|\sec x| \geq 1$. At which points in the picture do we have $\cos x = \sec x$? At which points do we have $\cos x = -\sec x$?

In Exercises 13–16, graph each pair of functions on the same set of axes. Adjust the viewing rectangle as necessary so that the graphs are shown for at least two periods.

13. $Y_1 = 0.6 \sin(x/2)$
$Y_2 = 0.6 \csc(x/2)$

14. $Y_1 = -1.2 \sin(\pi x/3)$
$Y_2 = -1.2 \csc(\pi x/3)$

15. $Y_1 = 3 \cos\left(2x - \frac{\pi}{6}\right)$
$Y_2 = 3 \sec\left(2x - \frac{\pi}{6}\right)$

16. $Y_1 = -1.5 \cos\left(\pi x - \frac{\pi}{6}\right)$
$Y_2 = -1.5 \sec\left(\pi x - \frac{\pi}{6}\right)$

17. Graph the two functions $y = \tan^2 x$ and $y = \sec^2 x - 1$. What do you observe? What does this demonstrate?

18. Graph the two functions $y = \cot^2 x$ and $y = \csc^2 x - 1$. What do you observe? What does this demonstrate?

19. **(a)** Graph the two functions $y = \tan x$ and $y = x$ in the standard viewing rectangle. Observe that $\tan x$ and

x are very close to one another when x is close to zero. In fact, the approximation $\tan x \approx x$ is often used in applications when it is known that x is close to zero.

(b) To obtain a better view of $y = \tan x$ and $y = x$ near the origin, adjust the viewing rectangle so that x extends from $-\pi/2$ to $\pi/2$ and y extends from -2 to 2. Again, note that when x is close to zero, the values of $\tan x$ are indeed close to x. To see numerical evidence of this, complete the following table.

x	0.000123	0.01	0.05	0.1	0.2	0.3	0.4	0.5
tan x								

(c) If you study *Taylor polynomials* in calculus, you'll see that an even better approximation to $\tan x$ is

$$\tan x \approx x + \tfrac{1}{3}x^3 \quad \text{when } x \text{ is close to } 0$$

Add the graph of $y = x + \frac{1}{3}x^3$ to the picture that you obtained in part (b). Describe what you see.

(d) To see numerical evidence of how well $\tan x$ is approximated by $x + \frac{1}{3}x^3$, add a third row to the table you worked out in part (b); in this third row, show the values for $x + \frac{1}{3}x^3$. When you've completed the table, note that the new values are much closer to $\tan x$ than were the values of x.

20. **(a)** Graph the equations $y = \tan x$ and $y = 2$ in the standard viewing rectangle.

(b) Use the graph to give a rough estimate for the smallest positive root of the equation $\tan x = 2$.

Answer: $x \approx 1$

(c) Use the graphing utility to determine the root more accurately, say, through the first four decimal places.

(d) Let r denote the root that you determined in part (c). Is the number $r + \pi$ also a root of the equation $\tan x = 2$?

21. **(a)** Graph the equations $y = \tan x$ and $y = x$ in the standard viewing rectangle. Use the graph to give a rough estimate for the smallest positive root of the equation $\tan x = x$.

Answer: Something between 4 and 5, call it 4.5

(b) Use the graphing utility to determine the root more accurately, say, through the first four decimal places.

(c) Let r denote the root that you determined in part (b). Is the number $r + \pi$ also a root of the equation $\tan x = x$?

CHAPTER SEVEN SUMMARY OF PRINCIPAL FORMULAS AND TERMS

FORMULAS OR TERMS	PAGE REFERENCE	COMMENTS
1. $\theta = \dfrac{s}{r}$	408	This equation defines the *radian measure* θ of an angle. We assume here that the vertex of the angle is placed at the center of a circle of radius r and that s is the length of the intercepted arc. See Figure 1 in Section 7.1.
2. $s = r\theta$	417	This formula expresses the arc length s on a circle in terms of the radius r and the radian measure θ of the central angle subtended by the arc.
3. $A = \frac{1}{2}r^2\theta$	417	This formula expresses the area of a sector of a circle in terms of the radius r and the radian measure θ of the central angle.
4. $\omega = \dfrac{\theta}{t}$ $v = \dfrac{d}{t}$ $v = r\omega$	420, 421	Suppose that a wheel rotates about its axis at a constant rate. If a radial line on the wheel turns through an angle of measure θ in time t, then the *angular speed* ω of the wheel is defined by $\omega = \theta/t$. If a point on the rotating wheel travels a distance d in time t, then the linear speed v of the point is defined by $v = d/t$. If the angular speed ω is expressed in radians per unit time, then the linear and angular speeds are related by $v = r\omega$.

FORMULAS OR TERMS	PAGE REFERENCE	COMMENTS
5. $\cos t$ $\sin t$ $\tan t$ $\sec t$ $\csc t$ $\cot t$	428	Let $P(x, y)$ denote the point on the unit circle such that the arc length from $(1, 0)$ is t. Or equivalently, let $P(x, y)$ denote the point where the terminal side of the angle with radian measure t intersects the unit circle. Then the six trigonometric functions of the real number t are defined as follows: $\cos t = x$ | $\sec t = \frac{1}{x}$ $(x \neq 0)$ $\sin t = y$ | $\csc t = \frac{1}{y}$ $(y \neq 0)$ $\tan t = \frac{y}{x}$ $(x \neq 0)$ | $\cot t = \frac{x}{y}$ $(y \neq 0)$
6. $\sec t = \frac{1}{\cos t}$ $\csc t = \frac{1}{\sin t}$ $\cot t = \frac{1}{\tan t}$ $\tan t = \frac{\sin t}{\cos t}$ $\cot t = \frac{\cos t}{\sin t}$	430	These five identities are direct consequences of the definitions of the trigonometric functions.
7. $\sin^2 t + \cos^2 t = 1$ $\tan^2 t + 1 = \sec^2 t$ $\cot^2 t + 1 = \csc^2 t$	430	These are the three *Pythagorean identities.*
8. $\cos(-t) = \cos t$ $\sin(-t) = -\sin t$ $\tan(-t) = -\tan t$	433	These three identities are sometimes referred to as the *opposite angle identities.* In each case, a trigonometric function of $-t$ is expressed in terms of that same function of t.
9. Periodic function	438	A function f is periodic if there is a positive number p such that the equation $f(x+p) = f(x)$ holds for all x in the domain of f. The smallest such number p is called the *period* of f. Important examples: The period of $y = \sin x$ is 2π; the period of $y = \cos x$ is 2π; the period of $y = \tan x$ is π.
10. Amplitude	439	Let m and M denote the smallest and the largest values, respectively, of the periodic function f. Then the *amplitude* of f is defined to be the number $\frac{1}{2}(M - m)$. Important examples: The amplitude for both $y = \sin x$ and $y = \cos x$ is 1; amplitude is not defined for the function $y = \tan x$.
11. Period $= \frac{2\pi}{B}$	458	This formula gives the period for the functions $y = A \sin(Bx - C)$ and $y = A \cos(Bx - C)$.
12. Phase shift $= \frac{C}{B}$	458	The phase shift serves as a guide in graphing functions of the form $y = A \sin(Bx - C)$ or $y = A \cos(Bx - C)$. For instance, to graph one complete cycle of $y = A \sin(Bx - C)$, first sketch one complete cycle of $y = A \sin Bx$, beginning at $x = 0$. Then draw a curve with exactly the same shape, but beginning at $x = C/B$ rather than $x = 0$. This will represent one cycle of $y = A \sin(Bx - C)$.

FORMULAS OR TERMS	PAGE REFERENCE	COMMENTS
13. Simple harmonic motion	462	Oscillation described by an equation of the form $y = A\sin(Bx - C)$ or $y = A\cos(Bx - C)$ is referred to as *simple harmonic motion.* The standard example for simple harmonic motion is the motion of a mass on the end of a spring (neglecting friction), as depicted in Figures 1 and 3 in Section 7.6.
14. Period, frequency, and amplitude for simple harmonic motion	463	The amplitude for simple harmonic motion is the maximum displacement from the equilibrium position. The period is the time required for one complete cycle of the motion. The frequency f, measured in cycles per second, is computed from the equation $f = 1/\text{period}$.

WRITING MATHEMATICS

1. Say whether the statement is TRUE or FALSE. Write out your reason or reasons in complete sentences. If you draw a diagram to accompany your writing, be sure that you clearly label any parts of the diagram to which you refer.
   (a) If t is a real number less than π, then $\cos t$ is positive.
   (b) The equation $1 + \tan^2 t = \sec^2 t$ is true for every real number t.
   (c) If t is a positive real number, then $\cos^2 t$ is a positive real number.
   (d) Between any two consecutive turning points, the graph of $y = \sin x$ is decreasing.
   (e) If a and b are real numbers and $\ln(\sin a) = \ln(\sin b)$, then $\sin a = \sin b$.
   (f) If a and b are real numbers and $\ln(\sin a) = \ln(\sin b)$, then $a = b$.
   (g) If a and b are real numbers and $\sin(\ln a) = \sin(\ln b)$, then $\ln a = \ln b$.
2. Let $f(x) = x + \sin x$. This function has an interesting property: the iterates can be used to obtain better and better approximations to the number π. The following table, for instance, shows the first five iterates of $x_0 = 0.8$. As you can see, the fifth iterate is indeed an excellent approximation to π. In fact, the fifth iterate, x_5, agrees with π through the first 15 decimal places. [As you'll see in part (a), this behavior is not peculiar to $x_0 = 0.8$. Any value for x_0 between 0 and π will produce similar results.]

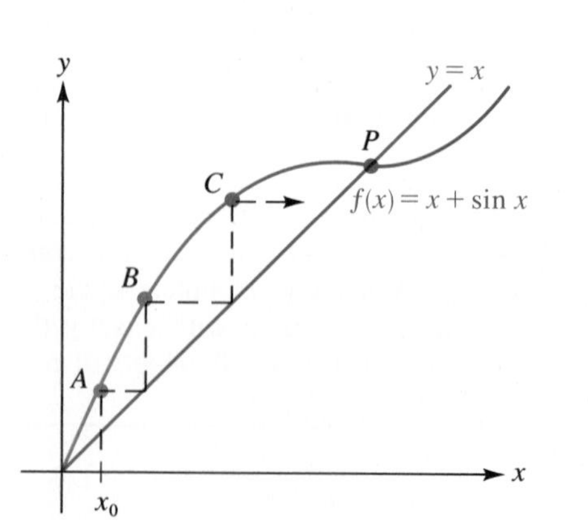

x_1	1.5173560908 . . .
x_2	2.5159285012 . . .
x_3	3.1015642481 . . .
x_4	3.1415819650 . . .
x_5	3.1415926535 . . .

Iterates of $x_0 = 0.8$ for the function $f(x) = x + \sin x$.

   (a) Use your calculator to set up three tables of iterates, similar to the given table, using $x_0 = 0.9$, $x_0 = 1.5$, and $x_0 = 2$. Describe the results.
   (b) Use the figure in the margin to explain why the successive iterates approach the number π.

CHAPTER SEVEN REVIEW EXERCISES

1. Evaluate the following: **(a)** $\sin(5\pi/3)$; **(b)** $\cot(11\pi/6)$.
2. Graph the function $y = 3\cos 3\pi x$ over the interval $-\frac{1}{3} \leq x \leq \frac{1}{3}$. Specify the x-intercepts and the coordinates of the highest point on the graph.
3. Simplify the following expression:

$$\cos t - \cos(-t) + \sin t - \sin(-t)$$

4. In the expression $1/\sqrt{4 - t^2}$, make the substitution $t = 2\cos x$ and simplify the result. Assume that $0 < x < \pi$.
5. Graph the equation $y = \sec(2\pi x - 3)$ for one complete cycle.
6. If $\sec t = -5/3$ and $\pi < t < \frac{3\pi}{2}$, compute $\cot t$.
7. Graph the function $y = -\sin(2x - \pi)$ for one complete period. Specify the amplitude, period, and phase shift.
8. Graph $y = -\frac{1}{2}\tan(\pi x/3)$ over one period.

In Exercises 9–20, evaluate each expression without using a calculator or tables.

9. $\cos \pi$
10. $\sin(-3\pi/2)$
11. $\csc(2\pi/3)$
12. $\tan(\pi/3)$
13. $\tan(11\pi/6)$
14. $\cos 0$
15. $\sin(\pi/6)$
16. $\sec(3\pi/4)$
17. $\cot(5\pi/4)$
18. $\tan(-7\pi/4)$
19. $\csc(-5\pi/6)$
20. $\sin^2(\pi/7) + \cos^2(\pi/7)$

Exercises 21–30 are calculator exercises. (Set your calculator to the radian mode.) In Exercises 21–26, where numerical answers are required, round your results to three decimal places.

21. Evaluate $\sin 1$.
22. Evaluate $\cos 2$.
23. Evaluate $\sin(3\pi/2)$.
24. Evaluate $\sin(0.78)$.
25. Evaluate $\sin(\sin 0.0123)$.
26. Evaluate $\sin[\sin(\sin 0.0123)]$.
27. Verify that $\sin^2 1776 + \cos^2 1776 = 1$.
28. Verify that $\sin 14 = 2\sin 7 \cos 7$.
29. Verify that $\cos(0.5) = \cos^2(0.25) - \sin^2(0.25)$.
30. Verify that $\cos(0.3) = \left[\frac{1}{2}(1 + \cos 0.6)\right]^{1/2}$.
31. In the expression $\sqrt{25 - x^2}$, make the substitution $x = 5\sin\theta$ $\left(0 < \theta < \frac{\pi}{2}\right)$ and simplify the result.
32. In the expression $(49 + x^2)^{1/2}$, make the substitution $x = 7\tan\theta$ $\left(0 < \theta < \frac{\pi}{2}\right)$ and simplify the result.
33. In the expression $(x^2 - 100)^{1/2}$, make the substitution $x = 10\sec\theta$ $\left(0 < \theta < \frac{\pi}{2}\right)$ and simplify the result.
34. In the expression $(x^2 - 4)^{-3/2}$, make the substitution $x = 2\sec\theta$ $\left(0 < \theta < \frac{\pi}{2}\right)$ and simplify the result.
35. In the expression $(x^2 + 5)^{-1/2}$, make the substitution $x = \sqrt{5}\tan\theta$ $\left(0 < \theta < \frac{\pi}{2}\right)$ and simplify the result.
36. If $\sin\theta = -5/13$ and $\pi < \theta < \frac{3\pi}{2}$, compute $\cos\theta$.
37. If $\cos\theta = 8/17$ and $\sin\theta$ is negative, where $0 < \theta < 2\pi$, compute $\tan\theta$.
38. If $\sec\theta = -25/7$ and $\pi < \theta < \frac{3\pi}{2}$, compute $\cot\theta$.
39. In the following figure, P is the center of the circle, which has radius $\sqrt{2}$ units. If the radian measure of angle BPA is θ, express the area of the shaded region in terms of θ. Simplify your answer as much as possible.

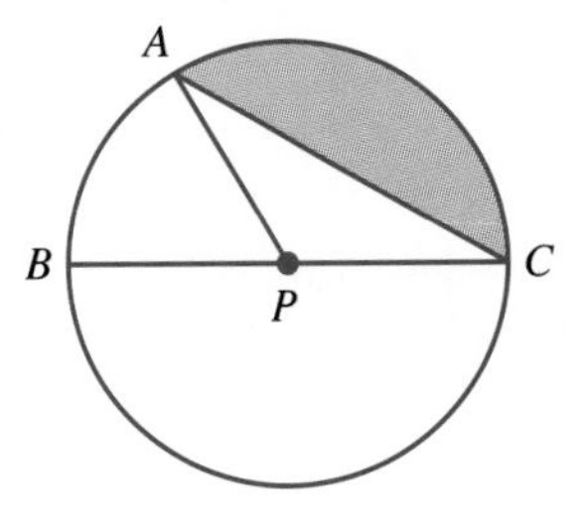

40. Express the area of the shaded region in the following figure in terms of r and θ. (Assume θ is in radians.)

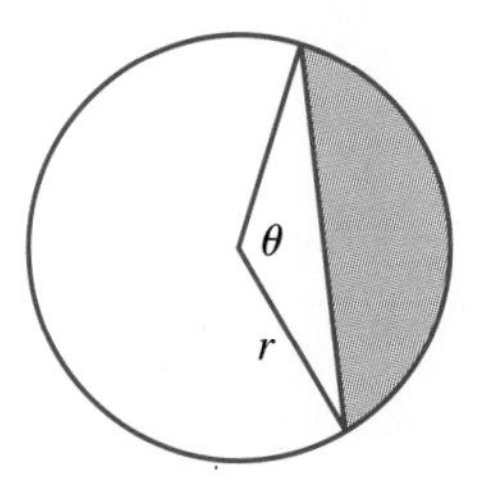

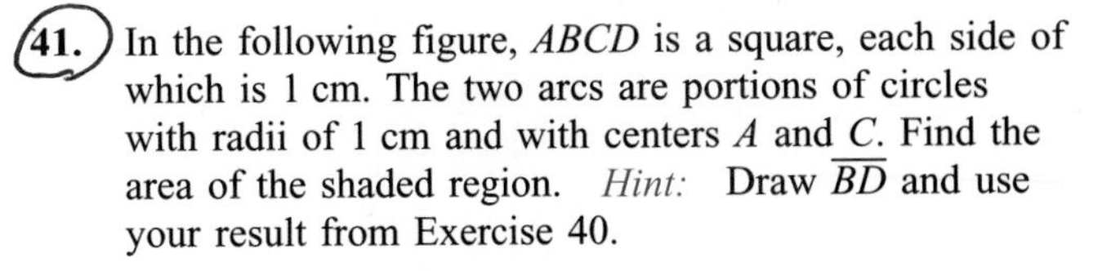

41. In the following figure, $ABCD$ is a square, each side of which is 1 cm. The two arcs are portions of circles with radii of 1 cm and with centers A and C. Find the area of the shaded region. *Hint:* Draw $\overline{BD}$ and use your result from Exercise 40.

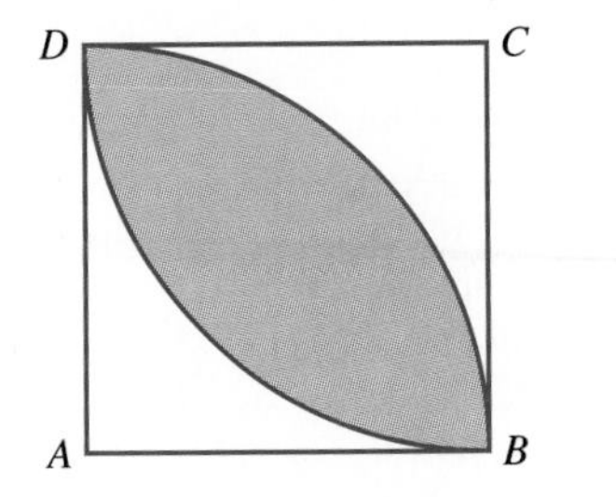

In Exercises 42–46, a function of the form $y = A \sin Bx$ or $y = A \cos Bx$ is graphed for one period. Determine the equation in each case. (Assume $B > 0$.)

42.

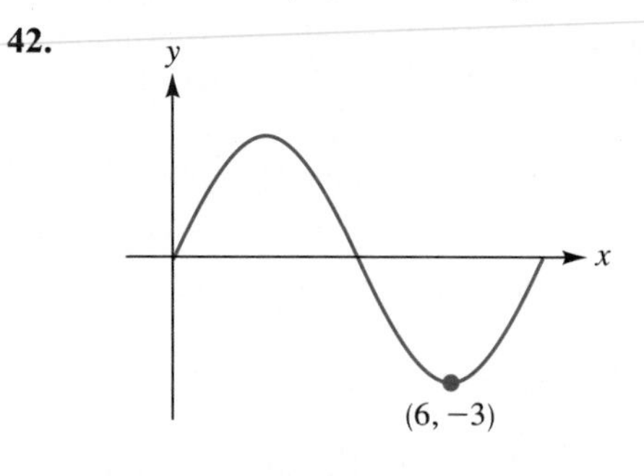

43.

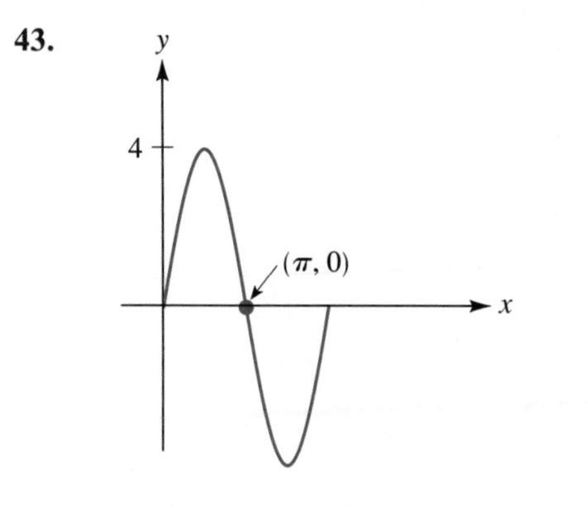

44.

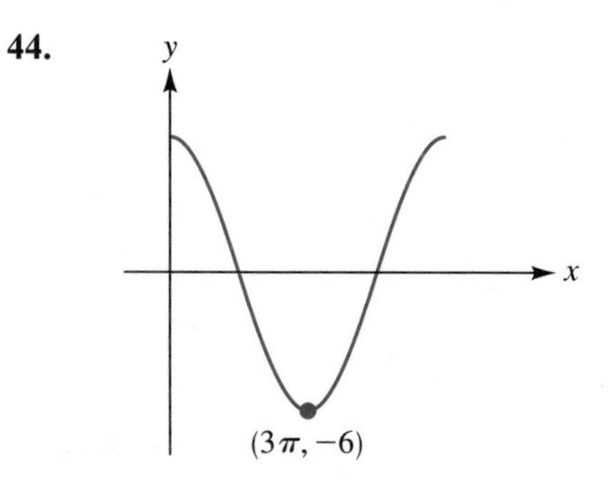

45.

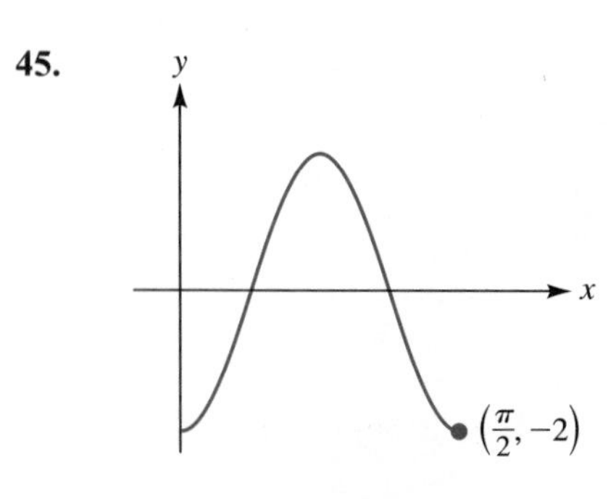

46.

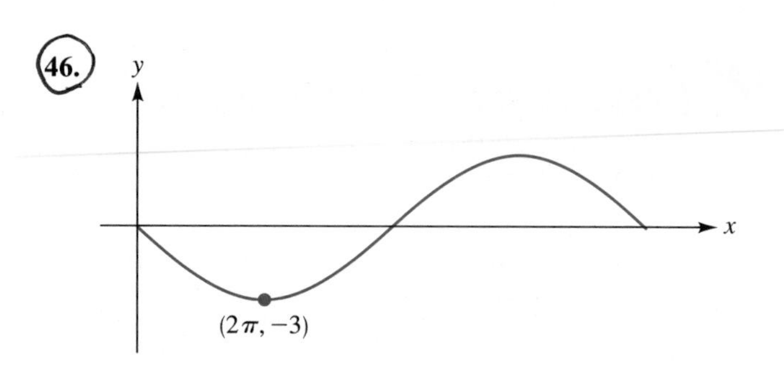

In Exercises 47–52, sketch the graph of each function for one complete cycle. In each case, specify the x-intercepts and the coordinates of the highest and lowest points on the graph.

47. $y = -3 \cos 4x$

48. $y = 2 \sin(3\pi x/4)$

49. $y = 2 \sin\left(\frac{\pi x}{2} - \frac{\pi}{4}\right)$

50. $y = -\sin(x - 1)$

51. $y = 3 \cos\left(\frac{\pi x}{3} - \frac{\pi}{3}\right)$

52. $y = -2 \cos(x + \pi)$

In Exercises 53–56, sketch the graph of each function for one period.

53. **(a)** $y = \tan(\pi x/4)$
(b) $y = \cot(\pi x/4)$

54. **(a)** $y = 2 \cot 2x$
(b) $y = 2 \tan 2x$

55. **(a)** $y = 3 \sec(x/4)$
(b) $y = 3 \csc(x/4)$

56. **(a)** $y = \sec \pi x$
(b) $y = \csc \pi x$

57. A mass on a tabletop is attached to a spring, as shown in the figure. The coordinate system has been chosen so that the equilibrium position of the mass corresponds to $s = 0$. Assume that the mass moves in simple harmonic motion described by the equation $s = -2.5 \cos(\pi t/8)$, where s is in cm and t is in seconds.

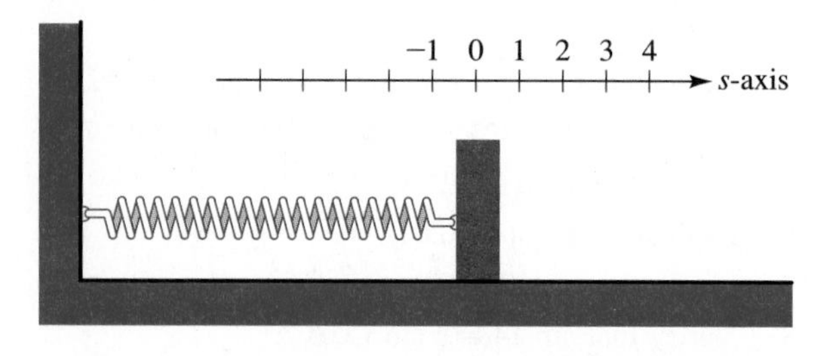

(a) Graph two complete cycles of the function $s = -2.5 \cos(\pi t/8)$ beginning at $t = 0$. Specify the amplitude, period, and frequency of the motion.
(b) Use the graph to determine the times in this interval at which the mass is farthest from the equilibrium position.
(c) When during this interval of time is the mass passing through the equilibrium position?

58. In the following figure, the arc is a portion of a circle with center O and radius 1. Use the figure to prove the following result: If α and β are acute angles (radian measure) with $\beta > \alpha$, then $\sin\beta - \sin\alpha < \beta - \alpha$. *Hint:* The area of the segment of the circle determined by chord $\overline{AB}$ is less than the area of the segment determined by chord $\overline{AC}$.

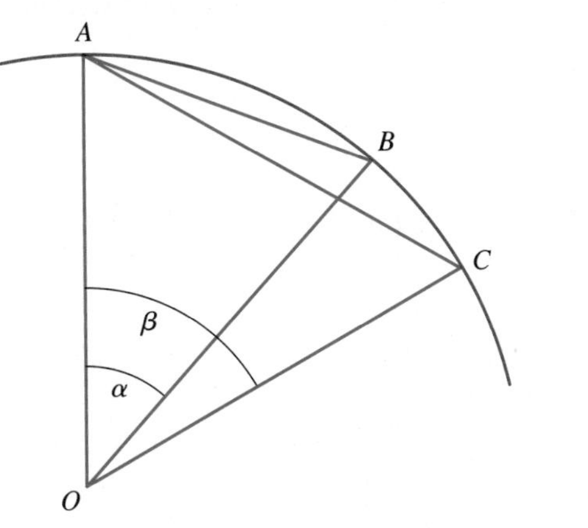

59. In the following figure, the arc is a portion of a circle with center O and radius 1, and $\angle OAE = \pi/2$. Use the figure to prove the following result: If α and β are acute angles (radian measure) with $\beta > \alpha$, then $\tan\beta - \tan\alpha > \beta - \alpha$. *Hint:* The area of region $ABDA$ is less than the area of region $ABCEDA$.

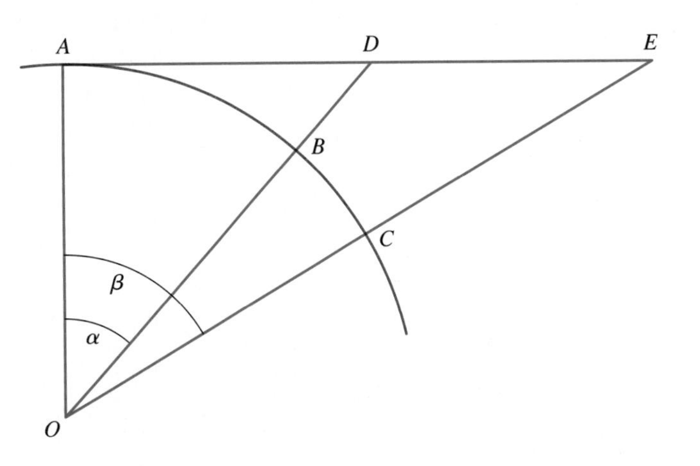

CHAPTER SEVEN TEST

1. Evaluate each expression without using a calculator or tables.
   (a) $\cos(4\pi/3)$ **(b)** $\csc(-5\pi/6)$ **(c)** $\sin^2(3\pi/4) + \cos^2(3\pi/4)$
2. In the expression $1/\sqrt{16 - t^2}$, make the substitution $t = 4\sin u$ and simplify the result. Assume that $0 < u < \frac{\pi}{2}$.
3. Graph the function $y = 0.5\sec(4\pi x - 1)$ for one complete cycle.
4. Graph the function $y = -\sin\left(3x - \frac{\pi}{4}\right)$ for one complete cycle. Specify the amplitude, period, and phase shift.
5. Graph the function $y = 3\tan(\pi x/4)$ on the interval $0 \leqslant x \leqslant 4$.
6. **(a)** Convert 175° to radian measure.
   (b) Convert 5 radians to degree measure.
7. Two points B and C are on a circle of radius 5 cm. The center of the circle is A, and angle BAC is 75°.
   (a) Find the length of the (shorter) arc of the circle from B to C.
   (b) Find the area of the (smaller) sector determined by angle BAC.
8. A wheel rotates about its axis with an angular speed of 25 revolutions/sec.
   (a) Find the angular speed of the wheel in radians/sec.
   (b) Find the linear speed of a point on the wheel that is 5 cm from the center.
9. Prove that the following equation is an identity:

$$\frac{\cot\theta}{1 + \tan(-\theta)} + \frac{\tan\theta}{1 + \cot(-\theta)} = \cot\theta + \tan\theta + 1$$

10. A point moves in simple harmonic motion along the x-axis. The x-coordinate of the point at time t is given by $x = 10\cos(\pi t/3)$, where t is in seconds and x is in centimeters.
    (a) Graph this function for two complete cycles, beginning at $t = 0$.
    (b) At what times during these two cycles is the point passing through the origin? At what times is the point farthest from the origin?
11. Evaluate each expression. (Show your work or supply reasons; *don't* use a calculator.)
    (a) $\sin^2 13 + \cos^2 13$
    (b) $\sin 5 + \sin(-5)$
    (c) $\tan 1 + \tan(-1 - 2\pi)$
12. The following figure shows the graph of $y = \sin x$ for $0 \leq x \leq 1.6$.

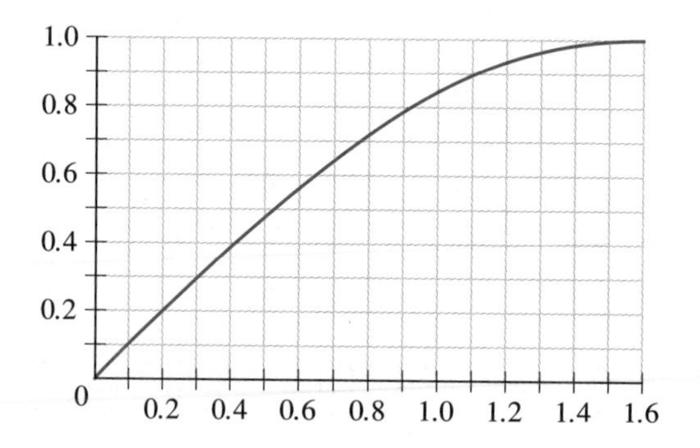

    (a) Use the graph to estimate, to the nearest tenth, a root of the equation $\sin x = 0.9$.
    (b) Use a calculator to obtain a more accurate value for the root in part (a). Round the answer to four decimal places.
    (c) Use the reference-angle concept and a calculator to find another root of the equation $\sin x = 0.9$ in the interval $0 \leq x \leq 2\pi$. Round the answer to four decimal places.

ANALYTICAL TRIGONOMETRY

INTRODUCTION

. . . through the improvements in algebraic symbolism . . . trigonometry became, in the 17th century, largely an analytic science, and as such it entered the field of higher mathematics.

David Eugene Smith in *History of Mathematics* (New York: Ginn and Company, 1925)

This chapter is devoted to some of the more algebraic (as opposed to geometric) portions of trigonometry. In Section 8.1 we develop six basic identities known as the *addition formulas*. Then, in the next two sections, we consider a number of identities that follow rather directly from these addition formulas. In Section 8.4 we return to a topic that was introduced briefly in the previous chapter: solving trigonometric equations. In solving many of these equations, we'll use the identities developed in the previous sections. We'll also make use of the inverse trigonometric functions that were introduced in the previous chapter (in Section 7.4). In the last section of this chapter, Section 8.5, we take a more careful look at the inverse trigonometric functions and their properties.

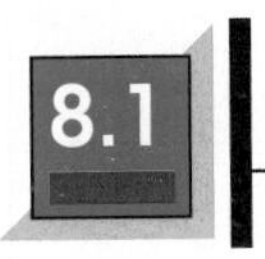

THE ADDITION FORMULAS

It has long been recognized that the addition formulas are the heart of trigonometry. Indeed, Professor Rademacher and others have shown that the entire body of trigonometry can be derived from the assumption that there exist functions S and C such that

1. $S(x-y) = S(x)C(y) - C(x)S(y)$
2. $C(x-y) = C(x)C(y) + S(x)S(y)$
3. $\lim_{x\to 0^+} \frac{S(x)}{x} = 1$

Professor Frederick H. Young in his article, "The Addition Formulas" from *The Mathematics Teacher,* vol. L (1957), pp. 45–48.

For any real numbers r, s, and t, it is always true that $r(s+t) = rs + rt$. This is the so called **distributive law** for real numbers. If f is a function, however, it is not true in general that $f(s+t) = f(s) + f(t)$. For example, consider the cosine function. It is not true in general that $\cos(s+t) = \cos s + \cos t$. For instance, with $s = \pi/6$ and $t = \pi/3$, we have

$$\cos\left(\frac{\pi}{6} + \frac{\pi}{3}\right) \stackrel{?}{=} \cos\frac{\pi}{6} + \cos\frac{\pi}{3}$$

$$\cos\frac{\pi}{2} \stackrel{?}{=} \cos\frac{\pi}{6} + \cos\frac{\pi}{3}$$

$$0 \stackrel{?}{=} \frac{\sqrt{3}}{2} + \frac{1}{2} \qquad \text{No!}$$

In this section, we will see just what $\cos(s+t)$ does equal. The correct formula for $\cos(s+t)$ is one of a group of important trigonometric identities called the **addition formulas.** We begin with the four addition formulas for sine and cosine.

THE ADDITION FORMULAS FOR SINE AND COSINE

$$\sin(s+t) = \sin s\ \cos t + \cos s\ \sin t$$
$$\sin(s-t) = \sin s\ \cos t - \cos s\ \sin t$$
$$\cos(s+t) = \cos s\ \cos t - \sin s\ \sin t$$
$$\cos(s-t) = \cos s\ \cos t + \sin s\ \sin t$$

Our strategy for deriving these formulas will be as follows. First we'll prove the fourth formula—this takes some effort. The other three formulas are then relatively easy to derive from the fourth one. As background for our discussion, you will need to recall that the distance d between two points (x_1, y_1) and (x_2, y_2) is given by $d = \sqrt{(x_2 - x_1)^2 + (y_2 - y_1)^2}$.

To prove the fourth formula in the box, we use Figure 1. The idea behind the proof is as follows.* We begin in Figure 1(a) with the unit circle and the angles s, t, and $s - t$. Then we rotate $\triangle OPQ$ about the origin until the point P coincides with the point $(1, 0)$, as indicated in Figure 1(b). Although this rotation changes the coordinates for the points P and Q, it certainly has no effect upon the length of the line segment $\overline{PQ}$. Thus, whether we calculate PQ using the coordinates in Figure 1(a) or those in Figure 1(b), the results must be the same. As you'll see, by equating the two expressions for PQ, we will obtain the required formula.

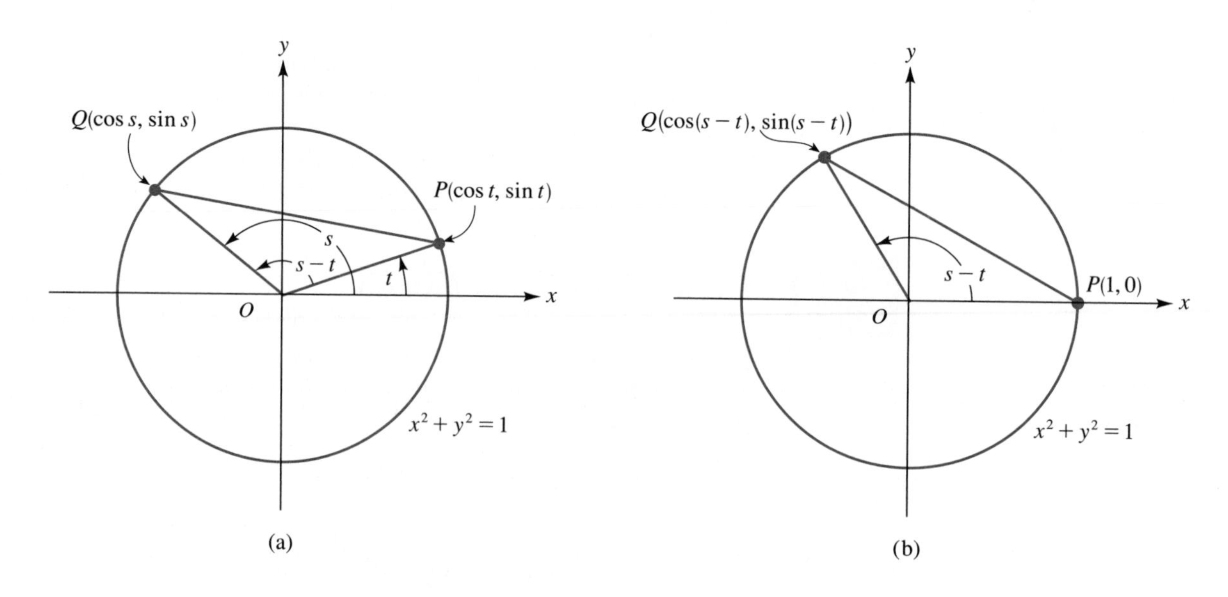

FIGURE 1

Applying the distance formula in Figure 1(a), we have

$$\begin{aligned} PQ &= \sqrt{(\cos s - \cos t)^2 + (\sin s - \sin t)^2} \\ &= \sqrt{\cos^2 s - 2\cos s\cos t + \cos^2 t + \sin^2 s - 2\sin s\sin t + \sin^2 t} \\ &= \sqrt{2 - 2\cos s\cos t - 2\sin s\sin t} \qquad \text{(Why?)} \qquad (1) \end{aligned}$$

Next, applying the distance formula in Figure 1(b), we have

$$\begin{aligned} PQ &= \sqrt{[\cos(s-t) - 1]^2 + [\sin(s-t) - 0]^2} \\ &= \sqrt{\cos^2(s-t) - 2\cos(s-t) + 1 + \sin^2(s-t)} \end{aligned}$$

The right-hand side of this last equation can be simplified using the fact that $\cos^2(s - t) + \sin^2(s - t) = 1$. We then have

$$PQ = \sqrt{2 - 2\cos(s-t)} \qquad (2)$$

*This idea for the proof can be traced back to the great French mathematician Augustin-Louis Cauchy (1789–1857).

From equations (1) and (2), we conclude that

$$\begin{aligned} 2 - 2\cos(s - t) &= 2 - 2\cos s \cos t - 2\sin s \sin t \\ -2\cos(s - t) &= -2\cos s \cos t - 2\sin s \sin t \\ \cos(s - t) &= \cos s \cos t + \sin s \sin t \end{aligned}$$

This completes the proof of the fourth addition formula. Before deriving the other three addition formulas, let's look at some applications of this result.

EXAMPLE 1 Simplify the expression $\cos 2\theta \cos \theta + \sin 2\theta \sin \theta$.

Solution According to the identity that we just proved, we have

$$\cos 2\theta \cos \theta + \sin 2\theta \sin \theta = \cos(2\theta - \theta) = \cos \theta$$

Thus, the given expression is equal to $\cos \theta$.

EXAMPLE 2 Simplify $\cos(\theta - \pi)$.

Solution We use the formula for $\cos(s - t)$ with s and t replaced by θ and π, respectively. This yields

$$\cos(s - t) = \cos s \cos t + \sin s \sin t$$

$$\updownarrow \quad \updownarrow$$

$$\begin{aligned} \cos(\theta - \pi) &= \cos \theta \cos \pi + \sin \theta \sin \pi \\ &= (\cos \theta)(-1) + (\sin \theta)(0) = -\cos \theta \end{aligned}$$

Thus, the required simplification is $\cos(\theta - \pi) = -\cos \theta$.

In Example 2, we found that $\cos(\theta - \pi) = -\cos \theta$. This type of identity is often referred to as a **reduction formula.** The next example develops two basic reduction formulas that we will need to use later in this section.

EXAMPLE 3 Prove the following identities:

(a) $\cos\left(\frac{\pi}{2} - \alpha\right) = \sin \alpha$; **(b)** $\sin\left(\frac{\pi}{2} - \beta\right) = \cos \beta$.

Solution

(a) $$\begin{aligned} \cos\left(\tfrac{\pi}{2} - \alpha\right) &= \cos \tfrac{\pi}{2} \cos \alpha + \sin \tfrac{\pi}{2} \sin \alpha \\ &= (0)\cos \alpha + (1)\sin \alpha = \sin \alpha \end{aligned}$$

This proves the identity. Incidentally, if we use degree measure instead of radian measure, this identity states that

$$\cos(90^\circ - \alpha) = \sin \alpha$$

as we saw earlier, in Section 6.2.

(b) Since the identity $\cos\left(\frac{\pi}{2} - \alpha\right) = \sin \alpha$ holds for all values of α, we can simply replace α by the quantity $\frac{\pi}{2} - \beta$ to obtain

$$\begin{aligned} \cos\left[\tfrac{\pi}{2} - \left(\tfrac{\pi}{2} - \beta\right)\right] &= \sin\left(\tfrac{\pi}{2} - \beta\right) \\ \cos\left(\tfrac{\pi}{2} - \tfrac{\pi}{2} + \beta\right) &= \sin\left(\tfrac{\pi}{2} - \beta\right) \\ \cos \beta &= \sin\left(\tfrac{\pi}{2} - \beta\right) \end{aligned}$$

This proves the identity.

The two identities in Example 3 are worth memorizing. For simplicity, we replace both α and β by the single θ to get

$$\cos\left(\frac{\pi}{2} - \theta\right) = \sin\theta \quad \text{and} \quad \sin\left(\frac{\pi}{2} - \theta\right) = \cos\theta$$

In Examples 2 and 3, radian measure was used. However, if you look back at Figure 1 and the derivation of the formula for $\cos(s - t)$, you will see that the derivation makes no reference, implicit or explicit, to a specific system of angle measurement. (For instance, the derivation does not involve the formula $s = r\theta$, which does require radian measure.) Thus, the formula for $\cos(s - t)$ is also valid when angles are measured in degrees. In the next example, we apply the formula in just such a case.

EXAMPLE 4 Use the formula for $\cos(s - t)$ to determine the exact value of $\cos 15°$.

Solution First observe that $15° = 45° - 30°$. Then we have

$$\begin{aligned}\cos 15° &= \cos(45° - 30°)\\ &= \cos 45° \cos 30° + \sin 45° \sin 30° \qquad \text{using the formula for } \cos(s-t) \text{ with } s = 45° \text{ and } t = 30°\\ &= \left(\frac{\sqrt{2}}{2}\right)\left(\frac{\sqrt{3}}{2}\right) + \left(\frac{\sqrt{2}}{2}\right)\left(\frac{1}{2}\right)\\ &= \frac{\sqrt{6}}{4} + \frac{\sqrt{2}}{4} = \frac{\sqrt{6}+\sqrt{2}}{4}\end{aligned}$$

Thus, the exact value of $\cos 15°$ is $(\sqrt{6} + \sqrt{2})/4$.

Now let's return to our derivations of the addition formulas. Using the fourth addition formula, we can easily derive the third formula as follows. In the formula

$$\cos(s - t) = \cos s \cos t + \sin s \sin t$$

we replace t by the quantity $-t$. This is permissible because the formula holds for all real numbers. We obtain

$$\cos[s - (-t)] = \cos s \cos(-t) + \sin s \sin(-t)$$

On the right-hand side of this equation, we can use the identities developed for $\cos(-t)$ and $\sin(-t)$ in Section 7.3. Doing this yields

$$\cos(s + t) = (\cos s)(\cos t) + (\sin s)(-\sin t)$$

which is equivalent to

$$\cos(s + t) = \cos s \cos t - \sin s \sin t$$

This is the third addition formula, as we wished to prove.

Next we derive the formula for $\sin(s + t)$. We have

$$\begin{aligned}\sin(s+t) &= \cos\left[\tfrac{\pi}{2} - (s+t)\right] && \text{replacing } \theta \text{ by } s+t \text{ in the identity } \sin\theta = \cos\left(\tfrac{\pi}{2} - \theta\right)\\ &= \cos\left[\left(\tfrac{\pi}{2} - s\right) - t\right]\\ &= \cos\left(\tfrac{\pi}{2} - s\right)\cos t + \sin\left(\tfrac{\pi}{2} - s\right)\sin t && \text{(Why?)}\\ &= \sin s\cos t + \cos s \sin t && \text{(Why?)}\end{aligned}$$

This proves the first addition formula.

Finally, we can use the first addition formula to prove the second one as follows:

$$\begin{aligned}\sin(s-t) &= \sin[s + (-t)]\\ &= \sin s\cos(-t) + \cos s\sin(-t)\\ &= (\sin s)(\cos t) + (\cos s)(-\sin t)\\ &= \sin s\cos t - \cos s\sin t\end{aligned}$$

This completes the proofs of the four addition formulas for sine and cosine.

EXAMPLE 5 If $\sin s = 3/5$ $\left(0 < s < \frac{\pi}{2}\right)$ and $\sin t = -\sqrt{3}/4$ $\left(\pi < t < \frac{3\pi}{2}\right)$, compute $\sin(s - t)$.

Solution $\quad \sin(s-t) = \underbrace{\sin s}_{\text{given as } 3/5} \cos t - \cos s \underbrace{\sin t}_{\text{given as } -\sqrt{3}/4} \qquad (3)$

In view of equation (3), we need to find only $\cos t$ and $\cos s$. These can be determined using the Pythagorean identity $\cos^2\theta = 1 - \sin^2\theta$. We have

$$\begin{aligned}\cos^2 t &= 1 - \sin^2 t\\ &= 1 - \left(\frac{-\sqrt{3}}{4}\right)^2 = 1 - \frac{3}{16} = \frac{13}{16}\end{aligned}$$

Therefore,

$$\cos t = \sqrt{13}/4 \quad \text{or} \quad \cos t = -\sqrt{13}/4$$

We choose the negative value here for cosine, since it is given that $\pi < t < \frac{3\pi}{2}$. Thus

$$\cos t = -\sqrt{13}/4$$

Similarly, to find $\cos s$, we have

$$\cos^2 s = 1 - \sin^2 s = 1 - \left(\frac{3}{5}\right)^2 = \frac{16}{25}$$

Therefore,

$$\cos s = 4/5 \qquad \cos s \text{ is positive, since } 0 < s < \tfrac{\pi}{2}$$

Finally, we substitute the values we've obtained for $\cos t$ and $\cos s$, along with the given data, back into equation (3). This yields

$$\sin(s-t) = \left(\frac{3}{5}\right)\left(\frac{-\sqrt{13}}{4}\right) - \left(\frac{4}{5}\right)\left(\frac{-\sqrt{3}}{4}\right)$$

$$= \frac{-3\sqrt{13}}{20} + \frac{4\sqrt{3}}{20} = \frac{-3\sqrt{13}+4\sqrt{3}}{20}$$

In the box that follows, we list the two addition formulas for the tangent function. As you'll see, these two formulas follow directly from the addition formulas for sine and cosine.

ADDITION FORMULAS FOR TANGENT

$$\tan(s+t) = \frac{\tan s + \tan t}{1 - \tan s \tan t}$$

$$\tan(s-t) = \frac{\tan s - \tan t}{1 + \tan s \tan t}$$

To prove the formula for $\tan(s+t)$, we begin with

$$\tan(s+t) = \frac{\sin(s+t)}{\cos(s+t)}$$

$$= \frac{\sin s \cos t + \cos s \sin t}{\cos s \cos t - \sin s \sin t} \tag{4}$$

Now we divide both numerator and denominator on the right-hand side of equation (4) by the quantity $\cos s \cos t$. This yields

$$\tan(s+t) = \frac{\dfrac{\sin s \cancel{\cos t}}{\cos s \cancel{\cos t}} + \dfrac{\cancel{\cos s}\sin t}{\cancel{\cos s}\cos t}}{\dfrac{\cancel{\cos s}\,\cancel{\cos t}}{\cancel{\cos s}\,\cancel{\cos t}} - \dfrac{\sin s \sin t}{\cos s \cos t}} = \frac{\tan s + \tan t}{1 - \tan s \tan t}$$

This proves the formula for $\tan(s+t)$. The formula for $\tan(s-t)$ can be deduced from this with the aid of the identity $\tan(-t) = -\tan t$, which was derived in Section 7.3. We have

$$\tan(s-t) = \tan[s+(-t)]$$

$$= \frac{\tan s + \tan(-t)}{1 - \tan s \tan(-t)} = \frac{\tan s + (-\tan t)}{1 - (\tan s)(-\tan t)}$$

$$= \frac{\tan s - \tan t}{1 + \tan s \tan t} \quad \text{as required}$$

EXAMPLE 6 Simplify the expression $\dfrac{\tan\frac{\pi}{9} + \tan\frac{2\pi}{9}}{1 - \tan\frac{\pi}{9}\tan\frac{2\pi}{9}}$.

Solution

$$\frac{\tan\frac{\pi}{9}+\tan\frac{2\pi}{9}}{1-\tan\frac{\pi}{9}\tan\frac{2\pi}{9}}=\tan\left(\frac{\pi}{9}+\frac{2\pi}{9}\right) \quad \text{using the formula for } \tan(s+t)$$

$$=\tan\frac{3\pi}{9}=\tan\frac{\pi}{3}=\sqrt{3}$$

EXAMPLE 7 Compute $\tan\frac{\pi}{12}$, using the fact that $\frac{\pi}{12}=\frac{\pi}{3}-\frac{\pi}{4}$.

Solution

$$\tan\frac{\pi}{12}=\tan\left(\frac{\pi}{3}-\frac{\pi}{4}\right)$$

$$=\frac{\tan\frac{\pi}{3}-\tan\frac{\pi}{4}}{1+\tan\frac{\pi}{3}\tan\frac{\pi}{4}} \quad \text{using the formula for } \tan(s-t) \text{ with } s=\pi/3 \text{ and } t=\pi/4$$

$$=\frac{\sqrt{3}-1}{1+\sqrt{3}(1)}$$

So, $\tan\frac{\pi}{12}=\frac{\sqrt{3}-1}{\sqrt{3}+1}$. We can write this answer in a more compact form by rationalizing the denominator. As you can check, the result is $\tan\frac{\pi}{12}=2-\sqrt{3}$.

EXERCISE SET 8.1

Do 2, 7, 11, 17, 25, 29, 41

A

In Exercises 1–10, use the addition formulas for sine and cosine to simplify the expression.

1. $\sin\theta\cos 2\theta+\cos\theta\sin 2\theta$
2. $\sin\frac{\pi}{6}\cos\frac{\pi}{3}+\cos\frac{\pi}{6}\sin\frac{\pi}{3}$
3. $\sin 3\theta\cos\theta-\cos 3\theta\sin\theta$
4. $\sin 110^\circ\cos 20^\circ-\cos 110^\circ\sin 20^\circ$
5. $\cos 2u\cos 3u-\sin 2u\sin 3u$
6. $\cos 2u\cos 3u+\sin 2u\sin 3u$
7. $\cos\frac{2\pi}{9}\cos\frac{\pi}{18}+\sin\frac{2\pi}{9}\sin\frac{\pi}{18}$
8. $\cos\frac{3\pi}{10}\cos\frac{\pi}{5}-\sin\frac{3\pi}{10}\sin\frac{\pi}{5}$
9. $\sin(A+B)\cos A-\cos(A+B)\sin A$
10. $\cos(s-t)\cos t-\sin(s-t)\sin t$

In Exercises 11–14, simplify each expression (as in Example 2).

11. $\sin\left(\theta-\frac{3\pi}{2}\right)$
12. $\cos\left(\frac{3\pi}{2}+\theta\right)$
13. $\cos(\theta+\pi)$
14. $\sin(\theta-\pi)$
15. Expand $\sin(t+2\pi)$ using the appropriate addition formula, and check to see that your answer agrees with the formula in the first box on page 434.
16. Follow the directions in Exercise 15, but use $\cos(t+2\pi)$.
17. Use the formula for $\cos(s+t)$ to compute the exact value of $\cos 75^\circ$.
18. Use the formula for $\sin(s-t)$ to compute the exact value of $\sin\frac{\pi}{12}$.
19. Use the formula for $\sin(s+t)$ to find $\sin\frac{7\pi}{12}$.
20. Determine the exact value of **(a)** $\sin 105^\circ$ and **(b)** $\cos 105^\circ$.

In Exercises 21–24, use the addition formulas for sine and cosine to simplify each expression.

21. $\sin\left(\frac{\pi}{4}+s\right)-\sin\left(\frac{\pi}{4}-s\right)$
22. $\sin\left(t+\frac{\pi}{6}\right)-\sin\left(t-\frac{\pi}{6}\right)$

23. $\cos\left(\frac{\pi}{3} - \theta\right) - \cos\left(\frac{\pi}{3} + \theta\right)$

24. $\cos\left(\theta - \frac{\pi}{4}\right) + \cos\left(\theta + \frac{\pi}{4}\right)$

In Exercises 25–28, compute the indicated quantity using the following data.

$$\sin\alpha = \frac{12}{13}, \text{ where } \frac{\pi}{2} < \alpha < \pi$$

$$\cos\beta = -\frac{3}{5}, \text{ where } \pi < \beta < \frac{3\pi}{2}$$

$$\cos\theta = \frac{7}{25}, \text{ where } -2\pi < \theta < -\frac{3\pi}{2}$$

25. **(a)** $\sin(\alpha+\beta)$ **(b)** $\cos(\alpha+\beta)$

26. **(a)** $\sin(\alpha-\beta)$ **(b)** $\cos(\alpha-\beta)$

27. **(a)** $\sin(\theta-\beta)$ **(b)** $\sin(\theta+\beta)$

28. **(a)** $\cos(\alpha+\theta)$ **(b)** $\cos(\alpha-\theta)$

29. Suppose that $\sin\theta = \frac{1}{5}$ and $0 < \theta < \frac{\pi}{2}$.
(a) Compute $\cos\theta$.
(b) Compute $\sin 2\theta$. *Hint:* $\sin 2\theta = \sin(\theta+\theta)$

30. Suppose that $\cos\theta = \frac{12}{13}$ and $\frac{3\pi}{2} < \theta < 2\pi$.
(a) Compute $\sin\theta$.
(b) Compute $\cos 2\theta$. *Hint:* $\cos 2\theta = \cos(\theta+\theta)$.

31. Given $\tan\theta = -\frac{2}{3}$, where $\frac{\pi}{2} < \theta < \pi$, and $\csc\beta = 2$, where $0 < \beta < \frac{\pi}{2}$, find $\sin(\theta+\beta)$ and $\cos(\beta-\theta)$.

32. Given $\sec s = \frac{5}{4}$, where $\sin s < 0$, and $\cot t = -1$, where $\frac{\pi}{2} < t < \pi$, find $\sin(s-t)$ and $\cos(s+t)$.

In Exercises 33–36, prove that each equation is an identity.

33. $\sin\left(t + \frac{\pi}{4}\right) = (\sin t + \cos t)/\sqrt{2}$

34. $\cos\left(t + \frac{\pi}{4}\right) = (\cos t - \sin t)/\sqrt{2}$

35. $\sin\left(t + \frac{\pi}{4}\right) + \cos\left(t + \frac{\pi}{4}\right) = \sqrt{2}\cos t$

36. $\sec(\alpha+\beta) = \dfrac{\sec\alpha\sec\beta}{1 - \tan\alpha\tan\beta}$

In Exercises 37–40, use the given information to compute $\tan(s+t)$ *and* $\tan(s-t)$.

37. $\tan s = 2$ and $\tan t = 3$

38. $\tan s = \frac{1}{2}$ and $\tan t = \frac{1}{3}$

39. $s = \frac{3\pi}{4}$ and $\tan t = -4$

40. $s = \frac{7\pi}{4}$ and $\tan t = -2$

In Exercises 41–46, use the addition formulas for tangent to simplify each expression.

41. $\dfrac{\tan t + \tan 2t}{1 - \tan t \tan 2t}$

42. $\dfrac{\tan\frac{\pi}{5} - \tan\frac{\pi}{30}}{1 + \tan\frac{\pi}{5}\tan\frac{\pi}{30}}$

43. $\dfrac{\tan 70^\circ - \tan 10^\circ}{1 + \tan 70^\circ \tan 10^\circ}$

44. $\dfrac{2\tan\frac{\pi}{12}}{1 - \tan^2\frac{\pi}{12}}$

45. $\dfrac{\tan(x-y) + \tan y}{1 - \tan(x-y)\tan y}$

46. $[\tan(\theta+\pi)][\tan(\theta-\pi)] + 1$

47. Compute $\tan\frac{7\pi}{12}$ and rationalize the answer.
Hint: $\frac{7\pi}{12} = \frac{\pi}{3} + \frac{\pi}{4}$

48. Compute $\tan 15^\circ$ using the fact that $15^\circ = 45^\circ - 30^\circ$. Then check that your answer is consistent with the result in Example 7.

B

In Exercises 49–58, prove that each equation is an identity.

49. $\dfrac{\sin(s+t)}{\cos s \cos t} = \tan s + \tan t$

50. $\dfrac{\cos(s-t)}{\cos s \sin t} = \cot t + \tan s$

51. $\cos(A-B) - \cos(A+B) = 2\sin A \sin B$

52. $\sin(A-B) + \sin(A+B) = 2\sin A\cos B$

53. $\cos(A+B)\cos(A-B) = \cos^2 A - \sin^2 B$

54. $\sin(A+B)\sin(A-B) = \cos^2 B - \cos^2 A$

55. $\cos(\alpha+\beta)\cos\beta + \sin(\alpha+\beta)\sin\beta = \cos\alpha$

56. $\cos\left(\theta + \frac{\pi}{4}\right) + \sin\left(\theta - \frac{\pi}{4}\right) = 0$

57. $\tan 2\theta = \dfrac{2\tan\theta}{1 - \tan^2\theta}$

58. $\tan\left(\frac{\pi}{4} + \theta\right) - \tan\left(\frac{\pi}{4} - \theta\right) = 2\tan 2\theta$
Hint: Use the addition formulas for tangent and the result in Exercise 57.

In Exercises 59–61, you are asked to derive expressions for the average rates of change of the functions $\sin x$, $\cos x$, *and* $\tan x$. *In each case, assume that the interval is* $[x, x+h]$. *(The results are used in calculus in the study of* derivatives.*)*

59. Let $f(x) = \sin x$. Show that

$$\frac{\Delta f}{\Delta x} = (\sin x)\left(\frac{\cos h - 1}{h}\right) + (\cos x)\left(\frac{\sin h}{h}\right)$$

60. Let $g(x) = \cos x$. Show that

$$\frac{\Delta g}{\Delta x} = (\cos x)\left(\frac{\cos h - 1}{h}\right) - (\sin x)\left(\frac{\sin h}{h}\right)$$

61. Let $T(x) = \tan x$. Show that

$$\frac{\Delta T}{\Delta x} = \frac{\tan h}{h} \cdot \frac{\sec^2 x}{1 - \tan h \tan x}$$

62. Let θ be the acute angle defined by the following figure.

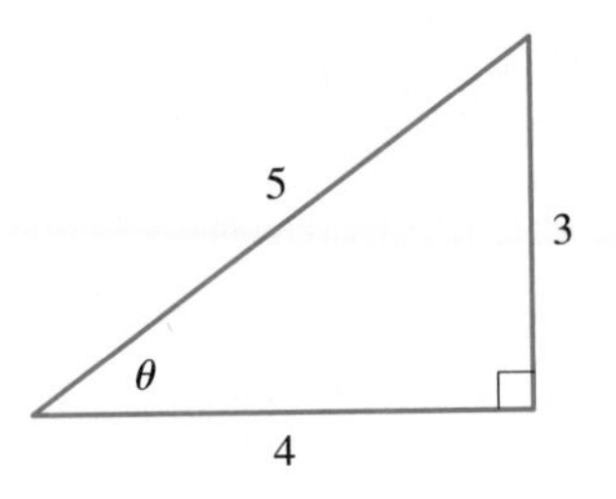

Use an addition formula and the figure to show that $5\sin(x+\theta) = 4\sin x + 3\cos x$.

63. Let a and b be positive constants, and let θ be the acute angle (in radian measure) defined by the following figure.

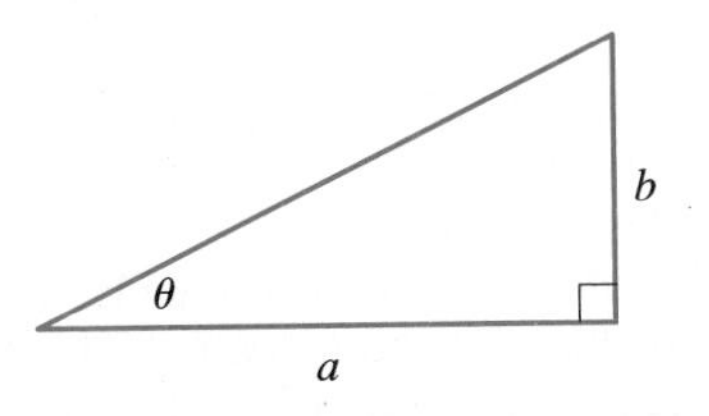

(a) Use an addition formula and the figure to show that $\sqrt{a^2+b^2}\sin(x+\theta) = a\sin x + b\cos x$.
(b) Use the result in part (a) to specify the maximum value of the function $f(x) = a\sin x + b\cos x$.

64. **(a)** Use an addition formula to show that $2\sin\left(x+\frac{\pi}{6}\right) = \cos x + \sqrt{3}\sin x$.
(b) Use the result in part (a) to graph the function $f(x) = \cos x + \sqrt{3}\sin x$ for one period.

65. **(a)** Use an addition formula to show that $\sqrt{2}\cos\left(x-\frac{\pi}{4}\right) = \cos x + \sin x$.
(b) Use the result in part (a) to graph the function $f(x) = \cos x + \sin x$ for one period.

66. Let A, B, and C be the angles of a triangle, so that $A+B+C=\pi$.
(a) Show that $\sin(A+B) = \sin C$.
(b) Show that $\cos(A+B) = -\cos C$.
(c) Show that $\tan(A+B) = -\tan C$.

67. Suppose that A, B, and C are the angles of a triangle, so that $A+B+C=\pi$. Show that

$$\cos^2 A + \cos^2 B + \cos^2 C + 2\cos A\cos B\cos C = 1$$

68. Prove that

$$\frac{\sin(\alpha-\beta)}{\cos\alpha\cos\beta} + \frac{\sin(\beta-\gamma)}{\cos\beta\cos\gamma} + \frac{\sin(\gamma-\alpha)}{\cos\gamma\cos\alpha} = 0$$

69. Suppose that $a^2+b^2=1$ and $c^2+d^2=1$. Prove that $|ac+bd| \le 1$. *Hint:* Let $a=\cos\theta$, $b=\sin\theta$, $c=\cos\phi$, and $d=\sin\phi$.

In Exercises 70–72, simplify the expression.

70. $\cos\left(\frac{\pi}{6}+t\right)\cos\left(\frac{\pi}{6}-t\right) - \sin\left(\frac{\pi}{6}+t\right)\sin\left(\frac{\pi}{6}-t\right)$
Hint: If your solution relies on four separate addition formulas, then you are doing this the hard way.

71. $\sin\left(\frac{\pi}{3}-t\right)\cos\left(\frac{\pi}{3}+t\right) + \cos\left(\frac{\pi}{3}-t\right)\sin\left(\frac{\pi}{3}+t\right)$

72. $\dfrac{\tan(A+2B) - \tan(A-2B)}{1+\tan(A+2B)\tan(A-2B)}$

73. If $\alpha+\beta=\frac{\pi}{4}$, show that $(1+\tan\alpha)(1+\tan\beta)=2$.

Exercises 74 and 75 outline simple geometric derivations of the formulas for $\sin(\alpha+\beta)$ and $\cos(\alpha+\beta)$ in the case where α and β are acute angles, with $\alpha+\beta<90°$. The exercises rely on the accompanying figures, which are constructed as follows. Begin, in Figure A, with $\alpha = \angle GAD$, $\beta = \angle HAG$, and $AH = 1$. Then, from H, draw perpendiculars to $\overline{AD}$ and to $\overline{AG}$, as shown in Figure B. Finally, draw $\overline{FE} \perp \overline{BH}$ and $\overline{FC} \perp \overline{AD}$.

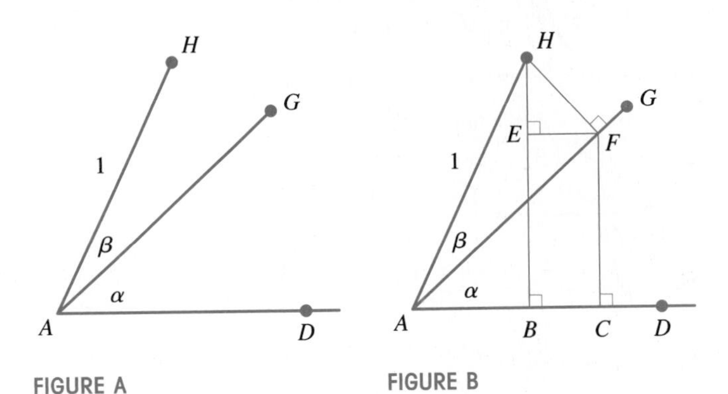

FIGURE A FIGURE B

74. *Formula for* $\sin(\alpha+\beta)$. Supply the reasons or steps behind each statement.
(a) $BH = \sin(\alpha+\beta)$ **(b)** $FH = \sin\beta$
(c) $\angle BHF = \alpha$
(d) $EH = \cos\alpha\sin\beta$
Hint: Use $\triangle EFH$ and the result in part (b).
(e) $AF = \cos\beta$ **(f)** $CF = \sin\alpha\cos\beta$
(g) $\sin(\alpha+\beta) = \sin\alpha\cos\beta + \cos\alpha\sin\beta$
Hint: $\sin(\alpha+\beta) = BH = EH + CF$

75. *Formula for* $\cos(\alpha+\beta)$. Supply the reasons or steps behind each statement.
(a) $\cos(\alpha+\beta) = AB$
(b) $AC = \cos\alpha\cos\beta$
Hint: Use $\triangle ACF$ and the result in Exercise 74(e).
(c) $EF = \sin\alpha\sin\beta$
(d) $\cos(\alpha+\beta) = \cos\alpha\cos\beta - \sin\alpha\sin\beta$
Hint: $AB = AC - BC$

76. Let S and C be two functions. Assume that the domain for both S and C is the set of all real numbers, and that

S and C satisfy the following two identities.

$$S(x-y) = S(x)C(y) - C(x)S(y) \quad (1)$$
$$C(x-y) = C(x)C(y) + S(x)S(y) \quad (2)$$

Also, suppose that the function S is not identically zero. That is,

$$S(x) \neq 0 \quad \text{for at least one real number } x \quad (3)$$

(a) Show that $S(0) = 0$. *Hint:* In identity (1), let $x = y$.
(b) Show that $C(0) = 1$. *Hint:* In identity (1), let $y = 0$.
(c) Explain (in complete sentences) why it was necessary to use condition (3) in the work for part (b).
(d) Prove the identity $[C(x)]^2 + [S(x)]^2 = 1$. *Hint:* In identity (2), let $y = x$.
(e) Show that C is an even function and S is an odd function. That is, prove the identities $C(-x) = C(x)$ and $S(-x) = -S(x)$. *Hint:* Write $-x$ as $0 - x$.

In Exercises 77–80, prove the identities.

77. $\dfrac{\sin(A+B)}{\sin(A-B)} = \dfrac{\tan A + \tan B}{\tan A - \tan B}$

78. $\dfrac{\cos(A+B)}{\cos(A-B)} = \dfrac{1 - \tan A \tan B}{1 + \tan A \tan B}$

79. $\cot(A+B) = \dfrac{\cot A \cot B - 1}{\cot A + \cot B}$

80. $\cot(A-B) = \dfrac{\cot A \cot B + 1}{\cot B - \cot A}$

81. Let $f(t) = \cos^2 t + \cos^2\left(t + \frac{2\pi}{3}\right) + \cos^2\left(t - \frac{2\pi}{3}\right)$.
(a) Complete the table. (Use a calculator.)

t	1	2	3	4
$f(t)$				

(b) On the basis of your results in part (a), make a conjecture about the function f. Prove that your conjecture is correct.

C

82. **(a)** Use your calculator to check that $\tan 50° - \tan 40° = 2 \tan 10°$.
(b) Prove that the equation in part (a) is indeed an identity.

83. If $\tan B = \dfrac{n \sin A \cos A}{1 - n \sin^2 A}$, show that $\tan(A-B) = (1-n)\tan A$.

84. If triangle ABC is not a right triangle, and $\cos A = \cos B \cos C$, show that $\tan B \tan C = 2$.

85. **(a)** The angles of a triangle are $A = 20°$, $B = 50°$, and $C = 110°$. Use your calculator to compute the sum $\tan A + \tan B + \tan C$ and then the product $\tan A \tan B \tan C$. What do you observe?
(b) The angles of a triangle are $\alpha = \pi/10$, $\beta = 3\pi/10$, and $\gamma = 3\pi/5$. Use your calculator to compute $\tan \alpha + \tan \beta + \tan \gamma$ and $\tan \alpha \tan \beta \tan \gamma$.
(c) If triangle ABC is not a right triangle, prove that $\tan A + \tan B + \tan C = \tan A \tan B \tan C$.

8.2 THE DOUBLE-ANGLE FORMULAS

Ptolemy (c. 150) knew substantially the sine of half an angle . . . and it is probable that Hipparchus (c. 140 B.C.) and certain that Varahamihira (c. 505) knew the relation that we express as $\sin \frac{\phi}{2} = \sqrt{(1 - \cos \phi)/2}$.

David Eugene Smith in *History of Mathematics* (New York: Ginn and Company, 1925)

There are a number of basic identities that follow from the addition formulas. In the following two boxes, we summarize some of the most useful of these.

THE DOUBLE-ANGLE FORMULAS

a. $\sin 2\theta = 2 \sin \theta \cos \theta$

b. $\cos 2\theta = \cos^2 \theta - \sin^2 \theta$

c. $\tan 2\theta = \dfrac{2 \tan \theta}{1 - \tan^2 \theta}$

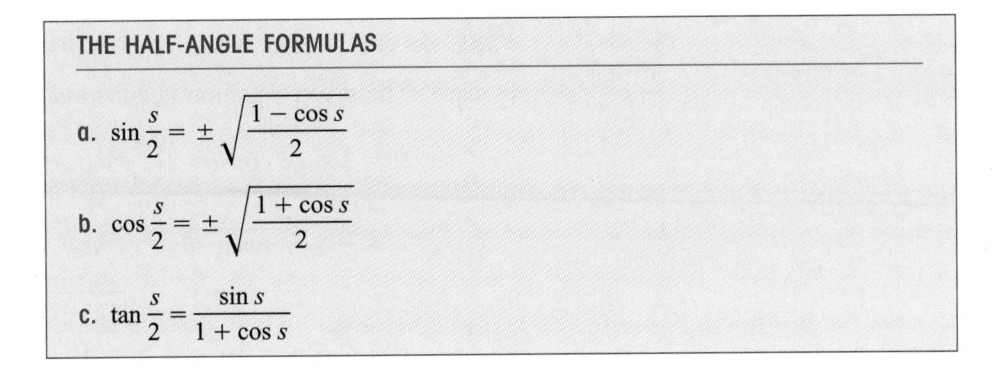

THE HALF-ANGLE FORMULAS

a. $\sin\dfrac{s}{2} = \pm\sqrt{\dfrac{1-\cos s}{2}}$

b. $\cos\dfrac{s}{2} = \pm\sqrt{\dfrac{1+\cos s}{2}}$

c. $\tan\dfrac{s}{2} = \dfrac{\sin s}{1+\cos s}$

The identities for $\sin 2\theta$, $\cos 2\theta$, and $\tan 2\theta$ are all derived in the same way: we replace 2θ by $(\theta + \theta)$ and use the appropriate addition formula. For instance, for $\sin 2\theta$ we have

$$\begin{aligned}\sin 2\theta &= \sin(\theta+\theta) = \sin\theta\cos\theta + \cos\theta\sin\theta \\ &= 2\sin\theta\cos\theta\end{aligned}$$

This establishes the formula for $\sin 2\theta$. (Exercise 33 asks you to carry out the corresponding derivations for $\cos 2\theta$ and $\tan 2\theta$.)

EXAMPLE 1 If $\sin\theta = 4/5$ and $\pi/2 < \theta < \pi$, find the quantities $\cos\theta$, $\sin 2\theta$, and $\cos 2\theta$.

Solution We have

$$\cos^2\theta = 1 - \sin^2\theta = 1 - \left(\frac{4}{5}\right)^2 = \frac{9}{25}$$

Consequently, $\cos\theta = 3/5$ or $\cos\theta = -3/5$. We want the negative value for the cosine here, since $\pi/2 < \theta < \pi$. Thus,

$$\cos\theta = -\frac{3}{5}$$

Now that we know the values of $\cos\theta$ and $\sin\theta$, the double-angle formulas can be used to determine $\sin 2\theta$ and $\cos 2\theta$. We have

$$\sin 2\theta = 2\sin\theta\cos\theta = 2\left(\frac{4}{5}\right)\left(-\frac{3}{5}\right) = -\frac{24}{25}$$

and

$$\begin{aligned}\cos 2\theta &= \cos^2\theta - \sin^2\theta \\ &= \left(-\frac{3}{5}\right)^2 - \left(\frac{4}{5}\right)^2 = \frac{9}{25} - \frac{16}{25} = -\frac{7}{25}\end{aligned}$$

The required values are therefore $\cos\theta = -3/5$, $\sin 2\theta = -24/25$, and $\cos 2\theta = -7/25$.

EXAMPLE 2 If $x = 4 \sin \theta$, $0 < \theta < \pi/2$, express $\sin 2\theta$ in terms of x.

Solution The given equation is equivalent to $\sin \theta = x/4$, so we have

$$\sin 2\theta = 2 \sin \theta \cos \theta = 2\left(\frac{x}{4}\right)\cos \theta$$

$$= \frac{x}{2}\sqrt{1 - \sin^2 \theta} \qquad \text{(Why is the positive root appropriate?)}$$

$$= \frac{x}{2}\sqrt{1 - \frac{x^2}{16}} = \frac{x}{2}\sqrt{\frac{16 - x^2}{16}} = \frac{x\sqrt{16 - x^2}}{8}$$

At the start of this section we listed formulas for $\cos 2\theta$ and for $\sin 2\theta$. There are also formulas for $\cos 3\theta$ and for $\sin 3\theta$:

$$\cos 3\theta = 4 \cos^3 \theta - 3 \cos \theta \qquad \sin 3\theta = 3 \sin \theta - 4 \sin^3 \theta$$

In the next example we derive the formula for $\cos 3\theta$, and Exercise 40 asks you to derive the formula for $\sin 3\theta$. Although these formulas needn't be memorized, they are useful. For instance, Exercise 64 in Section 8.4 shows how the formula for $\cos 3\theta$ can be used to solve certain types of cubic equations. *Historical note:* The identities for $\cos 3\theta$ and $\sin 3\theta$ are usually attributed to the French mathematician François Viète (1540–1603). Recent research, however, has shown that a geometric version of the formula for $\sin 3\theta$ was developed much earlier by the Persian mathematican and astronomer Jashmid al-Kāshi (d. 1429). (For background and details, see the article by Professor Farhad Riahi, "An Early Iterative Method for the Determination of sin 1°," in *The College Mathematics Journal,* vol. 26, January 1995, pp. 16–21.)

EXAMPLE 3 Prove the following identity:

$$\cos 3\theta = 4 \cos^3 \theta - 3 \cos \theta$$

Solution

$$\begin{aligned}\cos 3\theta &= \cos(2\theta + \theta)\\ &= \underbrace{\cos 2\theta}_{(\cos^2\theta - \sin^2\theta)} \cos \theta - \underbrace{\sin 2\theta}_{(2 \sin \theta \cos \theta)} \sin \theta\\ &= (\cos^2 \theta - \sin^2 \theta)\cos \theta - (2 \sin \theta \cos \theta)\sin \theta\\ &= \cos^3 \theta - \sin^2 \theta \cos \theta - 2 \sin^2 \theta \cos \theta\end{aligned}$$

Collecting like terms now gives us

$$\cos 3\theta = \cos^3 \theta - 3 \sin^2 \theta \cos \theta$$

Finally, we replace $\sin^2 \theta$ by the quantity $1 - \cos^2 \theta$. This yields

$$\begin{aligned}\cos 3\theta &= \cos^3 \theta - 3(1 - \cos^2 \theta)\cos \theta\\ &= \cos^3 \theta - 3 \cos \theta + 3 \cos^3 \theta\\ &= 4 \cos^3 \theta - 3 \cos \theta \qquad \text{as required}\end{aligned}$$

In the box that follows, we list several alternate ways of writing the formula for $\cos 2\theta$. Formulas (c) and (d) are quite useful in calculus.

EQUIVALENT FORMS OF THE FORMULA $\cos 2\theta = \cos^2\theta - \sin^2\theta$

a. $\cos 2\theta = 2\cos^2\theta - 1$ b. $\cos 2\theta = 1 - 2\sin^2\theta$

c. $\cos^2\theta = \dfrac{1 + \cos 2\theta}{2}$ d. $\sin^2\theta = \dfrac{1 - \cos 2\theta}{2}$

One way to prove identity (a) is as follows:

$$\begin{aligned}\cos 2\theta &= \cos^2\theta - \sin^2\theta\\ &= \cos^2\theta - (1 - \cos^2\theta)\\ &= \cos^2\theta - 1 + \cos^2\theta\\ &= 2\cos^2\theta - 1 \qquad \text{as required}\end{aligned}$$

In this last equation, if we add 1 to both sides and then divide by 2, the result is identity (c). (Verify this.) The proofs for (b) and (d) are similar; see Exercise 34.

EXAMPLE 4 Express $\cos^4 t$ in a form that does not involve powers of the trigonometric functions.

Solution

$$\begin{aligned}\cos^4 t &= (\cos^2 t)^2 = \left(\frac{1+\cos 2t}{2}\right)^2 && \text{using the formula for } \cos^2\theta\\ &= \frac{1 + 2\cos 2t + \cos^2 2t}{4}\\ &= \frac{1 + 2\cos 2t + \frac{1}{2}(1 + \cos 4t)}{4} && \text{using the formula for } \cos^2\theta \text{ with } \theta = 2t\end{aligned}$$

An easy way to simplify this last expression is to multiply both the numerator and the denominator by 2. As you should check for yourself, the final result is

$$\cos^4 t = \frac{3 + 4\cos 2t + \cos 4t}{8}$$

The last three formulas we are going to prove in this section are the **half-angle formulas:**

$$\cos\frac{s}{2} = \pm\sqrt{\frac{1+\cos s}{2}}$$

$$\sin\frac{s}{2} = \pm\sqrt{\frac{1-\cos s}{2}}$$

$$\tan\frac{s}{2} = \frac{\sin s}{1+\cos s}$$

To derive the formula for $\cos(s/2)$, we begin with one of the alternate forms of the cosine double-angle formula:

$$\cos^2\theta = \frac{1+\cos 2\theta}{2} \qquad \text{or} \qquad \cos\theta = \pm\sqrt{\frac{1+\cos 2\theta}{2}}$$

Since this identity holds for all values of θ, we may replace θ by $s/2$ to obtain

$$\cos\frac{s}{2} = \pm\sqrt{\frac{1+\cos 2(s/2)}{2}} = \pm\sqrt{\frac{1+\cos s}{2}}$$

This is the required formula for $\cos(s/2)$. To derive the formula for $\sin(s/2)$, we follow exactly the same procedure, except that we begin with the identity $\sin^2\theta = \frac{1}{2}(1-\cos 2\theta)$. [Exercise 34(c) asks you to complete the proof.] In both formulas, the sign before the radical is determined by the quadrant in which the angle or arc $s/2$ terminates.

EXAMPLE 5 Evaluate cos 105° using a half-angle formula.

Solution

$$\cos 105^\circ = \cos\frac{210^\circ}{2} = \pm\sqrt{\frac{1+\cos 210^\circ}{2}} \qquad \text{using the formula for } \cos(s/2) \text{ with } s = 210^\circ$$

$$= \pm\sqrt{\frac{1+\left(-\sqrt{3}/2\right)}{2}}$$

$$= \pm\sqrt{\frac{1-\left(\sqrt{3}/2\right)}{2}\cdot\frac{2}{2}} = \pm\sqrt{\frac{2-\sqrt{3}}{4}}$$

$$= \frac{\pm\sqrt{2-\sqrt{3}}}{2}$$

We want to choose the negative value here, since the terminal side of 105° lies in the second quadrant. Thus, we finally obtain

$$\cos 105^\circ = \frac{-\sqrt{2-\sqrt{3}}}{2}$$

Our last task is to establish the formula for $\tan(s/2)$. To do this, we first prove the equivalent identity, $\tan\theta = \dfrac{\sin 2\theta}{1+\cos 2\theta}$.

PROOF THAT $\tan\theta = \dfrac{\sin 2\theta}{1+\cos 2\theta}$

$$\frac{\sin 2\theta}{1+\cos 2\theta} = \frac{2\sin\theta\cos\theta}{2\cos^2\theta} \qquad \text{using the identity } \cos 2\theta = 2\cos^2\theta - 1 \text{ in the denominator}$$

$$= \frac{\sin\theta}{\cos\theta} = \tan\theta$$

If we now replace θ by $s/2$ in the identity $\tan\theta = \dfrac{\sin 2\theta}{1+\cos 2\theta}$, the result is

$$\tan\frac{s}{2} = \frac{\sin s}{1+\cos s}$$

This is the half-angle formula for the tangent.

We conclude this section with a summary of the principal trigonometric identities developed in this section and in Chapter 7. For completeness, the list also includes two sets of trigonometric identities that we did not discuss in this section. These are the so-called **product-to-sum formulas** and **sum-to-product formulas.** Proofs and applications of these formulas are discussed in the next section.

PROPERTY SUMMARY — PRINCIPAL TRIGONOMETRIC IDENTITIES

I. Consequences of the definitions

(a) $\csc\theta = \dfrac{1}{\sin\theta}$ **(b)** $\sec\theta = \dfrac{1}{\cos\theta}$ **(c)** $\cot\theta = \dfrac{1}{\tan\theta}$

(d) $\tan\theta = \dfrac{\sin\theta}{\cos\theta}$ **(e)** $\cot\theta = \dfrac{\cos\theta}{\sin\theta}$

II. The Pythagorean identities

(a) $\sin^2\theta + \cos^2\theta = 1$ **(b)** $\tan^2\theta + 1 = \sec^2\theta$ **(c)** $\cot^2\theta + 1 = \csc^2\theta$

III. The opposite-angle formulas

(a) $\sin(-\theta) = -\sin\theta$ **(b)** $\cos(-\theta) = \cos\theta$ **(c)** $\tan(-\theta) = -\tan\theta$

IV. The reduction formulas

(a) $\sin(\theta + 2\pi k) = \sin\theta$ **(b)** $\cos(\theta + 2\pi k) = \cos\theta$

(c) $\sin\left(\frac{\pi}{2} - \theta\right) = \cos\theta$ **(d)** $\cos\left(\frac{\pi}{2} - \theta\right) = \sin\theta$

V. The addition formulas

(a) $\sin(s + t) = \sin s \cos t + \cos s \sin t$ **(b)** $\sin(s - t) = \sin s \cos t - \cos s \sin t$

(c) $\cos(s + t) = \cos s \cos t - \sin s \sin t$ **(d)** $\cos(s - t) = \cos s \cos t + \sin s \sin t$

(e) $\tan(s + t) = \dfrac{\tan s + \tan t}{1 - \tan s \tan t}$ **(f)** $\tan(s - t) = \dfrac{\tan s - \tan t}{1 + \tan s \tan t}$

VI. The double-angle formulas

(a) $\sin 2\theta = 2\sin\theta\cos\theta$ **(b)** $\cos 2\theta = \cos^2\theta - \sin^2\theta$ **(c)** $\tan 2\theta = \dfrac{2\tan\theta}{1 - \tan^2\theta}$

VII. The half-angle formulas

(a) $\sin\dfrac{\theta}{2} = \pm\sqrt{\dfrac{1 - \cos\theta}{2}}$ **(b)** $\cos\dfrac{\theta}{2} = \pm\sqrt{\dfrac{1 + \cos\theta}{2}}$ **(c)** $\tan\dfrac{\theta}{2} = \dfrac{\sin\theta}{1 + \cos\theta}$

VIII. The product-to-sum formulas

(a) $\sin A \sin B = \frac{1}{2}[\cos(A - B) - \cos(A + B)]$ **(b)** $\sin A \cos B = \frac{1}{2}[\sin(A + B) + \sin(A - B)]$

(c) $\cos A \cos B = \frac{1}{2}[\cos(A + B) + \cos(A - B)]$

IX. The sum-to-product formulas

(a) $\sin\alpha + \sin\beta = 2\sin\dfrac{\alpha + \beta}{2}\cos\dfrac{\alpha - \beta}{2}$ **(b)** $\sin\alpha - \sin\beta = 2\cos\dfrac{\alpha + \beta}{2}\sin\dfrac{\alpha - \beta}{2}$

(c) $\cos\alpha + \cos\beta = 2\cos\dfrac{\alpha + \beta}{2}\cos\dfrac{\alpha - \beta}{2}$ **(d)** $\cos\alpha - \cos\beta = -2\sin\dfrac{\alpha + \beta}{2}\sin\dfrac{\alpha - \beta}{2}$

EXERCISE SET 8.2

A

In Exercises 1–8, use the given information to evaluate each expression.

1. $\cos\varphi = \frac{7}{25}$ $(0° < \varphi < 90°)$
   (a) $\sin 2\varphi$ **(b)** $\cos 2\varphi$ **(c)** $\tan 2\varphi$
2. $\cos\varphi = \frac{3}{5}$ $(0° < \varphi < 90°)$
   (a) $\sin 2\varphi$ **(b)** $\cos 2\varphi$ **(c)** $\tan 2\varphi$
3. $\tan u = -4$ $\left(\frac{3\pi}{2} < u < 2\pi\right)$
   (a) $\sin 2u$ **(b)** $\cos 2u$ **(c)** $\tan 2u$
4. $\cot s = 2$ $\left(\pi < s < \frac{3\pi}{2}\right)$
   (a) $\sin 2s$ **(b)** $\cos 2s$ **(c)** $\tan 2s$
5. $\sin\alpha = \sqrt{3}/2$ $(0° < \alpha < 90°)$
   (a) $\sin(\alpha/2)$ **(b)** $\cos(\alpha/2)$ **(c)** $\tan(\alpha/2)$
6. $\cos\beta = -\frac{1}{8}$ $(180° < \beta < 270°)$
   (a) $\sin(\beta/2)$ **(b)** $\cos(\beta/2)$ **(c)** $\tan(\beta/2)$
7. $\cos\theta = -\frac{7}{9}$ $\left(\frac{\pi}{2} < \theta < \pi\right)$
   (a) $\sin(\theta/2)$ **(b)** $\cos(\theta/2)$ **(c)** $\tan(\theta/2)$
8. $\cos\theta = \frac{12}{13}$ $\left(\frac{3\pi}{2} < \theta < 2\pi\right)$
   (a) $\sin(\theta/2)$ **(b)** $\cos(\theta/2)$ **(c)** $\tan(\theta/2)$

In Exercises 9–12, use the given information to compute each of the following: **(a)** $\sin 2\theta$; **(b)** $\cos 2\theta$; **(c)** $\sin(\theta/2)$; **(d)** $\cos(\theta/2)$.

9. $\sin\theta = \frac{3}{4}$ and $\frac{\pi}{2} < \theta < \pi$
10. $\cos\theta = \frac{2}{5}$ and $\frac{3\pi}{2} < \theta < 2\pi$
11. $\cos\theta = -\frac{1}{3}$ and $180° < \theta < 270°$
12. $\sin\theta = -\frac{1}{10}$ and $270° < \theta < 360°$

In Exercises 13–16, use an appropriate half-angle formula to evaluate each quantity.

13. **(a)** $\sin(\pi/12)$ **(b)** $\cos(\pi/12)$ **(c)** $\tan(\pi/12)$
14. **(a)** $\sin(\pi/8)$ **(b)** $\cos(\pi/8)$ **(c)** $\tan(\pi/8)$
15. **(a)** $\sin 105°$ **(b)** $\cos 105°$ **(c)** $\tan 105°$
16. **(a)** $\sin 165°$ **(b)** $\cos 165°$ **(c)** $\tan 165°$

In Exercises 17–24, refer to the two triangles and compute the quantities indicated.

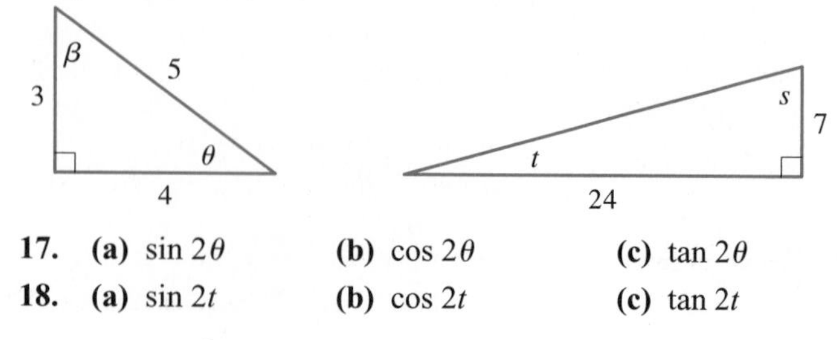

17. **(a)** $\sin 2\theta$ **(b)** $\cos 2\theta$ **(c)** $\tan 2\theta$
18. **(a)** $\sin 2t$ **(b)** $\cos 2t$ **(c)** $\tan 2t$
19. **(a)** $\sin 2\beta$ **(b)** $\cos 2\beta$ **(c)** $\tan 2\beta$
20. **(a)** $\sin 2s$ **(b)** $\cos 2s$ **(c)** $\tan 2s$
21. **(a)** $\sin(\theta/2)$ **(b)** $\cos(\theta/2)$ **(c)** $\tan(\theta/2)$
22. **(a)** $\sin(s/2)$ **(b)** $\cos(s/2)$ **(c)** $\tan(s/2)$
23. **(a)** $\sin(\beta/2)$ **(b)** $\cos(\beta/2)$ **(c)** $\tan(\beta/2)$
24. **(a)** $\sin(t/2)$ **(b)** $\cos(t/2)$ **(c)** $\tan(t/2)$

In Exercises 25–28, use the given information to express $\sin 2\theta$ and $\cos 2\theta$ in terms of x.

25. $x = 5\sin\theta$ $\left(0 < \theta < \frac{\pi}{2}\right)$
26. $x = \sqrt{2}\cos\theta$ $\left(0 < \theta < \frac{\pi}{2}\right)$
27. $x - 1 = 2\sin\theta$ $\left(0 < \theta < \frac{\pi}{2}\right)$
28. $x + 1 = 3\sin\theta$ $\left(\frac{\pi}{2} < \theta < \pi\right)$

In Exercises 29–32, express each quantity in a form that does not involve powers of the trigonometric functions (as in Example 4).

29. $\sin^4\theta$
30. $\cos^6\theta$
31. $\sin^4(\theta/2)$
32. $\sin^6(\theta/4)$
33. Prove each of the following double-angle formulas. *Hint:* As in the text, replace 2θ with $\theta + \theta$, and use an appropriate addition formula.
    (a) $\cos 2\theta = \cos^2\theta - \sin^2\theta$
    (b) $\tan 2\theta = \dfrac{2\tan\theta}{1 - \tan^2\theta}$
34. **(a)** Beginning with the identity $\cos 2\theta = \cos^2\theta - \sin^2\theta$, prove that $\cos 2\theta = 1 - 2\sin^2\theta$.
    (b) Using the result in part (a), prove that $\sin^2\theta = (1 - \cos 2\theta)/2$.
    (c) Derive the formula for $\sin(s/2)$ as follows: using the identity in part (b), replace θ with $s/2$, and then take square roots.

B

In Exercises 35–50, prove that the given equations are identities.

35. $\cos 2s = \dfrac{1 - \tan^2 s}{1 + \tan^2 s}$
36. $1 + \cos 2t = \cot t \sin 2t$
37. $\cos\theta = 2\cos^2(\theta/2) - 1$
38. $\dfrac{\sin 2\theta}{\sin\theta} - \dfrac{\cos 2\theta}{\cos\theta} = \sec\theta$
39. $\sin^4\theta = \dfrac{3 - 4\cos 2\theta + \cos 4\theta}{8}$
40. $\sin 3\theta = 3\sin\theta - 4\sin^3\theta$

41. $\sin 2\theta = \dfrac{2\tan\theta}{1+\tan^2\theta}$

42. $2\csc 2\theta = \dfrac{\csc^2\theta}{\cot\theta}$

43. $\sin 2\theta = 2\sin^3\theta\cos\theta + 2\sin\theta\cos^3\theta$

44. $\cot\theta = \dfrac{1+\cos 2\theta}{\sin 2\theta}$

45. $\dfrac{1+\tan(\theta/2)}{1-\tan(\theta/2)} = \tan\theta + \sec\theta$

46. $\tan\theta + \cot\theta = 2\csc 2\theta$

47. $2\sin^2(45^\circ - \theta) = 1 - \sin 2\theta$

48. $(\sin\theta - \cos\theta)^2 = 1 - \sin 2\theta$

49. $1 + \tan\theta\tan 2\theta = \tan 2\theta\cot\theta - 1$

50. $\tan\left(\frac{\pi}{4}+\theta\right) - \tan\left(\frac{\pi}{4}-\theta\right) = 2\tan 2\theta$

51. If $\tan\alpha = \frac{1}{11}$ and $\tan\beta = \frac{5}{6}$, find $\alpha + \beta$, given that $0 < \alpha < \frac{\pi}{2}$ and $0 < \beta < \frac{\pi}{2}$. *Hint:* Compute $\tan(\alpha + \beta)$.

52. Let $z = \tan\theta$. Show that

$$\cos 2\theta = \frac{1-z^2}{1+z^2} \quad \text{and} \quad \sin 2\theta = \frac{2z}{1+z^2}$$

53. **(a)** Use a calculator to verify that the value $x = \cos 20^\circ$ is a root of the cubic equation $8x^3 - 6x - 1 = 0$.
(b) Use the identity $\cos 3\theta = 4\cos^3\theta - 3\cos\theta$ (from Example 3 on page 494) to prove that $\cos 20^\circ$ is a root of the cubic equation $8x^3 - 6x - 1 = 0$.
Hint: In the given identity, substitute $\theta = 20^\circ$.

54. **(a)** Use a calculator to verify that the value $x = \sin\dfrac{5\pi}{18}$ is a root of the cubic equation $8x^3 - 6x + 1 = 0$.
(b) Use the identity $\sin 3\theta = 3\sin\theta - 4\sin^3\theta$ (from Exercise 40) to prove that $\sin\dfrac{5\pi}{18}$ is a root of the equation $8x^3 - 6x + 1 = 0$. *Hint:* In the identity, substitute $\theta = 5\pi/18$.

55. The following figure shows a semicircle with radius $AO = 1$.

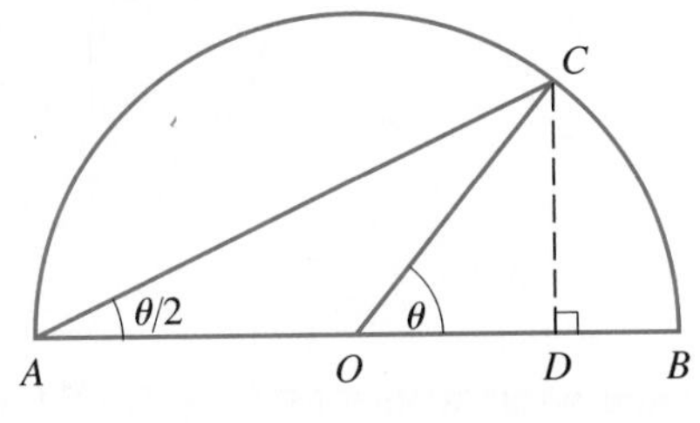

(a) Use the figure to derive the formula

$$\tan\frac{\theta}{2} = \frac{\sin\theta}{1+\cos\theta} \quad \left(0 < \theta < \frac{\pi}{2}\right)$$

Hint: Show that $CD = \sin\theta$ and $OD = \cos\theta$. Then look at right triangle ADC to find $\tan(\theta/2)$.
(b) Use the formula developed in part (a) to show that
(i) $\tan 15^\circ = \dfrac{1}{2+\sqrt{3}} = 2 - \sqrt{3}$;
(ii) $\tan(\pi/8) = \sqrt{2} - 1$.

56. In this exercise, we'll use the accompanying figure to prove the following identities:

$$\cos 2\theta = 2\cos^2\theta - 1$$
$$\sin 2\theta = 2\sin\theta\cos\theta$$
$$\cos 3\theta = 4\cos^3\theta - 3\cos\theta$$
$$\sin 3\theta = 3\sin\theta - 4\sin^3\theta$$

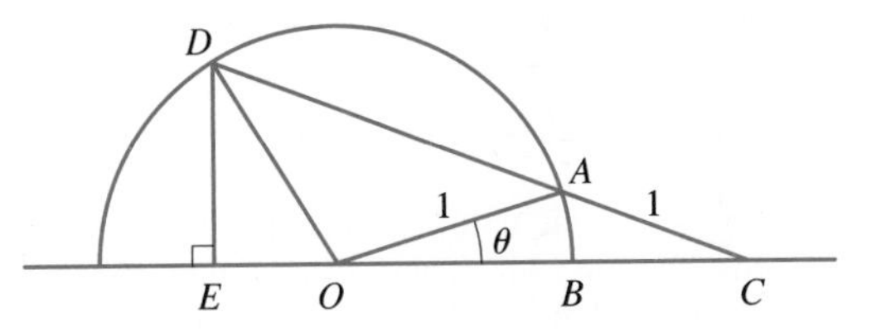

[The figure and the technique in this exercise are adapted from the article by Wayne Dancer, "Geometric Proofs of Multiple Angle Formulas," in *American Mathematical Monthly,* vol. 44 (1937), pp. 366–367.] The figure is constructed as follows. We start with $\angle AOB = \theta$ in standard position in the unit circle, as shown. The point C is chosen on the extended diameter such that $CA = 1$. Then $\overline{CA}$ is extended to meet the circle at D, and radius $\overline{DO}$ is drawn. Finally, from D, a perpendicular is drawn to the diameter, as shown.

Supply the reason or reasons that justify each of the following statements.

(a) $\angle ACO = \theta$
(b) $\angle DAO = 2\theta$ *Hint:* $\angle DAO$ is an exterior angle to $\triangle AOC$.
(c) $\angle ODA = 2\theta$
(d) $\angle DOE = 3\theta$
(e) From O, draw a perpendicular to $\overline{CD}$, meeting $\overline{CD}$ at F. From A, draw a perpendicular to $\overline{OC}$, meeting $\overline{OC}$ at G. Then $GC = OG = \cos\theta$ and $FA = DF = \cos 2\theta$.
(f) $\cos\theta = \dfrac{1+\cos 2\theta}{2\cos\theta}$ *Hint:* Use $\triangle CFO$.
(g) $\cos 2\theta = 2\cos^2\theta - 1$ *Hint:* In the equation in part (f), solve for $\cos 2\theta$.
(h) $OF = \sin 2\theta$
(i) $\sin 2\theta = 2\sin\theta\cos\theta$ *Hint:* Find $\sin\theta$ in $\triangle CFO$, and then solve the resulting equation for $\sin 2\theta$.
(j) $DC = 1 + 2\cos 2\theta$ and $EO = \cos 3\theta$

Exercise 56 continues on page 500

(k) $\cos\theta = \dfrac{2\cos\theta + \cos 3\theta}{1 + 2\cos 2\theta}$ *Hint:* Compute $\cos\theta$ in $\triangle CDE$ and then use part (j).

(l) $\cos 3\theta = 4\cos^3\theta - 3\cos\theta$ *Hint:* Use the results in parts (k) and (g).

(m) $DE = \sin 3\theta$

(n) $\sin 3\theta = 3\sin\theta - 4\sin^3\theta$ *Hint:* Compute $\sin\theta$ in $\triangle CDE$.

57. Prove the following identities involving products of cosines. *Suggestion:* In each case, begin with the right-hand side and use the double-angle formula for the sine.

(a) $\cos\theta\cos 2\theta = \dfrac{\sin 4\theta}{4\sin\theta}$

(b) $\cos\theta\cos 2\theta\cos 4\theta = \dfrac{\sin 8\theta}{8\sin\theta}$

(c) $\cos\theta\cos 2\theta\cos 4\theta\cos 8\theta = \dfrac{\sin 16\theta}{16\sin\theta}$

58. (a) Use your calculator to evaluate the expression $\cos 72^\circ \cos 144^\circ$. Then follow steps (b) through (d) to *prove* that $\cos 72^\circ \cos 144^\circ = -1/4$.

(b) Multiply the expression $\cos 72^\circ \cos 144^\circ$ by the quantity $(\sin 72^\circ)/(\sin 72^\circ)$, which equals 1. Show that the result can be written

$$\frac{\frac{1}{2}\sin 144^\circ \cos 144^\circ}{\sin 72^\circ}$$

(c) Explain why the expression obtained in part (b) is equal to

$$\frac{\frac{1}{4}\sin 288^\circ}{\sin 72^\circ}$$

(d) Use the reference-angle concept to explain why the expression in part (c) is equal to $-1/4$, as required.

59. (a) Use your calculator to evaluate the expression $\cos 72^\circ + \cos 144^\circ$. Then follow steps (b) through (d) to *prove* that $\cos 72^\circ + \cos 144^\circ = -1/2$.

(b) Use the observation

$$\cos 72^\circ + \cos 144^\circ = \cos(108^\circ - 36^\circ) + \cos(108^\circ + 36^\circ)$$

and the addition formulas for cosine to show that

$$\cos 72^\circ + \cos 144^\circ = 2\cos 108^\circ \cos 36^\circ$$

(c) Use the reference-angle concept to explain why $\cos 108^\circ \cos 36^\circ = \cos 72^\circ \cos 144^\circ$.

(d) Use parts (b) and (c) and the identity in Exercise 58(a) to conclude that $\cos 72^\circ + \cos 144^\circ = -1/2$, as required.

60. In the figure at the top of the next column, the points A_1, A_2, A_3, A_4, and A_5 are the vertices of a regular pentagon. Follow steps (a) through (c) to show that

$$(A_1A_2)(A_1A_3)(A_1A_4)(A_1A_5) = 5$$

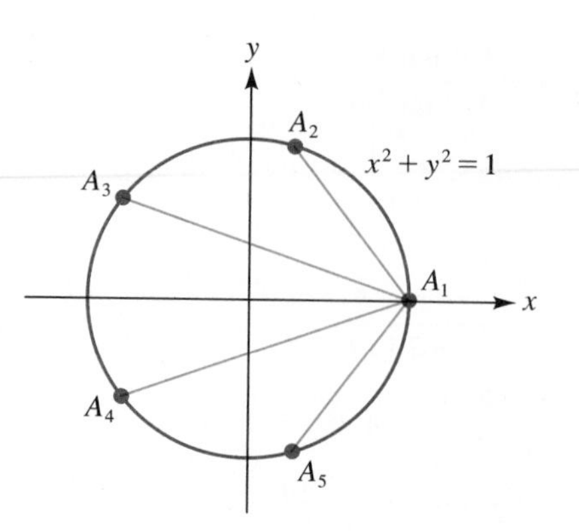

[This is a particular case of a general result due to Roger Cotes (1682–1716): Suppose that a regular n-gon is inscribed in the unit circle. Let the vertices of the n-gon be $A_1, A_2, A_3, \ldots, A_n$. Then the product $(A_1A_2)(A_1A_3)\cdots(A_1A_n)$ is equal to n, the number of sides of the polygon.]

(a) What are the coordinates of the points A_2, A_3, A_4, and A_5? (Give your answers in terms of sines and cosines.)

(b) Show that

$$(A_1A_2)(A_1A_3)(A_1A_4)(A_1A_5) = (2 - 2\cos 72^\circ)(2 - 2\cos 144^\circ)$$

(c) Show that the expression on the right-hand side of the equation in part (b) is equal to 5. *Hint:* Use the equations given in Exercises 58(a) and 59(a).

For Exercises 61 and 62, refer to the following figures. Figure A shows an equilateral triangle inscribed in the unit circle. Figure B shows a regular pentagon inscribed in the unit circle. In both figures, the coordinates of the point P are (x, 0).

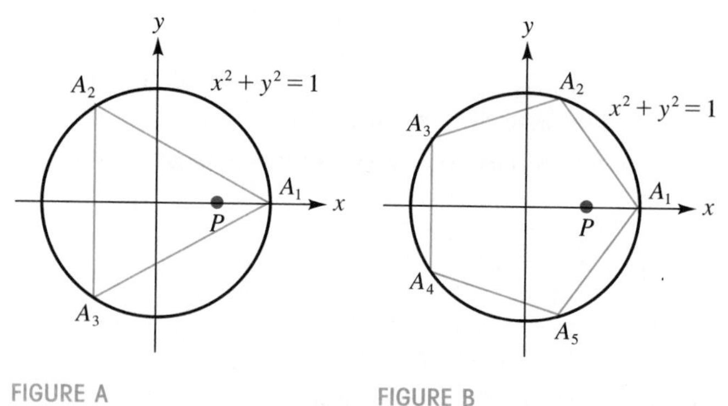

FIGURE A FIGURE B

61. For Figure A, show that, $(PA_1)(PA_2)(PA_3) = 1 - x^3$.

62. For Figure B, show that

$$(PA_1)(PA_2)(PA_3)(PA_4)(PA_5) = 1 - x^5$$

Hints: First find the coordinates of A_2 and A_3 in

terms of sines and cosines. Next, show that

$$(PA_1)(PA_2)(PA_3)(PA_4)(PA_5) = (1-x)[1-2x\cos 72° + x^2)(1-2x\cos 144° + x^2)$$

Then expand the expression within the brackets and simplify using the equations in Exercises 58(a) and 59(a).

Remark: The results in Exercises 61 and 62 are particular cases of the following theorem of Roger Cotes: Suppose that a regular n-gon $A_1A_2A_3 \cdots A_n$ is inscribed in the unit circle. Suppose that A_1 is the point $(1, 0)$ and that P is a point with coordinates $(x, 0)$, where $0 \leq x \leq 1$. Then the product $(PA_1)(PA_2)(PA_3) \cdots (PA_n)$ is equal to $1 - x^n$.

In Exercises 63 and 64, the results and the techniques are taken from the article by Zalman Usiskin, "Products of Sines," which appeared in The Two-Year College Mathematics Journal, *vol. 10 (1979), pp. 334–340.*

63. **(a)** Use your calculator to check that $\sin 18° \sin 54° = 1/4$.
(b) Supply reasons for each of the following steps to prove that the equation in part (a) is indeed correct.
(i) $\sin 72° = 2\sin 36° \cos 36° = 2 \sin 36° \sin 54°$
(ii) $\sin 72° = 4 \sin 18° \cos 18° \sin 54°$
$= 4 \sin 18° \sin 72° \sin 54°$
(iii) $1/4 = \sin 18° \sin 54°$

64. **(a)** Use your calculator to check that $\sin 10° \sin 50° \sin 70° = 1/8$.
(b) Prove that the equation in part (a) is indeed correct. *Hints:* Use the technique in Exercise 63; begin with the equation $\sin 80° = 2 \sin 40° \cos 40°$.

65. **(a)** Use two of the addition formulas from the previous section to show that

$$\cos(60° - \theta)\cos(60° + \theta) = (4\cos^2\theta - 3)/4$$

(b) Show that

$$\cos\theta\cos(60° - \theta)\cos(60° + \theta) = (\cos 3\theta)/4$$

Hint: Use the result in part (a) and the identity for $\cos 3\theta$ that was derived in Example 3 on page 494.
(c) Use the result in part (b) to show that $\cos 20° \cos 40° \cos 80° = 1/8$.
(d) Use a calculator to check that $\cos 20° \cos 40° \cos 80° = 1/8$.

66. **(a)** Use two of the addition formulas from the previous section to show that

$$\sin(60° - \theta)\ \sin(60° + \theta) = (3 - 4\sin^2\theta)/4$$

(b) Show that

$$\sin\theta\sin(60° - \theta)\sin(60° + \theta) = (\sin 3\theta)/4$$

Hint: Use the result in part (a) and the identity for $\sin 3\theta$ in Exercise 40.
(c) Use the result in part (b) to show that $\sin 20° \sin 40° \sin 80° = \sqrt{3}/8$.
(d) Use a calculator to check that $\sin 20° \sin 40° \sin 80° = \sqrt{3}/8$.

C

67. In this exercise we show that the irrational number $\cos\frac{2\pi}{7}$ is a root of the cubic equation

$$8x^3 + 4x^2 - 4x - 1 = 0.$$

(a) Prove the following two identities.

$$\cos 3\theta = 4\cos^3\theta - 3\cos\theta$$

$$\cos 4\theta = 8\cos^4\theta - 8\cos^2\theta + 1$$

(b) Let $\theta = 2\pi/7$. Use the reference-angle concept [not the formulas in part (a)] to explain why $\cos 3\theta = \cos 4\theta$.
(c) Now use the formulas in part (a) to show that if $\theta = 2\pi/7$, then

$$8\cos^4\theta - 4\cos^3\theta - 8\cos^2\theta + 3\cos\theta + 1 = 0$$

(d) Show that the equation in part (c) can be written

$$(\cos\theta - 1)(8\cos^3\theta + 4\cos^2\theta - 4\cos\theta - 1) = 0$$

Conclude that the value $x = \cos\frac{2\pi}{7}$ satisfies the equation $8x^3 + 4x^2 - 4x - 1 = 0$.
(e) Use your calculator (in the radian mode) to check that $x = \cos\frac{2\pi}{7}$ satisfies the cubic equation $8x^3 + 4x^2 - 4x - 1 = 0$. *Remark:* An interesting fact about the real number $\cos\frac{2\pi}{7}$ is that it cannot be expressed in terms of radicals within the real-number system.

68. *Calculation of* $\sin 18°$, $\cos 18°$, *and* $\sin 3°$.
(a) Prove that $\cos 3\theta = 4\cos^3\theta - 3\cos\theta$.
(b) Supply a reason for each statement.
(i) $\sin 36° = \cos 54°$
(ii) $2\sin 18° \cos 18° = 4\cos^3 18° - 3\cos 18°$
(iii) $2 \sin 18° = 4\cos^2 18° - 3$
(c) In equation (iii), replace $\cos^2 18°$ by $1 - \sin^2 18°$ and then solve the resulting equation for $\sin 18°$. Thus, show that $\sin 18° = \frac{1}{4}(\sqrt{5} - 1)$.
(d) Show that $\cos 18° = \frac{1}{4}\sqrt{10 + 2\sqrt{5}}$.
(e) Show that $\sin 3°$ is equal to

$$\tfrac{1}{16}\left[(\sqrt{5} - 1)(\sqrt{6} + \sqrt{2}) - 2(\sqrt{3} - 1)\sqrt{5 + \sqrt{5}}\,\right]$$

Hint: $3° = 18° - 15°$
(f) Use your calculator to check the results in parts (c), (d), and (e).

8.3 THE PRODUCT-TO-SUM AND SUM-TO-PRODUCT FORMULAS

In the summer of 1580, . . . [the Polish mathematician Paul Wittich] *went for a short time to Uraniborg to work with Tycho Brahe. He soon showed himself to be a skillful mathematician, for with Tycho he discovered—or, more precisely, rediscovered—the method of prosthaphaeresis, by which the products and quotients of trigonometric functions . . . can be replaced by simpler sums and differences.*

The method of prosthaphaeresis originated with Johann Werner [a German astronomer, mathematician, and geographer (1468–1522)], *who developed it in conjunction with the law of cosines for sides of a spherical triangle.*

Charles C. Gillipsie (ed.), *Dictionary of Scientific Biography,* Vol. XIV (New York: Charles Scribner's Sons, 1976)

The addition formulas for sine and cosine (from Section 8.1) can be used to establish identities concerning sums and products of sines and cosines. These identities are useful at times for simplifying expressions in trigonometry and in calculus. We begin with the three *product-to-sum formulas.*

THE PRODUCT-TO-SUM FORMULAS

$$\sin A \sin B = \frac{1}{2}[\cos(A - B) - \cos(A + B)]$$

$$\sin A \cos B = \frac{1}{2}[\sin(A - B) + \sin(A + B)]$$

$$\cos A \cos B = \frac{1}{2}[\cos(A - B) + \cos(A + B)]$$

In order to derive the first identity in the box, we write down the addition formulas for $\cos(A - B)$ and for $\cos(A + B)$:

$$\cos A \cos B + \sin A \sin B = \cos(A - B) \qquad (1)$$

$$\cos A \cos B - \sin A \sin B = \cos(A + B) \qquad (2)$$

If we subtract equation (2) from equation (1), we have

$$2 \sin A \sin B = \cos(A - B) - \cos(A + B)$$

Now, dividing both sides of this last equation by 2, we obtain the required identity:

$$\sin A \sin B = \frac{1}{2}[\cos(A - B) - \cos(A + B)]$$

The derivations of the remaining two product-to-sum formulas are entirely similar. (See Exercise 42.)

EXAMPLE 1 Use the formula for $\sin A \sin B$ to simplify the expression $\sin 15° \sin 75°$.

Solution In the identity for $\sin A \sin B$, we substitute the values $A = 15°$ and $B = 75°$. This yields

$$\begin{aligned}\sin 15° \sin 75° &= \frac{1}{2}[\cos(15° - 75°) - \cos(15° + 75°)]\\ &= \frac{1}{2}[\cos(-60°) - \cos(90°)]\\ &= \frac{1}{2}\left(\frac{1}{2} - 0\right) = \frac{1}{4}\end{aligned}$$

EXAMPLE 2 Convert the product $\sin 4x \cos 3x$ to a sum.

Solution Using the formula for $\sin A \cos B$ with $A = 4x$ and $B = 3x$, we have

$$\begin{aligned}\sin 4x \cos 3x &= \frac{1}{2}[\sin(4x - 3x) + \sin(4x + 3x)] \\ &= \frac{1}{2}[\sin x + \sin 7x] \\ &= \frac{1}{2}\sin x + \frac{1}{2}\sin 7x\end{aligned}$$

NOTE This result is *not* equal to $\sin(x/2) + \sin(7x/2)$.

EXAMPLE 3 Simplify the following expression.

$$\cos\left(s + \frac{\pi}{4}\right)\cos\left(s - \frac{\pi}{4}\right)$$

Solution In the formula

$$\cos A \cos B = \frac{1}{2}[\cos(A - B) + \cos(A + B)]$$

we will let $A = s + \frac{\pi}{4}$ and $B = s - \frac{\pi}{4}$. Notice that with these values for A and B, we have

$$A - B = \pi/2 \qquad \text{and} \qquad A + B = 2s$$

Consequently,

$$\begin{aligned}\cos\left(s + \frac{\pi}{4}\right)\cos\left(s - \frac{\pi}{4}\right) &= \frac{1}{2}\left(\cos\frac{\pi}{2} + \cos 2s\right) \\ &= \frac{1}{2}(0 + \cos 2s) = \frac{1}{2}\cos 2s\end{aligned}$$

As we saw in Examples 1 and 3, converting a product into a sum can produce in some instances a much simpler form of a given expression. Of course, there are times when it may be more useful to proceed in the other direction. That is, there are times when we want to convert a sum (or a difference) into a product. The following *sum-to-product formulas* are useful here.

THE SUM-TO-PRODUCT FORMULAS

$$\sin\alpha + \sin\beta = 2\sin\frac{\alpha+\beta}{2}\cos\frac{\alpha-\beta}{2}$$

$$\sin\alpha - \sin\beta = 2\cos\frac{\alpha+\beta}{2}\sin\frac{\alpha-\beta}{2}$$

$$\cos\alpha + \cos\beta = 2\cos\frac{\alpha+\beta}{2}\cos\frac{\alpha-\beta}{2}$$

$$\cos\alpha - \cos\beta = -2\sin\frac{\alpha+\beta}{2}\sin\frac{\alpha-\beta}{2}$$

Each of these formulas can be derived from one of the product-to-sum formulas that are listed on page 502. For instance, to derive the formula for $\cos \alpha - \cos \beta$, we begin with

$$\sin A \sin B = \frac{1}{2}[\cos(A - B) - \cos(A + B)] \tag{3}$$

If we let

$$A + B = \alpha \tag{4}$$

and

$$A - B = \beta \tag{5}$$

then we have

$$A = \frac{\alpha + \beta}{2} \quad \text{adding equations (4) and (5) and then dividing by 2}$$

and

$$B = \frac{\alpha - \beta}{2} \quad \text{subtracting equation (5) from equation (4) and then dividing by 2}$$

Now we use these last two equations and also equations (4) and (5) to substitute in equation (3). The result is

$$\sin \frac{\alpha + \beta}{2} \sin \frac{\alpha - \beta}{2} = \frac{1}{2}(\cos \beta - \cos \alpha)$$

or, after multiplying by 2,

$$2 \sin \frac{\alpha + \beta}{2} \sin \frac{\alpha - \beta}{2} = \cos \beta - \cos \alpha$$

This last equation is equivalent to

$$\cos \alpha - \cos \beta = -2 \sin \frac{\alpha + \beta}{2} \sin \frac{\alpha - \beta}{2}$$

This is the fourth sum-to-product formula, as required. (For the derivation of the remaining three sum-to-product formulas, see Exercises 43–45.)

EXAMPLE 4 Convert each expression to a product and simplify where possible: **(a)** $\sin 50° + \sin 70°$; **(b)** $\cos \theta - \cos 5\theta$.

Solution

(a) In the formula

$$\sin \alpha + \sin \beta = 2 \sin \frac{\alpha + \beta}{2} \cos \frac{\alpha - \beta}{2}$$

we set $\alpha = 50°$ and $\beta = 70°$. Then we have

$$\begin{aligned} \sin 50° + \sin 70° &= 2 \sin \frac{50° + 70°}{2} \cos \frac{50° - 70°}{2} \\ &= 2 \sin 60° \cos(-10°) \\ &= \sqrt{3} \cos 10° \quad \text{using the identity } \cos(-\theta) = \cos \theta \end{aligned}$$

(b) Using the formula

$$\cos \alpha - \cos \beta = -2 \sin \frac{\alpha+\beta}{2} \sin \frac{\alpha-\beta}{2}$$

we have

$$\begin{aligned}\cos \theta - \cos 5\theta &= -2 \sin \frac{\theta+5\theta}{2} \sin \frac{\theta-5\theta}{2} \\ &= -2 \sin 3\theta \sin(-2\theta) \\ &= 2 \sin 3\theta \sin 2\theta \qquad \text{using the identity } \sin(-t) = -\sin t\end{aligned}$$

EXAMPLE 5

(a) Use a sum-to-product formula to show that

$$\sin x + \cos x = \sqrt{2} \cos\left(x - \frac{\pi}{4}\right)$$

(b) Use the identity in part (a) to graph the function $y = \sin x + \cos x$.

Solution

(a) Using the identity $\cos x = \sin\left(\frac{\pi}{2} - x\right)$, we have

$$\begin{aligned}\sin x + \cos x &= \sin x + \sin\left(\frac{\pi}{2} - x\right) \\ &= 2 \sin \frac{x + \left(\frac{\pi}{2} - x\right)}{2} \cos \frac{x - \left(\frac{\pi}{2} - x\right)}{2} \\ &= 2 \sin \frac{\pi}{4} \cos\left(x - \frac{\pi}{4}\right) \qquad \text{Check the algebra.} \\ &= \sqrt{2} \cos\left(x - \frac{\pi}{4}\right)\end{aligned}$$

Actually, there is an easier way to obtain this identity. Expand $\cos\left(x - \frac{\pi}{4}\right)$ using an addition formula, then multiply the result by $\sqrt{2}$; the result will be $\cos x + \sin x$. We use the sum-to-product formula in this example because we want to begin with $\cos x + \sin x$ and work toward a form that can be graphed using the techniques of the previous chapter.

(b) In view of the identity established in part (a), we need to graph only the function $y = \sqrt{2} \cos\left(x - \frac{\pi}{4}\right)$. This is a function of the form $y = A \cos(Bx - C)$, which we studied in Section 7.5. In Figure 1 we show the graph of $y = \sqrt{2} \cos\left(x - \frac{\pi}{4}\right)$. You should verify for yourself (using the techniques of Section 7.5) that the amplitude, period, and phase shift indicated in Figure 1 are indeed correct.

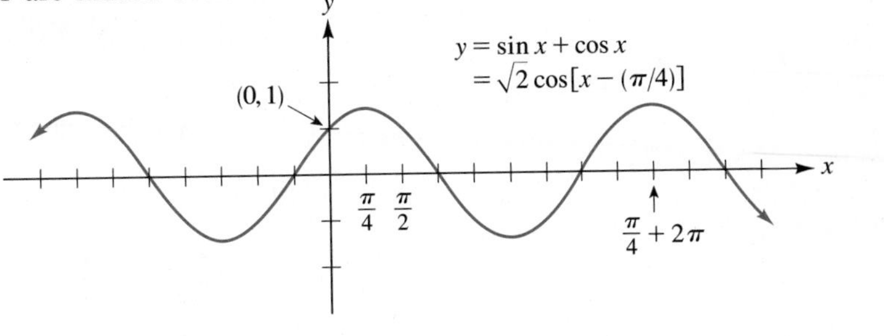

FIGURE 1
The graph of $y = \sin x + \cos x$ or $y = \sqrt{2} \cos(x - \pi/4)$.
Amplitude: $\sqrt{2}$
Period: 2π
Phase shift: $\pi/4$

EXERCISE SET 8.3

A

In Exercises 1–22, use a product-to-sum formula to convert each expression to a sum or difference. Simplify where possible.

1. $\cos 70^\circ \cos 20^\circ$
2. $\cos 50^\circ \cos 40^\circ$
3. $\sin 5^\circ \sin 85^\circ$
4. $\sin 130^\circ \sin 10^\circ$
5. $\sin 20^\circ \cos 10^\circ$
6. $\cos 18^\circ \sin 72^\circ$
7. $\cos \frac{\pi}{5} \cos \frac{4\pi}{5}$
8. $\cos \frac{5\pi}{12} \cos \frac{\pi}{12}$
9. $\sin \frac{2\pi}{7} \sin \frac{5\pi}{7}$
10. $\sin \frac{3\pi}{8} \sin \frac{\pi}{8}$
11. $\sin \frac{7\pi}{12} \cos \frac{\pi}{12}$
12. $\cos \frac{7\pi}{8} \sin \frac{\pi}{8}$
13. $\sin 3x \sin 4x$
14. $\cos 5x \cos 2x$
15. $\sin 6\theta \cos 5\theta$
16. $\sin \frac{2\theta}{3} \sin \frac{5\theta}{3}$
17. $\cos \frac{\theta}{2} \sin \frac{3\theta}{2}$
18. $\cos \frac{t}{4} \cos \frac{3t}{4}$
19. $\sin(2x + y) \sin(2x - y)$
20. $\cos\left(\theta + \frac{\pi}{6}\right) \cos\left(\theta - \frac{\pi}{6}\right)$
21. $\sin 2t \cos(s - t)$
22. $\cos(\alpha + 2\beta) \sin(2\alpha - \beta)$

In Exercises 23–34, convert each expression into a product and simplify where possible.

23. $\cos 35^\circ + \cos 55^\circ$
24. $\cos 50^\circ - \cos 10^\circ$
25. $\sin \frac{\pi}{5} - \sin \frac{3\pi}{10}$
26. $\sin \frac{\pi}{12} + \sin \frac{11\pi}{12}$
27. $\cos 5\theta - \cos 3\theta$
28. $\sin \frac{5\theta}{2} - \sin \frac{\theta}{2}$
29. $\sin 35^\circ + \cos 65^\circ$ *Hint:* Use the identity $\cos \theta = \sin(90^\circ - \theta)$.
30. $\cos \frac{3\pi}{8} - \sin \frac{\pi}{8}$ *Hint:* Use the identity $\cos \theta = \sin\left(\frac{\pi}{2} - \theta\right)$.
31. $\sin\left(\frac{\pi}{3} + 2\theta\right) - \sin\left(\frac{\pi}{3} - 2\theta\right)$
32. $\cos\left(\frac{5\pi}{12} + \theta\right) + \cos\left(\frac{\pi}{12} - \theta\right)$
33. $\dfrac{\cos \frac{5\pi}{12} + \sin \frac{5\pi}{12}}{\cos \frac{\pi}{12} - \sin \frac{\pi}{12}}$
34. $\dfrac{\sin 47^\circ + \cos 17^\circ}{\cos 47^\circ + \sin 17^\circ}$

In Exercises 35–38, prove that the equations are identities.

35. $\dfrac{\sin s + \sin t}{\cos s + \cos t} = \tan\left(\dfrac{s + t}{2}\right)$
36. $\dfrac{\cos[(n - 2)\theta] - \cos n\theta}{\sin[(n - 2)\theta] + \sin n\theta} = \tan \theta$
37. $\dfrac{\sin 2x + \sin 2y}{\cos 2x + \cos 2y} = \tan(x + y)$
38. $\sin(\theta + \phi) \sin(\theta - \phi) = \sin^2 \theta - \sin^2 \phi$

In Exercises 39 and 40, convert each sum to a product.

39. $\cos 7\theta + \cos 5\theta + \cos 3\theta + \cos \theta$
Hint: The given expression can be written $(\cos 7\theta + \cos 5\theta) + (\cos 3\theta + \cos \theta)$. After converting the quantities in parentheses to products, look for a common term to factor out.

40. $\sin 2\theta + \sin 4\theta + \sin 6\theta + \sin 8\theta$

41. **(a)** Express the quantity $\sqrt{2}\,[\sin(x/2) + \cos(x/2)]$ as a product.
(b) Use your result in part (a) to graph the function $f(x) = \sqrt{2}\,[\sin(x/2) + \cos(x/2)]$ for two complete cycles. Specify the amplitude, period, and phase shift.

B

42. **(a)** Derive the product-to-sum formula for $\cos A \cos B$. *Hint:* Add equations (1) and (2) in the text on page 502.
(b) Derive the product-to-sum formula for $\sin A \cos B$. *Hint:* Start by writing down the two addition formulas for $\sin(A + B)$ and $\sin(A - B)$, then add the two equations.

43. Derive the sum-to-product formula for $\cos \alpha + \cos \beta$. *Hint:* Follow the method used in the text to derive the formula for $\cos \alpha - \cos \beta$, but rather than beginning with equation (3), begin instead with $\cos A \cos B = \frac{1}{2}[\cos(A - B) + \cos(A + B)]$.

44. Derive the sum-to-product formula for $\sin \alpha + \sin \beta$. *Hint:* Follow the method used in the text to derive the formula for $\cos \alpha - \cos \beta$, but rather than beginning with equation (3), begin instead with $\sin A \cos B = \frac{1}{2}[\sin(A - B) + \sin(A + B)]$.

45. Derive the sum-to-product formula for $\sin \alpha - \sin \beta$. *Hint:* In the formula for $\sin \alpha + \sin \beta$ (which we derived in Exercise 44), replace each occurrence of β with $-\beta$.

46. For this exercise, follow steps (a) through (c) to show that

$$\cos \frac{\pi}{7} \cos \frac{2\pi}{7} \cos \frac{3\pi}{7} = \frac{1}{8}$$

(a) Start with the expression on the left side of the given equation and multiply both the numerator and the denominator by the quantity $\sin \frac{\pi}{7}$. Show that the resulting expression can be written

$$\frac{\frac{1}{2} \sin \frac{2\pi}{7} \cos \frac{2\pi}{7} \cos \frac{3\pi}{7}}{\sin \frac{\pi}{7}}$$

(b) Explain why the expression obtained in part (a) is equal to

$$\frac{\frac{1}{4}\sin\frac{4\pi}{7}\cos\frac{3\pi}{7}}{\sin\frac{\pi}{7}}$$

(c) Now use a product-to-sum formula to simplify the numerator of the expression in part (b). Then conclude that the original expression is indeed equal to $1/8$, as we wished to show.

47. The following problem appears in *Problems in Elementary Mathematics* by V. Lidsky et al. (Moscow: MIR Publishers, 1973).

Simplify the following expression:

$$\frac{1}{2\sin 10°} - 2\sin 70°$$

C

In Exercises 48–50, assume that $A + B + C = 180°$. *Prove that each equation is an identity.*

48. $\sin 2A + \sin 2B + \sin 2C = 4\sin A\sin B\sin C$

49. $\cos A + \cos B + \cos C = 1 + 4\sin(A/2)\sin(B/2)\sin(C/2)$

50. $\sin A + \sin B + \sin C = 4\cos(A/2)\cos(B/2)\cos(C/2)$

51. **(a)** The accompanying figure shows three triangles. For each triangle, use a calculator to verify that the sum of the cosines of the angles of the triangle is less than $3/2$.

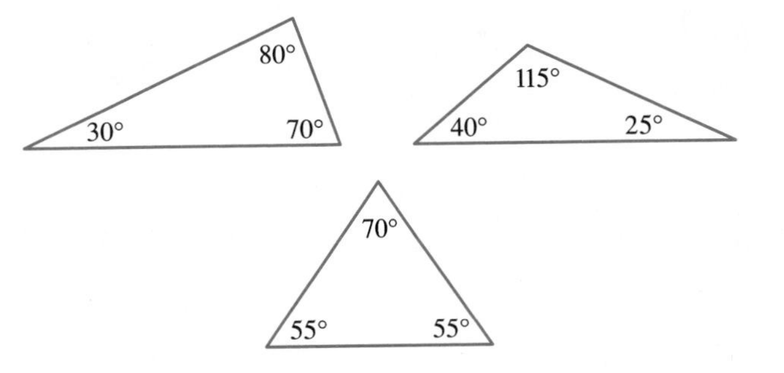

(b) What is the sum of the cosines of the three angles in an equilateral triangle?

(c) Let A, B, and C denote the three angles of a triangle, so that $A + B + C = 180°$. The following sequence of steps proves the inequality $\cos A + \cos B + \cos C \leq 3/2$. Supply the reasons or calculations that support each step. [The proof, by W. O. J. Moser, appeared in *The American Mathematical Monthly,* vol. 67 (1960), p. 695.]

(i) $\cos A + \cos B + \cos C$
$= 2\cos[(A+B)/2]\cos[(A-B)/2] + \cos C$

(ii) $\leq 2\cos[(A+B)/2] + \cos C$

(iii) $= 2\cos[(180° - C)/2] + \cos C$

(iv) $= 2\cos[90° - (C/2)] + \cos C$

(v) $= 2\sin(C/2) + (1 - 2\sin^2(C/2)$

(vi) $= (3/2) - 2[\sin(C/2) - (1/2)]^2$

(vii) $\leq 3/2$

8.4 TRIGONOMETRIC EQUATIONS

In this section, we consider some techniques for solving equations involving the trigonometric functions. As usual, by a **solution** or **root** of an equation we mean a value of the variable for which the equation becomes a true statement.

EXAMPLE 1 Consider the trigonometric equation $\sin x + \cos x = 1$. Is $x = \pi/4$ a solution? Is $x = \pi/2$ a solution?

Solution To see if the value $x = \pi/4$ satisfies the given equation, we write

$$\sin\frac{\pi}{4} + \cos\frac{\pi}{4} \stackrel{?}{=} 1$$

$$\frac{\sqrt{2}}{2} + \frac{\sqrt{2}}{2} \stackrel{?}{=} 1$$

$$\sqrt{2} \stackrel{?}{=} 1 \qquad \text{No.}$$

Thus, $x = \pi/4$ is not a solution. In a similar fashion, we can check to see if $x = \pi/2$ is a solution:

$$\sin\frac{\pi}{2} + \cos\frac{\pi}{2} \stackrel{?}{=} 1$$

$$1 + 0 \stackrel{?}{=} 1 \qquad \text{Yes.}$$

Thus $x = \pi/2$ is a solution.

The example that we've just concluded serves to remind us of the difference between a *conditional equation* and an *identity.* An identity is true for all values of the variable in its domain. For example, the equation $\sin^2 t + \cos^2 t = 1$ is an identity: it is true for every real number t. In contrast to this, a conditional equation is true only for some (or perhaps even none) of the values of the variable. The equation in Example 1 is a conditional equation; we saw that it is false when $x = \pi/4$ and true when $x = \pi/2$. The equation $\sin t = 2$ is an example of a conditional equation that has no solution. (Why?) The equations that we are going to solve in this section are conditional equations that involve the trigonometric functions. In general, there is no single technique that can be used to solve every trigonometric equation. In the examples that follow, we illustrate some of the more common approaches to solving trigonometric equations. As background for Example 2, you should review Example 2 in Section 7.4. (In that example, we obtained solutions for the trigonometric equation $\cos x = 0.8$.)

EXAMPLE 2 Consider the equation

$$\sin x = 1/2 \tag{1}$$

(a) Use your knowledge of the sine function (and not a calculator) to find all solutions of equation (1) in the open interval $(0, 2\pi)$.
(b) Use a calculator to find all solutions of equation (1) in the open interval $(0, 2\pi)$. Check that the answers are consistent with those in part (a).
(c) Find all real-number solutions of equation (1).

Solution

(a) First of all, we know from earlier work that $\sin(\pi/6) = 1/2$. So, one solution of equation (1) is certainly $x_1 = \pi/6$. To find another solution, we note that in Quadrant II, as well as in Quadrant I, $\sin x$ is positive. In Quadrant II, the angle with a reference angle of $\pi/6$ is $5\pi/6$; see Figure 1. Thus $\sin(5\pi/6) = 1/2$, and we conclude that $x_2 = 5\pi/6$ is also a solution of equation (1). Since $\sin x$ is negative in Quadrants III and IV, we needn't look there for solutions. In summary, then, the solutions of the equation $\sin x = 1/2$ in the open interval $(0, 2\pi)$ are $x_1 = \pi/6$ and $x_2 = 5\pi/6$.

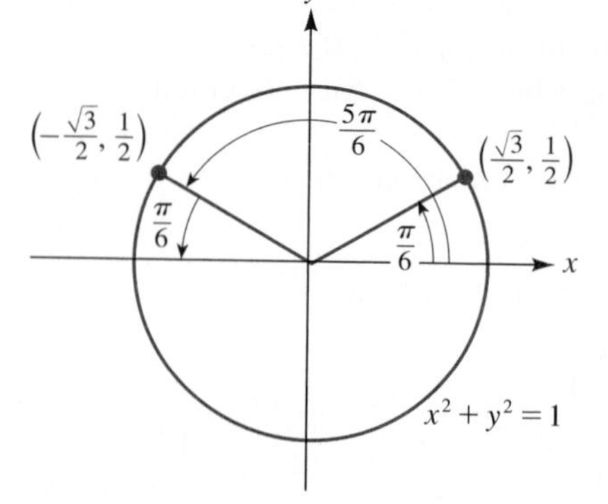

FIGURE 1

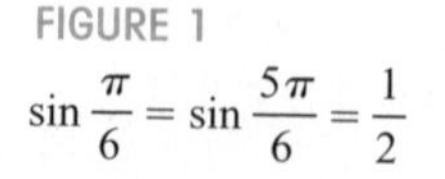
$\sin\frac{\pi}{6} = \sin\frac{5\pi}{6} = \frac{1}{2}$

(b) We can determine a root of equation (1) by using a calculator and the *inverse sine function,* denoted by $\sin^{-1}$. Applying the inverse sine function to both sides of equation (1) yields

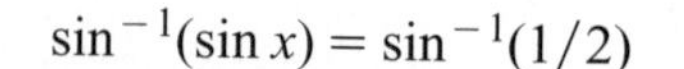

$$\sin^{-1}(\sin x) = \sin^{-1}(1/2)$$

and therefore

$$x = \sin^{-1}(1/2)$$

Now we use a calculator, set in the radian mode, to evaluate the expression $\sin^{-1}(1/2)$. As in Section 7.4 (where we used the inverse cosine function), we show keystrokes for the two basic types of calculators: input-before-function calculators and function-before-input calculators.

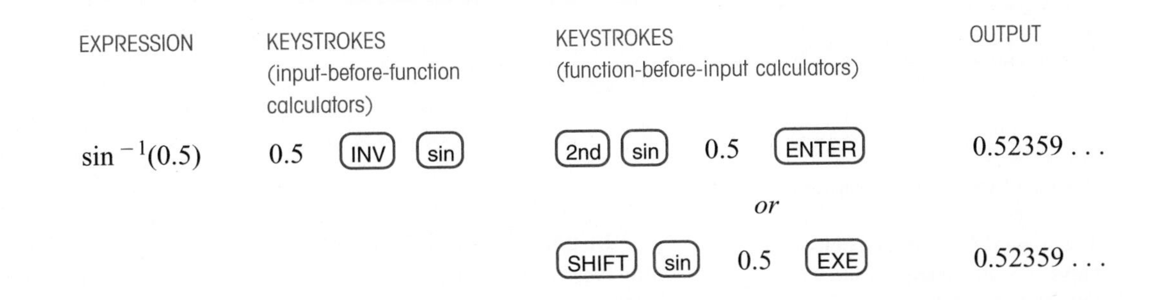

EXPRESSION	KEYSTROKES (input-before-function calculators)	KEYSTROKES (function-before-input calculators)	OUTPUT
$\sin^{-1}(0.5)$	0.5 INV sin	2nd sin 0.5 ENTER	0.52359 . . .
		or	
		SHIFT sin 0.5 EXE	0.52359 . . .

This gives us a root 0.52359. . . for equation (1). Indeed, this is the decimal approximation for the root $x_1 = \pi/6$ that we determined in part (a). Take a moment to confirm this fact; that is, use your calculator to verify that $\pi/6 = 0.52359 \ldots$.

As explained in part (a), there is another root, one in Quadrant II. Taking the calculator value 0.52359. . . for our reference angle, the second-quadrant root then is

$$\pi - 0.52359 \ldots = 2.61799 \ldots$$

This is the decimal approximation for the root $x_2 = 5\pi/6$ that we determined in part (a). (Use your calculator to verify that $5\pi/6 = 2.61799 \ldots$.)

(c) We can use the results in part (a), along with the fact that the sine function is periodic, to specify all real number solutions for the given equation. Since the period of the sine function is 2π, we have

$$\sin(x + 2k\pi) = \sin x \qquad \text{for every integer } k$$

Consequently, since $\sin\frac{\pi}{6} = \frac{1}{2}$, we know that $\sin\left(\frac{\pi}{6} + 2k\pi\right)$ is also equal to $1/2$. In other words, for every integer k, the quantity $\frac{\pi}{6} + 2k\pi$ is a solution of the given equation. Following the same reasoning, the quantity $\frac{5\pi}{6} + 2k\pi$ is also a solution of the given equation for every integer k. In summary, there are infinitely many real-number solutions of the equation $\sin x = 1/2$. These solutions are given by the expressions

$$\frac{\pi}{6} + 2k\pi \qquad \text{and} \qquad \frac{5\pi}{6} + 2k\pi \qquad \text{where } k \text{ is an integer}$$

Figure 2 (on the next page) shows some of these solutions; they are the x-coordinates of the points where the sine curve intersects the line $y = 1/2$.

GRAPHICAL PERSPECTIVE

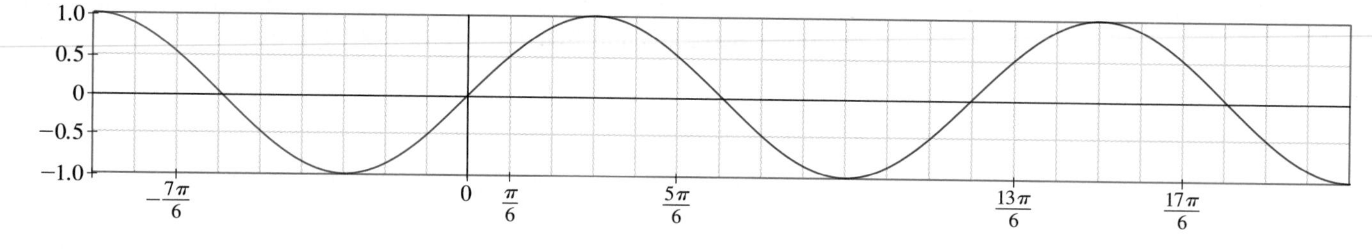

FIGURE 2

A view of the curve $y = \sin x$ for $-1.5\pi \leq x \leq 3.5\pi$.

The x-coordinates of the points where the curve $y = \sin x$ intersects the line $y = 1/2$ are solutions of the equation $\sin x = 1/2$. This view shows five of the solutions: $-7\pi/6$, $\pi/6$, $5\pi/6$, $13\pi/6$, and $17\pi/6$.

In part (b) of the example just concluded, we indicated how the inverse sine function can be used to solve a simple trigonometric equation. In the next section, the inverse sine function is analyzed in detail. But for our present work with this function, you need to know only the following definition:

$\sin^{-1}(x)$ denotes that number in the interval $[-\pi/2, \pi/2]$ whose sine is x.

As examples of this definition, we have

$\sin^{-1}(1/2) = \pi/6$	because $\sin\frac{\pi}{6} = \frac{1}{2}$ and $\frac{\pi}{6}$ is in the interval $[-\frac{\pi}{2}, \frac{\pi}{2}]$
$\sin^{-1}(1/2) \neq 5\pi/6$	Although $\sin\frac{5\pi}{6} = \frac{1}{2}$, the number $\frac{5\pi}{6}$ is not in the required interval $[-\frac{\pi}{2}, \frac{\pi}{2}]$.
$\sin^{-1}(-1) = -\pi/2$	because $\sin(-\frac{\pi}{2}) = -1$ and $-\frac{\pi}{2}$ is in the interval $[-\frac{\pi}{2}, \frac{\pi}{2}]$
$\sin^{-1}(-1) \neq 3\pi/2$	Although $\sin\frac{3\pi}{2} = -1$, the number $\frac{3\pi}{2}$ is not in the required interval $[-\frac{\pi}{2}, \frac{\pi}{2}]$.

For some of the exercises in this section, you'll need to use the inverse cosine function (which we worked with in Section 7.4) or the *inverse tangent function,* rather than the inverse sine function that we've just been discussing. For reference, we define the three functions in the box that follows. (All three functions are discussed at greater length in the next section.)

DEFINITION: Inverse Trigonometric Functions

NAME OF FUNCTION	ABBREVIATION	DEFINITION
Inverse cosine	$\cos^{-1}$	$\cos^{-1}(x)$ is the (unique) number in the interval $[0, \pi]$ whose cosine is x.
Inverse sine	$\sin^{-1}$	$\sin^{-1}(x)$ is the (unique) number in the interval $[-\frac{\pi}{2}, \frac{\pi}{2}]$ whose sine is x.
Inverse tangent	$\tan^{-1}$	$\tan^{-1}(x)$ is the (unique) number in the interval $(-\frac{\pi}{2}, \frac{\pi}{2})$ whose tangent is x.

The next two examples show how factoring can be used to solve a trigonometric equation.

Now we use a calculator, set in the radian mode, to evaluate the expression $\sin^{-1}(1/2)$. As in Section 7.4 (where we used the inverse cosine function), we show keystrokes for the two basic types of calculators: input-before-function calculators and function-before-input calculators.

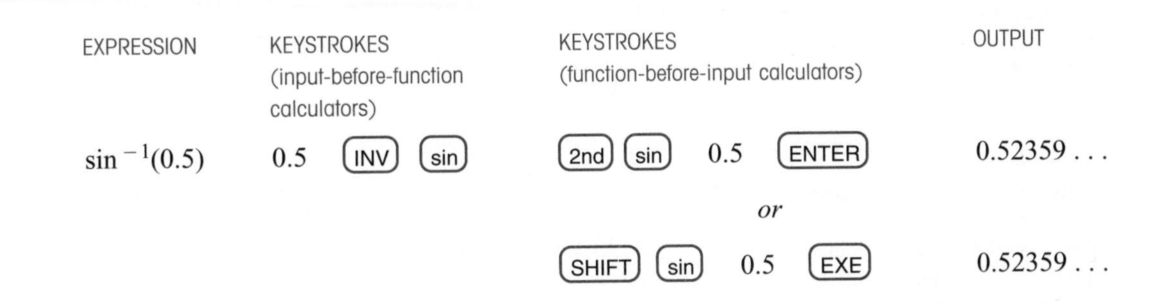

EXPRESSION	KEYSTROKES (input-before-function calculators)	KEYSTROKES (function-before-input calculators)	OUTPUT
$\sin^{-1}(0.5)$	0.5 INV sin	2nd sin 0.5 ENTER	0.52359 . . .
		or	
		SHIFT sin 0.5 EXE	0.52359 . . .

This gives us a root 0.52359. . . for equation (1). Indeed, this is the decimal approximation for the root $x_1 = \pi/6$ that we determined in part (a). Take a moment to confirm this fact; that is, use your calculator to verify that $\pi/6 = 0.52359 \ldots$.

As explained in part (a), there is another root, one in Quadrant II. Taking the calculator value 0.52359. . . for our reference angle, the second-quadrant root then is

$$\pi - 0.52359 \ldots = 2.61799 \ldots$$

This is the decimal approximation for the root $x_2 = 5\pi/6$ that we determined in part (a). (Use your calculator to verify that $5\pi/6 = 2.61799 \ldots$.)

(c) We can use the results in part (a), along with the fact that the sine function is periodic, to specify all real number solutions for the given equation. Since the period of the sine function is 2π, we have

$$\sin(x + 2k\pi) = \sin x \qquad \text{for every integer } k$$

Consequently, since $\sin\frac{\pi}{6} = \frac{1}{2}$, we know that $\sin\left(\frac{\pi}{6} + 2k\pi\right)$ is also equal to $1/2$. In other words, for every integer k, the quantity $\frac{\pi}{6} + 2k\pi$ is a solution of the given equation. Following the same reasoning, the quantity $\frac{5\pi}{6} + 2k\pi$ is also a solution of the given equation for every integer k. In summary, there are infinitely many real-number solutions of the equation $\sin x = 1/2$. These solutions are given by the expressions

$$\frac{\pi}{6} + 2k\pi \qquad \text{and} \qquad \frac{5\pi}{6} + 2k\pi \qquad \text{where } k \text{ is an integer}$$

Figure 2 (on the next page) shows some of these solutions; they are the x-coordinates of the points where the sine curve intersects the line $y = 1/2$.

GRAPHICAL PERSPECTIVE

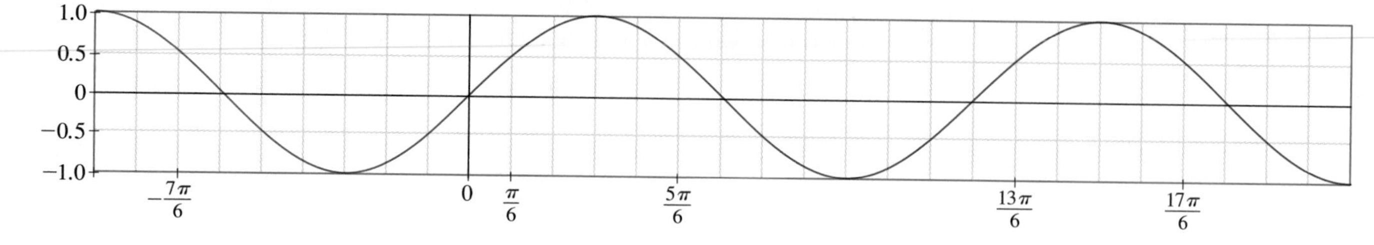

FIGURE 2
A view of the curve $y = \sin x$ for $-1.5\pi \leq x \leq 3.5\pi$.
The x-coordinates of the points where the curve $y = \sin x$ intersects the line $y = 1/2$ are solutions of the equation $\sin x = 1/2$. This view shows five of the solutions: $-7\pi/6$, $\pi/6$, $5\pi/6$, $13\pi/6$, and $17\pi/6$.

In part (b) of the example just concluded, we indicated how the inverse sine function can be used to solve a simple trigonometric equation. In the next section, the inverse sine function is analyzed in detail. But for our present work with this function, you need to know only the following definition:

$\sin^{-1}(x)$ denotes that number in the interval $[-\pi/2, \pi/2]$ whose sine is x.

As examples of this definition, we have

$\sin^{-1}(1/2) = \pi/6$ because $\sin\frac{\pi}{6} = \frac{1}{2}$ and $\frac{\pi}{6}$ is in the interval $[-\frac{\pi}{2}, \frac{\pi}{2}]$

$\sin^{-1}(1/2) \neq 5\pi/6$ Although $\sin\frac{5\pi}{6} = \frac{1}{2}$, the number $\frac{5\pi}{6}$ is not in the required interval $[-\frac{\pi}{2}, \frac{\pi}{2}]$.

$\sin^{-1}(-1) = -\pi/2$ because $\sin(-\frac{\pi}{2}) = -1$ and $-\frac{\pi}{2}$ is in the interval $[-\frac{\pi}{2}, \frac{\pi}{2}]$

$\sin^{-1}(-1) \neq 3\pi/2$ Although $\sin\frac{3\pi}{2} = -1$, the number $\frac{3\pi}{2}$ is not in the required interval $[-\frac{\pi}{2}, \frac{\pi}{2}]$.

For some of the exercises in this section, you'll need to use the inverse cosine function (which we worked with in Section 7.4) or the *inverse tangent function,* rather than the inverse sine function that we've just been discussing. For reference, we define the three functions in the box that follows. (All three functions are discussed at greater length in the next section.)

DEFINITION: Inverse Trigonometric Functions

NAME OF FUNCTION	ABBREVIATION	DEFINITION
Inverse cosine	$\cos^{-1}$	$\cos^{-1}(x)$ is the (unique) number in the interval $[0, \pi]$ whose cosine is x.
Inverse sine	$\sin^{-1}$	$\sin^{-1}(x)$ is the (unique) number in the interval $[-\frac{\pi}{2}, \frac{\pi}{2}]$ whose sine is x.
Inverse tangent	$\tan^{-1}$	$\tan^{-1}(x)$ is the (unique) number in the interval $(-\frac{\pi}{2}, \frac{\pi}{2})$ whose tangent is x.

The next two examples show how factoring can be used to solve a trigonometric equation.

EXAMPLE 3 Find all real-number solutions of the equation

$$\cos^2 x + \cos x - 2 = 0$$

Solution By factoring the expression on the left-hand side, we obtain

$$(\cos x + 2)(\cos x - 1) = 0$$

Therefore

$$\cos x = -2 \qquad \text{or} \qquad \cos x = 1$$

We discard the result $\cos x = -2$ because the value of $\cos x$ is never less than -1. From the equation $\cos x = 1$, we conclude that $x = 0$ is one solution. After $x = 0$, the next time we have $\cos x = 1$ is when $x = 2\pi$. (You can see this by looking at the graph of $y = \cos x$ or by considering the unit circle definition of the cosine.) In general, then, the solutions of the equation $\cos^2 x + \cos x - 2 = 0$ are given by $x = 2\pi k$, where k is an integer.

EXAMPLE 4 Find all solutions of the equation $\tan^2 \theta + \tan \theta - 6 = 0$ in the interval $[0, 2\pi]$. Round the final answers to three decimal places.

Solution By factoring, we have

$$(\tan \theta - 2)(\tan \theta + 3) = 0$$

and therefore

$$\tan \theta = 2 \qquad \text{or} \qquad \tan \theta = -3$$

Applying the inverse tangent function to both sides of the equation $\tan \theta = 2$ gives us

$$\tan^{-1}(\tan \theta) = \tan^{-1}(2)$$

and consequently

$$\begin{aligned} \theta &= \tan^{-1}(2) \\ &\approx 1.107 \qquad \text{using a calculator set in the radian mode} \end{aligned}$$

Now, since the period of the tangent function is π, we can find another root of the equation $\tan \theta = 2$ just by adding π to the root $\tan^{-1}(2)$. This gives us

$$\tan^{-1}(2) + \pi \approx 4.249 \qquad \text{using a calculator set in the radian mode}$$

Note that this root is also in the required interval $[0, 2\pi]$. However, adding π again to this value will take us out of the interval $[0, 2\pi]$. So let's turn to the other equation that we obtained through factoring: $\tan \theta = -3$. Applying the inverse tangent function to both sides of the equation $\tan \theta = -3$ gives us

$$\tan^{-1}(\tan \theta) = \tan^{-1}(-3)$$

and therefore

$$\begin{aligned} \theta &= \tan^{-1}(-3) \\ &= -1.2490457\ldots \qquad \text{using a calculator} \end{aligned}$$

Although this last value is a root of the equation $\tan \theta = -3$, it is not in the required interval $[0, 2\pi]$. To generate a root that does belong to this interval, we add π (the period of the tangent function) to the quantity $\tan^{-1}(-3)$ to obtain

$$\tan^{-1}(-3) + \pi \approx 1.893 \qquad \text{using a calculator}$$

Still another root can be obtained by again adding π:

$$\tan^{-1}(-3) + 2\pi \approx 5.034$$

Adding π again would take us out of the interval $[0, 2\pi]$, so we stop here. Let's summarize our results. In the interval $[0, 2\pi]$, there are four roots of the equation $\tan^2 \theta + \tan \theta - 6 = 0$. They are:

$$\begin{aligned} \tan^{-1}(2) &\approx 1.107 \\ \tan^{-1}(2) + \pi &\approx 4.249 \\ \tan^{-1}(-3) + \pi &\approx 1.893 \\ \tan^{-1}(-3) + 2\pi &\approx 5.034 \end{aligned}$$

GRAPHICAL PERSPECTIVE

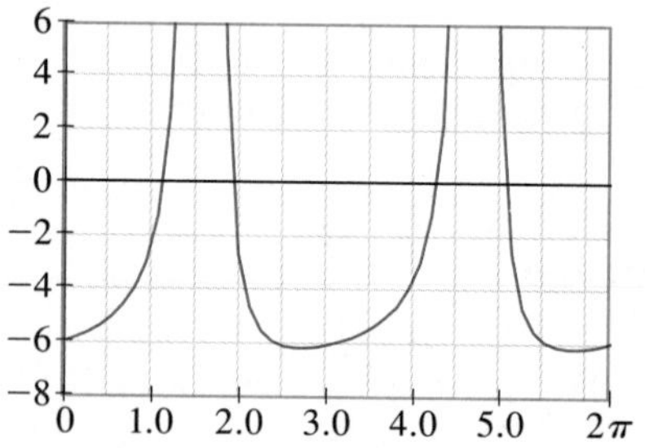

For a graphical check of these results, see Figure 3, which shows the graph of the equation $y = \tan^2 \theta + \tan \theta - 6$. The θ-intercepts of the graph are the roots of the equation $\tan^2 \theta + \tan \theta - 6 = 0$. In the figure, note that each intercept is consistent with one of the roots that we have determined. For instance, the largest intercept in Figure 3 is approximately 5.0. This corresponds to the root $\tan^{-1}(-3) + 2\pi \approx 5.034$.

FIGURE 3
A view of $y = \tan^2 \theta + \tan \theta - 6$ for $0 \leq \theta \leq 2\pi$.

In some equations, more than one trigonometric function is present. A common approach here is to express the various functions in terms of a single one. The next example demonstrates this technique.

EXAMPLE 5 Find all real-number solutions of the equation

$$3 \tan^2 x - \sec^2 x - 5 = 0$$

Solution We use the Pythagorean identity $\tan^2 x + 1 = \sec^2 x$ to substitute for $\sec^2 x$ in the given equation. This gives us

$$\begin{aligned} 3 \tan^2 x - (\tan^2 x + 1) - 5 &= 0 \\ 2 \tan^2 x - 6 &= 0 \\ 2 \tan^2 x &= 6 \\ \tan^2 x &= 3 \\ \tan x &= \pm\sqrt{3} \end{aligned}$$

Since the period of the tangent function is π, we need to find only those values of x between 0 and π that satisfy $\tan x = \pm\sqrt{3}$. The other solutions will then be obtained by adding multiples of π to these solutions. Now, in Quadrant I, the tangent is positive and we know that $\tan x = \sqrt{3}$ when $x = \pi/3$. In Quadrant II, $\tan x$ is negative and we know that $\tan 2\pi/3 = -\sqrt{3}$, since the reference angle for $2\pi/3$ is $\pi/3$. Thus, the solutions between 0 and π are $x = \pi/3$ and $x = 2\pi/3$. It follows that all of the solutions to the equation $3 \tan^2 x - \sec^2 x - 5 = 0$ are given by

$$x = \frac{\pi}{3} + \pi k \qquad \text{and} \qquad x = \frac{2\pi}{3} + \pi k \qquad \text{where } k \text{ is an integer}$$

The technique used in Example 5—expressing the various functions in terms of a single function—is most useful when it does not involve introducing a radical expression. For instance, consider the equation $\sin s + \cos s = 1$. Although we could begin by replacing $\cos s$ by the expression $\pm\sqrt{1 - \sin^2 s}$, it turns out to be easier in this situation to begin by squaring both sides of the given equation. This is done in the next example.

EXAMPLE 6 Find all solutions of the equation $\sin s + \cos s = 1$ satisfying $0° \leq s < 360°$.

Solution Squaring both sides of the equation yields

$$(\sin s + \cos s)^2 = 1^2$$
$$\sin^2 s + 2 \sin s \cos s + \cos^2 s = 1$$

(These add to 1: $\sin^2 s$ and $\cos^2 s$)

Consequently, we have

$$2 \sin s \cos s = 0$$

From this last equation, we conclude that $\sin s = 0$ or $\cos s = 0$. When $\sin s = 0$, we know that $s = 0°$ or $s = 180°$. And when $\cos s = 0$, we know that $s = 90°$ or $s = 270°$. Now we must go back and check which (if any) of these values is a solution to the *original* equation. This must be done whenever we square both sides in the process of solving an equation.

$s = 0°$	$\sin 0° + \cos 0° \stackrel{?}{=} 1$ $0 + 1 \stackrel{?}{=} 1$	True
$s = 90°$	$\sin 90° + \cos 90° \stackrel{?}{=} 1$ $1 + 0 \stackrel{?}{=} 1$	True
$s = 180°$	$\sin 180° + \cos 180° \stackrel{?}{=} 1$ $0 + (-1) \stackrel{?}{=} 1$	False
$s = 270°$	$\sin 270° + \cos 270° \stackrel{?}{=} 1$ $-1 + 0 \stackrel{?}{=} 1$	False

We conclude that the only solutions of the equation $\sin s + \cos s = 1$ on the interval $0° \leq s < 360°$ are $s = 0°$ and $s = 90°$.

In the example that follows, we consider an equation that involves a multiple of the unknown angle.

EXAMPLE 7 Solve the equation $\sin 3x = 1$ on the interval $0 \leq x \leq 2\pi$.

Solution We know that $\sin(\pi/2) = 1$. Thus, one solution can be found by writing $3x = \pi/2$, from which we conclude that $x = \pi/6$. We can look for other solutions in the required interval by writing, more generally,

$$3x = \frac{\pi}{2} + 2\pi k$$
$$= \frac{\pi + 4\pi k}{2}$$

and therefore

$$x = \frac{\pi + 4\pi k}{6}$$

Thus, when $k = 1$, we obtain

$$x = \frac{\pi + 4\pi(1)}{6} = \frac{5\pi}{6}$$

With $k = 2$, we have

$$x = \frac{\pi + 4\pi(2)}{6} = \frac{9\pi}{6} = \frac{3\pi}{2}$$

And with $k = 3$, we have

$$x = \frac{\pi + 4\pi(3)}{6} = \frac{13\pi}{6} \qquad \text{which is greater than } 2\pi$$

We conclude that the solutions of $\sin 3x = 1$ on the interval $0 \leq x \leq 2\pi$ are $\pi/6$, $5\pi/6$, and $3\pi/2$.

In Example 7, notice that we did not need to make use of a formula for $\sin 3x$, even though the expression $\sin 3x$ did appear in the given equation. In the next example, however, we do make use of the identity $\sin 2x = 2 \sin x \cos x$.

EXAMPLE 8 Solve the equation $\sin x \cos x = 1$.

Solution We could begin by squaring both sides. (Exercise 59 will ask you to use this approach.) However, with the double-angle formula for sine in mind, we can proceed instead as follows. We multiply both sides of the given equation by 2. This yields

$$2 \sin x \cos x = 2$$

and, consequently,

$$\sin 2x = 2 \qquad \text{using the double-angle formula}$$

This last equation has no solution, since the value of the sine function never exceeds 1. Thus, the equation $\sin x \cos x = 1$ has no solution.

For the last example in this section, we look at another case in which a calculator is required.

EXAMPLE 9 Find all angles θ in the interval $0° \leq \theta \leq 360°$ satisfying the equation $2 \sin \theta = \cos \theta$.

Solution We first want to rewrite the given equation using a single function, rather than both sine and cosine. The easiest way to do this is to divide both sides by $\cos \theta$. Nothing is lost here in assuming $\cos \theta \neq 0$. (If $\cos \theta$ were zero, then θ would be 90° or 270°; but neither of those angles is a solution of the given equation.) Dividing both sides of the given equation by $\cos \theta$, we have

$$\frac{2 \sin \theta}{\cos \theta} = \frac{\cos \theta}{\cos \theta}$$

$$2 \tan \theta = 1$$

$$\tan \theta = 0.5$$

From experience, we know that none of the angles with which we are familiar (the multiples of 30° and 45°) has a value of 0.5 for the tangent. Thus, as in Example 4, we use the inverse tangent function and a calculator. First, applying the inverse tangent function to both sides of the equation $\tan\theta = 0.5$, we have

$$\tan^{-1}(\tan\theta) = \tan^{-1}(0.5)$$

and therefore

$$\theta = \tan^{-1}(0.5)$$

Now, according to the statement of the problem, our answers are to be in degrees, not radians. So, before using the calculator to evaluate the expression $\tan^{-1}(0.5)$, we first set the calculator in the degree mode. Then, as you should check for yourself, we obtain

$$\theta = \tan^{-1}(0.5) \approx 26.6° \qquad \text{using a calculator set in the degree mode}$$

To find another angle θ satisfying the equation $\tan\theta = 0.5$, we use the fact that the period of the tangent is π radians, which is 180°. Thus, a second solution (rounded to one decimal place) is

$$26.6° + 180° = 206.6°$$

Adding another 180° to this last value will take us out of the specified interval $0° \leq \theta \leq 360°$, so we stop here. In summary, there are two angles between 0° and 360° satisfying the given equation $2\sin\theta = \cos\theta$. They are 26.6° and 206.6° (rounded to one decimal place). For a graphical check of these results, see Figure 4, which shows the graphs of the equations $y = 2\sin\theta$ and $y = \cos\theta$ for $0° \leq \theta \leq 360°$. (In Figure 4, we are viewing the horizontal axis as the θ-axis, where θ is measured in degrees.) The graphs intersect at two points. The θ-coordinates of these points are solutions of the equation $2\sin\theta = \cos\theta$. The figure shows that at one of the intersection points, θ is slightly less than 30°, while at the other intersection point, θ is slightly less than 210°. These observations are consistent with the numerical values we've obtained for the roots.

GRAPHICAL PERSPECTIVE

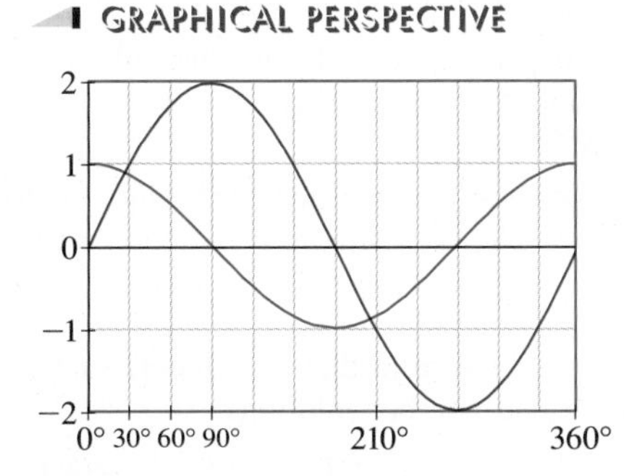

FIGURE 4
Graphs of $y = 2\sin\theta$ and $y = \cos\theta$ for $0° \leq \theta \leq 360°$.

Do 5, 7, 15, 19, 21, 33, 41

EXERCISE SET 8.4

A

1. Is $\theta = \pi/2$ a solution of $2\cos^2\theta - 3\cos\theta = 0$?
2. Is $x = 15°$ a solution of $(\sqrt{3}/3)\cos 2x + \sin 2x = 1$?
3. Is $x = 3\pi/4$ a solution of $\tan^2 x - 3\tan x + 2 = 0$?
4. Is $t = 2\pi/3$ a solution of $2\sin t + 2\cos t = \sqrt{3} - 1$?

In Exercises 5–22, determine all solutions of the given equations. Express your answers using radian measure.

5. $\sin\theta = \sqrt{3}/2$
6. $\sin\theta = \sqrt{2}/2$
7. $\sin\theta = -1/2$
8. $\sin\theta + (\sqrt{2}/2) = 0$
9. $\cos\theta = -1$
10. $\cos\theta = 1/2$
11. $\tan\theta = \sqrt{3}$
12. $\tan\theta + (\sqrt{3}/3) = 0$
13. $\tan x = 0$
14. $2\sin^2 x - 3\sin x + 1 = 0$
15. $2\cos^2\theta + \cos\theta = 0$
16. $\sin^2 x - \sin x - 6 = 0$
17. $\cos^2 t\sin t - \sin t = 0$
18. $\cos\theta + 2\sec\theta = -3$
19. $2\cos^2 x - \sin x - 1 = 0$
20. $2\cot^2 x + \csc^2 x - 2 = 0$
21. $\sqrt{3}\sin t - \sqrt{1 + \sin^2 t} = 0$
22. $\sec\alpha + \tan\alpha = \sqrt{3}$

In Exercises 23–36, use a calculator to find all solutions in the interval $(0, 2\pi)$. Round the answers to two decimal places.

23. $\cos x = 0.184$

24. $\cos t = -0.567$

25. $\sin x = 1/\sqrt{5}$

26. $\sin t = -0.301$

27. $\tan x = 6$

28. $\tan t = -5.25$

29. $\sin t = 5 \cos t$

30. $\sin x \cos x = 0.035$

31. $\sec t = 2.24$ *Hint:* The equation is equivalent to $\cos t = 1/(2.24)$.

32. $\cot x = -3.27$

33. $\tan^2 x + \tan x - 12 = 0$

34. $15 \sin^2 x - 26 \sin x + 8 = 0$

35. $16 \sin^3 x - 12 \sin^2 x + 36 \sin x - 27 = 0$
Hint: Factor by grouping.

36. $3 \cos^3 x - 9 \cos^2 x + \cos x - 3 = 0$
Hint: Factor by grouping.

In Exercises 37–42, find all solutions in the interval $0° \leq \theta \leq 360°$. Where necessary, use a calculator and round to one decimal place.

37. $\sin \theta = 1/4$

38. $\cos \theta = -4/5$

39. $9 \tan^2 \theta - 16 = 0$

40. $5 \sin^2 \theta + 13 \sin \theta - 6 = 0$

41. $\cos^2 \theta - \cos \theta - 1 = 0$
Hint: You'll need to use the quadratic formula.

42. $\tan^2 \theta - \tan \theta - 1 = 0$

In Exercises 43–50, determine all of the solutions in the interval $0° \leq \theta < 360°$.

43. $\cos 3\theta = 1$

44. $\tan 2\theta = -1$

45. $\sin 3\theta = -\sqrt{2}/2$

46. $\sin(\theta/2) = \frac{1}{2}$

47. $\sin \theta = \cos(\theta/2)$
Hint: $\sin \theta = 2 \sin(\theta/2) \cos(\theta/2)$

48. $2 \sin^2 \theta - \cos 2\theta = 0$

49. $\sin 2\theta = \sqrt{3} \cos 2\theta$ *Hint:* Divide by $\cos 2\theta$.

50. $\sin 2\theta = -2 \cos \theta$

B

51. Find all solutions of the equation $\tan 3x - \tan x = 0$ in the interval $0 \leq x < 2\pi$. *Hint:* Write $\tan 3x = \tan(2x + x)$ and use the addition formula for tangent.

52. Find all solutions of the equation $2 \sin x = 1 - \cos x$ in the interval $0° \leq x < 360°$. Use a calculator and round the answer(s) to one decimal place.

53. Find all solutions of the equation $\cos(x/2) = 1 + \cos x$ in the interval $0 \leq x < 2\pi$.

54. Find all solutions of the equation

$$\sin 3x \cos x + \cos 3x \sin x = \sqrt{3}/2$$

in the interval $0 < x < 2\pi$.

55. Find all real numbers θ for which $\sec 4\theta + 2 \sin 4\theta = 0$.

56. Consider the equation $\sin^2 x - \cos^2 x = 7/25$.
(a) Solve the equation for $\cos x$.
(b) Find all of the solutions of the equation satisfying $0 < x < \pi$.
(c) Solve the original equation by means of a double-angle formula and use the result to check your answers in part (b).

57. Find a solution of the equation $4 \sin \theta - 3 \cos \theta = 2$ in the interval $0° < \theta < 90°$. *Hint:* Add $3 \cos \theta$ to both sides, then square.

58. Find all solutions of the equation

$$\sin^3 \theta \cos \theta - \sin \theta \cos^3 \theta = -1/4$$

in the interval $0 < \theta < \pi$. *Hint:* Factor the left-hand side, then use the double-angle formulas.

59. Consider the equation $\sin x \cos x = 1$.
(a) Square both sides and then replace $\cos^2 x$ by $1 - \sin^2 x$. Show that the resulting equation can be written $\sin^4 x - \sin^2 x + 1 = 0$.
(b) Show that the equation $\sin^4 x - \sin^2 x + 1 = 0$ has no solutions. Conclude from this that the original equation has no real-number solutions.

60. The accompanying figure shows a portion of the graph of the periodic function

$$f(x) = \sin(1 + \sin x)$$

Find the x-coordinates of the turning points P, Q, and R. Round the answers to three decimal places.
Hint: The x-coordinate of Q is halfway between the x-coordinates of P and R.

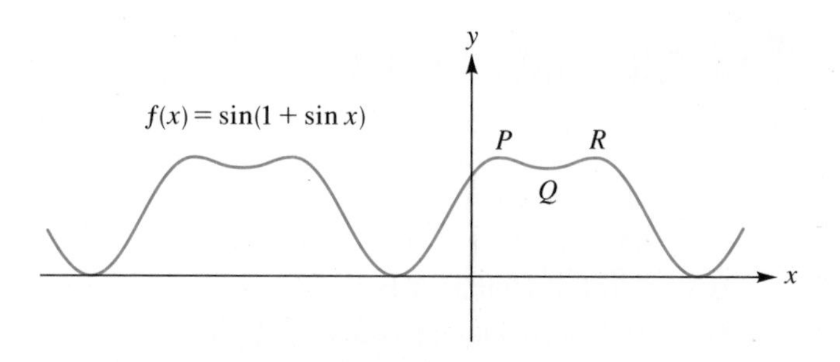

61. **(a)** Find the smallest solution of $\cos x = 0.412$ in the interval $(1000, \infty)$. (Round the answer to three decimal places.)
(b) Find the smallest solution of $\cos x = -0.412$ in the interval $(1000, \infty)$.

62. The accompanying figure shows a graph of the function $f(x) = x + 0.4 \sin(2\pi x)$ on the interval $0 \leqslant x \leqslant 1.5$. (Functions of this form occur in mathematical biology in the study of rhythmic behaviors, such as heartbeat.) Using calculus, it can be shown that the x-coordinates of the turning points of this function are found by solving the equation

$$1 + 0.8\pi \cos(2\pi x) = 0$$

Use this fact to find the x-coordinates of the turning points P, Q, and R in the figure. Round the answers to three decimal places.

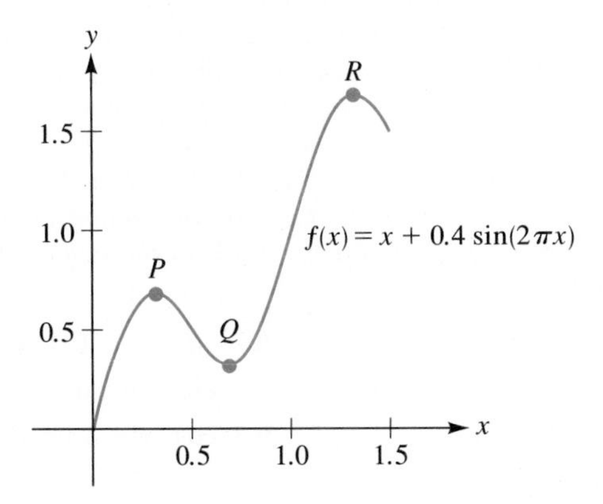

63. (As background for this exercise, you need to know the definitions of *fixed point* and *iterate* from Section 4.3.) The following figure shows a view of the function $f(x) = x + 0.4 \sin(2\pi x)$ along with the line $y = x$.

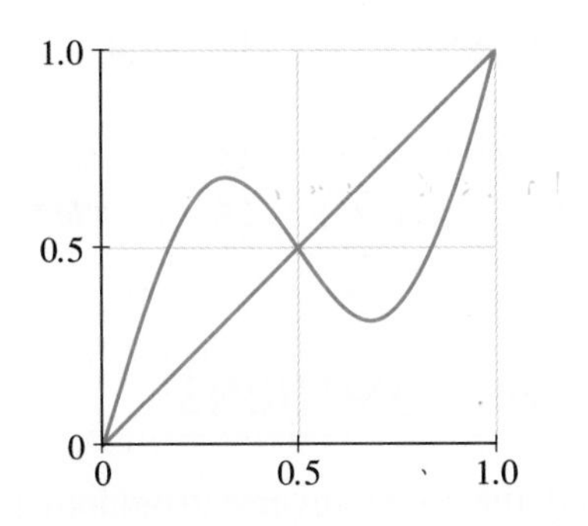

(a) From the figure, it appears that the value $x = 0.5$ is a fixed point for the function f. Confirm this algebraically by solving the equation $f(x) = x$ on the interval $0 \leqslant x \leqslant 1$.

(b) Compute the first ten iterates of $x_0 = 0.25$ under the function f. Round the answers to four decimal places. Describe any patterns that you detect. Are the iterates approaching the fixed point $x = 0.5$?

(c) Follow part (b) using $x_0 = 0.45$.

C

64. In this exercise, you will see how certain cubic equations can be solved by using the following identity (which we proved in Example 3 in Section 8.2):

$$4 \cos^3 \theta - 3 \cos \theta = \cos 3\theta \tag{1}$$

For example, suppose that we wish to solve the equation

$$8x^3 - 6x - 1 = 0 \tag{2}$$

To transform this equation into a form where the stated identity is useful, we make the substitution $x = a \cos \theta$, where a is a constant to be determined. With this substitution, equation (2) can be written

$$8a^3 \cos^3 \theta - 6a \cos \theta = 1 \tag{3}$$

In equation (3), the coefficient of $\cos^3 \theta$ is $8a^3$. Since we want this coefficient to be 4 [as it is in equation (1)], we divide both sides of equation (3) by $2a^3$ to obtain

$$4 \cos^3 \theta - \frac{3}{a^2} \cos \theta = \frac{1}{2a^3} \tag{4}$$

Next, a comparison of equations (4) and (1) leads us to require that $3/a^2 = 3$. Thus, $a = \pm 1$. For convenience, we choose $a = 1$; equation (4) then becomes

$$4 \cos^3 \theta - 3 \cos \theta = 1/2 \tag{5}$$

Comparing equation (5) with the identity in (1) leads us to the equation

$$\cos 3\theta = 1/2$$

As you can check, the solutions here are of the form

$$\theta = 20^\circ + 120k^\circ \qquad \text{and} \qquad \theta = 100^\circ + 120k^\circ$$

Thus,

$$x = \cos(20^\circ + 120k^\circ) \qquad \text{and} \qquad x = \cos(100^\circ + 120k^\circ)$$

Now, however, as you can again check, only three of the angles yield distinct values for $\cos \theta$, namely, $\theta = 20^\circ$, $\theta = 140^\circ$, and $\theta = 260^\circ$. Thus, the solutions of the equation $8x^3 - 6x - 1 = 0$ are given by $x = \cos 20^\circ$, $x = \cos 140^\circ$, and $x = \cos 260^\circ$.

Use the method just described to solve the following equations.

(a) $x^3 - 3x + 1 = 0$

Answers: $2 \cos 40^\circ$, $-2 \cos 20^\circ$, $2 \cos 80^\circ$

(b) $x^3 - 36x - 72 = 0$

(c) $x^3 - 6x + 4 = 0$ *Answers:* 2, $-1 \pm \sqrt{3}$

(d) $x^3 - 7x - 7 = 0$ (Round your answers to three decimal places.)

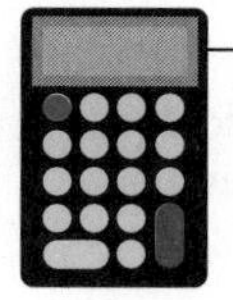

GRAPHING UTILITY EXERCISES FOR SECTION 8.4

1. **(a)** Graph the equation $y = \cos^2 x + \cos x - 2$. Use a viewing rectangle extending from 0 to 10 in the x-direction and from -3 to 3 in the y-direction.
   (b) The picture obtained in part (a) indicates that the graph has an x-intercept between 6 and 7. By zooming in on this x-intercept, or by using a solve key, obtain the first two decimal places of this intercept.
   (c) According to Example 3, what is the exact value of the x-intercept that you approximated in part (b)?
2. **(a)** Graph the equation $y = \tan^2 x + \tan x - 6$ for $0 \leq x \leq 2\pi$.
   (b) By zooming in on each x-intercept of the graph in part (a), estimate the roots of the equation $\tan^2 x + \tan x - 6 = 0$. Then check that your estimates are consistent with the values obtained in Example 4.
3. **(a)** Graph the function $y = \sin(\pi x/180)$ in the standard viewing rectangle. Describe what you see.
   (b) Change the viewing rectangle so that x extends from 0 to 360. The graph that you obtain is the graph of $y = \sin x$, where x is now measured in degrees.
   (c) According to Example 6, there are two solutions of the equation $\sin s + \cos s = 1$ in the interval $0° \leq s < 360°$. Confirm this by graphing the two equations.
   $$y = \sin(\pi x/180) + \cos(\pi x/180)$$
   and
   $$y = 1 \quad \text{for } 0 \leq x \leq 360$$
   and noting that there are two intersection points. (In Example 6, the intersection point at $x = 360$ is excluded.)
   (d) In Example 6 we saw that 90° is a solution of the equation $\sin s + \cos s = 1$ in the interval $0° \leq s < 360°$. Confirm this by zooming in on the appropriate intersection point in part (c) and checking the x-coordinate.

In Exercises 4–25, solve the equations on the interval $[0, 2\pi]$ as follows. Graph the expression on each side of the equation and then zoom in on the intersection points until you are certain of the first three decimal places in each answer. For instance, for Exercise 4, when you graph the two equations $y = \cos x$ and $y = 0.623$ on the interval $[0, 2\pi]$, you'll see that there are two intersection points. The x-coordinates of these points are roots of the equation $\cos x = 0.623$.

4. $\cos x = 0.623$
5. $\sin x = -0.438$
6. $\cos x = \tan x$
7. $\cos x = \tan 2x$
8. $\cos^2 x = 2 \sin x$
9. $\cos^3 x = 2 \sin x$
10. $\cos 2x + 1 = \cos(2x + 1)$
11. $\cos 2x + 0.9 = \cos(2x + 1)$
12. $\cos(x/2) = \cos(x^2/2)$
13. $\cos(x/2) = \cos(x^2/12)$
14. $2 \sin x - 3 \cos x = \tan(x/4)$
15. $2 \sin x + 3 \cos x = \tan(x/4)$
16. $\sqrt{x} = \tan x$
17. $x^2 = \tan x$
18. $\sin(\cos x) = \sin x$
19. $\cos(\sin x) = \sin x$
20. $\tan x = x$
21. $\cos x = x$
22. $\sin^3 x + \cos^3 x = 0.5$
23. $\sin^3 x - \cos^3 x = 0.5$
24. $1 - \tan^2 x = 2 \sin(x/5)$
25. $1 - \tan^2 x = 2 \sin(x/6)$

THE INVERSE TRIGONOMETRIC FUNCTIONS

The notation $\cos^{-1} \theta$ must not be understood to signify $\frac{1}{\cos \theta}$.

John Herschel in *Philosophical Transactions of London,* 1813

In previous sections, we introduced the three inverse trigonometric functions $y = \sin^{-1} x$, $y = \cos^{-1} x$, and $y = \tan^{-1} x$. Because we were using these functions as tools to solve equations, our emphasis there was more computational than theoretical in nature. Now we want to take a second look at these functions, this time from a more conceptual point of view. As background for this work, you should be familiar with the material on inverse functions in Section 3.5.

As indicated in Figure 1(a), the sine function in not one-to-one. Therefore there is no inverse function. However, let us now consider the **restricted sine function:**

$$y = \sin x \qquad \left(-\tfrac{\pi}{2} \leq x \leq \tfrac{\pi}{2}\right)$$

As indicated in Figure 1(b), the restricted sine function *is* one-to-one, and therefore the inverse function does exist in this case. We refer to the inverse of the restricted sine function as the **inverse sine function.**

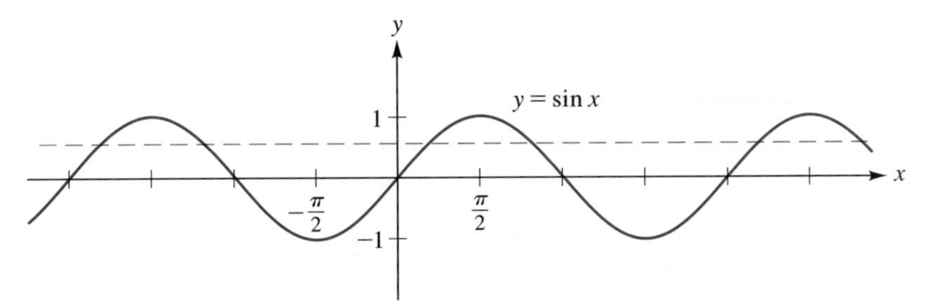

(a) The horizontal line test shows that the sine function is not one-to-one.

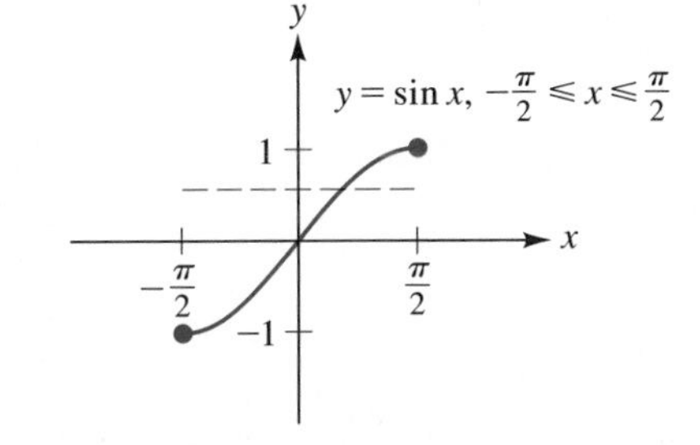

(b) By restricting the domain of the sine function to the closed interval $[-\pi/2, \pi/2]$, we obtain the restricted sine function. The horizontal line test shows that the restricted sine function is one-to-one.

FIGURE 1

Two notations are commonly used to denote the inverse sine function:

$$y = \sin^{-1} x \qquad \text{and} \qquad y = \arcsin x$$

Initially, at least, we will use the notation $y = \sin^{-1} x$. The graph of $y = \sin^{-1} x$ is easily obtained using the fact that the graph of a function and its inverse are reflections of one another about the line $y = x$. Figure 2(a) shows the graph of the restricted sine function and its inverse, $y = \sin^{-1} x$. Figure 2(b) shows the graph of $y = \sin^{-1} x$ alone. From Figure 2(b), we see that the domain of $y = \sin^{-1} x$ is the interval $[-1, 1]$, while the range is $[-\frac{\pi}{2}, \frac{\pi}{2}]$.

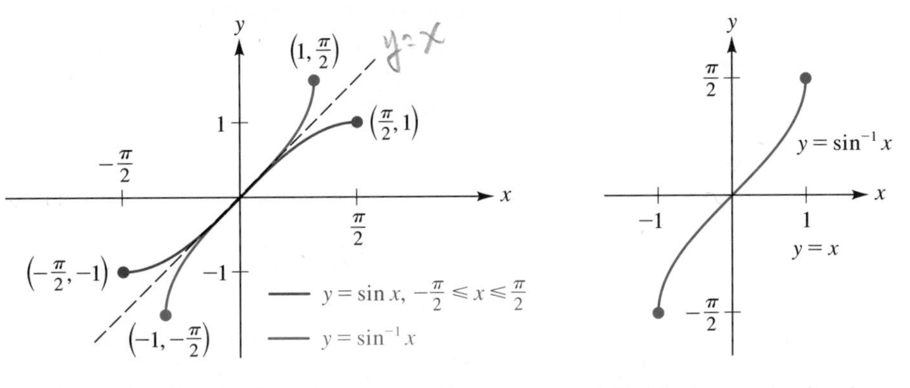

FIGURE 2 (a) The graphs of a function and its inverse are mirror images of one another about the line $y = x$. (b) The inverse sine function.

Values of $\sin^{-1} x$ are computed according to the following rule.

$\sin^{-1} x$ is that number in the interval $[-\pi/2, \pi/2]$ whose sine is x.

To see why this is so, we begin with the restricted sine function:

$$y = \sin x \qquad \left(-\frac{\pi}{2} \leq x \leq \frac{\pi}{2}\right)$$

As explained in Section 3.5, the inverse function is obtained by interchanging x and y (the inputs and the outputs). So, for the inverse function, we have

$$x = \sin y \qquad \left(-\tfrac{\pi}{2} \leq y \leq \tfrac{\pi}{2}\right) \tag{1}$$

Equation (1) tells us that y is the number in the interval $\left[-\frac{\pi}{2}, \frac{\pi}{2}\right]$ whose sine is x. This is what we wished to show.

EXAMPLE 1 Evaluate: **(a)** $\sin^{-1}(1/2)$; **(b)** $\sin^{-1}(-1/2)$.

Solution

(a) $\sin^{-1}(1/2)$ is that number in the interval $\left[-\frac{\pi}{2}, \frac{\pi}{2}\right]$ whose sine is $1/2$. Since $\sin(\pi/6) = 1/2$, we conclude that $\sin^{-1}(1/2) = \pi/6$.

(b) Since $\sin(-\pi/6) = -1/2$ and $-\pi/6$ is in the interval $\left[-\frac{\pi}{2}, \frac{\pi}{2}\right]$, we conclude that $\sin^{-1}(-1/2) = -\pi/6$.

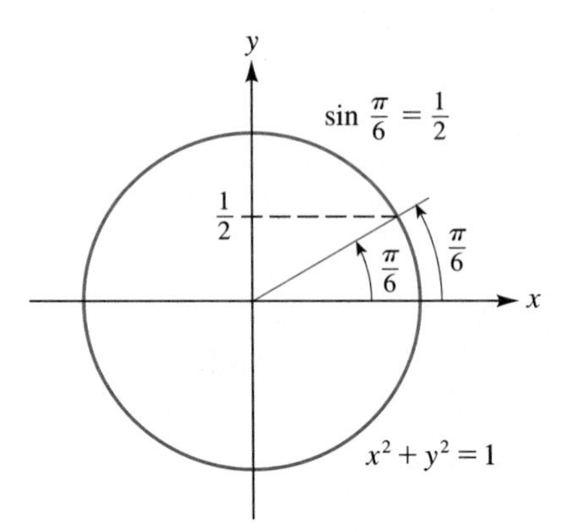

FIGURE 3

We can use the result in Example 1(a) to see why the notation *arcsin* is used for the inverse sine function. With the arcsin notation, the result in Example 1(a) can be written

$$\arcsin \frac{1}{2} = \frac{\pi}{6}$$

Now, as Figure 3 indicates, the arc length with a sine of $1/2$ is $\pi/6$. This is the idea behind the arcsin notation.

EXAMPLE 2 Evaluate $\arcsin(3/4)$. Round the answer to three decimal places.

Solution The quantity $\arcsin(3/4)$ is that number (or arc length, or angle in radians) in the interval $\left[-\frac{\pi}{2}, \frac{\pi}{2}\right]$ whose sine is $3/4$. Since we're not familiar with an angle with a sine of $3/4$, we use a calculator. As in previous sections, we show keystrokes for the two basic types of calculators: input-before-function calculators and function-before-input calculators.

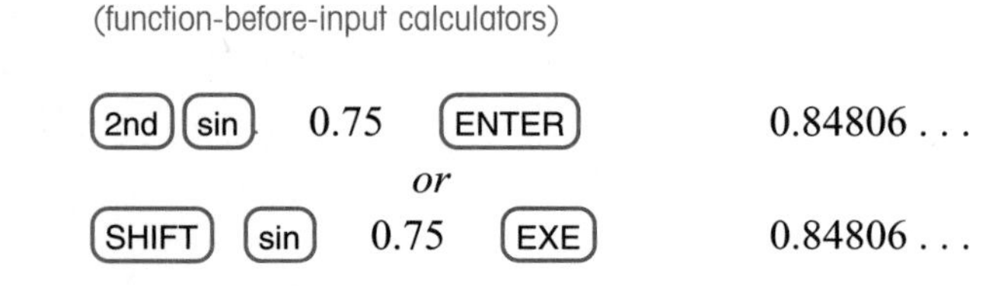

EXPRESSION	KEYSTROKES (input-before-function calculators)	KEYSTROKES (function-before-input calculators)	OUTPUT
arcsin(0.75)	0.75 [INV] [sin]	[2nd] [sin] 0.75 [ENTER]	0.84806 . . .
		or	
		[SHIFT] [sin] 0.75 [EXE]	0.84806 . . .

Now, rounding to three decimal places, our result is

$$\arcsin(3/4) \approx 0.848$$

EXAMPLE 3 Show that the following expressions are not equal:

$$\sin^{-1} 0 \qquad \text{and} \qquad \frac{1}{\sin 0}$$

Solution The quantity $\sin^{-1} 0$ is that number in the interval $\left[-\frac{\pi}{2}, \frac{\pi}{2}\right]$ whose sine is 0. Since $\sin 0 = 0$, we conclude that

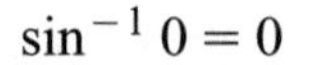

$$\sin^{-1} 0 = 0$$

On the other hand, since $\sin 0 = 0$, the expression $1/(\sin 0)$ is not even defined. Thus, the two given expressions certainly are not equal.

If f and f^{-1} are any pair of inverse functions, then by definition,

$$f[f^{-1}(x)] = x \qquad \text{for every } x \text{ in the domain of } f^{-1}$$

and

$$f^{-1}[f(x)] = x \qquad \text{for every } x \text{ in the domain of } f$$

Applying these facts to the restricted sine function, $y = \sin x$, and its inverse, $y = \sin^{-1} x$, we obtain the following two basic identities.

$\sin(\sin^{-1} x) = x$	for every x in the interval $[-1, 1]$
$\sin^{-1}(\sin x) = x$	for every x in the interval $\left[-\frac{\pi}{2}, \frac{\pi}{2}\right]$

The following example indicates that the domain restrictions accompanying these two identities cannot be ignored.

EXAMPLE 4 Compute each quantity that is defined.

(a) $\sin^{-1}\left(\sin \frac{\pi}{4}\right)$ **(b)** $\sin^{-1}(\sin \pi)$

(c) $\sin(\sin^{-1} 2)$ **(d)** $\sin\left[\sin^{-1}\left(-1/\sqrt{5}\right)\right]$

Solution

(a) Since $\pi/4$ lies in the domain of the restricted sine function, the identity $\sin^{-1}(\sin x) = x$ is applicable here. Thus,

$$\sin^{-1}\left(\sin \frac{\pi}{4}\right) = \frac{\pi}{4}$$

CHECK $\sin \frac{\pi}{4} = \sqrt{2}/2$. Therefore $\sin^{-1}\left(\sin \frac{\pi}{4}\right) = \sin^{-1}\left(\sqrt{2}/2\right) = \pi/4$.

(b) The number π is not in the domain of the restricted sine function, so the identity $\sin^{-1}(\sin x) = x$ does not apply in this case. However, since $\sin \pi = 0$, we have

$$\sin^{-1}(\sin \pi) = \sin^{-1} 0 = 0$$

Thus, $\sin^{-1}(\sin \pi)$ is equal to 0, not π.

(c) The number 2 is not in the domain of the inverse sine function. Thus, the expression $\sin(\sin^{-1} 2)$ is undefined.

(d) The identity $\sin(\sin^{-1} x) = x$ is applicable here. (Why?) Thus,

$$\sin\left[\sin^{-1}\left(-1/\sqrt{5}\right)\right] = -1/\sqrt{5}$$

EXAMPLE 5 If $\sin \theta = x/3$ and $0 < \theta < \frac{\pi}{2}$, express the quantity $\theta - \sin 2\theta$ as a function of x.

Solution The given conditions tell us that $\theta = \sin^{-1}(x/3)$. That expresses θ in terms of x. To express $\sin 2\theta$ in terms of x, we have

$$\begin{aligned}\sin 2\theta &= 2\sin\theta\cos\theta = 2(\sin\theta)\sqrt{1-\sin^2\theta}\\ &= 2\left(\frac{x}{3}\right)\sqrt{1-\frac{x^2}{9}} = \frac{2x}{3}\sqrt{\frac{9-x^2}{9}}\\ &= \frac{2x\sqrt{9-x^2}}{9}\end{aligned}$$

Now, combining the results, we have

$$\theta - \sin 2\theta = \sin^{-1}\left(\frac{x}{3}\right) - \frac{2x\sqrt{9-x^2}}{9}$$

We turn now to the inverse cosine function. We'll define it by following the same general procedure that we used to define the inverse sine. We begin by defining the **restricted cosine function:**

$$y = \cos x \qquad (0 \leq x \leq \pi)$$

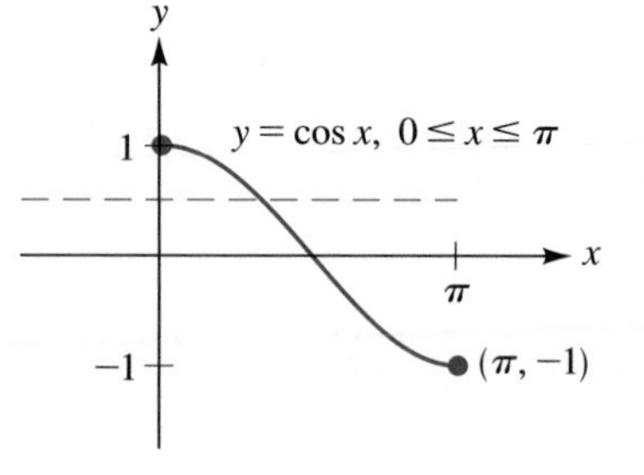

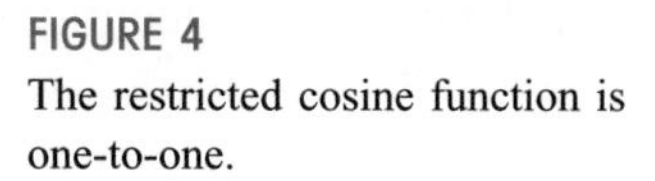

FIGURE 4
The restricted cosine function is one-to-one.

As indicated by the horizontal line test in Figure 4, the restricted cosine function is one-to-one, and therefore the inverse function exists. We denote the **inverse cosine function** by

$$y = \cos^{-1} x \qquad \text{or} \qquad y = \arccos x$$

The graph of $y = \cos^{-1} x$ is obtained by reflecting the graph of the restricted cosine function about the line $y = x$. Figure 5(a) displays the graph of the restricted cosine function along with $y = \cos^{-1} x$. Figure 5(b) shows the graph of $y = \cos^{-1} x$ alone. From Figure 5(b), we see that the domain of $y = \cos^{-1} x$ is the interval $[-1, 1]$, while the range is $[0, \pi]$.

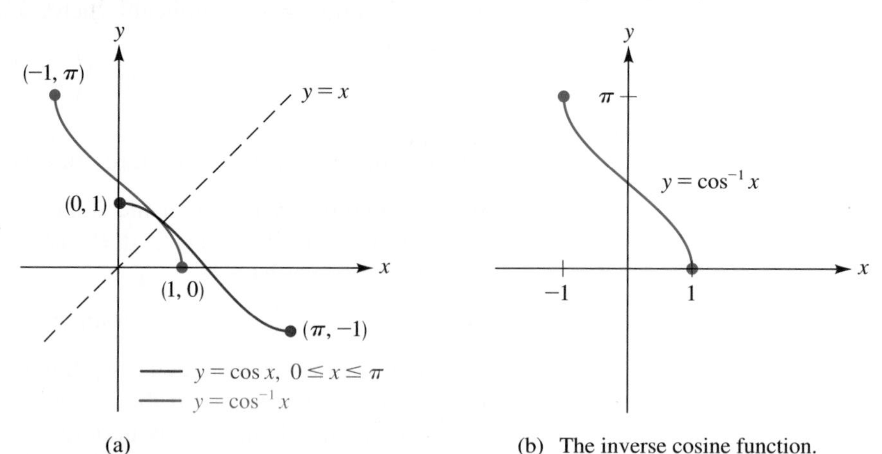

FIGURE 5 (a) (b) The inverse cosine function.

As we mentioned previously, the defining equations for inverse functions are

$$f[f^{-1}(x)] = x \qquad \text{for every } x \text{ in the domain of } f^{-1}$$

and

$$f^{-1}[f(x)] = x \qquad \text{for every } x \text{ in the domain of } f$$

In the particular case of the restricted cosine function and the inverse cosine function, these two identities read as follows.

$\cos(\cos^{-1} x) = x$	for every x in the interval $[-1, 1]$
$\cos^{-1}(\cos x) = x$	for every x in the interval $[0, \pi]$

As stated in the previous section, the values of $\cos^{-1} x$ are computed according to the following rule.

> $\cos^{-1} x$ is that number in the interval $[0, \pi]$ whose cosine is x.

To see why this is so, we begin with the restricted cosine function:

$$y = \cos x \qquad (0 \leq x \leq \pi)$$

Then for the inverse function, we interchange x and y (the inputs and the outputs) to obtain

$$x = \cos y \qquad (0 \leq y \leq \pi)$$

This last equation tells us that y is that number in the interval $[0, \pi]$ whose cosine is x. This is what we wished to show.

EXAMPLE 6 Compute each of the following:
(a) $\cos^{-1}(0)$; **(b)** $\arccos\left(\cos \frac{2}{5}\right)$.

Solution
(a) $\cos^{-1}(0)$ is that number in the interval $[0, \pi]$ whose cosine is 0. Since $\cos(\pi/2) = 0$, we have

$$\cos^{-1}(0) = \pi/2$$

(b) Since the number $2/5$ is in the domain of the restricted cosine function, the identity $\arccos(\cos x) = x$ is applicable. Thus,

$$\arccos\left(\cos \frac{2}{5}\right) = \frac{2}{5}$$

EXAMPLE 7 Show that $\sin(\cos^{-1} x) = \sqrt{1 - x^2}$ for $-1 \leq x \leq 1$.

Solution We use the identity $\sin y = \sqrt{1 - \cos^2 y}$, which is valid for $0 \leq y \leq \pi$. Substituting $\cos^{-1} x$ for y in this identity, we obtain

$$\begin{aligned} \sin(\cos^{-1} x) &= \sqrt{1 - [\cos(\cos^{-1} x)]^2} \\ &= \sqrt{1 - x^2} \qquad \text{as required} \end{aligned}$$

Before leaving this example, we point out an alternate method of solution that is useful when the restriction on x is $0 < x < 1$. In this case, we let $\theta = \cos^{-1} x$. Then θ is the radian measure of the acute angle whose cosine is x and we can sketch θ as shown in Figure 6. The sides of the triangle in Figure 6

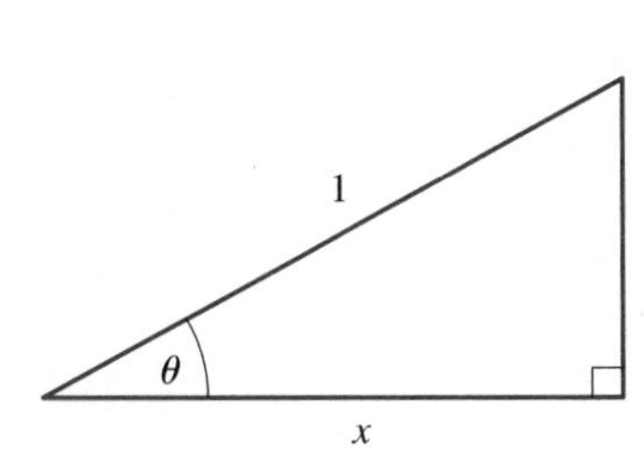

FIGURE 6

have been labeled in such a way that $\cos\theta = x$. Then, by the Pythagorean theorem, we find that the third side of the triangle is $\sqrt{1-x^2}$. We therefore have

$$\sin\theta = \frac{\text{opposite}}{\text{hypotenuse}} = \frac{\sqrt{1-x^2}}{1} = \sqrt{1-x^2}$$

Just as there is a basic identity connecting $\sin x$ and $\cos x$, namely, $\sin^2 x + \cos^2 x = 1$, there is also an identity connecting $\sin^{-1} x$ and $\cos^{-1} x$:

$$\sin^{-1} x + \cos^{-1} x = \frac{\pi}{2} \quad \text{for every } x \text{ in the closed interval } [-1, 1]$$

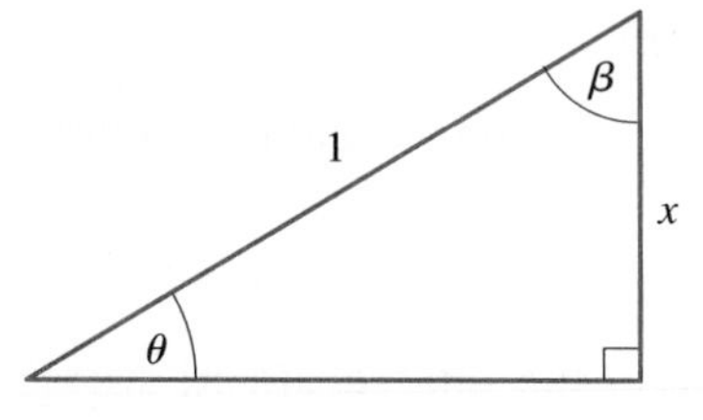

FIGURE 7

We can use Figure 7 to see why this identity is valid when $0 < x < 1$. (Exercise 55 shows you how to establish this identity for all values of x from -1 to 1.) In Figure 7, assume that θ and β are the radian measures of the indicated acute angles. Then we have

$$\sin\theta = \frac{x}{1} = x \qquad \text{and} \qquad \cos\beta = \frac{x}{1} = x$$

and therefore

$$\theta = \sin^{-1} x \qquad \text{and} \qquad \beta = \cos^{-1} x$$

But, from Figure 7, we know that $\theta + \beta = \pi/2$. Consequently, we have $\sin^{-1} x + \cos^{-1} x = \pi/2$, as required.

Now let us turn to the definition of the inverse tangent function. We begin by defining the **restricted tangent function:**

$$y = \tan x \qquad \left(-\frac{\pi}{2} < x < \frac{\pi}{2}\right)$$

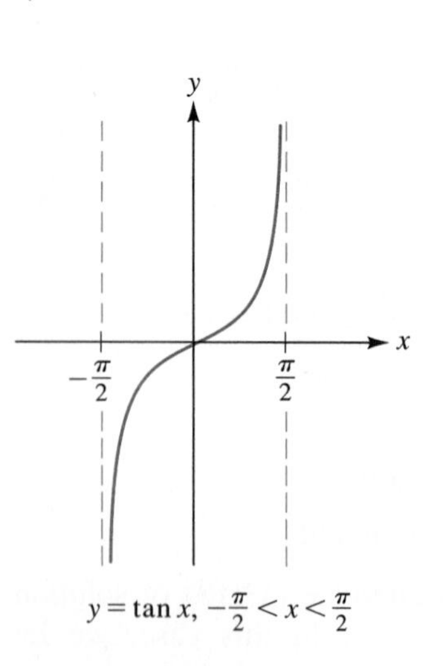

FIGURE 8
The restricted tangent function.

Figure 8 shows the graph of the restricted tangent function. As you can check by applying the horizontal line test, the restricted tangent function is one-to-one. This tells us that the inverse function exists. The two common notations for this **inverse tangent function** are

$$y = \tan^{-1} x \qquad \text{or} \qquad y = \arctan x$$

The graph of $y = \tan^{-1} x$ is obtained in the same way we obtained the graphs of the inverse sine and the inverse cosine. That is, we use the fact that the graphs of a function and its inverse are mirror images of one another about the line $y = x$. In Figure 9(a), we've reflected the restricted tangent function about the line $y = x$ to obtain the graph of $y = \tan^{-1} x$. In Figure 9(b) we show the graph of $y = \tan^{-1} x$ alone.

From the graph in Figure 9(b), we can see that the domain of the inverse tangent function is the set of all real numbers, and the range is the open interval $\left(-\frac{\pi}{2}, \frac{\pi}{2}\right)$. Furthermore, if we follow exactly the same line of reasoning used previously for the inverse sine and the inverse cosine, we obtain the following properties of the inverse tangent function.

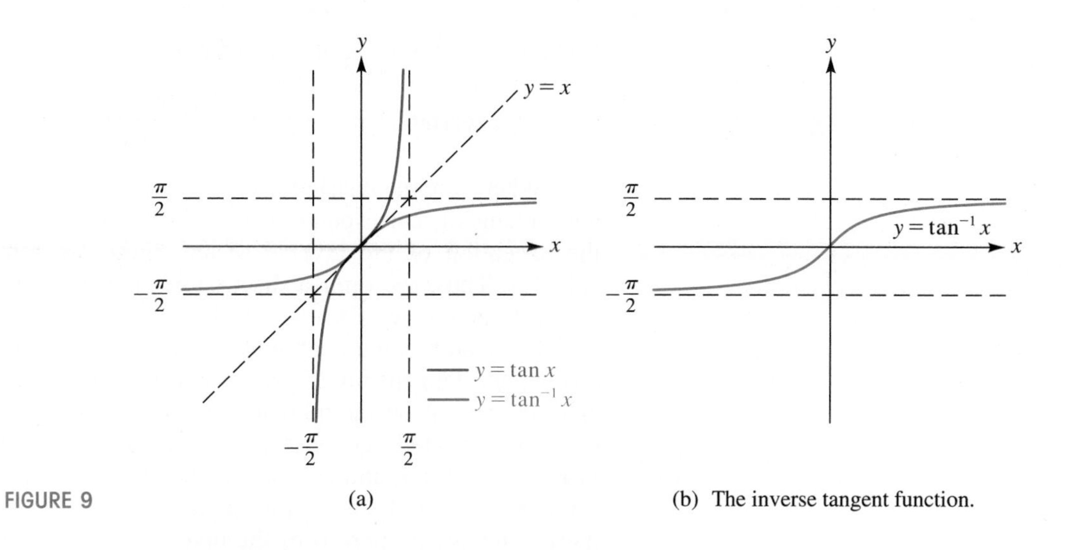

FIGURE 9 (a) (b) The inverse tangent function.

> $\tan^{-1} x$ is that number in the interval $\left(-\frac{\pi}{2}, \frac{\pi}{2}\right)$ whose tangent is x.
>
> $\tan(\tan^{-1} x) = x$ for every real number x
>
> $\tan^{-1}(\tan x) = x$ for every x in the open interval $\left(-\frac{\pi}{2}, \frac{\pi}{2}\right)$

EXAMPLE 8 Evaluate: **(a)** $\tan^{-1}(-1)$; **(b)** $\tan\left(\tan^{-1}\sqrt{5}\right)$.

Solution

(a) The quantity $\tan^{-1}(-1)$ is that number in the interval $\left(-\frac{\pi}{2}, \frac{\pi}{2}\right)$ whose tangent is -1. Since $\tan(-\pi/4) = -1$, we have $\tan^{-1}(-1) = -\pi/4$.

(b) The identity $\tan(\tan^{-1} x) = x$ holds for all real numbers x. We therefore have $\tan\left(\tan^{-1}\sqrt{5}\right) = \sqrt{5}$.

EXAMPLE 9 If $\tan\theta = x/3$ and $0 < \theta < \frac{\pi}{2}$, express the quantity $\theta - \tan 2\theta$ as a function of x.

Solution The given equation tells us that $\theta = \tan^{-1}(x/3)$. That expresses θ in terms of x. To express $\tan 2\theta$ in terms of x, we have

$$\tan 2\theta = \frac{2\tan\theta}{1 - \tan^2\theta} = \frac{2(x/3)}{1 - (x/3)^2}$$

$$= \frac{2x/3}{1 - (x^2/9)} = \frac{6x}{9 - x^2} \quad \text{multiplying numerator and denominator by 9}$$

Now, combining the results, we have $\theta - \tan 2\theta = \tan^{-1}\dfrac{x}{3} - \dfrac{6x}{9 - x^2}$.

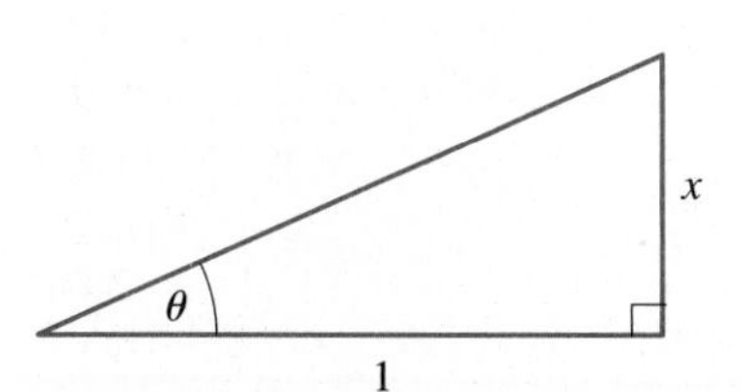

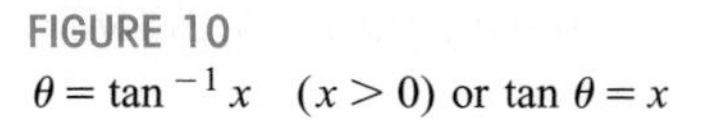

FIGURE 10
$\theta = \tan^{-1} x \quad (x > 0)$ or $\tan\theta = x$

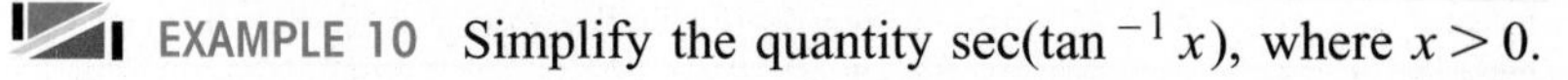

EXAMPLE 10 Simplify the quantity $\sec(\tan^{-1} x)$, where $x > 0$.

Solution We let $\theta = \tan^{-1} x$ or, equivalently, $\tan\theta = x$. Now, as shown in Figure 10, we sketch a right triangle with an angle θ (in radians) whose tangent is x. The Pythagorean theorem tells us that the length of the hypotenuse in this triangle is $\sqrt{1+x^2}$. Consequently, we have

$$\sec(\tan^{-1} x) = \sec\theta = \frac{\text{hypotenuse}}{\text{adjacent}} = \frac{\sqrt{1+x^2}}{1} = \sqrt{1+x^2}$$

When suitable restrictions are placed on the domains of the secant, cosecant, and cotangent, corresponding inverse functions can be defined. However, with the exception of the inverse secant, these are rarely, if ever, encountered in calculus. Thus, we omit a discussion of these functions here. (For the inverse secant function, see Exercise 65.)

Let's take a moment now to review our work in this section. The sine, cosine, and tangent functions are not one-to-one functions. However, with suitable restrictions on the domains, we do obtain one-to-one functions. The graphs of these restricted versions of the sine, cosine, and tangent functions are shown in Figures 1(b), 4, and 8, respectively. It is the inverses of these restricted functions that are the focus in this section. In the three boxes that follow, we summarize the key properties of the inverse sine, the inverse cosine, and the inverse tangent.

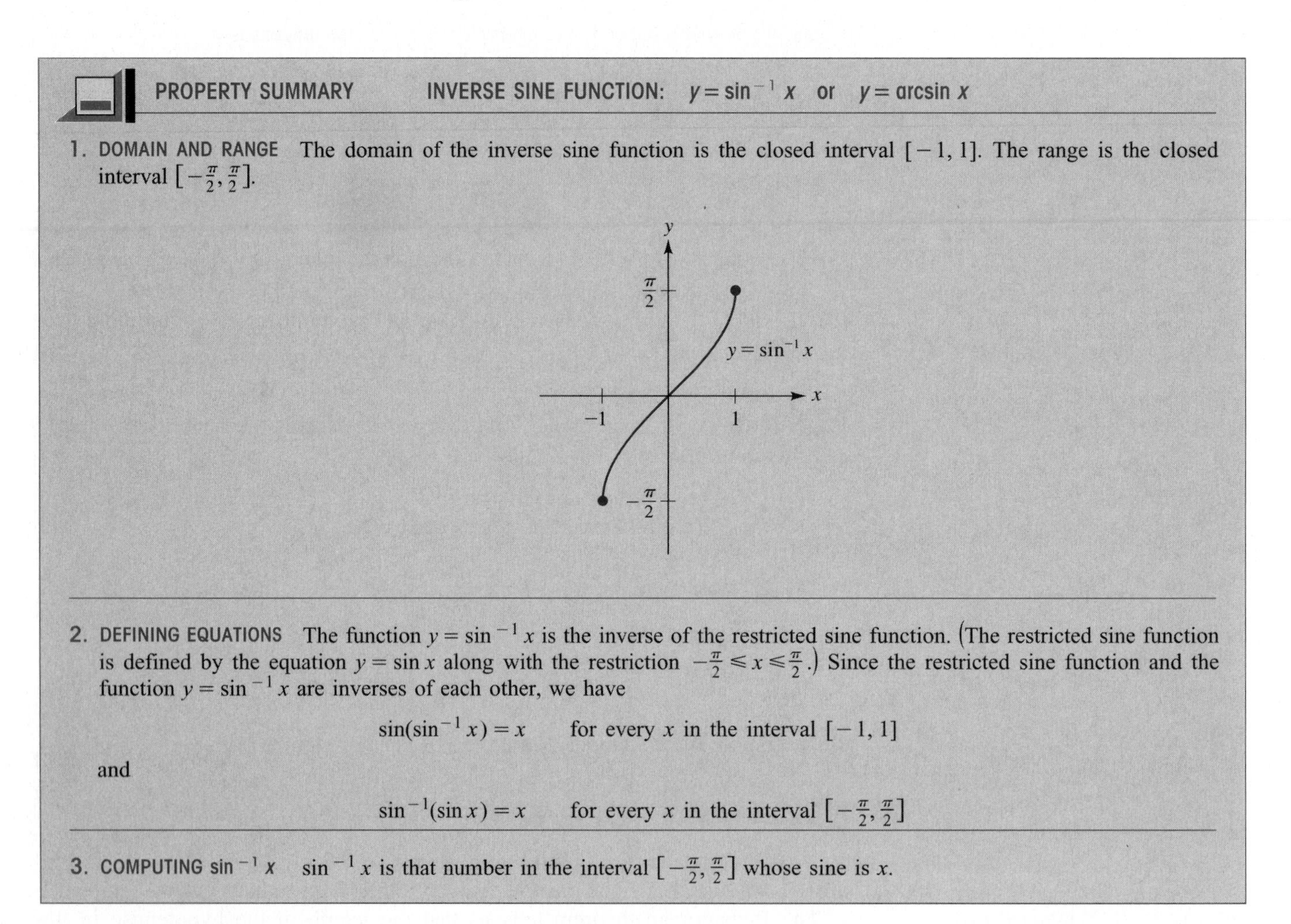

PROPERTY SUMMARY — INVERSE SINE FUNCTION: $y = \sin^{-1} x$ or $y = \arcsin x$

1. **DOMAIN AND RANGE** The domain of the inverse sine function is the closed interval $[-1, 1]$. The range is the closed interval $[-\frac{\pi}{2}, \frac{\pi}{2}]$.

2. **DEFINING EQUATIONS** The function $y = \sin^{-1} x$ is the inverse of the restricted sine function. (The restricted sine function is defined by the equation $y = \sin x$ along with the restriction $-\frac{\pi}{2} \le x \le \frac{\pi}{2}$.) Since the restricted sine function and the function $y = \sin^{-1} x$ are inverses of each other, we have

$$\sin(\sin^{-1} x) = x \qquad \text{for every } x \text{ in the interval } [-1, 1]$$

and

$$\sin^{-1}(\sin x) = x \qquad \text{for every } x \text{ in the interval } [-\tfrac{\pi}{2}, \tfrac{\pi}{2}]$$

3. **COMPUTING $\sin^{-1} x$** $\sin^{-1} x$ is that number in the interval $[-\frac{\pi}{2}, \frac{\pi}{2}]$ whose sine is x.

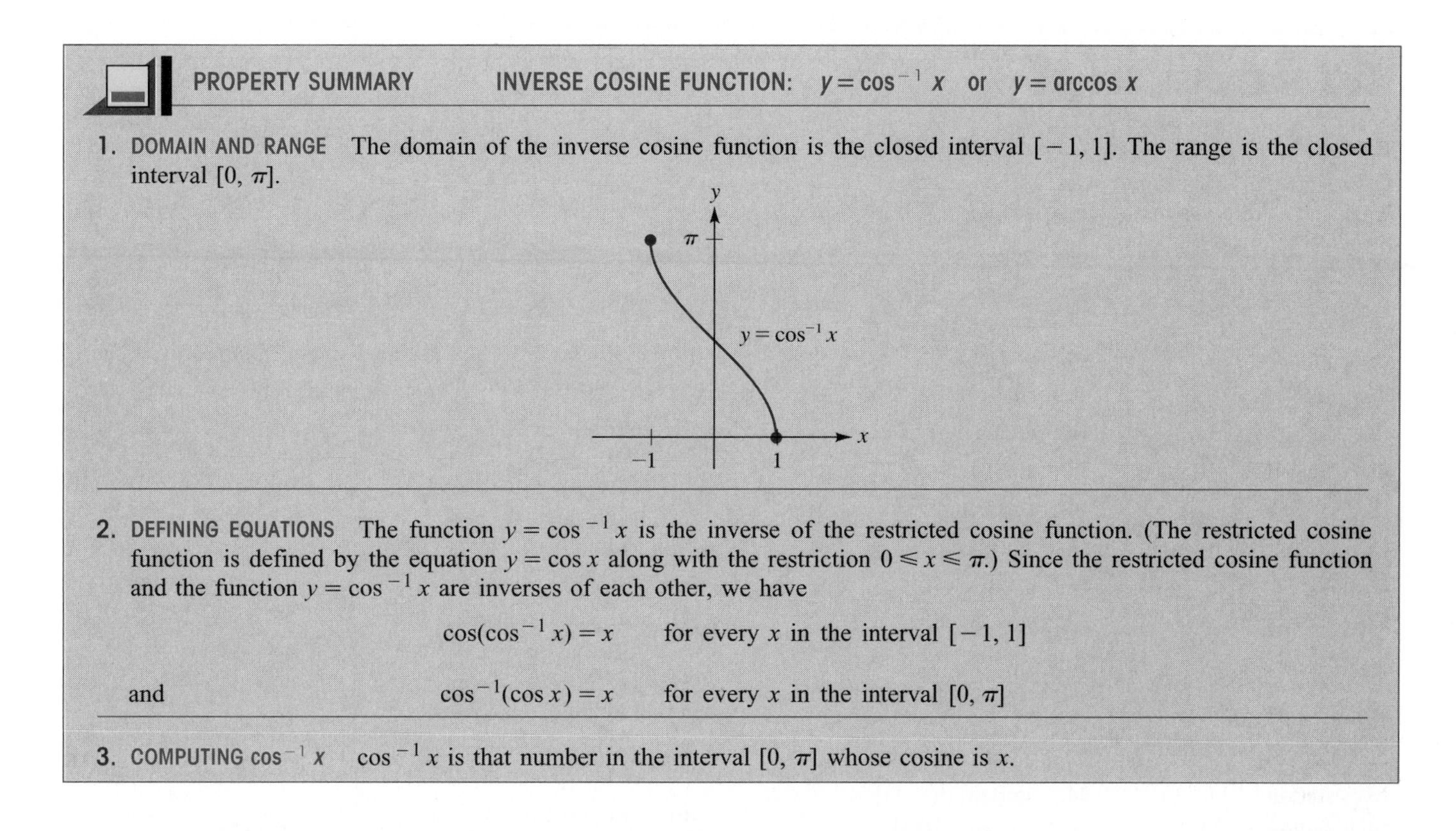

PROPERTY SUMMARY **INVERSE COSINE FUNCTION: $y = \cos^{-1} x$ or $y = \arccos x$**

1. **DOMAIN AND RANGE** The domain of the inverse cosine function is the closed interval $[-1, 1]$. The range is the closed interval $[0, \pi]$.

2. **DEFINING EQUATIONS** The function $y = \cos^{-1} x$ is the inverse of the restricted cosine function. (The restricted cosine function is defined by the equation $y = \cos x$ along with the restriction $0 \leq x \leq \pi$.) Since the restricted cosine function and the function $y = \cos^{-1} x$ are inverses of each other, we have

$$\cos(\cos^{-1} x) = x \qquad \text{for every } x \text{ in the interval } [-1, 1]$$

and

$$\cos^{-1}(\cos x) = x \qquad \text{for every } x \text{ in the interval } [0, \pi]$$

3. **COMPUTING $\cos^{-1} x$** $\cos^{-1} x$ is that number in the interval $[0, \pi]$ whose cosine is x.

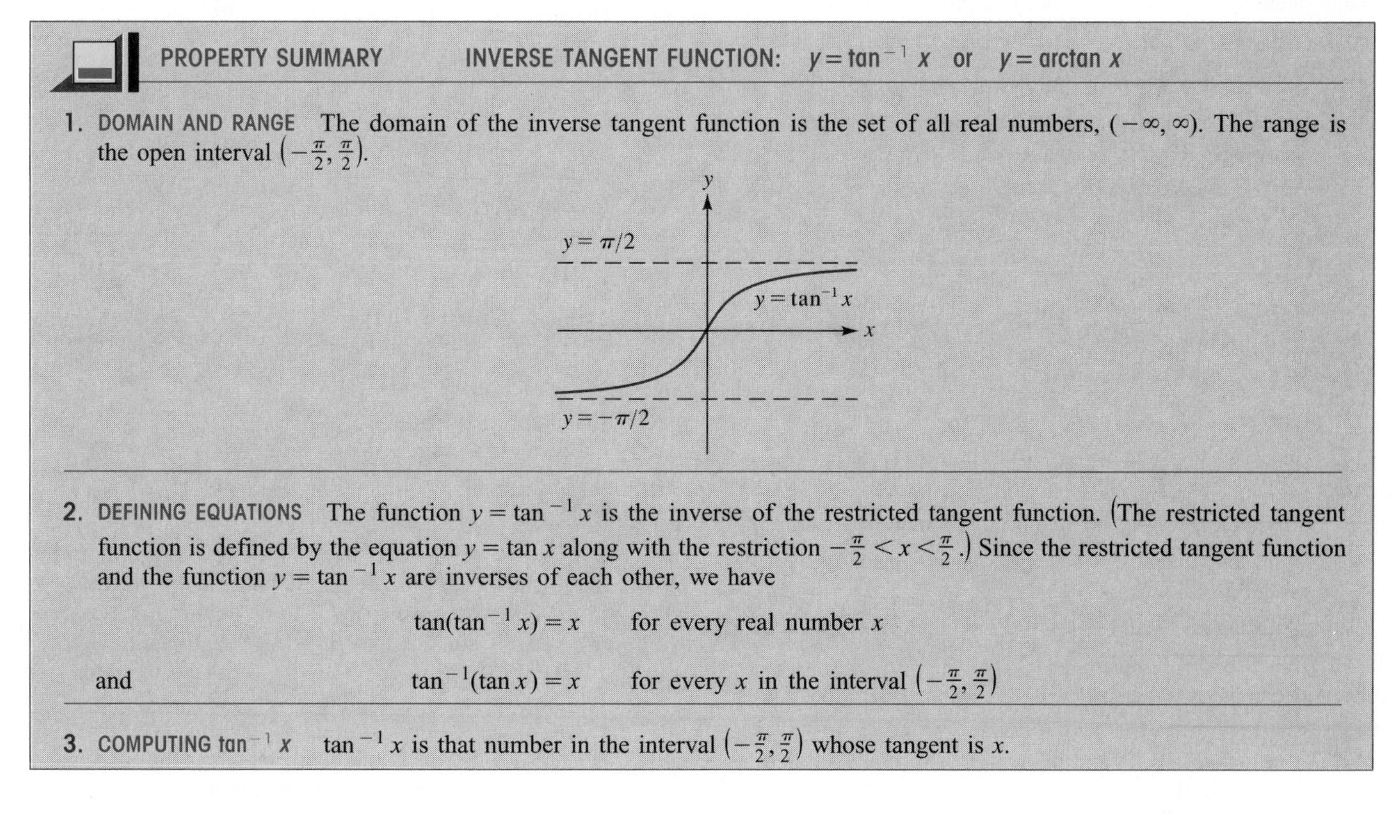

PROPERTY SUMMARY **INVERSE TANGENT FUNCTION: $y = \tan^{-1} x$ or $y = \arctan x$**

1. **DOMAIN AND RANGE** The domain of the inverse tangent function is the set of all real numbers, $(-\infty, \infty)$. The range is the open interval $\left(-\frac{\pi}{2}, \frac{\pi}{2}\right)$.

2. **DEFINING EQUATIONS** The function $y = \tan^{-1} x$ is the inverse of the restricted tangent function. (The restricted tangent function is defined by the equation $y = \tan x$ along with the restriction $-\frac{\pi}{2} < x < \frac{\pi}{2}$.) Since the restricted tangent function and the function $y = \tan^{-1} x$ are inverses of each other, we have

$$\tan(\tan^{-1} x) = x \qquad \text{for every real number } x$$

and

$$\tan^{-1}(\tan x) = x \qquad \text{for every } x \text{ in the interval } \left(-\frac{\pi}{2}, \frac{\pi}{2}\right)$$

3. **COMPUTING $\tan^{-1} x$** $\tan^{-1} x$ is that number in the interval $\left(-\frac{\pi}{2}, \frac{\pi}{2}\right)$ whose tangent is x.

EXERCISE SET 8.5

Do 1, 3, 9, 13, 15, 11, 21, 29

A

In Exercises 1–20, evaluate each of the quantities that is defined, but do not use a calculator or tables. If a quantity is undefined, say so.

1. $\sin^{-1}(\sqrt{3}/2)$
2. $\cos^{-1}(-1)$
3. $\tan^{-1}\sqrt{3}$
4. $\arccos(-\sqrt{2}/2)$
5. $\arctan(-1/\sqrt{3})$
6. $\arcsin(-1)$
7. $\tan^{-1}1$
8. $\sin^{-1}0$
9. $\cos^{-1}2\pi$
10. $\arctan 0$
11. $\sin[\sin^{-1}(\frac{1}{4})]$
12. $\cos[\cos^{-1}(\frac{4}{3})]$
13. $\cos[\cos^{-1}(\frac{3}{4})]$
14. $\tan(\arctan 3\pi)$
15. $\arctan[\tan(-\frac{\pi}{7})]$
16. $\sin(\arcsin 2)$
17. $\arcsin(\sin\frac{\pi}{2})$
18. $\arccos(\cos\frac{\pi}{8})$
19. $\arccos(\cos 2\pi)$
20. $\sin^{-1}(\sin\frac{3\pi}{2})$

In Exercises 21–30, evaluate the given quantities without using a calculator or tables.

21. $\tan[\sin^{-1}(\frac{4}{5})]$
22. $\cos(\arcsin\frac{2}{7})$
23. $\sin(\tan^{-1}1)$
24. $\sin[\tan^{-1}(-1)]$
25. $\tan(\arccos\frac{5}{13})$
26. $\cos[\sin^{-1}(\frac{2}{3})]$
27. $\cos(\arctan\sqrt{3})$
28. $\sin[\cos^{-1}(\frac{1}{3})]$
29. $\sin[\arccos(-\frac{1}{3})]$
30. $\tan(\arcsin\frac{20}{29})$

31. Use a calculator to evaluate each of the following quantities. Express your answers to two decimal places without rounding them.
(a) $\sin^{-1}(\frac{3}{4})$
(b) $\cos^{-1}(\frac{2}{3})$
(c) $\tan^{-1}\pi$
(d) $\tan^{-1}(\tan^{-1}\pi)$

In Exercises 32–34, evaluate the given expressions without using a calculator or tables.

32. $\csc[\sin^{-1}(\frac{1}{2}) - \cos^{-1}(\frac{1}{2})]$

33. $\sec[\cos^{-1}(\sqrt{2}/2) + \sin^{-1}(-1)]$

34. $\cot[\cos^{-1}(-1/2) + \cos^{-1}(0) + \tan^{-1}(1/\sqrt{3})]$

35. Show that $\cos(\sin^{-1}x) = \sqrt{1-x^2}$ for $-1 \le x \le 1$.
Suggestion: Use the method of Example 7 in the text.

36. If $\sin\theta = 2x$ and $0 < \theta < \frac{\pi}{2}$, express $\theta + \cos 2\theta$ as a function of x.

37. If $\sin\theta = 3x/2$ and $0 < \theta < \frac{\pi}{2}$, express $\frac{1}{4}\theta - \sin 2\theta$ as a function of x.

38. If $\cos\theta = x - 1$ and $0 < \theta < \frac{\pi}{2}$, express $2\theta - \cos 2\theta$ as a function of x.

39. If $\tan\theta = \frac{1}{2}(x-1)$ and $0 < \theta < \frac{\pi}{2}$, express $\theta - \cos\theta$ as a function of x.

40. If $\tan\theta = \frac{1}{3}(x+1)$ and $0 < \theta < \frac{\pi}{2}$, express $2\theta + \tan 2\theta$ as a function of x.

B

In Exercises 41–50, use the graphing techniques from Section 3.3 to sketch the graph of each function.

41. **(a)** $y = -\sin^{-1}x$
(b) $y = \sin^{-1}(-x)$
(c) $y = -\sin^{-1}(-x)$

42. **(a)** $y = -\cos^{-1}x$
(b) $y = \cos^{-1}(-x)$
(c) $y = -\cos^{-1}(-x)$

43. **(a)** $f(x) = \arccos(x+1)$
(b) $g(x) = \arccos x + \frac{\pi}{2}$

44. **(a)** $F(x) = \arcsin(x-1)$
(b) $G(x) = \arcsin x - \frac{\pi}{4}$

45. **(a)** $y = \arcsin(2-x) + \frac{\pi}{2}$
(b) $y = -\arcsin(2-x) + \frac{\pi}{2}$

46. **(a)** $y = \cos^{-1}(1-x)$
(b) $y = -\cos^{-1}(1-x) - \pi$

47. **(a)** $y = -\tan^{-1}x$
(b) $y = \tan^{-1}(-x)$
(c) $y = -\tan^{-1}(-x)$

48. **(a)** $y = \arctan(x+3)$
(b) $y = \arctan x + \frac{\pi}{2}$
(c) $y = -\arctan(-x) + \frac{\pi}{2}$

49. $f(x) = -\arctan(1-x) - \frac{\pi}{2}$

50. $y = \tan^{-1}(2-x) + \frac{\pi}{4}$

51. Evaluate $\sin(2\tan^{-1}4)$.
Hint: $\sin 2\theta = 2\sin\theta\cos\theta$.

52. Evaluate $\cos[2\sin^{-1}(\frac{5}{13})]$.

53. Evaluate $\sin(\arccos\frac{3}{5} - \arctan\frac{7}{13})$. *Suggestion:* Use the formula for $\sin(x-y)$.

54. Show that $\sin[\sin^{-1}(\frac{1}{3}) + \sin^{-1}(\frac{1}{4})] = (\sqrt{15} + 2\sqrt{2})/12$.

55. In the text we showed that $\sin^{-1}x + \cos^{-1}x = \pi/2$ for x in the open interval (0, 1). Follow steps (a) and (b) to show that this identity actually holds for every x in the closed interval $[-1, 1]$.
(a) Let $\alpha = \sin^{-1}x$ and $\beta = \cos^{-1}x$. Explain why $-\frac{\pi}{2} \le \alpha + \beta \le \frac{3\pi}{2}$. *Hint:* What are the ranges of the inverse sine and the inverse cosine functions?
(b) Use the addition formula for sine to show that $\sin(\alpha + \beta) = 1$. Conclude [with the help of part (a)] that $\alpha + \beta = \pi/2$, as required.

In Exercises 56–58, solve the given equations.

56. $\cos^{-1} t = \sin^{-1} t$ *Hint:* Compute the cosine of both sides.

57. $\sin^{-1}(3t - 2) = \sin^{-1} t - \cos^{-1} t$

58. $\sin^{-1}(\sin^{-1} x) = \sin^{-1}(\cos^{-1} x)$

59. Show that $\arctan x + \arctan y = \arctan \frac{x+y}{1-xy}$ when x and y are positive and $xy < 1$.

60. Use the identity in Exercise 59 to show that the following equations are correct.

(a) $\arctan \frac{1}{2} + \arctan \frac{1}{3} = \pi/4$

(b) $\arctan \frac{1}{4} + \arctan \frac{3}{5} = \pi/4$

(c) $2 \arctan \frac{1}{3} + \arctan \frac{1}{7} = \pi/4$

Hint: $2 \arctan \frac{1}{3} = \arctan \frac{1}{3} + \arctan \frac{1}{3}$

(d) $\arctan \frac{1}{2} + \arctan \frac{1}{5} + \arctan \frac{1}{8} = \pi/4$

In Exercises 61–63, solve the equations.

61. $2 \tan^{-1} x = \tan^{-1} \frac{1}{4x}$ *Hint:* Compute the tangent of both sides.

62. $\tan^{-1} x = \cos^{-1} x$

63. $2 \tan^{-1} \sqrt{t - t^2} = \tan^{-1} t + \tan^{-1}(1 - t)$

64. (a) As you can see in the accompanying figure, the graphs of $y = \cos^{-1} x$ and $y = \tan^{-1} x$ intersect at a point in Quadrant I. By solving the equation $\cos^{-1} x = \tan^{-1} x$, show that the x-coordinate of this intersection point is given by

$$x = \sqrt{\frac{\sqrt{5} - 1}{2}}$$

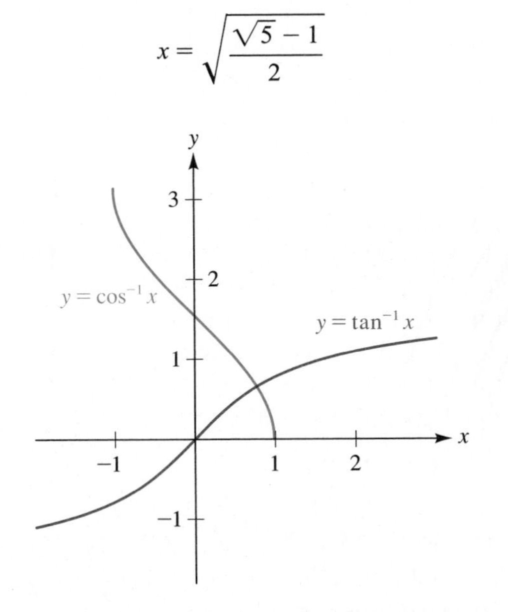

(b) Use the result in part (a) along with your calculator to specify the coordinates of the intersection point. Express both coordinates as decimals, rounded to three places.

65. (a) Assume that the domain of the **restricted secant function** is $[0, \frac{\pi}{2}) \cup [\pi, \frac{3\pi}{2})$. Sketch the restricted secant function and note that it is one-to-one.

(b) The **inverse secant function** can be defined as follows:

> $\sec^{-1} x$ is the unique number in the set $[0, \frac{\pi}{2}) \cup [\pi, \frac{3\pi}{2})$ whose secant is x.

Using the fact that this function is the inverse of the restricted secant function, sketch the graph of $y = \sec^{-1} x$.

(c) Evaluate $\sec^{-1}(2/\sqrt{3})$ and $\sec^{-1}(-2/\sqrt{3})$.

(d) Evaluate $\sec^{-1}(\sqrt{2})$ and $\sec^{-1}(-\sqrt{2})$.

(e) Evaluate $\sec(\sec^{-1} 2)$ and $\sec^{-1}(\sec 0)$.

66. Evaluate each of the following quantities. (The inverse secant function is defined in the previous exercise.)

(a) $\cos[\sec^{-1}(\frac{4}{3})]$ **(b)** $\sin(\sec^{-1} 4)$

(c) $\tan[\sec^{-1}(-3)]$ **(d)** $\cot[\sec^{-1}(\frac{13}{12})]$

In Exercises 67 and 68, we make use of some simple geometric figures to evaluate sums involving the inverse tangent function. The idea here is taken from the note by Professor Edward M. Harris entitled Behold! Sums of Arctan *(The College Mathematics Journal, vol. 18 no. 2, p. 141).*

67. In this exercise, we use the following figure to show that $\tan^{-1}(\frac{2}{3}) + \tan^{-1}(\frac{1}{5}) = \frac{\pi}{4}$

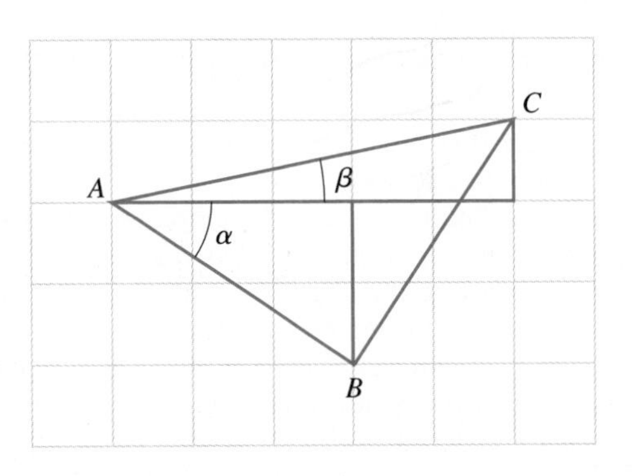

(a) Show that $\angle ABC = \pi/2$. *Hint:* Compute the slopes of $\overline{AB}$ and $\overline{BC}$. (Assume that the grid lines are marked off at one-unit intervals.)

(b) Show that $AB = BC$ and conclude that $\angle BAC = \pi/4$.

(c) Now explain why $\alpha = \tan^{-1}(\frac{2}{3})$, $\beta = \tan^{-1}(\frac{1}{5})$, and $\tan^{-1}(\frac{2}{3}) + \tan^{-1}(\frac{1}{5}) = \pi/4$.

68. In this exercise, we use the accompanying figure to show that $\tan^{-1} 1 + \tan^{-1} 2 + \tan^{-1} 3 = \pi$.

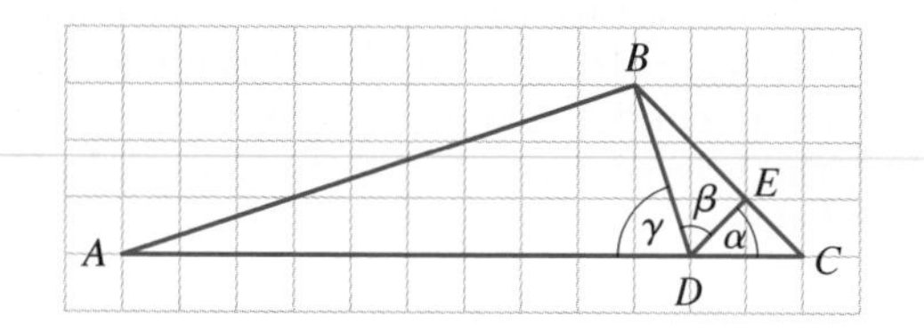

(a) By computing slopes, show that $\overline{DE} \perp \overline{BC}$ and $\overline{AB} \perp \overline{BD}$. (Assume that the grid lines are at one-unit intervals.)

(b) Determine the following lengths: *DE, CE, BE, AB,* and *BD*.

(c) Show that $\tan\alpha = 1$, $\tan\beta = 2$, and $\tan\gamma = 3$. Then explain why $\tan^{-1} 1 + \tan^{-1} 2 + \tan^{-1} 3 = \pi$.

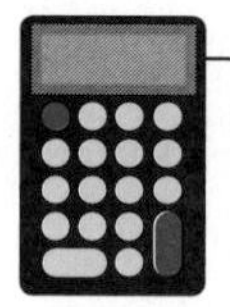

GRAPHING UTILITY EXERCISES FOR SECTION 8.5

1. (a) Graph the function $y = \arcsin x$ using a viewing rectangle that extends from -2 to 2 in both the x- and the y-directions. Then use the graphing utility to estimate the maximum and the minimum values of the function.

(b) What are the exact maximum and minimum values of the function $y = \arcsin x$? (If you need help answering this, refer to the graph on page 526.) Check that your estimates in part (a) are consistent with these results.

2. Using the viewing rectangle specified in Exercise 1, graph the function $f(x) = \sin(\arcsin x)$. What do you observe? What identity does this demonstrate?

3. Using the viewing rectangle specified in Exercise 1, graph the function $g(x) = \sin^{-1} x + \cos^{-1} x$. What do you observe? What is the exact value for the y-intercept of the graph? What identity does this demonstrate?

4. (a) In the standard viewing rectangle, graph the function $y = \arctan x$.

(b) According to the text, the graph in part (a) has two horizontal asymptotes. What are the equations for these two asymptotes? Add the graphs of the two asymptotes to the picture obtained in part (a). Finally, to emphasize the fact that the two lines are indeed asymptotes, change the viewing rectangle so that x extends from -50 to 50. What do you observe?

5. In Example 7 we proved the identity $\sin(\cos^{-1} x) = \sqrt{1 - x^2}$ for $-1 \leq x \leq 1$. Demonstrate this identity visually by graphing the two functions $y = \sin(\cos^{-1} x)$ and $y = \sqrt{1 - x^2}$ and noting that the graphs appear to be identical.

6. Demonstrate the identity $\cos(\sin^{-1} x) = \sqrt{1 - x^2}$ for $-1 \leq x \leq 1$ by graphing the two functions $y = \cos(\sin^{-1} x)$ and $y = \sqrt{1 - x^2}$ and noting that the graphs appear to be identical.

7. In Example 10 we found that $\sec(\tan^{-1} x) = \sqrt{1 + x^2}$ for $x > 0$. Actually, this identity is valid for all real numbers. Demonstrate this visually by graphing the two functions $y = \sec(\tan^{-1} x)$ and $y = \sqrt{1 + x^2}$.

In Exercises 8–23, use graphs to determine if there are solutions for each equation in the interval [0, 1]. *If there are solutions, use the graphing utility to find them accurately to two decimal places.*

8. $\cos^{-1} x = \tan^{-1} x$ **9.** $x = \arccos x$

10. $\cos^{-1} x = x^2$

11. (a) $1.3\left(x - \frac{1}{2}\right)^2 = \cos^{-1} x$
(b) $1.4\left(x - \frac{1}{2}\right)^2 = \cos^{-1} x$

12. $\tan^{-1} x = \sin 3x$

13. (a) $\arccos x = 2 \sin 3x$
(b) $\arccos x = 2 \sin 4x$

14. (a) $\sin(2.3x) = \arctan x$
(b) $\sin(2.2x) = \arctan x$

15. (a) $1/(\tan^{-1} x + \sin^{-1} x) = \sin 2x$
(b) $1/(\tan^{-1} x + \sin^{-1} x) = \sin 3x$

16. (a) $1/(\sin^{-1} x + \cos^{-1} x) = 4x^3$
(b) $1/(\sin^{-1} x + \cos^{-1} x) = 5x^3$

17. $\sin^{-1} x = \cos^{-1} x$

18. $\sin^{-1} x = \sin^{-1}(\sin^{-1} x)$

19. $\cos^{-1} x = \cos^{-1}(\cos^{-1} x)$

20. $\cos^{-1}(\sin^{-1} x) = \sin^{-1}(\cos^{-1} x)$

21. $1/(\sin x) = \sin^{-1} x$

22. $1/(\cos x) = \cos^{-1} x$

23. $1/(\tan x) = \tan^{-1} x$

CHAPTER EIGHT SUMMARY OF PRINCIPAL FORMULAS AND TERMS

FORMULAS OR TERMS	PAGE REFERENCE	COMMENTS
1. $\sin(s+t)=\sin s\cos t+\cos s\sin t$ $\sin(s-t)=\sin s\cos t-\cos s\sin t$ $\cos(s+t)=\cos s\cos t-\sin s\sin t$ $\cos(s-t)=\cos s\cos t+\sin s\sin t$	483	These identities are referred to as the *addition formulas* for sine and cosine.
2. $\cos\left(\frac{\pi}{2}-\theta\right)=\sin\theta$ $\sin\left(\frac{\pi}{2}-\theta\right)=\cos\theta$	486	These two identities, called *reduction formulas,* hold for all real numbers θ. In terms of angles, the identities state that the sine of an angle is equal to the cosine of its complement.
3. $\tan(s+t)=\dfrac{\tan s+\tan t}{1-\tan s\tan t}$ $\tan(s-t)=\dfrac{\tan s-\tan t}{1+\tan s\tan t}$	488	These are the *addition formulas* for tangent.
4. $\sin 2\theta=2\sin\theta\cos\theta$ $\cos 2\theta=\cos^2\theta-\sin^2\theta$ $\tan 2\theta=\dfrac{2\tan\theta}{1-\tan^2\theta}$	492	These are the *double-angle formulas.* There are four other forms of the double-angle formula for cosine. They appear in the box on page 495.
5. $\sin\dfrac{s}{2}=\pm\sqrt{\dfrac{1-\cos s}{2}}$ $\cos\dfrac{s}{2}=\pm\sqrt{\dfrac{1+\cos s}{2}}$ $\tan\dfrac{s}{2}=\dfrac{\sin s}{1+\cos s}$	493	These three identities are referred to as the *half-angle formulas.* In the case of the half-angle formulas for sine and cosine, the sign before the radical is determined by the quadrant in which the angle or arc terminates.
6. $\sin A\sin B=\frac{1}{2}[\cos(A-B)-\cos(A+B)]$ $\sin A\cos B=\frac{1}{2}[\sin(A-B)+\sin(A+B)]$ $\cos A\cos B=\frac{1}{2}[\cos(A-B)+\cos(A+B)]$	502	These *product-to-sum* identities are derived from the addition formulas for sine and cosine.
7. $\sin\alpha+\sin\beta=2\sin\dfrac{\alpha+\beta}{2}\cos\dfrac{\alpha-\beta}{2}$ $\sin\alpha-\sin\beta=2\cos\dfrac{\alpha+\beta}{2}\sin\dfrac{\alpha-\beta}{2}$ $\cos\alpha+\cos\beta=2\cos\dfrac{\alpha+\beta}{2}\cos\dfrac{\alpha-\beta}{2}$ $\cos\alpha-\cos\beta=-2\sin\dfrac{\alpha+\beta}{2}\sin\dfrac{\alpha-\beta}{2}$	503	These are the *sum-to-product* identities.
8. The restricted sine function	518	The domain of the function $y=\sin x$ is the set of all real numbers. By allowing inputs only from the closed interval $[-\pi/2, \pi/2]$, we obtain the restricted sine function. (See Figure 1 on page 519.) The motivation for restricting the domain is that the restricted sine function is one-to-one, and therefore the inverse function exists.

FORMULAS OR TERMS	PAGE REFERENCE	COMMENTS
9. $\sin^{-1} x$	519	$\sin^{-1} x$ is that number in the interval $[-\pi/2, \pi/2]$ whose sine is x. An alternate but entirely equivalent form of this expression is $\arcsin x$. The basic properties of the inverse sine function are summarized in the box on page 526.
10. The restricted cosine function	522	The domain of the function $y = \cos x$ is the set of all real numbers. By allowing inputs only from the closed interval $[0, \pi]$, we obtain the restricted cosine function. (See Figure 4 on page 522.) The motivation for restricting the domain is that the restricted cosine function is one-to-one, and therefore the inverse function exists.
11. $\cos^{-1} x$	522, 523	$\cos^{-1} x$ is that number in the interval $[0, \pi]$ whose cosine is x. An alternate but entirely equivalent form of this expression is $\arccos x$. The basic properties of the inverse cosine function are summarized in the box on page 527.
12. The restricted tangent function	524	The tangent function, $y = \tan x$, is not one-to-one. However, by allowing inputs only from the open interval $(-\pi/2, \pi/2)$, we obtain the restricted tangent function, which is one-to-one. (See Figure 8 on page 524.) Since the restricted tangent function is one-to-one, the inverse function exists.
13. $\tan^{-1} x$	524, 525	$\tan^{-1} x$ is that number in the interval $(-\pi/2, \pi/2)$ whose tangent is x. An alternate but entirely equivalent form of this expression is $\arctan x$. The basic properties of the inverse tangent function are summarized on page 527.

WRITING MATHEMATICS

1. Say whether the statement is TRUE or FALSE. Write out your reason or reasons in complete sentences. If you draw a diagram to accompany your writing, be sure that you clearly label any parts of the diagram to which you refer.
   (a) The equation $\tan^2 t + 1 = \sec^2 t$ is true for every real number t.
   (b) There is no real number x satisfying the equation $\cos(\frac{\pi}{4} + x) = 2$.
   (c) There is no real number x satisfying the equation $\cos(\frac{\pi}{4} + x) = \cos x$.
   (d) For every number x in the closed interval $[-1, 1]$, we have $\sin^{-1} x = 1/\sin x$.
   (e) There is no real number x for which $\sin^{-1} x = 1/\sin x$. *Hint:* Draw a careful sketch of the graphs of the inverse sine function and the cosecant function on the interval $0 < x \leq 1$.
   (f) The equation $\sin(x + y) = \sin x \cos y + \cos x \sin y$ holds for all real numbers x and y.
   (g) The equation $\tan(x + y) = \dfrac{\tan x + \tan y}{1 - \tan x \tan y}$ holds for all real numbers x and y.
2. There is a formula for calculating the angle between two given lines in the x-y plane. The derivation of this formula relies on the identity for

$\tan(s - t)$. Find a book on analytic geometry in the library. Look up this formula, find out how it is derived, and how to use it. You can work with a classmate or your instructor. Then on your own, write a summary of what you have learned. Include an example (like the ones in this precalculus text) explaining how the formula is applied in a specific case.

3. This exercise consists of two parts. In the first part you are going to follow some simple instructions to construct a regular pentagon. In the second part, you'll write a paper explaining why the construction is valid.

(a) Use the following instructions to make a poster showing a regular pentagon inscribed in the unit circle. The poster should also show the steps used in the construction.

> Draw a unit circle. Label the origin O; let B and C denote the points $(1, 0)$ and $(0, 1)$, respectively; and let A denote the point where the bisector of $\angle OCB$ meets the x-axis. Now, from A, draw a line segment straight up to the unit circle, meeting the unit circle at a point P. Then the line segment joining B and P will be one side of a regular pentagon inscribed in the unit circle.

Remark: This is perhaps the simplest geometric construction known for the regular pentagon. The ancient Greek mathematicians had a more complicated method. The construction given here was discovered (only) a century ago by H. W. Richmond; it appeared in the *Quarterly Journal of Mathematics,* vol. 26 (1893), pp. 296–297.

(b) With a group of classmates or your instructor, work out the details in the following terse justification for the construction in part (a). Then, on your own, carefully write out the justification in full detail. This will involve a mixture of English sentences and equations, much like the exposition in this textbook. At each of the main steps, be sure to tell the reader where you are going and what that step will accomplish.

> We want to show $\angle POB = 72°$. Let $\angle OCB = \theta$. From right triangle OCB we get $\sin\theta = 2\sqrt{5}/5$ and $\cos\theta = \sqrt{5}/5$. Using these values in the half-angle formula for tangent then yields (after simplifying) $\tan(\theta/2) = (\sqrt{5} - 1)/2$. Next, from right triangle OCA we have $OA = \frac{1}{2}\tan(\theta/2)$. From these last two equations, we conclude that $OA = (\sqrt{5} - 1)/4$. However, according to Exercise 68(c) in Section 8.2, $\sin 18° = (\sqrt{5} - 1)/4$. It now follows that $OA = \cos 72°$, and therefore $\angle POB = 72°$.

(In working out the details, you can assume the result from Exercise 68(c) in Section 8.2.)

CHAPTER EIGHT REVIEW EXERCISES

In Exercises 1–32, prove that the equations are identities.

1. $\cot(x + y) = \dfrac{\cot x \cot y - 1}{\cot x + \cot y}$

2. $\cos 2x = \dfrac{1 - \tan^2 x}{1 + \tan^2 x}$

3. $\sin 2x = \dfrac{2\tan x}{1 + \tan^2 x}$

4. $\sin^2 x - \sin^2 y = \sin(x + y)\sin(x - y)$

5. $\tan^2 x - \tan^2 y = \dfrac{\sin(x + y)\sin(x - y)}{\cos^2 x \cos^2 y}$

6. $2 \csc 2x = \sec x \csc x$
7. $(\sin x)[\tan(x/2) + \cot(x/2)] = 2$
8. $\tan\left(x + \frac{\pi}{4}\right) = (1 + \tan x)/(1 - \tan x)$
9. $\tan\left(\frac{\pi}{4} + x\right) - \tan\left(\frac{\pi}{4} - x\right) = 2 \tan 2x$
10. $\dfrac{\cot x - 1}{\cot x + 1} = \dfrac{1 - \sin 2x}{\cos 2x}$
11. $2 \sin\left(\dfrac{\pi}{4} - \dfrac{x}{2}\right) \cos\left(\dfrac{\pi}{4} - \dfrac{x}{2}\right) = \cos x$
12. $\dfrac{\tan(x + y) - \tan y}{1 + \tan(x + y) \tan y} = \tan x$
13. $\dfrac{1 - \tan\left(\frac{1}{4}\pi - t\right)}{1 + \tan\left(\frac{1}{4}\pi - t\right)} = \tan t$
14. $\tan\left(\frac{1}{4}\pi + \frac{1}{2}t\right) = \tan t + \sec t$
15. $\dfrac{\tan(\alpha - \beta) + \tan \beta}{1 - \tan(\alpha - \beta) \tan \beta} = \tan \alpha$
16. $(1 + \tan \theta)\left[1 + \tan\left(\frac{1}{4}\pi - \theta\right)\right] = 2$
17. $\tan 3\theta = \dfrac{3t - t^3}{1 - 3t^2}$, where $t = \tan \theta$
18. $\sin\left(t + \frac{2}{3}\pi\right) \cos\left(t - \frac{2}{3}\pi\right) + \cos\left(t + \frac{2}{3}\pi\right) \sin\left(t - \frac{2}{3}\pi\right) = \sin 2t$
19. $\tan 2x + \sec 2x = \dfrac{\cos x + \sin x}{\cos x - \sin x}$
20. $\cos^4 x - \sin^4 x = \cos 2x$
21. $2 \sin x + \sin 2x = \dfrac{2 \sin^3 x}{1 - \cos x}$
22. $1 + \tan x \tan(x/2) = \sec x$
23. $\tan \dfrac{x}{2} = \dfrac{1 - \cos x + \sin x}{1 + \cos x + \sin x}$
24. $\dfrac{\sin 3x}{\sin x} - \dfrac{\cos 3x}{\cos x} = 2$
25. $\sin(x + y) \cos y - \cos(x + y) \sin y = \sin x$
26. $\dfrac{\sin x + \sin 2x}{\cos x - \cos 2x} = \cot \dfrac{x}{2}$
27. $\dfrac{1 - \tan^2(x/2)}{1 + \tan^2(x/2)} = \cos x$
28. $4 \sin(x/4) \cos(x/4) \cos(x/2) = \sin x$
29. $\sin 4x = 4 \sin x \cos x - 8 \sin^3 x \cos x$
30. $\cos 4x = 8 \cos^4 x - 8 \cos^2 x + 1$
31. $\sin 5x = 16 \sin^5 x - 20 \sin^3 x + 5 \sin x$
32. $\cos 5x = 16 \cos^5 x - 20 \cos^3 x + 5 \cos x$

In Exercises 33–41, establish the identities by applying the sum-to-product formulas.

33. $\sin 80° - \sin 20° = \cos 50°$
34. $\sin 65° + \sin 25° = \sqrt{2} \cos 20°$
35. $\dfrac{\cos x - \cos 3x}{\sin x + \sin 3x} = \tan x$
36. $\dfrac{\sin 3° + \sin 33°}{\cos 3° + \cos 33°} = \tan 18°$
37. $\sin(5\pi/12) + \sin(\pi/12) = \sqrt{6}/2$
38. $\cos 10° - \sin 10° = \sqrt{2} \sin 35°$
39. $\dfrac{\cos 3y + \cos(2x - 3y)}{\sin 3y + \sin(2x - 3y)} = \cot x$
40. $\dfrac{\sin 10° - \sin 50°}{\cos 50° - \cos 10°} = \sqrt{3}$
41. $\dfrac{\sin 40° - \sin 20°}{\cos 20° - \cos 40°} = \dfrac{\sin 10° - \sin 50°}{\cos 50° - \cos 10°}$
42. Suppose that a and b are in the open interval $(0, \pi/2)$ and $a \neq b$. If $\sin a + \sin b = \cos a + \cos b$, show that $a + b = \pi/2$. *Hint:* Begin with the sum-to-product formulas.
43. Refer to the figure. Using calculus, it can be shown that the area of the first-quadrant region under the curve $y = 1/(1 + x^2)$ from $x = 0$ to $x = a$ is given by the expression $\tan^{-1} a$. Use this fact to carry out the following calculations.

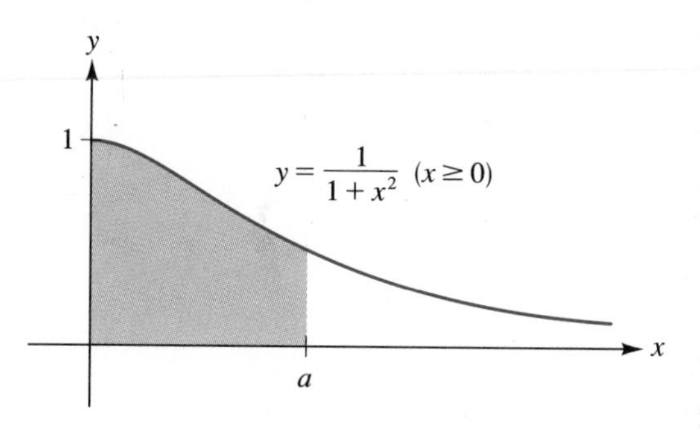

   (a) Find the area of the first-quadrant region under the curve $y = 1/(1 + x^2)$ from $x = 0$ to $x = 1$.
   (b) Find a value of a so that the area of the first-quadrant region under the curve from $x = 0$ to $x = a$ is: (i) 1.5; (ii) 1.56; (iii) 1.57. In each case, round your answer to the nearest integer.
44. The figure at the top of the next column shows the graph of the curve $y = 1/\sqrt{1 - x^2}$ for $0 \leq x < 1$. Using calculus it can be shown that the area of the first-quadrant region bounded by this curve, the coordinate axes, and the line $x = a$ is given by $\sin^{-1} a$. Use this fact to carry out the following calculations.
   (a) Find the area of the first-quadrant region under this curve from $x = 0$ to $x = \sqrt{2}/2$. Give both the exact form of the answer and a calculator approximation rounded to three decimal places.

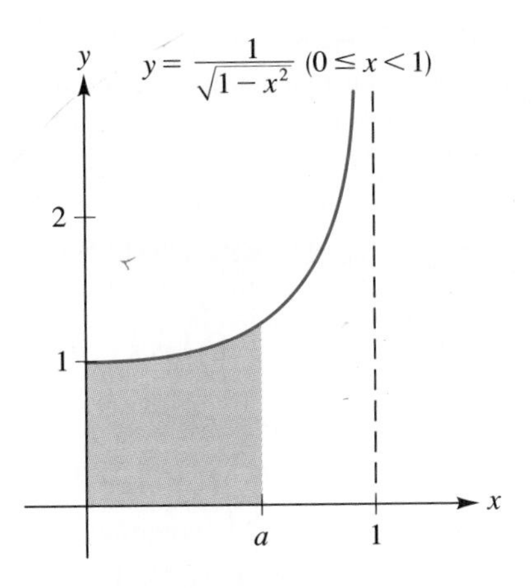

(b) Find a value of a so that the area of the first-quadrant region under the curve from $x = 0$ to $x = a$ is 1.5. Round your answer to three decimal places.

In Exercises 45–61, find all solutions of each equation in the interval $[0, 2\pi)$. *In cases where a calculator is necessary, round the answers to two decimal places.*

45. $\tan x = 4.26$

46. $\tan x = -4.26$

47. $\csc x = 2.24$

48. $\sin(\sin x) = \pi/6$

49. $\tan^2 x - 3 = 0$

50. $\cot^2 x - \cot x = 0$

51. $1 + \sin x = \cos x$

52. $2 \sin 3x - \sqrt{3} = 0$

53. $\sin x - \cos 2x + 1 = 0$

54. $\sin x + \sin 2x = 0$

55. $3 \csc x - 4 \sin x = 0$

56. $2 \sin^2 x + \sin x - 1 = 0$

57. $2 \sin^4 x - 3 \sin^2 x + 1 = 0$

58. $\sec^2 x - \sec x - 2 = 0$

59. $\sin^4 x + \cos^4 x = 5/8$

60. $4 \sin^2 2x + \cos 2x - 2 \cos^2 x - 2 = 0$

61. $\cot x + \csc x + \sec x = \tan x$ *Suggestion:* Using sines and cosines, the given equation becomes $\cos^2 x - \sin^2 x + \cos x + \sin x = 0$, which can be factored.

62. If A and B both are solutions of the equation $a \cos x + b \sin x = c$, show that $\tan[\frac{1}{2}(A + B)] = b/a$. *Hint:* The given information yields two equations. After subtracting one of those equations from the other and rearranging, you will have $\frac{\cos A - \cos B}{\sin A - \sin B} = -\frac{b}{a}$. Now use the sum-to-product formulas.

63. Evaluate $\cos \tan^{-1} \sin \tan^{-1}(\sqrt{2}/2)$.

In Exercises 64–87, evaluate each expression (without using a calculator or tables).

64. $\cos^{-1}(-\sqrt{2}/2)$

65. $\arctan(\sqrt{3}/3)$

66. $\sin^{-1} 0$

67. $\arcsin \frac{1}{2}$

68. $\arctan \sqrt{3}$

69. $\cos^{-1}(\frac{1}{2})$

70. $\tan^{-1}(-1)$

71. $\cos^{-1}(-\frac{1}{2})$

72. $\sin(\sin^{-1} 1)$

73. $\cos[\cos^{-1}(\frac{2}{7})]$

74. $\sin[\arccos(-\frac{1}{2})]$

75. $\sin[\tan^{-1}(-1)]$

76. $\cot[\cos^{-1}(\frac{1}{2})]$

77. $\sec[\cos^{-1}(\sqrt{2}/3)]$

78. $\sin(\frac{3\pi}{2} + \arccos \frac{3}{5})$

79. $\tan[\frac{\pi}{4} + \sin^{-1}(\frac{5}{13})]$

80. $\sin[2 \sin^{-1}(\frac{4}{5})]$

81. $\tan(2 \tan^{-1} 2)$

82. $\sin^{-1}(\sin \frac{\pi}{7})$

83. $\cos[\frac{1}{2} \cos^{-1}(\frac{4}{5})]$

84. $\sec^{-1}(-2/\sqrt{3})$

85. $\sec^{-1}(-\sqrt{2})$

86. $\sin[\sec^{-1}(\frac{29}{20})]$

87. $\sec[\sec^{-1}(\sqrt{6})]$

88. In this exercise we investigate the relationship between the variables x and θ in the accompanying figure. (Assume that θ is in radians.)

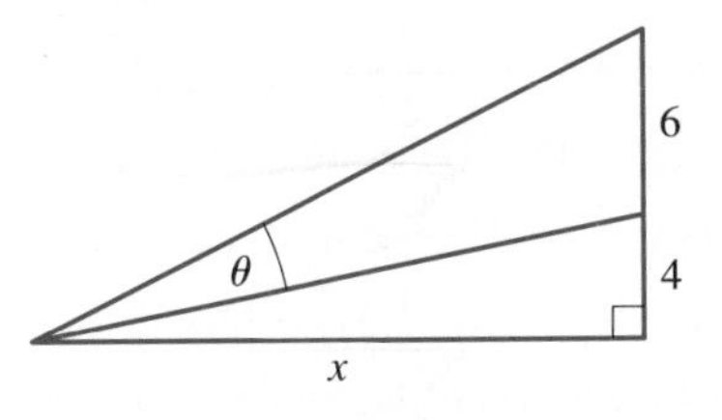

(a) Refer to the figure. Show that

$$\theta = \tan^{-1}(10/x) - \tan^{-1}(4/x)$$

(b) Use your calculator to complete the following table. (Round the results to two decimal places.) Which x-value in the table yields the largest value for θ?

x	0.1	1	2	3	10	100
θ						

(c) As indicated in the following graph (on the next page), the value of x that makes θ as large as possible is a number between 5 and 10, closer to 5 than to 10. Using calculus, it can be shown that this value of x is, in fact, $2\sqrt{10}$. Use your calculator to evaluate $2\sqrt{10}$; check that the result is consistent with the given graph. What is the corresponding value of θ in this case? Also, give the coordinates of the highest point on the accompanying graph. Round both coordinates to two decimal places.

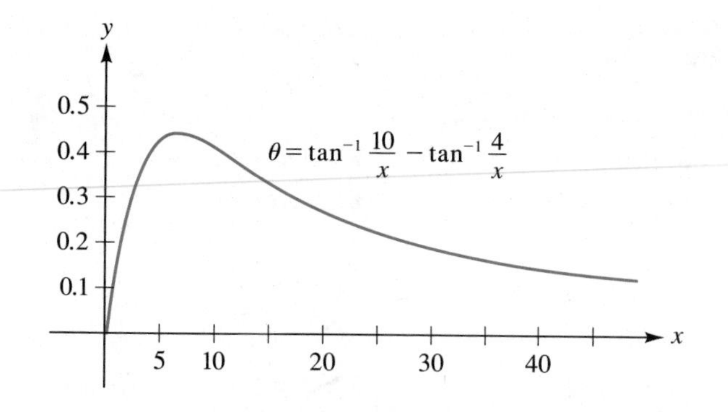

(d) What is the degree measure for the angle θ obtained in part (c)? Round the answer to one decimal place.

In Exercises 89–96; show that each equation is an identity.

89. $\tan(\tan^{-1} x + \tan^{-1} y) = (x + y)/(1 - xy)$

90. $\tan^{-1}(x/\sqrt{1 - x^2}) = \sin^{-1} x$

91. $\sin(2 \arctan x) = 2x/(1 + x^2)$

92. $\cos(2 \cos^{-1} x) = 2x^2 - 1$

93. $\sin[\frac{1}{2} \sin^{-1}(x^2)] = \sqrt{\frac{1}{2} - \frac{1}{2}\sqrt{1 - x^4}}$

94. $\tan^{-1}(\frac{1}{3}) + \tan^{-1}(\frac{1}{5}) = \tan^{-1}(\frac{4}{7})$

95. $\arcsin(4\sqrt{41}/41) + \arcsin(\sqrt{82}/82) = \pi/4$

96. $\tan[\sin^{-1}(\frac{1}{3}) + \cos^{-1}(\frac{1}{2})] = \frac{1}{5}(8\sqrt{2} + 9\sqrt{3})$

97. **(a)** Use a calculator to compute the quantity $\cos 20° \cos 40° \cos 60° \cos 80°$. Give your answer to as many decimal places as is shown on your calculator.

(b) Now use a product-to-sum formula to *prove* that the display on your calculator is the exact value of the given expression, not an approximation.

98. In this exercise, we will use the accompanying figure to derive the half-angle formula for sine:

$$\sin\frac{\theta}{2} = \pm\sqrt{\tfrac{1}{2}(1 - \cos\theta)}$$

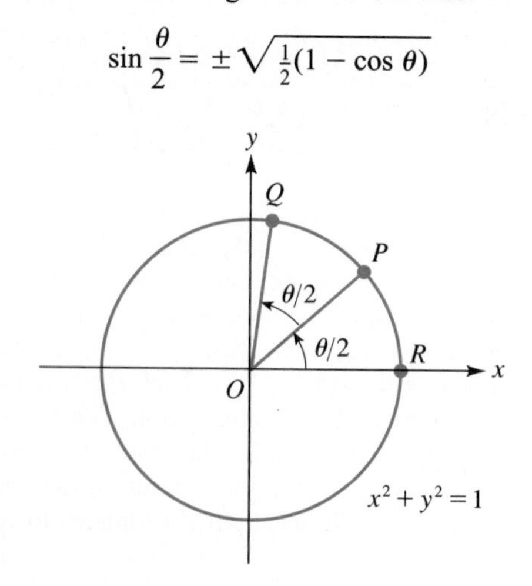

Our derivation will make use of the formula for the distance between two points and the identity $\sin^2 t + \cos^2 t = 1$. However, we will not rely on an addition formula for sine, as we did in Section 8.2

(a) Explain why the coordinates of P and Q are $P(\cos\frac{\theta}{2}, \sin\frac{\theta}{2})$ and $Q(\cos\theta, \sin\theta)$.

(b) Use the distance formula to show that

$$(PQ)^2 = 2 - 2\cos\tfrac{\theta}{2}\cos\theta - 2\sin\tfrac{\theta}{2}\sin\theta$$

and

$$(PR)^2 = 2 - 2\cos\tfrac{\theta}{2}$$

(c) Explain why $\triangle POR$ is congruent to $\triangle QOP$.

(d) From part (c), it follows that $(PQ)^2 = (PR)^2$. By equating the expressions for $(PQ)^2$ and $(PR)^2$ [obtained in part (b)], show that

$$\sin\tfrac{\theta}{2}\sin\theta = (\cos\tfrac{\theta}{2})(1 - \cos\theta)$$

(e) Square both sides of the equation obtained in part (d); then replace $\cos^2(\theta/2)$ by $1 - \sin^2(\theta/2)$ and show that the resulting equation can be written

$$[\sin^2(\theta/2)](2 - 2\cos\theta) = (1 - \cos\theta)^2$$

(f) Solve the equation in part (e) for the quantity $\sin\frac{\theta}{2}$. You should obtain

$$\sin\tfrac{\theta}{2} = \pm\sqrt{\tfrac{1}{2}(1 - \cos\theta)} \quad \text{as required}$$

In Exercises 99–101, prove that the equations are identities. (These identities appear in Trigonometry and Double Algebra, *by August DeMorgan, published in 1849.)*

99. $\sin 2\theta = 2/(\cot\theta + \tan\theta)$

100. $\sin 2\theta = \dfrac{\tan(45° + \theta) - \tan(45° - \theta)}{\tan(45° + \theta) + \tan(45° - \theta)}$

101. $\cos 2\theta = 1/(1 + \tan 2\theta \tan\theta)$

102. Prove the following two identities. These identities were given by the Swiss mathematician Leonhard Euler in 1748.

(a) $\tan 2\theta = \dfrac{2\tan\theta}{1 - \tan^2\theta}$ **(b)** $\cot 2\theta = \dfrac{\cot\theta - \tan\theta}{2}$

103. Prove the following two identities. These identities were given by the Swiss mathematician Johann Heinrich Lambert in 1765.

(a) $\sin 2\theta = \dfrac{2\tan\theta}{1 + \tan^2\theta}$ **(b)** $\cos 2\theta = \dfrac{1 - \tan^2\theta}{1 + \tan^2\theta}$

104. Prove the following identity, which was essentially given by the German mathematician Johann Müller (known as "Regiomontanus") around 1464.

$$\frac{\sin A + \sin B}{\sin A - \sin B} = \frac{\tan[\frac{1}{2}(A + B)]}{\tan[\frac{1}{2}(A - B)]}$$

105. Prove the following identity, given by the Austrian mathematician George Joachim Rhaeticus in 1569.

$$\cos n\theta = \cos[(n-2)\theta] - 2\sin\theta \sin[(n-1)\theta]$$

106. Suppose that $x + \frac{1}{x} = 2\cos\theta$.

(a) Show that $x^2 + \frac{1}{x^2} = 2\cos 2\theta$.

(b) Show that $x^3 + \frac{1}{x^3} = 2\cos 3\theta$.

CHAPTER EIGHT TEST

Do

1. Use an appropriate addition formula to simplify the expression $\sin(\theta + \frac{3\pi}{2})$.
2. Compute $\cos 2t$ given that $\sin t = -2\sqrt{5}/5$ and $\frac{3\pi}{2} < t < 2\pi$.
3. Compute $\tan(\theta/2)$ given that $\cos\theta = -5/13$ and $\pi < \theta < \frac{3\pi}{2}$.
4. Use a calculator to find all solutions of the equation $\sin x = 3\cos x$ in the interval $(0, 2\pi)$.
5. Find all solutions of the equation
$$2\sin^2 x + 7\sin x + 3 = 0$$
on the interval $0 \leq x \leq 2\pi$.
6. If $\cos\alpha = 2/\sqrt{5}$ $(\frac{3\pi}{2} < \alpha < 2\pi)$ and $\sin\beta = 4/5$ $(\frac{\pi}{2} < \beta < \pi)$, compute $\sin(\beta - \alpha)$.
7. Find all solutions of the equation $\sin(x + 30°) = \sqrt{3}\sin x$ on the interval $0° < x < 90°$.
8. If $\csc\theta = -3$ and $\pi < \theta < \frac{3\pi}{2}$, compute $\sin(\theta/2)$.
9. On the same set of axes, sketch the graphs of the restricted sine function and the function $y = \sin^{-1} x$. Specify the domain and the range for each function.
10. Compute each of the following quantities
(a) $\sin^{-1}[\sin(\pi/10)]$ **(b)** $\sin^{-1}(\sin 2\pi)$
11. Compute $\cos(\arcsin\frac{3}{4})$.
12. Prove that the following equation is an identity:
$$\tan\left(\frac{\pi}{4} + \frac{\theta}{2}\right) = \frac{1 + \cos\theta + \sin\theta}{1 + \cos\theta - \sin\theta}$$
13. Use a product-to-sum formula to simplify the expression $\sin(7\pi/24)\cos(\pi/24)$.
14. Use the sum-to-product formulas to simplify the expression
$$\frac{\sin 3\theta + \sin 5\theta}{\cos 3\theta + \cos 5\theta}.$$
15. Simplify each of the following expressions.
(a) $\sec[\arctan\sqrt{x^2 - 1}]$ (Assume that $x > 1$.)
(b) $\sin[\sec^{-1}(5/3) + \tan^{-1}(3/4)]$
16. Sketch a graph of the function $y = \tan^{-1} x$ and specify the domain and the range.

ADDITIONAL TOPICS IN TRIGONOMETRY

The subject of trigonometry is an excellent example of a branch of mathematics . . . which was motivated by both practical and intellectual interests—surveying, map-making, and navigation on the one hand, and curiosity about the size of the universe on the other. With it the Alexandrian mathematicians triangulated the universe and rendered precise their knowledge about the Earth and the heavens.

Morris Klein in *Mathematics in Western Culture* (New York: Oxford University Press, 1953)

INTRODUCTION

In this chapter we develop six topics that require a background in basic trigonometry. The first two of these topics, discussed in Section 9.1, are the law of sines and the law of cosines. These laws relate the angles and the lengths of the sides in any triangle. The next two sections, 9.2 and 9.3, introduce the important topic of vectors, first from a geometric standpoint, then from an algebraic standpoint. In Sections 9.4 through 9.6 we expand upon some of the ideas in Chapters 2 and 3 on graphs and equations. The topics presented here are parametric equations and polar coordinates.

THE LAW OF SINES AND THE LAW OF COSINES

The ratio of the sides of a triangle to each other is the same as the ratio of the sines of the opposite angles.

Bartholomaus Pitiscus (1561–1613) in his text, *Trigonometriae sive de dimensione triangulorum libri quinque.* This text was first published in Frankfort, Germany, in 1595, and according to several historians of mathematics, it was the first satisfactory textbook on trigonometry.

In this section we discuss two formulas relating the sides and the angles in any triangle: the *law of sines* and the *law of cosines.* These formulas can be used to determine an unknown side or angle using given information about the triangle. As you will see, which formula to apply in a particular case depends upon what data are initially given. To help you keep track of this, we'll use the following notations from elementary geometry.

NOTATION	EXPLANATION
SAA	One side and two angles are given.
SSA	Two sides and an angle opposite one of those sides are given.
SAS	Two sides and the included angle are given.
SSS	Three sides of the triangle are given.

Also, we will often follow the convention of denoting the angles of a triangle by A, B, and C and the lengths of the corresponding opposite sides by a, b, and c; see Figure 1. With this notation, we are ready to state the law of sines.

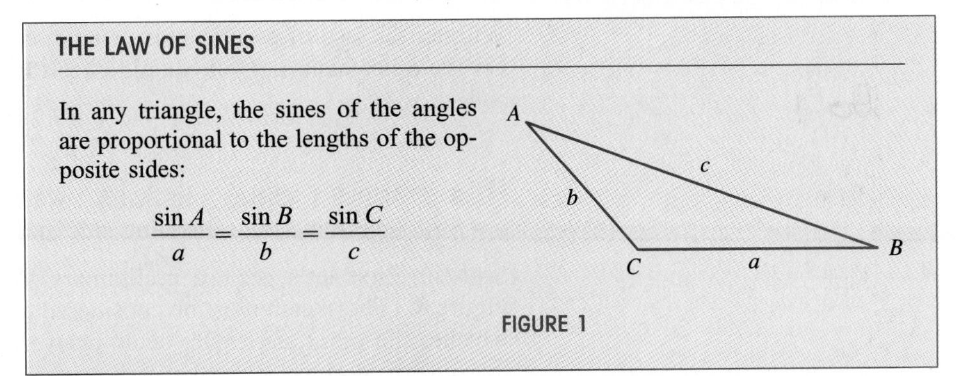

THE LAW OF SINES

In any triangle, the sines of the angles are proportional to the lengths of the opposite sides:

$$\frac{\sin A}{a} = \frac{\sin B}{b} = \frac{\sin C}{c}$$

FIGURE 1

The proof of the law of sines is easy. We use the following result from Section 6.4: The area of any triangle is equal to half the product of two sides times the sine of the included angle. Thus, with reference to Figure 1, we have

$$\tfrac{1}{2}bc \sin A = \tfrac{1}{2}ac \sin B = \tfrac{1}{2}ab \sin C$$

since each of these three expressions equals the area of triangle ABC. Now we just multiply through by the quantity $2/abc$ to obtain

$$\frac{\sin A}{a} = \frac{\sin B}{b} = \frac{\sin C}{c}$$

which completes the proof.

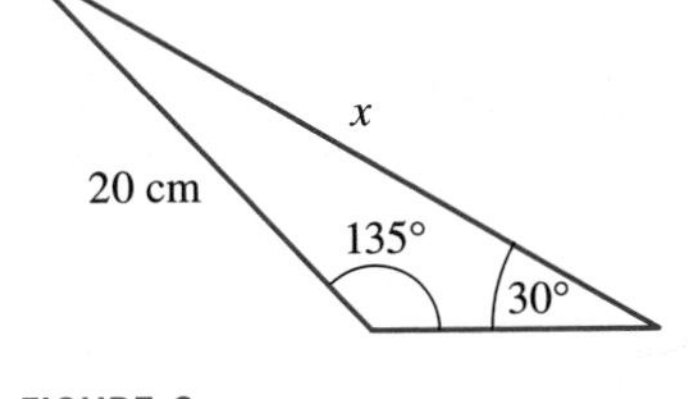

FIGURE 2

EXAMPLE 1 (SAA) Find the length x in Figure 2.

Solution We can determine x directly by applying the law of sines. We have

$$\frac{\sin 30^\circ}{20} = \frac{\sin 135^\circ}{x}$$

Length of side opposite the 30° angle (20) — Length of side opposite the 135° angle (x)

$$x \sin 30^\circ = 20 \sin 135^\circ$$
$$x(1/2) = 20(\sqrt{2}/2)$$
$$x = 20\sqrt{2} \text{ cm}$$

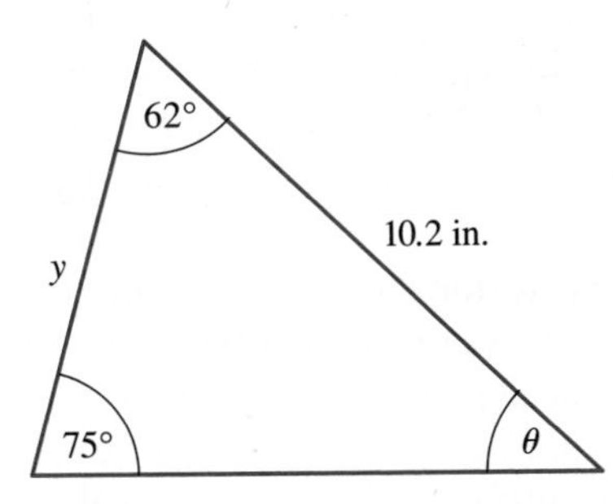

FIGURE 3

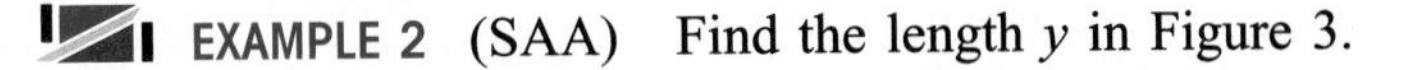

EXAMPLE 2 (SAA) Find the length y in Figure 3.

Solution First we need to determine the angle θ in Figure 3. Since the sum of the angles in any triangle is 180°, we have

$$\theta = 180^\circ - (75^\circ + 62^\circ) = 43^\circ$$

Now, using the law of sines, we obtain

$$\frac{\sin 75^\circ}{10.2} = \frac{\sin 43^\circ}{y}$$
$$y \sin 75^\circ = (10.2)\sin 43^\circ$$
$$y = \frac{(10.2)\sin 43^\circ}{\sin 75^\circ}$$

Without the use of a calculator or tables, this is the final form for the answer. On the other hand (as you should check for yourself), a calculator yields

$$y \approx 7.2 \text{ in.}$$

EXAMPLE 3 (SSA) In $\triangle ABC$, we are given $\angle C = 45°$, $b = 4\sqrt{2}$ ft, and $c = 8$ ft. Determine the remaining side and angles.

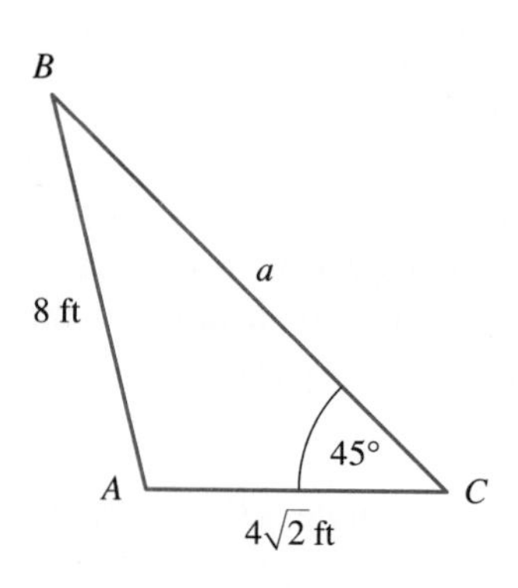

FIGURE 4

Solution First let's draw a preliminary sketch conveying the given data; see Figure 4. (The sketch must be considered tentative. At the outset, we don't know whether the other angles are acute or even whether the given data are compatible.) To find angle B, we have (according to the law of sines)

$$\frac{\sin B}{4\sqrt{2}} = \frac{\sin 45°}{8}$$

and therefore

$$8 \sin B = 4\sqrt{2} \sin 45° = 4\sqrt{2}\left(\frac{\sqrt{2}}{2}\right) = 4$$

$$\sin B = \frac{1}{2}$$

From our previous work, we know that one possibility for B is 30°, since $\sin 30° = 1/2$. However, there is another possibility. Since the reference angle for 150° is 30°, we know that $\sin 150°$ is also equal to $1/2$. Which angle do we want? For the problem at hand, this is easy to answer. Since angle C is given as 45°, angle B cannot equal 150°, for the sum of 45° and 150° exceeds 180°. We conclude that

$$\angle B = 30°$$

Next, since $\angle B = 30°$ and $\angle C = 45°$, we have

$$\angle A = 180° - (30° + 45°) = 105°$$

Finally, we use the law of sines to find a. From the equation $\dfrac{\sin A}{a} = \dfrac{\sin C}{c}$, we have $a \sin C = c \sin A$, and therefore

$$a = \frac{c \sin A}{\sin C} = \frac{8 \sin 105°}{\sin 45°} = \frac{8 \sin 105°}{1/\sqrt{2}}$$

$$= 8\sqrt{2} \sin 105° \text{ ft} \approx 10.9 \text{ ft}$$

In the preceding example, two possibilities arose for the angle B: both 30° and 150°. However, it turned out that the value 150° was incompatible with the given information in the problem. In Exercise 13 at the end of this section, you will see a case in which both of two possibilities are compatible with the given data. This results in two distinct solutions to the problem. In contrast to this, Exercise 11(a) shows a case in which there is no triangle fulfilling the given conditions. For these reasons, the case SSA is sometimes referred to as the **ambiguous case.**

Now we turn to the law of cosines.

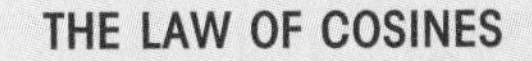

THE LAW OF COSINES

In any triangle, the square of the length of any side equals the sum of the squares of the lengths of the other two sides minus twice the product of the lengths of those other two sides times the cosine of their included angle.

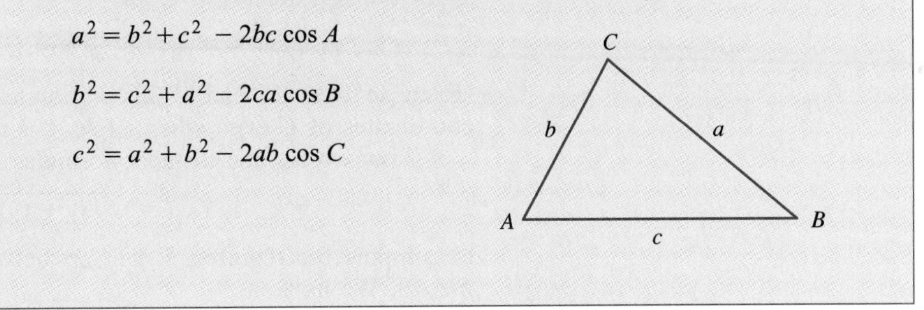

$$a^2 = b^2 + c^2 - 2bc \cos A$$

$$b^2 = c^2 + a^2 - 2ca \cos B$$

$$c^2 = a^2 + b^2 - 2ab \cos C$$

Before looking at a proof of this law, we make two preliminary comments. First, it is important to understand that the three equations in the box all follow the same pattern. For example, look at the first equation:

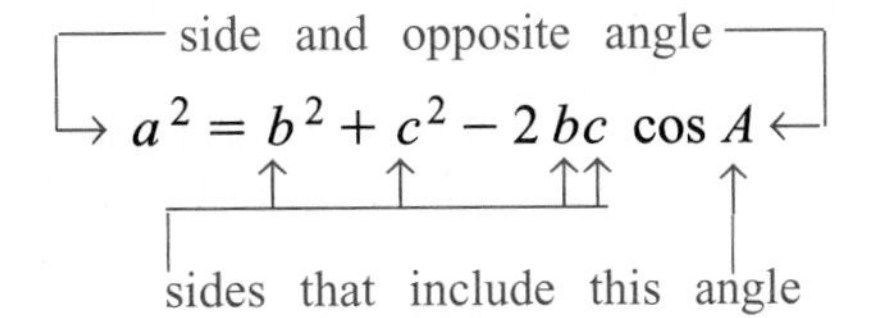

Now check for yourself that the other two equations also follow this pattern. It is the pattern that is important here; after all, not every triangle is labeled *ABC*.

The second observation is that the law of cosines is a generalization of the Pythagorean theorem. In fact, look what happens to the equation

$$a^2 = b^2 + c^2 - 2bc \cos A$$

when angle A is a right angle:

$$a^2 = b^2 + c^2 - 2bc \underbrace{\cos 90^\circ}_{0}$$

$$a^2 = b^2 + c^2 \qquad \text{which is the Pythagorean theorem}$$

Now let us prove the law of cosines:

$$a^2 = b^2 + c^2 - 2bc \cos A$$

(The other two equations can be proved in the same way. Indeed, just relabeling the figure would suffice.) The proof that we give uses coordinate geometry in a very nice way to complement the trigonometry. We begin by placing angle A in standard position, as indicated in Figure 5. (So, in the figure, angle A is then identified with angle CAB.) Then if u and v denote the lengths indicated in Figure 5, the coordinates of C are (u, v) and we have

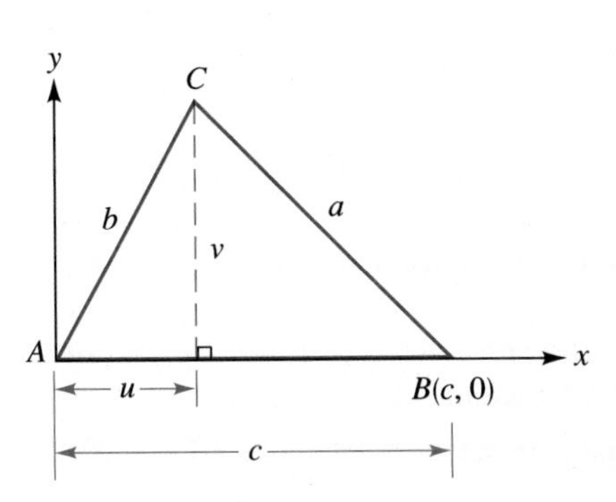

FIGURE 5

$$\cos A = \frac{\text{adjacent}}{\text{hypotenuse}} = \frac{u}{b} \qquad \text{and therefore} \qquad u = b \cos A$$

Similarly, we have

$$\sin A = \frac{\text{opposite}}{\text{hypotenuse}} = \frac{v}{b} \qquad \text{and therefore} \qquad v = b \sin A$$

Thus, the coordinates of C are

$$(b \cos A, b \sin A)$$

(Exercise 44 at the end of this section asks you to check that these represent the coordinates of C even when angle A is not acute.)

Now we use the distance formula,

$$d = \sqrt{(x_2 - x_1)^2 + (y_2 - y_1)^2}$$

to compute the required distance a between the points $C(b \cos A, b \sin A)$ and $B(c, 0)$. We have

$$\begin{aligned} a &= \sqrt{(b \cos A - c)^2 + (b \sin A - 0)^2} \\ &= \sqrt{b^2 \cos^2 A - 2bc \cos A + c^2 + b^2 \sin^2 A} \\ &= \sqrt{b^2 \underbrace{(\cos^2 A + \sin^2 A)}_{1} - 2bc \cos A + c^2} \\ &= \sqrt{b^2 + c^2 - 2bc \cos A} \end{aligned}$$

Squaring both sides of this last equation gives us the law of cosines, as we set out to prove.

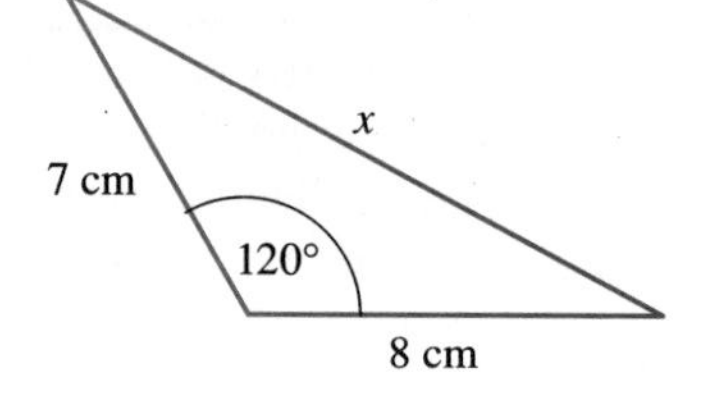

FIGURE 6

EXAMPLE 4 (SAS) Compute the length x in Figure 6.

Solution The law of cosines is directly applicable. We have

$$\begin{aligned} x^2 &= 7^2 + 8^2 - 2(7)(8)\cos 120^\circ \\ &= 49 + 64 - 112\left(-\tfrac{1}{2}\right) = 169 \\ x &= \sqrt{169} = 13 \text{ cm} \end{aligned}$$

If the equation $a^2 = b^2 + c^2 - 2bc \cos A$ is solved for $\cos A$, the result is

$$\cos A = \frac{b^2 + c^2 - a^2}{2bc}$$

This expresses the cosine of an angle in a triangle in terms of the lengths of the sides. In a similar fashion, we obtain the corresponding formulas

$$\cos B = \frac{c^2 + a^2 - b^2}{2ca} \qquad \text{and} \qquad \cos C = \frac{a^2 + b^2 - c^2}{2ab}$$

These alternative forms for the law of cosines are used in the next example.

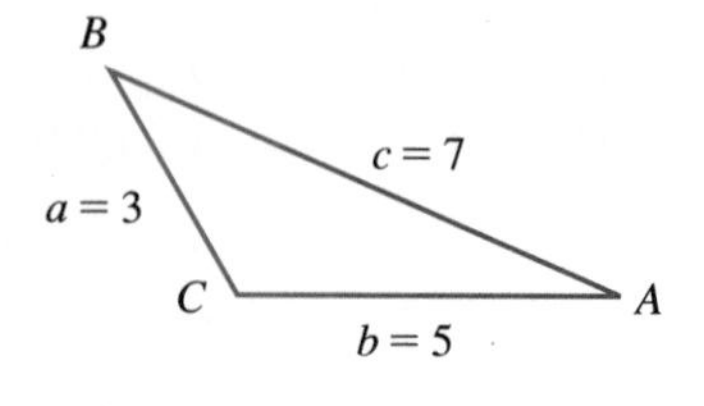

FIGURE 7

EXAMPLE 5 (SSS) In triangle ABC, the sides are $a = 3$ units, $b = 5$ units, and $c = 7$ units. Find the angles. (Use degree measure.)

Solution Figure 7 summarizes the given data. We have

$$\cos A = \frac{b^2 + c^2 - a^2}{2bc} = \frac{5^2 + 7^2 - 3^2}{2(5)(7)} = \frac{65}{70} = \frac{13}{14}$$

Now that we know $\cos A = 13/14$, we can find $\angle A$ by using a calculator to compute $\cos^{-1}(13/14)$. Since we are required to give the answer in degree measure, we first set the calculator to the degree mode. Then we obtain

$$\cos^{-1}(13/14) \approx 21.7867893 \qquad \text{using a calculator set in the degree mode}$$

and therefore

$$\angle A \approx 21.8^\circ \qquad \text{rounding to one decimal place}$$

In a similar manner, we have

$$\cos B = \frac{c^2 + a^2 - b^2}{2ca} = \frac{7^2 + 3^2 - 5^2}{2(7)(3)} = \frac{33}{42} = \frac{11}{14}$$

So, $\cos B = 11/14$ and a calculator then yields $\angle B \approx 38.2132107^\circ \approx 38.2^\circ$.

At this point, we can find $\angle C$ in either of two ways. Each has its advantage. The first way is to begin by computing $\cos C$ in the same way that we found $\cos A$ and $\cos B$. As you can check, the result is $\cos C = -1/2$. A calculator is not needed in this case. We know from previous work that $\angle C$ must be 120°. The second method that can be used relies on the fact that the sum of the angles in a triangle is 180°. Thus, we have

$$\begin{aligned} \angle C &= 180^\circ - \angle A - \angle B \\ &\approx 180^\circ - 21.7867893^\circ - 38.2132107^\circ \\ &\approx 120.0^\circ \end{aligned}$$

This way is quicker than the first method. The disadvantage, however, is that we know that $\angle C$ is only approximately 120°. The first method (using the cosine law) is longer, but it tells us that $\angle C$ is exactly 120°. In summary, then, the three required angles are

$$\angle A \approx 21.8^\circ \qquad \angle B \approx 38.2^\circ \qquad \angle C = 120^\circ$$

We conclude this section with an example indicating how the law of sines and the law of cosines are used in navigation. In this example, you'll see the term *bearing* used in specifying the location of one point relative to another. To explain this term, we refer to Figure 8.

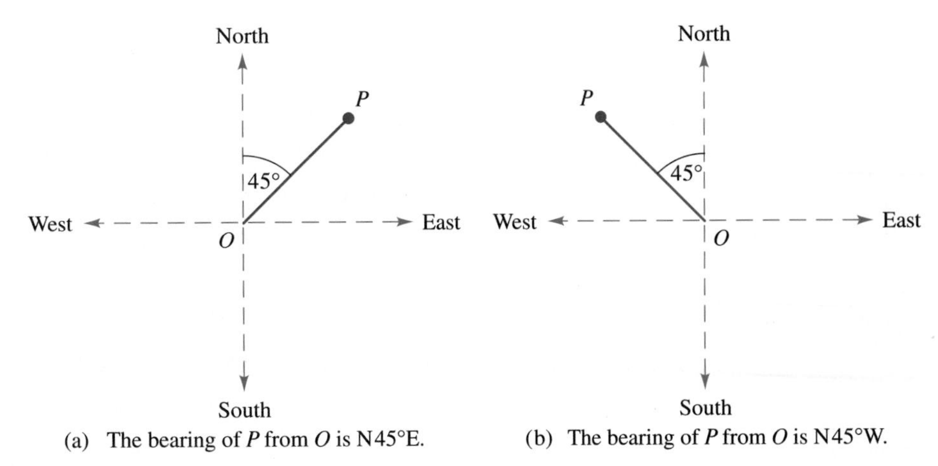

FIGURE 8 (a) The bearing of P from O is N45°E. (b) The bearing of P from O is N45°W.

In Figure 8(a), the bearing of P from O is N45°E (read "north, 45° east"). This bearing tells us the *acute* angle between line segment $\overline{OP}$ and the north–south line through O. In Figure 8(b), the bearing of P from O is N45°W (read "north, 45° west"). Again, note that the bearing gives us the acute angle between line segment $\overline{OP}$ and the north–south line through O. Figure 9 provides additional examples of this terminology.

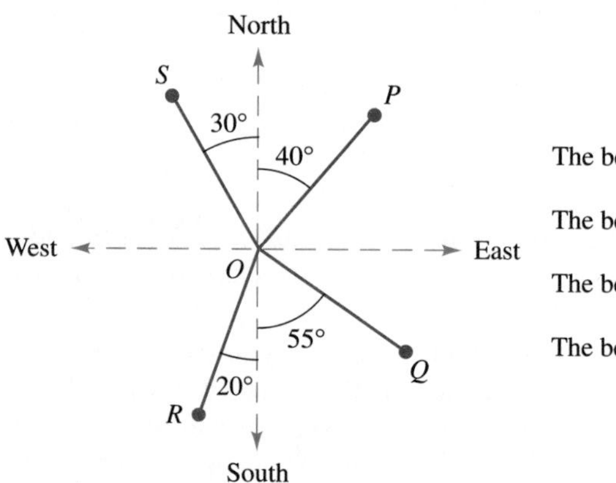

The bearing of P from O is N40°E.

The bearing of Q from O is S55°E.

The bearing of R from O is S20°W.

The bearing of S from O is N30°W.

FIGURE 9
The bearing is specified by means of the acute angle measured from the north–south line.

EXAMPLE 6 A small fire is sighted from ranger stations A and B. The bearing of the fire from station A is N35°E, and the bearing of the fire from station B is N49°W. Station A is 1.3 miles due west of station B.

(a) How far is the fire from each ranger station?

(b) At fire station C, which is 1.5 miles from A, there is a helicopter that can be used to drop water on the fire. If the bearing of C from A is S42°E, find the distance from C to the fire, and find the bearing of the fire from C.

Solution

(a) In Figure 10 we have sketched the situation involving ranger stations A and B and the fire (denoted by F).

We compute the angles of $\triangle ABF$ as follows.

$$\angle FAB = 90° - 35° = 55° \qquad \angle FBA = 90° - 49° = 41°$$
$$\angle F = 180° - (55° + 41°) = 84°$$

We can now use the law of sines in $\triangle ABF$ to compute AF and BF.

$$\frac{\sin(\angle F)}{AB} = \frac{\sin(\angle FBA)}{AF}$$

and therefore

$$\frac{\sin 84°}{1.3} = \frac{\sin 41°}{AF}$$

or

$$AF = \frac{(1.3)\sin 41°}{\sin 84°}$$

$$\frac{\sin(\angle F)}{AB} = \frac{\sin(\angle FAB)}{BF}$$

and therefore

$$\frac{\sin 84°}{1.3} = \frac{\sin 55°}{BF}$$

or

$$BF = \frac{(1.3)\sin 55°}{\sin 84°}$$

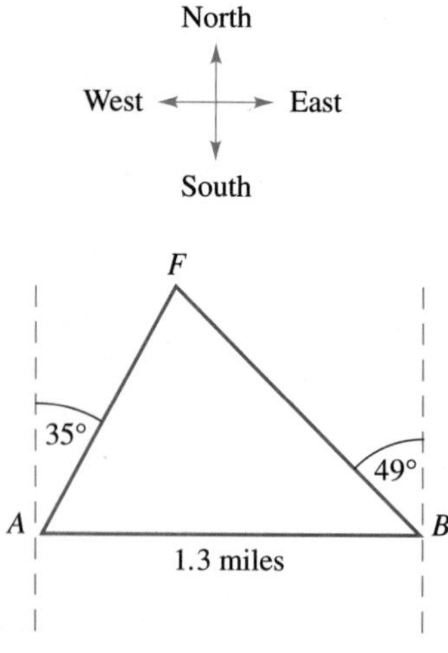

FIGURE 10

We use a calculator to evaluate these expressions for AF and BF. As you should check for yourself, the results (rounded to the nearest tenth of a mile) are

$$AF \approx 0.9 \text{ miles} \quad \text{and} \quad BF \approx 1.1 \text{ miles}$$

(b) We draw a sketch of the situation, as shown in Figure 11. In Figure 11, we can compute CF, the distance from the helicopter to the fire, using the law of cosines in $\triangle CAF$. First, note that $\angle CAF = 48° + 55° = 103°$. So we have

$$CF = \sqrt{AC^2 + AF^2 - 2 \cdot AC \cdot AF \cdot \cos 103°}$$

$$= \sqrt{1.5^2 + \left(\frac{1.3 \sin 41°}{\sin 84°}\right)^2 - 2(1.5)\left(\frac{1.3 \sin 41°}{\sin 84°}\right)\cos 103°}$$

$$\approx 1.9 \text{ miles} \qquad \text{using a calculator and rounding to one decimal place}$$

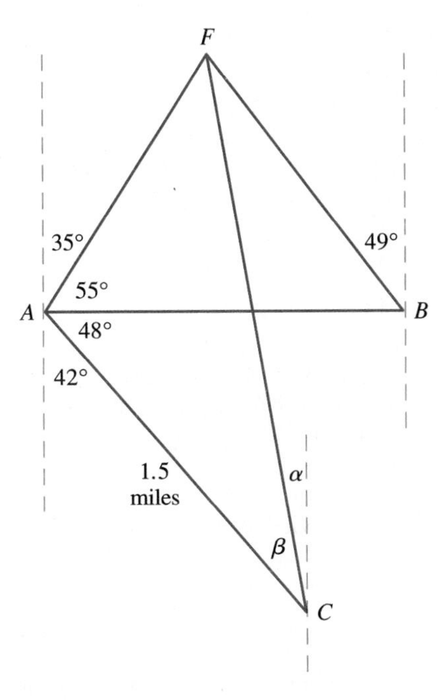

FIGURE 11

In order to find the bearing of the fire at F from the fire station at C, we need to determine the angle α in Figure 11. First we find the angle β. Using the law of sines in $\triangle CAF$, we have

$$\frac{\sin \beta}{AF} = \frac{\sin 103°}{CF}$$

and therefore

$$\sin \beta = \frac{AF \cdot \sin 103°}{CF}$$

Before using the inverse sine function and a calculator to compute β, we note from Figure 11 that β is an acute angle (because β is an angle in $\triangle CAF$, and in that triangle, $\angle CAF$ is greater than 90°). So in this particular application of the law of sines, there is no ambiguity. We have then

$$\beta = \sin^{-1}\left(\frac{AF \cdot \sin 103°}{CF}\right)$$

Now, on the right-hand side of this last equation, we substitute the expressions we obtained previously for AF and CF; then, using a calculator set in the degree mode, we obtain

$$\beta \approx 26° \qquad \text{rounding to the nearest degree}$$

(You should verify this calculator value for yourself.) Now that we know β, we can determine the bearing of the fire from station C. From Figure 11, we have

$$\alpha + \beta = 42° \qquad \text{(Why?)}$$

and therefore

$$\alpha = 42° - \beta = 42° - 26° = 16° \qquad \text{to the nearest degree}$$

In summary now, fire station C is approximately 1.9 miles from the fire, and the bearing of the fire from station C is N16°W.

EXERCISE SET 9.1

A

In Exercises 1–8, assume that the vertices and the lengths of the sides of a triangle are labeled as in Figure 1 in the text. For Exercises 1–4, leave your answers in terms of radicals or the trigonometric functions; that is, don't use a calculator. In Exercises 5–8, use a calculator and round your final answers to one decimal place.

1. If $\angle A = 60°$, $\angle B = 45°$, and $BC = 12$ cm, find AC.
2. If $\angle A = 30°$, $\angle B = 135°$, and $BC = 4$ cm, find AC.
3. If $\angle B = 100°$, $\angle C = 30°$, and $AB = 10$ cm, find BC.
4. If $\angle A = \angle B = 35°$, and $AB = 16$ cm, find AC and BC.
5. If $\angle A = 36°$, $\angle B = 50°$, and $b = 12.61$ cm, find a and c.
6. If $\angle B = 81°$, $\angle C = 55°$, and $b = 6.24$ cm, find c and a.
7. If $a = 29.45$ cm, $b = 30.12$ cm, and $\angle B = 66°$, find the remaining side and angles of the triangle.
8. If $a = 52.15$ cm, $c = 42.90$ cm, and $\angle A = 125°$, find the remaining side and angles of the triangle.

In Exercises 9 and 10, use degree measure for your answers. In parts (c) and (d), use a calculator and round the results to one decimal place.

9. **(a)** In $\triangle ABC$, $\sin B = \sqrt{2}/2$. What are the possible values for $\angle B$?
   (b) In $\triangle DEF$, $\cos E = \sqrt{2}/2$. What are the possible values for $\angle E$?
   (c) In $\triangle GHI$, $\sin H = 1/4$. What are the possible values for $\angle H$?
   (d) In $\triangle JKL$, $\cos K = -2/3$. What are the possible values for $\angle K$?
10. **(a)** In $\triangle ABC$, $\sin B = \sqrt{3}/2$. What are the possible values for $\angle B$?
    (b) In $\triangle DEF$, $\cos E = -\sqrt{3}/2$. What are the possible values for $\angle E$?
    (c) In $\triangle GHI$, $\sin H = 2/9$. What are the possible values for $\angle H$?
    (d) In $\triangle JKL$, $\cos K = 2/3$. What are the possible values for $\angle K$?
11. **(a)** Show that there is no triangle satisfying the conditions $a = 2.0$ ft, $b = 6.0$ ft, and $\angle A = 23.1°$
    Hint: Try computing $\sin B$ using the law of sines.
    (b) If $a = 2.0$ ft, $b = 3.0$ ft, $\angle A = 23.1°$, and $\angle B$ is obtuse, show that $c = 1.1$ ft.
12. **(a)** Show that there is no triangle with $a = 2$, $b = 3$, and $\angle A = 42°$.
    (b) Is there any triangle in which $a = 2$, $b = 3$, and $\angle A = 41°$?
13. Let $b = 1$, $a = \sqrt{2}$, and $\angle B = 30°$.
    (a) Use the law of sines to show that $\sin A = \sqrt{2}/2$. Conclude that $\angle A = 45°$ or $\angle A = 135°$.
    (b) Assuming that $\angle A = 45°$, determine the remaining parts of $\triangle ABC$.
    (c) Assuming that $\angle A = 135°$, determine the remaining parts of $\triangle ABC$.
    (d) Find the areas of the two triangles.
14. Let $a = 30$, $b = 36$, and $\angle A = 20°$.
    (a) Show that $\angle B = 24.23°$ or $\angle B = 155.77°$ (rounding to two decimal places).
    (b) Determine the remaining parts for each of the two possible triangles. Round your final results to two decimal places. [However, in your calculations, do *not* work with rounded values. (Why?)]
    (c) Find the areas of the two triangles.
15. Find the lengths a, b, c, and d in the following figure. Leave your answers in terms of trigonometric functions (rather than using a calculator).

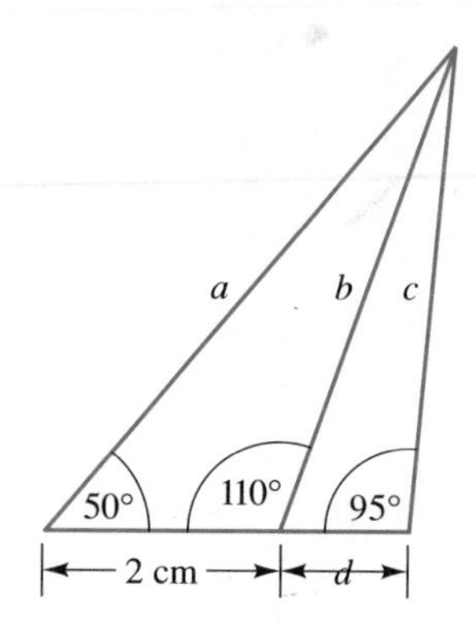

16. In the following figure, $\overline{PQR}$ is a straight line segment. Find the distance PR. Round your final answer to two decimal places.

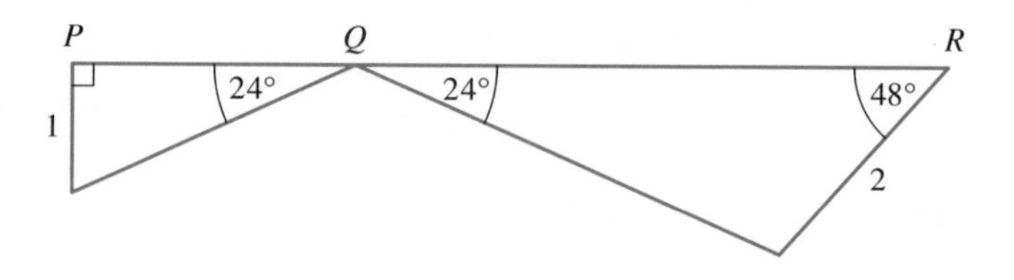

17. Two points P and Q are on opposite sides of a river (see the sketch). From P to another point R on the same side is 300 ft. Angles PRQ and RPQ are found to be 20° and 120°, respectively. Compute the distance from P to Q, across the river. (Round your answer to the nearest foot.)

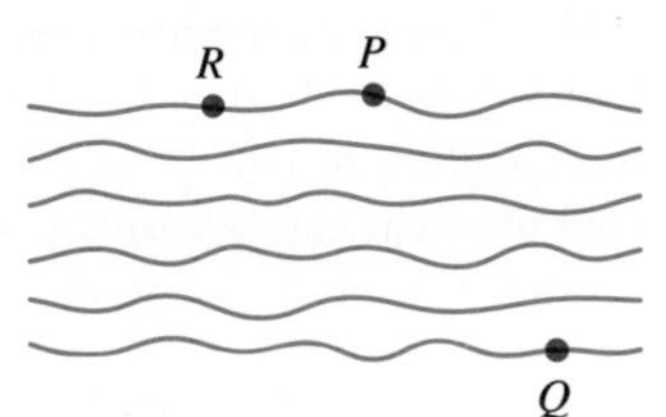

Figure for Exercise 17

18. Determine the angle θ in the accompanying figure. Round your answer to two decimal places.

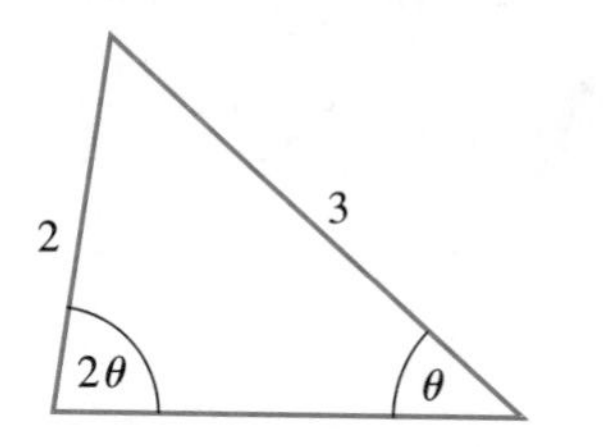

In Exercises 19–22, use the law of cosines to determine the length x in each figure. For Exercises 19 and 20, leave your answers in terms of radicals. In Exercises 21 and 22, use a calculator and round the answers to one decimal place.

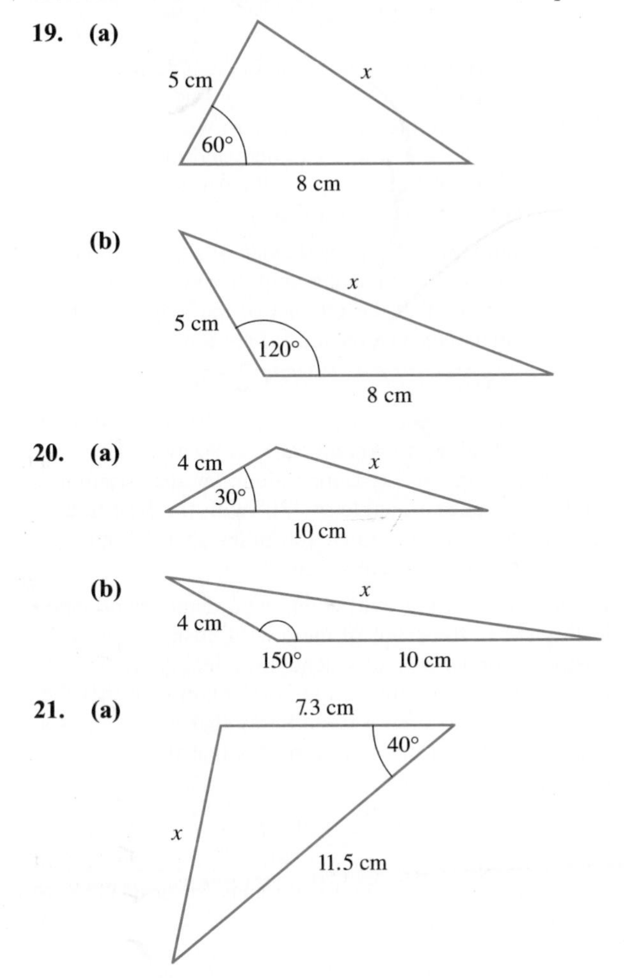

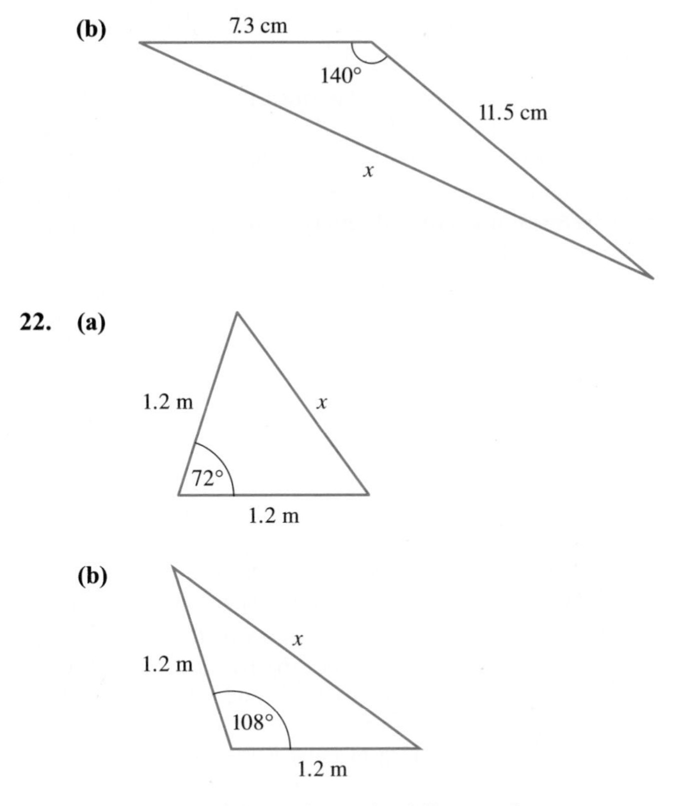

In Exercises 23 and 24, refer to the following figure.

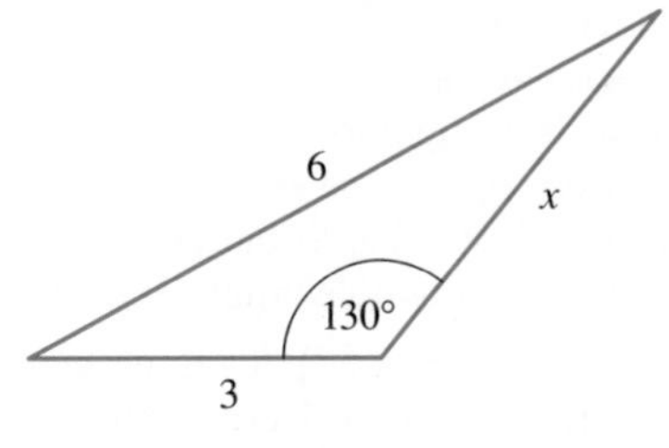

23. In applying the law of cosines to the figure, a student incorrectly writes $x^2 = 3^2 + 6^2 - 2(3)(6)\cos 130°$. Why is this incorrect? What is the correct equation?

24. In applying the law of cosines to the figure, a student writes $6^2 = 3^2 + x^2 + 6x \cos 50°$. Why is this correct?

In Exercises 25 and 26, use the given information to find the cosine of each angle in $\triangle ABC$.

25. $a = 6$ cm, $b = 7$ cm, $c = 10$ cm

26. $a = 17$ cm, $b = 8$ cm, $c = 15$ cm (For this particular triangle, you can check your answers, because there is an alternate method of solution that does not require the law of cosines.)

In Exercises 27–30, compute each angle of the given triangle. Where necessary, use a calculator and round to one decimal place.

27. $a = 7, b = 8, c = 13$

28. $a = 33, b = 7, c = 37$

29. $a = b = 2/\sqrt{3}, c = 2$

30. $a = 36, b = 77, c = 85$

In Exercises 31–34, round each answer to one decimal place.

31. A regular pentagon is inscribed in a circle of radius 1 unit. Find the perimeter of the pentagon. *Hint:* First find the length of a side using the law of cosines.

32. Find the perimeter of a regular nine-sided polygon inscribed in a circle of radius 4 cm. (See the hint for Exercise 31.)

33. In $\triangle ABC$, $\angle A = 40°$, $b = 6.1$ cm, and $c = 3.2$ cm.
(a) Find a using the law of cosines.
(b) Find $\angle C$ using the law of sines.
(c) Find $\angle B$.

34. In parallelogram $ABCD$, you are given $AB = 6$ in., $AD = 4$ in., and $\angle A = 40°$. Find the length of each diagonal.

35. Town B is 26 miles from town A at a bearing of S15°W. Town C is 54 miles from town A at a bearing of S7°E. Compute the distance from town B to town C. Round your final answer to the nearest mile.

36. Town C is 5 miles due east of town D. Town E is 12 miles from town C at a bearing (from C) of N52°E.
(a) How far apart are towns D and E? (Round to the nearest one-half mile.)
(b) Find the bearing of town E from town D. (Round the angle to the nearest degree.)

37. An airplane crashes in a lake and is spotted by observers at lighthouses A and B along the coast. Lighthouse B is 1.50 miles due east of lighthouse A. The bearing of the airplane from lighthouse A is S20°E; the bearing of the plane from lighthouse B is S42°W. Find the distance from each lighthouse to the crash site. (Round your final answers to two decimal places.)

38. (Continuation of Exercise 37) A rescue boat is in the lake, three-fourths of a mile from lighthouse B, and at a bearing of S35°E from lighthouse B.
(a) Find the distance from the rescue boat to the airplane. Express your answer using miles and feet, with the portion in feet rounded to the nearest ten feet.
(b) Find the bearing of the plane from the rescue boat. (Your answer should have the form of Sθ°W. Round θ to two decimal places.)

39. (Refer to the following figure.) When the Sun is viewed from the Earth, it subtends an angle of $\theta = 32'$ $\left(= \frac{32}{60} \text{ degree}\right)$. Assuming that the distance d from the Earth to the Sun is 92,690,000 miles, use the law of cosines to compute the diameter D of the Sun. Round the answer to the nearest ten thousand miles.

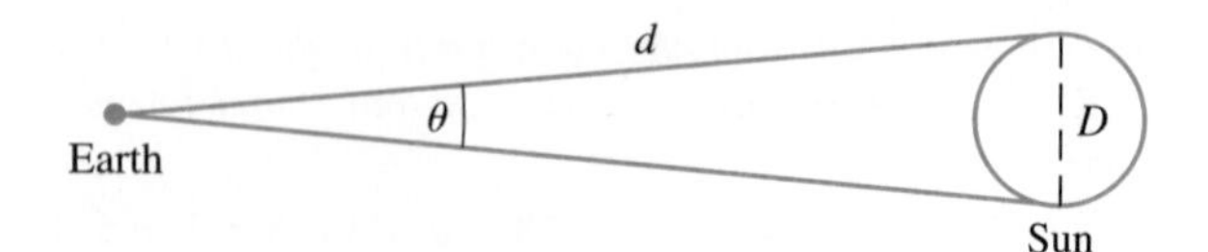

40. Compute the lengths CD and CE in the accompanying figure. Round the final answers to two decimal places.

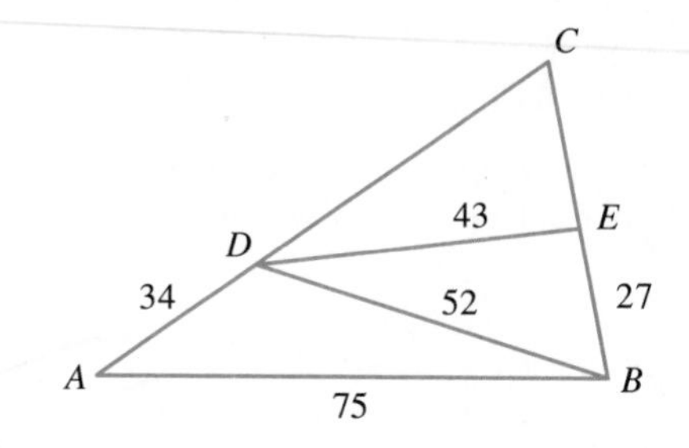

B

41. **(a)** Let m and n be positive numbers, with $m > n$. Furthermore, suppose that in triangle ABC the lengths a, b, and c are given by

$$a = 2mn + n^2 \qquad b = m^2 - n^2$$
$$c = m^2 + n^2 + mn$$

Show that $\cos C = -1/2$ and conclude that $\angle C = 120°$.

(b) Give an example of a triangle in which the lengths of the sides are whole numbers and one of the angles is 120°. (Specify the three sides; you needn't find the other angles.)

42. If the lengths of two adjacent sides of a parallelogram are a and b, and if the acute angle formed by these two sides is θ, show that the product of the lengths of the two diagonals is given by the expression

$$\sqrt{(a^2 + b^2)^2 - 4a^2b^2 \cos^2 \theta}$$

43. Two trains leave the railroad station at noon. The first train travels along a straight track at 90 mph. The second train travels at 75 mph along another straight track that makes an angle of 130° with the first track. At what time are the trains 400 miles apart? Round your answer to the nearest minute.

44. In this exercise, you will complete a detail mentioned in the text in the proof of the law of cosines. Let the positive numbers u and v denote the lengths indicated in the figure at the top of the next column, so that the coordinates of C are $(-u, v)$. Show that $u = -b \cos A$ and $v = b \sin A$. Conclude from this that the coordinates of C are

$$(b \cos A, b \sin A)$$

Hint: Use the right-triangle definitions for cosine and sine along with the addition formulas for $\cos(180° - \theta)$ and $\sin(180° - \theta)$.

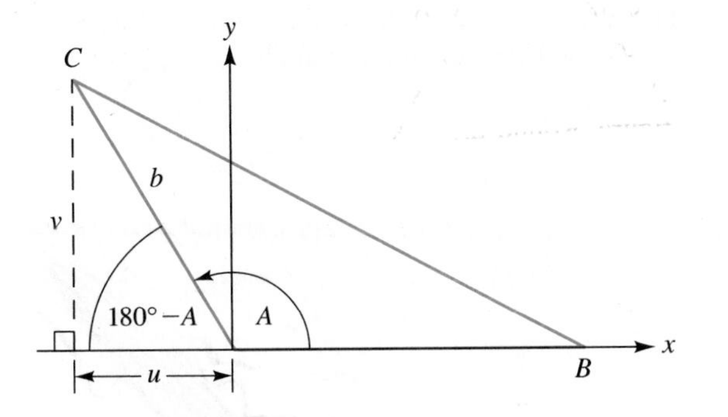

45. In the following figure, $ABCD$ is a square, $AB = 1$, and $\angle EAB = \angle EBA = 15°$. Show that $\triangle CDE$ is equilateral. *Hint:* First use the law of sines to find AE. Then use the law of cosines to find DE.

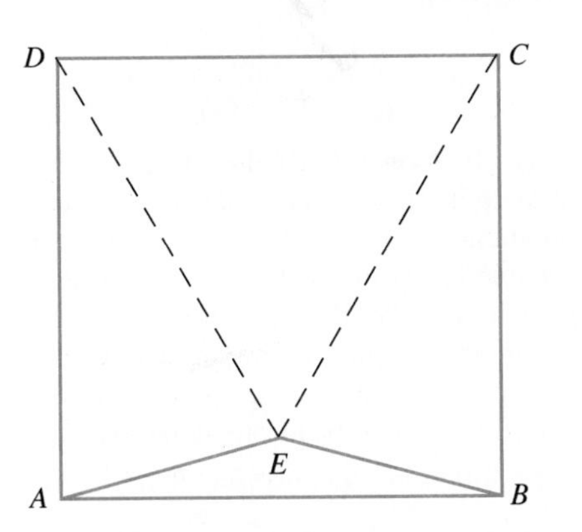

46. Use steps (a) through (c) to show that the area of any triangle ABC is given by the following expression:

$$\frac{a^2 - b^2}{2} \cdot \frac{\sin A \sin B}{\sin(A - B)}$$

(a) Use the law of sines to show that $(a^2 - b^2)\sin A \sin B = ab(\sin^2 A - \sin^2 B)$.

(b) Prove the trigonometric identity $\sin(A - B)\sin(A + B) = \sin^2 A - \sin^2 B$.

(c) Use the results in parts (a) and (b) to show that $\frac{a^2 - b^2}{2} \cdot \frac{\sin A \sin B}{\sin(A - B)} = \frac{1}{2}ab \sin C$, which is the area of $\triangle ABC$, as required.

As background for Exercises 47 and 48, refer to the figure at the top of the next column. The smaller circle in the figure is the ***inscribed circle*** *for $\triangle ABC$. Each side of $\triangle ABC$ is tangent to the inscribed circle. The larger circle is the* ***circumscribed circle*** *for $\triangle ABC$. The circumscribed circle is the circle passing through the three vertices of the triangle.*

In Exercises 47 and 48, you will derive expressions for the radii of the circumscribed circle and the inscribed circle for $\triangle ABC$. In these exercises, assume as given the following two results from geometry:

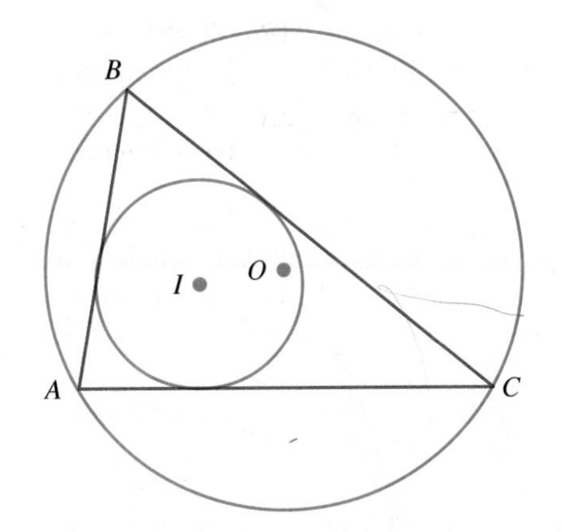

i. *The three angle bisectors of the angles of a triangle meet in a point. This point (labeled I in the figure) is the center of the inscribed circle.*

ii. *The perpendicular bisectors of the sides of a triangle meet in a point. This point (labeled O in the figure) is the center of the circumscribed circle.*

47. The following figure shows the circumscribed circle for $\triangle ABC$. The point O is the center of the circle and $\mathcal{R}$ is the radius.

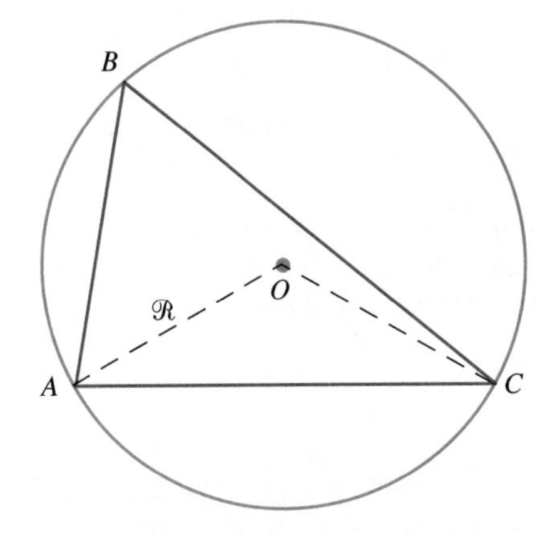

In this exercise, you will derive the formulas for the radius $\mathcal{R}$ of the circumscribed circle for $\triangle ABC$:

$$\mathcal{R} = \frac{a}{2 \sin A} = \frac{b}{2 \sin B} = \frac{c}{2 \sin C}$$

(a) According to a theorem from geometry, the measure of $\angle AOC$ is twice the measure of $\angle B$. What theorem is this? (State the theorem using complete sentences.)

(b) Draw a perpendicular from O to $\overline{AC}$, meeting $\overline{AC}$ at T. Explain why $AT = TC = b/2$. (As usual, b denotes the length of the side opposite angle B.)

(c) Explain why $\triangle ATO$ is congruent to $\triangle CTO$.

(d) Use the results in parts (a) and (c) to show that $\angle COT = \angle B$.

(e) Use the result in part (d) to show that $\mathcal{R} = b/(2 \sin B)$. From this, we can conclude (by the law of sines) that $\mathcal{R} = \dfrac{a}{2 \sin A} = \dfrac{b}{2 \sin B} = \dfrac{c}{2 \sin C}$, as required.

(f) In the figure that we used for parts (a) through (e), the center of the circle falls within $\triangle ABC$. Draw a figure in which the center lies outside of $\triangle ABC$ and prove that the equation $\mathcal{R} = b/(2 \sin B)$ is true in this case, too.

48. In this exercise you will show that the radius r of the inscribed circle for $\triangle ABC$ is given by

$$r = 4\mathcal{R} \sin \frac{A}{2} \sin \frac{B}{2} \sin \frac{C}{2}$$

In the following figure, the inscribed circle (with center I) is tangent to side $\overline{AC}$ at the point D.

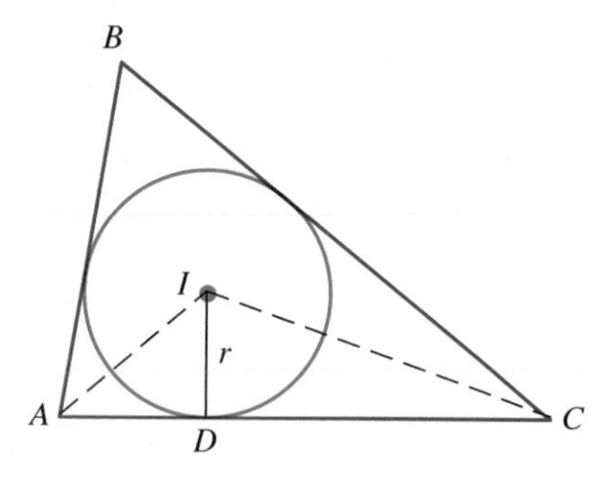

(a) According to a theorem from geometry, $\overline{ID}$ is perpendicular to $\overline{AC}$. What theorem is this? (State the theorem using complete sentences.)

(b) Show that $AD = r \cot(A/2)$ and $DC = r \cot(C/2)$.

(c) Use the results in part (b) to show that $r = \dfrac{b}{\cot(A/2) + \cot(C/2)}$. *Hint:* From the figure, we have $b = AD + DC$. (As usual, the letter b denotes the length AC in $\triangle ABC$.)

(d) Show that $r = \dfrac{b \sin(A/2) \sin(C/2)}{\sin[(A + C)/2]}$. *Hint:* Begin with the expression for r in part (c) and convert to sines and cosines. Then simplify and use an addition formula.

(e) Show that $r = \dfrac{b \sin(A/2) \sin(C/2)}{\cos(B/2)}$. *Hint:* Begin with the expression for r in part (d) and use the fact that $A + B + C = 180°$.

(f) From Exercise 47, we have $b = 2\mathcal{R} \sin B$. Use this to substitute for b in part (e). Show that the resulting equation can be written

$r = 4\mathcal{R} \sin \dfrac{A}{2} \sin \dfrac{B}{2} \sin \dfrac{C}{2}$, as required.

49. The following figure shows a quadrilateral with sides a, b, c, and d inscribed in a circle.

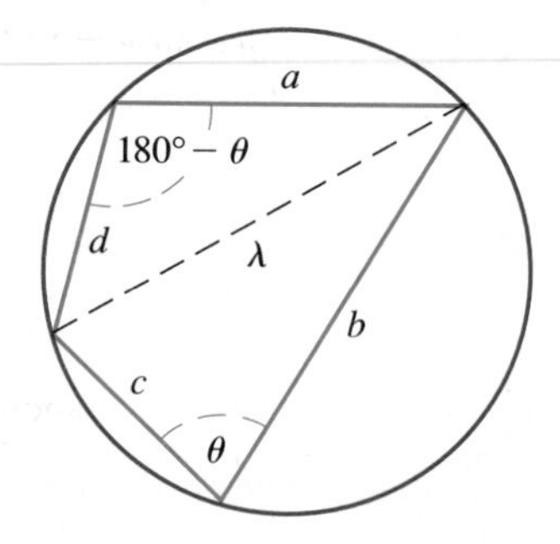

If λ denotes the length of the diagonal indicated in the figure, prove that

$$\lambda^2 = \frac{(ab + cd)(ac + bd)}{bc + ad}$$

This result is known as **Brahmagupta's theorem.** It is named after its discoverer, a seventh-century Hindu mathematician. *Hint:* Assume as given the theorem from geometry stating that when a quadrilateral is inscribed in a circle, the opposite angles are supplementary. Apply the law of cosines in both of the triangles in the figure to obtain expressions for λ^2. Then eliminate $\cos \theta$ from one equation.

50. Prove the following identity for $\triangle ABC$:

$$\frac{\cos A}{a} + \frac{\cos B}{b} + \frac{\cos C}{c} = \frac{a^2 + b^2 + c^2}{2abc}$$

Suggestion: Use the law of cosines to substitute for a^2, for b^2, and for c^2 in the numerator of the expression on the right-hand side.

51. In $\triangle ABC$, suppose that $a^4 + b^4 + c^4 = 2(a^2 + b^2)c^2$. Find $\angle C$. (There are two answers.) *Hint:* Solve the given equation for c^2.

52. In this section we have seen that the cosines of the angles in a triangle can be expressed in terms of the lengths of the sides. For instance, for $\cos A$ in $\triangle ABC$, we obtained $\cos A = (b^2 + c^2 - a^2)/2bc$. This exercise shows how to derive corresponding expressions for the sines of the angles. For ease of notation in this exercise, let us agree to use the letter T to denote the following quantity:

$$T = 2(a^2b^2 + b^2c^2 + c^2a^2) - (a^4 + b^4 + c^4)$$

Then the sines of the angles in $\triangle ABC$ are given by

$$\sin A = \frac{\sqrt{T}}{2bc} \qquad \sin B = \frac{\sqrt{T}}{2ac} \qquad \sin C = \frac{\sqrt{T}}{2ab}$$

In the steps that follow, we'll derive the first of these

three formulas, the derivations for the other two being entirely similar.

(a) In $\triangle ABC$, why is the positive root always appropriate in the formula $\sin A = \sqrt{1-\cos^2 A}$?

(b) In the formula in part (a), replace $\cos A$ by $(b^2 + c^2 - a^2)/2bc$ and show that the result can be written

$$\sin A = \frac{\sqrt{4b^2c^2 - (b^2 + c^2 - a^2)^2}}{2bc}$$

(c) On the right-hand side of the equation in part (b), carry out the indicated multiplication. After combining like terms, you should obtain $\sin A = \sqrt{T}/2bc$, as required.

53. In the two easy steps that follow, we derive the law of sines by using the formulas obtained in Exercise 52. (Since the formulas in Exercise 52 were obtained using the law of cosines, we are, in essence, showing how to derive the law of sines from the law of cosines.)

(a) Use the formulas in Exercise 52 to check that each of the three fractions $(\sin A)/a$, $(\sin B)/b$, and $(\sin C)/c$ is equal to $\sqrt{T}/2abc$.

(b) Conclude from part (a) that $(\sin A)/a = (\sin B)/b = (\sin C)/c$.

54. In this exercise you are going to use the law of cosines and the law of sines to determine the area of the shaded equilateral triangle in Figure A. Begin by labeling points, as shown in Figure B.

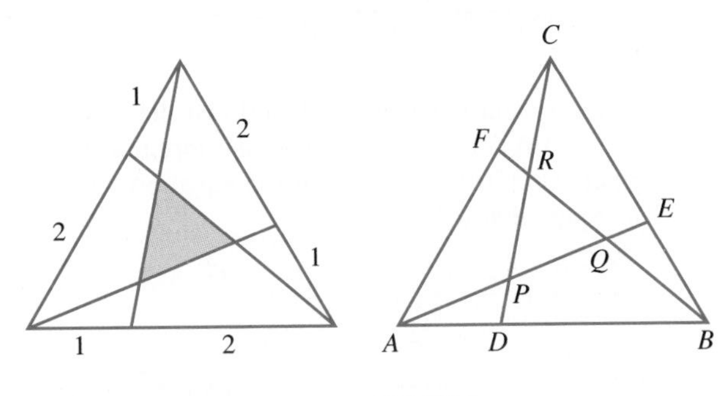

FIGURE A FIGURE B

(a) Apply the law of cosines in $\triangle ABE$ to show that $AE = \sqrt{7}$.

(b) Apply the law of sines in $\triangle ABE$ to show that $\sin(\angle AEB) = (3\sqrt{3})/(2\sqrt{7})$.

(c) Apply the law of sines in $\triangle CFB$ to show that $\sin(\angle FBC) = (\sqrt{3})/(2\sqrt{7})$. *Hint:* By symmetry, $BF = AE$.

(d) Apply the law of sines in $\triangle QEB$ to show that $QE = 1/\sqrt{7}$.

(e) Apply the law of sines in $\triangle QEB$ to show that $QB = 3/\sqrt{7}$.

(f) Show that $PQ = 3\sqrt{7}/7$.
Hint: $PQ = AE - (QE + AP)$, and, by symmetry, $AP = QB$.

(g) Use the result in part (f) to find the area of equilateral triangle PQR.

55. This exercise is adapted from a problem proposed by Professor Norman Schaumberger in the May 1990 issue of *The College Mathematics Journal.* In the accompanying figure, radius $OA = 1$ and $\angle AFC = 60°$. Follow steps (a) through (f) to prove that

$$AD \cdot AB - AE \cdot AC = BC$$

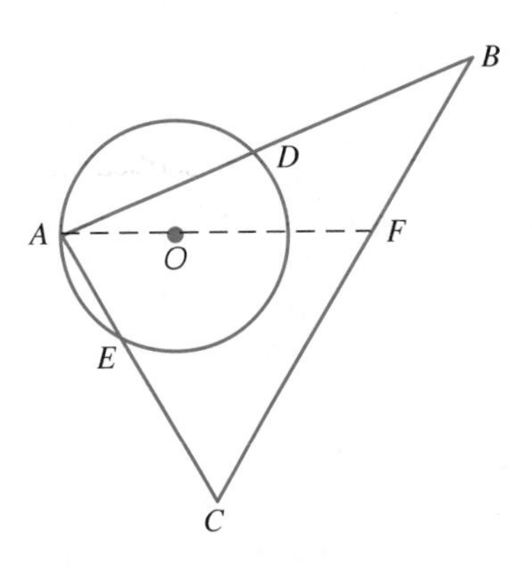

(a) Let $\angle BAF = \alpha$; show that $AD = 2\cos\alpha$.
Hint: In isosceles triangle AOD, drop a perpendicular from O to side $\overline{AD}$.

(b) Let $\angle CAF = \beta$; show that $AE = 2\cos\beta$.

(c) Explain why $\angle B = 60° - \alpha$ and $\angle C = 120° - \beta$.

(d) Using the law of sines, show that

$$AB = \frac{BC\sin(120° - \beta)}{\sin(\alpha + \beta)}$$

and

$$AC = \frac{BC\sin(60° - \alpha)}{\sin(\alpha + \beta)}$$

(e) Using the results in parts (a), (b), and (d), verify that

$$AD \cdot AB - AE \cdot AC = BC \cdot \left[\frac{2\cos\alpha\sin(120° - \beta) - 2\cos\beta\sin(60° - \alpha)}{\sin(\alpha + \beta)}\right]$$

(f) To complete the proof, you need to show that the quantity in brackets in part (e) is equal to 1. In other words, you want to show that

$$2\cos\alpha\sin(120° - \beta) - 2\cos\beta\sin(60° - \alpha) = \sin(\alpha + \beta)$$

Use the addition formulas for sine to prove that this last equation is indeed an identity.

C

56. *Heron's formula* Approximately 2000 years ago, Heron of Alexandria derived a formula for the area of a triangle in terms of the lengths of the sides. A more modern derivation of Heron's formula is indicated in the steps that follow.

(a) Use the expression for $\sin A$ in Exercise 52(b) to show that

$$\sin^2 A = \frac{(a-b+c)(a+b-c)(b+c-a)(b+c+a)}{4b^2c^2}$$

Hint: Use difference-of-squares factoring repeatedly.

(b) Let s denote one-half of the perimeter of $\triangle ABC$. That is, let $s = \frac{1}{2}(a+b+c)$. Using this notation (which is due to Euler), verify that

(i) $a+b+c = 2s$
(ii) $-a+b+c = 2(s-a)$
(iii) $a-b+c = 2(s-b)$
(iv) $a+b-c = 2(s-c)$

Then, using this notation and the result in part (a), show that

$$\sin A = \frac{2\sqrt{s(s-a)(s-b)(s-c)}}{bc}$$

Note: Since $\sin A$ is positive (why?), the positive root is appropriate here.

(c) Use the result in part (b) and the formula area $\triangle ABC = \frac{1}{2}bc \sin A$ to conclude that

$$\text{area } \triangle ABC = \sqrt{s(s-a)(s-b)(s-c)}$$

This is Heron's formula. For historical background and a purely geometric proof, see *An Introduction to the History of Mathematics,* 6th ed., by Howard Eves (Philadelphia: Saunders College Publishing, 1990), pp. 178 and 194.

57. In this exercise, you will derive a formula for the length of an angle bisector in a triangle. (The formula will be needed in Exercise 58.) Let f denote the length of the bisector of angle C in $\triangle ABC$, as shown in the following figure.

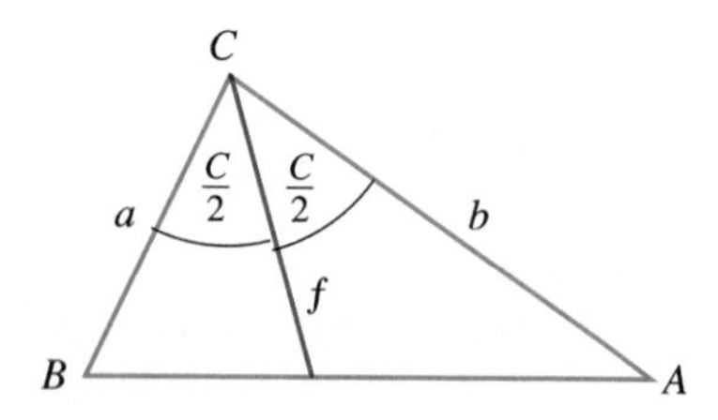

(a) Explain why

$$\frac{1}{2}af \sin\frac{C}{2} + \frac{1}{2}bf \sin\frac{C}{2} = \frac{1}{2}ab \sin C$$

Hint: Use areas.

(b) Show that $f = \dfrac{2ab\cos(C/2)}{a+b}$.

(c) By the law of cosines, $\cos C = (a^2+b^2-c^2)/2ab$. Use this to show that

$$\cos\frac{C}{2} = \frac{1}{2}\sqrt{\frac{(a+b-c)(a+b+c)}{ab}}$$

(d) Show that the length of the angle bisector in terms of the sides is given by

$$f = \frac{\sqrt{ab}}{a+b}\sqrt{(a+b-c)(a+b+c)}$$

58. In triangle XYZ (in the accompanying figure), $\overline{SX}$ bisects angle ZXY and $\overline{TY}$ bisects angle ZYX. In this exercise, you are going to prove the following theorem, known as the **Steiner–Lehmus theorem:** If the lengths of the angle bisectors $\overline{SX}$ and $\overline{TY}$ are equal, then $\triangle XYZ$ is isosceles (with $XZ = YZ$).

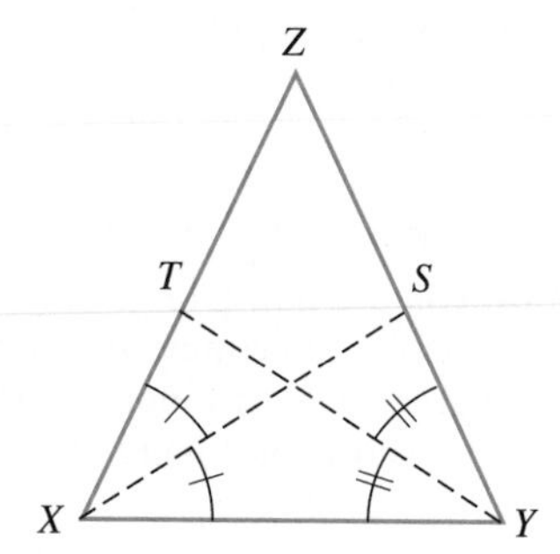

(a) Let x, y, and z denote the lengths of the sides $\overline{YZ}$, $\overline{XZ}$, and $\overline{XY}$, respectively. Use the formula in Exercise 57(d) to show that the equation $TY = SX$ is equivalent to

$$\frac{\sqrt{xz}}{x+z}\sqrt{(x+z-y)(x+z+y)} = \frac{\sqrt{yz}}{y+z}\sqrt{(y+z-x)(y+z+x)} \quad (1)$$

(b) What common factors do you see on both sides of equation (1)? Divide both sides of equation (1) by those common factors. You should obtain

$$\frac{\sqrt{x}}{x+z}\sqrt{x+z-y} = \frac{\sqrt{y}}{y+z}\sqrt{y+z-x} \quad (2)$$

(c) Clear equation (2) of fractions, and then square both sides. After combining like terms and then grouping, the equation can be written

$$(3x^2yz - 3xy^2z) + (x^3y - xy^3) + (x^2z^2 - y^2z^2) + (xz^3 - yz^3) = 0 \quad (3)$$

(d) Show that equation (3) can be written

$$(x-y)[3xyz + xy(x+y) + x^2(x+y) + z^3] = 0$$

Now notice that the quantity in brackets in this last equation must be positive. (Why?) Consequently, $x - y = 0$, and so $x = y$, as required.

Remark: This theorem has a fascinating history, beginning in 1840 when C. L. Lehmus first proposed the theorem to the great Swiss geometer Jacob Steiner (1796–1863). For background (and much shorter proofs!), see either of the following references: *Scientific American,* vol. 204 (1961), pp. 166–168; *American Mathematical Monthly,* vol. 70 (1963), pp. 79–80.

59. Show that

$$\text{area } \triangle ABC = \frac{a^2 \sin 2B + b^2 \sin 2A}{4}$$

60. For any triangle ABC, show that

$$\frac{\sin(A-B)}{\sin(A+B)} = \frac{a^2-b^2}{c^2}$$

9.2 VECTORS IN THE PLANE, A GEOMETRIC APPROACH

The idea of a parallelogram of velocities may be found in various ancient Greek authors, and the concept of a parallelogram of forces was not uncommon in the sixteenth and seventeenth centuries.

Michael J. Crowe in *A History of Vector Analysis* (Notre Dame, Ind.: University of Notre Dame Press, 1967)

Vector notation is compact. If we can express a law of physics in vector form we usually find it easier to understand and to manipulate mathematically.

David Halliday and Robert Resnick in *Fundamentals of Physics,* 3rd ed. (New York: John Wiley and Sons, Inc., 1988)

Certain quantities, such as temperature, length, and mass, can be specified by means of a single number (assuming that a system of units has been agreed on). We call these quantities **scalars.** On the other hand, quantities such as force and velocity are characterized by both a *magnitude* (a positive number) and a *direction*. We call these quantities **vectors.**

Geometrically, a vector is a directed line segment or arrow. The vector in Figure 1, for instance, represents a wind velocity of 5 mph from the west. The length of this vector represents the magnitude of the wind velocity, while the direction of the vector indicates the direction of the wind velocity. As another example, the vector in Figure 2 represents a force acting on an object: the magnitude of the force is 3 pounds, and the force acts at an angle of 135° with the horizontal.

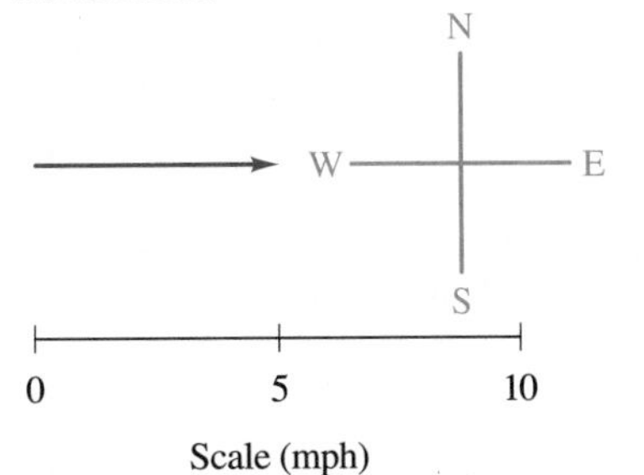

FIGURE 1
A vector representing a wind velocity of 5 mph from the west.

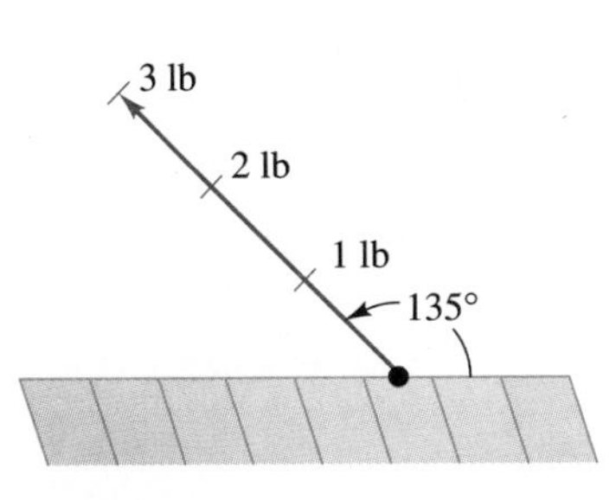

FIGURE 2
A vector representing a force of 3 lb acting at an angle of 135° with the horizontal.

In a moment, we are going to discuss the important concept of vector addition, but first let us agree on some matters of notation. Suppose that we have a vector drawn from a point P to a point Q, as shown in Figure 3. The point P in Figure 3 is called the **initial point** of the vector, and Q is the **terminal point.** We can denote this vector by

$$\overrightarrow{PQ}$$

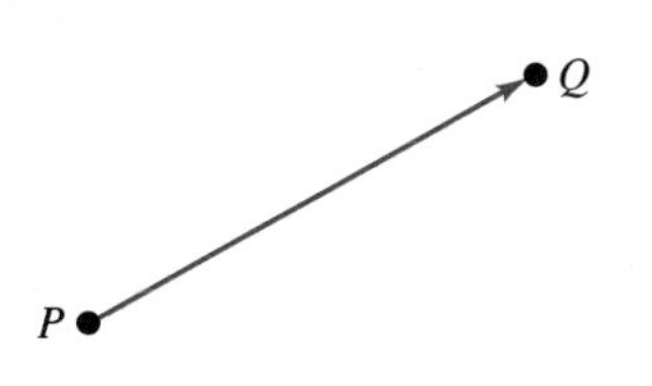

FIGURE 3

The length or **magnitude** of the vector $\overrightarrow{PQ}$ is denoted by $|\overrightarrow{PQ}|$. On the printed page, vectors are often indicated by boldface letters, such as **a**, **A**, and **v**.

A word about notation. As you know, the notation (a, b) can denote either a point in the x-y plane or an open interval on the number line. In each instance, however, there is usually no danger of confusion; the context makes it clear which meaning is intended. Now we have a similar situation with the notation for the length of a vector. As has perhaps already occurred to you, the same vertical bars that we are using to denote the length of a vector are also used, in another context, to denote the absolute value of a real number. Again, it will be clear from the context which meaning is intended. Some books avoid this situation by using double bars to indicate the length of a vector: $\|\mathbf{v}\|$. In this text, we use the notation $|\mathbf{v}|$ simply because that is the one found in most calculus books.

If two vectors **a** and **b** have the same length and the same direction, we say that they are *equal* and we write $\mathbf{a} = \mathbf{b}$; see Figure 4. Notice that our definition for vector equality involves magnitude and direction, but not location. Thus, when it is convenient to do so, we are free to move a given vector to another location, provided we do not alter the magnitude or the direction.

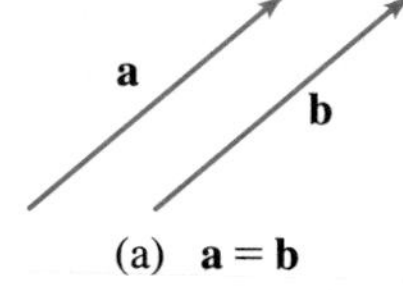

(a) $\mathbf{a} = \mathbf{b}$

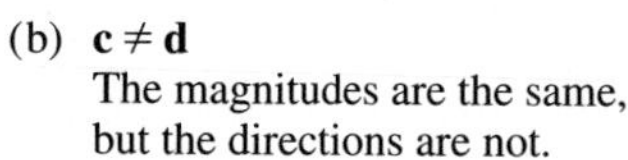

(b) $\mathbf{c} \neq \mathbf{d}$
The magnitudes are the same, but the directions are not.

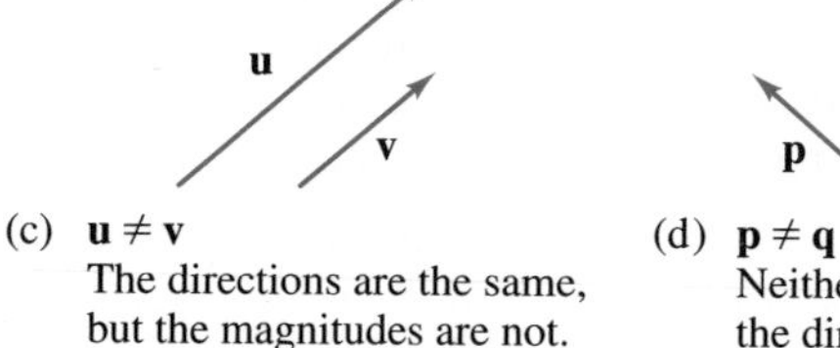

(c) $\mathbf{u} \neq \mathbf{v}$
The directions are the same, but the magnitudes are not.

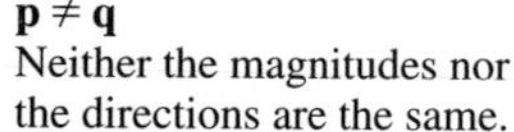

(d) $\mathbf{p} \neq \mathbf{q}$
Neither the magnitudes nor the directions are the same.

FIGURE 4

As motivation for the definition of vector addition, let's suppose that an object moves from a point P to a point Q. Then we can represent this *displacement* by the vector $\overrightarrow{PQ}$. (Indeed, the word *vector* is derived from the Latin *vectus,* meaning "carried.") Now suppose that after moving from P to Q, the object moves from Q to R. Then, as you can see in Figure 5, the net effect is a displacement from P to R. We say in this case that the vector $\overrightarrow{PR}$ is the **sum** or **resultant** of the vectors $\overrightarrow{PQ}$ and $\overrightarrow{QR}$, and we write

$$\overrightarrow{PQ} + \overrightarrow{QR} = \overrightarrow{PR}$$

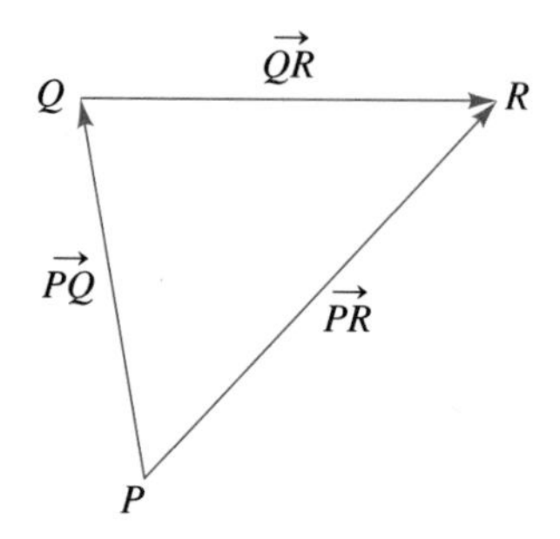

FIGURE 5

These ideas are formalized in the definition that follows.

DEFINITION: Vector Addition

Let **u** and **v** be two vectors. Position **v** (without changing its magnitude or direction) so that its initial point coincides with the terminal point of **u**, as in Figure 6(a). Then, as indicated in Figure 6(b), the vector $\mathbf{u} + \mathbf{v}$ is the directed line segment from the initial point of **u** to the terminal point of **v**. The vector $\mathbf{u} + \mathbf{v}$ is called the **sum** or **resultant** of **u** and **v**.

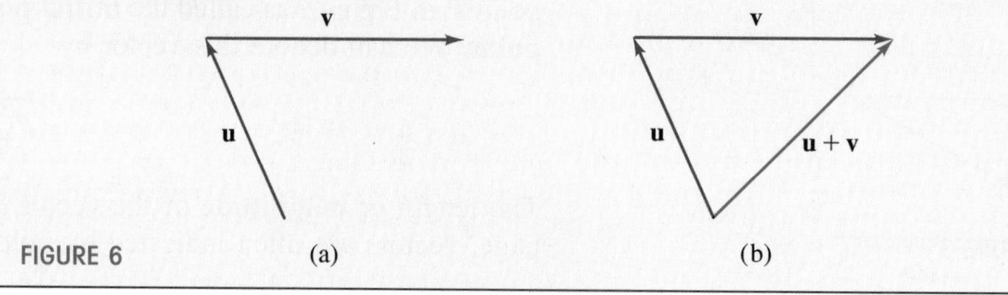

FIGURE 6 (a) (b)

EXAMPLE 1 Referring to Figure 7, **(a)** determine the initial and terminal points of $\mathbf{u} + \mathbf{v}$; and **(b)** compute $|\mathbf{u} + \mathbf{v}|$.

FIGURE 7

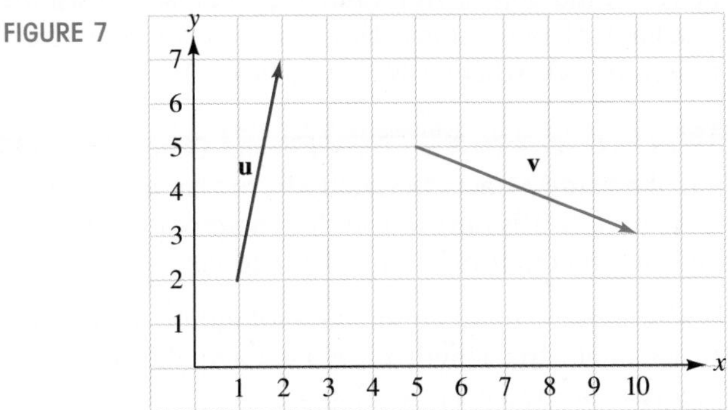

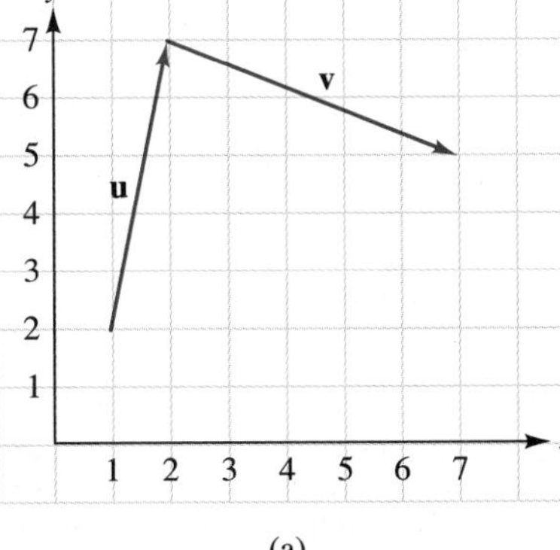

(a)

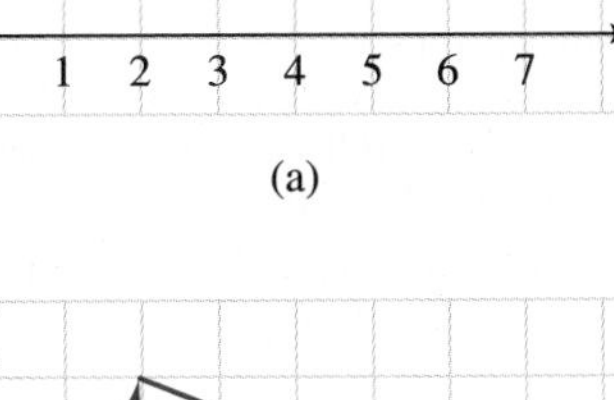

(b)

FIGURE 8

Solution

(a) According to the definition, we first need to move $\mathbf{v}$ (without changing its length or direction) so that its initial point coincides with the terminal point of $\mathbf{u}$. From Figure 7, we see that this can be accomplished by moving each point of $\mathbf{v}$ three units in the negative x-direction and two units in the positive y-direction. Figure 8(a) shows the new location of $\mathbf{v}$, and Figure 8(b) indicates the sum $\mathbf{u} + \mathbf{v}$. From Figure 8(b), we see that the initial and terminal points of $\mathbf{u} + \mathbf{v}$ are $(1, 2)$ and $(7, 5)$, respectively.

(b) We can use the distance formula to determine $|\mathbf{u} + \mathbf{v}|$. Using the points $(1, 2)$ and $(7, 5)$ that were obtained in part (a), we have

$$|\mathbf{u} + \mathbf{v}| = \sqrt{(7-1)^2 + (5-2)^2} = \sqrt{45} = \sqrt{9 \cdot 5} = 3\sqrt{5}$$

One important consequence of our definition for vector addition is that this operation is *commutative*. That is, for any two vectors $\mathbf{u}$ and $\mathbf{v}$, we have

$$\mathbf{u} + \mathbf{v} = \mathbf{v} + \mathbf{u}$$

Figure 9 indicates why this is so. Vector addition can also be carried out by using the **parallelogram law.** In Figure 9, to determine $\mathbf{u} + \mathbf{v}$, position $\mathbf{u}$ and $\mathbf{v}$ so that their initial points coincide. Then, as indicated in Figure 10, the vector $\mathbf{u} + \mathbf{v}$ is the directed diagonal of the parallelogram determined by $\mathbf{u}$ and $\mathbf{v}$.

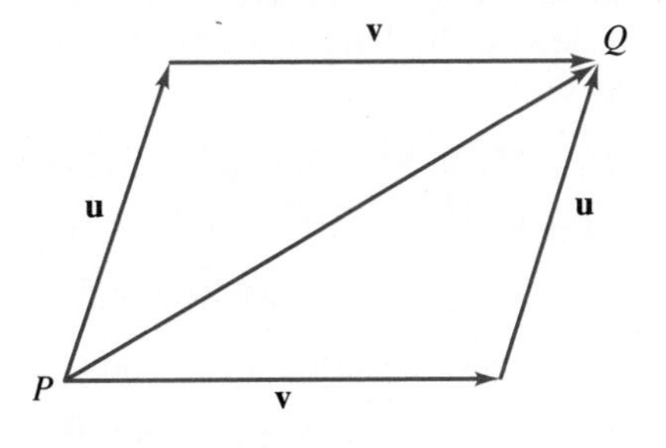

FIGURE 9
The upper triangle shows the sum $\mathbf{u} + \mathbf{v}$, while the lower triangle shows the sum $\mathbf{v} + \mathbf{u}$. Since in both cases the resultant is $\overrightarrow{PQ}$, it follows that $\mathbf{u} + \mathbf{v} = \mathbf{v} + \mathbf{u}$.

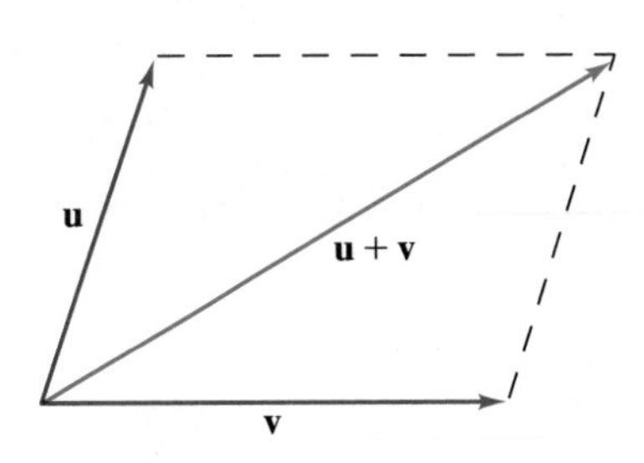

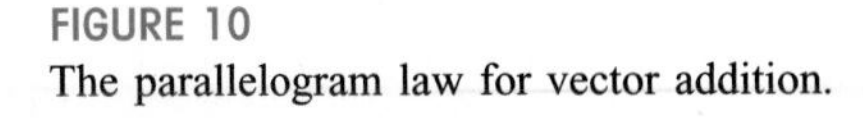

FIGURE 10
The parallelogram law for vector addition.

It is a fact—and it has been verified experimentally—that if two forces **F** and **G** act on an object, the net effect is the same as if just the resultant **F** + **G** acted on the object. In Example 2, we use the parallelogram law to compute the resultant of two forces. Note that the units of force used in this example are **newtons** (N), where 1 N $\approx$ 0.2248 lb.

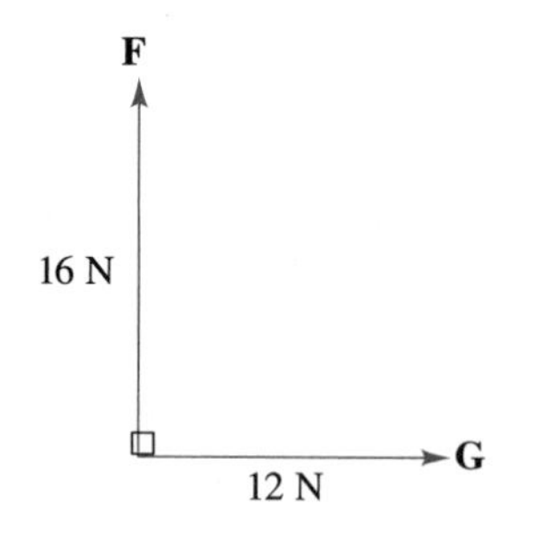

FIGURE 11

EXAMPLE 2 Two forces **F** and **G** act on an object. As indicated in Figure 11, the force **G** acts horizontally to the right with a magnitude of 12 N, while **F** acts vertically upward with a magnitude of 16 N. Determine the magnitude and the direction of the resultant force.

Solution We complete the parallelogram, as shown in Figure 12. Now we need to calculate the length of **F** + **G** and the angle θ. Applying the Pythagorean theorem in Figure 12, we have

$$|\mathbf{F}+\mathbf{G}| = \sqrt{12^2+16^2} = \sqrt{144+256} = \sqrt{400} = 20$$

Also from Figure 12, we have

$$\tan\theta = \frac{16}{12} = \frac{4}{3}$$

Consequently,

$$\theta = \tan^{-1}\frac{4}{3} \approx 53.1^\circ \qquad \text{using a calculator set in the degree mode}$$

Now we can summarize our results. The magnitude of **F** + **G** is 20 N, and the angle θ between **F** + **G** and the horizontal is (approximately) 53.1°.

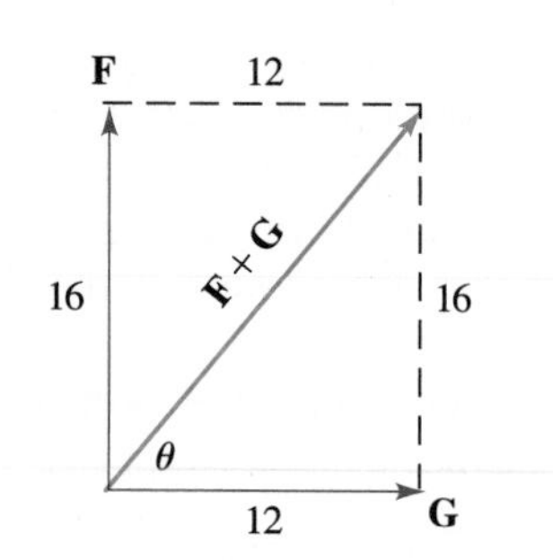

FIGURE 12

In Example 2, we determined the resultant for two perpendicular forces. The next example shows how to compute the resultant when the forces are not perpendicular. Our calculations will make use of both the law of sines and the law of cosines.

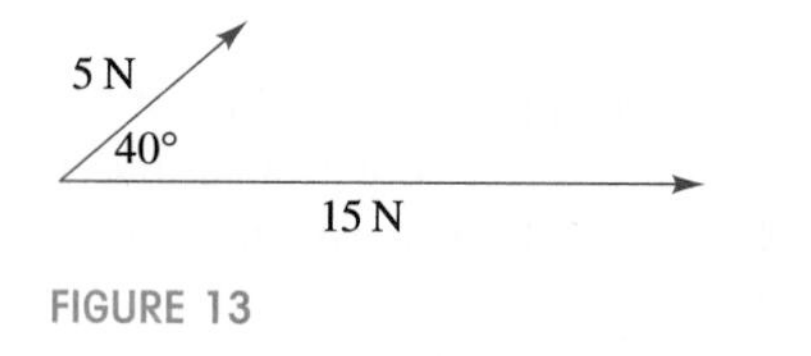

FIGURE 13

EXAMPLE 3 Determine the resultant of the two forces in Figure 13. (Round the answers to one decimal place.)

Solution As in the previous example, we complete the parallelogram. In Figure 14, the angle in the lower right-hand corner of the parallelogram is 140°. This is because the sum of two adjacent angles in any parallelogram is always 180°. Letting d denote the length of the diagonal in Figure 14, we can use the law of cosines to write

$$\begin{aligned} d^2 &= 15^2 + 5^2 - 2(15)(5)\cos 140^\circ \\ d &= \sqrt{250 - 150\cos 140^\circ} = \sqrt{250 + 150\cos 40^\circ} \qquad \text{(Why?)} \\ &= \sqrt{25(10 + 6\cos 40^\circ)} = 5\sqrt{10 + 6\cos 40^\circ} \\ &\approx 19.1 \qquad \text{using a calculator} \end{aligned}$$

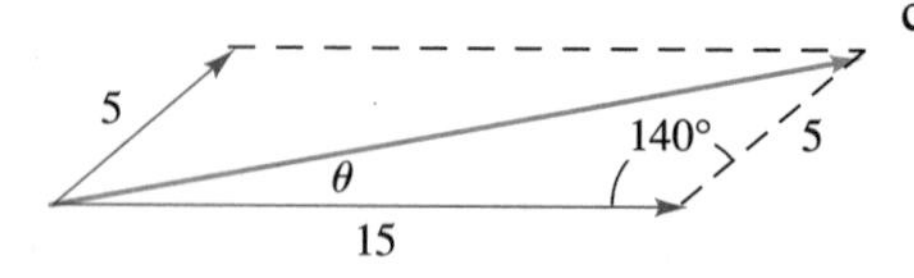

FIGURE 14

So the magnitude of the resultant is 19.1 N (to one decimal place). To specify the direction of the resultant, we need to determine the angle θ in Figure 14. Using the law of sines, we have

$$\frac{\sin\theta}{5} = \frac{\sin 140^\circ}{d}$$

and, consequently,

$$\sin\theta = \frac{5\sin 40^\circ}{d} \qquad \text{(Why?)}$$

$$= \frac{5\sin 40^\circ}{5\sqrt{10+6\cos 40^\circ}} = \frac{\sin 40^\circ}{\sqrt{10+6\cos 40^\circ}}$$

Using a calculator now, we obtain

$$\theta = \sin^{-1}\left(\frac{\sin 40^\circ}{\sqrt{10+6\cos 40^\circ}}\right) \approx 9.7^\circ$$

In summary, the magnitude of the resultant force is about 19.1 N, and the angle θ between the resultant and the 15 N force is approximately 9.7°.

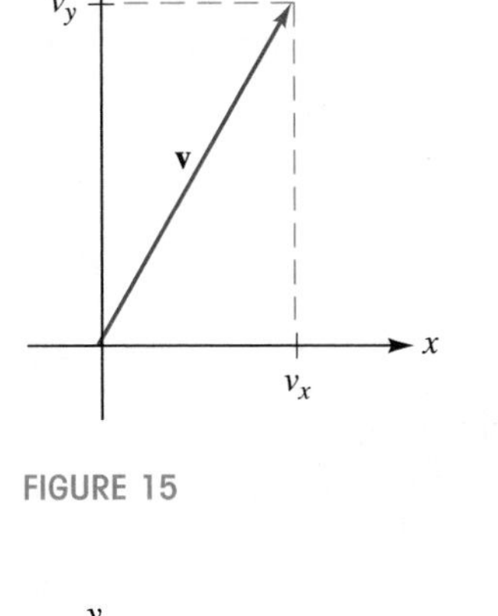

FIGURE 15

As background for the next example, we introduce the notion of *components* of a vector. (You will see this concept again in the next section, but in a more algebraic context.) Suppose that the initial point of a vector **v** is located at the origin of a rectangular coordinate system, as shown in Figure 15. Now suppose we draw perpendiculars from the terminal point of **v** to the axes, as indicated by the blue dashed lines in Figure 15. Then the coordinates v_x and v_y are called the **components** of the vector **v** in the x- and y-directions, respectively. For an example involving components, refer back to Figure 12. The horizontal component of the vector **F** + **G** is 12 N, and the vertical component is 16 N.

EXAMPLE 4 Determine the horizontal and vertical components of the velocity vector **v** in Figure 16.

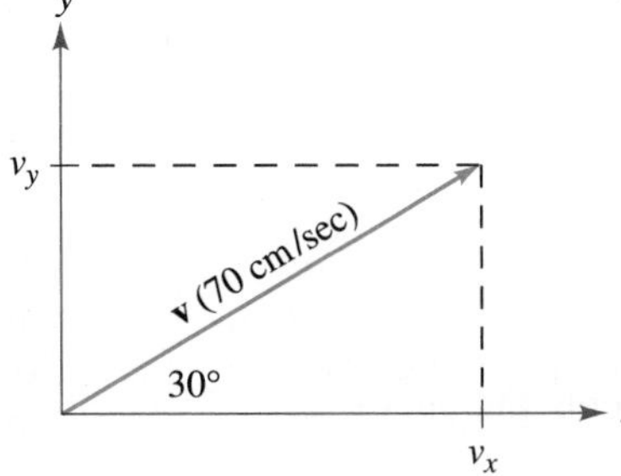

FIGURE 16

Solution From Figure 16, we can write

$$\cos 30^\circ = \frac{\text{adjacent}}{\text{hypotenuse}} = \frac{v_x}{70}$$

and, consequently,

$$v_x = (\cos 30^\circ)(70) = \frac{\sqrt{3}}{2}(70)$$

$$= 35\sqrt{3} \approx 61 \text{ cm/sec} \qquad \text{to two significant digits}$$

Similarly, we have

$$\sin 30^\circ = \frac{v_y}{70}$$

and therefore

$$v_y = (\sin 30^\circ)(70) = 35 \text{ cm/sec}$$

In summary, now, the x-component of the velocity is about 61 cm/sec, and the y-component is 35 cm/sec.

Our last example in this section will indicate how vectors are used in navigation. First, however, let's introduce some terminology. Suppose that an airplane has a **heading** of due east. This means that the airplane is pointed due east, and if there were no wind effects, the plane would indeed travel due east

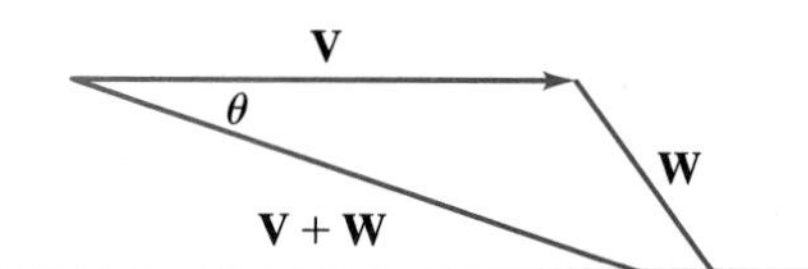

FIGURE 17

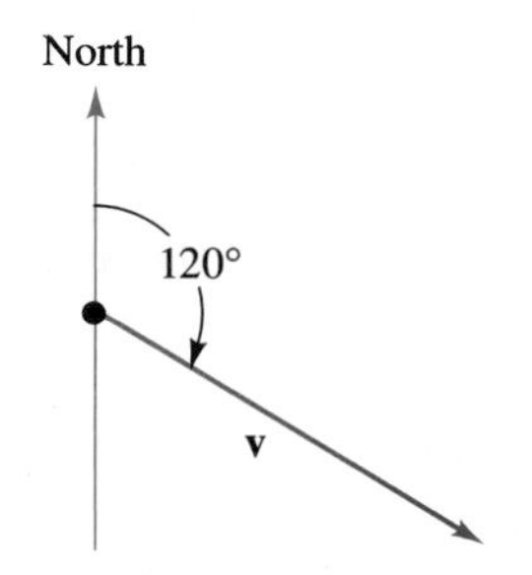

FIGURE 18

with respect to the ground. The **air speed** is the speed of the airplane relative to the air, whereas the **ground speed** is the plane's speed relative to the ground. Again, if there were no wind effects, then the air speed and the ground speed would be equal. Now suppose that the heading and air speed of an airplane are represented by the velocity vector **V** in Figure 17. (The direction of **V** is the heading; the magnitude of **V** is the air speed.) Also suppose that the wind velocity is represented by the vector **W** in Figure 17. Then the vector sum $\mathbf{V} + \mathbf{W}$ represents the actual velocity of the plane with respect to the ground. The direction of $\mathbf{V} + \mathbf{W}$ is called the **course**; it is the direction in which the airplane is moving with respect to the ground. The magnitude of $\mathbf{V} + \mathbf{W}$ is the ground speed (which was defined previously). The angle θ in Figure 17 is called, naturally enough, the **drift angle.**

In navigation, directions are given in terms of the angle measured clockwise from true north. For example, the direction of the velocity vector **v** in Figure 18 is 120°.

EXAMPLE 5 [Refer to Figure 19(a).] The heading and air speed of an airplane are 60° and 250 mph, respectively. If the wind is 40 mph from 150°, find the ground speed, the drift angle, and the course.

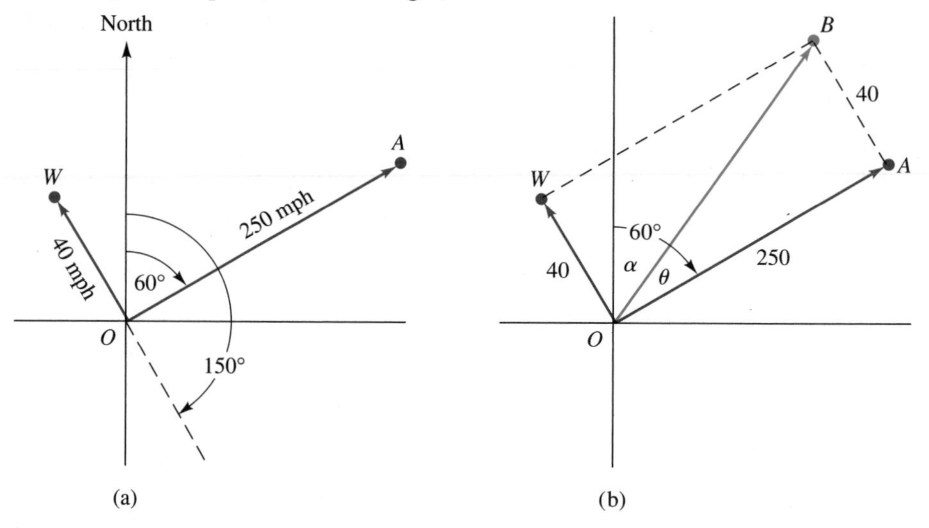

FIGURE 19

Solution In Figure 19(a), the vector $\overrightarrow{OA}$ represents the air speed of 250 mph and the heading of 60°. The vector $\overrightarrow{OW}$ represents a wind of 40 mph from 150°. Also, angle $WOA = 90°$. (Why?) In Figure 19(b), we have completed the parallelogram to obtain the vector sum $\overrightarrow{OA} + \overrightarrow{OW} = \overrightarrow{OB}$. The length of $\overrightarrow{OB}$ represents the ground speed, and θ is the drift angle. Because triangle BOA is a right triangle, we have

$$\tan\theta = \frac{40}{250} = \frac{4}{25} \qquad\qquad |\overrightarrow{OB}| = \sqrt{250^2 + 40^2}$$

$$\theta \approx 9.1° \qquad\qquad \approx 253.2$$

From these calculations, we conclude that the ground speed is approximately 253.2 mph and the drift angle is about 9.1°. We still need to compute the course (that is, the direction of vector $\overrightarrow{OB}$). From Figure 19(b), we have

$$\alpha = 60° - \theta \approx 60° - 9.1° = 50.9°$$

Thus, the course is 50.9°, to one decimal place. (We are using the convention, mentioned previously, that directions are given in terms of the angle measured clockwise from true north.)

EXERCISE SET 9.2

A

In Exercises 1–26, assume that the coordinates of the points P, Q, R, S, and O are as follows:

$P(-1, 3)$ $Q(4, 6)$ $R(4, 3)$ $S(5, 9)$ $O(0,0)$

For each exercise, draw the indicated vector (using graph paper) and compute its magnitude. In Exercises 7–20, compute the sums using the definition given on page 554. In Exercises 21–26, use the parallelogram law to compute the sums.

1. $\overrightarrow{PQ}$
2. $\overrightarrow{QP}$
3. $\overrightarrow{SQ}$
4. $\overrightarrow{QS}$
5. $\overrightarrow{OP}$
6. $\overrightarrow{PO}$
7. $\overrightarrow{PQ} + \overrightarrow{QS}$
8. $\overrightarrow{SQ} + \overrightarrow{QP}$
9. $\overrightarrow{OP} + \overrightarrow{PQ}$
10. $\overrightarrow{OS} + \overrightarrow{SQ}$
11. $(\overrightarrow{OS} + \overrightarrow{SQ}) + \overrightarrow{QP}$
12. $(\overrightarrow{OS} + \overrightarrow{SP}) + \overrightarrow{PR}$
13. $\overrightarrow{OP} + \overrightarrow{QS}$
14. $\overrightarrow{QS} + \overrightarrow{PO}$
15. $\overrightarrow{SR} + \overrightarrow{PO}$
16. $\overrightarrow{OS} + \overrightarrow{QO}$
17. $\overrightarrow{OP} + \overrightarrow{RQ}$
18. $\overrightarrow{OP} + \overrightarrow{QR}$
19. $\overrightarrow{SQ} + \overrightarrow{RO}$
20. $\overrightarrow{SQ} + \overrightarrow{OR}$
21. $\overrightarrow{OP} + \overrightarrow{OR}$
22. $\overrightarrow{OP} + \overrightarrow{OQ}$
23. $\overrightarrow{RP} + \overrightarrow{RS}$
24. $\overrightarrow{QP} + \overrightarrow{QR}$
25. $\overrightarrow{SO} + \overrightarrow{SQ}$
26. $\overrightarrow{SQ} + \overrightarrow{SR}$

In Exercises 27–32, the vectors **F** *and* **G** *denote two forces that act on an object:* **G** *acts horizontally to the right, and* **F** *acts vertically upward. In each case, use the information that is given to compute* $|\mathbf{F} + \mathbf{G}|$ *and* θ*, where* θ *is the angle between* **G** *and the resultant.*

27. $|\mathbf{F}| = 4$ N, $|\mathbf{G}| = 5$ N
28. $|\mathbf{F}| = 15$ N, $|\mathbf{G}| = 6$ N
29. $|\mathbf{F}| = |\mathbf{G}| = 9$ N
30. $|\mathbf{F}| = 28$ N, $|\mathbf{G}| = 1$ N
31. $|\mathbf{F}| = 3.22$ N, $|\mathbf{G}| = 7.21$ N
32. $|\mathbf{F}| = 4.06$ N, $|\mathbf{G}| = 26.83$ N

In Exercises 33–38, the vectors **F** *and* **G** *represent two forces acting on an object, as indicated in the following figure. In each case, use the given information to compute (to two decimal places) the magnitude and direction of the resultant. (Give the direction of the resultant by specifying the angle* θ *between* **F** *and the resultant.)*

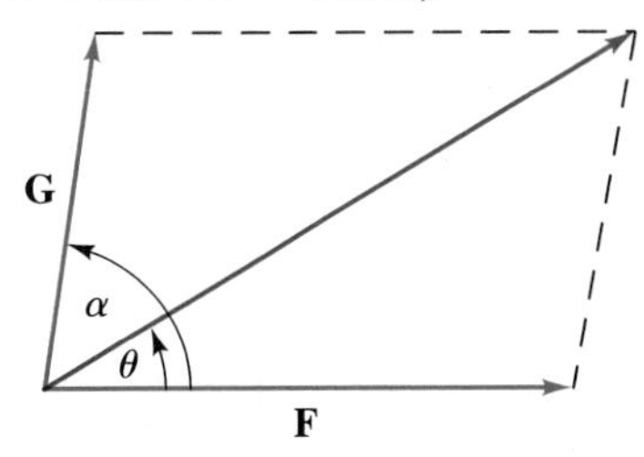

33. $|\mathbf{F}| = 5$ N, $|\mathbf{G}| = 4$ N, $\alpha = 80°$
34. $|\mathbf{F}| = 8$ N, $|\mathbf{G}| = 10$ N, $\alpha = 60°$
35. $|\mathbf{F}| = 16$ N, $|\mathbf{G}| = 25$ N, $\alpha = 35°$
36. $|\mathbf{F}| = 4.24$ N, $|\mathbf{G}| = 9.01$ N, $\alpha = 45°$
37. $|\mathbf{F}| = 50$ N, $|\mathbf{G}| = 25$ N, $\alpha = 130°$
38. $|\mathbf{F}| = 1.26$ N, $|\mathbf{G}| = 2.31$ N, $\alpha = 160°$

In Exercises 39–46, the initial point for each vector is the origin, and θ *denotes the angle (measured counterclockwise) from the x-axis to the vector. In each case, compute the horizontal and vertical components of the given vector. (Round your answers to two decimal places.)*

39. The magnitude of **V** is 16 cm/sec, and $\theta = 30°$.
40. The magnitude of **V** is 40 cm/sec, and $\theta = 60°$.
41. The magnitude of **F** is 14 N, and $\theta = 75°$.
42. The magnitude of **F** is 23.12 N, and $\theta = 52°$.
43. The magnitude of **V** is 1 cm/sec, and $\theta = 135°$.
44. The magnitude of **V** is 12 cm/sec, and $\theta = 120°$.
45. The magnitude of **F** is 1.25 N, and $\theta = 145°$.
46. The magnitude of **F** is 6.34 N, and $\theta = 175°$.

In Exercises 47–50, use the given flight data to compute the ground speed, the drift angle, and the course. (Round your answers to two decimal places.)

47. The heading and air speed are 30° and 300 mph, respectively; the wind is 25 mph from 120°.

48. The heading and air speed are 45° and 275 mph, respectively; the wind is 50 mph from 135°.

49. The heading and air speed are 100° and 290 mph, respectively; the wind is 45 mph from 190°.

50. The heading and air speed are 90° and 220 mph, respectively; the wind is 80 mph from **(a)** 180°; **(b)** 90°.

B

51. A block weighing 12 lb rests on an inclined plane, as indicated in Figure A. Determine the components of the weight perpendicular to and parallel to the plane. Round your answers to two decimal places.
Hint: In Figure B, the component of the weight perpendicular to the plane is $|\overrightarrow{OR}|$; the component parallel to the plane is $|\overrightarrow{OP}|$. Why does angle QOR equal 35°?

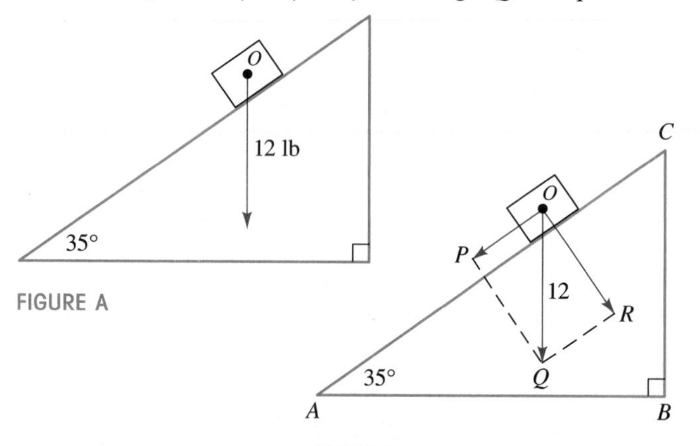

FIGURE A

FIGURE B

In Exercises 52 and 53, you are given the weight of a block on an inclined plane, along with the angle θ that the inclined plane makes with the horizontal. In each case, determine the components of the weight perpendicular to and parallel to the plane. (Round your answers to two decimal places where necessary.)

52. 15 lb; $\theta = 30°$

53. 12 lb; $\theta = 10°$

54. A block rests on an inclined plane that makes an angle of 20° with the horizontal. The component of the weight parallel to the plane is 34.2 lb.
(a) Determine the weight of the block. (Round your answer to one decimal place.)
(b) Determine the component of the weight perpendicular to the plane. (Round your answer to one decimal place.)

55. In Section 9.3, we will see that vector addition is associative. That is, for any three vectors **A, B,** and **C,** we have $(\mathbf{A} + \mathbf{B}) + \mathbf{C} = \mathbf{A} + (\mathbf{B} + \mathbf{C})$. In this exercise, you are going to check that this property holds in a particular case. Let **A, B,** and **C** be the vectors with initial and terminal points as follows:

Vector	Initial Point	Terminal Point
A	$(-1, 2)$	$(2, 4)$
B	$(1, 2)$	$(3, 0)$
C	$(6, 2)$	$(4, -3)$

(a) Use the definition of vector addition on page 554 to determine the initial and terminal points of $(\mathbf{A} + \mathbf{B}) + \mathbf{C}$. *Suggestion:* Use graph paper.
(b) Use the definition of vector addition to determine the initial and terminal points of $\mathbf{A} + (\mathbf{B} + \mathbf{C})$. [Your answers should agree with those in part (a).]

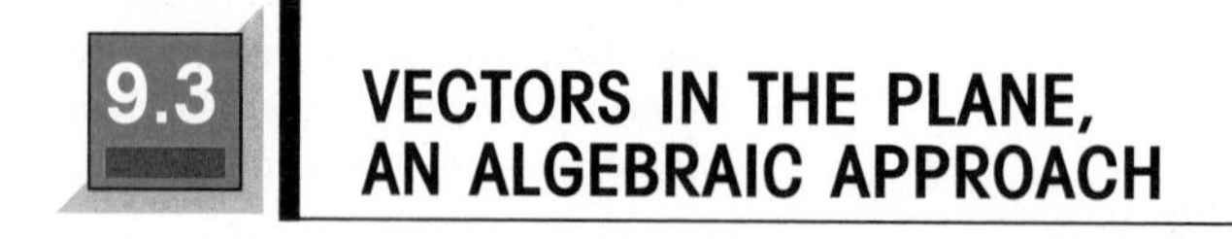

9.3 VECTORS IN THE PLANE, AN ALGEBRAIC APPROACH

A great many of the mathematical ideas that apply to physics and engineering are collected in the concept of vector spaces. This branch of mathematics has applications in such practical problems as calculating the vibrations of bridges and airplane wings. Logical extensions to spaces of infinitely many dimensions are widely used in modern theoretical physics as well as in many branches of mathematics itself.

From *The Mathematical Sciences*, edited by the Committee on Support of Research in the Mathematical Sciences with the collaboration of George Boehm (Cambridge, Mass.: The M.I.T. Press, 1969)

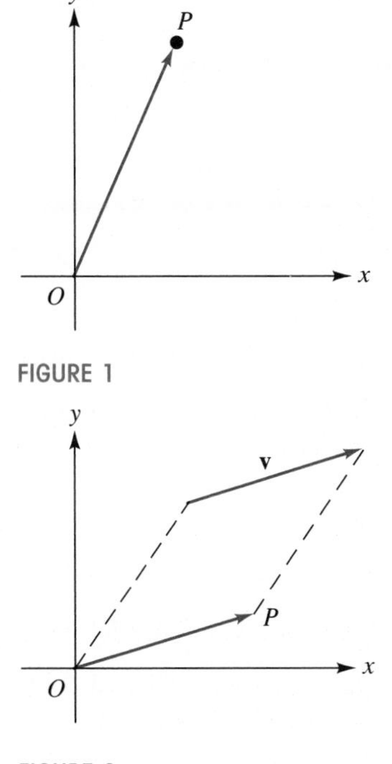

FIGURE 1

FIGURE 2

The geometric concept of a vector in the plane can be recast in an algebraic setting. This is useful both for computational purposes and (as our opening quotation implies) for more advanced work.

Consider an x-y coordinate system and a vector $\overrightarrow{OP}$ with initial point the origin, as shown in Figure 1. We call $\overrightarrow{OP}$ the **position vector** (or **radius vector**) of the point P. Most of our work in this section will involve such position vectors. There is no loss of generality in focusing on these types of vectors, for, as indicated in Figure 2, each vector $\mathbf{v}$ in the plane is equal to a unique position vector $\overrightarrow{OP}$.

If the coordinates of the point P are (a, b), we call a and b the **components** of the vector $\overrightarrow{OP}$, and we use the notation

$$\langle a, b\rangle$$

to denote this vector (see Figure 3). The number a is the **horizontal component** or ***x*-component** of the vector; b is the **vertical component** or ***y*-component.**

In the previous section, we said that two vectors are equal provided they have the same length and the same direction. For vectors $\langle a, b\rangle$ and $\langle c, d\rangle$, this implies that

$$\langle a, b\rangle = \langle c, d\rangle \qquad \text{if and only if} \qquad a = c \quad \text{and} \quad b = d$$

(If you've studied complex numbers in a previous course, notice the simiarlity here. Two complex numbers are equal provided their corresponding real and imaginary parts are equal; two vectors are equal provided their corresponding components are equal.)

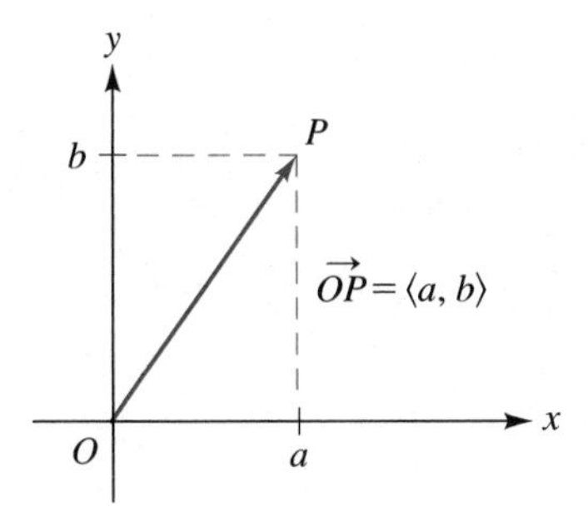

FIGURE 3

The vector $\overrightarrow{OP}$ is denoted by $\langle a, b\rangle$. The coordinates a and b are the components of the vector.

It's easy to calculate the length of a vector $\mathbf{v}$ when its components are given. Suppose that $\mathbf{v} = \langle v_1, v_2\rangle$. Applying the Pythagorean theorem in Figure 4, we have

$$|\mathbf{v}|^2 = v_1^2 + v_2^2$$

and, consequently,

$$|\mathbf{v}| = \sqrt{v_1^2 + v_2^2}$$

Although Figure 4 shows the point (v_1, v_2) in Quadrant I, you can check for yourself that the same formula results when (v_1, v_2) is located in any of the other three quadrants. We therefore have the following general formula.

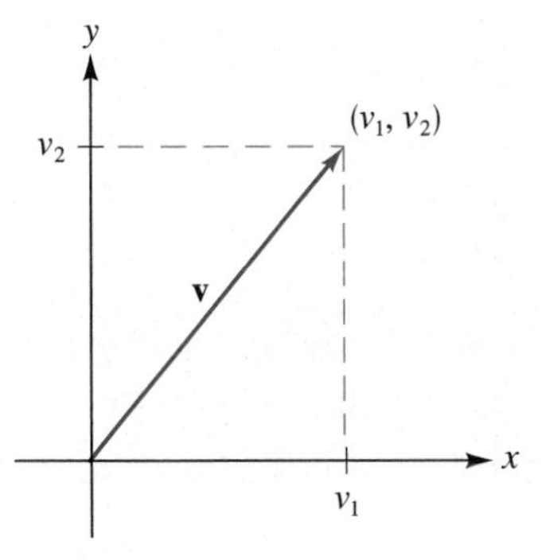

FIGURE 4

THE LENGTH OF A VECTOR

If $\mathbf{v} = \langle v_1, v_2\rangle$, then

$$|\mathbf{v}| = \sqrt{v_1^2 + v_2^2}$$

EXAMPLE 1 Compute the length of the vector $\mathbf{v} = \langle 2, -4\rangle$.

Solution

$$|\mathbf{v}| = \sqrt{v_1^2 + v_2^2} = \sqrt{2^2 + (-4)^2} = \sqrt{20} = 2\sqrt{5}$$

Even if the initial point of a vector **v** is not the origin, we can still find the *components* of **v** by determining the components of the equivalent position vector. The formula that we are going to derive for this is as follows.

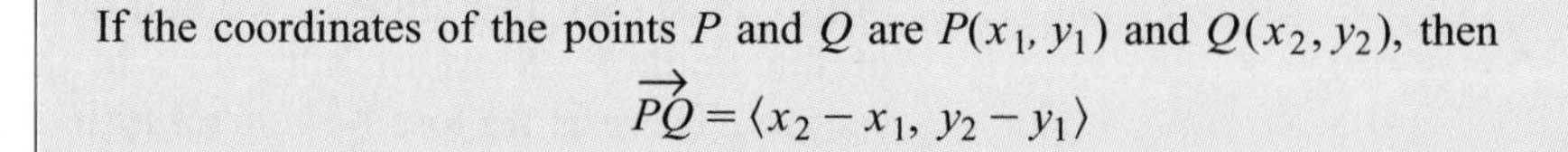

If the coordinates of the points P and Q are $P(x_1, y_1)$ and $Q(x_2, y_2)$, then

$$\overrightarrow{PQ} = \langle x_2 - x_1, y_2 - y_1 \rangle$$

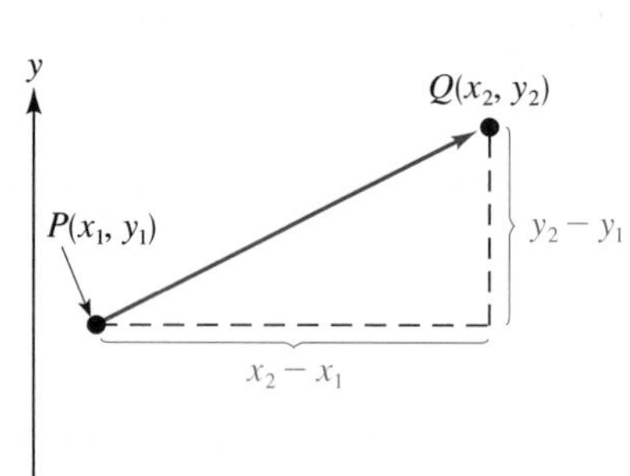

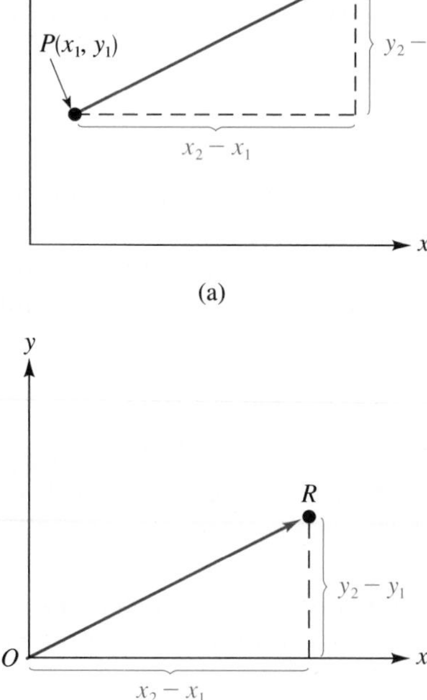

FIGURE 5

To derive this formula, we first construct the right triangle shown in Figure 5(a). Now we let R denote the point $(x_2 - x_1, y_2 - y_1)$ and draw the position vector $\overrightarrow{OR}$ shown in Figure 5(b). Since the right triangles in Figures 5(a) and 5(b) are congruent and have corresponding legs that are parallel, we have

$$\overrightarrow{PQ} = \overrightarrow{OR} = \langle x_2 - x_1, y_2 - y_1 \rangle$$

as required. We can summarize this result as follows. For any vector **v**, we have

x-component of **v**: (x-coordinate of terminal point) − (x-coordinate of initial point)

y-component of **v**: (y-coordinate of terminal point) − (y-coordinate of initial point)

EXAMPLE 2 Let P and Q be the points $P(3, 1)$ and $Q(7, 3)$. Find the components of $\overrightarrow{PQ}$.

Solution

$$\begin{aligned} x\text{-component of } \overrightarrow{PQ} &= x\text{-coordinate of } Q - x\text{-coordinate of } P \\ &= 7 - 3 = 4 \end{aligned}$$

$$\begin{aligned} y\text{-component of } \overrightarrow{PQ} &= y\text{-coordinate of } Q - y\text{-coordinate of } P \\ &= 3 - 1 = 2 \end{aligned}$$

Consequently, $\overrightarrow{PQ} = \langle 4, 2 \rangle$

Vector addition is particularly simple to carry out when the vectors are in component form. Indeed, we can use the parallelogram law to verify the following result.

THEOREM: Vector Addition

If $\mathbf{u} = \langle u_1, u_2 \rangle$ and $\mathbf{v} = \langle v_1, v_2 \rangle$, then $\mathbf{u} + \mathbf{v} = \langle u_1 + v_1, u_2 + v_2 \rangle$.

This theorem tells us that vector addition can be carried out *componentwise*; in other words, to add two vectors, just add the corresponding components. For example,

$$\langle 1, 2 \rangle + \langle 3, 7 \rangle = \langle 4, 9 \rangle$$

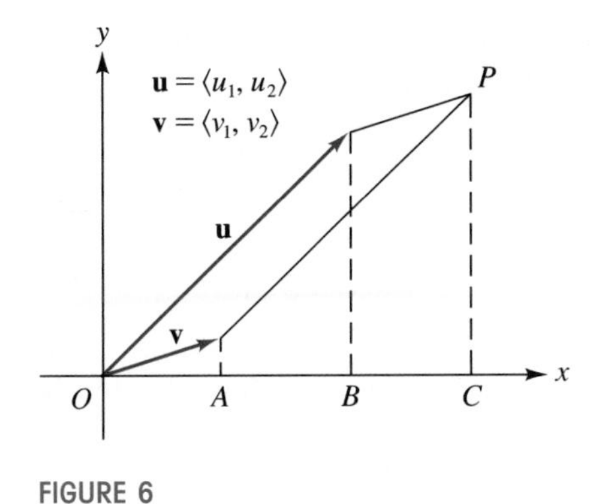

FIGURE 6

To see why this theorem is valid, consider Figure 6, where we have completed the parallelogram. (*Caution:* In reading the derivation that follows, don't confuse notation such as OA with $\overrightarrow{OA}$; recall that OA denotes the length of the line segment $\overline{OA}$.) Since the x-component of $\mathbf{u}$ is u_1, we have

$$OB = u_1$$

Also,

$$BC = v_1 \qquad \text{(Why?)}$$

Therefore,

$$OC = OB + BC = u_1 + v_1$$

But OC is the x-component of the vector $\overrightarrow{OP}$ and, by the parallelogram law, $\overrightarrow{OP} = \mathbf{u} + \mathbf{v}$. In other words, the x-component of $\mathbf{u} + \mathbf{v}$ is $u_1 + v_1$, as we wished to show. The fact that the y-component of $\mathbf{u} + \mathbf{v}$ is $u_2 + v_2$ is proved in a similar fashion.

The vector $\langle 0, 0\rangle$ is called the **zero vector,** and it is denoted by **0.** Notice that for any vector $\mathbf{v} = \langle v_1, v_2\rangle$, we have

$$\mathbf{v} + \mathbf{0} = \mathbf{v} \qquad \text{because} \qquad \langle v_1, v_2\rangle + \langle 0, 0\rangle = \langle v_1, v_2\rangle$$

and

$$\mathbf{0} + \mathbf{v} = \mathbf{v} \qquad \text{because} \qquad \langle 0, 0\rangle + \langle v_1, v_2\rangle = \langle v_1, v_2\rangle$$

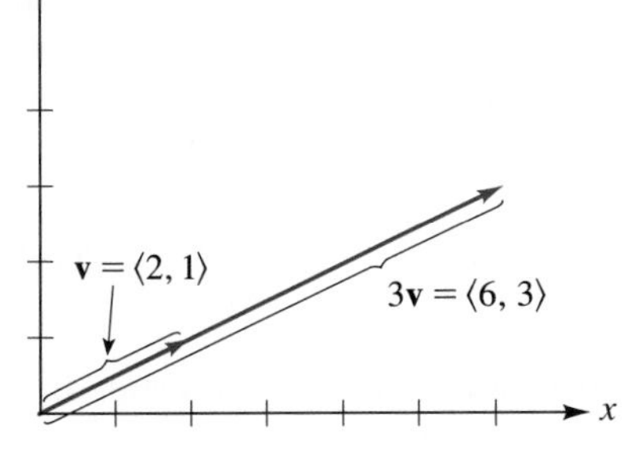

(a) The vectors **v** and 3**v** have the same direction. The length of 3**v** is three times that of **v**.

So for the operation of vector addition, the zero vector plays the same role as does the real number zero in addition of real numbers. There are, in fact, several other ways in which vector addition resembles ordinary addition of real numbers. We'll return to this point again near the end of this section.

In the box that follows, we define an operation called **scalar multiplication,** in which a vector is "multiplied" by a real number (a **scalar**) to obtain another vector.

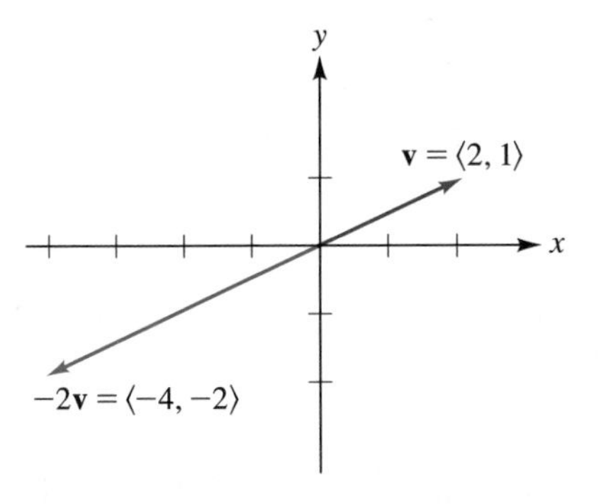

(b) The vectors **v** and −2**v** have opposite directions. The length of −2**v** is twice that of **v**.

FIGURE 7

DEFINITION: Scalar Multiplication

For each real number k and each vector $\mathbf{v} = \langle x, y\rangle$, we define a vector $k\mathbf{v}$ by the equation

$$k\mathbf{v} = k\langle x, y\rangle = \langle kx, ky\rangle$$

EXAMPLES

If $\mathbf{v} = \langle 2, 1\rangle$, then

$$2\mathbf{v} = \langle 4, 2\rangle \quad 3\mathbf{v} = \langle 6, 3\rangle$$

$$0\mathbf{v} = \langle 0, 0\rangle = \mathbf{0}$$

$$-1\mathbf{v} = \langle -2, -1\rangle$$

In geometric terms, the length of $k\mathbf{v}$ is $|k|$ times the length of $\mathbf{v}$. The vectors $\mathbf{v}$ and $k\mathbf{v}$ have the same direction if $k > 0$ and opposite directions if $k < 0$. For example, let $\mathbf{v} = \langle 2, 1\rangle$. In Figure 7(a) we show the vectors $\mathbf{v}$ and $3\mathbf{v}$, while in Figure 7(b) we show $\mathbf{v}$ and $-2\mathbf{v}$.

EXAMPLE 3 Let $\mathbf{v} = \langle 3, 4\rangle$ and $\mathbf{w} = \langle -1, 2\rangle$. Compute each of the following: **(a)** $\mathbf{v} + \mathbf{w}$; **(b)** $-2\mathbf{v} + 3\mathbf{w}$; **(c)** $|-2\mathbf{v} + 3\mathbf{w}|$.

Solution

(a) $\mathbf{v} + \mathbf{w} = \langle 3, 4\rangle + \langle -1, 2\rangle = \langle 3 - 1, 4 + 2\rangle = \langle 2, 6\rangle$

(b) $-2\mathbf{v} + 3\mathbf{w} = -2\langle 3, 4\rangle + 3\langle -1, 2\rangle = \langle -6, -8\rangle + \langle -3, 6\rangle$
$= \langle -9, -2\rangle$

(c) $|-2\mathbf{v} + 3\mathbf{w}| = |\langle -9, -2\rangle| = \sqrt{(-9)^2 + (-2)^2} = \sqrt{85}$

For each vector $\mathbf{v}$, we define a vector $-\mathbf{v}$, called the **negative** of $\mathbf{v}$, by the equation

$$-\mathbf{v} = -1\mathbf{v}$$

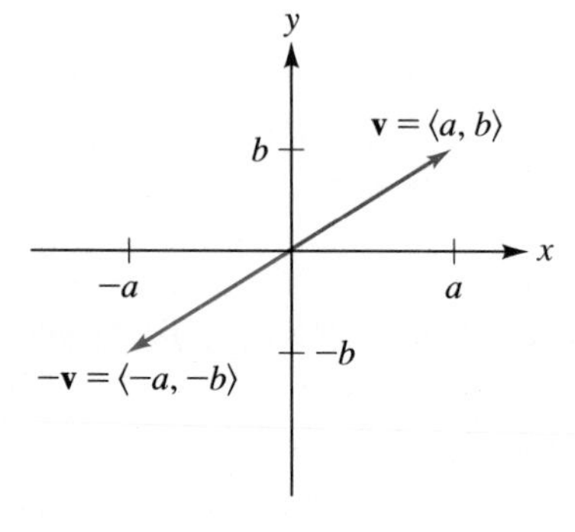

FIGURE 8

Thus, if $\mathbf{v} = \langle a, b\rangle$, then $-\mathbf{v} = \langle -a, -b\rangle$. As indicated in Figure 8, the vectors $\mathbf{v}$ and $-\mathbf{v}$ have the same length but opposite directions.

We can use the ideas in the preceding paragraph to define **vector subtraction.** Given two vectors $\mathbf{u}$ and $\mathbf{v}$, we define a vector $\mathbf{u} - \mathbf{v}$ by the equation

$$\mathbf{u} - \mathbf{v} = \mathbf{u} + (-\mathbf{v}) \tag{1}$$

First let's see what equation (1) is saying in terms of components, then we will indicate a simple geometric interpretation of vector subtraction.

If $\mathbf{u} = \langle u_1, u_2\rangle$ and $\mathbf{v} = \langle v_1, v_2\rangle$, then equation (1) tells us that

$$\begin{aligned}\mathbf{u} - \mathbf{v} &= \langle u_1, u_2\rangle + \langle -v_1, -v_2\rangle \\ &= \langle u_1 + (-v_1), u_2 + (-v_2)\rangle\end{aligned}$$

That is,

$$\mathbf{u} - \mathbf{v} = \langle u_1 - v_1, u_2 - v_2\rangle \tag{2}$$

In other words, to subtract two vectors, just subtract the corresponding components.

Vector subtraction can be interpreted geometrically. According to the formula in the first box on page 562, the right-hand side of equation (2) represents a vector drawn from the terminal point of $\mathbf{v}$ to the terminal point of $\mathbf{u}$. Figure 9 summarizes this fact, and Figure 10 provides a geometric comparison of vector addition and vector subtraction.

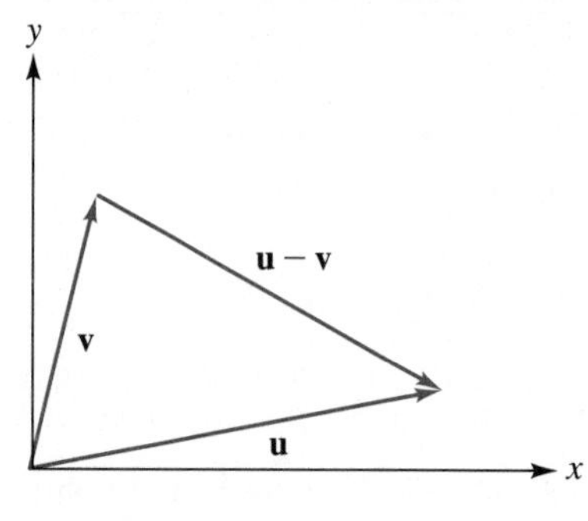

FIGURE 9

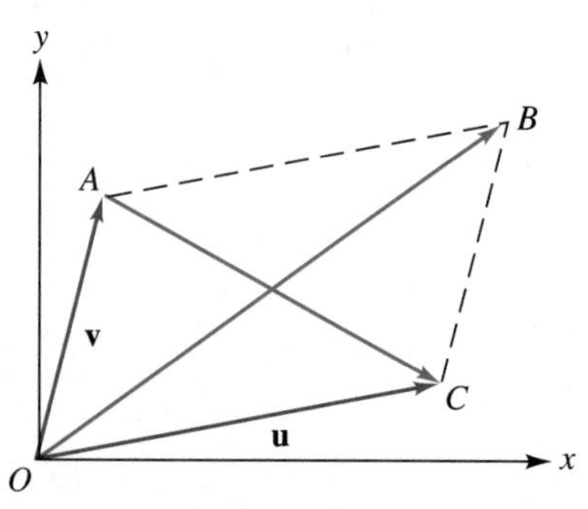

FIGURE 10
$\mathbf{u} + \mathbf{v}$ and $\mathbf{u} - \mathbf{v}$ are the directed diagonals of parallelogram $OABC$: $\mathbf{u} + \mathbf{v} = \overrightarrow{OB}$ and $\mathbf{u} - \mathbf{v} = \overrightarrow{AC}$.

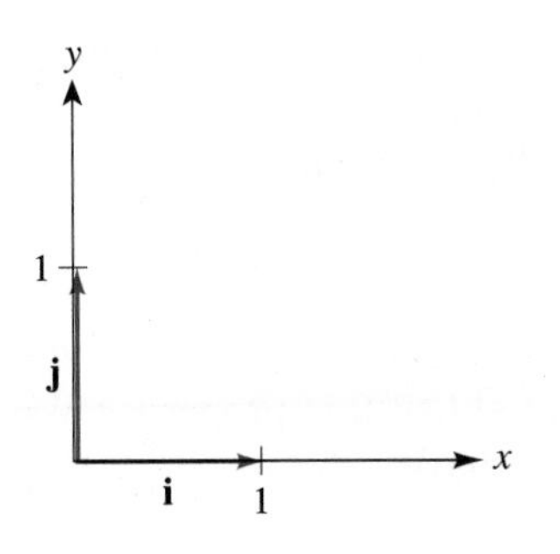

FIGURE 11

EXAMPLE 4 Let $\mathbf{u} = \langle 5, 3 \rangle$ and $\mathbf{v} = \langle -1, 2 \rangle$. Compute $3\mathbf{u} - \mathbf{v}$.

Solution
$$\begin{aligned} 3\mathbf{u} - \mathbf{v} &= 3\langle 5, 3 \rangle - \langle -1, 2 \rangle \\ &= \langle 15, 9 \rangle - \langle -1, 2 \rangle = \langle 16, 7 \rangle \end{aligned}$$

The next three examples deal with unit vectors. By definition, any vector of length 1 is called a **unit vector.** Two particularly useful unit vectors are

$$\mathbf{i} = \langle 1, 0 \rangle \quad \text{and} \quad \mathbf{j} = \langle 0, 1 \rangle$$

These are shown in Figure 11.

Any vector $\mathbf{v} = \langle x, y \rangle$ can be uniquely expressed in terms of the unit vectors **i** and **j** as follows:

$$\langle x, y \rangle = x\mathbf{i} + y\mathbf{j} \tag{3}$$

To verify equation (3), we have

$$\begin{aligned} x\mathbf{i} + y\mathbf{j} &= x\langle 1, 0 \rangle + y\langle 0, 1 \rangle \\ &= \langle x, 0 \rangle + \langle 0, y \rangle = \langle x, y \rangle \end{aligned}$$

EXAMPLE 5

(a) Express the vector $\langle 3, -7 \rangle$ in terms of the unit vectors **i** and **j**.
(b) Express the vector $\mathbf{v} = -4\mathbf{i} + 5\mathbf{j}$ in component form.

Solution
(a) Using equation (3), we can write

$$\langle 3, -7 \rangle = 3\mathbf{i} + (-7)\mathbf{j} = 3\mathbf{i} - 7\mathbf{j}$$

(b)
$$\begin{aligned} \mathbf{v} = -4\mathbf{i} + 5\mathbf{j} &= -4\langle 1, 0 \rangle + 5\langle 0, 1 \rangle \\ &= \langle -4, 0 \rangle + \langle 0, 5 \rangle = \langle -4, 5 \rangle \end{aligned}$$

Thus, the component form of **v** is $\langle -4, 5 \rangle$.

EXAMPLE 6 Find a unit vector **u** that has the same direction as the vector $\mathbf{v} = \langle 3, 4 \rangle$.

Solution First, let's determine the length of **v**:

$$|\mathbf{v}| = \sqrt{3^2 + 4^2} = \sqrt{25} = 5$$

So we want a vector whose length is one-fifth that of **v** and whose direction is the same as that of **v**. Such a vector is

$$\mathbf{u} = \tfrac{1}{5}\mathbf{v} = \tfrac{1}{5}\langle 3, 4 \rangle = \left\langle \tfrac{3}{5}, \tfrac{4}{5} \right\rangle$$

(You should check for yourself now that the length of this vector is 1.)

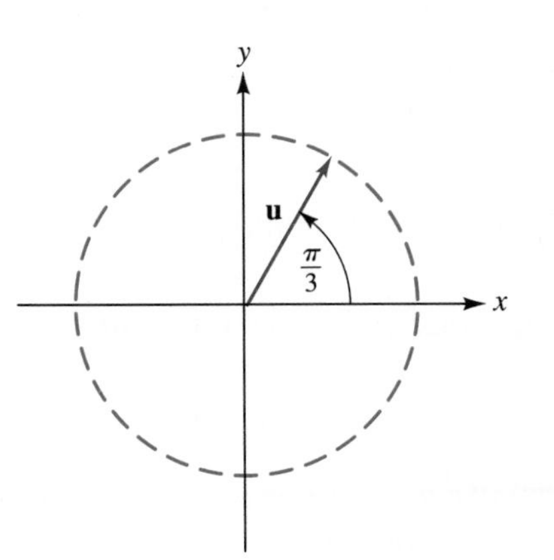

FIGURE 12

EXAMPLE 7 The angle from the positive x-axis to the unit vector **u** is $\pi/3$, as indicated in Figure 12. Determine the components of **u**.

Solution Let P denote the terminal point of **u**. Since P lies on the unit circle, the coordinates of P are by definition $\left(\cos\frac{\pi}{3}, \sin\frac{\pi}{3}\right)$. We therefore have

$$\mathbf{u} = \left\langle \frac{1}{2}, \frac{\sqrt{3}}{2} \right\rangle$$

It was mentioned earlier that there are several ways in which vector addition resembles ordinary addition of real numbers. In the previous section, for example, we saw that vector addition is commutative. That is, for any two vectors **u** and **v**,

$$\mathbf{u} + \mathbf{v} = \mathbf{v} + \mathbf{u}$$

By using components, we can easily verify that this property holds. (In the previous section, we used a geometric argument to establish this property.) We begin by letting $\mathbf{u} = \langle u_1, u_2 \rangle$ and $\mathbf{v} = \langle v_1, v_2 \rangle$. Then we have

$$\begin{aligned}\mathbf{u} + \mathbf{v} &= \langle u_1, u_2 \rangle + \langle v_1, v_2 \rangle \\ &= \langle u_1 + v_1, u_2 + v_2 \rangle \\ &= \langle v_1 + u_1, v_2 + u_2 \rangle \qquad \text{Addition of real numbers is commutative.} \\ &= \langle v_1, v_2 \rangle + \langle u_1, u_2 \rangle \\ &= \mathbf{v} + \mathbf{u}\end{aligned}$$

which is what we wanted to show.

There are a number of other properties of vector addition and scalar multiplication that can be proved in a similar fashion. In the following box, we list a particular collection of these properties, known as the **vector space properties.** (Exercises 55–58 ask that you verify these properties by using components, just as we did for the commutative property.)

PROPERTIES OF VECTOR ADDITION AND SCALAR MULTIPLICATION

For all vectors **u**, **v**, and **w**, and for all scalars (real numbers) a and b, the following properties hold.

1. $\mathbf{u} + (\mathbf{v} + \mathbf{w}) = (\mathbf{u} + \mathbf{v}) + \mathbf{w}$
2. $\mathbf{0} + \mathbf{v} = \mathbf{v} + \mathbf{0} = \mathbf{v}$
3. $\mathbf{v} + (-\mathbf{v}) = \mathbf{0}$
4. $\mathbf{u} + \mathbf{v} = \mathbf{v} + \mathbf{u}$
5. $a(\mathbf{u} + \mathbf{v}) = a\mathbf{u} + a\mathbf{v}$
6. $(a + b)\mathbf{v} = a\mathbf{v} + b\mathbf{v}$
7. $(ab)\mathbf{v} = a(b\mathbf{v})$
8. $1\mathbf{v} = \mathbf{v}$

EXERCISE SET 9.3

A

In Exercises 1–6, sketch each vector in an x-y coordinate system, and compute the length of the vector.

1. $\langle 4, 3 \rangle$ **2.** $\langle 5, 12 \rangle$ **3.** $\langle -4, 2 \rangle$

4. $\langle -6, -6 \rangle$ **5.** $\langle \frac{3}{4}, -\frac{1}{2} \rangle$ **6.** $\langle -3, 0 \rangle$

In Exercises 7–12, the coordinates of two points P and Q are given. In each case, determine the components of the vector $\overrightarrow{PQ}$. Write your answers in the form $\langle a, b \rangle$.

7. $P(2, 3)$ and $Q(3, 7)$

8. $P(5, 1)$ and $Q(4, 9)$

9. $P(-2, -3)$ and $Q(-3, -2)$

10. $P(0, -4)$ and $Q(0, -8)$

11. $P(-5, 1)$ and $Q(3, -4)$

12. $P(1, 0)$ and $Q(0, 1)$

In Exercises 13–32, assume that the vectors **a**, **b**, **c**, *and* **d** *are defined as follows:*

$$\mathbf{a} = \langle 2, 3 \rangle \qquad \mathbf{b} = \langle 5, 4 \rangle \qquad \mathbf{c} = \langle 6, -1 \rangle \qquad \mathbf{d} = \langle -2, 0 \rangle$$

Compute each of the indicated quantities.

13. $\mathbf{a} + \mathbf{b}$ **14.** $\mathbf{c} + \mathbf{d}$

15. $2\mathbf{a} + 4\mathbf{b}$ **16.** $-2\mathbf{c} + 2\mathbf{d}$

17. $|\mathbf{b}+\mathbf{c}|$

18. $|5\mathbf{b}+5\mathbf{c}|$

19. $|\mathbf{a}+\mathbf{c}|-|\mathbf{a}|-|\mathbf{c}|$

20. $1/|\mathbf{d}|$

21. $\mathbf{a}+(\mathbf{b}+\mathbf{c})$

22. $(\mathbf{a}+\mathbf{b})+\mathbf{c}$

23. $3\mathbf{a}+4\mathbf{a}$

24. $|4\mathbf{b}+5\mathbf{b}|$

25. $\mathbf{a}-\mathbf{b}$

26. $\mathbf{b}-\mathbf{c}$

27. $3\mathbf{b}-4\mathbf{d}$

28. $\dfrac{1}{|3\mathbf{b}-4\mathbf{d}|}(3\mathbf{b}-4\mathbf{a})$

29. $\mathbf{a}-(\mathbf{b}+\mathbf{c})$

30. $(\mathbf{a}-\mathbf{b})-\mathbf{c}$

31. $|\mathbf{c}+\mathbf{d}|^2-|\mathbf{c}-\mathbf{d}|^2$

32. $|\mathbf{a}+\mathbf{b}|^2+|\mathbf{a}-\mathbf{b}|^2-2|\mathbf{a}|^2-2|\mathbf{b}|^2$

In Exercises 33–38, express each vector in terms of the unit vectors **i** *and* **j**.

33. $\langle 3, 8\rangle$

34. $\langle 4, -2\rangle$

35. $\langle -8, -6\rangle$

36. $\langle -9, 0\rangle$

37. $3\langle 5, 3\rangle + 2\langle 2, 7\rangle$

38. $|\langle 12, 5\rangle|\langle 3, 4\rangle + |\langle 3, 4\rangle|\langle 12, 5\rangle$

In Exercises 39–42, express each vector in the form $\langle \mathbf{a}, \mathbf{b}\rangle$.

39. $\mathbf{i}+\mathbf{j}$

40. $\mathbf{i}-2\mathbf{j}$

41. $5\mathbf{i}-4\mathbf{j}$

42. $\dfrac{1}{|\mathbf{i}+\mathbf{j}|}(\mathbf{i}+\mathbf{j})$

In Exercises 43–48, find a unit vector having the same direction as the given vector.

43. $\langle 4, 8\rangle$

44. $\langle -3, 3\rangle$

45. $\langle 6, -3\rangle$

46. $\langle -12, 5\rangle$

47. $8\mathbf{i}-9\mathbf{j}$

48. $\langle 7, 3\rangle - \mathbf{i} + \mathbf{j}$

In Exercises 49–54, you are given an angle θ *measured counterclockwise from the positive x-axis to a unit vector* $\mathbf{u} = \langle u_1, u_2\rangle$. *In each case, determine the components* u_1 *and* u_2.

49. $\theta = \pi/6$

50. $\theta = \pi/4$

51. $\theta = 2\pi/3$

52. $\theta = 3\pi/4$

53. $\theta = 5\pi/6$

54. $\theta = 3\pi/2$

B

In Exercises 55–58, let $\mathbf{u} = \langle u_1, u_2\rangle$, $\mathbf{v} = \langle v_1, v_2\rangle$, *and* $\mathbf{w} = \langle w_1, w_2\rangle$. *Refer to the box on page 566.*

55. Verify Properties 1 and 2.

56. Verify Properties 3 and 4.

57. Verify Properties 5 and 6.

58. Verify Properties 7 and 8.

In Exercises 59–75, we study the dot product of two vectors. Given two vectors $\mathbf{A} = \langle x_1, y_1\rangle$ *and* $\mathbf{B} = \langle x_2, y_2\rangle$, *we define the* ***dot product*** $\mathbf{A}\cdot\mathbf{B}$ *as follows.*

$$\mathbf{A}\cdot\mathbf{B} = x_1x_2 + y_1y_2$$

For example, if $\mathbf{A} = \langle 3, 4\rangle$ *and* $\mathbf{B} = \langle -2, 5\rangle$, *then* $\mathbf{A}\cdot\mathbf{B} = (3)(-2) + (4)(5) = 14$. *Notice that the dot product of two vectors is a real number. For this reason, the dot product is also known as the* ***scalar product.*** *For Exercises 59–61, the vectors* **u**, **v**, *and* **w** are defined as follows:

$$\mathbf{u} = \langle -4, 5\rangle \qquad \mathbf{v} = \langle 3, 4\rangle \qquad \mathbf{w} = \langle 2, -5\rangle$$

59. **(a)** Compute $\mathbf{u}\cdot\mathbf{v}$ and $\mathbf{v}\cdot\mathbf{u}$.
(b) Compute $\mathbf{v}\cdot\mathbf{w}$ and $\mathbf{w}\cdot\mathbf{v}$.
(c) Show that for any two vectors **A** and **B**, we have $\mathbf{A}\cdot\mathbf{B} = \mathbf{B}\cdot\mathbf{A}$. That is, show that the dot product is commutative. *Hint:* Let $\mathbf{A} = \langle x_1, y_1\rangle$ and let $\mathbf{B} = \langle x_2, y_2\rangle$.

60. **(a)** Compute $\mathbf{v}+\mathbf{w}$.
(b) Compute $\mathbf{u}\cdot(\mathbf{v}+\mathbf{w})$.
(c) Compute $\mathbf{u}\cdot\mathbf{v}+\mathbf{u}\cdot\mathbf{w}$.
(d) Show that for any three vectors **A**, **B**, and **C**, we have $\mathbf{A}\cdot(\mathbf{B}+\mathbf{C}) = \mathbf{A}\cdot\mathbf{B}+\mathbf{A}\cdot\mathbf{C}$.

61. **(a)** Compute $\mathbf{v}\cdot\mathbf{v}$ and $|\mathbf{v}|^2$.
(b) Compute $\mathbf{w}\cdot\mathbf{w}$ and $|\mathbf{w}|^2$.

62. Show that for any vector **A**, we always have $|\mathbf{A}|^2 = \mathbf{A}\cdot\mathbf{A}$. That is, the square of the length of a vector is equal to the dot product of the vector with itself. *Hint:* Let $\mathbf{A} = \langle x, y\rangle$.

Let θ *(where* $0 \leq \theta \leq \pi$*) denote the angle between the two nonzero vectors* **A** *and* **B**. *Then it can be shown that the cosine of* θ *is given by the formula*

$$\cos\theta = \frac{\mathbf{A}\cdot\mathbf{B}}{|\mathbf{A}|\,|\mathbf{B}|}$$

(See Exercise 75 for the derivation of this result.) In Exercises 63–68, use this formula to find the cosine of the angle between the given pair of vectors. Also, in each case use a calculator to compute the angle. Express the angle using degrees and using radians. Round the values to two decimal places.

63. $\mathbf{A} = \langle 4, 1\rangle$ and $\mathbf{B} = \langle 2, 6\rangle$

64. $\mathbf{A} = \langle 3, -1\rangle$ and $\mathbf{B} = \langle -2, 5\rangle$

65. $\mathbf{A} = \langle 5, 6\rangle$ and $\mathbf{B} = \langle -3, -7\rangle$

66. $\mathbf{A} = \langle 3, 0\rangle$ and $\mathbf{B} = \langle 1, 4\rangle$

67. **(a)** $\mathbf{A} = \langle -8, 2\rangle$ and $\mathbf{B} = \langle 1, -3\rangle$
(b) $\mathbf{A} = \langle -8, 2\rangle$ and $\mathbf{B} = \langle -1, 3\rangle$

68. **(a)** $\mathbf{A} = \langle 7, 12\rangle$ and $\mathbf{B} = \langle 1, 2\rangle$
(b) $\mathbf{A} = \langle 7, 12\rangle$ and $\mathbf{B} = \langle -1, -2\rangle$

69. **(a)** Compute the cosine of the angle between the vectors $\langle 2, 5\rangle$ and $\langle -5, 2\rangle$.
(b) What can you conclude from your answer in part (a)?
(c) Draw a sketch to check your conclusion in part (b).

70. Follow Exercise 69, but use the vectors $\langle 6, -8\rangle$ and $\langle -4, -3\rangle$.

71. Suppose that $\mathbf{A}$ and $\mathbf{B}$ are nonzero vectors and $\mathbf{A} \cdot \mathbf{B} = 0$. Explain why $\mathbf{A}$ and $\mathbf{B}$ are perpendicular.

72. Find a value for t such that the vectors $\langle 15, -3 \rangle$ and $\langle -4, t \rangle$ are perpendicular.

73. Find a unit vector that is perpendicular to the vector $\langle -12, 5 \rangle$, (There are two answers.)

C

74. Suppose that $\mathbf{A}$ and $\mathbf{B}$ are nonzero vectors such that $|\mathbf{A} + \mathbf{B}|^2 = |\mathbf{A}|^2 + |\mathbf{B}|^2$. Show that $\mathbf{A}$ and $\mathbf{B}$ are perpendicular. *Hint:* Let $\mathbf{A} = \langle x_1, y_1 \rangle$ and let $\mathbf{B} = \langle x_2, y_2 \rangle$. After making some calculations, you should be able to apply the result in Exercise 71.

75. Refer to the figure at the top of the next column. In this exercise we are going to derive the following formula for the cosine of the angle θ between two nonzero vectors
$\mathbf{A}$ and $\mathbf{B}$:

$$\cos \theta = \frac{\mathbf{A} \cdot \mathbf{B}}{|\mathbf{A}|\,|\mathbf{B}|}$$

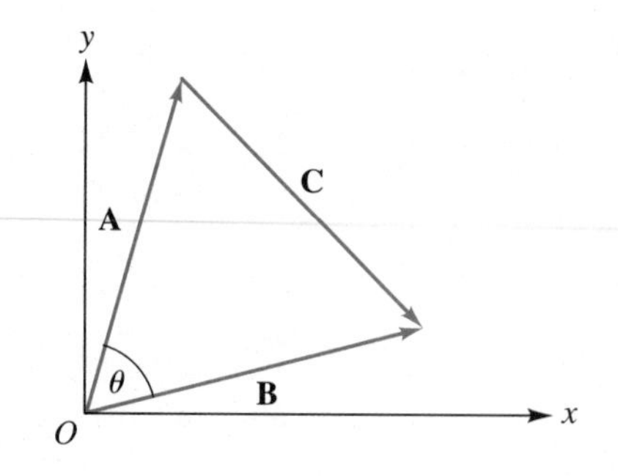

(a) Let $\mathbf{A} = \langle x_1, y_1 \rangle$ and $\mathbf{B} = \langle x_2, y_2 \rangle$. Compute the length of the vector $\mathbf{C}$ in the figure using the fact that $\mathbf{C} = \langle x_2 - x_1, y_2 - y_1 \rangle$.

(b) Using your result in part (a), show that

$$|\mathbf{C}|^2 = |\mathbf{A}|^2 + |\mathbf{B}|^2 - 2\mathbf{A} \cdot \mathbf{B}$$

(c) According to the law of cosines, we have

$$|\mathbf{C}|^2 = |\mathbf{A}|^2 + |\mathbf{B}|^2 - 2|\mathbf{A}|\,|\mathbf{B}| \cos \theta$$

Set this expression for $|\mathbf{C}|^2$ equal to the expression obtained in part (b), and then solve for $\cos \theta$ to obtain the required formula.

9.4 PARAMETRIC EQUATIONS

We will introduce the idea of parametric equations through a simple example. Suppose that we have a point $P(x, y)$ that moves in the x-y plane, and that the x- and y-coordinates of P at time t (in seconds) are given by the following pair of equations.

$$x = 2t \qquad \text{and} \qquad y = \tfrac{1}{2}t^2 \qquad (t \geq 0)$$

This pair of *parametric equations* tells us the location of the point P at any time t. For instance, when $t = 1$, we have

$$x = 2t = 2(1) = 2 \qquad \text{and} \qquad y = \tfrac{1}{2}t^2 = \tfrac{1}{2}(1)^2 = \tfrac{1}{2}$$

In other words, after one second, the coordinates of P are $\left(2, \frac{1}{2}\right)$. Let's see where P is after two seconds. We have

$$x = 2t = 2(2) = 4 \qquad \text{and} \qquad y = \tfrac{1}{2}t^2 = \tfrac{1}{2}(2)^2 = 2$$

So, after two seconds, the location of the point P is (4, 2). In Table 1, we have computed the values of x and y (and hence the location of P) for integral values of t running from 0 to 5. Figure 1(a) shows the points that are determined. In Figure 1(b) we have joined the points with a smooth curve. [If we were not certain of the pattern emerging in Figure 1(a), we could have computed additional points using fractional values for t.]

TABLE 1

t	x	y
0	0	0
1	2	$\frac{1}{2}$
2	4	2
3	6	$\frac{9}{2}$
4	8	8
5	10	$\frac{25}{2}$

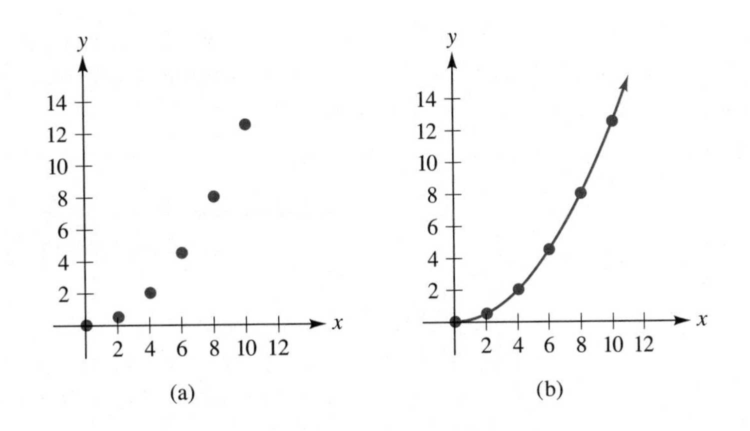

FIGURE 1
$x = 2t,\ y = \frac{1}{2}t^2 \quad (t \geqslant 0)$

The curve in Figure 1(b) appears to be a portion of a parabola. We can confirm this as follows. From the equation $x = 2t$, we obtain $t = x/2$. Now we use this result to substitute for t in the equation $y = \frac{1}{2}t^2$. This yields

$$y = \frac{1}{2}\left(\frac{x}{2}\right)^2 = \frac{1}{8}x^2$$

We conclude from this that the curve in Figure 1(b) is a portion of the parabola $y = \frac{1}{8}t^2$. The equations $x = 2t$ and $y = \frac{1}{2}t^2$ (with the restriction $t \geqslant 0$) are the **parametric equations** for the curve. The variable t is called the **parameter.** In our initial example, the parameter t represented time. In other cases the parameter might have an interpretation as a slope of a certain line, or as an angle. And in still other cases, there may be no immediate physical interpretation for the parameter.

Before looking at additional examples, we point out two advantages in using parametric equations to describe curves. First, by restricting the values of the parameter (as we did in our initial example), we can focus on specific portions of a curve. Second, parametric equations let us think of a curve as a path traced out by a moving point; as the parameter t increases, a definite direction of motion is established.

In physics, parametric equations are often used to describe the motion of an object. Suppose, for example, that a ball is thrown from a height of 6 feet, with an initial speed of 88 ft/sec and at an angle of 35° with the horizontal, as shown in Figure 2. Then (neglecting air resistance and spin), it can be shown that the parametric equations for the path of the ball are

$$x = (88 \cos 35°)t \tag{1}$$
$$y = 6 + (88 \sin 35°)t - 16t^2 \tag{2}$$

In these equations, x and y are measured in feet and t is in seconds, with $t = 0$ corresponding to the instant the ball is thrown.

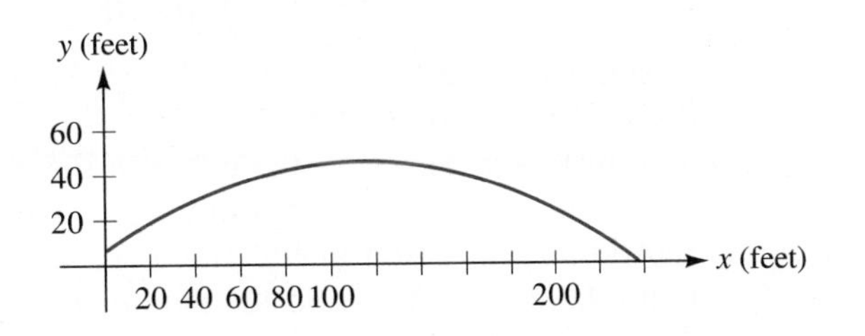

FIGURE 2
The path of a ball thrown from a height of 6 ft, with an initial speed of 88 ft/s, at an angle of 35° with the horizontal.

With equations (1) and (2) we can calculate the location of the ball at any time t. For instance, to determine the location when $t = 1$ second, we substitute $t = 1$ in the equations to obtain

$$x = (88 \cos 35°)(1) \approx 72.1 \text{ ft}$$

and

$$y = 6 + (88 \sin 35°)(1) - 16(1)^2 \approx 40.5 \text{ ft}$$

So, after one second, the ball has traveled a horizontal distance of approximately 72.1 ft, and the height of the ball is approximately 40.5 ft. In Figure 3 we display this information along with the results for similar calculations corresponding to $t = 2$ and $t = 3$ seconds. [You should use equations (1) and (2) and your calculator to check the results in Figure 3 for yourself.]

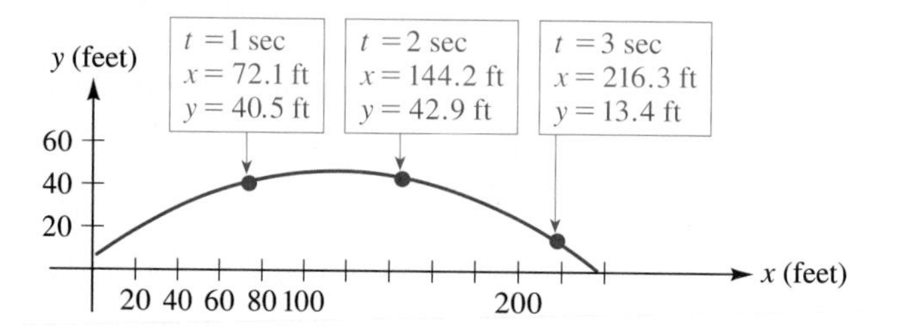

FIGURE 3
The location of the ball after 1, 2, and 3 seconds.

From Figure 3, we can see that the total horizontal distance traveled by the ball in flight is something between 220 ft and 240 ft. Using the parametric equations for the path of the ball, we can determine this distance exactly. We can also find out how long the ball is in the air. When the ball does hit the ground, its y-coordinate is zero. Replacing y by zero in equation (2), we have

$$0 = 6 + (88 \sin 35°)t - 16t^2$$

This is a quadratic equation in the variable t. To solve for t, we use the quadratic formula

$$t = \frac{-b \pm \sqrt{b^2 - 4ac}}{2a}$$

with the following values for a, b, and c:

$$a = -16 \qquad b = 88 \sin 35° \qquad c = 6$$

As Exercise 19 asks you to check, the results are $t \approx 3.27$ and $t \approx -0.11$. We discard the negative root here because in equations (1) and (2) and in Figure 3, t represents time, with $t = 0$ corresponding to the instant that the ball is thrown. Now, using $t = 3.27$, we can compute x:

$$x = (88 \cos 35°)t \approx (88 \cos 35°)(3.27) \approx 235.7 \text{ ft}$$

In summary, the ball is in the air for about 3.3 seconds, and the total horizontal distance is approximately 236 feet.

Just as in Figure 1, the curve in Figures 2 and 3 is a parabola. Exercise 20 asks you to verify this by solving equation (1) for t in terms of x and then substituting the result in equation (2). This process of obtaining an explicit equation relating x and y from the parametric equations is referred to as *eliminating*

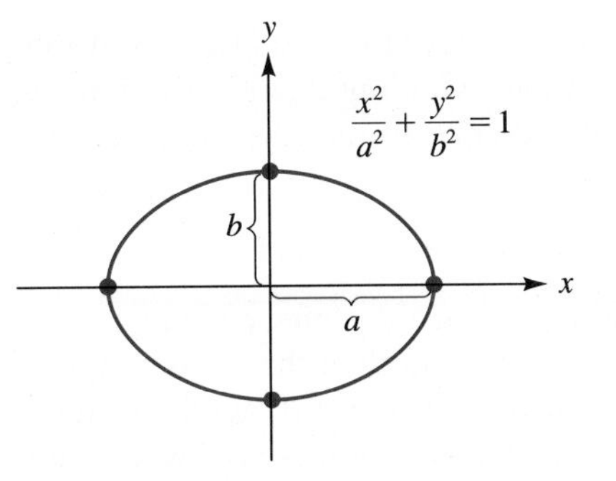

FIGURE 4
The ellipse $(x^2/a^2)+(y^2/b^2)=1$ is symmetric about both coordinate axes. The intercepts of the ellipse are $x=\pm a$ and $y=\pm b$.

the parameter. The next example shows a technique that is often useful in this context. As background for this example, we point out that the graph of an equation of the form $\frac{x^2}{a^2}+\frac{y^2}{b^2}=1$ is an *ellipse*, as indicated in Figure 4. (We will study this curve in detail in Chapter 11.)

EXAMPLE 1 The parametric equations of an ellipse are

$$x=6\cos t \quad \text{and} \quad y=3\sin t \qquad (0\leq 2t\leq 2\pi)$$

(a) Eliminate the parameter t to obtain an x-y equation for the curve.
(b) Graph the ellipse and indicate the points corresponding to $t=0$, $\pi/2$, π, $3\pi/2$, and 2π. As t increases, what is the direction of travel along the curve?

Solution
(a) So that we can apply the identity $\cos^2 t+\sin^2 t=1$, we divide the first equation by 6 and the second by 3. This gives us

$$\frac{x}{6}=\cos t \quad \text{and} \quad \frac{y}{3}=\sin t$$

Squaring and then adding these two equations yields

$$\left(\frac{x}{6}\right)^2+\left(\frac{y}{3}\right)^2=\cos^2 t+\sin^2 t=1$$

or

$$\frac{x^2}{6^2}+\frac{y^2}{3^2}=1$$

This is the x-y equation for the ellipse.
(b) Figure 5 shows the graph of the ellipse $(x^2/6^2)+(y^2/3^2)=1$. When $t=0$, the parametric equations yield

$$x=6\cos 0=6 \quad \text{and} \quad y=3\sin 0=0$$

Thus, with $t=0$, we obtain the point (6, 0) on the ellipse, as indicated in Figure 5. The points corresponding to $t=\pi/2$, π, $3\pi/2$, and 2π are obtained similarly. Note that as t increases, the direction of travel around the ellipse is counterclockwise, and that when $t=2\pi$, we are back to the starting point (6, 0).

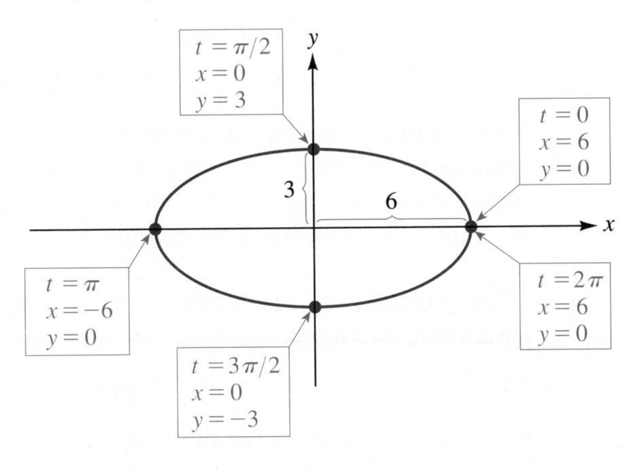

FIGURE 5
Parametric equations for this ellipse are $x=6\cos t$ and $y=3\sin t$. The x-y equation is $(x^2/6^2)+(y^2/3^2)=1$.

The parametric equations $x = 6 \cos t$ and $y = 3 \sin t$ that we graphed in Figure 5 are by no means the only parametric equations yielding that ellipse. Indeed, there are numerous parametric equations that give rise to the same ellipse. One such pair is

$$x = 6 \sin t \qquad \text{and} \qquad y = 3 \cos t \qquad (0 \leq t \leq 2\pi)$$

As you can check for yourself, these equations lead to the same x-y equation as before, namely, $(x^2/6^2) + (y^2/3^2) = 1$. The difference now is that in tracing out the curve from the parametric equations, we start (when $t = 0$) with the point $(0, 3)$ on the ellipse and we travel clockwise rather than counterclockwise. Similarly, the equations

$$x = 6 \sin 2t \qquad \text{and} \qquad y = 3 \cos 2t$$

also produce the ellipse. In this case, as you can check for yourself, as t runs from 0 to 2π, we make two complete trips around the ellipse. If we think of t as time, then the parametric equations $x = 6 \sin 2t$ and $y = 3 \cos 2t$ describe a point traveling around the ellipse $(x^2/6^2) + (y^2/3^2) = 1$ twice as fast as would be the case with the equations $x = 6 \sin t$ and $y = 3 \cos t$.

In Example 1 we saw that the parametric equations $x = 6 \cos t$ and $y = 3 \sin t$ represent an ellipse. The same method we used in Example 1 can be used to establish the following general results.

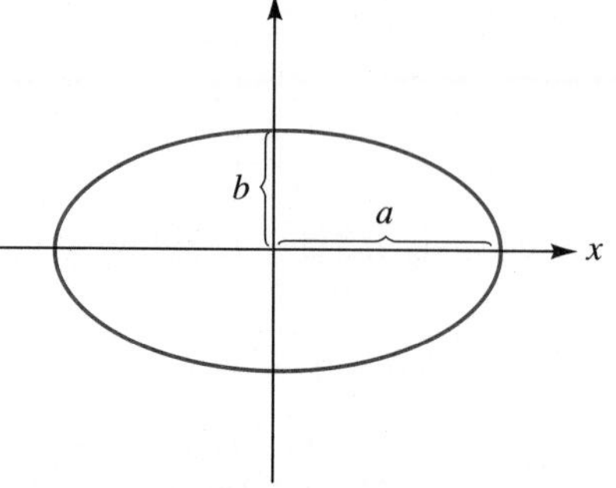

FIGURE 6
$x = a \cos t$ and $y = b \sin t$
or
$x = a \sin t$ and $y = b \cos t$

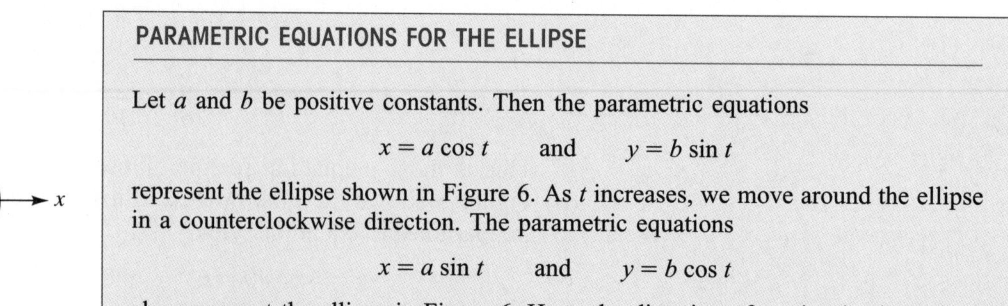

PARAMETRIC EQUATIONS FOR THE ELLIPSE

Let a and b be positive constants. Then the parametric equations

$$x = a \cos t \qquad \text{and} \qquad y = b \sin t$$

represent the ellipse shown in Figure 6. As t increases, we move around the ellipse in a counterclockwise direction. The parametric equations

$$x = a \sin t \qquad \text{and} \qquad y = b \cos t$$

also represent the ellipse in Figure 6. Here, the direction of motion is clockwise.

There is a particular case involving the parametric equations $x = a \cos t$ and $y = b \sin t$ that deserves mention. When a and b are equal, we have $x = a \cos t$ and $y = a \sin t$, from which we deduce that

$$\begin{aligned} x^2 + y^2 &= a^2 \cos^2 t + a^2 \sin^2 t \\ &= a^2(\cos^2 t + \sin^2 t) = a^2 \end{aligned}$$

Thus, with $a = b$, the parametric equations describe a circle of radius a. And specializing further still, if $a = b = 1$, we obtain $x = \cos t$ and $y = \sin t$ as parametric equations for the unit circle. This agrees with our unit circle definitions for sine and cosine back in Section 6.4.

EXAMPLE 2 The position of a point $P(x, y)$ at time t is given by the parametric equations

$$x = 1 + 2t \qquad \text{and} \qquad y = 2 + 4t \qquad (t \geq 0)$$

Find the x-y equation for the path traced out by the point P.

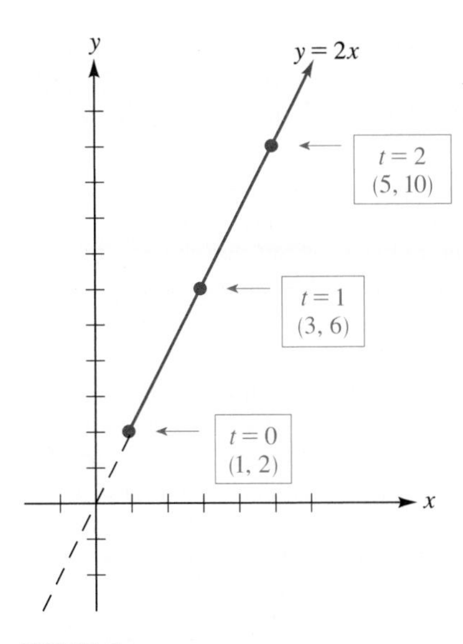

FIGURE 7
The parametric equations $x = 1 + 2t$ and $y = 2 + 4t$, with the restriction $t \geqslant 0$, describe the portion of the line $y = 2x$ to the right of and including the point $(1, 2)$.

Solution From the parametric equation for x, we obtain $t = (x - 1)/2$. Using this to substitute for t in the second parametric equation, we have

$$y = 2 + 4\left(\frac{x-1}{2}\right) = 2 + 2(x - 1) = 2x$$

This tells us that the point P moves along the line $y = 2x$. However, the entire line is not traced out, but only a portion of it. This is because of the original restriction $t \geqslant 0$. To see which portion of the line is described by the equations, we can successively let $t = 0$, 1, and 2 in the parametric equations. As you can check, the points obtained are $(1, 2)$, $(3, 6)$, and $(5, 10)$. So, as t increases, we move to the right along the line $y = 2x$, starting from the point $(1, 2)$; see Figure 7.

One of the recurrent techniques that we have used for analyzing parametric equations in this section has been that of eliminating the parameter. It is important to note, however, that it is not always a simple matter to eliminate the parameter. And indeed, in some cases it may not even be possible. When this occurs we have several techniques available. We can set up a table with t, x, and y and plot points; we can use the techniques of calculus; or we can use a graphing utility to obtain the graph. Even when we use a graphing utility, however, the techniques of calculus are helpful in understanding why the graph looks as it does.

EXERCISE SET 9.4

A

In Exercises 1–6, you are given the parametric equations of a curve and a value for the parameter t. Find the coordinates of the point on the curve corresponding to the given value of t.

1. $x = 2 - 4t,\ y = 3 - 5t;\ t = 0$
2. $x = 3 - t^2,\ y = 4 + t^3;\ t = -1$
3. $x = 5 \cos t,\ y = 2 \sin t;\ t = \pi/6$
4. $x = 4 \cos 2t,\ y = 6 \sin 2t;\ t = \pi/3$
5. $x = 3 \sin^3 t,\ y = 3 \cos^3 t;\ t = \pi/4$
6. $x = \sin t - \sin 2t,\ y = \cos t + \cos 2t;\ t = 2\pi/3$

In Exercises 7–16, graph the parametric equations after eliminating the parameter t. Specify the direction on the curve corresponding to increasing values of t.

7. $x = t + 1,\ y = t^2$
8. $x = 2t - 1,\ y = t^2 - 1$
9. $x = t^2 - 1,\ y = t + 1$
10. $x = t - 4,\ y = |t|$
11. $x = 5 \cos t,\ y = 2 \sin t$
12. $x = 2 \sin t,\ y = 3 \cos t$
13. $x = 4 \cos 2t,\ y = 6 \sin 2t$
14. $x = 2 \cos(t/2),\ y = \sin(t/2)$
15. **(a)** $x = 2 \cos t,\ y = 2 \sin t$
    (b) $x = 4 \cos t,\ y = 2 \sin t$
16. **(a)** $x = 3 \sin t,\ y = 3 \cos t$
    (b) $x = 5 \sin t,\ y = 3 \cos t$
17. The following figure shows the parametric equations and the path for a ball thrown from a height of 5 ft, with an initial speed of 100 ft/sec and at an angle of 70° with the horizontal.

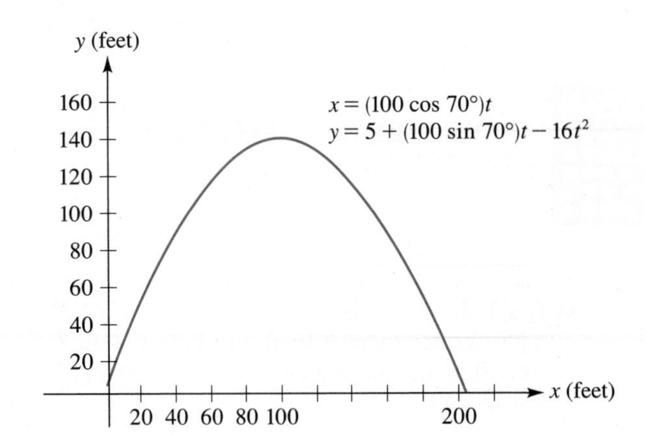

(a) Compute the x- and y-coordinates of the ball when $t = 1$, 2, and 3 seconds. (Round the answers to one decimal place.)

(b) How long is the ball in flight? (Round the answer to two decimal places.) What is the total horizontal distance traveled by the ball before it lands? (Round to the nearest foot.) Check that your answer is consistent with the figure.

18. The figure shows the parametric equations and the path for a ball thrown from a height of 5 ft, with an initial speed of 100 ft/s and at an angle of 45° with the horizontal. (So except for the initial angle, the data is the same as in Exercise 17.)

(a) Compute the x- and y-coordinates of the ball when $t = 1$, 2, and 3 seconds. (Round the answers to one decimal place.)

(b) How long is the ball in flight? (Round the answer to two decimal places.) What is the total horizontal distance traveled by the ball before it lands? (Round to the nearest foot.)

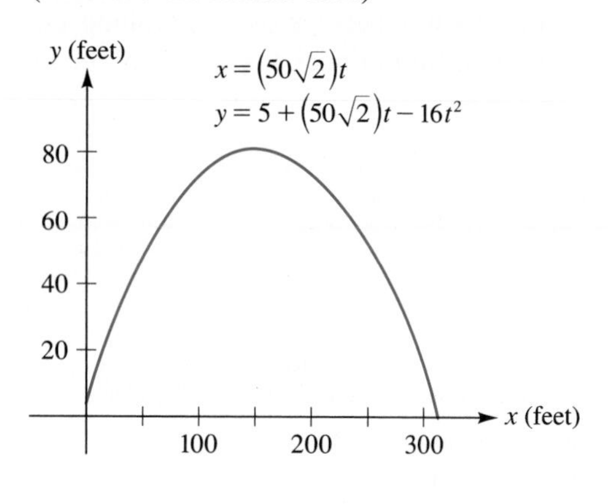

19. In the text we said that the solutions of the quadratic equation $0 = 6 + (88 \sin 35°)t - 16t^2$ are $t \approx 3.27$ and $t \approx -0.11$. Use the quadratic formula and your calculator to verify these results.

20. Refer to parametric equations (1) and (2) in the text. By eliminating the parameter t, show that the equations describe a parabola. (That is, show that the resulting x-y equation has the form $y = ax^2 + bx + c$.)

B

21. The curve in the accompanying figure is called an *astroid.* (It is also known as a *hypocycloid of four cusps.*) A pair of parametric equations for the curve is $x = \cos^3 t$, $y = \sin^3 t$. By eliminating the parameter t, find the x-y equation for the curve. *Hint:* In each equation, raise both sides to the two-thirds power.

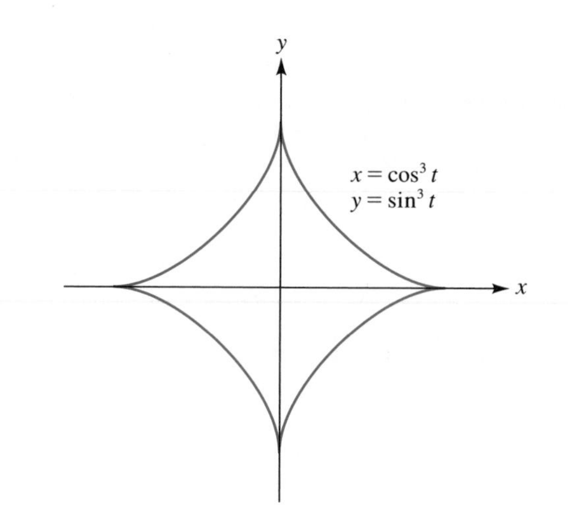

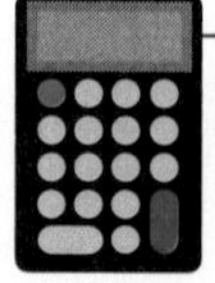

GRAPHING UTILITY EXERCISES FOR SECTION 9.4

1. Use the standard viewing rectangle for this exercise.

(a) Graph the parametric equations $x = 2t$, $y = 0.5t^2$ with t-values running from 0 to 1; from 0 to 3; from 0 to 4. Compare the results (in a complete sentence or two).

(b) Graph the parametric equations given in part (a) using t-values running from -5 to 5. Compare your picture to the graph shown in Figure 1(b) on page 569. What are the restrictions on t in Figure 1(b)?

2. **(a)** Graph the parametric equations $x = (88 \cos 35°)t$, $y = 6 + (88 \sin 35°)t - 16t^2$. Use settings that will give you a picture similar to Figure 3 on page 570.
(b) Use the graphing utility to estimate the x-intercept of the graph. Check that your answer is consistent with the value determined in the text.

3. Graph each of the following pairs of parametric equations. According to the text on page 572, the graphs should all appear identical. (If you are using a graphing calculator, set it to the radian mode.)
(a) $x = 6 \cos t,\ y = 3 \sin t \quad (0 \leq t \leq 2\pi)$
(b) $x = 6 \sin t,\ y = 3 \cos t \quad (0 \leq t \leq 2\pi)$
(c) $x = 6 \sin 2t,\ y = 3 \cos 2t \quad (0 \leq t \leq \pi)$

In Exercises 4–20, graph the parametric equations using the given range for the parameter t. In each case, begin with the standard viewing rectangle and then make adjustments, as necessary, so that the graph utilizes as much of the viewing screen as possible. For example, in graphing the circle given by $x = \cos t$ and $y = \sin t$, it would be natural to choose a viewing rectangle extending from -1 to 1 in both the x- and the y-directions.

4. **(a)** $x = 3t + 2,\ y = 3t - 2, \quad (-2 \leq t \leq 2)$
(b) $x = 3t + 2,\ y = 3t - 2, \quad (-3 \leq t \leq 3)$

5. $x = \ln(3t + 2),\ y = \ln(3t - 2)$ using:
(a) $2/3 < t \leq 1$
(b) $2/3 < t \leq 10$
(c) $2/3 < t \leq 100$
(d) $2/3 < t \leq 1000$

6. $x = 4 \cos t,\ y = 3 \sin t, \quad 0 \leq t \leq 2\pi$ *(ellipse)*

7. $x = 4 \cos t,\ y = -3 \sin t, \quad 0 \leq t \leq 2\pi$ (the same as the ellipse in Exercise 6, but traced out in the opposite direction)

8. $x = 4 \cos t,\ y = 3 \sin t, \quad 0 \leq t \leq \pi/2$ (one-quarter of an ellipse)

9. $x = 4 \cos^3 t,\ y = 4 \sin^3 t, \quad 0 \leq t \leq 2\pi$ [This curve is known as the *astroid* or *hypocycloid of four cusps*. It was first studied by the Danish astronomer Olof Roemer (1644–1710) and by the Swiss mathematician James Bernoulli (1654–1705).]

10. $x = 2 \cos t + \cos 2t,\ y = 2 \sin t - \sin 2t, \quad 0 \leq t \leq 2\pi$ [This curve is the *deltoid*. It was first studied by the Swiss mathematician Leonhard Euler (1707–1783).]

11. $x = \dfrac{\cos t}{1 + \sin^2 t},\ y = \dfrac{\sin t \cos t}{1 + \sin^2 t}, \quad 0 \leq t \leq 2\pi$ [This curve is the *lemniscate of Bernoulli*. The Swiss mathematician James Bernoulli (1654–1705) studied the curve and took the name *lemniscate* from the Greek *lemniskos*, meaning "ribbon."]

12. $x = 2 \tan t,\ y = 2 \cos^2 t, \quad 0 \leq t \leq 2\pi$ *Remark:* If you eliminate the parameter t, you'll find that the Cartesian form of the curve is $y = 8/(x^2 + 4)$. (Verify this last statement, first algebraically, then graphically.) The curve is known as the *witch of Agnesi*, named after the Italian mathematician and scientist Maria Gaetana Agnesi (1718–1799). The word "witch" in the name of the curve is the result of a mistranslation from Italian to English. In Agnesi's time, the curve was known as *la versiera*, an Italian name with a Latin root meaning "to turn." In translation, the word *versiera* was confused with another Italian word *avversiera*, which means "wife of the devil" or "witch."

13. $x = \dfrac{t^2 - 1}{3t^2 + 1},\ y = \dfrac{t(t^2 - 1)}{3t^2 + 1}, \quad (-\infty < t < \infty)$
Hint: Use $-10 \leq t \leq 10$. [This curve is the *folium of Descartes*. The word *folium* means "leaf." When Descartes drew the curve in 1638, he did not use negative values for coordinates. Thus, he obtained only the first-quadrant portion of the curve, which resembles a leaf or loop. The complete graph of the curve was first given by the Dutch mathematician and scientist Christiaan Huygens in 1692. Huygens' graph is reproduced on page 68 in John Stillwell's *Mathematics and Its History* (New York: Springer Verlag, 1989).]

14. $x = 3t^2,\ y = 2t^3, \quad -2 \leq t \leq 2$ *(semi-cubical parabola)*

15. $x = \sec t,\ y = 2 \tan t, \quad 0 \leq t \leq 2\pi$ *(hyperbola)*

16. $x = 3(t^2 - 3),\ y = t(t^2 - 3), \quad -3 \leq t \leq 3$ *(Tschirnhausen's cubic)*

17. $x = 8 \cos t + 2 \cos 4t,\ y = 8 \sin t - 2 \sin 4t,$ $0 \leq t \leq 2\pi$ *(hypocycloid with five cusps)*

18. $x = 8 \cos t + \cos 8t,\ y = 8 \sin t - \sin 8t,$ $0 \leq t \leq 2\pi$ *(hypocycloid with nine cusps)*

19. $x = \cos t + t \sin t,\ y = \sin t - t \cos t$ *(involute of a circle)*
(a) $-\pi \leq t \leq \pi$
(b) $-2\pi \leq t \leq 2\pi$
(c) $-4\pi \leq t \leq 4\pi$

20. $x = \sin(0.8t + \pi),\ y = \sin t, \quad 0 \leq t \leq 10\pi$ *(Bowditch curve)*

NOTE Historical background and information about many of the curves mentioned on this page (and throughout Section 9.6) is available on the Internet through the World Wide Web. The address is

http://www-groups.dcs.st-and.ac.uk/~history/Curves/

9.5 INTRODUCTION TO POLAR COORDINATES

[Jacob Bernoulli (1654–1705)] *was one of the first to use polar coordinates in a general manner, and not simply for spiral shaped curves.*

Florian Cajori in *A History of Mathematics,* 4th ed. (New York: Chelsea Publishing Company, 1985)

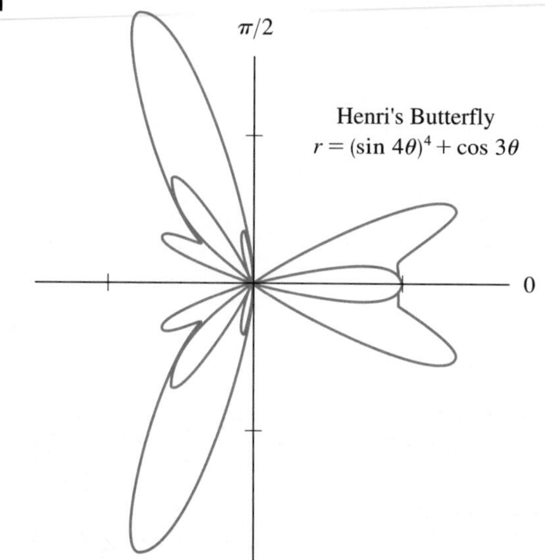

FIGURE 1
This polar coordinate graph was discovered by Henri Berger, a student in one of the author's mathematics classes at UCLA in Spring of 1988.

Up until now, we have always specified the location of a point in the plane by means of a rectangular coordinate system. In this section, we introduce another coordinate system that can be used to locate points in the plane. This is the system of **polar coordinates.** We begin by drawing a half-line or ray emanating from a fixed point O. The fixed point O is called the **pole** or **origin,** and the half-line is called the **polar axis.** The polar axis is usually depicted as being horizontal and extending to the right, as indicated in Figure 2(a). Now let P be any point in the plane. As indicated in Figure 2(b), we initially let r denote the distance from O to P and we let θ denote the angle measured from the polar axis to $\overline{OP}$. (Just as in our earlier work with angles in standard position, we take the measure of θ to be positive if the rotation is counterclockwise and negative if it is clockwise.) Then the ordered pair (r, θ) serves to locate the point P with respect to the pole and the polar axis. We refer to r and θ as polar coordinates of P, and we write $P(r, \theta)$ to indicate that P is the point with polar coordinates (r, θ).

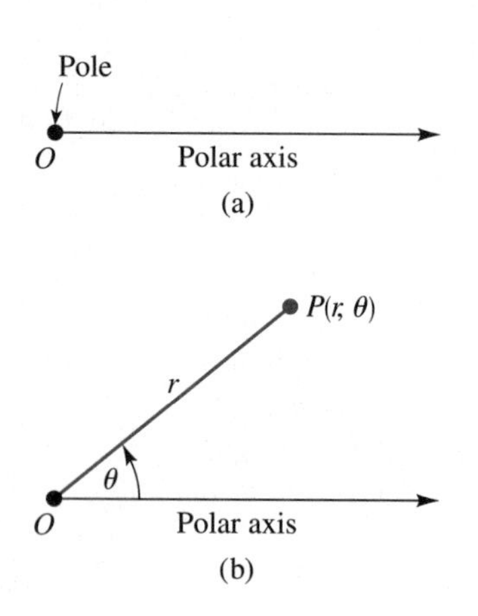

FIGURE 2

Plotting points in polar coordinates is facilitated by the use of polar coordinate graph paper, such as that shown in Figure 3(a) (on the next page). Figure 3(b) shows the points with polar coordinates $A(2, \pi/6)$, $B(3, 2\pi/3)$, and $C(4, 0)$.

There is a minor complication that arises in using polar rather than rectangular coordinates. Consider, for example, the point $C(4, 0)$ in Figure 3(b). This point could just as well have been labeled with the coordinates $(4, 2\pi)$, or $(4, 2k\pi)$ for any integral value of k. Similarly, the coordinates (r, θ) and $(r, \theta + 2k\pi)$ represent the same point for all integral values of k. This is in marked contrast to the situation with rectangular coordinates, where the coordinate representation of each point is unique. Also, what polar coordinates should we assign to the origin? For if $r = 0$ in Figure 2(b), we cannot really define an angle θ. The convention agreed on to cover this case is that the coordinates $(0, \theta)$ denote the origin for all values of θ. Finally, we point out that in working with polar coordinates, it is sometimes useful to let r take on negative values.

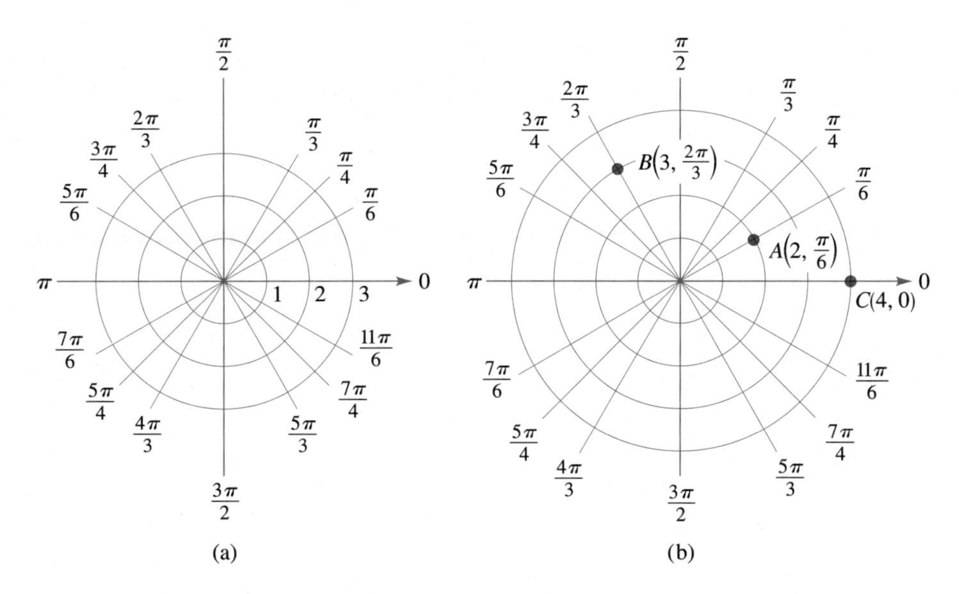

FIGURE 3

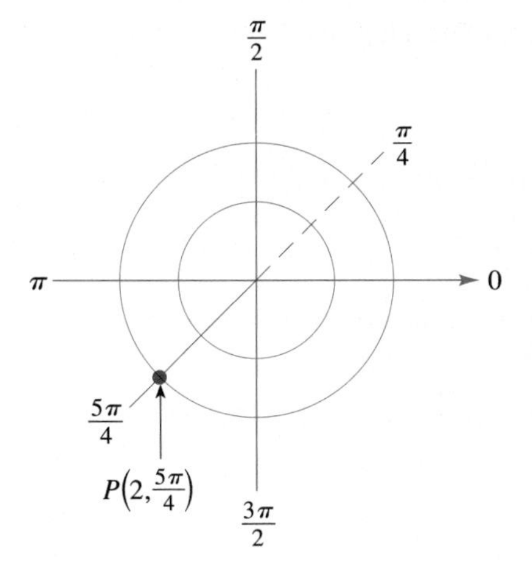

FIGURE 4

For example, consider the point $P(2, 5\pi/4)$ in Figure 4. The coordinates $(2, 5\pi/4)$ indicate that to reach P from the origin, we go 2 units in the direction $5\pi/4$. Alternately, we can describe this as -2 units in the $\pi/4$ direction. (Refer again to Figure 4.) For reasons such as this, we will adhere to the convention that the polar coordinates (r, θ) and $(-r, \theta + \pi)$ represent the same point. (Since r can now, in fact, be positive, negative, or zero, r is referred to as a *directed distance.*)

It is often useful to consider both rectangular and polar coordinates simultaneously. To do this, we draw the two coordinate systems so that the origins coincide and the positive x-axis coincides with the polar axis; see Figure 5. Suppose now that a point P, other than the origin, has rectangular coordinates (x, y) and polar coordinates (r, θ), as indicated in Figure 5. We wish to find equations relating the two sets of coordinates. From Figure 5, we see that

$$x^2 + y^2 = r^2$$

$$\sin\theta = \frac{y}{r} \qquad \cos\theta = \frac{x}{r} \qquad \tan\theta = \frac{y}{x}$$

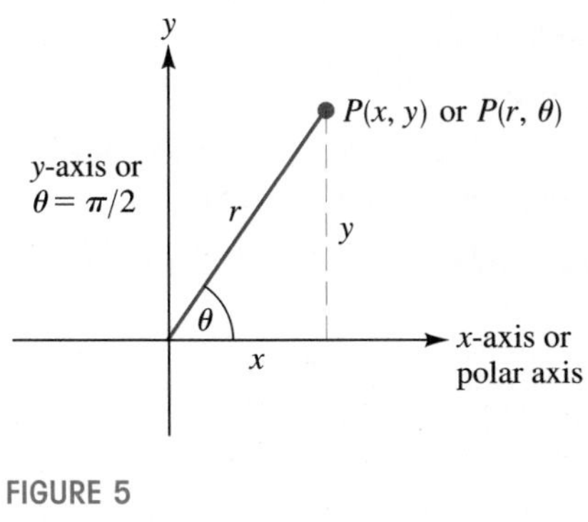

FIGURE 5

Although Figure 5 displays the point P in the first quadrant, it can be shown that these same equations hold when P is in any quadrant. For reference we summarize these equations as follows.

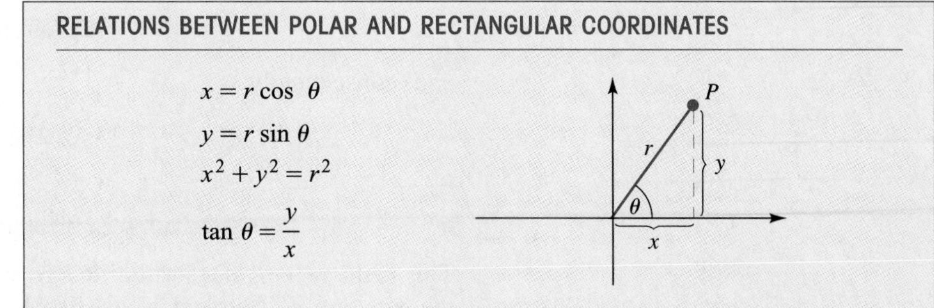

RELATIONS BETWEEN POLAR AND RECTANGULAR COORDINATES

$$x = r\cos\theta$$

$$y = r\sin\theta$$

$$x^2 + y^2 = r^2$$

$$\tan\theta = \frac{y}{x}$$

EXAMPLE 1 The polar coordinates of a point are $(5, \pi/6)$. Find the rectangular coordinates.

Solution We are given that $r = 5$ and $\theta = \pi/6$. Thus,

$$x = r\cos\theta = 5\cos\frac{\pi}{6} = 5\left(\frac{\sqrt{3}}{2}\right)$$

and

$$y = r\sin\theta = 5\sin\frac{\pi}{6} = 5\left(\frac{1}{2}\right)$$

The rectangular coordinates are therefore $\left(\frac{5}{2}\sqrt{3}, \frac{5}{2}\right)$.

The definition of the graph of an equation in polar coordinates is similar to the corresponding definition for rectangular coordinates. The **graph** of an equation in polar coordinates is the set of all points (r, θ) with coordinates that satisfy the given equation. It is often the case that the equation of a curve is simpler in one coordinate system than in another. The next two examples show instances of this.

EXAMPLE 2 Convert each polar equation to rectangular form:
(a) $r = \cos\theta + 2\sin\theta$; **(b)** $r^2 = \sin 2\theta$.

Solution

(a) In view of the transformation equations $x = r\cos\theta$ and $y = r\sin\theta$, we multiply both sides of the given equation by r to obtain

$$r^2 = r\cos\theta + 2r\sin\theta$$

and therefore

$$x^2 + y^2 = x + 2y \qquad \text{or} \qquad x^2 - x + y^2 - 2y = 0$$

This is the rectangular form of the given equation.
Question for review: What is the graph of this last equation?

(b) Using the double-angle formula for $\sin 2\theta$, we have

$$r^2 = 2\sin\theta\cos\theta$$

Now, in order to obtain the expressions $r\sin\theta$ and $r\cos\theta$ on the right-hand side of the equation, we multiply both sides by r^2. This yields

$$r^4 = 2(r\sin\theta)(r\cos\theta)$$

and, consequently,

$$(x^2 + y^2)^2 = 2yx$$

or

$$x^4 + 2x^2y^2 + y^4 - 2xy = 0$$

This is the rectangular form of the given equation. Notice how much simpler the equation is in its polar coordinate form.

EXAMPLE 3 Convert the rectangular equation $x^2 + y^2 + ax = a\sqrt{x^2 + y^2}$ to polar form, expressing r as a function of θ. Assume that a is a constant.

Solution Using the relations $x^2 + y^2 = r^2$ and $x = r \cos \theta$, we obtain

$$r^2 + ar \cos \theta = ar$$

Notice that this equation is satisfied by $r = 0$. In other words, the graph of this equation will pass through the origin. This is consistent with the fact that the original equation is satisfied when x and y are both zero. Now assume for the moment that $r \neq 0$. Then we can divide both sides of the last equation by r to obtain

$$r + a \cos \theta = a \quad \text{or} \quad r = a - a \cos \theta$$

This expresses r as a function of θ, as required. Note that when $\theta = 0$, we obtain

$$r = a - a(1) = 0$$

That is, nothing has been lost in dividing through by r; the graph will still pass through the origin.

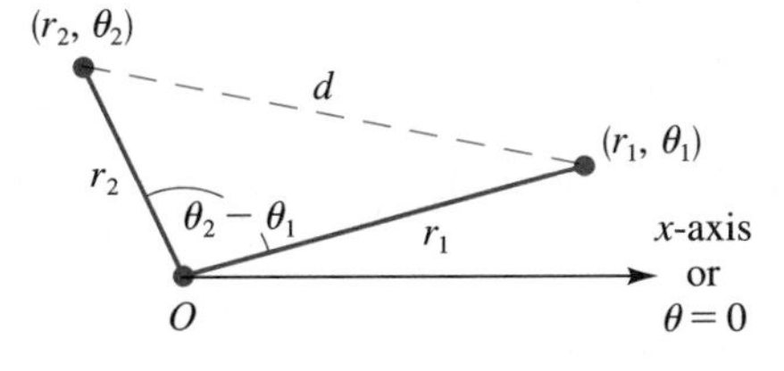

FIGURE 6

When we work in an x-y coordinate system, one of the basic tools is the formula for the distance between two points: $d = \sqrt{(x_2 - x_1)^2 + (y_2 - y_1)^2}$. There is a corresponding distance formula for use with polar coordinates. In Figure 6, we let d denote the distance between the points with polar coordinates (r_1, θ_1) and (r_2, θ_2). Then, applying the law of cosines, we have $d^2 = r_1^2 + r_2^2 - 2r_1r_2 \cos(\theta_2 - \theta_1)$. This is the distance formula in polar coordinates. For ease of reference, we repeat this result in the box that follows.

PROPERTY SUMMARY

DISTANCE FORMULA FOR POLAR COORDINATES

Let d denote the distance between two points with polar coordinates (r_1, θ_1) and (r_2, θ_2). Then

$$d^2 = r_1^2 + r_2^2 - 2r_1r_2 \cos(\theta_2 - \theta_1)$$

EXAMPLE 4 Compute the distance between the points with polar coordinates $(2, 5\pi/6)$ and $(4, \pi/6)$.

Solution We use the distance formula, taking the points (r_1, θ_1) and (r_2, θ_2) to be $(2, 5\pi/6)$ and $(4, \pi/6)$, respectively. This yields

$$\begin{aligned} d^2 &= r_1^2 + r_2^2 - 2r_1r_2 \cos(\theta_2 - \theta_1) \\ &= 2^2 + 4^2 - 2(2)(4)\cos\left(\frac{\pi}{6} - \frac{5\pi}{6}\right) \\ &= 20 - 16 \cos\left(-\frac{2\pi}{3}\right) = 20 - 16\left(-\frac{1}{2}\right) = 28 \end{aligned}$$

$$d = \sqrt{28} = 2\sqrt{7}$$

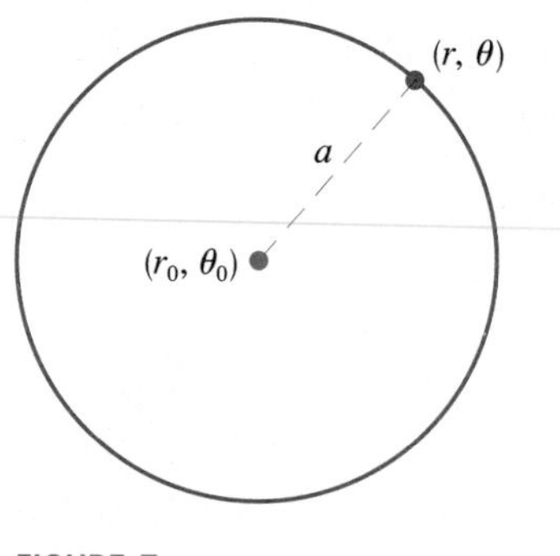

FIGURE 7

You should check for yourself that the same result is obtained if we choose (r_1, θ_1) and (r_2, θ_2) to be $(4, \pi/6)$ and $(2, 5\pi/6)$, respectively. This is because, in general, $\cos(\theta_2 - \theta_1) = \cos(\theta_1 - \theta_2)$. (Why?)

The distance formula can be used to find an equation for a circle. Suppose that the radius of the circle is a and that the polar coordinates of the center are (r_0, θ_0). As indicated in Figure 7, we let (r, θ) denote an arbitrary point on the circle. Then, since the distance between the points (r, θ) and (r_0, θ_0) is a, the radius of the circle, we have

$$r^2 + r_0^2 - 2rr_0 \cos(\theta - \theta_0) = a^2 \tag{1}$$

This is the required equation. There is a special case of this equation that is worth noting. If the center of the circle is located at the origin, then $r_0 = 0$ and equation (1) becomes simply $r^2 = a^2$. Therefore

$$r = a \qquad \text{or} \qquad r = -a$$

The graph of each of these equations is a circle of radius a, with center at the origin.

EXAMPLE 5 Determine a polar equation for the circle satisfying the given conditions.

(a) The radius is 2 and the polar coordinates of the center are $(4, \pi/5)$.
(b) The radius is 5 and the center is the origin.

Solution

(a) Using equation (1), we have

$$r^2 + 4^2 - 2r(4)\cos\left(\theta - \frac{\pi}{5}\right) = 2^2$$

and therefore

$$r^2 - 8r\cos\left(\theta - \frac{\pi}{5}\right) = -12$$

(b) A polar equation for a circle of radius a and with center at the origin is $r = a$. So, if the radius is 5, the required equation is $r = 5$. (The equation $r = -5$ would yield the same graph.)

We conclude this section by discussing equations of lines in polar coordinates. There are two cases to consider: lines that pass through the origin and lines that do not pass through the origin. First, consider a line through the origin, and suppose that the line makes an angle θ_0 with the positive x-axis; see Figure 8. The polar equation of this line is simply

$$\theta = \theta_0$$

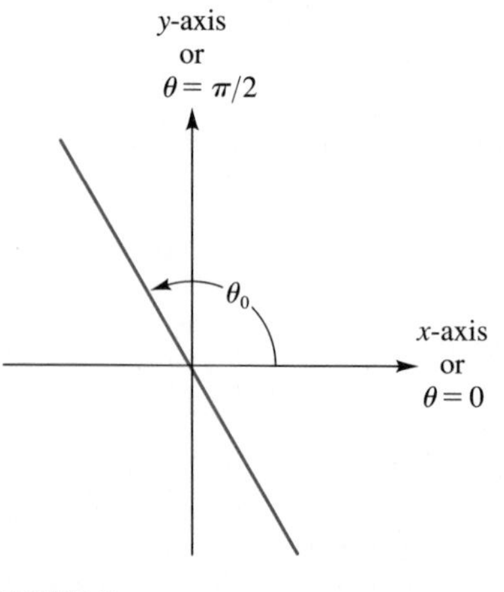

FIGURE 8

Notice that this equation poses no restrictions on r. In other words, for every real number r, the point with polar coordinates (r, θ_0) lies on this line.

EXAMPLE 6 Graph the line with polar equation $\theta = \pi/3$, and locate the points on this line for which $r = 1, 2$, and -1.

Solution See Figure 9.

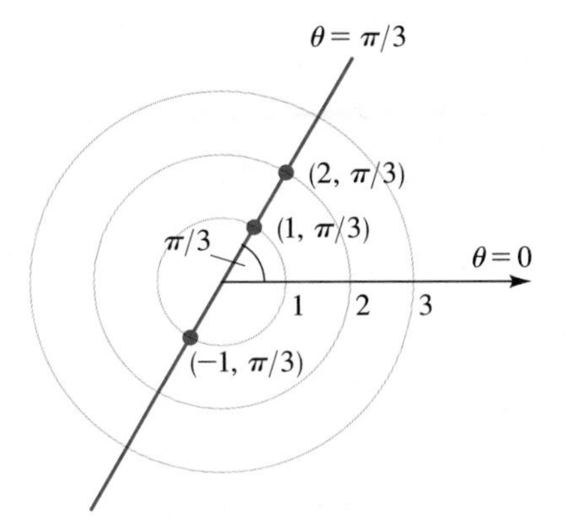

FIGURE 9

Now let us determine the polar equation for a line $\mathscr{L}$ that does not pass through the origin. As indicated in Figure 10, we assume that the perpendicular distance from the origin to $\mathscr{L}$ is d, and that the polar coordinates of the point N (in Figure 10) are (d, α). The point $P(r, \theta)$ in Figure 10 denotes an arbitrary point on line $\mathscr{L}$. We want to find an equation relating r and θ.

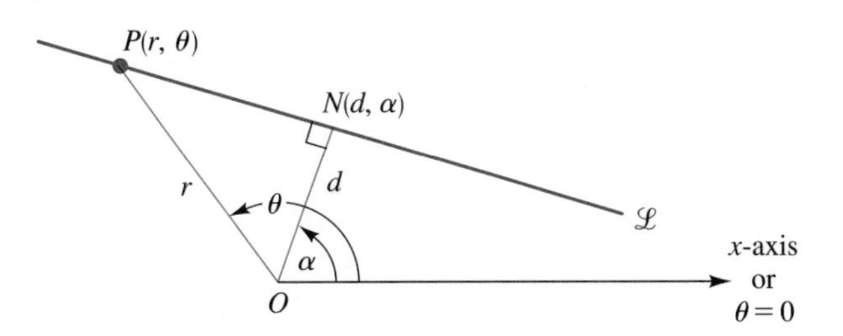

FIGURE 10

In right triangle ONP, note that $\angle PON = \theta - \alpha$, and therefore

$$\cos(\theta - \alpha) = \frac{d}{r} \qquad \text{or} \qquad r\cos(\theta - \alpha) = d$$

This is the polar equation of the line $\mathscr{L}$. For reference, in the box that follows, we summarize our results about lines.

PROPERTY SUMMARY **POLAR EQUATIONS FOR LINES**

1. LINE THROUGH THE ORIGIN
   Suppose that a line passes through the origin and makes an angle θ_0 with the positive x-axis (as indicated in Figure 8 on page 580). Then an equation for the line is $\theta = \theta_0$.
2. LINE NOT PASSING THROUGH THE ORIGIN
   Suppose that the perpendicular distance from the origin to the line is d and that the point (d, α) is the foot of the perpendicular from the origin to the line (as indicated in Figure 10). Then a polar equation for the line is $r\cos(\theta - \alpha) = d$.

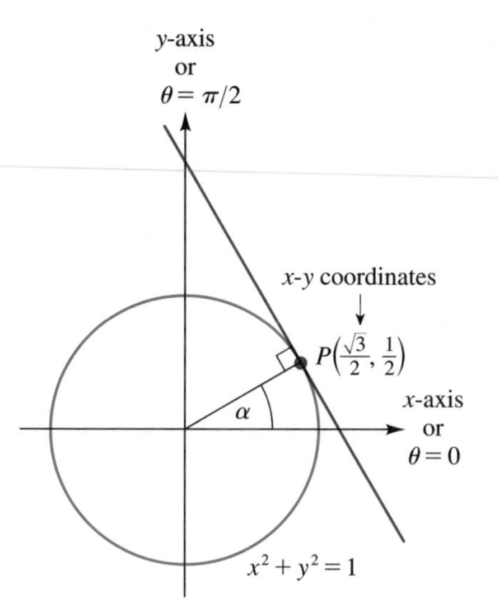

FIGURE 11

EXAMPLE 7 Figure 11 shows a line tangent to the unit circle at the point P. The x-y coordinates of P are $\left(\sqrt{3}/2, \frac{1}{2}\right)$.

(a) What are the polar coordinates of the point P?
(b) Find the polar equation for the tangent line.
(c) Use the polar equation to find the x- and y-intercepts of the tangent line.

Solution

(a) Using the unit-circle definition for cosine, we have $\cos\alpha = \sqrt{3}/2$ and, consequently, $\alpha = \pi/6$. The polar coordinates of P are therefore $(1, \pi/6)$.

(b) In the general equation $r\cos(\theta - \alpha) = d$, we use the values $\alpha = \pi/6$ and $d = 1$ to obtain

$$r\cos\left(\theta - \frac{\pi}{6}\right) = 1 \tag{2}$$

This is the polar equation for the tangent line.

(c) The x-axis corresponds to $\theta = 0$. Substituting $\theta = 0$ in equation (2) gives us $r\cos\left(0 - \dfrac{\pi}{6}\right) = 1$, and therefore

$$r\cos\frac{\pi}{6} = 1 \qquad \text{(Why?)}$$

$$r = \frac{1}{\sqrt{3}/2} = \frac{2}{\sqrt{3}} = \frac{2\sqrt{3}}{3}$$

This tells us that the tangent line meets the x-axis at the point with polar coordinates $\left(2\sqrt{3}/3, 0\right)$. So, the x-intercept of the line is $2\sqrt{3}/3$. For the y-intercept, we set $\theta = \pi/2$ in equation (2). As you should check for yourself, this yields $r = 2$. Thus, the y-intercept of the tangent line is 2.

EXERCISE SET 9.5

A

In Exercises 1–3, convert the given polar coordinates to rectangular coordinates.

1. **(a)** $(3, 2\pi/3)$ **(b)** $(4, 11\pi/6)$ **(c)** $(4, -\pi/6)$
2. **(a)** $(5, \pi/4)$ **(b)** $(-5, \pi/4)$ **(c)** $(-5, -\pi/4)$
3. **(a)** $(1, \pi/2)$ **(b)** $(1, 5\pi/2)$ **(c)** $(-1, \pi/8)$

In Exercises 4–6, convert the given rectangular coordinates to polar coordinates. Express your answers in such a way that r is nonnegative and $0 \leq \theta < 2\pi$.

4. $\left(3, \sqrt{3}\right)$
5. $(-1, -1)$
6. $(0, -2)$

In Exercises 7–16, convert to rectangular form.

7. $r = 2\cos\theta$
8. $2\sin\theta - 3\cos\theta = r$
9. $r = \tan\theta$
10. $r = 4$
11. $r = 3\cos 2\theta$
12. $r = 4\sin 2\theta$
13. $r^2 = 8/(2 - \sin^2\theta)$
14. $r^2 = 1/(3 + \cos^2\theta)$
15. $r\cos\left(\theta - \frac{\pi}{6}\right) = 2$
16. $r\sin\left(\theta + \frac{\pi}{4}\right) = 6$

In Exercises 17–24, convert to polar form.

17. $3x - 4y = 2$
18. $x^2 + y^2 = 25$
19. $y^2 = x^3$
20. $y = x^2$
21. $2xy = 1$

22. $x^2 + 4x + y^2 + 4y = 0$

23. $9x^2 + y^2 = 9$

24. $x^2(x^2 + y^2) = y^2$

In Exercises 25–30, you are given a polar equation and its graph. Use the equation to determine polar coordinates of the points labeled with capital letters. (For the r-values, where necessary, use a calculator and round to two decimal places.)

25. $r = \dfrac{4}{1 + \sin\theta}$

26. $r = \dfrac{2}{\cos\left(\theta - \frac{\pi}{3}\right)}$

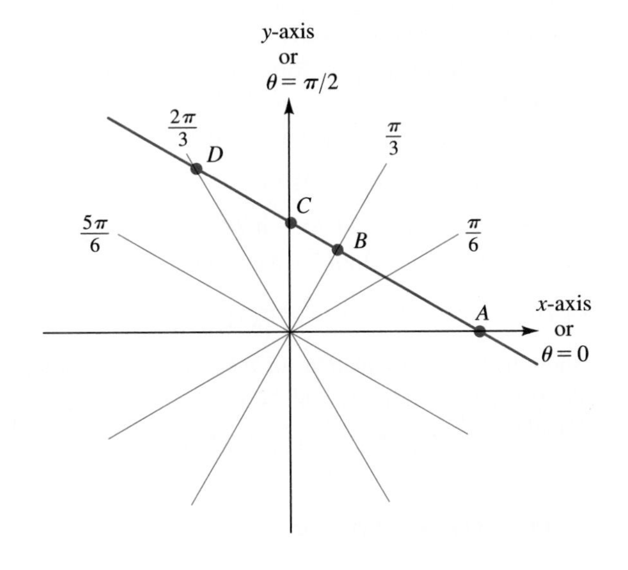

27. $r = 2\cos 2\theta$

28. $r = -1 - \sin\theta$

29. $r = e^{\theta/6}$

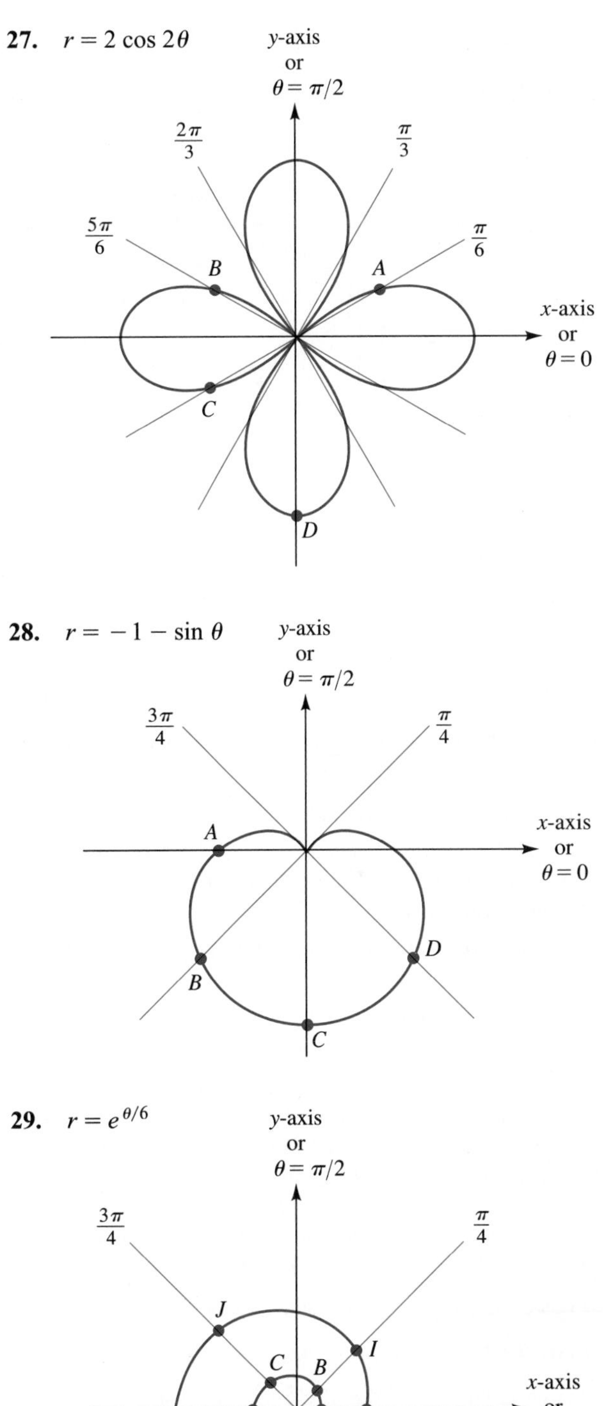

30. $r^2 = \cos 2\theta$

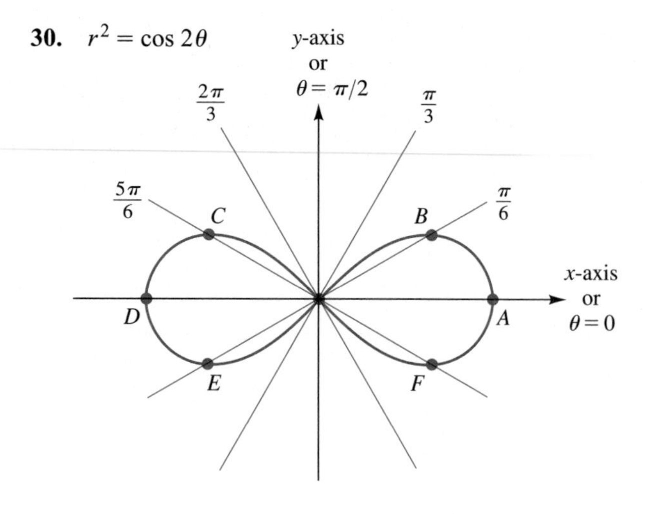

In Exercises 31–34, compute the distance between the given points. (The coordinates are polar coordinates.)

31. $(2, 2\pi/3)$ and $(4, \pi/6)$ **32.** $(4, \pi)$ and $(3, 7\pi/4)$

33. $(4, 4\pi/3)$ and $(1, 0)$ **34.** $(3, 5\pi/6)$ and $(5, 5\pi/3)$

In Exercises 35–38, determine a polar equation for the circle satisfying the given conditions.

35. The radius is 2 and the polar coordinates for the center are: **(a)** $(4, 0)$; **(b)** $(4, 2\pi/3)$; **(c)** $(0, 0)$.

36. The radius is $\sqrt{6}$ and the polar coordinates of the center are: **(a)** $(2, \pi)$; **(b)** $(2, 3\pi/4)$; **(c)** $(0, 0)$.

37. The radius is 1 and the polar coordinates of the center are: **(a)** $(1, 3\pi/2)$; **(b)** $(1, \pi/4)$.

38. The radius is 6 and the polar coordinates of the center are: **(a)** $(-3, 5\pi/4)$; **(b)** $(-2, \pi/4)$.

In Exercises 39–42, the polar equation of a line is given. In each case: **(a)** *specify the perpendicular distance from the origin to the line;* **(b)** *determine the polar coordinates of the points on the line corresponding to $\theta = 0$ and $\theta = \frac{\pi}{2}$;* **(c)** *specify the polar coordinates of the foot of the perpendicular from the origin to the line; and* **(d)** *use the results in parts (a), (b), and (c) to sketch the line.*

39. $r\cos\left(\theta - \frac{\pi}{6}\right) = 2$ **40.** $r\cos\left(\theta - \frac{\pi}{4}\right) = 1$

41. $r\cos\left(\theta + \frac{2\pi}{3}\right) = 4$ **42.** $r\cos\left(\theta - \frac{\pi}{2}\right) = \sqrt{2}$

43. A line is tangent to the circle $x^2 + y^2 = 4$ at a point P with x-y coordinates $\left(-\sqrt{3}, 1\right)$.
(a) What are the polar coordinates of P?
(b) Find a polar equation for the tangent line.
(c) Use the polar equation to determine the x- and y-intercepts of the line.

44. Follow Exercise 43 using the circle $x^2 + y^2 = 36$ and the point P with x-y coordinates $\left(-3, -3\sqrt{3}\right)$.

Notice that the rectangluar form of the polar equation $r\cos\theta = a$ is $x = a$. Thus, the graph of the polar equation $r\cos\theta = a$ is a vertical line with an x-intercept of a. Similarly, the graph of the polar equation $r\sin\theta = b$ is a horizontal line with a y-intercept of b. In Exercises 45 and 46, use these observations to graph the polar equations.

45. **(a)** $r\cos\theta = 3$ **(b)** $r\sin\theta = 3$

46. **(a)** $r\cos\theta = -2$ **(b)** $r\sin\theta = 4$

B

47. The accompanying figure shows the graph of a line $\mathscr{L}$.

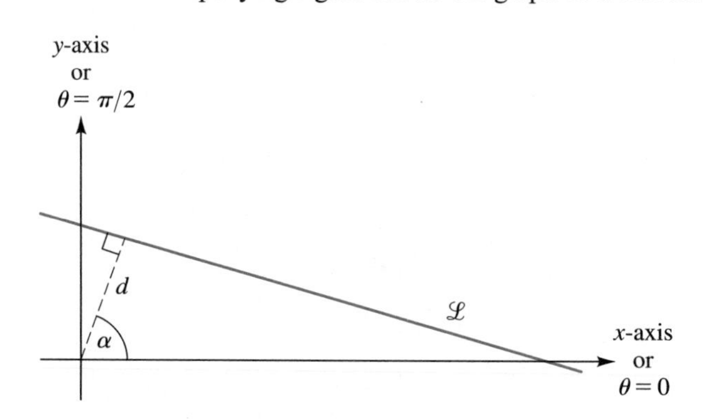

(a) According to the text, what is the polar equation for $\mathscr{L}$?
(b) By converting the equation in part (a) to rectangular form, show that the x-y equation for $\mathscr{L}$ is $x\cos\alpha + y\sin\alpha = d$. (This equation is called the **normal form** for a line.) *Hint:* First use one of the addition formulas for cosine.

48. By converting the polar equation

$$r = a\cos\theta + b\sin\theta$$

to rectangular form, show that the graph is a circle, and find the center and the radius.

49. Show that the rectangular form of the equation $r = a\sin 3\theta$ is $(x^2 + y^2)^2 = ay(3x^2 - y^2)$.

50. Show that the rectangular form of the equation

$$r = ab/(1 - a\cos\theta) \quad (a < 1)$$

is $(1 - a^2)x^2 + y^2 - 2a^2bx - a^2b^2 = 0$.

51. Show that the polar form of the equation $\left(x^2/a^2\right) - \left(y^2/b^2\right) = 1$ is $r^2 = \dfrac{a^2b^2}{(b^2\cos^2\theta - a^2\sin^2\theta)}$.

52. Show that a polar equation for the line passing through the two points $A(r_1, \theta_1)$ and $B(r_2, \theta_2)$ is $rr_1\sin(\theta - \theta_1) - rr_2\sin(\theta - \theta_2) + r_1r_2\sin(\theta_1 - \theta_2) = 0$. *Hint:* Use the following figure and the observation that

$$\text{area } \triangle OAB = (\text{area } \triangle OBP) + (\text{area } \triangle OPA)$$

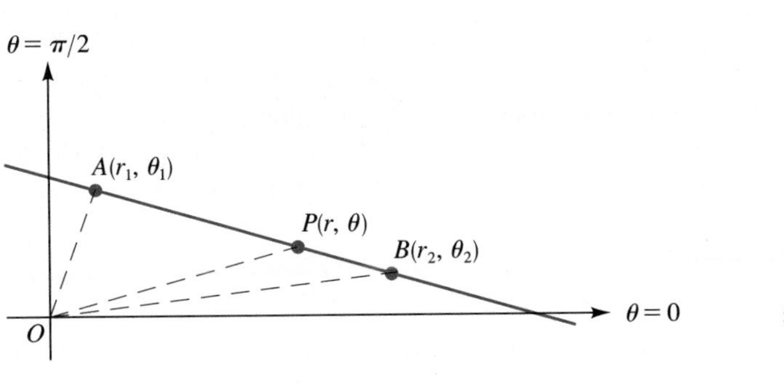

53. The following figure shows an equilateral triangle inscribed in the unit circle.

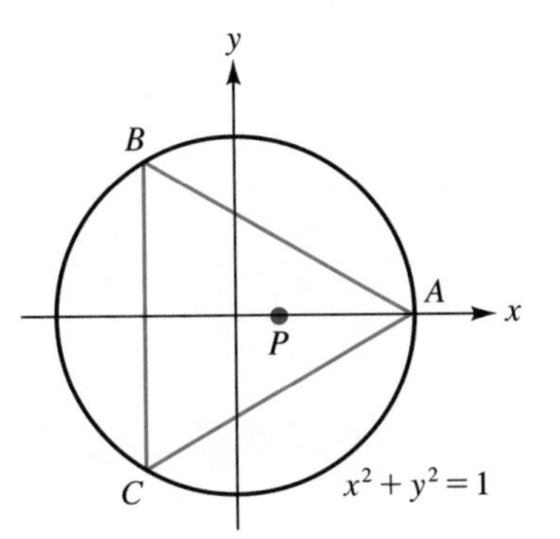

(a) Specify polar coordinates for the points A, B, and C.

(b) In the figure, P denotes an arbitrary point between the origin and A on the x-axis. Let the polar coordinates of P be $(r, 0)$. Use the distance formula for polar coordinates to show that

$$(PA)(PB)(PC) = 1 - r^3$$

[This result is due to the English mathematician Roger Cotes (1682–1716).]

54. The following figure shows an equilateral triangle inscribed in the unit circle.

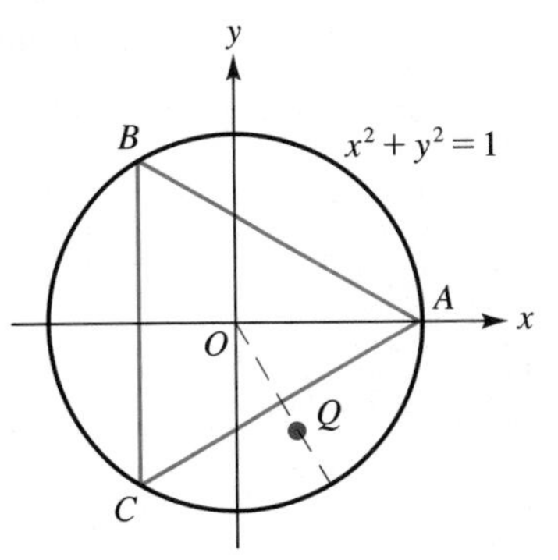

The point Q lies on the bisector of $\angle AOC$, and the distance from O to Q is r. Use the distance formula for polar coordinates to show that

$$(QA)(QB)(QC) = 1 + r^3$$

(This result also is due to Roger Cotes.)

CURVES IN POLAR COORDINATES

In principle, the graph of any polar equation $r = f(\theta)$ can be obtained by setting up a table and plotting a sufficient number of points. Indeed, this is just the way a graphing calculator or a computer operates. Figure 1, for example, shows a

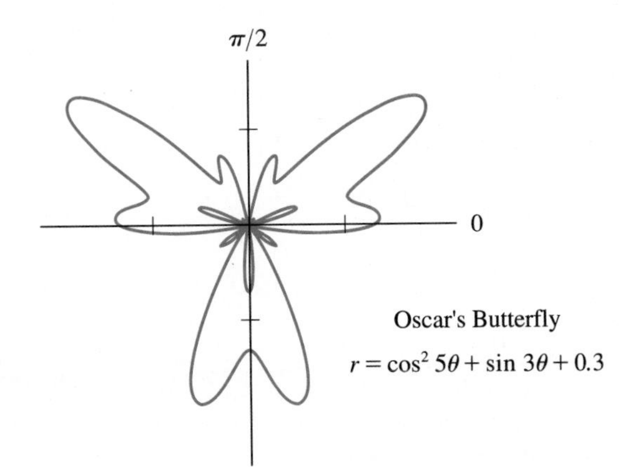

FIGURE 1
This polar coordinate graph was discovered by Oscar Ramirez, a precalculus student at UCLA in the Fall of 1991.

computer-generated graph of the polar equation $r = \cos^2 5\theta + \sin 3\theta + 0.3$. In this section, we'll concentrate on a few basic types of polar equations and their graphs. With the exception of Example 1, the polar equations that we consider involve trigonometric functions. In order to *understand* why the graphs look as they do, we will often use symmetry and the basic properties of the sine and cosine functions.

EXAMPLE 1 Graph the polar equation $r = \theta/\pi$ for $\theta \geq 0$.

Solution The equation shows that as θ increases, so does r. Geometrically, this means that as we plot points, moving counterclockwise with increasing θ, the points will be farther and farther from the origin. In Table 1, we have computed values for r corresponding to some convenient values of θ. In Figure 2, we've plotted the points in the table and connected them with a smooth curve. The curve is known as the **spiral of Archimedes.**

TABLE 1

θ	0	$\frac{\pi}{4}$	$\frac{\pi}{2}$	$\frac{3\pi}{4}$	π	$\frac{5\pi}{4}$	$\frac{3\pi}{2}$	$\frac{7\pi}{4}$	2π	$\frac{5\pi}{2}$	3π	4π
r	0	$\frac{1}{4}$	$\frac{1}{2}$	$\frac{3}{4}$	1	$\frac{5}{4}$	$\frac{3}{2}$	$\frac{7}{4}$	2	$\frac{5}{2}$	3	4

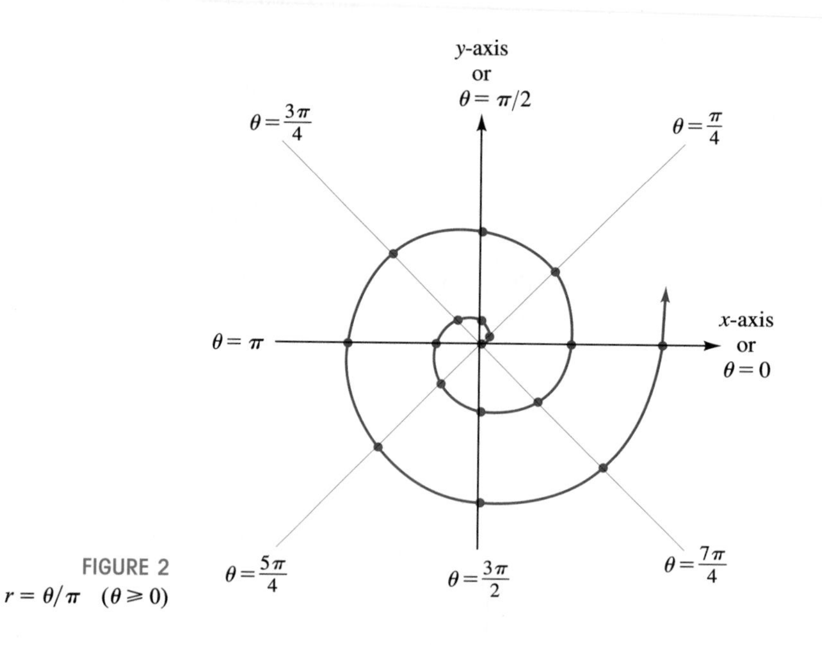

FIGURE 2
$r = \theta/\pi \quad (\theta \geq 0)$

We saw in Chapters 2 and 3 that symmetry considerations can often be used to lessen the amount of work involved in graphing equations. This is also true for polar equations. In the box that follows, we list four tests for symmetry in polar coordinates.

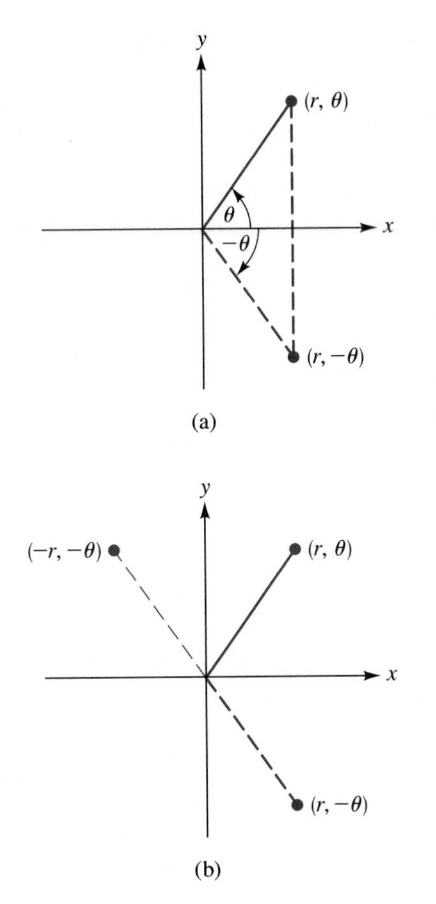

FIGURE 3

SYMMETRY TESTS IN POLAR COORDINATES

If the following substitution yields an equivalent equation,	then the graph is symmetric about:
1. replacing θ with $-\theta$,	the x-axis;
2. replacing r and θ with $-r$ and $-\theta$, respectively,	the y-axis;
3. replacing θ with $\pi-\theta$,	the y-axis;
4. replacing r with $-r$,	the origin.

CAUTION As opposed to the case with the x-y symmetry tests (in Chapter 2), if a polar equation fails a symmetry test, then it is still possible for the graph to possess the indicated symmetry. This is a consequence of the fact that the polar coordinates of a point are not unique. (For an example, see Example 3.)

The validity of the first test follows from the fact that the points (r, θ) and $(r, -\theta)$ are reflections of each other about the x-axis; see Figure 3(a). The validity of test 2 follows from the fact that the points (r, θ) and $(-r, -\theta)$ are reflections of each other about the y-axis, as indicated in Figure 3(b). The other tests can be justified in a similar manner.

EXAMPLE 2 Graph the polar equation $r = 1 - \cos\theta$.

Solution We know (from Section 7.3) that the cosine function satisfies the identity $\cos(-\theta) = \cos\theta$. So, according to symmetry test 1, the graph of $r = 1 - \cos\theta$ will be symmetric about the x-axis. In Figure 4(a), we've set up a table (using a calculator as necessary) with some convenient values of θ running from 0 to π. Plotting the points in the table leads to the curve shown in Figure 4(a). Reflecting this curve in the x-axis, we obtain the graph in Figure 4(b). The

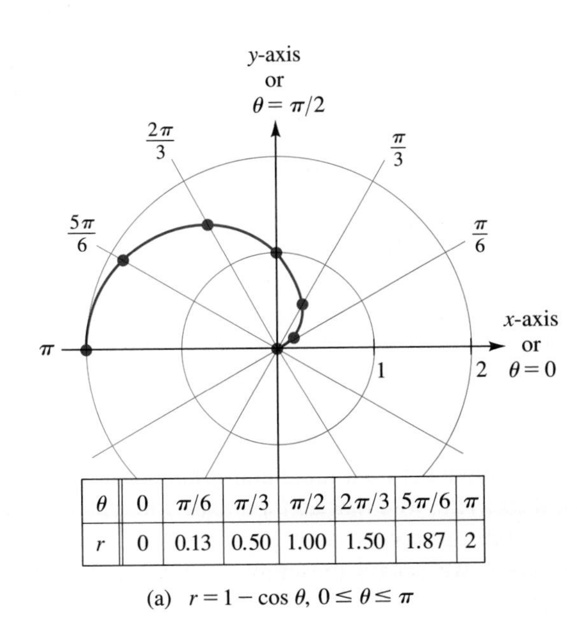

θ	0	$\pi/6$	$\pi/3$	$\pi/2$	$2\pi/3$	$5\pi/6$	π
r	0	0.13	0.50	1.00	1.50	1.87	2

(a) $r = 1 - \cos\theta,\ 0 \le \theta \le \pi$

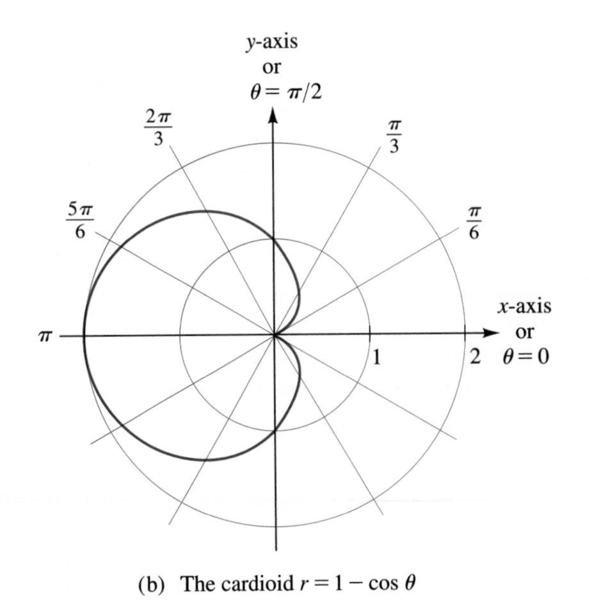

(b) The cardioid $r = 1 - \cos\theta$

FIGURE 4

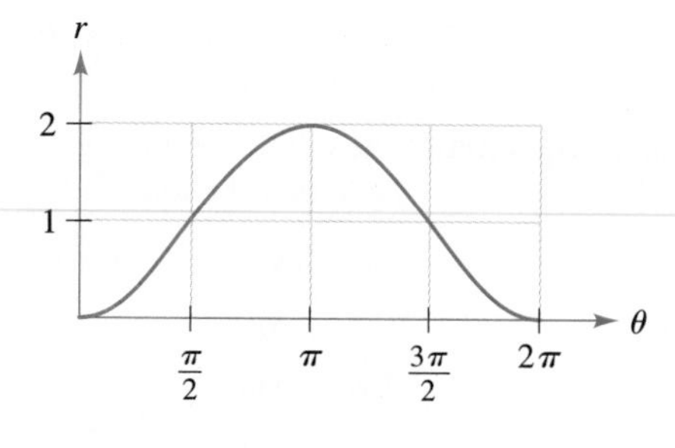

FIGURE 5
The graph of $r = 1 - \cos\theta$ in a rectangular θ-r coordinate system.

fact that the period of the cosine function is 2π implies that we need not consider values of θ beyond 2π. (Why?) Thus, the curve in Figure 4(b) is the required graph of $r = 1 - \cos\theta$. The curve is a **cardioid.**

In graphing polar equations, just as in graphing Cartesian *x*-*y* equations, there is always the question of whether we have plotted a sufficient number of points to reveal the essential features of the graph. As a check on our work in Figure 4(a), we can graph the equation $r = 1 - \cos\theta$ in a *rectangular θ-r* coordinate system, as shown in Figure 5. (This is easy to do using reflection and translation.) From the graph in Figure 5, we see that as θ increases from 0 to π, the values of r increase from 0 to 2. Now, recall that in polar coordinates, r represents the directed distance from the origin. So, on the polar graph, as we go from $\theta = 0$ to π, the points we obtain should be farther and farther from the origin, starting with $r = 0$ and ending with $r = 2$. The polar graph in Figure 4(a) is consistent with these observations.

EXAMPLE 3 Graph the polar curve $r = 1 + 2\sin\theta$.

Solution Using symmetry test 3, we replace θ by $\pi - \theta$ in the given equation:

$$r = 1 + 2\sin(\pi - \theta) = 1 + 2(\sin\pi\cos\theta - \cos\pi\sin\theta)$$
$$= 1 + 2\sin\theta \qquad \text{because } \sin\pi = 0 \text{ and } \cos\pi = -1$$

This shows that the graph is symmetric about the y-axis. (Note that symmetry test 2 fails for this equation, yet the graph is symmetric about the y-axis.)

Next, to see what to expect in polar coordinates, we first sketch the graph of the given equation in a rectangular θ-r coordinate system; see Figure 6. In Figure 6, notice that the maximum value of $|r|$ is 3. This implies that the required polar graph is contained within a circle of radius 3 about the origin. (We'll refer to this sketch several more times in the course of the solution.)

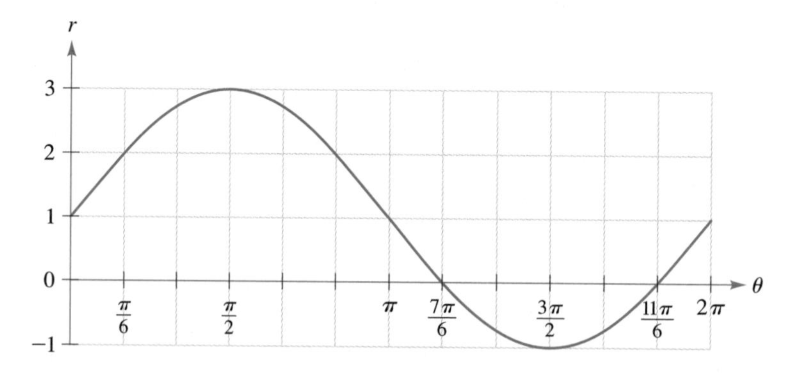

FIGURE 6
The graph of $r = 1 + 2\sin\theta$ in a rectangular θ-r coordinate system.

In Figure 7(a), we have set up a table and graphed the given polar equation for θ running from 0 to $\pi/2$. Note that as θ increases from 0 to $\pi/2$, the r-values increase from 1 to 3 (in agreement with Figure 6). By reflecting the curve in Figure 7(a) in the y-axis, we obtain the graph of $r = 1 + 2\sin\theta$ for $0 \leqslant \theta \leqslant \pi$; see Figure 7(b).

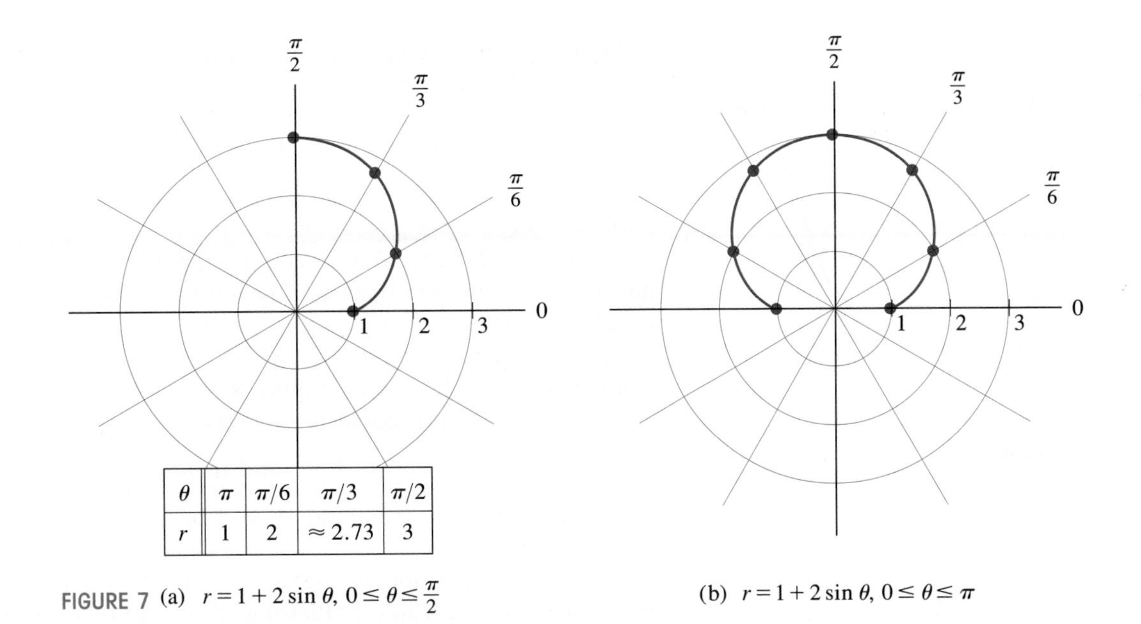

θ	π	$\pi/6$	$\pi/3$	$\pi/2$
r	1	2	≈ 2.73	3

FIGURE 7 (a) $r = 1 + 2\sin\theta,\ 0 \leq \theta \leq \frac{\pi}{2}$ (b) $r = 1 + 2\sin\theta,\ 0 \leq \theta \leq \pi$

Next, let's consider the values of r as θ increases from π to $3\pi/2$. Figure 6 tells us what to expect on this interval:

> As θ increases from π to $7\pi/6$, the r-values decrease from 1 to 0; as θ increases from $7\pi/6$ to $3\pi/2$, the r-values decrease from 0 to -1. However (in view of the convention regarding negative r-values), this will mean that the points on the polar graph are, in fact, moving farther and farther from the origin as θ increases from $7\pi/6$ to $3\pi/2$.

In Figure 8(a), we've set up a table for $\pi \leq \theta \leq \frac{3\pi}{2}$ and added the corresponding points from Figure 7(b). Finally, again taking into account the y-axis symmetry, we sketch the final graph for $0 \leq \theta \leq 2\pi$, as indicated in Figure 8(b). The curve

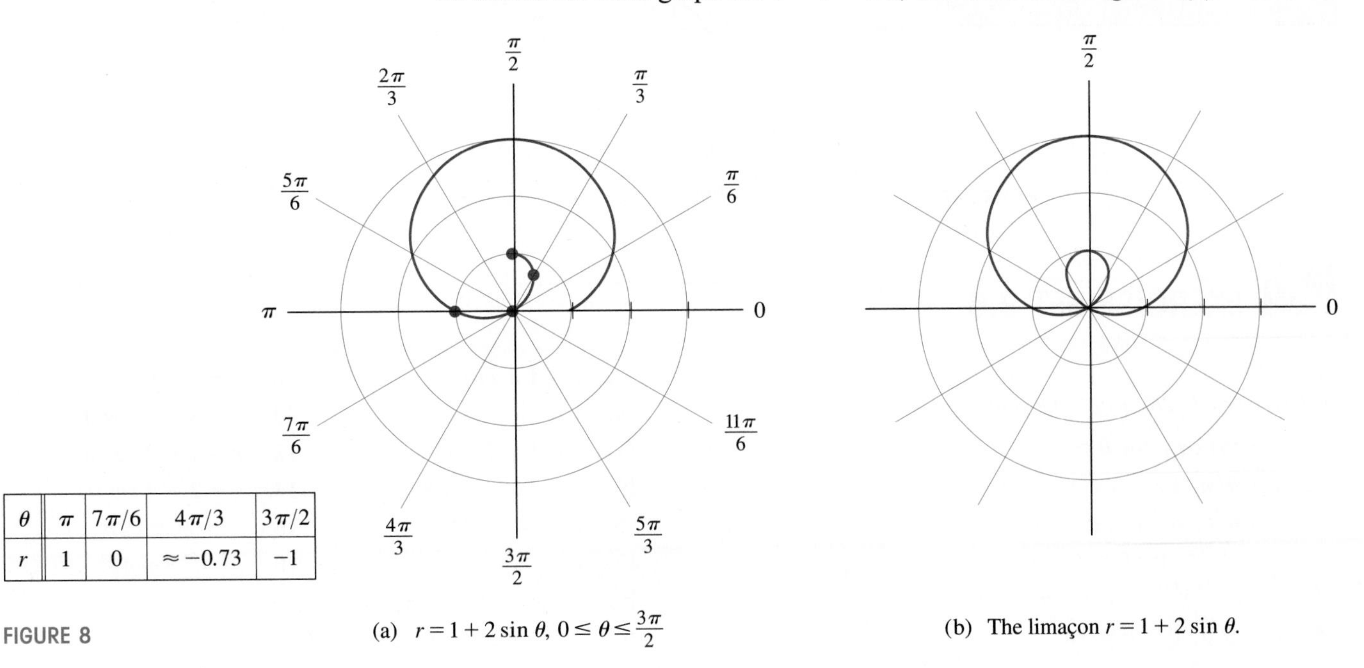

θ	π	$7\pi/6$	$4\pi/3$	$3\pi/2$
r	1	0	≈ -0.73	-1

FIGURE 8 (a) $r = 1 + 2\sin\theta,\ 0 \leq \theta \leq \frac{3\pi}{2}$ (b) The limaçon $r = 1 + 2\sin\theta$.

in Figure 8(b) is a **limaçon.** (Not all limaçons have inner loops. See, for example, Exercises 23 and 24.)

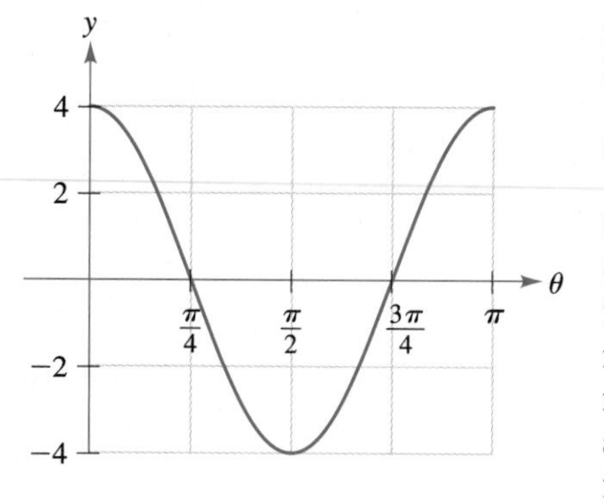

FIGURE 9
$y = 4\cos 2\theta$

EXAMPLE 4 Graph the polar equation $r^2 = 4\cos 2\theta$.

Solution As a guide for graphing the given polar equation, we first sketch the curve $y = 4\cos 2\theta$ in a rectangular θ-y coordinate system. As indicated in Figure 9, the period of $4\cos 2\theta$ is π. So, in graphing the given polar equation, we will not need to consider values of θ beyond π. Figure 9 also shows that $4\cos 2\theta$ is negative on the interval $\frac{\pi}{4} < \theta < \frac{3\pi}{4}$. The value of r^2, however, cannot be negative. Consequently, we do not need to consider values of θ in the open interval from $\pi/4$ to $3\pi/4$. Finally, Figure 9 shows that the absolute value of $4\cos 2\theta$ is always less than or equal to 4. Consequently, for the polar graph, we have

$$r^2 \leq 4 \qquad \text{and therefore} \qquad |r| \leq 2$$

This tells us that the polar graph will be contained within a circle of radius 2 about the origin.

For the polar graph, we begin by computing a table of values with θ running from 0 to $\pi/4$; see Table 2. Plotting the points in Table 2 and joining them with a smooth curve leads to the graph shown in Figure 10(a). Then, rather than setting up another table with θ running from $3\pi/4$ to π, we rely on symmetry to complete the graph. According to the first two symmetry tests, the curve is symmetric about both the x-axis and the y-axis. Thus, we obtain the graph shown in Figure 10(b). The curve is called a **lemniscate.**

TABLE 2

θ	$r = \pm 2\sqrt{\cos 2\theta}$
0	± 2
$\pi/12$	± 1.86
$\pi/8$	± 1.68
$\pi/6$	± 1.41
$\pi/4$	0

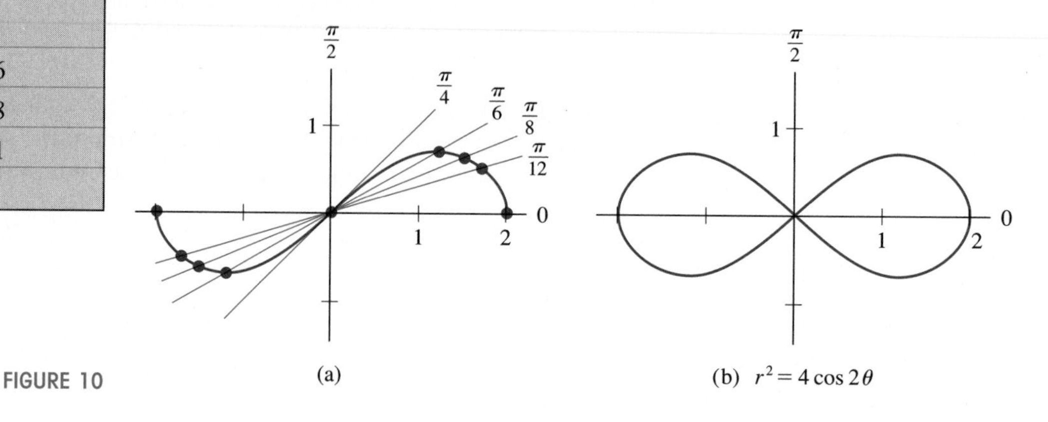

FIGURE 10 (a) (b) $r^2 = 4\cos 2\theta$

EXERCISE SET 9.6

A

In Exercises 1–26, graph the polar equations.

1. $r = \theta/(2\pi)$, for $\theta \geq 0$
2. $r = \theta/\pi$, for $-4\pi \leq \theta \leq 0$
3. $r = \ln \theta$, for $1 \leq \theta \leq 3\pi$
4. $r = e^{\theta/2\pi}$, for $0 \leq \theta \leq 2\pi$
5. $r = e^{-\theta/(2\pi)}$, for $0 \leq \theta \leq 2\pi$
6. $r = \sqrt{\theta}$, $0 \leq \theta \leq 3\pi$
7. $r = 1 + \cos \theta$
8. $r = 1 - \sin \theta$
9. $r = 2 - 2\sin \theta$
10. $r = 3 + 3\cos \theta$
11. $r = 1 - 2\sin \theta$
12. $r = 1 + 2\cos \theta$
13. $r = 2 + 4\cos \theta$
14. $r = 2 - 4\sin \theta$
15. $r^2 = 4\sin 2\theta$
16. $r^2 = 9\cos 2\theta$
17. $r^2 = \cos 4\theta$
18. $r^2 = 3\sin 4\theta$
19. $r = \cos 2\theta$ (*four-leafed rose*)
20. $r = 2\sin 2\theta$ (*four-leafed rose*)

21. $r = \sin 3\theta$ *(three-leafed rose)*

22. $r = 2\cos 5\theta$ *(five-leafed rose)*

23. $r = 4 + 2\sin\theta$ *(limaçon with no inner loop)*

24. $r = 1.5 - \cos\theta$ *(limaçon with no inner loop)*

25. $r = 8\tan\theta$ *(kappa curve)*

26. $r = \csc\theta + 2$ *(conchoid of Nicomedes)*

In Exercises 27 and 28, match one of the graphs with each equation. Hint: *Compute the values of r when* $\theta = 0$, $\pi/2$, π, *and* $3\pi/2$.

27. **(a)** $r = 3 + 3\sin\theta$ **(b)** $r = 3 - 3\sin\theta$
(c) $r = 3 - 3\cos\theta$ **(d)** $r = 3 + 3\cos\theta$

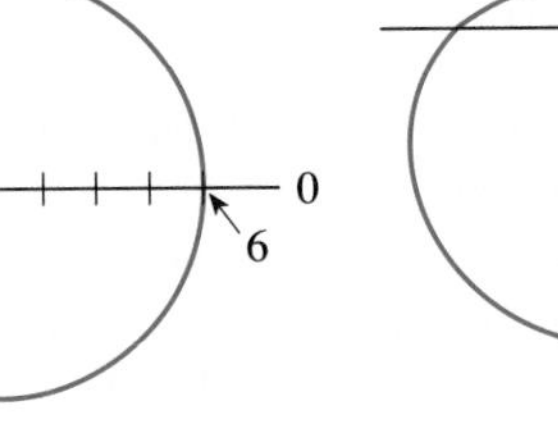

GRAPH A GRAPH B

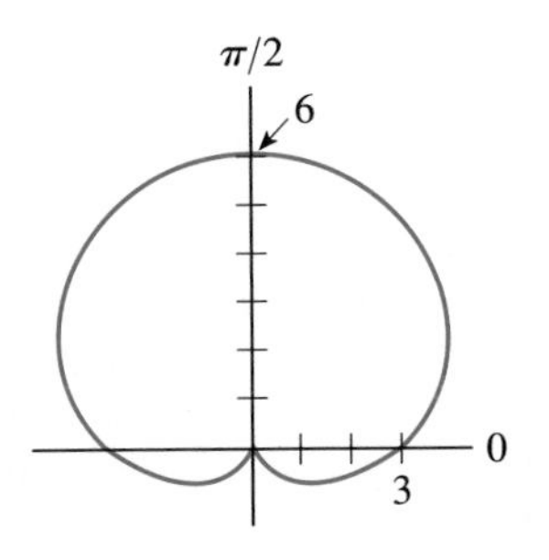

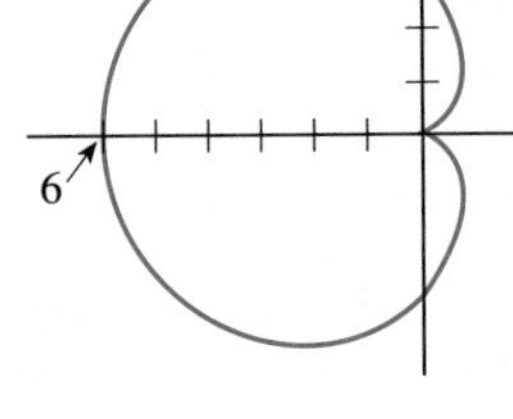

GRAPH C GRAPH D

28. **(a)** $r = 3 + 2\cos\theta$ **(b)** $r = 2 + 3\cos\theta$
(c) $r = 2 - 3\cos\theta$ **(d)** $r = 3 - 2\cos\theta$

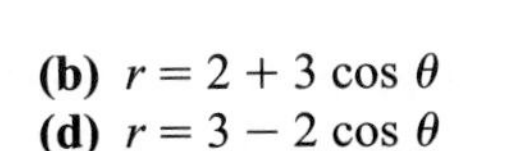

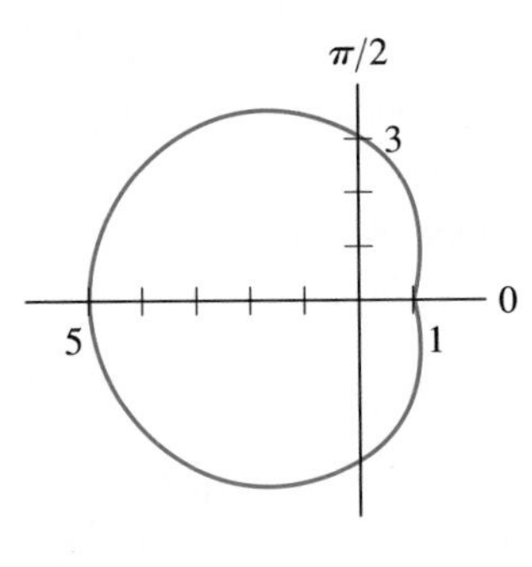

GRAPH A GRAPH B

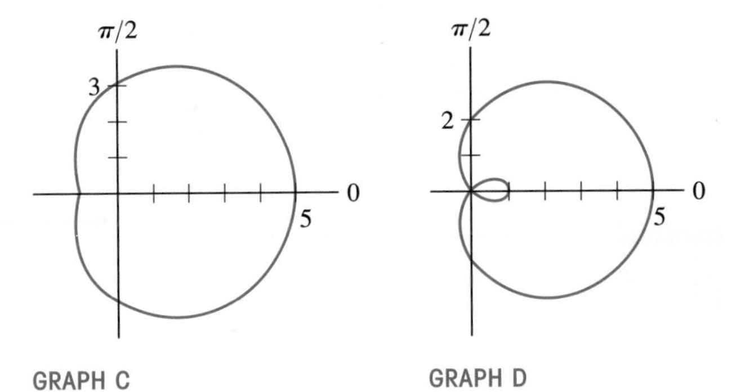

GRAPH C GRAPH D

B

29. The following figure shows a graph of the *logarithmic spiral* $\ln r = a\theta$ for $0 \leqslant \theta \leqslant 2\pi$. (In this equation, a denotes a positive constant.)

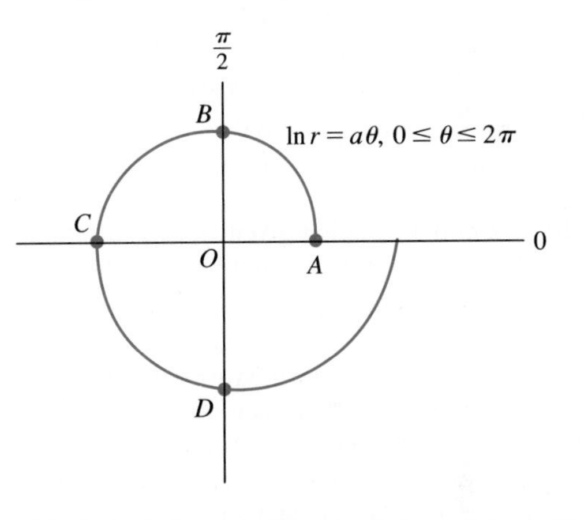

(a) Find the polar coordinates of the points A, B, C, and D, and then show that

$$\frac{OD}{OC} = \frac{OC}{OB} = \frac{OB}{OA} = e^{a\pi/2}$$

(b) Show that $\angle ABC$ and $\angle BCD$ are right angles. *Hint:* Use rectangular coordinates to compute slopes.

30. **(a)** Graph the polar curves $r = 2\cos\theta - 1$ and $r = 2\cos\theta + 1$. What do you observe?

(b) Part (a) shows that algebraically nonequivalent polar equations may have identical graphs. (This is another consequence of the fact that the polar coordinates of a point are not unique.) Show that both equations in part (a) can be written $(x^2 - 2x + y^2)^2 - x^2 - y^2 = 0$.

31. Follow Exercise 30 using the two polar equations $r = \cos(\theta/2)$ and $r = \sin(\theta/2)$.

32. Let k denote a positive constant, and let F_1 and F_2 denote the points with rectangular coordinates $(-k, 0)$ and $(k, 0)$, respectively. A curve known as the *lemniscate of Bernoulli* is defined as the set of points $P(x, y)$ such that $(F_1P) \times (F_2P) = k^2$.

(a) Show that the rectangular equation of the curve is $(x^2 + y^2)^2 = 2k^2(x^2 - y^2)$.

(b) Show that the polar equation is

$$r^2 = 2k^2 \cos 2\theta$$

(c) Graph the equation $r^2 = 2k^2 \cos 2\theta$.

GRAPHING UTILITY EXERCISES FOR SECTION 9.6

1. In the same picture, graph the four polar equations $r = 2$, $r = 4$, $r = 6$, and $r = 8$. Describe the graphs.

2. In the same picture, graph the polar equations $r = 1$ and $r \cos[\theta - (\pi/6)] = 1$. What do you observe? Check that your result is consistent with Figure 11 on page 582. *Hint:* Write the equation

$$r \cos[\theta - (\pi/6)] = 1 \text{ as } r = \frac{1}{\cos[\theta - (\pi/6)]}.$$

3. **(a)** Use one of the polar symmetry tests to show that the graph of $r = \cos^2 \theta - 2 \cos \theta$ is symmetric about the x-axis.

(b) Graph the equation given in part (a) and note that the curve is indeed symmetric about the x-axis.

4. As background for this exercise, read the caution note at the bottom of the box on page 587.

(a) Graph the polar equation $r = \sin^2 \theta - 2 \sin \theta$. What type of symmetry do you observe?

(b) Check (algebraically) that the equation $r = \sin^2 \theta - 2 \sin \theta$ satisfies symmetry test 3 on page 587, but not symmetry test 2.

In Exercises 5–14, graph the polar equations.

5. **(a)** $r = \cos 2\theta$ (*four-leafed rose*)
(b) $r = \sin 2\theta$

6. **(a)** $r = \cos 4\theta$ (*eight-leafed rose*)
(b) $r = \sin 4\theta$

7. **(a)** $r = \cos 3\theta$ (*three-leafed rose*)
(b) $r = \sin 3\theta$

8. **(a)** $r = \cos 5\theta$ (*five-leafed rose*)
(b) $r = \sin 5\theta$

9. $r = 4(1 + \cos \theta)$ (*cardioid*)

10. $r = \theta$ (*spiral of Archimedes*) *Suggestion:* Use a viewing rectangle extending from -30 to 30 in both the x- and the y-directions. Let θ run from 0 to 2π, then from 0 to 4π, and finally from 0 to 8π.

11. $r = 1/\theta$ (*hyperbolic spiral*) *Suggestion:* Use a viewing rectangle extending from -1 to 1 in both the x- and the y-directions. Let θ run from 0 to 2π, then from 0 to 4π, and finally from 0 to 8π.

12. $r = 1/\sqrt{\theta}$ (*spiral of Cotes,* or *lituus*) *Suggestion:* Use the guidelines in Exercise 11.

13. $r = \sqrt{\theta}$ (*spiral of Fermat*) *Suggestion:* Use a viewing rectangle extending from -5 to 5 in both the x- and the y-directions. Let θ run from θ to 2π, then from 0 to 4π, and finally from 0 to 8π.

14. $r^2 = 4 \cos 2\theta$ (*lemniscate of Bernoulli*) *Hint:* The equation is equivalent to the two equations $r = \pm 2\sqrt{\cos 2\theta}$. You'll find, however, that you need to consider only one of these.

15. The *limaçon of Pascal* is defined by the equation $r = a \cos \theta + b$. [The curve is named after Étienne Pascal (1588–1640), the father of Blaise Pascal. According to mathematics historian Howard Eves, however, the curve is misnamed; it appears earlier in the writings of Albrecht Dürer (1471–1528).] When $a = b$, the curve is a cardioid; when $a = 0$ the curve is a circle. Graph the following limaçons.

(a) $r = 2 \cos \theta + 1$ **(b)** $r = 2 \cos \theta - 1$
(c) $r = \cos \theta + 2$ **(d)** $r = \cos \theta - 2$
(e) $r = 2 \cos \theta + 2$ **(f)** $r = \cos \theta$

16. Graph the *cissoid of Diocles,* $r = 10 \sin \theta \tan \theta$. [The Greek mathematician Diocles (ca 180 B.C.) used the geometric properties of this curve to solve the problem of duplicating the cube. For an explanation of the problem and historical background, see the text by Howard Eves, *An Introduction to the History of Mathematics,* 6th ed., pp. 109–127 (Philadelphia: Saunders College Publishing, 1990).]

17. The polar curve $r = \cos^4(\theta/4)$ has a property that is difficult to detect without the aid of a graphing utility. Graph the equation in the standard viewing rectangle. Note that the curve appears to have a simple inner loop to the left of the origin. Now zoom in on the origin. What do you observe?

In Exercises 18–20, graph the polar curves.

18. **(a)** $r = 2|\sin \theta|^{\sin \theta}$ **(b)** $r = 2|\sin \theta|^{\sin(\theta/2)}$

19. $r = (\sin \theta)/\theta, \quad 0 < \theta \leq 4\pi$

20. Graph the polar curve $r = 1.5 \sin\left(\frac{30\theta}{17} + \frac{\pi}{30}\right) + 0.5$. Use a viewing rectangle extending from -2 to 2 in both the x- and the y-directions. Let θ run from 0 to 34π.

CHAPTER NINE SUMMARY OF PRINCIPAL FORMULAS AND NOTATION

FORMULAS OR NOTATIONS	PAGE REFERENCE	COMMENT
1. $\dfrac{\sin A}{a} = \dfrac{\sin B}{b} = \dfrac{\sin C}{c}$	539	This is the *law of sines*. In words it states that in any triangle, the sines of the angles are proportional to the lengths of the opposite sides.
2. $a^2 = b^2 + c^2 - 2bc \cos A$	541	This is the *law of cosines*. We can use it to calculate the third side of a triangle when we are given two sides and the included angle. The formula can also be used to calculate the angles of a triangle when we know the three sides, as in Example 5 on page 542.
3. Vector	553 561	Geometrically, a vector in the plane is a directed line segment. Vectors can be used to represent quantities that have both magnitude and direction, such as force and velocity. For examples, see Figures 1 and 2 in Section 9.2. Algebraically, a vector in the plane is an ordered pair of real numbers, denoted by $\langle x, y\rangle$; see Figure 3 in Section 9.3. The numbers x and y are called the *components* of the vector $\langle x, y\rangle$.
4. Vector equality	554 561	Geometrically, two vectors are said to be equal provided they have the same length and the same direction. In terms of components, this means that two vectors are equal if and only if their corresponding components are equal.
5. Vector addition	554 555 562	Geometrically, two vectors can be added by using the parallelogram law, as in Figure 10 in Section 9.2. Algebraically, this is equivalent to the following componentwise formula for vector addition: $$\langle a, b\rangle + \langle c, d\rangle = \langle a + c, b + d\rangle$$
6. $\lvert\mathbf{v}\rvert = \sqrt{v_1^2 + v_2^2}$	554 561	The expression on the left-hand side of the equation denotes the *length* or *magnitude* of the vector $\mathbf{v}$. The expression on the right-hand side of the equation tells us how to compute the length of $\mathbf{v}$ in terms of its respective x- and y-components v_1 and v_2.
7. $\overrightarrow{PQ} = \langle x_2 - x_1, y_2 - y_1\rangle$	562	This formula gives the components of the vector $\overrightarrow{PQ}$, where P and Q are the points $P(x_1, y_1)$ and $Q(x_2, y_2)$; see Figure 5 in Section 9.3 for a derivation of this formula.
8. The zero vector	563	The zero vector, denoted by $\mathbf{0}$, is the vector $\langle 0, 0\rangle$. The zero vector plays the same role in vector addition as does the real number zero in addition of real numbers. For any vector $\mathbf{v}$, we have $$\mathbf{0} + \mathbf{v} = \mathbf{v} + \mathbf{0} = \mathbf{v}$$
9. Scalar multiplication	563	For each real number k and each vector $\mathbf{v} = \langle x, y\rangle$, the vector $k\mathbf{v}$ is defined by the equation $$k\mathbf{v} = \langle kx, ky\rangle$$ Geometrically, the length of $k\mathbf{v}$ is $\lvert k\rvert$ times the length of $\mathbf{v}$. The operation that forms the vector $k\mathbf{v}$ from the scalar k and the vector $\mathbf{v}$ is called scalar multiplication.
10. $\mathbf{i} = \langle 1, 0\rangle$ $\mathbf{j} = \langle 0, 1\rangle$	565	These two equations define the *unit vectors* $\mathbf{i}$ and $\mathbf{j}$; see Figure 11 in Section 9.3. Any vector $\langle x, y\rangle$ can be expressed in terms of the unit vectors $\mathbf{i}$ and $\mathbf{j}$ as follows: $$\langle x, y\rangle = x\mathbf{i} + y\mathbf{j}$$

FORMULAS OR NOTATIONS	PAGE REFERENCE	COMMENT
11. Parametric equations	568 569	Suppose the x- and y-coordinates of the points on a curve are expressed as functions of another variable t: $x = f(t)$ and $y = g(t)$. These equations are parametric equations of the curve, and t is called a parameter. An example: A pair of parametric equations for the ellipse $(x^2/a^2) + (y^2/b^2) = 1$ is $x = a \cos t$ and $y = b \sin t$.
12. $x = r \cos \theta$ $y = r \sin \theta$ $x^2 + y^2 = r^2$ $\tan \theta = y/x$	577	These formulas relate the rectangular coordinates of a point (x, y) to its polar coordinates (r, θ). See the boxed figure on page 577.
13. $d^2 = r_1^2 + r_2^2 - 2r_1r_2 \cos(\theta_2 - \theta_1)$	579	This is the formula for the (square of the) distance d between the points with polar coordinates (r_1, θ_1) and (r_2, θ_2).
14. $r^2 + r_0^2 - 2rr_0 \cos(\theta - \theta_0) = a^2$	580	This is a polar equation for the circle with center (r_0, θ_0) and radius a.
15. $r \cos(\theta - \alpha) = d$	581	This is a polar equation for a line not passing through the origin. In the equation, d and α are constants; d is the perpendicular distance from the origin to the line, and the (polar) point (d, α) is the foot of the perpendicular from the origin to the line.

WRITING MATHEMATICS

A regular feature in *Mathematics Magazine* is the "Proof Without Words" section. The idea is to present a proof of a well-known theorem in such a way that the entire proof is "obvious" from a well-chosen picture and at most several equations. Of course, what is obvious to a mathematician may not be so obvious to her or his students (as you well know).

The following clever proof without words for the law of cosines was developed by Professor Sidney H. Kung of Jacksonville University. (It was published in *Mathematics Magazine,* vol. 63, no. 5, December 1990.) Study the proof (or discuss it with friends) until you see how it works. Then, in a paragraph or two, write out the details of the proof as if you were explaining it to a classmate. (To facilitate your explanation, you will probably need to label some of the points in the given figure.)

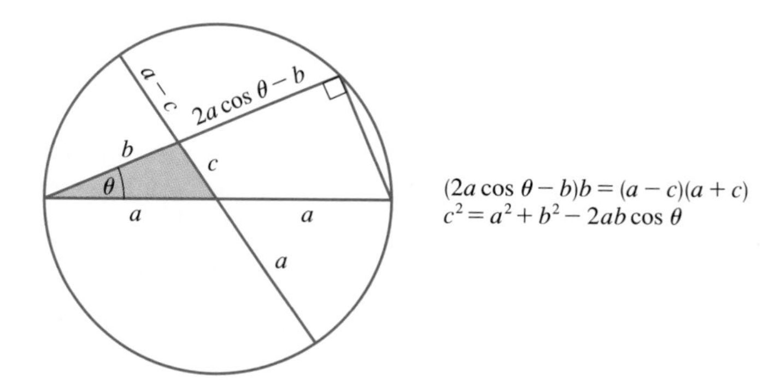

Proof without words for the law of cosines, $\theta < 90°$ (by Professor Sidney H. Kung).

CHAPTER NINE REVIEW EXERCISES

In Exercises 1–8, the lengths of the three sides of a triangle are denoted by a, b, and c; the angles opposite these sides are A, B, and C, respectively. In each exercise, use the given data to find the remaining sides and angles. Use a calculator and round the final answers to one decimal place.

1. $\angle A = 40°$, $\angle B = 85°$, $c = 16$ cm
2. $\angle C = 84°$, $\angle B = 16°$, $a = 9$ cm
3. **(a)** $a = 8$ cm, $b = 9$ cm, $\angle A = 52°$, $\angle B < 90°$
   (b) Use the data in part (a), but assume $\angle B > 90°$.
4. **(a)** $a = 6.25$ cm, $b = 9.44$ cm, $\angle A = 12°$, $\angle B < 90°$
   (b) Use the data in part (a), but assume $\angle B > 90°$.
5. $a = 18$ cm, $b = 14$ cm, $\angle C = 24°$
6. $a = 32.16$ cm, $b = 50.12$ cm, $\angle C = 156°$
7. $a = 4$ cm, $b = 7$ cm, $c = 9$ cm
8. $a = 12.61$ cm, $b = 19.01$ cm, $c = 14.14$ cm

In Exercises 9–18, refer to the following figure. In each case, determine the indicated quantity. Use a calculator and round your result to two decimal places.

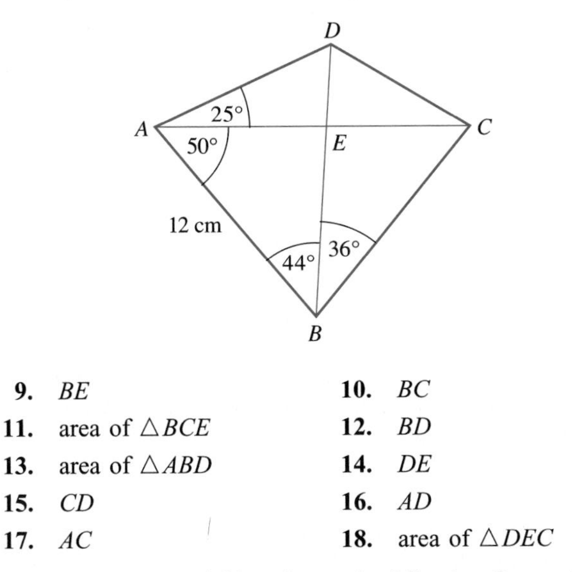

9. BE
10. BC
11. area of $\triangle BCE$
12. BD
13. area of $\triangle ABD$
14. DE
15. CD
16. AD
17. AC
18. area of $\triangle DEC$

In Exercises 19 and 20, refer to the following figure, in which $\overline{BD}$ bisects $\angle ABC$.

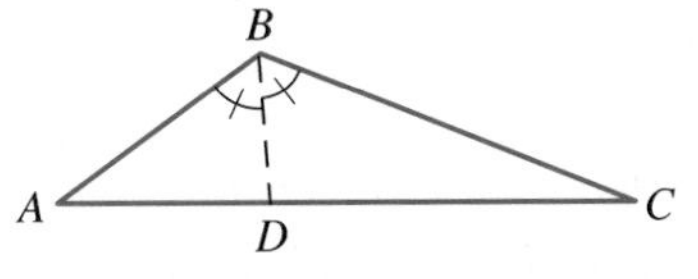

19. If $AB = 24$ cm, $BC = 40$ cm, and $AC = 56$ cm, find the length of the angle bisector $\overline{BD}$.
20. If $AB = 105$ cm, $BC = 120$ cm, and $AC = 195$ cm, find the length of the angle bisector $\overline{BD}$.

In Exercises 21 and 22, refer to the following figure, in which $AD = DC$.

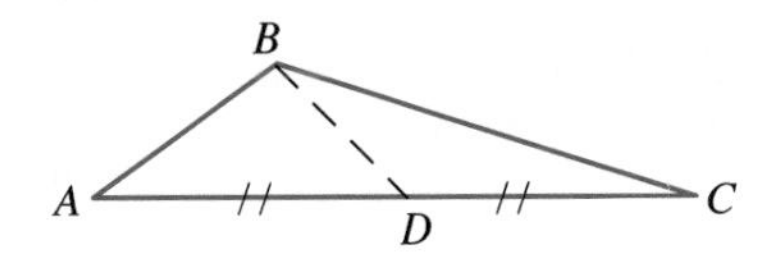

21. If $AB = 12$ cm, $BC = 26$ cm, and $AD = DC = 17$ cm, find BD.
22. If $AB = 16$ cm, $BC = 22$ cm, and $AD = DC = 17$ cm, find BD.
23. In $\triangle ABC$, $a = 4$, $b = 5$, and $c = 6$.
   (a) Find $\cos A$ and $\cos C$.
   (b) Using the results in part (a), show that angle C is twice angle A.
24. In $\triangle ABC$, suppose that angle C is twice angle A. Show that $ab = c^2 - a^2$.

In Exercises 25–30, refer to the framework shown in the accompanying figure. (Assume that the figure is drawn to scale.) In the figure, $\overline{AC}$ is parallel to $\overline{ED}$ and $BE = BD$. In each case, compute the indicated quantity. Round the final answers to two decimal places.

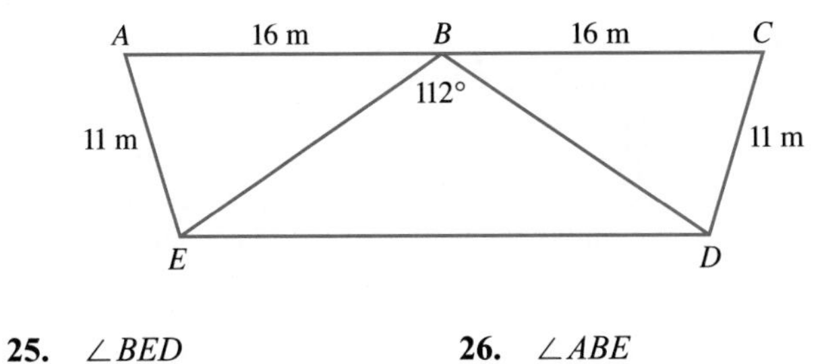

25. $\angle BED$
26. $\angle ABE$
27. $\angle AEB$
28. $\angle BAE$
29. BE
30. DE
31. The given figure shows a regular pentagon inscribed in a circle. The radius of the circle is 10 cm. Determine the following lengths, rounding the answers to two decimal places.
   (a) one side of the pentagon; **(b)** AC.

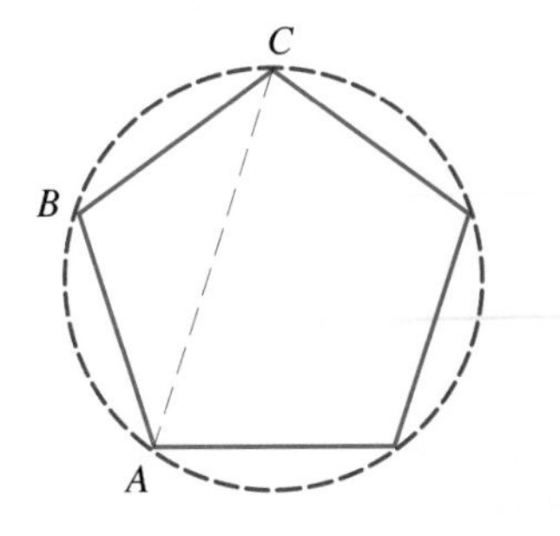

32. A pilot on a training flight is supposed to leave the airport and fly for 100 miles in the direction N40°E. Then he is supposed to turn around and fly directly back to the airport. By error, however, he makes the 100-mile return trip in the direction S25°W. How far is he now from the airport, and in what direction must he fly? (Round the answers to one decimal place.)

33. The following figures show the same circle of diameter D inscribed first in an equilateral triangle and then in a regular hexagon. Let a denote the length of a side of the triangle and let b denote the length of a side of the hexagon. Show that $ab = D^2$.

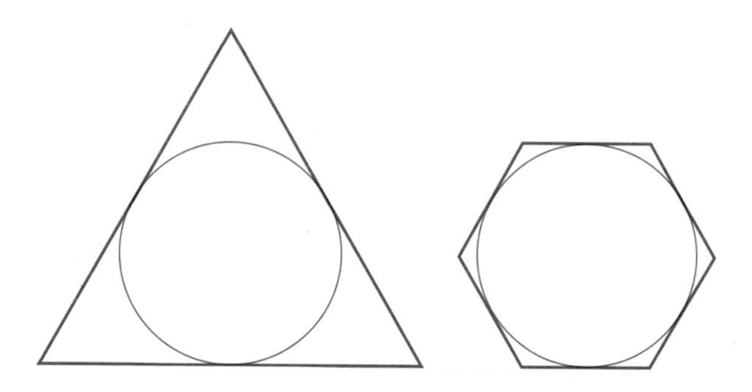

34. Suppose that $\triangle ABC$ is a right triangle with the right angle at C. Use the law of cosines to prove the following statements.

(a) The square of the distance from C to the midpoint of the hypotenuse is equal to one-fourth the square of the hypotenuse.

(b) The sum of the squares of the distances from C to the two points that trisect the hypotenuse is equal to five-ninths the square of the hypotenuse.

(c) Let P, Q, and R be points on the hypotenuse such that $AP = PQ = QR = RB$. Derive a result similar to your results in parts (a) and (b) for the sum of the squares of the distances from C to P, Q, and R.

35. The perimeter and the area of the triangle in the figure are 20 cm and $10\sqrt{3}$ cm^2, respectively. Find a and b. (Assume $a < b$.)

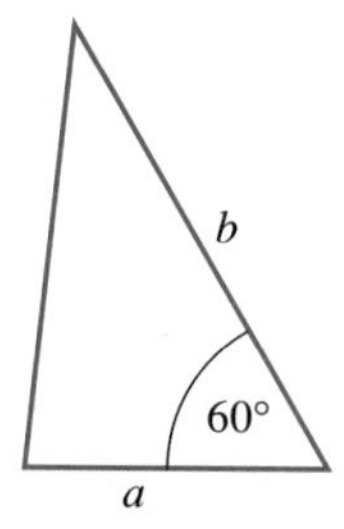

36. In $\triangle ABC$, suppose that $\angle B - \angle A = 90°$.

(a) Show that $c^2 = (b^2 - a^2)^2/(a^2 + b^2)$. *Hint:* Let P be the point on $\overline{AC}$ such that $\angle CBP = 90°$ and $\angle PBA = \angle A$. Show that $CP = (a^2 + b^2)/2b$.

(b) Show that $\dfrac{2}{c^2} = \dfrac{1}{(b+a)^2} + \dfrac{1}{(b-a)^2}$.

37. Two forces **F** and **G** act on an object, **G** horizontally to the right with a magnitude of 15 N, and **F** vertically upward with a magnitude of 20 N. Determine the magnitude and direction of the resultant force. (Use a calculator to determine the angle between the horizontal and the resultant; round the result to one decimal place.)

38. Determine the resultant of the two forces in the accompanying figure. (Use a calculator, and round the values you obtain for the magnitude and direction to one decimal place.)

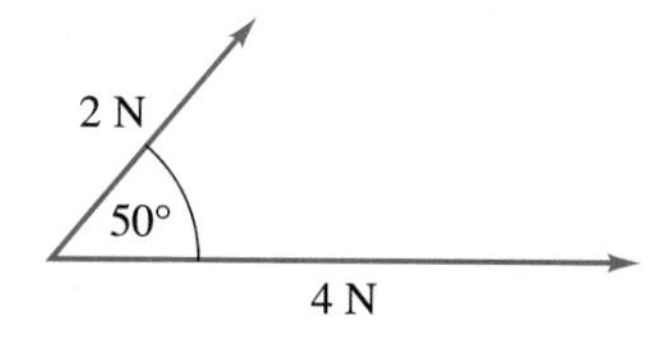

39. Determine the horizontal and vertical components of the velocity vector **v** in the following figure. (Use a calculator and round your answers to one decimal place.)

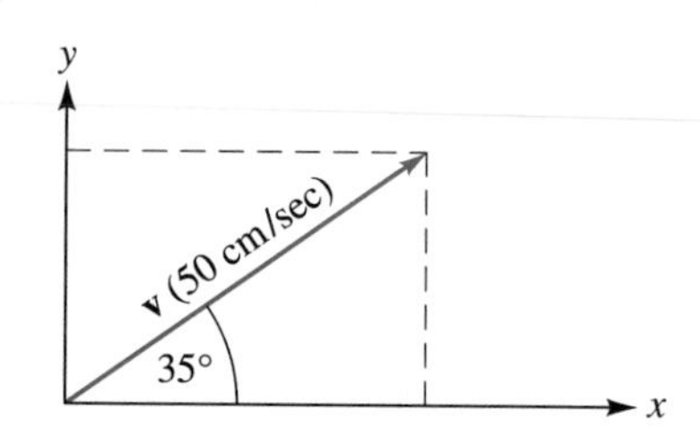

40. The heading and air speed of an airplane are 50° and 220 mph, respectively. If the wind is 60 mph from 140°, find the ground speed, the drift angle, and the course. (Use a calculator and round your answers to one decimal place.)

41. A block rests on an inclined plane that makes an angle of 24° with the horizontal. The component of the weight parallel to the plane is 14.8 lb. Determine the weight of the block and the component of the weight perpendicular to the plane. (Use a calculator and round your answers to one decimal place.)

42. Find the length of the vector $\langle 20, 99\rangle$.

43. For which values of b will the vectors $\langle 2, 6\rangle$ and $\langle -5, b\rangle$ have the same length?

44. Given the points $A(2, 6)$ and $B(-7, 4)$, find the components of the following vectors. Write your answers in the form $\langle x, y\rangle$.

(a) $\overrightarrow{AB}$ **(b)** $\overrightarrow{BA}$ **(c)** $3\overrightarrow{AB}$ **(d)** $\dfrac{1}{|\overrightarrow{AB}|}\overrightarrow{AB}$

In Exercises 45–56, compute each of the indicated quantities, given that the vectors **a, b, c,** *and* **d** *are defined as follows:*

$$\mathbf{a} = \langle 3, 5\rangle \quad \mathbf{b} = \langle 7, 4\rangle \quad \mathbf{c} = \langle 2, -1\rangle \quad \mathbf{d} = \langle 0, 3\rangle$$

45. $\mathbf{a} + \mathbf{b}$ **46.** $\mathbf{b} - \mathbf{d}$

47. $3\mathbf{c} + 2\mathbf{a}$ **48.** $|\mathbf{a}|$

49. $|\mathbf{b} + \mathbf{d}|^2 - |\mathbf{b} - \mathbf{d}|^2$ **50.** $\mathbf{a} + (\mathbf{b} + \mathbf{c})$

51. $(\mathbf{a} + \mathbf{b}) + \mathbf{c}$ **52.** $\mathbf{a} - (\mathbf{b} - \mathbf{c})$

53. $(\mathbf{a} - \mathbf{b}) - \mathbf{c}$

54. $|\mathbf{a} + \mathbf{b}|^2 + |\mathbf{a} - \mathbf{b}|^2 - 2|\mathbf{a}|^2 - 2|\mathbf{b}|^2$

55. $4\mathbf{c} + 2\mathbf{a} - 3\mathbf{b}$ **56.** $|4\mathbf{c} + 2\mathbf{a} - 3\mathbf{b}|$

57. Express the vector $\langle 7, -6\rangle$ in terms of the unit vectors **i** and **j**.

58. Express the vector $4\mathbf{i} - 6\mathbf{j}$ in the form $\langle a, b\rangle$.

59. Find a unit vector having the same direction as $\langle 6, 4\rangle$.

60. Find two unit vectors that are perpendicular to $\langle \cos\theta, \sin\theta\rangle$. Simplify the components as much as possible.

In Exercises 61 and 62, compute the distance between the points with the given polar coordinates. Use a calculator and round the final answers to two decimal places.

61. $(3, \pi/12)$ and $(2, 17\pi/18)$ **62.** $(1, 1)$ and $(3, 2)$

In Exercises 63 and 64, determine a polar equation for the circle satisfying the given conditions.

63. The radius is 3 and the polar coordinates of the center are $(5, \pi/6)$.

64. The radius is 1 and the polar coordinates of the center are $(-3, \pi/4)$.

In Exercises 65 and 66, the polar equation of a line is given. In each case, **(a)** *specify the perpendicular distance from the origin to the line;* **(b)** *determine the polar coordinates of the points on the line corresponding to $\theta = 0$ and $\theta = \pi/2$;* **(c)** *specify the polar coordinates of the foot of the perpendicular from the origin to the line; and* **(d)** *sketch the line.*

65. $r\cos\left(\theta - \frac{\pi}{3}\right) = 3$ **66.** $r\cos(\theta - 1) = \sqrt{5}$

In Exercises 67–74, graph the polar equations.

67. **(a)** $r = 2 - 2\cos\theta$ **(b)** $r = 2 - 2\sin\theta$

68. **(a)** $r^2 = 4\cos 2\theta$ **(b)** $r^2 = 4\sin 2\theta$

69. **(a)** $r = 2\cos\theta - 1$ **(b)** $r = 2\sin\theta - 1$

70. **(a)** $r = 2\cos\theta$ **(b)** $r = 2\cos\theta + 1$

71. **(a)** $r = 4\sin 2\theta$ **(b)** $r = 4\cos 2\theta$

72. **(a)** $r = 4\sin 3\theta$ **(b)** $r = 4\cos 3\theta$

73. **(a)** $r = 1 + 2\sin(\theta/2)$ **(b)** $r = 1 - 2\cos(\theta/2)$

74. **(a)** $r = (1.5)^\theta \quad (\theta \geqslant 0)$ **(b)** $r = (1.5)^\theta \quad (\theta \leqslant 0)$

In Exercises 75–80, graph the curve (or line) determined by the parametric equations. Indicate the direction of travel along the curve as t increases.

75. $x = 3 - 5t,\ y = 1 + t$ **76.** $x = t - 1,\ y = 3t$

77. $x = 3\sin t,\ y = 6\cos t$

78. $x = 2\cos t,\ y = 2\sin t$

79. $x = 4\sec t,\ y = 3\tan t$

80. $x = 1 + 3\cos t,\ y = 2 + 4\sin t$

81. In the accompanying figure, P and Q are two points on the polar curve $r = (\sin\theta)/\theta$ such that $\angle AOP = \angle POQ$. Follow steps (a) through (f) to prove that $\overline{PQ} \perp \overline{OQ}$.

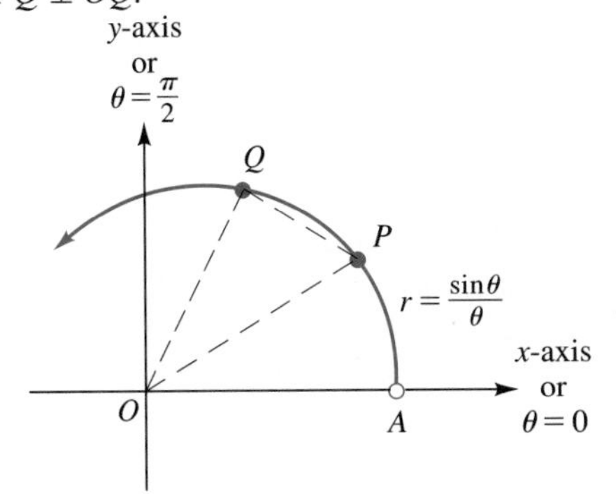

(a) Let $\alpha = \angle AOP = \angle POQ$. What are the polar coordinates of P and Q in terms of α?

(b) What are the rectangular coordinates of P and Q in terms of α?

(c) Using the coordinates determined in part (b), show that the slope of $\overline{OQ}$ is $\tan 2\alpha$.

(d) Using the coordinates determined in part (b), show that the slope of $\overline{PQ}$ is

$$\frac{\sin^2 2\alpha - 2\sin^2\alpha}{\sin 2\alpha\cos 2\alpha - 2\sin\alpha\cos\alpha}$$

(e) Show that the expression in part (d) can be simplified to $-\cot 2\alpha$. *Hint:* In the numerator, replace $\sin^2 2\alpha$ by $1 - \cos^2 2\alpha$.

(f) Use the results in parts (c) and (e) to explain why $\overline{PQ} \perp \overline{OQ}$.

82. In this exercise you will use the following figure to derive the *Mollweide formula*:

$$\frac{a + b}{c} = \frac{\cos\frac{1}{2}(A - B)}{\sin\frac{1}{2}C}$$

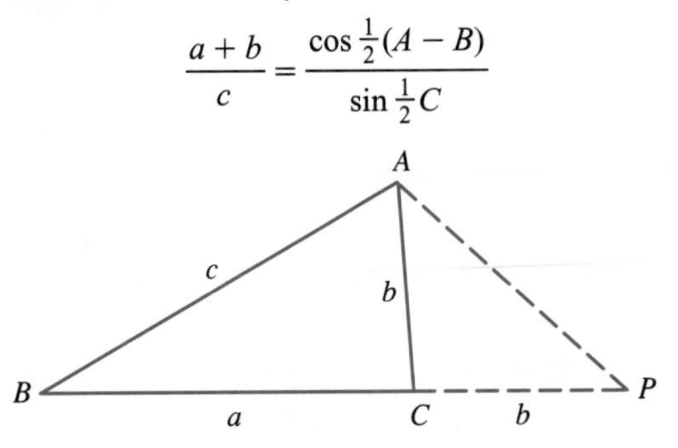

The figure is constructed as follows. Starting with $\triangle ABC$, extend side $\overline{BC}$ so that $CP = b$, as shown. Then draw line segment $\overline{AP}$.

(a) Show that $\angle APC = \angle PAC = \frac{1}{2}C$.

(b) Show that $\angle BAP = 90° + \frac{1}{2}(A - B)$. *Hint:* Start with the fact that $\angle BAP = A + \frac{1}{2}C$.

(c) Use the law of sines now in $\triangle ABP$ to show that

$$\frac{\sin[90° + \frac{1}{2}(A - B)]}{a + b} = \frac{\sin \frac{1}{2}C}{c}$$

(d) Use the result in part (c) to obtain the required identity.

83. In this exercise, we use the law of sines to deduce the **law of tangents** for $\triangle ABC$:

$$\frac{a - b}{a + b} = \frac{\tan \frac{1}{2}(A - B)}{\tan \frac{1}{2}(A + B)}$$

This law was given by the Danish physician and mathematician Thomas Fink in his text *Geometria Rotundi*, published in Basel in 1583. (Our use of the terms "tangent" and "secant" is also due to Fink.)

(a) Suppose that a, b, x, and y are real numbers such that $\dfrac{a}{b} = \dfrac{x}{y}$. Verify that $\dfrac{a - b}{a + b} = \dfrac{x - y}{x + y}$.

(b) Use the law of sines and the result in part (a) to show that

$$\frac{a - b}{a + b} = \frac{\sin A - \sin B}{\sin A + \sin B}$$

(c) Use the result in part (a) and the sum-to-product formulas from Section 8.3 to complete the derivation of the law of tangents.

84. The following figure shows a sector of the unit circle $x^2 + y^2 = 1$. The central angle for the sector is α (a constant). The point $P(\cos\theta, \sin\theta)$ denotes an arbitrary point on the arc of the sector. From P, perpendiculars are drawn to the sides of the sector, meeting the sides at Q and R, as shown. Show that the distance from Q to R does not depend upon θ.

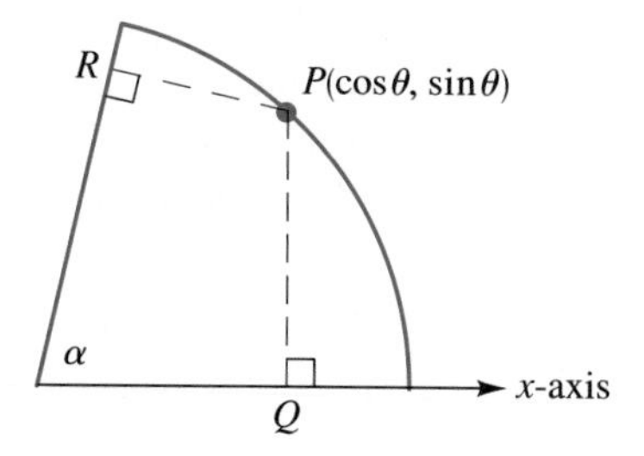

CHAPTER NINE TEST

1. In triangle ABC, let $A = 120°$, $b = 5$ cm, and $c = 3$ cm. Find a.
2. The sides of a triangle are 2 cm, 3 cm, and 4 cm. Determine the cosine of the angle opposite the longest side. On the basis of your answer, explain whether or not the angle opposite the longest side is an acute angle.
3. Two of the angles in a triangle are 30° and 45°. If the side opposite the 45° angle is $20\sqrt{2}$ cm, find the side opposite the 30° angle.
4. Each side of the square $STUV$ is 8 cm long. The point P lies on diagonal $\overline{SU}$ such that $SP = 2$ cm. Find the distance from P to V.
5. In $\triangle ABC$, $b = 5.8$ cm, $c = 3.2$ cm, and $A = 27°$. Find the remaining sides and angles of the triangle.
6. Two forces **F** and **G** act on an object. The force **G** acts horizontally with a magnitude of 2 N and **F** acts vertically upward with a magnitude of 4 N.

   (a) Find the magnitude of the resultant.

   (b) Find $\tan\theta$, where θ is the angle between **G** and the resultant.

7. Two forces act on an object, as shown in the figure at the left.

   (a) Find the magnitude of the resultant. (Leave your answer in terms of radicals and the trigonometric functions.)

   (b) Find $\sin\theta$, where θ is the angle between the 12 N force and the resultant.

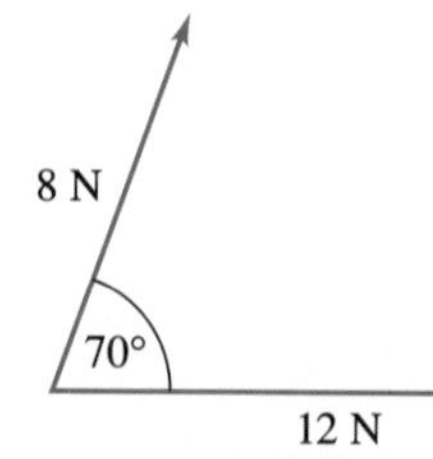

8. The heading and air speed of an airplane are 40° and 300 mph, respectively. If the wind is 50 mph from 130°, find the ground speed and the tangent of the drift angle. (Leave your answer in terms of radicals and the trigonometric functions.)

9. Let $\mathbf{A} = \langle 2, 4 \rangle$, $\mathbf{B} = \langle 3, -1 \rangle$, and $\mathbf{C} = \langle 4, -4 \rangle$.
   (a) Find $2\mathbf{A} + 3\mathbf{B}$.
   (b) Find $|2\mathbf{A} + 3\mathbf{B}|$.
   (c) Express $\mathbf{C} - \mathbf{B}$ in terms of $\mathbf{i}$ and $\mathbf{j}$.

10. Let P and Q be the points $(4, 5)$ and $(-7, 2)$, respectively. Find a unit vector having the same direction as $\overrightarrow{PQ}$.

11. Convert the polar equation $r^2 = \cos 2\theta$ to rectangular form.

12. Graph the equation $r = 2(1 - \cos \theta)$.

13. Given the parametric equations $x = 4 \sin t$ and $y = 2 \cos t$, eliminate the parameter t and sketch the graph.

14. Compute the distance between the two points with polar coordinates $(4, 10\pi/21)$ and $(1, \pi/7)$.

15. Find a polar equation for the circle with center (in polar coordinates) $(5, \pi/2)$ and with radius 2. Does the (polar) point $(2, \pi/6)$ lie on this circle?

16. Refer to the following figure.

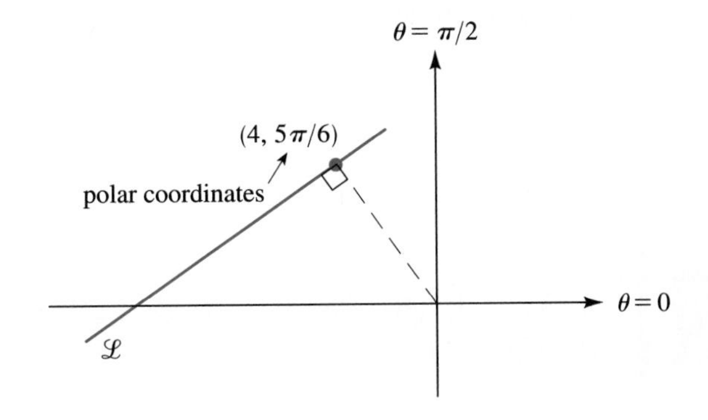

   (a) Determine a polar equation for the line $\mathscr{L}$ in the figure.
   (b) Use an addition formula to show that the polar equation of the line can be written $-r\sqrt{3} \cos \theta + r \sin \theta - 8 = 0$.

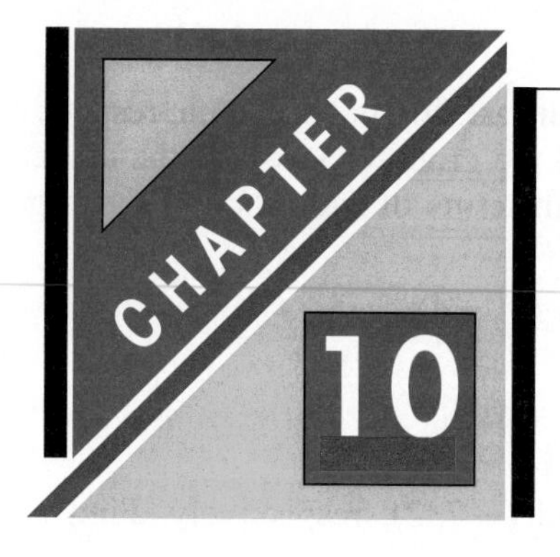

SYSTEMS OF EQUATIONS

INTRODUCTION

Many problems in a variety of disciplines give rise to ... [linear] systems. For example, in physics, in order to find the currents in an electrical circuit containing known voltages and resistances, a system of linear equations must be solved. In chemistry, the balancing of chemical equations requires the solution of a system of linear equations. And in economics, the Leontief input–output model reduces problems concerning the production and consumption of goods to systems of linear equations.

Leslie Hogben in *Elementary Linear Algebra* (St. Paul: West Publishing Company, 1987)

In this chapter we consider systems of equations. Roughly speaking, a system of equations is just a collection of equations with a common set of unknowns. In solving such systems, we try to find values for the unknowns that simultaneously satisfy each of the equations in the system. In Section 10.1 we review two techniques that are often presented in intermediate algebra for solving systems involving two linear equations in two unknowns. An important technique for solving larger systems of equations is developed in Section 10.2; this technique is known as *Gaussian elimination.* After that, in Section 10.3, we introduce matrices as a tool for reducing the amount of bookkeeping involved in Gaussian elimination. Two additional techniques for solving systems of linear equations are explained in Sections 10.4 and 10.5. The technique in Section 10.4 uses the idea of an inverse matrix; the technique in Section 10.5 involves determinants and Cramer's rule. Section 10.6 contains a brief discussion of nonlinear systems of equations. In the last section of this chapter, Section 10.7, we consider systems of inequalities.

SYSTEMS OF TWO LINEAR EQUATIONS IN TWO UNKNOWNS

Both in theory and in applications, it's often necessary to solve two equations in two unknowns. You may have been introduced to the idea of simultaneous equations in a previous course in algebra; however, to put matters on a firm foundation, we begin here with the basic definitions. By a **linear equation in two variables** we mean an equation of the form

$$ax + by = c$$

where the constants a and b are not both zero. The two variables needn't always be denoted by the letters x and y, of course; it is the *form* of the equation that matters. Table 1 displays some examples.

TABLE 1

Equations in Two Variables	Is It Linear? Yes	No
$3x - 8y = 12$	√	
$-s + 4t = 0$	√	
$2x - 3y^2 = 1$		√
$y = 4 - 2x$	√	
$\frac{4}{u} + \frac{5}{v} = 3$		√

An ordered pair of numbers (x_0, y_0) is said to be a **solution of the linear equation** $ax + by = c$ provided we obtain a true statement when we replace x and y in the equation by x_0 and y_0, respectively. For example, the ordered pair (3, 2) is a solution of the equation $x - y = 1$, since $3 - 2 = 1$. On the other hand, (2, 3) is not a solution of $x - y = 1$, since $2 - 3 \neq 1$.

Now consider a **system** of two linear equations in two unknowns:

$$\begin{cases} ax + by = c \\ dx + ey = f \end{cases}$$

If we can find an ordered pair that is a solution to both equations, then we say that ordered pair is a **solution of the system.** Sometimes, to emphasize the fact that a solution must satisfy both equations, we refer to the system as a pair of **simultaneous equations.** A system that has at least one solution is said to be **consistent.** If there are no solutions, the system is **inconsistent.**

EXAMPLE 1 Consider the system

$$\begin{cases} x + \ \ y = 2 \\ 2x - 3y = 9 \end{cases}$$

(a) Is (1, 1) a solution of the system?
(b) Is (3, −1) a solution of the system?

Solution

(a) Although (1, 1) is a solution of the first equation, it is not a solution of the system because it does not satisfy the second equation. (Check this for yourself.)

(b) (3, −1) satisfies the first equation:

$$3 + (-1) \stackrel{?}{=} 2 \qquad \text{True}$$

(3, −1) satisfies the second equation:

$$2(3) - 3(-1) \stackrel{?}{=} 9$$
$$6 + 3 \stackrel{?}{=} 9 \qquad \text{True}$$

Since (3, −1) satisfies both equations, it is a solution of the system.

We can gain an important perspective on systems of linear equations by looking at Example 1 in graphical terms. Table 2 shows how each of the state-

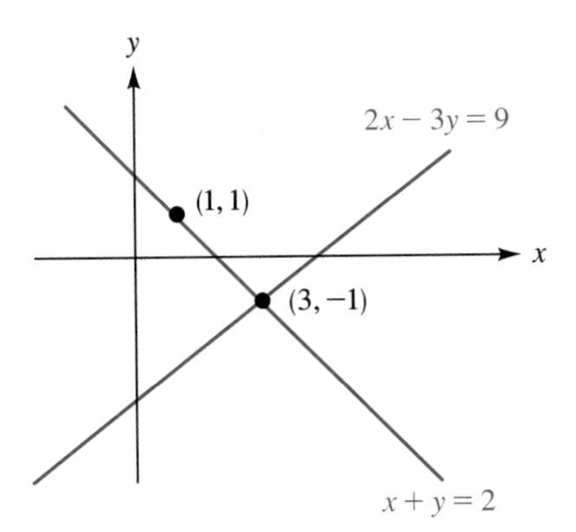

FIGURE 1

TABLE 2

Algebraic Idea	Corresponding Geometric Idea
1. The ordered pair (1, 1) is a solution of the equation $x + y = 2$.	1. The point (1, 1) lies on the line $x + y = 2$. See Figure 1.
2. The ordered pair (1, 1) is not a solution of the equation $2x - 3y = 9$.	2. The point (1, 1) does not lie on the line $2x - 3y = 9$. See Figure 1.
3. The ordered pair (3, −1) is a solution of the system $\begin{cases} x + \ \ y = 2 \\ 2x - 3y = 9 \end{cases}$	3. The point (3, −1) lies on both of the lines $x + y = 2$ and $2x - 3y = 9$. See Figure 1.

ments in that example can be rephrased using the geometric ideas with which we are already familiar.

In Example 1, we verified that $(3, -1)$ is a solution of the system

$$\begin{cases} x + \ \ y = 2 \\ 2x - 3y = 9 \end{cases}$$

Are there any other solutions of this particular system? No: Figure 1 shows us that there are no other solutions, since $(3, -1)$ is clearly the only point common to both lines. In a moment we'll look at two important methods for solving systems of linear equations in two unknowns. But even before we consider these methods, we can say something about the solutions of linear systems.

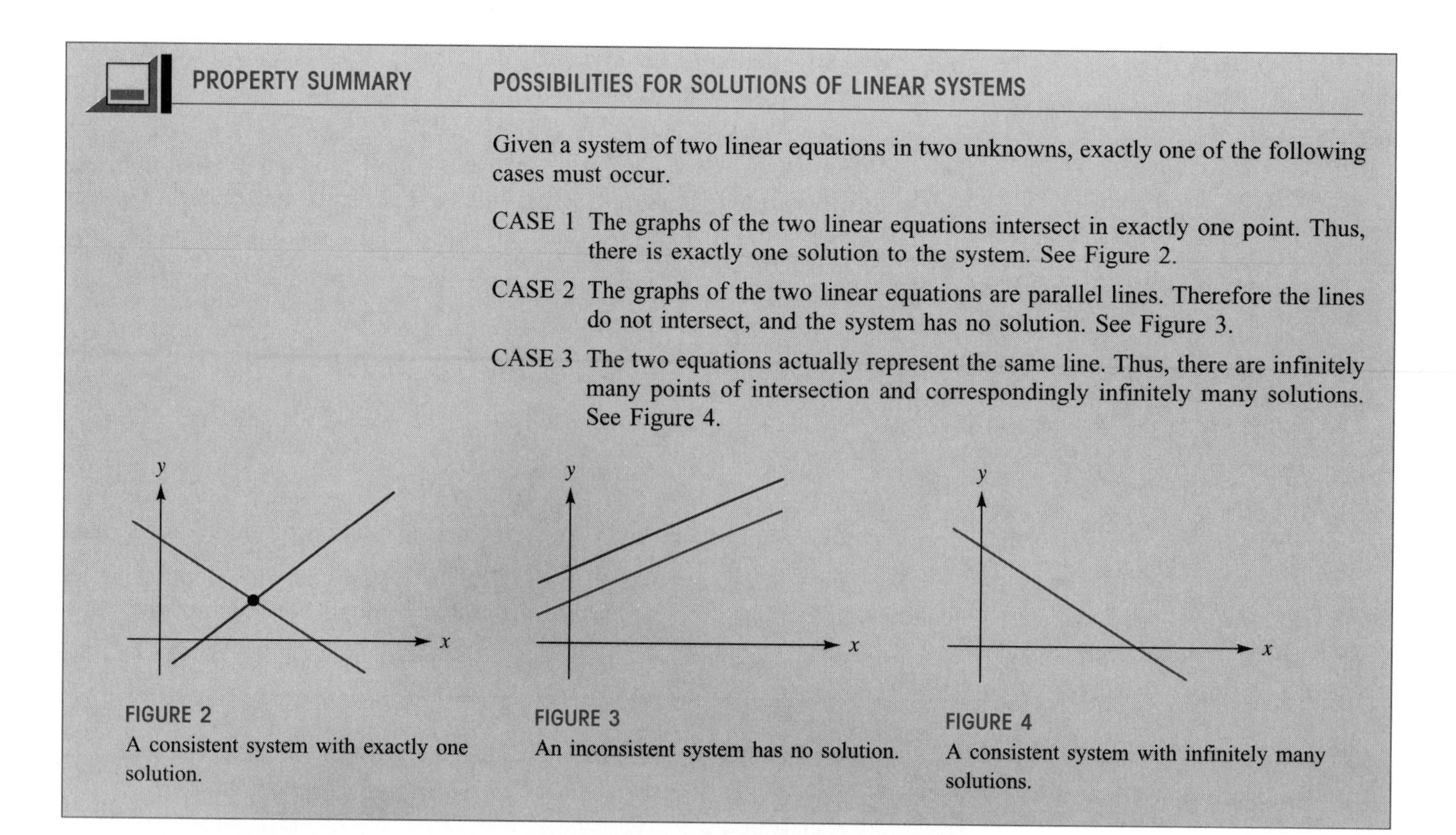

PROPERTY SUMMARY

POSSIBILITIES FOR SOLUTIONS OF LINEAR SYSTEMS

Given a system of two linear equations in two unknowns, exactly one of the following cases must occur.

CASE 1 The graphs of the two linear equations intersect in exactly one point. Thus, there is exactly one solution to the system. See Figure 2.

CASE 2 The graphs of the two linear equations are parallel lines. Therefore the lines do not intersect, and the system has no solution. See Figure 3.

CASE 3 The two equations actually represent the same line. Thus, there are infinitely many points of intersection and correspondingly infinitely many solutions. See Figure 4.

FIGURE 2
A consistent system with exactly one solution.

FIGURE 3
An inconsistent system has no solution.

FIGURE 4
A consistent system with infinitely many solutions.

We are going to review two methods from basic algebra for solving systems of two linear equations in two unknowns. These methods are the **substitution method** and the **addition–subtraction method.** We'll begin by demonstrating the substitution method. Consider the system

$$\begin{cases} 3x + 2y = 17 & (1) \\ 4x - 5y = -8 & (2) \end{cases}$$

We first choose one of the two equations and then use it to express one of the variables in terms of the other. In the case at hand, neither equation appears particularly simpler than the other, so let's just start with the first equation and solve for x in terms of y. We have

$$\begin{aligned} 3x &= 17 - 2y \\ x &= \tfrac{1}{3}(17 - 2y) \end{aligned} \tag{3}$$

Now we use equation (3) to substitute for x in the equation that we have not yet used, namely, equation (2). This yields

$$\begin{aligned} 4\left[\tfrac{1}{3}(17 - 2y)\right] - 5y &= -8 \\ 4(17 - 2y) - 15y &= -24 \qquad \text{multiplying by 3} \\ -23y &= -92 \\ y &= 4 \end{aligned}$$

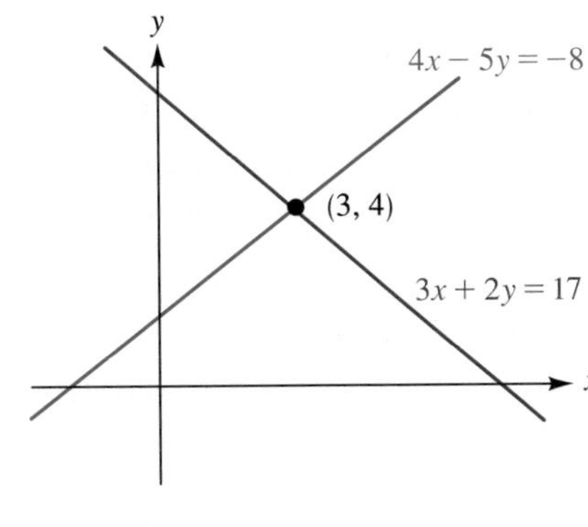

FIGURE 5

The value $y = 4$ that we have just obtained can now be used in equation (3) to find x. Replacing y with 4 in equation (3) yields

$$x = \tfrac{1}{3}[17 - 2(4)] = \tfrac{1}{3}(9) = 3$$

We have now found that $x = 3$ and $y = 4$. As you can easily check, this pair of values indeed satisfies both of the original equations. We write our solution as the ordered pair (3, 4). Figure 5 summarizes the situation. It shows that the system is consistent and that (3, 4) is the only solution.

Generally speaking, it is not necessary to graph the equations in a given system in order to decide whether the system is consistent. Rather, this information will emerge as you attempt to follow an algebraic method of solution. Examples 2 and 3 will illustrate this.

EXAMPLE 2 Solve the system

$$\begin{cases} \tfrac{3}{2}x - 3y = -9 \\ x - 2y = 4 \end{cases}$$

Solution We use the substitution method. Since it is easy to solve the second equation for x, we begin there:

$$\begin{aligned} x - 2y &= 4 \\ x &= 4 + 2y \end{aligned}$$

Now we substitute this result in the first equation of our system to obtain

$$\begin{aligned} \tfrac{3}{2}(4 + 2y) - 3y &= -9 \\ 6 + 3y - 3y &= -9 \\ 6 &= -9 \qquad \text{False} \end{aligned}$$

Since the substitution process leads us to this obviously false statement, we conclude that the given system has no solution; that is, the system is inconsistent.

QUESTION What can you say about the graphs of the two given equations?

EXAMPLE 3 Solve the system

$$\begin{cases} 3x + 4y = 12 \\ 2y = 6 - \frac{3}{2}x \end{cases}$$

Solution We use the method of substitution. Since it is easy to solve the second equation for y, we begin there:

$$2y = 6 - \tfrac{3}{2}x \qquad \text{and therefore} \qquad y = 3 - \tfrac{3}{4}x$$

Now we use this result to substitute for y in the first equation of the original system. The result is

$$\begin{aligned} 3x + 4\left(3 - \tfrac{3}{4}x\right) &= 12 \\ 3x + 12 - 3x &= 12 \\ 3x - 3x &= 12 - 12 \\ 0 &= 0 \qquad \text{Always true} \end{aligned}$$

This last identity imposes no restrictions on x. Graphically speaking, this says that our two lines intersect for every value of x. In other words, the two lines coincide. We could have foreseen this initially had we solved both equations for y. As you can verify, the result in both cases is

$$y = -\tfrac{3}{4}x + 3$$

Every point on this line yields a solution to our system of equations. In summary, then, our system is consistent and the solutions to the system have the form $(x, -\frac{3}{4}x + 3)$, where x can be any real number. For instance, when $x = 0$, we obtain the solution $(0, 3)$. When $x = 1$, we obtain the solution $\left(1, \frac{9}{4}\right)$. The idea here is that for *each* value of x we obtain a solution; thus, there are infinitely many solutions.

Now let's turn to the addition–subtraction method of solving systems of equations. By way of example, consider the system

$$\begin{cases} 2x + 3y = 5 \\ 4x - 3y = -1 \end{cases}$$

Notice that if we add these two equations, the result is an equation involving only the unknown x:

$$\begin{aligned} 6x &= 4 \\ x &= 4/6 = 2/3 \end{aligned}$$

There are now several ways in which the corresponding value of y can be obtained. As you can easily check, substituting the value $x = 2/3$ in either of the original equations leads to the result $y = 11/9$.

Another way to find y is by multiplying both sides of the first equation by -2. (You'll see why in a moment.) We display the work this way:

$$\begin{aligned} 2x + 3y &= 5 \quad \xrightarrow{\text{Multiply by } -2} \quad -4x - 6y = -10 \\ 4x - 3y &= -1 \quad \xrightarrow{\text{No change}} \quad 4x - 3y = -1 \end{aligned}$$

Adding the last two equations then gives us

$$-9y = -11$$
$$y = 11/9$$

The required solution is therefore $(2/3, 11/9)$.

In the previous example, we were able to find x directly by adding the two equations. As the next example shows, it may be necessary first to multiply both sides of each equation by an appropriate constant.

EXAMPLE 4 Solve the system

$$\begin{cases} 5x - 3y = 4 \\ 2x + 4y = 1 \end{cases}$$

Solution To eliminate x, we could multiply the second equation by $5/2$ and then subtract the resulting equation from the first equation. However, to avoid working with fractions, we proceed as follows.

$$5x - 3y = 4 \xrightarrow{\text{Multiply by 2}} 10x - 6y = 8 \quad (4)$$
$$2x + 4y = 1 \xrightarrow{\text{Multiply by 5}} 10x + 20y = 5 \quad (5)$$

Subtracting equation (5) from equation (4) then yields

$$-26y = 3$$
$$y = -3/26$$

To find x, we return to the original system and work in a similar manner:

$$5x - 3y = 4 \xrightarrow{\text{Multiply by 4}} 20x - 12y = 16$$
$$2x + 4y = 1 \xrightarrow{\text{Multiply by 3}} 6x + 12y = 3$$

Upon adding the last two equations, we obtain

$$26x = 19$$
$$x = 19/26$$

The solution of the given system of equations is therefore $(19/26, -3/26)$.

We conclude this section with some problems that can be solved using simultaneous equations.

EXAMPLE 5 Determine the constants b and c so that the parabola $y = x^2 + bx + c$ passes through the points $(-3, 1)$ and $(1, -2)$.

Solution Since the point $(-3, 1)$ lies on the curve $y = x^2 + bx + c$, the coordinates must satisfy the equation. Thus we have

$$1 = (-3)^2 + b(-3) + c$$
$$-8 = -3b + c \quad (6)$$

This gives us one equation in two unknowns. We need another equation involving b and c. Since the point $(1, -2)$ also lies on the graph of $y = x^2 + bx + c$, we must have

$$-2 = 1^2 + b(1) + c$$
$$-3 = b + c \qquad (7)$$

Rewriting equations (6) and (7), we have the system

$$\begin{cases} -3b + c = -8 & (8) \\ b + c = -3 & (9) \end{cases}$$

Subtracting equation (9) from (8) then yields

$$-4b = -5 \qquad \text{and therefore} \qquad b = 5/4$$

One way to obtain the corresponding value of c is to replace b by $5/4$ in equation (9). This yields

$$\frac{5}{4} + c = -3$$
$$c = -3 - \frac{5}{4} = -\frac{17}{4}$$

The required values of b and c are therefore

$$b = 5/4, \qquad c = -17/4$$

(Exercise 58 at the end of this section asks you to check that the parabola $y = x^2 + \frac{5}{4}x - \frac{17}{4}$ indeed passes through the given points.)

EXAMPLE 6 Find the area of the triangular region bounded by the lines $y = 3x$, $y = -\frac{1}{2}x + 7$, and the x-axis; see Figure 6.

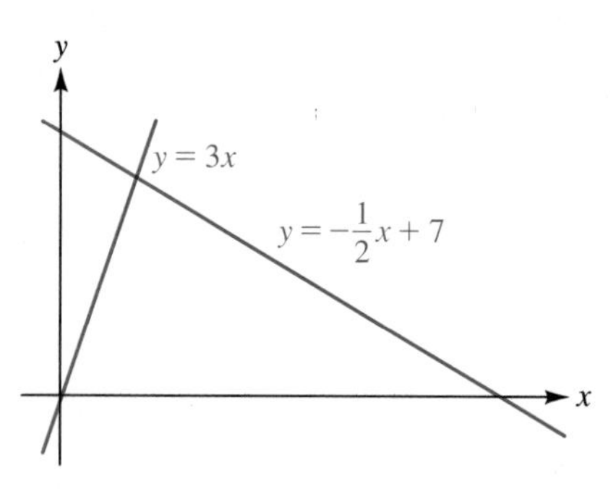

FIGURE 6

Solution To compute the area of this triangle, we need to know the base and the height. To determine the base, we first compute the x-intercept of the line $y = -\frac{1}{2}x + 7$:

$$0 = -\tfrac{1}{2}x + 7$$
$$\tfrac{1}{2}x = 7$$
$$x = 14$$

The height of the triangle is the y-coordinate of the intersection point of the lines $y = 3x$ and $y = -\frac{1}{2}x + 7$ (see Figure 6). We can use the substitution method to solve this pair of simultaneous equations for y. From the equation $y = 3x$, we obtain $x = \frac{1}{3}y$. Then, substituting this in the equation $y = -\frac{1}{2}x + 7$ yields

$$y = -\frac{1}{2}\left(\frac{1}{3}y\right) + 7$$
$$6y = -y + 42$$
$$7y = 42$$
$$y = 6$$

Now that we know that the base of the triangle is 14 units and the height is 6 units, we can compute the required area:

$$A = \frac{1}{2}bh = \frac{1}{2}(14)(6) = 42 \text{ square units}$$

In the next example, we solve a mixture problem using a system of two equations in two unknowns.

EXAMPLE 7 Suppose that a chemistry student can obtain two acid solutions from the stockroom. The first solution is 20% acid and the second solution is 45% acid. (The percentages are by volume.) How many milliliters of each solution should the student mix together to obtain 100 ml of a 30% acid solution?

Solution We begin by assigning letters to denote the required quantities.

Let x denote the number of milliliters of the 20% solution to be used.
Let y denote the number of milliliters of the 45% solution to be used.

We summarize the data in Table 3. Since the final mixture must total 100 ml,

TABLE 3

Type of Solution	Number of ml	Percent of Acid	Total Acid (ml)
First solution (20% acid)	x	20	$(0.20)x$
Second solution (45% acid)	y	45	$(0.45)y$
Mixture	$x+y$	30	$(0.30)(x+y)$

we have the equation

$$x + y = 100 \tag{10}$$

This gives us one equation in two unknowns. However, we need a second equation. Looking at the data in the right-hand column of Table 3, we can write

$$\underbrace{0.20x}_{\substack{\text{Amount of acid in}\\ x \text{ ml of the 20\% solution}}} + \underbrace{0.45y}_{\substack{\text{Amount of acid in}\\ y \text{ ml of the 45\% solution}}} = \underbrace{(0.30)(x+y)}_{\substack{\text{Amount of acid in}\\ \text{the final mixture}}}$$

Thus,

$$\begin{aligned} 0.20x + 0.45y &= 0.30(x+y) \\ 20x + 45y &= 30(x+y) \\ 4x + 9y &= 6(x+y) = 6x + 6y \\ -2x + 3y &= 0 \end{aligned} \tag{11}$$

Equations (10) and (11) can be solved by either the substitution method or the addition method. As Exercise 59 at the end of this section asks you to show, the results are

$$x = 60 \text{ ml} \qquad \text{and} \qquad y = 40 \text{ ml}$$

EXERCISE SET 10.1

A

1. Which of the following are linear equations in two variables?
   (a) $3x + 3y = 10$ (b) $2x + 4xy + 3y = 1$
   (c) $u - v = 1$ (c) $x = 2y + 6$
2. Which of the following are linear equations in two variables?
   (a) $y = x$ (b) $y = x^2$
   (c) $\frac{4}{x} - \frac{3}{y} = -1$ (d) $2w + 8z = -4w + 3$
3. Is (5, 1) a solution of the following system?
$$\begin{cases} 2x - 8y = 2 \\ 3x + 7y = 22 \end{cases}$$
4. Is (14, −2) a solution of the following system?
$$\begin{cases} x + y = 12 \\ x - y = 4 \end{cases}$$
5. Is (0, −4) a solution of the following system?
$$\begin{cases} \frac{1}{6}x + \frac{1}{2}y = -2 \\ \frac{2}{3}x + \frac{3}{4}y = 2 \end{cases}$$
6. Is (12, −8) a solution of the system in Exercise 5?
7. Is (3, −2) a solution of the following system?
$$\begin{cases} \frac{2}{7}x - \frac{1}{5}y = \frac{44}{35} \\ \frac{1}{3}x - \frac{5}{4}y = \frac{7}{2} \end{cases}$$

In Exercises 8–18, use the substitution method to find all solutions of each system.

8. $\begin{cases} 4x - y = 7 \\ -2x + 3y = 9 \end{cases}$
9. $\begin{cases} 3x - 2y = -19 \\ x + 4y = -4 \end{cases}$
10. $\begin{cases} 6x - 2y = -3 \\ 5x + 3y = 4 \end{cases}$
11. $\begin{cases} 4x + 2y = 3 \\ 10x + 4y = 1 \end{cases}$
12. $\begin{cases} \frac{3}{2}x - 5y = 1 \\ x + \frac{3}{4}y = -1 \end{cases}$
13. $\begin{cases} 13x - 8y = -3 \\ -7x + 2y = 0 \end{cases}$
14. $\begin{cases} 4x + 6y = 3 \\ -6x - 9y = -\frac{9}{2} \end{cases}$
15. $\begin{cases} -\frac{2}{5}x + \frac{1}{4}y = 3 \\ \frac{1}{4}x - \frac{2}{5}y = -3 \end{cases}$
16. $\begin{cases} 0.02x - 0.03y = 1.06 \\ 0.75x + 0.50y = -0.01 \end{cases}$
17. $\begin{cases} \sqrt{2}x - \sqrt{3}y = \sqrt{3} \\ \sqrt{3}x - \sqrt{8}y = \sqrt{2} \end{cases}$
18. $\begin{cases} 7x - 3y = -12 \\ \frac{14}{3}x - 2y = 2 \end{cases}$

In Exercises 19–28, use the addition–subtraction method to find all solutions of each system of equations.

19. $\begin{cases} 5x + 6y = 4 \\ 2x - 3y = -3 \end{cases}$
20. $\begin{cases} -8x + y = -2 \\ 4x - 3y = 1 \end{cases}$
21. $\begin{cases} 4x + 13y = -5 \\ 2x - 54y = -1 \end{cases}$
22. $\begin{cases} 16x - 3y = 100 \\ 16x + 10y = 10 \end{cases}$
23. $\begin{cases} \frac{1}{4}x - \frac{1}{3}y = 4 \\ \frac{2}{7}x - \frac{1}{7}y = \frac{1}{10} \end{cases}$

   Suggestion for Exercise 23: First clear both equations of fractions.

24. $\begin{cases} 2.1x - 3.5y = 1.2 \\ 1.4x + 2.6y = 1.1 \end{cases}$
25. $\begin{cases} 8x + 16y = 5 \\ 2x + 5y = \frac{5}{4} \end{cases}$
26. $\begin{cases} 8x + 16y = 5 \\ 2x + 4y = 1 \end{cases}$
27. $\begin{cases} 125x - 40y = 45 \\ \frac{1}{10}x + \frac{1}{10}y = \frac{3}{10} \end{cases}$
28. $\begin{cases} \sqrt{6}x - \sqrt{3}y = 3\sqrt{2} - \sqrt{3} \\ \sqrt{2}x - \sqrt{5}y = \sqrt{6} + \sqrt{5} \end{cases}$
29. Find b and c, given that the parabola $y = x^2 + bx + c$ passes through (0, 4) and (2, 14).
30. Determine the constants a and b, given that the parabola $y = ax^2 + bx + 1$ passes through (−1, 11) and (3, 1).
31. Determine the constants A and B, given that the line $Ax + By = 2$ passes through the points (−4, 5) and (7, −9).
32. Find the area of the triangular region in the first quadrant bounded by the lines $y = 5x$, $y = -3x + 6$, and the x-axis.
33. Find the area of the triangular region in the first quadrant bounded by the x-axis and the lines $y = 2x - 5$ and $y = -\frac{1}{2}x + 3$.
34. Find the area of the triangular region in the first quadrant bounded by the y-axis and the lines $y = 2x + 2$ and $y = -\frac{3}{2}x + 9$.
35. A student in a chemistry laboratory has access to two acid solutions. The first solution is 10% acid and the second is 35% acid. (The percentages are by volume.) How many cubic centimeters of each should she mix together to obtain 200 cm^3 of a 25% acid solution?
36. One salt solution is 15% salt and another is 20% salt. How many cubic centimeters of each solution must be mixed to obtain 50 cm^3 of a 16% salt solution?
37. A shopkeeper has two types of coffee beans on hand. One type sells for \$5.20/lb, the other for \$5.80/lb.

How many pounds of each type must be mixed to produce 16 lb of a blend that sells for \$5.50/lb?

38. A certain alloy contains 10% tin and 30% copper. (The percentages are by weight.) How many pounds of tin and how many pounds of copper must be melted with 1000 lb of the given alloy to yield a new alloy containing 20% tin and 35% copper?

B

39. Find x and y in terms of a and b:

$$\begin{cases} \dfrac{x}{a} + \dfrac{y}{b} = 1 \\ \dfrac{x}{b} + \dfrac{y}{a} = 1 \end{cases}$$

Does your solution impose any conditions on a and b?

40. Solve the following system for x and y in terms of a and b, where $a \neq b$:

$$\begin{cases} ax + by = 1/a \\ b^2x + a^2y = 1 \end{cases}$$

41. Solve the following system for x and y in terms of a and b, where $a \neq b$:

$$\begin{cases} ax + a^2y = 1 \\ bx + b^2y = 1 \end{cases}$$

Does your solution impose any additional conditions on a and b?

42. Solve the following system for s and t:

$$\begin{cases} \dfrac{3}{s} - \dfrac{4}{t} = 2 \\ \dfrac{5}{s} + \dfrac{1}{t} = -3 \end{cases}$$

Hint: Make the substitutions $1/s = x$ and $1/t = y$ in order to obtain a system of two linear equations.

43. Solve the following system for s and t:

$$\begin{cases} \dfrac{1}{2s} - \dfrac{1}{2t} = -10 \\ \dfrac{2}{s} + \dfrac{3}{t} = 5 \end{cases}$$

(Use the hint in Exercise 42.)

In Exercises 44–53, find all solutions of the given systems. For Exercises 47–52, use a calculator and round the final answers to two decimal places.

44. $\begin{cases} 0.5x - 0.8y = 0.3 \\ 0.4x - 0.1y = 0.9 \end{cases}$

45. $\begin{cases} \dfrac{2w-1}{3} + \dfrac{z+2}{4} = 4 \\ \dfrac{w+3}{2} - \dfrac{w-z}{3} = 3 \end{cases}$

46. $\dfrac{x-y}{2} = \dfrac{x+y}{3} = 1$

47. $\begin{cases} 1.03x - 2.54y = 5.47 \\ 3.85x + 4.29y = -1.84 \end{cases}$

48. $\begin{cases} 2.39x + 8.16y = -2.83 \\ 1.01x + 2.98y = 4.41 \end{cases}$

49. $\begin{cases} 2\ln x - 5\ln y = 11 \\ \ln x + \ln y = -5 \end{cases}$

50. $\begin{cases} 3\ln x + \ln y = 3 \\ 4\ln x - 6\ln y = -7 \end{cases}$

Hint: Let $u = \ln x$ and $v = \ln y$.

51. $\begin{cases} e^x - 3e^y = 2 \\ 3e^x + e^y = 16 \end{cases}$

52. $\begin{cases} e^x + 2e^y = 4 \\ \frac{1}{2}e^x - e^y = 0 \end{cases}$

Hint: Let $u = e^x$ and $v = e^y$.

53. $\begin{cases} 4\sqrt{x^2 - 3x} - 3\sqrt{y^2 + 6y} = -4 \\ \frac{1}{2}\left(\sqrt{x^2 - 3x} + \sqrt{y^2 + 6y}\right) = 3 \end{cases}$

Hint: Let $u = \sqrt{x^2 - 3x}$ and $v = \sqrt{y^2 + 6y}$.

54. The sum of two numbers is 64. Twice the larger number plus five times the smaller number is 20. Find the two numbers. (Let x denote the larger number and let y denote the smaller number.)

55. In a two-digit number, the sum of the digits is 14. Twice the tens digit exceeds the units digit by one. Find the number.

56. The sum of the digits in a two-digit number is 14. Furthermore, the number itself is 2 greater than 11 times the tens digit. Find the number.

57. The perimeter of a rectangle is 34 in. If the length is 2 in. more than twice the width, find the length and the width.

58. Verify that the parabola $y = x^2 + \frac{5}{4}x - \frac{17}{4}$ indeed passes through $(-3, 1)$ and $(1, -2)$, as stated in the text.

59. Consider the following system from Example 7:

$$\begin{cases} x + y = 100 \\ -2x + 3y = 0 \end{cases}$$

(a) Solve this system using the method of substitution.
Answer: $(60, 40)$

(b) Solve the system using the addition–subtraction method.

60. Consider the following system:

$$\begin{cases} 2x - 5y = -2 \\ 3x + 4y = 5 \end{cases}$$

(a) Solve the system using the substitution method.

(b) Solve the system using the addition–subtraction method. *Answer:* $\left(\frac{17}{23}, \frac{16}{23}\right)$

(c) Verify that your solution indeed satisfies both equations of the system.

61. Solve the following system for x and y:

$$\begin{cases} by = x + ab \\ cy = x + ac \end{cases}$$

(Assume that a, b, and c are constants and that $b \neq c$.)

62. Solve for x and y in terms of a, b, c, d, e, and f:

$$\begin{cases} ax + by = c \\ dx + ey = f \end{cases}$$

(Assume that $ae - bd \neq 0$.)

63. Solve the following system for x and y in terms of a and b, where $a \neq b$:

$$\begin{cases} \dfrac{a}{bx} + \dfrac{b}{ay} = a + b \\ \dfrac{b}{x} + \dfrac{a}{y} = a^2 + b^2 \end{cases}$$

C

64. Solve the following system for x and y in terms of a and b, where $ab \neq -1$:

$$\begin{cases} \dfrac{x + y - 1}{x - y + 1} = a \\ \dfrac{y - x + 1}{x - y + 1} = ab \end{cases}$$

Answer: $x = (a + 1)/(ab + 1)$
$y = a(b + 1)/(ab + 1)$

65. Given that the lines $7x + 5y = 4$, $x + ky = 3$, and $5x + y + k = 0$ are concurrent (pass through a common point), what are the possible values for k?

66. Solve the following system for x and y in terms of a and b:

$$\begin{cases} (a + b)x + (a^2 + b^2)y = a^3 + b^3 \\ (a - b)x + (a^2 - b^2)y = a^3 - b^3 \end{cases}$$

(Assume that a and b are nonzero and that $a \neq \pm b$.)

67. The vertices of triangle ABC are $A(2a, 0)$, $B(2b, 0)$, and $C(0, 2)$.

(a) Show that the equations of the sides are $x + by = 2b$, $x + ay = 2a$, and $y = 0$.

(b) Show that the equations of the medians are $x + (2a - b)y = 2a$, $x + (2b - a)y = 2b$, and $2x + (a + b)y = 2(a + b)$.

(c) Show that all three medians intersect at the point $(\frac{2}{3}(a + b), \frac{2}{3})$.

(d) Show that the equations of the altitudes are $bx - y = 2ab$, $ax - y = 2ab$, and $x = 0$.

(e) Show that all three altitudes intersect at the point $(0, -2ab)$.

(f) Show that the equations of the perpendicular bisectors of the sides are $x = a + b$, $bx - y = b^2 - 1$, and $ax - y = a^2 - 1$.

(g) Show that all three perpendicular bisectors intersect at the point $(a + b, ab + 1)$.

(h) Show that the three points found in parts (c), (e), and (g) all lie on a line. This is the **Euler line** of the triangle.

GRAPHING UTILITY EXERCISES FOR SECTION 10.1

In Exercises 1–4, you are given a system of two linear equations. By graphing the pair of equations, determine which one of the three cases described in Figures 2 through 4 (on page 602) applies. (You're not being asked to solve the system.)

1. $\begin{cases} 3x + 7y = 10 \\ 6x - 3y = 1 \end{cases}$

2. $\begin{cases} y = \sqrt{3}(1 - 3x)/3 \\ \sqrt{3}y + 3x - 1 = 0 \end{cases}$

3. $\begin{cases} 5y = 10.5x - 25.5 \\ 21x = 50 + 10y \end{cases}$

4. $\begin{cases} 2y = x - 18 \\ y = 0.4x + 1 \end{cases}$

*In Exercises 5–8, you are given a system of two linear equations. **(a)** By graphing the pair of equations, estimate the solution for the system. (Zoom in on the intersection point until you are sure about the first two decimal places for each coordinate.) **(b)** Use an algebraic method to find the solution. Check that your answers are consistent with the graphical estimates obtained in part (a).*

5. $\begin{cases} y = -8x + 6 \\ x = 8 + y \end{cases}$

6. $\begin{cases} 7x + 4y = 35 \\ x - 9y = 35 \end{cases}$

7. $\begin{cases} 0.35x - 0.41y = 4 \\ 3.02x - 1.27y = -3 \end{cases}$

8. $\begin{cases} \frac{1}{3}x + \frac{1}{4}y = \frac{1}{5} \\ \frac{1}{6}x + \frac{1}{7}y = \frac{1}{8} \end{cases}$

9. **(a)** Graph the two lines $3y - 4x = 0$ and $x + y = 7$.
 (b) Given that the two lines in part (a) intersect at a point with integer coordinates, use the graph to specify the coordinates. Then check your result by using one of the algebraic solution techniques from this section.

10. *(Continuation of Exercise 9)* Let P denote the point that you determined in Exercise 9(b), let Q denote the x-intercept of the line $x + y = 7$, and let O denote the origin. The figure at the top of the next column shows the two lines that you graphed in Exercise 9(a) and also the circle passing through the three points P, Q, and O. The circle is called the *circumcircle* for $\triangle PQO$.

 Use the following two facts to find the equation of the circumcircle:
 (i) The radius of the circle is $\sqrt{12.5}$.

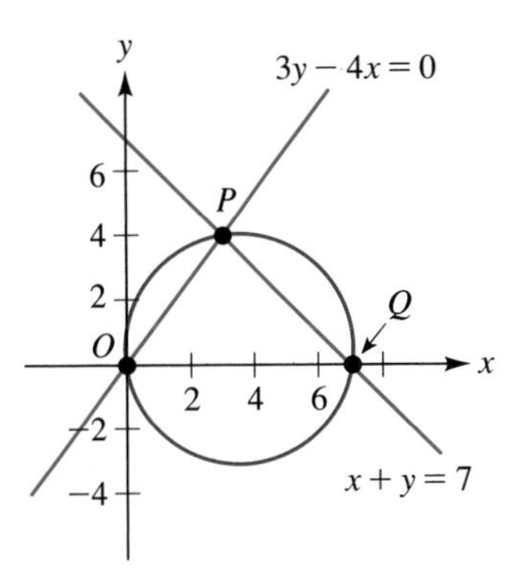

 (ii) The center of the circle is found by solving the system

$$\begin{cases} 6x + 8y = 25 \\ x - y = 3 \end{cases}$$

 Check your answer by graphing the circle along with the two lines given in Exercise 9(a). Does your circle pass through all three points P, Q, and O?

10.2 GAUSSIAN ELIMINATION

A method of solution is perfect if we can foresee from the start, and even prove, that following that method we shall obtain our aim.

Gottfried Wilhelm von Leibniz (1646–1716)

... the first electronic computer, the ABC, named after its designers Atanasoff and Berry, was built specifically to solve systems of 29 equations in 29 unknowns, a formidable task without the aid of a computer.

Angela B. Shiflet in *Discrete Mathematics for Computer Science* (St. Paul: West Publishing Company, 1987)

In the previous section, we solved systems of linear equations in two unknowns. In this section, we introduce the technique known as Gaussian elimination for solving systems of linear equations in which there are more than two unknowns.*

As a first example, consider the following system of three linear equations in the three unknowns x, y, and z:

$$\begin{cases} 3x + 2y - z = -3 \\ \qquad 5y - 2z = 2 \\ \qquad\qquad 5z = 20 \end{cases}$$

This system is easy to solve using the process of *back-substitution.* Dividing both sides of the third equation by 5 yields $z = 4$. Then, substituting $z = 4$ back into the second equation gives us

$$5y - 2(4) = 2$$
$$5y = 10$$
$$y = 2$$

*The technique is named after Carl Friedrich Gauss (1777–1855). Early in the nineteenth century, Gauss used this technique (and introduced the method of least squares for minimizing errors) in analyzing the orbit of the asteroid Pallas. However, the essentials of Gaussian elimination were in existence long before Gauss's time. Indeed, a version of the method appears in the Chinese text *Chui-Chang Suan-Shu* ("Nine Chapters on the Mathematical Art"), written approximately two thousand years ago.

Finally, substituting the values $z = 4$ and $y = 2$ back into the first equation yields

$$\begin{aligned} 3x + 2(2) - 4 &= -3 \\ 3x &= -3 \\ x &= -1 \end{aligned}$$

We have now found that $x = -1$, $y = 2$, and $z = 4$. If you go back and check, you will find that these values indeed satisfy all three equations in the given system. Furthermore, the algebra we've just carried out shows that these are the only possible values for x, y, and z satisfying all three equations. We summarize by saying that the **ordered triple** $(-1, 2, 4)$ is the solution of the given system.

The system that we just considered was easy to solve (using back-substitution) because of the special form in which it was written. This form is called **upper-triangular form.** Although the following definition of upper-triangular form refers to systems with three unknowns, the same type of definition can be given for systems with any number of unknowns. Table 1 displays examples of systems in upper-triangular form.

UPPER-TRIANGULAR FORM (THREE VARIABLES)

A system of linear equations in x, y, and z is said to be in **upper-triangular form** provided x appears in no equation after the first and y appears in no equation after the second. (It is possible that y may not even appear in the second equation.)

TABLE 1 Examples of Systems in Upper-Triangular Form

2 Unknowns: x, y	3 Unknowns: x, y, z	4 Unknowns: x, y, z, t
$\begin{cases} 3x + 5y = 7 \\ 8y = 5 \end{cases}$	$\begin{cases} 4x - 3y + 2z = -5 \\ 7y + z = 9 \\ -4z = 3 \end{cases}$	$\begin{cases} x - y + z - 4t = 1 \\ 3y - 2z + t = -1 \\ 3z - 5t = 4 \\ 6t = 7 \end{cases}$
	$\begin{cases} 15x - 2y + z = 1 \\ 3z = -8 \end{cases}$	$\begin{cases} 2x + y + 2z - t = -3 \\ 4z + 3t = 1 \\ 5t = 6 \end{cases}$
		$\begin{cases} 8x + 3y - z + t = 2 \\ 2y + z - 4t = 1 \end{cases}$

When we were solving linear systems of equations with two unknowns in the previous section, we observed that there were three possibilities: a unique solution, infinitely many solutions, and no solution. As the next three examples indicate, the situation is similar when dealing with larger systems.

EXAMPLE 1 Find all solutions of the system

$$\begin{cases} x + y + 2z = 2 \\ 3y - 4z = -5 \\ 6z = 3 \end{cases}$$

Solution The system is in upper-triangular form, so we can use back-substitution. Dividing the third equation by 6 yields $z = 1/2$. Substituting this value for z back into the second equation then yields

$$\begin{aligned} 3y - 4(1/2) &= -5 \\ 3y &= -3 \\ y &= -1 \end{aligned}$$

Now, substituting the values $z = 1/2$ and $y = -1$ back into the first equation, we obtain

$$\begin{aligned} x + (-1) + 2(1/2) &= 2 \\ x &= 2 \end{aligned}$$

As you can easily check, the values $x = 2$, $y = -1$, and $z = 1/2$ indeed satisfy all three equations. Furthermore, the algebra we've just carried out shows that these are the only possible values for x, y, and z satisfying all three equations. We summarize by saying that the unique solution to our system is the ordered triple $(2, -1, 1/2)$.

EXAMPLE 2 Find all solutions of the system

$$\begin{cases} -2x + y + 3z = 6 \\ \qquad\quad 2z = 10 \end{cases}$$

Solution Again, the system is in upper-triangular form, and we use back-substitution. Solving the second equation for z yields $z = 5$. Then replacing z by 5 in the first equation gives us

$$\begin{aligned} -2x + y + 3(5) &= 6 \\ -2x + y &= -9 \\ y &= 2x - 9 \end{aligned}$$

At this point, we've made use of both equations in the given system. There is no third equation to provide additional restrictions on x, y, or z. We know from the previous section that the equation $y = 2x - 9$ has infinitely many solutions, all of the form

$$(x, 2x - 9) \qquad \text{where } x \text{ is a real number}$$

It follows, then, that there are infinitely many solutions to the given system and they may be written

$$(x, 2x - 9, 5) \qquad \text{where } x \text{ is a real number}$$

For instance, choosing in succession $x = 0$, $x = 1$, and $x = 2$ yields the solutions $(0, -9, 5)$, $(1, -7, 5)$, and $(2, -5, 5)$. (We remark in passing that any linear system in upper-triangular form in which the number of unknowns exceeds the number of equations will always have infinitely many solutions.)

EXAMPLE 3 Find all solutions of the system

$$\begin{cases} 4x - 7y + 3z = 1 \\ 3x + \ \ y - 2z = 4 \\ 4x - 7y + 3z = 6 \end{cases}$$

(Note that the system is not in upper-triangular form.)

Solution Look at the left-hand sides of the first and third equations: they are identical. Thus, if there were values for x, y, and z that satisfied both equations, it would follow that $1 = 6$, which is clearly impossible. We conclude that the given system has no solutions.

As Examples 1 and 2 have demonstrated, systems in upper-triangular form can be readily solved. In view of this, it would be useful to have a technique for converting a given system into an equivalent system in upper-triangular form. (The expression **equivalent system** means a system with exactly the same set of solutions as the original system.) **Gaussian elimination** is one such technique. We will demonstrate this technique in Examples 4, 5, and 6. In using Gaussian elimination, we will rely on what are called the three **elementary operations,** listed in the box that follows. These are operations that, when performed on an equation in a system, produce an equivalent system.

THE ELEMENTARY OPERATIONS

1. Multiply both sides of an equation by a nonzero constant.
2. Interchange the order in which two equations of a system are listed.
3. To one equation add a multiple of another equation in the system.

EXAMPLE 4 Find all solutions of the system

$$\begin{cases} x + 2y + z = 3 \\ 2x + y + z = 16 \\ x + y + 2z = 9 \end{cases}$$

Solution First we want to eliminate x from the second and third equations. To eliminate x from the second equation, we add to it -2 times the first equation. The result is the equivalent system

$$\begin{cases} x + 2y + z = 3 \\ -3y - z = 10 \\ x + y + 2z = 9 \end{cases}$$

To eliminate x from the third equation, we add to it -1 times the first equation. The result is the equivalent system

$$\begin{cases} x + 2y + z = 3 \\ -3y - z = 10 \\ -y + z = 6 \end{cases}$$

Now to bring the system into upper-triangular form, we need to eliminate y from the third equation. We could do this by adding $-1/3$ times the second equation to the third equation. However, to avoid working with fractions as long as possible, we proceed instead to interchange the second and third equations to obtain the equivalent system

$$\begin{cases} x + 2y + z = 3 \\ -y + z = 6 \\ -3y - z = 10 \end{cases}$$

Now we add -3 times the second equation to the last equation to obtain the equivalent system

$$\begin{cases} x + 2y + z = 3 \\ \quad -y + z = 6 \\ \qquad -4z = -8 \end{cases}$$

The system is now in upper-triangular form, and back-substitution yields, in turn, $z = 2$, $y = -4$, and $x = 9$. (Check this for yourself.) The required solution is therefore $(9, -4, 2)$.

In Table 2 we list some convenient abbreviations that are used in describing the elementary operations. You should plan on using these abbreviations yourself: they'll make it simpler for you (and your instructor) to check your work. In Table 2, the notation E_i stands for the ith equation in a system. For instance, for the initial system in Example 4, the symbol E_1 denotes the first equation: $x + 2y + z = 3$.

TABLE 2 ABBREVIATIONS FOR THE ELEMENTARY OPERATIONS

Abbreviation	Explanation
1. cE_i	Multiply both sides of the ith equation by c.
2. $E_i \leftrightarrow E_j$	Interchange the ith and jth equations.
3. $cE_i + E_j$	Add c times the ith equation to the jth equation.

EXAMPLE 5 Solve the system

$$\begin{cases} 4x - 3y + 2z = 40 \\ 5x + 9y - 7z = 47 \\ 9x + 8y - 3z = 97 \end{cases}$$

Solution

$$\begin{cases} 4x - 3y + 2z = 40 \\ 5x + 9y - 7z = 47 \\ 9x + 8y - 3z = 97 \end{cases} \xrightarrow{(-1)E_2 + E_1} \begin{cases} -x - 12y + 9z = -7 \\ 5x + 9y - 7z = 47 \\ 9x + 8y - 3z = 97 \end{cases}$$

adding -1 times the second equation to the first equation

$$\xrightarrow[9E_1 + E_3]{5E_1 + E_2} \begin{cases} -x - 12y + 9z = -7 \\ \quad -51y + 38z = 12 \\ \quad -100y + 78z = 34 \end{cases}$$

$$\xrightarrow{\frac{1}{2}E_3} \begin{cases} -x - 12y + 9z = -7 \\ \quad -51y + 38z = 12 \\ \quad -50y + 39z = 17 \end{cases}$$

$$\xrightarrow{(-1)E_3 + E_2} \begin{cases} -x - 12y + 9z = -7 \\ \quad -y - z = -5 \\ \quad -50y + 39z = 17 \end{cases}$$

to allow working with smaller but integral coefficients

$$\xrightarrow{-50E_2 + E_3} \begin{cases} -x - 12y + 9z = -7 \\ -y - z = -5 \\ 89z = 267 \end{cases}$$

The system is now in upper-triangular form. Solving the third equation, we obtain $z = 3$. Substituting this value back into the second equation yields $y = 2$. (Check this for yourself.) Finally, substituting $z = 3$ and $y = 2$ back into the first equation yields $x = 10$. (Again, check this for yourself.) The solution to the system is therefore $(10, 2, 3)$.

EXAMPLE 6 Solve the system

$$\begin{cases} x + 2y + 4z = 0 \\ x + 3y + 9z = 0 \end{cases}$$

Solution This system is similar to the one in Example 2 in that there are fewer equations than there are unknowns. By subtracting the first equation from the second, we readily obtain an equivalent system in upper-triangular form:

$$\begin{cases} x + 2y + 4z = 0 \\ y + 5z = 0 \end{cases}$$

Although the system is now in upper-triangular form, notice that the second equation does not determine y or z uniquely; that is, there are infinitely many number pairs (y, z) satisfying the second equation. We can solve the second equation for y in terms of z; the result is $y = -5z$. Now we replace y with $-5z$ in the first equation to obtain

$$x + 2(-5z) + 4z = 0 \qquad \text{or} \qquad x = 6z$$

At this point we've used both of the equations in the system to express x and y in terms of z. Furthermore, there is no third equation in the system to provide additional restrictions on x, y, or z. We therefore conclude that the given system has infinitely many solutions. These solutions have the form

$$(6z, -5z, z) \qquad \text{where } z \text{ is any real number}$$

EXAMPLE 7 Solve the system

$$\begin{cases} x - 4y + z = 3 \\ 3x + 5y - 2z = -1 \\ 7x + 6y - 3z = 2 \end{cases}$$

Solution

$$\begin{cases} x - 4y + z = 3 \\ 3x + 5y - 2z = -1 \\ 7x + 6y - 3z = 2 \end{cases} \xrightarrow[-7E_1 + E_3]{-3E_1 + E_2} \begin{cases} x - 4y + z = 3 \\ 17y - 5z = -10 \\ 34y - 10z = -19 \end{cases}$$

$$\xrightarrow{-2E_2 + E_3} \begin{cases} x - 4y + z = 3 \\ 17y - 5z = -10 \\ 0 = 1 \end{cases}$$

From the third equation in this last system, we conclude that this system, and consequently the original system, has no solution. (Reason: If there *were* values for x, y, and z satisfying the original system, then it would follow that $0 = 1$, which is clearly impossible.)

EXERCISE SET 10.2

A

In Exercises 1–10, the systems of linear equations are in upper-triangular form. Find all solutions of each system.

1. $\begin{cases} 2x + y + z = -9 \\ 3y - 2z = -4 \\ 8z = -8 \end{cases}$

2. $\begin{cases} -3x + 7y + 2z = -19 \\ y + z = 1 \\ -2z = -2 \end{cases}$

3. $\begin{cases} 8x + 5y + 3z = 1 \\ 3y + 4z = 2 \\ 5z = 3 \end{cases}$

4. $\begin{cases} 2x + 7z = -4 \\ 5y - 3z = 6 \\ 6z = 18 \end{cases}$

5. $\begin{cases} -4x + 5y = 0 \\ 3y + 2z = 1 \\ 3z = -1 \end{cases}$

6. $\begin{cases} 3x - 2y + z = 4 \\ 3z = 9 \end{cases}$

7. $\begin{cases} -x + 8y + 3z = 0 \\ 2z = 0 \end{cases}$

8. $\begin{cases} -x + y + z + w = 9 \\ 2y - z - w = 9 \\ 3z + 2w = 1 \\ 11w = 22 \end{cases}$

9. $\begin{cases} 2x + 3y + z + w = -6 \\ y + 3z - 4w = 23 \\ 6z - 5w = 31 \\ -2w = 10 \end{cases}$

10. $\begin{cases} 7x - y - z + w = 3 \\ 2y - 3z - 4w = -2 \\ 3w = 6 \end{cases}$

In Exercises 11–30, find all solutions of each system.

11. $\begin{cases} x + y + z = 12 \\ 2x - y - z = -1 \\ 3x + 2y + z = 22 \end{cases}$

12. $\begin{cases} A + B = 0 \\ -6A - B + C = 2 \\ 9A - 6B + 2C = 1 \end{cases}$

Answer: $\left(-\frac{3}{25}, \frac{3}{25}, \frac{7}{5}\right)$

13. $\begin{cases} 2x - 3y + 2z = 4 \\ 4x + 2y + 3z = 7 \\ 5x + 4y + 2z = 7 \end{cases}$

14. $\begin{cases} x + 2z = 5 \\ y - 30z = -16 \\ x - 2y + 4z = 8 \end{cases}$

15. $\begin{cases} 3x + 3y - 2z = 13 \\ 6x + 2y - 5z = 13 \\ 7x + 5y - 3z = 26 \end{cases}$

16. $\begin{cases} 2x + 5y - 3z = 4 \\ 4x - 3y + 2z = 9 \\ 5x + 6y - 2z = 18 \end{cases}$

17. $\begin{cases} x + y + z = 1 \\ -2x + y + z = -2 \\ 3x + 6y + 6z = 5 \end{cases}$

18. $\begin{cases} 7x + 5y - 7z = -10 \\ 2x + y + z = 7 \\ x + y - 3z = -8 \end{cases}$

19. $\begin{cases} 2x - y + z = -1 \\ x + 3y - 2z = 2 \\ -5x + 6y - 5z = 5 \end{cases}$

20. $\begin{cases} -2x + 2y - z = 0 \\ 3x - 4y + z = 1 \\ 5x - 8y + z = 4 \end{cases}$

21. $\begin{cases} 2x - y + z = 4 \\ x + 3y + 2z = -1 \\ 7x + 5z = 11 \end{cases}$

22. $\begin{cases} 3x + y - z = 10 \\ 8x - y - 6z = -3 \\ 5x - 2y - 5z = 1 \end{cases}$

23. $\begin{cases} x + y + z + w = 4 \\ x - 2y - z - w = 3 \\ 2x - y + z - w = 2 \\ x - y + 2z - 2w = -7 \end{cases}$

24. $\begin{cases} x + y - 3z + 2w = 0 \\ -2x - 2y + 6z + w = -5 \\ -x + 3y + 3z + 3w = -5 \\ 2x + y - 3z - w = 4 \end{cases}$

25. $\begin{cases} 2x + 3y + 2z = 5 \\ x + 4y - 3z = 1 \end{cases}$

26. $\begin{cases} 4x - y - 3z = 2 \\ 6x + 5y - z = 0 \end{cases}$

27. $\begin{cases} x - 2y - 2z + 2w = -10 \\ 3x + 4y - z - 3w = 11 \\ -4x - 3y - 3z + 8w = -21 \end{cases}$

28. $\begin{cases} 2x + y + z + w = 1 \\ x + 3y - 3z - 3w = 0 \\ -3x - 4y + 2z + 2w = -1 \end{cases}$

29. $\begin{cases} 4x - 2y + 3z = -2 \\ 6y - 4z = 6 \end{cases}$

30. $\begin{cases} 6x - 2y - 5z + w = 3 \\ 5x - y + z + 5w = 4 \\ -4y - 7z - 9w = -5 \end{cases}$

B

31. **(a)** Find constants A, B, and C so that the parabola $y = Ax^2 + Bx + C$ passes through the three points $(1, -2)$, $(-1, 0)$, and $(2, 3)$.
(b) Find constants a, b, and c so that the curve $y = x^3 + ax^2 + bx + c$ passes through the three points given in part (a).

32. Follow Exercise 31 using the three points $(1, -1)$, $(2, -2)$, and $(3, 3)$.

33. A parabola $y = ax^2 + bx + c$ passes through the three points $(-1, 1)$, $(1, -2)$, and $(4, 1)$. Find the x-intercepts and the vertex of the parabola.

34. Suppose that the height of an object as a function of time is given by $f(t) = at^2 + bt + c$, where t is time in seconds, $f(t)$ is the height in feet at time t, and a, b, and c are certain constants. If, after 1, 2, and 3 seconds, the corresponding heights are 184 ft, 136 ft, and 56 ft, respectively, find the time at which the object is at ground level (height = 0 ft).

35. Find the equation of the circle passing through the points $(-1, -5)$, $(1, 2)$, and $(-2, 0)$. Write your answer in the form $Ax^2 + Ay^2 + Bx + Cy + D = 0$.

36. Find the equation of a circle that passes through the origin and the points $(8, -4)$ and $7, -1)$. Specify the center and the radius.

37. Solve the following system for x, y, and z:

$$\begin{cases} e^x + e^y - 2e^z = 2a \\ e^x + 2e^y - 4e^z = 3a \\ \frac{1}{2}e^x - 3e^y + e^z = -5a \end{cases}$$

(Assume that $a > 0$.) *Hint:* Let $A = e^x$, $B = e^y$, and $C = e^z$. Solve the resulting linear system for A, B, and C.

38. Solve the following system for α, β, and γ:

$$\begin{cases} \ln \alpha - \ln \beta - \ln \gamma = 2 \\ 3 \ln \alpha + 5 \ln \beta - 2 \ln \gamma = 1 \\ 2 \ln \alpha - 4 \ln \beta + \ln \gamma = 2 \end{cases}$$

39. The following figure displays three circles that are mutually tangent. The line segments joining the centers have lengths a, b, and c, as shown. Let r_1, r_2, and r_3 denote the radii of the circles, as indicated in the figure. Show that

$$r_1 = \frac{a + c - b}{2} \qquad r_2 = \frac{a + b - c}{2}$$

$$r_3 = \frac{b + c - a}{2}$$

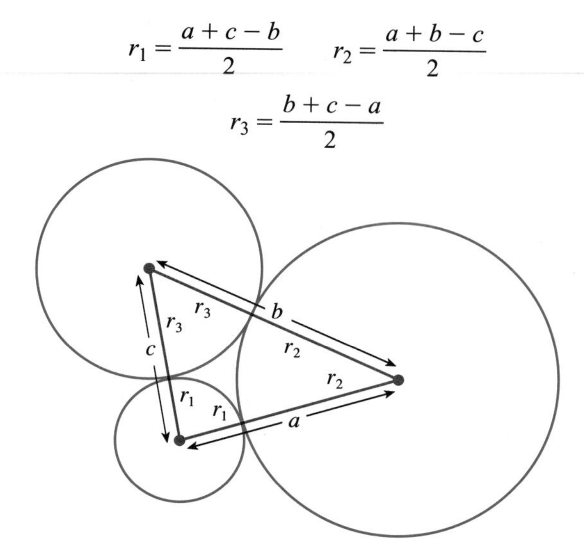

40. Consider the system

$$\begin{cases} \lambda x + y + z = a \\ x + \lambda y + z = b \\ x + y + \lambda z = c \end{cases}$$

(a) Assuming that the value of the constant λ is neither 1 nor -2, find all solutions of the system.
(b) If $\lambda = -2$, how must a, b, and c be related for the system to have solutions? Find these solutions.
(c) If $\lambda = 1$, how must a, b, and c be related for the system to have solutions? Find these solutions.

C

The following exercise appears in Algebra for Colleges and Schools *by H. S. Hall and S. R. Knight, revised by F. L. Sevenoak (New York: The Macmillan Company, 1906).*

41. A, B, and C are three towns forming a triangle. A man has to walk from one to the next, ride thence to the next, and drive thence to his starting point. He can walk, ride, and drive a mile in a, b, and c minutes, respectively. If he starts from B he takes $a + c - b$ hours, if he starts from C he takes $b + a - c$ hours, and if he starts from A he takes $c + b - a$ hours. Find the length of the circuit. [Assume that the circuit from A to B to C is counterclockwise.]

MATRICES

Arthur Cayley (1821–1895) and James Joseph Sylvester (1814–1897), two English mathematicians, invented the matrix ... in the 1850s.... The operations of addition and multiplication of matrices were later defined and the algebra of matrices was then developed. In 1925, Werner Heisenberg, a German physicist, used matrices in developing his theory of quantum mechanics, extending the role of matrices from algebra to the area of applied mathematics.

John K. Luedeman and Stanley M. Lukawecki in *Elementary Linear Algebra* (St. Paul: West Publishing Co., 1986)

Recall that in Gaussian Elimination, row operations are used to change the coefficient matrix to an upper triangular matrix. The solution is then found by back substitution, starting from the last equation in the reduced system.

Steven C. Althoen and Renate McLaughlin in "Gauss–Jordan Reduction: A Brief History," *American Mathematical Monthly*, vol. 94 (1987), pp. 130–142

As you saw in the previous section, there can be a good deal of bookkeeping involved in using Gaussian elimination to solve systems of equations. We can lighten the load somewhat by using a **matrix** (pl.: **matrices**), which is simply a rectangular array of numbers enclosed in parentheses or brackets. Here are three examples:

$$\begin{pmatrix} 2 & 3 \\ -5 & 4 \end{pmatrix} \qquad \begin{pmatrix} -6 & 0 & 1 & \frac{1}{4} \\ \frac{2}{3} & 1 & 5 & 8 \end{pmatrix} \qquad \begin{pmatrix} \pi & 0 & 0 \\ 0 & 1 & 9 \\ -1 & -2 & 3 \\ -4 & 8 & 6 \end{pmatrix}$$

The particular numbers constituting a matrix are called its **entries** or **elements.** For instance, the entries in the matrix $\begin{pmatrix} 2 & 3 \\ -5 & 4 \end{pmatrix}$ are the four numbers 2, 3, -5, and 4. In this section, the entries will always be real numbers. However, it is also possible to consider matrices in which some or all of the entries are nonreal complex numbers.

It is convenient to agree on a standard system for labeling the rows and columns of a matrix. The rows are numbered from top to bottom and the columns from left to right, as indicated in the following example:

$$\begin{matrix} & & \text{column 1} & \text{column 2} \\ & & \downarrow & \downarrow \\ \text{row 1} & \rightarrow & \\ \text{row 2} & \rightarrow & \\ \text{row 3} & \rightarrow & \end{matrix} \begin{pmatrix} 5 & -3 \\ 0 & 2 \\ -1 & 16 \end{pmatrix}$$

We express the **size,** or **dimension,** of a matrix by specifying the number of rows and the number of columns, in that order. For instance, we would say that the matrix

$$\begin{pmatrix} 5 & -3 \\ 0 & 2 \\ -1 & 16 \end{pmatrix}$$

is a 3×2 (read "3 by 2") matrix, not 2×3. The following example will help fix in your mind the terminology that we have introduced.

EXAMPLE 1 Consider the matrix

$$\begin{pmatrix} 1 & 3 & 5 \\ 7 & 9 & 11 \end{pmatrix}$$

(a) List the entries.
(b) What is the size of the matrix?
(c) Which element is in the second row, third column?

Solution

(a) The entries are 1, 3, 5, 7, 9, and 11.
(b) Since there are two rows and three columns, this is a 2×3 matrix.
(c) To locate the element in the second row, third column, we draw lines through the second row and the third column and see where they intersect:

$$\begin{pmatrix} 1 & 3 & 5 \\ 7 & 9 & 11 \end{pmatrix}$$

Thus, the entry in row 2, column 3 is 11.

There is a natural way to use matrices to describe and solve systems of linear equations. Consider, for example, the following system of linear equations in **standard form** (with the x-, y-, and z-terms lined up on the lefthand side and the constant terms on the right-hand side of each equation):

$$\begin{cases} x + 2y - 3z = 4 \\ 3x \qquad\; + z = 5 \\ -x - 3y + 4z = 0 \end{cases}$$

The **coefficient matrix** of this system is the matrix

$$\begin{pmatrix} 1 & 2 & -3 \\ 3 & 0 & 1 \\ -1 & -3 & 4 \end{pmatrix}$$

As the name implies, the coefficient matrix of the system is the matrix whose entries are the coefficients of x, y, and z, written in the same relative positions as they appear in the system. Notice the zero appearing in the second row, second column of the matrix. It is there because the coefficient of y in the second equation is in fact zero. The **augmented matrix** of the system of equations considered here is

$$\left(\begin{array}{rrr|r} 1 & 2 & -3 & 4 \\ 3 & 0 & 1 & 5 \\ -1 & -3 & 4 & 0 \end{array}\right)$$

As you can see, the augmented matrix is formed by *augmenting* the coefficient matrix with the column of constant terms taken from the right-hand side of the given system of equations. To help relate the augmented matrix to the original system of equations, we will write the augmented matrix this way:

$$\left(\begin{array}{ccc|c} 1 & 2 & -3 & 4 \\ 3 & 0 & 1 & 5 \\ -1 & -3 & 4 & 0 \end{array}\right)$$

EXAMPLE 2 Write the coefficient matrix and the augmented matrix for the system

$$\begin{cases} 8x - 2y + z = 1 \\ 3x - 4z + y = 2 \\ 12y - 3z - 6 = 0 \end{cases}$$

Solution First we write the system in standard form, with the x-, y-, and z-terms lined up and the constant terms on the right. This yields

$$\begin{cases} 8x - 2y + z = 1 \\ 3x + y - 4z = 2 \\ 12y - 3z = 6 \end{cases}$$

The coefficient matrix is then

$$\begin{pmatrix} 8 & -2 & 1 \\ 3 & 1 & -4 \\ 0 & 12 & -3 \end{pmatrix}$$

while the augmented matrix is

$$\left(\begin{array}{ccc|c} 8 & -2 & 1 & 1 \\ 3 & 1 & -4 & 2 \\ 0 & 12 & -3 & 6 \end{array}\right)$$

In the previous section we used the three elementary operations in solving systems of linear equations. In Table 1 we express these operations in the language of matrices. The matrix operations are called the **elementary row operations.**

TABLE 1

Elementary Operations for a System of Linear Equations	Corresponding Elementary Row Operations for a Matrix
1. Multiply both sides of an equation by a nonzero constant.	1′. Multiply each entry in a given row by a nonzero constant.
2. Interchange two equations.	2′. Interchange two rows.
3. To one equation, add a multiple of another equation.	3′. To one row, add a multiple of another row.

Table 2 displays an example of each elementary row operation. In the table, notice that the notation for describing these operations is essentially the same as that introduced in the previous section. For example, in the previous section, $10E_1$ indicated that the first equation in a system was multiplied by 10. Now, $10R_1$ indicates that each entry in the first *row* of a matrix is multiplied by 10.

TABLE 2

Examples of the Elementary Row Operations	Comments
$\begin{pmatrix} 1 & 2 & 3 \\ 4 & 5 & 6 \\ 7 & 8 & 9 \end{pmatrix} \xrightarrow{10R_1} \begin{pmatrix} 10 & 20 & 30 \\ 4 & 5 & 6 \\ 7 & 8 & 9 \end{pmatrix}$	Multiply each entry in row 1 by 10.
$\begin{pmatrix} 1 & 2 & 3 \\ 4 & 5 & 6 \\ 7 & 8 & 9 \end{pmatrix} \xrightarrow{R_2 \leftrightarrow R_3} \begin{pmatrix} 1 & 2 & 3 \\ 7 & 8 & 9 \\ 4 & 5 & 6 \end{pmatrix}$	Interchange rows 2 and 3.
$\begin{pmatrix} 1 & 2 & 3 \\ 4 & 5 & 6 \\ 7 & 8 & 9 \end{pmatrix} \xrightarrow{-4R_1 + R_2} \begin{pmatrix} 1 & 2 & 3 \\ 0 & -3 & -6 \\ 7 & 8 & 9 \end{pmatrix}$	To each entry in row 2, add -4 times the corresponding entry in row 1.

We are now ready to use matrices to solve systems of equations. In the example that follows, we'll use the same system of equations used in Example 5 of the previous section. *Suggestion:* After reading the next example, carefully compare each step with the corresponding one taken in Example 5 of the previous section.

EXAMPLE 3 Solve the system

$$\begin{cases} 4x - 3y + 2z = 40 \\ 5x + 9y - 7z = 47 \\ 9x + 8y - 3z = 97 \end{cases}$$

Solution

$$\left(\begin{array}{ccc:c} 4 & -3 & 2 & 40 \\ 5 & 9 & -7 & 47 \\ 9 & 8 & -3 & 97 \end{array}\right) \xrightarrow{(-1)R_2 + R_1} \left(\begin{array}{ccc:c} -1 & -12 & 9 & -7 \\ 5 & 9 & -7 & 47 \\ 9 & 8 & -3 & 97 \end{array}\right)$$

$$\xrightarrow[9R_1 + R_3]{5R_1 + R_2} \left(\begin{array}{ccc:c} -1 & -12 & 9 & -7 \\ 0 & -51 & 38 & 12 \\ 0 & -100 & 78 & 34 \end{array}\right)$$

$$\xrightarrow{\frac{1}{2}R_3} \left(\begin{array}{ccc:c} -1 & -12 & 9 & -7 \\ 0 & -51 & 38 & 12 \\ 0 & -50 & 39 & 17 \end{array}\right)$$

$$\xrightarrow{(-1)R_3 + R_2} \left(\begin{array}{rrr|r} -1 & -12 & 9 & -7 \\ 0 & -1 & -1 & -5 \\ 0 & -50 & 39 & 17 \end{array}\right)$$

$$\xrightarrow{-50R_2 + R_3} \left(\begin{array}{rrr|r} -1 & -12 & 9 & -7 \\ 0 & -1 & -1 & -5 \\ 0 & 0 & 89 & 267 \end{array}\right)$$

This last augmented matrix represents a system of equations in upper-triangular form:

$$\begin{cases} -x - 12y + 9z = -7 \\ \quad\;\; -y - z = -5 \\ \qquad\qquad 89z = 267 \end{cases}$$

As you should now check for yourself, this yields the values $z = 3$, then $y = 2$, then $x = 10$. The solution of the original system is therefore $(10, 2, 3)$.

For the remainder of this section (and in part of the next) we are going to study matrices without referring to systems of equations. As motivation for this, we point out that matrices are essential tools in many fields of study. For example, a knowledge of matrices and their properties is needed for work in computer graphics. To begin, we need to say what it means for two matrices to be equal.

DEFINITION: Equality of Matrices

Two matrices are equal provided they have the same size (same number of rows, same number of columns) and the corresponding entries are equal.

EXAMPLES

$$\begin{pmatrix} 2 & 3 \\ 4 & 5 \end{pmatrix} = \begin{pmatrix} 2 & 3 \\ 4 & 5 \end{pmatrix}$$

$$\begin{pmatrix} 2 & 3 & 0 \\ 4 & 5 & 0 \end{pmatrix} \neq \begin{pmatrix} 2 & 3 \\ 4 & 5 \end{pmatrix}$$

$$\begin{pmatrix} 2 & 3 \\ 4 & 5 \end{pmatrix} \neq \begin{pmatrix} 2 & 3 \\ 5 & 4 \end{pmatrix}$$

Now we can define matrix addition and subtraction. These operations are defined only between matrices of the same size.

DEFINITION: Matrix Addition

To add (or subtract) two matrices of the same size, add (or subtract) the corresponding entries.

EXAMPLES

$$\begin{pmatrix} 2 & 3 \\ -1 & 4 \end{pmatrix} + \begin{pmatrix} 6 & 1 \\ 0 & -4 \end{pmatrix} = \begin{pmatrix} 8 & 4 \\ -1 & 0 \end{pmatrix}$$

$$\begin{pmatrix} 2 & 3 \\ -1 & 4 \\ 9 & 10 \end{pmatrix} - \begin{pmatrix} 6 & 1 \\ 0 & -4 \\ 3 & 2 \end{pmatrix} = \begin{pmatrix} -4 & 2 \\ -1 & 8 \\ 6 & 8 \end{pmatrix}$$

Many properties of the real numbers also apply to matrices. For instance, matrix addition is *commutative*:

$$A + B = B + A \qquad \text{where } A \text{ and } B \text{ are matrices of the same size}$$

Matrix addition is also *associative*:

$$A + (B + C) = (A + B) + C \qquad \text{where } A, B, \text{ and } C \text{ are matrices of the same size}$$

In the next example, we verify these properties in two specific instances.

EXAMPLE 4 Let $A = \begin{pmatrix} 1 & 2 \\ 3 & 4 \end{pmatrix}$, $B = \begin{pmatrix} 0 & -5 \\ 8 & -1 \end{pmatrix}$, and $C = \begin{pmatrix} 6 & 7 \\ 8 & 9 \end{pmatrix}$.

(a) Show that $A + C = C + A$.
(b) Show that $A + (B + C) = (A + B) + C$.

Solution

(a) $A + C = \begin{pmatrix} 1 & 2 \\ 3 & 4 \end{pmatrix} + \begin{pmatrix} 6 & 7 \\ 8 & 9 \end{pmatrix} = \begin{pmatrix} 7 & 9 \\ 11 & 13 \end{pmatrix}$

$$C + A = \begin{pmatrix} 6 & 7 \\ 8 & 9 \end{pmatrix} + \begin{pmatrix} 1 & 2 \\ 3 & 4 \end{pmatrix} = \begin{pmatrix} 7 & 9 \\ 11 & 13 \end{pmatrix}$$

This shows that $A + C = C + A$, since both $A + C$ and $C + A$ represent the matrix $\begin{pmatrix} 7 & 9 \\ 11 & 13 \end{pmatrix}$.

(b) First we compute $A + (B + C)$:

$$\begin{aligned} A + (B + C) &= \begin{pmatrix} 1 & 2 \\ 3 & 4 \end{pmatrix} + \left[\begin{pmatrix} 0 & -5 \\ 8 & -1 \end{pmatrix} + \begin{pmatrix} 6 & 7 \\ 8 & 9 \end{pmatrix}\right] \\ &= \begin{pmatrix} 1 & 2 \\ 3 & 4 \end{pmatrix} + \begin{pmatrix} 6 & 2 \\ 16 & 8 \end{pmatrix} \\ &= \begin{pmatrix} 7 & 4 \\ 19 & 12 \end{pmatrix} \end{aligned}$$

Next we compute $(A + B) + C$:

$$\begin{aligned} (A + B) + C &= \left[\begin{pmatrix} 1 & 2 \\ 3 & 4 \end{pmatrix} + \begin{pmatrix} 0 & -5 \\ 8 & -1 \end{pmatrix}\right] + \begin{pmatrix} 6 & 7 \\ 8 & 9 \end{pmatrix} \\ &= \begin{pmatrix} 1 & -3 \\ 11 & 3 \end{pmatrix} + \begin{pmatrix} 6 & 7 \\ 8 & 9 \end{pmatrix} \\ &= \begin{pmatrix} 7 & 4 \\ 19 & 12 \end{pmatrix} \end{aligned}$$

We conclude from these calculations that $A + (B + C) = (A + B) + C$, since both sides of that equation represent the matrix $\begin{pmatrix} 7 & 4 \\ 19 & 12 \end{pmatrix}$.

We will now define an operation on matrices called **scalar multiplication.** First of all, the word *scalar* here just means *real number,* so we are talking about multiplying a matrix by a real number. (In more advanced work, nonreal complex scalars are also considered.)

DEFINITION: Scalar Multiplication

To multiply a matrix by a scalar, multiply each entry in the matrix by that scalar.

EXAMPLES

$$2\begin{pmatrix} 5 & 9 & 0 \\ -1 & 2 & 3 \end{pmatrix} = \begin{pmatrix} 10 & 18 & 0 \\ -2 & 4 & 6 \end{pmatrix}$$

$$1\begin{pmatrix} 1 & 2 \\ 3 & 4 \end{pmatrix} = \begin{pmatrix} 1 & 2 \\ 3 & 4 \end{pmatrix}$$

There are two simple but useful properties of scalar multiplication that are worth noting at this point. We'll omit the proofs of these two properties; however, Example 5 does ask us to verify them for a particular case.

PROPERTY SUMMARY — PROPERTIES OF SCALAR MULTIPLICATION

1. $c(kM) = (ck)M$ for all scalars c and k and any matrix M
2. $c(M + N) = cM + cN$, where c is any scalar and M and N are any matrices of the same size

EXAMPLE 5 Let $c = 2$, $k = 3$, $M = \begin{pmatrix} 1 & 2 \\ 3 & 4 \end{pmatrix}$, and $N = \begin{pmatrix} 5 & 6 \\ 7 & 8 \end{pmatrix}$.

(a) Show that $c(kM) = (ck)M$.
(b) Show that $c(M + N) = cM = cN$.

Solution

(a) $$c(kM) = 2\left[3\begin{pmatrix} 1 & 2 \\ 3 & 4 \end{pmatrix}\right]$$

$$= 2\begin{pmatrix} 3 & 6 \\ 9 & 12 \end{pmatrix} = \begin{pmatrix} 6 & 12 \\ 18 & 24 \end{pmatrix}$$

$$(ck)M = (2 \cdot 3)\begin{pmatrix} 1 & 2 \\ 3 & 4 \end{pmatrix}$$

$$= 6\begin{pmatrix} 1 & 2 \\ 3 & 4 \end{pmatrix} = \begin{pmatrix} 6 & 12 \\ 18 & 24 \end{pmatrix}$$

Thus, $c(kM) = (ck)M$, since in both cases the result is $\begin{pmatrix} 6 & 12 \\ 18 & 24 \end{pmatrix}$.

(b) $c(M+N) = 2\left[\begin{pmatrix} 1 & 2 \\ 3 & 4 \end{pmatrix} + \begin{pmatrix} 5 & 6 \\ 7 & 8 \end{pmatrix}\right]$

$$= 2\begin{pmatrix} 6 & 8 \\ 10 & 12 \end{pmatrix} = \begin{pmatrix} 12 & 16 \\ 20 & 24 \end{pmatrix}$$

$$cM + cN = 2\begin{pmatrix} 1 & 2 \\ 3 & 4 \end{pmatrix} + 2\begin{pmatrix} 5 & 6 \\ 7 & 8 \end{pmatrix}$$

$$= \begin{pmatrix} 2 & 4 \\ 6 & 8 \end{pmatrix} + \begin{pmatrix} 10 & 12 \\ 14 & 16 \end{pmatrix} = \begin{pmatrix} 12 & 16 \\ 20 & 24 \end{pmatrix}$$

Thus, $c(M+N) = cM + cN$, since both sides equal $\begin{pmatrix} 12 & 16 \\ 20 & 24 \end{pmatrix}$.

A matrix with zeros for all of its entries plays the same role in matrix addition as does the number zero in ordinary addition of real numbers. For instance, in the case of 2×2 matrices, we have

$$\begin{pmatrix} a & b \\ c & d \end{pmatrix} + \begin{pmatrix} 0 & 0 \\ 0 & 0 \end{pmatrix} = \begin{pmatrix} a & b \\ c & d \end{pmatrix}$$

and

$$\begin{pmatrix} 0 & 0 \\ 0 & 0 \end{pmatrix} + \begin{pmatrix} a & b \\ c & d \end{pmatrix} = \begin{pmatrix} a & b \\ c & d \end{pmatrix}$$

for all real numbers a, b, c, and d. The matrix $\begin{pmatrix} 0 & 0 \\ 0 & 0 \end{pmatrix}$ is called the **additive identity** for 2×2 matrices. Similarly, any matrix with all zero entries is the additive theory for matrices of that size. It is sometimes convenient to denote an additive identity matrix by a boldface zero: **0.** With this notation, we can write

$$A + \mathbf{0} = \mathbf{0} + A = A \qquad \text{for any matrix } A$$

For this matrix equation, it is understood that the size of the matrix $\mathbf{0}$ is the same as the size of A. With this notation, we also have

$$A - A = \mathbf{0} \qquad \text{for any matrix } A$$

Our last topic in this section is matrix multiplication. We will begin with the simplest case and then work up to the more general situation. By convention, a matrix with only one row is called a **row vector.** Examples of row vectors are

$$(2 \quad 13), \qquad (-1 \quad 4 \quad 3), \qquad \text{and} \qquad (0 \quad 0 \quad 0 \quad 1)$$

Similarly, a matrix with only one column is called a **column vector.** Examples are

$$\begin{pmatrix} 2 \\ 13 \end{pmatrix}, \quad \begin{pmatrix} -1 \\ 4 \\ 3 \end{pmatrix}, \quad \text{and} \quad \begin{pmatrix} 0 \\ 0 \\ 0 \\ 1 \end{pmatrix}$$

The following definition tells us how to multiply a row vector and a column vector when they have the same number of entries.

DEFINITION: The Inner Product of a Row Vector and a Column Vector

Let A be a row vector and B a column vector, and assume that the number of columns in A is the same as the number of rows in B. Then the **inner product** $A \cdot B$ is defined to be the number obtained by multiplying the corresponding entries and then adding the products.

EXAMPLES

$$(1 \quad 2 \quad 3) \cdot \begin{pmatrix} 4 \\ 5 \\ 6 \end{pmatrix} = 1 \cdot 4 + 2 \cdot 5 + 3 \cdot 6 = 32$$

$$(1 \quad 2) \cdot \begin{pmatrix} 4 \\ 5 \\ 6 \end{pmatrix} \quad \text{is not defined}$$

$$(1 \quad 2 \quad 3) \cdot \begin{pmatrix} 4 \\ 5 \end{pmatrix} \quad \text{is not defined}$$

An important observation here is that the end result of taking the inner product is always just a number. The definition of matrix product that we now give depends on this observation.

DEFINITION: The Product of Two Matrices

Let A and B be two matrices, and assume that the number of columns in A is the same as the number of rows in B. Then the **product matrix** AB is computed according to the following rule:

The entry in the ith row and the jth column of AB is the inner product of the ith row of A with the jth column of B.

The matrix AB will have as many rows as A and as many columns as B.

As an example of matrix multiplication, we will compute the product AB, where $A = \begin{pmatrix} 1 & 2 \\ 3 & 4 \end{pmatrix}$ and $B = \begin{pmatrix} 5 & 6 & 0 \\ 7 & 8 & 1 \end{pmatrix}$. In other words, we will compute $\begin{pmatrix} 1 & 2 \\ 3 & 4 \end{pmatrix}\begin{pmatrix} 5 & 6 & 0 \\ 7 & 8 & 1 \end{pmatrix}$. Before we attempt to carry out the calculations of any matrix multiplication, however, we should check on two points.

1. Is the product defined? That is, does the number of columns in A equal the number of rows in B? In this case, yes; the common number is 2.
2. What is the size of the product? According to the definition, the product AB will have as many rows as A and as many columns as B. Thus, the size of AB will be 2×3.

Schematically, then, the situation looks like this:

$$\begin{pmatrix} 1 & 2 \\ 3 & 4 \end{pmatrix}\begin{pmatrix} 5 & 6 & 0 \\ 7 & 8 & 1 \end{pmatrix} = \begin{pmatrix} ? & ? & ? \\ ? & ? & ? \end{pmatrix}$$

We have six positions to fill. The computations are presented in Table 3. Reading from the table, we see that our result is

$$AB = \begin{pmatrix} 1 & 2 \\ 3 & 4 \end{pmatrix}\begin{pmatrix} 5 & 6 & 0 \\ 7 & 8 & 1 \end{pmatrix} = \begin{pmatrix} 19 & 22 & 2 \\ 43 & 50 & 4 \end{pmatrix}$$

TABLE 3

Position	How to Compute	Computation
row 1, column 1	inner product of row 1 and column 1 $\begin{pmatrix} 1 & 2 \\ 3 & 4 \end{pmatrix}\begin{pmatrix} 5 & 6 & 0 \\ 7 & 8 & 1 \end{pmatrix}$	$1 \cdot 5 + 2 \cdot 7 = 19$
row 1, column 2	inner product of row 1 and column 2 $\begin{pmatrix} 1 & 2 \\ 3 & 4 \end{pmatrix}\begin{pmatrix} 5 & 6 & 0 \\ 7 & 8 & 1 \end{pmatrix}$	$1 \cdot 6 + 2 \cdot 8 = 22$
row 1, column 3	inner product of row 1 and column 3 $\begin{pmatrix} 1 & 2 \\ 3 & 4 \end{pmatrix}\begin{pmatrix} 5 & 6 & 0 \\ 7 & 8 & 1 \end{pmatrix}$	$1 \cdot 0 + 2 \cdot 1 = 2$
row 2, column 1	inner product of row 2 and column 1 $\begin{pmatrix} 1 & 2 \\ 3 & 4 \end{pmatrix}\begin{pmatrix} 5 & 6 & 0 \\ 7 & 8 & 1 \end{pmatrix}$	$3 \cdot 5 + 4 \cdot 7 = 43$
row 2, column 2	inner product of row 2 and column 2 $\begin{pmatrix} 1 & 2 \\ 3 & 4 \end{pmatrix}\begin{pmatrix} 5 & 6 & 0 \\ 7 & 8 & 1 \end{pmatrix}$	$3 \cdot 6 + 4 \cdot 8 = 50$
row 2, column 3	inner product of row 2 and column 3 $\begin{pmatrix} 1 & 2 \\ 3 & 4 \end{pmatrix}\begin{pmatrix} 5 & 6 & 0 \\ 7 & 8 & 1 \end{pmatrix}$	$3 \cdot 0 + 4 \cdot 1 = 4$

EXAMPLE 6 Let $A = \begin{pmatrix} 1 & 2 \\ 3 & 4 \end{pmatrix}$ and $B = \begin{pmatrix} 5 & 6 \\ 7 & 8 \end{pmatrix}$. By computing AB and then BA, show that $AB \neq BA$. This shows that, in general, matrix multiplication is not commutative.

Solution

$$AB = \begin{pmatrix} 1 & 2 \\ 3 & 4 \end{pmatrix}\begin{pmatrix} 5 & 6 \\ 7 & 8 \end{pmatrix} = \begin{pmatrix} 1 \cdot 5 + 2 \cdot 7 & 1 \cdot 6 + 2 \cdot 8 \\ 3 \cdot 5 + 4 \cdot 7 & 3 \cdot 6 + 4 \cdot 8 \end{pmatrix} = \begin{pmatrix} 19 & 22 \\ 43 & 50 \end{pmatrix}$$

$$BA = \begin{pmatrix} 5 & 6 \\ 7 & 8 \end{pmatrix}\begin{pmatrix} 1 & 2 \\ 3 & 4 \end{pmatrix} = \begin{pmatrix} 5 \cdot 1 + 6 \cdot 3 & 5 \cdot 2 + 6 \cdot 4 \\ 7 \cdot 1 + 8 \cdot 3 & 7 \cdot 2 + 8 \cdot 4 \end{pmatrix} = \begin{pmatrix} 23 & 34 \\ 31 & 46 \end{pmatrix}$$

Comparing the two matrices AB and BA, we conclude that $AB \neq BA$.

EXERCISE SET 10.3

A

In Exercises 1–4, specify the size of each matrix.

1. **(a)** $\begin{pmatrix} -4 & 0 & 5 \\ 2 & 8 & -1 \end{pmatrix}$ **(b)** $\begin{pmatrix} 7 & 1 \\ 4 & -3 \\ 0 & 0 \end{pmatrix}$

2. **(a)** $\begin{pmatrix} 1 & 0 \\ 0 & -1 \end{pmatrix}$ **(b)** $\begin{pmatrix} 1 \\ 6 \\ 8 \\ 1 \end{pmatrix}$

3. $\begin{pmatrix} 1 & a & b & c \\ a & 1 & 0 & a \\ b & 0 & 1 & b \\ c & a & b & 1 \\ 0 & 0 & 0 & 1 \end{pmatrix}$ **4.** $(-3 \quad 1 \quad 6 \quad 0)$

In Exercises 5–8, write the coefficient matrix and the augmented matrix for each system.

5. $\begin{cases} 2x + 3y + 4z = 10 \\ 5x + 6y + 7z = 9 \\ 8x + 9y + 10z = 8 \end{cases}$ **6.** $\begin{cases} 5x - y + z = 0 \\ 4y + 2z = 1 \\ 3x + y + z = -1 \end{cases}$

7. $\begin{cases} x + z + w = -1 \\ x + y + 2w = 0 \\ y + z + w = 1 \\ 2x - y - z = 2 \end{cases}$ **8.** $\begin{cases} 8x - 8y = 5 \\ x - y + z = 1 \end{cases}$

In Exercises 9–22, use matrices to solve each system of equations.

9. $\begin{cases} x - y + 2z = 7 \\ 3x + 2y - z = -10 \\ -x + 3y + z = -2 \end{cases}$ **10.** $\begin{cases} 2x - 3y + 4z = 14 \\ 3x - 2y + 2z = 12 \\ 4x + 5y - 5z = 16 \end{cases}$

11. $\begin{cases} x + z = -2 \\ -3x + 2y = 17 \\ x - y - z = -9 \end{cases}$ **12.** $\begin{cases} 5x + y + 10z = 23 \\ 4x + 2y - 10z = 76 \\ 3x - 4y = 18 \end{cases}$

13. $\begin{cases} x + y + z = -4 \\ 2x - 3y + z = -1 \\ 4x + 2y - 3z = 33 \end{cases}$ **14.** $\begin{cases} 2x + 3y - 4z = 7 \\ x - y + z = -\frac{3}{2} \\ 6x - 5y - 2z = -7 \end{cases}$

15. $\begin{cases} 3x - 2y + 6z = 0 \\ x + 3y + 20z = 15 \\ 10x - 11y - 10z = -9 \end{cases}$

16. $\begin{cases} 3A - 3B + C = 4 \\ 6A + 9B - 3C = -7 \\ A - 2B - 2C = -3 \end{cases}$

17. $\begin{cases} 4x - 3y + 3z = 2 \\ 5x + y - 4z = 1 \\ 9x - 2y - z = 3 \end{cases}$

18. $\begin{cases} 6x + y - z = -1 \\ -3x + 2y + 2z = 2 \\ 5y + 3z = 1 \end{cases}$

19. $\begin{cases} x - y + z + w = 6 \\ x + y - z + w = 4 \\ x + y + z - w = -2 \\ -x + y + z + w = 0 \end{cases}$

20. $\begin{cases} x + 2y - z - 2w = 5 \\ 2x + y + 2z + w = -7 \\ -2x - y - 3z - 2w = 10 \\ z + w = -3 \end{cases}$

21. $\begin{cases} 15A + 14B + 26C = 1 \\ 18A + 17B + 32C = -1 \\ 21A + 20B + 38C = 0 \end{cases}$

22. $\begin{cases} A + B + C + D + E = -1 \\ 3A - 2B - 2C + 3D + 2E = 13 \\ 3C + 4D - 4E = 7 \\ 5A - 4B + E = 30 \\ C - 2E = 3 \end{cases}$

In Exercises 23–50, the matrices A, B, C, D, E, F, and G are defined as follows:

$$A = \begin{pmatrix} 2 & 3 \\ -1 & 4 \end{pmatrix} \quad B = \begin{pmatrix} 1 & -1 \\ 3 & 0 \end{pmatrix} \quad C = \begin{pmatrix} 1 & 0 \\ 0 & 1 \end{pmatrix}$$

$$D = \begin{pmatrix} -1 & 2 & 3 \\ 4 & 0 & 5 \end{pmatrix} \quad E = \begin{pmatrix} 2 & 1 \\ 8 & -1 \\ 6 & 5 \end{pmatrix}$$

$$F = \begin{pmatrix} 5 & -1 \\ -4 & 0 \\ 2 & 3 \end{pmatrix} \quad G = \begin{pmatrix} 0 & 0 \\ 0 & 0 \\ 0 & 0 \end{pmatrix}$$

In each exercise, carry out the indicated matrix operations if they are defined. If an operation is not defined, say so.

23. $A + B$ **24.** $A - B$ **25.** $2A + 2B$
26. $2(A + B)$ **27.** AB **28.** BA
29. AC **30.** CA **31.** $3D + E$
32. $E + F$ **33.** $2F - 3G$ **34.** DE
35. ED **36.** DF **37.** FD
38. $A + D$ **39.** $G + A$ **40.** DG
41. GD **42.** $(A + B) + C$ **43.** $A + (B + C)$
44. CD **45.** DC **46.** $5E - 3F$
47. $A^2(= AA)$ **48.** A^2A **49.** AA^2
50. C^2
51. Let

$$A=\begin{pmatrix}-1 & 3 & 4\\ 3 & 2 & -3\\ 9 & 1 & 6\end{pmatrix} \quad B=\begin{pmatrix}7 & 0 & 1\\ 0 & 0 & 3\\ -1 & 2 & 4\end{pmatrix}$$

$$C=\begin{pmatrix}4 & 6 & 1\\ 2 & 1 & 3\\ -1 & -1 & 2\end{pmatrix}$$

(a) Compute $A(B+C)$. **(b)** Compute $AB+AC$.
(c) Compute $(AB)C$. **(d)** Compute $A(BC)$.

B

52. Let $A=\begin{pmatrix}1 & 2\\ 3 & 4\end{pmatrix}$ and $B=\begin{pmatrix}5 & 6\\ 7 & 8\end{pmatrix}$. Let A^2 and B^2 denote the matrix products AA and BB, respectively. Compute each of the following.
(a) $(A+B)(A+B)$ **(b)** $A^2+2AB+B^2$
(c) $A^2+AB+BA+B^2$

53. Let $A=\begin{pmatrix}3 & 5\\ 7 & 9\end{pmatrix}$ and $B=\begin{pmatrix}2 & 4\\ 6 & 8\end{pmatrix}$. Compute each of the following.
(a) A^2-B^2 **(b)** $(A-B)(A+B)$
(c) $(A+B)(A-B)$ **(d)** $A^2+AB-BA-B^2$

54. Let

$$A=\begin{pmatrix}1 & 0\\ 0 & 1\end{pmatrix} \quad B=\begin{pmatrix}1 & 0\\ 0 & -1\end{pmatrix}$$

$$C=\begin{pmatrix}-1 & 0\\ 0 & 1\end{pmatrix} \quad D=\begin{pmatrix}-1 & 0\\ 0 & -1\end{pmatrix}$$

Complete the following multiplication table.

	A	B	C	D
A				
B			D	
C				
D				

Hint: In the second row, third column, D is the proper entry because (as you can check) $BC=D$.

55. In this exercise, let's agree to write the coordinates (x, y) of a point in the plane as the 2×1 matrix $\begin{pmatrix}x\\ y\end{pmatrix}$.

(a) Let $A=\begin{pmatrix}1 & 0\\ 0 & -1\end{pmatrix}$ and $Z=\begin{pmatrix}x\\ y\end{pmatrix}$. Compute the matrix AZ. After computing AZ, observe that it represents the point obtained by reflecting $\begin{pmatrix}x\\ y\end{pmatrix}$ about the x-axis.

(b) Let $B=\begin{pmatrix}-1 & 0\\ 0 & 1\end{pmatrix}$ and $Z=\begin{pmatrix}x\\ y\end{pmatrix}$. Compute the matrix BZ. After computing BZ, observe that it represents the point obtained by reflecting $\begin{pmatrix}x\\ y\end{pmatrix}$ about the y-axis.

(c) Let A, B, and Z represent the matrices defined in parts (a) and (b). Compute the matrix $(AB)Z$, and then interpret it in terms of reflection about the axes.

56. In this exercise, we continue to explore some of the connections between matrices and geometry. As in Exercise 55, we will use 2×1 matrices to specify the coordinates of points in the plane. Let P, S, and T be the matrices defined as follows:

$$P=\begin{pmatrix}\cos x\\ \sin x\end{pmatrix} \quad S=\begin{pmatrix}\cos\theta & -\sin\theta\\ \sin\theta & \cos\theta\end{pmatrix}$$

$$T=\begin{pmatrix}\cos\beta & -\sin\beta\\ \sin\beta & \cos\beta\end{pmatrix}$$

Notice that the point P lies on the unit circle.

(a) Compute the matrix SP. After computing SP, observe that it represents the point on the unit circle obtained by rotating P (about the origin) through an angle θ.

(b) Show that

$$ST=TS=\begin{pmatrix}\cos(\theta+\beta) & -\sin(\theta+\beta)\\ \sin(\theta+\beta) & \cos(\theta+\beta)\end{pmatrix}$$

(c) Compose $(ST)P$? What is the angle through which P is rotated?

57. A function f is defined as follows.The domain of f is the set of all 2×2 matrices (with real entries). If $A=\begin{pmatrix}a & b\\ c & d\end{pmatrix}$, then $f(A)=ad-bc$.

(a) Let $A=\begin{pmatrix}1 & 2\\ 3 & 4\end{pmatrix}$ and $B=\begin{pmatrix}3 & -1\\ 5 & 8\end{pmatrix}$. Compute $f(A), f(B)$, and $f(AB)$. Is it true, in this case, that $f(A)\cdot f(B)=f(AB)$?

(b) Let $A=\begin{pmatrix}a & b\\ c & d\end{pmatrix}$ and $B=\begin{pmatrix}e & f\\ g & h\end{pmatrix}$. Show that $f(A)\cdot f(B)=f(AB)$.

58. The **trace** of a 2×2 matrix $\begin{pmatrix}a & b\\ c & d\end{pmatrix}$ is defined by

$$\operatorname{tr}\begin{pmatrix}a & b\\ c & d\end{pmatrix}=a+d$$

(a) If $A=\begin{pmatrix}1 & 2\\ 3 & 4\end{pmatrix}$ and $B=\begin{pmatrix}5 & 6\\ 7 & 8\end{pmatrix}$, verify that $\operatorname{tr}(A+B)=\operatorname{tr}A+\operatorname{tr}B$.

(b) If $A=\begin{pmatrix}a & b\\ c & d\end{pmatrix}$ and $B=\begin{pmatrix}e & f\\ g & h\end{pmatrix}$, show that $\operatorname{tr}(A+B)=\operatorname{tr}A+\operatorname{tr}B$.

59. Let $A = \begin{pmatrix} a & b \\ c & d \end{pmatrix}$. The **transpose** of A is the matrix denoted by A^{T} and defined by $A^{\mathrm{T}} = \begin{pmatrix} a & c \\ b & d \end{pmatrix}$. In other words, A^{T} is obtained by switching the columns and rows of A. Show that the following equations hold for all 2×2 matrices A and B.
(a) $(A+B)^{\mathrm{T}} = A^{\mathrm{T}} + B^{\mathrm{T}}$ **(b)** $(A^{\mathrm{T}})^{\mathrm{T}} = A$
(c) $(AB)^{\mathrm{T}} = B^{\mathrm{T}}A^{\mathrm{T}}$

60. Find an example of two 2×2 matrices A and B for which $AB = \mathbf{0}$ but neither A nor B is $\mathbf{0}$.

THE INVERSE OF A SQUARE MATRIX

We have noted Cayley's work in analytic geometry, especially in connection with the use of determinants; but [Arthur] *Cayley* [1821–1895] *also was one of the first men to study matrices, another instance of the British concern for form and structure in algebra.*

Carl B. Boyer in *A History of Mathematics,* 2nd ed., revised by Uta C. Merzback (New York: John Wiley and Sons, 1991)

A matrix that has the same number of rows and columns is called a **square matrix.** So, two examples of square matrices are

$$A = \begin{pmatrix} 1 & 2 \\ 3 & 4 \end{pmatrix} \quad \text{and} \quad B = \begin{pmatrix} -5 & 6 & 7 \\ \frac{1}{2} & 0 & 1 \\ 8 & 4 & -3 \end{pmatrix}$$

The matrix A is said to be a square matrix of **order two** (or, more simply, a 2×2 matrix); B is a square matrix of **order three** (that is, a 3×3 matrix). We will first present the concepts and techniques of this section in terms of square matrices of order two. After that, we'll show how the ideas carry over to larger square matrices.

We begin by defining a special matrix I_2:

$$I_2 = \begin{pmatrix} 1 & 0 \\ 0 & 1 \end{pmatrix}$$

The matrix I_2 has the following property: For every 2×2 matrix $A = \begin{pmatrix} a & b \\ c & d \end{pmatrix}$, we have (as you can easily verify)

$$\begin{pmatrix} a & b \\ c & d \end{pmatrix}\begin{pmatrix} 1 & 0 \\ 0 & 1 \end{pmatrix} = \begin{pmatrix} a & b \\ c & d \end{pmatrix} \quad \text{and} \quad \begin{pmatrix} 1 & 0 \\ 0 & 1 \end{pmatrix}\begin{pmatrix} a & b \\ c & d \end{pmatrix} = \begin{pmatrix} a & b \\ c & d \end{pmatrix}$$

In other words,

$$AI_2 = A \quad \text{and} \quad I_2A = A$$

This shows that the matrix I_2 plays the same role in the multiplication of 2×2 matrices as does the number 1 in the multiplication of real numbers. For this reason, I_2 is called the **multiplicative identity** for square matrices of order two.

If we have two real numbers a and b such that $ab = 1$, then we say that a and b are (multiplicative) inverses. In the box that follows we apply this same terminology to matrices.

> **DEFINITION: The Inverse of a 2×2 Matrix**
>
> If A and B are 2×2 matrices such that
>
> $$AB = I_2 \quad \text{and} \quad BA = I_2$$
>
> then A and B are said to be **inverses** of one another.

EXAMPLE 1 Find the inverse of $A = \begin{pmatrix} 1 & 2 \\ 3 & 4 \end{pmatrix}$.

Solution We need to find numbers a, b, c, and d such that the following two matrix equations are valid:

$$\begin{pmatrix} 1 & 2 \\ 3 & 4 \end{pmatrix}\begin{pmatrix} a & b \\ c & d \end{pmatrix} = \begin{pmatrix} 1 & 0 \\ 0 & 1 \end{pmatrix} \tag{1}$$

$$\begin{pmatrix} a & b \\ c & d \end{pmatrix}\begin{pmatrix} 1 & 2 \\ 3 & 4 \end{pmatrix} = \begin{pmatrix} 1 & 0 \\ 0 & 1 \end{pmatrix} \tag{2}$$

From equation (1) we have

$$\begin{pmatrix} a + 2c & b + 2d \\ 3a + 4c & 3b + 4d \end{pmatrix} = \begin{pmatrix} 1 & 0 \\ 0 & 1 \end{pmatrix}$$

For this last equation to be valid, the corresponding entries in the two matrices must be equal. (Why?) Consequently, we obtain four equations, two involving a and c, and two involving b and d:

$$\begin{cases} a + 2c = 1 \\ 3a + 4c = 0 \end{cases} \qquad \begin{cases} b + 2d = 0 \\ 3b + 4d = 1 \end{cases}$$

As Exercise 32(a) asks you to show, the solution to the first system is $a = -2$ and $c = 3/2$, and for the second system, $b = 1$ and $d = -1/2$. Furthermore (as you can check), these same values are obtained for a, b, c and d if we begin with matrix equation (2) rather than (1). Thus, the required inverse matrix is

$$\begin{pmatrix} -2 & 1 \\ \frac{3}{2} & -\frac{1}{2} \end{pmatrix}$$

Exercise 32(b) asks you to carry out the matrix multiplication to confirm that we indeed have

$$\begin{pmatrix} 1 & 2 \\ 3 & 4 \end{pmatrix}\begin{pmatrix} -2 & 1 \\ \frac{3}{2} & -\frac{1}{2} \end{pmatrix} = \begin{pmatrix} -2 & 1 \\ \frac{3}{2} & -\frac{1}{2} \end{pmatrix}\begin{pmatrix} 1 & 2 \\ 3 & 4 \end{pmatrix}$$

$$= \begin{pmatrix} 1 & 0 \\ 0 & 1 \end{pmatrix}$$

In Example 1 we found the inverse of a square matrix by solving systems of equations. However, as you've seen in previous sections, there are systems of equations that do not have solutions. In the present context, this implies that there are matrices that do not have inverses. For example (as Exercise 33 asks you to show), the matrix $\begin{pmatrix} 2 & 5 \\ 6 & 15 \end{pmatrix}$ does not have an inverse. A matrix that does not have an inverse is called a **singular matrix.** If a matrix does possess an inverse, we say that the matrix is **nonsingular.** So, according to the result in Example 1, the matrix $\begin{pmatrix} 1 & 2 \\ 3 & 4 \end{pmatrix}$ is nonsingular.

It can be proven that if a square matrix has an inverse, then that inverse is unique. (In other words, you can never find two different inverses for the same matrix.) The inverse of a nonsingular matrix A is denoted by A^{-1} (read *A inverse*).

There are several methods that can be used to compute the inverse of a matrix (or to show that there is no inverse). One method involves working with systems of equations, as in Example 1. A more efficient method, however, is the following. (We'll demonstrate the method with an example, but we won't give a formal proof of its validity.)

Suppose that we want to compute the inverse (if it exists) for the matrix

$$A = \begin{pmatrix} 6 & 2 \\ 13 & 4 \end{pmatrix}$$

We begin by writing a larger matrix, formed by merging A and I_2 as follows:

$$\left(\begin{array}{cc:cc} 6 & 2 & 1 & 0 \\ 13 & 4 & 0 & 1 \end{array}\right)$$

A ⟶ (left block) ↑ ; ↑ (right block) ⟵ I_2

A convenient abbreviation for this matrix is $(A \,\vdots\, I_2)$. Now we carry out elementary row operations on this matrix until one of the following two situations occurs:

1. Either the matrix takes on the form

$$\left(\begin{array}{cc:cc} 1 & 0 & a & b \\ 0 & 1 & c & d \end{array}\right)$$

I_2 ↑ (left block)

In this case, $A^{-1} = \begin{pmatrix} a & b \\ c & d \end{pmatrix}$.

2. Or, one of the rows to the left of the dashed line consists entirely of zeros. In this case, A^{-1} does not exist.

The calculations for our example then run as follows.

$$\left(\begin{array}{cc:cc} 6 & 2 & 1 & 0 \\ 13 & 4 & 0 & 1 \end{array}\right) \xrightarrow{-2R_1 + R_2} \left(\begin{array}{cc:cc} 6 & 2 & 1 & 0 \\ 1 & 0 & -2 & 1 \end{array}\right)$$

$$\xrightarrow{R_1 \leftrightarrow R_2} \left(\begin{array}{cc:cc} 1 & 0 & -2 & 1 \\ 6 & 2 & 1 & 0 \end{array}\right)$$

$$\xrightarrow{-6R_1 + R_2} \left(\begin{array}{cc:cc} 1 & 0 & -2 & 1 \\ 0 & 2 & 13 & -6 \end{array}\right)$$

$$\xrightarrow{\frac{1}{2}R_2} \left(\begin{array}{cc:cc} 1 & 0 & -2 & 1 \\ 0 & 1 & \frac{13}{2} & -3 \end{array}\right)$$

The inverse matrix can now be read off:

$$A^{-1} = \begin{pmatrix} -2 & 1 \\ \frac{13}{2} & -3 \end{pmatrix}$$

(You should verify for yourself that we indeed have $AA^{-1} = I_2$ and $A^{-1}A = I_2$.)

Inverse matrices can be used to solve certain systems of equations in which the number of unknowns is the same as the number of equations. Before explaining how this works, we describe how a system of equations can be written in matrix form. Consider the system

$$\begin{cases} x + 2y = 8 \\ 3x + 4y = 6 \end{cases} \tag{3}$$

As defined in the previous section, the coefficient matrix A for this system is

$$A = \begin{pmatrix} 1 & 2 \\ 3 & 4 \end{pmatrix}$$

Now we define two matrices X and B:

$$X = \begin{pmatrix} x \\ y \end{pmatrix} \qquad B = \begin{pmatrix} 8 \\ 6 \end{pmatrix}$$

Then system (3) can be written as a single matrix equation:

$$AX = B \tag{4}$$

To see why this is so, we expand equation (4) to obtain

$$\begin{pmatrix} 1 & 2 \\ 3 & 4 \end{pmatrix}\begin{pmatrix} x \\ y \end{pmatrix} = \begin{pmatrix} 8 \\ 6 \end{pmatrix} \qquad \text{using the definitions of } A, X, \text{ and } B$$

$$\begin{pmatrix} x + 2y \\ 3x + 4y \end{pmatrix} = \begin{pmatrix} 8 \\ 6 \end{pmatrix} \qquad \text{carrying out the matrix multiplication}$$

By equating the corresponding entries of the matrices in this last equation, we obtain $x + 2y = 8$ and $3x + 4y = 6$, as given initially in system (3).

EXAMPLE 2 Use an inverse matrix to solve the system

$$\begin{cases} x + 2y = 8 \\ 3x + 4y = 6 \end{cases}$$

Solution As explained just prior to this example, the matrix form for this system is

$$AX = B \tag{5}$$

where

$$A = \begin{pmatrix} 1 & 2 \\ 3 & 4 \end{pmatrix} \qquad X = \begin{pmatrix} x \\ y \end{pmatrix} \qquad B = \begin{pmatrix} 8 \\ 6 \end{pmatrix}$$

From Example 1, we know that

$$A^{-1} = \begin{pmatrix} -2 & 1 \\ \frac{3}{2} & -\frac{1}{2} \end{pmatrix}$$

Now we multiply both sides of equation (5) by A^{-1} to obtain

$$A^{-1}(AX) = A^{-1}B$$
$$(A^{-1}A)X = A^{-1}B \qquad \text{Matrix multiplication is associative.}$$
$$I_2X = A^{-1}B$$
$$X = A^{-1}B \qquad \text{(Why?)}$$

Substituting the actual matrices into this last equation, we have

$$\begin{pmatrix} x \\ y \end{pmatrix} = \begin{pmatrix} -2 & 1 \\ \frac{3}{2} & -\frac{1}{2} \end{pmatrix}\begin{pmatrix} 8 \\ 6 \end{pmatrix} = \begin{pmatrix} -10 \\ 9 \end{pmatrix}$$

Therefore $x = -10$ and $y = 9$, as required.

All of the ideas we have discussed for square matrices of order two can be carried over directly to larger square matrices. It is easy to check that the following matrices, I_3 and I_4, are the multiplicative identities for 3×3 and 4×4 matrices, respectively:

$$I_3 = \begin{pmatrix} 1 & 0 & 0 \\ 0 & 1 & 0 \\ 0 & 0 & 1 \end{pmatrix} \qquad I_4 = \begin{pmatrix} 1 & 0 & 0 & 0 \\ 0 & 1 & 0 & 0 \\ 0 & 0 & 1 & 0 \\ 0 & 0 & 0 & 1 \end{pmatrix}$$

These identity matrices are described by saying that they have ones down the **main diagonal** and zeros everywhere else. (Larger identity matrices can be defined following this same pattern.) Sometimes, when it is clear from the context or as a matter of convenience, we'll omit the subscript and denote the appropriately sized identity matrix simply by I. (This is done in the box that follows.)

PROPERTY SUMMARY **THE INVERSE OF A SQUARE MATRIX**

1. Suppose that A and B are square matrices of the same size, and let I denote the identity matrix of that size. Then A and B are said to be **inverses** of one another provided

$$AB = I \quad \text{and} \quad BA = I$$

2. It can be shown that every square matrix has at most one inverse.
3. A square matrix that has an inverse is said to be **nonsingular** (or **invertible**). A square matrix that does not have an inverse is called **singular.**
4. If the matrix A is nonsingular, then the inverse of A is denoted by A^{-1}. In this case, we have

$$AA^{-1} = I \quad \text{and} \quad A^{-1}A = I$$

EXAMPLE 3 Let

$$A = \begin{pmatrix} 5 & 0 & 2 \\ 2 & 2 & 1 \\ -3 & 1 & -1 \end{pmatrix}$$

Use the elementary row operations to compute A^{-1}, if it exists.

Solution Following the method that we described for the 2×2 case, we first write down the matrix $(A \mid I_3)$ formed by merging A and I_3:

$$\left(\begin{array}{rrr|rrr} 5 & 0 & 2 & 1 & 0 & 0 \\ 2 & 2 & 1 & 0 & 1 & 0 \\ -3 & 1 & -1 & 0 & 0 & 1 \end{array}\right)$$

Now we carry out the elementary row operations, trying to obtain I_3 to the left of the dashed line. We have

$$\left(\begin{array}{rrr|rrr} 5 & 0 & 2 & 1 & 0 & 0 \\ 2 & 2 & 1 & 0 & 1 & 0 \\ -3 & 1 & -1 & 0 & 0 & 1 \end{array}\right) \xrightarrow{-2R_2+R_1} \left(\begin{array}{rrr|rrr} 1 & -4 & 0 & 1 & -2 & 0 \\ 2 & 2 & 1 & 0 & 1 & 0 \\ -3 & 1 & -1 & 0 & 0 & 1 \end{array}\right)$$

$$\xrightarrow[3R_1+R_3]{-2R_1+R_2} \left(\begin{array}{rrr|rrr} 1 & -4 & 0 & 1 & -2 & 0 \\ 0 & 10 & 1 & -2 & 5 & 0 \\ 0 & -11 & -1 & 3 & -6 & 1 \end{array}\right) \xrightarrow{1R_3+R_2} \left(\begin{array}{rrr|rrr} 1 & -4 & 0 & 1 & -2 & 0 \\ 0 & -1 & 0 & 1 & -1 & 1 \\ 0 & -11 & -1 & 3 & -6 & 1 \end{array}\right)$$

$$\xrightarrow{(-1)R_2} \left(\begin{array}{rrr|rrr} 1 & -4 & 0 & 1 & -2 & 0 \\ 0 & 1 & 0 & -1 & 1 & -1 \\ 0 & -11 & -1 & 3 & -6 & 1 \end{array}\right) \xrightarrow[11R_2+R_3]{4R_2+R_1} \left(\begin{array}{rrr|rrr} 1 & 0 & 0 & -3 & 2 & -4 \\ 0 & 1 & 0 & -1 & 1 & -1 \\ 0 & 0 & -1 & -8 & 5 & -10 \end{array}\right)$$

$$\xrightarrow{(-1)R_3} \left(\begin{array}{rrr|rrr} 1 & 0 & 0 & -3 & 2 & -4 \\ 0 & 1 & 0 & -1 & 1 & -1 \\ 0 & 0 & 1 & 8 & -5 & 10 \end{array}\right)$$

The required inverse is therefore

$$A^{-1} = \begin{pmatrix} -3 & 2 & -4 \\ -1 & 1 & -1 \\ 8 & -5 & 10 \end{pmatrix}$$

EXERCISE SET 10.4

A

In Exercises 1–4, the matrices A, B, C, and D are defined as follows.

$$A = \begin{pmatrix} 4 & -1 \\ -5 & 2 \end{pmatrix} \qquad B = \begin{pmatrix} \frac{1}{2} & 5 \\ 3 & 1 \end{pmatrix}$$

$$C = \begin{pmatrix} 3 & 0 & -2 \\ 0 & 5 & 6 \\ 1 & 4 & -7 \end{pmatrix} \qquad D = \begin{pmatrix} 1 & 2 & 3 \\ 4 & 5 & 6 \\ 7 & 8 & 9 \end{pmatrix}$$

1. Compute AI_2 and I_2A to verify that $AI_2 = I_2A = A$.
2. Compute BI_2 and I_2B to verify that $BI_2 = I_2B = B$.
3. Compute CI_3 and I_3C to verify that $CI_3 = I_3C = C$.
4. Compute DI_3 and I_3D to verify that $DI_3 = I_3D = D$.

In Exercises 5–12, compute A^{-1}, if it exists, using the method of Example 1.

5. $A = \begin{pmatrix} 7 & 9 \\ 4 & 5 \end{pmatrix}$
6. $A = \begin{pmatrix} 3 & -8 \\ 2 & -5 \end{pmatrix}$
7. $A = \begin{pmatrix} -3 & 1 \\ 5 & 6 \end{pmatrix}$
8. $A = \begin{pmatrix} -4 & 0 \\ 9 & 3 \end{pmatrix}$
9. $A = \begin{pmatrix} -2 & 3 \\ -4 & 6 \end{pmatrix}$
10. $A = \begin{pmatrix} \frac{5}{3} & -2 \\ -\frac{2}{3} & 1 \end{pmatrix}$

11. $A = \begin{pmatrix} \frac{1}{3} & \frac{1}{3} \\ -\frac{1}{9} & \frac{2}{9} \end{pmatrix}$

12. $A = \begin{pmatrix} -3 & 7 \\ 12 & -28 \end{pmatrix}$

In Exercises 13–26, compute the inverse matrix, if it exists, using elementary row operations (as shown in Example 3).

13. $\begin{pmatrix} 2 & 1 \\ 3 & 2 \end{pmatrix}$

14. $\begin{pmatrix} -6 & 5 \\ 18 & -15 \end{pmatrix}$

15. $\begin{pmatrix} 0 & -11 \\ 1 & 6 \end{pmatrix}$

16. $\begin{pmatrix} -2 & 13 \\ -4 & 25 \end{pmatrix}$

17. $\begin{pmatrix} \frac{2}{3} & -\frac{1}{4} \\ -8 & 3 \end{pmatrix}$

18. $\begin{pmatrix} -\frac{2}{5} & \frac{1}{3} \\ -6 & 5 \end{pmatrix}$

19. $\begin{pmatrix} -5 & 4 & -3 \\ 10 & -7 & 6 \\ 8 & -6 & 5 \end{pmatrix}$

20. $\begin{pmatrix} 1 & 0 & -2 \\ -3 & -1 & 6 \\ 2 & 1 & -5 \end{pmatrix}$

21. $\begin{pmatrix} 1 & 2 & -1 \\ 0 & 3 & 0 \\ -4 & 0 & 5 \end{pmatrix}$

22. $\begin{pmatrix} 1 & -4 & -8 \\ 1 & 2 & 5 \\ 1 & 1 & 3 \end{pmatrix}$

23. $\begin{pmatrix} -7 & 5 & 3 \\ 3 & -2 & -2 \\ 3 & -2 & -1 \end{pmatrix}$

24. $\begin{pmatrix} 2 & -1 & -1 \\ 1 & 0 & -1 \\ -2 & 1 & 2 \end{pmatrix}$

25. $\begin{pmatrix} 1 & 2 & 3 \\ 4 & 5 & 6 \\ 7 & 8 & 9 \end{pmatrix}$

26. $\begin{pmatrix} 2 & 1 & 3 \\ 4 & 5 & -7 \\ 2 & 1 & 3 \end{pmatrix}$

27. If $A = \begin{pmatrix} 3 & 8 \\ 4 & 11 \end{pmatrix}$, then $A^{-1} = \begin{pmatrix} 11 & -8 \\ -4 & 3 \end{pmatrix}$. Use this fact and the method of Example 2 to solve the following systems.

(a) $\begin{cases} 3x + 8y = 5 \\ 4x + 11y = 7 \end{cases}$ **(b)** $\begin{cases} 3x + 8y = -12 \\ 4x + 11y = 0 \end{cases}$

28. If $A = \begin{pmatrix} 3 & -7 \\ 4 & -9 \end{pmatrix}$, then $A^{-1} = \begin{pmatrix} -9 & 7 \\ -4 & 3 \end{pmatrix}$. Use this fact and the method of Example 2 to solve the following systems.

(a) $\begin{cases} 3x - 7y = 30 \\ 4x - 9y = 39 \end{cases}$ **(b)** $\begin{cases} 3x - 7y = -45 \\ 4x - 9y = -71 \end{cases}$

29. The inverse of the matrix

$$A = \begin{pmatrix} 3 & 2 & 6 \\ 1 & 1 & 2 \\ 2 & 2 & 5 \end{pmatrix} \quad \text{is} \quad A^{-1} = \begin{pmatrix} 1 & 2 & -2 \\ -1 & 3 & 0 \\ 0 & -2 & 1 \end{pmatrix}$$

Use this fact to solve the following systems.

(a) $\begin{cases} 3x + 2y + 6z = 28 \\ x + y + 2z = 9 \\ 2x + 2y + 5z = 22 \end{cases}$ **(b)** $\begin{cases} 3x + 2y + 6z = -7 \\ x + y + 2z = -2 \\ 2x + 2y + 5z = -6 \end{cases}$

30. The inverse of the matrix

$$A = \begin{pmatrix} 1 & -1 & 1 \\ 2 & -3 & 2 \\ -4 & 6 & 1 \end{pmatrix} \quad \text{is} \quad A^{-1} = \begin{pmatrix} 3 & -\frac{7}{5} & -\frac{1}{5} \\ 2 & -1 & 0 \\ 0 & \frac{2}{5} & \frac{1}{5} \end{pmatrix}$$

Use this fact to solve the following systems.

(a) $\begin{cases} x - y + z = 5 \\ 2x - 3y + 2z = -15 \\ -4x + 6y + z = 25 \end{cases}$

(b) $\begin{cases} x - y + z = 1 \\ 2x - 3y + 2z = -2 \\ -4x + 6y + z = 0 \end{cases}$

B

31. Let $A = \begin{pmatrix} 1 & -6 & 3 \\ 2 & -7 & 3 \\ 4 & -12 & 5 \end{pmatrix}$.

(a) Compute the matrix product AA. What do you observe?

(b) Use the result in part (a) to solve the following system.

$$\begin{cases} x - 6y + 3z = 19/2 \\ 2x - 7y + 3z = 11 \\ 4x - 12y + 5z = 19 \end{cases}$$

32. **(a)** Solve the following two systems and then check to see that your results agree with those given in Example 1.

$$\begin{cases} a + 2c = 1 \\ 3a + 4c = 0 \end{cases} \qquad \begin{cases} b + 2d = 0 \\ 3b + 4d = 1 \end{cases}$$

(b) At the end of Example 1, it is asserted that

$$\begin{pmatrix} 1 & 2 \\ 3 & 4 \end{pmatrix}\begin{pmatrix} -2 & 1 \\ \frac{3}{2} & -\frac{1}{2} \end{pmatrix} = \begin{pmatrix} -2 & 1 \\ \frac{3}{2} & -\frac{1}{2} \end{pmatrix}\begin{pmatrix} 1 & 2 \\ 3 & 4 \end{pmatrix} = \begin{pmatrix} 1 & 0 \\ 0 & 1 \end{pmatrix}$$

Carry out the indicated matrix multiplications to verify that these equations are valid.

33. Use the technique in Example 1 to show that the matrix $\begin{pmatrix} 2 & 5 \\ 6 & 15 \end{pmatrix}$ does not have an inverse.

34. Use the elementary row operations (as in Example 3) to find the inverse of the following matrix.

$$\begin{pmatrix} 1 & 1 & 1 & 1 \\ 1 & 2 & 3 & 4 \\ 1 & 3 & 6 & 10 \\ 1 & 4 & 10 & 20 \end{pmatrix}$$

35. Let $A = \begin{pmatrix} 2 & 3 \\ 4 & 5 \end{pmatrix}$ and $B = \begin{pmatrix} 7 & 8 \\ 6 & 7 \end{pmatrix}$.
(a) Compute A^{-1}, B^{-1}, and $B^{-1}A^{-1}$.
(b) Compute $(AB)^{-1}$. What do you observe?

36. Let $A = \begin{pmatrix} a & b \\ c & d \end{pmatrix}$. Compute A^{-1}. (Assume that $ad - bc \neq 0$.)

GRAPHING UTILITY EXERCISES FOR SECTIONS 10.3 AND 10.4

In Exercises 1–33, the matrices A, B, C, D, E, F, and G are defined as follows.

$$A = \begin{pmatrix} 9 & 5 \\ 7 & 4 \end{pmatrix} \quad B = \begin{pmatrix} 2 & -9 \\ 1 & -4 \end{pmatrix} \quad C = \begin{pmatrix} 3 & 2 & 6 \\ 1 & 1 & 2 \\ 2 & 2 & 5 \end{pmatrix}$$

$$D = \begin{pmatrix} 9 & 4 & 4 \\ 2 & 2 & 1 \\ -3 & 1 & -1 \end{pmatrix} \quad E = \begin{pmatrix} 5 & 4 & 1 \\ -1 & 2 & 9 \\ 0 & 8 & -6 \end{pmatrix}$$

$$F = \begin{pmatrix} -2 & 0 \\ 3 & 1 \\ 5 & 2 \end{pmatrix} \quad G = \begin{pmatrix} 3 & -2 & -1 \\ 7 & -4 & 0 \end{pmatrix}$$

In each exercise, carry out the indicated matrix operations. If an operation is undefined, say so.

1. AB **2.** BA **3.** AE
4. AC **5.** EC **6.** CA
7. $A(A+B)$ **8.** $(A+B)A$ **9.** $AA+AB$
10. DD **11.** EA **12.** AF
13. FA **14.** CD **15.** DC
16. $C(D-E)$ **17.** $CD-CE$ **18.** CF
19. CG **20.** AG **21.** GA
22. A^{-1} **23.** B^{-1} **24.** $(AB)^{-1}$
25. $(BA)^{-1}$ **26.** $B^{-1}A^{-1}$ **27.** $A^{-1}B^{-1}$
28. C^{-1} **29.** G^{-1} **30.** F^{-1}
31. $(DC)^{-1}$ **32.** $D^{-1}C^{-1}$ **33.** $(FG)^{-1}$

In Exercises 34–38: **(a)** *compute the inverse of the coefficient matrix for the system; and* **(b)** *use matrix methods to solve the system. In cases where the final answer involves decimals, round to three decimal places.*

34. **(a)** $\begin{cases} x + 4y = 7 \\ 2x + 7y = 12 \end{cases}$ **(b)** $\begin{cases} x + 4y = -3 \\ 2x + 7y = 0 \end{cases}$

35. **(a)** $\begin{cases} 8x - 5y = -13 \\ 3x + 4y = 48 \end{cases}$ **(b)** $\begin{cases} 8x - 5y = 5 \\ 3x + 4y = -2 \end{cases}$

36. **(a)** $\begin{cases} x + 2y + 2z = 3 \\ 3x + y = -1 \\ x + y + z = 12 \end{cases}$ **(b)** $\begin{cases} x + 2y + 2z = 0 \\ 3x + y = 6 \\ x + y + z = 1 \end{cases}$

37. **(a)** $\begin{cases} 5x - 2y - 2z = 15 \\ 3x + y = 4 \\ x + y + z = -4 \end{cases}$ **(b)** $\begin{cases} x + 2y + 2z = 0 \\ 3x + y = 6 \\ x + y + z = 1 \end{cases}$

38. **(a)** $\begin{cases} 2x + 3y + z + w = 3 \\ 6x + 6y - 5z - 2w = 15 \\ x - y + z + \frac{1}{6}w = -3 \\ 4x + 9y + 3z + 2w = -3 \end{cases}$

(b) $\begin{cases} 2x + 3y + z + w = 0 \\ 6x + 6y - 5z - 2w = 0 \\ x - y + z + \frac{1}{6}w = 5 \\ 4x + 9y + 3z + 2w = 1 \end{cases}$

10.5 DETERMINANTS AND CRAMER'S RULE

The idea of the determinant ... dates back essentially to Leibniz (1693), the Swiss mathematician Gabriel Cramer (1750), and Lagrange (1773); the name is due to Cauchy (1812).
Y. Mikami has pointed out that the Japanese mathematician Seki Kōwa had the idea of a determinant sometime before 1683.

Dirk J. Struik in *A Concise History of Mathematics,* 4th ed. (New York: Dover Publications, 1987)

As you saw in the previous section, square matrices and their inverses can be used to solve certain systems of equations. In this section, we are going to associate a number with each square matrix. This number is called the **determinant** of the matrix. As you'll see, this too has an application in solving systems of equations.

The determinant of a matrix A can be denoted by det A or $|A|$. Determinants are also denoted simply by replacing the parentheses of matrix notation with vertical lines. Thus, three examples of determinants are

$$\begin{vmatrix} 1 & 2 \\ 3 & 4 \end{vmatrix} \qquad \begin{vmatrix} 1 & 2 & 3 \\ 4 & 5 & 6 \\ 7 & 8 & 9 \end{vmatrix} \qquad \begin{vmatrix} 3 & 7 & 8 & 9 \\ 5 & 6 & 4 & 3 \\ -9 & 9 & 0 & 1 \\ 1 & 3 & -2 & 1 \end{vmatrix}$$

A determinant with n rows and n columns is said to be an ***n*th-order determinant.** Therefore, the determinants we've just written are, respectively, second-, third-, and fourth-order determinants. As with matrices, we speak of the numbers in a determinant as its **entries.** We also number the rows and the columns of a determinant as we do with matrices. However, unlike matrices, each determinant has a numerical value. The value of a second-order determinant is defined as follows.

DEFINITION: 2 × 2 Determinant

$$\begin{vmatrix} a & b \\ c & d \end{vmatrix} = ad - bc$$

Table 1 illustrates how this definition is used to evaluate, or *expand,* a second-order determinant.

TABLE 1

Determinant	Value of Determinant
$\begin{vmatrix} 3 & 7 \\ 5 & 10 \end{vmatrix}$	$\begin{vmatrix} 3 & 7 \\ 5 & 10 \end{vmatrix} = 3(10) - 7(5) = 30 - 35 = -5$
$\begin{vmatrix} 3 & -7 \\ 5 & 10 \end{vmatrix}$	$\begin{vmatrix} 3 & -7 \\ 5 & 10 \end{vmatrix} = 3(10) - (-7)(5) = 30 + 35 = 65$
$\begin{vmatrix} a & a^3 \\ 1 & a^2 \end{vmatrix}$	$\begin{vmatrix} a & a^3 \\ 1 & a^2 \end{vmatrix} = a(a^2) - a^3(1) = a^3 - a^3 = 0$

In general, the value of an nth-order determinant ($n > 2$) is defined in terms of certain determinants of order $n - 1$. For instance, the value of a third-order determinant is defined in terms of second-order determinants. We'll use the following example to introduce the necessary terminology here:

$$\begin{vmatrix} 8 & 3 & 5 \\ 2 & 4 & 6 \\ 9 & 1 & 7 \end{vmatrix}$$

Pick a given entry—say, 8—and imagine crossing out all entries occupying the same row and the same column as 8.

$$\begin{vmatrix} 8 & 3 & 5 \\ 2 & 4 & 6 \\ 9 & 1 & 7 \end{vmatrix}$$

Now we are left with the second-order determinant $\begin{vmatrix} 4 & 6 \\ 1 & 7 \end{vmatrix}$. This second-order determinant is called the **minor** of the entry 8. Similarly, to find the minor of the entry 6 in the original determinant, imagine crossing out all entries that occupy the same row and the same column as 6.

$$\begin{vmatrix} 8 & 3 & 5 \\ 2 & 4 & 6 \\ 9 & 1 & 7 \end{vmatrix}$$

We are left with the second-order determinant $\begin{vmatrix} 8 & 3 \\ 9 & 1 \end{vmatrix}$, which by definition is the minor of the entry 6. In the same manner, *the minor of any element is the determinant obtained by crossing out the entries occupying the same row and column as the given element.*

Closely related to the minor of an entry in a determinant is the cofactor of that entry. The **cofactor** of an entry is defined as the minor multiplied by $+1$ or -1, according to the scheme displayed in Figure 1.

$$\begin{vmatrix} + & - & + \\ - & + & - \\ + & - & + \end{vmatrix}$$

FIGURE 1

After looking at an example, we'll give a more formal rule for computing cofactors, one that will not rely on a figure and that will also apply to larger determinants.

EXAMPLE 1 Consider the determinant

$$\begin{vmatrix} 1 & 2 & 3 \\ 4 & 5 & 6 \\ 7 & 8 & 9 \end{vmatrix}$$

Compute the minor and the cofactor of the entry 4.

Solution By definition, we have

$$\text{minor of } 4 = \begin{vmatrix} 2 & 3 \\ 8 & 9 \end{vmatrix} = 18 - 24 = -6$$

Thus, the minor of the entry 4 is -6. To compute the cofactor of 4, we first notice that 4 is located in the second row and first column of the given determinant. Upon checking the corresponding position in Figure 1, we see a negative sign. Therefore, we have by definition

$$\begin{aligned} \text{cofactor of } 4 &= (-1)(\text{minor of } 4) \\ &= (-1)(-6) = 6 \end{aligned}$$

The cofactor of 4 is therefore 6.

The following rule tells us how cofactors can be computed without relying on Figure 1. For reference, we also restate the definition of a minor. (You should verify for yourself that this rule yields results that are consistent with Figure 1.)

DEFINITION: Minors and Cofactors

The **minor** of an entry b in a determinant is the determinant formed by suppressing the entries in the row and in the column in which b appears.

Suppose that the entry b is in the ith row and the jth column. Then the **cofactor** of b is given by the expression

$$(-1)^{i+j}(\text{minor of } b)$$

We are now prepared to state the definition that tells us how to evaluate a third-order determinant. Actually, as you'll see later, the definition is quite general and may be applied to determinants of any size.

DEFINITION: The Value of a Determinant

Multiply each entry in the first row of the determinant by its cofactor and then add the results. The value of the determinant is defined to be this sum.

To see how this definition is used, let's evaluate the determinant

$$\begin{vmatrix} 1 & 2 & 3 \\ 4 & 5 & 6 \\ 7 & 8 & 9 \end{vmatrix}$$

The definition tells us to multiply each entry in the first row by its cofactor and then add the results. Carrying out this procedure, we have

$$\begin{aligned} \begin{vmatrix} 1 & 2 & 3 \\ 4 & 5 & 6 \\ 7 & 8 & 9 \end{vmatrix} &= 1\begin{vmatrix} 5 & 6 \\ 8 & 9 \end{vmatrix} - 2\begin{vmatrix} 4 & 6 \\ 7 & 9 \end{vmatrix} + 3\begin{vmatrix} 4 & 5 \\ 7 & 8 \end{vmatrix} \\ &= 1(45-48) - 2(36-42) + 3(32-35) \\ &= 0 \qquad \text{(Check the arithmetic!)} \end{aligned}$$

So the value of this particular determinant is zero. The procedure we've used here is referred to as **expanding the determinant along its first row.** The following theorem (stated here without proof) tells us that the value of a determinant can be obtained by expanding along any row or column; the results are the same in all cases.

THEOREM:

Select any row or any column in a determinant and multiply each element in that row or column by its cofactor. Then add the results. The number obtained will be the value of the determinant. (In other words, the number obtained will be the same as that obtained by expanding the determinant along its first row.)

According to this theorem, we could have evaluated the determinant

$$\begin{vmatrix} 1 & 2 & 3 \\ 4 & 5 & 6 \\ 7 & 8 & 9 \end{vmatrix}$$

by expanding it along any row or any column. Let's expand it along the second column and check to see that the result agrees with the value we obtained earlier. Expanding along the second column, we have

$$\begin{aligned}\begin{vmatrix} 1 & 2 & 3 \\ 4 & 5 & 6 \\ 7 & 8 & 9 \end{vmatrix} &= -2\begin{vmatrix} 4 & 6 \\ 7 & 9 \end{vmatrix} + 5\begin{vmatrix} 1 & 3 \\ 7 & 9 \end{vmatrix} - 8\begin{vmatrix} 1 & 3 \\ 4 & 6 \end{vmatrix} \\ &= -2(36-42) + 5(9-21) - 8(6-12) \\ &= -2(-6) + 5(-12) - 8(-6) \\ &= 12 - 60 + 48 = 0 \qquad \text{as obtained previously}\end{aligned}$$

We would have obtained the same result had we chosen to begin with any other row or column. (Exercise 13 at the end of this section asks you to verify this.)

There are three basic rules that make it easier to evaluate determinants. These are summarized in the box that follows. (Suggestions for proving these can be found in the exercises.)

PROPERTY SUMMARY — RULES FOR MANIPULATING DETERMINANTS

EXAMPLES

1. If each entry in a given row is multiplied by the constant k, then the value of the determinant is multiplied by k. This is true for columns, also.

$$10\begin{vmatrix} 1 & 3 & 4 \\ 1 & 2 & 3 \\ 4 & 5 & 6 \end{vmatrix} = \begin{vmatrix} 10 & 30 & 40 \\ 1 & 2 & 3 \\ 4 & 5 & 6 \end{vmatrix}$$

$$k\begin{vmatrix} a & b & c \\ d & e & f \\ g & h & i \end{vmatrix} = \begin{vmatrix} a & kb & c \\ d & ke & f \\ g & kh & i \end{vmatrix}$$

2. If a multiple of one row is added to another row, the value of the determinant is not changed. This applies to columns, also.

$$\begin{vmatrix} a & b & c \\ d & e & f \\ g & h & i \end{vmatrix} = \begin{vmatrix} a & b & c \\ d+ka & e+kb & f+kc \\ g & h & i \end{vmatrix}$$

3. If two rows are interchanged, then the value of the determinant is multiplied by -1. This applies to columns, also.

$$\begin{vmatrix} 1 & 2 & 3 \\ 4 & 5 & 6 \\ a & b & c \end{vmatrix} = -\begin{vmatrix} 4 & 5 & 6 \\ 1 & 2 & 3 \\ a & b & c \end{vmatrix}$$

EXAMPLE 2 Evaluate the determinant $\begin{vmatrix} 15 & 14 & 26 \\ 18 & 17 & 32 \\ 21 & 20 & 42 \end{vmatrix}$.

Solution

$$\begin{vmatrix} 15 & 14 & 26 \\ 18 & 17 & 32 \\ 21 & 20 & 42 \end{vmatrix} = (3\times 2)\begin{vmatrix} 5 & 14 & 13 \\ 6 & 17 & 16 \\ 7 & 20 & 21 \end{vmatrix}$$

using Rule 1 to factor 3 from the first column and 2 from the third column

$$= 6\begin{vmatrix} 5 & 1 & 13 \\ 6 & 1 & 16 \\ 7 & -1 & 21 \end{vmatrix}$$

using Rule 2 to subtract the third column from the second column

$$= 6\begin{vmatrix} 12 & 0 & 34 \\ 13 & 0 & 37 \\ 7 & -1 & 21 \end{vmatrix}$$

using Rule 2 to add the third row to the first and second rows

$$= 6\left[-\left(-1\begin{vmatrix} 12 & 34 \\ 13 & 37 \end{vmatrix}\right)\right]$$ expanding the determinant along the second column

$$= 6\begin{vmatrix} 12 & 34 \\ 1 & 3 \end{vmatrix}$$ using Rule 2 to subtract the first row from the second row

$$= 6(36 - 34) = 12$$

The value of the given determinant is therefore 12. Notice the general strategy. We used Rules 1 and 2 until one column (or one row) contained two zeros. At that point, it is a simple matter to expand the determinant along that column (or row).

EXAMPLE 3 Show that

$$\begin{vmatrix} 1 & 1 & 1 \\ a & b & c \\ a^2 & b^2 & c^2 \end{vmatrix} = (b-a)(c-a)(c-b)$$

Solution

$$\begin{vmatrix} 1 & 1 & 1 \\ a & b & c \\ a^2 & b^2 & c^2 \end{vmatrix} = \begin{vmatrix} 1 & 0 & 0 \\ a & b-a & c-a \\ a^2 & b^2-a^2 & c^2-a^2 \end{vmatrix}$$ using Rule 2 to subtract the first column from the second and third columns

$$= \begin{vmatrix} 1 & 0 & 0 \\ a & b-a & c-a \\ a^2 & (b-a)(b+a) & (c-a)(c+a) \end{vmatrix}$$

$$= (b-a)(c-a)\begin{vmatrix} 1 & 0 & 0 \\ a & 1 & 1 \\ a^2 & b+a & c+a \end{vmatrix}$$ using Rule 1 to factor $b-a$ from the second column and $c-a$ from the third column

$$= (b-a)(c-a)[(c+a)-(b+a)]$$ expanding along first row

$$= (b-a)(c-a)(c-b)$$

The definition that we gave for third-order determinants can also be applied to define fourth-order (or larger) determinants. Consider, for example, the fourth-order determinant A given by

$$A = \begin{vmatrix} 25 & 40 & 5 & 10 \\ 9 & 0 & 3 & 6 \\ -2 & 3 & 11 & -17 \\ -3 & 4 & 7 & 2 \end{vmatrix}$$

By definition, we can evaluate this determinant by selecting the first row, multiplying each entry by its cofactor, and then adding the results. This yields

$$A = 25\begin{vmatrix} 0 & 3 & 6 \\ 3 & 11 & -17 \\ 4 & 7 & 2 \end{vmatrix} - 40\begin{vmatrix} 9 & 3 & 6 \\ -2 & 11 & -17 \\ -3 & 7 & 2 \end{vmatrix}$$

$$+ 5\begin{vmatrix} 9 & 0 & 6 \\ -2 & 3 & -17 \\ -3 & 4 & 2 \end{vmatrix} - 10\begin{vmatrix} 9 & 0 & 3 \\ -2 & 3 & 11 \\ -3 & 4 & 7 \end{vmatrix}$$

The problem is now reduced to evaluating four third-order determinants. If we had instead expanded down the second column, the ensuing work would be somewhat less because of the zero in that column. Nevertheless, there would still be three third-order determinants to evaluate.

Because of the amount of computation involved, we in fact rarely evaluate fourth-order determinants directly from the definition. Instead, we use the three rules (given just before Example 2) to simplify the determinant first. In the example that follows, we show how this is done. (We will accept the fact that the three rules are applicable in evaluating determinants of any size.)

EXAMPLE 4 Evaluate the determinant A given by

$$A = \begin{vmatrix} 25 & 40 & 5 & 10 \\ 9 & 0 & 3 & 6 \\ -2 & 3 & 11 & -17 \\ -3 & 4 & 7 & 2 \end{vmatrix}$$

Solution

$$A = 5 \times 3 \begin{vmatrix} 5 & 8 & 1 & 2 \\ 3 & 0 & 1 & 2 \\ -2 & 3 & 11 & -17 \\ -3 & 4 & 7 & 2 \end{vmatrix} = 15 \begin{vmatrix} 2 & 8 & 0 & 0 \\ 3 & 0 & 1 & 2 \\ -35 & 3 & 0 & -39 \\ -24 & 4 & 0 & -12 \end{vmatrix}$$

Factor 5 from the first row and 3 from the second row.

Subtract the second row from the first. Subtract 11 times the second row from the third. Subtract 7 times the second row from the fourth.

$$= 15 \times 2 \times 4 \begin{vmatrix} 1 & 4 & 0 & 0 \\ 3 & 0 & 1 & 2 \\ -35 & 3 & 0 & -39 \\ -6 & 1 & 0 & -3 \end{vmatrix} = 120 \cdot (-1) \begin{vmatrix} 1 & 4 & 0 \\ -35 & 3 & -39 \\ -6 & 1 & -3 \end{vmatrix}$$

Factor 2 from the first row and 4 from the fourth row.

Expand the determinant along the third column.

$$= -120 \cdot (-3) \begin{vmatrix} 1 & 4 & 0 \\ -35 & 3 & 13 \\ -6 & 1 & 1 \end{vmatrix} = 360 \begin{vmatrix} 1 & 0 & 0 \\ -35 & 143 & 13 \\ -6 & 25 & 1 \end{vmatrix}$$

Factor -3 from the third column.

Subtract 4 times the first column from the second.

$$= 360 \begin{vmatrix} 143 & 13 \\ 25 & 1 \end{vmatrix} = 360 \begin{vmatrix} -182 & 0 \\ 25 & 1 \end{vmatrix}$$

Subtract 13 times the second row from the first row.

$$= 360(-182) = -65520$$

Determinants can be used to solve certain systems of linear equations in which there are as many unknowns as there are equations. In the box that follows, we state **Cramer's rule** for solving a system of three linear equations in three unknowns.* A more general but entirely similar version of Cramer's rule holds for n equations in n unknowns.

CRAMER'S RULE

Consider the system

$$\begin{cases} a_1x + b_1y + c_1z = d_1 \\ a_2x + b_2y + c_2z = d_2 \\ a_3x + b_3y + c_3z = d_3 \end{cases}$$

Let the four determinants D, D_x, D_y, and D_z be defined as follows:

$$D = \begin{vmatrix} a_1 & b_1 & c_1 \\ a_2 & b_2 & c_2 \\ a_3 & b_3 & c_3 \end{vmatrix} \qquad D_x = \begin{vmatrix} d_1 & b_1 & c_1 \\ d_2 & b_2 & c_2 \\ d_3 & b_3 & c_3 \end{vmatrix}$$

$$D_y = \begin{vmatrix} a_1 & d_1 & c_1 \\ a_2 & d_2 & c_2 \\ a_3 & d_3 & c_3 \end{vmatrix} \qquad D_z = \begin{vmatrix} a_1 & b_1 & d_1 \\ a_2 & b_2 & d_2 \\ a_3 & b_3 & d_3 \end{vmatrix}$$

Then if $D \neq 0$, the unique values of x, y, and z satisfying the system are given by

$$x = \frac{D_x}{D} \qquad y = \frac{D_y}{D} \qquad z = \frac{D_z}{D}$$

[If $D = 0$, the solutions (if any) can be found using Gaussian elimination.]

Notice that the determinant D in Cramer's rule is just the determinant of the coefficient matrix of the given system. If you replace the first column of D with the column of numbers on the right side of the given system, you obtain D_x. The determinants D_y and D_z are obtained similarly.

Before actually proving Cramer's rule, let's take a look at how it's applied.

EXAMPLE 5 Use Cramer's rule to find all solutions of the following system of equations:

$$\begin{cases} 2x + 2y - 3z = -20 \\ x - 4y + z = 6 \\ 4x - y + 2z = -1 \end{cases}$$

Solution First we list the determinants D, D_x, D_y, and D_z:

$$D = \begin{vmatrix} 2 & 2 & -3 \\ 1 & -4 & 1 \\ 4 & -1 & 2 \end{vmatrix} \qquad D_x = \begin{vmatrix} -20 & 2 & -3 \\ 6 & -4 & 1 \\ -1 & -1 & 2 \end{vmatrix}$$

$$D_y = \begin{vmatrix} 2 & -20 & -3 \\ 1 & 6 & 1 \\ 4 & -1 & 2 \end{vmatrix} \qquad D_z = \begin{vmatrix} 2 & 2 & -20 \\ 1 & -4 & 6 \\ 4 & -1 & -1 \end{vmatrix}$$

*The rule is named after one of its discoverers, the Swiss mathematician Gabriel Cramer (1704–1752).

The calculations for evaluating D begin as follows:

$$D = \begin{vmatrix} 2 & 2 & -3 \\ 1 & -4 & 1 \\ 4 & -1 & 2 \end{vmatrix} \underset{\uparrow}{=} \begin{vmatrix} 0 & 10 & -5 \\ 1 & -4 & 1 \\ 0 & 15 & -2 \end{vmatrix}$$

Subtract twice the second row from the first.
Subtract 4 times the second row from the third.

Since we now have two zeros in the first column, it is an easy matter to expand D along that column to obtain

$$\begin{aligned} D &= -1\begin{vmatrix} 10 & -5 \\ 15 & -2 \end{vmatrix} \\ &= -1[-20-(-75)] = -1(55) = -55 \end{aligned}$$

The value of D is therefore -55. (Since this value is nonzero, Cramer's rule does apply.) As Exercise 31 at the end of this section asks you to verify, the values of the other three determinants are

$$D_x = 144 \qquad D_y = 61 \qquad D_z = -230$$

By Cramer's rule, then, the unique values of x, y, and z that satisfy the system are

$$x = \frac{D_x}{D} = \frac{144}{-55} = -\frac{144}{55} \qquad y = \frac{D_y}{D} = \frac{61}{-55} = -\frac{61}{55} \qquad z = \frac{D_z}{D} = \frac{-230}{-55} = \frac{46}{11}$$

One way we can prove Cramer's rule is to use Gaussian elimination to solve the system

$$\begin{cases} a_1x + b_1y + c_1z = d_1 \\ a_2x + b_2y + c_2z = d_2 \\ a_3x + b_3y + c_3z = d_3 \end{cases} \tag{1}$$

A much shorter and simpler proof, however, has been found by D. E. Whitford and M. S. Klamkin.* This is the proof we give here; it makes effective use of the rules employed in this section for manipulating determinants.

Consider the system of equations (1) and assume that $D \neq 0$. We will show that if x, y, and z satisfy the system, then in fact $x = D_x/D$, with similar equations giving y and z. (Exercise 64 at the end of this section then shows how to check that these values indeed satisfy the given system.) We have

$$D_x = \begin{vmatrix} d_1 & b_1 & c_1 \\ d_2 & b_2 & c_2 \\ d_3 & b_3 & c_3 \end{vmatrix} \quad \text{by definition}$$

$$= \begin{vmatrix} (a_1x + b_1y + c_1z) & b_1 & c_1 \\ (a_2x + b_2y + c_2z) & b_2 & c_2 \\ (a_3x + b_3y + c_3z) & b_3 & c_3 \end{vmatrix} \quad \text{using the equations in (1) to substitute for } d_1, d_2, \text{ and } d_3$$

*The proof was published in the *American Mathematical Monthly,* vol. **60** (1953); pp. 186–187.

$$= \begin{vmatrix} a_1x & b_1 & c_1 \\ a_2x & b_2 & c_2 \\ a_3x & b_3 & c_3 \end{vmatrix}$$ subtracting y times the second column as well as z times the third column from the first column

$$= x \begin{vmatrix} a_1 & b_1 & c_1 \\ a_2 & b_2 & c_2 \\ a_3 & b_3 & c_3 \end{vmatrix}$$ factoring x out of the first column

$$= xD$$ by definition

We now have $D_x = xD$, which is equivalent to $x = D_x/D$, as required. The formulas for y and z are obtained similarly.

EXERCISE SET 10.5

A

In Exercises 1–6, evaluate the determinants.

1. **(a)** $\begin{vmatrix} 2 & -17 \\ 1 & 6 \end{vmatrix}$ **(b)** $\begin{vmatrix} 1 & 6 \\ 2 & -17 \end{vmatrix}$

2. **(a)** $\begin{vmatrix} 1 & 0 \\ 0 & 1 \end{vmatrix}$ **(b)** $\begin{vmatrix} 0 & 1 \\ 0 & 1 \end{vmatrix}$

3. **(a)** $\begin{vmatrix} 5 & 7 \\ 500 & 700 \end{vmatrix}$ **(b)** $\begin{vmatrix} 5 & 500 \\ 7 & 700 \end{vmatrix}$

4. **(a)** $\begin{vmatrix} -8 & -3 \\ 4 & -5 \end{vmatrix}$ **(b)** $\begin{vmatrix} -3 & -8 \\ -5 & 4 \end{vmatrix}$

5. $\begin{vmatrix} \sqrt{2}-1 & \sqrt{2} \\ \sqrt{2} & \sqrt{2}+1 \end{vmatrix}$

6. $\begin{vmatrix} \sqrt{3}+\sqrt{2} & 1+\sqrt{5} \\ 1-\sqrt{5} & \sqrt{3}-\sqrt{2} \end{vmatrix}$

In Exercises 7–12, refer to the following determinant:

$$\begin{vmatrix} -6 & 3 & 8 \\ 5 & -4 & 1 \\ 10 & 9 & -10 \end{vmatrix}$$

7. Evaluate the minor of the entry 3.

8. Evaluate the cofactor of the entry 3.

9. Evaluate the minor of -10.

10. Evaluate the cofactor of -10.

11. **(a)** Multiply each entry in the first row by its minor and find the sum of the results.
(b) Multiply each entry in the first row by its cofactor and find the sum of the results.
(c) Which gives you the value of the determinant, part (a) or part (b)?

12. **(a)** Multiply each entry in the first column by its cofactor and find the sum of the results.
(b) Follow the same instructions as in part (a), but use the second column.
(c) Follow the same instructions as in part (a), but use the third column.

13. Let $A = \begin{vmatrix} 1 & 2 & 3 \\ 4 & 5 & 6 \\ 7 & 8 & 9 \end{vmatrix}$. Evaluate A by expanding it along **(a)** the second row; **(b)** the third row; **(c)** the first column; **(d)** the third column.

In Exercises 14–21, evaluate the determinants.

14. $\begin{vmatrix} 1 & 2 & -1 \\ 2 & -1 & 1 \\ 4 & 0 & 2 \end{vmatrix}$

15. $\begin{vmatrix} 5 & 10 & 15 \\ 1 & 2 & 3 \\ -9 & 11 & 7 \end{vmatrix}$

16. $\begin{vmatrix} 8 & 4 & 2 \\ 3 & 9 & 3 \\ -2 & 8 & 6 \end{vmatrix}$

17. $\begin{vmatrix} 1 & 2 & -3 \\ 4 & 5 & -9 \\ 0 & 0 & 1 \end{vmatrix}$

18. $\begin{vmatrix} 3 & 0 & 0 \\ 0 & 19 & 0 \\ 0 & 0 & 10 \end{vmatrix}$

19. $\begin{vmatrix} -6 & -8 & 18 \\ 25 & 12 & 15 \\ -9 & 4 & 13 \end{vmatrix}$

20. $\begin{vmatrix} 23 & 0 & 47 \\ -37 & 0 & 18 \\ 14 & 0 & 25 \end{vmatrix}$

21. $\begin{vmatrix} 16 & 0 & -64 \\ -8 & 15 & -12 \\ 30 & -20 & 10 \end{vmatrix}$

22. Use the method illustrated in Example 3 to show that

$$\begin{vmatrix} 1 & 1 & 1 \\ a & b & c \\ a^3 & b^3 & c^3 \end{vmatrix} = (b-a)(c-a)(c^2-b^2+ac-ab)$$
$$= (b-a)(c-a)(c-b)(a+b+c)$$

23. Use the method shown in Example 3 to express the determinant $\begin{vmatrix} 1 & x & x^2 \\ 1 & y & y^2 \\ 1 & z & z^2 \end{vmatrix}$ as a product.

24. Show that

$$\begin{vmatrix} 1 & 1 & 1 \\ a^2 & b^2 & c^2 \\ a^3 & b^3 & c^3 \end{vmatrix} = (b-a)(c-a)(bc^2 - b^2c + ac^2 - ab^2)$$

$$= (b-a)(c-a)(c-b)(bc+ac+ab)$$

25. Simplify the determinant $\begin{vmatrix} 1 & 1 & 1 \\ 1 & 1+x & 1 \\ 1 & 1 & 1+y \end{vmatrix}$.

26. Use the method shown in Example 3 to express the following determinant as a product of three factors:

$$\begin{vmatrix} 1 & 1 & 1 \\ a & b & c \\ bc & ca & ab \end{vmatrix}$$

In Exercises 27–30, evaluate the determinants.

27. $\begin{vmatrix} 1 & -1 & 0 & 2 \\ 0 & 1 & -1 & 0 \\ 2 & 1 & 0 & -1 \\ -2 & 2 & 1 & 1 \end{vmatrix}$

28. $\begin{vmatrix} 3 & -2 & 3 & 4 \\ 1 & 4 & -3 & 2 \\ 6 & 3 & -6 & -3 \\ -1 & 0 & 1 & 5 \end{vmatrix}$

29. $\begin{vmatrix} 2 & 0 & 0 & 0 \\ 0 & 3 & 0 & 0 \\ 0 & 0 & 4 & 0 \\ 0 & 0 & 0 & 5 \end{vmatrix}$ **30.** $\begin{vmatrix} 7 & -8 & 1 & 2 \\ 21 & 4 & 3 & -1 \\ -35 & 8 & 3 & -2 \\ 14 & 16 & 0 & 1 \end{vmatrix}$

31. Verify the following statements (from Example 5).

(a) $\begin{vmatrix} -20 & 2 & -3 \\ 6 & -4 & 1 \\ -1 & -1 & 2 \end{vmatrix} = 144$

(b) $\begin{vmatrix} 2 & -20 & -3 \\ 1 & 6 & 1 \\ 4 & -1 & 2 \end{vmatrix} = 61$

(c) $\begin{vmatrix} 2 & 2 & -20 \\ 1 & -4 & 6 \\ 4 & -1 & -1 \end{vmatrix} = -230$

32. Consider the following system:

$$\begin{cases} x + y + 3z = 0 \\ x + 2y + 5z = 0 \\ x - 4y - 8z = 0 \end{cases}$$

(a) Without doing any calculations, find one obvious solution of this system.
(b) Calculate the determinant D.
(c) List all solutions of this system.

In Exercises 33–42, use Cramer's rule to solve those systems for which $D \neq 0$. In cases where $D = 0$, use Gaussian elimination or matrix methods.

33. $\begin{cases} 3x + 4y - z = 5 \\ x - 3y + 2z = 2 \\ 5x - 6z = -7 \end{cases}$ **34.** $\begin{cases} 3A - B - 4C = 3 \\ A + 2B - 3C = 9 \\ 2A - B + 2C = -8 \end{cases}$

35. $\begin{cases} 3x + 2y - z = -6 \\ 2x - 3y - 4z = -11 \\ x + y + z = 5 \end{cases}$ **36.** $\begin{cases} 5x - 3y - z = 16 \\ 2x + y - 3z = 5 \\ 3x - 2y + 2z = 5 \end{cases}$

37. $\begin{cases} 2x + 5y + 2z = 0 \\ 3x - y - 4z = 0 \\ x + 2y - 3z = 0 \end{cases}$ **38.** $\begin{cases} 4u + 3v - 2w = 14 \\ u + 2v - 3w = 6 \\ 2u - v + 4w = 2 \end{cases}$

39. $\begin{cases} 12x - 11z = 13 \\ 6x + 6y - 4z = 26 \\ 6x + 2y - 5z = 13 \end{cases}$

40. $\begin{cases} 3x + 4y + 2z = 1 \\ 4x + 6y + 2z = 7 \\ 2x + 3y + z = 11 \end{cases}$

41. $\begin{cases} x + y + z + w = -7 \\ x - y + z - w = -11 \\ 2x - 2y - 3z - 3w = 26 \\ 3x + 2y + z - w = -9 \end{cases}$

42. $\begin{cases} 2A - B - 3C + 2D = -2 \\ A - 2B + C - 3D = 4 \\ 3A - 4B + 2C - 4D = 12 \\ 2A + 3B - C - 2D = -4 \end{cases}$

43. Find all values of x for which

$$\begin{vmatrix} x-4 & 0 & 0 \\ 0 & x+4 & 0 \\ 0 & 0 & x+1 \end{vmatrix} = 0$$

44. Find all values of x for which

$$\begin{vmatrix} 1 & x & x^2 \\ 1 & 1 & 1 \\ 4 & 5 & 0 \end{vmatrix} = 0$$

45. By expanding the determinant $\begin{vmatrix} a & b & c \\ a & b & c \\ d & e & f \end{vmatrix}$ down the first column, show that its value is zero.

46. By expanding the determinant $\begin{vmatrix} ka & kb & kc \\ d & e & f \\ g & h & i \end{vmatrix}$ along its first row, show that it is equal to $k \begin{vmatrix} a & b & c \\ d & e & f \\ g & h & i \end{vmatrix}$.

B

47. Show that

$$\begin{vmatrix} a_1+A_1 & b_1 & c_1 \\ a_2+A_2 & b_2 & c_2 \\ a_3+A_3 & b_3 & c_3 \end{vmatrix} = \begin{vmatrix} a_1 & b_1 & c_1 \\ a_2 & b_2 & c_2 \\ a_3 & b_3 & c_3 \end{vmatrix} + \begin{vmatrix} A_1 & b_1 & c_1 \\ A_2 & b_2 & c_2 \\ A_3 & b_3 & c_3 \end{vmatrix}$$

48. Show that

$$\begin{vmatrix} a_1 & b_1 & c_1 \\ a_2 & b_2 & c_2 \\ a_3 & b_3 & c_3 \end{vmatrix} = \begin{vmatrix} a_1+kb_1 & b_1 & c_1 \\ a_2+kb_2 & b_2 & c_2 \\ a_3+kb_3 & b_3 & c_3 \end{vmatrix}$$

49. By expanding each determinant along a row or column, show that

$$\begin{vmatrix} a_1 & b_1 & c_1 \\ a_2 & b_2 & c_2 \\ a_3 & b_3 & c_3 \end{vmatrix} = -\begin{vmatrix} a_2 & b_2 & c_2 \\ a_1 & b_1 & c_1 \\ a_3 & b_3 & c_3 \end{vmatrix}$$

50. Solve for x in terms of a, b, and c:

$$\begin{vmatrix} a & a & x \\ c & c & c \\ b & x & b \end{vmatrix} = 0 \qquad (c \neq 0)$$

51. Show that

$$\begin{vmatrix} 1+a & 1 & 1 & 1 \\ 1 & 1+b & 1 & 1 \\ 1 & 1 & 1+c & 1 \\ 1 & 1 & 1 & 1+d \end{vmatrix} = abcd\left(\frac{1}{a}+\frac{1}{b}+\frac{1}{c}+\frac{1}{d}+1\right)$$

52. Show that

$$\begin{vmatrix} 1 & a & a_2 \\ a^2 & 1 & a \\ a & a^2 & 1 \end{vmatrix} = (a^3-1)^2$$

53. Consider the determinant $D = \begin{vmatrix} a_1 & b_1 & c_1 \\ a_2 & b_2 & c_2 \\ a_3 & b_3 & c_3 \end{vmatrix}$. Let A_1 denote the cofactor of a_1, let B_1 denote the cofactor of b_1, and so on. Prove that

$$\begin{vmatrix} B_2 & C_2 \\ B_3 & C_3 \end{vmatrix} = a_1 D$$

54. Show that

$$\begin{vmatrix} 1 & bc & b+c \\ 1 & ca & c+a \\ 1 & ab & a+b \end{vmatrix} = (b-c)(c-a)(a-b)$$

55. Find the value of each determinant.

(a) $\begin{vmatrix} 1 & 0 & 0 \\ x & 1 & 0 \\ x & y & 1 \end{vmatrix}$ **(b)** $\begin{vmatrix} 1 & 0 & 0 & 0 \\ x & 1 & 0 & 0 \\ x & y & 1 & 0 \\ x & y & z & 1 \end{vmatrix}$

56. Evaluate the determinant

$$\begin{vmatrix} a & b & c \\ a & a+b & a+b+c \\ a & 2a+b & 3a+2b+c \end{vmatrix}$$

57. Show that

$$\begin{vmatrix} 1 & a & a & a \\ 1 & b & a & a \\ 1 & a & b & a \\ 1 & a & a & b \end{vmatrix} = (b-a)^3$$

58. Show that

$$\begin{vmatrix} a & 1 & 1 & 1 \\ 1 & a & 1 & 1 \\ 1 & 1 & a & 1 \\ 1 & 1 & 1 & a \end{vmatrix} = (a-1)^3(a+3)$$

59. Solve the following system for x, y, and z:

$$\begin{cases} ax + by + cz = k \\ a^2x + b^2y + c^2z = k^2 \\ a^3x + b^3y + c^3z = k^3 \end{cases}$$

(Assume that the values of a, b, and c are all distinct and all nonzero.)

60. Show that the equation

$$\begin{vmatrix} x & y & 1 \\ x_1 & y_1 & 1 \\ 1 & m & 0 \end{vmatrix} = 0$$

represents a line that has slope m and passes through the point (x_1, y_1).

For Exercise 61, use the fact that the equation of a line passing through (x_1, y_1) and (x_2, y_2) can be written

$$\begin{vmatrix} x & y & 1 \\ x_1 & y_1 & 1 \\ x_2 & y_2 & 1 \end{vmatrix} = 0$$

For Exercise 62, use the fact that the equation of a circle passing through (x_1, y_1), (x_2, y_2), and (x_3, y_3) can be written

$$\begin{vmatrix} x^2+y^2 & x & y & 1 \\ x_1^2+y_1^2 & x_1 & y_1 & 1 \\ x_2^2+y_2^2 & x_2 & y_2 & 1 \\ x_3^2+y_3^2 & x_3 & y_3 & 1 \end{vmatrix} = 0$$

61. Find the equation of the line passing through $(-3, -1)$ and $(2, 9)$. Write the answer in the form $y = mx + b$.

62. Find the equation of the circle passing through $(7, 0)$, $(5, -6)$, and $(-1, -4)$. Write the answer in the form $(x - h)^2 + (y - k)^2 = r^2$.

C

63. Show that the area of the triangle in Figure A is $\frac{1}{2}\begin{vmatrix} a & b \\ c & d \end{vmatrix}$. *Hint:* Figure B indicates how the required area can be found using a rectangle and three right triangles.

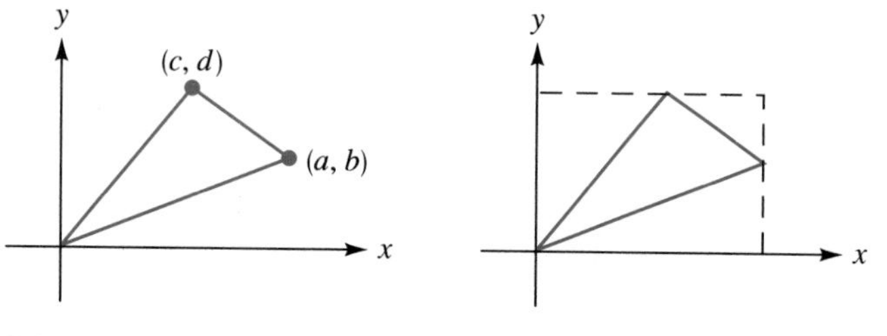

FIGURE A

FIGURE B

64. This exercise completes the derivation of Cramer's rule given in the text. Using the same notation, and assuming $D \neq 0$, we need to show that the values $x = D_x/D$, $y = D_y/D$, and $z = D_z/D$ satisfy the equations in (1) on page 646. We will show that these values satisfy the first equation in (1), the verification for the other equations being entirely similar.

(a) Check that substituting the values $x = D_x/D$, $y = D_y/D$, and $z = D_z/D$ in the first equation of (1) yields an equation equivalent to

$$a_1D_x + b_1D_y + c_1D_z - d_1D = 0$$

(b) Show that the equation in part (a) can be written

$$a_1\begin{vmatrix} b_1 & c_1 & d_1 \\ b_2 & c_2 & d_2 \\ b_3 & c_3 & d_3 \end{vmatrix} - b_1\begin{vmatrix} a_1 & c_1 & d_1 \\ a_2 & c_2 & d_2 \\ a_3 & c_3 & d_3 \end{vmatrix} + c_1\begin{vmatrix} a_1 & b_1 & d_1 \\ a_2 & b_2 & d_2 \\ a_3 & b_3 & d_3 \end{vmatrix} - d_1\begin{vmatrix} a_1 & b_1 & c_1 \\ a_2 & b_2 & c_2 \\ a_3 & b_3 & c_3 \end{vmatrix} = 0$$

(c) Show that the equation in part (b) can be written

$$\begin{vmatrix} a_1 & b_1 & c_1 & d_1 \\ a_1 & b_1 & c_1 & d_1 \\ a_2 & b_2 & c_2 & d_2 \\ a_3 & b_3 & c_3 & d_3 \end{vmatrix} = 0$$

(d) Now explain why the equation in (c) indeed holds.

65. Let D denote the determinant of the matrix $\begin{pmatrix} a & b \\ c & d \end{pmatrix}$.

(a) Show that the inverse of this matrix is $\frac{1}{D}\begin{pmatrix} d & -b \\ -c & a \end{pmatrix}$. Assume that $ad - bc \neq 0$.

(b) Use the result in part (a) to find the inverse of the matrix $\begin{pmatrix} -6 & 7 \\ 1 & 9 \end{pmatrix}$.

66. Assuming that $p + q + r = 0$, solve the following equation for x:

$$\begin{vmatrix} p - x & r & q \\ r & q - x & p \\ q & p & r - x \end{vmatrix} = 0$$

Hint: After adding certain rows or columns, the quantity $p + q + r$ will appear.

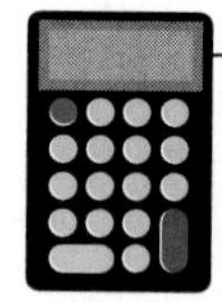

GRAPHING UTILITY EXERCISES FOR SECTION 10.5

1. According to the text, the value of the determinant

$$\begin{vmatrix} 1 & 2 & 3 \\ 4 & 5 & 6 \\ 7 & 8 & 9 \end{vmatrix}$$

is zero. Use a graphing utility to obtain this result.

2. Evaluate the determinant $\begin{vmatrix} 15 & 14 & 26 \\ 18 & 17 & 32 \\ 21 & 20 & 42 \end{vmatrix}$. Check your answer by referring to Example 2.

3. Evaluate the determinant

$$\begin{vmatrix} 25 & 40 & 5 & 10 \\ 9 & 0 & 3 & 6 \\ -2 & 3 & 11 & -17 \\ -3 & 4 & 7 & 2 \end{vmatrix}$$

Check your answer by referring to Example 4.

4. According to Example 3 in the text, the following

equation is an identity:

$$\begin{vmatrix} 1 & 1 & 1 \\ a & b & c \\ a^2 & b^2 & c^2 \end{vmatrix} = (b-a)(c-a)(c-b)$$

Evaluate each side of this equation using the values $a = \sqrt{5}$, $b = \sqrt{6}$, and $c = \sqrt{7}$, and check to see that the results indeed agree.

5. Consider the two determinants

$$\begin{vmatrix} 1 & 2 & 3 \\ -7 & -4 & 5 \\ 9 & 2 & 6 \end{vmatrix} \quad \text{and} \quad \begin{vmatrix} 10 & 20 & 30 \\ -7 & -4 & 5 \\ 9 & 2 & 6 \end{vmatrix}$$

(a) According to Item 1 in the Property Summary Box on page 642, how are the values of these determinants related?

(b) Evaluate each determinant to verify your answer in part (a).

6. Consider the two determinants

$$\begin{vmatrix} \sqrt{2} & \sqrt{5} & \sqrt{7} \\ -6 & 10 & 0 \\ 8 & -1 & -6 \end{vmatrix} \quad \text{and} \quad \begin{vmatrix} -6 & 10 & 0 \\ \sqrt{2} & \sqrt{5} & \sqrt{7} \\ 8 & -1 & -6 \end{vmatrix}.$$

(a) According to Item 3 in the Property Summary Box on page 642, how are the values of these determinants related?

(b) Evaluate each determinant to verify your answer in part (a).

7. Consider the system

$$\begin{cases} 3x - 3y + 4z = 11 \\ 4x + 8y - 5z = -6 \\ -2x - 7y - z = 0 \end{cases}$$

(a) Compute each of the determinants D, D_x, D_y, and D_z.

(b) Use Cramer's rule to solve the given system of equations. Round your answers to two decimal places.

8. Consider the system

$$\begin{cases} 0.2x - 3.1y + 4.4z - 1.2w = 5.3 \\ -6.1x + 1.1y - 1.9z - 1.9w = -3.2 \\ 0.8x + 0.3y + 3.7z + 3.0w = 10.1 \\ 1.0x - 5.7y - 2.6z - 5.4w = -6.6 \end{cases}$$

(a) Compute each of the determinants D, D_x, D_y, D_z, and D_w.

(b) Use Cramer's rule to solve the given system of equations. Round your answers to one decimal place.

10.6 NONLINEAR SYSTEMS OF EQUATIONS

In the previous sections of this chapter, we looked at several techniques for solving systems of linear equations. In particular, we saw that if solutions exist, they can always be found by Gaussian elimination. In the present section, we consider **nonlinear systems** of equations, that is, systems in which at least one of the equations is not linear. There is no single technique that serves to solve all nonlinear systems. However, simple substitution will often suffice. The work in this section focuses on examples showing some of the more common approaches. In all of the examples and in the exercises, we will be concerned exclusively with solutions (x, y) in which both x and y are real numbers. In Example 1, the system consists of one linear equation and one quadratic equation. Such a system can always be solved by substitution.

EXAMPLE 1 Find all solutions (x, y) of the following system, where x and y are real numbers:

$$\begin{cases} 2x + y = 1 \\ y = 4 - x^2 \end{cases}$$

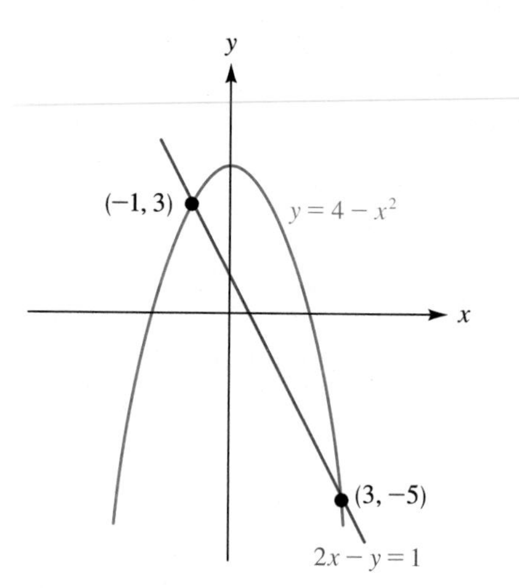

FIGURE 1

Solution We use the second equation to substitute for y in the first equation. This will yield an equation with only one unknown:

$$2x + (4 - x^2) = 1$$
$$-x^2 + 2x + 3 = 0$$
$$x^2 - 2x - 3 = 0$$
$$(x - 3)(x + 1) = 0$$

$x - 3 = 0$	$x + 1 = 0$
$x = 3$	$x = -1$

The values $x = 3$ and $x = -1$ can now be substituted back into either of the original equations. Substituting $x = 3$ in the equation $y = 4 - x^2$ yields $y = -5$. Similarly, substituting $x = -1$ in the equation $y = 4 - x^2$ gives us $y = 3$. We have now obtained two ordered pairs, $(3, -5)$ and $(-1, 3)$. As you can easily check, both of these are solutions of the given system. Figure 1 displays the graphical interpretation of this result. The line $2x + y = 1$ intersects the parabola $y = 4 - x^2$ at the points $(-1, 3)$ and $(3, -5)$.

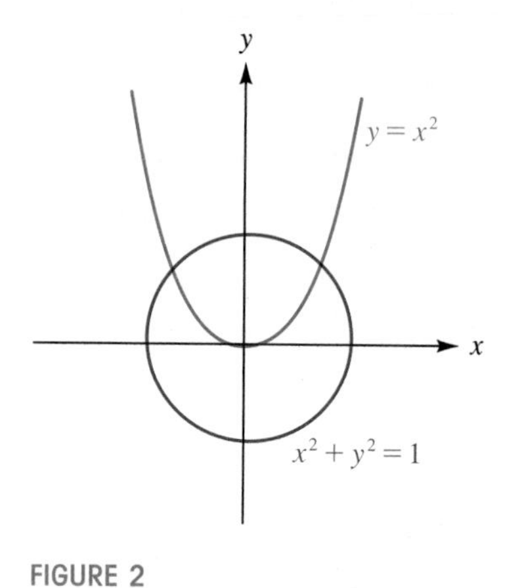

FIGURE 2

EXAMPLE 2 Where do the graphs of the parabola $y = x^2$ and the circle $x^2 + y^2 = 1$ intersect? See Figure 2.

Solution The system that we wish to solve is

$$\begin{cases} y = x^2 & (1) \\ x^2 + y^2 = 1 & (2) \end{cases}$$

In view of equation (1), we can replace the x^2-term of equation (2) by y. Doing this yields

$$y + y^2 = 1$$

or

$$y^2 + y - 1 = 0$$

This last equation can be solved by using the quadratic formula with $a = 1$, $b = 1$, and $c = -1$. As you can check, the results are

$$y = \frac{-1 + \sqrt{5}}{2} \quad \text{and} \quad y = \frac{-1 - \sqrt{5}}{2}$$

However, from Figure 2, it is clear that the y-coordinate at each intersection point is positive. Therefore, we discard the negative number $y = -\frac{1}{2}(1 + \sqrt{5})$ from further consideration in this context. Substituting the positive number $y = \frac{1}{2}(-1 + \sqrt{5})$ back in the equation $y = x^2$ then gives us

$$x^2 = (-1 + \sqrt{5})/2$$

Therefore,

$$x = \pm\sqrt{(-1 + \sqrt{5})/2}$$

By choosing the positive square root, we obtain the x-coordinate for the intersection point in the first quadrant. That point is therefore

$$\left(\sqrt{(-1+\sqrt{5})/2}, (-1+\sqrt{5})/2\right) \approx (0.79, 0.62) \qquad \text{using a calculator}$$

Similarly, the negative square root yields the x-coordinate for the intersection point in the second quadrant. That point is

$$\left(-\sqrt{(-1+\sqrt{5})/2}, (-1+\sqrt{5})/2\right) \approx (-0.79, 0.62) \qquad \text{using a calculator}$$

We have now found the two intersection points, as required.

EXAMPLE 3 Find all solutions (x, y) of the following system, where x and y are real numbers:

$$\begin{cases} xy = 1 & (3) \\ y = 3x + 1 & (4) \end{cases}$$

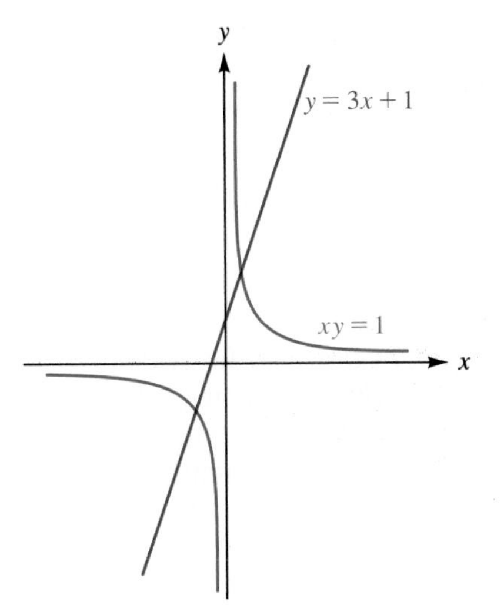

FIGURE 3

Solution Since these equations are easy to graph, we do so, because that will tell us something about the required solutions. As Figure 3 indicates, there are two intersection points, one in the first quadrant, the other in the third quadrant. One way to begin now would be to solve equation (3) for one unknown in terms of the other. However, to avoid introducing fractions at the outset, let's use equation (4) to substitute for y in equation (3). This yields

$$x(3x+1) = 1$$

or

$$3x^2 + x - 1 = 0$$

This last equation can be solved by using the quadratic formula. As you should verify, the solutions are

$$x = (-1+\sqrt{13})/6 \qquad \text{and} \qquad x = (-1-\sqrt{13})/6$$

The corresponding y-values can now be obtained by substituting for x in either of the given equations. We will substitute in equation (4). (Exercise 23 at the end of this section asks you to substitute in equation (3) as well and then to show that the very different-looking answers obtained in that way are in fact equal to those found here.) Substituting $x = (-1+\sqrt{13})/6$ into the equation $y = 3x + 1$ gives us

$$\begin{aligned} y &= 3\left(\frac{-1+\sqrt{13}}{6}\right) + 1 \\ &= \frac{-1+\sqrt{13}}{2} + \frac{2}{2} = \frac{1+\sqrt{13}}{2} \end{aligned}$$

Thus, one of the intersection points is

$$\left((-1+\sqrt{13})/6), (1+\sqrt{13})/2\right)$$

Notice that this must be the first-quadrant point of intersection, since both coordinates are positive. The intersection point in the third quadrant is obtained in exactly the same manner. As you can check, substituting $x = -\frac{1}{6}(1+\sqrt{13})$ in the equation $y = 3x + 1$ yields $y = \frac{1}{2}(1 - \sqrt{13})$. Thus, the other intersection point is

$$\left(-(1 + \sqrt{13})/6, (1 - \sqrt{13})/2\right)$$

As Figure 3 indicates, there are no other solutions.

EXAMPLE 4 Find all real numbers x and y satisfying the system of equations

$$\begin{cases} y = \sqrt{x} \\ (x + 2)^2 + y^2 = 1 \end{cases}$$

Solution We use the first equation to substitute for y in the second equation. This yields

$$(x + 2)^2 + (\sqrt{x})^2 = 1$$
$$x^2 + 4x + 4 + x = 1$$
$$x^2 + 5x + 3 = 0$$

Using the quadratic formula to solve this last equation, we have

$$x = \frac{-5 \pm \sqrt{5^2 - 4(1)(3)}}{2(1)} = \frac{-5 \pm \sqrt{13}}{2}$$

The two values of x are thus

$$(-5 + \sqrt{13})/2 \quad \text{and} \quad (-5 - \sqrt{13})/2$$

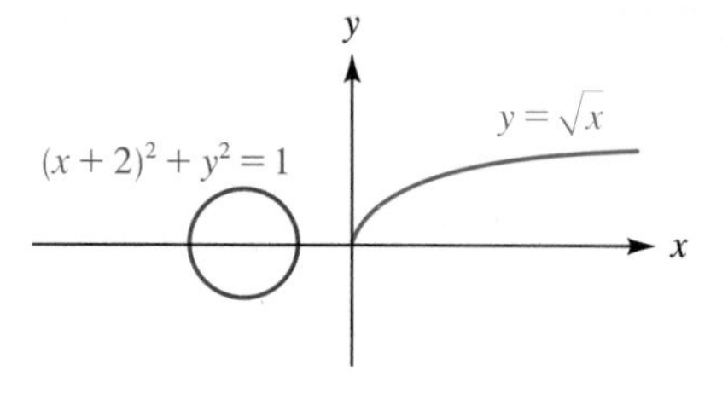

FIGURE 4

However, notice that both of these quantities are negative. (The second is obviously negative; without using a calculator, can you explain why the first is negative?) Thus, neither of these quantities is an appropriate x-input in the equation $y = \sqrt{x}$, since we are looking for y-values that are real numbers. We conclude from this that there are no pairs of real numbers x and y satisfying the given system. In geometric terms, this means that the two graphs do not intersect; see Figure 4.

In the next example, we look at a system that can be reduced to a linear system through appropriate substitutions.

EXAMPLE 5 Solve the system

$$\begin{cases} \dfrac{2}{x^2} - \dfrac{3}{y^2} = -6 \\[2ex] \dfrac{3}{x^2} + \dfrac{4}{y^2} = 59 \end{cases}$$

Solution Let $u = 1/x^2$ and $v = 1/y^2$, so that the system becomes

$$\begin{cases} 2u - 3v = -6 \\ 3u + 4v = 59 \end{cases}$$

This is now a linear system. As Exercise 46 asks you to show, the solution is $u = 9$, $v = 8$. In view of the definitions of u and v, then, we have

$\frac{1}{x^2} = 9$	$\frac{1}{y^2} = 8$
$x^2 = 1/9$	$y^2 = 1/8$
$x = +1/3$	$y = \pm 1/\sqrt{8} = \pm 1/(2\sqrt{2})$
	$= \pm\sqrt{2}/4$

This gives us four possible solutions for the original system:

$$\left(\frac{1}{3}, \frac{\sqrt{2}}{4}\right) \quad \left(\frac{1}{3}, -\frac{\sqrt{2}}{4}\right) \quad \left(-\frac{1}{3}, \frac{\sqrt{2}}{4}\right) \quad \left(-\frac{1}{3}, -\frac{\sqrt{2}}{4}\right)$$

As you can check, all four of these pairs satisfy the given system.

EXAMPLE 6 Determine all solutions (x, y) of the following system, where x and y are real numbers:

$$\begin{cases} y = 3^x & (5) \\ y = 3^{2x} - 2 & (6) \end{cases}$$

Solution We'll use the substitution method. First we rewrite equation (6) as

$$y = (3^x)^2 - 2 \qquad (7)$$

Now, in view of equation (5), we can replace 3^x with y in equation (7) to obtain

$$y = y^2 - 2$$
$$0 = y^2 - y - 2$$
$$0 = (y + 1)(y - 2)$$

From this last equation, we see that $y = -1$ or $y = 2$. With $y = -1$, equation (5) becomes $-1 = 3^x$, contrary to the fact that 3^x is positive for all real numbers x. Thus, we discard the case where $y = -1$. On the other hand, if $y = 2$, equation (5) becomes

$$2 = 3^x$$

We can solve this exponential equation by taking the logarithm of both sides. Using base e logarithms, we have

$$\ln 2 = \ln 3^x = x \ln 3$$

and, consequently,

$$x = \frac{\ln 2}{\ln 3} \qquad \text{CAUTION: } \frac{\ln 2}{\ln 3} \neq \ln 2 - \ln 3$$

We've now found that $x = (\ln 2)/(\ln 3)$ and $y = 2$. Figure 5 displays a graphical interpretation of this result. (Using a calculator for the x-coordinate, we find that $x \approx 0.6$, which is consistent with Figure 5.)

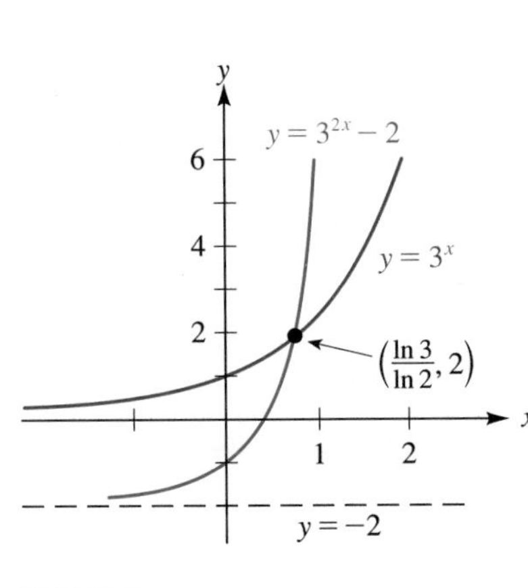

FIGURE 5

EXERCISE SET 10.6

A

In Exercises 1–22, find all solutions (x, y) of the given systems, where x and y are real numbers.

1. $\begin{cases} y = 3x \\ y = x^2 \end{cases}$

2. $\begin{cases} y = x + 3 \\ y = 9 - x^2 \end{cases}$

3. $\begin{cases} x^2 + y^2 = 25 \\ 24y = x^2 \end{cases}$

4. $\begin{cases} 3x + 4y = 12 \\ x^2 - y + 1 = 0 \end{cases}$

5. $\begin{cases} xy = 1 \\ y = -x^2 \end{cases}$

6. $\begin{cases} x + 2y = 0 \\ xy = -2 \end{cases}$

7. $\begin{cases} 2x^2 + y^2 = 17 \\ x^2 + 2y^2 = 22 \end{cases}$

8. $\begin{cases} x - 2y = 1 \\ y^2 - x^2 = 3 \end{cases}$

9. $\begin{cases} y = 1 - x^2 \\ y = x^2 - 1 \end{cases}$

10. $\begin{cases} xy = 4 \\ y = 4x \end{cases}$

11. $\begin{cases} xy = 4 \\ y = 4x + 1 \end{cases}$

12. $\begin{cases} \dfrac{2}{x^2} + \dfrac{5}{y^2} = 3 \\ \dfrac{3}{x^2} - \dfrac{2}{y^2} = 1 \end{cases}$

13. $\begin{cases} \dfrac{1}{x^2} - \dfrac{3}{y^2} = 14 \\ \dfrac{2}{x^2} + \dfrac{1}{y^2} = 35 \end{cases}$

14. $\begin{cases} y = -\sqrt{x} \\ (x - 3)^2 + y^2 = 4 \end{cases}$

15. $\begin{cases} y = -\sqrt{x - 1} \\ (x - 3)^2 + y^2 = 4 \end{cases}$

16. $\begin{cases} y = -\sqrt{x - 6} \\ (x - 3)^2 + y^2 = 4 \end{cases}$

17. $\begin{cases} y = 2^x \\ y = 2^{2x} - 12 \end{cases}$

18. $\begin{cases} y = e^{4x} \\ y = e^{2x} + 6 \end{cases}$

19. $\begin{cases} 2(\log_{10} x)^2 - (\log_{10} y)^2 = -1 \\ 4(\log_{10} x)^2 - 3(\log_{10} y)^2 = -11 \end{cases}$

20. $\begin{cases} y = \log_2(x + 1) \\ y = 5 - \log_2(x - 3) \end{cases}$

21. $\begin{cases} 2^x \cdot 3^y = 4 \\ x + y = 5 \end{cases}$

22. $\begin{cases} a^{2x} + a^{2y} = 10 \\ a^{x+y} = 4 \end{cases} \quad (a > 0)$

    Hint: Use the substitutions $a^x = t$ and $a^y = u$.

23. Let $x = (-1 + \sqrt{13})/6$. Using this x-value, show that the equations $y = 3x + 1$ and $y = 1/x$ yield the same y-value.

24. A sketch shows that the line $y = 100x$ intersects the parabola $y = x^2$ at the origin. Are there any other intersection points? If so, find them. If not, explain why not.

25. Solve the following system for x and y:

$$\begin{cases} ax + by = 2 \\ abxy = 1 \end{cases}$$

(Assume that neither a nor b is zero.)

B

26. Let a, b, and c be constants (with $a \neq 0$), and consider the system

$$\begin{cases} y = ax^2 + bx + c \\ y = k \end{cases}$$

For which value of k (in terms of a, b, and c) will the system have exactly one solution? What is that solution? What is the relationship between the solution you've found and the graph of $y = ax^2 + bx + c$?

27. Find all solutions of the system

$$\begin{cases} x^3 + y^3 = 3473 \\ x + y = 23 \end{cases}$$

28. Solve the following system for x, y, and z:

$$\begin{cases} yz = p^2 \\ zx = q^2 \\ xy = r^2 \end{cases}$$

(Assume that p, q, and r are positive constants.)

29. If the diagonal of a rectangle has length d and the perimeter of the rectangle is $2p$, express the lengths of the sides in terms of d and p.

30. Solve the following system for x and y:

$$\begin{cases} xy + pq = 2px \\ x^2y^2 + p^2q^2 = 2q^2y^2 \end{cases} \quad (pq \neq 0)$$

Hint: Square the first equation, then subtract the second. This results in an equation that can be written $(2px + qy)(px - qy) = 0$.

31. Solve the following system for u and v:

$$\begin{cases} u^2 - v^2 = 9 \\ \sqrt{u + v} + \sqrt{u - v} = 4 \end{cases}$$

Hint: Square the second equation.

32. Solve for x and y in terms of p, q, a, and b:

$$\begin{cases} q^{\ln x} = p^{\ln y} \\ (px)^{\ln a} = (qy)^{\ln b} \end{cases}$$

(Assume all constants and variables are positive.)

33. The accompanying figure shows the graph of a function $N = N_0 e^{kt}$. In each case, determine the constants N_0 and k so that the graph passes through **(a)** (2, 3) and (8, 24); **(b)** (1/2, 1) and (4, 10).

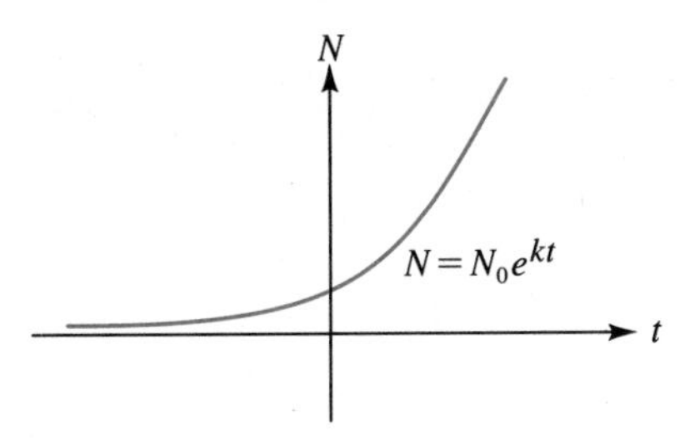

34. [As background for this exercise, first work Exercise 33(a).] Measurements (begun in 1958, by Charles D. Keeling of the Scripps Institution of Oceanography) show that the concentration of carbon dioxide in the atmosphere has increased exponentially over the years 1958–1988. In 1958, the average carbon dioxide concentration was $N = 315$ ppm (parts per million), and in 1988 it was $N = 351$ ppm.

(a) Use this data to determine the constants k and N_0 in the growth law $N = N_0 e^{kt}$. *Hint:* The curve $N = N_0 e^{kt}$ passes through the two points (1958, 315) and (1988, 351). Use this to obtain a system of two equations in two unknowns.

(b) Stephen H. Schneider in his article "The Changing Climate" [*Scientific American,* vol. 261 (1989), pp. 70–79] estimates that, under certain circumstances, the concentration of carbon dioxide in the atmosphere could reach 600 ppm by the year 2080. Use your results in part (a) to make a projection for the year 2080. Is your projection higher or lower than Dr. Schneider's?

35. Solve the following system for x, y, and z in terms of p, q, and r.

$$\begin{cases} x(x+y+z) = p^2 \\ y(x+y+z) = q^2 \\ z(x+y+z) = r^2 \end{cases}$$

(Assume that p, q, and r are nonzero.) *Hint:* Denote $x + y + z$ by w; then add the three equations.

36. **(a)** Find the points where the line $y = -2x - 2$ intersects the parabola $y = \frac{1}{2}x^2$.

(b) On the same set of axes, sketch the line $y = -2x - 2$ and the parabola $y = \frac{1}{2}x^2$. Be certain that your sketch is consistent with the results obtained in part (a).

37. If a right triangle has area 180 cm^2 and hypotenuse 41 cm, find the lengths of the two legs.

38. The sum of two numbers is 8, while their product is -128. What are the two numbers?

39. If a rectangle has perimeter 46 cm and area 60 cm^2, find the length and the width.

40. Find all right triangles for which the perimeter is 24 units and the area is 24 square units.

41. Solve the following system for x and y using the substitution method:

$$\begin{cases} x^2 + y^2 = 5 & (1) \\ xy = 2 & (2) \end{cases}$$

42. The substitution method in Exercise 41 leads to a quadratic equation. Here is an alternative approach to solving that system; this approach leads to linear equations. Multiply equation (2) by 2 and add the resulting equation to equation (1). Now take square roots to conclude that $x + y = \pm 3$. Next multiply equation (2) by 2 and subtract the resulting equation from equation (1). Take square roots to conclude that $x - y = \pm 1$. You now have the following four linear systems, each of which can be solved (with almost no work) by the addition–subtraction method.

$$\begin{cases} x + y = 3 \\ x - y = 1 \end{cases} \qquad \begin{cases} x + y = 3 \\ x - y = -1 \end{cases}$$

$$\begin{cases} x + y = -3 \\ x - y = 1 \end{cases} \qquad \begin{cases} x + y = -3 \\ x - y = -1 \end{cases}$$

Solve these systems and compare your results with those obtained in Exercise 41.

43. Solve the following system using the method explained in Exercise 42.

$$\begin{cases} x^2 + y^2 = 7 \\ xy = 3 \end{cases}$$

44. Solve the following system using the substitution method:

$$\begin{cases} 3xy - 4x^2 = 2 \\ -5x^2 + 3y^2 = 7 \end{cases}$$

(Begin by solving the first equation for y.)

45. Here is an alternative approach for solving the system in Exercise 44. Let $y = mx$, where m is a constant to be determined. Replace y with mx in both equations of the system to obtain the following pair of equations:

$$x^2(3m - 4) = 2 \qquad (1)$$
$$x^2(-5 + 3m^2) = 7 \qquad (2)$$

Now divide equation (2) by equation (1). After clearing fractions and simplifying, you can write the resulting equation as $2m^2 - 7m + 6 = 0$. Solve this last equation by factoring. The values of m can then be used in equation (1) to determine values for x. In each case, the corresponding y-values are determined by the equation $y = mx$.

46. Solve the following system. (You should obtain $u = 9$ and $v = 8$, as stated in Example 5.)

$$\begin{cases} 2u - 3v = -6 \\ 3u + 4v = 59 \end{cases}$$

47. Solve the following system for x and y:

$$\begin{cases} xy + pq = 2px \\ x^2y^2 + p^2q^2 = 2q^2y^2 \end{cases}$$

Hint: Square the first equation, then subtract the second. This results in an equation that can be factored.

C

48. Solve the following system for x and y:

$$\begin{cases} \dfrac{1}{x^2} + \dfrac{1}{xy} = \dfrac{1}{a^2} \\ \dfrac{1}{y^2} + \dfrac{1}{xy} = \dfrac{1}{b^2} \end{cases}$$

(Assume that a and b are positive.)

49. Solve the following system for x and y:

$$\begin{cases} x^4 = y^6 \\ \ln \dfrac{x}{y} = \dfrac{\ln x}{\ln y} \end{cases}$$

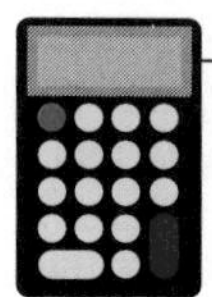

GRAPHING UTILITY EXERCISES FOR SECTION 10.6

In Exercises 1–14, you are given a nonlinear system consisting of two equations in two unknowns. ***(a)*** *Solve each system by graphing both equations and estimating the coordinates of the points where the two graphs intersect. Zoom in on the intersection points until you are sure of the first three decimal places in each coordinate.* ***(b)*** *For Exercises 1–8, also use an algebraic method to find the solutions. Round the final answers to four decimal places. Check to see that your results are consistent with the graphical estimates obtained in part (a).*

1. $\begin{cases} y = x - 5 \\ y = -x^2 + 2 \end{cases}$

2. $\begin{cases} y = x^2 - 1 \\ y = -2x^4 + 3 \end{cases}$

3. $\begin{cases} x^2 + y^2 = 1 \\ y = 3x \end{cases}$

4. $\begin{cases} x^2 + y^2 = 1 \\ y = 3x^2 \end{cases}$

5. $\begin{cases} y = \sqrt{x+1} + 1 \\ 3x + 4y = 12 \end{cases}$

6. $\begin{cases} y = \sqrt{x} \\ y = 2x^3 \end{cases}$

7. $\begin{cases} y = 4^{2x} \\ y = 4^x + 3 \end{cases}$

8. $\begin{cases} y = \ln x \\ y = 1 + \ln(x - 5) \end{cases}$

9. $\begin{cases} y = e^{x/2} \\ y = x^2 \end{cases}$

10. $\begin{cases} x^2 - 2x + y = 0 \\ y = 3x^3 - x^2 - 10x \end{cases}$

11. $\begin{cases} y = \sqrt{x} - 1 \\ y = \ln x \end{cases}$

Hint: There are two solutions.

12. $\begin{cases} y = \sqrt[3]{x} \\ y = \ln x \end{cases}$

Hint: There are two solutions.

13. $\begin{cases} y = x^3 \\ y = (e^x + e^{-x})/2 \end{cases}$

Hint: There are two solutions.

14. $\begin{cases} y = x^3 \\ y = (e^x - e^{-x})/2 \end{cases}$

Hint: There are four solutions; use symmetry to cut the amount of work in half.

10.7 SYSTEMS OF INEQUALITIES

How typesetters frequently improvised signs by the use of forms primarily intended for other purposes is illustrated in the inequality signs appearing in 1743 in the papers of Euler. The radical sign √ is turned into the position > for "greater than" and into < for "less than."

Florian Cajori in *A History of Mathematical Notations, vol. II* (La Salle, Illinois: The Open Court Publishing Company, 1929)

In the first section of this chapter, we solved systems of equations in two unknowns. Now we are going to consider systems of inequalities in two unknowns. The techniques we develop are used in calculus when discussing functions of two variables and in business and economics in the study of linear programming.

Let a, b, and c denote real numbers, and assume that a and b are not both zero. Then all of the following are called **linear inequalities:**

$$ax + by + c < 0 \qquad ax + by + c > 0$$
$$ax + by + c \leqslant 0 \qquad ax + by + c \geqslant 0$$

An ordered pair of numbers (x_0, y_0) is said to be a **solution** of a given inequality (linear or not) provided we obtain a true statement upon substituting x_0 and y_0 for x and y, respectively. For instance, $(-1, 1)$ is a solution of the inequality $2x + 3y - 6 < 0$, since substitution yields

$$2(-1) + 3(1) - 6 < 0$$
$$-5 < 0 \qquad \text{True}$$

Like a linear equation in two unknowns, a linear inequality has infinitely many solutions. For this reason, we often represent the solutions graphically. When we do this, we say that we are **graphing the inequality.**

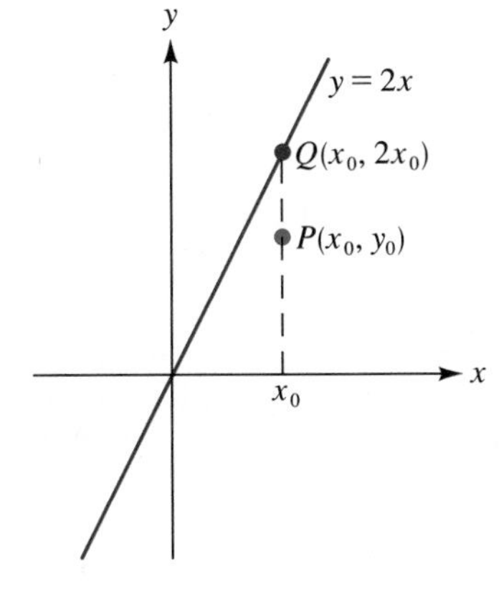

FIGURE 1

For example, let's graph the inequality $y < 2x$. First we observe that the coordinates of the points on the line $y = 2x$ do not, by definition, satisfy this inequality. So it remains to consider points above the line and points below the line. We will in fact show that the required graph consists of all points that lie below the line. Take any point $P(x_0, y_0)$ not on the line $y = 2x$. Let $Q(x_0, 2x_0)$ be the point on $y = 2x$ with the same first coordinate as P. Then, as indicated in Figure 1, the point P lies below the line if and only if the y-coordinate of P is less than the y-coordinate of Q. In other words, (x_0, y_0) lies below the line if and only if $y_0 < 2x_0$. This last statement is equivalent to saying that (x_0, y_0) satisfies the inequality $y < 2x$. This shows that the graph of $y < 2x$ consists of all points below the line $y = 2x$ (see Figure 2). The broken line in Figure 2 indicates that the points on $y = 2x$ are not part of the required graph. If the original inequality had been $y \leqslant 2x$, then we would use a solid line rather than a broken one, indicating that the line is included in the graph. And if the original inequality had been $y > 2x$, the graph would be the region above the line.

Just as the graph of $y < 2x$ is the region below the line $y = 2x$, it is true in general that the graph of $y < f(x)$ is the region below the graph of the function f. For example, Figures 3 and 4 display the graphs of $y < x^2$ and $y \geqslant x^2$.

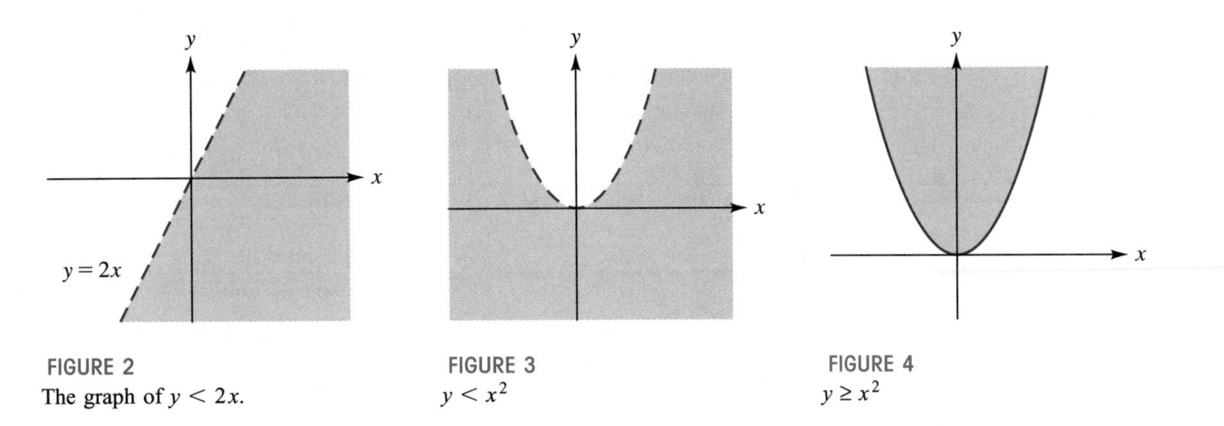

FIGURE 2 The graph of $y < 2x$.

FIGURE 3 $y < x^2$

FIGURE 4 $y \geq x^2$

Example 1 summarizes the technique developed so far for graphing an inequality. Following this example, we will point out a useful alternative method.

EXAMPLE 1 Graph the inequality $4x - 3y \leq 12$.

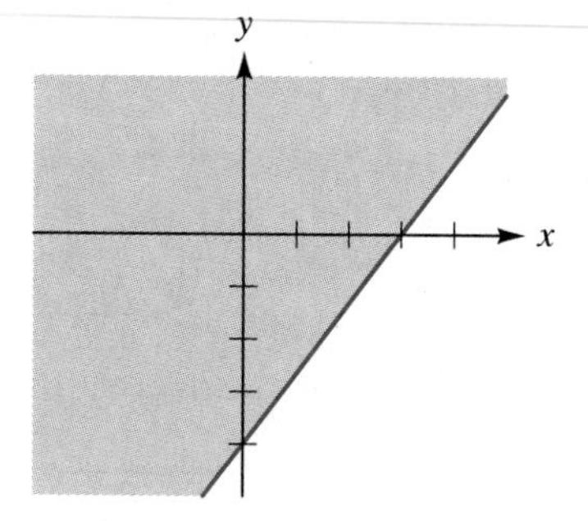

FIGURE 5
The graph of $4x - 3y \leq 12$.

Solution The graph will include the line $4x - 3y = 12$ and either the region above the line or the region below it. To decide which region, we solve the inequality for y:

$$4x - 3y \leq 12$$
$$-3y \leq -4x + 12$$
$$y \geq \frac{4}{3}x - 4 \qquad \text{Multiplying or dividing by a negative number reverses an inequality.}$$

This last inequality tells us that we want the region above the line, as well as the line itself. Figure 5 displays the required graph.

There is another method that we can use in Example 1 to determine which side of the line we want. This method involves a **test point.** We pick any convenient point that is not on the line $4x - 3y = 12$. Then we test to see if this point satisfies the given inequality. For example, let's pick the point (0, 0). Substituting these coordinates in the inequality $4x - 3y \leq 12$ yields

$$4(0) - 3(0) \overset{?}{\leq} 12$$
$$0 \overset{?}{\leq} 12 \qquad \text{True}$$

We conclude from this that the required side of the line includes the point (0, 0). In agreement with Figure 5, we see this is the region above the line.

Next we discuss *systems* of inequalities in two unknowns. As with systems of equations, a **solution** of a system of inequalities is an ordered pair (x_0, y_0) that satisfies all of the inequalities in the system. As a first example, let's graph the points that satisfy the following nonlinear system.

$$\begin{cases} y - x^2 \geq 0 \\ x^2 + y^2 < 1 \end{cases}$$

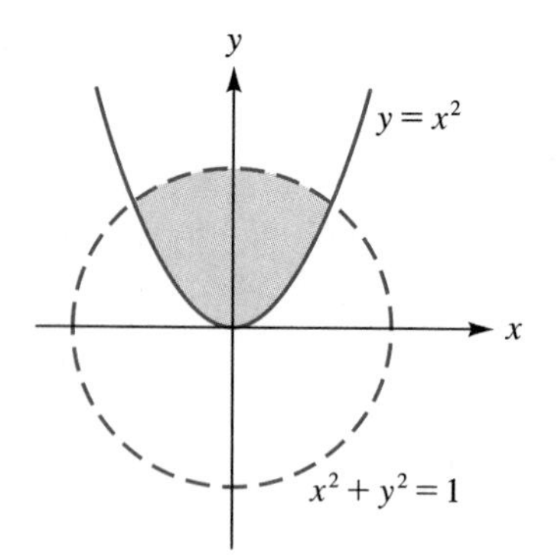

FIGURE 6

By writing the first inequality as $y \geq x^2$, we see that it describes the set of points on or above the parabola $y = x^2$. For the second inequality, we must decide whether it describes the points inside or outside the circle $x^2 + y^2 = 1$. One way to do this is to choose (0, 0) as a test point. Substituting the values $x = 0$ and $y = 0$ in the inequality $x^2 + y^2 < 1$ yields the true statement $0^2 + 0^2 < 1$. Since (0, 0) lies within the circle and satisfies the inequality, we conclude that the inequality $x^2 + y^2 < 1$ describes the set of all points within the circle. Now we put our information together. We wish to graph the points that lie on or above the parabola $y = x^2$ but within the circle $x^2 + y^2 = 1$. This is the shaded region shown in Figure 6.

EXAMPLE 2 Graph the system

$$\begin{cases} -x + 3y \leq 12 \\ x + y \leq 8 \\ x \geq 0 \\ y \geq 0 \end{cases}$$

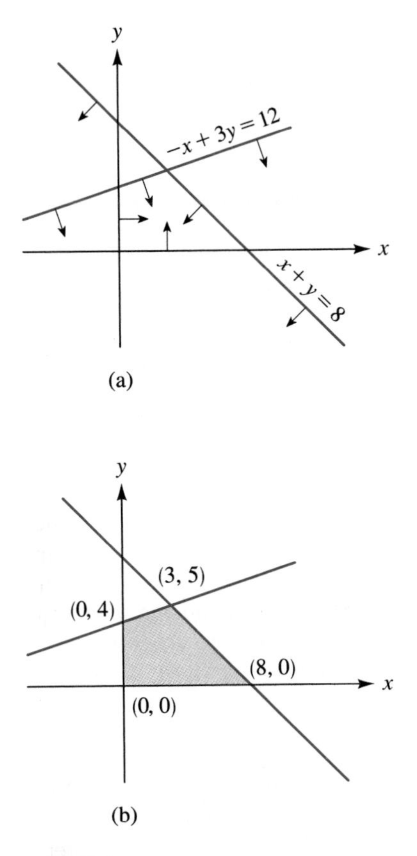

FIGURE 7

Solution Solving the first inequality for y, we have

$$3y \leqslant x + 12$$
$$y \leqslant \tfrac{1}{3}x + 4$$

Thus, the first inequality is satisfied by the points on or below the line $-x + 3y = 12$. Similarly, by solving the second inequality for y, we see that it describes the set of points on or below the line $x + y = 8$. The third inequality, $x \geqslant 0$, describes the points on or to the right of the y-axis. Similarly, the fourth inequality, $y \geqslant 0$, describes the points on or above the x-axis. We summarize these statements in Figure 7(a). The arrows indicate which side of each line we wish to consider.

Finally, Figure 7(b) shows the graph of the given system. The coordinates of each point in the shaded region satisfy all four of the given inequalities. The coordinates (3, 5) in Figure 7(b) were found by solving the system of linear equations $-x + 3y = 12$ and $x + y = 8$. (You'll need to carry out this work as part of Exercise 25.)

We can use Figure 7(b) to introduce some terminology that is useful in describing sets of points in the plane. A **vertex** of a region is a corner, or point, where two adjacent bounding sides meet. Thus, the vertices of the shaded region in Figure 7(b) are the four points (0, 0), (8, 0), (3, 5), and (0, 4). The shaded region in Figure 7(b) is **convex.** This means that, given any two points in that region, the straight line segment joining these two points lies wholly within the region. Figure 8(a) displays another example of a convex set, while Figure 8(b) shows a set that is not convex. The shaded region in Figure 8(b) is also an example of a **bounded region.** By this we mean that the region can be wholly contained within some (sufficiently large) circle. Perhaps the simplest example of a region that is not bounded is the entire x-y plane itself. Figure 9 shows another example of an unbounded region.

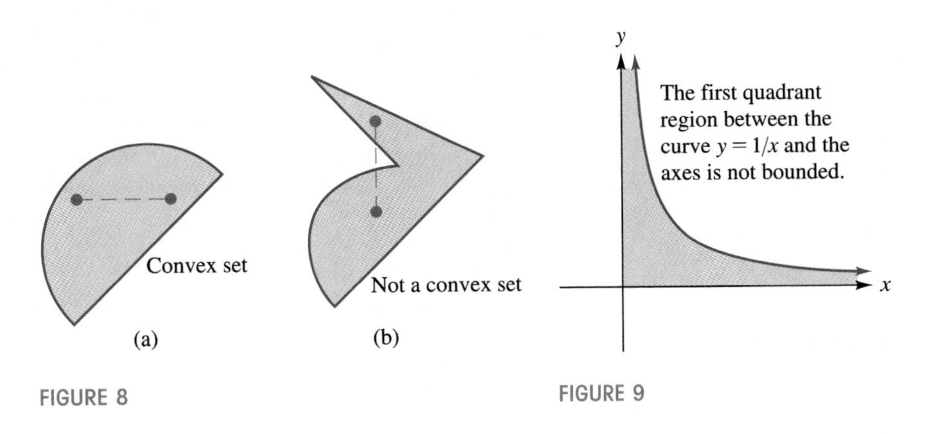

FIGURE 8

FIGURE 9

EXERCISE SET 10.7

A

In Exercises 1 and 2, decide whether or not the ordered pairs are solutions of the given inequality.

1. $4x - 6y + 3 \geqslant 0$: **(a)** $(1, 2)$; **(b)** $(0, \frac{1}{2})$.

2. $5x + 2y < 1$: **(a)** $(-1, 3)$; **(b)** $(0, 0)$.

In Exercises 3–16, graph the given inequalities.

3. $2x - 3y > 6$

4. $2x - 3y < 6$

5. $2x - 3y \geq 6$

6. $2x - 3y \leq 6$

7. $x - y < 0$

8. $y \leq \frac{1}{2}x - 1$

9. $x \geq 1$

10. $y < 0$

11. $x > 0$

12. $y \leq \sqrt{x}$

13. $y > x^3 + 1$

14. $y \leq |x - 2|$

15. $x^2 + y^2 \geq 25$

16. $y \geq e^x - 1$

In Exercises 17–22, graph the systems of inequalities.

17. $\begin{cases} y \leq x^2 \\ x^2 + y^2 \leq 1 \end{cases}$

18. $\begin{cases} y < x \\ x^2 + y^2 < 1 \end{cases}$

19. $\begin{cases} y \geq 1 \\ y \leq |x| \end{cases}$

20. $\begin{cases} x \geq 0 \\ y \geq 0 \\ y < \sqrt{x} \\ x \leq 4 \end{cases}$

21. $\begin{cases} x \geq 0 \\ y \geq 0 \\ y \leq 1 - x^2 \end{cases}$

22. $\begin{cases} y < 2x \\ y > \frac{1}{2}x \end{cases}$

In Exercises 23–34, graph the systems of linear inequalities. In each case specify the vertices. Is the region convex? Is the region bounded?

23. $\begin{cases} y \leq x + 5 \\ y \leq -2x + 14 \\ x \geq 0 \\ y \geq 0 \end{cases}$

24. $\begin{cases} y \geq x + 5 \\ y \geq -2x + 14 \\ x \geq 0 \\ y \geq 0 \end{cases}$

25. $\begin{cases} -x + 3y \leq 12 \\ x + y \leq 8 \\ x \geq 0 \\ y \geq 0 \end{cases}$

26. $\begin{cases} y \geq 2x \\ y \geq -x + 6 \end{cases}$

27. $\begin{cases} 0 \leq 2x - y + 3 \\ x + 3y \leq 23 \\ 5x + y \leq 45 \end{cases}$

28. $\begin{cases} 0 \leq 2x - y + 3 \\ x + 3y \leq 23 \\ 5x + y \leq 45 \\ x \geq 0 \\ y \geq 0 \end{cases}$

29. $\begin{cases} 5x + 6y < 30 \\ y > 0 \end{cases}$

30. $\begin{cases} 5x + 6y < 30 \\ x > 0 \end{cases}$

31. $\begin{cases} 5x + 6y < 30 \\ x > 0 \\ y > 0 \end{cases}$

32. $\begin{cases} 2x + 3y \geq 6 \\ 2x + 3y \leq 12 \end{cases}$

33. $\begin{cases} x \geq 0 \\ y \geq 0 \\ 20 - x \geq 0 \\ 30 - y \geq 0 \\ x + y \leq 40 \\ x + y \geq 35 \end{cases}$

34. $\begin{cases} x \geq 0 \\ y \geq 0 \\ 3x - y + 1 \geq 0 \\ 0 \leq x - y + 3 \\ y \leq 5 \\ x \leq \frac{1}{3}(17 - y) \\ x \leq \frac{1}{2}(y + 8) \end{cases}$

B

35. Graph the following system of inequalities and specify the vertices.

$$\begin{cases} x \geq 0 \\ y \geq e^x \end{cases} \qquad y \leq e^{-x} + 1$$

A formula such as

$$f(x,y) = \sqrt{2x - y + 1}$$

defines a function of two variables. *The inputs for such a function are ordered pairs* (x,y) *of real numbers. For example, using the ordered pair* $(3, 5)$ *as an input, we have*

$$f(3, 5) = \sqrt{2(3) - 5 + 1} = \sqrt{2}$$

So the input $(3, 5)$ *yields an output of* $\sqrt{2}$*. We define the domain for this function just as we did in Chapter 4: The* domain *is the set of all inputs that yield real-number outputs. For instance, the ordered pair* $(1, 4)$ *is not in the domain of the function we have been discussing, because (as you should check for yourself)* $f(1, 4) = \sqrt{-1}$*, which is not a real number. We can determine the domain of the function in equation (1) by requiring that the quantity under the radical sign be nonnegative. Thus, we require that* $2x - y + 1 \geq 0$ *and, consequently,* $y \leq 2x + 1$*. (Check this.) The following figure shows the graph of this inequality; the domain of our function is the set of ordered pairs making up the graph. In Exercises 36–41, follow a similar procedure and sketch the domain of the given function.*

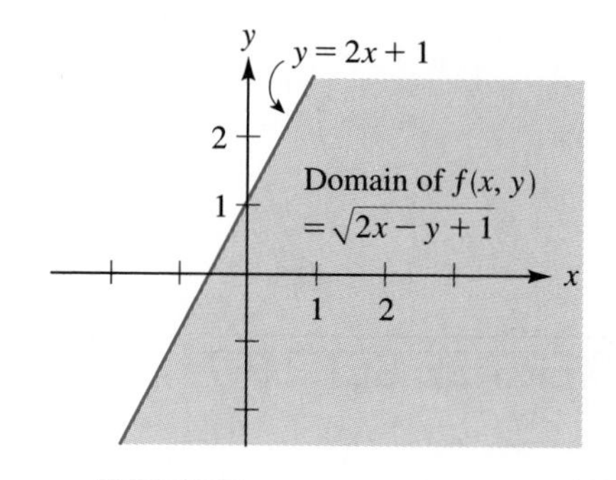

36. $f(x,y) = \sqrt{x + y + 2}$

37. $f(x,y) = \sqrt{x^2 + y^2 - 1}$

38. $g(x,y) = \sqrt{25 - x^2 - y^2}$

39. $g(x,y) = \ln(x^2 - y)$

40. $h(x,y) = \ln(xy)$

41. $h(x,y) = \sqrt{x} + \sqrt{y}$

CHAPTER TEN SUMMARY OF PRINCIPAL TERMS

TERMS	PAGE REFERENCE	COMMENT
1. Linear equation in two variables	600	A linear equation in two variables is an equation of the form $ax + by = c$, where a, b, and c are constants and x and y are variables or unknowns. Similarly, a linear equation in three variables is an equation of the form $ax + by + cz = d$.
2. Solution of a linear equation	600	A solution of the linear equation $ax + by = c$ is an ordered pair of numbers (x_0,y_0) such that $ax_0 + by_0 = c$. Similarly, a solution of the linear equation $ax + by + cz = d$ is an ordered triple of numbers (x_0,y_0,z_0) such that $ax_0 + by_0 + cz_0 = d$.
3. Consistent system; inconsistent system	601	A system of equations is consistent if it has at least one solution; otherwise, the system is inconsistent. See Figures 2, 3, and 4 on page 602 for a geometric interpretation of these terms.
4. Upper-triangular form (three variables)	612	A system of linear equations in x, y, and z is said to be in upper-triangular form if x appears in no equation after the first and y appears in no equation after the second. This definition can be extended to include systems with any number of unknowns. See Table 1 in Section 10.2 for examples of systems that are in upper-triangular form. When a system is in upper-triangular form, it is a simple matter to obtain the solutions; see, for instance, Examples 1 and 2 in Section 10.2.
5. Gaussian elimination	611, 614	This is a technique for converting a system of equations to upper-triangular form. See Examples 4 and 5 in Section 10.2.
6. Elementary operations	614	These are operations that can be performed on an equation in a system without altering the solution set. See the box on page 614 for a list of these operations.
7. Matrix	619	A matrix is a rectangular array of numbers, enclosed in parentheses or brackets. The numbers constituting the rectangular array are called the entries or the elements in the matrix. The size or dimension of a matrix is expressed by specifying the number of rows and the number of columns in that order. For examples of this terminology, see Example 1 in Section 10.3.
8. Matrix equality	623	Two matrices are said to be equal provided they are the same size and their corresponding entries are equal.
9. Matrix addition and subtraction	623	To add two matrices of the same size, add the corresponding entries. Similarly, to subtract two matrices of the same size, subtract the corresponding entries.
10. Matrix multiplication	627	Let A and B be two matrices. The matrix product AB is defined only when the number of columns in A is the same as the number of rows in B. In this case, the matrix AB will have as many rows as A and as many columns as B. The entry in the ith row and jth column of AB is the number formed as follows: multiply the corresponding entries in the ith row of A and the jth column of B, then add the results.
11. Square matrix	631	A matrix that has the same number of rows and columns is called a square matrix. An $n \times n$ square matrix is said to be an nth-order square matrix.

TERMS	PAGE REFERENCE	COMMENT
12. Multiplicative identity matrix	631, 635	For the set of $n \times n$ matrices, the multiplicative identity matrix I_n is the $n \times n$ matrix with ones down the main diagonal (upper left corner to bottom right corner) and zeros everywhere else. For example, $I_2 = \begin{pmatrix} 1 & 0 \\ 0 & 1 \end{pmatrix}$ and $I_3 = \begin{pmatrix} 1 & 0 & 0 \\ 0 & 1 & 0 \\ 0 & 0 & 1 \end{pmatrix}$ For every $n \times n$ matrix A, we have $AI_n = A$ and $I_nA = A$.
13. Inverse matrix	631, 635	Given an $n \times n$ matrix A, the inverse matrix (if it exists) is denoted by A^{-1}. The defining equations for A^{-1} are $AA^{-1} = I$ and $A^{-1}A = I$. Here, I stands for the multiplicative identity matrix that is the same size as A. If a square matrix has an inverse, it is said to be nonsingular; if there is no inverse, then the matrix is singular.
14. Minor	640, 641	The minor of an entry b in a determinant is the determinant formed by suppressing the entries in the row and column in which b appears.
15. Cofactor	640, 641	If an entry b appears in the ith row and the jth column of a determinant, then the cofactor of b is computed by multiplying the minor of b times the number $(-1)^{i+j}$.
16. Determinant	639, 641	The value of the 2×2 determinant $\begin{vmatrix} a & b \\ c & d \end{vmatrix}$ is defined to be $ad - bc$. For larger determinants, the value can be found as follows. Select any row or column and multiply each entry in that row or column by its cofactor, then add the results. The resulting sum is the value of the determinant. It can be shown that this value is independent of the particular row or column that is chosen.
17. Cramer's rule	645	This rule yields the solutions of certain systems of linear equations in terms of determinants. For a statement of the rule, see page 645. Example 5 on pages 645–646 shows how the rule is applied. The proof of Cramer's rule begins on page 646.
18. Linear inequality in two variables	659	A linear inequality in two variables is any one of the four types of inequalities that result when the equal sign in the equation $ax + by = c$ is replaced by one of the four symbols $<, \leqslant, >, \geqslant$. For any of these linear inequalities, a solution is an ordered pair of numbers (x_0, y_0) with the property that a true statement is obtained when x and y (in the inequality) are replaced by x_0 and y_0, respectively.
19. Vertex	661	A vertex of a region in the x-y plane is a corner or point where two adjacent bounding sides meet.
20. Convex region	661	A region in the x-y plane is convex if, given any two points in the region, the line segment joining those points lies wholly within the region.
21. Bounded region	661	A region in the x-y plane is bounded if it can be completely contained within some (sufficiently large) circle.

WRITING MATHEMATICS

On your own or with a group of classmates, complete the following exercise. Then (strictly on your own) write out a detailed solution. This will involve a combination of English composition and algebra (much like the exposition in this textbook). At each stage, be sure to tell the reader (in complete sentences) where you are headed and why each of the main steps is necessary.

Three integers are said to form a **Pythagorean triple** if the square of the largest is equal to the sum of the squares of the other two. For example, 3, 4, and 5 form a Pythagorean triple because $3^2 + 4^2 = 5^2$. Do you know any other Pythagorean triples? This exercise shows how to develop expressions for generating an infinite number of Pythagorean triples.

(a) Refer to the figure. Suppose that the coordinates of the point P are $(a/c, b/c)$, where a, b, and c are integers. Explain why a, b, and c form a Pythagorean triple.

(b) Let m denote the slope of the line passing through P and $(0, -1)$. Using the methods of Section 10.6, show that the coordinates of the point P are

$$\left(\frac{2m}{m^2+1}, \frac{m^2-1}{m^2+1}\right)$$

(c) Use the results in parts (a) and (b) to explain why the three numbers $2m$, $m^2 - 1$, and $m^2 + 1$ form a Pythagorean triple. Then complete the following table to obtain examples of Pythagorean triples.

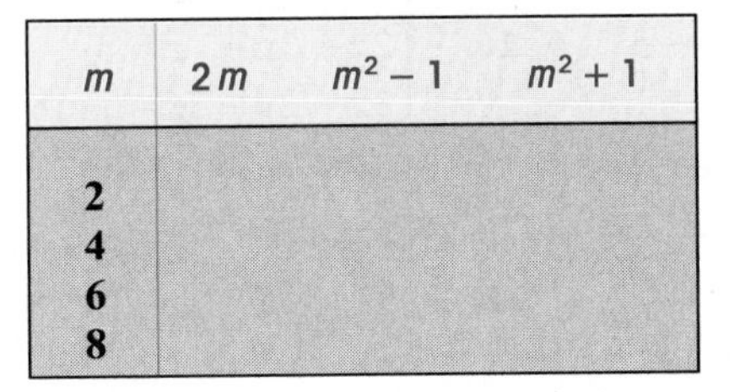

m	$2m$	m^2-1	m^2+1
2			
4			
6			
8			

CHAPTER TEN REVIEW EXERCISES

In Exercises 1–38, solve each system of equations. If there are no solutions in a particular case, say so. In cases where there are literal (rather than numerical) coefficients, specify any restrictions that your solutions impose on those coefficients.

1. $\begin{cases} x + y = -2 \\ x - y = 8 \end{cases}$

2. $\begin{cases} x - y = 1 \\ x + y = 5 \end{cases}$

3. $\begin{cases} 2x + y = 2 \\ x + 2y = 7 \end{cases}$

4. $\begin{cases} 3x + 2y = 6 \\ 5x + 4y = 4 \end{cases}$

5. $\begin{cases} 7x + 2y = 9 \\ 4x + 5y = 63 \end{cases}$

6. $\begin{cases} \dfrac{x}{2} + \dfrac{y}{3} = 9 \\ \dfrac{x}{5} - \dfrac{y}{2} = -4 \end{cases}$

7. $\begin{cases} 2x - \dfrac{y}{2} = -8 \\ \dfrac{x}{3} + \dfrac{y}{8} = -1 \end{cases}$

8. $\begin{cases} 3x - 14y - 1 = 0 \\ -6x + 28y - 3 = 0 \end{cases}$

9. $\begin{cases} 3x + 5y - 1 = 0 \\ 9x - 10y - 8 = 0 \end{cases}$

10. $\begin{cases} 9x + 15y - 1 = 0 \\ 6x + 10y + 1 = 0 \end{cases}$

11. $\begin{cases} \frac{2}{3}x = -\frac{1}{2}y - 12 \\ \dfrac{x}{2} = y + 2 \end{cases}$

12. $\begin{cases} 0.1x + 0.2y = -5 \\ -0.2x - 0.5y = 13 \end{cases}$

13. $\begin{cases} \dfrac{1}{x} + \dfrac{1}{y} = -1 \\ \dfrac{2}{x} + \dfrac{5}{y} = -14 \end{cases}$

14. $\begin{cases} \dfrac{2}{x} + \dfrac{15}{y} = -9 \\ \dfrac{1}{x} + \dfrac{10}{y} = -2 \end{cases}$

15. $\begin{cases} ax + (1-a)y = 1 \\ (1-a)x + y = 0 \end{cases}$

16. $\begin{cases} ax - by - 1 = 0 \\ (a-1)x + by + 2 = 0 \end{cases}$

17. $\begin{cases} 2x - y = 3a^2 - 1 \\ 2y + x = 2 - a^2 \end{cases}$

18. $\begin{cases} 3ax + 2by = 3a^2 - ab + 2b^2 \\ 3bx + 2ay = 2a^2 + 5ab - 3b^2 \end{cases}$

19. $\begin{cases} 5x - y = 4a^2 - 6b^2 \\ 2x + 3y = 5a^2 + b^2 \end{cases}$

20. $\begin{cases} \dfrac{2b}{x} - \dfrac{3}{y} = 7ab \\ \dfrac{4a}{x} + \dfrac{5a}{by} = 3a^2 \end{cases}$

21. $\begin{cases} px - qy = q^2 \\ qx + py = p^2 \end{cases}$

22. $\begin{cases} x - y = \dfrac{a-b}{a+b} \\ x + y = 1 \end{cases}$

23. $\begin{cases} \dfrac{4a}{x} - \dfrac{3b}{y} = a - 7b \\ \dfrac{3a^2}{x} - \dfrac{2b^2}{y} = (3a+b)(a-2b) \end{cases}$

24. $\begin{cases} 6b^2x^{-1/2} - 5a^2y^{-1/2} - a^2b^2 = 0 \\ 2b^2x^{-1/2} + a^2y^{-1/2} - 3a^2b^2 = 0 \end{cases}$

Hint: Let $u = x^{-1/2}$ and $v = y^{-1/2}$.

25. $\begin{cases} x + y + z = 9 \\ x - y - z = -5 \\ 2x + y - 2z = -1 \end{cases}$

26. $\begin{cases} x - 4y + 2z = 9 \\ 2x + y + z = 3 \\ 3x - 2y - 3z = -18 \end{cases}$

27. $\begin{cases} 4x - 4y + z = 4 \\ 2x + 3y + 3z = -8 \\ x + y + z = -3 \end{cases}$

28. $\begin{cases} x - 8y + z = 1 \\ 5x + 16y + 3z = 3 \\ 4x - 4y - 4z = -4 \end{cases}$

29. $\begin{cases} -2x + y + z = 1 \\ x - 2y + z = -2 \\ x + y - 2z = 4 \end{cases}$

30. $\begin{cases} -x + y + z = 1 \\ x - y + z = -1 \\ x + y - z = 1 \end{cases}$

31. $\begin{cases} 4x + 2y - 3z = 15 \\ 2x + y + 3z = 3 \end{cases}$

32. $\begin{cases} 3x + 2y + 17z = 1 \\ x + 2y + 3z = 3 \end{cases}$

33. $\begin{cases} x + 2y - 3z = -2 \\ 2x - y + z = 1 \\ 3x - 4y + 5z = 1 \end{cases}$

34. $\begin{cases} 9x + y + z = 0 \\ -3x + y - z = 0 \\ 3x - 5y + 3z = 0 \end{cases}$

35. $\begin{cases} x + y + z = a + b \\ 2x - y + 2z = -a + 5b \\ x - 2y + z = -2a + 4b \end{cases}$

36. $\begin{cases} ax + by - 2az = 4ab + 2b^2 \\ x + y + z = 4a + 2b \\ bx + ay + 4az = 5a^2 + b^2 \end{cases}$

37. $\begin{cases} x + y + z + w = 8 \\ 3x + 3y - z - w = 20 \\ 4x - y - z + 2w = 18 \\ 2x + 5y + 5z - 5w = 8 \end{cases}$

38. $\begin{cases} x - 2y + 3z + w = 1 \\ x + y + z + w = 5 \\ 2x + 3y + 2z - w = 3 \\ 3x + y - z + 2w = 4 \end{cases}$

In Exercises 39–46, evaluate each of the determinants.

39. $\begin{vmatrix} 1 & 5 \\ -6 & 4 \end{vmatrix}$

40. $\begin{vmatrix} \frac{1}{6} & 1 \\ 2 & 12 \end{vmatrix}$

41. $\begin{vmatrix} 4 & 0 & 3 \\ -2 & 1 & 5 \\ 0 & 2 & -1 \end{vmatrix}$

42. $\begin{vmatrix} 2 & 6 & 4 \\ 6 & 18 & 24 \\ 15 & 5 & -10 \end{vmatrix}$

43. $\begin{vmatrix} 1 & 5 & 7 \\ 1 & 5 & 7 \\ 17 & 19 & 21 \end{vmatrix}$

44. $\begin{vmatrix} 0 & 2 & 4 & 0 \\ 4 & 0 & 6 & 2 \\ 0 & 0 & 1 & 1 \\ 14 & 7 & 1 & 0 \end{vmatrix}$

45. $\begin{vmatrix} 1 & 0 & 0 & 0 \\ 0 & 2 & 0 & 0 \\ 0 & 0 & 3 & 0 \\ 0 & 0 & 0 & 4 \end{vmatrix}$

46. $\begin{vmatrix} 1 & a & b & c \\ 0 & 2 & d & e \\ 0 & 0 & 3 & f \\ 0 & 0 & 0 & 4 \end{vmatrix}$

47. Show that $\begin{vmatrix} a & b & c \\ b & c & a \\ c & a & b \end{vmatrix} = 3abc - a^3 - b^3 - c^3$.

48. Show that $\begin{vmatrix} 1 & 1 & 1 \\ 1 & 1+x & 1 \\ 1 & 1 & 1+x^2 \end{vmatrix} = x^3$.

49. Show that

$$\begin{vmatrix} a^2 + x & b & c & d \\ -b & 1 & 0 & 0 \\ -c & 0 & 1 & 0 \\ -d & 0 & 0 & 1 \end{vmatrix} = a^2 + b^2 + c^2 + d^2 + x$$

50. In a two-digit number, the sum of the digits is 11. Four times the units digit exceeds the tens digit by 4. Find the number.

51. Determine constants a and b so that the parabola $y = ax^2 + bx - 1$ passes through the points $(-2, 5)$ and $(2, 9)$.

52. Find two numbers whose sum and difference are 52 and 10, respectively.

53. The vertices of triangle ABC are $A(-2, 0)$, $B(4, 0)$, and $C(0, 6)$. Let A_1, B_1, and C_1 denote the midpoints of sides $\overline{BC}$, $\overline{AC}$, and $\overline{AB}$, respectively.

(a) Find the point where the line segments $\overline{AA_1}$ and $\overline{BB_1}$ intersect. *Answer:* $(2/3, 2)$

(b) Follow the instructions given in part (a) using $\overline{BB_1}$ and $\overline{CC_1}$.

(c) Follow the instructions given in part (a) using $\overline{AA_1}$ and $\overline{CC_1}$.

(d) Let P denote the point $(\frac{2}{3}, 2)$ that you found in part (a). Compute each of the following ratios: AP/PA_1, BP/PB_1, CP/PC_1. What do you observe?

54. An **altitude** of a triangle is a line segment drawn from a vertex perpendicular to the opposite side. Suppose that the vertices of triangle ABC are as given in Exercise 53. Find the intersection point for each pair of altitudes. What do you observe about the three answers?

55. This exercise appears in *Plane and Solid Analytic Geometry,* by W. F. Osgood and W. G. Graustein (New York: Macmillan, 1920): Let P be any point (a, a) of the line $x - y = 0$, other than the origin. Through P draw two lines, of arbitrary slopes m_1 and m_2, intersecting the x-axis in A_1 and A_2, and the y-axis in B_1 and B_2, respectively. Prove that the lines $\overline{A_1B_2}$ and $\overline{A_2B_1}$ will, in general, meet on the line $x + y = 0$.

56. Determine constants h, k, and r so that the circle $(x - h)^2 + (y - k)^2 = r^2$ passes through the three points $(0, 0)$, $(0, 1)$ and $(1, 0)$.

57. The vertices of a triangle are the points of intersection of the lines $y = x - 1$, $y = -x - 2$, and $y = 2x + 3$. Find the equation of the circle passing through these three intersection points.

58. The vertices of triangle ABC are $A(0, 0)$, $B(3, 0)$, and $C(0, 4)$.

(a) Find the center and the radius of the circle that passes though A, B, and C. This circle is called the **circumcircle** for triangle ABC.

(b) The figure shows the **inscribed circle** for triangle ABC; this is the circle that is tangent to all three sides of the triangle. Find the center and the radius of this circle using the following two facts.

(i) The center of the inscribed circle is the common intersection point of the three angle bisectors of the triangle.

(ii) The line that bisects angle B has slope $-1/2$.

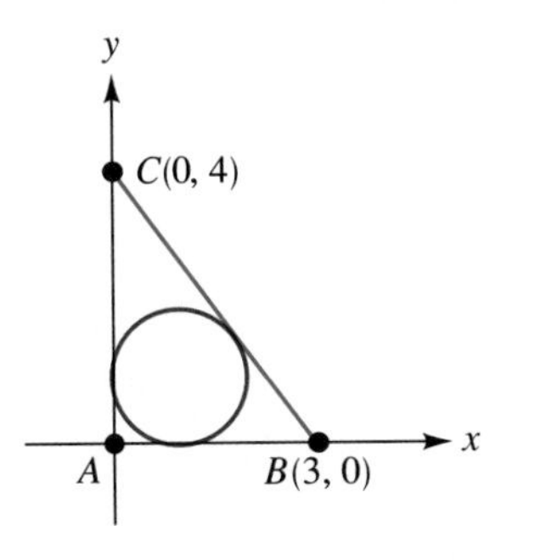

(c) Verify the following statement for triangle ABC. (The statement actually holds for any triangle; the result is known as *Euler's theorem.*) Let R and r denote the radii of the circles in parts (a) and (b), respectively. Then the distance d between the centers of the circles satisfies the equation

$$d^2 = R^2 - 2rR$$

(d) Verify the following statement for triangle ABC (which in fact holds for any triangle): The area of triangle ABC is equal to one-half the product of the perimeter and the radius of the inscribed circle.

(e) Verify the following statement for triangle ABC (which in fact holds for any triangle): The sum of the reciprocals of the lengths of the altitudes is equal to the reciprocal of the radius of the inscribed circle.

In Exercises 59–82, the matrices A, B, C, D, E, and F are defined as follows:

$$A = \begin{pmatrix} 3 & -2 \\ 1 & 5 \end{pmatrix} \qquad B = \begin{pmatrix} 2 & 1 \\ 1 & 8 \end{pmatrix}$$

$$C = \begin{pmatrix} -1 & 0 \\ 0 & -1 \end{pmatrix} \qquad D = \begin{pmatrix} -4 & 0 & 6 \\ 1 & 3 & 2 \end{pmatrix}$$

$$E = \begin{pmatrix} 3 & -1 \\ 4 & 1 \\ -5 & 9 \end{pmatrix} \qquad F = \begin{pmatrix} 2 & 6 \\ 5 & 3 \\ 5 & 8 \end{pmatrix}$$

In each exercise, carry out the indicated matrix operations if they are defined. If an operation is not defined, say so.

59. $2A + 2B$

60. $2(A + B)$

61. $4B$

62. $B + 4$

63. AB

64. BA

65. $AB - BA$

66. $B + E$

67. $B + C$

68. $A(B + C)$

69. $AB + AC$

70. $(B + C)A$

71. $BA + CA$

72. $D + E$

73. DE

74. $(EE)D$

75. $E(ED)$

76. $E + F$

77. EF

78. $3E - 2F$

79. $(A + B) + C$

80. $A + (B + C)$

81. $(AB)C$

82. $A(BC)$

83. For a square matrix A, the notation A^2 means AA. Similarly, A^3 means AAA. If $A = \begin{pmatrix} 1 & 1 \\ 0 & 1 \end{pmatrix}$, verify that

$$A^2 = \begin{pmatrix} 1 & 2 \\ 0 & 1 \end{pmatrix} \quad \text{and} \quad A^3 = \begin{pmatrix} 1 & 3 \\ 0 & 1 \end{pmatrix}$$

84. Let A be the matrix $\begin{pmatrix} 0 & 0 & 0 \\ a & 0 & 0 \\ b & c & 0 \end{pmatrix}$. Compute A^2 and A^3.

For Exercises 85–88, in each case compute the inverse of the matrix in part (a), and then use that inverse to solve the system of equations in part (b).

85. (a) $\begin{pmatrix} 1 & 5 \\ 2 & 9 \end{pmatrix}$ (b) $\begin{cases} x + 5y = 3 \\ 2x + 9y = -4 \end{cases}$

86. (a) $\begin{pmatrix} 5 & -4 \\ 14 & -11 \end{pmatrix}$ (b) $\begin{cases} 5x - 4y = 2 \\ 14x - 11y = 5 \end{cases}$

87. (a) $\begin{pmatrix} 1 & -2 & 3 \\ 2 & -5 & 10 \\ -1 & 2 & -2 \end{pmatrix}$ (b) $\begin{cases} x - 2y + 3z = -2 \\ 2x - 5y + 10z = -3 \\ -x + 2y - 2z = 6 \end{cases}$

88. (a) $\begin{pmatrix} -1 & 4 & 2 \\ -3 & 10 & 5 \\ 0 & 3 & 1 \end{pmatrix}$

(b) $\begin{cases} -x + 4y + 2z = 8 \\ -3x + 10y + 5z = 0 \\ 3y + z = -12 \end{cases}$

In Exercises 89 and 90, find the inverse of each matrix.

89. $\begin{pmatrix} 5 & 3 & 6 & -7 \\ 3 & -4 & 0 & -9 \\ 0 & 1 & -1 & -1 \\ 2 & 2 & 3 & -2 \end{pmatrix}$

90. $\begin{pmatrix} 1 & -1 & 0 & -3 \\ 5 & -2 & 3 & -11 \\ 2 & 3 & 2 & -3 \\ 4 & 5 & 5 & -5 \end{pmatrix}$

In Exercises 91–98, compute D, D_x, D_y, D_z (and D_w where appropriate) for each system of equations. Use Cramer's rule to solve the systems in which $D \neq 0$. If $D = 0$, solve the system using Gaussian elimination or matrix methods.

91. $\begin{cases} 2x - y + z = 1 \\ 3x + 2y + 2z = 0 \\ x - 5y - 3z = -2 \end{cases}$

92. $\begin{cases} x + 2y - z = -1 \\ 2x - 3y + 3z = 3 \\ 2x + 3y + z = 1 \end{cases}$

93. $\begin{cases} x + 2y + 3z = -1 \\ 4x + 5y + 6z = 2 \\ 7x + 8y + 9z = -3 \end{cases}$

94. $\begin{cases} 3x + 2y - 2z = 0 \\ 2x + 3y - z = 0 \\ 8x + 7y - 5z = 0 \end{cases}$

95. $\begin{cases} 3x + 2y - 2z = 1 \\ 2x + 3y - z = -2 \\ 8x + 7y - 5z = 0 \end{cases}$

96. $\begin{cases} x + y + z + w = 5 \\ x - y - z + w = 3 \\ 2x + 3y + 3z + 2w = 21 \\ 4z - 3w = -7 \end{cases}$

97. $\begin{cases} 2x - y + z + 3w = 15 \\ x + 2y + 2w = 12 \\ 3y + 3z + 4w = 12 \\ -4x + y - 4z = -11 \end{cases}$

98. $\begin{cases} x + y + z = (a + b)^2 \\ \dfrac{bx}{a} + \dfrac{ay}{b} - z = 0 \\ x + y - z = (a - b)^2 \end{cases}$

In Exercises 99–112, find all solutions (x, y) for each system, where x and y are real numbers.

99. $\begin{cases} y = 6x \\ y = x^2 \end{cases}$

100. $\begin{cases} y = 4x \\ y = x^3 \end{cases}$

101. $\begin{cases} y = 9 - x^2 \\ y = x^2 - 9 \end{cases}$

102. $\begin{cases} x^3 - y = 0 \\ xy - 16 = 0 \end{cases}$

103. $\begin{cases} x^2 - y^2 = 9 \\ x^2 + y^2 = 16 \end{cases}$

104. $\begin{cases} 2x + 3y = 6 \\ y = \sqrt{x + 1} \end{cases}$

105. $\begin{cases} x^2 + y^2 = 1 \\ y = \sqrt{x} \end{cases}$

106. $\begin{cases} \dfrac{x}{11} + \dfrac{y}{12} = 2 \\ \dfrac{xy}{132} = 1 \end{cases}$

107. $\begin{cases} x^2 + y^2 = 1 \\ y = 2x^2 \end{cases}$

108. $\begin{cases} x^2 - 3xy + y^2 = -11 \\ 2x^2 + xy - y^2 = 8 \end{cases}$

109. $\begin{cases} x^2 + 2xy + 3y^2 = 68 \\ 3x^2 - xy + y^2 = 18 \end{cases}$

110. $\begin{cases} \dfrac{x^2}{a^2} + \dfrac{y^2}{b^2} = 1 \\ \dfrac{x^2}{b^2} + \dfrac{y^2}{a^2} = 1 \end{cases}$ $(a > b > 0)$

111. $\begin{cases} 2(x - 3)^2 - (y + 1)^2 = -1 \\ -3(x - 3)^2 + 2(y + 1)^2 = 6 \end{cases}$

Hint: Let $u = x - 3$ and $v = y + 1$.

112. $\begin{cases} x^4 = y - 1 \\ y - 3x^2 + 1 = 0 \end{cases}$

Exercises 113–118 appear (in German) in an algebra text by Leonhard Euler, first published in 1770. The English versions given here are taken from the translated version, Elements of Algebra, 5th ed., *by Leonhard Euler (London: Longman, Orme, and Co., 1840). [This, in turn, has been reprinted by Springer-Verlag (New York, 1984).]*

113. Required two numbers, whose sum may be s, and their proportion as a to b.

Answer: $\dfrac{as}{a+b}$ and $\dfrac{bs}{a+b}$

114. The sum $2a$, and the sum of the squares $2b$, of two numbers being given; to find the numbers.

Answer: $a-\sqrt{b-a^2}$ and $a+\sqrt{b-a^2}$

115. To find three numbers, so that [the sum of] one-half of the first, one-third of the second, and one-quarter of the third, shall be equal to 62; one-third of the first, one-quarter of the second, and one-fifth of the third, equal to 47; and one-quarter of the first, one-fifth of the second, and one-sixth of the third, equal to 38. Answer: 24, 60, 120

116. Required two numbers, whose product may be 105, and whose squares [when added] may together make 274.

117. Required two numbers, whose product may be m, and the sum of the squares n $(n \geqslant 2m)$.

118. Required two numbers such that their sum, their product, and the difference of their squares may all be equal.

In Exercises 119–124, graph each system of inequalities and specify whether the region is convex or bounded.

119. $\begin{cases} x^2+y^2 \geqslant 1 \\ y-4x \leqslant 0 \\ y-x \geqslant 0 \\ x \geqslant 0 \\ y \geqslant 0 \end{cases}$

120. $\begin{cases} x^2+y^2 \leqslant 1 \\ x \geqslant 0 \\ y \geqslant 0 \end{cases}$

121. $\begin{cases} y-\sqrt{x} \leqslant 0 \\ y \geqslant 0 \\ x \geqslant 1 \\ x-4 \leqslant 0 \end{cases}$

122. $\begin{cases} y-|x| \leqslant 0 \\ x+1 \geqslant 0 \\ x-1 \leqslant 0 \\ y+1 \geqslant 0 \end{cases}$

123. $\begin{cases} y \leqslant 1/x \\ y \geqslant 0 \\ x \geqslant 1 \end{cases}$

124. $\begin{cases} y-100x \leqslant 0 \\ y-x^2 \geqslant 0 \\ x \geqslant 0 \end{cases}$

CHAPTER 10 TEST

1. Determine all solutions of the system

$$\begin{cases} 3x+4y=12 \\ y=x^2+2x+3 \end{cases}$$

2. Find all solutions of the system

$$\begin{cases} x-2y=13 \\ 3x+5y=-16 \end{cases}$$

3. **(a)** Find all solutions of the following system using Gaussian elimination:

$$\begin{cases} x+4y-z=0 \\ 3x+y+z=-1 \\ 4x-4y+5z=-7 \end{cases}$$

(b) Compute D, D_x, D_y, and D_z for the system in part (a). Then check your answer in part (a) using Cramer's rule.

4. Suppose that the matrices A and B are defined as follows:

$$A=\begin{pmatrix} 1 & -3 \\ 2 & -1 \end{pmatrix} \qquad B=\begin{pmatrix} 0 & 4 \\ 1 & 3 \end{pmatrix}$$

(a) Compute $2A-B$. **(b)** Compute BA.

5. Determine the area of the triangular region in the first quadrant that is bounded by the x-axis and the lines $y = 2x$ and $y = -x + 6$.
6. Find all solutions of the system

$$\begin{cases} \dfrac{1}{2x} + \dfrac{1}{3y} = 10 \\[2ex] -\dfrac{5}{x} - \dfrac{4}{y} = -4 \end{cases}$$

7. Specify the coefficient matrix for the system

$$\begin{cases} x + y - z = -1 \\ 2x - y + 2z = 11 \\ x - 2y + z = 10 \end{cases}$$

Also specify the augmented matrix for this system.

8. Use matrix methods to find all solutions of the system displayed in the previous problem.
9. Find the equation of a line that passes through the point of intersection of the lines $x + y = 11$ and $3x + 2y = 7$ and that is perpendicular to the line $2x - 4y = 7$.
10. Consider the determinant

$$\begin{vmatrix} 2 & 3 & -1 \\ 0 & 1 & 4 \\ 5 & -2 & 6 \end{vmatrix}$$

(a) What is the minor of the entry in the third row, second column?
(b) What is the cofactor of the entry in the third row, second column?

11. Evaluate the determinant

$$\begin{vmatrix} 4 & -5 & 0 \\ -8 & 10 & 7 \\ 16 & 20 & 14 \end{vmatrix}$$

12. Find all solutions of the system

$$\begin{cases} x^2 + y^2 = 15 \\ xy = 5 \end{cases}$$

13. Find the solutions of the system

$$\begin{cases} A + 2B + 3C = 1 \\ 2A - B - C = 2 \end{cases}$$

14. **(a)** Determine the inverse of the following matrix:

$$\begin{pmatrix} 10 & -2 & 5 \\ 6 & -1 & 4 \\ 1 & 0 & 1 \end{pmatrix}$$

(b) Use the inverse matrix determined in part (a) to solve the following system:

$$\begin{cases} 10u - 2v + 5w = -1 \\ 6u - v + 4w = -2 \\ u + w = 3 \end{cases}$$

15. Graph the inequality $5x - 6y \geqslant 30$.
16. Determine constants P and Q so that the parabola $y = Px^2 + Qx - 5$ passes through the two points $(-2, -1)$ and $(-1, -2)$.
17. Graph the inequality $(x - 2)^2 + y^2 > 1$. Is the solution set bounded? Is it convex?
18. Graph the following system of inequalities and specify the vertices:

$$\begin{cases} x \geqslant 0 \\ y \geqslant 0 \\ 2y - x \leqslant 14 \\ x + 3y \leqslant 36 \\ 9x + y \leqslant 99 \end{cases}$$

19. Use Cramer's rule to solve the following system for x and y in terms of a and b. Assume $a \neq b$. Simplify your answers.

$$\begin{cases} ax + by = a^2 \\ bx + ay = b^2 \end{cases}$$

20. Given that the inverse of the matrix $\begin{pmatrix} 5 & 0 & 2 \\ 2 & 2 & 1 \\ -3 & 1 & -1 \end{pmatrix}$ is the matrix $\begin{pmatrix} -3 & 2 & -4 \\ -1 & 1 & -1 \\ 8 & -5 & 10 \end{pmatrix}$, solve the following system of equations:

$$\begin{cases} 5 \ln x + 2 \ln z = 3 \\ 2 \ln x + 2 \ln y + \ln z = -1 \\ -3 \ln x + \ln y - \ln z = 2 \end{cases}$$

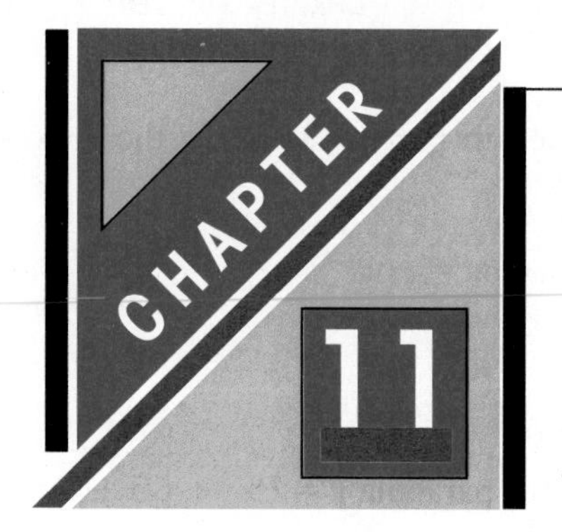

ANALYTIC GEOMETRY

The Greeks knew the properties of the curves given by cutting a cone with a plane—the ellipse, the parabola and hyperbola. Kepler discovered by analysis of astronomical observations, and Newton proved mathematically . . . that the planets move in ellipses. The geometry of Ancient Greece thus became the cornerstone of modern astronomy.

John Lighton Synge (1897–1987)

Navigator, lay in a conic section flight path to the cloud's center.

Captain James T. Kirk in *Star Trek, The Motion Picture*

Circle
Ellipse
Hyperbola
Line L
Parabola
Circle

INTRODUCTION

In this chapter we study analytic geometry, with an emphasis on the *conic sections.* The **conic sections** (or **conics,** for short) are the curves formed when a plane intersects the surface of a right cylindrical cone. As indicated in the figure in the margin, these curves are the circle, the ellipse, the hyperbola, and the parabola.

The study of the conics dates back over 2000 years to ancient Greece, where Apollonius of Perga (262–190 B.C.) wrote an eight-volume treatise on the subject. However, until the seventeenth century, the conic sections were studied only as a portion of pure (as opposed to applied) mathematics. Then, in the seventeenth century, it was discovered that the conic sections were crucial in expressing some of the most important laws of nature. This is essentially the observation that is made in the opening quotation by the physicist J. L. Synge.

11.1 THE BASIC EQUATIONS

As background for the work in this chapter, you need to be familiar with the following results from Chapter 2.

1. The distance formula: $d = \sqrt{(x_2 - x_1)^2 + (y_2 - y_1)^2}$
2. The equation for a circle: $(x - h)^2 + (y - k)^2 = r^2$
3. The slope of a line: $m = \dfrac{y_2 - y_1}{x_2 - x_1}$
4. The point–slope formula: $y - y_1 = m(x - x_1)$
5. The slope–intercept formula: $y = mx + b$
6. The condition for two nonvertical lines to be parallel: $m_1 = m_2$
7. The condition for two nonvertical lines to be perpendicular: $m_1 m_2 = -1$
8. The midpoint formula: $(x_0, y_0) = \left(\dfrac{x_1 + x_2}{2}, \dfrac{y_1 + y_2}{2}\right)$

In this section we are going to develop two additional results to supplement those we just listed. The first of these results concerns the slope of a line. We begin by defining the **angle of inclination** (or, simply the **inclination**) of a line to be the angle θ measured counterclockwise from the positive side or positive direction of the x-axis to the line; see Figure 1. If θ denotes the angle of inclination, we always have $0° \leq \theta < 180°$ if θ is measured in degrees and $0 \leq \theta < \pi$ if θ is in radians. (Notice that when $\theta = 0°$, the line is horizontal.)

As you might suspect, there is a simple relationship between the angle of inclination and the slope of a line. We state this relationship in the box that follows.

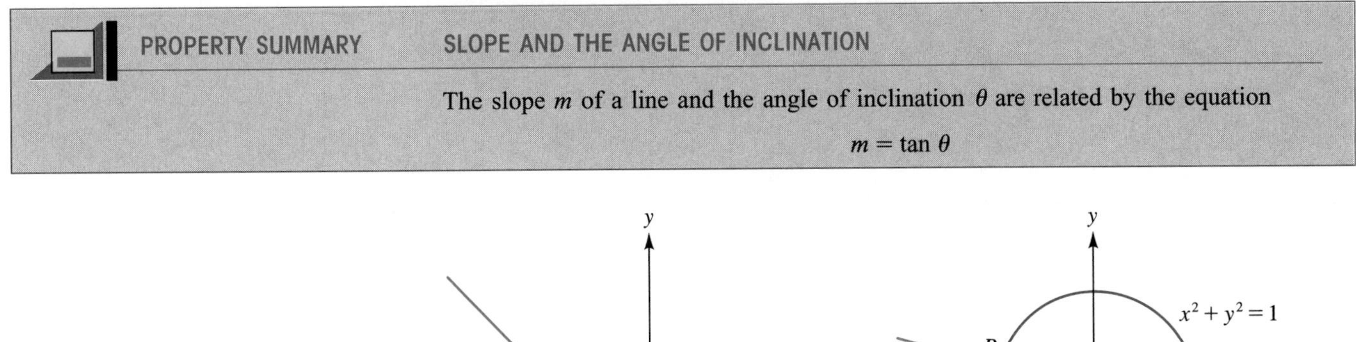

PROPERTY SUMMARY

SLOPE AND THE ANGLE OF INCLINATION

The slope m of a line and the angle of inclination θ are related by the equation

$$m = \tan\theta$$

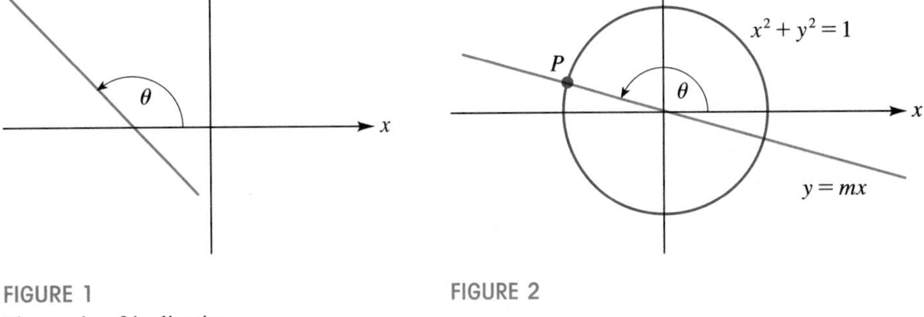

FIGURE 1
The angle of inclination.

FIGURE 2

The formula $m = \tan\theta$ provides a useful connection between elementary coordinate geometry and trigonometry. To derive this formula, we can work with the line $y = mx$ rather than $y = mx + b$ (because these lines are parallel and therefore have equal angles of inclination). As indicated in Figure 2, we let P denote the point where the line $y = mx$ intersects the unit circle. Then the coordinates of P are $(\cos\theta, \sin\theta)$, and we can compute the slope m using the point P and the origin:

$$m = \frac{y_2 - y_1}{x_2 - x_1} = \frac{\sin\theta - 0}{\cos\theta - 0} = \frac{\sin\theta}{\cos\theta} = \tan\theta \qquad \text{as required}$$

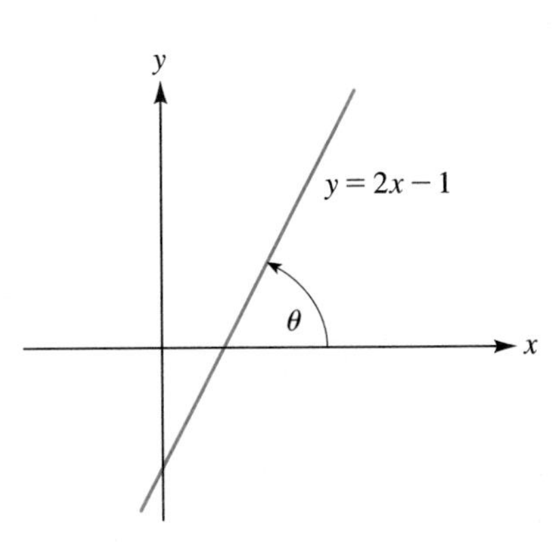

FIGURE 3

EXAMPLE 1 Determine the acute angle θ between the x-axis and the line $y = 2x - 1$; see Figure 3. Express the answer in degrees, rounded to one decimal place.

Solution From the equation $y = 2x - 1$, we read directly that $m = 2$. Then, since $\tan\theta = m$, we conclude that

$$\tan\theta = 2$$

and therefore

$$\theta = \tan^{-1} 2$$

$$\theta \approx 63.4° \qquad \text{using a calculator set in the degree mode}$$

In Section 2.1, we reviewed the formula for the distance between two points. Now we consider another kind of distance formula, one that gives the distance d from a point to a line. As indicated in Figure 4, distance in this context means the *shortest* distance, which is the perpendicular distance. In the box that follows, we show two equivalent forms for this formula. Although the second form is more widely known, the first is just as useful and somewhat simpler to derive.

FIGURE 4

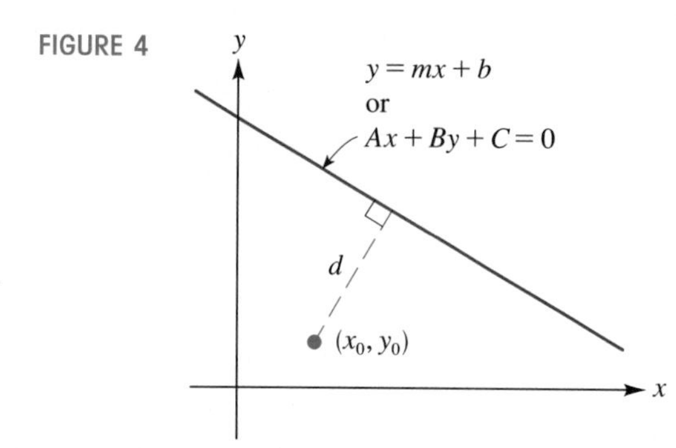

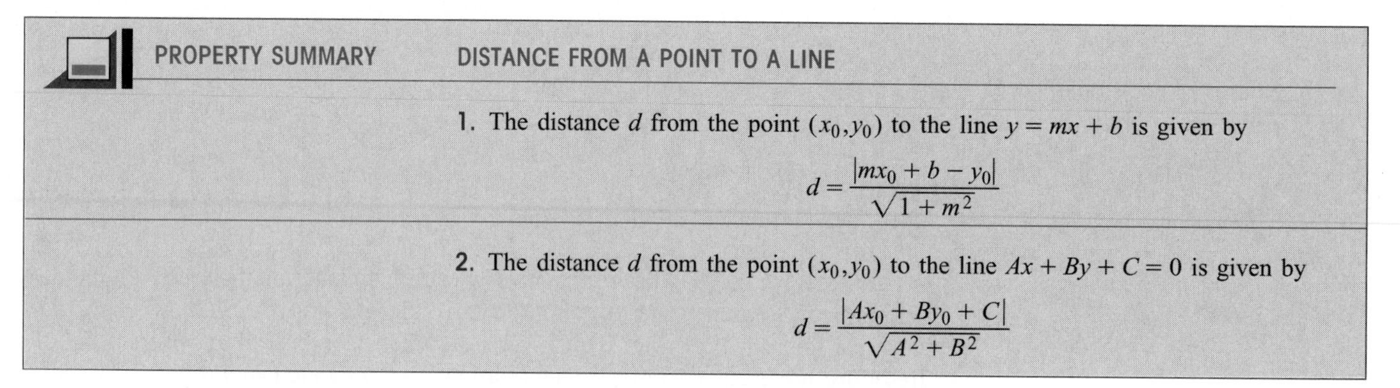

PROPERTY SUMMARY **DISTANCE FROM A POINT TO A LINE**

1. The distance d from the point (x_0, y_0) to the line $y = mx + b$ is given by

$$d = \frac{|mx_0 + b - y_0|}{\sqrt{1 + m^2}}$$

2. The distance d from the point (x_0, y_0) to the line $Ax + By + C = 0$ is given by

$$d = \frac{|Ax_0 + By_0 + C|}{\sqrt{A^2 + B^2}}$$

We will derive the first formula in the box with the aid of Figure 5. (Each dashed line in the figure is parallel to the x- or y-axis.)

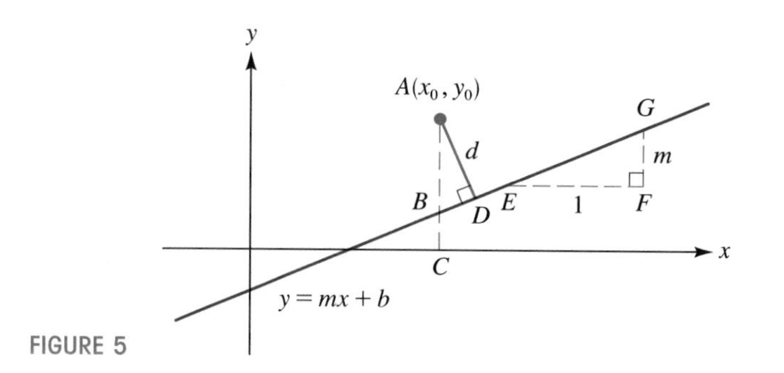

FIGURE 5

In Figure 5, $\triangle ABD$ is similar to $\triangle EGF$. (Exercise 44 asks you to verify this.) So we have

$$\frac{AB}{EG} = \frac{AD}{EF} \qquad \text{corresponding sides of similar triangles are proportional}$$

and therefore

$$\frac{AB}{\sqrt{1+m^2}} = \frac{d}{1} \qquad \text{using the Pythagorean theorem and the fact that } EF = 1$$

or

$$d = \frac{AB}{\sqrt{1+m^2}} \tag{1}$$

Next, from Figure 5 we have

$$\begin{aligned} AB &= AC - BC \\ &= y_0 - (mx_0 + b) \qquad \text{(Why?)} \end{aligned}$$

Now we can use this last equation to substitute for AB in equation (1). This yields

$$d = \frac{y_0 - (mx_0 + b)}{\sqrt{1+m^2}}$$

For the general case (in which the point and line may not be situated as in Figure 5), we need to use the absolute value of the quantity in the numerator, to assure that AB and d are nonnegative. We then have

$$d = \frac{|y_0 - (mx_0 + b)|}{\sqrt{1+m^2}} = \frac{|mx_0 + b - y_0|}{\sqrt{1+m^2}} \qquad \text{as required}$$

EXAMPLE 2

(a) Find the distance from the point $(-3, 1)$ to the line $y = -2x + 7$; see Figure 6.
(b) Find the equation of the circle that has center $(-3, 1)$ and that is tangent to the line $y = -2x + 7$.

Solution
(a) To find the distance from the point $(-3, 1)$ to the line $y = -2x + 7$, we use the formula $d = |mx_0 + b - y_0|/\sqrt{1+m^2}$ with $x_0 = -3$, $y_0 = 1$, $m = -2$, and $b = 7$. This yields

$$d = \frac{|(-2)(-3) + 7 - 1|}{\sqrt{1+(-2)^2}} = \frac{12}{\sqrt{5}} = \frac{12\sqrt{5}}{5} \text{ units}$$

FIGURE 6

(b) The equation of a circle with center (h, k) and radius of length r is $(x-h)^2 + (y-k)^2 = r^2$. We are given here that (h, k) is $(-3, 1)$. Furthermore, the distance determined in part (a) is the length of the radius. (A theorem from geometry tells us that the radius drawn to the point of tangency is perpendicular to the tangent.) Thus, the equation of the required circle is

$$[x - (-3)]^2 + (y-1)^2 = \left(12\sqrt{5}/5\right)^2$$

or

$$(x+3)^2 + (y-1)^2 = 144/5$$

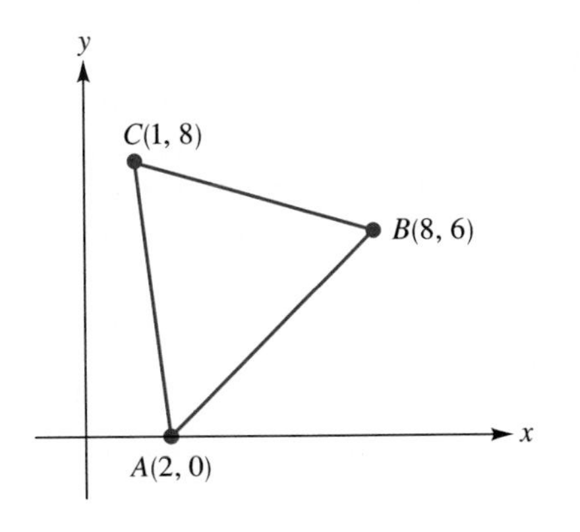

FIGURE 7

EXAMPLE 3 Find the area of triangle ABC in Figure 7.

Solution The area of any triangle is one-half the product of the base and the height. Let's view $\overline{AB}$ as the base. Then, using the formula for the distance between two points, we have

$$\begin{aligned} AB &= \sqrt{(x_2-x_1)^2+(y_2-y_1)^2} \\ &= \sqrt{(8-2)^2+(6-0)^2} \\ &= \sqrt{36+36} = \sqrt{2\times 36} = 6\sqrt{2} \text{ units} \end{aligned}$$

With $\overline{AB}$ as the base, the height of the triangle is the perpendicular distance from C to $\overline{AB}$. To compute that distance, we first need the equation of the line through A and B. The slope of the line is

$$m = \frac{y_2-y_1}{x_2-x_1} = \frac{6-0}{8-2} = \frac{6}{6} = 1$$

Then, using the point (2, 0) and the slope $m = 1$, we have

$$\begin{aligned} y - y_1 &= m(x-x_1) \\ y - 0 &= 1(x-2) \end{aligned}$$

and therefore

$$y = x - 2$$

We can now compute the height of the triangle by finding the perpendicular distance from $C(1, 8)$ to $y = x - 2$:

$$\begin{aligned} \text{height} &= \frac{|mx_0+b-y_0|}{\sqrt{1+m^2}} \\ &= \frac{|1(1)+(-2)-8|}{\sqrt{1+1^2}} = \frac{|-9|}{\sqrt{2}} = \frac{9}{\sqrt{2}} \text{ units} \end{aligned}$$

Now we're ready to compute the area of the triangle, since we know the base and the height. We have

$$\begin{aligned} \text{area} &= \frac{1}{2}(6\sqrt{2})\frac{9}{\sqrt{2}} \\ &= 27 \text{ square units} \end{aligned}$$

EXAMPLE 4 From the point (8, 1), a line is drawn tangent to the circle $x^2 + y^2 = 20$, as shown in Figure 8. Find the slope of this tangent line.

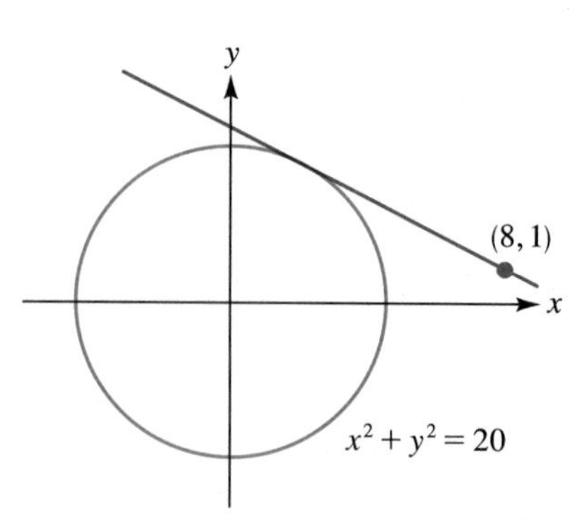

FIGURE 8

Solution This is a problem in which the direct approach is not the simplest. The direct approach would be first to determine the coordinates of the point of tangency. Then, using those coordinates along with (8, 1), the required slope could be computed. As it turns out, however, the coordinates of the point of tangency are not very easy to determine. In fact, one of the advantages of the following method is that those coordinates need not be found.

In Figure 8, let m denote the slope of the tangent line. Because the tangent line passes through (8, 1), we can write its equation

$$\begin{aligned} y - y_1 &= m(x-x_1) \\ y - 1 &= m(x-8) \qquad \text{or} \qquad y = mx - 8m + 1 \end{aligned}$$

Now, since the distance from the origin to the tangent line is $\sqrt{20}$ units (the radius of the circle), we have

$$\begin{aligned}\sqrt{20} &= \frac{|mx_0 + b - y_0|}{\sqrt{1+m^2}}\\ &= \frac{|m(0) - 8m + 1 - 0|}{\sqrt{1+m^2}} \qquad \text{using } x_0 = 0 = y_0 \text{ and } b = -8m+1\\ &= \frac{|-8m+1|}{\sqrt{1+m^2}}\end{aligned}$$

To solve this equation for m, we square both sides to obtain

$$20 = \frac{64m^2 - 16m + 1}{1+m^2}$$

or

$$\begin{aligned}20(1+m^2) &= 64m^2 - 16m + 1\\ 0 &= 44m^2 - 16m - 19\\ 0 &= (22m - 19)(2m+1)\end{aligned}$$

From this we see that the two roots of the equation are $19/22$ and $-1/2$. Because the slope of the tangent line specified in Figure 8 is negative, we choose the value $m = -1/2$; this is the required slope.

The last formula that we consider in this section is

$$d = \frac{|Ax_0 + By_0 + C|}{\sqrt{A^2+B^2}}$$

As we pointed out earlier, this formula gives the distance from the point (x_0, y_0) to the line $Ax + By + C = 0$. To derive the formula, first note that the slope and the y-intercept of the line $Ax + By + C = 0$ are

$$m = -\frac{A}{B} \qquad \text{and} \qquad b = -\frac{C}{B}$$

So we have

$$d = \frac{|mx_0 + b - y_0|}{\sqrt{1+m^2}} = \frac{|(-A/B)x_0 + (-C/B) - y_0|}{\sqrt{1+(-A/B)^2}}$$

Now, to complete the derivation, we need to show that when this last expression is simplified, the result is $|Ax_0 + By_0 + C|/\sqrt{A^2+B^2}$. Exercise 43 asks you to carry out the details.

EXAMPLE 5 Use the formula $d = |Ax_0 + By_0 + C|/\sqrt{A^2+B^2}$ to compute the distance from the point $(-3, 1)$ to the line $y = -2x + 7$.

NOTE In Example 2, we computed this quantity using the distance formula, $d = |mx_0 + b - y_0|/\sqrt{1+m^2}$.

Solution First, we write the given equation $y = -2x + 7$ in the form $Ax + By + C = 0$:

$$2x + y - 7 = 0$$

From this we see that $A = 2$, $B = 1$, and $C = -7$. Thus, we have

$$d = \frac{|Ax_0 + By_0 + C|}{\sqrt{A^2 + B^2}}$$

$$= \frac{|2(-3) + 1(1) + (-7)|}{\sqrt{2^2 + 1^2}} = \frac{12}{\sqrt{5}} = \frac{12\sqrt{5}}{5}$$

The required distance is therefore $12\sqrt{5}/5$ units, as we obtained previously in Example 2(a).

EXERCISE SET 11.1

A

Exercises 1–12 are review exercises. To solve these problems, you will need to utilize the formulas listed at the beginning of this section.

1. Find the distance between the points $(-5, -6)$ and $(3, -1)$.

2. Find the equation of the line that passes through $(2, -4)$ and is parallel to the line $3x - y = 1$. Write your answer in the form $y = mx + b$.

3. Find the equation of a line that is perpendicular to the line $4x - 5y - 20 = 0$ and has the same y-intercept as the line $x - y + 1 = 0$. Write your answer in the form $Ax + By + C = 0$.

4. Find the equation of the line passing through the points $(6, 3)$ and $(1, 0)$. Write your answer in the form $y = mx + b$.

5. Find the equation of the line that is the perpendicular bisector of the line segment joining the points $(2, 1)$ and $(6, 7)$. Write your answer in the form $Ax + By + C = 0$.

6. Find the area of the circle $(x - 12)^2 + (y + \sqrt{5})^2 = 49$.

7. Find the x- and y-intercepts of the circle with center $(1, 0)$ and radius 5.

8. Find the equation of the line that has a positive slope and is tangent to the circle $(x - 1)^2 + (y - 1)^2 = 4$ at one of its y-intercepts. Write your answer in the form $y = mx + b$.

9. Suppose that the coordinates of A, B, and C are $A(1, 2)$, $B(6, 1)$, and $C(7, 8)$. Find the equation of the line passing through C and through the midpoint of the line segment $\overline{AB}$. Write your answer in the form $ax + by + c = 0$.

10. Find the equation of the line passing through the point $(-4, 0)$ and through the point of intersection of the lines $2x - y + 1 = 0$ and $3x + y - 16 = 0$. Write your answer in the form $y = mx + b$.

11. Find the perimeter of $\triangle ABC$ in the following figure.

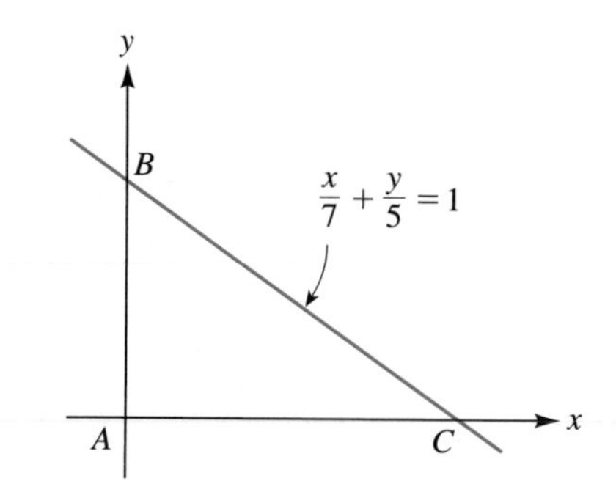

12. Find the sum of the x- and y-intercepts of the line

$$(\csc^2 \alpha)x + (\sec^2 \alpha)y = 1$$

(Assume that α is a constant.)

In Exercises 13–16, determine the angle of inclination of each line. Express the answer in both radians and degrees. In cases where a calculator is necessary, round the answer to two decimal places.

13. $y = \sqrt{3}x + 4$

14. $x + \sqrt{3}y - 2 = 0$

15. **(a)** $y = 5x + 1$
(b) $y = -5x + 1$

16. **(a)** $3x - y - 3 = 0$
(b) $3x + y - 3 = 0$

In Exercises 17–20, find the distance from the point to the line using **(a)** *the formula* $d = |mx_0 + b - y_0|/\sqrt{1 + m^2}$; **(b)** *the formula* $d = |Ax_0 + By_0 + C|/\sqrt{A^2 + B^2}$.

17. $(1, 4)$; $y = x - 2$

18. $(-2, -3)$; $y = -4x + 1$

19. $(-3, 5)$; $4x + 5y + 6 = 0$

20. $(0, -3)$; $3x - 2y = 1$

21. **(a)** Find the equation of the circle that has center $(-2, -3)$ and is tangent to the line $2x + 3y = 6$.
(b) Find the radius of the circle that has center $(1, 3)$ and is tangent to the line $y = \frac{1}{2}x + 5$.

22. Find the area of the triangle with vertices (3, 1), (−2, 7), and (6, 2). *Hint:* Use the method shown in Example 3.

23. Find the area of the quadrilateral $ABCD$ with vertices $A(0, 0)$, $B(8, 2)$, $C(4, 7)$, and $D(1, 6)$. *Suggestion:* Draw a diagonal and use the method shown in Example 3 for the two resulting triangles.

24. From the point (7, −1), tangent lines are drawn to the circle $(x-4)^2+(y-3)^2=4$. Find the slopes of these lines.

25. From the point (0, −5), tangent lines are drawn to the circle $(x-3)^2+y^2=4$. Find the slope of each tangent.

26. Find the distance between the two parallel lines $y=2x-1$ and $y=2x+4$. *Hint:* Draw a sketch; then find the distance from the origin to each line.

27. Find the distance between the two parallel lines $3x+4y=12$ and $3x+4y=24$.

28. Find the equation of the line that passes through (3, 2) and whose x- and y-intercepts are equal. (There are two answers.)

29. Find the equation of the line that passes through the point (2, 6) in such a way that the segment of the line cut off between the axes is bisected by the point (2, 6).

30. Find the equation of the line whose angle of inclination is 60° and whose distance from the origin is four units. (There are two answers.)

B

31. Find the equation of the angle bisector in the accompanying figure. *Hint:* Let (x, y) be a point on the angle bisector. Then (x, y) is equidistant from the two given lines.

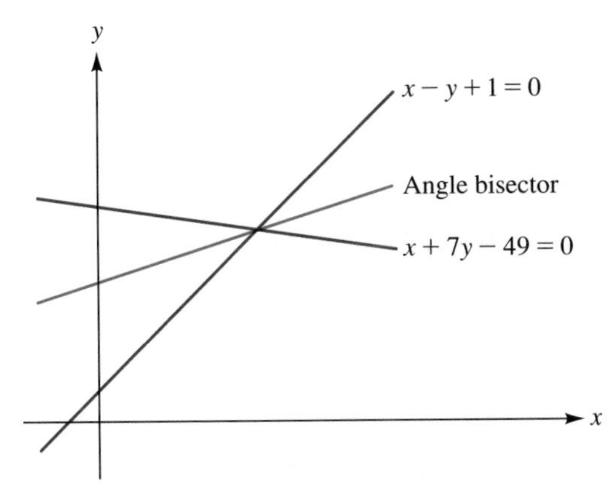

32. Find the center and the radius of the circle that passes through the points (−2, 7), (0, 1), and (2, −1).

33. **(a)** Find the center and the radius of the circle passing through the points $A(-12, 1)$, $B(2, 1)$ and $C(0, 7)$.
(b) Let R denote the radius of the circle in part (a). In $\triangle ABC$, let a, b, and c be the lengths of the sides opposite angles A, B, and C, respectively. Show that the area of $\triangle ABC$ is equal to $abc/4R$.

34. (Continuation of Exercise 33.)
(a) Let H denote the point where the altitudes of $\triangle ABC$ intersect. Find the coordinates of H.
(b) Let d denote the distance from H to the center of the circle in Exercise 33(a). Show that
$$d^2=9R^2-(a^2+b^2+c^2)$$

35. Suppose the line $x-7y+44=0$ intersects the circle $x^2-4x+y^2-6y=12$ at points P and Q. Find the length of the chord $\overline{PQ}$.

36. The point (1, −2) is the midpoint of a chord of the circle $x^2-4x+y^2+2y=15$. Find the length of the chord.

37. Show that the product of the distances from the point $(0, c)$ to the lines $ax+y=0$ and $x+by=0$ is
$$\frac{|bc^2|}{\sqrt{a^2+a^2b^2+b^2+1}}$$

38. Suppose that the point (x_0, y_0) lies on the circle $x^2+y^2=a^2$. Show that the equation of the line tangent to the circle at (x_0, y_0) is $x_0x+y_0y=a^2$.

39. The vertices of $\triangle ABC$ are $A(0, 0)$, $B(8, 0)$, and $C(8, 6)$.
(a) Find the equations of the three lines that bisect the angles in $\triangle ABC$. *Hint:* Make use of the identity $\tan(\theta/2)=(\sin\theta)/(1+\cos\theta)$.
(b) Find the points where each pair of angle bisectors intersect. What do you observe?

40. Show that the equations of the lines with slope m that are tangent to the circle $x^2+y^2=a^2$ are
$$y=mx+a\sqrt{1+m^2} \quad\text{and}\quad y=mx-a\sqrt{1+m^2}$$

41. The point (x, y) is equidistant from the point $(0, 1/4)$ and the line $y=-1/4$. Show that x and y satisfy the equation $y=x^2$.

42. The point (x_0, y_0) is equidistant from the line $x+2y=0$ and the point (3, 1). Find (and simplify) an equation relating x_0 and y_0.

43. **(a)** Find the slope m and the y-intercept b of the line $Ax+By+C=0$
(b) Use the formula $d=|mx_0+b-y_0|/\sqrt{1+m^2}$ to show that the distance from the point (x_0, y_0) to the line $Ax+By+C=0$ is given by
$$d=\frac{|Ax_0+By_0+C|}{\sqrt{A^2+B^2}}$$

44. Refer to Figure 5 on page 674. Show that $\triangle ABD$ is similar to $\triangle EGF$.

C

45. Show that the distance of the point (x_1, y_1) from the line passing through the two points (x_2, y_2) and (x_3, y_3) is given by $d = |D|/\sqrt{(x_2 - x_3)^2 + (y_2 - y_3)^2}$, where

$$D = \begin{vmatrix} x_1 & y_1 & 1 \\ x_2 & y_2 & 1 \\ x_3 & y_3 & 1 \end{vmatrix}$$

46. Let (a_1, b_1), (a_2, b_2), and (a_3, b_3) be three noncollinear points. Show that the equation of the circle passing through these three points is

$$\begin{vmatrix} x^2 + y^2 & x & y & 1 \\ a_1^2 + b_1^2 & a_1 & b_1 & 1 \\ a_2^2 + b_2^2 & a_2 & b_2 & 1 \\ a_3^2 + b_3^2 & a_3 & b_3 & 1 \end{vmatrix} = 0$$

47. Find the equation of the circle that passes through the points (6, 3) and $(-4, -3)$ and that has its center on the line $y = 2x - 7$.

48. Let a be a positive number and suppose that the coordinates of points P and Q are $P(a \cos \theta, a \sin \theta)$ and $Q(a \cos \beta, a \sin \beta)$. Show that the distance from the origin to the line passing through P and Q is

$$a \left| \cos\left(\frac{\theta - \beta}{2}\right) \right|$$

49. Find the equation of a circle that has radius 5 and is tangent to the line $2x + 3y = 26$ at the point (4, 6). Write your answer in standard form. (There are two answers.)

50. Find the equation of the circle passing through $(2, -1)$ and tangent to the line $y = 2x + 1$ at (1, 3). Write your answer in standard form.

51. For the last exercise in this section, you will prove an interesting property of the curve $x^3 + y^3 = 6xy$. This curve, known as the *folium of Descartes,* is shown in Figure A. Figure B and the accompanying caption indicate the property that we will establish.

(a) Suppose that the slope of the line segment $\overline{OQ}$ is t. By solving the system of equations

$$\begin{cases} y = tx \\ x^3 + y^3 = 6xy \end{cases}$$

show that the coordinates of the point Q are $x = 6t/(1 + t^3)$ and $y = 6t^2/(1 + t^3)$.

(b) Show that the slope of the line joining the points P and Q is $(t^2 - t - 1)/(t^2 + t - 1)$.

(c) Suppose that the slope of the line segment $\overline{OR}$ is u. By repeating the procedure used in parts (a) and (b), you'll find that the slope of the line joining the points P and R is $(u^2 - u - 1)/(u^2 + u - 1)$. Now, since the points P, Q, and R are collinear (lie on the same line), it must be the case that

$$\frac{t^2 - t - 1}{t^2 + t - 1} = \frac{u^2 - u - 1}{u^2 + u - 1}$$

Working from this equation, show that $tu = -1$. This shows that $\angle ROQ$ is a right angle, as required.

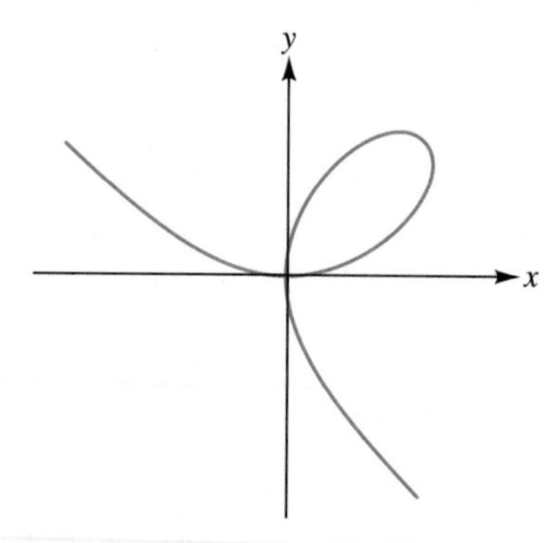

The Folium of Descartes, $x^3 + y^3 = 6xy$

FIGURE A

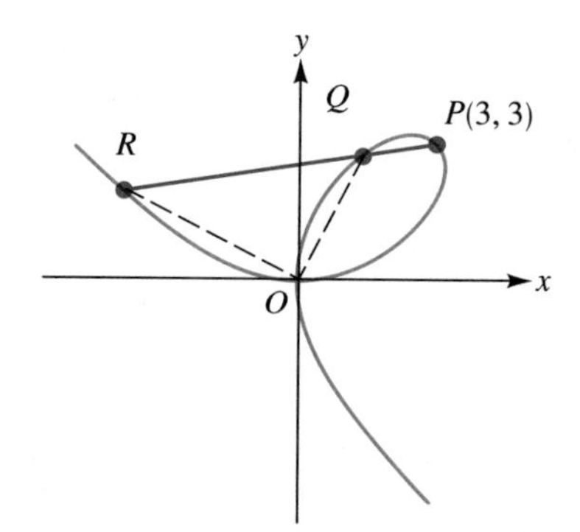

A property of the folium: Suppose that a line through the point $P(3, 3)$ meets the folium again at points Q and R. Then $\angle ROQ$ is a right angle.

FIGURE B

11.2 THE PARABOLA

In Section 4.2, we saw that the graph of a quadratic function $y = ax^2 + bx + c$ is a symmetric U-shaped curve called a *parabola.* In this section, we give a more general definition of the parabola, a definition that emphasizes the geometric properties of the curve.

> **DEFINITION: The Parabola**
>
> A **parabola** is the set of all points in the plane equally distant from a fixed line and a fixed point not on the line. The fixed line is called the **directrix;** the fixed point is called the **focus.**

Let us initially suppose that the focus of the parabola is the point $(0, p)$ and the directrix is the line $y = -p$. We will assume throughout this section that p is positive. To understand the geometric content of the definition of the parabola, the special graph paper displayed in Figure 1(a) is useful. The common center of the concentric circles in Figure 1(a) is the focus $(0, p)$. Thus, all the points on a given circle are at a fixed distance from the focus. The radii of the circles increase in increments of p units. Similarly, the broken horizontal lines in the figure are drawn at intervals that are multiples of p units from the directrix $y = -p$. By considering the points where the circles intersect the horizontal lines, we can find a number of points equally distant from the focus $(0, p)$ and the directrix $y = -p$; see Figure 1(b).

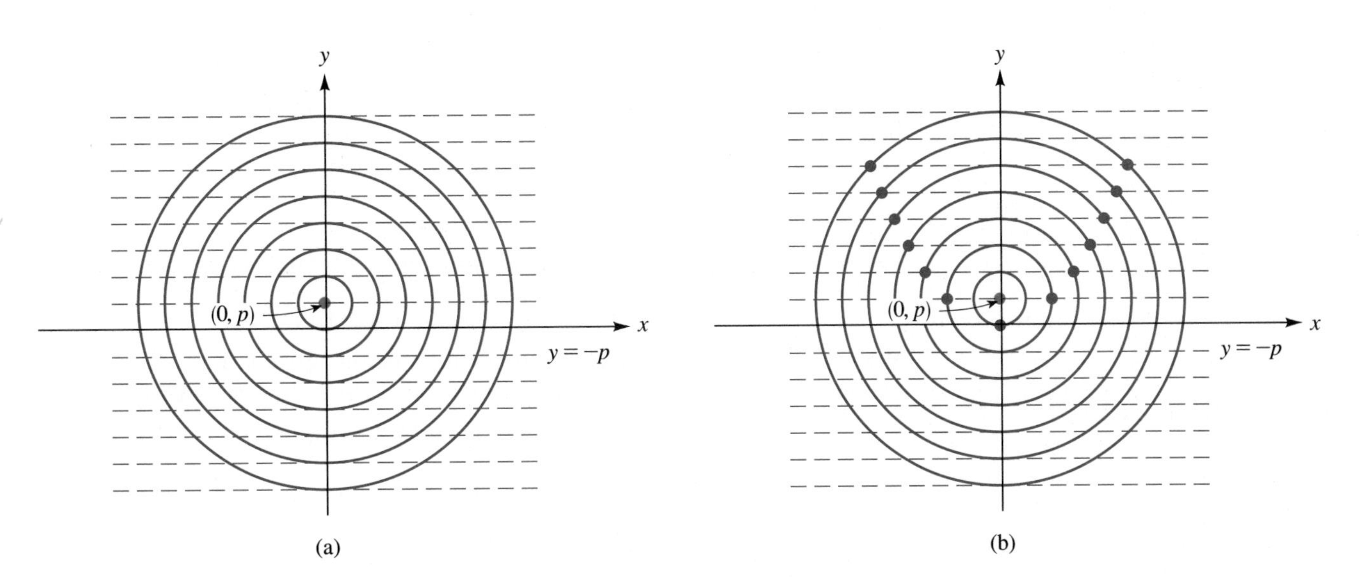

FIGURE 1

The graphical method we have just described lets us locate many points equally distant from the focus $(0, p)$ and the directrix $y = -p$. Figure 1(b) shows that the points on the parabola are symmetric about a line, in this case the y-axis. Also, by studying the figure, you should be able to convince yourself that in this case there can be no points below the x-axis that satisfy the stated condition. However, to describe the required set of points completely, and to show that our new definition is consistent with the old one, Figure 1(b) is inadequate. We need to bring algebraic methods to bear on the problem. Thus, let d_1 denote the distance from the point $P(x, y)$ to the focus $(0, p)$ and let d_2 denote the distance from $P(x, y)$ to the directrix $y = -p$, as shown in Figure 2. The distance d_1 is then

$$d_1 = \sqrt{(x-0)^2 + (y-p)^2} = \sqrt{x^2 + y^2 - 2py + p^2}$$

The distance d_2 in Figure 2 is just the vertical distance between the points P and Q. Thus,

$$\begin{aligned} d_2 &= (y\text{-coordinate of } P) - (y\text{-coordinate of } Q) \\ &= y - (-p) = y + p \end{aligned}$$

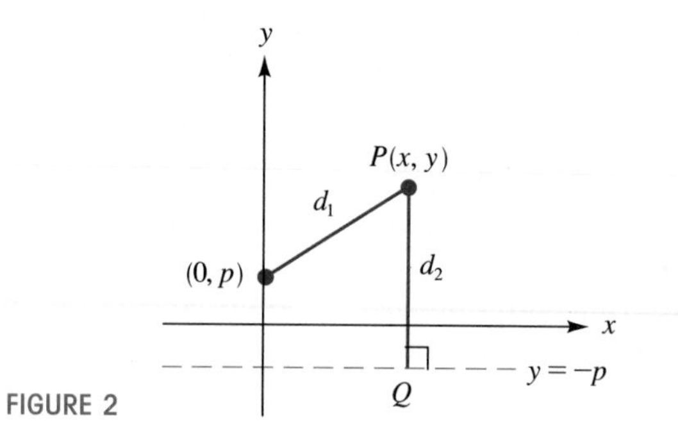

FIGURE 2

(Absolute value signs are unnecessary here for, as noted earlier, P cannot lie below the x-axis.) The condition that P be equally distant from the focus $(0, p)$ and the directrix $y = -p$ can be expressed by the equation

$$d_1 = d_2$$

Using the expressions we've found for d_1 and d_2, this last equation becomes

$$\sqrt{x^2 + y^2 - 2py + p^2} = y + p$$

We can obtain a simpler but equivalent equation by squaring both sides. (Two nonnegative quantities are equal if and only if their squares are equal.) Thus,

$$x^2 + y^2 - 2py + p^2 = y^2 + 2py + p^2$$

After combining like terms, this equation becomes

$$x^2 = 4py$$

This is the equation of a parabola with focus $(0, p)$ and directrix $y = -p$. In the box that follows, we summarize the properties of the parabola $x^2 = 4py$. As Figure 3(b) indicates, the terminology we've introduced applies equally well to an arbitrary parabola for which the axis of symmetry is not necessarily parallel to one of the coordinate axes and the vertex is not necessarily the origin.

PROPERTY SUMMARY THE PARABOLA

1. The **parabola** is the set of points equidistant from a fixed line called the **directrix** and a fixed point, not on the line, called the **focus.**
2. The **axis** of a parabola is the line drawn through the focus and perpendicular to the directrix.
3. The **vertex** of a parabola is the point where the parabola intersects its axis. The vertex is located halfway between the focus and the directrix. See Figure 3.

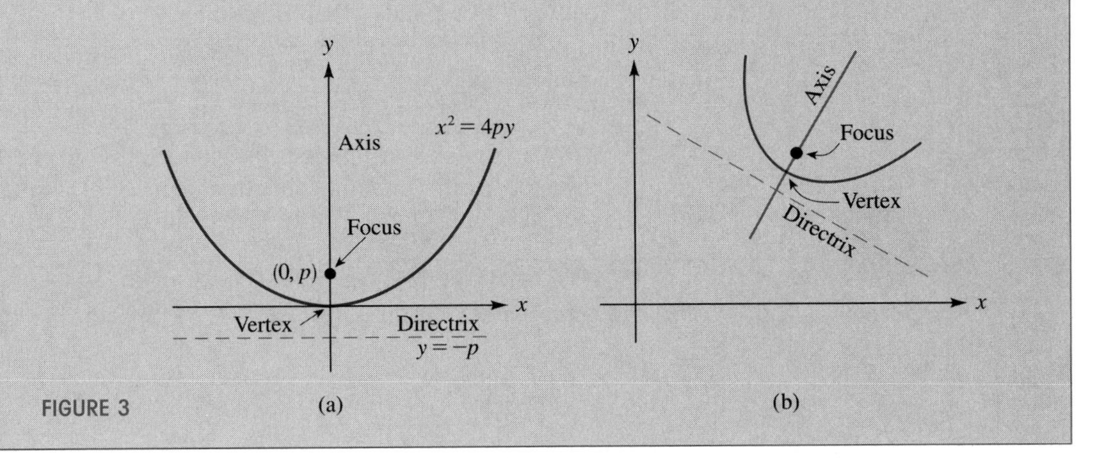

FIGURE 3

EXAMPLE 1

(a) Refer to Figure 4. Determine the focus and the directrix of the parabola $x^2 = 16y$.

(b) As indicated in Figure 4, the point Q is on the parabola $x^2 = 16y$ and the x-coordinate of Q is -12. Find the y-coordinate of Q.

(c) Verify that the point Q is equidistant from the focus and the directrix of the parabola.

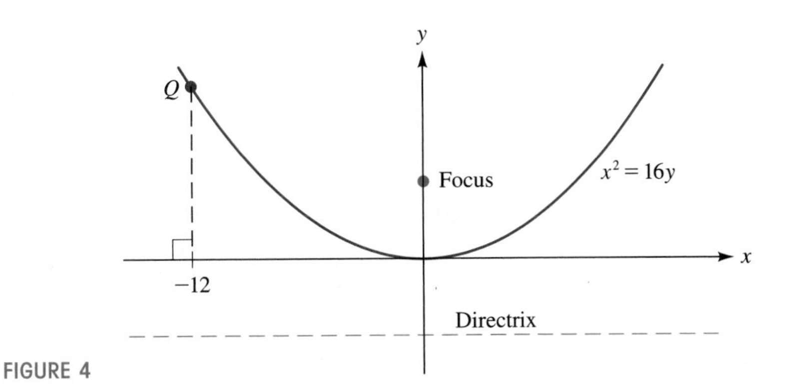

FIGURE 4

Solution

(a) We know that for the basic parabola $x^2 = 4py$, the focus is the point $(0, p)$ and the directrix is the line $y = -p$. Comparing the given equation $x^2 = 16y$

with $x^2 = 4py$, we see that $4p = 16$ and therefore $p = 4$. Thus, the focus of the parabola $x^2 = 16y$ is (0, 4) and the directrix is $y = -4$.

(b) We are given that the x-coordinate of Q is -12. Substituting $x = -12$ in the given equation $x^2 = 16y$ yields $(-12)^2 = 16y$. Therefore $y = 144/16 = 9$, and the coordinates of Q are $(-12, 9)$.

(c) In part (b) we found that the y-coordinate of Q is 9. From part (a) we know that the equation of the directrix is $y = -4$. Thus, the vertical distance from Q down to the directrix is $9 - (-4) = 13$. Next, to calculate the distance from $Q(-12, 9)$ to the focus (0, 4), we can use the formula for the distance between two points. This yields

$$\begin{aligned} d &= \sqrt{(x_2 - x_1)^2 + (y_2 - y_1)^2} \\ &= \sqrt{(-12 - 0)^2 + (9 - 4)^2} = \sqrt{144 + 25} = \sqrt{169} = 13 \end{aligned}$$

We've now shown that the distance from Q to the focus is the same as the distance from Q to the directrix, namely, 13 units. In other words, Q is indeed equidistant from the focus and the directrix of the parabola.

EXAMPLE 2 Determine the equation of the parabola in Figure 5, given that the curve passes through the point (3, 5). Specify the focus and the directrix.

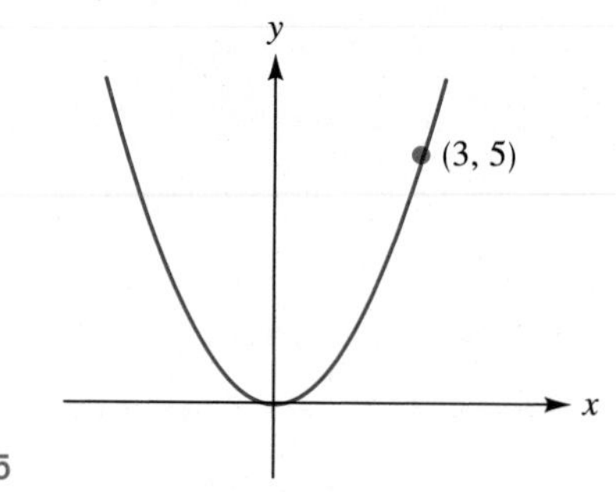

FIGURE 5

Solution The general equation for a parabola in this position is $x^2 = 4py$. Since the point (3, 5) lies on the curve, its coordinates must satisfy the equation $x^2 = 4py$. Thus,

$$\begin{aligned} 3^2 &= 4p(5) \\ 9 &= 20p \\ 9/20 &= p \end{aligned}$$

With $p = 9/20$, the equation $x^2 = 4py$ becomes

$$x^2 = 4(9/20)y$$

or

$$x^2 = (9/5)y \qquad \text{as required}$$

Furthermore, since $p = 9/20$, the focus is $(0, 9/20)$ and the directrix is $y = -9/20$.

We have seen that the equation of a parabola with focus $(0, p)$ and directrix $y = -p$ is $x^2 = 4py$. By following the same method, we can obtain general equations for parabolas with other orientations. The basic results are summarized in Figure 6.

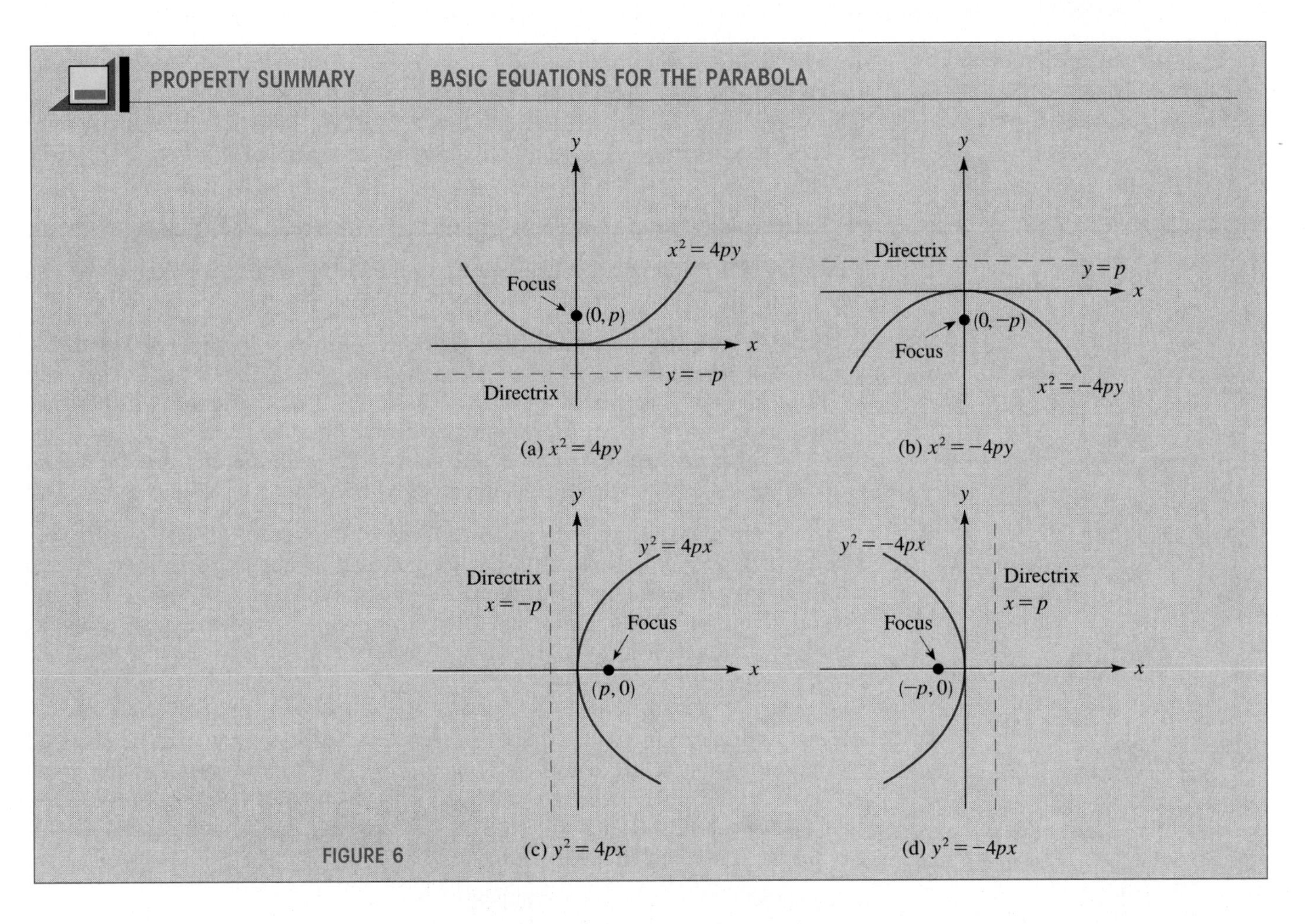

FIGURE 6

A **chord** of a parabola is a straight line segment joining any two points on the curve. If the chord passes through the focus, it is called a **focal chord.** For purposes of graphing, it is useful to know the length of the focal chord perpendicular to the axis of the parabola. This is the length of the horizontal line segment $\overline{AB}$ in Figure 7 and the vertical line segment $\overline{A'B'}$ in Figure 8. We will call this length the **focal width.**

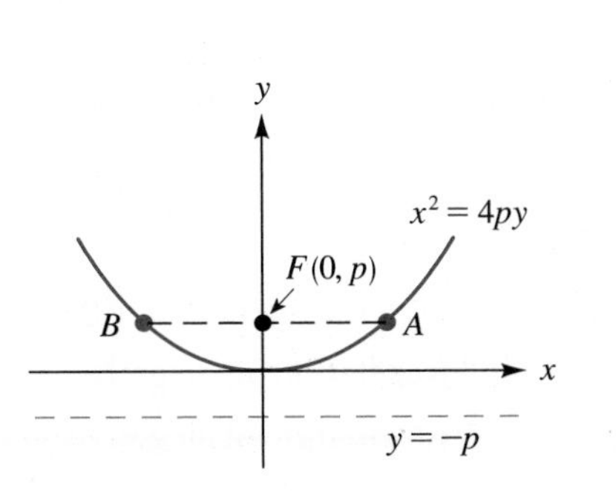

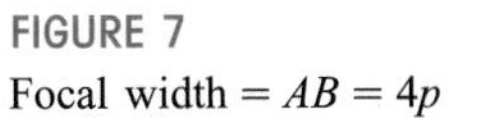
FIGURE 7
Focal width = $AB = 4p$

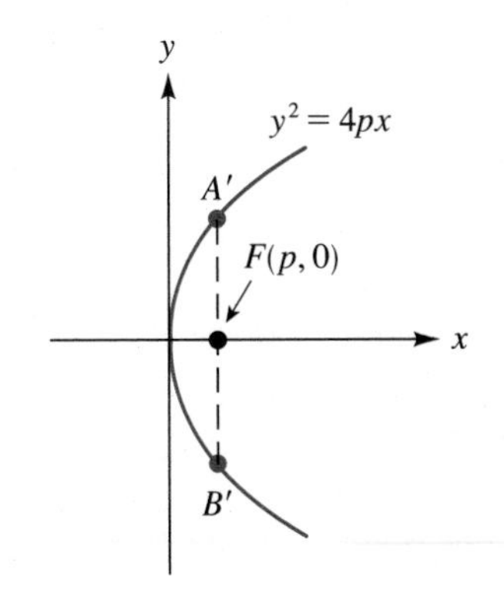

FIGURE 8
Focal width = $A'B' = 4p$

In Figure 7, the distance from A to F is the same as the distance from A to the line $y = -p$. (Why?) But the distance from A to the line $y = -p$ is $2p$. Therefore, $AF = 2p$ and AB is twice this, or $4p$. We have shown that the focal width of the parabola is $4p$. In other words, given a parabola $x^2 = 4py$, the width at its focus is $4p$, the coefficient of y. In the same way, the length of the focal chord $\overline{A'B'}$ in Figure 8 is also $4p$, the coefficient of x in that case.

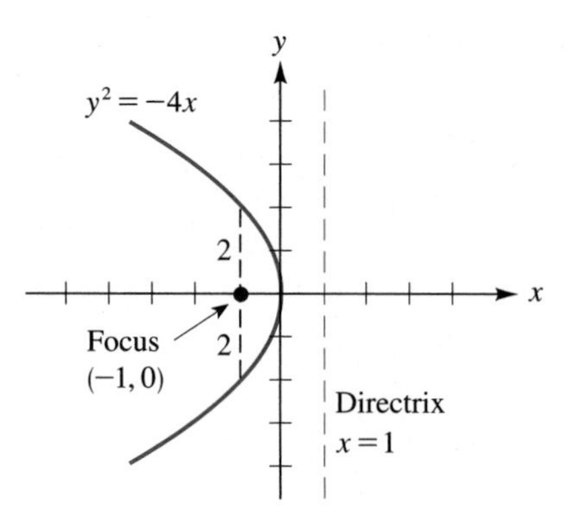

FIGURE 9

EXAMPLE 3 Find the focus and the directrix of the parabola $y^2 = -4x$, and sketch the graph.

Solution Comparing the basic equation $y^2 = -4px$ [in Figure 6(d)] with the equation at hand, we see that $4p = 4$ and thus $p = 1$. The focus is therefore $(-1, 0)$, and the directrix is $x = 1$. The basic form of the graph will be as in Figure 6(d). For purposes of graphing, we note that the focal width is 4 (the absolute value of the coefficient of x). This, along with the fact that the vertex is $(0, 0)$, gives us enough information to draw the graph; see Figure 9.

In Examples 1 through 3 and in Figure 6, the vertex of each parabola is located at the origin. Now we want to consider parabolas that are translated (shifted) from this standard position. As background for this, we review and generalize the results about translation in Section 3.3. Consider, as an example, the two equations

$$y = (x - 1)^2 \qquad \text{and} \qquad y = x^2 + 1$$

As we saw in Section 3.3, the graphs of these two equations can each be obtained by translating the graph of $y = x^2$. To graph $y = (x - 1)^2$, we translate the graph of $y = x^2$ to the right 1 unit; to graph $y = x^2 + 1$, we translate $y = x^2$ up 1 unit.

To see the underlying pattern here, let's rewrite the equation $y = x^2 + 1$ as $y - 1 = x^2$. Then the situation is this:

EQUATION	HOW GRAPH IS OBTAINED
$y = (x - 1)^2$	translation in positive x-direction
$y - 1 = x^2$	translation in positive y-direction

OBSERVATION *In the equation $y = x^2$, the effect of replacing x with $x - 1$ is to translate the graph one unit in the positive x-direction. Similarly, replacing y with $y - 1$ translates the graph one unit in the positive y-direction.*

As a second example before we generalize, consider the equations

$$y = (x + 1)^2 \qquad \text{and} \qquad y = x^2 - 1$$

The first equation involves a translation in the negative x-direction; the second involves a translation in the negative y-direction. As before, to see the underlying pattern, we rewrite the second equation as $y + 1 = x^2$. Now we have the following situation:

EQUATION	HOW GRAPH IS OBTAINED
$y = (x + 1)^2$	translation in negative x-direction
$y + 1 = x^2$	translation in negative y-direction

OBSERVATION *In the equation $y = x^2$, the effect of replacing x with $x + 1$ is to translate the graph one unit in the negative x-direction. Similarly, replacing y with $y + 1$ translates the graph one unit in the negative y-direction.*

Both of the examples that we have just considered are specific instances of the following basic result. (The result is valid whether or not the given equation and graph represent a function.)

PROPERTY SUMMARY **TRANSLATION AND COORDINATES**

Suppose that we have an equation that determines a graph in the x-y plane, and let h and k denote positive numbers. Then, replacing x with $x-h$ or $x+h$, or replacing y with $y-k$ or $y+k$, has the following effects on the graph of the original equation.

REPLACEMENT	RESULTING TRANSLATION
1. x replaced with $x-h$	h units in the positive x-direction
2. y replaced with $y-k$	k units in the positive y-direction
3. x replaced with $x+h$	h units in the negative x-direction
4. y replaced with $y+k$	k units in the negative y-direction

EXAMPLE 4 Graph the parabola $(y+1)^2 = -4(x-2)$. Specify the vertex, the focus, the directrix, and the axis of symmetry.

Solution The given equation is obtained from $y^2 = -4x$ (which we graphed in the previous example) by replacing x and y with $x-2$ and $y+1$, respectively. So the required graph is obtained by translating the parabola in Figure 9 to the right 2 units and down 1 unit. In particular, this means that the vertex moves from $(0, 0)$ to $(2, -1)$; the focus moves from $(-1, 0)$ to $(1, -1)$; the directrix moves from $x = 1$ to $x = 3$; and the axis of symmetry moves from $y = 0$ (which is the x-axis) to $y = -1$. The required graph is shown in Figure 10.

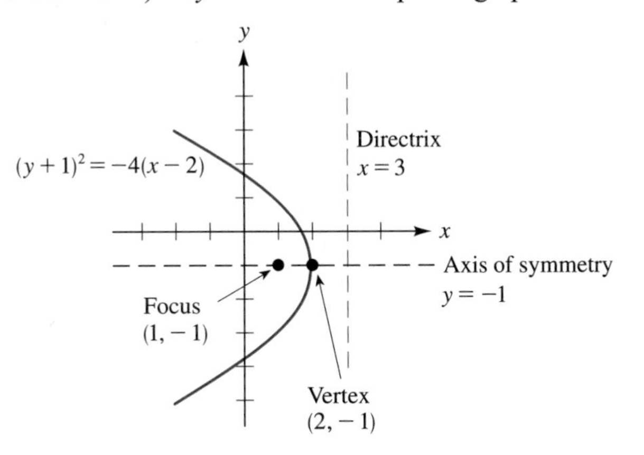

FIGURE 10

EXAMPLE 5 Graph the parabola $2y^2 - 4y - x + 5 = 0$, and specify each of the following: vertex, focus, directrix, axis of symmetry, and focal width.

Solution Just as we did in Section 4.2, we use the technique of completing the square:

$$2(y^2 - 2y \qquad) = x - 5$$

$$2(y^2 - 2y + 1) = x - 5 + 2 \qquad \text{adding 2 to both sides}$$

$$(y-1)^2 = \frac{1}{2}(x-3)$$

The graph of this last equation is obtained by translating the graph of $y^2 = \frac{1}{2}x$ "right 3, up 1." This moves the vertex from (0, 0) to (3, 1). Now, for $y^2 = \frac{1}{2}x$, the focus and directrix are determined by setting $4p = 1/2$. Therefore $p = 1/8$, and consequently the focus and directrix of $y^2 = \frac{1}{2}x$ are $(1/8, 0)$ and $x = -1/8$, respectively. Thus, the focus of the translated curve is $(3\frac{1}{8}, 1)$ and the directrix is $x = 2\frac{7}{8}$. Figure 11(a) shows the graph of $y^2 = \frac{1}{2}x$, and Figure 11(b) shows the translated graph. (You should check for yourself that the information accompanying Figure 11(b) is correct.)

FIGURE 11

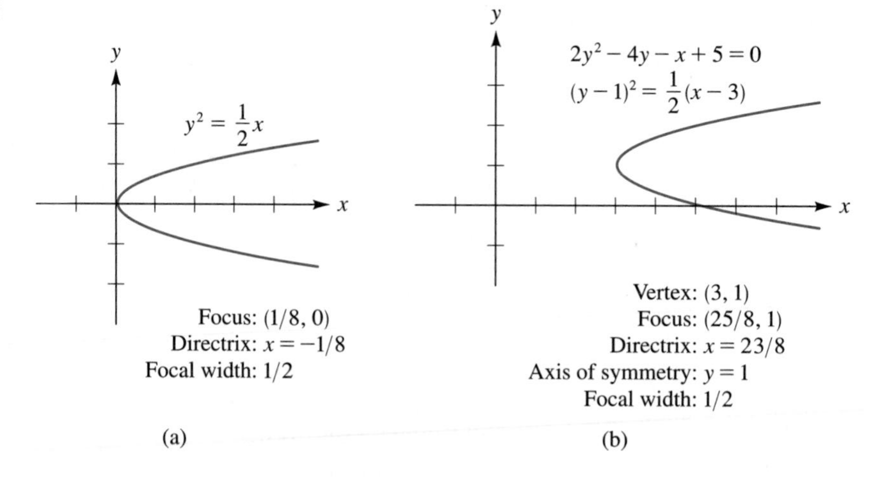

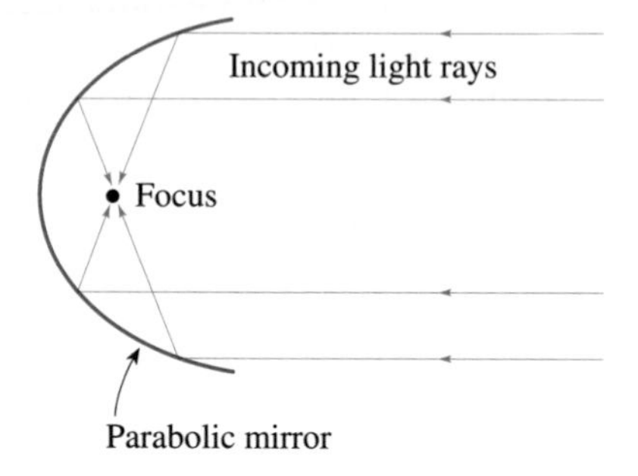

FIGURE 12

There are numerous applications of the parabola in the sciences. Figure 12 shows a cross section of a parabolic mirror in a telescope. As indicated in Figure 12, light rays coming in parallel to the axis of the parabola are reflected through the focus. (The word *focus* comes from a Latin word meaning "fireplace.") Parabolic reflectors are also used in communication and surveillance systems, in radio telescopes, and in automobile headlights. In an automobile headlight, the bulb is located at the focus. To diagram the parabolic mirror of an automobile headlight, just reverse the directions of the arrows in Figure 12. [For a more complete description of the uses of the parabola, see the article, "The Standup Conic Presents: The Parabola and Applications" by Lee Whitt in *The Journal of Undergraduate Mathematics and its Applications,* vol. III, no. 3 (1982)].

EXERCISE SET 11.2

A

In Exercises 1–22, graph the parabolas. In each case, specify the focus, the directrix, and the focal width. For Exercises 13–22, also specify the vertex.

1. $x^2 = 4y$
2. $x^2 = 16y$
3. $y^2 = -8x$
4. $y^2 = 12x$
5. $x^2 = -20y$
6. $x^2 - y = 0$
7. $y^2 + 28x = 0$
8. $4y^2 + x = 0$
9. $x^2 = 6y$
10. $y^2 = -10x$
11. $4x^2 = 7y$
12. $3y^2 = 4x$
13. $y^2 - 6y - 4x + 17 = 0$
14. $y^2 + 2y + 8x + 17 = 0$
15. $x^2 - 8x - y + 18 = 0$
16. $x^2 + 6y + 18 = 0$
17. $y^2 + 2y - x + 1 = 0$
18. $2y^2 - x + 1 = 0$
19. $2x^2 - 12x - y + 18 = 0$
20. $y + \sqrt{2} = (x - 2\sqrt{2})^2$
21. $2x^2 - 16x - y + 33 = 0$
22. $\frac{1}{4}y^2 - y - x + 1 = 0$

B

For Exercises 23 and 24, refer to the figure below.

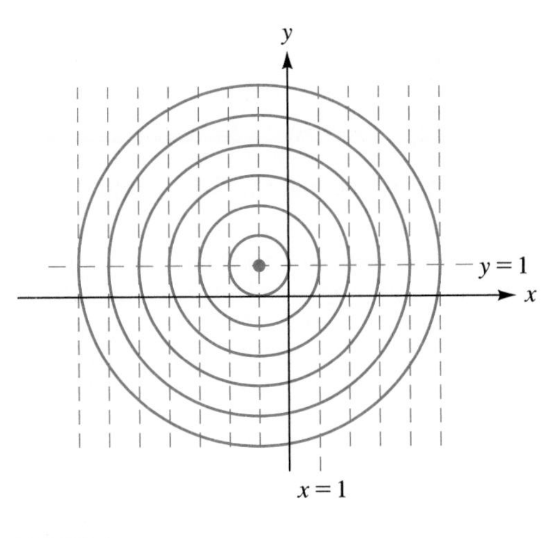

FIGURE A

23. Make a photocopy of Figure A. Then, in your copy of Figure A, indicate (by drawing dots) eleven points that are equidistant from the point $(-1, 1)$ and the line $x = 1$. What is the equation of the line of symmetry for the set of dots?

24. The eleven dots that you located in Exercise 23 are part of a parabola. Find the equation of that parabola and sketch its graph. Specify the vertex, the focus and directrix, and the focal width.

For Exercises 25–30, find the equation of the parabola satisfying the given conditions. In each case, assume that the vertex is at the origin.

25. The focus is $(0, 3)$.

26. The directrix is $y - 8 = 0$.

27. The directrix is $x + 32 = 0$.

28. The focus lies on the y-axis, and the parabola passes through the point $(7, -10)$.

29. The parabola is symmetric about the x-axis, the x-coordinate of the focus is negative, and the length of the focal chord perpendicular to the x-axis is 9.

30. The focus is the smaller of the two x-intercepts of the circle $x^2 - 8x + y^2 - 6y + 9 = 0$.

31. Let P denote the point $(8, 8)$ on the parabola $x^2 = 8y$, and let $\overline{PQ}$ be a focal chord.

(a) Find the equation of the line through the point $(8, 8)$ and the focus.

(b) Find the coordinates of Q.

(c) Find the length of $\overline{PQ}$.

(d) Find the equation of the circle with this focal chord as a diameter.

(e) Show that the circle determined in part (d) intersects the directrix of the parabola in only one point. Conclude from this that the directrix is tangent to the circle. Draw a sketch of the situation.

32. The following figure shows a parabolic cross section of a reflecting mirror. Find the distance from the vertex of the parabola to the focus. *Hint:* Choose a convenient coordinate system.

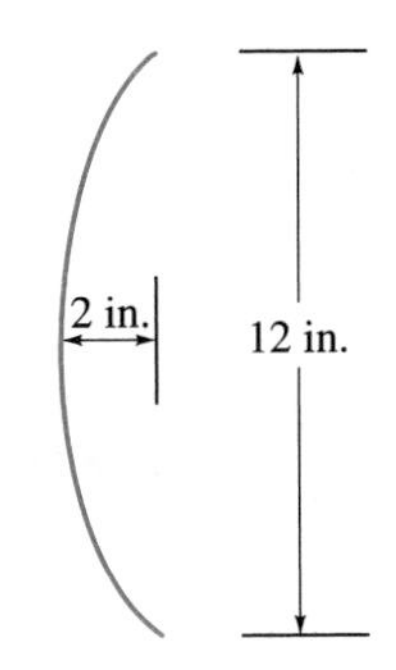

33. An arch is in the shape of a parabola with a vertical axis. The arch is 15 ft high at the center and 40 ft wide at the base. At what height above the base is the width 20 ft? *Hint:* Choose a convenient coordinate system in which the vertex of the parabola is at the origin.

34. The following figure depicts the cable of a suspension bridge. The cable is in the form of a parabola with a vertical axis. The horizontal highway on the bridge is 300 ft long. The longest of the vertical supporting wires is 100 ft; the shortest is 40 ft. Find the length of a supporting wire that is 50 ft from the middle.

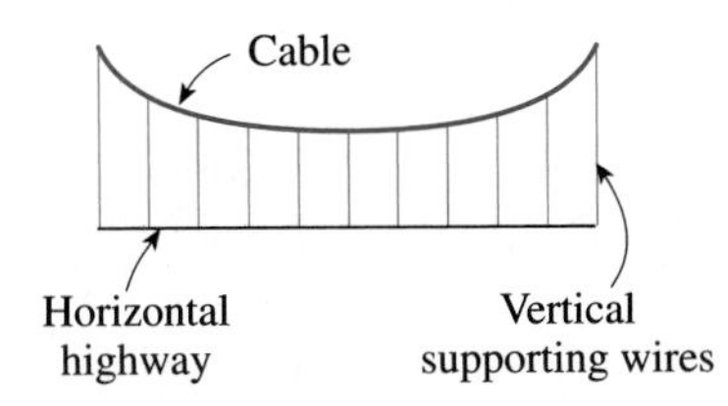

35. The segments $\overline{AA'}$ and $\overline{BB'}$ are focal chords of the parabola $x^2 = 2y$. The coordinates of A and B are $(4, 8)$ and $(-2, 2)$, respectively.

(a) Find the equation of the line through A and B'.

(b) Find the equation of the line through B and A'.

(c) Show that the two lines you have found intersect at a point on the directrix.

36. Let $\overline{PQ}$ be the horizontal focal chord of the parabola $x^2 = 8y$. Let R denote the point where the directrix of the parabola meets the y-axis. Show that $\overline{PR}$ is perpendicular to $\overline{QR}$.

37. Suppose $\overline{PQ}$ is a focal chord of the parabola $y = x^2$ and that the coordinates of P are $(2, 4)$.
(a) Find the coordinates of Q.
(b) Find the coordinates of M, the midpoint of $\overline{PQ}$.
(c) A perpendicular is drawn from M to the y-axis, meeting the y-axis at S. Also, a line perpendicular to the focal chord is drawn through M, meeting the y-axis at T. Find ST and verify that it is equal to one-half the focal width of the parabola.

38. Let F be the focus of the parabola $x^2 = 8y$, and let P denote the point on the parabola with coordinates $(8, 8)$. Let $\overline{PQ}$ be a focal chord. If V denotes the vertex of the parabola, verify that

$$PF \cdot FQ = VF \cdot PQ$$

39. In the following figure, $\triangle OAB$ is equilateral and $\overline{AB}$ is parallel to the x-axis. Find the length of a side and the area of triangle OAB.

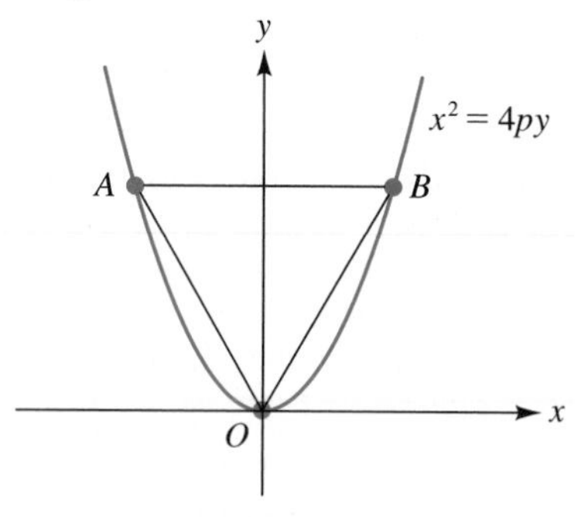

C

40. If $\overline{PQ}$ is a focal chord of the parabola $x^2 = 4py$ and the coordinates of P are (x_0, y_0), show that the coordinates of Q are

$$\left(\frac{-4p^2}{x_0}, \frac{p^2}{y_0}\right)$$

41. If $\overline{PQ}$ is a focal chord of the parabola $y^2 = 4px$ and the coordinates of P are (x_0, y_0), show that the coordinates of Q are

$$\left(\frac{p^2}{x_0}, \frac{-4p^2}{y_0}\right)$$

42. Let F and V denote the focus and the vertex, respectively, of the parabola $x^2 = 4py$. If $\overline{PQ}$ is a focal chord of the parabola, show that

$$PF \cdot FQ = VF \cdot PQ$$

43. Let $\overline{PQ}$ be a focal chord of the parabola $y^2 = 4px$, and let M be the midpoint of $\overline{PQ}$. A perpendicular is drawn from M to the x-axis, meeting the x-axis at S. Also from M, a line segment is drawn that is perpendicular to $\overline{PQ}$ and that meets the x-axis at T. Show that the length of $\overline{ST}$ is one-half the focal width of the parabola.

44. Let $\overline{AB}$ be a chord (not necessarily a focal chord) of the parabola $y^2 = 4px$, and suppose that $\overline{AB}$ subtends a right angle at the vertex. (In other words, $\angle AOB = 90°$, where O is the origin in this case.) Find the x-intercept of the segment $\overline{AB}$. What is surprising about this result? *Hint:* Begin by writing the coordinates of A and B as $A(a^2/4p, a)$ and $B(b^2/4p, b)$.

45. Let $\overline{PQ}$ be a focal chord of the parabola $x^2 = 4py$. Complete the following steps to prove that the circle with $\overline{PQ}$ as a diameter is tangent to the directrix of the parabola. Let the coordinates of P be (x_0, y_0).
(a) Show that the coordinates of Q are

$$\left(\frac{-4p^2}{x_0}, \frac{p^2}{y_0}\right)$$

(b) Show that the midpoint of $\overline{PQ}$ is

$$\left(\frac{x_0^2 - 4p^2}{2x_0}, \frac{y_0^2 + p^2}{2y_0}\right)$$

(c) Show that the length of $\overline{PQ}$ is $(y_0 + p)^2/y_0$. *Suggestion:* This can be done using the formula for the distance between two points, but the following is simpler. Let F be the focus. Then $PQ = PF + FQ$. Now, both PF and FQ can be determined by using the definition of the parabola rather than the distance formula.
(d) Show that the distance from the center of the circle to the directrix equals the radius of the circle. How does this complete the proof?

46. P and Q are two points on the parabola $y^2 = 4px$, the coordinates of P are (a, b), and the slope of $\overline{PQ}$ is m. Find the y-coordinate of the midpoint of $\overline{PQ}$. *Hint for checking:* The final answer is independent of both a and b.

GRAPHING UTILITY EXERCISES FOR SECTION 11.2

In Exercises 1–4, graph the parabolas. These graphs provide examples of the four basic orientations shown in Figure 6 on page 685.

1. $x^2 = 8y$ *Hint:* Enter this as $\frac{1}{8}x^2$.
2. $x^2 = -8y$
3. $y^2 = 8x$
   Hint: Enter this as $\sqrt{8x}$ and $-\sqrt{8x}$.
4. $y^2 = -8x$
5. In this exercise we graph the parabola $4x + y^2 + 2y - 7 = 0$.
   (a) By completing the square, show that the equation can be written $(y + 1)^2 = -4(x - 2)$.
   (b) Solving the equation in part (a) for y gives us the two functions $y = -1 \pm \sqrt{-4(x-2)}$. Graph these two functions and check that your graph is consistent with Figure 9 on page 686.
6. **(a)** Complete the square to determine the vertex of the parabola $y = 2x^2 - 16x + 33$.
   (b) Graph this function. Check that the general appearance of the graph you obtain is consistent with your result in part (a).
   (c) Use the graphing utility to estimate the coordinates of the vertex. How close are your estimates to the values determined in part (a)?

Exercises 7–10 deal with lines that are tangent to a parabola. To determine the equations for the tangent lines in these problems, use the following result, which was known (in a purely geometric form) to Archimedes.

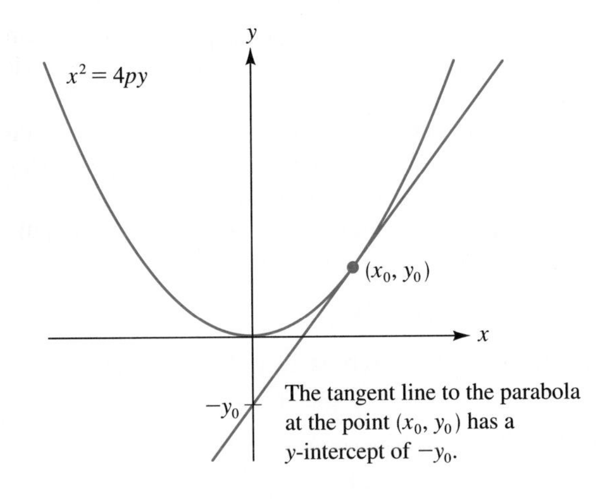

The tangent line to the parabola at the point (x_0, y_0) has a y-intercept of $-y_0$.

7. As a first example, consider the parabola $x^2 = 8y$. Note that the point (4, 2) lies on this parabola. (Why?) According to the statement accompanying the figure, the tangent to the parabola at (4, 2) will have a y-intercept of $y_0 = -2$. Use this fact to find the equation of the tangent line. Then use the graphing utility to graph the parabola and the tangent line.
8. Graph the parabola $x^2 = 2y$ and the tangent to the curve at the point $(-2, 2)$.
9. Graph the parabola $x^2 = y$ and the tangent to the curve at the point $(-3, 9)$.
10. In this exercise, we use the graphing utility to verify a particular case of the following general result:
    Suppose $\overline{PQ}$ is a focal chord of a parabola. Then the tangents at P and Q are perpendicular to each other, and they intersect at a point on the directrix.
    (a) The line segment $\overline{PQ}$ is a focal chord of the parabola $x^2 = 8y$ and the coordinates of P are (8, 8). Find the coordinates of Q. Also find the equation of the directrix.
    (b) Find the equations of the tangent lines at P and Q.
    (c) Graph the parabola, the two tangent lines, and the directrix. If your calculations are correct, the tangent lines will meet at a point on the directrix and they will be perpendicular.

In Exercises 11–14, graph the equations and estimate the coordinates of the intersection points. Also, in Exercises 11 and 12, use the techniques of Section 10.6 to determine the exact coordinates of the intersection points.

11. $x^2 = 5y$; $y^2 = 5x$
12. $(y + 1)^2 = \frac{1}{2}(x - 1)$; $2x + y = 5$
13. $y^2 = 0.25x$; $y = \ln x$
14. $x^2 = \frac{1}{3}y$; $y = e^x$

TANGENTS TO PARABOLAS (OPTIONAL)

Many of the more important properties of the parabola relate to the tangent to the curve. For instance, there is a close connection between the optical property illustrated in Figure 12 on page 688 and the tangents to the parabola. (See Exercise 19 at the end of this section for details.) In general, the techniques of calculus are required to deal with tangents to curves. However, for curves with equations and graphs as simple as the parabola, the methods of algebra are often adequate. The following discussion shows how we can determine the tangent to a parabola without using calculus.

To begin with, we need a definition for the tangent to a parabola. As motivation, recall from geometry that a tangent line to a circle is defined to be a line that intersects the circle in exactly one point (see Figure 1). However, this definition is not quite adequate for the parabola. For instance, the y-axis intersects the parabola $y = x^2$ in exactly one point, but surely it is not tangent to the curve. With this in mind, we adopt the definition in the following box for the tangent to a parabola. Along with this definition, we make the assumption that through each point P on a parabola, there is only one tangent line that can be drawn.

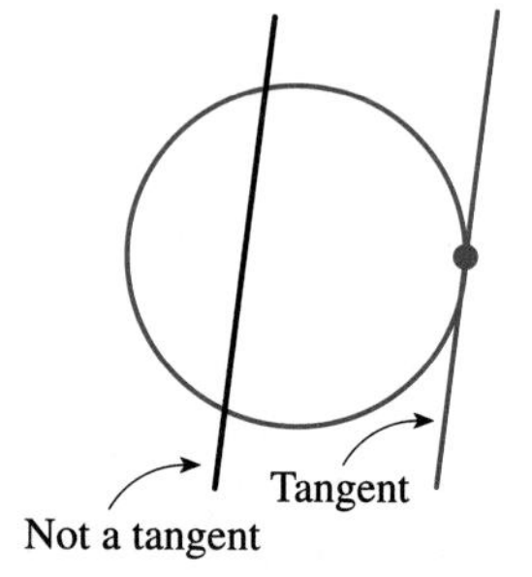

FIGURE 1

DEFINITION: Tangent to a Parabola

Let P be a point on a parabola. Then a line through P is said to be **tangent** to the parabola at P provided that the line intersects the parabola only at P and the line is not parallel to the axis of the parabola.

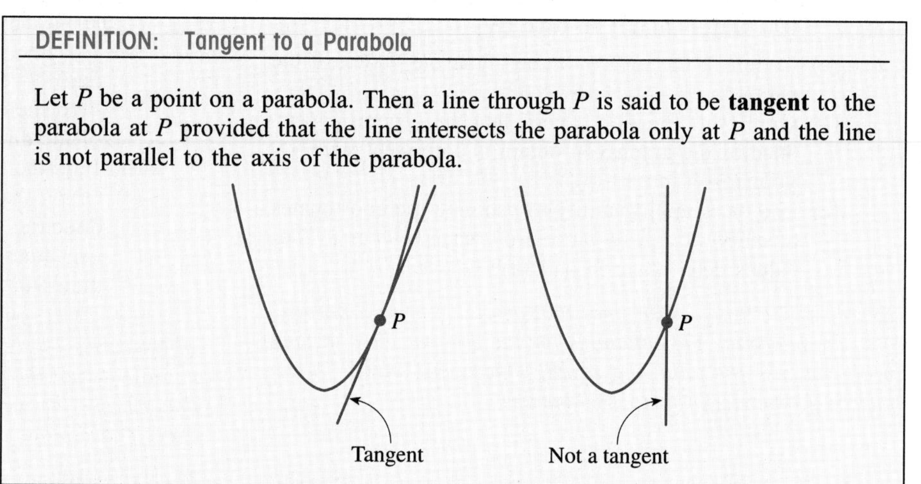

We'll demonstrate how to find tangents to parabolas by means of an example. The method used here can be used for any parabola. Suppose that we want to find the equation of the tangent to the parabola $x^2 = 4y$ at the point (2, 1) on the curve (see Figure 2). Let m denote the slope of the tangent line. Since the line must pass through the point (2, 1), its equation is:

$$y - 1 = m(x - 2)$$

This is the tangent line, so it intersects the parabola in only one point, namely, (2, 1). Algebraically, this means that the ordered pair (2, 1) is the only solution of the following system of equations:

$$\begin{cases} y - 1 = m(x - 2) & (1) \\ x^2 = 4y & (2) \end{cases}$$

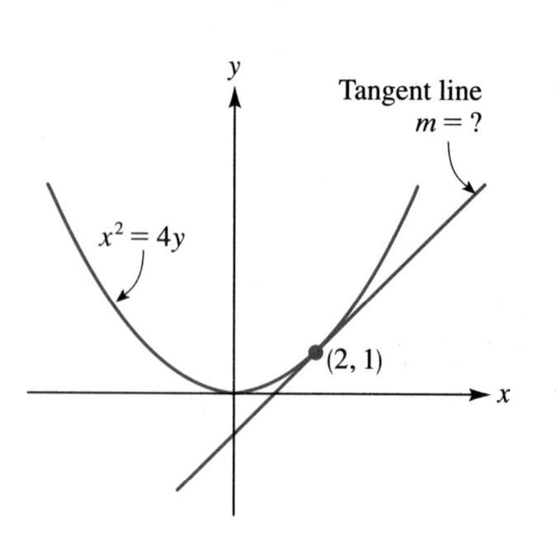

FIGURE 2

The strategy now is to solve this system of equations in terms of m. Then we'll reconcile the results with the fact that (2, 1) is known to be the unique solution. This will allow us to determine m. From equation (2), we have $y = x^2/4$. Then, substituting for y in equation (1), we obtain

$$\frac{x^2}{4} - 1 = m(x - 2)$$

$$x^2 - 4 = 4m(x - 2) \quad \text{multiplying both sides by 4}$$

$$(x - 2)(x + 2) = 4m(x - 2) \quad \text{factoring}$$

$$(x - 2)(x + 2) - 4m(x - 2) = 0 \quad \text{subtracting } 4m(x - 2) \text{ from both sides}$$

Observe now that the quantity $x - 2$ appears twice on the left-hand side of the last equation, so we can factor it out to obtain

$$(x - 2)[(x + 2) - 4m] = 0$$

$$x - 2 = 0 \quad \Big| \quad x + 2 - 4m = 0$$

$$x = 2 \quad \Big| \quad x = 4m - 2$$

Now look at these two x-values that we've just determined. The first value, $x = 2$, yields no new information, since we knew from the start that this was the x-coordinate of the intersection point. However, the second value, $x = 4m - 2$, must also equal 2, since the system has but one solution. Thus we have

$$4m - 2 = 2$$

$$4m = 4$$

$$m = 1$$

Therefore, the slope of the tangent line is 1. Substituting this value of m in equation (1) yields

$$y - 1 = 1(x - 2)$$

or

$$y = x - 1 \quad \text{as required}$$

EXERCISE SET 11.3

A

In Exercises 1–8, use the method shown in the text to find the equation of the tangent to the parabola at the given point. In each case, include a sketch with your answer.

1. $x^2 = y$; $(2, 4)$

2. $x^2 = -2y$; $(2, -2)$

3. $x^2 = 8y$; $(4, 2)$

4. $x^2 = 12y$; $(6, 3)$

5. $x^2 = -y$; $(-3, -9)$

6. $x^2 = -6y$; $(\sqrt{6}, -1)$

7. $y^2 = 4x$; $(1, 2)$

8. $y^2 = -8x$; $(-8, -8)$

B

In Exercises 9–11, find the slope of the tangent to the curve at the indicated point. (Use the method shown in the text for parabolas.)

9. $y = \sqrt{x}$; $(4, 2)$
Hint: $x - 4 = (\sqrt{x} - 2)(\sqrt{x} + 2)$

10. $y = 1/x$; $(3, 1/3)$

11. $y = x^3$; $(2, 8)$

12. Consider the parabola $x^2 = 4y$, and let (x_0, y_0) denote a point on the parabola in the first quadrant.
(a) Find the y-intercept of the line tangent to the parabola at the point on the parabola where $y_0 = 1$.
(b) Repeat part (a) using $y_0 = 2$.
(c) Repeat part (a) using $y_0 = 3$.
(d) On the basis of your results in parts (a)–(c), make a conjecture about the y-intercept of the line that is tangent to the parabola at the point (x_0, y_0) on the curve. Verify your conjecture by computing this y-intercept.

C

13. Let (x_0, y_0) be a point on the parabola $x^2 = 4py$. Using the method explained in the text, show that the equation of the line tangent to the parabola at (x_0, y_0) is

$$y = \frac{x_0}{2p}x - y_0$$

Thus, the y-intercept of the line tangent to $x^2 = 4py$ at (x_0, y_0) is just $-y_0$.

14. Let (x_0, y_0) be a point on the parabola $y^2 = 4px$. Show that the equation of the line tangent to the parabola at (x_0, y_0) is

$$y = \frac{2p}{y_0}x + \frac{y_0}{2}$$

Show that the x-intercept of this line is $-x_0$.

Exercises 15–22 contain results about tangents to parabolas. In working these problems, you'll find it convenient to use the facts developed in Exercises 13 and 14. Also assume, as given, the results about focal chords in Exercises 40 and 41 of Exercise Set 11.2. In some of the problems, reference is made to the normal line. *The* normal line *or* normal *to a parabola at the point (x_0, y_0) on the parabola is defined as the line through (x_0, y_0) that is perpendicular to the tangent at (x_0, y_0).*

15. Let $\overline{PQ}$ be a focal chord of the parabola $x^2 = 4py$.
(a) Show that the tangents to the parabola at P and Q are perpendicular to each other.
(b) Show that the tangents to the parabola at P and Q intersect at a point on the directrix.
(c) Let D be the intersection point of the tangents at P and Q. Show that the line segment from D to the focus is perpendicular to $\overline{PQ}$.

16. Let $P(x_0, y_0)$ be a point [other than (0, 0)] on the parabola $y^2 = 4px$. Let A be the point where the normal line to the parabola at P meets the axis of the parabola. Let B be the point where the line drawn from P perpendicular to the axis of the parabola meets the axis. Show that $AB = 2p$.

17. Let $P(x_0, y_0)$ be a point on the parabola $y^2 = 4px$. Let A be the point where the normal line to the parabola at P meets the axis of the parabola. Let F be the focus of the parabola. If a line is drawn from A perpendicular to $\overline{FP}$, meeting $\overline{FP}$ at Z, show that $ZP = 2p$.

18. Let $\overline{AB}$ be a chord (not necessarily a focal chord) of the parabola $x^2 = 4py$. Let M be the midpoint of the chord and let C be the point where the tangents at A and B intersect.
(a) Show that $\overline{MC}$ is parallel to the axis of the parabola.
(b) If D is the point where the parabola meets $\overline{MC}$, show that $CD = DM$.

19. In this exercise, we prove the *reflection property* of parabolas. The following figure shows a line tangent to the parabola $y^2 = 4px$ at $P(x_0, y_0)$. The dashed line through H and P is parallel to the axis of the parabola. We wish to prove that $\alpha = \beta$. That is, we wish to prove the reflection property: the angle of incidence equals the angle of reflection.

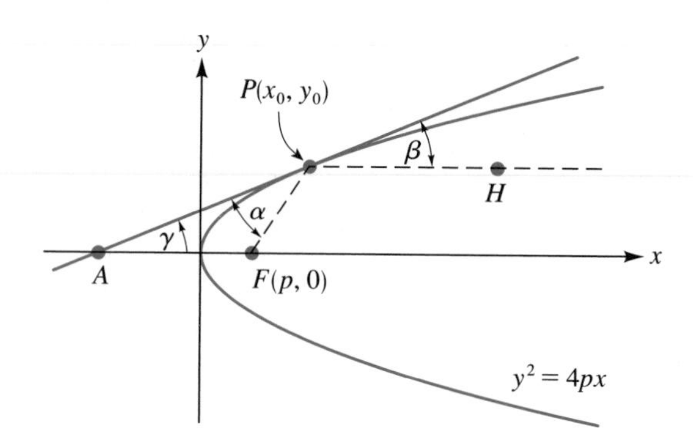

(a) Show that $FA = x_0 + p = FP$. *Hint:* Regarding FP, it is easier to rely on the definition of a parabola than on the distance formula.
(b) Why does $\alpha = \gamma$? Why does $\gamma = \beta$?
(c) Conclude that $\alpha = \beta$, as required.

20. From a point $P(x_0, y_0)$ on the parabola $x^2 = 4py$, a tangent line is drawn meeting the axis of the parabola at A. From the focus F, a line is drawn perpendicular to $\overline{AP}$, meeting $\overline{AP}$ at B. Finally, from P, a line drawn perpendicular to the directrix meets the directrix at C.
(a) Show that B lies on the line that is tangent to the parabola at the vertex.
(b) Show that B is the midpoint of $\overline{FC}$.

21. Verify that the point $A(3, 2)$ lies on the line that is tangent to the parabola $x^2 = 4y$ at $P(4, 4)$. Let F be the focus of the parabola. A perpendicular is drawn from A

to $\overline{FP}$, meeting $\overline{FP}$ at B. Also from A, a perpendicular is drawn to the directrix, meeting the directrix at C. Show that $FB = FC$. (This result is known as *Adams's theorem;* it holds for any parabola and any point on a tangent line.)

22. The segment $\overline{AB}$ is a focal chord of the parabola $y^2 = 4x$ and the coordinates of A are $(4, 4)$. Normals drawn through A and B meet the parabola again at A' and B', respectively. Prove that $\overline{A'B'}$ is three times as long as $\overline{AB}$.

THE ELLIPSE

We have here apparently [in the work of Anthemius of Tralles (a sixth-century Greek architect and mathematician)] *the first mention of the construction of an ellipse by means of a string stretched tight round the foci.*

Sir Thomas Heath in *A History of Greek Mathematics,* Vol. II (Oxford: The Clarendon Press, 1921)

The heavenly motions are nothing but a continuous song for several voices, to be perceived by the intellect, not by the ear.

Johannes Kepler (1571–1630)

The true orbit of Mars was even less of a circle than the Earth's. It took almost two years for Kepler to realize that its orbital shape is that of an ellipse. An ellipse is the shape of a circle when viewed at an angle.

Phillip Flower in *Understanding the Universe* (St. Paul: West Publishing Company, 1990)

In this section we discuss the symmetric, oval-shaped curve known as the *ellipse.* As Kepler discovered and Newton later proved, this is the curve described by the planets in their motions around the sun.

> **DEFINITION: The Ellipse**
>
> An **ellipse** is the set of all points in the plane, the sum of whose distances from two fixed points is constant. Each fixed point is called a **focus** (plural: **foci**).

Subsequently we will derive an equation describing the ellipse just as we found an equation for the parabola in Section 11.2. But first let's consider some rather immediate consequences of the definition. In fact, we can learn a great deal about the ellipse even before we derive its equation.

There is a simple mechanical method for constructing an ellipse that arises directly from the definition of the curve. Mark the given foci, say F_1 and F_2, on a drawing board and insert thumbtacks at those points. Now take a piece of string that is longer than the distance from F_1 to F_2, and tie the ends of the string to the tacks. Next, pull the string taut with a pencil point and touch the pencil point to the drawing board. Then if you move the pencil while keeping the string taut, the curve traced out will be an ellipse, as indicated in Figure 1. The reason the curve is an ellipse is that for each point on the curve, the sum of the distances from the foci is constant, the constant being the length of the string. By actually carrying out this construction for yourself several times, each time varying the distance between the foci or the length of the string, you can learn a great deal about the ellipse. For instance, when the distance between the foci is small compared to the length of the string, the ellipse begins to resemble

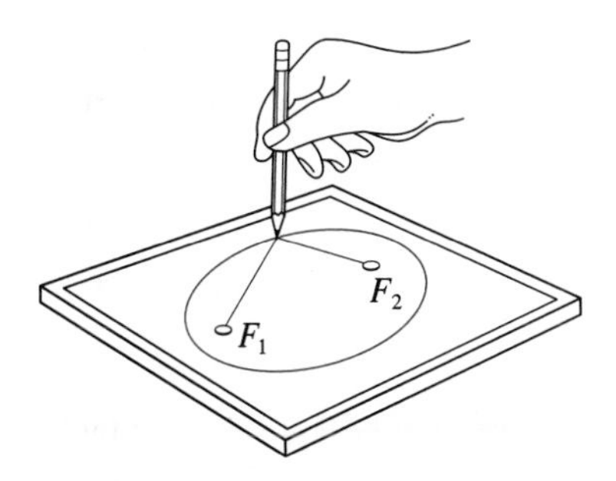

FIGURE 1

a circle, as in Figure 2. On the other hand, when the distance between the foci is nearly equal to the length of the string, the ellipse becomes relatively flat, as in Figure 3.

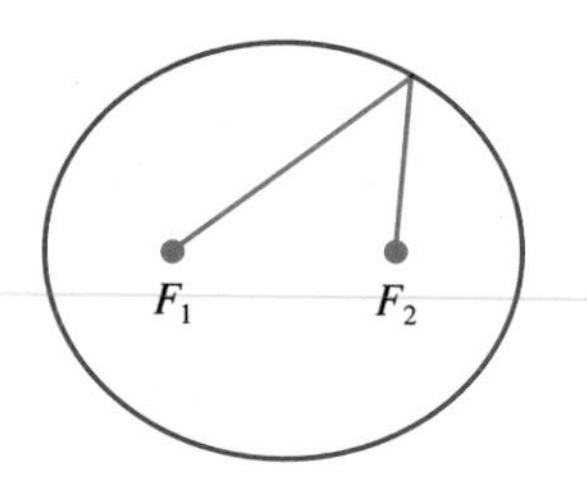

FIGURE 2
When the distance between the foci is small compared to the length of the string, the ellipse resembles a circle.

FIGURE 3
When the distance between the foci is nearly equal to the length of the string, the ellipse is relatively flat.

We now derive one of the standard forms for the equation of an ellipse. As indicated in Figure 4, we assume that the foci are $F_1(-c, 0)$ and $F_2(c, 0)$ and that the sum of the distances from the foci to a point $P(x, y)$ on the ellipse is $2a$. Since the point P lies on the ellipse, we have

$$F_1P + F_2P = 2a$$

and therefore

$$\sqrt{(x+c)^2 + (y-0)^2} + \sqrt{(x-c)^2 + (y-0)^2} = 2a$$

or

$$\sqrt{(x-c)^2 + y^2} = 2a - \sqrt{(x+c)^2 + y^2} \tag{1}$$

Now, by following a straightforward but lengthy process of squaring and simplifying (as outlined in detail in Exercise 40), we find that equation (1) becomes

$$(a^2 - c^2)x^2 + a^2y^2 = a^2(a^2 - c^2) \tag{2}$$

To write equation (2) in a more symmetric form, we define the positive number b by the equation

$$b^2 = a^2 - c^2 \tag{3}$$

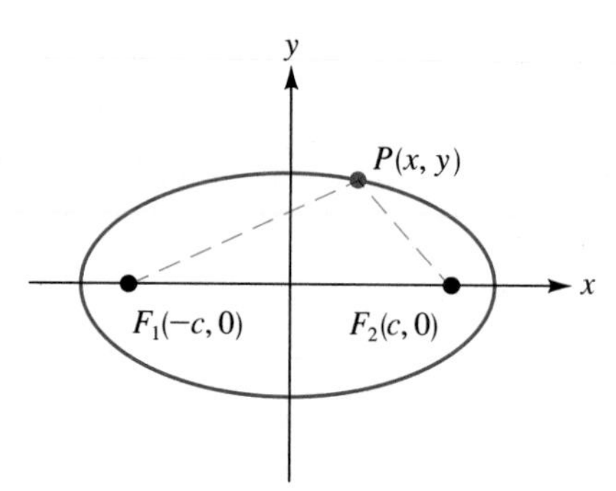

FIGURE 4
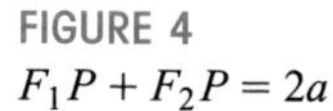
$F_1P + F_2P = 2a$

NOTE For this definition to make sense, we need to know that the right-hand side of equation (3) is positive. (See Exercise 41 at the end of this section for details.)

Finally, using equation (3) to substitute for $a^2 - c^2$ in equation (2), we obtain

$$b^2x^2 + a^2y^2 = a^2b^2$$

which can be written

$$\frac{x^2}{a^2} + \frac{y^2}{b^2} = 1 \qquad (a > b) \tag{4}$$

We have now shown that the coordinates of each point on the ellipse satisfy equation (4). Conversely, it can be shown that if x and y satisfy equation (4), then the point (x, y) indeed lies on the ellipse. We refer to equation (4) as the **standard form** for the equation of an ellipse with foci $(-c, 0)$ and $(c, 0)$. For an ellipse in this form, it will always be the case that a is greater than b; this follows from equation (3).

For purposes of graphing, we want to know the intercepts of the ellipse. To find the x-intercepts, we set $y = 0$ in equation (4) to obtain

$$\frac{x^2}{a^2} = 1$$

$$x^2 = a^2 \quad \text{or} \quad x = \pm a$$

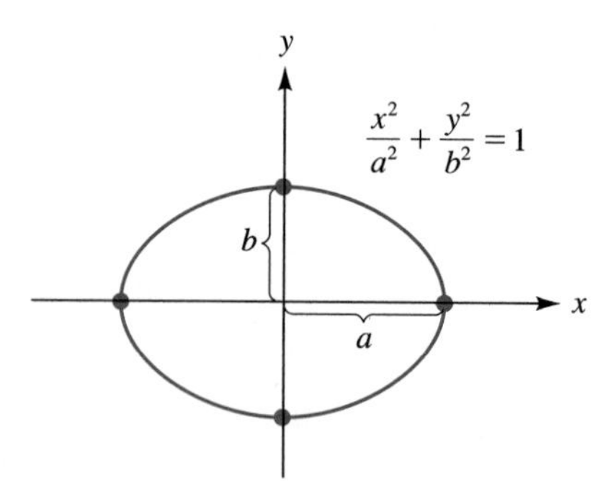

FIGURE 5
The intercepts of the ellipse $(x^2/a^2) + (y^2/b^2) = 1$ are $x = \pm a$ and $y = \pm b$.

The x-intercepts are therefore a and $-a$. In a similar fashion, you can check that the y-intercepts are b and $-b$. Also (according to the symmetry tests in Section 2.4), note that the graph of equation (4) must be symmetric about both coordinate axes. Figure 5 shows the graph of the ellipse $(x^2/a^2) + (y^2/b^2) = 1$.

(Calculator exercises at the end of this section will help to convince you that the general shape of the curve in Figure 5 is correct.)

In the box that follows, we define several terms that are useful in describing and analyzing the ellipse.

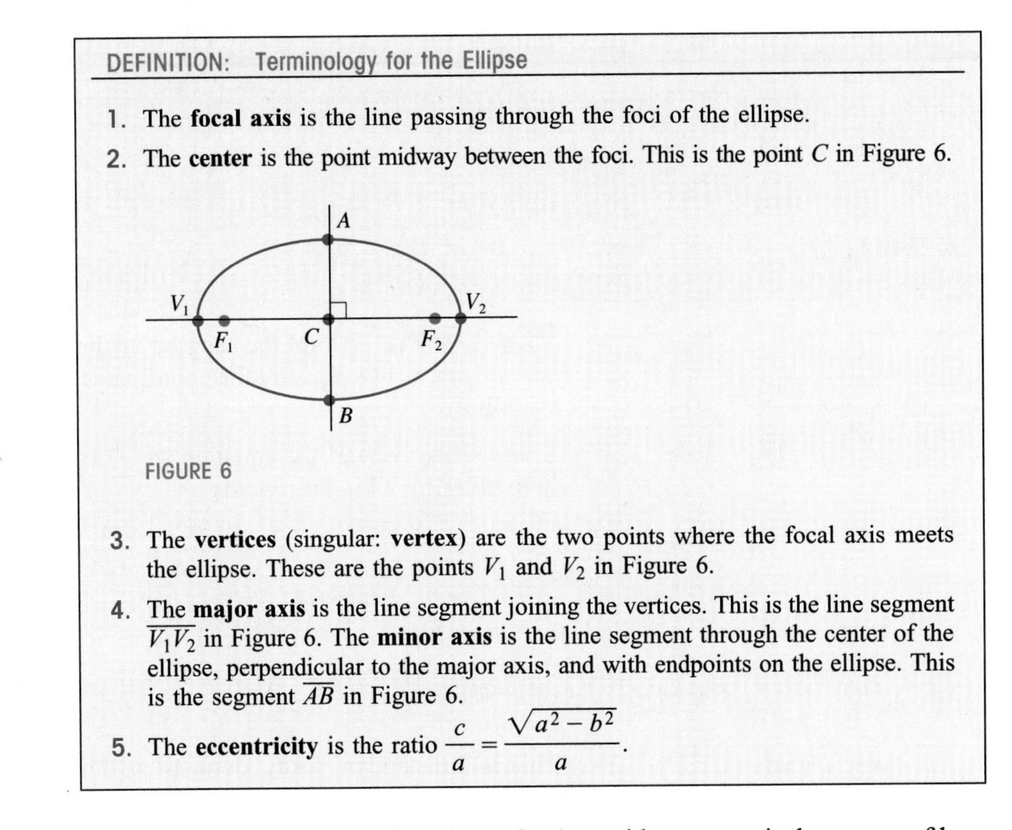

DEFINITION: Terminology for the Ellipse

1. The **focal axis** is the line passing through the foci of the ellipse.
2. The **center** is the point midway between the foci. This is the point C in Figure 6.

FIGURE 6

3. The **vertices** (singular: **vertex**) are the two points where the focal axis meets the ellipse. These are the points V_1 and V_2 in Figure 6.
4. The **major axis** is the line segment joining the vertices. This is the line segment $\overline{V_1V_2}$ in Figure 6. The **minor axis** is the line segment through the center of the ellipse, perpendicular to the major axis, and with endpoints on the ellipse. This is the segment $\overline{AB}$ in Figure 6.
5. The **eccentricity** is the ratio $\dfrac{c}{a} = \dfrac{\sqrt{a^2 - b^2}}{a}$.

The eccentricity (as defined in the box) provides a numerical measure of how much the ellipse deviates from being a circle. As Figure 7 indicates, the closer the eccentricity is to zero, the more the ellipse resembles a circle. In the other direction, as the eccentricity approaches 1, the ellipse becomes increasingly flat.

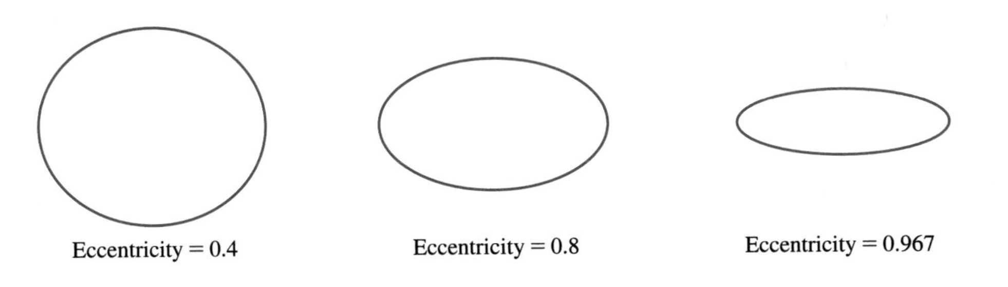

FIGURE 7
Eccentricity is a number between 0 and 1 that describes the shape of an ellipse. The narrowest ellipse in this figure has the same proportions as the orbit of Halley's comet. By way of contrast, the eccentricity of Earth's orbit is 0.0017; if an ellipse with this eccentricity were included in Figure 7, it would appear indistinguishable from a circle.

In the next box, we summarize our discussion up to this point. (We use the letter e to denote the eccentricity; this is the conventional choice, even though the same letter is used with a very different meaning in connection with exponential functions and logarithms.)

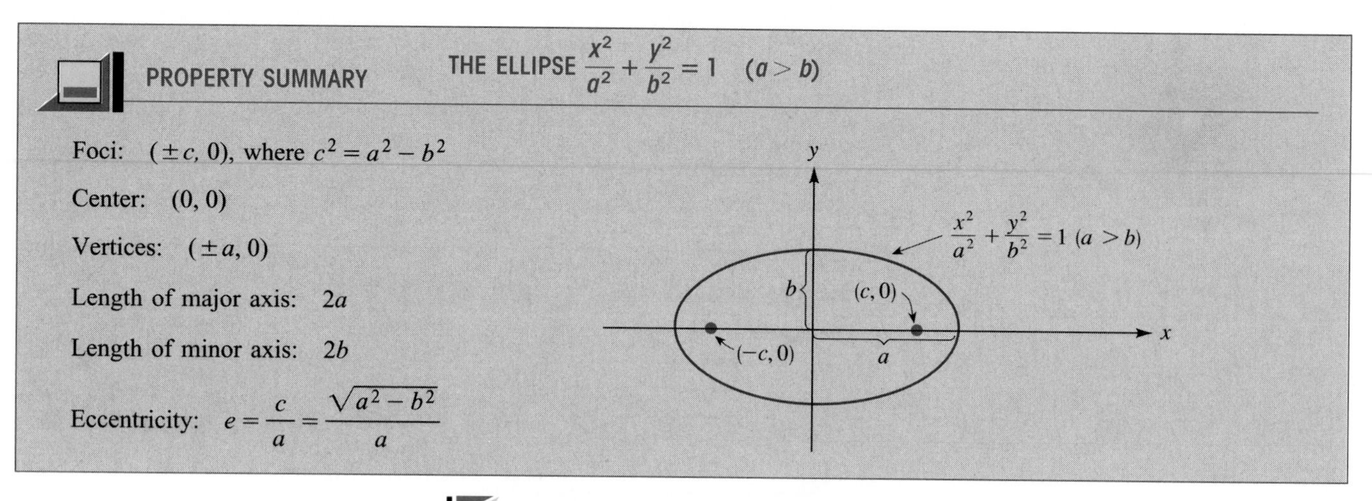

PROPERTY SUMMARY THE ELLIPSE $\frac{x^2}{a^2} + \frac{y^2}{b^2} = 1 \quad (a > b)$

Foci: $(\pm c, 0)$, where $c^2 = a^2 - b^2$

Center: $(0, 0)$

Vertices: $(\pm a, 0)$

Length of major axis: $2a$

Length of minor axis: $2b$

Eccentricity: $e = \frac{c}{a} = \frac{\sqrt{a^2 - b^2}}{a}$

EXAMPLE 1 Find the lengths of the major and minor axes of the ellipse $9x^2 + 16y^2 = 144$. Specify the coordinates of the foci and the eccentricity. Graph the ellipse.

Solution To convert the equation $9x^2 + 16y^2 = 144$ to standard form, we divide both sides by 144. This yields

$$\frac{9x^2}{144} + \frac{16y^2}{144} = \frac{144}{144}$$

$$\frac{x^2}{16} + \frac{y^2}{9} = 1$$

$$\frac{x^2}{4^2} + \frac{y^2}{3^2} = 1$$

This is the standard form. Comparing this equation with $(x^2/a^2) + (y^2/b^2) = 1$, we see that $a = 4$ and $b = 3$. Thus, the major and minor axes are 8 and 6 units, respectively. Next, to determine the foci, we use the equation $c^2 = a^2 - b^2$. We have

$$c^2 = 4^2 - 3^2 = 7 \qquad \text{and therefore} \qquad c = \sqrt{7}$$

(We choose the positive square root because $c > 0$.) It follows that the coordinates of the foci are $(-\sqrt{7}, 0)$ and $(\sqrt{7}, 0)$. We now calculate the eccentricity using the formula $e = c/a$. This yields $e = \sqrt{7}/4$. Figure 8 shows the graph along with the required information.

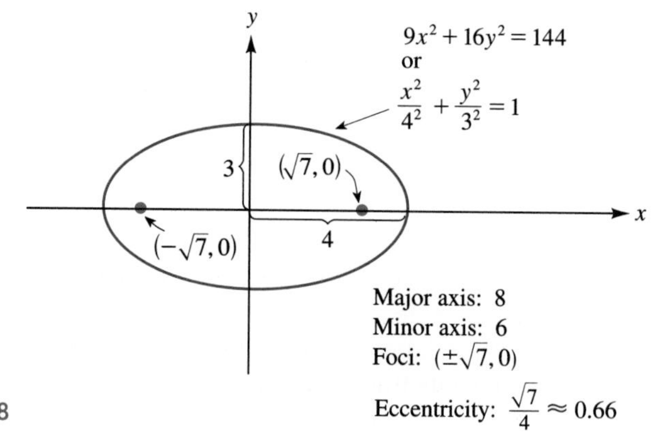

FIGURE 8

EXAMPLE 2 The foci of an ellipse are $(-1, 0)$ and $(1, 0)$, and the eccentricity is $1/3$. Find the equation of the ellipse (in standard form), and specify the lengths of the major and minor axes.

Solution Since the foci are $(\pm 1, 0)$, we have $c = 1$. Using the equation $e = c/a$ with $e = 1/3$ and $c = 1$, we obtain

$$\frac{1}{3} = \frac{1}{a} \qquad \text{and therefore} \qquad a = 3$$

Recall now that b^2 is defined by the equation $b^2 = a^2 - c^2$. In view of this, we have

$$b^2 = 3^2 - 1^2 = 8$$
$$b = \sqrt{8} = 2\sqrt{2}$$

The equation of the ellipse in standard form is therefore

$$\frac{x^2}{3^2} + \frac{y^2}{(2\sqrt{2})^2} = 1$$

Furthermore, since $a = 3$ and $b = 2\sqrt{2}$, the lengths of the major and minor axes are 6 and $4\sqrt{2}$ units, respectively.

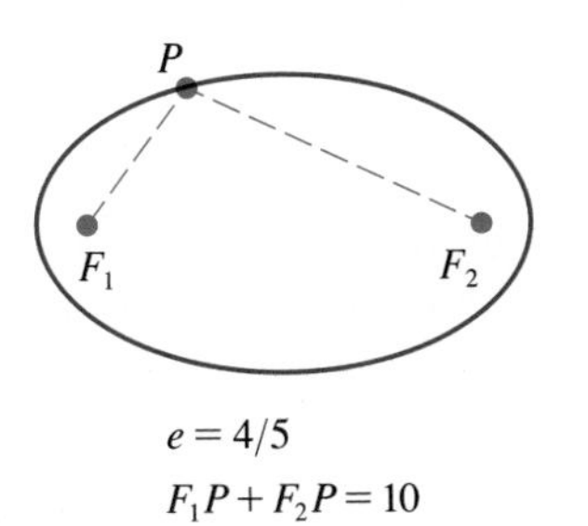

$e = 4/5$
$F_1P + F_2P = 10$

FIGURE 9

EXAMPLE 3 The eccentricity of an ellipse is $4/5$, and the sum of the distances from a point P on the ellipse to the foci is 10 units. Compute the distance between the foci F_1 and F_2. See Figure 9.

Solution We are required to find the distance between the foci F_1 and F_2. By definition, this is the quantity $2c$. Since the sum of the distances from a point on the ellipse to the foci is 10 units, we have

$$2a = 10 \qquad \text{by definition of } 2a$$
$$a = 5$$

Now we substitute the values $a = 5$ and $e = 4/5$ in the formula $e = c/a$:

$$\frac{4}{5} = \frac{c}{5} \qquad \text{and therefore} \qquad c = 4$$

It now follows that the required distance $2c$ is 8 units.

Suppose now that we translate the graph of $(x^2/a^2) + (y^2/b^2) = 1$ by h units in the x-direction and k units in the y-direction, as shown in Figure 10. Then the equation of the translated ellipse is

$$\frac{(x-h)^2}{a^2} + \frac{(y-k)^2}{b^2} = 1 \tag{5}$$

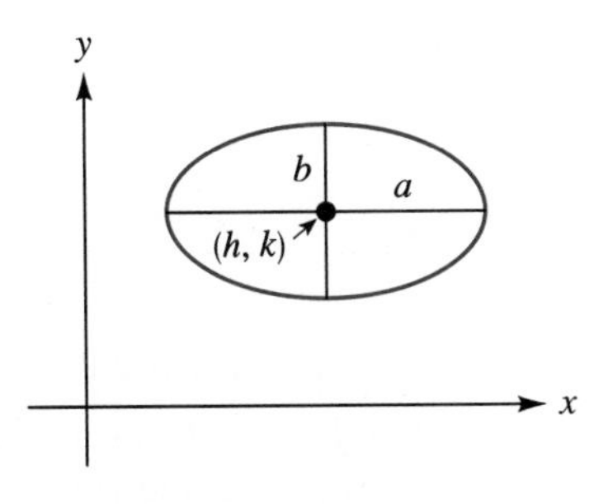

FIGURE 10

Equation (5) is another **standard form** for the equation of an ellipse. As indicated in Figure 10, the center of this ellipse is the point (h, k). In the next example, we use the technique of completing the square to convert an equation for an ellipse to standard form. Once the equation is in standard form, the graph is readily obtained.

EXAMPLE 4 Determine the center, foci, and eccentricity of the ellipse

$$4x^2 + 9y^2 - 8x - 54y + 49 = 0$$

Graph the ellipse.

Solution We will convert the given equation to standard form by using the technique of completing the square. We have

$$4x^2 - 8x + 9y^2 - 54y = -49$$
$$4(x^2 - 2x) + 9(y^2 - 6y) = -49$$
$$4(x^2 - 2x + 1) + 9(y^2 - 6y + 9) = -49 + 4(1) + 9(9)$$
$$4(x-1)^2 + 9(y-3)^2 = 36$$

Dividing this last equation by 36, we obtain

$$\frac{(x-1)^2}{9} + \frac{(y-3)^2}{4} = 1$$

or

$$\frac{(x-1)^2}{3^2} + \frac{(y-3)^2}{2^2} = 1$$

This last equation represents an ellipse with center at (1, 3) and with $a = 3$ and $b = 2$. We can calculate c using the formula $c^2 = a^2 - b^2$:

$$c^2 = 3^2 - 2^2 = 5 \qquad \text{and therefore} \qquad c = \sqrt{5}$$

Since the center of this ellipse is (1, 3), the foci are therefore $(1 + \sqrt{5}, 3)$ and $(1 - \sqrt{5}, 3)$. Finally, the eccentricity is c/a, which is $\sqrt{5}/3$. Figure 11 shows the graph of this ellipse.

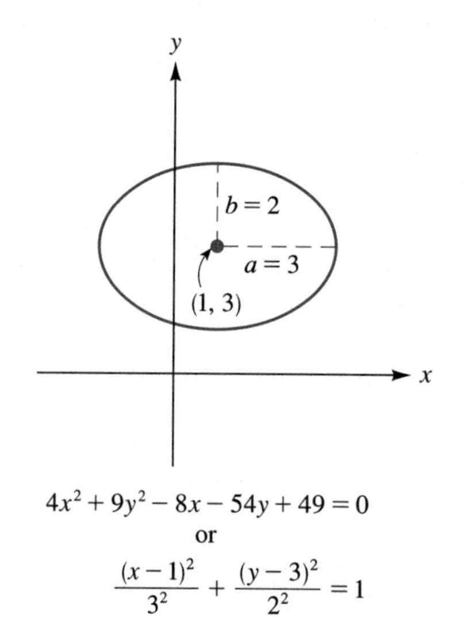

$$4x^2 + 9y^2 - 8x - 54y + 49 = 0$$
or
$$\frac{(x-1)^2}{3^2} + \frac{(y-3)^2}{2^2} = 1$$

FIGURE 11

In developing the equation $(x^2/a^2) + (y^2/b^2) = 1$, we assumed that the foci of the ellipse were located on the x-axis at the points $(-c, 0)$ and $(c, 0)$. If instead the foci are located on the y-axis at the points $(0, c)$ and $(0, -c)$, then the same method we used in the previous case will show the equation of the ellipse to be

$$\frac{x^2}{b^2} + \frac{y^2}{a^2} = 1 \qquad (a > b)$$

We still assume that $2a$ represents the sum of the distances from a point on the ellipse to the foci. In the box that follows, we summarize the situation for the ellipse with foci $(0, c)$ and $(0, -c)$.

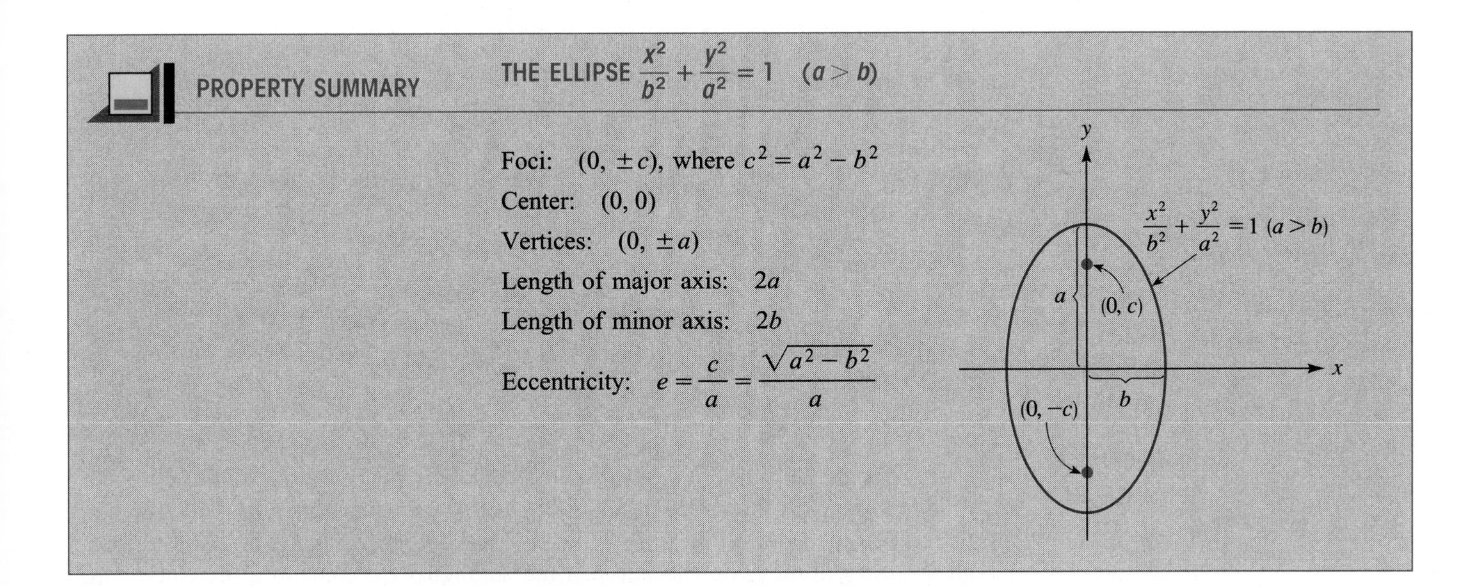

PROPERTY SUMMARY

THE ELLIPSE $\frac{x^2}{b^2} + \frac{y^2}{a^2} = 1 \quad (a > b)$

Foci: $(0, \pm c)$, where $c^2 = a^2 - b^2$

Center: $(0, 0)$

Vertices: $(0, \pm a)$

Length of major axis: $2a$

Length of minor axis: $2b$

Eccentricity: $e = \frac{c}{a} = \frac{\sqrt{a^2 - b^2}}{a}$

We now have two standard forms for the ellipse:

$$\text{Foci on the } x\text{-axis at } (\pm c, 0): \quad \frac{x^2}{a^2} + \frac{y^2}{b^2} = 1 \qquad (a > b)$$

$$\text{Foci on the } y\text{-axis at } (0, \pm c): \quad \frac{x^2}{b^2} + \frac{y^2}{a^2} = 1 \qquad (a > b)$$

Because a is greater than b in both standard forms, it is always easy to determine by inspection whether the foci lie on the x-axis or the y-axis. Consider, for instance, the equation $(x^2/5^2) + (y^2/6^2) = 1$. In this case, since $6 > 5$, we have $a = 6$. And since 6^2 appears under y^2, we conclude that the foci lie on the y-axis.

EXAMPLE 5 The point $(5, 3)$ lies on an ellipse with vertices $(0, \pm 2\sqrt{21})$. Find the equation of the ellipse. Write the answer both in standard form and in the form $Ax^2 + By^2 = C$.

Solution Since the vertices are $(0, \pm 2\sqrt{21})$, the standard form for the equation in this case is $(x^2/b^2) + (y^2/a^2) = 1$. Furthermore, in view of the coordinates of the vertices, we have $a = 2\sqrt{21}$. Therefore,

$$\frac{x^2}{b^2} + \frac{y^2}{84} = 1$$

Now, since the point (5, 3) lies on the ellipse, its coordinates must satisfy this last equation. We thus have

$$\frac{5^2}{b^2} + \frac{3^2}{84} = 1$$

$$\frac{25}{b^2} + \frac{3}{28} = 1$$

$$\frac{25}{b^2} = \frac{25}{28}$$

From this last equation, we see that $b^2 = 28$ and therefore

$$b = \sqrt{28} = 2\sqrt{7}$$

Now that we've determined a and b, we can write the equation of the ellipse in standard form. It is

$$\frac{x^2}{(2\sqrt{7})^2} + \frac{y^2}{(2\sqrt{21})^2} = 1$$

As you should verify for yourself, this equation can also be written in the equivalent form

$$3x^2 + y^2 = 84$$

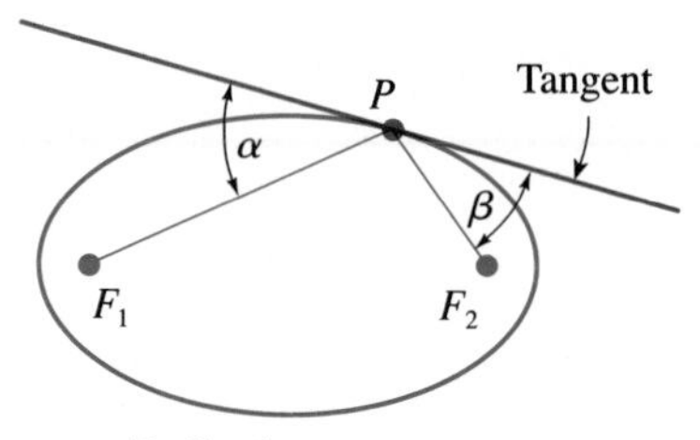

Reflection property of the ellipse: $\alpha = \beta$

FIGURE 12

As with the parabola, many of the interesting properties of the ellipse are related to tangent lines. We define a **tangent to an ellipse** as a line that intersects the ellipse in exactly one point. Figure 12 shows a line tangent to an ellipse at an arbitrary point P on the curve. Line segments $\overline{F_1P}$ and $\overline{F_2P}$ are drawn from the foci to the point of tangency. These two segments are called **focal radii.** One of the most basic properties of the ellipse is the **reflection property.** According to this property, the focal radii drawn to the point of tangency make equal angles with the tangent. (The steps for proving this fact are outlined in detail in Exercise 65 at the end of this section. For a similar result concerning the parabola, see Exercise 19 in the previous section.)

If $P(x_1, y_1)$ is a point on the ellipse $(x^2/a^2) + (y^2/b^2) = 1$, then the equation of the line tangent to the ellipse at $P(x_1, y_1)$ is

$$\frac{x_1x}{a^2} + \frac{y_1y}{b^2} = 1$$

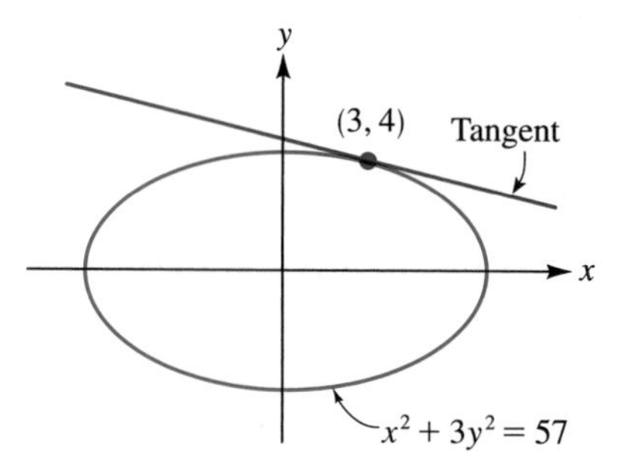

FIGURE 13

This equation is easy to remember because it so closely resembles the equation of the ellipse itself. However, notice that the equation $(x_1x/a^2) + (y_1y/b^2) = 1$ is indeed linear, since x_1, y_1, a, and b all denote constants. This equation for the tangent to an ellipse can be derived using the same technique we employed with the parabola in the previous section. (See Exercise 64 at the end of this section for the details.) The two examples that follow show how this equation is used.

EXAMPLE 6 Find the equation of the line that is tangent to the ellipse $x^2 + 3y^2 = 57$ at the point (3, 4) on the ellipse; see Figure 13. Write the answer in the form $y = mx + b$.

Solution We know that the equation of the line tangent to the ellipse at the point (x_1, y_1) is $(x_1x/a^2)+(y_1y/b^2)=1$. We are given that $x_1=3$ and $y_1=4$. Thus we need to determine a^2 and b^2. To do that, we convert the equation $x^2+3y^2=57$ to standard form by dividing through by 57. This yields

$$\frac{x^2}{57}+\frac{y^2}{19}=1$$

It follows that $a^2=57$ and $b^2=19$. The equation of the tangent line then becomes

$$\frac{3x}{57}+\frac{4y}{19}=1$$

$$\frac{x}{19}+\frac{4y}{19}=1$$

$$x+4y=19$$

Solving for y, we obtain

$$y=-\frac{1}{4}x+\frac{19}{4}$$

EXAMPLE 7 Show that the slope of the tangent line in Figure 14 is equal to the eccentricity of the ellipse, $e=c/a$.

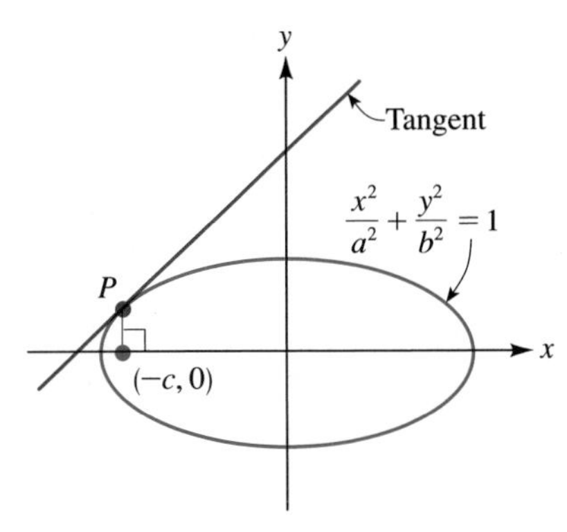

FIGURE 14

Solution We first determine the coordinates of the point P in Figure 14. Evidently, the x-coordinate of P is $-c$. Substituting this value for x in the equation of the ellipse yields

$$\frac{c^2}{a^2}+\frac{y^2}{b^2}=1$$

To solve this equation for y, let's first clear the equation of fractions by multiplying by a^2b^2. This yields

$$b^2c^2+a^2y^2=a^2b^2$$

$$a^2y^2=a^2b^2-b^2c^2=b^2(a^2-c^2)$$

$$=b^2(b^2) \qquad \text{by definition of } b^2$$

$$=b^4$$

$$y^2=\frac{b^4}{a^2}$$

$$y=\frac{b^2}{a}$$

The coordinates of P are therefore $(-c, b^2/a)$. Substituting these coordinates for x_1 and y_1 in the equation $(x_1x/a^2)+(y_1y/b^2)=1$, we obtain

$$-\frac{cx}{a^2}+\frac{(b^2/a)y}{b^2}=1$$

$$-\frac{cx}{a^2}+\frac{y}{a}=1$$

Therefore,

$$\frac{y}{a} = \frac{c}{a^2}x + 1 \qquad \text{and so} \qquad y = \frac{c}{a}x + a$$

The slope of the line is c/a, the coefficient of x. This is what we wanted to show. ◢

We conclude this section by mentioning a few of the applications of the ellipse. Each planet in the solar system moves in an elliptical orbit with the sun at one focus. Some gears in machines are elliptical rather than circular. (In certain brands of racing bikes, this is true of one of the gears in front.) As with the parabola, the reflection property of the ellipse has applications in optics and acoustics. As indicated in Figure 15, a light ray or sound emitted from one focus of an ellipse is always reflected through the other focus. This property is used in the design of "whispering galleries." In these rooms (with elliptical cross sections), a person standing at focus F_2 can hear a whisper from focus F_1 while others closer to F_1 might hear nothing. Statuary Hall in the Capitol building in Washington, D.C., is a whispering gallery. This idea is used in a modern medical device known as the *lithotripter,* in which high-energy sound waves are used to break up kidney stones. The patient is positioned with the kidney stone at one focus in an elliptical water bath while sound waves are emitted from the other focus.

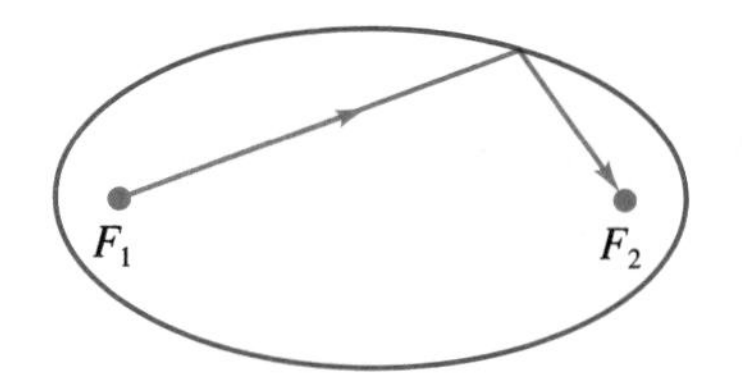

FIGURE 15
A consequence of the reflection property of the ellipse. A light ray or sound emitted from one focus is reflected through the other focus.

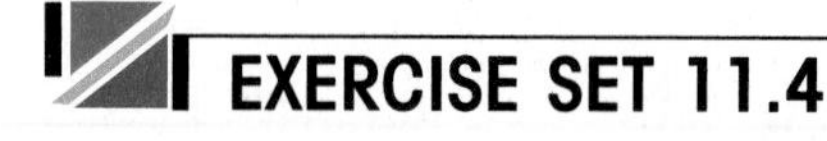

EXERCISE SET 11.4

A

In Exercises 1–24, graph the ellipses. In each case, specify the lengths of the major and minor axes, the foci, and the eccentricity. For Exercises 13–24, also specify the center of the ellipse.

1. $4x^2 + 9y^2 = 36$
2. $4x^2 + 25y^2 = 100$
3. $x^2 + 16y^2 = 16$
4. $9x^2 + 25y^2 = 225$
5. $x^2 + 2y^2 = 2$
6. $2x^2 + 3y^2 = 3$
7. $16x^2 + 9y^2 = 144$
8. $25x^2 + y^2 = 25$
9. $15x^2 + 3y^2 = 5$
10. $9x^2 + y^2 = 4$
11. $2x^2 + y^2 = 4$
12. $36x^2 + 25y^2 = 400$
13. $\dfrac{(x-5)^2}{5^2} + \dfrac{(y+1)^2}{3^2} = 1$
14. $\dfrac{(x-1)^2}{2^2} + \dfrac{(y+4)^2}{3^2} = 1$
15. $\dfrac{(x-1)^2}{1^2} + \dfrac{(y-2)^2}{2^2} = 1$
16. $\dfrac{x^2}{4^2} + \dfrac{(y-3)^2}{2^2} = 1$
17. $\dfrac{(x+3)^2}{3^2} + \dfrac{y^2}{1^2} = 1$
18. $\dfrac{(x-2)^2}{2^2} + \dfrac{(y-2)^2}{2^2} = 1$
19. $3x^2 + 4y^2 - 6x + 16y + 7 = 0$
20. $16x^2 + 64x + 9y^2 - 54y + 1 = 0$
21. $5x^2 + 3y^2 - 40x - 36y + 188 = 0$
22. $x^2 + 16y^2 - 160y + 384 = 0$
23. $16x^2 + 25y^2 - 64x - 100y + 564 = 0$
24. $4x^2 + 4y^2 - 32x + 32y + 127 = 0$

In Exercises 25–32, find the equation of the ellipse satisfying the given conditions. Write the answer both in standard form and in the form $Ax^2 + By^2 = C$.

25. Foci $(\pm 3, 0)$; vertices $(\pm 5, 0)$
26. Foci $(0, \pm 1)$; vertices $(0, \pm 4)$
27. Vertices $(\pm 4, 0)$; eccentricity $1/4$
28. Foci $(0, \pm 2)$; endpoints of the minor axis $(\pm 5, 0)$
29. Foci $(0, \pm 2)$; endpoints of the major axis $(0, \pm 5)$
30. Endpoints of the major axis $(\pm 10, 0)$; endpoints of the minor axes $(0, \pm 4)$
31. Center at the origin; vertices on the x-axis; length of major axis twice the length of minor axis; $(1, \sqrt{2})$ lies on the ellipse

32. Eccentricity 3/5; one endpoint of the minor axis $(-8, 0)$; center at the origin

33. Find the equation of the tangent to the ellipse $x^2 + 3y^2 = 76$ at each of the given points. Write your answers in the form $y = mx + b$.
(a) $(8, 2)$ **(b)** $(-7, 3)$ **(c)** $(1, -5)$

34. **(a)** Find the equation of the line tangent to the ellipse $x^2 + 3y^2 = 84$ at the point $(3, 5)$ on the ellipse. Write your answer in the form $y = mx + b$.
(b) Repeat part (a), but at the point $(-3, -5)$ on the ellipse.
(c) Are the lines determined in (a) and (b) parallel?

35. A line is drawn tangent to the ellipse $x^2 + 3y^2 = 52$ at the point $(2, 4)$ on the ellipse.
(a) Find the equation of this tangent line.
(b) Find the area of the first-quadrant triangle bounded by the axes and this tangent line.

36. Tangent lines are drawn to the ellipse $x^2 + 3y^2 = 12$ at the points $(3, -1)$ and $(-3, -1)$ on the ellipse.
(a) Find the equation of each tangent line. Write your answers in the form $y = mx + b$.
(b) Find the point where the tangent lines intersect.

37. Let F_1 and F_2 be the foci of the ellipse $9x^2 + 25y^2 = 225$. Let P be the point $(1, 6\sqrt{6}/5)$.
(a) Show that P lies on the ellipse.
(b) Find the equation of the line that is tangent to the ellipse at P. Write your answer in the form $Ax + By + C = 0$.
(c) Let d_1 denote the distance from F_1 to the tangent line determined in part (b). Similarly let d_2 denote the distance from F_2 to this tangent line. Compute d_1 and d_2.
(d) Verify that $d_1 d_2 = b^2$, where b is half the length of the minor axis.

38. Solve the equation $(x^2/3^2) + (y^2/2^2) = 1$ for y to obtain $y = \pm\frac{1}{3}\sqrt{36 - 4x^2}$. Then complete the following table and use the results to graph the given equation. (Use the fact that the graph must be symmetric about the y-axis.)

x	0	0.5	1.0	1.5	2.0	2.5	3
y	± 2						0

39. Solve the equation $(x^2/1^2) + (y^2/4^2) = 1$ for y to obtain $y = \pm 4\sqrt{1 - x^2}$. Then complete the following tables and use the results to graph the given equation. (Use the fact that the graph must be symmetric about the y-axis.)

x	0	0.1	0.2	0.3	0.4	0.5
y	± 4					

x	0.6	0.7	0.8	0.9	1.0
y					0

B

40. This exercise outlines the steps needed to complete the derivation of the equation $(x^2/a^2) + (y^2/b^2) = 1$.
(a) Square both sides of equation (1) on page 696. After simplifying, you should obtain
$$a\sqrt{(x + c)^2 + y^2} = a^2 + xc$$
(b) Square both sides of the equation in part (a). Show that the result can be written
$$a^2x^2 - x^2c^2 + a^2y^2 = a^4 - a^2c^2$$
(c) Verify that the equation in part (b) is equivalent to
$$(a^2 - c^2)x^2 + a^2y^2 = a^2(a^2 - c^2)$$
(d) Using the equation in part (c), replace the quantity $a^2 - c^2$ with b^2. Then show that the resulting equation can be rewritten $(x^2/a^2) + (y^2/b^2) = 1$, as required.

41. In the text we defined the positive number b^2 by the equation $b^2 = a^2 - c^2$. For this definition to make sense, we need to show that the quantity $a^2 - c^2$ is positive. This can be done as follows. First, recall that in any triangle, the sum of the lengths of two sides is always greater than the length of the third side. Now apply this fact to triangle F_1PF_2 in Figure 4 to show that $2a > 2c$. Conclude from this that $a^2 - c^2 > 0$, as required.

42. A line is drawn tangent to the ellipse $(x^2/a^2) + (y^2/b^2) = 1$ at the point (x_1, y_1) on the ellipse. Let P and Q denote the points where the tangent meets the y- and x-axes, respectively. Show that the midpoint of PQ is $(a^2/2x_1, b^2/2y_1)$.

43. A *normal* to an ellipse is a line drawn perpendicular to the tangent at the point of tangency. Show that the equation of the normal to the ellipse $(x^2/a^2) + (y^2/b^2) = 1$ at the point (x_1, y_1) can be written
$$a^2y_1x - b^2x_1y = (a^2 - b^2)x_1y_1$$

44. Let (x_1, y_1) be any point on the ellipse $(x^2/a^2) + (y^2/b^2) = 1$ $(a > b)$ other than one of the endpoints of the major or minor axis. Show that the normal at (x_1, y_1) does not pass through the origin. *Hint:* Find the y-intercept of the normal.

45. Find the points of intersection of the ellipses

$$\frac{x^2}{a^2}+\frac{y^2}{b^2}=1 \quad \text{and} \quad \frac{x^2}{b^2}+\frac{y^2}{a^2}=1 \quad (a>b)$$

Include a sketch with your answer.

46. Let F_1 and F_2 denote the foci of the ellipse $(x^2/a^2)+(y^2/b^2)=1$ $(a>b)$. Suppose that P is one of the endpoints of the minor axis and angle F_1PF_2 is a right angle. Compute the eccentricity of the ellipse.

47. Let $P(x_1, y_1)$ be a point on the ellipse $(x^2/a^2)+(y^2/b^2)=1$. Let N be the point where the normal through P meets the x-axis, and let F be the focus $(-c, 0)$. Show that $FN/FP=e$, where e denotes the eccentricity.

48. Let $P(x_1, y_1)$ be a point on the ellipse $b^2x^2+a^2y^2=a^2b^2$. Suppose that the tangent to the ellipse at P meets the y-axis at A and the x-axis at B. If $AP=PB$, what are x_1 and y_1 (in terms of a and b)?

49. **(a)** Verify that the points $A(5, 1)$, $B(4, -2)$, and $C(-1, 3)$ all lie on the ellipse $x^2+3y^2=28$.
(b) Find a point D on the ellipse such that $\overline{CD}$ is parallel to $\overline{AB}$.
(c) If O denotes the center of the ellipse, show that the triangles OAC and OBD have equal areas. *Suggestion:* In computing the areas, the formula given in Exercise 14(b) of Exercise Set 2.1 is useful.

50. Find the points of intersection of the parabola $y=x^2$ and the ellipse $b^2x^2+a^2y^2=a^2b^2$.

51. Recall that the two line segments joining a point on the ellipse to the foci are called *focal radii*. These are the segments $\overline{F_1P}$ and $\overline{F_2P}$ in the following figure.

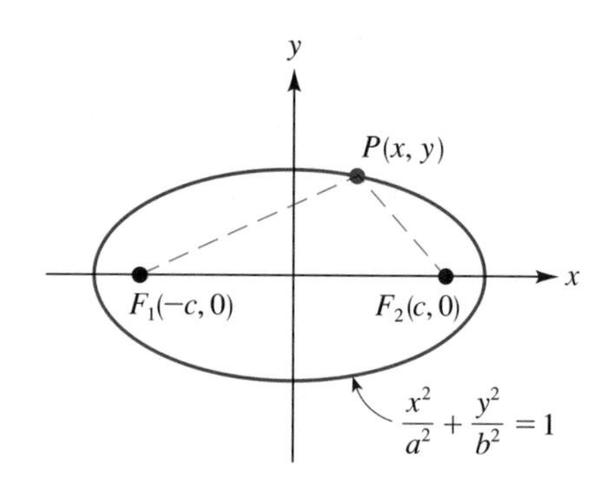

(a) Show that $F_1P=a+ex$. *Hint:* If you try to do this from scratch, it can involve a rather lengthy calculation. Begin instead with the equation $a\sqrt{(x+c)^2+y^2}=a^2+xc$ [from Exercise 40(a)], and divide both sides by a.
(b) Show that $F_2P=a-ex$. *Hint:* Make use of the result in part (a), together with the fact that $F_1P+F_2P=2a$, by definition.

52. Find the coordinates of a point P in the first quadrant on the ellipse $9x^2+25y^2=225$ such that $\angle F_2PF_1$ is a right angle.

53. The accompanying figure shows the two tangent lines drawn from the point (h, k) to the ellipse $b^2x^2+a^2y^2=a^2b^2$. Follow steps (a) through (d) to show that the equation of the line passing through the two points of tangency is $b^2hx+a^2ky=a^2b^2$.

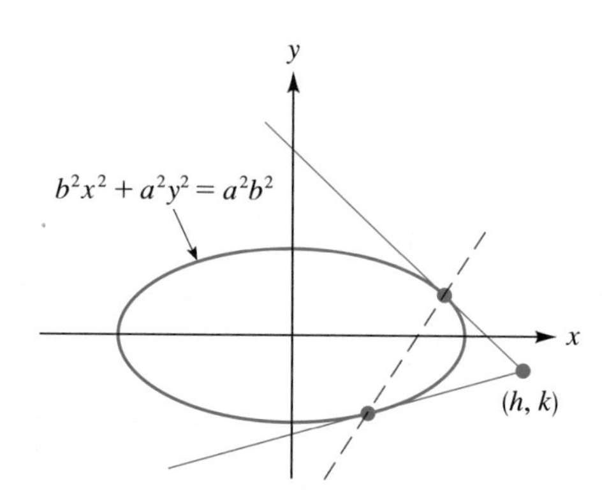

(a) Let (x_1, y_1) be one of the points of tangency. Check that the equation of the tangent line through this point is $b^2x_1x+a^2y_1y=a^2b^2$.
(b) Using the result in part (a), explain why $b^2x_1h+a^2y_1k=a^2b^2$.
(c) In a similar fashion, show that $b^2x_2h+a^2y_2k=a^2b^2$, where (x_2, y_2) is the other point of tangency.
(d) The equation $b^2hx+a^2ky=a^2b^2$ represents a line. Explain why this line must pass through the points (x_1, y_1) and (x_2, y_2).

C

54. The normal to the ellipse $b^2x^2+a^2y^2=a^2b^2$ at $P(x_1, y_1)$ meets the x-axis at A and the y-axis at B. Show that

$$PA \cdot PB = F_1P \cdot F_2P$$

where, as usual, F_1 and F_2 are the foci.

55. In the following figure, $P(x, y)$ denotes an arbitrary point on the ellipse $(x^2/a^2)+(y^2/b^2)=1$, and R is the foot of the perpendicular drawn from P to the line $x=a/e$. Show that $F_2P/PR=e$. The line $x=a/e$ is called a *directrix* of the ellipse. *Hint:* This result is easy to prove if you make use of the formula for F_2P in Exercise 51(b).

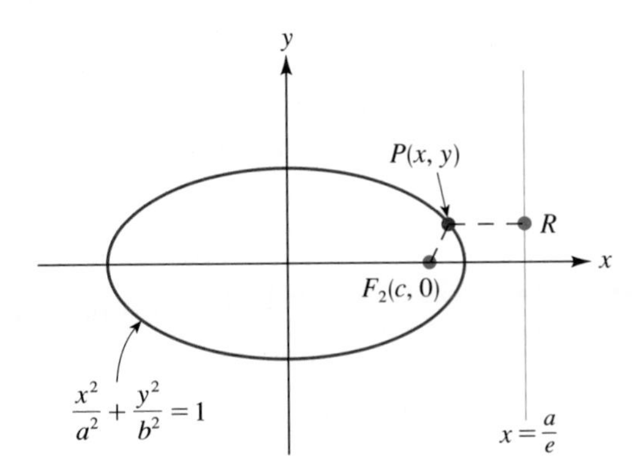

56. Let $P(x, y)$ be a point on the ellipse $(x^2/a^2)+(y^2/b^2)=1$. Let Q be the foot of the perpendicular drawn from P to the line $x=-a/e$. Let F_1 be the focus located at $(-c, 0)$. Show that $F_1P/PQ=e$. The line $x=-a/e$ is called a *directrix* of the ellipse.

57. Let D denote the distance from the center of the ellipse $(x^2/a^2)+(y^2/b^2)=1$ to the directrix $x=a/e$, and let d denote the distance from the focus $F_2(c, 0)$ to the directrix. Show that $D/d=a^2/b^2$.

58. At the point $P(x_1, y_1)$ on the ellipse $(x^2/a^2)+(y^2/b^2)=1$, a tangent line is drawn that meets the directrix $x=a/e$ at Q. If F denotes the focus located at $(c, 0)$, show that angle PFQ is a right angle.

59. **(a)** Show that the point $(c, b^2/a)$ lies on the ellipse $(x^2/a^2)+(y^2/b^2)=1$, where c has its usual meaning.
(b) A tangent is drawn to the ellipse $(x^2/a^2)+(y^2/b^2)=1$ at the point $(c, b^2/a)$. Find the point where this tangent intersects the directrix $x=a/e$.

60. The *auxiliary circle* of an ellipse is the circle centered at the center of the ellipse and with radius half the length of the major axis of the ellipse.
(a) Find the equation of the auxiliary circle of the ellipse $x^2+3y^2=12$. Sketch the circle and the ellipse.
(b) Verify that the point $P(3, 1)$ lies on the ellipse.
(c) Find the equation of the line tangent to the ellipse at P.
(d) Let T be the point where the perpendicular drawn from a focus to the tangent meets the tangent. Show that T lies on the auxiliary circle.

61. The points P and Q have the same x-coordinate. The point P lies on the circle $x^2+y^2=a^2$, and Q lies on the ellipse $(x^2/a^2)+(y^2/b^2)=1$. Show that the tangent to the circle at P meets the tangent to the ellipse at Q at a point on the x-axis. (Assume that P and Q are distinct, and that neither point is an endpoint of the major or minor axis of the ellipse.)

62. A line segment that passes through the center of an ellipse is called a *diameter*.
(a) If (x_1, y_1) is an endpoint of a diameter of the ellipse $(x^2/a^2)+(y^2/b^2)=1$, show that the other endpoint is $(-x_1, -y_1)$.
(b) If (x_1, y_1) is an endpoint of a diameter of the ellipse $(x^2/a^2)+(y^2/b^2)=1$, show that the length of the diameter is $2\sqrt{(a^2-b^2)x_1^2+a^2b^2}/a$.

63. Let $\overline{PQ}$ be a diameter of the ellipse $(x^2/a^2)+(y^2/b^2)=1$. Show that the tangents to the ellipse at P and Q are parallel.

64. This exercise outlines the steps required to show that the equation of the tangent to the ellipse $(x^2/a^2)+(y^2/b^2)=1$ at the point (x_1, y_1) on the ellipse is $(x_1x/a^2)+(y_1y/b^2)=1$.
(a) Show that the equation $(x^2/a^2)+(y^2/b^2)=1$ is equivalent to

$$b^2x^2+a^2y^2=a^2b^2 \tag{1}$$

Conclude that (x_1, y_1) lies on the ellipse if and only if

$$b^2x_1^2+a^2y_1^2=a^2b^2 \tag{2}$$

(b) Subtract equation (2) from equation (1) to show that

$$b^2(x^2-x_1^2)+a^2(y^2-y_1^2)=0 \tag{3}$$

Equation (3) is equivalent to equation (1) provided only that (x_1, y_1) lies on the ellipse. In the following steps, we will find the algebra much simpler if we use equation (3) to represent the ellipse, rather than the equivalent and perhaps more familiar equation (1).
(c) Let the equation of the line tangent to the ellipse at (x_1, y_1) be

$$y-y_1=m(x-x_1)$$

Explain why the following system of equations must have exactly one solution, namely, (x_1, y_1):

$$\begin{cases} b^2(x^2-x_1^2)+a^2(y^2-y_1^2)=0 & (4)\\ y-y_1=m(x-x_1) & (5)\end{cases}$$

(d) Solve equation (5) for y and then substitute for y in equation (4) to obtain

$$b^2(x^2-x_1^2)+a^2m^2(x-x_1)^2+2a^2my_1(x-x_1)=0 \tag{6}$$

(e) Show that equation (6) can be written

$$(x-x_1)[b^2(x+x_1)+a^2m^2(x-x_1)+2a^2my_1]=0 \tag{7}$$

(f) Equation (7) is a quadratic equation in x, but as pointed out earlier, $x=x_1$ must be the only solution. (That is, $x = x_1$ is a double root.) Thus, the factor in brackets must equal zero when x is replaced by x_1. Use this observation to show that

$$m = -\frac{b^2 x_1}{a^2 y_1}$$

This represents the slope of the line tangent to the ellipse at (x_1, y_1).

(g) Using this value for m, show that equation (5) becomes

$$b^2 x_1 x + a^2 y_1 y = b^2 x_1^2 + a^2 y_1^2 \qquad (8)$$

(h) Now use equation (2) to show that equation (8) can be written

$$\frac{x_1 x}{a^2} + \frac{y_1 y}{b^2} = 1$$

which is what we set out to show.

65. This exercise outlines a proof of the reflection property of the ellipse. We want to prove that right triangles F_1AP and F_2BP in the accompanying figure are similar.

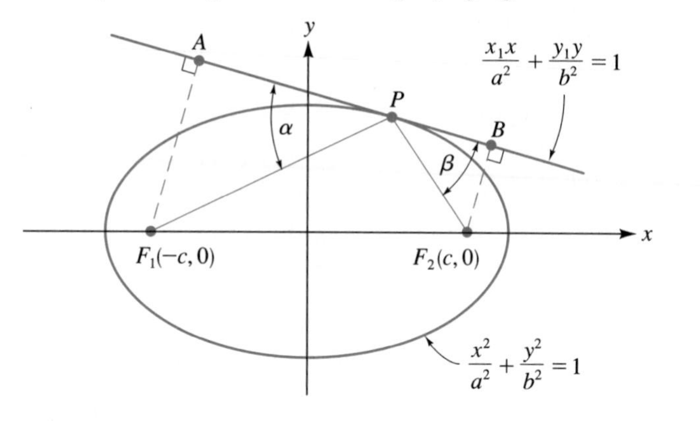

This will imply that $\alpha = \beta$, as required, since corresponding angles in similar triangles are equal. To show that the right triangles F_1AP and F_2BP are similar, it is enough to demonstrate that

$$\frac{F_1A}{F_1P} = \frac{F_2B}{F_2P}$$

(a) The quantity F_1A is the distance from the point $(-c, 0)$ to the line $(x_1x/a^2) + (y_1y/b^2) = 1$. Verify that this distance is given by

$$F_1A = \frac{\left|\frac{x_1 c}{a^2} + 1\right|}{\sqrt{A^2 + B^2}}$$

where $A = x_1/a^2$ and $B = y_1/b^2$.

(b) In the preceding expression for F_1A, replace c by ae. Then, by making use of the expression for F_1P obtained in Exercise 51(a), show that

$$F_1A = \frac{F_1P}{a\sqrt{A^2 + B^2}}$$

(c) Show similarly that

$$F_2B = \frac{F_2P}{a\sqrt{A^2 + B^2}}$$

(d) Now use the expressions for F_1A and F_2B obtained in parts (b) and (c) to verify that

$$\frac{F_1A}{F_1P} = \frac{F_2B}{F_2P}$$

This implies that right triangles F_1AP and F_2BP are similar and, consequently, that $\alpha = \beta$, as required.

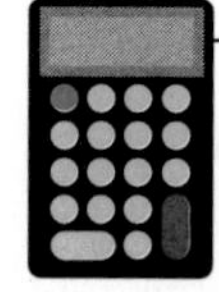

GRAPHING UTILITY EXERCISES FOR SECTION 11.4

1. On page 697 in the text was the statement that calculator exercises at the end of that section would help to convince you that the general shape of the ellipse shown in Figure 5 (on page 696) is correct. Now let's make that "graphing utility exercises" rather than merely "calculator exercises."

(a) Take the equation $(x^2/3^2) + (y^2/2^2) = 1$, as given in Exercise 38 on page 705, and solve for y to obtain

$$y = \pm\tfrac{1}{3}\sqrt{36 - 4x^2}$$

Now graph these two functions. Which types of symmetry does the ellipse possess?

(b) Take the equation $(x^2/1^2) + (y^2/4^2) = 1$, as given in Exercise 39 on page 705, and solve for y to obtain

$$y = \pm 4\sqrt{1 - x^2}$$

Graph these two functions. Which types of symmetry does the ellipse possess?

2. (a) Solve the equation $4x^2 + 9y^2 = 36$ for y, and then graph this ellipse (as in Exercise 1).

(b) If you translate the ellipse in part (a) to the right 1 unit and up 3 units, what is the equation of the new ellipse?

(c) Solve the equation obtained in part (b) for y, and check that the result can be written

$$y = \pm\tfrac{1}{3}\sqrt{36 - 4(x-1)^2} + 3$$

(d) Graph the two functions given in part (c). Use the graph to estimate the center of the ellipse. Check that your result in part (b) is consistent with this estimate.

3. In this exercise you will graph four ellipses. Each ellipse will have the same major axis, but the eccentricities will vary from 0 to 0.8. (Actually, you'll graph only the top half of each ellipse, the portion corresponding to $y \geq 0$.)

(a) The equation $b^2x^2 + 16y^2 = 16b^2$ represents an ellipse. Show that the length of the horizontal axis of this ellipse is 8.

(b) Check that the following values for b^2 yield the indicated eccentricities.

b^2	16	13.44	10.24	5.76
Eccentricity	0	0.4	0.6	0.8

(c) Solve the equation $b^2x^2 + 16y^2 = 16b^2$ for y. Show that the result can be written

$$y = \pm(0.25)\sqrt{b^2(16 - x^2)}$$

Now graph the (top half) of the four ellipses corresponding to the four values of b^2 in the table. Be sure to use the same scale and same size unit on both axes so that the top of the ellipse with eccentricity 0 indeed appears to be a semicircle. If you have done things correctly, your graphs should be similar to those shown in the following figure.

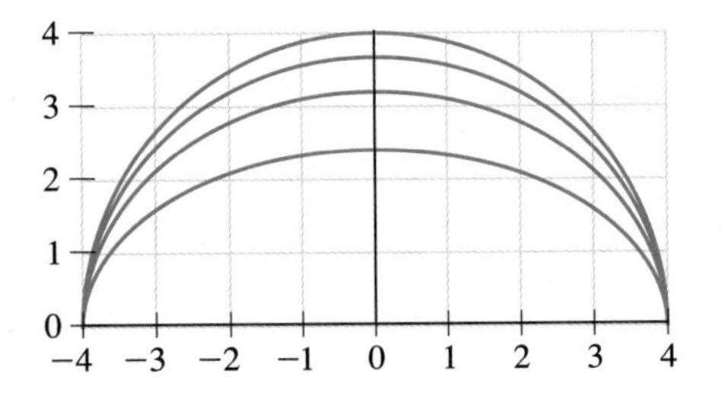

4. The *auxiliary circle* of an ellipse is defined to be the circle with diameter the same as the major axis of the ellipse. Determine the equation of the auxiliary circle for the ellipse $9x^2 + 25y^2 = 225$, and then graph the ellipse and the circle.

In Exercises 5 and 6, determine the equation of the tangent to the ellipse at the given point, then graph the ellipse and the tangent line.

5. $3x^2 + y^2 = 12$; $(1, -3)$ **6.** $x^2 + 3y^2 = 28$; $(4, 2)$

7. (As background for this exercise, you need to have completed Exercises 4 and 5.) In this exercise you're going to illustrate an interesting result concerning an ellipse and its auxiliary circle.

(a) Graph the ellipse $x^2 + 3y^2 = 12$ for $y \geq 0$.

(b) Find the values of a, b, and c for this ellipse.

(c) Find the equation of the auxiliary circle for this ellipse. Add the graph of the top half of this circle to your picture from part (a).

(d) Verify (algebraically) that the point $P(3, 1)$ lies on the ellipse. Then find the equation of the tangent line to the ellipse at P. Next, add the graph of this tangent line to the picture that you obtained in part (c).

(e) In part (b) you determined the value of c. Find the equation of the line passing through $(-c, 0)$ and perpendicular to the tangent in part (d). Then add the graph of this line to your picture. If you've done things correctly, you should obtain a figure similar to the following one. This figure provides an example of this general result: The line through the focus and perpendicular to the tangent meets the tangent at a point on the auxiliary circle.

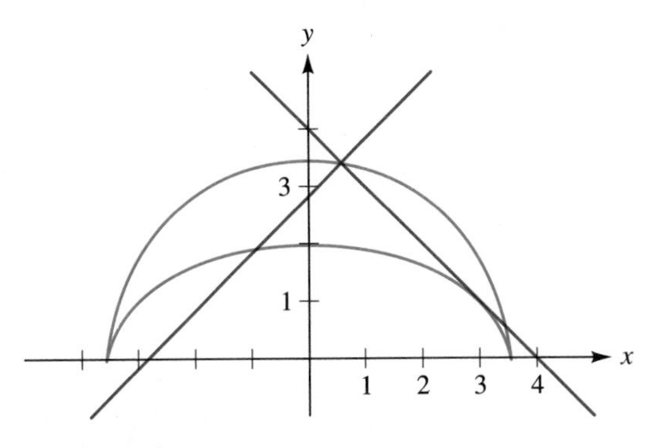

FIGURE FOR PROBLEM 7
The line through the focus and perpendicular to a tangent meets the tangent at a point on the auxiliary circle.

11.5 THE HYPERBOLA

The ellipse is the general shape of any closed orbit. . . . It is also possible to have orbits that are not closed, whose shapes are represented by open curves. Even if the two bodies are not bound together by their mutual gravitational attraction, their gravitational attraction for each other still affects their relative motion. Then the general shape of their orbit is a hyperbola.

Theodore P. Snow in *The Dynamic Universe: An Introduction to Astronomy,* 4th ed. (St. Paul: West Publishing Company, 1990)

. . . Menaechmus [c. 350 B.C.] *is reputed to have discovered the curves that were later known as the ellipse, the parabola, and the hyperbola.*

Carl B. Boyer in *A History of Mathematics,* 2nd ed., revised by Uta C. Merzbach (New York: John Wiley and Sons, 1991)

In the previous section, we defined an ellipse as the set of points P such that the sum of the distances from P to two fixed points is constant. By considering the difference instead of the sum, we are led to the definition of the hyperbola.

DEFINITION: The Hyperbola

A **hyperbola** is the set of all points in the plane, the difference of whose distances from two fixed points is a positive constant. Each fixed point is called a **focus.**

As with the ellipse, we label the foci F_1 and F_2. Before obtaining an equation for the hyperbola, we can see the general features of the curve by using two sets of concentric circles, with centers F_1 and F_2, to locate points satisfying the definition of a hyperbola. In Figure 1, we've plotted a number of points P such that either $F_1P - F_2P = 3$ or $F_2P - F_1P = 3$. By joining these points, we obtain the graph of the hyperbola shown in Figure 1.

Unlike the parabola or the ellipse, the hyperbola is composed of two distinct

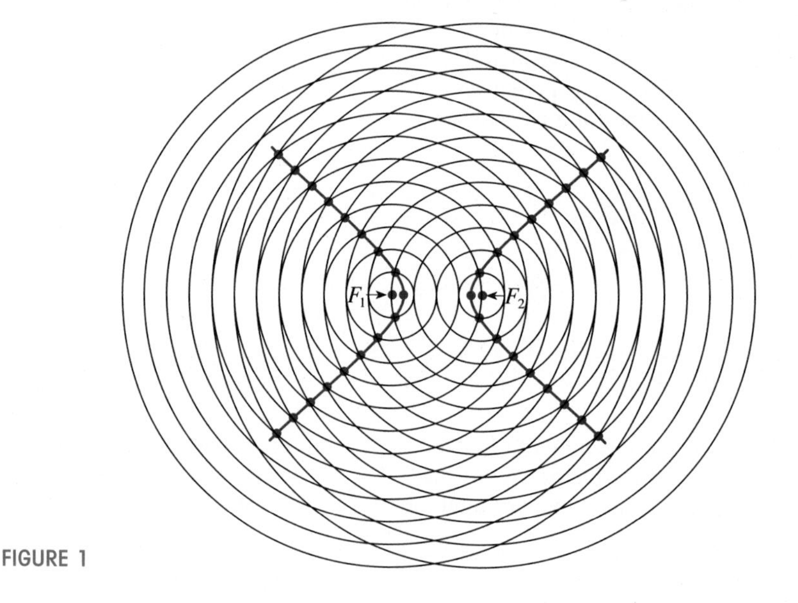

FIGURE 1

parts, or **branches.** As you can check, the left branch in Figure 1 corresponds to the equation $F_2P - F_1P = 3$, while the right branch corresponds to the equation $F_1P - F_2P = 3$. Figure 1 also reveals that the hyperbola possesses two types of symmetry. First, it is symmetric about the line passing through the two foci

F_1 and F_2; this line is referred to as the **focal axis** of the hyperbola. Second, the hyperbola is symmetric about the line that is the perpendicular bisector of the segment $\overline{F_1F_2}$.

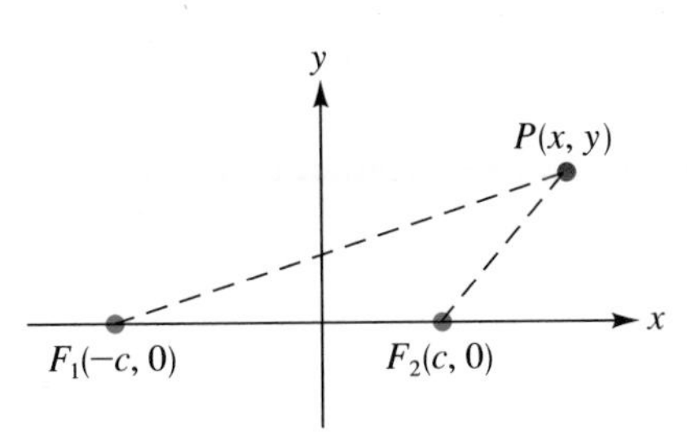

FIGURE 2

To derive an equation for the hyperbola, let us initially assume that the foci are located at the points $F_1(-c, 0)$ and $F_2(c, 0)$, as indicated in Figure 2. We will use $2a$ to denote the positive constant referred to in the definition of the hyperbola. By definition, then, $P(x, y)$ lies on the hyperbola if and only if

$$|F_1P - F_2P| = 2a$$

or, equivalently,

$$F_1P - F_2P = \pm 2a$$

If we use the formula for the distance between two points, this last equation becomes

$$\sqrt{(x+c)^2 + (y-0)^2} - \sqrt{(x-c)^2 + (y-0)^2} = \pm 2a$$

or

$$\sqrt{(x+c)^2 + y^2} - \sqrt{(x-c)^2 + y^2} = \pm 2a$$

We can simplify this equation by carrying out the same procedure that we used for the ellipse in the previous section. As Exercise 42 asks you to verify, the resulting equation is

$$(c^2 - a^2)x^2 - a^2y^2 = a^2(c^2 - a^2) \tag{1}$$

Before further simplifying equation (1), we point out that the quantity $c^2 - a^2$ [which appears twice in equation (1)] is positive. To see why this is so, refer back to Figure 2. In triangle F_1F_2P (as in any triangle), the length of any side is less than the sum of the lengths of the other two sides. Thus

$$F_1P < F_1F_2 + F_2P \qquad \text{and} \qquad F_2P < F_1F_2 + F_1P$$

and therefore

$$F_1P - F_2P < F_1F_2 \qquad \text{and} \qquad F_2P - F_1P < F_1F_2$$

These last two equations tell us that

$$|F_1P - F_2P| < F_1F_2$$

Therefore, in view of the definitions of $2a$ and $2c$, we have

$$0 < 2a < 2c \qquad \text{or} \qquad 0 < a < c$$

This last inequality tells us that $c^2 - a^2$ is positive, as we wished to show.

Now, since $c^2 - a^2$ is positive, we can define the positive number b by the equation

$$b^2 = c^2 - a^2$$

With this notation, equation (1) becomes

$$b^2x^2 - a^2y^2 = a^2b^2$$

Dividing by a^2b^2, we obtain

$$\frac{x^2}{a^2} - \frac{y^2}{b^2} = 1 \tag{2}$$

We have now shown that the coordinates of every point on the hyperbola satisfy equation (2). Conversely, it can be shown that if the coordinates of a point satisfy equation (2), then the point satisfies the original definition of the hyperbola. Equation (2) is the **standard form** for the equation of a hyperbola with foci $F_1(-c, 0)$ and $F_2(c, 0)$.

The intercepts of the hyperbola are readily obtained from equation (2). To find the x-intercepts, we set y equal to zero to obtain

$$\frac{x^2}{a^2} = 1$$

$$x^2 = a^2 \qquad \text{or} \qquad x = \pm a$$

Thus, the hyperbola crosses the x-axis at the points $(-a, 0)$ and $(a, 0)$. On the other hand, the curve does not cross the y-axis, for if we set x equal to zero in equation (2), we obtain $-y^2/b^2 = 1$ or

$$y^2 = -b^2 \qquad (b > 0)$$

Since the square of any real number y is nonnegative, this last equation has no solution. Therefore, the graph does not cross the y-axis. Finally, let us note that (according to the symmetry tests in Section 2.4) the graph of equation (2) must be symmetric about both coordinate axes.

Before graphing the hyperbola $(x^2/a^2) - (y^2/b^2) = 1$, we point out the important fact that the two lines $y = (b/a)x$ and $y = -(b/a)x$ are asymptotes for the curve. We can see why as follows. First we solve equation (2) for y:

$$\frac{x^2}{a^2} - \frac{y^2}{b^2} = 1$$

$$b^2x^2 - a^2y^2 = a^2b^2 \qquad \text{multiplying both sides by } a^2b^2$$

$$-a^2y^2 = a^2b^2 - b^2x^2$$

$$y^2 = \frac{b^2x^2 - a^2b^2}{a^2} = \frac{b^2(x^2 - a^2)}{a^2}$$

$$y = \pm\frac{b}{a}\sqrt{x^2 - a^2} \tag{3}$$

Now, as x grows arbitrarily large, the value of the quantity $\sqrt{x^2 - a^2}$ becomes closer and closer to x itself. Table 1 provides some empirical evidence for this statement in the case when $a = 5$. (A formal proof of the statement properly belongs to calculus.) In summary, then, we have the approximation $\sqrt{x^2 - a^2} \approx x$ as x grows arbitrarily large. So, in view of equation (3), we have

TABLE 1

x	$\sqrt{x^2 - 5^2}$
100	99.875
1000	999.987
10000	9999.999

$$y = \pm\frac{b}{a}\sqrt{x^2 - a^2} \approx \pm\frac{b}{a}x \qquad \text{as } x \text{ grows arbitrarily large}$$

In other words, the two lines $y = \pm(b/a)x$ are asymptotes for the hyperbola.

A simple way to sketch the two asymptotes and then graph the hyperbola is as follows. First draw the rectangle with vertices (a, b), $(-a, b)$, $(-a, -b)$, and $(a, -b)$, as indicated in Figure 3(a). The slopes of the diagonals in this rectangle are b/a and $-b/a$. Thus by extending these diagonals as in Figure 3(b), we obtain the two asymptotes $y = \pm(b/a)x$. Now, since the x-intercepts of the hyperbola are a and $-a$, we can sketch the curve as shown in Figure 3(c).

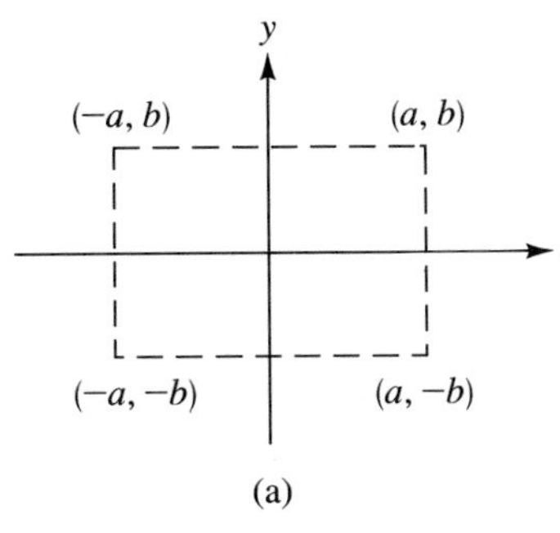

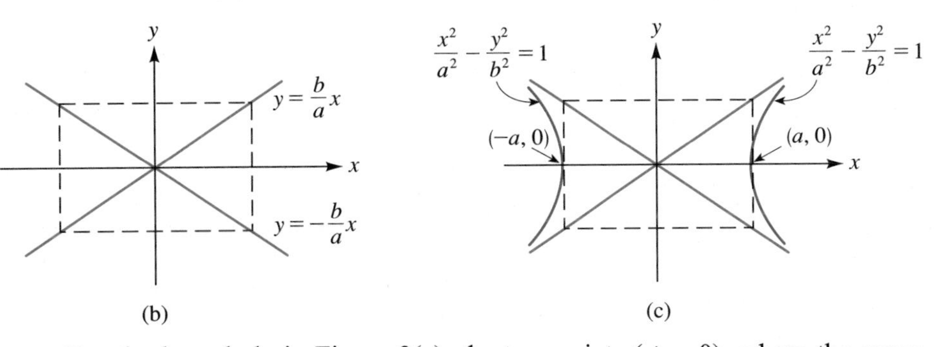

FIGURE 3
Steps in graphing the hyperbola $(x^2/a^2) - (y^2/b^2) = 1$ and its asymptotes.

For the hyperbola in Figure 3(c), the two points $(\pm a, 0)$, where the curve meets the x-axis, are referred to as **vertices.** The midpoint of the line segment joining the two vertices is called the **center** of the hyperbola. (Equivalently, we can define the center as the point of intersection of the two asymptotes.) For the hyperbola in Figure 3(c), the center coincides with the origin. The line segment joining the vertices of a hyperbola is the **transverse axis** of the hyperbola. For reference, in the box that follows we summarize our work up to this point on the hyperbola. Several new terms describing the hyperbola are also given in the box.

PROPERTY SUMMARY

THE HYPERBOLA $\frac{x^2}{a^2} - \frac{y^2}{b^2} = 1$

1. The **foci** are the points $F_1(-c, 0)$ and $F_2(c, 0)$. The hyperbola is the set of points P such that $|F_1P - F_2P| = 2a$.
2. The **focal axis** is the line passing through the foci.
3. The **vertices** are the points at which the hyperbola intersects its focal axis. In Figure 4, these are the two points $V_1(-a, 0)$ and $V_2(a, 0)$.

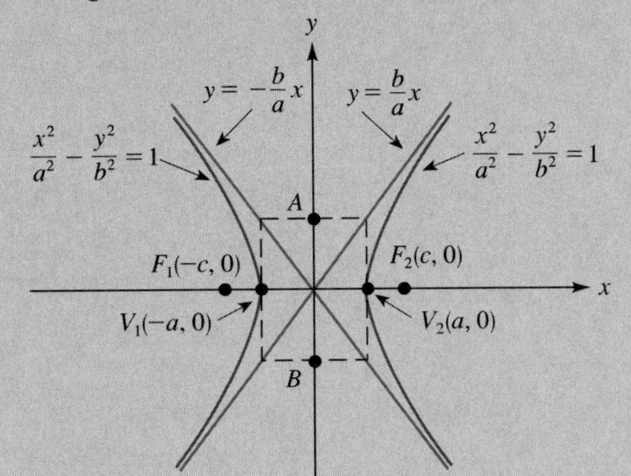

FIGURE 4

4. The **center** is the point on the focal axis midway between the foci. The center of the hyperbola in Figure 4 is the origin.
5. The **transverse axis** is the line segment joining the two vertices. In Figure 4, the length of the transverse axis $\overline{V_1V_2}$ is $2a$.
6. The **conjugate axis** is the line segment perpendicular to the transverse axis, passing through the center and extending a distance b on either side of the center. In Figure 4, this is the segment $\overline{AB}$.
7. The **eccentricity** e is defined by $e = c/a$, where $c^2 = a^2 + b^2$.

In general, if A, B, and C are positive numbers, then the graph of an equation of the form

$$Ax^2 - By^2 = C$$

will be a hyperbola of the type shown in Figure 4. Example 1 shows why this is so.

EXAMPLE 1 Graph the hyperbola $16x^2 - 9y^2 = 144$ after determining the following: vertices, foci, eccentricity, lengths of the transverse and conjugate axes, and asymptotes.

Solution First we convert the given equation to standard form by dividing both sides by 144. This yields

$$\frac{x^2}{9} - \frac{y^2}{16} = 1 \qquad \text{or} \qquad \frac{x^2}{3^2} - \frac{y^2}{4^2} = 1$$

By comparing this with the equation $(x^2/a^2) - (y^2/b^2) = 1$, we see that $a = 3$ and $b = 4$. The value of c can be determined by using the equation $c^2 = a^2 + b^2$. We have

$$c^2 = 3^2 + 4^2 = 25 \qquad \text{and therefore} \qquad c = 5$$

Now that we know the values of a, b, and c, we can list the required information:

Vertices: $(\pm 3, 0)$ Length of transverse axis $(= 2a)$: 6
Foci: $(\pm 5, 0)$ Length of conjugate axis $(= 2b)$: 8
Eccentricity: $\frac{5}{3}$ Asymptotes: $y = \pm \frac{4}{3}x$

The graph of the hyperbola is shown in Figure 5.

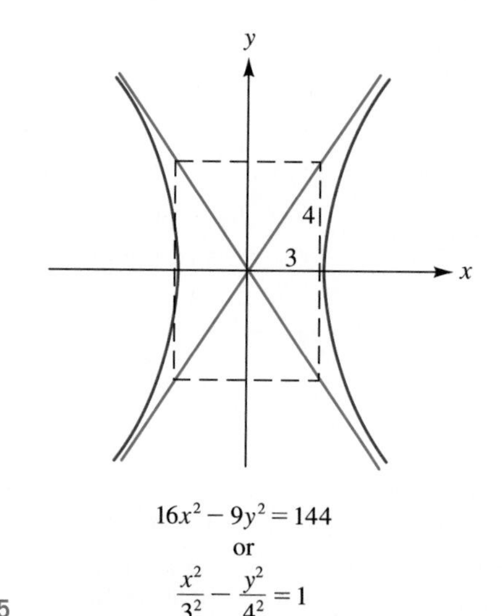

FIGURE 5

We can use the same method that we used to derive the equation for the hyperbola with foci $(\pm c, 0)$ when the foci are instead located on the y-axis at the points $(0, \pm c)$. The equation of the hyperbola in this case is

$$\frac{y^2}{a^2} - \frac{x^2}{b^2} = 1$$

We summarize the basic properties of this hyperbola in the following box.

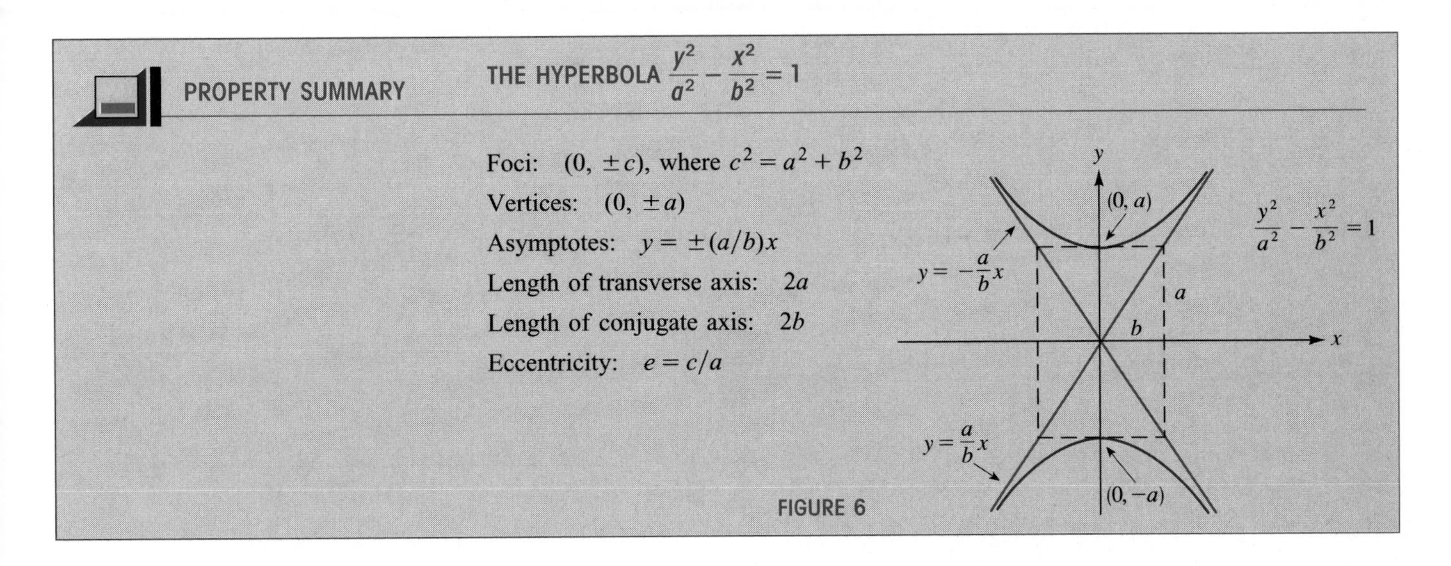

FIGURE 6

EXAMPLE 2 Use the technique of completing the square to show that the graph of the following equation is a hyperbola:

$$9y^2 - 54y - 25x^2 + 200x - 544 = 0$$

Graph the hyperbola and specify the center, the vertices, the foci, the length of the transverse axis, and the equations of the asymptotes.

Solution

$$9(y^2 - 6y \qquad) - 25(x^2 - 8x \qquad) = 544 \qquad \text{factoring}$$
$$9(y^2 - 6y + 9) - 25(x^2 - 8x + 16) = 544 + 81 - 400 \qquad \text{completing the squares}$$
$$9(y - 3)^2 - 25(x - 4)^2 = 225$$
$$\frac{(y - 3)^2}{5^2} - \frac{(x - 4)^2}{3^2} = 1 \qquad \text{dividing by 225}$$

Now, the graph of this last equation is obtained by translating the graph of the equation

$$\frac{y^2}{5^2} - \frac{x^2}{3^2} = 1 \tag{4}$$

to the right 4 units and up 3 units. So first we analyze the graph of equation (4). The general form of the graph is shown in Figure 6 in the preceding Property Summary Box. In this case, we have $a = 5$, $b = 3$, and

$$c = \sqrt{a^2 + b^2} = \sqrt{5^2 + 3^2} = \sqrt{34} \ (\approx 5.8)$$

Consequently, the vertices [for equation (4)] are $(0, \pm 5)$; the foci are $(0, \pm\sqrt{34})$; and the asymptotes are $y = \pm\frac{5}{3}x$. Figure 7 shows the graph of this hyperbola. Finally, by translating the graph in Figure 7 to the right 4 units and up 3 units, we obtain the graph of the original equation, as shown in Figure 8. You should verify for yourself that the information accompanying Figure 8 is correct.

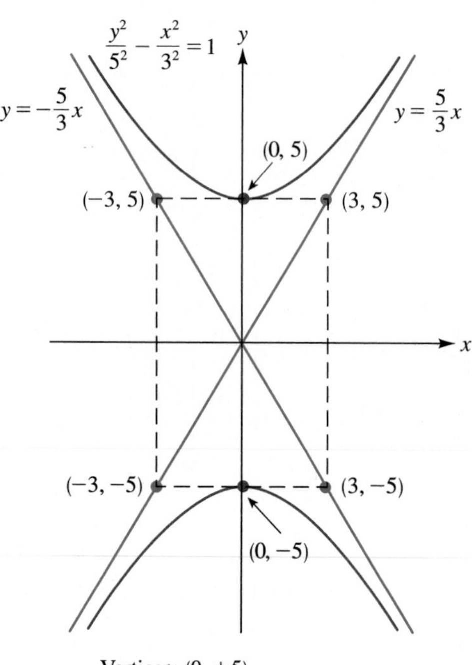

FIGURE 7

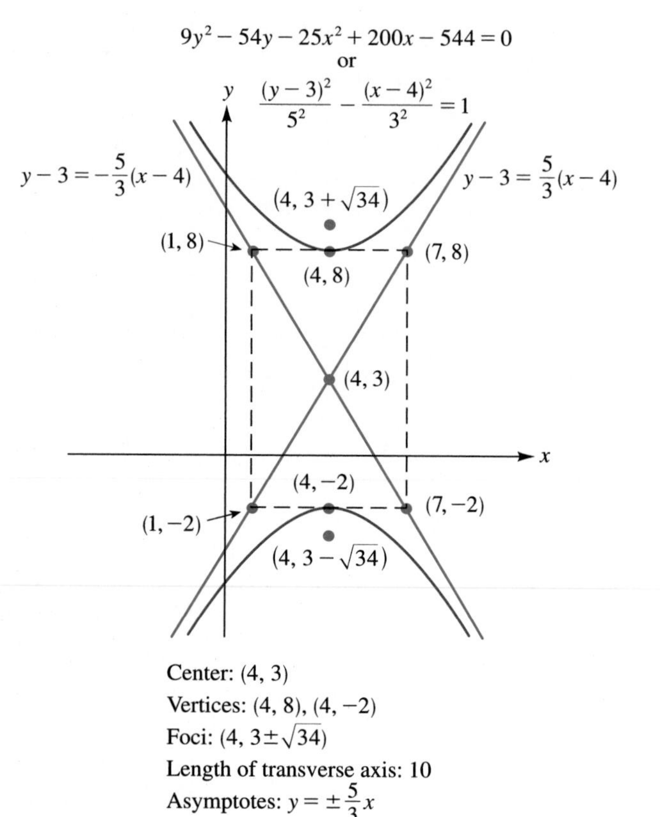

FIGURE 8

If you reread the example we have just completed, you'll see that it was not necessary to know in advance that the given equation represented a hyperbola. Rather, this fact emerged naturally after we completed the square. Indeed, completing the square is a useful technique for identifying the graph of any equation of the form

$$Ax^2 + Cy^2 + Dx + Ey + F = 0$$

EXAMPLE 3 Identify the graph of the equation

$$4x^2 - 32x - y^2 + 2y + 63 = 0$$

Solution As before, we complete the squares:

$$4(x^2 - 8x) - (y^2 - 2y) = -63$$

$$4(x^2 - 8x + 16) - (y^2 - 2y + 1) = -63 + 64 - 1$$

$$4(x - 4)^2 - (y - 1)^2 = 0$$

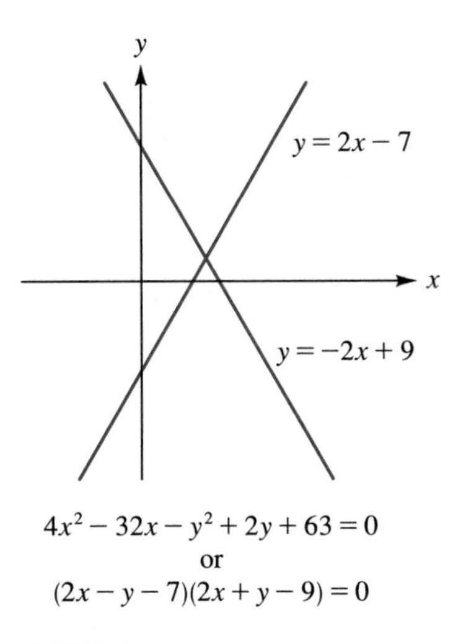

$4x^2 - 32x - y^2 + 2y + 63 = 0$
or
$(2x - y - 7)(2x + y - 9) = 0$

FIGURE 9

Since the right-hand side of this last equation is zero, dividing both sides by 4 will not bring the equation into one of the standard forms. Indeed, if we factor the left-hand side of the equation as a difference of two squares, we obtain

$$[2(x-4)-(y-1)][2(x-4)+(y-1)]=0$$

$$(2x-y-7)(2x+y-9)=0$$

$$\begin{array}{r|r} 2x-y-7=0 & 2x+y-9=0 \\ -y=-2x+7 & y=-2x+9 \\ y=2x-7 & \end{array}$$

Thus the given equation is equivalent to the two linear equations $y = 2x - 7$ and $y = -2x + 9$. These two lines together constitute the graph. See Figure 9.

The two lines that we graphed in Figure 9 are actually the asymptotes for the hyperbola $4(x-4)^2 - (y-1)^2 = 1$. (Verify this for yourself.) For that reason, the graph in Figure 9 is referred to as a **degenerate hyperbola.** There are other cases similar to this that can arise in graphing equations of the form $Ax^2 + Cy^2 + Dx + Ey + F = 0$. For instance, as you can check for yourself by completing the squares, the graph of the equation

$$x^2 - 2x + 4y^2 - 16y + 17 = 0$$

consists of the single point $(1, 2)$. We refer to the graph in this case as a **degenerate ellipse.** Similarly, as you can check by completing the squares, the equation $x^2 - 2x + 4y^2 - 16y + 18 = 0$ has no graph; there are no points with coordinates that satisfy the equation.

We can obtain the equation of a tangent line to a hyperbola using the same ideas employed for the parabola and the ellipse in the previous sections. The result is this: The equation of the tangent to the hyperbola $(x^2/a^2) - (y^2/b^2) = 1$ at the point (x_1, y_1) on the curve is

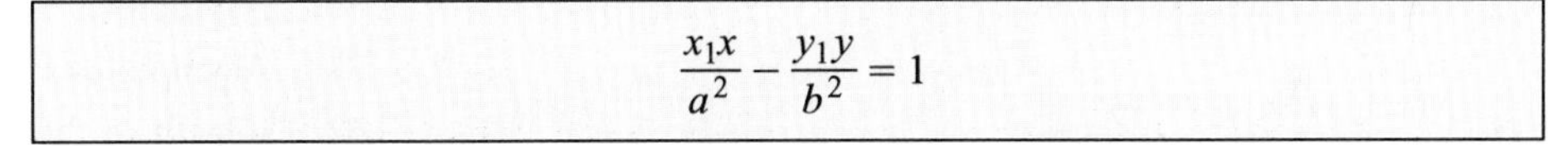

$$\frac{x_1 x}{a^2} - \frac{y_1 y}{b^2} = 1$$

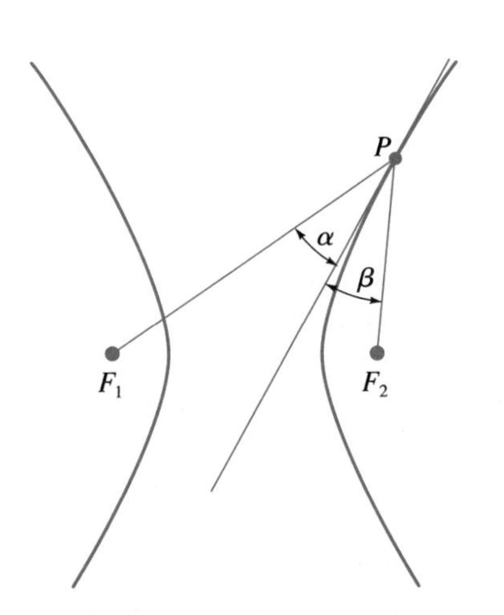

FIGURE 10
Reflection property of the hyperbola: The tangent at P bisects the angle formed by the focal radii drawn to P, so $\alpha = \beta$.

(See Exercise 53 at the end of this section for an outline of the derivation.) As with the parabola and the ellipse, many of the interesting properties of the hyperbola are related to the tangent lines. For instance, the hyperbola has a reflection property that is similar to the reflection properties of the parabola and the ellipse. To state this property, we first define a **focal radius** of a hyperbola as a line segment drawn from a focus to a point on the hyperbola. Then the **reflection property** of the hyperbola is that the tangent line bisects the angle formed by the focal radii drawn to the point of tangency, as indicated in Figure 10.

We conclude this section by listing several applications of the hyperbola. Some comets have hyperbolic orbits. Unlike Halley's comet, which has an elliptical orbit, these comets pass through the solar system once and never return. The Cassegrain telescope (invented by the Frenchman Sieur Cassegrain in 1672) uses both a hyperbolic mirror and a parabolic mirror. The Hubble Space Telescope, launched into orbit in April 1990, utilizes a Cassegrain telescope. The hyperbola is also used in some navigation systems. In the LORAN (LOng RAnge

Navigation) system, an airplane or a ship at a point P receives radio signals that are transmitted simultaneously from two locations, F_1 and F_2. The time difference between the two signals is converted to a difference in distances: $F_1P - F_2P$. This locates the ship along one branch of a hyperbola. Then, using data from a second set of signals, the ship is located along a second hyperbola. The intersection of the two hyperbolas then determines the location of the airplane or ship.

EXERCISE SET 11.5

A

For Exercises 1–24, graph the hyperbolas. In each case in which the hyperbola is nondegenerate, specify the following: vertices, foci, lengths of transverse and conjugate axes, eccentricity, and equations of the asymptotes. In Exercises 11–24, also specify the centers.

1. $x^2 - 4y^2 = 4$

2. $y^2 - x^2 = 1$

3. $y^2 - 4x^2 = 4$

4. $25x^2 - 9y^2 = 225$

5. $16x^2 - 25y^2 = 400$

6. $9x^2 - y^2 = 36$

7. $2y^2 - 3x^2 = 1$

8. $x^2 - y^2 = 9$

9. $4y^2 - 25x^2 = 100$

10. $x^2 - 3y^2 = 3$

11. $\dfrac{(x-5)^2}{5^2} - \dfrac{(y+1)^2}{3^2} = 1$

12. $\dfrac{(x-5)^2}{3^2} - \dfrac{(y+1)^2}{5^2} = 1$

13. $\dfrac{(y-2)^2}{2^2} - \dfrac{(x-1)^2}{1^2} = 1$

14. $\dfrac{(y-3)^2}{2^2} - \dfrac{x^2}{1^2} = 1$

15. $\dfrac{(x+3)^2}{4^2} - \dfrac{(y-4)^2}{4^2} = 1$

16. $\dfrac{(x+1)^2}{5^2} - \dfrac{(y-2)^2}{3^2} = 1$

17. $x^2 - y^2 + 2y - 5 = 0$

18. $16x^2 - 32x - 9y^2 + 90y - 353 = 0$

19. $x^2 - y^2 - 4x + 2y - 6 = 0$

20. $x^2 - 8x - y^2 + 8y - 25 = 0$

21. $y^2 - 25x^2 + 8y - 9 = 0$

22. $9y^2 - 18y - 4x^2 - 16x - 43 = 0$

23. $x^2 + 7x - y^2 - y + 12 = 0$

24. $9x^2 + 9x - 16y^2 + 4y + 2 = 0$

25. Let $P(x, y)$ be a point in the first quadrant on the hyperbola $(x^2/2^2) - (y^2/1^2) = 1$. Let Q be the point in the first quadrant with the same x-coordinate as P and lying on an asymptote to the hyperbola. Show that $PQ = (x - \sqrt{x^2 - 4})/2$.

26. The distance PQ in Exercise 25 represents the vertical distance between the hyperbola and the asymptote. Complete the following table to see numerical evidence that this separation distance approaches zero as x gets larger and larger. (Round each entry to one significant digit.)

x	10	50	100	500	1000	10000
PQ						

In Exercises 27–36, determine the equation of the hyperbola satisfying the given conditions. Write each answer in the form $Ax^2 - By^2 = C$ or in the form $Ay^2 - Bx^2 = C$.

27. Foci $(\pm 4, 0)$; vertices $(\pm 1, 0)$

28. Foci $(0, \pm 5)$; vertices $(0, \pm 3)$

29. Asymptotes $y = \pm \frac{1}{2}x$; vertices $(\pm 2, 0)$

30. Asymptotes $y = \pm x$; foci $(0, \pm 1)$

31. Asymptotes $y = \pm\sqrt{10}x/5$; foci $(\pm\sqrt{7}, 0)$

32. Length of the transverse axis 6; eccentricity $4/3$; center $(0, 0)$; focal axis horizontal

33. Vertices $(0, \pm 7)$; graph passes through the point $(1, 9)$

34. Eccentricity 2; foci $(\pm 1, 0)$

35. Length of the transverse axis 6; length of the conjugate axis 2; foci on the y-axis; center at the origin

36. Asymptotes $y = \pm 2x$; graph passes through $(1, \sqrt{3})$

37. Show that the two asymptotes of the hyperbola $x^2 - y^2 = 16$ are perpendicular to each other.

38. **(a)** Verify that the point $P(6, 4\sqrt{3})$ lies on the hyperbola $16x^2 - 9y^2 = 144$.
(b) In Example 1, we found that the foci of this hyperbola were $F_1(-5, 0)$ and $F_2(5, 0)$. Compute the lengths F_1P and F_2P, where P is the point $(6, 4\sqrt{3})$.
(c) Verify that $|F_1P - F_2P| = 2a$.

39. **(a)** Verify that the point $P(5, 6)$ lies on the hyperbola $5y^2 - 4x^2 = 80$.
(b) Find the foci.
(c) Compute the lengths of the line segments $\overline{F_1P}$ and $\overline{F_2P}$, where P is the point $(5, 6)$.
(d) Verify that $|F_1P - F_2P| = 2a$.

B

40. **(a)** Let e_1 denote the eccentricity of the hyperbola $(x^2/4^2) = (y^2/3^2) = 1$, and let e_2 denote the eccentricity of the hyperbola $(x^2/3^2) - (y^2/4^2) = 1$. Verify that $e_1^2e_2^2 = e_1^2 + e_2^2$.
(b) Let e_1 and e_2 denote the eccentricities of the two hyperbolas $(x^2/A^2) - (y^2/B^2) = 1$ and $(y^2/B^2) - (x^2/A^2) = 1$, respectively. Verify that $e_1^2e_2^2 = e_1^2 + e_2^2$.

41. **(a)** If the hyperbola $(x^2/a^2) - (y^2/b^2) = 1$ has perpendicular asymptotes, show that $a = b$. What is the eccentricity in this case?
(b) Show that the asymptotes of the hyperbola $(x^2/a^2) - (y^2/a^2) = 1$ are perpendicular. What is the eccentricity of this hyperbola?

42. Derive equation (1) in this section from the equation that precedes it.

43. Let $P(x, y)$ be a point on the right-hand branch of the hyperbola $(x^2/a^2) - (y^2/b^2) = 1$. As usual, let F_2 denote the focus located at $(c, 0)$. The following steps outline a proof of the fact that the length of the line segment $\overline{F_2P}$ in this case is given by $F_2P = xe - a$.
(a) Explain why

$$\sqrt{(x + c)^2 + y^2} - \sqrt{(x - c)^2 + y^2} = 2a$$

(b) In the preceding equation, add the quantity $\sqrt{(x - c)^2 + y^2}$ to both sides and then square both sides. Show that the result can be written

$$xc - a^2 = a\sqrt{(x - c)^2 + y^2}$$

or
$$xc - a^2 = a(F_2P)$$

(c) Divide both sides of the preceding equation by a to show that $xe - a = F_2P$, as required.

44. Let $P(x, y)$ be a point on the right-hand branch of the hyperbola $(x^2/a^2) - (y^2/b^2) = 1$. As usual, let F_1 denote the focus located at $(-c, 0)$. Show that $F_1P = xe + a$. *Hint:* Use the result in Exercise 43 along with the fact that the right-hand branch is defined by the equation $F_1P - F_2P = 2a$.

45. Let P be a point on the right-hand branch of the hyperbola $x^2 - y^2 = k^2$. If d denotes the distance from P to the center of the hyperbola, show that

$$d^2 = (F_1P)(F_2P)$$

46. Let $P(x, y)$ be a point on the right-hand branch of the hyperbola $(x^2/a^2) - (y^2/b^2) = 1$. From P, a line segment is drawn perpendicular to the line $x = a/e$, meeting this line at D. If F_2 denotes (as usual) the focus located at $(c, 0)$, show that

$$\frac{F_2P}{PD} = e$$

The line $x = a/e$ is called the *directrix* of the hyperbola corresponding to the focus F_2. *Hint:* Use the expression for F_2P developed in Exercise 43.

47. Consider the hyperbola $(x^2/a^2) - (y^2/b^2) = 1$, and let D be the point where the asymptote $y = (b/a)x$ meets the directrix $x = a/e$. If F denotes the focus corresponding to this directrix, show that the line segment $\overline{FD}$ is perpendicular to the asymptote.

48. Consider the hyperbola $(x^2/a^2) - (y^2/b^2) = 1$ and suppose that a perpendicular is drawn from the focus $F_2(c, 0)$ to the asymptote $y = (b/a)x$, meeting the asymptote at A.
(a) Show that $F_2A = b$.
(b) Show that the distance from the center of the hyperbola to A is equal to a.

49. By solving the system

$$\begin{cases} y = \frac{4}{3}x - 1 \\ 16x^2 - 9y^2 = 144 \end{cases}$$

show that the line and the hyperbola intersect in exactly one point. Draw a sketch of the situation. This demonstrates that a line that intersects a hyperbola in exactly one point does not have to be a tangent line.

In Exercises 50–52, find the equation of the line that is tangent to the hyperbola at the given point. Write your answer in the form $y = mx + b$.

50. $x^2 - 4y^2 = 16$; $(5, 3/2)$

51. $3x^2 - y^2 = 12$; $(4, 6)$

52. $16x^2 - 25y^2 = 400$; $(10, 4\sqrt{3})$

C

53. We define a *tangent line to a hyperbola* as a line that is not parallel to an asymptote and that intersects the hyperbola in exactly one point. Show that the equation of the line tangent to the hyperbola $(x^2/a^2) - (y^2/b^2) = 1$ at the point (x_1, y_1) on the curve is

$$\frac{x_1x}{a^2} - \frac{y_1y}{b^2} = 1$$

Hint: Allow for signs, but follow exactly the same steps as were supplied in Exercise 64 of Exercise Set 11.4, where we found the tangent to the ellipse. You should find that the slope in the present case is $m = (b^2x_1/a^2y_1)$. Explain why this slope cannot equal the slope of an asymptote as long as (x_1, y_1) is on the hyperbola.

54. (The results in this exercise will be used in Exercise 56 to prove the reflection property of the hyperbola.) Let $P(x_1, y_1)$ be a point on the right-hand branch of the hyperbola $(x^2/a^2) - (y^2/b^2) = 1$. Suppose that the tangent to the hyperbola at P meets the x-axis at A.
(a) Show that the coordinates of A are $(a^2/x_1, 0)$.
(b) Show that $F_1A = a(x_1e + a)/x_1$.
(c) Show that $F_2A = a(x_1e - a)/x_1$.

55. (The result in this exercise will be used in Exercise 56 to prove the reflection property of the hyperbola.) Refer to the following figure. If $F_1P/F_1A = F_2P/F_2A$, show that $\alpha = \beta$.

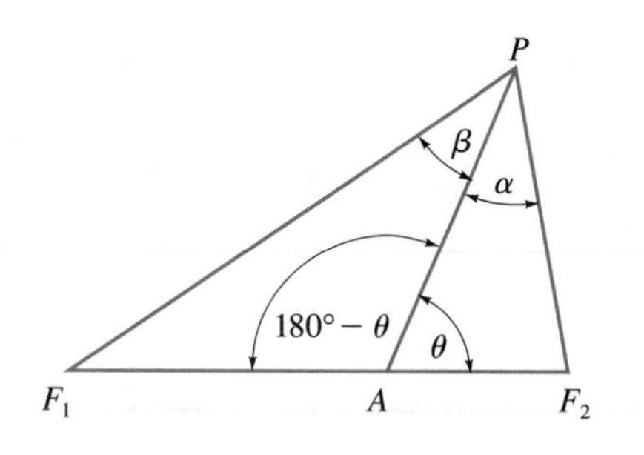

Hint: Use the law of sines in triangle AF_2P to show that $\dfrac{F_2P}{F_2A} = \dfrac{\sin\theta}{\sin\alpha}$. Then carry out the same procedure in triangle AF_1P to show that $\dfrac{F_1P}{F_1A} = \dfrac{\sin\theta}{\sin\beta}$.

56. According to the reflection property of the hyperbola, the tangent to the hyperbola at P bisects the angle formed by the focal radii at P. This exercise outlines a proof of this property. The figure (at the top of the next column) shows a point $P(x_1, y_1)$ on the right-hand branch of the hyperbola $(x^2/a^2) - (y^2/b^2) = 1$. The tangent at P cuts the x-axis at A.
(a) Show that $F_1P/F_1A = x_1/a$. *Hint:* This is easy if you use the expressions for F_1P and F_1A developed in Exercises 44 and 54, respectively.
(b) Show that $F_2P/F_2A = x_1/a$. *Hint:* Use the expressions for F_2P and F_2A developed in Exercises 43 and 54, respectively.
(c) Use the results in parts (a) and (b) along with the result of Exercise 55 to conclude that angle F_1PA equals angle F_2PA, as required.

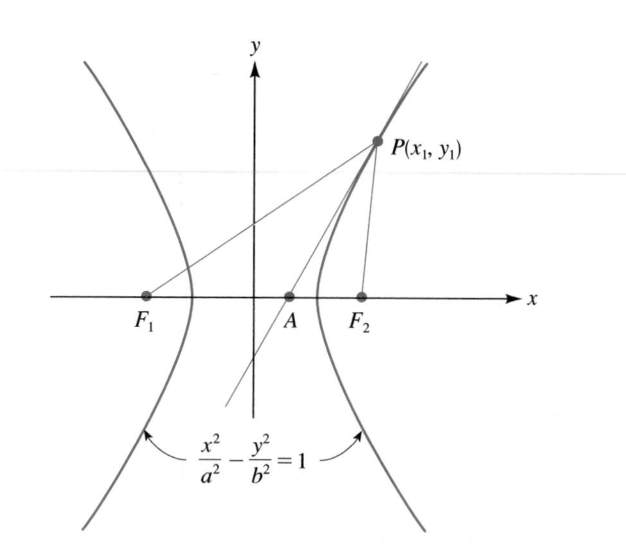

57. The *normal line* to a hyperbola at a point P on the hyperbola is the line through P that is perpendicular to the tangent at P. If the coordinates of P are (x_1, y_1), show that the equation of the normal line is

$$a^2y_1x + b^2x_1y = x_1y_1(a^2 + b^2)$$

58. Suppose the point $P(x_1, y_1)$ is on the hyperbola $(x^2/a^2) - (y^2/b^2) = 1$. A tangent is drawn at P, meeting the x-axis at A and the y-axis at B. Also, perpendiculars are drawn from P to the x- and y-axes, meeting these axes at C and D, respectively. If O denotes the origin, show that: **(a)** $OA \cdot OC = a^2$; and **(b)** $OB \cdot OD = b^2$.

59. Suppose the point $P(x_1, y_1)$ is in the first quadrant on the hyperbola $(x^2/a^2) - (y^2/b^2) = 1$. The normal line through P meets the x- and y-axes at Q and R, respectively. If O denotes the origin, show that the area of $\triangle OQR$ is $\dfrac{x_1y_1(a^2 + b^2)^2}{2a^2b^2}$.

60. Let $P(x_1, y_1)$ be a point in the first quadrant on the hyperbola $(x^2/a^2) - (y^2/b^2) = 1$. Let D be the point where the tangent at P meets the line $x = a/e$. Show that $\angle DF_2P$ is a right angle.

61. The tangent to the hyperbola $(x^2/a^2) - (y^2/b^2) = 1$ at the point P meets the asymptotes at A and B. Show that P is the midpoint of $\overline{AB}$.

62. At the point P on the hyperbola $(x^2/a^2) - (y^2/b^2) = 1$, a tangent line is drawn that meets the lines $x = a$ and $x = -a$ at S and T, respectively. Show that the circle with $\overline{ST}$ as a diameter passes through the two foci of the hyperbola.

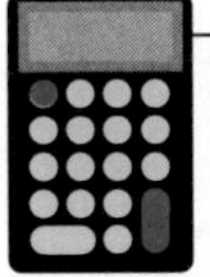

GRAPHING UTILITY EXERCISES FOR SECTION 11.5

1. In this exercise you are going to look at the graph of the hyperbola $16x^2 - 9y^2 = 144$ from two perspectives.
   (a) Solve the given equation for y, then graph the two resulting functions in the standard viewing rectangle.
   (b) Determine the equations of the asymptotes. Add the graphs of the asymptotes to your picture from part (a).
   (c) Looking at your picture from part (b), you can see that the hyperbola seems to be moving closer and closer to its asymptotes as $|x|$ gets large. To see more dramatic evidence of this, change the viewing rectangle so that both x and y extend from -100 to 100. At this scale, the hyperbola is virtually indistinguishable from its asymptotes.
2. In this exercise we graph the hyperbola $\dfrac{(y-3)^2}{5^2} - \dfrac{(x-4)^2}{3^2} = 1$.
   (a) Solve the equation for y to obtain $y = 3 \pm 5\sqrt{1 + \frac{1}{9}(x-4)^2}$.
   (b) In the standard viewing rectangle, graph the two equations that you obtained in part (a). Then, for a better view, adjust the viewing rectangle so that both x and y extend from -20 to 20.
   (c) Show that the equations of the asymptotes are $y = \pm\frac{5}{3}(x-4) + 3$. Add the graphs of these asymptotes to your picture from part (b). Check to see that your result is consistent with what you see in Figure 8 on page 716.
3. In this exercise we graph a hyperbola in which the axes of the curve are not parallel to the coordinate axes. The equation is $x^2 + 4xy - 2y^2 = 6$.
   (a) Use the quadratic formula to solve the equation for y in terms of x. Show that the result can be written $y = x \pm \frac{1}{2}\sqrt{6x^2 - 12}$.
   (b) Graph the two equations obtained in part (a). Use the standard viewing rectangle.
   (c) It can be shown that the equations of the asymptotes are $y = (1 \pm 0.5\sqrt{6})x$. Add the graphs of these asymptotes to the picture that you obtained in part (b).
   (d) Change the viewing rectangle so that both x and y extend from -50 to 50. What do you observe?
4. Use the technique indicated in Exercise 3 to graph the hyperbola $x^2 + xy - 2y^2 = 1$. Use a viewing rectangle that extends from -20 to 20 in both the x- and the y-directions.

11.6 THE FOCUS–DIRECTRIX PROPERTY OF CONICS

Here then [in the *Conics,* written by Apollonius of Perga (ca. 262–ca. 190 B.C.)] *we have the properties of the three curves expressed in the precise language of the Pythagorean application of areas, and the curves are named accordingly:* parabola *(παραβολη) where the rectangle is exactly* applied [equal], hyperbola *(υπερβολη) where it* exceeds, *and* ellipse *(ελλειψιζ) where it* falls short.

Sir Thomas Heath in *A History of Greek Mathematics,* Vol. II (Oxford: The Clarendon Press, 1921)

The most important contribution which Pappus [ca. 300] *made to our knowledge of the conics was his publication of the focus, directrix, eccentricity theorem.*

Julian Lowell Coolidge in *A History of the Conic Sections and Quadric Surfaces* (London: Oxford University Press, 1946)

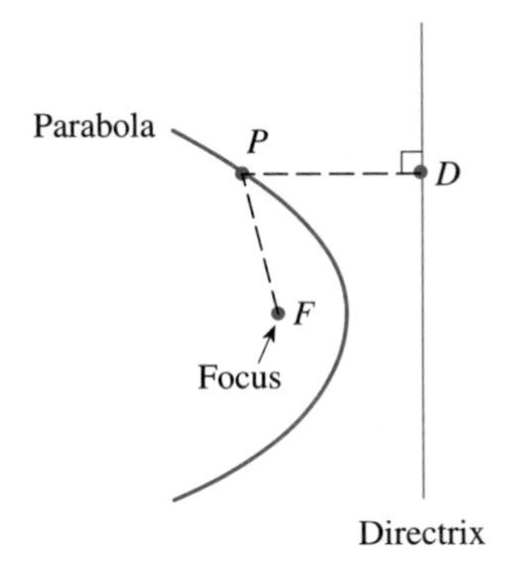

FIGURE 1

We begin by recalling the focus–directrix property that we used in defining the parabola. A point P is on a parabola if and only if the distance from P to the focus is equal to the distance from P to the directrix. For the parabola in Figure 1, this means that for each point P on the parabola, we have $FP = PD$

or, equivalently,

$$\frac{FP}{PD} = 1$$

The ellipse and the hyperbola also can be characterized by focus–directrix properties. We'll begin with the ellipse. To help us in subsequent computations, we need to know the lengths of the line segments $\overline{F_1P}$ and $\overline{F_2P}$ in Figure 2.

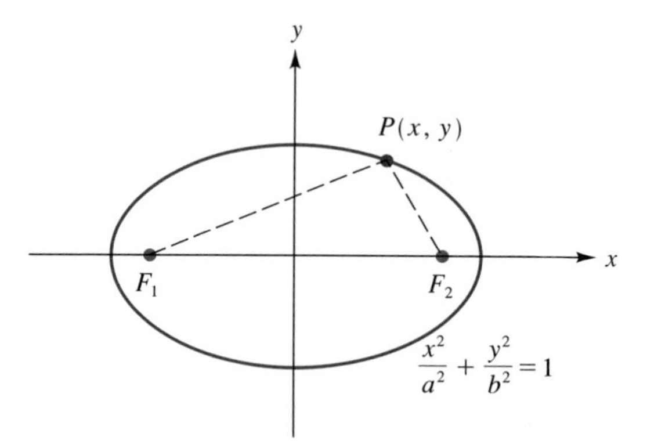

FIGURE 2
The segments drawn from a point on the ellipse to the foci are called focal radii. Here $\overline{F_1P}$ and $\overline{F_2P}$ are the focal radii.

These line segments joining a point on the ellipse to the foci are called **focal radii.** In the box that follows, we give a formula for the length of each focal radius. In the formulas, $e(= c/a)$ denotes the eccentricity of the ellipse. (For the derivation of these formulas, see Exercise 15 at the end of this section.)

THE FOCAL RADII OF THE ELLIPSE $\frac{x^2}{a^2} + \frac{y^2}{b^2} = 1$

As indicated in Figure 2, let $P(x, y)$ be a point on the ellipse. Then the lengths of the focal radii are

$$F_1P = a + ex \qquad \text{and} \qquad F_2P = a - ex$$

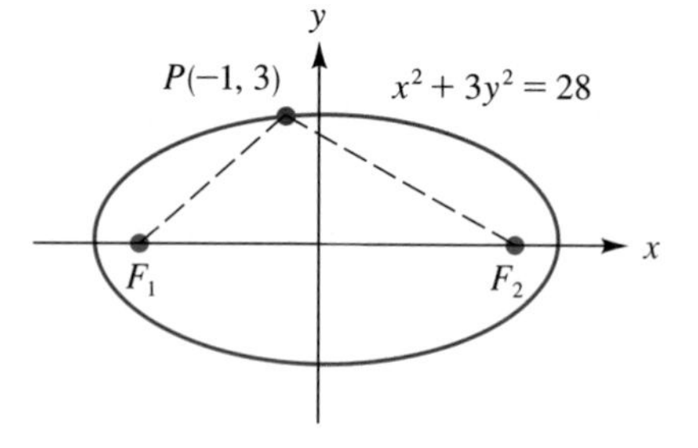

FIGURE 3

EXAMPLE 1 Figure 3 shows the ellipse $x^2 + 3y^2 = 28$. Compute the lengths of the focal radii drawn to the point $P(-1, 3)$.

Solution To apply the formulas in the box, we need to know the values of a and e for the given ellipse. (We already know that $x = -1$; that is given.) As you should verify for yourself using the techniques of Section 11.4, we have $a = 2\sqrt{7}$ and $e = \sqrt{6}/3$. Therefore

$$\begin{aligned} F_1P &= a + ex \\ &= 2\sqrt{7} + \left(\frac{\sqrt{6}}{3}\right)(-1) = \frac{6\sqrt{7} - \sqrt{6}}{3} \end{aligned}$$

and

$$\begin{aligned} F_2P &= a - ex \\ &= 2\sqrt{7} - \left(\frac{\sqrt{6}}{3}\right)(-1) = \frac{6\sqrt{7} + \sqrt{6}}{3} \end{aligned}$$

In Figure 4 we show the ellipse $(x^2/a^2)+(y^2/b^2)=1$ and the vertical line $x=a/e$. The line $x=a/e$ is a **directrix** of the ellipse. In Figure 4, F_2P is the distance from the focus F_2 to a point P on the ellipse, and PD is the distance from P to the directrix $x=a/e$. We are going to prove the following remarkable *focus–directrix property of the ellipse:* For any point $P(x, y)$ on the ellipse, the ratio of F_2P to PD is equal to e, the eccentricity of the ellipse. To prove this, we proceed as follows:

$$\frac{F_2P}{PD}=\frac{a-ex}{PD} \qquad \text{using our formula for } F_2P$$

$$=\frac{a-ex}{\dfrac{a}{e}-x} \qquad \text{using Figure 4}$$

$$=\frac{e(a-ex)}{a-ex} \qquad \text{multiplying both numerator and denominator by } e$$

$$=e$$

So, $F_2P/PD=e$, as we wished to show.

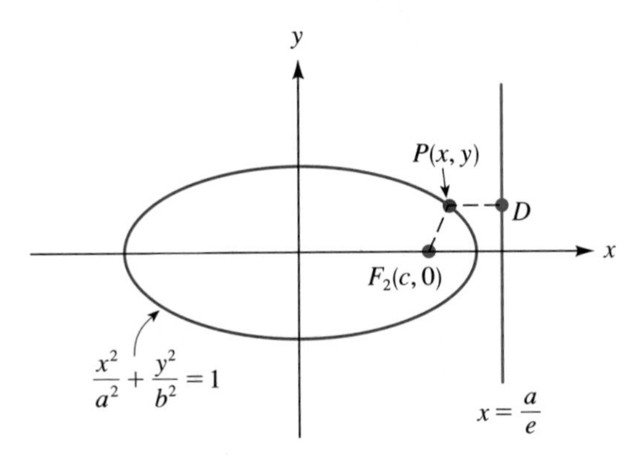

FIGURE 4

The directrix $x=a/e$ is associated with the focus $F_2(c, 0)$. What about the focus $F_1(-c, 0)$? From the symmetry of the ellipse, it follows that the line $x=-a/e$ is the directrix associated with this focus. More specifically, referring to Figure 5, not only do we have $F_2P/PD=e$, we also have $F_1P/PE=e$.

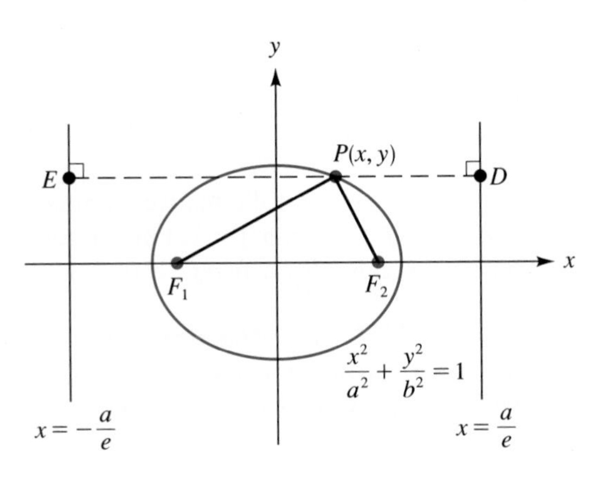

FIGURE 5
For every point P on the ellipse, we have $F_1P/PE=e$ and $F_2P/PD=e$.

We have seen that each point P on the ellipse in Figure 5 satisfies the following two equations:

$$\frac{F_1P}{PE} = e \tag{1}$$

$$\frac{F_2P}{PD} = e \tag{2}$$

Now, conversely, suppose that the point $P(x, y)$ in Figure 6 satisfies equations (1) and (2). We are going to show that $P(x, y)$ must, in fact, lie on the ellipse $(x^2/a^2) + (y^2/b^2) = 1$.*

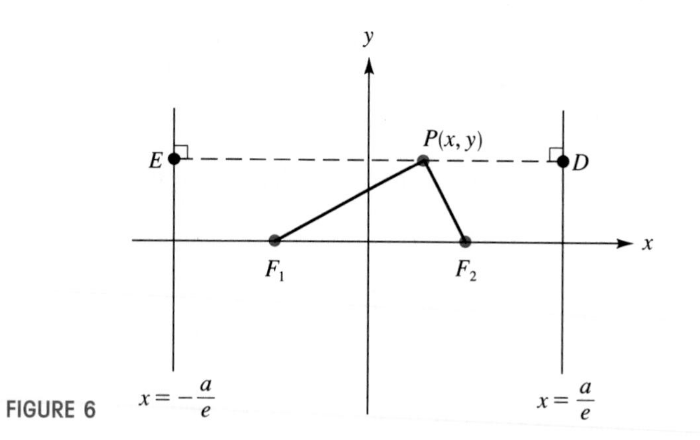

FIGURE 6

In Figure 6, the distance from E to D is $2a/e$. Therefore

$$PE + PD = 2a/e$$

and, consequently,

$$\frac{F_1P}{e} + \frac{F_2P}{e} = \frac{2a}{e} \qquad \text{using equations (1) and (2) to substitute for } PE \text{ and } PD$$

or

$$F_1P + F_2P = 2a \qquad \text{multiplying both sides by } e$$

Thus, by definition, P lies on the ellipse $(x^2/a^2) + (y^2/b^2) = 1$, as we wished to show.

In the box that follows, we summarize the focus–directrix properties of the ellipse $(x^2/a^2) + (y^2/b^2) = 1$. For reference, we also include the original defining property of the ellipse from Section 11.4.

*The short proof given here was communicated to the author by Professor Ray Redheffer. For a proof using equation (2) but not equation (1), see Exercise 16.

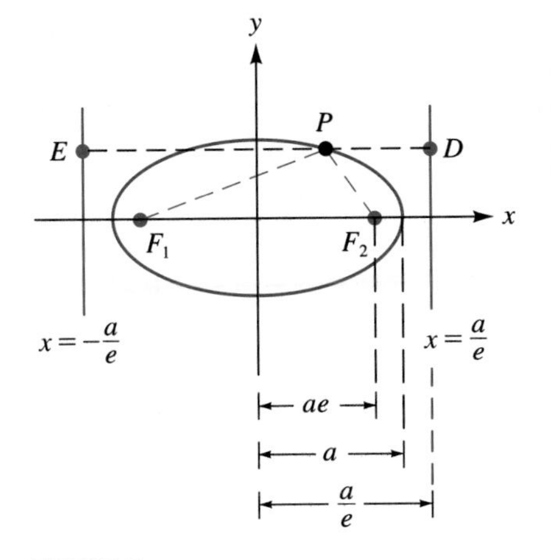

FIGURE 7
The ellipse $(x^2/a^2)+(y^2/b^2)=1$ with foci $F_1(-c, 0)$ and $F_2(c, 0)$ and directrices $x = \pm a/e$.

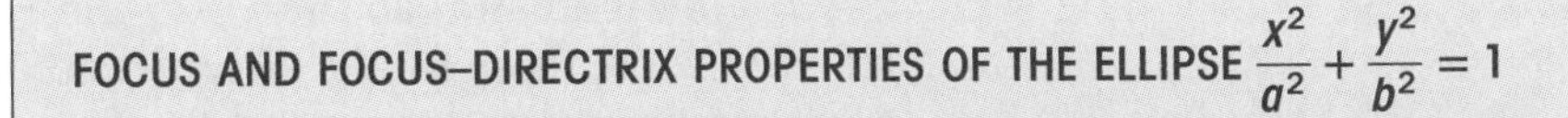

FOCUS AND FOCUS–DIRECTRIX PROPERTIES OF THE ELLIPSE $\dfrac{x^2}{a^2}+\dfrac{y^2}{b^2}=1$

Refer to Figure 7.

1. A point P is on the ellipse if and only if the sum of the distances from P to the foci $F_1(-c, 0)$ and $F_2(c, 0)$ is $2a$.
2. A point P is on the ellipse if and only if
$$\frac{F_2P}{PD}=e$$
where F_2P is the distance from P to the focus F_2, PD is the distance from P to the directrix $x=a/e$, and e $(=c/a)$ is the eccentricity of the ellipse.
3. A point P is on the ellipse if and only if
$$\frac{F_1P}{PE}=e$$
where F_1P is the distance from P to the focus F_1 and PE is the distance from P to the directrix $x=-a/e$.

EXAMPLE 2 The foci of an ellipse are $(\pm 3, 0)$ and the directrix corresponding to the focus $(3, 0)$ is $x=5$. Find the equation of the ellipse. Write the answer in the form $Ax^2+By^2=C$.

Solution Using the given data (and referring to Figure 7), we have

$$ae=3 \qquad \text{and} \qquad a/e=5$$

From the first equation we obtain $e=3/a$. Substituting this value of e in the second equation yields

$$\frac{a}{3/a}=5 \qquad \text{and therefore} \qquad a^2=15$$

Now we can calculate b^2 by using the relation $b^2=a^2-c^2$ and the given information $c=3$:

$$b^2=a^2-c^2=15-9=6$$

Substituting the values we've obtained for a^2 and b^2 in the equation $(x^2/a^2)+(y^2/b^2)=1$ yields

$$\frac{x^2}{15}+\frac{y^2}{6}=1$$

or

$$2x^2+5y^2=30 \qquad \text{multiplying both sides by 30}$$

The focus–directrix property can be developed for the hyperbola in just the same way that we have proceeded for the ellipse. In fact, the algebra is so similar that we shall omit the details here and simply summarize the results. The hyperbola $(x^2/a^2)-(y^2/b^2)=1$ has two directrices: they are the vertical lines

$x = a/e$ in Figure 8(a) and $x = -a/e$ in Figure 8(b). Notice that these equations are identical to those for the directrices of the ellipse. In the box following Figure 8, we summarize the focus and the focus–directrix properties of the hyperbola.

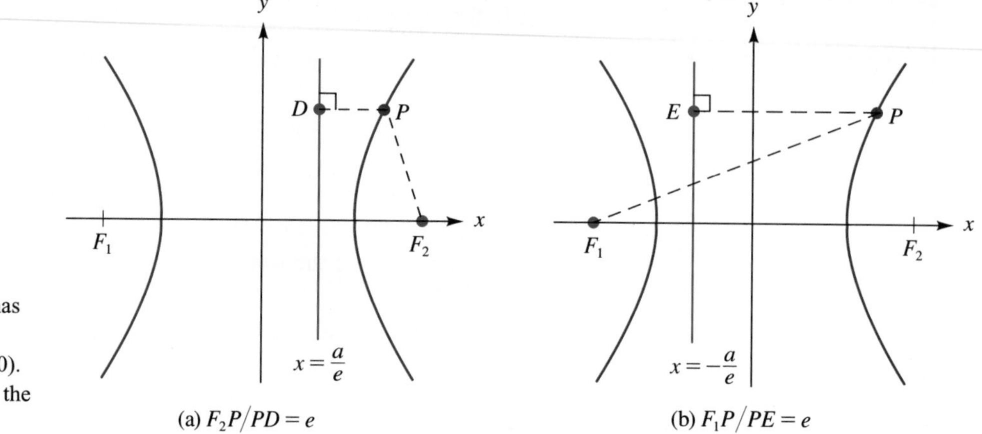

FIGURE 8
The hyperbola $(x^2/a^2) - (y^2/b^2) = 1$ has two directrices. The directrix $x = a/e$ corresponds to the focus $F_2(c, 0)$. The directrix $x = -a/e$ corresponds to the focus $F_1(-c, 0)$.

FOCUS AND FOCUS–DIRECTRIX PROPERTIES OF THE HYPERBOLA $\frac{x^2}{a^2} - \frac{y^2}{b^2} = 1$

Refer to Figure 8.

1. A point P is on the hyperbola if and only if the difference of the distances from P to the foci $F_1(-c, 0)$ and $F_2(c, 0)$ is $2a$.
2. A point P is on the hyperbola if and only if

$$\frac{F_2P}{PD} = e$$

where F_2P is the distance from P to the focus F_2, PD is the distance from P to the directrix $x = a/e$, and e $(= c/a)$ is the eccentricity of the hyperbola.

3. A point P is on the hyperbola if and only if

$$\frac{F_1P}{PE} = e$$

where F_1P is the distance from P to the focus F_1 and PE is the distance from P to the directrix $x = -a/e$.

EXAMPLE 3 Determine the foci, the eccentricity, and the directrices for the hyperbola $9x^2 - 16y^2 = 144$.

Solution To convert the given equation to standard form, we divide both sides by 144. This yields

$$\frac{x^2}{4^2} - \frac{y^2}{3^2} = 1$$

So, we have $a = 4$, $b = 3$, and

$$c = \sqrt{a^2 + b^2} = \sqrt{4^2 + 3^2} = 5$$

The foci are therefore $(\pm 5, 0)$, and the eccentricity is

$$e = \frac{c}{a} = \frac{5}{4}$$

The directrices are

$$x = \pm \frac{a}{e} = \pm \frac{4}{5/4} = \pm \frac{16}{5}$$

EXAMPLE 4 Figure 9 shows the graph of the hyperbola $9x^2 - 16y^2 = 144$. (This is the hyperbola discussed in the previous example.)

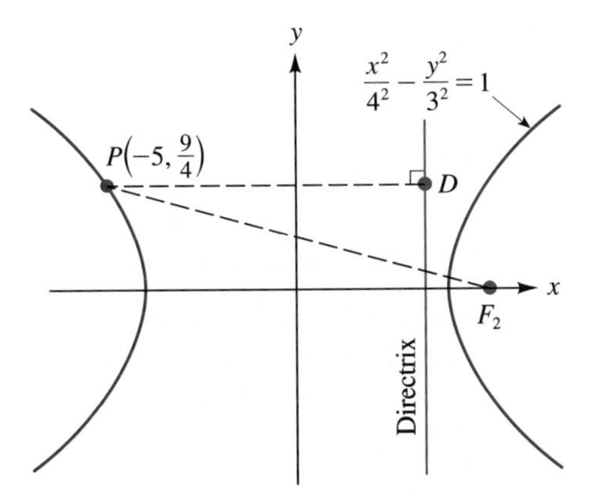

FIGURE 9

(a) Verify that the point $P(-5, 9/4)$ lies on this hyperbola.
(b) Compute F_2D and PD, and verify that

$$\frac{F_2P}{PD} = e$$

Solution
(a) Substituting the values $x = -5$ and $y = 9/4$ in the equation of the hyperbola yields

$$9(-5)^2 - 16(9/4)^2 = 144$$

or

$$225 - 81 = 144$$

Since this last equation is correct, we conclude that the point $P(-5, 9/4)$ indeed lies on the hyperbola.
(b) From the previous example, we know that the coordinates of the focus F_2 are $(5, 0)$. So, using the distance formula, we have

$$F_2P = \sqrt{(5-(-5))^2 + \left(0 - \frac{9}{4}\right)^2}$$

$$= \sqrt{100 + \frac{81}{16}} = \sqrt{\frac{1681}{16}} = \frac{41}{4}$$

Next, we use the fact (from Example 3) that the equation of the directrix in Figure 9 is $x = 16/5$. Thus, in Figure 9,

$$PD = |-5| + \frac{16}{5} = \frac{41}{5}$$

Finally, we compute the ratio F_2P/PD:

$$\frac{F_2P}{PD} = \frac{41/4}{41/5} = \frac{5}{4}$$

This is the same number that we obtained for the eccentricity in Example 3. So, we have verified in this case that the ratio of F_2P to PD is equal to the eccentricity.

The focus–directrix property provides a unified approach to the parabola, the ellipse, and the hyperbola. For both the ellipse and the hyperbola, we've seen that for any point P on the curve, we have

$$\frac{\text{distance from } P \text{ to a focus}}{\text{distance from } P \text{ to the corresponding directrix}} = e \tag{3}$$

where e is a positive constant, the eccentricity of the curve. For the ellipse we have $e < 1$, and for the hyperbola we have $e > 1$. Equation (3) also holds for the parabola if we agree on the convention that the eccentricity of a parabola is 1. This is because with $e = 1$, equation (3) tells us that the distance from P to the focus equals the distance from P to the directrix, which is the defining condition for the parabola. The following theorem summarizes these remarks.

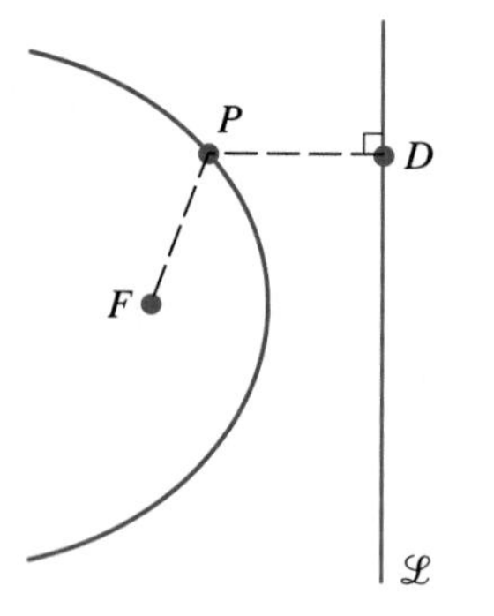

FIGURE 10

THEOREM: THE FOCUS-DIRECTRIX PROPERTY OF CONICS

Refer to Figure 10. Let $\mathcal{L}$ be a fixed line, F a fixed point, and e a positive constant. Consider the set of points P satisfying the condition $FP/PD = e$. (The point D is defined in the figure.) Then

(a) If $e = 1$, the set of points is a *parabola* with focus F and directrix $\mathcal{L}$.
(b) If $0 < e < 1$, the set of points is an *ellipse* with focus F, corresponding directrix $\mathcal{L}$, and eccentricity e.
(c) If $e > 1$, the set of points is a *hyperbola* with focus F, corresponding directrix $\mathcal{L}$, and eccentricity e.

NOTE In our development in this section, we've always considered cases in which the directrix is vertical, for simplicity. However, the preceding theorem is valid for any orientation of the directrix $\mathcal{L}$.

EXERCISE SET 11.6

A

In Exercises 1–4, you are given an ellipse and a point P on the ellipse. Find F_1P and F_2P, the lengths of the focal radii.

1. $x^2 + 3y^2 = 76;\ P(-8, 2)$

2. $x^2 + 3y^2 = 57;\ P(3, -4)$

3. $(x^2/15^2) + (y^2/5^2) = 1;\ P(9, 4)$

4. $2x^2 + 3y^2 = 14;\ P(-1, -2)$

In Exercises 5–10, determine the foci, the eccentricity, and the directrices for each ellipse and hyperbola.

5. **(a)** $(x^2/4^2) + (y^2/3^2) = 1$
(b) $(x^2/4^2) - (y^2/3^2) = 1$

6. **(a)** $x^2 + 4y^2 = 1$
(b) $x^2 - 4y^2 = 1$

7. **(a)** $12x^2 + 13y^2 = 156$
(b) $12x^2 - 13y^2 = 156$

8. **(a)** $x^2 + 2y^2 = 2$
(b) $x^2 - 2y^2 = 2$

9. **(a)** $25x^2 + 36y^2 = 900$
(b) $25x^2 - 36y^2 = 900$

10. **(a)** $4x^2 + 25y^2 = 100$
(b) $4x^2 - 25y^2 = 100$

In Exercises 11 and 12, use the given information to find the equation of the ellipse. Write the answer in the form $Ax^2 + By^2 = C$.

11. The foci are $(\pm 1, 0)$ and the directrices are $x = \pm 4$.

12. The foci are $(\pm\sqrt{3}, 0)$ and the eccentricity is 2/3.

In Exercises 13 and 14, use the given information to find the equation of the hyperbola. Write the answer in the form $Ax^2 - By^2 = C$.

13. The foci are $(\pm 2, 0)$ and the directrices are $x = \pm 1$.

14. The foci are $(\pm 3, 0)$ and the eccentricity is 2.

B

15. In this exercise we show that the focal radii of the ellipse $(x^2/a^2) + (y^2/b^2) = 1$ are $F_1P = a + ex$ and $F_2P = a - ex$. The method used here, which avoids the use of radicals, appears in the eighteenth-century text *Traité analytique des sections coniques* (Paris: 1707) by the Marquis de l'Hospital (1661–1704). (For another method, one that does use radicals, see Exercise 51 in Section 11.4.) For convenience, let $d_1 = F_1P$ and $d_2 = F_2P$, as indicated in the accompanying figure.

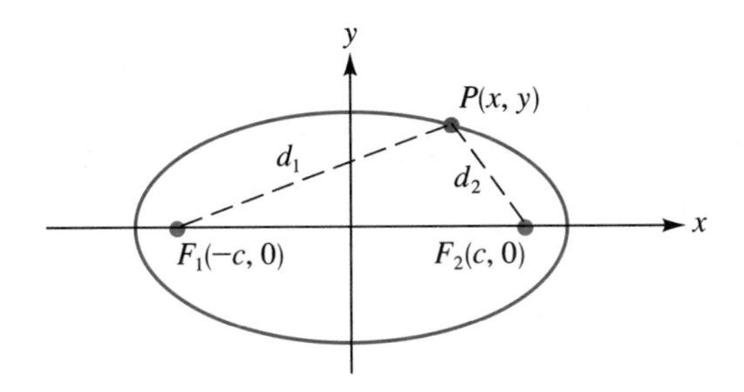

(a) Using the distance formula, verify that

$$d_1^2 = (x + c)^2 + y^2 \quad \text{and} \quad d_2^2 = (x - c)^2 + y^2$$

(b) Use the two equations in part (a) to show that $d_1^2 - d_2^2 = 4cx$.

(c) Explain why $d_1 + d_2 = 2a$.

(d) Factor the left-hand side of the equation in part (b) and substitute for one of the factors using the equation in part (c). Show that the result can be written $d_1 - d_2 = 2cx/a$.

(e) Add the equations in parts (c) and (d). Show that the resulting equation can be written $d_1 = a + ex$, as required.

(f) Use the equation in part (c) and the result in part (e) to show that $d_2 = a - ex$.

16. In this exercise we show that if the point $P(x, y)$ in the accompanying figure satisfies the condition

$$\frac{F_2P}{PD} = e$$

then, in fact, the point $P(x, y)$ lies on the ellipse $(x^2/a^2) + (y^2/b^2) = 1$.

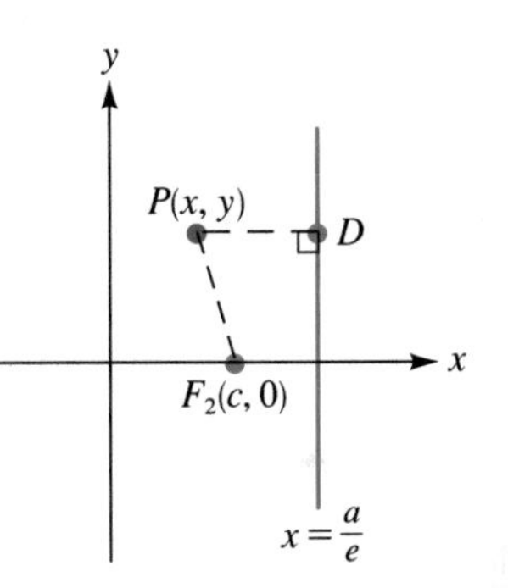

(a) From the given equation we have $(F_2P)^2 = e^2(PD)^2$. Use the distance formula and the figure to deduce from this equation that

$$(x - c)^2 + y^2 = e^2\left(\frac{a}{e} - x\right)^2$$

(b) In the equation in part (a), replace e with c/a. After carrying out the indicated operations and simplifying, show that the equation can be written

$$a^2x^2 - c^2x^2 + a^2y^2 = a^4 - a^2c^2$$

(c) The equation in part (b) is equivalent to $(a^2 - c^2)x^2 + a^2y^2 = a^2(a^2 - c^2)$. Now replace the quantity $a^2 - c^2$ by b^2 and show that the resulting equation can be written $(x^2/a^2) + (y^2/b^2) = 1$; thus, P lies on the ellipse, as we wished to show.

11.7 THE CONICS IN POLAR COORDINATES

The use of polar coordinates permits a unified treatment of the conic sections, and it is the polar coordinate equations for these curves that are used in celestial mechanics.

Professor Bernice Kastner in her text *Space Mathematics,* published in 1985 by NASA (National Aeronautics and Space Administration)

As indicated in the opening quotation, the polar coordinate equations of the parabola, ellipse, and hyperbola are useful in applications. We will develop these equations by using the focus–directrix property of the conics that we discussed in Section 11.6. Suppose that we have a conic with focus F, directrix $\mathcal{L}$, and eccentricity e. Then, as shown in Figure 1, we can set up a polar coordinate system in which the focus of the conic is the origin or pole, and the directrix is perpendicular to the polar axis. In Figure 1, we've used d to denote the distance from the focus to the directrix.

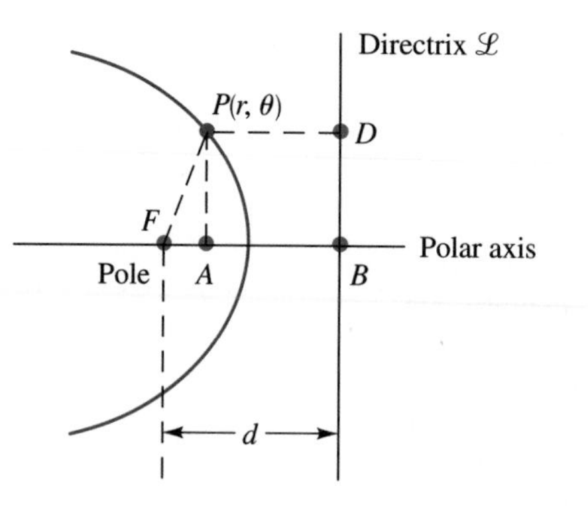

FIGURE 1
A conic with focus F and directrix $\mathcal{L}$.

Using the focus–directrix property for the conic in Figure 1, we have

$$\frac{FP}{PD} = e \qquad \text{and therefore} \qquad \frac{r}{FB - FA} = e$$

In this last equation, we can replace FB by d and FA by $r \cos \theta$. (Why?) This yields

$$\frac{r}{d - r \cos \theta} = e$$

Now, as Exercise 22 asks you to verify, when we solve this equation for r, we obtain

$$r = \frac{ed}{1 + e \cos \theta}$$

This is the polar form for the equation of the conic in Figure 1, in which the directrix is vertical and to the right of the focus. By the same technique, we can obtain similar equations when the directrix is to the left of the focus and when the directrix is horizontal. We summarize the results in the following box.

POLAR EQUATIONS OF THE CONICS

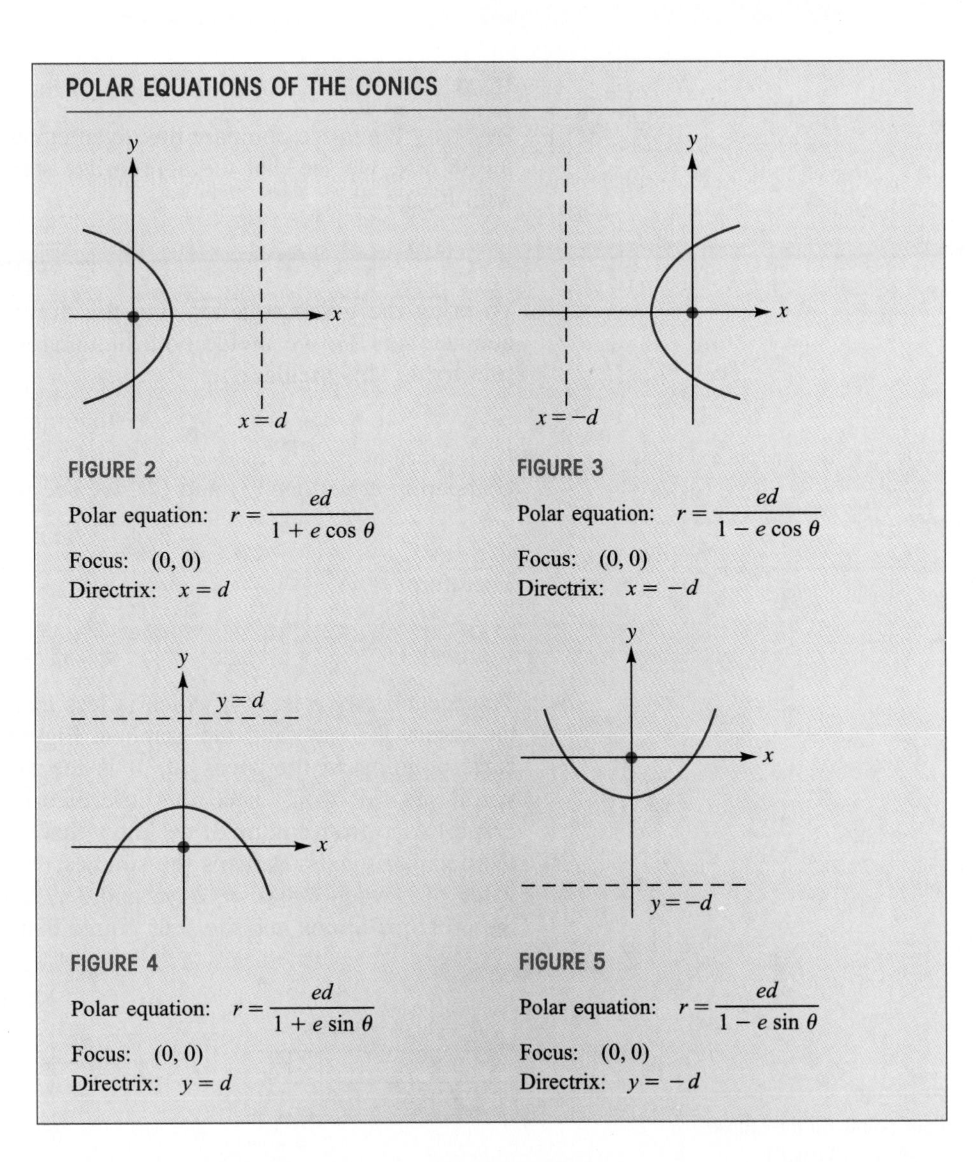

FIGURE 2

Polar equation: $r = \dfrac{ed}{1 + e\cos\theta}$

Focus: (0, 0)

Directrix: $x = d$

FIGURE 3

Polar equation: $r = \dfrac{ed}{1 - e\cos\theta}$

Focus: (0, 0)

Directrix: $x = -d$

FIGURE 4

Polar equation: $r = \dfrac{ed}{1 + e\sin\theta}$

Focus: (0, 0)

Directrix: $y = d$

FIGURE 5

Polar equation: $r = \dfrac{ed}{1 - e\sin\theta}$

Focus: (0, 0)

Directrix: $y = -d$

For the example that follows and for the exercises at the end of this section, it will be convenient to have formulas that express b in terms of e and a for the ellipse and the hyperbola. For the ellipse, we have

$$b^2 = a^2 - c^2 = a^2 - (ae)^2 = a^2(1 - e^2)$$

and therefore

$$b = a\sqrt{1 - e^2} \qquad \text{for the ellipse} \tag{1}$$

Similarly, for the hyperbola, we have

$$b^2 = c^2 - a^2 = (ae)^2 - a^2 = a^2(e^2 - 1)$$

and therefore

$$b = a\sqrt{e^2 - 1} \qquad \text{for the hyperbola} \tag{2}$$

EXAMPLE 1 Sketch the graph of the conic $r = 8/(4 - 3\cos\theta)$.

Solution When we compare the given equation with the four basic types shown in the box, we see that the appropriate standard equation is the one associated with Figure 3:

$$r = \frac{ed}{1 - e\cos\theta} \tag{3}$$

To bring the given equation into this form (in which the first term in the denominator is 1), we divide both numerator and denominator on the right-hand side by 4. This yields

$$r = \frac{2}{1 - \frac{3}{4}\cos\theta} \tag{4}$$

Comparing equations (3) and (4), we see that

$$e = 3/4 \qquad \text{and} \qquad ed = 2$$

Therefore

$$d = \frac{2}{e} = \frac{2}{3/4} = \frac{8}{3}$$

The eccentricity e is $3/4$, which is less than 1, so the conic is an ellipse. From the result $d = 8/3$ (and the graph in Figure 3), we conclude that the directrix corresponding to the focus $(0, 0)$ is the vertical line $x = -8/3$. (Actually, as you'll see, we won't need this information about the directrix in drawing the graph.) Also from Figure 3, we know that the major axis of the ellipse lies along the polar or x-axis. Perhaps the simplest way to proceed now is to compute the value of r when $\theta = 0$, $\pi/2$, π, and $3\pi/2$. In Figure 6 we show the results of these computations and the four points that are determined.

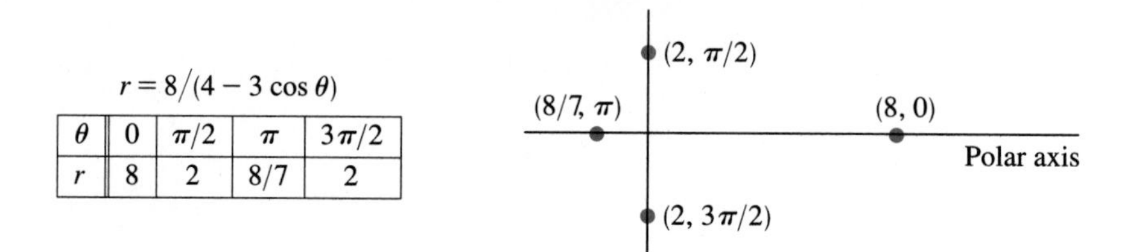

$r = 8/(4 - 3\cos\theta)$

θ	0	$\pi/2$	π	$3\pi/2$
r	8	2	8/7	2

FIGURE 6
Four points on the ellipse $r = 8/(4 - 3\cos\theta)$.

Since the major axis of this ellipse lies along the polar axis, the length of the major axis is

$$2a = 8 + \frac{8}{7} = \frac{64}{7} \qquad \text{and therefore} \qquad a = \frac{32}{7}$$

For the x-coordinate of the center of the ellipse, we want the number (on the x-axis) that is halfway between $-8/7$ and 8. As you can check (by averaging the two numbers), this x-coordinate is $24/7$.

The last piece of information that we need for drawing an accurate sketch is the value of b for this ellipse. Using equation (1), we have

$$\begin{aligned} b &= a\sqrt{1 - e^2} \\ &= \frac{32}{7}\sqrt{1 - \left(\frac{3}{4}\right)^2} = \frac{32}{7}\sqrt{\frac{7}{16}} = \frac{8}{7}\sqrt{7} \end{aligned}$$

With a calculator, we find that $b \approx 3.02$. We can now draw the graph, as shown in Figure 7.

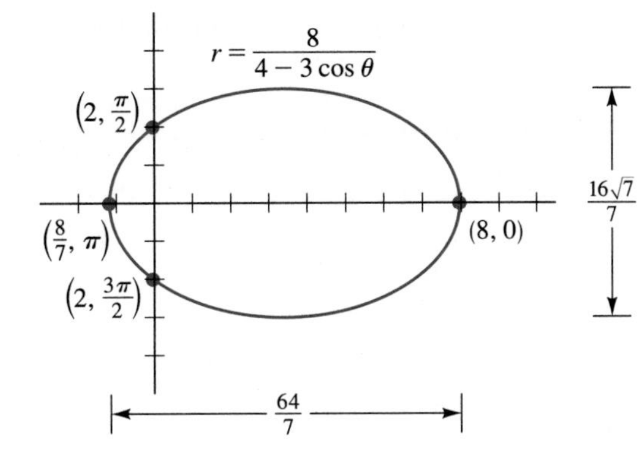

FIGURE 7

EXAMPLE 2 Find the polar equation for the parabola in Figure 8. The focus of the parabola is (0, 0) and the (rectangular) equation of the directrix is $y = -9/4$.

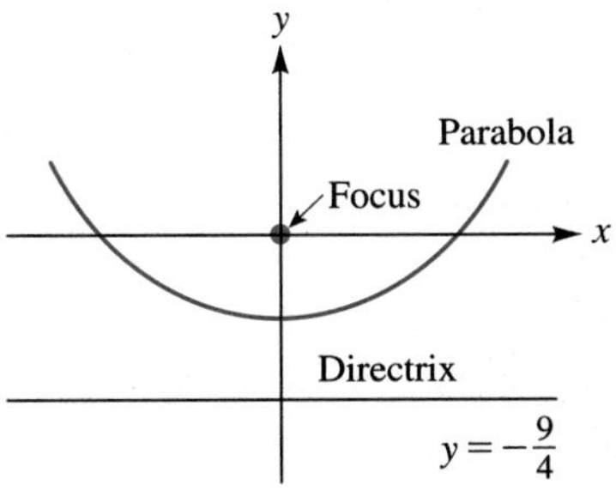

y
Parabola
Focus
x
Directrix
$y = -\frac{9}{4}$

FIGURE 8

Solution The parabola in Figure 8 is a conic of the type shown in Figure 5, so the required equation must be of the form

$$r = \frac{ed}{1 - e\sin\theta}$$

For a parabola, the eccentricity e is 1. From Figure 8, we see that the distance d from the focus to the directrix is 9/4. Using these values for e and d, the equation becomes

$$r = \frac{1(9/4)}{1 - (1)\sin\theta}$$

or

$$r = \frac{9}{4 - 4\sin\theta} \qquad \text{multiplying both numerator and denominator by 4}$$

EXAMPLE 3 Graph the conic $r = 9/(4 + 5\cos\theta)$.

Solution To bring this equation into standard form, we divide both the numerator and the denominator of the fraction by 4. This yields

$$r = \frac{9/4}{1 + (5/4)\cos\theta}$$

Comparing this with the standard form

$$r = \frac{ed}{1 + e\cos\theta}$$

we see that the eccentricity e is 5/4, which is greater than 1, so the conic is a hyperbola. Figure 2 on page 731 shows us the general form for one branch of this hyperbola. From Figure 2, we see that the transverse axis of the hyperbola must lie along the polar or x-axis. Now, just as we did in Example 1, we compute the values of r corresponding to $\theta = 0$, $\pi/2$, π, and $3\pi/2$. Figure 9(a) shows the four points that are obtained. The two points $(1, 0)$ and $(-9, \pi)$ both lie on the transverse axis of the hyperbola, so they must be the vertices. This allows us to draw the rough sketch in Figure 9(b).

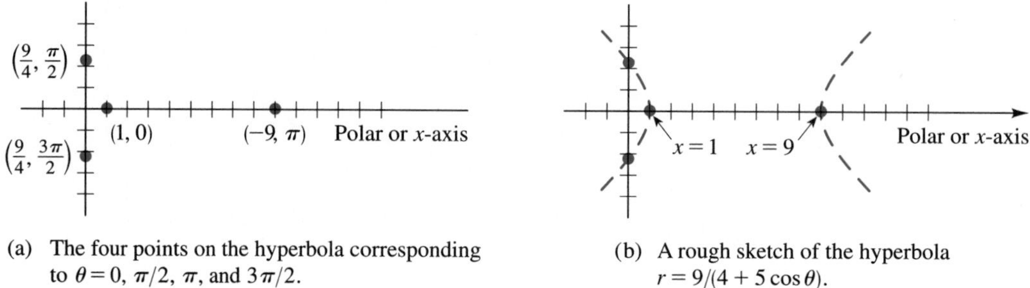

FIGURE 9

(a) The four points on the hyperbola corresponding to $\theta = 0$, $\pi/2$, π, and $3\pi/2$.

(b) A rough sketch of the hyperbola $r = 9/(4 + 5\cos\theta)$.

For a more accurate drawing of the hyperbola, we need to determine the values of a and b. Using Figure 9(b), we have

$$2a = 9 - 1 = 8 \qquad \text{and therefore} \qquad a = 4$$

The center of the hyperbola lies on the x-axis, halfway between the points with rectangular coordinates $(1, 0)$ and $(9, 0)$. Thus the center is $(5, 0)$ and consequently $c = 5$. (We are using the fact that the origin is a focus.) Now that we know the values of a and c, we can calculate b:

$$b^2 = c^2 - a^2 = 5^2 - 4^2 = 9 \qquad \text{and therefore} \qquad b = 3$$

(Another way to calculate b is to use the formula on page 731: $b = a\sqrt{e^2 - 1}$). In summary, then, we have the following information to use in drawing the hyperbola in the x-y coordinate system: the rectangular coordinates of the vertices are $(1, 0)$ and $(9, 0)$; the center is $(5, 0)$; and the values of a and b are 4 and 3, respectively. This allows us to draw the graph as shown in Figure 10.

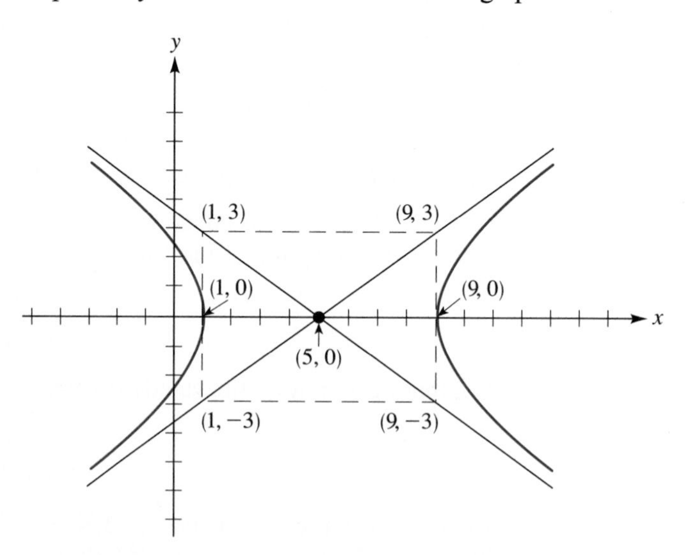

FIGURE 10

The hyperbola $r = 9/(4 + 5\cos\theta)$: the left-hand focus is located at the origin and the coordinates shown are x-y coordinates.

EXERCISE SET 11.7

A

In Exercises 1 and 2, graph each ellipse. Specify the eccentricity, the center, and the endpoints of the major and minor axes.

1. **(a)** $r = \dfrac{6}{3 + 2\cos\theta}$ **(b)** $r = \dfrac{6}{3 - 2\cos\theta}$

2. **(a)** $r = \dfrac{12}{5 + 3\sin\theta}$ **(b)** $r = \dfrac{12}{5 - 3\sin\theta}$

In Exercises 3 and 4, graph each parabola. Specify the (rectangular) coordinates of the vertex and the equation of the directrix.

3. **(a)** $r = \dfrac{5}{2 + 2\cos\theta}$ **(b)** $r = \dfrac{5}{2 - 2\cos\theta}$

4. **(a)** $r = \dfrac{2}{1 + \sin\theta}$ **(b)** $r = \dfrac{2}{1 - \sin\theta}$

In Exercises 5 and 6, graph each hyperbola. Specify the eccentricity, the center, and the values of a, b, and c.

5. **(a)** $r = \dfrac{3}{2 + 4\cos\theta}$ **(b)** $r = \dfrac{3}{2 - 4\cos\theta}$

6. **(a)** $r = \dfrac{3}{3 + 4\sin\theta}$ **(b)** $r = \dfrac{3}{3 - 4\sin\theta}$

In Exercises 7–18, graph each conic section. If the conic is a parabola, specify (using rectangular coordinates) the vertex and the directrix. If the conic is an ellipse, specify the center, the eccentricity, and the lengths of the major and minor axes. If the conic is a hyperbola, specify the center, the eccentricity, and the lengths of the transverse and conjugate axes.

7. $r = \dfrac{24}{2 - 3\cos\theta}$

8. $r = \dfrac{16}{10 + 5\sin\theta}$

9. $r = \dfrac{8}{5 + 3\sin\theta}$

10. $r = \dfrac{9}{1 + 2\cos\theta}$

11. $r = \dfrac{12}{5 - 5\sin\theta}$

12. $r = \dfrac{5}{3 - 2\sin\theta}$

13. $r = \dfrac{12}{7 + 5\cos\theta}$

14. $r = \dfrac{5}{3 + 3\cos\theta}$

15. $r = \dfrac{4}{5 + 5\sin\theta}$

16. $r = \dfrac{14}{7 - 8\cos\theta}$

17. $r = \dfrac{9}{1 - 2\cos\theta}$

18. $r = \dfrac{2}{\sqrt{2} - \sqrt{2}\sin\theta}$

B

*For Exercises 19–21, refer to the following figure. The line segment $\overline{PQ}$ is called a **focal chord** of the conic.*

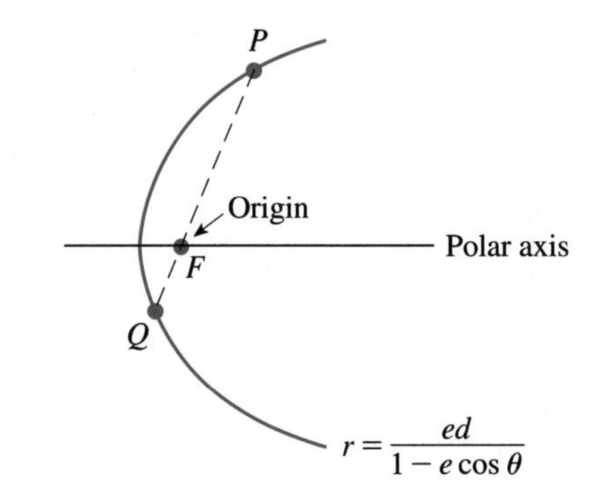

19. Show that $\dfrac{1}{FP} + \dfrac{1}{FQ} = \dfrac{2}{ed}$. What is remarkable about this result? *Hint:* Denote the polar coordinates of P by (r, θ), where $r = ed/(1 - e\cos\theta)$. Now, what are the polar coordinates of Q?

20. If the coordinates of P are (r, θ), show that

$$PQ = \frac{2ed}{1 - e^2\cos^2\theta}$$

Hint: $PQ = FP + FQ$.

21. In the given figure (preceding Exercise 19), suppose that we draw a focal chord $\overline{AB}$ that is perpendicular to $\overline{PQ}$. Show that the sum $\dfrac{1}{PQ} + \dfrac{1}{AB}$ is a constant.

22. Solve the equation $r/(d - r\cos\theta) = e$ for r. [As stated in the text, you should obtain $r = ed/(1 + e\cos\theta)$.]

ROTATION OF AXES

In Section 11.5 we saw that the equation

$$Ax^2 + Cy^2 + Dx + Ey + F = 0$$

can represent, in general, one of three curves: a parabola, an ellipse, or a hyperbola. (We use the phrase ''in general'' here to allow for the so-called degenerate cases.) In the present section, we will find that the second-degree equation

$$Ax^2 + Bxy + Cy^2 + Dx + Ey + F = 0 \tag{1}$$

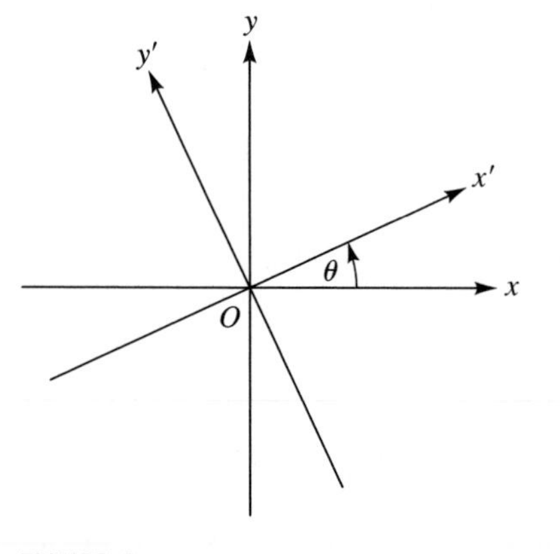

FIGURE 1

also represents, in general, one of these three curves. The difference now is that due to the xy-term in equation (1), the axes of the curves will no longer be parallel or perpendicular to the x- and y-axes. To study the curves defined by equation (1), it is useful first to introduce the technique known as *rotation of axes.*

Suppose that the x- and y-axes are rotated through a positive angle θ to yield a new x'-y' coordinate system, as shown in Figure 1. This procedure is referred to as a **rotation of axes.** We wish to obtain formulas relating the old and new coordinates. Let P be a given point with coordinates (x, y) in the original coordinate system and (x', y') in the new coordinate system. In Figure 2, let r denote the distance OP and α the angle measured from the x-axis to $\overline{OP}$. From Figure 2, we have

$$\cos \alpha = \frac{\text{adjacent}}{\text{hypotenuse}} = \frac{x}{r} \quad \text{and} \quad \sin \alpha = \frac{\text{opposite}}{\text{hypotenuse}} = \frac{y}{r}$$

Thus,

$$x = r \cos \alpha \quad \text{and} \quad y = r \sin \alpha \tag{2}$$

Again from Figure 2, we have

$$\cos(\alpha - \theta) = \frac{x'}{r} \quad \text{and} \quad \sin(\alpha - \theta) = \frac{y'}{r}$$

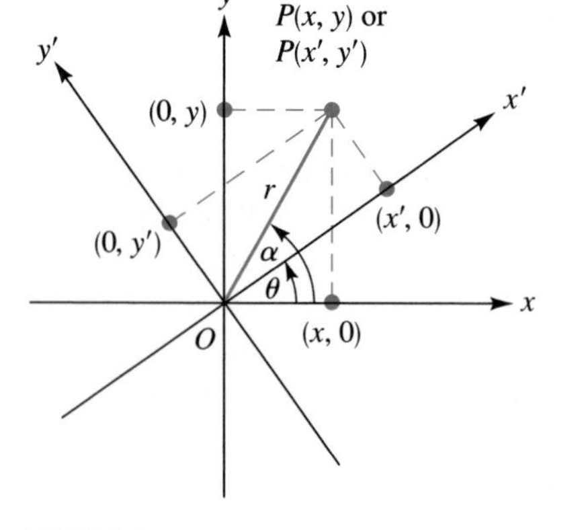

FIGURE 2

Thus,

$$x' = r \cos(\alpha - \theta) \quad \text{and} \quad y' = r \sin(\alpha - \theta)$$

or

$$x' = r \cos \alpha \cos \theta + r \sin \alpha \sin \theta$$

and

$$y' = r \sin \alpha \cos \theta - r \cos \alpha \sin \theta$$

With the aid of equations (2), this last pair of equations can be rewritten

$$x' = x \cos \theta + y \sin \theta \quad \text{and} \quad y' = -x \sin \theta + y \cos \theta$$

These two equations tell us how to express the new coordinates (x', y') in terms of the original coordinates (x, y) and the angle of rotation θ. On the other hand, it is also useful to express x and y in terms of x', y', and θ. This can be accomplished by treating the two equations we've just derived as a system of two equations in the unknowns x and y. As Exercise 41 at the end of this section asks you to verify, the results of solving this system for x and y are $x = x' \cos\theta - y' \sin\theta$ and $y = x' \sin\theta + y' \cos\theta$.

FORMULAS FOR THE ROTATION OF AXES

$$\begin{cases} x = x' \cos\theta - y' \sin\theta \\ y = x' \sin\theta + y' \cos\theta \end{cases} \qquad \begin{cases} x' = x \cos\theta + y \sin\theta \\ y' = -x \sin\theta + y \cos\theta \end{cases}$$

EXAMPLE 1 Suppose that the angle of rotation from the x-axis to the x'-axis is 45°. If the coordinates of a point P are (2, 0) with respect to the x'-y' coordinate system, what are the coordinates of P with respect to the x-y system?

Solution Substitute the values $x' = 2$, $y' = 0$, and $\theta = 45°$ in the formulas

$$\begin{cases} x = x' \cos\theta - y' \sin\theta \\ y = x' \sin\theta + y' \cos\theta \end{cases}$$

This yields

$$\begin{aligned} x &= 2\cos 45° - 0 \sin 45° \\ &= 2(\sqrt{2}/2) = \sqrt{2} \end{aligned} \qquad \begin{aligned} y &= 2\sin 45° + 0\cos 45° \\ &= 2(\sqrt{2}/2) = \sqrt{2} \end{aligned}$$

Thus, the coordinates of P in the x-y system are $(\sqrt{2}, \sqrt{2})$.

EXAMPLE 2 Suppose that the angle of rotation from the x-axis to the x'-axis is 45°. Write the equation $xy = 1$ in terms of the x'-y' coordinate system and then sketch the graph of this equation.

Solution With $\theta = 45°$, the rotation formulas for x and y become

$$\begin{cases} x = x' \cos 45° - y' \sin 45° \\ y = x' \sin 45° + y' \cos 45° \end{cases}$$

Thus we have

$$x = x'\left(\frac{\sqrt{2}}{2}\right) - y'\left(\frac{\sqrt{2}}{2}\right) = \frac{\sqrt{2}}{2}(x' - y')$$

and

$$y = x'\left(\frac{\sqrt{2}}{2}\right) + y'\left(\frac{\sqrt{2}}{2}\right) = \frac{\sqrt{2}}{2}(x' + y')$$

If we now substitute these expressions for x and y in the equation $xy = 1$, we obtain

$$\begin{aligned} \left[\frac{\sqrt{2}}{2}(x' - y')\right]\left[\frac{\sqrt{2}}{2}(x' + y')\right] &= 1 \\ \frac{1}{2}[(x')^2 - (y')^2] &= 1 \\ \frac{(x')^2}{(\sqrt{2})^2} - \frac{(y')^2}{(\sqrt{2})^2} &= 1 \end{aligned}$$

This last equation represents a hyperbola in the x'-y' coordinate system. With respect to this x'-y' system, the hyperbola can be analyzed using the techniques developed in Section 11.5. The results (as you should verify for yourself) are as follows:

$$xy = 1$$

or

$$\frac{(x')^2}{(\sqrt{2})^2} - \frac{(y')^2}{(\sqrt{2})^2} = 1$$

Focal axis: x'-axis
Center: origin
Vertices: $(\pm\sqrt{2}, 0)$
Foci: $(\pm 2, 0)$
Asymptotes: $y' = \pm x'$

These specifications are in terms of the x'-y' coordinate system.

As noted, the preceding specifications are in terms of the x'-y' coordinate system. However, since the original equation, $xy = 1$, is given in terms of the x-y system, we would like to express these specifications in terms of the same x-y coordinate system. This can be done by the method shown in Example 1. The results are as follows;

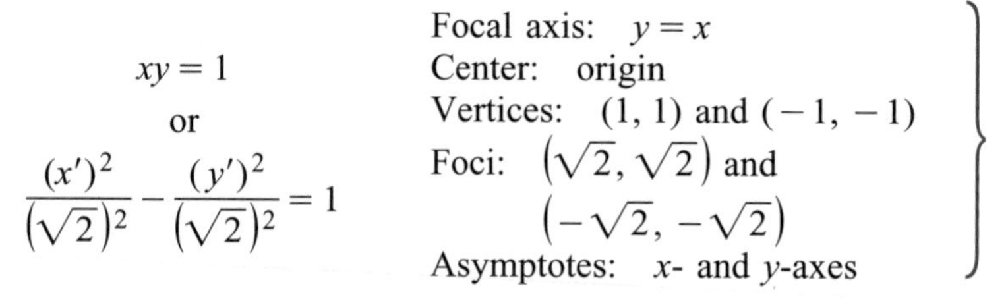

$$xy = 1$$

or

$$\frac{(x')^2}{(\sqrt{2})^2} - \frac{(y')^2}{(\sqrt{2})^2} = 1$$

Focal axis: $y = x$
Center: origin
Vertices: $(1, 1)$ and $(-1, -1)$
Foci: $(\sqrt{2}, \sqrt{2})$ and $(-\sqrt{2}, -\sqrt{2})$
Asymptotes: x- and y-axes

These specifications are in terms of the x-y coordinate system.

Figure 3 displays the graph of this hyperbola.

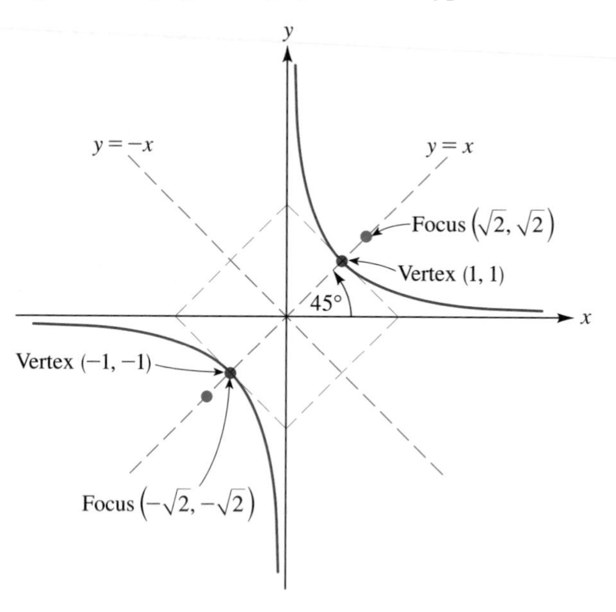

FIGURE 3
$xy = 1$

In Example 2, we saw that a rotation of 45° reduced the given equation to one of the standard forms with which we are already familiar. Now let us consider the situation in greater generality. We begin with the second-degree equation

$$Ax^2 + Bxy + Cy^2 + Dx + Ey + F = 0 \qquad (B \neq 0) \tag{3}$$

If we rotate the axes through an angle θ, equation (3) will, after some simplification, take on the form

$$A'(x')^2 + B'x'y' + C'(y')^2 + D'x' + E'y' + F' = 0 \tag{4}$$

for certain constants A', B', C', D', E', and F'. We wish to determine an angle of rotation θ for which $B' = 0$. The reason we want to do this is that if $B' = 0$, we will be able to analyze equation (4) using the techniques of the previous sections. We begin with the rotation formulas:

$$x = x' \cos \theta - y' \sin \theta$$
$$y = x' \sin \theta + y' \cos \theta$$

Substituting these expressions for x and y in equation (3) yields

$$\begin{aligned} A(x' \cos \theta - y' \sin \theta)^2 &+ B(x' \cos \theta - y' \sin \theta)(x' \sin \theta + y' \cos \theta) \\ &+ C(x' \sin \theta + y' \cos \theta)^2 + D(x' \cos \theta - y' \sin \theta) \\ &+ E(x' \sin \theta + y' \cos \theta) + F = 0 \end{aligned}$$

We can simplify this equation by performing the indicated operations and then collecting like terms. As Exercise 44 asks you to verify, the resulting equation is

$$A'(x')^2 + B'x'y' + C'(y')^2 + D'x' + E'y' + F' = 0 \tag{5}$$

where

$$\begin{aligned} A' &= A \cos^2 \theta + B \sin \theta \cos \theta + C \sin^2 \theta \\ B' &= 2(C - A)\sin \theta \cos \theta + B(\cos^2 \theta - \sin^2 \theta) \\ C' &= A \sin^2 \theta - B \sin \theta \cos \theta + C \cos^2 \theta \\ D' &= D \cos \theta + E \sin \theta \\ E' &= E \cos \theta - D \sin \theta \\ F' &= F \end{aligned}$$

Thus, B' will be zero provided that

$$2(C - A)\sin \theta \cos \theta + B(\cos^2 \theta - \sin^2 \theta) = 0$$

By using the double-angle formulas, we can write this last equation as

$$(C - A)\sin 2\theta + B \cos 2\theta = 0$$

or

$$B \cos 2\theta = (A - C)\sin 2\theta$$

Now, dividing both sides by $B \sin 2\theta$, we obtain

$$\frac{\cos 2\theta}{\sin 2\theta} = \frac{A - C}{B}$$

or

$$\cot 2\theta = \frac{A - C}{B}$$

We have now shown that if θ satisfies the condition $\cot 2\theta = (A - C)/B$, then equation (5) will contain no $x'y'$-term. Although we will not prove it here, it can be shown that there is always a value of θ in the range $0° < \theta < 90°$ for which $\cot 2\theta = (A - C)/B$. In subsequent examples, we will always choose θ in this range.

EXAMPLE 3 Graph the equation $2x^2 + \sqrt{3}xy + y^2 = 2$.

Solution We first rotate the axes through an angle θ so that the new equation will contain no $x'y'$-term. To choose an appropriate value of θ, we require that

$$\cot 2\theta = \frac{A - C}{B} = \frac{2 - 1}{\sqrt{3}} = \frac{1}{\sqrt{3}}$$

Thus, $\cot 2\theta = 1/\sqrt{3}$, from which we conclude that $2\theta = 60°$, or $\theta = 30°$. With this value of θ, the rotation formulas become

$$x = x'\left(\frac{\sqrt{3}}{2}\right) - y'\left(\frac{1}{2}\right) \quad \text{and} \quad y = x'\left(\frac{1}{2}\right) + y'\left(\frac{\sqrt{3}}{2}\right)$$

Now we use these formulas to substitute for x and y in the given equation. This yields

$$2\left(\frac{x'\sqrt{3}}{2} - \frac{y'}{2}\right)^2 + \sqrt{3}\left(\frac{x'\sqrt{3}}{2} - \frac{y'}{2}\right)\left(\frac{x'}{2} + \frac{y'\sqrt{3}}{2}\right) + \left(\frac{x'}{2} + \frac{y'\sqrt{3}}{2}\right)^2 = 2$$

After simplification, this last equation becomes

$$(y')^2 + 5(x')^2 = 4$$

or

$$\frac{(y')^2}{2^2} + \frac{(x')^2}{(2/\sqrt{5})^2} = 1$$

We recognize this as the equation of an ellipse in the x'-y' coordinate system. The focal axis is the y'-axis. The values of a and b are 2 and $2/\sqrt{5}$, respectively. The ellipse can now be sketched as in Figure 4.

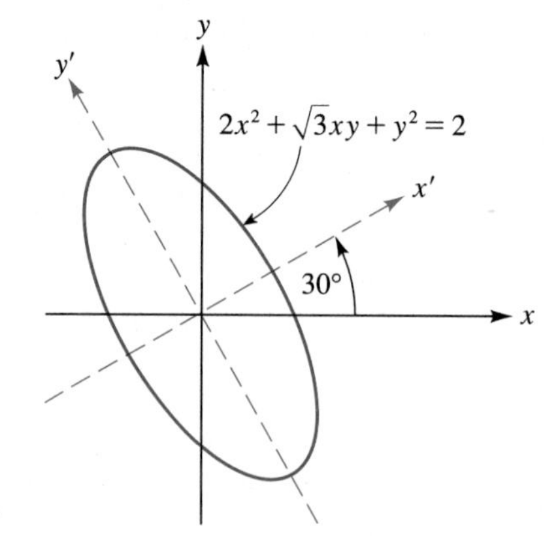

FIGURE 4

In Example 3 we were able to determine the angle θ directly, since we recognized the quantity $1/\sqrt{3}$ as the value of $\cot 60°$. However, this is the exception rather than the rule. In most problems, the value of $(A - C)/B$ is not so easily identified as the cotangent of one of the more familiar angles. The next examples demonstrate a technique that can be used in such cases. The technique relies on the following three trigonometric identities:

1. $\sec^2 \beta = 1 + \tan^2 \beta$
2. $\sin \theta = \sqrt{\dfrac{1 - \cos 2\theta}{2}}$
3. $\cos \theta = \sqrt{\dfrac{1 + \cos 2\theta}{2}}$

The positive square roots are appropriate, since $0° < \theta < 90°$.

EXAMPLE 4 Graph the equation $x^2 + 4xy - 2y^2 = 6$.

Solution We have $A = 1$, $B = 4$, and $C = -2$. Therefore,

$$\cot 2\theta = \frac{A - C}{B} = \frac{1 - (-2)}{4} = \frac{3}{4}$$

Since $\cot 2\theta = 3/4$, it follows that $\tan 2\theta = 4/3$. Therefore

$$\begin{aligned}\sec^2 2\theta &= 1 + \left(\frac{4}{3}\right)^2 \qquad \text{(Why?)}\\ &= \frac{9}{9} + \frac{16}{9} = \frac{25}{9}\end{aligned}$$

Thus

$$\sec 2\theta = \pm\frac{5}{3}$$

At this point, we need to decide whether the positive or negative sign is appropriate. Since we are assuming that $0° < \theta < 90°$, the angle 2θ must lie in either Quadrant I or Quadrant II. To decide which, we note that the value determined for $\cot 2\theta$ was positive. That rules out the possibility that 2θ might lie in Quadrant II. We conclude that in this case, $0° < 2\theta < 90°$. Therefore, the sign of $\sec 2\theta$ must be positive, and we have

$$\sec 2\theta = \frac{5}{3} \qquad \text{or} \qquad \cos 2\theta = \frac{3}{5}$$

The values of $\sin \theta$ and $\cos \theta$ can now be obtained as follows:

$$\begin{aligned}\sin \theta &= \sqrt{\frac{1 - \cos 2\theta}{2}}\\ &= \sqrt{\frac{1 - \frac{3}{5}}{2}}\\ &= \frac{1}{\sqrt{5}} \qquad \text{after simplifying}\end{aligned}$$

$$\begin{aligned}\cos \theta &= \sqrt{\frac{1 + \cos 2\theta}{2}}\\ &= \sqrt{\frac{1 + \frac{3}{5}}{2}}\\ &= \frac{2}{\sqrt{5}} \qquad \text{after simplifying}\end{aligned}$$

With these values for sin θ and cos θ, the rotation formulas become

$$x = x'\left(\frac{2}{\sqrt{5}}\right) - y'\left(\frac{1}{\sqrt{5}}\right) = \frac{1}{\sqrt{5}}(2x' - y')$$

and

$$y = x'\left(\frac{1}{\sqrt{5}}\right) + y'\left(\frac{2}{\sqrt{5}}\right) = \frac{1}{\sqrt{5}}(x' + 2y')$$

We now substitute these expressions for x and y in the original equation, $x^2 + 4xy - 2y^2 = 6$. This yields

$$\left(\frac{2x' - y'}{\sqrt{5}}\right)^2 + 4\left(\frac{2x' - y'}{\sqrt{5}}\right)\left(\frac{x' + 2y'}{\sqrt{5}}\right) - 2\left(\frac{x' + 2y'}{\sqrt{5}}\right)^2 = 6$$

As Exercise 42 asks you to verify, this equation can be simplified to

$$2(x')^2 - 3(y')^2 = 6$$

or

$$\frac{(x')^2}{(\sqrt{3})^2} - \frac{(y')^2}{(\sqrt{2})^2} = 1$$

This last equation represents a hyperbola with center at the origin of the x'-y' coordinate system and with a focal axis that coincides with the x'-axis. We can sketch the hyperbola using the methods of Section 11.5, but it is first necessary to know the angle θ between the x- and x'-axes. We have

$$\sin \theta = 1/\sqrt{5}$$

$$\theta = \sin^{-1}(1/\sqrt{5})$$

$$\theta \approx 27^\circ \quad \text{using a calculator set in the degree mode}$$

Figure 5 shows the required graph.

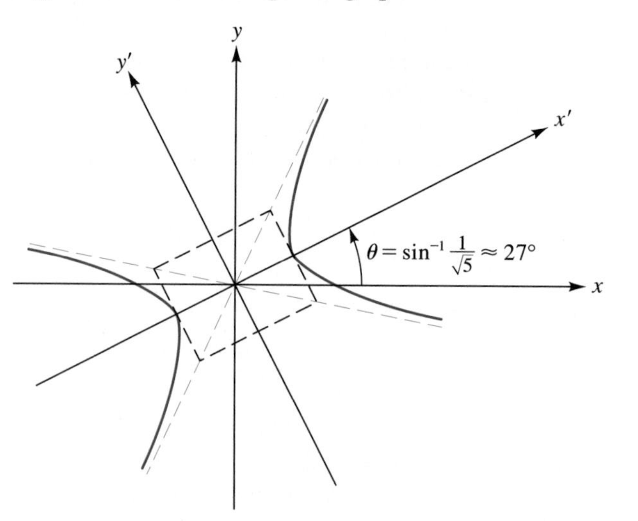

FIGURE 5
$x^2 + 4xy - 2y^2 = 6$

In Examples 3 and 4, we graphed equations of the form

$$Ax^2 + Bxy + Cy^2 + F = 0$$

The technique used in those examples is equally effective in graphing equations of the form $Ax^2 + Bxy + Cy^2 + Dx + Ey + F = 0$, in which the x- and y-terms are present. This is demonstrated in Example 5. Since the general technique employed in Example 5 is the same as in the previous examples, we will merely outline the procedure and the results in the solution, leaving the detailed calculations to Exercise 43 at the end of this section.

EXAMPLE 5 Graph the equation $16x^2 - 24xy + 9y^2 + 110x - 20y + 100 = 0$.

Outline of Solution

$$\cot 2\theta = \frac{A - C}{B} = -\frac{7}{24}$$

Now, proceeding as in the last example, we find that $\cos 2\theta = -7/25$, $\cos\theta = 3/5$, and $\sin\theta = 4/5$. Thus, the rotation formulas become

$$x = x'\left(\frac{3}{5}\right) - y'\left(\frac{4}{5}\right) = \frac{1}{5}(3x' - 4y')$$

and

$$y = x'\left(\frac{4}{5}\right) + y'\left(\frac{3}{5}\right) = \frac{1}{5}(4x' + 3y')$$

Next we substitute for x and y in the given equation. After straightforward but lengthy computations, we obtain

$$(y')^2 + 2x' - 4y' + 4 = 0$$

We graphed equations of this form in Section 11.2 by completing the square. Using that technique here, we have

$$\begin{aligned}(y')^2 - 4y' &= -2x' - 4\\ (y')^2 - 4y' + 4 &= -2x' - 4 + 4\\ (y' - 2)^2 &= -2x'\end{aligned}$$

This is the equation of a parabola. With respect to the x'-y' system, the vertex is $(0, 2)$ and the axis of the parabola is the line $y' = 2$. In terms of the x-y system, the vertex is $\left(-\frac{8}{5}, \frac{6}{5}\right)$ and the axis of the parabola is the line $-4x + 3y = 10$. Finally, the angle of rotation is $\theta = \sin^{-1}\left(\frac{4}{5}\right)$. Using a calculator, we find this to be approximately 53°. The required graph is shown in Figure 6.

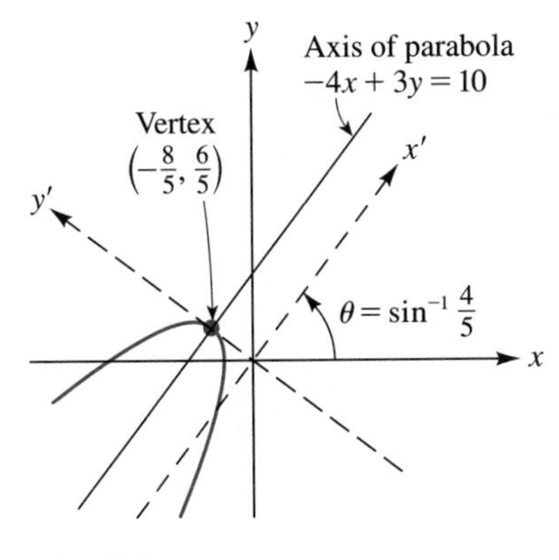

FIGURE 6
$16x^2 - 24xy + 9y^2 + 110x - 20y + 100 = 0$

EXERCISE SET 11.8

A

In Exercises 1–3, an angle of rotation is specified, followed by the coordinates of a point in the x'-y' system. Find the coordinates of each point with respect to the x-y system.

1. $\theta = 30°$; $(x', y') = (\sqrt{3}, 2)$
2. $\theta = 60°$; $(x', y') = (-1, 1)$
3. $\theta = 45°$; $(x', y') = (\sqrt{2}, -\sqrt{2})$

In Exercises 4–6, an angle of rotation is specified, followed by the coordinates of a point in the x-y system. Find the coordinates of each point with respect to the x'-y' system.

4. $\theta = 45°$; $(x, y) = (0, -2)$

5. $\theta = \sin^{-1}(\frac{5}{13})$; $(x, y) = (-3, 1)$
6. $\theta = 15°$; $(x, y) = (1, 0)$

In Exercises 7–14, find $\sin\theta$ *and* $\cos\theta$, *where* θ *is the (acute) angle of rotation that eliminates the* $x'y'$*-term.* Note: *You are not asked to graph the equation.*

7. $25x^2 - 24xy + 18y^2 + 1 = 0$
8. $x^2 + 24xy + 8y^2 - 8 = 0$
9. $x^2 - 24xy + 8y^2 - 8 = 0$
10. $220x^2 + 119xy + 100y^2 = 0$
11. $x^2 - 2\sqrt{3}xy - y^2 = 3$
12. $5x^2 + 12xy - 4 = 0$
13. $161xy - 240y^2 - 1 = 0$
14. $4x^2 - 5xy + 4y^2 + 2 = 0$
15. Suppose that the angle of rotation is 45°. Write the equation $2xy = 9$ in terms of the x'-y' coordinate system and then graph the equation.
16. Suppose that the angle of rotation is 45°. Write the equation $5x^2 - 6xy + 5y^2 + 16 = 0$ in terms of the x'-y' system.

In Exercises 17–40, graph the equations.

17. $7x^2 + 8xy + y^2 - 1 = 0$
18. $2x^2 - \sqrt{3}xy + y^2 - 20 = 0$
19. $x^2 + 4xy + 4y^2 = 1$
20. $x^2 + 4xy + 4y^2 = 0$
21. $9x^2 - 24xy + 16y^2 - 400x - 300y = 0$
22. $8x^2 + 12xy + 13y^2 = 34$
23. $4xy + 3y^2 + 4x + 6y = 1$
24. $x^2 - 2xy + y^2 + x - y = 0$
25. $3x^2 - 2xy + 3y^2 - 6\sqrt{2}x + 2\sqrt{2}y + 4 = 0$
26. $x^2 + 3xy + y^2 = 1$
27. $(x - y)^2 = 8(y - 6)$
28. $4x^2 - 4xy + y^2 - 4x + 2y + 1 = 0$
29. $3x^2 + 4xy + 6y^2 = 7$
30. $x^2 + 2\sqrt{3}xy + 3y^2 + 12\sqrt{3}x - 12y - 24 = 0$
31. $17x^2 - 12xy + 8y^2 - 80 = 0$
32. $7x^2 - 2\sqrt{3}xy + 5y^2 = 32$
33. $3xy - 4y^2 + 18 = 0$
34. $x^2 + y^2 = 2xy + 4x + 4y - 8$
35. $(x + y)^2 + 4\sqrt{2}(x - y) = 0$
36. $41x^2 - 24xy + 9y^2 = 3$
37. $3x^2 - \sqrt{15}xy + 2y^2 = 3$
38. $3x^2 + 10xy + 3y^2 - 2\sqrt{2}x + 2\sqrt{2}y - 10 = 0$
39. $3x^2 - 2xy + 3y^2 + 2 = 0$
40. $(x + y)(x + y + 1) = 2$

B

41. Solve for x and y:

$$\begin{cases} (\cos\theta)x + (\sin\theta)y = x' \\ (-\sin\theta)x + (\cos\theta)y = y' \end{cases}$$

Answer: $x = x'\cos\theta - y'\sin\theta$, $y = x'\sin\theta + y'\cos\theta$

42. Simplify the equation:

$$\left(\frac{2x' - y'}{\sqrt{5}}\right)^2 + 4\left(\frac{2x' - y'}{\sqrt{5}}\right)\left(\frac{x' + 2y'}{\sqrt{5}}\right) - 2\left(\frac{x' + 2y'}{\sqrt{5}}\right)^2 = 6$$

Answer: $2(x')^2 - 3(y')^2 = 6$

43. Refer to Example 5 in the text.
    (a) Show that $\cos 2\theta = -7/25$.
    (b) Show that $\cos\theta = 3/5$ and $\sin\theta = 4/5$.
    (c) Make the substitutions $x = \frac{1}{5}(3x' - 4y')$ and $y = \frac{1}{5}(4x' + 3y')$ in the given equation $16x^2 - 24xy + 9y^2 + 110x - 20y + 100 = 0$ and show that the resulting equation simplifies to $(y')^2 + 2x' - 4y' + 4 = 0$.
44. Make the substitutions $x = x'\cos\theta - y'\sin\theta$ and $y = x'\sin\theta + y'\cos\theta$ in the equation $Ax^2 + Bxy + Cy^2 + Dx + Ey + F = 0$ and show that the result is

$$A'(x')^2 + B'x'y' + C'(y')^2 + D'x' + E'y' + F' = 0$$

where

$$A' = A\cos^2\theta + B\sin\theta\cos\theta + C\sin^2\theta$$
$$B' = 2(C - A)\sin\theta\cos\theta + B(\cos^2\theta - \sin^2\theta)$$
$$C' = A\sin^2\theta - B\sin\theta\cos\theta + C\cos^2\theta$$
$$D' = D\cos\theta + E\sin\theta$$
$$E' = E\cos\theta - D\sin\theta$$
$$F' = F$$

45. (Refer to Exercise 44.) Show that $A + C = A' + C'$.
46. Complete the following steps to derive the equation $(B')^2 - 4A'C' = B^2 - 4AC$.
    (a) Show that $A' - C' = (A - C)\cos 2\theta + B\sin 2\theta$.
    (b) Show that $B' = B\cos 2\theta - (A - C)\sin 2\theta$.
    (c) Square the equations in parts (a) and (b), then add the two resulting equations to show that $(A' - C')^2 + (B')^2 = (A - C)^2 + B^2$.
    (d) Square the equation given in Exercise 45, then subtract the result from the equation in part (c). The result can be written $(B')^2 - 4A'C' = B^2 - 4AC$, as required.

CHAPTER ELEVEN SUMMARY OF PRINCIPAL TERMS AND FORMULAS

TERMS AND FORMULAS	PAGE REFERENCE	COMMENT
1. Conic sections	672	These are the curves that are formed when a plane intersects the surface of a right circular cone. As indicated in the figure at the beginning of the chapter, these curves are the circle, the ellipse, the hyperbola, and the parabola.
2. Angle of inclination	672	The angle of inclination of a line is the angle between the x-axis and the line, measured counterclockwise from the positive side or positive direction of the x-axis to the line.
3. $m = \tan\theta$	673	The slope of a line is equal to the tangent of the angle of inclination.
4. $d = \dfrac{\lvert mx_0 + b - y_0\rvert}{\sqrt{1+m^2}}$	674	This is a formula for the (perpendicular) distance d from the point (x_0, y_0) to the line $y = mx + b$.
5. $d = \dfrac{\lvert Ax_0 + By_0 + C\rvert}{\sqrt{A^2+B^2}}$	674, 677	This is a formula for the (perpendicular) distance d from the point (x_0, y_0) to the line $Ax + By + C = 0$.
6. Parabola	681	A parabola is the set of all points in the plane equally distant from a fixed line and a fixed point not on the line. The fixed line is called the *directrix,* and the fixed point is called the *focus.*
7. Axis (of a parabola)	683	This is the line that is drawn through the focus of the parabola, perpendicular to the directrix.
8. Vertex (of a parabola)	683	This is the point at which the parabola intersects its axis.
9. Focal chord (of a parabola)	685	A focal chord of a parabola is a line segment that passes through the focus and has endpoints on the parabola.
10. Focal width (of a parabola)	686	The focal width of a parabola is the length of the focal chord that is perpendicular to the axis of the parabola. For a given value of p, the two parabolas $x^2 = 4py$ and $y^2 = 4px$ have the same focal width; it is $4p$.
11. Tangent line (to a parabola)	692	A line that is not parallel to the axis of a parabola is tangent to the parabola provided that it intersects the parabola in exactly one point.
12. Ellipse	695	An ellipse is the set of all points in the plane, the sum of whose distances from two fixed points is constant. Each fixed point is called a *focus* of the ellipse.
13. Eccentricity (of an ellipse)	697	The eccentricity is a number that measures how much the ellipse deviates from being a circle. See Figure 7 in Section 11.4. The eccentricity e is defined by the formula $e = c/a$, where c and a are defined by the following conventions. The distance between the foci is denoted by $2c$. The sum of the distances from a point on the ellipse to the two foci is denoted by $2a$.
14. Focal axis (of an ellipse)	697	This is the line passing through the two foci.
15. Center (of an ellipse)	697	This is the midpoint of the line segment joining the foci.
16. Vertices (of an ellipse)	697	The two points at which an ellipse meets its focal axis are called the vertices of the ellipse.
17. Major axis (of an ellipse)	697	This is the line segment joining the two vertices of the ellipse.
18. Minor axis (of an ellipse)	697	This is the line segment through the center of the ellipse, perpendicular to the major axis, and with endpoints on the ellipse.

TERMS AND FORMULAS	PAGE REFERENCE	COMMENT
19. $\frac{x^2}{a^2} + \frac{y^2}{b^2} = 1 \quad (a > b)$	696	This is the standard form for the equation of an ellipse with foci $(\pm c, 0)$.
$\frac{x^2}{b^2} + \frac{y^2}{a^2} = 1 \quad (a > b)$	701	This is the standard form for the equation of an ellipse with foci $(0, \pm c)$.
20. Tangent line to an ellipse	702	A tangent to an ellipse is a line that intersects the ellipse in exactly one point. See Figure 12 in Section 11.4.
21. $\frac{x_1 x}{a^2} + \frac{y_1 y}{b^2} = 1$	702	This is the equation of the line that is tangent to the ellipse $(x^2/a^2) + (y^2/b^2) = 1$ at the point (x_1, y_1) on the ellipse.
22. Hyperbola	710	A hyperbola is the set of all points in the plane, the difference of whose distances from two fixed points is a positive constant. The two fixed points are the *foci,* and the line passing through the foci is the *focal axis.*
23. $\frac{x^2}{a^2} - \frac{y^2}{b^2} = 1$	711	This is the standard form for the equation of a hyperbola with foci $(\pm c, 0)$.
$\frac{y^2}{a^2} - \frac{x^2}{b^2} = 1$	715	This is the standard form for the equation of a hyperbola with foci $(0, \pm c)$.
24. Asymptote	712	A line is said to be an asymptote for a curve if the distance between the line and the curve approaches zero as we move farther and farther out along the line. The asymptotes for the hyperbola $(x^2/a^2) - (y^2/b^2) = 1$ are the two lines $y = \pm(b/a)x$. For the hyperbola $(y^2/a^2) - (x^2/b^2) = 1$, the asymptotes are $y = \pm(a/b)x$.
25. Focal axis (of a hyperbola)	713	This is the line passing through the foci.
26. Vertices (of a hyperbola)	713	The two points at which the hyperbola intersects its focal axis are called vertices.
27. Center (of a hyperbola)	713	This is the point on the focal axis midway between the foci.
28. Transverse axis	713	This is the line segment joining the two vertices of a hyperbola.
29. Conjugate axis	713	This is the line segment perpendicular to the transverse axis of the hyperbola, passing through the center and extending a distance $b(=\sqrt{c^2 - a^2})$ on either side of the center.
30. Eccentricity (of a hyperbola)	713	For both of the standard forms for the hyperbola, the eccentricity e is defined by $e = c/a$.
31. $\frac{x_1 x}{a^2} - \frac{y_1 y}{b^2} = 1$	717	This is the equation of the tangent to the hyperbola $(x^2/a^2) - (y^2/b^2) = 1$ at the point (x_1, y_1) on the curve. (See Exercise 53 in Exercise Set 11.5 for the definition of a tangent to a hyperbola.)
32. Focus–directrix property of conics	728	Refer to Figure 10 on page 728. Let $\mathscr{L}$ be a fixed line, F a fixed point, and e a positive constant. Consider the set of points P satisfying the condition $FP/PD = e$, where D is defined by Figure 10. Then: **(a)** If $e = 1$ the set of points is a *parabola* with focus F and directrix $\mathscr{L}$; **(b)** If $0 < e < 1$, the set of points is an *ellipse* with focus F, corresponding directrix $\mathscr{L}$, and eccentricity e; **(c)** If $e > 1$, the set of points is a *hyperbola* with focus F, corresponding directrix $\mathscr{L}$, and eccentricity e.

TERMS AND FORMULAS	PAGE REFERENCE	COMMENT
33. $r = \dfrac{ed}{1 \pm e \cos \theta}$ $r = \dfrac{ed}{1 \pm e \sin \theta}$	730, 731	These are polar equations of conics with the focus at the pole or origin, as indicated in Figures 2–5 on page 731. The eccentricity is e, and d is the distance from the focus to the directrix.
34. $\begin{cases} x = x' \cos \theta - y' \sin \theta \\ y = x' \sin \theta + y' \cos \theta \end{cases}$ $\begin{cases} x' = x \cos \theta + y \sin \theta \\ y' = -x \sin \theta + y \cos \theta \end{cases}$	736, 737	These are the formulas that relate the coordinates of a point in the x-y system to the coordinates in the rotated x'-y' system. See Figures 1 and 2 in Section 11.8.
35. $\cot 2\theta = \dfrac{A - C}{B}$	739	This formula determines an angle of rotation θ. When the equation $Ax^2 + Bxy + Cy^2 + Dx + Ey + F = 0 \quad (B \neq 0)$ is written in the x'-y' system, the resulting equation does not contain an $x'y'$-term. The graph can then be analyzed by means of the technique of completing the square.

WRITING MATHEMATICS

Write out your answers to Questions 1 and 2 in complete sentences. If you draw a diagram to accompany your writing, or if you use equations, be sure that you clearly label any elements to which you refer.

1. Refer to Figure 1 on page 695. If the thumbtacks are 3 in. apart, what length of string should be used to produce an ellipse with eccentricity 2/3?
2. Refer to Figure 1 on page 695 to explain each of the following.
   (a) When the eccentricity of an ellipse is close to 1, the ellipse is very flat.
   (b) When the eccentricity of an ellipse is close to 0, the ellipse resembles a circle.
3. Investigate the geometric significance of the eccentricity of a hyperbola by completing the three steps that follow. Then, write a report telling what you have done, what patterns you have observed, and what relationship you have found between the eccentricity and the shape of the hyperbola.
   (a) Use the definition of the eccentricity e to show that $e = \sqrt{1 + (b/a)^2}$. (This shows that the eccentricity is always greater than 1.)
   (b) Compute the eccentricity for each of the following hyperbolas. (Use a calculator.)
      (i) $(0.0201)x^2 - y^2 = 0.0201$
      (ii) $3x^2 - y^2 = 3$
      (iii) $8x^2 - y^2 = 8$
      (iv) $15x^2 - y^2 = 15$
      (v) $99x^2 - y^2 = 99$
   (c) On the same set of axes, sketch the Quadrant I portion of each of the hyperbolas in part (b).

CHAPTER ELEVEN REVIEW EXERCISES

In Exercises 1–15, refer to the following figure and show that the given statements are correct.

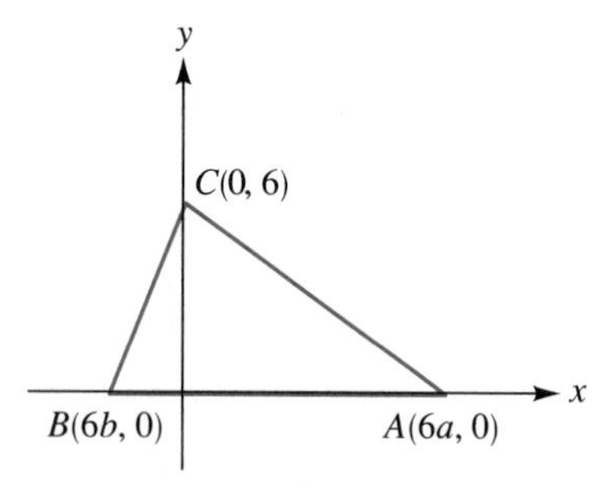

1. The equations of the lines forming the sides of $\triangle ABC$ are $x + by = 6b$, $x + ay = 6a$, and $y = 0$.
2. The equations of the lines forming the medians of $\triangle ABC$ are $2x + (a+b)y = 6(a+b)$, $x - (a-2b)y = 6b$, and $x - (b-2a)y = 6a$. (A *median* is a line segment drawn from a vertex of a triangle to the midpoint of the opposite side.)
3. Each pair of medians of $\triangle ABC$ intersect at the point $G(2a + 2b, 2)$. (The point G is called the *centroid* of $\triangle ABC$.)
4. The equations of the lines forming the altitudes of $\triangle ABC$ are $y = ax - 6ab$, $y = bx - 6ab$, and $x = 0$. (An *altitude* is a line segment drawn from a vertex to the opposite side, perpendicular to that side.)
5. Each pair of altitudes intersect at the point $H(0, -6ab)$. (The point H is the *orthocenter* of $\triangle ABC$.)
6. The equations of the perpendicular bisectors of the sides of $\triangle ABC$ are $x = 3a + 3b$, $bx - y = 3b^2 - 3$, and $ax - y = 3a^2 - 3$.
7. Each pair of perpendicular bisectors intersect at the point $(O(3a + 3b, 3ab + 3)$. (The point O is called the *circumcenter* of $\triangle ABC$.)
8. The distance from the circumcenter O to each vertex is $3\sqrt{(a^2+1)(b^2+1)}$. (This distance, denoted by R, is the *circumradius* of $\triangle ABC$. Note that the circle with center O and radius R passes through the points A, B, and C.)
9. In $\triangle ABC$, let p, q, and r denote the lengths BC, AC, and AB, respectively. Then the area of $\triangle ABC$ is $pqr/4R$.
10. $AH^2 + BC^2 = 4(OA)^2$
11. $OH^2 = 9R^2 - (p^2 + q^2 + r^2)$ *Hint:* See Exercise 9.
12. $GH^2 = 4R^2 - \frac{4}{9}(p^2 + q^2 + r^2)$
13. $HA^2 + HB^2 + HC^2 = 12R^2 - (p^2 + q^2 + r^2)$
14. The points H, G, and O are collinear. (The line through these three points is the *Euler line* of $\triangle ABC$.)
15. $GH = 2(GO)$

In Exercises 16–18, find the angle of inclination for each line. Use a calculator to express your answers in degrees. (Round to one decimal place.)

16. $y = 4x - 3$ **17.** $2x + 3y = 6$ **18.** $y = 2x$
19. Find the distance from the point $(-1, -3)$ to the line $5x + 6y = 30$.
20. Find the distance from the point $(2, 1)$ to the line $y = \frac{1}{2}x + 4$.
21. The vertices of an equilateral triangle are $(\pm 6, 0)$ and $(0, 6\sqrt{3})$. Verify that the sum of the three distances from the point $(1, 2)$ to the sides of the triangle is equal to the height of the triangle. (It can be shown that, for any point inside an equilateral triangle, the sum of the distances to the sides is equal to the height of the triangle. This is *Viviani's theorem.*)
22. A tangent line is drawn from the point $(-12, -1)$ to the circle $x^2 + y^2 = 20$. Find the slope of this line, given that its y-intercept is positive.

In Exercises 23–26, find the equation of the parabola satisfying the given conditions. In each case, assume the vertex is $(0, 0)$.

23. **(a)** The focus is $(4, 0)$. **(b)** The focus is $(0, 4)$.
24. The focus lies on the x-axis, and the curve passes through the point $(3, 1)$.
25. The parabola is symmetric about the y-axis, the y-coordinate of the focus is positive, and the length of the focal chord perpendicular to the y-axis is 12.
26. The focus of the parabola is the center of the circle $x^2 - 8x + y^2 + 15 = 0$.

In Exercises 27–30, find the equation of the ellipse satisfying the given conditions. Write your answers in the form $Ax^2 + By^2 = C$.

27. Foci $(\pm 2, 0)$; endpoints of the major axis $(\pm 8, 0)$
28. Foci $(0, \pm 1)$; endpoints of the minor axis $(\pm 4, 0)$
29. Eccentricity $4/5$; one end of the minor axis at $(-6, 0)$; center at the origin
30. For any point P on a certain ellipse, the sum of the distances from $(1, 2)$ and $(-1, -2)$ is 12. Find the equation of the ellipse. *Hint:* Use the distance formula and the definition of an ellipse.

In Exercises 31–34, find the equation of the hyperbola satisfying the given conditions. Write each answer in the form $Ax^2 - By^2 = C$ or in the form $Ay^2 - Bx^2 = C$.

31. Foci $(\pm 6, 0)$; vertices $(\pm 2, 0)$

32. Asymptotes $y = \pm 2x$; foci $(0, \pm 3)$

33. Eccentricity 4; foci $(\pm 3, 0)$

34. Length of the transverse axis 3; eccentricity 5/4; center (0, 0); focal axis horizontal

In Exercises 35–39, graph the parabolas, and in each case specify the vertex, the focus, the directrix, and the focal width.

35. $x^2 = 10y$

36. $x^2 = 5y$

37. $x^2 = -12(y - 3)$

38. $x^2 = -8(y + 1)$

39. $(y - 1)^2 = -4(x - 1)$

In Exercises 40–45, graph the ellipses, and in each case specify the center, the foci, the lengths of the major and minor axes, and the eccentricity.

40. $x^2 + 2y^2 = 4$

41. $4x^2 + 9y^2 = 144$

42. $49x^2 + 9y^2 = 441$

43. $9x^2 + y^2 = 9$

44. $\dfrac{(x-1)^2}{5^2} + \dfrac{(y+2)^2}{3^2} = 1$

45. $\dfrac{(x+3)^2}{3^2} + \dfrac{y^2}{3^2} = 1$

In Exercises 46–51, graph the hyperbolas. In each case specify the center, the vertices, the foci, the equations of the asymptotes, and the eccentricity.

46. $x^2 - 2y^2 = 4$

47. $4x^2 - 9y^2 = 144$

48. $49y^2 - 9x^2 = 441$

49. $9y^2 - x^2 = 9$

50. $\dfrac{(x-1)^2}{5^2} - \dfrac{(y+2)^2}{3^2} = 1$

51. $\dfrac{(y+3)^2}{3^2} - \dfrac{x^2}{3^2} = 1$

In Exercises 52–65, use the technique of completing the square to graph the given equation. If the graph is a parabola, specify the vertex, axis, focus, and directrix. If the graph is an ellipse, specify the center, foci, and lengths of the major and minor axes. If the graph is a hyperbola, specify the center, vertices, foci, and equations of the asymptotes. Finally, if the equation has no graph, say so.

52. $3x^2 + 4y^2 - 6x + 16y + 7 = 0$

53. $y^2 - 16x - 8y + 80 = 0$

54. $y^2 + 4x + 2y - 15 = 0$

55. $16x^2 + 64x + 9y^2 - 54y + 1 = 0$

56. $16x^2 - 32x - 9y^2 + 90y - 353 = 0$

57. $x^2 + 6x - 12y + 33 = 0$

58. $5x^2 + 3y^2 - 40x - 36y + 188 = 0$

59. $x^2 - y^2 - 4x + 2y - 6 = 0$

60. $9x^2 - 90x - 16y^2 + 32y + 209 = 0$

61. $x^2 + 2y - 12 = 0$

62. $y^2 - 25x^2 + 8y - 9 = 0$

63. $x^2 + 16y^2 - 160y + 384 = 0$

64. $16x^2 + 25y^2 - 64x - 100y + 564 = 0$

65. $16x^2 - 25y^2 - 64x + 100y - 36 = 0$

66. Let F_1 and F_2 denote the foci of the hyperbola $5x^2 - 4y^2 = 80$.
(a) Verify that the point P with coordinates (6, 5) lies on the hyperbola.
(b) Compute the quantity $(F_1P - F_2P)^2$.

67. Show that the coordinates of the vertex of the parabola $Ax^2 + Dx + Ey + F = 0$ are given by

$$x = -\frac{D}{2A} \quad \text{and} \quad y = \frac{D^2 - 4AF}{4AE}$$

68. If the equation $Ax^2 + Cy^2 + Dx + Ey + F = 0$ represents an ellipse or a hyperbola, show that the center is the point $(-D/2A, -E/2C)$.

69. The figure shows the parabola $x^2 = 4py$ and a circle with center at the origin and diameter $3p$. If V and F denote the vertex and focus of the parabola, respectively, show that the common chord of the circle and parabola bisects the line segment $\overline{VF}$.

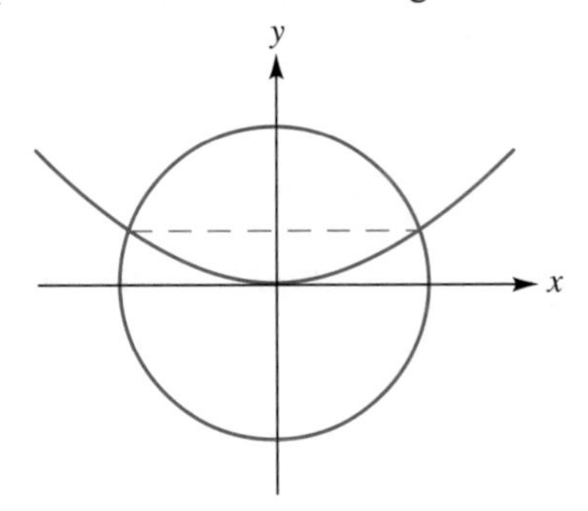

70. The following figure shows an ellipse and a parabola. As indicated in the figure, the curves are symmetric about the x-axis and they both have an x-intercept of 5.

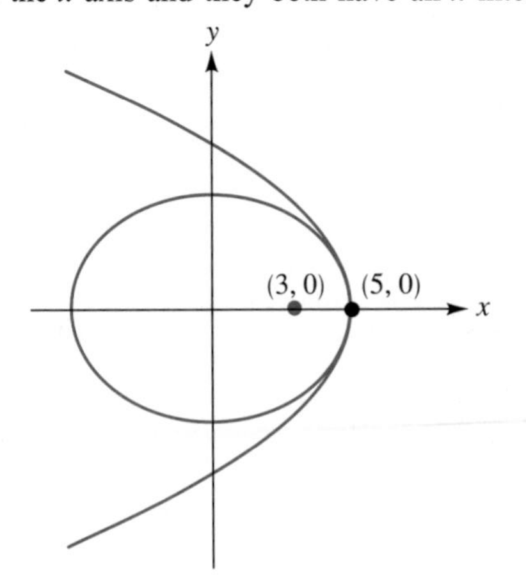

Find the equation of the ellipse and the parabola, given that the point (3, 0) is a focus for both curves.

71. Show that the equation of a line tangent to the circle $(x-h)^2+(y-k)^2=r^2$ at the point (a, b) on the circle is

$$(a-h)(x-h)+(b-k)(y-k)=r^2$$

72. The area A of an ellipse $(x^2/a^2)+(y^2/b^2)=1$ is given by the formula $A=\pi ab$. Use this formula to compute the area of the ellipse $5x^2+6y^2=60$.

73. As you know, there is a simple expression for the circumference of a circle of radius a, namely, $2\pi a$. However, there is no similar type of elementary expression for the circumference of an ellipse. (The circumference of an ellipse can be computed to as many decimal places as required using the methods of calculus.) Nevertheless, there are some interesting elementary formulas that allow us to approximate the circumference of an ellipse quite closely. Three such formulas follow, along with the names of their discoverers and the dates of discovery. Each formula yields an approximate value for the circumference of the ellipse $(x^2/a^2)+(y^2/b^2)=1$.

$$C_1=\pi\left[a+b+\tfrac{1}{2}(\sqrt{a}-\sqrt{b})^2\right]$$

Giuseppe Peano, 1887

$$C_2=\pi\left[3(a+b)-\sqrt{(a+3b)(3a+b)}\right]$$

Srinivasa Ramanujan, 1914

$$C_3=\frac{\pi}{2}\left[a+b+\sqrt{2(a^2+b^2)}\right]$$

R. A. Johnson, 1930

Use these formulas to complete the following table of approximations for the circumference of the ellipse $(x^2/5^2)+(y^2/3^2)=1$.

	Approximation Obtained	Percentage Error
C_1		
C_2		
C_3		

Round the values of C_1, C_2, and C_3 to six decimal places. To complete the right-hand column of the table, you need two facts. First, the actual circumference of the ellipse, rounded to six decimal places, is 25.526999. Second, percentage error in an approximation is given by

$$\frac{|\text{true value} - \text{approximation}|}{\text{true value}} \times 100$$

Round the percentage errors to two significant digits. Which of the three approximations is the best?

74. In the following figure, V is the vertex of the parabola $y=ax^2+bx+c$.

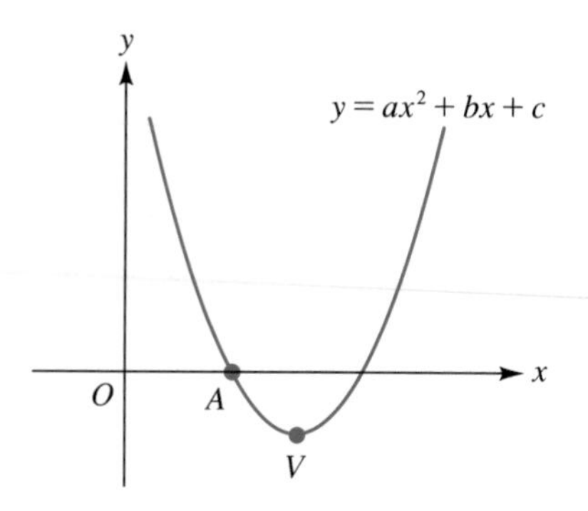

If r_1 and r_2 are the roots of the equation $ax^2+bx+c=0$, show that

$$VO^2-VA^2=r_1r_2$$

CHAPTER ELEVEN TEST

1. Find the focus and the directrix of the parabola $y^2=-12x$, and sketch the graph.
2. Graph the hyperbola $x^2-4y^2=4$. Specify the foci and the asymptotes.
3. **(a)** Determine an angle of rotation θ so that there is no $x'y'$-term present when the equation

$$x^2+2\sqrt{3}xy+3y^2-12\sqrt{3}x+12y=0$$

is transformed to the x'-y' coordinate system.

(b) Graph the equation

$$x^2 + 2\sqrt{3}xy + 3y^2 - 12\sqrt{3}x + 12y = 0$$

4. Determine the angle of inclination for the line $y = (1/\sqrt{3})x - 4$.
5. The foci of an ellipse are $(0, \pm 2)$ and the eccentricity is $1/2$. Determine the equation of the ellipse. Write your answer in standard form.
6. Tangents are drawn from the point $(-4, 0)$ to the circle $x^2 + y^2 = 1$. Find the slopes of the tangents.
7. The x-intercept of a line is 2, and its angle of inclination is 60°. Find the equation of the line. Write your answer in the form $Ax + By + C = 0$.
8. Determine the equation of the hyperbola with foci $(\pm 2, 0)$ and with asymptotes $y = \pm(1/\sqrt{3})x$. Write your answer in standard form.
9. Let F_1 and F_2 denote the foci of the hyperbola $5x^2 - 4y^2 = 80$.
   (a) Verify that the point $P(6, 5)$ lies on the hyperbola.
   (b) Compute the quantity $(F_1P - F_2P)^2$.
10. Graph the ellipse $4x^2 + 25y^2 = 100$. Specify the foci and the lengths of the major and minor axes.
11. Compute the distance from the point $(-1, 0)$ to the line $2x - y - 1 = 0$.
12. Graph the equation $16x^2 + y^2 - 64x + 2y + 65 = 0$.
13. Graph the equation $\dfrac{(x+4)^2}{3^2} - \dfrac{(y-4)^2}{1^2} = 1$.
14. Graph the equation $r = 9/(5 - 4\cos\theta)$. Which type of conic is this?
15. Graph the parabola $(x - 1)^2 = 8(y - 2)$. Specify the focal width and the vertex.
16. Consider the ellipse $(x^2/6^2) + (y^2/5^2) = 1$.
    (a) What are the equations of the directrices?
    (b) If P is a point on the ellipse in Quadrant I such that the x-coordinate of P is 3, compute F_1P and F_2P, the lengths of the focal radii.
17. Determine the equation of the line that is tangent to the ellipse $x^2 + 3y^2 = 52$ at the point $(-2, 4)$. Write your answer in the form $y = mx + b$.
18. Find the equation of the line that is tangent to the parabola $x^2 = 2y$ at the point $(4, 8)$. Write your answer in the form $y = mx + b$.

ROOTS OF POLYNOMIAL EQUATIONS

It is necessary that I make some general statements concerning the nature of equations. . . .

René Descartes (1596–1650) in *La Géométrie* (1637)

This polynomial can be derived using Hückel's molecular orbital theory. The roots of the polynomial represent the allowed energy levels of the pi electrons.

Jun-ichi Aihara in his article, "Why Aromatic Compounds Are Stable," *Scientific American,* vol. 266 (March 1992), p. 65

INTRODUCTION

As suggested by the opening quotations, the business of solving polynomial equations is an old subject that still has contemporary applications at the cutting edge of technology and science. In this chapter we continue the work begun in Section 1.4 on solving polynomial equations. The first two sections of the chapter are preparatory. In Section 12.1 we introduce the *complex-number system,* and in Section 12.2 we discuss the division process for polynomials. Section 12.3 presents two basic tools of the trade that are used in studying polynomial equations. These are the *remainder theorem* and the *factor theorem.* These two theorems are then applied repeatedly in Section 12.4 to develop fundamental results concerning polynomial equations. You can view much of the material in this section as a kind of generalization of what you already know about quadratic equations. However, don't look for a general formula that is similar to the quadratic formula, one that can be used to solve every polynomial equation. (The historical note at the end of Section 12.4 will tell you more about this.) The next two sections of the chapter (12.5 and 12.6) present additional results that are used in actually solving polynomial equations or in determining the nature of their solutions. In the last two sections of the chapter we discuss *partial fractions,* a topic needed for parts of calculus. As you'll see, the theoretical underpinnings for partial fractions have all been developed earlier in this chapter. You'll also find that the work with partial fractions requires some of the techniques for solving systems of equations that you studied in Chapter 10.

THE COMPLEX NUMBER SYSTEM

In the following I shall denote the expression $\sqrt{-1}$ by the letter i so that ii = −1.

Leonhard Euler in a paper presented to the Saint Petersburg Academy in 1777.

In elementary algebra, complex numbers appear as expressions $a + bi$, where a and b are ordinary real numbers and $i^2 = -1$. . . . Complex numbers are manipulated by the usual rules of algebra, with the convention that i^2 is to be replaced by -1 whenever it occurs.

Ralph Boas in *Invitation to Complex Analysis* (New York: Random House, 1987)

The preceding quotation from Professor Boas summarizes the basic approach we will follow in this section. Near the end of the section, after you've become accustomed to working with complex numbers, we'll present a formal list of some of the basic definitions and properties that can be used to develop the subject more rigorously, as is required in more advanced courses.

When we solve equations in this chapter, you'll see instances in which the real-number system proves to be inadequate. In particular, since the square of a real number is never negative, there is no real number x such that $x^2 = -1$. To overcome this inconvenience, mathematicians define the symbol i by the equation

$$i^2 = -1$$

For reasons that are more historical than mathematical, i is referred to as the **imaginary unit.** This name is unfortunate in a sense, because to an engineer or a mathematician, i is neither less "real" nor less tangible than any real number. Having said this, however, we do have to admit that i does not belong to the real-number system.

Algebraically, we operate with the symbol i as if it were any letter in a polynomial expression. However, when we see i^2, we must remember to replace it by -1. Here are four sample calculations involving i.

1. $3i + 2i = 5i$
2. $-2i^2 + 6i = -2(-1) + 6i = 2 + 6i$
3. $(-i)^2 = i^2 = -1$
4. $0i = 0$

An expression of the form $a + bi$, where a and b are real numbers, is called a **complex number.*** Thus, four examples of complex numbers are:

$2 + 3i$

$4 + (-5)i$ (usually written $4 - 5i$)

$1 - \sqrt{2}i$ (also written $1 - i\sqrt{2}$)

$\frac{1}{2} + \frac{3}{2}i$ $\left(\text{also written } \frac{1 + 3i}{2}\right)$

Given a complex number $a + bi$, we say that a is the **real part** of $a + bi$, and b is the **imaginary part** of $a + bi$. For example, the real part of $3 - 4i$ is 3, and the imaginary part is -4. Observe that both the real part and the imaginary part of a complex number are themselves real numbers.

We define the notion of *equality* for complex numbers in terms of their real and imaginary parts. Two complex numbers are said to be **equal** if their corresponding real and imaginary parts are equal. We can write this definition symbolically as follows:

$$a + bi = c + di \quad \text{if and only if} \quad a = c \quad \text{and} \quad b = d$$

*The term *complex number* is attributed to Carl Friedrich Gauss (1777–1855). The term *imaginary number* originated with René Descartes (1596–1650).

EXAMPLE 1 Determine the real numbers c and d such that $10 + 4i = 2c + di$.

Solution Equating the real parts of the two complex numbers gives us $2c = 10$, and therefore $c = 5$. Similarly, equating the imaginary parts yields $d = 4$. These are the required values for c and d.

As the next example indicates, addition, subtraction, and multiplication of complex numbers are carried out using the usual rules of algebra, with the understanding (as mentioned before) that i^2 is always to be replaced with -1. (We'll discuss division of complex numbers subsequently.)

EXAMPLE 2 Let $z = 2 + 5i$ and $w = 3 - 4i$. Compute each of the following: **(a)** $w + z$; **(b)** $3z$; **(c)** $w - 3z$; **(d)** zw; **(e)** wz.

Solution

(a) $w + z = (3 - 4i) + (2 + 5i) = (3 + 2) + (-4i + 5i) = 5 + i$

(b) $3z = 3(2 + 5i) = 6 + 15i$

(c) $w - 3z = (3 - 4i) - (6 + 15i) = (3 - 6) + (-4i - 15i) = -3 - 19i$

(d) $zw = (2 + 5i)(3 - 4i) = 6 - 8i + 15i - 20i^2$
$= 6 + 7i - 20(-1) = 26 + 7i$

(e) $wz = (3 - 4i)(2 + 5i) = 6 + 15i - 8i - 20i^2$
$= 6 + 7i - 20(-1) = 26 + 7i$

If you look over the result of part (a) in Example 2, you can see that the sum is obtained simply by adding the corresponding real and imaginary parts of the given numbers. Likewise in part (c), the difference is obtained by subtracting the corresponding real and imaginary parts. In part (d), however, notice that the product is *not* obtained in a similar fashion. (Exercise 64 provides some perspective on this.) Finally, notice that the results in parts (d) and (e) are identical. In fact, it can be shown that for any two complex numbers z and w, we always have $zw = wz$. In other words, just as with real numbers, multiplication of complex numbers is *commutative.* Furthermore, along the same lines, it can be shown that all of the properties of real numbers listed in Appendix A.3 continue to hold for complex numbers. We'll return to this point at the end of this section and in the exercises.

As background for the discussion of division of complex numbers, we introduce the notion of a *complex conjugate,* or simply a *conjugate.**

DEFINITION: Complex Conjugate

Let $z = a + bi$. The **complex conjugate** of z, denoted by $\bar{z}$, is defined by

$$\bar{z} = a - bi$$

EXAMPLES

If $z = 3 + 4i$, then $\bar{z} = 3 - 4i$.
If $w = 9 - 2i$, then $\overline{w} = 9 + 2i$.

*The term "conjugates" *(conjuguées)* was introduced by the nineteenth-century French mathematician A. L. Cauchy in his text *Cours d'Analyse Algébrique* (Paris: 1821).

EXAMPLE 3

(a) If $z = 6 - 3i$, compute $z\bar{z}$. **(b)** If $w = a + bi$, compute $w\bar{w}$.

Solution

(a)
$$\begin{aligned} z\bar{z} &= (6 - 3i)(6 + 3i) \\ &= 36 + 18i - 18i - 9i^2 \\ &= 36 + 9 = 45 \end{aligned}$$

(b)
$$\begin{aligned} w\bar{w} &= (a + bi)(a - bi) \\ &= a^2 - abi + abi - b^2i^2 \\ &= a^2 + b^2 \end{aligned}$$

NOTE The result in part (b) shows that the product of a complex number and its conjugate is always a real number.

Quotients of complex numbers are easy to compute using conjugates. Suppose, for example, that we wish to compute the quotient

$$\frac{5 - 2i}{3 + 4i}$$

To do this, we take the conjugate of the denominator, namely, $3 - 4i$, and then multiply the given fraction by $\frac{3 - 4i}{3 - 4i}$, which equals 1. This yields

$$\begin{aligned} \frac{5 - 2i}{3 + 4i} &= \frac{5 - 2i}{3 + 4i} \cdot \frac{3 - 4i}{3 - 4i} \\ &= \frac{15 - 26i + 8i^2}{9 - 16i^2} \\ &= \frac{7 - 26i}{25} && \text{since } 8i^2 = -8 \text{ and } -16i^2 = 16 \\ &= \frac{7}{25} - \frac{26}{25}i && \text{as required} \end{aligned}$$

In the box that follows, we summarize our procedure for computing quotients. The condition $w \neq 0$ means that w is any complex number other than $0 + 0i$.

PROCEDURE FOR COMPUTING QUOTIENTS

Let z and w be two complex numbers, $w \neq 0$. Then z/w is computed as follows:

$$\frac{z}{w} = \frac{z}{w} \cdot \frac{\bar{w}}{\bar{w}}$$

EXAMPLE 4

Let $z = 3 + 4i$ and $w = 1 - 2i$. Compute each of the following quotients: **(a)** $\frac{1}{z}$; **(b)** $\frac{z}{w}$.

Solution

(a)
$$\begin{aligned} \frac{1}{z} &= \frac{1}{z} \cdot \frac{\bar{z}}{\bar{z}} = \frac{1}{3 + 4i} \cdot \frac{3 - 4i}{3 - 4i} \\ &= \frac{3 - 4i}{9 - 16i^2} = \frac{3 - 4i}{25} \\ &= \frac{3}{25} - \frac{4}{25}i \end{aligned}$$

(b)
$$\begin{aligned} \frac{z}{w} &= \frac{z}{w} \cdot \frac{\bar{w}}{\bar{w}} = \frac{3 + 4i}{1 - 2i} \cdot \frac{1 + 2i}{1 + 2i} \\ &= \frac{3 + 10i + 8i^2}{1 - 4i^2} = \frac{-5 + 10i}{5} \\ &= -1 + 2i \end{aligned}$$

We began this section by defining i with the equation $i^2 = -1$. This can be rewritten

$$i = \sqrt{-1}$$

provided that we agree to certain conventions regarding principal square roots and negative numbers. In dealing with the principal square root of a negative real number, say, $\sqrt{-5}$, we shall write

$$\sqrt{-5} = \sqrt{(-1)(5)} = \sqrt{-1}\sqrt{5} = i\sqrt{5}$$

In other words, we are allowing the use of the rule $\sqrt{ab} = \sqrt{a}\sqrt{b}$ when a is -1 and b is a positive real number. However, the rule $\sqrt{ab} = \sqrt{a}\sqrt{b}$ *cannot* be used when both a and b are negative. If that were allowed, we could write

$$1 = (-1)(-1)$$

and then $\qquad \sqrt{1} = \sqrt{(-1)(-1)} = \sqrt{-1}\sqrt{-1} = (i)(i)$

Consequently, $\qquad 1 = i^2 \qquad$ and therefore $\qquad 1 = -1$

which is a contradiction. Again, the point here is that the rule $\sqrt{ab} = \sqrt{a}\sqrt{b}$ cannot be applied when both a and b are negative.

EXAMPLE 5 Simplify: **(a)** i^4; **(b)** i^{101}; **(c)** $\sqrt{-12} + \sqrt{-27}$; **(d)** $\sqrt{-9}\sqrt{-4}$.

Solution

(a) We make use of the defining equation for i, which is $i^2 = -1$. Thus we have

$$i^4 = (i^2)^2 = (-1)^2 = 1$$

(The result, $i^4 = 1$, is worth remembering.)

(b) $i^{101} = i^{100}i = (i^4)^{25}i = 1^{25}i = i$

(c) $\sqrt{-12} + \sqrt{-27} = \sqrt{12}\sqrt{-1} + \sqrt{27}\sqrt{-1}$
$$= \sqrt{4}\sqrt{3}\,i + \sqrt{9}\sqrt{3}\,i$$
$$= 2\sqrt{3}\,i + 3\sqrt{3}\,i = 5\sqrt{3}\,i$$

(d) $\sqrt{-9}\sqrt{-4} = (3i)(2i) = 6i^2 = -6$

NOTE $\sqrt{-9}\sqrt{-4} \neq \sqrt{36}$; why?

We have developed the complex-number system in this section for the same reason that complex numbers were developed historically: they're needed to solve polynomial equations. Consider, for example, the quadratic equation

$$ax^2 + bx + c = 0$$

where a, b, and c are real numbers and $a \neq 0$. In Section 1.4, we noted that if the discriminant $b^2 - 4ac$ is negative, then the equation has no real roots. As the next example indicates, such equations do have two complex-number roots. Furthermore (assuming the coefficients a, b, and c in the equation are real numbers) these roots always turn out to be complex conjugates.

EXAMPLE 6 Solve the quadratic equation $x^2 - 4x + 6 = 0$.

Solution We use the quadratic formula with $a = 1$, $b = -4$, and $c = 6$:

$$x = \frac{-b \pm \sqrt{b^2 - 4ac}}{2a} = \frac{4 \pm \sqrt{16 - 4(1)(6)}}{2(1)}$$

$$= \frac{4 \pm \sqrt{-8}}{2} = \frac{4 \pm (2\sqrt{2})i}{2} \qquad \text{Check the algebra.}$$

$$= 2 \pm \sqrt{2}i$$

In summary, the two roots of the quadratic equation are the complex numbers $2 + \sqrt{2}i$ and $2 - \sqrt{2}i$. These numbers are complex conjugates.

NOTE When you write answers such as these on paper, be sure you make it clear that the symbol i is outside (not inside) the radical sign. To emphasize this distinction, sometimes it's helpful to write the roots in the form $2 + i\sqrt{2}$ and $2 - i\sqrt{2}$.

Near the beginning of this section, we mentioned that we'd eventually present a formal list of some of the basic definitions and properties that can be used to develop the subject more rigorously. These are given in the two boxes that follow. As you'll see in some of the exercises, these definitions and properties are indeed consistent with our work in this section and with the properties of real numbers listed in Appendix A.3.

DEFINITION: Addition, Subtraction, Multiplication, and Division for Complex Numbers

Let $z = a + bi$ and $w = c + di$. Then $z + w$, $z - w$, and zw are defined as follows.

1. $z + w = (a + bi) + (c + di) = (a + c) + (b + d)i$
2. $z - w = (a + bi) - (c + di) = (a - c) + (b - d)i$
3. $zw = (a + bi)(c + di) = (ac - bd) + (ad + bc)i$

Furthermore, if $w \neq 0$, then z/w is defined as follows:

4. $\dfrac{z}{w} = \dfrac{a + bi}{c + di} \cdot \dfrac{c - di}{c - di} = \left(\dfrac{ac + bd}{c^2 + d^2}\right) + \left(\dfrac{bc - ad}{c^2 + d^2}\right)i$

PROPERTY SUMMARY **PROPERTIES OF COMPLEX CONJUGATES**

If $z = a + bi$, the complex conjugate, $\bar{z} = a - bi$, has the following properties.

1. $\bar{\bar{z}} = z$
2. $z = \bar{z}$ if and only if z is a real number
3. $\overline{z + w} = \bar{z} + \bar{w}$; $\overline{z - w} = \bar{z} - \bar{w}$
4. $\overline{zw} = \bar{z}\,\bar{w}$; $\dfrac{\bar{z}}{\bar{w}} = \overline{\left(\dfrac{z}{w}\right)}$
5. $(\bar{z})^n = \overline{z^n}$ for each natural number n

EXERCISE SET 12.1

A

1. Complete the table.

i^2	i^3	i^4	i^5	i^6	i^7	i^8
-1						

2. Simplify the following expression, and write the answer in the form $a + bi$.

$$1 + 3i - 5i^2 + 4 - 2i - i^3$$

For Exercises 3 and 4, specify the real and imaginary parts of each complex number.

3. **(a)** $4 + 5i$ **(b)** $4 - 5i$ **(c)** $\frac{1}{2} - i$ **(d)** $16i$

4. **(a)** $-2 + \sqrt{7}i$ **(b)** $1 + 5^{1/3}i$ **(c)** $-3i$ **(d)** 0

5. Determine the real numbers c and d such that

$$8 - 3i = 2c + di$$

6. Determine the real numbers a and b such that

$$27 - 64i = a^3 - b^3i$$

7. Simplify each of the following.
(a) $(5 - 6i) + (9 + 2i)$ **(b)** $(5 - 6i) - (9 + 2i)$

8. If $z = 1 + 4i$, compute $z - 10i$.

9. Compute each of the following.
(a) $(3 - 4i)(5 + i)$ **(b)** $(5 + i)(3 - 4i)$ **(c)** $\frac{3 - 4i}{5 + i}$ **(d)** $\frac{5 + i}{3 - 4i}$

10. Compute each of the following.
(a) $(2 + 7i)(2 - 7i)$ **(b)** $\frac{-1 + 3i}{2 + 7i}$ **(c)** $\frac{1}{2 + 7i}$ **(d)** $\frac{1}{2 + 7i} \cdot (-1 + 3i)$

In Exercises 11–36, evaluate each expression using the values $z = 2 + 3i$, $w = 9 - 4i$, and $w_1 = -7 - i$.

11. **(a)** $z + w$ **(b)** $\overline{z} + w$ **(c)** $z + \overline{z}$

12. **(a)** $\overline{z} + \overline{w}$ **(b)** $\overline{(z + w)}$ **(c)** $w - \overline{w}$

13. $(z + w) + w_1$
14. $z + (w + w_1)$
15. zw
16. wz
17. $z\overline{z}$
18. $w\overline{w}$
19. $z(ww_1)$
20. $(zw)w_1$
21. $z(w + w_1)$
22. $zw + zw_1$
23. $z^2 - w^2$
24. $(z - w)(z + w)$
25. $(zw)^2$
26. z^2w^2
27. z^3
28. z^4
29. z/w
30. w/z
31. $\overline{z}/\overline{w}$
32. $\overline{(z/w)}$
33. $z/\overline{z}$
34. $\overline{z}/z$
35. $(w - \overline{w})/(2i)$
36. $(w + \overline{w})/2$

For Exercises 37–40, compute each quotient.

37. $\frac{i}{5 + i}$
38. $\frac{1 - i\sqrt{3}}{1 + i\sqrt{3}}$
39. $\frac{1}{i}$
40. $\frac{i + i^2}{i^3 + i^4}$

In Exercises 41–48, simplify each expression.

41. $\sqrt{-49} + \sqrt{-9} + \sqrt{-4}$
42. $\sqrt{-25} + i$
43. $\sqrt{-20} - 3\sqrt{-45} + \sqrt{-80}$
44. $\sqrt{-4}\sqrt{-4}$
45. $1 + \sqrt{-36}\sqrt{-36}$
46. $i - \sqrt{-100}$
47. $3\sqrt{-128} - 4\sqrt{-18}$
48. $64 + \sqrt{-64}\sqrt{-64}$

In Exercises 49–56: **(a)** *compute the discriminant of the quadratic and note that it is negative (and therefore the equation has no real-number roots);* **(b)** *use the quadratic formula to obtain the two complex-conjugate roots of each equation.*

49. $x^2 - x + 1 = 0$
50. $x^2 - 6x + 12 = 0$
51. $5z^2 + 2z + 2 = 0$
52. $-10z^2 + 4z - 2 = 0$
53. $2z^2 + 3z + 4 = 0$
54. $3z^2 - 7z + 5 = 0$
55. $\frac{1}{6}z^2 - \frac{1}{4}z + 1 = 0$
56. $\frac{1}{2}z^2 + 2z + \frac{9}{4} = 0$

57. **(a)** Evaluate the expression $x^2 - 4x + 6$ when $x = 2 - i\sqrt{2}$.
(b) How does your result in part (a) relate to Example 6 in this section?

58. **(a)** Find the roots of the quadratic equation $x^2 - 2x + 5 = 0$.
(b) Compute the product of the two roots that you obtained in part (a).
(c) Check your answer in part (b) by applying the result given in Exercise 115(b) in Exercise Set 1.4.

59. Let $z = a + bi$ and $w = c + di$. Compute each of the following quantities, and then check that your results agree with the definitions in the first box on page 757.

(a) $z + w$ **(b)** $z - w$ **(c)** zw **(d)** $\frac{z}{w}$

B

60. Show that $\left(\frac{-1+i\sqrt{3}}{2}\right)^2 + \left(\frac{-1-i\sqrt{3}}{2}\right)^2 = -1.$

61. Let $z = \frac{-1+i\sqrt{3}}{2}$ and $w = \frac{-1-i\sqrt{3}}{2}$. Verify the following statements.
(a) $z^3 = 1$ and $w^3 = 1$
(b) $zw = 1$
(c) $z = w^2$ and $w = z^2$
(d) $(1 - z + z^2)(1 + z - z^2) = 4$

62. Let $z = a + bi$ and $w = c + di$.
(a) Show that $\overline{\overline{z}} = z$.
(b) Show that $\overline{(z + w)} = \overline{z} + \overline{w}$.

63. Show that the complex number 0 $(=0 + 0i)$ has the following properties.
(a) $0 + z = z$ and $z + 0 = z$, for all complex numbers z. *Hint:* Let $z = a + bi$.
(b) $0 \cdot z = 0$ and $z \cdot 0 = 0$, for all complex numbers z.

64. This exercise indicates one of the reasons why multiplication of complex numbers is not carried out simply by multiplying the corresponding real and imaginary parts of the numbers. (Recall that addition and subtraction *are* carried out in this manner.) Suppose for the moment that we were to define multiplication in this seemingly less complicated way:

$$(a + bi)(c + di) = ac + (bd)i \qquad (*)$$

(a) Compute $(2 + 3i)(5 + 4i)$, assuming that multiplication is defined by $(*)$.
(b) Still assuming that multiplication is defined by $(*)$, find two complex numbers z and w such that $z \neq 0$, $w \neq 0$, but $zw = 0$ (where 0 denotes the complex number $0 + 0i$).

Now notice that the result in part (b) is contrary to our expectation or desire that the product of two nonzero numbers be nonzero, as is the case for real numbers. On the other hand, it can be shown that when multiplication is carried out as described in the text, then the product of two complex numbers is nonzero if and only if both factors are nonzero.

65. **(a)** Show that addition of complex numbers is commutative. That is, show that $z + w = w + z$ for all complex numbers z and w. *Hint:* Let $z = a + bi$ and $w = c + di$.
(b) Show that multiplication of complex numbers is commutative. That is, show that $zw = wz$ for all complex numbers z and w.

66. Let $z = a + bi$.
(a) show that $(\overline{z})^2 = \overline{z^2}$.
(b) Show that $(\overline{z})^3 = \overline{z^3}$.

In Exercises 67–70, find all roots of each equation. Hints: First, factor by grouping. In Exercises 67 and 68, each equation has three roots; in Exercise 69, the equation has six roots; in Exercise 70, there are five roots.

67. $x^3 - 3x^2 + 4x - 12 = 0$

68. $2x^3 + 4x^2 + 3x + 6 = 0$

69. $x^6 - 9x^4 + 16x^2 - 144 = 0$

70. $x^5 + 4x^3 + 8x^2 + 32 = 0$

C

71. Let a and b be real numbers. Find the real and imaginary parts of the quantity

$$\frac{a+bi}{a-bi} + \frac{a-bi}{a+bi}$$

72. Find the real and imaginary parts of the quantity

$$\left(\frac{a+bi}{a-bi}\right)^2 - \left(\frac{a-bi}{a+bi}\right)^2$$

73. Find the real part of $\frac{(a+bi)^2}{a-bi} - \frac{(a-bi)^2}{a+bi}$.

74. **(a)** Let $\alpha = \left(\frac{\sqrt{a^2+b^2}+a}{2}\right)^{1/2}$

and let $\beta = \left(\frac{\sqrt{a^2+b^2}-a}{2}\right)^{1/2}$

Show that the square of the complex number $\alpha + \beta i$ is $a + bi$.
(b) Use the result in part (a) to find a complex number z such that $z^2 = i$.
(c) Use the result in part (a) to find a complex number z such that $z^2 = -7 + 24i$.

One of the basic properties of real numbers is that if a product is equal to zero, then at least one of the factors is zero. Exercises 75 and 76 show that this property also holds for complex numbers. For Exercises 75 and 76, assume that $zw = 0$, where $z = a + bi$ and $w = c + di$.

75. If $a \neq 0$, prove that $w = 0$. (That is, prove that $c = d = 0$.)

76. If $b \neq 0$, prove that $w = 0$. (That is, prove that $c = d = 0$.)

DIVISION OF POLYNOMIALS

Although the process of long division for polynomials is often taught in elementary algebra courses, it usually does not receive sufficient emphasis there. As with ordinary long division for numbers, the process is best learned by first watching someone do examples and then practicing on your own. The terms *quotient, remainder, divisor,* and *dividend* will be used here in the same way they are used in ordinary division of numbers. For instance, when 7 is divided by 2, the quotient is 3 and the remainder is 1. We write this

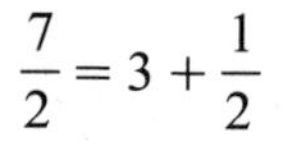

$$\frac{7}{2} = 3 + \frac{1}{2}$$

or, equivalently,

$$7 = 2 \times 3 + 1$$

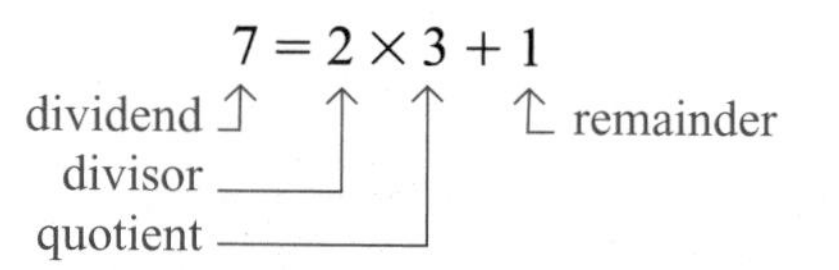

The process of long division for polynomials follows the same four-step cycle used in ordinary long division of numbers: divide, multiply, subtract, bring down. As a first example, we divide $2x^2 - 7x + 8$ by $x - 2$. Notice that in setting up the division, we write both the dividend and the divisor in decreasing powers of x.

$$\begin{array}{r|l} & \quad 2x \;\; - 3 \\ \hline x - 2 & 2x^2 - 7x + 8 \\ & \underline{2x^2 - 4x} \\ & \qquad\; -3x + 8 \\ & \qquad\; \underline{-3x + 6} \\ & \qquad\qquad\quad 2 \end{array}$$

1. Divide the first term of the dividend by the first term of the divisor: $\dfrac{2x^2}{x} = 2x$. The result becomes the first term of the quotient, as shown.
2. Multiply the divisor $x - 2$ by the term $2x$ obtained in the previous step. This yields the quantity $2x^2 - 4x$, which is written below the dividend, as shown.
3. From the quantity $2x^2 - 7x$ in the dividend, subtract the quantity $2x^2 - 4x$. This yields $-3x$.
4. Bring down the $+8$ in the dividend, as shown. The resulting quantity, $-3x + 8$, is now treated as the dividend and the entire process is repeated.

We've now found that when $2x^2 - 7x + 8$ is divided by $x - 2$, the quotient is $2x - 3$ and the remainder is 2. This is summarized by writing either

$$\frac{2x^2 - 7x + 8}{x - 2} = 2x - 3 + \frac{2}{x - 2} \tag{1}$$

or, after multiplying through by $x - 2$,

$$\underbrace{2x^2 - 7x + 8}_{\text{dividend}} = \underbrace{(x - 2)}_{\text{divisor}}\underbrace{(2x - 3)}_{\text{quotient}} + \underbrace{2}_{\text{remainder}} \tag{2}$$

There are two observations to be made here. First, notice that equation (2) is valid for all real numbers x, whereas equation (1) carries the implicit restriction that x may not equal 2. For this reason, we often prefer to write our results in the form of equation (2). Second, notice that the degree of the remainder is less than the degree of the divisor. This is very similar to the situation with ordinary division of positive integers, where the remainder is always less than the divisor.

As another example of the long division process, we divide $3x^4 - 2x^3 + 2$ by $x^2 - 1$. Notice in what follows that we have inserted the terms in the divisor and dividend that have coefficients of zero. These terms act as place holders.

$$\begin{array}{r|l}
 & 3x^2 - 2x + 3 \\ \hline
x^2 + 0x - 1 & 3x^4 - 2x^3 + 0x^2 + 0x + 2 \\
 & \underline{3x^4 + 0x^3 - 3x^2} \\
 & \quad -2x^3 + 3x^2 + 0x \\
 & \quad \underline{-2x^3 + 0x^2 + 2x} \\
 & \qquad\qquad 3x^2 - 2x + 2 \\
 & \qquad\qquad \underline{3x^2 + 0x - 3} \\
 & \qquad\qquad\qquad -2x + 5
\end{array}$$

We can write this result as

$$\frac{3x^4 - 2x^3 + 2}{x^2 - 1} = 3x^2 - 2x + 3 + \frac{-2x + 5}{x^2 - 1}$$

or, multiplying through by $x^2 - 1$,

$$\underbrace{3x^4 - 2x^3 + 2}_{\text{dividend}} = \underbrace{(x^2 - 1)}_{\text{divisor}}\underbrace{(3x^2 - 2x + 3)}_{\text{quotient}} + \underbrace{(-2x + 5)}_{\text{remainder}}$$

Notice that this last equation holds for all values of x, whereas the previous equation carries the restrictions that x may be neither 1 nor -1. Also, as in our previous example, observe that the degree of the remainder is less than that of the divisor.

There is a theorem, commonly referred to as the **division algorithm,** that summarizes rather nicely the key results of the long division process. We state the theorem here without proof.

THE DIVISION ALGORITHM

Let $p(x)$ and $d(x)$ be polynomials, and assume that $d(x)$ is not the zero polynomial. Then there are unique polynomials $q(x)$ and $R(x)$ such that

$$p(x) = d(x) \cdot q(x) + R(x)$$

where either $R(x)$ is the zero polynomial or the degree of $R(x)$ is less than the degree of $d(x)$.

The polynomials $p(x)$, $d(x)$, $q(x)$, and $R(x)$ are referred to, respectively, as the *dividend, divisor, quotient,* and *remainder.* When $R(x) = 0$, we have $p(x) = d(x) \cdot q(x)$ and we say that $d(x)$ and $q(x)$ are **factors** of $p(x)$. Also, since $d(x)$ is not the zero polynomial, notice that the equation $p(x) = d(x) \cdot q(x) + R(x)$ implies that the degree of $q(x)$ is less than or equal to the degree of $p(x)$. (Why?)

EXAMPLE 1 Let $p(x) = x^3 + 2x^2 - 4$ and $d(x) = x - 3$. Use the long division process to find the polynomials $q(x)$ and $R(x)$ such that

$$p(x) = d(x) \cdot q(x) + R(x)$$

where either $R(x) = 0$ or the degree of $R(x)$ is less than the degree of $d(x)$.

Solution After inserting the term $0x$ in the dividend $p(x)$, we use long division to divide $p(x)$ by $d(x)$:

$$\begin{array}{r|l} & x^2 + 5x \ + 15 \\ \hline x - 3 & x^3 + 2x^2 + \ 0x - \ 4 \\ & \underline{x^3 - 3x^2} \\ & \qquad 5x^2 + \ 0x \\ & \qquad \underline{5x^2 - 15x} \\ & \qquad\qquad 15x - \ 4 \\ & \qquad\qquad \underline{15x - 45} \\ & \qquad\qquad\qquad 41 \end{array}$$

We now have

$$\underbrace{x^3 + 2x^2 - 4}_{p(x)} = \underbrace{(x - 3)}_{d(x)}\underbrace{(x^2 + 5x + 15)}_{q(x)} + \underbrace{41}_{R(x)}$$

Thus, $q(x) = x^2 + 5x + 15$ and $R(x) = 41$. Notice that the degree of $R(x)$ is less than the degree of $d(x)$.

The long division procedure for polynomials can be streamlined when the divisor is of the form $x - r$. This shortened version, known as **synthetic division,** will be useful in subsequent sections when we are solving polynomial equations.

We can explain the idea behind synthetic division by using the long division carried out in Example 1. The basic idea is that in the long division process, it is the *coefficients* of the various polynomials that carry all the necessary information. In our example, for instance, the quotient and remainder can be abbreviated by writing down a sequence of four numbers:

$$1 \quad 5 \quad 15 \quad 41$$

By studying the long division process, you will find that these numbers are obtained through the following four steps:

STEP 1 Write down the first coefficient of the dividend. This will be the first coefficient of the quotient.

Result $\boxed{1}$

STEP 2 Multiply the 1 obtained in the previous step by the -3 in the divisor. Then subtract the result from the second coefficient of the dividend:

$$-3 \times 1 = -3 \qquad 2 - (-3) = 5 \qquad \textit{Result } \boxed{5}$$

STEP 3 Multiply the 5 obtained in the previous step by the -3 in the divisor. Then subtract the result from the third coefficient of the dividend:

$$-3 \times 5 = -15 \qquad 0 - (-15) = 15 \qquad \textit{Result } \boxed{15}$$

STEP 4 Multiply the 15 obtained in the previous step by the -3 in the divisor. Then subtract the result from the fourth coefficient of the dividend:

$$-3 \times 15 = -45 \qquad -4 - (-45) = 41 \qquad \textit{Result } \boxed{41}$$

A convenient format for setting up this process involves writing the constant term of the divisor and the coefficients of the dividend as follows:

$$\begin{array}{r|rrrr} -3 & 1 & 2 & 0 & -4 \\ \hline & & & & \end{array}$$

Now, using this format, let's again go through the four steps we have just described:

STEP 1 Bring down the 1.

$$\begin{array}{r|rrrr} -3 & 1 & 2 & 0 & -4 \\ & & & & \\ \hline & 1 & & & \end{array}$$

STEP 2 $-3 \times 1 = -3$
$2 - (-3) = 5$

$$\begin{array}{r|rrrr} -3 & 1 & 2 & 0 & -4 \\ & & -3 & & \\ \hline & 1 & 5 & & \end{array}$$

STEP 3 $-3 \times 5 = -15$
$0 - (-15) = 15$

$$\begin{array}{r|rrrr} -3 & 1 & 2 & 0 & -4 \\ & & -3 & -15 & \\ \hline & 1 & 5 & 15 & \end{array}$$

STEP 4 $-3 \times 15 = -45$
$-4 - (-45) = 41$

$$\begin{array}{r|rrrr} -3 & 1 & 2 & 0 & -4 \\ & & -3 & -15 & -45 \\ \hline & 1 & 5 & 15 & 41 \end{array}$$

Although we have now obtained the required sequence of numbers, 1 5 15 41, there is one further simplification that can be made. In Steps 2 through 4, we can add instead of subtract if we use 3 instead of -3 in the initial format. (You will see the motivation for this in the next section when we discuss the remainder theorem.) With this change, let us now summarize the technique of synthetic division using the example with which we've been working. The method is applicable for any polynomial division where the divisor has the form $x - a$.

TO DIVIDE $x^3 + 2x^2 - 4$ BY $x - 3$ USING SYNTHETIC DIVISION

		COMMENTS
Format	$\begin{array}{r\|rrrr} 3 & 1 & 2 & 0 & -4 \\ \hline & & & & \end{array}$	Since the divisor is $x - 3$, the format begins with 3. The coefficients from the dividend are written in the order corresponding to decreasing powers of x. A zero coefficient is inserted as a place holder.
Procedure	$\begin{array}{r\|rrrr} 3 & 1 & 2 & 0 & -4 \\ & & 3 & 15 & 45 \\ \hline & 1 & 5 & 15 & 41 \end{array}$	**STEP 1** Bring down the 1. **STEP 2** $3 \times 1 = 3;\ 2 + 3 = 5$ **STEP 3** $3 \times 5 = 15;\ 0 + 15 = 15$ **STEP 4** $3 \times 15 = 45;\ -4 + 45 = 41$
Answer	Quotient: $x^2 + 5x + 15$ Remainder: 41	The degree of the first term in the quotient is one less than the degree of the first term of the dividend.

EXAMPLE 2 Use synthetic division to divide $x^3 - 6x + 4$ by $x - 2$.

Solution

$$\begin{array}{r|rrrr} 2 & 1 & 0 & -6 & 4 \\ & & 2 & 4 & -4 \\ \hline & 1 & 2 & -2 & 0 \end{array}$$

Looking at the third row of numbers in the synthetic division we've carried out, we see that the quotient is $x^2 + 2x - 2$ and the remainder is 0. In other words, both $x - 2$ and $x^2 + 2x - 2$ are factors of $x^3 - 6x + 4$, and we have

$$x^3 - 6x + 4 = (x - 2)(x^2 + 2x - 2)$$

EXAMPLE 3 Use synthetic division to divide $x^5 - a^5$ by $x - a$.

Solution

$$\begin{array}{r|rrrrrr} a & 1 & 0 & 0 & 0 & 0 & -a^5 \\ & & a & a^2 & a^3 & a^4 & a^5 \\ \hline & 1 & a & a^2 & a^3 & a^4 & 0 \end{array}$$

As before, we read off the quotient and remainder from the third row of numbers. The quotient is

$$x^4 + ax^3 + a^2x^2 + a^3x + a^4$$

and the remainder is zero. So, we have

$$\underbrace{x^5 - a^5}_{\text{dividend}} = \underbrace{(x - a)}_{\text{divisor}}\underbrace{(x^4 + ax^3 + a^2x^2 + a^3x + a^4)}_{\text{quotient}} \quad \underbrace{+\,0}_{\text{remainder}}$$

or

$$x^5 - a^5 = (x - a)(x^4 + ax^3 + a^2x^2 + a^3x + a^4)$$

The last equation in Example 3 tells us how to factor $x^5 - a^5$. One factor is $x - a$. Notice the pattern in the second factor, $x^4 + ax^3 + a^2x^2 + a^3x + a^4$.

first term: a^0x^{5-1}
second term: a^1x^{5-2}
third term: a^2x^{5-3}
fourth term: a^3x^{5-4}
fifth term: a^4x^{5-5}

In the same way that we've found a factorization for $x^5 - a^5$, we can find a factorization for $x^n - a^n$ for any positive integer $n \geqslant 2$. In each case, the first factor is $x - a$, while the second factor follows the same pattern just described for $x^5 - a^5$. We state the general result in the box that follows. The result can be proved by using *mathematical induction* (discussed in a later chapter) or by using the *remainder theorem* (discussed in the next section).

FACTORIZATION OF $x^n - a^n$

$$x^n - a^n = (x - a)(x^{n-1} + ax^{n-2} + a^2x^{n-3} + \cdots + a^{n-1})$$

In our development of synthetic division, we assumed that the form of the divisor was $x - r$. The next example shows what to do when the form of the divisor is $x + r$.

EXAMPLE 4 Use synthetic division to divide $x^4 - 2x^3 + 5x^2 - 4x + 3$ by $x + 1$.

Solution We first need to write the divisor $x + 1$ in the form $x - r$. We have

$$x + 1 = x - (-1)$$

In other words, r is -1, and this is the value we use to set up the synthetic division format. The format then is

$$\begin{array}{r|rrrrr} -1 & 1 & -2 & 5 & -4 & 3 \\ & & & & & \\ \hline & & & & & \end{array}$$

Now we carry out the synthetic division procedure:

$$\begin{array}{r|rrrrr} -1 & 1 & -2 & 5 & -4 & 3 \\ & & -1 & 3 & -8 & 12 \\ \hline & 1 & -3 & 8 & -12 & 15 \end{array}$$

The quotient is therefore $x^3 - 3x^2 + 8x - 12$, and the remainder is 15. We can summarize this result by writing

$$x^4 - 2x^3 + 5x^2 - 4x + 3 = (x + 1)(x^3 - 3x^2 + 8x - 12) + 15$$

(Notice that the degree of the remainder is less than the degree of the divisor, in agreement with the division algorithm.)

EXERCISE SET 12.2

A

In Exercises 1–20, use long division to find the quotients and the remainders. Also, write each answer in the form $p(x) = d(x) \cdot q(x) + R(x)$, as in equation (2) in the text.

1. $\dfrac{x^2 - 8x + 4}{x - 3}$
2. $\dfrac{x^3 - 4x^2 + x - 2}{x - 5}$
3. $\dfrac{x^2 - 6x - 2}{x + 5}$
4. $\dfrac{3x^2 + 4x - 1}{x - 1}$
5. $\dfrac{6x^3 - 2x + 3}{2x + 1}$
6. $\dfrac{x^4 - 4x^3 + 6x^2 - 4x + 1}{x - 1}$
7. $\dfrac{x^5 + 2}{x + 3}$
8. $\dfrac{4x^3 - x^2 + 8x - 1}{x^2 - x + 1}$
9. $\dfrac{x^6 - 64}{x - 2}$
10. $\dfrac{x^6 + 64}{x - 2}$
11. $\dfrac{5x^4 - 3x^2 + 2}{x^2 - 3x + 5}$
12. $\dfrac{8x^6 - 36x^4 + 54x^2 - 27}{2x^2 - 3}$
13. $\dfrac{3y^3 - 4y^2 - 3}{y^2 + 5y + 2}$
14. $\dfrac{4y^4 - y^3 + 2y - 1}{2y^2 - 3y - 4}$
15. $\dfrac{t^4 - 4t^3 + 4t^2 - 16}{t^2 - 2t + 4}$
16. $\dfrac{2t^5 - 6t^4 - t^2 + 2t + 3}{t^3 - 2}$
17. $\dfrac{z^5 - 1}{z - 1}$
18. $\dfrac{1 + z + z^2 + z^3}{1 + z + z^2}$
19. $\dfrac{ax^2 + bx + c}{x - r}$
20. $\dfrac{ax^3 + bx^2 + cx + d}{x - r}$

In Exercises 21–40, use synthetic division to find the quotients and remainders. Also, in each case, write the result of the division in the form $p(x) = d(x) \cdot q(x) + R(x)$, as in equation (2) in the text.

21. $\dfrac{x^2 - 6x - 2}{x - 5}$
22. $\dfrac{3x^2 + 4x - 1}{x - 1}$
23. $\dfrac{4x^2 - x - 5}{x + 1}$
24. $\dfrac{x^2 - 1}{x + 2}$

25. $\dfrac{6x^3 - 5x^2 + 2x + 1}{x - 4}$

26. $\dfrac{x^4 - 4x^3 + 6x^2 - 4x + 1}{x - 1}$

27. $\dfrac{x^3 - 1}{x - 2}$

28. $\dfrac{x^3 - 8}{x - 2}$

29. $\dfrac{x^5 - 1}{x + 2}$

30. $\dfrac{x^3 - 8x^2 - 1}{x + 3}$

31. $\dfrac{x^4 - 6x^3 + 2}{x + 4}$

32. $\dfrac{3x^3 - 2x^2 + x + 1}{x - \frac{1}{2}}$

33. $\dfrac{x^3 - 4x^2 - 3x + 6}{x - 10}$

34. $\dfrac{1 + 3x + 3x^2 + x^3}{x + 1}$

35. $\dfrac{x^3 - x^2}{x + 5}$

36. $\dfrac{5x^4 - 4x^3 + 3x^2 - 2x + 1}{x + \frac{1}{2}}$

37. $\dfrac{14 - 27x - 27x^2 + 54x^3}{x - \frac{2}{3}}$

38. $\dfrac{14 - 27x - 27x^2 + 54x^3}{x + \frac{2}{3}}$

39. $\dfrac{x^4 + 3x^2 + 12}{x - 3}$

40. **(a)** $\dfrac{x^4 - 16}{x - 2}$ **(b)** $\dfrac{x^4 + 16}{x + 2}$

In Exercises 41–44, each expression has the form $x^n - a^n$. Write each expression as a product of two factors (as in the box on page 764).

41. $x^5 - 32$

42. $y^6 - 1$

43. $z^4 - 81$

44. $x^7 - y^7$

B

In Exercises 45–48, use synthetic division to determine the quotient $q(x)$ and the remainder $R(x)$ in each case.

45. $\dfrac{6x^2 - 8x + 1}{3x - 4}$ *Hint:* Divide both numerator and denominator by 3. (Why?)

46. $\dfrac{4x^3 + 6x^2 - 6x - 5}{2x - 3}$

47. $\dfrac{6x^3 + 1}{2x + 1}$

48. $\dfrac{5x^3 - 3x^2 + 1}{3x + 1}$

49. When $x^3 + kx + 1$ is divided by $x + 1$, the remainder is -4. Find k.

50. **(a)** Show that when $x^3 + kx + 6$ is divided by $x + 3$, the remainder is $-21 - 3k$.
(b) Determine a value of k such that $x + 3$ will be a factor of $x^3 + kx + 6$.

51. When $x^2 + 2px - 3q^2$ is divided by $x - p$, the remainder is zero. Show that $p^2 = q^2$.

52. Given that $x - 3$ is a factor of $x^3 - 2x^2 - 4x + 3$, solve the equation $x^3 - 2x^2 - 4x + 3 = 0$.

The process of synthetic division applies equally well when some or all of the coefficients are nonreal complex numbers. In Exercises 53–56, use synthetic division to determine the quotient $q(x)$ and the remainder $R(x)$ in each case.

53. $\dfrac{x^2 - 4x + 1}{x - i}$

54. $\dfrac{x^3 - 2x^2 - 4}{x - 3i}$

55. $\dfrac{x^2 - 2x + 2}{x - (1 + i)}$

56. $\dfrac{x^3 - x^2 + 4x - 4}{x + 2i}$

57. Given that the identity $f(x) = d(x) \cdot q(x) + R(x)$ holds for the following polynomials, evaluate $f(\sqrt{3})$.
Hint (of sorts): There's an easy way and a tedious way.

$f(x) = 2x^5 + 5x^4 - 8x^3 + 7x^2 - 9 \quad d(x) = x^2 - 3$

$q(x) = 2x^3 + 5x^2 - 2x + 22 \quad R(x) = -6x + 57$

58. Given that the identity $f(t) = d(t) \cdot q(t) + R(t)$ holds for the following polynomials, evaluate $f(4)$.

$f(t) = t^5 - 3t^4 + 2t^3 - 5t^2 + 6t - 7 \quad d(t) = t - 4$

$q(t) = t^4 + t^3 + 6t^2 + 19t + 82 \quad R(t) = 321$

59. Find the remainder when $t^5 - 5a^4t + 4a^5$ is divided by $t - a$.

60. When $f(x)$ is divided by $(x - a)(x - b)$, the remainder is $Ax + B$. Apply the division algorithm to show that

$$A = \frac{f(a) - f(b)}{a - b} \quad \text{and} \quad B = \frac{bf(a) - af(b)}{b - a}$$

ROOTS OF POLYNOMIAL EQUATIONS: THE REMAINDER THEOREM AND THE FACTOR THEOREM

Descartes recommended [in his *La Géométrie* of 1637] *that all terms* [of an equation] *should be taken to one side and equated with zero. Though he was not the first to suggest this, he was the earliest writer to realize the advantage to be gained. He pointed out that a polynomial* $f(x)$ *was divisible by* $(x - a)$ *if and only if* a *was a root of* $f(x)$.

David M. Burton in *The History of Mathematics, An Introduction,* 2d ed. (Dubuque, Iowa: Wm. C. Brown Publishers, 1991)

The techniques for solving polynomial equations of degree 2 were discussed in Section 1.4. Now we want to extend those ideas. Our focus in this section and in the remainder of the chapter is on solving polynomial equations of any degree, that is, equations of the form

$$f(x) = a_n x^n + a_{n-1}x^{n-1} + \cdots + a_1 x + a_0 = 0 \qquad (1)$$

Here, as in Chapter 1, a **root** or **solution** of equation (1) is a number r that when substituted for x leads to a true statement. Thus, r is a root of equation (1) provided $f(r) = 0$. We also refer to the number r in this case as a **zero** of the function f.

EXAMPLE 1

(a) Is -3 a zero of the function f defined by $f(x) = x^4 + x^2 - 6$?

(b) Is $\sqrt{2}$ a root of the equation $x^4 + x^2 - 6 = 0$?

Solution

(a) By definition, -3 will be a zero of f if $f(-3) = 0$. We have

$$f(-3) = (-3)^4 + (-3)^2 - 6 = 81 + 9 - 6 = 84 \neq 0$$

Thus, -3 is not a zero of the function f.

(b) To check if $\sqrt{2}$ is a root of the given equation, we have

$$(\sqrt{2})^4 + (\sqrt{2})^2 - 6 \stackrel{?}{=} 0$$

$$4 + 2 - 6 \stackrel{?}{=} 0$$

$$0 \stackrel{?}{=} 0 \qquad \text{True}$$

Thus, $\sqrt{2}$ is a root of the equation $x^4 + x^2 - 6 = 0$.

There are cases in which a root of an equation is what we call a **repeated root.** Consider, for instance, the equation $x(x - 1)(x - 1) = 0$. We have

$$x(x-1)(x-1) = 0$$

$$x = 0 \quad \Big| \quad x - 1 = 0,\; x = 1 \quad \Big| \quad x - 1 = 0,\; x = 1$$

The roots of the equation are therefore 0, 1, and 1. The repeated root here is $x = 1$. We say in this case that 1 is a **double root** or, equivalently, that 1 is a **root of multiplicity 2.** More generally, if a root is repeated k times, we call it a **root of multiplicity *k*.**

EXAMPLE 2 State the multiplicity of each root of the equation

$$(x - 4)^2(x - 5)^3 = 0$$

Solution We have $(x - 4)(x - 4)(x - 5)(x - 5)(x - 5) = 0$. By setting each factor equal to zero, we obtain the roots 4, 4, 5, 5, and 5. From this we see that the root 4 has multiplicity 2, while the root 5 has multiplicity 3. Notice that it is not really necessary to write out all the factors as we did here; the exponents of the factors in the original equation give us the required multiplicities.

There are two simple but important theorems that will form the basis for much of our subsequent work with polynomials: the *remainder theorem* and the *factor theorem.* We begin with a statement of the remainder theorem.

THE REMAINDER THEOREM

When a polynomial $f(x)$ is divided by $x - r$, the remainder is $f(r)$.

Before turning to a proof of the remainder theorem, let's see what the theorem is saying in two particular cases. First, suppose that we divide the polynomial $f(x) = 2x^2 - 3x + 4$ by $x - 1$. Then according to the remainder theorem, the remainder in this case should be the number $f(1)$. Let's check:

$$\begin{array}{r|rrr} 1 & 2 & -3 & 4 \\ & & 2 & -1 \\ \hline & 2 & -1 & \textcircled{3} \end{array} \qquad \begin{aligned} f(x) &= 2x^2 - 3x + 4 \\ f(1) &= 2(1)^2 - 3(1) + 4 \\ f(1) &= \textcircled{3} \end{aligned}$$

As the calculations show, the remainder is indeed equal to $f(1)$. As a second example, let us divide the polynomial $g(x) = ax^2 + bx + c$ by $x - r$. According to the remainder theorem, the remainder should be $g(r)$. Again, let us check:

$$\begin{array}{r|lll} r & a & b & c \\ & & ar & ar^2 + br \\ \hline & a & ar + b & \boxed{ar^2 + br + c} \end{array} \qquad \begin{aligned} g(x) &= ax^2 + bx + c \\ \\ g(r) &= \boxed{ar^2 + br + c} \end{aligned}$$

The calculations show that the remainder is equal to $g(r)$, as we wished to check.

In the example just concluded, we verified that the remainder theorem holds for any quadratic polynomial $g(x) = ax^2 + bx + c$. A general proof of the remainder theorem can easily be given along these same lines. The only drawback is that it becomes slightly cumbersome to carry out the synthetic division process when the dividend is

$$a_n x^n + a_{n-1}x^{n-1} + \cdots + a_1x + a_0$$

For this reason, mathematicians often prefer to base the proof of the remainder theorem on the division algorithm given in the previous section. This is the path we will follow here.

To prove the remainder theorem, we must show that when the polynomial $f(x)$ is divided by $x - r$, the remainder is $f(r)$. Now, according to the division algorithm, we can write

$$f(x) = (x - r) \cdot q(x) + R(x) \tag{2}$$

In this identity, either $R(x)$ is the zero polynomial or the degree of $R(x)$ is less than that of $x - r$. But since the degree of $x - r$ is 1, we must have in this case that the degree of $R(x)$ is zero. Thus, in *either* case, the remainder $R(x)$ is a constant. Denoting this constant by c, we can rewrite equation (2) as

$$f(x) = (x - r) \cdot q(x) + c$$

Now if we set $x = r$ in this identity, we obtain

$$f(r) = (r - r) \cdot q(r) + c = c$$

We have now shown that $f(r) = c$. But by definition, c is the remainder $R(x)$. Thus, $f(r) = R(x)$. This proves the remainder theorem.

EXAMPLE 3 Let $f(x) = 2x^3 - 5x^2 + x - 6$.

(a) Use the remainder theorem to evaluate $f(3)$.
(b) Is $x - 3$ a factor of $f(x) = 2x^3 - 5x^2 + x - 6$?

Solution

(a) According to the remainder theorem, $f(3)$ is the remainder when $f(x)$ is divided by $x - 3$. Using synthetic division, we have

$$\begin{array}{r|rrrr} 3 & 2 & -5 & 1 & -6 \\ & & 6 & 3 & 12 \\ \hline & 2 & 1 & 4 & 6 \end{array}$$

The remainder is 6, and therefore $f(3) = 6$.

(b) By definition, $x - 3$ is a factor of $f(x)$ if we obtain a zero remainder when $f(x)$ is divided by $x - 3$. But from our work in part (a), we know that the remainder is 6, not 0, so $x - 3$ is not a factor of $f(x)$.

EXAMPLE 4 In the previous section it was stated, but not proved, that $x - a$ is a factor of $f(x) = x^n - a^n$. Use the remainder theorem to prove this fact.

Solution We need to show that when $f(x) = x^n - a^n$ is divided by $x - a$, the remainder is zero. By the remainder theorem, the remainder is equal to $f(a)$, which is easy to find:

$$f(a) = a^n - a^n = 0$$

Thus, the remainder is zero, and $x - a$ is a factor of $x^n - a^n$, as required.

From our experience with quadratic equations, we know that there is a close connection between factoring a quadratic polynomial $f(x)$ and solving the polynomial equation $f(x) = 0$. The factor theorem states this relationship between roots and factors in a precise form. Furthermore, the factor theorem tells us that this relationship holds for polynomials of all degrees, not just quadratics.

THE FACTOR THEOREM

Let $f(x)$ be a polynomial. If $f(r) = 0$, then $x - r$ is a factor of $f(x)$. Conversely, if $x - r$ is a factor of $f(x)$, then $f(r) = 0$.

In terms of roots, we can summarize the factor theorem by saying that r is a root of the equation $f(x) = 0$ if and only if $x - r$ is a factor of $f(x)$. To prove the factor theorem, let us begin by assuming that $f(r) = 0$. We want to show that $x - r$ is a factor of $f(x)$. Now, according to the remainder theorem, if $f(x)$ is divided by $x - r$, the remainder is $f(r)$. So we can write

$$f(x) = (x - r) \cdot q(x) + f(r) \qquad \text{for some polynomial } q(x)$$

But since $f(r)$ is zero, this equation becomes

$$f(x) = (x - r) \cdot q(x)$$

This last equation tells us that $x - r$ is a factor of $f(x)$, as we wished to prove.

Now, conversely, let us assume that $x - r$ is a factor of $f(x)$. We want to show that $f(r) = 0$. Since $x - r$ is a factor of $f(x)$, we can write

$$f(x) = (x - r) \cdot q(x) \qquad \text{for some polynomial } q(x)$$

If we now let $x = r$ in this last equation, we obtain

$$f(r) = (r - r) \cdot q(r) = 0$$

as we wished to show.

The example that follows indicates how the factor theorem can be used to solve equations.

EXAMPLE 5 Solve the equation $x^3 - 2x + 1 = 0$, given that one root is $x = 1$.

Solution Since $x = 1$ is a root, the factor theorem tells us that $x - 1$ is a factor of $x^3 - 2x + 1$. In other words,

$$x^3 - 2x + 1 = (x - 1) \cdot q(x) \qquad \text{for some polynomial } q(x)$$

To determine this other factor $q(x)$, we divide $x^3 - 2x + 1$ by $x - 1$:

$$\begin{array}{r|rrrr} 1 & 1 & 0 & -2 & 1 \\ & & 1 & 1 & -1 \\ \hline & 1 & 1 & -1 & 0 \end{array}$$

Thus, $q(x) = x^2 + x - 1$ and we have the factorization

$$x^3 - 2x + 1 = (x - 1)(x^2 + x - 1)$$

Using this identity, the original equation becomes

$$(x - 1)(x^2 + x - 1) = 0$$

Now the problem is reduced to solving the linear equation $x - 1 = 0$ and the quadratic equation $x^2 + x - 1 = 0$. The linear equation yields $x = 1$, but (from the statement of the problem) we already know that 1 is a root. So if we are to find any additional roots, they must come from the equation $x^2 + x - 1 = 0$. Using the quadratic formula, we obtain

$$x = \frac{-1 \pm \sqrt{(1)^2 - 4(1)(-1)}}{2(1)} = \frac{-1 \pm \sqrt{5}}{2}$$

We now have three roots of the cubic equation $x^3 - 2x + 1 = 0$. They are 1, $(-1 + \sqrt{5})/2$, and $(-1 - \sqrt{5})/2$. As you will see in the next section, a cubic equation can have at most three roots. So in this case, we have determined all the roots; that is, we have solved the equation.

Before going on to other examples, let's take a moment to summarize the technique we used in Example 5. We want to solve a polynomial equation $f(x) = 0$, given that one root is $x = r$. Since r is a root, the factor theorem tells us that $x - r$ is a factor of $f(x)$. Then, with the aid of synthetic division, we obtain a factorization

$$f(x) = (x - r) \cdot q(x) \qquad \text{for some polynomial } q(x)$$

This gives rise to the two equations $x - r = 0$ and $q(x) = 0$. Since the first of these only reasserts that r is a root, we try to solve the second equation, $q(x) = 0$. We refer to the equation $q(x) = 0$ as the **reduced equation.** Example 5 showed you the idea behind this terminology; the degree of $q(x)$ is one less than that of $f(x)$. If, as in Example 5, the reduced equation happens to be a quadratic equation, then we can always determine the remaining roots by factoring or by the quadratic formula. In subsequent sections, we will look at techniques that are helpful in cases where $q(x)$ is not quadratic.

EXAMPLE 6 Solve the equation $x^4 + 2x^3 - 7x^2 - 20x - 12 = 0$, given that $x = 3$ and $x = -2$ are roots.

Solution Since $x = 3$ is a root, the factor theorem tells us that $x - 3$ is a factor of the polynomial $x^4 + 2x^3 - 7x^2 - 20x - 12$. That is,

$$x^4 + 2x^3 - 7x^2 - 20x - 12 = (x - 3) \cdot q(x) \qquad \text{for some polynomial } q(x)$$

As in Example 5, we can find $q(x)$ by synthetic division. We have

$$\begin{array}{r|rrrrr} 3 & 1 & 2 & -7 & -20 & -12 \\ & & 3 & 15 & 24 & 12 \\ \hline & 1 & 5 & 8 & 4 & 0 \end{array}$$

Thus, $q(x) = x^3 + 5x^2 + 8x + 4$, and our original equation is equivalent to

$$(x - 3)(x^3 + 5x^2 + 8x + 4) = 0$$

Now, $x = -2$ is also a root of this equation. But $x = -2$ surely is not a root of the equation $x - 3 = 0$, so it must be a root of the reduced equation $x^3 + 5x^2 + 8x + 4 = 0$. Again the factor theorem is applicable. Since $x = -2$ is a root of the reduced equation, $x + 2$ must be a factor of $x^3 + 5x^2 + 8x + 4$. That is,

$$x^3 + 5x^2 + 8x + 4 = (x + 2) \cdot p(x) \qquad \text{for some polynomial } p(x)$$

We can use synthetic division to determine $p(x)$:

$$\begin{array}{r|rrrr} -2 & 1 & 5 & 8 & 4 \\ & & -2 & -6 & -4 \\ \hline & 1 & 3 & 2 & 0 \end{array}$$

Thus, $p(x) = x^2 + 3x + 2$, and our reduced equation can be written

$$(x + 2)(x^2 + 3x + 2) = 0$$

This gives rise to a second reduced equation.

$$x^2 + 3x + 2 = 0$$

In this case, the roots are readily obtained by factoring. We have

$$x^2 + 3x + 2 = 0$$
$$(x + 2)(x + 1) = 0$$
$$x + 2 = 0 \quad \Big| \quad x + 1 = 0$$
$$x = -2 \quad \Big| \quad x = -1$$

Now, of the two roots we've just found, -2 happens to be one of the roots that we were initially given in the statement of the problem. On the other hand, -1 is a distinct additional root. In summary, then, we have three distinct roots: 3, -2, and -1, where the root -2 has multiplicity 2. As you will see in the next section, a fourth-degree equation can have at most four (not necessarily distinct) roots. So in the case at hand, we have found all the roots of the given equation.

EXAMPLE 7 In each case, find a polynomial equation $f(x) = 0$ satisfying the given conditions. If there is no such equation, say so.

(a) The numbers -1, 4, and 5 are roots.
(b) A factor of $f(x)$ is $x - 3$, and -4 is a root of multiplicity 2.
(c) The degree of f is 4, the number -5 is a root of multiplicity 3, and 6 is a root of multiplicity 2.

Solution

(a) If $f(x)$ is any polynomial containing the factors $x + 1$, $x - 4$, and $x - 5$, then the equation $f(x) = 0$ will certainly be satisfied with $x = -1$, $x = 4$, or $x = 5$. The simplest polynomial equation in this case is therefore

$$(x + 1)(x - 4)(x - 5) = 0$$

This is a polynomial equation with the required roots. If required, we can carry out the multiplication on the left-hand side of the equation. As you can check, this yields $x^3 - 8x^2 + 11x + 20 = 0$.

(b) According to the factor theorem, since -4 is a root, $x + 4$ must be a factor of $f(x)$. In fact, since -4 is a root of multiplicity 2, the quantity $(x + 4)^2$ must be a factor of $f(x)$. The following equation therefore satisfies the given conditions:

$$(x - 3)(x + 4)^2 = 0$$

(c) We are given two roots, one with multiplicity 3, the other with multiplicity 2. Thus the degree of $f(x)$ must be at least $3 + 2 = 5$. (Why?) But this then contradicts the given condition that the degree of f should be 4. Consequently, there is no polynomial equation that satisfies the given conditions.

EXERCISE SET 12.3

A

In Exercises 1–6, determine whether the given value for the variable is a root of the equation.

1. $12x - 8 = 112$; $x = 10$
2. $12x^2 - x - 20 = 0$; $x = 5/4$
3. $x^2 - 2x - 4 = 0$; $x = 1 - \sqrt{5}$
4. $1 - x + x^2 - x^3 = 0$; $x = -1$
5. $2x^2 - 3x + 1 = 0$; $x = 1/2$

6. $(x-1)(x-2)(x-3)=0;\ x=4$

In Exercises 7–14, determine whether the given value is a zero of the function.

7. $f(x)=3x-2;\ x=2/3$

8. $g(x)=1+x^2;\ x=-1$

9. $h(x)=5x^3-x^2+2x+8;\ x=-1$

10. $F(x)=-2x^5+3x^4+8x^3;\ x=0$

11. $f(t)=1+2t+t^3-t^5;\ t=2$

12. $f(t)=1+2t+t^3-t^5;\ t=\sqrt{2}$

13. $f(x)=2x^3-3x+1$: **(a)** $x=(\sqrt{3}-1)/2$; **(b)** $x=(\sqrt{3}+1)/2$

14. $g(x)=x^4+8x^3+9x^2-8x-10$: **(a)** $x=1$; **(b)** $x=\sqrt{6}-4$; **(c)** $x=\sqrt{6}+4$

15. List the distinct roots of each of the following equations. In the case of a repeated root, give its multiplicity.
(a) $(x-1)(x-2)^3(x-3)=0$
(b) $(x-1)(x-1)(x-1)=0$
(c) $(x-5)^6(x+1)^4=0$
(d) $x^5(x-1)=0$

16. In this exercise, we verify that the remainder theorem is valid for the cubic polynomial

$$g(x)=ax^3+bx^2+cx+d$$

(a) Compute $g(r)$.
(b) Using synthetic division, divide $g(x)$ by $x-r$. Check that the remainder you obtain is the same as the answer in part (a).

In Exercises 17–24, use the remainder theorem (as in Example 3) to evaluate $f(x)$ for the given value of x.

17. $f(x)=4x^3-6x^2+x-5;\ x=-3$

18. $f(x)=2x^3-x-4;\ x=4$

19. $f(x)=6x^4+5x^3-8x^2-10x-3;\ x=1/2$

20. $f(x)=x^5-x^4-x^3-x^2-x-1;\ x=-2$

21. $f(x)=x^2+3x-4;\ x=-\sqrt{2}$

22. $f(x)=-3x^3+8x^2-12;\ x=5$

23. $f(x)=\frac{1}{2}x^3-5x^2-13x-10;\ x=12$

24. $f(x)=x^7-7x^6+5x^4+1;\ x=-3$

In Exercises 25–38, you are given a polynomial equation and one or more roots. Solve each equation using the method shown in Examples 5 and 6. To help you decide if you have found all the roots in each case, you may rely on the following theorem, discussed in the next section: A polynomial equation of degree n has at most n (not necessarily distinct) roots.

25. $x^3-4x^2-9x+36=0;\ -3$ is a root.

26. $x^3+7x^2+11x+5=0;\ -1$ is a root.

27. $x^3+x^2-7x+5=0;\ 1$ is a root.

28. $x^3+8x^2-3x-24=0;\ -8$ is a root.

29. $3x^3-5x^2-16x+12=0;\ -2$ is a root.

30. $2x^3-5x^2-46x+24=0;\ 6$ is a root.

31. $2x^3+x^2-5x-3=0;\ -3/2$ is a root.

32. $6x^4-19x^3-25x^2+18x+8=0;\ 4$ and $-1/3$ are roots.

33. $x^4-15x^3+75x^2-125x=0;\ 5$ is a root.

34. $2x^3+5x^2-8x-20=0;\ 2$ is a root.

35. $x^4+2x^3-23x^2-24x+144=0;\ -4$ and 3 are roots.

36. $6x^5+5x^4-29x^3-25x^2-5x=0;\ \sqrt{5}$ and $-1/3$ are roots.

37. $x^3+7x^2-19x-9=0;\ -9$ is a root.

38. $4x^5-15x^4+8x^3+19x^2-12x-4=0;\ 1, 2,$ and -1 are roots.

In Exercises 39 and 40, use the given information to solve each equation.

39. $f(x)=4x^4-12x^3+5x^2+6x+1=0;\ 2x^2-3x-1$ is a factor of $f(x)$.

40. $f(x)=3x^5-3x^4-x^3-24x^2+24x+8=0;$ $3x^2-3x-1$ is a factor of $f(x)$.

For Exercises 41 and 42, refer to the following computer-generated tables for the functions f and g.

$f(t)=t^3-4t+3$

t	$f(t)$
0.500	1.125000
0.750	0.421875
1.000	0.000000
1.250	−0.046875
1.500	0.375000

$g(t)=t^5+2t^4+t^3+2t^2-t-2$

t	$g(t)$
−3.00	−89.0
−2.50	−22.15625
−2.25	−7.4228515625
−2.00	0.0
−1.75	2.8603515625

41. **(a)** What is the remainder when $f(t)$ is divided by $t-\frac{1}{2}$?
(b) What is the remainder when $f(t)$ is divided by $t-1.25$?
(c) Specify a linear factor of $f(t)$.
(d) Solve the equation $f(t)=0$.

42. **(a)** What is the remainder when $g(t)$ is divided by $t+3$?
(b) What is the remainder when $g(t)$ is divided by $t+\frac{5}{2}$?
(c) Specify a linear factor of $g(t)$.

In Exercises 43–50, find a polynomial equation $f(x) = 0$ satisfying the given conditions. If no such equation is possible, state this.

43. Degree 3; the coefficient of x^3 is 1; three roots are 3, -4, and 5

44. Degree 3; the coefficients are integers; $1/2$, $2/5$, and $-3/4$ are roots

45. Degree 3; -1 is a root of multiplicity two; $x + 6$ is a factor of $f(x)$

46. Degree 3; 4 is a root of multiplicity two; -1 is a root of multiplicity two

47. Degree 4; $1/2$ is a root of multiplicity three; $x^2 - 3x - 4$ is a factor of $f(x)$

48. Degree 3; -2 and -3 are roots

49. Degree 4; $x^2 - 3x + 1$ and $x + 6$ are factors of $f(x)$

50. Degree 4; the coefficients are integers; $1/2$ is a root of multiplicity two; $2x^2 - 4x - 1$ is a factor of $f(x)$

In Exercises 51–54, use the remainder theorem (as in Example 3) to evaluate $f(x)$ for the given value of x. Use a calculator, and round your answers to two decimal places.

51. $f(x) = x^3 - 3x^2 + 12x + 9$; $x = 1.16$

52. $f(x) = x^3 - 2x - 5$:
(a) $x = 2.09$ **(b)** $x = 2.094$ **(c)** $x = 2.0945$

53. $f(x) = x^3 - 5x - 2$: **(a)** $x = 2.41$; **(b)** $x = 2.42$

54. $f(x) = x^4 - 2x^3 - 5x^2 + 10x - 3$:
(a) $x = -2.3$ **(b)** $x = -2.302$
(c) $x = -2.30277$

55. A computer was used to generate the following information about the function $f(x) = x^3 + x^2 - 18x + 10$.

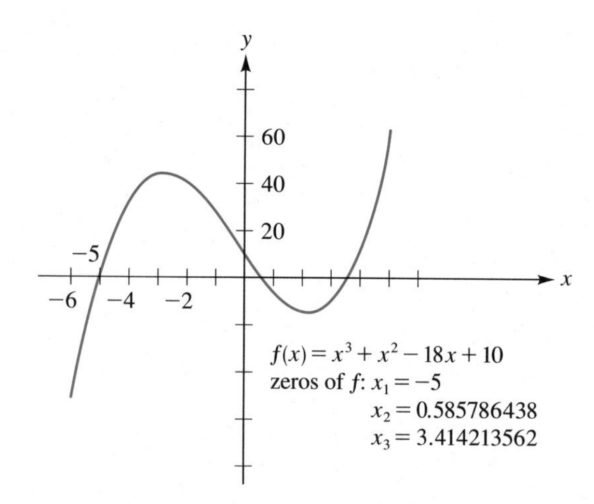

(a) Determine the exact values for the zeros x_2 and x_3.
(b) Use your calculator to evaluate the answers in part (a), and check that they are consistent with the computer values given in the figure.

56. A computer was used to generate the following information about the function

$$g(x) = -2x^3 + 7x^2 - 2x - 6$$

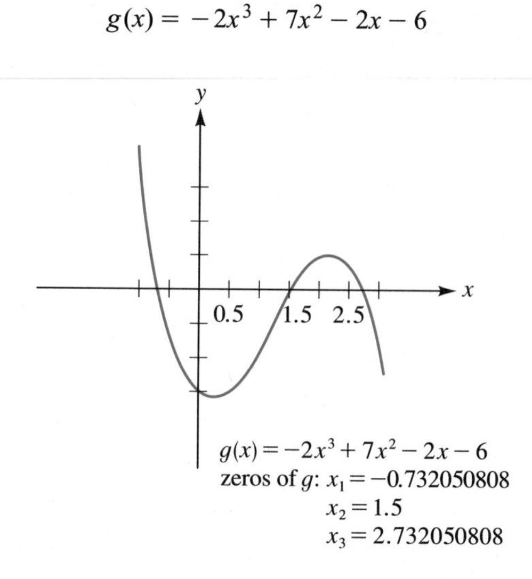

(a) Determine the exact values for the zeros x_1 and x_3.
(b) Use your calculator to evaluate the answers in part (a), and check that they are consistent with the computer values given in the figure.

B

In Exercises 57–60, determine whether the given value is a zero of the function.

57. $f(x) = \frac{1}{2}x^2 + bx + c$; $x = -b - \sqrt{b^2 - 2c}$

58. $Q(x) = ax^2 + bx + c$; $x = \dfrac{-b + \sqrt{b^2 - 4ac}}{2a}$

Hint: Look before you leap!

59. $F(x) = 2x^4 + 4x + 1$; $x = \left(-\sqrt{2} + \sqrt{2\sqrt{2} - 2}\right)/2$

60. $f(x) = x^3 - 3x^2 + 3x - 3$: **(a)** $x = \sqrt[3]{2} - 1$;
(b) $x = \sqrt[3]{2} + 1$

61. Determine values for a and b such that $x - 1$ is a factor of both $x^3 + x^2 + ax + b$ and $x^3 - x^2 - ax + b$.

62. Determine a quadratic equation with the given roots:
(a) $a/b, -b/a$; **(b)** $-a + 2\sqrt{2b}, -a - 2\sqrt{2b}$

63. One root of the equation $x^2 + bx + 1 = 0$ is twice the other; find b. (There are two answers.)

64. Determine a value for a such that one root of the equation $ax^2 + x - 1 = 0$ is five times the other.

C

65. Solve the equation $x^3 - 12x + 16 = 0$, given that one of the roots has multiplicity 2.

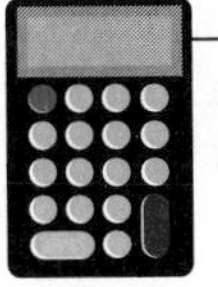

GRAPHING UTILITY EXERCISES FOR SECTION 12.3

In Exercises 1–4, you are given a polynomial equation and a real number b. Use a graph to show that the number b is not a root of the equation. [Do this by graphing the corresponding polynomial function in an appropriate viewing rectangle and noting that b is not an x-intercept for the graph.]

1. $x^3 + x^2 - 3x - 3 = 0$; $b = 2$
2. $x^3 + x^2 - 5 = 0$; $b = 1.5$
3. $12x^2 + x - 12 = 0$; $b = 0.9$
4. $5x^3 - 30x - 4 = 0$; $b = 2.5$

In Exercises 5–8, you are given a polynomial equation $f(x) = 0$ and a root r. **(a)** *Graph the corresponding polynomial function $y = f(x)$.* **(b)** *Which x-intercept corresponds to the given root r? (To answer this, you'll first need to compute a decimal approximation for r.)* **(c)** *Zoom in several times on the x-intercept that you indicated in part (b). Use the graphing utility to obtain an approximate value for the x-coordinate. Note that the value is consistent with the value for r that you computed in part (b).* **(d)** *Verify empirically that r is a root of the equation by using your calculator or computer to evaluate $f(r)$.* **(e)** *Prove that r is a root by doing algebra (with paper and pencil) to evaluate $f(r)$.*

5. $x^3 - 9x - 10 = 0$; $r = 1 + \sqrt{6}$
6. $2x^3 - 6x^2 - x = 0$; $r = (3 - \sqrt{11})/2$
7. $x^4 - 2x^3 - 3x^2 - 2x - 4 = 0$; $r = 1 - \sqrt{5}$
8. $x^6 - 7x^3 + 12$; $r = \sqrt[3]{4}$
9. **(a)** Graph the function $f(x) = x^3 - 2x + 1$. How many x-intercepts are there?
   (b) By using a ZOOM key or a SOLVE key, estimate the roots of the equation $x^3 - 2x + 1 = 0$. [The roots are the x-intercepts of the graph in part (a).]
   (c) Check that your answers in part (b) are consistent with those obtained in Example 5 of this section.
10. **(a)** Graph the function $f(x) = x^4 + 2x^3 - 7x^2 - 20x - 12$. How many x-intercepts are there?
    (b) By using a ZOOM key or a SOLVE key, estimate the roots of the equation
    $$x^4 + 2x^3 - 7x^2 - 20x - 12 = 0$$
    (c) Check that your answers in part (b) are consistent with those obtained in Example 6 of this section.
11. **(a)** In the standard viewing rectangle, graph the function $f(x) = x^4 + x^2 - 6$. Note that the viewing rectangle shows there are (at least) two x-intercepts. Zoom out several times to convince yourself that the graph has only two x-intercepts.
    (b) Return to the standard viewing rectangle. Note from the graph that there is an x-intercept between 1 and 2. By using the ZOOM or the SOLVE key, estimate this intercept.
    (c) Review Example 1(b) in the text. Then explain how your work in part (b) of this exercise relates to Example 1(b).
12. **(a)** In the standard viewing rectangle, graph the four equations
    $$y = x^n - 2^n, \quad (n = 2, 3, 4, 5)$$
    (b) Switch to a viewing rectangle that extends from -5 to 5 in the x-direction and from -40 to 20 in the y-direction. Name one property shared by all four graphs.
    (c) How does your answer in part (b) relate to the result in Example 4 of this section.

12.4 THE FUNDAMENTAL THEOREM OF ALGEBRA

One of the most intriguing problems was the question of the number of roots of an equation, which brought in negative and imaginary quantities, and led to the conclusion, in the work of Girard and Descartes, that an equation of degree n can have no more than n roots. The more precise statement, that an equation of degree n always has one root, and hence always has n roots (allowing for multiple roots), became known as the fundamental theorem of algebra. *After several attempts by D'Alembert, Euler, and others, the proof was finally given by Gauss in 1799.*

From *A Source Book in Mathematics, 1200–1800,* D. J. Struik, ed. (Princeton: Princeton University Press, 1986)

Every equation of algebra has as many solutions as the exponent of the highest term indicates.

Albert Girard, 1629, as quoted by David M. Burton in *The History of Mathematics, An Introduction,* 2d ed. (Dubuque, Iowa: Wm. C. Brown Publishers, 1991)

. . . we arrive from all the investigations explained above at the rigorous proof of the theorem that every integral rational algebraic function of one variable can be decomposed into real factors of the first and second degree.

Carl Friedrich Gauss in his doctoral dissertation (1799)

Does every polynomial equation (of degree at least 1) have a root? Or, to put the question another way, is it possible to write a polynomial equation that has no solution? Certainly, if we consider only real roots (i.e., roots that are real numbers), then it is easy to specify an equation with no real roots:

$$x^2 = -1$$

This equation has no real roots because the square of a real number is never negative. On the other hand, if we expand our base of operations from the real-number system to the complex-number system, then $x^2 = -1$ does indeed have a root. In fact, both i and $-i$ are roots in this case.

As it turns out, the situation just described for the equation $x^2 = -1$ holds quite generally. That is, within the complex-number system, every polynomial equation of degree at least 1 has at least one root. This is the substance of a remarkable theorem that was first proved by the great mathematician Carl Friedrich Gauss in 1799. (Gauss was only 22 years old at the time.) Although there are many fundamental theorems in algebra, Gauss's result has come to be known as *the* **fundamental theorem of algebra.**

THE FUNDAMENTAL THEOREM OF ALGEBRA

Every polynomial equation of the form

$$a_nx^n + a_{n-1}x^{n-1} + \cdots + a_1x + a_0 = 0 \qquad (n \geq 1,\ a_n \neq 0)$$

has at least one root within the complex number system. (This root may be a real number.)

Although most of this section deals with polynomials with real coefficients, Gauss's theorem still applies in cases in which some or all of the coefficients are nonreal complex numbers. The proof of Gauss's theorem is usually given in the post-calculus course called complex variables.

The fundamental theorem of algebra asserts that every polynomial equation of degree at least 1 has a root. There are two initial observations to be made here. First, notice that the theorem says nothing about actually finding the root. Second, notice that the theorem deals only with polynomial equations. Indeed, it is easy to specify a nonpolynomial equation that does not have a root. Such an equation is $1/x = 0$. The expression on the left-hand side of this equation can never be zero because the numerator is 1.

EXAMPLE 1 Which of the following equations has at least one root?

(a) $x^3 - 17x^2 + 6x - 1 = 0$ **(b)** $\sqrt{2}x^{47} - \pi x^{25} + \sqrt{3} = 0$
(c) $x^2 - 2ix + (3 + i) = 0$

Solution All three equations are polynomial equations, so, according to the fundamental theorem of algebra, each equation has at least one root.

In Section 1.4 we used factoring as a tool for solving quadratic equations. The next theorem, a consequence of the fundamental theorem of algebra, tells us that (in principle, at least) any polynomial of degree n can be factored into a product of n linear factors. In proving the **linear factors theorem,** we will need to use the factor theorem. If you reread the proof of that theorem in the previous section, you will see that it makes no difference whether the number r appearing in the factor $x - r$ is a real number or a nonreal complex number. Thus, the factor theorem is valid in either case.

THE LINEAR FACTORS THEOREM

Let $f(x) = a_n x^n + a_{n-1}x^{n-1} + \cdots + a_1 x + a_0$, where $n \geqslant 1$ and $a_n \neq 0$. Then $f(x)$ can be expressed as a product of n linear factors:

$$f(x) = a_n(x - r_1)(x - r_2) \cdots (x - r_n)$$

(The complex numbers r_k appearing in these factors are not necessarily all distinct, and some or all of the r_k may be real numbers.)

PROOF OF THE LINEAR FACTORS THEOREM According to the fundamental theorem of algebra, the equation $f(x) = 0$ has a root; let's call this root r_1. By the factor theorem, $x - r_1$ is a factor of $f(x)$, and we can write

$$f(x) = (x - r_1) \cdot Q_1(x)$$

for some polynomial $Q_1(x)$ that has degree $n - 1$ and leading coefficient a_n. If the degree of $Q_1(x)$ happens to be zero, we're done. On the other hand, if the degree of $Q_1(x)$ is at least 1, another application of the fundamental theorem of algebra followed by the factor theorem gives us

$$Q_1(x) = (x - r_2) \cdot Q_2(x)$$

where the degree of $Q_2(x)$ is $n - 2$ and the leading coefficient of $Q_2(x)$ is a_n. We now have

$$f(x) = (x - r_1)(x - r_2) \cdot Q_2(x)$$

We continue this process until the quotient is $Q_n(x) = a_n$. As a result, we obtain

$$\begin{aligned} f(x) &= (x - r_1)(x - r_2) \cdots (x - r_n)a_n \\ &= a_n(x - r_1)(x - r_2) \cdots (x - r_n) \end{aligned}$$

as we wished to show.

The linear factors theorem tells us that any polynomial can be expressed as a product of linear factors. The theorem gives us no information, however, as to how those factors can actually be obtained. The next example demonstrates a case in which the factors are readily obtainable; this is always the case with quadratic polynomials.

EXAMPLE 2 Express each of the following second-degree polynomials in the form $a_n(x - r_1)(x - r_2)$: **(a)** $3x^2 - 5x - 2$; **(b)** $x^2 - 4x + 5$.

Solution

(a) A factorization for $3x^2 - 5x - 2$ can be found by simple trial and error. We have

$$3x^2 - 5x - 2 = (3x + 1)(x - 2)$$

We now write the factor $3x+1$ as $3(x+\frac{1}{3})$. This, in turn, can be written $3[x-(-\frac{1}{3})]$. The final factorization is then

$$3x^2-5x-2=3[x-(-\tfrac{1}{3})](x-2)$$

(b) From the factor theorem, or from our more elementary work with quadratic equations, we know that if r_1 and r_2 are the roots of the equation $x^2-4x+5=0$, then $x-r_1$ and $x-r_2$ are the factors of x^2-4x+5. That is, $x^2-4x+5=(x-r_1)(x-r_2)$. The values for r_1 and r_2 in this case are readily obtained by using the quadratic formula. As you can check, the results are

$$r_1=2+i \qquad \text{and} \qquad r_2=2-i$$

The required factorization is therefore

$$x^2-4x+5=[x-(2+i)][x-(2-i)]$$

Using the linear factors theorem, we can show that every polynomial equation of degree $n \geqslant 1$ has exactly n roots. To help you follow the reasoning, we make two preliminary comments. First, we agree that a root of multiplicity k will be counted as k roots. For example, although the third-degree equation $(x-1)(x-4)^2=0$ has only two distinct roots, namely 1 and 4, it has three roots *if* we agree to count the repeated root 4 two times. The second preliminary comment concerns the *zero-product property,* which states that $pq=0$ if and only if $p=0$ or $q=0$. When we stated this in Section 1.4, we were working within the real-number system. However, the property is also valid within the complex-number system. (See Exercises 75 and 76 in Section 12.1.) Now let's state and prove our theorem.

THEOREM

Every polynomial equation of degree $n \geqslant 1$ has exactly n roots, where a root of multiplicity k is counted k times.

PROOF OF THE THEOREM Using the linear factors theorem, we can write the nth-degree polynomial equation $f(x)=a_nx^n+a_{n-1}x^{n-1}+\cdots+a_1x+a_0=0$ as

$$f(x)=a_n(x-r_1)(x-r_2)\cdots(x-r_n)=0 \qquad (a_n \neq 0) \tag{1}$$

By the factor theorem, each of the numbers $r_1, r_2, \ldots, r_n$ is a root. Some of these numbers may in fact be equal; in other words, we may have repeated roots in this list. In any case, if we agree to count a root of multiplicity k as k roots, then we obtain exactly n roots from the list $r_1, r_2, \ldots, r_n$. Furthermore, the equation $f(x)=0$ can have no other roots, as we now show. Suppose that r is any number distinct from all the numbers $r_1, r_2, \ldots, r_n$. Replacing x with r in equation (1) yields

$$f(r)=a_n(r-r_1)(r-r_2)\cdots(r-r_n)$$

But the expression on the right-hand side of this last equation cannot be zero, because none of the factors is zero. Thus, $f(r)$ is not zero, and so r is not a root. This completes the proof of the theorem.

TABLE 1

Root	Multiplicity
3	2
−2	1
0	2

EXAMPLE 3 Find a polynomial $f(x)$ with leading coefficient 1 such that the equation $f(x) = 0$ has only those roots specified in Table 1. What is the degree of this polynomial?

Solution The expressions $(x - 3)^2$, $x + 2$, and $(x - 0)^2$ all must appear as factors of $f(x)$ and, furthermore, no other linear factor can appear. The form of $f(x)$ is therefore

$$f(x) = a_n(x - 3)^2(x + 2)(x - 0)^2$$

Since the leading coefficient a_n is to be 1, we can rewrite this last equation as

$$f(x) = x^2(x - 3)^2(x + 2)$$

This is the required polynomial. The degree here is 5. This can be seen either by multiplying out the factors or by simply adding the multiplicities of the roots in Table 1.

EXAMPLE 4 Find a quadratic function f that has zeros of 3 and 5 and a graph that passes through the point $(2, -9)$.

Solution The general form of a quadratic function with 3 and 5 as zeros is $f(x) = a_n(x - 3)(x - 5)$. Since the graph passes through $(2, -9)$, we have

$$-9 = a_n(2 - 3)(2 - 5) = a_n(3)$$
$$-3 = a_n$$

The required function is therefore

$$f(x) = -3(x - 3)(x - 5)$$

If we wish, we can carry out the multiplication and rewrite this as

$$f(x) = -3x^2 + 24x - 45$$

The next example shows how we can use the factored form of a polynomial $f(x)$ to determine the relationships between the coefficients of the polynomial and the roots of the equation $f(x) = 0$.

EXAMPLE 5 Let r_1 and r_2 be the roots of the equation $x^2 + bx + c = 0$. Show that

$$r_1r_2 = c \qquad \text{and} \qquad r_1 + r_2 = -b$$

Solution Since r_1 and r_2 are the roots of the equation $x^2 + bx + c = 0$, we have the identity

$$x^2 + bx + c = (x - r_1)(x - r_2)$$

After multiplying out the right-hand side, we can rewrite this identity as

$$x^2 + bx + c = x^2 - (r_1 + r_2)x + r_1r_2$$

By equating coefficients, we readily obtain $r_1r_2 = c$ and $r_1 + r_2 = -b$, as required.

The technique used in Example 5 can be used to obtain similar relationships between the roots and the coefficients of polynomial equations of any given degree. In Table 2, for instance, we show the relationships obtained in Example 5, along with the corresponding relationships that can be derived for a cubic equation. (Exercise 31 at the end of this section asks you to verify the results for the cubic equation.)

TABLE 2

Equation	Roots	Relationships between Roots and Coefficients
$x^2 + bx + c = 0$	r_1, r_2	$r_1 + r_2 = -b$ $r_1 r_2 = c$
$x^3 + bx^2 + cx + d = 0$	r_1, r_2, r_3	$r_1 + r_2 + r_3 = -b$ $r_1 r_2 + r_2 r_3 + r_3 r_1 = c$ $r_1 r_2 r_3 = -d$

We conclude this section with some remarks concerning the solving of polynomial equations by formulas. You know that the roots of the quadratic equation $ax^2 + bx + c = 0$ are given by the formula

$$x = \frac{-b \pm \sqrt{b^2 - 4ac}}{2a}$$

The question is, are there similar formulas for the solutions of higher-degree equations? By "similar" we mean a formula involving the coefficients and radicals. To answer this question, we look at a bit of history. As early as 1700 B.C., Babylonian mathematicians were able to solve quadratic equations. This is clear from the study of the clay tablets with cuneiform numerals that archeologists have found. The ancient Greeks also were able to solve quadratic equations. Like the Babylonians, the Greeks worked without the aid of algebra as we know it. The mathematicians of ancient Greece used geometric constructions to solve equations. Of course, since all quantities were interpreted geometrically, negative roots were never considered.

The general quadratic formula was known to the Moslem mathematicians sometime before A.D. 1000. For the next 500 years, mathematicians searched for, but did not discover, a formula to solve the general cubic equation. Indeed, in 1494, Luca Pacioli stated in his text *Summa di Arithmetica* that the general cubic equation could not be solved by the algebraic techniques then available. All of this was to change, however, within the next several decades.

Around 1515, the Italian mathematician Scipione del Ferro solved the cubic equation $x^3 + px + q = 0$ using algebraic techniques. This essentially constituted a solution of the seemingly more general equation $x^3 + bx^2 + cx + d = 0$. The reason for this is that if we make the substitution $x = y - b/3$ in the latter equation, the result is a cubic equation with no y^2-term. By 1540, the Italian mathematician Ludovico Ferrari had solved the general fourth-degree equation. Actually, at that time in Renaissance Italy, there was considerable controversy as to exactly who discovered the various formulas first. Details of the dispute can be found in any text on the history of mathematics. But for our purposes here, the point is simply that by the middle of the sixteenth century, all polynomial equations of degree 4 or less could be solved by the formulas that had been discovered. The common feature of these formulas was that they involved the coefficients, the four basic operations of arithmetic, and various radicals. For example, a formula for the solution of the equation $x^3 + px + q = 0$ is as follows:

$$x = \sqrt[3]{\frac{-q}{2} + \sqrt{\frac{q^2}{4} + \frac{p^3}{27}}} + \sqrt[3]{\frac{-q}{2} - \sqrt{\frac{q^2}{4} + \frac{p^3}{27}}}$$

To get some idea of the practical difficulties inherent in computing with this formula, try using it to show that $x = -2$ is a root of the cubic equation $x^3 + 4x + 16 = 0$.

For more than 200 years after the cubic and quartic (fourth-degree) equations had been solved, mathematicians continued to search for a formula that would yield the solutions of the general fifth-degree equation. The first breakthrough, if it can be called that, occurred in 1770, when the French mathematician Joseph Louis Lagrange found a technique that served to unify and summarize all of the previous methods used for the equations of degrees 2, 3, and 4. However, Lagrange then showed that his technique could not work in the case of the general fifth-degree equation. While we will not describe the details of Lagrange's work here, it is worth pointing out that he relied on the types of relationships between roots and coefficients that we looked at in Table 2 of this section.

Finally, in 1828, the Norwegian mathematician Niels Henrik Abel proved that for the general polynomial equation of degree 5 or higher, there could be no formula yielding the solutions in terms of the coefficients and radicals. This is not to say that such equations do not possess solutions. In fact they must, as we saw earlier in this section. It is just that we cannot in every case express the solutions in terms of the coefficients and radicals. For example, it can be shown that the equation $x^5 - 6x + 3 = 0$ has a real root between 0 and 1, but this number cannot be expressed in terms of the coefficients and radicals. (We can, however, compute the root to as many decimal places as we wish, as you will see in the next section.) In 1830, the French mathematician Evariste Galois completed matters by giving conditions for determining exactly which polynomial equations can be solved in terms of coefficients and radicals.

NOTE Historical background on the history of polynomial equations and the fundamental theorem of algebra is available on the Internet through the Web site maintained by David E. Joyce. The address is

http://aleph0.clarku.edu:80/~djoyce/mathhist/subjects.html

Once at this Web site, choose "The history of algebra."

EXERCISE SET 12.4

A

According to the fundamental theorem of algebra, which of the equations in Exercises 1 and 2 have at least one root?

1. **(a)** $x^5 - 14x^4 + 8x + 53 = 0$
(b) $(4.17)x^3 + (2.06)x^2 + (0.01)x + 1.23 = 0$
(c) $ix^2 + (2 + 3i)x - 17 = 0$
(d) $x^{2.1} + 3x^{0.3} + 1 = 0$

2. **(a)** $\sqrt{3}x^{17} + \sqrt{2}x^{13} + \sqrt{5} = 0$
(b) $17x^{\sqrt{3}} + 13x^{\sqrt{2}} + \sqrt{5} = 0$
(c) $1/(x^2 + 1) = 0$
(d) $2^{3x} - 2^x - 1 = 0$

In Exercises 3–10, express each polynomial in the form $a_n(x - r_1)(x - r_2) \cdots (x - r_n)$.

3. $x^2 - 2x - 3$
4. $x^3 - 2x^2 - 3x$
5. $4x^2 + 23x - 6$
6. $6x^2 + x - 12$
7. $x^2 - 5$
8. $x^2 + 5$
9. $x^2 - 10x + 26$
10. $x^3 + 2x^2 - 3x - 6$

In Exercises 11–16, find a polynomial $f(x)$ with leading coefficient 1 such that the equation $f(x) = 0$ has the given roots and no others. If the degree of $f(x)$ is 7 or more, express $f(x)$ in factored form; otherwise, express $f(x)$ in the form $a_nx^n + a_{n-1}x^{n-1} + \cdots + a_1x + a_0$.

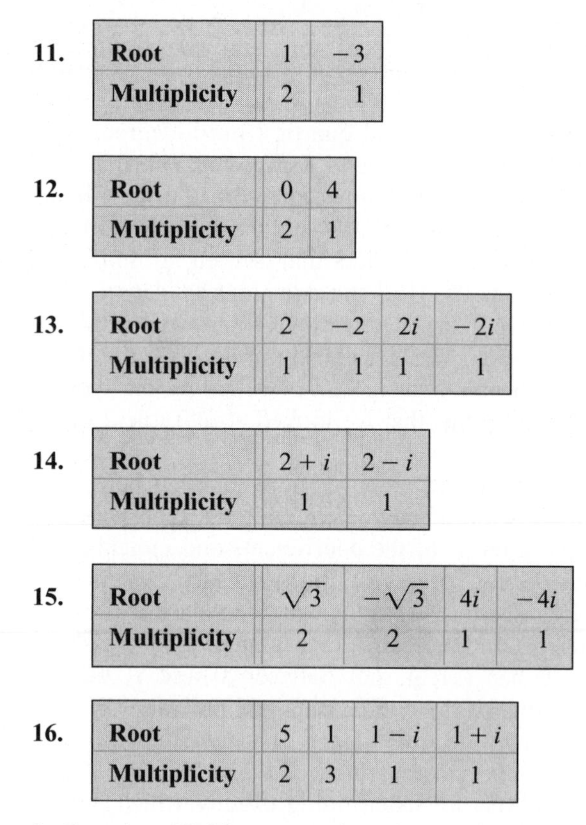

11.

Root	1	−3
Multiplicity	2	1

12.

Root	0	4
Multiplicity	2	1

13.

Root	2	−2	$2i$	$-2i$
Multiplicity	1	1	1	1

14.

Root	$2+i$	$2-i$
Multiplicity	1	1

15.

Root	$\sqrt{3}$	$-\sqrt{3}$	$4i$	$-4i$
Multiplicity	2	2	1	1

16.

Root	5	1	$1-i$	$1+i$
Multiplicity	2	3	1	1

In Exercises 17–20, express the polynomial $f(x)$ in the form $a_nx^n + a_{n-1}x^{n-1} + \cdots + a_1x + a_0$.

17. Find a quadratic function that has zeros -4 and 9 and a graph that passes through the point (3, 5).

18. Find a quadratic function that has a maximum value of 2 and that has -2 and 4 as zeros.

19. Find a third-degree polynomial function that has zeros -5, 2, and 3 and a graph that passes through the point (0, 1).

20. Find a fourth-degree polynomial function that has zeros $\sqrt{2}$, $-\sqrt{2}$, 1, and -1 and a graph that passes through $(2, -20)$.

In Exercises 21–28, find a quadratic equation with the given roots and no others. Write your answers in the form $Ax^2 + Bx + C = 0$. Suggestion: Make use of Table 2.

21. $r_1 = -i$, $r_2 = -\sqrt{3}$

22. $r_1 = 1 + i\sqrt{3}$, $r_2 = 1 - i\sqrt{3}$

23. $r_1 = 9$, $r_2 = -6$

24. $r_1 = 5$, $r_2 = 3/4$

25. $r_1 = 1 + \sqrt{5}$, $r_2 = 1 - \sqrt{5}$

26. $r_1 = 6 - 5i$, $r_2 = 6 + 5i$

27. $r_1 = a + \sqrt{b}$, $r_2 = a - \sqrt{b}$ $(b > 0)$

28. $r_1 = a + bi$, $r_2 = a - bi$

B

29. Express the polynomial $x^4 + 64$ as a product of four linear factors. *Hint:* First add and subtract the quantity $16x^2$ so that you can use the difference-of-squares factoring formula.

30. Suppose that p and q are positive integers with $p > q$. Find a quadratic equation with integer coefficients and roots $\sqrt{p}/(\sqrt{p} \pm \sqrt{p-q})$.

31. Let r_1, r_2, and r_3 be the roots of the equation $x^3 + bx^2 + cx + d = 0$. Use the method shown in Example 5 to verify the following relationships:

$$r_1 + r_2 + r_3 = -b$$
$$r_1r_2 + r_2r_3 + r_3r_1 = c$$
$$r_1r_2r_3 = -d$$

32. Let r_1, r_2, r_3, and r_4 be the roots of the equation $x^4 + bx^3 + cx^2 + dx + e = 0$. Use the method shown in Example 5 to prove the following facts:

$$r_1 + r_2 + r_3 + r_4 = -b$$
$$r_1r_2 + r_2r_3 + r_3r_4 + r_4r_1 + r_2r_4 + r_3r_1 = c$$
$$r_1r_2r_3 + r_2r_3r_4 + r_3r_4r_1 + r_4r_1r_2 = -d$$
$$r_1r_2r_3r_4 = e$$

33. Solve the equation $x^3 - 4x^2 - 9x + 36 = 0$, given that the sum of two of the roots is 0. *Suggestion:* Use Table 2.

34. Solve the equation $x^3 - 75x + 250 = 0$, given that two of the roots are equal. *Suggestion:* Use Table 2.

35. For this exercise, assume as given the following trigonometric identity:

$$\tan 3\theta = (\tan^3\theta - 3\tan\theta)/(3\tan^2\theta - 1)$$

(a) Use the given identity to show that the number $\tan 15°$ is a root of the cubic equation

$$x^3 - 3x^2 - 3x + 1 = 0 \qquad (1)$$

(b) Use a calculator to check that $\tan 15°$ indeed appears to be a root of equation (1).

(c) Factor (by grouping) the left-hand side of equation (1). Conclude that $\tan 15°$ is a root of the reduced equation

$$x^2 - 4x + 1 = 0 \qquad (2)$$

(d) The work in parts (a) and (c) shows that the number tan 15° is a root of equation (2). By following the same technique, show that the number tan 75° is also a root of equation (2).

(e) Use Table 2 in this section to evaluate each of the following quantities. Then use a calculator to check your answers.

(i) tan 15° + tan 75°

(ii) tan 15° tan 75°

(f) Use the quadratic formula to solve equation (2). Which root is tan 15° and which is tan 75°?

36. For this exercise, assume as given the following two identities which were derived in Exercise Set 8.2 (Exercises 58 and 59).

$$(\cos 72°)(\cos 144°) = -1/4$$

and

$$\cos 72° + \cos 144° = -1/2$$

(a) Find a quadratic equation, with integer coefficients, whose roots are the numbers cos 72° and cos 144°. *Hint:* Use the result in Example 5 or Table 2.

(b) Use the quadratic formula to solve the equation in part (a). Conclude that

$$\cos 72° = \tfrac{1}{4}(-1 + \sqrt{5})$$

and

$$\cos 144° = \tfrac{1}{4}(-1 - \sqrt{5})$$

(c) Use a calculator to check the results in part (b).

37. (a) In the trigonometric identity $\cos 3\theta = 4\cos^3\theta - 3\cos\theta$, make the substitution $\theta = 20°$ and conclude that the number cos 20° is a root of the equation $8x^3 - 6x - 1 = 0$.

(b) Follow the method in part (a) to show that each of the numbers cos 100° and cos 140° is a root of $8x^3 - 6x - 1 = 0$.

(c) Use Table 2 to explain why

$$(\cos 20°)(\cos 100°)(\cos 140°) = 1/8$$

(d) Use the result in part (c) and the reference angle concept to show that

$$(\cos 20°)(\cos 40°)(\cos 80°) = 1/8$$

(e) Use a calculator to check the result in part (d).

C

38. (a) Let r_1, r_2, r_3, and r_4 be four real roots of the equation $x^4 + ax^2 + bx + c = 0$. Show that

$$r_1 + r_2 + r_3 + r_4 = 0$$

Hint: Use the first formula in Exercise 32.

(b) Suppose a circle intersects the parabola $y = x^2$ in the points $(x_1, y_1), \ldots, (x_4, y_4)$. Show that

$$x_1 + x_2 + x_3 + x_4 = 0$$

Hint: Use the results in part (a).

39. (a) Let r_1, r_2, and r_3 be three distinct numbers that are roots of the equation $f(x) = Ax^2 + Bx + C = 0$. Show that $f(x) = 0$ for all values of x. *Hint:* You need to show that $A = B = C = 0$. First show that both A and B are zero as follows. If either A or B were nonzero, then the equation $f(x) = 0$ would be a polynomial equation of degree at most 2 with three distinct roots. Why is that impossible?

(b) Use the result in part (a) to prove the following identity:

$$\frac{a^2 - x^2}{(a-b)(a-c)} + \frac{b^2 - x^2}{(b-c)(b-a)} + \frac{c^2 - x^2}{(c-a)(c-b)} - 1 = 0$$

Hint: Let $f(x)$ denote the quadratic expression on the left-hand side of the equation. Compute $f(a)$, $f(b)$, and $f(c)$.

40. Prove that the following equation is an identity:

$$\frac{(x-a)(x-b)c^2}{(c-a)(c-b)} + \frac{(x-b)(x-c)a^2}{(a-b)(a-c)} + \frac{(x-c)(x-a)b^2}{(b-c)(b-a)} - x^2 = 0$$

Hint: Use the result in Exercise 39(a).

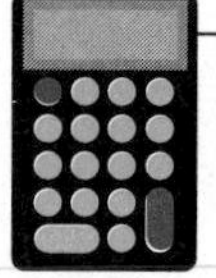

GRAPHING UTILITY EXERCISES FOR SECTION 12.4

In each of Exercises 1–8, you are given a polynomial equation $f(x) = 0$. According to the fundamental theorem of algebra, each of these equations has at least one root. However, the fundamental theorem does not tell you whether the equation has any real-number roots. Use a graph to determine whether the equation has at least one real root. Note: You are not being asked to solve the equation. [Use the fact that a root of the equation $f(x) = 0$ is an x-intercept for the graph of the function $y = f(x)$.] In Exercises 1 and 2, check your response by computing the discriminant for the quadratic.

1. $x^2 - 3x + 2.26 = 0$

2. $x^2 - 2x - 290 = 0$

3. $x^3 - 3x^2 + 3 = 0$

4. $x^4 - 3x^2 + 3 = 0$

5. $x^4 + x^3 + x^2 + x + 1 = 0$

6. $x^5 + x^3 + x^2 + x + 1 = 0$

7. $0.2x^3 + 4.4x^2 - 109x - 1 = 0$

8. $x^4 - \sqrt{35}x^2 + 2.79\pi = 0$

9. In his book *Ars Magna* (*The Great Art*), the Renaissance mathematician Gerolamo Cardano (1501–1576) gave the follow formula for a root of the equation $x^3 + ax = b$:

$$x = \sqrt[3]{\frac{b}{2} + \sqrt{\frac{b^2}{4} + \frac{a^3}{27}}} - \sqrt[3]{-\frac{b}{2} + \sqrt{\frac{b^2}{4} + \frac{a^3}{27}}}$$

(a) Use this formula and your calculator to compute a root of the cubic equation $x^3 + 3x = 76$.

(b) Use a graph to check the answer in part (a). That is, graph the function $y = x^3 + 3x - 76$, and note the x-intercept. (Use a viewing rectangle that extends from -10 to 10 in the x-direction and from -200 to 200 in the y-direction.) Also check the answer simply by substituting it in the equation $x^3 + 3x = 76$.

10. Consider the cubic equation $x^3 = 15x + 4$, which was solved by Rafael Bombelli in his text *L'Algebra parte maggiore del arithmetica* (Bologna: 1572). Bombelli was among the first to use complex numbers to obtain real-number solutions.

(a) Rewrite the equation in the form $x^3 - 15x = 4$, and apply Cardano's formula (given in the previous exercise). Show that this yields

$$x = \sqrt[3]{2 + 11i} - \sqrt[3]{-2 + 11i}$$

(b) Using paper and pencil (or a computer algebra system), show that

$$(2 + i)^3 = 2 + 11i \quad \text{and} \quad (-2 + i)^3 = -2 + 11i$$

(c) Use the results in part (b) to simplify the root given in part (a). You should obtain $x = 4$.

(d) Use a graph to check visually that $x = 4$ is a root of $x^3 = 15x + 4$. (That is, graph the function $y = x^3 - 15x - 4$ and note that $x = 4$ appears to be one of the x-intercepts.) Also, check, algebraically that the value $x = 4$ satisfies the equation.

(e) Given that $x = 4$ is a root of the equation $x^3 = 15x + 4$, use the techniques from Section 12.3 to determine the remaining two roots. Check your answers by using a graphing utility to determine the x-intercepts of the graph in part (d).

12.5 RATIONAL AND IRRATIONAL ROOTS

Of the two leading commentators on Descartes in the 17th century, the first (1649) was his warm personal friend, Florimond de Beaune, an officeholder at Blois. He also wrote on algebra, being one of the first to treat scientifically of the superior and inferior limits of the roots of a numerical equation.

David Eugene Smith in *History of Mathematics,* Vol. 1 (New York: Ginn and Co., 1923)

Jacques Peletier (1517–1582), a French man of letters, poet, and mathematician, had observed as early as 1558, that the root of an equation is a divisor of the last term.

Florian Cajori in *A History of Mathematics,* 4th ed. (New York: Chelsea Publishing Co., 1985)

As we saw in the previous section, not every polynomial equation has a real root. Furthermore, even if a polynomial equation does possess a real root, that root isn't necessarily a rational number. (The equation $x^2 = 2$ provides a simple example.) If a polynomial equation with integer coefficients does have a rational root, however, we can find that root by applying the **rational roots theorem,** which we now state.

> **THE RATIONAL ROOTS THEOREM**
>
> Consider the polynomial equation
>
> $$a_n x^n + a_{n-1}x^{n-1} + \cdots + a_1 x + a_0 = 0 \qquad (n \geqslant 1,\ a_n \neq 0)$$
>
> and suppose that all the coefficients are integers. Let p/q be a rational number, where p and q have no common factors other than ± 1. If p/q is a root of the equation, then p is a factor of a_0 and q is a factor of a_n.

A proof of the rational roots theorem is outlined in Exercise 39 at the end of this section. For the moment, though, let's just see why the theorem is plausible. Suppose that the two rational numbers a/b and c/d are the roots of a certain quadratic equation. Then, from our experience with quadratics (or by the linear factors theorem), we know that the equation can be written in the form

$$k\left(x - \frac{a}{b}\right)\left(x - \frac{c}{d}\right) = 0 \tag{1}$$

where k is a constant. Now, as Exercise 28 will ask you to check, if we carry out the multiplication and clear of fractions, equation (1) becomes

$$(kbd)x^2 - (kad + kbc)x + kac = 0 \tag{2}$$

Observe that a and c (the numerators of the two roots) are factors of the constant term kac in equation (2), just as the rational roots theorem asserts. Furthermore, b and d (the denominators of the roots) are factors of the coefficient of the x^2-term in equation (2), again as the theorem asserts.

The following example shows how the rational roots theorem can be used to solve a polynomial equation.

EXAMPLE 1 Find the rational roots (if any) of the equation $2x^3 - x^2 - 9x - 4 = 0$. Then solve the equation.

Solution First we list the factors of a_0, the factors of a_n, and the possibilities for rational roots:

factors of $a_0 = -4$: $\pm 1,\ \pm 2,\ \pm 4$

factors of $a_3 = 2$: $\pm 1,\ \pm 2$

possible rational roots: $\pm \dfrac{1}{1},\ \pm \dfrac{1}{2},\ \pm \dfrac{2}{1},\ \pm \dfrac{4}{1}$

Now we can use synthetic division to test whether or not any of these possibilities is a root. (A zero remainder will tell us that we have a root.) As you can check, the first three possibilities (1, -1, and $1/2$) are not roots. However, using $-1/2$, we have

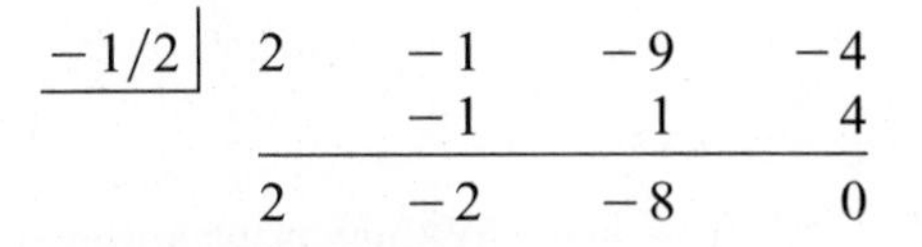

$-1/2$	2	-1	-9	-4
		-1	1	4
	2	-2	-8	0

Thus, $x = -1/2$ is a root. We could now continue to check the remaining possibilities in this same manner. At this point, however, it is simpler to consider the reduced equation $2x^2 - 2x - 8 = 0$, or $x^2 - x - 4 = 0$. Since this is a quadratic equation, it can be solved directly. We have

$$x = \frac{-(-1) \pm \sqrt{(-1)^2 - 4(1)(-4)}}{2(1)} = \frac{1 \pm \sqrt{17}}{2}$$

We have now determined three distinct roots. Since the degree of the original equation is 3, there can be no other roots. We conclude that $x = -1/2$ is the only rational root. The three roots of the equation are $-1/2$ and $(1 \pm \sqrt{17})/2$.

As Example 1 indicates, the number of possibilities for rational roots can be relatively large, even for rather simple equations. The next theorem we develop allows us to reduce the number of possibilities. We say that a real number B is an **upper bound** for the roots of an equation if every real root is less than or equal to B. Similarly, a real number b is a **lower bound** if every real root is greater than or equal to b. The following theorem tells us how synthetic division can be used in determining upper and lower bounds for roots.

THE UPPER AND LOWER BOUND THEOREM FOR REAL ROOTS

Consider the polynomial equation

$$f(x) = a_n x^n + a_{n-1} x^{n-1} + \cdots + a_1 x + a_0 = 0$$

where all of the coefficients are real numbers and a_n is positive.

1. If we use synthetic division to divide $f(x)$ by $x - B$, where $B > 0$, and we obtain a third row containing no negative numbers, then B is an upper bound for the real roots of $f(x) = 0$.
2. If we use synthetic division to divide $f(x)$ by $x - b$, where $b < 0$, and we obtain a third row in which the numbers are alternately positive and negative, then b is a lower bound for the real roots of $f(x) = 0$. (In determining whether the signs alternate in the third row, zeros are ignored.)

CAUTION A number may fail the lower bound test but still be a lower bound. For an example, see Exercise 55.

We will prove the first part of this theorem. A proof of the second part can be developed along similar lines. To prove the first part of the theorem, we use the division algorithm to write

$$f(x) = (x - B) \cdot Q(x) + R \tag{3}$$

The remainder R here is a constant that may be zero. To show that B is an upper bound, we must show that any number greater than B is not a root. Toward this end, let p be a number that is greater than B. Note that p must be positive, since B is positive. Then with $x = p$, equation (3) becomes

$$f(p) = (p - B) \cdot Q(p) + R \tag{4}$$

We are now going to show that the right-hand side of equation (4) is positive. This will tell us that p is not a root. First, look at the factor $(p - B)$. This is positive, since p is greater than B. Next consider $Q(p)$. By hypothesis, the coefficients of $Q(x)$ are all nonnegative. Furthermore, the leading coefficient of $Q(x)$ is a_n, which is positive. Since p is also positive, it follows that $Q(p)$ must be positive. Finally, the number R is nonnegative because, in the synthetic division of $f(x)$ by $x - B$, all the numbers in the third row are nonnegative. It now follows that the right-hand side of equation (4) is positive. Consequently, $f(p)$ is not zero and p is not a root of the equation $f(x) = 0$. This is what we wished to show.

EXAMPLE 2 Determine the rational roots, or show that none exist, for the equation

$$\frac{1}{4}x^4 - \frac{3}{4}x^3 + \frac{17}{4}x^2 + 4x + 5 = 0$$

Solution We will use the rational roots theorem along with the upper and lower bound theorem. First of all, if we are to apply the rational roots theorem, then our equation must have integer coefficients. In view of this, we multiply both sides of the given equation by 4 to obtain

$$x^4 - 3x^3 + 17x^2 + 16x + 20 = 0$$

As in the previous example, we list the factors of a_0, the factors of a_n, and the possibilities for rational roots:

factors of $a_0 = 20$: $\pm 1, \pm 2, \pm 4, \pm 5, \pm 10, \pm 20$

factors of $a_4 = 1$: ± 1

possible rational roots: $\pm 1, \pm 2, \pm 4, \pm 5, \pm 10, \pm 20$

Our strategy here will be to first check for positive roots, beginning with 1 and working upward. The checks for $x = 1$, $x = 2$, and $x = 4$ are as follows:

1	1	−3	17	16	20
		1	−2	15	31
	1	−2	15	31	51

2	1	−3	17	16	20
		2	−2	30	92
	1	−1	15	46	112

4	1	−3	17	16	20
		4	4	84	400
	1	1	21	100	420

As you can see, none of the remainders here is zero. However, notice that in the division corresponding to $x = 4$, all the numbers appearing in the third row are nonnegative. It therefore follows that 4 is an upper bound for the roots of the given equation. In view of this, we needn't bother to check the remaining values $x = 5$, $x = 10$, and $x = 20$, since none of those can be roots. At this point we can conclude that the given equation has no positive rational roots.

Next we check for negative rational roots, beginning with -1 and working downward (if necessary). Checking $x = -1$, we have

−1	1	−3	17	16	20
		−1	4	−21	5
	1	−4	21	−5	25

Two conclusions can be drawn from this synthetic division. First, $x = -1$ is not a root of the equation. Second, -1 is a lower bound for the roots because the signs in the third row of the synthetic division alternate. This means that we needn't bother to check any of the numbers -2, -4, -5, -15, and -20: none of them can be roots, since they are all less than -1.

Let's summarize our results. We have shown that the given equation has no positive rational roots and no negative rational roots. Furthermore, by inspection we see that zero is not a root of the equation. Thus, the given equation possesses no rational roots.

We conclude this section by demonstrating a method for approximating irrational roots. The method depends on the **location theorem.**

THE LOCATION THEOREM

Let $f(x)$ be a polynomial, all of whose coefficients are real numbers. If a and b are real numbers such that $f(a)$ and $f(b)$ have opposite signs, then the equation $f(x) = 0$ has at least one real root between a and b.

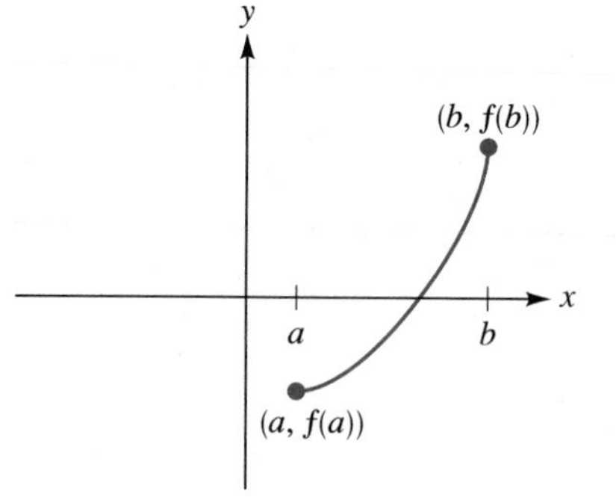

FIGURE 1

Figure 1 indicates why this theorem is plausible. If the point $(a, f(a))$ lies below the x-axis and $(b, f(b))$ lies above the x-axis, then it certainly seems that the graph of f must cross the x-axis at some point x_0 between a and b. At this intercept we have $f(x_0) = 0$; that is, x_0 is a root of the equation $f(x) = 0$. (The location theorem is a special case of the *intermediate value theorem;* the proof is usually discussed in calculus courses.)

Our technique for approximating (or "locating") irrational roots uses the **method of successive approximations.** We will demonstrate this method in Example 3.

EXAMPLE 3 The following equation has exactly one positive root. Locate this root between successive hundredths.

$$f(x) = x^3 + 2x - 4 = 0$$

Solution First we need to find two numbers a and b such that $f(a)$ and $f(b)$ have opposite signs. By inspection (or by trial and error), we find that $f(1) = -1$ and $f(2) = 8$. Thus, according to the location theorem, the equation $f(x) = 0$ has a real root in the interval $(1, 2)$.

Now that we have located the root between successive integers, we can locate it between successive tenths. We compute $f(1.0)$, $f(1.1)$, $f(1.2)$, and so on, up through $f(2.0)$ if necessary, until we find a sign change. As you can verify, using synthetic division and a calculator, the results are as follows:

$$f(1.0) = -1 \qquad \left.\begin{array}{l} f(1.1) = -0.469 \\ f(1.2) = 0.128 \end{array}\right\} \leftarrow \text{sign change}$$

This shows that the root lies between 1.1 and 1.2.

Having located the root between successive tenths, we follow a similar process to locate the root between successive hundredths. Using synthetic division and a calculator, we obtain these results:

$$\begin{array}{lll} f(1.10) = -0.469 & f(1.13) \approx -0.297 & f(1.16) \approx -0.119 \\ f(1.11) \approx -0.412 & f(1.14) \approx -0.238 & f(1.17) \approx -0.058 \\ f(1.12) \approx -0.355 & f(1.15) \approx -0.179 & f(1.18) \approx \ \ 0.003 \end{array} \left.\right\} \leftarrow \text{sign change}$$

This shows that the root lies between 1.17 and 1.18, so we've located the root between successive hundredths, as required.

The procedure described in Example 3 could be continued to yield closer and closer approximations for the required root. We note in passing that such calculations are easily handled with the aid of a programmable calculator.

EXERCISE SET 12.5

A

In Exercises 1–6, list the possibilities for rational roots.

1. $4x^3 - 9x^2 - 15x + 3 = 0$
2. $x^4 - x^3 + 10x^2 - 24 = 0$
3. $8x^5 - x^2 + 9 = 0$
4. $18x^4 - 10x^3 + x^2 - 4 = 0$
5. $\frac{2}{3}x^3 - x^2 - 5x + 2 = 0$
6. $\frac{1}{2}x^4 - 5x^3 + \frac{4}{3}x^2 + 8x - \frac{1}{3} = 0$

In Exercises 7–12, show that each equation has no rational roots.

7. $x^3 - 3x + 1 = 0$
8. $x^3 + 8x^2 - 1 = 0$
9. $x^3 + x^2 - x + 1 = 0$
10. $x^4 + 4x^3 + 4x^2 - 16 = 0$
11. $12x^4 - x^2 - 6 = 0$
12. $4x^5 - x^4 - x^3 - x^2 + x - 8 = 0$

For Exercises 13–25, find the rational roots of each equation and then solve the equation. (Use the rational roots theorem and the upper and lower bound theorem, as in Example 2.)

13. $x^3 + 3x^2 - x - 3 = 0$
14. $2x^3 - 5x^2 - 3x + 9 = 0$
15. $4x^3 + x^2 - 20x - 5 = 0$
16. $3x^3 - 16x^2 + 17x - 4 = 0$
17. $9x^3 + 18x^2 + 11x + 2 = 0$
18. $4x^3 - 10x^2 - 25x + 4 = 0$
19. $x^4 + x^3 - 25x^2 - x + 24 = 0$
20. $10x^4 + 107x^3 + 301x^2 + 171x + 23 = 0$
21. $x^4 - 4x^3 + 6x^2 - 4x + 1 = 0$
22. $24x^3 - 46x^2 + 29x - 6 = 0$
23. $x^3 - \frac{5}{2}x^2 - 23x + 12 = 0$
24. $x^3 - \frac{17}{3}x^2 - \frac{10}{3}x + 8 = 0$
25. $2x^4 - \frac{9}{10}x^3 - \frac{29}{10}x^2 + \frac{27}{20}x - \frac{3}{20} = 0$

In Exercises 26 and 27, determine integral upper and lower bounds for the real roots of the equations. (Follow the method used within the solution of Example 2.)

26. **(a)** $x^3 + 2x^2 - 5x + 20 = 0$
    (b) $x^5 - 3x^2 + 100 = 0$
27. **(a)** $5x^4 - 10x - 12 = 0$
    (b) $3x^4 - 4x^3 + 5x^2 - 2x - 4 = 0$
    (c) $2x^4 - 7x^3 - 5x^2 + 28x - 12 = 0$
28. Referring to equation (1) in this section, multiply out the left-hand side, then clear the equation of fractions. Check that your result agrees with equation (2).

In Exercises 29–34, each equation has exactly one positive root. In each case, locate the root between successive hundredths. Use a calculator.

29. $x^3 + x - 1 = 0$
30. $x^3 - 2x - 5 = 0$
31. $x^5 - 200 = 0$
32. $x^3 - 3x^2 + 3x - 26 = 0$
33. $x^3 - 8x^2 + 21x - 22 = 0$
34. $2x^4 - x^3 - 12x^2 - 16x - 8 = 0$

In Exercises 35–38, each equation has exactly one negative root. In each case, use a calculator to locate the root between successive hundredths.

35. $x^3 + x^2 - 2x + 1 = 0$
36. $x^5 + 100 = 0$
37. $x^3 + 2x^2 + 2x + 101 = 0$
38. $x^4 + 4x^3 - 6x^2 - 8x - 3 = 0$

B

39. This exercise outlines a proof of the rational roots theorem. At one point in the proof, we will need to rely on the following fact, which is proved in courses on number theory.

FACT FROM NUMBER THEORY Suppose that A, B, and C are integers and that A is a factor of the number BC. If A has no factor in common with C (other than ± 1), then A must be a factor of B.

(a) Let $A = 2$, $B = 8$, and $C = 5$. Verify that the fact from number theory is correct here.
(b) Let $A = 20$, $B = 8$, and $C = 5$. Note that A is a factor of BC, but A is not a factor of B. Why doesn't this contradict the fact from number theory?
(c) Now we're ready to prove the rational roots theorem. We begin with a polynomial equation with integer coefficients:

$$a_n x^n + a_{n-1}x^{n-1} + \cdots + a_1 x + a_0 = 0 \qquad (n \geqslant 1,\ a_n \neq 0)$$

We assume that the rational number p/q is a root of the equation and that p and q have no common factors other than 1. Why is the following equation now true?

$$a_n\left(\frac{p}{q}\right)^n + a_{n-1}\left(\frac{p}{q}\right)^{n-1} + \cdots + a_1\left(\frac{p}{q}\right) + a_0 = 0$$

(d) Show that the last equation in part (c) can be written

$$p(a_n p^{n-1} + a_{n-1}qp^{n-2} + \cdots + a_1 q^{n-1}) = -a_0 q^n$$

Since p is a factor of the left-hand side of this last equation, p must also be a factor of the right-hand side. That is, p must be a factor of $a_0 q^n$. But since p and q have no common factors, neither do p and q^n. Our fact from number theory now tells us that p must be a factor of a_0, as we wished to show. (The proof that q is a factor of a_n is carried out in a similar manner.)

40. The location theorem asserts that the polynomial equation $f(x) = 0$ has a root in the open interval (a, b) whenever $f(a)$ and $f(b)$ have unlike signs. If $f(a)$ and $f(b)$ have the same sign, can the equation $f(x) = 0$ have a root between a and b? *Hint:* Look at the graph of $f(x) = x^2 - 2x + 1$ with $a = 0$ and $b = 2$.

Use a calculator for Exercises 41–43. Round your answers to two decimal places.

41. On the same set of axes, sketch the graphs of $y = x^3$ and $y = 1 - 3x$. Find the x-coordinate of the point where the graphs intersect.

42. On the same set of axes, sketch the graphs of $y = x^3$ and $y = x + 1$. Find the x-coordinate of the point where the graphs intersect.

43. Find the x-coordinate of the point where the curves $y = x^2 - 1$ and $y = \sqrt{x}$ meet.

44. In a note that appeared in *The Two-Year College Mathematics Journal* [vol. 12 (1981), pp. 334–336], Professors Warren Page and Leo Chosid explain how the process of testing for rational roots can be shortened. In essence, their result is as follows. Suppose that we have a polynomial with integer coefficients and we are testing for a possible root p/q. Then, if a noninteger is generated at any point in the synthetic division process, p/q cannot be a root of the polynomial. For example, suppose we want to know if 4/3 is a root of $6x^4 - 10x^3 + 2x^2 - 9x + 8 = 0$. The first few steps of the synthetic division are as follows.

$$\begin{array}{c|ccccc} \frac{4}{3} & 6 & -10 & 2 & -9 & 8 \\ & & 8 & -\frac{8}{3} & & \\ \hline & 6 & -2 & & & \end{array}$$

Since the noninteger $-8/3$ has been generated in the synthetic division process, the process can be stopped; 4/3 is not a root of the polynomial. Use this idea to shorten your work in testing to see if the numbers 3/4, 1/8, and $-3/2$ are roots of the equation

$$8x^5 - 5x^4 + 3x^2 - 2x - 6 = 0$$

45. In a note that appeared in *The College Mathematics Journal* [vol. 20 (1989), pp. 139–141], Professor Don Redmond proved the following interesting result.

Consider the polynomial equation $f(x) = a_n x^n + a_{n-1}x^{n-1} + \cdots + a_1 x + a_0 = 0$, and suppose that the degree of $f(x)$ is at least 2 and that all of the coefficients are integers. If the three numbers a_0, a_n, and $f(1)$ are all odd, then the given equation has no rational roots.

Use this result to show that the following equations have no rational roots.
(a) $9x^5 - 8x^4 + 3x^2 - 2x + 27 = 0$
(b) $5x^5 + 5x^4 - 11x^2 - 3x - 25 = 0$

46. **(a)** Use a calculator to verify that the number tan 9° appears to be a root of the following equation.

$$x^4 - 4x^3 - 14x^2 - 4x + 1 = 0 \qquad (1)$$

In parts (b) through (d) of this exercise, you will *prove* that tan 9° is indeed a root and that tan 9° is irrational.

(b) Use the trigonometric identity

$$\tan 5\theta = \frac{\tan^5\theta - 10\tan^3\theta + 5\tan\theta}{5\tan^4\theta - 10\tan^2\theta + 1}$$

to show that the number $x = \tan 9°$ is a root of the fifth-degree equation

$$x^5 - 5x^4 - 10x^3 + 10x^2 + 5x - 1 = 0 \qquad (2)$$

Hint: In the given trigonometric identity, substitute $\theta = 9°$.

(c) List the possibilities for the rational roots of equation (2). Then use synthetic division and the remainder theorem to show that there is only one rational root. What is the reduced equation in this case?

(d) Use your work in parts (b) and (c) to explain (in complete sentences) why the number tan 9° is an irrational root of equation (1).

47. As background for this exercise you need to have worked Exercise 46.

(a) Follow exactly the same method used in parts (b) through (d) of Exercise 46 to show that the number $-\tan 27°$ is an irrational root of equation (1) in Exercise 46.

(b) From Exercise 46 and part (a) of this exercise, we know that both of the numbers tan 9° and $-\tan 27°$ are roots of equation (1) in Exercise 46. By following the same method, it can also be shown that the numbers $-\tan 63°$ and tan 81° are roots of equation (1). Assuming this fact, along with the results in Exercise 32 on page 782, evaluate each of the following quantities, then use a calculator to check your results.

(i) $\tan 9° \tan 27° \tan 63° \tan 81°$

(ii) $\tan 9° - \tan 27° - \tan 63° + \tan 81°$

48. (a) Let $\theta = 2\pi/7$. Use the reference angle concept to explain why $\cos 3\theta = \cos 4\theta$, then use your calculator to confirm the result.

(b) For this portion of the exercise, assume as given the following two trigonometric identities:

$$\cos 3\theta = 4\cos^3\theta - 3\cos\theta$$
$$\cos 4\theta = 8\cos^4\theta - 8\cos^2\theta + 1$$

Use these identities and the result in part (a) to show that $\cos(2\pi/7)$ is a root of the equation

$$8x^4 - 4x^3 - 8x^2 + 3x + 1 = 0 \qquad (1)$$

(c) List the prossibilities for the rational roots of equation (1). Then use synthetic division and the remainder theorem to show that there is only one rational root. Check that the reduced equation in this case is

$$8x^3 + 4x^2 - 4x - 1 = 0 \qquad (2)$$

(d) The work in parts (a) through (c) shows that the number $\cos(2\pi/7)$ is a root of equation (2). By following the same technique, it can be shown that the numbers $\cos(4\pi/7)$ and $\cos(6\pi/7)$ also are roots of equation (2). Use this fact, along with Table 2 in Section 12.4, to evaluate each of the following quantities. Then use a calculator to check your answers.

(i) $\cos\frac{2\pi}{7}\cos\frac{4\pi}{7}\cos\frac{6\pi}{7}$

(ii) $\cos\frac{2\pi}{7} + \cos\frac{4\pi}{7} + \cos\frac{6\pi}{7}$

C

In Exercises 49–53, you need to know that a **prime number** *is a positive integer greater than* 1 *with no factors other than itself and* 1. *Thus, the first seven prime numbers are* 2, 3, 5, 7, 11, 13, *and* 17.

49. Find all prime numbers p for which the equation $x^2 + x - p = 0$ has a rational root.

50. Find all prime numbers p for which the equation $x^3 + x^2 + x - p = 0$ has at least one rational root. For each value of p that you find, find the corresponding *real* roots of the equation.

51. Consider the equation $x^3 + px - q = 0$, where p and q are prime numbers. Observe that there are only four possible rational roots here: 1, -1, q, and $-q$.

(a) Show that if $x = 1$ is a root, then we must have $q = 3$ and $p = 2$. What are the remaining roots in this case?

(b) Show that none of the numbers -1, q, and $-q$ can be a root of the equation. *Hint:* For each case, assume the contrary and deduce a contradiction.

52. Consider the equation $x^2 + x - pq = 0$, where p and q are prime numbers. If this equation has rational roots, show that these roots must be -3 and 2. *Suggestion:* The possible rational roots are ± 1, $\pm p$, $\pm q$, and $\pm pq$. In each case, assume that the given number is a root and see where that leads.

53. If p and q are prime numbers, show that the equation $x^3 + px - pq = 0$ has no rational roots.

54. Find all integral values of b for which the equation $x^3 - b^2x^2 + 3bx - 4 = 0$ has a rational root.

55. Let $f(x) = x^3 + 3x^2 - x - 3$.

(a) Factor $f(x)$ by using the basic factoring techniques in Section 1.3.

(b) Sketch the graph of $f(x) = x^3 + 3x^2 - x - 3$. Note that -3 is a lower bound for the roots.

(c) Show that the number -3 fails the lower bound test. This shows that a number may fail the lower bound test and yet be a lower bound. (We say that the lower bound test provides a *sufficient* but not a *necessary* condition for a lower bound.)

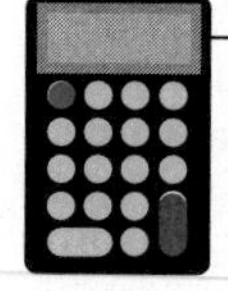

GRAPHING UTILITY EXERCISES FOR SECTION 12.5

In Exercises 1 and 2, you are given a polynomial equation $f(x) = 0$. In parentheses following each equation is the example number in which that equation was considered. For each equation $f(x) = 0$, carry out the following steps. **(a)** *Graph the corresponding function $y = f(x)$.* **(b)** *Use the graphing utility to determine (or at least approximate) the x-intercepts of the graph.* [*These x-intercepts are the roots of the equation $f(x) = 0$.*] **(c)** *Check that the values obtained in part (b) are consistent with the roots determined in the text.*

1. $2x^3 - x^2 - 9x - 4 = 0$ (Example 1)

2. $x^3 + 2x - 4 = 0$ (Example 3)

In Exercises 3–8, you are given a polynomial equation $f(x) = 0$. **(a)** *Use a graph to determine each real root between successive integers.* **(b)** *Use the method of Example 3 to locate each root between successive: tenths; hundreds; thousands. (Use the graphing utility to set up the appropriate tables.)* **(c)** *Check your answers in part (b) either by using a* SOLVE *key or by zooming in on the x-intercepts.*

3. $-x^3 + 2x + 2 = 0$

4. $-x^3 + 3x - 1 = 0$

5. $x^3 + 11x^2 - 1 = 0$

6. $x^3 - 11x^2 - 1 = 0$

7. $x^4 - 48x - 2 = 0$

8. $-\frac{1}{720}x^6 + \frac{1}{24}x^4 - \frac{1}{2}x^2 + 1 = 0$

9. Let $f(x) = \frac{1}{4}x^4 - \frac{3}{4}x^3 + \frac{17}{4}x^2 + 4x + 5$. According to Example 2, the equation $f(x) = 0$ has no rational roots. Use a graph to find out if the equation has any real roots.

In Exercises 10–12, you are given a polynomial equation $f(x) = 0$. By graphing the equation $y = f(x)$, find a closed interval $[c, d]$ that satisfies both of the following conditions: the number c is the largest integer that is a lower bound for the roots; and the number d is the smallest integer that is an upper bound for the roots.

10. $x^4 - 6x^3 - 16x^2 + 64x - 6 = 0$

11. $x^3 - 16.25x^2 + 3.875x + 2 = 0$

12. $x^4 + 24x^3 + 12x^2 - 6x - 3 = 0$

12.6 CONJUGATE ROOTS AND DESCARTES'S RULE OF SIGNS

An equation can have as many true [positive] *roots as it contains changes of sign, from plus to minus or from minus to plus; and as many false* [negative] *roots as the number of times two plus signs or two minus signs are found in succession.*

René Descartes (1637)

We now proceed to investigate a remarkable theorem, implicit in the work of [Thomas] *Harriot* [1560–1621] *but first used explicitly by Descartes (1637), which limits the number of positive or negative roots of an equation. . . .*

Remarkable as the Harriot–Descartes's Rule of Signs is, it still leaves uncertainty as to the exact number of real roots in an equation: it only gives an upper limit to them. The problem of finding an exact test . . . was finally solved in 1829 by [Jacques Charles François] *Sturm.*

H. W. Turnbull in *Theory of Equations* (Edinburgh: Oliver and Boyd, 1939)

As you know from earlier work involving quadratic equations with real coefficients, when nonreal complex roots occur, they occur in conjugate pairs. For instance, as you can check by means of the quadratic formula, the roots of the equation $x^2 - 2x + 5 = 0$ are $1 + 2i$ and $1 - 2i$. The **conjugate roots theorem** tells us that the situation is the same for all polynomial equations with real coefficients.

THE CONJUGATE ROOTS THEOREM

Let $f(x)$ be a polynomial, all of whose coefficients are real numbers. Suppose that $a + bi$ is a root of the equation $f(x) = 0$, where a and b are real and $b \neq 0$. Then $a - bi$ is also a root of the equation.

To prove the conjugate roots theorem, we use four of the properties of complex conjugates listed in Section 12.1:

Property 1: $\bar{z}_1\bar{z}_2 = \overline{z_1 z_2}$
Property 2: $(\bar{z})^m = \overline{z^m}$
Property 3: $\bar{r} = r$ for every real number r
Property 4: $\bar{z}_1 + \bar{z}_2 = \overline{z_1 + z_2}$

Using these properties, we can prove the theorem as follows. We begin with a polynomial with real coefficients:

$$f(x) = a_n x^n + a_{n-1}x^{n-1} + \cdots + a_1 x + a_0$$

We must show that if $z = a + bi$ is a root of $f(x) = 0$, then $\bar{z} = a - bi$ is also a root. We have

$$\begin{aligned}
f(\bar{z}) &= a_n(\bar{z})^n + a_{n-1}(\bar{z})^{n-1} + \cdots + a_1\bar{z} + a_0 \\
&= \bar{a}_n\overline{z^n} + \bar{a}_{n-1}\overline{z^{n-1}} + \cdots + \bar{a}_1\bar{z} + \bar{a}_0 && \text{Properties 3 and 2} \\
&= \overline{a_n z^n} + \overline{a_{n-1}z^{n-1}} + \cdots + \overline{a_1 z} + \bar{a}_0 && \text{Property 1} \\
&= \overline{a_n z^n + a_{n-1}z^{n-1} + \cdots + a_1 z + a_0} && \text{Property 4} \\
&= \overline{f(z)} = \bar{0} && f(z) = 0, \text{ since } z \text{ is a root} \\
&= 0 && \text{Property 3}
\end{aligned}$$

We have now shown that $f(\bar{z}) = 0$, given that $f(z) = 0$. Thus, $\bar{z}$ is a root, as we wished to show.

Although the conjugate roots theorem concerns nonreal complex roots, it can nevertheless be used to obtain information about real roots, as the next two examples demonstrate.

EXAMPLE 1 Solve the equation $f(x) = 2x^4 - 3x^3 + 12x^2 + 22x - 60 = 0$, given that one root is $1 + 3i$.

Solution Since all of the coefficients of $f(x)$ are real numbers, we know that the conjugate of $1 + 3i$ must also be a root. Thus, $1 + 3i$ and $1 - 3i$ are roots, from which it follows that $[x - (1 + 3i)]$ and $[x - (1 - 3i)]$ are factors of $f(x)$. As you can check, the product of these two factors is $x^2 - 2x - 10$. Thus we must have

$$f(x) = (x^2 - 2x + 10) \cdot Q(x)$$

for some polynomial $Q(x)$. We compute $Q(x)$ using long division:

$$\begin{array}{r|l}
 & 2x^2 + x - 6 \\
\hline
x^2 - 2x + 10 & 2x^4 - 3x^3 + 12x^2 + 22x - 60 \\
 & \underline{2x^4 - 4x^3 + 20x^2} \\
 & \qquad x^3 - 8x^2 + 22x \\
 & \qquad \underline{x^3 - 2x^2 + 10x} \\
 & \qquad\qquad -6x^2 + 12x - 60 \\
 & \qquad\qquad \underline{-6x^2 + 12x - 60} \\
 & \qquad\qquad\qquad\qquad 0
\end{array}$$

Thus, $Q(x) = 2x^2 + x - 6$, and the original equation becomes

$$f(x) = (x^2 - 2x + 10)(2x^2 + x - 6) = 0$$

We can now find any additional roots by solving the equation $2x^2 + x - 6 = 0$. We have

$$2x^2 + x - 6 = 0$$
$$(2x - 3)(x + 2) = 0$$

$$2x - 3 = 0 \quad\Big|\quad x + 2 = 0$$
$$x = 3/2 \quad\Big|\quad x = -2$$

We now have four distinct roots of the original equation: $1 + 3i$, $1 - 3i$, $3/2$, and -2. Since the degree of the equation is 4, there can be no other roots.

EXAMPLE 2 Show that the equation $x^3 - 2x^2 + x - 1 = 0$ has at least one irrational root.

Solution Allowing for multiple roots, the equation has three roots. We are going to show that at least one of these three roots must be irrational. For convenience, we break the argument into two cases.

CASE 1 *All three roots are real numbers.*
Applying the rational roots theorem to the given equation shows that there are no rational roots. (Verify this for yourself.) Thus, in this case, all three real roots must be irrational.

CASE 2 *At least one root is a nonreal complex number.*
If we have one nonreal complex root $a + bi$, then we know that its conjugate, $a - bi$, is also a root. The third root then must be a real number. (Why?) As we pointed out in Case 1, however, the given equation has no rational roots. Thus, the real root must be irrational.

There is a theorem, similar to the conjugate roots theorem, that tells us about irrational roots of the form $a + b\sqrt{c}$. As background for this theorem, let's look at two preliminary examples. First, we consider the equation $x^2 - 2x - 5 = 0$. As you can check, the roots in this case are $1 + \sqrt{6}$ and $1 - \sqrt{6}$. However, it is not true in general that irrational roots such as these always occur in pairs. Consider as a second example the quadratic equation

$$(x + 2)(x - \sqrt{3}) = 0$$

or

$$x^2 + (2 - \sqrt{3})x - 2\sqrt{3} = 0$$

Here, one of the roots is $\sqrt{3}$, yet $-\sqrt{3}$ is not a root. This type of behavior can occur in polynomial equations where not all of the coefficients are rational. On the other hand, when the coefficients are all rational, we do have the following theorem. (See Exercise 41 at the end of this section for a proof.)

THEOREM

Let $f(x)$ be a polynomial in which all the coefficients are rational. Suppose that $a + b\sqrt{c}$ is a root of the equation $f(x) = 0$, where a, b, and c are rational and $\sqrt{c}$ is irrational. Then $a - b\sqrt{c}$ is also a root of the equation.

EXAMPLE 3 Find a quadratic equation with rational coefficients and a leading coefficient of 1 such that one of the roots is $r_1 = 4 + 5\sqrt{3}$.

Solution If one root is $r_1 = 4 + 5\sqrt{3}$, then the other is $r_2 = 4 - 5\sqrt{3}$. We denote the required equation by $x^2 + bx + c = 0$. Thus, according to Table 2 in Section 12.4, we have

$$b = -(r_1 + r_2) = -[(4 + 5\sqrt{3}) + (4 - 5\sqrt{3})] = -8$$

and

$$c = r_1 r_2 = (4 + 5\sqrt{3})(4 - 5\sqrt{3}) = 16 - 75 = -59$$

The required equation is therefore $x^2 - 8x - 59 = 0$. This answer can also be obtained without using the table. Since the roots are $4 \pm 5\sqrt{3}$, we can write the required equation as $[x - (4 + 5\sqrt{3})][x - (4 - 5\sqrt{3})] = 0$. As you can now check by multiplying out the two factors, this equation is equivalent to $x^2 - 8x - 59 = 0$, as obtained previously.

We conclude this section with a discussion of **Descartes's rule of signs.** This rule, published by Descartes in 1637, provides us with information about the types of roots an equation can have, even before we attempt to solve the equation. In order to state Descartes's rule of signs, we first explain what is meant by a variation in sign in a polynomial with real coefficients. Suppose that $f(x)$ is a polynomial with real coefficients, written in descending (or ascending) powers of x. For example, let $f(x) = 2x^3 - 4x^2 - 3x + 1$. Then we say that there is a **variation in sign** if two successive coefficients have opposite signs. In the case of $f(x) = 2x^3 - 4x^2 - 3x + 1$, there are two variations in sign, the first occurring as we go from 2 to -4 and the second occurring as we go from -3 to 1. In looking for variations in sign, we ignore terms with zero coefficients. Table 1 shows a few more examples of how we count variations in sign.

TABLE 1

Polynomial	Number of Variations in Sign
$x^2 + 4x$	0
$-3x^5 + x^2 + 1$	1
$x^3 + 3x^2 - x + 6$	2

We now state Descartes's rule of signs and look at some examples. Although the proof of this theorem is not difficult, it is rather lengthy, and we shall omit it here.

DESCARTES'S RULE OF SIGNS

Let $f(x)$ be a polynomial, all of whose coefficients are real numbers, and consider the equation $f(x) = 0$. Then:

a. The number of positive roots either is equal to the number of variations in sign of $f(x)$ or is less than that by an even integer.
b. The number of negative roots either is equal to the number of variations in sign of $f(-x)$ or is less than that by an even integer.

EXAMPLE 4 Use Descartes's rule of signs to obtain information regarding the roots of the equation $x^3 + 8x + 5 = 0$.

Solution Let $f(x) = x^3 + 8x + 5$. Then, since there are no variations in sign for $f(x)$, we see from part (a) of Descartes's rule that the given equation has no positive roots. Next we compute $f(-x)$ to learn about the possibilities for negative roots: we have $f(-x) = -x^3 - 8x + 5$. So $f(-x)$ has one sign change, and consequently [by part (b) of Descartes's rule] the original equation has one negative root. Furthermore, notice that zero is not a root of the equation. Thus, the equation has only one real root, a negative root. Since the equation has a total of three roots, we can conclude that we have one negative root and two nonreal complex roots. The two nonreal roots will be complex conjugates.

EXAMPLE 5 Use Descartes's rule to obtain information regarding the roots of the equation $x^4 + 3x^2 - 7x - 5 = 0$.

Solution Let $f(x) = x^4 + 3x^2 - 7x - 5$. Then $f(x)$ has one variation in sign, so according to part (a) of Descartes's rule, the equation has one positive root. That leaves three roots still to be accounted for, since the degree of the equation is 4. We have $f(-x) = x^4 + 3x^2 + 7x - 5$. Since $f(-x)$ has one sign change, we know from part (b) of Descartes's rule that the equation has one negative root. Noting now that zero is not a root, we conclude that the two remaining roots must be nonreal complex roots. In summary, then, the equation has one positive root, one negative root, and two nonreal complex (conjugate) roots.

EXAMPLE 6 Use Descartes's rule to obtain information regarding the roots of the equation $f(x) = x^3 - x^2 + 3x + 2 = 0$.

Solution Since $f(x)$ has two variations in sign, the given equation has either two positive roots or no positive roots. To see how many negative roots are possible, we compute

$$\begin{aligned} f(-x) &= (-x)^3 - (-x)^2 + 3(-x) + 2 \\ &= -x^3 - x^2 - 3x + 2 \end{aligned}$$

Since $f(-x)$ has one variation in sign, we conclude from part (b) of Descartes's rule that the equation has exactly one negative root. In summary, then, there are two possibilities:

either: one negative root and two positive roots;

or: one negative root and two nonreal complex roots.

By using Descartes's rule in Examples 4 and 5, we were able to determine the exact numbers of positive and negative roots for the given equations. As Example 6 indicates, however, there are cases in which a direct application of Descartes's rule provides several distinct possibilities for the types of roots, rather than a single definitive result. (Exercises 55 and 56 in the Review Exercises for this chapter illustrate a technique that is sometimes useful in gaining additional information from Descartes's rule. In particular, Exercise 56 will show you that the equation in Example 6 has no positive roots.)

EXERCISE SET 12.6

A

In Exercises 1–16, an equation is given, followed by one or more roots of the equation. In each case, determine the remaining roots.

1. $x^2 - 14x + 53 = 0;\ x = 7 - 2i$
2. $x^2 - x - \frac{1535}{4} = 0;\ x = \frac{1}{2} + 8\sqrt{6}$
3. $x^3 - 13x^2 + 59x - 87 = 0;\ x = 5 + 2i$
4. $x^4 - 10x^3 + 30x^2 - 10x - 51 = 0;\ x = 4 + i$
5. $x^4 + 10x^3 + 38x^2 + 66x + 45 = 0;\ x = -2 + i$

6. $2x^3 + 11x^2 + 30x - 18 = 0;\ x = -3 - 3i$

7. $4x^3 - 47x^2 + 232x + 61 = 0;\ x = 6 - 5i$

8. $9x^4 + 18x^3 + 20x^2 - 32x - 64 = 0;\ x = -1 + \sqrt{3}i$

9. $4x^4 - 32x^3 + 81x^2 - 72x + 162 = 0;\ x = 4 + \sqrt{2}i$

10. $2x^4 - 17x^3 + 137x^2 - 57x - 65 = 0;\ x = 4 - 7i$

11. $x^4 - 22x^3 + 140x^2 - 128x - 416 = 0;\ x = 10 + 2i$

12. $4x^4 - 8x^3 + 24x^2 - 20x + 25 = 0;\ x = (1 + 3i)/2$

13. $15x^3 - 16x^2 + 9x - 2 = 0;\ x = (1 + \sqrt{2}i)/3$

14. $x^5 - 5x^4 + 30x^3 + 18x^2 + 92x - 136 = 0;$
$x = -1 + i\sqrt{3}, x = 3 - 5i$

15. $x^7 - 3x^6 - 4x^5 + 30x^4 + 27x^3 - 13x^2 - 64x + 26 = 0;$
$x = 3 - 2i,\ x = -1 + i,\ x = 1$

16. $x^6 - 2x^5 - 2x^4 + 2x^3 + 2x + 1 = 0;\ x = 1 + \sqrt{2}$

In Exercises 17–20, find a quadratic equation with rational coefficients, one of whose roots is the given number. Write your answer so that the coefficient of x^2 is 1. Use either of the methods shown in Example 3.

17. $r_1 = 1 + \sqrt{6}$ **18.** $r_1 = 2 - \sqrt{3}$

19. $r_1 = (2 + \sqrt{10})/3$ **20.** $r_1 = \frac{1}{2} + \frac{1}{4}\sqrt{5}$

In Exercises 21–36, use Descartes's rule of signs to obtain information regarding the roots of the equations.

21. $x^3 + 5 = 0$ **22.** $x^4 + x^2 + 1 = 0$

23. $2x^5 + 3x + 4 = 0$ **24.** $x^3 + 8x - 3 = 0$

25. $5x^4 + 2x - 7 = 0$ **26.** $x^3 - 4x^2 + x - 1 = 0$

27. $x^3 - 4x^2 - x - 1 = 0$

28. $x^8 + 4x^6 + 3x^4 + 2x^2 + 5 = 0$

29. $3x^8 + x^6 - 2x^2 - 4 = 0$

30. $12x^4 - 5x^3 - 7x^2 - 4 = 0$

31. $x^9 - 2 = 0$

32. $x^9 + 2 = 0$

33. $x^8 - 2 = 0$

34. $x^8 + 2 = 0$

35. $x^6 + x^2 - x - 1 = 0$

36. $x^7 + x^2 - x - 1 = 0$

B

37. Consider the equation $x^4 + cx^2 + dx - e = 0$, where c, d, and e are positive. Show that the equation has one positive root, one negative root, and two nonreal complex roots.

38. Consider the equation $x^n - 1 = 0$.
(a) Show that the equation has $n - 2$ nonreal complex roots when n is even.
(b) How many nonreal complex roots are there when n is odd?

39. Find the polynomial $f(x)$ of lowest degree with rational coefficients and with leading coefficient 1, such that $\sqrt{3} + 2i$ is a root of the equation $f(x) = 0$.

40. Find a quadratic polynomial $f(x)$ with integer coefficients, such that $(2 + \sqrt{3})/(2 - \sqrt{3})$ is a root of the equation $f(x) = 0$. *Hint:* First rationalize the given root.

C

41. Let $f(x)$ be a polynomial, with rational coefficients. Suppose that $a + b\sqrt{c}$ is a root of $f(x) = 0$, where a, b, and c are rational and $\sqrt{c}$ is irrational. Complete the following steps to prove that $a - b\sqrt{c}$ is also a root of the equation $f(x) = 0$.
(a) If $b = 0$, we're done. Why?
(b) (From now on we'll assume that $b \neq 0$.) Let $d(x) = [x - (a + b\sqrt{c})][x - (a - b\sqrt{c})]$. Explain why $d(a + b\sqrt{c}) = 0$.
(c) Verify that $d(x) = (x - a)^2 - b^2c$. Thus, $d(x)$ is a quadratic polynomial with rational coefficients.
(d) Now suppose that we use the long division process to divide the polynomial $f(x)$ by the quadratic polynomial $d(x)$. We'll obtain a quotient $Q(x)$ and a remainder. Since the degree of $d(x)$ is 2, our remainder will be of degree 1 or less. In other words, the general form of this remainder will be $Cx + D$. Furthermore, C and D will have to be rational, because all of the coefficients in $f(x)$ and in $d(x)$ are rational. In summary, we have the identity

$$f(x) = d(x) \cdot Q(x) + (Cx + D)$$

Now make the substitution $x = a + b\sqrt{c}$ in this identity, and conclude that $C = D = 0$.
(e) Using the result in part (d), we have

$$f(x) = [x - (a + b\sqrt{c})][x - (a - b\sqrt{c})] \cdot Q(x)$$

Let $x = a - b\sqrt{c}$ in this last identity and conclude that $a - b\sqrt{c}$ is a root of $f(x) = 0$, as required.

42. Find a polynomial $f(x)$ with integer coefficients such that one root of $f(x) = 0$ is $x = \sqrt{2} + \sqrt[3]{2}$.

43. As you know, if a quadratic equation $f(x) = x^2 + Bx + C = 0$ has two real roots, there is a geometric interpretation: the roots are the x-intercepts for the graph of f. Less well known is the fact that there is a geometric interpretation when the quadratic equation $f(x) = x^2 + Bx + C = 0$ has nonreal complex roots. If these roots are $a \pm bi$, where a and b are real numbers, show that the coordinates of the vertex for the graph of f are (a, b^2). *Hint:* Start by using Table 2 in Section 12.4 to show that $f(x)$ can be written $x^2 - 2ax + (a^2 + b^2)$.

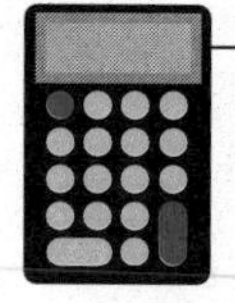

GRAPHING UTILITY EXERCISES FOR SECTION 12.6

1. Let $f(x) = 2x^4 - 3x^3 + 12x^2 + 22x - 60$.
   (a) Use Descartes's rule to verify that the equation $f(x) = 0$ has one negative root.
   (b) Use Descartes's rule to verify that the equation $f(x) = 0$ has either one or three positive roots.
   (c) Graph the equation $y = f(x)$. Use the graph to say which of the two cases in part (b) actually holds.
   (d) Use the graph to estimate the real roots of the equation $f(x) = 0$. Check that your answers are consistent with the values obtained in Example 1.
2. Let $f(x) = x^3 - 2x^2 + x - 1$.
   (a) Without using a graphing utility, explain (in complete sentences) why the equation $f(x) = 0$ must have either three real roots or only one real root. (If you get stuck, see Example 2 in the text.)
   (b) Use a graph to demonstrate that the equation $f(x) = 0$ has, in fact, only one real root.
   (c) According to Example 2, the root (or x-intercept) that you observed in part (b) is an irrational number. Use the graphing utility to obtain an approximation for this root.
3. Let $f(x) = x^3 + 8x + 5$.
   (a) Use Descartes's rule to explain in complete sentences why the equation $f(x) = 0$ has no positive roots and exactly one negative root. (If you need help, see Example 4 in the text.)
   (b) Use a graph to confirm the results in part (a). That is, graph $y = f(x)$ and note that there is but one x-intercept and it is negative.
   (c) Use a graphing utility to compute the root of $x^3 + 8x + 5 = 0$. (Do this either by repeatedly zooming in on the x-intercept of the graph or by using a SOLVE key.)
   (d) Use the general formula given on page 781 to compute the root of $x^3 + 8x + 5 = 0$. Check that the answer agrees with the value that you obtained in part (c).
4. Let $f(x) = x^3 - x^2 + 3x + 2$.
   (a) Use Descartes's rule to explain in complete sentences why the equation $f(x) = 0$ has
   either: one negative root and two positive roots;
   or: one negative root and two nonreal complex roots.
   (If you need help, review Example 6 in the text.)
   (b) Use a graph to determine which of the two possibilities in part (a) is actually the case.
   (c) Use a graphing utility to compute the real root(s) of the equation $f(x) = 0$.

12.7 INTRODUCTION TO PARTIAL FRACTIONS

In elementary Algebra, a group of fractions connected by the signs of addition and subtraction is reduced to a more simple form by being collected into one single fraction whose denominator is the lowest common denominator of the given fractions. But the converse process of separating a fraction into a group of simpler, or partial, *fractions is often required.*

H. S. Hall and S. R. Knight, in *Higher Algebra* (London: Macmillan and Co., 1946). (This classic text was first published in 1887.)

In calculus, there are times when it is helpful to express a given function in terms of simpler functions. In Section 3.4 we learned that composition of functions provided one way to do this. For example, if $R(x) = 1/(x^2 - 1)$, then, as you can easily check, $R(x)$ can be expressed as a composition of two simpler functions as follows:

$$R(x) = f(g(x)) \qquad \text{where } g(x) = x^2 - 1 \quad \text{and} \quad f(x) = 1/x$$

In this section, we introduce another way that certain types of functions can be broken down into simpler functions. Instead of using composition of functions, now we'll be using the sum and difference of functions. We can again use as an example $R(x) = 1/(x^2 - 1)$. This time $R(x)$ can be expressed as a *difference* of two simpler expressions as follows:

$$\frac{1}{x^2 - 1} = \frac{1}{2(x - 1)} - \frac{1}{2(x + 1)}$$

You can check this result for yourself by using the least common denominator to combine the two fractions on the right-hand side of the equation.

Our basic goal in this and the next section is to be able to write a given fractional expression as a sum or difference of two or more simpler fractions. For instance, in Example 1, we will be given the fraction $\frac{2x-3}{(x-1)(x+1)}$, and we will be asked to find constants A and B so that

$$\frac{2x-3}{(x-1)(x+1)} = \frac{A}{x-1} + \frac{B}{x+1} \tag{1}$$

When A and B are determined, the right-hand side of equation (1) is called the **partial fraction decomposition** of the given fraction. The adjective *partial* is used because each denominator on the right-hand side of equation (1) is a *part* of the denominator on the other side of the equation.

One of the basic tools that can be used in finding partial fractions is supplied by the following theorem, which we state here without proof.

EQUATING-THE-COEFFICIENTS THEOREM

Suppose that $P(x)$ and $Q(x)$ are polynomials such that $P(x) = Q(x)$ for all x. Then the corresponding coefficients of the two polynomials are equal.

The next two statements supply examples of what this theorem is saying.

1. If $ax + b = 5x - 11$ for all x, then $a = 5$ and $b = -11$.
2. If $px^3 + qx^2 + rx + s = x^2 + 2$ for all x, then $p = 0$, $q = 1$, $r = 0$, and $s = 2$.

In Example 1, we apply the theorem to obtain a partial fraction decomposition.

EXAMPLE 1 Determine constants A and B so that the following equation is an identity:

$$\frac{2x-3}{(x-1)(x+1)} = \frac{A}{x-1} + \frac{B}{x+1} \tag{2}$$

Solution First, to clear equation (2) of fractions, we multiply both sides by the least common denominator $(x-1)(x+1)$. This yields

$$\begin{aligned} 2x - 3 &= A(x+1) + B(x-1) \\ &= Ax + A + Bx - B \\ &= (A+B)x + (A-B) \end{aligned}$$

So we have

$$2x - 3 = (A+B)x + (A-B)$$

Now, since this last equation is supposed to be an identity, we can use our equating-the-coefficients theorem to obtain the two equations

$$\begin{cases} A + B = 2 \\ A - B = -3 \end{cases}$$

Adding these two equations gives us $2A = -1$ and therefore $A = -1/2$. Subtracting the two equations yields $2B = 5$, and consequently $B = 5/2$. We've now found that $A = -1/2$ and $B = 5/2$, as required. The partial fraction decomposition is therefore

$$\frac{2x-3}{(x-1)(x+1)} = \frac{-1/2}{x-1} + \frac{5/2}{x+1} = \frac{-1}{2(x-1)} + \frac{5}{2(x+1)}$$

You should check for yourself that combining the two fractions on the right-hand side of this last equation indeed yields $\frac{2x-3}{(x-1)(x+1)}$.

In Example 1, we found the partial fraction decomposition by solving a system of two linear equations. There is a shortcut that is often useful in such problems. As before, we start by multiplying both sides of equation (2) by the common denominator $(x-1)(x+1)$ to obtain

$$2x - 3 = A(x+1) + B(x-1) \tag{3}$$

Now, equation (3) is an identity; in particular it must hold when $x = 1$. (You're about to see why we've singled out $x = 1$.) Substituting $x = 1$ in equation (3) gives us

$$2(1) - 3 = A(1+1) + B(1-1)$$

or

$$-1 = 2A$$

and consequently

$$A = -1/2 \qquad \text{as obtained previously.}$$

The value for B is obtained similarly. Go back to equation (3), and this time let $x = -1$. This gives us

$$2(-1) - 3 = A(-1+1) + B(-1-1)$$

or

$$-5 = -2B \qquad \text{and therefore} \qquad B = 5/2$$

Again, this agrees with the result obtained previously.

EXAMPLE 2 Determine constants A and B so that the following equation is an identity:

$$\frac{x}{(x+4)^2} = \frac{A}{x+4} + \frac{B}{(x+4)^2}$$

Solution We'll show two methods: first, the method that we used in Example 1, where we equated the coefficients; and second, the shortcut method explained after Example 1. For ease of reference, we'll call this shortcut method the *convenient-values method.*

FIRST METHOD *(equating coefficients)* As in Example 1, we start by multiplying both sides of the given identity by the least common denominator. Here, the least common denominator is $(x+4)^2$, and we obtain

$$x = A(x+4) + B = Ax + 4A + B$$

That is,

$$x = Ax + (4A + B)$$

Now, by equating coefficients, we obtain the system of equations

$$\begin{cases} A = 1 \\ 4A + B = 0 \end{cases}$$

The first of these two equations gives us the value of A directly. Substituting $A = 1$ in the second equation then yields $4(1) + B = 0$, and therefore $B = -4$. In summary, $A = 1$, $B = -4$, and the partial fraction decomposition is

$$\frac{x}{(x+4)^2} = \frac{1}{x+4} - \frac{4}{(x+4)^2}$$

You should check for yourself that combining the two fractions on the right-hand side of this last equation indeed produces the fraction on the left-hand side of the equation.

SECOND METHOD ***(convenient values)*** As before, we start by multiplying both sides of the given identity by $(x+4)^2$ to obtain

$$x = A(x+4) + B \tag{4}$$

Letting $x = -4$ in identity (4) gives us

$$-4 = A(-4+4) + B$$

or

$$B = -4 \qquad \text{as obtained previously}$$

With the value $B = -4$, identity (4) reads

$$x = A(x+4) - 4 \tag{5}$$

At this point, substituting any value for x (other than -4, which we've already exploited) will produce the required value for A. Using $x = 0$ then in equation (5) yields

$$0 = A(0+4) - 4 \qquad \text{or} \qquad -4A = -4$$

and therefore

$$A = 1 \qquad \text{as obtained previously}$$

In the examples up to this point, the convenient-values method appears to be somewhat more efficient (that is, shorter) than equating the coefficients. (In Example 2, perhaps it's a toss-up.) In the next example, we again show our two methods of solution. You can decide for yourself which method you prefer. In the convenient-values method, you'll see that one of the values we choose to substitute in the identity is a complex number. For purposes of completeness, we mention in passing the following theorem from post-calculus mathematics that justifies this technique.

If p and q are polynomial functions such that $p(x) = q(x)$ for all real numbers x, then $p(z) = q(z)$ for all complex numbers z.

As we've just said, we're stating this result only for the purposes of completeness; you certainly don't need to memorize this theorem or know any more about it to work the exercises.

EXAMPLE 3 Determine real numbers A, B, and C so that the following equation is an identity:

$$\frac{7x^2 - 9x + 29}{(x-2)(x^2+9)} = \frac{A}{x-2} + \frac{Bx + C}{x^2+9}$$

Solution

FIRST METHOD ***(equating coefficients)*** The least common denominator is $(x-2)(x^2+9)$. Multiplying both sides of the given identity by this common denominator yields

$$\begin{aligned} 7x^2 - 9x + 29 &= A(x^2+9) + (Bx + C)(x-2) \\ &= Ax^2 + 9A + Bx^2 - 2Bx + Cx - 2C \\ &= (A+B)x^2 + (-2B + C)x + (9A - 2C) \end{aligned}$$

Equating coefficients now gives us a system of three equations in three unknowns:

$$\begin{cases} A + B = 7 \\ -2B + C = -9 \\ 9A - 2C = 29 \end{cases}$$

Exercise 23 asks you to solve this system to obtain $A = 3$, $B = 4$, and $C = -1$. The partial fraction decomposition is therefore

$$\frac{7x^2 - 9x + 29}{(x-2)(x^2+9)} = \frac{3}{x-2} + \frac{4x - 1}{x^2+9}$$

SECOND METHOD ***(convenient values)*** Going back to the original identity and multiplying both sides by the least common denominator yields (as in the first method)

$$7x^2 - 9x + 29 = A(x^2+9) + (Bx + C)(x-2) \tag{6}$$

Because of the factor $x - 2$ on the right-hand side of identity (6), we choose the convenient value $x = 2$. Substituting $x = 2$ in equation (6) gives us

$$7(2^2) - 9(2) + 29 = A(2^2 + 9) + 0$$

or

$$39 = 13A \qquad \text{and therefore} \qquad A = 3$$

Next, we want to choose another value for x, one that will cause the factor $x^2 + 9$ to be zero. Thus we choose $x = 3i$. (The other root of the equation $x^2 + 9 = 0$, namely, $x = -3i$, would also be an acceptable choice here.) Making the substitution $x = 3i$ (along with $x^2 = -9$) in identity (6) yields

$$\begin{aligned} 7(-9) - 9(3i) + 29 &= A(-9+9) + [B(3i) + C](3i - 2) \\ -34 - 27i &= -9B - 6Bi + 3Ci - 2C \\ &= (-9B - 2C) + (-6B + 3C)i \end{aligned}$$

Equating the real parts and equating the imaginary parts from both sides of this last equation gives the equations

$$-9B - 2C = -34 \qquad (7)$$

and

$$-6B + 3C = -27$$

or

$$-2B + C = -9 \qquad \text{dividing by 3} \qquad (8)$$

As Exercise 24 asks you to check, the solution to the system consisting of equations (7) and (8) is $B = 4$ and $C = -1$. In summary, then, we have $A = 3$, $B = 4$, and $C = -1$, as obtained previously.

EXERCISE SET 12.7

A

In Exercises 1–22, determine the constants (denoted by capital letters) so that each equation is an identity. For Exercises 1–6, do each problem in two ways: **(a)** *use the equating-the-coefficients theorem, as in Example 1; and* **(b)** *use the convenient-values method that was explained after Example 1. For the remainder of the exercises, use either method (or a combination).*

1. $\dfrac{7x-6}{(x-2)(x+2)} = \dfrac{A}{x-2} + \dfrac{B}{x+2}$

2. $\dfrac{5x+27}{(x+3)(x-3)} = \dfrac{A}{x+3} + \dfrac{B}{x-3}$

3. $\dfrac{6x-25}{(2x+5)(2x-5)} = \dfrac{A}{2x+5} + \dfrac{B}{2x-5}$

4. $\dfrac{x}{(4x+3)(4x-3)} = \dfrac{A}{4x+3} + \dfrac{B}{4x-3}$

5. $\dfrac{1}{(x+1)(3x-1)} = \dfrac{A}{x+1} + \dfrac{B}{3x-1}$

6. $\dfrac{1-x}{(4x+3)(2x-1)} = \dfrac{A}{4x+3} + \dfrac{B}{2x-1}$

7. $\dfrac{8x+3}{(x+3)^2} = \dfrac{A}{x+3} + \dfrac{B}{(x+3)^2}$

8. $\dfrac{7x}{(x-5)^2} = \dfrac{A}{x-5} + \dfrac{B}{(x-5)^2}$

9. $\dfrac{6-x}{(5x+4)^2} = \dfrac{A}{5x+4} + \dfrac{B}{(5x+4)^2}$

10. $\dfrac{30x-17}{(6x-1)^2} = \dfrac{A}{6x-1} + \dfrac{B}{(6x-1)^2}$

11. $\dfrac{3x^2+7x-2}{(x-1)(x^2+1)} = \dfrac{A}{x-1} + \dfrac{Bx+C}{x^2+1}$

12. $\dfrac{15x^2-35x+77}{(2x-5)(x^2+3)} = \dfrac{A}{2x-5} + \dfrac{Bx+C}{(x^2+3)}$

13. $\dfrac{x^2+1}{(x+1)(x^2+4)} = \dfrac{A}{x+1} + \dfrac{Bx+C}{x^2+4}$

14. $\dfrac{x-7}{x(x^2+2)} = \dfrac{A}{x} + \dfrac{Bx+C}{x^2+2}$

15. $\dfrac{1}{x(x^2-x+1)} = \dfrac{A}{x} + \dfrac{Bx+C}{x^2-x+1}$

16. $\dfrac{2x^2-11x-6}{(x+2)(x^2-2x+4)} = \dfrac{A}{x+2} + \dfrac{Bx+C}{x^2-2x+4}$

17. $\dfrac{3x^2-2}{(x-2)(x+1)(x-1)} = \dfrac{A}{x-2} + \dfrac{B}{x+1} + \dfrac{C}{x-1}$

18. $\dfrac{1}{(x-1)(x-2)(x-3)} = \dfrac{A}{x-1} + \dfrac{B}{x-2} + \dfrac{C}{x-3}$

19. $\dfrac{4x^2-47x+133}{(x-6)^3} = \dfrac{A}{x-6} + \dfrac{B}{(x-6)^2} + \dfrac{C}{(x-6)^3}$

20. $\dfrac{x^2+2x}{(x+1)^3} = \dfrac{A}{x+1} + \dfrac{B}{(x+1)^2} + \dfrac{C}{(x+1)^3}$

21. $\dfrac{x^2-2}{(x^2+2)^2} = \dfrac{Ax+B}{x^2+2} + \dfrac{Cx+D}{(x^2+2)^2}$

22. $\dfrac{x^3+2x^2}{(x^2+3)^2} = \dfrac{Ax+B}{x^2+3} + \dfrac{Cx+D}{(x^2+3)^2}$

23. Solve the following system of equations. (As indicated in Example 3, you should obtain $A = 3$, $B = 4$, and $C = -1$.)

$$\begin{cases} A + B = 7 \\ -2B + C = -9 \\ 9A - 2C = 29 \end{cases}$$

24. Solve the following system of equations. (As indicated in the text, you should obtain $B = 4$ and $C = -1$.)

$$\begin{cases} -9B - 2C = -34 \\ -2B + C = -9 \end{cases}$$

Exercises 25 and 26 provide practice using the convenient-values method with complex numbers. As background, you should review Example 3.

25. In this exercise you'll use the convenient-values method to determine real numbers A, B, and C such that the following equation is an identity:

$$\frac{x+1}{x(x^2+4)} = \frac{A}{x} + \frac{Bx+C}{x^2+4}$$

(a) Multiplying both sides of the given identity by the least common denominator yields

$$x + 1 = A(x^2+4) + (Bx+C)x \qquad (1)$$

Determine A by substituting $x = 0$ in identity (1).

(b) In identity (1), substitute $x = 2i$ and show that the resulting equation can be written

$$2i + 1 = -4B + 2iC \qquad (2)$$

(c) Determine B and C by equating real and imaginary parts in equation (2).

26. In this exercise you'll use the convenient-values method to determine real numbers A, B, C, and D such that the following equation is an identity:

$$\frac{x^2+x+1}{(x^2+1)(x^2+16)} = \frac{Ax+B}{x^2+1} + \frac{Cx+D}{x^2+16}$$

(a) Multiplying both sides of the given identity by the least common denominator yields

$$x^2 + x + 1 = (Ax+B)(x^2+16) + (Cx+D)(x^2+1) \qquad (3)$$

In identity (3), substitute $x = 4i$ and show that the resulting equation can be written

$$4i - 15 = -60Ci - 15D$$

Determine C and D by equating real and imaginary parts in this last equation.

(b) In identity (3), substitute $x = i$ and show that the resulting equation can be written

$$i = 15Ai + 15B$$

Determine A and B by equating real and imaginary parts in this last equation.

B

In Exercises 27 and 28, the equating-the-coefficients theorem is used as a tool to factor polynomials.

27. Use the equating-the-coefficients theorem to find a factorization of $x^4 - x^3 + x^2 - x + 1$ of the form

$$x^4 - x^3 + x^2 - x + 1 = (x^2 + bx + 1)(x^2 + cx + 1)$$

28. Use the equating-the-coefficients theorem to determine constants b, c, and d in the following factorization of a fifth-degree polynomial:

$$x^5 - x^4 + x^3 - x^2 + x - 1 = (x^2 + bx + 1)(x^3 + cx^2 + dx - 1)$$

(Assume $b > 0$.)

C

29. Let $f(x) = x^6 - x^5 + x^4 - x^3 + x^2 - x + 1$. Use the equating-the-coefficients theorem to show that a factorization with the following form is impossible:

$$f(x) = (x^3 + ax^2 + bx + 1)(x^3 + cx^2 + dx + 1)$$

12.8 MORE ABOUT PARTIAL FRACTIONS

If the denominator of a rational function has two relatively prime factors, then this rational function can be expressed as the sum of two fractions whose denominators are equal to the two factors.

Leonhard Euler (1707–1783) in *Introductio in analysis infinitorum* (1748)

In the examples and exercises in the previous section, you were told what the general form for each partial fraction should look like, and then the required constants were computed. In calculus, however, when a partial fraction decomposition is needed, you're usually not specifically told what the general form for the partial fractions should be. Rather, you're expected to know this. In this section we list the necessary guidelines. But first, consider the following example indicating why these guidelines are needed.

Suppose, for instance, that we want to find the partial fraction decomposition for $1/[x(x^2+1)]$, and we guess that the general form is

$$\frac{1}{x(x^2+1)} = \frac{A}{x} + \frac{B}{x^2+1} \tag{1}$$

Multiplying both sides of this last equation by the least common denominator gives us

$$\begin{aligned} 1 &= A(x^2+1) + Bx \\ &= Ax^2 + A + Bx \\ &= Ax^2 + Bx + A \end{aligned}$$

Equating coefficients now, as we did in the previous section, yields $A = 0$, $B = 0$, and $A = 1$. The two equations $A = 0$ and $A = 1$ are contradictory. The only way out is to conclude that there is no partial fraction decomposition of the form shown in equation (1). This example demonstrates the need for guidelines for setting up partial fraction decompositions.

We begin with a few definitions. Suppose that we have two polynomials $p(x)$ and $q(x)$. Following the terminology in Section 4.7, we call an expression of the form $p(x)/q(x)$ a **rational expression.** A rational expression is said to be **proper** if the degree of the numerator is less than the degree of the denominator. So, each of the following is a proper rational expression:

$$\frac{4x-5}{2x^2+3x-1} \qquad \frac{x^3-2x-5}{x^5-9x^3-4} \qquad \frac{x^2-5}{(x+1)(x+2)(x+3)}$$

An **improper** rational expression is one in which the degree of the numerator is greater than or equal to the degree of the denominator. Examples of improper rational expressions are $(x^2+1)/(x-1)$ and $(x^3-4)/(x^3+2x^2+1)$. All of the partial fraction guidelines that we are about to discuss pertain to proper rational expressions. (Near the end of this section, we'll explain what to do for an improper rational expression.)

The final term that we need to define is *irreducible quadratic polynomial.* The definition is given in the box that follows.

DEFINITION: Irreducible Quadratic Polynomial

Let $f(x)$ be a quadratic polynomial, all of whose coefficients are real. Then $f(x)$ is said to be **irreducible** provided the equation $f(x) = 0$ has no real roots. This is equivalent to saying that $f(x)$ cannot be factored into the form $(ax + b)(cx + d)$, where a, b, c, and d are real numbers.

As examples, note that $x^2 + 5$ is irreducible, but $x^2 - 5$ is not irreducible.

The first step in determining a partial fraction decomposition is to factor the denominator. In the previous section, the denominators were already factored for you. If you glance back at the previous section, you'll see that the denominators contained one or more of the following types of factors, and no others:

$$\begin{aligned} \text{Linear factors:} &\quad ax + b \\ \text{Powers of linear factors:} &\quad (ax + b)^n, \quad n \geq 2 \\ \text{Irreducible quadratic factors:} &\quad ax^2 + bx + c \\ \text{Powers of irreducible quadratic factors:} &\quad (ax^2 + bx + c)^n, \quad n \geq 2 \end{aligned}$$

A remarkable theorem tells us that, in fact, *every* polynomial with real coefficients can be factored (over the real numbers) using only the types of factors we've just listed. We state this theorem in the box that follows.

LINEAR AND QUADRATIC FACTORS THEOREM

Let $f(x)$ be a polynomial, all of whose coefficients are real numbers. Then $f(x)$ can be factored (over the real numbers) into a product of linear and/or irreducible quadratic factors.

To see a familiar example, consider the cubic polynomial $x^3 - 1$. From our work on factoring in Section 1.3, we know that

$$x^3 - 1 = (x - 1)(x^2 + x + 1)$$

So in this case, there is one linear factor and one irreducible quadratic factor. (You should check for yourself, using the discriminant, that the quadratic factor is indeed irreducible.) However, not every polynomial, of course, is so easily factored. For instance, although the theorem tells us that the polynomial $x^3 + x + 1$ *can* be factored into linear and/or irreducible quadratic factors, it doesn't tell us how to find those factors. Indeed, that can be a very difficult job, quite beyond this course. For the examples and exercises in this section (and in any calculus text), you'll be able to find the factors using factoring techniques from Section 1.3 or from this chapter. At the end of this section we show how the linear and quadratic factors theorem follows from our earlier work in this chapter.

For all of the guidelines that we give, we assume that we're starting with a proper rational expression $p(x)/q(x)$ and that $q(x)$ has been factored into linear and/or irreducible quadratic factors. We also assume that $p(x)$ and $q(x)$ have no common factors; that is, the fraction $p(x)/q(x)$ has been reduced to lowest terms. The first two guidelines tell us what to do when $q(x)$ contains a linear factor $ax + b$ or a power of this factor.

GUIDELINES FOR THE PARTIAL FRACTIONS SETUP: LINEAR FACTORS

(a) If the denominator contains a linear factor $ax + b$ and no higher power of this factor, then the partial fraction setup must contain a term $A_1/(ax + b)$, where A_1 is a constant to be determined.

(b) [This encompasses part (a).] More generally, for each factor in the denominator of the form $(ax + b)^n$, the partial fraction setup must contain the following sum of n fractions:

$$\frac{A_1}{ax+b} + \frac{A_2}{(ax+b)^2} + \cdots + \frac{A_n}{(ax+b)^n}$$

where $A_1, A_2, \ldots, A_n$ are constants to be determined.

EXAMPLE 1 Determine the form of the partial fraction decomposition:

(a) $\dfrac{5x+1}{4x^2-9}$; **(b)** $\dfrac{x^2}{x^3+2x^2-5x-10}$.

Solution

(a) The denominator can be factored using difference-of-squares factoring:

$$4x^2 - 9 = (2x-3)(2x+3)$$

Thus, the denominator contains two distinct linear factors. So, according to the first guideline in the box preceding this example, the form of the partial fraction decomposition is

$$\frac{5x+1}{4x^2-9} = \frac{A_1}{2x-3} + \frac{A_2}{2x+3}$$

(b) The denominator can be factored by grouping (as in Section 1.3):

$$\begin{aligned} x^3 + 2x^2 - 5x - 10 &= (x^3 - 5x) + (2x^2 - 10) \\ &= x(x^2-5) + 2(x^2-5) \\ &= (x^2-5)(x+2) \qquad (2) \end{aligned}$$

We're not finished with the factoring yet! If this had been a factoring problem from Section 1.3, equation (2) would indeed be our final result. That's because in Section 1.3 we were restricted to factors with integer (or possibly rational) coefficients. But here the restriction is removed, and we can factor $x^2 - 5$ as a difference of squares:

$$x^2 - 5 = \left(x - \sqrt{5}\right)\left(x + \sqrt{5}\right)$$

Combining this last result with equation (2), our final factorization looks this way:

$$x^3 + 2x^2 - 5x - 10 = \left(x - \sqrt{5}\right)\left(x + \sqrt{5}\right)(x+2)$$

Thus, the denominator contains three distinct linear factors, and the form of the partial fraction decomposition is

$$\frac{x^2}{x^3+2x^2-5x-10} = \frac{A_1}{x-\sqrt{5}} + \frac{A_2}{x+\sqrt{5}} + \frac{A_3}{x+2}$$

EXAMPLE 2 Determine the form of the partial fraction decomposition for $\dfrac{6x-1}{x^3+2x^2-5x-6}$.

Solution In Example 1(a), we factored a denominator quite similar to this one by grouping the terms. As you can check, however, that won't work here. The next strategy then is to apply the rational roots theorem to the polynomial equation

$$x^3+2x^2-5x-6=0 \tag{3}$$

As Exercise 35 asks you to show, this leads to three rational roots: -1, 2, and -3. Consequently, we have a factorization into three distinct linear factors:

$$x^3+2x^2-5x-6=(x+1)(x-2)(x+3)$$

The form of the partial fraction decomposition must therefore be

$$\frac{6x-1}{x^3+2x^2-5x-6}=\frac{A_1}{x+1}+\frac{A_2}{x-2}+\frac{A_3}{x+3}$$

In Examples 1 and 2, none of the linear factors in the denominators are repeated. The next example shows cases in which there are repeated linear factors.

EXAMPLE 3 Determine the form of the partial fraction decomposition for each of the following:

(a) $\dfrac{5x-1}{x^2-6x+9}$; **(b)** $\dfrac{2x^4+x+1}{x^5+3x^4+3x^3+x^2}$.

Solution

(a) The denominator is a perfect square:

$$x^2-6x+9=(x-3)^2$$

So, according to the second guideline in the box preceding Example 1, the form of the partial fraction decomposition is

$$\frac{5x-1}{x^2-6x+9}=\frac{A_1}{x-3}+\frac{A_2}{(x-3)^2}$$

(b) As explained in Section 1.3, the first step in factoring is to look for a common term. In this case there is one: it is x^2, and we have

$$x^5+3x^4+3x^3+x^2=x^2(x^3+3x^2+3x+1)$$

Now what? Regarding the second factor on the right-hand side of this last equation, we could try to factor it by grouping. As you can check, that does work. It's more direct, however, to observe that this factor is actually a perfect cube: $x^3+3x^2+3x+1=(x+1)^3$. Putting things together then, we have the final factorization

$$x^5+3x^4+3x^3+x^2=x^2(x+1)^3$$

In summary, there are two repeated linear factors: the factor x occurs twice and the factor $x+1$ occurs three times. The form of the partial fraction decomposition is therefore

$$\frac{2x^4+x+1}{x^5+3x^4+3x^3+x^2}=\frac{A_1}{x}+\frac{A_2}{x^2}+\frac{B_1}{x+1}+\frac{B_2}{(x+1)^2}+\frac{B_3}{(x+1)^3}$$

The next two guidelines apply in cases where the denominator contains one or more irreducible quadratic factors. After you read these guidelines, compare them to those that we listed previously for linear factors. You'll see that almost everything is the same. The only difference, in fact, is this: for the quadratic factors, the form of the numerator is $Ax + B$ rather than just A.

GUIDELINES FOR THE PARTIAL FRACTIONS SETUP: IRREDUCIBLE QUADRATIC FACTORS

(a) If the denominator contains an irreducible quadratic factor $ax^2 + bx + c$ and no higher power of this factor, then the partial fractions setup must contain a term $\frac{A_1x+B_1}{ax^2+bx+c}$, where A_1 and B_1 are constants to be determined.

(b) [This encompasses part (a).] More generally, for each factor in the denominator of the form $(ax^2 + bx + c)^n$, where $ax^2 + bx + c$ is irreducible, the partial fractions setup must contain the following sum of n fractions:

$$\frac{A_1x+B_1}{ax^2+bx+c}+\frac{A_2x+B_2}{(ax^2+bx+c)^2}+\cdots+\frac{A_nx+B_n}{(ax^2+bx+c)^n}$$

where the A_i and B_i are constants to be determined.

EXAMPLE 4 Determine the partial fraction decomposition for $\frac{x^3+4x^2+1}{x^4-16}$.

Solution Our first job is to factor the denominator. Using difference-of-squares factoring, we have

$$\begin{aligned} x^4-16 &= (x^2-4)(x^2+4) \\ &= (x-2)(x+2)(x^2+4) \end{aligned}$$

So there are two linear factors and one irreducible quadratic factor, none of which is repeated. The form of the partial fraction decomposition is therefore

$$\frac{x^3+4x^2+1}{(x-2)(x+2)(x^2+4)}=\frac{A}{x-2}+\frac{B}{x+2}+\frac{Cx+D}{x^2+4}$$

(For ease of reading and writing, we are denoting the constants by A, B, C, and D instead of, say A_1, A_2, B_1, and C_1, respectively.) Multiplying both sides of this equation by the common denominator gives

$$x^3+4x^2+1=A(x+2)(x^2+4)+B(x-2)(x^2+4)+(Cx+D)(x-2)(x+2) \qquad (4)$$

Substituting $x = 2$ in identity (4) yields

$$\begin{aligned} 2^3+4(2^2)+1 &= A(2+2)(2^2+4)+0+0 \\ 25 &= 32A \\ A &= 25/32 \end{aligned}$$

Similarly, letting $x = -2$ in identity (4) gives us

$$(-2)^3 + 4(-2)^2 + 1 = 0 + B(-4)(8) + 0$$
$$9 = -32B$$
$$B = -9/32$$

At this point, we've found A and B, but we still need C and D. If we substitute the values we've just found for A and B in equation (4), we have

$$x^3 + 4x^2 + 1 = \tfrac{25}{32}(x+2)(x^2+4) - \tfrac{9}{32}(x-2)(x^2+4) + (Cx+D)(x-2)(x+2) \quad (5)$$

Observe now that in identity (5), letting $x = 0$ will yield an equation involving D alone. Exercise 36(a) asks you to follow through with the arithmetic and algebra to obtain $D = 15/8$. Finally, in identity (5), we replace D by $15/8$ and make the substitution $x = 1$. As Exercise 36(b) asks you to verify, the end result is $C = 1/2$. Putting everything together now, the required partial fraction decomposition is

$$\frac{x^3 + 4x^2 + 1}{x^4 - 16} = \frac{\frac{25}{32}}{x-2} + \frac{-\frac{9}{32}}{x+2} + \frac{\frac{1}{2}x + \frac{15}{8}}{x^2+4}$$
$$= \frac{25}{32(x-2)} - \frac{9}{32(x+2)} + \frac{4x+15}{8(x^2+4)}$$

EXAMPLE 5 Determine the partial fraction decomposition for the following expression

$$\frac{3x^3 - x^2 + 7x - 3}{x^4 + 6x^2 + 9}$$

Solution The denominator factors as a perfect square:

$$x^4 + 6x^2 + 9 = (x^2+3)^2$$

Since the repeated factor $x^2 + 3$ is irreducible, our guidelines tell us that the form of the partial fraction decomposition is

$$\frac{3x^3 - x^2 + 7x - 3}{(x^2+3)^2} = \frac{Ax+B}{x^2+3} + \frac{Cx+D}{(x^2+3)^2}$$

Multiplying both sides of this identity by $(x^2+3)^2$ gives us

$$3x^3 - x^2 + 7x - 3 = (Ax+B)(x^2+3) + (Cx+D) \quad (6)$$

At this point we can use either of the two techniques discussed in the previous section, the equating-the-coefficients technique or the convenient-values technique. As Exercise 37 asks you to show, the results are $A = 3$, $B = -1$, $C = -2$, and $D = 0$. The partial fraction decomposition then is

$$\frac{3x^3 - x^2 + 7x - 3}{x^4 + 6x^2 + 9} = \frac{3x-1}{x^2+3} - \frac{2x}{(x^2+3)^2}$$

Each of the examples we've considered in this and the previous section involved proper rational expressions. For improper rational expressions, we can first use long division to express the improper fraction in the general form

(polynomial) + (proper rational expression)

Then the techniques we've developed can be applied to the second term in this sum. For instance, suppose we're given the improper rational expression $(2x^3 + 4x^2 - 15x - 36)/(x^2 - 9)$. As you can check for yourself by using long division, we obtain

$$\frac{2x^3 + 4x^2 - 15x - 36}{x^2 - 9} = \underbrace{2x + 4}_{\text{polynomial}} + \underbrace{\frac{3x}{x^2 - 9}}_{\text{proper rational expression}}$$

At this stage, a partial fraction decomposition can be worked out for the proper rational expression $3x/(x^2 - 9)$. As Exercise 38(b) asks you to check, the result is

$$\frac{3x}{x^2 - 9} = \frac{3}{2(x - 3)} + \frac{3}{2(x + 3)}$$

So, our final decomposition of the given improper rational fraction is

$$\frac{2x^3 + 4x^2 - 15x - 36}{x^2 - 9} = 2x + 4 + \frac{3}{2(x - 3)} + \frac{3}{2(x + 3)}$$

The linear and quadratic factors theorem that we've discussed in this section is a direct consequence of two theorems that appeared earlier in the chapter: the linear factors theorem and the conjugate roots theorem. Here's why: If we start with a polynomial $f(x)$, the linear factors theorem says that we can decompose $f(x)$ into linear factors:

$$f(x) = (x - c_1)(x - c_2) \cdots (x - c_n) \tag{7}$$

Now, it's possible that some, or even all, of these c_i may be nonreal complex numbers. For instance, suppose that c_1 is a nonreal complex number, with $c_1 = a + bi$. Then, assuming that all of the coefficients of $f(x)$ are real numbers, the conjugate roots theorem tells us that in one of the other factors, call it $x - c_2$ for simplicity, we must have $c_2 = a - bi$. Now look what happens when we compute the product of these two factors:

$$\begin{aligned}(x - c_1)(x - c_2) &= [x - (a + bi)][x - (a - bi)] \\ &= x^2 - (a - bi)x - (a + bi)x + (a + bi)(a - bi) \\ &= x^2 - 2ax + (a^2 + b^2) \qquad \text{(Check the algebra.)}\end{aligned}$$

As Exercise 45 asks you to check, this last quadratic polynomial is irreducible, that is, it has no real roots. In summary, the two linear factors $x - c_1$ and $x - c_2$ that contain nonreal complex numbers give rise to one irreducible quadratic factor with real coefficients. So, similarly, after pairing up any other linear factors in equation (7) that contain nonreal complex numbers, what are we left with? The right-hand side of equation (7) will contain only linear factors and/or irreducible quadratic factors, all with real coefficients, just as the linear and quadratic factors theorem asserts.

EXERCISE SET 12.8

A

In Exercises 1–4, determine if the given quadratic polynomial is irreducible. [Recall from the text that a quadratic polynomial $f(x)$ is irreducible if the equation $f(x) = 0$ has no real roots.]

1. **(a)** $x^2 - 16$ **(b)** $x^2 + 16$

2. **(a)** $x^2 + 17$ **(b)** $x^2 - 17$

3. **(a)** $x^2 + 3x - 4$ **(b)** $x^2 + 3x + 4$

4. **(a)** $24x^2 + x - 3$ **(b)** $x^2 + 24x + 144$

In Exercises 5–16: **(a)** *factor the denominator of the given rational expression;* **(b)** *determine the form of the partial fraction decomposition for the given rational expression; and* **(c)** *determine the values of the constants in the partial fraction decomposition that you gave in part (b). To help you in spotting errors, use the fact that in part (c), each of the required constants turns out to be an integer.*

5. $\dfrac{11x + 30}{x^2 - 100}$

6. $\dfrac{x + 18}{x^2 - 36}$

7. $\dfrac{8x - 2\sqrt{5}}{x^2 - 5}$

8. $\dfrac{2\sqrt{11}}{x^2 - 11}$

9. $\dfrac{7x + 39}{x^2 - x - 6}$

10. $\dfrac{19x - 15}{4x^2 - 5x}$

11. $\dfrac{3x^2 + 17x - 38}{x^3 - 3x^2 - 4x + 12}$

12. $\dfrac{16x^2 + 9x - 2}{3x^3 + x^2 - 2x}$

13. $\dfrac{5x^2 + 2x + 5}{x^3 + x^2 + x}$

14. $\dfrac{x^2 - 3x - 1}{x^3 - x^2 + 2x - 2}$

15. $\dfrac{2x^3 + 5x - 4}{x^4 + 2x^2 + 1}$

16. $\dfrac{11x^3 + 35x - 7}{x^4 + 6x^2 + 9}$

In Exercises 17–34, determine the partial fraction decomposition for each of the given rational expressions. Hint: In Exercises 17 and 18, use the rational roots theorem to factor the denominator.

17. $\dfrac{x^2 + 2}{x^3 - 3x^2 - 16x - 12}$

18. $\dfrac{1}{x^3 + x^2 - 10x + 8}$

19. $\dfrac{5 - x}{6x^2 - 19x + 15}$

20. $\dfrac{2x}{32x^2 - 12x + 1}$

21. $\dfrac{2x + 1}{x^3 - 5x}$

22. $\dfrac{2x + 1}{x^3 + 5x}$

23. $\dfrac{x^3 + 2}{x^4 + 8x^2 + 16}$

24. $\dfrac{x^3 + 2}{x^4 - 8x^2 + 16}$

25. $\dfrac{x^3 + x - 3}{x^4 - 15x^3 + 75x^2 - 125x}$

26. $\dfrac{4x^2}{2x^3 - 5x^2 - 4x + 3}$

27. $\dfrac{1}{x^3 - 1}$

28. $\dfrac{x}{x^3 + 8}$

29. $\dfrac{7x^3 + 11x^2 - x - 2}{x^4 + 2x^3 + x^2}$

30. $\dfrac{4x - 5}{x^4 + 2x^3 + x^2 - 1}$
Hint: Review the factoring in Exercise 29.

31. $\dfrac{x^3 - 5}{x^4 - 81}$

32. $\dfrac{x + 1}{x^4 - 16}$

33. $\dfrac{1}{x^4 + x^3 + 2x^2 + x + 1}$ Hint: To factor the denominator, replace $2x^2$ with $x^2 + x^2$ and group as follows: $(x^4 + x^3 + x^2) + (x^2 + x + 1)$.

34. $\dfrac{x^3 + 4x^2 + 16x}{x^4 + 64}$ Hint: To factor the denominator, add and subtract the term $16x^2$.

35. Use the rational roots theorem and the remainder theorem to determine the roots of the equation $x^3 + 2x^2 - 5x - 6 = 0$. (This is to verify a statement made in Example 2.)

36. This exercise completes two details mentioned in Example 4.
(a) In identity (5) in the text, let $x = 0$ to obtain an equation involving D alone, then solve the equation. You should obtain $D = 15/8$.
(b) In identity (5), we replaced D by $15/8$ and made the substitution $x = 1$. Check that this leads to the result $C = 1/2$.

37. This exercise completes calculations mentioned in Example 5.
(a) Show that identity (6) in the text leads to the following system of equations:

$$\begin{cases} A = 3 \\ B = -1 \\ 3A + C = 7 \\ 3B + D = -3 \end{cases}$$

(b) From the system in part (a), we have $A = 3$ and $B = -1$, which agrees with the values given in Example 5. Now determine C and D and check that your answers agree with the values given in Example 5.

38. This exercise completes the discussion of improper rational expressions in this section.

(a) Use long division to obtain the following result:

$$\frac{2x^3 + 4x^2 - 15x - 36}{x^2 - 9} = (2x + 4) + \frac{3x}{x^2 - 9}$$

(b) Find constants A and B such that $3x/(x^2 - 9) = A/(x - 3) + B/(x + 3)$. (According to the text, you should obtain $A = B = 3/2$.)

In Exercises 39–44, you are given an improper rational expression. First, use long division to rewrite the expression in the form

(polynomial) + (proper rational expression).

Next, obtain the partial fraction decomposition for the proper rational expression. Finally, rewrite the given improper rational expression in the form

(polynomial) + (partial fractions)

39. $\dfrac{6x^3 - 16x^2 - 13x + 25}{x^2 - 4x + 3}$

40. $\dfrac{2x^5 - 11x^4 - 4x^3 + 53x^2 - 24x - 5}{2x^3 + x^2 - 10x - 5}$

41. $\dfrac{x^5 - 10x^4 + 36x^3 - 55x^2 + 32x + 1}{x^4 - 6x^3 + 12x^2 - 8x}$

42. $\dfrac{x^6 + 3x^5 + 9x^3 + 26x^2 + 3x + 8}{x^3 + 8}$

43. $\dfrac{x^6 + 2x^5 + 5x^4 - x^2 - 2x - 4}{x^4 - 1}$

44. $\dfrac{2x^7 + 3x^6 - 2x^4 - 4x^3 + 2}{x^4 + 2x^3 + x^2 - 1}$

Hint: After the long division, you can factor the denominator by writing it as $x^2(x^2 + 2x + 1) - 1^2$ and then using the difference-of-squares technique.

45. This exercise completes a detail mentioned in the text in the derivation of the linear and quadratic factors theorem. Let a and b be real numbers with $b \neq 0$. Show that the quadratic polynomial $x^2 - 2ax + a^2 + b^2$ is irreducible.

B

In Exercises 46–52, determine the partial fraction decomposition for each of the given expressions.

46. $\dfrac{1}{(x - a)(x - b)}$ $(a \neq b)$ **47.** $\dfrac{px + q}{(x - a)(x - b)}$ $(a \neq b)$

48. $\dfrac{1}{(x - a)(x + a)}$ $(a \neq 0)$

49. $\dfrac{px + q}{(x - a)(x + a)}$ $(a \neq 0)$

50. $\dfrac{x^2 + px + q}{(x - a)(x - b)(x - c)}$

(Assume that a, b, and c are all unequal.)

51. $\dfrac{1}{(1 - ax)(1 - bx)(1 - cx)}$ (Assume that a, b, and c all are nonzero and all unequal.)

52. **(a)** $\dfrac{x^2 - 1}{(x^2 + 1)^3}$ **(b)** $\dfrac{x^5 - 1}{(x^2 + 1)^3}$

C

53. Find the partial fraction decomposition: $\dfrac{1}{x^4 + 1}$.

Hint: To factor the denominator, add and subtract a term.

54. **(a)** Determine the general form for the partial fraction decomposition of $1/(x^4 - x^3 + x^2 - x + 1)$. (Note that you are not required to find the numerical values for each constant.) *Hint:* See Exercise 27 in Section 12.7.

(b) Determine the general form for the partial fraction decomposition of $1/(x^5 - 1)$.

CHAPTER TWELVE SUMMARY OF PRINCIPAL TERMS AND NOTATION

TERMS OR NOTATIONS	PAGE REFERENCE	COMMENT
1. $i^2 = -1$	753	This is the defining equation for the symbol i. The quantity i is not a real number.
2. Complex numbers	753	A complex number is an expression of the form $a + bi$, where a and b are real numbers. With the exception of division, complex numbers are combined using the usual rules of algebra, along with the convention $i^2 = -1$. Division is carried out using complex conjugates (defined in Item 3), as indicated in the first box on page 757.
3. If $z = c + di$, then $\bar{z} = c - di$.	754	$\bar{z}$ is called the complex conjugate of z.
4. The division algorithm	761	This is a theorem that summarizes the results of the long division process for polynomials. Suppose that $p(x)$ and $d(x)$ are polynomials and $d(x)$ is not the zero polynomial. Then according to the division algorithm, there are unique polynomials $q(x)$ and $R(x)$ such that $$p(x) = d(x) \cdot q(x) + R(x)$$ where either $R(x)$ is the zero polynomial or the degree of $R(x)$ is less than the degree of $d(x)$.
5. Root, solution, zero	767	A root, or solution, of a polynomial equation $f(x) = 0$ is a number r such that $f(r) = 0$. The root r is also called a zero of the function f.
6. The remainder theorem	768	The remainder theorem asserts that when a polynomial $f(x)$ is divided by $x - r$, the remainder is $f(r)$. Example 3 in Section 12.3 shows how this theorem can be used with synthetic division to evaluate a polynomial.
7. The factor theorem	769	The factor theorem makes two statements about a polynomial $f(x)$. First, if $f(r) = 0$, then $x - r$ is a factor of $f(x)$. And second, if $x - r$ is a factor of $f(x)$, then $f(r) = 0$.
8. The fundamental theorem of algebra	776	Let $f(x)$ be a polynomial of degree 1 or greater. The fundamental theorem of algebra asserts that the equation $f(x) = 0$ has at least one root among the complex numbers. (See Example 1 in Section 12.4.)
9. The linear factors theorem	777	Let $f(x) = a_n x^n + a_{n-1}x^{n-1} + \cdots + a_1 x + a_0$, where $n \geqslant 1$ and $a_n \neq 0$. Then this theorem asserts that $f(x)$ can be expressed as a product of n linear factors: $$f(x) = a_n(x - r_1)(x - r_2) \cdots (x - r_n)$$ (The complex numbers $r_1, r_2, \ldots, r_n$ are not necessarily all distinct, and some or all of them may be real numbers.) On page 778, the linear factors theorem is used to prove that every polynomial equation of degree $n \geqslant 1$ has exactly n roots, where a root of multiplicity k is counted k times.
10. The rational roots theorem	784	Given a polynomial equation, this theorem tells us which rational numbers are candidates for roots of the equation. For a statement of the theorem, see page 785. A proof of the theorem is outlined in Exercise 39, Exercise Set 12.5. For an example of how the theorem is applied, see Example 1 in Section 12.5.

TERMS OR NOTATIONS	PAGE REFERENCE	COMMENT
11. Upper bound for roots; lower bound for roots	786	A real number B is an upper bound for the roots of an equation if every real root is less than or equal to B. Similarly, a real number b is a lower bound for the roots if every real root is greater than or equal to b.
12. The upper and lower bound theorem for real roots	786	This theorem tells how synthetic division can be used in determining upper and lower bounds for roots of equations. For the statement and proof of the theorem, see page 786. For a demonstration of how the theorem is applied, see Example 2 in Section 12.5.
13. The location theorem	788	Let $f(x)$ be a polynomial with real coefficients. If a and b are real numbers such that $f(a)$ and $f(b)$ have opposite signs, then the equation $f(x) = 0$ has at least one root between a and b. To see how this theorem is applied, see Example 3 in Section 12.5.
14. The conjugate roots theorem	792	Let $f(x)$ be a polynomial with real coefficients, and suppose that $a + bi$ is a root of the equation $f(x) = 0$, where a and b are real numbers and $b \neq 0$. Then this theorem asserts that $a - bi$ is also a root of the equation. (In other words, for polynomial equations in which all of the coefficients are real numbers, when complex nonreal roots occur, they occur in conjugate pairs.) For illustrations of how this theorem is applied, see Examples 1 and 2 in Section 12.6.
15. Variation in sign	795	Suppose that $f(x)$ is a polynomial with real coefficients, written in descending or ascending powers of x. Then a variation in sign occurs whenever two successive coefficients have opposite signs. For examples, see Table 1 in Section 12.6.
16. Descartes's rule of signs	795	Let $f(x)$ be a polynomial, all of whose coefficients are real numbers, and consider the equation $f(x) = 0$. Then, according to Descartes's rule: **a.** The number of positive roots either is equal to the number of variations in sign of $f(x)$ or is less than that by an even integer. **b.** The number of negative roots either is equal to the number of variations in sign of $f(-x)$ or is less than that by an even integer. Examples 4–6 in Section 12.6 show how Descartes's rule is applied.
17. Proper rational expression	805	Let $p(x)$ and $q(x)$ be polynomials such that the degree of $p(x)$ is less than the degree of $q(x)$. Then the fraction $p(x)/q(x)$ is called a proper rational expression.
18. Irreducible quadratic polynomial	806	Let $ax^2 + bx + c$ be a quadratic polynomial, all of whose coefficients are real numbers. The polynomial is irreducible provided the equation $ax^2 + bx + c = 0$ has no real-number roots.
19. The linear and quadratic factors theorem	806	Let $f(x)$ be a polynomial, all of whose coefficients are real numbers. Then $f(x)$ can be factored (over the real numbers) into a product of linear and/or irreducible quadratic factors.
20. Partial fraction decomposition	806	A proper rational expression $p(x)/q(x)$ can be decomposed into a sum of simpler partial fractions. The denominators of the partial fractions are built from the linear and/or irreducible quadratic factors of the original denominator $q(x)$. For details, see the boxes on pages 807 and 809.

WRITING MATHEMATICS

I. Discuss each of the following statements with a classmate and decide whether it is true or false. Then (on your own), write out the reason (or reasons) for each decision.

1. Every equation has a root.
2. Every polynomial equation of degree 4 has four distinct roots.
3. No cubic equation can have a root of multiplicity 4.
4. The degree of the polynomial $x(x-1)(x-2)(x-3)$ is 3.
5. The degree of the polynomial $6(x+1)^2(x-5)^4$ is 6.
6. Every polynomial of degree n, where $n \geq 1$, can be written in the form $(x-r_1)(x-r_2)\cdots(x-r_n)$.
7. The sum of the roots of the polynomial equation $x^2 - px + q = 0$ is p.
8. The product of the roots of the polynomial equation $2x^3 - x^2 + 3x - 1 = 0$ is 1.
9. Although a polynomial equation of degree n may have n distinct roots, the fundamental theorem of algebra tells us how to find only one of the roots.
10. Every polynomial equation of degree $n \geq 1$ has at least one real root.
11. If all of the coefficients in a polynomial equation are real, then at least one of the roots must be real.
12. Every cubic equation with roots $\sqrt{5}$, $\sqrt{6}$, and $\sqrt{7}$ can be written in the form
$$a_n(x-\sqrt{5})(x-\sqrt{6})(x-\sqrt{7}) = 0$$
13. Every polynomial equation of degree ≥ 1 has at least one root.
14. According to the rational roots theorem, there are four possibilities for the rational roots of the equation $x^5 + 6x^2 - 2 = 0$.
15. According to the rational roots theorem, there are only two possibilities for the rational roots of the equation $\frac{1}{3}x^3 - 5x + 1 = 0$.
16. The sum of the roots of the equation $x^2 - 12x + 16 = 0$ is -12.
17. According to the location theorem, if $f(x)$ is a polynomial and $f(a)$ and $f(b)$ have the same sign, then the equation $f(x) = 0$ has at least one root between a and b.
18. According to Descartes's rule, the equation $x^7 - 4x + 3 = 0$ has two positive roots.

II. The following identities display two distinct factorizations of the fourth-degree polynomial $x^4 - 3x^2 + 1$:

$$x^4 - 3x^2 + 1 = (x^2 + x - 1)(x^2 - x - 1) \quad (1)$$

$$x^4 - 3x^2 + 1 = (x^2 + \sqrt{5}x + 1)(x^2 - \sqrt{5}x + 1) \quad (2)$$

(a) First, convince yourself that both identity (1) and identity (2) are valid. (You can do this by carrying out the multiplication on the right-hand side of each identity. Alternatively, you might find a clever technique for factoring the left-hand side of each identity.)

(b) Working with a classmate, explain why the two distinct factorizations shown in part (a) do *not* lead to two essentially different partial fraction decompositions for the rational expression $1/(x^4 - 3x^2 + 1)$. (To get to the bottom of this, you may actually need to work out the partial fractions in each case.)

(c) On your own, write a report on what you've done in parts (a) and (b). Be sure to give particular attention to any points at which you got stuck and how you worked through those points.

CHAPTER TWELVE REVIEW EXERCISES

In Exercises 1 and 2, you are given polynomials $p(x)$ and $d(x)$. In each case, use synthetic division to determine polynomials $q(x)$ and $R(x)$ such that

$$p(x) = d(x) \cdot q(x) + R(x)$$

where either $R(x) = 0$ or the degree of $R(x)$ is less than the degree of $d(x)$.

1. $p(x) = x^4 + 3x^3 - x^2 - 5x + 1$; $d(x) = x + 2$

2. $p(x) = 4x^4 + 2x + 1$; $d(x) = x - 2$

In Exercises 3–8, use synthetic division to find the quotients and the remainders.

3. $\dfrac{x^4 - 2x^2 + 8}{x - 3}$

4. $\dfrac{x^3 - 1}{x - 2}$

5. $\dfrac{2x^3 - 5x^2 - 6x - 3}{x + 4}$

6. $\dfrac{x^2 + x - 3\sqrt{2}}{x - \sqrt{2}}$

7. $\dfrac{5x^2 - 19x - 4}{x + 0.2}$

8. $\dfrac{x^3 - 3a^2x^2 - 4a^4x + 9a^6}{x - a^2}$

In Exercises 9–16, use synthetic division and the remainder theorem to find the indicated values of the functions. Use a calculator for Exercises 15 and 16.

9. $f(x) = x^5 - 10x + 4$; $f(10)$

10. $f(x) = x^4 + 2x^3 - x$; $f(-2)$

11. $f(x) = x^3 - 10x^2 + x - 1$; $f(1/10)$

12. $f(x) = x^4 - 2a^2x^2 + 3a^3x - a^4$; $f(-a)$

13. $f(x) = x^3 + 3x^2 + 3x + 1$; $f(a - 1)$

14. $f(x) = x^3 - 1$; $f(1.1)$

15. $f(x) = x^4 + 4x^3 - 6x^2 - 8x - 2$;
(a) $f(-0.3)$ (Round the result to 2 decimal places.)
(b) $f(-0.39)$ (Round the result to 3 decimal places.)
(c) $f(-0.394)$ (Round the result to 5 decimal places.)

16. $f(-4.907378)$, where f is the function in Exercise 15. (Round the result to three decimal places.)

17. Find a value for a such that 3 is a root of the equation $x^3 - 4x^2 - ax - 6 = 0$.

18. For which values of b will -1 be a root of the equation $x^3 + 2b^2x^2 + x - 48 = 0$?

19. For which values of a will $x - 1$ be a factor of the polynomial $a^2x^3 + 3ax^2 + 2$?

20. Use synthetic division to verify that $\sqrt{2} - 1$ is a root of the equation $x^6 + 14x^3 - 1 = 0$.

21. Let $f(x) = ax^3 + bx^2 + cx + d$ and suppose that r is a root of the equation $f(x) = 0$.
(a) Show that $r - h$ is a root of $f(x + h) = 0$.
(b) Show that $-r$ is a root of the equation $f(-x) = 0$.
(c) Show that kr is a root of the equation $f(x/k) = 0$.

22. Suppose that r is a root of the equation $a_2x^2 + a_1x + a_0 = 0$. Show that mr is a root of the quadratic equation $a_2x^2 + ma_1x + m^2a_0 = 0$.

In Exercises 23–28, list the possibilities for the rational roots of the equations.

23. $x^5 - 12x^3 + x - 18 = 0$

24. $x^5 - 12x^3 + x - 17 = 0$

25. $2x^4 - 125x^3 + 3x^2 - 8 = 0$

26. $\frac{3}{5}x^3 - 8x^2 - \frac{1}{2}x + \frac{3}{2} = 0$

27. $x^3 + x - p = 0$, where p is a prime number

28. $x^3 + x - pq = 0$, where both p and q are prime numbers

In Exercises 29–36, each equation has at least one rational root. Solve the equations. Suggestion: *Use the upper and lower bound theorem to eliminate some of the possibilities for rational roots.*

29. $2x^3 + x^2 - 7x - 6 = 0$

30. $x^3 + 6x^2 - 8x - 7 = 0$

31. $2x^3 - x^2 - 14x + 10 = 0$

32. $2x^3 + 12x^2 + 13x + 15 = 0$

33. $\frac{3}{2}x^3 + \frac{1}{2}x^2 + \frac{1}{2}x - 1 = 0$

34. $x^4 - 2x^3 - 13x^2 + 38x - 24 = 0$

35. $x^5 + x^4 - 14x^3 - 14x^2 + 49x + 49 = 0$

36. $8x^5 + 12x^4 + 14x^3 + 13x^2 + 6x + 1 = 0$

37. Solve the equation $x^3 - 9x^2 + 24x - 20 = 0$, using the fact that one of the roots has multiplicity 2.

38. One root of the equation $x^2 + kx + 2k = 0 \quad (k \neq 0)$ is twice the other. Find k and find the roots of the equation.

39. State each of the following theorems.
(a) The division algorithm
(b) The remainder theorem
(c) The factor theorem
(d) The fundamental theorem of algebra

40. Find a quadratic equation with roots $a - \sqrt{a^2 - 1}$ and $a + \sqrt{a^2 - 1}$, where $a > 1$.

In Exercises 41–44, write each polynomial in the form

$$a_n(x - r_1)(x - r_2) \cdots (x - r_n)$$

41. $6x^2 + 7x - 20$ **42.** $x^2 + x - 1$

43. $x^4 - 4x^3 + 5x - 20$ **44.** $x^4 - 4x^2 - 5$

Each of Exercises 45–48 gives an equation, followed by one or more roots. Solve the equation.

45. $x^3 - 7x^2 + 25x - 39 = 0$; $x = 2 - 3i$

46. $x^3 + 6x^2 - 24x + 160 = 0$; $x = 2 + 2i\sqrt{3}$

47. $x^4 - 2x^3 - 4x^2 + 14x - 21 = 0$; $x = 1 + i\sqrt{2}$

48. $x^5 + x^4 - x^3 + x^2 + x - 1 = 0$; $x = (1 + i\sqrt{3})/2$, $x = (-1 - \sqrt{5})/2$

In Exercises 49–54, use Descartes's rule of signs to obtain information regarding the roots of the equations.

49. $x^3 + 8x - 7 = 0$ **50.** $3x^4 + x^2 + 4x - 2 = 0$

51. $x^3 + 3x + 1 = 0$ **52.** $2x^6 + 3x^2 + 6 = 0$

53. $x^4 - 10 = 0$ **54.** $x^4 + 5x^2 - x + 2 = 0$

55. Consider the equation $x^3 + x^2 + x + 1 = 0$.
(a) Use Descartes's rule to show that the equation has either one or three negative roots.
(b) Now show that the equation cannot have three negative roots. *Hint:* Multiply both sides of the equation by $x - 1$. Then simplify the left-hand side and reapply Descartes's rule to the new equation.
(c) Actually, the original equation can be solved using the basic factoring techniques discussed in Section 1.3. Solve the equation in this manner.

56. Use Descartes's rule to show that the equation $x^3 - x^2 + 3x + 2 = 0$ has no positive roots. *Hint:* Multiply both sides of the equation by $x + 1$ and apply Descartes's rule to the resulting equation.

57. Let P be the point in the first quadrant where the curve $y = x^3$ intersects the circle $x^2 + y^2 = 1$. Locate the x-coordinate of P within successive hundredths.

58. Let P be the point in the first quadrant where the parabola $y = 4 - x^2$ intersects the curve $y = x^3$. Locate the x-coordinate of P within successive hundredths.

59. Consider the equation $x^3 - 36x - 84 = 0$.
(a) Use Descartes's rule to check that this equation has exactly one positive root.
(b) Use the upper and lower bound theorem to show that 7 is an upper bound for the positive root.
(c) Using a calculator, locate the positive root within successive hundredths.

60. Consider the equation $x^3 - 3x + 1 = 0$.
(a) Use Descartes's rule to check that this equation has exactly one negative root.
(b) Use the upper and lower bound theorem to show that -2 is a lower bound for the negative root.
(c) Using a calculator, locate the negative root within successive hundredths.

In Exercises 61–64, find polynomial equations that have integer coefficients and the given values as roots.

61. $4 - \sqrt{5}$

62. $a + b$ and $a - b$, where a and b are integers

63. $6 - 2i$ and $\sqrt{5}$

64. $\dfrac{5 + \sqrt{6}}{5 - \sqrt{6}}$ *Hint:* First rationalize the expression.

65. Find a fourth-degree polynomial equation with integer coefficients, such that $x = 1 + \sqrt{2} + \sqrt{3}$ is a root. *Hint:* Begin by writing the given relationship as $x - 1 = \sqrt{2} + \sqrt{3}$; then square both sides.

66. Find a cubic equation with integer coefficients, such that $x = 1 + \sqrt[3]{2}$ is a root. Is $1 - \sqrt[3]{2}$ also a root of the equation?

In Exercises 67–70, first determine the zeros of each function; then sketch the graph.

67. $y = x^3 - 2x^2 - 3x$ **68.** $y = x^4 + 3x^3 + 3x^2 + x$

69. $y = x^4 - 4x^2$ **70.** $y = x^3 + 6x^2 + 5x - 12$

For Exercises 71–78, carry out the indicated operations and express your answer in the standard form $a + bi$.

71. $(3 - 2i)(3 + 2i) + (1 + 3i)^2$

72. $2i(1 + i)^2$

73. $(1 + i\sqrt{2})(1 - i\sqrt{2}) + (\sqrt{2} + i)(\sqrt{2} - i)$

74. $(2 + 3i)/(1 + i)$ **75.** $(3 - i\sqrt{3})/(3 + i\sqrt{3})$

76. $\dfrac{1 + i}{1 - i} + \dfrac{1 - i}{1 + i}$

77. $-\sqrt{-2}\sqrt{-9} + \sqrt{-8} - \sqrt{-72}$

78. $\dfrac{\sqrt{-4} - \sqrt{-3}\sqrt{-3}}{\sqrt{-100}}$

79. The **real part** of a complex number z is denoted by $\operatorname{Re}(z)$. For instance, $\operatorname{Re}(2 + 5i) = 2$. Show that for any complex number z, we have $\operatorname{Re}(z) = \frac{1}{2}(z + \bar{z})$.
Hint: Let $z = a + bi$.

80. The **imaginary part** of a complex number z is denoted by $\operatorname{Im}(z)$. For instance, $\operatorname{Im}(2 + 5i) = 5$. Show that for any complex number z, we have $\operatorname{Im}(z) = \dfrac{1}{2i}(z - \bar{z})$.

81. The **absolute value** of the complex number $a + bi$ is defined by $|a + bi| = \sqrt{a^2 + b^2}$.
(a) Compute $|6 + 2i|$ and $|6 - 2i|$.
(b) As you know, the absolute value of the real number -3 is 3. Now write -3 in the form $-3 + 0i$ and compute its absolute value using the new definition. (The point here is to observe that the two results agree.)
(c) Let $z = a + bi$. Show that $z\bar{z} = |z|^2$.

In Exercises 82–86, verify that the formulas are correct by carrying out the operations indicated on the left-hand side of the equations. (This list of formulas appears in A Treatise on Algebra, *by George Peacock, published in 1845.)*

82. $\dfrac{1}{a + bi} + \dfrac{1}{a - bi} = \dfrac{2a}{a^2 + b^2}$

83. $\dfrac{1}{a - bi} - \dfrac{1}{a + bi} = \dfrac{2bi}{a^2 + b^2}$

84. $\dfrac{a + bi}{a - bi} + \dfrac{a - bi}{a + bi} = \dfrac{2(a^2 - b^2)}{a^2 + b^2}$

85. $\dfrac{a + bi}{a - bi} - \dfrac{a - bi}{a + bi} = \dfrac{4abi}{a^2 + b^2}$

86. $\dfrac{a + bi}{c + di} + \dfrac{a - bi}{c - di} = \dfrac{2(ac + bd)}{c^2 + d^2}$

In Exercises 87–92, determine the partial fraction decomposition for each expression.

87. $\dfrac{2x - 1}{100 - x^2}$

88. $\dfrac{x}{x^2 - 20}$

89. $\dfrac{1}{x^3 + 2x^2 + x}$

90. $\dfrac{1}{x^4 + x^2}$

91. $\dfrac{x^3 + 2}{x^4 + 6x^2 + 9}$

92. $\dfrac{4x^3 - 3x^2 + 24x - 15}{x^4 + 11x^2 + 30}$

93. **(a)** Let $f(x) = x^4 - 2x^3 + x^2 - 1$. Use the equating-the-coefficients theorem to find a factorization for $f(x)$ of the form $f(x) = (x^2 + bx + 1)(x^2 + cx - 1)$.
(b) Find the partial fraction decomposition for $x^3/(x^4 - 2x^3 + x^2 - 1)$.

94. Let $f(x) = x^4 - 7x^2 + 1$.
(a) Use the equating-the-coefficients theorem to find a factorization for $f(x)$ of the form $f(x) = (x^2 + bx - 1)(x^2 + cx - 1)$.
(b) Are the quadratic factors that you found in part (a) irreducible?
(c) Use the result in part (a) to obtain a partial fraction decomposition for $1/(x^4 - 7x^2 + 1)$.
(d) Use the equating-the-coefficients theorem to find a factorization for $f(x)$ of the form $f(x) = (x^2 + Bx + 1)(x^2 + Cx + 1)$.
(e) Are the quadratic factors that you found in part (d) irreducible?
(f) Use the result in part (d) to obtain a partial fraction decomposition for $1/(x^4 - 7x^2 + 1)$. Do you obtain the same decomposition that you found in part (c), or is this a new decomposition?

CHAPTER TWELVE TEST

1. Let $f(x) = 6x^4 - 5x^3 + 7x^2 - 2x - 2$. Make use of the remainder theorem and synthetic division to compute $f(1/2)$.
2. Solve the equation $x^3 + x^2 - 11x - 15 = 0$, given that one of the roots is -3.
3. List the possibilities for the rational roots of the equation $2x^5 - 4x^3 + x - 6 = 0$.
4. Find a quadratic function with zeros 1 and -8 and with a y-intercept of -24.
5. Use synthetic division to divide $4x^3 + x^2 - 8x + 3$ by $x + 1$.
6. **(a)** State the factor theorem.
   (b) State the fundamental theorem of algebra.
   (c) State the linear and quadratic factors theorem.
7. **(a)** The equation $x^3 - 2x^2 - 1 = 0$ has just one positive root. Use the upper and lower bound theorem to determine the smallest integer that is an upper bound for the root.
   (b) Locate the root between successive tenths. (Use a calculator.)
8. Solve the equation $x^5 - 6x^4 + 11x^3 + 16x^2 - 50x + 52 = 0$, given that two of the roots are $1 + i$ and $3 - 2i$.
9. Let $p(x) = x^4 + 2x^3 - x + 6$ and $d(x) = x^2 + 1$. Find polynomials $q(x)$ and $R(x)$ such that $p(x) = d(x) \cdot q(x) + R(x)$.
10. Express the polynomial $2x^2 - 6x + 5$ in the factored form $a_n(x - r_1)(x - r_2)$.
11. Consider the equation $x^4 - x^3 + 24 = 0$.
    (a) List the possibilities for rational roots.
    (b) Use the upper and lower bound theorem to show that 2 is an upper bound for the roots.
    (c) In view of parts (a) and (b), what possibilities now remain for positive rational roots?
    (d) Which (if any) of the possibilities in part (c) are actually roots?
12. **(a)** Find the rational roots of the cubic equation $2x^3 - x^2 - x - 3 = 0$.
    (b) Find all solutions of the equation in part (a).
13. Use Descartes's rule of signs to obtain information regarding the roots of the following equation: $3x^4 + x^2 - 5x - 1 = 0$.
14. Find a cubic polynomial $f(x)$ with integer coefficients, such that $1 - 3i$ and -2 are roots of the equation $f(x) = 0$.
15. Find a polynomial $f(x)$ with leading coefficient 1, such that the equation $f(x) = 0$ has the following roots and no other:

Root	Multiplicity
2	1
$3i$	3
$1 + \sqrt{2}$	2

Write your answer in the form $a_n(x - r_1)(x - r_2) \cdots (x - r_n)$.

16. Simplify the following expression and write the answer in the form $a + bi$:

$$(4 + 2i)(4 - 2i) + \frac{3 + i}{1 + 2i}$$

17. Consider the quadratic equation $2x^2 - 4x + 3 = 0$.

(a) Compute the discriminant. What does this say about the roots of the equation?

(b) Find the roots of the equation.

In Exercises 18–20, determine the partial fraction decomposition for each expression.

18. $\dfrac{3x - 1}{x^3 - 16x}$

19. $\dfrac{1}{x^3 - x^2 + 3x - 3}$

20. $\dfrac{4x^2 - 15x + 20}{x^3 - 4x^2 + 4x}$

ADDITIONAL TOPICS IN ALGEBRA

INTRODUCTION

... in the ordinary treatise on the elements of algebra, these [additional] *topics are either completely omitted or treated carelessly. For this reason, I am certain that the material I have gathered in this book is quite sufficient to remedy that defect.*

Leonhard Euler (1707–1783) in his classic text, *Introductio in analysis infinitorum* (1748)

A strong background in algebra is an important prerequisite for courses in calculus and in probability and statistics. In this final chapter we develop several additional topics that help provide a foundation in those areas of study. We begin in Section 13.1 with the principle of mathematical induction. This gives us a framework for proving statements about the natural numbers. In Section 13.2 we discuss the binomial theorem, which is used to analyze and expand expressions of the form $(a + b)^n$. As you'll see, the proof of the binomial theorem uses mathematical induction. Section 13.3 introduces the related (but distinct) concepts of sequences and series. Then in the next two sections (Sections 13.4 and 13.5), we study arithmetic and geometric sequences and series. The last topic introduced in Section 13.5 concerns finding sums of infinite geometric series. In a sense, this is an appropriate topic for our last chapter, for it is closely related to the idea of a *limit,* which is the starting point for calculus. Finally, in the last section of this chapter (Section 13.6) we introduce DeMoivre's theorem.

13.1 MATHEMATICAL INDUCTION

Is mathematics an experimental science? The answer to this question is both yes and no, as the following example illustrates. Consider the problem of determining a formula for the sum of the first n odd natural numbers:

$$1 + 3 + 5 + \cdots + (2n - 1)$$

We begin by doing some calculations in the hope that this may shed some light on the problem. Table 1 shows the results of calculating the sum of the first n odd natural numbers for values of n ranging from 1 to 5. Upon inspecting the table, we observe that each sum in the right-hand column is the square of the corresponding entry in the left-hand column. For instance, for $n = 5$, we see that

$$\underbrace{1 + 3 + 5 + 7 + 9}_{\text{five terms}} = 5^2$$

Now let's try the next case, where $n = 6$, and see if the pattern persists. That is, we want to know if it is true that

$$\underbrace{1 + 3 + 5 + 7 + 9 + 11}_{\text{six terms}} = 6^2$$

TABLE 1

n	$1 + 3 + 5 + \cdots + (2n - 1)$
1	$1 = 1$
2	$1 + 3 = 4$
3	$1 + 3 + 5 = 9$
4	$1 + 3 + 5 + 7 = 16$
5	$1 + 3 + 5 + 7 + 9 = 25$

TABLE 2

n	2^n	$(n + 1)^2$
1	2	4
2	4	9
3	8	16
4	16	25
5	32	36

As you can easily check, this last equation is true. Thus, based on the experimental (or empirical) evidence, we are led to the following conjecture:

CONJECTURE The sum of the first n odd natural numbers is n^2. That is, $1 + 3 + 5 + \cdots + (2n - 1) = n^2$, for each natural number n.

At this point, the ''law'' we've discovered is indeed really only a conjecture. After all, we've checked it only for values of n ranging from 1 to 6. It is conceivable at this point (although we may feel it is unlikely) that the conjecture is false for certain values of n. For the conjecture to be useful, we must be able to prove that it holds without exception for *all* natural numbers n. In fact, we will subsequently prove that this conjecture is valid. But before explaining the method of proof to be used, let's look at one more example.

Again, let n denote a natural number. Then consider the following question: Which quantity is the larger, 2^n or $(n + 1)^2$? As before, we begin by doing some calculations. This is the experimental stage of our work. According to Table 2, the quantity $(n + 1)^2$ is larger than 2^n for each value of n up through $n = 5$. Thus, we make the following conjecture:

CONJECTURE $(n + 1)^2 > 2^n$ for all natural numbers n.

Again, we note that this is only a conjecture at this point. Indeed, if we try the case where $n = 6$, we find that the pattern does not persist. That is, when $n = 6$, we find that 2^n is 64, while $(n + 1)^2$ is only 49. So in this example, the conjecture is not true in general; we have found a value of n for which it fails.

The preceding examples show that experimentation does have a place in mathematics, but we must be careful with the results. When experimentation leads to a conjecture, proof is required before the conjecture can be viewed as a valid law. For the remainder of this section, we shall discuss one such method of proof—*mathematical induction.*

In order to state the principle of mathematical induction, we first introduce some notation. Suppose that for each natural number n we have a statement P_n to be proved. Consider, for instance, the first conjecture we arrived at:

$$1 + 3 + 5 + \cdots + (2n - 1) = n^2$$

If we denote this statement by P_n, then we have, for example, that

P_1 is the statement that $1 = 1^2$
P_2 is the statement that $1 + 3 = 2^2$
P_3 is the statement that $1 + 3 + 5 = 3^2$

With this notation, we can now state the **principle of mathematical induction.**

THE PRINCIPLE OF MATHEMATICAL INDUCTION

Suppose that for each natural number n, we have a statement P_n for which the following two conditions hold:

1. P_1 is true.
2. For each natural number k, if P_k is true, then P_{k+1} is true.

Then all of the statements are true; that is, P_n is true for all natural numbers n.

The idea behind mathematical induction is a simple one. Think of each statement P_n as the rung of a ladder to be climbed. Then we make the analogy shown in Table 3.

TABLE 3

Mathematical Induction	Ladder Analogy
Hypotheses { **1.** P_1 is true. **2.** If P_k is true, then P_{k+1} is true, for any k.	Hypotheses { **1′.** You can reach the first rung. **2′.** If you are on the kth rung, you can reach the $(k+1)$st rung, for any k.
Conclusion { **3.** P_n is true for all n.	Conclusion { **3′.** You can climb the entire ladder.

According to the principle of mathematical induction, we can prove that a statement or formula P_n is true for all n if we carry out the following two steps:

STEP 1 Show that P_1 is true.
STEP 2 Assume that P_k is true, and on the basis of this assumption, show that P_{k+1} is true.

In Step 2, the assumption that P_k is true is referred to as the **induction hypothesis.** (In computer science, Step 1 is sometimes referred to as the **initialization step.**) Now let's turn to some examples of proof by mathematical induction.

EXAMPLE 1 Use mathematical induction to prove that

$$1 + 3 + 5 + \cdots + (2n - 1) = n^2$$

for all natural numbers n.

Solution Let P_n denote the statement that $1 + 3 + 5 + \cdots + (2n - 1) = n^2$. Then we want to show that P_n is true for all natural numbers n.

STEP 1 We must check that P_1 is true. But P_1 is just the statement that $1 = 1^2$, which is true.

STEP 2 Assuming that P_k is true, we must show that P_{k+1} is true. Thus, we assume that

$$1 + 3 + 5 + \cdots + (2k - 1) = k^2 \qquad (1)$$

That is the induction hypothesis. We must now show that

$$1 + 3 + 5 + \cdots + (2k - 1) + [2(k + 1) - 1] = (k + 1)^2 \qquad (2)$$

In order to derive equation (2) from equation (1), we add the quantity $[2(k + 1) - 1]$ to both sides of equation (1). (The motivation for this stems from the observation that the left-hand sides of equations (1) and (2) differ only by the quantity $[2(k + 1) - 1]$.) We obtain

$$\begin{aligned} 1 + 3 + 5 + \cdots + (2k - 1) + [2(k + 1) - 1] &= k^2 + [2(k + 1) - 1] \\ &= k^2 + 2k + 1 \\ &= (k + 1)^2 \end{aligned}$$

This last equation is what we wanted to show. And having now carried out Steps 1 and 2, we conclude by the principle of mathematical induction that P_n is true for all natural numbers n.

EXAMPLE 2 Use mathematical induction to prove that

$$2^3 + 4^3 + 6^3 + \cdots + (2n)^3 = 2n^2(n+1)^2$$

for all natural numbers n.

Solution Let P_n denote the statement that

$$2^3 + 4^3 + 6^3 + \cdots + (2n)^3 = 2n^2(n+1)^2$$

Then we want to show that P_n is true for all natural numbers n.

STEP 1 We must check that P_1 is true, where P_1 is the statement that

$$2^3 = 2(1^2)(1+1)^2 \qquad \text{or} \qquad 8 = 8$$

Thus, P_1 is true.

STEP 2 Assuming that P_k is true, we must show that P_{k+1} is true. Thus, we assume that

$$2^3 + 4^3 + 6^3 + \cdots + (2k)^3 = 2k^2(k+1)^2 \tag{3}$$

We must now show that

$$2^3 + 4^3 + 6^3 + \cdots + (2k)^3 + [2(k+1)]^3 = 2(k+1)^2(k+2)^2 \tag{4}$$

Adding $2(k+1)]^3$ to both sides of equation (3) yields

$$\begin{aligned} 2^3 + 4^3 + 6^3 + \cdots + (2k)^3 + [2(k+1)]^3 &= 2k^2(k+1)^2 + [2(k+1)]^3 \\ &= 2k^2(k+1)^2 + 8(k+1)^3 \\ &= 2(k+1)^2[k^2 + 4(k+1)] \\ &= 2(k+1)^2(k^2 + 4k + 4) \\ &= 2(k+1)^2(k+2)^2 \end{aligned}$$

We have now derived equation (4) from equation (3), as we wished to do. Having carried out Steps 1 and 2, we conclude by the principle of mathematical induction that P_n is true for all natural numbers n.

TABLE 4

n	$2^{2n}-1$
1	$3\ (=3 \cdot 1)$
2	$15\ (=3 \cdot 5)$
3	$63\ (=3 \cdot 21)$
4	$255\ (=3 \cdot 85)$

EXAMPLE 3 As indicated in Table 4, the number 3 is a factor of $2^{2n} - 1$ when $n = 1$, 2, 3, and 4. Use mathematical induction to show that 3 is a factor of $2^{2n} - 1$ for all natural numbers n.

Solution Let P_n denote the statement that 3 is a factor of $2^{2n} - 1$. We want to show that P_n is true for all natural numbers n.

STEP 1 We must check that P_1 is true. But P_1 in this case is just the statement that 3 is a factor of $2^{2(1)} - 1$; that is, 3 is a factor of 3, which is surely true.

STEP 2 Assuming that P_k is true, we must show that P_{k+1} is true. Thus, we assume that

$$3 \text{ is a factor of } 2^{2k} - 1 \tag{5}$$

and we must show that

$$3 \text{ is a factor of } 2^{2(k+1)} - 1$$

The strategy here will be to rewrite the expression $2^{2(k+1)} - 1$ in such a way that the induction hypothesis, statement (5), can be applied. We have

$$\begin{aligned} 2^{2(k+1)} - 1 &= 2^{2k+2} - 1 \\ &= 2^2 \cdot 2^{2k} - 1 \\ &= 4 \cdot 2^{2k} - 4 + 3 \\ 2^{2(k+1)} - 1 &= 4(2^{2k} - 1) + 3 \end{aligned} \tag{6}$$

Now, look at the right-hand side of equation (6). By the induction hypothesis, 3 is a factor of $2^{2k} - 1$. Thus, 3 is a factor of $4(2^{2k} - 1)$, from which it certainly follows that 3 is a factor of $4(2^{2k} - 1) + 3$. In summary, then, 3 is a factor of the right-hand side of equation (6). Consequently, 3 must be a factor of the left-hand side of equation (6), which is what we wished to show.

Having now completed Steps 1 and 2, we conclude by the principle of mathematical induction that P_n is true for all natural numbers n. In other words, 3 is a factor of $2^{2n} - 1$ for all natural numbers n.

There are instances in which a given statement P_n is false for certain initial values of n but true thereafter. An example of this is provided by the statement

$$2^n > (n+1)^2$$

As you can easily check, this statement is false for $n = 1$, 2, 3, 4, and 5. But, as Example 4 shows, the statement is true for $n \geqslant 6$. In Example 4, we adapt the principle of mathematical induction by beginning in Step 1 with a consideration of P_6 rather than P_1.

EXAMPLE 4 Use mathematical induction to prove that

$$2^n > (n+1)^2 \qquad \text{for all natural numbers } n \geqslant 6$$

Solution

STEP 1 We must first check that P_6 is true. But P_6 is simply the assertion that

$$2^6 > (6+1)^2 \qquad \text{or} \qquad 64 > 49$$

Thus, P_6 is true.

STEP 2 Assuming that P_k is true, where $k \geqslant 6$, we must show that P_{k+1} is true. Thus, we assume that

$$2^k > (k+1)^2 \qquad \text{where } k \geqslant 6 \tag{7}$$

We must show that

$$2^{k+1} > (k+2)^2$$

Multiplying both sides of inequality (7) by 2 gives us

$$2(2^k) > 2(k+1)^2 = 2k^2 + 4k + 2$$

This can be rewritten

$$2^{k+1} > k^2 + 4k + (k^2 + 2)$$

However, since $k \geqslant 6$, it is certainly true that

$$k^2 + 2 > 4$$

We therefore have

$$2^{k+1} > k^2 + 4k + 4 \quad \text{or} \quad 2^{k+1} > (k+2)^2$$

as we wished to show.

Having now completed Steps 1 and 2, we conclude that P_n is true for all natural numbers $n \geq 6$.

EXERCISE SET 13.1

A

In Exercises 1–18, use the principle of mathematical induction to show that the statements are true for all natural numbers.

1. $1 + 2 + 3 + \cdots + n = n(n+1)/2$

2. $2 + 4 + 6 + \cdots + 2n = n(n+1)$

3. $1 + 4 + 7 + \cdots + (3n-2) = n\,(3n-1)/2$

4. $5 + 9 + 13 + \cdots + (4n+1) = n(2n+3)$

5. $1^2 + 2^2 + 3^2 + \cdots + n^2 = n(n+1)(2n+1)/6$

6. $2^2 + 4^2 + 6^2 + \cdots + (2n)^2 = 2n(n+1)(2n+1)/3$

7. $1^2 + 3^2 + 5^2 + \cdots + (2n-1)^2 = n(2n-1)(2n+1)/3$

8. $2 + 2^2 + 2^3 + \cdots + 2^n = 2^{n+1} - 2$

9. $3 + 3^2 + 3^3 + \cdots + 3^n = (3^{n+1} - 3)/2$

10. $e^x + e^{2x} + e^{3x} + \cdots + e^{nx} = \dfrac{e^{(n+1)x} - e^x}{e^x - 1} \quad (x \neq 0)$

11. $1^3 + 2^3 + 3^3 + \cdots + n^3 = [n(n+1)/2]^2$

12. $2^3 + 4^3 + 6^3 + \cdots + (2n)^3 = 2n^2(n+1)^2$

13. $1^3 + 3^3 + 5^3 + \cdots + (2n-1)^3 = n^2(2n^2 - 1)$

14. $1 \cdot 2 + 3 \cdot 4 + 5 \cdot 6 + \cdots + (2n-1)(2n)$
$= n(n+1)(4n-1)/3$

15. $1 \cdot 3 + 3 \cdot 5 + 5 \cdot 7 + \cdots + (2n-1)(2n+1)$
$= n(4n^2 + 6n - 1)/3$

16. $\dfrac{1}{1 \times 3} + \dfrac{1}{2 \times 4} + \dfrac{1}{3 \times 5} + \cdots + \dfrac{1}{n(n+2)}$
$= \dfrac{n(3n+5)}{4(n+1)(n+2)}$

17. $1 + \dfrac{3}{2} + \dfrac{5}{2^2} + \dfrac{7}{2^3} + \cdots + \dfrac{2n-1}{2^{n-1}} = 6 - \dfrac{2n+3}{2^{n-1}}$

18. $1 + 2 \cdot 2 + 3 \cdot 2^2 + 4 \cdot 2^3 + \cdots + n \cdot 2^{n-1}$
$= (n-1)2^n + 1$

19. Show that $n \leq 2^{n-1}$ for all natural numbers n.

20. Show that 3 is a factor of $n^3 + 2n$ for all natural numbers n.

21. Show that $n^2 + 4 < (n+1)^2$ for all natural numbers $n \geq 2$.

22. Show that $n^3 > (n+1)^2$ for all natural numbers $n \geq 3$.

In Exercises 23–26, prove that the statement is true for all natural numbers in the specified range. Use a calculator to carry out Step 1.

23. $(1.5)^n > 2n;\ n \geq 7$

24. $(1.25)^n > n;\ n \geq 11$

25. $(1.1)^n > n;\ n \geq 39$

26. $(1.1)^n > 5n;\ n \geq 60$

B

27. Let $f(n) = \dfrac{1}{1 \cdot 2} + \dfrac{1}{2 \cdot 3} + \dfrac{1}{3 \cdot 4} + \cdots + \dfrac{1}{n(n+1)}$.

(a) Complete the following table.

n	1	2	3	4	5
$f(n)$					

(b) On the basis of the results in the table, what would you guess to be the value of $f(6)$? Compute $f(6)$ to see if this is correct.

(c) Make a conjecture about the value of $f(n)$, and prove it using mathematical induction.

28. Let $f(n) = \dfrac{1}{1 \times 3} + \dfrac{1}{3 \times 5} + \dfrac{1}{5 \times 7} + \cdots + \dfrac{1}{(2n-1)(2n+1)}$.

(a) Complete the following table.

n	1	2	3	4
$f(n)$				

(b) On the basis of the results in the table, what would you guess to be the value of $f(5)$? Compute $f(5)$ to see if your guess is correct.

(c) Make a conjecture about the value of $f(n)$, and prove it using mathematical induction.

29. Suppose that a function f satisfies the following conditions:

$$f(1) = 1$$
$$f(n) = f(n-1) + 2\sqrt{f(n-1)} + 1 \quad (n \geq 2)$$

(a) Complete the table.

n	1	2	3	4	5
$f(n)$					

(b) On the basis of the results in the table, what would you guess to be the value of $f(6)$? Compute $f(6)$ to see if your guess is correct.

(c) Make a conjecture about the value of $f(n)$ when n is a natural number, and prove the conjecture using mathematical induction.

30. This exercise demonstrates the necessity of carrying out both Step 1 and Step 2 before considering an induction proof valid.

(a) Let P_n denote the statement that $n^2 + 1$ is even. Check that P_1 is true. Then give an example showing that P_n is not true for all n.

(b) Let Q_n denote the statement that $n^2 + n$ is odd. Show that Step 2 of an induction proof can be completed in this case, but not Step 1.

31. A *prime number* is a natural number that has no factors other than itself and 1. For technical reasons, 1 is not considered a prime. Thus, the list of the first seven primes looks like this: 2, 3, 4, 7, 11, 13, 17. Let P_n be the statement that $n^2 + n + 11$ is prime. Check that P_n is true for all values of n less than 10. Check that P_{10} is false.

32. Prove that if $x \neq 1$,

$$1 + 2x + 3x^2 + \cdots + nx^{n-1} = \frac{1 - x^n}{(1-x)^2} - \frac{nx^n}{1-x}$$

for all natural numbers n.

33. If $r \neq 1$, show that $1 + r + r^2 + \cdots + r^{n-1} = \dfrac{(r^n - 1)}{(r - 1)}$ for all natural numbers n.

34. Use mathematical induction to show that

$$x^n - 1 = (x-1)(1 + x + x^2 + \cdots + x^{n-1})$$

for all natural numbers n.

35. Prove that 5 is a factor of $n^5 - n$ for all natural numbers $n \geq 2$.

36. Prove that 4 is a factor of $5^n + 3$ for all natural numbers n.

37. Prove that 5 is a factor of $2^{2n+1} + 3^{2n+1}$ for all nonnegative integers n.

38. Prove that 8 is a factor of $3^{2n} - 1$ for all natural numbers n.

39. Prove that 3 is a factor of $2^{n+1} + (-1)^n$ for all nonnegative integers n.

40. Prove that 6 is a factor of $n^3 + 3n^2 + 2n$ for all natural numbers n.

41. Use mathematical induction to show that $x - y$ is a factor of $x^n - y^n$ for all natural numbers n.
Suggestion for Step 2: Verify and then use the fact that $x^{k+1} - y^{k+1} = x^k(x - y) + (x^k - y^k)y$.

In Exercises 42 and 43, use mathematical induction to prove that the formulas hold for all natural numbers n.

42. $\log_{10}(a_1 a_2 \cdots a_n) = \log_{10} a_1 + \log_{10} a_2 + \cdots + \log_{10} a_n$

43. $(1 + p)^n \geq 1 + np$, where $p > -1$

13.2 THE BINOMIAL THEOREM

A mathematician, like a painter or a poet, is a maker of patterns.

G. E. Hardy (1877–1947)

[Blaise Pascal] *made numerous discoveries relating to this array and set them forth in his* Traité du triangle arithmétique, *published posthumously in 1665, and among these was essentially our present Binomial Theorem for positive integral exponents.*

David Eugene Smith in *History of Mathematics,* Vol. II (New York: Ginn and Co., 1925)

If you look back at Section 1.3, you will see that two of the special products listed there are

$$(a + b)^2 = a^2 + 2ab + b^2$$

and

$$(a + b)^3 = a^3 + 3a^2b + 3ab^2 + b^3$$

Our present goal is to develop a general formula, known as the *binomial theorem,* for expanding any product of the form $(a + b)^n$, when n is a natural number.

We begin by looking for patterns in the expansion of $(a + b)^n$. To do this, let's list the expansions of $(a + b)^n$ for $n = 1, 2, 3, 4$, and 5. (Exercises 1 and 2 ask you to verify these results simply by repeated multiplication.)

$$
\begin{aligned}
(a+b)^1 &= a + b\\
(a+b)^2 &= a^2 + 2ab + b^2\\
(a+b)^3 &= a^3 + 3a^2b + 3ab^2 + b^3\\
(a+b)^4 &= a^4 + 4a^3b + 6a^2b^2 + 4ab^3 + b^4\\
(a+b)^5 &= a^5 + 5a^4b + 10a^3b^2 + 10a^2b^3 + 5ab^4 + b^5
\end{aligned}
$$

After surveying these results, we note the following patterns.

PROPERTY SUMMARY **PATTERNS OBSERVED IN $(a + b)^n$ for $n = 1, 2, 3, 4, 5$**

GENERAL STATEMENT	EXAMPLE
There are $n + 1$ terms.	There are 4 $(= 3 + 1)$ terms in the expansion of $(a + b)^3$.
The expansion begins with a^n and ends with b^n	$(a + b)^3$ begins with a^3 and ends with b^3.
The sum of the exponents in each term is n.	The sum of the exponents in each term of $(a + b)^3$ is 3.
The exponents of a decrease by 1 from term to term.	$(a+b)^3 = a^{\textcircled{3}} + 3a^{\textcircled{2}}b + 3a^{\textcircled{1}}b^2 + a^{\textcircled{0}}b^3$
The exponents of b increase by 1 from term to term.	$(a+b)^3 = a^3b^{\textcircled{0}} + 3a^2b^{\textcircled{1}} + 3ab^{\textcircled{2}} + b^{\textcircled{3}}$
When n is even, the coefficients are symmetric about the middle term.	The sequence of coefficients for $(a + b)^4$ is 1, 4, 6, 4, 1.
When n is odd, the coefficients are symmetric about the two middle terms.	The sequence of coefficients for $(a + b)^5$ is 1, 5, 10, 10, 5, 1.

The patterns we have just observed for $(a + b)^n$ persist for all natural numbers n. (This will follow from the binomial theorem, which is proved at the end of this section.) Thus, for example, the form of $(a + b)^6$ must be as follows. (Each question mark denotes a coefficient to be determined.)

$$(a+b)^6 = a^6 + ?a^5b + ?a^4b^2 + ?a^3b^3 + ?a^2b^4 + ?ab^5 + b^6$$

The problem now is to find the proper coefficient for each term. To do this, we need to discover additional patterns in the expansion of $(a + b)^n$.

We have already written out the expansions of $(a + b)^n$ for values of n ranging from 1 to 5. Let us now write only the coefficients appearing in those expansions. The resulting triangular array of numbers is known as **Pascal's triangle.*** For reasons of symmetry, we begin with $(a + b)^0$ rather than $(a + b)^1$.

$(a+b)^0$ 1

$(a+b)^1$ 1 1

$(a+b)^2$ 1 2 1

$(a+b)^3$ 1 3 3 1

$(a+b)^4$ 1 4 6 4 1

$(a+b)^5$ 1 5 10 10 5 1

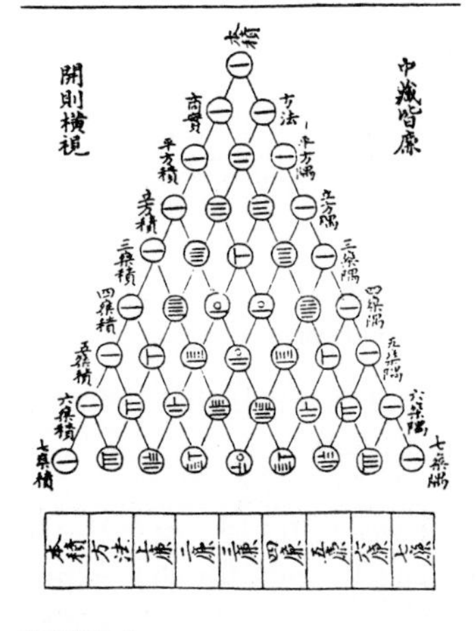

FIGURE 1
"Pascal's" triangle, by Chu-Shi-Kie, A.D. 1303.

*The array is named after Blaise Pascal, a seventeenth-century French mathematician and philosopher. However, as Figure 1 indicates, the Pascal triangle was known to Chinese mathematicians centuries earlier.

The key observation regarding Pascal's triangle is this: *Each entry in the array* (other than the 1's along the sides) *is the sum of the two numbers diagonally above it.* For instance, the 6 that appears in the fifth row is the sum of the two 3's diagonally above it. Using this observation, we can form as many additional rows as we please. The coefficients for $(a+b)^n$ will then appear in the $(n+1)$st row of the array.* For instance, to obtain the row corresponding to $(a+b)^6$, we have

$$\begin{array}{lccccccccccccc} \text{sixth row. } (a+b)^5: & & 1 & \searrow\swarrow & 5 & \searrow\swarrow & 10 & \searrow\swarrow & 10 & \searrow\swarrow & 5 & \searrow\swarrow & 1 & \\ & & & \downarrow & & \downarrow & & \downarrow & & \downarrow & & \downarrow & & \\ \text{seventh row, } (a+b)^6: & 1 & & 6 & & 15 & & 20 & & 15 & & 6 & & 1 \end{array}$$

Thus, the sequence of coefficients for $(a+b)^6$ is 1, 6, 15, 20, 15, 6, 1. This answers the question raised earlier about the expansion of $(a+b)^6$. We have

$$(a+b)^6 = a^6 + 6a^5b + 15a^4b^2 + 20a^3b^3 + 15a^2b^4 + 6ab^5 + b^6$$

For analytical work or for larger values of the exponent n, it is inefficient to rely on Pascal's triangle. For this reason, we point out another pattern in the expansions of $(a+b)^n$.

> In the expansion of $(a+b)^n$, the coefficient of any term after the first can be generated as follows: from the *previous* term, multiply the coefficient by the exponent of a and then divide by the number of that previous term.

To see how this method is used, let's compute the second, third, and fourth coefficients in the expansion of $(a+b)^6$. To compute the coefficient of the second term, we go back to the first term, which is a^6. We have

$$\text{coefficient of second term} = \frac{1 \cdot 6}{1} = 6$$

(1: coefficient of first term; 6: exponent of a in first term; denominator 1: number of first term)

Thus the second term is $6a^5b$ and, consequently, we have

$$\text{coefficient of third term} = \frac{6 \cdot 5}{2} = 15$$

(6: coefficient of second term; 5: exponent of a in second term; denominator 2: number of second term)

Continuing now with this method, you should check for yourself that the coefficient of the fourth term in the expansion of $(a+b)^6$ is 20.

NOTE We now know that the first four coefficients are 1, 6, 15, and 20. By symmetry, it follows that the complete sequence of coefficients for this expan-

*That these numbers actually are the appropriate coefficients follows from the binomial theorem, which is proved at the end of this section.

sion is 1, 6, 15, 20, 15, 6, 1. No additional calculation of coefficients is necessary.

EXAMPLE 1 Expand $(2x - y^2)^7$.

Solution First we write the expression of $(a + b)^7$ using the method explained just prior to this example, or using Pascal's triangle. As you should check for yourself, the expansion is

$$(a + b)^7 = a^7 + 7a^6b + 21a^5b^2 + 35a^4b^3 + 35a^3b^4 + 21a^2b^5 + 7ab^6 + b^7$$

Now we make the substitutions $a = 2x$ and $b = -y^2$. This yields

$$\begin{aligned}[2x + (-y^2)]^7 &= (2x)^7 + 7(2x)^6(-y^2) + 21(2x)^5(-y^2)^2 \\ &\quad + 35(2x)^4(-y^2)^3 + 35(2x)^3(-y^2)^4 \\ &\quad + 21(2x)^2(-y^2)^5 + 7(2x)(-y^2)^6 + (-y^2)^7 \\ &= 128x^7 - 448x^6y^2 + 672x^5y^4 - 560x^4y^6 + 280x^3y^8 \\ &\quad - 84x^2y^{10} + 14xy^{12} - y^{14}\end{aligned}$$

This is the required expansion. Notice how the signs alternate in the final answer; this is characteristic of all expressions of the form $(a - b)^n$.

In preparation for the binomial theorem, we introduce two notations that are used not only in connection with the binomial theorem, but in many other areas of mathematics as well. The first of these notations is $n!$ (read "n factorial").

> **DEFINITION: The Factorial Symbol**
>
> $n! = 1 \cdot 2 \cdot 3 \cdots n$
> where n is a natural number
>
> $0! = 1$
>
> **EXAMPLES**
>
> $3! = 1 \cdot 2 \cdot 3 = 6$
>
> $\dfrac{6!}{4!} = \dfrac{6 \cdot 5 \cdot 4 \cdot 3 \cdot 2 \cdot 1}{4 \cdot 3 \cdot 2 \cdot 1}$
>
> $= 6 \cdot 5 = 30$

EXAMPLE 2 Simplify the expression $\dfrac{(n + 1)!}{n - 1)!}$.

Solution

$$\frac{(n + 1)!}{(n - 1)!} = \frac{(n + 1) \cdot n \cdot (n - 1)!}{(n - 1)!} = (n + 1) \cdot n = n^2 + n$$

The second notation that we introduce in preparation for the binomial theorem is $\dbinom{n}{k}$. This notation is read "n choose k," because it can be shown that $\dbinom{n}{k}$ is equal to the number of ways of choosing a subset of k elements from a set with n elements.

DEFINITION: The Binomial Coefficient $\binom{n}{k}$

Let n and k be nonnegative integers, with $k \leq n$. Then the binomial coefficient $\binom{n}{k}$ is defined by

$$\binom{n}{k} = \frac{n!}{k!(n-k)!}$$

EXAMPLE

$$\binom{5}{2} = \frac{5!}{2!(5-2)!} = \frac{5!}{2!3!} = \frac{5 \cdot 4 \cdot 3 \cdot 2 \cdot 1}{(2 \cdot 1)(3 \cdot 2 \cdot 1)} = \frac{5 \cdot 4}{2 \cdot 1} = 10$$

The binomial coefficients are so named because they are indeed the coefficients in the expansion of $(a+b)^n$. More precisely, the relationship is this: The coefficients in the expansion of $(a+b)^n$ are the $n+1$ numbers

$$\binom{n}{0}, \binom{n}{1}, \binom{n}{2}, \ldots, \binom{n}{n}$$

Subsequently, we will see why this statement is true. For now, however, let's look at an example. Consider the binomial coefficients $\binom{3}{0}$, $\binom{3}{1}$, $\binom{3}{2}$, and $\binom{3}{3}$. According to our statement, these four quantities should be the coefficients in the expansion of $(a+b)^3$. Let's check:

$$\binom{3}{0} = \frac{3!}{0!(3-0)!} = \frac{3!}{1(3!)} = 1$$

$$\binom{3}{1} = \frac{3!}{1!(3-1)!} = \frac{3 \cdot 2 \cdot 1}{1(2 \cdot 1)} = 3$$

$$\binom{3}{2} = \frac{3!}{2!(3-2)!} = \frac{3 \cdot 2 \cdot 1}{(2 \cdot 1)1} = 3$$

$$\binom{3}{3} = \frac{3!}{3!(3-3)!} = \frac{3!}{3!0!} = 1$$

The values of $\binom{3}{0}$, $\binom{3}{1}$, $\binom{3}{2}$, and $\binom{3}{3}$ are thus 1, 3, 3, and 1, respectively. But these last four numbers are indeed the coefficients in the expression of $(a+b)^3$, as we wished to check.

We are now in a position to state the binomial theorem, after which we will look at several applications. Finally, at the end of this section, we will use mathematical induction to prove the theorem. In the statement of the theorem that follows, we are assuming that the exponent n is a natural number.

THE BINOMIAL THEOREM

$$(a+b)^n = \binom{n}{0}a^n + \binom{n}{1}a^{n-1}b + \binom{n}{2}a^{n-2}b^2 + \dots + \binom{n}{n-1}ab^{n-1} + \binom{n}{n}b^n$$

One of the uses of the binomial theorem is in identifying specific terms in an expansion without computing the entire expansion. This is particularly helpful when the exponent n is relatively large. Looking back at the statement of the binomial theorem, there are three observations we can make. First, the coefficient of the rth term is $\binom{n}{r-1}$. For instance, the coefficient of the third term is $\binom{n}{3-1} = \binom{n}{2}$. The second observation is that the exponent for a in the rth term is $n-(r-1)$. For instance, the exponent for a in the third term is $n-(3-1) = n-2$. Finally, we observe that the exponent for b in the rth term is $r-1$, the same quantity that appears in the lower position of the corresponding binomial coefficient. For instance, the exponent for b in the third term is $r-1 = 3-1 = 2$. We summarize these three observations with the following statement.

The rth term in the expansion of $(a+b)^n$ is

$$\binom{n}{r-1}a^{n-r+1}b^{r-1}$$

EXAMPLE 3 Find the fifteenth term in the expansion of $\left(x^2 - \frac{1}{x}\right)^{18}$.

Solution Using the values $r = 15$, $n = 18$, $a = x^2$, and $b = -1/x$, we have

$$\begin{aligned}
\binom{n}{r-1}a^{n-r+1}b^{r-1} &= \binom{18}{15-1}(x^2)^{18-15+1}\left(\frac{-1}{x}\right)^{15-1} \\
&= \binom{18}{14}x^8 \cdot \frac{1}{x^{14}} \\
&= \frac{18 \cdot 17 \cdot 16 \cdot 15 \cdot (14!)}{14!(4 \cdot 3 \cdot 2 \cdot 1)}x^{-6} \\
&= \frac{18 \cdot 17 \cdot 16 \cdot 15}{4 \cdot 3 \cdot 2 \cdot 1}x^{-6} \\
&= 3060x^{-6} \qquad \text{as required}
\end{aligned}$$

EXAMPLE 4 Find the coefficient of the term containing x^4 in the expansion of $(x + y^2)^{30}$.

Solution Again we use the fact that the rth term in the expansion of $(a+b)^n$ is $\binom{n}{r-1}a^{n-r+1}b^{r-1}$. In this case, n is 30 and x plays the role of a. The exponent for x is then $n - r + 1$ or $30 - r + 1$. To see when this exponent is 4, we write

$$30 - r + 1 = 4 \qquad \text{and therefore} \qquad r = 27$$

The required coefficient is therefore $\binom{30}{27-1}$. We then have

$$\binom{30}{26} = \frac{30!}{26!(30-26)!} = \frac{30 \cdot 29 \cdot 28 \cdot 27}{4 \cdot 3 \cdot 2 \cdot 1}$$

After carrying out the indicated arithmetic, we find that $\binom{30}{26} = 27{,}405$. This is the required coefficient.

EXAMPLE 5 Find the coefficient of the term containing a^9 in the expansion of $(a + 2\sqrt{a})^{10}$.

Solution The rth term in this expansion is

$$\binom{10}{r-1} a^{10-r+1}(2\sqrt{a})^{r-1}$$

We can rewrite this as

$$\binom{10}{r-1} a^{10-r+1}(2^{r-1})(a^{1/2})^{r-1}$$

or

$$2^{r-1}\binom{10}{r-1} a^{10-r+1+(r-1)/2}$$

This shows that the general form of the coefficient we wish to find is $2^{r-1}\binom{10}{r-1}$. We now need to determine r when the exponent of a is 9. Thus, we require that

$$10 - r + 1 + \frac{r-1}{2} = 9$$

$$-r + \frac{r-1}{2} = -2$$

$$-2r + r - 1 = -4 \qquad \text{multiplying both sides by 2}$$

$$r = 3$$

The required coefficient is now obtained by substituting $r = 3$ in the expression $2^{r-1}\binom{10}{r-1}$. Thus, the required coefficient is

$$2^2\binom{10}{2} = \frac{4 \cdot 10!}{2!(10-2)!} = 2 \cdot 10 \cdot 9 = 180$$

There are three simple identities involving the binomial coefficients that will simplify our proof of the binomial theorem.

IDENTITY 1 $\binom{r}{0} = 1$ for all nonnegative integers r

IDENTITY 2 $\binom{r}{r} = 1$ for all nonnegative integers r

IDENTITY 3 $\binom{k}{r} + \binom{k}{r-1} = \binom{k+1}{r}$ for all natural numbers k and r with $r \leq k$

All three of these identities can be proved directly from the definitions of the binomial coefficients, without the need for mathematical induction. The proofs of the first two are straightforward, and we omit them here. The proof of the third identity runs as follows:

$$\binom{k}{r} + \binom{k}{r-1} = \frac{k!}{r!(k-r)!} + \frac{k!}{(r-1)!(k-r+1)!}$$
$$= \frac{k!}{r(r-1)!(k-r)!} + \frac{k!}{(r-1)!(k-r+1)(k-r)!}$$

Now, the common denominator on the right-hand side of the last equation is $r(r-1)!(k-r+1)(k-r)!$ Thus, we have

$$\binom{k}{r} + \binom{k}{r-1} = \frac{k!(k-r+1)}{r(r-1)!(k-r+1)(k-r)!} + \frac{k!r}{r(r-1)!(k-r+1)(k-r)!}$$
$$= \frac{k!(k-r+1) + k!r}{r(r-1)!(k-r+1)(k-r)!}$$
$$= \frac{k!(k-r+1+r)}{r(r-1)!(k-r+1)(k-r)!} = \frac{k!(k+1)}{r!(k-r+1)!}$$
$$= \frac{(k+1)!}{r![(k+1)-r]!}$$
$$= \binom{k+1}{r} \quad \text{as required}$$

Taken together, the three identities show why the $(n+1)$st row of Pascal's triangle consists of the numbers $\binom{n}{0}, \binom{n}{1}, \binom{n}{2}, \ldots, \binom{n}{n}$. Identities 1 and 2 tell us that this row of numbers begins and ends with 1. Identity 3 is just a statement of the fact that each entry in the row, other than the initial and final 1, is generated by adding the two entries diagonally above it.

We conclude this section by using mathematical induction to prove the binomial theorem. The statement P_n that we wish to prove for all natural numbers n is this:

$$(a+b)^n = \binom{n}{0}a^n + \binom{n}{1}a^{n-1}b + \cdots + \binom{n}{n-1}ab^{n-1} + \binom{n}{n}b^n$$

First, we check that P_1 is true. The statement P_1 asserts that

$$(a+b)^1 = \binom{1}{0}a^1 + \binom{1}{1}a^0b$$

However, in view of Identities 1 and 2, this last equation becomes

$$(a+b)^1 = 1 \cdot a + 1 \cdot b$$

which is surely true. Now let's assume that P_k is true and, on the basis of this assumption, show that P_{k+1} is true. The statement P_k is

$$(a+b)^k = \binom{k}{0}a^k + \binom{k}{1}a^{k-1}b + \cdots + \binom{k}{k-1}ab^{k-1} + \binom{k}{k}b^k$$

Multiplying both sides of this equation by the quantity $(a+b)$ yields

$$\begin{aligned}
(a+b)^{k+1} &= (a+b)\left[\binom{k}{0}a^k + \binom{k}{1}a^{k-1}b + \cdots + \binom{k}{k-1}ab^{k-1} + \binom{k}{k}b^k\right] \\
&= a\left[\binom{k}{0}a^k + \binom{k}{1}a^{k-1}b + \cdots + \binom{k}{k-1}ab^{k-1} + \binom{k}{k}b^k\right] \\
&\qquad + b\left[\binom{k}{0}a^k + \binom{k}{1}a^{k-1}b + \cdots + \binom{k}{k-1}ab^{k-1} + \binom{k}{k}b^k\right] \\
&= \binom{k}{0}a^{k+1} + \binom{k}{1}a^kb + \cdots + \binom{k}{k-1}a^2b^{k-1} + \binom{k}{k}ab^k \\
&\qquad + \binom{k}{0}a^kb + \binom{k}{1}a^{k-1}b^2 + \cdots + \binom{k}{k-1}ab^k + \binom{k}{k}b^{k+1} \\
&= \binom{k}{0}a^{k+1} + \left[\binom{k}{1} + \binom{k}{0}\right]a^kb + \cdots + \left[\binom{k}{k} + \binom{k}{k-1}\right]ab^k + \binom{k}{k}b^{k+1}
\end{aligned}$$

We can now make some substitutions on the right-hand side of this last equation. The initial binomial coefficient $\binom{k}{0}$ can be replaced by $\binom{k+1}{0}$, because both are equal to 1, according to Identity 1. Similarly, the binomial coefficient $\binom{k}{k}$ appearing at the end of the equation can be replaced by $\binom{k+1}{k+1}$, since both are equal to 1 according to Identity 2. Finally, we can use Identity 3 to simplify each of the sums in the brackets. We obtain

$$(a+b)^{k+1} = \binom{k+1}{0}a^{k+1} + \binom{k+1}{1}a^kb + \cdots + \binom{k+1}{k}ab^k + \binom{k+1}{k+1}b^{k+1}$$

But this last equation is just the statement P_{k+1}; that is, we have derived P_{k+1} from P_k, as we wished to do. The induction proof is now complete.

EXERCISE SET 13.2

A

In Exercises 1 and 2, verify each statement directly, without using the techniques developed in this section.

1. **(a)** $(a+b)^2 = a^2 + 2ab + b^2$
(b) $(a+b)^3 = a^3 + 3a^2b + 3ab^2 + b^3$
Hint: $(a+b)^3 = (a+b)(a+b)^2$

2. **(a)** $(a+b)^4 = a^4 + 4a^3b + 6a^2b^2 + 4ab^3 + b^4$
Hint: Use the result in Exercise 1(b) and the fact that $(a+b)^4 = (a+b)(a+b)^3$.
(b) $(a+b)^5 = a^5 + 5a^4b + 10a^3b^2 + 10a^2b^3 + 5ab^4 + b^5$

In Exercises 3–28, carry out the indicated expansions.

3. $(a+b)^9$
4. $(a-b)^9$
5. $(2A+B)^3$
6. $(1+2x)^6$
7. $(1-2x)^6$
8. $(3x^2-y)^5$
9. $(\sqrt{x}+\sqrt{y})^4$
10. $(\sqrt{x}-\sqrt{y})^4$
11. $(x^2+y^2)^5$
12. $(5A-B^2)^3$
13. $[1-(1/x)]^6$
14. $(3x+y^2)^4$
15. $[(x/2)-(y/3)]^3$
16. $(1-z^2)^7$
17. $(ab^2+c)^7$
18. $[x-(1/x)]^8$
19. $(x+\sqrt{2})^8$
20. $(4A-\frac{1}{2})^5$
21. $(\sqrt{2}-1)^3$
22. $(1+\sqrt{5})^4$
23. $(\sqrt{2}+\sqrt{3})^5$
24. $(\frac{1}{2}-2a)^6$
25. $(2\sqrt[3]{2}-\sqrt[3]{4})^3$
26. $(x+y+1)^4$
Suggestion: Rewrite the expression as $[(x+y)+1]^4$.
27. $(x^2-2x-1)^5$
Suggestion: Rewrite the expression as $[x^2-(2x+1)]^5$.
28. $[x^2-2x-(1/x)]^6$

In Exercises 29–38, evaluate or simplify each expression.

29. $5!$
30. **(a)** $3!+2!$ **(b)** $(3+2)!$
31. $\binom{7}{3}\binom{3}{2}$
32. $\dfrac{20!}{18!}$
33. **(a)** $\binom{5}{3}$ **(b)** $\binom{5}{4}$
34. **(a)** $\binom{7}{7}$ **(b)** $\binom{7}{0}$
35. $\dfrac{(n+2)!}{n!}$
36. $\dfrac{n[(n-2)!]}{(n+1)!}$
37. $\binom{6}{4}+\binom{6}{3}-\binom{7}{4}$
38. $(3!)!+(3!)^2$

39. Find the fifteenth term in the expansion of $(a+b)^{16}$.

40. Find the third term in the expansion of $(a-b)^{30}$.

41. Find the one hundredth term in the expansion of $(1+x)^{100}$.

42. Find the twenty-third term in the expansion of $[x-(1/x^2)]^{25}$.

43. Find the coefficient of the term containing a^4 in the expansion of $(\sqrt{a}-\sqrt{x})^{10}$.

44. Find the coefficient of the term containing a^4 in the expansion of $(3a-5x)^{12}$.

45. Find the coefficient of the term containing y^8 in the expansion of $[(x/2)-4y]^9$.

46. Find the coefficient of the term containing x^6 in the expansion of $[x^2-(1/x)]^{12}$.

47. Find the coefficient of the term containing x^3 in the expansion of $(1-\sqrt{x})^8$.

48. Find the coefficient of the term containing a^8 in the expansion of $[a-(2/\sqrt{a})]^{14}$.

49. Find the term that does not contain A in the expansion of $[(1/A)+3A^2]^{12}$.

50. Find the coefficient of B^{-10} in the expansion of $[(B^2/2)-(3/B^3)]^{10}$.

B

51. Show that the coefficient of x^n in the expansion of $(1+x)^{2n}$ is $(2n)!/(n!)^2$.

52. Find n so that the coefficients of the eleventh and thirteenth terms in $(1+x)^n$ are the same.

53 (a) Complete the following table.

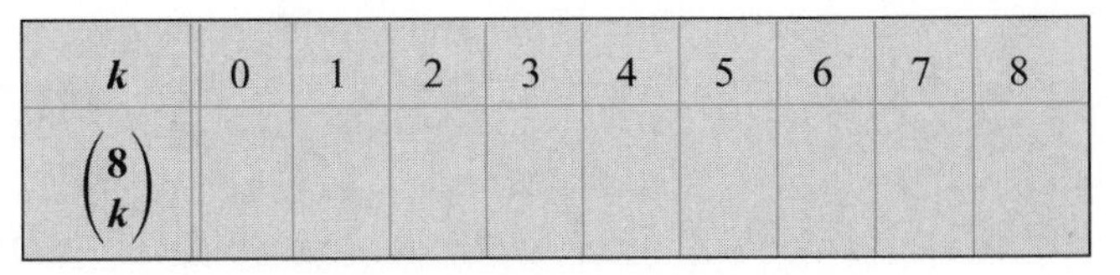

k	0	1	2	3	4	5	6	7	8
$\binom{8}{k}$									

(b) Use the results in part (a) to verify that
$$\binom{8}{0}+\binom{8}{1}+\binom{8}{2}+\cdots+\binom{8}{8}=2^8.$$

(c) By taking $a=b=1$ in the expansion of $(a+b)^n$, show that $\binom{n}{0}+\binom{n}{1}+\binom{n}{2}+\cdots+\binom{n}{n}=2^n$.

C

54. Two real numbers A and B are defined by $A=\sqrt[99]{99!}$ and $B=\sqrt[100]{100!}$. Which number is larger, A or B?
Hint: Compare A^{9900} and B^{9900}.

55. This exercise outlines a proof of the identity

$$\binom{n}{0}^2 + \binom{n}{1}^2 + \binom{n}{2}^2 + \cdots + \binom{n}{n}^2 = \binom{2n}{n}.$$

(a) Verify that

$$(1+x)^n\left(1+\frac{1}{x}\right)^n = \frac{(1+x)^{2n}}{x^n} \qquad (1)$$

(This requires only basic algebra, not the binomial theorem.)

(b) Show that the coefficient of the term independent of x on the right side of equation (1) is $\binom{2n}{n}$.

(c) Use the binomial theorem to expand $(1+x)^n$. Then show that the coefficient of the term independent of x on the left side of equation (1) is

$$\binom{n}{0}^2 + \binom{n}{1}^2 + \cdots + \binom{n}{n}^2.$$

13.3 INTRODUCTION TO SEQUENCES AND SERIES

Students of calculus do not always understand that infinite series are primarily tools for the study of functions.

George F. Simmons in *Calculus with Analytic Geometry (New York: McGraw-Hill, 1985)*

This section and the next two sections in this chapter deal with numerical sequences. We will begin with a somewhat informal definition of this concept. Then, after looking at some examples and terminology, we'll point out how the function concept is involved in defining a sequence. A **numerical sequence** is an ordered list of numbers. Here are four examples:

Example A: $1, \sqrt{2}, 10$

Example B: $2, 4, 6, 8, \ldots$

Example C: $1, \frac{1}{2}, \frac{1}{4}, \frac{1}{8}, \ldots$

Example D: $1, 1, 1, 1, \ldots$

The individual entries in a numerical sequence are called the **terms** of the sequence. In this chapter, the terms in each sequence will always be real numbers so, for convenience, we will drop the adjective *numerical* and refer simply to *sequences*. (It is worth pointing out, however, that in more advanced courses, sequences are studied in which the individual terms are functions.) Any sequence possessing only a finite number of terms is called a **finite sequence.** Thus the sequence in Example A is a finite sequence. On the other hand, Examples B, C, and D are examples of what we call **infinite sequences;** each contains infinitely many terms. As Example D indicates, it is not necessary that all the terms in a sequence be distinct. In this chapter, all the sequences we discuss will be infinite sequences.

In Examples B, C, and D, the three dots are read "and so on." In using this notation, we are assuming that it is clear what the subsequent terms of the sequence are. Toward this end, we often specify a formula for the nth term in a sequence. Example B in this case would appear this way:

$$2, 4, 6, 8, \cdots 2n, \cdots$$

A letter with subscripts is often used to denote the various terms in a sequence. For instance, if we denote the sequence in Example B by $a_1, a_2, a_3, \cdots$, then we have $a_1 = 2$, $a_2 = 4$, $a_3 = 6$, and, in general, $a_n = 2n$. Of course, there is nothing special about the letter a in this context; any other letter would do just as well.

EXAMPLE 1 Consider the sequence $a_1, a_2, a_3, \cdots$, in which the nth term a_n is given by

$$a_n = \frac{n}{n+1}$$

Compute the first three terms of the sequence, as well as the one thousandth term.

Solution To obtain the first term, we replace n by 1 in the given formula. This yields

$$a_1 = \frac{1}{1+1} = \frac{1}{2}$$

The other terms are similarly obtained. We have

$$a_2 = \frac{2}{2+1} = \frac{2}{3}$$
$$a_3 = \frac{3}{3+1} = \frac{3}{4}$$
$$a_{1000} = \frac{1000}{1000+1} = \frac{1000}{1001}$$

A word about notation: In the example we just concluded, we denoted the sequence by $a_1, a_2, a_3, \ldots$. In this case the subscripts begin with 1 and run through the natural numbers. The next example shows a case where the subscripts start with 0 rather than 1. (There is no new concept here; as we've said, it's just a matter of notation.)

EXAMPLE 2 Consider the sequence $b_0, b_1, b_2, \cdots$, in which the general term is given by

$$b_n = (-10)^n \quad \text{for } n \geqslant 0$$

Compute the sum of the first three terms of this sequence.

Solution We are asked to compute the sum $b_0 + b_1 + b_2$. Using the given formula for b_n, we have

$$b_0 = (-10)^0 = 1$$
$$b_1 = (-10)^1 = -10$$
$$b_2 = (-10)^2 = 100$$

Therefore $b_0 + b_1 + b_2 = 1 + (-10) + 100 = 91$.

In Examples 1 and 2, we were given a formula for the general term in each sequence. The next example shows a different way of specifying a sequence: each term after the first (or after the first several) is defined in terms of what has come before. This is an example of a **recursive definition.** Recursive definitions are particularly useful in computer programming.

EXAMPLE 3 Compute the first three terms of the sequence $b_1, b_2, b_3, \cdots$, which is defined recursively by

$$b_1 = 4$$
$$b_n = 2(b_{n-1} - 1) \quad \text{for } n \geqslant 2$$

Solution We are given the first term: $b_1 = 4$. To find b_2, we replace n by 2 in the formula $b_n = 2(b_{n-1} - 1)$ to obtain

$$b_2 = 2(b_1 - 1) = 2(4 - 1) = 6$$

Thus $b_2 = 6$. Next we use this value of b_2 in the formula $b_n = 2(b_{n-1} - 1)$ to obtain b_3. Replacing n by 3 in this formula yields

$$b_3 = 2(b_2 - 1) = 2(6 - 1) = 10$$

We have now found the first three terms of the sequence: $b_1 = 4$, $b_2 = 6$, and $b_3 = 10$.

If you think about the central idea behind the concept of a sequence, you can see that a function is involved: For each input n, we have an output a_n. As indicated in the previous examples, the inputs n may be the natural numbers 1, 2, 3, ⋯, or they may be the nonnegative integers 0, 1, 2, 3,⋯. For these reasons, the formal definition of a sequence is stated as follows.

DEFINITION: Sequence

A sequence is a function whose domain is either the set of natural numbers or the set of nonnegative integers.

If we denote the function by f for the moment, then $f(1)$ denotes what we have been calling a_1. Similarly, $f(2) = a_2$, $f(3) = a_3$, and, in general, $f(n) = a_n$. So the sequence notation with subscripts is just another variety of function notation. Furthermore, since a sequence is a function, we can draw a graph, as indicated in the next example.

TABLE 1

n	0	1	2	3
x_n	1	$\frac{1}{2}$	$\frac{1}{4}$	$\frac{1}{8}$

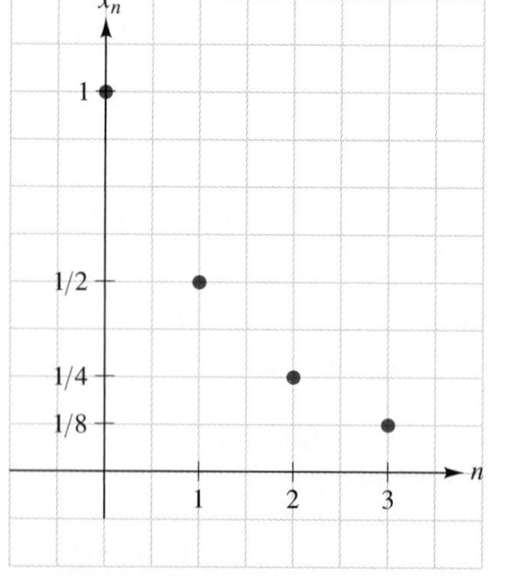

FIGURE 1

Graph of the sequence $x_n = \frac{1}{2^n}$ for $n = 0, 1, 2$, and 3.

EXAMPLE 4 Consider the sequence defined by

$$x_n = \frac{1}{2^n}, \quad \text{for } n \geqslant 0$$

Graph this sequence for $n = 0, 1, 2$, and 3.

Solution First we compute the required values of x_n, as shown in Table 1. Then, locating the inputs n on the horizontal axis and the outputs x_n on the vertical axis, we obtain the graph shown in Figure 1.

We will often be interested in the sum of certain terms of a sequence. Consider, for example, the sequence

$$10, 20, 30, 40, \cdots$$

in which the nth term is $10n$. The sum of the first four terms in this sequence is

$$10 + 20 + 30 + 40 = 100$$

More generally, the sum of the first n terms in this sequence is indicated by

$$10 + 20 + 30 + \cdots + 10n$$

This expression is an example of a **finite series,** which simply means a sum of a finite number of terms.

We can indicate the sum of the first n terms of the sequence $a_1, a_2, a_3, \cdots$ by

$$a_1 + a_2 + a_3 + \cdots + a_n$$

Another way to indicate this sum uses what is called **sigma notation,** which we now introduce. The capital Greek letter sigma is written Σ. We define the notation $\sum_{k=1}^{n} a_k$ by the equation

$$\sum_{k=1}^{n} a_k = a_1 + a_2 + a_3 + \cdots + a_n$$

For example, $\sum_{k=1}^{3} a_k$ stands for the sum $a_1 + a_2 + a_3$, the idea in this case being to replace the subscript k successively by 1, 2, and 3 and then add the results.

For a more concrete example, let's evaluate $\sum_{k=1}^{4} k^2$. We have

$$\sum_{k=1}^{4} k^2 = 1^2 + 2^2 + 3^3 + 4^2$$
$$= 1 + 4 + 9 + 16 = 30$$

There is nothing special about the choice of the letter k in the expression $\sum_{k=1}^{4} k^2$. For instance, we could equally well write

$$\sum_{j=1}^{4} j^2 = 1^2 + 2^2 + 3^2 + 4^2 = 30$$

The letter k in the expression $\sum_{k=1}^{4} k^2$ is called the **index of summation.** Similarly, the letter j appearing in $\sum_{j=1}^{4} j^2$ is the index of summation in that case. As we have seen, the choice of the letter used for the index of summation has no effect on the value of the indicated sum. For this reason, the index of summation is referred to as a *dummy variable.* The next two examples provide further practice with the sigma notation.

EXAMPLE 5 Express each of the following sums without sigma notation.

(a) $\sum_{k=1}^{3} (3k-2)^2$ **(b)** $\sum_{i=1}^{4} ix^{i-1}$ **(c)** $\sum_{j=1}^{5} (a_{j+1} - a_j)$

Solution

(a) The notation $\sum_{k=1}^{3} (3k-2)^2$ directs us to replace k successively by 1, 2, and 3 in the expression $(3k-2)^2$ and then to add the results. We thus obtain

$$\sum_{k=1}^{3} (3k-2)^2 = 1^2 + 4^2 + 7^2 = 66$$

(b) The notation $\sum_{i=1}^{4} ix^{i-1}$ directs us to replace i successively by 1, 2, 3, and 4 in the expression ix^{i-1} and then to add the results. We have

$$\sum_{i=1}^{4} ix^{i-1} = 1x^0 + 2x^1 + 3x^2 + 4x^3$$
$$= 1 + 2x + 3x^2 + 4x^3$$

(c) To expand $\sum_{j=1}^{5} (a_{j+1} - a_j)$, we replace j successively by 1, 2, 3, 4, and 5 in the expression $a_{j+1} - a_j$ and then add. We obtain

$$\sum_{j=1}^{5} (a_{j+1} - a_j) = (a_2 - a_1) + (a_3 - a_2) + (a_4 - a_3) + (a_5 - a_4) + (a_6 - a_5)$$
$$= a_2 - a_1 + a_3 - a_2 + a_4 - a_3 + a_5 - a_4 + a_6 - a_5$$

Combining like terms, we have

$$\sum_{j=1}^{5} (a_{j+1} - a_j) = a_6 - a_1$$

Sums such as $\sum_{j=1}^{5} (a_{j+1} - a_j)$ are known as **collapsing** or **telescoping sums.**

EXAMPLE 6 Use sigma notation to rewrite each sum.

(a) $\frac{x}{1!} + \frac{x^2}{2!} + \frac{x^3}{3!} + \cdots + \frac{x^{12}}{12!}$ **(b)** $\frac{x}{2!} + \frac{x^2}{3!} + \frac{x^3}{4!} + \cdots + \frac{x^n}{(n+1)!}$

Solution

(a) Since the exponents on x run from 1 to 12, we choose a dummy variable, say k, running from 1 to 12. Also, we notice that if x^k is the numerator of a given term in the sum, then $k!$ is the corresponding denominator. Consequently, the sum can be written

$$\frac{x}{1!} + \frac{x^2}{2!} + \frac{x^3}{3!} + \cdots + \frac{x^{12}}{12!} = \sum_{k=1}^{12} \frac{x^k}{k!}$$

(b) Since the exponents on x run from 1 to n, we choose a dummy variable, say k, running from 1 to n. (Note that both of the letters n and x would be inappropriate here as dummy variables.) Also, we notice that if the numerator of a given term in the sum is x^k, then the corresponding denominator is $(k+1)!$. Thus, the given sum can be written

$$\frac{x}{2!} + \frac{x^2}{3!} + \frac{x^3}{4!} + \cdots + \frac{x^n}{(n+1)!} = \sum_{k=1}^{n} \frac{x^k}{(k+1)!}$$

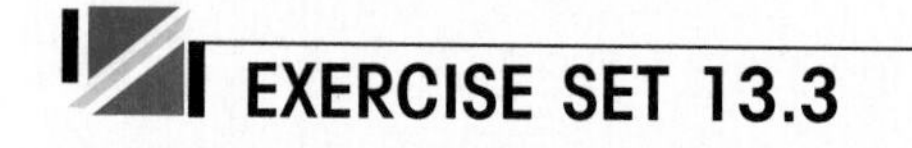

EXERCISE SET 13.3

A

In Exercises 1–14, compute the first four terms in each sequence. For Exercises 1–10, assume that each sequence is defined for $n \geq 1$; in Exercises 11–14, assume that each sequence is defined for $n \geq 0$.

1. $a_n = n/(n+1)$ **2.** $a_n = 1/n^2$

3. $b_n = (-1)^n$

4. $b_n = (n+1)^2$

5. $c_n = 2^{-n}$

6. $c_n = (-1)^n(2^n)$

7. $x_n = 3n$

8. $x_n = (n-1)^{n+1}$

9. $y_n = [1 + (1/n)]^n$

10. $y_n = [(-1)^n]\sqrt{n}$

11. $a_n = (n-1)/(n+1)$

12. $a_n = (-1)^n/(n+2)$

13. $b_n = (-2)^{n+1}/(n+1)^2$

14. $b_n = (n+2)^{-n}$

In Exercises 15–22, the sequences are defined recursively. Compute the first five terms in each sequence.

15. $a_1 = 1;\ a_n = (1 + a_{n-1})^2,\ n \geqslant 2$

16. $a_1 = 2;\ a_n = \sqrt{a_{n-1}^2 + 1},\ n \geqslant 2$

17. $a_1 = 2;\ a_2 = 2;\ a_n = a_{n-1}a_{n-2},\ n \geqslant 3$

18. $F_1 = 1;\ F_2 = 1;\ F_n = F_{n-1} + F_{n-2},\ n \geqslant 3$

19. $a_1 = 1;\ a_{n+1} = na_n,\ n \geqslant 1$

20. $a_1 = 1;\ a_2 = 2;\ a_n = a_{n-1}/a_{n-2},\ n \geqslant 3$

21. $a_1 = 0;\ a_n = 2^{a_{n-1}},\ n \geqslant 2$

22. $a_1 = 0;\ a_2 = 1;\ a_n = (a_{n-1} + a_{n-2})/2,\ n \geqslant 3$

In Exercises 23–28, find the sum of the first five terms of the sequence. (In Exercises 23–27, assume that the sequences are defined for $n \geqslant 1$.)

23. $a_n = 2^n$

24. $b_n = 2^{-n}$

25. $a_n = n^2 - n$

26. $b_n = (n-1)!$

27. $a_n = (-1)^n/n!$

28. $a_1 = 1;\ a_n = \dfrac{1}{n-1} - \dfrac{1}{n+1},\quad n \geqslant 2$

29. Find the sum of the first five terms of the sequence that is defined recursively by $a_1 = 1;\ a_2 = 2;\ a_n = a_{n-1}^2 + a_{n-2}^2,\ n \geqslant 3$.

30. Find the sum of the first four terms of the sequence defined recursively by $a_1 = 2,\ a_n = n/a_{n-1},\ n \geqslant 2$.

31. Find the sum of the first four terms of the sequence defined recursively by $a_1 = 2;\ a_n = (a_{n-1})^2,\ n \geqslant 2$.

32. Find the sum of the first six terms of the sequence defined recursively by $a_1 = 1;\ a_2 = 2;\ a_{n+1} = a_na_{n-1},\ n \geqslant 2$.

In Exercises 33–38, graph the sequences for the indicated values of n.

33. $a_n = (-1)^n;\ n = 1, 2, 3, 4$

34. $a_n = 2n - 1;\ n = 1, 2, 3$

35. $a_n = 1/n;\ n = 1, 2, 3, 4$

36. $a_n = (-1)^n/n;\ n = 1, 2, 3, 4$

37. $a_n = (n-1)/(n+1);\ n = 1, 2, 3$

38. $a_n = [1 + (1/n)]^n;\ n = 1, 2, 3$

In Exercises 39–50, express each of the sums without using sigma notation. Simplify your answers where possible.

39. $\sum_{k=1}^{3} (k-1)$

40. $\sum_{k=1}^{5} k$

41. $\sum_{k=4}^{5} k^2$

42. $\sum_{k=2}^{6} (1-2k)$

43. $\sum_{n=1}^{3} x^n$

44. $\sum_{n=1}^{3} (n-1)x^{n-2}$

45. $\sum_{n=1}^{4} \dfrac{1}{n}$

46. $\sum_{n=0}^{4} 3^n$

47. $\sum_{j=1}^{9} \log_{10}\left(\dfrac{j}{j+1}\right)$

48. $\sum_{j=2}^{5} \log_{10} j$

49. $\sum_{j=1}^{6} \left(\dfrac{1}{j} - \dfrac{1}{j+1}\right)$

50. $\sum_{j=1}^{5} (x^{j+1} - x^j)$

In Exercises 51–60, rewrite the sums using sigma notation.

51. $5 + 5^2 + 5^3 + 5^4$

52. $5 + 5^2 + 5^3 + \cdots + 5^n$

53. $x + x^2 + x^3 + x^4 + x^5 + x^6$

54. $x + 2x^2 + 3x^3 + 4x^4 + 5x^5 + 6x^6$

55. $\frac{1}{1} + \frac{1}{2} + \frac{1}{3} + \cdots + \frac{1}{12}$

56. $\dfrac{1}{1} + \dfrac{1}{2} + \dfrac{1}{3} + \cdots + \dfrac{1}{n}$

57. $2 - 2^2 + 2^3 - 2^4 + 2^5$

58. $\binom{10}{3} + \binom{10}{4} + \binom{10}{5} + \cdots + \binom{10}{10}$

59. $1 - 2 + 3 - 4 + 5$

60. $\frac{1}{2} - \frac{1}{4} + \frac{1}{8} - \frac{1}{16} + \frac{1}{32} - \frac{1}{64} + \frac{1}{128}$

B

61. A sequence is defined recursively as follows: $s_1 = 5$ and $s_n = \sqrt{s_{n-1}}$ for $n \geqslant 2$.
   (a) Compute the first eight terms of this sequence. (Use a calculator; for the answers, round to three decimal places.) What do you observe about the answers?
   (b) Use a calculator to compute s_{15}. (Round the answer to six decimal places.)
   (c) In view of your work in parts (a) and (b), what number do you think would be a very close approximation to s_{100}?

62. Follow Exercise 61 but take s_1 to be 500 rather than 5.

63. A sequence is defined recursively as follows: $s_1 = 0.7$ and $s_n = (s_{n-1})^2$ for $n \geqslant 2$.
   (a) Compute the first six terms of this sequence. (Use a calculator; for the answers, round to five decimal places.) What do you observe about the answers?
   (b) Use a calculator to compute s_{10}. (Report the answer as shown on your calculator screen.)

(c) In view of your work in parts (a) and (b), what number do you think would be a very close approximation to s_{100}?

64. A sequence is defined recursively as follows:

$$x_1 = 1 \quad \text{and} \quad x_n = \frac{x_{n-1}}{1 + x_{n-1}} \quad \text{for } n > 1$$

(a) Complete the following table:

n	1	2	3	4
x_n				

(b) On the basis of the results in the table, make a guess about the value of x_5, then compute x_5 to see if your guess is correct.

(c) The sequence given at the start of this exercise is defined recursively. Make a conjecture about a simpler way to define this sequence, then use mathematical induction to prove that your conjecture is correct.

Exercises 65–73 involve the ***Fibonacci sequence,*** *which is defined recursively as follows:*

$$F_1 = 1; \quad F_2 = 1; \quad F_{n+2} = F_n + F_{n+1} \quad \text{for } n \geq 1$$

This sequence was first studied by the Italian mathematician and merchant Leonardo of Pisa (c. 1170–1240), better known as Fibonacci ("son of Bonaccio").

65. **(a)** Complete the following table for the first ten terms of the Fibonacci sequence.

F_1	F_2	F_3	F_4	F_5	F_6	F_7	F_8	F_9	F_{10}
1	1								

(b) Given that $F_{20} = 6765$ and $F_{21} = 10946$, compute F_{22} and F_{19}.

(c) Given that $F_{29} = 514229$ and $F_{31} = 1346269$, compute F_{30}.

66. **(a)** Use the table in Exercise 65(a) to verify that $F_8 + F_9 + F_{10} = 2(F_8 + F_9)$.

(b) Show that $F_{100} + F_{101} + F_{102} = 2(F_{100} + F_{101})$.

(c) Show that $F_n + F_{n+1} + F_{n+2} = 2(F_n + F_{n+1})$ for all natural numbers $n \geq 3$. *Hint:* Mathematical induction is not needed.

(d) Does the identity in part (c) hold for either of the values $n = 1$ or $n = 2$?

In Exercises 67–69: **(a)** Verify that the given equation holds for $n = 1$, $n = 2$, and $n = 3$; **(b)** Use mathematical induction to show that the equation holds for all natural numbers.

67. $F_1 + F_2 + F_3 + \cdots + F_n = F_{n+2} - 1$

68. $F_1^2 + F_2^2 + F_3^2 + \cdots + F_n^2 = F_n F_{n+1}$

69. $F_{n+1}^2 = F_n F_{n+2} + (-1)^n$ *Hint for part (b):* Add $F_{k+1}F_{k+2}$ to both sides of the equation in the induction hypothesis. Then factor F_{k+1} from the left-hand side and factor F_{k+2} from the first two terms on the right-hand side.

70. Use mathematical induction to prove that $F_n \geq n$ for all natural numbers $n \geq 5$.

71. We've seen that the Fibonacci sequence is defined recursively; each term after the second is the sum of the previous two terms. There is, in fact, an explicit formula for the nth term of the Fibonacci sequence:

$$F_n = \frac{(1+\sqrt{5})^n - (1-\sqrt{5})^n}{2^n\sqrt{5}}$$

This formula was discovered by the French-English mathematician Abraham deMoivre more than 500 years after Fibonacci first introduced the sequence in 1202 in his book *Liber Abaci.* (For a proof of this formula, see Exercises 72 and 73.)

(a) Using algebra (and not your calculator), check that this formula gives the right answers for F_1 and F_2.

(b) Use this formula and your calculator to compute F_{24} and F_{25}.

(c) Use the formula and your calculator to compute F_{26}. Then check your answer by using the results in part (b).

72. (This result will be used in Exercise 73.) Suppose that x is a real number such that $x^2 = x + 1$. Use mathematical induction to prove that

$$x^n = F_n x + F_{n-1} \quad \text{for } n \geq 2$$

Hint: You can carry out the induction proof without solving the given quadratic equation.

73. In this exercise, we use the result in Exercise 72 to derive the following formula for the nth Fibonacci number:

$$F_n = \frac{(1+\sqrt{5})^n - (1-\sqrt{5})^n}{2^n\sqrt{5}} \tag{1}$$

The clever method used here was discovered by Erwin Just; it appeared in *Mathematics Magazine,* vol. 44 (1971), p. 199.

Let α and β denote the roots of the quadratic equation $x^2 = x + 1$. Then, according to Exercise 72, for $n \geq 2$ we have

$$\alpha^n = F_n\alpha + F_{n-1} \tag{2}$$

and

$$\beta^n = F_n\beta + F_{n-1} \tag{3}$$

(a) Subtract equation (3) from equation (2) to show that

$$F_n = (\alpha^n - \beta^n)/(\alpha - \beta) \quad \text{for } n \geq 2 \tag{4}$$

(b) Use the quadratic formula to show that the roots of the equation $x^2 = x + 1$ are given by
$\alpha = (1 + \sqrt{5})/2$ and $\beta = (1 - \sqrt{5})/2$.

(c) In equation (4), substitute for α and β using the values obtained in part (b). Show that this leads to equation (1).

(d) The work in parts (a) through (c) shows that equation (1) holds for $n \geqslant 2$. Now complete the derivation by checking that equation (1) also holds for $n = 1$.

74. This exercise requires a knowledge of matrix multiplication. Use mathematical induction to show that

$$\begin{pmatrix} 1 & 1 \\ 1 & 0 \end{pmatrix}^n = \begin{pmatrix} F_{n+1} & F_n \\ F_n & F_{n-1} \end{pmatrix} \quad \text{for } n \geqslant 2$$

13.4 ARITHMETIC SEQUENCES AND SERIES

One of the most natural ways to generate a sequence is to begin with a fixed number a and then repeatedly add a fixed constant d. This yields the sequence

$$a,\ a+d,\ a+2d,\ a+3d,\ \cdots \tag{1}$$

Such a sequence is called an **arithmetic sequence** or **arithmetic progression.** Notice that the difference between any two consecutive terms is the constant d. We call d the **common difference.** Here are several examples of arithmetic sequences:

Example A: $1, 2, 3, \cdots$

Example B: $3, 7, 11, 15, \cdots$

Example C: $10, 5, 0, -5, \cdots$

In Example A, the first term is $a = 1$ and the common difference is $d = 1$. For Example B we have $a = 3$. The value of d in this example is found by subtracting any two consecutive terms; thus $d = 4$. Finally, in Example C we have $a = 10$ and $d = -5$. Notice when the common difference is negative, the terms of the sequence decrease.

There is a simple formula for the nth term in an arithmetic sequence. In arithmetic sequence (1), notice that

$$\begin{aligned} a_1 &= a + 0d \\ a_2 &= a + 1d \\ a_3 &= a + 2d \\ a_4 &= a + 3d \end{aligned}$$

Following this pattern, it appears that the formula for a_n should be given by $a_n = a + (n-1)d$. Indeed, this is the correct formula, and Exercise 32 asks you to verify it using mathematical induction.

THE nth TERM OF AN ARITHMETIC SEQUENCE

The nth term of an arithmetic sequence $a, a+d, a+2d, \cdots$ is given by

$$a_n = a + (n-1)d$$

EXAMPLE 1 Determine the one hundredth term of the arithmetic sequence

$$7, 10, 13, 16, \cdots$$

Solution The first term is $a = 7$ and the common difference is $d = 3$. Substituting these values in the formula $a_n = a + (n - 1)d$ yields

$$a_n = 7 + (n - 1)3 = 3n + 4$$

To find the 100th term, we replace n by 100 in this last equation to obtain

$$a_{100} = 3(100) + 4 = 304 \qquad \text{as required}$$

EXAMPLE 2 Determine the arithmetic sequence in which the second term is -2 and the eighth term is 40.

Solution We are given that the second term is -2. Using this information in the formula $a_n = a + (n - 1)d$, we have

$$-2 = a + (2 - 1)d = a + d$$

This gives us one equation in two unknowns. We are also given that the eighth term is 40. Therefore,

$$40 = a + (8 - 1)d = a + 7d$$

We now have a system of two equations in two unknowns:

$$\begin{cases} -2 = a + d \\ 40 = a + 7d \end{cases}$$

Subtracting the first equation from the second gives us

$$42 = 6d \qquad \text{or} \qquad 7 = d$$

To find a, we replace d by 7 in the first equation of the system. This yields

$$-2 = a + 7 \qquad \text{or} \qquad -9 = a$$

We have now determined the sequence, since we know that the first term is -9 and the common difference is 7. The first four terms of the sequence are -9, -2, 5, and 12.

Next we would like to derive a formula for the sum of the first n terms of an arithmetic sequence. Such a sum is referred to as an **arithmetic series.** If we use S_n to denote the required sum, we have

$$S_n = a + (a + d) + \cdots + [a + (n - 2)d] + [a + (n - 1)d] \tag{2}$$

Of course, we must obtain the same sum if we add the terms from right to left rather than left to right. That is, we must have

$$S_n = [a + (n - 1)d] + [a + (n - 2)d] + \cdots + (a + d) + a \tag{3}$$

Let's now add equations (2) and (3). Adding the left-hand sides is easy; we obtain $2S_n$. Now we add the corresponding terms on the right-hand sides. For the first term, we have

$$\underset{\substack{\text{first term}\\\text{in equation (2)}}}{\underset{\uparrow}{a}} + \underset{\substack{\text{first term}\\\text{in equation (3)}}}{\underbrace{[a + (n - 1)d]}} = 2a + (n - 1)d$$

Next we add the second terms:

$$\underset{\substack{\text{second term}\\\text{in equation (2)}}}{\underset{\uparrow}{(a + d)}} + \underset{\substack{\text{second term}\\\text{in equation (3)}}}{\underbrace{[a + (n - 2)d]}} = 2a + d + (n - 2)d = 2a + (n - 1)d$$

Notice that the sum of the second terms is again $2a + (n - 1)d$, the same quantity we arrived at with the first terms. As you can check, this pattern continues all the way through to the last terms. For instance,

$$\underset{\substack{\text{last term}\\\text{in equation (2)}}}{\underbrace{[a + (n - 1)d]}} + \underset{\substack{\text{last term}\\\text{in equation (3)}}}{\underset{\uparrow}{a}} = 2a + (n - 1)d$$

We conclude from these observations that by adding the right-hand sides of equations (2) and (3), the quantity $2a + (n - 1)d$ is added a total of n times. Therefore,

$$2S_n = n[2a + (n - 1)d]$$

$$S_n = \frac{n}{2}[2a + (n - 1)d]$$

This gives us the desired formula for the sum of the first n terms in an arithmetic sequence. There is an alternate form of this formula, which now follows rather quickly:

$$\begin{aligned} S_n &= \frac{n}{2}[2a + (n - 1)d] \\ &= \frac{n}{2}\{a + [a + (n - 1)d]\} \\ &= \frac{n}{2}(a + a_n) = n\left(\frac{a + a_n}{2}\right) \end{aligned}$$

This last equation is easy to remember. It says that the sum of an arithmetic series is obtained by averaging the first and last terms and then multiplying this average by n, the number of terms. For reference, we summarize both formulas as follows.

FORMULAS FOR THE SUM OF AN ARITHMETIC SERIES

$$S_n = \frac{n}{2}(2a + (n - 1)d]$$

$$S_n = n\left(\frac{a + a_n}{2}\right)$$

EXAMPLE 3 Find the sum of the first 30 terms of the arithmetic sequence 2, 6, 10, 14, $\cdots$.

Solution We have $a = 2$, $d = 4$, and $n = 30$. Substituting these values in the formula $S_n = (n/2)[2a + (n-1)d]$ then yields

$$\begin{aligned} S_{30} &= \frac{30}{2}[2(2) + (30-1)4] \\ &= 15[4 + 29(4)] \\ &= 1800 \qquad \text{(Check the arithmetic)} \end{aligned}$$

EXAMPLE 4 In a certain arithmetic sequence, the first term is 6 and the fortieth term is 71. Find the sum of the first 40 terms and also the common difference for the sequence.

Solution We have $a = 6$ and $a_{40} = 71$. Using these values in the formula $S_n = n(a + a_n)/2$ yields

$$S_{40} = \frac{40}{2}(6 + 71) = 20(77) = 1540$$

The sum of the first 40 terms is thus 1540. The value of d can now be found by using the formula $a_n = a + (n-1)d$. We have $a_{40} = a + (40-1)d$, and therefore

$$\begin{aligned} 71 &= 6 + 39d \\ 39d &= 65 \\ d &= \frac{65}{39} = \frac{5}{3} \end{aligned}$$

The required value of d is therefore $5/3$.

EXAMPLE 5 Show that $\sum_{k=1}^{50} (3k-2)$ represents an arithmetic series, and compute the sum.

Solution There are two different ways to see that $\sum_{k=1}^{50} (3k-2)$ is an arithmetic series. One way is simply to write out the first few terms and look at the pattern. We have

$$\sum_{k=1}^{50} (3k-2) = 1 + 4 + 7 + 10 + \cdots + 148$$

From this it is clear that we are indeed summing the terms in an arithmetic sequence in which $d = 3$ and $a = 1$.

Another, more formal way to show that $\sum_{k=1}^{50} (3k-2)$ represents an arithmetic series is to prove that the difference between successive terms in the indicated sum is a constant. Now, the form of a typical term in this sum is $3k - 2$. Thus, the form of the next term must be $[3(k+1) - 2]$. The difference between these terms is then

$$[3(k+1)-2]-(3k-2)=3k+3-2-3k+2=3$$

The difference therefore is constant, as we wished to show.

To evaluate $\sum_{k=1}^{50}(3k-2)$, we can use either of our two formulas for the sum of an arithmetic series. Using the formula $S_n=(n/2)[2a+(n-1)d]$, we obtain

$$S_{50}=\frac{50}{2}[2(1)+49(3)]=25(149)=3725$$

Thus, the required sum is 3725. You should check for yourself that the same value is obtained using the formula $S_n=n(a+a_n)/2$.

EXERCISE SET 13.4

A

1. Find the common difference d for each of the following arithmetic sequences.
   (a) $1, 3, 5, 7, \cdots$ **(b)** $10, 6, 2, -2, \cdots$
   (c) $2/3, 1, 4/3, 5/3, \cdots$
   (d) $1, 1+\sqrt{2}, 1+2\sqrt{2}, 1+3\sqrt{2}, \cdots$
2. Which of the following are arithmetic sequences?
   (a) $2, 4, 8, 16, \cdots$ (b) $5, 9, 13, 17, \cdots$
   (c) $3, 11/5, 7/5, 3/5, \cdots$
   (d) $-1, -1, -1, -1, \cdots$
   (e) $-1, 1, -1, 1, \cdots$

In Exercises 3–8, find the indicated term in each sequence.

3. $10, 21, 32, 43, \cdots; a_{12}$ 4. $7, 2, -3, -8, \cdots; a_{20}$
5. $6, 11, 16, 21, \cdots; a_{100}$
6. $2/5, 4/5, 6/5, 8/5, \cdots; a_{30}$
7. $-1, 0, 1, 2, \cdots; a_{1000}$
8. $42, 1, -40, -81, \cdots; a_{15}$
9. The 4th term in an arithmetic sequence is -6, and the 10th term is 5. Find the common difference and the first term.
10. The 5th term in an arithmetic sequence is $1/2$, and the 20th term is $7/8$. Find the first three terms of the sequence.
11. The 60th term in an arithmetic sequence is 105, and the common difference is 5. Find the first term.
12. Find the common difference in an arithmetic sequence in which $a_{10}-a_{20}=70$.
13. Find the common difference in an arithmetic sequence in which $a_{15}-a_7=-1$.
14. Find the sum of the first 16 terms in the sequence 2, 11, 20, 29, $\cdots$.
15. Find the sum of the first 1000 terms in the sequence 1, 2, 3, 4, $\cdots$.
16. Find the sum of the first 50 terms in an arithmetic series that has 1st term -8 and 50th term 139.
17. Find the sum: $\frac{\pi}{3}+\frac{2\pi}{3}+\pi+\frac{4\pi}{3}+\cdots+\frac{13\pi}{3}$.
18. Find the sum: $\frac{1}{e}+\frac{3}{e}+\frac{5}{e}+\cdots+\frac{21}{e}$.
19. Determine the first term of an arithmetic sequence in which the common difference is 5 and the sum of the first 38 terms is 3534.
20. The sum of the first 12 terms in an arithmetic sequence is 156. What is the sum of the first and twelfth terms?
21. In a certain arithmetic sequence, the 1st term is 4 and the 16th term is -100. Find the sum of the first 16 terms and also the common difference for the sequence.
22. The 5th and 50th terms of an arithmetic sequence are 3 and 30, respectively. Find the sum of the first 10 terms.
23. The 8th term in an arithmetic sequence is 5, and the sum of the first 10 terms is 20. Find the common difference and the first term of the sequence.

In Exercises 24–26, find each sum.

24. $\sum_{i=1}^{10}(2i-1)=1+3+5+\cdots+19$
25. $\sum_{k=1}^{20}(4k+3)$ 26. $\sum_{n=5}^{100}(2n-1)$
27. The sum of three consecutive terms in an arithmetic sequence is 30, and their product is 360. Find the three

terms. *Suggestion:* Let x denote the *middle* term and d the common difference.

28. The sum of three consecutive terms in an arithmetic sequence is 21, and the sum of their squares is 197. Find the three terms.

29. The sum of three consecutive terms in an arithmetic sequence is 6, and the sum of their cubes is 132. Find the three terms.

30. In a certain arithmetic sequence, $a = -4$ and $d = 6$. If $S_n = 570$, find n.

31. Let $a_1 = 1/(1+\sqrt{2})$, $a_2 = -1$, and $a_3 = 1/(1-\sqrt{2})$.
(a) Show that $a_2 - a_1 = a_3 - a_2$.
(b) Find the sum of the first six terms in the arithmetic sequence

$$\frac{1}{1+\sqrt{2}}, \; -1, \; \frac{1}{1-\sqrt{2}}, \dots$$

32. Using mathematical induction, prove that the nth term of the sequence $a,\ a+d,\ a+2d,\ \dots$ is given by

$$a_n = a + (n-1)d$$

B

33. Let b denote a positive constant. Find the sum of the first n terms in the sequence

$$\frac{1}{1+\sqrt{b}}, \frac{1}{1-b}, \frac{1}{1-\sqrt{b}}, \dots$$

34. The sum of the first n terms in a certain arithmetic sequencen is given by $S_n = 3n^2 - n$. Show that the rth term is given by $a_r = 6r - 4$.

35. Let $a_1, a_2, a_3, \dots$ be an arithmetic sequence, and let S_k denote the sum of the first k terms. If $S_n/S_m = n^2/m^2$, show that

$$\frac{a_n}{a_m} = \frac{2n-1}{2m-1}$$

36. If the common difference in an arithmetic sequence is twice the first term, show that

$$\frac{S_n}{S_m} = \frac{n^2}{m^2}$$

37. The lengths of the sides of a right triangle form three consecutive terms in an arithmetic sequence. Show that the triangle is similar to the 3-4-5 right triangle.

38. Suppose that $1/a$, $1/b$, and $1/c$ are three consecutive terms in an arithmetic sequence. Show that:
(a) $\dfrac{a}{c} = \dfrac{a-b}{b-c}$; **(b)** $b = \dfrac{2ac}{a+c}$.

39. Suppose that a, b, and c are three positive numbers with $a > c > 2b$. If $1/a$, $1/b$, and $1/c$ are consecutive terms in an arithmetic sequence, show that

$$\ln(a+c) + \ln(a-2b+c) = 2\ln(a-c)$$

GEOMETRIC SEQUENCES AND SERIES

A **geometric sequence** or **geometric progression** is a sequence of the form

$$a,\ ar,\ ar^2,\ ar^3,\ \dots \quad \text{where } a \text{ and } r \text{ are constants}$$

As you can see, each term after the first in a geometric sequence is obtained by multiplying the previous term by r. The number r is called the **common ratio,** since the ratio of any term to the previous one is always r. For instance, the ratio of the fourth term to the third is $ar^3/ar^2 = r$. Here are two examples of geometric sequences:

$$1, \frac{1}{2}, \frac{1}{4}, \frac{1}{8}, \dots$$

$$10, -100, 1000, -10{,}000, \dots$$

In the first example, we have $a = 1$ and $r = 1/2$; in the second example, we have $a = 10$ and $r = -10$.

EXAMPLE 1 In a certain geometric sequence, the first term is 2, the third term is 3, and the common ratio is negative. Find the second term.

Solution Let x denote the second term, so that the sequence begins

$$2, x, 3, \cdots$$

By definition, the ratios $3/x$ and $x/2$ must be equal. Thus, we have

$$\frac{3}{x} = \frac{x}{2}$$
$$x^2 = 6$$
$$x = \pm\sqrt{6}$$

Now, the second term must be negative, since the first term is positive and the common ratio is negative. Thus, the second term is $x = -\sqrt{6}$.

TABLE 1

n	a_n
1	ar^0
2	ar^1
3	ar^2
4	ar^3
⋮	⋮

The formula for the nth term of a geometric sequence is easily deduced by considering Table 1. The table indicates that the exponent on r is one less than the value of n in each case. On the basis of this observation, it appears that the nth term must be given by $a_n = ar^{n-1}$. Indeed, it can be shown by mathematical induction that this formula does hold for all natural numbers n. (Exercise 30 asks you to carry out the proof.) We summarize this result as follows:

nth TERM OF A GEOMETRIC SEQUENCE

The nth term of the geometric sequence, $a, ar, ar^2, \cdots$ is given by

$$a_n = ar^{n-1}$$

EXAMPLE 2 Find the seventh term in the geometric sequence 2, 6, 18,⋯.

Solution We can find the common ratio r by dividing the second term by the first. Thus, $r = 3$. Now, using $a = 2$, $r = 3$, and $n = 7$ in the formula $a_n = ar^{n-1}$, we have

$$a_7 = 2(3)^6 = 2(729) = 1458$$

The seventh term of the sequence is therefore 1458.

Suppose that we begin with a geometric sequence $a, ar, ar^2, \cdots$, in which $r \neq 1$. If we add the first n terms and denote the sum by S_n, we have

$$S_n = a + ar + ar^2 + \cdots + ar^{n-2} + ar^{n-1} \qquad (1)$$

This sum is called a **finite geometric series.** We would like to find a formula for S_n. To do this, we multiply equation (1) by r to obtain

$$rS_n = ar + ar^2 + ar^3 + \cdots + ar^{n-1} + ar^n \qquad (2)$$

We now subtract equation (2) from equation (1). This yields (after combining like terms)

$$S_n - rS_n = a - ar^n$$
$$S_n(1 - r) = a(1 - r^n)$$
$$S_n = \frac{a(1 - r^n)}{1 - r} \qquad (r \neq 1)$$

This is the formula for the sum of a finite geometric series. We summarize this result in the box that follows.

FORMULA FOR THE SUM OF A GEOMETRIC SERIES

Let S_n denote the sum $a + ar + ar^2 + \cdots + ar^{n-1}$, and assume that $r \neq 1$. Then

$$S_n = \frac{a(1 - r^n)}{1 - r}$$

EXAMPLE 3 Evaluate the sum $\frac{1}{2^1} + \frac{1}{2^2} + \frac{1}{2^3} + \cdots + \frac{1}{2^{10}}$.

Solution This is a finite geometric series with $a = 1/2$, $r = 1/2$, and $n = 10$. Using these values in the formula for S_n yields

$$S_{10} = \frac{\frac{1}{2}[1 - (\frac{1}{2})^{10}]}{1 - \frac{1}{2}} = \frac{\frac{1}{2}[1 - \frac{1}{1024}]}{\frac{1}{2}} = 1 - \frac{1}{1024} = \frac{1023}{1024}$$

We would now like to attach a meaning to certain expressions of the form

$$a + ar + ar^2 + \cdots$$

Such an expression is called an **infinite geometric series.** The three dots indicate (intuitively at least) that the additions are to be carried out indefinitely, without end. To see how to proceed here, let's start by looking at some examples involving finite geometric series. In particular, we'll consider the series

$$\frac{1}{2^1} + \frac{1}{2^2} + \frac{1}{2^3} + \cdots + \frac{1}{2^n}$$

for increasing values of n. The idea is to look for a pattern as n grows ever larger. Let $S_1 = \frac{1}{2}$, $S_2 = \frac{1}{2^1} + \frac{1}{2^2}$, $S_3 = \frac{1}{2^1} + \frac{1}{2^2} + \frac{1}{2^3}$, and, in general,

$$S_n = \frac{1}{2^1} + \frac{1}{2^2} + \cdots + \frac{1}{2^n}$$

Then we can compute S_n for any given value of n by means of the formula for the sum of a finite geometric series. From Table 2, which displays the results of these calculations, it seems clear that as n grows larger and larger, the value of S_n grows ever closer to 1. More precisely (but leaving the details for calculus), it can be shown that the value of S_n can be made as close to 1 as we please, provided only that n is sufficiently large. For this reason, we say that the *sum of the infinite geometric series* $\frac{1}{2^1} + \frac{1}{2^2} + \frac{1}{2^3} + \cdots$ is 1. That is,

$$\frac{1}{2^1} + \frac{1}{2^2} + \frac{1}{2^3} + \cdots = 1$$

TABLE 2

n	$S_n = \frac{1}{2} + \frac{1}{2^2} + \cdots + \frac{1}{2^n}$
1	0.5
2	0.75
5	0.96875
10	0.999023437···
15	0.999969482···
20	0.999999046···
25	0.999999970···

We can arrive at this last result another way. First we compute the sum of the *finite* geometric series $\frac{1}{2^1} + \frac{1}{2^2} + \cdots + \frac{1}{2^n}$. As you can check, the result is

$$S_n = 1 - (1/2)^n$$

Now, as n grows larger and larger, the value of $(1/2)^n$ gets closer and closer to zero. Thus, as n grows ever larger, the value of S_n will more and more resemble $1 - 0$, or 1.

Now let's repeat our reasoning to obtain a formula for the sum of the infinite geometric series

$$a + ar + ar^2 + \cdots \qquad \text{where } |r| < 1$$

First we consider the finite geometric series

$$a + ar + ar^2 + \cdots + ar^{n-1}$$

The sum S_n in this case is

$$S_n = \frac{a(1 - r^n)}{1 - r}$$

We want to know how S_n behaves as n grows ever larger. This is where the assumption $|r| < 1$ is crucial. Just as $(1/2)^n$ approaches zero as n grows larger and larger, so will r^n approach zero as n grows larger and larger. Thus, as n grows ever larger, the sum S_n will more and more resemble

$$\frac{a(1-0)}{1-r} = \frac{a}{1-r}$$

For this reason, we say that the sum of the infinite geometric series is $a/(1 - r)$. We will make free use of this result in the subsequent examples. However, a more rigorous development of infinite series properly belongs to calculus.

FORMULA FOR THE SUM OF AN INFINITE GEOMETRIC SERIES

Suppose that $|r| < 1$. Then the sum S of the infinite geometric series $a + ar + ar^2 + \cdots$ is given by

$$S = \frac{a}{1-r}$$

EXAMPLE 4 Find the sum of the infinite geometric series $1 + \frac{2}{3} + \frac{4}{9} + \cdots$.

Solution In this case, we have $a = 1$ and $r = 2/3$. Thus,

$$S = \frac{a}{1-r} = \frac{1}{1-\frac{2}{3}} = \frac{1}{\frac{1}{3}} = 3$$

The sum of the series is 3.

EXAMPLE 5 Find a fraction equivalent to the repeating decimal $0.2\overline{35}$.

Solution Let $S = 0.2\overline{35}$. Then we have

$$\begin{aligned} S &= 0.2353535\cdots \\ &= \frac{2}{10} + \frac{35}{1000} + \frac{35}{100{,}000} + \frac{35}{10{,}000{,}000} + \cdots \end{aligned}$$

Now, the expression following 2/10 on the right-hand side of this last equation is an infinite geometric series in which $a = 35/1000$ and $r = 1/100$. Thus,

$$S = \frac{2}{10} + \frac{a}{1-r}$$

$$= \frac{2}{10} + \frac{\frac{35}{1000}}{1 - \frac{1}{100}} = \frac{233}{990} \qquad \text{(Check the arithmetic!)}$$

The given decimal is therefore equivalent to 233/990.

EXERCISE SET 13.5

A

1. Find the 2nd term in a geometric sequence in which the 1st term is 9, the 3rd term is 4, and the common ratio is positive.
2. Find the 5th term in a geometric sequence in which the 4th term is 4, the 6th term is 6, and the common ratio is negative.
3. The product of the first three terms in a geometric sequence is 8000. If the 1st term is 4, find the 2nd and 3rd terms.

In Exercises 4–8, find the indicated term of the given geometric sequence.

4. $9, 81, 729, \cdots; a_7$
5. $-1, 1, -1, 1, \cdots; a_{100}$
6. $1/2, 1/4, 1/8, \cdots; a_9$
7. $2/3, 4/9, 8/27, \cdots; a_8$
8. $1, -\sqrt{2}, 2, \cdots; a_6$
9. Find the common ratio in a geometric sequence in which the first term is 1 and the seventh term is 4096.
10. Find the first term in a geometric sequence in which the common ratio is 4/3 and the tenth term is 16/9.
11. Find the sum of the first ten terms of the sequence 7, 14, 28, $\cdots$.
12. Find the sum of the first five terms of the sequence $-1/2, 3/10, -9/50, \cdots$.
13. Find the sum: $1 + \sqrt{2} + 2 + \cdots + 32$.
14. Find the sum of the first 12 terms in the sequence $-4, -2, -1, \cdots$.

In Exercises 15–17, evaluate each sum.

15. $\sum_{k=1}^{6} \left(\frac{3}{2}\right)^k$
16. $\sum_{k=1}^{6} \left(\frac{2}{3}\right)^{k+1}$
17. $\sum_{k=2}^{6} \left(\frac{1}{10}\right)^k$

In Exercises 18–22, determine the sum of each infinite geometric series.

18. $\frac{1}{4} + \frac{1}{4^2} + \frac{1}{4^3} + \cdots$
19. $\frac{2}{3} - \frac{4}{9} + \frac{8}{27} - \cdots$
20. $\frac{9}{10} + \frac{9}{100} + \frac{9}{1000} + \cdots$
21. $1 + \frac{1}{1.01} + \frac{1}{(1.01)^2} + \cdots$
22. $-1 - \frac{1}{\sqrt{2}} - \frac{1}{2} - \cdots$

In Exercises 23–27, express each repeating decimal as a fraction.

23. $0.555\cdots$
24. $0.\overline{47}$
25. $0.\overline{123}$
26. $0.050505\cdots$
27. $0.4\overline{32}$

B

28. The lengths of the sides in a right triangle form three consecutive terms of a geometric sequence. Find the common ratio of the sequence. (There are two distinct answers.)
29. The product of three consecutive terms in a geometric sequence is -1000, and their sum is 15. Find the common ratio. (There are two answers.) *Suggestion:* Denote the terms by a/r, a, and ar.
30. Use mathematical induction to prove that the nth term of the geometric sequence $a, ar, ar^2, \cdots$ is ar^{n-1}.
31. Show that the sum of the following infinite geometric series is 3/2:
$$\frac{\sqrt{3}}{\sqrt{3}+1} + \frac{\sqrt{3}}{\sqrt{3}+3} + \cdots$$
32. Let A_1 denote the area of an equilateral triangle, each side of which is 1 unit long. A second equilateral triangle is formed by joining the midpoints of the sides of the first triangle. Let A_2 denote the area of this second triangle. This process is then repeated to form a third triangle with area A_3; and so on. Find the sum of the areas: $A_1 + A_2 + A_3 + \cdots$.
33. Let $a_1, a_2, a_3, \cdots$ be a geometric sequence such that $r \neq 1$. Let $S = a_1 + a_2 + a_3 + \cdots + a_n$, and let $T = \frac{1}{a_1} + \frac{1}{a_2} + \cdots + \frac{1}{a_n}$. Show that $\frac{S}{T} = a_1 a_n$.

34. Suppose that a, b, and c are three consecutive terms in a geometric sequence. Show that $\frac{1}{a+b}$, $\frac{1}{2b}$, and $\frac{1}{c+b}$ are three consecutive terms in an arithmetic sequence.

35. A ball is dropped from a height of 6 ft. Assuming that on each bounce the ball rebounds to one-third of its previous height, find the total distance traveled by the ball.

13.6 DEMOIVRE'S THEOREM

Go to Mr. DeMoivre; he knows these things better than I do.

Isaac Newton [according to Boyer's *A History of Mathematics* (New York: John Wiley and Sons, Inc., 1968)]

In this section, we explore one of the many important connections between trigonometry and the complex number system. We begin with an observation that was first made in the year 1797 by the Norwegian surveyor and mathematician Caspar Wessel. He realized, essentially, that the complex numbers can be visualized as points in the x-y plane by identifying the complex number $a + bi$ with the point (a, b). In this context, we often refer to the x-y plane as the **complex plane,** and we refer to the complex numbers as points in this plane.

EXAMPLE 1 Plot the point $2 + 3i$ in the complex plane.

Solution The complex number $2 + 3i$ is identified with the point $(2, 3)$. See Figure 1.

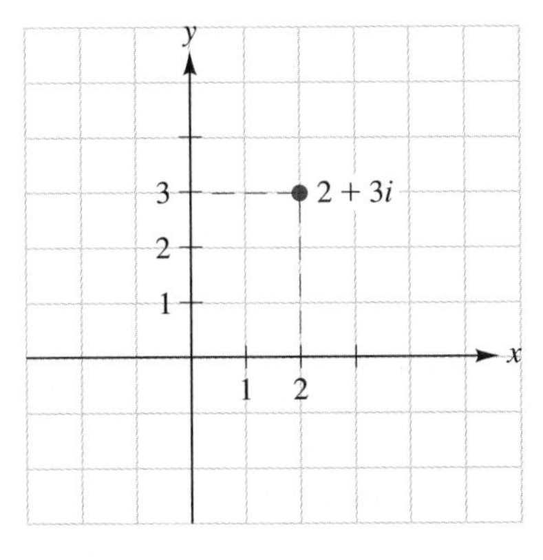

FIGURE 1

As indicated in Figure 2, the distance from the origin to the point $a + bi$ is denoted by r. We call the distance r the **modulus** of the complex number $a + bi$. The angle θ in Figure 2 (measured counterclockwise from the positive x-axis) is referred to as the **argument** of the complex number $a + bi$. (Using the terminology of Section 11.7, r and θ are the **polar coordinates** of the point $a + bi$.)

From Figure 2, we have the following three equations relating the quantities a, b, r, and θ. (Although Figure 2 shows $a + bi$ in the first quadrant, the equations remain valid for the other quadrants as well.)

$$r = \sqrt{a^2 + b^2} \tag{1}$$

$$a = r\cos\theta \tag{2}$$

$$b = r\sin\theta \tag{3}$$

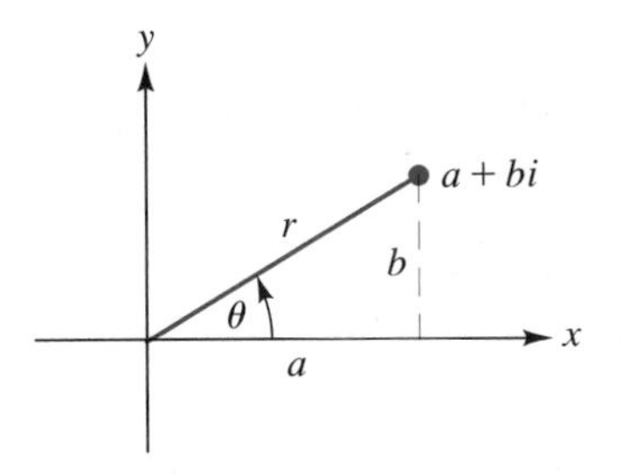

FIGURE 2

If we have a complex number $z = a + bi$, we can use equations (2) and (3) to write

$$z = (r\cos\theta) + (r\sin\theta)i = r(\cos\theta + i\sin\theta)$$

That is,

$$z = r(\cos\theta + i\sin\theta) \tag{4}$$

The expression that appears on the right-hand side of equation (4) is called the **trigonometric** (or **polar**) **form** of the complex number z. In contrast, the expression $a + bi$ is referred to as the **rectangular form** of the complex number z.

EXAMPLE 2 Express the complex number $z = 3\left(\cos\frac{\pi}{3} + i\sin\frac{\pi}{3}\right)$ in rectangular form.

Solution $z = 3\left(\cos \frac{\pi}{3} + i \sin \frac{\pi}{3}\right)$
$= 3\left(\frac{1}{2} + \frac{1}{2}\sqrt{3}i\right) = \frac{3}{2} + \frac{3}{2}\sqrt{3}i$

The rectangular form is therefore $\frac{3}{2} + \frac{3}{2}\sqrt{3}i$.

EXAMPLE 3 Find the trigonometric form for the complex number $-\sqrt{2} + i\sqrt{2}$.

Solution We are asked to write the given number in the form $r(\cos \theta + i \sin \theta)$, so we need to find r and θ. Using equation (1) and the values $a = -\sqrt{2}$, $b = \sqrt{2}$, we have

$$r = \sqrt{(-\sqrt{2})^2 + (\sqrt{2})^2} = \sqrt{2+2} = \sqrt{4} = 2$$

Now that we know r, we can use equations (2) and (3) to determine θ. From equation (2), we obtain

$$\cos \theta = \frac{a}{r} = \frac{-\sqrt{2}}{2} \tag{5}$$

Similarly, equation (3) gives us

$$\sin \theta = \frac{b}{r} = \frac{\sqrt{2}}{2} \tag{6}$$

One angle satisfying both of equations (5) and (6) is $\theta = 3\pi/4$. (There are other angles, and we'll return to this point in a moment.) In summary, then, we have $r = 2$ and $\theta = 3\pi/4$, so the required trigonometric form is

$$2\left(\cos \frac{3\pi}{4} + i \sin \frac{3\pi}{4}\right)$$

In the example we just completed, we noted that $\theta = 3\pi/4$ was only one angle satisfying the conditions $\cos \theta = -\sqrt{2}/2$ and $\sin \theta = \sqrt{2}/2$. Another such angle is $\frac{3\pi}{4} + 2\pi$. Indeed, any angle of the form $\frac{3\pi}{4} + 2\pi k$, where k is an integer, would do just as well. The upshot of this is that θ, the argument of a complex number, is not uniquely determined. In Example 3, we followed a common convention in converting to trigonometric form: we picked θ in the interval $0 \leq \theta < 2\pi$. Furthermore, although it won't cause us any difficulties in this section, you might also note that the argument θ is undefined for the complex number $0 + 0i$. (Why?)

We are now ready to derive a formula that will make it easy to multiply two complex numbers in trigonometric form. Suppose that the two complex numbers are

$$r(\cos \alpha + i \sin \alpha) \qquad \text{and} \qquad R(\cos \beta + i \sin \beta)$$

Then their product is

$$\begin{aligned} &rR[(\cos \alpha + i \sin \alpha)(\cos \beta + i \sin \beta)] \\ &\quad = rR[(\cos \alpha \cos \beta - \sin \alpha \sin \beta) + i(\sin \alpha \cos \beta + \cos \alpha \sin \beta)] \\ &\quad = rR[\cos(\alpha + \beta) + i \sin(\alpha + \beta)] \qquad \text{using the addition formulas from Section 8.1} \end{aligned}$$

Notice that the modulus of the product is rR, which is the product of the two original moduli. Also, the argument is $\alpha + \beta$, which is the *sum* of the two original arguments. So, to multiply two complex numbers, just multiply the moduli and add the arguments. There is a similar rule for obtaining the quotient of two complex numbers: Divide their moduli and subtract the arguments. These two rules are stated more precisely in the box that follows. (For a proof of the division rule, see Exercise 75.)

Let $z = r(\cos\alpha + i\sin\alpha)$ and $w = R(\cos\beta + i\sin\beta)$. Then

$$zw = rR[\cos(\alpha+\beta) + i\sin(\alpha+\beta)]$$

Also, if $R \neq 0$, then

$$\frac{z}{w} = \frac{r}{R}[\cos(\alpha-\beta) + i\sin(\alpha-\beta)]$$

EXAMPLE 4 Let $z = 8\left(\cos\frac{5\pi}{3} + i\sin\frac{5\pi}{3}\right)$ and $w = 4\left(\cos\frac{2\pi}{3} + i\sin\frac{2\pi}{3}\right)$. Compute **(a)** zw and **(b)** z/w. Express each answer in both trigonometric and rectangular form.

Solution

(a) $zw = (8)(4)\left[\cos\left(\frac{5\pi}{3} + \frac{2\pi}{3}\right) + i\sin\left(\frac{5\pi}{3} + \frac{2\pi}{3}\right)\right]$

$= 32\left(\cos\frac{7\pi}{3} + i\sin\frac{7\pi}{3}\right) = 32\left(\cos\frac{\pi}{3} + i\sin\frac{\pi}{3}\right)$ trigonometric form

$= 32\left(\frac{1}{2} + \frac{1}{2}\sqrt{3}i\right) = 16 + 16\sqrt{3}i$ rectangular form

(b) $\frac{z}{w} = \frac{8}{4}\left[\cos\left(\frac{5\pi}{3} - \frac{2\pi}{3}\right) + i\sin\left(\frac{5\pi}{3} - \frac{2\pi}{3}\right)\right]$

$= 2(\cos\pi + i\sin\pi)$ trigonometric form

$= 2(-1 + i\cdot 0) = -2$ rectangular form

EXAMPLE 5 Compute z^2, where $z = r(\cos\theta + i\sin\theta)$.

Solution $z^2 = [r(\cos\theta + i\sin\theta)][r(\cos\theta + i\sin\theta)]$

$= r^2[\cos(\theta+\theta) + i\sin(\theta+\theta)]$

$= r^2(\cos 2\theta + i\sin 2\theta)$

The result in Example 5 is a particular case of an important theorem attributed to Abraham DeMoivre (1667–1754). In the box that follows, we state *DeMoivre's theorem.* (The theorem can be proved using mathematical induction, which we discussed in Section 13.1.)

DEMOIVRE'S THEOREM

Let n be a natural number. Then

$$[r(\cos\theta + i\sin\theta)]^n = r^n(\cos n\theta + i\sin n\theta)$$

EXAMPLE 6 Use DeMoivre's theorem to compute $\left(-\sqrt{2} + i\sqrt{2}\right)^5$. Express your answer in rectangular form.

Solution In Example 3, we saw that the trigonometric form of $-\sqrt{2} + i\sqrt{2}$ is given by

$$-\sqrt{2} + i\sqrt{2} = 2\left(\cos \tfrac{3\pi}{4} + i \sin \tfrac{3\pi}{4}\right)$$

Therefore,

$$\begin{aligned}(-\sqrt{2} + i\sqrt{2})^5 &= 2^5\left(\cos \tfrac{15\pi}{4} + i \sin \tfrac{15\pi}{4}\right)\\ &= 32\left[\tfrac{1}{2}\sqrt{2} + i\left(-\tfrac{1}{2}\sqrt{2}\right)\right]\\ &= 16\sqrt{2} - 16\sqrt{2}i\end{aligned}$$

The next two examples show how DeMoivre's theorem is used in computing roots. If n is a natural number and $z^n = w$, then we say that z is an ***n*th root** of w. The work in the examples also relies on the following observation about equality between nonzero complex numbers in trigonometric form: If $r(\cos\theta + i\sin\theta) = R(\cos A + i \sin A)$, then $r = R$ and $\theta = A + 2\pi k$, where k is an integer.

EXAMPLE 7 Find the cube roots of $8i$.

Solution First we express $8i$ in trigonometric form. As can be seen from Figure 3,

$$8i = 8\left(\cos \tfrac{\pi}{2} + i \sin \tfrac{\pi}{2}\right)$$

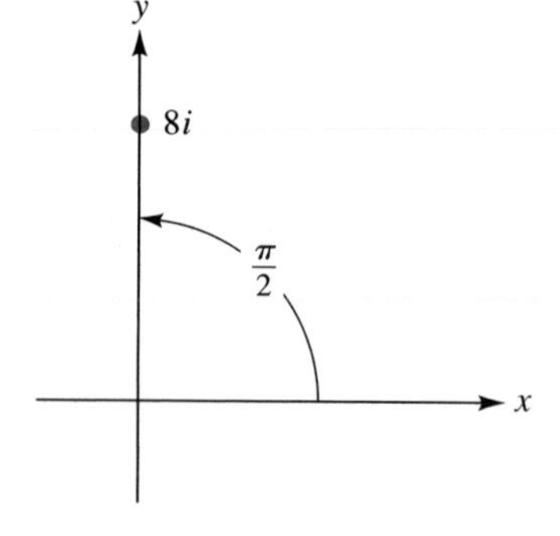

FIGURE 3

Now we let $z = r(\cos\theta + i\sin\theta)$ denote a cube root of $8i$. Then the equation $z^3 = 8i$ becomes

$$r^3(\cos 3\theta + i \sin 3\theta) = 8\left(\cos \tfrac{\pi}{2} + i \sin \tfrac{\pi}{2}\right) \tag{7}$$

From equation (7) we conclude that $r^3 = 8$ and, consequently, that $r = 2$. Also from equation (7), we have

$$3\theta = \tfrac{\pi}{2} + 2\pi k$$

or

$$\theta = \tfrac{\pi}{6} + \tfrac{2\pi k}{3} \qquad \text{dividing by 3} \tag{8}$$

If we let $k = 0$, equation (8) yields $\theta = \pi/6$. Thus, one of the cube roots of $8i$ is

$$2\left(\cos \tfrac{\pi}{6} + i \sin \tfrac{\pi}{6}\right) = 2\left(\tfrac{1}{2}\sqrt{3} + i \cdot \tfrac{1}{2}\right) = \sqrt{3} + i$$

Next we let $k = 1$ in equation (8). As you can check, this yields $\theta = 5\pi/6$. So another cube root of $8i$ is

$$2\left(\cos \tfrac{5\pi}{6} + i \sin \tfrac{5\pi}{6}\right) = 2\left(-\tfrac{1}{2}\sqrt{3} + i \cdot \tfrac{1}{2}\right) = -\sqrt{3} + i$$

Similarly, using $k = 2$ in equation (8) yields $\theta = 3\pi/2$. (Verify this.) Consequently, a third cube root is

$$2\left(\cos \tfrac{3\pi}{2} + i \sin \tfrac{3\pi}{2}\right) = 2[0 + i(-1)] = -2i$$

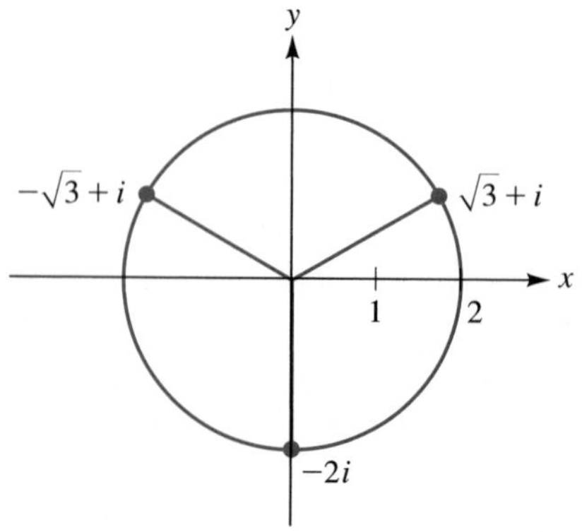

We have now found three distinct cube roots of $8i$. If we were to continue the process, using $k = 3$ for example, we would find that no additional roots are obtained in this manner. We therefore conclude that there are exactly three cube roots of $8i$. We have plotted these cube roots in Figure 4. Notice that the points lie equally spaced on a circle of radius 2 about the origin.

FIGURE 4

When we used DeMoivre's theorem in Example 7, we found that the number $8i$ had three distinct cube roots. Along these same lines, it is true in general that any nonzero number $a + bi$ possesses exactly n distinct nth roots. This is true even if $b = 0$. In the next example, for instance, we compute the five fifth roots of 2.

EXAMPLE 8 Compute the five fifth roots of 2.

Solution We will follow the procedure we used in Example 7. In trigonometric form, the number 2 becomes $2(\cos 0 + i \sin 0)$. Now we let $z = r(\cos\theta + i\sin\theta)$ denote a fifth root of 2. Then the equation $z^5 = 2$ becomes

$$r^5(\cos 5\theta + i \sin 5\theta) = 2(\cos 0 + i \sin 0) \qquad (9)$$

From equation (9), we see that $r^5 = 2$ and, consequently, $r = 2^{1/5}$. Also from equation (9), we have

$$5\theta = 0 + 2\pi k \qquad \text{or} \qquad \theta = 2\pi k/5$$

Using the values $k = 0$, 1, 2, 3, and 4 in succession, we obtain the following results:

k	0	1	2	3	4
θ	0	$2\pi/5$	$4\pi/5$	$6\pi/5$	$8\pi/5$

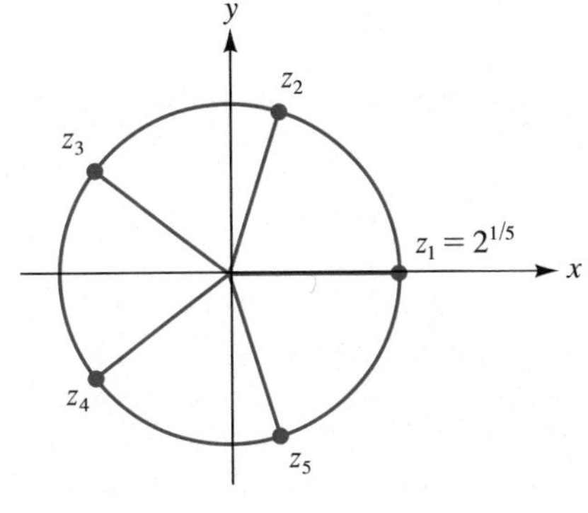

FIGURE 5

The five fifth roots of 2 are therefore

$$z_1 = 2^{1/5}(\cos 0 + i \sin 0) = 2^{1/5}$$
$$z_2 = 2^{1/5}\left(\cos \tfrac{2\pi}{5} + i \sin \tfrac{2\pi}{5}\right)$$
$$z_3 = 2^{1/5}\left(\cos \tfrac{4\pi}{5} + i \sin \tfrac{4\pi}{5}\right)$$
$$z_4 = 2^{1/5}\left(\cos \tfrac{6\pi}{5} + i \sin \tfrac{6\pi}{5}\right)$$
$$z_5 = 2^{1/5}\left(\cos \tfrac{8\pi}{5} + i \sin \tfrac{8\pi}{5}\right)$$

In Figure 5, we have plotted these five fifth roots. Notice that the points are equally spaced on a circle of radius $2^{1/5}$ about the origin.

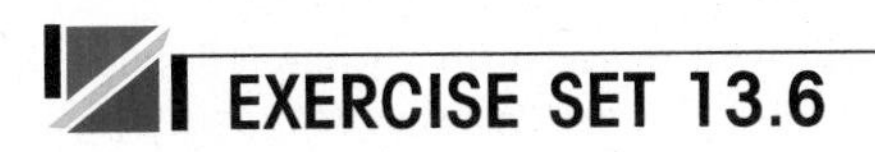

EXERCISE SET 13.6

A

In Exercises 1–8, plot each point in the complex plane.

1. $4 + 2i$ **2.** $4 - 2i$ **3.** $-5 + i$
4. $-3 - 5i$ **5.** $1 - 4i$ **6.** i
7. $-i$ **8.** $1\ (= 1 + 0i)$

In Exercises 9–18, convert each complex number to rectangular form.

9. $2\left(\cos \frac{1}{4}\pi + i \sin \frac{1}{4}\pi\right)$ **10.** $6\left(\cos \frac{5}{3}\pi + i \sin \frac{5}{3}\pi\right)$
11. $4\left(\cos \frac{5}{6}\pi + i \sin \frac{5}{6}\pi\right)$ **12.** $3\left(\cos \frac{3}{2}\pi + i \sin \frac{3}{2}\pi\right)$
13. $\sqrt{2}(\cos 225° + i \sin 225°)$
14. $\frac{1}{2}(\cos 240° + i \sin 240°)$
15. $\sqrt{3}\left(\cos \frac{1}{2}\pi + i \sin \frac{1}{2}\pi\right)$ **16.** $5(\cos \pi + i \sin \pi)$
17. $4(\cos 75° + i \sin 75°)$ *Hint:* Use the addition formulas from Section 8.1 to evaluate cos 75° and sin 75°.
18. $2\left(\cos \frac{1}{8}\pi + i \sin \frac{1}{8}\pi\right)$ *Hint:* Use the half-angle formulas from Section 8.2 to evaluate $\cos(\pi/8)$ and $\sin(\pi/8)$.

In Exercises 19–28, convert from rectangular to trigonometric form. (In each case, choose an argument θ such that $0 \leqslant \theta < 2\pi$.)

19. $\frac{1}{2}\sqrt{3} + \frac{1}{2}i$ **20.** $\sqrt{2} + \sqrt{2}i$
21. $-1 + \sqrt{3}i$ **22.** -4

23. $-2\sqrt{3} - 2i$
24. $-3\sqrt{2} - 3\sqrt{2}i$
25. $-6i$
26. $-4 - 4\sqrt{3}i$
27. $\frac{1}{4}\sqrt{3} - \frac{1}{4}i$
28. 16

In Exercises 29–54, carry out the indicated operations. Express your results in rectangular form for those cases in which the trigonometric functions are readily evaluated without tables or a calculator.

29. $2(\cos 22° + i \sin 22°) \times 3(\cos 38° + i \sin 38°)$
30. $4(\cos 5° + i \sin 5°) \times 6(\cos 130° + i \sin 130°)$
31. $\sqrt{2}\left(\cos \frac{1}{3}\pi + i \sin \frac{1}{3}\pi\right) \times \sqrt{2}\left(\cos \frac{4}{3}\pi + i \sin \frac{4}{3}\pi\right)$
32. $\left(\cos \frac{1}{5}\pi + i \sin \frac{1}{5}\pi\right) \times \left(\cos \frac{1}{20}\pi + i \sin \frac{1}{20}\pi\right)$
33. $3\left(\cos \frac{1}{7}\pi + i \sin \frac{1}{7}\pi\right) \times \sqrt{2}\left(\cos \frac{1}{7}\pi + i \sin \frac{1}{7}\pi\right)$
34. $\sqrt{3}(\cos 3° + i \sin 3°) \times \sqrt{3}(\cos 38° + i \sin 38°)$
35. $6(\cos 50° + i \sin 50°) \div 2(\cos 5° + i \sin 5°)$
36. $\sqrt{3}(\cos 140° + i \sin 140°) \div 3(\cos 5° + i \sin 5°)$
37. $2^{4/3}\left(\cos \frac{5}{12}\pi + i \sin \frac{5}{12}\pi\right) \div 2^{1/3}\left(\cos \frac{1}{4}\pi + i \sin \frac{1}{4}\pi\right)$
38. $\sqrt{6}\left(\cos \frac{16}{9}\pi + i \sin \frac{16}{9}\pi\right) \div \sqrt{2}\left(\cos \frac{1}{9}\pi + i \sin \frac{1}{9}\pi\right)$
39. $\left(\cos \frac{2}{5}\pi + i \sin \frac{2}{5}\pi\right) \div \left(\cos \frac{2}{5}\pi + i \sin \frac{2}{5}\pi\right)$
40. $\left(\cos \frac{2}{5}\pi + i \sin \frac{2}{5}\pi\right) \div \left(\cos \frac{1}{10}\pi + i \sin \frac{1}{10}\pi\right)$
41. $\left[3\left(\cos \frac{1}{3}\pi + i \sin \frac{1}{3}\pi\right)\right]^5$
42. $\left[\sqrt{2}\left(\cos \frac{5}{6}\pi + i \sin \frac{5}{6}\pi\right)\right]^4$
43. $\left[\frac{1}{2}\left(\cos \frac{1}{24}\pi + i \sin \frac{1}{24}\pi\right)\right]^6$
44. $[\sqrt{3}(\cos 70° + i \sin 70°)]^3$
45. $[2^{1/5}(\cos 63° + i \sin 63°)]^{10}$
46. $\left[2\left(\cos \frac{1}{5}\pi + i \sin \frac{1}{5}\pi\right)\right]^3$
47. $2(\cos 200° + i \sin 200°) \times \sqrt{2}(\cos 20° + i \sin 20°) \times \frac{1}{2}(\cos 5° + i \sin 5°)$
48. $\left[\dfrac{\cos(\pi/8) + i \sin(\pi/8)}{\cos(-\pi/8) + i \sin(-\pi/8)}\right]^5$
49. $\left[\frac{1}{2}(1 - \sqrt{3}i)\right]^5$ *Hint:* Convert to trigonometric form.
50. $(1 - i)^3$
51. $(-2 - 2i)^5$
52. $\left(-\frac{1}{2} + \frac{1}{2}\sqrt{3}i\right)^6$
53. $\left(-2\sqrt{3} - 2i\right)^4$
54. $(1 + i)^{16}$

In Exercises 55–61, use DeMoivre's theorem to find the indicated roots. Express the results in rectangular form.

55. Cube roots of $-27i$
56. Cube roots of 2
57. Eighth roots of 1
58. Square roots of i
59. Cube roots of 64
60. Square roots of $-\frac{1}{2} - \frac{1}{2}\sqrt{3}i$
61. Sixth roots of 729

Use a calculator to complete Exercises 62–65.

62. Compute $(9 + 9i)^6$.
63. Compute $(7 - 7i)^8$.
64. Compute the cube roots of $1 + 2i$. Express your answers in rectangular form, with the real and imaginary parts rounded to two decimal places.
65. Compute the fifth roots of i. Express your answers in rectangular form, with the real and imaginary parts rounded to two decimal places.

B

In Exercises 66–68, find the indicated roots. Express the results in rectangular form.

66. Find the fourth roots of i. *Hint:* Use the half-angle formulas from Section 8.2.
67. Find the fourth roots of $8 - 8\sqrt{3}i$. *Hint:* Use the addition formulas or the half-angle formulas.
68. Find the square roots of $7 + 24i$. *Hint:* You'll need to use the half-angle formulas from Section 8.2.
69. **(a)** Compute the three cube roots of 1.
    (b) Let z_1, z_2, and z_3 denote the three cube roots of 1. Verify that $z_1 + z_2 + z_3 = 0$ and also that $z_1z_2 + z_2z_3 + z_3z_1 = 0$.
70. **(a)** Compute the four fourth roots of 1.
    (b) Verify that the sum of these four fourth roots is 0.
71. Evaluate $\left(-\frac{1}{2} + \frac{1}{2}\sqrt{3}i\right)^5 + \left(-\frac{1}{2} - \frac{1}{2}\sqrt{3}i\right)^5$. *Hint:* Use DeMoivre's theorem.
72. Show that $\left(-\frac{1}{2} + \frac{1}{2}\sqrt{3}i\right)^6 + \left(-\frac{1}{2} - \frac{1}{2}\sqrt{3}i\right)^6 = 2$.
73. Compute $(\cos \theta + i \sin \theta)(\cos \theta - i \sin \theta)$.
74. In the identity $(\cos \theta + i \sin \theta)^2 = \cos 2\theta + i \sin 2\theta$, carry out the multiplication on the left-hand side of the equation. Then equate the corresponding real parts and imaginary parts from each side of the equation that results. What do you obtain?
75. Show that

$$\frac{r(\cos \alpha + i \sin \alpha)}{R(\cos \beta + i \sin \beta)} = \frac{r}{R}[\cos(\alpha - \beta) + i \sin(\alpha - \beta)]$$

Suggestion: Begin with the quantity on the left side and multiply it by $(\cos \beta - i \sin \beta)/(\cos \beta - i \sin \beta)$.

76. Show that

$$1 + \cos \theta + i \sin \theta = 2 \cos\left(\tfrac{\theta}{2}\right)\left[\cos\left(\tfrac{\theta}{2}\right) + i \sin\left(\tfrac{\theta}{2}\right)\right]$$

77. If $z = r(\cos \theta + i \sin \theta)$, show that

$$\frac{1}{z} = \frac{1}{r}(\cos \theta - i \sin \theta)$$

Hint: $1/z = [1(\cos 0 + i \sin 0)]/[r(\cos \theta + i \sin \theta)]$.

78. Show that $\dfrac{1 + \sin \theta + i \cos \theta}{1 + \sin \theta - i \cos \theta} = \sin \theta + i \cos \theta$.
Assume that $\theta \neq \frac{3}{2}\pi + 2\pi k$, where k is an integer.
Hint: Work with the left-hand side; first "rationalize"

the denominator by multiplying by the quantity $\dfrac{(1 + \sin\theta) + i\cos\theta}{(1 + \sin\theta) + i\cos\theta}$.

79. If $w + \dfrac{1}{w} = 2\cos\theta$, show that $w = \cos\theta \pm i\sin\theta$.

Hint: Use the quadratic formula to solve the given equation for w.

80. If $w + \dfrac{1}{w} = 2\cos\theta$, show that $w^n + \dfrac{1}{w^n} = 2\cos n\theta$.

Hint: Use the results in Exercises 79 and 77.

CHAPTER THIRTEEN SUMMARY OF PRINCIPAL TERMS AND FORMULAS

TERMS OR FORMULAS	PAGE REFERENCE	COMMENT
1. The principle of mathematical induction	823	This principle can be stated as follows. Suppose that for each natural number n we have a statement P_n. Suppose P_1 is true. Also suppose that P_{k+1} is true whenever P_k is true. Then according to the principle of mathematical induction, all of the statements are true; that is, P_n is true for all natural numbers n. The principle of mathematical induction has the status of an axiom; that is, we accept its validity without proof.
2. Pascal's triangle	829	Pascal's triangle refers to the triangular array of numbers displayed on page 829. Additional rows can be added to the triangle according to the following rule: Each entry in the array (other than the 1's along the sides) is the sum of the two numbers diagonally above it. The numbers in the nth row of Pascal's triangle are the coefficients of the terms in the expansion of $(a + b)^{n-1}$.
3. $n!$ (read: n factorial)	831	This denotes the product of the first n natural numbers. For example, $4! = (4)(3)(2)(1) = 24$.
4. $\binom{n}{k}$ (read: n choose k)	832	Let n and k be nonnegative integers with $k \leq n$. Then the *binomial coefficient* $\binom{n}{k}$ is defined by $$\binom{n}{k} = \frac{n!}{k!(n-k)!}$$
5. The binomial theorem	833	The binomial theorem is a formula that allows us to analyze and expand expressions of the form $(a + b)^n$. If we use the *sigma notation,* then the statement of the binomial theorem that appears on page 833 can be abbreviated to read $$(a + b)^n = \sum_{k=0}^{n} \binom{n}{k} a^{n-k} b^k$$
6. The complex plane	855	Refer to Figure 2 on page 855. A complex number $a + bi$ can be identified with the point (a, b) in the x-y plane. In this context, the x-y plane is referred to as the complex plane. The distance r in the figure is the *modulus* of the complex number $a + bi$; the angle θ is an *argument* of $a + bi$.

TERMS OR FORMULAS	PAGE REFERENCE	COMMENT
7. $z = r(\cos\theta + i\sin\theta)$	855	The expression on the right-hand side is the *trigonometric form* of the complex number $z = a + bi$. (The expression $a + bi$ is the *rectangular form* of the complex number z.) The trigonometric form and the rectangular form are related by the following equations: $r = \sqrt{a^2 + b^2}$ $\quad a = r\cos\theta \quad b = r\sin\theta$
8. DeMoivre's theorem	857	Let n be a natural number and let $z = r(\cos\theta + i\sin\theta)$. Then DeMoivre's theorem states that $z^n = r^n(\cos n\theta + i\sin n\theta)$ For applications of this theorem, see Examples 6, 7, and 8 in Section 13.6.

WRITING MATHEMATICS

1. A sequence is defined recursively as follows: $a_0 = \pi$; $a_{n+1} = \dfrac{a_n}{a_n + 1}$ for $n \geqslant 0$. Compute a_0, a_1, and a_2. Describe (in a complete sentence or two) what pattern you observe. Make a conjecture about a general formula for a_n. Then use mathematical induction to prove that the formula is valid. (As in the examples in the text, use complete sentences in writing and explaining the proof.)
2. A sequence is defined recursively as follows.
$$a_1 = 3/2$$
$$a_{n+1} = a_n^2 - 2a_n + 2 \quad \text{for } n \geqslant 1$$
Use mathematical induction to show that $1 < a_n < 2$ for all natural numbers n.
3. Study the procedure on page 851 to obtain the formula for the sum of a finite geometric series. Work with a classmate or your instructor to apply the procedure to find the sum of the series $1 + 2\sin\theta + 3\sin^2\theta + 4\sin^3\theta + \cdots + n\sin^{n-1}\theta$. Then, on your own, write a paragraph or two explaining in detail what you've done.
4. Study the following purported ''proof'' by mathematical induction, and explain the fallacy.

   Let P_n denote the following statement: In any set of n numbers, the numbers are all equal.

   STEP 1 P_1 is clearly true.
   STEP 2 Assume that P_k is true. We must show that P_{k+1} is true. That is, given a set of $k + 1$ numbers $a_1, a_2, a_3, \ldots, a_{k+1}$, we must show that $a_1 = a_2 = a_3 = \cdots = a_{k+1}$. By the inductive hypothesis, however, we have $a_1 = a_2 = a_3 = \cdots = a_k$ and also $a_2 = a_3 = a_4 = \cdots = a_{k+1}$. These last two sets of equations imply that $a_1 = a_2 = a_3 = \cdots = a_k = a_{k+1}$, as we wished to show. Having now completed Steps 1 and 2, we conclude that P_n is true for every n.

CHAPTER THIRTEEN REVIEW EXERCISES

In Exercises 1–10, use the principle of mathematical induction to show that the statements are true for all natural numbers.

1. $5 + 10 + 15 + \cdots + 5n = \frac{5}{2}n(n+1)$

2. $10 + 10^2 + 10^3 + \cdots + 10^n = \frac{10}{9}(10^n - 1)$

3. $1 \cdot 2 + 2 \cdot 3 + 3 \cdot 4 + \cdots + n(n+1) = \frac{1}{3}n(n+1)(n+2)$

4. $\frac{1}{2} + \frac{2}{2^2} + \frac{3}{2^3} + \cdots + \frac{n}{2^n} = 2 - \frac{2+n}{2^n}$

5. $1 + 3 \cdot 2 + 5 \cdot 2^2 + 7 \cdot 2^3 + \cdots + (2n-1) \cdot 2^{n-1} = 3 + (2n-3) \cdot 2^n$

6. $\frac{1}{1 \cdot 4} + \frac{1}{4 \cdot 7} + \frac{1}{7 \cdot 10} + \cdots + \frac{1}{(3n-2)(3n+1)} = \frac{n}{3n+1}$

7. $1 + 2^2 \cdot 2 + 3^2 \cdot 2^2 + 4^2 \cdot 2^3 + \cdots + n^2 \cdot 2^{n-1} = (n^2 - 2n + 3)2^n - 3$

8. 9 is a factor of $n^3 + (n+1)^3 + (n+2)^3$.

9. 3 is a factor of $7^n - 1$.

10. 8 is a factor of $9^n - 1$.

In Exercises 11–20, expand the given expressions.

11. $(3a + b^2)^4$

12. $(5a - 2b)^3$

13. $(x + \sqrt{x})^4$

14. $(1 - \sqrt{3})^6$

15. $(x^2 - 2y^2)^5$

16. $\left(\frac{1}{a} + \frac{2}{b}\right)^3$

17. $\left(1 + \frac{1}{x}\right)^5$

18. $\left(x^3 + \frac{1}{x^2}\right)^6$

19. $(a\sqrt{b} - b\sqrt{a})^4$

20. $(x^{-2} + y^{5/2})^8$

21. Find the fifth term in the expansion of $(3x + y^2)^5$.

22. Find the eighth term in the expansion of $(2x - y)^9$.

23. Find the coefficient of the term containing a^5 in the expansion of $(a - 2b)^7$.

24. Find the coefficient of the term containing b^8 in the expansion of $\left(2a - \frac{b}{3}\right)^{10}$.

25. Find the coefficient of the term containing x^3 in the expansion of $(1 + \sqrt{x})^8$.

26. Expand $(1 + \sqrt{x} + x)^6$. *Suggestion:* Rewrite the expression as $[(1 + \sqrt{x}) + x]^6$.

In Exercises 27–32, verify each assertion by computing the indicated binomial coefficients.

27. $\binom{2}{0}^2 + \binom{2}{1}^2 + \binom{2}{2}^2 = \binom{4}{2}$

28. $\binom{3}{0}^2 + \binom{3}{1}^2 + \binom{3}{2}^2 + \binom{3}{3}^2 = \binom{6}{3}$

29. $\binom{4}{0}^2 + \binom{4}{1}^2 + \binom{4}{2}^2 + \binom{4}{3}^2 + \binom{4}{4}^2 = \binom{8}{4}$

30. $\binom{2}{0} + \binom{2}{1} + \binom{2}{2} = 2^2$

31. $\binom{3}{0} + \binom{3}{1} + \binom{3}{2} + \binom{3}{3} = 2^3$

32. $\binom{4}{0} + \binom{4}{1} + \binom{4}{2} + \binom{4}{3} + \binom{4}{4} = 2^4$

In Exercises 33–38, compute the first four terms in each sequence. Also, in Exercises 33–35, graph the sequences for $n = 1, 2, 3,$ and 4.

33. $a_n = \frac{2n}{n+1}$

34. $a_n = \frac{3n-2}{3n+2}$

35. $a_n = (-1)^n\left(1 - \frac{1}{n+1}\right)$

36. $a_0 = 4;\ a_n = 2a_{n-1},\ n \geqslant 1$

37. $a_0 = -3;\ a_n = 4a_{n-1},\ n \geqslant 1$

38. $a_0 = 1;\ a_1 = 2;\ a_n = 3a_{n-1} + 2a_{n-2},\ n \geqslant 2$

In Exercises 39 and 40, evaluate each sum.

39. **(a)** $\sum_{k=1}^{3} (-1)^k(2k+1)$ **(b)** $\sum_{k=0}^{8} \left(\frac{1}{k+1} - \frac{1}{k+2}\right)$

40. **(a)** $\sum_{j=1}^{4} \frac{(-1)^j}{j}$ **(b)** $\sum_{n=1}^{5} \left(\frac{1}{n} - \frac{1}{n+1}\right)$

In Exercises 41 and 42, rewrite each sum using sigma notation.

41. $\frac{5}{3} + \frac{5}{3^2} + \frac{5}{3^3} + \frac{5}{3^4} + \frac{5}{3^5}$

42. $\frac{1}{2} - \frac{2}{2^2} + \frac{3}{2^3} - \frac{4}{2^4} + \frac{5}{2^5} - \frac{6}{2^6}$

In Exercises 43–46, find the indicated term in each sequence.

43. 5, 9, 13, 17, ⋯; a_{18}

44. 5, 9/2, 4, 7/2, ⋯; a_{20}

45. 10, 5, 5/2, 5/4, ⋯; a_{12}

46. $\sqrt{2}+1,\ 1,\ \sqrt{2}-1,\ 3-2\sqrt{2},\ \cdots;\ a_{10}$

47. Determine the sum of the first 12 terms of an arithmetic sequence in which the first term is 8 and the twelfth term is 43/2.

48. Find the sum of the first 45 terms in the sequence 10, 29/3, 28/3, 9, ⋯.

49. Find the sum of the first 10 terms in the sequence 7, 70, 700, ⋯.

50. Find the sum of the first 12 terms in the sequence 1/3, −2/9, 4/27, −8/81, ⋯.

51. In a certain geometric sequence, the third term is 4 and the fifth term is 10. Find the sixth term, given that the common ratio is negative.

52. For a certain infinite geometric series, the first term is 2 and the sum is 18/11. Find the common ratio r.

In Exercises 53–56, find the sum of each infinite geometric series.

53. $\frac{3}{5}+\frac{3}{25}+\frac{3}{125}+\cdots$

54. $\frac{7}{10}+\frac{7}{100}+\frac{7}{1000}+\cdots$

55. $\frac{1}{9}-\frac{1}{81}+\frac{1}{729}-\cdots$

56. $1+\frac{1}{1+\sqrt{2}}+\frac{1}{(1+\sqrt{2})^2}+\cdots$

57. Find a fraction equivalent of $0.\overline{45}$.

58. Find a fraction equivalent to $0.2\overline{13}$.

In Exercises 59–62, verify each equation using the formula for the sum of an arithmetic series:

$$S_n=\frac{n}{2}[2a+(n-1d]$$

(The formulas given in Exercises 59–62 appear in Elements of Algebra *by Leonhard Euler, first published in 1770.)*

59. $1+2+3+\cdots+n=n+n(n-1)/2$

60. $1+3+5+\cdots$ (to n terms) $=n+2n(n-1)/2$

61. $1+4+7+\cdots$ (to n terms) $=n+3n(n-1)/2$

62. $1+5+9+\cdots$ (to n terms) $=n+4n(n-1)/2$

63. In this exercise, you will use the following (remarkably simple) formula for approximating sums of powers of integers:

$$1^k+2^k+3^k+\cdots+n^k\approx\frac{\left(n+\frac{1}{2}\right)^{k+1}}{k+1} \qquad (1)$$

[This formula appears in the article, "Sums of Powers of Integers" by B. L. Burrows and R. F. Talbot, published in the *American Mathematical Monthly,* vol. 91 (1984), p. 394.]

(a) Use formula (1) to estimate the sum $1^2+2^2+3^2+\cdots+50^2$. Round your answer to the nearest integer.

(b) Compute the exact value of the sum in part (a) using the formula $\sum_{k=1}^{n}k^2=\frac{1}{6}n(n+1)(2n+1)$. (This formula can be proved using mathematical induction.) Then compute the percent error for the approximation obtained in part (a). The percent error is given by

$$\frac{|\text{actual value}-\text{approximate value}|}{\text{actual value}}\times 100$$

(c) Use formula (1) to estimate the sum $1^4+2^4+3^4+\cdots+200^4$. Round your answer to six significant digits.

(d) The following formula for the sum $1^4+2^4+\cdots+n^4$ can be proved using mathematical induction:

$$\sum_{k=1}^{n}k^4=\frac{n(n+1)(2n+1)(3n^2+3n-1)}{30}$$

Use this formula to compute the sum in part (c). Round your answer to six significant digits. Then use this result to compute the percent error for the approximation in part (c).

64. According to *Stirling's formula* [named after James Stirling (1692–1770)], the quantity $n!$ can be approximated as follows:

$$n!\approx\sqrt{2\pi n}\left(\frac{n}{e}\right)^n$$

In this formula, e is the constant 2.718... (discussed in Section 5.2). Use a calculator to complete the following table. Round your answers to five significant digits. As you will see, the numbers in the right-hand column approach 1 as n increases. This shows that, in a certain sense, the approximation improves as n increases.

n	$n!$	$\sqrt{2\pi n}\,(n/e)^n$	$\frac{n!}{\sqrt{2\pi n}\,(n/e)^n}$
10			
20			
30			
40			
50			
60			
65			

65. The nonzero numbers a, b, and c are consecutive terms in a geometric sequence, and $a + b + c = 70$. Furthermore, $4a$, $5b$, and $4c$ are consecutive terms in an arithmetic sequence. Find a, b, and c.

66. The nonzero numbers a, b, and c are consecutive terms in a geometric sequence, and a, $2b$, and c are consecutive terms in an arithmetic sequence. Show that the common ratio in the geometric sequence must be either $2 + \sqrt{3}$ or $2 - \sqrt{3}$.

67. If the numbers $\dfrac{1}{b+c}$, $\dfrac{1}{c+a}$, and $\dfrac{1}{a+b}$ are consecutive terms in an arithmetic sequence, show that a^2, b^2, and c^2 are also consecutive terms in an arithmetic sequence.

68. If a, b, and c are consecutive terms in a geometric sequence, prove that $\dfrac{1}{a+b}$, $\dfrac{1}{2b}$, and $\dfrac{1}{c+b}$ are consecutive terms in an arithmetic sequence.

69. **(a)** Find a value for x such that $3 + x$, $4 + x$, and $5 + x$ are consecutive terms in a geometric sequence.
(b) Given three numbers a, b, and c, find a value for x (in terms of a, b, and c) such that $a + x$, $b + x$, and $c + x$ are consecutive terms in a geometric sequence.

70. If $\ln(A + C) + \ln(A + C - 2B) = 2 \ln(A - C)$, show that $1/A$, $1/B$, and $1/C$ are consecutive terms in an arithmetic sequence.

71. Let $a_1, a_2, a_3, \cdots$ be an arithmetic sequence with common difference d, and let r ($\neq 1$) be a real number. In this exercise we develop a formula for the sum of the series

$$a_1 + ra_2 + r^2a_3 + \cdots + r^{n-1}a_n$$

The method we use here is essentially the same as the method used in the text to derive the formula for the sum of a finite geometric series. So, as background for this exercise, you should review the derivation on page 851 for the sum of a geometric series.

(a) Let S denote the required sum. Show that

$$\begin{aligned} S - rS &= a_1 + rd + r^2d + \cdots + r^{n-1}d - r^na_n \\ &= a_1 + \frac{d(r - r^n)}{1 - r} - r^na_n \end{aligned}$$

(b) Show that $S = \dfrac{a_1 - r^na_n}{1 - r} + \dfrac{d(r - r^n)}{(1 - r)^2}$.

72. Use your calculator and the formula in Exercise 71(b) to find the sum of each of the following series.
(a) $1 + 2 \cdot 2 + 2^2 \cdot 3 + 2^3 \cdot 4 + \cdots + 2^{13} \cdot 14$
(b) $2 + 4 \cdot 5 + 4^2 \cdot 8 + 4^3 \cdot 11 + \cdots + 4^6 \cdot 20$
(c) $3 - \dfrac{1}{2} \cdot 5 + \dfrac{1}{2^2} \cdot 7 - \dfrac{1}{2^3} \cdot 9 + \cdots +$ (to 10 terms)

In Exercises 73–76, convert each complex number to rectangular form.

73. $3\left(\cos \frac{1}{3}\pi + i \sin \frac{1}{3}\pi\right)$ **74.** $\cos \frac{1}{6}\pi + i \sin \frac{1}{6}\pi$

75. $2^{1/4}\left(\cos \frac{7}{4}\pi + i \sin \frac{7}{4}\pi\right)$

76. $5\left[\cos\left(-\frac{1}{4}\pi\right) + i \sin\left(-\frac{1}{4}\pi\right)\right]$

In Exercises 77–80, express the complex numbers in trigonometric form.

77. $\frac{1}{2}\left(1 + \sqrt{3}i\right)$ **78.** $3i$

79. $-3\sqrt{2} - 3\sqrt{2}i$ **80.** $2\sqrt{3} - 2i$

In Exercises 81–88, carry out the indicated operations. Express your results in rectangular form for those cases in which the trigonometric functions are readily evaluated without tables or a calculator.

81. $5\left(\cos \frac{1}{7}\pi + i \sin \frac{1}{7}\pi\right) \times 2\left(\cos \frac{3}{28}\pi + i \sin \frac{3}{28}\pi\right)$

82. $4\left(\cos \frac{1}{12}\pi + i \sin \frac{1}{12}\pi\right) \times 3\left(\cos \frac{1}{12}\pi + i \sin \frac{1}{12}\pi\right)$

83. $8\left(\cos \frac{1}{12}\pi + i \sin \frac{1}{12}\pi\right) \div 4\left(\cos \frac{1}{3}\pi + i \sin \frac{1}{3}\pi\right)$

84. $4(\cos 32^\circ + i \sin 32^\circ) \div 2^{1/2}(\cos 2^\circ + i \sin 2^\circ)$

85. $\left[3^{1/4}\left(\cos \frac{1}{36}\pi + i \sin \frac{1}{36}\pi\right)\right]^{12}$

86. $\left[2\left(\cos \frac{2}{15}\pi + i \sin \frac{2}{15}\pi\right)\right]^5$

87. $\left(\sqrt{3} + i\right)^{10}$ **88.** $\left(\sqrt{2} - \sqrt{2}i\right)^{15}$

In Exercises 89–92, use DeMoivre's theorem to find the indicated roots. Express your results in rectangular form.

89. Sixth roots of 1 **90.** Cube roots of $-64i$

91. Square roots of $\sqrt{2} - \sqrt{2}i$

92. Fourth roots of $1 + \sqrt{3}i$

93. Find the five fifth roots of $1 + i$. Use a calculator to express the roots in rectangular form. (Round each decimal to two places in the final answer.)

94. For our last exercise in the last chapter of this book, we take the following problem from the classic text by Isaac Todhunter, *Plane Trigonometry for the Use of Colleges and Schools,* 5th ed. (London: Macmillan & Co., 1874):

> If the cotangents of the angles of a triangle be in arithmetical progression, the squares of the sides will also be in arithmetical progression.

Hint: Write the given information as an equation involving sines and cosines. Use the formulas on page 542 to express the cosines in terms of the lengths of the sides. For the sines, use the formulas in Exercise 52 on page 550. [For a shorter proof, see page 68 in the January 1995 issue of *The College Mathematics Journal.* Although the solution given there applies to the converse of the present problem, it can easily be adjusted to fit this problem. (As is usual with all math journal articles, you'll find that there are alegbra steps that you will need to fill in for yourself.)]

CHAPTER THIRTEEN TEST

1. Use the principle of mathematical induction to show that the following formula is valid for all natural numbers n:
$$1^2 + 2^2 + 3^2 + \cdots + n^2 = \frac{n(n+1)(2n+1)}{6}.$$
2. Express each of the following sums without using sigma notation, and then evaluate each sum.
   (a) $\sum_{k=0}^{2} (10k - 1)$ **(b)** $\sum_{k=1}^{3} (-1)^k k^2$
3. **(a)** Write the formula for the sum S_n of a finite geometric series.
   (b) Evaluate the sum $\frac{3}{2} + \frac{3^2}{2^2} + \frac{3^3}{2^3} + \cdots + \frac{3^{10}}{2^{10}}$.
4. **(a)** Determine the coefficient of the term containing a^3 in the expansion of $(a - 2b^3)^{11}$.
   (b) Find the fifth term of the expansion in part (a).
5. Expand the expression $(3x^2 + y^3)^5$.
6. Determine the sum of the first 12 terms of an arithmetic sequence in which the first term is 8 and the twelfth term is $43/2$.
7. Find the sum of the following infinite geometric series:
$$\frac{7}{10} + \frac{7}{100} + \frac{7}{1000} + \cdots.$$
8. A sequence is defined recursively as follows: $a_1 = 1$; $a_2 = 1$; and $a_n = (a_{n-1})^2 + a_{n-2}$ for $n \geqslant 3$. Determine the fourth and fifth terms in this sequence.
9. In a certain geometric sequence, the third term is 4 and the fifth term is 10. Find the sixth term, given that the common ratio is negative.
10. What is the twentieth term in the arithmetic sequence $-61, -46, -31, \ldots$?
11. Find the rectangular form for the complex number $z = 2\left(\cos\frac{2}{3}\pi + i\sin\frac{2}{3}\pi\right)$.
12. Find the trigonometric form of the complex number $\sqrt{2} - \sqrt{2}i$.
13. Let $z = 3\left(\cos\frac{2\pi}{9} + i\sin\frac{2\pi}{9}\right)$ and $w = 5\left(\cos\frac{\pi}{9} + i\sin\frac{\pi}{9}\right)$. Compute the product zw and express your answer in rectangular form.
14. Compute the cube roots of $64i$.

APPENDIX

A.1 USING A GRAPHING UTILITY

INTRODUCTION

Throughout this text you will find exercise sections labeled *Graphing Utility Exercises.* The term **graphing utility** refers to either a graphics calculator or a computer with software for graphing and analyzing functions. Although the details in the use of these graphing utilities differ, all share certain common features. In this appendix we discuss entering (typing in) equations and changing the viewing rectangle and the scale for graphs. As you work through the examples and exercises in this appendix, you should refer as necessary to the owner's manual or documentation that came with your graphing utility. Another useful resource is the *Graphing Utilities Manual* that is available with this text. This supplement contains keystroke-level instructions for the general procedures discussed in this appendix.

GRAPHING EQUATIONS Graphing utilities utilize the function concept for graphing. Given an equation in the form $y = f(x)$, the utility computes pairs of points (x, y) and then plots each point, one at a time, on the screen. The utility is not "smart." For example, it does not know ahead of time that the graph of the equation $2x - 3y = 12$ will be a line, or that $x^2 + y^2 = 25$ will be a circle. Indeed for equations such as these, you will first need to solve for y before entering the function. The next two examples show instances of this.

EXAMPLE 1 Graph the equation $2x - 3y = 12$.

Solution Before entering the equation in the graphing utility, you first need to solve for y:

$$\begin{aligned} 2x - 3y &= 12 \\ -3y &= -2x + 12 \\ y &= \tfrac{2}{3}x - 4 \end{aligned}$$

Now enter the equation into the graphing utility:

$$Y = (2/3)X - 4$$

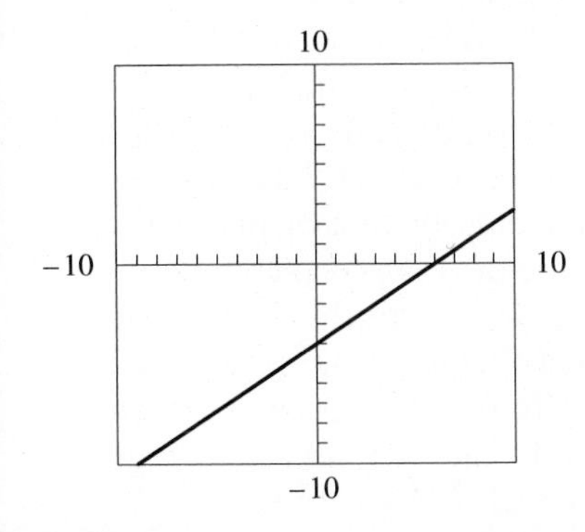

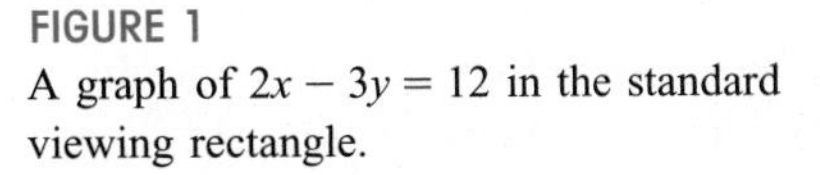

FIGURE 1
A graph of $2x - 3y = 12$ in the standard viewing rectangle.

NOTE For entering this equation, some graphing utilities require that you use a multiplication symbol (*) and type the equation as

$$Y = (2/3) * X - 4$$

On the other hand, if your graphing utility understands algebraic symbolism, you can save a few keystrokes and simply enter the function as

$$Y = 2X/3 - 4$$

With the equation solved for y, you can use a graph command in the graphing utility to obtain a graph. For many graphing utilities, the default view or the **standard viewing rectangle** is $-10 \leq x \leq 10$ and $-10 \leq y \leq 10$. This is the view shown in Figure 1.

EXAMPLE 2 Graph the equation $x^2 + y^2 = 25$.

Solution According to Chapter 2, the graph of this equation is a circle with center (0, 0) and radius 5. In order for the graphing utility to generate a graph, you first need to solve the given equation for y in terms of x:

$$x^2 + y^2 = 25$$
$$y^2 = 25 - x^2$$
$$y = \pm\sqrt{25 - x^2}$$

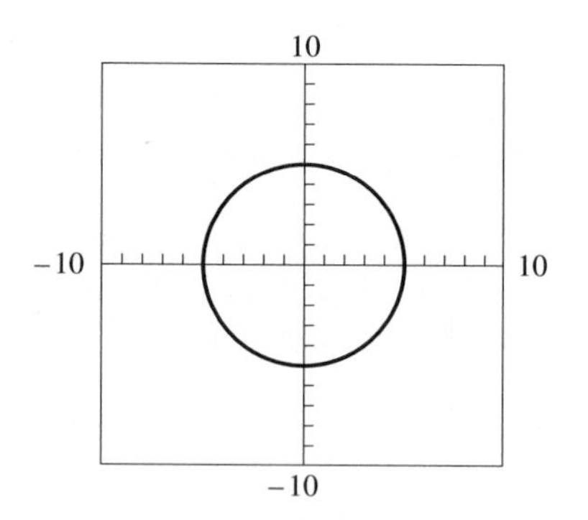

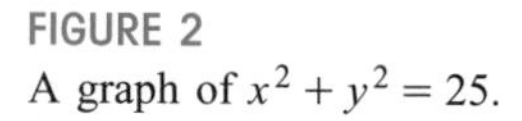

FIGURE 2
A graph of $x^2 + y^2 = 25$.

Note that there are two separate equations here, $y = \sqrt{25 - x^2}$ and $y = -\sqrt{25 - x^2}$. Enter both of these into your graphing utility, consulting the owner's manual as necessary for entering the square roots and exponents. The equations that follow use the carat symbol (ˆ) to indicate that the next number is an exponent.

$$Y_1 = \sqrt{25 - X\hat{}2} \qquad Y_2 = -\sqrt{25 - X\hat{}2}$$

If you use the standard viewing rectangle now, you should obtain a graph similar to that shown in Figure 2. If the graph that you obtain looks more like an oval than a circle, then you need to consult the owner's manual to find out how to display the true proportions in a graph. For instance, some graphing calculators have a ''squaring'' feature within the ZOOM menu that will automatically adjust the viewing rectangle to show true proportions.

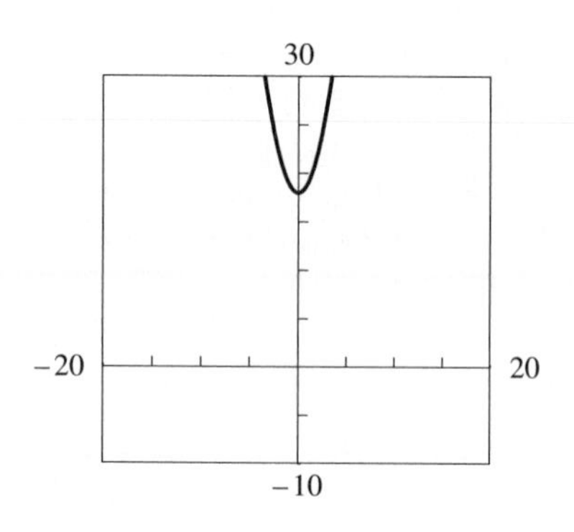

FIGURE 3(a)
A first view of $y = x^2 + 18$.

CHANGING THE SCALE OF THE GRAPHS Graphing utilities allow you to modify the viewing rectangle. Graphing calculators have menus that allow you to enter Xmin, Xmax, Ymin, and Ymax, as well as Xscl (for X scale) and Yscl. The Xscl and Yscl entries determine what each tick mark on the axes represents. Computer software applications usually require that you supply an interval of values to use on each axis. Furthermore, many utilities support a zoom feature that allows for instant adjustment of the view. The next two examples illustrate these features.

EXAMPLE 3 Graph the equation $y = x^2 + 18$.

Solution Begin by entering the function $Y = X\hat{}2 + 18$. Try to graph this in the standard viewing rectangle; you'll see that no graph appears. This is because the y-numbers (given by $x^2 + 18$) are always greater than or equal to 18. This is beyond the y-boundary of the standard viewing rectangle, which extends only to 10 in the y-direction. At this point some experimenting is called for. Adjust the viewing rectangle so that x extends from -20 to 20 and y extends from -10 to 30. (If you have settings for Xscl and Yscl, set each to 5 so that fewer tick marks are shown.) With these settings, you should obtain a display similar to the one shown in Figure 3(a). The view in Figure 3(a) is acceptable because it shows all of the essential features of the graph. However, much of the viewing screen in Figure 3(a) is not utilized. For a view that fills more of the screen, try adjusting the viewing rectangle so that x extends from -5 to 5 and y extends from 0 to 50. You should obtain a graph similar to the one shown in Figure 3(b). [In Figure 3(b), we've set Xscl and Yscl to 1 and 5, respectively.]

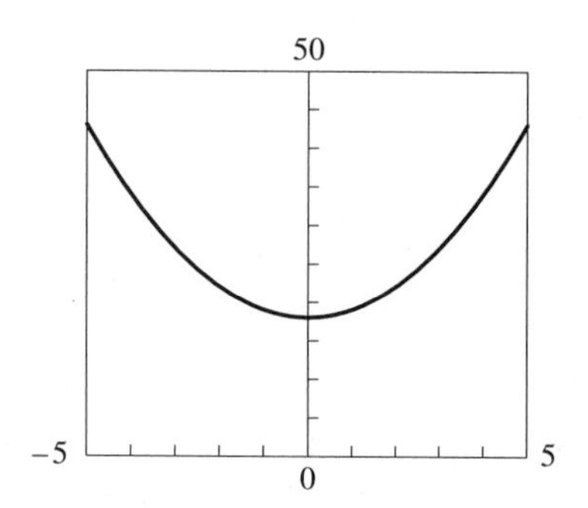

FIGURE 3(b)
Another view of $y = x^2 + 18$.

CAUTION The two graphs in Figure 3 appear to be quite different from one another, even though they both represent the same equation. Thus, if you obtain a

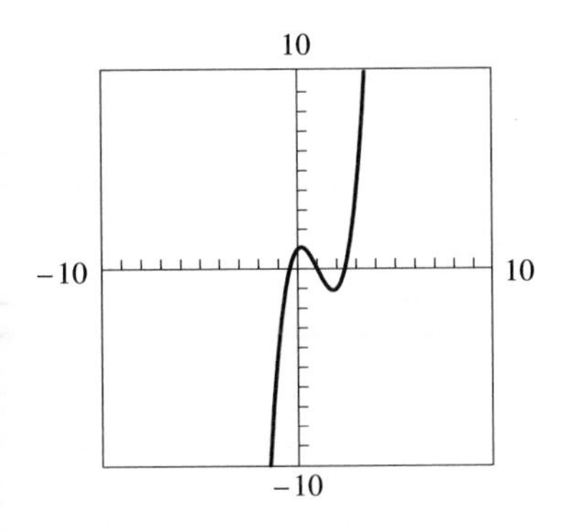

FIGURE 4(a)
$y = x^3 - 3x^2 + x + 1$
$-10 \leq x \leq 10$; Xscl = 1
$-10 \leq y \leq 10$; Yscl = 1

graph from a graphing utility and then use it to sketch the curve by hand on paper, in addition to the equation, you must also include the specifications for the particular viewing rectangle you've chosen.

EXAMPLE 4 Graph the equation $y = x^3 - 3x^2 + x + 1$ and estimate (to one decimal place) the largest x-intercept.

Solution Begin by entering the function Y = X^3 − 3X^2 + X + 1 and graphing it in the standard viewing rectangle. You should obtain a display similar to that shown in Figure 4(a). From the graph in Figure 4(a), we see that the largest x-intercept is between $x = 2$ and $x = 3$. With Figure 4(a) in mind, then, adjust the viewing rectangle so that x extends from 2 to 3 with Xscl = 0.1 and y extends from -1.5 to 1.5 with Yscl = 0.5. Figure 4(b) shows the resulting graph. From the figure it is clear that the x-intercept we are looking for is approximately 2.4.

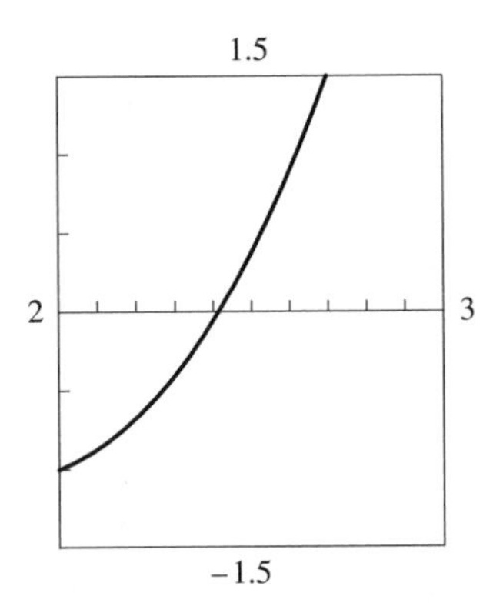

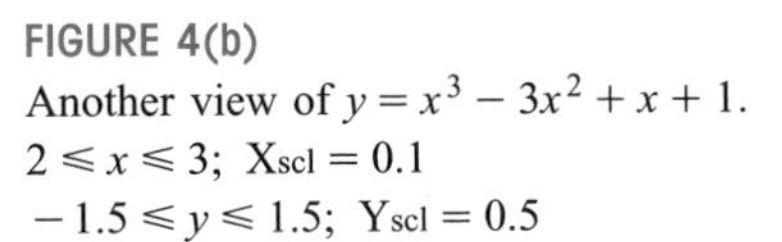
FIGURE 4(b)
Another view of $y = x^3 - 3x^2 + x + 1$.
$2 \leq x \leq 3$; Xscl = 0.1
$-1.5 \leq y \leq 1.5$; Yscl = 0.5

SOLVING EQUATIONS Graphing utilities can be used to solve equations. The methods they use can vary greatly depending on the utility. Some computer programs allow you to enter the equation directly, and they return exact values for answers. Others allow you to enter intervals in which to search for a solution, and they return very accurate estimates of those solutions. Some graphing calculators have a "solver" feature, which again yields accurate estimates of the solutions. All graphing calculators have a "trace" feature, which allows you to trace along a curve and then zoom in until the x-intercept can be accurately approximated. In the solution of the next example, we simply give the results that can be obtained in this manner. Check your owner's manual to determine which of the results (approximate or exact) are available with your graphing utility.

EXAMPLE 5 Solve the equation $x^3 - 3x^2 + x + 1 = 0$. Approximate your answers to three decimal places.

Solution You entered and graphed the function $y = x^3 - 3x^2 + x + 1$ in Example 4. The solutions to the equation are the x-intercepts of the graph. The exact solutions are $1 - \sqrt{2}$, $1 + \sqrt{2}$, and 1. The approximate solutions are -0.414, 2.414, and 1.000.

If your graphing utility provides alternate methods for solving equations (most do), experiment with them and become familiar with each one. Often a less exact approximation method is adequate for a particular problem.

MAXIMUM AND MINIMUM VALUES Graphing utilities can be used to find maximum and minimum values, or turning points, which appear as high and low points on a graph. Many graphing utilities provide advanced features for this purpose (such as numerical derivatives), but tracing and zooming in on the curve near these points is often sufficient to determine the coordinates.

EXAMPLE 6 Estimate (to one decimal place) the coordinates of the high and low points of $y = x^3 - 3x^2 + x + 1$.

Solution You entered and graphed this function in Example 4. Using tracing and zooming features (or more advanced features if available), the high point is found

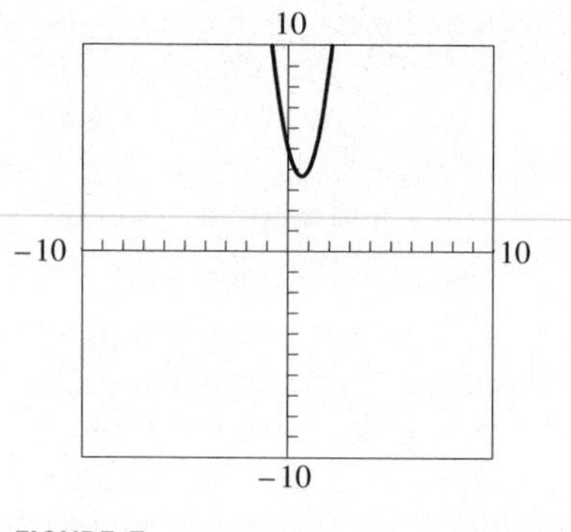

FIGURE 5
The graph of $y = 3x^2 - 4x + 5$.

to be approximately (0.2, 1.1) and the low point is approximately (1.8, −1.1). Verify these results for yourself.

EXAMPLE 7 Find the vertex of the parabola $y = 3x^2 - 4x + 5$. Approximate your coordinates to two decimal places.

Solution Begin by entering the function.

$$Y = 3X\hat{}2 - 4X + 5$$

Using the standard viewing rectangle, the graph should look similar to that shown in Figure 5. Using tracing and zooming features, the low point is approximately (0.67, 3.67). Verify this. The exact coordinates of this vertex are $\left(\frac{2}{3}, \frac{11}{3}\right)$.

This appendix has provided you with a brief introduction to just a few of the many features of graphing utilities. Experiment and consult your owner's manual to discover other features as you work the graphing utility exercises in this text.

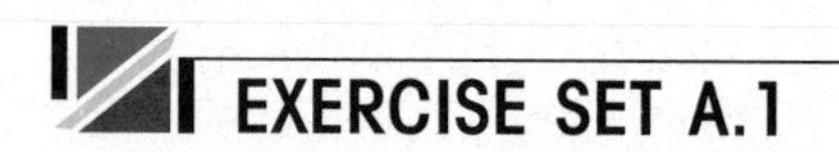

EXERCISE SET A.1

In Exercises 1–14, use a graphing utility to sketch the graph of the equation; use the standard viewing rectangle: $-10 \leq x \leq 10$ and $-10 \leq y \leq 10$.

1. $y = 3x - 5$
2. $y = -2x + 3$
3. $y = -\frac{3}{4}x + 2$
4. $y = \frac{2}{3}x - 5$
5. $3x - 2y = 7$
6. $-2x + 3y = 5$
7. $y = 2x^2 - 6$
8. $y = -3x^2 + 8$
9. $y = x^2 - 6x + 5$
10. $y = -x^2 - 4x + 2$
11. $x^2 + y^2 = 36$
12. $x^2 + y^2 = 81$
13. $(x - 2)^2 + (y + 1)^2 = 49$
14. $(x + 3)^2 + (y - 4)^2 = 25$

In Exercises 15–22, use a graphing utility to sketch the graph of the equation. Make appropriate adjustments to the viewing rectangle so that the essential features of the graph are clear. In Exercises 17–20, be sure your graph shows true proportions.

15. $y = x^2 + 32$
16. $y = -16 - x^2$
17. $x^2 + y^2 = 0.1$
18. $x^2 + y^2 = 0.25$
19. $(x - 1)^2 + (y + 1)^2 = 1$
20. $(x + 1)^2 + (y - 2)^2 = 1$
21. $y = 2x^3 + x^2$
22. $y = 2x^2 + x^3$

In Exercises 23–30, use a graphing utility to find the solutions to each equation. Approximate your answers to three decimal places. Include a sketch with your answer.

23. $x^2 + 2x = 4$
24. $x^2 - 2x = 4$
25. $x^3 - 4x^2 - 5x = 0$
26. $x^3 - 5x^2 - x + 5 = 0$
27. $x^3 + 5x^2 + 5x + 1 = 0$
28. $x^3 - 8x + 8 = 0$
29. $x^4 + 3x^3 - 5x^2 - 9x + 10 = 0$
30. $x^4 - x^3 - 8x^2 + 3x + 9 = 0$

In Exercises 31–34, use a graphing utility to find the vertex of each parabola. Approximate the coordinates to two decimal places. Include a sketch with your answer.

31. $y = 2x^2 - 5x + 7$
32. $y = 3x^2 + 8x - 9$
33. $y = 9x^2 - 7x - 13$
34. $y = -6x^2 + 11x - 5$

In Exercises 35–40, estimate (to one decimal place) the coordinates of the turning points for each curve. Include a sketch with your answer.

35. $y = 2x^3 + 4x^2 - 8x - 14$
36. $y = 2x^3 - 4x^2 - 8x - 14$
37. $y = x^3 - 3x^2 - x - 7$
38. $y = x^3 + 3x^2 + x - 7$
39. $y = x^4 - 2x^3 - 5x^2 + 6x + 14$
40. $y = 2x^4 - 3x^3 - 6x^2 + 8x - 5$

A.2 SIGNIFICANT DIGITS AND CALCULATORS

Many of the numbers that we use in scientific work and in daily life are approximations. In some cases the approximations arise because the numbers are obtained through measurements or experiments. Consider, for example, the following statement from an astronomy textbook:

The diameter of the Moon is 3476 km.

We interpret this statement as meaning that the actual diameter D is closer to 3476 km than it is to either 3475 km or 3477 km. In other words,

$$3475.5 \text{ km} \leq D \leq 3476.5 \text{ km}$$

The interval [3475.5, 3476.5] in this example provides information about the accuracy of the measurement. Another way to indicate accuracy in an approximation is by specifying the number of *significant digits* it contains. The measurement 3476 km has four significant digits. In general, the number of significant digits in a given number is found as follows.

SIGNIFICANT DIGITS

The number of significant digits in a given number is determined by counting the digits from left to right, beginning with the left-most nonzero digit.

EXAMPLES

Number	Number of Significant Digits
1.43	3
0.52	2
0.05	1
4837	4
4837.0	5

Numbers obtained through measurements are not the only source of approximations in scientific work. For example, to five significant digits, we have the following approximation for the irrational number π:

$$\pi \approx 3.1416$$

This statement tells us that π is closer to 3.1416 than it is to either 3.1415 or 3.1417. In other words,

$$3.14155 \leq \pi \leq 3.14165$$

Table 1 provides some additional examples of the ideas we've introduced.

TABLE 1

Number	Number of Significant Digits	Range of Measurement
37	2	[36.5, 37.5]
37.0	3	[36.95, 37.05]
268.1	4	[268.05, 268.15]
1.036	4	[1.0355, 1.0365]
0.036	2	[0.0355, 0.0365]

There is an ambiguity involving zero that can arise in counting significant digits. Suppose that someone measures the width w of a rectangle and reports the result as 30 cm. How many significant digits are there? If the value 30 cm was obtained by measuring to the nearest 10 cm, then only the digit 3 is significant, and we can conclude only that the width w lies in the range $25 \text{ cm} \leq w \leq 35$ cm. On the other hand, if the 30 cm was obtained by measuring to the nearest 1 cm, then both the digits 3 and 0 are significant and we have $29.5 \text{ cm} \leq w \leq 30.5$ cm.

By using **scientific notation** we can avoid the type of ambiguity discussed in the previous paragraph. A number written in the form

$$b \times 10^n \qquad \text{where } 1 \leq b < 10 \text{ and } n \text{ is an integer}$$

is said to be expressed in scientific notation. For the example in the previous paragraph, then, we would write

$$w = 3 \times 10^1 \text{ cm} \qquad \text{if the measurement is to the nearest 10 cm}$$

and

$$w = 3.0 \times 10^1 \text{ cm} \qquad \text{if the measurement is to the nearest 1 cm}$$

As the figures in Table 2 indicate, for a number $b \times 10^n$ in scientific notation, the number of significant digits is just the number of digits in b. (This is one of the advantages in using scientific notation; the number of significant digits, and hence the accuracy of the measurement, is readily apparent.)

TABLE 2

Measurement	Number of Significant Digits	Range of Measurement
Mass of the earth:		
6×10^{27} g	1	$[5.5 \times 10^{27}$ g, 6.5×10^{27} g$]$
6.0×10^{27} g	2	$[5.95 \times 10^{27}$ g, 6.05×10^{27} g$]$
5.974×10^{27} g	4	$[5.9735 \times 10^{27}$ g, 5.9745×10^{27} g$]$
Mass of a proton:		
1.67×10^{-24} g	3	$[1.665 \times 10^{-24}$ g, 1.675×10^{-24} g$]$

There are numerous exercises in this text in which a calculator either is required or is extremely useful. Although you could work many of these exercises using tables instead of a calculator, this author recommends that you purchase a scientific calculator. (A graphing calculator is better still, but considerably more expensive.) ''Scientific calculator'' is a generic term describing a calculator with (at least) the following features or functions beyond the usual arithmetic functions:

1. Memory
2. Scientific notation
3. Powers and roots
4. Logarithms (base ten and base e)
5. Trigonometric functions
6. Inverse functions

Because of the variety and differences in the calculators that are available, no specific instructions are provided in this section of the appendix for operating a calculator. When you buy a calculator, read the owner's manual carefully and work through some of the examples in it. In general, learn to use the memory capabilities of the calculator so that, as far as possible, you don't need to write down the results of the intermediate steps in a given calculation.

Many of the numerical exercises in the text ask that you round the answers to a specified number of decimal places. Our rules for rounding are as follows.

RULES FOR ROUNDING A NUMBER (WITH MORE THAN n DECIMAL PLACES) TO n DECIMAL PLACES

1. If the digit in the $(n+1)$st decimal place is greater than 5, increase the digit in the nth place by 1. If the digit in the $(n+1)$st place is less than 5, leave the nth digit unchanged.
2. If the digit in the $(n+1)$st decimal place is 5 and there is at least one nonzero digit to the right of this 5, increase the digit in the nth decimal place by 1.
3. If the digit in the $(n+1)$st decimal place is 5 and there are no nonzero digits to the right of this 5, then increase the digit in the nth decimal place by 1 only if this results in an even digit.

The examples in Table 3 illustrate the use of these rules.

TABLE 3

Number	Rounded to One Decimal Place	Rounded to Three Decimal Places
4.3742	4.4	4.374
2.0515	2.1	2.052
2.9925	3.0	2.992

These same rules can be adapted for rounding a result to a specified number of significant digits. As examples of this, we have

2347 rounded to two significant digits is $2300 = 2.3 \times 10^3$

2347 rounded to three significant digits is $2350 = 2.35 \times 10^3$

975 rounded to two significant digits is $980 = 9.8 \times 10^2$

0.985 rounded to two significant digits is $0.98 = 9.8 \times 10^{-1}$

In calculator exercises that ask you to round your answers, it's important that you postpone rounding until the final calculation is carried out. For example, suppose that you are required to determine the hypotenuse x of the right triangle in Figure 1 to two significant digits. Using the Pythagorean theorem, we have

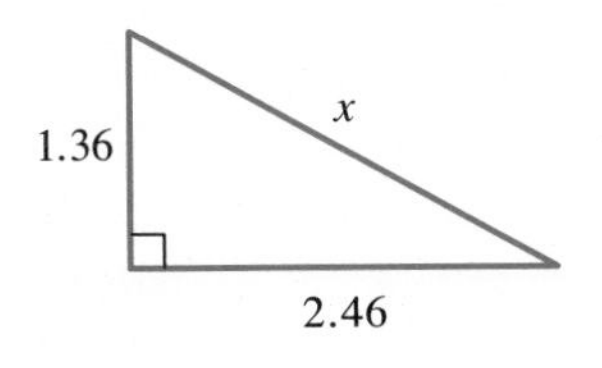

FIGURE 1

$$x = \sqrt{(1.36)^2 + (2.46)^2}$$

$$= 2.8 \quad \text{using a calculator and rounding the final result to two significant digits}$$

TABLE 4

Number	Range of Measurement
1.36	[1.355, 1.365]
2.46	[2.455, 2.465]

On the other hand, if we first round each of the given lengths to two significant digits, we obtain

$$x = \sqrt{(1.4)^2 + (2.5)^2}$$
$$= 2.9 \quad \text{to two significant digits}$$

This last result is inappropriate and we can see why as follows. As Table 4 shows, the maximum possible values for the sides are 1.365 and 2.465, respectively.

Thus, the maximum possible value for the hypotenuse must be

$$\sqrt{(1.365)^2 + (2.465)^2} = 2.817\ldots \quad \text{calculator display}$$
$$= 2.8 \quad \text{to two significant digits}$$

This shows that the value 2.9 is indeed inappropriate, as we stated previously.

An error often made by people working with calculators and approximations is to report a final answer with a greater degree of accuracy than the data warrant. Consider, for example, the right triangle in Figure 2. Using the Pythagorean theorem and a calculator with an eight-digit display, we obtain

$$h = 3.6055513 \text{ cm}$$

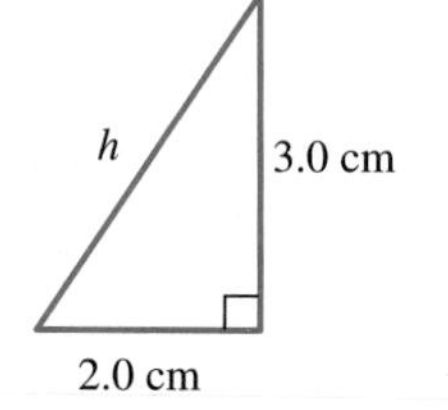

FIGURE 2

This value for h is inappropriate, since common sense tells us that the answer should be no more accurate than the data used to obtain that answer. In particular, since the given sides of the triangle apparently were measured only to the nearest tenth of a centimeter, we certainly should not expect any improvement in accuracy for the resulting value of the hypotenuse. An appropriate form for the value of h here would be $h = 3.6$ cm. In general, for calculator exercises in this text that do not specify a required number of decimal places or significant digits in the final results, you should use the following guidelines.

GUIDELINES FOR COMPUTING WITH APPROXIMATIONS

1. *For adding and subtracting:* Round the final result so that it contains only as many decimal places as there are in the data with the fewest decimal places.
2. *For multiplying and dividing:* Round the final result so that it contains only as many significant digits as there are in the data with the fewest significant digits.
3. *For powers and roots:* In computing a power or a root of a real number b, round the result so that it contains as many significant digits as there are in b.

A.3 PROPERTIES OF THE REAL NUMBERS

I do not like × as a symbol for multiplication, as it is easily confounded with x; . . . often I simply relate two quantities by an interposed dot . . . .

G. W. Leibniz in a letter dated July 29, 1698

Today's familiar plus and minus signs were first used in 15th-century Germany as warehouse marks. They indicated when a container held something that weighed over or under a certain standard weight.

Martin Gardner in "Mathematical Games" (*Scientific American,* June 1977)

In this section we will first list the basic properties for the real number system. After that we'll summarize some of the more down-to-earth implications of those properties, such as the procedures for working with signed numbers and fractions. At the end of the section, we present a more theoretical treatment of some of these ideas.

The set of real numbers is **closed** with respect to the operations of addition and multiplication. This just means that when we add or multiply two real numbers, the result (that is, the **sum** or the **product**) is again a real number. Some of the other most basic properties and definitions for the real-number system are listed in the following box. In the box, the lowercase letters a, b, and c denote arbitrary real numbers.

PROPERTY SUMMARY	SOME FUNDAMENTAL PROPERTIES OF THE REAL NUMBERS
Commutative properties	$a + b = b + a \qquad ab = ba$
Associative properties	$a + (b + c) = (a + b) + c \qquad a(bc) = (ab)c$
Identity properties	**1.** There is a unique real number 0 (called **zero** or the **additive identity**) such that $a + 0 = a$ and $0 + a = a$ **2.** There is a unique real number 1 (called **one** or the **multiplicative identity**) such that $a \cdot 1 = a$ and $1 \cdot a = a$
Inverse properties	**1.** For each real number a there is a real number $-a$ (called the **additive inverse** of a or the **negative** of a) such that $a + (-a) = 0$ and $(-a) + a = 0$ **2.** For each real number $a \neq 0$, there is a real number denoted by $\frac{1}{a}$ (or $1/a$ or a^{-1}) and called the **multiplicative inverse** or **reciprocal** of a, such that $a \cdot \frac{1}{a} = 1$ and $\frac{1}{a} \cdot a = 1$
Distributive properties	$a(b + c) = ab + ac \qquad (b + c)a = ba + ca$

On reading this list of properties for the first time, many students ask the natural question, "Why do we even bother to list such obvious properties?" One reason is that all the other laws of arithmetic and algebra (including the "not-so-obvious" ones) can be derived from our rather short list. For example, as you'll see near the end of this section, the rule $0 \cdot a = 0$ can be proved using the distributive property, as can the rule that the product of two negative numbers is a positive number. The distributive properties indicate in a sense the way in which multiplication and addition interact with each other. More specifically, each of the following examples relies on the distributive properties.

EXAMPLE	REMARK
$12 \times 103 = 12(100 + 3)$ $= 1200 + 36 = 1236$	The distributive property can be used as an aid in mental calculations.
$3a^2(a + 2b) = 3a^2 \cdot a + 3a^2 \cdot 2b$ $= 3a^3 + 6a^2b$	Here an application of the distributive property is followed by a simplification.
$A(c + d) = Ac + Ad$	This is a direct application of the distributive property. In the next example we are going to replace A with the quantity $(a + b)$.
$(a + b)(c + d) = (a + b)c + (a + b)d$ $= ac + bc + ad + bd$	The distributive properties are used three times here.
$(x + 3)(x + 6) = (x + 3)x + (x + 3)6$ $= x^2 + 3x + 6x + 18$ $= x^2 + 9x + 18$	This is similar to the previous example. Examples of this type are discussed in greater detail in Section 1.6.

The distributive and associative properties can both be generalized. The **generalized distributive property** asserts that

$$a(b_1 + b_2 + \ldots + b_n) = ab_1 + ab_2 + \ldots + ab_n$$

The **generalized associative property** tells us that we can group terms as we please in computing a sum or a product. For instance, we have

$$\begin{aligned} 17 + 47 + 53 + 80 &= 17 + [(47 + 53) + 80] \\ &= 17 + [100 + 80] \\ &= 17 + 180 = 197 \end{aligned}$$

EXAMPLE 1 Identify the property of the real numbers that justifies each statement.

(a) $(17 \times 50) \times 2 = 17 \times (50 \times 2) = 17 \times 100 = 1700$
(b) Since π and $\sqrt{2}$ are real numbers, so is $\pi + \sqrt{2}$.
(c) $3 + y^2 = y^2 + 3$
(d) $[y + (z + 1)]x = yx + (z + 1)x$
(e) $(z + 1) + [-(z + 1)] = 0$
(f) $(a^2 + 1) \cdot \left(\dfrac{1}{a^2 + 1}\right) = 1$

Solution
(a) The associative property of multiplication
(b) The closure property with respect to addition
(c) The commutative property of addition
(d) The distributive property
(e) The additive inverse property
(f) The multiplicative inverse property

If you look back at the box on page A-11, you'll see that the operations of subtraction and division are not mentioned. This is because addition and multiplication are actually the more fundamental operations. In fact, we can define subtraction and division in terms of addition and multiplication as follows.

DEFINITIONS: Subtraction and Division

Subtraction and division are defined in terms of addition and multiplication, respectively, as follows.

$$a - b = a + (-b)$$

$$a \div b = a \cdot \left(\frac{1}{b}\right) = ab^{-1} \qquad \text{provided that } b \neq 0$$

Division by zero is not defined. To see why this must be the case, suppose just for the moment that division by zero were defined. Then we would have $1/0 = k$ for some real number k. But then, multiplying both sides of the equation by 0 would yield $1 = k \cdot 0$; that is, $1 = 0$, a contradiction. In summary, if we wish the rules of arithmetic and algebra to be consistent, division by zero must remain undefined.

In basic algebra you were introduced to a number of procedures that were useful in working with algebraic expressions. In the three boxes that follow, we summarize the properties of the real numbers that are behind those procedures you learned. In the statements of these properties, the lowercase letters a, b, c, and d are used to denote arbitrary real numbers. In the properties of fractions, we are assuming that none of the denominators is zero.

PROPERTY SUMMARY — PROPERTIES OF ZERO

PROPERTY	EXAMPLE
1. $a \cdot 0 = 0$	$(x^3 - \pi) \cdot 0 = 0$
2. If $\frac{a}{b} = 0$, then $a = 0$.	If $\frac{x^2 + x - 1}{2x} = 0$, then $x^2 + x - 1 = 0$.
3. If $ab = 0$, then $a = 0$ or $b = 0$.	If $5x = 0$ then $x = 0$.

PROPERTY SUMMARY — PROPERTIES OF NEGATIVES

PROPERTY	EXAMPLE
1. $-(-a) = a$	$-[-(x+1)] = x + 1$
2. $(-a)b = -(ab) = a(-b)$	$(-3)4 = -12 = 3(-4)$
3. $(-a)(-b) = ab$	$[-(x+y)][-(x+y)] = (x+y)^2$
4. $-a = (-1)a$	$-(x+y-z) = -1(x+y-z) = -x-y+z$

PROPERTY SUMMARY — PROPERTIES OF FRACTIONS

(The denominators in each case are assumed to be nonzero.)

PROPERTY	EXAMPLE
1. $\frac{a}{b} = \frac{c}{d}$ if and only if $ad = bc$	$\frac{8}{12} = \frac{2}{3}$ since $8 \cdot 3 = 12 \cdot 2$
2. $\frac{a}{b} = \frac{ac}{bc}$	$\frac{4}{7} = \frac{4 \cdot 5}{7 \cdot 5} = \frac{(4 \cdot 5)\pi}{(7 \cdot 5)\pi}$
3. $\frac{a}{b} \pm \frac{c}{b} = \frac{a \pm c}{b}$	$\frac{10}{3} + \frac{3}{3} = \frac{13}{3}$; $\frac{10}{3} - \frac{3}{3} = \frac{7}{3}$
4. $\frac{a}{b} \pm \frac{c}{d} = \frac{ad \pm bc}{bd}$	$\frac{2}{3} + \frac{3}{4} = \frac{2 \cdot 4 + 3 \cdot 3}{3 \cdot 4} = \frac{17}{12}$; $\frac{2}{3} - \frac{3}{4} = \frac{2 \cdot 4 - 3 \cdot 3}{3 \cdot 4} = \frac{-1}{12}$
5. $\frac{a}{b} \cdot \frac{c}{d} = \frac{ac}{bd}$	$\frac{5}{4} \cdot \frac{7}{9} = \frac{35}{36}$
6. $\frac{a}{b} \div \frac{c}{d} = \frac{a}{b} \cdot \frac{d}{c} = \frac{ad}{bc}$	$\frac{3}{5} \div \frac{2}{7} = \frac{3}{5} \cdot \frac{7}{2} = \frac{21}{10}$; $1 \div \frac{1}{10} = \frac{1}{1} \cdot \frac{10}{1} = 10$
7. $-\frac{a}{b} = \frac{-a}{b} = \frac{a}{-b}$	$-\frac{3}{4} = \frac{-3}{4} = \frac{3}{-4}$; $-\frac{x-1}{x+1} = \frac{-(x-1)}{x+1} = \frac{x-1}{-(x+1)}$
8. $\frac{a-b}{b-a} = -1$	$\frac{7-x}{x-7} = -1$; $\frac{x^2-y^2}{y^2-x^2} = -1$

For the next example, you'll need to recall the definition of least common multiple. The **least common multiple (l.c.m.)** is the smallest positive integer that is a multiple of each of the given integers. *Examples:* The l.c.m. of 2 and 3 is 6; the l.c.m. of 3, 5, and 6 is 30.

EXAMPLE 2 Simplify the following expression (that is, write it as a single fraction):

$$\frac{\frac{5}{4} + \frac{7}{3}}{\frac{1}{6} + \frac{3}{2}}$$

Solution The individual denominators are 4, 3, 6, and 2. Since the least common multiple of these numbers is 12, we multiple the entire original fraction by 12/12 (which is equal to 1) to obtain

$$\frac{12}{12} \cdot \frac{\frac{5}{4} + \frac{7}{3}}{\frac{1}{6} + \frac{3}{2}} = \frac{12\left(\frac{5}{4} + \frac{7}{3}\right)}{12\left(\frac{1}{6} + \frac{3}{2}\right)} = \frac{15 + 28}{2 + 18} = \frac{43}{20}$$

The properties of fractions that we have listed in the preceding Property Summary Box are used throughout the book when we work with fractional expressions. To help you gain some additional perspective here, we list several common types of errors to avoid when working with fractions.

ERRORS TO AVOID

ERROR	CORRECTION	COMMENT
$\frac{1}{x}+\frac{1}{y} \neq \frac{1}{x+y}$	$\frac{1}{x}+\frac{1}{y}=\frac{1(y)+x(1)}{xy}=\frac{x+y}{xy}$	Add fractions with unlike denominators using Property 4 on page A-14.
$\frac{a}{2}+\frac{3}{7} \neq \frac{a+3}{2+7}$	$\frac{a}{2}+\frac{3}{7}=\frac{a(7)+2(3)}{(2)(7)}=\frac{7a+6}{14}$	Again, use Property 4.
$(1/a)+3 \neq \frac{1}{a+3}$	$\frac{1}{a}+\frac{3}{1}=\frac{1(1)+a(3)}{a(1)}=\frac{1+3a}{a}$	Writing $1/a$ and 3 as $\frac{1}{a}$ and $\frac{3}{1}$, respectively, makes it clear that Property 4 applies.
$6(x+y) \neq 6x+y$	$6(x+y)=6x+6y$	According to the distributive property, the factor 6 must be distributed over *both* terms of the sum.
$6(xy) \neq (6x)(6y)$	$6(xy)=(6x)y=6xy$	The distributive law does not apply.

Why is zero times any number equal to zero? Why does "negative times negative equal positive"? We conclude this section by showing how the fundamental properties listed in the box on page A-11 are used to prove these statements.

THEOREM

Let a and b be real numbers. Then

(a) $a \cdot 0 = 0$

(b) $-a = (-1)a$

(c) $-(-a) = a$

(d) $a(-b) = -ab$

(e) $(-a)(-b) = ab$

PROOF OF PART (a)

$$a \cdot 0 = a \cdot (0+0) \quad \text{additive identity property}$$

$$a \cdot 0 = a \cdot 0 + a \cdot 0 \quad \text{distributive property}$$

Now since $a \cdot 0$ is a real number, it has an additive inverse, $-(a \cdot 0)$. Adding this to both sides of the last equation, we obtain

$$a \cdot 0 + [-(a \cdot 0)] = (a \cdot 0 + a \cdot 0) + [-(a \cdot 0)]$$

$$a \cdot 0 + [-(a \cdot 0)] = a \cdot 0 + \{a \cdot 0 + [-(a \cdot 0)]\} \quad \text{associative property of addition}$$

$$0 = a \cdot 0 + 0 \quad \text{additive inverse property}$$

$$0 = a \cdot 0 \quad \text{additive identity property}$$

Thus $a \cdot 0 = 0$, as we wished to show.

PROOF OF PART (b)

$0 = 0 \cdot a$	using part (a) and the commutative property of multiplication
$= [1 + (-1)]a$	additive inverse property
$= 1 \cdot a + (-1)a$	distributive property
$= a + (-1)a$	multiplicative identity property

Now, by adding $-a$ to both sides of this last equation, we obtain

$-a + 0 = -a + [a + (-1)a]$	
$-a = (-a + a) + (-1)a$	additive identity property and associative property of addition
$-a = 0 + (-1)a$	additive inverse property
$-a = (-1)a$	additive identity property

This last equation asserts that $-a = (-1)a$, as we wished to show.

PROOF OF PART (c)

$-(-a) + (-a) = 0$	additive inverse property

By adding a to both sides of this last equation, we obtain

$[-(-a) + (-a)] + a = 0 + a$	
$-(-a) + (-a + a) = a$	associative property of addition and additive identity property
$-(-a) + 0 = a$	additive inverse property
$-(-a) = a$	additive identity property

This last equation states that $-(-a) = a$, as we wished to show.

PROOF OF PART (d)

$a(-b) = a[(-1)b]$	using part (b)
$= [a(-1)]b$	associative property of multiplication
$= [(-1)a]b$	commutative property of multiplication
$= (-1)(ab)$	associative property of multiplication
$= -(ab)$	using part (b)

Thus $a(-b) = -ab$, as we wished to show.

PROOF OF PART (e)

$(-a)(-b) = -[(-a)b]$	using part (d)
$= -[b(-a)]$	commutative property of multiplication
$= -[-(ba)]$	using part (d)
$= ba$	using part (c)
$= ab$	commutative property of multiplication

We've now shown that $(-a)(-b) = ab$, as required.

EXERCISE SET A.3

A

For Exercises 1–10, use the distributive properties to carry out the indicated multiplication.

1. $8 \times 104 = 8(100 + 4) = ?$

2. $17 \times 102 = 17(100 + 2) = ?$

3. $45 \times 98 = 45[100 + (-2)] = ?$

4. **(a)** $2a[3a + (-1)]$ **(b)** $2a(3a - 4)$ **(c)** $2a(3a + b - 4)$

5. **(a)** $(A + B)(A + B)$ **(b)** $(A + B)(A - B)$ **(c)** $(A - B)(A - B)$

6. **(a)** $(A + B)(C + D)$ **(b)** $(A + B)(2A + 3B)$ **(c)** $(A + B)(2A - 3B)$

7. **(a)** $(x + 2y)(x + 2y)$ **(b)** $(x + 2y)(x - 2y)$ **(c)** $(x - 2y)(x - 2y)$

8. **(a)** $(x + 3)(x + 3)$ **(b)** $(x + 3)(x - 3)$ **(c)** $(x^2 + 3)(x^2 - 3)$

9. **(a)** $A(x^2 - x + 1)$ **(b)** $(x + 1)(x^2 - x + 1)$

10. **(a)** $(2x + 3)(x - 1)$ **(b)** $(3x + 2)(x - 1)$

In Exercises 11–24, identify the property or properties of the real numbers that justify each statement.

11. $x + 3y = 3y + x$

12. $x(3y) = (3y)x$

13. $\frac{a^2}{b^2} + \left(-\frac{a^2}{b^2}\right) = 0$

14. $6(x + 8) = 6x + 48$

15. $(x + 1) + y = x + (1 + y)$

16. $(x + y) \cdot \left(\frac{1}{x + y}\right) = 1$

17. $(x + y)[(x + y)(x - y)] = (x + y)^2(x - y)$

18. $(\sqrt{2})(1) = \sqrt{2}$

19. $(5 + \pi) + 0 = 5 + \pi$

20. $[(x + y) + (x^2 + y^2)]x = (x + y)x + (x^2 + y^2)x$

21. Since $\sqrt{7}$ and $\sqrt{\pi}$ are real numbers, so is $\sqrt{7} + \sqrt{\pi}$.

22. $y(y + y^2 + y^3) = y^2 + y^3 + y^4$

23. $(x + 2)(x^2 + 4) = (x + 2)x^2 + (x + 2)4$

24. Since $\frac{1}{\sqrt{7}}$ and $\sqrt{5}$ are real numbers, so is $\left(\frac{1}{\sqrt{7}}\right)(\sqrt{5})$.

For Exercises 25–34, give numerical examples to show that the purported "rules" are not, in general, valid.

25. $\frac{1}{b} + \frac{1}{c} = \frac{1}{b + c}$

26. $\frac{1}{a} - \frac{1}{b} = \frac{1}{a - b}$

27. $3(x + y) = 3x + y$

28. $4(a - b) = 4a - b$

29. $x^2(x + y) = x^3 + y$

30. $a(b^2 - c^2) = ab^2 - c^2$

31. $3(xy) = (3x)(3y)$

32. $5(bc) = (5b)(5c)$

33. $a - (b - c) = a - b - c$

34. $a - (b + c) = a - b + c$

In Exercises 35–60, simplify the given expressions.

35. $1 + 2(-3)(4)$

36. $(97)(-98)(99)(0)$

37. $4 - (-3 + 5)$

38. $(-1)^4$

39. $-[(-1)^4]$

40. -1^4

41. $1 - \{1 - [-(-1 - 1)]\}$

42. $4^2 - (-4)^3$

43. $\frac{7}{5} + \frac{2}{3}$

44. $\frac{3}{4} - \frac{4}{3}$

45. $\frac{7}{x} - \frac{12}{x}$

46. $\frac{a + 1}{a^2 + 1} + \frac{a - 1}{a^2 + 1}$

47. $\frac{x^2 + y^2}{x + y} - \frac{x^2 - y^2}{x + y}$

48. $\frac{7}{12} \cdot \frac{5}{4}$

49. $\frac{7}{12} \cdot \left(\frac{-7}{3}\right)$

50. $1 \div \frac{1}{3}$

51. $\frac{2}{5} \div \frac{7}{3}$

52. $1 \div \frac{2}{3} + 1$

53. $1 \div \left(\frac{2}{3} + 1\right)$

54. $\frac{\frac{4}{5} + \frac{1}{2}}{\frac{5}{2}}$

55. **(a)** $\frac{\frac{4}{3} + \frac{1}{2}}{\frac{11}{6} + \frac{5}{4}}$ **(b)** $\frac{\frac{4}{3} - \frac{1}{2}}{\frac{1}{2} - \frac{4}{3}}$

56. **(a)** $\frac{b - a}{a - b}$ **(b)** $\frac{x - 7}{7 - x}$ **(c)** $\frac{3 - x^4}{x^4 - 3}$

57. $\frac{1 - (x + y)}{x + y - 1}$

58. $\frac{\frac{1}{5} - \frac{1}{4}}{\frac{1}{5} + \frac{1}{4}}$

59. $\frac{\frac{1}{2} - \frac{1}{3} + \frac{1}{4}}{\frac{1}{2} + \frac{1}{3} - \frac{1}{4}}$

60. $1 + \cfrac{1}{1 - \cfrac{1}{1 - \frac{1}{5}}}$

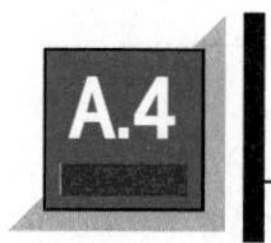

$\sqrt{2}$ IS IRRATIONAL

The proof is by reductio ad absurdum, *and* reductio ad absurdum, *which Euclid loved so much, is one of a mathematician's finest weapons.*

G. H. Hardy (1877–1947)

We will use an *indirect proof* to show that the square root of 2 is an irrational number. The strategy is as follows:

1. We suppose that $\sqrt{2}$ is a rational number.
2. Using (1) and the usual rules of logic and algebra, we derive a contradiction.
3. On the basis of the contradiction in (2), we conclude that the supposition in (1) is untenable; that is, we conclude that $\sqrt{2}$ is irrational.

In carrying out the proof, we'll assume as known the following three statements:

If x is an even natural number, then $x = 2k$ for some natural number k.

Any rational number can be written in the form a/b, where the integers a and b have no common integral factors other than ± 1. (In other words, any fraction can be reduced to lowest terms.)

If x is a natural number and x^2 is even, then x is even.

Our indirect proof now proceeds as follows. Suppose that $\sqrt{2}$ is a rational number. Then we can write

$$\sqrt{2} = \frac{a}{b} \qquad \text{where } a \text{ and } b \text{ are natural numbers with no common factor other than 1} \tag{1}$$

Since both sides of equation (1) are positive, we can square both sides to obtain the equivalent equation

$$2 = \frac{a^2}{b^2}$$

or

$$2b^2 = a^2 \tag{2}$$

Since the left-hand side of equation (2) is an even number, the right-hand side must also be even. But if a^2 is even then a is even, and so

$$a = 2k \qquad \text{for some natural number } k$$

Using this last equation to substitute for a in equation (2), we have

$$2b^2 = (2k)^2 = 4k^2$$

or

$$b^2 = 2k^2$$

Hence (reasoning as before) b^2 is even, and therefore b is even. But then we have that both b and a are even, contrary to our hypothesis that b and a have no common factor other than 1. We conclude from this that equation (1) cannot hold; that is, there is no rational number a/b such that $\sqrt{2} = a/b$. Thus $\sqrt{2}$ is irrational, as we wished to prove.

A.5 REVIEW OF INTEGER EXPONENTS AND *n*TH ROOTS

In algebra, you became familiar with the definition and basic properties of the **exponential notation** a^n, where a is a real number and n is a natural number. Since that definition and those properties are so fundamental, we repeat them here in the two boxes that follow.

DEFINITION 1: Base and Exponent

Given a real number a and a natural number n, we define a^n by

$$a^n = \underbrace{a \cdot a \cdot a \cdots a}_{n \text{ factors}}$$

In the expression a^n, the number a is the **base** and n is the **exponent** or **power** to which the base is raised.

PROPERTY SUMMARY — PROPERTIES OF EXPONENTS

PROPERTY	EXAMPLES
1. $a^m a^n = a^{m+n}$	$a^5 a^6 = a^{11}$; $(x+1)(x+1)^2 = (x+1)^3$
2. $(a^m)^n = a^{mn}$	$(2^3)^4 = 2^{12}$; $[(x+1)^2]^3 = (x+1)^6$
3. $\dfrac{a^m}{a^n} = \begin{cases} a^{m-n} & \text{if } m > n \\ \dfrac{1}{a^{n-m}} & \text{if } m < n \\ 1 & \text{if } m = n \end{cases}$	$\dfrac{a^6}{a^2} = a^4$; $\dfrac{a^2}{a^6} = \dfrac{1}{a^4}$; $\dfrac{a^5}{a^5} = 1$
4. $(ab)^m = a^m b^m$; $\left(\dfrac{a}{b}\right)^m = \dfrac{a^m}{b^m}$	$(2x^2)^3 = 2^3 \cdot (x^2)^3 = 8x^6$; $\left(\dfrac{x^2}{y^3}\right)^4 = \dfrac{x^8}{y^{12}}$

Each of these properties is a direct consequence of the definition of a^n. For instance, according to the first property, we have $a^2 a^3 = a^5$. To verify that this is indeed correct, we note that

$$a^2 a^3 = (aa)(aaa) = a^5$$

Now we want to extend our definition of a^n to allow for exponents that are integers but not necessarily natural numbers. We begin by defining a^0.

DEFINITION 2: Zero Exponent

For any nonzero real number a,

$$a^0 = 1$$

(0^0 is not defined.)

EXAMPLES

(a) $2^0 = 1$

(b) $(-\pi)^0 = 1$

(c) $\left(\dfrac{3}{1 + a^2 + b^2}\right)^0 = 1$

It's easy to see the motivation for defining a^0 to be 1. Assuming that the exponent zero is to have the same properties as exponents that are natural numbers, we can write

$$a^0 a^n = a^{0+n}$$

That is,

$$a^0 a^n = a^n$$

Now we divide both sides of this last equation by a^n to obtain $a^0 = 1$, which agrees with our definition.

Our next definition (see the box that follows) assigns a meaning to the expression a^{-n} when n is a natural number. Again, it's easy to see the motivation for this definition. We have

$$a^n a^{-n} = a^{n+(-n)} = a^0 = 1$$

That is,

$$a^n a^{-n} = 1$$

Now we divide both sides of this last equation by a^n to obtain $a^{-n} = 1/a^n$, in agreement with Definition 3.

DEFINITION 3: Negative Exponent

For any nonzero real number a and natural number n,

$$a^{-n} = \frac{1}{a^n}$$

EXAMPLES

(a) $2^{-1} = \dfrac{1}{2^1} = \dfrac{1}{2}$

(b) $\left(\dfrac{1}{10}\right)^{-1} = \dfrac{1}{(\frac{1}{10})^1} = 10$

(c) $x^{-2} = \dfrac{1}{x^2}$

(d) $(a^2b)^{-3} = \dfrac{1}{(a^2b)^3} = \dfrac{1}{a^6b^3}$

(e) $\dfrac{1}{2^{-3}} = \dfrac{1}{1/2^3} = 2^3 = 8$

It can be shown that the four properties of exponents that we listed earlier continue to hold now for all integer exponents. We make use of this fact in the next three examples.

EXAMPLE 1 Simplify the following expression. Write the answer in such a way that only positive exponents appear.

$$(a^2b^3)^2(a^5b)^{-1}$$

Solution

FIRST METHOD

$$(a^2b^3)^2(a^5b)^{-1} = (a^2b^3)^2 \cdot \frac{1}{a^5b} = \frac{a^4b^6}{a^5b} = \frac{b^{6-1}}{a^{5-4}} = \frac{b^5}{a}$$

ALTERNATE METHOD

$$(a^2b^3)^2(a^5b)^{-1} = (a^4b^6)(a^{-5}b^{-1}) = a^{4-5}b^{6-1} = a^{-1}b^5 = \frac{b^5}{a} \quad \text{as obtained previously}$$

EXAMPLE 2 Simplify the following expression, writing the answer so that negative exponents are not used.

$$\left(\frac{a^{-5}b^2c^0}{a^3b^{-1}}\right)^3$$

Solution We show two solutions. The first makes immediate use of the property $(a^m)^n = a^{mn}$. In the second solution we begin by working within the parentheses.

FIRST SOLUTION

$$\left(\frac{a^{-5}b^2c^0}{a^3b^{-1}}\right)^3 = \frac{a^{-15}b^6}{a^9b^{-3}} = \frac{b^{6-(-3)}}{a^{9-(-15)}} = \frac{b^9}{a^{24}}$$

ALTERNATE SOLUTION

$$\left(\frac{a^{-5}b^2c^0}{a^3b^{-1}}\right)^3 = \left(\frac{b^3}{a^8}\right)^3 = \frac{b^9}{a^{24}}$$

Now that we've completed our review of integer exponents, we are ready to consider roots. Here is the basic definition.

DEFINITION 4: *n*th Roots

Let n be a natural number. If a and b are real numbers and

$$a^n = b$$

then we say that a is an ***n*th root** of b. When $n = 2$ and when $n = 3$, we refer to the roots as **square roots** and **cube roots,** respectively.

EXAMPLES

Both 3 and -3 are square roots of 9 because $3^2 = 9$ and $(-3)^2 = 9$.

Both 2 and -2 are fourth roots of 16 because $2^4 = 16$ and $(-2)^4 = 16$.

2 is a cube root of 8 because $2^3 = 8$.

-3 is a fifth root of -243 because $(-3)^5 = -243$.

As the examples in the box suggest, square roots, fourth roots, and all *even* roots of positive numbers occur in pairs, one positive and one negative. In these cases, we use the notation $\sqrt[n]{b}$ to denote the positive, or **principal,** nth root of b. As examples of this notation, we can write

$\sqrt[4]{81} = 3$ (The principal fourth root of 81 is 3.)

$\sqrt[4]{81} \neq -3$ (The principal fourth root of 81 is not -3.)

$-\sqrt[4]{81} = -3$ (The negative of the principal fourth root of 81 is -3.)

The symbol $\sqrt{}$ is called a **radical sign,** and the number within the radical sign is the **radicand.** The natural number n used in the notation $\sqrt[n]{}$ is called the **index** of the radical. For square roots, as you know from basic algebra, we suppress the index and simply write $\sqrt{}$ rather than $\sqrt[2]{}$. So, for example, $\sqrt{25} = 5$.

On the other hand, as we saw in the examples, cube roots, fifth roots, and, in fact, all *odd* roots occur singly, not in pairs. In these cases, we again use the notation $\sqrt[n]{b}$ for the nth root. The definition and examples in the following box summarize our discussion up to this point.

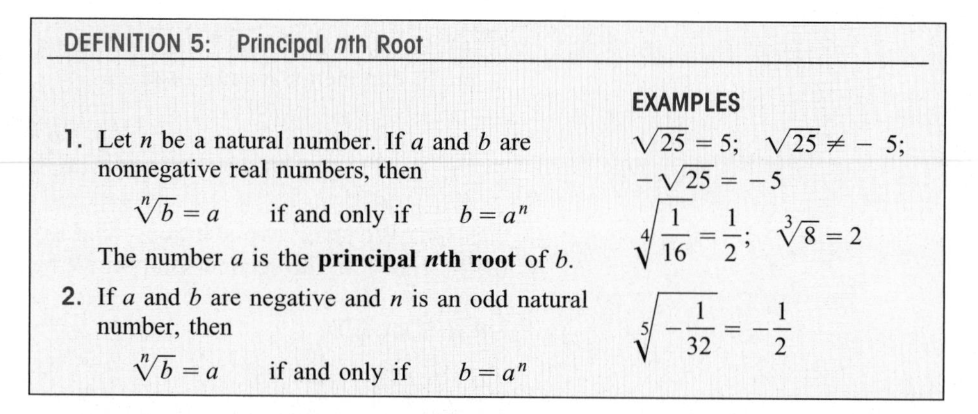

DEFINITION 5: Principal nth Root

1. Let n be a natural number. If a and b are nonnegative real numbers, then

$$\sqrt[n]{b} = a \quad \text{if and only if} \quad b = a^n$$

The number a is the **principal nth root** of b.

2. If a and b are negative and n is an odd natural number, then

$$\sqrt[n]{b} = a \quad \text{if and only if} \quad b = a^n$$

EXAMPLES

$\sqrt{25} = 5$; $\sqrt{25} \neq -5$; $-\sqrt{25} = -5$

$\sqrt[4]{\frac{1}{16}} = \frac{1}{2}$; $\sqrt[3]{8} = 2$

$\sqrt[5]{-\frac{1}{32}} = -\frac{1}{2}$

There are five properties of nth roots that are frequently used in simplifying certain expressions. The first four are similar to the properties of square roots that are developed in elementary algebra. For reference, we list these properties side by side in the following box. Property 5 is listed here only for the sake of completeness; we'll postpone discussing it until Appendix A.6.

PROPERTY SUMMARY

PROPERTIES OF nTH ROOTS	CORRESPONDING PROPERTIES FOR SQUARE ROOTS
Suppose that x and y are real numbers and that m and n are natural numbers. Then each of the following properties holds, provided only that the expressions on both sides of the equation are defined (and so represent real numbers).	
1. $(\sqrt[n]{x})^n = x$	$(\sqrt{x})^2 = x$
2. $\sqrt[n]{xy} = \sqrt[n]{x}\sqrt[n]{y}$	$\sqrt{xy} = \sqrt{x}\sqrt{y}$
3. $\sqrt[n]{\dfrac{x}{y}} = \dfrac{\sqrt[n]{x}}{\sqrt[n]{y}}$	$\sqrt{\dfrac{x}{y}} = \dfrac{\sqrt{x}}{\sqrt{y}}$
4. n even: $\sqrt[n]{x^n} = \lvert x \rvert$ n odd: $\sqrt[n]{x^n} = x$	$\sqrt{x^2} = \lvert x \rvert$
5. $\sqrt[m]{\sqrt[n]{x}} = \sqrt[mn]{x}$	

These properties are used in simplifying expressions involving nth roots. In general, we try to factor the expression under the radical so that one factor is the largest perfect nth power that we can find. Then we apply Property 2 or 3. For instance, the expression $\sqrt{72}$ is simplified as follows:

$$\sqrt{72} = \sqrt{(36)(2)} = \sqrt{36}\sqrt{2} = 6\sqrt{2}$$

In this procedure, we began by factoring 72 as (36)(2). Note that 36 is the largest factor of 72 that is a perfect square. If we were to begin instead with a different factorization, say, $72 = (9)(8)$, we can still arrive at the same answer, but it would take longer. (Check this for yourself.) As another example, let's simplify $\sqrt[3]{40}$. First, what (if any) is the largest perfect-cube factor of 40? Since the first few perfect cubes are

$$1^3 = 1 \qquad 2^3 = 8 \qquad 3^3 = 27 \qquad 4^3 = 64$$

we see that 8 is a perfect-cube factor of 40, and we write

$$\sqrt[3]{40} = \sqrt[3]{(8)(5)} = \sqrt[3]{8}\sqrt[3]{5} = 2\sqrt[3]{5}$$

EXAMPLE 3 Simplify: **(a)** $\sqrt{12} + \sqrt{75}$; **(b)** $\sqrt{\dfrac{162}{49}}$.

Solution

(a) $\sqrt{12} + \sqrt{75} = \sqrt{(4)(3)} + \sqrt{(25)(3)}$
$= \sqrt{4}\sqrt{3} + \sqrt{25}\sqrt{3}$
$= 2\sqrt{3} + 5\sqrt{3} = (2+5)\sqrt{3}$
$= 7\sqrt{3}$

(b) $\sqrt{\dfrac{162}{49}} = \dfrac{\sqrt{162}}{\sqrt{49}} = \dfrac{\sqrt{81}\sqrt{2}}{7} = \dfrac{9\sqrt{2}}{7}$

EXAMPLE 4 Simplify: $\sqrt[3]{16} + \sqrt[3]{250} - \sqrt[3]{128}$.

Solution $\sqrt[3]{16} + \sqrt[3]{250} - \sqrt[3]{128} = \sqrt[3]{8}\sqrt[3]{2} + \sqrt[3]{125}\sqrt[3]{2} - \sqrt[3]{64}\sqrt[3]{2}$
$= 2\sqrt[3]{2} + 5\sqrt[3]{2} - 4\sqrt[3]{2}$
$= 3\sqrt[3]{2}$

EXAMPLE 5 Simplify each of the following expressions by removing the largest possible perfect-square or perfect-cube factor from within the radical:

(a) $\sqrt{8x^2}$;

(b) $\sqrt{8x^2}$, where $x \geqslant 0$;

(c) $\sqrt{18a^7}$, where $a \geqslant 0$;

(d) $\sqrt[3]{16y^5}$.

Solution

(a) $\sqrt{8x^2} = \sqrt{(4)(2)(x^2)} = \sqrt{4}\sqrt{2}\sqrt{x^2} = 2\sqrt{2}\,|x|$

(b) $\sqrt{8x^2} = \sqrt{4}\sqrt{2}\sqrt{x^2}$
$= 2\sqrt{2}x$ $\qquad \sqrt{x^2} = x$ because $x \geqslant 0$

(c) $\sqrt{18a^7} = \sqrt{(9a^6)(2a)} = \sqrt{9a^6}\sqrt{2a}$
$= 3a^3\sqrt{2a}$

(d) $\sqrt[3]{16y^5} = \sqrt[3]{8y^3}\sqrt[3]{2y^2} = 2y\sqrt[3]{2y^2}$

There are times when it is convenient to rewrite fractions involving radicals in alternate forms. Suppose, for example, that we want to rewrite the fraction $5/\sqrt{3}$ in an equivalent form not involving a radical in the denominator. This is called **rationalizing the denominator.** The procedure here is to multiply by 1 in this way:

$$\frac{5}{\sqrt{3}} = \frac{5}{\sqrt{3}} \cdot 1 = \frac{5}{\sqrt{3}} \cdot \frac{\sqrt{3}}{\sqrt{3}} = \frac{5\sqrt{3}}{3}$$

That is, $\dfrac{5}{\sqrt{3}} = \dfrac{5\sqrt{3}}{3}$, as required.

To rationalize a denominator of the form $a + \sqrt{b}$, we need to multiply the fraction not by $\sqrt{b}/\sqrt{b}$, but rather by $(a - \sqrt{b})/(a - \sqrt{b})$ $(= 1)$. To see why this is necessary, notice that

$$(a + \sqrt{b})\sqrt{b} = a\sqrt{b} + b$$

which still contains a radical, whereas

$$\begin{aligned}(a + \sqrt{b})(a - \sqrt{b}) &= a^2 - a\sqrt{b} + a\sqrt{b} - b \\ &= a^2 - b\end{aligned}$$

which is free of radicals. Similarly, to rationalize a denominator of the form $a - \sqrt{b}$, we multiply the fraction by $(a + \sqrt{b})/(a + \sqrt{b})$. (The quantities $a + \sqrt{b}$ and $a - \sqrt{b}$ are said to be **conjugates** of each other.) The next two examples make use of these ideas.

EXAMPLE 6 Simplify: $\dfrac{1}{\sqrt{2}} - 3\sqrt{50}$.

Solution First, we rationalize the denominator in the fraction $1/\sqrt{2}$:

$$\frac{1}{\sqrt{2}} = \frac{1}{\sqrt{2}} \cdot 1 = \frac{1}{\sqrt{2}} \cdot \frac{\sqrt{2}}{\sqrt{2}} = \frac{\sqrt{2}}{2}$$

Next, we simplify the expression $3\sqrt{50}$:

$$3\sqrt{50} = 3\sqrt{(25)(2)} = 3\sqrt{25}\sqrt{2} = (3)(5)\sqrt{2} = 15\sqrt{2}$$

Now, putting things together, we have

$$\begin{aligned}\frac{1}{\sqrt{2}} - 3\sqrt{50} &= \frac{\sqrt{2}}{2} - 15\sqrt{2} = \frac{\sqrt{2}}{2} - \frac{30\sqrt{2}}{2} \\ &= \frac{\sqrt{2} - 30\sqrt{2}}{2} = \frac{(1 - 30)\sqrt{2}}{2} \\ &= \frac{-29\sqrt{2}}{2}\end{aligned}$$

EXAMPLE 7 Rationalize the denominator in the expression $\dfrac{4}{2 + \sqrt{3}}$.

Solution We multiply by 1, writing it as $\dfrac{2 - \sqrt{3}}{2 - \sqrt{3}}$.

$$\begin{aligned}\frac{4}{2 + \sqrt{3}} \cdot 1 &= \frac{4}{2 + \sqrt{3}} \cdot \frac{2 - \sqrt{3}}{2 - \sqrt{3}} \\ &= \frac{4(2 - \sqrt{3})}{4 - (\sqrt{3})^2} = \frac{4(2 - \sqrt{3})}{4 - 3} = \frac{8 - 4\sqrt{3}}{1} \\ &= 8 - 4\sqrt{3}\end{aligned}$$

NOTE Check for yourself that multiplying the original fraction by $\dfrac{2 + \sqrt{3}}{2 + \sqrt{3}}$ does *not* eliminate radicals in the denominator.

In the next example, we are asked to rationalize the numerator rather than the denominator. This is useful at times in calculus.

EXAMPLE 8 Rationalize the *numerator:* $\dfrac{\sqrt{x}-\sqrt{3}}{x-3}$, where $x \geq 0$, $x \neq 3$.

Solution

$$\begin{aligned}\frac{\sqrt{x}-\sqrt{3}}{x-3}\cdot 1 &= \frac{\sqrt{x}-\sqrt{3}}{x-3}\cdot\frac{\sqrt{x}+\sqrt{3}}{\sqrt{x}+\sqrt{3}}\\ &= \frac{(\sqrt{x})^2-(\sqrt{3})^2}{(x-3)(\sqrt{x}+\sqrt{3})} = \frac{x-3}{(x-3)(\sqrt{x}+\sqrt{3})}\\ &= \frac{1}{\sqrt{x}+\sqrt{3}}\end{aligned}$$

The strategy for rationalizing numerators or denominators involving nth roots is similar to that used for square roots. To rationalize a numerator or a denominator involving an nth root, we multiply that numerator or denominator by a factor that yields a product that itself is a perfect nth power. The next example displays two instances of this.

EXAMPLE 9

(a) Rationalize the denominator: $\dfrac{6}{\sqrt[3]{7}}$.

(b) Rationalize the denominator: $\dfrac{ab}{\sqrt[4]{a^2b^3}}$, where $a > 0$, $b > 0$.

Solution

(a)
$$\begin{aligned}\frac{6}{\sqrt[3]{7}}\cdot 1 &= \frac{6}{\sqrt[3]{7}}\cdot\frac{\sqrt[3]{7^2}}{\sqrt[3]{7^2}}\\ &= \frac{6\sqrt[3]{7^2}}{\sqrt[3]{7^3}}\\ &= \frac{6\sqrt[3]{49}}{7} \quad \text{as required}\end{aligned}$$

(b)
$$\begin{aligned}\frac{ab}{\sqrt[4]{a^2b^3}}\cdot 1 &= \frac{ab}{\sqrt[4]{a^2b^3}}\cdot\frac{\sqrt[4]{a^2b}}{\sqrt[4]{a^2b}}\\ &= \frac{ab\sqrt[4]{a^2b}}{\sqrt[4]{a^4b^4}} = \frac{ab\sqrt[4]{a^2b}}{ab} = \sqrt[4]{a^2b}\end{aligned}$$

A.6 REVIEW OF RATIONAL EXPONENTS

. . . Nicole Oresme, a bishop in Normandy (about 1323–1382), first conceived the notion of fractional powers, afterwards rediscovered by [Simon] *Stevin* [1548–1620] . . .

Florian Cajori in *A History of Elementary Mathematics* (New York: Macmillan Company, 1917)

We can use the concept of an nth root to give a meaning to fractional exponents that is useful and, at the same time, consistent with our earlier work. First, by way of motivation, suppose that we want to assign a value to $5^{1/3}$. Assuming that the usual properties of exponents continue to apply here, we can write

$$(5^{1/3})^3 = 5^1 = 5$$

That is,

$$(5^{1/3})^3 = 5$$

or

$$5^{1/3} = \sqrt[3]{5}$$

By replacing 5 and 3 with b and n, respectively, we can see that we want to define $b^{1/n}$ to mean $\sqrt[n]{b}$. Also, by thinking of $b^{m/n}$ as $(b^{1/n})^m$, we see that the definition for $b^{m/n}$ ought to be $(\sqrt[n]{b})^m$. These definitions are formalized in the box that follows.

DEFINITION: Rational Exponents

	EXAMPLES
1. Let b denote a real number and n a natural number. We define $b^{1/n}$ by $$b^{1/n} = \sqrt[n]{b}$$ (If n is even, we require that $b \geq 0$.)	$4^{1/2} = \sqrt{4} = 2$ $(-8)^{1/3} = \sqrt[3]{-8} = -2$
2. Let m/n be a rational number reduced to lowest terms. Assume that n is positive and that $\sqrt[n]{b}$ exists. Then, $$b^{m/n} = (\sqrt[n]{b})^m$$ or equivalently, $$b^{m/n} = \sqrt[n]{b^m}$$	$8^{2/3} = (\sqrt[3]{8})^2 = 2^2 = 4$ or, equivalently, $8^{2/3} = \sqrt[3]{8^2} = \sqrt[3]{64} = 4$

It can be shown that the four properties of exponents that we listed in Appendix A.5 continue to hold for rational exponents in general. In fact, we'll take this for granted rather than follow the lengthy argument needed for its verification. We will also assume that these properties apply to irrational exponents. So, for instance, we have

$$\left(2^{\sqrt{5}}\right)^{\sqrt{5}} = 2^5 = 32$$

(The definition of irrational exponents is discussed in Section 5.1.) In the next three examples, we display the basic techniques for working with rational exponents.

EXAMPLE 1 Simplify each of the following quantities. Express the answers using positive exponents. If an expression does not represent a real number, say so.

(a) $49^{1/2}$ **(b)** $-49^{1/2}$ **(c)** $(-49)^{1/2}$ **(d)** $49^{-1/2}$

Solution

(a) $49^{1/2}=\sqrt{49}=7$

(b) $-49^{1/2}=-(49^{1/2})=-\sqrt{49}=-7$

(c) The quantity $(-49)^{1/2}$ does not represent a real number because there is no real number x such that $x^2=-49$.

(d) $49^{-1/2}=\sqrt{49^{-1}}=\sqrt{1/49}=\sqrt{1}/\sqrt{49}=1/7$

Alternatively, we have

$49^{-1/2}=(49^{1/2})^{-1}=7^{-1}=1/7$

EXAMPLE 2 Simplify each of the following. Write the answers using positive exponents. (Assume that $a>0$.)

(a) $(5a^{2/3})(4a^{3/4})$

(b) $\sqrt[5]{\dfrac{16a^{1/3}}{a^{1/4}}}$

(c) $(x^2+1)^{1/5}(x^2+1)^{4/5}$

Solution

(a)
$$(5a^{2/3})(4a^{3/4})=20a^{(2/3)+(3/4)}$$
$$=20a^{17/12} \qquad \text{because } \frac{2}{3}+\frac{3}{4}=\frac{17}{12}$$

(b)
$$\sqrt[5]{\frac{16a^{1/3}}{a^{1/4}}}=\left(\frac{16a^{1/3}}{a^{1/4}}\right)^{1/5}$$
$$=(16a^{1/12})^{1/5} \qquad \text{because } \frac{1}{3}-\frac{1}{4}=\frac{1}{12}$$
$$=16^{1/5}a^{1/60}$$

(c) $(x^2+1)^{1/5}(x^2+1)^{4/5}=(x^2+1)^1=x^2+1$

EXAMPLE 3 Simplify: **(a)** $32^{-2/5}$; **(b)** $(-8)^{4/3}$.

Solution

(a) $32^{-2/5}=(\sqrt[5]{32})^{-2}=2^{-2}=\dfrac{1}{2^2}=\dfrac{1}{4}$

Alternatively, we have

$32^{-2/5}=(2^5)^{-2/5}=2^{-2}=\dfrac{1}{2^2}=\dfrac{1}{4}$

(b) $(-8)^{4/3}=(\sqrt[3]{-8})^4=(-2)^4=16$

Alternatively, we can write

$(-8)^{4/3}=[(-2)^3]^{4/3}=(-2)^4=16$

Rational exponents can be used to simplify certain expressions containing radicals. For example, one of the properties of nth roots that we listed but did not discuss in the previous section is $\sqrt[m]{\sqrt[n]{x}}=\sqrt[mn]{x}$. Using exponents, it is easy to verify this property. We have

$$\sqrt[m]{\sqrt[n]{x}} = (x^{1/n})^{1/m}$$
$$= x^{1/mn} = \sqrt[mn]{x} \qquad \text{as we wished to show}$$

EXAMPLE 4 Consider the expression $\sqrt{x}\sqrt[3]{y^2}$, where x and y are positive.

(a) Rewrite the expression using rational exponents.
(b) Rewrite the expression using only one radical sign.

Solution

(a) $\sqrt{x}\sqrt[3]{y^2} = x^{1/2}y^{2/3}$

(b) $\sqrt{x}\sqrt[3]{y^2} = x^{1/2}y^{2/3}$

$$= x^{3/6}y^{4/6} \qquad \text{rewriting the fractions using a common denominator}$$
$$= \sqrt[6]{x^3}\sqrt[6]{y^4} = \sqrt[6]{x^3y^4}$$

EXAMPLE 5 Rewrite the following expression using rational exponents (assume that x, y, and z are positive):

$$\sqrt{\frac{\sqrt[3]{x}\sqrt[4]{y^3}}{\sqrt[5]{z^4}}}$$

Solution

$$\sqrt{\frac{\sqrt[3]{x}\sqrt[4]{y^3}}{\sqrt[5]{z^4}}} = \left(\frac{\sqrt[3]{x}\sqrt[4]{y^3}}{\sqrt[5]{z^4}}\right)^{1/2} = \left(\frac{x^{1/3}y^{3/4}}{z^{4/5}}\right)^{1/2}$$
$$= \frac{x^{1/6}y^{3/8}}{z^{2/5}} = x^{1/6}y^{3/8}z^{-2/5}$$

TABLES

TABLE 1 EXPONENTIAL FUNCTIONS

x	e^x	e^{-x}	x	e^x	e^{-x}
0.00	1.0000	1.0000	1.5	4.4817	0.2231
0.01	1.0101	0.9901	1.6	4.9530	0.2019
0.02	1.0202	0.9802	1.7	5.4739	0.1827
0.03	1.0305	0.9704	1.8	6.0496	0.1653
0.04	1.0408	0.9608	1.9	6.6859	0.1496
0.05	1.0513	0.9512	2.0	7.3891	0.1353
0.06	1.0618	0.9418	2.1	8.1662	0.1225
0.07	1.0725	0.9324	2.2	9.0250	0.1108
0.08	1.0833	0.9231	2.3	9.9742	0.1003
0.09	1.0942	0.9139	2.4	11.023	0.0907
0.10	1.1052	0.9048	2.5	12.182	0.0821
0.11	1.1163	0.8958	2.6	13.464	0.0743
0.12	1.1275	0.8869	2.7	14.880	0.0672
0.13	1.1388	0.8781	2.8	16.445	0.0608
0.14	1.1503	0.8694	2.9	18.174	0.0550
0.15	1.1618	0.8607	3.0	20.086	0.0498
0.16	1.1735	0.8521	3.1	22.198	0.0450
0.17	1.1853	0.8437	3.2	24.533	0.0408
0.18	1.1972	0.8353	3.3	27.113	0.0369
0.19	1.2092	0.8270	3.4	29.964	0.0334
0.20	1.2214	0.8187	3.5	33.115	0.0302
0.21	1.2337	0.8106	3.6	36.598	0.0273
0.22	1.2461	0.8025	3.7	40.447	0.0247
0.23	1.2586	0.7945	3.8	44.701	0.0224
0.24	1.2712	0.7866	3.9	49.402	0.0202
0.25	1.2840	0.7788	4.0	54.598	0.0183
0.30	1.3499	0.7408	4.1	60.340	0.0166
0.35	1.4191	0.7047	4.2	66.686	0.0150
0.40	1.4918	0.6703	4.3	73.700	0.0136
0.45	1.5683	0.6376	4.4	81.451	0.0123
0.50	1.6487	0.6065	4.5	90.017	0.0111
0.55	1.7333	0.5769	4.6	99.484	0.0101
0.60	1.8221	0.5488	4.7	109.95	0.0091
0.65	1.9155	0.5220	4.8	121.51	0.0082
0.70	2.0138	0.4966	4.9	134.29	0.0074
0.75	2.1170	0.4724	5.0	148.41	0.0067
0.80	2.2255	0.4493	5.5	244.69	0.0041
0.85	2.3396	0.4274	6.0	403.43	0.0025
0.90	2.4596	0.4066	6.5	665.14	0.0015
0.95	2.5857	0.3867	7.0	1096.6	0.0009
1.0	2.7183	0.3679	7.5	1808.0	0.0006
1.1	3.0042	0.3329	8.0	2981.0	0.0003
1.2	3.3201	0.3012	8.5	4914.8	0.0002
1.3	3.6693	0.2725	9.0	8103.1	0.0001
1.4	4.0552	0.2466	10.0	22026	0.00005

TABLE 2 LOGARITHMS TO THE BASE TEN

x	0	1	2	3	4	5	6	7	8	9
1.0	.0000	.0043	.0866	.0128	.0170	.0212	.0253	.0294	.0334	.0374
1.1	.0414	.0453	.0492	.0531	.0569	.0607	.0645	.0682	.0719	.0755
1.2	.0792	.0828	.0864	.0899	.0934	.0969	.1004	.1038	.1072	.1106
1.3	.1139	.1173	.1206	.1239	.1271	.1303	.1335	.1367	.1399	.1430
1.4	.1461	.1492	.1523	.1553	.1584	.1614	.1644	.1673	.1703	.1732
1.5	.1761	.1790	.1818	.1847	.1875	.1093	.1931	.1959	.1987	.2014
1.6	.2041	.2068	.2095	.2122	.2148	.2175	.2201	.2227	.2253	.2279
1.7	.2304	.2330	.2355	.2380	.2405	.2430	.2455	.2480	.2504	.2529
1.8	.2553	.2577	.2601	.2625	.2648	.2672	.2695	.2718	.2742	.2765
1.9	.2788	.2810	.2833	.2856	.2878	.2900	.2923	.2945	.2967	.2989
2.0	.3010	.3032	.3054	.3075	.3096	.3118	.3139	.3160	.3181	.3201
2.1	.3222	.3243	.3263	.3284	.3304	.3324	.3345	.3365	.3385	.3404
2.2	.2424	.3444	.3464	.3483	.3502	.3522	.3541	.3560	.3579	.3598
2.3	.3617	.3636	.3655	.3674	.3692	.3711	.3729	.3747	.3766	.3784
2.4	.3802	.3820	.3838	.3856	.3874	.3892	.3909	.3927	.3945	.3962
2.5	.3979	.3997	.4014	.4031	.4048	.4065	.4082	.4099	.4116	.4133
2.6	.4150	.4166	.4183	.4200	.4216	.4232	.4249	.4265	.4281	.4298
2.7	.4314	.4330	.4346	.4362	.4378	.4393	.4409	.4425	.4440	.4456
2.8	.4472	.4487	.4502	.4518	.4533	.4548	.4564	.4579	.4594	.4609
2.9	.4624	.4639	.4654	.4669	.4683	.4698	.4713	.4728	.4742	.4757
3.0	.4771	.4786	.4800	.4814	.4829	.4843	.4857	.4871	.4886	.4900
3.1	.4914	.4928	.4942	.4955	.4969	.4983	.4997	.5011	.5024	.5038
3.2	.5051	.5065	.5079	.5092	.5105	.5119	.5132	.5145	.5159	.5172
3.3	.5185	.5198	.5211	.5224	.5237	.5250	.5263	.5276	.5289	.5302
3.4	.5315	.5328	.5340	.5353	.5366	.5378	.5391	.5403	.5416	.5428
3.5	.5441	.5453	.5465	.5478	.5490	.5502	.5514	.5527	.5539	.5551
3.6	.5563	.5575	.5587	.5599	.5611	.5623	.5635	.5647	.5658	.5670
3.7	.5682	.5694	.5705	.5717	.5729	.5740	.5752	.5763	.5775	.5786
3.8	.5798	.5809	.5821	.5832	.5843	.5855	.5866	.5877	.5888	.5899
3.9	.5911	.5922	.5933	.5944	.5955	.5966	.5977	.5988	.5999	.6010
4.0	.6021	.6031	.6042	.6053	.6064	.6075	.6085	.6096	.6107	.6117
4.1	.6128	.6138	.6149	.6160	.6170	.6180	.6191	.6201	.6212	.6222
4.2	.6232	.6243	.6253	.6263	.6274	.6284	.6294	.6304	.6314	.6325
4.3	.6335	.6345	.6355	.6365	.6375	.6385	.6395	.6405	.6415	.6425
4.4	.6435	.6444	.6454	.6464	.6474	.6484	.6493	.6503	.6513	.6522
4.5	.6532	.6542	.6551	.6561	.6571	.6580	.6590	.6599	.6609	.6618
4.6	.6628	.6637	.6646	.6656	.6665	.6675	.6684	.6693	.6702	.6712
4.7	.6721	.6730	.6739	.6749	.6758	.6767	.6776	.6785	.6794	.6803
4.8	.6812	.6821	.6830	.6839	.6848	.6857	.6866	.6875	.6884	.6893
4.9	.6902	.6911	.6920	.6928	.6937	.6946	.6955	.6964	.6972	.6981
5.0	.6990	.6998	.7007	.7016	.7024	.7033	.7042	.7050	.7059	.7067
5.1	.7076	.7084	.7093	.7101	.7110	.7118	.7126	.7135	.7143	.7152
5.2	.7160	.7168	.7177	.7185	.7193	.7202	.7210	.7218	.7226	.7235
5.3	.7243	.7251	.7259	.7267	.7275	.7284	.7292	.7300	.7308	.7316
5.4	.7324	.7332	.7340	.7348	.7356	.7364	.7372	.7380	.7388	.7396

TABLE 2 LOGARITHMS TO THE BASE TEN *(continued)*

x	0	1	2	3	4	5	6	7	8	9
5.5	.7404	.7412	.7419	.7427	.7435	.7443	.7451	.7459	.7466	.7474
5.6	.7482	.7490	.7497	.7505	.7513	.7520	.7528	.7536	.7543	.7551
5.7	.7559	.7566	.7574	.7582	.7589	.7597	.7604	.7612	.7619	.7627
5.8	.7634	.7642	.7649	.7657	.7664	.7672	.7679	.7686	.7694	.7701
5.9	.7709	.7716	.7723	.7731	.7738	.7745	.7752	.7760	.7767	.7774
6.0	.7782	.7789	.7796	.7803	.7810	.7818	.7825	.7832	.7839	.7846
6.1	.7853	.7860	.7868	.7875	.7882	.7889	.7896	.7903	.7910	.7917
6.2	.7924	.7931	7938	.7945	.7952	.7959	.7966	.7973	.7980	.7987
6.3	.7993	.8000	.8007	.8014	.8021	.8028	.8035	.8041	.8048	.8055
6.4	.8062	.8069	.8075	.8082	.8089	.8096	.8102	.8109	.8116	.8122
6.5	.8129	.8136	.8142	.8149	.8156	.8162	.8169	.8176	.8182	.8189
6.6	.8195	.8202	.8209	.8215	.8222	.8288	.8235	.8241	.8248	.8254
6.7	.8261	.8267	.8274	.8280	.8287	.8293	.8299	.8306	.8312	.8319
6.8	.8325	.8331	.8338	.8344	.8351	.8357	.8363	.8370	.8376	.8382
6.9	.8388	.8395	.8401	.8407	.8414	.8420	.8426	.8432	.8439	.8445
7.0	.8451	.8457	.8463	.8470	.8476	.8482	.8488	.8494	.8500	.8506
7.1	.8513	.8519	.8525	.8531	.8537	.8543	.8549	.8555	.8561	.8567
7.2	.8573	.8579	.8585	.8591	.8597	.8603	.8609	.8615	.8621	.8627
7.3	.8633	.8639	.8645	.8651	.8657	.8663	.8669	.8675	.8681	.8686
7.4	.8692	.8698	.8704	.8710	.8716	.8722	.8727	.8733	.8739	.8745
7.5	.8751	.8756	.8762	.8768	.8774	.8779	.8785	.8791	.8797	.8802
7.6	.8808	.8814	.8820	.8825	.8831	.8837	.8842	.8848	.8854	.8859
7.7	.8865	.8871	.8876	.8882	.8887	.8893	.8899	.8904	.8910	.8915
7.8	.8921	.8927	.8932	.8938	.8943	.8949	.8954	.8960	.8965	.8971
7.9	.8976	.8982	.8987	.8993	.8998	.9004	.9009	.9015	.9020	.9025
8.0	.9031	.9036	.9042	.9047	.9053	.9058	.9063	.9069	.9074	.9079
8.1	.9085	.9090	.9096	.9101	.9106	.9112	.9117	.9122	.9128	.9133
8.2	.9138	.9143	.9149	.9154	.9259	.9165	.9170	.9175	.9180	.9186
8.3	.9191	.9196	.9201	.9206	.9212	.9217	.9222	.9227	.9232	.9238
8.4	.9243	.9248	.9253	.9258	.9263	.9269	.9274	.9279	.9284	.9289
8.5	.9294	.9299	.9304	.9309	.9315	.9320	.9325	.9330	.9335	.9340
8.6	.9345	.9350	.9355	.9360	.9365	.9370	.9375	.9380	.9385	.9390
8.7	.9395	.9400	.9405	.9410	.9415	.9420	.9425	.9430	.9435	.9440
8.8	.9445	.9450	.9455	.9460	.9465	.9469	.9474	.9479	.9484	.9489
8.9	.9494	.9499	.9504	.9509	.9513	.9518	.9523	.9528	.9533	.9538
9.0	.9542	.9547	.9552	.9557	.9562	.9566	.9571	.9576	.9581	.9586
9.1	.9590	.9595	.9600	.9605	.9609	.9614	.9619	.9624	.9628	.9633
9.2	.9638	.9643	.9647	.9652	.9657	.9661	.9666	.9671	.9675	.9680
9.3	.9685	.9689	.9694	.9699	.9703	.9708	.9713	.9717	.9722	.9727
9.4	.9731	.9736	.9741	.9745	.9750	.9754	.9759	.9763	.9768	.9773
9.5	.9777	.9782	.9786	.9791	.9795	.9800	.9805	.9809	.9814	.9818
9.6	.9823	.9827	.9823	.9836	.9841	.9845	.9850	.9854	.9859	.9863
9.7	.9868	.9872	.9877	.9881	.9886	.9890	.9894	.9899	.9903	.9908
9.8	.9912	.9917	.9921	.9926	.9930	.9934	.9939	.9943	.9948	.9952
9.9	.9956	.9961	.9965	.9969	.9974	.9978	.9983	.9987	.9991	.9996

TABLE 3 NATURAL LOGARITHMS

n	$\ln n$	n	$\ln n$	n	$\ln n$
		4.5	1.5041	9.0	2.1972
0.1	−2.3026	4.6	1.5261	9.1	2.2083
0.2	−1.6904	4.7	1.5476	9.2	2.2192
0.3	−1.2040	4.8	1.5686	9.3	2.2300
0.4	−0.9163	4.9	1.5892	9.4	2.2407
0.5	−0.6931	5.0	1.6094	9.5	2.2513
0.6	−0.5108	5.1	1.6292	9.6	2.2618
0.7	−0.3567	5.2	1.6487	9.7	2.2721
0.8	−0.2231	5.3	1.6677	9.8	2.2824
0.9	−0.1054	5.4	1.6864	9.9	2.2925
1.0	0.0000	5.5	1.7047	10	2.3026
1.1	0.0953	5.6	1.7228	11	2.3979
1.2	0.1823	5.7	1.7405	12	2.4849
1.3	0.2624	5.8	1.7579	13	2.5649
1.4	0.3365	5.9	1.7750	14	2.6391
1.5	0.4055	6.0	1.7918	15	2.7081
1.6	0.4700	6.1	1.8083	16	2.7726
1.7	0.5306	6.2	1.8245	17	2.8332
1.8	0.5878	6.3	1.8405	18	2.8904
1.9	0.6419	6.4	1.8563	19	2.9444
2.0	0.6931	6.5	1.8718	20	2.9957
2.1	0.7419	6.6	1.8871	25	3.2189
2.2	0.7885	6.7	1.9021	30	3.4012
2.3	0.8329	6.8	1.9169	35	3.5553
2.4	0.8755	6.9	1.9315	40	3.6889
2.5	0.9163	7.0	1.9459	45	3.8067
2.6	0.9555	7.1	1.9601	50	3.9120
2.7	0.9933	7.2	1.9741	55	4.0073
2.8	1.0296	7.3	1.9879	60	4.0943
2.9	1.0647	7.4	2.0015	65	4.1774
3.0	1.0986	7.5	2.0149	70	4.2485
3.1	1.1314	7.6	2.0281	75	4.3175
3.2	1.1632	7.7	2.0412	80	4.3820
3.3	1.1939	7.8	2.0541	85	4.4427
3.4	1.2238	7.9	2.0669	90	4.4998
3.5	1.2528	8.0	2.0794	95	4.5539
3.6	1.2809	8.1	2.0919	100	4.6052
3.7	1.3083	8.2	2.1041		
3.8	1.3350	8.3	2.1163		
3.9	1.3610	8.4	2.1282		
4.0	1.3863	8.5	2.1401		
4.1	1.4110	8.6	2.1518		
4.2	1.4351	8.7	2.1633		
4.3	1.4586	8.8	2.1748		
4.4	1.4816	8.9	2.1861		

TABLE 4 VALUES OF TRIGONOMETRIC FUNCTIONS

Angle θ									
Degrees	Radians	sin θ	csc θ	tan θ	cot θ	sec θ	cos θ		
0°00′	.0000	.0000	No value	.0000	No value	1.000	1.0000	1.5708	90°00′
10	029	029	343.8	029	343.8	000	000	679	50
20	058	058	171.9	058	171.9	000	000	650	40
30	087	087	114.6	087	114.6	000	1.0000	621	30
40	116	116	85.95	116	85.94	000	.9999	592	20
50	145	145	68.76	145	68.75	000	999	563	10
1°00′	.0175	.0175	57.30	.0175	57.29	1.000	.9998	1.5533	89°00′
10	204	204	49.11	204	49.10	000	998	504	50
20	233	233	42.98	233	42.96	000	997	475	40
30	262	262	38.20	262	38.19	000	997	446	30
40	291	291	34.38	291	34.37	000	996	417	20
50	320	320	31.26	320	31.24	001	995	388	10
2°00′	.0349	.0349	28.65	.0349	28.64	1.001	.9994	1.5359	88°00′
10	378	378	26.45	378	26.43	001	993	330	50
20	407	407	24.56	407	24.54	001	992	301	40
30	436	436	22.93	437	22.90	001	990	272	30
40	465	465	21.49	466	21.47	001	989	243	20
50	495	494	20.23	495	20.21	001	988	213	10
3°00′	.0524	.0523	19.11	.0524	19.08	1.001	.9986	1.5184	87°00′
10	553	552	18.10	553	18.07	002	985	155	50
20	582	581	17.20	582	17.17	002	983	126	40
30	611	610	16.38	612	16.35	002	981	097	30
40	640	640	15.64	641	15.60	002	980	068	20
50	669	669	14.96	670	14.92	002	978	039	10
4°00′	.0698	.0698	14.34	.0699	14.30	1.002	.9976	1.5010	86°00′
10	727	727	13.76	729	13.73	003	974	981	50
20	756	756	13.23	758	13.20	003	971	952	40
30	785	785	12.75	787	12.71	003	969	923	30
40	814	814	12.29	816	12.25	003	967	893	20
50	844	843	11.87	846	11.83	004	964	864	10
5°00′	.0873	.0872	11.47	.0875	11.43	1.004	.9962	1.4835	85°00′
10	902	901	11.10	904	11.06	004	959	806	50
20	931	929	10.76	934	10.71	004	957	777	40
30	960	958	10.43	963	10.39	005	954	748	30
40	.0989	.0987	10.13	.0992	10.08	005	951	719	20
50	.1018	.1016	9.839	.1022	9.788	005	948	690	10
6°00′	.1047	.1045	9.567	.1051	9.514	1.006	.9945	1.4661	84°00′
10	076	074	9.309	080	9.255	006	942	632	50
20	105	103	9.065	110	9.010	006	939	603	40
30	134	132	8.834	139	8.777	006	936	573	30
40	164	161	8.614	169	8.556	007	932	544	20
50	193	190	8.405	198	8.345	007	929	515	10
7°00′	.1222	.1219	8.206	.1228	8.144	1.008	.9925	1.4486	83°00′
		cos θ	sec θ	cot θ	tan θ	csc θ	sin θ	Radians	Degrees
								Angle θ	

TABLE 4 VALUES OF TRIGONOMETRIC FUNCTIONS *(continued)*

Angle θ									
Degrees	Radians	sin θ	csc θ	tan θ	cot θ	sec θ	cos θ		
7°00′	.1222	.1219	8.206	.1228	8.144	1.008	.9925	1.4486	83°00′
10	251	248	8.016	257	7.953	008	922	457	50
20	280	276	7.834	287	7.770	008	918	428	40
30	309	305	7.661	317	7.596	009	914	399	30
40	338	334	7.496	346	7.429	009	911	370	20
50	367	363	7.337	376	7.269	009	907	341	10
8°00′	.1396	.1392	7.185	.1405	7.115	1.010	.9903	1.4312	82°00′
10	425	421	7.040	435	6.968	010	899	283	50
20	454	449	6.900	465	827	011	894	254	40
30	484	478	765	495	691	011	890	224	30
40	513	507	636	524	561	012	886	195	20
50	542	536	512	554	435	012	881	166	10
9°00′	.1571	.1564	6.392	.1584	6.314	1.012	.9877	1.4137	81°00′
10	600	593	277	614	197	013	872	108	50
20	629	622	166	644	6.084	013	868	079	40
30	658	650	6.059	673	5.976	014	863	050	30
40	687	679	5.955	703	871	014	858	1.4021	20
50	716	708	855	733	769	015	853	1.3992	10
10°00′	.1745	.1736	5.759	.1763	5.671	1.015	.9848	1.3963	80°00′
10	774	765	665	793	576	016	843	934	50
20	804	794	575	823	485	016	838	904	40
30	833	822	487	853	396	017	833	875	30
40	862	851	403	883	309	018	827	846	20
50	891	880	320	914	226	018	822	817	10
11°00′	.1920	.1908	5.241	.1944	5.145	1.019	.9816	1.3788	79°00′
10	949	937	164	.1974	5.066	019	811	759	50
20	.1978	965	089	.2004	4.989	020	805	730	40
30	.2007	.1994	5.016	035	915	020	799	701	30
40	036	.2022	4.945	065	843	021	793	672	20
50	065	051	876	095	773	022	787	643	10
12°00′	.2094	.2079	4.810	.2126	4.705	1.022	.9781	1.3614	78°00′
10	123	108	745	156	638	023	775	584	50
20	153	136	682	186	574	024	769	555	40
30	182	164	620	217	511	024	763	526	30
40	211	193	560	247	449	025	757	497	20
50	240	221	502	278	390	026	750	468	10
13°00′	.2269	.2250	4.445	.2309	4.331	1.026	.9744	1.3439	77°00′
10	298	278	390	339	275	027	737	410	50
20	327	306	336	370	219	028	730	381	40
30	356	334	284	401	165	028	724	352	30
40	385	363	232	432	113	029	717	323	20
50	414	391	182	462	061	030	710	294	10
14°00′	.2443	.2419	4.134	.2493	4.011	1.031	.9703	1.3265	76°00′
		cos θ	sec θ	cot θ	tan θ	csc θ	sin θ	Radians	Degrees
								Angle θ	

TABLE 4 VALUES OF TRIGONOMETRIC FUNCTIONS *(continued)*

Angle θ									
Degrees	Radians	sin θ	csc θ	tan θ	cot θ	sec θ	cos θ		
14°00′	.2443	.2419	4.134	.2493	4.011	1.031	.9703	1.3265	76°00′
10	473	447	086	524	3.962	031	696	235	50
20	502	476	4.039	555	914	032	689	206	40
30	531	504	3.994	586	867	033	681	177	30
40	560	532	950	617	821	034	674	148	20
50	589	560	906	648	776	034	667	119	10
15°00′	.2618	.2588	3.864	.2679	3.732	1.035	.9659	1.3090	75°00′
10	647	616	822	711	689	036	652	061	50
20	676	644	782	742	647	037	644	032	40
30	705	672	742	773	606	038	636	1.3003	30
40	734	700	703	805	566	039	628	1.2974	20
50	763	728	665	836	526	039	621	945	10
16°00′	.2793	.2756	3.628	.2867	3.487	1.040	.9613	1.2915	74°00′
10	822	784	592	899	450	041	605	886	50
20	851	812	556	931	412	042	596	857	40
30	880	840	521	962	376	043	588	828	30
40	909	868	487	.2944	340	044	580	799	20
50	938	896	453	.3026	305	045	572	770	10
17°00′	.2967	.2924	3.420	.3057	3.271	1.046	.9563	1.2741	73°00′
10	.2996	952	388	089	237	047	555	712	50
20	.3025	.2979	357	121	204	048	546	683	40
30	054	.3007	326	153	172	048	537	654	30
40	083	035	295	185	140	049	528	625	20
50	113	062	265	217	108	050	520	595	10
18°00′	.3142	.3090	3.236	.3249	3.078	1.051	.9511	1.2566	72°00′
10	171	118	207	281	047	052	502	537	50
20	200	145	179	314	3.018	053	492	508	40
30	229	173	152	346	2.989	054	483	479	30
40	258	201	124	378	960	056	474	450	20
50	287	228	098	411	932	057	465	421	10
19°00′	.3316	.3256	3.072	.3443	2.904	1.058	.9455	1.2392	71°00′
10	345	283	046	476	877	059	446	363	50
20	374	311	3.021	508	850	060	436	334	40
30	403	338	2.996	541	824	061	426	305	30
40	432	365	971	574	798	062	417	275	20
50	462	393	947	607	773	063	407	246	10
20°00′	.3491	.3420	2.924	.3640	2.747	1.064	.9397	1.2217	70°00′
10	520	448	901	673	723	065	387	188	50
20	549	475	878	706	699	066	377	159	40
30	578	502	855	739	675	068	367	130	30
40	607	529	833	772	651	069	356	101	20
50	636	557	812	805	628	070	346	072	10
21°00′	.3665	.3584	2.790	.3839	2.605	1.071	.9336	1.2043	69°00′
		cos θ	sec θ	cot θ	tan θ	csc θ	sin θ	Radians	Degrees
								Angle θ	

TABLE 4 VALUES OF TRIGONOMETRIC FUNCTIONS *(continued)*

Angle θ Degrees	Radians	sin θ	csc θ	tan θ	cot θ	sec θ	cos θ		
21°00′	.3665	.3584	2.790	.3839	2.605	1.071	.9336	1.2043	69°00′
10	694	611	769	872	583	072	325	1.2014	50
20	723	638	749	906	560	074	315	1.1985	40
30	752	665	729	939	539	075	304	956	30
40	782	692	709	.3973	517	076	293	926	20
50	811	719	689	.4006	496	077	283	897	10
22°00′	.3840	.3746	2.669	.4040	2.475	1.079	.9272	1.1868	68°00′
10	869	773	650	074	455	080	261	839	50
20	898	800	632	108	434	081	250	810	40
30	927	827	613	142	414	082	239	781	30
40	956	854	595	176	394	084	228	752	20
50	985	881	577	210	375	085	216	723	10
23°00′	.4014	.3907	2.559	.4245	2.356	1.086	.9205	1.1694	67°00′
10	043	934	542	279	337	088	194	665	50
20	072	961	525	314	318	089	182	636	40
30	102	.3987	508	348	300	090	171	606	30
40	131	.4014	491	383	282	092	159	577	20
50	160	041	475	417	264	093	147	548	10
24°00′	.4189	.4067	2.459	.4452	2.246	1.095	.9135	1.1519	66°00′
10	218	094	443	487	229	096	124	490	50
20	247	120	427	522	211	097	112	461	40
30	276	147	411	557	194	099	100	432	30
40	305	173	396	592	177	100	088	403	20
50	334	200	381	628	161	102	075	374	10
25°00′	.4363	.4226	2.366	.4663	2.145	1.103	.9063	1.1345	65°00′
10	392	253	352	699	128	105	051	316	50
20	422	279	337	734	112	106	038	286	40
30	451	305	323	770	097	108	026	257	30
40	480	331	309	806	081	109	013	228	20
50	509	358	295	841	066	111	.9001	199	10
26°00′	.4538	.4384	2.281	.4877	2.050	1.113	.8988	1.1170	64°00′
10	567	410	268	913	035	114	975	141	50
20	596	436	254	950	020	116	962	112	40
30	625	462	241	.4986	2.006	117	949	083	30
40	654	488	228	.5022	1.991	119	936	054	20
50	683	514	215	059	977	121	923	1.1025	10
27°00′	.4712	.4540	2.203	.5095	1.963	1.122	.8910	1.0996	63°00′
10	741	566	190	132	949	124	897	966	50
20	771	592	178	169	935	126	884	937	40
30	800	617	166	206	921	127	870	908	30
40	829	643	154	243	907	129	857	879	20
50	858	669	142	280	894	131	843	850	10
28°00′	.4887	.4695	2.130	.5317	1.881	1.133	.8829	1.0821	62°00′
		cos θ	sec θ	cot θ	tan θ	csc θ	sin θ	Radians	Degrees Angle θ

TABLE 4 VALUES OF TRIGONOMETRIC FUNCTIONS *(continued)*

Angle θ									
Degrees	Radians	sin θ	csc θ	tan θ	cot θ	sec θ	cos θ		
28°00′	.4887	.4695	2.130	.5317	1.881	1.133	.8829	1.0821	62°00′
10	916	720	118	354	868	134	816	792	50
20	945	746	107	392	855	136	802	763	40
30	.4974	772	096	430	842	138	788	734	30
40	.5003	797	085	467	829	140	774	705	20
50	032	823	074	505	816	142	760	676	10
29°00′	.5061	.4848	2.063	.5543	1.804	1.143	.8746	1.0647	61°00′
10	091	874	052	581	792	145	732	617	50
20	120	899	041	619	780	147	718	588	40
30	149	924	031	658	767	149	704	559	30
40	178	950	020	696	756	151	689	530	20
50	207	.4975	010	735	744	153	675	501	10
30°00′	.5236	.5000	2.000	.5774	1.732	1.155	.8660	1.0472	60°00′
10	265	025	1.990	812	720	157	646	443	50
20	294	050	980	851	709	159	631	414	40
30	323	075	970	890	698	161	616	385	30
40	352	100	961	930	686	163	601	356	20
50	381	125	951	.5969	675	165	587	327	10
31°00′	.5411	.5150	1.942	.6009	1.664	1.167	.8572	1.0297	59°00′
10	440	175	932	048	653	169	557	268	50
20	469	200	923	088	643	171	542	239	40
30	498	225	914	128	632	173	526	210	30
40	527	250	905	168	621	175	511	181	20
50	556	275	896	208	611	177	496	152	10
32°00′	.5585	.5299	1.887	.6249	1.600	1.179	.8480	1.0123	58°00′
10	614	324	878	289	590	181	465	094	50
20	643	348	870	330	580	184	450	065	40
30	672	373	861	371	570	186	434	036	30
40	701	398	853	412	560	188	418	1.0007	20
50	730	422	844	453	550	190	403	.9977	10
33°00′	.5760	.5446	1.836	.6494	1.540	1.192	.8387	.9948	57°00′
10	789	471	828	536	530	195	371	919	50
20	818	495	820	577	520	197	355	890	40
30	847	519	812	619	511	199	339	861	30
40	876	544	804	661	501	202	323	832	20
50	905	568	796	703	492	204	307	803	10
34°00′	.5934	.5592	1.788	.6745	1.483	1.206	.8290	.9774	56°00′
10	963	616	781	787	473	209	274	745	50
20	.5992	640	773	830	464	211	258	716	40
30	.6021	644	766	873	455	213	241	687	30
40	050	688	758	916	446	216	225	657	20
50	080	712	751	.6959	437	218	208	628	10
35°00′	.6109	.5736	1.743	.7002	1.428	1.221	.8192	.9599	55°00′
		cos θ	sec θ	cot θ	tan θ	csc θ	sin θ	Radians	Degrees
								Angle θ	

TABLE 4 VALUES OF TRIGONOMETRIC FUNCTIONS *(continued)*

Angle θ									
Degrees	Radians	sin θ	csc θ	tan θ	cot θ	sec θ	cos θ		
35°00′	.6109	.5736	1.743	.7002	1.428	1.221	.8192	.9599	55°00′
10	138	760	736	046	419	223	175	570	50
20	167	783	729	089	411	226	158	541	40
30	196	807	722	133	402	228	141	512	30
40	225	831	715	177	393	231	124	483	20
50	254	854	708	221	385	233	107	454	10
36°00′	.6283	.5878	1.701	.7265	1.376	1.236	.8090	.9425	54°00′
10	312	901	695	310	368	239	073	396	50
20	341	925	688	355	360	241	056	367	40
30	370	948	681	400	351	244	039	338	30
40	400	972	675	445	343	247	021	308	20
50	429	.5995	668	490	335	249	.8004	279	10
37°00′	.6458	.6018	1.662	.7536	1.327	1.252	.7986	.9250	53°00′
10	487	041	655	581	319	255	969	221	50
20	516	065	649	627	311	258	951	192	40
30	545	088	643	673	303	260	934	163	30
40	574	111	636	720	295	263	916	134	20
50	603	134	630	766	288	266	898	105	10
38°00′	.6632	.6157	1.624	.7813	1.280	1.269	.7880	.9076	52°00′
10	661	180	618	860	272	272	862	047	50
20	690	202	612	907	265	275	844	.9018	40
30	720	225	606	.7954	257	278	826	.8988	30
40	749	248	601	.8002	250	281	808	959	20
50	778	271	595	050	242	284	790	930	10
39°00′	.6807	.6293	1.589	.8098	1.235	1.287	.7771	.8901	51°00′
10	836	316	583	146	228	290	753	872	50
20	865	338	578	195	220	293	735	843	40
30	894	361	572	243	213	296	716	814	30
40	923	383	567	292	206	299	698	785	20
50	952	406	561	342	199	302	679	756	10
40°00′	.6981	.6428	1.556	.8391	1.192	1.305	.7660	.8727	50°00′
10	.7010	450	550	441	185	309	642	698	50
20	039	472	545	491	178	312	623	668	40
30	069	494	540	541	171	315	604	639	30
40	098	517	535	591	164	318	585	610	20
50	127	539	529	642	157	322	566	581	10
41°00′	.7156	.6561	1.524	.8693	1.150	1.325	.7547	.8552	49°00′
10	185	583	519	744	144	328	528	523	50
20	214	604	514	796	137	332	509	494	40
30	243	626	509	847	130	335	490	465	30
40	272	648	504	899	124	339	470	436	20
50	301	670	499	.8952	117	342	451	407	10
42°00′	.7330	.6691	1.494	.9004	1.111	1.346	.7431	.8378	48°00′
		cos θ	sec θ	cot θ	tan θ	csc θ	sin θ	Radians	Degrees
								Angle θ	

TABLE 4 VALUES OF TRIGONOMETRIC FUNCTIONS *(continued)*

Angle θ Degrees	Radians	sin θ	csc θ	tan θ	cot θ	sec θ	cos θ		
42°00′	.7330	.6691	1.494	.9004	1.111	1.346	.7431	.8378	48°00′
10	359	713	490	057	104	349	412	348	50
20	389	734	485	110	098	353	392	319	40
30	418	756	480	163	091	356	373	290	30
40	447	777	476	217	085	360	353	261	20
50	476	799	471	271	079	364	333	232	10
43°00′	.7505	.6820	1.466	.9325	1.072	1.367	.7314	.8203	47°00′
10	534	841	462	380	066	371	294	174	50
20	563	862	457	435	060	375	274	145	40
30	592	884	453	490	054	379	254	116	30
40	621	905	448	545	048	382	234	087	20
50	650	926	444	601	042	386	214	058	10
44°00′	.7679	.6947	1.440	.9657	1.036	1.390	.7193	.8029	46°00′
10	709	967	435	713	030	394	173	.7999	50
20	738	.6988	431	770	024	398	153	970	40
30	767	.7009	427	827	018	402	133	941	30
40	796	030	423	884	012	406	112	912	20
50	825	050	418	.9942	006	410	092	883	10
45°00′	.7854	.7071	1.414	1.000	1.000	1.414	.7071	.7854	45°00′
		cos θ	sec θ	cot θ	tan θ	csc θ	sin θ	Radians	Degrees Angle θ

ANSWERS TO SELECTED EXERCISES

CHAPTER ONE

EXERCISE SET 1.1

1. (a) natural number, integer, rational number **(b)** integer, rational number **3. (a)** rational number **(b)** irrational number **5. (a)** natural number, integer, rational number **(b)** rational number **7. (a)** rational number **(b)** rational number **9.** irrational number **11.** irrational number

13.
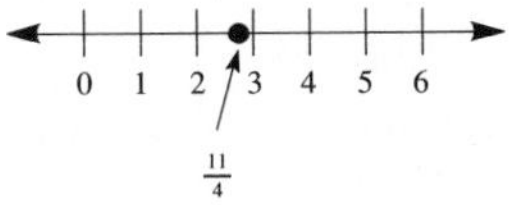

15.
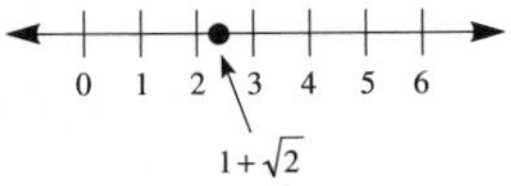

17.
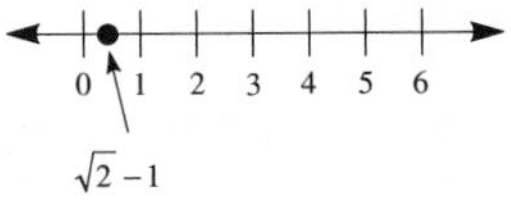

19.
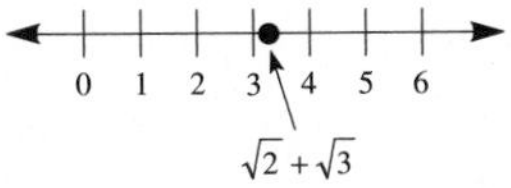

21.
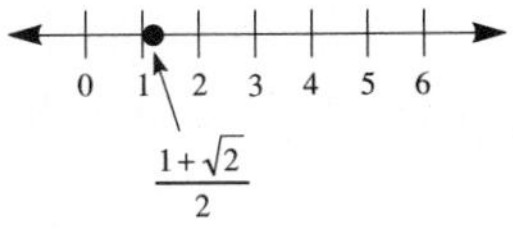

23.
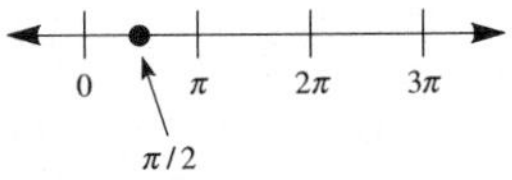

25.
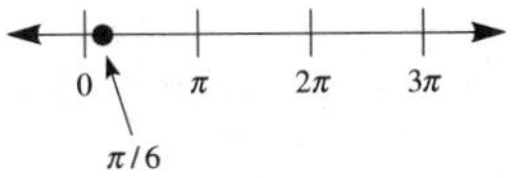

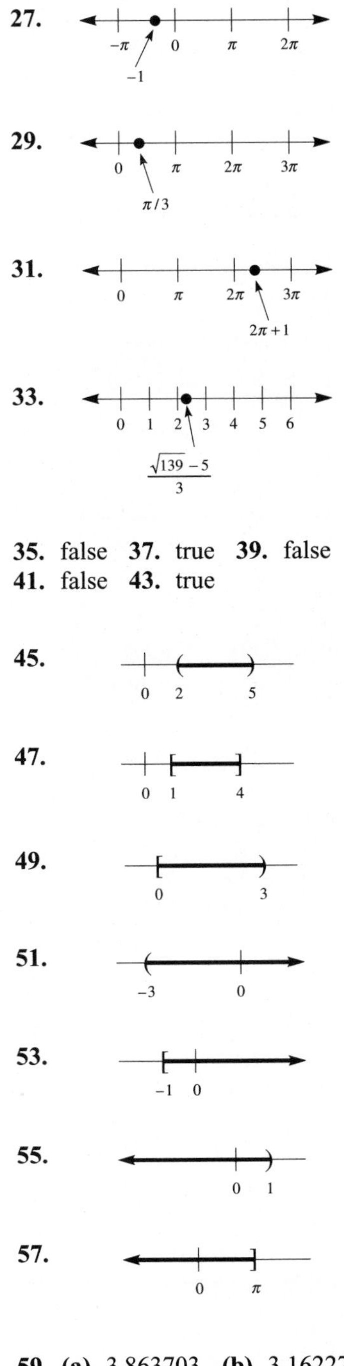

27.
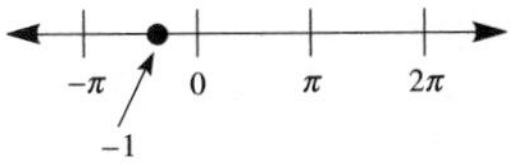

29.
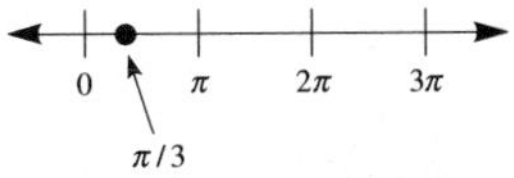

31.
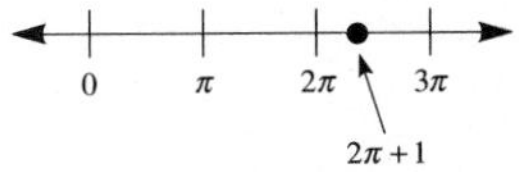

33.
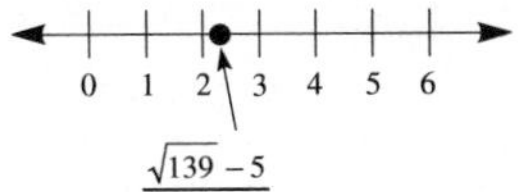

35. false **37.** true **39.** false **41.** false **43.** true

45.
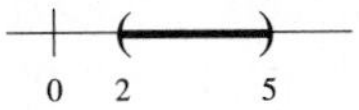

47.
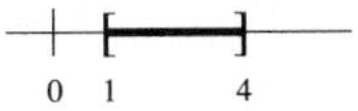

49.
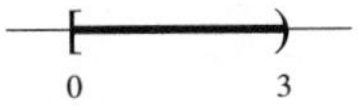

51.
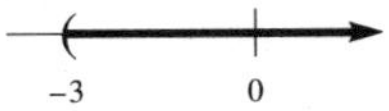

53.
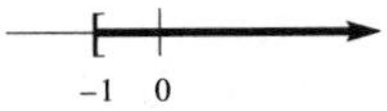

55.
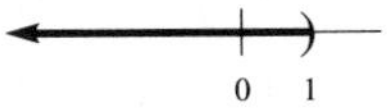

57.
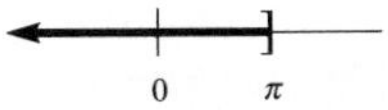

59. (a) 3.863703 **(b)** 3.162277 **(c)** 1.847759 **61. (a)** one decimal place **(b)** two decimal places **(c)** six decimal places **(d)** nine decimal places **63. (a)** $a=\sqrt{2}$ and $b=\sqrt{8}$ (Other answers are possible.) **(b)** $a=\sqrt{2}$ and $b=\sqrt{3}$ (Other answers are possible.) **65. (a)** $2^{1/2}=\sqrt{2}$ (Other answers are possible.) **(b)** $(\sqrt{2})^2=2$ (Other answers are possible.)

EXERCISE SET 1.2

1. 3 **3.** 6 **5.** 2 **7.** 0 **9.** 0 **11.** 17 **13.** 1 **15.** 0 **17.** 25 **19.** -1 **21.** 0 **23.** 3 **25.** $\sqrt{2}-2$ **27.** $x-3$ **29.** t^2+1 **31.** $\sqrt{3}+4$ **33.** $-2x+7$ **35.** 1 **37.** $3x+11$ **39.** $|x-4|=8$ **41.** $|x-1|=1/2$ **43.** $|x-1|\geqslant 1/2$ **45.** $|y+4|<1$ **47.** $|y|<3$ **49.** $|x^2-a^2|<M$

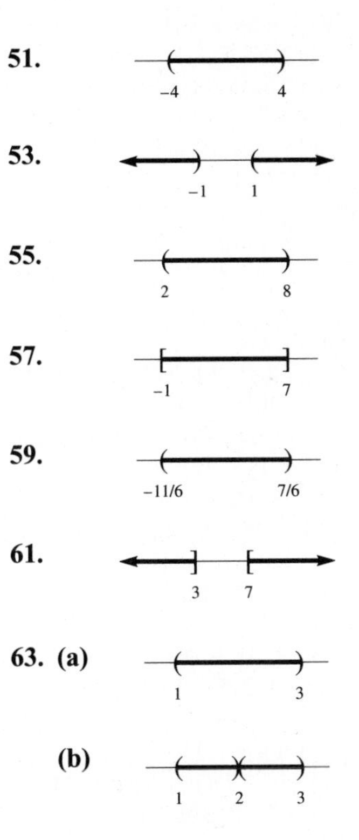

51.
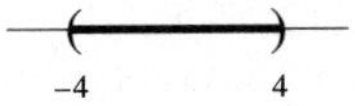

53.
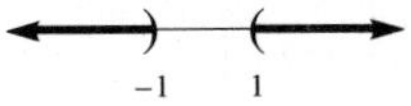

55.
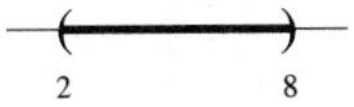

57.
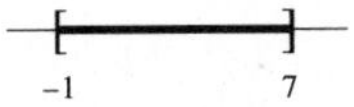

59.
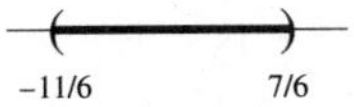

61.
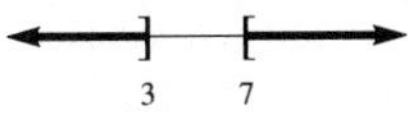

63. (a)
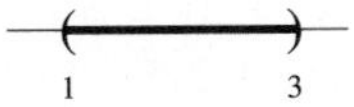

(b)
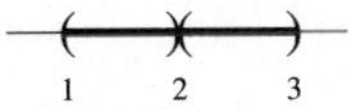

(c) The interval in (b) does not include 2.

EXERCISE SET 1.3

1. (a) all real numbers **(b)** all nonnegative real numbers **3. (a)** all real numbers **(b)** all real numbers
5. (a) all nonnegative real numbers
(b) all positive real numbers except 1
7. (a) degree: 0; coefficents: 4
(b) degree: 3; coefficients: 4
(c) degree: 6; coefficients: 1, 4, -1, 2
11. (a) $x^2 - y^2$ **(b)** $x^4 - 25$
13. (a) $A^2 - 16$
(b) $a^2 + 2ab + b^2 - 16$
15. (a) $x^2 - 16x + 64$
(b) $4x^4 - 20x^2 + 25$
17. (a) $x + 2\sqrt{xy} + y$
(b) $2x + y - 2\sqrt{x^2 + xy}$
19. (a) $a^3 + 3a^2 + 3a + 1$
(b) $27x^6 - 54x^4a^2 + 36x^2a^4 - 8a^6$
21. (a) $x^3 + 1$ **(b)** $x^6 + 1$
23. (a) $(x + 8)(x - 8)$ **(b)** $7x^2(x^2 + 2)$
(c) $z(11 + z)(11 - z)$
(d) $(ab + c)(ab - c)$
25. (a) $(x + 3)(x - 1)$
(b) $(x - 3)(x + 1)$ **(c)** irreducible
(d) $(-x + 3)(x + 1)$ or $-(x - 3)(x + 1)$
27. (a) $(x + 1)(x^2 - x + 1)$
(b) $(x + 6)(x^2 - 6x + 36)$
(c) $8(5 - x^2)(25 + 5x^2 + x^4)$
(d) $(4ax - 5)(16a^2x^2 + 20ax + 25)$
29. (a) $(12 + x)(12 - x)$
(b) irreducible **(c)** $(9 + y)(15 - y)$
31. (a) $h^3(1 + h)(1 - h)$
(b) $h^3(10 + h)(10 - h)$
(c) $(h + 1)^3(11 + h)(9 - h)$
33. (a) $(x - 8)(x - 5)$ **(b)** irreducible
35. (a) $(x + 9)(x - 4)$
(b) $(x - 9)(x - 4)$
37. (a) $(3x + 2)(x - 8)$ **(b)** irreducible
39. (a) $(3x - 1)(2x + 5)$
(b) $(6x + 5)(x - 1)$ **41. (a)** $(t^2 + 1)^2$
(b) $(t + 1)^2(t - 1)^2$ **(c)** irreducible
43. (a) $x(4x^2 - 20x - 25)$
(b) $x(2x - 5)^2$ **45. (a)** $(a - c)(b + a)$
(b) $(u + v - y)(x + u + v)$
47. $(xz + t)(xz + y)$ **49.** $(a^2 - 2b^2c^2)^2$
51. irreducible
53. $(x + 4)(x^2 - 4x + 16)$
55. $x(x^2 + 3xy + 3y^2)$
57. $(x - y)(x^2 + xy + y^2 + 1)$
59. (a) $(p^2 + 1)(p + 1)(p - 1)$
(b) $(p^4 + 1)(p^2 + 1)(p + 1)(p - 1)$
61. $(x + 1)^3$ **63.** irreducible
65. $\left(\frac{5}{4} + c\right)\left(\frac{5}{4} - c\right)$
67. $\left(z^2 + \frac{9}{4}\right)\left(z + \frac{3}{2}\right)\left(z - \frac{3}{2}\right)$
69. $\left(\dfrac{5}{mn} - 1\right)\left(\dfrac{25}{m^2n^2} + \dfrac{5}{mn} + 1\right)$
71. $\left(\frac{1}{2}x + y\right)^2$
73. $(x - a)(8x - 8a + 1)(8x - 8a - 1)$
75. $(ax + b)(x + 1)$ **77.** $-x(x + 1)^{1/2}$
79. $x(x + 1)^{-3/2}$ **81.** $-\frac{2}{3}x(2x + 3)^{1/2}$
83. $(A + B)^3$
85. $x(a^2 + x^2)^{-3/2}(2a^2 + x^2)$
87. $(y^2 - pq)(y - p)(y - q)$

89. (a)

x	$\dfrac{x^2 - 16}{x - 4}$
3.9	7.9
3.99	7.99
3.999	7.999
3.9999	7.9999
3.99999	7.99999

x	$\dfrac{x^2 - 16}{x - 4}$
4.1	8.1
4.01	8.01
4.001	8.001
4.0001	8.0001
4.00001	8.00001

(b) 8 **(c)** $\dfrac{x^2 - 16}{x - 4} = x + 4$, if $x \neq 4$, so $x + 4$ will approach 8.
91. $(x^2 + 4x + 8)(x^2 - 4x + 8)$
93. $2(x - t)(x + y + z + t)$
95. $3(b - a)(c - a)(c - b)$

EXERCISE SET 1.4

1. yes **3.** no **5.** no
7. $x = 6$ or $x = -1$ **9.** $x = \pm 10$
11. $x = \frac{6}{5}$ **13.** $z = -\frac{1}{5}$ or $z = \frac{3}{2}$
15. $x = 1$ or $x = -3$
17. $x = \frac{1}{2}$ or $x = 6$
19. $x = 0, -13$, or 12
21. $y = \frac{1}{2}$ or $y = \pm\frac{1}{2}\sqrt{3}$
23. $y = 1$ or $y = \pm\sqrt{2}$
25. $x = -1$ or $x = -9$
27. $x = \left(1 \pm \sqrt{21}\right)/2$
29. $x = \left(-3 \pm \sqrt{41}\right)/4$
31. $x = -\frac{1}{6}$ or $x = -\frac{5}{2}$
33. $x = \left(1 \pm \sqrt{41}\right)/4$
35. $x = \left(6 \pm \sqrt{42}\right)/6$
37. $x \approx -1.47$ or $x \approx -1.53$
39. $x = -67$ or $x = -89$
41. $x = \frac{2}{3}$ or $x = 1$
43. $x = \left(-2 \pm \sqrt{13}\right)/3$
45. $x = \pm 2\sqrt{6}$ **47.** $x = \left(1 \pm \sqrt{5}\right)/2$
49. $x = \left(3 \pm \sqrt{15}\right)/3$
51. $x = \sqrt{5}/2$ or $x = -2\sqrt{5}/5$
53. $x = 1/2y$ or $x = 1/y$
55. $x = 0$ or $x = p + q$
57. $x = -5a/4$ or $x = 4a/3$
59. $r = \left(-h \pm \sqrt{h^2 + 40}\right)/2$
61. $t = 0$ or $t = v_0/16$
63. $x = 0, -2b$, or b **65.** two real roots **67.** two real roots **69.** one real root **71.** two real roots **73.** $k = 36$
75. $k = \pm 2\sqrt{5}$ **77. (a)** $x = 24$
(b) no real number solutions
79. $x = -1$ **81.** $x = \pm 2$ or $x = \pm 3$
83. $x = -4$ or $x = -\frac{20}{9}$
85. no real number solutions
87. no real number solutions
89. $x = \frac{5}{2}$ or $x = -4$
91. $x = \frac{1}{3}$ or $x = \frac{1}{2}$ **93.** $x = 1$
95. $x = -3 \pm 2\sqrt{2}$ **97.** Any value of x (except $x = -1$ or $x = -3$) is a solution to the equation.
99. $x = a + b$
101. $x = a^2b^2/(2a + b)^2$
105. 1 meter **107. (a)** 266 people will have caught the flu. **(b)** 8 days
109. $L/4$ inches and $3L/4$ inches
111. (b) $k = -\frac{1}{2}$ **(c)** $y = \pm\frac{\sqrt{5}}{2}$, $x = -\frac{1}{2} \pm \frac{\sqrt{5}}{2}$ **113.** $A = 0$ and $B = 0$; $A = 1$ and $B = -2$

CHAPTER 1 REVIEW EXERCISES

1. $(a - 4b)(a + 4b)$
3. $(1 - a)(7 + 4a + a^2)$ **5.** $x(ax + b)^2$
7. $(4x + 1)(2x + 1)$
9. $a^4x^4(1 - xa)(1 + xa)(1 + x^2a^2)$
11. $(2 + a)^3$ **13.** $z^3(4xy + 1)(xy - 1)$
15. $(1 - x)(1 + x + x^2)(1 + x)(1 - x + x^2)$
17. $(ax + b - 2abx)(ax + b + 2abx)$
19. $(a - b)(a + b + c + ab)$
21. $81x^2 - y^2$ **23.** $4a^2b^2 - 12ab + 9$

25. $8a^3 + 36a^2 + 54a + 27$
27. $1 - 27a^3$ **29.** $x - y$
31. $x + 3x^{2/3}y^{1/3} + 3x^{1/3}y^{2/3} + y$
33. $x^4 - 6x^3 + 11x^2 - 6x + 1$
35. $|x - 6| = 2$ **37.** $|a - b| = 3$
39. $|x| > 10$ **41.** $\sqrt{6} - 2$
43. $x^4 + x^2 + 1$
45. **(a)** $-2x + 5$ **(b)** 1 **(c)** $2x - 5$

47.

3 5

49.

-5 0

51.

-1 0

53.

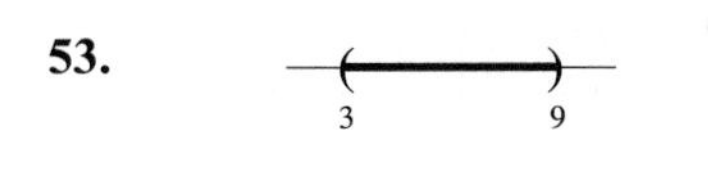

3 9

55.

-2 0

57. **(a)**

-1 4 9

(b)

-1 9

59.

x	$1 + \frac{x^2}{2} + \frac{3x^4}{8}$	$1/\sqrt{1 - x^2}$
0.1	1.00504	1.00504
0.01	1.00005	1.00005

61. $x = \frac{1}{3}$ or $x = -\frac{1}{2}$
63. $x = -6$ or $x = 4$ **65.** $x = -1$ or $x = 10$ **67.** $x = \frac{2}{3}$ or $x = 3$
69. $x = \pm\sqrt{1 + \sqrt{3}}$ or $x = 0$
71. $x = -7$ **73.** $x = \frac{1}{4}$ **75.** $x = 10$
77. $x = 2$ **79.** $k = \frac{6}{5}$
81. $k = 0$ or $k = 1$
83. **(a)** $t = \frac{1}{4}$ sec (on the way up); $t = \frac{15}{4}$ sec (on the way down)
(b) 4 seconds **85.** $v_0 = 16(t_1 + t_2)$

CHAPTER 1 TEST

1. **(a)** $(6 - x)(12 + x^2)$ **(b)** irreducible
2. **(a)** irreducible
(b) $z(3xy - 1)(2xy + 1)$
3. $-3x(1 - x^2)^{-3/2}(2x^2 + x - 2)$
4. $(3a + 2b)(3x - 2y)$
5. $8a^3 + 36a^2b^2 + 54ab^4 + 27b^6$
6. 1 **7.** **(a)** $\left(\frac{39}{10}, \frac{41}{10}\right)$ **(b)** $[-2, \infty)$
8. **(a)** $x = -5$ or $x = 1$
(b) $x = -2 \pm \sqrt{5}$
9. $x = 2$ **10.** $x = 0, -\frac{3}{2}$, or 1
11. $y = -3$ or $y = \pm 2\sqrt{2}$
12. $k = (3 + \sqrt{13})/2 \approx 3.30$

CHAPTER TWO

EXERCISE SET 2.1

1. **3.** **(a)**

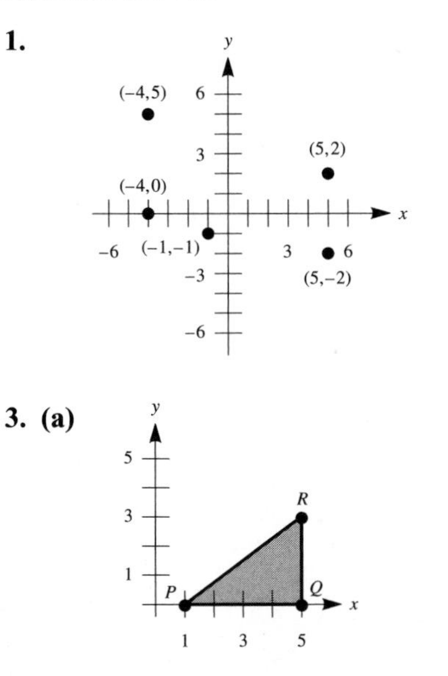

(b) 6 square units **5.** **(a)** 5 **(b)** 13
7. **(a)** 10 **(b)** 9 **9.** 4
11. **(a)** $(4, 1/2)$ **(b)** $(-6, 7)$
13. **(a)** yes **(b)** yes **(c)** no
15. 0; the three points are collinear.
17. center: (3, 1); radius: 5; y-intercepts: 5, -3 **19.** center: (0, 0); radius: $\sqrt[4]{2}$; y-intercepts: $\pm\sqrt[4]{2}$
21. center: $(-4, 3)$; radius: 1; no y-intercepts **23.** center: $\left(-3, \frac{1}{3}\right)$; radius: $\sqrt{2}$; no y-intercepts
25. **(a)** $y = \frac{4}{5}$ **(b)** $y = -\frac{4}{5}$
27. **(a)** $x = -\frac{\sqrt{2}}{2}$ **(b)** $x = \frac{\sqrt{2}}{2}$

29. **(a)** $y = \frac{\sqrt{5}}{3}$ **(b)** $y = -\frac{\sqrt{5}}{3}$
31. **(a)** $y \approx 0.758$ **(b)** $x \approx 0.758$
33. $(x - 1)^2 + (y - 1)^2 = 34$
35. **(a)** $(6, 5)$ **(b)** $\left(\frac{1}{2}, -\frac{3}{2}\right)$
(c) $(1, -4)$ **37.** $(11, -8)$

39. **(a)**

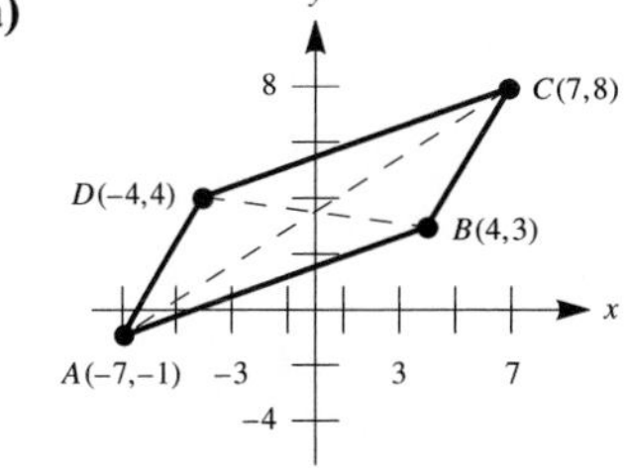

(b) AC: $(0, 7/2)$; BD: $(0, 7/2)$
(c) The diagonals of a parallelogram bisect each other. **41.** **(a)** 128 **(b)** 96
(c) 4/3

43. $y = -3 \pm \sqrt{4 - (x + 5)^2}$
(a) **(b)** **(c)**

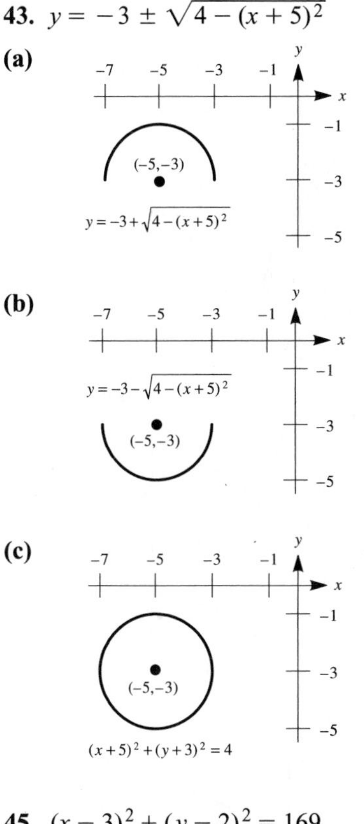

45. $(x - 3)^2 + (y - 2)^2 = 169$
47. $(x - 3)^2 + (y - 5)^2 = 9$
49. $a = \sqrt{2}$; $b = \sqrt{3}$; $c = 2$; $d = \sqrt{5}$; $e = \sqrt{6}$; $f = \sqrt{7}$; $g = 2\sqrt{2}$
51. $(x - 3)^2 + (y - 2)^2 = 13$

53. $t = 7, -3$
55. (b) $M_{\overline{OB}} = \left(\frac{a+b}{2}, \frac{c}{2}\right)$; $M_{\overline{AC}} = \left(\frac{a+b}{2}, \frac{c}{2}\right)$
63. (a) Each side has a length of c and each angle is 90°. **(b)** $b - a$ **(c)** $a^2 + b^2 = c^2$
65. (c) $u = 5$, $v = 12$, $w = 13$; area = perimeter = 30

GRAPHING UTILITY EXERCISES FOR SECTION 2.1

1.

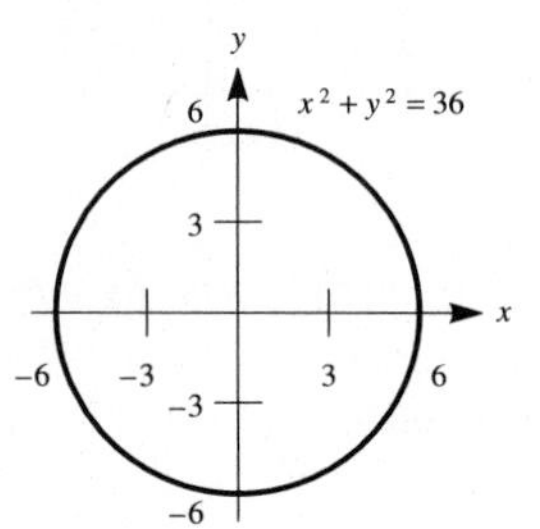

3.

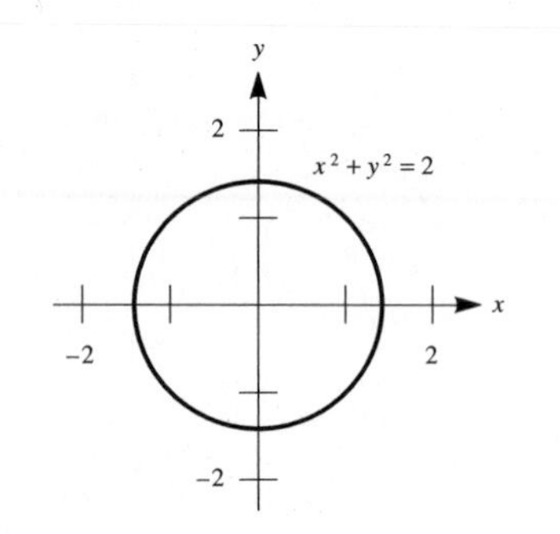

5.

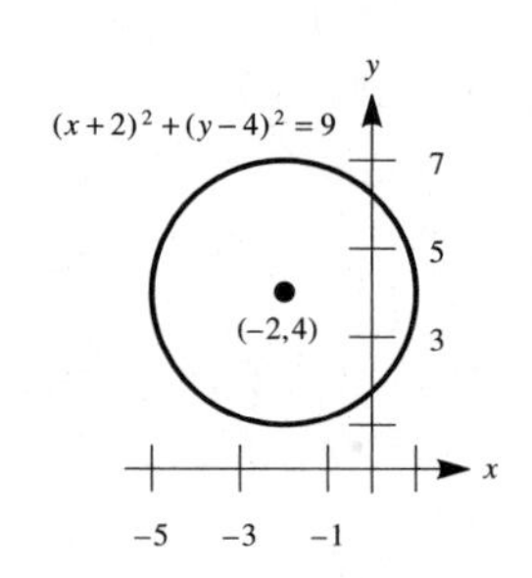

7.

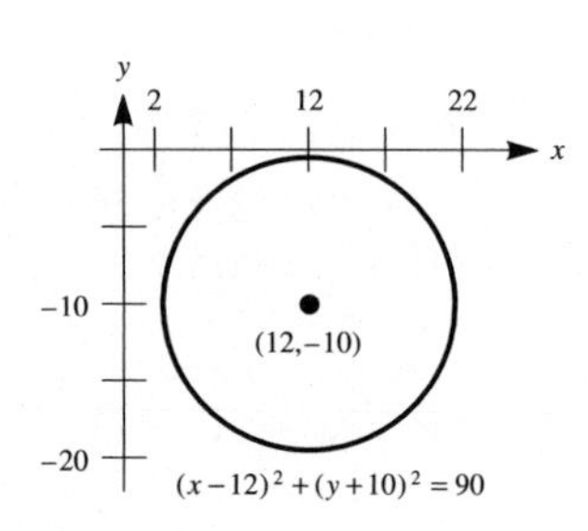

9. (a)

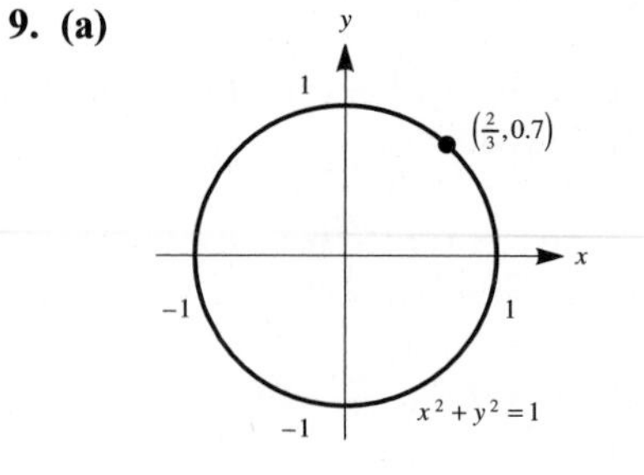

(b) $y = \sqrt{5}/3 \approx 0.745$

11. (a)

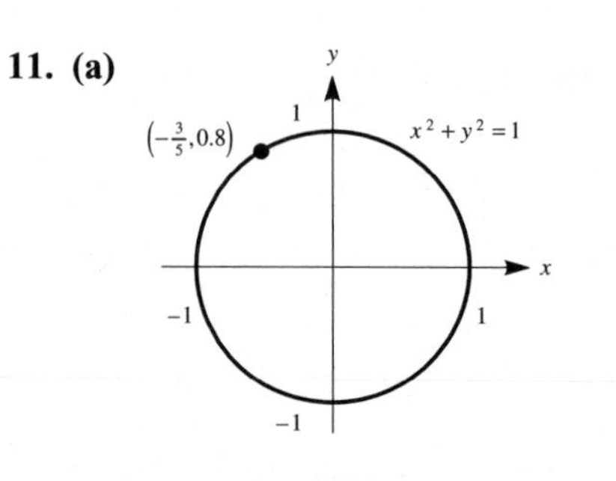

(b) $y = 4/5 = 0.8$

13. (a)

(b) $y = -\sqrt{0.51} \approx -0.714$

15. (a)

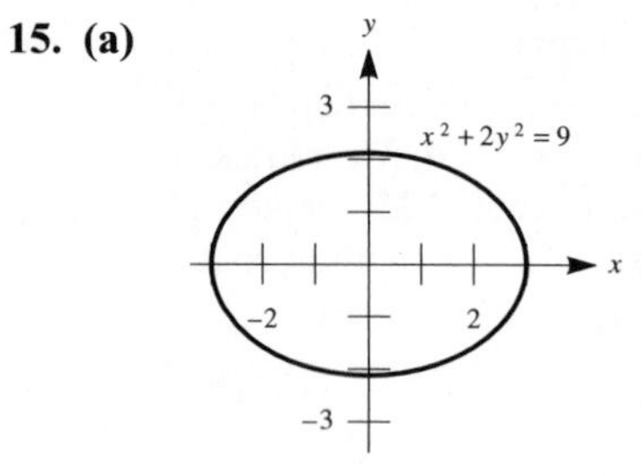

(b) The curves intersect at $(-3, 0)$ and $(3, 0)$.

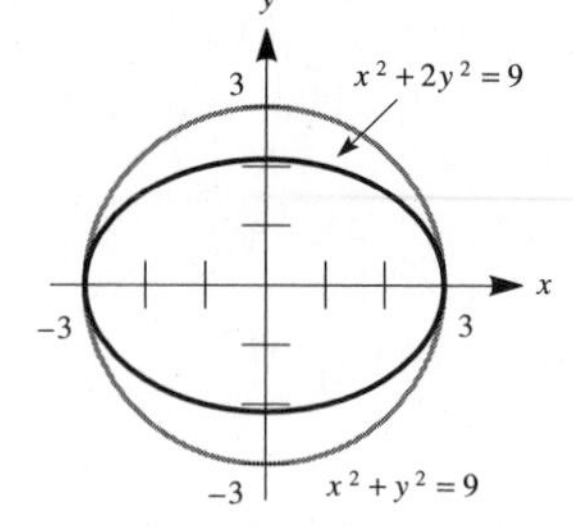

17. (a)

(b)

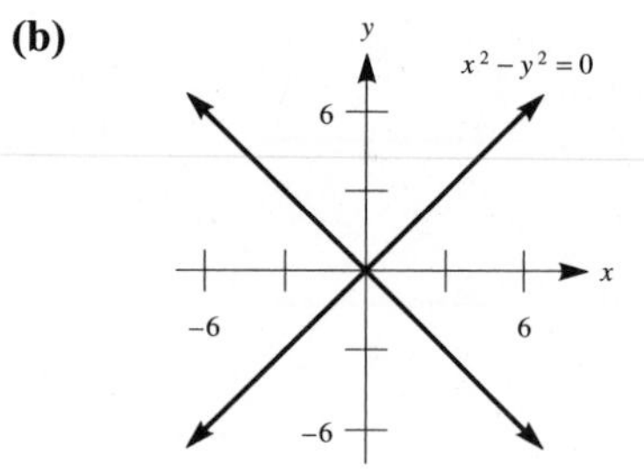

(c)

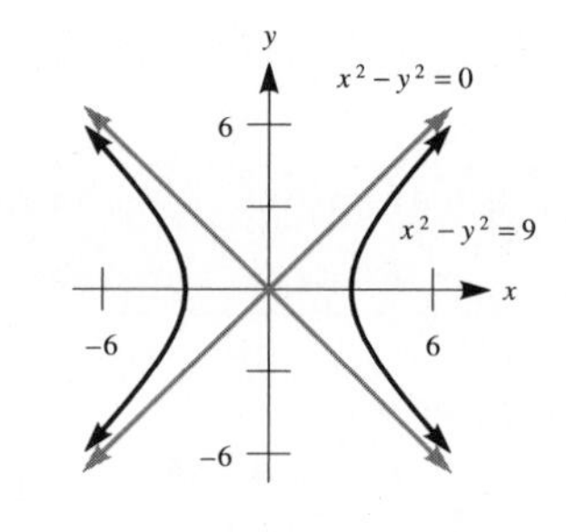

19.

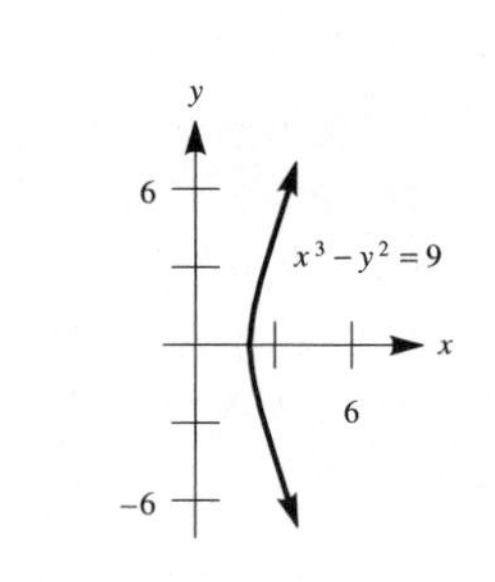

21. **(a,b)**

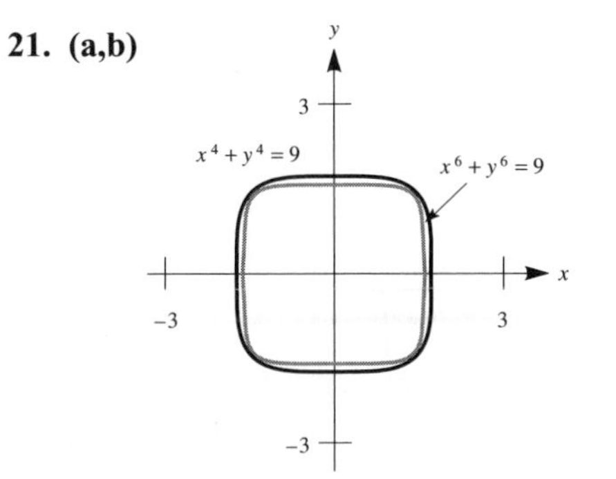

(c)

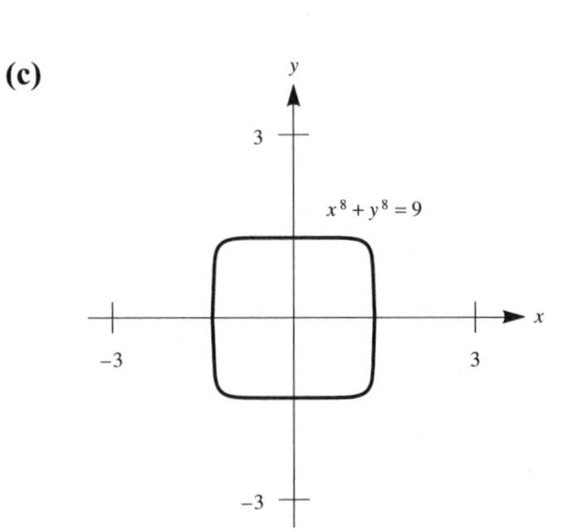

23.

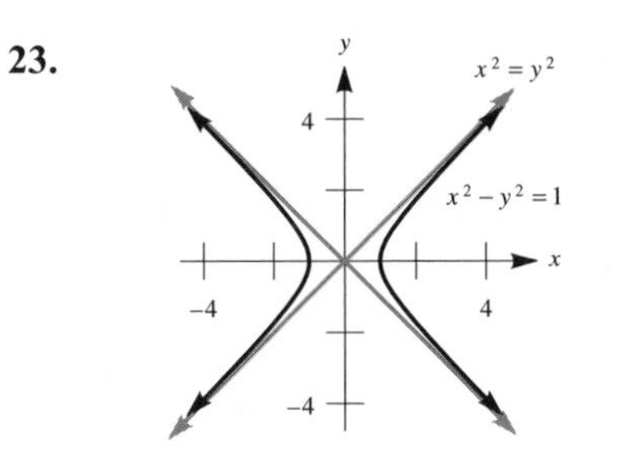

EXERCISE SET 2.2

1. no **3.** no **5.** yes **7.** $a=-9/2$
9. $a=4/121$ **11.** $a=3$
13. domain: $x \neq 4$

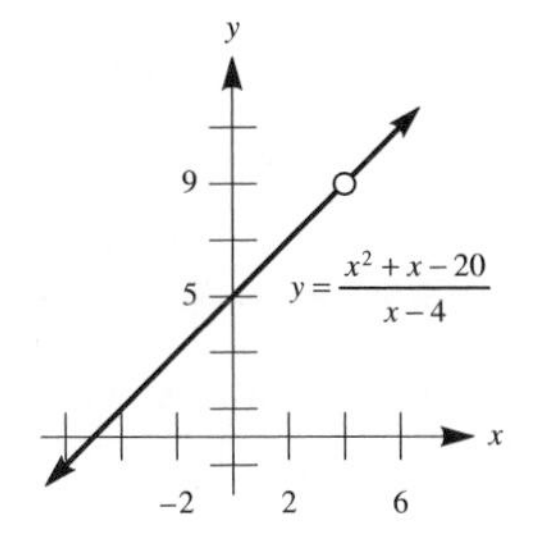

15. domain: $x \neq 3$

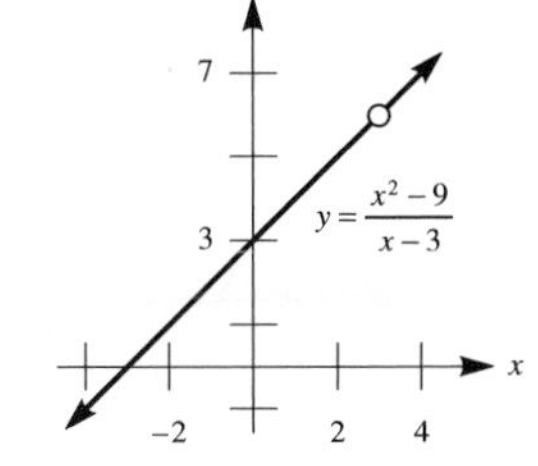

17. domain: $x \neq \pm 1$

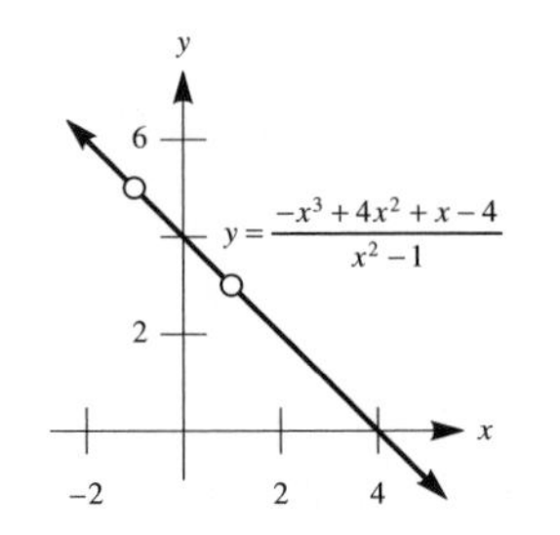

19. x-intercept: 4; y-intercept: 3

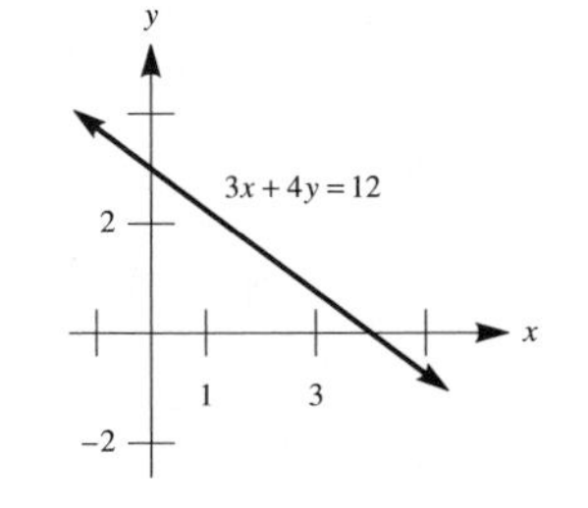

21. x-intercept: 2; y-intercept: -4

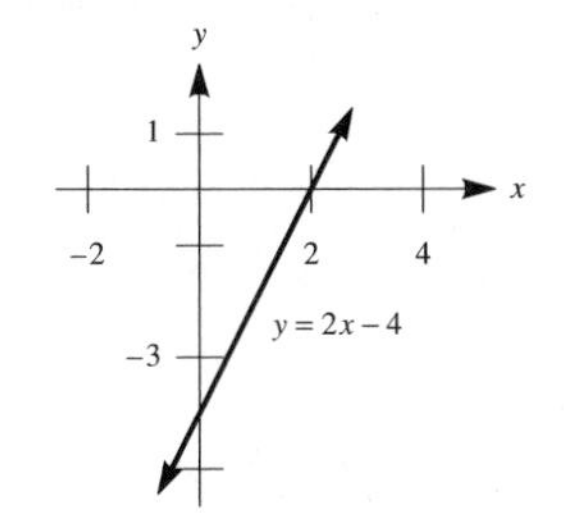

23. x-intercept: 1; y-intercept: 1

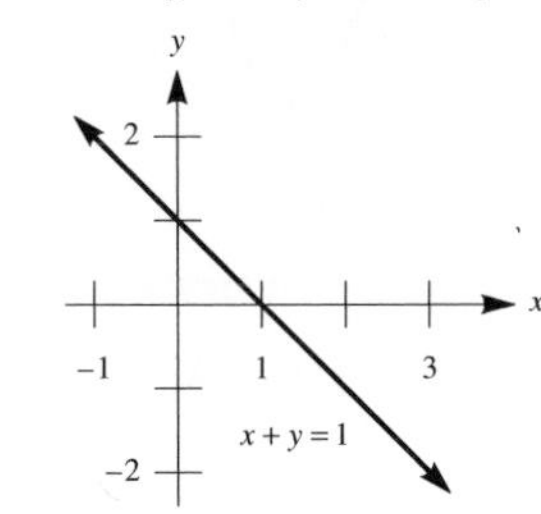

25. **(a)** x-intercepts: -1, -2; y-intercept: 2 **(b)** no x-intercepts; y-intercept: 3 **27.** **(a)** x-intercepts: $(-1 \pm \sqrt{5})/2$; y-intercept: -1 **(b)** no x-intercepts; y-intercept: 1
29. x-intercepts: 0, $-1-2\sqrt{3} \approx -4.46$, $-1+2\sqrt{3} \approx 2.46$; y-intercept: 0
31. x-intercept: $-8/3$; y-intercepts: -2, 6
33. x-intercepts: $2-2\sqrt{2} \approx -0.83$, $2+2\sqrt{2} \approx 4.83$; y-intercepts: $1-\sqrt{5} \approx -1.24$, $1+\sqrt{5} \approx 3.24$
35. **(a)**

x	-3	-2	-1	0	1	2	3
$y=\lvert x\rvert$	3	2	1	0	1	2	3

(b)

x	-3	-2	-1	0	1	2	3
$y=x^2$	9	4	1	0	1	4	9

(c)

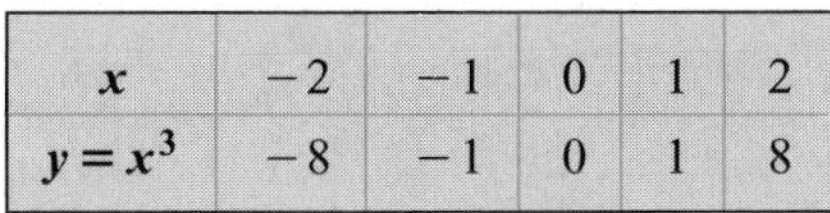

x	-2	-1	0	1	2
$y=x^3$	-8	-1	0	1	8

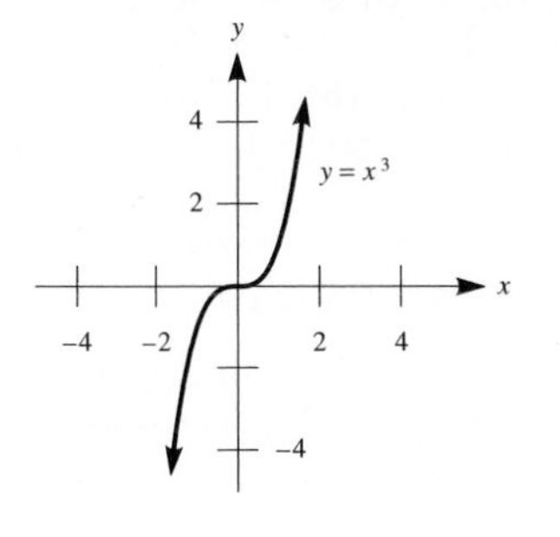

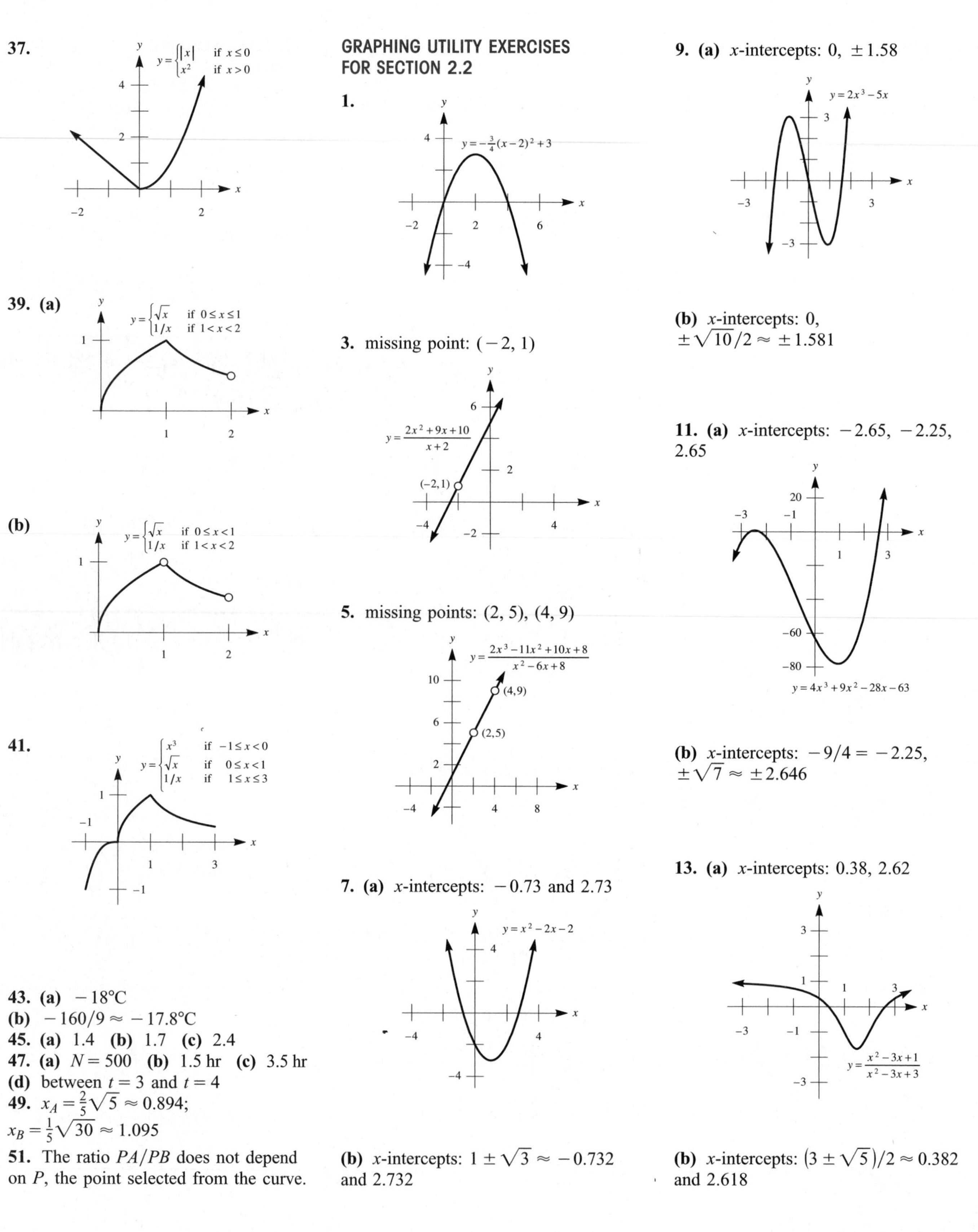

37.

39. (a)

(b)

41.

43. (a) $-18°$C
(b) $-160/9 \approx -17.8°$C
45. (a) 1.4 **(b)** 1.7 **(c)** 2.4
47. (a) $N = 500$ **(b)** 1.5 hr **(c)** 3.5 hr
(d) between $t = 3$ and $t = 4$
49. $x_A = \frac{2}{5}\sqrt{5} \approx 0.894$;
$x_B = \frac{1}{5}\sqrt{30} \approx 1.095$
51. The ratio PA/PB does not depend on P, the point selected from the curve.

GRAPHING UTILITY EXERCISES FOR SECTION 2.2

1.

3. missing point: $(-2, 1)$

5. missing points: (2, 5), (4, 9)

7. (a) x-intercepts: -0.73 and 2.73

(b) x-intercepts: $1 \pm \sqrt{3} \approx -0.732$ and 2.732

9. (a) x-intercepts: 0, ± 1.58

(b) x-intercepts: 0, $\pm\sqrt{10}/2 \approx \pm 1.581$

11. (a) x-intercepts: -2.65, -2.25, 2.65

(b) x-intercepts: $-9/4 = -2.25$, $\pm\sqrt{7} \approx \pm 2.646$

13. (a) x-intercepts: 0.38, 2.62

(b) x-intercepts: $(3 \pm \sqrt{5})/2 \approx 0.382$ and 2.618

15. x-intercepts: -1.879, 0.347, 1.532

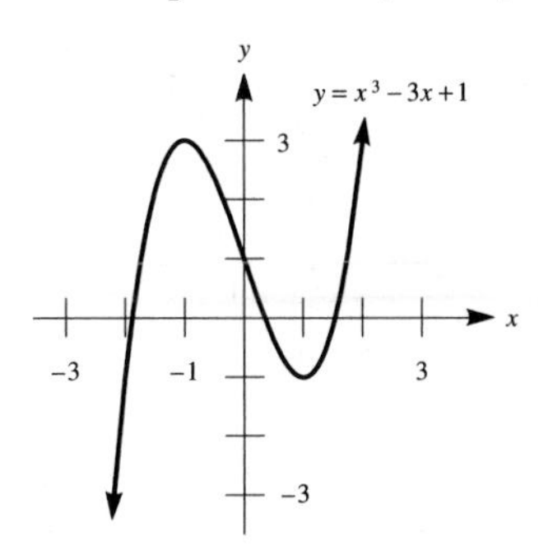

17. x-intercepts: -0.815, 0.875, 5.998

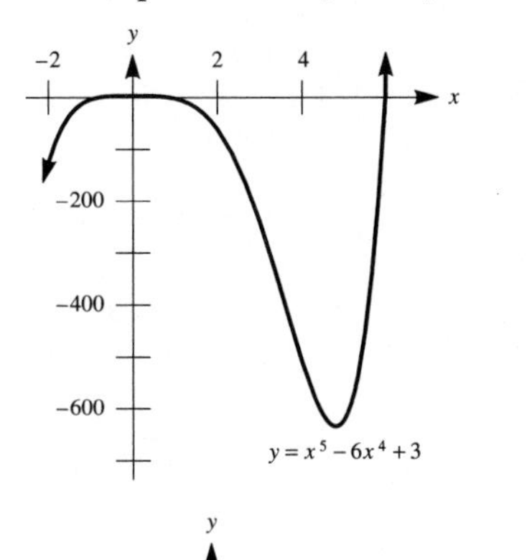

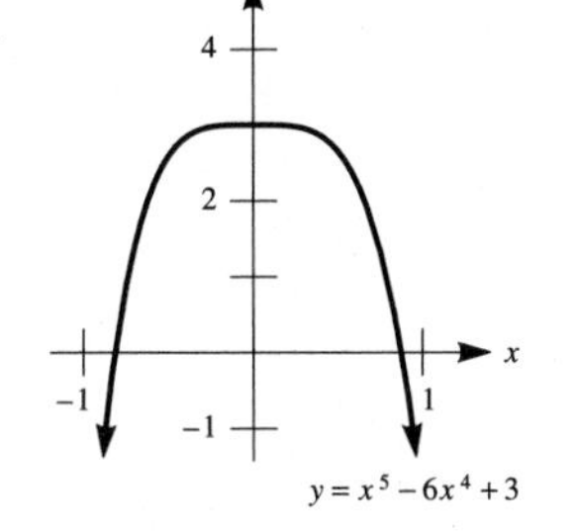

19. (a)

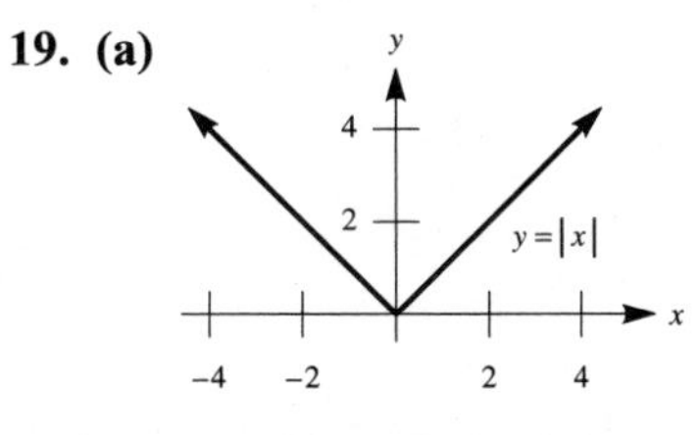

(b)

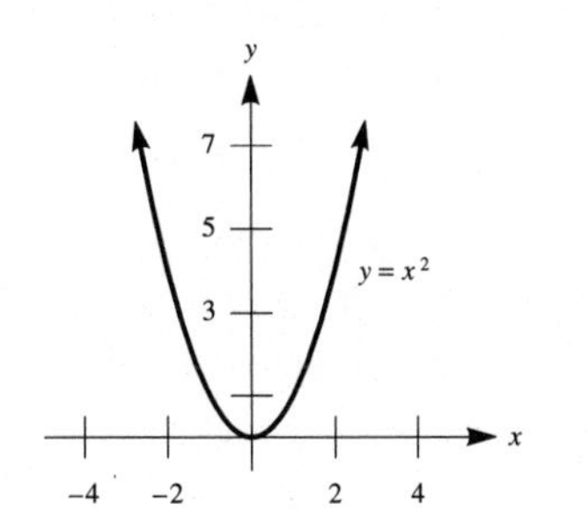

(c)

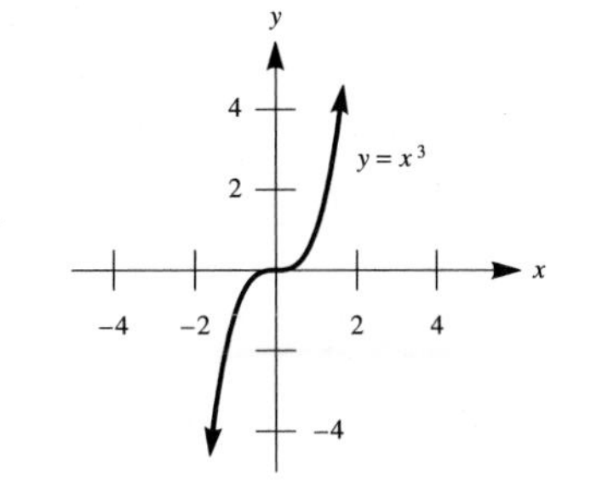

21. (a)

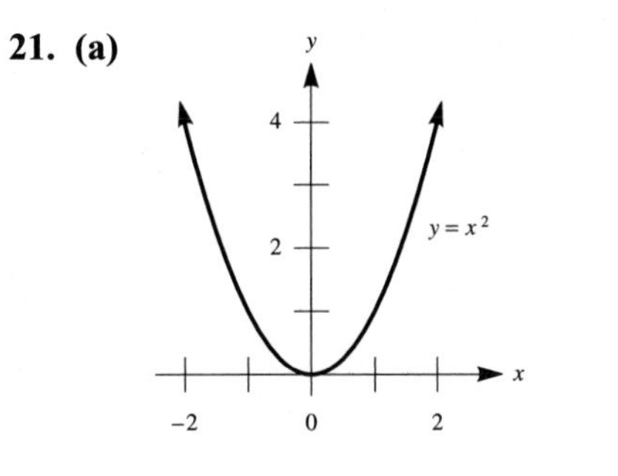

(b)

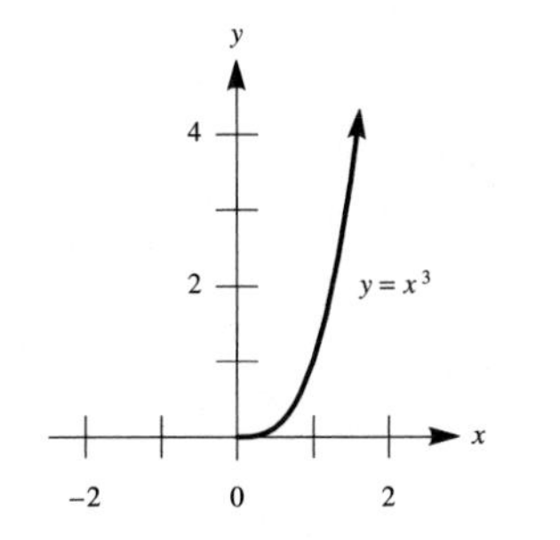

(c) $(0, 0)$ and $(1, 1)$; $x^2 > x^3$ on the interval $(0, 1)$

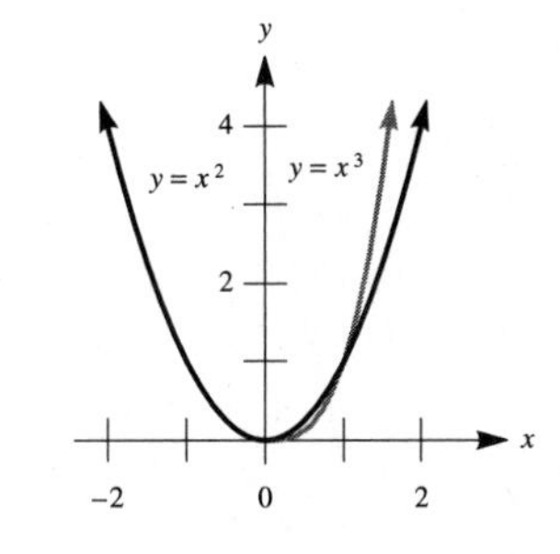

23.

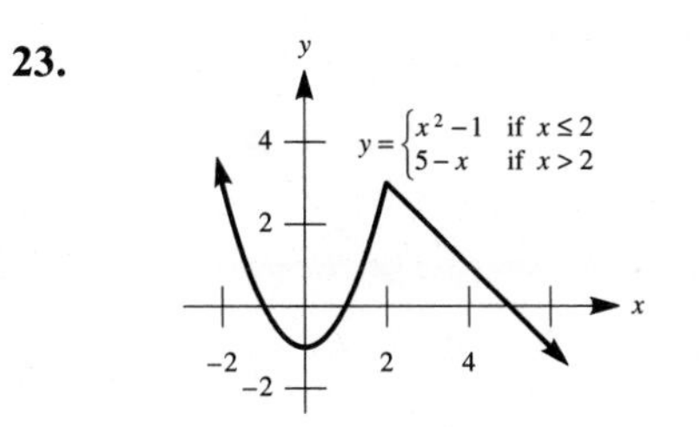

25.

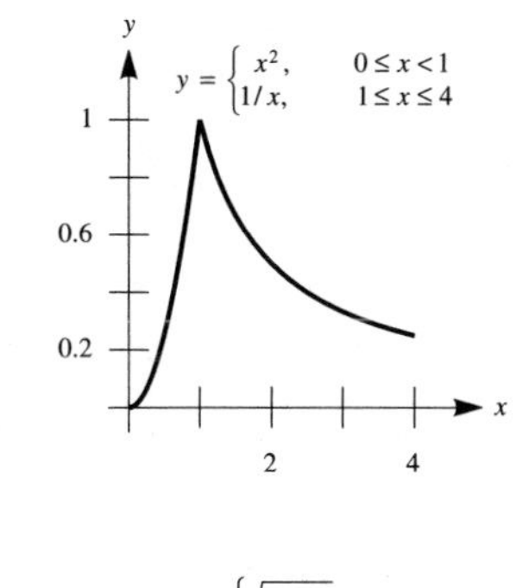

27.

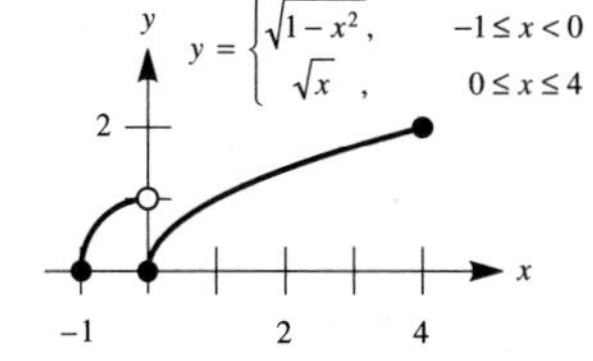

29.

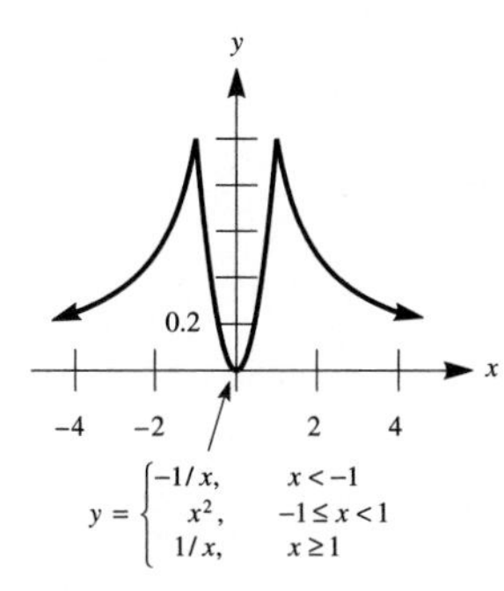

31.

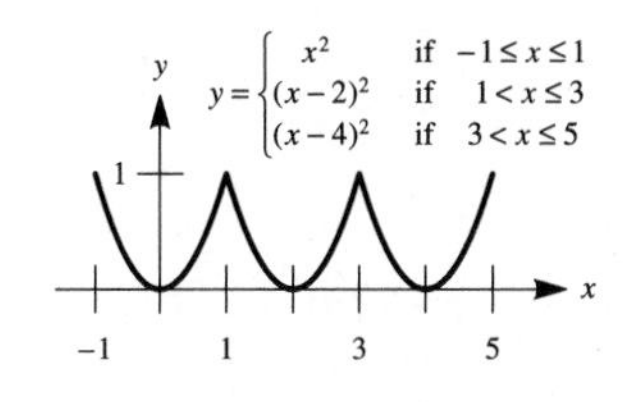

EXERCISE SET 2.3

1. (a) -2 **(b)** 3 **(c)** $-7/3$ **(d)** 3

3. (a) slope $= 1$

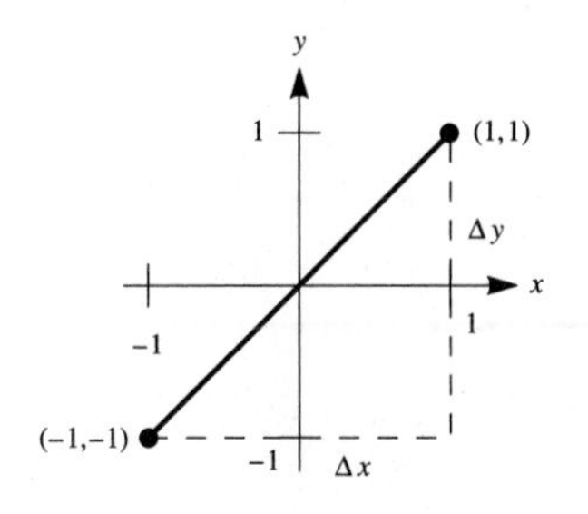

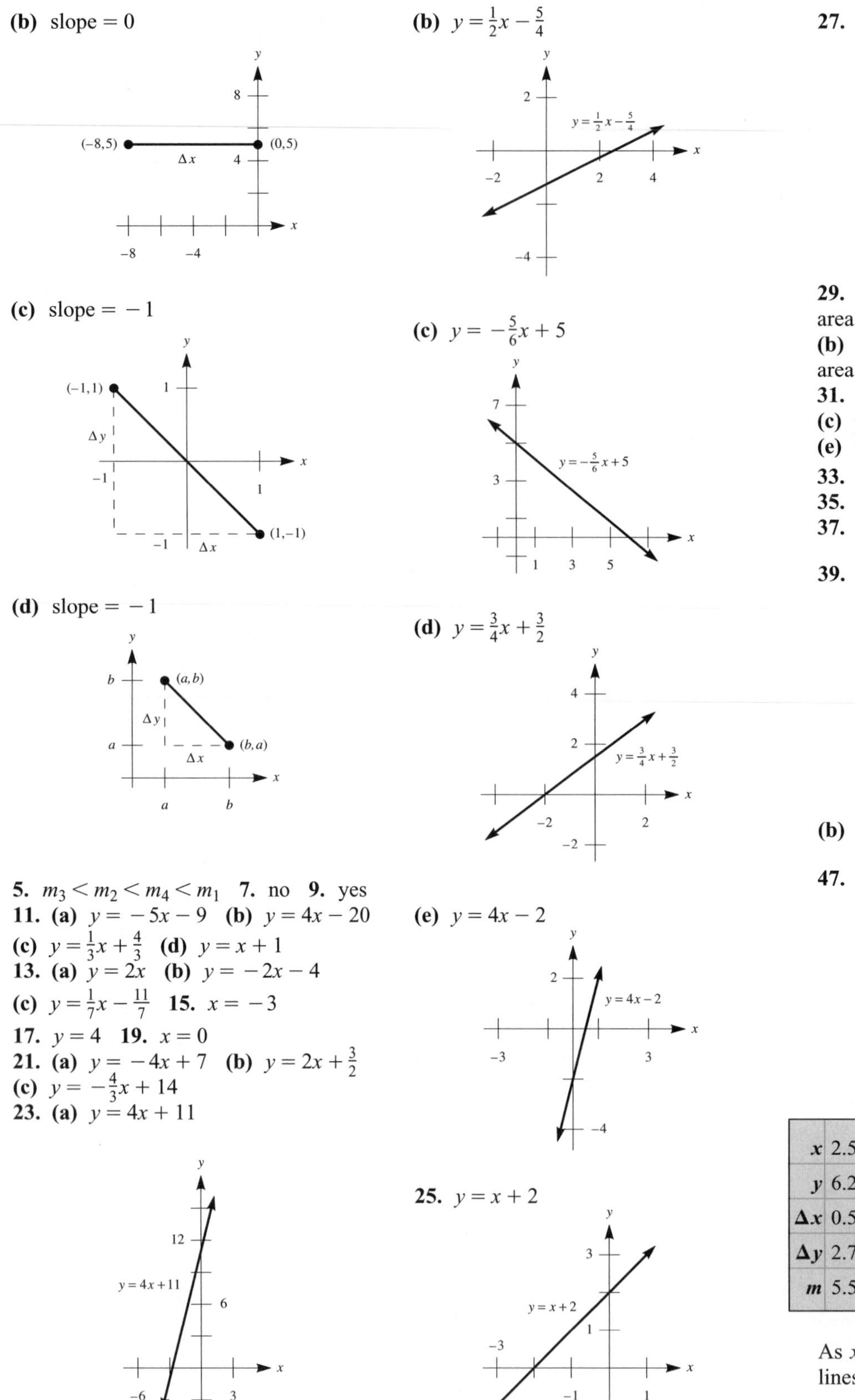

(b) slope $= 0$

(c) slope $= -1$

(d) slope $= -1$

5. $m_3 < m_2 < m_4 < m_1$ **7.** no **9.** yes
11. **(a)** $y = -5x - 9$ **(b)** $y = 4x - 20$
(c) $y = \frac{1}{3}x + \frac{4}{3}$ **(d)** $y = x + 1$
13. **(a)** $y = 2x$ **(b)** $y = -2x - 4$
(c) $y = \frac{1}{7}x - \frac{11}{7}$ **15.** $x = -3$
17. $y = 4$ **19.** $x = 0$
21. **(a)** $y = -4x + 7$ **(b)** $y = 2x + \frac{3}{2}$
(c) $y = -\frac{4}{3}x + 14$
23. **(a)** $y = 4x + 11$

(b) $y = \frac{1}{2}x - \frac{5}{4}$

(c) $y = -\frac{5}{6}x + 5$

(d) $y = \frac{3}{4}x + \frac{3}{2}$

(e) $y = 4x - 2$

25. $y = x + 2$

27. $y - 4 = 0$

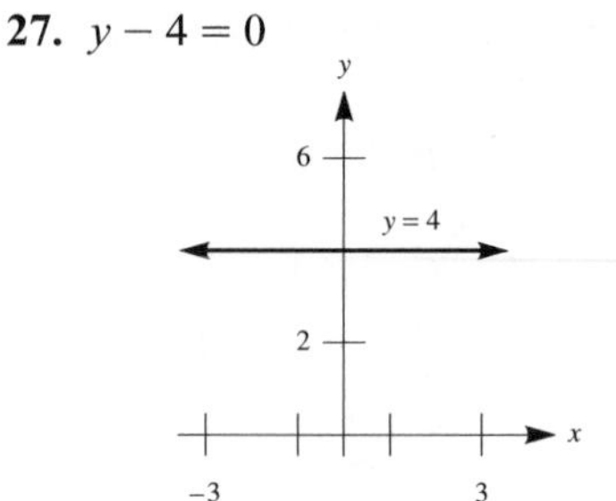

29. **(a)** x-intercept: 5; y-intercept: 3; area $= 15/2$; perimeter $= 8 + \sqrt{34}$
(b) x-intercept: 5; y-intercept: -3; area $= 15/2$; perimeter $= 8 + \sqrt{34}$
31. **(a)** neither **(b)** parallel
(c) perpendicular **(d)** perpendicular
(e) neither **(f)** parallel
33. $y = \frac{2}{5}x + \frac{12}{5}$; $2x - 5y + 12 = 0$
35. $y = -\frac{4}{3}x + \frac{16}{3}$; $4x + 3y - 16 = 0$
37. $y = \frac{3}{5}x + 11$; $3x - 5y + 55 = 0$

39. **(a)**

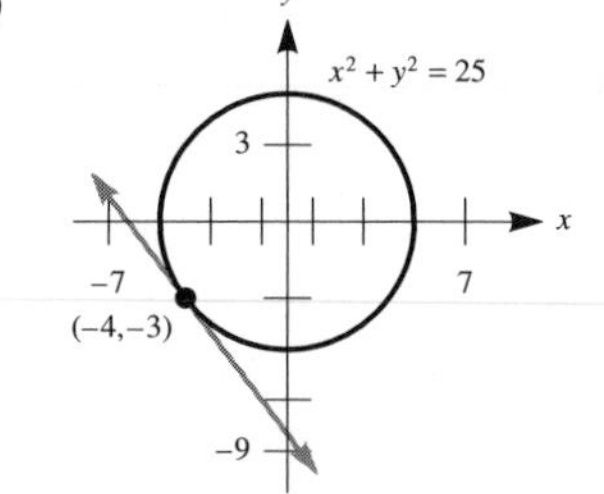

(b) $y = -\frac{4}{3}x - \frac{25}{3}$

47.

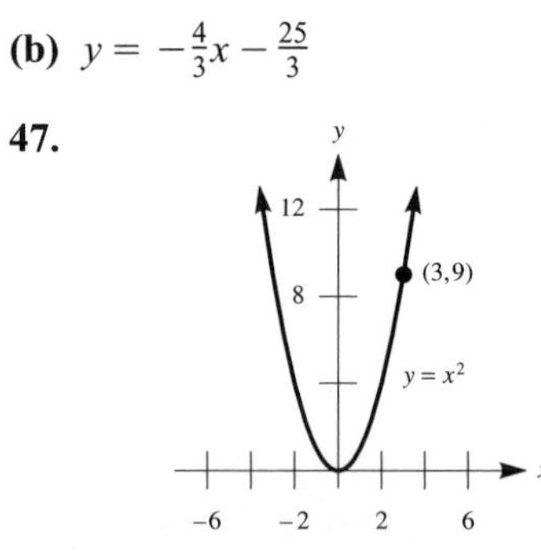

x	2.5	2.9	2.99	2.999	2.9999
y	6.25	8.41	8.9401	8.994001	8.99940001
Δx	0.5	0.1	0.01	0.001	0.0001
Δy	2.75	0.59	0.0599	0.005999	0.00059999
m	5.5	5.9	5.99	5.999	5.9999

As x approaches 3, the slope of these lines approaches 6. The slope of the tangent line is 6. **49.** $\left(-\frac{1}{2}, -\frac{1}{8}\right)$

51. $\left(8, \frac{1}{8}\right)$ **53.** 44.1 sq. units

55. (a)

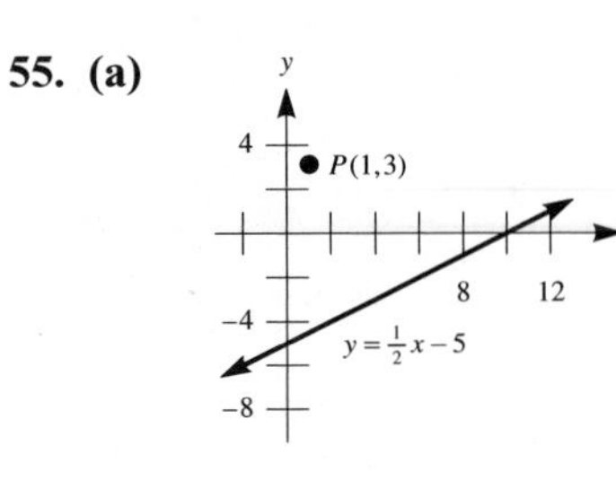

(b) $y = -2x + 5$ **(c)** $(4, -3)$
(d) $3\sqrt{5}$ **59. (a)** $x = 410 - \frac{2}{3}P$
(b) 230 units **(c)** \$307.50/unit
63. In calculating slope, either point can be chosen as the starting point.
65. (a) $y = 2x$; **(b)** $y = \sqrt{mn}\,x$

GRAPHING UTILITY EXERCISES FOR SECTION 2.3

1. (a) $y = x$; $y = 2x$; $y = 3x$; $y = 10x$ As the positive slopes increase, the lines become steeper.

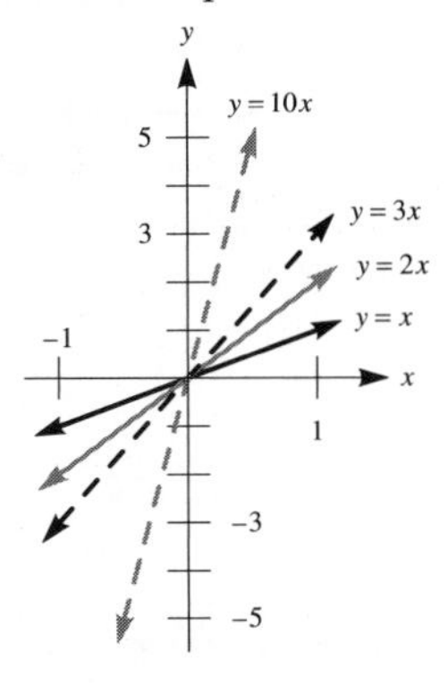

(b) $y = -x$; $y = -2x$; $y = -3x$; $y = -10x$ As the negative slopes decrease, the lines become steeper.

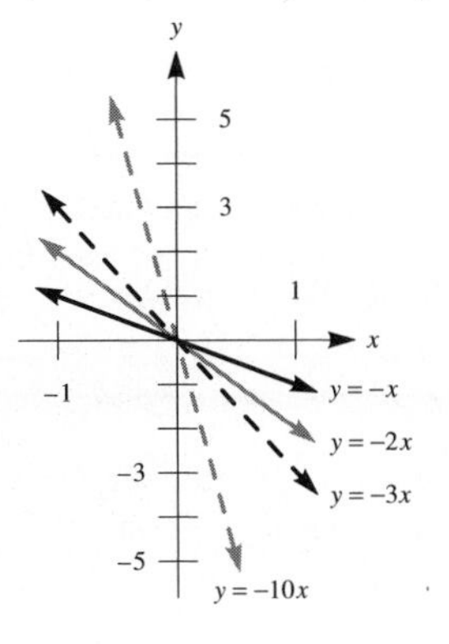

3.

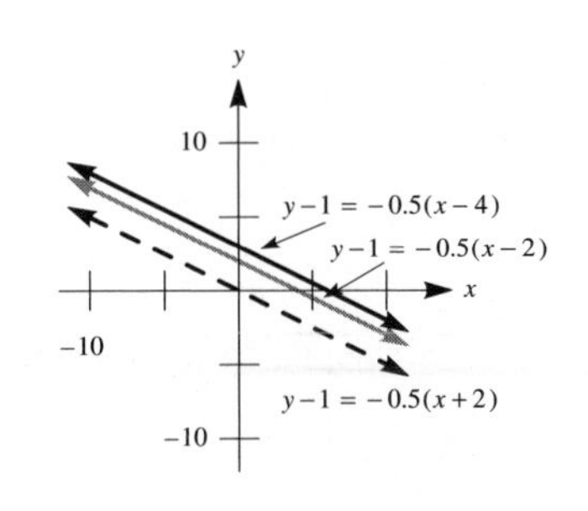

5. (a) $y = -\frac{3}{2}x + 3$; x-intercept: 2; y-intercept: 3

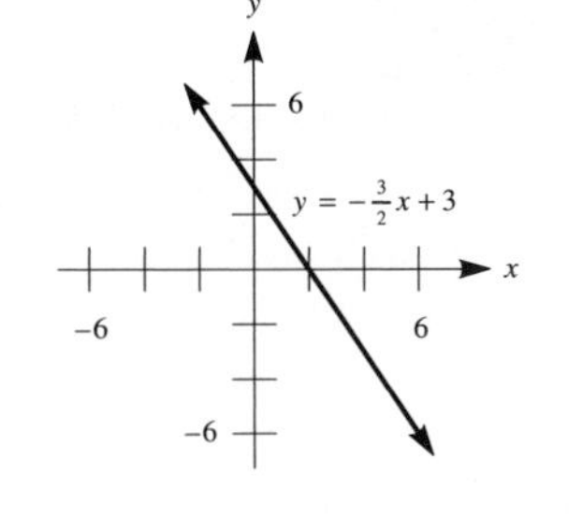

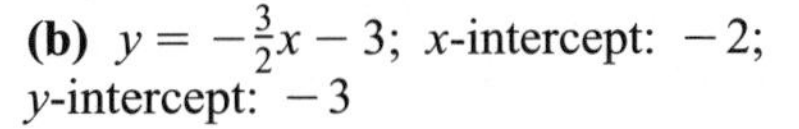

(b) $y = -\frac{3}{2}x - 3$; x-intercept: -2; y-intercept: -3

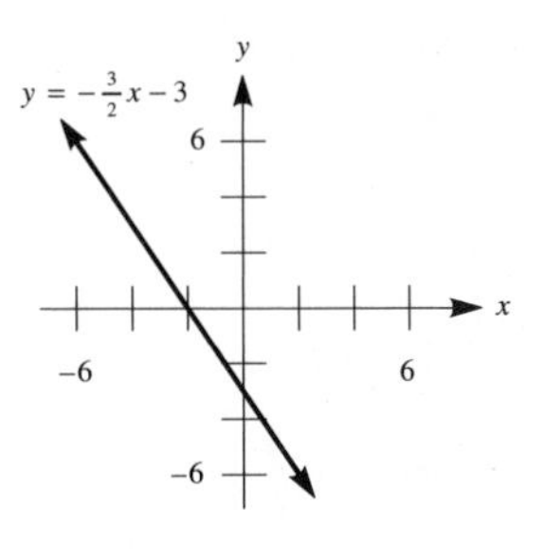

(c) $y = -\frac{5}{6}x + 5$; x-intercept: 6; y-intercept: 5

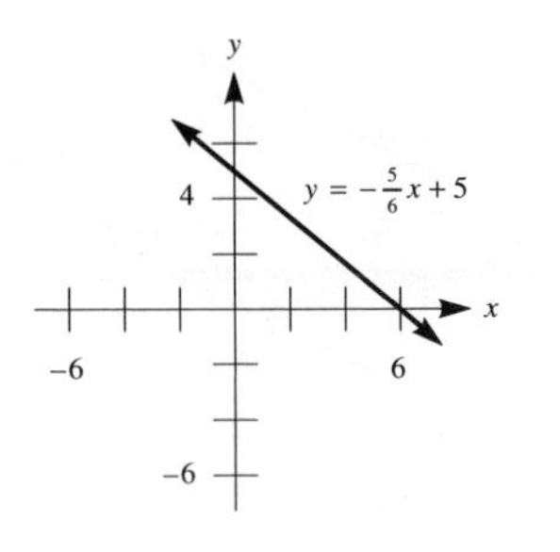

(d) $y = -\frac{5}{6}x - 5$; x-intercept: -6; y-intercept: -5

(e) A line with an x-intercept of a and a y-intercept of b.

7. $y = \frac{1}{2}x + 5$

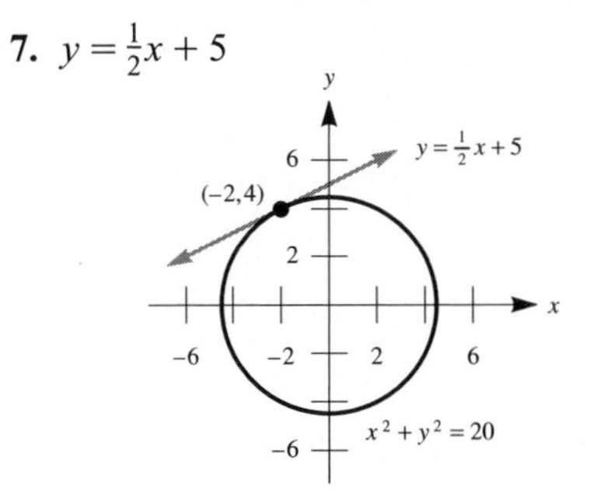

9. $y = -\frac{8}{15}x + \frac{73}{5}$

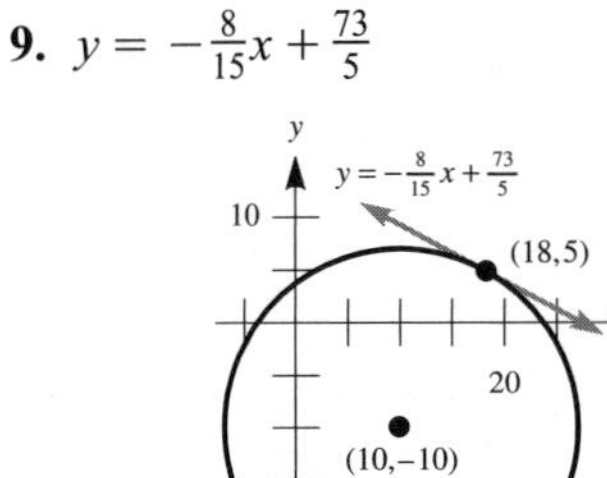

11. $y = -\frac{3}{4}x$

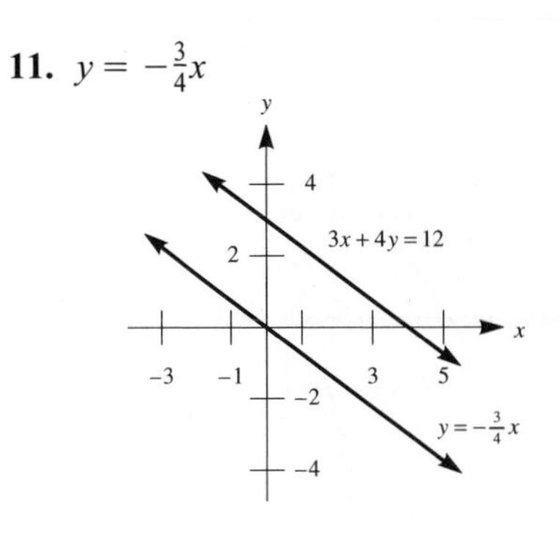

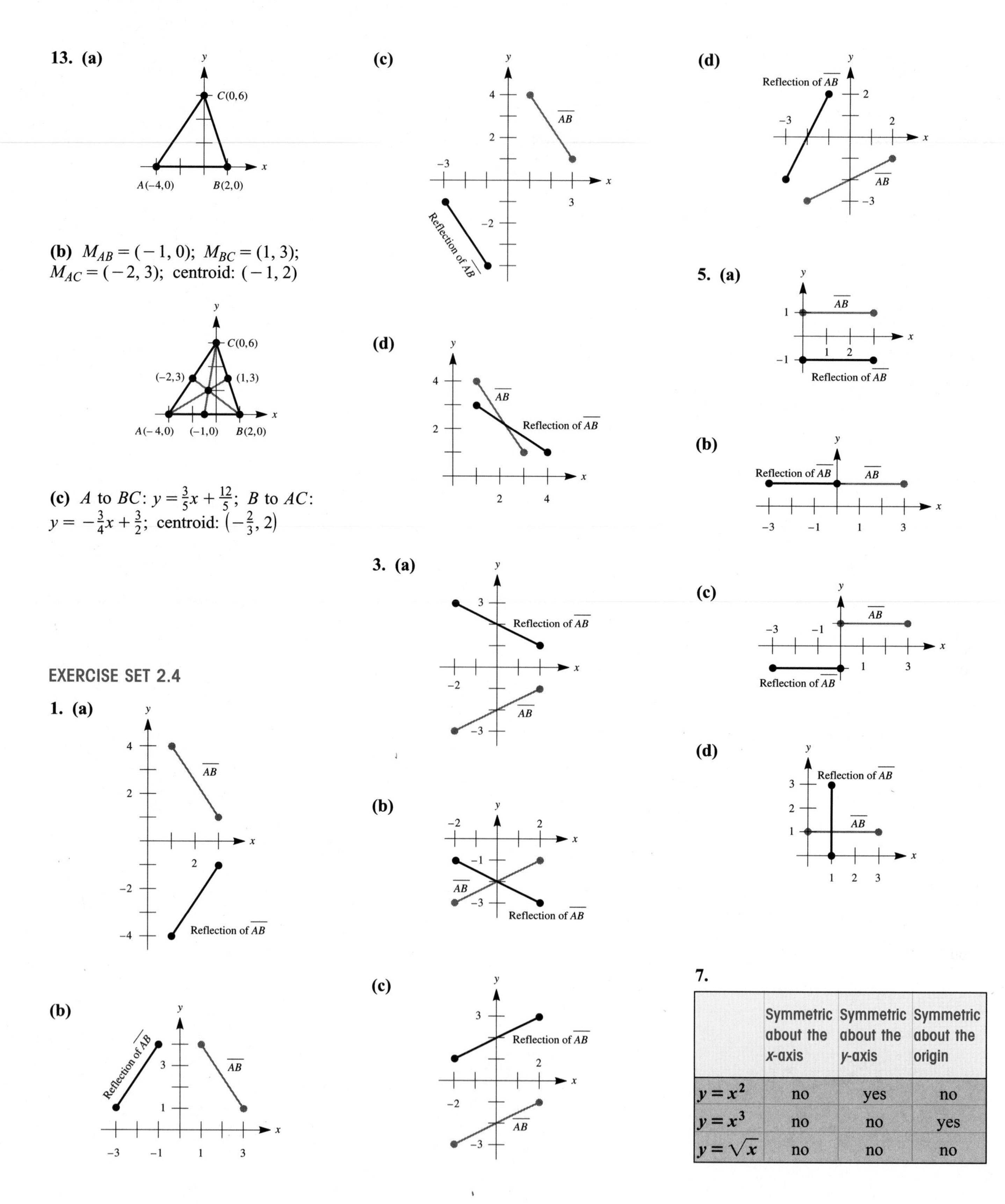

	Symmetric about the x-axis	Symmetric about the y-axis	Symmetric about the origin
$y = x^2$	no	yes	no
$y = x^3$	no	no	yes
$y = \sqrt{x}$	no	no	no

9. x-intercepts: 2, -2; y-intercept: 4; symmetric about the y-axis

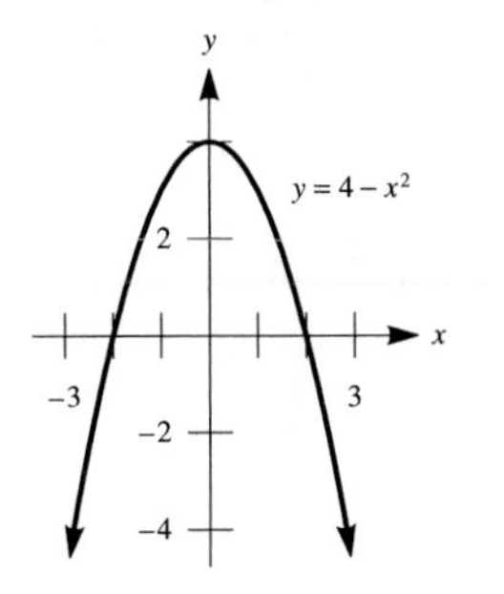

11. no x- or y-intercepts; symmetric about the origin

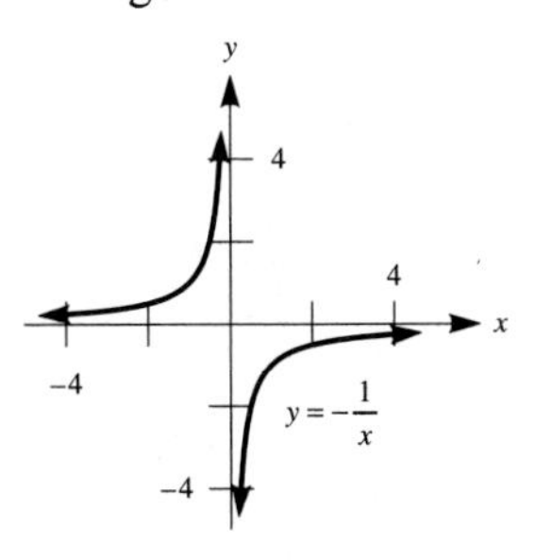

13. x- and y-intercepts: 0; symmetric about the y-axis

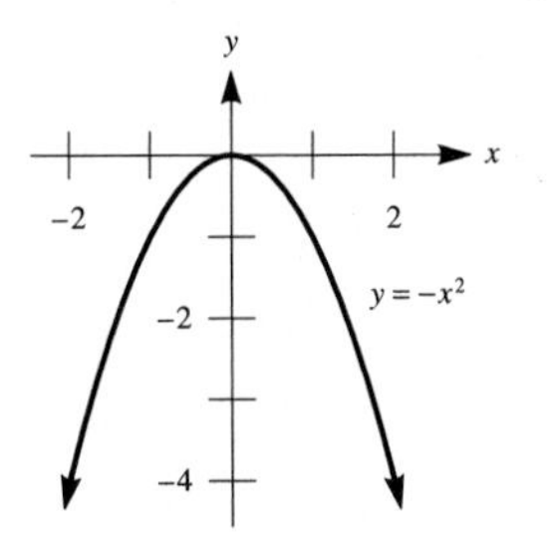

15. no x- or y-intercepts; symmetric about the origin

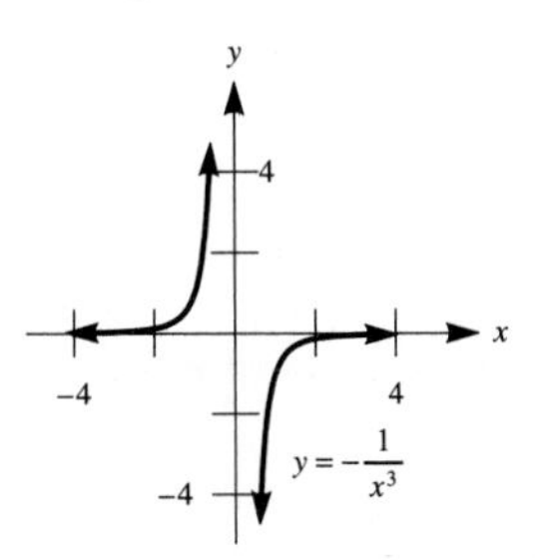

17. x- and y-intercepts: 0; symmetric about the y-axis

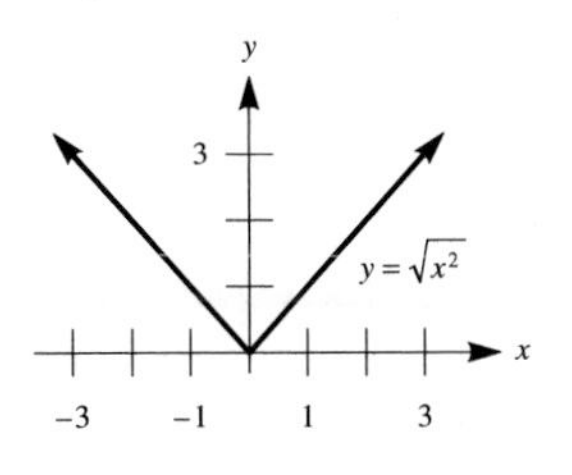

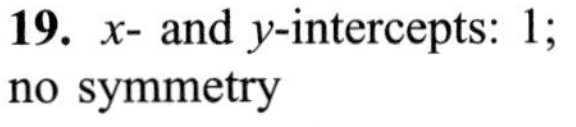

19. x- and y-intercepts: 1; no symmetry

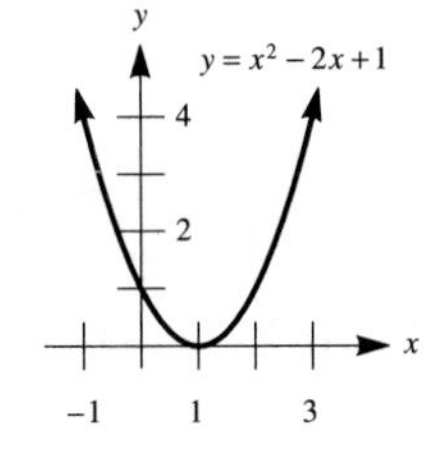

21. x-intercept: 2; no y-intercept; symmetric about the x-axis

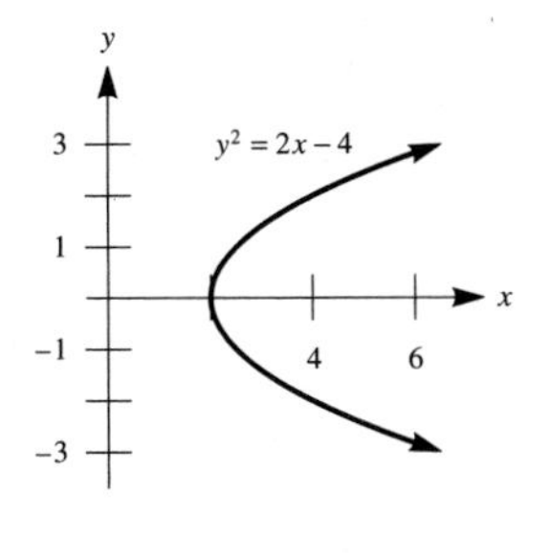

23. x-intercepts: $(-1 \pm \sqrt{33})/4$; y-intercept: -4; no symmetry

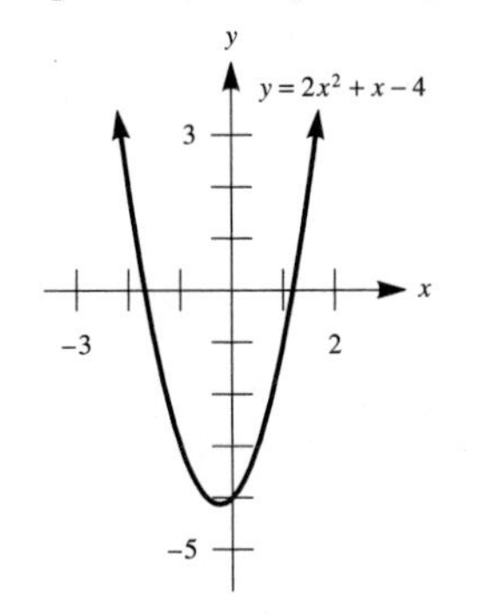

25. (a) x- and y-intercepts: 2; no symmetry

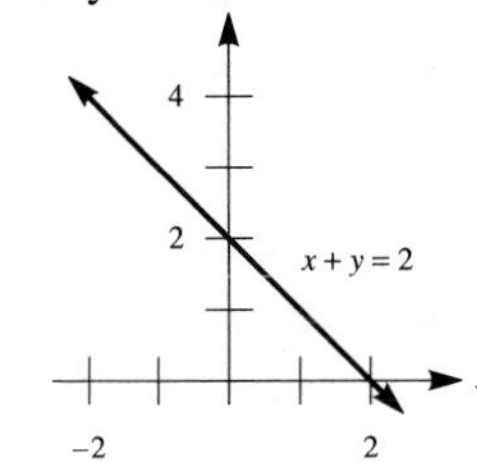

(b) x-intercepts: ± 2; y-intercept: 2; symmetric about the y-axis

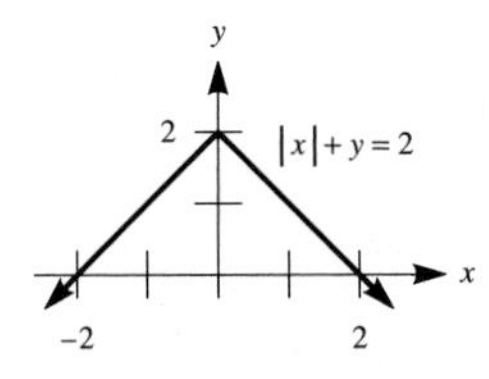

27. (a) x-intercepts: $8/3$; y-intercepts: -2, 2 **(b)** $[8/3, \infty)$ **(c)** The graph of $y = |\frac{3}{4}x - 2|$ can be obtained by reflecting $y = \frac{3}{4}x - 2$ about the x-axis for the interval $(-\infty, 8/3)$. For the interval $[8/3, \infty)$, no reflection is necessary. **33.** $y = 2x - 6$ **35.** x-intercepts: 0, $\pm 3\sqrt{3} \approx \pm 5.2$; y-intercept: 0; symmetric with respect to the origin

x	0	0.5	1	1.5	2
y	0	−13.4	−26	−37.1	−46

x	2.5	3	3.5	4	4.5
y	−51.9	−54	−51.6	−44	−30.4

x	5	5.5	6
y	−10	17.9	54

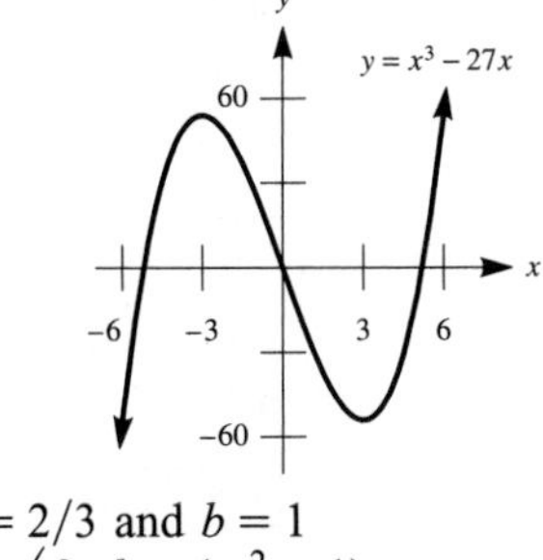

37. $m = 2/3$ and $b = 1$

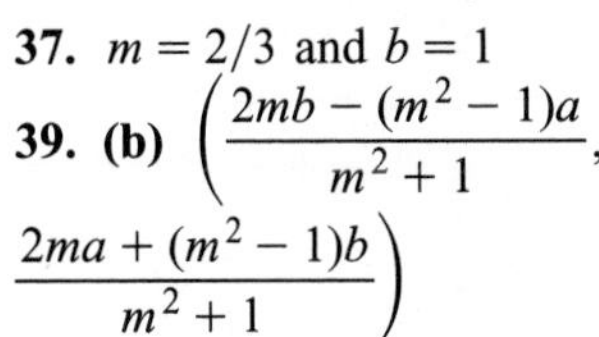

39. (b) $\left(\dfrac{2mb - (m^2 - 1)a}{m^2 + 1}, \dfrac{2ma + (m^2 - 1)b}{m^2 + 1} \right)$

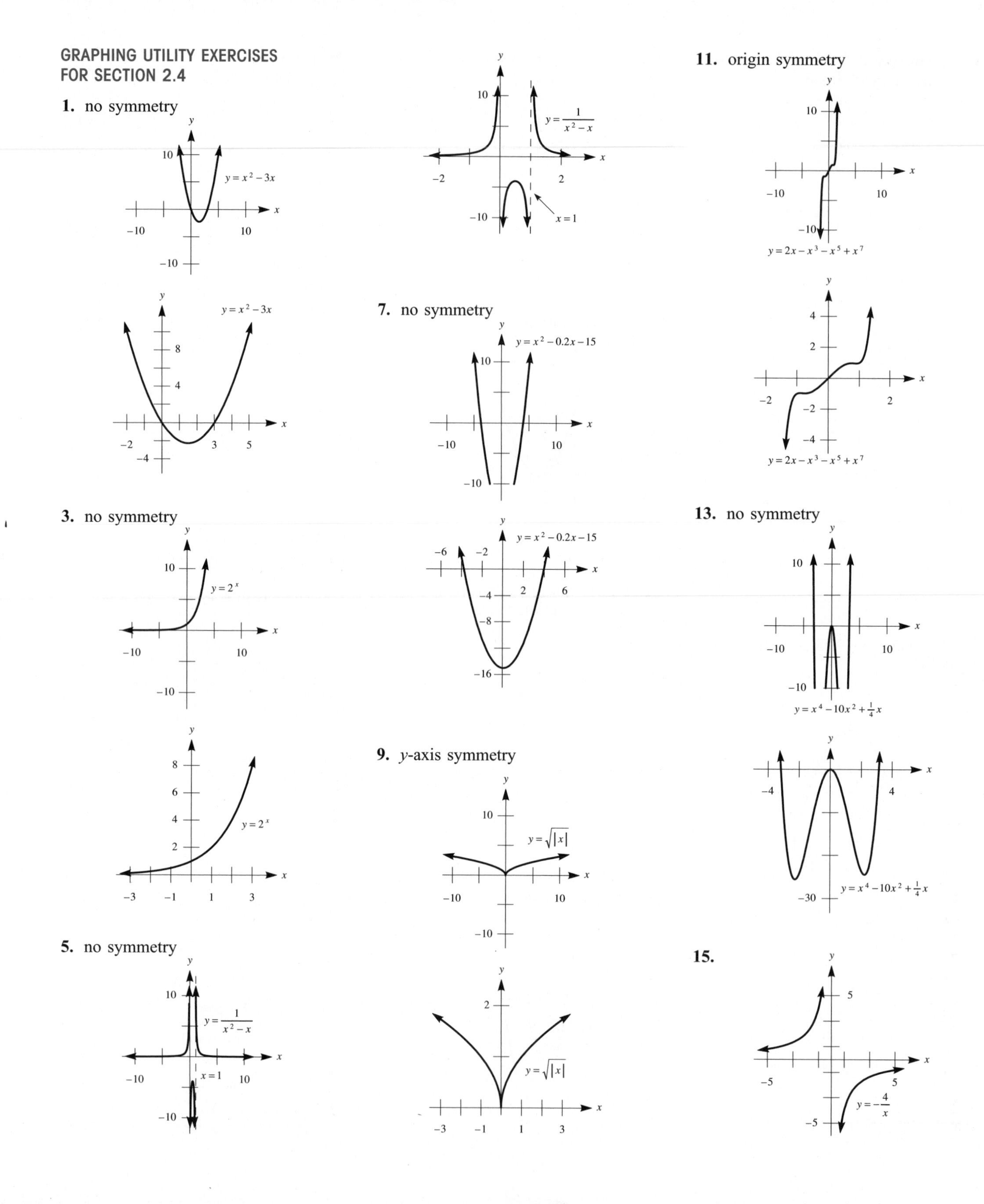
GRAPHING UTILITY EXERCISES FOR SECTION 2.4
1. no symmetry
$y = x^2 - 3x$
$y = x^2 - 3x$
3. no symmetry
$y = 2^x$
$y = 2^x$
5. no symmetry
$y = \frac{1}{x^2 - x}$
$x = 1$
$y = \frac{1}{x^2 - x}$
$x = 1$
7. no symmetry
$y = x^2 - 0.2x - 15$
$y = x^2 - 0.2x - 15$
9. y-axis symmetry
$y = \sqrt{|x|}$
$y = \sqrt{|x|}$
11. origin symmetry
$y = 2x - x^3 - x^5 + x^7$
$y = 2x - x^3 - x^5 + x^7$
13. no symmetry
$y = x^4 - 10x^2 + \frac{1}{4}x$
$y = x^4 - 10x^2 + \frac{1}{4}x$
15.
$y = -\frac{4}{x}$

15.

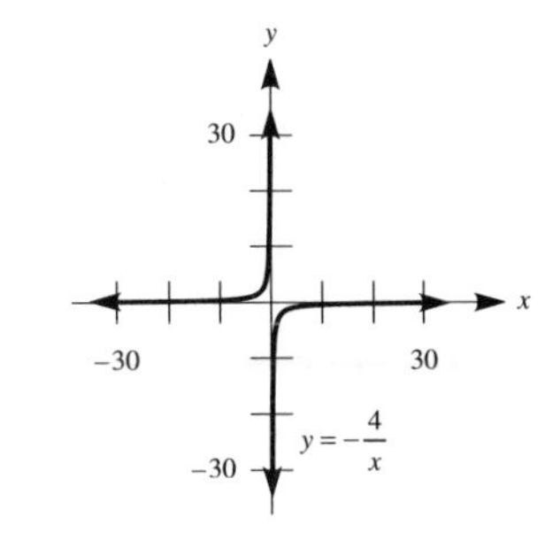

17. (a)

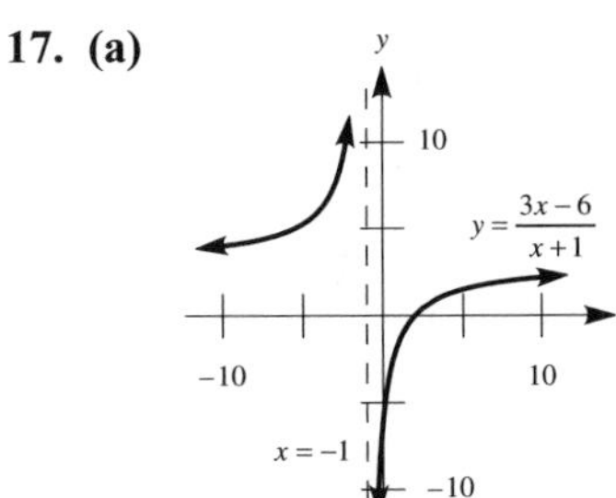

(b)

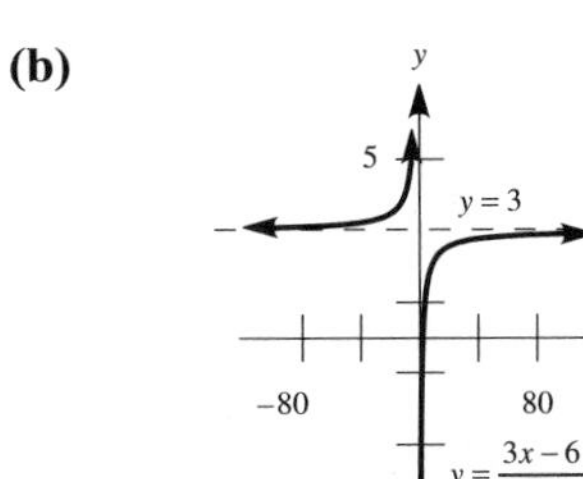

(c)

x	10	100	1000
$\frac{3x-6}{x+1}$	2.18182	2.91089	2.99101

x	10,000	100,000
$\frac{3x-6}{x+1}$	2.99910	2.99991

x	−10	−100	−1000
$\frac{3x-6}{x+1}$	4	3.09091	3.00901

x	−10,000	−100,000
$\frac{3x-6}{x+1}$	3.00090	3.00009

(d)

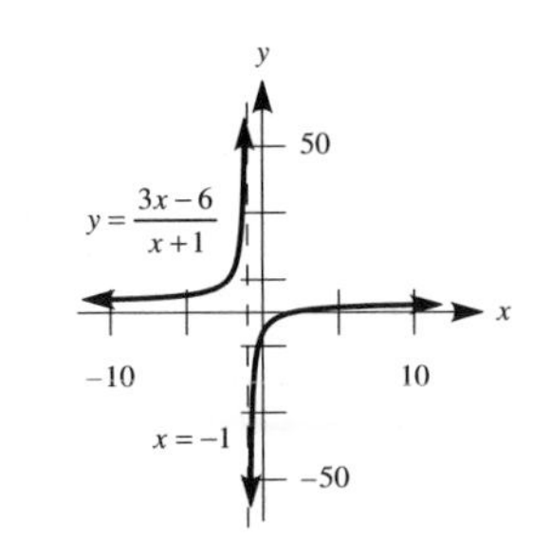

EXERCISE SET 2.5

1. $(-\infty, -1)$ **3.** $[1/3, \infty)$
5. $(-\infty, -9/2)$ **7.** $[-6, \infty)$
9. $(-\infty, 5/2)$ **11.** $[-17/23, \infty)$
13. $[4, 6]$ **15.** $[-1/2, 1]$
17. $(3.98, 3.998)$ **19. (a)** $[-\frac{1}{2}, \frac{1}{2}]$
(b) $(-\infty, -1/2) \cup (1/2, \infty)$
21. (a) $(-\infty, 0) \cup (0, \infty)$ **(b)** no solution **23. (a)** $(-\infty, 3)$ **(b)** $(1, 3)$
(c) $(-\infty, 1) \cup (3, \infty)$ **25. (a)** $[-4, \infty)$
(b) $[-4, 6]$ **(c)** $(-\infty, -4) \cup (6, \infty)$
27. (a) $(a-c, \infty)$ **(b)** $(a-c, a+c)$
(c) $(-\infty, a-c] \cup [a+c, \infty)$
29. $(-10, 14)$ **31.** $(-11, 1)$
33. (a) $(-2h, h)$ **(b)** $(h, -2h)$
35. (a) $[a, \infty)$ **(b)** $a \approx 0.3$
(c) $[2/7, \infty)$; $a = 2/7$ **37. (a)** $[a, b]$
(b) $a \approx 0.1$, $b \approx 0.6$
(c) $[\frac{1}{8}, \frac{5}{8}]$; $a = 1/8$, $b = 5/8$
39. $-175° \leq C \leq 125°$
41. $-13° \leq F \leq 887°$
43. $9200° \leq F \leq 11700°$
45. (a)

a	b	$\sqrt{ab}$ (G.M.)	$\frac{a+b}{2}$ (A.M.)	Which is larger, G.M. or A.M.?
1	2	1.4142	1.5	A.M.
1	3	1.7320	2.0	A.M.
1	4	2.0000	2.5	A.M.
2	3	2.4495	2.5	A.M.
3	4	3.4641	3.5	A.M.
5	10	7.0711	7.5	A.M.
9	10	9.4868	9.5	A.M.
99	100	99.4987	99.5	A.M.
999	1000	999.4999	999.5	A.M.

47. $(0, 3)$

49. (a)

x	y	$\frac{x}{y} + \frac{y}{x}$	True or False: $\frac{x}{y} + \frac{y}{x} \geq 2$
1	1	2.0000	true
2	3	2.1667	true
3	5	2.2667	true
4	7	2.3214	true
5	9	2.3556	true
9	10	2.0111	true
49	50	2.0004	true
99	100	2.0001	true

53. (c) $x = \frac{15}{2}$; $\frac{15}{2}$ ft by $\frac{15}{2}$ ft

EXERCISE SET 2.6

1. (a) $(-\infty, -1] \cup [4, \infty)$ **(b)** $[-1, 4]$
3. (a) no solution **(b)** $(-\infty, \infty)$
5. (a) $[-1, 1] \cup [3, \infty)$
(b) $(-\infty, -1) \cup (1, 3)$ **7.** $(-3, 2)$
9. $(-\infty, 2) \cup (9, \infty)$
11. $(-\infty, 4] \cup [5, \infty)$
13. $(-\infty, -4] \cup [4, \infty)$ **15.** no solution **17.** $(-7, -6) \cup (0, \infty)$
19. $[-15, 0] \cup [15, \infty)$ **21.** $(-\infty, \infty)$
23. $(-\infty, -3/4) \cup (-2/3, 0)$
25. $\left(-\infty, \frac{-1-\sqrt{5}}{2}\right) \cup \left(\frac{-1+\sqrt{5}}{2}, \infty\right)$
27. $[4-\sqrt{14}, 4+\sqrt{14}]$
29. $[-4, -3] \cup [1, \infty)$
31. $(-\infty, -6) \cup (-5, -4)$
33. $(-\infty, -1/3) \cup (1/3, 2) \cup (2, \infty)$
35. $(-\infty, -2/3] \cup \{3, -1/2\}$
37. $(-\infty, -\sqrt{5}] \cup [-2, 2] \cup [\sqrt{5}, \infty)$
39. $(-3, 3) \cup (4, \infty)$
41. $(-\sqrt{14}, -2\sqrt{2}) \cup (2\sqrt{2}, \sqrt{14})$
43. $\left(-\infty, \frac{3-\sqrt{11}}{2}\right) \cup \left(\frac{3+\sqrt{11}}{2}, 4\right)$
45. $(-2, -1) \cup (1, \infty)$ **47.** $(-1, 1]$
49. $(-\infty, 3/2) \cup [2, \infty)$
51. $(-\infty, -1) \cup (0, 9)$
53. $(-\frac{7}{2}, -\frac{4}{3}) \cup (-1, 0) \cup (1, \infty)$

55. $(-\infty, -1)$ **57.** $(-1, 0)$
59. $(-5, \infty)$ **61.** $[-1, 1) \cup (2, 4]$
63. $(-1, 2-\sqrt{5}) \cup (1, 2+\sqrt{5})$
65. $(-3/2, 0)$ **67.** $(-\infty, -1) \cup (0, \infty)$
69. **(a)** $(-\infty, -1] \cup [5, \infty)$
(b) $(-\infty, -1) \cup (5, \infty)$
71. $(-\infty, -2] \cup [2, \infty)$
73. $(-\infty, 0) \cup (1/2, \infty)$
75. $\left(\frac{1}{5}, \frac{4}{5}\right)$ **77.** $(0, 2)$

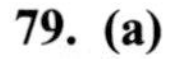

79. **(a)**

x	1	2	5
$\dfrac{x^2+1000}{2x^2+x}$	333.67	100.40	18.64

x	10	100	200	10,000
$\dfrac{x^2+1000}{2x^2+x}$	5.24	0.55	0.51	0.50

(b) $(2000, \infty)$ **(c)** $n = 2001$

81. $\left(-\infty, \dfrac{3a+5b}{8}\right)$

GRAPHING UTILITY EXERCISES FOR SECTION 2.6

1. $y > 0$ on the set $(-1, 0) \cup (3, \infty)$

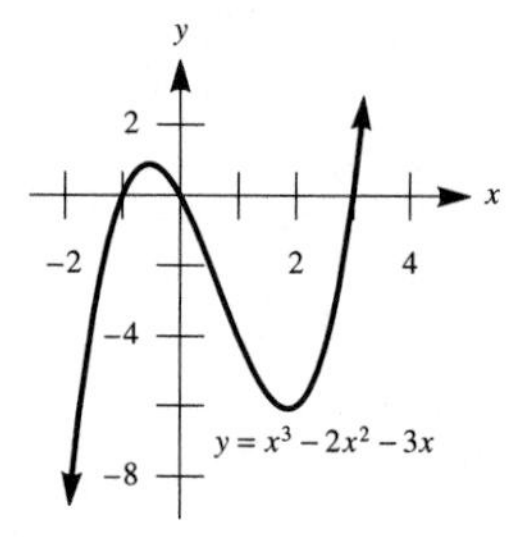

3. **(a)**

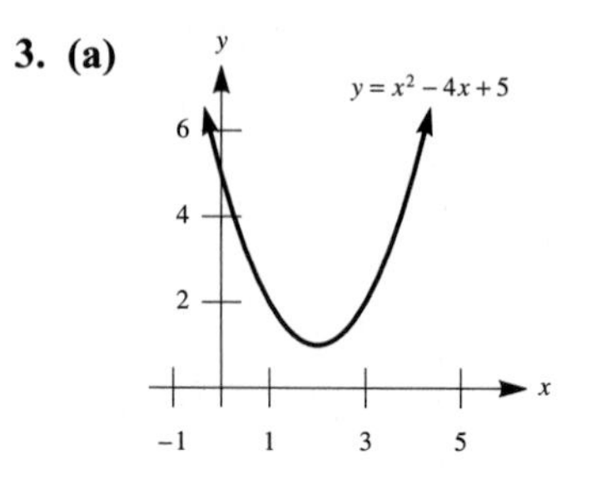

(b) $(-\infty, \infty)$ **(c)** no solution

5. $[0.697, 4.303]$; $\left[\dfrac{5-\sqrt{13}}{2}, \dfrac{5+\sqrt{13}}{2}\right]$

7. $(-\infty, -0.186] \cup [2.686, \infty)$; $\left(-\infty, \dfrac{5-\sqrt{33}}{4}\right] \cup \left[\dfrac{5+\sqrt{33}}{4}, \infty\right)$

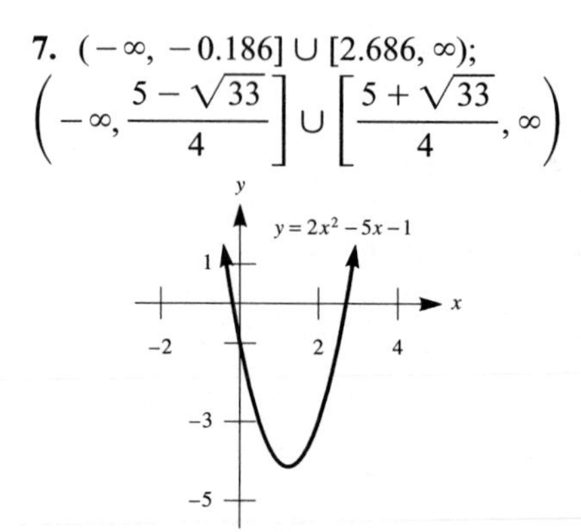

9. $(-0.329, 24.329)$; $(12-2\sqrt{38}, 12+2\sqrt{38})$

11. $(-\infty, -1.554) \cup (1.554, \infty)$; $\left(-\infty, -\sqrt{1+\sqrt{2}}\right) \cup \left(\sqrt{1+\sqrt{2}}, \infty\right)$

13. $[-2.236, 2.236]$; $[-\sqrt{5}, \sqrt{5}]$

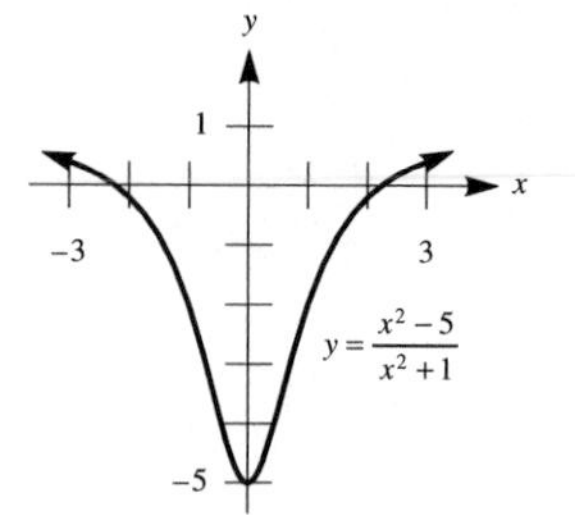

15. $(-2.236, 2.236)$; $(-\sqrt{5}, \sqrt{5})$

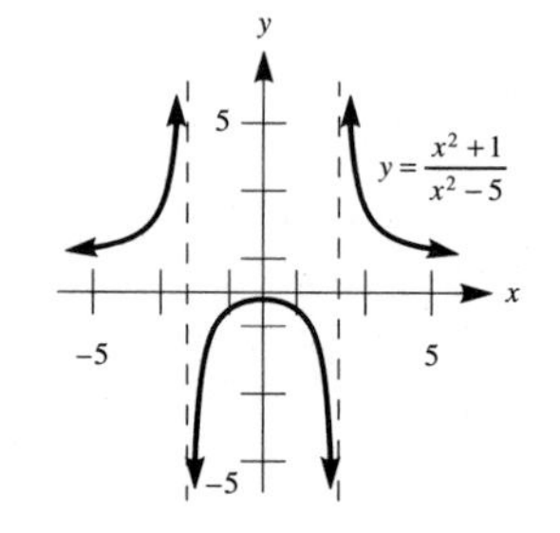

17. $[-0.453, \infty)$

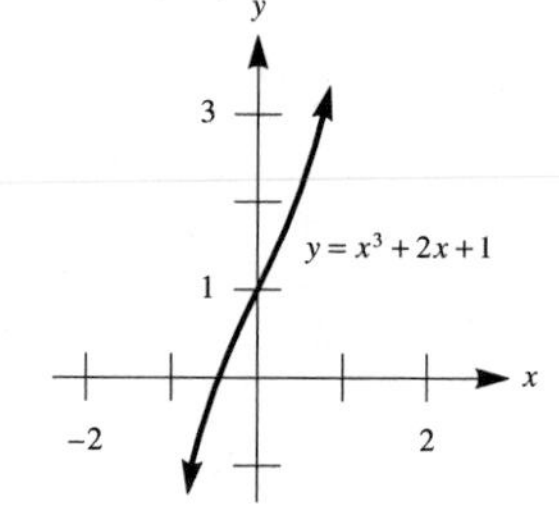

19. $(-\infty, 0.544) \cup (1, \infty)$

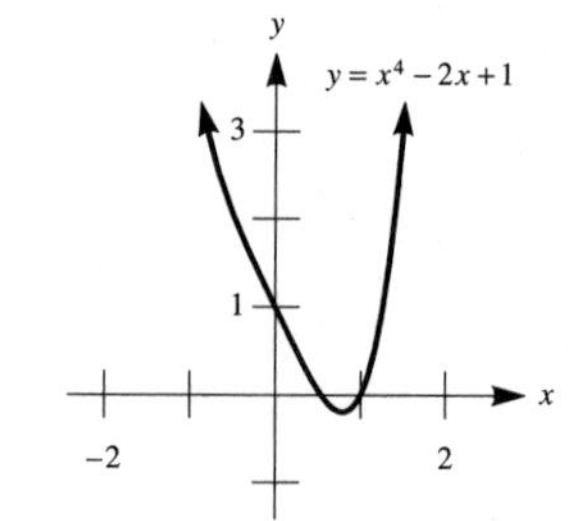

21. $(-3.079, 0) \cup (3.079, \infty)$

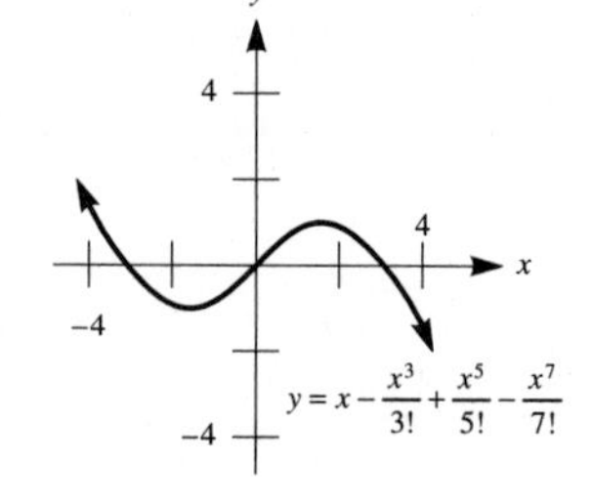

23. $(-\infty, -3.618) \cup (-1.382, -0.854] \cup [5.854, \infty)$

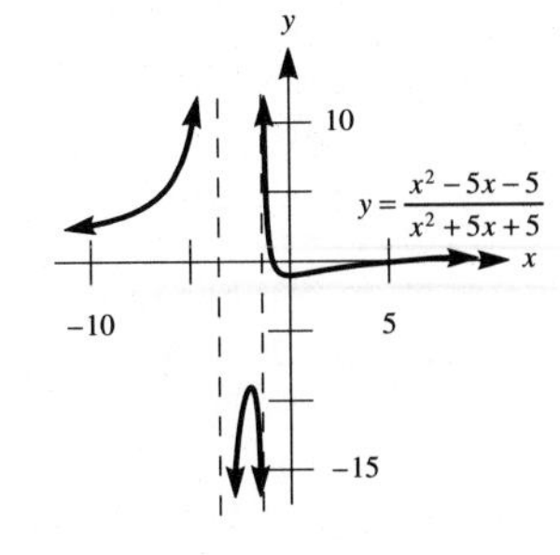

CHAPTER 2 REVIEW EXERCISES

1. $y = -2x - 6$ **3.** $y = \frac{1}{4}x - \frac{5}{2}$
5. $y = 2x + 8$
7. $y = -2$ **9.** $y = x + 1$
11. $9x - 4y + 14 = 0$ **13.** $2x + y = 0$
15. $3x - 4y = 0$ **17.** $x + y - 1 = 0$ or $x - 2y - 4 = 0$ **19.** 13 **21.** (15, 6)
23. y-axis **25.** origin **27.** origin
29. origin **31.** origin **33.** x-axis
35. x-axis, y-axis, origin, $y = x$
37. x-axis **39.** x-axis **41.** $\left(-\frac{27}{13}, \frac{31}{13}\right)$ and $(3, -1)$ **43.** $(-2, 4)$ and $(1, 1)$

45.

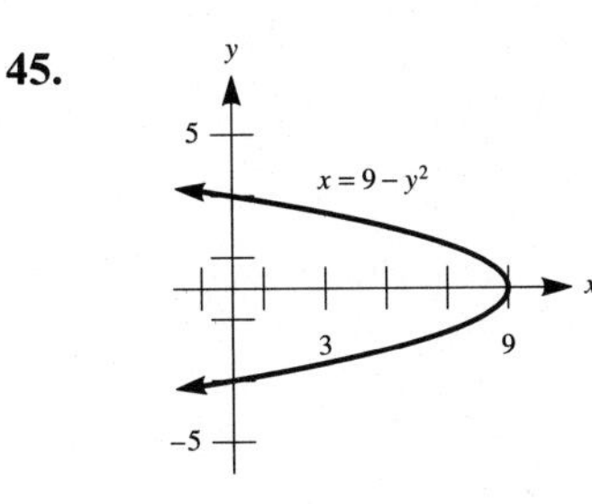

47.

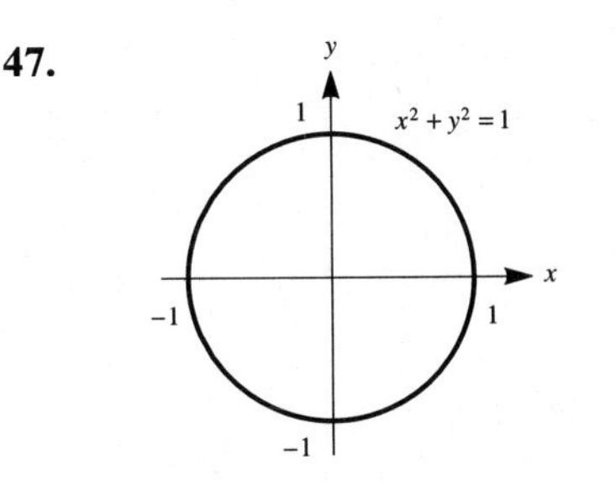

49.

51.

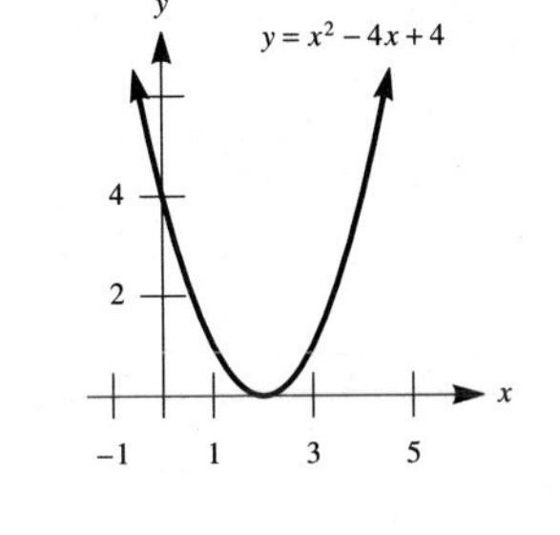

53.

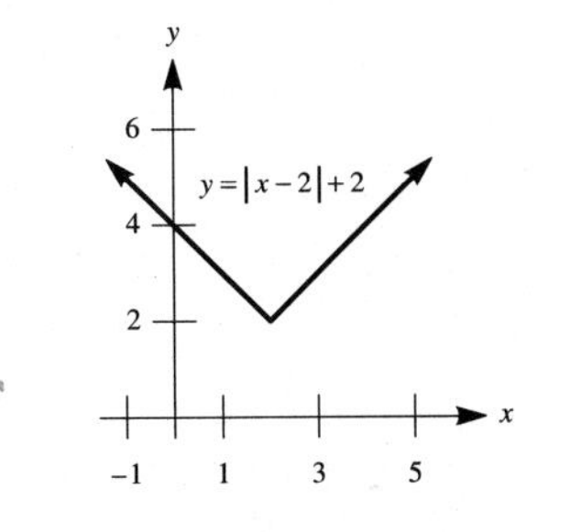

55.

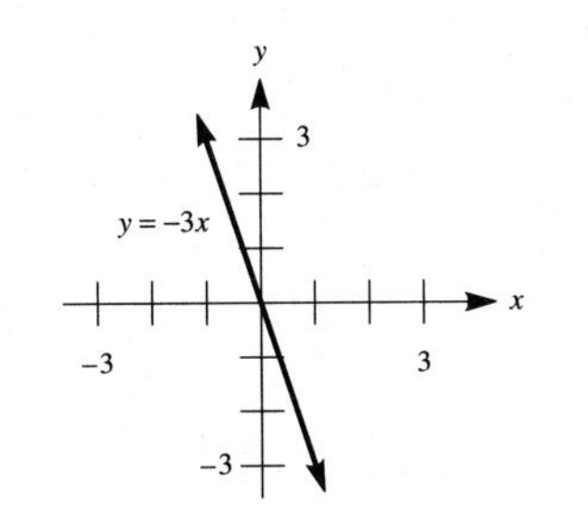

57.

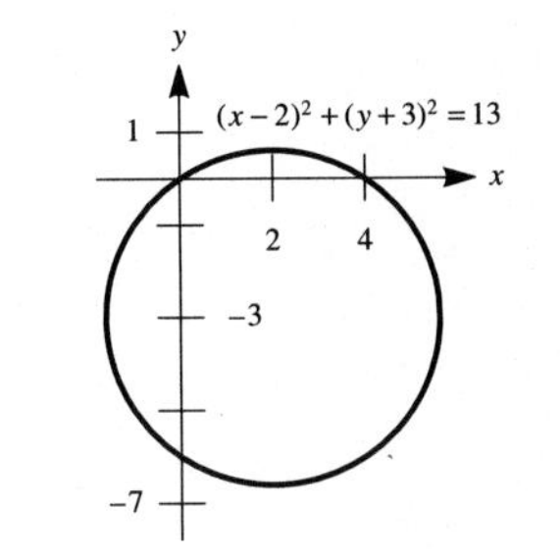

59.

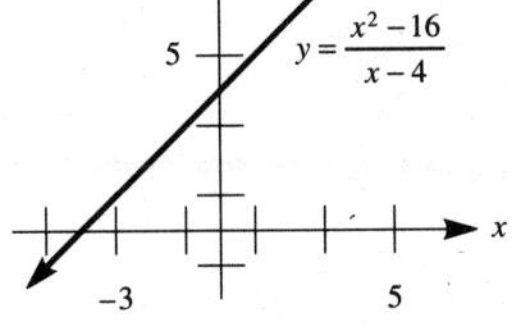

61.

63. $t = 19$ **65.** $m = x^2 - 2x + 4$
67. $MA = MB = MC = \sqrt{b^2 + c^2}$; they are all the same. **69. (a)** $\sqrt{130}$ **(b)** $-11/3$ **(c)** $\left(\frac{7}{2}, -\frac{1}{2}\right)$ **71. (a)** 2 **(b)** $\sqrt{3}/3$ **(c)** $(0, 0)$ **73.** $49/6$
75. (a) $\left(\frac{5}{3}, \frac{11}{3}\right)$; they are all the same point. **(b)** $\left(\frac{2a}{3} + \frac{2b}{3}, \frac{2c}{3}\right)$; all medians of a triangle intersect at a point that is 2/3 of the distance from each vertex to the midpoint of the opposite side. **77.** $(-\infty, 7]$ **79.** $(-2, 1)$
81. $\left[-\frac{1}{2}, \frac{1}{2}\right]$ **83.** $\left(-\frac{41}{10}, -\frac{39}{10}\right)$
85. $(-\infty, -2] \cup [3, \infty)$ **87.** $(-8, 5)$
89. $\left(3 - \sqrt{10}, 3 + \sqrt{10}\right)$
91. $(-5, -3) \cup (3, 5)$
93. $(-2, \infty)$ **95.** $(-\infty, -1) \cup [1, 9]$
97. $(-\infty, -1/2) \cup [1/6, \infty)$
99. $(0, 25]$ **101.** $\left[-\frac{1}{2}, \frac{1}{2}\right]$ **103.** $6\sqrt{2}$
107. $13\sqrt{5}/5$ **109.** $17\sqrt{13}/13$
111. (a) $(x-1)^2 + (y-1)^2 = 1$
(b) $\overline{AT}$: $y = \frac{9}{8}x$; $\overline{BU}$: $y = -\frac{1}{4}x + 1$; $\overline{CS}$: $y = -3x + 3$ **(c)** $\left(\frac{8}{11}, \frac{9}{11}\right)$; the point is the same.

CHAPTER 2 TEST

1. $y = 5x - 7$ **2.** $y = \frac{6}{5}x - \frac{17}{5}$
3. $y = -3x$ **4.** $y = \frac{4}{5}x$ **5.** (3, 9)
6. (a) origin **(b)** y-axis **(c)** x-axis
7. x-intercept: 5; y-intercept: -3

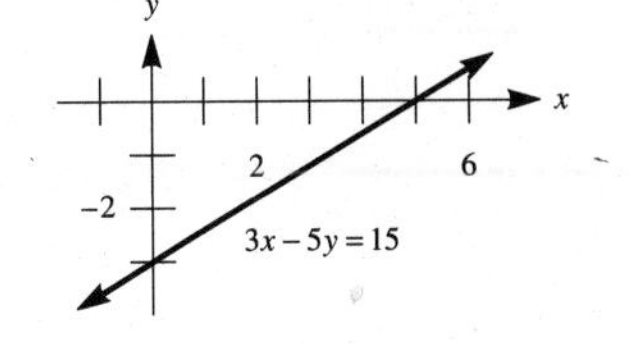

8. x-intercepts: $(5 \pm \sqrt{59})/2$; y-intercepts: $(-3 \pm \sqrt{43})/2$

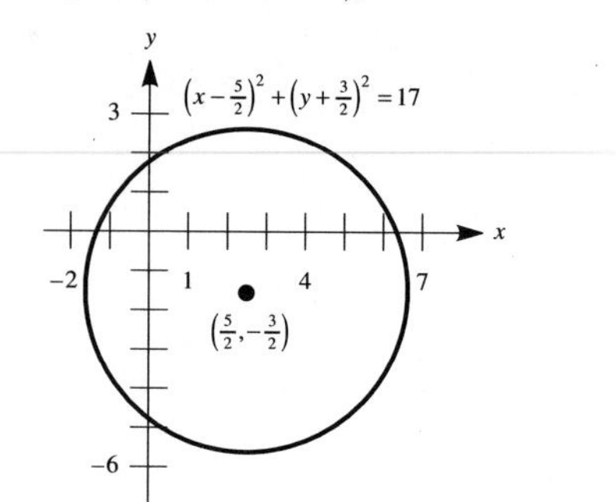

9. no x-intercept; y-intercept: 3

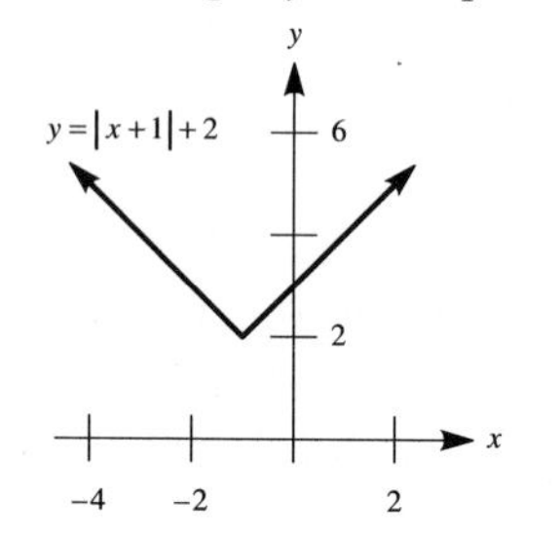

10. $t=1$ **11.** $(-4, 16)$ and $(3, 9)$
12. yes

13. (a)

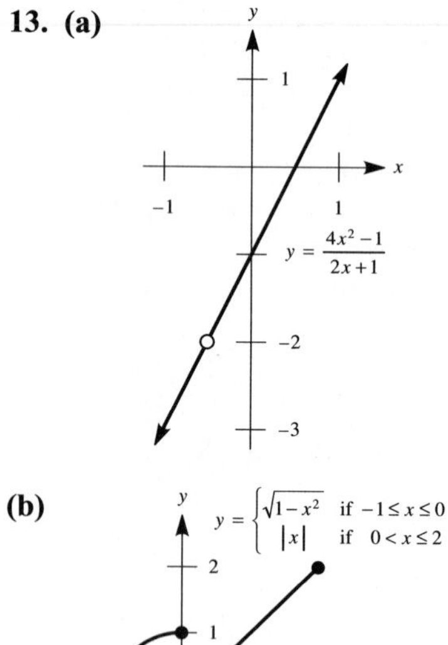

(b)

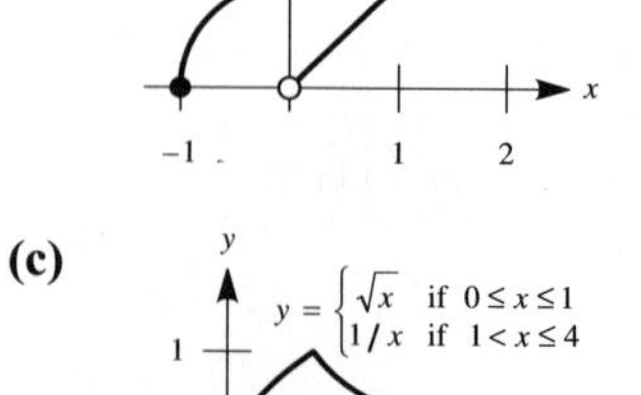

(c)

$y=\begin{cases}\sqrt{x} & \text{if } 0\le x\le 1\\ 1/x & \text{if } 1<x\le 4\end{cases}$

14. (a)

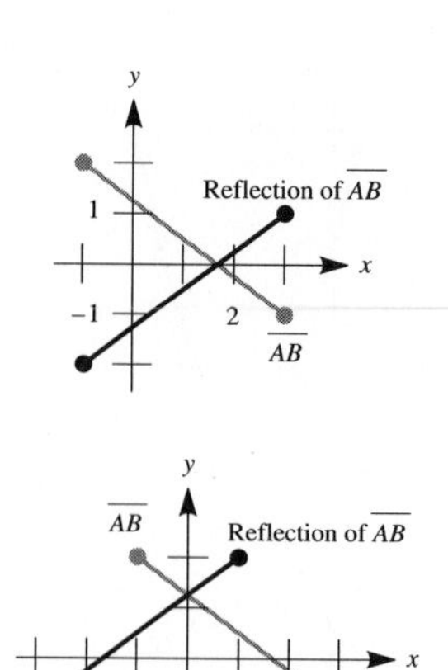

(b)

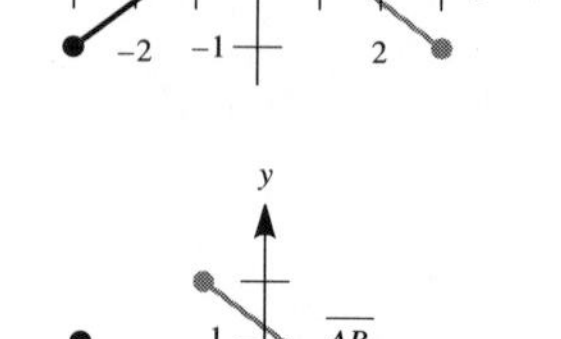

(c)

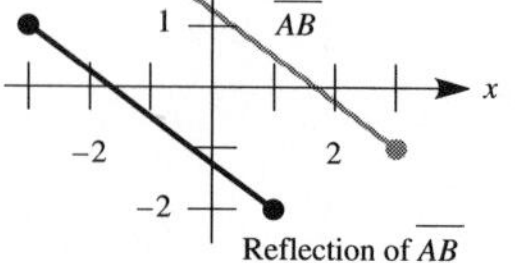

(d)

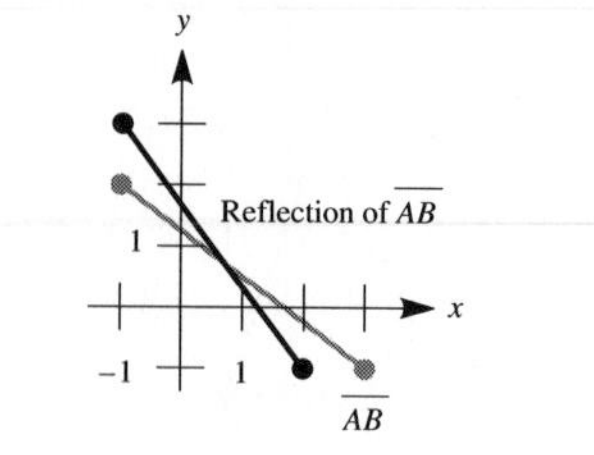

15. $\frac{121}{5}$ sq. units **16.** $(-\infty, 3]$
17. $\left(\frac{27}{10}, \frac{31}{10}\right)$ **18.** $[-8, \infty)$
19. $\left(-2, \frac{-3-\sqrt{3}}{3}\right] \cup \left(-1, \frac{-3+\sqrt{3}}{3}\right] \cup (0, \infty)$ **20.** $\left[\frac{8}{3}, 4\right]$

CHAPTER THREE

EXERCISE SET 3.1

1. (a) $(-\infty, \infty)$
(b) $(-\infty, 1/5) \cup (1/5, \infty)$
(c) $(-\infty, 1/5]$ **(d)** $(-\infty, \infty)$
3. (a) $(-\infty, \infty)$
(b) $(-\infty, -3) \cup (-3, 3) \cup (3, \infty)$
(c) $(-\infty, -3] \cup [3, \infty)$ **(d)** $(-\infty, \infty)$
5. (a) $(-\infty, \infty)$
(b) $(-\infty, 3) \cup (3, 5) \cup (5, \infty)$
(c) $(-\infty, 3] \cup [5, \infty)$ **(d)** $(-\infty, \infty)$
7. (a) $(-\infty, -3) \cup (-3, \infty)$
(b) $(-\infty, -3) \cup [2, \infty)$
(c) $(-\infty, -3) \cup (-3, \infty)$
9. (a) $(-\infty, \infty)$ **(b)** $(-\infty, \infty)$
(c) $[-2, 1] \cup [3, \infty)$
11. $(-\infty, 1) \cup (1, \infty)$
13. domain: $(-\infty, \infty)$; range: $(-\infty, \infty)$
15. domain: $(-\infty, \infty)$; range: $(-\infty, \infty)$
17. domain: $(-\infty, 6) \cup (6, \infty)$; range: $(-\infty, 4/3) \cup (4/3, \infty)$
19. (a) domain: $(-\infty, 5) \cup (5, \infty)$; range: $(-\infty, 1) \cup (1, \infty)$
(b) domain: $(-\infty, \sqrt[3]{5}) \cup (\sqrt[3]{5}, \infty)$; range: $(-\infty, 1) \cup (1, \infty)$
21. domain: $(-\infty, \infty)$; range: $[4, \infty)$
23. domain: $(-\infty, 0) \cup (0, \infty)$; range: $(-\infty, a) \cup (a, \infty)$ **25.** f, g, F, H
27. (a) f: $\{1, 2, 3\}$; g: $\{2, 3\}$; F: $\{1\}$; H: $\{1, 2\}$ **(b)** g: $\{i, j\}$; F: $\{i, j\}$; G: $\{k\}$ **29. (a)** $y=(x-3)^2$
(b) $y=x^2-3$ **(c)** $y=(3x)^2$
(d) $y=3x^2$ **31. (a)** -1 **(b)** 1
(c) 5 **(d)** $-5/4$ **(e)** z^2-3z+1
(f) x^2-x-1 **(g)** a^2-a-1
(h) x^2+3x+1 **(i)** 1 **(j)** $4-3\sqrt{3}$
(k) $1-\sqrt{2}$ **(l)** 2 **33. (a)** $12x^2$
(b) $6x^2$ **(c)** $3x^4$ **(d)** $9x^4$ **(e)** $\frac{3}{4}x^2$
(f) $\frac{3}{2}x^2$ **35. (a)** 1 **(b)** -7 **(c)** -3
(d) $-7/18$ **(e)** $-2x^2-4x-1$
(f) $1-2x^2-4xh-2h^2$
(g) $-4xh-2h^2$ **(h)** $-4x-2h$
37. (a) domain: $(-\infty, 2) \cup (2, \infty)$; range: $(-\infty, 2) \cup (2, \infty)$ **(b)** 1/2
(c) 0 **(d)** 1 **(e)** $(2x^2-1)/(x^2-2)$
(f) $(2-x)/(1-2x)$
(g) $(2a-1)/(a-2)$
(h) $(2x-3)/(x-3)$
39. (a) $d(1)=80$; $d(3/2)=108$; $d(2)=128$; $d(t_0)=-16t_0^2+96t_0$
(b) $t=0, 6$ **(c)** $t=(12 \pm \sqrt{143})/4$
41. $g(3)=1$; $g(x+4)=|x|$ **43.** 8
45. (a) $2x+h$ **(b)** $4x+2h$
(c) $6x^2+6xh+2h^2$ **47.** $4x+2h-3$
49. $-3/x(x+h)$ **51.** $4x+4a$
53. $x+a-2$ **55.** $4x^2+4ax+4a^2$
57. $-5(x+a)/a^2x^2$
59. (a) $-1/(x-1)(a-1)$
(b) $-1/2(x-1)$
(c) $-1/(x-1)(x+h-1)$
(d) $-1/2(2+h)$ **61. (a)** $x_0=11/5$
(b) $x_0=\pm 2$ **(c)** $x_0=0, 1$
(d) $x_0=3/2, -1$
63. (a) \$125.51 **(b)** \$363.76

65. (a)

n	2	3	4	5
$g(n)$	1.4142	1.4422	1.4142	1.3797

n	6	7	8
$g(n)$	1.3480	1.3205	1.2968

(b) 15
67. $a = 1$ and $b = 2$ (Other answers are possible.) **69.** $a = 2$ (Other answers are possible.) **71.** n
75. symmetric about the y-axis
79. (a) $f(a) = 0$; $f(2a) = 1/3$; $f(3a) = 1/2$; no **83.** $a = 1/2$ and $b = -3/2$ **89.** $k = -1/3$ **91. (a)** 0 **(b)** 1 **(c)** 2 **(d)** 6 **(e)** -1 **(f)** -2 **(g)** -6 **(h)** $1/2$ **93.** 0
95. (a) x **(b)** $22/7$ **97.** G is not a function, since it assigns more than one value to some inputs x. Define G to assign the closest prime number greater than or equal to x. **99.** $P(1) = 17$; $P(2) = 19$; $P(3) = 23$; $P(4) = 29$; $x = 17$ **101.** $g(g(a)) = a$; $g(g(g(a))) = b$
103. $k = 2$

EXERCISE SET 3.2

1. $y = \sqrt{3} \approx 1.732$
3. $y = \sqrt{5}/5 \approx 0.447$
5. $P(4, \sqrt[3]{4}) \approx P(4, 1.587)$; $Q(\sqrt[3]{4}, \sqrt[3]{4}) \approx Q(1.587, 1.587)$; $R(\sqrt[3]{4}, \sqrt[9]{4}) \approx R(1.587, 1.167)$
7. $P(\sqrt[3]{3}, 3 - 3\sqrt[3]{3}) \approx P(1.442, -1.327)$; $Q(3 - 3\sqrt[3]{3}, 3 - 3\sqrt[3]{3}) \approx Q(-1.327, -1.327)$; $R(3 - 3\sqrt[3]{3}, -63 - 72\sqrt[3]{3} + 81\sqrt[3]{9}) \approx R(-1.327, 1.645)$
9. domain: $[-4, 2]$; range: $[-3, 3]$
11. domain: $[-3, 4]$; range: $[-2, 2]$
13. domain: $[-4, -1) \cup (-1, 4]$; range: $[-2, 3)$ **15.** domain: $[-4, 3]$; range: $\{2\}$ **17. (a)** 1 **(b)** -3 **(c)** no **(d)** $x = 2$ **(e)** -1
19. (a) positive **(b)** $f(-2) = 4$; $f(1) = 1$; $f(2) = 2$; $f(3) = 0$ **(c)** $f(2)$ **(d)** -3 **(e)** 3 **(f)** domain: $[-2, 4]$; range: $[-2, 4]$ **21. (a)** $g(-2)$ **(b)** 5 **(c)** $f(2) - g(2)$ **(d)** $x = -2, 3$ **(e)** 4 is in the range of f

23.

Function	$\lvert x \rvert$	x^2	x^3
Domain	$(-\infty, \infty)$	$(-\infty, \infty)$	$(-\infty, \infty)$
Range	$[0, \infty)$	$[0, \infty)$	$(-\infty, \infty)$
Turning Point	(0, 0)	(0, 0)	none
Maximum Value	none	none	none
Minimum Value	0	0	none
Interval(s) where Increasing	$(0, \infty)$	$(0, \infty)$	$(-\infty, \infty)$
Interval(s) where Decreasing	$(-\infty, 0)$	$(-\infty, 0)$	none

25. (a) $[-1, 1]$ **(b)** 1 (occurring at $x = 1$) **(c)** -1 (occurring at $x = 3$) **(d)** $[0, 1]$ and $[3, 4]$ **(e)** $[1, 3]$
27. (a) $[-3, 0]$ **(b)** 0 (occurring at $x = 0$ and $x = 4$) **(c)** -3 (occurring at $x = 2$) **(d)** $[2, 4]$ **(e)** $[0, 2]$
29. (a) function **(b)** not a function **(c)** not a function **(d)** function

31.

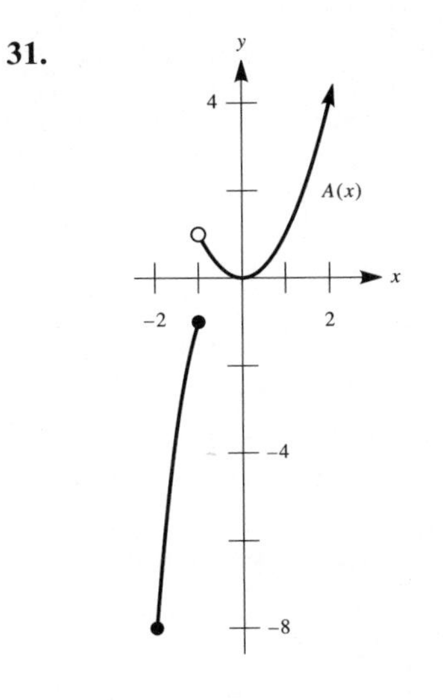

33.

35. (a)

(b)

37.

39.

41. domain: $(-\infty, \infty)$; range: $\{3\}$
43. 7; they are the same **45.** 10
47. 0 **49.** 2 **51.** $-3/ab$
53. $-2(a^2 + ab + b^2)$
55. (a) $\frac{1}{3}$ °C/min **(b)** $\frac{4}{3}$ °C/min **(c)** $\frac{1}{2}$ °C/min **57.** $b = 5$
59. (a)

Interval	[1, 1.1]	[1, 1.01]	[1, 1.001]
$\Delta s/\Delta t$	33.6	32.16	32.016

Interval	[1, 1.0001]	[1, 1.00001]
$\Delta s/\Delta t$	32.0016	32.00016

(b) 32 ft/sec

61. **(a)** 4 °F/hr **(b)** -2 °F/hr
(c) 0 °F/hr **(d)** 1 °F/hr
63. **(a)** 1983–1984 period
(b) 1983–1984: 6; 1984–1985: 1
(c) negative **(d)** -5.7
(e) 1982–1992: 3.62; 1991–1992: 6.4; greater over the 1991–1992 period

65. **(a)** $2x + h$
(b)

h	2	1	0.1	0.01
$\Delta f/\Delta x$ on the interval $[1, 1 + h]$	4	3	2.1	2.01

h	0.001	0.0001
$\Delta f/\Delta x$ on the interval $[1, 1 + h]$	2.001	2.0001

h	-2	-1	-0.1	-0.01
$\Delta f/\Delta x$ on the interval $[1, 1 + h]$	0	1	1.9	1.99

h	-0.001	-0.0001
$\Delta f/\Delta x$ on the interval $[1, 1 + h]$	1.999	1.9999

(c) 2 **(d)** $y = 2x - 1$

67. **(a)**

t	0	0.25	0.5	0.75	1	1.25	1.5
$S(t)$	0	0.25	0.5	0.75	1	1	1

t	1.75	2	2.25	2.5	2.75	3
$S(t)$	1	1	1	1	1	1

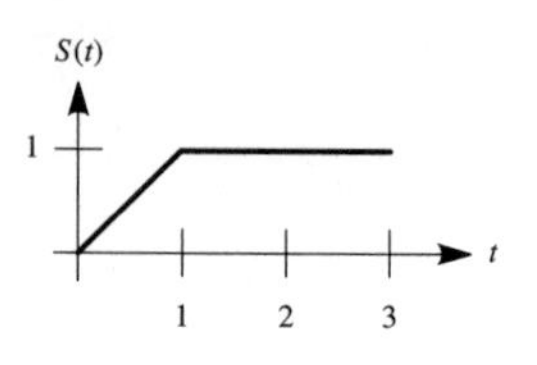

(b)

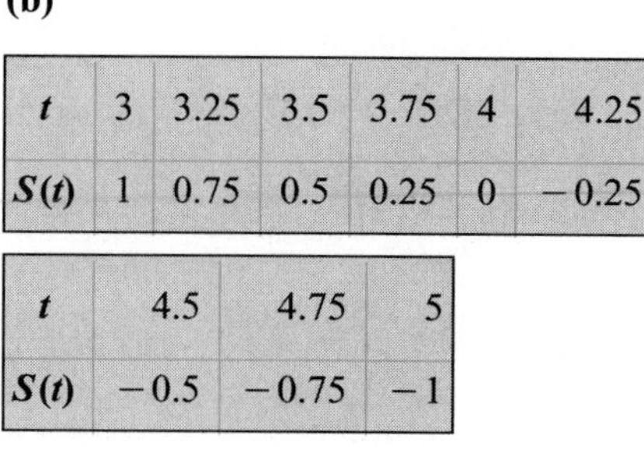

t	3	3.25	3.5	3.75	4	4.25
$S(t)$	1	0.75	0.5	0.25	0	-0.25

t	4.5	4.75	5
$S(t)$	-0.5	-0.75	-1

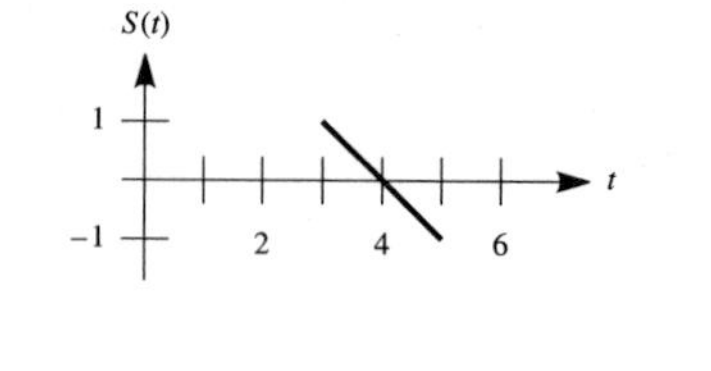

(c)

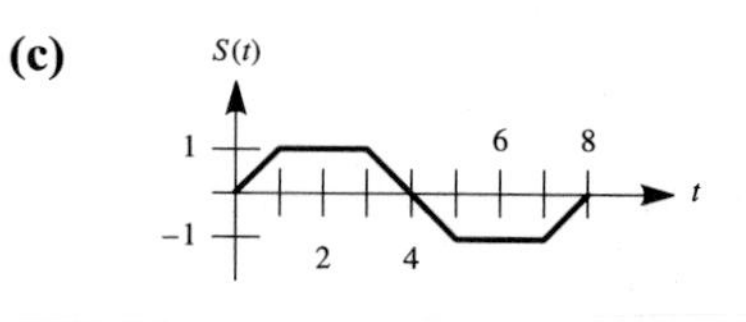

(d) The graph is identical to that of part (c) except for the t-values.

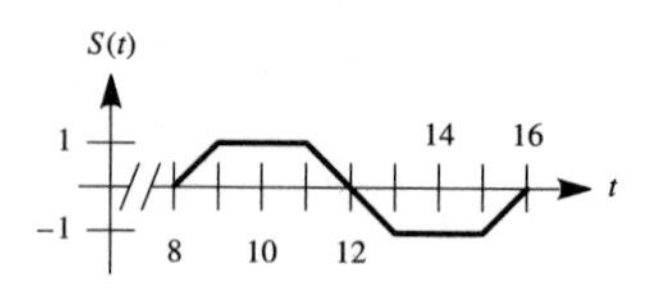

69. **(a)**

x	0	0.1	0.5	0.9	1.0
$[x]$	0	0	0	0	1

(b)

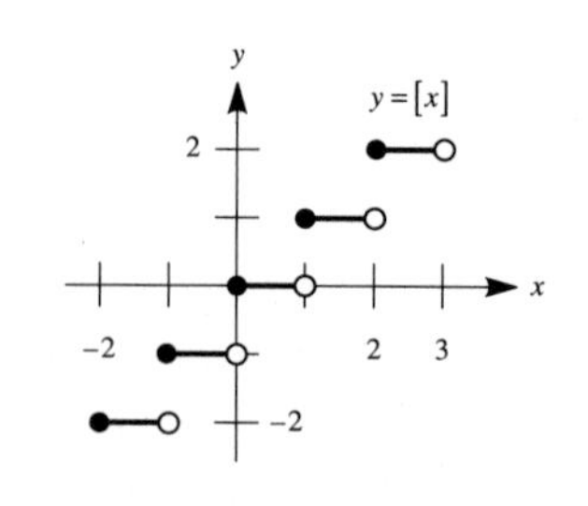

(c) all integers

GRAPHING UTILITY EXERCISES FOR SECTION 3.2

1. **(a)** two turning points

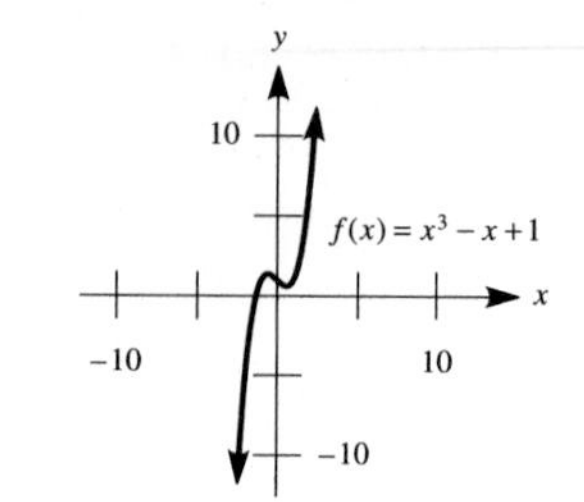

(b)

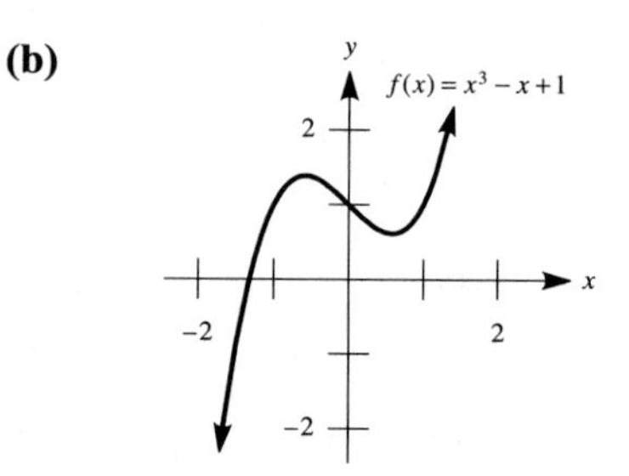

(c) $x \approx \pm 0.568$
(d) $x = \pm 1/\sqrt{3} \approx \pm 0.577$
(e) increasing: $(-\infty, -1/\sqrt{3})$ and $(1/\sqrt{3}, \infty)$;
decreasing: $(-1/\sqrt{3}, 1/\sqrt{3})$
3. turning points: $x \approx 1.2, 2.0$;
increasing: $(-\infty, 6/5)$ and $(2, \infty)$;
decreasing: $(6/5, 2)$

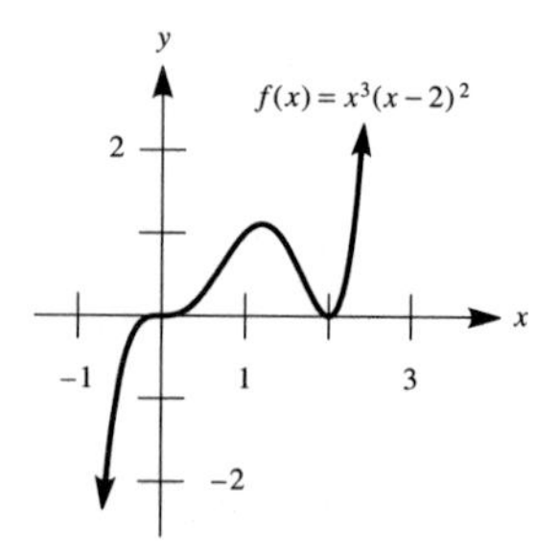

5. turning point: $x \approx 0.630$; increasing: $(-\infty, \sqrt[3]{-0.5})$ and $(\sqrt[3]{-0.5}, \sqrt[3]{0.25})$;
decreasing: $(\sqrt[3]{0.25}, \infty)$

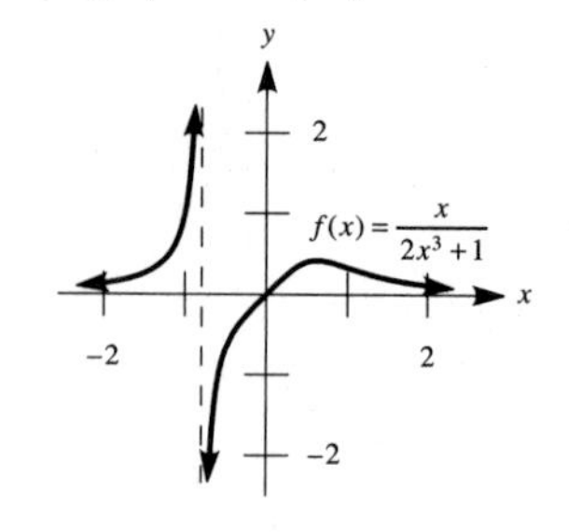

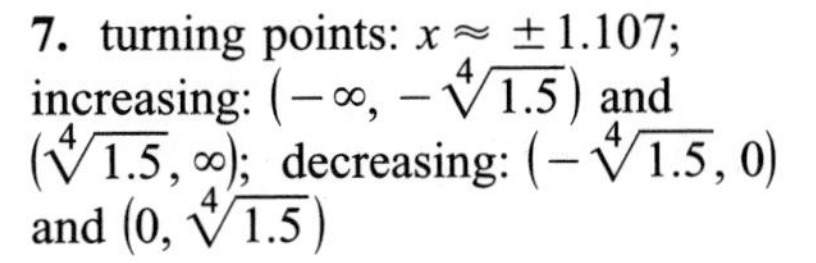

7. turning points: $x \approx \pm 1.107$; increasing: $(-\infty, -\sqrt[4]{1.5})$ and $(\sqrt[4]{1.5}, \infty)$; decreasing: $(-\sqrt[4]{1.5}, 0)$ and $(0, \sqrt[4]{1.5})$

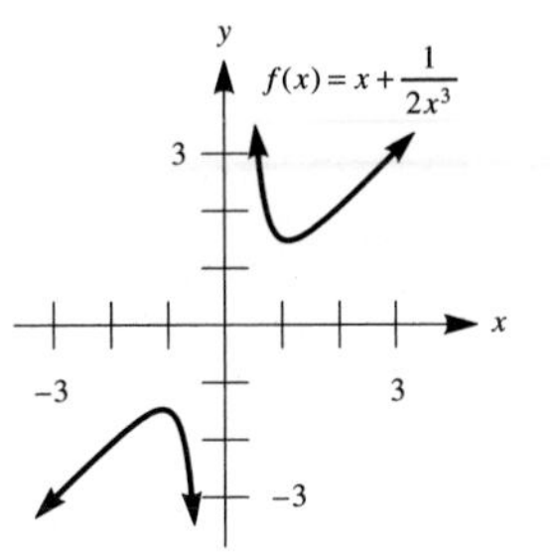

9. (a) $(-\infty, \infty)$
(b)

y
3
$f(x) = \sqrt[3]{x}$
-3
x
3
-3

EXERCISE SET 3.3

1. (a) C **(b)** F **(c)** I **(d)** A **(e)** J **(f)** K **(g)** D **(h)** B **(i)** E **(j)** H **(k)** G

3.

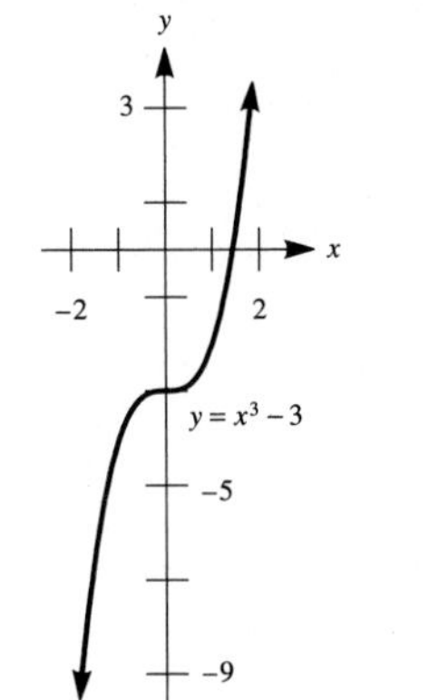

5.

y
8
$y = (x+4)^2$
4
x
-8
-4

7.

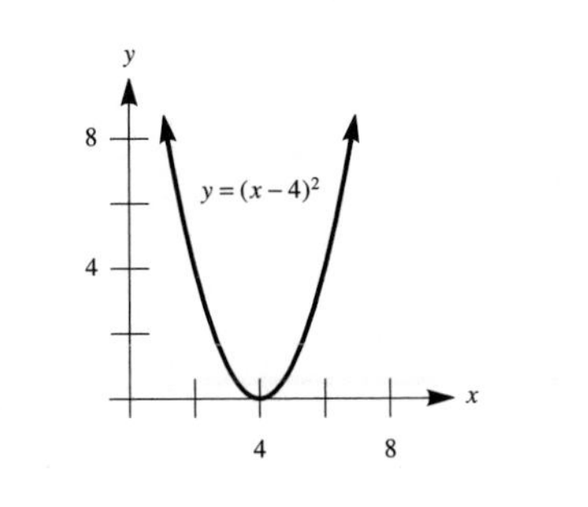

9.

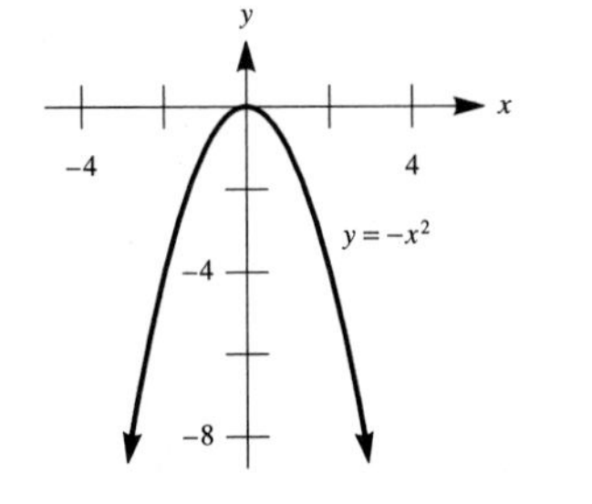

11.

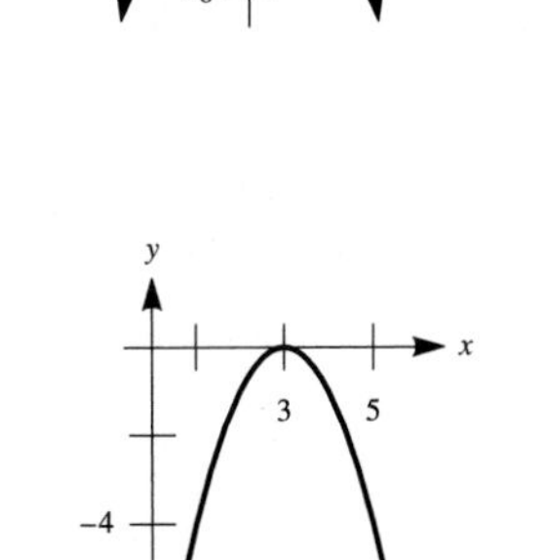

13.

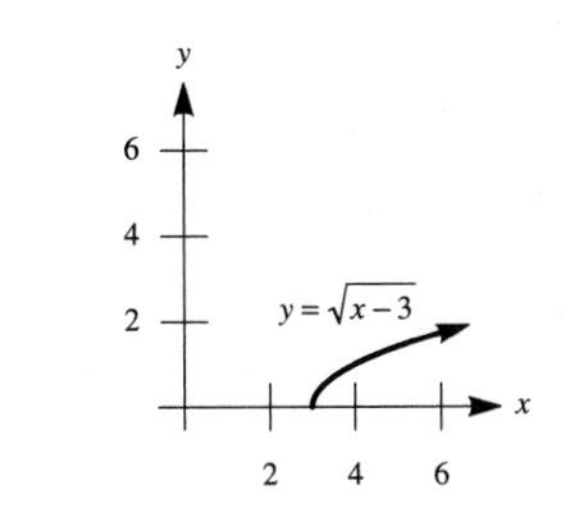

15.

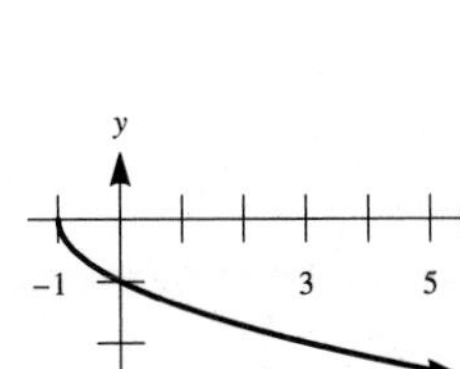

17.

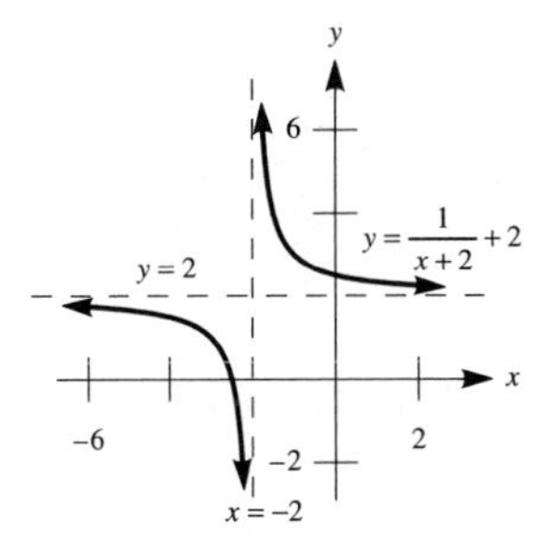

19.

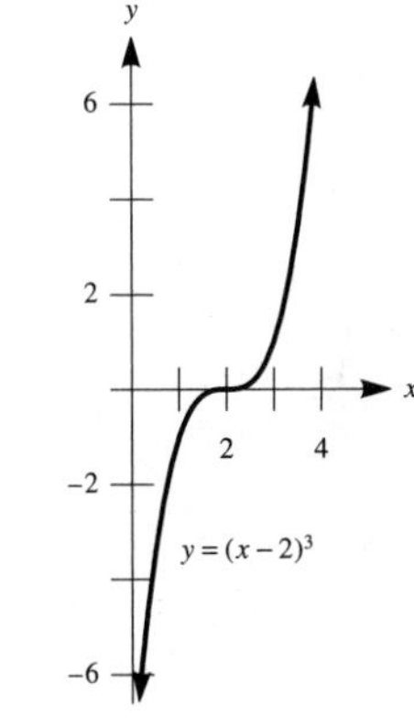

21.

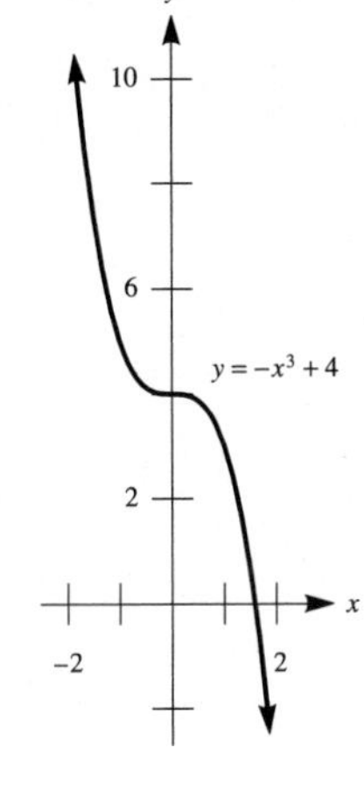

23. (a)

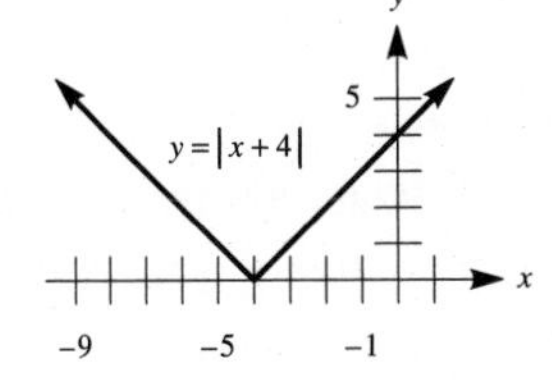

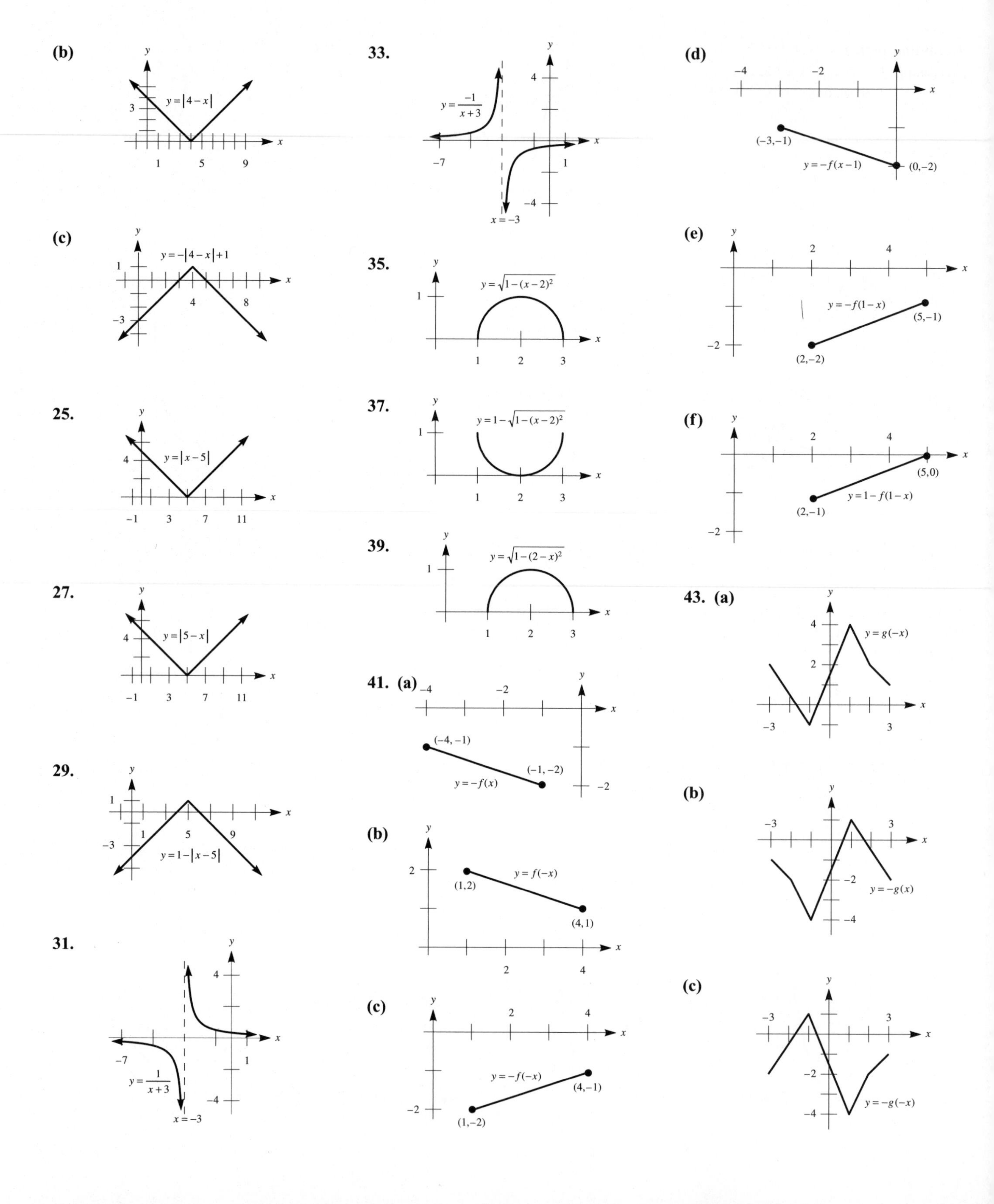
(b)
y = |4 − x|
(c)
y = −|4 − x| + 1
25.
y = |x − 5|
27.
y = |5 − x|
29.
y = 1 − |x − 5|
31.
y = 1/(x + 3)
x = −3
33.
y = −1/(x + 3)
x = −3
35.
y = √(1 − (x − 2)²)
37.
y = 1 − √(1 − (x − 2)²)
39.
y = √(1 − (2 − x)²)
41. (a)
(−4, −1)
(−1, −2)
y = −f(x)
(b)
y = f(−x)
(1,2)
(4,1)
(c)
y = −f(−x)
(4,−1)
(1,−2)
(d)
(−3,−1)
y = −f(x − 1)
(0,−2)
(e)
y = −f(1 − x)
(5,−1)
(2,−2)
(f)
(5,0)
y = 1 − f(1 − x)
(2,−1)
43. (a)
y = g(−x)
(b)
y = −g(x)
(c)
y = −g(−x)

45. **(a)**

x	x^2	$x^2 - 1$	$x^2 + 1$
0	0	−1	1
±1	1	0	2
±2	4	3	5
±3	9	8	10

(b) The graphs are vertical displacements of each other.

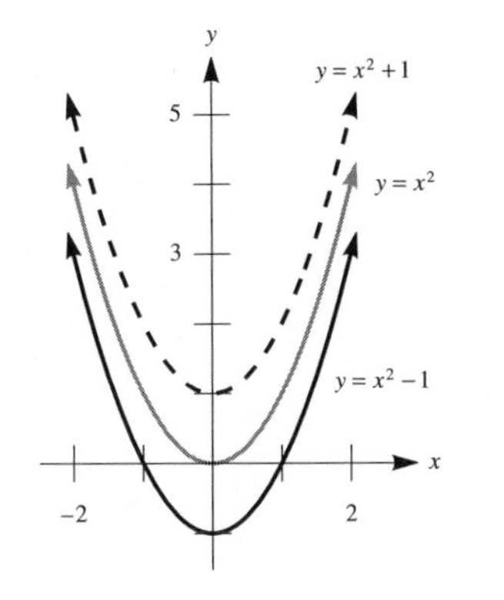

47. **(a)**

x	$\sqrt{x}$	$-\sqrt{x}$
0	0.0	0.0
1	1.0	−1.0
2	1.4	−1.4
3	1.7	−1.7
4	2.0	−2.0
5	2.2	−2.2

(b) The graphs are reflections across the x-axis.

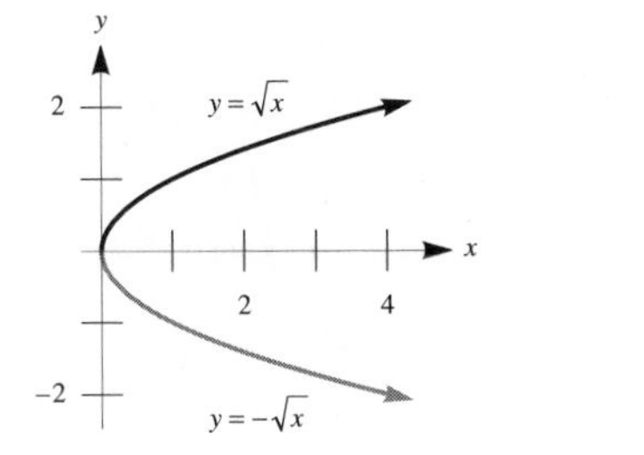

49. **(a)**

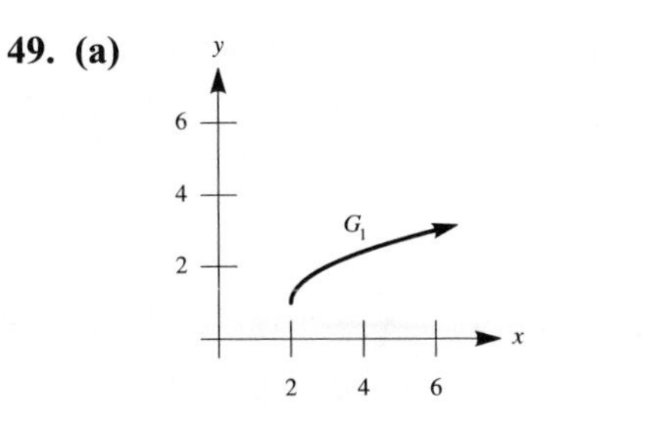

(b)

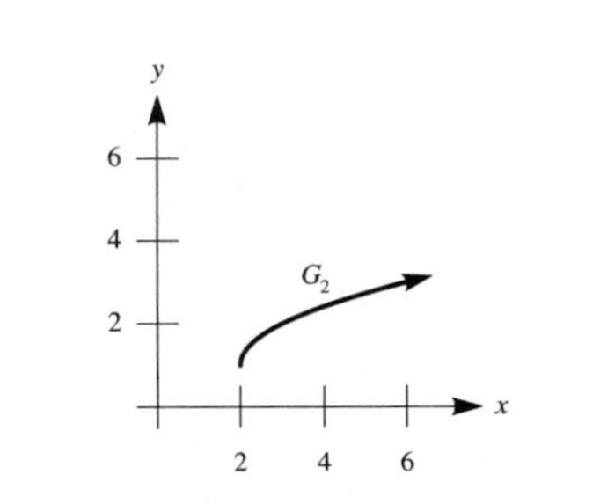

51. **(a)**

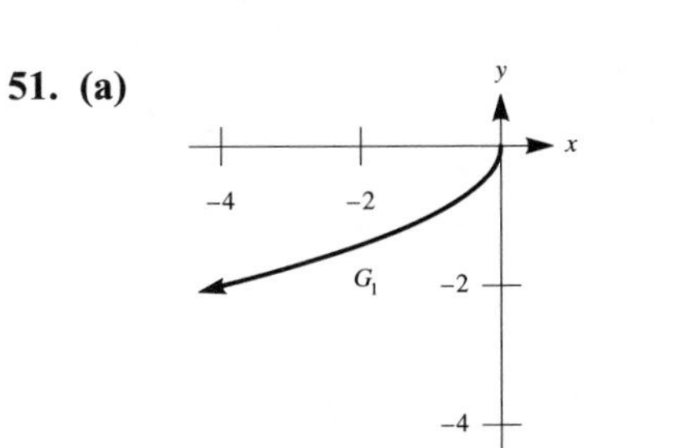

(b)

53. **(a)**

(b)

55. **(a)**

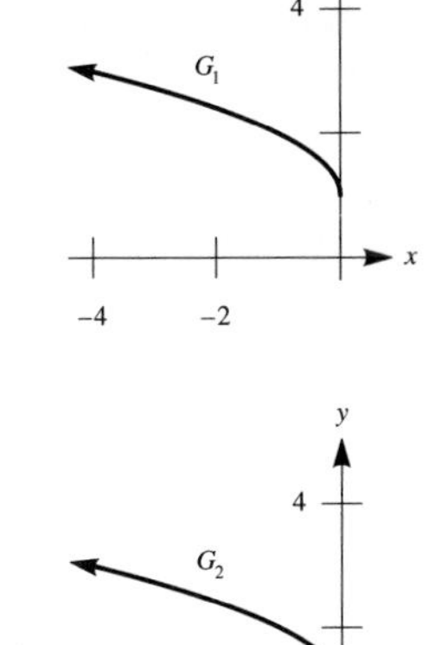

(b)

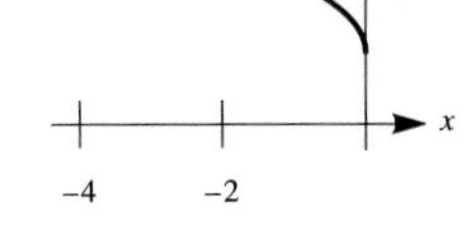

57. **(a)**

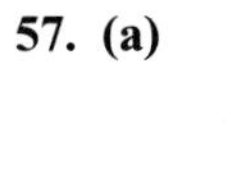

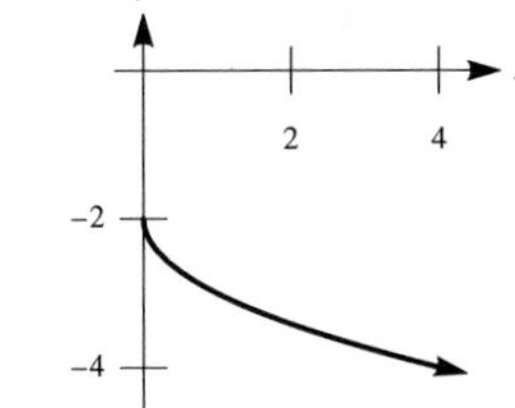

(c) $y = -\sqrt{x} - 2$

59. **(a)**

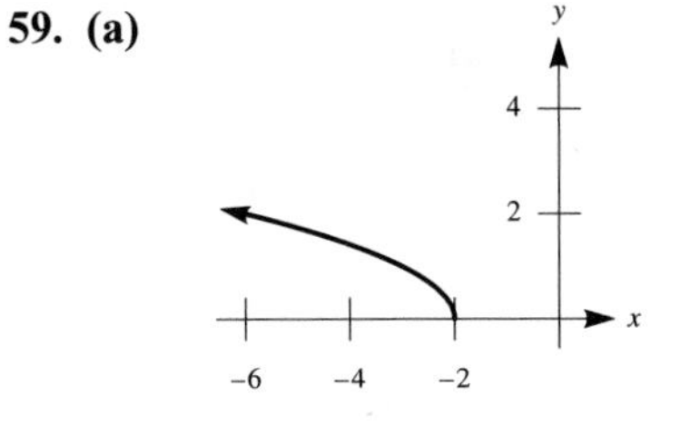

(c) $y = \sqrt{-x - 2}$

61. **(a)** $(-a, b + 2)$ **(b)** $(-a, -b + 2)$ **(c)** $(a + 3, -b)$ **(d)** $(a - 1, 1 - b)$ **(e)** $(-a + 1, b)$ **(f)** $(-a + 1, -b + 1)$

63.

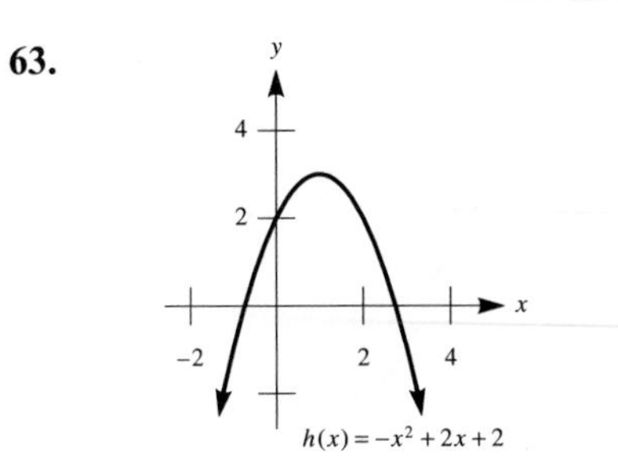

65. (a) Replacing x by $-x$ does not change the equation. **67. (a)** even **(b)** even **(c)** neither **(d)** neither **(e)** odd **69.** $(-\infty, -1)$

GRAPHING UTILITY EXERCISES FOR SECTION 3.3

1. (a)

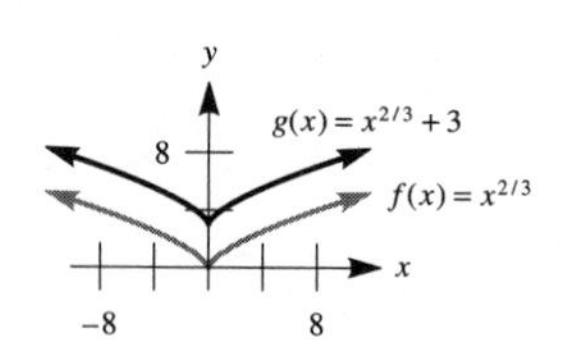

(b)

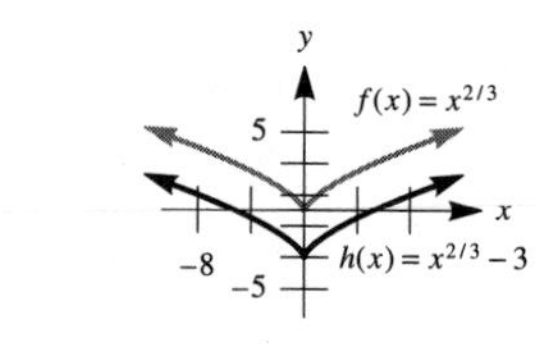

3. (a)

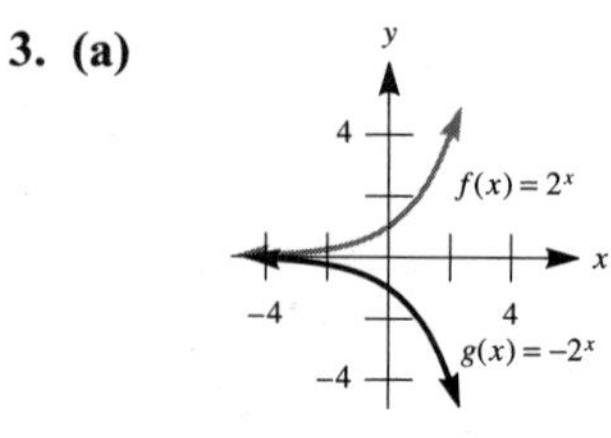

(b)

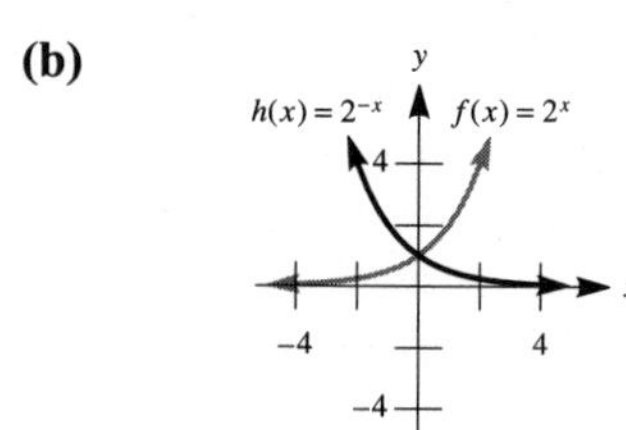

5. (a)

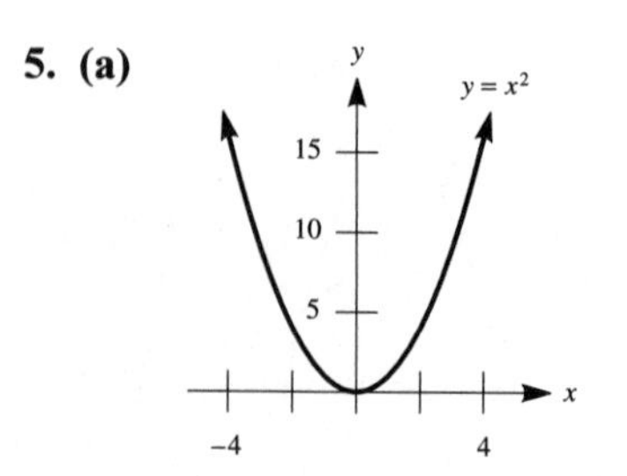

(b) Translate $y = x^2$ to the left 3 units.

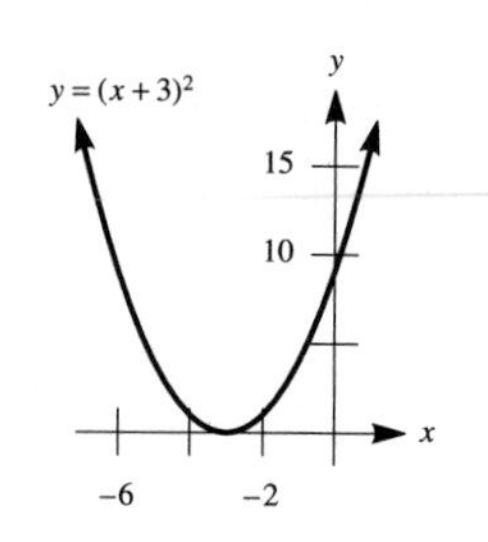

(c) Translate $y = x^2$ to the left 3 units, then reflect it across the x-axis.

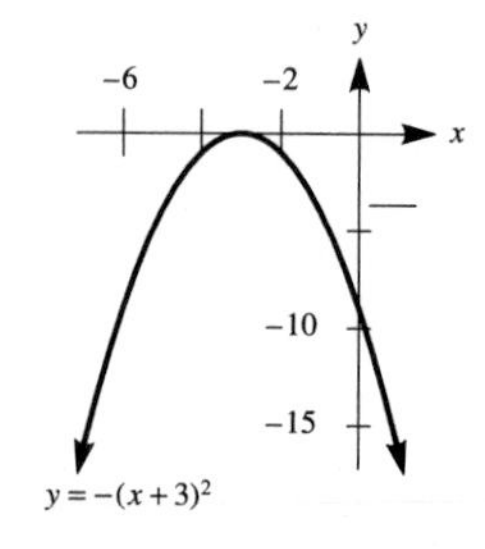

(d) Translate $y = x^2$ to the left 3 units, then reflect it across the y-axis.

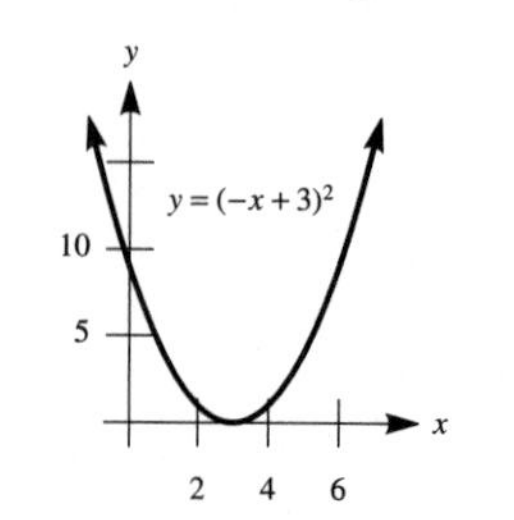

7. (a)

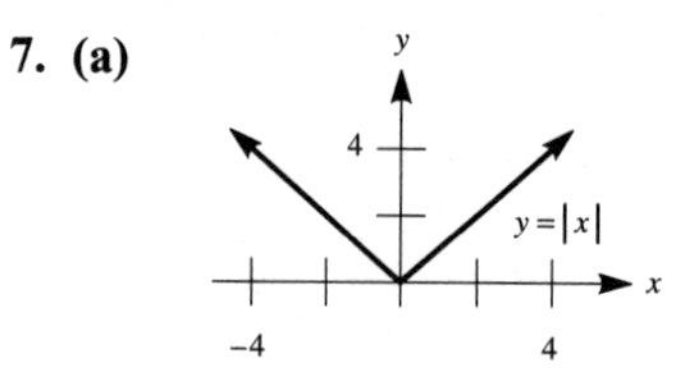

(b) Translate $y = |x|$ to the left 4 units.

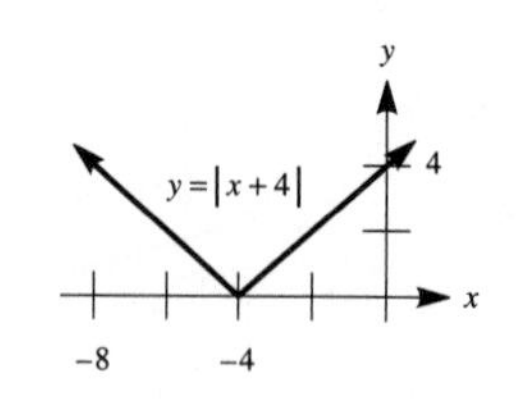

(c) Translate $y = |x|$ to the left 4 units, then reflect it across the y-axis.

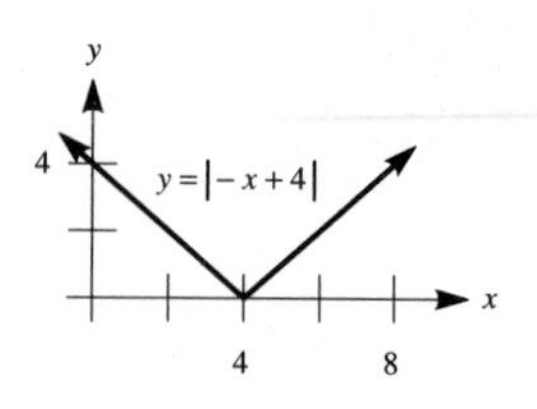

(d) Translate $y = |x|$ to the left 4 units, then reflect it across the y-axis, then translate it down 2 units.

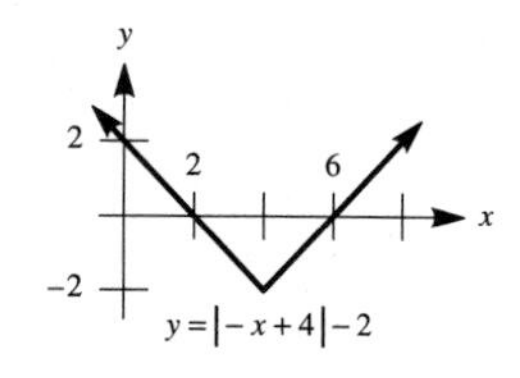

9. (a)

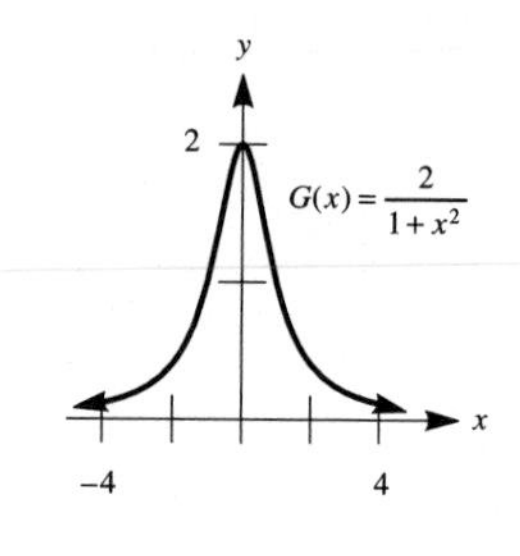

(b) Translate $G(x)$ to the left 3 units.

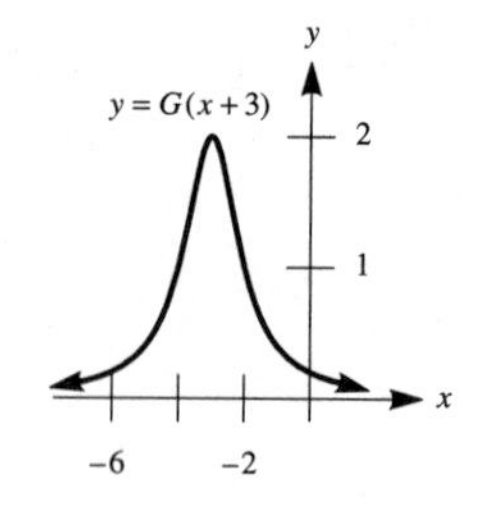

(c) Translate $G(x)$ to the right 3 units.

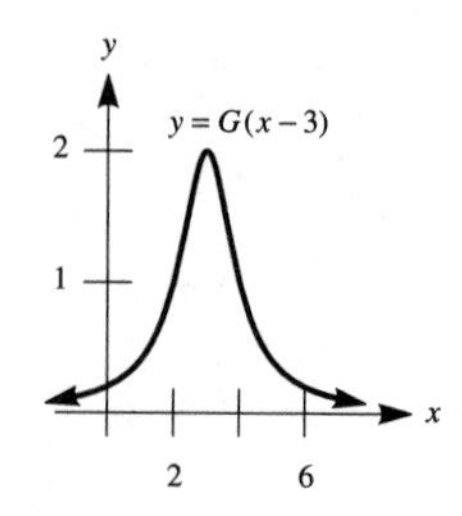

(d) Reflect $G(x)$ across the x-axis, then translate it up 2 units.

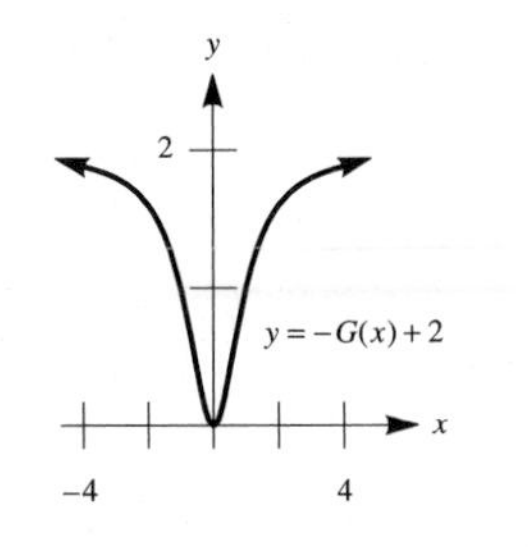

EXERCISE SET 3.4

1. (a) $x^2 - x - 7$ **(b)** $-x^2 + 5x + 5$ **(c)** 5 **3. (a)** $x^2 - 2x - 8$ **(b)** $-x^2 + 2x + 8$ **5. (a)** $4x - 2$ **(b)** $4x - 2$ **(c)** -4
7. (a) $\dfrac{-x^4 + 22x^2 - 4x - 80}{2x^3 - x^2 - 18x + 9}$
(b) $-80/9$
9. (a) $-x^5 + 9x^3 + 2x^2 - 18$ **(b)** $-x^5 + 9x^3 + 2x^2 - 18$ **(c)** -24
11. (a) $-6x - 14$ **(b)** -74 **(c)** $-6x - 7$ **(d)** -67
13. (a) $(f \circ g)(x) = -x$; $(f \circ g)(-2) = 2$; $(g \circ f)(x) = 2 - x$; $(g \circ f)(-2) = 4$
(b) $(f \circ g)(x) = 9x^2 - 3x - 6$; $(f \circ g)(-2) = 36$; $(g \circ f)(x) = -3x^2 + 9x + 14$; $(g \circ f)(-2) = -16$
(c) $(f \circ g)(x) = (1 - x^4)/3$; $(f \circ g)(-2) = -5$; $(g \circ f)(x) = 1 - x^4/81$; $(g \circ f)(-2) = 65/81$
(d) $(f \circ g)(x) = 2^{x^2+1}$; $(f \circ g)(-2) = 32$; $(g \circ f)(x) = 2^{2x} + 1$; $(g \circ f)(-2) = 17/16$
(e) $(f \circ g)(x) = 3x^5 - 4x^2$; $(f \circ g)(-2) = -112$; $(g \circ f)(x) = 3x^5 - 4x^2$; $(g \circ f)(-2) = -112$
(f) $(f \circ g)(x) = x$; $(f \circ g)(-2) = -2$; $(g \circ f)(x) = x$; $(g \circ f)(-2) = -2$
15. (a) $(-x + 7)/6x$ **(b)** $(-t + 7)/6t$ **(c)** $5/12$ **(d)** $(1 - 6x)/7$ **(e)** $(1 - 6y)/7$ **(f)** $-11/7$
17. (a) $M(7) = 13/5$; $M[M(7)] = 7$ **(b)** x **(c)** 7 **19. (a)** 1 **(b)** -3 **(c)** -1 **(d)** 2 **(e)** 2 **(f)** -3
21. (a) $(T \circ I)(x) = 4x^3 - 3x^2 + 6x - 1$; $(I \circ T)(x) = 4x^3 - 3x^2 + 6x - 1$
(b) $(G \circ I)(x) = ax^2 + bx + c$; $(I \circ G)(x) = ax^2 + bx + c$
(c) Given any function $f(x)$ and given $I(x) = x$, $(f \circ I)(x) = f(x)$ and $(I \circ f)(x) = f(x)$.
23.

x	0	1	2	3	4
$(f \circ g)(x)$	1	3	2	undef.	2

x	-1	0	1	2	3	4
$(g \circ f)(x)$	0	0	3	4	2	undef.

25. (a)

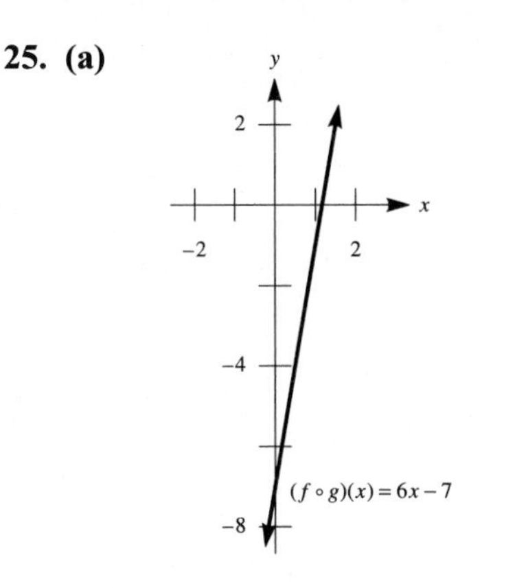

(b)

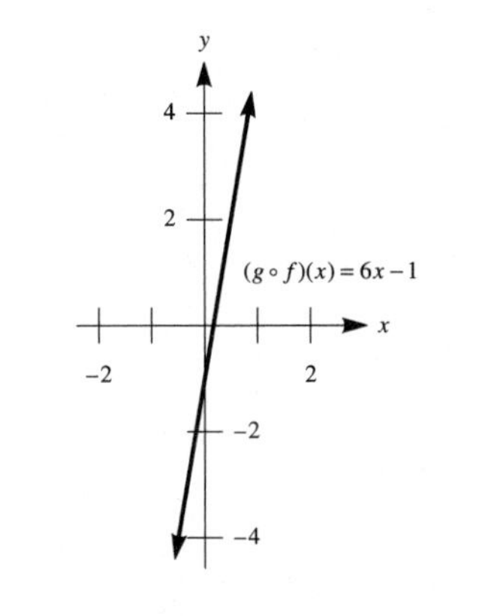

27. (a) domain: $[0, \infty)$; range: $[-3, \infty)$

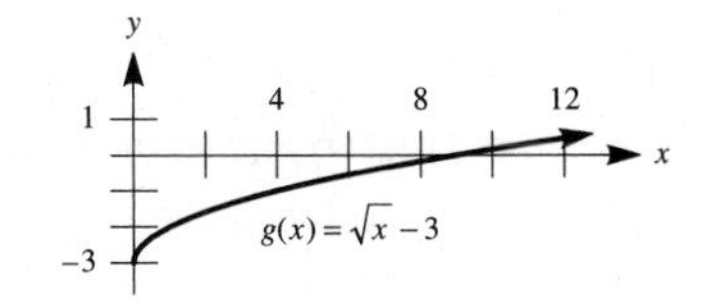

(b) domain: $(-\infty, \infty)$; range: $(-\infty, \infty)$

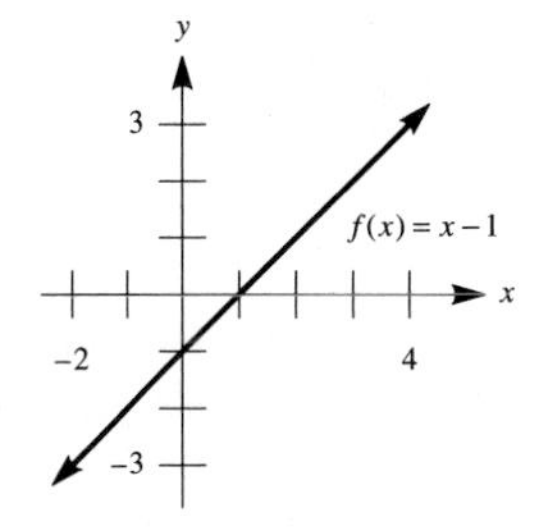

(c) domain: $[0, \infty)$; range: $[-4, \infty)$

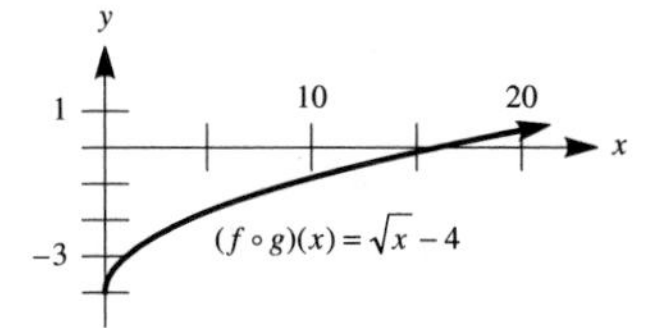

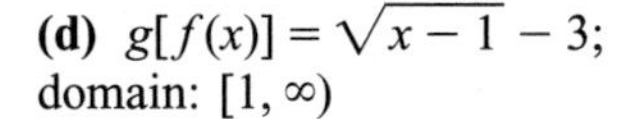
(d) $g[f(x)] = \sqrt{x - 1} - 3$; domain: $[1, \infty)$

(e)

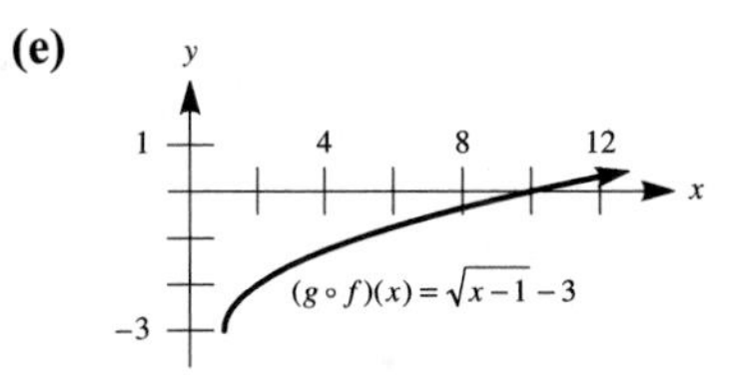

29. $f(x) = x^4$ and $g(x) = 3x - 1$; $C(x) = (f \circ g)(x)$ **31. (a)** $f(x) = \sqrt[3]{x}$ and $g(x) = 3x + 4$; $F(x) = (f \circ g)(x)$
(b) $f(x) = |x|$ and $g(x) = 2x - 3$; $G(x) = (f \circ g)(x)$
(c) $f(x) = x^5$ and $g(x) = ax + b$; $H(x) = (f \circ g)(x)$ **(d)** $f(x) = 1/x$ and $g(x) = \sqrt{x}$; $T(x) = (f \circ g)(x)$
33. (a) $(b \circ c)(x)$ **(b)** $(a \circ d)(x)$ **(c)** $(c \circ d)(x)$ **(d)** $(c \circ b)(x)$ **(e)** $(c \circ a)(x)$ **(f)** $(a \circ c)(x)$ **(g)** $(b \circ d)(x) = (d \circ b)(x)$
35. (a) 2, 4, 8, 16, 32, 64 **(b)** 0, 0, 0, 0, 0, 0 **(c)** $-2, -4, -8, -16, -32, -64$
37. (a) $-3, -5, -9, -17, -33, -65$ **(b)** $-1, -1, -1, -1, -1, -1$ **(c)** 3, 7, 15, 31, 63, 127
39. (a) 0.81, 0.656, 0.430, 0.185, 0.034, 0.001 **(b)** 1, 1, 1, 1, 1, 1 **(c)** 1.21, 1.464, 2.144, 4.595, 21.114, 445.792 **41.** 0.316, 0.562, 0.750, 0.866

43. $(C \circ f)(t) = 2\pi/(t^2 + 1)$; $\dfrac{\pi}{5}$ ft
45. **(a)** $(C \circ f)(t) = 100 + 450t - 25t^2$
(b) \$1225 **(c)** no
47. $f(x) = (x + 6)/4$
49. $a = -1/2$ and $b = 1/2$
51. **(a)** $2x + 2a - 2$ **(b)** $4x + 4a - 4$
53. **(a)** $x^2/2 + 1$ **(b)** $(x + 1)^2/2$
(c) $x^2/4 + 1$ **(d)** $(x + 1)^2/4$
(e) $(x^2 + 1)/2$ **55.** **(a)** $(g \circ f \circ h)(x)$
(b) $(h \circ g \circ f)(x)$ **(c)** $(f \circ g \circ h)(x)$
(d) $(h \circ f \circ g)(x)$ **57.** **(a)** $f(1) = 4$; $f(2) = 1$; $f(3) = 10$; $f(4) = 2$; $f(5) = 16$; $f(6) = 3$ **(b)** 4, 2, 1
(c) 10, 5, 16, 8, 4, 2, 1
(d) $x_0 = 2$: 2, 1; $x_0 = 4$: 4, 2, 1;
$x_0 = 5$: 5, 16, 8, 4, 2, 1;
$x_0 = 6$: 6, 3, 10, 5, 16, 8, 4, 2, 1;
$x_0 = 7$: 7, 22, 11, 34, 17, 52, 26, 13, 40, 20, 10, 5, 16, 8, 4, 2, 1

GRAPHING UTILITY EXERCISES FOR SECTION 3.4

1.

x	0.1	0.2	0.3
$f(g(x))$	194.672	148.176	109.744
$g(f(x))$	4.992	4.936	4.784

3.

x	7	8	9
$f(g(x))$	1614	2218	2918
$g(f(x))$	−571	−747	−947

5. **(a)**

t	0	0.5	1	1.5	2	2.5
A	707	737	804	903	1034	1199

t	3	3.5	4	4.5	5
A	1402	1648	1940	2284	2685

(b) 800 m^2 **(c)** 707 m^2; 3 hr
(d) 0 to 2.5: 196.8 m^2/hr;
2.5 to 5: 594.3 m^2/hr
The area is increasing faster over the interval from $t = 2.5$ to $t = 5$.
7. **(a)** 3, 2.259259259, 1.963308018, 1.914212754, 1.912932041, 1.912931183 (repeats)
The iterates converge to 1.912931183.
(b) $\sqrt[3]{7} \approx 1.912931183$; they are the same. **(c)** The fifth iterate agrees through the first three decimal places. The sixth iterate agrees through the first eight decimal places.

EXERCISE SET 3.5

3. 7 **5.** **(a)** 4 **(b)** −1 **(c)** $\sqrt{2}$
(d) $t + 1$ **7.** **(a)** $f^{-1}(x) = (x - 1)/2$
(b) $f^{-1}(5) = 2$; $1/f(5) = 1/11$
9. **(a)** $f^{-1}(x) = (x + 1)/3$

(c)

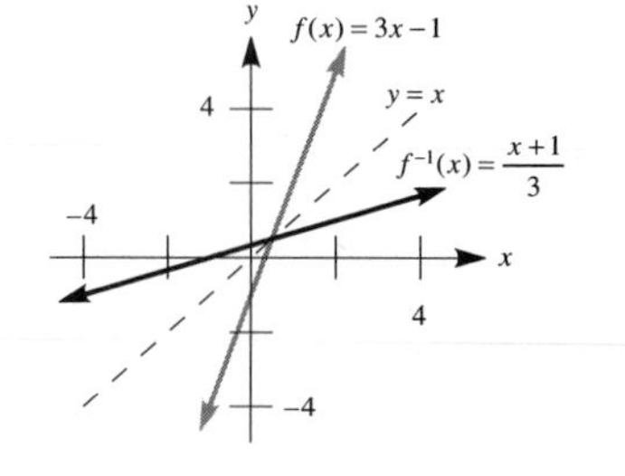

11. **(a)** $f^{-1}(x) = x^2 + 1$ for $x \geqslant 0$

(c)

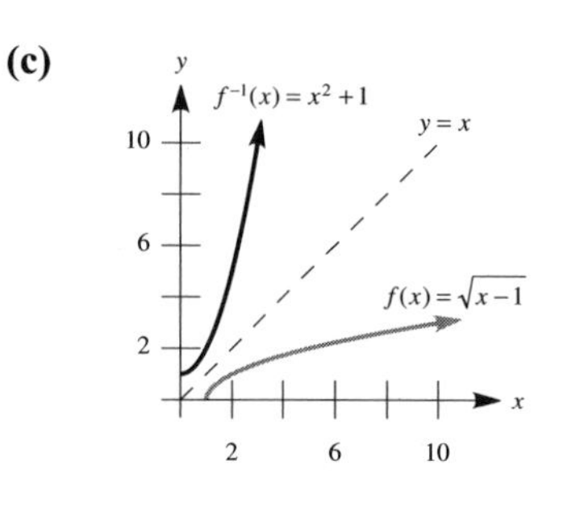

13. **(a)** domain: $(-\infty, 3) \cup (3, \infty)$; range: $(-\infty, 1) \cup (1, \infty)$
(b) $f^{-1}(x) = (3x + 2)/(x - 1)$
(c) domain: $(-\infty, 1) \cup (1, \infty)$; range: $(-\infty, 3) \cup (3, \infty)$; the domain of f is equal to the range of f^{-1}, and the range of f is equal to the domain of f^{-1}. **15.** $f^{-1}(x) = \sqrt[3]{(x - 1)/2}$

17. $f[f^{-1}(x)] = f^{-1}[f(x)] = x$

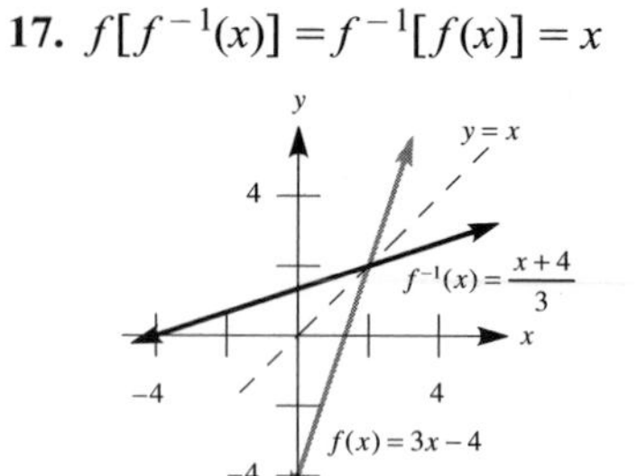

19. **(a)** $f^{-1}(x) = \sqrt[3]{x + 1} + 3$

(b)

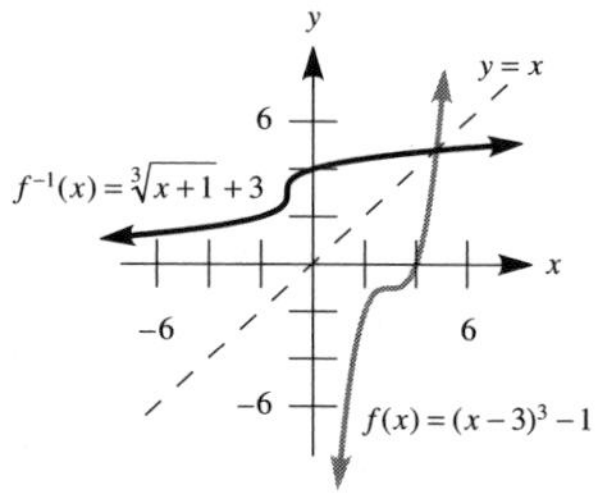

21. **(a)**

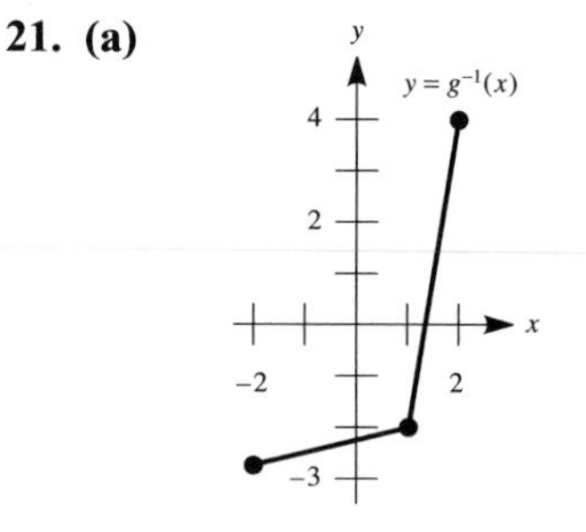

(b)

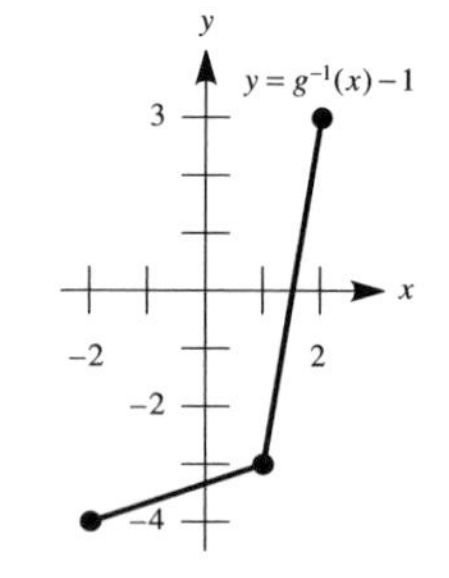

(c)

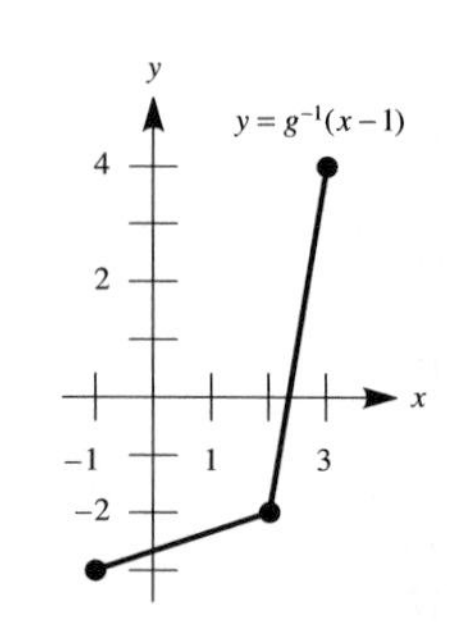

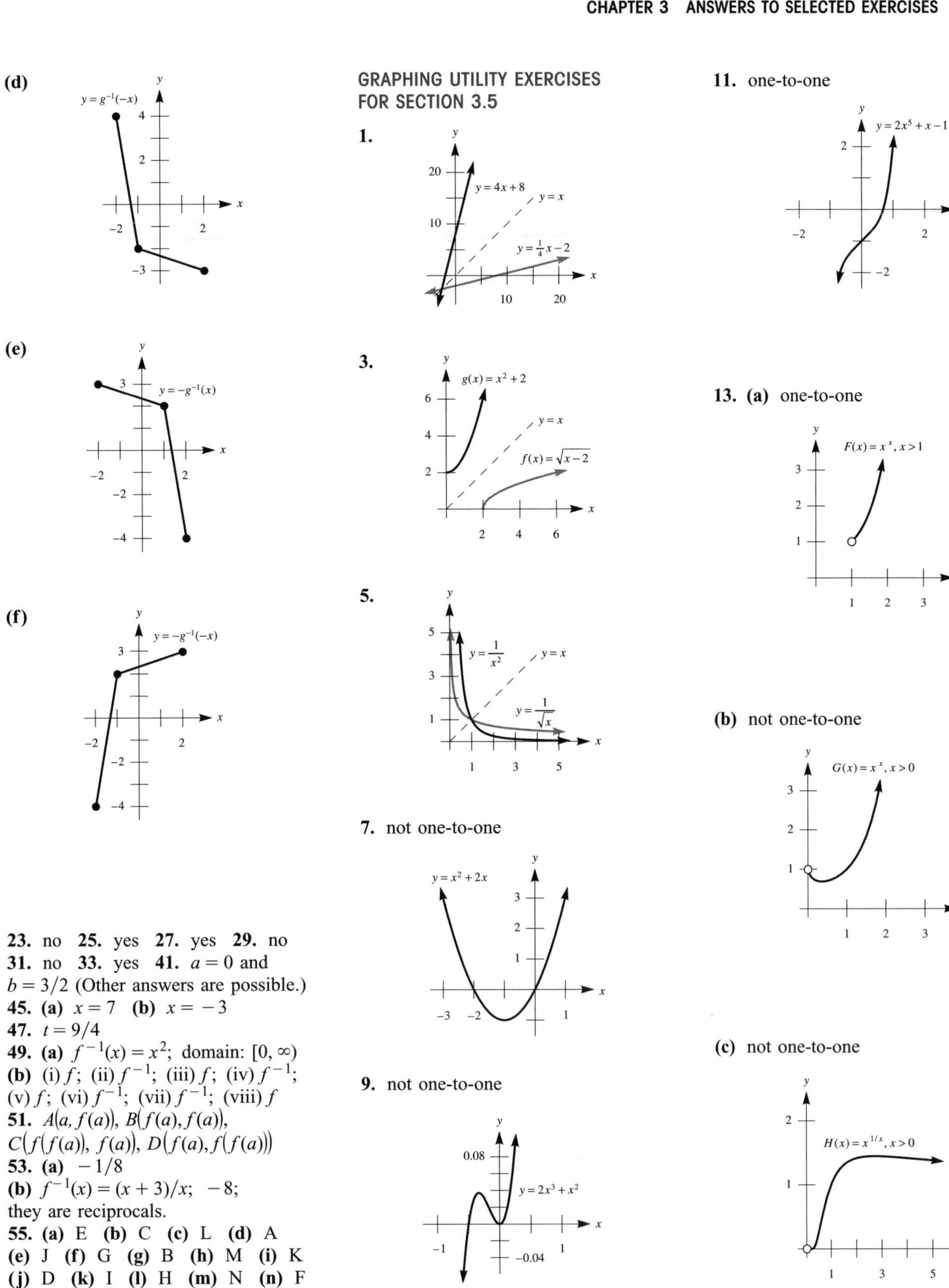

23. no **25.** yes **27.** yes **29.** no
31. no **33.** yes **41.** $a = 0$ and $b = 3/2$ (Other answers are possible.)
45. (a) $x = 7$ **(b)** $x = -3$
47. $t = 9/4$
49. (a) $f^{-1}(x) = x^2$; domain: $[0, \infty)$
(b) (i) f; (ii) f^{-1}; (iii) f; (iv) f^{-1}; (v) f; (vi) f^{-1}; (vii) f^{-1}; (viii) f
51. $A(a, f(a))$, $B(f(a), f(a))$, $C(f(f(a)), f(a))$, $D(f(a), f(f(a)))$
53. (a) $-1/8$
(b) $f^{-1}(x) = (x + 3)/x$; -8; they are reciprocals.
55. (a) E **(b)** C **(c)** L **(d)** A **(e)** J **(f)** G **(g)** B **(h)** M **(i)** K **(j)** D **(k)** I **(l)** H **(m)** N **(n)** F

15. $F^{-1}(x) = -\sqrt{1-x^2}$ with domain $0 \leqslant x \leqslant 1$

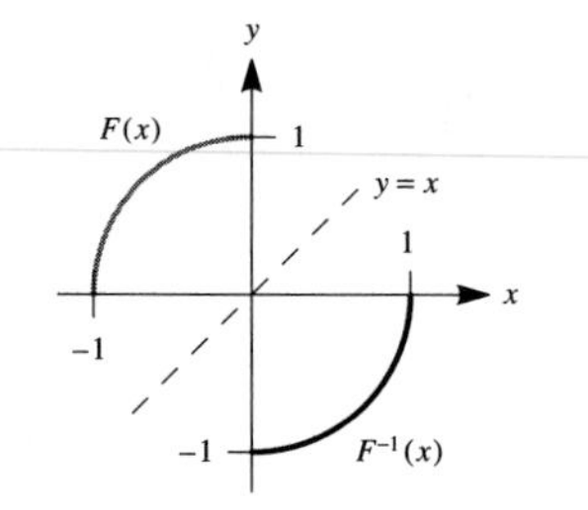

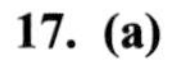
17. (a)

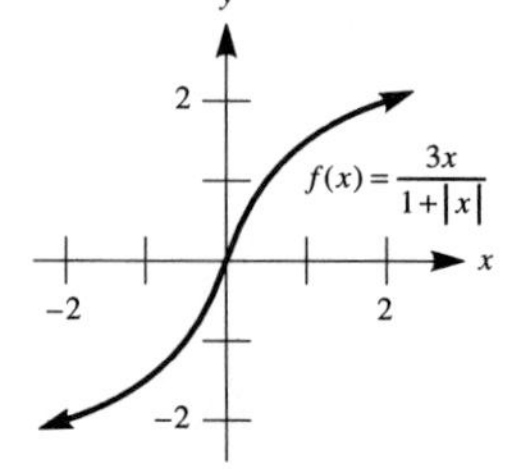

(b) $f^{-1}(x) = \begin{cases} x/(3-x) & \text{if } \quad 0 \leqslant x < 3 \\ x/(3+x) & \text{if } -3 < x < 0 \end{cases}$

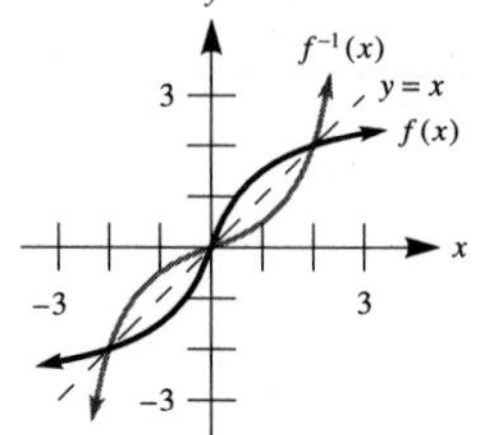

CHAPTER 3 REVIEW EXERCISES

1. (a) $(-\infty, 3]$
(b) $(-\infty, 1/2) \cup (1/2, \infty)$
3. yes **5. (a)** $-1/ax$
(b) $1 - 4x - 2h$
7. (a) $g^{-1}(x) = 1/(3x+5)$

(b)

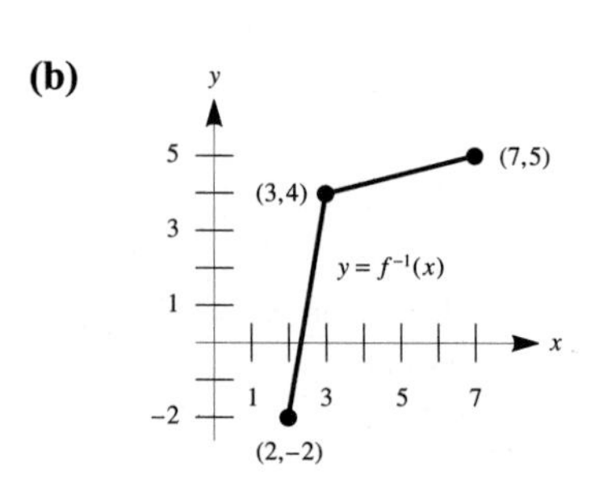

9. (a) x-intercepts: $-5, 1$; y-intercept: -1

(b) x-intercept: -1; y-intercept: $-1/2$

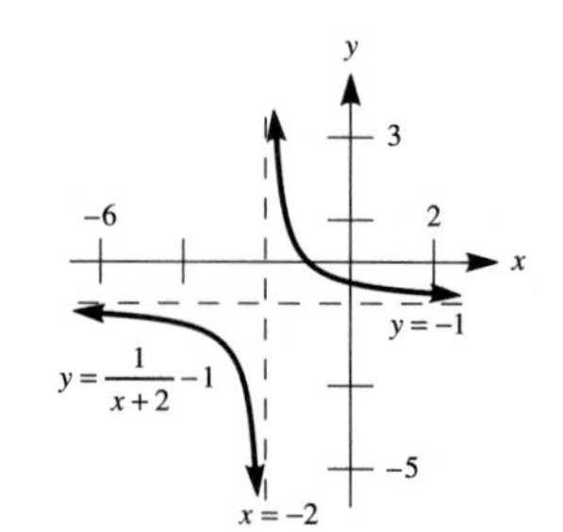

11. (a) 5 **(b)** $7 - 4\sqrt{2}$
13. $m(h) = 10 + h$

15.

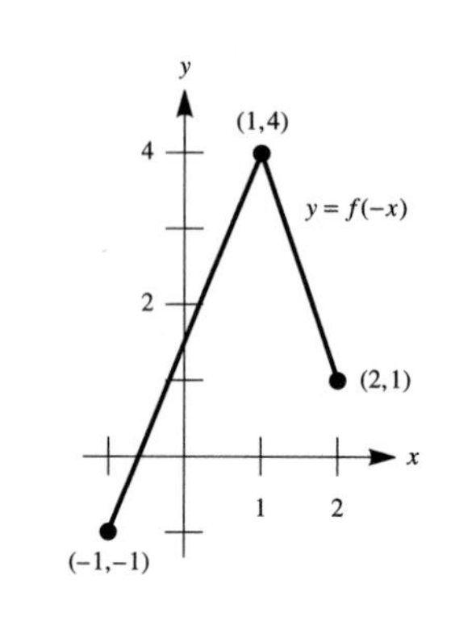

17. (a) 2.5, 3.25, 3.625
(b) The y-coordinate of point A is $f(1) = 2.5$; the y-coordinate of point B is $f(2.5) = 3.25$; and the y-coordinate of point C is $f(3.25) = 3.625$. **(c)** $y = 4$
(d) 3.813, 3.906, 3.953, 3.977, 3.988, 3.994, 3.997; these iterates are approaching 4.
19. (a) 0.4, 1.2, 1.4 **(b)** 0.45, 1.20, 1.48 **(c)** $y = (1+\sqrt{5})/2$
(d) 0.447, 1.203, 1.484, 1.576, 1.605, 1.614, 1.617, 1.618 (repeated); $(1+\sqrt{5})/2 \approx 1.618$

21. x-intercepts: $1 \pm \sqrt{2}$; y-intercept: 1

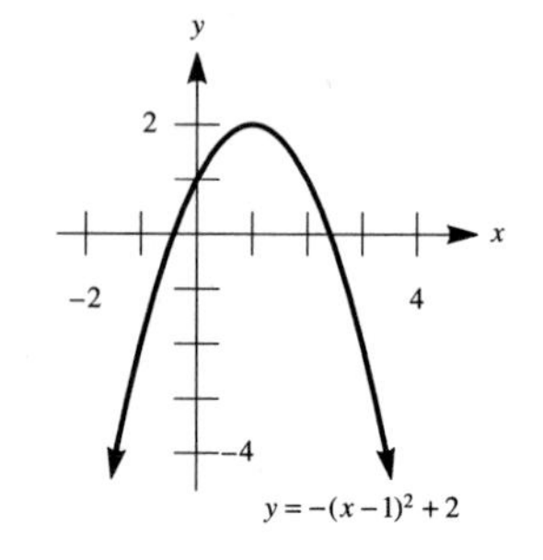

23. no x-intercepts; y-intercept: 1

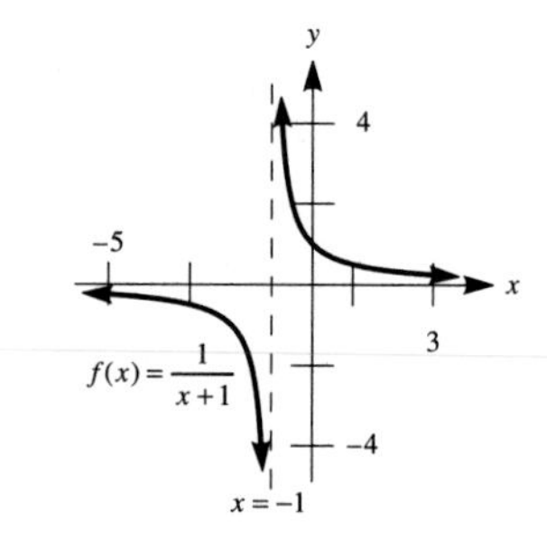

25. x-intercept: -3; y-intercept: 3

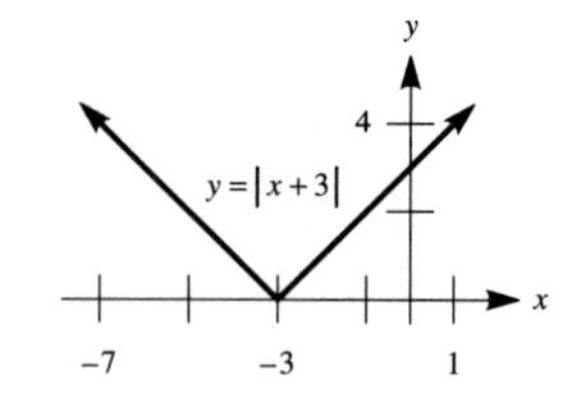

27. x-intercepts: ± 1; y-intercept: 1

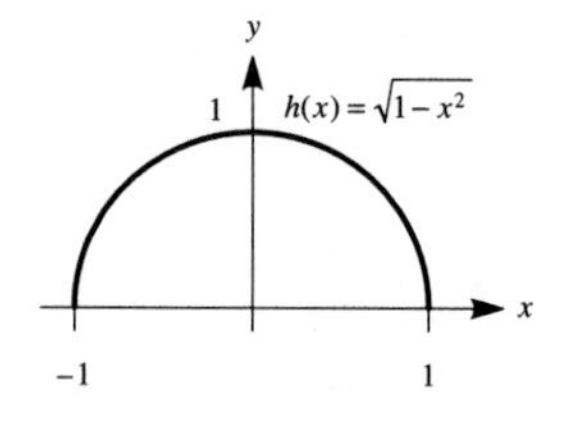

29. x-intercept: 0; y-intercept: 0

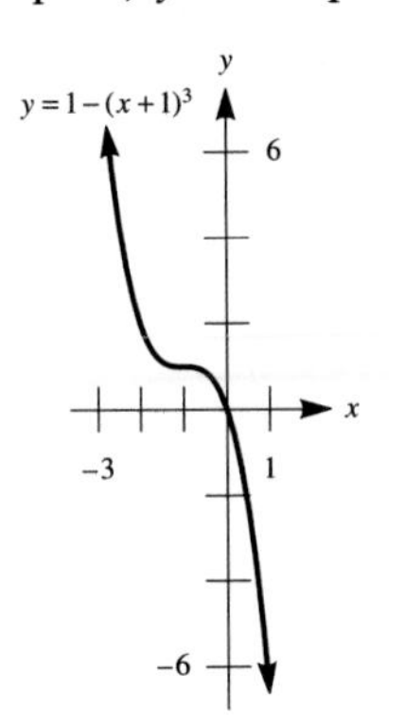

31. x-intercept: 0; y-intercept: 0

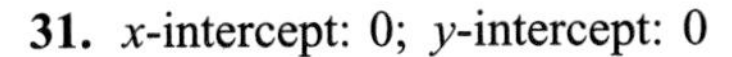

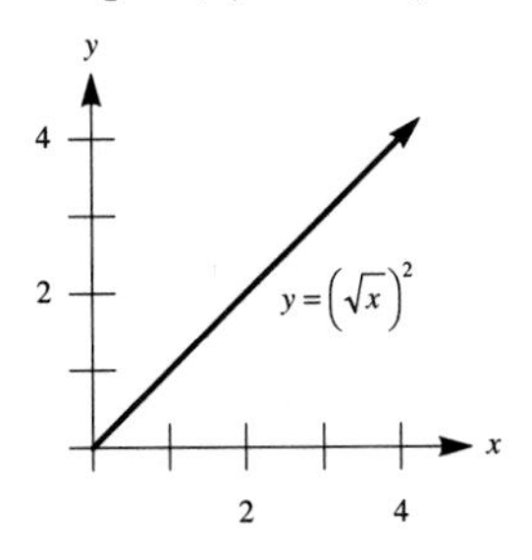

33. x-intercept: 1; no y-intercept

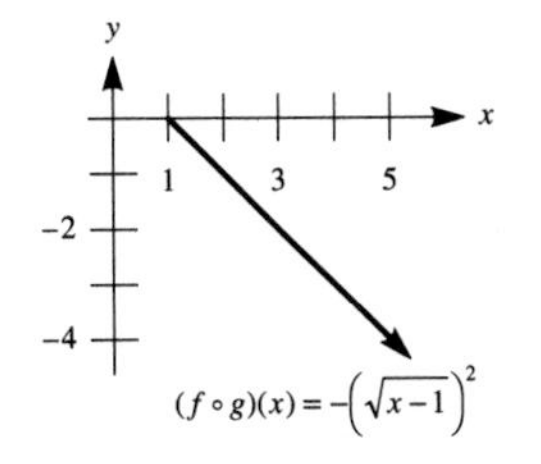

35. x-intercepts: -1, 0; y-intercept: 0

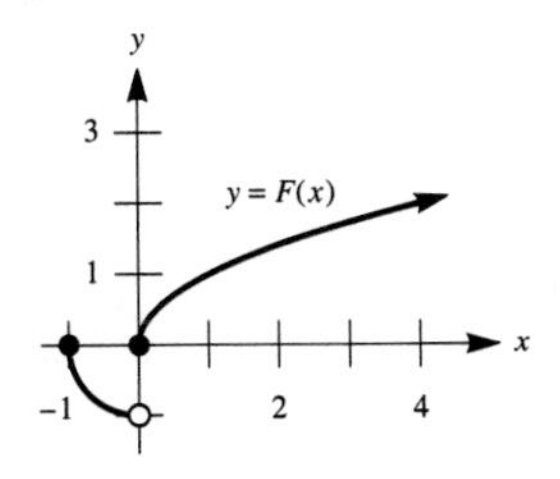

37. no x- or y-intercepts

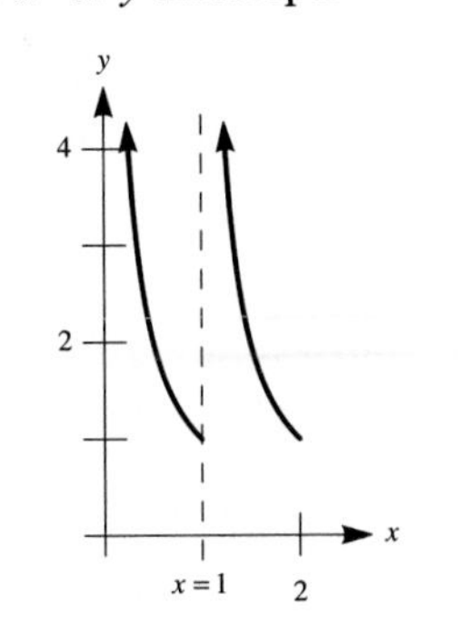

39. x-intercept: $1/2$; y-intercept: -1

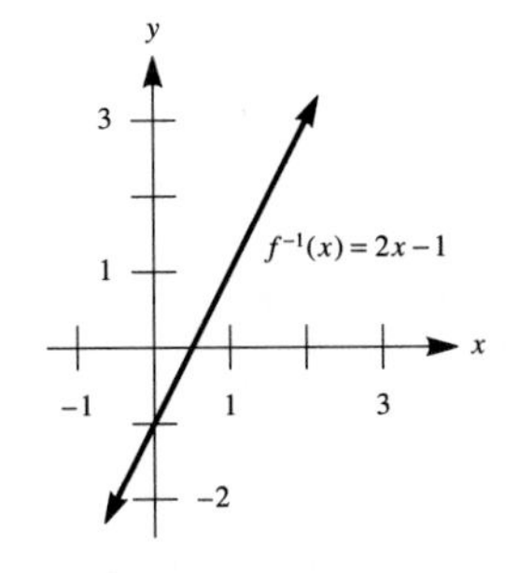

41. x-intercept: 0; y-intercept: 0

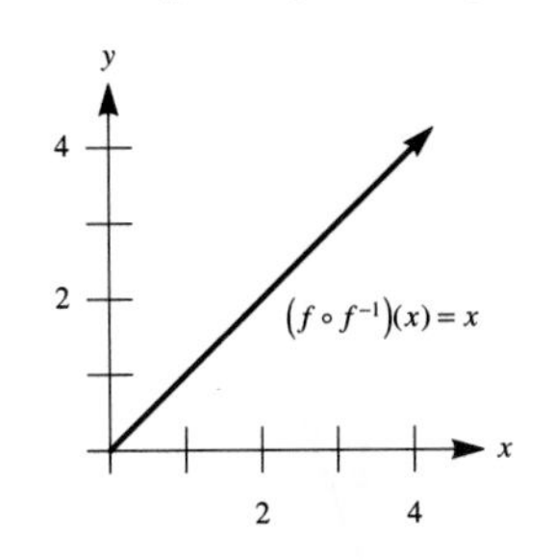

43. $(-\infty, -3) \cup (-3, 3) \cup (3, \infty)$
45. $(-\infty, 4]$ **47.** $(-\infty, \infty)$
49. $(-\infty, -1] \cup [3, \infty)$
51. $(-\infty, 0) \cup (0, \infty)$
53. $(-\infty, 1/3) \cup (1/3, \infty)$
55. $(-\infty, 0) \cup (0, \infty)$
57. $(-\infty, 2) \cup (2, \infty)$ **59.** $(f \circ g)(x)$
61. $(G \circ g)(x)$ **63.** $(g \circ f \circ G)(x)$
65. $(g \circ G \circ G)(x)$ **67.** 12 **69.** $-9/19$
71. $t^2 + t$ **73.** $x^2 - 5x + 6$ **75.** -4
77. 6 **79.** $x^4 - x^2$
81. $-2x^3 + 3x^2 - x$ **83.** $4x^2 - 2x$
85. $-2x^2 + 2x + 1$
87. $(2x + 2)/(2x - 5)$
89. $2x + h - 1$
91. $F^{-1}(x) = (4x + 3)/(1 - x)$

93. x for $x \neq -4$ **95.** x
97. $g^{-1}(-x) = (1 + x)/2$ **99.** $22/7$
101. negative **103.** -1 **105.** -1
107. $(0, -2)$ and $(5, 1)$ **109.** $[-6, 0]$ and $[5, 8]$ **111.** 0 (occurring at $x = 2$)
113. no **115.** $x = 4$ **117.** **(a)** $x = 10$ **(b)** $x = 0$ **119.** **(a)** 5 **(b)** -3 **(c)** 4 **(d)** $1/4$ **121.** $(f \circ f)(10)$ is larger **123.** $[0, 4]$ **125.** 5
127. $(1, 3) \cup (6, 10)$ **129.** $(4, 7)$

CHAPTER 3 TEST

1. $(-\infty, -1] \cup [6, \infty)$
2. $(-\infty, 2/3) \cup (2/3, \infty)$
3. **(a)** $2x^2 - 2x - 2$ **(b)** $2x^2 - 5x + 2$ **(c)** 54 **4.** $-2/at$ **5.** $4x + 2h - 5$
6. $g^{-1}(x) = -x/(6x + 4)$ **7.** $-8/5$
8. **(a)** x-intercepts: 4, 2; y-intercept: -2

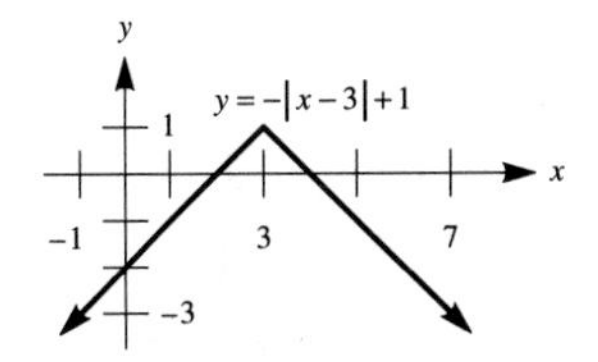

(b) x-intercept: $-5/2$; y-intercept: $-5/3$

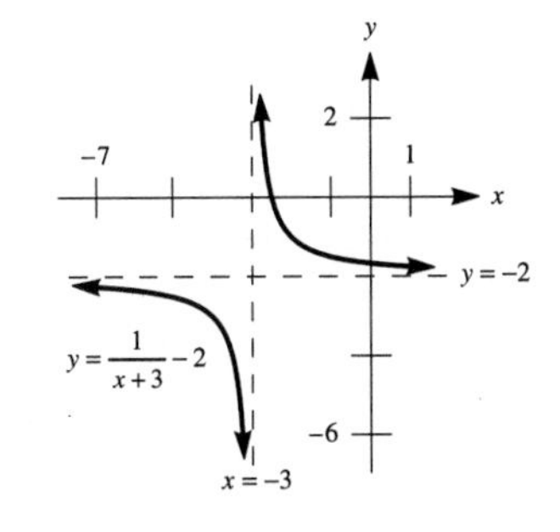

9. **(a)** $[-3, 1]$ **(b)** $(-2, 1)$ **(c)** -3 (which occurs at $x = 2$) **(d)** 1 (which occurs at $x = -2$) **(e)** $(-2, 2)$ **(f)** $-1/3$
10. **(a)** $23/4$ **(b)** $12 - 7\sqrt{3}$

11. domain: $(-\infty, -1) \cup (-1, \infty)$

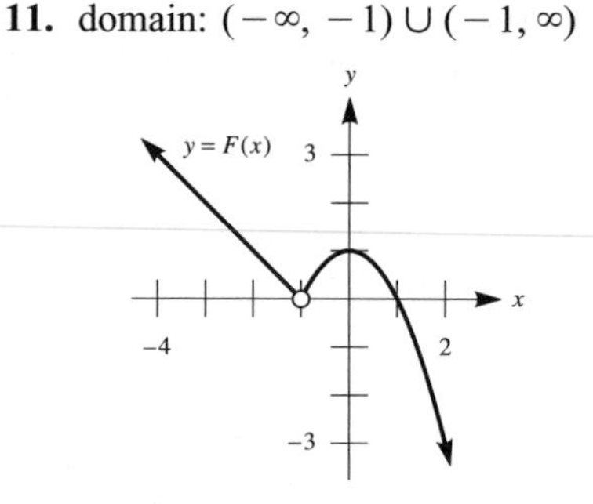

12. $P(-1, \sqrt{2}) \approx P(-1, 1.41)$;
$Q(\sqrt{2}, \sqrt{\sqrt{2}+3}) \approx Q(1.41, 2.41)$
14. (a) 0.4, 0.8, 0.4, 0.8, 0.4, 0.8
(b) 0.4, 0.8, 0.4, 0.8, 0.4, 0.8; yes

15.

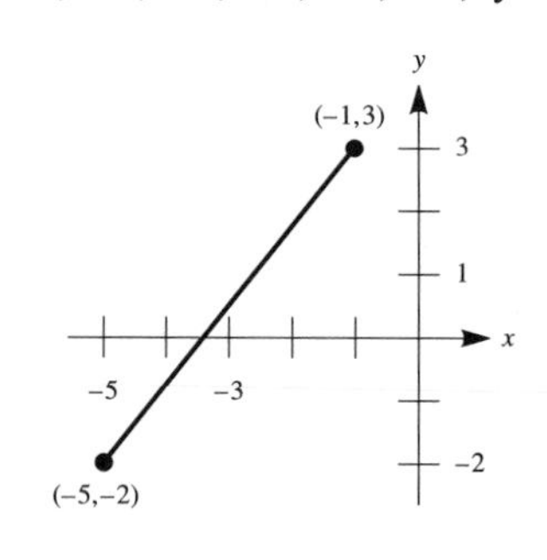

16. $t = 1$ **17.** $b = 10$
18. (a) $\dfrac{\Delta F}{\Delta t} = 0.16t_1 - 0.8$
(b) -0.16 °C/hr

CHAPTER FOUR

EXERCISE SET 4.1

1. $f(x) = \frac{2}{3}x + \frac{2}{3}$ **3.** $f(x) = \sqrt{2}x$
5. $f(x) = x - \frac{7}{2}$ **7.** $f(x) = \sqrt{3}$
9. $f(x) = \frac{1}{2}x - 2$ **11.** yes
13. $V(t) = -2375t + 20000$
15. (a) $V(t) = -12000t + 60000$
(b)

End of Year	Yearly Depreciation	Accumulated Depreciation	Value V
0	0	0	60,000
1	12,000	12,000	48,000
2	12,000	24,000	36,000
3	12,000	36,000	24,000
4	12,000	48,000	12,000
5	12,000	60,000	0

17. (a) \$530 **(b)** \$538 **(c)** \$8/fan
19. (a) 50 **(b)** \$50/player
(c) The answers are the same.
21. (a) $\frac{4}{5}$ ft/sec **(b)** 0 cm/sec
(c) 8 mph **23. (a)** B is traveling faster.
(b) A is farther to the right.
(c) $t = 8$ sec
25. (a) $y = 369669.5x - 708277846$
(b) 27,364,000 **(c)** too low **(d)** 8.1%
27. (a) $y = 295554.3x - 575450553$
(b) 12,703,000 **(c)** too low **(d)** 1.8%
29. (a) $y = 4.92x - 35.55$
(b) 102.21 cm **(c)** 4.2%
(d) 33.33 cm; too low; 8.3% error
(e) 377.73 cm; too high; 48.4% error

31. (a)

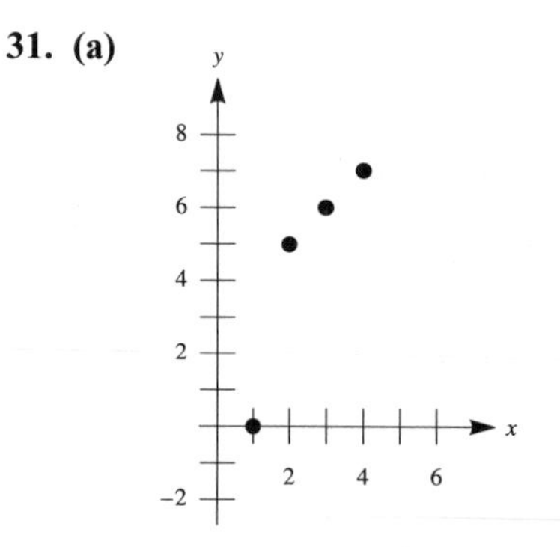

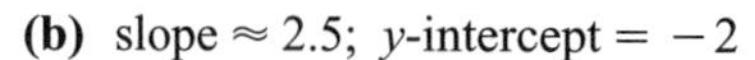
(b) slope ≈ 2.5; y-intercept $= -2$

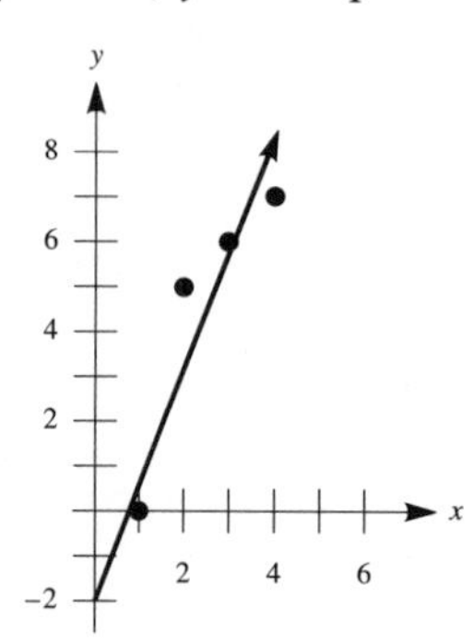

(c)

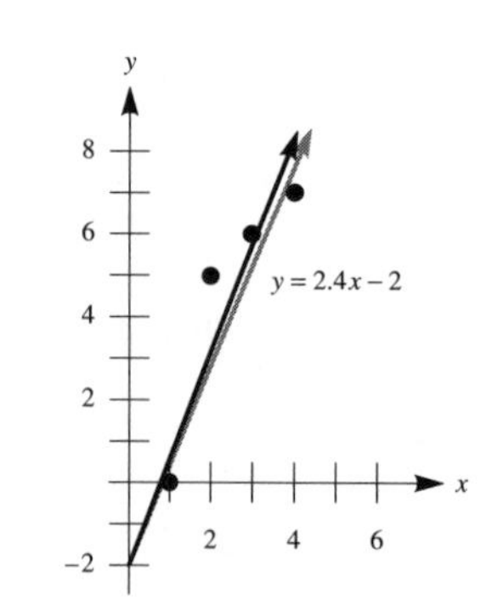

33. (a) yes
(b) $f^{-1}(x) = \dfrac{x + 71238863.429}{37546.068}$
(c) the year 2004
35. (a) 3.6174 billion dollars **(b)** no
37. (a) 3:58.0; too high
(b) 3:57.6; too high
39. (a) $x = 3$

(b)

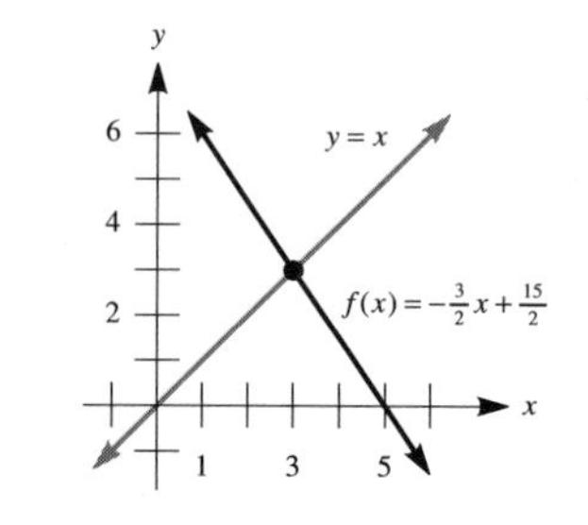

(c) 3, 3, 3, 3, 3
41. (a) $x = 3$ **(b)** 2, 2.5, 2.75, 2.875, 2.9375 **(c)** 2, 2.5, 2.75, 2.875, 2.9375, 2.9688, 2.9844, 2.9922, 2.9961, 2.9980, 2.9990, 2.9995 **(d)** The iterates are approaching the number 3, which is the fixed point of the function f.
43. (a) A **(b)** A^2 **(c)** AC **(d)** AC
45. (a) $1/AC$ **(b)** $1/AC$
49. $f(x) = 3x + 1$
51. (c) $m = 2.4$, $b = -2$
53. $y = 2.4x - 0.4$
55. $y = 0.084x + 37.241$
57. (a) $f(x) = \sqrt{2}x + (-1 + \sqrt{2})$ and $f(x) = -\sqrt{2}x + (-1 - \sqrt{2})$
(b) $f(x) = \sqrt[3]{2}x + (-1 + \sqrt[3]{2})$
59. yes; $f(x) = x$

GRAPHING UTILITY EXERCISES FOR SECTION 4.1

1. (a)

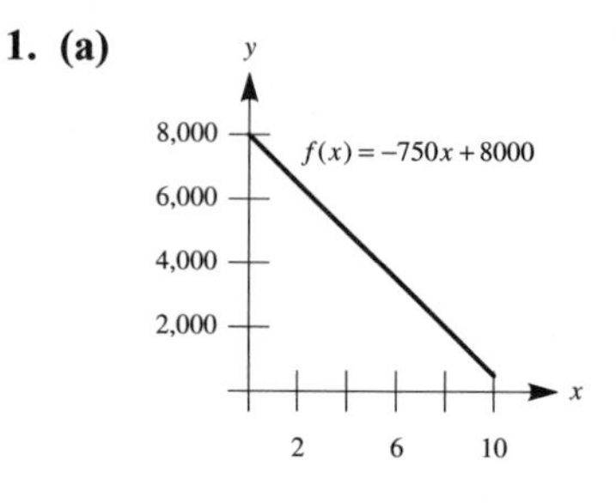

(b) 4250; the same

3. (a)

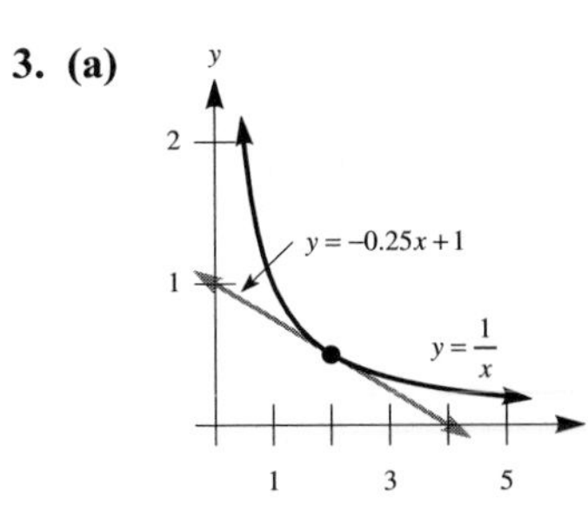

(b)

x	1.9	1.99	1.999
$1/x$	0.5263	0.502513	0.5002501
$-0.25x+1$	0.525	0.5025	0.50025

x	2.1	2.01	2.001
$1/x$	0.4762	0.497512	0.4997501
$-0.25x+1$	0.475	0.4975	0.49975

5. (a) $y = 1.0808x - 2080.1169$
(b) 73.918 trillion cubic feet
(c) yes; too low **(d)** The world reserves are equivalent to 68 years of natural gas production.

EXERCISE SET 4.2

1. vertex: $(-2, 0)$; axis of symmetry: $x = -2$; minimum value: 0; x-intercept: -2; y-intercept: 4

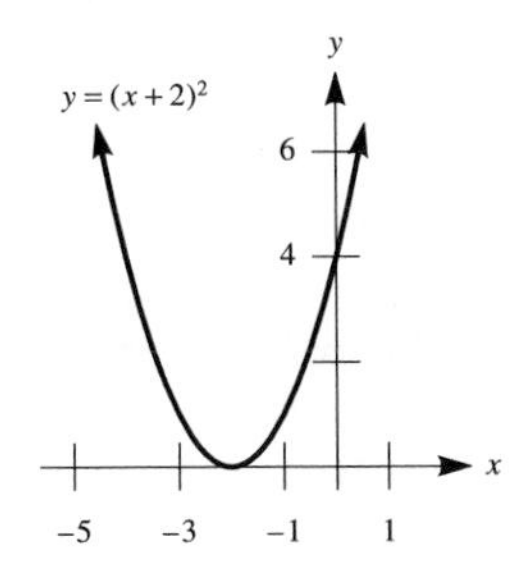

3. vertex: $(-2, 0)$; axis of symmetry: $x = -2$; minimum value: 0; x-intercept: -2; y-intercept: 8

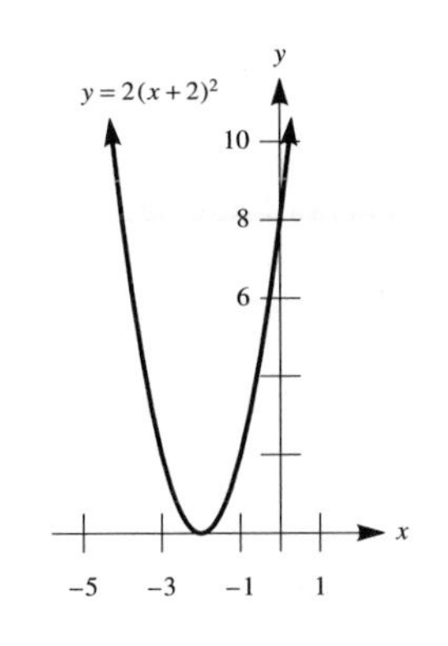

5. vertex: $(-2, 4)$; axis of symmetry: $x = -2$; maximum value: 4; x-intercepts: $-2 \pm \sqrt{2}$; y-intercept: -4

7. vertex: $(2, -4)$; axis of symmetry: $x = 2$; minimum value: -4; x-intercepts: 0 and 4; y-intercept: 0

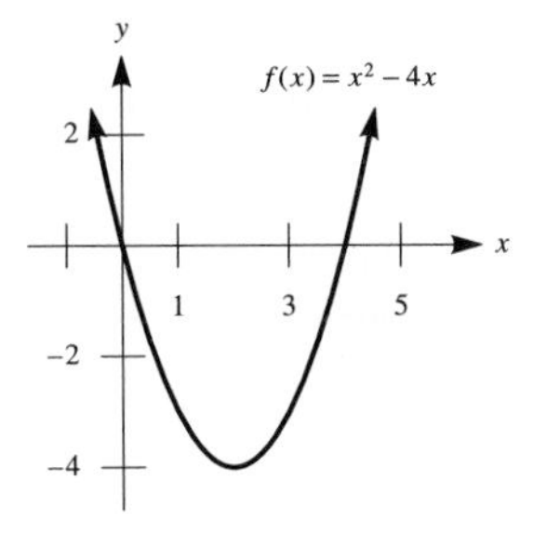

9. vertex: $(0, 1)$; axis of symmetry: $x = 0$; maximum value: 1; x-intercepts: ± 1; y-intercept: 1

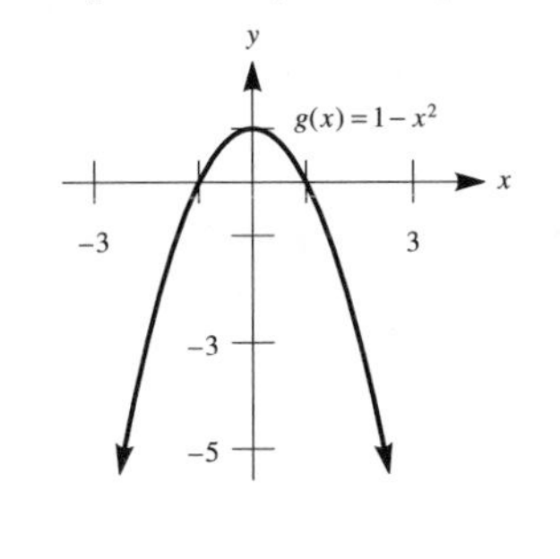

11. vertex: $(1, -4)$; axis of symmetry: $x = 1$; minimum value: -4; x-intercepts: 3 and -1; y-intercept: -3

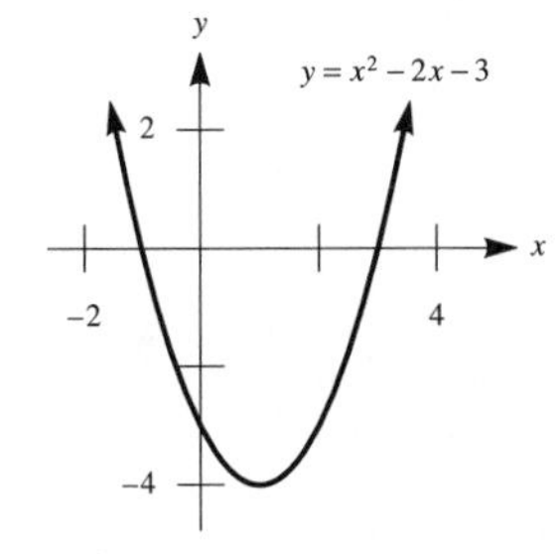

13. vertex: $(3, 11)$; axis of symmetry: $x = 3$; maximum value: 11; x-intercepts: $3 \pm \sqrt{11}$; y-intercept: 2

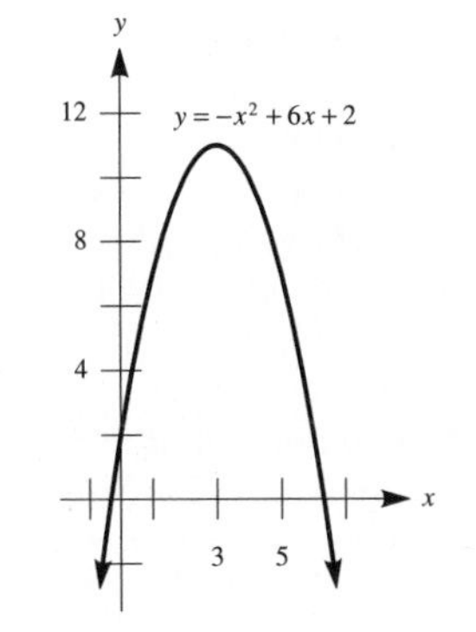

15. vertex: (0, 0); axis of symmetry: $t = 0$; minimum value: 0; t-intercept: 0; s-intercept: 0

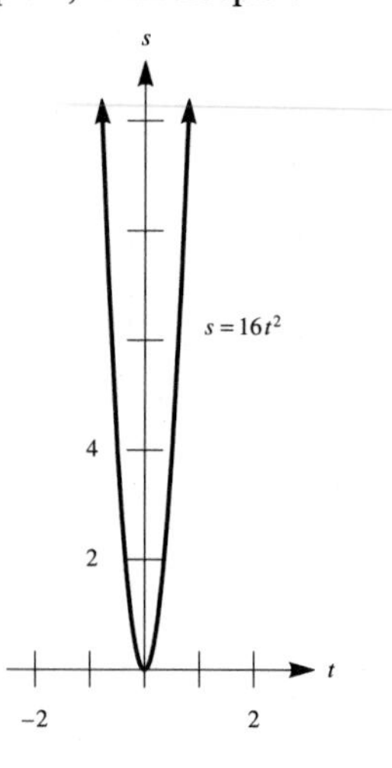

17. vertex: $\left(\frac{1}{6}, \frac{9}{4}\right)$; axis of symmetry: $t = 1/6$; maximum value: 9/4; t-intercepts: $-1/3$ and $2/3$; s-intercept: 2

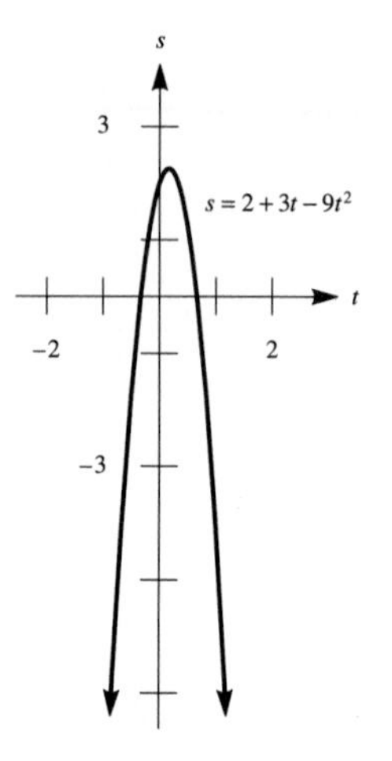

19. $x = 1$; minimum
21. $x = 3/2$; maximum
23. $x = 0$; minimum
25. minimum value: -13
27. maximum value: 25/8
29. maximum value: 1000
31. 5 units **33.** quadratic
35. neither **37.** linear
39. (a) minimum: 8 at $x = 3$
(b) minimum: 4 at $x = 3$
(c) minimum: 64 at $x = \pm\sqrt{3}$
41. (a) maximum: 4 at $x = 2$
(b) maximum: $2\sqrt[3]{2}$ at $x = 2$
(c) maximum: 16 at $x = \pm\sqrt{2}$
43. $ax_1 + ax_2 + b$ **49. (a)** The y-coordinate of point A is $f(x_0)$, the first iterate of x_0. The x-coordinate of point B is $f(x_0)$, so the y-coordinate of point B is $f(f(x_0))$, the second iterate of x_0. The x-coordinate of point C is $f(f(x_0))$, so the y-coordinate of point C is $f(f(f(x_0)))$, the third iterate of x_0. The x-coordinate of point D is $f(f(f(x_0)))$, so the y-coordinate of point D is $f(f(f(f(x_0))))$, the fourth iterate of x_0.
(b) -0.56, -0.2864, -0.518, -0.332 **51.** $y = -\frac{1}{2}(x-2)^2 + 2$
53. $y = \frac{1}{4}(x-3)^2 - 1$
55. $(x+3)^2 + (y+4)^2 = 25$ **59.** $b = 4$

GRAPHING UTILITY EXERCISES FOR SECTION 4.2

1. (a) As the coefficient of x^2 increases from 1 to 8, the graph narrows.

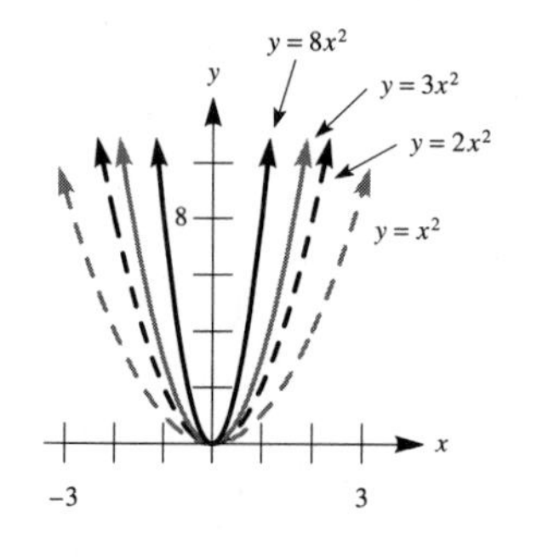

(b) The graph should be narrower than the others.

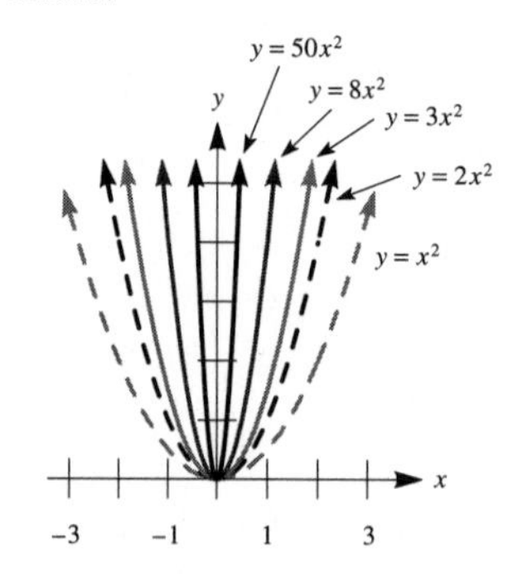

3. (a)

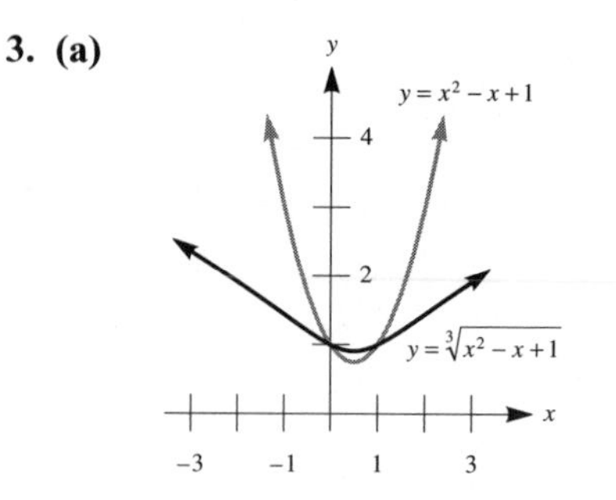

(b) 0.909
5. (a) $y = 330.8t + 218.8$
(b) 1542 thousand cases for 1990; 2534 thousand cases for 1993; too low
(c) $y = 62t^2 + 144.8t + 280.8$; 1852 thousand cases for 1990; 4332 thousand cases for 1993. The 1990 projection is closer to the actual value; more accurate; too low

7. (a) $y = 10.02x - 19598.75$; $y = -2.54x^2 + 10114.14x - 10068145.15$
(b)

	1991	1992
Estimate for world energy production (In quadrillion Btu) using linear model	351.07	361.09
Estimate for world energy production (In quadrillion Btu) using quadratic model	341.85	339.17

(c) quadratic model
(d) quadratic model

EXERCISE SET 4.3

1. $x = 1$ **3.** no fixed points
5. $x = -1$ and $x = 5$ **7.** $t = 1$
9. $t = -3$ and $t = 4$ **11.** $x = 0$ and $x = 4/9$ **13.** $u = -2$ and $u = 1$
15. $x = 10$ **17.** $x = -3$ and $x = -1$
19. (b) fourth iterate
(c) twelfth iterate

21. **(a)**

	x_1	x_2	x_3	x_4
From graph	1.7	0.8	1.4	1.0
From calculator	1.72	0.796	1.443	0.990

	x_5	x_6	x_7	x_8
From graph	1.3	1.1	1.2	1.1
From calculator	1.307	1.085	1.240	1.132

(b) $x = 20/17 \approx 1.176$
(c) eighth iterate

23. **(a)**

	x_1	x_2	x_3	x_4	x_5
From graph	0.36	0.92	0.29	0.82	0.58
From calculator	0.36	0.922	0.289	0.822	0.585

	x_6	x_7	x_8	x_9
From graph	0.97	0.11	0.40	0.96
From calculator	0.971	0.113	0.402	0.962

(b) $x_1 = (5 - \sqrt{5})/8$; $x_2 = (5 + \sqrt{5})/8$; $x_3 = (5 - \sqrt{5})/8$; $x_4 = (5 + \sqrt{5})/8$: the pattern is alternating iterate values.
(c) $x_1 \approx 0.344$; $x_2 \approx 0.903$; $x_3 \approx 0.352$; $x_4 \approx 0.912$; $x_5 \approx 0.320$; $x_6 \approx 0.871$; $x_7 \approx 0.450$; $x_8 \approx 0.990$; $x_9 \approx 0.039$; $x_{10} \approx 0.151$; more like that in part (a)

25. **(a)**

	x_1	x_2	x_3	x_4
From graph	0.6	0.8	0.4	0.8
From calculator	0.6	0.8	0.4	0.8

	x_5	x_6	x_7	x_8
From graph	0.4	0.8	0.4	0.8
From calculator	0.4	0.8	0.4	0.8

The pattern that emerges is alternating iterate values.
(b) $x_1 = 0.5$; $x_2 = 1.0$; $x_3 = x_4 = \cdots = x_8 = 0.0$: the third iterate and all remaining iterates are zero.
(c) $x_1 = -2$; $x_2 = -4$; $x_3 = -8$; $x_4 = -16$; $x_5 = -32$; $x_6 = -64$; $x_7 = -128$; $x_8 = -256$; $x_9 = -512$; $x_{10} = -1024$: the iterates are negative powers of 2.

27. **(a)** $x_{21} \approx 0.6632$; $x_{22} \approx 0.6477$; $x_{23} \approx 0.6617$; $x_{24} \approx 0.6492$; $x_{25} \approx 0.6605$
(b)

n	20	21	22
Number of fish after n breeding seasons	323	332	324

n	23	24	25
Number of fish after n breeding seasons	331	325	330

(c) $x_1 = 0.0675$; $x_2 \approx 0.04721$; $x_3 \approx 0.03373$; $x_4 \approx 0.02445$; $x_5 \approx 0.01789$: the iterates are approaching zero. Eventually the population will decrease to zero.

29. **(a)**

n	0	1	2	3
x_n	0.1	0.279	0.6236	0.7276
Number of fish after n breeding seasons	50	140	312	364

n	4	5	6	7
x_n	0.6143	0.7345	0.6046	0.7411
Number of fish after n breeding seasons	307	367	302	371

n	8	9	10
x_n	0.5948	0.7471	0.5857
Number of fish after n breeding seasons	297	374	293

n	21	22	23
x_n	0.7633	0.5601	0.7638
Number of fish after n breeding seasons	382	280	382

n	24	25	26
x_n	0.5592	0.7641	0.5587
Number of fish after n breeding seasons	280	382	279

(b) $x = 21/31 \approx 0.6774$
(c) $a \approx 0.7646$; $b \approx 0.5580$
(d) 382 fish and 279 fish

31. **(a)** $x_0 = c$: d, c, d, c, d, c; $x_0 = d$: c, d, c, d, c, d; the iterates are alternating values. **(b)** 0.8, 0.4, 0.8, 0.4, 0.8, 0.4 **(c)** The work in part (b) shows that $\{0.4, 0.8\}$ is a 2-cycle for the function T. **(d)** $x_1 = 0.8$; $x_2 = 0.4$

35. **(a)** $a = 6/7$ and $b = 3/7$
(b) $P(\frac{3}{7}, \frac{3}{7})$, $Q(\frac{3}{7}, \frac{6}{7})$, $R(\frac{6}{7}, \frac{6}{7})$, $S(\frac{6}{7}, \frac{3}{7})$

37. **(a)** $a = 1$ and $b = -2$
(b) $a = (-1 + \sqrt{4c - 3})/2$, $b = (-1 - \sqrt{4c - 3})/2$

GRAPHING UTILITY EXERCISES FOR SECTION 4.3

1. **(a)** one fixed point

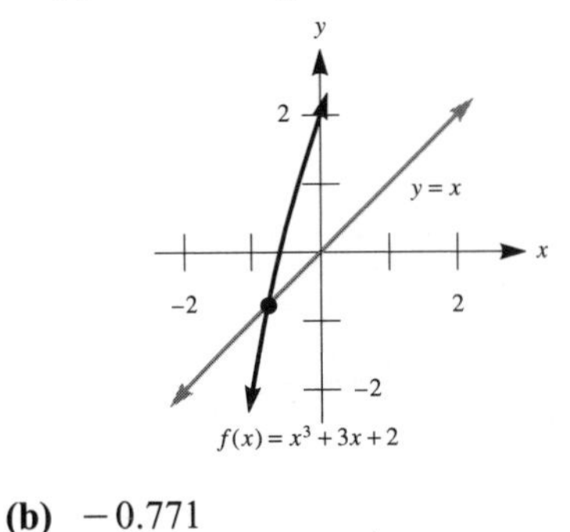

(b) -0.771

3. **(a)** three fixed points

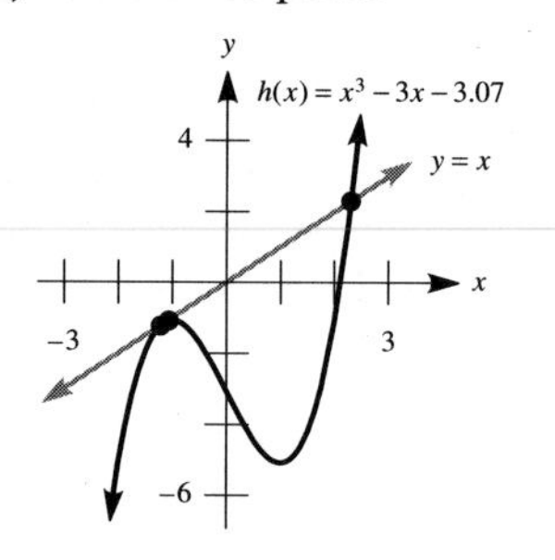

(b) -1.206, -1.103, 2.309

5. **(a)** two fixed points

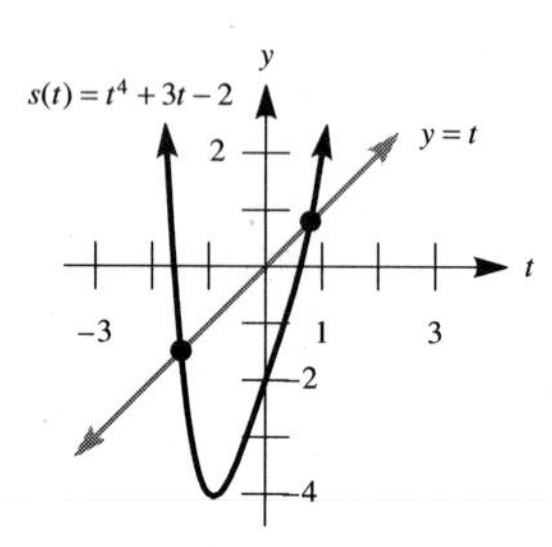

(b) -1.495, 0.798

7. **(a)** two fixed points

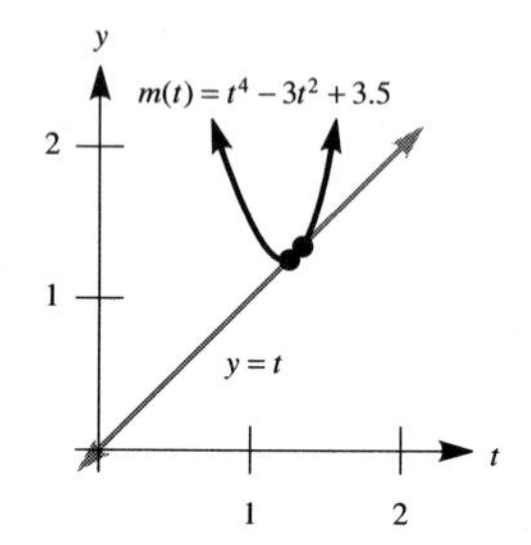

(b) 1.256, 1.344

9. **(a)** four fixed points

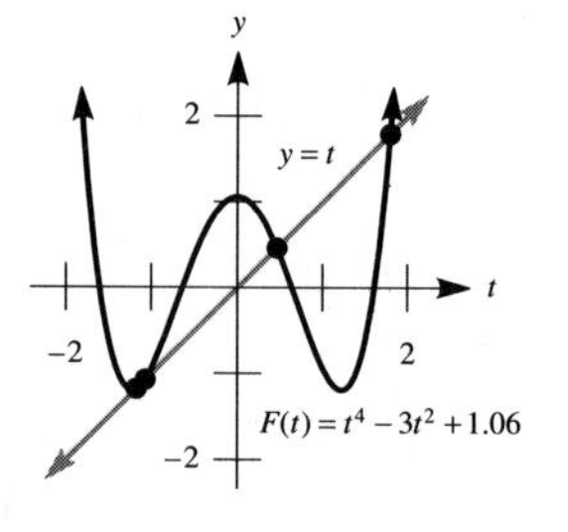

(b) -1.177, -1.083, 0.463, 1.797

11. **(a)** two fixed points

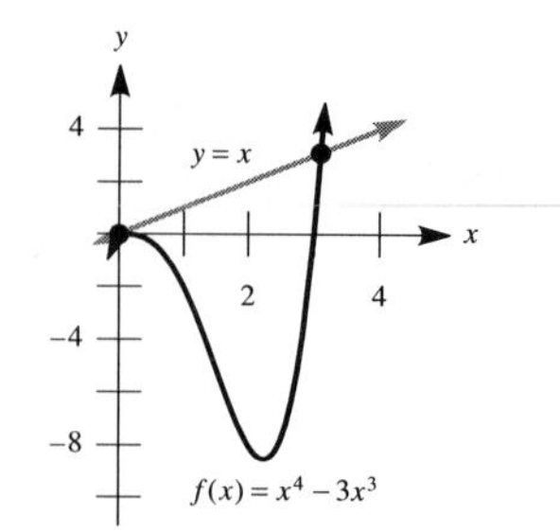

(b) 0, 3.104

13. **(a)**

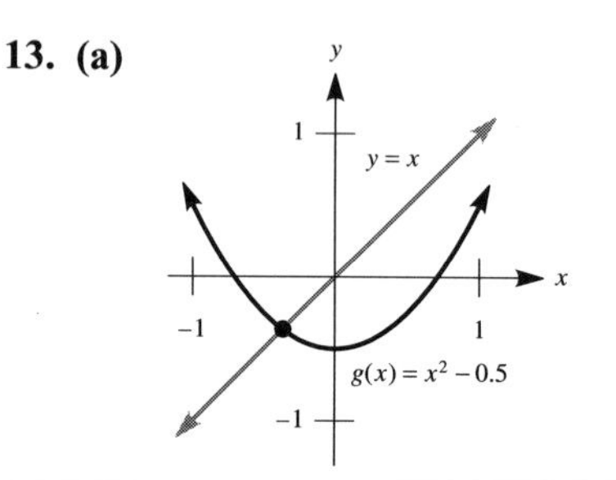

(b) $(1 - \sqrt{3})/2 \approx -0.36602540$
(c) **(i)** 12th iterate; yes
(ii) 25th iterate; no

EXERCISE SET 4.4

1. **(a)** $A(x) = 8x - x^2$
(b) $P(x) = 2x + 170/x$
3. **(a)** $D(x) = \sqrt{x^4 + 3x^2 + 1}$
(b) $m(x) = (x^2 + 1)/x$
5. **(a)** $A(y) = \pi y^2/4$
(b) $A(y) = \pi^2 y^2/16$
7. **(a)** $P(x) = 16x - x^2$
(b) $S(x) = 2x^2 - 32x + 256$
(c) $D(x) = x^3 - (16 - x)^3$ or $D(x) = (16 - x)^3 - x^3$
(d) $A(x) = 8$; the average does not depend on what the two numbers are.
9. $R(x) = -\frac{1}{4}x^2 + 8x$
11. **(a)**

x	1	2	3	4
$P(x)$	17.88	19.49	20.83	21.86

x	5	6	7
$P(x)$	22.49	22.58	21.75

(b) 22.58; $x = 6$ **(c)** 22.63
13. **(a)** $h(s) = \sqrt{3}s$
(b) $A(s) = \sqrt{3}s^2$ **(c)** $4\sqrt{3}$ cm
(d) $(25\sqrt{3}/4)$ in.^{2}
15. $V(r) = 2\pi r^3$ **17.** **(a)** $h(r) = 12/r^2$
(b) $S(r) = 2\pi r^2 + 24\pi/r$
19. $V(S) = S\sqrt{S\pi}/6\pi$
21. $A(x) = \frac{1}{2}x\sqrt{400 - x^2}$
23. $d(x) = (x + 4)\sqrt{x^2 + 25}/x$
25. $x = 25$; $L = 25$

x	5	10	20	24	24.8
$A(x)$	225	400	600	624	624.96

x	24.9	25	25.1	25.2	45
$A(x)$	624.99	625	624.99	624.96	225

27. **(a)**
$x = 2$

x	1	2	3	4
A	7.5	12	10.5	0

$x = 2.25$

x	1.75	2.00	2.25
A	11.3203	12.0000	12.3047

x	2.50	2.75
A	12.1875	11.6016

$x = 2.30$

x	2.15	2.20	2.25
A	12.2308	12.2760	12.3047

x	2.30	2.35
A	12.3165	12.3111

(b) $x = 2.30$; $A = 12.3168$
29. $A(x) = \frac{17}{144}x^2 - \frac{1}{3}x + \frac{1}{2}$
31. **(a)** $V(r) = \sqrt{3}\pi r^3/3$
(b) $S(r) = 2\pi r^2$
33. **(a)** $r(h) = 3h/\sqrt{h^2 - 9}$
(b) $h(r) = 3r/\sqrt{r^2 - 9}$
35. $A(x) = (4x^2 + \pi(14 - x)^2)/16\pi$
37. $A(r) = r(1 - 4\pi r)/4$
39. $A(x) = \pi x^2/3$

41. **(a)** $V(x) = 4x^3 - 28x^2 + 48x$
(b)

x (in.)	0	0.5	1.0	1.5	2.0	2.5	3.0
volume (in.³)	0	17.5	24	22.5	16	7.5	0

(c) $x = 1.0$
(d)

x (in.)	0.8	0.9	1.0	1.1	1.2	1.3	1.4
volume (in.³)	22.5	23.4	24	24.2	24.2	23.9	23.3

(e) $x = 1.1$
43. **(a)** $A(r) = 32r - 2r^2 - \pi r^2/2$
(b) downward; yes
45. **(a)** $y(s) = 3s/\sqrt{1 - s^2}$
(b) $s(y) = y/\sqrt{y^2 + 9}$
(c) $z(s) = 3/\sqrt{1 - s^2}$
(d) $s(z) = \sqrt{z^2 - 9}/z$
47. **(a)** $m(a) = (a^2 + 1)/a$
49. $A(x) = (8x - x^2)/4$
51. $A(m) = (2m^2 - 8m + 8)/(m^2 - 4m)$
53. $A = (ma - b)^2/(-2m)$

GRAPHING UTILITY EXERCISES FOR SECTION 4.4

1.

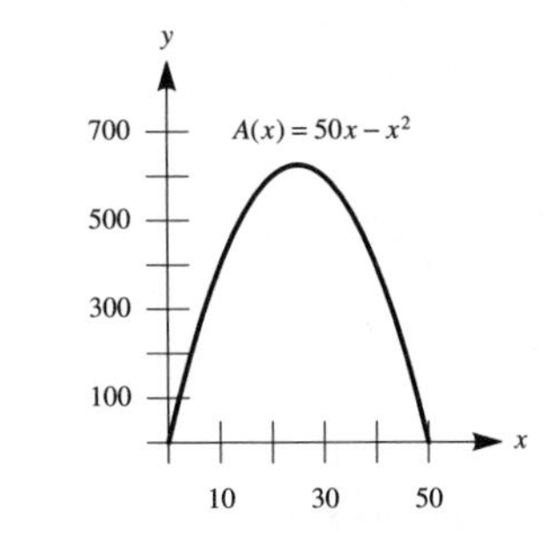

(a) yes **(b)** one **(c)** yes **(d)** no

3.

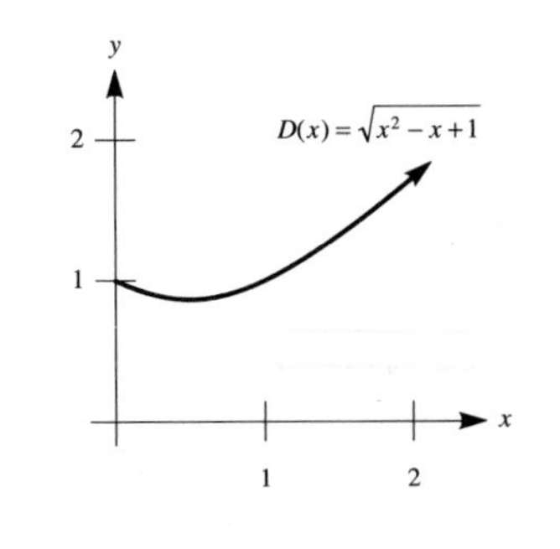

(a) no **(b)** one **(c)** no **(d)** yes

5.

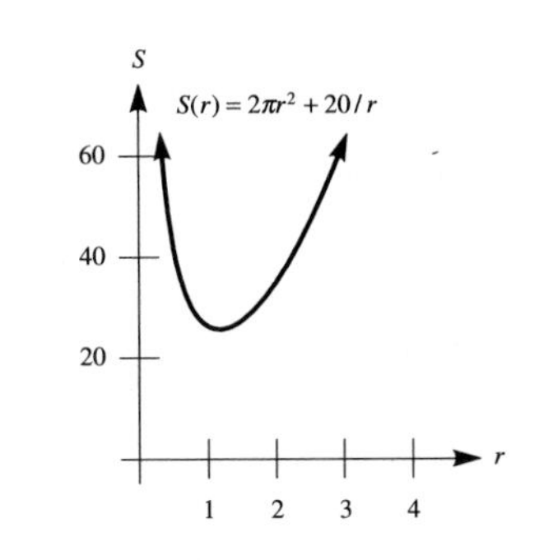

(a) no **(b)** one **(c)** no **(d)** yes

7. **(a)**

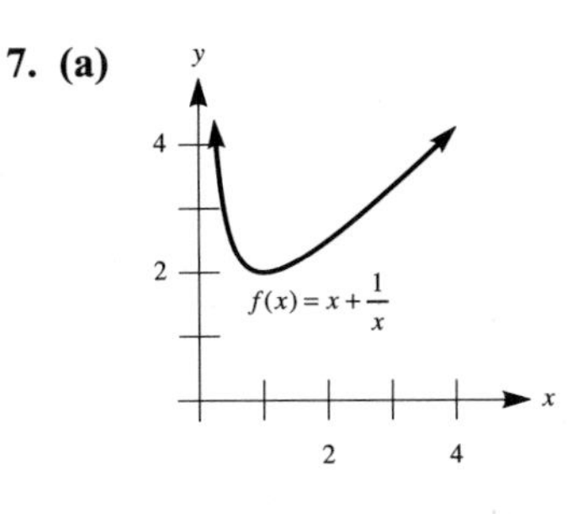

(a) no **(b)** one **(c)** no **(d)** yes

(b)

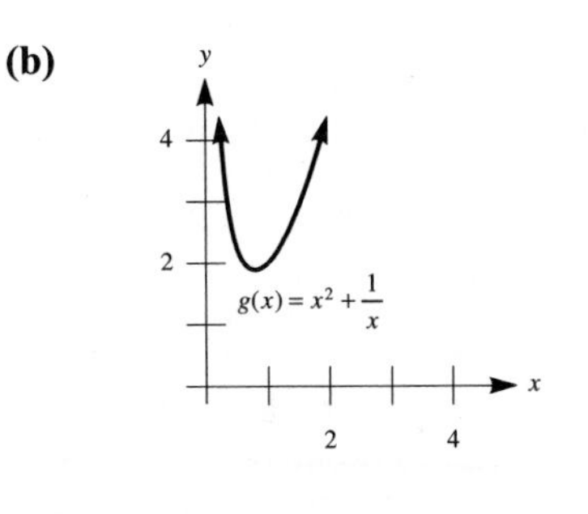

(a) no **(b)** one **(c)** no **(d)** yes

9. **(a)** $P(x) = -x^2 + \sqrt{11}x$
(b) maximum value

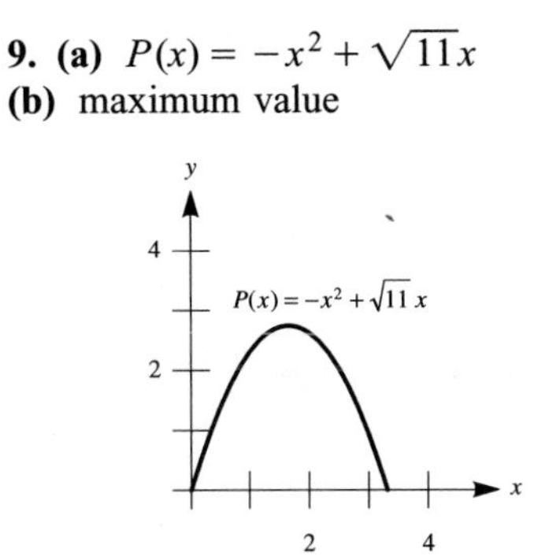

EXERCISE SET 4.5

1. 25/4 **3.** 1/2 **5.** $\frac{25}{4}$ m by $\frac{25}{4}$ m
7. 1250 in.² **9.** **(a)** 18 **(b)** 23/4
(c) 47/8 **(d)** 95/16
11. **(a)** $t = 1$: 16 ft; $t = 3/2$: 12 ft
(b) 16 ft; 1 sec
(c) $t = \frac{7}{4}$ sec or $t = \frac{1}{4}$ sec
13. point: $\left(\frac{7}{2}, \frac{2 + \sqrt{6}}{2}\right)$;
distance: $\sqrt{7}/2$
15. **(a)** 1/2 **(b)** 1/4
17. 125 ft by 250 ft **19.** $x = 40$
21. $x = 60$; maximum revenue: \$900;
$p = \$15$ **23.** **(a)** 36/13
(b) $6\sqrt{13}/13$; this is the square root of the answer to part (a).
25. **(a)** 225/2 **27.** $x = \sqrt{2}/2$;
$A = 1/2$ **29.** 49/12
33. 100 yd by 150 yd
37. **(a)** $p(x) = -2x + 500$ **(b)** revenue is \$31,250 when price is \$250
39. $t = \pm\sqrt{2}/2$ **41.** $t = \pm\sqrt{3}$
43. **(a)** Since $x^2 - x + 1 = \left(x - \frac{1}{2}\right)^2 + \frac{3}{4}$, the quantity must be positive.
We know that $\left(x - \frac{1}{2}\right)^2 \geq 0$,
so $\left(x - \frac{1}{2}\right)^2 + \frac{3}{4} \geq \frac{3}{4} > 0$.
(b) $D = -3 < 0$, so there are no x-intercepts (since no real solutions exist). Because the graph is a parabola opening up and there are no x-intercepts, the y-values must be positive for all values of x.
45. **(a)** $A(x) = \frac{4 + \pi}{16\pi}x^2 - 2x + 16$
(b) $x = 16\pi/(4 + \pi)$ **(c)** $\pi/4$
47. **(a)** $A(x) = \frac{4 + \pi}{16\pi}x^2 - \frac{L}{8}x + \frac{L^2}{16}$
(b) $x = \pi L/(4 + \pi)$ **(c)** $\pi/4$
49. 2 **53.** 2 **57.** **(a)** 1.143

GRAPHING UTILITY EXERCISES FOR SECTION 4.5

1. (a) $C(x) = x + 6400/x$
(b) 80 ft (at \$1/ft) and 40 ft (at \$2/ft)

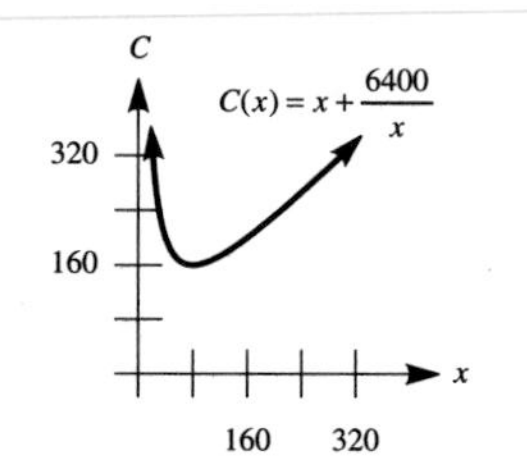

3. (a) $A(x) = 34 + 2x + 60/x$
(b) 7.48 in. by 7.48 in.

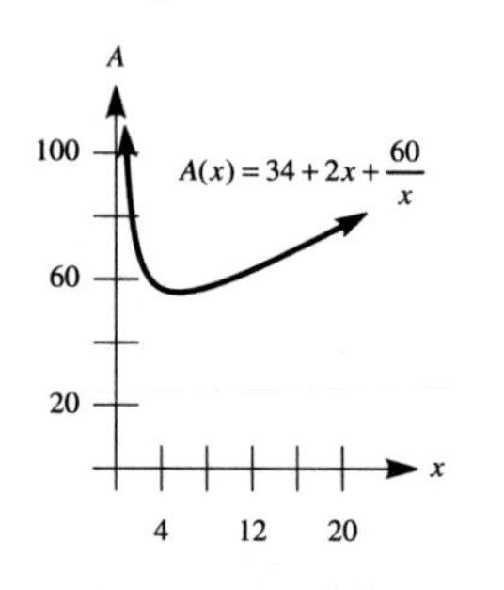

5. (a) $d(x) = \sqrt{x^4 + x^2 - 6x + 9}$
(b) (1, 1)

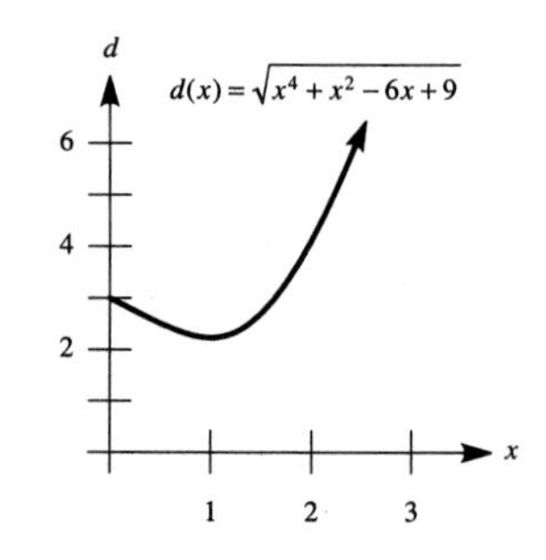

7. (a) $V(r) = 18\pi r^2 - \pi r^3$
(b) 2714 cubic units

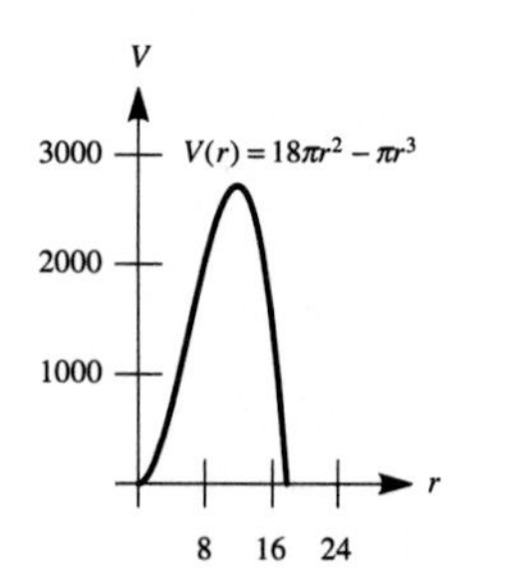

9. (a) $C(x) = 400x + 1500000/x$
(b) 61.2 ft by 81.7 ft

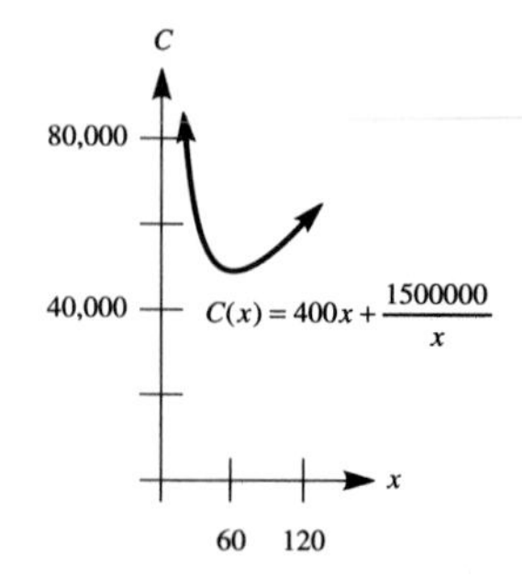

11. (a) $A(x) = 15x - \left(\frac{1}{2} + \frac{\pi}{8}\right)x^2$
(b) 8.40 ft by 4.20 ft

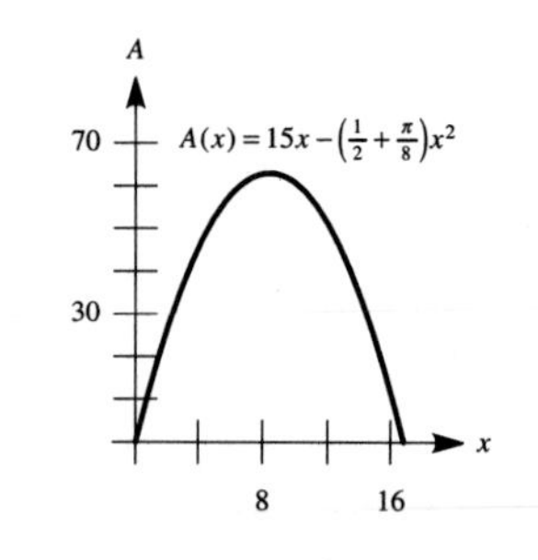

(c) The maximum value is $450/(4 + \pi) \approx 63.01$ when $x = 60/(4 + \pi) \approx 8.40$.

EXERCISE SET 4.6

1. This graph has 4 turning points, but a polynomial of degree 3 can have at most 2 turning points. **3.** As $|x|$ gets very large, the function should be similar to $f(x) = a_3x^3$. But then $f(x)$ does not have a parabolic shape like the given graph. **5.** As $|x|$ gets very large, with x negative, then the graph should resemble $2x^5$. But the y-values of $2x^5$ are always negative when x is negative, contrary to the given graph. **7.** This graph has a corner, which cannot occur in the graph of a polynomial function.

9. no x-intercepts; y-intercept: 5

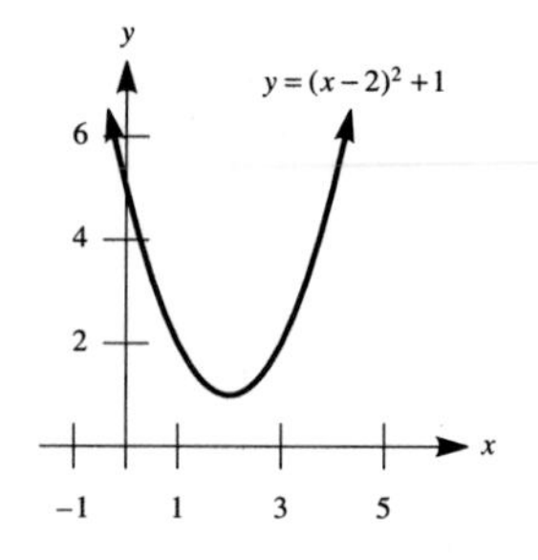

11. x-intercept: 1; y-intercept: -1

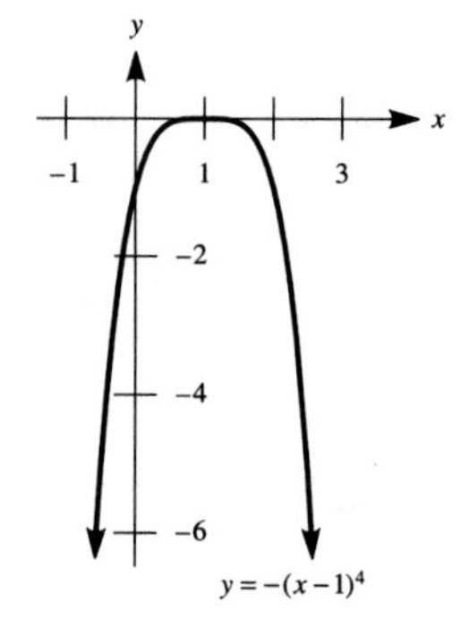

13. x-intercept: $4 + \sqrt[3]{2}$; y-intercept: -66

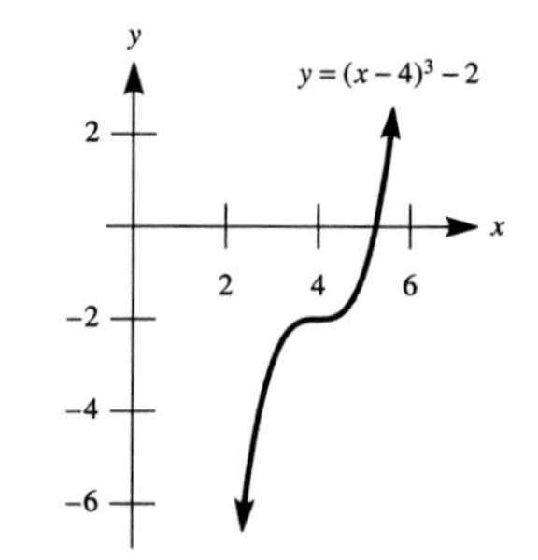

15. x-intercept: -5; y-intercept: -1250

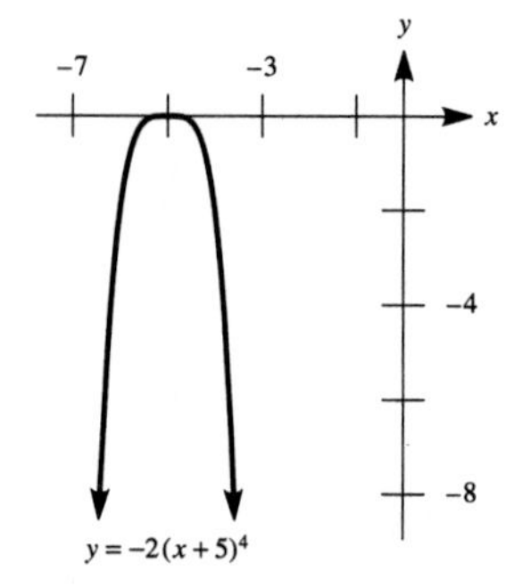

17. x-intercept: -1; y-intercept: $1/2$

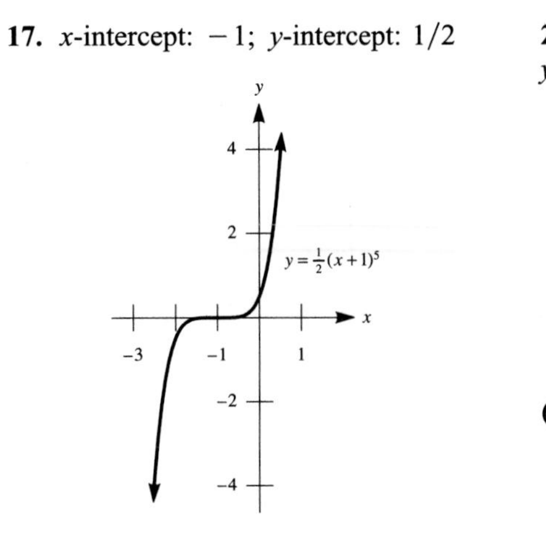

19. x- and y-intercepts: 0

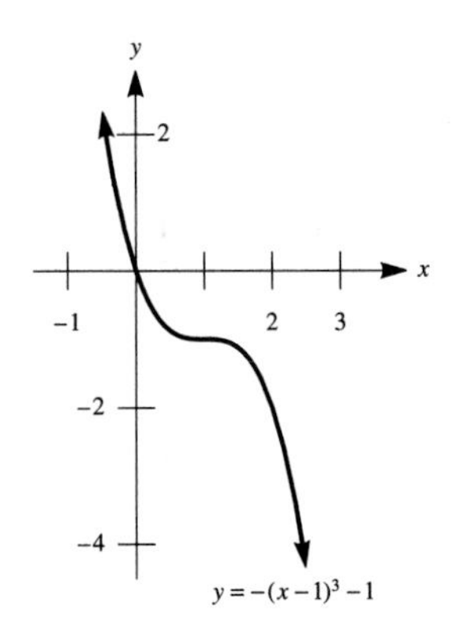

21. (a) x-intercepts: 2, 1, -1;
y-intercept: 2

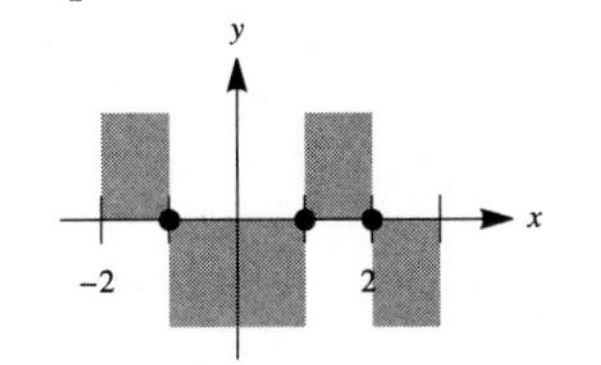

(b)

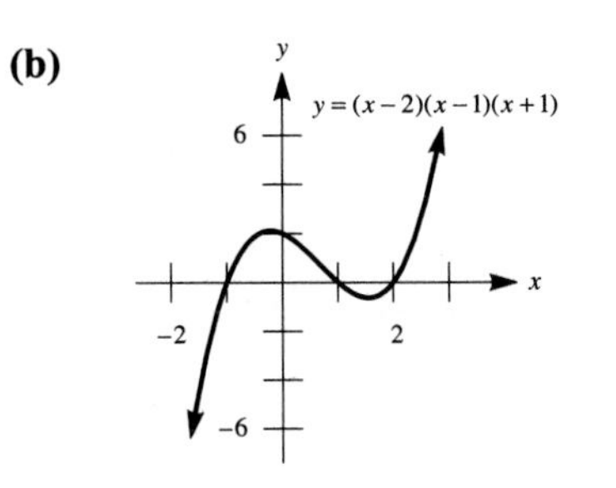

23. (a) x-intercepts: 0, 2, 1;
y-intercept: 0

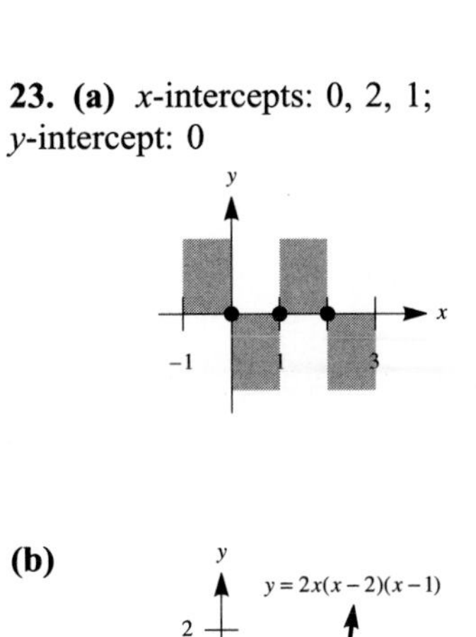

(b)

$y = 2x(x-2)(x-1)$

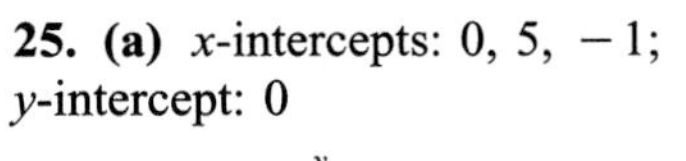

25. (a) x-intercepts: 0, 5, -1;
y-intercept: 0

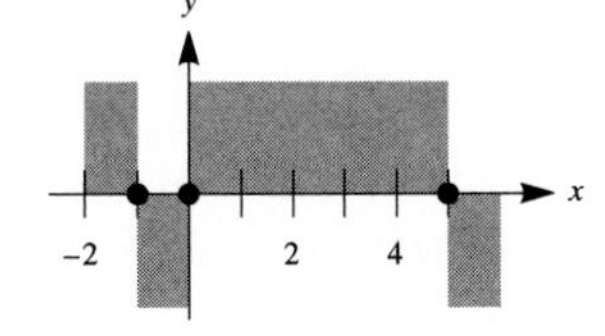

(b)

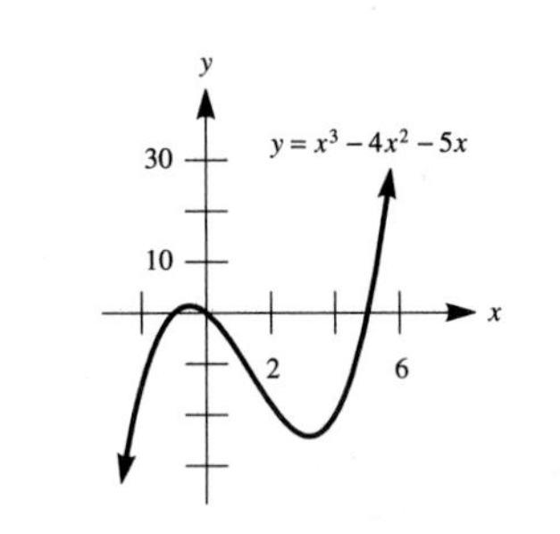

27. (a) x-intercepts: -3, -2, 2;
y-intercept: -12

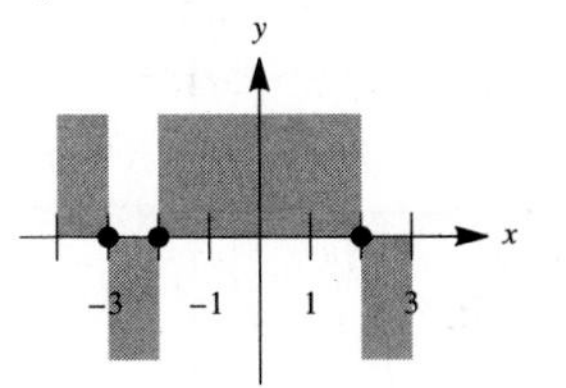

(b)

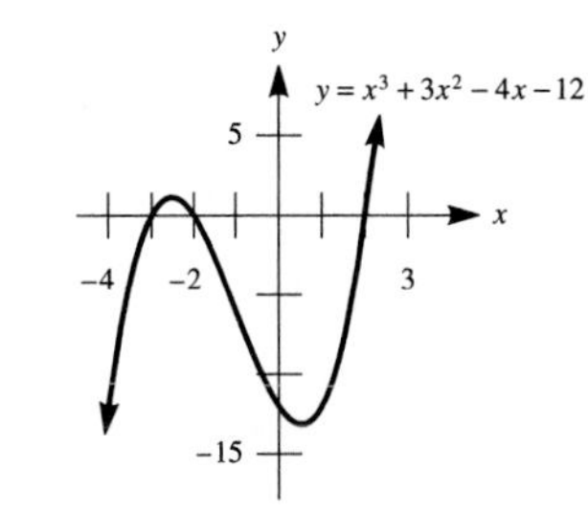

29. (a) x-intercepts: 0, -2;
y-intercept: 0

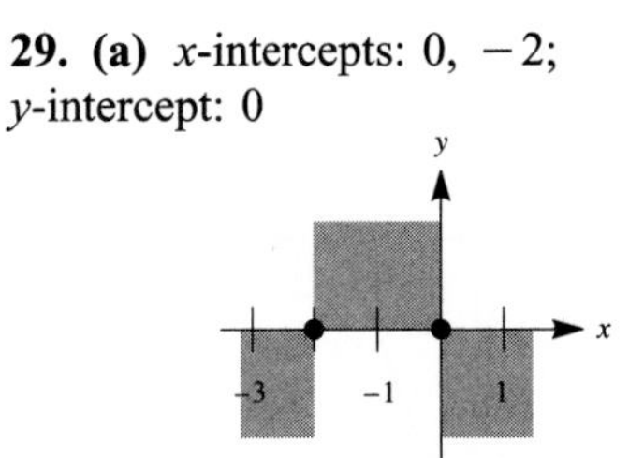

(b)

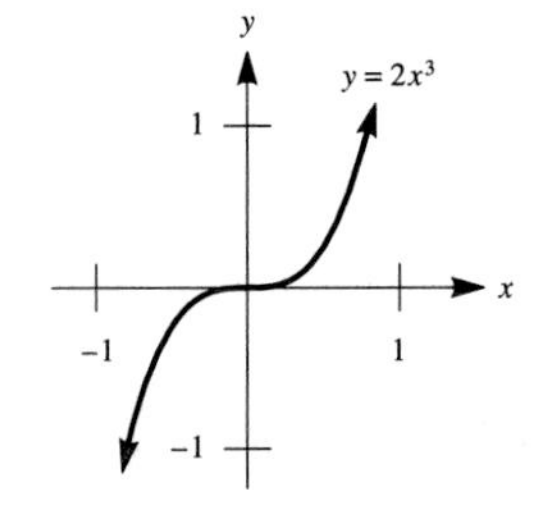

(c)

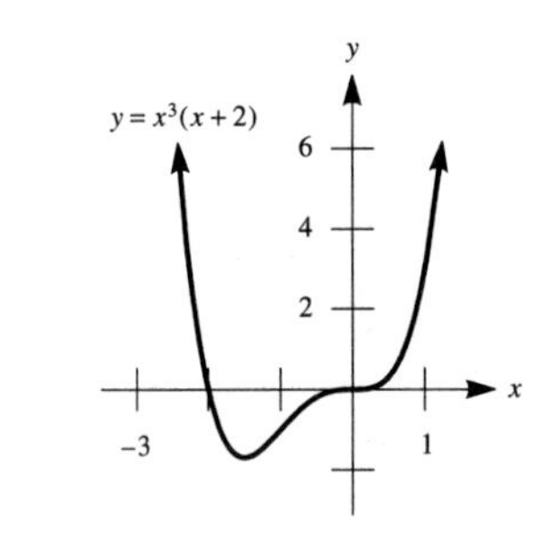

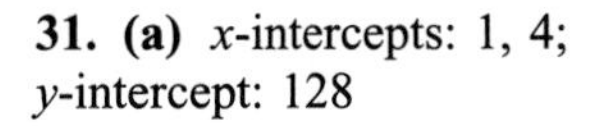

31. (a) x-intercepts: 1, 4;
y-intercept: 128

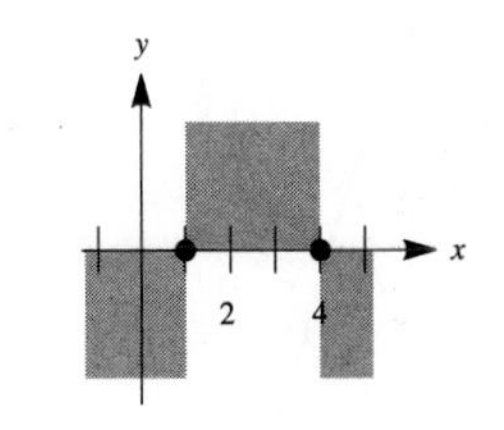

(b)

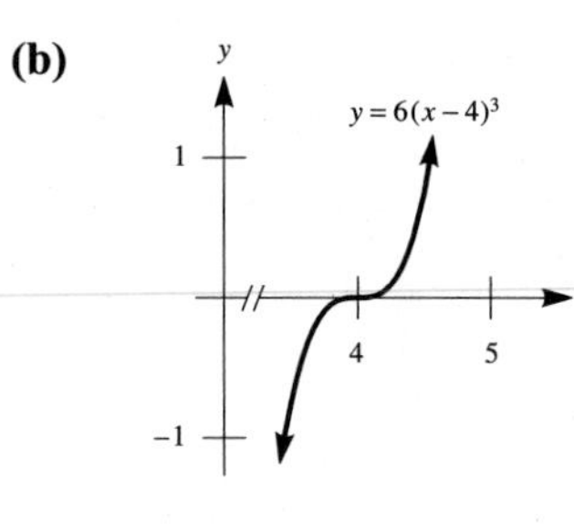

(c)

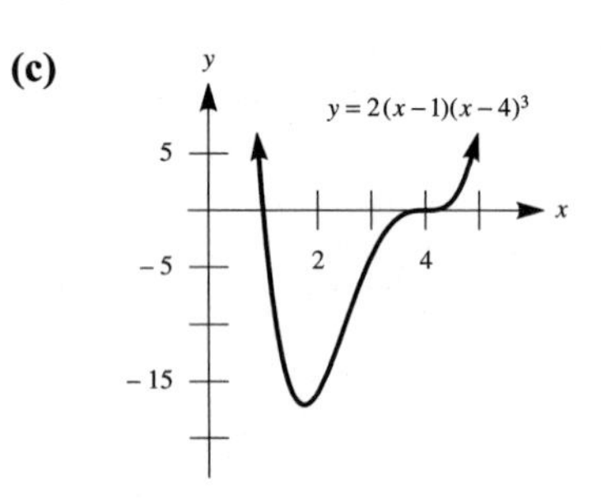

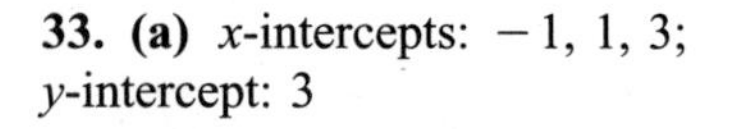

33. (a) x-intercepts: $-1, 1, 3$; y-intercept: 3

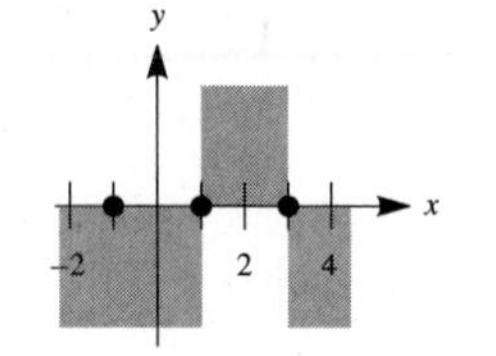

(b)

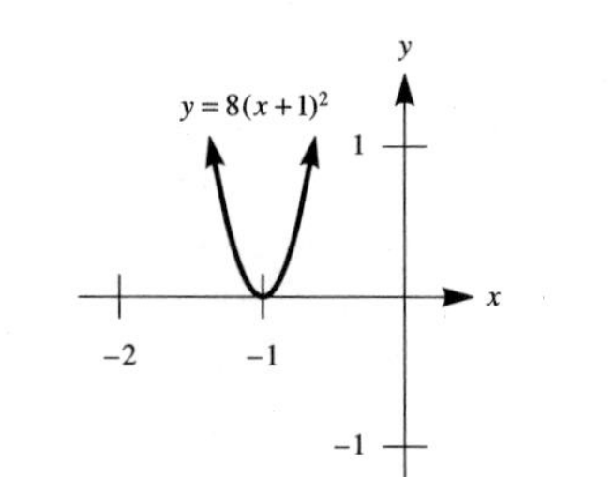

(c)

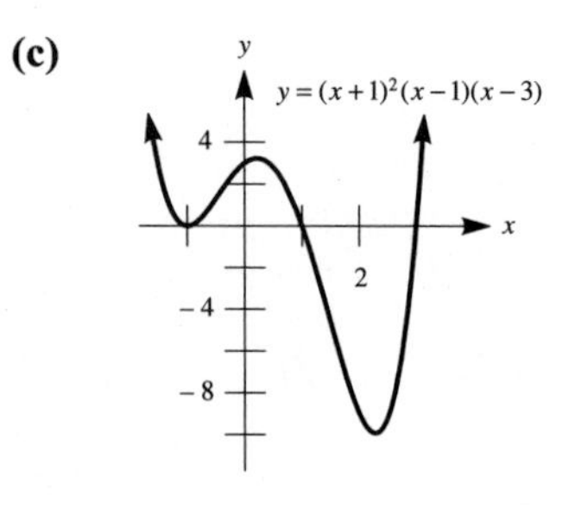

35. (a) x-intercepts: $0, 4, -2$; y-intercept: 0

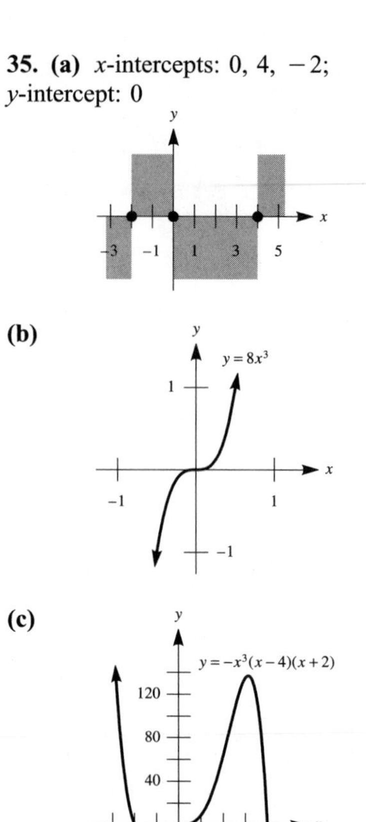

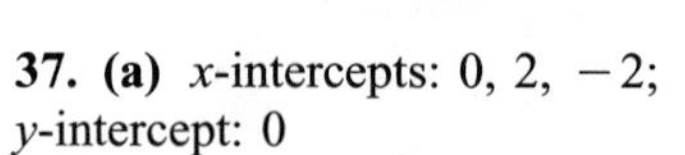

37. (a) x-intercepts: $0, 2, -2$; y-intercept: 0

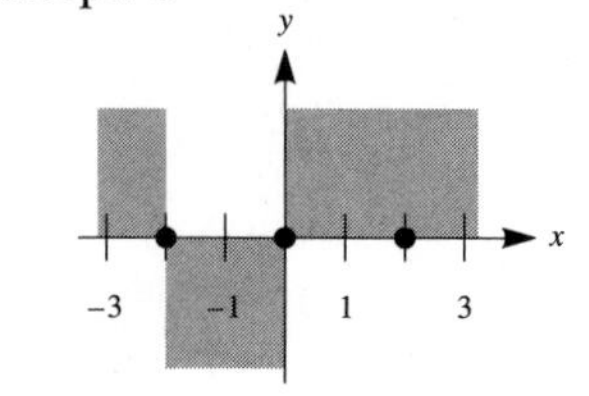

(b)

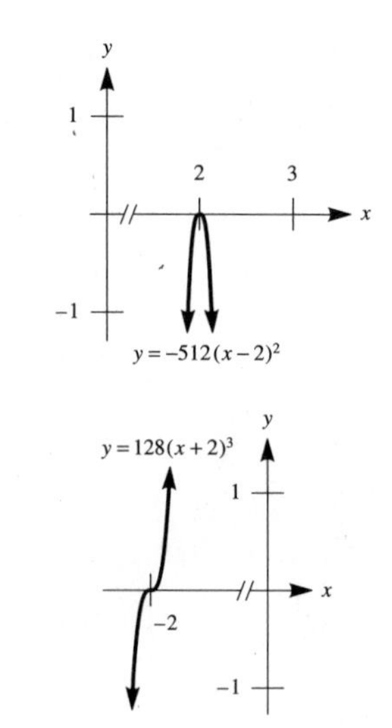

(c)

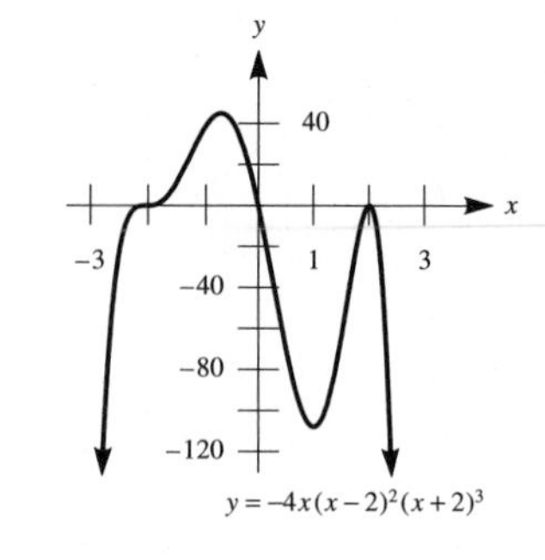

39. $x = (3 \pm \sqrt{29})/2 \approx -1.193$ and 4.193

41. $x = -6$ and $\pm\sqrt{3} \approx \pm 1.732$

43. $x = 0$ and $-\sqrt[3]{4} \approx -1.587$

45. From left to right, they are $f(x) = x$, $g(x) = x^2$, $h(x) = x^3$, $F(x) = x^4$, $G(x) = x^5$, and $H(x) = x^6$.

47. $[0, 0.68)$ **49.** no **51.** yes, at the point (100, 100)

53. (a)

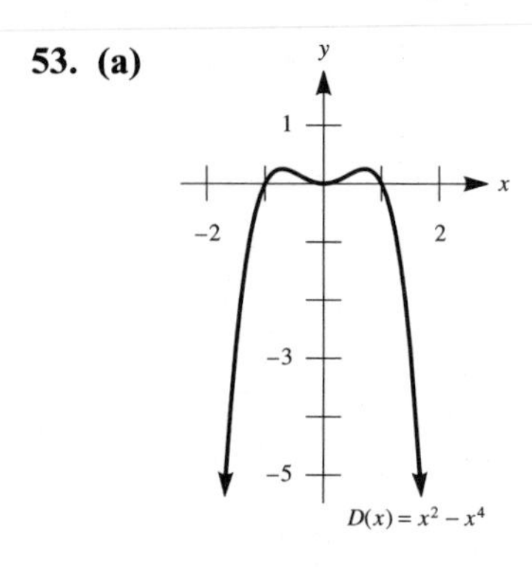

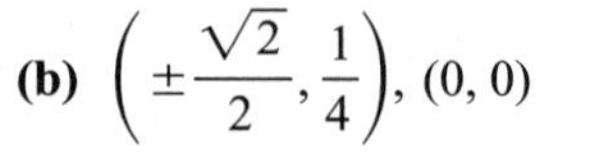

(b) $\left(\pm\dfrac{\sqrt{2}}{2}, \dfrac{1}{4}\right)$, (0, 0)

(c) $\frac{1}{4}$

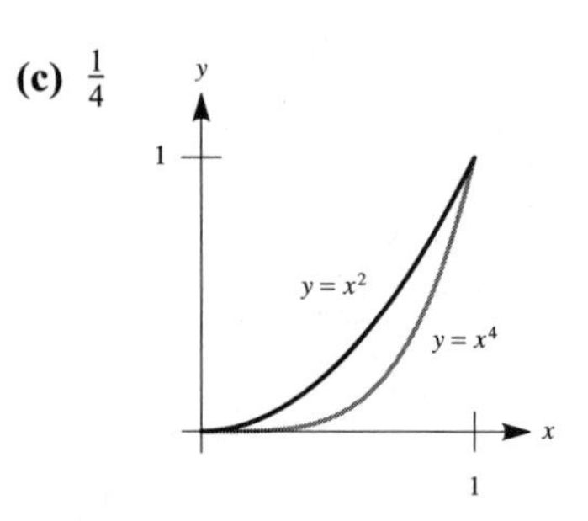

55. (c) (0, 6)

(d)

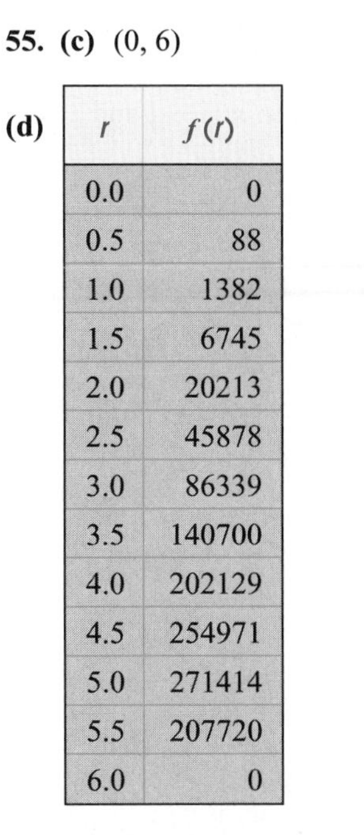

r	$f(r)$
0.0	0
0.5	88
1.0	1382
1.5	6745
2.0	20213
2.5	45878
3.0	86339
3.5	140700
4.0	202129
4.5	254971
5.0	271414
5.5	207720
6.0	0

maximum volume $\approx$ 522 cm^3

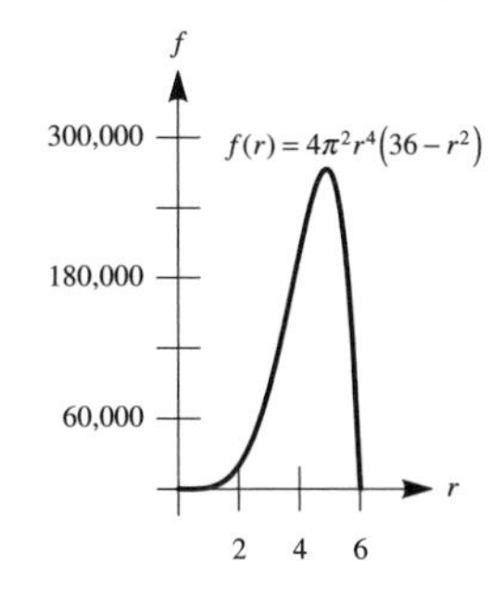

GRAPHING UTILITY EXERCISES FOR SECTION 4.6

1. (a)

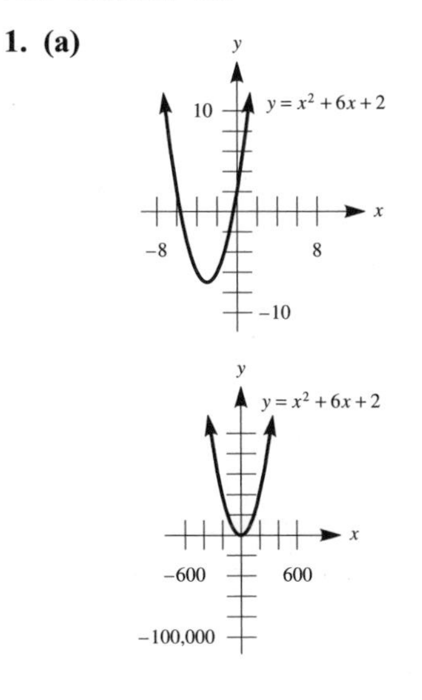

(b)

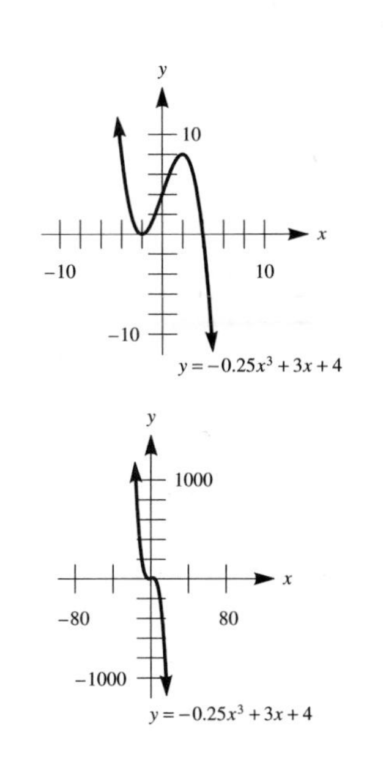

(c)

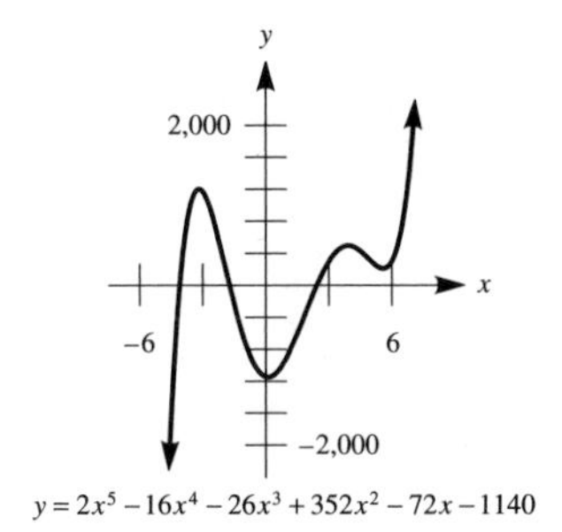

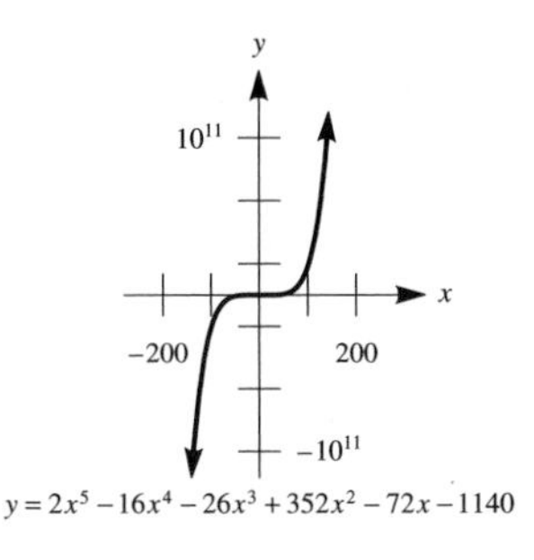

3. (1.738, 5.250) **5.** 81.872 in^3

EXERCISE SET 4.7

1. domain: $(-\infty, 3) \cup (3, \infty)$; x-intercept: -5; y-intercept: $-5/4$
3. domain: $(-\infty, -2) \cup (-2, 3) \cup (3, \infty)$; x-intercepts: 9, -1; y-intercept: $3/2$
5. domain: $(-\infty, 0) \cup (0, \infty)$; x-intercepts: -2, 2, 1; no y-intercepts

7. no x-intercept; y-intercept: $1/4$; horizontal asymptote: $y = 0$; vertical asymptote: $x = -4$

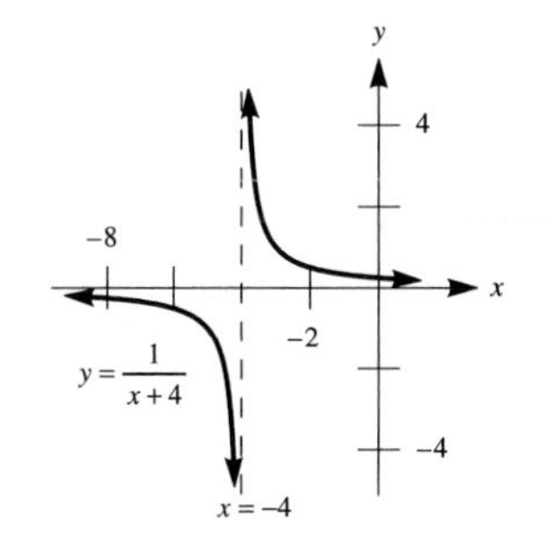

9. no x-intercept; y-intercept: $3/2$; horizontal asymptote: $y = 0$; vertical asymptote: $x = -2$

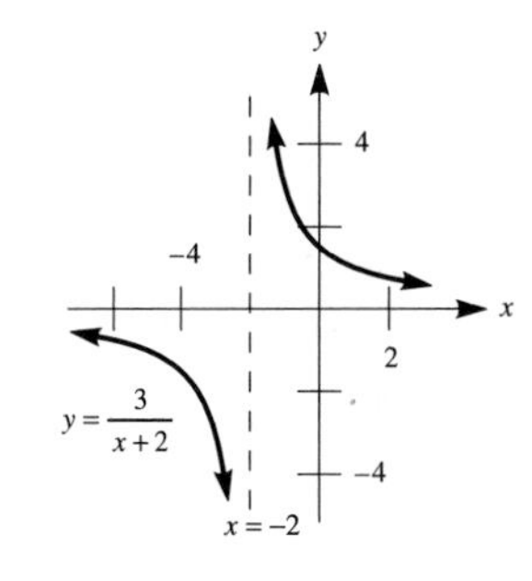

11. no x-intercept; y-intercept: $2/3$; horizontal asymptote: $y = 0$; vertical asymptote: $x = 3$

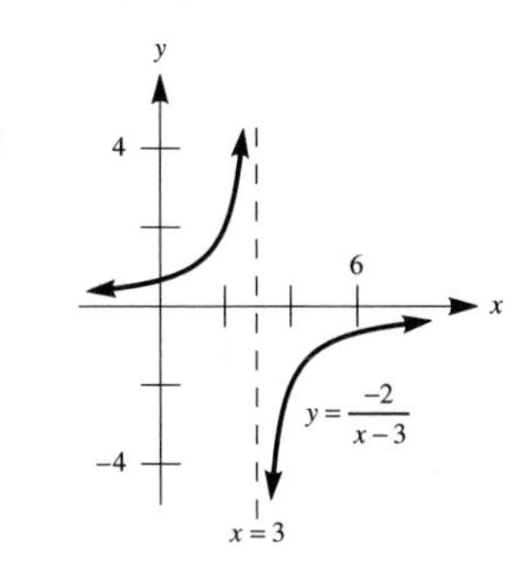

13. x-intercept: 3; y-intercept: 3; horizontal asymptote: $y = 1$; vertical asymptote: $x = 1$

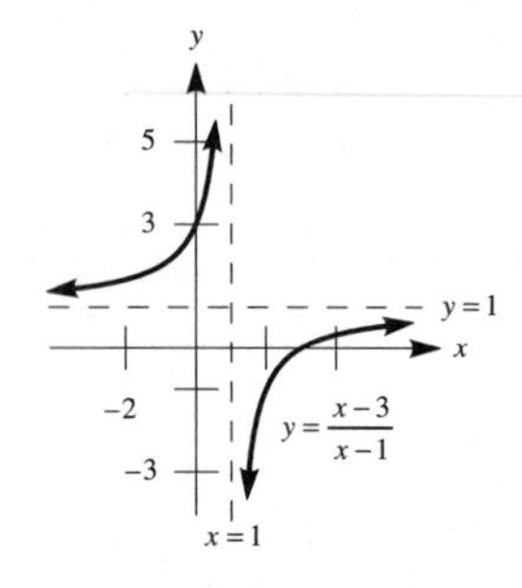

15. x-intercept: $1/2$; y-intercept: -2; horizontal asymptote: $y = 2$; vertical asymptote: $x = -1/2$

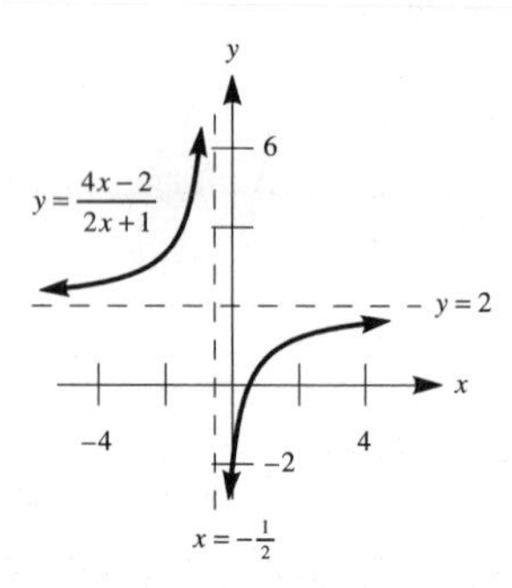

17. no x-intercept; y-intercept: $1/4$; horizontal asymptote: $y = 0$; vertical asymptote: $x = 2$

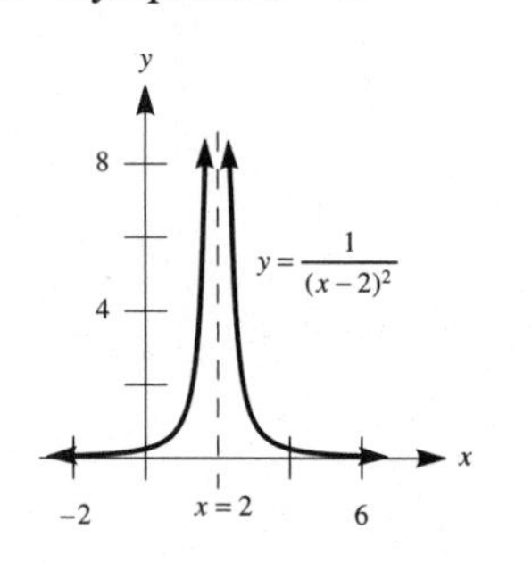

19. no x-intercept; y-intercept: 3; horizontal asymptote: $y = 0$; vertical asymptote: $x = -1$

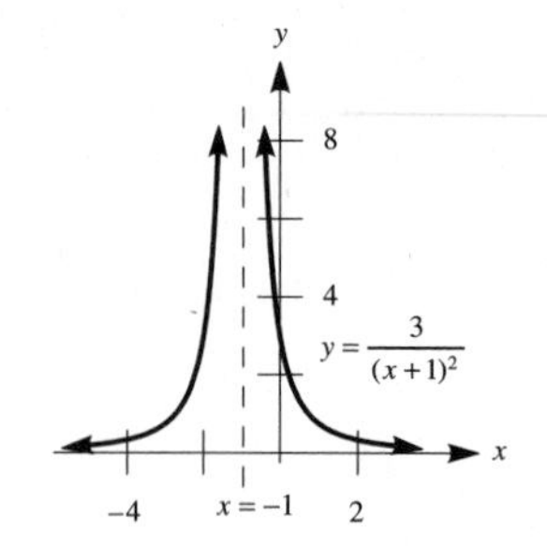

21. no x-intercept; y-intercept: $1/8$; horizontal asymptote: $y = 0$; vertical asymptote: $x = -2$

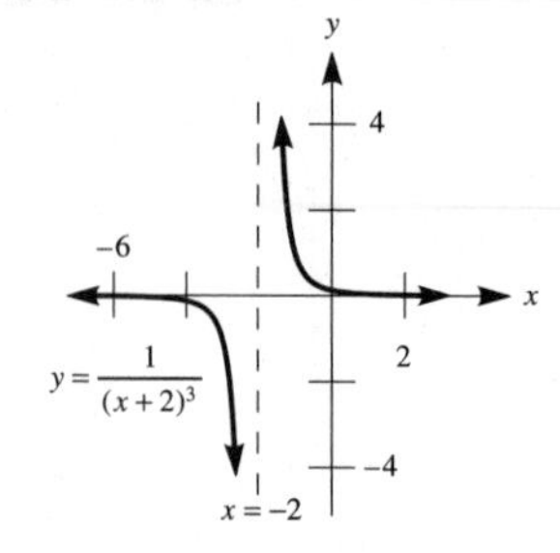

23. no x-intercept; y-intercept: $-4/125$; horizontal asymptote: $y = 0$; vertical asymptote: $x = -5$

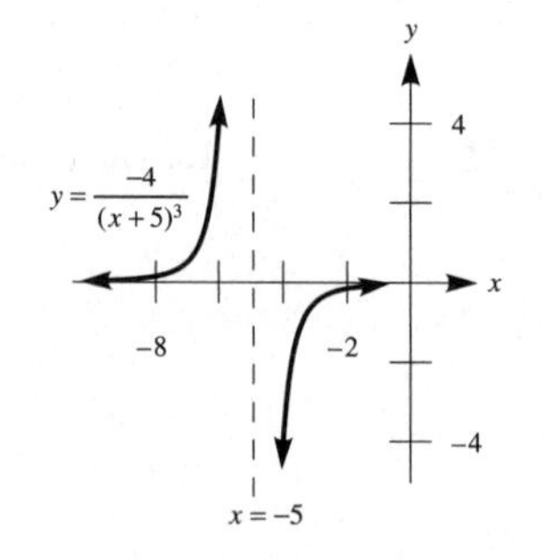

25. x- and y-intercepts: 0; horizontal asymptote: $y = 0$; vertical asymptotes: $x = -2$ and $x = 2$

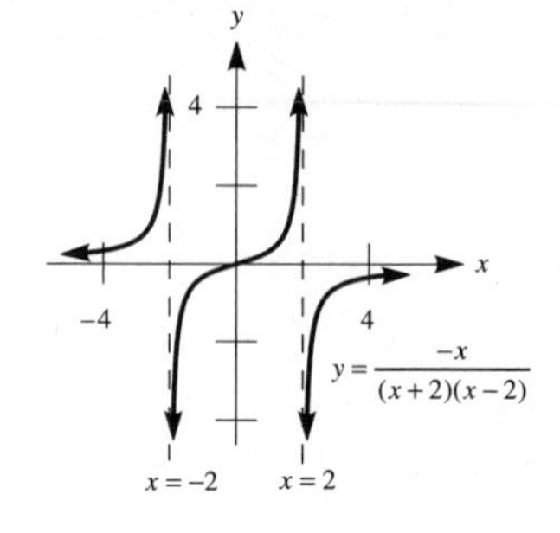

27. (a) x-intercept: 0; y-intercept: 0; horizontal asymptote: $y = 0$; vertical asymptotes: $x = 1$ and $x = -3$

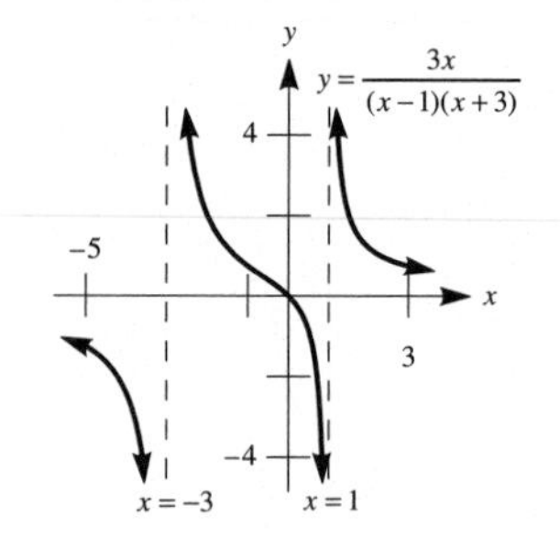

(b) x-intercept: 0; y-intercept: 0; horizontal asymptote: $y = 3$; vertical asymptotes: $x = 1$ and $x = -3$

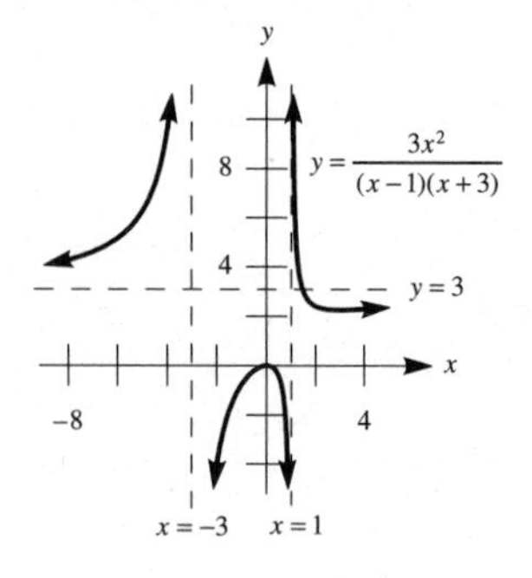

29. (a) x-intercepts: 2, 4;
no y-intercept;
horizontal asymptote: $y = 1$;
vertical asymptotes: $x = 0$ and $x = 1$

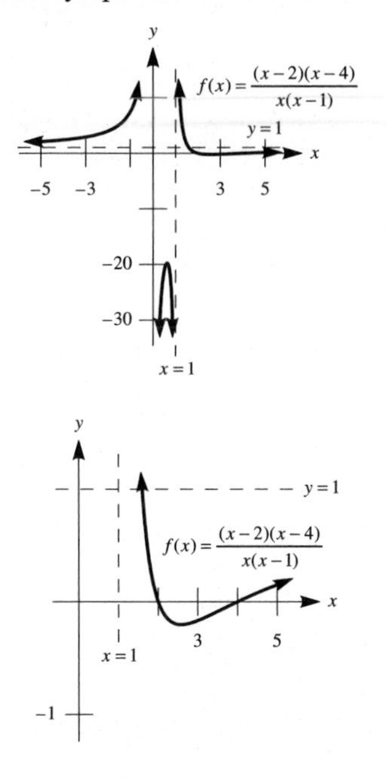

(b) x-intercepts: 2, 4; no y-intercept;
horizontal asymptote: $y = 1$;
vertical asymptotes: $x = 0$ and $x = 3$

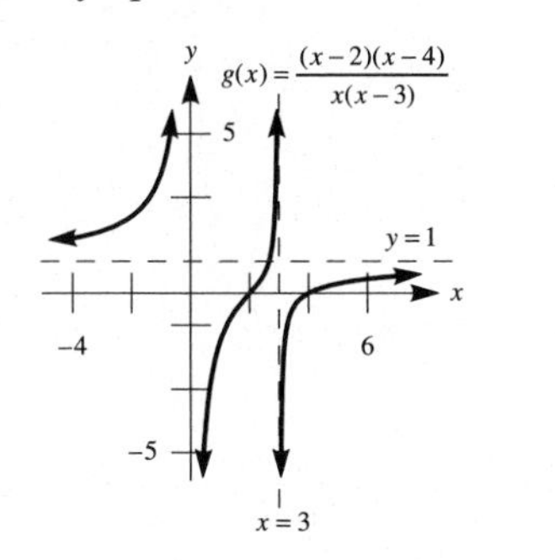

31. crosses at $x = 11/2$

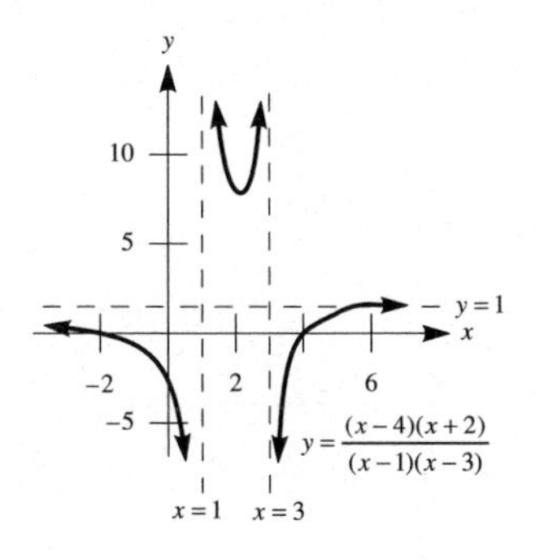

33. crosses at $x = 1/3$

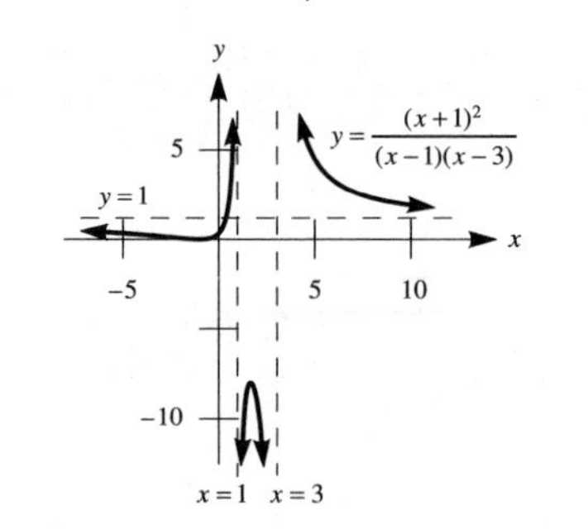

37. (a)

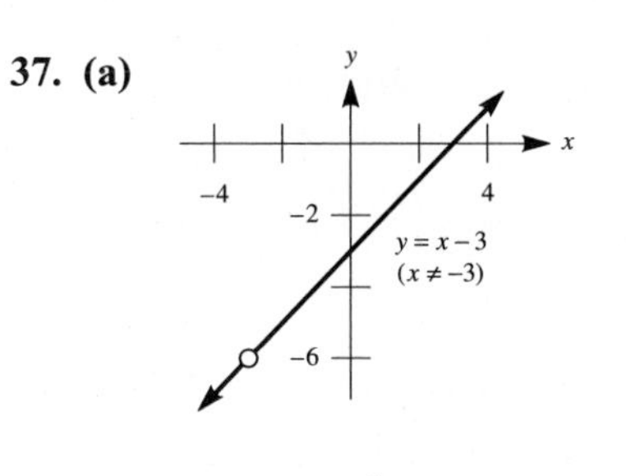

(b)

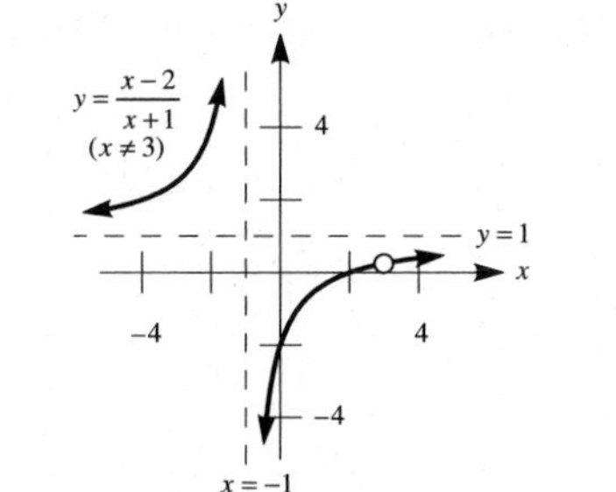

(c)

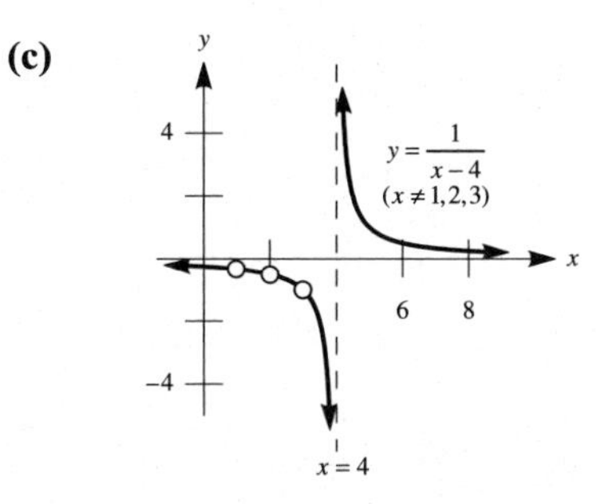

39. turning point: $(-3, -1/12)$

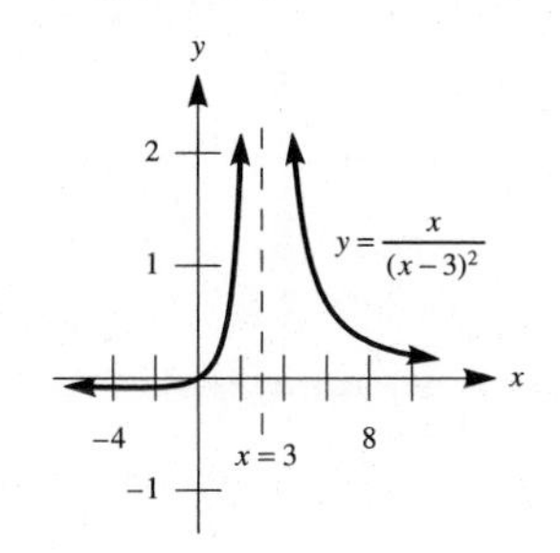

41. (b)

x	$x + 4$	$\frac{x^2 + x - 6}{x - 3}$
10	14	14.8571
100	104	104.0619
1000	1004	1004.0060

x	$x + 4$	$\frac{x^2 + x - 6}{x - 3}$
−10	−6	−6.4615
−100	−96	−96.0583
−1000	−996	−996.0600

(c) vertical asymptote: $x = 3$;
x-intercepts: $-3, 2$; y-intercept: 2

(d)

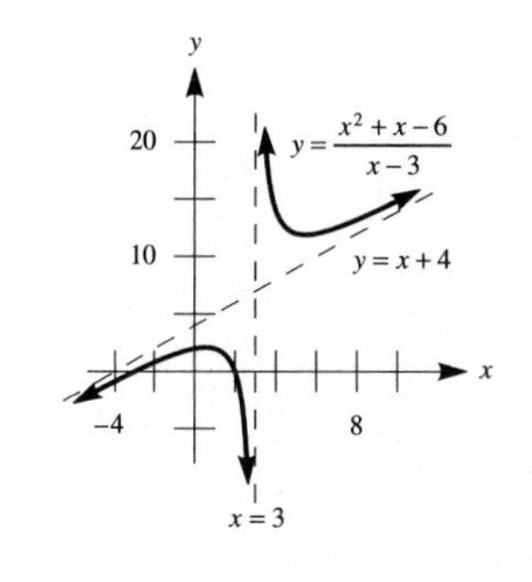

(e) $(3 + \sqrt{6}, 7 + 2\sqrt{6})$ and $(3 - \sqrt{6}, 7 - 2\sqrt{6})$

43.

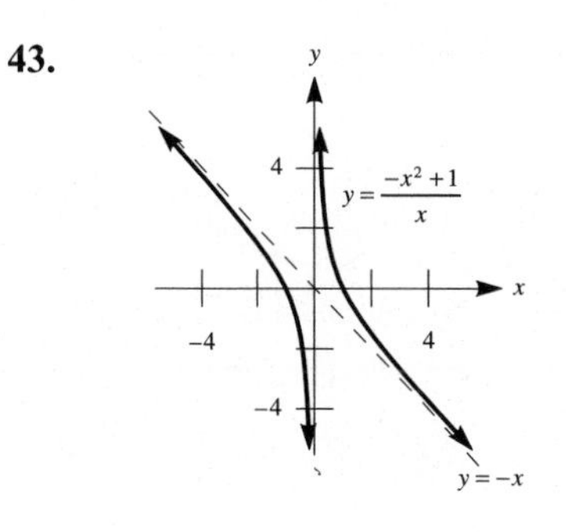

GRAPHING UTILITY EXERCISES FOR SECTION 4.7

1. $m \approx -0.5$

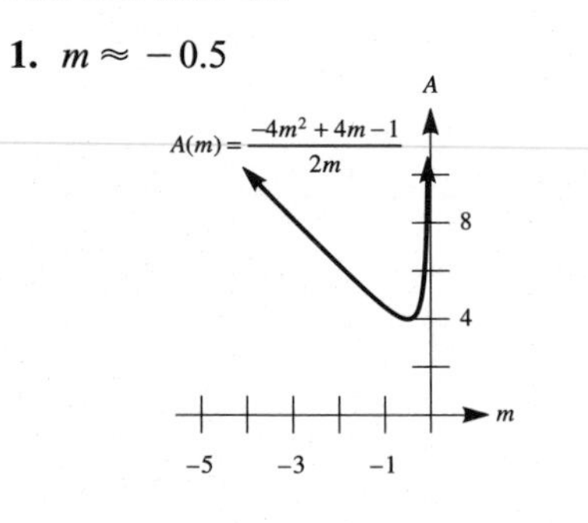

3. $P(1, 1)$

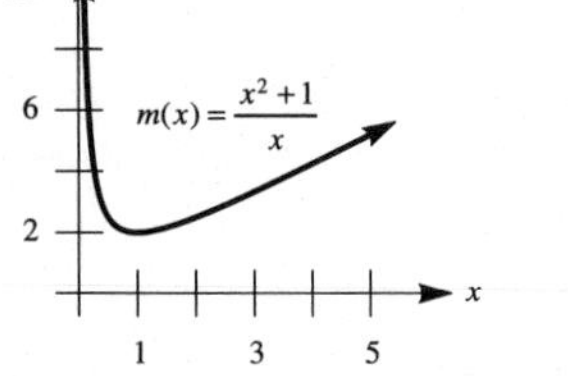

The line appears to be tangent to the parabola at (1, 1).

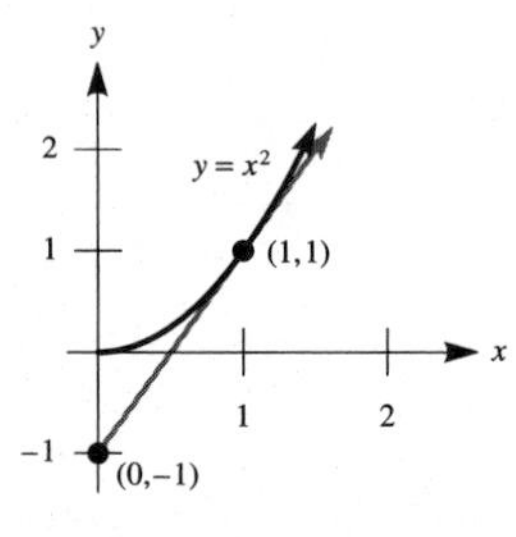

5. (a) $P(x) = (2x^2 + 12)/x$
(b) $x \approx 2.45$; length = 2.45 m

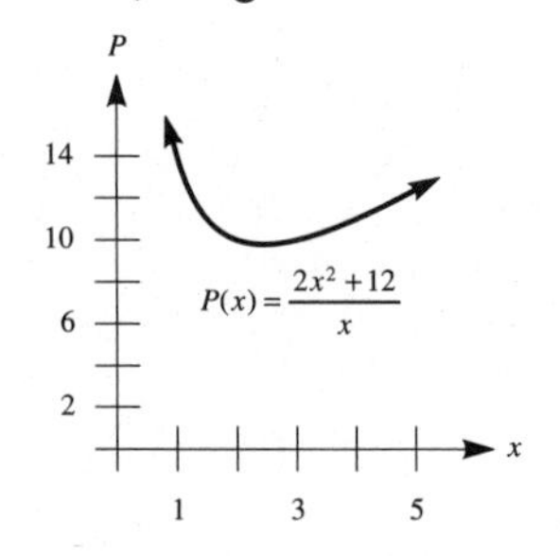

7. 9.73

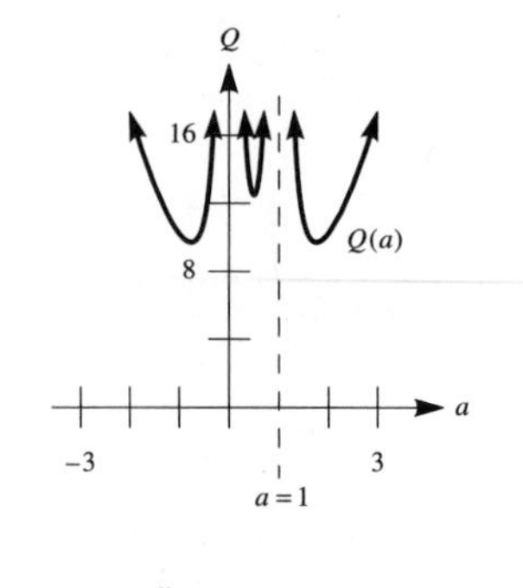

9. (a)

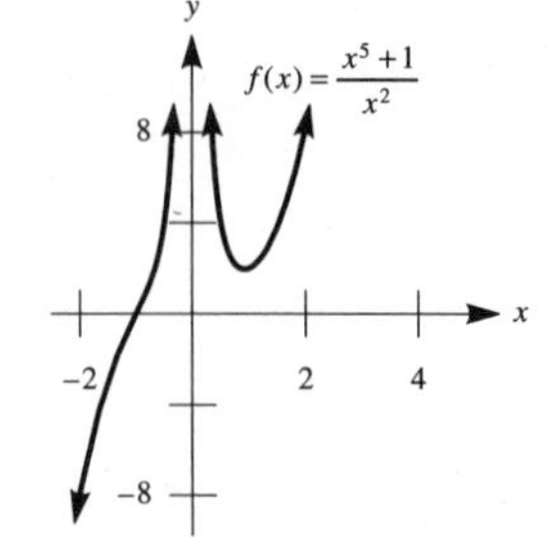

(b) As $|x|$ increases, the curve $f(x)$ approaches the curve $y = x^3$.

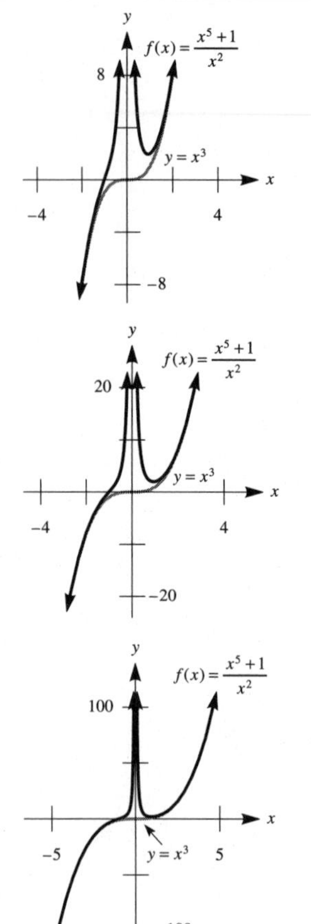

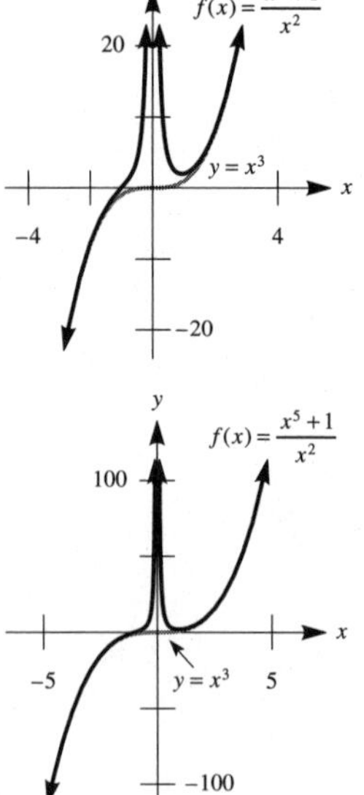

(c)

x	5	10	50	100	500
d	0.04	0.01	0.0004	0.0001	0.000004

x	−5	−10	−50	−100	−500
d	0.04	0.01	0.0004	0.0001	0.000004

(d) As $|x|$ increases, the quantity $1/x^2$ approaches 0, and thus $f(x) \approx x^3$ when $|x|$ gets very large.

CHAPTER 4 REVIEW EXERCISES

1. −5 **3.** \$20,000

5.

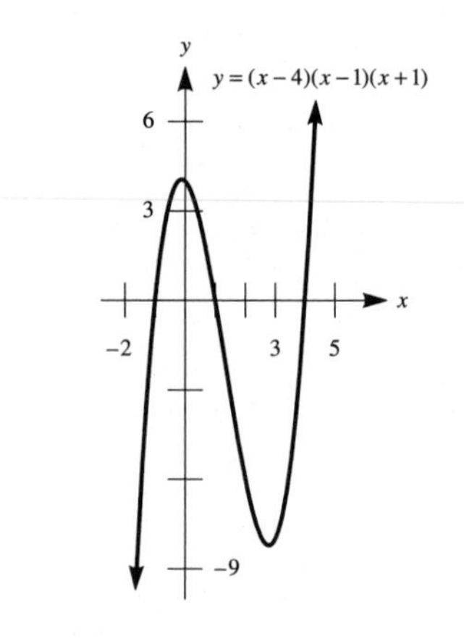

7. $V(t) = -180t + 1000$
9. x-intercept: $-5/3$; y-intercept: $5/2$; horizontal asymptote: $y = 3$; vertical asymptote: $x = -2$

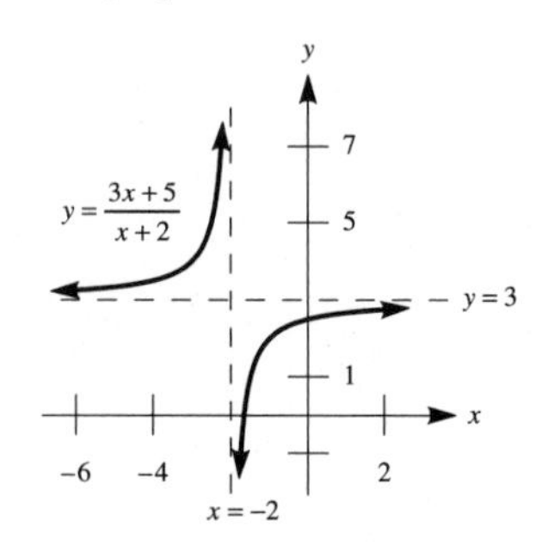

11.

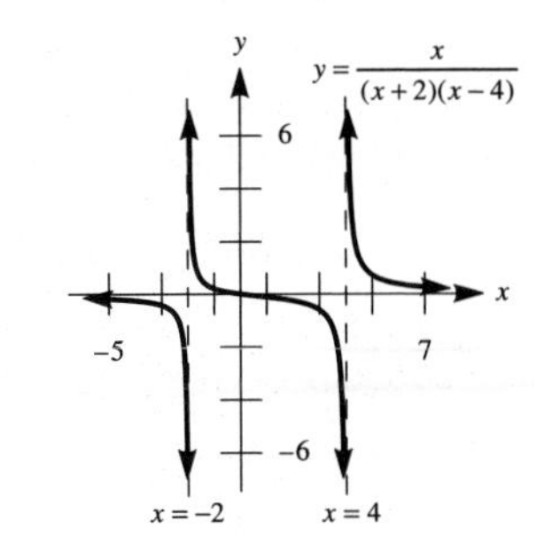

13. $P(w) = 2w + 2\sqrt{144 - \pi^2 w^2}/\pi$
15. $f(x) = x + 2$ **17.** $f(x) = \frac{3}{8}x - \frac{5}{2}$
19. $f(x) = -\frac{3}{4}x + \frac{11}{4}$

21. vertex: $(-1, -4)$;
x-intercepts: $1, -3$; y-intercept: -3

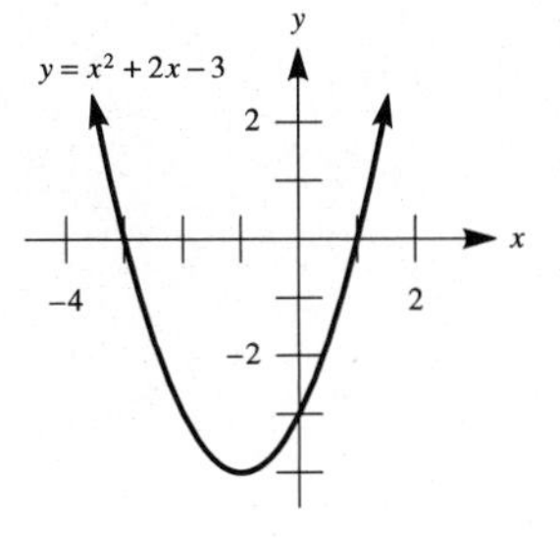

23. vertex: $(\sqrt{3}, 6)$; x-intercepts:
$\sqrt{3} + \sqrt{6}, \sqrt{3} - \sqrt{6}$; y-intercept: 3

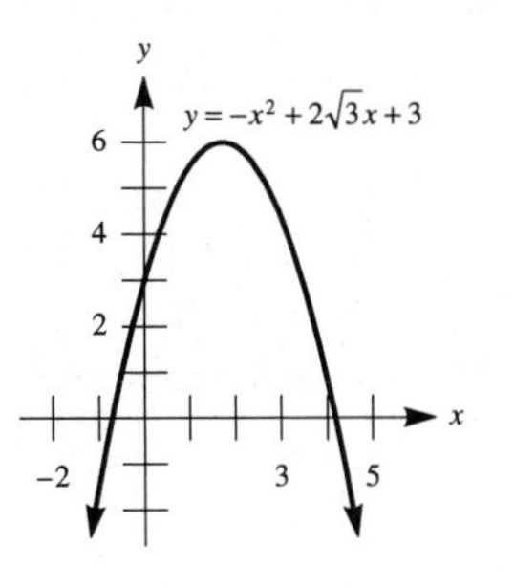

25. vertex: (2, 12); x-intercepts: 0, 4;
y-intercept: 0

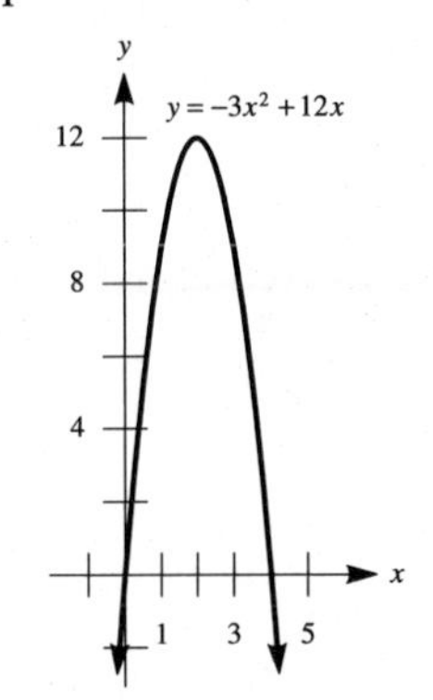

27. 5 **29.** 3/4
31. (a) maximum height: $\frac{v_0^2}{64}$ ft;
$t = \frac{v_0}{32}$ sec **(b)** $t = \frac{v_0}{16}$ sec
33. (a) $d = \sqrt{x^4 - 3x^2 + 4}$
(b) $\left(\frac{-\sqrt{6}}{2}, \frac{3}{2}\right)$ **35.** $b = \frac{17}{3}, -11$
37. 1 **39.** $\frac{225}{4}\ \text{cm}^2$ **41.** $a = 1$
43. $x = 400$ units; $p = \$80$
45. (a) graph: 3, $4\frac{1}{4}$, $5\frac{1}{2}$, $6\frac{1}{4}$, $6\frac{3}{4}$;
calculator: 3, 4.243, 5.402, 6.294, 6.903
(b) $(9 + 3\sqrt{5})/2 \approx 7.854$

49. x-intercepts: $-4, 2$; y-intercept: -8

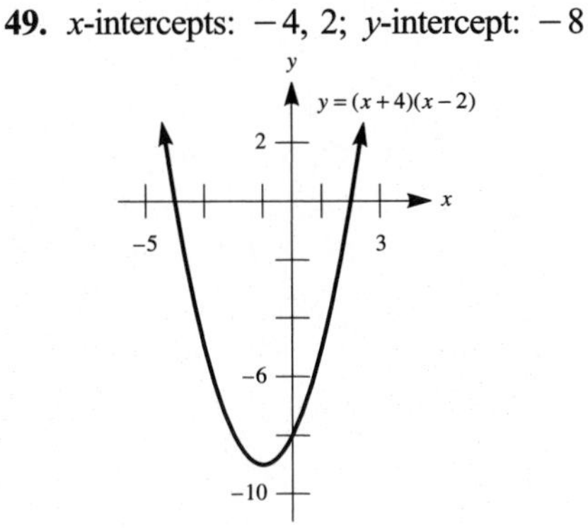

51. x-intercept: -5; y-intercept: -125

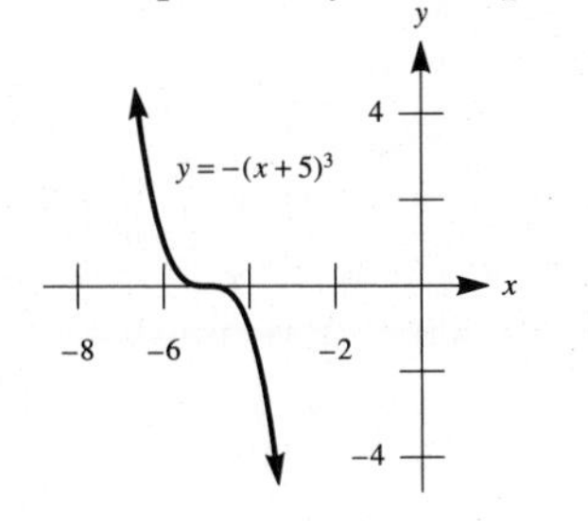

53. x-intercepts: $-1, 0$; y-intercept: 0

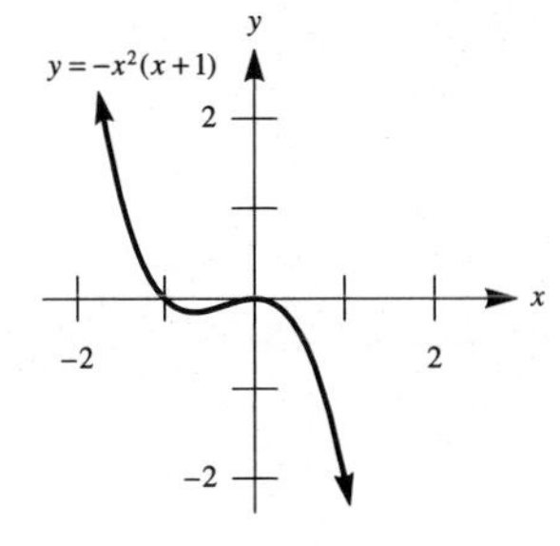

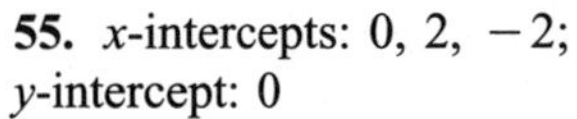
55. x-intercepts: $0, 2, -2$;
y-intercept: 0

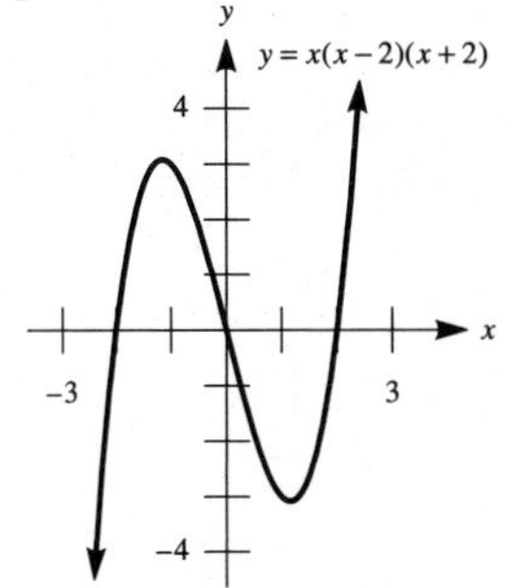

57. x-intercept: $-1/3$; no y-intercept;
horizontal asymptote: $y = 3$;
vertical asymptote: $x = 0$

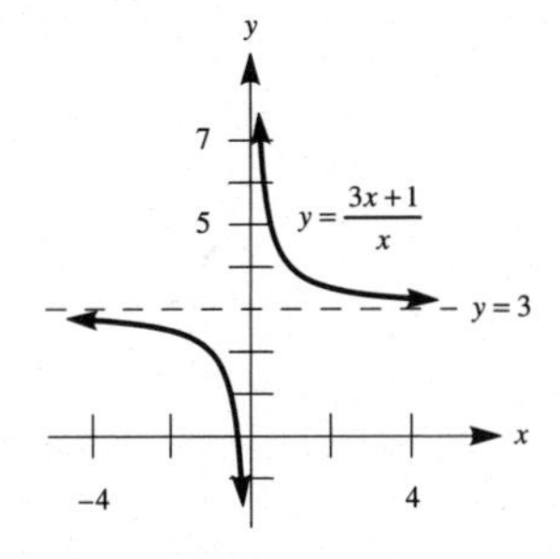

59. no x-intercept; y-intercept: -1;
horizontal asymptote: $y = 0$;
vertical asymptote: $x = 1$

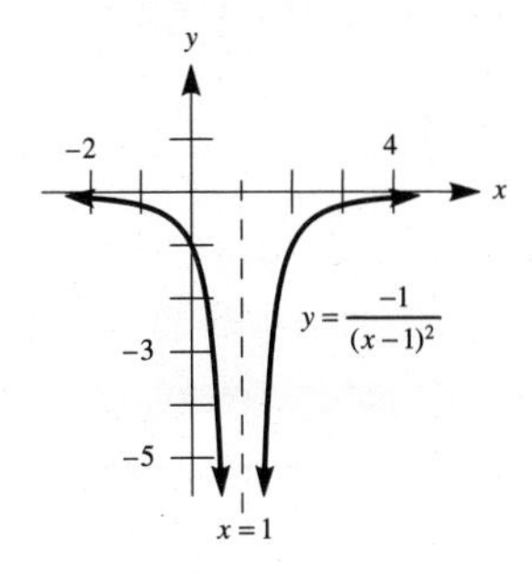

61. x-intercept: 2; y-intercept: 2/3; horizontal asymptote: $y = 1$; vertical asymptote: $x = 3$

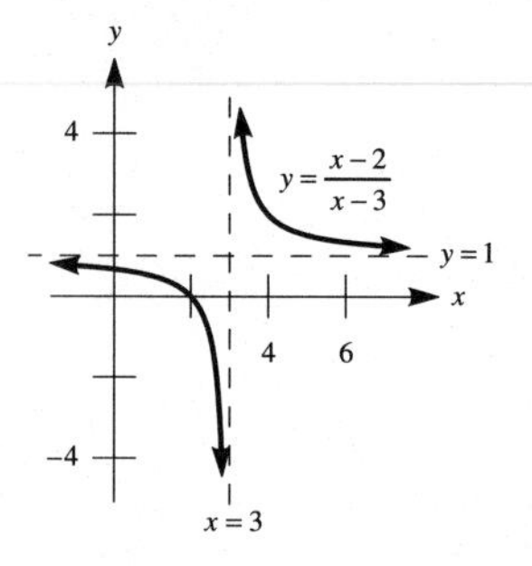

63. x-intercept: 1; y-intercept: 1/4; horizontal asymptote: $y = 1$; vertical asymptote: $x = 2$

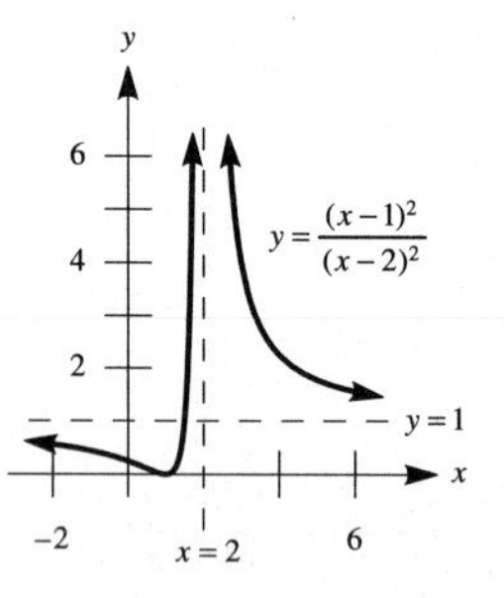

65. **(a)** 1 **(b)** $\sqrt{13}$ **(c)** $b = \pm\sqrt{2}/2$ **67.** $k = 6$
69. $(-\infty, 4 - 2\sqrt{3}] \cup [4 + 2\sqrt{3}, \infty)$
71. $A(m) = m/2$
73. $A(x) = (1 - x)\sqrt{1 - x^2}$
75. **(a)** $x_0 = -b/2a$ **(b)** The graph of f is symmetric about the vertical line $x = -b/2a$, which passes through the vertex.

CHAPTER 4 TEST

1. $-18/7$ **2.** **(a)** increasing: $(-\infty, 1)$; maximum: 2 **(b)** $t = 0$
3. **(a)** x-intercepts: 3, -4; y-intercept: -48

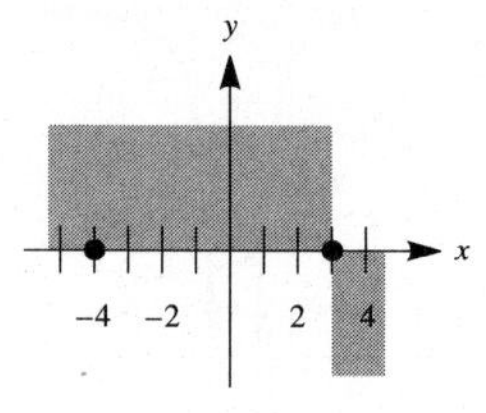

(b)

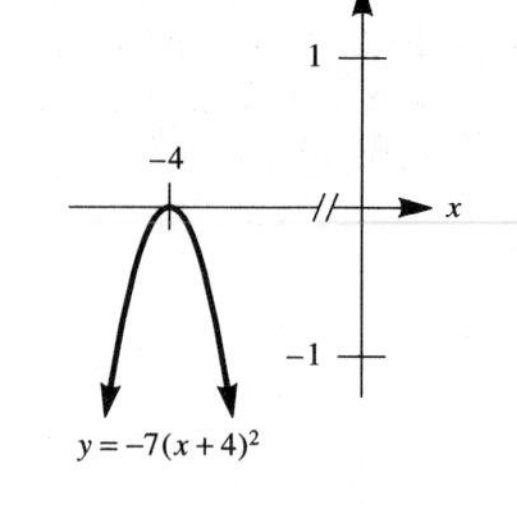

(c)

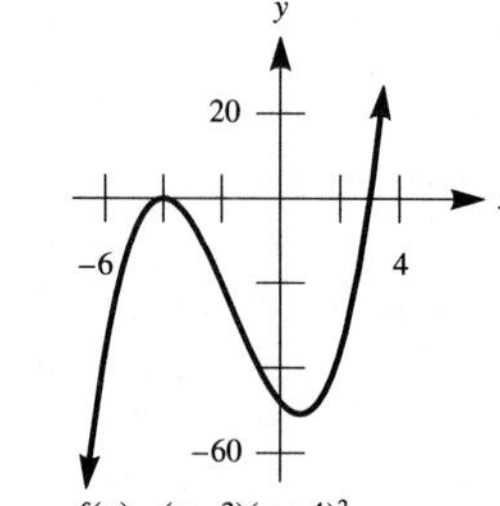

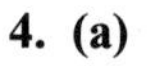

4. **(a)**

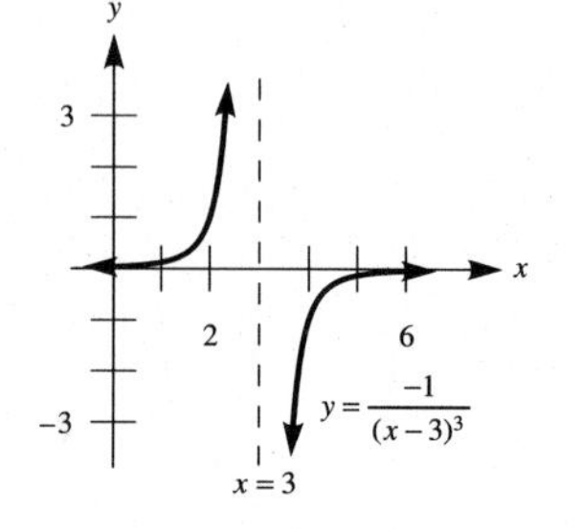

(b)

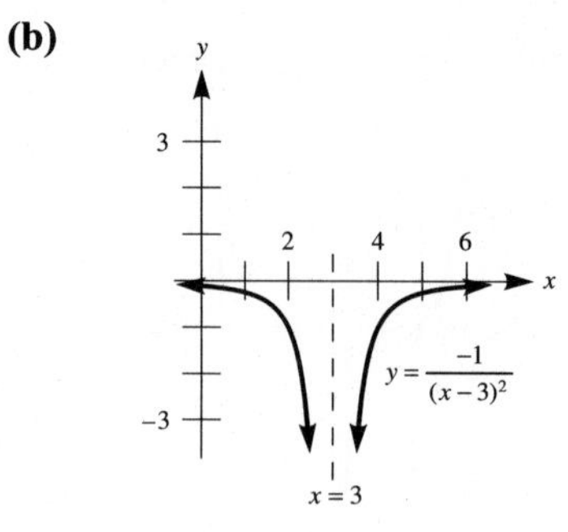

5. turning point: $\left(\frac{7}{2}, \frac{73}{4}\right)$; x-intercepts: $(7 \pm \sqrt{73})/2$; y-intercept: 6; axis of symmetry: $x = 7/2$

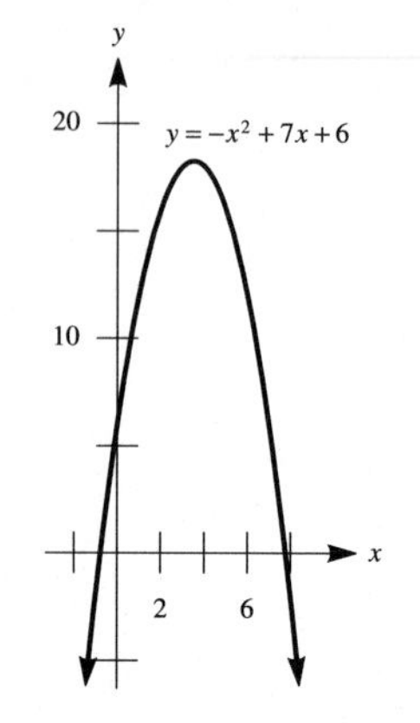

6. revenue = \$9600, price = \$40/unit
7. $V(t) = -1325t + 14000$
8. **(a)** $x = 0, 1/2$ **(b)** no fixed points
9. **(a)**

	x_1	x_2	x_3
From graph	0.56	0.28	0.52
From calculator	0.56	0.286	0.518

	x_4	x_5	x_6
From graph	0.33	0.49	0.36
From calculator	0.332	0.490	0.360

(b) $(-5 - \sqrt{85})/10 \approx -1.4220$ and $(-5 + \sqrt{85})/10 \approx 0.4220$: the iterates are approaching 0.4220.
(c) -0.4, 0.44, 0.4064, 0.4348, 0.4109, 0.4311; yes
(d) -3.4, -10.96, -119.52, -14285, -204055878, -4.16×10^{16}; no

10. x-intercept: 3; y-intercept: $-27/2$

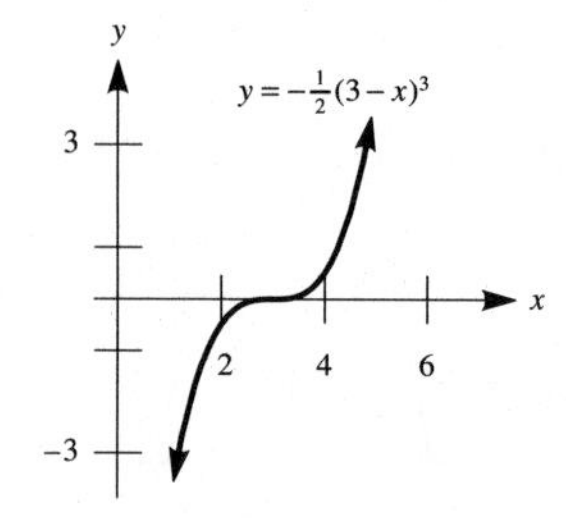

11. x-intercept: $3/2$; y-intercept: -3; vertical asymptote: $x = -1$; horizontal asymptote: $y = 2$

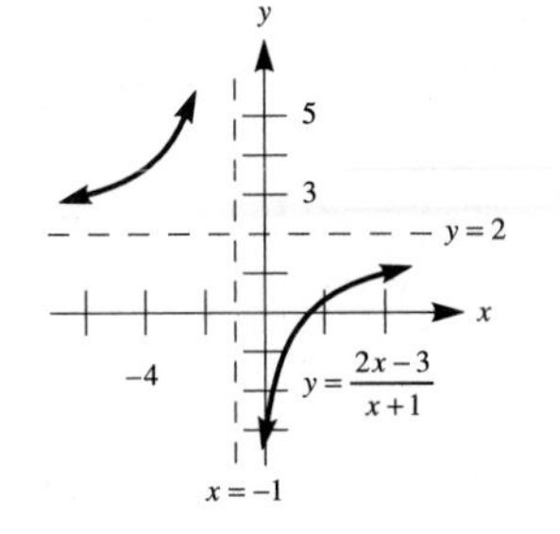

12. (a) $L(x) = \sqrt{10x^2 - 22x + 17}$
(b) $x = 1.1$
13. (a) vertical asymptotes: $x = 3$ and $x = -3$; horizontal asymptote: $y = 1$

(b)

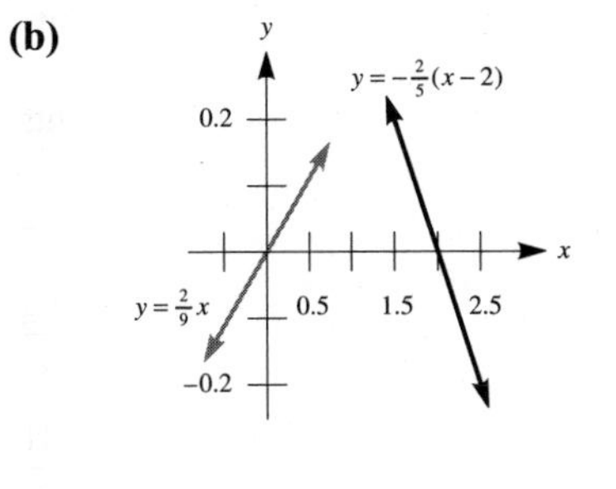

(c)

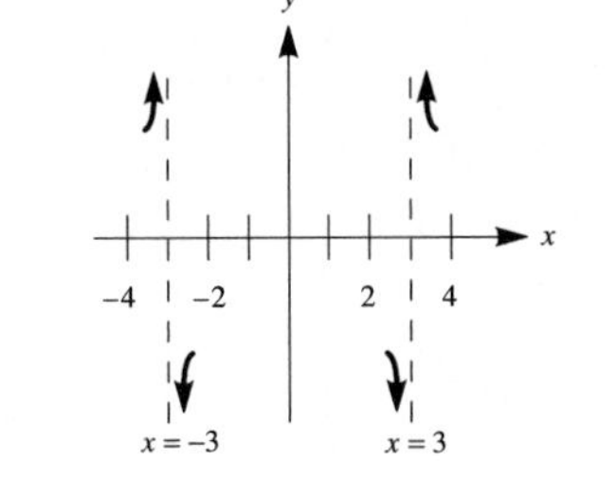

(d)

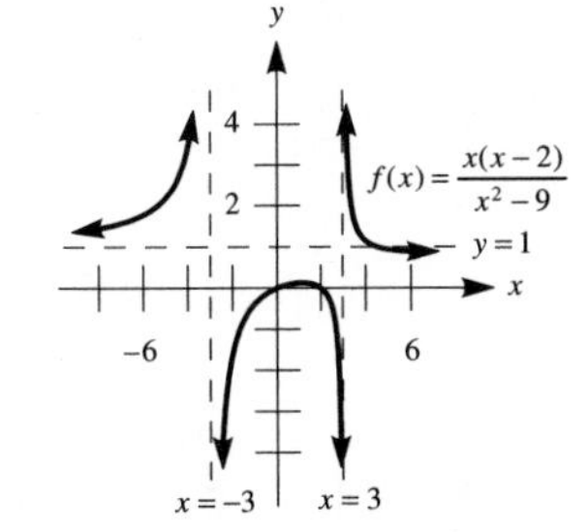

14. $A(w) = w\sqrt{64 - w^2}/2$
15. (a) As $|x|$ increases for x positive, $-x^3$ increases in the negative direction, which the pictured function does not.

(b) This graph has four turning points, and a polynomial function with highest-degree term $-x^3$ can have at most two turning points.

16. (a)

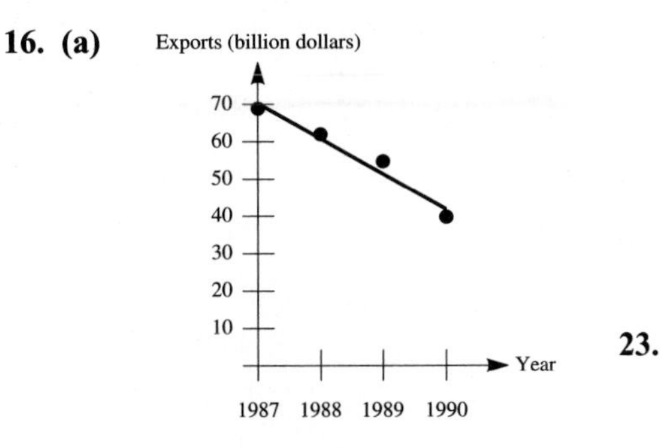

(b) \$32.6 billion; too high

CHAPTER FIVE

EXERCISE SET 5.1

1. (a) $2^{30} \approx 10^9$ **(b)** $2^{50} \approx 10^{15}$
3. 125 **5.** 16 **7.** 8 **9.** $5^{\sqrt{2}}$
11. (a) $x = 3$ **(b)** $t = 3/2$
(c) $y = 1/4$ **(d)** $z = 5/2$
13. $(-\infty, \infty)$ **15.** $(-\infty, \infty)$

17.

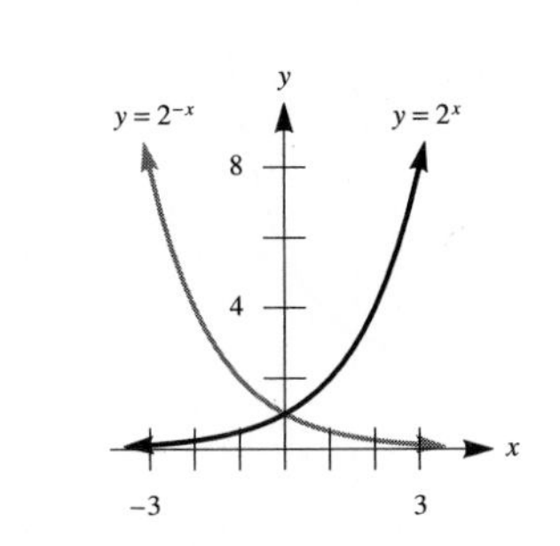

19.

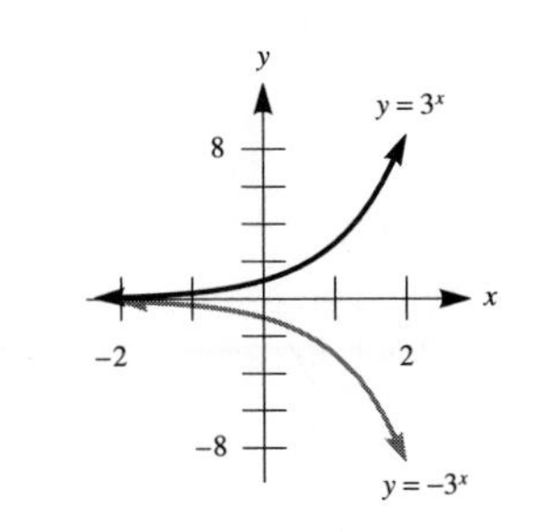

21.

23.

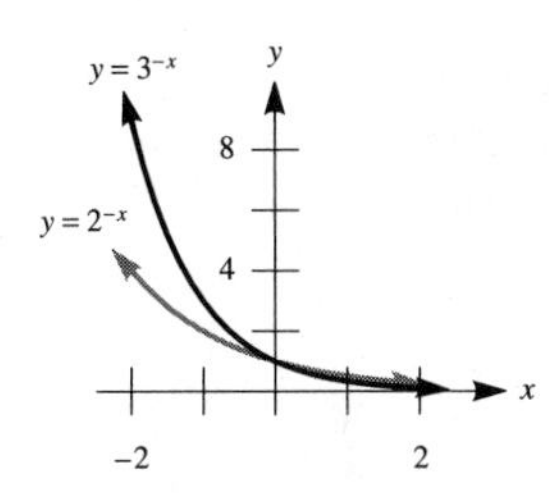

25. domain: $(-\infty, \infty)$; range: $(-\infty, 1)$; x- and y-intercepts: 0; asymptote: $y = 1$

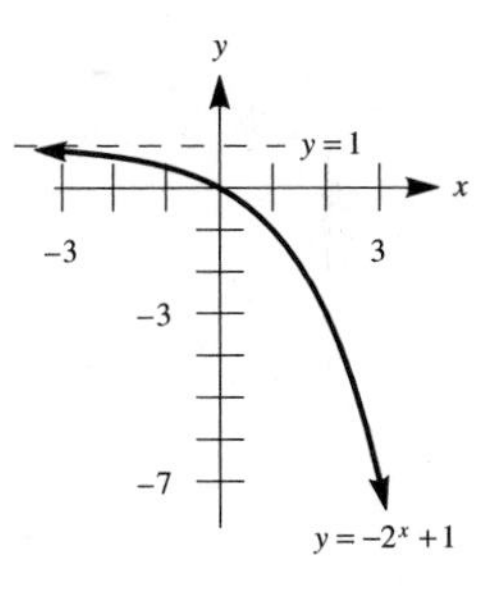

27. domain: $(-\infty, \infty)$; range: $(1, \infty)$; no x-intercept; y-intercept: 2; asymptote: $y = 1$

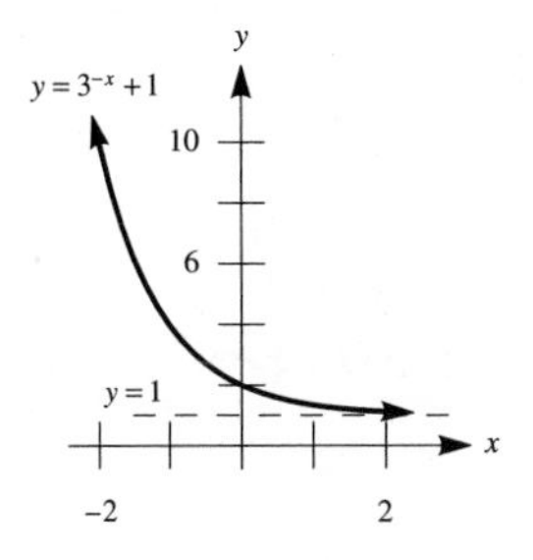

29. domain: $(-\infty, \infty)$; range: $(0, \infty)$; no x-intercept; y-intercept: $1/2$; asymptote: $y = 0$

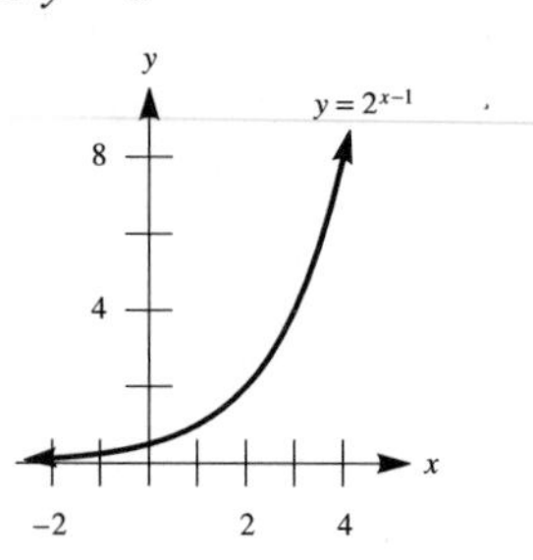

31. domain: $(-\infty, \infty)$; range: $(1, \infty)$; no x-intercept; y-intercept: 4; asymptote: $y = 1$

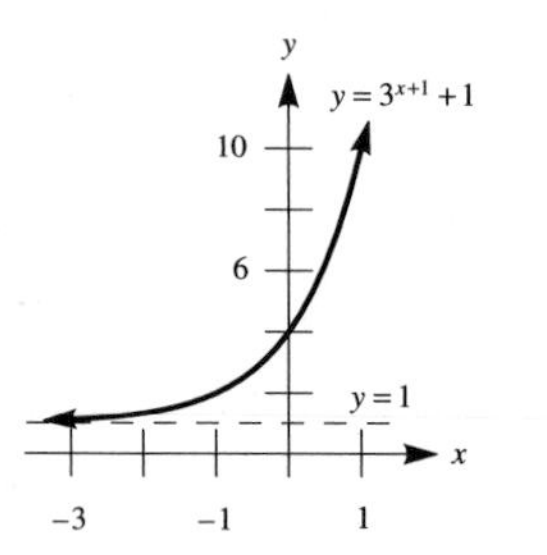

33. $x = -1/3$ **35.** $x = 3/2, 1$

37. (a)

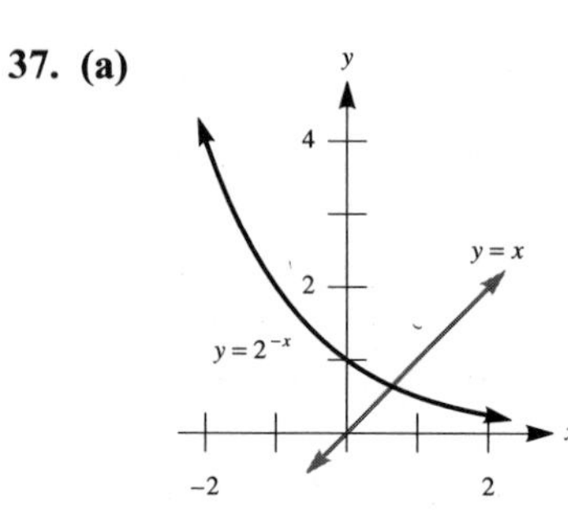

(b) between 0 and 1
(c) $(13 - \sqrt{137})/2 \approx 0.648$
(d) two decimal places

41. (a)

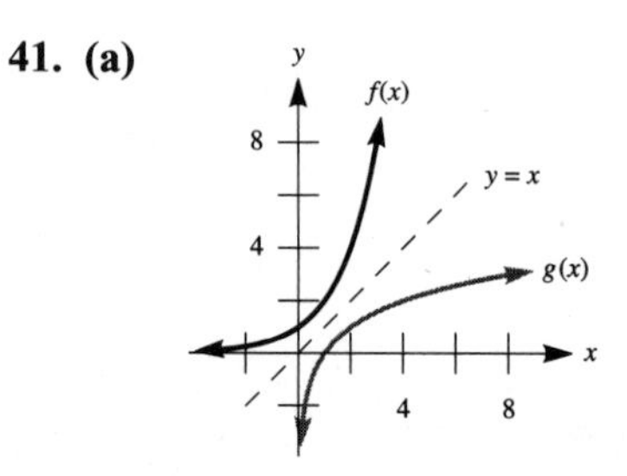

(b) domain: $(0, \infty)$; range: $(-\infty, \infty)$; x-intercept: 1; no y-intercept; asymptote: $x = 0$
43. (a) 1.4 **(b)** 1.41
45. (a) 1.5 **(b)** 1.52
47. (a) 1.7 **(b)** 1.73
49. (a) 1.6 **(b)** 1.62
51. $x \approx 0.3$ **53.** $x \approx 0.7$

55.

x	$\log_{10} x$
1	0.00
2	0.30
3	0.48
4	0.60
5	0.70
6	0.78
7	0.85
8	0.90
9	0.95
10	1.00

GRAPHING UTILITY EXERCISES FOR SECTION 5.1

1. $x = 1.5$

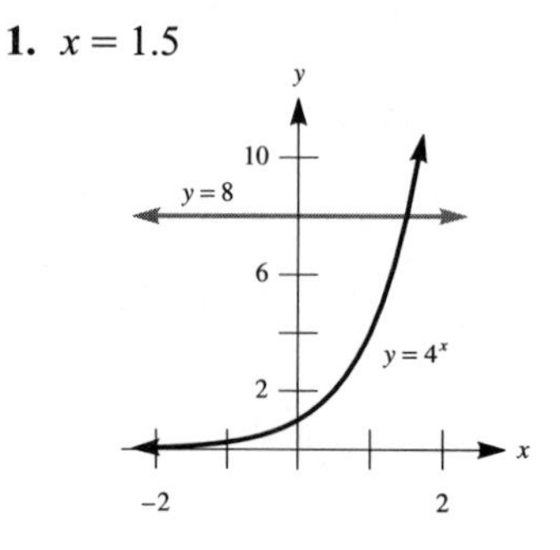

3. $x = -1$ and $x = 1$; they are the solutions to $x^2 2^x - 2^x = 0$.

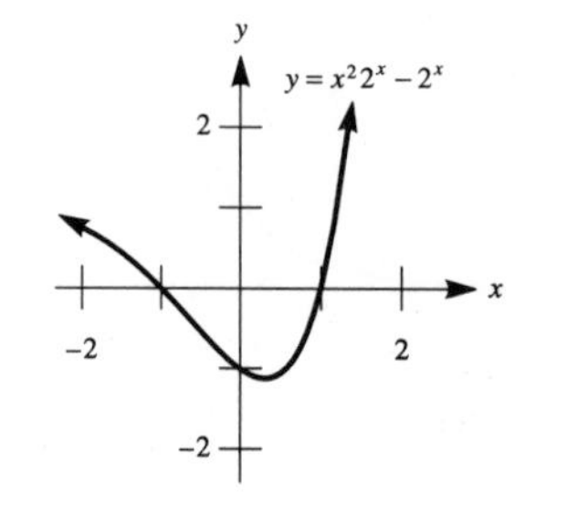

5. (a) no x-intercept; y-intercept: $-1/9$; asymptote: $y = 0$

(b)

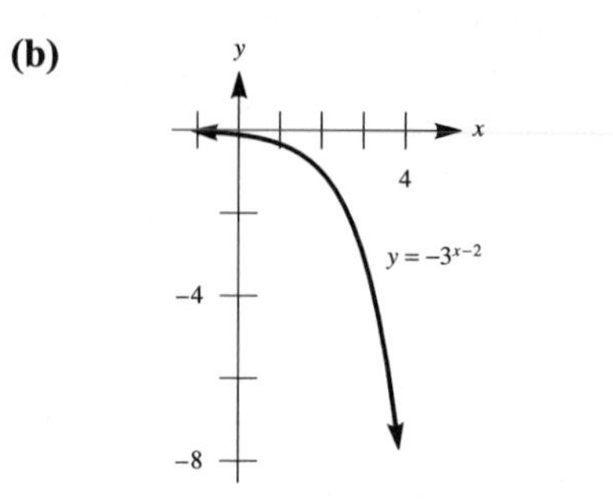

7. (a) x-intercept: -1; y-intercept: -3; asymptote: $y = -4$

(b)

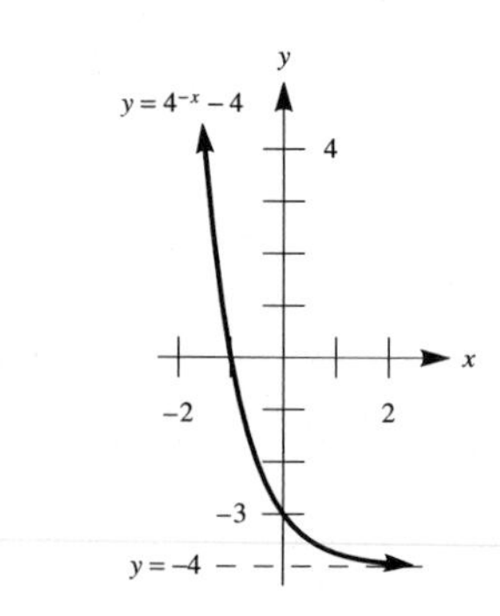

9. (a) no x-intercept; y-intercept: $1/10$; asymptote: $y = 0$

(b)

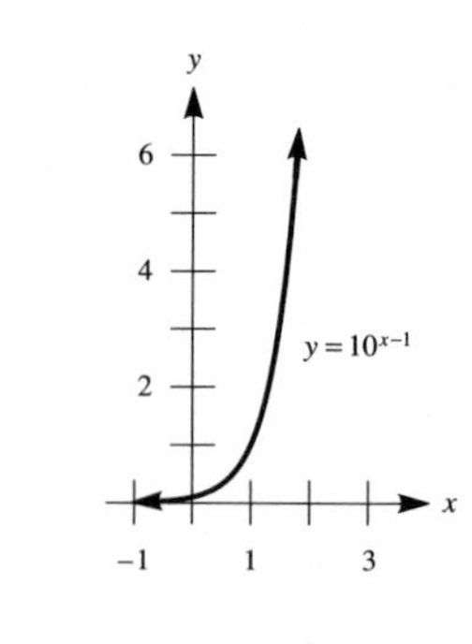

11. (a)

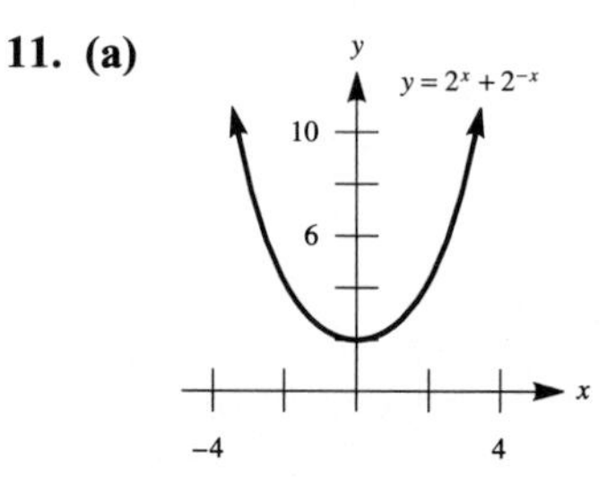

(b)

(c)

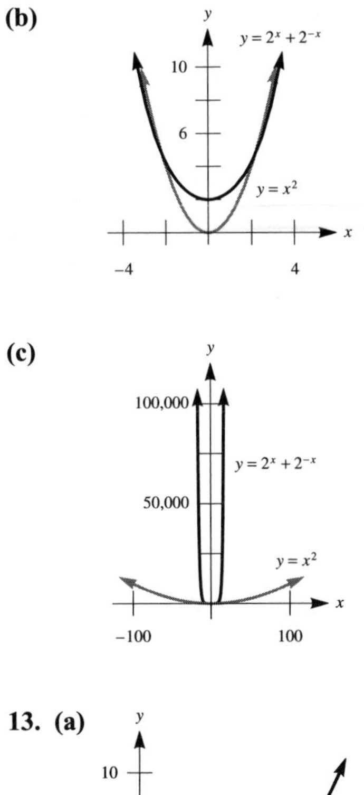

13. (a)

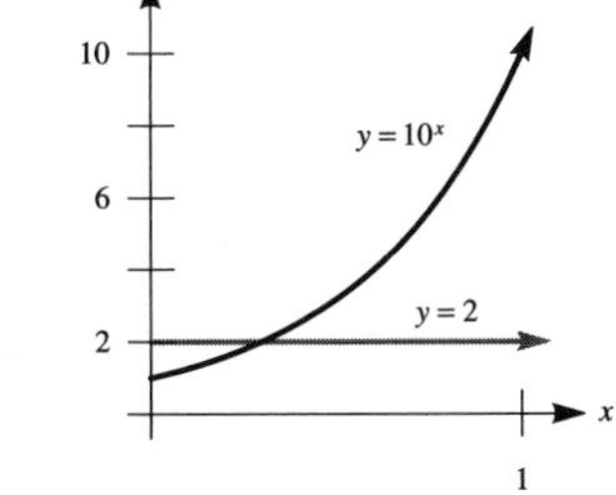

(b) The point (0.3, 2) is the approximate point of intersection, and thus $10^{0.3} \approx 2$.

EXERCISE SET 5.2

1. domain: $(-\infty, \infty)$; range: $(0, \infty)$; no x-intercept; y-intercept: 1; asymptote: $y = 0$

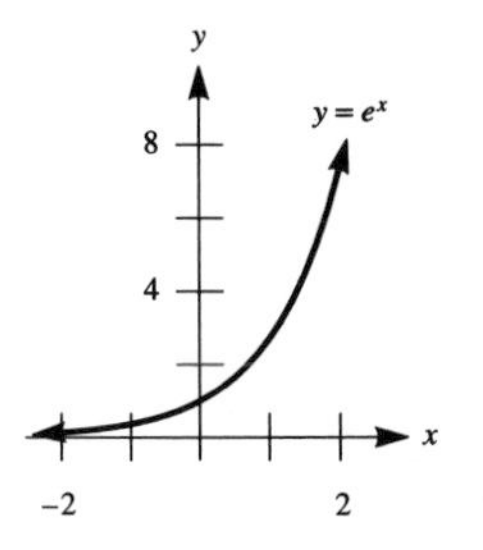

3. domain: $(-\infty, \infty)$; range: $(-\infty, 0)$; no x-intercept; y-intercept: -1; asymptote: $y = 0$

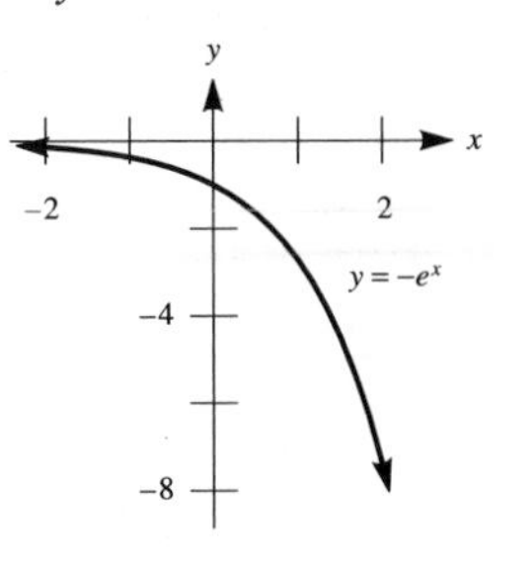

5. domain: $(-\infty, \infty)$; range: $(1, \infty)$; no x-intercept; y-intercept: 2; asymptote: $y = 1$

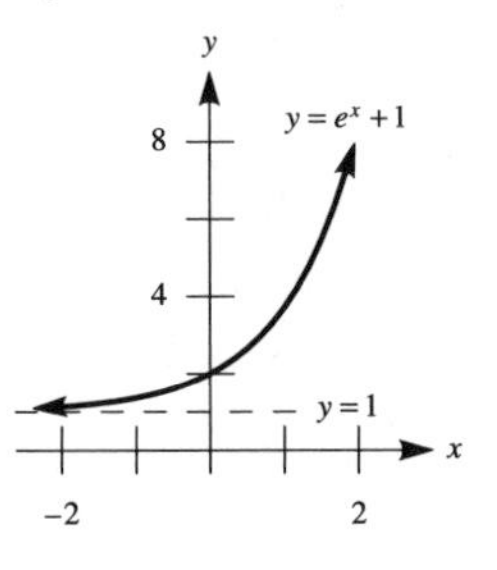

7. domain: $(-\infty, \infty)$; range: $(1, \infty)$; no x-intercept; y-intercept: $e + 1$; asymptote: $y = 1$

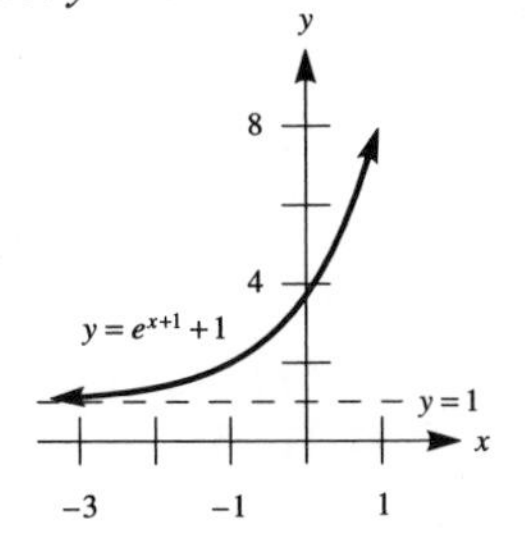

9. domain: $(-\infty, \infty)$; range: $(-\infty, 0)$; no x-intercept; y-intercept: $-1/e^2$; asymptote: $y = 0$

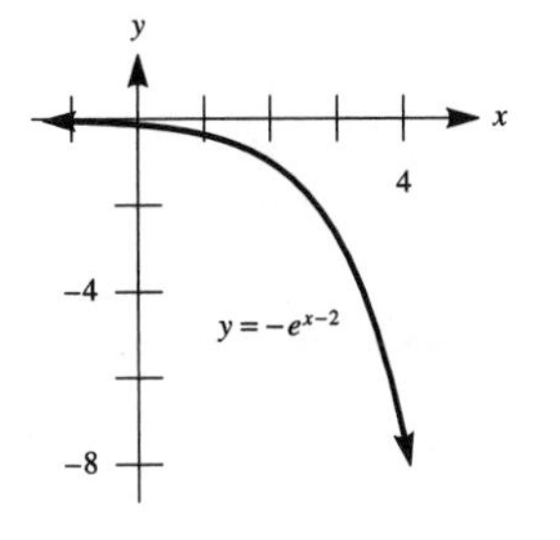

11. domain: $(-\infty, \infty)$; range: $(-\infty, e)$; x-intercept: 1; y-intercept: $e - 1$; asymptote: $y = e$

13.

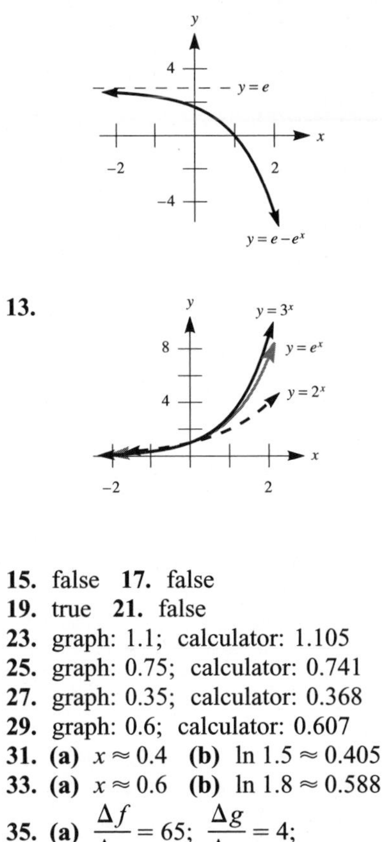

15. false **17.** false
19. true **21.** false
23. graph: 1.1; calculator: 1.105
25. graph: 0.75; calculator: 0.741
27. graph: 0.35; calculator: 0.368
29. graph: 0.6; calculator: 0.607
31. **(a)** $x \approx 0.4$ **(b)** $\ln 1.5 \approx 0.405$
33. **(a)** $x \approx 0.6$ **(b)** $\ln 1.8 \approx 0.588$
35. **(a)** $\dfrac{\Delta f}{\Delta x} = 65$; $\dfrac{\Delta g}{\Delta x} = 4$; $\dfrac{\Delta h}{\Delta x} \approx 12.70$

(b) $\dfrac{\Delta f}{\Delta x} = 2465$; $\dfrac{\Delta g}{\Delta x} = 256$; $\dfrac{\Delta h}{\Delta x} \approx 5122.13$

(c) $\dfrac{\Delta f}{\Delta x} = 12{,}209$; $\dfrac{\Delta g}{\Delta x} = 16{,}384$; $\dfrac{\Delta h}{\Delta x} \approx 2{,}066{,}413$

37. **(a)** e^π is larger **(b)** e^π is larger
39. **(a)**

	x_1	x_2	x_3
From graph	0.37	0.70	0.50
From calculator	0.3679	0.6922	0.5005

	x_4	x_5
From graph	0.60	0.55
From calculator	0.6062	0.5454

(b) seventh iterate; the eighth iterate does not begin with 56 in the first two decimal places; the ninth iterate does have 56 in the first two decimal places
(c) no
41. (a) $(-\infty, \infty)$ **(b)** 1
(c) symmetric about the y-axis

(d)

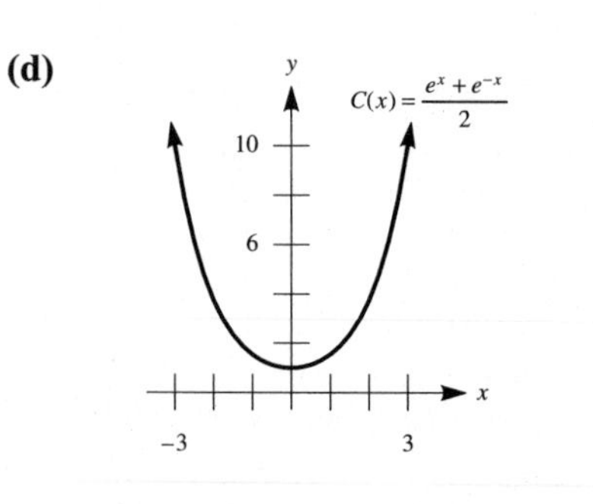

43. (a)

x	1000	10^4	10^5
$\left(1 + \frac{2}{x}\right)^x$	7.374312	7.387579	7.388908

x	10^6	10^7
$\left(1 + \frac{2}{x}\right)^x$	7.389041	7.389055

(b) $e^2 \approx 7.389056$; the values in the table are approaching e^2

45. (a)

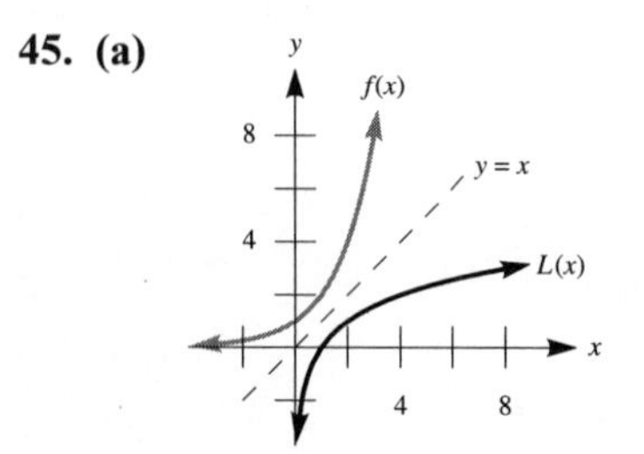

(b) domain: $(0, \infty)$; range: $(-\infty,\infty)$; x-intercept: 1; asymptote: $x = 0$
(c) **(i)** x-intercept: 1; asymptote: $x = 0$

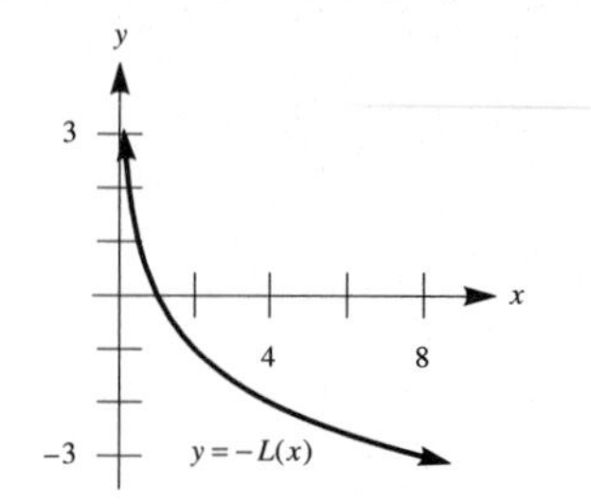

(ii) x-intercept: -1; asymptote: $x = 0$

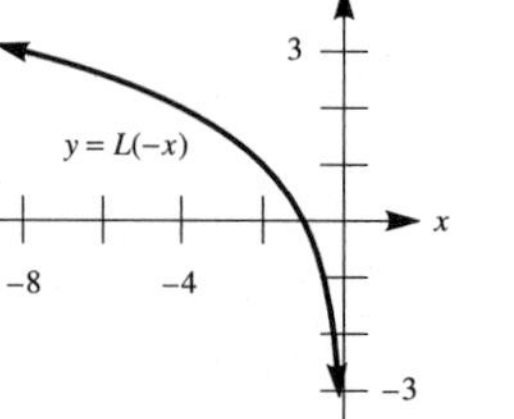

(iii) x-intercept: 2; asymptote: $x = 1$

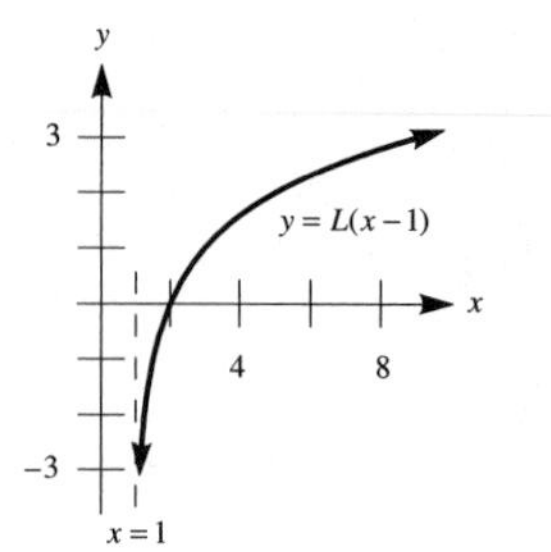

47. A, D, E, G **49.** B, D, E, G
51. A, D, F, G, H **53.** A, D, F, G, H

GRAPHING UTILITY EXERCISES FOR SECTION 5.2

1. (a)

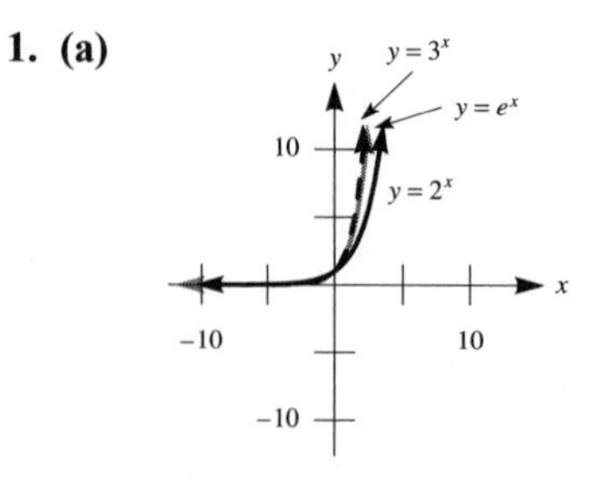

(b)

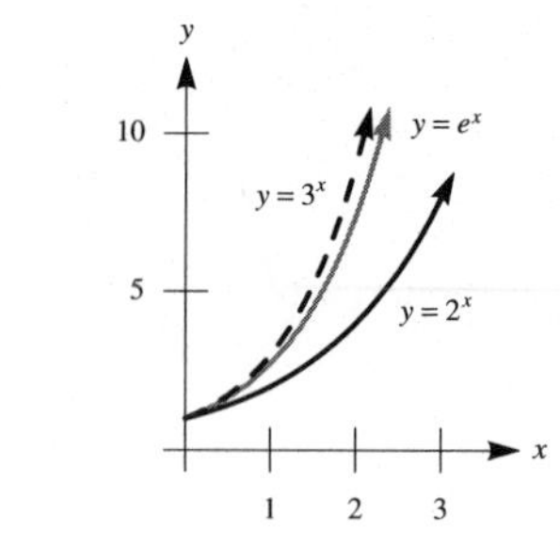

(c)

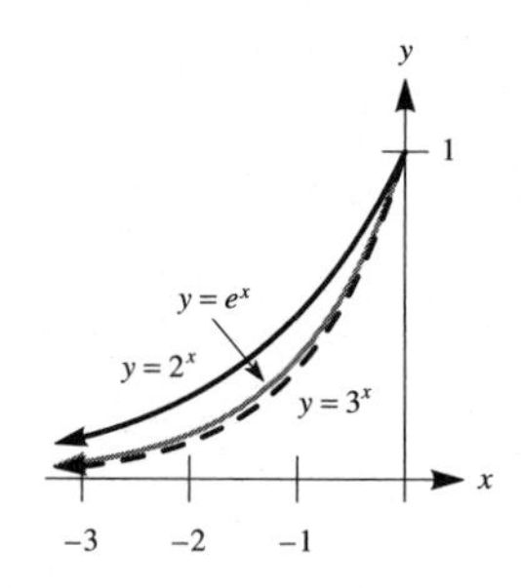

3. a reflection of $y = e^x$ across the y-axis

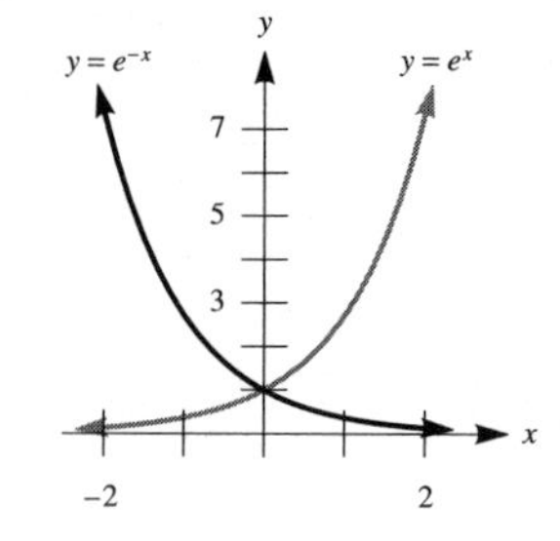

5. a reflection of $y = e^x$ across the x- and y-axes

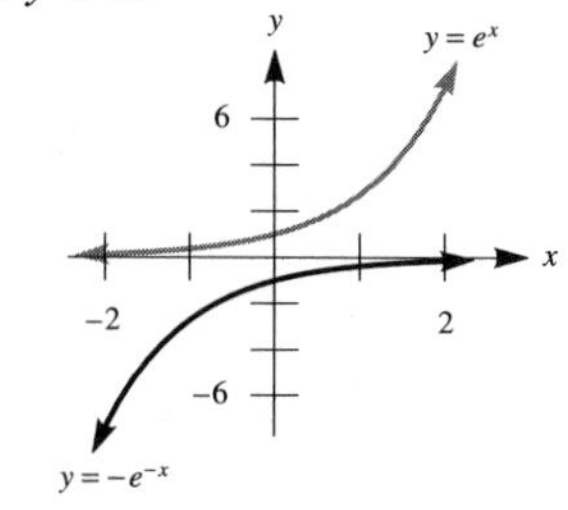

7. (a) a translation of $y = e^x$ to the right 1 unit

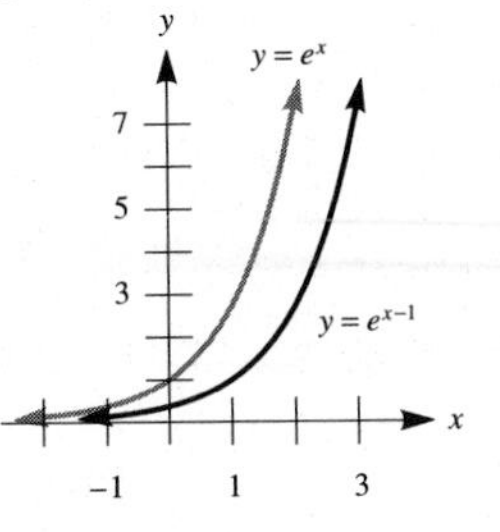

(b) a translation of $y = e^x$ to the right 1 unit, then a reflection across the y-axis

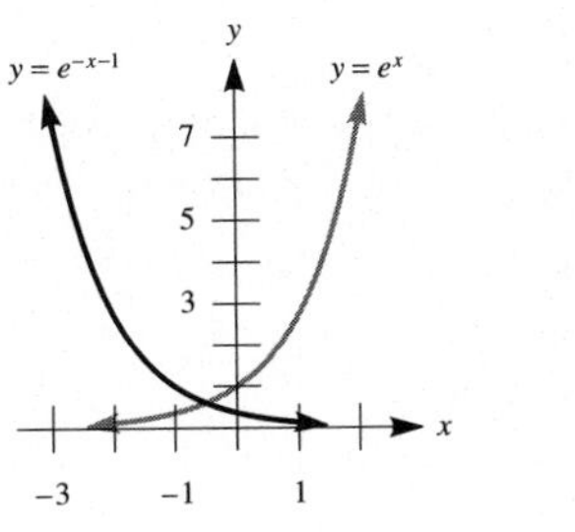

9. (a) y-intercept: 1; as k increases, the graph of $y = e^{x/k}$ flattens.

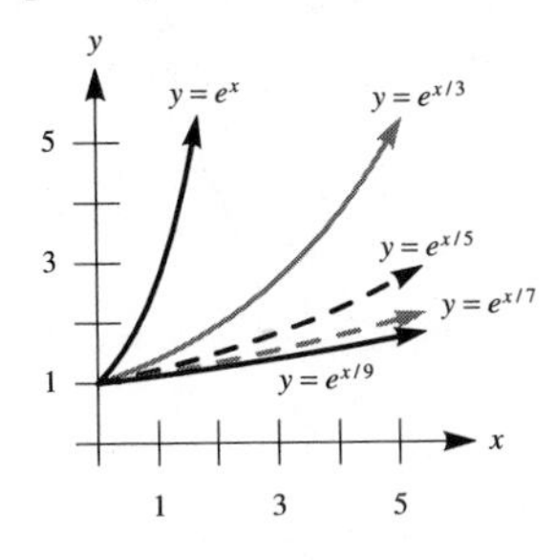

(b) It should be much flatter than the others.

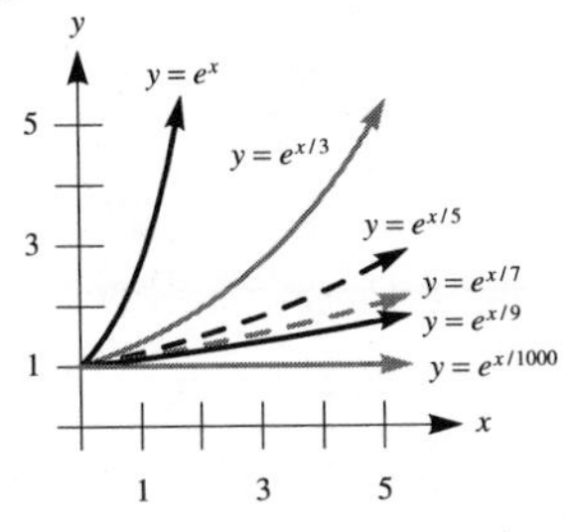

(c) The graph takes on the more familiar exponential shape.

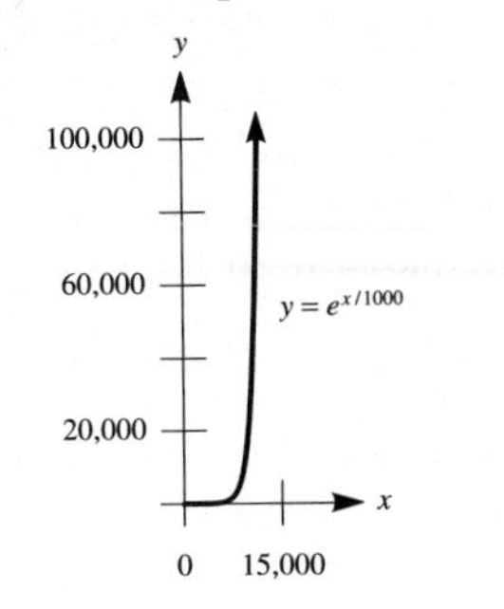

11. (a)

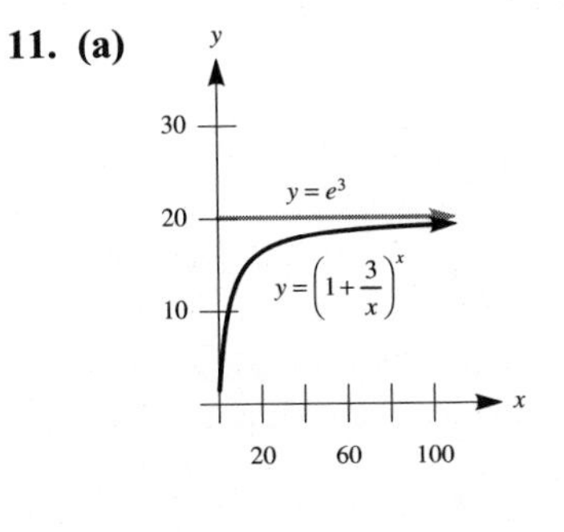

(b)

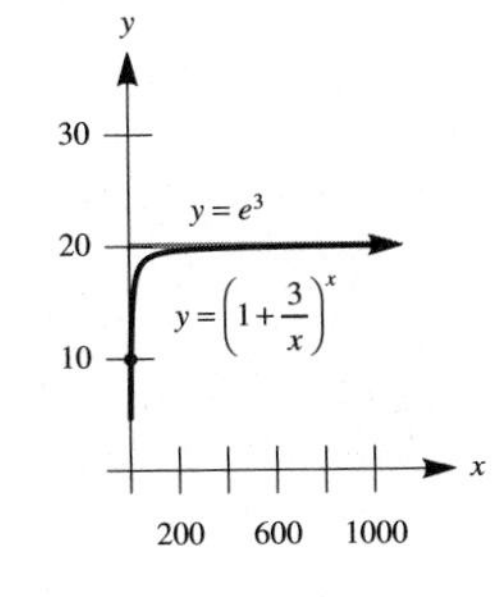

EXERCISE SET 5.3

1. (a) no **(b)** yes **(c)** yes
3. (a) $(4x + 1)/(2 - 3x)$
(b) $(3x + 4)/(2x - 1)$ **(c)** $1/2$
(d) -4 **5.** $(-1, 3)$ and $(6, -1)$
7. (a) $\log_3 9 = 2$ **(b)** $\log_{10} 1000 = 3$
(c) $\log_7 343 = 3$ **(d)** $\log_2 \sqrt{2} = 1/2$
9. (a) $2^5 = 32$ **(b)** $10^0 = 1$
(c) $e^{1/2} = \sqrt{e}$ **(d)** $3^{-4} = 1/81$
(e) $t^v = u$ **11.** $\log_5 30$
13. (a) $3/2$ **(b)** $-5/2$ **(c)** $3/2$
15. (a) $x = 1/16$ **(b)** $x = e^{-2} \approx 0.14$
17. (a) $(0, \infty)$ **(b)** $\left(-\infty, 3/4\right)$
(c) $(-\infty, 0) \cup (0, \infty)$ **(d)** $(0, \infty)$
(e) $(-\infty, -5) \cup (5, \infty)$
19. A: $(0, 1)$; B: $(1, 0)$; C: $(4, 2)$; D: $(2, 4)$

21. (a) domain: $(0, \infty)$; range: $(-\infty, \infty)$; x-intercept: 1; no y-intercept; asymptote: $x = 0$

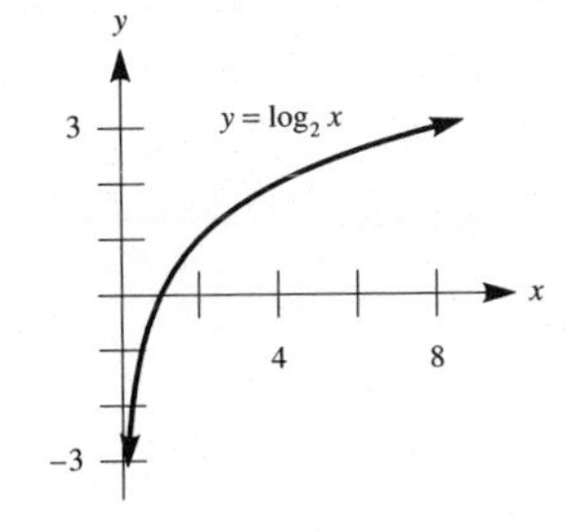

(b) domain: $(0, \infty)$; range: $(-\infty, \infty)$; x-intercept: 1; no y-intercept; asymptote: $x = 0$

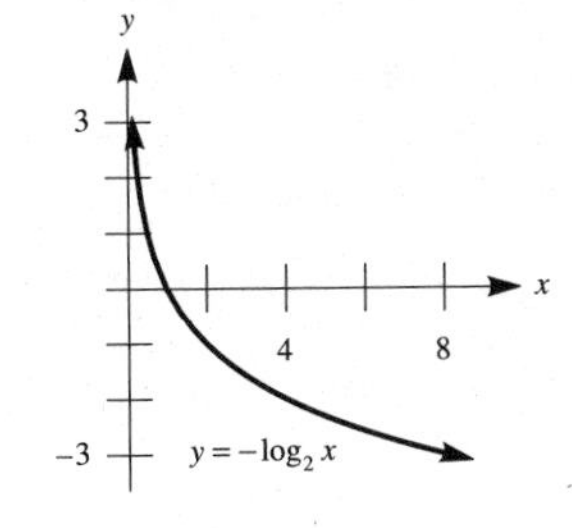

(c) domain: $(-\infty, 0)$; range: $(-\infty, \infty)$; x-intercept: -1; no y-intercept; asymptote: $x = 0$

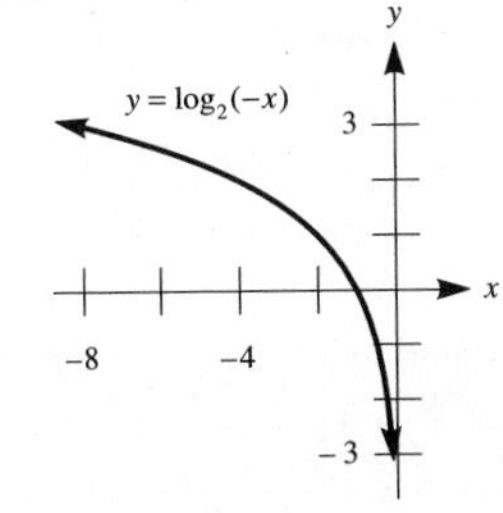

(d) domain: $(-\infty, 0)$; range: $(-\infty, \infty)$; x-intercept: -1; no y-intercept; asymptote: $x = 0$

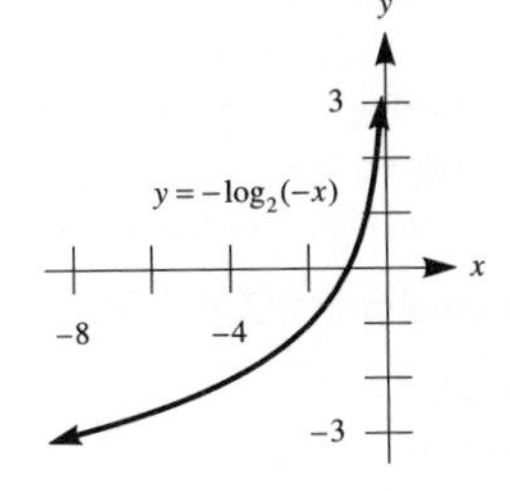

23. domain: $(2, \infty)$; range: $(-\infty, \infty)$; x-intercept: 5; no y-intercept; asymptote: $x = 2$

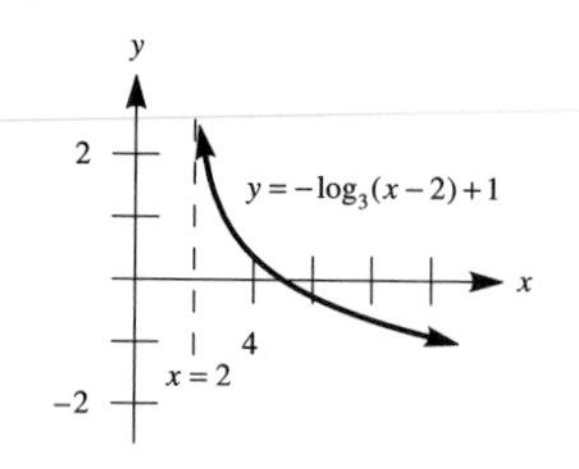

25. domain: $(-e, \infty)$; range: $(-\infty, \infty)$; x-intercept: $-e + 1$; y-intercept: 1; asymptote: $x = -e$

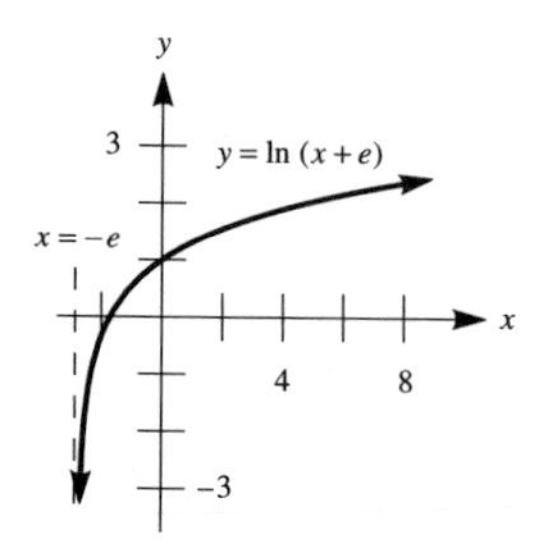

27. **(a)** 4 **(b)** -1 **(c)** $1/2$
29. $x = \log_{10} 25 \approx 1.40$
31. $x = \pm\sqrt{1 + \log_{10} 4} \approx \pm 1.27$
33. $t = (-3 + \ln 10)/2 \approx -0.35$
35. $t = (1 - \ln 12.405)/4 \approx -0.38$
37. **(a)** No, the value is not exact, because $-\log(0.4) \approx 0.39794$.

(b)

	x_1	x_2	x_3
From graph	0.22	0.65	0.20
From calculator	0.2218	0.6539	0.1845

	x_4	x_5	x_6
From graph	0.75	0.15	0.85
From calculator	0.7341	0.1342	0.8721

(c) $x_7 \approx 0.0594$; $x_8 \approx 1.2260$; $x_9 \approx -0.0885$; no
(d) $\log(-0.0885)$ does not exist.
39. $x \approx 1.61$; 0.035% error
41. **(a)** $A(1.6, 0.470)$
(b) $B(0.470, 0.470)$
(c) $C(0.470, -0.755)$
(d) $D(-0.755, -0.755)$
(e) $E(-0.755, -0.281)$
43. A: $g(x) = \ln x$; B: $f(x) = \sqrt{x}$
45. A: $g(x) = \ln x$; B: $f(x) = \sqrt[10]{x}$
47. Since $\log_3 3 = 1$ while $\log_2 3 > 1$, the upper curve must be $y = \log_2 x$ and the lower curve must be $y = \log_3 x$.

49. x-intercept: e; asymptote: $x = 0$

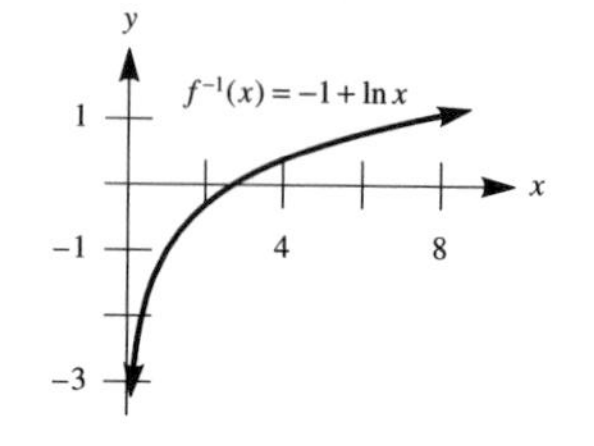

51. The area of the shaded region is less than that of the rectangle (defined by dashed lines on the graph), which has an area of two square units.

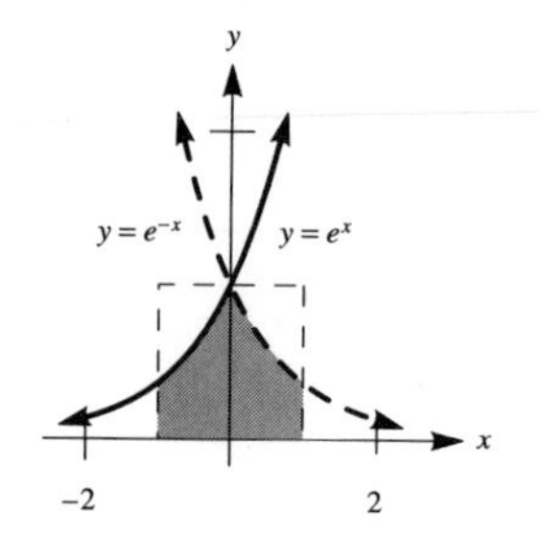

53. $x = \ln 6$ **55.** $x \approx 10^{30}$
57. **(a)** 3.5; acid **(b)** 0; acid
59. **(a)** $P(10) = 4$, $P(18) = 7$, $P(19) = 8$
(b)

x	$P(x)$	$\dfrac{x}{\ln x}$	$\dfrac{P(x)}{x/\ln x}$
10^2	25	22	1.151
10^4	1229	1086	1.132
10^6	78498	72382	1.084
10^8	5761455	5428681	1.061
10^9	50847534	48254942	1.054
10^{10}	455052512	434294482	1.048

(c)

x	$P(x)$	$\dfrac{x}{\ln x - 1.08366}$
10^2	25	28
10^4	1229	1231
10^6	78498	78543
10^8	5761455	5768004
10^9	50847534	50917519
10^{10}	455052512	455743004

x	$\dfrac{P(x)}{x/(\ln x - 1.08366)}$
10^2	0.8804
10^4	0.9988
10^6	0.9994
10^8	0.9989
10^9	0.9986
10^{10}	0.9985

61. A, D, E, H **63.** B, D, E, H
65. A, D, E, H **67.** A, D, E, H
69. $n = 27$

71.

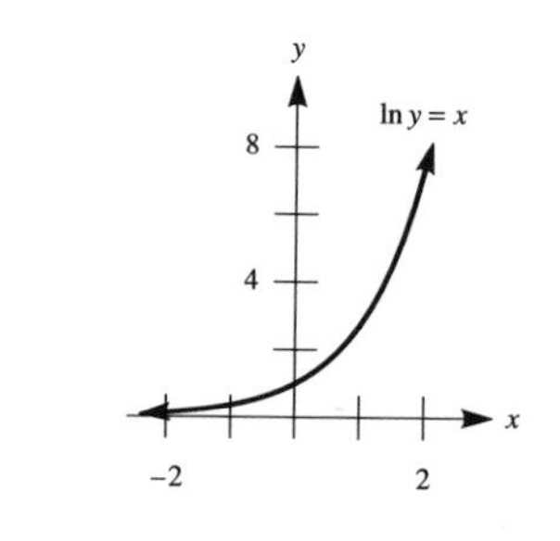

GRAPHING UTILITY EXERCISES FOR SECTION 5.3

1. **(a)** symmetric about the line $y = x$

(b)

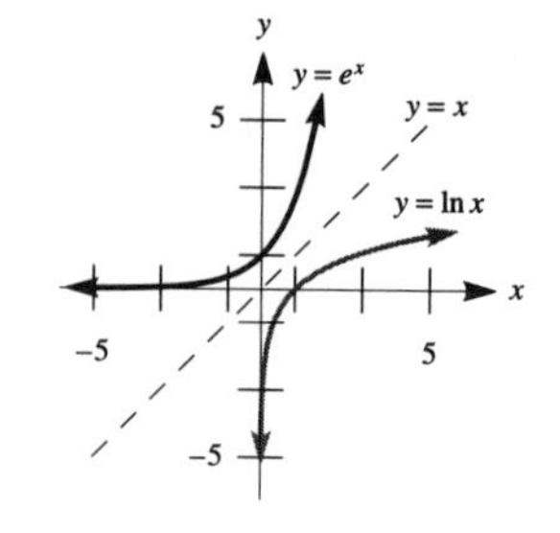

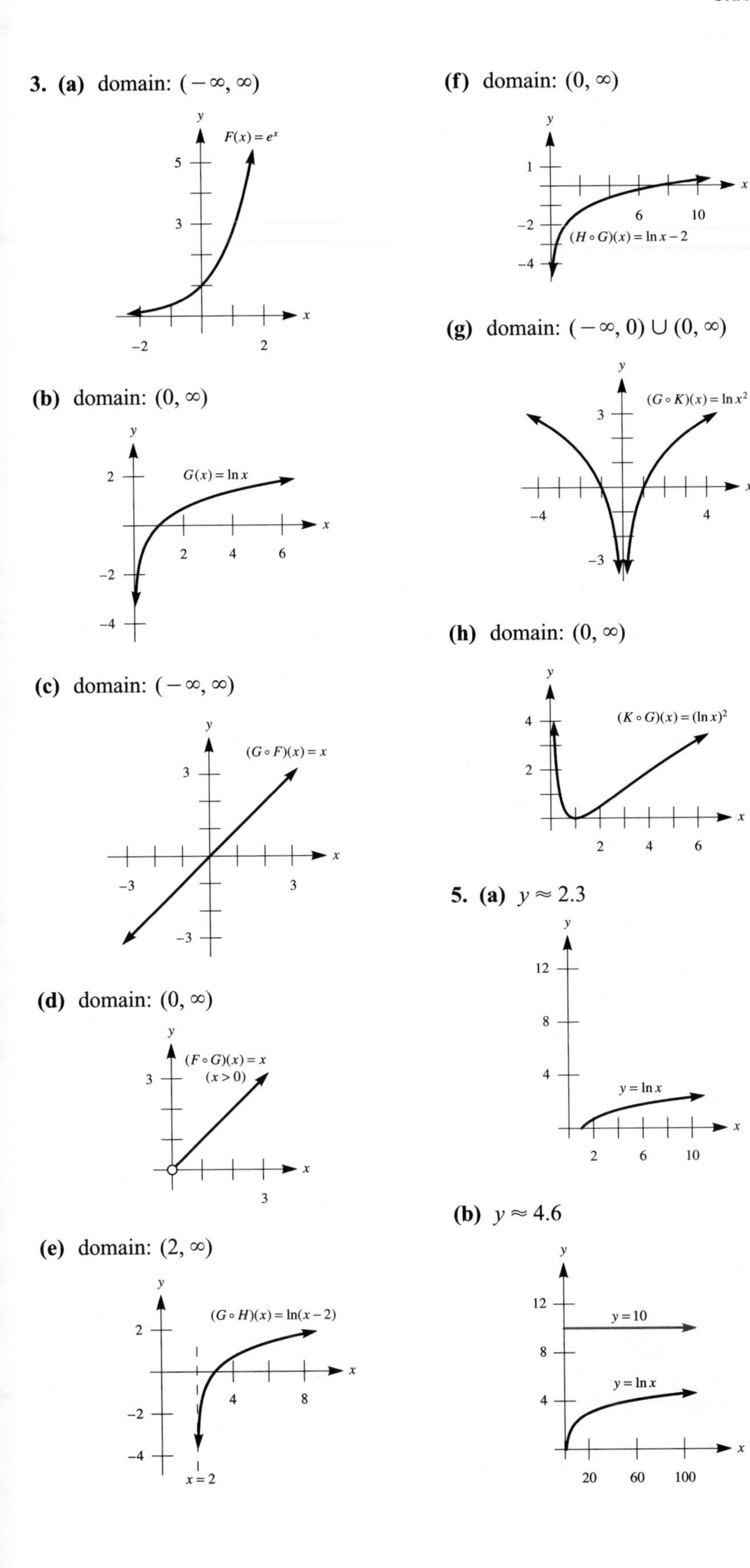

3. (a) domain: $(-\infty, \infty)$

(b) domain: $(0, \infty)$

(c) domain: $(-\infty, \infty)$

(d) domain: $(0, \infty)$

(e) domain: $(2, \infty)$

(f) domain: $(0, \infty)$

(g) domain: $(-\infty, 0) \cup (0, \infty)$

(h) domain: $(0, \infty)$

5. (a) $y \approx 2.3$

(b) $y \approx 4.6$

(c) $y \approx 6.9$

(d) $y \approx 9.2$

(e) $x = e^{10} \approx 22026$

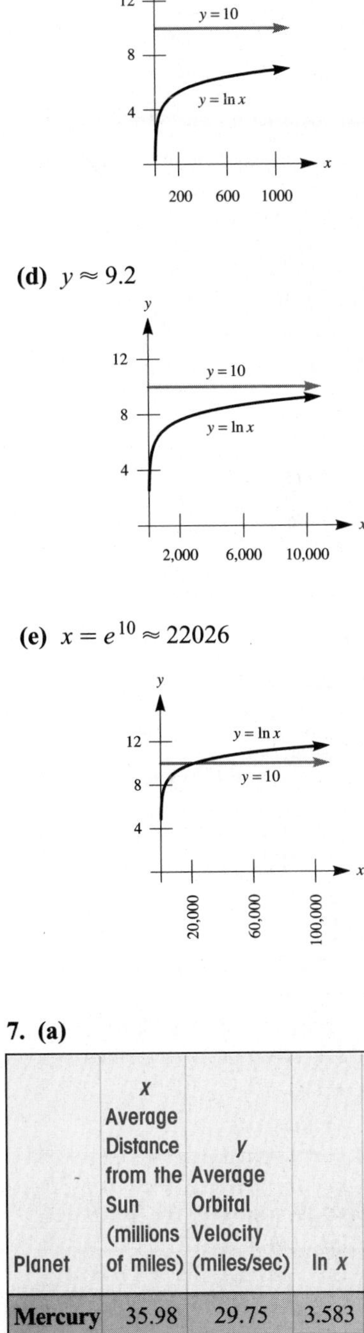

7. (a)

Planet	x Average Distance from the Sun (millions of miles)	y Average Orbital Velocity (miles/sec)	ln x	ln y
Mercury	35.98	29.75	3.583	3.393
Venus	67.08	21.76	4.206	3.080
Earth	92.96	18.51	4.532	2.918
Mars	141.64	14.99	4.953	2.707

(b) $\ln y = -0.500 \ln x + 5.184$

(d)

Planet	x Average Distance from the Sun (millions of miles)	y Average Orbital Velocity (miles/sec) from Equation (2)	y Average Orbital Velocity (miles/sec) from Observation
Jupiter	483.63	8.11	8.12
Saturn	888.22	5.99	5.99
Uranus	1786.55	4.22	4.23
Neptune	2799.06	3.37	3.38
Pluto	3700.75	2.93	2.95

41. (a) $x = (\ln 5)/(\ln 2) \approx 2.32$

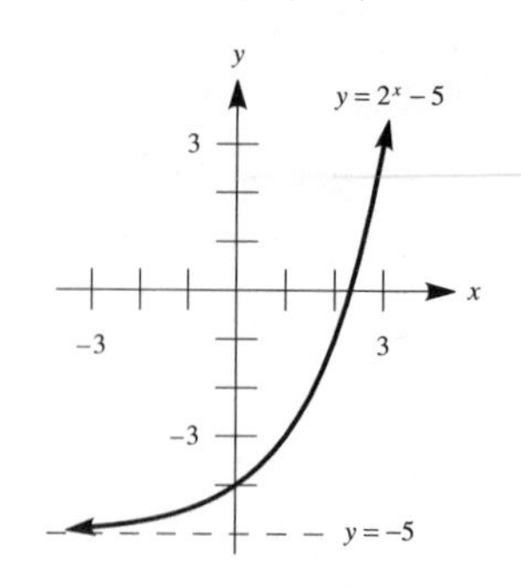

(b) $x = (2 \ln 5)/(\ln 2) \approx 4.64$

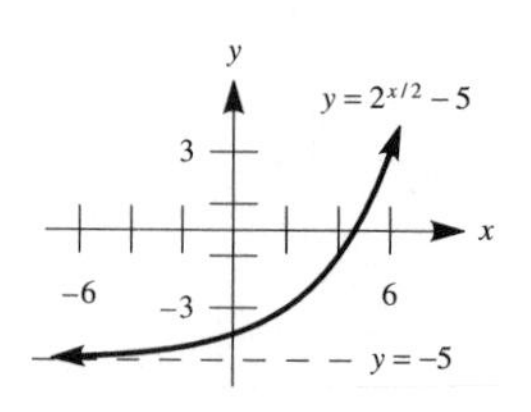

EXERCISE SET 5.4

1. 1 **3.** 1/2 **5.** 4 **7.** 0
9. 4 **11.** $\log_{10} 60$ **13.** $\log_5 20$
15. (a) $\ln 6$ **(b)** $\ln(3/16384)$
17. $\log_b\left[\dfrac{4(1+x)^3}{(1-x)^{3/2}}\right]$
19. $\log_{10}\left[\dfrac{27\sqrt{x+1}}{(x^2+1)^6}\right]$
21. (a) $2 \log_{10} x - \log_{10}(1+x^2)$
(b) $2 \ln x - \frac{1}{2}\ln(1+x^2)$
23. (a) $\frac{1}{2}\log_{10}(3+x) + \frac{1}{2}\log_{10}(3-x)$
(b) $\frac{1}{2}\ln(2+x) + \frac{1}{2}\ln(2-x) - \ln(x-1)$ $-\frac{3}{2}\ln(x+1)$
25. (a) $\frac{1}{2}\log_b x - \frac{1}{2}$
(b) $\ln(1+x^2) + \ln(1+x^4) + \ln(1+x^6)$
27. (a) $A+B$ **(b)** $-A-B$ **(c)** $3B$
(d) $-3B$ **29. (a)** $C-B$ **(b)** $B-C$
(c) $C-2B$ **(d)** $C-4A$
31. (a) $1/B$ **(b)** $(A+C+1)/B$
33. (a) $A/(B+1)$ **(b)** $(B+C)/(B+1)$
35. (a) 1 **(b)** 1
37. (a) $a+2b+3c$ **(b)** $1+\frac{1}{2}a$
(c) $\frac{1}{2}(1+a+b+c)$
(d) $1+a-\frac{1}{2}b-\frac{1}{2}c$
39. (a) $1+t$ **(b)** $u-t$
(c) $\frac{3}{2}t+\frac{1}{2}u-1$ **(d)** $2+t+\frac{1}{2}u$

43. $x = (\ln 5 - \ln 2 + 1)/2$
45. $x = (\ln 13)/(\ln 2)$ **47.** $x = 1/(\ln 10)$
49. $x = \pm\sqrt{1 + \dfrac{\log 12}{\log 3}} \approx \pm 1.806$
51. $(\log_{10} 5)/(\log_{10} 2)$
53. $(\log_{10} 3)/(\log_{10} e)$
55. $(\log_{10} 2)/(\log_{10} b)$
57. $(\ln 6)/(\ln 10)$ **59.** $1/(\ln 10)$
61. $[\ln(\ln x) - \ln(\ln 10)]/(\ln 10)$
63. (a) true **(b)** true **(c)** true
(d) false **(e)** true **(f)** false **(g)** true
(h) false **(i)** true **(j)** false
(k) false **(l)** true **(m)** true
65. (f) 2345.6 **(g)** 0.123456
67.

x	0.1	0.05
$\ln(1+x)$	0.095310	0.048790

x	0.005	0.0005
$\ln(1+x)$	0.004987	0.000499

69. $a = 4$, $b = -(\ln 243)/2.5$
71. (a) $x = (\log_{10} 3)/(\log_{10} 2)$
(b) $(\log_{10} 3)/(\log_{10} 2) \approx 1.585$; $(\ln 3)/(\ln 2) \approx 1.585$ **75.** x^3
79. (c) $\log_\pi 2 + \dfrac{1}{\log_\pi 2}$ is larger.
85. (b) Both sides are -0.470.

EXERCISE SET 5.5

1. $x = (\ln 3)/(2 \ln 3 - \ln 5) \approx 1.869$
3. $x = e^{e^{1.5}} \approx 88.384$ **5.** $x = \pm 8$
7. $x = (3 - \sqrt{809})/4 \approx -6.361$; $x = (3 + \sqrt{809})/4 \approx 7.861$
9. $x = \log 2 \approx 0.301$ **11. (a)** $x > 0$
(b) $e^{-\sqrt{3}} \approx 0.177$; $e^{\sqrt{3}} \approx 5.652$; 1
13. (a) $x > 0$ **(b)** $x = 6^{1/5} \approx 1.431$
15. $x > 0$ **17.** $x = \sqrt{3} \approx 1.732$
19. no real solutions
21. no real solutions **23.** $x = 7/3$
25. (a) no real solutions **(b)** $x = 0$
(c) $x = \ln 3 \approx 1.099$
(d) $x = \ln(1+\sqrt{5}) \approx 1.174$
27. $x = \ln[(1+\sqrt{5})/2] \approx 0.481$
29. $x = \dfrac{3 \ln 5}{5 \ln 2 - \ln 3 - \ln 5} \approx 6.372$
31. $x = 2$ **33.** $x = 1/2$ **35.** $x = 3$
37. $x = 203/99$ **39.** $x = 3$ **41.** $x = 7$
43. (a) $x = 10^y/[3(10^y) - 1]$
(b) $x = (1-y)/2$ **45.** $x \leqslant -1$
47. $x \geqslant \ln(43/4) \approx 2.375$
49. $x < (2 - e^2)/5 \approx -1.078$
51. $x \geqslant (\ln 100 - 2) \approx 2.605$
53. all real numbers
55. $-1 < x < 1/2$
57. $x \leqslant -1$ or $x \geqslant 5$ **59.** $0 < x < 1$
61. (a) $(4, \infty)$ **(b)** $4 < x \leqslant 7$
63. $0 < x < 3$
65. (a) Domain for $y = \log_2(2x-1)$: $x > 1/2$; domain for $y = \log_4 x$: $x > 0$
(b) $x > 1$
67. $x = e^{-4/3} \approx 0.264$; $x = e^2 \approx 7.389$
69. $x > 0$ **71.** $x = \beta^{-1/\alpha}$
73. $x = (1/k)\ln(y/A)$
75. $x = (-1/k)\ln[(a-y)/by]$
81. Since $e > 0$ and $2 - e < 2$, thus $\dfrac{2-e}{3} < \dfrac{2}{3}$.
83. $(-3, -\sqrt{6}) \cup (\sqrt{6}, 3)$
85. $x = 1$ and $x = 2$
87. $x = \ln(a-b)/\ln(a+b)$
89. $x = \pm b/(a+b)$

GRAPHING UTILITY EXERCISES FOR SECTION 5.5

1. (a)

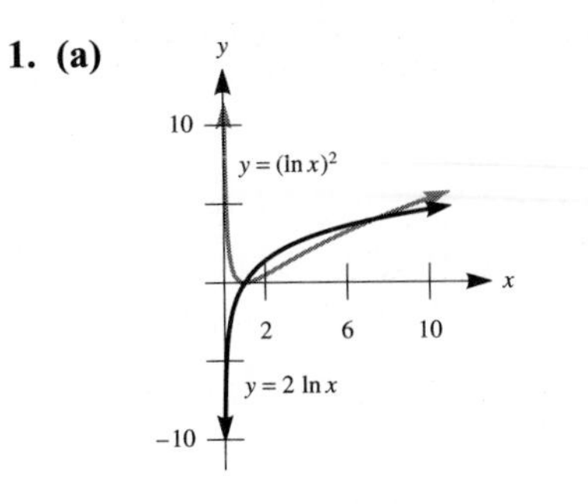

(b) $x \approx 7.39$ **(c)** $x = e^2 \approx 7.39$
3. (a) $x \approx -1.355$

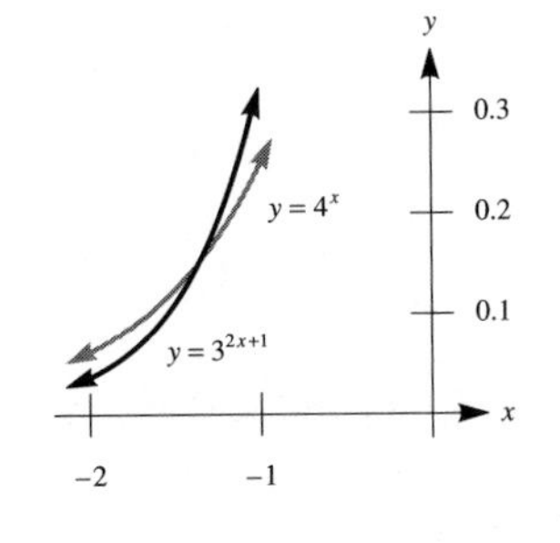

(b) $x = (\ln 3)/(\ln 4 - 2 \ln 3) \approx -1.355$
(c) $x \approx -1.355$

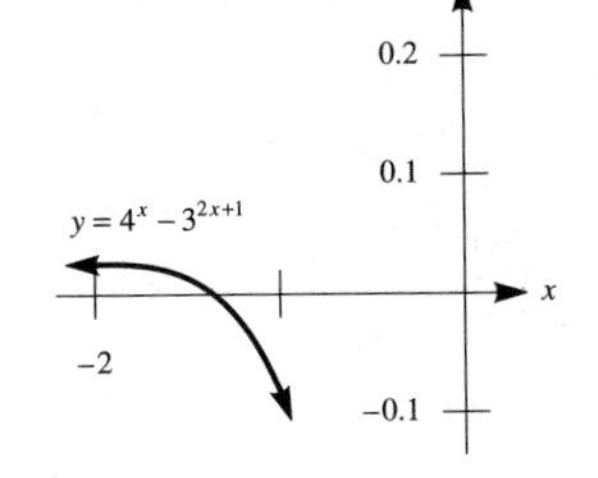

5. (a)

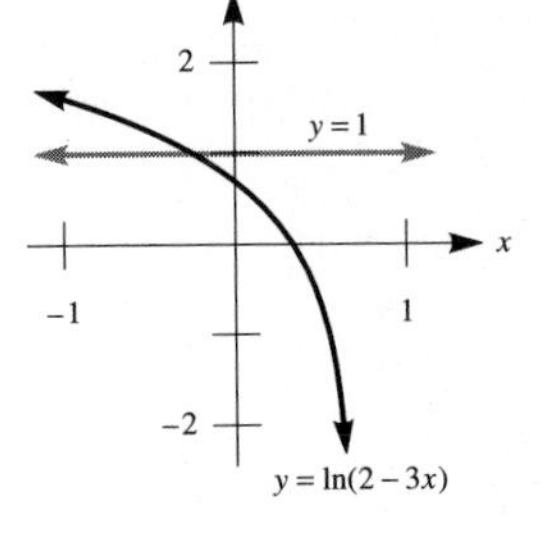

(b) The x-coordinate of the intersection point is the left endpoint of the solution set from Example 8(a) in the text.
(c) $x \approx -0.24$
(d) $x = (2 - e)/3 \approx -0.24$

EXERCISE SET 5.6

1. \$1009.98 **3.** 8.45% **5.** \$767.27
7. (a) \$3869.68 **(b)** \$4006.39
9. 13 quarters **11.** \$3487.50
13. \$2610.23 **15.** 5.83%
17. the 6% investment
19. (a) 14 yr **(b)** 13.86 yr **(c)** 1.01%
21. \$26.5 trillion **23. (a)** 14 yr

(b)

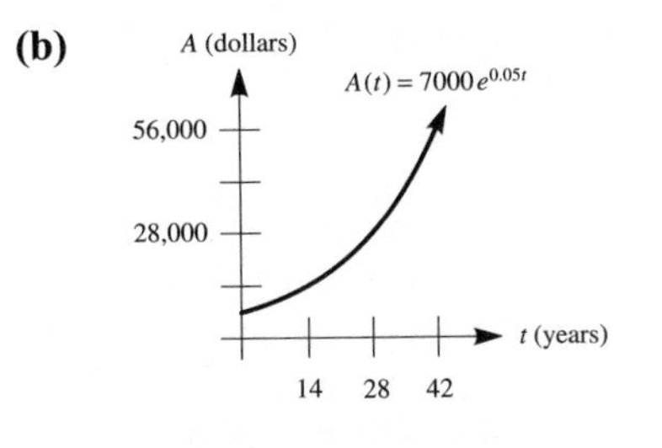

25. (a)

t (years)	1	2	3
A (account #1)	2081	2166	2254
A (account #2)	2082	2167	2255

t (years)	4	10
A (account #1)	2345	2978
A (account #2)	2347	2984

(b) account #1: 17.4 yr; account #2: 17.3 yr
(c) \$88/yr **(d)** \$88/yr
27. (a) the quadratic model
(b)

	1991	1992
Producer Price Index (logarithmic model)	116.5	117.9
Producer Price Index (quadratic model)	116.2	116.7
Producer Price Index (power model)	117.0	118.7
Producer Price Index (actual)	116.5	117.2

(c) logarithmic model **(d)** quadratic model **(e)** logarithmic: 127.0; quadratic: 107.1; power: 129.8; highest: power model; lowest: quadratic model **(f)** the quadratic model

EXERCISE SET 5.7

1. (a) $k \approx 0.3209$ **(b)** 9951 bacteria
(c) $t \approx 5.0$ hr **3.** $k \approx 0.1769$
5.

Region	1995 Population (billions)	Percent of Population in 1995
World	5.702	100
More dev.	1.169	20.5
Less dev.	4.533	79.5

Region	Relative Growth Rate (percent/yr)	Year 2000 Population (billions)
World	1.5	6.146
More dev.	0.2	1.181
Less dev.	1.9	4.985

Region	Percent of World Population in 2000
World	100
More dev.	19.22
Less dev.	81.11

7. (a)

Country	1995 Population (millions)	Relative Growth Rate (percent/yr)
Iraq	20.6	3.7
United Kingdom	58.6	0.2

Country	Year 2000 Population (millions)	Percent Increase in Population
Iraq	24.8	20.4
United Kingdom	59.2	1.0

(b) the year 2025

9. (a) Cyprus: 0.8 million; Gaza: 1.0 million **(b)** the year 2025; Gaza: 3.2 million
11. (a) $k \approx 0.0269$ **(b)** $k \approx 0.0041$
13. (a) $k \approx 0.0200$ **(b)** 170,853,155 **(c)** slower **15. (a)** 0.684% **(b)** 0.955% **(c)** 21,731,258 **(d)** higher
17. (a)

Region	1990 Population (millions)	Growth Rate (%)	2025 Population
North America	275.2	0.7	351.6
Soviet Union	291.3	0.7	372.2
Europe	499.5	0.2	535.7
Nigeria	113.3	3.1	335.3

(b) 222.0 million
(c) North America: 76.4 million; Soviet Union: 80.9 million; Europe: 36.2 million; combined: 193.5 million **(d)** yes
19. (a)

t (sec)	0	550	1100	1650	2200
$\mathcal{N}$ (g)	8	4	2	1	0.5

(b)

t (yr)	0	4.9×10^9	9.8×10^9
$\mathcal{N}$ (g)	10	5	2.5

t (yr)	14.7×10^9	19.6×10^9
$\mathcal{N}$ (g)	1.25	0.625

21. 0.55 g **23. (a)** 4.29 g **(b)** 79 hr

25. (a)

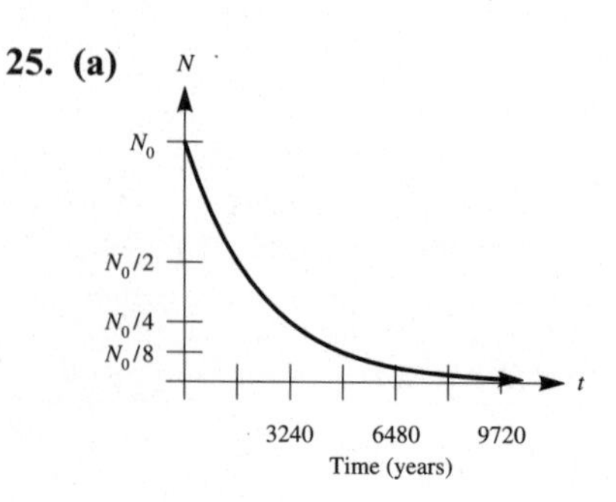

(b)

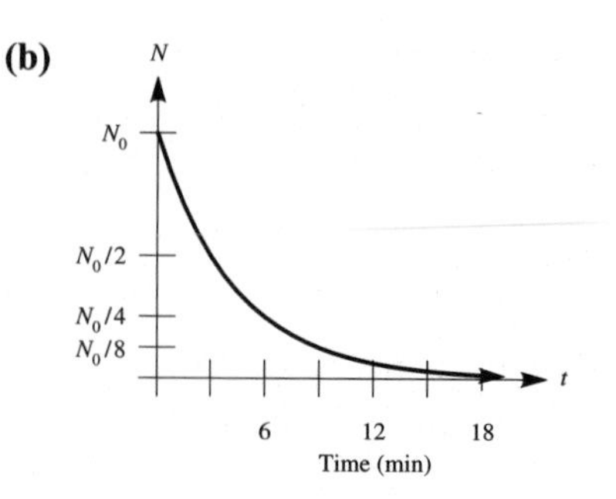

27. (a) 1.53 g **(b)** 43 yr
29. (a) 73,000 yr **(b)** 10 half-lives **(c)** 10 half-lives
31. (a) $k \approx -0.0248$ **(b)** 279 yr **(c)** 280 yr **33. (b)** 14.4 sec **(c)** 0.05 g **35.** 0.02799997 oz
37. (a) the year 2020 **(b)** the year 2032; the carrying capacity will be reached 12 years later.
39. (a) \$7212 million; lower **(b)** \$27.653 billion **(c)** $y = (2216.445)e^{0.1722t}$ **(d)** \$7399 million; closer
41. (b) 31 yr **(c)** 43 yr
43. (a) the year 2013 **(b)** the year 2036 **(c)** the year 2020
45. (a) the year 2002 **(b)** the year 2015 **(c)** the year 2003
47. (a) (i) 55 yr **(ii)** 27 yr **(b) (i)** 201 yr **(ii)** 132 yr **(c)** 447 yr
51. 4.181 billion years old
53. 15,505 years old
55. 8000 years; older
57. 1992 years old; the years 10–5 B.C.; they fit the historical range.
59. (a)

	$N(-1)$	$N(0)$	$N(1)$
From graph	0.25	0.5	1.0
From calculator	0.176	0.444	1.014

	$N(4)$	$N(5)$
From graph	3.5	3.75
From calculator	3.489	3.795

(b) $N(10) \approx 3.99854773$; $N(15) \approx 3.99999021$; $N(20) \approx 3.99999993$ **(c)** $t \approx 3$ **(d)** $t \approx 3.178$ **(e)** $N^{-1}(t) = \ln[8t/(4-t)]$ **(f)** 3.178
61. (a) $N(t) = 50/(1 + 207.33e^{-2.17t})$ **(b)** $N(1) \approx 2.0$; $N(3) \approx 38.2$; $N(4) \approx 48.3$; $N(5) \approx 49.8$ **(c)** 1 day 19.5 hr

CHAPTER 5 REVIEW EXERCISES

1. $\log_5 126$ is larger **3.** $\dfrac{4 \ln 1.5}{\ln 1.25}$ hr

5. not one-to-one

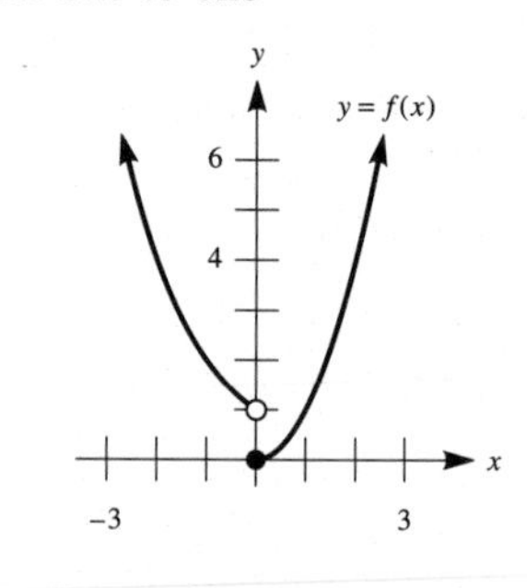

7. $x = (e+1)/(e-1)$

9. $y = e^x$: domain: $(-\infty, \infty)$; range: $(0, \infty)$
$y = \ln x$: domain: $(0, \infty)$; range: $(-\infty, \infty)$

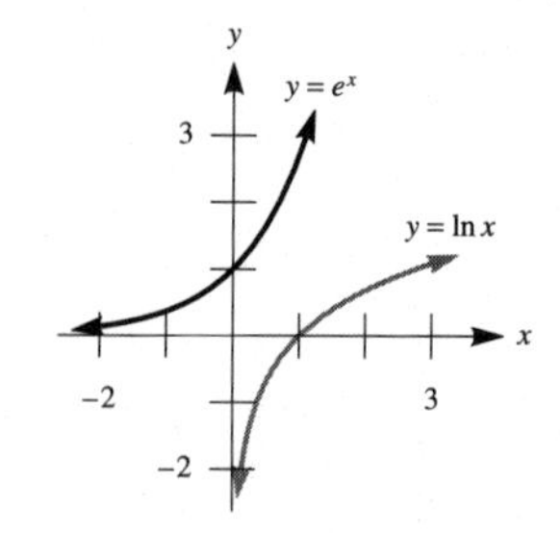

11. $-3/2$ **13.** $k \approx -0.05$
15. $x = 2 - \ln(12/5)$ **17.** the year 2015

19. 7 yr

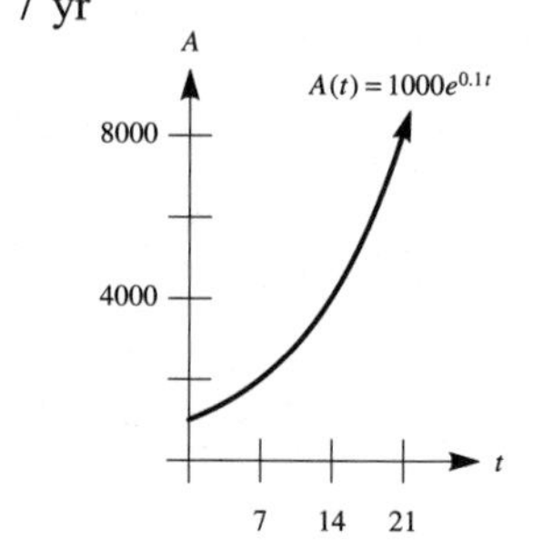

21. horizontal asymptote: $y = 0$; no vertical asymptote; no x-intercept; y-intercept: 1

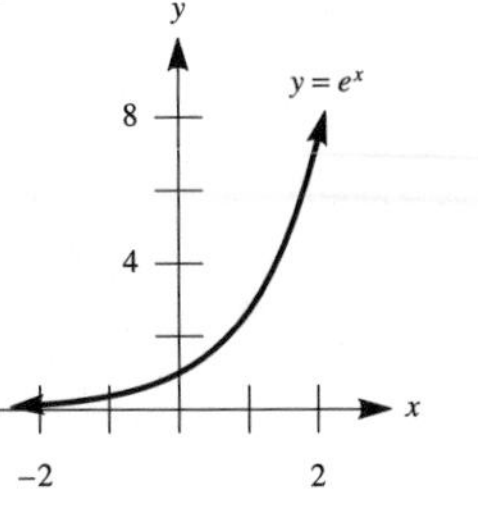

23. no horizontal asymptote; vertical asymptote: $x = 0$; x-intercept: 1; no y-intercept

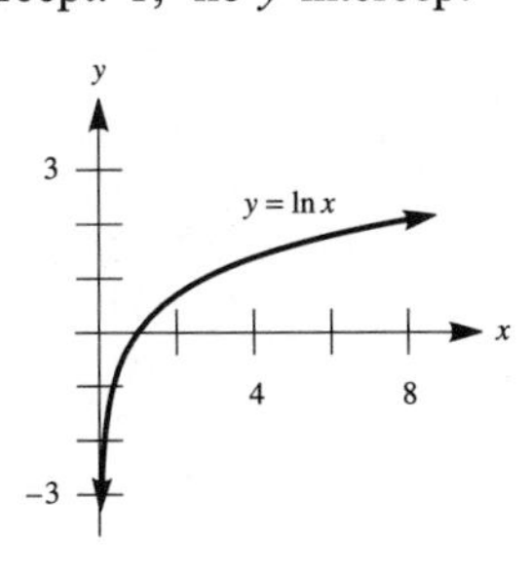

25. horizontal asymptote: $y = 1$; no vertical asymptote; no x-intercept; y-intercept: 3

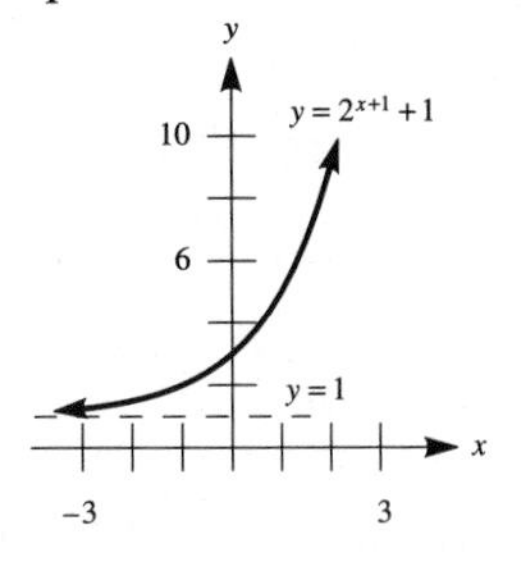

27. horizontal asymptote: $y = 0$; no vertical asymptote; no x-intercept; y-intercept: 1

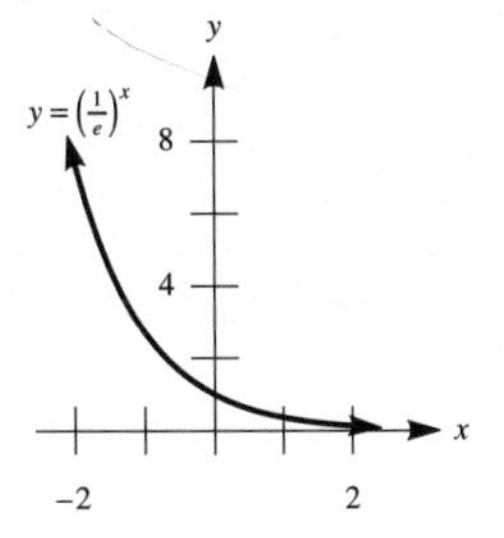

29. horizontal asymptote: $y = 1$; no vertical asymptote; no x-intercept; y-intercept: $e + 1$

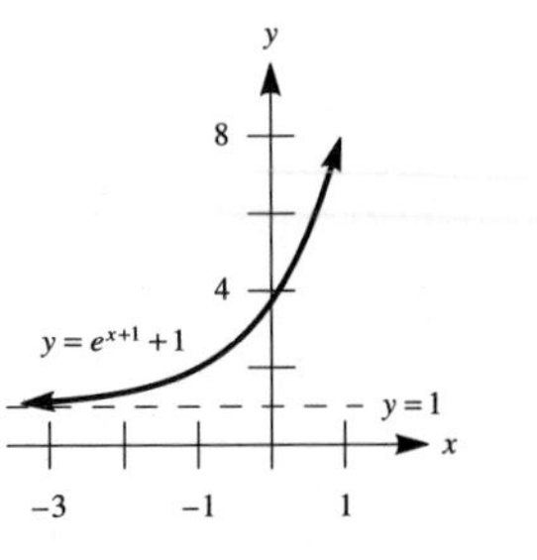

31. no asymptotes; x- and y-intercepts: 0

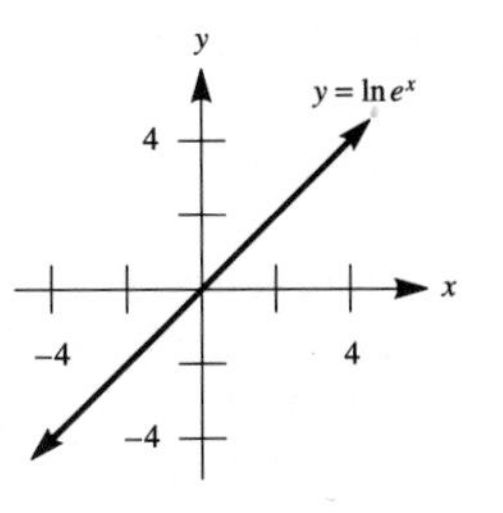

33. $x = 4$ **35.** $x = 3$ **37.** $x = 2/3$
39. $x = \sqrt[3]{3}$
41. $x = (1 - 2 \ln 3)/10$
43. $x = 200/99$ **45.** $x = 2$ **47.** $x > 0$
49. $x = 1$ **51.** $1/2$ **53.** $1/5$
55. -1 **57.** 16 **59.** 4 **61.** 2
63. 2 **65.** $9/14$
67. $2a + 3b + \dfrac{c}{2}$ **69.** $8a + 4b$
71. between 2 and 3 **73.** between 2 and 3 **75.** between -2 and -3

77. (a) Quadrant III

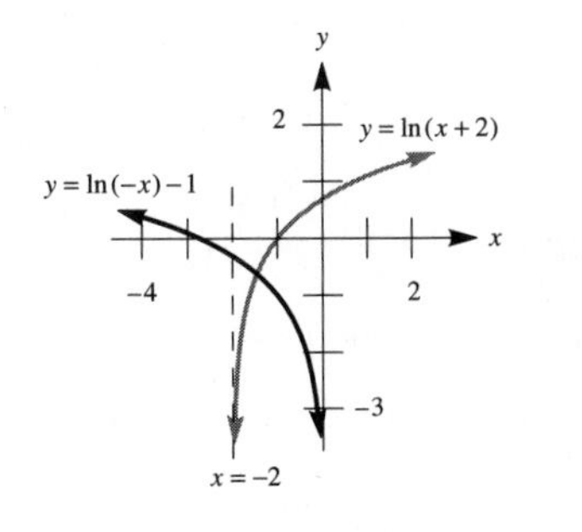

(b) $x = -2e/(e + 1) \approx -1.46$
79. $k = \ln(1/2)/T$ **81.** 6.25%
83. $d \ln(c/b)/\ln(1/2)$ days
85. $\log_{10} 2$ **87.** $\ln 10$ **89.** $\ln x^a y^b$

91. $\frac{1}{2} \ln(x - 3) + \frac{1}{2} \ln(x + 4)$
93. $3 \log_{10} x - \frac{1}{2} \log_{10}(1 + x)$
95. $\frac{1}{3} \log_{10} x - \frac{2}{3}$
97. $3 \ln(1 + 2e) - 3 \ln(1 - 2e)$
99. $(\ln 2) \Big/ \left[\ln\left(1 + \dfrac{R}{100}\right) \right]$ yr
101. 9.92%
103. $(100 \ln 2)/R$ yr **105. (a)** $7\frac{3}{4}$ yr
(b) $(\ln n) \Big/ \left[\ln\left(1 + \dfrac{R}{400}\right) \right]$ yr
107. (a) $(0, \infty)$ **(b)** $[1, \infty)$
109. (a) $(-\infty, -3) \cup (-3, 5) \cup (5, \infty)$
(b) $(-\infty, -3) \cup (5, \infty)$
111. $(-\infty, -1) \cup (1, \infty)$ **113.** -0.7
115. -2.2 **117.** 4.3 **119.** 2.7
121. 1.6 **123. (a)** 1.0986
(b) 0.00112% **(c)** $\ln \sqrt{3} \approx 0.5493$; $\ln 9 \approx 2.1972$; $\ln(1/3) \approx -1.0986$

CHAPTER 5 TEST

1. domain: $(-\infty, \infty)$; range: $(-3, \infty)$; x-intercept: $-(\ln 3)/(\ln 2)$; y-intercept: -2; asymptote: $y = -3$

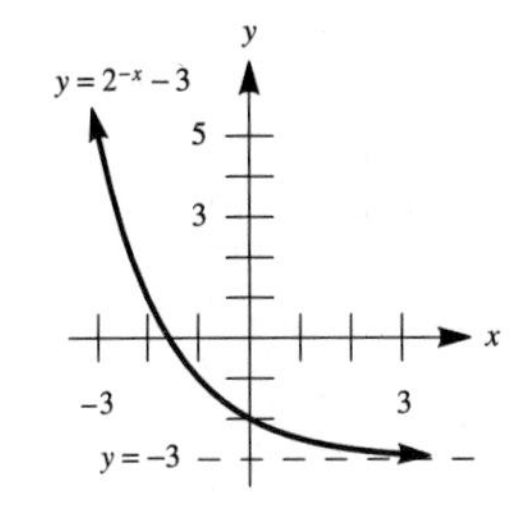

2. $\left(\ln \frac{5}{3}\right)/(\ln 1.033) \approx 16$ hr
3. $\log_2 17$ is larger **4.** $(\ln 15)/(\ln 2)$
5. $2^{40} \approx 10^{12}$
6. 4 yr
7. (a) $(0, \infty)$

(b)

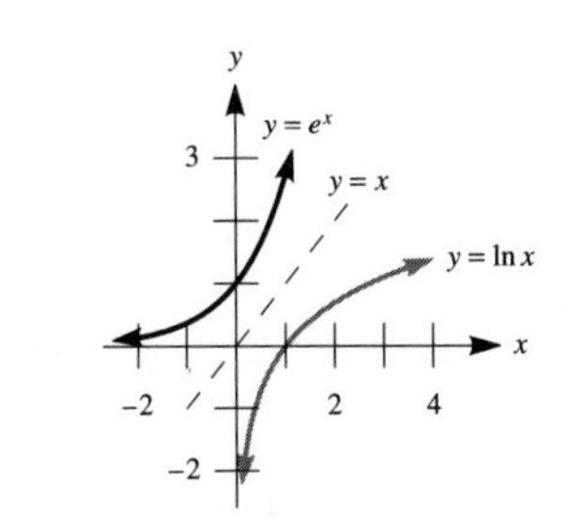

8. $3a - \frac{1}{2}b$ **9. (a)** $-1/2$ **(b)** -1
10. $x = 0, -2, 2$
11. (a) $k = (\ln \frac{1}{2})/4 \approx -0.1733$
(b) 0.35 g
12. $\ln[x^2/(x^2+1)^{1/3}]$
13. no solution
14. $g^{-1}(x) = 10^x + 1$; range: $(1, \infty)$
15. (a) 12 yr

(b)

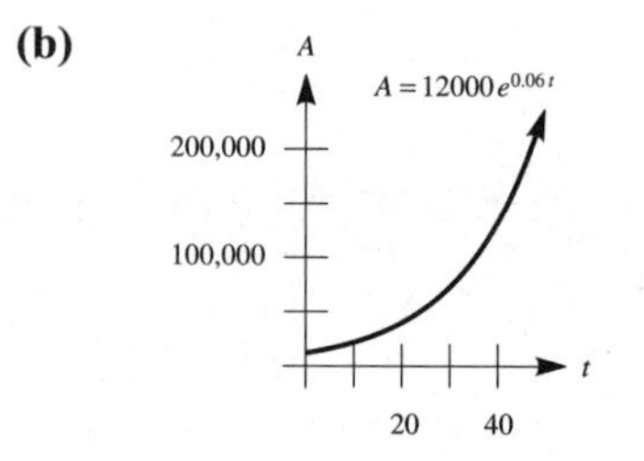

16. (a) $x > 0$ **(b)** $x = 1, e^2$

17. (a)

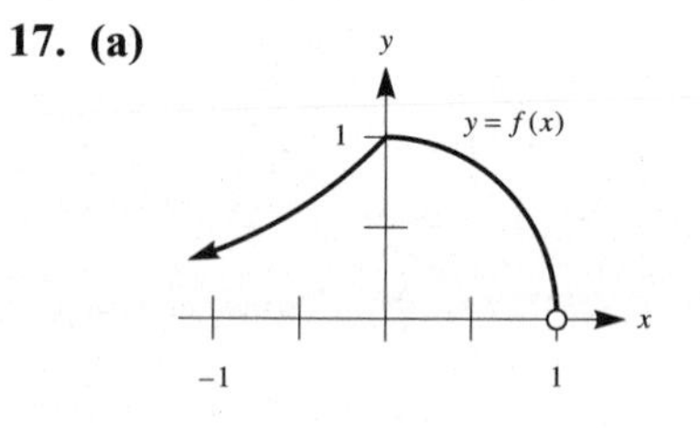

(b) no
18. $x = 4$ **19.** $x = \ln(3/2) \approx 0.405$
20. (a) $x > (\ln 1.6)/(\ln 0.3) \approx -0.39$
(b) $3 < x \leq 4$

CHAPTER SIX

EXERCISE SET 6.1

1. (a) $\sin\theta = \frac{15}{17}$, $\cos\theta = \frac{8}{17}$, $\tan\theta = \frac{15}{8}$, $\cot\theta = \frac{8}{15}$, $\sec\theta = \frac{17}{8}$, $\csc\theta = \frac{17}{15}$ **(b)** $\sin\beta = \frac{8}{17}$, $\cos\beta = \frac{15}{17}$, $\tan\beta = \frac{8}{15}$, $\cot\beta = \frac{15}{8}$, $\sec\beta = \frac{17}{15}$, $\csc\beta = \frac{17}{8}$
3. (a) $\sin\theta = \sqrt{5}/5$, $\cos\theta = 2\sqrt{5}/5$, $\tan\theta = \frac{1}{2}$, $\cot\theta = 2$, $\sec\theta = \sqrt{5}/2$, $\csc\theta = \sqrt{5}$ **(b)** $\sin\beta = 2\sqrt{5}/5$, $\cos\beta = \sqrt{5}/5$, $\tan\beta = 2$, $\cot\beta = \frac{1}{2}$, $\sec\beta = \sqrt{5}$, $\csc\beta = \sqrt{5}/2$
5. (a) $\cos A = 3\sqrt{13}/13$, $\sin A = 2\sqrt{13}/13$, $\tan A = \frac{2}{3}$
(b) $\sec B = \sqrt{13}/2$, $\csc B = \sqrt{13}/3$, $\cot B = \frac{2}{3}$
7. $\sin B = \frac{12}{13}$, $\cos B = \frac{5}{13}$, $\tan B = \frac{12}{5}$, $\cot B = \frac{5}{12}$, $\sec B = \frac{13}{5}$, $\csc B = \frac{13}{12}$
9. (a) $\sin B = \frac{4}{5}$, $\cos A = \frac{4}{5}$
(b) $\sin A = \frac{3}{5}$, $\cos B = \frac{3}{5}$
(c) $(\tan A)(\tan B) = 1$
11. (a) $\cos A = \frac{24}{25}$, $\sin A = \frac{7}{25}$, $\tan A = \frac{7}{24}$
(b) $\cos B = \frac{7}{25}$, $\sin B = \frac{24}{25}$, $\tan B = \frac{24}{7}$
(c) $(\tan A)(\tan B) = 1$
13. $\sin\theta \approx 0.906$, $\cos\theta \approx 0.423$, $\tan\theta \approx 2.145$ **15.** $\sin\theta \approx 0.623$, $\cos\theta \approx 0.783$, $\tan\theta \approx 0.795$
17. $\sin\theta \approx 0.985$, $\cos\theta \approx 0.173$, $\tan\theta \approx 5.706$ **19.** $\sec\theta \approx 1.064$, $\csc\theta \approx 2.924$, $\cot\theta \approx 2.747$
21. $\sec\theta \approx 1.049$, $\csc\theta \approx 3.326$, $\cot\theta \approx 3.172$ **23.** $\sec\theta \approx 1.000$, $\csc\theta \approx 57.299$, $\cot\theta \approx 57.290$
37. (a) $\cos 30° \approx 0.8660254038$, $\cos 45° \approx 0.7071067812$
(b) $\cos 30° \approx 0.8660254038$, $\cos 45° \approx 0.7071067812$
39. false **41.** true **43.** true **45.** true
47. (a) Since $RC < QB < PA$, we have $\sin 20° < \sin 40° < \sin 60°$.
(b) $\sin 20° \approx 0.3420$, $\sin 40° \approx 0.6428$, $\sin 60° \approx 0.8660$
49. (a) $\cos\theta$ **(b)** $\sec\beta$
53. $\sin 3° \approx 0.0523359562$, $\sin 6° \approx 0.1045284633$, $\sin 9° \approx 0.1564344650$, $\sin 12° \approx 0.2079116908$, $\sin 15° \approx 0.2588190451$, $\sin 18° \approx 0.3090169944$

EXERCISE SET 6.2

1. (a) $11SC$ **(b)** $11 \sin\theta\cos\theta$
3. (a) $-8C^3S$ **(b)** $-8\cos^3\theta\sin\theta$
5. (a) $1 + 2T + T^2$
(b) $1 + 2\tan\theta + \tan^2\theta$
7. (a) $T^2 + T - 6$
(b) $\tan^2\theta + \tan\theta - 6$
9. (a) -1 **(b)** -1 **11. (a)** $\dfrac{CS+2}{S}$
(b) $\dfrac{\cos A \sin A + 2}{\sin A}$
13. (a) $(T-1)(T+9)$
(b) $(\tan\beta - 1)(\tan\beta + 9)$
15. (a) $(2C+1)(2C-1)$
(b) $(2\cos B + 1)(2\cos B - 1)$
17. (a) $3ST^2(3ST+2)$
(b) $3\sec B\tan^2 B(3\sec B\tan B + 2)$
19. $\cos\theta = \sqrt{7}/4$, $\tan\theta = 3\sqrt{7}/7$, $\cot\theta = \sqrt{7}/3$, $\sec\theta = 4\sqrt{7}/7$, $\csc\theta = \frac{4}{3}$ **21.** $\sin\beta = \sqrt{22}/5$, $\tan\beta = \sqrt{66}/3$, $\cot\beta = \sqrt{66}/22$, $\sec\beta = 5\sqrt{3}/3$, $\csc\beta = 5\sqrt{22}/22$
23. $\cos A = \frac{12}{13}$, $\tan A = \frac{5}{12}$, $\cot A = \frac{12}{5}$, $\sec A = \frac{13}{12}$, $\csc A = \frac{13}{5}$
25. $\sin B = \frac{4}{5}$, $\cos B = \frac{3}{5}$, $\cot B = \frac{3}{4}$, $\sec B = \frac{5}{3}$, $\csc B = \frac{5}{4}$
27. $\sin C = \sqrt{5}/3$, $\cos C = \frac{2}{3}$, $\tan C = \sqrt{5}/2$, $\cot C = 2\sqrt{5}/5$, $\csc C = 3\sqrt{5}/5$ **29.** $\sin\alpha = \sqrt{3}/2$, $\cos\alpha = \frac{1}{2}$, $\tan\alpha = \sqrt{3}$, $\sec\alpha = 2$, $\csc\alpha = 2\sqrt{3}/3$
31. $\sin\theta = 0.887$, $\tan\theta \approx 1.916$, $\cot\theta \approx 0.522$, $\sec\theta \approx 2.162$, $\csc\theta \approx 1.128$ **33.** $\sin\theta \approx 0.768$, $\cos\theta \approx 0.640$, $\cot\theta \approx 0.833$, $\sec\theta \approx 1.562$, $\csc\theta \approx 1.302$
35. $\sin A + \cos A$ **37.** $\csc\theta$
39. $\cos^2 B$ **41.** $\cos A + 4$
43. $2\csc\theta$ **45.** 0 **47.** 1 **49.** $\tan\theta$
51. -1 **53.** $\sin^2\theta$ **55.** $-2\cos\theta - 1$

EXERCISE SET 6.3

1. $BC = 30$ cm, $AC = 30\sqrt{3}$ cm
3. $AB = \dfrac{32\sqrt{3}}{3}$ cm, $BC = \dfrac{16\sqrt{3}}{3}$ cm
5. $AC \approx 11.5$ cm, $BC \approx 9.6$ cm
7. (a) 15.59 ft **(b)** 9 ft
9. 34 million miles **11.** 141.1 m
13. (a) $h \approx 27.3$ ft **(b)** 906.9 ft^2
15. 1.5 in^2
17. $\frac{7}{2}\sin(360°/7) \approx 2.736$ square units
19. $\pi - 2\sqrt{2} \approx 0.313$ square units
21. $\sqrt{3} \approx 1.732$ square units
25. 10,660 ft **27.** 136 m
29. $BD = 18(\sqrt{3} - 1)$ cm
31. (a) $\angle BOA = 90° - \theta$, $\angle OAB = \theta$, $\angle BAP = 90° - \theta$, $\angle BPA = \theta$
(b) $AO = \sin\theta$, $AP = \cos\theta$, $OB = \sin^2\theta$, $BP = \cos^2\theta$
33. (a) $BC = \dfrac{5}{\sin\theta}$ **(b)** $AB = \dfrac{4}{\cos\theta}$
(c) $AC = 4\sec\theta + 5\csc\theta$

35. (a) $DE = \sin\theta$ **(b)** $OE = \cos\theta$
(c) $CF = \tan\theta$ **(d)** $OC = \sec\theta$
(e) $AB = \cot\theta$ **(f)** $OB = \csc\theta$
37. (b) 1080 miles
39. (b)

n	5	10	50	100
A_n	2.38	2.94	3.1333	3.1395

n	1,000	5,000	10,000
A_n	3.141572	3.1415918	3.1415924

(c) As n gets larger, A_n approaches the area of the circle, which is π.

EXERCISE SET 6.4

1. (a) reference angle: 70°

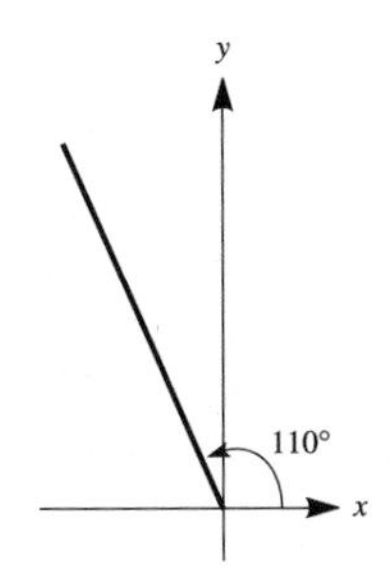

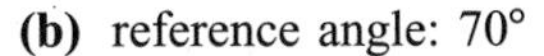
(b) reference angle: 70°

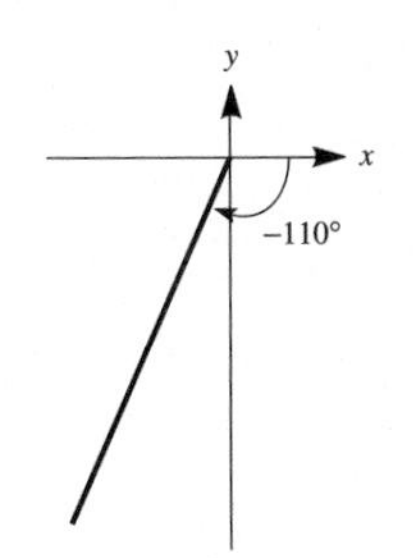

3. (a) reference angle: 20°

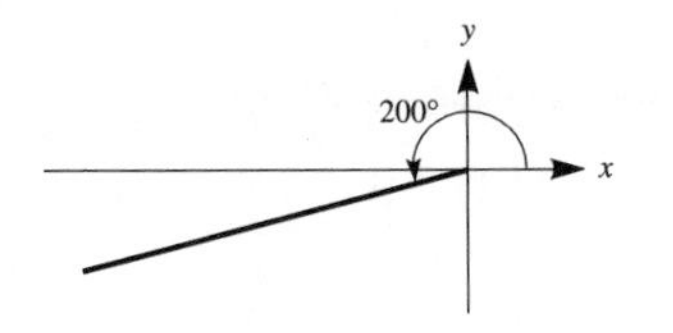

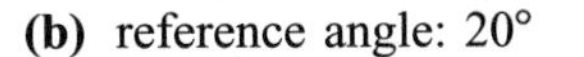
(b) reference angle: 20°

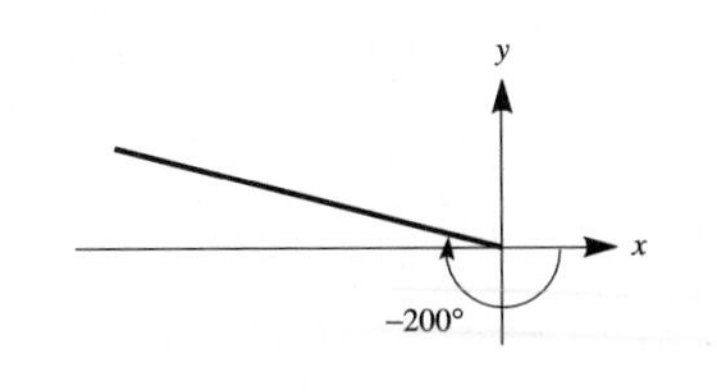

5. (a) reference angle: 60°

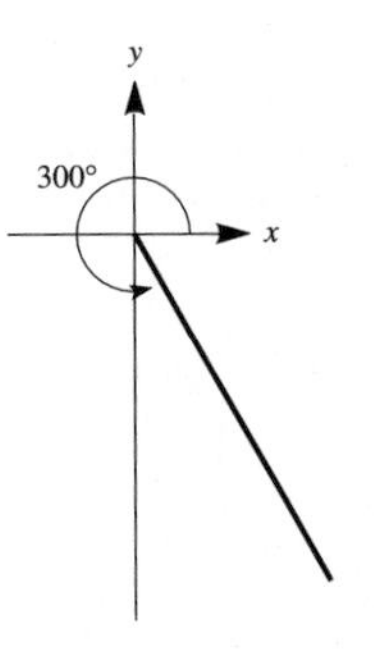

(b) reference angle: 60°

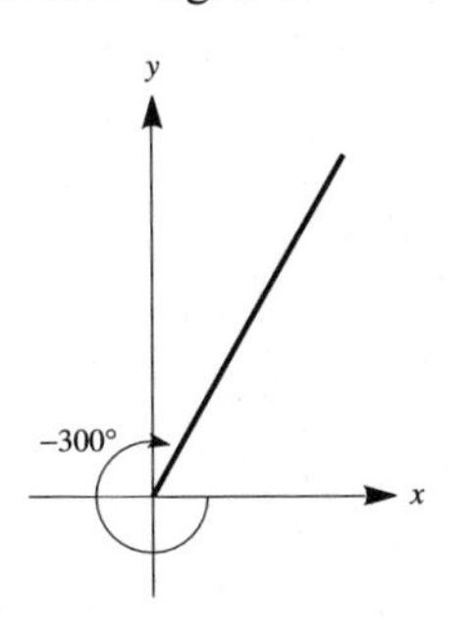

7. (a) reference angle: 60°

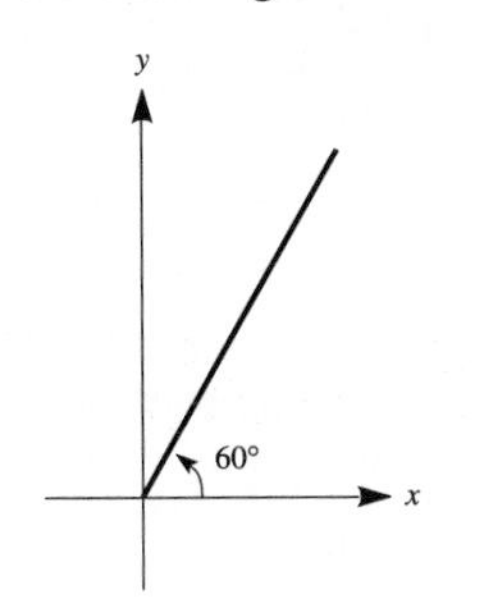

(b) reference angle: 60°

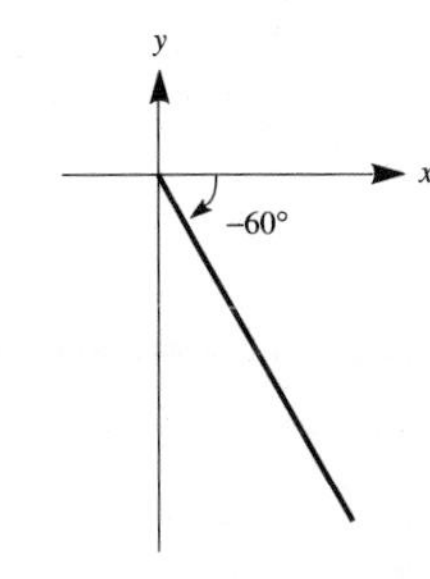

9. $\cos 270° = 0$, $\sec 270°$ is undefined, $\sin 270° = -1$, $\csc 270° = -1$, $\tan 270°$ is undefined, $\cot 270° = 0$
11. $\cos(-270°) = 0$, $\sec(-270°)$ is undefined, $\sin(-270°) = 1$, $\csc(-270°) = 1$, $\tan(-270°)$ is undefined, $\cot(-270°) = 0$
13. $\cos 810° = 0$, $\sec 810°$ is undefined, $\sin 810° = 1$, $\csc 810° = 1$, $\tan 810°$ is undefined, $\cot 810° = 0$
15. $\sin 10° \approx 0.2$, $\sin(-10°) \approx -0.2$; $\sin 10° \approx 0.17$, $\sin(-10°) \approx -0.17$
17. $\cos 80° \approx 0.2$, $\cos(-80°) \approx 0.2$; $\cos 80° \approx 0.17$, $\cos(-80°) \approx 0.17$
19. $\sin 120° \approx 0.9$, $\sin(-120°) \approx -0.9$; $\sin 120° \approx 0.87$, $\sin(-120°) \approx -0.87$
21. $\sin 150° = 0.5$, $\sin(-150°) = -0.5$; $\sin 150° = 0.5$, $\sin(-150°) = -0.5$
23. $\cos 220° \approx -0.8$, $\cos(-220°) \approx -0.8$; $\cos 220° \approx -0.77$, $\cos(-220°) \approx -0.77$
25. $\cos 310° \approx 0.6$, $\cos(-310°) \approx 0.6$; $\cos 310° \approx 0.64$, $\cos(-310°) \approx 0.64$
27. $\sin(40° + 360°) \approx 0.6$; $\sin(40° + 360°) \approx 0.64$
29. $\sin 70°$ is larger; $\sin 70° \approx 0.94$, $\cos 70° \approx 0.34$
31. $\cos 160°$ is larger; $\cos 170° \approx -0.98$, $\cos 160° \approx -0.94$
33. $\cos 290°$ is larger; $\cos 280° \approx 0.17$, $\cos 290° \approx 0.34$
35. $\sin 10°$ is larger; $\sin 10° \approx 0.17$, $\sin(-10°) \approx -0.17$
37. $\sin 80°$ is larger; $\sin 80° \approx 0.98$, $\sin 110° \approx 0.94$
39. $\cos(-110°)$ is larger; $\sin(-80°) \approx -0.98$, $\cos(-110°) \approx -0.34$
41. (a) $\cos 315° = \sqrt{2}/2$
(b) $\cos(-315°) = \sqrt{2}/2$

(c) $\sin 315° = -\sqrt{2}/2$
(d) $\sin(-315°) = \sqrt{2}/2$
43. (a) $\cos 300° = \frac{1}{2}$
(b) $\cos(-300°) = \frac{1}{2}$
(c) $\sin 300° = -\sqrt{3}/2$
(d) $\sin(-300°) = \sqrt{3}/2$
45. (a) $\cos 210° = -\sqrt{3}/2$
(b) $\cos(-210°) = -\sqrt{3}/2$
(c) $\sin 210° = -\frac{1}{2}$ **(d)** $\sin(-210°) = \frac{1}{2}$
47. (a) $\cos 390° = \sqrt{3}/2$
(b) $\cos(-390°) = \sqrt{3}/2$
(c) $\sin 390° = \frac{1}{2}$ **(d)** $\sin(-390°) = -\frac{1}{2}$
49. (a) $\sec 600° = -2$
(b) $\csc(-600°) = 2\sqrt{3}/3$
(c) $\tan 600° = \sqrt{3}$
(d) $\cot(-600°) = -\sqrt{3}/3$

51.

θ	$\sin\theta$	$\cos\theta$	$\tan\theta$
0°	0	1	0
30°	$\frac{1}{2}$	$\sqrt{3}/2$	$\sqrt{3}/3$
45°	$\sqrt{2}/2$	$\sqrt{2}/2$	1
60°	$\sqrt{3}/2$	$\frac{1}{2}$	$\sqrt{3}$
90°	1	0	undefined
120°	$\sqrt{3}/2$	$-\frac{1}{2}$	$-\sqrt{3}$
135°	$\sqrt{2}/2$	$-\sqrt{2}/2$	-1
150°	$\frac{1}{2}$	$-\sqrt{3}/2$	$-\sqrt{3}/3$
180°	0	-1	0

53. $\dfrac{35\sqrt{3}}{4}$ cm^2 **55.** 302.7 cm^2
57. $108\sqrt{3} \approx 187.06$ cm^2

59. (a)

	Terminal Side of Angle θ Lies in			
	Quadrant I	Quadrant II	Quadrant III	Quadrant IV
$\sin\theta$	positive	positive	negative	negative
$\cos\theta$	positive	negative	negative	positive
$\tan\theta$	positive	negative	positive	negative

61. (a)

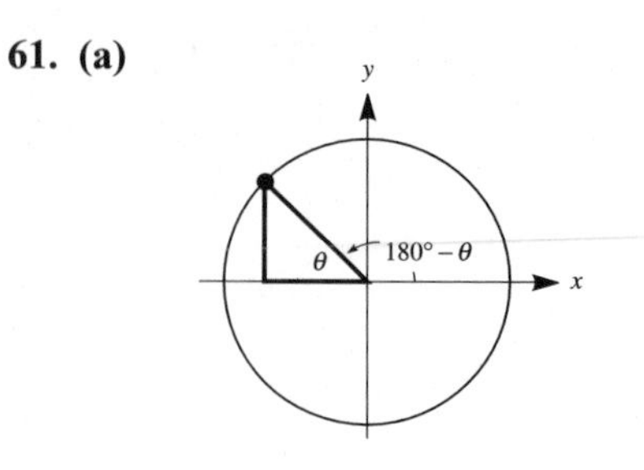

(b) θ **(c)** Because $180° - \theta$ lies in the second quadrant, where all y-coordinates are positive, $\sin(180° - \theta) = \sin\theta$.
63. $A(\theta) = \sin\theta$
65. (b) $0° < \theta < 180°$
67. $(0°, 180°)$ **69. (c)** $(0°, 180°)$
71. (d) $\sin 75° = (\sqrt{2} + \sqrt{6})/4$
(f) $\sin 105° \approx 0.9659$, but $\sin 45° + \sin 60° \approx 1.5731$.

EXERCISE SET 6.5

1. (a) $\cos\theta = -2\sqrt{6}/5$, $\tan\theta = -\sqrt{6}/12$, $\cot\theta = -2\sqrt{6}$, $\sec\theta = -5\sqrt{6}/12$, $\csc\theta = 5$
(b) $\cos\theta = -2\sqrt{6}/5$, $\tan\theta = \sqrt{6}/12$, $\cot\theta = 2\sqrt{6}$, $\sec\theta = -5\sqrt{6}/12$, $\csc\theta = -5$
3. (a) $\sin\theta = \frac{12}{13}$, $\tan\theta = \frac{12}{5}$, $\cot\theta = \frac{5}{12}$, $\sec\theta = \frac{13}{5}$, $\csc\theta = \frac{13}{12}$
(b) $\sin\theta = -\frac{12}{13}$, $\tan\theta = \frac{12}{5}$, $\cot\theta = \frac{5}{12}$, $\sec\theta = -\frac{13}{5}$, $\csc\theta = -\frac{13}{12}$
5. $\sin A = -\frac{1}{3}$, $\cos A = 2\sqrt{2}/3$, $\tan A = -\sqrt{2}/4$, $\cot A = -2\sqrt{2}$, $\sec A = 3\sqrt{2}/4$
7. $\cos B = -\frac{2}{3}$, $\sin B = -\sqrt{5}/3$, $\tan B = \sqrt{5}/2$, $\cot B = 2\sqrt{5}/5$, $\csc B = -3\sqrt{5}/5$
9. $\sin\theta = -\sqrt{9 - t^2}/3$, $\tan\theta = -\sqrt{9 - t^2}/t$, $\cot\theta = -t\sqrt{9 - t^2}/(9 - t^2)$, $\sec\theta = 3/t$, $\csc\theta = -3\sqrt{9 - t^2}/(9 - t^2)$
11. $\cos\theta = -\sqrt{1 - 9u^2}$, $\tan\theta = 3u\sqrt{1 - 9u^2}/(1 - 9u^2)$, $\cot\theta = \sqrt{1 - 9u^2}/3u$, $\sec\theta = -\sqrt{1 - 9u^2}/(1 - 9u^2)$, $\csc\theta = -1/3u$
13. $\sin\theta = \sqrt{9 - 3u^2}/3$, $\tan\theta = \sqrt{3 - u^2}/u$, $\cot\theta = u\sqrt{3 - u^2}/(3 - u^2)$, $\sec\theta = \sqrt{3}/u$, $\csc\theta = \sqrt{9 - 3u^2}/(3 - u^2)$
37. (a) not an identity **(b)** identity
45. (a) $(\cos\theta - \sin\theta)(1 + \cos\theta\sin\theta)$

CHAPTER 6 REVIEW EXERCISES

1. $\sqrt{2}/2$ **3.** $-\sqrt{3}$ **5.** -2 **7.** -1
9. $\sqrt{2}/2$ **11.** 1 **13.** $-2\sqrt{3}/3$
15. 2 **17.** 1
19. $a = \frac{1}{2}$, $c = \sqrt{3}/2$ **21.** $b = \frac{35}{2}$
23. $\sin A = \sqrt{55}/8$, $\cot A = 3\sqrt{55}/55$
25. 5 square units
27. $\sin^2 A + \cos^2 B = \frac{1}{5}$
29. $a = 16\sqrt{3}/3$
31. $c = 9$; area $= 9\sqrt{19}/4$
35. $\cos\theta = \dfrac{p^4 - q^4}{p^4 + q^4}$, $\tan\theta = \dfrac{2p^2q^2}{p^4 - q^4}$
37. $\sin A\cos A$ **39.** $\sin^2 A$
41. $\cos A + \sin A$ **43.** $\sin A\cos A$
45. $2\sin A\cos A$
47. $\cos 10° \approx 0.9848$
49. $\sin 10° \approx 0.1736$
51. $\sin\theta = \frac{4}{5}$, $\tan\theta = \frac{4}{3}$
53. $\tan\theta = -\frac{24}{7}$ **55.** $\cot\theta = \frac{5}{12}$
57. $\tan(90° - \theta) = \dfrac{5t\sqrt{1 - 25t^2}}{1 - 25t^2}$
59. $\sin\theta = (2\sqrt{3} - \sqrt{6})/6$
63. $3600\tan 35° \approx 2521$ cm^2
79. (a) $PN = \sin\theta$ **(b)** $ON = \cos\theta$
(c) $PT = \tan\theta$ **(d)** $OT = \sec\theta$
(e) $NA = 1 - \cos\theta$
(f) $NT = \sin\theta\tan\theta$
81. $A(\theta) = (1 - \cos\theta)^2/\sin\theta\cos\theta$

CHAPTER 6 TEST

1. (a) $\sqrt{3}/3$ **(b)** $\sqrt{2}$ **(c)** 1 **(d)** 0
2. (a) negative **(b)** negative **(c)** positive
3. (a) 1 **(b)** -1 **(c)** 0
4. $(2\cot\theta + 3)(\cot\theta + 4)$ **5.** 18 cm^2
6. (a) $\sqrt{2}/2$ **(b)** $-\sqrt{3}/3$ **(c)** -2
7. $\cos\theta = -2\sqrt{5}/5$, $\tan\theta = \frac{1}{2}$

8. $\cos\theta = \sqrt{9-u^2}/3$, $\cot\theta = -\sqrt{9-u^2}/u$
9. $-2\cos\theta - 1$
10. $\frac{81}{2}\sin 40° \approx 26.033 \text{ m}^2$
11. 1 **12.** $CD = 50(\tan 55° - \tan 25°)$
13. (a) sin 85° is larger. **(b)** cos 5° is larger. **(c)** tan 185° is larger.
14. (a) $AD = 2.75\cos 20°$, $CD = 2.75\sin 20°$, $DB = 3.25 - 2.75\cos 20°$ **(b)** $CB \approx 1.15$ cm

CHAPTER SEVEN

EXERCISE SET 7.1

1. $\theta = 2.5$ radians **3.** $\theta = 2$ radians
5. (a) $\frac{\pi}{4} \approx 0.79$ radians **(b)** $\frac{\pi}{2} \approx 1.57$ radians **(c)** $\frac{3\pi}{4} \approx 2.36$ radians
7. (a) 0 radians **(b)** $2\pi \approx 6.28$ radians **(c)** $\frac{5\pi}{2} \approx 7.85$ radians
9. (a) 15° **(b)** 30° **(c)** 45°
11. (a) 60° **(b)** 300° **(c)** 720°
13. (a) 114.59° **(b)** 171.89° **(c)** 565.49° **15.** smaller
17. $30° = \pi/6$ radians; $45° = \pi/4$ radians; $60° = \pi/3$ radians; $120° = 2\pi/3$ radians; $135° = 3\pi/4$ radians; $150° = 5\pi/6$ radians
19.

θ	$\sin\theta$	$\cos\theta$	$\tan\theta$
0	0	1	0
$\pi/2$	1	0	undefined
π	0	-1	0
$3\pi/2$	-1	0	undefined
2π	0	1	0

21. $0.5 < \cos 1 < 0.6$; $0.8 < \sin 1 < 0.9$; $\cos 1 \approx 0.54$; $\sin 1 \approx 0.84$
23. $0.5 < \cos(-1) < 0.6$; $-0.9 < \sin(-1) < -0.8$; $\cos(-1) \approx 0.54$; $\sin(-1) \approx -0.84$
25. $-0.7 < \cos 4 < -0.6$; $-0.8 < \sin 4 < -0.7$; $\cos 4 \approx -0.65$; $\sin 4 \approx -0.76$
27. $-0.7 < \cos(-4) < -0.6$; $0.7 < \sin(-4) < 0.8$; $\cos(-4) \approx -0.65$; $\sin(-4) \approx 0.76$
29. $0.8 < \sin(1+2\pi) < 0.9$; $\sin(1+2\pi) \approx 0.84$
31. (a) $-\sqrt{3}/2$ **(b)** $-\sqrt{3}/2$ **(c)** $\frac{1}{2}$ **(d)** $-\frac{1}{2}$ **33. (a)** $-\frac{1}{2}$ **(b)** $-\frac{1}{2}$ **(c)** $-\sqrt{3}/2$ **(d)** $\sqrt{3}/2$
35. (a) $-\sqrt{2}/2$ **(b)** $-\sqrt{2}/2$ **(c)** $-\sqrt{2}/2$ **(d)** $\sqrt{2}/2$
37. (a) -2 **(b)** $2\sqrt{3}/3$ **(c)** $\sqrt{3}$ **(d)** $-\sqrt{3}/3$
39. (a) $-2\sqrt{3}/3$ **(b)** -2 **(c)** $-\sqrt{3}/3$ **(d)** $\sqrt{3}$
41. approximately 57.30° **43.** $x = 0$

EXERCISE SET 7.2

1. $s = 4\pi$ ft **3.** $s = \pi/2$ cm
5. (a) $A = 12\pi \text{ cm}^2 \approx 37.70 \text{ cm}^2$ **(b)** $A = 50\pi/9 \text{ m}^2 \approx 17.45 \text{ m}^2$ **(c)** $A = 72\pi/5 \text{ m}^2 \approx 45.24 \text{ m}^2$ **(d)** $A = 1.296\pi \text{ cm}^2 \approx 4.07 \text{ cm}^2$
7. $2\pi/5$ radians
9. (a) $\left(10 + \frac{5\pi}{6}\right)$ in. ≈ 12.62 in. **(b)** $25\pi/12 \text{ in}^2 \approx 6.54 \text{ in}^2$
11. $25.2 \text{ cm}^2 - 18\sin 1.4 \text{ cm}^2 \approx 7.46 \text{ cm}^2$
13. (a) 12π radians/sec **(b)** 144π cm/sec **(c)** 72π cm/sec
15. (a) 6π radians/sec **(b)** 150π cm/sec **(c)** 75π cm/sec
17. (a) $50\pi/3$ radians/sec **(b)** 750π cm/sec **(c)** 375π cm/sec
19. (a) 0.000073 radians/sec **(b)** 1040 mph
21. (a) 200π radians/min **(b)** 2000π cm/min **(c)** $\frac{1000}{3}\pi$ radians/min **(d)** $\frac{500}{3}$ rpm
23. 4930 miles **25.** 1470 miles
27. 2690 miles **29. (a)** $P = \left(\frac{2\pi}{3} + 1\right)s$

(b) $A = \left(\frac{\pi}{3} - \frac{\sqrt{3}}{4}\right)s^2$

31. (a) $A_{ABC} = \left(\frac{\pi}{3} - \frac{\sqrt{3}}{4}\right)s^2$; $A_{ADF} = \left(\frac{\pi}{12} - \frac{\sqrt{3}}{16}\right)s^2$; $\frac{A_{ADF}}{A_{ABC}} = \frac{1}{4}$

(b) $A_{DBE} = \left(\frac{\pi}{8} - \frac{\sqrt{3}}{8}\right)s^2$

(c) $A_{DEF} = \left(\frac{\sqrt{3}}{4} - \frac{\pi}{8}\right)s^2$

(d) $A_{BEC} = \left(\frac{\pi}{12} - \frac{\sqrt{3}}{8}\right)s^2$

33. (b) $\frac{1}{2}\left(\pi - \sqrt{2}\right) \approx 0.86$
35. (a) $A(\cos\theta, \sin\theta)$
37. (a) $r(\theta) = 12/(2+\theta)$ **(b)** $A(\theta) = 72\theta/(2+\theta)^2$; no **(c)** $\theta(r) = (12/r) - 2$ **(d)** $A(r) = 6r - r^2$; yes **(e)** maximum $= 9 \text{ cm}^2$ when $r = 3$ cm, $\theta = 2$ radians

EXERCISE SET 7.3

1. (a) $\sqrt{3}/2$ **(b)** $\sqrt{3}/2$ **(c)** $-\frac{1}{2}$ **(d)** $\frac{1}{2}$ **3. (a)** $\sqrt{3}/2$ **(b)** $\sqrt{3}/2$ **(c)** $\frac{1}{2}$ **(d)** $-\frac{1}{2}$
5. (a) $-\sqrt{2}/2$ **(b)** $-\sqrt{2}/2$ **(c)** $-\sqrt{2}/2$ **(d)** $\sqrt{2}/2$ **7. (a)** 2 **(b)** $2\sqrt{3}/3$ **(c)** $-\sqrt{3}$ **(d)** $\sqrt{3}/3$
9. (a) $t = \pi/2, 3\pi/2, 5\pi/2$, and $7\pi/2$ (Other answers are possible.) **(b)** $t = -\pi/2, -3\pi/2, -5\pi/2$, and $-7\pi/2$ (Other answers are possible.)
11. (a) $\sin 2.06 \approx 0.88$; $\cos 2.06 \approx -0.47$; $\tan 2.06 \approx -1.88$; $\sec 2.06 \approx -2.13$; $\csc 2.06 \approx 1.13$; $\cot 2.06 \approx -0.53$
(b) $\sin(-2.06) \approx -0.88$; $\cos(-2.06) \approx -0.47$; $\tan(-2.06) \approx 1.88$; $\sec(-2.06) \approx -2.13$; $\csc(-2.06) \approx -1.13$; $\cot(-2.06) \approx 0.53$
13. (a) $\sin(\pi/6) \approx 0.50$; $\cos(\pi/6) \approx 0.87$; $\tan(\pi/6) \approx 1.73$; $\sec(\pi/6) \approx 1.15$; $\csc(\pi/6) \approx 2.00$; $\cot(\pi/6) \approx 0.58$
(b) $\sin\left(\frac{\pi}{6} + 2\pi\right) \approx 0.50$; $\cos\left(\frac{\pi}{6} + 2\pi\right) \approx 0.87$; $\tan\left(\frac{\pi}{6} + 2\pi\right) \approx 1.73$; $\sec\left(\frac{\pi}{6} + 2\pi\right) \approx 1.15$; $\csc\left(\frac{\pi}{6} + 2\pi\right) \approx 2.00$; $\cot\left(\frac{\pi}{6} + 2\pi\right) \approx 0.58$

25. $\cos t = -\frac{4}{5}$; $\tan t = \frac{3}{4}$
27. $\tan t = -\sqrt{39}/13$
29. $\sec\alpha = \frac{13}{5}$; $\cos\alpha = \frac{5}{13}$; $\sin\alpha = \frac{12}{13}$
35. $(\sqrt{7}\cos\theta)/7$ **37. (a)** $-\frac{2}{3}$
(b) $\frac{1}{4}$ **(c)** $\frac{1}{5}$ **(d)** $-\frac{1}{5}$
39. (a) $-(1+2\sqrt{2})/3$ **(b)** 1
41. (a) $\sqrt{2}/2$ **(b)** $\sqrt{3}/2$ **(c)** 1
43. $\cos^2 t$ **45.** $\sin^2\theta$ **55.** $\frac{13}{31}$
57. Assuming that the radius of the circle is 1, the coordinates of the point labeled (x, y) are $(\cos t, \sin t)$, and the coordinates of the point labeled $(-x, -y)$ are $(\cos(t+\pi), \sin(t+\pi))$. So $y = \sin t$ and $-y = \sin(t+\pi)$, from which it follows that $\sin(t+\pi) = -\sin t$. Similarly, $-x = \cos(t+\pi)$ and $x = \cos t$, from which it follows that $\cos(t+\pi) = -\cos t$. Since $t - \pi$ results in the same intersection point with the unit circle as $t + \pi$, identities (ii) and (iv) follow in a similar manner.

61. (a)

t	0.2	0.4	0.6	0.8
$f(t)$	219.07	50.53	19.70	9.55

t	1.0	1.2	1.4
$f(t)$	6.14	7.98	33.88

(b) The smallest output is 6.14, which occurs at $t = 1.0$. **(d)** $\tan^2 t + 9\cot^2 t = (\tan t - 3\cot t)^2 + 6 \geqslant 0 + 6 \geqslant 6$
(e) The answer from part (b) is consistent with this result.
63. (a) Each side is 0. **(b)** Each side is $(2-\sqrt{2})/2$. **(c)** no

65.

t	$1 - \frac{1}{2}t^2$	$\cos\theta$
0.02	0.9998	0.999800
0.05	0.99875	0.998750
0.1	0.995	0.995004
0.2	0.980	0.980067
0.3	0.955	0.955336

67.

x	$\frac{1}{3}x^3 + x$	$\frac{2}{15}x^5 + \frac{1}{3}x^3 + x$	$\tan x$
0.1	0.100333	0.100335	0.100335
0.2	0.202667	0.202709	0.202710
0.3	0.309	0.309324	0.309336
0.4	0.421333	0.422699	0.422793
0.5	0.541667	0.545833	0.546302

69. $P{:}\sin(\pi/12) \approx 0.259$; $Q{:}\frac{1}{2} = 0.5$; $R{:}\sqrt{2}/2 \approx 0.707$; $S{:}\sqrt{3}/2 \approx 0.866$; $T{:}\sin(5\pi/12) \approx 0.966$

EXERCISE SET 7.4

1. period: 2; amplitude: 1
3. period: 4; amplitude: 6
5. period: 4; amplitude: 2
7. period: 6; amplitude: $\frac{3}{2}$
9. $(-7\pi/2, 1) \approx (-10.996, 1)$
11. $(5\pi/2, 1) \approx (7.854, 1)$
13. $(-4\pi, 0) \approx (-12.566, 0)$
15. $(-3\pi, 0) \approx (-9.425, 0)$
17. $(-\pi, 0) \approx (-3.142, 0)$
19. increasing **21.** decreasing
23. $(9\pi/2, 0) \approx (14.137, 0)$
25. $(-4\pi, 1) \approx (-12.566, 1)$
27. $(\pi/2, 0) \approx (1.571, 0)$
29. $(4\pi, 1) \approx (12.566, 1)$
31. $(-5\pi/2, 0) \approx (-7.854, 0)$
33. decreasing **35.** increasing
37. (a) 0.8 **(b)** 0.7954 **(c)** 5.4878 **(d)** 2.3462 and 3.9370 **39. (a)** 1.15 **(b)** 1.1593 **(c)** 5.1239 **(d)** 1.9823 and 4.3009 **41. (a)** 0.9 **(b)** 0.9273 **(c)** 5.3559 **(d)** 2.2143 and 4.0689 **43. (a)** 0.1 **(b)** 0.1002 **(c)** 3.0414 **(d)** 3.2418 and 6.1830 **45. (a)** 0.4 **(b)** 0.4115 **(c)** 2.7301 **(d)** 3.5531 and 5.8717 **47. (a)** 1.0 **(b)** 1.0160 **(c)** 2.1256 **(d)** 4.1576 and 5.2672
49. (a) $C(\cos\theta, \sin\theta)$
51. (a) 0 **(b)** 0 **53.** $(\pi/2, \pi)$

55. (a)

	x_1	x_2	x_3
From graph	1.0	0.55	0.85
From calculator	0.99500	0.54450	0.85539

	x_4	x_5	x_6
From graph	0.65	0.80	0.70
From calculator	0.65593	0.79248	0.70208

	x_7
From graph	0.75
From calculator	0.76350

(b)

n	0	1	2	3
Number of fish after n breeding seasons	50	498	272	428

n	4	5	6	7
Number of fish after n breeding seasons	328	396	351	382

(c) 370 fish

GRAPHING UTILITY EXERCISES FOR SECTION 7.4

1. (a) x-intercept: π

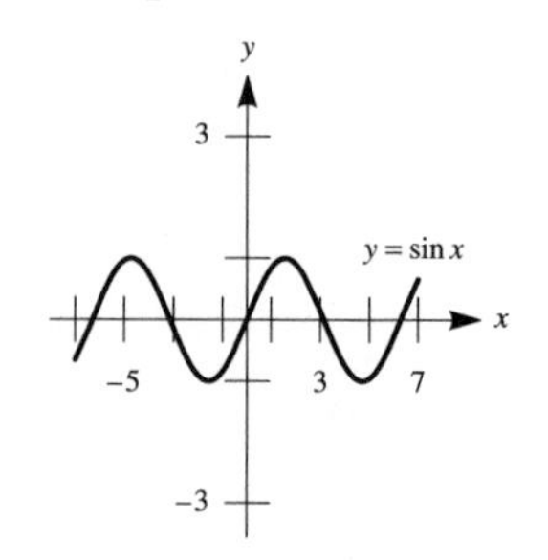

(b) four turning points; x-coordinate: $\pi/2$
(c) five turning points; x-coordinate: 2π

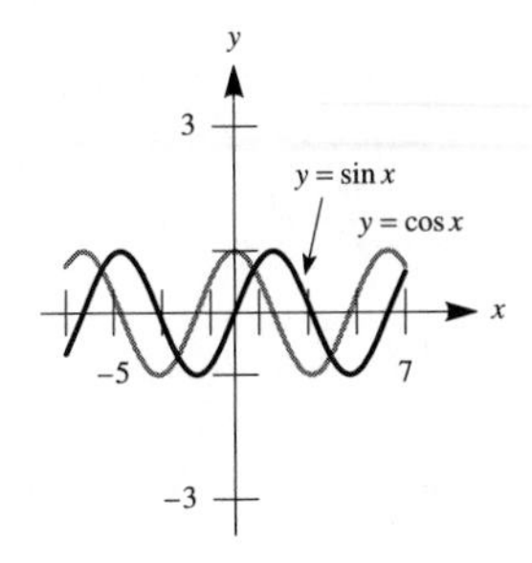

(d) $\pi/2$ units to the left

(e)

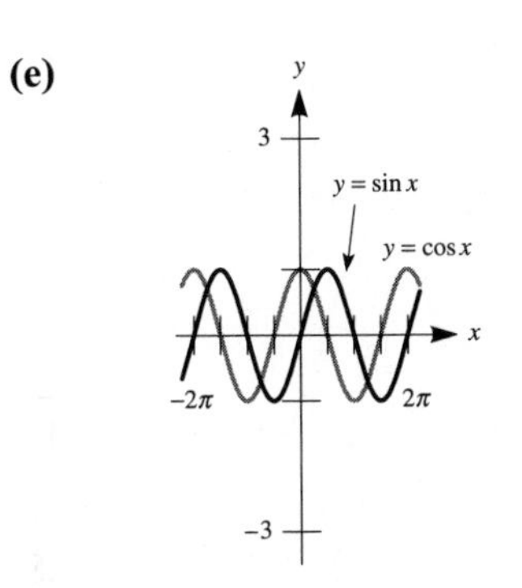

3. (a)

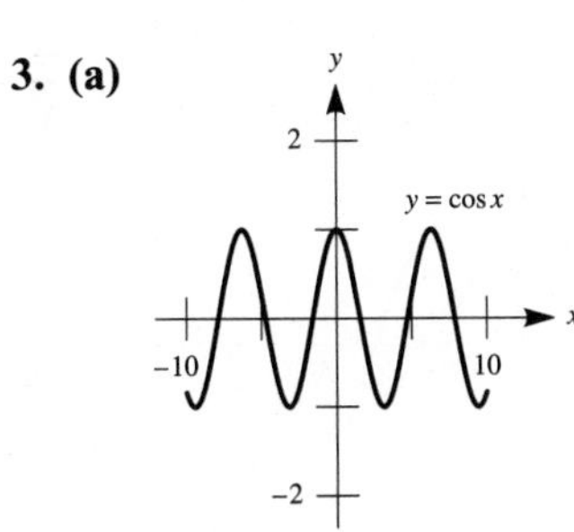

(b) In the vicinity of $x = 0$, the graphs of $y = \cos x$ and $y = 1 - 0.5x^2$ are very similar.

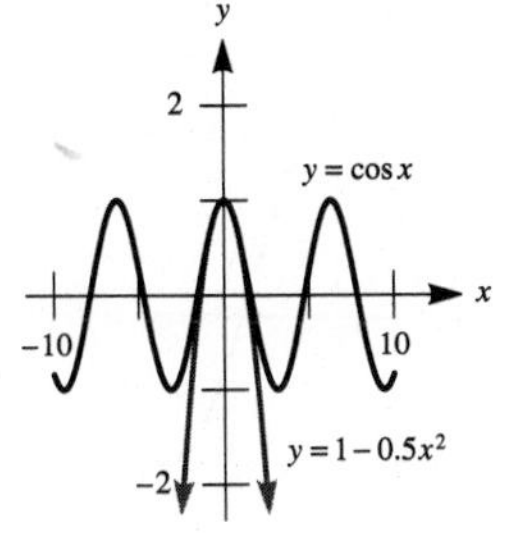

(c)

x	1	0.5	0.1
$\cos x$	0.54	0.8776	0.9950
$1 - 0.5x^2$	0.50	0.8750	0.9950

x	0.01	0.001
$\cos x$	0.99995	0.9999995
$1 - 0.5x^2$	0.99995	0.9999995

5. intersection points: $x \approx 2.498$ and $x \approx 3.785$

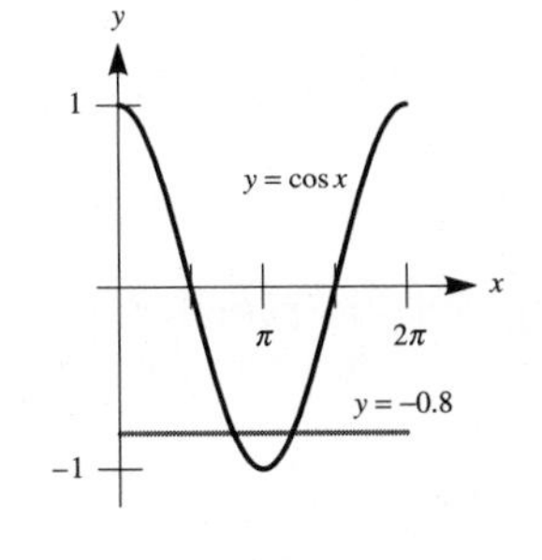

7. (a) $x \approx 0.7574$; $x \approx 2.3842$

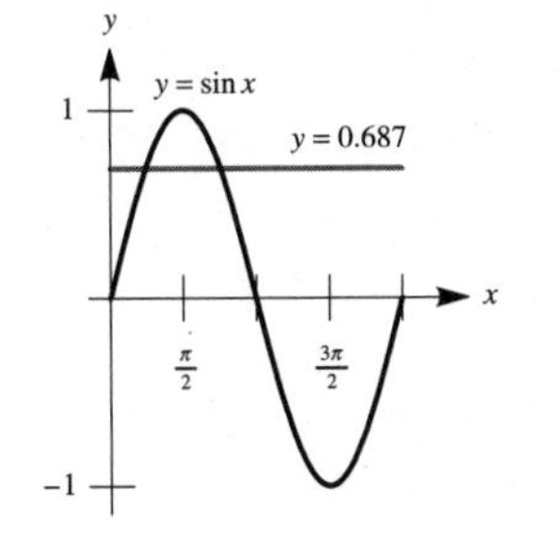

(b) $x \approx 3.8989$; $x \approx 5.5258$

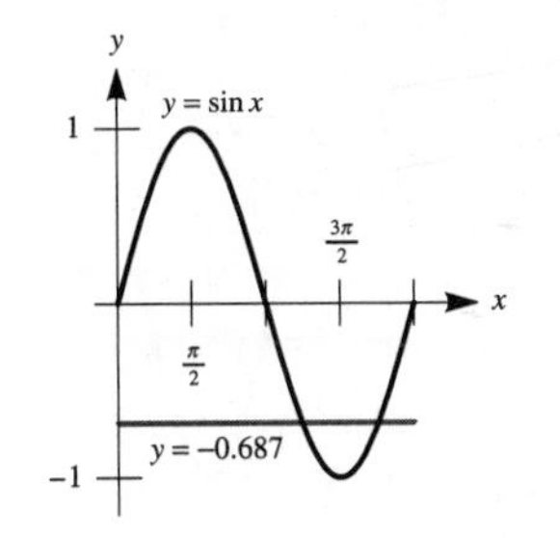

EXERCISE SET 7.5

1. (a) amplitude: 2; period: 2π; x-intercepts: 0, π, 2π; increasing: $(0, \pi/2)$ and $(3\pi/2, 2\pi)$

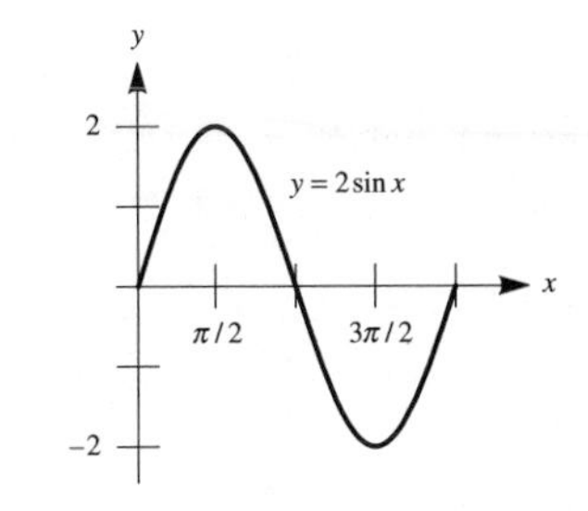

(b) amplitude: 1; period: π; x-intercepts: 0, $\pi/2$, π; increasing: $(\pi/4, 3\pi/4)$

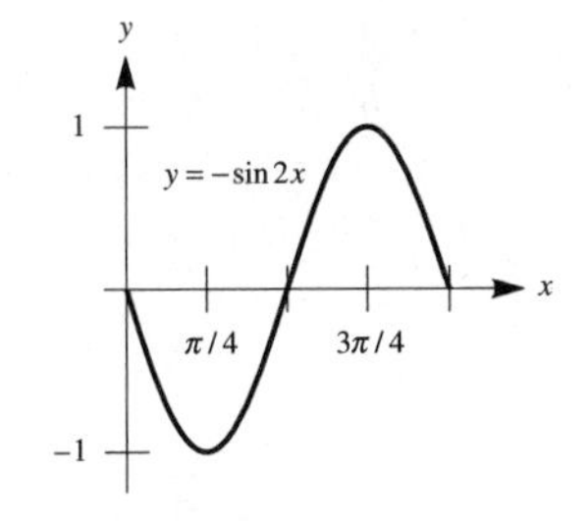

3. (a) amplitude: 1; period: π; x-intercepts: $\pi/4$, $3\pi/4$; increasing: $(\pi/2, \pi)$

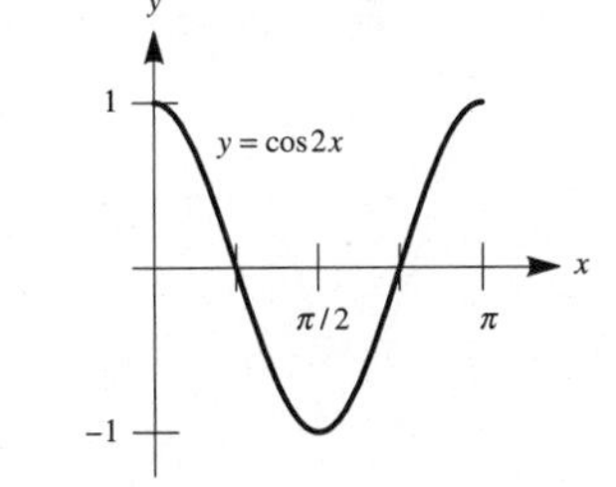

(b) amplitude: 2; period: π; x-intercepts: $\pi/4$, $3\pi/4$; increasing: $(\pi/2, \pi)$

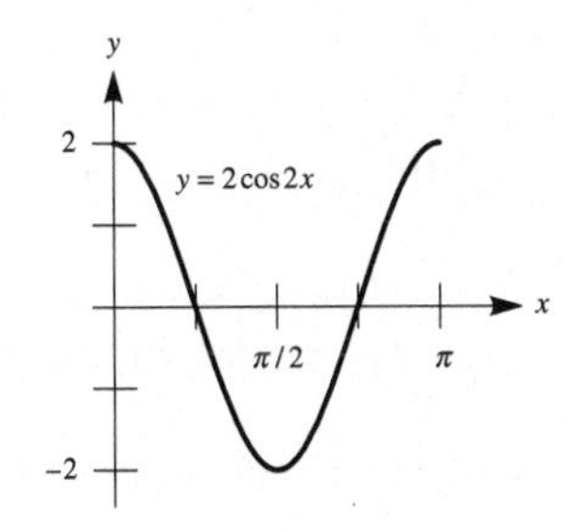

5. (a) amplitude: 3; period: 4; x-intercepts: 0, 2, 4; increasing: (0, 1) and (3, 4)

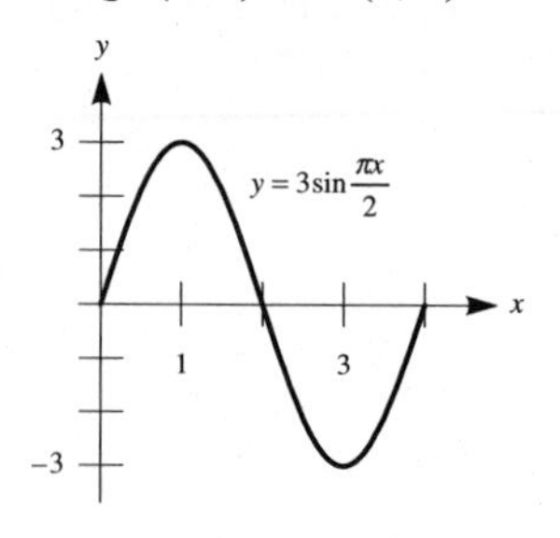

(b) amplitude: 3; period: 4; x-intercepts: 0, 2, 4; increasing: (1, 3)

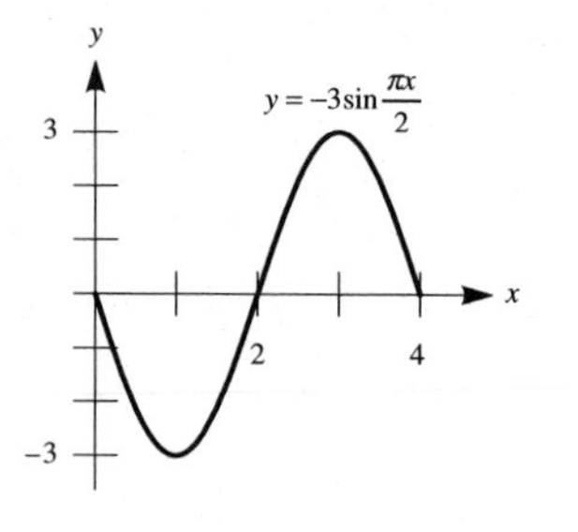

7. (a) amplitude: 1; period: 1; x-intercepts: $\frac{1}{4}$, $\frac{3}{4}$; increasing: $\left(\frac{1}{2}, 1\right)$

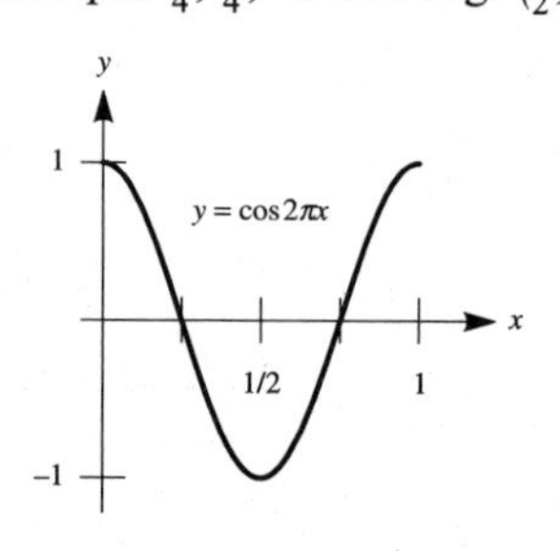

(b) amplitude: 4; period: 1; x-intercepts: $\frac{1}{4}$, $\frac{3}{4}$; increasing: $\left(0, \frac{1}{2}\right)$

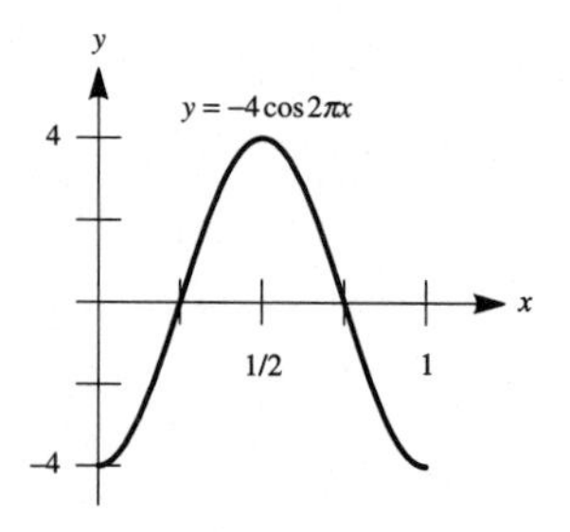

9. amplitude: 1; period: π; x-intercept: $3\pi/4$; increasing: $(0, \pi/4)$ and $(3\pi/4, \pi)$

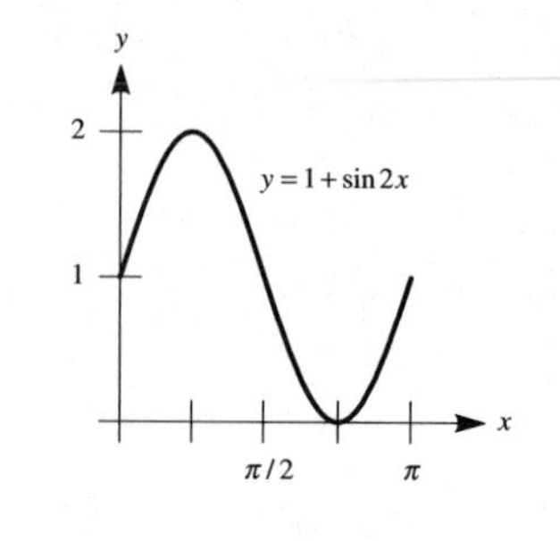

11. amplitude: 1; period: 6; x-intercepts: 0, 6; increasing: (0, 3)

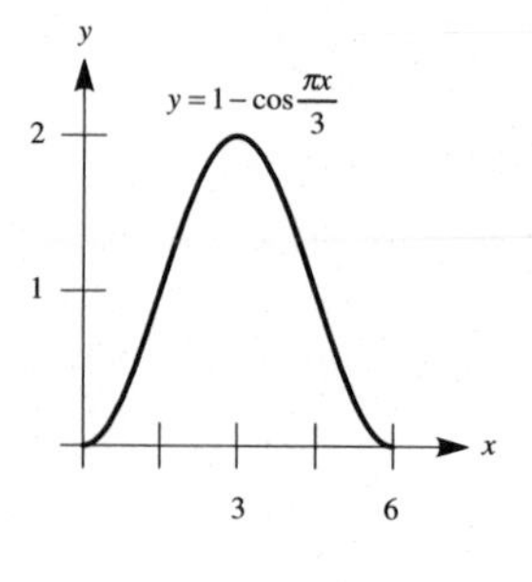

13. amplitude: 1; period: 2π; phase shift: $\pi/6$; x-intercepts: $\pi/6$, $7\pi/6$, $13\pi/6$; high point: $(2\pi/3, 1)$; low point: $(5\pi/3, -1)$

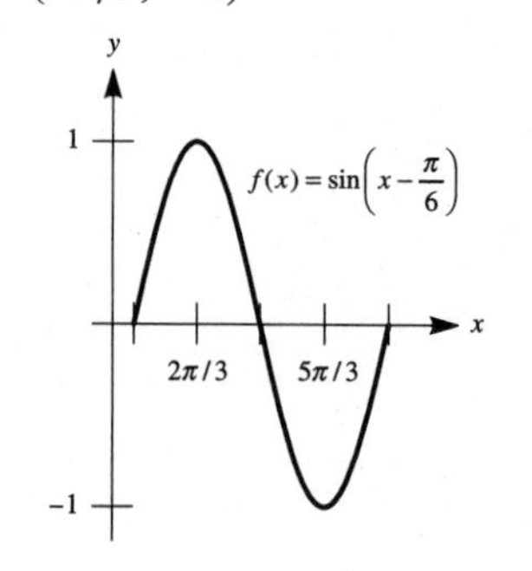

15. amplitude: 1; period: 2π; phase shift: $-\pi/4$; x-intercepts: $\pi/4$, $5\pi/4$; high point: $(3\pi/4, 1)$; low points: $(-\pi/4, -1)$ and $(7\pi/4, -1)$

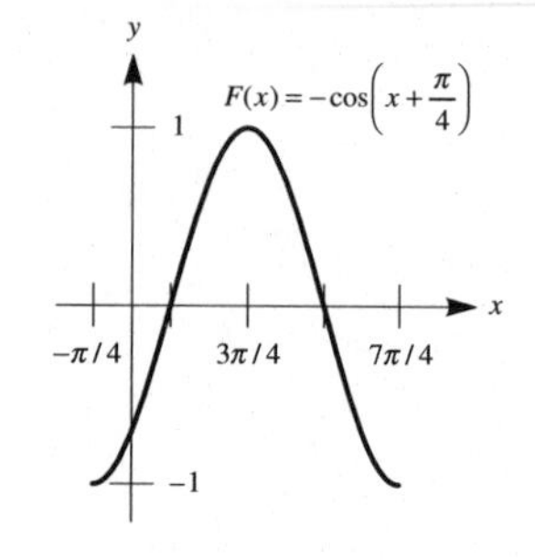

17. amplitude: 1; period: π; phase shift: $\pi/4$; x-intercepts: $\pi/4$, $3\pi/4$, $5\pi/4$; high point: $(\pi/2, 1)$; low point: $(\pi, -1)$

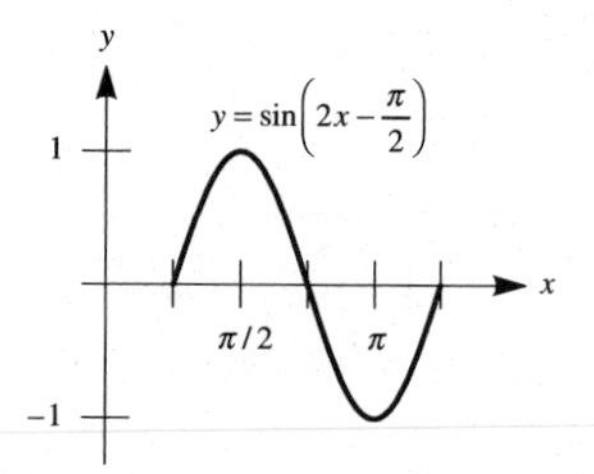

19. amplitude: 1; period: π; phase shift: $\pi/2$; x-intercepts: $3\pi/4$, $5\pi/4$; high points: $(\pi/2, 1)$ and $(3\pi/2, 1)$; low point: $(\pi, -1)$

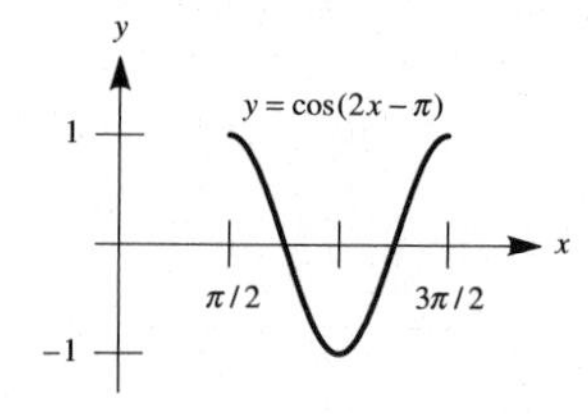

21. amplitude: 3; period: 4π; phase shift: $-\pi/3$; x-intercepts: $-\pi/3$, $5\pi/3$, $11\pi/3$; high point: $(2\pi/3, 3)$; low point: $(8\pi/3, -3)$

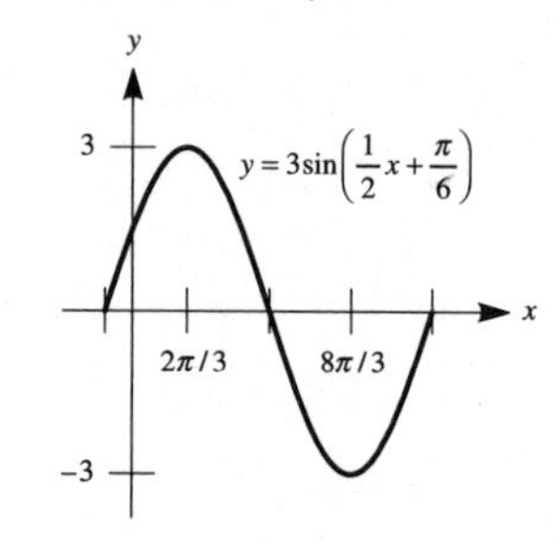

23. amplitude: 4; period: $2\pi/3$; phase shift: $\pi/12$; x-intercepts: $\pi/4$, $7\pi/12$; high points: $(\pi/12, 4)$ and $(3\pi/4, 4)$; low point: $(5\pi/12, -4)$

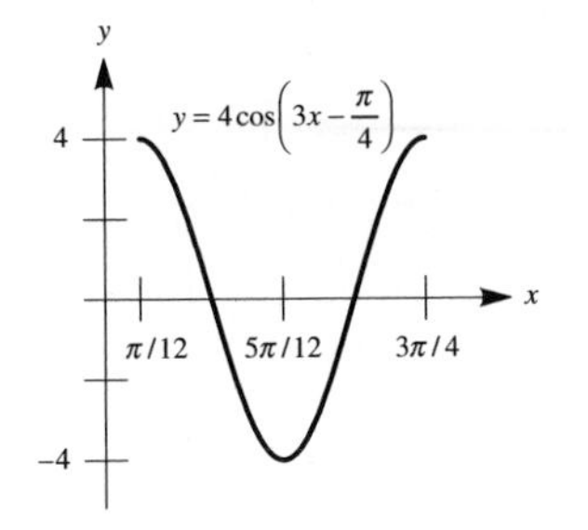

25. amplitude: $\frac{1}{2}$; period: 4; phase shift: 2π; x-intercepts: 2π, $2\pi + 2$, $2\pi + 4$; high point: $\left(2\pi + 1, \frac{1}{2}\right)$; low point: $\left(2\pi + 3, -\frac{1}{2}\right)$

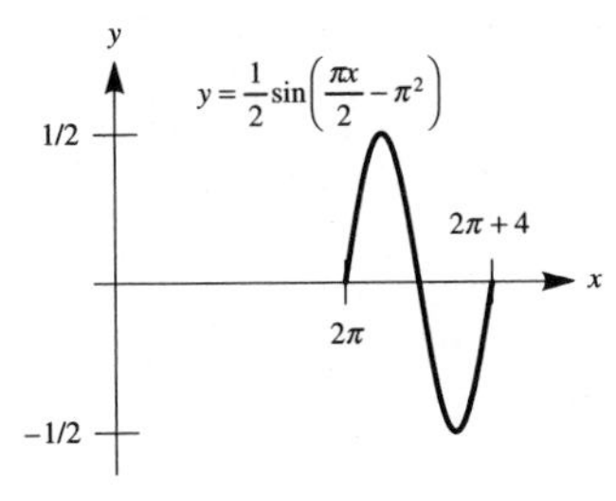

27. amplitude: 1; period: π; phase shift: $\pi/6$; x-intercepts: $\pi/6$ and $7\pi/6$; high point: $(2\pi/3, 2)$; low points: $(\pi/6, 0)$ and $(7\pi/6, 0)$

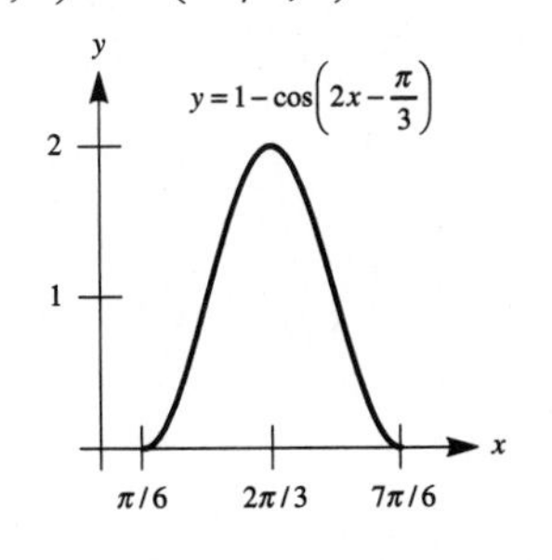

29. $y = 2\sin\frac{1}{2}x$ **31.** $y = -3\sin\pi x$
33. $y = -4\cos\frac{1}{5}x$

35.

37.

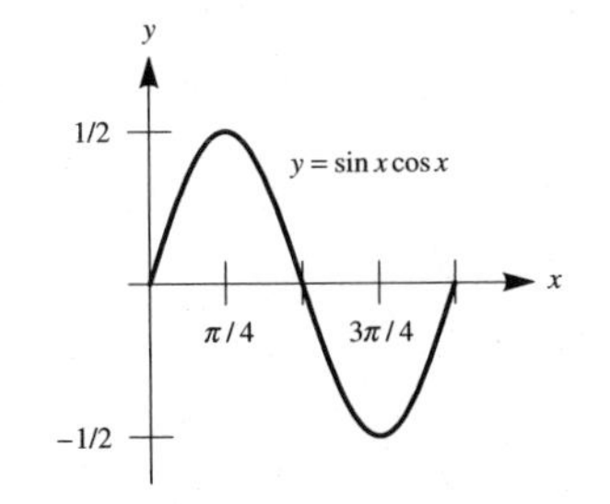

GRAPHING UTILITY EXERCISES FOR SECTION 7.5

1. $y = \sin x$: amplitude $= 1$, period $= 2\pi$; $y = 2\sin x$: amplitude $= 2$, period $= 2\pi$; $y = 3\sin x$: amplitude $= 3$, period $= 2\pi$; $y = 4\sin x$: amplitude $= 4$, period $= 2\pi$

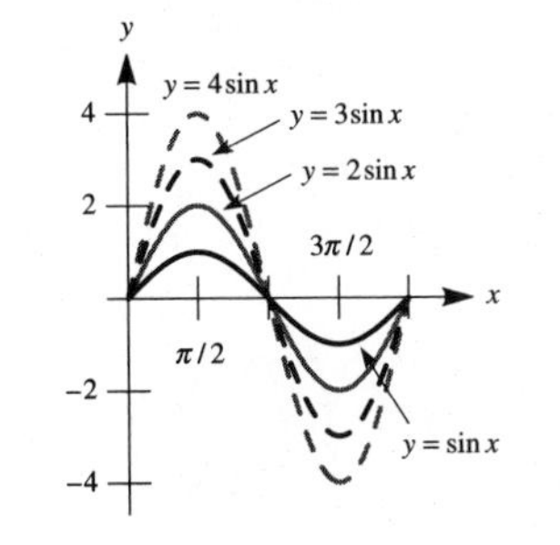

3. (a) $y = 2\sin\pi x$: amplitude $= 2$, period $= 2$; $y = \sin 2\pi x$: amplitude $= 1$, period $= 1$
(b)

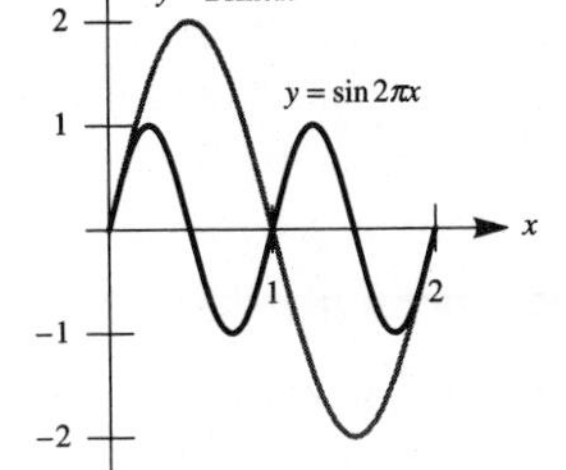

5. (a) amplitude: 2.5; period: $\frac{2}{3}$; phase shift: $-4/(3\pi)$

(b)

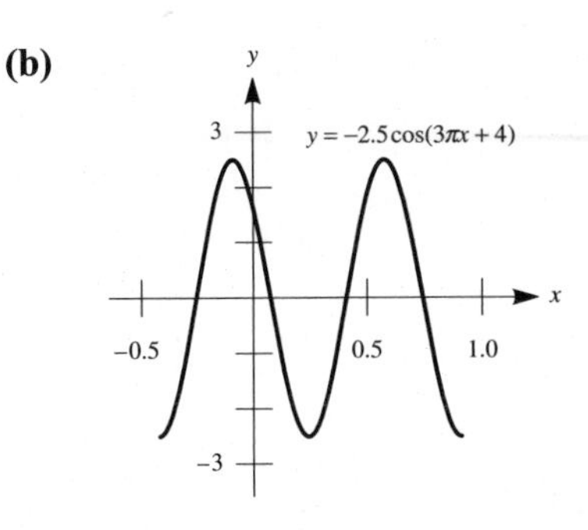

(c) high points: $(-0.09, 2.5)$ and $(0.58, 2.5)$; low points: $(-0.42, -2.5)$, $(0.24, -2.5)$, and $(0.91, -2.5)$
(d) high points: $\left(-\frac{4}{3\pi} + \frac{1}{3}, 2.5\right)$ and $\left(-\frac{4}{3\pi} + 1, 2.5\right)$; low points: $\left(-\frac{4}{3\pi}, -2.5\right)$, $\left(-\frac{4}{3\pi} + \frac{2}{3}, -2.5\right)$, and $\left(-\frac{4}{3\pi} + \frac{4}{3}, -2.5\right)$

7. (a) amplitude: 2.5; period: 6; phase shift: $-12/\pi$

(b)

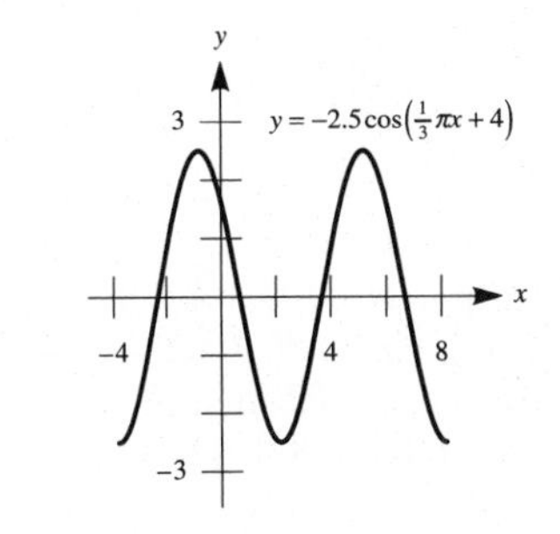

(c) high points: $(-0.82, 2.5)$ and $(5.18, 2.5)$; low points: $(-3.82, -2.5)$, $(2.18, -2.5)$, and $(8.18, -2.5)$
(d) high points: $\left(-\frac{12}{\pi} + 3, 2.5\right)$ and $\left(-\frac{12}{\pi} + 9, 2.5\right)$; low points: $\left(-\frac{12}{\pi}, -2.5\right)$, $\left(-\frac{12}{\pi} + 6, -2.5\right)$, and $\left(-\frac{12}{\pi} + 12, -2.5\right)$

9. (a) amplitude: 1; period: 4π; phase shift: -1.5

(b)

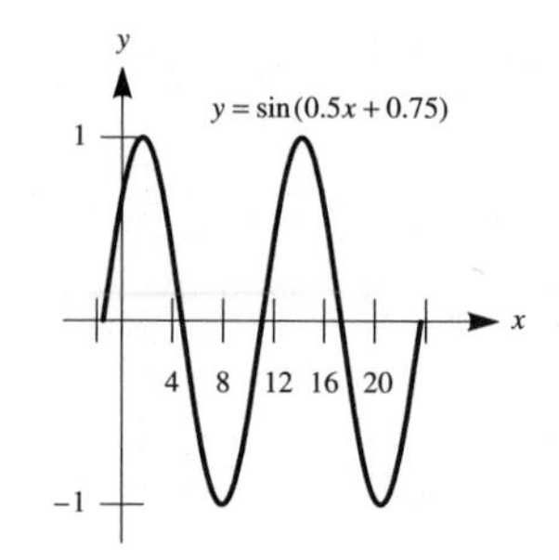

(c) high points: (1.64, 1) and (14.21, 1); low points: (7.92, −1) and (20.49, −1)
(d) high points: $(-1.5+\pi, 1)$ and $(-1.5+5\pi, 1)$; low points: $(-1.5+3\pi, -1)$ and $(-1.5+7\pi, -1)$
11. (a) amplitude: 0.02; period: 200; phase shift: 400

(b)

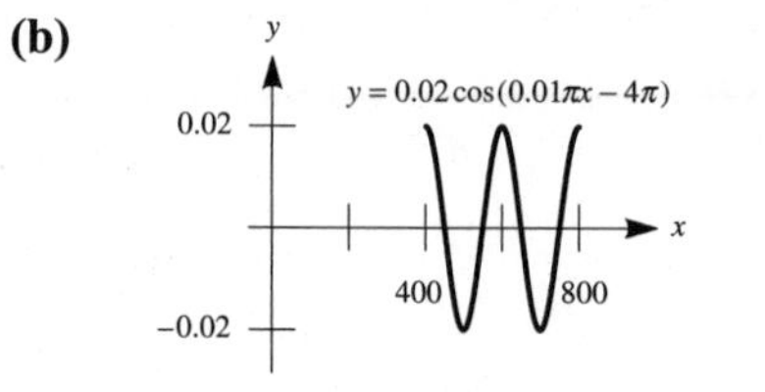

(c) high points: (400, 0.02), (600, 0.02), and (800, 0.02); low points: (500, −0.02) and (700, −0.02)
(d) high points: (400, 0.02), (600, 0.02), and (800, 0.02); low points: (500, −0.02) and (700, −0.02)
13. (a) Although the two graphs have the same period, the second graph appears to have a smaller amplitude.

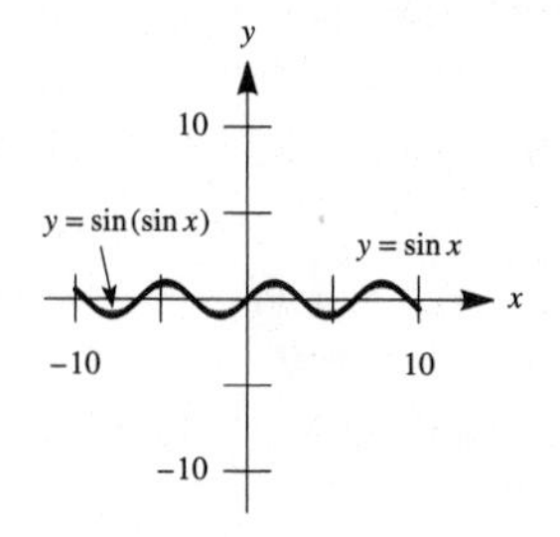

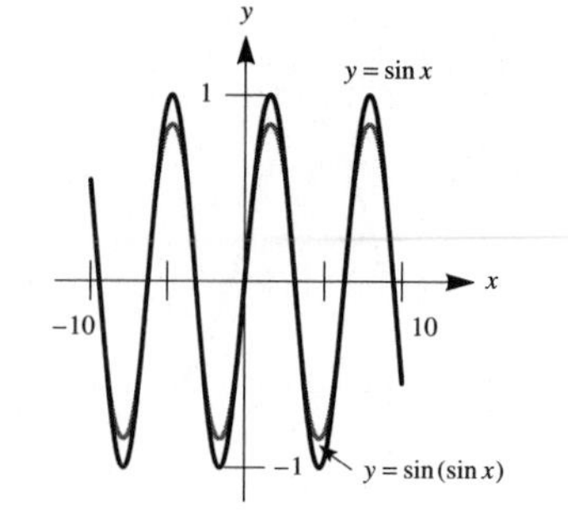

(b) 0.84 **(c)** The amplitude is $\sin\left(\sin\frac{\pi}{2}\right) = \sin 1 \approx 0.84$.

15. (a)

(b)

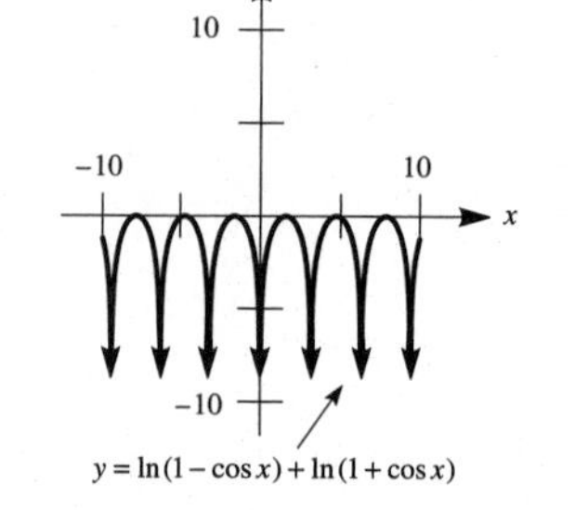

(c) Since $\ln(1-\cos x) + \ln(1+\cos x) = \ln(\sin^2 x)$, using properties of logarithms, the two graphs are identical.

EXERCISE SET 7.6

1. (a) $t = 0$: $s = 4$ cm; $t = 0.5$: $s \approx 2.83$ cm; $t = 1$: $s = 0$ cm; $t = 2$: $s = -4$ cm **(b)** amplitude: 4 cm; period: 4 sec; frequency: $\frac{1}{4}$ cps

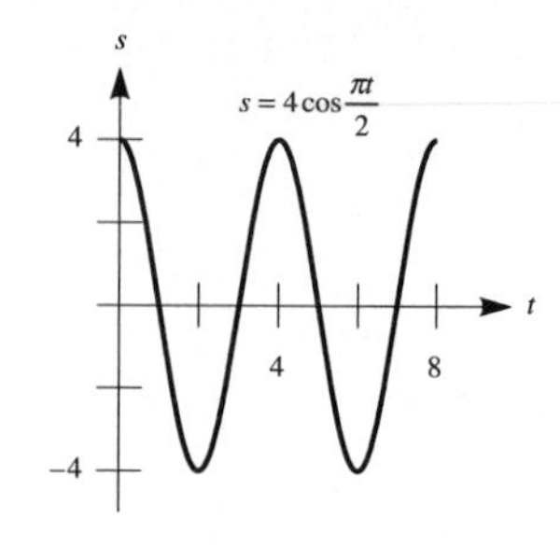

(c) $t = 0$, $t = 2$, $t = 4$, $t = 6$, and $t = 8$ sec **(d)** $t = 1$, $t = 3$, $t = 5$, and $t = 7$ sec **(e)** $2 < t < 4$ and $6 < t < 8$
3. (a) amplitude: 3 ft; period: 6 sec; frequency: $\frac{1}{6}$ cps

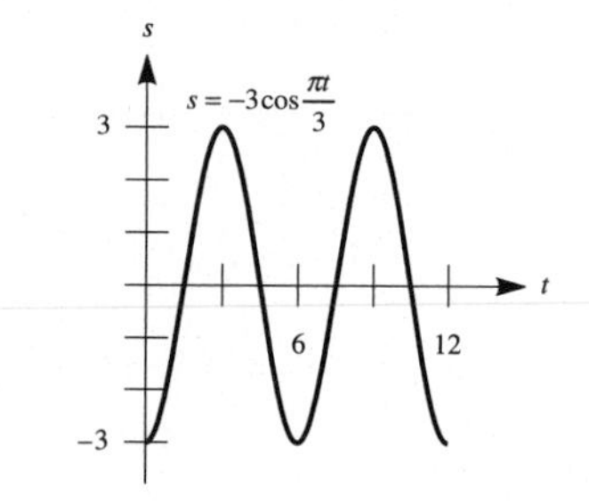

(b) $0 < t < 3$ and $6 < t < 9$
(c) $3 < t < 6$ and $9 < t < 12$

(d)

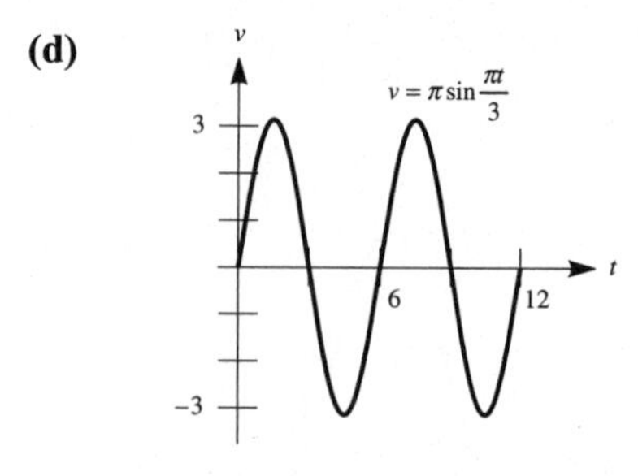

(e) $t = 0$, $t = 3$, $t = 6$, $t = 9$, and $t = 12$ sec; the mass (s-coordinate) is at −3, 3, −3, 3, and −3 ft, respectively.
(f) $t = 1.5$ and $t = 7.5$ sec; the mass is at 0 ft. **(g)** $t = 4.5$ and $t = 10.5$ sec; the mass is at 0 ft

(h)

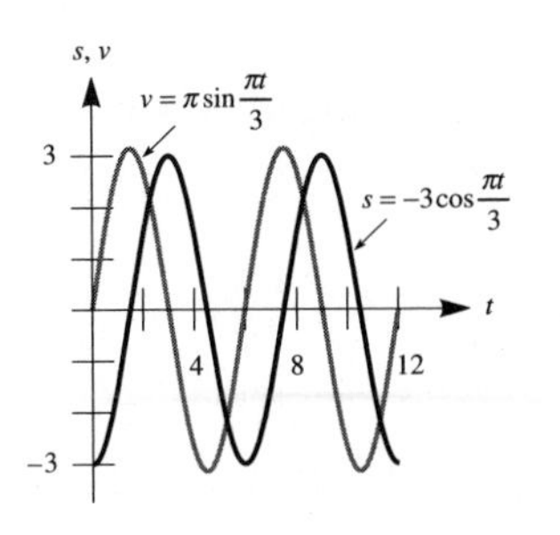

5. (a) amplitude: 170 volts; frequency: 60 cps

(b)

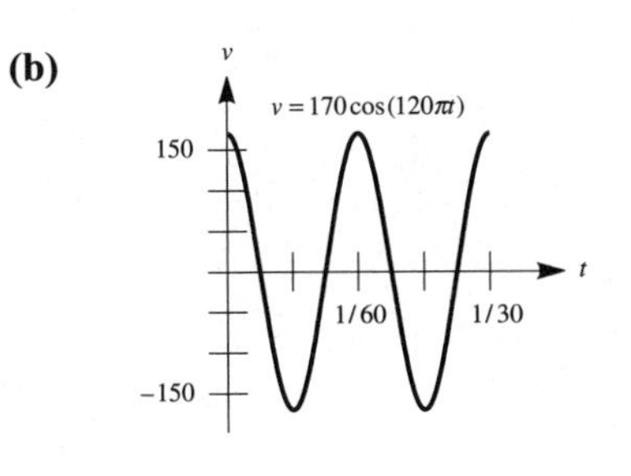

(c) $t = \frac{1}{60}$ sec and $t = \frac{1}{30}$ sec

7. (a)

***t* (sec)**	0	1	2	3
***θ* (radians)**	0	$\pi/3$	$2\pi/3$	π

***t* (sec)**	4	5	6	7
θ (radians)	$4\pi/3$	$5\pi/3$	2π	$7\pi/3$

(b) $1, \frac{1}{2}, -\frac{1}{2}, -1, -\frac{1}{2}, \frac{1}{2}, 1$, and $\frac{1}{2}$

(c)

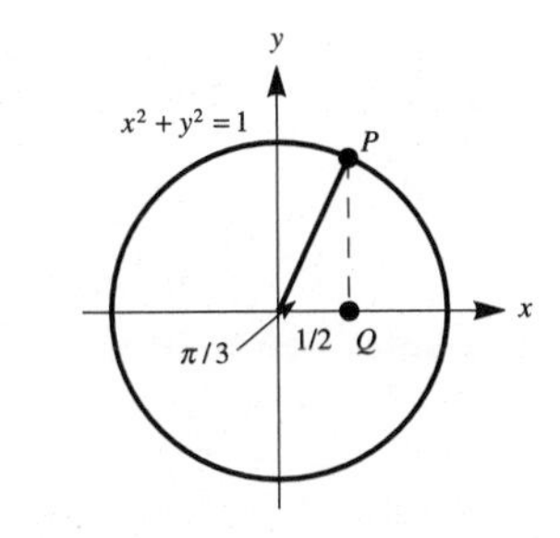

(d) $t = 2$ sec

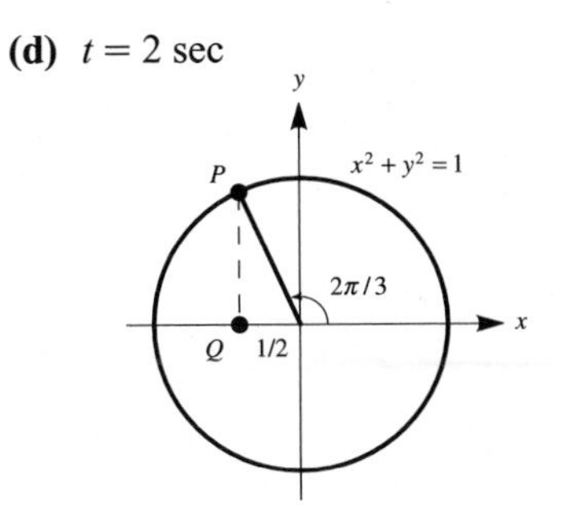

$t = 3$ sec

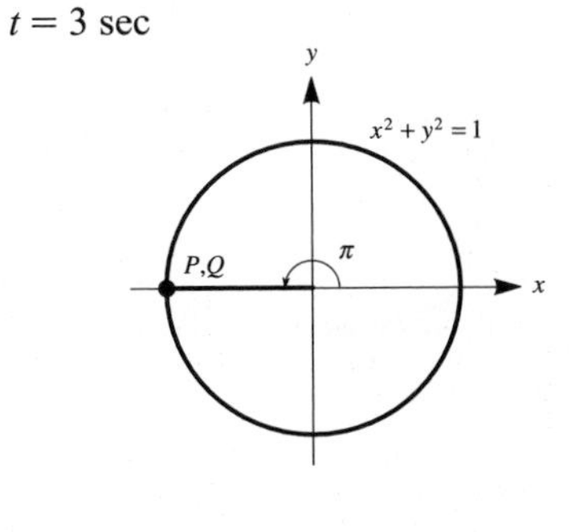

$t = 4$ sec

(e) The x-coordinate of Q is the same as the x-coordinate of P, which is $\cos \theta$. **(f)** amplitude: 1; period: 6; frequency: $\frac{1}{6}$

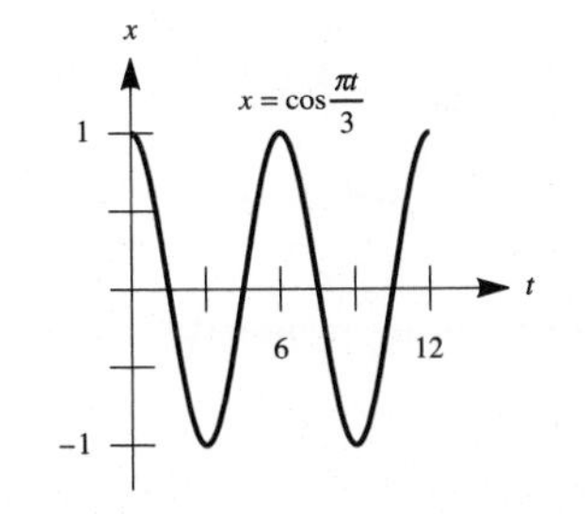

(g)

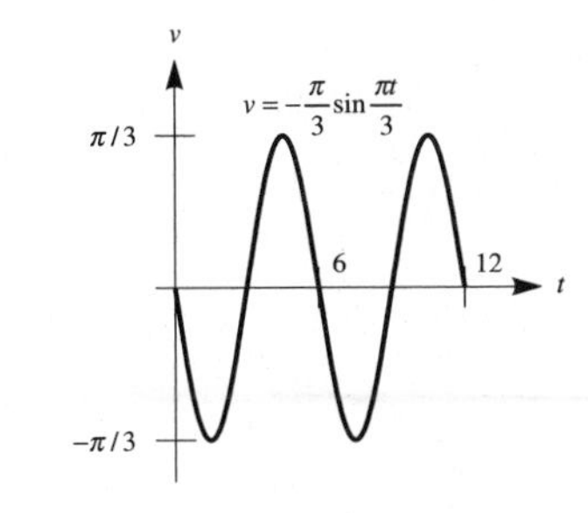

(h) $t = 0$, 3, 6, 9 and 12 sec; x-coordinates: 1, -1, 1, -1 and 1, respectively **(i)** $t = 4.5$ sec and $t = 10.5$ sec; Q is located at the origin. **(j)** $t = 1.5$ sec and $t = 7.5$ sec; Q is located at the origin.

EXERCISE SET 7.7

1. (a) x-intercept: $-\pi/4$; y-intercept: 1; asymptotes: $x = -3\pi/4$ and $x = \pi/4$

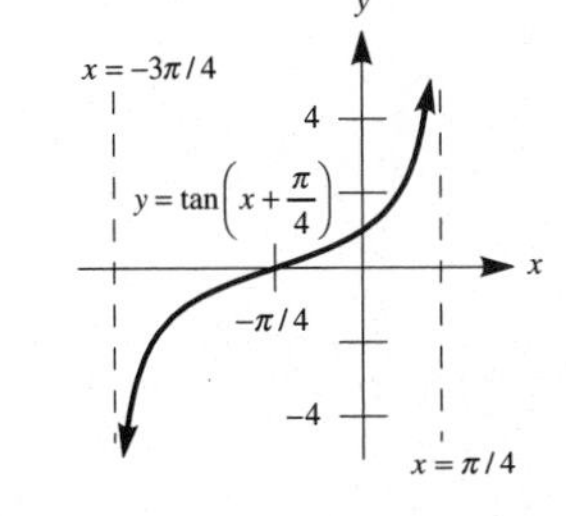

(b) x-intercept: $-\pi/4$; y-intercept: -1; asymptotes: $x = -3\pi/4$ and $x = \pi/4$

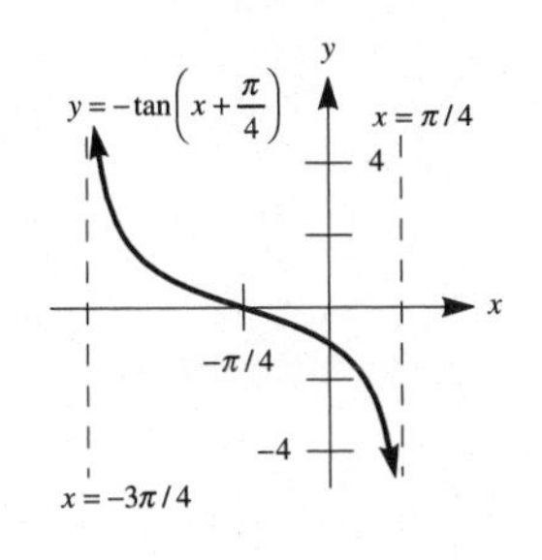

3. (a) x- and y-intercepts: 0; asymptotes: $x = -3\pi/2$ and $x = 3\pi/2$

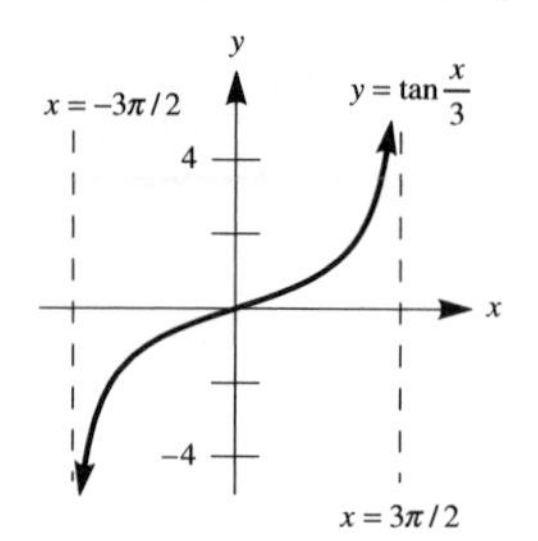

(b) x- and y-intercepts: 0; asymptotes: $x = -3\pi/2$ and $x = 3\pi/2$

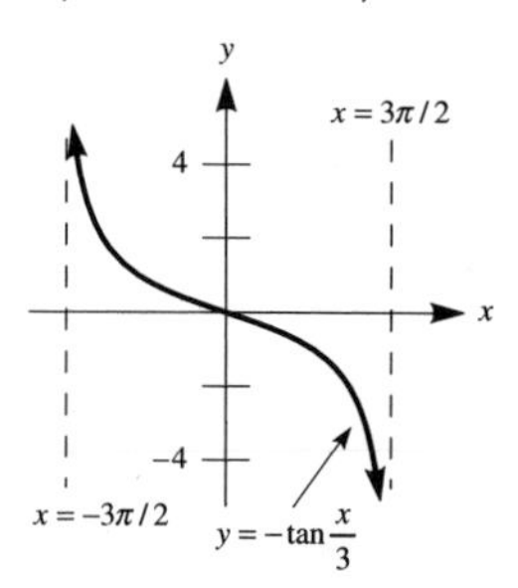

5. x- and y-intercepts: 0; asymptotes: $x = -1$ and $x = 1$

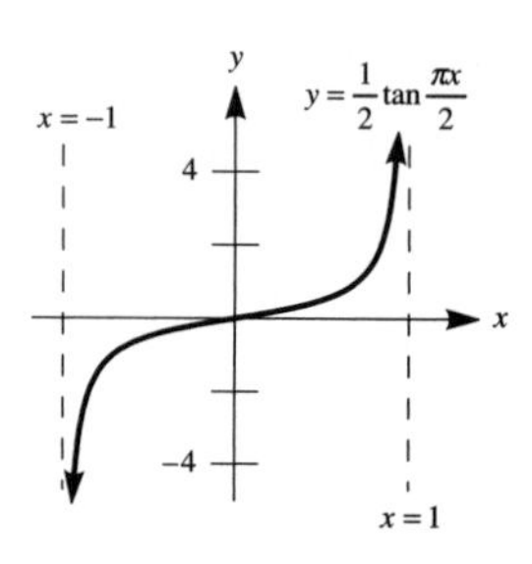

7. x-intercept: 1; no y-intercept; asymptotes: $x = 0$ and $x = 2$

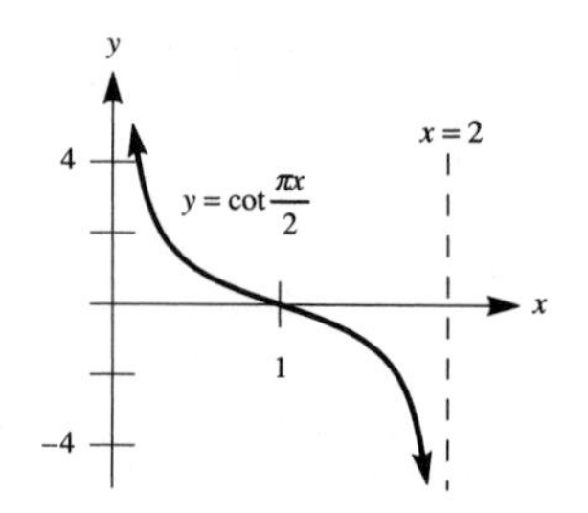

9. x-intercept: $3\pi/4$; y-intercept: 1; asymptotes: $x = \pi/4$ and $x = 5\pi/4$

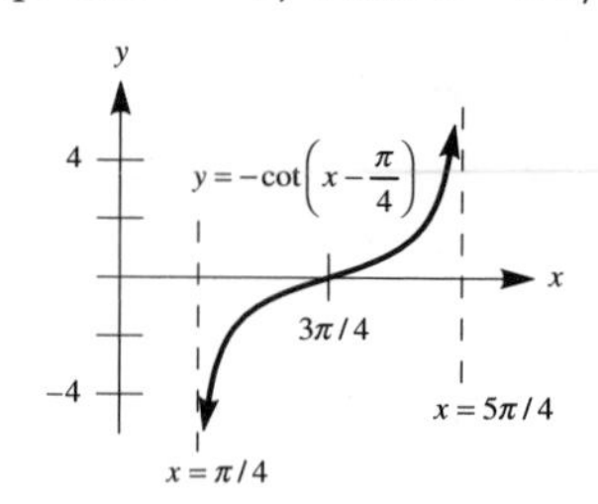

11. x-intercept: $\pi/4$; no y-intercept; asymptotes: $x = 0$ and $x = \pi/2$

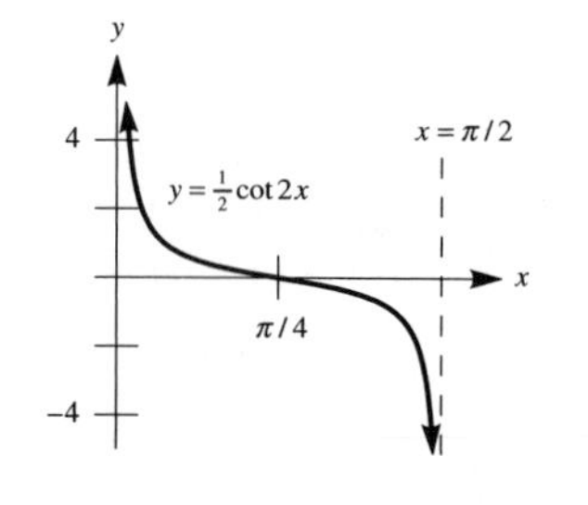

13. no x-intercept; y-intercept: $-\sqrt{2}$; asymptotes: $x = -3\pi/4$, $x = \pi/4$, and $x = 5\pi/4$

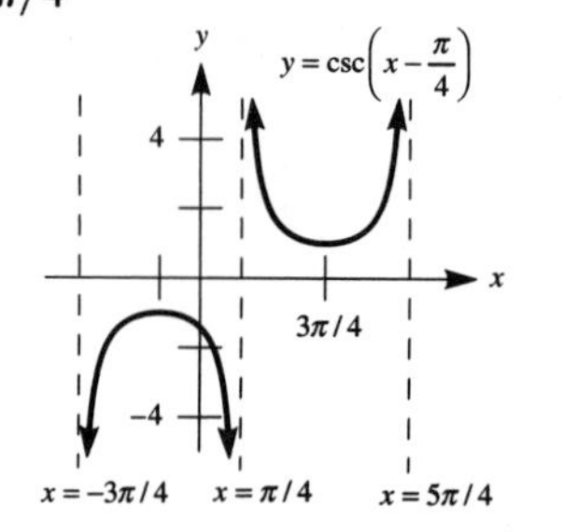

15. no x- or y-intercepts; asymptotes: $x = -2\pi$, $x = 0$, and $x = 2\pi$

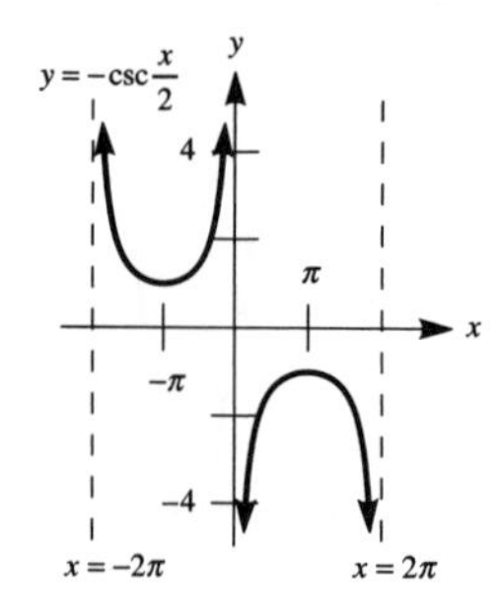

17. no x- or y-intercepts; asymptotes: $x = -1$, $x = 0$, and $x = 1$

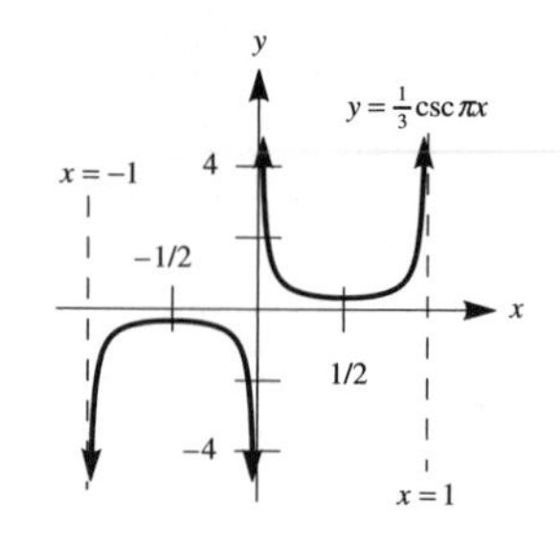

19. no x-intercept; y-intercept: -1; asymptotes: $x = -\pi/2$, $x = \pi/2$, and $x = 3\pi/2$

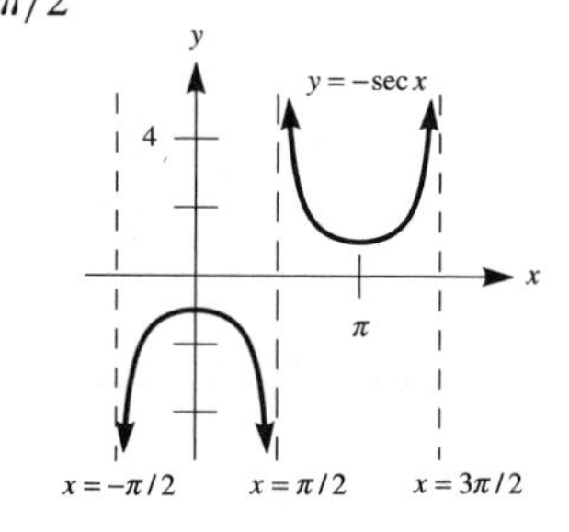

21. no x-intercept; y-intercept: -1; asymptotes: $x = \pi/2$, $x = 3\pi/2$, and $x = 5\pi/2$

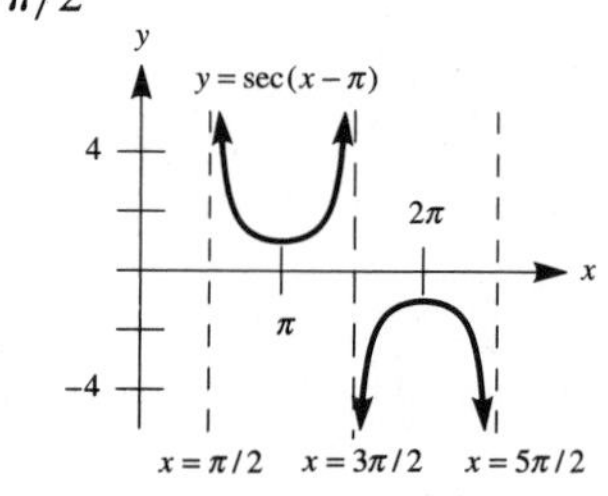

23. no x-intercept; y-intercept: 3; asymptotes: $x = -1$, $x = 1$, and $x = 3$

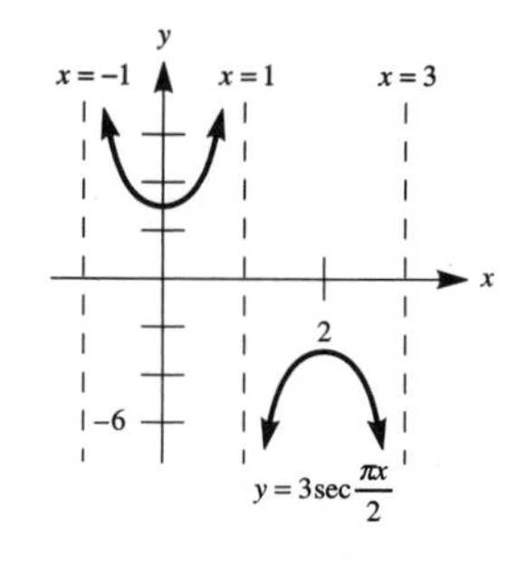

25. (a) x-intercepts: $-\frac{11}{18}$, $-\frac{5}{18}$, $\frac{1}{18}$, $\frac{7}{18}$ and $\frac{13}{18}$; y-intercept: -1; no asymptotes

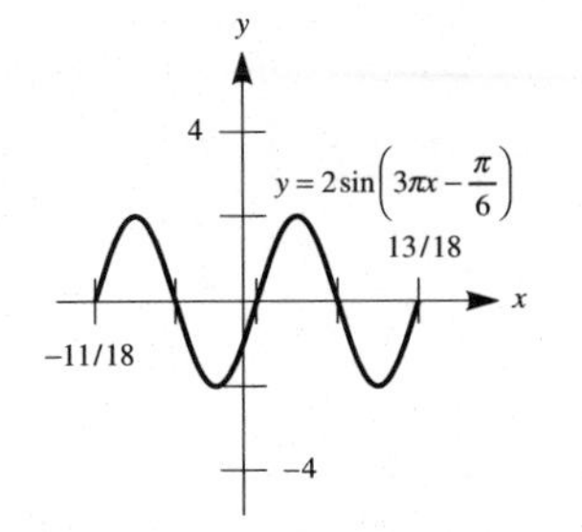

(b) no x-intercepts; y-intercept: -4; asymptotes: $x = -\frac{11}{18}$, $x = -\frac{5}{18}$, $x = \frac{1}{18}$, $x = \frac{7}{18}$, and $x = \frac{13}{18}$

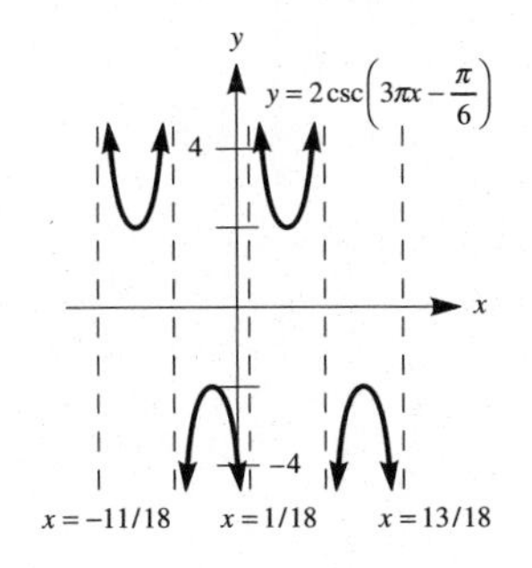

27. (a) x-intercepts: $-\frac{5}{8}$, $-\frac{1}{8}$, $\frac{3}{8}$, and $\frac{7}{8}$; y-intercept: $-3\sqrt{2}/2 \approx -2.12$; no asymptotes

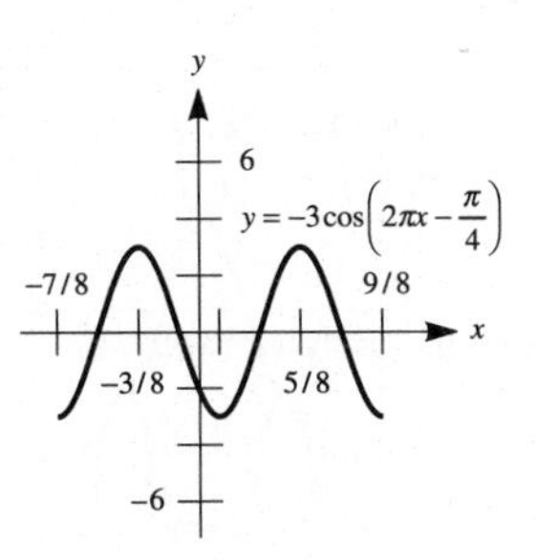

(b) no x-intercepts; y-intercept: $-3\sqrt{2}$; asymptotes: $x = -\frac{5}{8}$, $x = -\frac{1}{8}$, $x = \frac{3}{8}$, and $x = \frac{7}{8}$

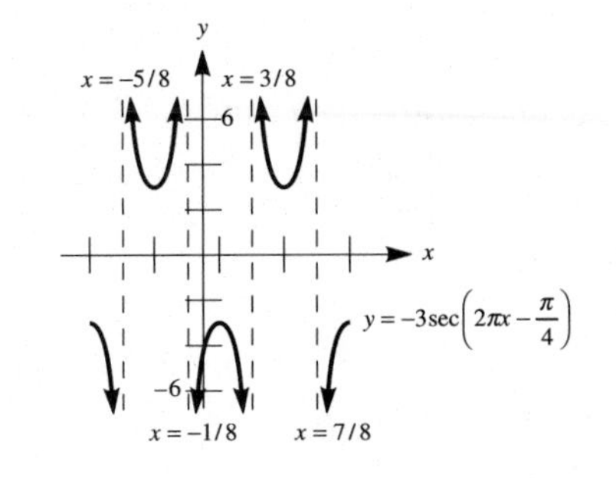

29. (a)

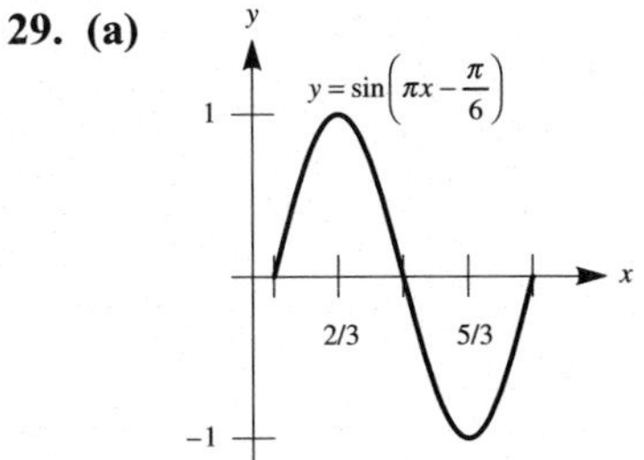

(b)

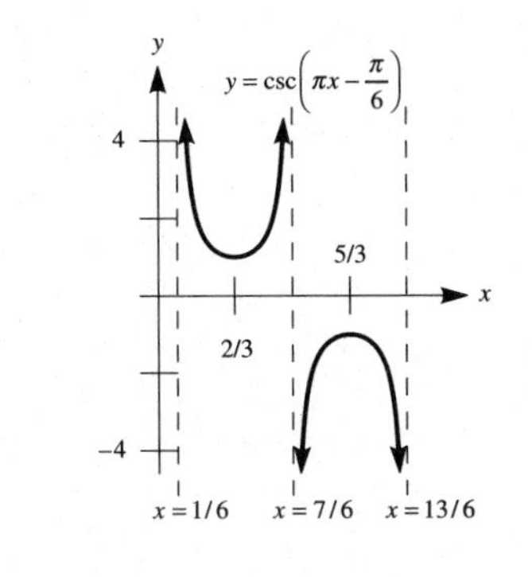

31. (a)

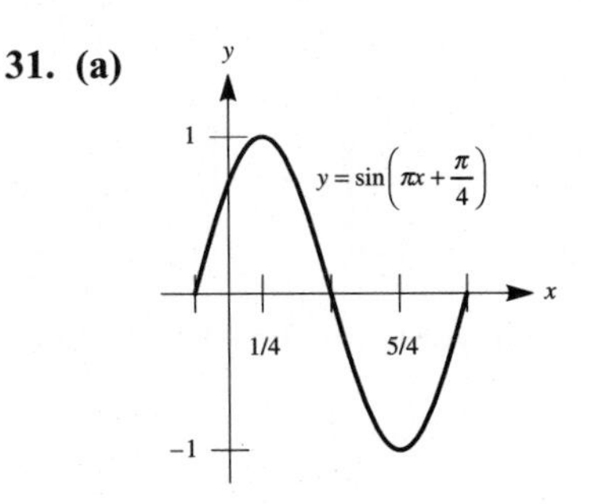

(b)

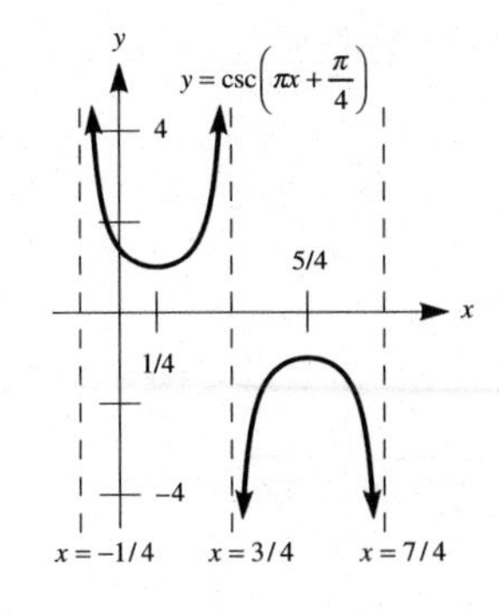

33.

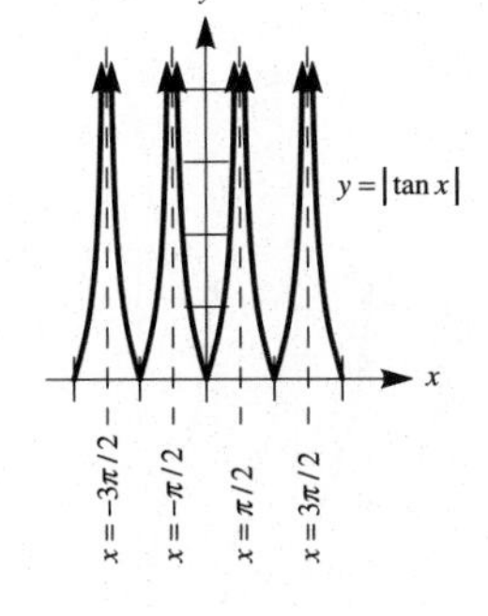

35.

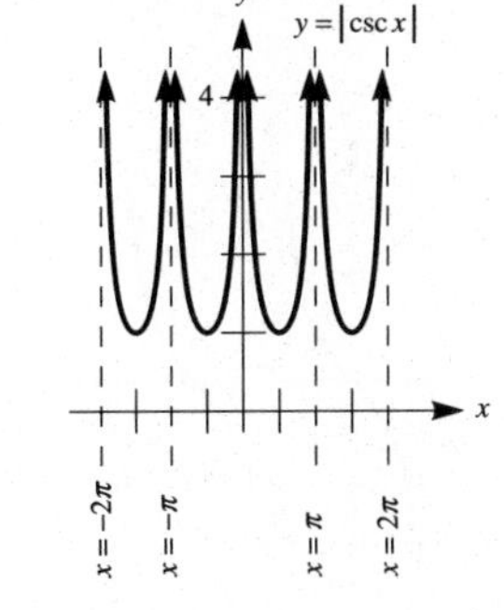

37. (a) Since P and Q are both points on the unit circle, the coordinates are $P(\cos s, \sin s)$ and $Q\left(\cos\left(s - \frac{\pi}{2}\right), \sin\left(s - \frac{\pi}{2}\right)\right)$.
(b) Since $\triangle OAP$ is congruent to $\triangle OBQ$ (labeling the third vertex B), $OA = OB$ and $AP = BQ$. Because the y-coordinate at Q is negative, we have concluded what was required.

GRAPHING UTILITY EXERCISES FOR SECTION 7.7

1. x-intercepts: $-\pi$, 0, π

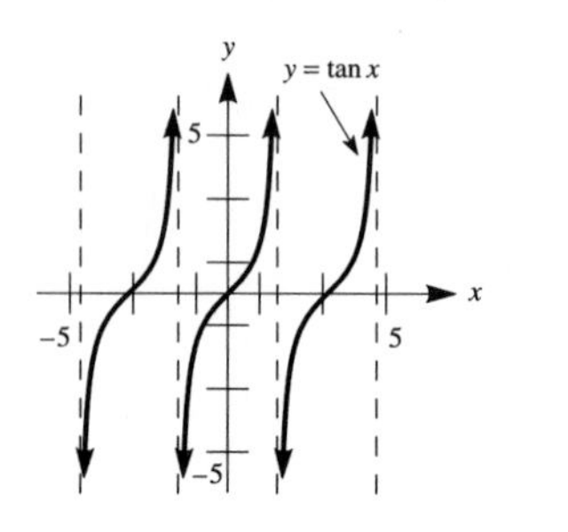

3.

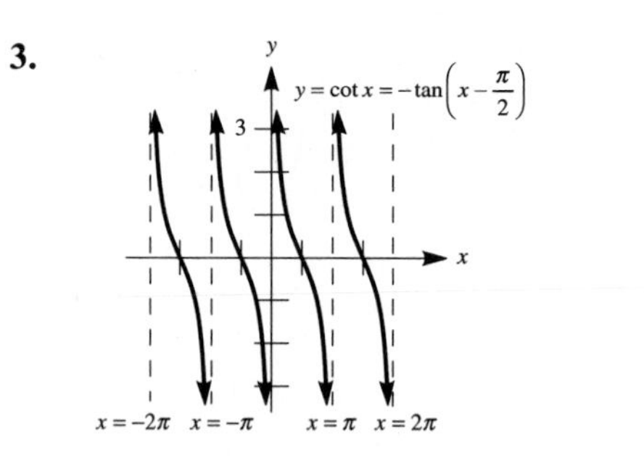

5. (a)

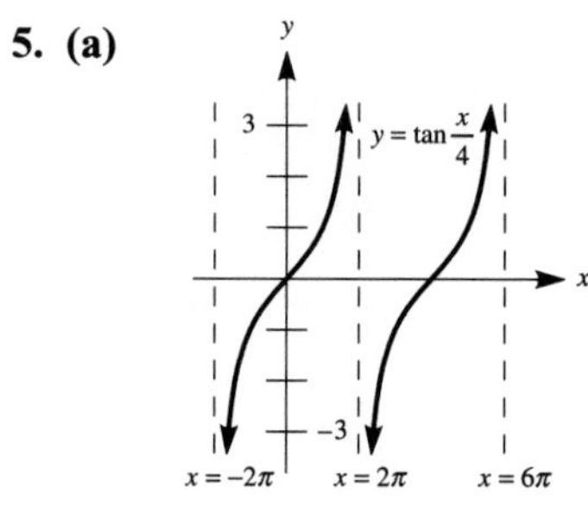

(b)

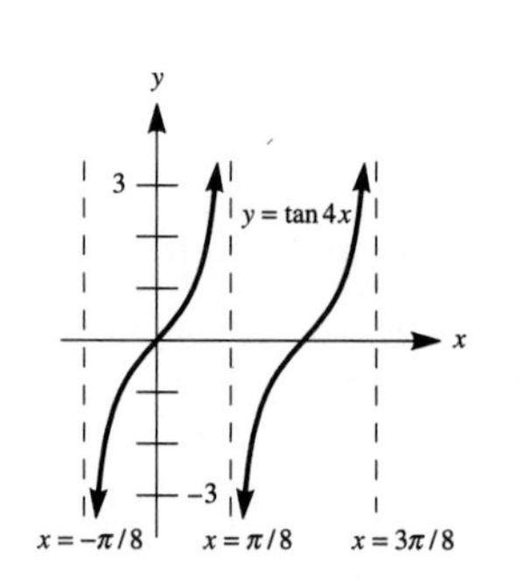

7. (a)

(b)

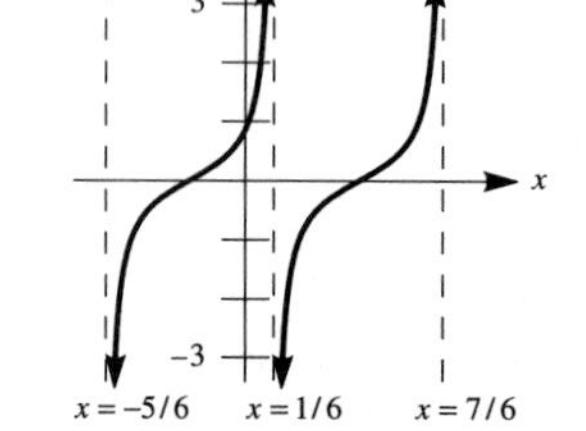

(c)

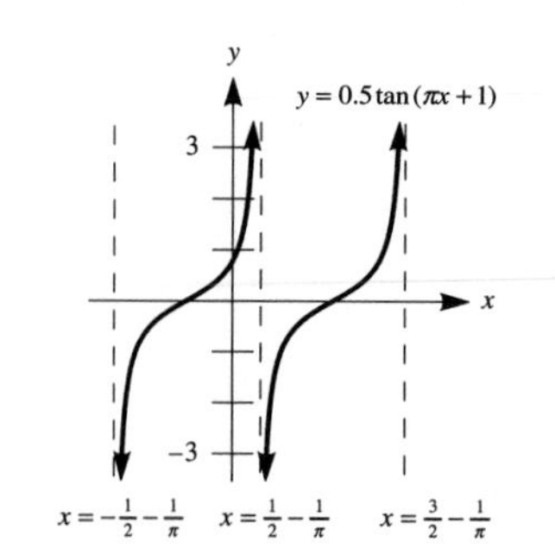

9. (a)

(b)

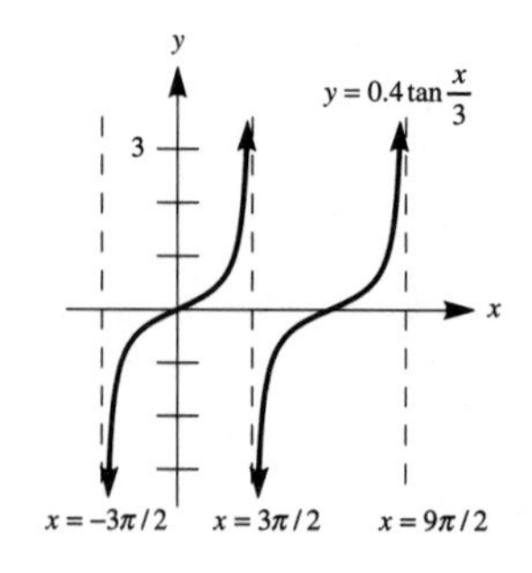

(c)

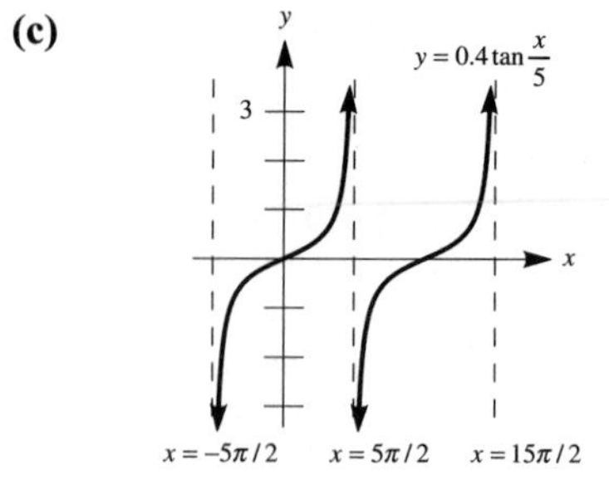

11. $\sin x = \csc x$ at odd multiples of $\pi/2$, such as $-3\pi/2$, $-\pi/2$, $\pi/2$, and $3\pi/2$; there are no points in which $\sin x = -\csc x$.

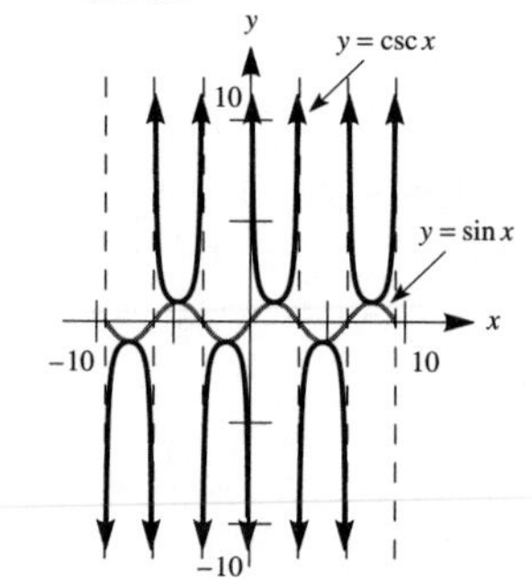

13.

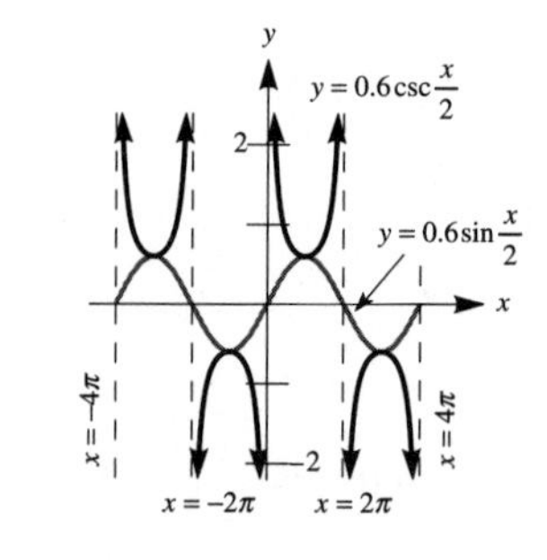

15.

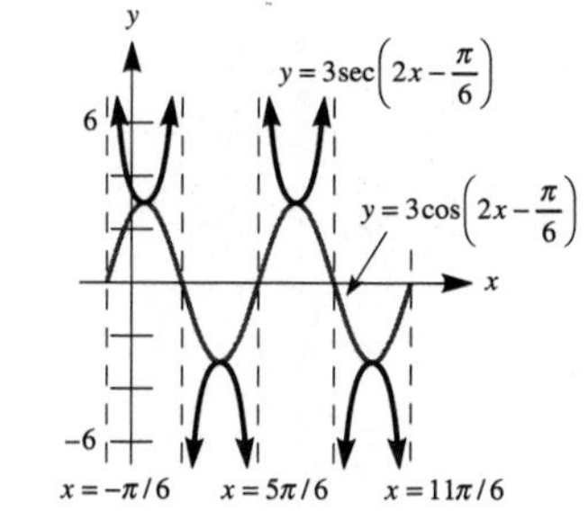

17. The two graphs are identical; this demonstrates that $\tan^2 x = \sec^2 x - 1$

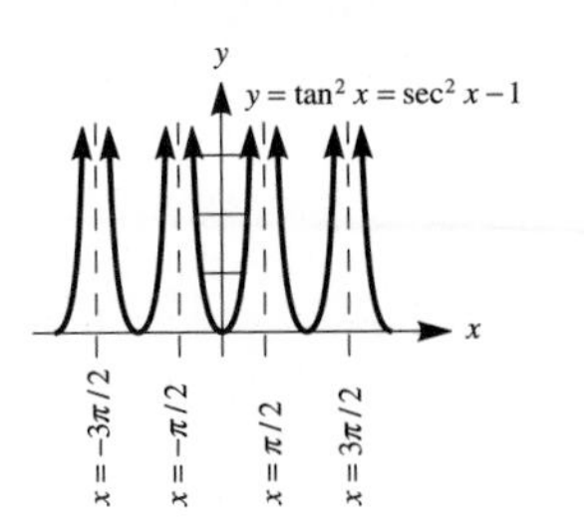

19. (a)

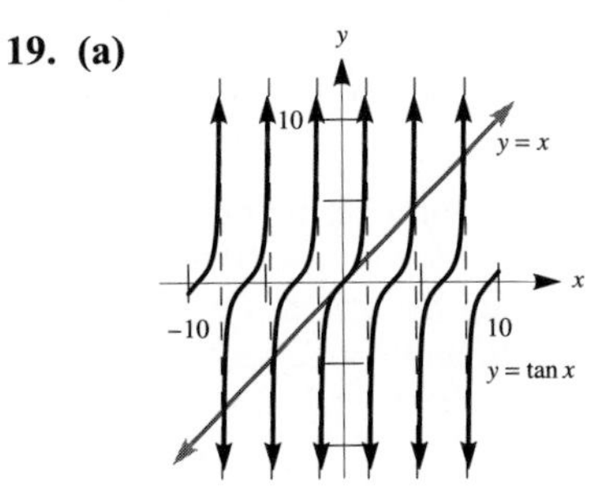

(b)

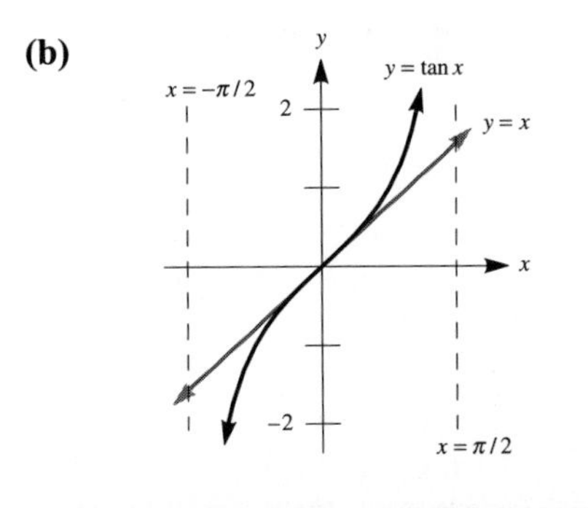

x	0.000123	0.01	0.05	0.1
$\tan x$	0.000123	0.010000	0.050042	0.100335
x	0.2	0.3	0.4	0.5
$\tan x$	0.202710	0.309336	0.422793	0.546302

(c) $y = x + \frac{1}{3}x^3$ is an even better approximation to $y = \tan x$

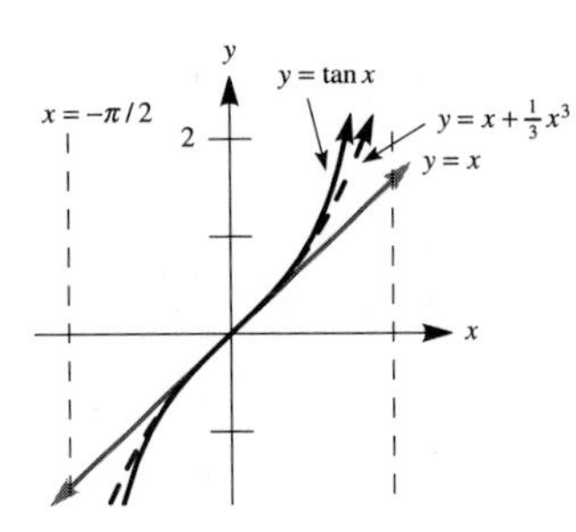

(d)

x	0.000123	0.01	0.05	0.1
$\tan x$	0.000123	0.010000	0.050042	0.100335
$x + \frac{x^3}{3}$	0.000123	0.010000	0.050042	0.100333
x	0.2	0.3	0.4	0.5
$\tan x$	0.202710	0.309336	0.422793	0.546302
$x + \frac{x^3}{3}$	0.202667	0.309	0.421333	0.541667

21. (a) They intersect at $x \approx 4.5$.

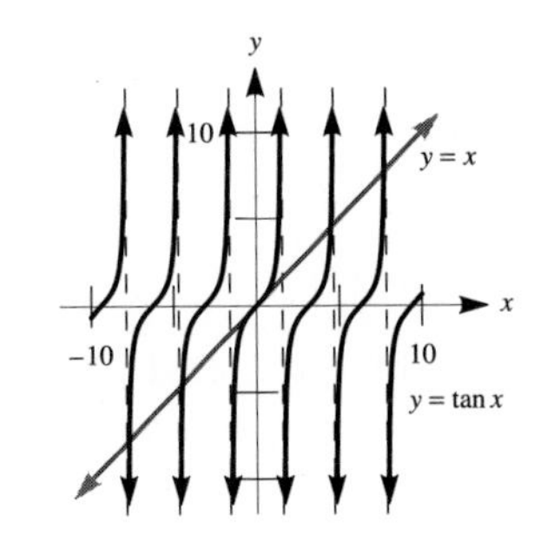

(b) $x \approx 4.4934$ **(c)** no

CHAPTER 7 REVIEW EXERCISES

1. (a) $-\sqrt{3}/2$ **(b)** $-\sqrt{3}$
3. $2 \sin t$

5.

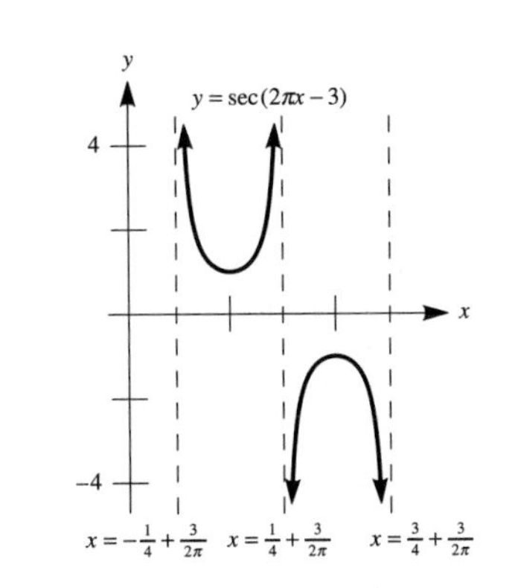

7. amplitude: 1; period: π; phase shift: $\pi/2$

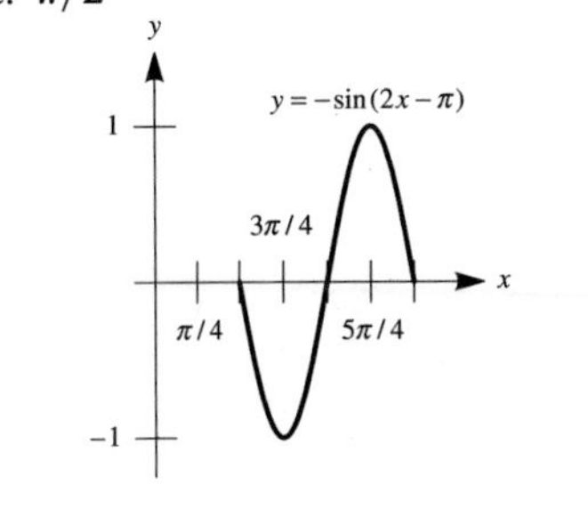

9. -1 **11.** $2\sqrt{3}/3$ **13.** $-\sqrt{3}/3$
15. $\frac{1}{2}$ **17.** 1 **19.** -2 **21.** 0.841
23. -1 **25.** 0.0123 **31.** $5 \cos \theta$
33. $10 \tan \theta$ **35.** $(\sqrt{5}/5)\cos \theta$
37. $-\frac{15}{8}$ **39.** $A(\theta) = \pi - \theta - \sin \theta$
41. $\frac{1}{2}(\pi - 2)\ \text{cm}^2$ **43.** $y = 4 \sin x$
45. $y = -2 \cos 4x$ **47.** x-intercepts: $\pi/8$ and $3\pi/8$; high point: $(\pi/4, 3)$; low points: $(0, -3)$ and $(\pi/2, -3)$

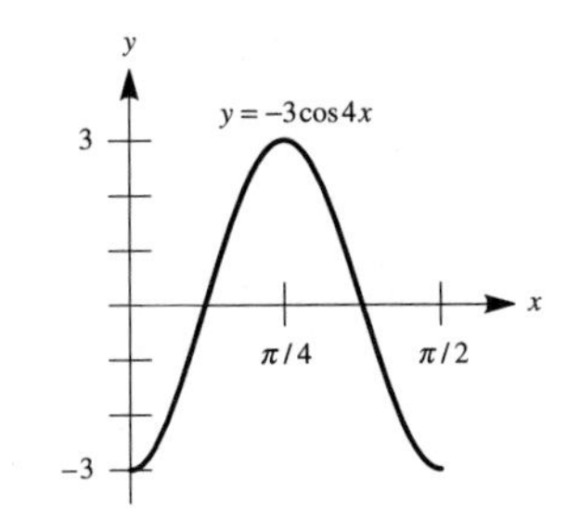

49. x-intercepts: $\frac{1}{2}$, $\frac{5}{2}$, and $\frac{9}{2}$; high point: $(\frac{3}{2}, 2)$; low point: $(\frac{7}{2}, -2)$

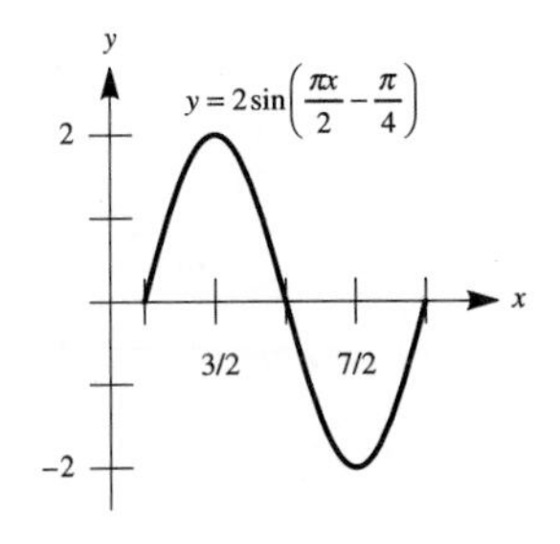

51. x-intercepts: $\frac{5}{2}$ and $\frac{11}{2}$; high points: (1, 3) and (7, 3); low point: (4, −3)

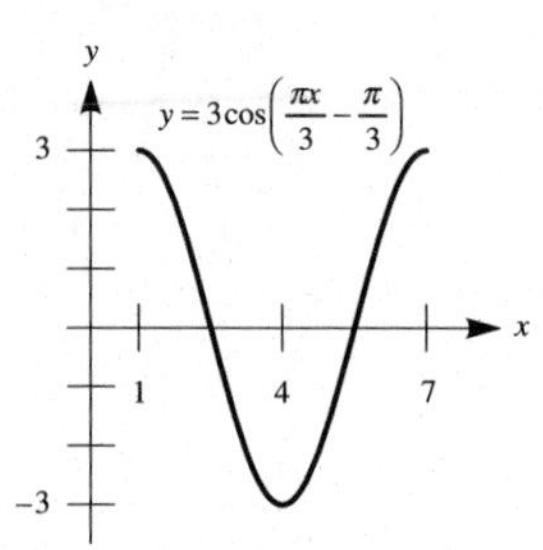

53. (a)

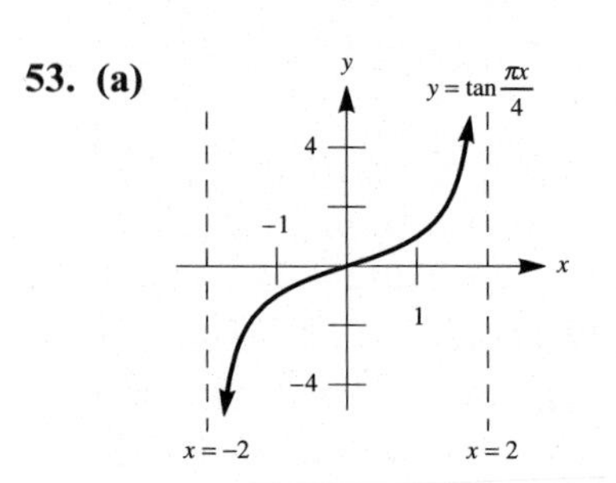

(b)

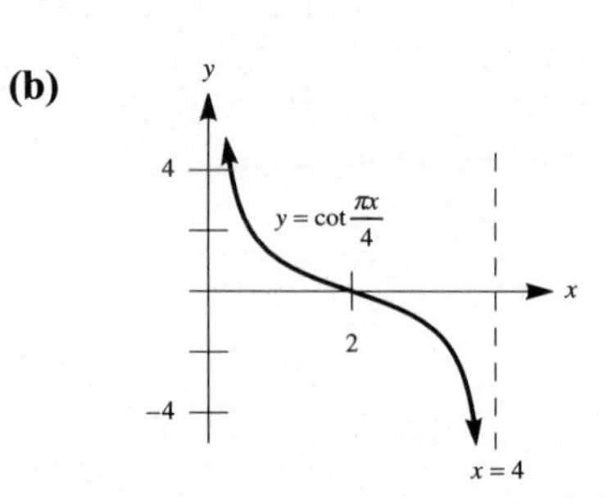

55. (a)

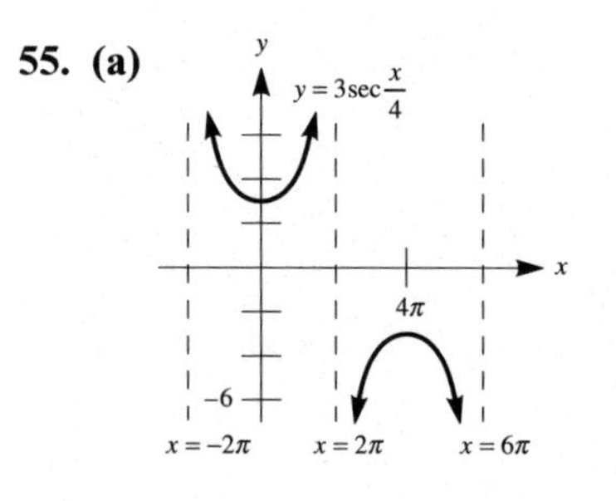

(b)

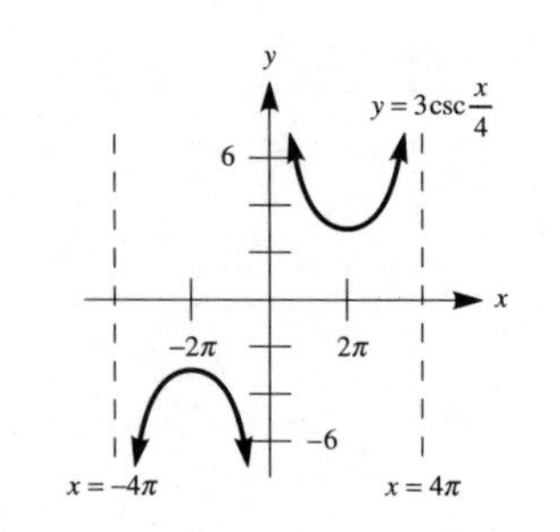

57. (a) amplitude: 2.5 cm; period: 16 sec; frequency: $\frac{1}{16}$ cps

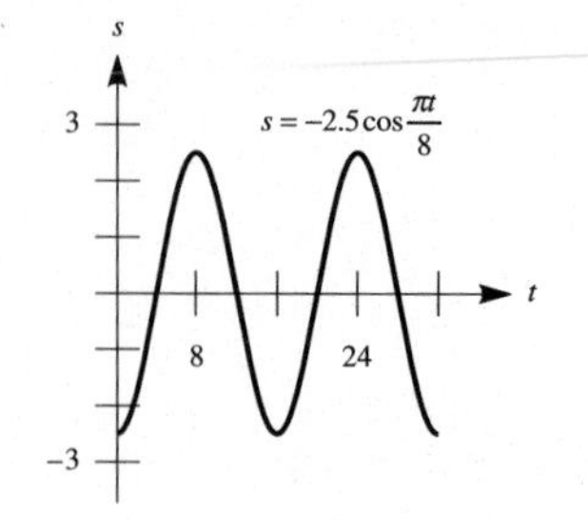

(b) $t = 0$, $t = 8$, $t = 16$, $t = 24$, and $t = 32$ sec **(c)** $t = 4$, $t = 12$, $t = 20$, and $t = 28$ sec

CHAPTER 7 TEST

1. (a) $-\frac{1}{2}$ **(b)** -2 **(c)** 1 **2.** $\frac{1}{4}\csc u$

3.

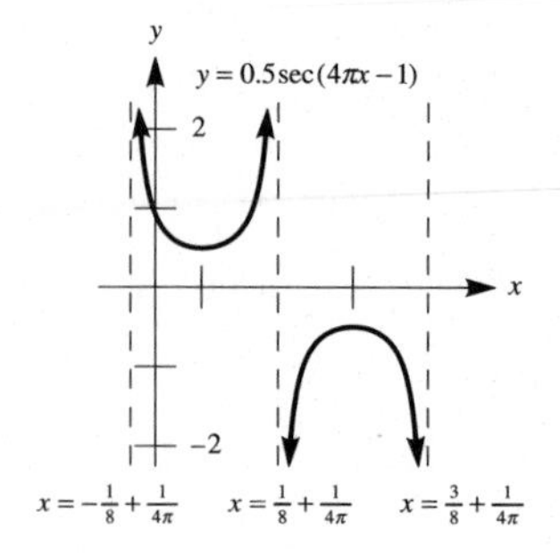

4. amplitude: 1; period: $2\pi/3$; phase shift: $\pi/12$

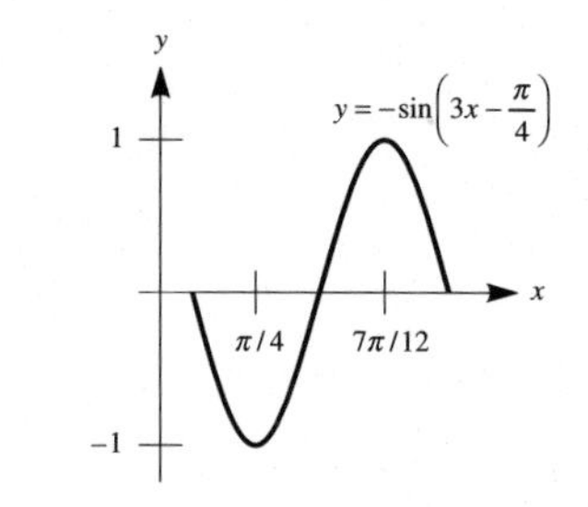

5.

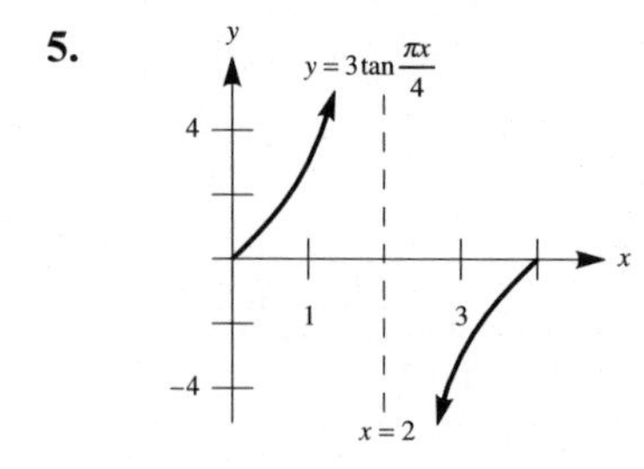

6. (a) $\frac{35}{36}\pi$ radians **(b)** $900°/\pi$
7. (a) $25\pi/12$ cm **(b)** $125\pi/24$ cm^2
8. (a) 50π radians/sec
(b) 250π cm/sec

10. (a)

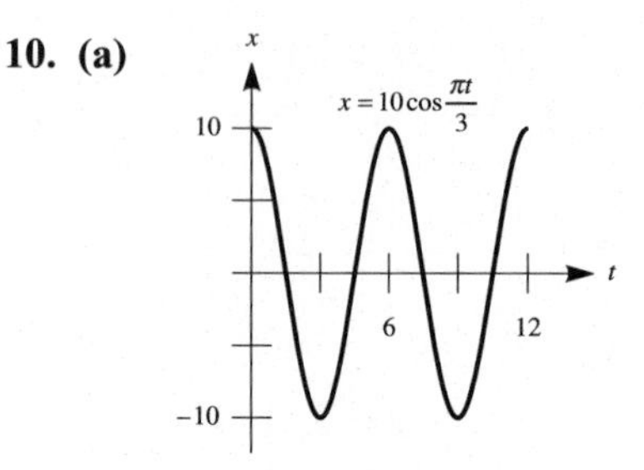

(b) passing through the origin: $t = 1.5$, $t = 4.5$, $t = 7.5$, and $t = 10.5$ sec; farthest from the origin: $t = 0$, $t = 3$, $t = 6$, $t = 9$, and $t = 12$ sec
11. (a) 1 **(b)** 0 **(c)** 0
12. (a) 1.1 **(b)** 1.1198 **(c)** 2.0218

CHAPTER EIGHT

EXERCISE SET 8.1

1. $\sin 3\theta$ **3.** $\sin 2\theta$ **5.** $\cos 5u$
7. $\sqrt{3}/2$ **9.** $\sin B$ **11.** $\cos\theta$
13. $-\cos\theta$ **15.** $\sin t$
17. $(\sqrt{6} - \sqrt{2})/4$
19. $(\sqrt{6} + \sqrt{2})/4$
21. $\sqrt{2}\sin s$ **23.** $\sqrt{3}\sin\theta$
25. (a) $-\frac{16}{65}$ **(b)** $\frac{63}{65}$
27. (a) $-\frac{44}{125}$ **(b)** $-\frac{4}{5}$
29. (a) $2\sqrt{6}/5$ **(b)** $4\sqrt{6}/25$
31. $\sin(\theta + \beta) = (2\sqrt{39} - 3\sqrt{13})/26$; $\cos(\beta - \theta) = (2\sqrt{13} - 3\sqrt{39})/26$
37. $\tan(s + t) = -1$; $\tan(s - t) = -\frac{1}{7}$
39. $\tan(s + t) = \frac{5}{3}$; $\tan(s - t) = \frac{3}{5}$
41. $\tan 3t$ **43.** $\sqrt{3}$ **45.** $\tan x$
47. $-2 - \sqrt{3}$ **63. (b)** $\sqrt{a^2 + b^2}$

65. (b)

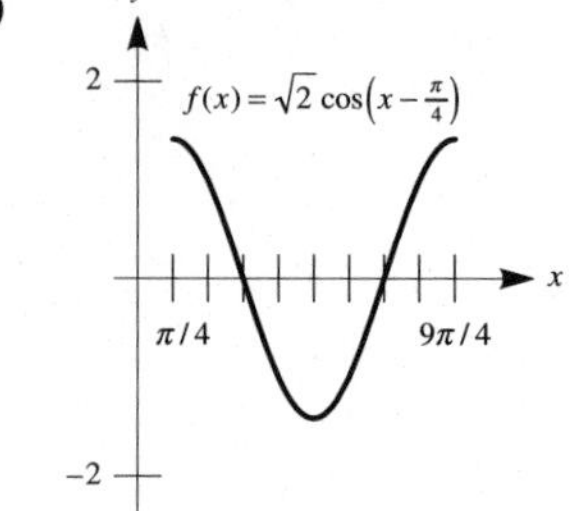

71. $\sqrt{3}/2$
75. (a) Using $\triangle ABH$, $\cos(\alpha+\beta) = AB/1$, so $\cos(\alpha+\beta) = AB$. **(b)** Using $\triangle ACF$, $\cos\alpha = AC/AF = AC/\cos\beta$, so $AC = \cos\alpha\cos\beta$. **(c)** Using $\triangle EFH$, $\sin(\angle EHF) = EF/HF$. But $\angle EHF = \alpha$, and $HF = \sin\beta$, so $\sin\alpha = EF/\sin\beta$, and thus $EF = \sin\alpha\sin\beta$. **(d)** From part (a), $\cos(\alpha+\beta) = AB = AC - BC$. But $AC = \cos\alpha\cos\beta$ from part (b), and $BC = EF = \sin\alpha\sin\beta$ from part (c), so $\cos(\alpha+\beta) = \cos\alpha\cos\beta - \sin\alpha\sin\beta$.

81. (a)

t	1	2	3	4
$f(t)$	1.5	1.5	1.5	1.5

(b) $f(t) = 1.5$
85. (a) The two values are equal: $\tan A + \tan B + \tan C \approx -1.1918$; $\tan A \tan B \tan C \approx -1.1918$.
(b) $\tan\alpha + \tan\beta + \tan\gamma \approx -1.3764$; $\tan\alpha\tan\beta\tan\gamma \approx -1.3764$

EXERCISE SET 8.2

1. (a) $\frac{336}{625}$ **(b)** $-\frac{527}{625}$ **(c)** $-\frac{336}{527}$
3. (a) $-\frac{8}{17}$ **(b)** $-\frac{15}{17}$ **(c)** $\frac{8}{15}$
5. (a) $\frac{1}{2}$ **(b)** $\sqrt{3}/2$ **(c)** $\sqrt{3}/3$
7. (a) $2\sqrt{2}/3$ **(b)** $\frac{1}{3}$ **(c)** $2\sqrt{2}$
9. (a) $-3\sqrt{7}/8$ **(b)** $-\frac{1}{8}$
(c) $\sqrt{8+2\sqrt{7}}/4$
(d) $\sqrt{8-2\sqrt{7}}/4$
11. (a) $4\sqrt{2}/9$ **(b)** $-\frac{7}{9}$ **(c)** $\sqrt{6}/3$
(d) $-\sqrt{3}/3$ **13. (a)** $\sqrt{2-\sqrt{3}}/2$
(b) $\sqrt{2+\sqrt{3}}/2$ **(c)** $2-\sqrt{3}$
15. (a) $\sqrt{2+\sqrt{3}}/2$
(b) $-\sqrt{2-\sqrt{3}}/2$ **(c)** $-2-\sqrt{3}$
17. (a) $\frac{24}{25}$ **(b)** $\frac{7}{25}$ **(c)** $\frac{24}{7}$
19. (a) $\frac{24}{25}$ **(b)** $-\frac{7}{25}$ **(c)** $-\frac{24}{7}$
21. (a) $\sqrt{10}/10$ **(b)** $3\sqrt{10}/10$ **(c)** $\frac{1}{3}$
23. (a) $\sqrt{5}/5$ **(b)** $2\sqrt{5}/5$ **(c)** $\frac{1}{2}$
25. $\sin 2\theta = 2x\sqrt{25-x^2}/25$; $\cos 2\theta = (25-2x^2)/25$
27. $\sin 2\theta = (x-1)\sqrt{3+2x-x^2}/2$; $\cos 2\theta = (1+2x-x^2)/2$
29. $\sin^4\theta = (3 - 4\cos 2\theta + \cos 4\theta)/8$
31. $\sin^4(\theta/2) = (3 - 4\cos\theta + \cos 2\theta)/8$

51. $\alpha+\beta = \pi/4$ **59. (a)** -0.5
(c) $\cos 108^\circ \cos 36^\circ = (-\cos 72^\circ)(-\cos 144^\circ) = \cos 72^\circ \cos 144^\circ$
63. (b) (i) This is true because of the double-angle formula for sine and the fact that $\cos 36^\circ = \sin 54^\circ$. **(ii)** This is true because of the double-angle formula for sine and the fact that $\cos 18^\circ = \sin 72^\circ$. **(iii)** Dividing each side by $4\sin 72^\circ$ produces this result.

EXERCISE SET 8.3

1. $\frac{1}{2}\cos 50^\circ$ **3.** $\frac{1}{2}\cos 80^\circ$
5. $\frac{1}{2}\sin 10^\circ + \frac{1}{2}$ **7.** $\frac{1}{2}\cos(3\pi/5) - \frac{1}{2}$
9. $\frac{1}{2}\cos(3\pi/7) + \frac{1}{2}$ **11.** $\frac{1}{2} + \frac{\sqrt{3}}{4}$
13. $\frac{1}{2}\cos x - \frac{1}{2}\cos 7x$
15. $\frac{1}{2}\sin\theta + \frac{1}{2}\sin 11\theta$
17. $\frac{1}{2}\sin\theta + \frac{1}{2}\sin 2\theta$
19. $\frac{1}{2}\cos 2y - \frac{1}{2}\cos 4x$
21. $\frac{1}{2}\sin(3t-s) + \frac{1}{2}\sin(t+s)$
23. $\sqrt{2}\cos 10^\circ$ **25.** $-\sqrt{2}\sin(\pi/20)$
27. $-2\sin 4\theta \sin\theta$ **29.** $\cos 5^\circ$
31. $\sin 2\theta$ **33.** $\sqrt{3}$
39. $4\cos\theta\cos 4\theta\cos 2\theta$
41. (a) $2\cos\left(\frac{x}{2} - \frac{\pi}{4}\right)$
(b) amplitude: 2; period: 4π; phase shift: $\pi/2$

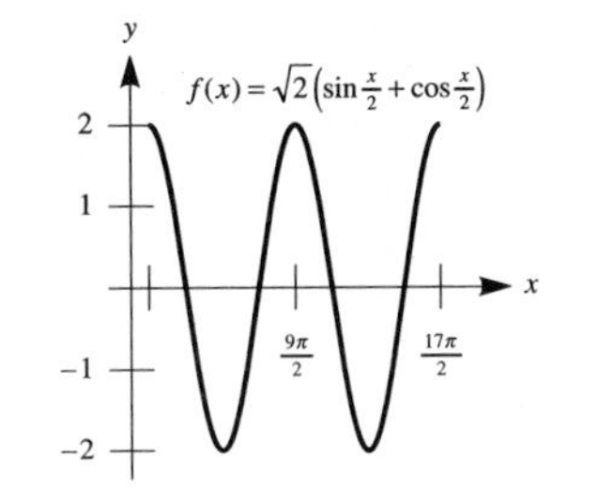

47. 1
51. (a) $\cos 30^\circ + \cos 70^\circ + \cos 80^\circ \approx 1.38$; $\cos 40^\circ + \cos 25^\circ + \cos 115^\circ \approx 1.25$; $\cos 55^\circ + \cos 55^\circ + \cos 70^\circ \approx 1.49$
(b) $\frac{3}{2}$ **(c) (i)** This is the sum-to-product formula for $\cos A + \cos B$. **(ii)** This is true because $\cos(A-B)/2 \leqslant 1$. **(iii)** This is true because $A + B = 180^\circ - C$. **(iv)** This is just division by 2. **(v)** The identities used are $\cos(90^\circ - \theta) = \sin\theta$ and $\cos\theta = 1 - 2\sin^2(\theta/2)$.
(vi) Multiplying this expression out shows they are equal. **(vii)** Since $2\left[\sin(C/2) - \frac{1}{2}\right]^2 \geqslant 0$, the expression is at most $\frac{3}{2}$.

EXERCISE SET 8.4

1. yes **3.** no **5.** $\theta = (\pi/3) + 2\pi k$ or $\theta = (2\pi/3) + 2\pi k$, where k is any integer **7.** $\theta = (7\pi/6) + 2\pi k$ or $\theta = (11\pi/6) + 2\pi k$, where k is any integer **9.** $\theta = \pi + 2\pi k$, where k is any integer **11.** $\theta = (\pi/3) + \pi k$, where k is any integer
13. $x = \pi k$, where k is any integer
15. $\theta = (\pi/2) + \pi k$, $\theta = (2\pi/3) + 2\pi k$ or $\theta = (4\pi/3) + 2\pi k$, where k is any integer **17.** $t = \pi k$, where k is any integer **19.** $x = (\pi/6) + 2\pi k$, $x = (5\pi/6) + 2\pi k$ or $x = (3\pi/2) + 2\pi k$, where k is any integer
21. $t = (\pi/4) + 2\pi k$ or $t = (3\pi/4) + 2\pi k$, where k is any integer
23. $x \approx 1.39, 4.90$ **25.** $x \approx 0.46, 2.68$
27. $x \approx 1.41, 4.55$ **29.** $t \approx 1.37, 4.51$
31. $t \approx 1.11, 5.18$
33. $x \approx 1.25, 1.82, 4.39, 4.96$
35. $x \approx 0.85, 2.29$
37. $\theta \approx 14.5^\circ, 165.5^\circ$
39. $\theta \approx 53.1^\circ, 126.9^\circ, 233.1^\circ, 306.9^\circ$
41. $\theta \approx 128.2^\circ, 231.8^\circ$
43. $\theta = 0^\circ, 120^\circ, 240^\circ$
45. $\theta = 75^\circ, 105^\circ, 195^\circ, 225^\circ, 315^\circ, 345^\circ$ **47.** $\theta = 60^\circ, 180^\circ, 300^\circ$
49. $\theta = 30^\circ, 120^\circ, 210^\circ, 300^\circ$
51. $x = 0$ or $x = \pi$
53. $x = 2\pi/3$ or $x = \pi$
55. $\theta = (3\pi/16) + k\pi/4$, where k is any integer **57.** $\theta = 60.45^\circ$
61. (a) $x \approx 1000.173$
(b) $x \approx 1001.022$
63. (b) The iterates are alternating between 0.3180 and 0.6820. They are not approaching the fixed point $x = 0.5$.

x	x_1	x_2	x_3	x_4
$f(x)$	0.65	0.3264	0.6812	0.3180

x	x_5	x_6	x_7	x_8
$f(x)$	0.6820	0.3180	0.6820	0.3180

x	x_9	x_{10}
$f(x)$	0.6820	0.3180

(c) The iterates are alternating between 0.3180 and 0.6820. They are not approaching the fixed point $x = 0.5$.

x	x_1	x_2	x_3	x_4
$f(x)$	0.5736	0.3951	0.6400	0.3318

x	x_5	x_6	x_7	x_8
$f(x)$	0.6801	0.3181	0.6820	0.3180

x	x_9	x_{10}
$f(x)$	0.6820	0.3180

GRAPHING UTILITY EXERCISES FOR SECTION 8.4

1. (a)

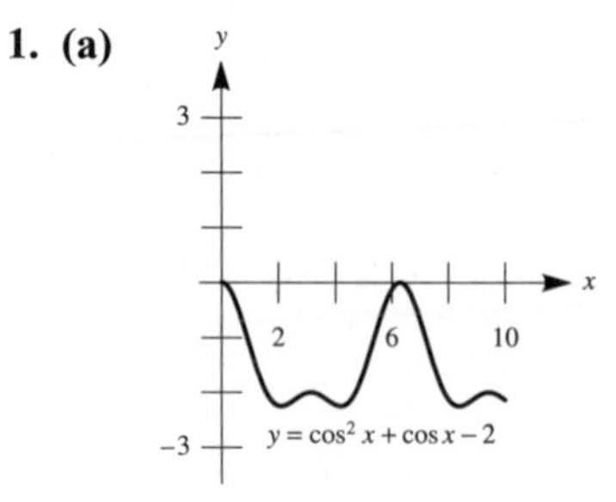

(b) $x \approx 6.28$ **(c)** $x = 2\pi$

3. (a) The graph is a sine function with amplitude 1 and period 360°.

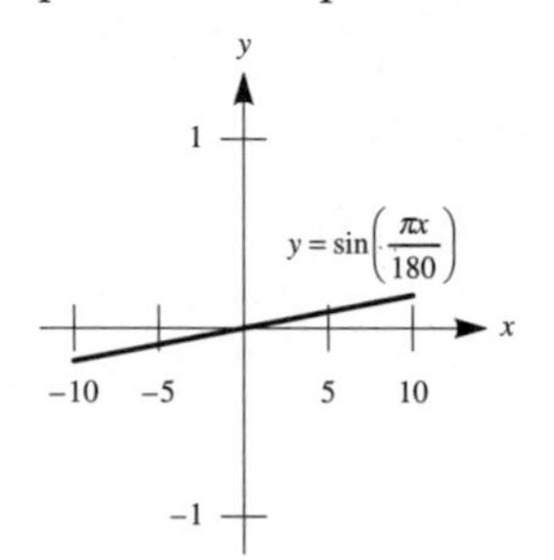

(b)

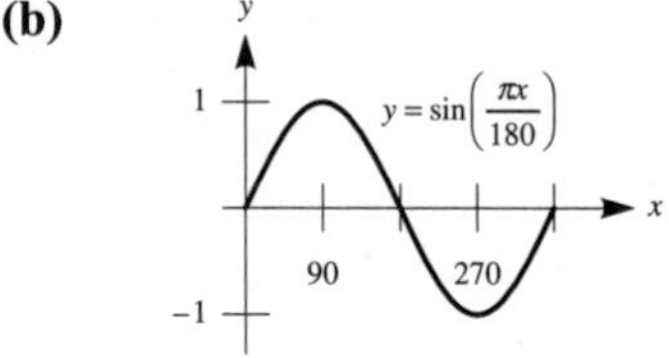

(c)

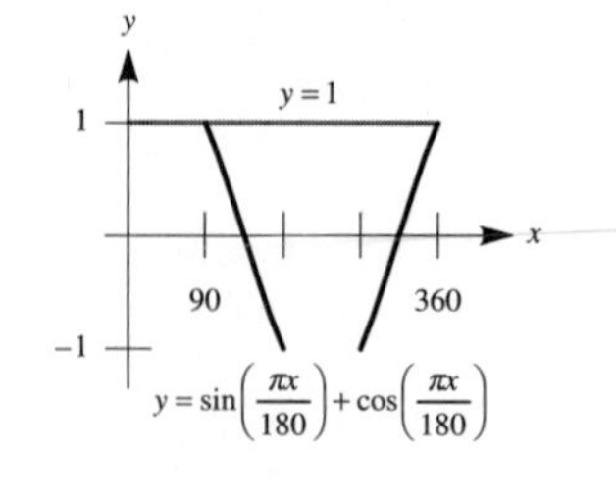

5. $x \approx 3.595, 5.830$

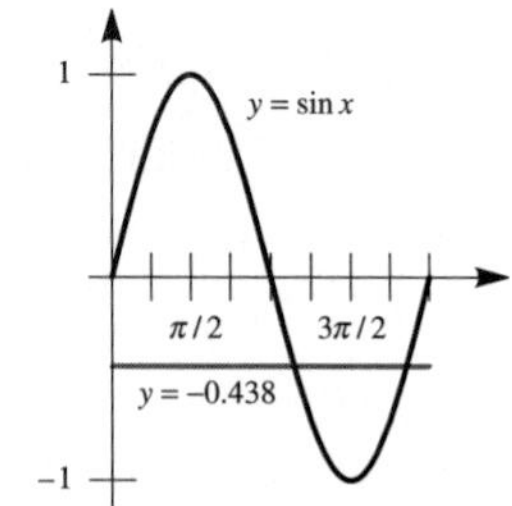
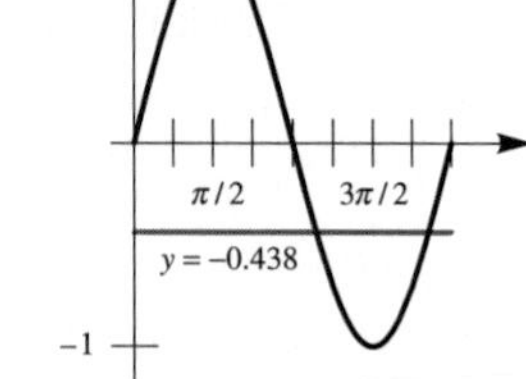

7. $x \approx 0.375, 1.571, 2.767, 4.712$

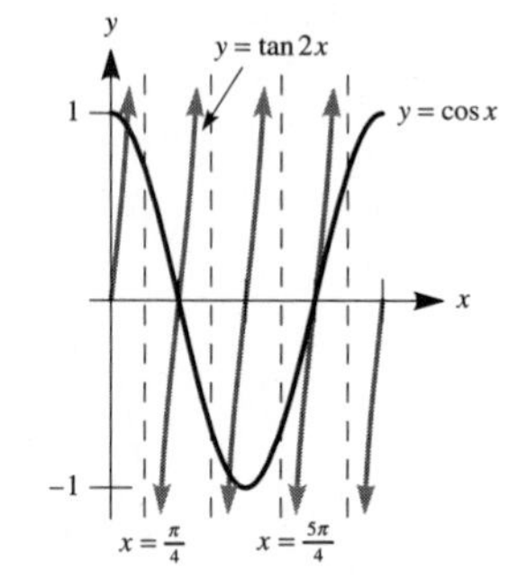

9. $x \approx 0.401, 3.542$

11. $x \approx 2.282, 5.424$

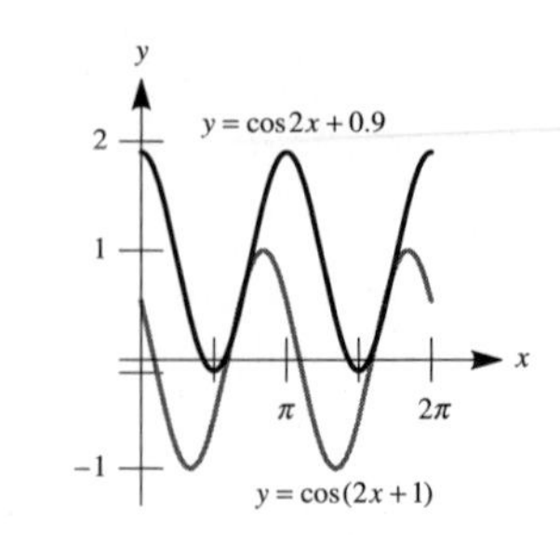

13. $x = 0, 6;$ $x \approx 6.187$

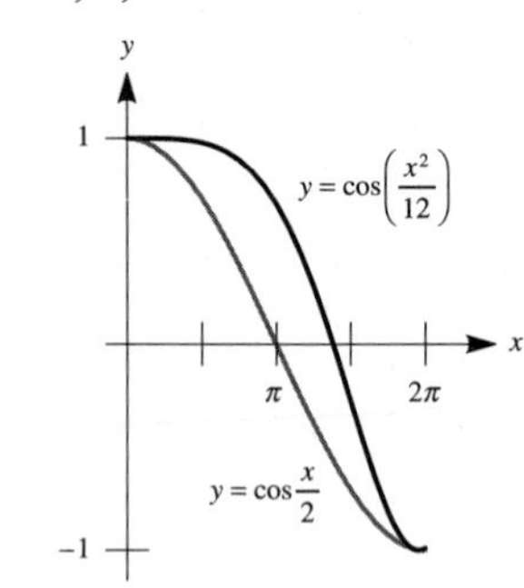

15. $x \approx 2.006$

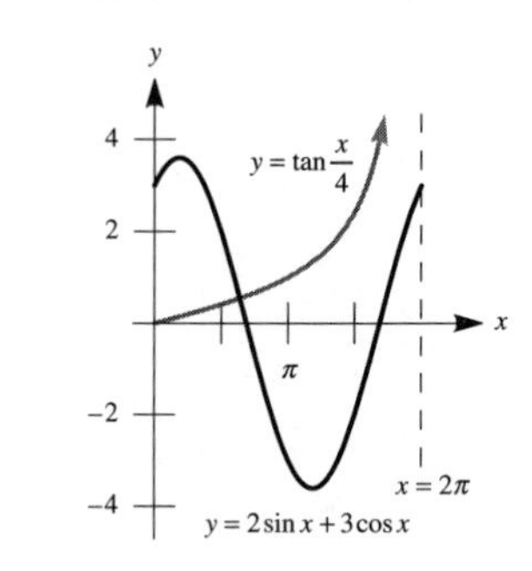

17. $x = 0$ and $x \approx 4.667$

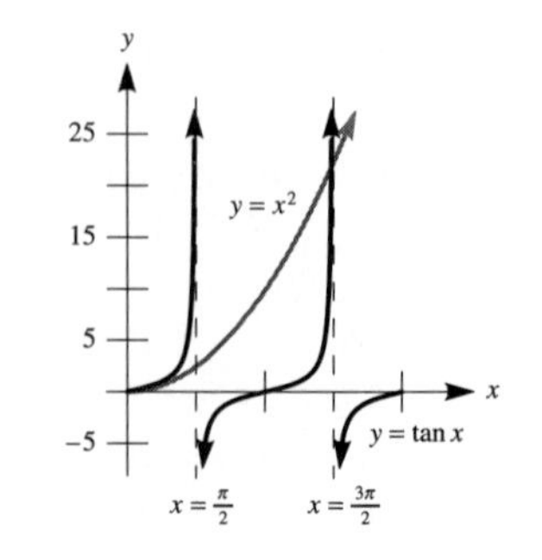

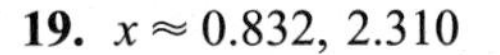

19. $x \approx 0.832, 2.310$

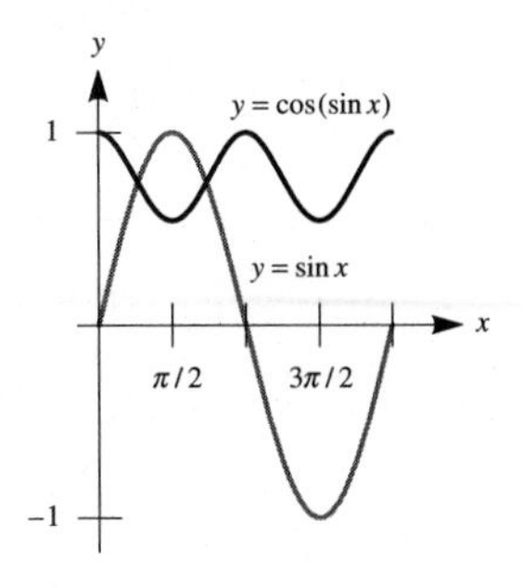

21. $x \approx 0.739$

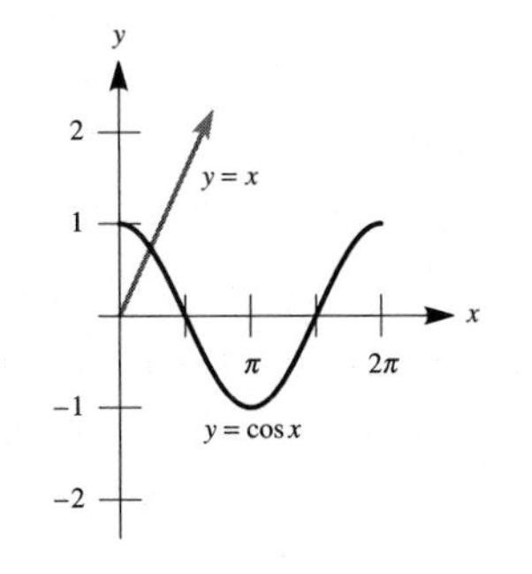

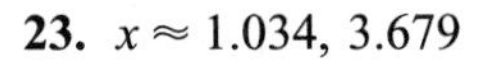

23. $x \approx 1.034, 3.679$

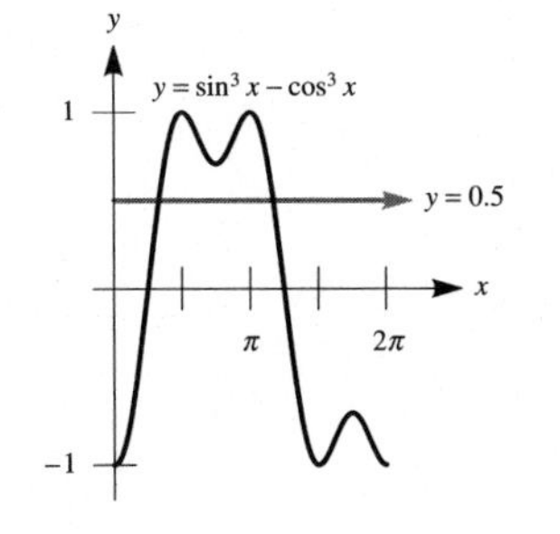

25. $x \approx 0.717, 2.864, 3.142$

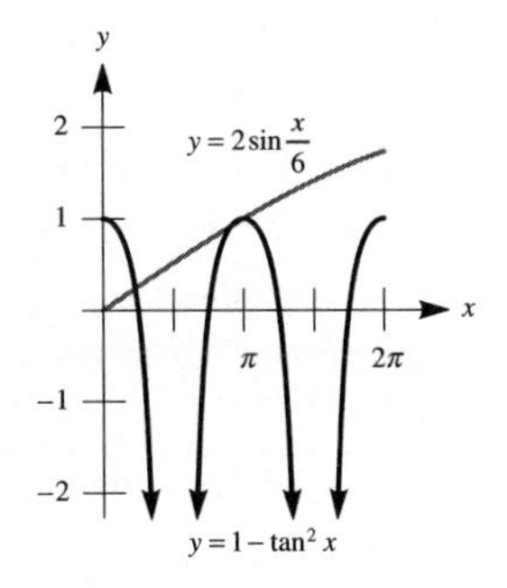

EXERCISE SET 8.5

1. $\pi/3$ **3.** $\pi/3$ **5.** $-\pi/6$ **7.** $\pi/4$
9. undefined **11.** $\frac{1}{4}$ **13.** $\frac{3}{4}$
15. $-\pi/7$ **17.** $\pi/2$ **19.** 0 **21.** $\frac{4}{3}$
23. $\sqrt{2}/2$ **25.** $\frac{12}{5}$ **27.** $\frac{1}{2}$
29. $2\sqrt{2}/3$
31. **(a)** 0.84 radians or 48.59°
(b) 0.84 radians or 48.19°
(c) 1.26 radians or 72.34°
(d) 0.90 radians or 51.57°
33. $\sqrt{2}$ **35.** $\sqrt{1-x^2}$
37. $\frac{\theta}{4} - \sin 2\theta =$
$\frac{1}{4}\sin^{-1}\left(\frac{3x}{2}\right) - \frac{3x\sqrt{4-9x^2}}{2}$
39. $\theta - \cos\theta =$
$\tan^{-1}\left(\frac{x-1}{2}\right) - \frac{2}{\sqrt{5-2x+x^2}}$

41. **(a)**

$y = -\sin^{-1} x$

(b)

$y = \sin^{-1}(-x)$

(c)

$y = -\sin^{-1}(-x)$

43. **(a)**

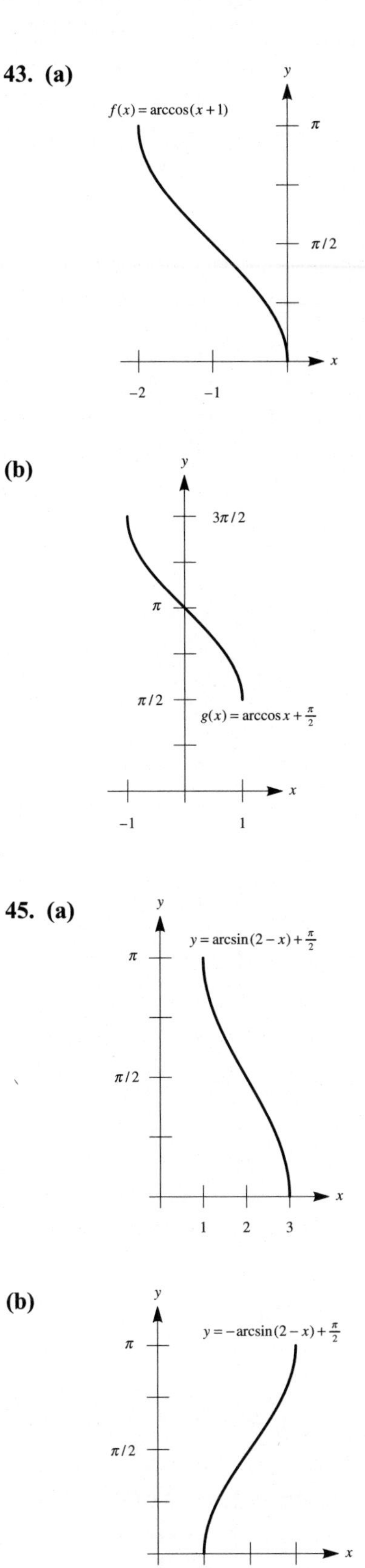

(b)

45. **(a)**

(b)

47. (a)

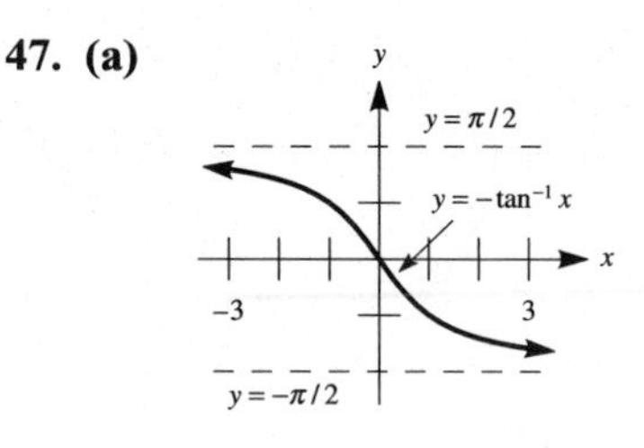

(b)

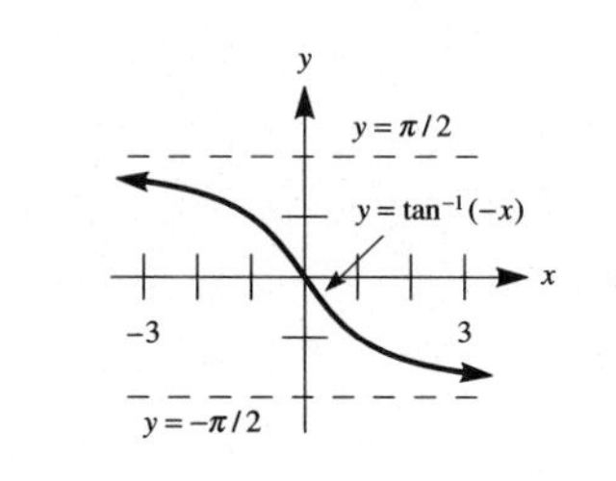

(c)

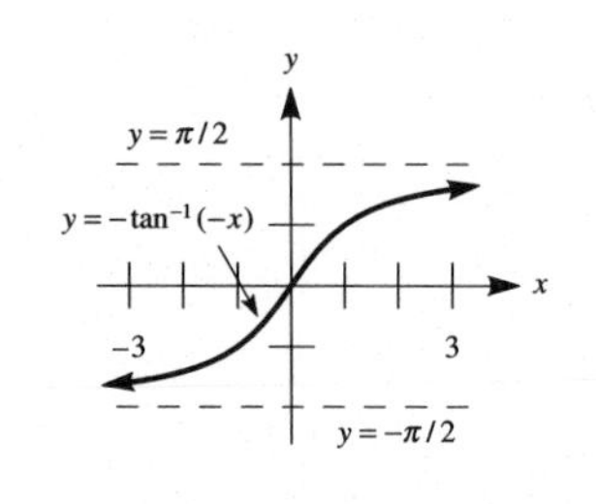

49.

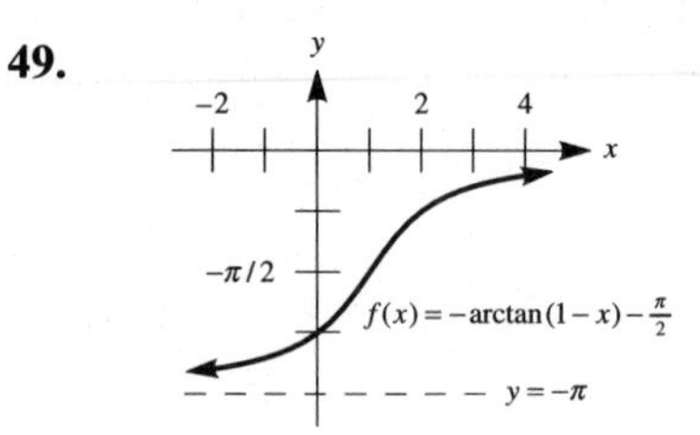

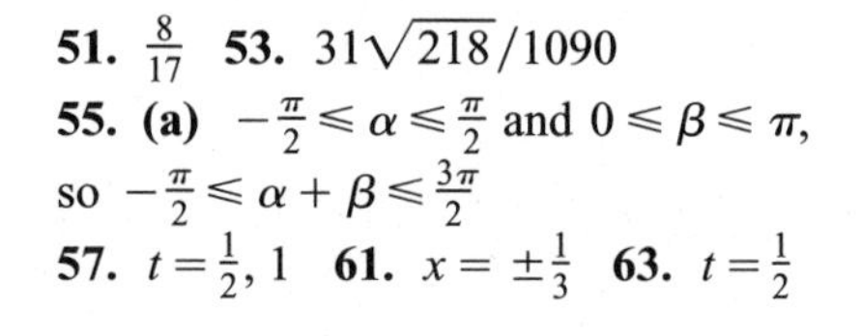

51. $\frac{8}{17}$ **53.** $31\sqrt{218}/1090$

55. (a) $-\frac{\pi}{2} \leq \alpha \leq \frac{\pi}{2}$ and $0 \leq \beta \leq \pi$, so $-\frac{\pi}{2} \leq \alpha + \beta \leq \frac{3\pi}{2}$

57. $t = \frac{1}{2}, 1$ **61.** $x = \pm\frac{1}{3}$ **63.** $t = \frac{1}{2}$

65. (a)

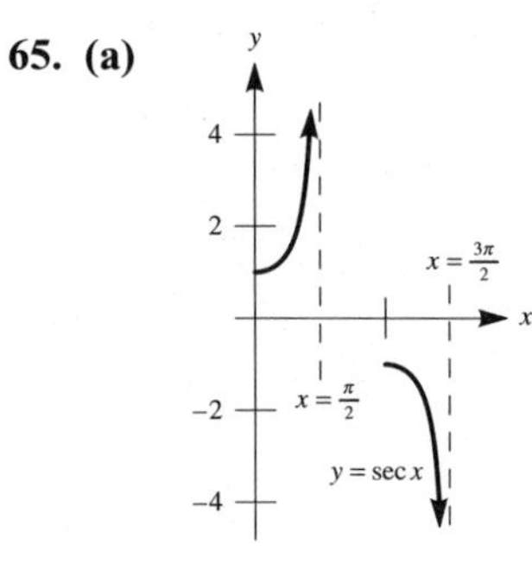

(b)

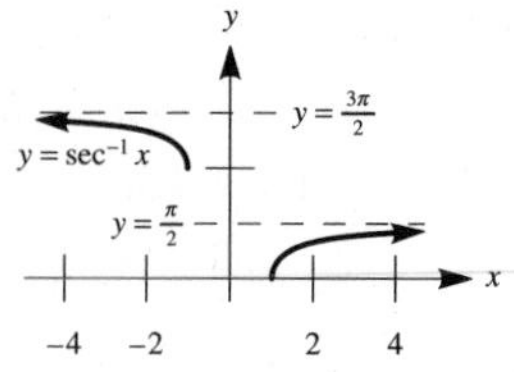

(c) $\sec^{-1}(2/\sqrt{3}) = \pi/6$; $\sec^{-1}(-2/\sqrt{3}) = 7\pi/6$

(d) $\sec^{-1}(\sqrt{2}) = \pi/4$; $\sec^{-1}(-\sqrt{2}) = 5\pi/4$

(e) $\sec(\sec^{-1} 2) = 2$; $\sec^{-1}(\sec 0) = 0$

GRAPHING UTILITY EXERCISES FOR SECTION 8.5

1. (a) maximum value: 1.57 (when $x = 1$); minimum value: -1.57 (when $x = -1$)

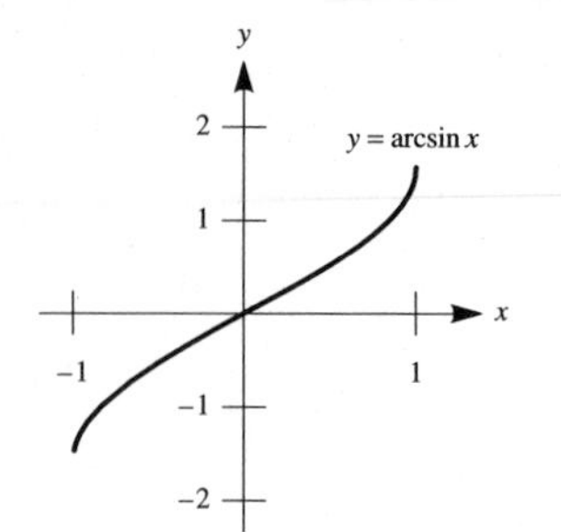

(b) maximum value: $\pi/2$ (when $x = 1$); minimum value: $-\pi/2$ (when $x = -1$)

3. The graph is a horizontal line with a y-intercept of $\pi/2$. This demonstrates the identity $\sin^{-1} x + \cos^{-1} x = \pi/2$, for $-1 \leq x \leq 1$.

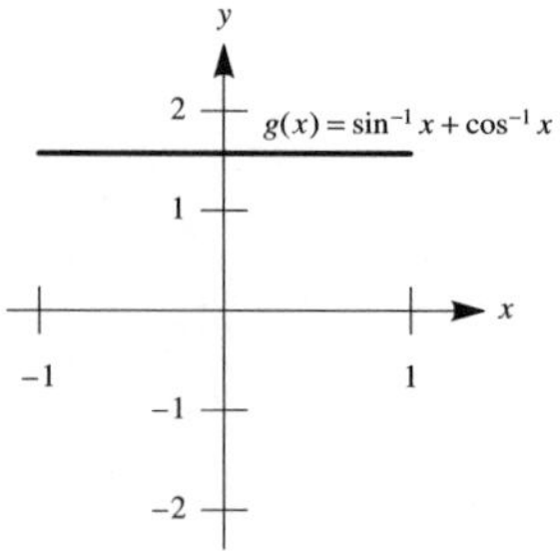

5.

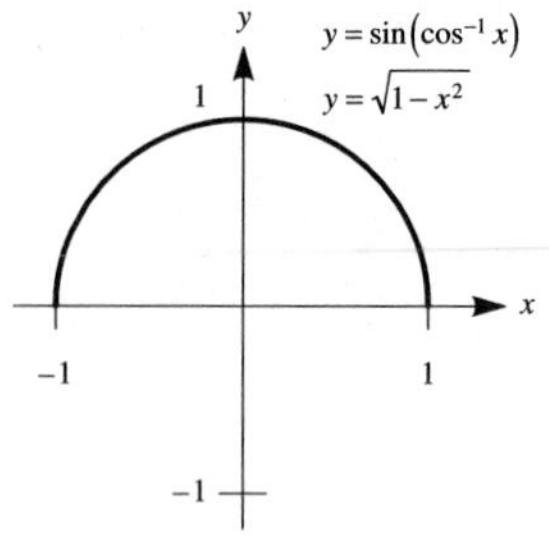

7.

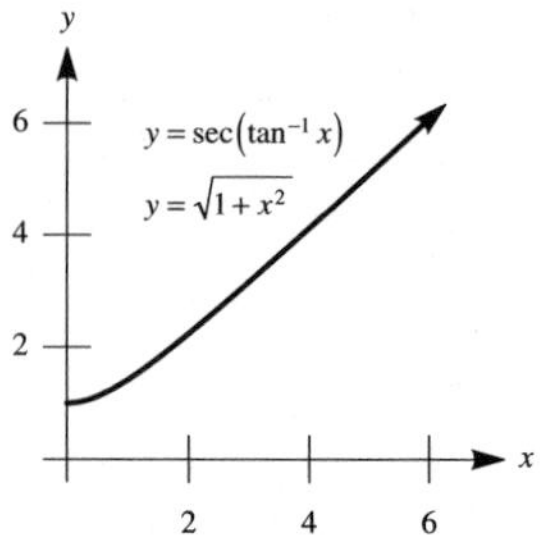

9. $x \approx 0.74$

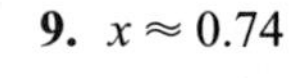

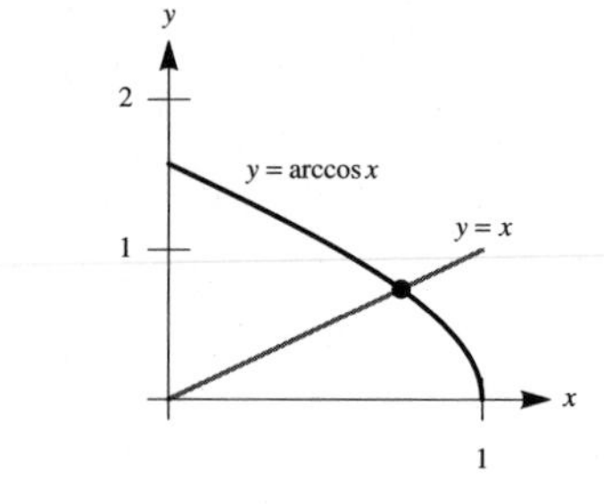

11. (a) $x \approx 0.96$

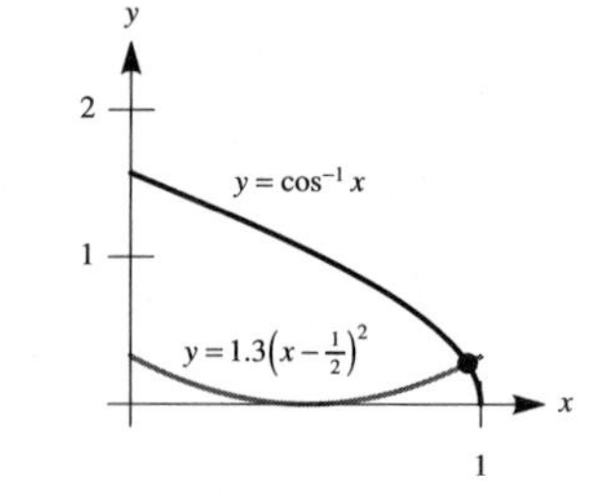

(b) $x \approx 0.96$

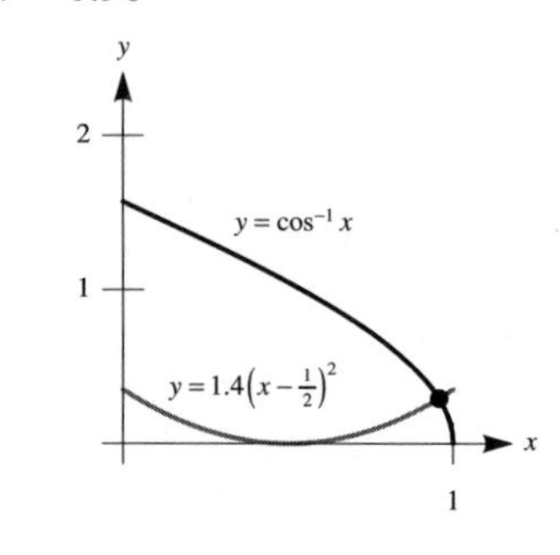

13. **(a)** $x \approx 0.24$

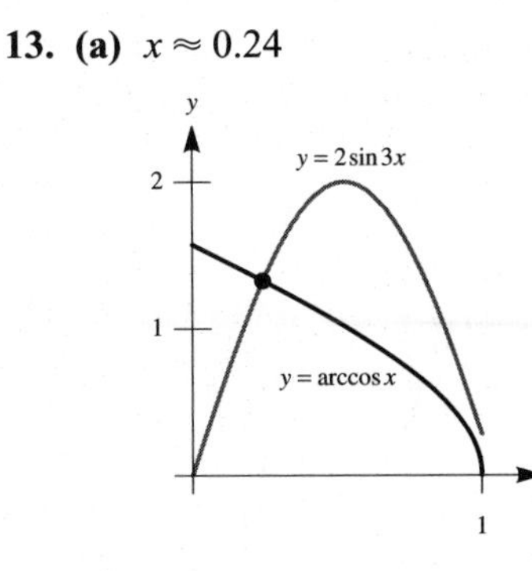

(b) $x \approx 0.19$ and $x \approx 0.68$

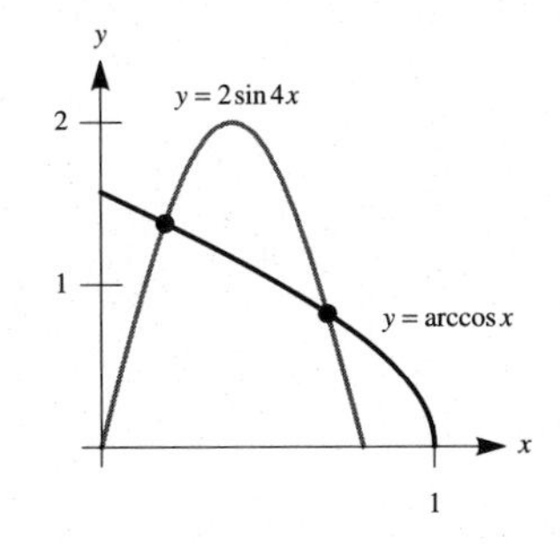

15. **(a)** $x \approx 0.56$

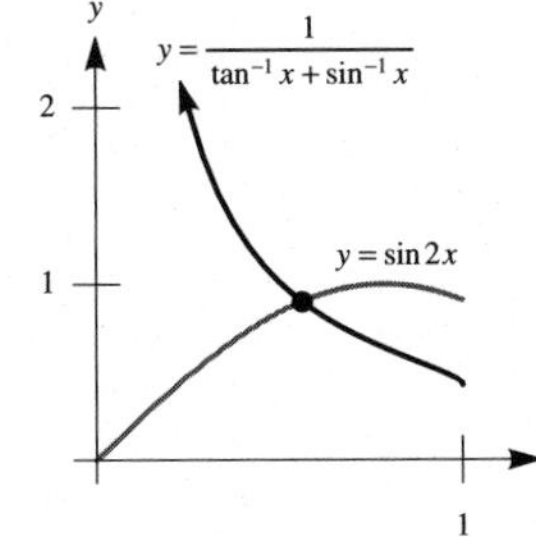

(b) $x \approx 0.51$ and $x \approx 0.84$

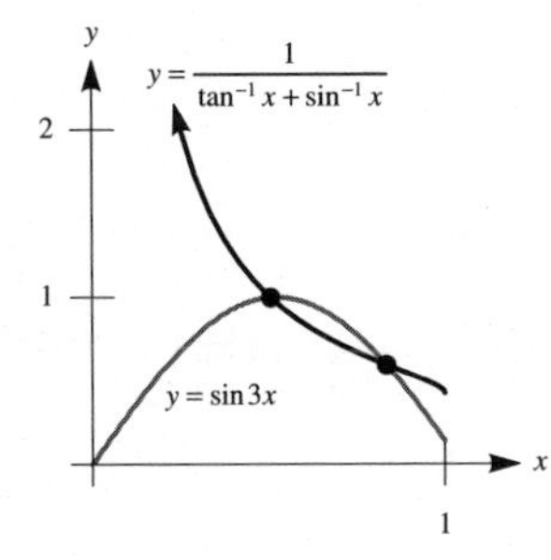

17. $x \approx 0.71$

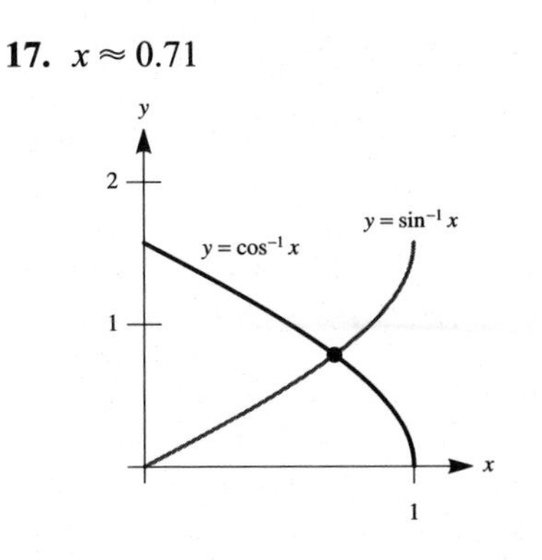

19. $x \approx 0.74$

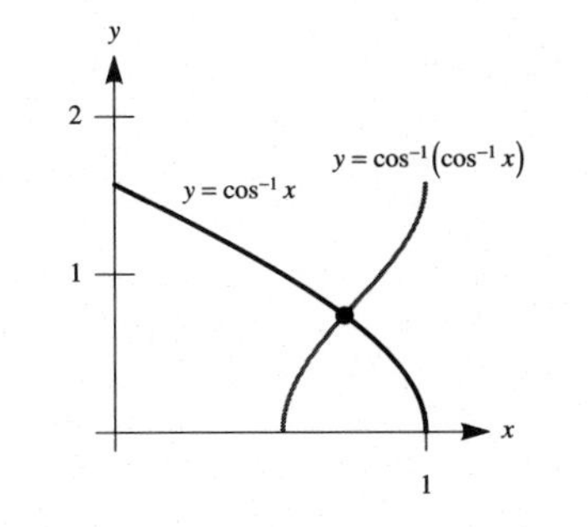

21. $x \approx 0.94$

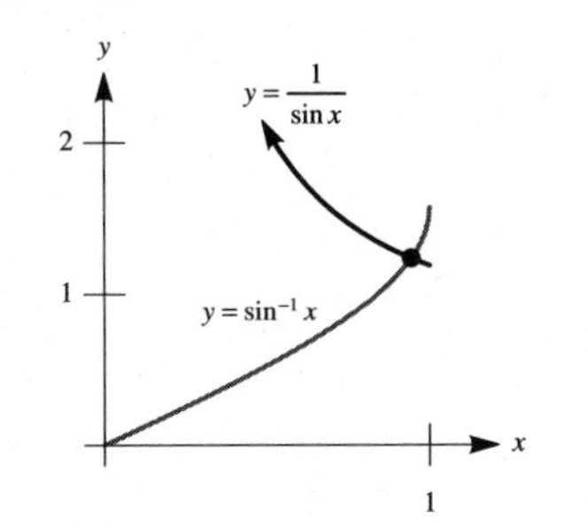

23. $x \approx 0.93$

CHAPTER 8 REVIEW EXERCISES

43. **(a)** $\pi/4$ **(b (i)** $a \approx 14$ **(ii)** $a \approx 93$ **(iii)** $a \approx 1256$

45. $x \approx 1.34$ or 4.48
47. $x \approx 0.46$ or 2.68
49. $x = \pi/3, 2\pi/3, 4\pi/3$, or $5\pi/3$
51. $x = 0$ or $3\pi/2$
53. $x = 0, \pi, 7\pi/6$, or $11\pi/6$
55. $x = \pi/3, 2\pi/3, 4\pi/3$, or $5\pi/3$
57. $x = \pi/4, \pi/2, 3\pi/4, 5\pi/4, 3\pi/2$, or $7\pi/4$ **59.** $x = \pi/6, \pi/3, 2\pi/3, 5\pi/6, 7\pi/6, 4\pi/3, 5\pi/3$, or $11\pi/6$
61. $x = 3\pi/4$ or $7\pi/4$
63. $\sqrt{3}/2$ **65.** $\pi/6$ **67.** $\pi/6$
69. $\pi/3$ **71.** $2\pi/3$ **73.** $\frac{2}{7}$
75. $-\sqrt{2}/2$ **77.** $3\sqrt{2}/2$ **79.** $\frac{17}{7}$
81. $-\frac{4}{3}$ **83.** $3\sqrt{10}/10$ **85.** $3\pi/4$
87. $\sqrt{6}$ **97.** **(a)** 0.0625

CHAPTER 8 TEST

1. $-\cos\theta$ **2.** $-\frac{3}{5}$ **3.** $-\frac{3}{2}$
4. $x \approx 1.25, 4.39$
5. $x = 7\pi/6, 11\pi/6$ **6.** $\sqrt{5}/5$
7. $x = 30°$ **8.** $\sqrt{18 + 12\sqrt{2}}/6$
9. restricted sine function: domain $= [-\pi/2, \pi/2]$, range $= [-1, 1]$; $y = \sin^{-1} x$: domain $= [-1, 1]$, range $= [-\pi/2, \pi/2]$

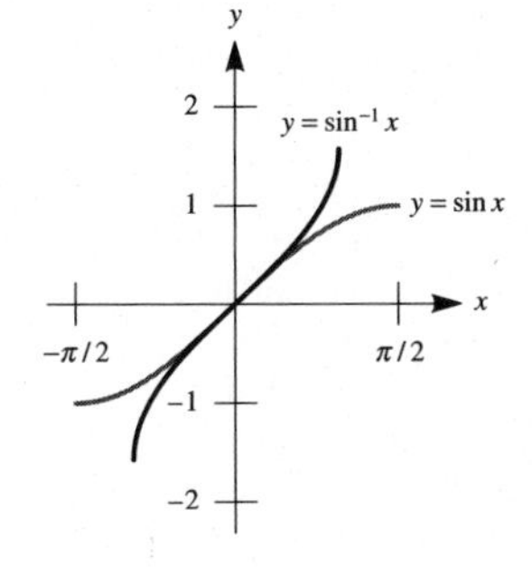

10. **(a)** $\pi/10$ **(b)** 0 **11.** $\sqrt{7}/4$
13. $(\sqrt{3} + \sqrt{2})/4$ **14.** $\tan 4\theta$
15. **(a)** x **(b)** 1
16. domain: $(-\infty, \infty)$; range: $(-\pi/2, \pi/2)$

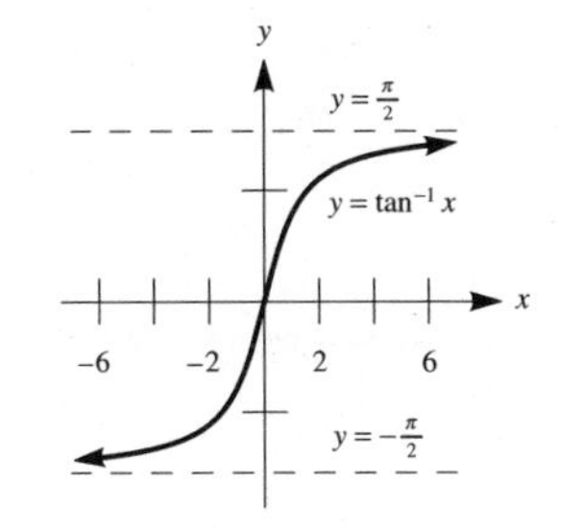

CHAPTER NINE

EXERCISE SET 9.1

1. $4\sqrt{6}$ cm **3.** $20 \sin 50°$ cm
5. $a \approx 9.7$ cm; $c \approx 16.4$ cm
7. $A \approx 63.3°$; $C \approx 50.7°$; $c \approx 25.5$ cm
9. **(a)** $45°$ or $135°$ **(b)** $45°$
(c) $14.5°$ or $165.5°$ **(d)** $131.8°$
11. **(a)** $\sin B \approx 1.18 > 1$
13. **(b)** $\angle C = 105°$; $c \approx 1.93$
(c) $\angle C = 15°$; $c \approx 0.52$
(d) $A_1 \approx 0.68$; $A_2 \approx 0.18$
15. $a = \dfrac{2 \sin 70°}{\sin 20°}$ cm; $b = \dfrac{2 \sin 50°}{\sin 20°}$ cm;
$c = \dfrac{2 \sin 50° \sin 70°}{\sin 20° \sin 85°}$ cm;
$d = \dfrac{2 \sin 50° \sin 15°}{\sin 20° \sin 85°}$ cm **17.** 160 ft
19. **(a)** $x = 7$ cm **(b)** $x = \sqrt{129}$ cm
21. **(a)** $x \approx 7.5$ cm **(b)** $x \approx 17.7$ cm
23. x is not the side opposite the $130°$ angle; $6^2 = x^2 + 3^2 - 2(x)(3)\cos 130°$
25. $\cos A = \frac{113}{140}$; $\cos B = \frac{29}{40}$; $\cos C = -\frac{5}{28}$
27. $A \approx 27.8°$; $B \approx 32.2°$; $C \approx 120°$
29. $A = 30°$; $B = 30°$; $C = 120°$
31. 5.9 units **33.** **(a)** $a \approx 4.2$ cm
(b) $C \approx 29.3°$ **(c)** $B \approx 110.7°$
35. approximately 31 miles
37. lighthouse A: 1.26 miles; lighthouse B: 1.60 miles
39. $D \approx 860{,}000$ miles
41. **(b)** $a = 5$, $b = 3$, $c = 7$ (using $m = 2$ and $n = 1$) **43.** 2:40 P.M.

47. **(f)**

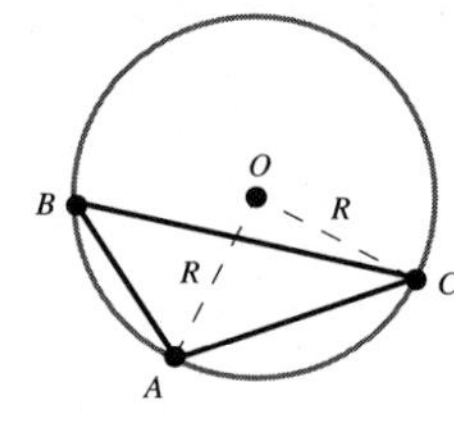

51. $C = 45°$ or $135°$

EXERCISE SET 9.2

1. $|\overrightarrow{PQ}| = \sqrt{34}$

3. $|\overrightarrow{SQ}| = \sqrt{10}$

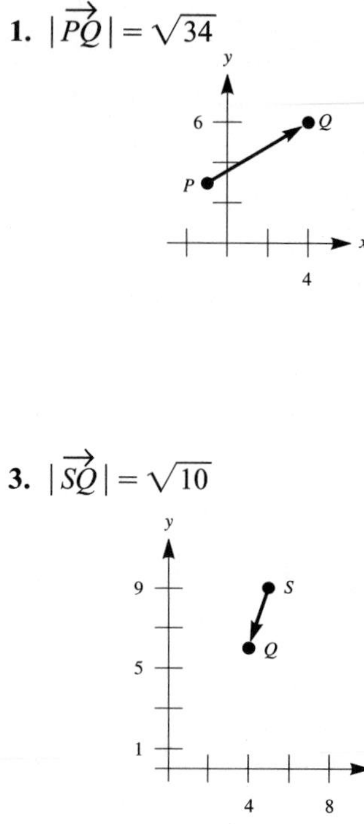

5. $|\overrightarrow{OP}| = \sqrt{10}$

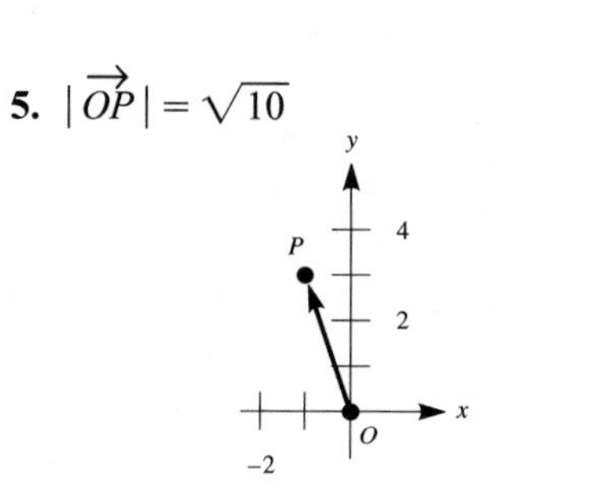

7. $|\overrightarrow{PQ} + \overrightarrow{QS}| = 6\sqrt{2}$

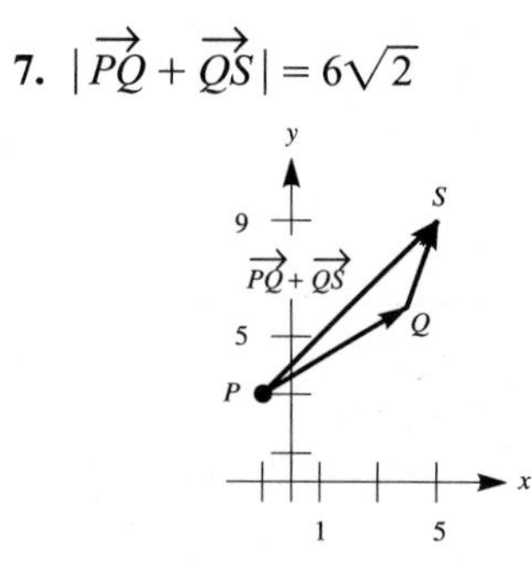

9. $|\overrightarrow{OP} + \overrightarrow{PQ}| = 2\sqrt{13}$

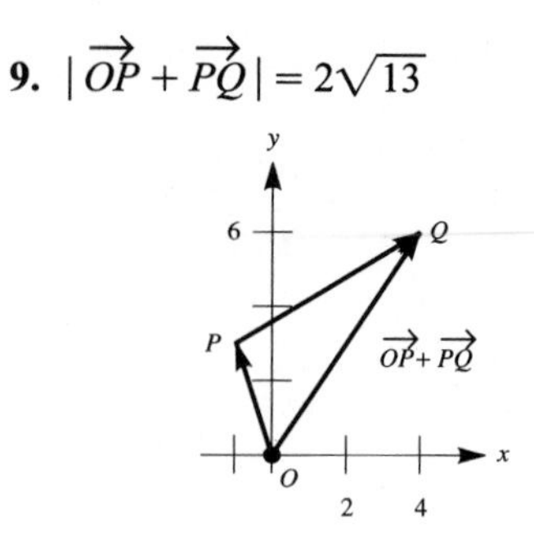

11. $|\overrightarrow{OS} + \overrightarrow{SQ} + \overrightarrow{QP}| = \sqrt{10}$

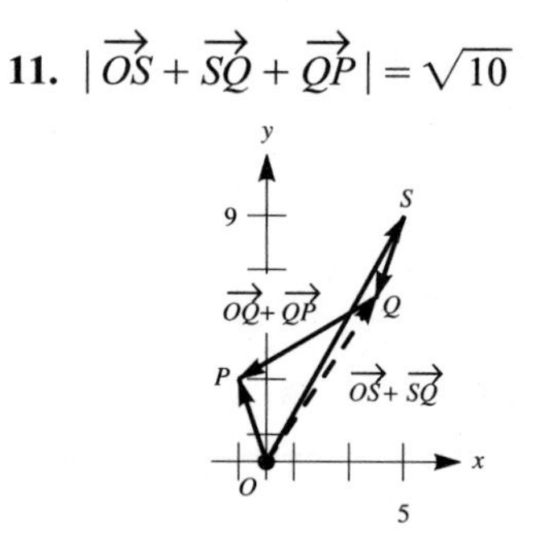

13. $|\overrightarrow{OP} + \overrightarrow{QS}| = 6$

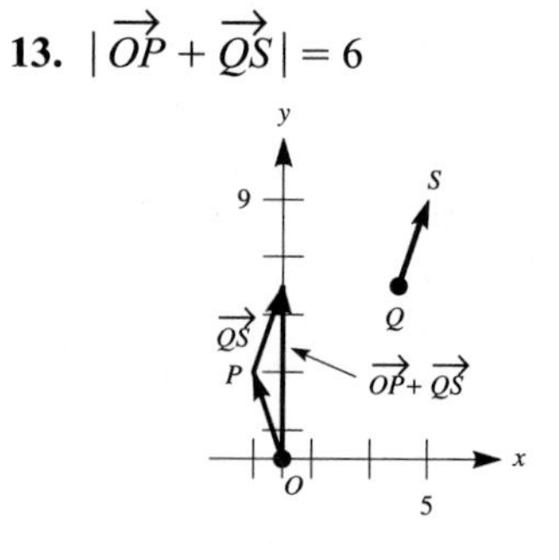

15. $|\overrightarrow{SR} + \overrightarrow{PO}| = 9$

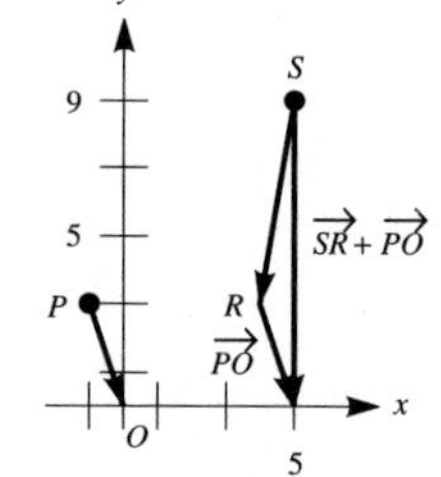

17. $|\overrightarrow{OP} + \overrightarrow{RQ}| = \sqrt{37}$

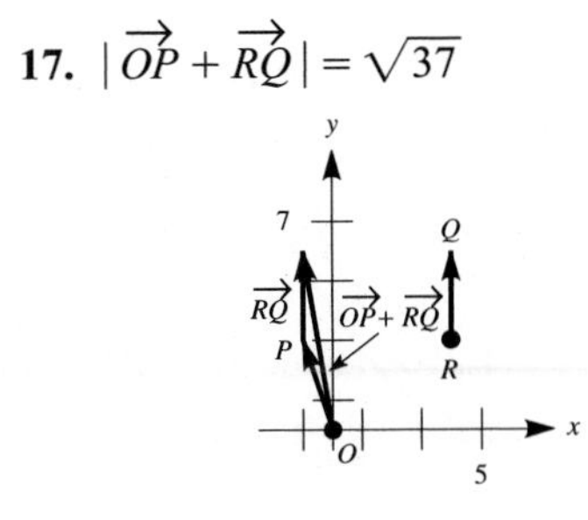

19. $|\overrightarrow{SQ} + \overrightarrow{RO}| = \sqrt{61}$

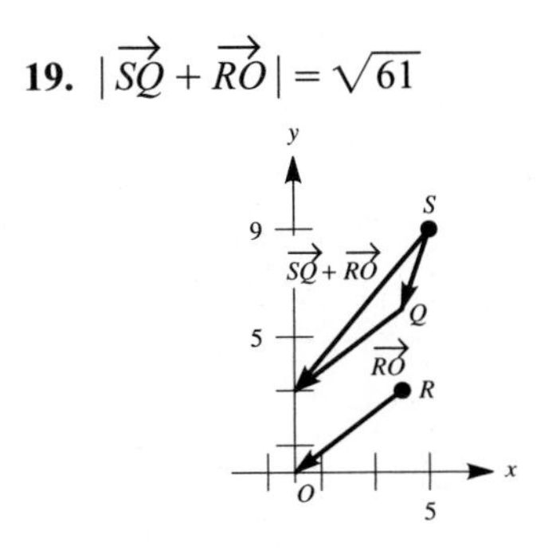

21. $|\overrightarrow{OP} + \overrightarrow{OR}| = 3\sqrt{5}$

23. $|\overrightarrow{RP} + \overrightarrow{RS}| = 2\sqrt{13}$

25. $|\overrightarrow{SO} + \overrightarrow{SQ}| = 6\sqrt{5}$

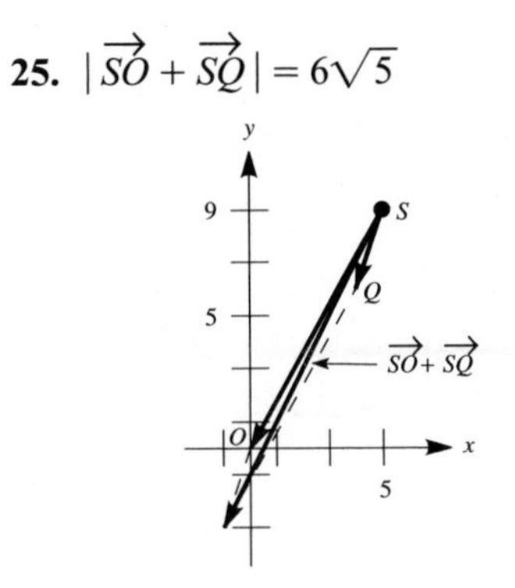

27. $|\mathbf{F} + \mathbf{G}| = \sqrt{41}$ N; $\theta \approx 38.7°$
29. $|\mathbf{F} + \mathbf{G}| = 9\sqrt{2}$ N; $\theta = 45°$
31. $|\mathbf{F} + \mathbf{G}| \approx 7.90$ N; $\theta \approx 24.1°$
33. $|\mathbf{F} + \mathbf{G}| \approx 6.92$ N; $\theta \approx 34.67°$
35. $|\mathbf{F} + \mathbf{G}| \approx 39.20$ N; $\theta \approx 21.46°$
37. $|\mathbf{F} + \mathbf{G}| \approx 38.96$ N; $\theta \approx 29.44°$
39. $V_x \approx 13.86$ cm/sec; $V_y = 8$ cm/sec
41. $F_x \approx 3.62$ N; $F_y \approx 13.52$ N
43. $V_x \approx -0.71$ cm/sec; $V_y \approx 0.71$ cm/sec
45. $F_x \approx -1.02$ N; $F_y \approx 0.72$ N
47. ground speed: 301.04 mph; drift angle: 4.76°; course: 25.24°
49. ground speed: 293.47 mph; drift angle: 8.82°; course: 91.18°
51. perpendicular: 9.83 lb; parallel: 6.88 lb **53.** perpendicular: 11.82 lb; parallel: 2.08 lb **55.** **(a)** initial point: $(-1, 2)$; terminal point: $(2, -3)$
(b) initial point: $(-1, 2)$; terminal point: $(2, -3)$

EXERCISE SET 9.3

1. length = 5

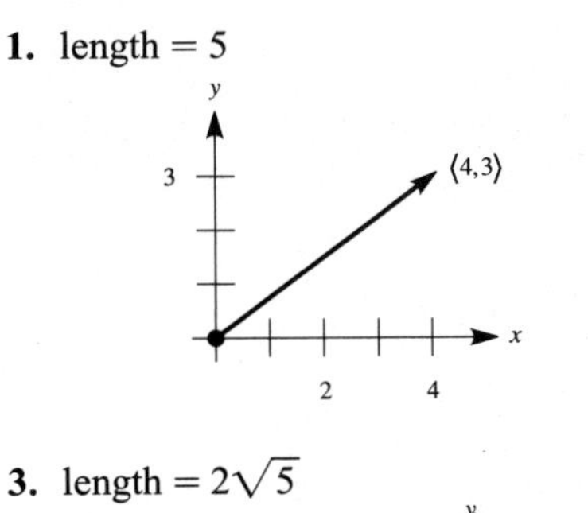

3. length = $2\sqrt{5}$

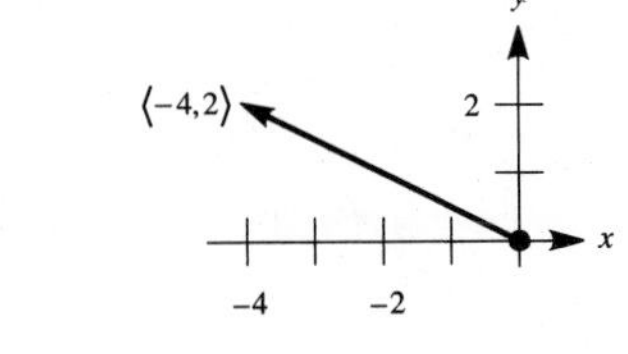

5. length = $\sqrt{13}/4$

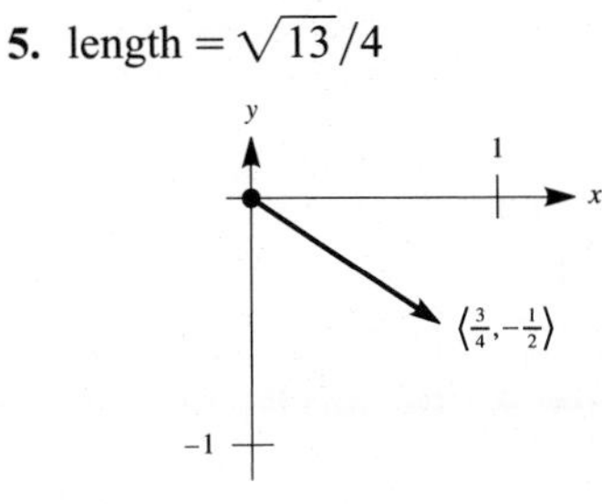

7. $\langle 1, 4 \rangle$ **9.** $\langle -1, 1 \rangle$ **11.** $\langle 8, -5 \rangle$
13. $\langle 7, 7 \rangle$ **15.** $\langle 24, 22 \rangle$ **17.** $\sqrt{130}$
19. $2\sqrt{17} - \sqrt{13} - \sqrt{37}$
21. $\langle 13, 6 \rangle$ **23.** $\langle 14, 21 \rangle$
25. $\langle -3, -1 \rangle$ **27.** $\langle 23, 12 \rangle$
29. $\langle -9, 0 \rangle$ **31.** -48 **33.** $3\mathbf{i} + 8\mathbf{j}$
35. $-8\mathbf{i} - 6\mathbf{j}$ **37.** $19\mathbf{i} + 23\mathbf{j}$
39. $\langle 1, 1 \rangle$ **41.** $\langle 5, -4 \rangle$
43. $\langle \sqrt{5}/5, 2\sqrt{5}/5 \rangle$
45. $\langle 2\sqrt{5}/5, -\sqrt{5}/5 \rangle$
47. $(8\sqrt{145}/145)\mathbf{i} - (9\sqrt{145}/145)\mathbf{j}$
49. $u_1 = \sqrt{3}/2$; $u_2 = \frac{1}{2}$
51. $u_1 = -\frac{1}{2}$; $u_2 = \sqrt{3}/2$
53. $u_1 = -\sqrt{3}/2$; $u_2 = \frac{1}{2}$
59. **(a)** $\mathbf{u} \cdot \mathbf{v} = 8$; $\mathbf{v} \cdot \mathbf{u} = 8$
(b) $\mathbf{v} \cdot \mathbf{w} = -14$; $\mathbf{w} \cdot \mathbf{v} = -14$
61. **(a)** $\mathbf{v} \cdot \mathbf{v} = 25$; $|\mathbf{v}|^2 = 25$
(b) $\mathbf{w} \cdot \mathbf{w} = 29$; $|\mathbf{w}|^2 = 29$
63. $\cos\theta = 7/\sqrt{170}$; $\theta \approx 57.53°$ or $\theta \approx 1.00$ radian
65. $\cos\theta = -57/\sqrt{3538}$; $\theta \approx 163.39°$ or $\theta \approx 2.85$ radians
67. **(a)** $\cos\theta = -7/\sqrt{170}$; $\theta \approx 122.47°$ or $\theta \approx 2.14$ radians
(b) $\cos\theta = 7/\sqrt{170}$; $\theta \approx 57.53°$ or $\theta \approx 1.00$ radian
69. **(a)** $\cos\theta = 0$
(b) The vectors are perpendicular.

(c)

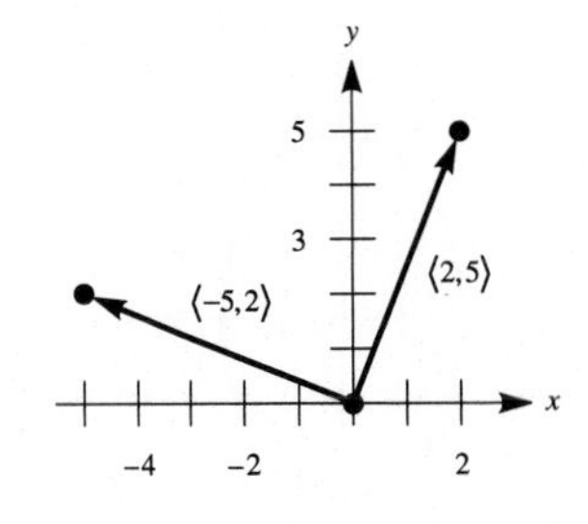

71. $\mathbf{A}\cdot\mathbf{B}=0$ implies $\cos\theta=0$ and thus $\theta=90°$, so the vectors are perpendicular.
73. $\left\langle\frac{5}{13},\frac{12}{13}\right\rangle$ and $\left\langle-\frac{5}{13},-\frac{12}{13}\right\rangle$
75. (a) $|\mathbf{C}|=\sqrt{x_1^2+y_1^2+x_2^2+y_2^2-2x_1x_2-2y_1y_2}$
(c) $\cos\theta=\mathbf{A}\cdot\mathbf{B}/|\mathbf{A}||\mathbf{B}|$

EXERCISE SET 9.4

1. $(2, 3)$ **3.** $(5\sqrt{3}/2, 1)$
5. $(3\sqrt{2}/4, 3\sqrt{2}/4)$

7.

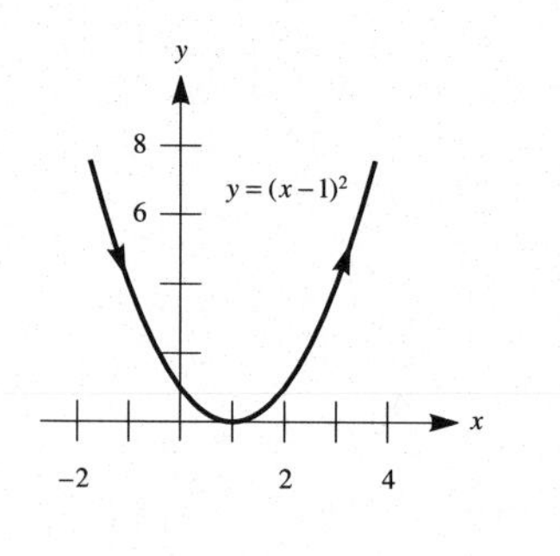

9.

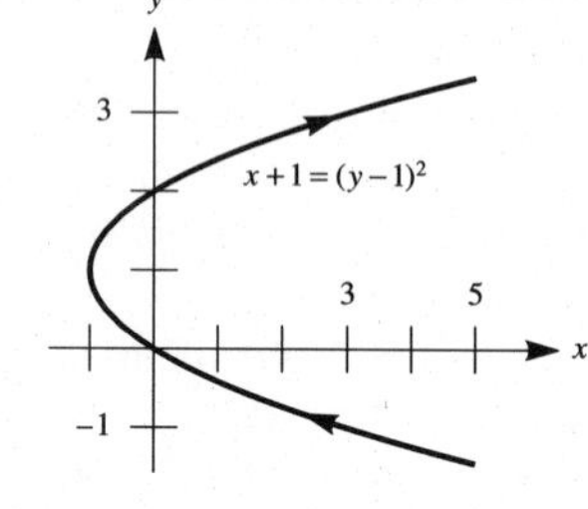

11.

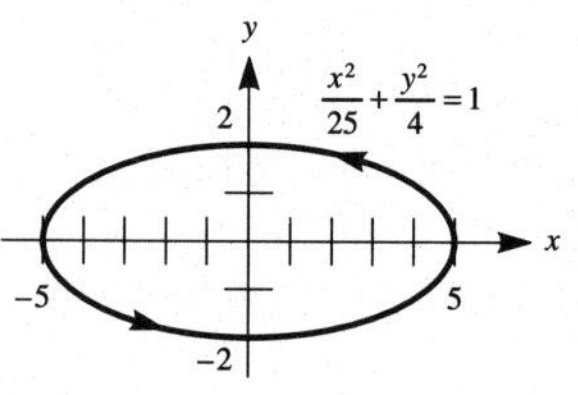

13.

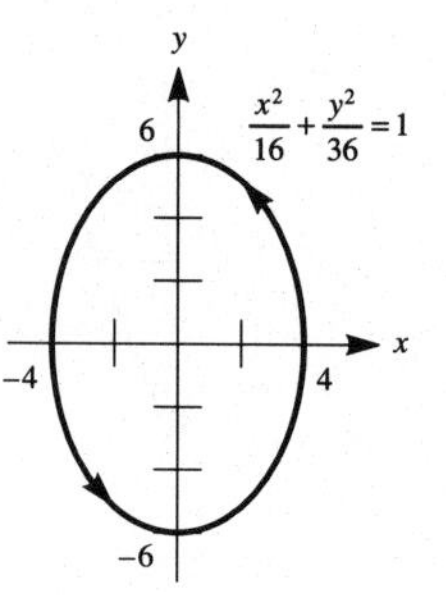

15. (a)

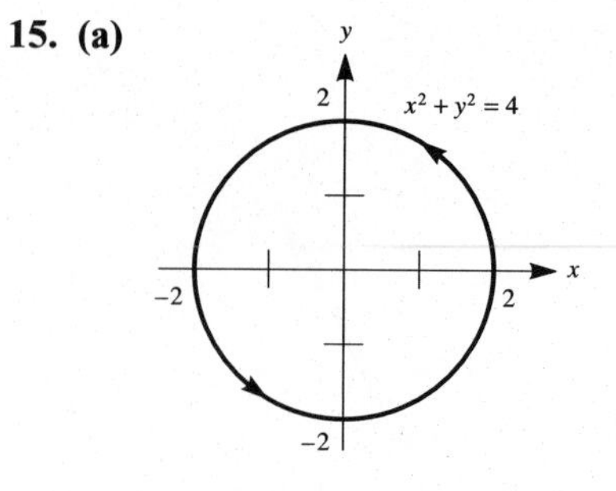

(b)

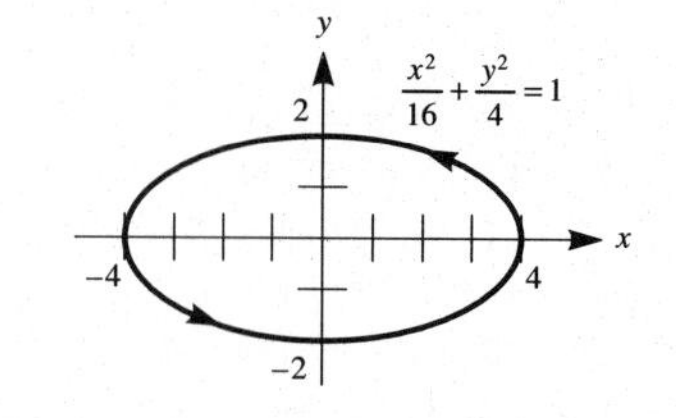

17. (a) $t=1$: $x\approx 34.2$, $y\approx 83.0$; $t=2$: $x\approx 68.4$, $y\approx 128.9$; $t=3$: $x\approx 102.6$, $y\approx 142.9$
(b) 5.93 seconds; 203 feet
21. $x^{2/3}+y^{2/3}=1$

GRAPHING UTILITY EXERCISES FOR SECTION 9.4

1. (a) $0\leqslant t\leqslant 1$

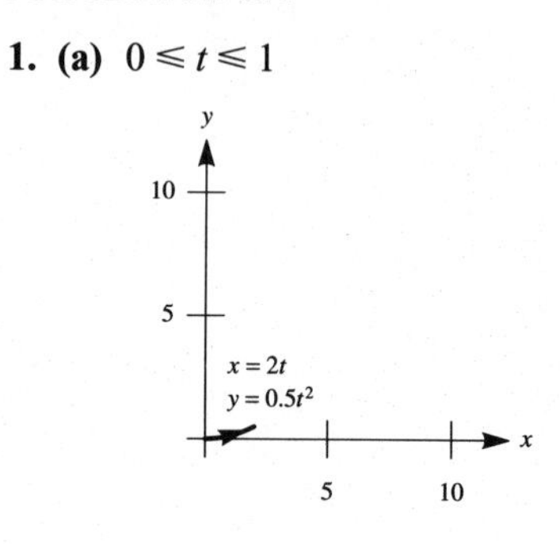

$0\leqslant t\leqslant 3$

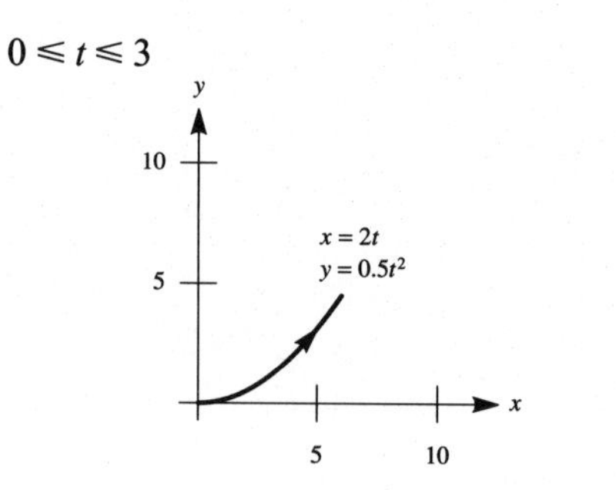

$0\leqslant t\leqslant 4$

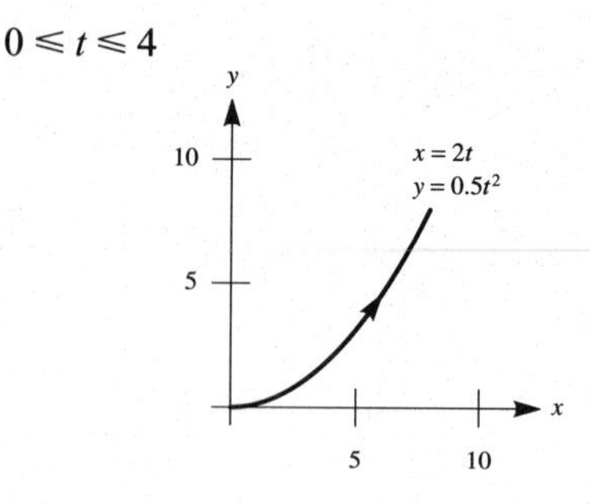

As the interval for t gets larger, the curve resembles a parabola.
(b) $-5\leqslant t\leqslant 5$

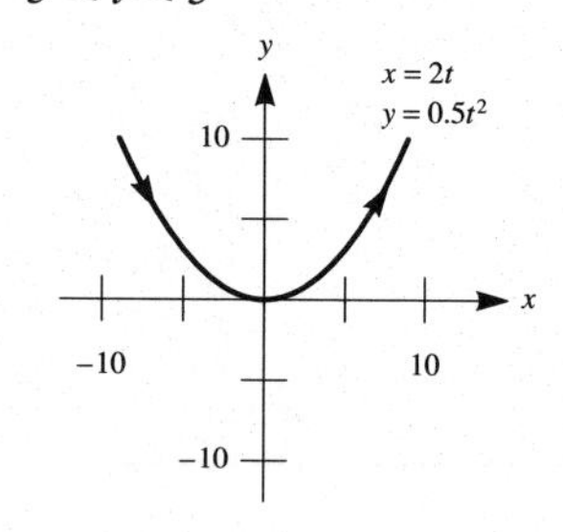

The restrictions on t in Figure 1(b) are $0\leqslant t\leqslant 5$.

3. (a)

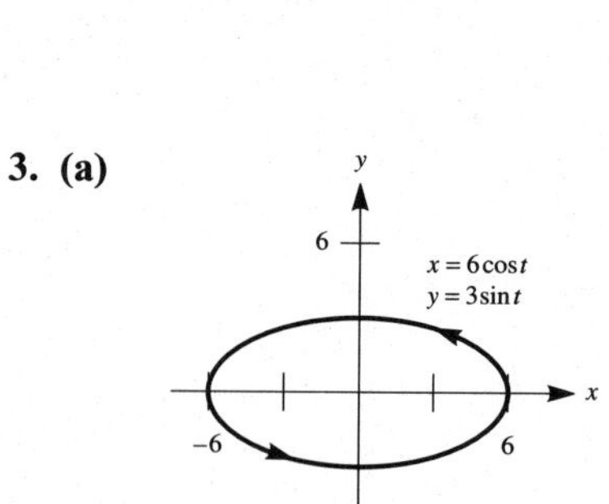

(b)

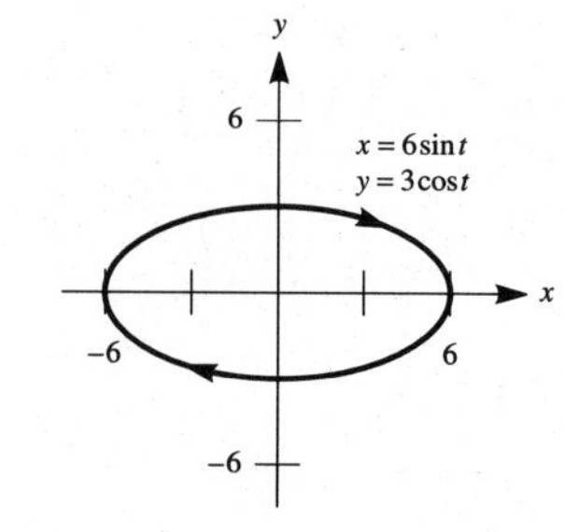

(c)

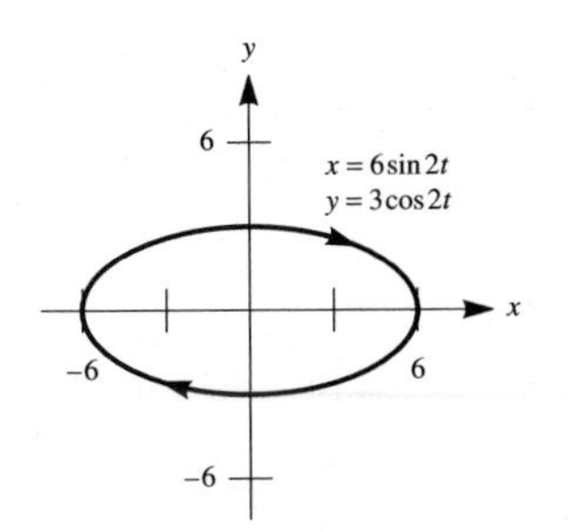

5. (a)

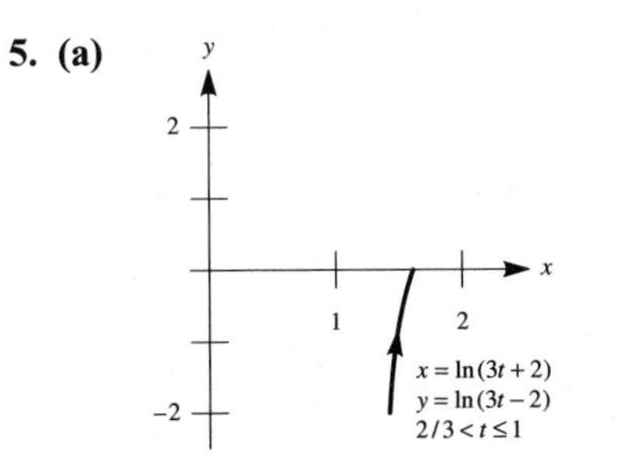

(b)

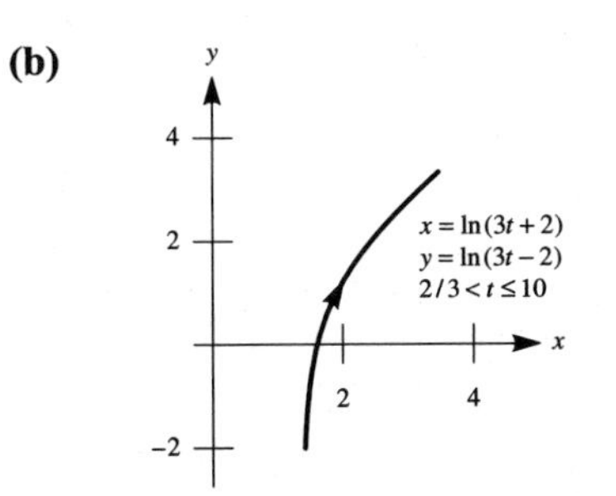

(c)

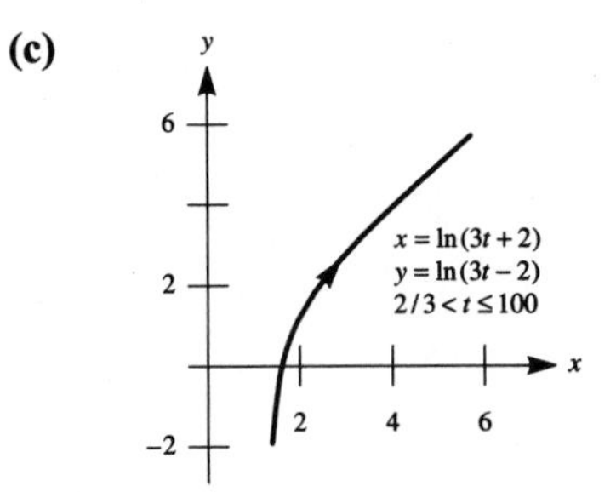

(d)

$x = \ln(3t+2)$
$y = \ln(3t-2)$
$2/3 < t \leq 1000$

7.

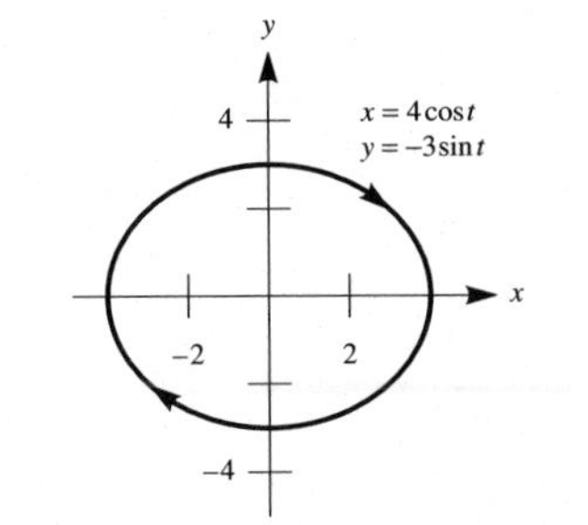

9.

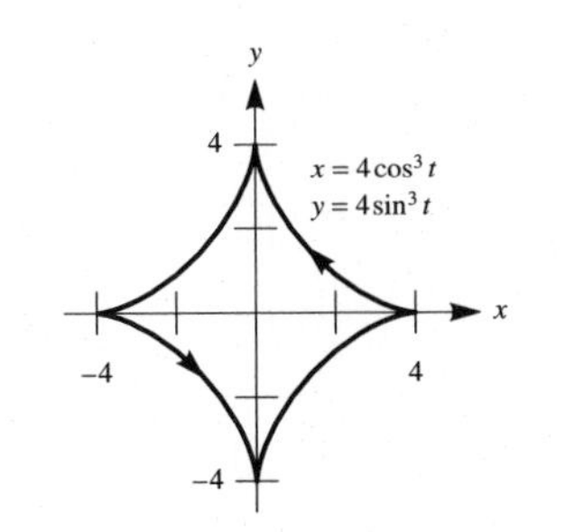

11.

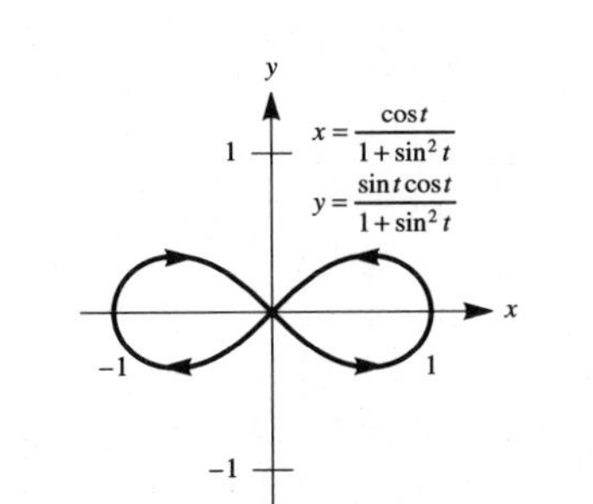

13.

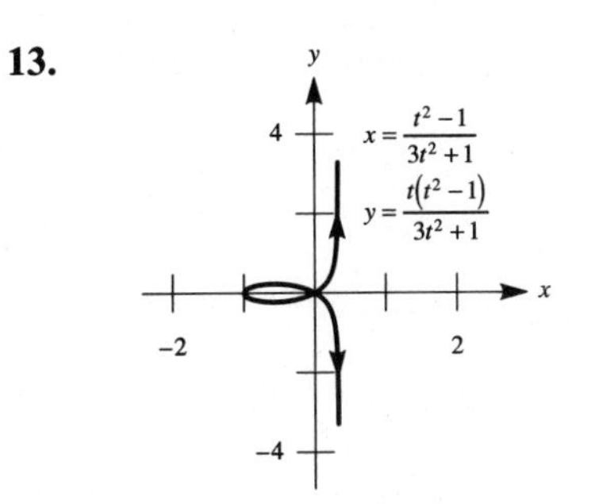

15.

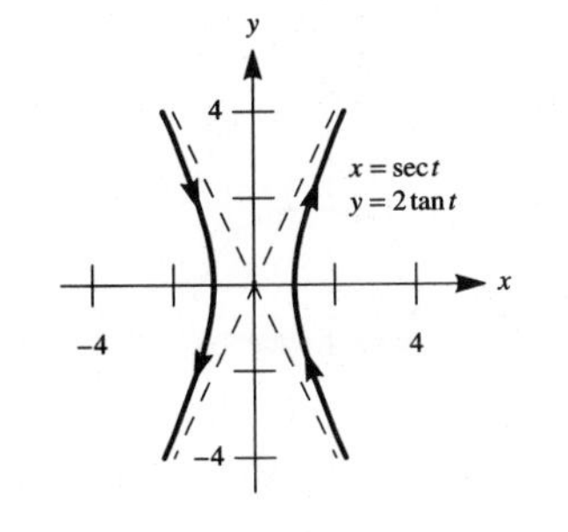

17.

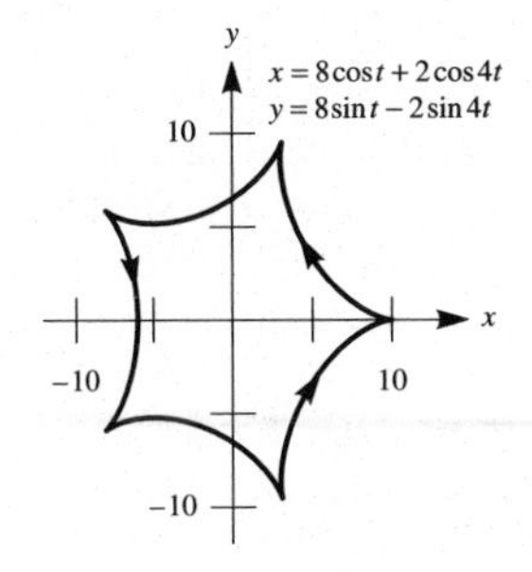

19. (a)

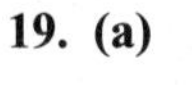

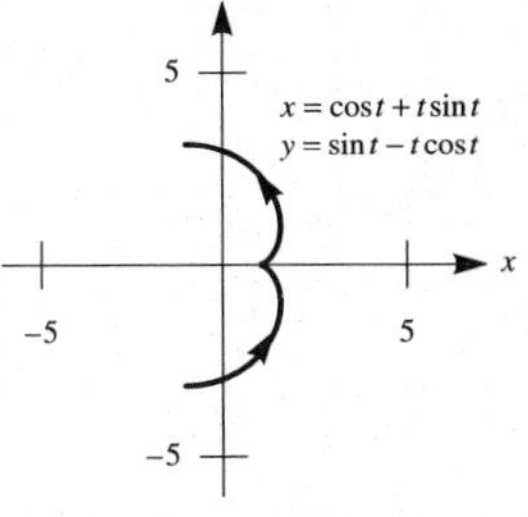

(b)

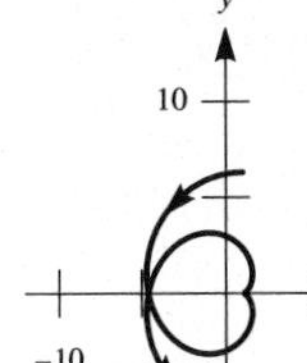

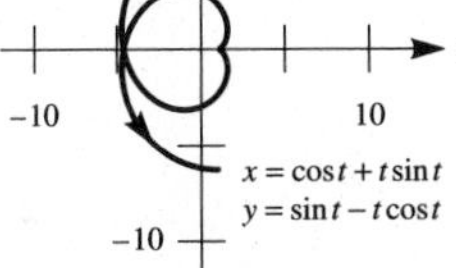

(c)

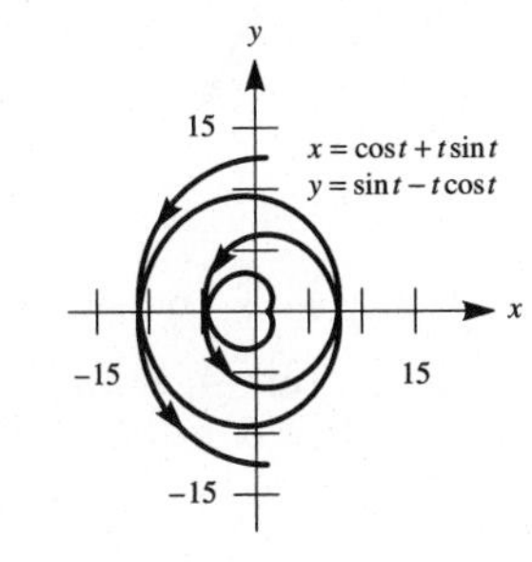

EXERCISE SET 9.5

1. (a) $\left(-\frac{3}{2}, 3\sqrt{3}/2\right)$ **(b)** $\left(2\sqrt{3}, -2\right)$
(c) $\left(2\sqrt{3}, -2\right)$
3. (a) $(0, 1)$ **(b)** $(0, 1)$
(c) $\left(-\sqrt{2+\sqrt{2}}/2, -\sqrt{2-\sqrt{2}}/2\right)$
5. $\left(\sqrt{2}, 5\pi/4\right)$ **7.** $(x-1)^2 + y^2 = 1$
9. $x^4 + x^2y^2 - y^2 = 0$
11. $(x^2+y^2)^3 = 9(x^2-y^2)^2$
13. $x^2/4 + y^2/8 = 1$

15. $y = -\sqrt{3}x + 4$
17. $r = 2/(3\cos\theta - 4\sin\theta)$
19. $r = \tan^2\theta \sec\theta$ **21.** $r^2 = \csc 2\theta$
23. $r^2 = 9/(9\cos^2\theta + \sin^2\theta)$
25. $A\left(\frac{8}{3}, \pi/6\right)$, $B\left(\frac{8}{3}, 5\pi/6\right)$, $C(4, \pi)$, $D(8, 7\pi/6)$
27. $A(1, \pi/6)$, $B(1, 5\pi/6)$, $C(1, 7\pi/6)$, $D(-2, \pi/2)$
29. $A(1, 0)$, $B(1.14, \pi/4)$, $C(1.48, 3\pi/4)$, $D(1.69, \pi)$, $E(1.92, 5\pi/4)$, $F(2.19, 3\pi/2)$, $G(2.50, 7\pi/4)$, $H(2.85, 2\pi)$, $I(3.25, 9\pi/4)$, $J(4.22, 11\pi/4)$, $K(5.48, 13\pi/4)$
31. $2\sqrt{5}$ **33.** $\sqrt{21}$
35. **(a)** $r^2 - 8r\cos\theta = -12$
(b) $r^2 - 8r\cos\left(\theta - \frac{2\pi}{3}\right) = -12$
(c) $r = 2$
37. **(a)** $r = 2\cos\left(\theta - \frac{3\pi}{2}\right)$
(b) $r = 2\cos\left(\theta - \frac{\pi}{4}\right)$
39. **(a)** 2 **(b)** $(4\sqrt{3}/3, 0)$ and $(4, \pi/2)$ **(c)** $(2, \pi/6)$

(d)

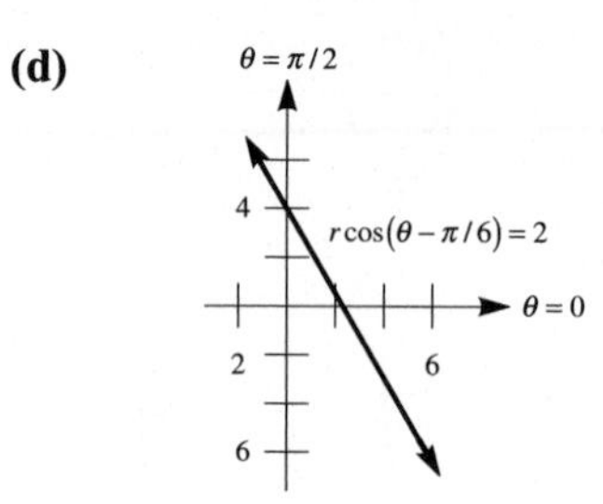

41. **(a)** 4 **(b)** $(-8, 0)$ and $\left(-\frac{8}{3}\sqrt{3}, \pi/2\right)$ **(c)** $(4, -2\pi/3)$

(d)

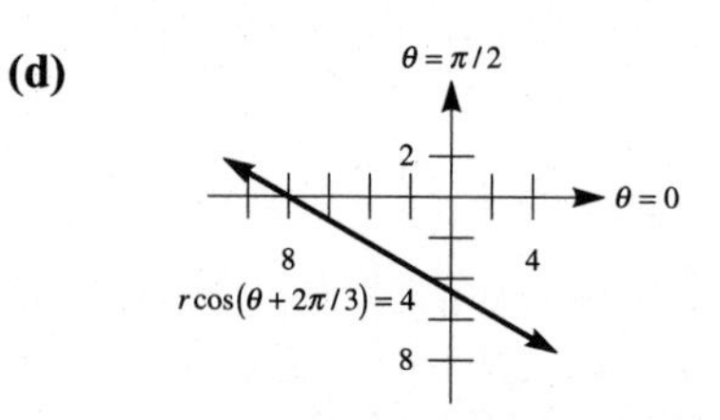

43. **(a)** $(2, 5\pi/6)$
(b) $r\cos\left(\theta - \frac{5\pi}{6}\right) = 2$
(c) x-intercept: $-4\sqrt{3}/3$; y-intercept: 4

45. **(a)**

(b)

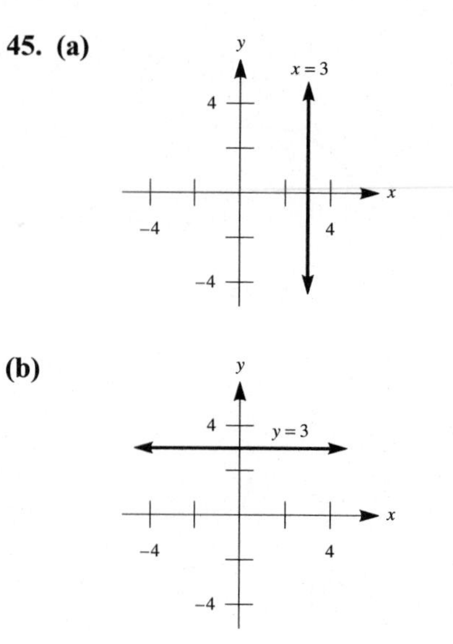

47. **(a)** $r\cos(\theta - \alpha) = d$
53. **(a)** $A(1, 0)$, $B(1, 2\pi/3)$, $C(1, 4\pi/3)$

EXERCISE SET 9.6

1.

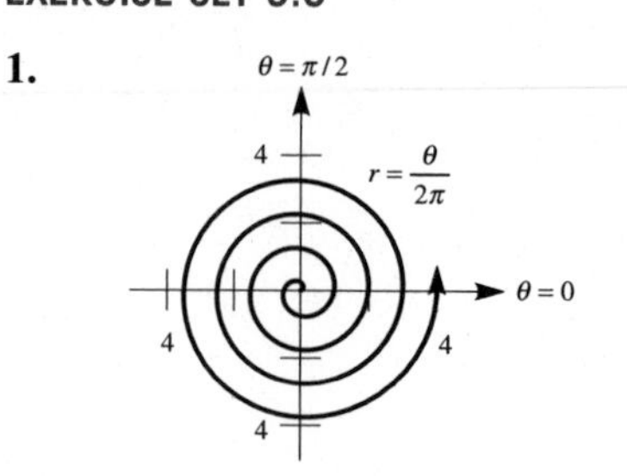

3.

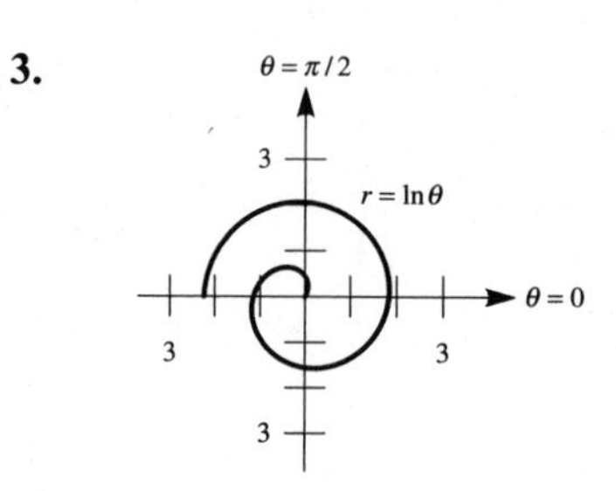

5.

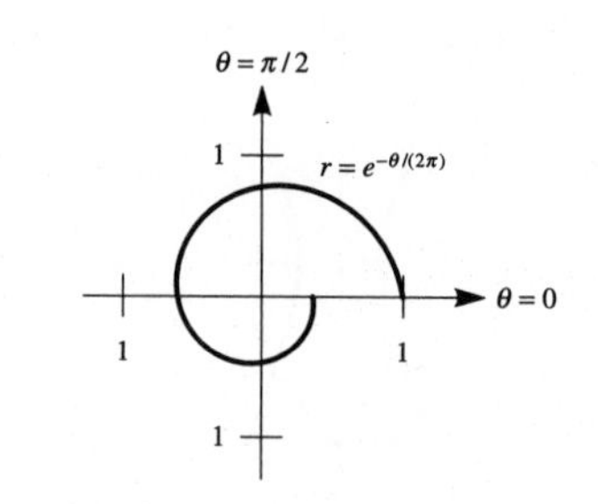

7.

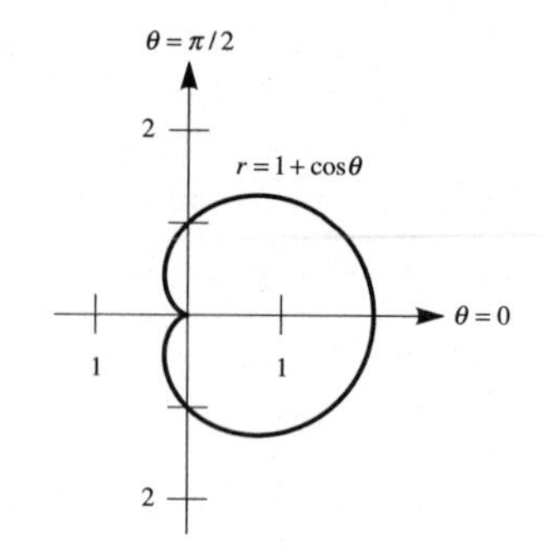

9.

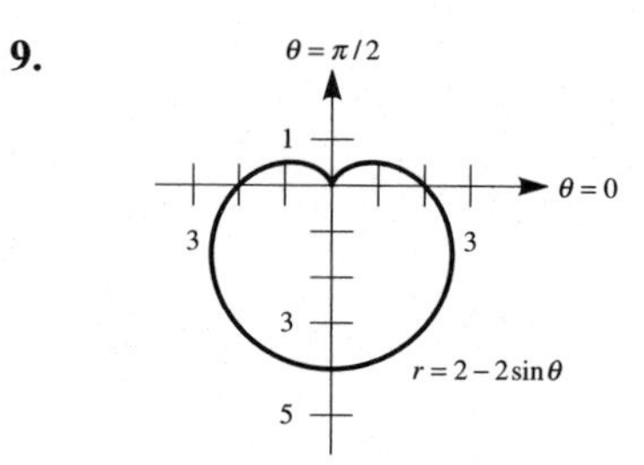

11.

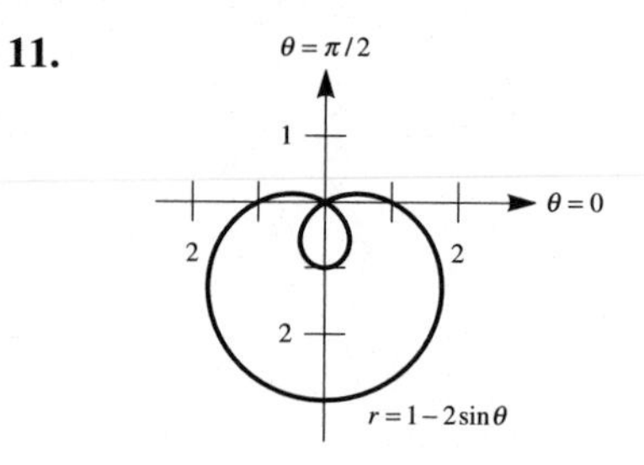

13.

θ = π/2
4
r = 2 + 4 cos θ
θ = 0
2
4
4

15.

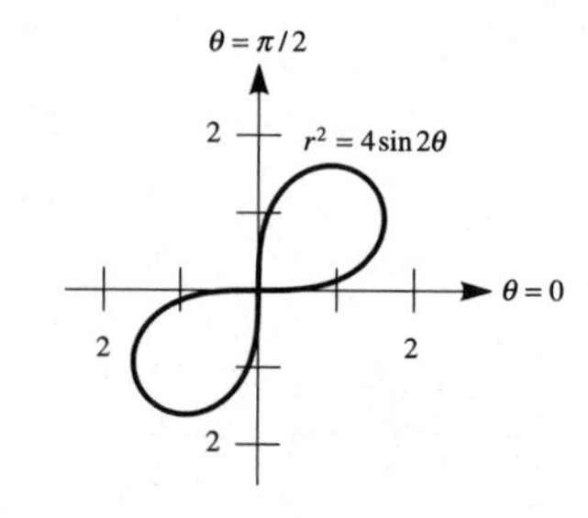

17.

19.

31. (a) The curves are identical.

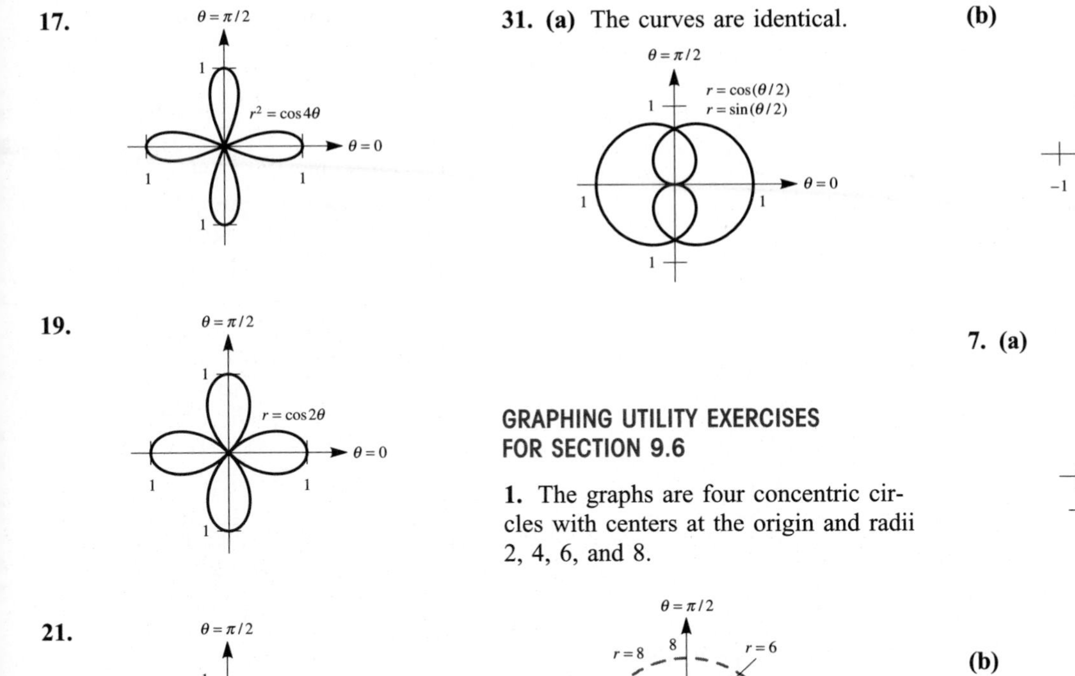

21.

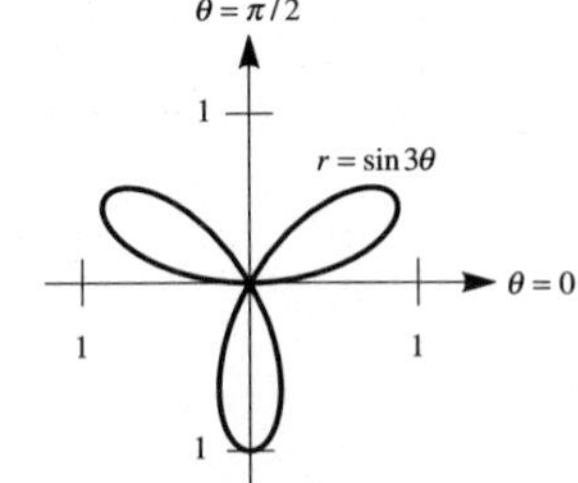

23.

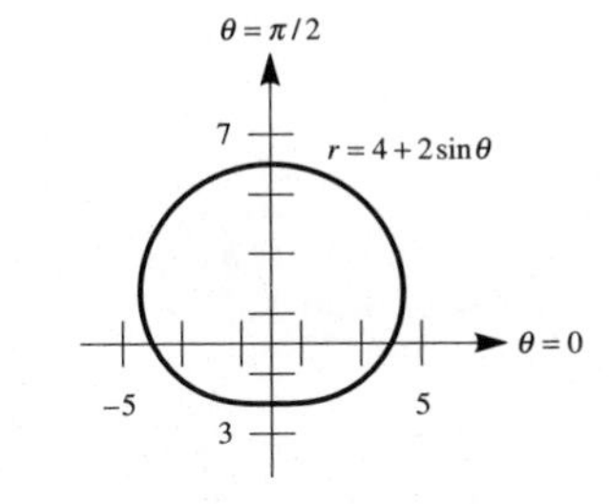

25.

$\theta = \pi/2$ $r = 8\tan\theta$ $\theta = 0$

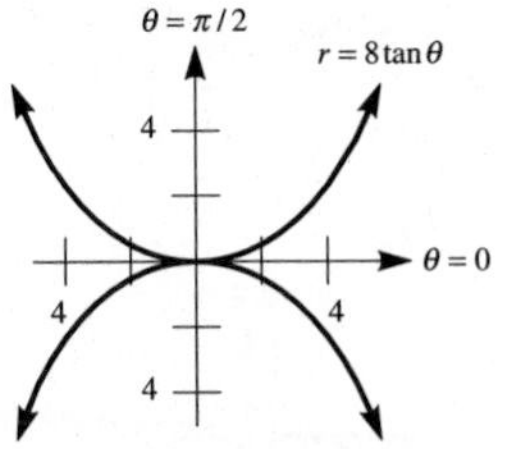

27. (a) C **(b)** B **(c)** D **(d)** A

29. (a) $A(1, 0)$; $B(e^{a\pi/2}, \pi/2)$; $C(e^{a\pi}, \pi)$; $D(e^{3a\pi/2}, 3\pi/2)$

GRAPHING UTILITY EXERCISES FOR SECTION 9.6

1. The graphs are four concentric circles with centers at the origin and radii 2, 4, 6, and 8.

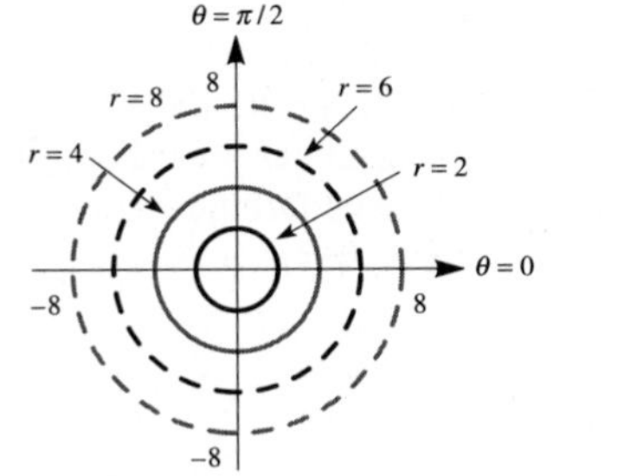

3. (b)

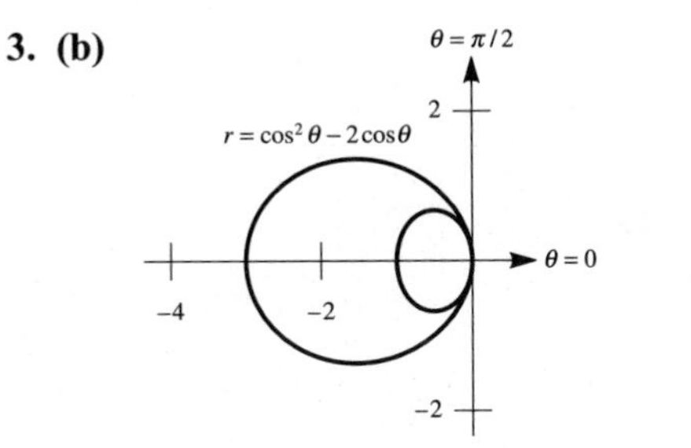

5. (a)

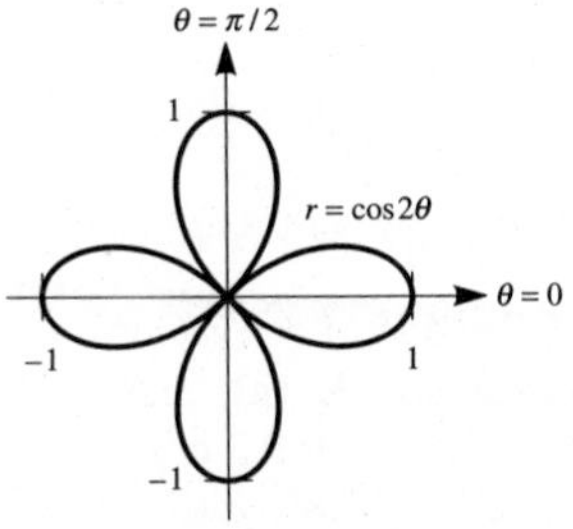

$\theta = \pi/2$ $r = \cos 2\theta$ $\theta = 0$

(b)

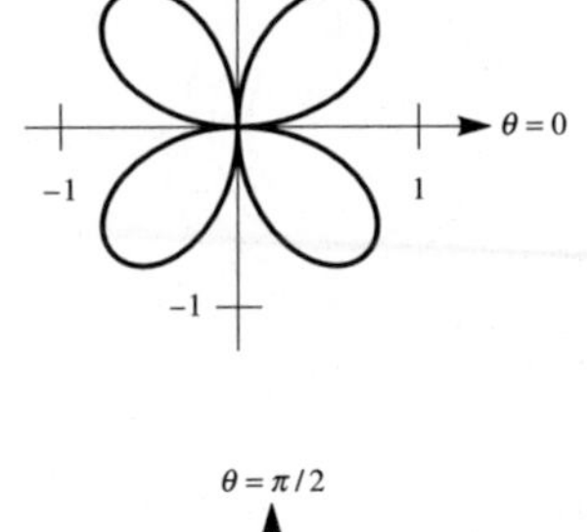

7. (a)

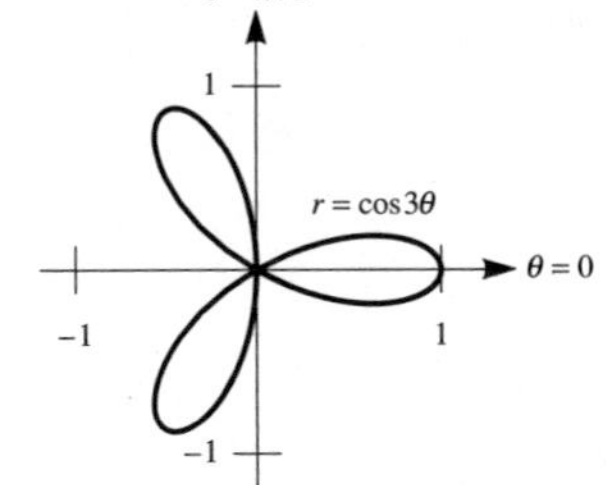

(b)

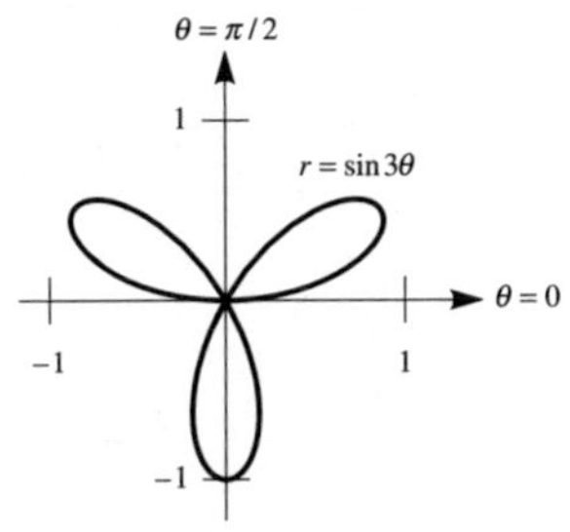

9.

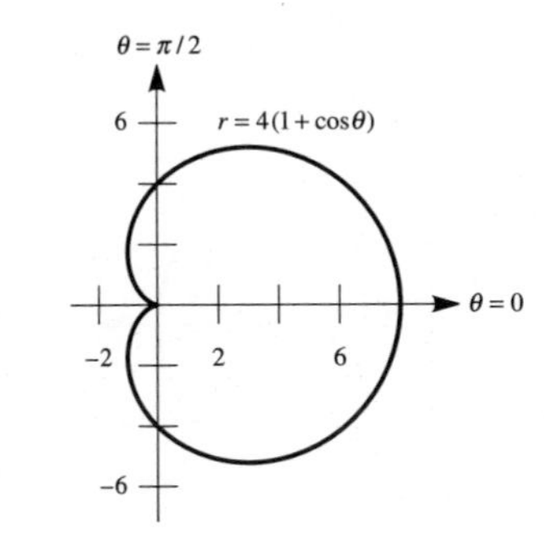

11.

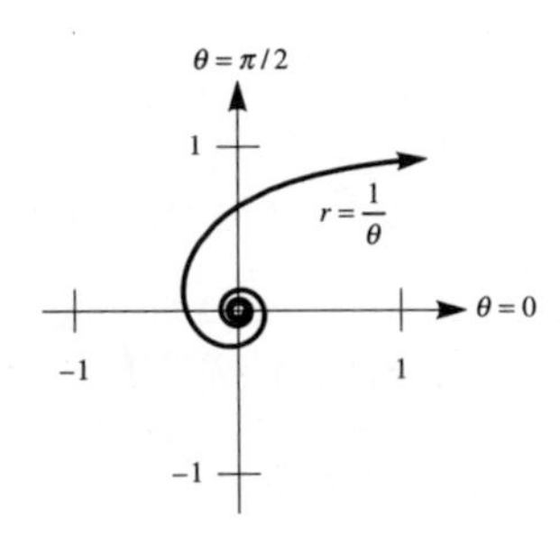

13.

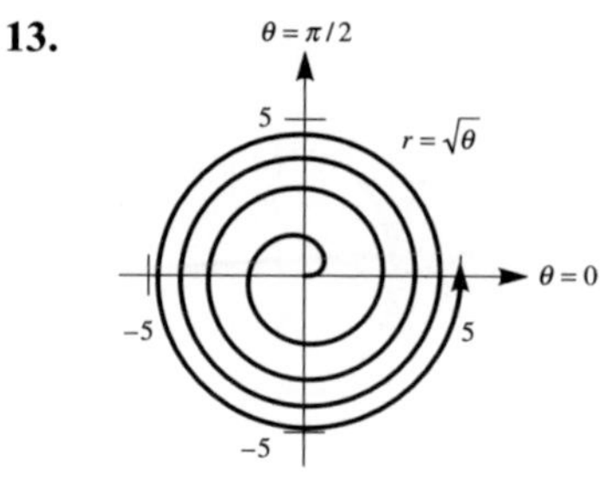

15. (a)

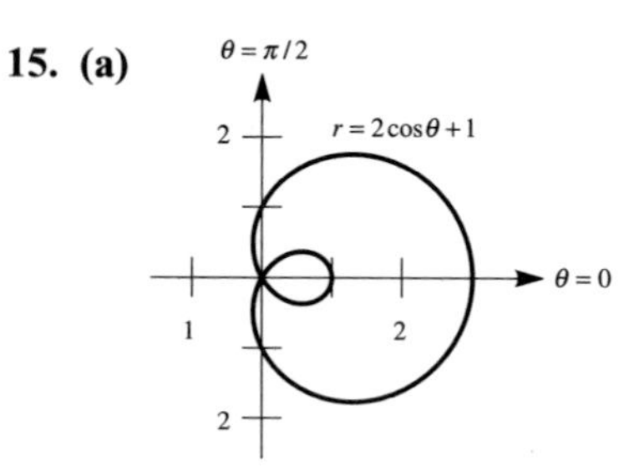

(b)

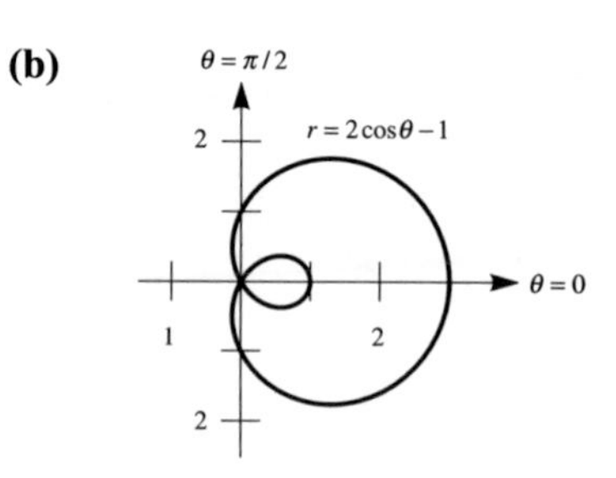

(c)

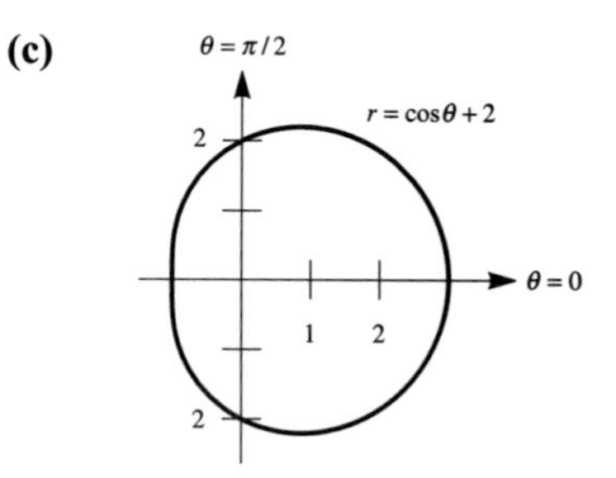

(d)

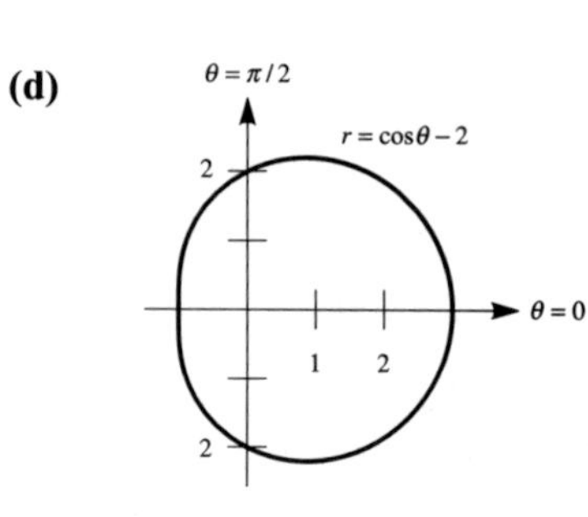

(e)

(f)

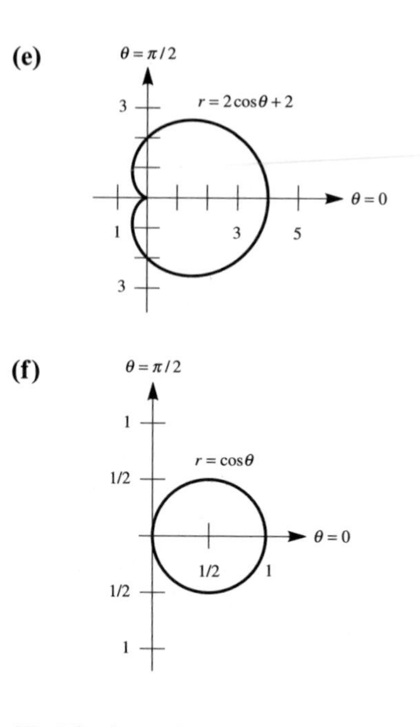

17. The inner loop near the origin is not simple but rather a cardioid-type shape that passes through both the first and fourth quadrants.

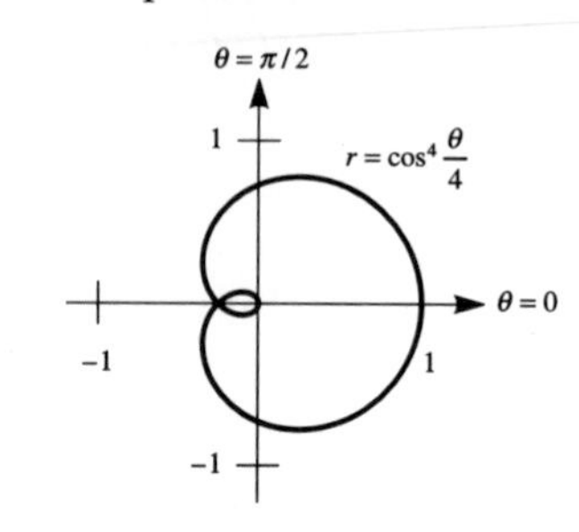

19.

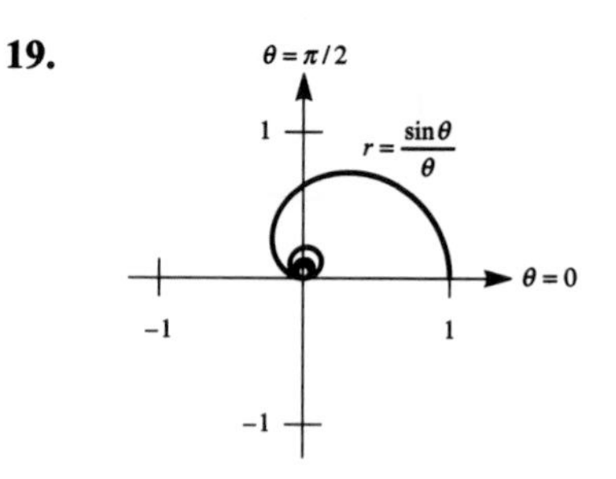

CHAPTER 9 REVIEW EXERCISES

1. $\angle C = 55°$, $a \approx 12.6$ cm, $b \approx 19.5$ cm **3. (a)** $\angle B \approx 62.4°$, $\angle C \approx 65.6°$, $c \approx 9.2$ cm **(b)** $\angle B \approx 117.6°$, $\angle C \approx 10.4°$, $c \approx 1.8$ cm **5.** $c \approx 7.7$ cm, $\angle A \approx 108.5°$, $\angle B \approx 47.5°$
7. $\angle C \approx 106.6°$, $\angle B \approx 48.2°$, $\angle A \approx 25.2°$ **9.** 9.21 cm **11.** 32.48 cm^2
13. 55.23 cm^2 **15.** 7.89 cm
17. 15.43 cm **19.** 15 cm **21.** 11 cm
23. (a) $\cos A = \frac{3}{4}$, $\cos C = \frac{1}{8}$ **25.** 36°
27. 121.24° **29.** 7.24 m
31. (a) 11.76 cm **(b)** 19.02 cm
35. $a = 5$, $b = 8$
37. $|\mathbf{R}| = 25$ N, $\theta \approx 53.1°$
39. $v_x \approx 41.0$ cm/sec, $v_y \approx 28.7$ cm/sec
41. $|\mathbf{W}| \approx 36.4$ lb, $|\mathbf{W}_p| \approx 33.2$ lb
43. $b = \pm\sqrt{15}$ **45.** $\langle 10, 9 \rangle$
47. $\langle 12, 7 \rangle$ **49.** 48 **51.** $\langle 12, 8 \rangle$
53. $\langle -6, 2 \rangle$ **55.** $\langle -7, -6 \rangle$
57. $7\mathbf{i} - 6\mathbf{j}$
59. $\langle 3\sqrt{13}/13, 2\sqrt{13}/13 \rangle$ **61.** 4.89
63. $r^2 - 10r\cos\left(\theta - \frac{\pi}{6}\right) = -16$
65. (a) 3 **(b)** $(6, 0)$, $\left(2\sqrt{3}, \frac{\pi}{2}\right)$
(c) $(3, \pi/3)$

(d)

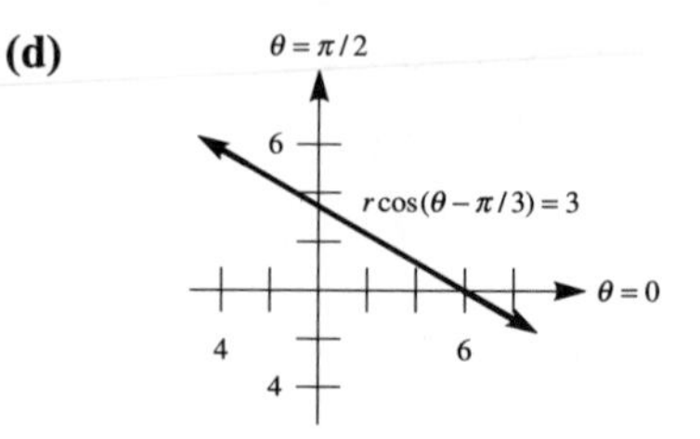

67. (a)

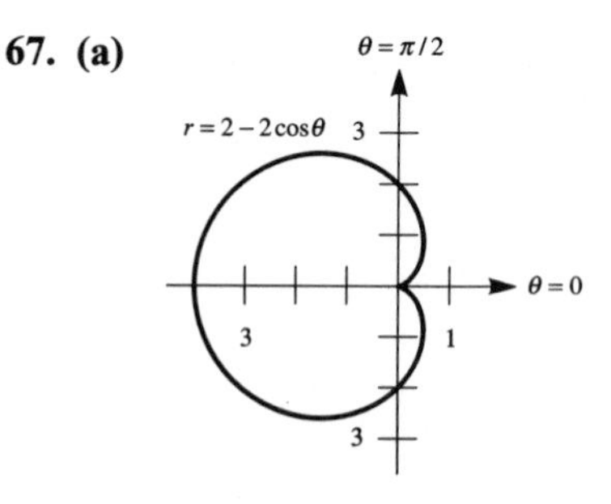

(b)

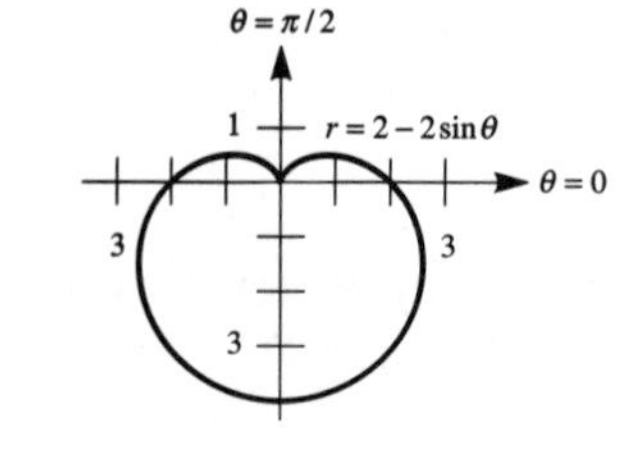

69. (a)

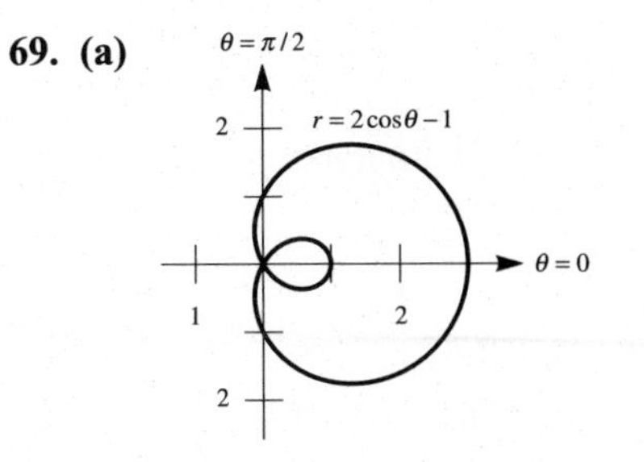

(b)

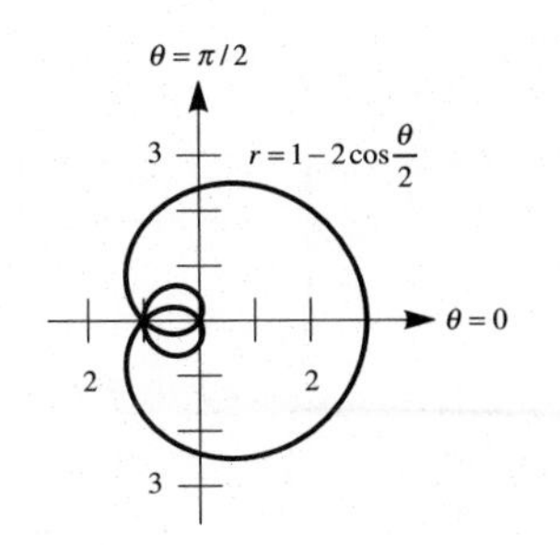

(b)

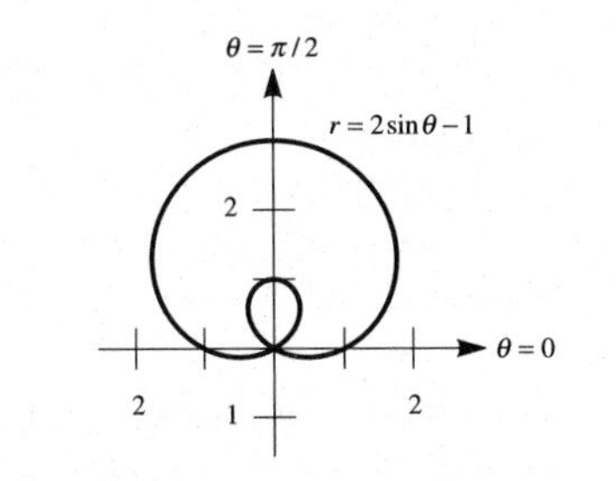

71. (a)

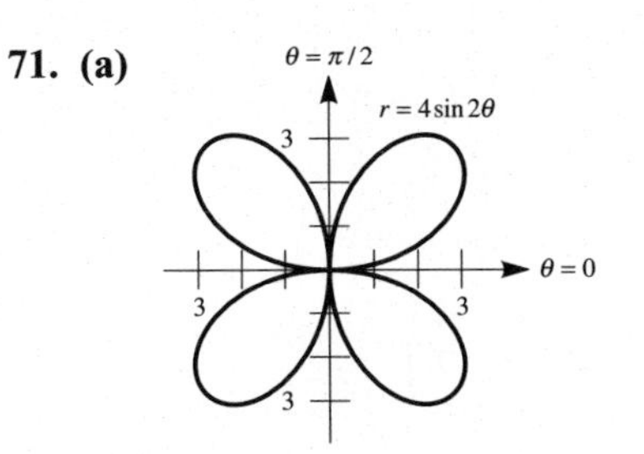

(b)

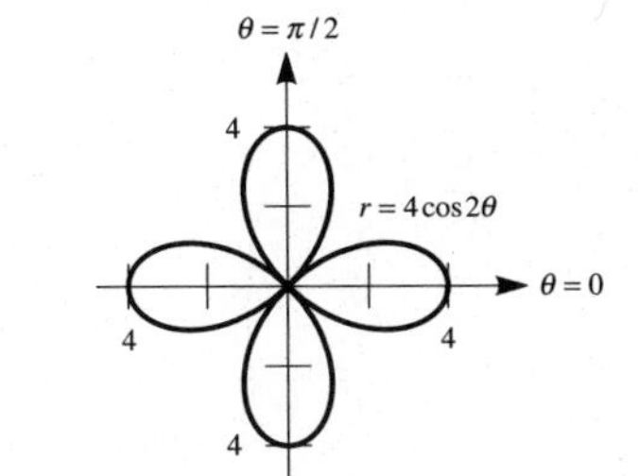

73. (a)

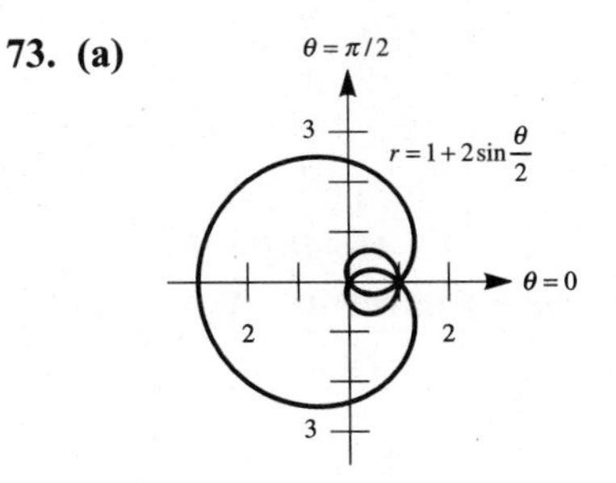

75.

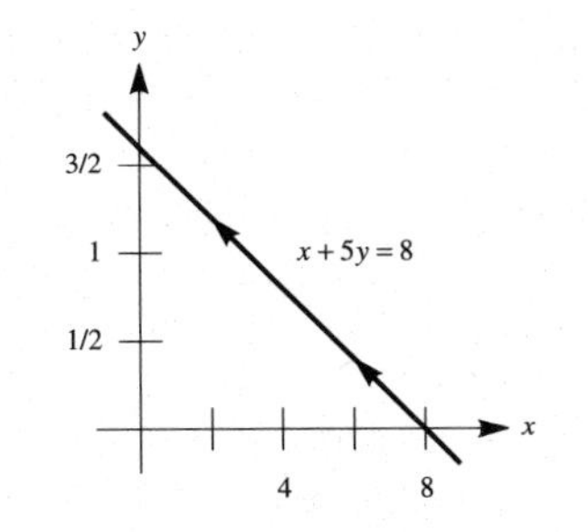

77.

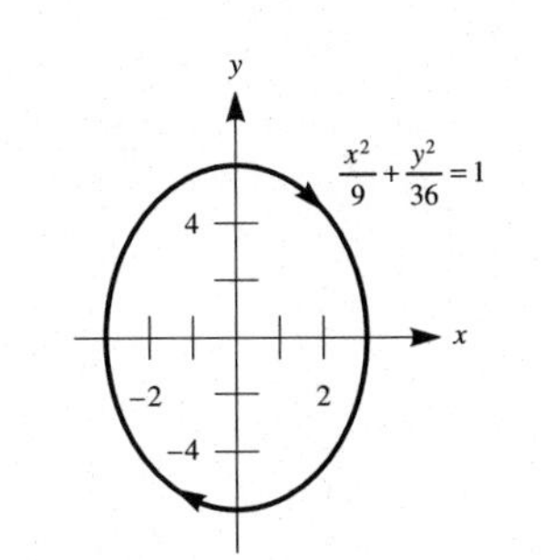

79.

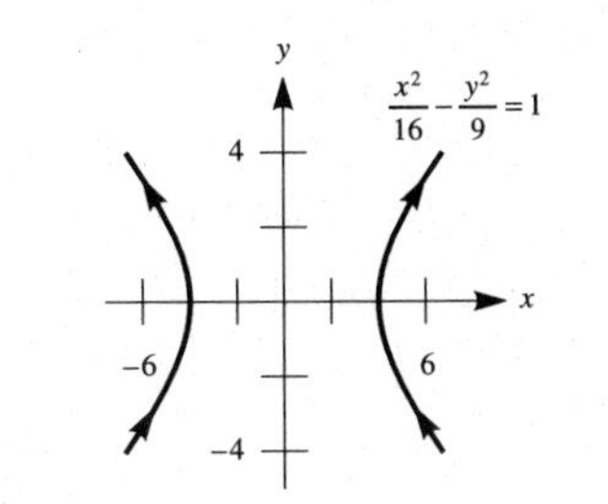

81. (a) $P\left(\dfrac{\sin \alpha}{\alpha}, \alpha\right)$, $Q\left(\dfrac{\sin 2\alpha}{2\alpha}, 2\alpha\right)$

(b) $P\left(\dfrac{\sin \alpha \cos \alpha}{\alpha}, \dfrac{\sin^2 \alpha}{\alpha}\right)$, $Q\left(\dfrac{\sin 2\alpha \cos 2\alpha}{2\alpha}, \dfrac{\sin^2 2\alpha}{2\alpha}\right)$

(f) The product of the slopes of $\overline{PQ}$ and $\overline{OQ}$ is -1.

CHAPTER 9 TEST

1. $a = 7$ cm **2.** If θ is the angle opposite the 4 cm side, then $\cos \theta = -\frac{1}{4} < 0$, so θ must be obtuse (not acute). **3.** 20 cm

4. $2\sqrt{17 - 4\sqrt{2}}$ cm
5. $a \approx 3.3$ cm, $C \approx 26.2°$, $B \approx 126.8°$
6. (a) $2\sqrt{5}$ N **(b)** $\tan \theta = 2$
7. (a) $4\sqrt{13 - 12 \cos 110°}$ N
(b) $\sin \theta = \dfrac{2 \sin 110°}{\sqrt{13 - 12 \cos 110°}}$
8. ground speed $= 50\sqrt{37}$ mph, $\tan \theta = \frac{1}{6}$
9. (a) $\langle 13, 5 \rangle$ **(b)** $\sqrt{194}$ **(c)** $\mathbf{i} - 3\mathbf{j}$
10. $\langle -11\sqrt{130}/130, -3\sqrt{130}/130 \rangle$
11. $(x^2 + y^2)^2 = x^2 - y^2$

12.

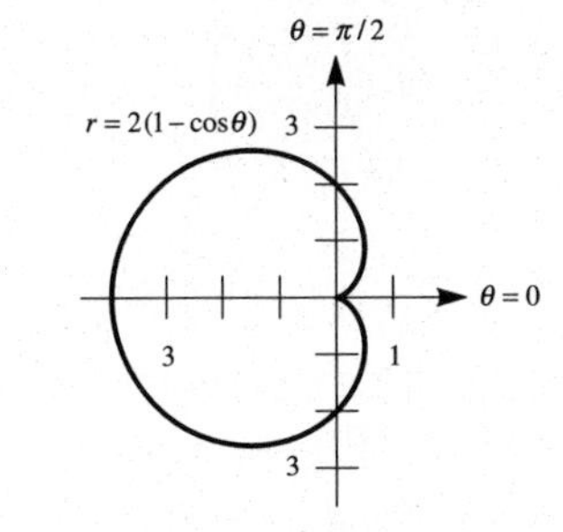

13.

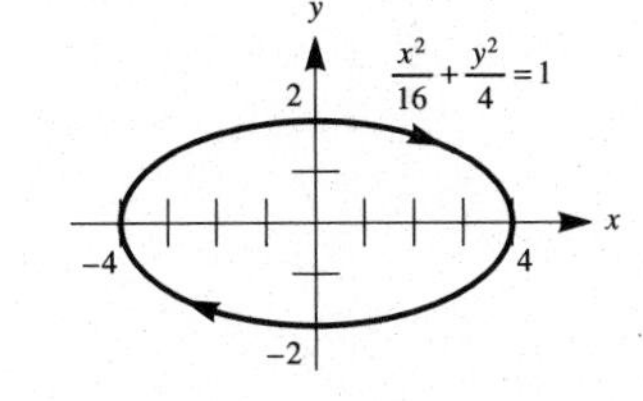

14. $\sqrt{13}$
15. $r^2 - 10r \cos\left(\theta - \frac{\pi}{2}\right) = -21$; no
16. (a) $r \cos\left(\theta - \frac{5\pi}{6}\right) = 4$

CHAPTER TEN

EXERCISE SET 10.1

1. (a) yes **(b)** no **(c)** yes **(d)** yes **3.** yes **5.** no **7.** yes **9.** $\left(-6, \frac{1}{2}\right)$ **11.** $\left(-\frac{5}{2}, \frac{13}{2}\right)$ **13.** $\left(\frac{1}{5}, \frac{7}{10}\right)$ **15.** $\left(-\frac{60}{13}, \frac{60}{13}\right)$ **17.** $\left(\sqrt{6}, 1\right)$ **19.** $\left(-\frac{2}{9}, \frac{23}{27}\right)$ **21.** $\left(-\frac{283}{242}, -\frac{3}{121}\right)$

23. $\left(-\frac{226}{25}, -\frac{939}{50}\right)$ **25.** $\left(\frac{5}{8}, 0\right)$
27. $(1, 2)$ **29.** $b = 3$, $c = 4$
31. $A = -28$, $B = -22$
33. $\frac{49}{20}$ sq. units **35.** 80 cc of the 10% solution, 120 cc of the 35% solution
37. 8 lb of \$5.20 coffee, 8 lb of \$5.80 coffee
39. $\left(\dfrac{ab}{a+b}, \dfrac{ab}{a+b}\right)$; yes: $a \neq \pm b$
41. $\left(\dfrac{a+b}{ab}, \dfrac{-1}{ab}\right)$; yes: $a \neq 0$, $b \neq 0$
43. $\left(-\frac{1}{11}, \frac{1}{9}\right)$ **45.** $(5, 2)$
47. $(1.32, -1.62)$
49. $(0.14, 0.05)$ **51.** $(1.61, 0)$
53. $(4, -8)$, $(4, 2)$, $(-1, -8)$, $(-1, 2)$
55. 59
57. length = 12 in.; width = 5 in.
59. (a) $(60, 40)$ **(b)** $(60, 40)$
61. $(0, a)$ **63.** $\left(\dfrac{1}{b}, \dfrac{1}{a}\right)$
65. $k = -29/7, 2$

GRAPHING UTILITY EXERCISES FOR SECTION 10.1

1. consistent, with exactly one solution

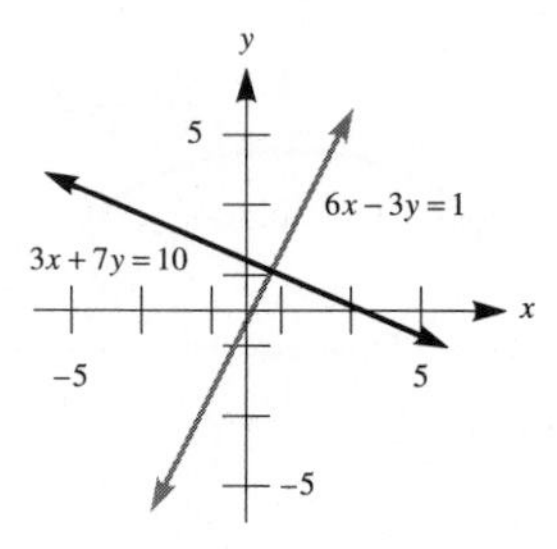

3. inconsistent, with no solution

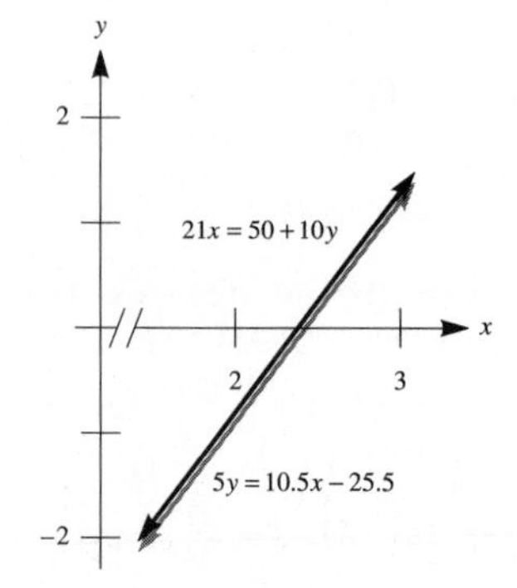

5. (a) $(1.56, -6.44)$

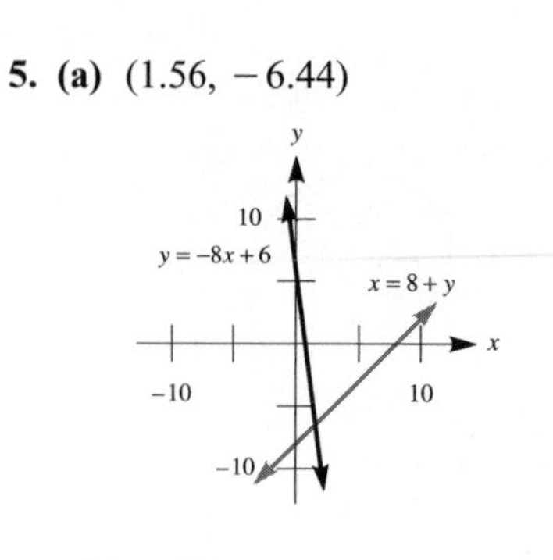

(b) $\left(\frac{14}{9}, -\frac{58}{9}\right)$

7. (a) $(-7.95, -16.54)$

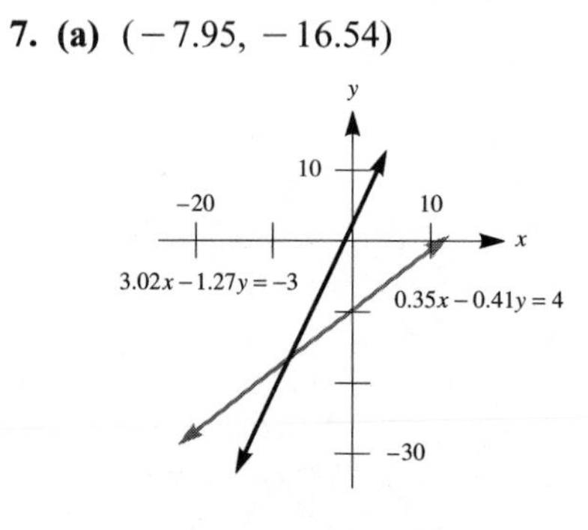

(b) $\left(-\frac{63100}{7937}, -\frac{131300}{7937}\right)$

9. (a)

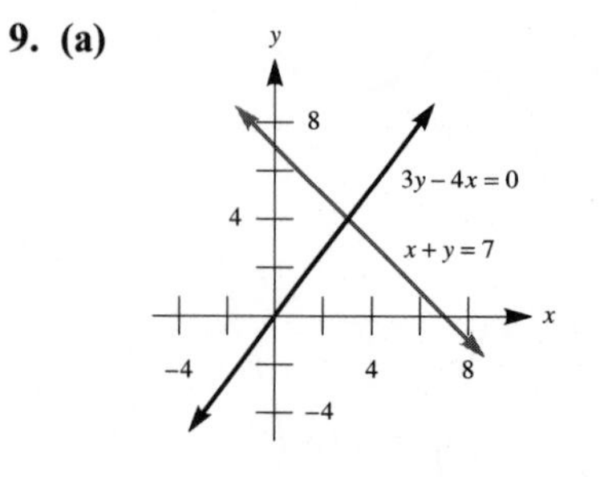

(b) $(3, 4)$

EXERCISE SET 10.2

1. $(-3, -2, -1)$ **3.** $\left(-\frac{1}{60}, -\frac{2}{15}, \frac{3}{5}\right)$
5. $\left(\frac{25}{36}, \frac{5}{9}, -\frac{1}{3}\right)$ **7.** $\left(x, \frac{x}{8}, 0\right)$, where x is any real number **9.** $(-1, 0, 1, -5)$
11. $\left(\frac{11}{3}, \frac{8}{3}, \frac{17}{3}\right)$ **13.** $(1, 0, 1)$
15. $(2, 3, 1)$
17. no solution (inconsistent)
19. $\left(\dfrac{z+1}{-7}, \dfrac{5(z+1)}{7}, z\right)$, where z is any real number
21. $\left(\dfrac{11-5z}{7}, \dfrac{-3z-6}{7}, z\right)$, where z is any real number
23. $(4, 1, -3, 2)$
25. $\left(\dfrac{17-17z}{5}, \dfrac{8z-3}{5}, z\right)$, where z is any real number
27. $\left(\dfrac{12+10w}{11}, \dfrac{146+19w}{55}, \dfrac{61w+159}{55}, w\right)$, where w is any real number
29. $\left(-\dfrac{5z}{12}, \dfrac{2z+3}{3}, z\right)$, where z is any real number
31. (a) $A = 2$, $B = -1$, $C = -3$
(b) $a = 0$, $b = -2$, $c = -1$
33. x-intercepts: $(3 \pm \sqrt{17})/2$; vertex: $\left(\frac{3}{2}, -\frac{17}{8}\right)$
35. $17x^2 + 17y^2 - 49x + 65y - 166 = 0$
37. $x = \ln a$; $y = \ln 2a$; $z = \ln(a/2)$
41. 60 miles

EXERCISE SET 10.3

1. (a) 2×3 **(b)** 3×2 **3.** 5×4
5. coefficient matrix:
$$\begin{pmatrix} 2 & 3 & 4 \\ 5 & 6 & 7 \\ 8 & 9 & 10 \end{pmatrix};$$
augmented matrix:
$$\begin{pmatrix} 2 & 3 & 4 & 10 \\ 5 & 6 & 7 & 9 \\ 8 & 9 & 10 & 8 \end{pmatrix}$$
7. coefficient matrix:
$$\begin{pmatrix} 1 & 0 & 1 & 1 \\ 1 & 1 & 0 & 2 \\ 0 & 1 & 1 & 1 \\ 2 & -1 & -1 & 0 \end{pmatrix};$$
augmented matrix:
$$\begin{pmatrix} 1 & 0 & 1 & 1 & -1 \\ 1 & 1 & 0 & 2 & 0 \\ 0 & 1 & 1 & 1 & 1 \\ 2 & -1 & -1 & 0 & 2 \end{pmatrix}$$
9. $(-1, -2, 3)$ **11.** $(-5, 1, 3)$
13. $(3, 0, -7)$ **15.** $(8, 9, -1)$
17. $\left(\dfrac{9z+5}{19}, \dfrac{31z-6}{19}, z\right)$ for any real number z
19. $(2, -1, 0, 3)$

21. no solution (inconsistent)

23. $\begin{pmatrix} 3 & 2 \\ 2 & 4 \end{pmatrix}$ **25.** $\begin{pmatrix} 6 & 4 \\ 4 & 8 \end{pmatrix}$

27. $\begin{pmatrix} 11 & -2 \\ 11 & 1 \end{pmatrix}$ **29.** $\begin{pmatrix} 2 & 3 \\ -1 & 4 \end{pmatrix}$

31. not defined **33.** $\begin{pmatrix} 10 & -2 \\ -8 & 0 \\ 4 & 6 \end{pmatrix}$

35. $\begin{pmatrix} 2 & 4 & 11 \\ -12 & 16 & 19 \\ 14 & 12 & 43 \end{pmatrix}$

37. $\begin{pmatrix} -9 & 10 & 10 \\ 4 & -8 & -12 \\ 10 & 4 & 21 \end{pmatrix}$

39. not defined **41.** $\begin{pmatrix} 0 & 0 & 0 \\ 0 & 0 & 0 \\ 0 & 0 & 0 \end{pmatrix}$

43. $\begin{pmatrix} 4 & 2 \\ 2 & 5 \end{pmatrix}$ **45.** not defined

47. $\begin{pmatrix} 1 & 18 \\ -6 & 13 \end{pmatrix}$ **49.** $\begin{pmatrix} -16 & 75 \\ -25 & 34 \end{pmatrix}$

51. (a) $\begin{pmatrix} -13 & 1 & 40 \\ 43 & 17 & 0 \\ 89 & 61 & 60 \end{pmatrix}$

(b) $\begin{pmatrix} -13 & 1 & 40 \\ 43 & 17 & 0 \\ 89 & 61 & 60 \end{pmatrix}$

(c) $\begin{pmatrix} -52 & -82 & 61 \\ 87 & 141 & 0 \\ 216 & 318 & 165 \end{pmatrix}$

(d) $\begin{pmatrix} -52 & -82 & 61 \\ 87 & 141 & 0 \\ 216 & 318 & 165 \end{pmatrix}$

53. (a) $\begin{pmatrix} 16 & 20 \\ 24 & 28 \end{pmatrix}$ **(b)** $\begin{pmatrix} 18 & 26 \\ 18 & 26 \end{pmatrix}$

(c) $\begin{pmatrix} 14 & 14 \\ 30 & 30 \end{pmatrix}$ **(d)** $\begin{pmatrix} 18 & 26 \\ 18 & 26 \end{pmatrix}$

55. (a) $\begin{pmatrix} x \\ -y \end{pmatrix}$ **(b)** $\begin{pmatrix} -x \\ y \end{pmatrix}$

(c) $\begin{pmatrix} -x \\ -y \end{pmatrix}$; reflection about the origin

57. (a) $f(A) = -2$; $f(B) = 29$; $f(AB) = -58$; yes

EXERCISE SET 10.4

5. $A^{-1} = \begin{pmatrix} -5 & 9 \\ 4 & -7 \end{pmatrix}$

7. $A^{-1} = \begin{pmatrix} -\frac{6}{23} & \frac{1}{23} \\ \frac{5}{23} & \frac{3}{23} \end{pmatrix}$

9. A^{-1} does not exist.

11. $A^{-1} = \begin{pmatrix} 2 & -3 \\ 1 & 3 \end{pmatrix}$

13. $A^{-1} = \begin{pmatrix} 2 & -1 \\ -3 & 2 \end{pmatrix}$

15. $A^{-1} = \begin{pmatrix} \frac{6}{11} & 1 \\ -\frac{1}{11} & 0 \end{pmatrix}$

17. A^{-1} does not exist.

19. $A^{-1} = \begin{pmatrix} -1 & 2 & -3 \\ 2 & 1 & 0 \\ 4 & -2 & 5 \end{pmatrix}$

21. $A^{-1} = \begin{pmatrix} 5 & -\frac{10}{3} & 1 \\ 0 & \frac{1}{3} & 0 \\ 4 & -\frac{8}{3} & 1 \end{pmatrix}$

23. $A^{-1} = \begin{pmatrix} 2 & 1 & 4 \\ 3 & 2 & 5 \\ 0 & -1 & 1 \end{pmatrix}$

25. A^{-1} does not exist.

27. (a) $x = -1$ and $y = 1$
(b) $x = -132$ and $y = 48$

29. (a) $x = 2$, $y = -1$, $z = 4$
(b) $x = 1$, $y = 1$, $z = -2$

31. (a) $AA = \begin{pmatrix} 1 & 0 & 0 \\ 0 & 1 & 0 \\ 0 & 0 & 1 \end{pmatrix}$; $A^{-1} = A$

(b) $x = 1/2$, $y = -1$, $z = 1$

35. (a) $A^{-1} = \begin{pmatrix} -\frac{5}{2} & \frac{3}{2} \\ 2 & -1 \end{pmatrix}$;

$B^{-1} = \begin{pmatrix} 7 & -8 \\ -6 & 7 \end{pmatrix}$;

$B^{-1}A^{-1} = \begin{pmatrix} -\frac{67}{2} & \frac{37}{2} \\ 29 & -16 \end{pmatrix}$

(b) $(AB)^{-1} = \begin{pmatrix} -\frac{67}{2} & \frac{37}{2} \\ 29 & -16 \end{pmatrix}$;

it's the same as $B^{-1}A^{-1}$.

GRAPHING UTILITY EXERCISES FOR SECTIONS 10.3 AND 10.4

1. $\begin{pmatrix} 23 & -101 \\ 18 & -79 \end{pmatrix}$ **3.** undefined

5. $\begin{pmatrix} 21 & 16 & 43 \\ 17 & 18 & 43 \\ -4 & -4 & -14 \end{pmatrix}$

7. $\begin{pmatrix} 139 & -36 \\ 109 & -28 \end{pmatrix}$ **9.** $\begin{pmatrix} 139 & -36 \\ 109 & -28 \end{pmatrix}$

11. undefined **13.** $\begin{pmatrix} -18 & -10 \\ 34 & 19 \\ 59 & 33 \end{pmatrix}$

15. $\begin{pmatrix} 39 & 30 & 82 \\ 10 & 8 & 21 \\ -10 & -7 & -21 \end{pmatrix}$

17. $\begin{pmatrix} 0 & -42 & 23 \\ 1 & -14 & 5 \\ -1 & -35 & 15 \end{pmatrix}$

19. undefined **21.** undefined

23. $\begin{pmatrix} -4 & 9 \\ -1 & 2 \end{pmatrix}$ **25.** $\begin{pmatrix} -11 & 26 \\ 19 & -45 \end{pmatrix}$

27. $\begin{pmatrix} -11 & 26 \\ 19 & -45 \end{pmatrix}$

29. G^{-1} does not exist.

31. $\begin{pmatrix} -21 & 56 & -26 \\ 0 & 1 & 1 \\ 10 & -27 & 12 \end{pmatrix}$

33. $(FG)^{-1}$ does not exist.

35. $A^{-1} = \begin{pmatrix} 0.085 & 0.106 \\ -0.064 & 0.170 \end{pmatrix}$

(a) (4, 9) **(b)** (0.213, −0.660)

37. $A^{-1} = \begin{pmatrix} 0.143 & 0 & 0.286 \\ -0.429 & 1 & -0.857 \\ 0.286 & -1 & 1.571 \end{pmatrix}$

(a) (1, 1, −6)
(b) (0.286, 5.143, −4.429)

EXERCISE SET 10.5

1. (a) 29 **(b)** −29 **3. (a)** 0 **(b)** 0
5. −1 **7.** −60 **9.** 9
11. (a) 314 **(b)** 674 **(c)** part (b)
13. (a) 0 **(b)** 0 **(c)** 0 **(d)** 0
15. 0 **17.** −3 **19.** 6848 **21.** 17120
23. $(y - x)(z - x)(z - y)$ **25.** xy
27. 20 **29.** 120 **33.** (1, 1, 2)
35. (2, −3, 6) **37.** (0, 0, 0)
39. $\left(13 - \frac{11}{3}y, y, 13 - 4y\right)$ for any real number y
41. (1, 0, −10, 2)
43. $x = 4$, $x = -4$, $x = -1$
55. (a) 1 **(b)** 1

59. $\left(\dfrac{k(k-b)(k-c)}{a(a-b)(a-c)}, \dfrac{k(k-a)(k-c)}{b(b-a)(b-c)}, \dfrac{k(k-a)(k-b)}{c(c-a)(c-b)}\right)$
61. $y = 2x + 5$
65. (b) $\begin{pmatrix} -\frac{9}{61} & \frac{7}{61} \\ \frac{1}{61} & \frac{6}{61} \end{pmatrix}$

GRAPHING UTILITY EXERCISES FOR SECTION 10.5

3. -65520 **5. (a)** The second determinant should be 10 times the first determinant. **(b)** 206, 2060
7. (a) $D = -219$, $D_x = -287$, $D_y = 124$, $D_z = -294$
(b) $(1.31, -0.57, 1.34)$

EXERCISE SET 10.6

1. (0, 0), (3, 9)
3. $(2\sqrt{6}, 1), (-2\sqrt{6}, 1)$
5. $(-1, -1)$
7. $(2, 3), (2, -3), (-2, 3), (-2, -3)$
9. $(1, 0), (-1, 0)$
11. $\left(\dfrac{-1+\sqrt{65}}{8}, \dfrac{1+\sqrt{65}}{2}\right)$, $\left(\dfrac{-1-\sqrt{65}}{8}, \dfrac{1-\sqrt{65}}{2}\right)$
13. $(\sqrt{17}/17, 1), (\sqrt{17}/17, -1), (-\sqrt{17}/17, 1), (-\sqrt{17}/17, -1)$
15. $(1, 0), (4, -\sqrt{3})$ **17.** (2, 4)
19. (100, 1000), (100, 1/1000), (1/100, 1000), (1/100, 1/1000)
21. $\left(\dfrac{2\ln 2 - 5\ln 3}{\ln 2 - \ln 3}, \dfrac{3\ln 2}{\ln 2 - \ln 3}\right)$
25. $\left(\dfrac{1}{a}, \dfrac{1}{b}\right)$ **27.** (9, 14), (14, 9)
29. dimensions: $\dfrac{p - \sqrt{2d^2 - p^2}}{2}$ by $\dfrac{p - \sqrt{2d^2 - p^2}}{2}$
31. $(5, 4), (5, -4)$
33. (a) $k = (\ln 8)/6$; $N_0 = 3/2$
(b) $k = \frac{2}{7}\ln 10$; $N_0 = 10^{-1/7}$
35. $\left(\dfrac{p^2}{A}, \dfrac{q^2}{A}, \dfrac{r^2}{A}\right)$, $\left(\dfrac{-p^2}{A}, \dfrac{-q^2}{A}, \dfrac{-r^2}{A}\right)$, where $A = \sqrt{p^2 + q^2 + r^2}$
37. 9 cm and 40 cm
39. width: 3 cm; length: 20 cm
41. $(1, 2), (-1, -2), (2, 1), (-2, -1)$
43. $\left(\dfrac{1+\sqrt{13}}{2}, \dfrac{-1+\sqrt{13}}{2}\right)$, $\left(\dfrac{-1+\sqrt{13}}{2}, \dfrac{1+\sqrt{13}}{2}\right)$, $\left(\dfrac{1-\sqrt{13}}{2}, \dfrac{-1-\sqrt{13}}{2}\right)$, $\left(\dfrac{-1-\sqrt{13}}{2}, \dfrac{1-\sqrt{13}}{2}\right)$
45. $(2, 3), (-2, -3), (1, 2), (-1, -2)$
47. (q, p), $\left(\dfrac{-1+\sqrt{3}}{2}q, (1-\sqrt{3})p\right)$, $\left(\dfrac{-1-\sqrt{3}}{2}q, (1+\sqrt{3})p\right)$
49. $(e^{9/2}, e^3)$

GRAPHING UTILITY EXERCISES FOR SECTION 10.6

1. (a) $(2.193, -2.807)$, $(-3.193, -8.193)$

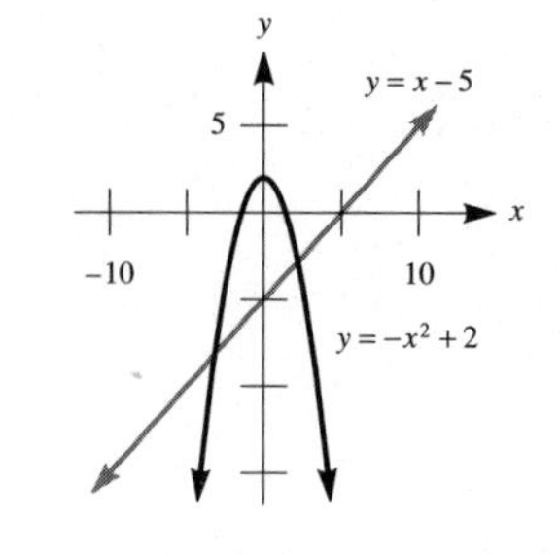

(b) $\left(\dfrac{-1+\sqrt{29}}{2}, \dfrac{-11+\sqrt{29}}{2}\right) \approx (2.1926, -2.8074)$; $\left(\dfrac{-1-\sqrt{29}}{2}, \dfrac{-11-\sqrt{29}}{2}\right) \approx (-3.1926, -8.1926)$

3. (a) $(0.316, 0.949), (-0.316, -0.949)$

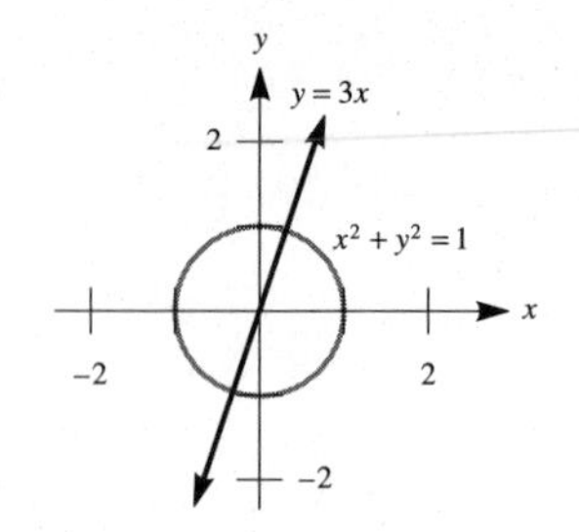

(b) $\left(\dfrac{\sqrt{10}}{10}, \dfrac{3\sqrt{10}}{10}\right) \approx (0.3162, 0.9487)$; $\left(-\dfrac{\sqrt{10}}{10}, -\dfrac{3\sqrt{10}}{10}\right) \approx (-0.3162, -0.9487)$

5. (a) $(0.852, 2.361)$

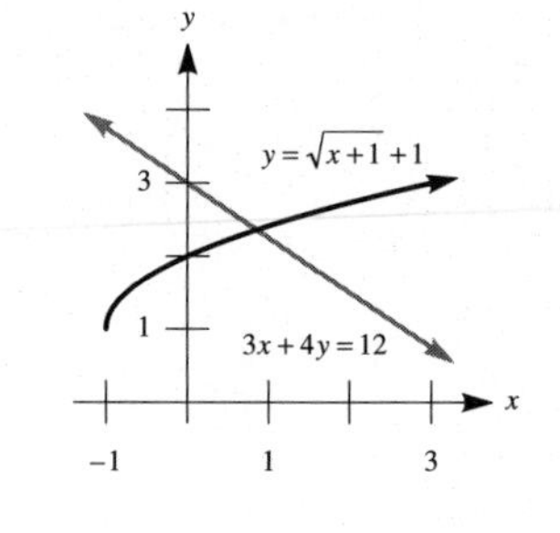

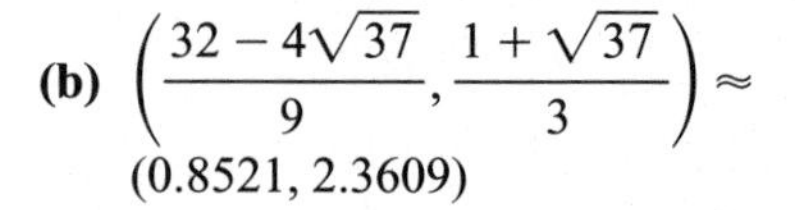

(b) $\left(\dfrac{32 - 4\sqrt{37}}{9}, \dfrac{1+\sqrt{37}}{3}\right) \approx (0.8521, 2.3609)$

7. (a) $(0.602, 5.303)$

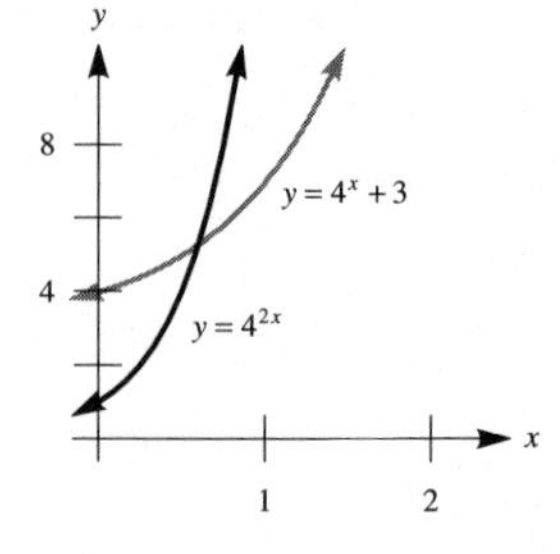

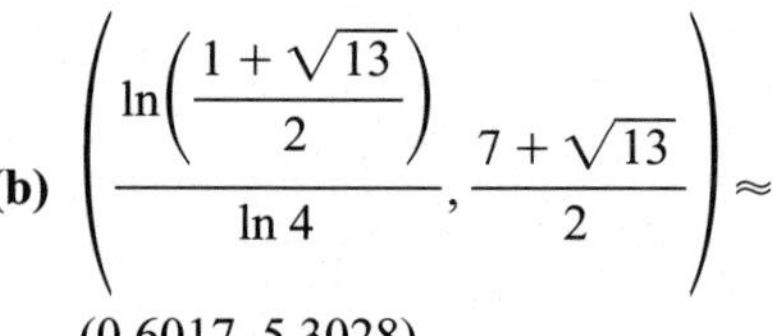

(b) $\left(\dfrac{\ln\left(\dfrac{1+\sqrt{13}}{2}\right)}{\ln 4}, \dfrac{7+\sqrt{13}}{2}\right) \approx (0.6017, 5.3028)$

9. $(-0.816, 0.665)$, $(1.430, 2.044)$

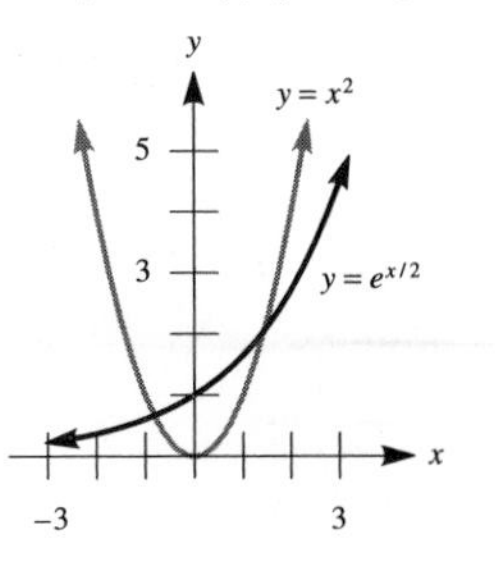

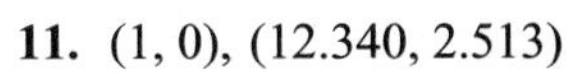

11. $(1, 0)$, $(12.340, 2.513)$

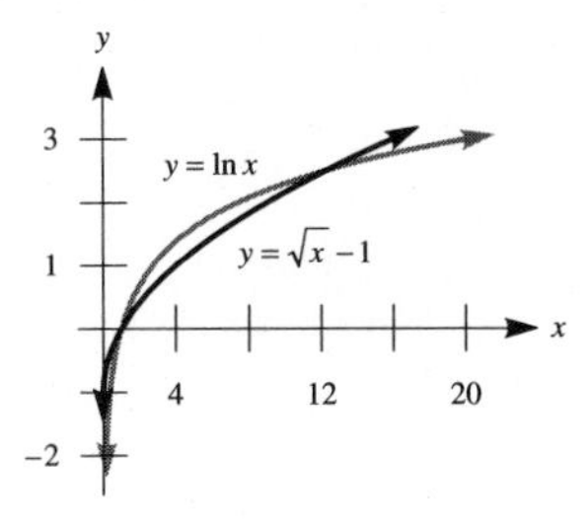

13. $(1.229, 1.855)$, $(6.135, 230.949)$

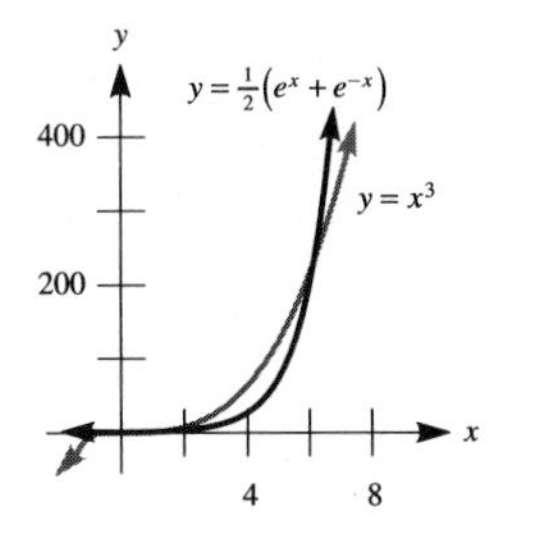

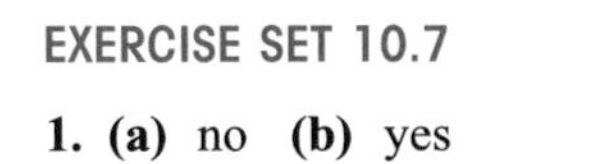

EXERCISE SET 10.7

1. **(a)** no **(b)** yes

3.

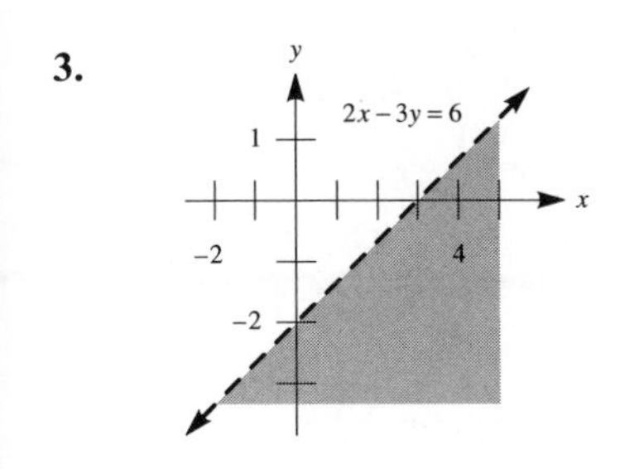

5.

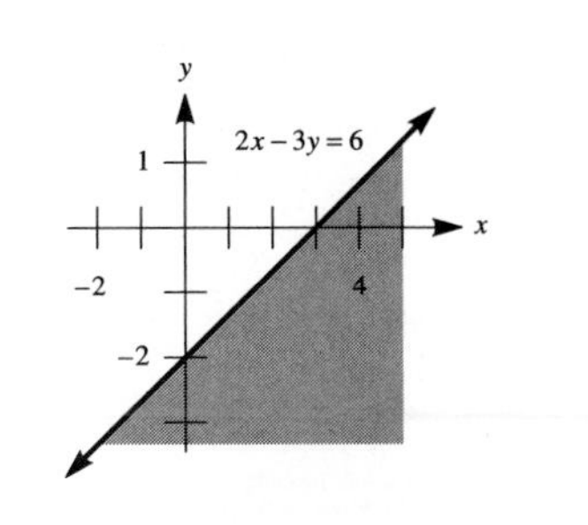

7.

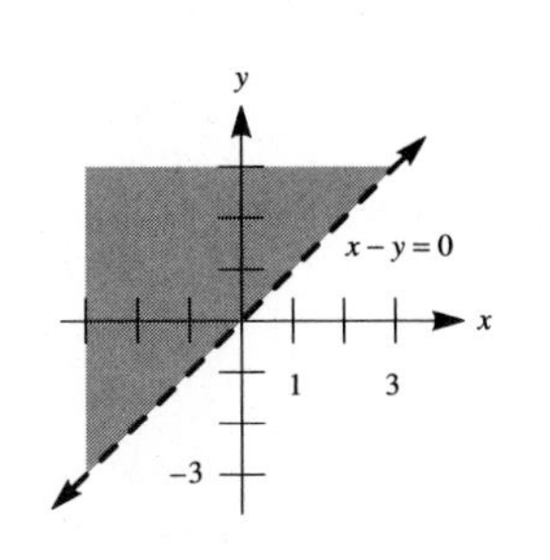

9.

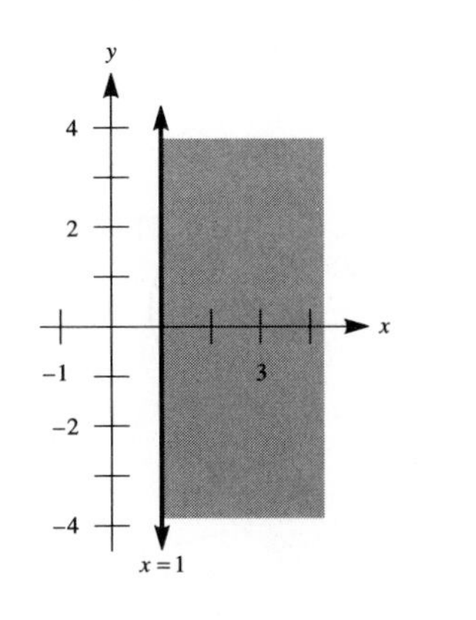

11.

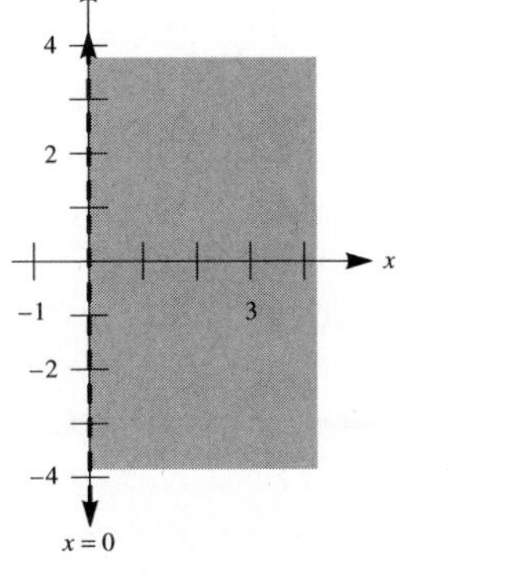

13.

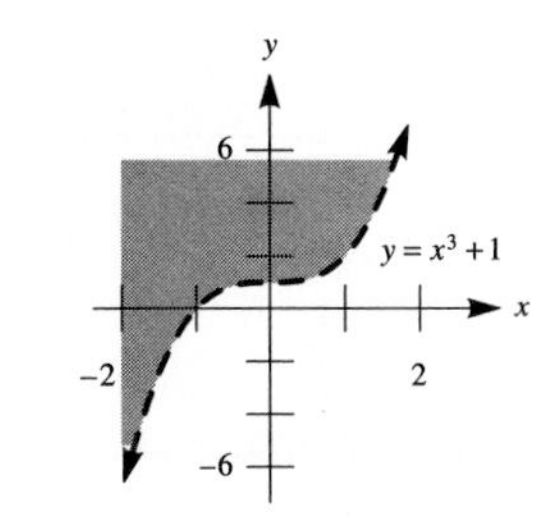

15.

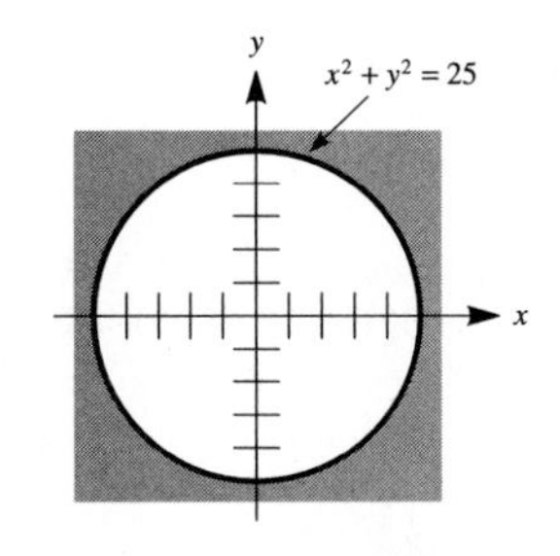

17.

19.

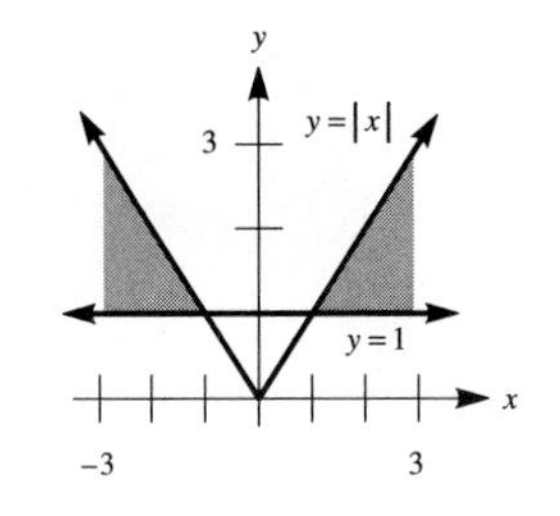

21.

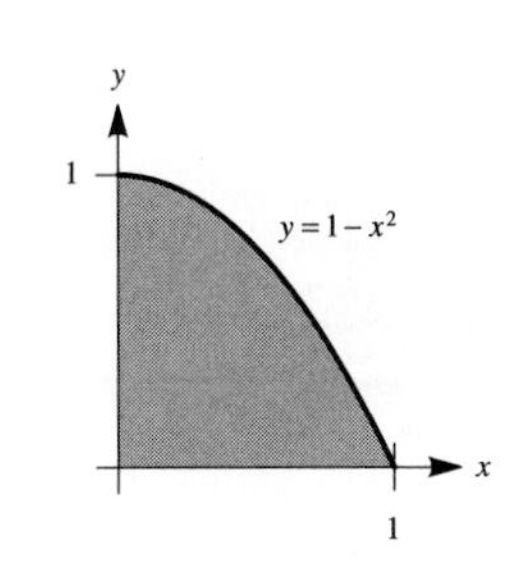

23. convex: yes; bounded: yes; vertices: (0, 0), (7, 0), (3, 8), (0, 5)

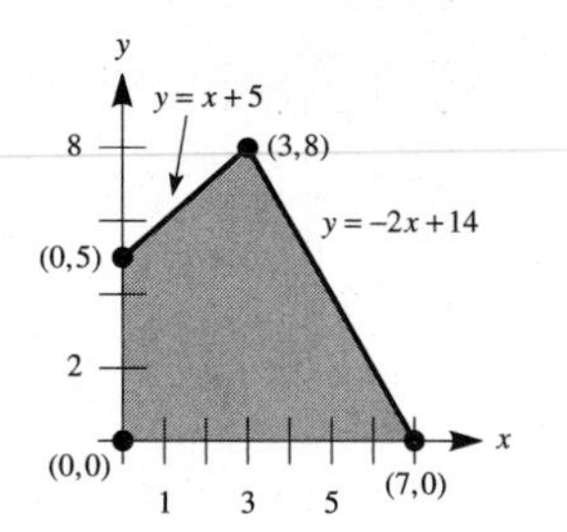

25. convex: yes; bounded: yes; vertices: (0, 0), (0, 4), (3, 5), (8, 0)

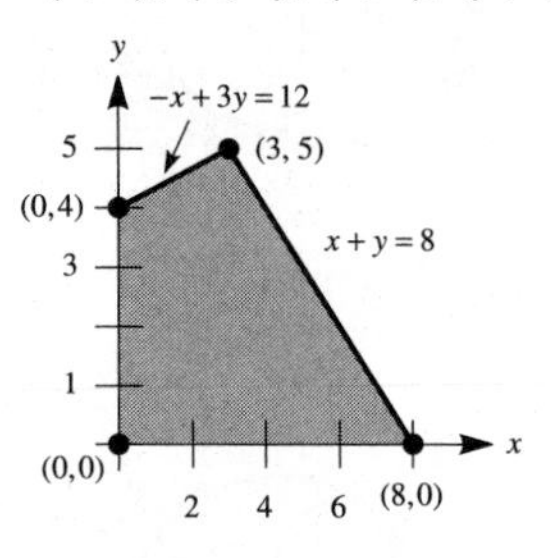

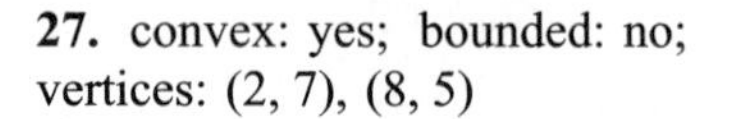
27. convex: yes; bounded: no; vertices: (2, 7), (8, 5)

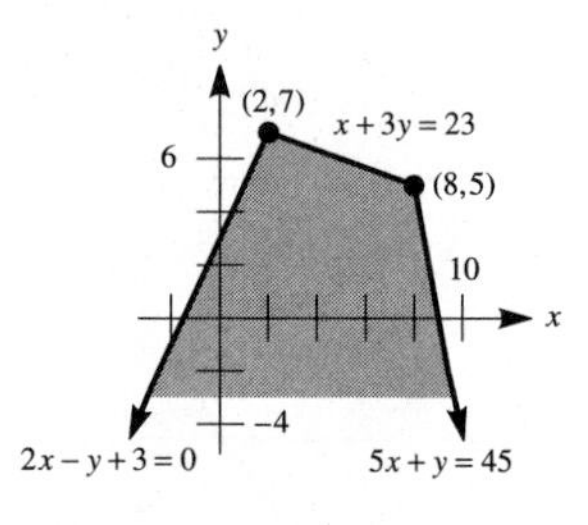

29. convex: yes; bounded: no; vertex: (6, 0)

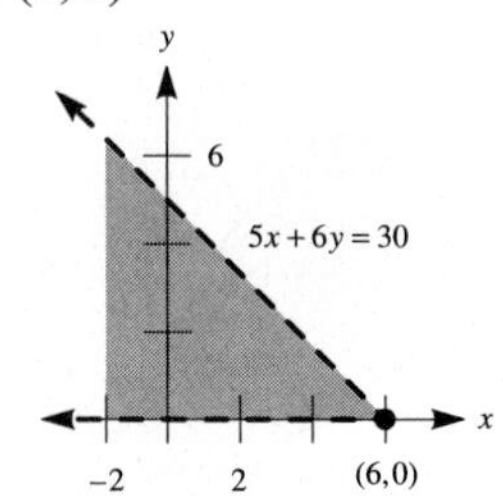

31. convex: yes; bounded: yes; vertices: (0, 0), (0, 5), (6, 0)

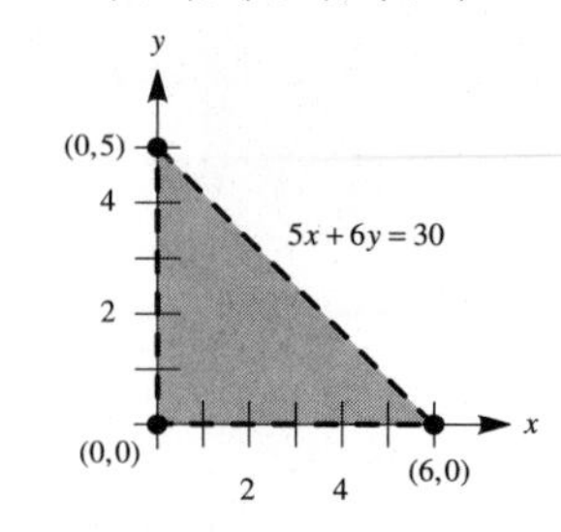

33. convex: yes; bounded: yes; vertices: (5, 30), (10, 30), (20, 15), (20, 20)

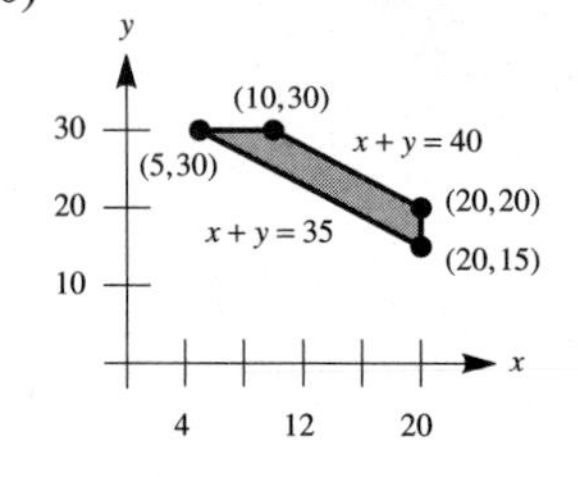

35. vertices: (0, 1), (0, 2), $\left(\ln\frac{1+\sqrt{5}}{2}, \frac{1+\sqrt{5}}{2}\right)$

37.

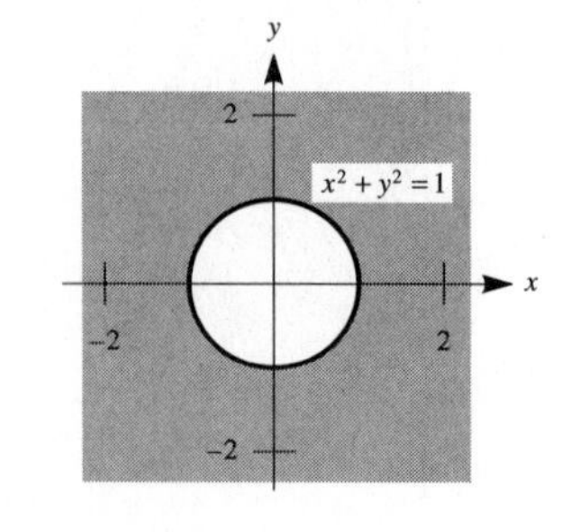

39.

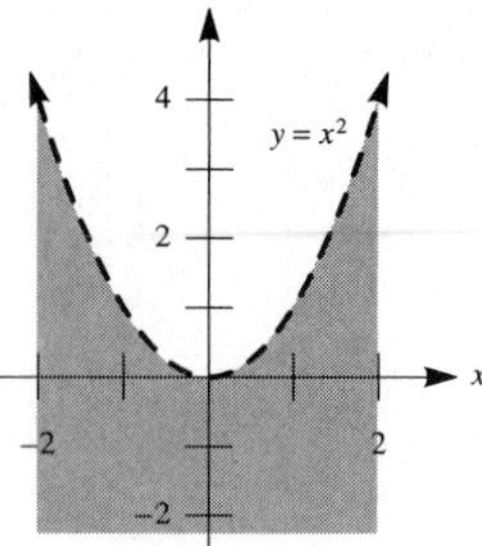

41.

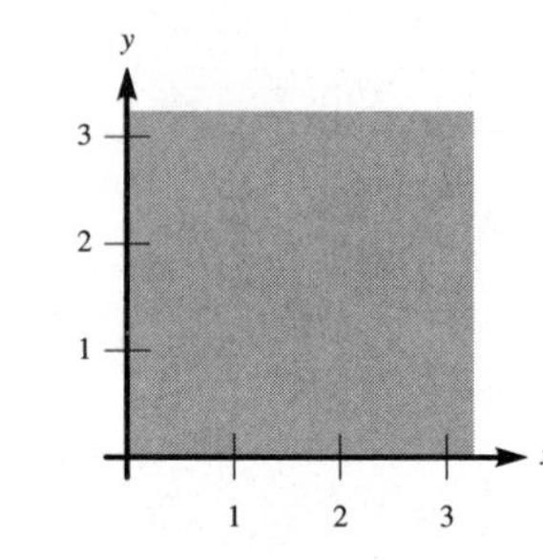

CHAPTER 10 REVIEW EXERCISES

1. $(3, -5)$ **3.** $(-1, 4)$ **5.** $(-3, 15)$ **7.** $\left(-\frac{18}{5}, \frac{8}{5}\right)$ **9.** $\left(\frac{2}{3}, -\frac{1}{5}\right)$ **11.** $(-12, -8)$ **13.** $\left(\frac{1}{3}, -\frac{1}{4}\right)$ **15.** $\left(\frac{-1}{a^2-3a+1}, \frac{1-a}{a^2-3a+1}\right)$, where $a \neq \frac{3 \pm \sqrt{5}}{2}$ **17.** $(a^2, 1-a^2)$ **19.** (a^2-b^2, a^2+b^2) **21.** $\left(\frac{pq(p+q)}{p^2+q^2}, \frac{p^3-q^3}{p^2+q^2}\right)$, where p and q are not both zero **23.** $\left(\frac{a}{a-b}, \frac{b}{a+b}\right)$, where $ab \neq 0$ and $a \neq \pm b$ **25.** $(2, 3, 4)$ **27.** $(-1, -2, 0)$ **29.** no solution (inconsistent) **31.** $(x, 6-2x, -1)$ for any real number x **33.** no solution (inconsistent) **35.** $(2b-z, a-b, z)$ for any real number z **37.** $(4, 3, -1, 2)$ **39.** 34 **41.** -56 **43.** 0 **45.** 24 **51.** $a = 2, b = 1$ **53.** **(a)** $(2/3, 2)$ **(b)** $(2/3, 2)$ **(c)** $(2/3, 2)$ **(d)** Each ratio is 2. **57.** $\left(x+\frac{17}{6}\right)^2 + \left(y+\frac{8}{3}\right)^2 = \frac{245}{36}$ **59.** $\begin{pmatrix} 10 & -2 \\ 4 & 26 \end{pmatrix}$ **61.** $\begin{pmatrix} 8 & 4 \\ 4 & 32 \end{pmatrix}$

63. $\begin{pmatrix} 4 & -13 \\ 7 & 41 \end{pmatrix}$ **65.** $\begin{pmatrix} -3 & -14 \\ -4 & 3 \end{pmatrix}$

67. $\begin{pmatrix} 1 & 1 \\ 1 & 7 \end{pmatrix}$ **69.** $\begin{pmatrix} 1 & -11 \\ 6 & 36 \end{pmatrix}$

71. $\begin{pmatrix} 4 & 3 \\ 10 & 33 \end{pmatrix}$ **73.** $\begin{pmatrix} -42 & 58 \\ 5 & 20 \end{pmatrix}$

75. undefined **77.** undefined

79. $\begin{pmatrix} 4 & -1 \\ 2 & 12 \end{pmatrix}$ **81.** $\begin{pmatrix} -4 & 13 \\ -7 & -41 \end{pmatrix}$

85. **(a)** $\begin{pmatrix} -9 & 5 \\ 2 & -1 \end{pmatrix}$ **(b)** $(-47, 10)$

87. **(a)** $\begin{pmatrix} 10 & -2 & 5 \\ 6 & -1 & 4 \\ 1 & 0 & 1 \end{pmatrix}$ **(b)** $(16, 15, 4)$

89. $\begin{pmatrix} -65 & 20 & 9 & 133 \\ 3 & -1 & 0 & -6 \\ 26 & -8 & -4 & -53 \\ -23 & 7 & 3 & 47 \end{pmatrix}$

91. $D = -20$, $D_x = 12$, $D_y = 13$, $D_z = -31$; $\left(-\frac{3}{5}, -\frac{13}{20}, \frac{31}{20}\right)$

93. $D = 0$; no solution (inconsistent)

95. $D = 0$; $(-5 - 4y, y, -8 - 5y)$ for any real number y

97. $D = 149$, $D_x = 596$, $D_y = 149$, $D_z = -149$, $D_w = 447$; $(4, 1, -1, 3)$

99. $(0, 0)$, $(6, 36)$ **101.** $(3, 0)$, $(-3, 0)$

103. $\left(\frac{5\sqrt{2}}{2}, \frac{\sqrt{14}}{2}\right)$, $\left(-\frac{5\sqrt{2}}{2}, \frac{\sqrt{14}}{2}\right)$, $\left(\frac{5\sqrt{2}}{2}, -\frac{\sqrt{14}}{2}\right)$, $\left(-\frac{5\sqrt{2}}{2}, -\frac{\sqrt{14}}{2}\right)$

105. $\left(\frac{-1+\sqrt{5}}{2}, \sqrt{\frac{-1+\sqrt{5}}{2}}\right)$ or $\left(\frac{-1+\sqrt{5}}{2}, \frac{\sqrt{-2+2\sqrt{5}}}{2}\right)$

107. $\left(\frac{\sqrt{-2+2\sqrt{17}}}{4}, \frac{-1+\sqrt{17}}{4}\right)$, $\left(\frac{-\sqrt{-2+2\sqrt{17}}}{4}, \frac{-1+\sqrt{17}}{4}\right)$

109. $(\sqrt{2}, 3\sqrt{2})$, $(-\sqrt{2}, -3\sqrt{2})$, $\left(\frac{7\sqrt{22}}{33}, \frac{31\sqrt{22}}{33}\right)$, $\left(\frac{-7\sqrt{22}}{33}, \frac{-31\sqrt{22}}{33}\right)$

111. $(5, 2)$, $(5, -4)$, $(1, 2)$, $(1, -4)$

113. $\frac{as}{a+b}$ and $\frac{bs}{a+b}$

115. 24, 60, 120

117. $\frac{\sqrt{n+2m}+\sqrt{n-2m}}{2}$ and $\frac{\sqrt{n+2m}-\sqrt{n-2m}}{2}$; $\frac{\sqrt{n-2m}-\sqrt{n+2m}}{2}$ and $\frac{-\sqrt{n-2m}-\sqrt{n+2m}}{2}$

119. convex: no; bounded: no

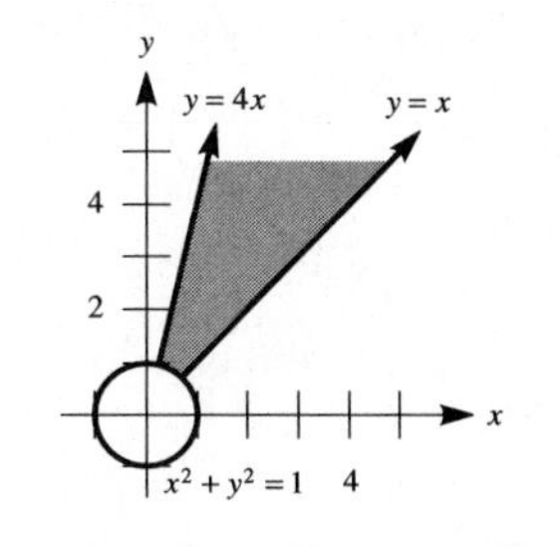

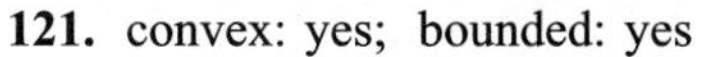

121. convex: yes; bounded: yes

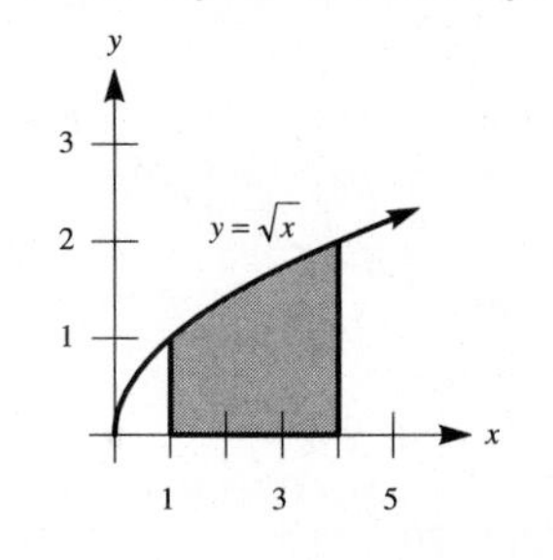

123. convex: no; bounded: no

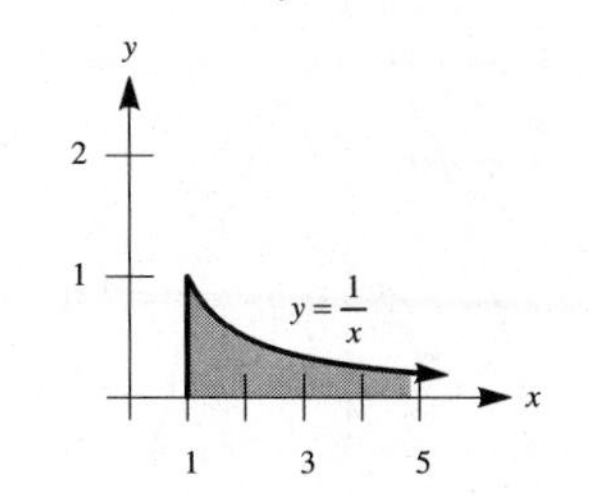

CHAPTER 10 TEST

1. $(0, 3)$, $\left(-\frac{11}{4}, \frac{81}{16}\right)$ **2.** $(3, -5)$

3. **(a)** $(1, -1, -3)$ **(b)** $D = -19$, $D_x = -19$, $D_y = 19$, $D_z = 57$; $(1, -1, -3)$

4. **(a)** $\begin{pmatrix} 2 & -10 \\ 3 & -5 \end{pmatrix}$ **(b)** $\begin{pmatrix} 8 & -4 \\ 7 & -6 \end{pmatrix}$

5. 12 sq. units **6.** $\left(\frac{1}{116}, -\frac{1}{144}\right)$

7. coefficient matrix: $\begin{pmatrix} 1 & 1 & -1 \\ 2 & -1 & 2 \\ 1 & -2 & 1 \end{pmatrix}$; augmented matrix: $\begin{pmatrix} 1 & 1 & -1 & -1 \\ 2 & -1 & 2 & 11 \\ 1 & -2 & 1 & 10 \end{pmatrix}$

8. $(3, -3, 1)$ **9.** $y = -2x - 4$

10. **(a)** 8 **(b)** -8 **11.** -1120

12. $\left(\frac{5+\sqrt{5}}{2}, \frac{5-\sqrt{5}}{2}\right)$, $\left(\frac{5-\sqrt{5}}{2}, \frac{5+\sqrt{5}}{2}\right)$, $\left(\frac{-5+\sqrt{5}}{2}, \frac{-5-\sqrt{5}}{2}\right)$, $\left(\frac{-5-\sqrt{5}}{2}, \frac{-5+\sqrt{5}}{2}\right)$

13. $\left(1 - \frac{1}{5}C, -\frac{7}{5}C, C\right)$, where C is any real number

14. **(a)** $\begin{pmatrix} 1 & -2 & 3 \\ 2 & -5 & 10 \\ -1 & 2 & -2 \end{pmatrix}$ **(b)** $(12, 38, -9)$

15.

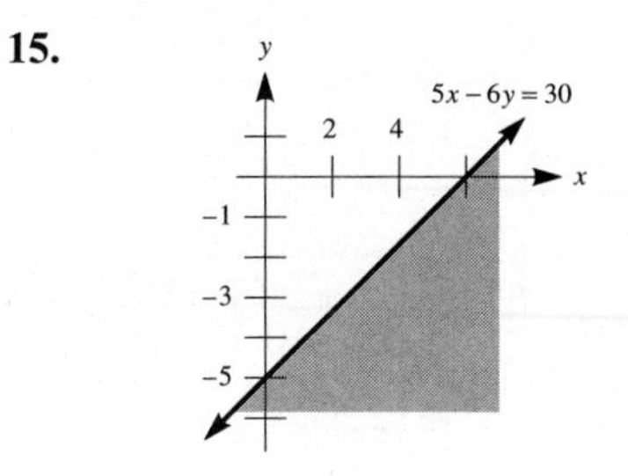

16. $P = -1$, $Q = -4$
17. bounded: no; convex: no

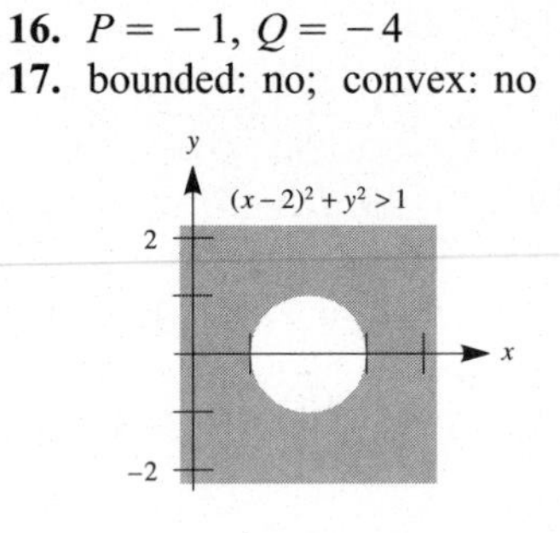

18. vertices: (0, 0), (0, 7), (6, 10), $\left(\frac{261}{26}, \frac{225}{26}\right)$, (11, 0)

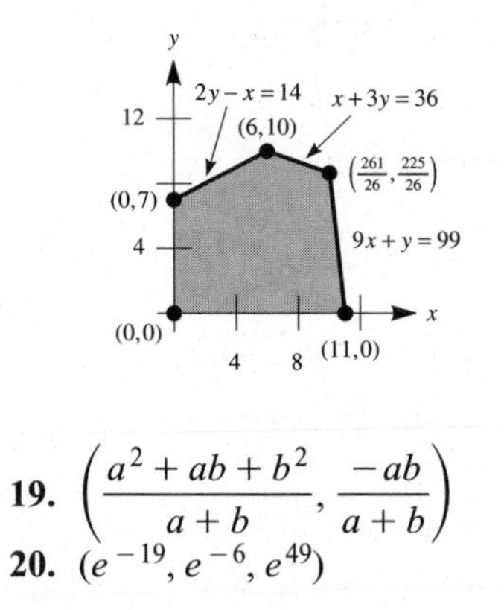

19. $\left(\dfrac{a^2 + ab + b^2}{a + b}, \dfrac{-ab}{a + b}\right)$
20. $(e^{-19}, e^{-6}, e^{49})$

43. (a) slope: $-\frac{A}{B}$; y-intercept: $-\frac{C}{B}$
47. $\left(x - \frac{26}{11}\right)^2 + \left(y + \frac{25}{11}\right)^2 = \frac{4964}{121}$
49. $\left(x - 4 - \dfrac{10\sqrt{13}}{13}\right)^2 + \left(y - 6 - \dfrac{15\sqrt{13}}{13}\right)^2 = 25$; $\left(x - 4 + \dfrac{10\sqrt{13}}{13}\right)^2 + \left(y - 6 + \dfrac{15\sqrt{13}}{13}\right)^2 = 25$

CHAPTER ELEVEN

EXERCISE SET 11.1

1. $\sqrt{89}$ **3.** $5x + 4y - 4 = 0$
5. $2x + 3y - 20 = 0$
7. x-intercepts: 6, -4; y-intercepts: $\pm 2\sqrt{6}$ **9.** $13x - 7y - 35 = 0$
11. $12 + \sqrt{74}$ **13.** $e = \frac{\pi}{3}$ or 60°
15. (a) $\theta = 1.37$ or 78.69°
(b) $\theta = 1.77$ or 101.31°
17. (a) $\dfrac{5\sqrt{2}}{2}$ **(b)** $\dfrac{5\sqrt{2}}{2}$
19. (a) $\dfrac{19\sqrt{41}}{41}$ **(b)** $\dfrac{19\sqrt{41}}{41}$
21. (a) $(x + 2)^2 + (y + 3)^2 = \frac{361}{13}$
(b) $\sqrt{5}$ **23.** $\frac{65}{2}$ **25.** $\dfrac{15 \pm 2\sqrt{30}}{5}$
27. $\frac{12}{5}$ **29.** $y = -3x + 12$
31. $y = \frac{1}{3}x + \frac{9}{2}$
33. (a) center: $(-5, 2)$; radius: $5\sqrt{2}$
35. $PQ = 5\sqrt{2}$ **39. (a)** A: $y = \frac{1}{3}x$; B: $y = -x + 8$; C: $y = 2x - 10$
(b) (6, 2)

EXERCISE SET 11.2

1. focus: (0, 1); directrix: $y = -1$; focal width: 4

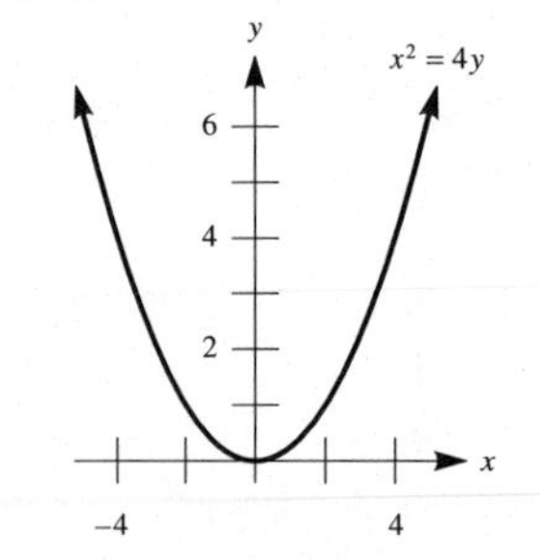

3. focus: $(-2, 0)$; directrix: $x = 2$; focal width: 8

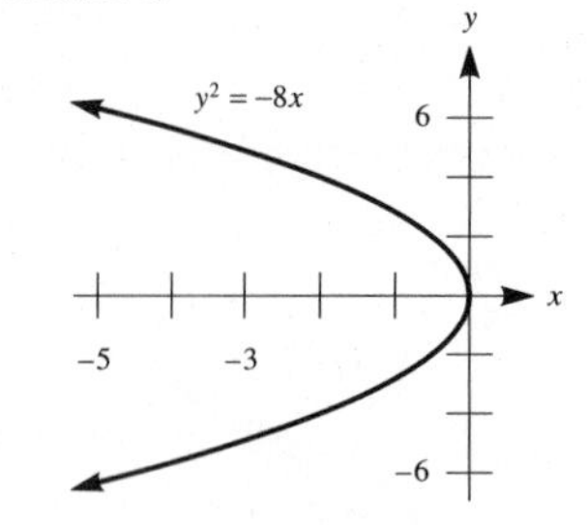

5. focus: $(0, -5)$; directrix: $y = 5$; focal width: 20

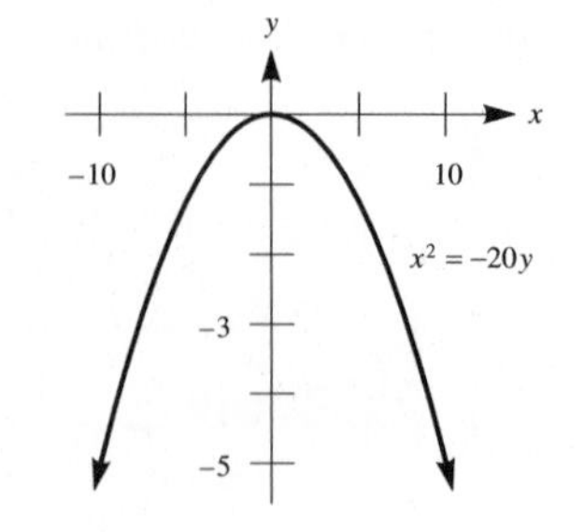

7. focus: $(-7, 0)$; directrix: $x = 7$; focal width: 28

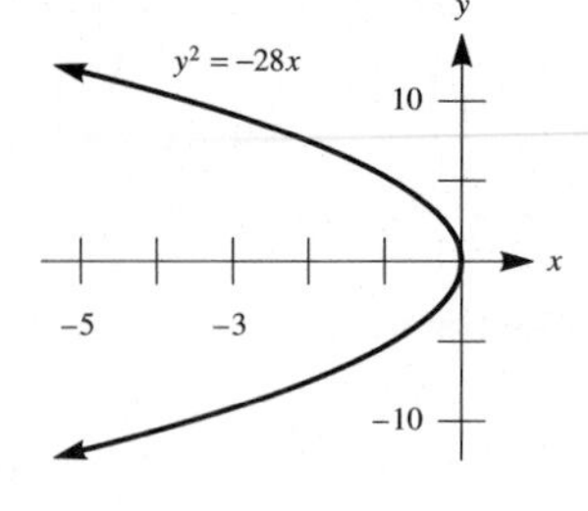

9. focus: $\left(0, \frac{3}{2}\right)$; directrix: $y = -\frac{3}{2}$; focal width: 6

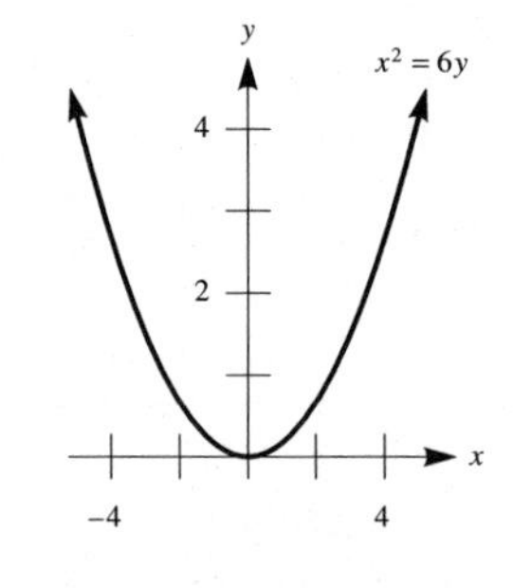

11. focus: $\left(0, \frac{7}{16}\right)$; directrix: $y = -\frac{7}{16}$; focal width: $\frac{7}{4}$

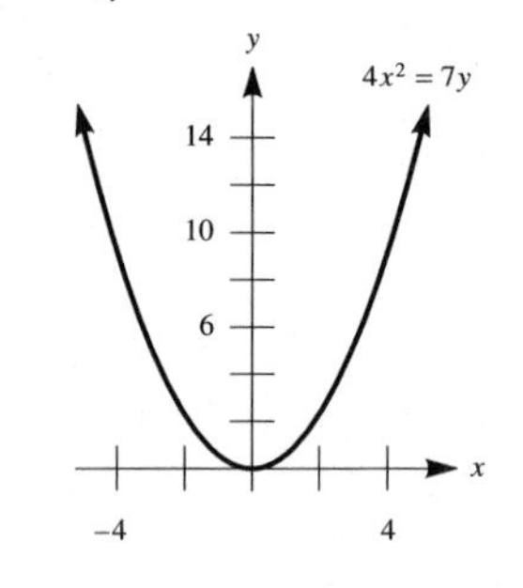

13. vertex: (2, 3); focus: (3, 3); directrix: $x = 1$; focal width: 4

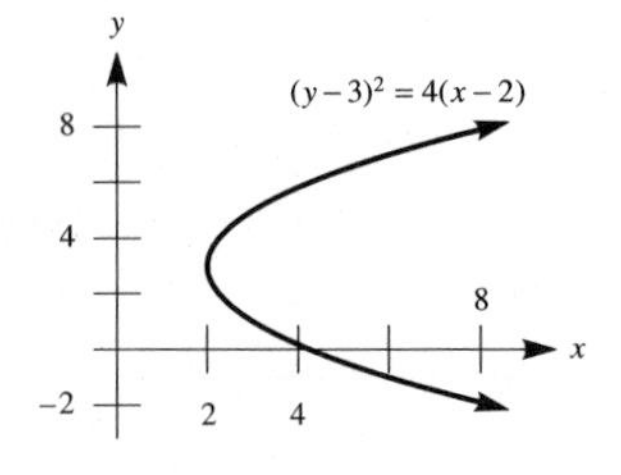

15. vertex: (4, 2); focus: $\left(4, \frac{9}{4}\right)$; directrix: $y = \frac{7}{4}$; focal width: 1

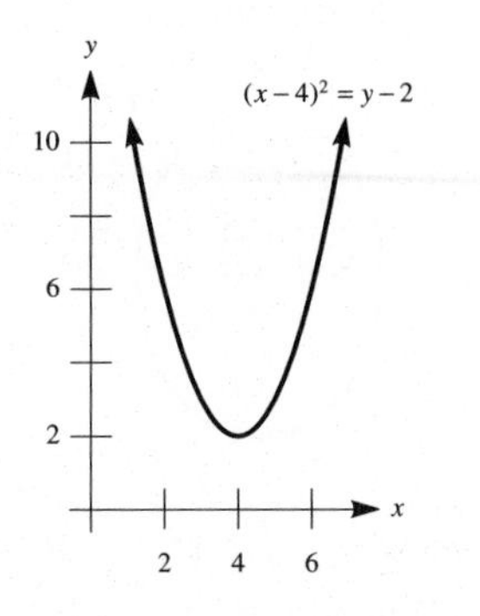

17. vertex: $(0, -1)$; focus: $\left(\frac{1}{4}, -1\right)$; directrix: $x = -\frac{1}{4}$; focal width: 1

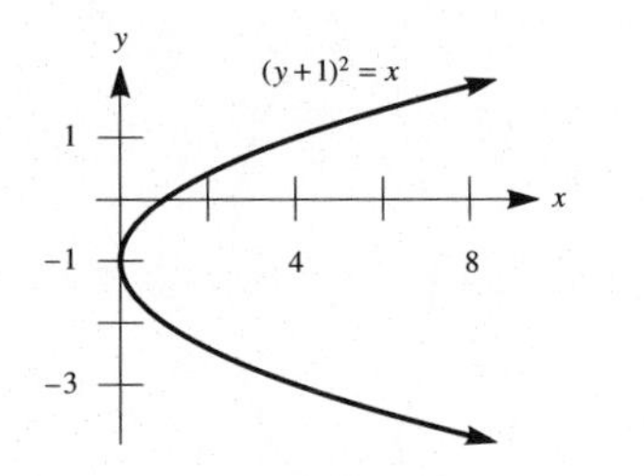

19. vertex: (3, 0); focus: $\left(3, \frac{1}{8}\right)$; directrix: $y = -\frac{1}{8}$; focal width: $\frac{1}{2}$

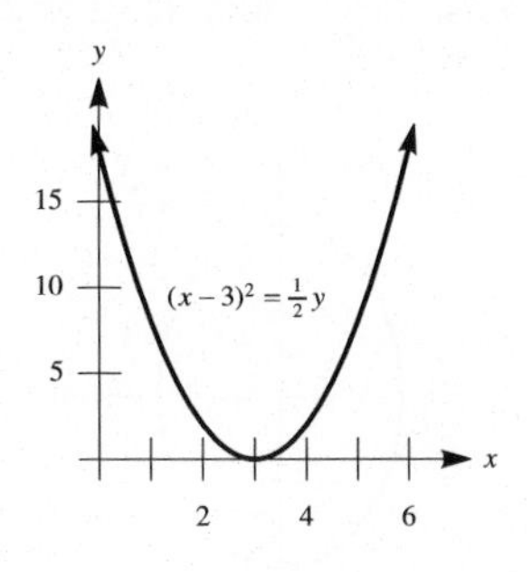

21. vertex: (4, 1); focus: $\left(4, \frac{9}{8}\right)$; directrix: $y = \frac{7}{8}$; focal width: $\frac{1}{2}$

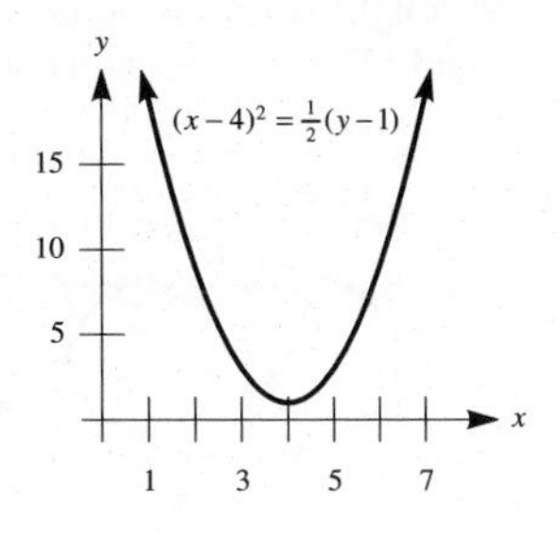

23. $y = 1$

25. $x^2 = 12y$ **27.** $y^2 = 128x$
29. $y^2 = -9x$
31. (a) $y = \frac{3}{4}x + 2$ **(b)** $\left(-2, \frac{1}{2}\right)$
(c) $\frac{25}{2}$ **(d)** $(x-3)^2 + \left(y - \frac{17}{4}\right)^2 = \frac{625}{16}$

(e)

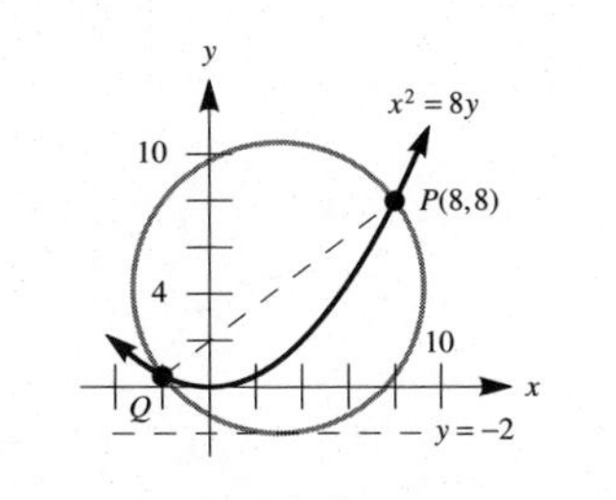

33. 11.25 ft **35. (a)** $y = \frac{9}{4}x - 1$
(b) $y = -\frac{9}{8}x - \frac{1}{4}$
37. (a) $\left(-\frac{1}{8}, \frac{1}{64}\right)$ **(b)** $\left(\frac{15}{16}, \frac{257}{128}\right)$ **(c)** $\frac{1}{2}$
39. length of side: $8\sqrt{3}p$ units; area: $48\sqrt{3}p^2$ sq. units

GRAPHING UTILITY EXERCISES FOR SECTION 11.2

1.

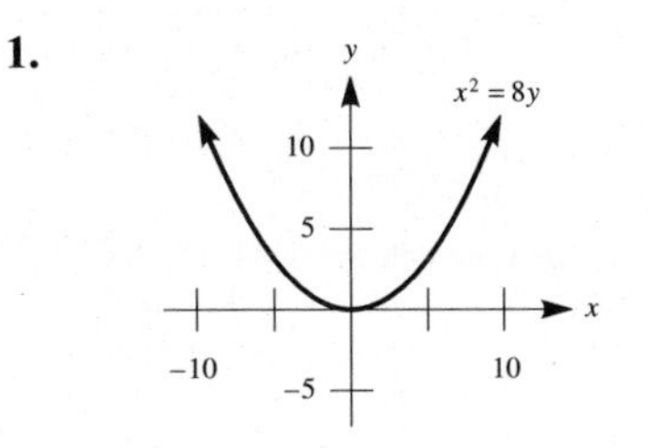

3.

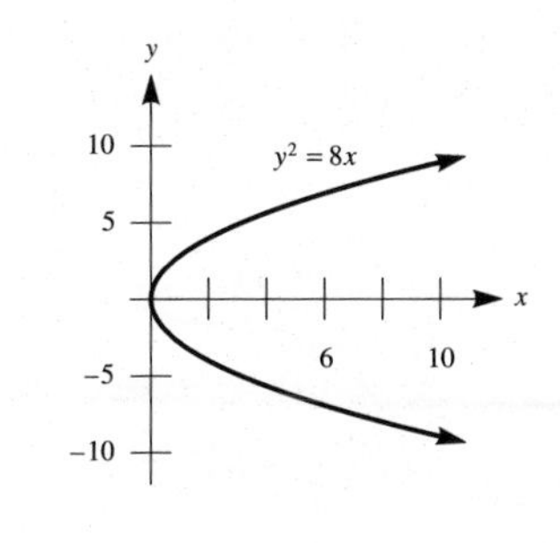

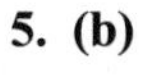

5. (b)

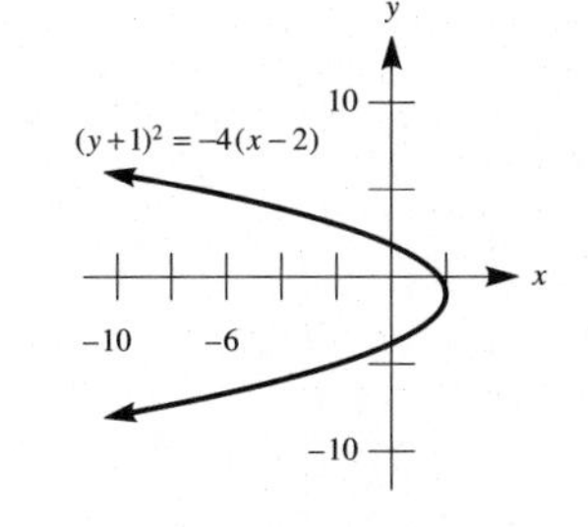

7. $y = x - 2$

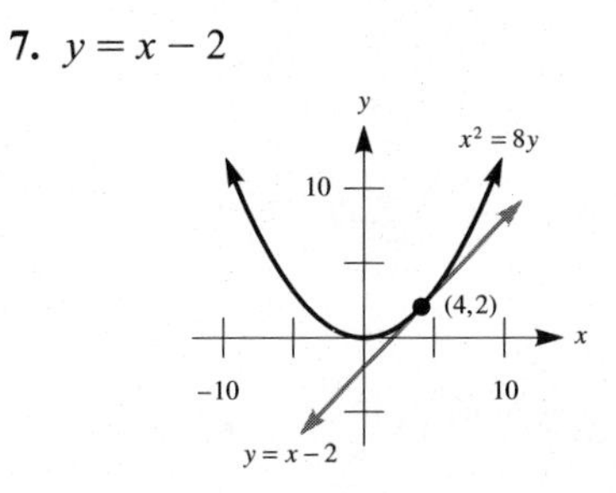

9.

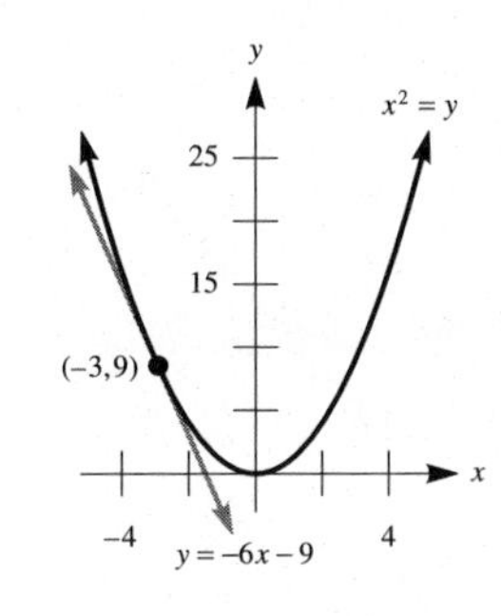

11. intersection points: (0, 0) and (5, 5)

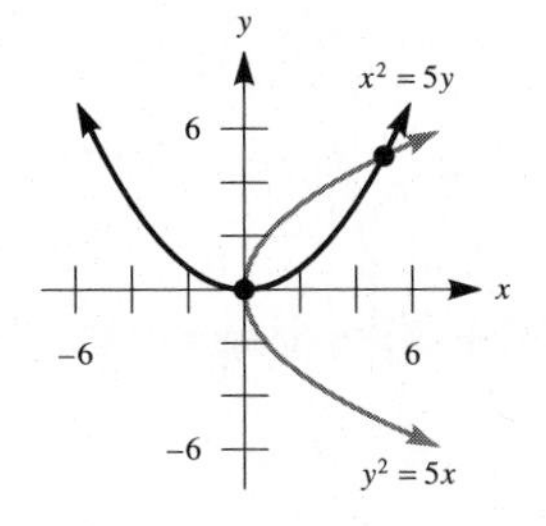

13. intersection points: (0.67, −0.41) and (2.04, 0.71)

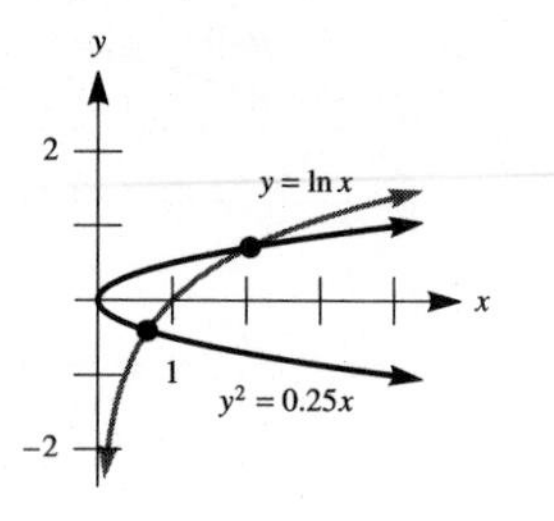

EXERCISE SET 11.3

1. $y = 4x - 4$

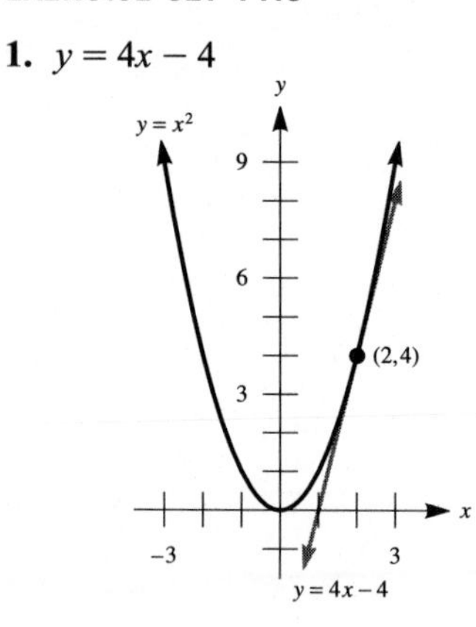

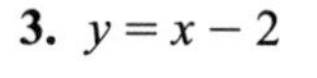

3. $y = x - 2$

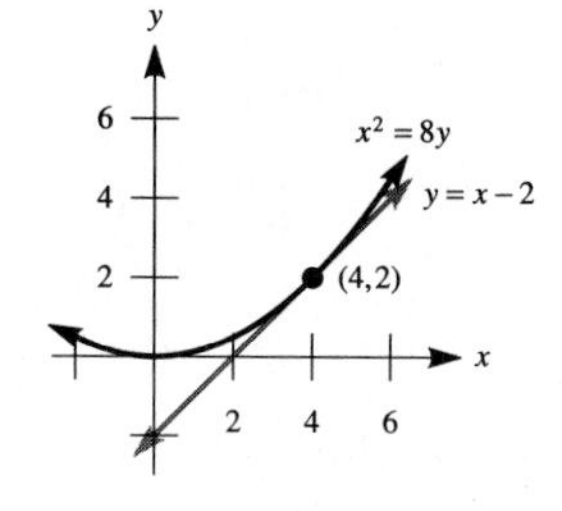

5. $y = 6x + 9$

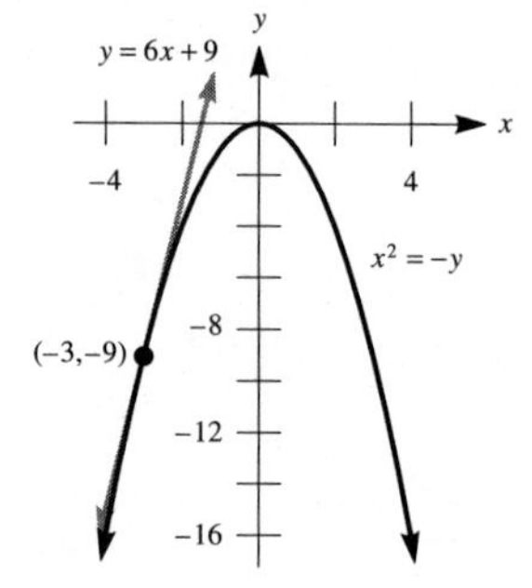

7. $y = x + 1$

9. $m = \frac{1}{4}$ **11.** $m = 12$

EXERCISE SET 11.4

1. length of major axis: 6; length of minor axis: 4; foci: $(\pm\sqrt{5}, 0)$; eccentricity: $\sqrt{5}/3$

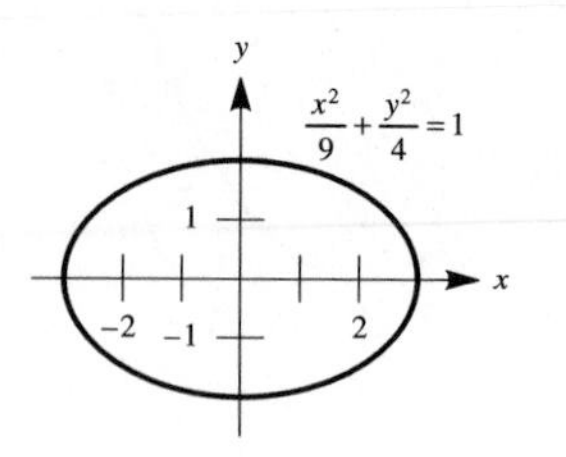

3. length of major axis: 8; length of minor axis: 2; foci: $(\pm\sqrt{15}, 0)$; eccentricity: $\sqrt{15}/4$

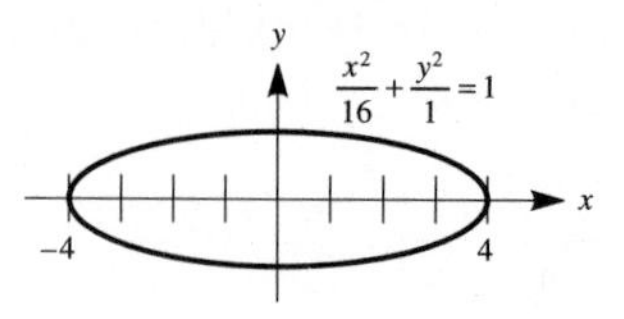

5. length of major axis: $2\sqrt{2}$; length of minor axis: 2; foci: $(\pm 1, 0)$; eccentricity: $\sqrt{2}/2$

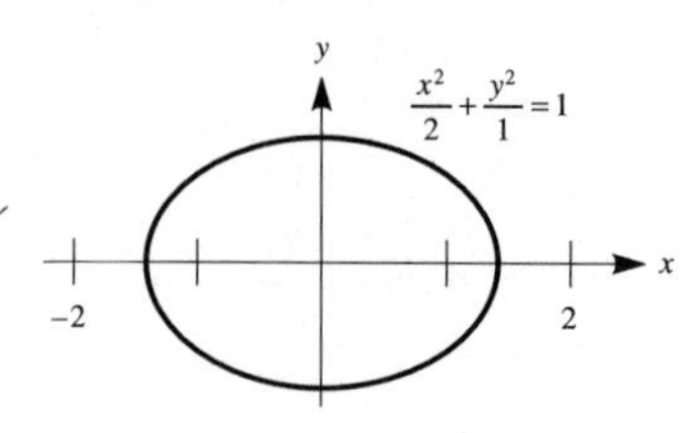

7. length of major axis: 8; length of minor axis: 6; foci: $(0, \pm\sqrt{7})$; eccentricity: $\sqrt{7}/4$

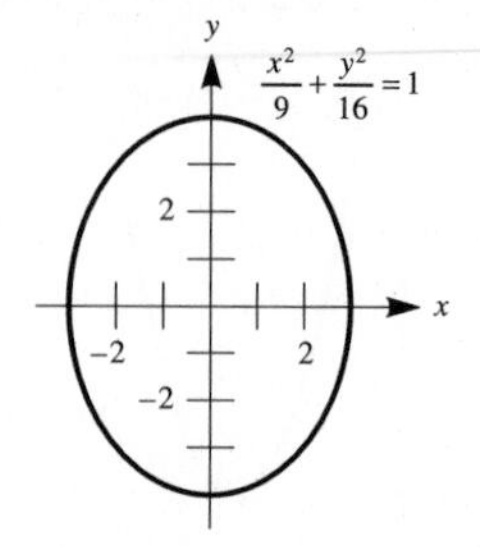

9. length of major axis: $2\sqrt{15}/3$; length of minor axis: $2\sqrt{3}/3$; foci: $(0, \pm 2\sqrt{3}/3)$; eccentricity: $2\sqrt{5}/5$

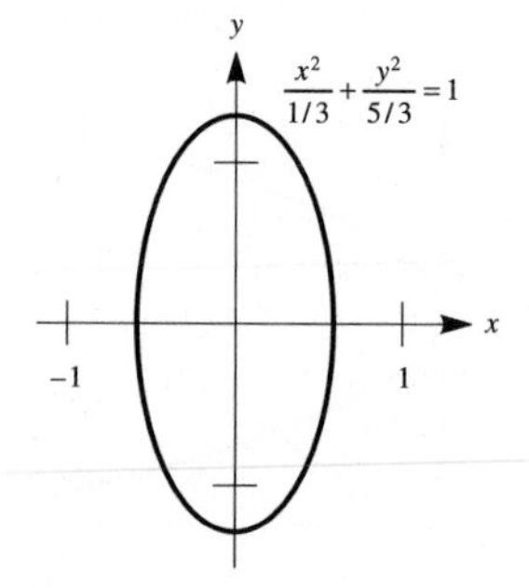

11. length of major axis: 4; length of minor axis: $2\sqrt{2}$; foci: $(0, \pm\sqrt{2})$; eccentricity: $\sqrt{2}/2$

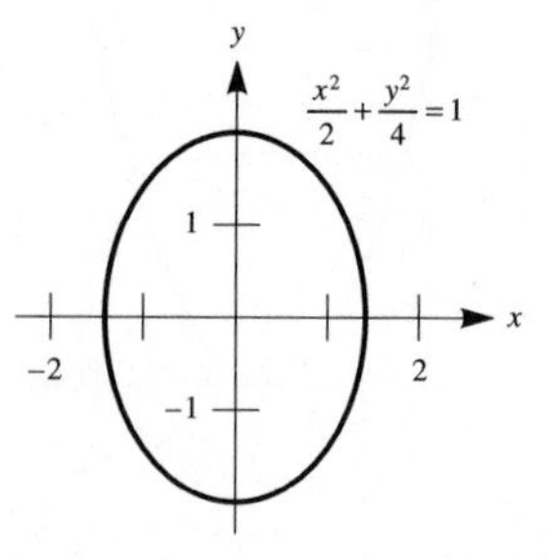

13. center: (5, −1); length of major axis: 10; length of minor axis: 6; foci: (9, −1) and (1, −1); eccentricity: 4/5

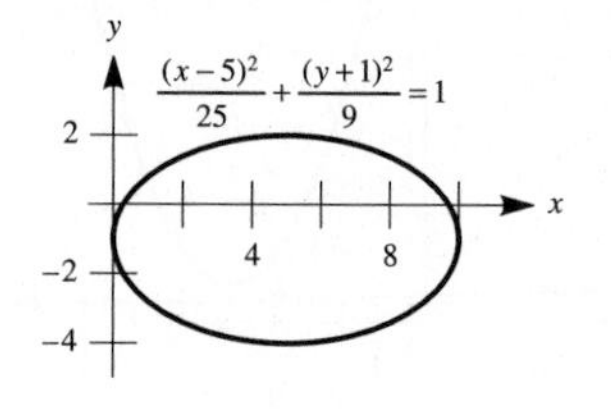

15. center: $(1, 2)$; length of major axis: 4; length of minor axis: 2; foci: $(1, 2 \pm \sqrt{3})$; eccentricity: $\sqrt{3}/2$

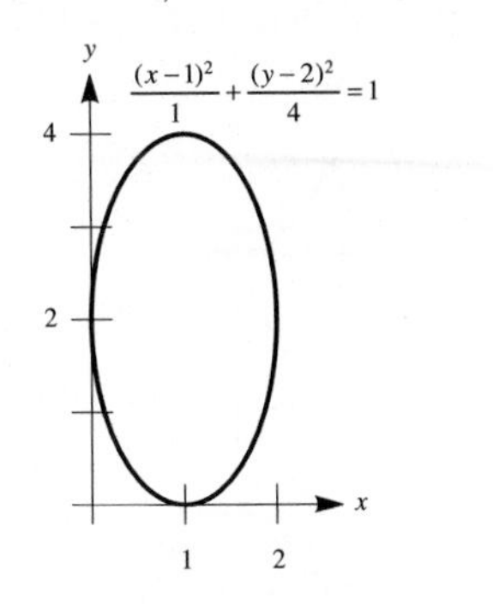

17. center: $(-3, 0)$; length of major axis: 6; length of minor axis: 2; foci: $(-3 \pm 2\sqrt{2}, 0)$; eccentricity: $2\sqrt{2}/3$

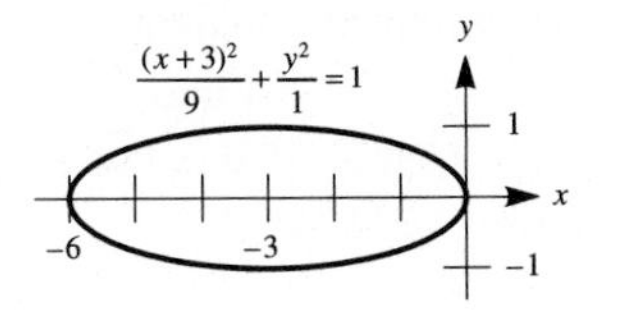

19. center: $(1, -2)$; length of major axis: 4; length of minor axis: $2\sqrt{3}$; foci: $(2, -2)$ and $(0, -2)$; eccentricity: $1/2$

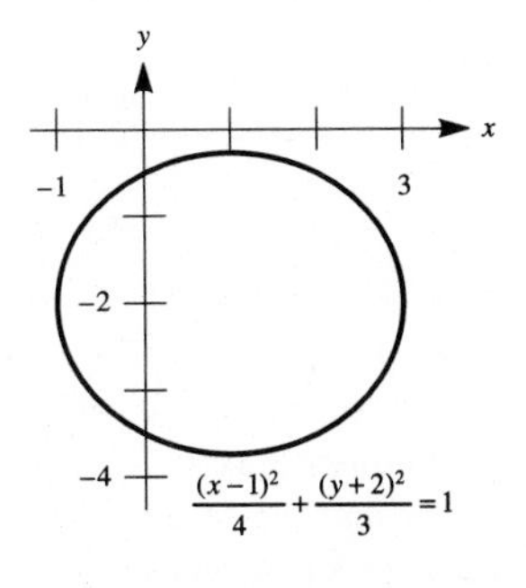

21. center: $(4, 6)$; degenerate ellipse (point)

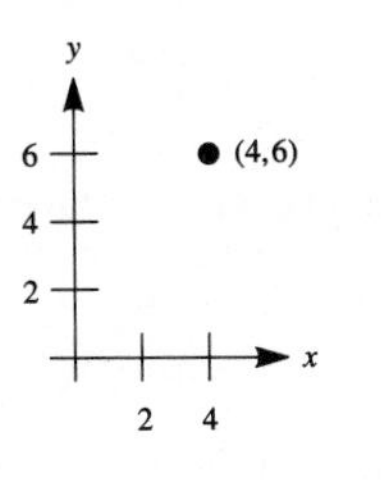

23. no graph

25. $\dfrac{x^2}{25} + \dfrac{y^2}{16} = 1$; $16x^2 + 25y^2 = 400$

27. $\dfrac{x^2}{16} + \dfrac{y^2}{15} = 1$; $15x^2 + 16y^2 = 240$

29. $\dfrac{x^2}{21} + \dfrac{y^2}{25} = 1$; $25x^2 + 21y^2 = 525$

31. $\dfrac{x^2}{9} + \dfrac{y^2}{9/4} = 1$; $x^2 + 4y^2 = 9$

33. **(a)** $y = -\frac{4}{3}x + \frac{38}{3}$ **(b)** $y = \frac{7}{9}x + \frac{76}{9}$ **(c)** $y = \frac{1}{15}x - \frac{76}{15}$

35. **(a)** $y = -6x + 26$ **(b)** $\frac{169}{3}$ sq. units

37. **(b)** $3x + 10\sqrt{6}y - 75 = 0$ **(c)** $d_1 = 87/\sqrt{609}$, $d_2 = 63/\sqrt{609}$

39.

x	0	0.1	0.2	0.3	0.4
y	± 4	± 3.98	± 3.92	± 3.82	± 3.67

x	0.5	0.6	0.7	0.8	0.9	1.0
y	± 3.46	± 3.2	± 2.86	± 2.4	± 1.74	0

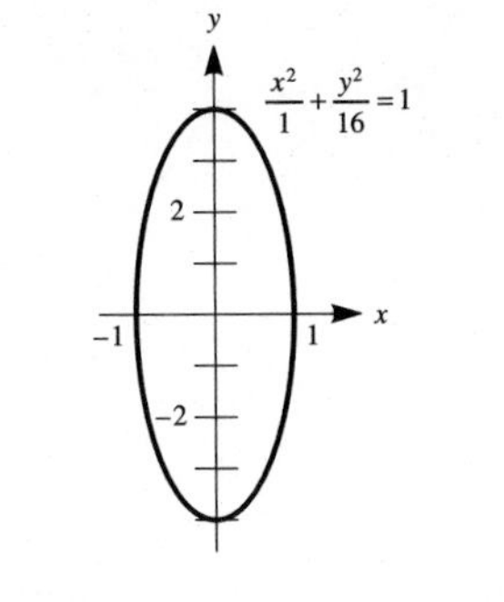

45. $\left(\dfrac{ab}{A}, \dfrac{ab}{A}\right), \left(\dfrac{ab}{A}, -\dfrac{ab}{A}\right), \left(-\dfrac{ab}{A}, \dfrac{ab}{A}\right), \left(-\dfrac{ab}{A}, -\dfrac{ab}{A}\right)$, where $A = \sqrt{a^2 + b^2}$

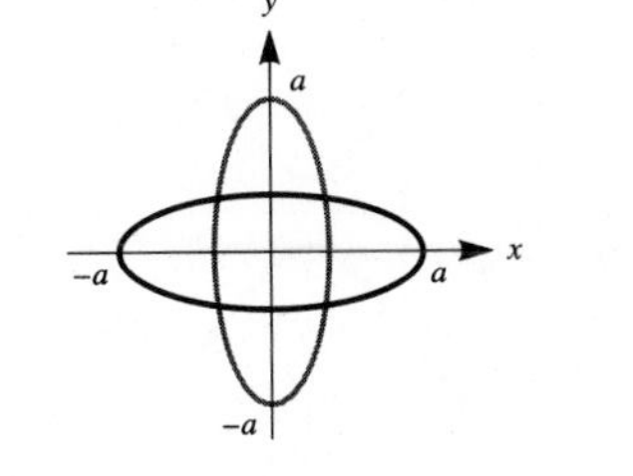

49. **(b)** $D\left(-\frac{20}{7}, -\frac{18}{7}\right)$

59. **(b)** $\left(\frac{a}{e}, 0\right)$

GRAPHING UTILITY EXERCISES FOR SECTION 11.4

1. **(a)** symmetry: x-axis, y-axis, origin

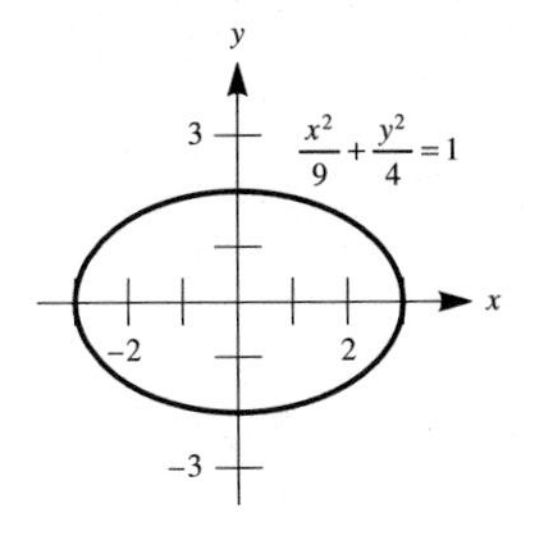

(b) symmetry: x-axis, y-axis, origin

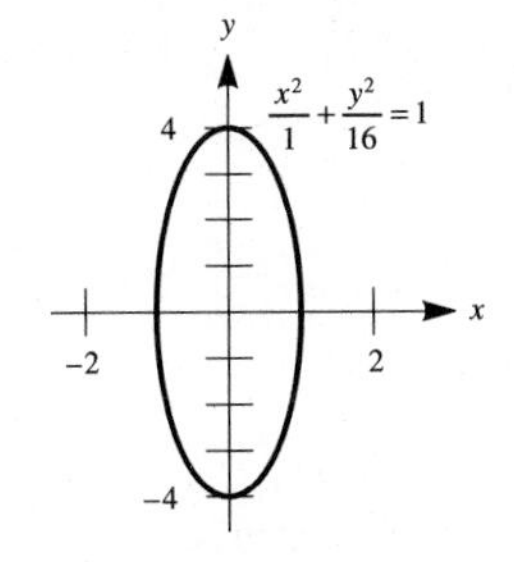

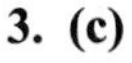
3. **(c)**

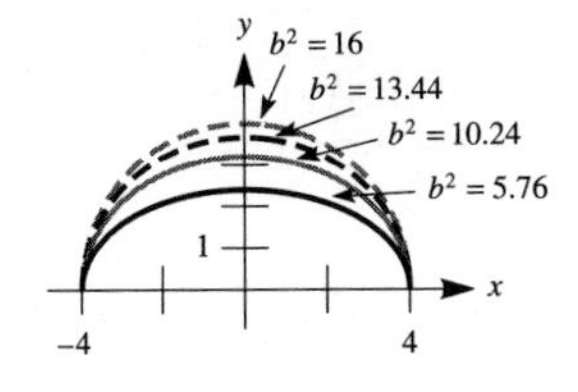

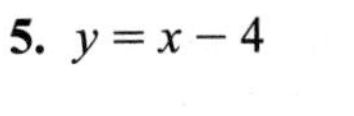
5. $y = x - 4$

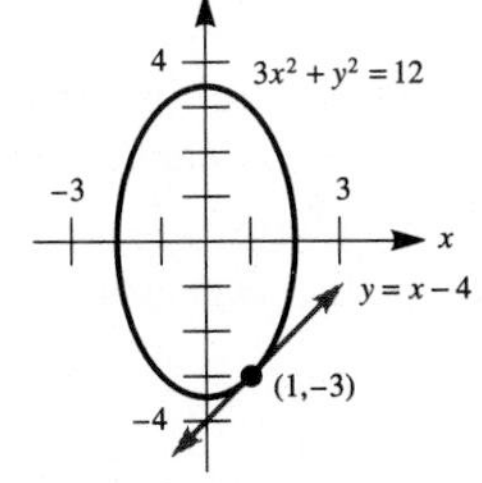

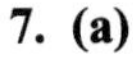
7. **(a)**

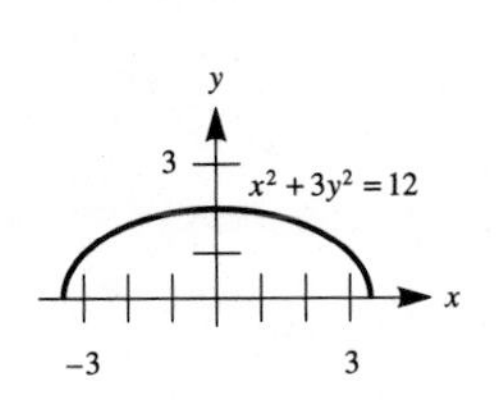

(b) $a = 2\sqrt{3}$, $b = 2$, $c = 2\sqrt{2}$
(c) $x^2 + y^2 = 12$

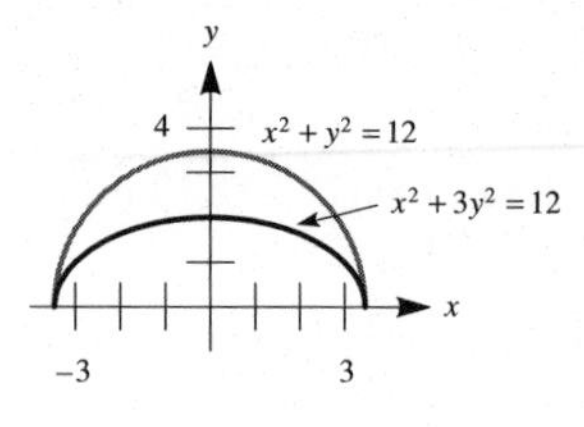

(d) $y = -x + 4$

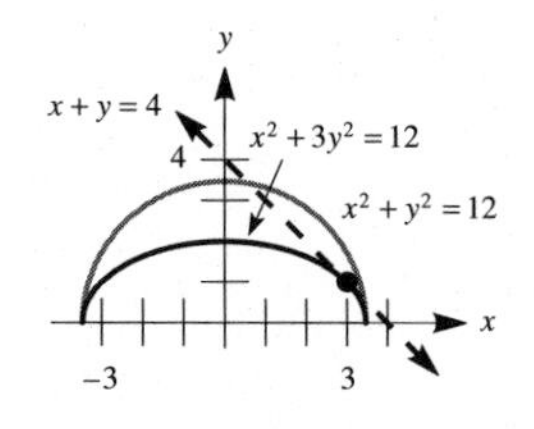

(e) $y = x + 2\sqrt{2}$

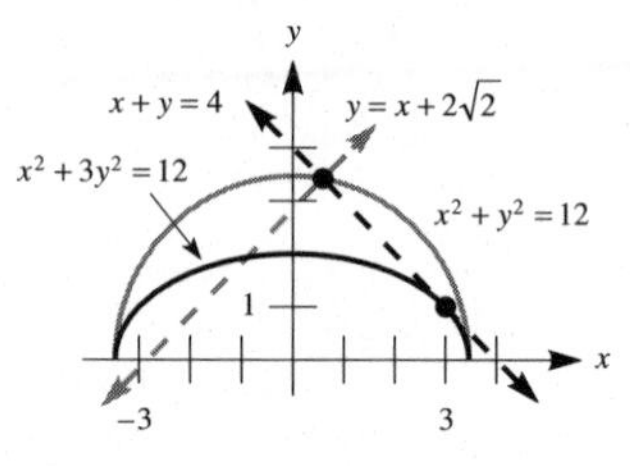

EXERCISE SET 11.5

1. vertices: $(\pm 2, 0)$; length of transverse axis: 4; length of conjugate axis: 2; asymptotes: $y = \pm\frac{1}{2}x$; foci: $\left(\pm\sqrt{5}, 0\right)$; eccentricity: $\sqrt{5}/2$

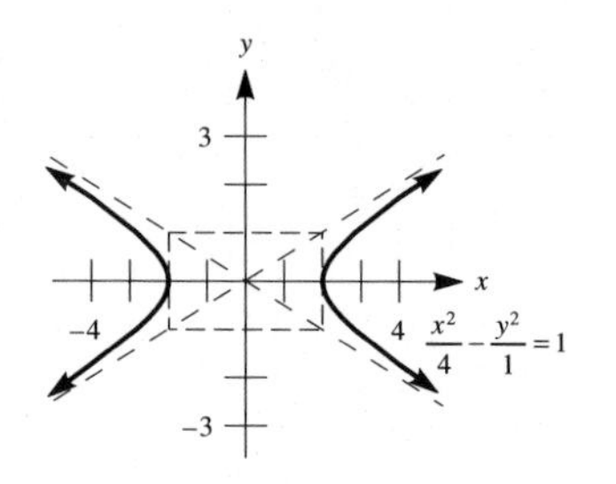

3. vertices: $(0, \pm 2)$; length of transverse axis: 4; length of conjugate axis: 2; asymptotes: $y = \pm 2x$; foci: $\left(0, \pm\sqrt{5}\right)$; eccentricity: $\sqrt{5}/2$

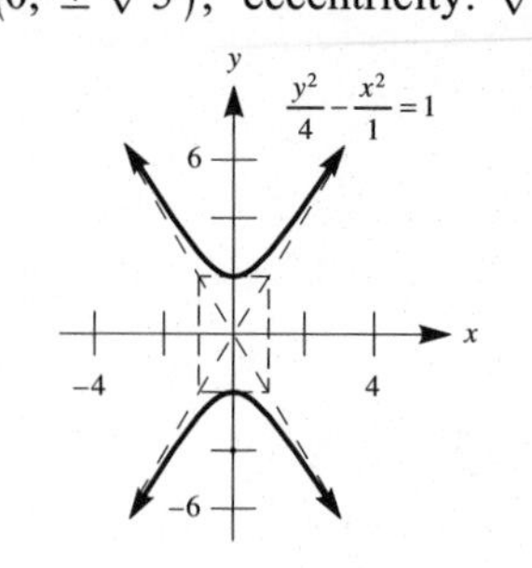

5. vertices: $(\pm 5, 0)$; length of transverse axis: 10; length of conjugate axis: 8; asymptotes: $y = \pm\frac{4}{5}x$; foci: $\left(\pm\sqrt{41}, 0\right)$; eccentricity: $\sqrt{41}/5$

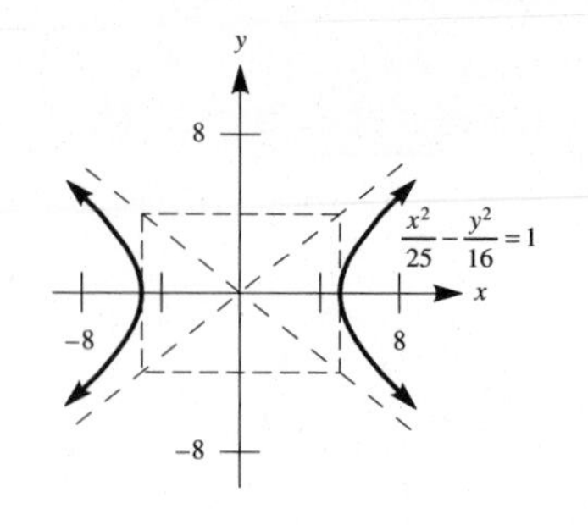

7. vertices: $\left(0, \pm\sqrt{2}/2\right)$; length of transverse axis: $\sqrt{2}$; length of conjugate axis: $2\sqrt{3}/3$; asymptotes: $y = \pm\dfrac{\sqrt{6}}{2}x$; foci: $\left(0, \pm\sqrt{30}/6\right)$; eccentricity: $\sqrt{15}/3$

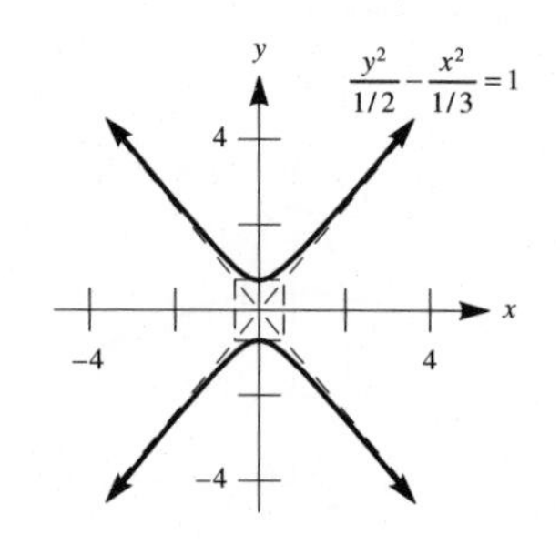

9. vertices: $(0, \pm 5)$; length of transverse axis: 10; length of conjugate axis: 4; asymptotes: $y = \pm\frac{5}{2}x$; foci: $\left(0, \pm\sqrt{29}\right)$; eccentricity: $\sqrt{29}/5$

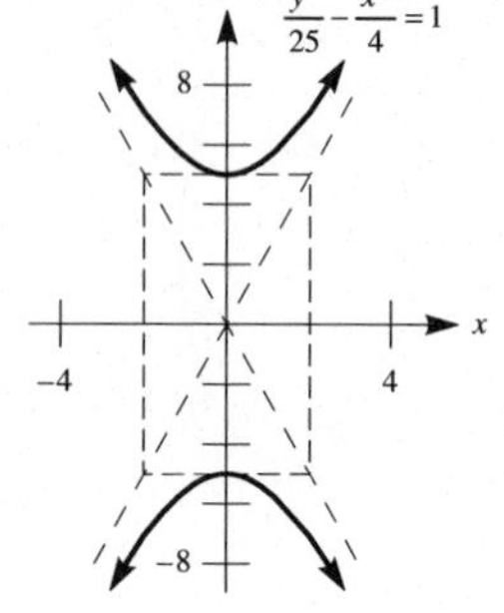

11. center: $(5, -1)$; vertices: $(10, -1)$ and $(0, -1)$; length of transverse axis: 10; length of conjugate axis: 6; asymptotes: $y = \frac{3}{5}x - 4$ and $y = -\frac{3}{5}x + 2$; foci: $\left(5 \pm \sqrt{34}, -1\right)$; eccentricity: $\sqrt{34}/5$

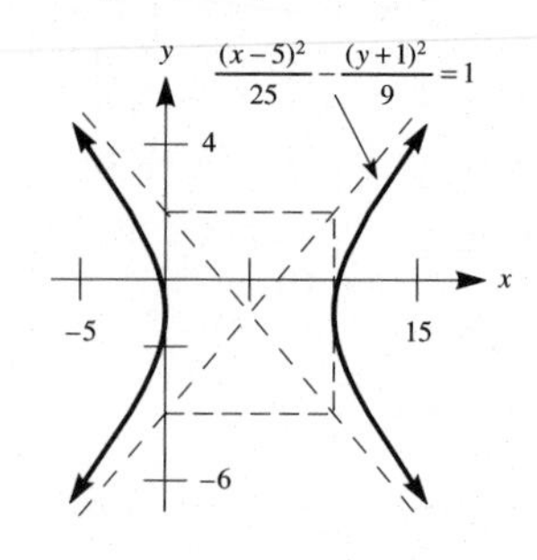

13. center: $(1, 2)$; vertices: $(1, 4)$ and $(1, 0)$; length of transverse axis: 4; length of conjugate axis: 2; asymptotes: $y = 2x$ and $y = -2x + 4$; foci: $\left(1, 2 \pm \sqrt{5}\right)$; eccentricity: $\sqrt{5}/2$

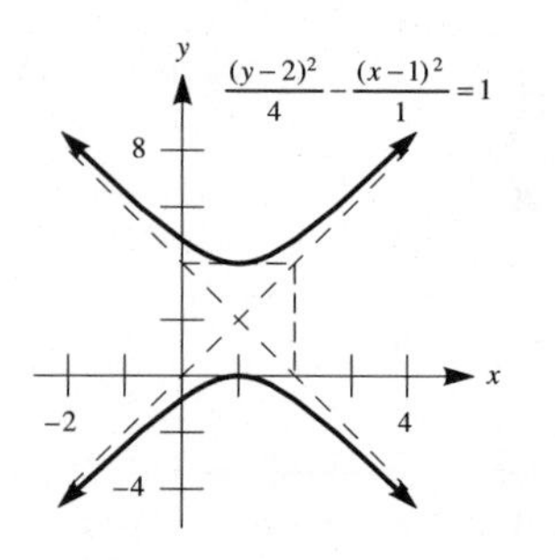

15. center: $(-3, 4)$; vertices: $(1, 4)$ and $(-7, 4)$; length of transverse axis: 8; length of conjugate axis: 8; asymptotes: $y = x + 7$ and $y = -x + 1$; foci: $\left(-3 \pm 4\sqrt{2}, 4\right)$; eccentricity: $\sqrt{2}$

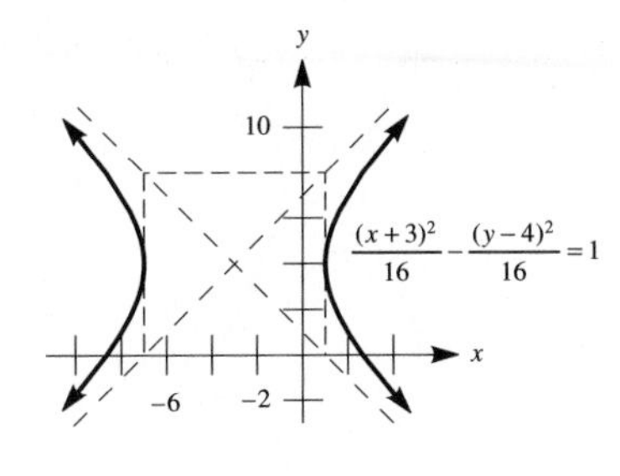

17. center: $(0, 1)$; vertices: $(2, 1)$ and $(-2, 1)$; lengths of both transverse and conjugate axes: 4; asymptotes: $y = x + 1$ and $y = -x + 1$; foci: $\left(\pm 2\sqrt{2}, 1\right)$; eccentricity: $\sqrt{2}$

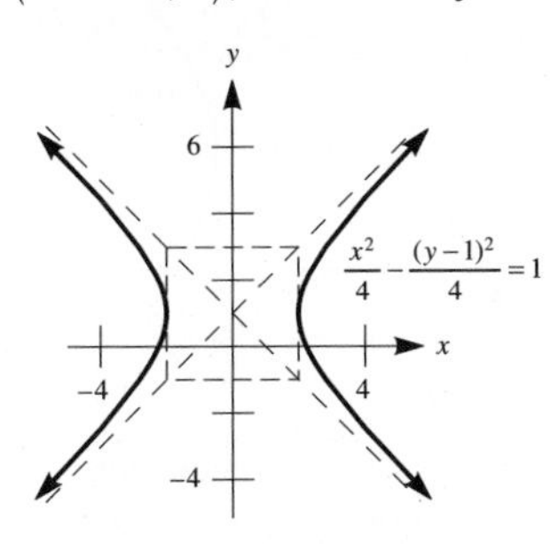

19. center: $(2, 1)$; vertices: $(5, 1)$ and $(-1, 1)$; lengths of both transverse and conjugate axes: 6; asymptotes: $y = x - 1$ and $y = -x + 3$; foci: $\left(2 \pm 3\sqrt{2}, 1\right)$; eccentricity: $\sqrt{2}$

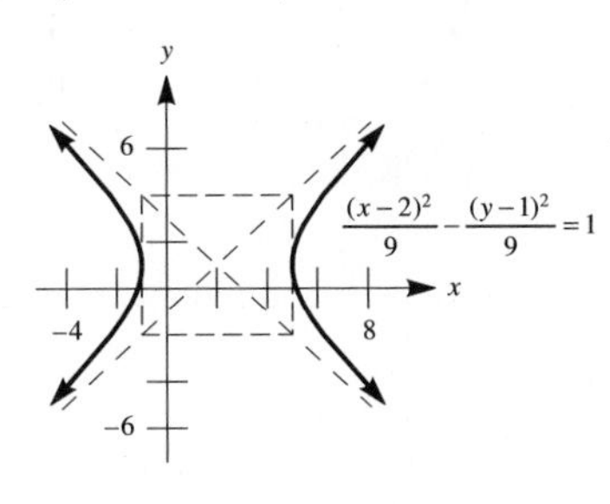

21. center: $(0, -4)$; vertices: $(0, 1)$ and $(0, -9)$; length of transverse axis: 10; length of conjugate axis: 2; asymptotes: $y = 5x - 4$ and $y = -5x - 4$; foci: $\left(0, -4 \pm \sqrt{26}\right)$; eccentricity: $\sqrt{26}/5$

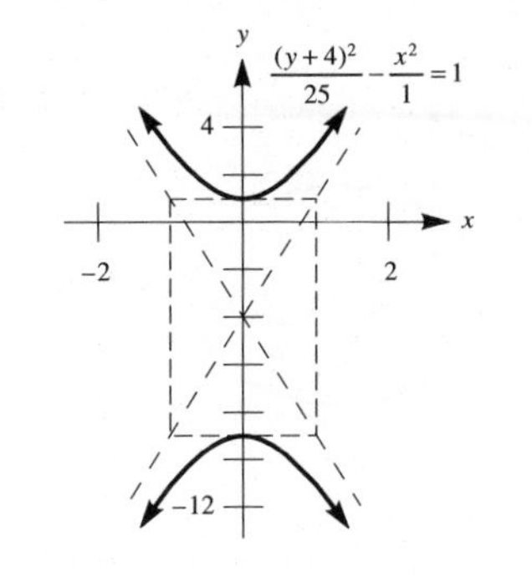

23. degenerate hyperbola

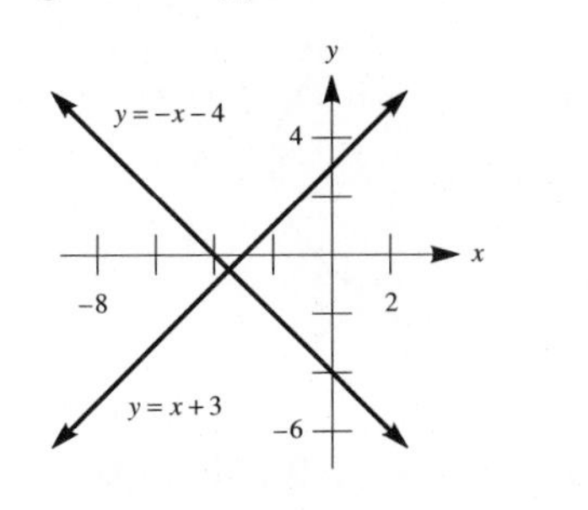

27. $15x^2 - y^2 = 15$ **29.** $x^2 - 4y^2 = 4$
31. $2x^2 - 5y^2 = 10$
33. $y^2 - 32x^2 = 49$
35. $y^2 - 9x^2 = 9$ **39.** **(b)** $(0, \pm 6)$
(c) $F_1P = 5$; $F_2P = 13$

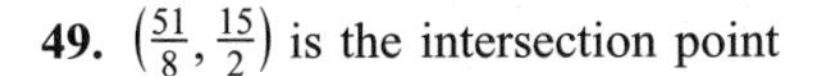
49. $\left(\frac{51}{8}, \frac{15}{2}\right)$ is the intersection point

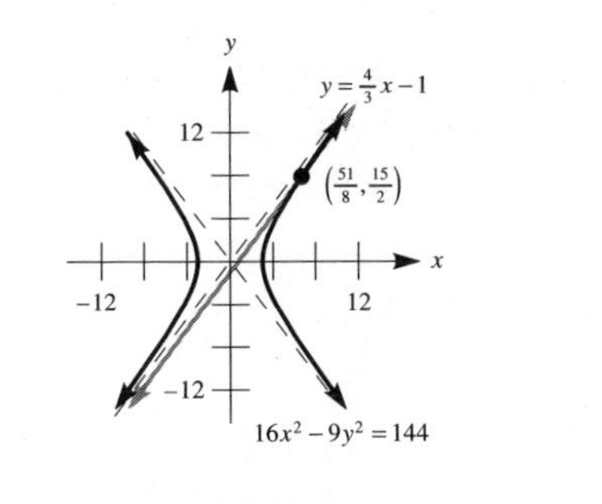

51. $y = 2x - 2$

GRAPHING UTILITY EXERCISES FOR SECTION 11.5

1. **(a)** $y = \pm\frac{1}{3}\sqrt{16x^2 - 144}$

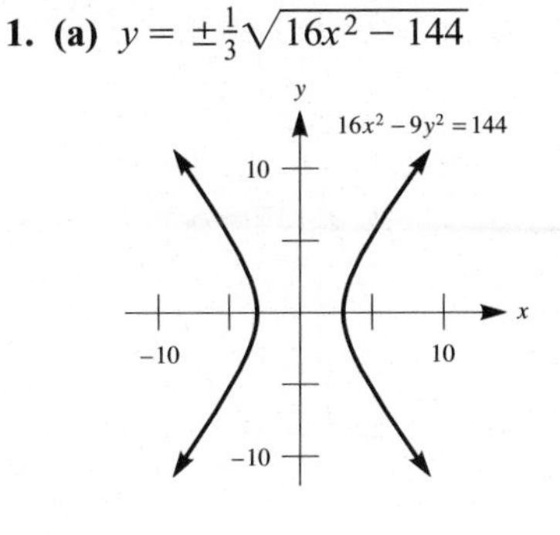

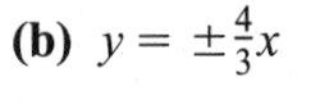
(b) $y = \pm\frac{4}{3}x$

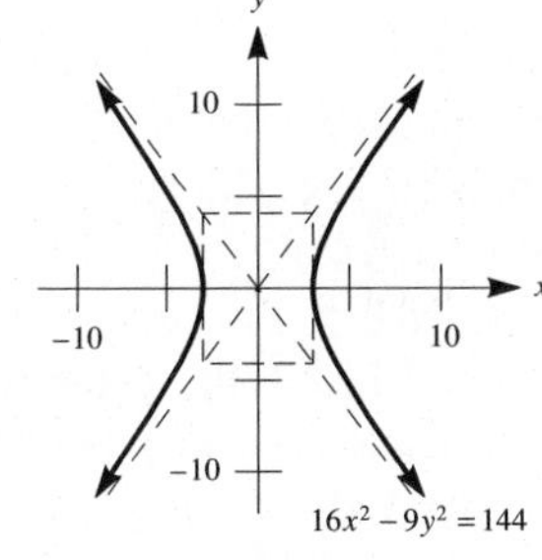

(c)

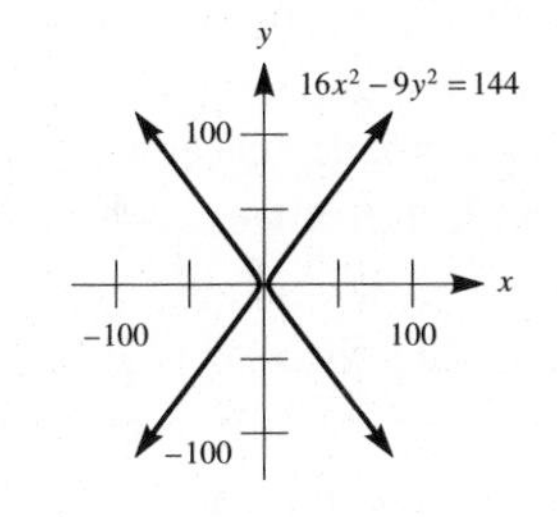

3. **(b)**

(c)

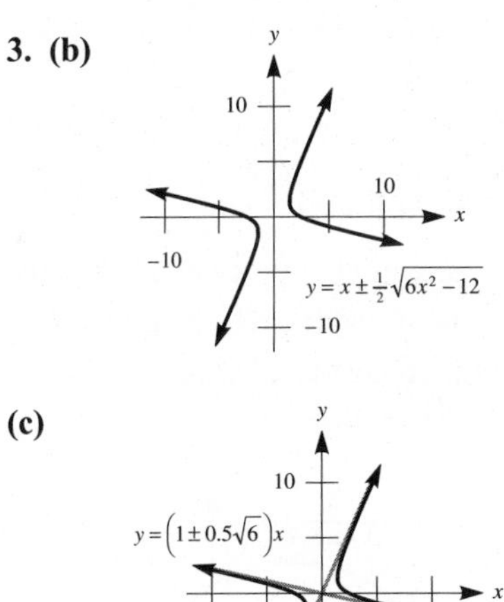

(d) The curve is indistinguishable from the asymptotes.

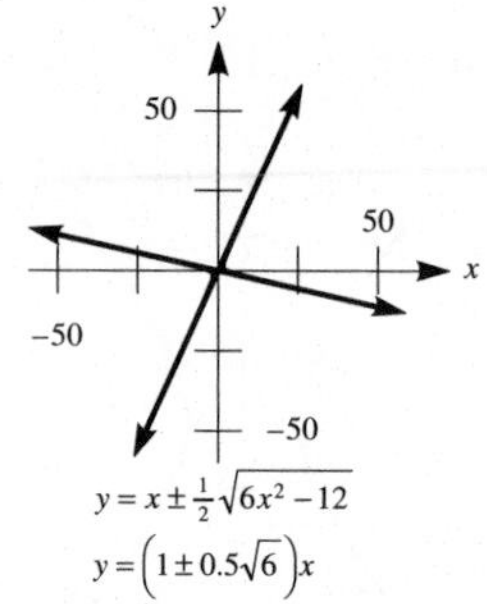

EXERCISE SET 11.6

1. $F_1P = \left(6\sqrt{19} - 8\sqrt{6}\right)/3$; $F_2P = \left(6\sqrt{19} + 8\sqrt{6}\right)/3$
3. $F_1P = 15 + 6\sqrt{2}$; $F_2P = 15 - 6\sqrt{2}$
5. (a) foci: $\left(\pm\sqrt{7}, 0\right)$; eccentricity: $\sqrt{7}/4$; directrices: $x = \pm 16\sqrt{7}/7$
(b) foci: $(\pm 5, 0)$; eccentricity: $5/4$; directrices: $x = \pm 16/5$
7. (a) foci: $(\pm 1, 0)$; eccentricity: $\sqrt{13}/13$; directrices: $x = \pm 13$
(b) foci: $(\pm 5, 0)$; eccentricity: $5\sqrt{13}/13$; directrices: $x = \pm 13/5$
9. (a) foci: $\left(\pm\sqrt{11}, 0\right)$; eccentricity: $\sqrt{11}/6$; directrices: $x = \pm 36\sqrt{11}/11$
(b) foci: $\left(\pm\sqrt{61}, 0\right)$; eccentricity: $\sqrt{61}/6$; directrices: $x = \pm 36\sqrt{61}/61$
11. $3x^2 + 4y^2 = 12$ **13.** $x^2 - y^2 = 2$

EXERCISE SET 11.7

1. (a) eccentricity: $\frac{2}{3}$; center: $\left(-\frac{12}{5}, 0\right)$; endpoints of major axis: $\left(\frac{6}{5}, 0\right)$ and $(-6, 0)$; endpoints of minor axis: $\left(-\frac{12}{5}, \pm 6\sqrt{5}/5\right)$

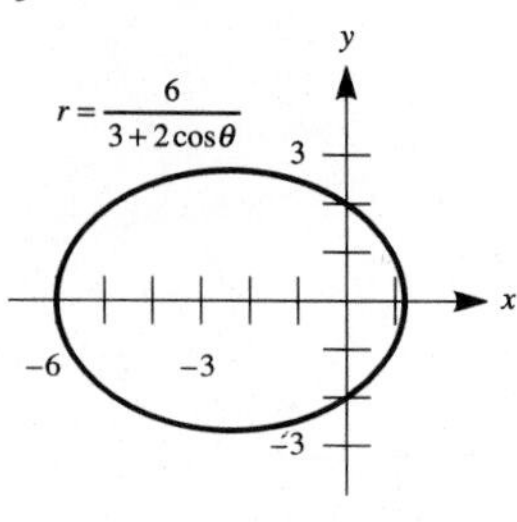

(b) eccentricity: $\frac{2}{3}$; center: $\left(\frac{12}{5}, 0\right)$; endpoints of major axis: $(6, 0)$ and $\left(-\frac{6}{5}, 0\right)$; endpoints of minor axis: $\left(\frac{12}{5}, \pm 6\sqrt{5}/5\right)$

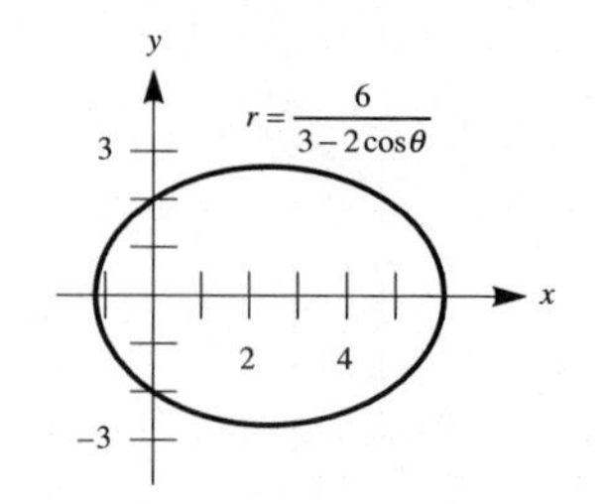

3. (a) vertex: $\left(\frac{5}{4}, 0\right)$; directrix: $x = \frac{5}{2}$

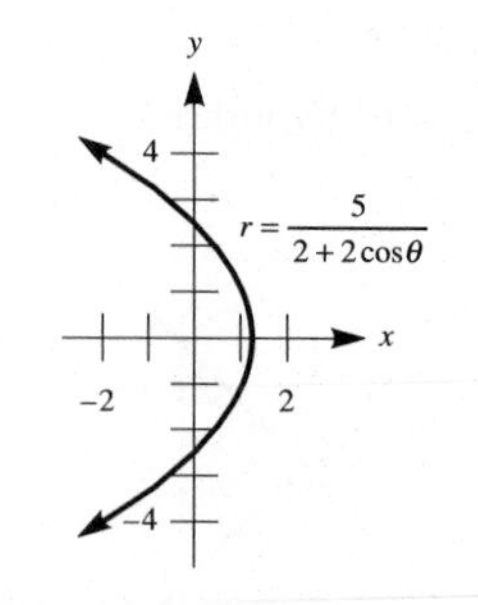

(b) vertex: $\left(-\frac{5}{4}, 0\right)$; directrix: $x = -\frac{5}{2}$

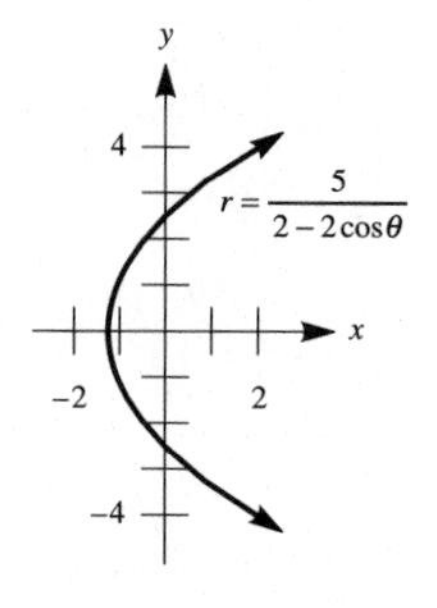

5. (a) eccentricity: 2; center: $(1, 0)$; $a = \frac{1}{2}$; $b = \frac{1}{2}\sqrt{3}$; $c = 1$

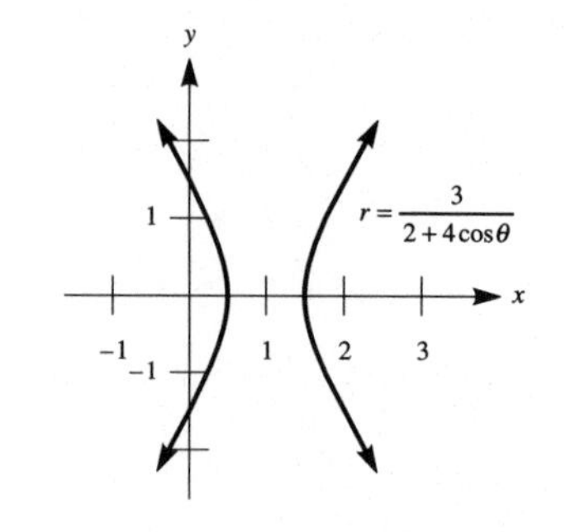

(b) eccentricity: 2; center: $(-1, 0)$; $a = \frac{1}{2}$; $b = \frac{1}{2}\sqrt{3}$; $c = 1$

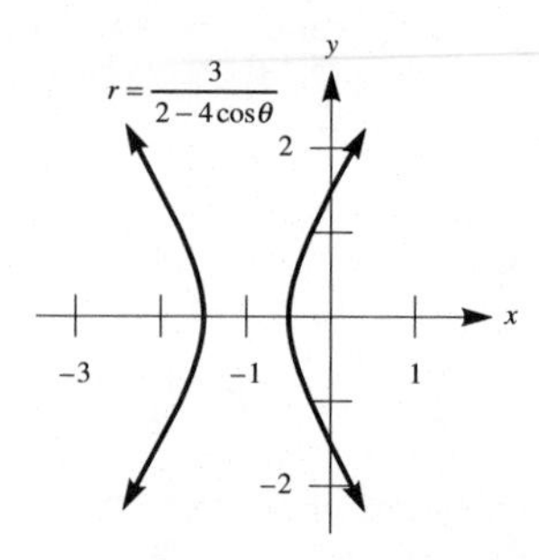

7. center: $\left(-\frac{72}{5}, 0\right)$; eccentricity: $\frac{3}{2}$; length of traverse axis: $\frac{96}{5}$; length of conjugate axis: $\frac{48}{5}\sqrt{5}$

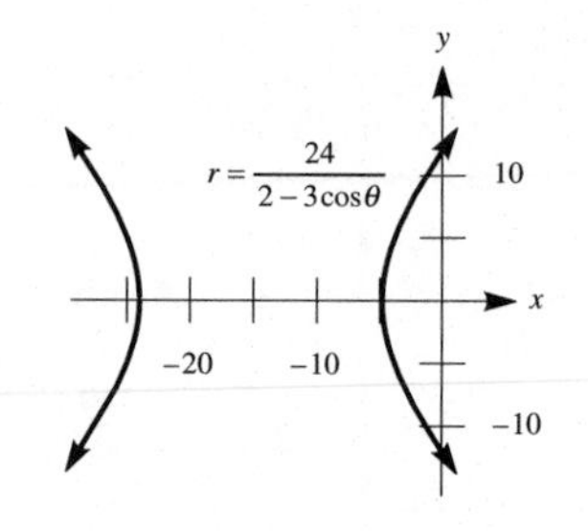

9. center: $\left(0, -\frac{3}{2}\right)$; eccentricity: $\frac{3}{5}$; length of major axis: 5; length of minor axis: 4

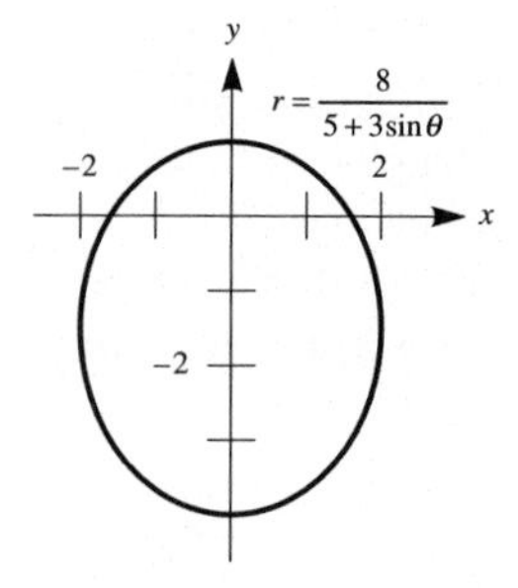

11. vertex: $\left(0, -\frac{6}{5}\right)$; directrix: $y = -\frac{12}{5}$

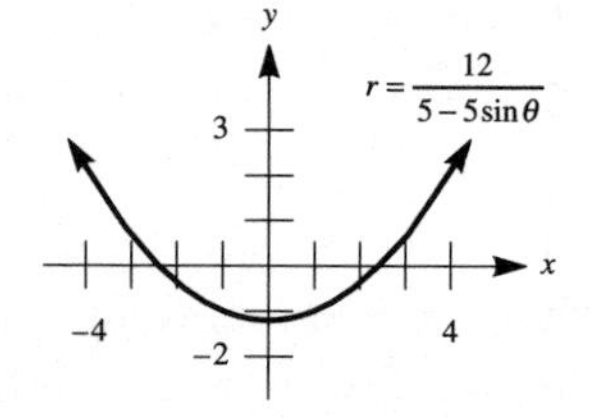

13. center: $\left(-\frac{5}{2}, 0\right)$; eccentricity: $\frac{5}{7}$; length of major axis: 7; length of minor axis: $2\sqrt{6}$

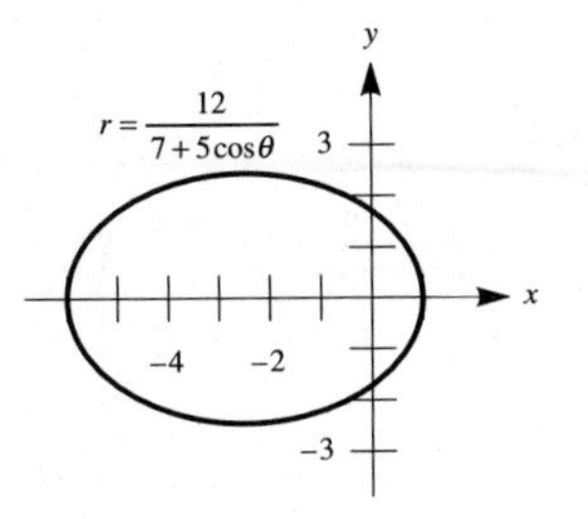

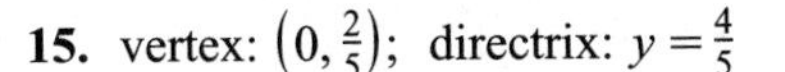
15. vertex: $\left(0, \frac{2}{5}\right)$; directrix: $y = \frac{4}{5}$

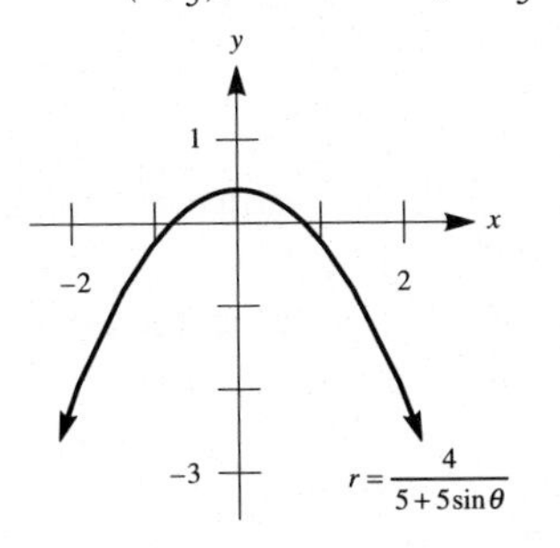

17. center: $(-6, 0)$; eccentricity: 2; length of transverse axis: 6; length of conjugate axis: $6\sqrt{3}$

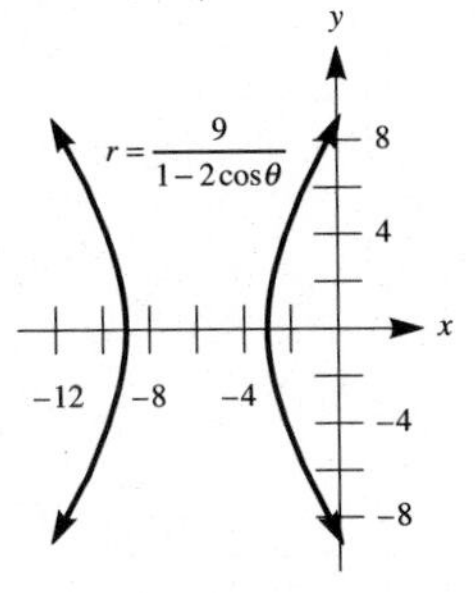

EXERCISE SET 11.8

1. $\left(\frac{1}{2}, 3\sqrt{3}/2\right)$ **3.** (2,0) **5.** $\left(-\frac{31}{13}, \frac{27}{13}\right)$
7. $\sin\theta = \frac{4}{5}$, $\cos\theta = \frac{3}{5}$
9. $\sin\theta = \frac{3}{5}$, $\cos\theta = \frac{4}{5}$
11. $\sin\theta = \sqrt{3}/2$, $\cos\theta = \frac{1}{2}$
13. $\sin\theta = \frac{7}{34}\sqrt{2}$, $\cos\theta = \frac{23}{34}\sqrt{2}$

15. $(x')^2 - (y')^2 = 9$

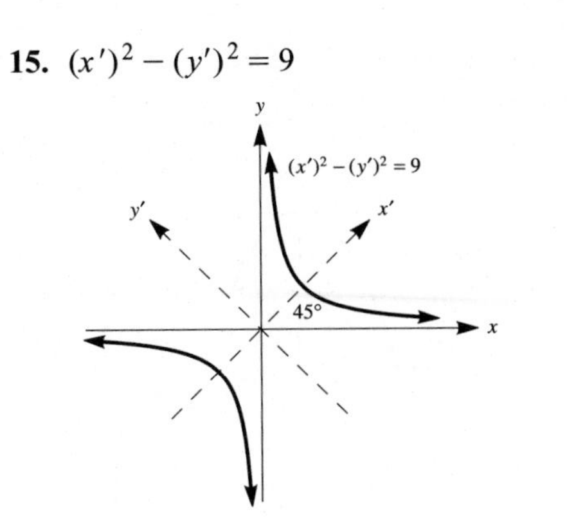

17.

$\dfrac{(x')^2}{1/9} - \dfrac{(y')^2}{1} = 1$

26.6°

19.

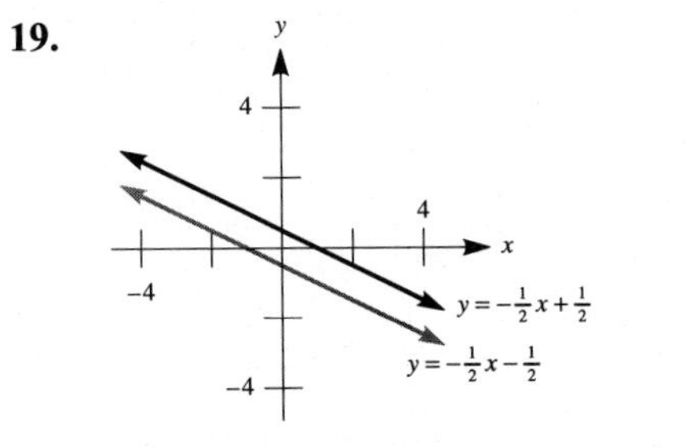

21.

36.9°

$(y')^2 = 20x'$

23.

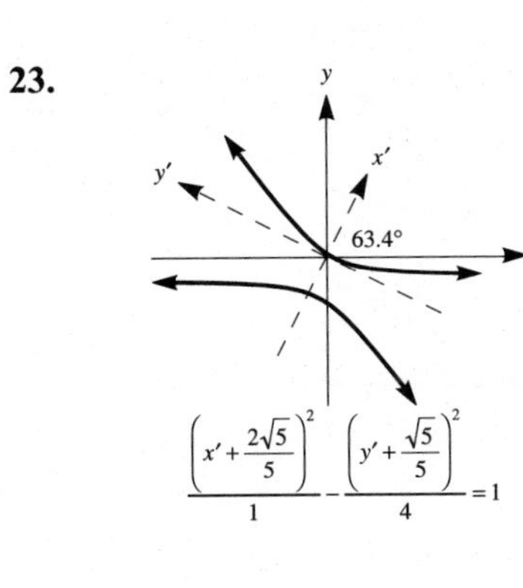

25.

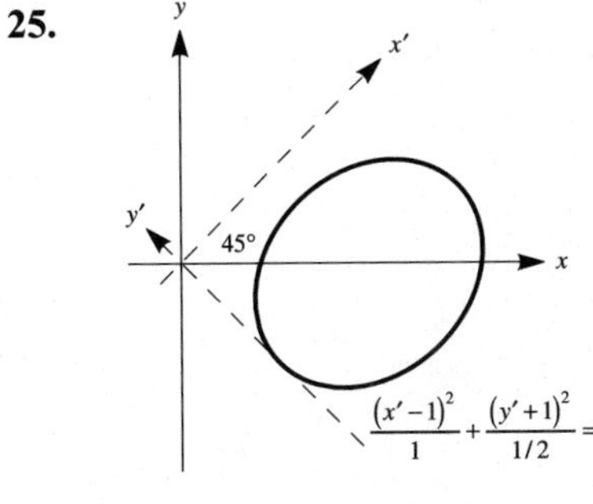

27.

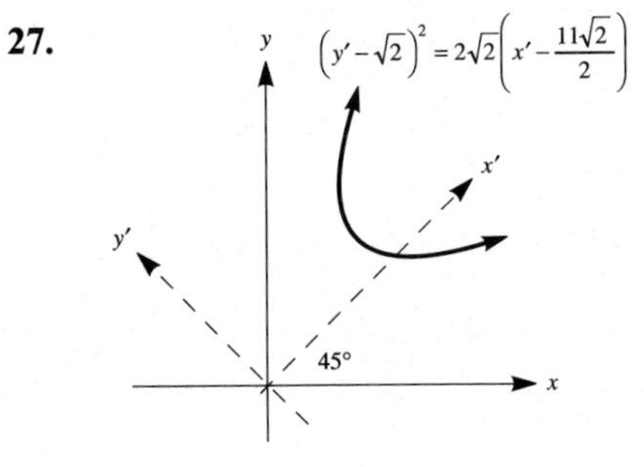

29.

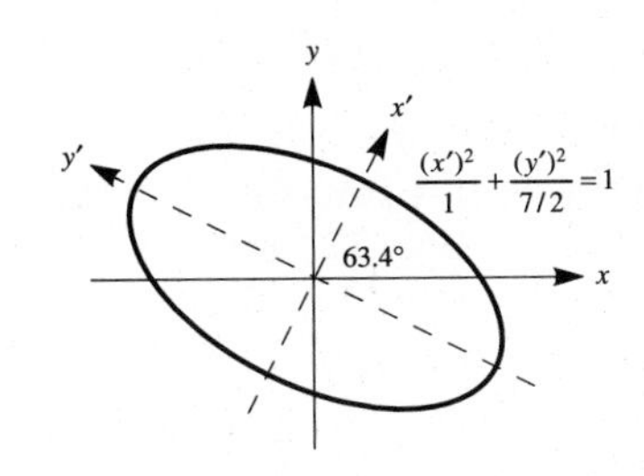

31.

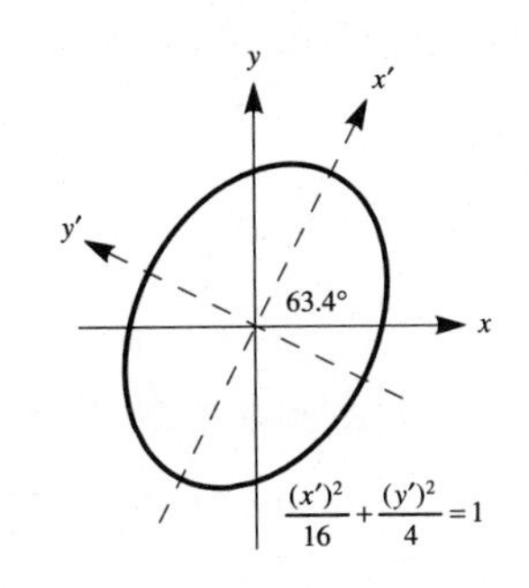

33.

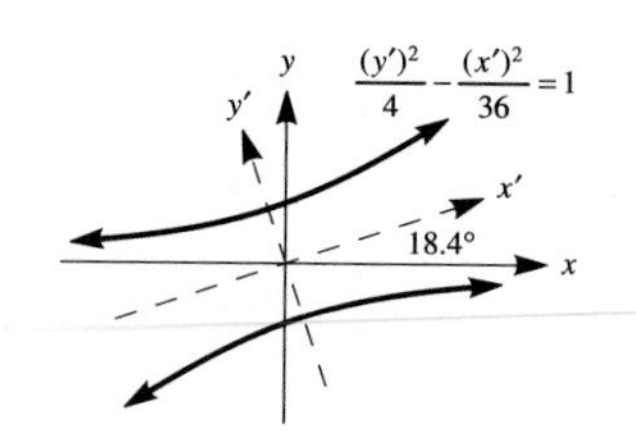

35.

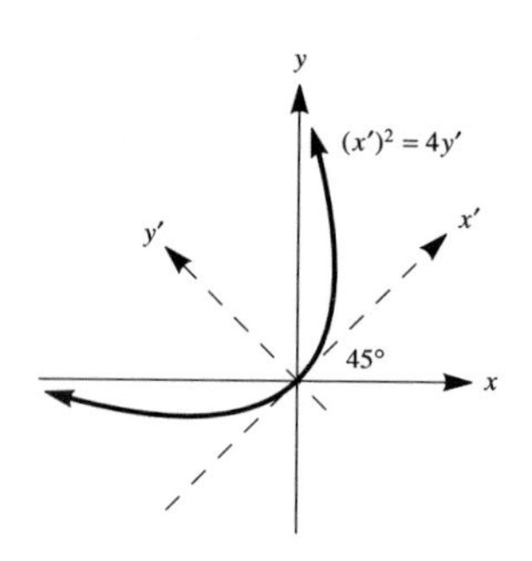

37.

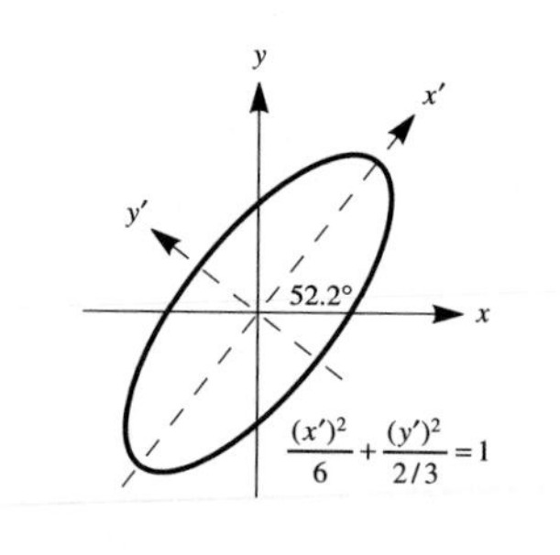

39. no graph

41. $x = x' \cos\theta - y' \sin\theta$; $y = x' \sin\theta + y' \cos\theta$

CHAPTER 11 REVIEW EXERCISES

17. 146.3° **19.** $53\sqrt{61}/61$

23. **(a)** $y^2 = 16x$ **(b)** $x^2 = 16y$

25. $x^2 = 12y$ **27.** $15x^2 + 16y^2 = 960$

29. $25x^2 + 9y^2 = 900$

31. $8x^2 - y^2 = 32$

33. $240x^2 - 16y^2 = 135$

35. vertex: (0, 0); focus: $(0, \frac{5}{2})$; directrix: $y = -\frac{5}{2}$; focal width: 10

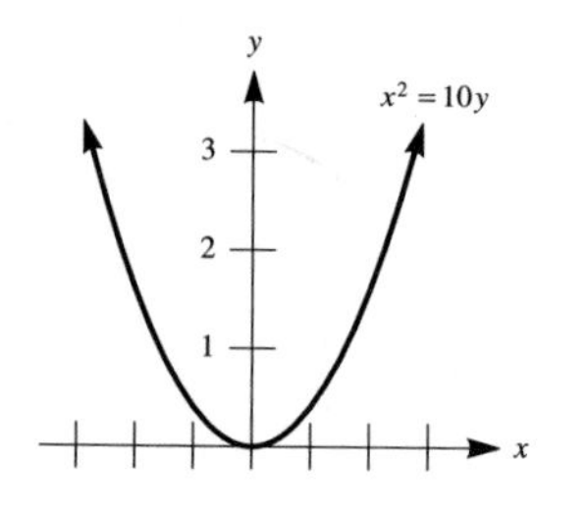

37. vertex: (0, 3); focus: (0, 0); directrix: $y = 6$; focal width: 12

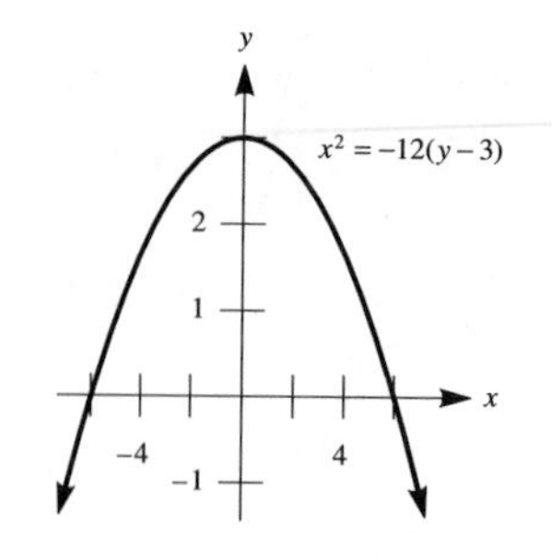

39. vertex: (1, 1); focus: (0, 1); directrix: $x = 2$; focal width: 4

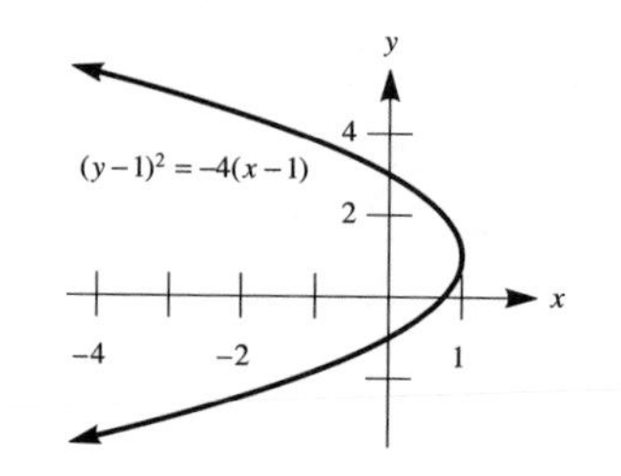

41. center: (0, 0); length of major axis: 12; length of minor axis: 8; foci: $(\pm 2\sqrt{5}, 0)$; eccentricity: $\sqrt{5}/3$

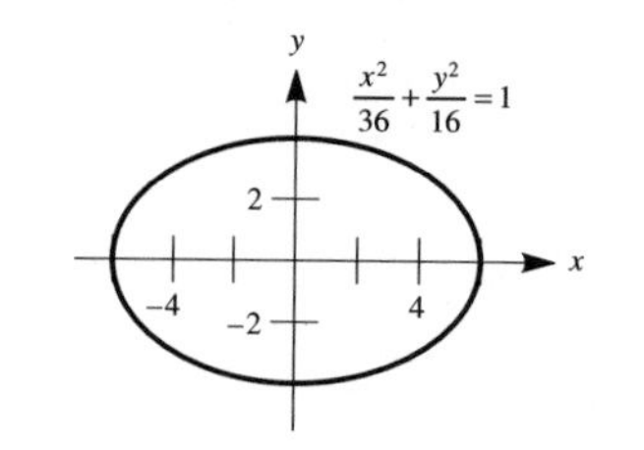

43. center: (0, 0); length of major axis: 6; length of minor axis: 2; foci: $(0, \pm 2\sqrt{2})$; eccentricity: $2\sqrt{2}/3$

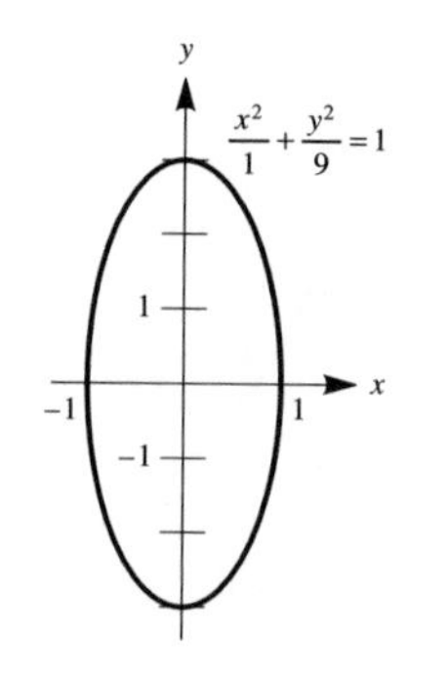

45. center: $(-3, 0)$; lengths of major and minor axes: 6; focus: $(-3, 0)$; eccentricity: 0

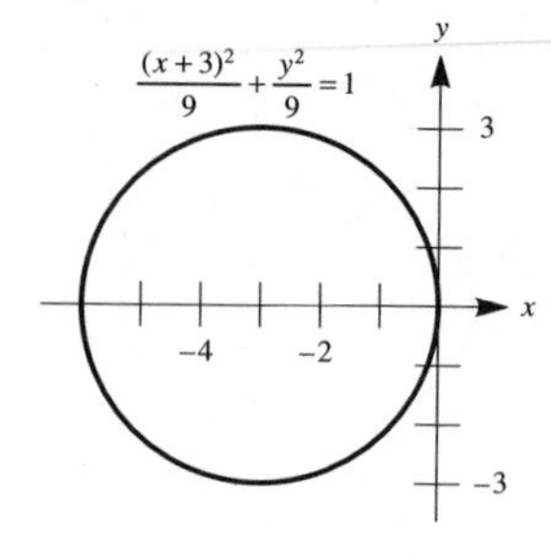

47. center: (0, 0); vertices: $(\pm 6, 0)$; asymptotes: $y = \pm\frac{2}{3}x$; foci: $(\pm 2\sqrt{13}, 0)$; eccentricity: $\sqrt{13}/3$

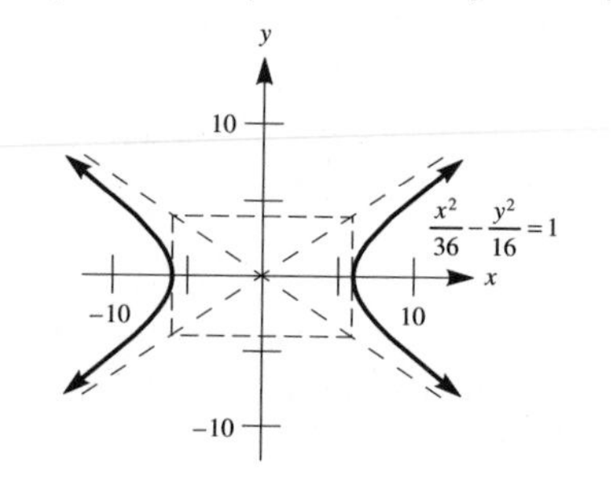

49. center: (0, 0); vertices: $(0, \pm 1)$; asymptotes: $y = \pm\frac{1}{3}x$; foci: $(0, \pm\sqrt{10})$; eccentricity: $\sqrt{10}$

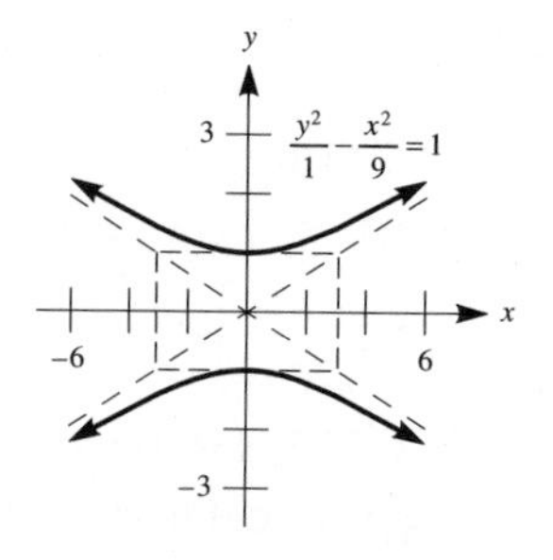

51. center: $(0, -3)$; vertices: $(0, 0)$ and $(0, -6)$; asymptotes: $y = x - 3$ and $y = -x - 3$; foci: $\left(0, -3 \pm 3\sqrt{2}\right)$; eccentricity: $\sqrt{2}$

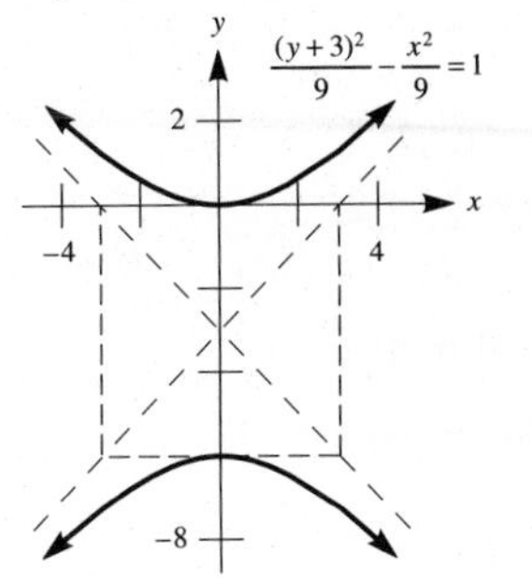

53. parabola: vertex: $(4, 4)$; axis of symmetry: $y = 4$; focus: $(8, 4)$; directrix: $x = 0$

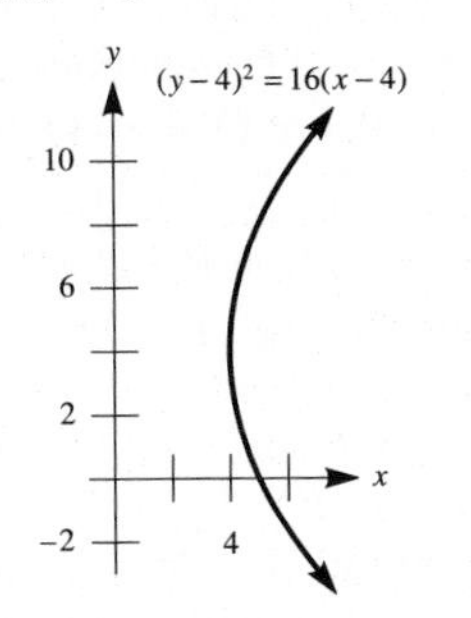

55. ellipse: center: $(-2, 3)$; length of major axis: 8; length of minor axis: 6; foci: $\left(-2, 3 \pm \sqrt{7}\right)$

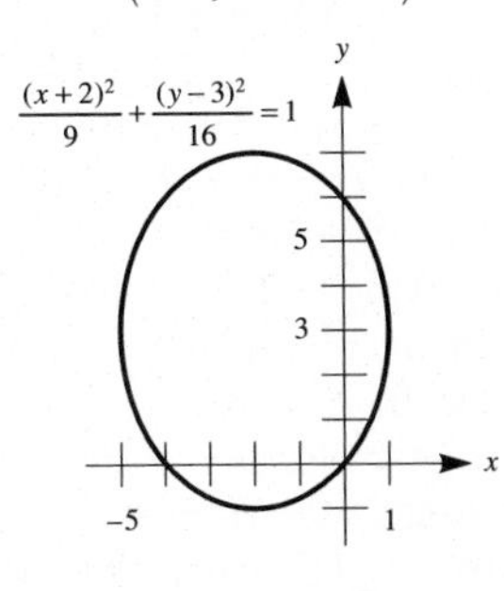

57. parabola: vertex: $(-3, 2)$; axis of symmetry: $x = -3$; focus: $(-3, 5)$; directrix: $y = -1$

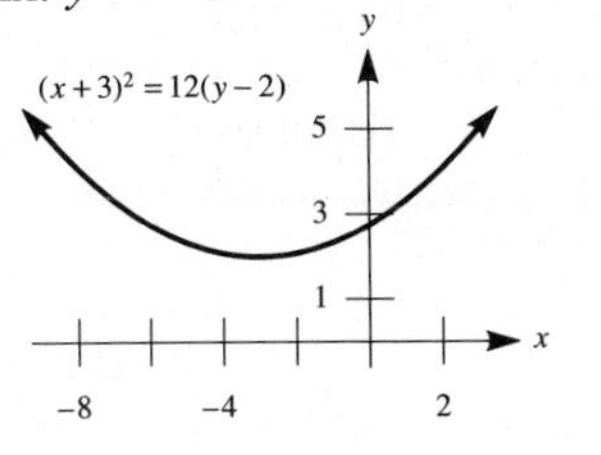

59. hyperbola: center: $(2, 1)$; vertices: $(5, 1)$ and $(-1, 1)$; asymptotes: $y = x - 1$ and $y = -x + 3$; foci: $\left(2 \pm 3\sqrt{2}, 1\right)$

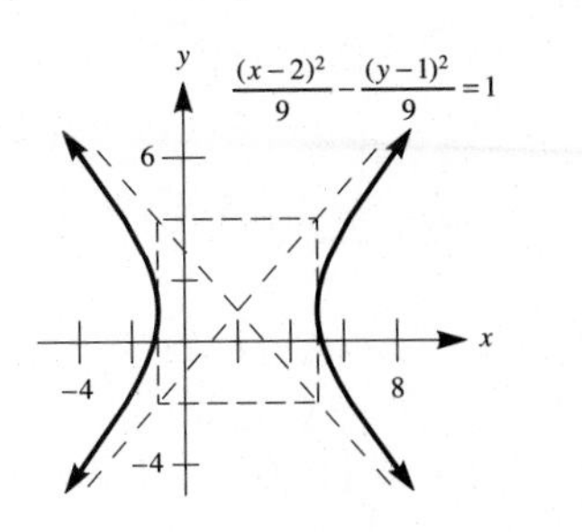

61. parabola: vertex: $(0, 6)$; axis of symmetry: $x = 0$; focus: $\left(0, \frac{11}{2}\right)$; directrix: $y = \frac{13}{2}$

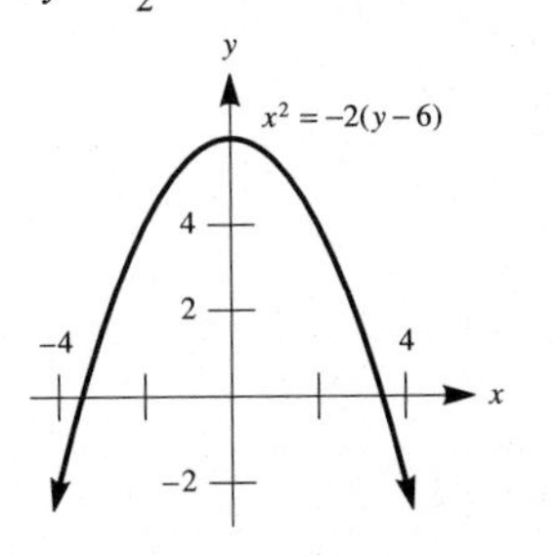

63. ellipse: center: $(0, 5)$; length of major axis: 8; length of minor axis: 2; foci: $\left(\pm\sqrt{15}, 5\right)$

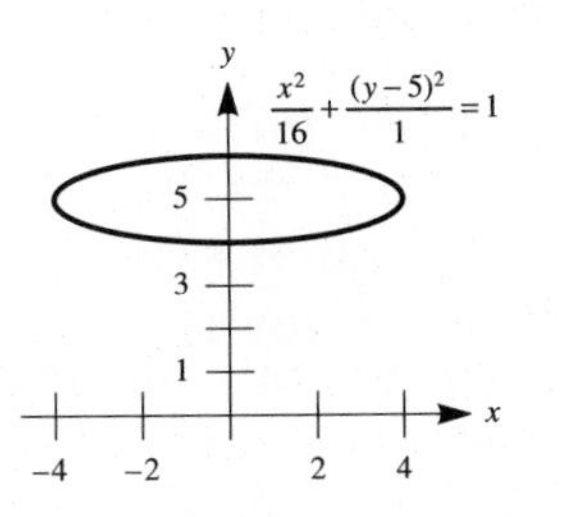

65. two lines: $4x - 5y = -2$ and $4x + 5y = 18$

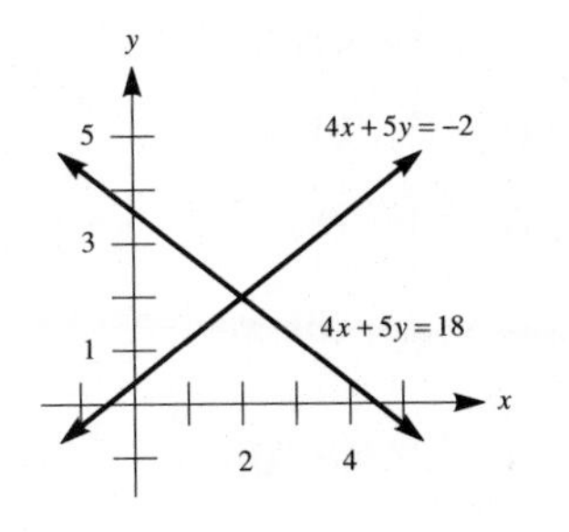

73. C_2 would be the best approximation

	Approximation Obtained	Percentage Error
C_1	25.531776	0.019
C_2	25.526986	0.000049
C_3	25.519489	0.029

CHAPTER 11 TEST

1. focus: $(-3, 0)$; directrix: $x = 3$

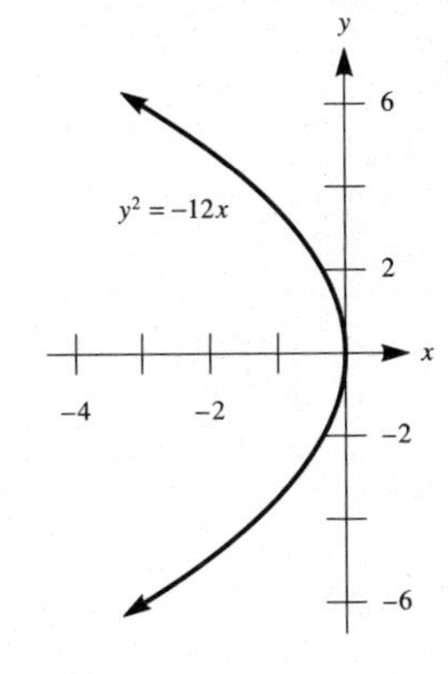

2. asymptotes: $y = \pm\frac{1}{2}x$; foci: $\left(\pm\sqrt{5}, 0\right)$

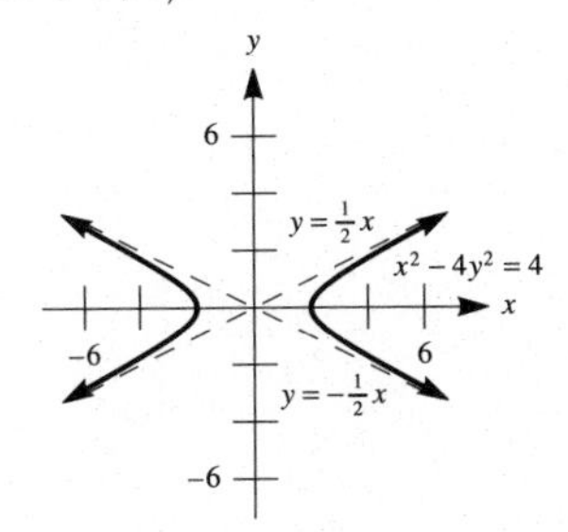

3. (a) $\theta = 60°$

(b)

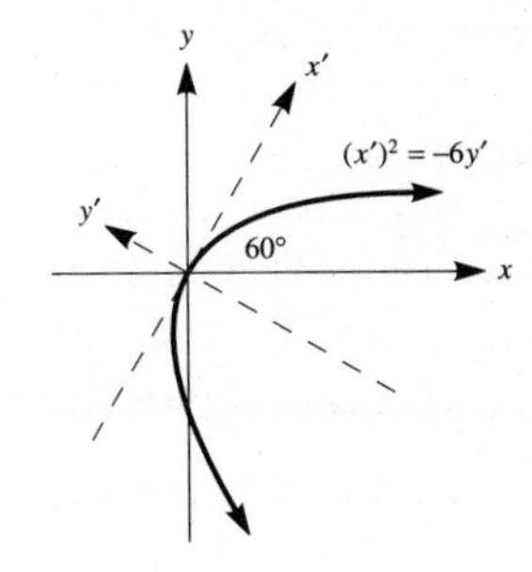

4. 30° **5.** $\dfrac{x^2}{12} + \dfrac{y^2}{16} = 1$

6. $m = \pm\dfrac{\sqrt{15}}{15}$

7. $\sqrt{3}x - y - 2\sqrt{3} = 0$

8. $\dfrac{x^2}{3} - \dfrac{y^2}{1} = 1$ **9. (b)** 64

10. length of major axis: 10; length of minor axis: 4; foci: $(\pm\sqrt{21}, 0)$

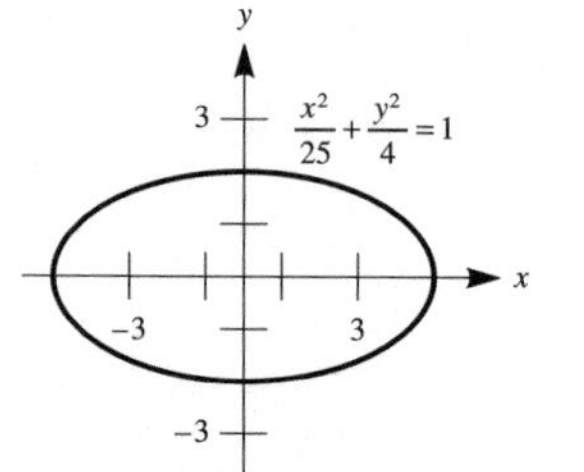

11. 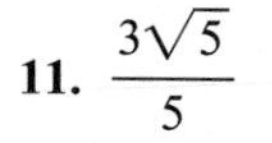 $\dfrac{3\sqrt{5}}{5}$

12.

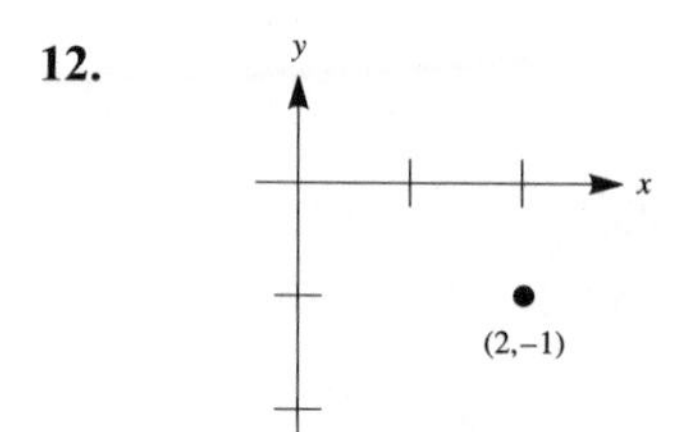

13.

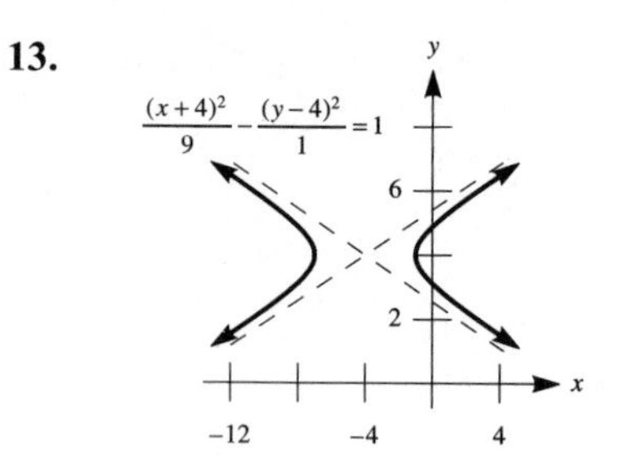

14. ellipse

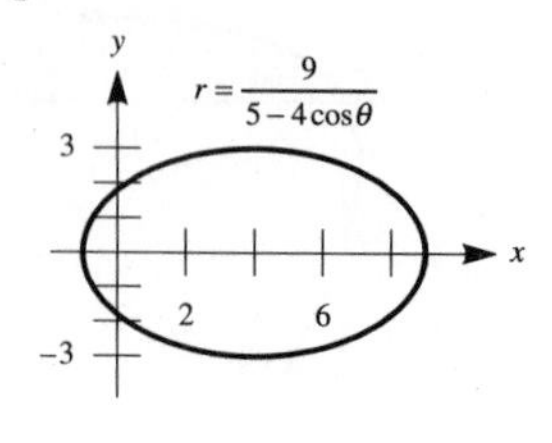

15. focal width: 8; vertex: (1, 2)

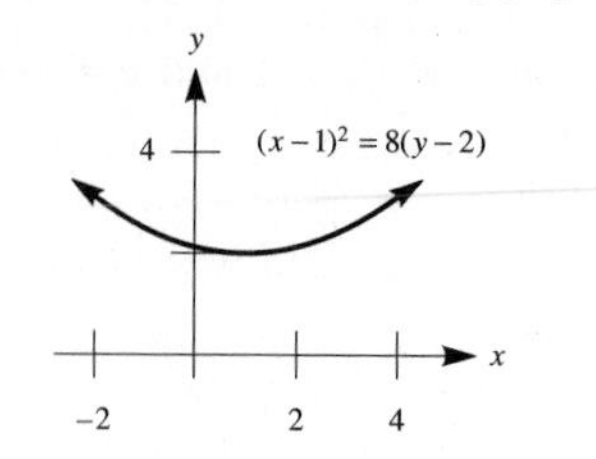

16. (a) $x = \pm 36\sqrt{11}/11$
(b) $F_1P = (12 + \sqrt{11})/2$;
$F_2P = (12 - \sqrt{11})/2$
17. $y = \frac{1}{6}x + \frac{13}{3}$ **18.** $y = 4x - 8$

CHAPTER TWELVE

EXERCISE SET 12.1

1.

i^2	i^3	i^4	i^5	i^6	i^7	i^8
-1	$-i$	1	i	-1	$-i$	1

3. (a) real part: 4; imaginary part: 5
(b) real part: 4; imaginary part: -5
(c) real part: $\frac{1}{2}$; imaginary part: -1
(d) real part: 0; imaginary part: 16
5. $c = 4$, $d = -3$
7. (a) $14 - 4i$ **(b)** $-4 - 8i$
9. (a) $19 - 17i$ **(b)** $19 - 17i$
(c) $\frac{11}{26} - \frac{23}{26}i$ **(d)** $\frac{11}{25} + \frac{23}{25}i$
11. (a) $11 - i$ **(b)** $11 - 7i$ **(c)** 4
13. $4 - 2i$ **15.** $30 + 19i$ **17.** 13
19. $-191 - 163i$ **21.** $19 - 4i$
23. $-70 + 84i$ **25.** $539 + 1140i$
27. $-46 + 9i$ **29.** $\frac{6}{97} + \frac{35}{97}i$
31. $\frac{6}{97} - \frac{35}{97}i$ **33.** $-\frac{5}{13} + \frac{12}{13}i$ **35.** -4
37. $\frac{1}{26} + \frac{5}{26}i$ **39.** $-i$ **41.** $12i$
43. $-3\sqrt{5}i$ **45.** -35 **47.** $12\sqrt{2}i$
49. (a) -3 **(b)** $x = \dfrac{1}{2} \pm \dfrac{\sqrt{3}}{2}i$
51. (a) -36 **(b)** $x = -\frac{1}{5} \pm \frac{3}{5}i$
53. (a) -23 **(b)** $x = -\dfrac{3}{4} \pm \dfrac{\sqrt{23}}{4}i$
55. (a) $-29/48$ **(b)** $x = \dfrac{3}{4} \pm \dfrac{\sqrt{87}}{4}i$

57. (a) 0 **(b)** This verifies the solution found in Example 6.
59. (a) $(a + c) + (b + d)i$
(b) $(a - c) + (b - d)i$
(c) $(ac - bd) + (bc + ad)i$
(d) $\dfrac{ac + bd}{c^2 + d^2} + \dfrac{bc - ad}{c^2 + d^2}i$
67. $x = 3, \pm 2i$
69. $x = \pm 3, \pm 2\sqrt{i}, \pm 2i\sqrt{i}$
71. real part: $\dfrac{2a^2 - 2b^2}{a^2 + b^2}$; imaginary part: 0 **73.** 0

EXERCISE SET 12.2

1. quotient: $x - 5$; remainder: -11; $x^2 - 8x + 4 = (x - 3)(x - 5) - 11$
3. quotient: $x - 11$; remainder: 53; $x^2 - 6x - 2 = (x + 5)(x - 11) + 53$
5. quotient: $3x^2 - \frac{3}{2}x - \frac{1}{4}$; remainder: $\frac{13}{4}$; $6x^3 - 2x + 3 = (2x + 1)\left(3x^2 - \frac{3}{2}x - \frac{1}{4}\right) + \frac{13}{4}$
7. quotient: $x^4 - 3x^3 + 9x^2 - 27x + 81$; remainder: -241; $x^5 + 2 = (x + 3)(x^4 - 3x^3 + 9x^2 - 27x + 81) - 241$
9. quotient: $x^5 + 2x^4 + 4x^3 + 8x^2 + 16x + 32$; remainder: 0; $x^6 - 64 = (x - 2)(x^5 + 2x^4 + 4x^3 + 8x^2 + 16x + 32) + 0$ **11.** quotient: $5x^2 + 15x + 17$; remainder: $-24x - 83$; $5x^4 - 3x^2 + 2 = (x^2 - 3x + 5)(5x^2 + 15x + 17) + (-24x - 83)$ **13.** quotient: $3y - 19$; remainder: $89y + 35$; $3y^3 - 4y^2 - 3 = (y^2 + 5y + 2)(3y - 19) + (89y + 35)$
15. quotient: $t^2 - 2t - 4$; remainder: 0; $t^4 - 4t^3 + 4t^2 - 16 = (t^2 - 2t + 4)(t^2 - 2t - 4) + 0$
17. quotient: $z^4 + z^3 + z^2 + z + 1$; remainder: 0; $z^5 - 1 = (z - 1)(z^4 + z^3 + z^2 + z + 1) + 0$
19. quotient: $ax + (b + ar)$; remainder: $c + r(b + ar) = ar^2 + br + c$; $ax^2 + bx + c = (x - r)[ax + (b + ar)] + (ar^2 + br + c)$
21. quotient: $x - 1$; remainder: -7; $x^2 - 6x - 2 = (x - 5)(x - 1) - 7$
23. quotient: $4x - 5$; remainder: 0; $4x^2 - x - 5 = (x + 1)(4x - 5) + 0$
25. quotient: $6x^2 + 19x + 78$; remainder: 313; $6x^3 - 5x^2 + 2x + 1 = (x - 4)(6x^2 + 19x + 78) + 313$
27. quotient: $x^2 + 2x + 4$; remainder: 7; $x^3 - 1 = (x - 2)(x^2 + 2x + 4) + 7$

29. quotient: $x^4 - 2x^3 + 4x^2 - 8x + 16$; remainder: -33; $x^5 - 1 = (x+2)(x^4 - 2x^3 + 4x^2 - 8x + 16) - 33$
31. quotient: $x^3 - 10x^2 + 40x - 160$; remainder: 642; $x^4 - 6x^3 + 2 = (x+4)(x^3 - 10x^2 + 40x - 160) + 642$
33. quotient: $x^2 + 6x + 57$; remainder: 576; $x^3 - 4x^2 - 3x + 6 = (x-10)(x^2 + 6x + 57) + 576$
35. quotient: $x^2 - 6x + 30$; remainder: -150; $x^3 - x^2 = (x+5)(x^2 - 6x + 30) - 150$
37. quotient: $54x^2 + 9x - 21$; remainder: 0; $54x^3 - 27x^2 - 27x + 14 = \left(x - \frac{2}{3}\right)(54x^2 + 9x - 21) + 0$
39. quotient: $x^3 + 3x^2 + 12x + 36$; remainder: 120; $x^4 + 3x^2 + 12 = (x-3)(x^3 + 3x^2 + 12x + 36) + 120$
41. $(x-2)(x^4 + 2x^3 + 4x^2 + 8x + 16)$
43. $(z-3)(z^3 + 3z^2 + 9z + 27)$
45. quotient: $2x$; remainder: 1
47. quotient: $3x^2 - \frac{3}{2}x + \frac{3}{4}$; remainder: $\frac{1}{4}$
49. $k = 4$ **53.** $q(x) = x + (-4 + i)$; $R(x) = -4i$ **55.** $q(x) = x + (-1 + i)$; $R(x) = 0$ **57.** $-6\sqrt{3} + 57$ **59.** 0
61. $a = -1, b = -1$
63. $b = \pm 3\sqrt{2}/2$
65. $x = 2$ (multiplicity 2), -4

GRAPHING UTILITY EXERCISES FOR SECTION 12.3

1. $f(2) = 3 \neq 0$

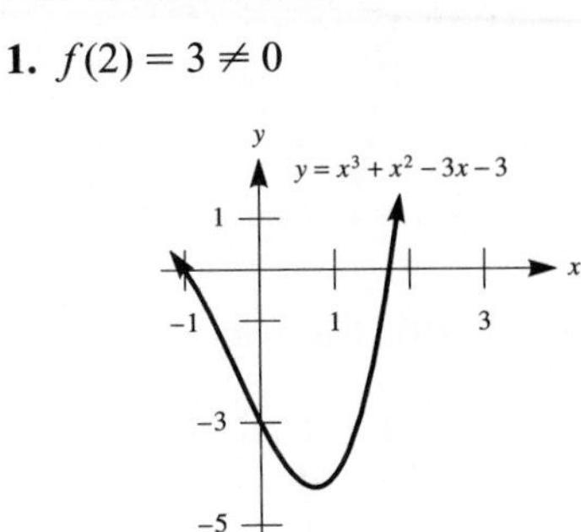

3. $f(0.9) = -1.38 \neq 0$

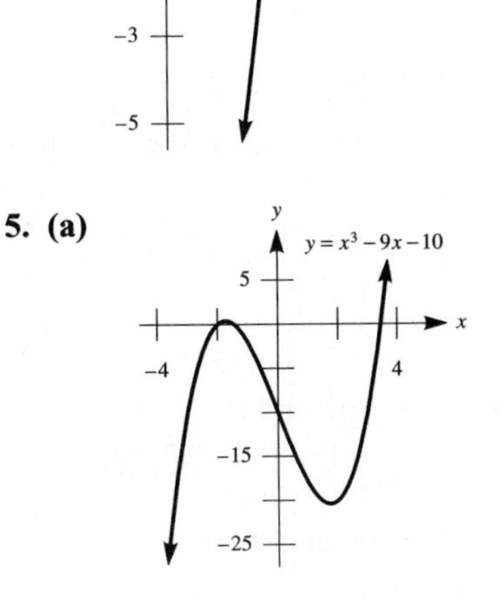

5. (a)

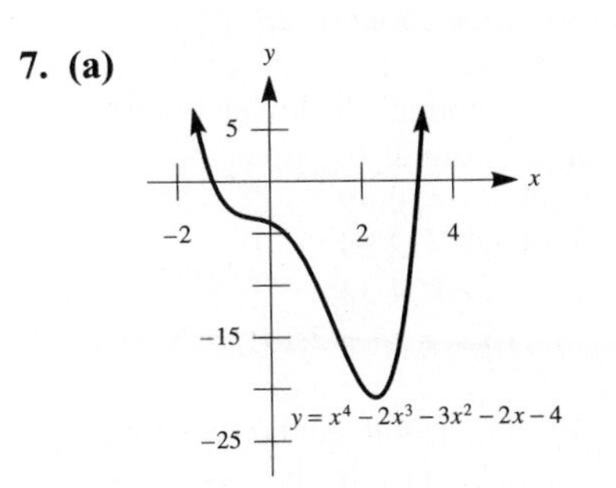

(b) $x \approx 3.449$ **(d)** $f(3.449) \approx 0$

7. (a)

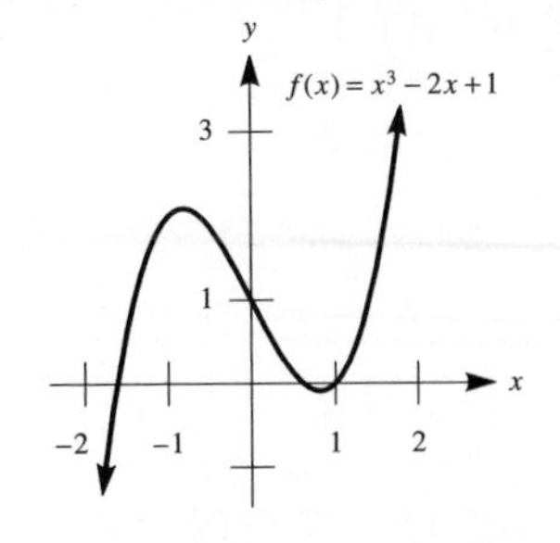

(b) $x \approx -1.236$ **(d)** $f(-1.236) \approx 0$

9. (a) three x-intercepts

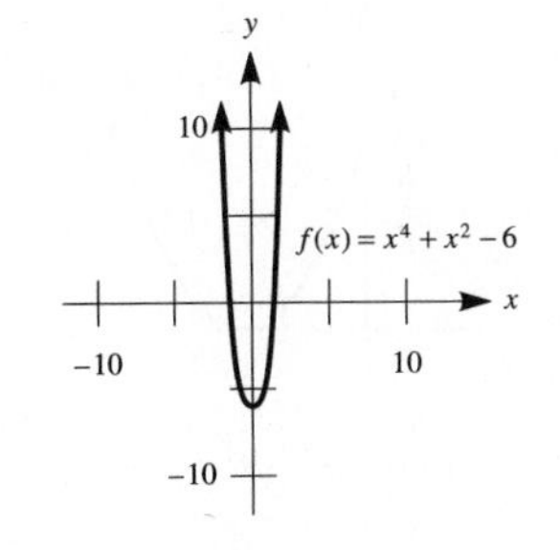

(b) $x \approx -1.618, 0.618, 1$ (exact)

11. (a) two x-intercepts

(b) $x \approx 1.414$ **(c)** Since $\sqrt{2} \approx 1.414$, our work verifies that $\sqrt{2}$ is a root of the equation $x^4 + x^2 - 6 = 0$.

EXERCISE SET 12.3

1. yes **3.** yes **5.** yes **7.** yes
9. yes **11.** no **13. (a)** yes **(b)** no
15. (a) 1, 2 (multiplicity 3), 3
(b) 1 (multiplicity 3)
(c) 5 (multiplicity 6), -1 (multiplicity 4)
(d) 0 (multiplicity 5), 1
17. -170 **19.** -9 **21.** $-3\sqrt{2} - 2$
23. -22 **25.** $\pm 3, 4$
27. $1, -1 \pm \sqrt{6}$ **29.** $-2, 2/3, 3$
31. $-3/2, \left(1 \pm \sqrt{5}\right)/2$ **33.** 0, 5
35. $-4, 3$ **37.** $-9, 1 \pm \sqrt{2}$
39. $x = \left(3 \pm \sqrt{17}\right)/4$ (each of multiplicity 2)
41. (a) 1.125 **(b)** -0.046875 **(c)** $t - 1$ **(d)** $t = 1, \left(-1 \pm \sqrt{13}\right)/2$
43. $x^3 - 4x^2 - 17x + 60 = 0$
45. $x^3 + 8x^2 + 13x + 6 = 0$
47. no such polynomial of degree 4 exists
49. $ax^4 + (3a + b)x^3 + (-17a + 3b)x^2 + (6a - 17b)x + 6b = 0$
51. 20.44 **53. (a)** -0.05 **(b)** 0.07
55. (a) $x_2 = 2 - \sqrt{2}$; $x_3 = 2 + \sqrt{2}$
57. yes **59.** yes

EXERCISE SET 12.4

1. (a) yes **(b)** yes **(c)** yes **(d)** no
3. $[x - (-1)](x - 3)$
5. $4\left(x - \frac{1}{4}\right)[x - (-6)]$
7. $\left[x - \left(-\sqrt{5}\right)\right]\left(x - \sqrt{5}\right)$
9. $[x - (5 + i)][x - (5 - i)]$
11. $f(x) = x^3 + x^2 - 5x + 3$
13. $f(x) = x^4 - 16$
15. $f(x) = x^6 + 10x^4 - 87x^2 + 144$
17. $f(x) = -\frac{5}{42}x^2 + \frac{25}{42}x + \frac{30}{7}$
19. $f(x) = \frac{1}{30}x^3 - \frac{19}{30}x + 1$
21. $x^2 + \left(i + \sqrt{3}\right)x + i\sqrt{3} = 0$
23. $x^2 - 3x - 54 = 0$
25. $x^2 - 2x - 4 = 0$
27. $x^2 - 2ax + a^2 - b = 0$
29. $x^4 + 64 = (x + 2 - 2i)(x + 2 + 2i)(x - 2 - 2i)(x - 2 + 2i)$
33. $x = 4, 3, -3$

35. (b) $\tan 15° \approx 0.2679$ is a root.
(c) $x^3 - 3x^2 - 3x + 1 = (x+1)(x^2 - 4x + 1)$
(e) (i) 4; **(ii)** 1
(f) $\tan 15° = 2 - \sqrt{3}$; $\tan 75° = 2 + \sqrt{3}$

GRAPHING UTILITY EXERCISES FOR SECTION 12.4

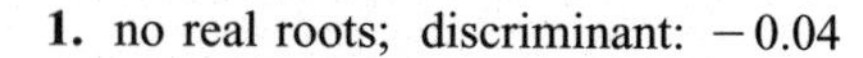

1. no real roots; discriminant: -0.04

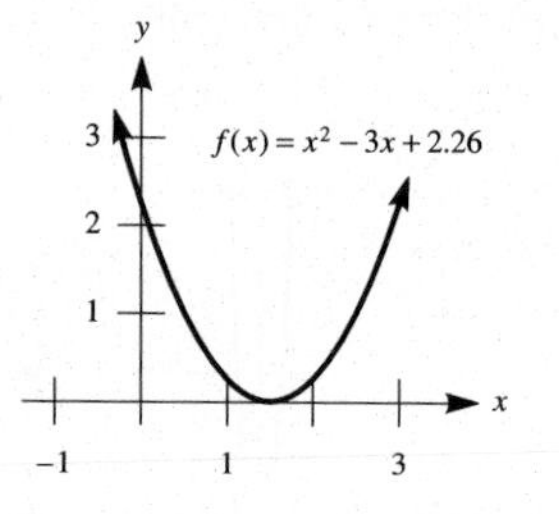

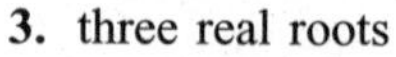

3. three real roots

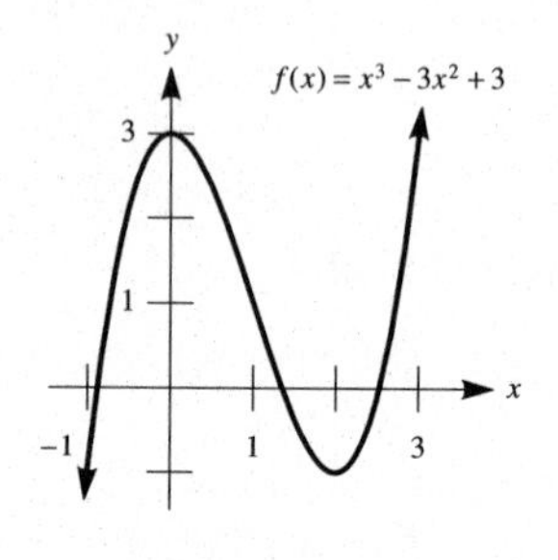

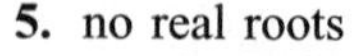

5. no real roots

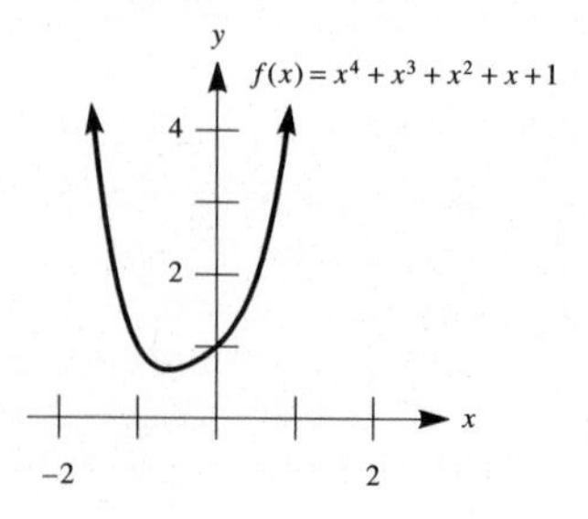

7. one real root

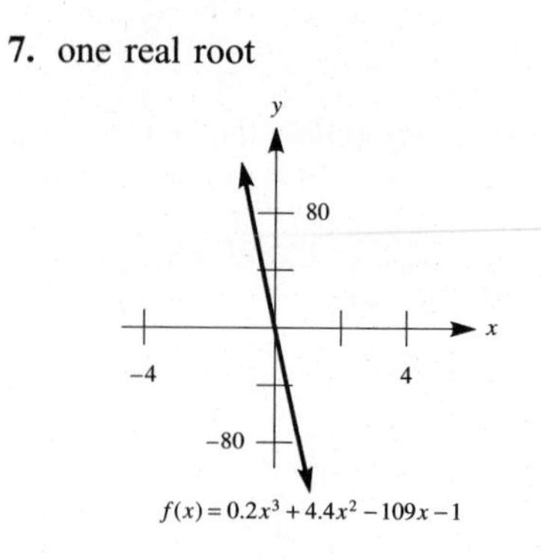

9. (a) $x = 4$ **(b)** The x-intercept is 4.

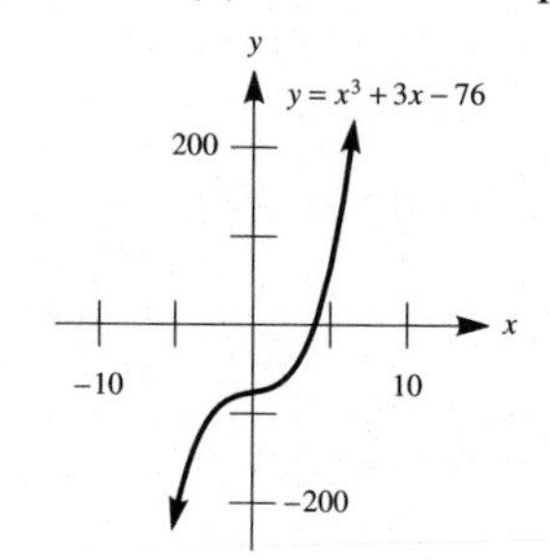

EXERCISE SET 12.5

1. $\pm 1, \pm\frac{1}{2}, \pm\frac{1}{4}, \pm 3, \pm\frac{3}{2}, \pm\frac{3}{4}$
3. $\pm 1, \pm\frac{1}{2}, \pm\frac{1}{4}, \pm\frac{1}{8}, \pm 3$ $\pm\frac{3}{2}, \pm\frac{3}{4}, \pm\frac{3}{8}, \pm 9, \pm\frac{9}{2}, \pm\frac{9}{4}, \pm\frac{9}{8}$
5. $\pm 1, \pm\frac{1}{2}, \pm 2, \pm 3, \pm\frac{3}{2}, \pm 6$
13. $x = 1, -1, -3$
15. $x = -\frac{1}{4}, -\sqrt{5}, \sqrt{5}$
17. $x = -1, -\frac{2}{3}, -\frac{1}{3}$
19. $x = 1, -1, (-1 + \sqrt{97})/2, (-1 - \sqrt{97})/2$
21. $x = 1$ (multiplicity 4)
23. $x = \frac{1}{2}, 6, -4$
25. $x = \frac{1}{4}, \frac{1}{5}, -\frac{\sqrt{6}}{2}, \frac{\sqrt{6}}{2}$
27. (a) upper bound: 2; lower bound: -1
(b) upper bound: 2; lower bound: -1
(c) upper bound: 6; lower bound: -2
29. $0.68 < r < 0.69$
31. $2.88 < r < 2.89$
33. $4.31 < r < 4.32$
35. $-2.15 < r < -2.14$
37. $-5.27 < r < -5.26$
39. (b) The result guarantees that A is a factor of B only in the case where A and C have no factor in common. Here A and C have a common factor of 5, so the result does not apply.
(c) Since $x = p/q$ is a root of the equation, this statement must be true.
41. $x = 0.32$

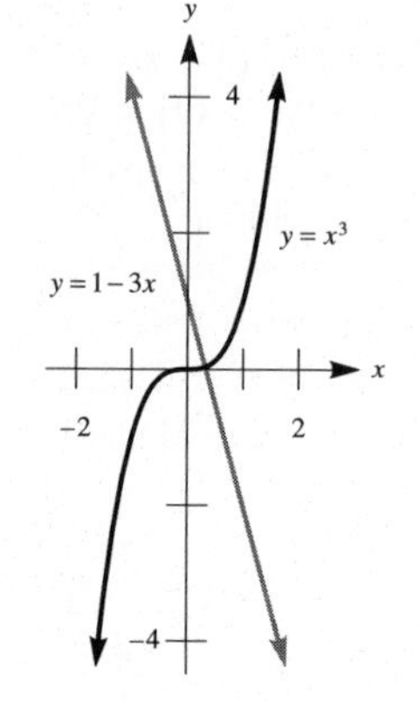

43. $x = 1.49$ **47. (b) (i)** 1 **(ii)** 4
49. $p = 2$
55. (a) $f(x) = (x+3)(x+1)(x-1)$

(b)

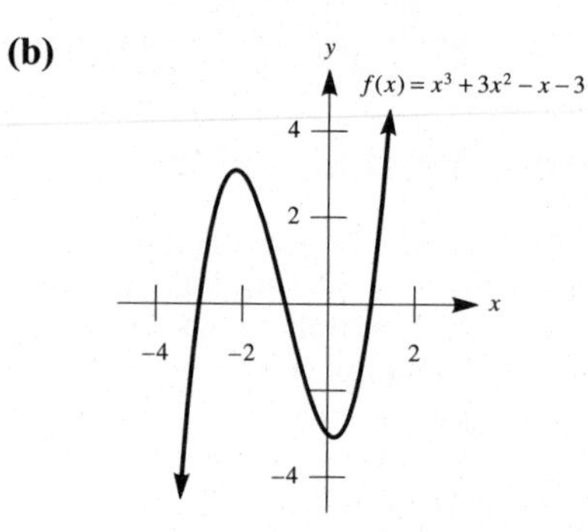

GRAPHING UTILITY EXERCISES FOR SECTION 12.5

1. (a)

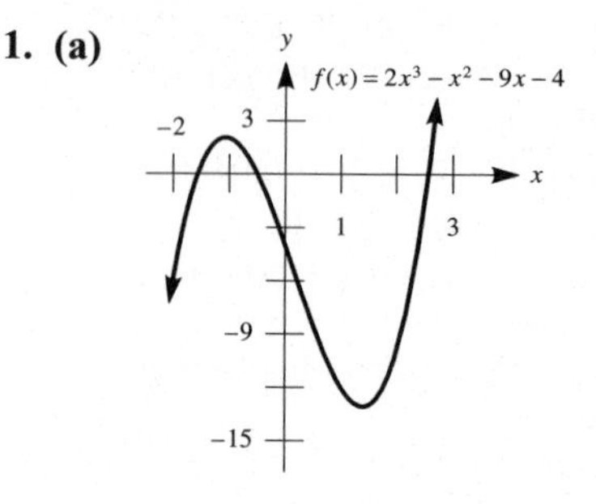

(b) x-intercepts: -0.5 (exact), $-1.562, 2.562$ (approximate)

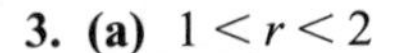

3. (a) $1 < r < 2$

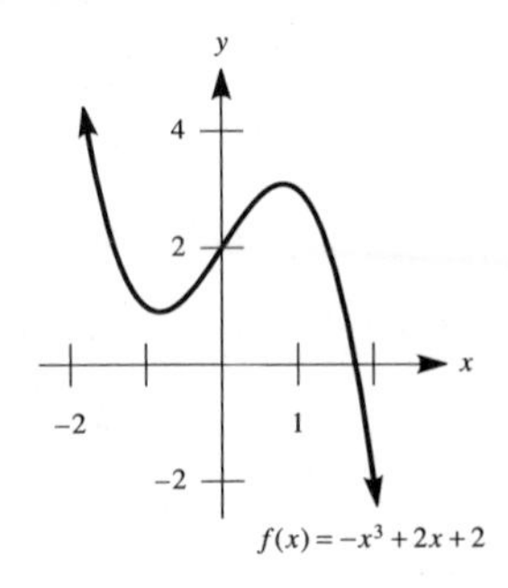

(b) tenths: $1.7 < r < 1.8$; hundredths: $1.76 < r < 1.77$; thousandths: $1.769 < r < 1.770$
(c) $r \approx 1.7693$
5. (a) $-11 < r_1 < -10$; $-1 < r_2 < 0$; $0 < r_3 < 1$

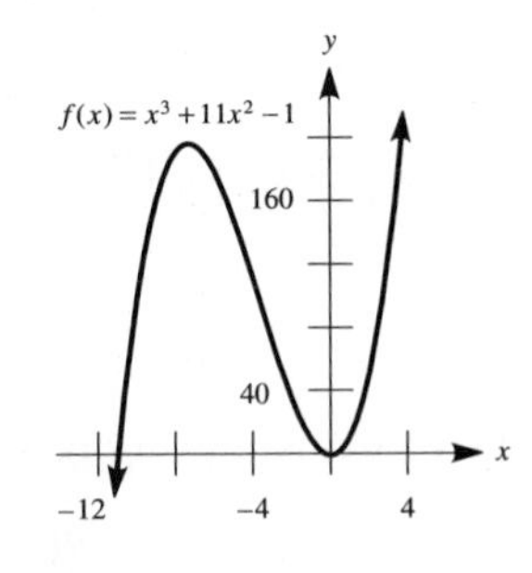

(b) tenths: $-11.0 < r_1 < -10.9$; $-0.4 < r_2 < -0.3$; $0.2 < r_3 < 0.3$; hundredths: $-11.00 < r_1 < -10.99$; $-0.31 < r_2 < -0.30$; $0.29 < r_3 < 0.30$; thousandths: $-10.992 < r_1 < -10.991$; $-0.306 < r_2 < -0.305$; $0.297 < r_3 < 0.298$
(c) $r_1 \approx -10.9917$; $r_2 \approx -0.3058$; $r_3 \approx 0.2975$
7. (a) $-1 < r_1 < 0$; $3 < r_2 < 4$

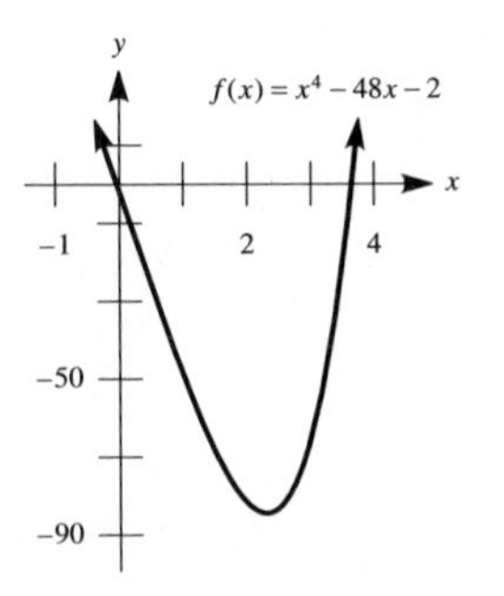

(b) tenths: $-0.1 < r_1 < 0.0$; $3.6 < r_2 < 3.7$; hundredths: $-0.05 < r_1 < -0.04$; $3.64 < r_2 < 3.65$; thousandths: $-0.042 < r_1 < -0.041$; $3.648 < r_2 < 3.649$
(c) $r_1 \approx -0.0417$; $r_2 \approx 3.6480$
9. no real roots

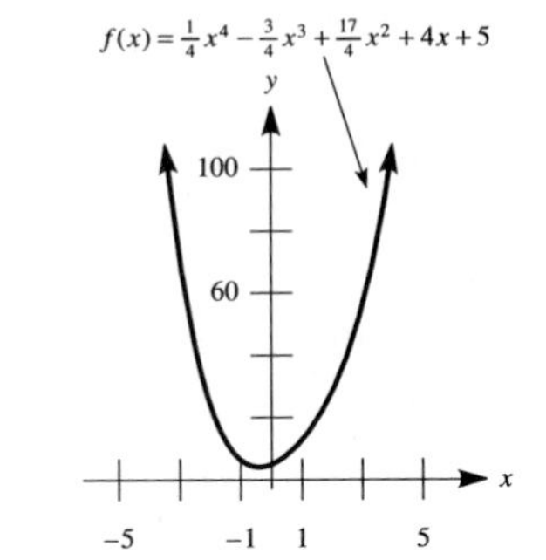

11. $[-1, 16]$

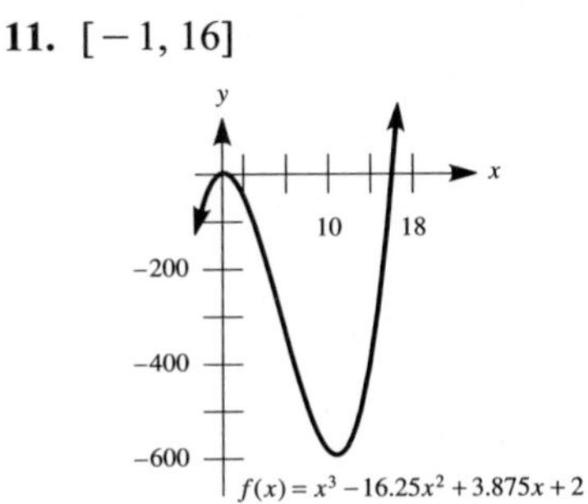

EXERCISE SET 12.6

1. $x = 7 + 2i$ **3.** $x = 5 - 2i$, 3
5. $x = -2 - i$, -3 (multiplicity 2)
7. $x = 6 + 5i$, $-\frac{1}{4}$
9. $x = 4 - \sqrt{2}i$, $-3i/2$, $3i/2$
11. $x = 10 - 2i$, $1 + \sqrt{5}$, $1 - \sqrt{5}$
13. $x = (1 - i\sqrt{2})/3$, $2/5$
15. $x = 3 + 2i$, $-1 - i$, $-1 + \sqrt{2}$, $-1 - \sqrt{2}$
17. $x^2 - 2x - 5 = 0$
19. $x^2 - \frac{4}{3}x - \frac{2}{3} = 0$
21. 2 complex roots and 1 negative real root
23. 4 complex roots and 1 negative real root
25. 2 complex roots, 1 positive real root, and 1 negative real root
27. either 1 positive real root and 2 negative real roots, or 1 positive real root and 2 complex roots
29. 1 positive real root, 1 negative real root, and 6 complex roots
31. 1 positive real root and 8 complex roots
33. 1 positive real root, 1 negative real root, and 6 complex roots
35. 1 positive real root, 1 negative real root, and 4 complex roots
37. 1 positive real root, 1 negative real root, and 2 complex roots
39. $f(x) = x^4 + 2x^2 + 49$

GRAPHING UTILITY EXERCISES FOR SECTION 12.6

1. (c) one positive real root

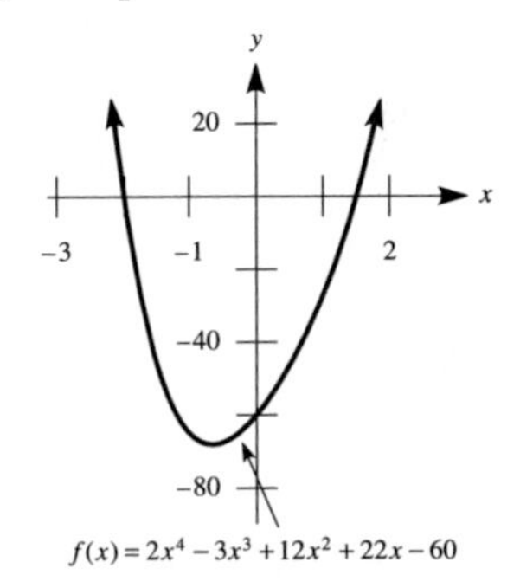

(d) $x = -2.000$ and 1.500
3. (a) Since $f(x) = x^3 + 8x + 5$ has no sign changes, there are no positive real roots. Since $f(-x) = -x^3 - 8x + 5$ has one sign change, there is exactly one negative real root.

(b)

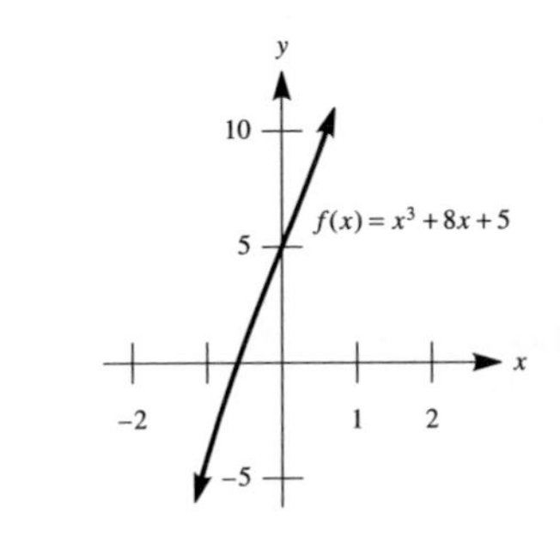

(c) $x \approx -0.5982$

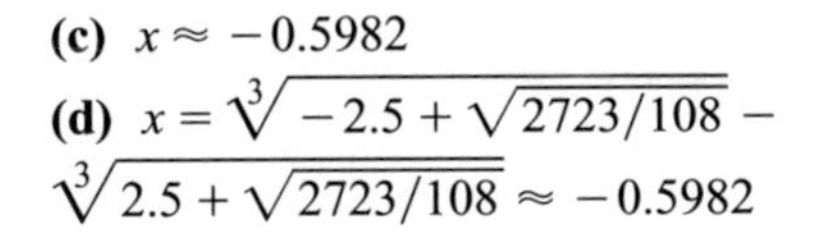

(d) $x = \sqrt[3]{-2.5 + \sqrt{2723/108}} - \sqrt[3]{2.5 + \sqrt{2723/108}} \approx -0.5982$

EXERCISE SET 12.7

1. (a) $A = 2, B = 5$ **(b)** $A = 2, B = 5$
3. (a) $A = 4, B = -1$
(b) $A = 4, B = -1$
5. (a) $A = -\frac{1}{4}, B = \frac{3}{4}$
(b) $A = -\frac{1}{4}, B = \frac{3}{4}$
7. $A = 8, B = -21$
9. $A = -\frac{1}{5}, B = \frac{34}{5}$
11. $A = 4, B = -1, C = 6$
13. $A = \frac{2}{5}, B = \frac{3}{5}, C = -\frac{3}{5}$
15. $A = 1, B = -1, C = 1$
17. $A = \frac{10}{3}, B = \frac{1}{6}, C = -\frac{1}{2}$
19. $A = 4, B = 1, C = -5$
21. $A = 0, B = 1, C = 0, D = -4$
23. $A = 3, B = 4, C = -1$
25. (a) $A = \frac{1}{4}$ **(c)** $C = 1, B = -\frac{1}{4}$
27. $\left(x^2 + \frac{-1+\sqrt{5}}{2}x + 1\right)\left(x^2 + \frac{-1-\sqrt{5}}{2}x + 1\right)$

EXERCISE SET 12.8

1. (a) no (reducible)
(b) yes (irreducible)
3. (a) no (reducible)
(b) yes (irreducible)
5. (a) $x^2 - 100 = (x+10)(x-10)$
(b) $\frac{11x+30}{x^2-100} = \frac{A}{x+10} + \frac{B}{x-10}$
(c) $A = 4, B = 7$
7. (a) $x^2 - 5 = (x+\sqrt{5})(x-\sqrt{5})$
(b) $\frac{8x - 2\sqrt{5}}{x^2-5} = \frac{A}{x+\sqrt{5}} + \frac{B}{x-\sqrt{5}}$
(c) $A = 5, B = 3$
9. (a) $x^2 - x - 6 = (x+2)(x-3)$
(b) $\frac{7x+39}{x^2-x-6} = \frac{A}{x+2} + \frac{B}{x-3}$
(c) $A = -5, B = 12$
11. (a) $x^3 - 3x^2 - 4x + 12 = (x-3)(x-2)(x+2)$
(b) $\frac{3x^2+17x-38}{x^3-3x^2-4x+12} = \frac{A}{x-3} + \frac{B}{x-2} + \frac{C}{x+2}$
(c) $A = 8, B = -2, C = -3$
13. (a) $x^3 + x^2 + x = x(x^2+x+1)$
(b) $\frac{5x^2+2x+5}{x^3+x^2+x} = \frac{A}{x} + \frac{Bx+C}{x^2+x+1}$
(c) $A = 5, B = 0, C = -3$
15. (a) $x^4 + 2x^2 + 1 = (x^2+1)^2$
(b) $\frac{2x^3+5x-4}{x^4+2x^2+1} = \frac{Ax+B}{x^2+1} + \frac{Cx+D}{(x^2+1)^2}$
(c) $A = 2, B = 0, C = 3, D = -4$
17. $\frac{x^2+2}{x^3-3x^2-16x-12} = \frac{19/28}{x-6} + \frac{-3/7}{x+1} + \frac{3/4}{x+2}$
19. $\frac{5-x}{6x^2-19x+15} = \frac{10}{3x-5} + \frac{-7}{2x-3}$
21. $\frac{2x+1}{x^3-5x} = \frac{-1/5}{x} + \frac{(1-2\sqrt{5})/10}{x+\sqrt{5}} + \frac{(1+2\sqrt{5})/10}{x-\sqrt{5}}$
23. $\frac{x^3+2}{x^4+8x^2+16} = \frac{x}{x^2+4} + \frac{-4x+2}{(x^2+4)^2}$
25. $\frac{x^3+x-3}{x^4-15x^3+75x^2-125x} = \frac{122/125}{x-5} + \frac{253/25}{(x-5)^2} + \frac{127/5}{(x-5)^3} + \frac{3/125}{x}$
27. $\frac{1}{x^3-1} = \frac{\frac{1}{3}}{x-1} + \frac{-\frac{1}{3}x - \frac{2}{3}}{x^2+x+1}$
29. $\frac{7x^3+11x^2-x-2}{x^4+2x^3+x^2} = \frac{3}{x} + \frac{-2}{x^2} + \frac{4}{x+1} + \frac{3}{(x+1)^2}$
31. $\frac{x^3-5}{x^4-81} = \frac{\frac{8}{27}}{x+3} + \frac{\frac{11}{54}}{x-3} + \frac{\frac{1}{2}x + \frac{5}{18}}{x^2+9}$
33. $\frac{1}{x^4+x^3+2x^2+x+1} = \frac{-x}{x^2+1} + \frac{x+1}{x^2+x+1}$
35. $x = 2, -3, -1$
37. (b) $C = -2, D = 0$
39. $\frac{6x^3-16x^2-13x+25}{x^2-4x+3} = 6x + 8 + \frac{-1}{x-1} + \frac{2}{x-3}$
41. $\frac{x^5-10x^4+36x^3-55x^2+32x+1}{x^4-6x^3+12x^2-8x} = x - 4 + \frac{-1/8}{x} + \frac{1/8}{x-2} + \frac{3/4}{(x-2)^2} + \frac{5/2}{(x-2)^3}$
43. $\frac{x^6+2x^5+5x^4-x^2-2x-4}{x^4-1} = x^2 + 2x + 5 + \frac{-\frac{1}{4}}{x+1} + \frac{\frac{1}{4}}{x-1} + \frac{\frac{1}{2}}{x^2+1}$
47. $\frac{px+q}{(x-a)(x-b)} = \frac{(pa+q)/(a-b)}{x-a} + \frac{(pb+q)/(b-a)}{x-b}$
49. $\frac{px+q}{(x-a)(x+a)} = \frac{(pa+q)/(2a)}{x-a} + \frac{(pa-q)/(2a)}{x+a}$
51. $\frac{1}{(1-ax)(1-bx)(1-cx)} = \frac{a^2/(a-b)(a-c)}{1-ax} + \frac{b^2/(b-a)(b-c)}{1-bx} + \frac{c^2/(c-a)(c-b)}{1-cx}$
53. $\frac{1}{x^4+1} = \frac{\frac{\sqrt{2}}{4}x + \frac{1}{2}}{x^2+\sqrt{2}x+1} + \frac{-\frac{\sqrt{2}}{4}x + \frac{1}{2}}{x^2-\sqrt{2}x+1}$

CHAPTER 12 REVIEW EXERCISES

1. $q(x) = x^3 + x^2 - 3x + 1$; $R(x) = -1$
3. quotient: $x^3 + 3x^2 + 7x + 21$; remainder: 71
5. quotient: $2x^2 - 13x + 46$; remainder: -187
7. quotient: $5x - 20$; remainder: 0
9. $f(10) = 99{,}904$
11. $f(1/10) = -999/1000$
13. $f(a-1) = a^3$
15. (a) $f(-0.3) \approx -0.24$
(b) $f(-0.39) \approx -0.007$
(c) $f(-0.394) \approx 0.00003$
17. $a = -5$ **19.** $a = -1, -2$
23. $\pm 1, \pm 2, \pm 3, \pm 6, \pm 9, \pm 18$
25. $\pm 1, \pm \frac{1}{2}, \pm 2, \pm 4, \pm 8$
27. $\pm p, \pm 1$ **29.** $x = 2, -\frac{3}{2}, -1$
31. $x = \frac{5}{2}, -1 + \sqrt{3}, -1 - \sqrt{3}$
33. $x = \frac{2}{3}, (-1 + i\sqrt{3})/2, (-1 - i\sqrt{3})/2$
35. $x = -1, -\sqrt{7}$ (multiplicity 2), $\sqrt{7}$ (multiplicity 2)
37. $x = 2$ (multiplicity 2), 5
39. (a) Let $p(x)$ and $d(x)$ be the polynomials where $d(x) \neq 0$. Then

there are unique polynomials $q(x)$ and $R(x)$ such that

$$p(x) = d(x) \cdot q(x) + R(x)$$

where either $R(x) = 0$ or the degree of $R(x)$ is less than the degree of $d(x)$.
(b) When a polynomial $f(x)$ is divided by $x - r$, the remainder is $f(r)$.
(c) Let $f(x)$ be a polynomial. If $f(r) = 0$, then $x - r$ is a factor of $f(x)$. Conversely, if $x - r$ is a factor of $f(x)$, then $f(r) = 0$.
(d) Every polynomial equation of the form

$$a_nx^n + a_{n-1}x^{n-1} + \cdots + a_1x + a_0 = 0 \quad (n \geq 1,\ a_n \neq 0)$$

has at least one root among the complex numbers. (This root may be a real number.)
41. $6\left(x - \frac{4}{3}\right)\left[x - \left(-\frac{5}{2}\right)\right]$
43. $(x - 4)\left(x - \left(-\sqrt[3]{5}\right)\right)$ $\left(x - \dfrac{\sqrt[3]{5} + i\sqrt{3\sqrt[3]{25}}}{2}\right)$ $\left(x - \dfrac{\sqrt[3]{5} - i\sqrt{3\sqrt[3]{25}}}{2}\right)$
45. $x = 2 - 3i,\ 2 + 3i,\ 3$
47. $x = 1 + i\sqrt{2},\ 1 - i\sqrt{2},\ \sqrt{7},\ -\sqrt{7}$
49. 1 positive real root and 2 complex roots **51.** 1 negative real root and 2 complex roots **53.** 1 positive real root, 1 negative real root, and 2 complex roots
55. **(c)** $x = -1,\ i,\ -i$
57. $0.82 < x < 0.83$
59. **(c)** $6.93 < x < 6.94$
61. $x^2 - 8x + 11 = 0$
63. $x^4 - 12x^3 + 35x^2 + 60x - 200 = 0$
65. $x^4 - 4x^3 - 4x^2 + 16x - 8 = 0$
67. zeros: 0, 3, -1

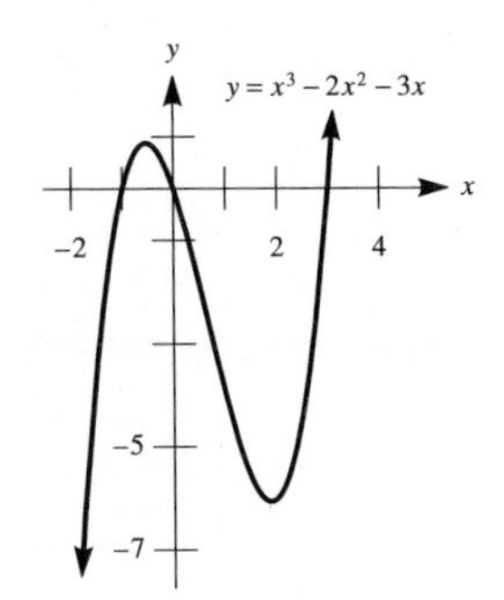

69. zeros: 0, -2, 2

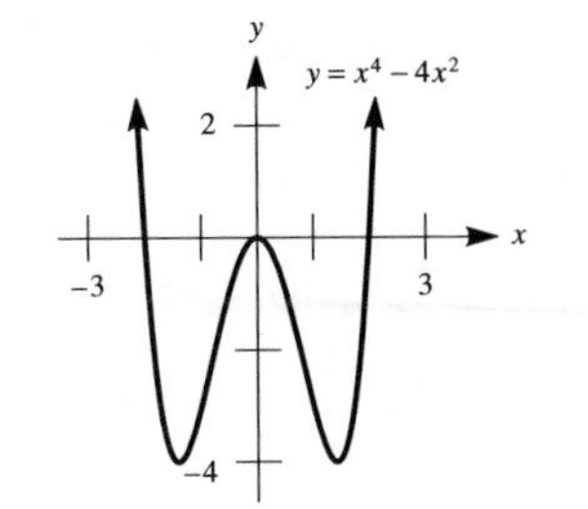

71. $5 + 6i$ **73.** 6 **75.** $\frac{1}{2} - \frac{\sqrt{3}}{2}i$
77. $3\sqrt{2} - 4\sqrt{2}i$
81. **(a)** $|6 + 2i| = |6 - 2i| = 2\sqrt{10}$
(b) 3
87. $\dfrac{2x - 1}{100 - x^2} = \dfrac{-21/20}{10 + x} + \dfrac{19/20}{10 - x}$
89. $\dfrac{1}{x^3 + 2x^2 + x} = \dfrac{1}{x} + \dfrac{-1}{x + 1} + \dfrac{-1}{(x + 1)^2}$
91. $\dfrac{x^3 + 2}{x^4 + 6x^2 + 9} = \dfrac{x}{x^2 + 3} + \dfrac{-3x + 2}{(x^2 + 3)^2}$
93. **(a)** $x^4 - 2x^3 + x^2 - 1 = (x^2 - x + 1)(x^2 - x - 1)$
(b) $\dfrac{x^3}{x^4 - 2x^3 + x^2 - 1} = \dfrac{\frac{1}{2}}{x^2 - x + 1} + \dfrac{x + \frac{1}{2}}{x^2 - x - 1}$

CHAPTER 12 TEST

1. $f\left(\frac{1}{2}\right) = -\frac{3}{2}$
2. $x = -3,\ 1 + \sqrt{6},\ 1 - \sqrt{6}$
3. $\pm 1,\ \pm\frac{1}{2},\ \pm 2,\ \pm 3,\ \pm\frac{3}{2},\ \pm 6$
4. $f(x) = 3x^2 + 21x - 24$
5. quotient: $4x^2 - 3x - 5$; remainder: 8
6. **(a)** Let $f(x)$ be a polynomial. If $f(r) = 0$, then $x - r$ is a factor of $f(x)$. Conversely, if $x - r$ is a factor of $f(x)$, then $f(r) = 0$.
(b) Every polynomial of the form

$$a_nx^n + a_{n-1}x^{n-1} + \cdots + a_1x + a_0 = 0 \quad (n \geq 1,\ a_n \neq 0)$$

has at least one root among the complex numbers. (This root may be a real number.)
(c) Let $f(x)$ be a polynomial, all of whose coefficients are real numbers. Then $f(x)$ can be factored (over the real numbers) into a product of linear and/or irreducible quadratic factors.
7. **(a)** 3 **(b)** $2.2 < x < 2.3$
8. $x = 1 \pm i,\ 3 \pm 2i,\ -2$
9. $q(x) = x^2 + 2x - 1,\ R(x) = -3x + 7$
10. $2\left[x - \left(\frac{3}{2} + \frac{1}{2}i\right)\right]\left[x - \left(\frac{3}{2} - \frac{1}{2}i\right)\right]$
11. **(a)** $\pm 1,\ \pm 2,\ \pm 3,\ \pm 4,\ \pm 6,\ \pm 8,\ \pm 12,\ \pm 24$ **(c)** $x = 1$ **(d)** none
12. **(a)** $x = 3/2$
(b) $x = 3/2,\ \left(-1 + i\sqrt{3}\right)/2,\ \left(-1 - i\sqrt{3}\right)/2$
13. 1 positive real root, 1 negative real root, and 2 complex roots
14. $x^3 + 6x + 20 = 0$
15. $f(x) = (x - 2)(x - 3i)^3\left[x - \left(1 + \sqrt{2}\right)\right]^2$
16. $21 - i$
17. **(a)** discriminant: -8; two complex conjugate roots
(b) $x = 1 \pm \left(\sqrt{2}/2\right)i$
18. $\dfrac{3x - 1}{x^3 - 16x} = \dfrac{\frac{1}{16}}{x} + \dfrac{-\frac{13}{32}}{x + 4} + \dfrac{\frac{11}{32}}{x - 4}$
19. $\dfrac{1}{x^3 - x^2 + 3x - 3} = \dfrac{\frac{1}{4}}{x - 1} + \dfrac{-\frac{1}{4}x - \frac{1}{4}}{x^2 + 3}$
20. $\dfrac{4x^2 - 15x + 20}{x^3 - 4x^2 + 4x} = \dfrac{5}{x} + \dfrac{-1}{x - 2} + \dfrac{3}{(x - 2)^2}$

CHAPTER THIRTEEN

EXERCISE SET 13.1

27. **(a)**

n	1	2	3	4	5
$f(n)$	$\frac{1}{2}$	$\frac{2}{3}$	$\frac{3}{4}$	$\frac{4}{5}$	$\frac{5}{6}$

(b) 6/7
(c) $f(n) = \dfrac{1}{1 \times 2} + \dfrac{1}{2 \times 3} + \cdots + \dfrac{1}{n(n + 1)} = \dfrac{n}{n + 1}$

29. **(a)**

n	1	2	3	4	5
$f(n)$	1	4	9	16	25

(b) 36 **(c)** $f(n) = n^2$

EXERCISE SET 13.2

3. $a^9 + 9a^8b + 36a^7b^2 + 84a^6b^3 + 126a^5b^4 + 126a^4b^5 + 84a^3b^6 + 36a^2b^7 + 9ab^8 + b^9$
5. $8A^3 + 12A^2B + 6AB^2 + B^3$
7. $1 - 12x + 60x^2 - 160x^3 + 240x^4 - 192x^5 + 64x^6$
9. $x^2 + 4x\sqrt{xy} + 6xy + 4y\sqrt{xy} + y^2$
11. $x^{10} + 5x^8y^2 + 10x^6y^4 + 10x^4y^6 + 5x^2y^8 + y^{10}$
13. $1 - \frac{6}{x} + \frac{15}{x^2} - \frac{20}{x^3} + \frac{15}{x^4} - \frac{6}{x^5} + \frac{1}{x^6}$
15. $\frac{x^3}{8} - \frac{x^2y}{4} + \frac{xy^2}{6} - \frac{y^3}{27}$
17. $a^7b^{14} + 7a^6b^{12}c + 21a^5b^{10}c^2 + 35a^4b^8c^3 + 35a^3b^6c^4 + 21a^2b^4c^5 + 7ab^2c^6 + c^7$
19. $x^8 + 8\sqrt{2}x^7 + 56x^6 + 112\sqrt{2}x^5 + 280x^4 + 224\sqrt{2}x^3 + 224x^2 + 64\sqrt{2}x + 16$
21. $5\sqrt{2} - 7$ **23.** $89\sqrt{3} + 109\sqrt{2}$
25. $12 - 24\sqrt[3]{2} + 12\sqrt[3]{4}$
27. $x^{10} - 10x^9 + 35x^8 - 40x^7 - 30x^6 + 68x^5 + 30x^4 - 40x^3 - 35x^2 - 10x - 1$
29. 120 **31.** 105 **33. (a)** 10 **(b)** 5
35. $n^2 + 3n + 2$ **37.** 0 **39.** $120a^2b^{14}$
41. $100x^{99}$ **43.** 45 **45.** 294912
47. 28 **49.** 40095 **51.** $\frac{(2n)!}{(n!)^2}$

53. (a)

k	0	1	2	3	4	5	6	7	8
$\binom{8}{k}$	1	8	28	56	70	56	28	8	1

EXERCISE SET 13.3

1. $\frac{1}{2}, \frac{2}{3}, \frac{3}{4}, \frac{4}{5}$ **3.** $-1, 1, -1, 1$
5. $\frac{1}{2}, \frac{1}{4}, \frac{1}{8}, \frac{1}{16}$ **7.** 3, 6, 9, 12
9. $2, \frac{9}{4}, \frac{64}{27}, \frac{625}{256}$ **11.** $-1, 0, \frac{1}{3}, \frac{1}{2}$
13. $-2, 1, -\frac{8}{9}, 1$
15. 1, 4, 25, 676, 458329
17. 2, 2, 4, 8, 32 **19.** 1, 1, 2, 6, 24
21. 0, 1, 2, 4, 16 **23.** 62 **25.** 40
27. $-\frac{19}{30}$ **29.** 903 **31.** 278

33.

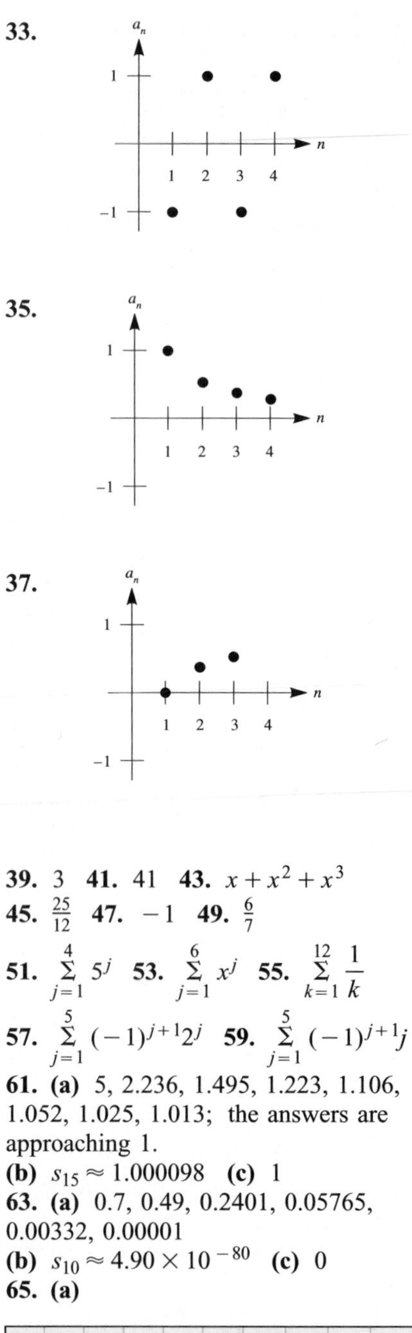

35.

37.

39. 3 **41.** 41 **43.** $x + x^2 + x^3$
45. $\frac{25}{12}$ **47.** -1 **49.** $\frac{6}{7}$
51. $\sum_{j=1}^{4} 5^j$ **53.** $\sum_{j=1}^{6} x^j$ **55.** $\sum_{k=1}^{12} \frac{1}{k}$
57. $\sum_{j=1}^{5} (-1)^{j+1}2^j$ **59.** $\sum_{j=1}^{5} (-1)^{j+1}j$
61. (a) 5, 2.236, 1.495, 1.223, 1.106, 1.052, 1.025, 1.013; the answers are approaching 1.
(b) $s_{15} \approx 1.000098$ **(c)** 1
63. (a) 0.7, 0.49, 0.2401, 0.05765, 0.00332, 0.00001
(b) $s_{10} \approx 4.90 \times 10^{-80}$ **(c)** 0
65. (a)

F_1	F_2	F_3	F_4	F_5	F_6	F_7	F_8	F_9	F_{10}
1	1	2	3	5	8	13	21	34	55

(b) $F_{22} = 17{,}711$; $F_{19} = 4181$
(c) $F_{30} = 832{,}040$
71. (a) $F_1 = 1$; $F_2 = 1$
(b) $F_{24} = 46{,}368$; $F_{25} = 75{,}025$
(c) $F_{26} = 121{,}393$

EXERCISE SET 13.4

1. (a) $d = 2$ **(b)** $d = -4$
(c) $d = \frac{1}{3}$ **(d)** $d = \sqrt{2}$
3. $a_{12} = 131$ **5.** $a_{100} = 501$
7. $a_{1000} = 998$ **9.** $d = \frac{11}{6}$, $a = -\frac{23}{2}$
11. $a = -190$ **13.** $d = -\frac{1}{8}$
15. 500,500 **17.** $91\pi/3$ **19.** $a_1 = \frac{1}{2}$
21. $S_{16} = -768$, $d = -\frac{104}{15}$
23. $d = \frac{6}{5}$, $a = -\frac{17}{5}$ **25.** 900
27. 2, 10, 18, or 18, 10, 2
29. $-1, 2, 5$ or $5, 2, -1$
31. (b) $-6 - 9\sqrt{2}$
33. $\frac{n}{2(1-b)}[2 + (n-3)\sqrt{b}]$

EXERCISE SET 13.5

1. 6 **3.** 20, 100 **5.** $a_{100} = 1$
7. $a_8 = \frac{256}{6561}$ **9.** $r = \pm 4$
11. 7161 **13.** $63 + 31\sqrt{2}$ **15.** $\frac{1995}{64}$
17. $\frac{11111}{1000000}$ or 0.011111
19. $\frac{2}{5}$ **21.** 101 **23.** $\frac{5}{9}$ **25.** $\frac{61}{495}$
27. $\frac{16}{37}$ **29.** $r = -\frac{1}{2}, -2$ **35.** 12 ft

EXERCISE SET 13.6

1.

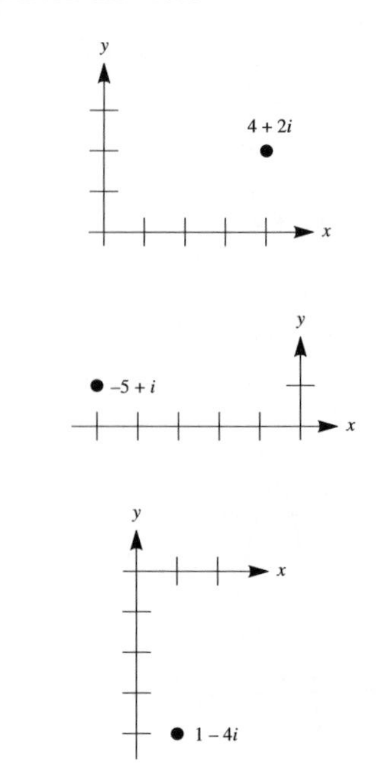

3.

5.

7.

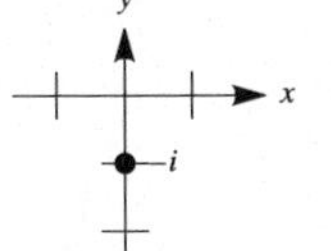

9. $\sqrt{2}+\sqrt{2}i$ **11.** $-2\sqrt{3}+2i$
13. $-1-i$ **15.** $\sqrt{3}i$
17. $(\sqrt{6}-\sqrt{2})+(\sqrt{6}+\sqrt{2})i$
19. $\cos(\pi/6)+i\sin(\pi/6)$
21. $2[\cos(2\pi/3)+i\sin(2\pi/3)]$
23. $4[\cos(7\pi/6)+i\sin(7\pi/6)]$
25. $6[\cos(3\pi/2)+i\sin(3\pi/2)]$
27. $\frac{1}{2}[\cos(11\pi/6)+i\sin(11\pi/6)]$
29. $3+3\sqrt{3}i$ **31.** $1-\sqrt{3}i$
33. $3\sqrt{2}\cos(2\pi/7)+[3\sqrt{2}\sin(2\pi/7)]i$
35. $\dfrac{3\sqrt{2}}{2}+\dfrac{3\sqrt{2}}{2}i$ **37.** $\sqrt{3}+i$
39. 1 **41.** $\dfrac{243}{2}-\dfrac{243\sqrt{3}}{2}i$
43. $\dfrac{\sqrt{2}}{128}+\dfrac{\sqrt{2}}{128}i$ **45.** $-4i$
47. $-1-i$ **49.** $\dfrac{1}{2}+\dfrac{\sqrt{3}}{2}i$
51. $128+128i$
53. $-128+128\sqrt{3}i$
55. $3i,\ -\frac{3}{2}\sqrt{3}-\frac{3}{2}i,\ \frac{3}{2}\sqrt{3}-\frac{3}{2}i$
57. $1,\ \frac{1}{2}\sqrt{2}+\frac{1}{2}\sqrt{2}i,\ i,$
$-\frac{1}{2}\sqrt{2}+\frac{1}{2}\sqrt{2}i,\ -1,$
$-\frac{1}{2}\sqrt{2}-\frac{1}{2}\sqrt{2}i,\ -i,\ \frac{1}{2}\sqrt{2}-\frac{1}{2}\sqrt{2}i$
59. $4,\ -2+2\sqrt{3}i,\ -2-2\sqrt{3}i$
61. $3,\ \frac{3}{2}+\frac{3}{2}\sqrt{3}i,\ -\frac{3}{2}+\frac{3}{2}\sqrt{3}i,\ -3,$
$-\frac{3}{2}-\frac{3}{2}\sqrt{3}i,\ \frac{3}{2}-\frac{3}{2}\sqrt{3}i$
63. 92,236,816
65. $0.95+0.31i,\ i,\ -0.95+0.31i,$
$-0.59-0.81i,\ 0.59-0.81i$
67. $\frac{1}{2}(\sqrt{6}-\sqrt{2})+\frac{1}{2}(\sqrt{6}+\sqrt{2})i,$
$-\frac{1}{2}(\sqrt{6}+\sqrt{2})+\frac{1}{2}(\sqrt{6}-\sqrt{2})i,$
$\frac{1}{2}(\sqrt{2}-\sqrt{6})-\frac{1}{2}(\sqrt{2}+\sqrt{6})i$, and
$\frac{1}{2}(\sqrt{2}+\sqrt{6})+\frac{1}{2}(\sqrt{2}-\sqrt{6})i$
69. **(a)** $1,\ -\frac{1}{2}+\frac{1}{2}\sqrt{3}i,\ -\frac{1}{2}-\frac{1}{2}\sqrt{3}i$
71. -1 **73.** 1

CHAPTER 13 REVIEW EXERCISES

11. $81a^4+108a^3b^2+54a^2b^4+12ab^6+b^8$
13. $x^4+4x^3\sqrt{x}+6x^3+4x^2\sqrt{x}+x^2$
15. $x^{10}-10x^8y^2+40x^6y^4-80x^4y^6+80x^2y^8-32y^{10}$
17. $1+\dfrac{5}{x}+\dfrac{10}{x^2}+\dfrac{10}{x^3}+\dfrac{5}{x^4}+\dfrac{1}{x^5}$
19. $a^4b^2-4a^3b^2\sqrt{ab}+6a^3b^3-4a^2b^3\sqrt{ab}+a^2b^4$
21. $15xy^8$ **23.** 84 **25.** 28
33. $1,\ \frac{4}{3},\ \frac{3}{2},\ \frac{8}{5}$

35. $-\frac{1}{2},\ \frac{2}{3},\ -\frac{3}{4},\ \frac{4}{5}$

37. $-3,\ -12,\ -48,\ -192$
39. **(a)** -5 **(b)** $\frac{9}{10}$ **41.** $\displaystyle\sum_{k=1}^{5}\frac{5}{3k}$
43. 73 **45.** $\frac{5}{1024}$ **47.** 177
49. 7,777,777,777 **51.** $-5\sqrt{10}$
53. $\frac{3}{4}$ **55.** $\frac{1}{10}$ **57.** $\frac{5}{11}$
63. **(a)** 42929 **(b)** 42925; 0.00932% error **(c)** 6.48040×10^{10}
(d) 6.48027×10^{10}; 2×10^{-3}% error
65. $a=40,\ b=20,\ c=10$ or $a=10,\ b=20,\ c=40$
69. **(a)** No such value for x exists.
(b) $x=(b^2-ac)/(a+c-2b)$
73. $\frac{3}{2}+\frac{3}{2}\sqrt{3}i$ **75.** $2^{-1/4}-2^{-1/4}i$
77. $1[\cos(\pi/3)+i\sin(\pi/3)]$
79. $6[\cos(5\pi/4)+i\sin(5\pi/4)]$
81. $5\sqrt{2}+5\sqrt{2}i$ **83.** $\sqrt{2}-\sqrt{2}i$
85. $\frac{27}{2}+\frac{27}{2}\sqrt{3}i$ **87.** $512-512\sqrt{3}i$
89. $1,\ \frac{1}{2}+\frac{1}{2}\sqrt{3}i,\ -\frac{1}{2}+\frac{1}{2}\sqrt{3}i,\ -1,$
$-\frac{1}{2}-\frac{1}{2}\sqrt{3}i$, and $\frac{1}{2}-\frac{1}{2}\sqrt{3}i$
91. $-\dfrac{\sqrt{4+2\sqrt{2}}}{2}+\dfrac{\sqrt{4-2\sqrt{2}}}{2}i$
and $\dfrac{\sqrt{4+2\sqrt{2}}}{2}-\dfrac{\sqrt{4-2\sqrt{2}}}{2}i$
93. $1.06+0.17i,\ 0.17+1.06i,$
$-0.95+0.49i,\ -0.76-0.76i$, and
$0.49-0.95i$

CHAPTER 13 TEST

2. **(a)** $-1+9+19=27$
(b) $-1+4-9=-6$
3. **(a)** $S_n=\dfrac{a(1-r^n)}{1-r}$ **(b)** $\frac{174075}{1024}$
4. **(a)** 42240 **(b)** $5280a^7b^{12}$
5. $243x^{10}+405x^8y^3+270x^6y^6+90x^4y^9+15x^2y^{12}+y^{15}$
6. 177 **7.** $\frac{7}{9}$ **8.** 5, 27 **9.** $-5\sqrt{10}$
10. 224 **11.** $-1+\sqrt{3}i$
12. $2\left[\cos\left(\frac{7\pi}{4}\right)+i\sin\left(\frac{7\pi}{4}\right)\right]$
13. $\frac{15}{2}+\frac{15}{2}\sqrt{3}i$ **14.** $2\sqrt{3}+2i$,
$-2\sqrt{3}+2i$, and $-4i$

APPENDIX

EXERCISE SET A.1

1.

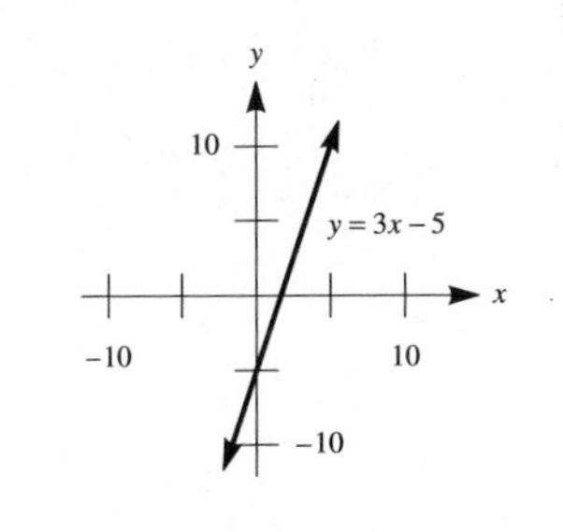

3.

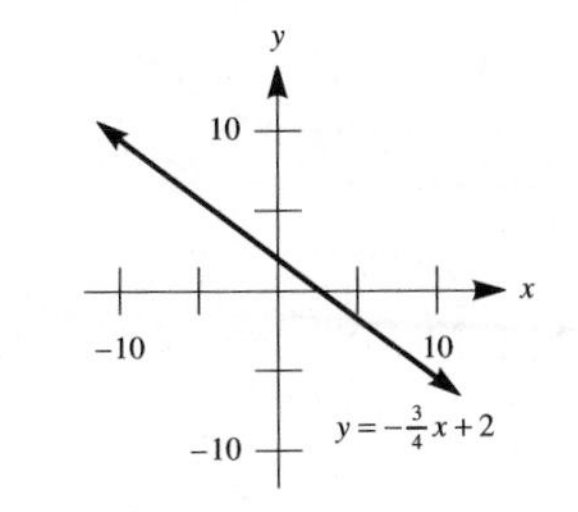

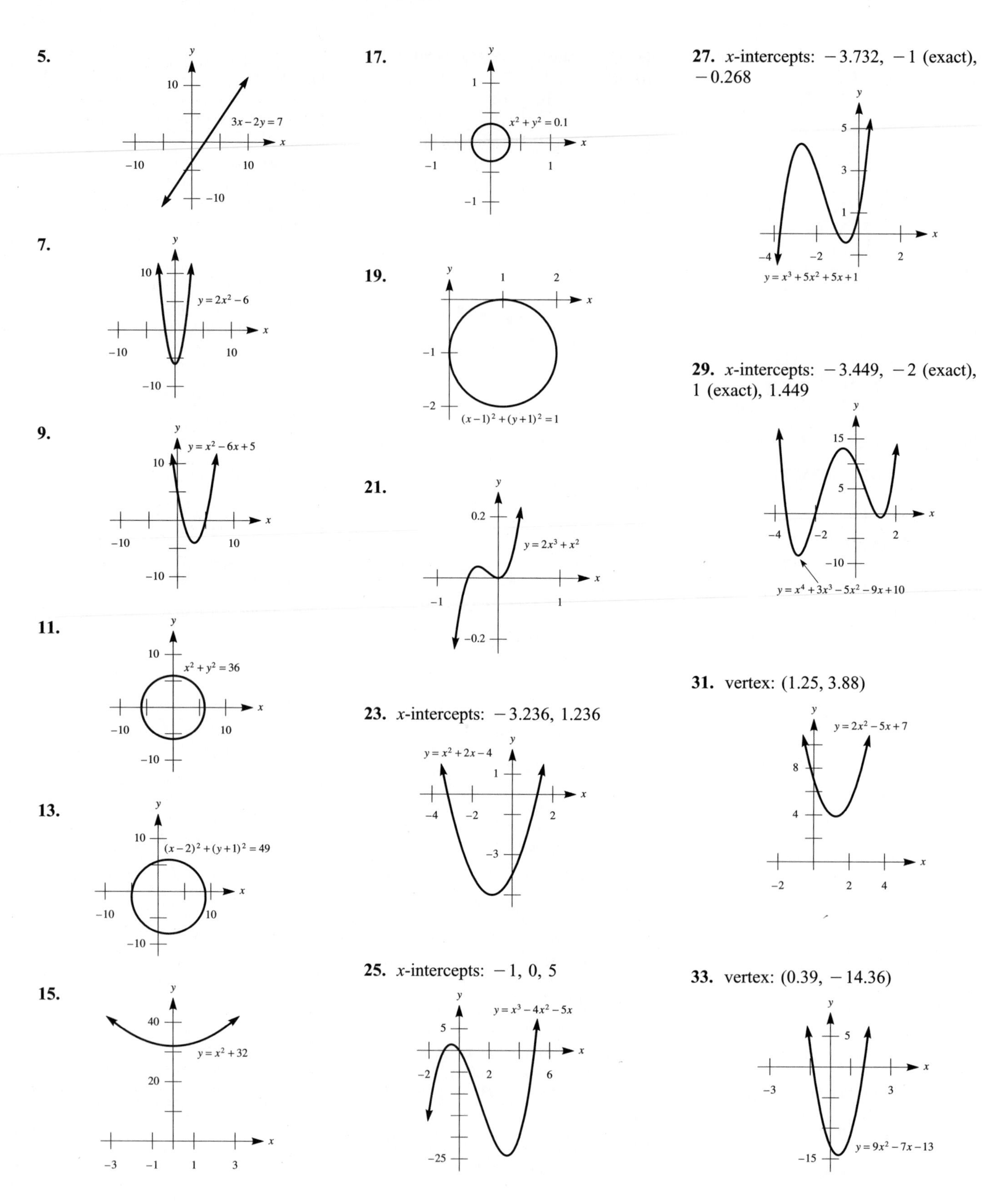

5.
$3x - 2y = 7$
7.
$y = 2x^2 - 6$
9.
$y = x^2 - 6x + 5$
11.
$x^2 + y^2 = 36$
13.
$(x-2)^2 + (y+1)^2 = 49$
15.
$y = x^2 + 32$
17.
$x^2 + y^2 = 0.1$
19.
$(x-1)^2 + (y+1)^2 = 1$
21.
$y = 2x^3 + x^2$
23. x-intercepts: -3.236, 1.236
$y = x^2 + 2x - 4$
25. x-intercepts: -1, 0, 5
$y = x^3 - 4x^2 - 5x$
27. x-intercepts: -3.732, -1 (exact), -0.268
$y = x^3 + 5x^2 + 5x + 1$
29. x-intercepts: -3.449, -2 (exact), 1 (exact), 1.449
$y = x^4 + 3x^3 - 5x^2 - 9x + 10$
31. vertex: (1.25, 3.88)
$y = 2x^2 - 5x + 7$
33. vertex: (0.39, -14.36)
$y = 9x^2 - 7x - 13$

35. high point: $(-2, 2)$; low point: $(0.7, -17.0)$

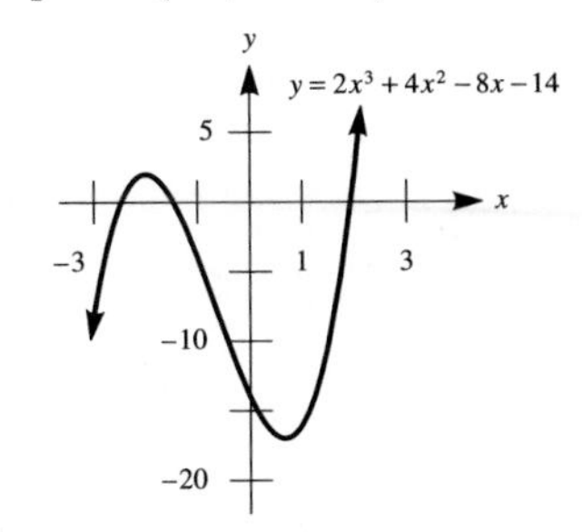

37. high point: $(-0.2, -6.9)$; low point: $(2.2, -13.1)$

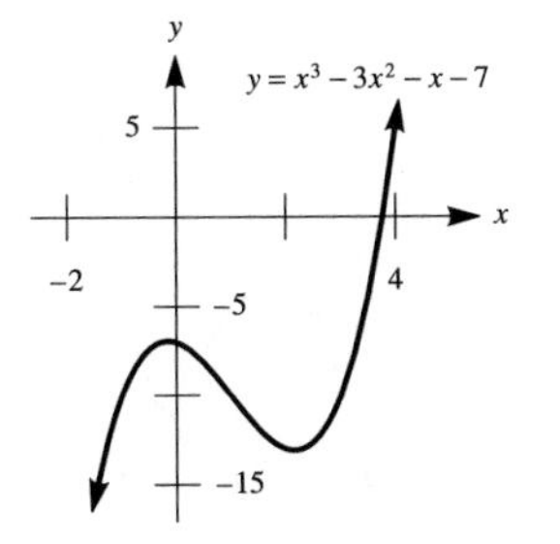

39. high point: $(0.5, 15.6)$; low points: $(-1.3, 5.0)$, $(2.3, 5.0)$

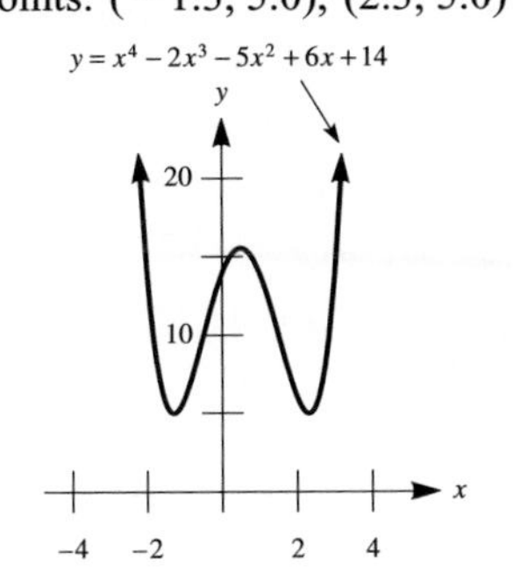

EXERCISE SET A.3

1. 832 **3.** 4410
5. (a) $A^2 + 2AB + B^2$ **(b)** $A^2 - B^2$ **(c)** $A^2 - 2AB + B^2$
7. (a) $x^2 + 4xy + 4y^2$ **(b)** $x^2 - 4y^2$ **(c)** $x^2 - 4xy + 4y^2$
9. (a) $Ax^2 - Ax + A$ **(b)** $x^3 + 1$
11. commutative property of addition
13. additive inverse property
15. associative property of addition
17. associative property of multiplication **19.** identity property of addition
21. closure property with respect to addition **23.** distributive property
25. $b = c = 1$ (Other answers are possible.) **27.** $x = 1$ and $y = 2$ (Other answers are possible.) **29.** $x = 2$ and $y = 1$ (Other answers are possible.)
31. $x = 2$ and $y = 3$ (Other answers are possible.) **33.** $a = 5$, $b = 3$, and $c = 1$ (Other answers are possible.)
35. -23 **37.** 2 **39.** -1 **41.** 2
43. $31/15$ **45.** $-5/x$
47. $2y^2/(x + y)$ **49.** $-49/36$
51. $6/35$ **53.** $3/5$
55. (a) $22/37$ **(b)** -1 **57.** -1
59. $5/7$ **61.** 16 **63.** 0
65. 1 **67.** 145

INDEX

A

B

C

D

E

F

G

H

M

N

O

P

FACTORING TECHNIQUES

TECHNIQUE	EXAMPLE OR FORMULA
Common factor	$3x^4 + 6x^3 - 12x^2 = 3x^2(x^2 + 2x - 4)$ $4(x^2 + 1) - x(x^2 + 1) = (x^2 + 1)(4 - x)$
Difference of squares	$x^2 - a^2 = (x - a)(x + a)$
Trial and error	$x^2 + 2x - 3 = (x + 3)(x - 1)$
Difference of cubes	$x^3 - a^3 = (x - a)(x^2 + ax + a^2)$
Sum of cubes	$x^3 + a^3 = (x + a)(x^2 - ax + a^2)$
Grouping	$x^3 - x^2 + x - 1 = (x^3 - x^2) + (x - 1)$ $= x^2(x - 1) + (x - 1) \cdot 1$ $= (x - 1)(x^2 + 1)$

FORMULAS FROM COORDINATE GEOMETRY

1. The distance formula: $d = \sqrt{(x_2 - x_1)^2 + (y_2 - y_1)^2}$
2. The equation for a circle: $(x - h)^2 + (y - k)^2 = r^2$
3. The definition of slope: $m = \dfrac{y_2 - y_1}{x_2 - x_1}$
4. The point-slope formula: $y - y_1 = m(x - x_1)$
5. The slope-intercept formula: $y = mx + b$
6. The condition for two nonvertical lines to be parallel: $m_1 = m_2$
7. The condition for two nonvertical lines to be perpendicular: $m_1 m_2 = -1$
8. The midpoint formula: $(x_0, y_0) = \left(\dfrac{x_1 + x_2}{2}, \dfrac{y_1 + y_2}{2}\right)$

PROPERTIES OF LOGARITHMS

In the following properties, b, P, and Q are positive and $b \neq 1$.

1. (a) $\log_b b = 1$
   (b) $\log_b 1 = 0$
2. $\log_b PQ = \log_b P + \log_b Q$
3. $\log_b (P/Q) = \log_b P - \log_b Q$
4. $\log_b P^n = n \log_b P$
5. $b^{\log_b P} = P$

THE GREEK ALPHABET

A	α	Alpha	H	η	Eta
B	β	Beta	Θ	θ	Theta
Γ	γ	Gamma	I	ι	Iota
Δ	δ	Delta	K	κ	Kappa
E	ϵ	Epsilon	Λ	λ	Lambda
Z	ζ	Zeta	M	μ	Mu
N	ν	Nu	T	τ	Tau
Ξ	ξ	Xi	Υ	υ	Upsilon
O	o	Omicron	Φ	ϕ	Phi
P	ρ	Rho	X	χ	Chi
Π	π	Pi	Ψ	ψ	Psi
Σ	σ	Sigma	Ω	ω	Omega

PRINCIPAL TRIGONOMETRIC IDENTITIES

I. CONSEQUENCES OF THE DEFINITIONS

(a) $\csc\theta = \dfrac{1}{\sin\theta}$ **(b)** $\sec\theta = \dfrac{1}{\cos\theta}$ **(c)** $\cot\theta = \dfrac{1}{\tan\theta}$ **(d)** $\tan\theta = \dfrac{\sin\theta}{\cos\theta}$ **(e)** $\cot\theta = \dfrac{\cos\theta}{\sin\theta}$

II. THE PYTHAGOREAN IDENTITIES

(a) $\sin^2\theta + \cos^2\theta = 1$ **(b)** $\tan^2\theta + 1 = \sec^2\theta$ **(c)** $\cot^2\theta + 1 = \csc^2\theta$

III. THE OPPOSITE-ANGLE FORMULAS

(a) $\sin(-\theta) = -\sin\theta$ **(b)** $\cos(-\theta) = \cos\theta$ **(c)** $\tan(-\theta) = -\tan\theta$

IV. THE REDUCTION FORMULAS

(a) $\sin(\theta + 2\pi k) = \sin\theta$ **(b)** $\cos(\theta + 2\pi k) = \cos\theta$

(c) $\sin\left(\dfrac{\pi}{2} - \theta\right) = \cos\theta$ **(d)** $\cos\left(\dfrac{\pi}{2} - \theta\right) = \sin\theta$

V. THE ADDITION FORMULAS

(a) $\sin(s + t) = \sin s\cos t + \cos s\sin t$ **(b)** $\sin(s - t) = \sin s\cos t - \cos s\sin t$

(c) $\cos(s + t) = \cos s\cos t - \sin s\sin t$ **(d)** $\cos(s - t) = \cos s\cos t + \sin s\sin t$

(e) $\tan(s + t) = \dfrac{\tan s + \tan t}{1 - \tan s\tan t}$ **(f)** $\tan(s - t) = \dfrac{\tan s - \tan t}{1 + \tan s\tan t}$

VI. THE DOUBLE-ANGLE FORMULAS

(a) $\sin 2\theta = 2\sin\theta\cos\theta$ **(b)** $\cos 2\theta = \cos^2\theta - \sin^2\theta$ **(c)** $\tan 2\theta = \dfrac{2\tan\theta}{1 - \tan^2\theta}$

VII. THE HALF-ANGLE FORMULAS

(a) $\sin\dfrac{\theta}{2} = \pm\sqrt{\dfrac{1 - \cos\theta}{2}}$ **(b)** $\cos\dfrac{\theta}{2} = \pm\sqrt{\dfrac{1 + \cos\theta}{2}}$ **(c)** $\tan\dfrac{\theta}{2} = \dfrac{\sin\theta}{1 + \cos\theta}$

VIII. THE PRODUCT-TO-SUM FORMULAS

(a) $\sin A\sin B = \frac{1}{2}[\cos(A - B) - \cos(A + B)]$ **(b)** $\sin A\cos B = \frac{1}{2}[\sin(A + B) + \sin(A - B)]$

(c) $\cos A\cos B = \frac{1}{2}[\cos(A + B) + \cos(A - B)]$

IX. THE SUM-TO-PRODUCT FORMULAS

(a) $\sin\alpha + \sin\beta = 2\sin\dfrac{\alpha + \beta}{2}\cos\dfrac{\alpha - \beta}{2}$ **(b)** $\sin\alpha - \sin\beta = 2\cos\dfrac{\alpha + \beta}{2}\sin\dfrac{\alpha - \beta}{2}$

(c) $\cos\alpha + \cos\beta = 2\cos\dfrac{\alpha + \beta}{2}\cos\dfrac{\alpha - \beta}{2}$ **(d)** $\cos\alpha - \cos\beta = -2\sin\dfrac{\alpha + \beta}{2}\sin\dfrac{\alpha - \beta}{2}$